CHILTON®

DaimlerChrysler Mechanical Service 2006 Edition

THOMSON
DELMAR LEARNING™

Australia • Canada • Mexico • Singapore • Spain • United Kingdom • United States

THOMSON
DELMAR LEARNING

Chilton®
DaimlerChrysler
Mechanical Service
2006 Edition

Vice President, Technology and Trades SBU:
Alar Elken

Executive Director, Professional Business Unit:
Gregory L. Clayton

Publisher, Professional Business Unit:
David Koontz

Marketing Director:
Beth A. Lutz

Marketing Specialist:
Brian McGrath

Marketing Coordinator:
Marissa Maiella

Production Director:
Mary Ellen Black

Production Manager:
Larry Main

Production Editor:
Elizabeth Hough

Editorial Assistant:
Christine Wade

Chilton Publishing Coordinator:
Paula Baillie

Editors:
Thomas A. Mellon
Richard J. Rivele
Jon Wallace

Cover Design:
Melinda Possinger

Printed in the United States of America

1 2 3 4 5 XX 07 06 05

For more information contact Thomson Delmar Learning Executive Woods 5 Maxwell Drive, PO Box 8007, Clifton Park, NY 12065-8007 Or find us on the World Wide Web at www.delmarlearning.com

ISBN: 1-4180-0600-9

ISSN: 1548-0887

NOTICE TO THE READER

Publisher does not warrant or guarantee any of the products described herein or perform any independent analysis in connection with any of the product information contained herein. Publisher does not assume, and expressly disclaims, any obligation to obtain and include information other than that provided to it by the manufacturer.

The reader is expressly warned to consider and adopt all safety precautions that might be indicated by the activities herein and to avoid all potential hazards. By following the instructions contained herein, the reader willingly assumes all risks in connection with such instructions.

The publisher makes no representation or warranties of any kind, including but not limited to, the warranties of fitness for particular purpose or merchantability, nor are any such representations implied with respect to the material set forth herein, and the publisher takes no responsibility with respect to such material. The publisher shall not be liable for any special, consequential, or exemplary damages resulting, in whole or part, from the readers' use of, or reliance upon, this material.

Table of Contents

Sections

Model Index

EDITORIAL POLICY

Manufacturer and Model Coverage

This manual does not cover every DaimlerChrysler model that is currently available on the market. Rather, the Chilton editorial staff makes judicious decisions as to which models warrant coverage, based on which vehicles are serviced by most technicians.

Model Year Information

Every effort is made to gather current data from the Original Vehicle Manufacturers (OEMs) when they publish it. However, OEMs release new model information throughout the year and may not adhere to the same schedule each year. As a result, not all models are equally current when each edition of this manual is published.

Although information in this manual is based on industry sources and is as complete as possible at the time of publication, some vehicle manufacturers may make changes which cannot be included here. While striving for total accuracy, the publisher cannot assume responsibility for any errors, changes, or omissions that may occur in the compilation of this data.

Safety Notice

Proper service and repair procedures are vital to the safe, reliable operation of all motor vehicles, as well as the personal safety of those performing the repairs. This manual outlines procedures for servicing and repairing vehicles using safe, effective methods. The procedures contain many NOTES, WARNINGS and CAUTIONS which should be followed along with standard safety procedures to reduce the possibility of personal injury or improper service which could damage the vehicle or compromise its safety.

Repair procedures, tools, parts, and technician skill and experience vary widely. It is not possible to anticipate all conceivable ways or conditions under which vehicles may be serviced, or to provide cautions for all possible hazards that may result. Standard and accepted safety precautions and equipment should be used when handling toxic or flammable fluids, and safety goggles or other protection should be used during cutting, grinding, chiseling, prying, or any other process that can cause material removal or projectiles.

Some procedures require using tools designed for a specific purpose. Before substituting another tool or procedure, you must be completely satisfied that neither your personal safety, nor the performance of the vehicle will be endangered.

LOCATING AND USING THE INFORMATION

Organization

To find where a particular model section or procedure is located, look in the Table of Contents. Main topics are listed with the page number on which they may be found. Following the main topics is an alphabetical listing of all of the procedures within the section and their page numbers.

Part Numbers

Part numbers listed in this book are not recommendations by the publisher for any product by brand name. They are references that can be used with interchanges manuals and aftermarket supplier catalogs to locate each brand supplier's discrete part number.

Special Tools

Special tools are recommended by the vehicle manufacturer to perform specific jobs. When necessary, special tools are referred to in the text by the part number of the tool manufacturer. These tools may be purchased, under the appropriate part number, from your local dealer or regional distributor, or an equivalent tool can be purchased locally from a tool supplier or parts outlet. Before substituting any tool for the one recommended, read the previous Safety Notice.

ACKNOWLEDGEMENT

The publisher would like to express appreciation to DaimlerChrysler for its assistance in producing this publication.

CHRYSLER AND DODGE

1

300 Series • Magnum

SPECIFICATION AND MAINTENANCE CHARTS

ENGINE AND VEHICLE IDENTIFICATION

Engine						
Code ①	Liters (cc)	Cu. In.	Cyl.	Fuel Sys.	Engine Type	Eng. Mfg.
R	2.7 (2736)	167	6	MPI	DOHC	DaimlerChrysler
G	3.5 (3507)	214	6	MPI	SOHC	DaimlerChrysler
H	5.7 (5654)	345	8	MPI	OHV	DaimlerChrysler

Model Year	
Code ②	Year
5	2005
6	2006

SMFI: Sequential Multi-port Fuel Injection

① 8th position of VIN

② 10th position of VIN

06009-300C-C01

GENERAL ENGINE SPECIFICATIONS

Year	Model	Engine Displacement Liters	Engine Series VIN	Net Horsepower @ rpm	Net Torque @ rpm (ft. lbs.)	Bore x Stroke (in.)	Com-pression Ratio	Oil Pressure @ rpm
2005	300, Magnum SE	2.7	R	190@6400	190@4000	3.386x3.091	9.67:1	45-105@3000
	300, Magnum SXT	3.5	G	250@6400	250@3800	3.780x3.189	10:01	45-105@3000
	300C, Magnum RT	5.7	H	340@5000	390@4000	3.910x3.580	9.6:1	25-110@3000

06009-300C-C02

GASOLINE ENGINE TUNE-UP SPECIFICATIONS

Year	Engine Displacement Liters	Engine VIN	Spark Plug Gap (in.)	Ignition Timing (deg.)	Fuel Pump (psi)	Idle Speed (rpm)	Valve Clearance	
							Intake	Exhaust
2005	2.7	R	0.048-0.058	①	53-63	②	HYD	HYD
	3.5	G	0.048-0.058	①	53-63	②	HYD	HYD
	5.7	H	0.045	①	53-63	②	HYD	HYD

NOTE: The Vehicle Emission Control Information (VECI) label often reflects specification changes made during production. The label figures must be used if they differ from those in this chart.

HYD: Hydraulic

① Ignition timing is controlled by the PCM and is not adjustable.

② Idle speed is controlled by the PCM and is not adjustable

06009-300C-C03

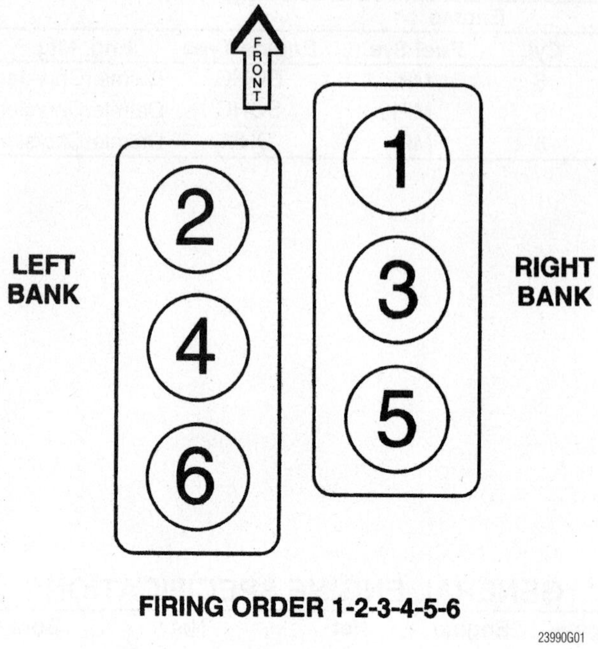

2.7L and 3.5L Engines
Firing order: 1–2–3–4–5–6
Distributorless ignition

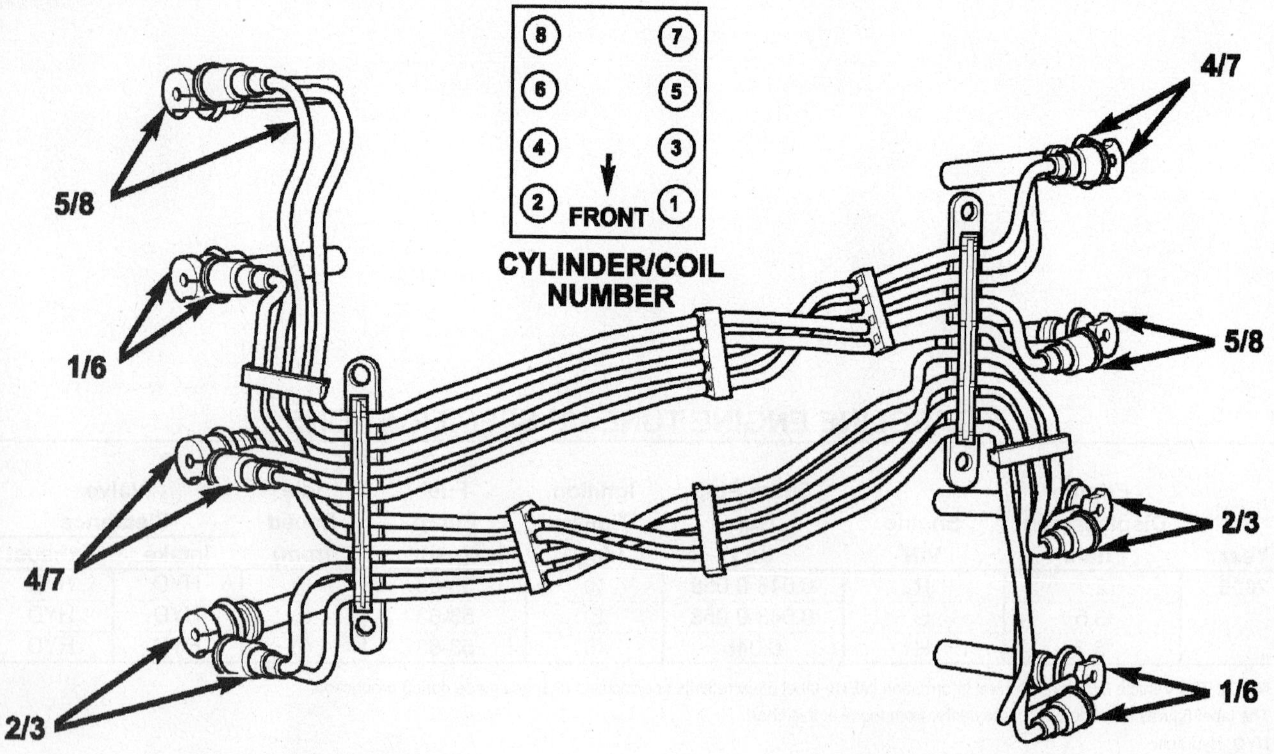

5.7L Engine
Firing order: 1–8–4–3–6–5–7–2
Distributorless ignition

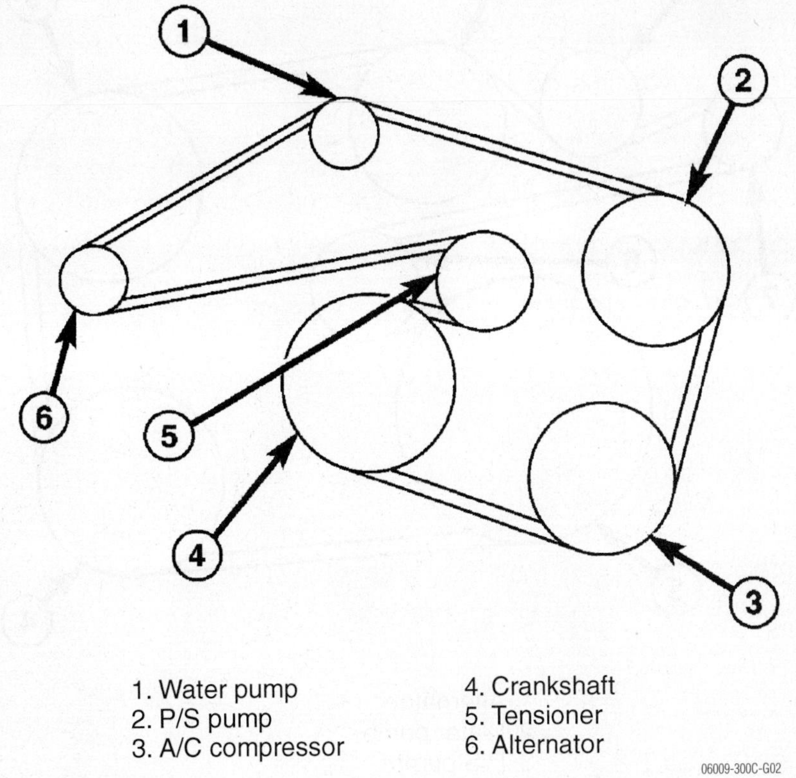

Accessory drive belt routing—2.7L engine

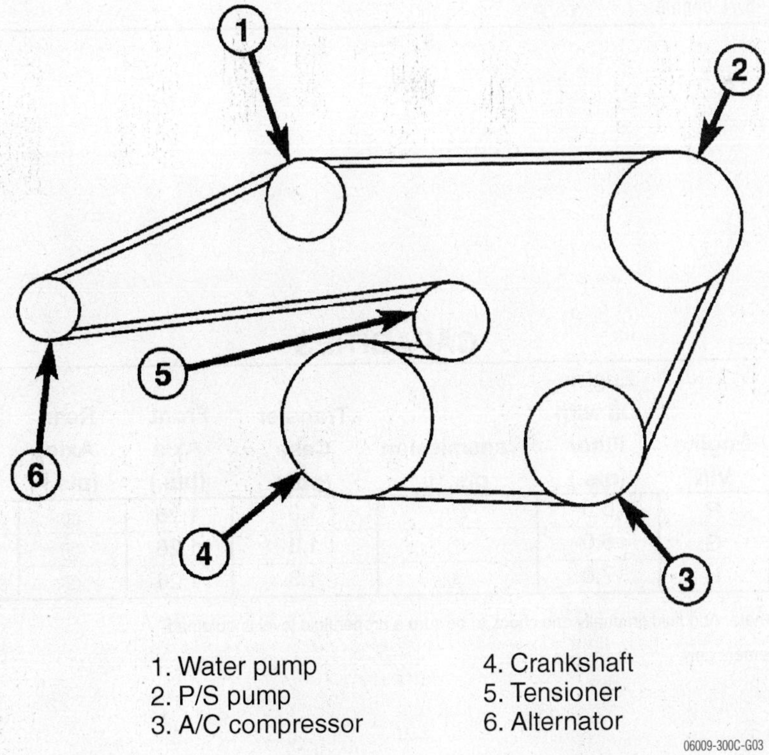

Accessory drive belt routing—3.5L engine

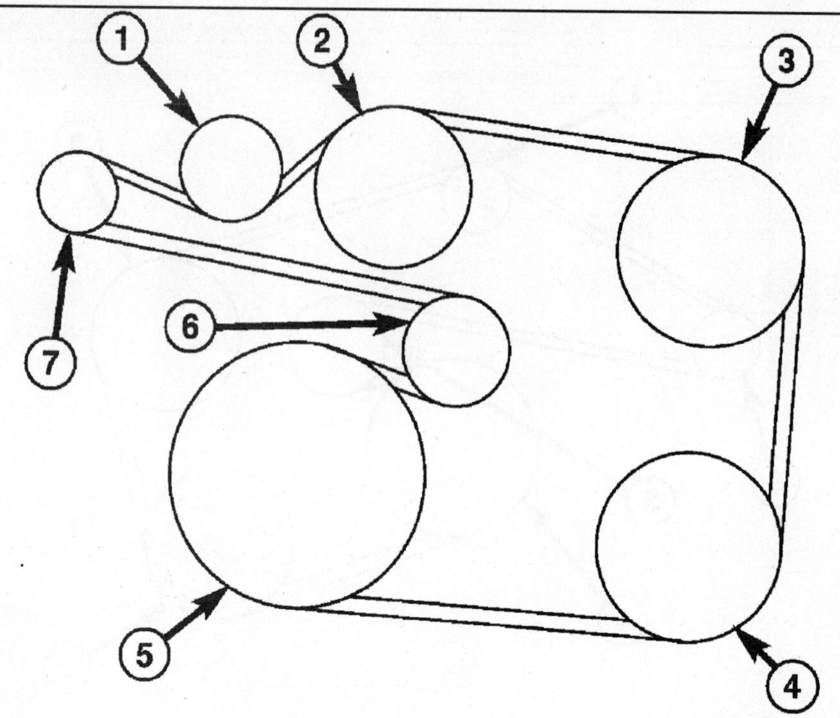

Accessory drive belt routing—5.7L engine

CAPACITIES

Year	Engine Displacement Liters	Engine VIN	Engine Oil with Filter (qts.)	Transmission pts. ①	Transfer Case (pts.)	Front Axle (pts.)	Rear Axle (pts.)	Fuel Tank (gal.)	Cooling System (qts.)
2005	2.7	R	6.0	②	1.3	1.26	③	18.0	9.7
	3.5	G	6.0	②	1.3	1.26	③	18.0	10.6
	5.7	H	7.0	②	1.3	1.26	③	19.0	14.6

NOTE: All capacities are approximate. Add fluid gradually and check to be sure a proper fluid level is obtained.

① For fluid drain and filter replacement only.

② NAG1: 10.6 pts.
42RLE: 8.0 pts.

③ 198mm axle: 3.0 pts.
210mm axle: 3.4 pts.

06009-300C-C04

VALVE SPECIFICATIONS

Year	Engine Displacement Liters	Engine VIN	Seat Angle (deg.)	Face Angle (deg.)	Spring Test Pressure (lbs. @ in.)	Spring Installed Height (in.)	Stem-to-Guide Clearance (in.) Intake	Stem-to-Guide Clearance (in.) Exhaust	Stem Diameter (in.) Intake	Stem Diameter (in.) Exhaust
2005	2.7	R	45-45.5	44.5-45.5	①	1.496	0.0009-0.0026	0.0020-0.0037	0.2337-0.2344	0.2326-0.2333
	3.5	G	45-45.5	44.5-45	②	1.496	0.0009-0.0026	0.0020-0.0037	0.2730-0.2737	0.2719-0.2726
	5.7	H	44.5-45	45-45.5	242@1.322	1.811	0.0008-0.0025	0.0019-0.0037	0.3120-0.3130	0.3110-0.3120

① Intake: 148-162@1.1417
Exhaust: 138-151@1.1811
② Intake: 130-144@1.239
Exhaust: 140-155@1.239

06009-300C-C05

CRANKSHAFT AND CONNECTING ROD SPECIFICATIONS

All measurements are given in inches.

Year	Engine Displacement Liters	Engine VIN	Crankshaft Main Brg. Journal Dia.	Crankshaft Main Brg. Oil Clearance	Crankshaft Shaft End-play	Crankshaft Thrust on No.	Connecting Rod Journal Diameter	Connecting Rod Oil Clearance	Connecting Rod Side Clearance
2005	2.7	R	2.4997-2.5004	0.0012-0.0022	0.0019-0.0108	3	2.1060-2.1067	0.0013-0.0027	0.0052-0.0150
	3.5	G	2.5190-2.5120	0.0014-0.0026	0.0040-0.0120	3	2.2282-2.2283	0.0010-0.0030	0.0153 max.
	5.7	H	2.5585-2.5595	0.0009-0.0020	0.0020-0.0110	3	2.1250-2.1260	0.0007-0.0023	0.0030-0.0137

06009-300C-C06

PISTON AND RING SPECIFICATIONS

All measurements are given in inches.

Year	Engine Displacement Liters	Engine VIN	Piston Clearance	Ring Gap Top Compression	Ring Gap Bottom Compression	Ring Gap Oil Control	Ring Side Clearance Top Compression	Ring Side Clearance Bottom Compression	Ring Side Clearance Oil Control
2005	2.7	R	0-0.0016	0.0080-0.0140	0.0146 0.0249	0.010-0.030	0.0013-0.0032	0.0016-0.0031	0.0022-0.0080
	3.5	G	0-0.0018	0.0080-0.0140	0.0078-0.0157	0.010-0.030	0.0016-0.0031	0.0016-0.0031	0.0015-0.0073
	5.7	H	0.0008-0.0019	0.0090-0.0149	0.0137-0.0236	0.006-0.026	0.0007-0.0026	0.0007-0.0022	0.0007-0.0091

06009-300C-C07

TORQUE SPECIFICATIONS

All readings in ft. lbs.

Year	Engine Displacement Liters	Engine VIN	Cylinder Head Bolts	Main Bearing Bolts	Rod Bearing Bolts	Crankshaft Damper Bolts	Flywheel Bolts	Manifold Intake	Manifold Exhaust	Spark Plugs	Oil Pan Drain Plug
2005	2.7	R	①	②	③	125	70	④	16	15	20
	3.5	G	⑤	②	③	70	70	⑥	16	20	20
	5.7	H	①	⑦	⑧	129	70	⑨	18	13	20

① See the text section

② First, tighten the main cap bolts to 15 ft. lbs. +90 degrees
Second, tighten the windasge tray bolts to 20 ft. lbs. +90 degrees
Third, tighten the tie (horizontal) bolts to 250 inch lbs.

③ 20 ft. lbs. + 90 degrees

④ Upper and Lower: 105 inch lbs.

⑤ Step 1: 45 ft. lbs.
Step 2: 65 ft. lbs.
Step 3: Verify 65 ft. lbs.
Step 4: + 90 degrees

⑥ Upper: 105 inch lbs.
Lower: 250 inch lbs.

⑦ First, install the main bearing cap bolts, and torque them, in sequence, to:
Step 1: 20 ft. lbs.
Step 2: +90 degrees
Then, install the crossbolts and torque them, in sequence, to 21 ft. lbs.

⑧ 15 ft. lbs. +90 degrees

⑨ 105 inch lbs.

06009-300C-C08

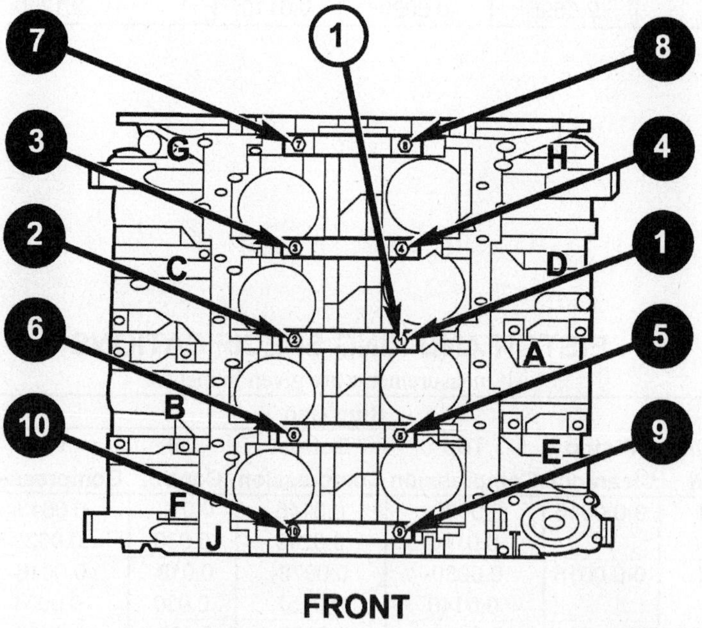

5.7L engine crankshaft torque sequence

WHEEL ALIGNMENT

			Caster		Camber		
Year	Model		Range (+/-Deg.)	Preferred Setting (Deg.)	Range (+/-Deg.)	Preferred Setting (Deg.)	Toe-in (in.)
2005	AWD	F	1.0	+5.60	0.50	0	0.20+/-0.20
		R	—	—	0.50	-0.75	0.20+/-0.30
	RWD	F	1.5	①	0.50	②	0.20+/-0.20
		R	—	—	0.50	-0.75	0.20+/-0.30

① Left +10.30
Right: +11.10

② Left: -0.10
Right: -0.20

06009-300C-C09

TIRE, WHEEL AND BALL JOINT SPECIFICATIONS

		OEM Tires		Tire Pressures (psi)		Wheel	Ball Joint	Wheel Lug
Year	Model	Standard	Optional	Front	Rear	Size	Play	Nut Torque
2005	300	P215/65TR17	P225/60HR18	①	①	std.: 7.0 opt: 7.5	0.059 in. max.	110
	Magnum SE	P215/65TR17	none	①	①	7.0	0.059 in. max.	110
	Magnum SXT	P225/60HR18	none	①	①	7.5	0.059 in. max.	110
	300C	P225/60HR18	none	①	①	7.5	0.059 in. max.	110
	Magnum RT	P225/60HR18	none	①	①	7.5	0.059 in. max.	110

NA: Information not available
OEM: Original Equipment Manufacturer
PSI: Pounds Per Square Inch
STD: Standard
OPT: Optional
① See placard on the vehicle

06009-300C-C10

BRAKE SPECIFICATIONS

All measurements in inches unless noted

			Brake Disc			Minimum	Brake Caliper	
Year	Model		Original Thickness	Minimum Thickness	Maximum Run-out	Lining Thickness	Bracket Bolts (ft. lbs.)	Mounting Pins (ft. lbs.)
2005	All	F	1.097-1.107	1.040	0.0014	NA	70	44
		R	①	②	0.0014	NA	85	44

NA: Information not available

① 17 inch brakes: 0.389-0.399
18 inch brakes: 0.861-0.871

② 17 inch brakes: 0.335
18 inch brakes: 0.807

06009-300C-C11

SCHEDULED MAINTENANCE INTERVALS

Chrysler 300 Series and Dodge Magnum

TO BE SERVICED	TYPE OF SERVICE	VEHICLE MILEAGE INTERVAL (x1000)												
		3	6	9	12	15	18	21	24	27	30	33	36	39
Engine oil & filter	R	✓	✓	✓	✓	✓	✓	✓	✓	✓	✓	✓	✓	✓
Tires	Rotate		✓		✓		✓		✓		✓		✓	
A/C filter	R					✓					✓			
Brake linings	I			✓			✓			✓			✓	
Engine air filter	S/I	✓	✓	✓	✓	✓	✓	✓	✓	✓	✓	✓	✓	✓
Engine air filter	R					✓					✓			
Spark plugs (5.7L)	R										✓			
PCV valve	S/I										✓			
Rear axle fluid	R	Every 48,000 miles												
Transfer case fluid	R	Every 48,000 miles												
Power steering fluid	R	Every 60,000 miles												
Auto. trans. fluid and filter	R	Every 60,000 miles												
Accessory drive belt	I/R	Every 72,000 miles												
Spark plugs (2.7L and 3.5L)	R	Every 100,000 miles												
Engine coolant	R	Every 102,000 miles												
Timing belt (3.5L)	R	Every 105,000 miles												

R: Replace S: Service I: Inspect Adj: Adjust

The above schedule is to be used if you drive under any of the following conditions:

Driving in temperatures under 32 degrees F

Stop and go traffic

Extensive engine idling

Driving in dusty conditions

Frequent trips under 10 miles

More than 50 % of your driving is in hot weather (90 deg. F) above 50 miles per hour

Trailer towing

Taxi, police or delivery service

If none of these conditions is met, double the maintenance intervals

06009-300C-C12

PRECAUTIONS

Before servicing any vehicle, please be sure to read all of the following precautions, which deal with personal safety, prevention of component damage, and important points to take into consideration when servicing a motor vehicle:

- Never open, service or drain the radiator or cooling system when the engine is hot; serious burns can occur from the steam and hot coolant.
- Observe all applicable safety precautions when working around fuel. Whenever servicing the fuel system, always work in a well-ventilated area. Do not allow fuel spray or vapors to come in contact with a spark, open flame, or excessive heat (a hot drop light, for example). Keep a dry chemical fire extinguisher near the work area. Always keep fuel in a container specifically designed for fuel storage; also, always properly seal fuel containers to avoid the possibility of fire or explosion. Refer to the additional fuel system precautions later in this section.
- Fuel injection systems often remain pressurized, even after the engine has been turned **OFF**. The fuel system pressure must be relieved before disconnecting any fuel lines. Failure to do so may result in fire and/or personal injury.
- Brake fluid often contains polyglycol ethers and polyglycols. Avoid contact with the eyes and wash your hands thoroughly after handling brake fluid. If you do get brake fluid in your eyes, flush your eyes with clean, running water for 15 minutes. If eye irritation persists, or if you have taken brake fluid internally, IMMEDIATELY seek medical assistance.
- The EPA warns that prolonged contact with used engine oil may cause a number of skin disorders, including cancer. You should make every effort to minimize your exposure to used engine oil. Protective gloves should be worn when changing oil. Wash your hands and any other exposed skin areas as soon as possible after exposure to used engine oil. Soap and water, or waterless hand cleaner should be used.
- All new vehicles are now equipped with an air bag system, often referred to as a Supplemental Restraint System (SRS) or Supplemental Inflatable Restraint (SIR) system. The system must be disabled before performing service on or around system components, steering column, instrument panel components, wiring and sensors. Failure to follow safety and disabling procedures could result in accidental air bag deployment, possible personal injury and unnecessary system repairs.
- Always wear safety goggles when working with, or around, the air bag system. When carrying a non-deployed air bag, be sure the bag and trim cover are pointed away from your body. When placing a non-deployed air bag on a work surface, always face the bag and trim cover upward, away from the surface. This will reduce the motion of the module if it is accidentally deployed. Refer to the additional air bag system precautions later in this section.
- Clean, high quality brake fluid from a sealed container is essential to the safe and proper operation of the brake system. You should always buy the correct type of brake fluid for your vehicle. If the brake fluid becomes contaminated, completely flush the system with new fluid. Never reuse any brake fluid. Any brake fluid that is removed from the system should be discarded. Also, do not allow any brake fluid to come in contact with a painted surface; it will damage the paint.
- Never operate the engine without the proper amount and type of engine oil; doing so WILL result in severe engine damage.
- Timing belt maintenance is extremely important. Many models utilize an interference-type, non-freewheeling engine. If the timing belt breaks, the valves in the cylinder head may strike the pistons, causing potentially serious (also time-consuming and expensive) engine damage. Refer to the maintenance interval charts for the recommended replacement interval for the timing belt, and to the timing belt section for belt replacement and inspection.
- Disconnecting the negative battery cable on some vehicles may interfere with the functions of the on-board computer system(s) and may require the computer to undergo a relearning process once the negative battery cable is reconnected.
- When servicing drum brakes, only disassemble and assemble one side at a time, leaving the remaining side intact for reference.
- Only an MVAC-trained, EPA-certified automotive technician should service the air conditioning system or its components.

ENGINE REPAIR

Alternator

REMOVAL

⁂ WARNING

Disconnect negative cable from battery before removing battery output wire (b+ wire) from alternator. Failure to do so can result in injury or damage to electrical system.

2.7L Engine

1. Before servicing the vehicle, refer to the precautions in the beginning of this section.
2. Disconnect negative battery cable.
3. Remove alternator drive belt.
4. Disconnect alternator field circuit plug.
5. Remove B+ terminal nut and wire.
6. Remove 2 lower mounting bolts.
7. Remove alternator.

To install:

8. Install alternator.
9. Install upper bolt mounting bolt.
10. Install 2 lower mounting bolts.
11. Tighten bolts to 65 Nm (48 ft. lbs.).
12. Install B+ terminal nut and wire. Tighten nut to 13 Nm (115 inch lbs.).
13. Connect alternator field circuit plug.
14. Install alternator drive belt.
15. Connect negative battery cable.

3.5L Engine

1. Before servicing the vehicle, refer to the precautions in the beginning of this section.
2. Disconnect negative battery cable.
3. Remove alternator drive belt.
4. Remove bracket bolts.
5. Remove upper mounting bolt.
6. Raise and support vehicle.
7. Remove middle splash pan.
8. Disconnect alternator field circuit plug.
9. Remove B+ terminal nut and wire.
10. Remove lower mounting bolts.
11. Remove alternator.

To install:

12. Install alternator.

13. Install lower mounting bolts. Loosen install the upper mounting bolt.
14. Tighten lower mounting bolts to 65 Nm (48 ft. lbs.).
15. Connect alternator field circuit plug.
16. Install B+ terminal nut and wire. Tighten nut to 13 Nm (115 inch lbs.).
17. Install middle splash pan.
18. Lower vehicle.
19. Remove loose installed upper bolt.
20. Install bracket and bracket bolt.
21. Install upper mounting bolt and bracket.
22. Tighten upper mounting bolt to 65 Nm (48 ft. lbs.).
23. Tighten Bracket bolt to 54 Nm (40 ft. lbs.).
24. Install alternator drive belt.
25. Connect negative battery cable.

5.7L Engine

1. Before servicing the vehicle, refer to the precautions in the beginning of this section.
2. Disconnect negative battery cable at battery.
3. Remove alternator drive belt.
4. Raise and support vehicle.
5. Unsnap plastic insulator cap from B+ output terminal.
6. Remove B+ terminal mounting nut at rear of alternator. Disconnect terminal from alternator.
7. Disconnect field wire connector at rear of alternator by pushing on connector tab.

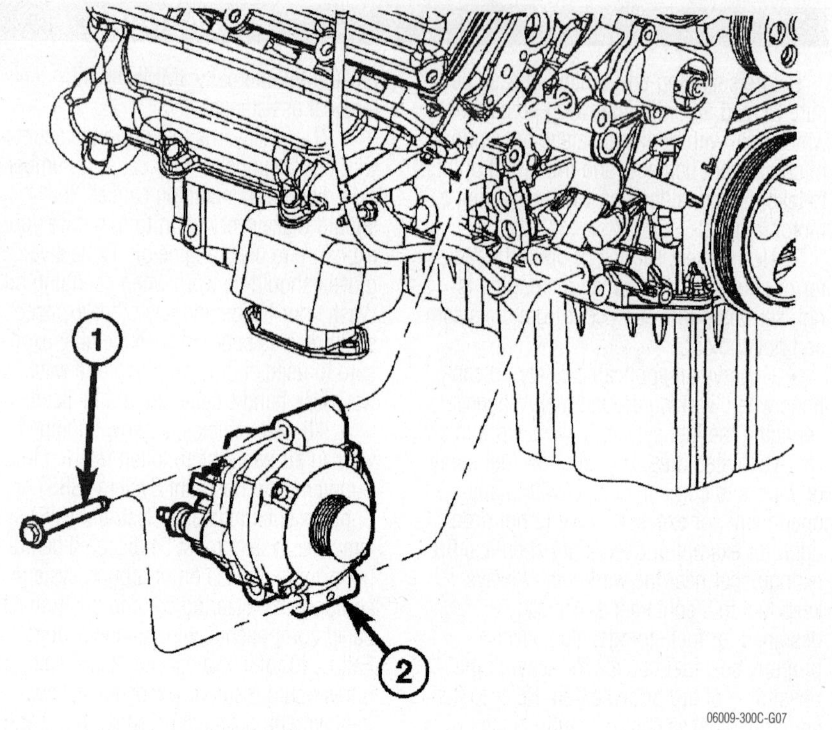

Alternator installation—5.7L engine

8. Remove alternator support bracket nut and bolt. Remove support bracket.
9. Remove 2 alternator mounting bolts.
10. Remove alternator from vehicle.

To install:

11. Position alternator to engine and install 2 mounting bolts.
12. Tighten bolts to 48 ft. lbs. (65 Nm).
13. Position support bracket to alternator and install bolt and nut. Tighten bolt/nut to 48 ft. lbs. (65 Nm).
14. Snap field wire connector into rear of alternator.
15. Install B+ terminal eyelet to alternator output stud.
16. Lower vehicle.

※※ WARNING

Never force a belt over a pulley rim using a screwdriver. The synthetic fiber of the belt can be damaged. When installing a serpentine accessory drive belt, the belt MUST be routed correctly. The water pump may be rotating in the wrong direction if the belt is installed incorrectly, causing the engine to overheat. Refer to belt routing label in engine compartment.

17. Install alternator drive belt.
18. Install negative battery cable to battery.

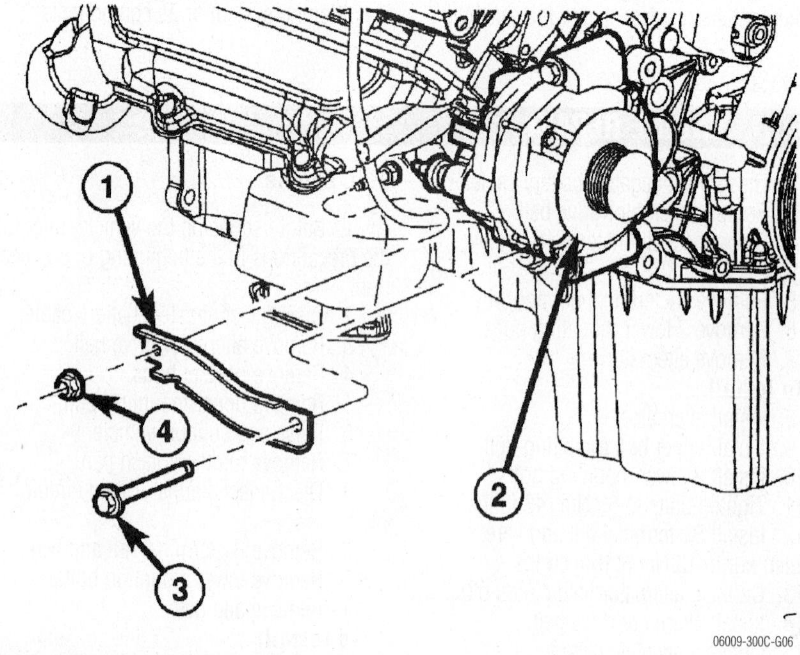

Alternator support bracket—5.7L engine

Engine Assembly

REMOVAL & INSTALLATION

2.7L Engine

1. Before servicing the vehicle, refer to the precautions in the beginning of this section.

2. Remove hood.
3. Release fuel pressure.
4. Disconnect negative battery cable located in the trunk.
5. Disconnect intake air temperature sensor electrical connector.
6. Remove air cleaner housing assembly.
7. Remove accessory drive belt.
8. Disconnect lines from power steering pump.
9. Remove power steering pump attaching bolts and set pump aside.
10. Recover air conditioning system using a suitable refrigerant recovery machine.
11. Disconnect the wire harness connector from the air conditioning compressor clutch coil connector.
12. Remove the nuts that secure the air conditioning suction line and air conditioning discharge line to the air conditioning compressor.
13. Disconnect the suction and discharge lines from the air conditioning compressor and remove and discard the dual plane seals.
14. Install plugs in, or tape over all of the opened refrigerant line fittings and the compressor ports.
15. Raise and support the vehicle.
16. Remove the front end splash shield.
17. Remove the bolts that secure the automatic transmission cooler line bracket and the air conditioning compressor to the cylinder block.
18. Position the cooler lines out of the way and remove the air conditioning compressor from the engine compartment.
19. Disconnect air conditioning compressor electrical connectors.
20. Remove air conditioning compressor.
21. Raise vehicle.
22. Remove lower splash shield bolts and remove splash shield.
23. Drain cooling system.
24. Drain the engine oil and remove the oil filter.
25. Disconnect down stream oxygen sensor connectors.
26. Disconnect exhaust pipes at the manifolds.
27. Disconnect starter electrical connectors.
28. Remove starter bolts and set starter aside.
29. Disconnect coolant pipe near starter from hose.
30. Disconnect crankshaft position sensor electrical connector and remove sensor.
31. Disconnect ground strap.
32. Disconnect oil pressure sensor electrical connector.
33. Remove 4 engine mount nuts.
34. Remove engine mount studs from mount.
35. Remove steering gear mounting bolts and lower steering gear for clearance.
36. Using a block of wood and a suitable lifting device under the oil pan, raise the engine.
37. Remove structural collar bolts.
38. Remove structural collar.
39. Remove lower bell housing bolts.
40. Lower engine and remove jackstand.
41. Mark flex plate to torque converter location.
42. Remove torque converter bolts.
43. Lower vehicle.
44. Remove wiper arms.
45. Remove wiper cowl.
46. Remove strut tower support.
47. Reposition purge solenoid.
48. Remove the intake manifold
49. Disconnect all vacuum lines and electrical connectors.
50. Disconnect all ground straps attaching to the engine.
51. Disconnect fuel line at fuel rail.
52. Disconnect coil, injector, capacitor, and knock sensor connectors.
53. Remove purge solenoid from bracket.
54. Remove the bolts from the EGR tube to exhaust manifold.
55. Remove the lower tube bolts to EGR valve and remove tube.
56. Remove upper tube bolts to EGR valve.
57. Remove the EGR valve mounting bolts.
58. Remove valve from vehicle. Clean mounting surface.
59. Remove upper bellhousing bolts and reposition electrical harness.
60. Attach lifting tool 8342A.
61. Support transmission with a block of wood and floor jack.
62. Hoist engine from engine compartment.

To install:

63. Attach special tool #8342A engine lifting fixture.
64. Lower engine into engine compartment.
65. Remove special tool #8342A.

⁂ WARNING

Do not tighten the transmission case to engine block bolts until all bolts have been hand started, and the two mating surfaces are completely joined together, as damage to cylinder block or transmission could occur.

66. Install bellhousing to cylinder block bolts and tighten to 68 Nm (50 ft. lbs.).
67. Install EGR valve.
68. Install the EGR valve mounting bolts.
69. Inspect rubber silicone seals on intake manifold end of EGR tube.
70. Install upper tube into the intake manifold, being careful that the silicone

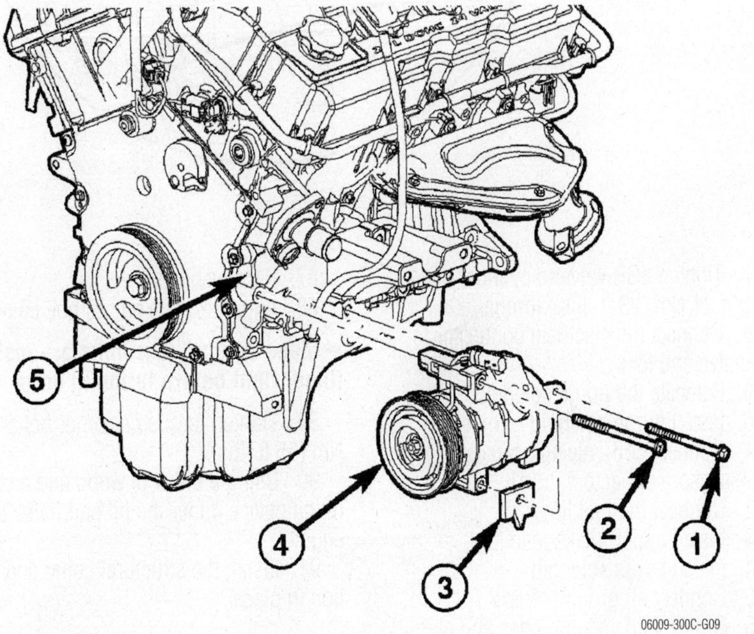

A/C compressor mounting—2.7L engine

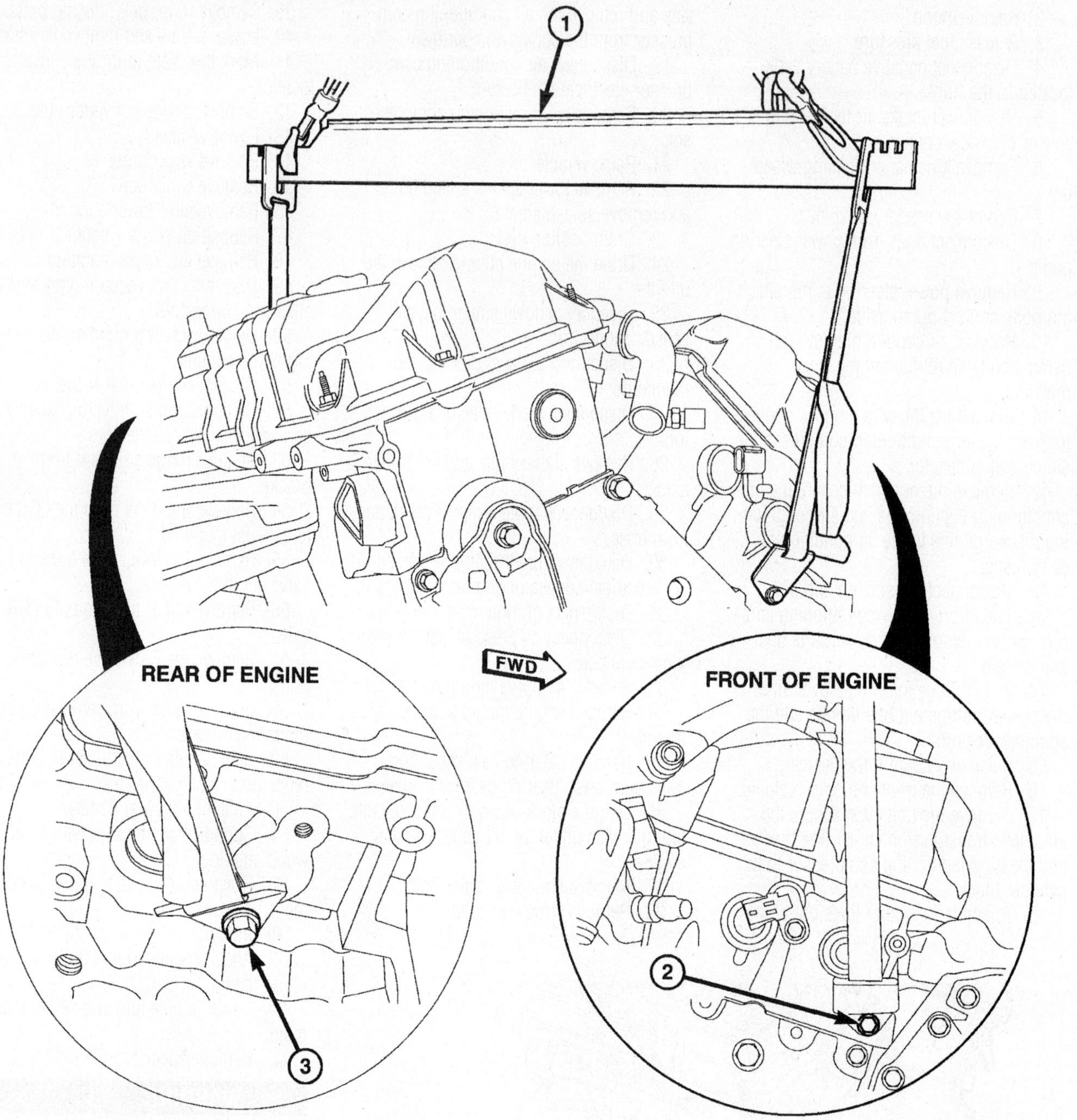

Engine lifting tool attached to the 2.7L engine

rubber seals are correctly installed and undamaged.

71. Install new gasket between the EGR valve and upper tube and install bolts.

72. Install the lower tube to exhaust manifold.

73. Install new gasket between the EGR valve and lower tube and install bolts.

74. Tighten the lower tube to EGR valve bolts to 11 Nm (95 inch lbs.) torque.

75. Tighten the lower tube to exhaust manifold bolts to 31 Nm (23 ft. lbs.) torque.

76. Tighten the upper tube to EGR valve bolts to 11 Nm (95 inch lbs.) torque.

77. Tighten EGR valve to cylinder head bolts to 31 Nm (23 ft. lbs.) torque.

78. Connect the electrical connector to EGR valve and lock.

79. Relocate the purge solenoid.

80. Install purge solenoid to bracket.

81. Connect coil, injector, capacitor, and knock sensor connectors.

82. Connect the fuel line.

83. Install upper intake manifold.

84. Install purge solenoid.

85. Connect all ground straps.

86. Connect all vacuum lines and electrical connectors.

87. Raise vehicle.

88. Align flexplate and torque converter.

➡Make sure all four bolts are installed finger tight before torquing bolts.

89. Tighten torque converter bolts to 75 Nm (55 ft. lbs.).

90. Using a block of wood and a suitable lifting device under the oil pan, raise the engine.

91. Install the structural collar and position in place.

➡Make sure that structural collar is flush with the oil pan and the transmission bell housing.

92. Finger tighten all bolts.
93. Tighten the structural collar to oil pan bolts to 55 Nm (40 ft. lbs.).
94. Tighten structural collar to transmission bolts to 55 Nm (40 ft. lbs.).
95. Align power steering pump with mounting holes on engine bracket.
96. Install three pump mounting bolts through access holes in pulley and engine bracket. Tighten bolts to 28 Nm (21 ft. lbs.) torque.

➡Always use a NEW O-ring on the end of the pressure hose.

97. Lubricate NEW O-ring on end of pressure hose with clean power steering fluid.
98. Install pressure hose to pump. Tighten pressure hose tube nut to 47 Nm (35 ft. lbs.) torque.
99. Install supply hose on pump. Install clamp securing hose in place.
100. Install serpentine drive belt.
101. Install air cleaner housing and inlet tube.

➡Be certain to check the refrigerant oil level if the air conditioning compressor is being replaced. Use only refrigerant oil of the type recommended for the air conditioning compressor in the vehicle.

102. Position the air conditioning compressor into the engine compartment.
103. Install the bolts that secure the air conditioning compressor and the automatic transmission cooler line bracket to the cylinder block. Tighten the bolts to 55 Nm (41 ft. lbs.).
104. Install the front end splash shields.
105. Lower the vehicle.
106. Remove the tape or plugs from the opened fittings on the air conditioning suction line and the air conditioning discharge line and the compressor ports.
107. Lubricate the new dual plane seals with clean refrigerant oil and install them onto the suction and the discharge line fittings. Use only the specified seals as they are made of a special material for the R-134a system. Use only refrigerant oil of the type recommended for the air conditioning compressor in the vehicle.
108. Install the suction and discharge lines onto the air conditioning compressor.
109. Install the nuts that secure the suction and discharge lines to the air conditioning compressor. Tighten the nuts to 23 Nm (17 ft. lbs.).
110. Connect the wire harness connector to the air conditioning compressor clutch coil connector.
111. Install the crankshaft position sensor and connect electrical connector.
112. Connect oil pressure sensor electrical connector.
113. Install starter, tighten bolts to 41 Nm (30 ft. lbs.)
114. Connect electrical connectors at starter.
115. Connect the coolant pipe to coolant hose.
116. Connect the exhaust pipes to the manifolds.
117. Connect the down stream oxygen sensor connectors.
118. Connect air conditioning compressor electrical connectors.
119. Install new oil filter.
120. Install lower splash shield and tighten retaining bolts.
121. Lower vehicle.
122. Install power steering pump.
123. Connect lines to power steering pump.
124. Install accessory drive belt.
125. Install air cleaner housing assembly.
126. Connect intake air temperature sensor connector.
127. Connect negative battery cable.
128. Fill coolant system.
129. Fill engine with correct grade of oil.
130. Install strut tower support.
131. Install cowl screen.
132. Install wiper arms.
133. Install the hood.
134. Fill the pump fluid reservoir to the proper level and let the fluid settle for at least two minutes.
135. Evacuate and charge air conditioning system
136. Start engine and check for leaks.
137. Turn steering wheel all the way to the left.
138. Raise the front wheels off the ground.
139. Slowly turn the steering wheel lock-to-lock 20 times with the engine off while checking the fluid level.

➡Vehicles with long return lines or oil coolers turn wheel 40 times.

140. Start the engine. With the engine idling maintain the fluid level.
141. Lower the front wheels and let the engine idle for two minutes.
142. Turn the steering wheel in both direction and verify power assist and quiet operation of the pump.

➡If the fluid is extremely foamy or milky looking, allow the vehicle to stand a few minutes and repeat the procedure.

3.5L Engine

1. Before servicing the vehicle, refer to the precautions in the beginning of this section.

➡Capture and store any residual fluid drainage, or leakage from ancillary components, in the appropriately marked containers.

2. Perform the fuel pressure release procedure.
3. Center and secure the steering wheel.
4. Disconnect negative battery cable.
5. Evacuate the air conditioning system.
6. Remove the hood.
7. Remove the windshield cowl assembly.
8. Raise and support the vehicle.
9. Remove the power steering pump and set it aside without disconnecting the lines.
10. Remove the lower engine close out panel.
11. Drain the cooling system.
12. Disconnect the lower radiator hose.
13. Remove the front drive axles and housing (if equipped with AWD).
14. Disconnect the alternator electrical connectors.
15. Separate the steering column coupling from the steering gear.
16. Remove the starter and spacer plate.
17. Disconnect the oil cooler hose at the oil cooler and remove the cooler hose retainer at the transmission.
18. Disconnect the transmission line bracket at the air conditioning compressor and allow the bolt to rest on the cradle.
19. Disconnect the ground strap at the right transmission housing.
20. Disconnect the engine block heater wiring connector and set aside (if equipped).
21. Remove the crankshaft position sensor.
22. Disconnect the left No.2 oxygen senor electrical connector and separate the exhaust manifold from the left exhaust pipe.
23. Disconnect the right No.2 oxygen senor electrical connector and separate the exhaust manifold from the right exhaust pipe.
24. Remove the flex plate inspection cover and torque converter bolts.
25. Remove the transmission housing to engine mounting bolts accessible.

26. Remove the engine mounting to cradle fasteners.
27. Lower the vehicle.
28. Remove the upper intake manifold.
29. Disconnect the heater hose and coolant reservoir hose from the rear coolant pipe.
30. Disconnect the oxygen sensor connector and the ground wire on the left cylinder head cover.
31. Disconnect the coolant temperature, cam position, oil pressure sensor electrical connectors.
32. Disconnect the left ignition coil and fuel injector harness connectors and position the wiring harness aside
33. Remove the right intake manifold support braces.
34. Disconnect the capacitor and ground strap from the right cylinder head cover.
35. Disconnect the oxygen sensor, knock, EGR, injector and ignition coil harness connectors and position the wiring harness aside.
36. Disconnect the engine wiring harness from the transmission housing and remove the remaining transmission housing bolts.
37. Connect the engine lifting bracket from special tool kit 8534B to the right rear of the cylinder head outer most bolt access hole.
38. Install a bolt into the inner most bolt access hole next to the engine lift bracket to assure lifting bracket positioning.
39. Connect an engine hoisting chain to the left timing chain cover engine lifting point and engine lift bracket.

**** WARNING**

While slowly separating the engine from the vehicle, constant checks must be made to assure proper positioning and that no damage to other components or wiring harnesses occur during separation.

➡As the engine is hoisted from the engine bay area, remove the loosened air conditioning compressor bolt retaining the transmission cooler lines and direct the lines aside.

40. Carefully remove the engine from the engine bay area.

To install:

**** WARNING**

Care must be taken when installing the engine to prevent pinching the power steering rack sensor with the left engine mount.

41. Install special tool 8534B, engine lifting bracket to the right rear of the cylinder head and install a bolt into the inner most bolt hole next to the bracket.
42. Connect an engine hoisting chain to the left timing chain cover lifting point and the engine lifting bracket.

**** WARNING**

Care must be taken when installing the engine to prevent pinching the power steering rack sensor with the left engine mount. Constantly check to assure proper positioning and no damage occurs to other components or harnesses until union is made.

➡As the engine is lowered into the engine bay area, install and hand tighten the air conditioning compressor bolts.

43. Carefully install the engine into the engine bay area and complete the union with the transmission.
44. Properly route the engine wiring harness behind the engine and tighten the accessible transmission bolts to 68 Nm (50 ft. lbs.).
45. Properly position the right engine harness and connect the oxygen sensor, knock sensor, EGR, fuel injector and ignition coil harness connectors.
46. Connect the capacitor and ground strap to the right cylinder head cover.
47. Install the right intake manifold support braces.
48. Properly position the left engine harness and connect the ignition coil and fuel injector harness connectors.
49. Connect the coolant temperature, cam position, oil pressure sensor harness connectors.
50. Connect the oxygen sensor, and the ground wire, to the left cylinder head cover.
51. Connect the heater hose and coolant reservoir hose to the rear coolant pipe.
52. Install the upper intake manifold.
53. Install the power steering pump. Install three pump mounting bolts through access holes in pulley and engine bracket. Tighten bolts to 28 Nm (21 ft. lbs.) torque.
54. Connect the upper radiator hose.
55. Raise and support the vehicle.
56. Install the engine mounting nuts and tighten to 101 Nm (75 ft. lbs.).
57. Install the remaining transmission housing bolts and tighten to 68 Nm (50 ft. lbs.).
58. Install the torque converter bolts. Tighten bolts to 75 Nm (55 ft. lbs.).
59. Install the torque converter inspection cover. Tighten bolts to 12 Nm (105 inch lbs.)
60. Connect the right exhaust pipe to the exhaust manifold and connect the oxygen sensor.
61. Install the front drive axle housing and front axle shafts (if equipped with AWD).
62. Connect the left exhaust pipe to the exhaust manifold and connect the oxygen sensor.
63. Install the crankshaft position sensor. Torque to 106 inch lbs. (12 Nm).
64. Connect the engine block heater (if equipped).
65. Connect the ground strap at the right transmission housing.
66. Tighten the air conditioning compressor bolts to 41 ft. lbs. (55 Nm).
67. Connect the oil cooler hose at the cooler and secure the cooler hose retainer at the transmission.
68. Connect the lower radiator hose.
69. Install the starter and spacer plate.
70. Align and connect the steering column coupler to the steering gear.
71. Connect the alternator electrical connectors.
72. Install new oil filter.
73. Install the lower engine close out panel.
74. Lower the vehicle.
75. Fill the cooling system to the proper level using the appropriate coolant.
76. Fill engine crankcase with proper oil to correct level.
77. Evacuate and recharge air conditioning.
78. Install the windshield wiper cowl assembly.
79. Install the hood.
80. Connect the negative battery cable.
81. Start engine and run until operating temperature is reached and radiator fans cycle.
82. Check all fluid levels and properly fill.

5.7L Engine

1. Before servicing the vehicle, refer to the precautions in the beginning of this section.
2. Remove the engine cover.
3. Perform the fuel system pressure release procedure.
4. Disconnect the battery negative cable.
5. Remove the air cleaner resonator and duct work as an assembly.
6. Remove the two push pins that secure the front cowl top panel to the right rear corner of the engine compartment.

7. Remove the front cowl top panel.
8. Remove the windshield wiper arms.
9. Remove the push-pin that secures each end of the cowl top panel to each front fender.
10. Disengage the two ¼-turn fasteners that secure the cowl top panel to the dash panel.
11. Remove the six push-pins that secure the cowl top panel to the strut tower support.
12. Disengage the integral retaining clips that secure the cowl top panel to the dash panel and remove the cowl panel from the engine compartment.
13. Drain cooling system.
14. Remove the accessory drive belt.
15. Remove upper radiator hose.
16. Disconnect cooling fan electrical connector.
17. Remove cooling fan mounting bolts.
18. Remove radiator cooling fan assembly from vehicle.
19. Remove the air conditioning compressor with the lines attached. Secure compressor out of the way.
20. Remove the alternator support bracket.
21. Remove alternator.
22. Remove the intake manifold and IAFM as an assembly.
23. Remove the ground wires from the rear of each cylinder head.
24. Disconnect the heater hoses.

➡It is not necessary to disconnect power steering hoses from pump, for power steering pump removal.

25. Remove the power steering pump and set aside.
26. Disconnect the fuel supply line.
27. Raise and support the vehicle on a hoist and drain the engine oil.
28. Remove the belly pan.
29. Remove engine front mount to frame nuts.
30. For AWD only: Mark the front driveshaft to the flange at both ends to ensure correct installation.
31. For AWD only: Remove front drive shaft fasteners from the differential and transfer case. Remove the driveshaft from the vehicle.
32. For AWD only: Remove the left and right front drive axles.
33. For AWD only: Remove the differential support bracket from the differential to the engine block.
34. For AWD only: Unbolt the differential from the oil pan.
35. For AWD only: Rotate the differential so the drive flange is facing forward and the oil pan side is facing up. Remove the differential out through the opening at the rear of the cradle.
36. For AWD only: Remove the intermediate shaft from the oil pan.
37. Disconnect the transmission oil cooler lines from their retainers at the oil pan bolts.
38. Disconnect exhaust pipe at manifolds.
39. Disconnect the starter wires. Remove starter motor.
40. Remove the torque converter access cover.
41. Remove drive plate to converter bolts.

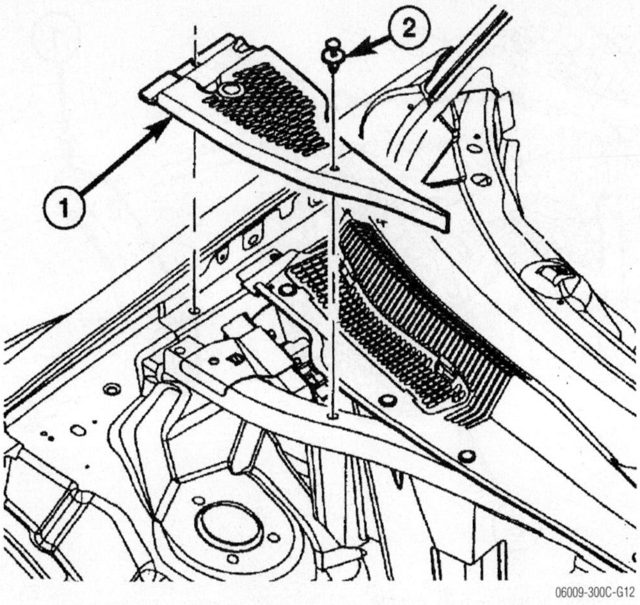

Remove the two push pins (2) that secure the front cowl top panel (1) to the right rear corner of the engine compartment

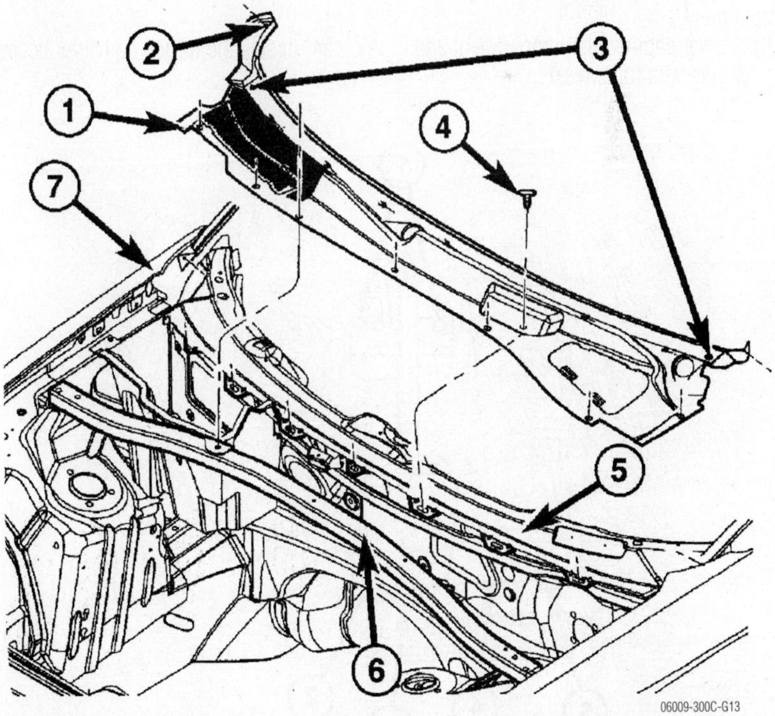

Remove the push-pin (2) that secures each end of the cowl top panel (1) to each front fender (7). Disengage the two ¼-turn fasteners that secure the cowl top panel to the dash panel. Remove the six push-pins (4) that secure the cowl top panel to the strut tower support (6)

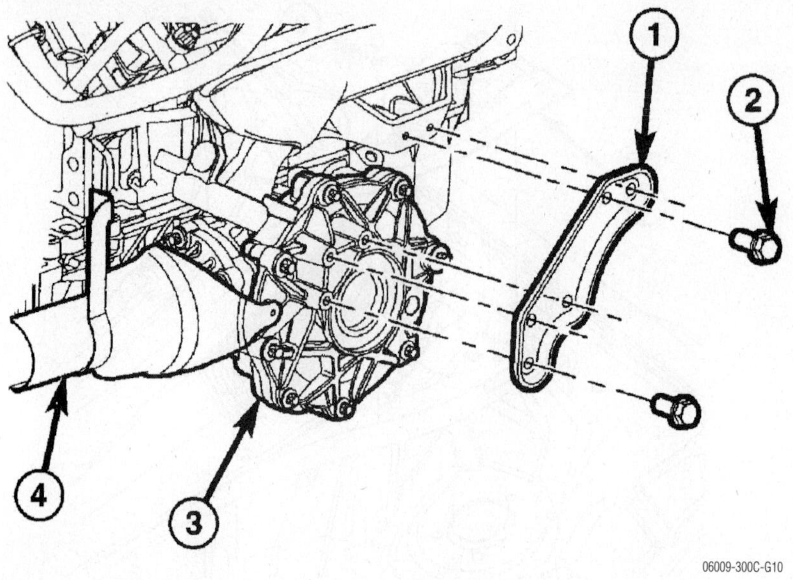

Differential support bracket—5.7L engine

42. Remove transmission bellhousing to engine block bolts.
43. Lower the vehicle.
44. Install engine lift fixture, special tool 8984 and 8984-UPD.
45. Separate engine from transmission, remove engine from vehicle, and install engine assembly on a repair stand.

To install:

46. Install engine lift fixture Special tool 8984 and 8984-UPD.
47. Position the engine in the engine compartment.
48. Lower engine into compartment and align engine with transmission.
49. Mate engine and transmission and install two transmission-to-engine block mounting bolts finger tight.
50. Lower engine assembly until engine mount studs rest in frame perches.
51. Install remaining transmission to engine block mounting bolts and tighten to 29 ft. lbs. (39 Nm).
52. Install and tighten engine mount to frame nuts.
53. Install drive plate to torque converter bolts. Torque to 31 ft. lbs. (42 Nm).
54. Install the torque converter access cover.

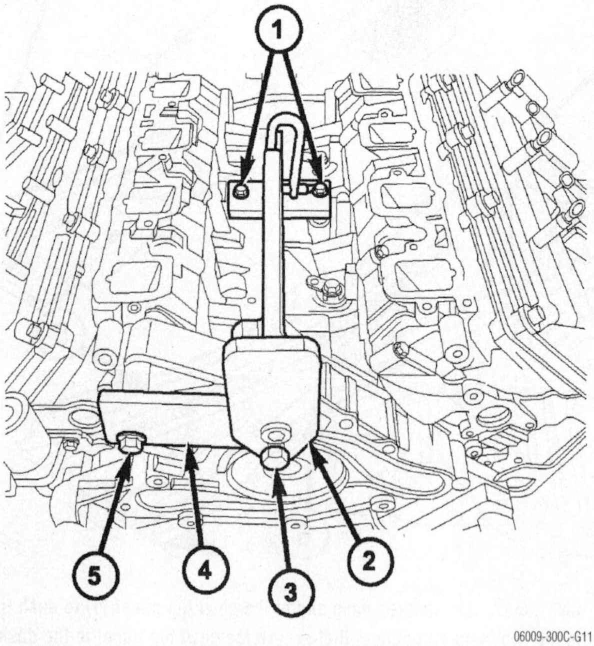

Install engine lift fixture (2), special tool 8984 and 8984-UPD (4). (1,3,5 are bolts)

55. Install the starter and connect the starter wires.
56. For AWD only: Install the intermediate shaft to the oil pan. Torque fasteners to 28 Nm (21 ft. lbs).
57. For AWD only: Install the front differential through the opening at the rear of the cradle, and attach to the oil pan.
58. For AWD only: Fasten the differential to the oil pan. Torque to 65 Nm (48 ft. lbs).
59. For AWD only: Install the differential support bracket to the differential and the engine block.
60. For AWD only: Install the left and right front drive axles.
61. For AWD only: Install front drive shaft into the differential and transfer case.
62. Install exhaust pipe to manifold.
63. Lower the vehicle.
64. Remove engine lift fixture, special tool 8984 and 8984-UPD.
65. Connect the fuel supply line.
66. Install the power steering pump. Torque to 21 ft. lbs. (28 Nm).
67. Connect the heater hoses.
68. Reconnect the ground wires to the rear of each cylinder head.
69. Install the intake manifold.
70. Install the alternator, alternator support bracket, and wire connections.
71. Install air conditioning compressor.
72. Install the accessory drive belt.
73. Position radiator cooling fan assembly in vehicle.
74. Install cooling fan mounting bolts. Tighten to 6 Nm (50 inch lbs.)
75. Connect cooling fan electrical connector.
76. Install upper radiator hose.
77. Connect the radiator lower hose.
78. Connect the transmission oil cooler lines to the radiator.
79. Connect the radiator upper hose.
80. Position the cowl top panel into the engine compartment.
81. Engage the integral retaining clips that secure the cowl top panel to the dash panel.
82. Install the six push-pins that secure the cowl top panel to the strut tower support.
83. Engage the ¼-turn fasteners that secure cowl top panel to the dash panel.
84. Install the push-pin that secures each end of the cowl top panel to each front fender.
85. Install the windshield wiper arms.
86. Position the front cowl top panel to the right rear corner of the engine compartment.
87. Install the two push-pins that secure the front cowl top panel.

88. Install the air cleaner resonator and duct work.
89. Add engine oil to crankcase.
90. Fill cooling system.
91. Install the engine cover.
92. Connect battery negative cable.
93. Start engine and inspect for leaks.
94. Install the belly pan.
95. Road test vehicle.

Water Pump

REMOVAL & INSTALLATION

2.7L Engine

1. Before servicing the vehicle, refer to the precautions in the beginning of this section.

⁂ CAUTION

Do not remove pressure cap with the system hot and pressurized. Serious burns from coolant can result.

2. Drain cooling system.
3. Disconnect negative battery cable.
4. Remove upper radiator hose.
5. Disconnect cooling fan electrical connector.
6. Remove cooling fan mounting bolts.
7. Remove radiator cooling fan assembly from vehicle.
8. Remove radiator fan assembly.
9. Remove accessory drive belt.

➡The water pump is driven by the primary timing chain.

10. Remove the timing chain and all chain guides.
11. Remove bolts attaching water pump to block.
12. Remove water pump and gasket.

To install

13. Clean all sealing surfaces.
14. Install water pump and gasket. Tighten mounting bolts to 12 Nm (105 inch lbs.).
15. Install timing chain guides and timing chain.
16. Install the accessory drive belts.
17. Position radiator cooling fan assembly in vehicle.
18. Install cooling fan mounting bolts. Tighten to 6 Nm (50 inch lbs.)
19. Connect cooling fan electrical connector.
20. Install upper radiator hose.
21. Fill cooling system.
22. Operate engine until it reaches normal operating temperature. Check cooling system and automatic transmission for correct fluid levels.

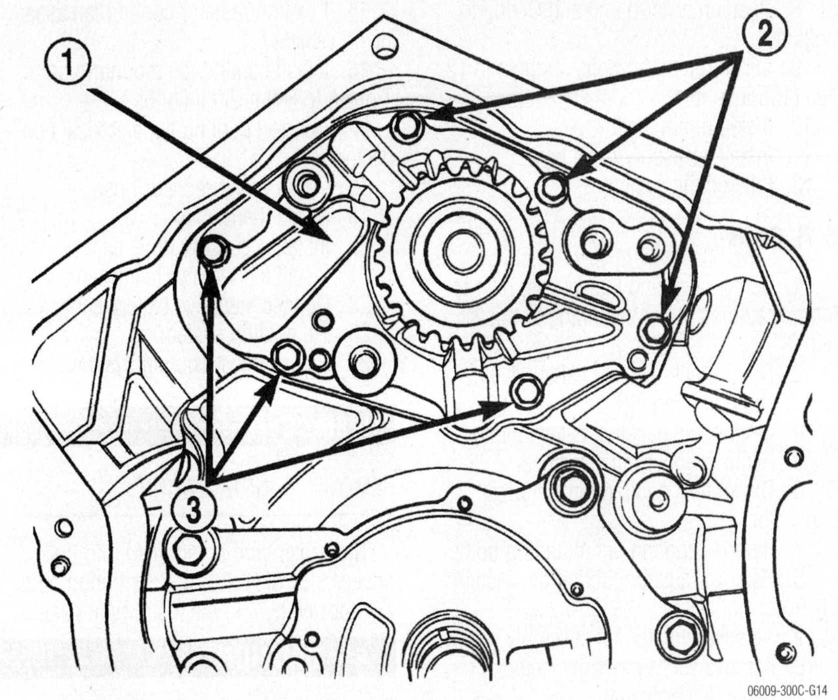

Remove bolts (2 and 3) attaching water pump (1) to block—2.7L engine

3.5L Engine

1. Before servicing the vehicle, refer to the precautions in the beginning of this section.

⁂ CAUTION

Do not remove pressure cap with the system hot and under pressure because serious burns from coolant can occur.

2. Drain cooling system.
3. Remove accessory drive belts.

➡The water pump is driven by the timing belt.

4. Remove engine timing belt.
5. Remove water pump mounting bolts. Note position of longer bolts for proper installation.
6. Remove water pump body from engine.

To install:

7. Clean all O-ring surfaces on front cover.

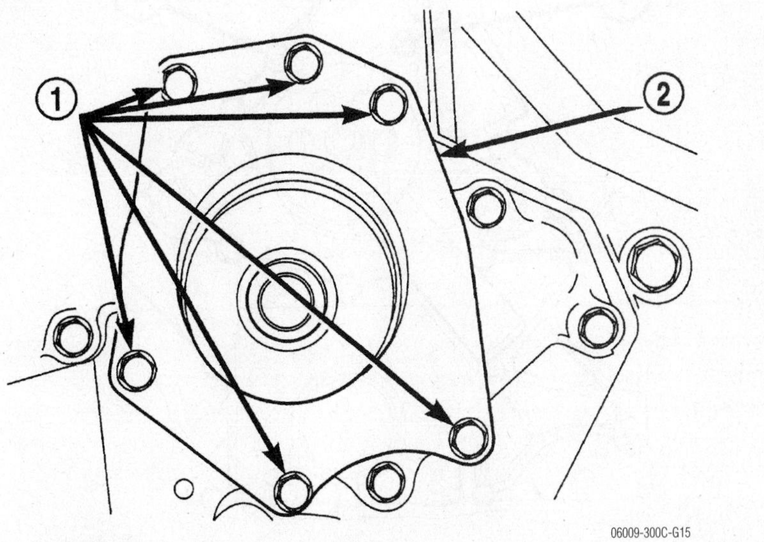

Water pump mounting. (1) bolts, (2) pump—3.5L engine

8. Position water pump and O-ring to engine.
9. Install mounting bolts. Tighten to 12 Nm (105 inch lbs.).
10. Install timing belt.
11. Install accessory drive belts.
12. Fill cooling system.

5.7L Engine

1. Before servicing the vehicle, refer to the precautions in the beginning of this section.
2. Disconnect negative battery cable.
3. Drain cooling system.
4. Disconnect negative battery cable.
5. Remove upper radiator hose.
6. Disconnect cooling fan electrical connector.
7. Remove cooling fan mounting bolts.
8. Remove radiator cooling fan assembly from vehicle.
9. Remove radiator fan assembly.
10. Remove accessory drive belt.
11. Remove thermostat.

➡The water pump mounting bolts are different lengths. Note the location of the water pump mounting bolts.

12. Remove water pump mounting bolts and remove water pump.

To install:

13. Install water pump and mounting bolts. Tighten mounting bolts to 28 Nm (20 ft. lbs.).
14. Make sure double ended bolt is in the proper location. Tighten double ended bolt to 28 Nm (20 ft. lbs.).
15. Position radiator cooling fan assembly in vehicle.
16. Install cooling fan mounting bolts. Tighten to 6 Nm (50 inch lbs.)
17. Connect cooling fan electrical connector.
18. Install upper radiator hose.
19. Install thermostat.
20. Install accessory drive belt.
21. Install the radiator fan assembly.
22. Connect negative battery cable.
23. Fill cooling system.
24. Pressure test cooling system.

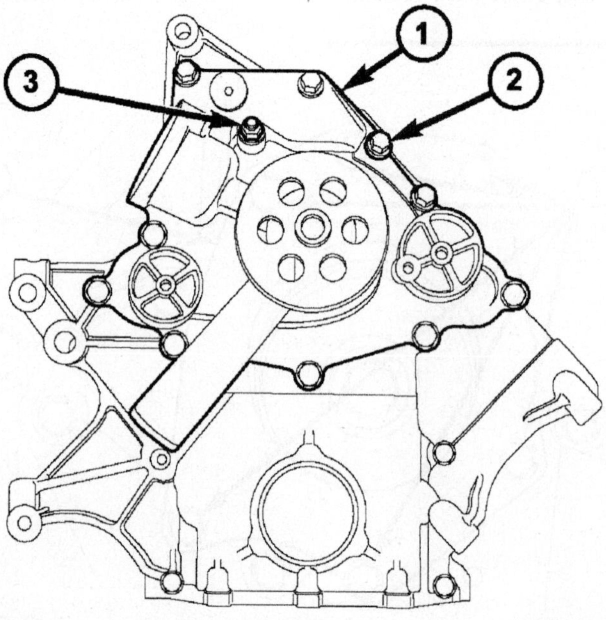

Water pump mounting (1) pump, (2) mounting bolts, (3) stud bolt

Heater Core

REMOVAL & INSTALLATION

The heater core is mounted into the driver's side of HVAC air distribution housing, located behind the instrument panel.

✲✲ CAUTION

On vehicles equipped with airbags, disable the airbag system before attempting any steering wheel, steering column, or instrument panel component diagnosis or service. Disconnect and isolate the negative battery (ground) cable, then wait two minutes for the airbag system capacitor to discharge before performing further diagnosis or service. This is the only sure way to disable the airbag system. Failure to take the proper precautions could result in accidental airbag deployment and possible personal injury or death.

1. Before servicing the vehicle, refer to the precautions in the beginning of this section.
2. Drain the engine cooling system
3. Disconnect and isolate the negative battery cable.
4. Carefully disconnect the heater hoses from the heater core tubes.
5. Remove the two push-pins (4) that secure the instrument panel silencer (1) to the instrument panel bracket (2).
6. Pull the instrument panel silencer rearward to disengage it from the brackets (3) located near the dash panel.
7. Remove the instrument panel silencer from the vehicle.
8. Remove the blend door actuator from the driver's side of the HVAC air distribution housing.
9. Remove the two screws that secure the flange to the front of the HVAC housing near the dash panel.
10. Remove the flange from the HVAC housing.

➡Take proper precautions to protect the carpeting from spilled engine coolant. Have absorbent toweling readily available to clean up any spills.

11. Remove the retaining clamps that secure the heater core tubes to the heater core.
12. Disconnect the heater core tubes from the heater core and remove and discard the O-ring seals.
13. Carefully pull the heater core tubes through the dash panel.
14. Install plugs in, or tape over the opened heater core ports.
15. Remove the screw that secures the heater core retaining bracket to the driver's side of the HVAC air distribution housing.
16. Remove the heater core retaining bracket from the air distribution housing.
17. Carefully pull the heater core out of the air distribution housing.

To install:

18. Carefully install the heater core into the driver's side of the HVAC air distribution housing.
19. Install heater core retaining bracket onto the air distribution housing.
20. Install the screw that secures the heater core retaining bracket onto the air distribution housing. Tighten the screw to 2.2 Nm (20 inch lbs.).
21. Remove the tape or plugs from the heater core ports.
22. Lubricate new rubber O-ring seals

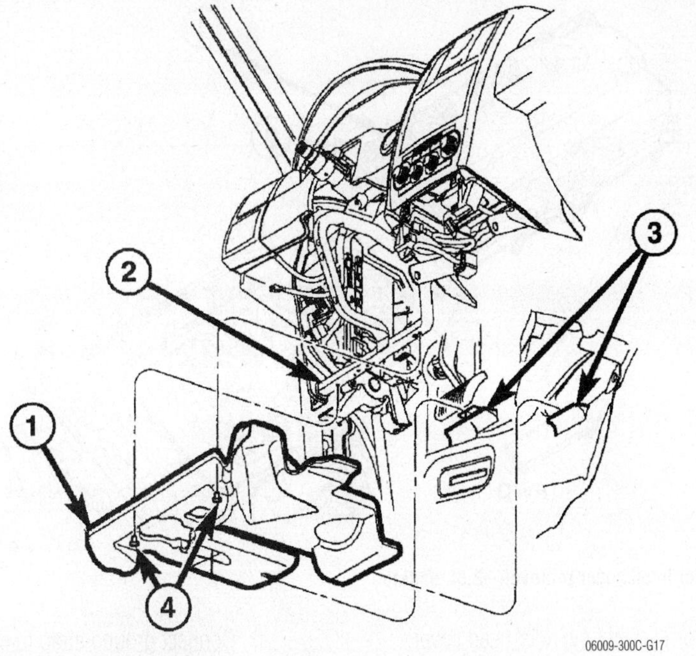

Driver's side instrument panel silencer

with clean engine coolant and install them onto the heater core tube fittings. Use only the specified O-rings as they are made of a special material for the engine cooling system.

23. Install the heater core tubes through the dash panel and onto the heater core.

24. Install the two retaining clamps that secure the heater core tubes to the heater core. Make sure that the clamps are installed correctly and securely.

25. Install the flange over the heater core tubes and onto the HVAC housing near the dash panel.

26. Install the two screws that secure the flange to the HVAC housing. Tighten the screws to 2.2 Nm (20 inch lbs.).

27. Install the blend door actuator to the driver's side of the air distribution housing.

28. Position the instrument panel silencer into the vehicle.

29. Install the instrument panel silencer above the brackets located near the dash panel.

30. Install the two push-pins that secure the instrument panel silencer to the instrument panel bracket.

31. Connect the heater hoses to the heater core tubes.

32. Connect the negative battery cable.

33. If the heater core is being replaced, flush the cooling system.

34. Refill the engine cooling system.

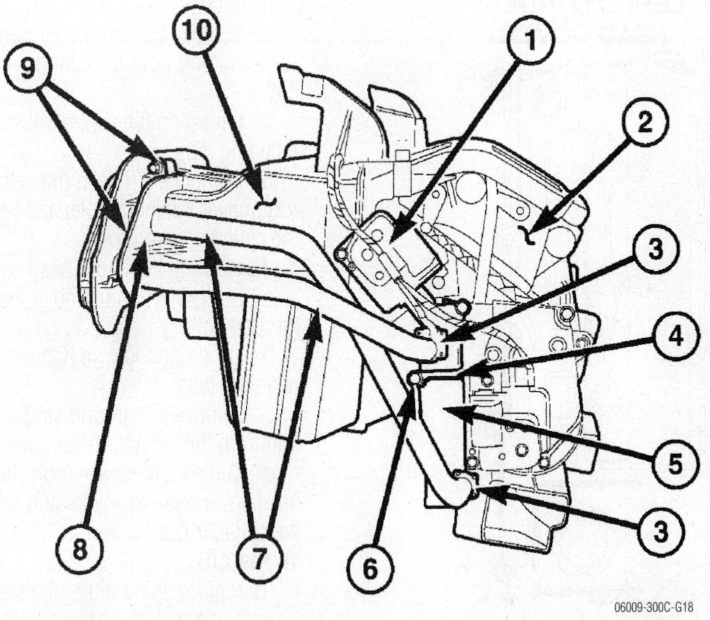

HVAC unit: (1) blend door actuator, (2) air distribution unit, (3) heater core tube retaining clamps, (4) heater core retaining bracket, (5) heater core, (6) screw, (7) heater core tubes, (8) flange, (9) screws (10) HVAC unit

Cylinder Head Covers

REMOVAL & INSTALLATION

2.7L Engine

1. Before servicing the vehicle, refer to the precautions in the beginning of this section.
2. For the left side cylinder head cover:
   a. Disconnect electrical connectors from ignition coils.
   b. Remove ground strap from cylinder head cover stud and disconnect capacitor connector. Reposition electrical harness.
   c. Disconnect electrical harness retaining clips from cylinder head cover studs. Reposition electrical harness.
   d. Remove make up air hose.
   e. Remove fastener attaching ignition coil capacitor.
   f. Remove ignition coils.
3. Loosen all cylinder head cover fasteners.

➡Cylinder head cover attaching bolts are captured to the cover.

⁂ WARNING

Make certain the double ended studs in the center of the cylinder head cover are loose before attempting to remove cover.

   a. Remove cylinder head cover.
4. For the right side cylinder head cover:
   a. Disconnect electrical connectors from ignition coils.
   b. Disconnect capacitor electrical connector.
   c. Remove PCV hose from cylinder head cover grommet.
   d. Remove ground strap from cylinder head cover stud.
   e. Disconnect electrical harness retaining clips from cylinder head cover studs. Reposition electrical harness.
   f. Remove fastener attaching ignition coil capacitor.

g. Remove ignition coils.
h. Remove foam insulator.
5. Loosen all cylinder head cover fasteners.

➡Cylinder head cover attaching bolts are captured to the cover.

**** WARNING**

Make certain the double ended studs in the center of the cylinder head cover are loose before attempting to remove cover.

6. Remove cylinder head cover.

To install:

7. For the left side cylinder head cover:
a. Clean cylinder head cover and both sealing surfaces. Inspect and replace gaskets as necessary.
b. Install cylinder head cover and hand start all fasteners. Verify that all double-ended studs are in the correct locations.
c. Tighten cylinder head cover attaching bolts and double-ended studs to 12 Nm (105 inch lbs.).
d. Install ignition coils.
e. Install ignition coil capacitor and fastener.
f. Install ground strap to cylinder head cover stud.
g. Connect all electrical connectors and harness clips.
h. Connect make up air hose.

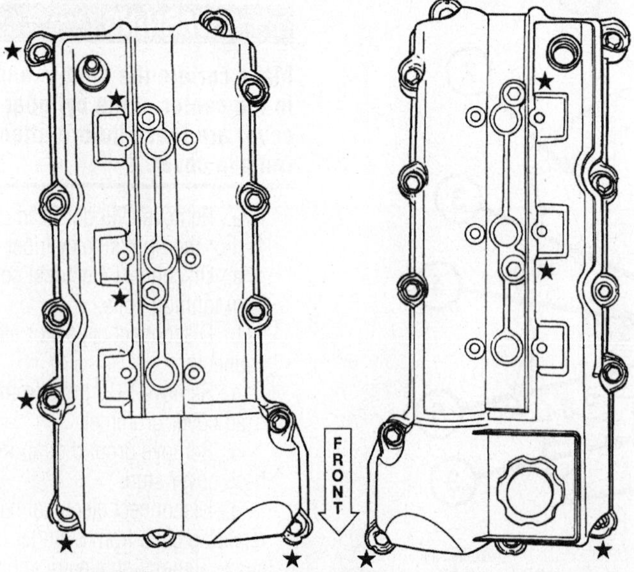

Cylinder head cover installation—2.7L engine

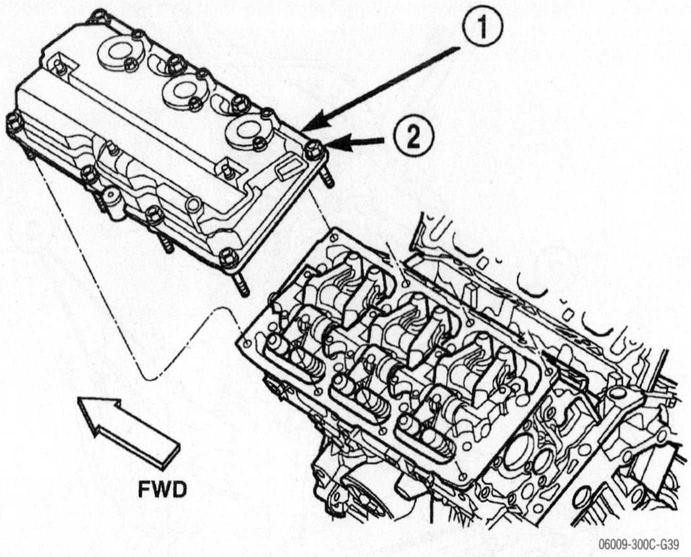

Cylinder head cover removal—3.5L engine

8. For the right cylinder head cover:
a. Clean cylinder head cover and both sealing surfaces. Inspect and replace gaskets as necessary.
b. Install cylinder head cover and hand start all fasteners. Verify that all double-ended studs are in the correct locations.
c. Tighten cylinder head cover attaching bolts and double-ended studs to 12 Nm (105 inch lbs.).
d. Install ignition coils.
e. Install ignition coil capacitor and fastener.
f. Connect ground strap to cylinder head cover stud.
g. Connect PCV hose to cylinder head cover grommet.
h. Connect all electrical connectors and harness clips.
i. Install foam insulator on top of cylinder head cover.

3.5L Engine

1. Before servicing the vehicle, refer to the precautions in the beginning of this section.
2. For the left cylinder head cover:
a. Disconnect and remove the three ignition coils.
b. Remove the ground strap/resistor retaining bolt from the cylinder head cover.
c. Lift up on the wire harness track retaining tabs.
d. Completely loosen the cylinder head cover retaining bolts and remove the cylinder head cover.
3. For the right cylinder head cover:
a. Disconnect and remove the three ignition coils.
b. Lift up on the wire harness track retaining tabs.
c. Remove the ground strap retaining bolt from the cylinder head cover.
d. Completely loosen the cylinder head cover retaining bolts and remove the cylinder head cover

To install:

4. To replace spark plug tube seals:
a. Using a suitable pry tool, carefully remove tube seals .
b. Position new seal with the part

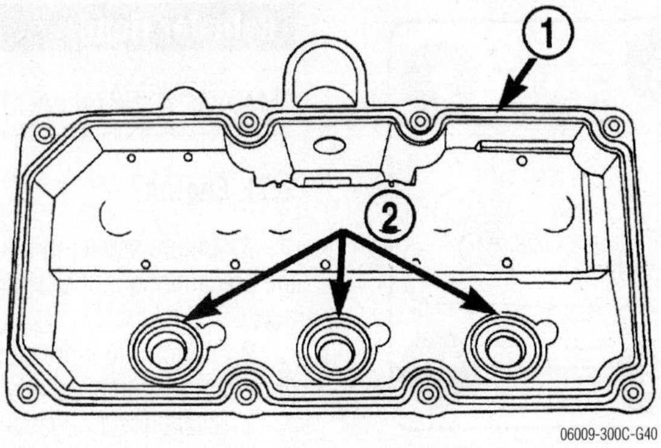

Cylinder head cover gasket (1) and seals (2)—3.5L engine

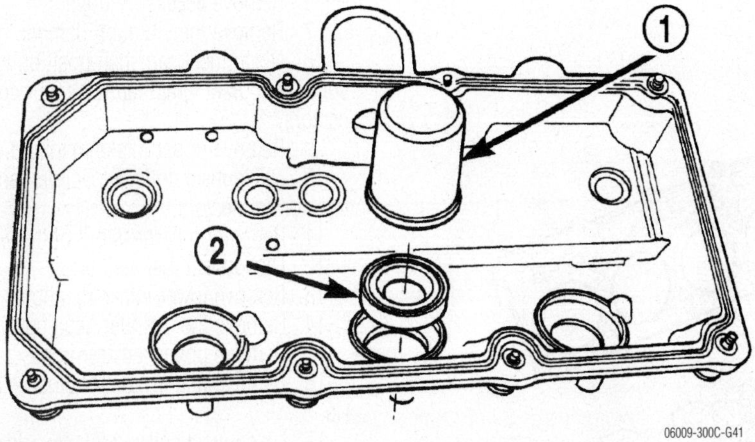

Spark plug tube seal (2) replacement using tool MD-998306 (1)—3.5L engine

number on seal facing cylinder head cover.

c. Install seals using Special Tool MD-998306.

5. For the right cylinder head cover:

a. Clean cylinder head and cover mating surfaces. Inspect and replace gasket and seals as necessary.

b. Install cylinder head cover bolts and tighten to 12 Nm (105 inch lbs.).

c. Install the ground strap retaining bolt to the cylinder head cover.

d. Install the wire harness track.

e. Install the ignition coils. Tighten mounting screws to 6.7 Nm (60 inch lbs.).

f. Connect the ignition coil electrical connectors.

6. For the left cylinder head cover:

a. Clean cylinder head and cover mating surfaces. Inspect and replace gasket and seals as necessary.

b. Install cylinder head cover and bolts. Tighten to 12 Nm (105 inch lbs.).

c. Position the wiring harness on the cylinder head cover.

d. Re-clip the wire harness track retaining tabs into the cover.

e. Install the ground strap/resistor retaining bolt onto the cylinder head cover.

f. Install the ignition coils. Tighten mounting screws to 6.7 Nm (60 inch lbs.)

5.7L Engine

1. Before servicing the vehicle, refer to the precautions in the beginning of this section.

2. For the right side cylinder head cover:

a. Disconnect electrical connectors from ignition coils.

b. Disconnect capacitor electrical connector.

c. Remove PCV hose from cylinder head cover grommet.

d. Remove ground strap from cylinder head cover stud.

e. Disconnect electrical harness retaining clips from cylinder head cover studs. Reposition electrical harness.

f. Remove fastener attaching ignition coil capacitor.

g. Remove ignition coils.

h. Remove foam insulator.

3. Loosen all cylinder head cover fasteners.

➡**Cylinder head cover attaching bolts are captured to the cover.**

⁂ WARNING

Make certain the double ended studs in the center of the cylinder head cover are loose before attempting to remove cover.

4. Remove cylinder head cover.

To install:

➡**The gasket may be used again, provided no cuts, tears, or deformation has occurred.**

⁂ WARNING

Do not use harsh cleaners to clean the cylinder head covers. Severe damage to covers may occur.

⁂ WARNING

DO NOT allow other components including the wire harness to rest on or against the engine cylinder head cover. Prolonged contact with other objects may wear a hole in the cylinder head cover.

5. Clean cylinder head cover and both sealing surfaces. Inspect and replace gasket as necessary.

6. Install cylinder head cover and hand start all fasteners. Verify that all double ended studs are in the correct location and install left and right ground straps.

➡**The ground straps must be installed in the same location as removed. The covers are machined to accept the ground straps in those locations only. The right hand ground strap is located on the front inboard stud. The left hand ground strap is located on the rear inboard stud.**

7. Tighten cylinder head cover bolts and double ended studs to 8 Nm (70 inch lbs). Begin torque sequence in the middle of head cover and torque bolts moving outward in a crisscross pattern from top to bottom.

8. Install ignition coil on plug, and torque fasteners to 12 Nm (105 inch lbs)

9. Connect, ignition coil electrical connectors.

10. Install PCV hose.

11. Connect battery negative cable.

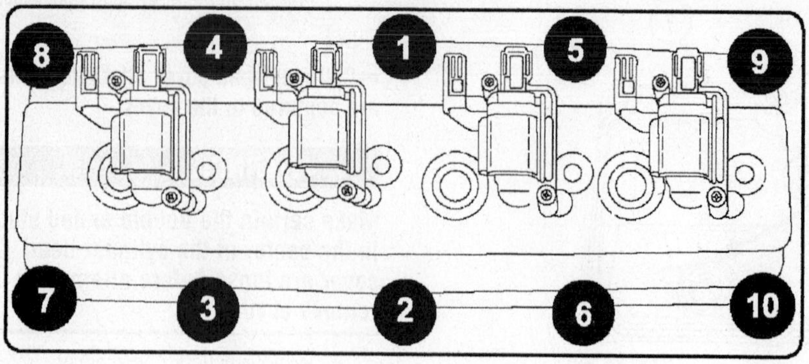

Cylinder head cover bolt loosening/tightening sequence—5.7L engine

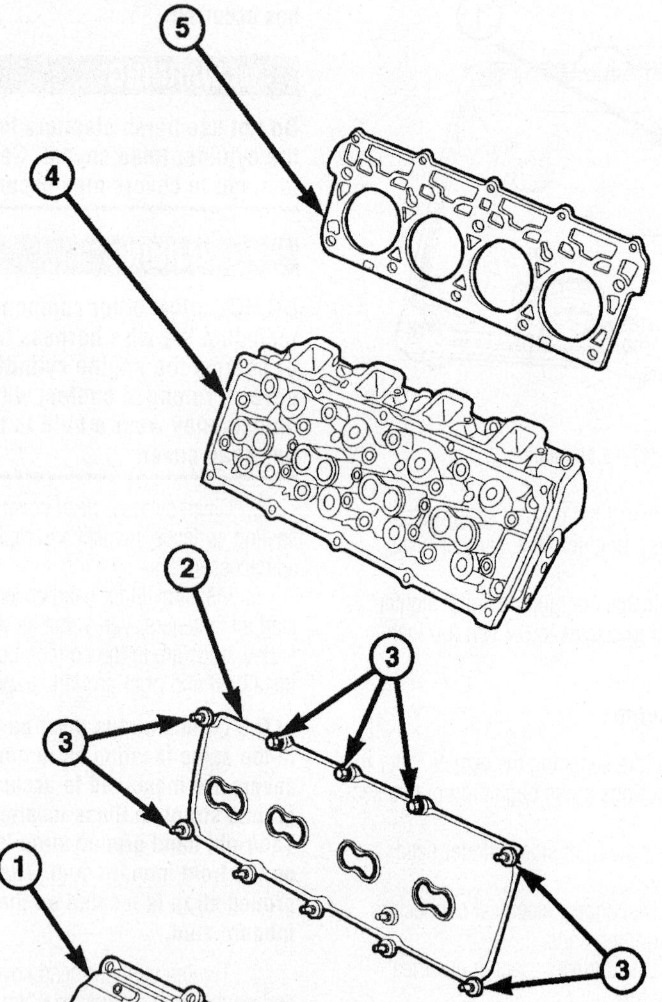

Cylinder head cover (1), cover gasket (2), studs (3), cylinder head (4), head gasket (5)—5.7L engine

Cylinder Head

REMOVAL & INSTALLATION

2.7L Engine

1. Before servicing the vehicle, refer to the precautions in the beginning of this section.
2. Perform fuel pressure release procedure before attempting any repairs.
3. Disconnect negative cable from battery.
4. Raise vehicle on hoist.
5. Drain cooling system.
6. Remove accessory drive belt.
7. Remove the vibration damper.
8. Disconnect camshaft position sensor, and coolant temperature sensor connectors.
9. Remove upper intake manifold.
10. Disconnect coils, capacitors, and injector connectors.
11. Reposition harness out of the way.
12. Disconnect fuel feed line.
13. Remove lower intake manifold.
14. Remove the cylinder head cover.
15. Remove upper radiator hose.
16. Remove upper radiator closure panels.
17. Disconnect cooling fan electrical connector.
18. Remove cooling fan mounting bolts.
19. Remove radiator cooling fan assembly from vehicle.
20. Raise vehicle.
21. Remove lower splash shield.
22. Remove lower radiator hose.
23. Remove lower condenser mount bolts.
24. Lower vehicle.
25. Remove upper radiator mounting brackets and bolts.
26. Remove upper condenser mounting bolts.
27. Separate condenser assembly from radiator.
28. Tilt radiator toward engine and remove radiator from vehicle.
29. Disconnect engine coolant temperature sensor connector.
30. Remove radiator upper hose at tube.
31. Remove heater hose from heater tube at rear of engine.
32. Disconnect heater tube from retaining clip at rear of engine.
33. Disconnect electrical connector from coolant temperature sensor.
34. Remove screws attaching heater tube to outlet connector.
35. Disengage heater tube from outlet

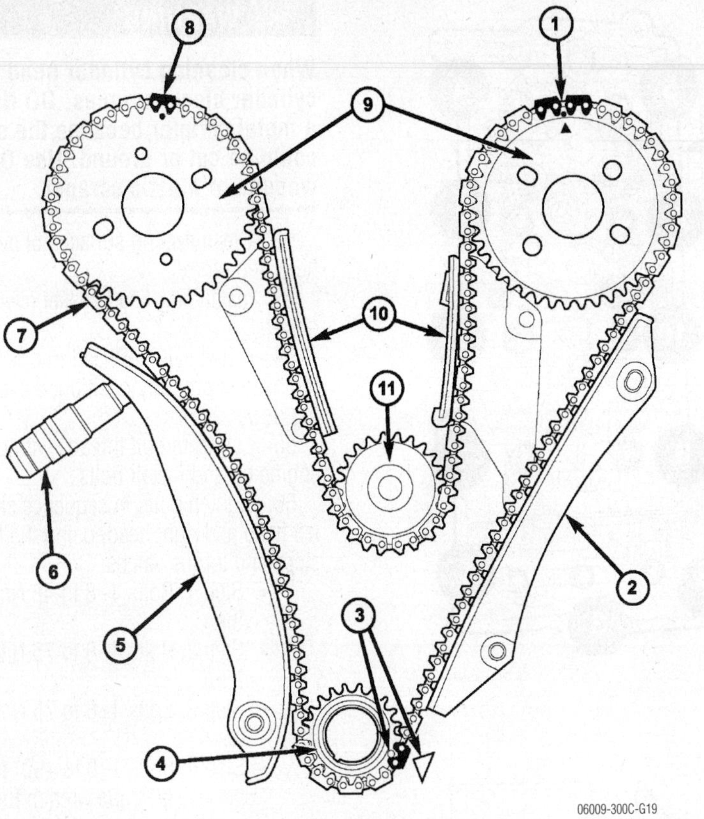

Rotate crankshaft until crankshaft sprocket timing mark aligns with timing mark on oil pump housing—2.7L engine

connector. To remove heater tube, move forward until the tube clears cylinder heads.

36. Remove coolant outlet.
37. Remove timing chain cover.
38. Rotate crankshaft until crankshaft sprocket timing mark aligns with timing mark on oil pump housing.
39. Remove primary timing chain.
40. Remove upper primary timing chain guides.
41. Remove camshaft bearing caps gradually in the sequence shown.
42. Remove camshafts and valvetrain components from cylinder head. Note component locations for reinstallation in original locations.
43. For left cylinder head removal:
    a. Remove fastener securing engine oil dipstick tube to cylinder head. Remove engine oil dipstick tube.
    b. Remove alternator.
44. For right cylinder head removal:
    a. Remove cylinder head ground strap.
    b. Disconnect EGR valve electrical connector and remove EGR valve from head.

✲✲ WARNING

Ensure cylinder head bolts 1–3 are removed before attempting the removal of cylinder head, as damage to cylinder head and/or block may occur.

45. Remove cylinder left head bolts and right head bolts in sequence shown.
46. Remove cylinder head(s).
47. Remove and discard cylinder head gasket.

To install:

To ensure engine gasket sealing, proper surface preparation must be performed, especially with the use of aluminum engine components and multi-layer steel cylinder head gaskets.

➡Multi-Layer Steel (MLS) head gaskets require a scratch free sealing surface. Remove all gasket material from cylinder head and block. Be careful not to gouge or scratch the aluminum head sealing surface. Clean all engine oil passages.

48. Before cleaning, check for leaks, damage and cracks.
49. Clean cylinder head and oil passages.
50. Check cylinder head for flatness.
51. Cylinder head must be flat within:
    - Standard dimension = less than 0.05 mm (0.002 inch.)
    - Service Limit = 0.2 mm (0.008 inch.)
    - Grinding Limit = Maximum of 0.2 mm (0.008 inch.) is permitted.

✲✲ WARNING

0.20 mm (0.008 in.) MAX is a combined total dimension of the stock removal limit from cylinder head and block top surface (deck) together.

➡The cylinder head bolts are tightened using a torque plus angle procedure. The bolts must be examined BEFORE reuse. If the threads are necked down the bolts must be replaced

Necking can be checked by holding a straight-edge against the threads. If all the threads do not contact the scale, the bolt must be replaced.

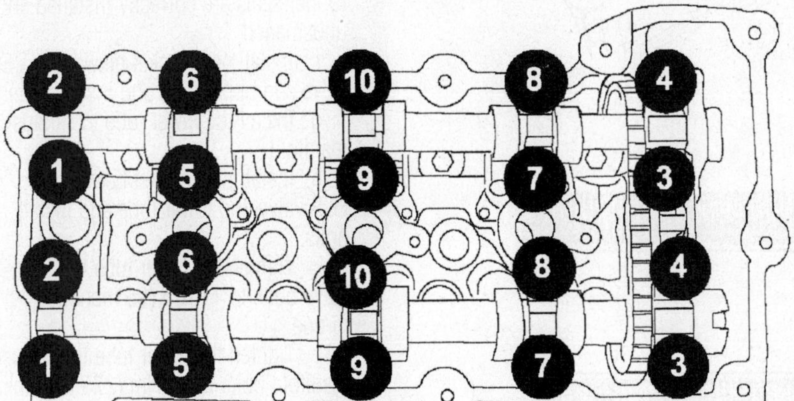

Camshaft bearing cap loosening sequence—2.7L engine

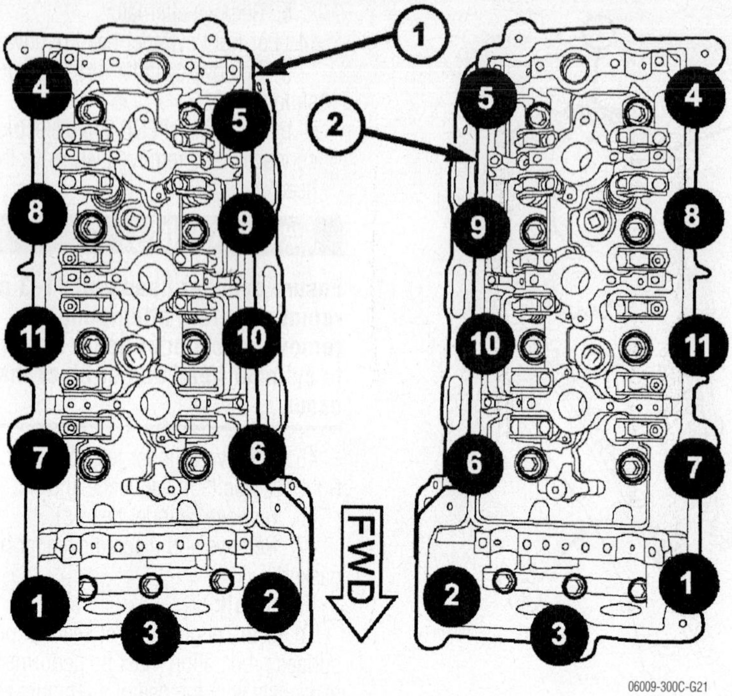

Cylinder head bolt loosening sequence—2.7L engine

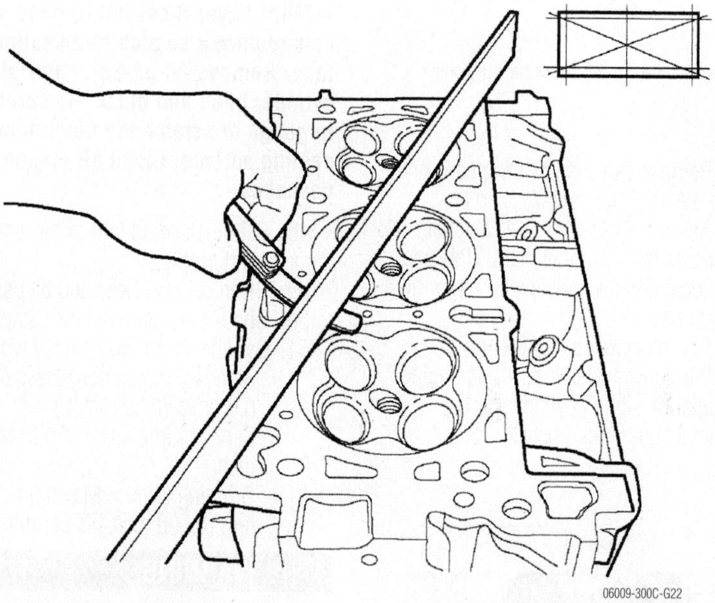

Check cylinder head for flatness

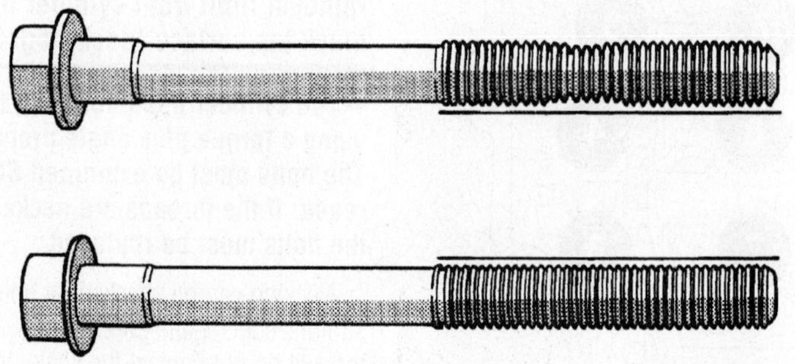

Check the head bolts for necking

** WARNING

When cleaning cylinder head and cylinder block surfaces, DO NOT use a metal scraper because the surfaces could be cut or ground. Use ONLY a wooden or plastic scraper.

52. Clean sealing surfaces of cylinder head and block.
53. Install new head gasket over locating dowels.
54. Install cylinder head to block, assuring head is properly positioned over locating dowels.
55. Lubricate bolt threads with clean engine oil and install bolts.
56. Tighten bolts in sequence shown for left head and right head, using the following steps and torque values:
    - Step 1: Bolts 1–8 to 48 Nm (35 ft. lbs.)
    - Step 2: Bolts 1–8 to 75 Nm (55 ft. lbs.)
    - Step 3: Bolts 1–8 to 75 Nm (55 ft. lbs.)
    - Step 4: Bolts 1–8 to +90° turn. Do not use a torque wrench for this step.
    - Step 5: Bolts 9–11 to 28 Nm (250 inch lbs.)
57. For left cylinder head installation:
    - Install engine oil dipstick tube.
    - Install alternator.
58. For right cylinder head installation:
    a. Install cylinder head ground strap.
    b. Clean mounting surface and install EGR valve.
    c. Install the EGR valve mounting bolts.
    d. Inspect rubber silicone seals on intake manifold end of EGR tube.
    e. Install upper tube into the intake manifold, being careful that the silicone rubber seals are correctly installed and undamaged..
    f. Install new gasket between the EGR valve and upper tube and install bolts.
    g. Install the lower tube to exhaust manifold.
    h. Install new gasket between the EGR valve and lower tube and install bolts.
    i. Tighten the lower tube to EGR valve bolts to 11 Nm (95 inch lbs.) torque.
    j. Tighten the lower tube to exhaust manifold bolts to 31 Nm (275 inch lbs.) torque.
    k. Tighten the upper tube to EGR valve bolts to 11 Nm (95 inch lbs.) torque.

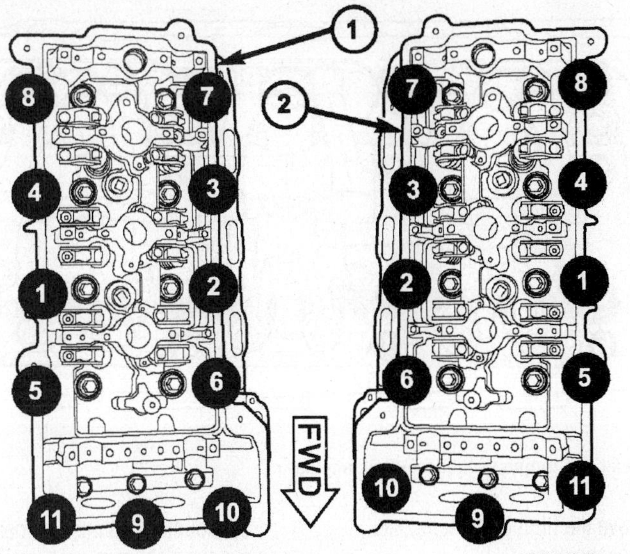

Cylinder head bolt torque sequence—2.7L engine

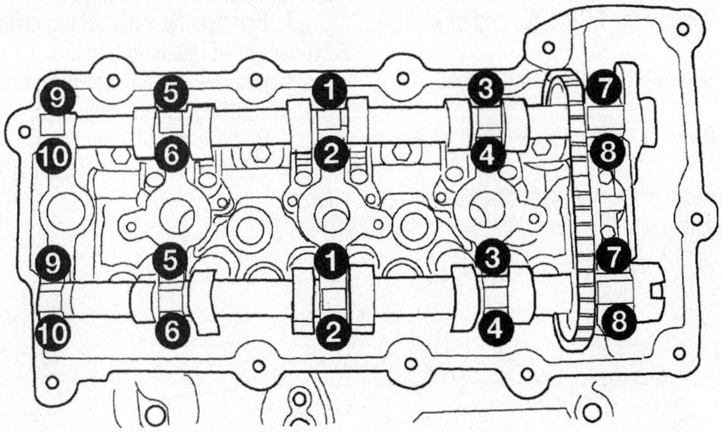

Camshaft bearing cap bolt torque sequence—2.7L engine

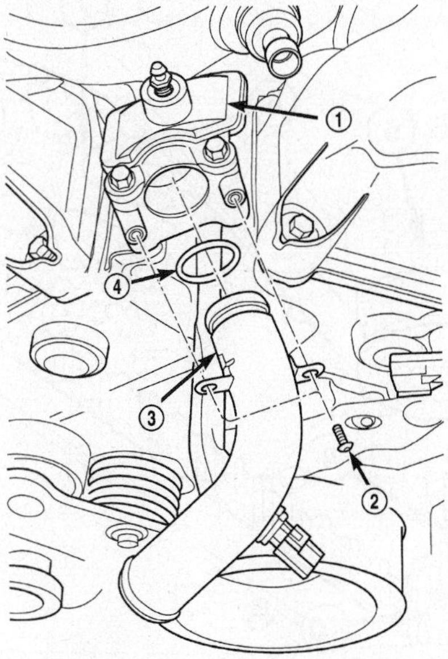

Coolant outlet installation—2.7L engine

l. Tighten EGR valve to cylinder head bolts to 31 Nm (275 inch lbs.) torque.

59. Install all valve train components and camshafts. Tighten camshaft bearing caps in sequence shown to 12 Nm (105 inch lbs.).
60. Install primary timing chain, guides and sprockets.
61. Inspect heater tube O-ring. Replace as necessary.
62. Lubricate O-ring with silicone type grease such as Mopar® dielectric grease.
63. Install the heater tube by inserting tube in between cylinder heads. Insert tube into outlet connector.
64. Attach heater tube to the retaining clip at rear of engine.
65. Install coolant outlet. Install attaching screws and tighten to 3 Nm (30 inch lbs.).
66. Position radiator into engine compartment. Seat the radiator assembly lower rubber isolators into the mounting holes in radiator lower support.
67. Install radiator mounting bracket and bolts. Tighten to 12 Nm (106 inch lbs.).
68. Position condenser on radiator and install upper mounting bolts. Tighten bolts to 6 Nm (50 inch lbs.).
69. Raise vehicle.
70. Install lower condenser mounting bolts. Tighten bolts to 10 Nm (88 inch lbs.)
71. Install lower radiator hose and clamp.
72. Lower vehicle.
73. Position radiator cooling fan assembly in vehicle.
74. Install cooling fan mounting bolts. Tighten to 6 Nm (50 inch lbs.)
75. Connect cooling fan electrical connector.
76. Install upper radiator upper hose. Align hose so it does not interfere with the accessory drive belt or engine. Position hose clamp so it will not interfere with the hood.
77. Install lower intake manifold.
78. Install the cylinder head cover.
79. Connect camshaft position sensor and coolant temperature sensor connectors.
80. Install timing chain cover.
81. Install crankshaft vibration damper.
82. Install upper intake manifold.
83. Connect oil pressure sensor connector.
84. Install accessory drive belt.
85. Fill cooling system.
86. Connect negative battery cable.
87. Start engine and check for leaks.

3.5L Engine

RIGHT SIDE

1. Before servicing the vehicle, refer to the precautions in the beginning of this section.
2. Perform the fuel relief procedure.
3. Disconnect the negative battery cable.
4. Drain cooling system.
5. Remove the upper intake manifold.
6. Remove the lower intake manifold.
7. Remove accessory drive belt.
8. Remove accessory drive belt idler pulley.
9. Remove the power steering mounting bolts and set the pump aside.
10. Raise the vehicle.
11. Remove crankshaft damper.
12. Remove lower outer timing belt cover bolts.
13. Remove front exhaust pipe to exhaust manifold mounting nuts.
14. Disconnect both oxygen sensor harness connectors on each side.
15. Lower vehicle.
16. Remove the remaining outer timing belt cover bolts and cover.
17. Rotate the engine to TDC and align timing belt marks.

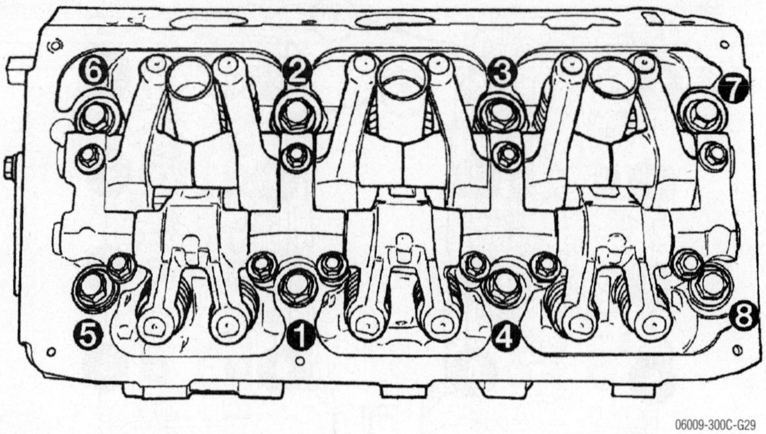

Cylinder head bolt torque sequence—3.5L engine

18. Remove the timing belt tensioner and reset the tensioner.
19. Remove the timing belt
20. Remove the right cylinder head cover to cylinder head ground strap and capacitor.
21. Remove the EGR valve and tube assembly.
22. Remove the right cylinder head cover.
23. Remove the right rocker arm assembly.
24. Remove the right rear camshaft thrust plate.
25. Counterhold the cam gear and remove the right cam gear retaining bolt.
26. Push the camshaft out of the back of the cylinder head approximately 3.5 inches and remove the right cam gear.
27. Remove the inner timing cover to cylinder head retaining bolts.
28. Remove the cylinder head bolts in REVERSE of tightening sequence.
29. Remove the cylinder head.
30. Clean and inspect all mating surfaces.

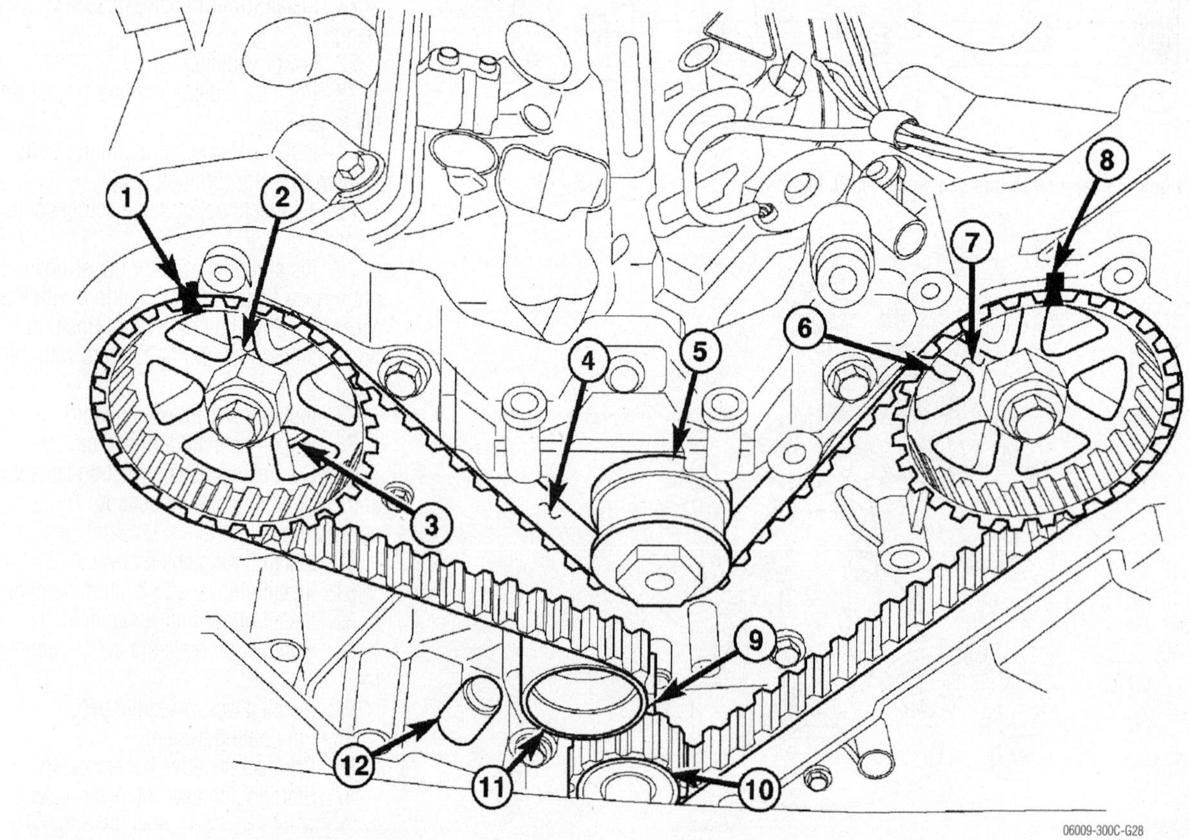

Rotate the engine to TDC and align timing belt marks (1, 8, 9)—3.5L engine

To install:

To ensure engine gasket sealing, proper surface preparation must be performed, especially with the use of aluminum engine components and multi-layer steel cylinder head gaskets.

➡Multi-Layer Steel (MLS) head gaskets require a scratch free sealing surface.

Remove all gasket material from cylinder head and block. Be careful not to gouge or scratch the aluminum head sealing surface. Clean all engine oil passages.

31. Before cleaning, check for leaks, damage and cracks.
32. Clean cylinder head and oil passages.
33. Check cylinder head for flatness.
34. Cylinder head must be flat within:
    - Standard dimension = less than 0.05 mm (0.002 inch.)
    - Service Limit = 0.2 mm (0.008 inch.)
    - Grinding Limit = Maximum of 0.2 mm (0.008 inch.) is permitted.

⁂ WARNING

0.20 mm (0.008 in.) MAX is a combined total dimension of the stock removal limit from cylinder head and block top surface (deck) together.

⁂ WARNING

The cylinder head gaskets are not interchangeable between cylinder heads and are clearly marked right or left.

➡The cylinder head bolts are tightened using a torque plus angle procedure. The bolts must be examined BEFORE reuse. If the threads are necked down the bolts must be replaced.

Necking can be checked by holding a scale or straight-edge against the threads. If all the threads do not contact the scale the bolt must be replaced.

⁂ WARNING

When cleaning cylinder head and cylinder block surfaces, DO NOT use a metal scraper because the surfaces could be cut or ground. Use ONLY a wooden or plastic scraper.

35. Clean sealing surfaces of cylinder head and block.

⁂ WARNING

Ensure that the correct head gaskets are used and are oriented correctly on cylinder block.

➡Before installing the cylinder head bolts, lubricate the threads with clean engine oil.

36. Install the cylinder head over locating dowels and finger tighten the head bolts.
37. Tighten the cylinder head bolts in the following sequence, using the 4 step torque-turn method. Tighten according to the following torque values:
    - Step 1: All to 61 Nm (45 ft. lbs.)
    - Step 2: All to 88 Nm (65 ft. lbs.)
    - Step 3: All (again) to 88 Nm (65 ft. lbs.)
    - Step 4: + 90° turn. Do not use a torque wrench for this step.
38. Bolt torque after 90° turn should be over 122 Nm (90 ft. lbs.) in the tightening direction. If not, replace the bolt.
39. Install the inner timing cover to cylinder head bolts. Tighten bolts to 54 Nm (40 ft. lbs.).
40. Install camshaft sprocket. Counterhold the camshaft sprocket gear and tighten the camshaft sprocket bolt to 102 Nm plus a 90° turn (75 lbs. ft. plus a 90° turn).
41. Install the rear camshaft thrust plate and seal and the EGR valve.
42. Rotate the camshaft gear to its alignment mark and check the left camshaft gear and crankshaft gear timing alignment marks.
43. Install the timing belt and tensioner.
44. Install the timing belt outer cover.
45. Install the power steering reservoir.
46. Install the vibration damper.
47. Install the accessory drive belt tensioner.
48. Install the accessory drive belt idler pulley.
49. Install the right exhaust manifold
50. Raise and support the vehicle.
51. Install the front exhaust pipe and connect the oxygen sensors.
52. Lower the vehicle.
53. Install the right rocker arm assembly.
54. Clean cylinder head and cover mating surfaces. Inspect and replace gasket and seals as necessary.
55. Install cylinder head cover bolts and tighten to 12 Nm (105 inch lbs.).
56. Install the ground strap retaining bolt to the cylinder head cover.
57. Install the wire harness track.
58. Install the ignition coils. Tighten mounting screws to 6.7 Nm (60 inch lbs.).
59. Connect the ignition coil electrical connectors.
60. Install the right cylinder head cover, ground strap and insulator.
61. Install lower intake manifold.
62. Install the fuel rail.
63. Install the upper intake manifold.
64. Connect the air cleaner element housing.
65. Fill the coolant system.
66. Connect the negative battery cable.

LEFT SIDE

1. Before servicing the vehicle, refer to the precautions in the beginning of this section.
2. Perform the fuel relief procedure.
3. Disconnect the negative battery cable.
4. Drain cooling system.
5. Remove the upper intake manifold.
6. Remove the lower intake manifold
7. Remove the accessory drive belt.
8. Remove the belt tensioner.
9. Remove the accessory drive idler pulley.
10. Remove the power steering mounting bolts and set pump aside.
11. Remove the crankshaft damper.
12. Remove the lower outer timing belt cover.
13. Raise and support the vehicle.
14. Remove the front exhaust pipe to exhaust manifold mounting nuts.
15. Disconnect both oxygen sensor harness connectors.
16. Lower the vehicle.
17. Remove the outer timing belt cover.
18. Rotate the engine to TDC and align the timing marks.
19. Remove the timing belt tensioner, timing belt, then reset tensioner.
20. Remove the left cylinder head cover to cylinder head ground strap.
21. Remove the left cylinder head cover.
22. Remove the left rocker arm assembly.
23. Remove the left camshaft thrust plate.
24. Counterhold the left cam gear and remove the cam gear retaining bolt.
25. Push the camshaft out of the back of the cylinder head approximately 3.5 inches and remove the cam gear. Remove the front timing belt housing to cylinder head bolts.
26. Remove the cylinder head bolts in REVERSE of tightening sequence.
27. Remove the cylinder head.
28. Clean and inspect all mating surfaces.

To install:

To ensure engine gasket sealing, proper surface preparation must be performed, especially with the use of aluminum engine components and multi-layer steel cylinder head gaskets.

➡Multi-Layer Steel (MLS) head gaskets require a scratch free sealing surface.

Remove all gasket material from cylinder head and block. Be careful not to gouge or scratch the aluminum head sealing surface. Clean all engine oil passages.

⁂ WARNING

The cylinder head gaskets are not interchangeable between cylinder heads and are clearly marked right or left.

➡The cylinder head bolts are tightened using a torque plus angle procedure. The bolts must be examined BEFORE reuse. If the threads are necked down the bolts must be replaced.

Necking can be checked by holding a scale or straight-edge against the threads. If all the threads do not contact the scale the bolt must be replaced.

⁂ WARNING

When cleaning cylinder head and cylinder block surfaces, DO NOT use a metal scraper because the surfaces could be cut or ground. Use ONLY a wooden or plastic scraper.

29. Clean sealing surfaces of cylinder head and block.

⁂ WARNING

Ensure that the correct head gaskets are used and are oriented correctly on cylinder block.

➡Before installing the cylinder head bolts, lubricate the threads with clean engine oil.

30. Install the cylinder head over locating dowels and finger tighten the head bolts.
31. Tighten the cylinder head bolts in the following sequence, using the 4 step torque-turn method. Tighten according to the following torque values:
    - Step 1: All to 61 Nm (45 ft. lbs.)
    - Step 2: All to 88 Nm (65 ft. lbs.)
    - Step 3: All (again) to 88 Nm (65 ft. lbs.)
    - Step 4: + 90° turn. Do not use a torque wrench for this step.
32. Bolt torque after 90° turn should be over 122 Nm (90 ft. lbs.) in the tightening direction. If not, replace the bolt.
33. Install the inner timing cover to cylinder head bolts. Tighten bolts to 54 N·(40 ft. lbs.).
34. Install camshaft sprocket. Counterhold the camshaft sprocket gear and tighten the camshaft sprocket bolt to 102 Nm plus a 90° turn (75 lbs. ft. plus a 90° turn).
35. Install the rear camshaft thrust plate and seal and the EGR valve.
36. Rotate the camshaft gear to its alignment mark and check the left camshaft gear and crankshaft gear timing alignment marks.
37. Install the timing belt and tensioner.
38. Install the timing belt outer cover.
39. Install the power steering reservoir.
40. Install the vibration damper.
41. Install the accessory drive belt tensioner.
42. Install the accessory drive belt idler pulley.
43. Install the right exhaust manifold.
44. Raise and support the vehicle.
45. Install the front exhaust pipe and connect the oxygen sensors.
46. Lower the vehicle.
47. Install the right rocker arm assembly.
48. Install the right cylinder head cover, ground strap and insulator.
49. Install lower intake manifold.
50. Install the fuel rail.
51. Install the upper intake manifold.
52. Connect the air cleaner element housing.
53. Fill the coolant system.
54. Connect the negative battery cable.

5.7L Engine

1. Before servicing the vehicle, refer to the precautions in the beginning of this section.
2. Perform the fuel system pressure release procedure.
3. Disconnect the fuel supply line.
4. Disconnect the battery negative cable.
5. Drain cooling system.
6. Remove the air cleaner resonator and duct work.
7. Remove closed crankcase ventilation system.
8. Disconnect the exhaust at the exhaust manifolds.
9. Disconnect the evaporation control system.
10. Disconnect heater hoses.
11. Remove the power steering pump.
12. Disconnect coil on plug connectors.

⁂ WARNING

The ground straps must be installed in the same location as removed. The covers are machined to accept the ground straps in those locations only.

13. Remove cylinder head cover.

➡The gasket may be used again, provided no cuts, tears, or deformation has occurred.

14. Remove intake manifold and throttle body as an assembly.
15. Remove rocker arm assemblies and push rods. Identify to ensure installation in original locations.
16. Remove the head bolts from each cylinder head, using the sequence provided, and remove cylinder heads. Discard the cylinder head gasket.

To install:

17. Clean all surfaces of cylinder block and cylinder heads.
18. Clean cylinder block front and rear gasket surfaces using a suitable solvent.
19. Inspect the cylinder head for out-of-flatness, using a straight-edge and a feeler gauge. If tolerances exceed 0.0508 mm (0.002 in.), replace the cylinder head.

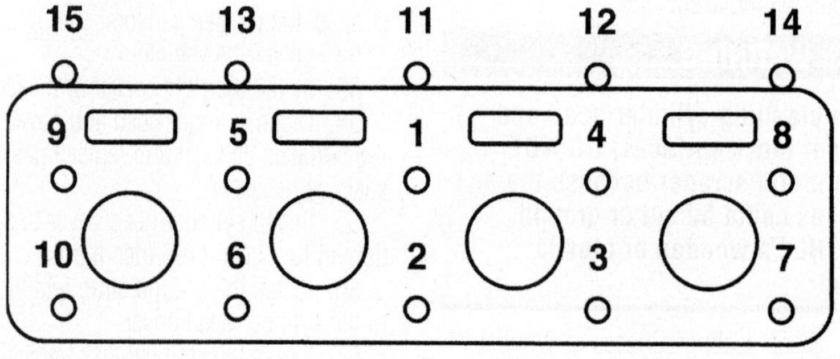

Cylinder head bolt removal/tightening sequence—5.7L engine

20. Inspect the valve seats for damage. Service the valve seats as necessary.

21. Inspect the valve guides for wear, cracks or looseness. If either condition exists, replace the cylinder head.

22. Inspect pushrods. Replace worn or bent pushrods.

23. Clean all surfaces of cylinder block and cylinder heads.

24. Clean cylinder block front and rear gasket surfaces using a suitable solvent.

** WARNING

The head gaskets are not interchangeable between left and right sides. They are marked "L" and "R" to indicate left and right sides.

** WARNING

The head gaskets are marked "TOP" to indicate which side goes up.

25. Position new cylinder head gaskets onto the cylinder block.

26. Position cylinder heads onto head gaskets and cylinder block.

27. Tighten the cylinder head bolts in three steps using the sequence provided:

- Step 1: Snug tighten M12 cylinder head bolts, in sequence, to 34 Nm (25 ft. lbs.) and M8 bolts to 20 Nm (15 ft. lbs.) torque.
- Step 2: Tighten M12 cylinder head bolts, in sequence, to 54 Nm (40 ft. lbs.) and verify M8 bolts to 20 Nm (15 ft. lbs.) torque.
- Step 3: Turn M12 cylinder head bolts, in sequence, 90 degrees and tighten M8 bolts to 34 Nm (25 ft. lbs.) torque.

28. Install pushrods and rocker arm assemblies in their original position, using pushrod retaining plate special tool 9070.

29. Install the intake manifold and throttle body assembly.

30. If required, adjust spark plugs to specifications. Install the plugs.

31. Connect the heater hoses.

32. Install the fuel supply line.

33. Install the power steering pump.

34. Install the drive belt.

35. Install the cylinder head cover.

36. Install ignition coil on plug, and torque fasteners to 12 Nm (105 inch lbs)

37. Connect, ignition coil electrical connectors.

38. Install PCV hose.

39. Connect the evaporation control system.

40. Install the air cleaner.

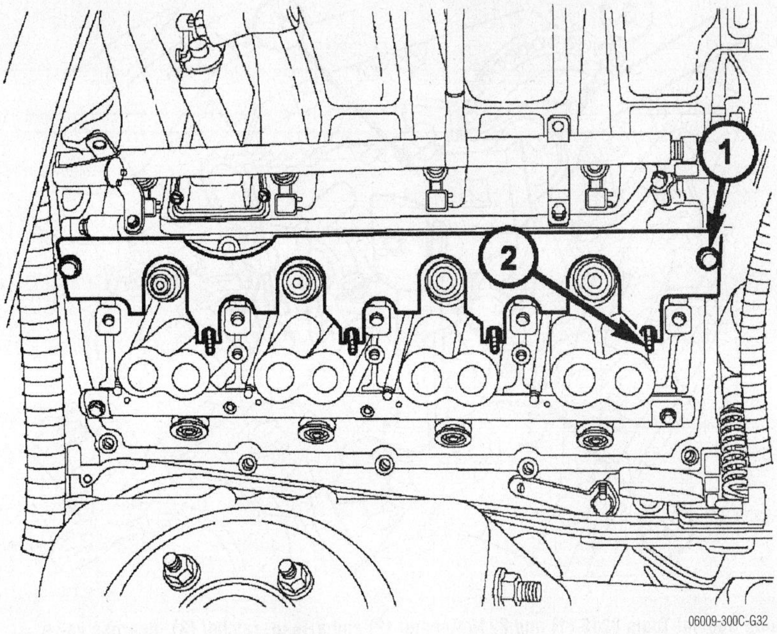

Pushrod retaining plate—5.7L engine

41. Fill cooling system.

42. Connect the negative cable to the battery.

43. Start engine check for leaks.

Rocker Arms/Shafts

REMOVAL & INSTALLATION

2.7L Engine

1. Before servicing the vehicle, refer to the precautions in the beginning of this section.

2. Disconnect negative battery cable located in trunk.

3. Remove upper intake manifold.

4. Remove the cylinder head cover.

** WARNING

Always rotate engine by turning the crankshaft. Failure to do so will result in valve and/or piston damage.

5. Rotate engine until the cam lobe is on its base circle (heel), on the rocker arm being removed.

** WARNING

Depress valve spring only enough to remove rocker arm.

6. Using Special Tools 8215 and 8216 Adaptor and a ⅜ in. drive ratchet, depress valve spring only enough to release tension on rocker arm.

7. Remove rocker arm from cylinder head.

** WARNING

If rocker arms are to be reused, identify position of rocker arms for reassembly in their original positions.

8. Repeat procedure for each rocker arm removed.

9. Inspect the rocker arm for wear or damage.

10. Inspect the cam follower assembly for wear or damage. Replace as necessary.

To install:

11. Lubricate rocker arms with clean engine oil before installation.

12. Rotate engine until cam lobe is on its base circle (heel) of rocker arm being installed.

13. Using Special Tools 8215-A, 8216-A adaptor, and a ⅜ in. drive ratchet, depress valve spring only enough to install rocker arm.

14. Install rocker arm in original position (if reused) over valve and lash adjuster. Release tension on valve spring.

➡Inspect rocker arm for proper engagement into lash adjuster and valve tip.

15. Repeat procedure for each rocker arm being installed.

16. Install the cylinder head cover.

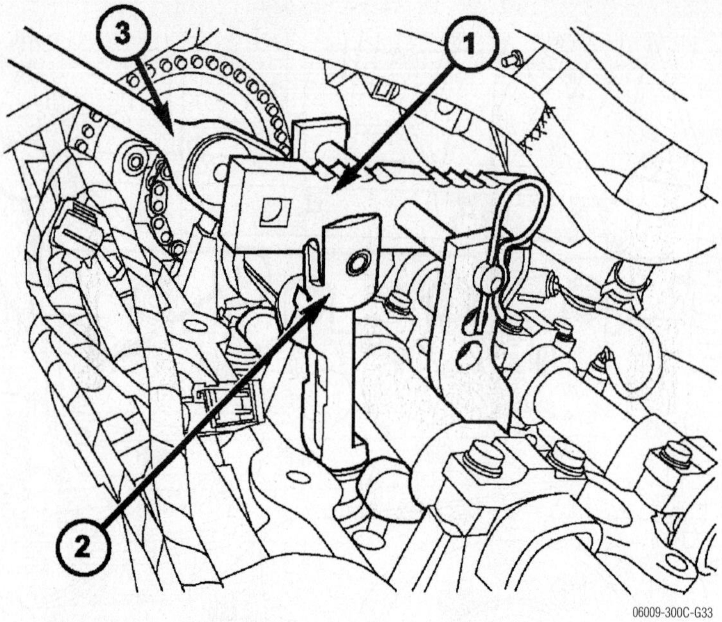

Using Special Tools 8215 (1) and 8216 Adaptor (2) and a 3/8 in. ratchet (3), depress valve spring only enough to release tension on rocker arm

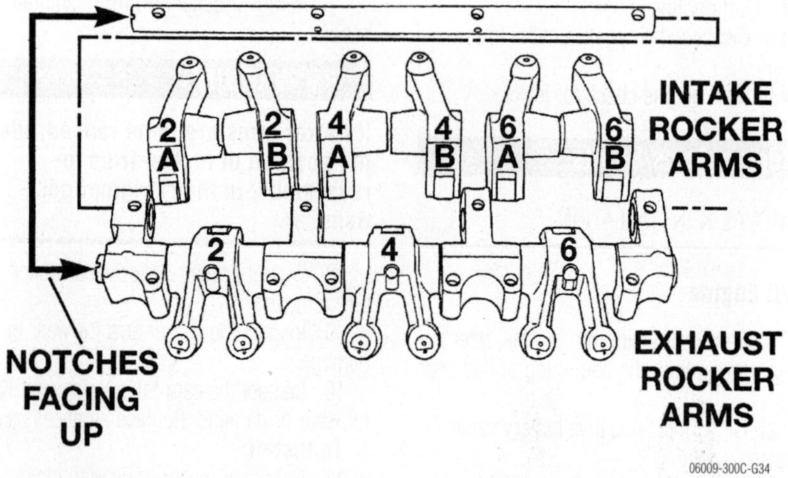

Rocker arm assemblies—3.5L engine

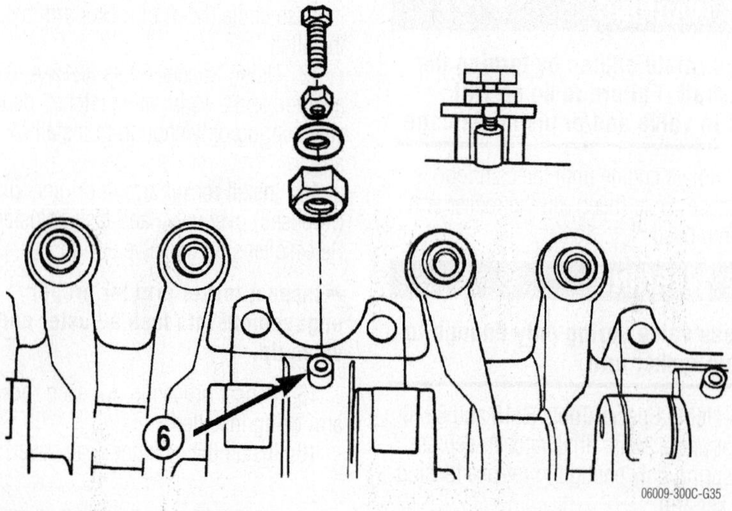

Dowel pin assembly—3.5L engine

3.5L Engine

REMOVAL

⁂ WARNING

The rocker arm shafts are hollow and are used as lubrication oil passages. The rocker arm and shaft assembly on the RIGHT side of the engine has an oil passage hole from the cylinder head to the third rocker shaft support. The rocker arm shaft assembly on the LEFT side of the engine has an oil passage hole from the cylinder head to the second rocker shaft support.

⁂ WARNING

Do not start or run engine with cylinder head cover removed from the engine. Damage or personal injury may occur.

1. Before servicing the vehicle, refer to the precautions in the beginning of this section.
2. Disconnect and isolate the negative battery cable.
3. Remove the upper intake manifold from the engine.
4. Cover lower intake manifold with a suitable cover during service.
5. Remove the cylinder head cover.
6. Identify the rocker arm assembly and rocker arms before disassembly.
7. Remove rocker arm assembly bolts.
8. Remove rocker arm assembly.

➡To prevent air ingestion into lash adjusters, avoid turning rocker arm assembly upside down.

⁂ WARNING

Do not allow rocker arm assembly to rest on lash adjusters, as damage may occur to lash adjusters and/or plastic retainers.

DISASSEMBLY

1. Before servicing the vehicle, refer to the precautions in the beginning of this section.
2. Remove dowel pin using a 4 mm screw, nut, spacer, and washer installed into the pin. Thread the screw into the pin, then loosen the nut on the screw. This will pull the dowel out of the shaft support. Do not reuse dowel pins. Remove the rocker arms and pedestals in order.
3. Check the rocker arm mounting por-

tion of the shafts for wear or damage. Replace if damaged or heavily worn.

4. Check shaft oil holes for clogging with small wire, clean as required.

INSPECTION

➡To prevent air ingestion into lash adjusters, avoid turning rocker arm assembly upside down.

⁂ WARNING

Do not allow rocker arm assembly to rest on lash adjusters, as damage may occur to lash adjuster and plastic retainer.

The intake and exhaust rocker arms are different. They should be identified before disassembling the assembly.

1. Check rocker arms for wear or damage:
   - Roller scuffing or wear
   - Shaft bore scuffing or wear
   - Swivel pad on lash adjuster missing or broken
   - Rocker arm showing signs of fatigue or cracking
   - Roller axle protruding from arm
2. Replace assembly as necessary if any rocker arms shows signs of wear.

ASSEMBLY

1. Before servicing the vehicle, refer to the precautions in the beginning of this section.

⁂ WARNING

Do Not remove lash adjuster from rocker arm assembly. Damage to the adjuster and rocker arm will result. New dowel pins must be installed when reassembling.

2. Install the rocker arms, and pedestals onto the shaft.
3. Install dowel pins. Dowel pins pass through the pedestal into the exhaust rocker shafts.

➡Dowel pins should be pressed in until they bottom-out against the rocker shaft in the pedestal.

➡Rocker arm and shaft assembly can be installed either prior to or after (preferred) cylinder head installation.

INSTALLATION

1. Before servicing the vehicle, refer to the precautions in the beginning of this section.

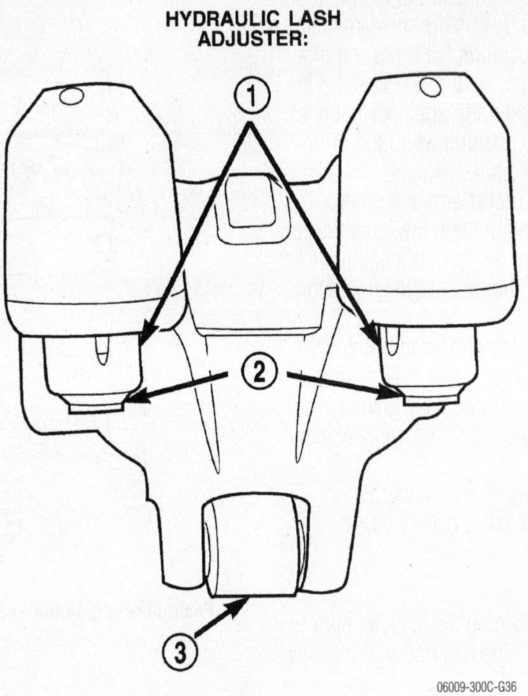

Hydraulic lash adjusters. (1) plunger, (2) contact pads, (3) roller—3.5L engine

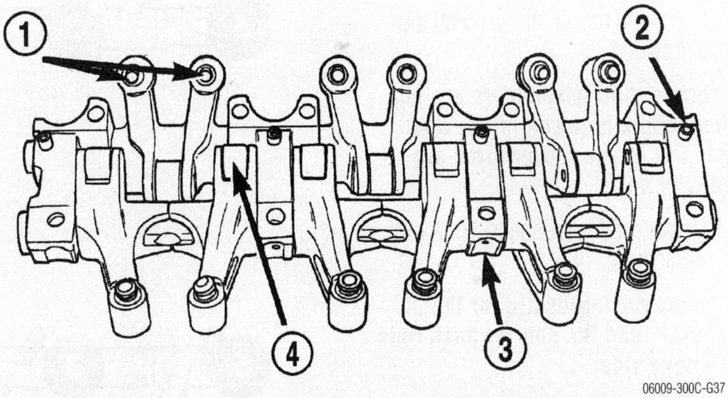

Rocker arms assembled. (1) contact pads, (2) dowel pins, (3) pedestals, (4) rollers—3.5L engine

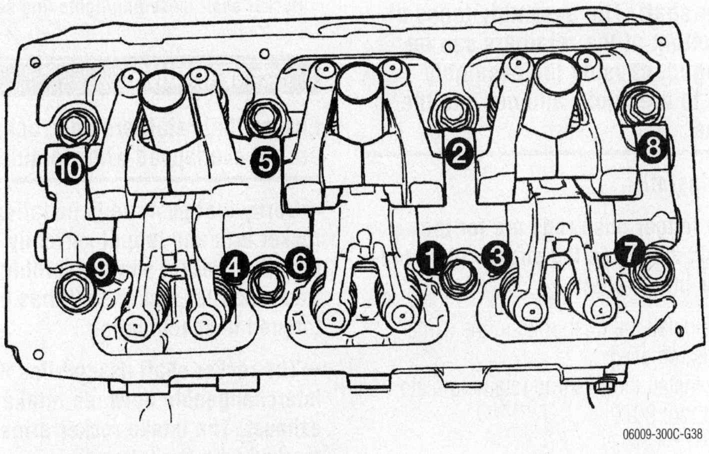

Rocker shaft bolt torque sequence—3.5L engine

2. Rotate camshaft gears to the position shown (1, 8, and 9) in. With the camshaft gears in these positions the lobes are in a neutral position (no load to the valve). This will allow the rocker arm shaft assembly to be tightened into position with little or no valve spring load on it.
3. Install the rocker arm and shaft assembly making sure that the identification marks face toward the front of engine for left head and toward the rear of the engine for right head.
4. Tighten rocker arm/shaft assembly bolts in sequence to 31 Nm (275 inch lbs.).
5. Install the cylinder head cover.
   a. Connect the ignition coil electrical connectors.
6. Install upper intake manifold.
7. Connect negative battery cable

5.7L Engine

1. Before servicing the vehicle, refer to the precautions in the beginning of this section.
2. Remove cylinder head cover.
3. Install pushrod retaining plate special tool 9070.
4. Loosen the rocker shafts using the sequence provided.

➡The rocker shaft assemblies are not interchangeable between intake and exhaust. The intake rocker arms are marked with an "I".

5. Remove the rocker shafts. Note location for reassembly.

➡The longer push rods are for the exhaust side, and the shorter push rods are for intake side.

6. Remove the pushrods. Note pushrod location for reassembly.

⁂ WARNING

Do not remove the retainers from the rocker shaft. The assembly tangs at the bottom of the retainers can be damaged, causing the assembly tangs to break off, and get into the engine.

To install:

➡The longer push rods are for the exhaust side, and the shorter push rods are for intake side.

7. Install the push rods in the same order as removed.
8. Install the pushrod retaining plate special tool 9070.

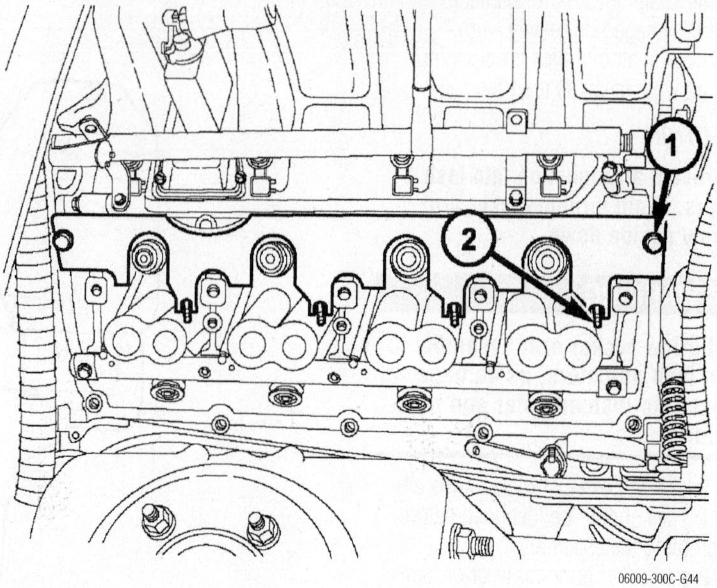

Pushrod retaining tool—5.7L engine

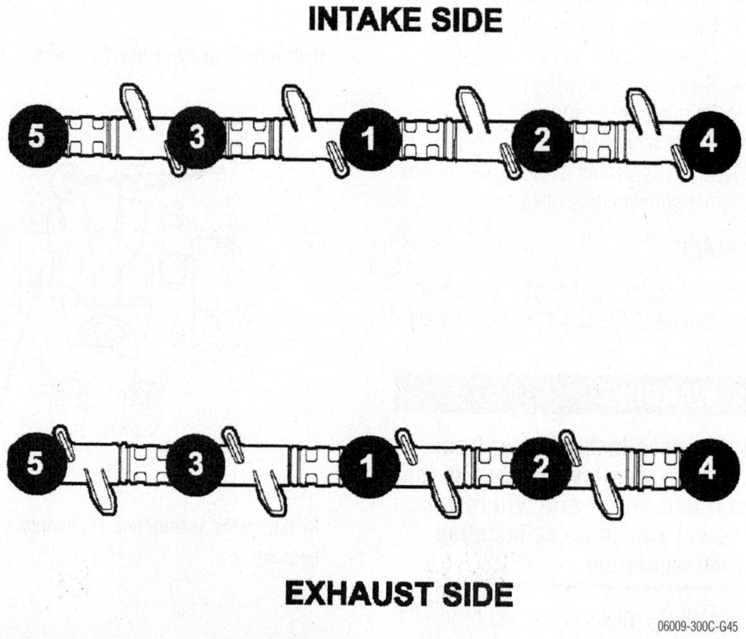

Rocker shaft loosening/tightening sequence—5.7L engine

⁂ WARNING

Ensure that retainers and rocker arms are not overlapped when torquing bolts.

➡Verify that pushrod is installed into rocker arm and tappet correctly while installing rocker shaft assembly. Recheck after rocker shaft has been torqued to specification.

➡The rocker shaft assemblies are not interchangeable between intake and exhaust. The intake rocker arms are marked with the letter "I".

9. Install rocker shaft assemblies in the same order as removed.
10. Tighten the rocker shaft bolts to 22 Nm (195 inch lbs.) torque, using the sequence provided.

⁂ WARNING

DO NOT rotate or crank the engine during or immediately after rocker arm installation. Allow the hydraulic roller tappets adequate time to bleed down (about 5 minutes).

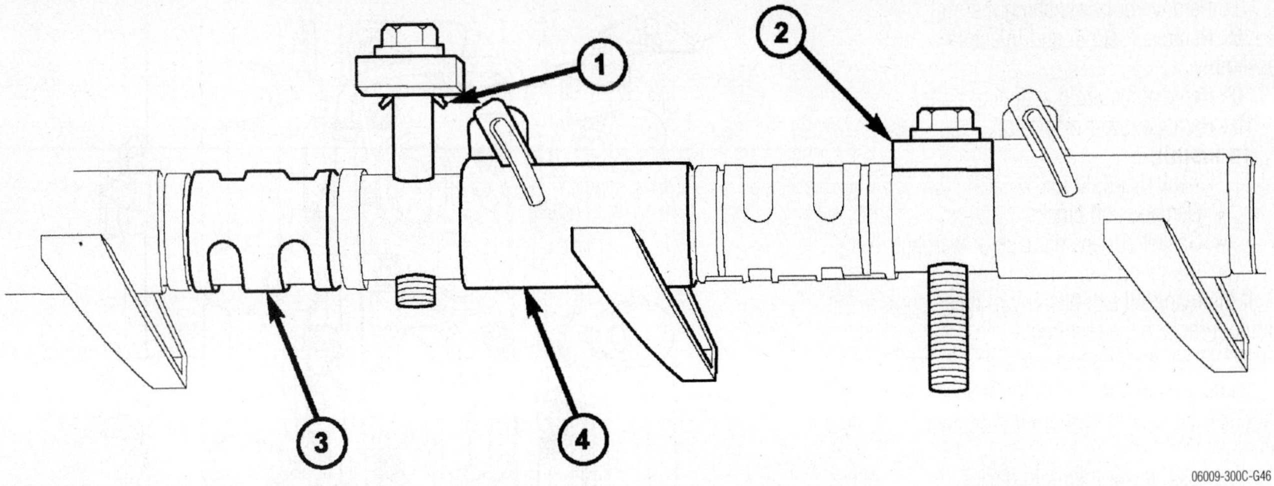

Assembly tangs (1), retainers (2), rocker arm (4)—5.7L engine

11. Remove pushrod retaining plate special tool 9070.
12. Install cylinder head cover.

Intake Manifold

REMOVAL & INSTALLATION

2.7L Engine

UPPER MANIFOLD

1. Before servicing the vehicle, refer to the precautions in the beginning of this section.
2. Disconnect negative battery cable.
3. Remove throttle body air inlet hose and air cleaner housing assembly.
4. Disconnect electrical connectors from the following components:
   - Manifold Absolute Pressure (MAP) Sensor
   - Electronic Throttle Control
   - Manifold Tuning Valve
5. Disconnect vapor purge hose, brake booster hose, positive crankcase ventilation (PCV) hose.
6. Remove manifold support brackets.
7. Remove manifold attaching bolts.
8. Remove upper manifold.
9. Remove foam insulator.
10. Check manifold for:
    - Damage and cracks
    - Gasket surface damage or warpage
    - Damaged or clogged EGR ports
11. If the manifold exhibits any damaged or warped conditions, replace the manifold. Clean EGR ports as necessary.

To install:

12. Clean and inspect sealing surfaces. Gaskets can be reused, if free of cuts or tears.

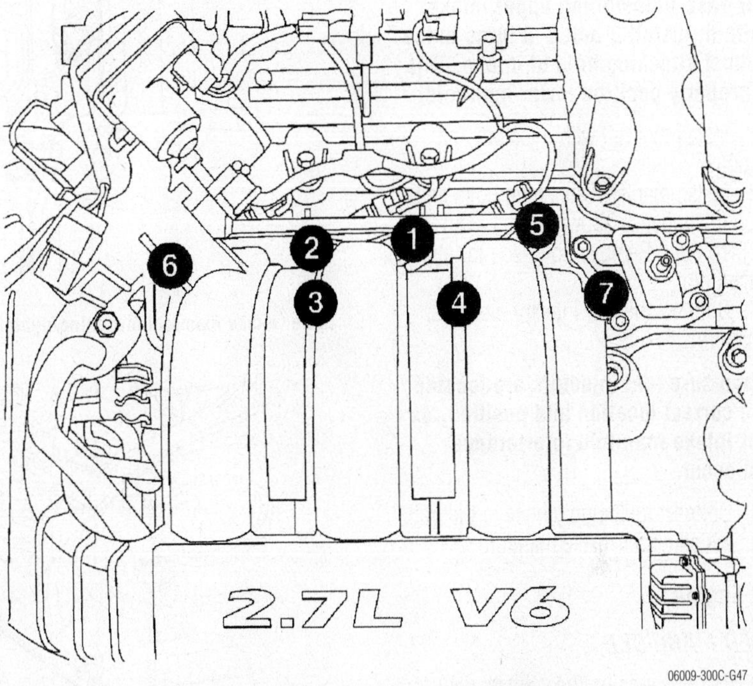

Upper intake manifold loosening/tightening sequence—2.7L engine

➡Make sure fuel injectors and wiring harnesses are in correct position to not interfere with upper manifold installation.

13. Install upper manifold gasket.
14. Position upper manifold onto lower manifold.
15. Install manifold attaching bolts and tighten in sequence shown in to 12 Nm (105 inch lbs.).
16. Connect PCV, brake booster, and vapor purge hoses.
17. Connect electrical connectors to the following components:
    - Manifold Absolute Pressure (MAP) Sensor
    - Electronic Throttle Control
    - Manifold Tuning Valve
18. Install throttle body air inlet hose and air cleaner housing assembly.

LOWER MANIFOLD

1. Before servicing the vehicle, refer to the precautions in the beginning of this section.
2. Release fuel system pressure.
3. Disconnect negative battery cable located in trunk.
4. Remove upper intake manifold.
5. Disconnect injector electrical connectors.
6. Disconnect fuel supply hose from fuel rail.

7. Remove bolts attaching fuel rail.
8. Remove fuel rail and injectors as an assembly.
9. Remove manifold attaching bolts.
10. Remove lower manifold.

To install:

11. Check manifold for:
- Damage and cracks
- Gasket surface damage or warpage
- Damaged fuel injector ports

If the manifold exhibits any of these conditions, replace the manifold.

12. Clean and inspect sealing surfaces of cylinder head and manifold. Gaskets can be reused provided they are free of cuts or tears.
13. Install lower manifold gasket.
14. Position manifold on cylinder head surfaces.

➡For ease of installing upper intake manifold, install a bolt 2–3 turns to the rearmost attaching hole of intake. This will properly position lower manifold.

15. Install fuel rail with injectors and start bolts.
16. Install manifold attaching bolts and tighten in sequence shown in to 12 Nm (105 inch lbs.). Remove bolt used for aligning manifold.
17. Connect the fuel injector electrical connectors.

➡Make sure fuel injectors are located in the correct location and position, as upper intake manifold interference could occur.

18. Connect fuel supply hose to fuel rail.
19. Install upper intake manifold.

3.5L Engine

UPPER MANIFOLD

1. Before servicing the vehicle, refer to the precautions in the beginning of this section.
2. Disconnect negative battery cable.
3. Disconnect the IAT sensor electrical connector.
4. Remove air inlet hose from the throttle body.
5. Disconnect the MAP sensor electrical connector.
6. Separate the engine electrical harness connectors from the intake manifold.
7. Disconnect the EGR tube, PCV, purge and power brake booster vacuum hoses from the upper intake manifold.
8. Disconnect the electronic throttle control electrical connector.
9. Remove the throttle bracket fasteners from the throttle body and cylinder head.

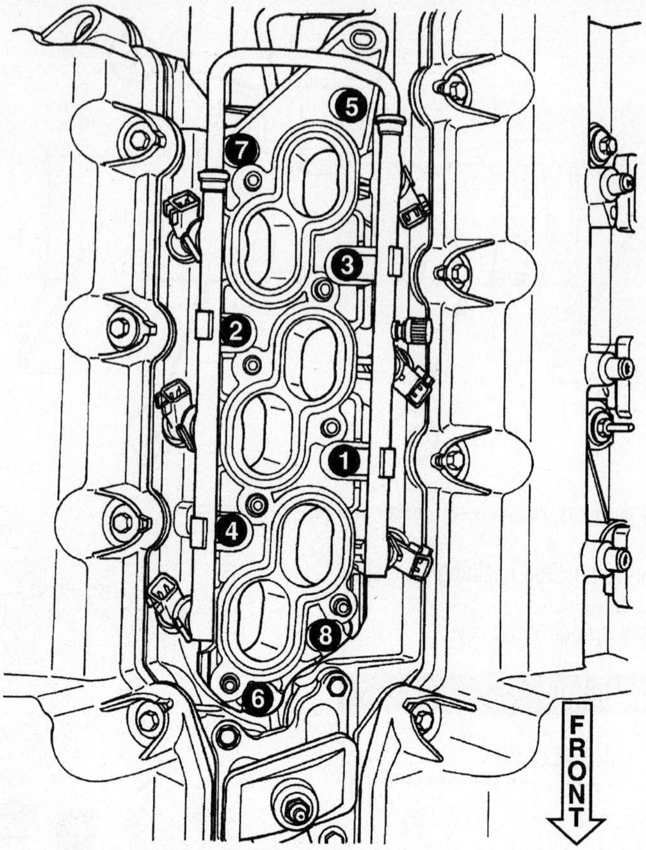

06009-300C-G48

Lower intake manifold loosening/tightening sequence—2.7L engine

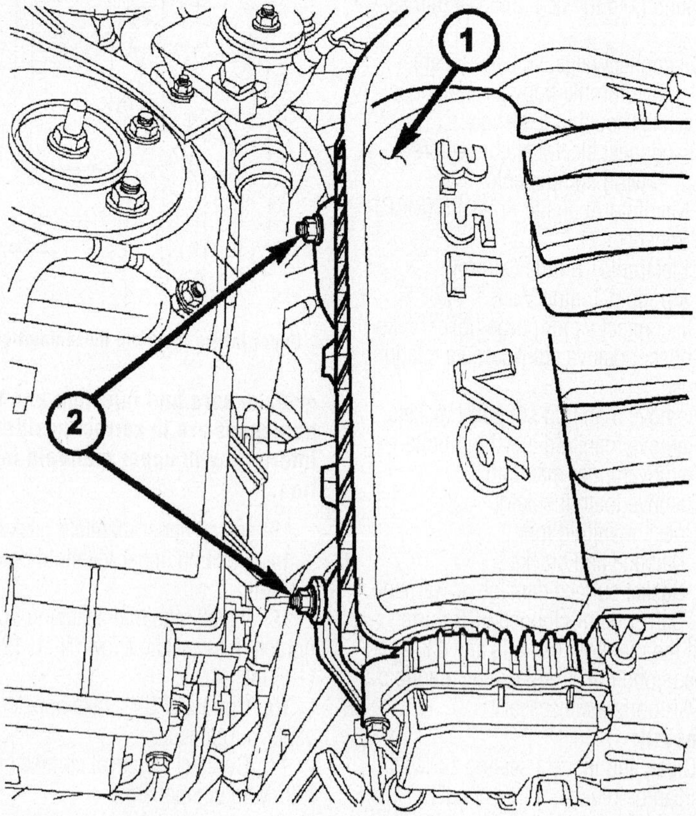

06009-300C-G48A

Right upper intake manifold (1) support brackets (2)—3.5L engine

10. Disconnect electrical connectors from the Manifold Tuning Valve (MTV) and Short Runner Valve.

11. Remove the right intake manifold support brackets.

12. Remove the upper intake manifold retaining bolts, insulation foam pad and manifold. Clean all gasket sealing surfaces.

To install:

13. Clean and inspect gasket sealing surfaces.

14. Position new gasket.

15. Install the upper intake manifold insulator foam.

16. Install the upper intake manifold. Tighten bolts to 12 Nm (105 inch lbs.) starting in the center working outward in a cross sequence pattern.

17. Install the right manifold support brackets. Tighten fasteners to 12 Nm (105 inch lbs.).

18. Install the throttle bracket. Tighten fasteners to 12 Nm (105 inch lbs.) at the throttle body and 28Nm (259 inch lbs.) at the cylinder head.

19. Connect the manifold tuning valve and short runner valve electrical connectors.

20. Connect the electronic throttle control harness connector.

21. Connect the engine electrical connectors to the intake manifold.

22. Connect the EGR tube, PCV, purge and power brake booster vacuum hoses to the intake manifold.

23. Connect the MAP sensor harness connector.

24. Install the inlet hose and connect the IAT sensor harness connector.

25. Connect negative battery cable.

LOWER MANIFOLD

1. Before servicing the vehicle, refer to the precautions in the beginning of this section.

2. Perform fuel pressure release procedure.

3. Drain the cooling system.

4. Disconnect the upper radiator hose from the thermostat housing.

5. Remove the upper intake manifold.

6. Reposition power steering fluid reservoir and bracket.

7. Disconnect the electrical connectors to fuel injectors and coolant temperature sensor.

8. Disconnect heater hose from the rear intake manifold.

9. Disconnect the coolant container hose at the rear intake manifold.

10. Disconnect the fuel supply hose from fuel rail.

11. Remove the bolts attaching fuel rail.

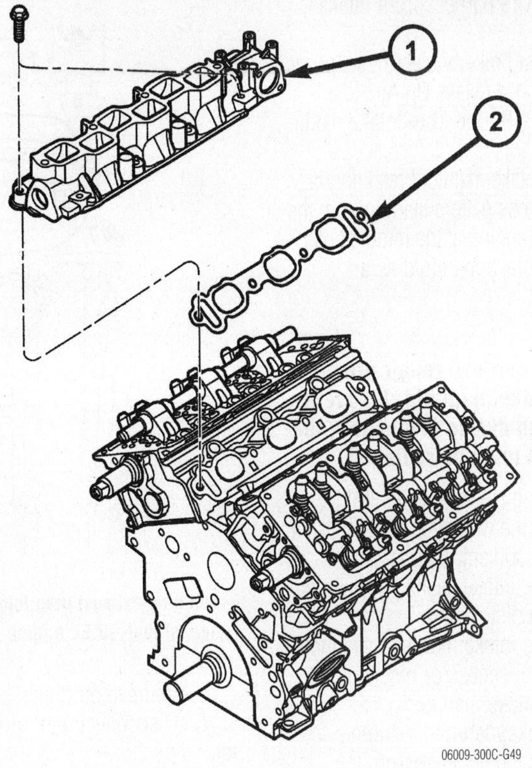

Lower intake manifold and gasket—3.5L engine

12. Remove fuel rail and injectors as an assembly.

13. Remove bolts attaching lower intake and remove intake manifold.

To install:

14. Clean all sealing surfaces.

15. Position new gaskets and intake manifold on cylinder head surfaces.

16. Install intake manifold bolts and gradually tighten in sequence shown until a torque of 28 Nm (250 inch lbs.) is obtained.

17. Install fuel rail and injectors as an assembly.

18. Connect fuel supply hose to fuel rail.

19. Connect heater hose to rear lower intake manifold.

20. Connect coolant container hose to the rear lower intake manifold.

21. Connect electrical connectors to fuel injectors and coolant temperature sensor.

22. Install power steering fluid reservoir and bracket.

23. Install upper intake manifold.

24. Connect the upper radiator hose to the thermostat housing.

25. Fill the cooling system.

5.7L Engine

1. Before servicing the vehicle, refer to the precautions in the beginning of this section.

2. Remove engine cover.

3. Bleed fuel system.

4. Disconnect negative cable from battery.

5. Remove air inlet hose.

6. Remove ignition wires from on top of intake manifold.

7. Disconnect electrical connectors for the following components:

- Manifold Absolute Pressure (MAP) Sensor
- Fuel Injectors
- ETC (Electric Throttle Control)

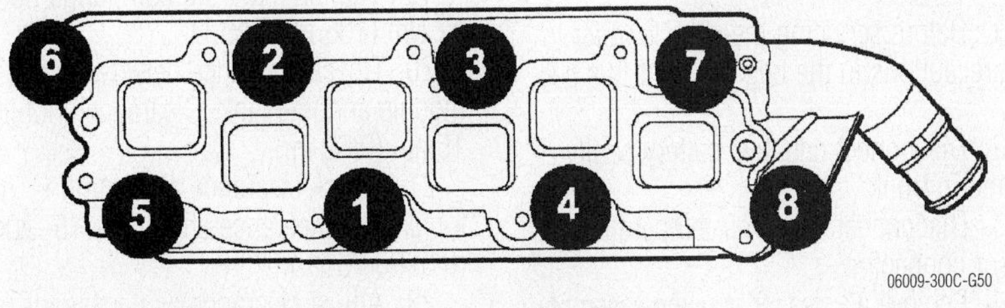

Lower intake manifold torque sequence—3.5L engine

8. Remove wire harness from intake manifold.

9. Disconnect brake booster hose, purge hose, and MUA hose (Make Up Air Hose).

10. Remove EGR tube from intake manifold.

11. Remove intake manifold retaining fasteners in a crisscross pattern starting from the outside bolts and ending at the middle bolts.

12. Remove intake manifold as an assembly.

To install:

➡There is no approved repair procedure for the intake manifold. If severe damage is found during inspection, the intake manifold must be replaced.

Before installing the intake manifold thoroughly clean the mating surfaces. Use a suitable cleaning solvent, then air dry.

13. Inspect the intake sealing surface for cracks, nicks and distortion.

14. Inspect the intake manifold vacuum hose fittings for looseness or blockage.

15. Position intake manifold.

16. Install intake manifold retaining bolts, and tighten in sequence from the middle bolts towards the outside in a crisscross pattern. Torque fasteners to 12 Nm (105 inch lbs.).

17. Install EGR tube.

18. Install wire harness on intake manifold.

19. Connect electrical connectors for the following components:

- Manifold Absolute Pressure (MAP) Sensor
- Fuel Injectors
- ETC (Electronic Throttle Control)

20. Install ignition wires.

21. Connect Brake booster hose, purge hose, and MUA hose (Make Up Air hose).

22. Install air inlet hose.

23. Connect negative cable to battery.

24. Install engine cover.

Exhaust Manifold

REMOVAL & INSTALLATION

2.7L Engine

LEFT SIDE

1. Before servicing the vehicle, refer to the precautions in the beginning of this section.

2. Disconnect negative battery cable located in trunk.

3. Disconnect intake air temperature sensor connector.

4. Remove air cleaner housing assembly.

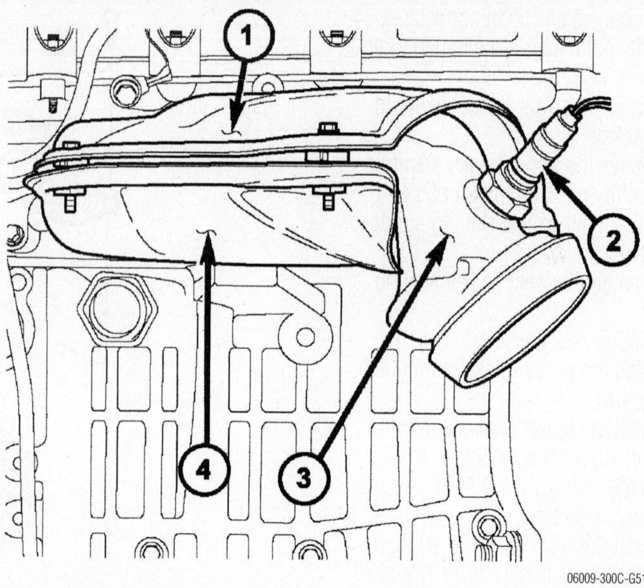

Left side exhaust manifold. (1) upper heat shield, (2) oxygen sensor, (3) manifold, (4) lower heat shield—3.5L engine

5. Remove oil dipstick tube.

6. Disconnect and remove oxygen sensor.

7. Remove upper heat shield and lower heat shield.

8. Raise vehicle on hoist.

9. Disconnect down stream oxygen sensor connector.

10. Remove exhaust pipe from manifold.

11. Remove exhaust manifold attaching bolts and remove manifold.

To install:

12. Inspect exhaust manifolds for damage or cracks.

13. Check manifold flatness.

14. Inspect the exhaust manifold gasket for obvious discoloration or distortion.

15. Check distortion of the cylinder head mounting surface with a straight-edge and thickness gauge.

16. Install exhaust manifold gasket and manifold. Tighten bolts starting at the center working outward to 23 Nm (200 inch lbs.).

17. Install manifold lower heat shield and upper heat shield. Tighten fasteners to 28 Nm (250 inch lbs.).

18. If removed, install the exhaust pipe bolts and cross brace. Tighten bolts to 55 Nm (40 ft. lbs.).

19. Tighten manifold ball flange nut to 12 Nm (106 inch lbs.).

20. Check clearance between exhaust module and fuel tank. Clearance should be 16mm (0.62 in.).

21. Check clearance at rear tunnel reinforcement. Clearance should be 15–20mm (0.59–0.78 in.).

22. Adjust clearance as necessary.

23. Tighten ball flange nuts to 34 Nm (25 ft. lbs.).

24. Connect down stream oxygen sensor connectors.

25. Lower vehicle.

26. Install oxygen sensor and connect electrical connector.

27. Connect negative battery cable.

RIGHT SIDE

1. Before servicing the vehicle, refer to the precautions in the beginning of this section.

2. Disconnect negative battery cable located in trunk.

3. Disconnect and remove upstream oxygen sensor.

4. Raise vehicle.

5. Remove EGR tube at EGR valve.

6. Remove EGR tube at manifold.

7. Disconnect down stream oxygen sensor electrical connector.

8. Disconnect exhaust pipe from manifold.

9. Remove upper heat shield and lower heat shield.

10. Remove manifold attaching bolts and remove manifold.

To install:

11. Inspect exhaust manifolds for damage or cracks.

12. Check manifold flatness.

13. Inspect the exhaust manifold gasket for obvious discoloration or distortion.

14. Check distortion of the cylinder head mounting surface with a straight-edge and thickness gauge.

15. Install exhaust manifold gasket and

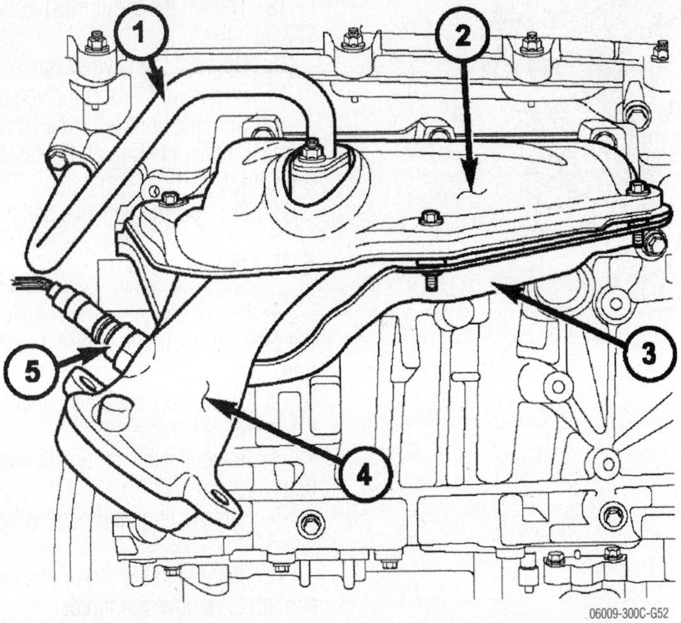

Right side exhaust manifold. (1) EGR tube, (2) upper heat shield, (3) lower heat shield, (4) manifold, (5) oxygen sensor—2.7L engine

manifold. Tighten bolts working from center outwards to 23 Nm (200 inch lbs.).

16. If removed, install bolts and cross brace. Tighten bolts to 55 Nm (40 ft. lbs.).
17. Tighten manifold ball flange nut to 12 Nm (106 inch lbs.).
18. Check clearance between exhaust module and fuel tank. Clearance should be 16mm (0.62 in.).
19. Check clearance at rear tunnel reinforcement. Clearance should be 15–20mm (0.59–0.78 in.).
20. Adjust clearance as necessary.
21. Tighten ball flange nuts to 34 Nm (25 ft. lbs.).
22. Connect oxygen sensor connectors.
23. Install lower heat shield and upper heat shield.
24. Install oxygen sensor and connect electrical connector.
25. Connect down stream oxygen sensor electrical connector.
26. Lower vehicle.
27. Install EGR tube using new gaskets. Tighten screws to 11 Nm (95 inch lbs.).
28. Connect negative battery cable.

3.5L Engine

LEFT SIDE

1. Before servicing the vehicle, refer to the precautions in the beginning of this section.
2. Disconnect and isolate the negative battery cable.
3. Raise and support the vehicle.
4. Separate the front exhaust pipe to manifold union.
5. Lower the vehicle.
6. Disconnect and remove the oxygen sensor from the exhaust manifold.
7. Remove the exhaust manifold shield retaining bolts, exhaust manifold, and discard gasket.

To install:

8. Inspect exhaust manifolds for damage or cracks.
9. Check manifold flatness.
10. Inspect the exhaust manifold gasket for obvious discoloration or distortion.

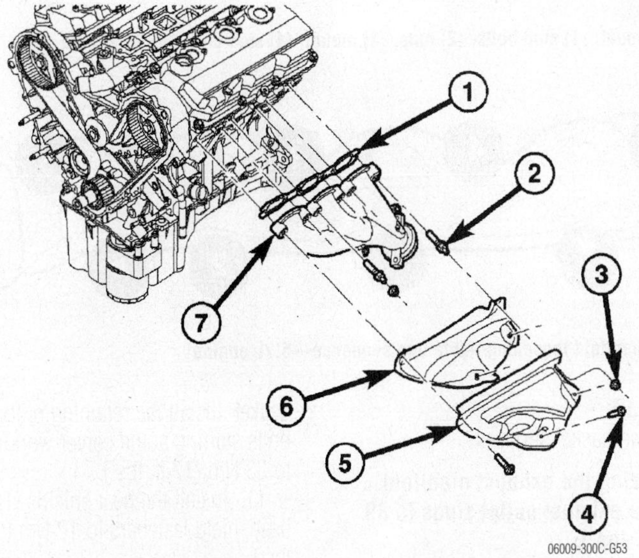

Left side exhaust manifold—3.5L engine

11. Check distortion of the cylinder head mounting surface with a straight-edge and thickness gauge.

➡If replacing the exhaust manifold, tighten the exhaust outlet studs to manifold to 40 Nm (29 ft. lbs.).

12. Position the exhaust manifold and gasket. Install the retaining bolts. Tighten 4 bolts starting at the center working outward to 23 Nm (17 ft. lbs.).
13. Install the exhaust manifold heat shields. Tighten the bolts to 12 Nm (105 inch lbs.).
14. Tighten the out most stud nuts to 8 Nm (73 inch lbs.).
15. Connect the oxygen sensor.
16. Raise and support the vehicle.
17. Connect the exhaust pipe to manifold union. Tighten the exhaust stud nuts to 34 Nm (25 ft. lbs.).
18. Connect the negative battery cable.

RIGHT SIDE

1. Before servicing the vehicle, refer to the precautions in the beginning of this section.
2. Disconnect the negative battery cable.
3. Disconnect the upstream oxygen sensor electrical connector.
4. Raise and support the vehicle.
5. Remove the exhaust manifold to exhaust pipe flange retaining bolts.
6. Lower the vehicle.
7. Remove the exhaust manifold heat shield and manifold.
8. Remove the oxygen sensor from the exhaust manifold.

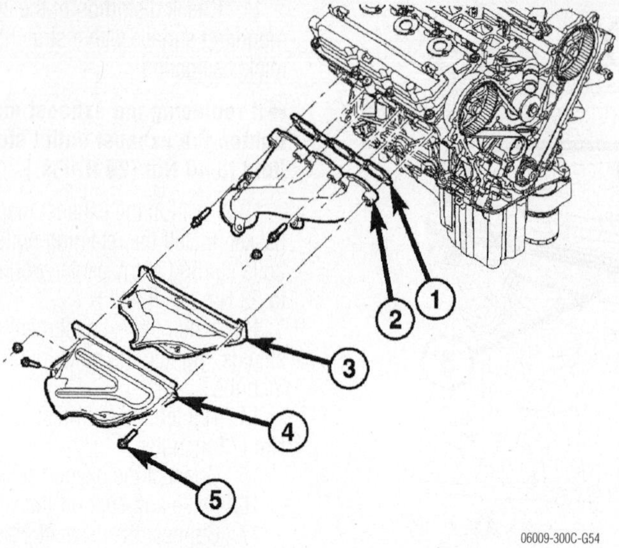

Right side exhaust manifold—3.5L engine

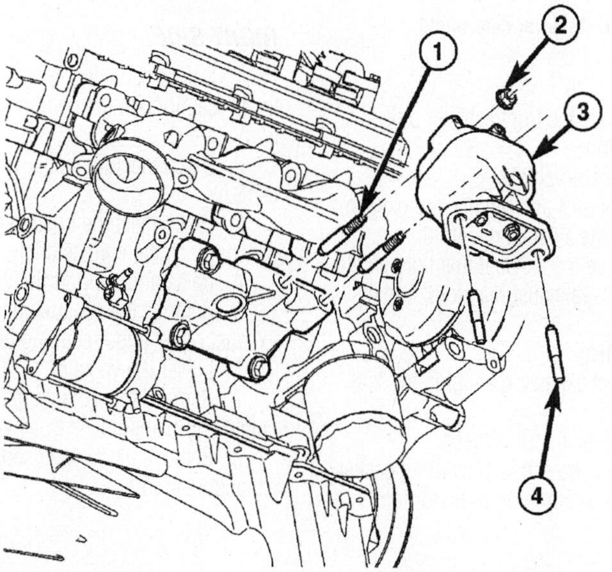

Engine mount. (1) stud bolts, (2) nuts, (3) mount, (4) stud bolts—5.7L engine

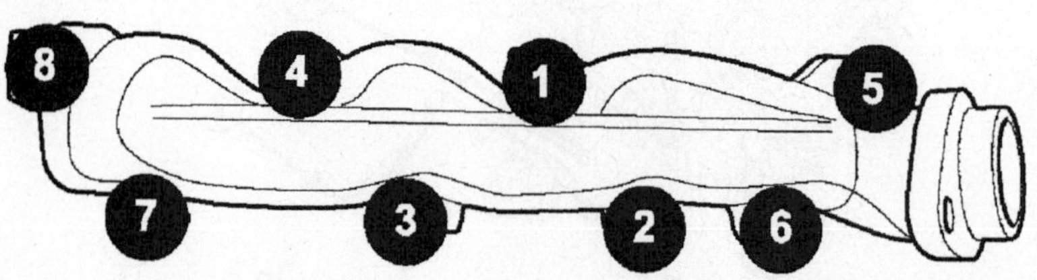

Exhaust manifold loosening/tightening sequence—5.7L engine

To install:

9. Clean gasket surfaces.

➡**If replacing the exhaust manifold, tighten the exhaust outlet studs to 39 Nm (29 ft. lbs.).**

10. Position the exhaust manifold and gasket. Install the retaining bolts. Tighten 4 bolts starting at the center working outward to 23 Nm (17 ft. lbs.).
11. Install the heat shields. Tighten the heat shield fasteners to 12 Nm (105 inch lbs.).
12. Tighten the 2 out most nuts to 8 Nm (73 inch lbs.).
13. Connect the oxygen sensor.
14. Raise and support the vehicle.
15. Connect the front exhaust pipe to exhaust manifold. Tighten the fasteners to 34 Nm (25 ft. lbs.).
16. Connect the negative battery cable.

5.7L Engine

1. Before servicing the vehicle, refer to the precautions in the beginning of this section.
2. Disconnect negative battery cable.
3. Raise vehicle.
4. Remove exhaust pipe to manifold bolts.
5. Remove engine mount to frame fasteners.
6. Using suitable jack, raise engine enough to remove manifolds.

**** WARNING**

Do not damage engine harness while raising the engine.

7. Remove the engine mount.
8. Remove heat shield.
9. Remove manifold bolts using sequence provided.
10. Remove manifold and gasket.

To install:

11. Clean mating surfaces on cylinder head and manifold. Wash with solvent and blow dry with compressed air.
12. Inspect manifold for cracks.
13. Inspect mating surfaces of manifold for flatness with a straight-edge. Gasket surfaces must be flat within 0.2 mm per 300 mm (0.008 inch per foot).
14. Install manifold gasket and manifold.
15. Install manifold bolts and tighten to 25 Nm (18 ft. lbs.).
16. Install heat shield and tighten nuts to 8 Nm (70 inch lbs.).
17. Install engine mounts. Torque to 70 ft. lbs. (95 Nm).
18. Lower engine.

**** WARNING**

Do not damage engine harness while lowering the engine.

19. Install and tighten right and left side engine mount to frame fasteners. Torque to 70 ft. lbs. (95 Nm).
20. Install exhaust flange to pipe bolts.
21. Lower vehicle.
22. Connect negative battery cable.

Structural Collar

REMOVAL & INSTALLATION

2.7L Engine

1. Before servicing the vehicle, refer to the precautions in the beginning of this section.
2. Disconnect negative battery cable located in trunk.
3. Raise vehicle on hoist.
4. Remove lower splash shield retaining bolts and splash shield.
5. Remove transmission mount bolts.
6. Place a transmission jack under the transmission pan and raise engine and transmission.
7. Remove bolts attaching structural collar to oil pan and transmission bellhousing.
8. Move structural collar to the right.
9. Remove structural collar out towards the rear.

To install:

10. Install structural collar from the right rear.

WARNING

The structural collar must be tightened using this service procedure, or damage to transmission case and/or oil pan may occur.

11. Position structural collar in place.
12. Finger tighten all bolts.

➡Make sure that structural collar is flush with the oil pan and the transmission bellhousing.

13. Install the vertical collar bolts to oil pan, pre-torque bolts to 1.1 Nm (10 inch lbs.).
14. Install the horizontal collar bolts to transmission and tighten to 55 Nm (40 ft. lbs.).
15. Starting with the center vertical bolts and working outward, final torque all bolts to 55 Nm (40 ft. lbs.).
16. Install lower splash shield and retaining bolts.
17. Lower vehicle.
18. Connect negative battery cable.

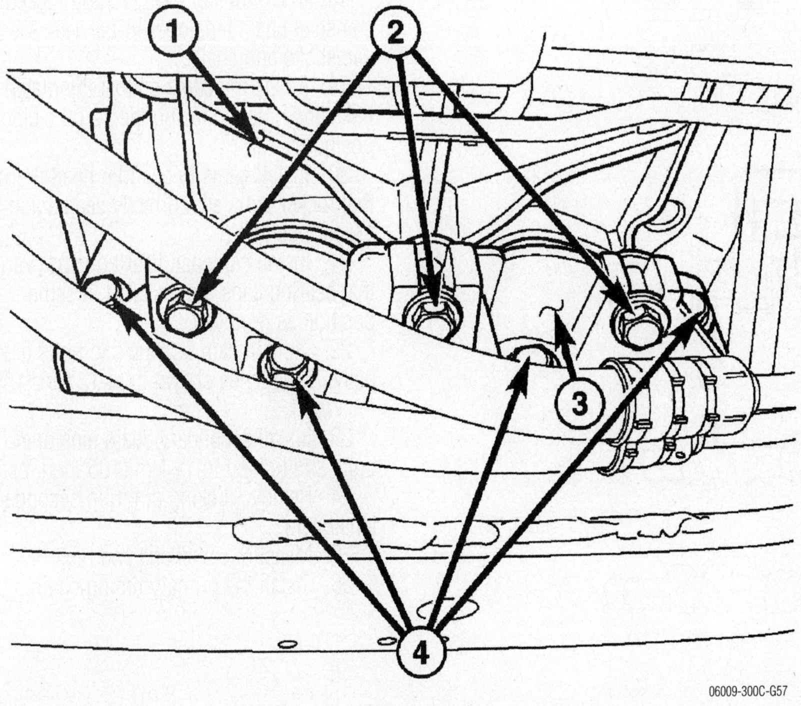

(3) structural collar, (1) oil pan, (2) vertical collar bolts, (4) horizontal bolts—2.7L engine

Camshaft and Valve Lifters

REMOVAL & INSTALLATION

2.7L Engine

1. Before servicing the vehicle, refer to the precautions in the beginning of this section.
2. Remove the primary timing chain.
3. Remove secondary chain tensioner mounting bolts.

➡Camshaft bearing caps have been marked during engine manufacturing. For example, number one exhaust camshaft bearing is marked "1E"

4. Slowly loosen camshaft bearing cap bolts in the order shown.
5. Remove camshaft bearing caps.
6. Remove intake camshaft, exhaust camshaft, secondary timing chain, and secondary timing chain tensioner together as an assembly.
7. Remove secondary timing chain tensioner and secondary timing chain from camshafts.
8. Remove rocker arms.

➡If lash adjusters and rocker arms are to be reused, always mark position for reassembly in their original positions.

9. Remove lash adjuster(s).

To install:

10. Inspect camshaft bearing journals for damage and binding. If journals are binding, check the cylinder head for damage. Also check cylinder head oil holes for clogging.
11. Inspect camshaft sprockets for excessive wear. Replace camshafts if necessary.
12. Check the cam lobe surfaces for abnormal wear and damage. Replace camshaft if defective. Measure the actual wear and replace, if out of limits—standard value is 0.0254 mm (0.001 in.); wear limit is 0.254 mm (0.010 in.).

WARNING

When the timing chain is removed and the cylinder heads are installed, DO NOT rotate the camshafts or crankshaft without first locating the proper crankshaft position. Failure to do so will result in valve and/or piston damage.

13. Install hydraulic lash adjuster making sure adjusters are at least partially full of oil. This can be verified by little or no plunger travel when lash adjuster is depressed.
14. Install rocker arm(s) and cylinder head covers.
15. Assemble camshaft chain on the cams. Verify that plated links are facing toward the front. Align the plated links to the dots on the camshaft sprockets.
16. If camshaft chain tensioner is already in the compressed and locked position, skip the next step.
17. When the camshaft chain tensioner is removed, it is necessary to compress and

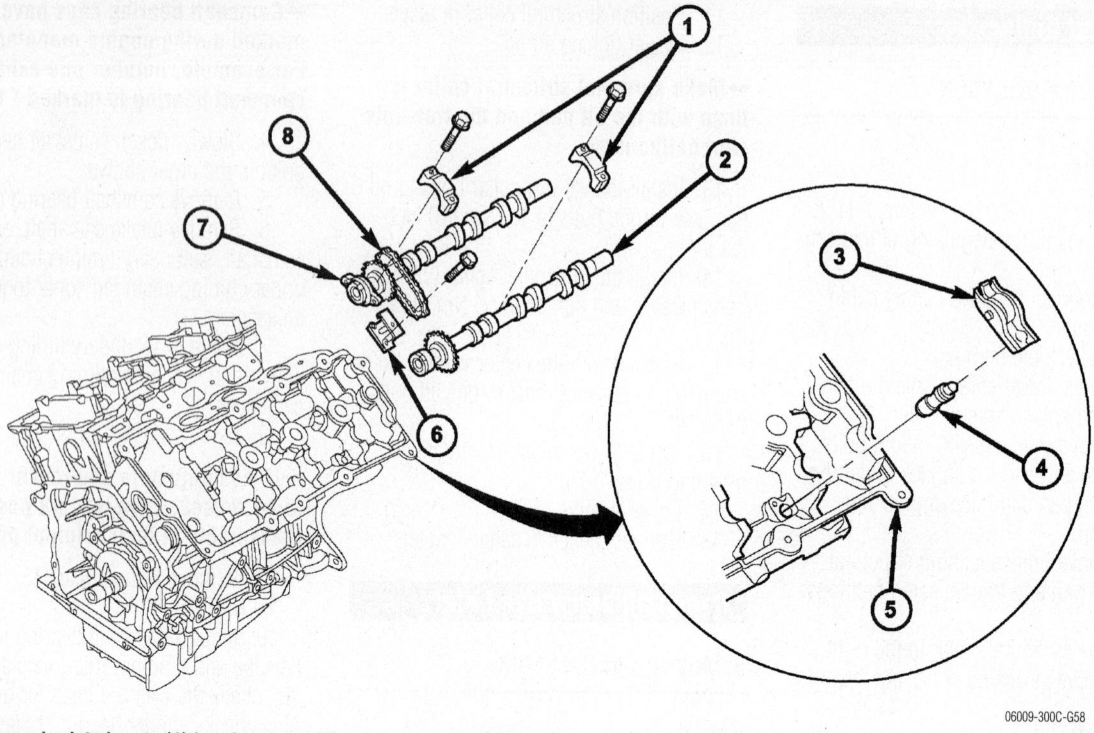

Camshafts and related parts (1) bearing caps, (2) exhaust camshaft (3) rocker arm, (4) lash adjuster, (5) head, (6) secondary timing chain tensioner, (7) intake camshaft—2.7L engine

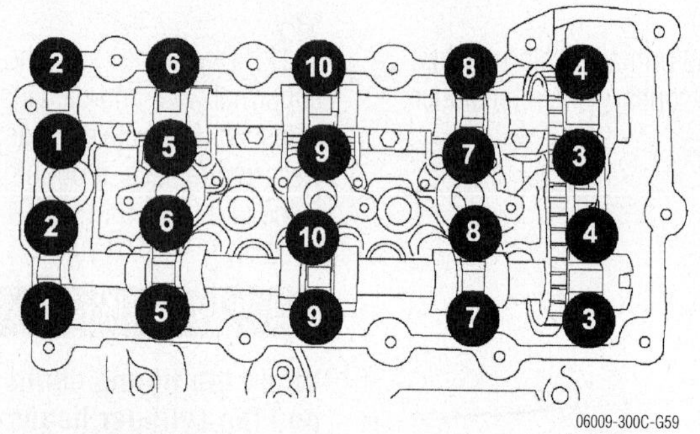

Camshaft bearing cap loosening sequence—2.7L engine

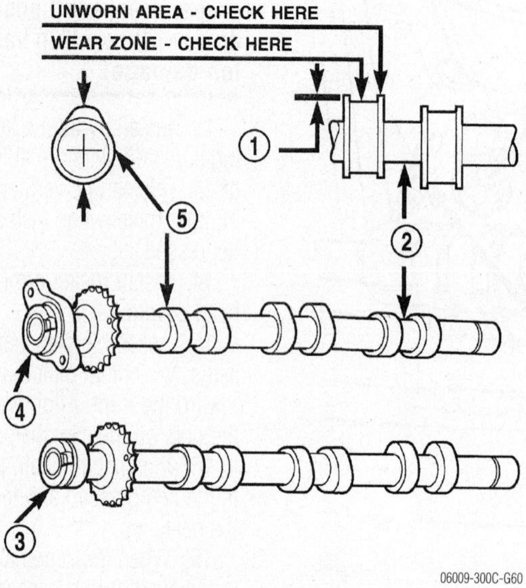

Camshaft inspection points—2.7L engine

lock the tensioner using the following procedures:

a. Place tensioner into a soft jaw vise.

b. SLOWLY compress tensioner until fabricated lock pin or the equivalent can be inserted into the locking holes.

c. Remove compressed and locked tensioner from the vise.

18. Insert the compressed and locked camshaft chain tensioner in-between the camshafts and chain.
19. Rotate the cams so that the plated links and dots are facing the 12:00 o'clock position.
20. Install cams to cylinder head. Verify that rocker arms are correctly seated and in proper positions.
21. Install camshaft bearing caps. Verify that bearing caps are installed in same position as removed.
22. Tighten cam bearing cap bolts gradually in sequence shown in to 12 Nm (105 inch lbs.).
23. Install secondary chain tensioner (2) bolts and tighten to 12 Nm (105 inch lbs.).
24. Remove locking pin from secondary tensioners.
25. Measure camshafts end play.
26. Install the primary timing chain.

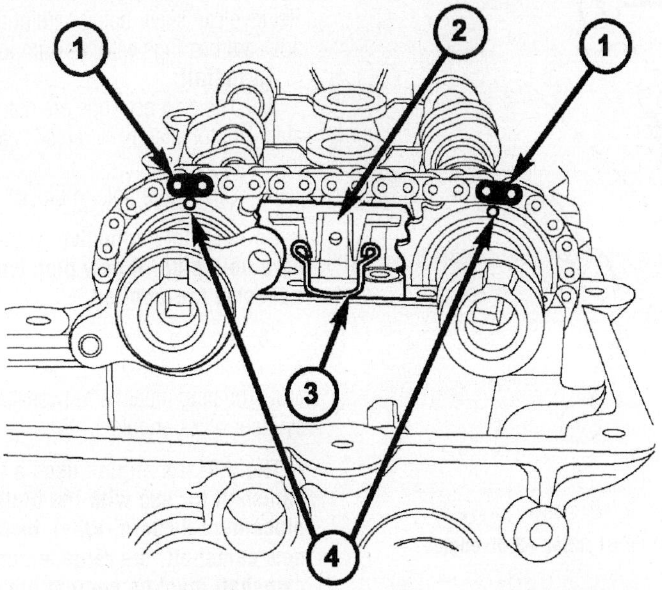

06009-300C-G61

Verify that plated links (1) are facing toward the front. Align the plated links (1) to the dots (4) on the camshaft sprockets—2.7L engine

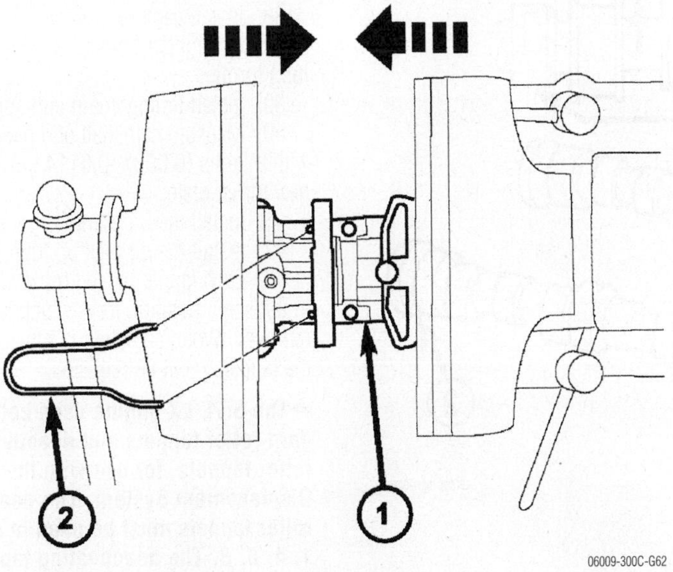

06009-300C-G62

Tensioner (1) and fabricated lock pin (2)—2.7L engine

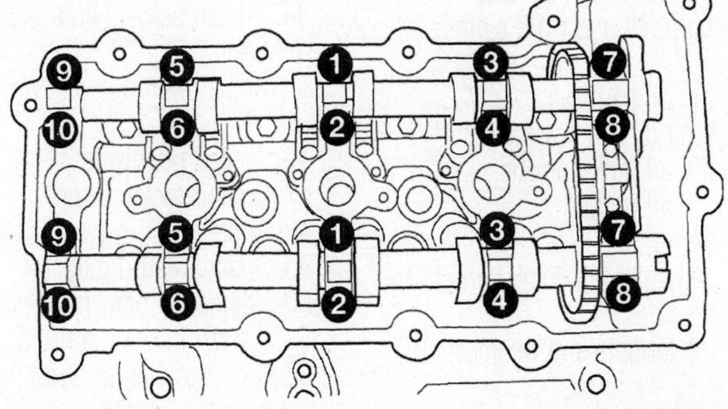

06009-300C-G63

Camshaft bearing torque sequence—2.7L engine

3.5L Engine

1. Before servicing the vehicle, refer to the precautions in the beginning of this section.

➡Camshafts are removed from the rear of each cylinder head.

2. Remove the cylinder head.

✲✲ WARNING

Care must be taken not to nick or scratch the journals when removing the camshaft.

3. Carefully remove the camshaft from the rear of the cylinder head.

To install:

4. Inspect camshaft bearing journals for damage and binding. If journals are binding, check the cylinder head for damage. Also check cylinder head oil holes for clogging.
5. Check the cam lobe and bearing surfaces for abnormal wear and damage. Replace camshaft if defective.

➡If camshaft is replaced due to lobe wear or damage, always replace the rocker arms.

6. Measure the lobe actual wear and replace camshaft if out of limit. Standard value is 0.0254 mm (0.001 in.), wear limit is 0.254 mm (0.010 in.).

➡Care must be taken not to scrape or nick the camshaft journals when installing the camshaft into position.

7. Lubricate camshaft bearing journals, camshaft lobes and camshaft seal with clean engine oil and install camshaft into cylinder head.
8. Install the cylinder head.

5.7L Engine

1. Before servicing the vehicle, refer to the precautions in the beginning of this section.
2. Remove the battery negative cable.
3. Remove the air cleaner assembly.
4. Drain coolant.
5. Remove the accessory drive belt.
6. Remove the alternator.
7. Remove the air conditioning compressor, and set aside. Do not disconnect the lines.
8. Remove upper radiator hose.
9. Remove upper radiator closure panels.
10. Disconnect cooling fan electrical connector.

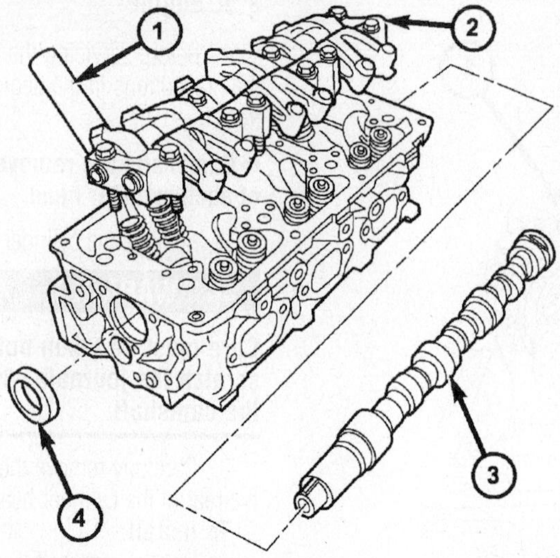

(1) rocker arm tool, (2) rocker arm assembly, (3) camshaft, (4) thrust collar—3.5L engine

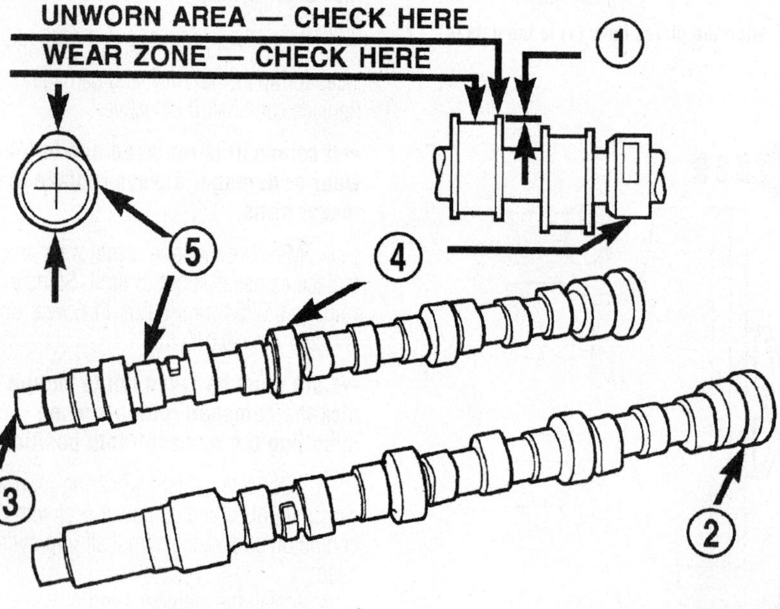

Camshaft inspection points—3.5L engine

11. Remove cooling fan mounting bolts.
12. Remove radiator cooling fan assembly from vehicle.
13. Raise vehicle.
14. Remove lower splash shield.
15. Remove lower radiator hose.
16. Remove lower condenser mount bolts.
17. Lower vehicle.
18. Remove upper radiator hose.
19. Remove upper radiator mounting brackets and bolts.
20. Remove upper condenser mounting bolts.
21. Separate condenser assembly from radiator.
22. Tilt radiator toward engine and remove radiator from vehicle.
23. Remove intake manifold.
24. Remove cylinder head covers.
25. Remove both left and right cylinder heads
26. Remove the oil pan.
27. Remove timing case cover.
28. Remove the oil pick up tube.
29. Remove the oil pump.
30. Remove timing chain.
31. Remove camshaft tensioner/thrust plate assembly.
32. Remove the tappets and retainer assembly.
33. Install a long bolt into front of camshaft to aid in removal of the camshaft. Remove camshaft, being careful not to damage cam bearings with the cam lobes.

To install:

34. The cam bearings are not serviceable. Do not attempt to replace cam bearings for any reason.
35. Clean core hole in block.

➡Do not apply adhesive to the new core hole plug. A new plug will have adhesive pre-applied.

36. Install a new core hole plug at the rear of camshaft, using suitable flat faced tool. The plug must be fully seated on the cylinder block shoulder.

➡The 5.7L LX engine uses a unique camshaft for use with the Multi Displacement System. When installing a new camshaft, the replacement camshaft must be compatible with the Multi Displacement System.

37. Lubricate camshaft lobes and camshaft bearing journals and insert the camshaft.
38. Install camshaft Tensioner plate assembly. Tighten bolts to 28 Nm (250 inch lbs.) torque.
39. Install timing chain and sprockets.
40. Measure camshaft end play. If not within limits (0.0031–0.0114 in.) install a new thrust plate.
41. Install the oil pump.
42. Install the oil pick up tube.
43. Each tappet reused must be installed in the same position from which it was removed. When camshaft is replaced, all of the tappets must be replaced.

➡The 5.7L LX engine uses both standard roller tappets and deactivating roller tappets, for use with the Multi Displacement System. The deactivating roller tappets must be used in cylinders 1, 4, 6, 8. The deactivating tappets can be identified by the two holes in the side of the tappet body, for the latching pins.

44. Install tappets and retaining yoke assembly.
45. Install both left and right cylinder heads.
46. Install pushrods.
47. Install rocker arms.
48. Install timing case cover.
49. Install the oil pan.
50. Install cylinder head covers.
51. Install intake manifold.
52. Position the air conditioning compressor on the engine.
53. Install the bolts that secure the air

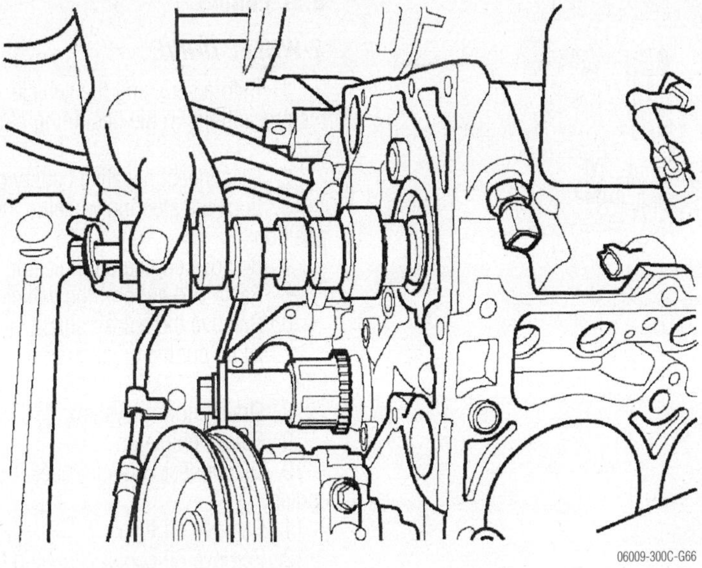

Camshaft removal/installation—5.7L engine

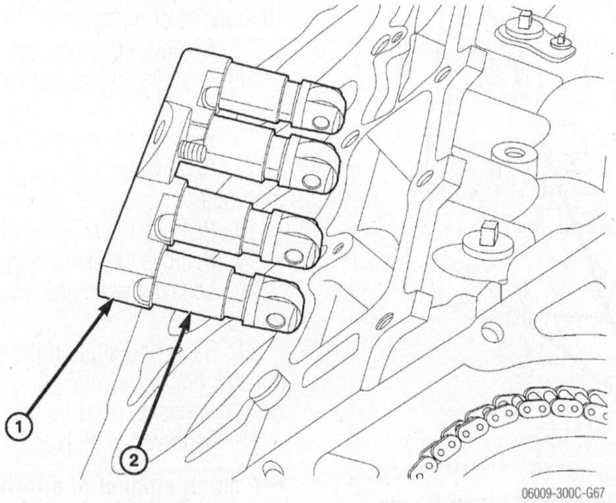

Lifter set—5.7L engine

conditioning compressor and the automatic transmission cooler line bracket to the cylinder block. Tighten the bolts to 55 Nm (41 ft. lbs.).

54. Install the air conditioning compressor.
55. Install the alternator.
56. Install the accessory drive belt.
57. Position radiator into engine compartment. Seat the radiator assembly lower rubber isolators into the mounting holes in radiator lower support.
58. Install radiator mounting bracket and bolts. Tighten to 12 Nm (106 inch lbs.).
59. Position condenser on radiator and install upper mounting bolts. Tighten bolts to 6 Nm (50 inch lbs.).
60. Raise vehicle.
61. Install lower condenser mounting bolts. Tighten bolts to 10 Nm (88 inch lbs.)
62. Install lower radiator hose and clamp.
63. Lower vehicle.
64. Position radiator cooling fan assembly in vehicle.
65. Install cooling fan mounting bolts. Tighten to 6 Nm (50 inch lbs.)
66. Connect cooling fan electrical connector.
67. Install upper radiator upper hose. Align hose so it does not interfere with the accessory drive belt or engine. Position hose clamp so it will not interfere with the hood.
68. Install the air cleaner assembly.
69. Install the battery negative cable.
70. Refill coolant.
71. Refill engine oil.
72. Start engine and check for leaks.

Oil Pan

REMOVAL & INSTALLATION

2.7L Engine

1. Before servicing the vehicle, refer to the precautions in the beginning of this section.
2. Disconnect negative battery cable located in trunk.
3. Remove engine oil indicator.
4. Raise vehicle on hoist.
5. Remove lower splash shield retaining bolts and splash shield.
6. Drain engine oil and remove oil filter.
7. Remove the steering coupler bolt and separate coupler from rack.
8. Disconnect power steering pressure switch electrical connector.
9. Remove steering rack mounting bolts.
10. Remove power steering line support from frame.
11. Reposition rack out of the way.
12. Remove structural collar.
13. Remove alternator mounting bracket to oil pan lower bolt.
14. Remove air conditioning compressor bracket to oil pan lower bolt.
15. Remove lower timing chain cover to oil pan bolts.

⁂ WARNING

Assure removal of the four lower timing cover bolts, as damage to the timing cover and/or oil pan may occur.

16. Remove oil pan attaching bolts. Remove oil pan and gasket.

To install:

17. Clean oil pan and sealing surfaces. Inspect timing chain cover gaskets. Replace as necessary.
18. Apply a ⅛ inch bead of Mopar® Engine RTV GEN II, or equivalent, to the front T-joints (oil pan gasket to timing cover gasket interface) and the rear T-joints (oil pan gasket to crankshaft rear oil seal retainer gasket interface).
19. Install oil pan gasket to block.

➡To prevent oil leaks at oil pan to timing chain cover, the following tightening sequence procedure must be performed.

20. Install oil pan and fasteners using the following tightening sequence:
    a. Install oil pan bolts and nuts finger tight only—just tight enough to com-

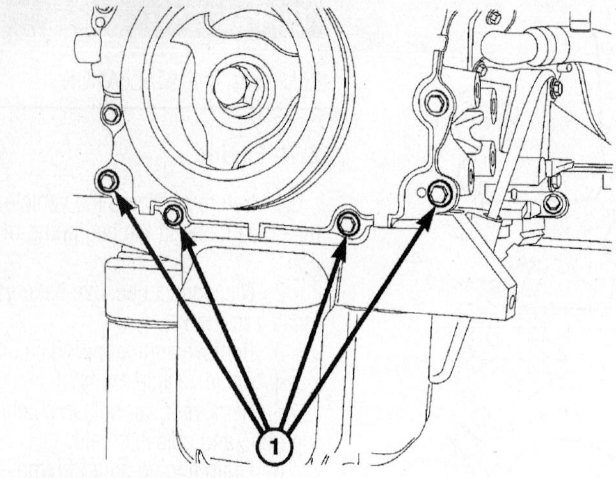

Lower timing chain cover-to-oil pan bolts—2.7L engine

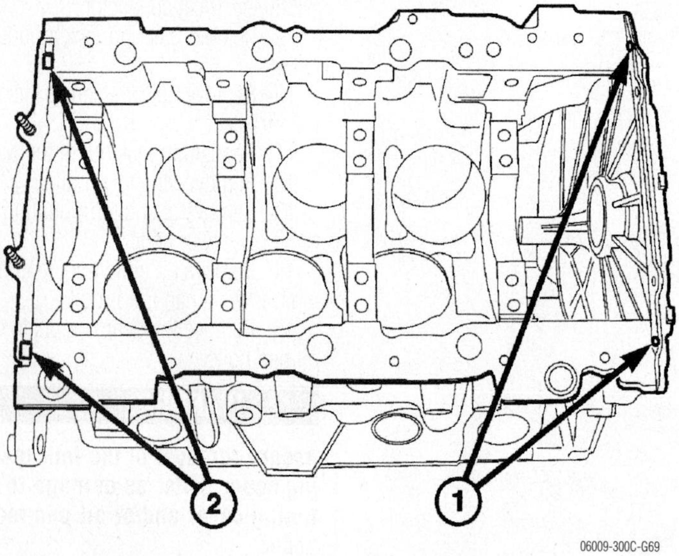

Apply a ⅛ inch bead of Mopar® Engine RTV GEN II, or equivalent, to the front T-joints (1) and the rear T-joints (2)—2.7L engine

press the gasket's rubber seal. Line up front of oil pan to be flush with front face of block.

b. Install lower timing chain cover bolts (1) and tighten to 12 Nm (105 inch lbs.).

c. Tighten oil pan bolts to 28 Nm (21 ft. lbs.).

d. Tighten oil pan nuts to 12 Nm (105 inch lbs.).

21. Install lower bolt attaching the air conditioning compressor to oil pan. Tighten bolt to 28 Nm (21 ft. lbs.).
22. Install lower bolt attaching the alternator bracket to oil pan. Tighten bolt to 28 Nm (21 ft. lbs.).
23. Install oil filter and drain plug.
24. Install structural collar.
25. Position steering gear in place and install mounting bolts. Tighten bolts to 70 Nm (95 ft. lbs.).
26. Install the steering coupling lower pinch bolt at the gear using a new bolt. Tighten to 30 Nm (22 ft. lbs.).
27. Connect electrical connector to steering gear.
28. Install power steering line support and tighten.
29. Install lower splash shield and retaining bolts.
30. Lower vehicle.
31. Install engine oil level indicator.
32. Fill engine crankcase with proper oil to correct level.
33. Connect negative battery cable.
34. Start engine and check for leaks.

3.5L Engine

2-WHEEL DRIVE

1. Before servicing the vehicle, refer to the precautions in the beginning of this section.
2. Disconnect negative battery cable.
3. Lock the steering wheel in the center position.
4. Remove engine oil indicator.
5. Raise and support the vehicle.
6. Remove the splash shield.
7. Drain engine oil and remove the oil filter.
8. Drain cooling system.
9. Raise vehicle on hoist.
10. Disconnect coolant hoses from oil cooler.
11. Remove oil filter.
12. Remove oil cooler attaching fastener from center of oil cooler.
13. Remove oil cooler.
14. Separate the steering column coupler from the steering gear.
15. Remove steering gear to cradle mounting bolts and suspend steering gear aside
16. Remove the flex plate access cover.
17. Remove the rear oil pan to transmission bolts.
18. Remove the two rear oil pan bolts.
19. Remove the remaining oil pan bolts.
20. Loosen the engine mount bolts at the cradle.
21. Raise and support the engine using a suitable floor jack with a block of wood at the transmission housing.
22. Remove the oil pan.

➡A small amount of oil will remain in the oil pan. Use care when removing the oil pan from the engine.

23. Clean all mating surfaces.

To install:

24. Clean oil pan and all gasket surfaces.
25. Apply a ⅛ inch bead of Mopar® Engine RTV GEN II, or equivalent, at the parting line of the oil pump housing and the rear seal retainer.
26. Install oil pan gasket to the engine block.
27. Install the oil pan while aligning the oil level indicator tube and attach fasteners finger tight.

➡Assure that the rear face of the oil pan is flush to the transmission bell housing when installing the oil pan.

28. Pre-torque the horizontal rear oil pan to transmission bolts to 1.4 Nm (12 inch lbs.).

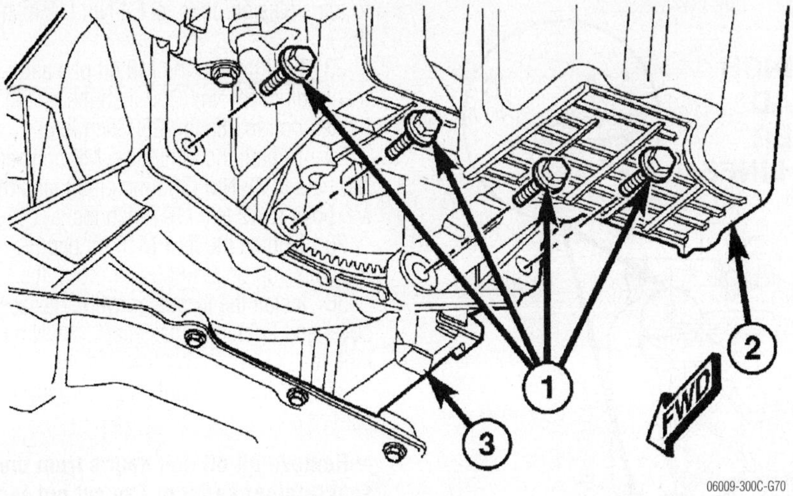

Rear oil pan (2) to transmission (3) bolts (1)—3.5L engine

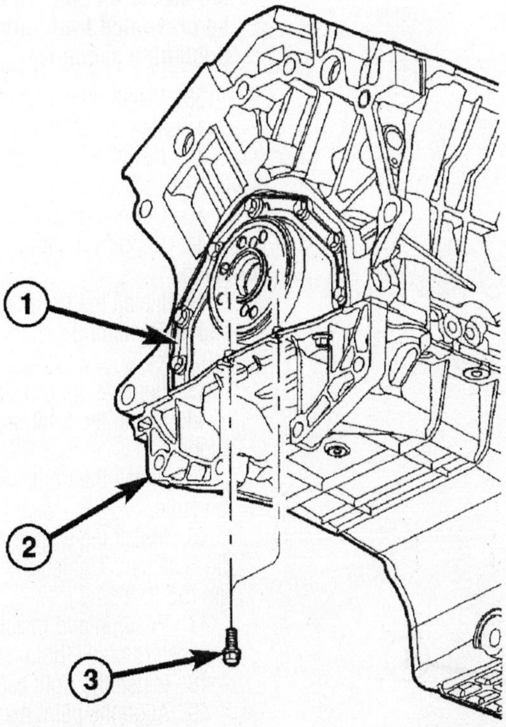

Rear oil pan bolts—3.5L engine

29. First tighten the M8 oil pan alignment bolt to 28 Nm (250 inch lbs.), then tighten bolt to 28 Nm (250 inch lbs.).
30. Tighten the remaining M8 bolts and M8 nuts to 28 Nm (250 inch lbs.), and the M6 bolts to 12 Nm (105 inch lbs.).
31. Tighten the four M10 oil pan to transmission bolts to 55 Nm (40 ft. lbs.).
32. Lower the engine and remove the lifting fixture. Tighten the engine mount to cradle fasteners to 75 Nm (55 ft. lbs.).
33. Install the flex plate inspection cover and tighten the fastener to 11 Nm (97 inch lbs.)
34. Position oil cooler to fitting on oil pan.

➡Remove all oil and debris from the seal retainer surface. The cut out section of the oil cooler seal retainer flange (top) must be aligned with the tab on the oil pan. The oil cooler must be prevented from turning during the tightening sequence.

35. Install oil cooler attaching fastener and tighten to 75 Nm (55 ft. lbs.).
36. Install oil filter and tighten to 12 Nm (106 inch lbs.).
37. Connect coolant hoses to oil cooler.
38. Fill cooling system.
39. Install the engine oil filter.
40. Install the steering gear to cradle bolts and tighten the fasteners to 70 Nm (95 ft. lbs.).
41. Connect the steering gear coupler and tighten the fastener to 30 Nm (22 ft. lbs.)
42. Attach the splash shield.
43. Lower the vehicle.
44. Install the engine oil indicator.
45. Fill engine crankcase with proper oil to correct level.
46. Connect negative battery cable.

ALL WHEEL DRIVE

1. Before servicing the vehicle, refer to the precautions in the beginning of this section.
2. Disconnect the negative battery cable.
3. Drain the cooling system.
4. Remove the oil level indicator tube fastener at the right exhaust manifold.
5. Raise and support the vehicle.
6. Drain engine oil and remove the oil filter.
7. Remove the left axle shaft.
8. Disconnect the power steering hoses from the power steering rack.
9. Remove the left front axle intermediate shaft support bracket and housing.
10. Remove the front drive shaft heat shield.
11. Paint mark and remove the front driveshaft.
12. Remove the right catalytic converter.
13. Separate the power steering hose from cradle hold down and position the hose aside.
14. Remove the lower engine mount to cradle fasteners.
15. Remove the front sway bar.
16. Raise and support the engine with a jackstand.
17. Remove the front axle housing support bracket.
18. Remove the front axle housing fasteners, rotate the housing position and remove the axle housing.
19. Lower the engine and remove the jackstand.
20. Disconnect coolant hoses from oil cooler.
21. Remove oil filter.
22. Remove oil cooler attaching fastener from center of oil cooler.
23. Remove oil cooler.
24. Separate the oil level indicator tube from the oil pan.
25. Remove the torque converter access cover.

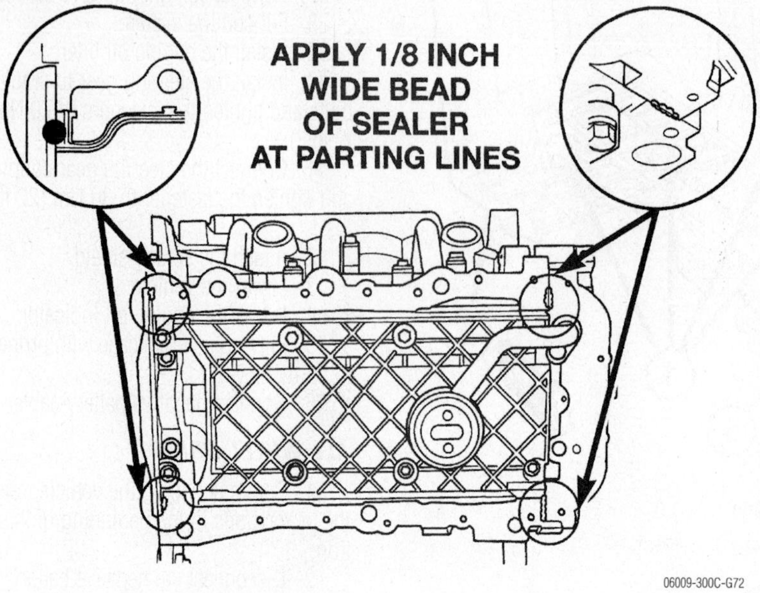

Oil pan sealer application—3.5L engine

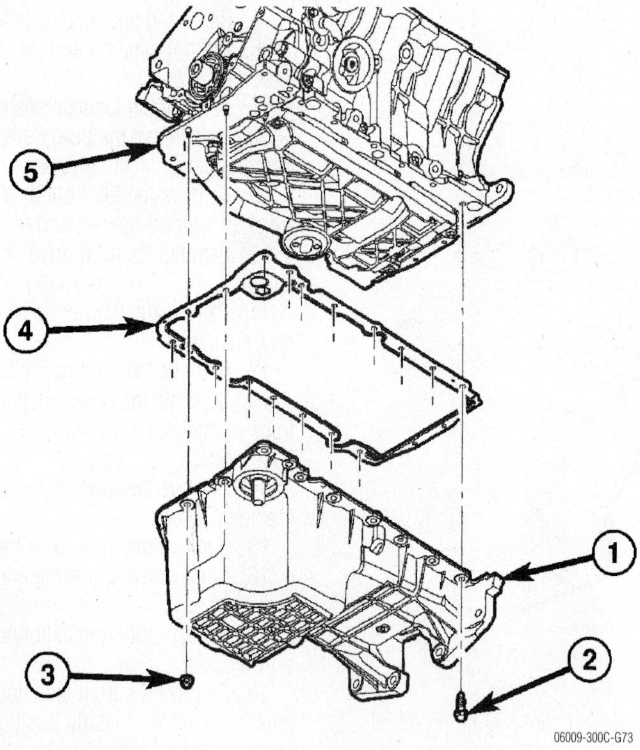

Oil pan and gasket installation—3.5L engine

26. Remove the oil pan.

To install:

27. Clean oil pan and all mating surfaces.
28. Replace the oil level indicator tube seal.
29. Apply a ⅛ inch bead of Mopar® Engine RTV GEN II, or equivalent, at the parting line of the oil pump housing and the rear seal retainer.
30. Install the oil pan gasket to the engine block.
31. Install the oil pan while aligning the oil level indicator tube and attach the fasteners finger tight.

➡Assure that the rear face of the oil pan is flush to the transmission bell housing when installing the oil pan.

32. Pre-torque the horizontal rear oil pan to transmission bolts to 1.4 Nm (12 inch lbs.)
33. First tighten the M8 oil pan alignment bolt to 28 Nm (250 inch lbs.), then tighten bolt to 28 Nm (250 inch lbs.).
34. Tighten the remaining M8 bolts and M8 nuts to 28 Nm (250 inch lbs.), and the M6 bolts to 12 Nm (105 Inch lbs.).
35. Tighten the four M10 oil pan to transmission bolts to 55 Nm (40 ft. lbs.).
36. Install the torque converter access cover and tighten the fastener to 11 Nm (97 inch lbs.).
37. Position oil cooler to fitting on oil pan.

➡Remove all oil and debris from the seal retainer surface. The cut out section of the oil cooler seal retainer flange (top) must be aligned with the tab on the oil pan. The oil cooler must be prevented from turning during the tightening sequence.

38. Install oil cooler attaching fastener and tighten to 75 Nm (55 ft. lbs.).
39. Install oil filter and tighten to 12 Nm (106 inch lbs.).
40. Connect coolant hoses to oil cooler.
41. Support the engine with a jackstand.
42. Install the front drive axle housing. Tighten mounting bolts to oil pan to 65 Nm (48 ft. lbs.).
43. Remove the jackstand.
44. Install the front drive axle housing support.
45. Install the front suspension sway bar to cradle.
46. Install the engine mount to front cradle fasteners. Tighten the fasteners to 75 Nm (55 ft. lbs.).
47. Position and install the power steering hose to cradle hold downs.
48. Install the right catalytic converter.
49. Align the paint marks and install the front drive shaft.
50. Install the front drive shaft heat shield.
51. Install the left front drive axle intermediate shaft housing and support bracket. Tighten fasteners to 25 Nm (19 ft. lbs.).
52. Connect the power steering hoses to the steering rack.
53. Install the left front drive axle.
54. Install the right front drive axle.
55. Lower the vehicle.
56. Install the oil level indicator tube fastener to the right exhaust manifold.
57. Fill engine crankcase with the proper oil to the correct level.
58. Fill the cooling system.
59. Fill the power steering.

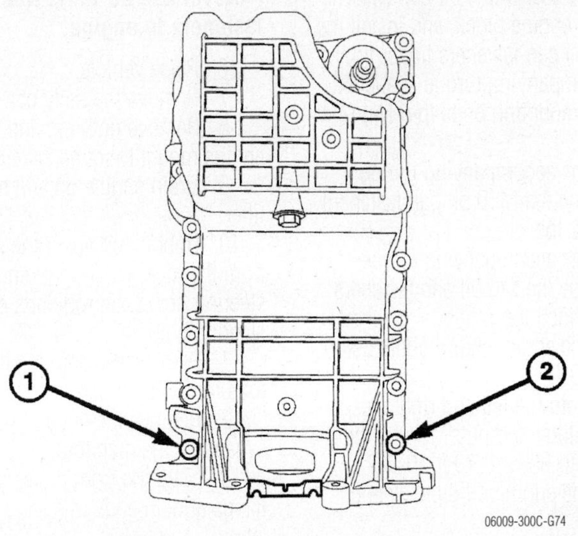

First tighten the M8 (1) oil pan alignment bolt to 28 Nm (250 inch lbs.), then tighten bolt (2) to 28 Nm (250 inch lbs.)—3.5L engine

60. Connect the negative battery cable.

5.7L Engine

2-WHEEL DRIVE

1. Before servicing the vehicle, refer to the precautions in the beginning of this section.
2. Disconnect the negative battery cable.
3. Remove the intake manifold.
4. Install engine lift fixture special tool 8984 and adapter 8984-UPD. See the Engine Removal & Installation procedure.

⁂ WARNING

Never use air tools when installing fasteners to engine.

5. Raise vehicle.
6. Remove the belly pan.
7. Drain engine oil and remove the oil filter.
8. Unbolt and lower the steering rack from the mounts. Do not remove power steering hoses, tie rod ends or disconnect steering column.
9. Remove both left and right side engine hydromount to frame nuts, and studs.
10. Remove the engine oil dipstick and tube from the oil pan.
11. Lower the vehicle.
12. Install engine support fixture special tool 8534. Do not use the third leg. See the Engine Removal & Installation procedure.
13. Raise engine using special tool 8534 and 8984 to provide clearance to remove oil pan.

➡Do not pry on oil pan or oil pan gasket. Gasket is integral to engine windage tray and does not come out with oil pan.

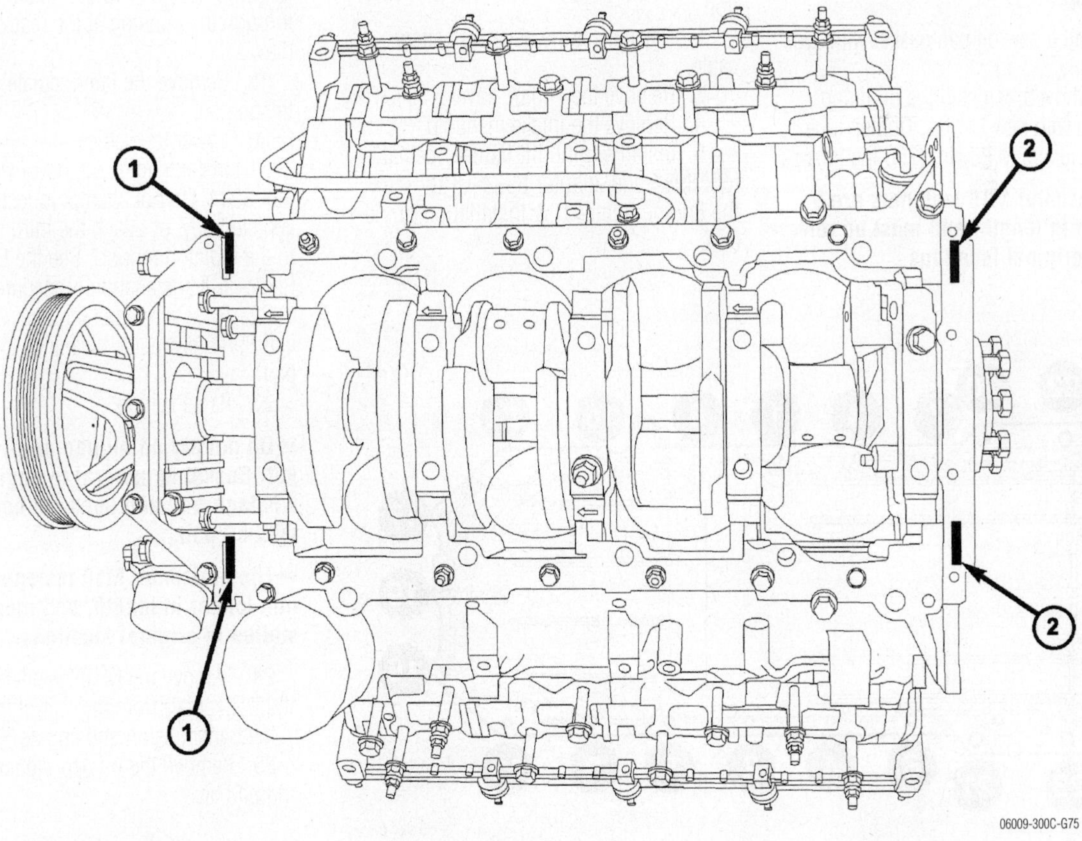

RTV sealer application—5.7L engine

➡The horizontal M10 fasteners are 5 mm longer in length, and must be reinstalled in original locations.

14. Remove the M10 fasteners (vertical and horizontal) from the rear of the oil pan to the transmission and engine.
15. Remove the oil pan mounting bolts and oil pan.

➡When the oil pan is removed, a new oil pan gasket/windage tray assembly must be installed. The old gasket cannot be reused.

16. Discard the integral windage tray and gasket and replace.

To install:

17. Clean the oil pan gasket mating surface of the block and oil pan.

➡Mopar® Engine RTV must be applied to the 4 T-joints, (area where front cover, rear retainer, block, and oil pan gasket meet). The bead of RTV should cover the bottom of the gasket. This area is approximately 4.5 mm x 25 mm in each of the 4 T-joint locations.

18. Apply Mopar® Engine RTV at the 4 T-joints.

➡When the oil pan is removed, a new oil pan gasket/windage tray assembly must be installed. The old gasket cannot be reused.

19. Install a new oil pan gasket/windage tray assembly.
20. If removed, reinstall the oil pump pickup tube with new O-ring. Tighten tube to pump fasteners to 28 Nm (250 inch lbs.).

➡The horizontal M10 fasteners are 5 mm longer in length, and must be reinstalled in original locations.

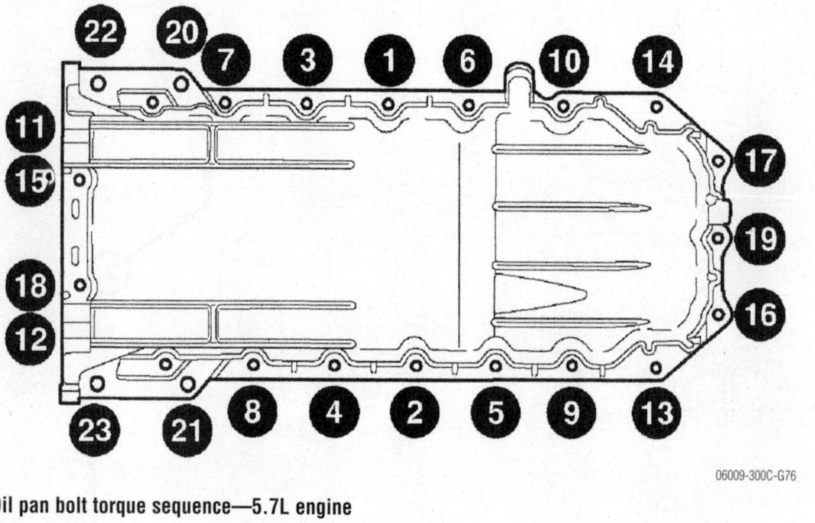

Oil pan bolt torque sequence—5.7L engine

21. Align the rear of the oil pan with the rear face of the engine block, and install the M10 and M6 oil pan fasteners finger tight. Using the accompanying torque sequence, torque the M6 mounting bolts to 5 Nm (44 inch lbs.).
22. Using the accompanying torque sequence, torque the M10 oil pan fasteners to 54 Nm (39 ft. lbs.).
23. Using the accompanying torque sequence, torque the M6 oil pan fasteners to 12 Nm (106 inch lbs.).
24. Lower the engine into mounts using special tool 8534.
25. Install both the left and right side engine mount studs and nuts. Torque the studs and nuts to 95 Nm (70 ft. lbs.).
26. Install the engine oil dipstick and tube.
27. Install the steering rack.
28. Remove special tool 8534.
29. Remove special tool 8984.
30. Install the intake manifold.
31. Fill engine oil.
32. Install oil filter, if removed.
33. Reconnect the negative battery cable.
34. Start engine and check for leaks.
35. Install the belly pan.

ALL WHEEL DRIVE

1. Before servicing the vehicle, refer to the precautions in the beginning of this section.
2. Disconnect the negative battery cable.
3. Remove the engine cover.
4. Remove the intake manifold.
5. Install engine lift fixture special tool 8984 and adapter 8984-UPD. See the Engine Removal & Installation procedure.

➡Never use air tools when installing fasteners to engine.

6. Raise vehicle.
7. Remove the belly pan.
8. Remove both left and right side engine mount to frame fasteners.
9. Drain engine oil and remove the oil filter.
10. Unbolt and lower the steering rack from the mounts. Do not remove power steering hoses, tie rod ends or disconnect steering column.
11. Remove the engine oil dipstick and tube from the oil pan.
12. Remove the left and right side exhaust at the manifolds.
13. Mark the front driveshaft to the flange at both ends to ensure correct installation.
14. Remove front driveshaft fasteners from the differential and transfer case. Remove the driveshaft from the vehicle.
15. Remove the left and right front drive axles.
16. Remove the differential support bracket from the differential to the engine block.
17. Unbolt the differential from the oil pan.
18. Rotate the differential so the drive flange is facing forward and the oil pan side is facing up. Remove the differential out through the opening at the rear of the cradle.
19. Remove the intermediate shaft from the oil pan.
20. Lower the vehicle.
21. Install engine support fixture special tool 8534. Do not attempt to fasten fixture to vehicle body, or attach the third support leg to the radiator support. See the Engine Removal & Installation procedure.
22. Raise engine using special tool 8534 and 8984 to provide clearance to remove oil pan.
23. Raise vehicle.

➡Do not pry on oil pan or oil pan gasket. Gasket is integral to engine windage tray and does not come out with oil pan.

➡The horizontal M10 fasteners are 5 mm longer in length, and must be reinstalled in original locations.

24. Remove the M10 fasteners (vertical and horizontal) from the rear of the oil pan to the transmission and engine.
25. Remove the oil pan mounting bolts and oil pan.

➡When the oil pan is removed, a new oil pan gasket/windage tray assembly must be installed. The old gasket cannot be reused.

26. Discard the integral windage tray and gasket and replace.

To install:

27. Clean the oil pan gasket mating surface of the block and oil pan.

➡Mopar® Engine RTV must be applied to the 4 T-joints, (area where front cover, rear retainer, block, and oil pan gasket meet). The bead of RTV should cover the bottom of the gasket. This area is approximately 4.5 mm x 25 mm in each of the 4 T-joint locations.

28. Apply Mopar® Engine RTV at the 4 T-joints.

➡When the oil pan is removed, a new oil pan gasket/windage tray assembly must be installed. The old gasket cannot be reused.

29. Install a new oil pan gasket/windage tray assembly.
30. If removed, reinstall the oil pump pickup tube with new O-ring. Tighten tube to pump fasteners to 28 Nm (250 inch lbs.).

➡The horizontal M10 fasteners are 5 mm longer in length, and must be reinstalled in original locations.

31. Align the rear of the oil pan with the rear face of the engine block, and install the M10 and M6 oil pan fasteners finger tight. Using the following torque sequence, torque the M6 mounting bolts to 5 Nm (44 inch lbs.).
32. Using the following torque sequence, torque the M10 oil pan fasteners to 54 Nm (39 ft. lbs.).
33. Using the following torque sequence, torque the M6 oil pan fasteners to 12 Nm (106 inch lbs.).
34. Install the intermediate shaft to the oil pan. Torque fasteners to 28 Nm (21 ft. lbs).
35. Install the front differential through the opening at the rear of the cradle, and attach to the oil pan.
36. Fasten the differential to the oil pan. Torque to 65 Nm (48 ft. lbs).
37. Install the differential support bracket to the differential and the engine block.
38. Install the left and right front drive axles.
39. Install front driveshaft into the differential and transfer case.
40. Install the engine oil dipstick and tube.
41. Install the steering gear.
42. Lower the engine into mounts using special tool 8534.
43. Install both the left and right side engine mount studs and nuts. Torque to 70 ft. lbs. (95 Nm).
44. Remove special tool 8534.
45. Remove special tool 8984.
46. Install the intake manifold.
47. Install the left and right side exhaust at the manifolds.
48. Fill engine oil.
49. Install oil filter, if removed.
50. Reconnect the negative battery cable.
51. Start engine and check for leaks.
52. Install the engine cover.

Oil Pump

REMOVAL & INSTALLATION

2.7L Engine

1. Before servicing the vehicle, refer to the precautions in the beginning of this section.
2. Remove crankshaft vibration damper.
3. Remove timing chain cover.
4. Remove timing chain and sprockets.
5. Remove oil pan.
6. Remove oil pick-up tube and O-ring.
7. Ensure that crankshaft position is at 60° ATDC of No.1 cylinder, or crankshaft sprocket mark aligns with mark on oil pump. This position will properly locate oil pump upon installation.
8. Remove oil pump attaching bolts.
9. Remove oil pump.

To install:

⁂ WARNING

Crankshaft position must be at 60° ATDC of No.1 cylinder before installing oil pump. This position will properly locate oil pump. If not properly located, severe damage to oil pump can occur.

10. Prime oil pump before installation by filling rotor cavity with engine oil.
11. If crankshaft has been rotated, it must be repositioned to 60° ATDC of No.1 cylinder prior to oil pump installation.
12. Install oil pump carefully over crankshaft and into position.
13. Install oil pump attaching bolts. Tighten bolts to 28 Nm (250 inch lbs.).
14. Install oil pick-up tube with new O-ring. Lubricate O-ring with clean engine oil before installation. Tighten attaching bolts to 28 Nm (250 inch lbs.).
15. Install oil pan.
16. Install timing chain and sprockets.
17. Install timing chain cover.
18. Install crankshaft vibration damper.
19. Fill crankcase with engine oil to correct level.

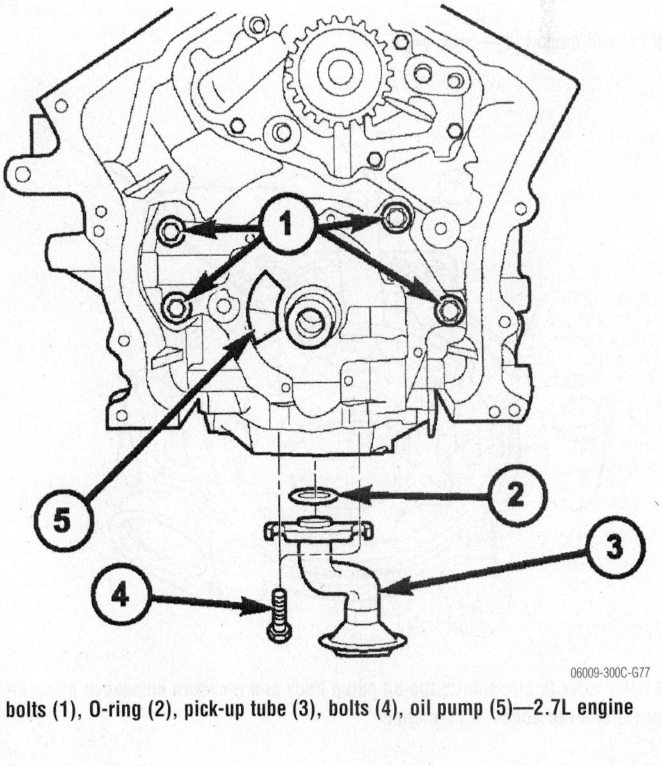

Oil pump bolts (1), O-ring (2), pick-up tube (3), bolts (4), oil pump (5)—2.7L engine

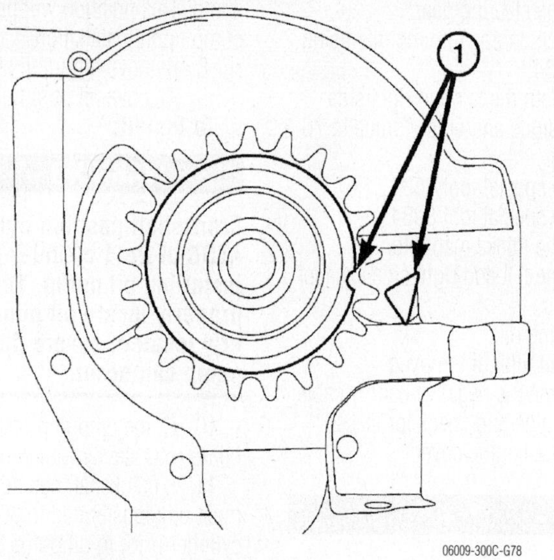

Crankshaft at 60 degrees ATDC—2.7L engine

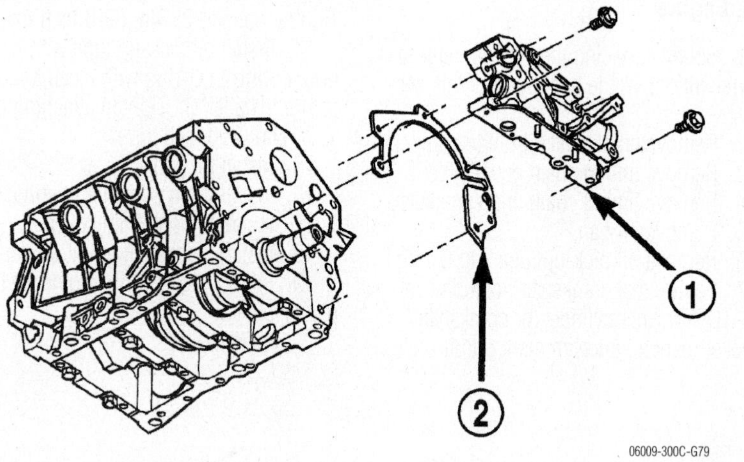

Oil pump (1) and gasket (2)—3.5L engine

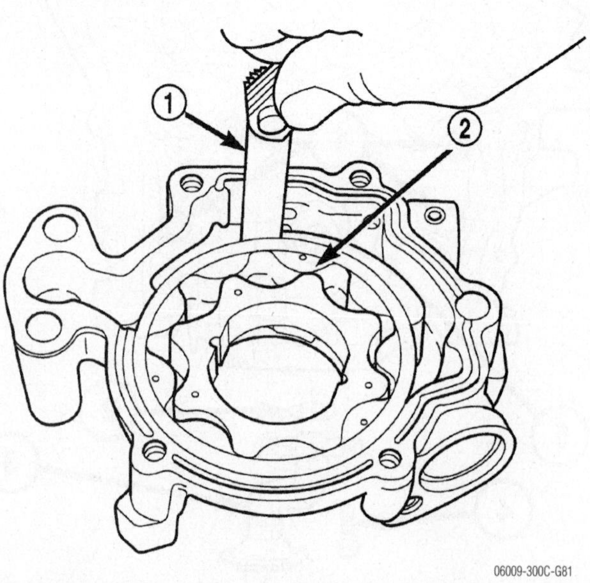

Press the outer rotor to one side of the oil pump body and measure clearance between the outer rotor (2) and the body—5.7L engine

3.5L Engine

1. Before servicing the vehicle, refer to the precautions in the beginning of this section.
2. Drain the cooling system.
3. Remove the timing belt.
4. Remove the crankshaft sprocket.
5. Remove the oil pan.
6. Remove the oil pickup tube.
7. Remove the oil pump fasteners. Remove the oil pump and gasket from engine.

To install:

8. Prime oil pump before installation by filling rotor cavity with clean engine oil.
9. Install oil pump and gasket carefully over the crankshaft. Position pump onto block and tighten bolts to 28 Nm (250 inch lbs.).
10. Install new O-ring on oil pickup tube.
11. Install oil pickup tube.
12. Install oil pan.
13. Install crankshaft sprocket.
14. Install timing belt.
15. Install the timing belt covers.
16. Install the crankshaft vibration damper. Torque to 70 ft. lbs. (95 Nm).
17. Install the accessory drive belt.
18. Fill the cooling system.
19. Fill engine crankcase with proper oil to the correct level.

5.7L Engine

1. Before servicing the vehicle, refer to the precautions in the beginning of this section.
2. Remove the oil pan and pick-up tube.
3. Remove the timing chain cover.
4. Remove the four bolts, and the oil pump.
5. Wash all parts in a suitable solvent.

➡Oil pump pressure relief valve and spring should not be removed from the oil pump. If these components are disassembled and or removed from the pump the entire oil pump assembly must be replaced.

6. Remove the pump cover.
7. Clean all parts thoroughly. Mating surface of the oil pump housing should be smooth. If the pump cover is scratched or grooved the oil pump assembly should be replaced.
8. Slide outer rotor into the body of the oil pump. Press the outer rotor to one side of the oil pump body and measure clearance between the outer rotor and the body. If the measurement is 0.235mm (0.009 in.)

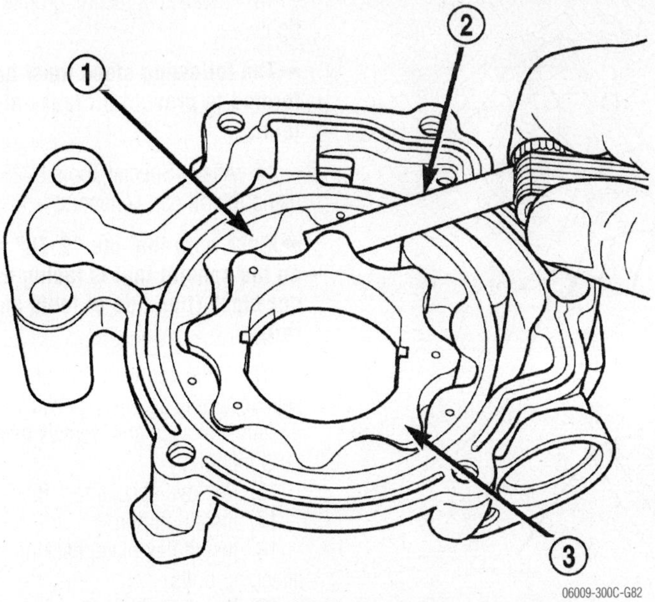

Measure the clearance between the inner (3) and outer rotors (1)—5.7L engine

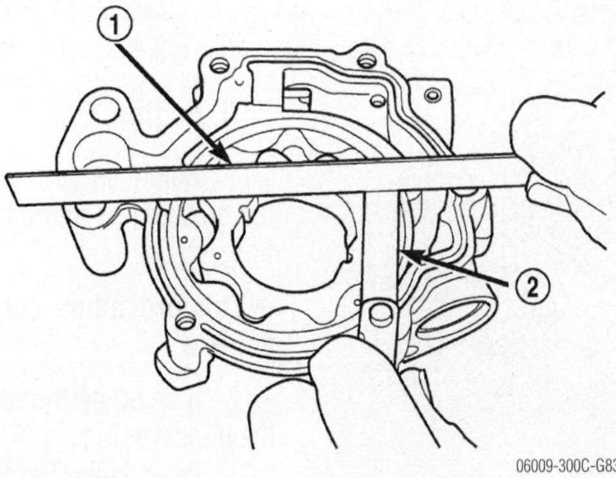

If a feeler gauge (2) of 0.095 mm (0.0038 in.) or greater can be inserted between the straight-edge and the rotors, the pump must be replaced—5.7L engine

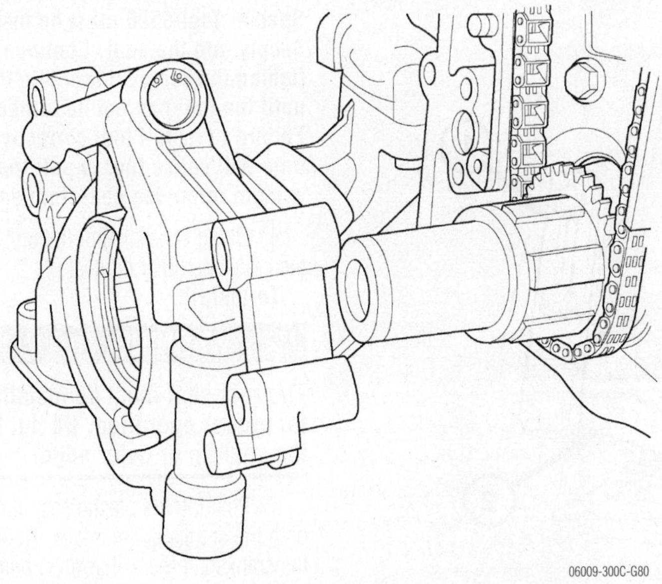

Oil pump—5.7L engine

or more the oil pump assembly must be replaced.

9. Install the inner rotor in the into the oil pump body. Measure the clearance between the inner and outer rotors. If the clearance between the rotors is 0.150 mm (0.006 in.) or more the oil pump assembly must be replaced.

10. Place a straight-edge across the body of the oil pump (between the bolt holes), if a feeler gauge of 0.095 mm (0.0038 in.) or greater can be inserted between the straight-edge and the rotors, the pump must be replaced.

11. Reinstall the pump cover. Torque fasteners to 15 Nm (132 inch lbs.).

To install:

12. Position the oil pump onto the crankshaft and install the 4 oil pump retaining bolts.

13. Tighten the oil pump retaining bolts to 28 Nm (250 inch lbs.).

14. Install the timing chain cover.

15. Install the pick-up tube and oil pan.

Rear Main Seal

REMOVAL & INSTALLATION

2.7L and 3.5L Engines

1. Before servicing the vehicle, refer to the precautions in the beginning of this section.
2. Disconnect negative battery cable.
3. Raise vehicle.
4. Remove splash shield retaining bolts and splash shield.
5. Remove the structural collar.
6. Remove transmission
7. Remove flex plate attaching bolts, backing plate and flex plate.
8. Remove oil pan.

➡The integrated stamped steel rear crankshaft seal is not interchangeable with the cast aluminum rear seal adapter and seal assembly.

9. Remove seal retainer attaching screws.
10. Remove crankshaft rear oil seal/adapter.

To install:

➡The integrated rear crankshaft seal is not interchangeable with the cast aluminum rear seal adapter and seal assembly.

11. Clean sealing surfaces.
12. Install seal assembly using special tool 6926-1.

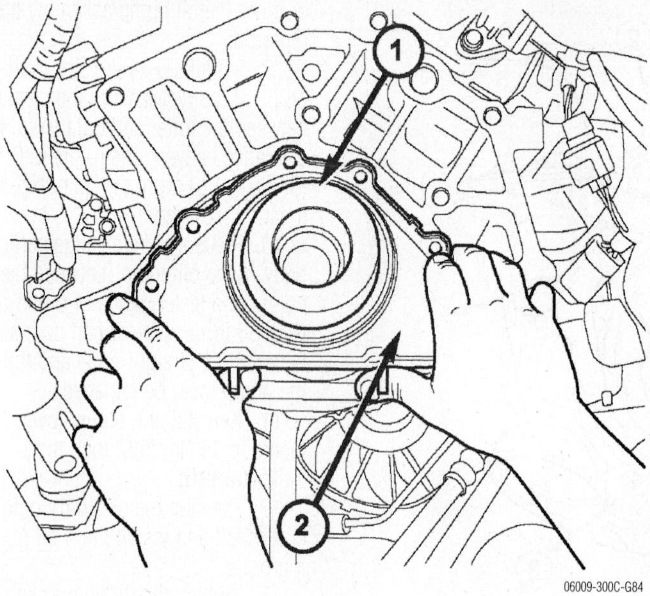

Rear main seal/retainer assembly—2.7L engine

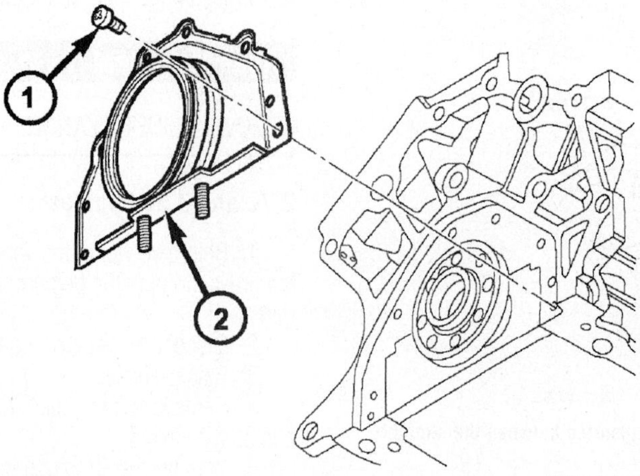

Install seal assembly (2) using special tool 6926-1(1)—2.7L engine

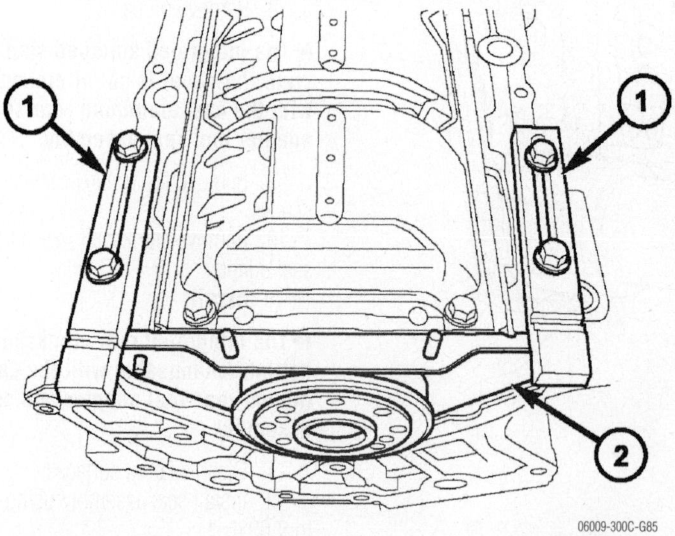

Attach Special Tools 8225 (1) to pan rail using the oil pan fasteners—2.7L engine

13. Install seal retaining bolts finger tight.

➡The following steps must be performed to prevent oil leaks at sealing joints.

14. Attach Special Tools 8225 to pan rail using the oil pan fasteners.

➡Make sure that the "2.7L" stamped on the special tool is facing the cylinder block (flat side of tools against pan rail).

15. While applying firm pressure to the seal assembly against Special Tools 8225, tighten seal assembly screws to 12 Nm (105 inch lbs.).
16. Remove special tool 8225.
17. Install oil pan.
18. Install flex plate, backing plate, and attaching bolts.
19. Install transmission.
20. Install lower splash shield retaining bolts and splash shield.
21. Lower vehicle.
22. Fill with oil.

5.7L Engine

1. Before servicing the vehicle, refer to the precautions in the beginning of this section.

➡This procedure can be performed in vehicle.

2. If being performed in vehicle, remove the transmission.
3. Remove the flexplate.

➡The crankshaft oil seal can not be reused after removal.

➡The crankshaft rear oil seal remover Special Tool 8506 must be installed deeply into the seal. Continue to tighten the removal tool into the seal until the tool can not be turned farther. Failure to install tool correctly the first time will cause tool to pull free of seal without removing seal from engine.

4. Using Special Tool 8506 (1), remove the crankshaft rear oil seal (2).

To install:

✱✱ WARNING

The rear seal must be installed dry for proper operation. Do not lubricate the seal lip or outer edge.

5. Position the plastic seal guide (2) onto the crankshaft rear face. Then position the crankshaft rear oil seal (3) onto the guide.

6. Using Special Tools 8349 Crankshaft Rear Oil Seal Installer (1) and C-4171 Driver Handle, with a hammer, tap the seal (3) into place. Continue to tap on the driver handle until the seal installer seats against the cylinder block crankshaft bore.
7. Install the flexplate.
8. Install the transmission.

Timing Belt Front Cover

REMOVAL & INSTALLATION

3.5L Engine

1. Before servicing the vehicle, refer to the precautions in the beginning of this section.
2. Perform fuel pressure release procedure.
3. Disconnect negative battery cable.
4. Remove accessory drive belt.
5. Remove accessory drive belt tensioner.
6. Remove bolts for power steering pump. Reposition power steering pump aside.
7. Raise vehicle on hoist.
8. Remove crankshaft damper bolt and remove the damper with a puller.
9. Remove the lower front timing belt cover fasteners.
10. Lower the vehicle.
11. Remove the upper timing belt cover bolts and remove front timing belt cover.

To install:

12. Exchange the accessory drive belt pulley if necessary. Tighten bolt to 61 Nm (45 ft. lbs.).
13. Install upper front timing belt cover.
14. Install lower timing belt front cover.
15. Tighten the timing cover bolts as follows:
    - M6 bolts: 12 Nm (105 inch lbs.)
    - M8 bolts: 28 Nm (250 inch lbs.)
    - M10 bolts: 54 Nm (40 ft. lbs.)
16. Install power steering pump fasteners. Tighten bolts to 23 Nm (200 inch lbs.).
17. Install crankshaft damper. Torque to 70 ft. lbs. (95 Nm).
18. Install accessory drive belt tensioner. Torque fastener to 28 Nm (250 inch lbs.).
19. Install accessory drive belt.
20. Lower vehicle.
21. Connect negative battery cable.

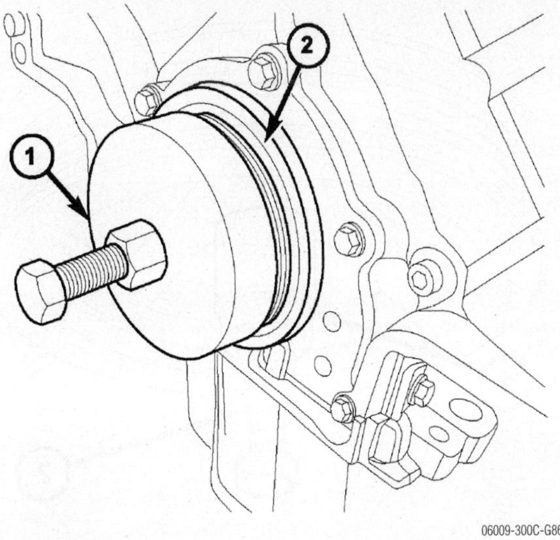

Rear main seal removal—5.7L engine

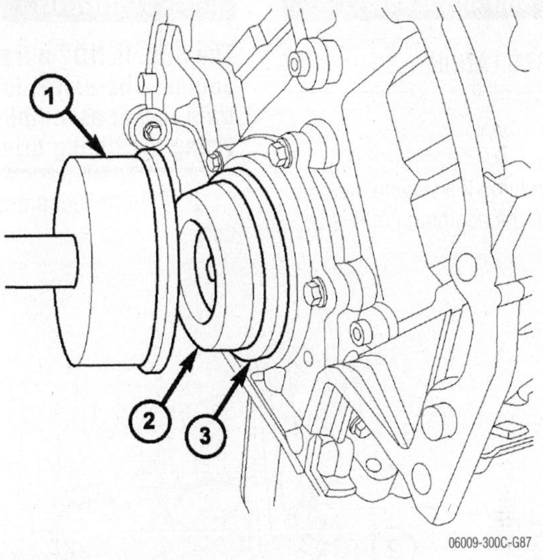

Rear main seal installation—5.7L engine

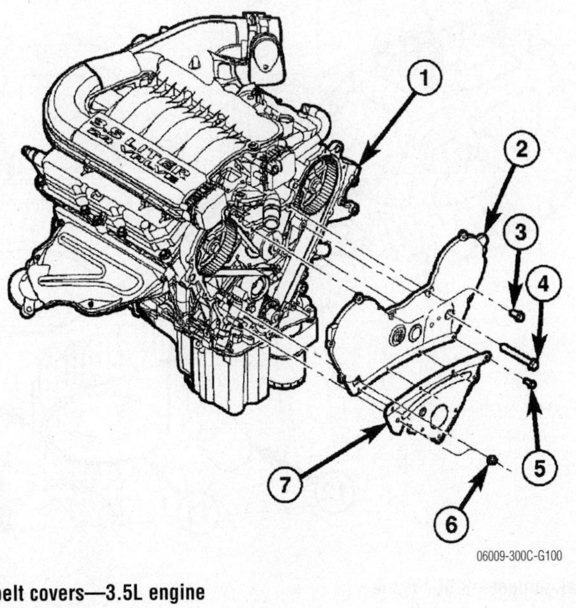

Front timing belt covers—3.5L engine

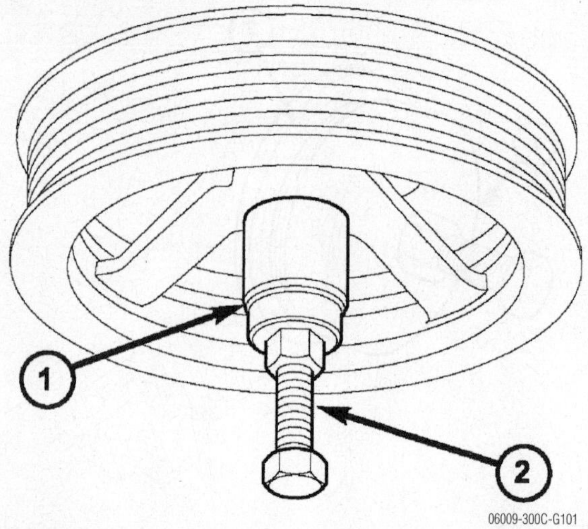

Damper installation—3.5L engine

Timing Belt and Sprockets

REMOVAL & INSTALLATION

3.5L Engine

1. Before servicing the vehicle, refer to the precautions in the beginning of this section.

WARNING

The 3.5L is NOT a freewheeling engine. Therefore, loosen the valve train rocker assemblies before servicing the timing drive.

2. Perform fuel pressure release procedure.
3. Disconnect negative battery cable.
4. Remove both cylinder head covers and loosen the rocker arm assemblies.
5. Remove the front timing belt cover.
6. Mark belt running direction, if timing belt is to be reused.

WARNING

When aligning timing marks, always rotate engine by turning the crankshaft. Failure to do so will result in valve and/or piston damage.

7. Rotate engine clockwise until crankshaft (10) mark aligns with the TDC mark on oil pump housing (9) and the camshaft sprocket (2, 7) timing marks (1, 8) are aligned with the marks on the rear cover.
8. Remove the timing belt tensioner (12) and remove timing belt.
9. Inspect the tensioner for fluid leakage.
10. Inspect the pivot and bolt for free movement, bearing grease leakage, and smooth rotation. If not rotating freely, replace the arm and pulley assembly.
11. When tensioner is removed from the engine it is necessary to compress the plunger into the tensioner body.

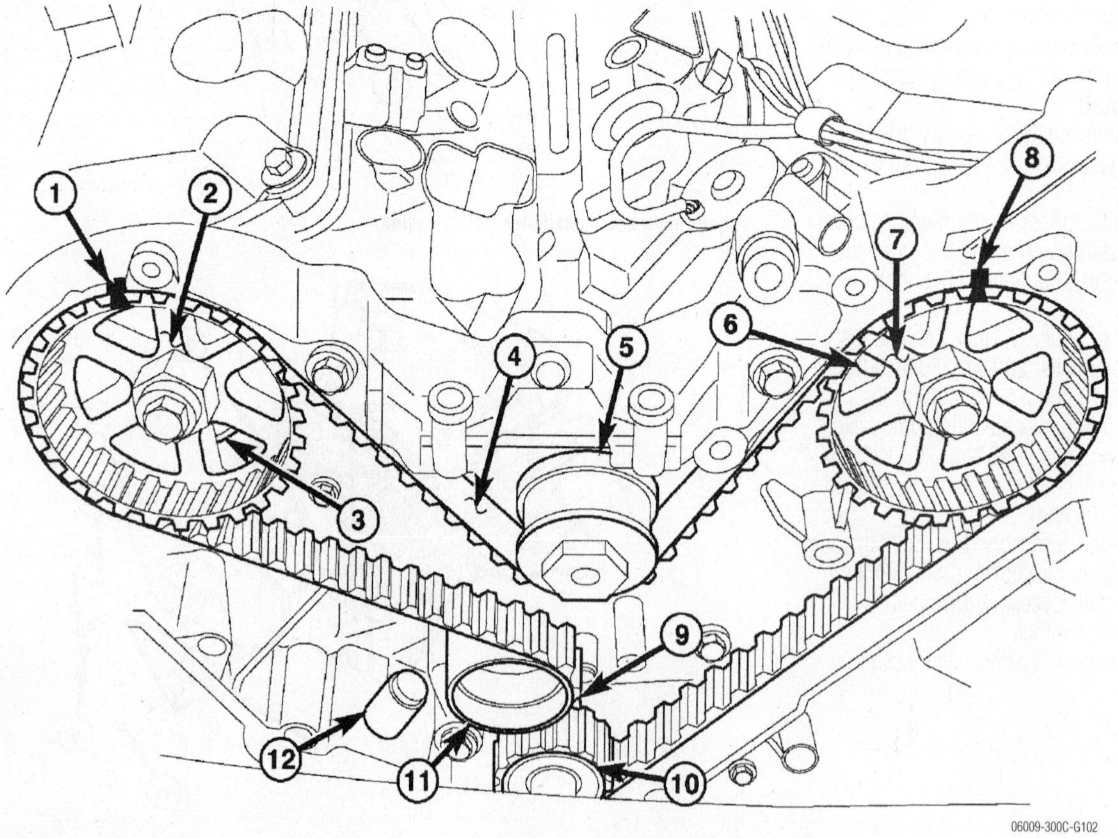

Timing mark alignment—3.5L engine

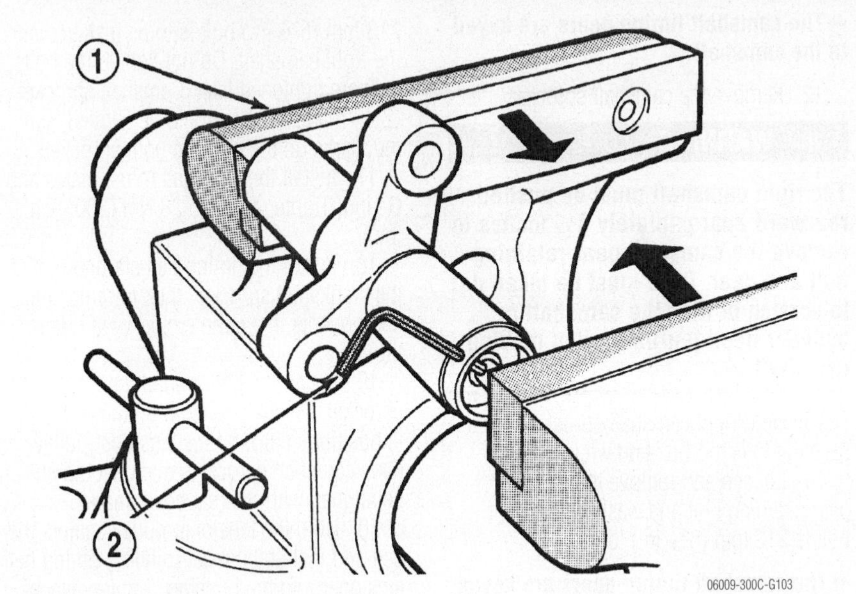

Position the tensioner in a vise—3.5L engine

➡Index the tensioner in the vise the same way it is installed on the engine. This ensures proper pin orientation when tensioner is installed on the engine.

a. Place the tensioner into a vise and SLOWLY compress the plunger. Total bleed down of tensioner should take about 5 minutes.

b. When plunger is compressed into the tensioner body install a pin through the body and plunger to retain plunger in place until tensioner is installed.

12. Inspect both sides of the timing belt. Replace belt if any of the following conditions exist:

a. Hardening of back rubber back side is glossy without resilience and leaves no indent when pressed with fingernail.

b. Cracks (4) on rubber back.

c. Cracks or peeling (1) of canvas.

d. Cracks on rib root.

e. Cracks on belt sides.

f. Missing teeth (2).

g. Abnormal wear (7) of belt sides. The sides are normal if they are sharp as if cut by a knife.

h. Vehicle mileage or time at component maintenance requirement.

13. If none of the above conditions are seen on the belt, the front timing belt cover can be installed.

To install:

WARNING

If camshafts have moved from the timing marks, always rotate camshaft towards the direction nearest to the timing marks. Do not turn camshafts a full revolution or damage to valves and/or pistons could result.

14. Align the crankshaft sprocket with the TDC mark on oil pump cover.

15. Align the camshaft sprockets timing reference marks with the marks on the rear cover.

16. Install the timing belt starting at the crankshaft sprocket going in a counterclockwise direction. Install the belt around the last sprocket. Maintain tension on the belt as it is positioned around the tensioner pulley.

➡If the camshaft gears have been removed it is only necessary to have the camshaft gear retaining bolts installed to a snug torque at this time.

17. Holding the tensioner pulley against the belt, install the tensioner into the housing and tighten to 28 Nm (250 inch lbs.). Each camshaft sprocket mark should remain aligned the cover marks.

18. When tensioner is in place pull retaining pin to allow the tensioner to extend to the pulley bracket.

19. Rotate crankshaft sprocket 2 revolutions and check the timing marks on the camshafts and crankshaft. The marks should line up within their respective locations. If marks do not line up, repeat procedure.

➡If camshaft gears have been removed and timing is correct, counterhold and tighten the camshaft gears to final torque specification.

20. Install the front timing belt cover.

21. Tighten the rocker arm assemblies and install the cylinder head covers.

22. Connect negative battery cable.

CAMSHAFT SPROCKETS

1. Before servicing the vehicle, refer to the precautions in the beginning of this section.

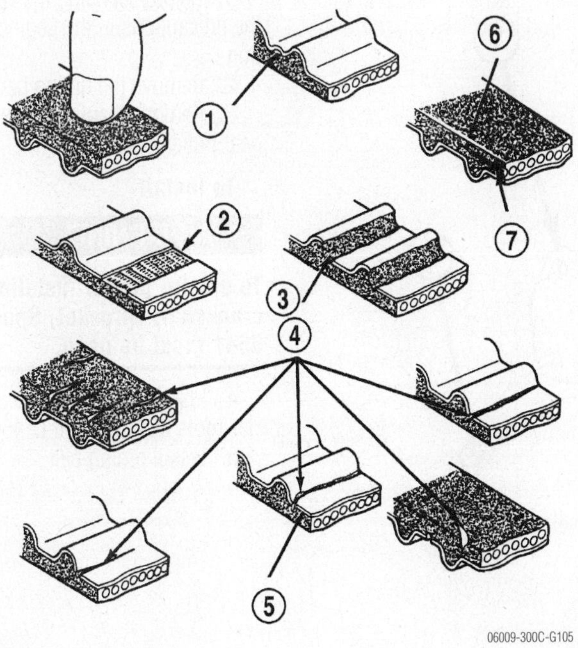

Timing belt inspection

✲✲ WARNING

The 3.5L engine is not a free-wheeling design. Therefore, care should be taken not to rotate the camshafts or crankshaft with the timing belt removed.

➡The camshaft timing gears are keyed to the camshaft.

2. Perform fuel pressure release procedure.
3. Remove front timing belt cover.
4. Position crankshaft sprocket to the TDC mark on the oil pump housing by turning crankshaft in the clockwise direction.
5. Install a dial indicator in number 1 cylinder to check TDC of the piston. Rotate the crankshaft until the piston is at exactly TDC.
6. Remove camshaft retainer/thrust plate from rear of right cylinder head.
7. Remove the right cylinder head cover.
8. Remove the right rocker arm assembly.
9. Remove the timing belt tensioner and timing belt.
10. Hold left camshaft sprocket with a 36 mm (1 7/16 in.) box end wrench.
11. Loosen and remove the camshaft gear retaining bolt and washer. The left bolt is 255 mm (10.0 in.) long.

➡The camshaft timing gears are keyed to the camshaft.

12. Remove the camshaft sprocket.

✲✲ WARNING

The right camshaft must be pushed rearward approximately 3½ inches to remove the camshaft gear retaining bolt and gear. Care must be taken not to scratch or nick the camshaft or cylinder head journals when moving camshaft.

13. Hold right camshaft sprocket with a 36 mm (1 7/16 in.) box end wrench.
14. Loosen and remove the camshaft gear retaining bolt and washer. The right bolt is 213 mm (8⅜ in.) long.

➡The camshaft timing gears are keyed to the camshaft.

15. Remove the camshaft sprocket.

To install:

✲✲ WARNING

The camshaft sprockets are keyed and not interchangeable from side to side because of the camshaft position sensor pick-up.

16. Install camshaft sprockets onto the camshafts. Install NEW sprocket attaching bolts into place. The 255 mm (10 in.) bolt is to be installed in the left camshaft and the 213 mm (8⅜ in.) bolt is to be installed into the right camshaft. Do not tighten the bolts; they are tightened later. Camshaft sprocket (2, 7) marks (1, 8) should be aligned with the marks on the cover at both sprockets.
17. Install the camshaft thrust plates and O-ring. Tighten bolts to 28 Nm (250 inch lbs.).
18. Install the timing belt starting first at the crankshaft sprocket, then to remaining components in a counterclockwise direction.
19. Install the belt around the last sprocket. Maintain tension on the belt as it is positioned around the tensioner pulley. Each camshaft sprockets mark should still be aligned with the rear cover marks.
20. Hold the tensioner pulley against the belt and install the reset (pinned) timing belt tensioner into the housing. Tighten attaching bolts to 28 Nm (250 inch lbs.).
21. Remove tensioner retaining pin to allow the tensioner to extend to the pulley bracket.
22. Using a dial indicator, position the number 1 piston at TDC.
23. Hold the camshaft sprocket hex with a 36 mm (1 7/16 in.) wrench and tighten the camshaft bolts to the following:
    - Right side: 102 Nm (75 ft. lbs.) +90° turn
    - Left side: 102 Nm (75 ft. lbs.) +90° turn
24. Remove dial indicator and install spark plug.
25. Install front timing belt cover.

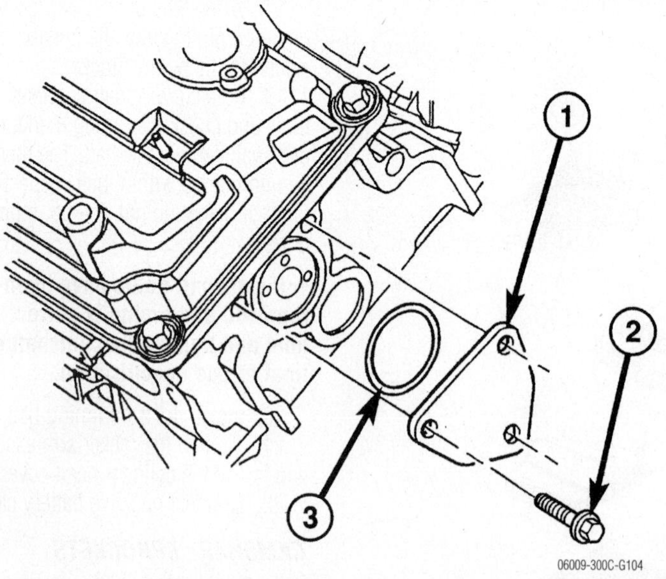

Camshaft retainer/thrust plate (1), bolts (2) and O-ring (3)—3.5L engine

CRANKSHAFT SPROCKET

1. Before servicing the vehicle, refer to the precautions in the beginning of this section.
2. Remove the timing belt.
3. Remove crankshaft sprocket using a gear puller.

To install

✲✲ WARNING

To ensure proper installation depth of crankshaft sprocket, Special Tool 6641 must be used.

4. Install crankshaft sprocket using Special Tools 6641 (1) and C-4685-C1.
5. Install timing belt.

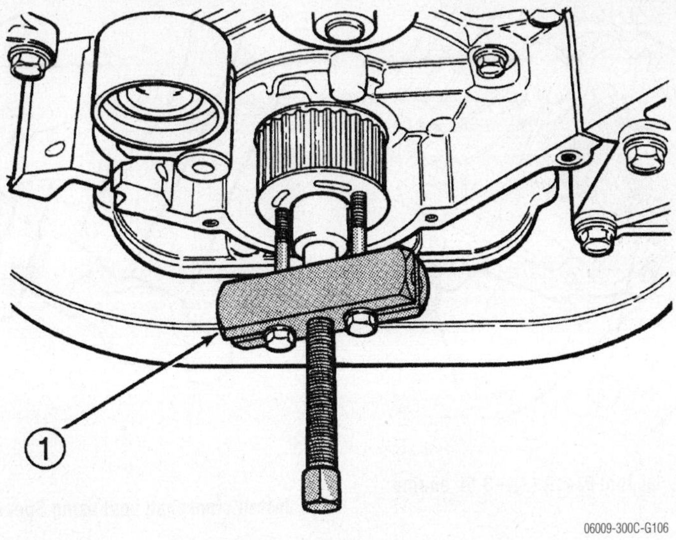

Remove crankshaft sprocket using a gear puller (1)—3.5L engine

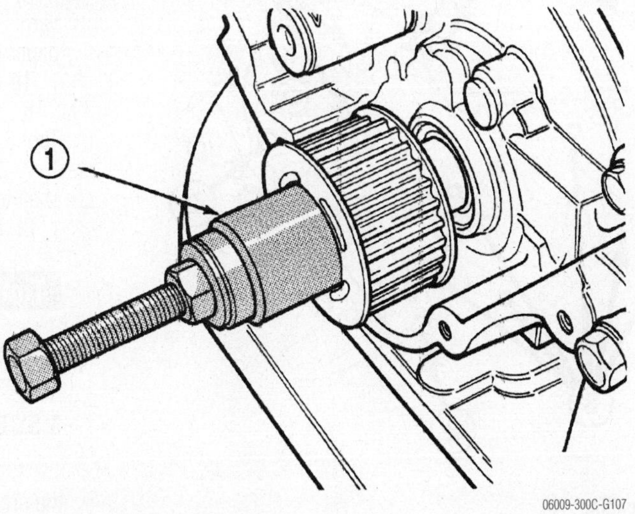

To ensure proper installation depth of crankshaft sprocket, Special Tool 6641 must be used—3.5L engine

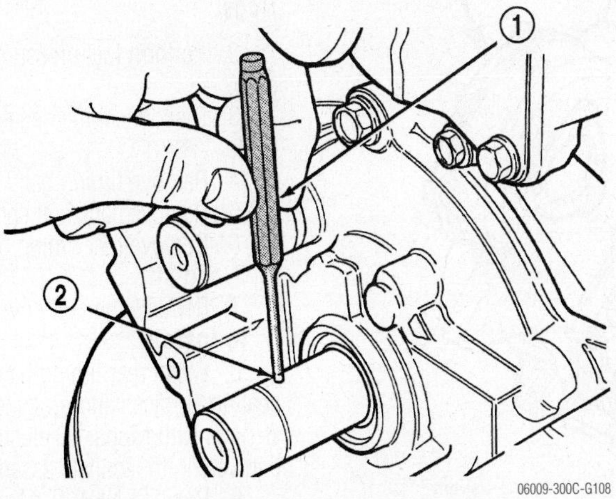

Tap the dowel pin (2) out of the crankshaft—3.5L engine

Crankshaft Front Seal

REMOVAL & INSTALLATION

3.5L Engine

For front seal service on the 2.7L and 5.7L engines, see Timing Chain Cover and Seal.

1. Before servicing the vehicle, refer to the precautions in the beginning of this section.
2. Remove the crankshaft sprocket.
3. Tap the dowel pin out of the crankshaft.
4. Remove crankshaft seal using Special Tool 6341A.

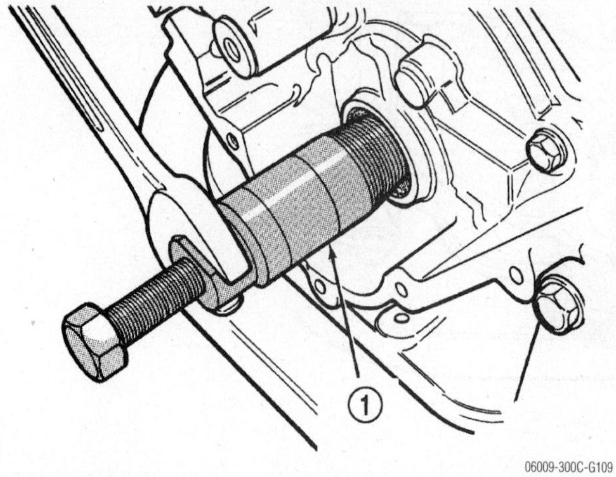

Remove crankshaft seal using Special Tool 6341A (1)—3.5L engine

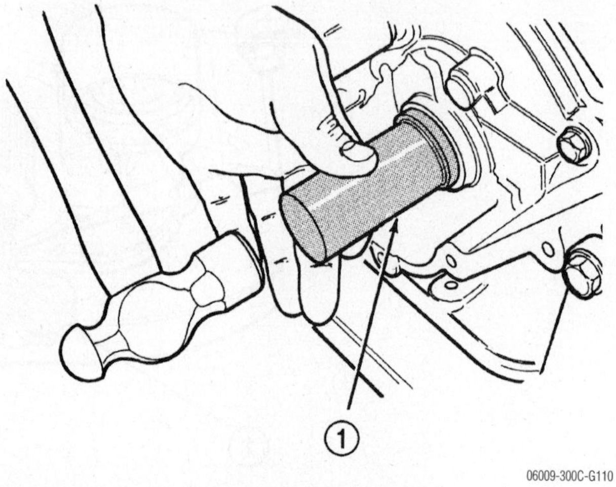

Install crankshaft seal using Special Tool 6342 (1)—3.5L engine

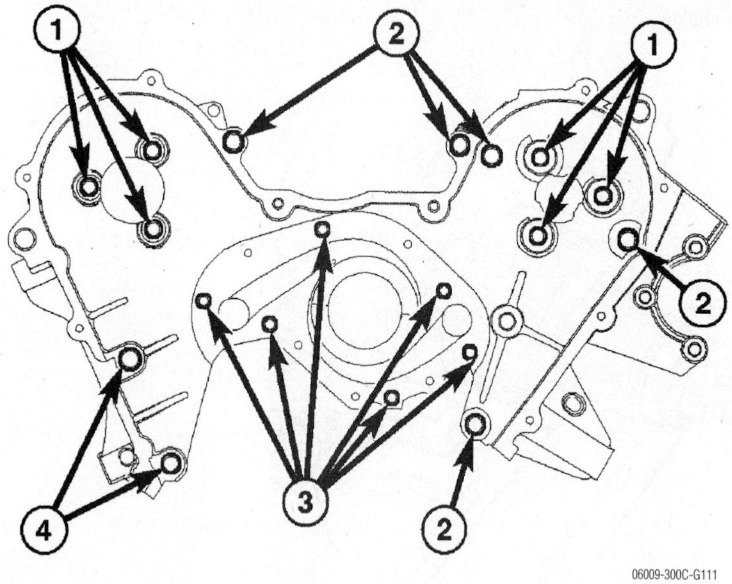

Rear timing belt cover bolts—3.5L engine

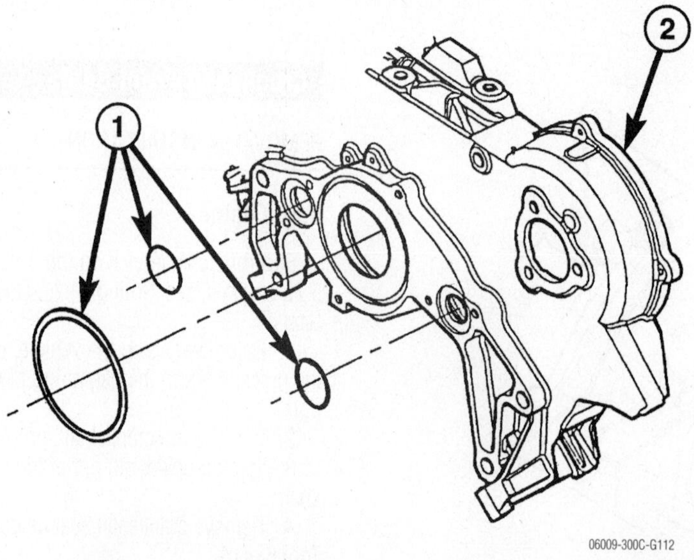

Rear timing belt cover O-rings—3.5L engine

➡Do not nick shaft seal surface or seal bore.

5. Shaft seal lip surface must be free of varnish, dirt or nicks. Polish with 400 grit paper if necessary.

To install:

6. Install crankshaft seal using Special Tool 6342.
7. Install the dowel pin into the crankshaft to 1.2 mm (0.047 in.) protrusion.
8. Install the crankshaft sprocket.

Timing Belt Rear Cover

REMOVAL & INSTALLATION

3.5L Engine

1. Before servicing the vehicle, refer to the precautions in the beginning of this section.

➡The rear timing belt cover has O-rings to seal the water pump passages to cylinder block. Do not reuse the O-rings.

2. Perform fuel pressure release procedure.
3. Disconnect the negative battery cable.
4. Remove timing belt.
5. Remove camshaft sprockets.
6. Remove rear timing belt cover bolts (1, 2, and 3).
7. Remove the rear cover.

To install:

8. Clean rear timing belt cover O-ring sealing surfaces and grooves. Lubricate new O-rings with Mopar® Dielectric Grease, or equivalent, to facilitate assembly.
9. Position NEW O-rings on cover.

10. Install rear timing belt cover. Tighten bolts to the following specified torque:
- M10 (2, 4): 54 Nm (40 ft. lbs.)
- M8 (1): 28 Nm (20 ft. lbs.)
- M6 (3): 12 Nm (105 inch lbs.)

11. Install camshaft sprockets.
12. Install timing belt.

Timing Chain Cover and Seal

REMOVAL & INSTALLATION

2.7L Engine

1. Before servicing the vehicle, refer to the precautions in the beginning of this section.
2. Disconnect negative battery cable.
3. Drain cooling system.
4. Remove coolant recovery bottle.
5. Raise vehicle on hoist.
6. Remove lower splash shield retaining bolts and splash shield.
7. Remove accessory drive belts.
8. Use special tool 9365 Damper Holder to hold damper while removing attaching bolt.
9. Remove damper by using Special Tools 8194 Insert and 8454 Puller.
10. Lower vehicle.
11. Remove timing chain cover bolts.
12. Remove timing chain cover.
13. Discard timing chain cover gasket.
14. Install Special Tool 8194, insert into crankshaft nose. Remove seal using Special Tool 6771, Remover.

To install:

15. Inspect and clean timing chain cover sealing surfaces.
16. Before installing timing cover gasket apply a ⅛ inch bead of Mopar® Engine RTV GEN II, or equivalent, to the parting lines between the oil pan and cylinder block.
17. Install timing cover and gasket. Tighten M10 cover bolts to 54 Nm (40 ft. lbs.) and M6 bolts to 12 Nm (105 inch lbs.).
18. Install new seal using Special Tools 6780-2 Sleeve, 6780-1 Installer, and 8179 Stud.
19. Install damper using Special Tools

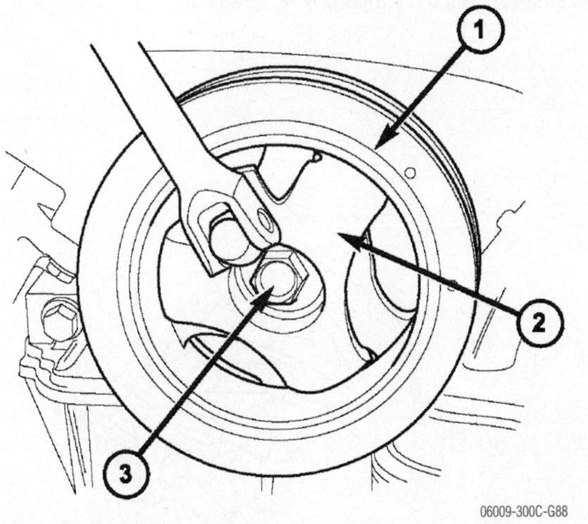

Use special tool 9365 Damper Holder (2) to hold damper (1) while removing attaching bolt (3)—2.7L engine

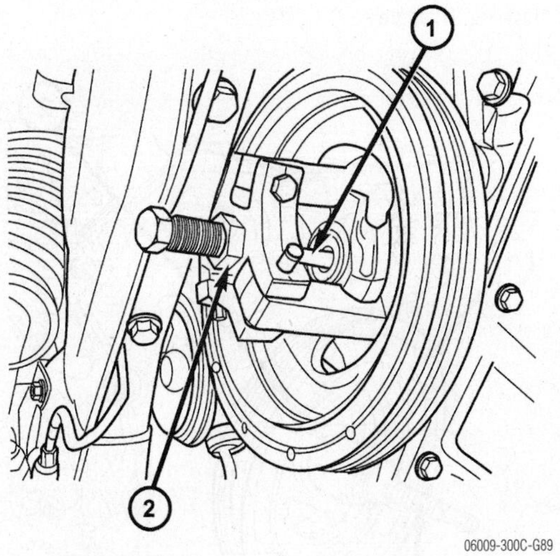

Remove damper by using Special Tools 8194 Insert (1) and 8454 Puller (2)—2.7L engine

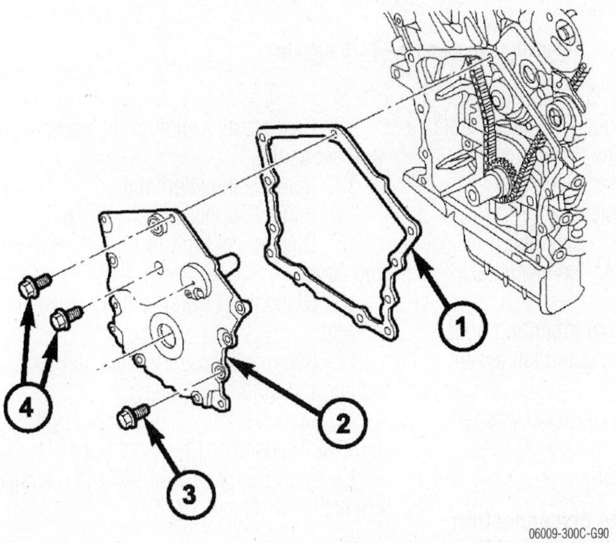

Timing chain cover gasket (1), cover (2), and bolts (3 & 4)—2.7L engine

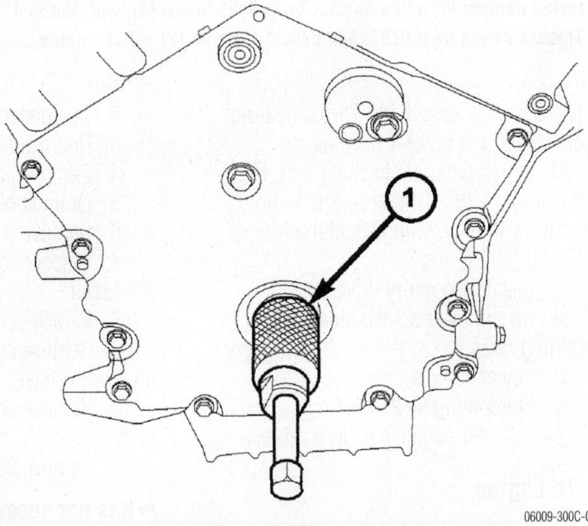

Install Special Tool 8194 (1), Insert into crankshaft nose to remove the seal—2.7L engine

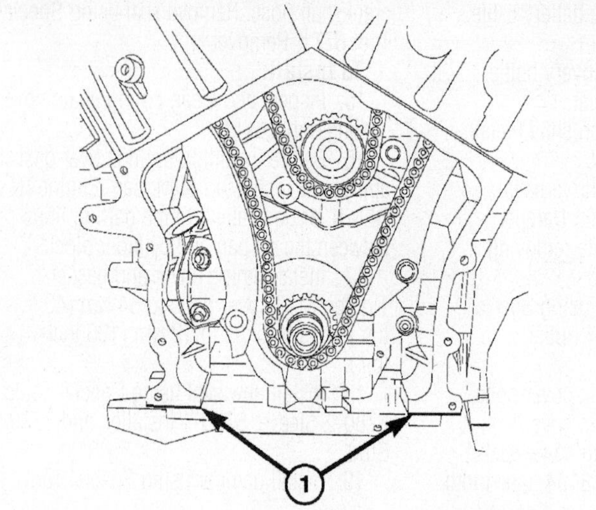

RTV application points at parting lines between the oil pan and cylinder block—2.7L engine

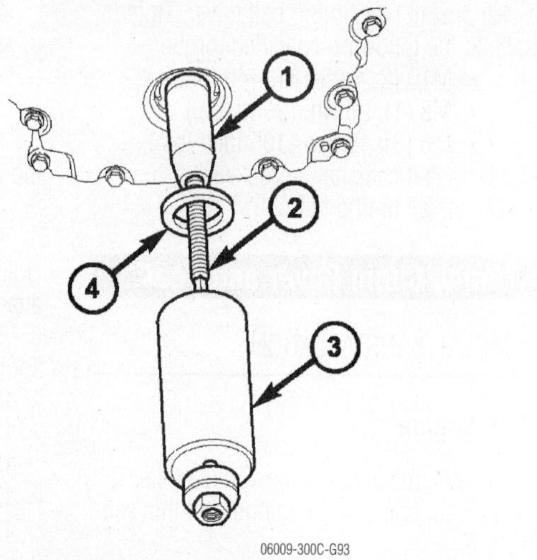

Install new seal (4) using Special Tools 6780-2 Sleeve (1), 6780-1 Installer (3), and 8179 Stud (2)—2.7L engine

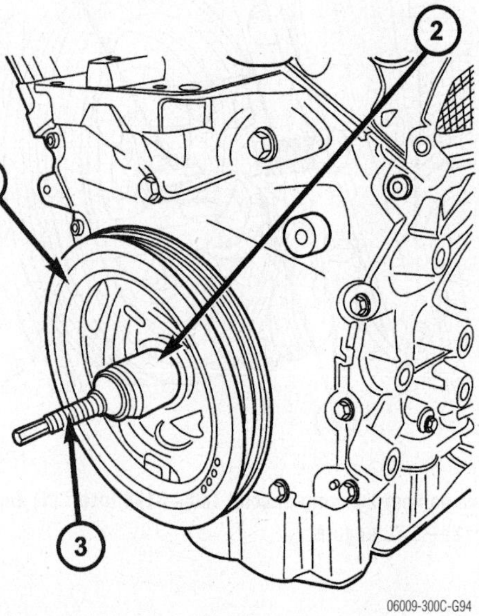

Install damper (1) using Special Tools 8179 Screw (3), with Nut and Thrust Bearing from 6792, and 6792-1 Installer (2)—2.7L engine

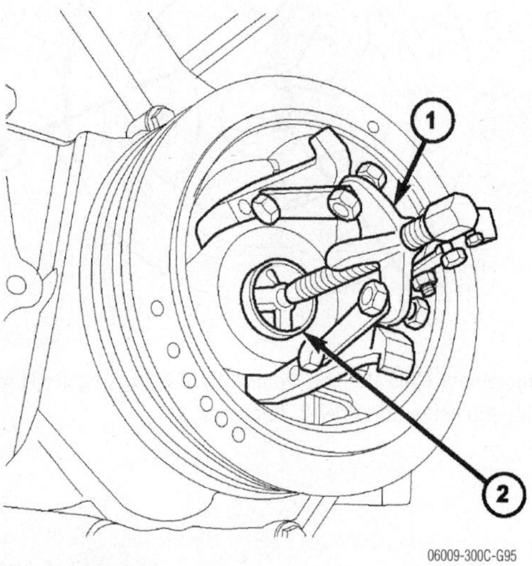

Removing the damper—5.7L engine

8179 Screw, with Nut and Thrust Bearing from 6792, and 6792-1 Installer.

20. Install damper attaching bolt. Use special tool 9365 damper holder to hold damper while tightening attaching bolt to 170 Nm (125 ft. lbs.).

21. Install accessory drive belts.

22. Install lower splash shield and retaining bolts.

23. Lower vehicle.

24. Fill cooling system.

25. Connect negative battery cable.

5.7L Engine

1. Before servicing the vehicle, refer to the precautions in the beginning of this section.
2. Disconnect the battery negative cable.
3. Remove the engine cover.
4. Remove air cleaner assembly.
5. Drain cooling system.
6. Remove accessory drive belt.
7. Disconnect cooling fan electrical connector.
8. Remove cooling fan mounting bolts.
9. Remove radiator cooling fan assembly from vehicle.
10. Remove coolant bottle and washer bottle.
11. Remove fan shroud.

➡It is not necessary to disconnect air conditioning lines or discharge refrigerant.

12. Remove air conditioning compressor and set aside.
13. Remove the alternator.
14. Remove upper radiator hose.
15. Disconnect both heater hoses at timing cover.
16. Disconnect lower radiator hose at engine.
17. Remove accessory drive belt tensioner and both idler pulleys.
18. Remove crankshaft damper bolt and remove the damper (2) with a puller (1).
19. Using Special Tool 9071 (1), remove crankshaft front seal (2).

➡Do not remove the hoses from the power steering pump.

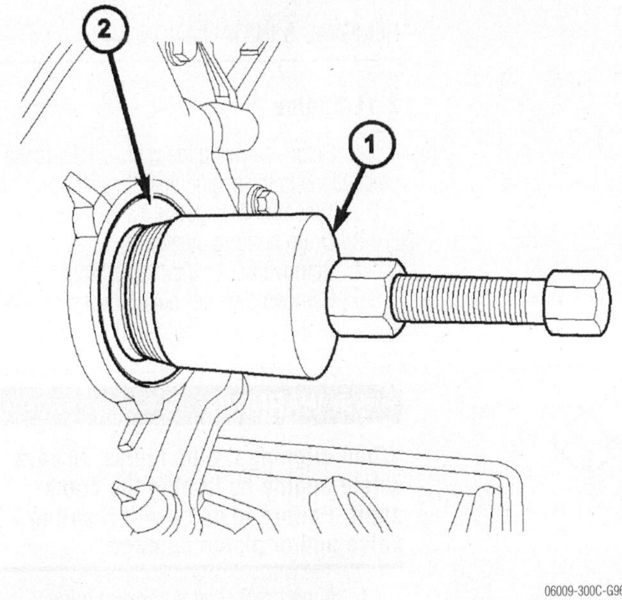

Removing the front seal—5.7L engine

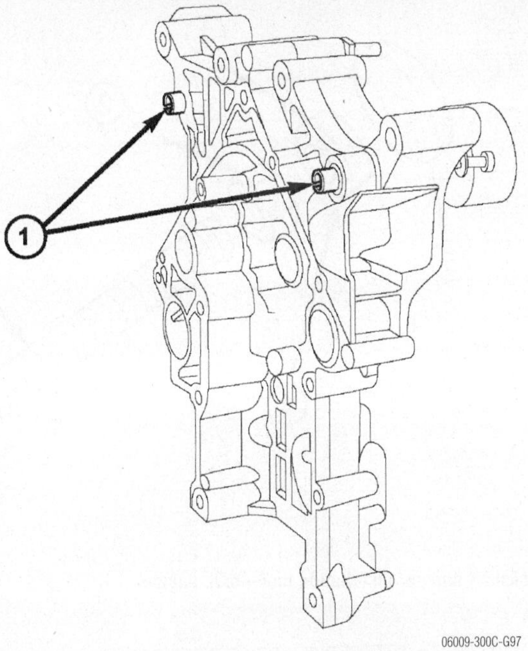

Timing cover slide bushings in place—5.7L engine

20. Remove power steering pump and set aside.
21. Remove the dipstick support bolt.
22. Drain the engine oil.
23. Remove the oil pan and pick up tube.

➡It is not necessary to remove water pump for timing cover removal.

24. Remove timing cover bolts and remove cover.
25. Verify that timing cover slide bushings (1) are located in timing cover.

To install:

26. Clean timing chain cover and block surface.

➡Always install a new gasket on timing cover.

27. Verify that the slide bushings are installed in timing cover.
28. Install cover and new gasket. Tighten fasteners to 28 Nm (250 inch lbs.).

➡The large lifting stud is torqued to 55 Nm (40 ft. lbs.).

29. Install the oil pan and pick up tube.
30. Install the air conditioning compressor.
31. Install the alternator.
32. Install power steering pump.
33. Install the dipstick support bolt.
34. Install the thermostat housing.

✱✱ WARNING

The front crankshaft seal must be installed dry. Do not apply lubricant to sealing lip or to outer edge.

35. Using Special Tool 8348 (1) and 8512A (2), install crankshaft front seal.

✱✱ WARNING

To prevent severe damage to the crankshaft, damper or Special Tool 8512-A, thoroughly clean the damper bore and the crankshaft nose before installing damper.

36. Slide damper onto crankshaft slightly.

✱✱ WARNING

Special Tool 8512-A, is assembled in a specific sequence. Failure to assemble this tool in this sequence can result in tool failure and severe damage to either the tool or the crankshaft.

37. Assemble Special Tool 8512-A as follows, the nut is threaded onto the shaft first (2). Then, the roller bearing (1) is placed onto the threaded rod (3). The hardened bearing surface of the bearing (1) MUST face the nut (2). Then, the hardened washer (5) slides onto the threaded rod (3). Once assembled coat the threaded rod's threads with Mopar® Nickel Anti-Seize or (Loctite® No. 771).

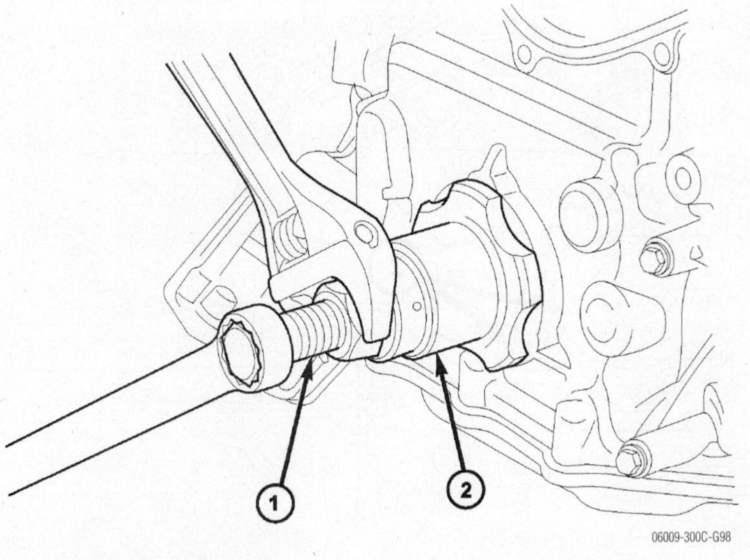

Front seal installation—5.7L engine

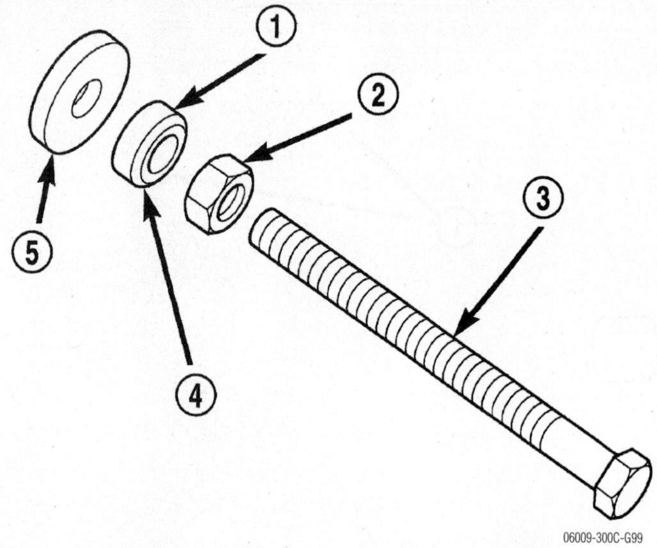

Crankshaft damper installation tool—5.7L engine

38. Using Special Tool 8512-A, press damper onto crankshaft.
39. Install then tighten crankshaft damper bolt to 176 Nm (129 ft. lbs.).
40. Install accessory drive belt tensioner assembly and both idler pulleys.
41. Install radiator lower hose.
42. Install both heater hoses.
43. Position radiator cooling fan assembly in vehicle.
44. Install cooling fan mounting bolts. Tighten to 6 Nm (50 inch lbs.)
45. Connect cooling fan electrical connector.
46. Install the fan and fan drive assembly
47. Install the accessory drive belt.
48. Install the coolant bottle and washer bottle.
49. Install the upper radiator hose.
50. Install the air cleaner assembly.
51. Fill cooling system.
52. Refill engine oil.
53. Connect the battery negative cable.
54. Install the engine cover.

Timing Chain and Sprockets

REMOVAL & INSTALLATION

2.7L Engine

1. Before servicing the vehicle, refer to the precautions in the beginning of this section.
2. Disconnect negative battery cable.
3. Drain cooling system.
4. Remove upper intake manifold.
5. Remove cylinder head covers, crankshaft vibration damper, and timing chain cover.

**** WARNING**

When aligning timing marks, always rotate engine by turning the crankshaft. Failure to do so will result in valve and/or piston damage.

6. Align crankshaft sprocket timing mark to mark on oil pump housing. The mark on oil pump housing is 60° ATDC of no.1 cylinder.

**** WARNING**

When the timing chain is removed and the cylinder heads are still installed, DO NOT rotate the camshafts or crankshaft without first locating the proper crankshaft position. Failure to do so will result in valve and/or piston damage.

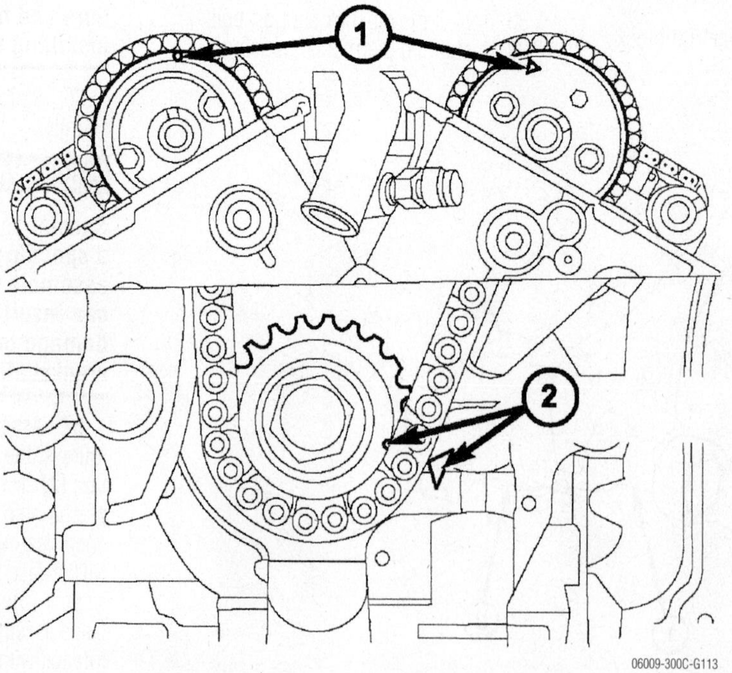

Align crankshaft sprocket timing mark to mark on oil pump housing. The mark on oil pump housing is 60° ATDC of no.1 cylinder—2.7L engine

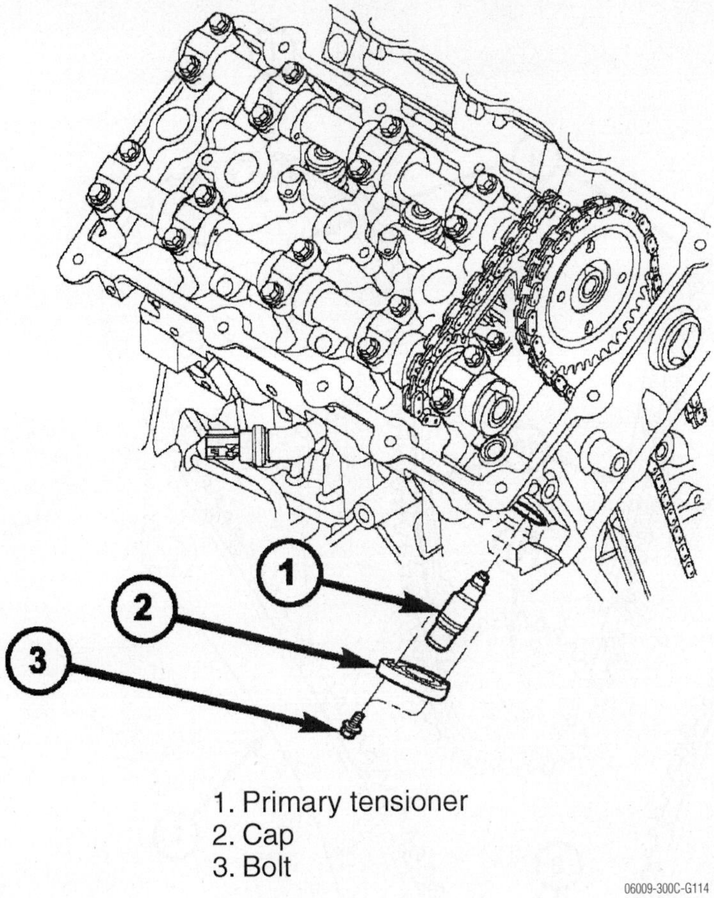

Primary timing chain tensioner removal—2.7L engine

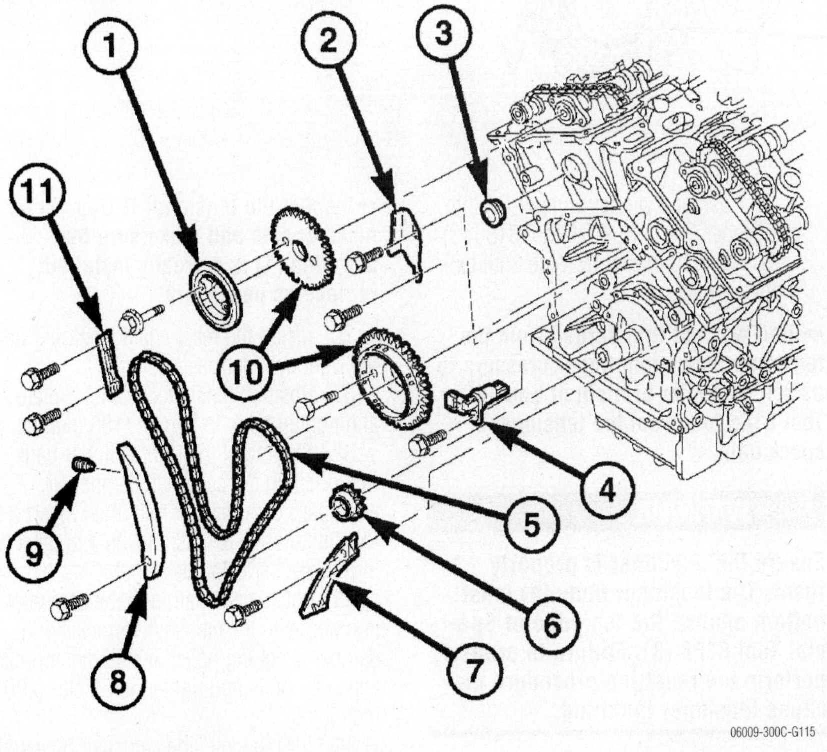

Timing chain and related parts—2.7L engine

7. Remove primary timing chain tensioner retainer cap and tensioner from right cylinder head.

8. Disconnect and remove camshaft position sensor (4) from left cylinder head.

9. Remove timing chain guide access plugs (3) from cylinder heads.

➡When camshaft sprocket bolts are removed, the camshafts will rotate in a clockwise direction.

10. Starting with the right camshaft sprocket, remove the sprocket attaching bolts. Remove camshaft damper (1) (if equipped) and sprocket.

11. Remove left side camshaft sprocket attaching bolts and remove sprocket.

12. Remove lower chain guide (7) and tensioner arm (8).

13. Remove the primary timing chain (5).

14. Remove crankshaft sprocket (6).

To install:

15. Inspect all sprockets (4, 9, and 11) and chain guides (2, 5,and 10). Replace if worn.

16. Install crankshaft sprocket.

17. If removed, install right and left side short chain guides (10). Tighten attaching bolts to 28 Nm (250 inch lbs.).

18. Align crankshaft sprocket timing mark to the mark on oil pump housing (3).

➡Lubricate timing chain and guides with engine oil before installation.

19. Place left side primary chain sprocket onto the chain so that the timing mark is located in-between the two (plated) timing links (1).

20. Lower the primary chain with left side sprocket through the left cylinder head opening.

➡The camshaft sprockets can be allowed to float on the camshaft hub during installation.

21. Loosely position left side camshaft sprocket over camshaft hub.

22. Align timing (plated) link to the crankshaft sprocket timing mark (3).

23. Position primary chain onto water pump drive sprocket (11).

24. Align right camshaft sprocket timing mark to the timing (plated) link on the timing chain (8) and loosely position over camshaft hub.

25. Verify that all chain timing (plated) links are properly aligned to the timing marks on all sprockets.

26. Install left side lower chain guide (2) and tensioner arm (5). Tighten attaching bolts to 28 Nm (250 inch lbs.).

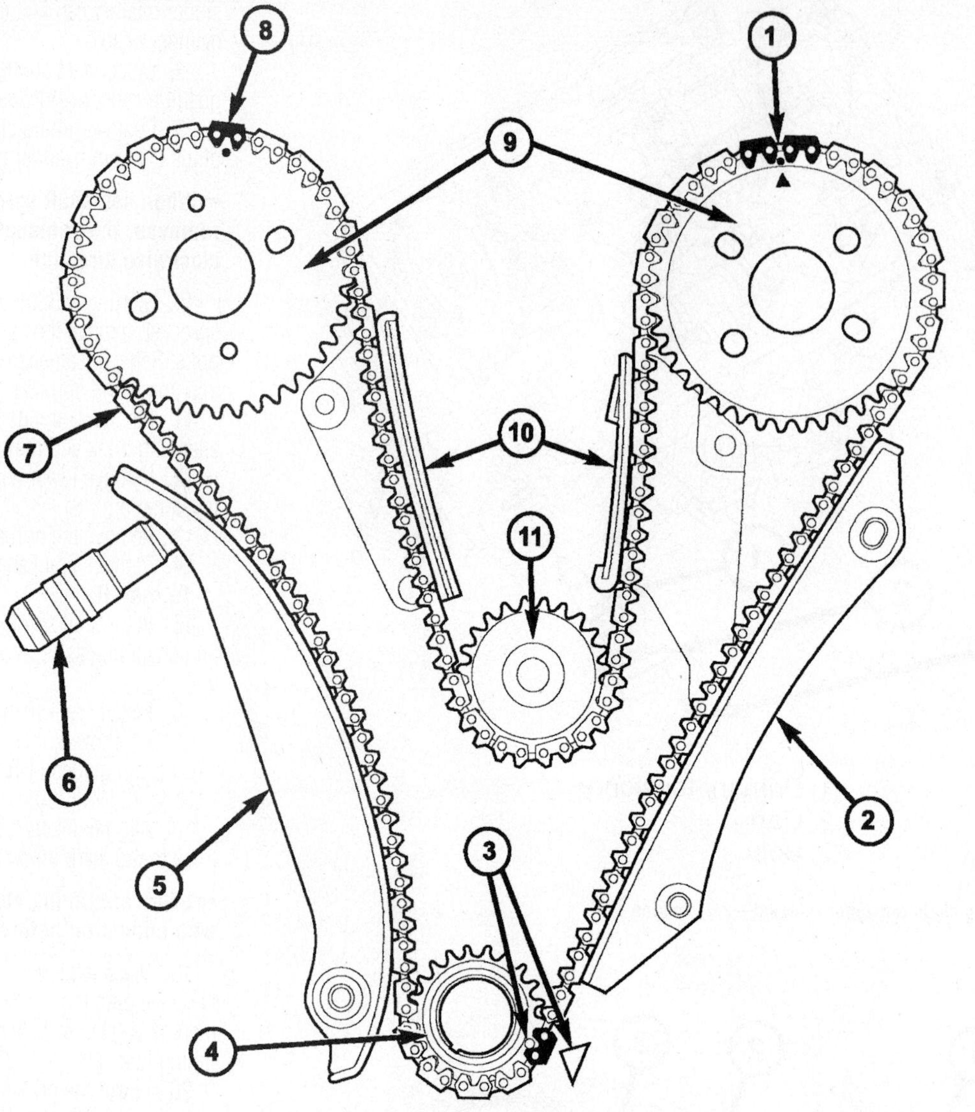

Timing chain installation—2.7L engine

➡Inspect O-ring on chain guide access plugs before installing. Replace O-ring as necessary.

27. Install chain guide access plugs to cylinder heads. Tighten plugs to 20 Nm (15 ft. lbs.).

➡To reset the primary timing chain tensioner, engine oil will first need to be purged from the tensioner.

28. Purge oil from timing chain tensioner using the following procedure:
    a. Place the check ball (2) end of tensioner into the shallow end of Special Tool 8186 (3).
    b. Using hand pressure, slowly depress tensioner until oil is purged from tensioner.
29. Reset timing chain tensioner using the following procedure:
    a. Position cylinder plunger (4) into the deeper end of Special Tool 8186 (3).
    b. Apply a downward force until tensioner is reset.

➡If oil was not first purged from the tensioner, use slight finger pressure to assist the center arm pin of Special Tool 8186 to unseat the tensioner's check ball.

⁂ WARNING

Ensure the tensioner is properly reset. The tensioner body (4) must bottom against the top edge of Special Tool 8186 (3). Failure to properly perform the resetting procedure may cause tensioner jamming.

➡Inspect the tensioner O-ring (2) for nicks or cuts and make sure the snapring (1) is correctly installed, replace as necessary.

30. Install the reset chain tensioner into the right cylinder head.
31. Position tensioner retaining plate and tighten bolts to 12 Nm (105 inch lbs.).
32. Starting at the right cylinder bank, first position the camshaft damper (if equipped) on camshaft hub, then insert a ⅜ in. square drive extension with a breaker bar into intake camshaft drive hub.
33. Rotate camshaft until the camshaft hub aligns to the camshaft sprocket and damper attaching holes. Install the sprocket attaching bolts and tighten to 28 Nm (250 inch lbs.).
34. Turn the left side camshaft by insert-

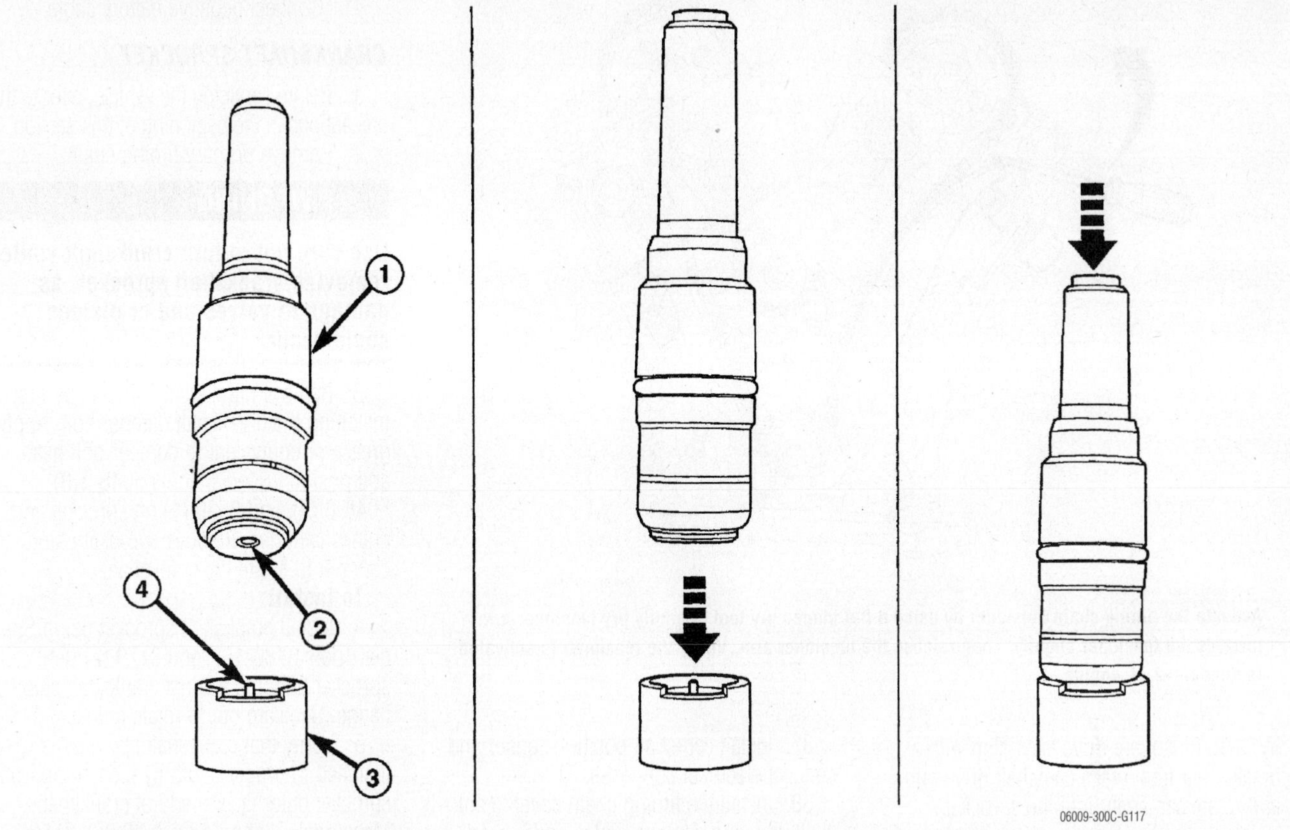

Purging oil from the tensioner—2.7L engine

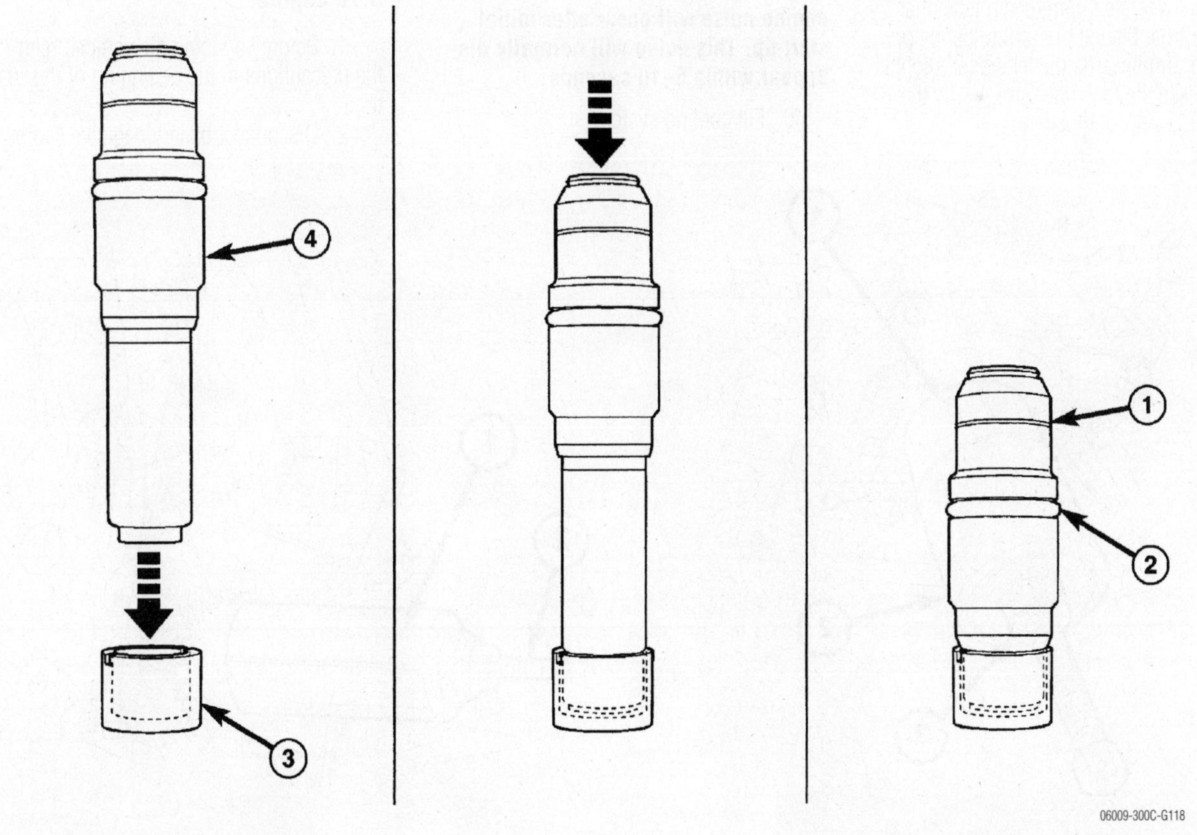

Reset the tensioner—2.7L engine

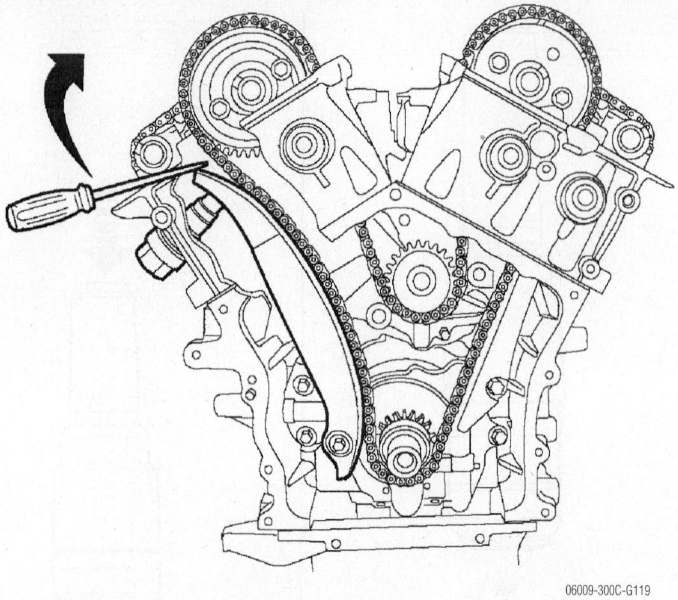

Activate the timing chain tensioner by using a flat bladed pry tool to gently pry tensioner arm towards the tensioner slightly. Then release the tensioner arm. Verify the tensioner is activated (extends)—2.7L engine

ing a ⅜ in. square drive extension with a breaker bar into intake camshaft drive hub and rotate camshaft until the sprocket attaching bolts can be installed. Tighten sprocket bolts to 28 Nm (250 inch lbs.).

35. Rotate engine slightly clockwise to remove timing chain slack, if necessary.

36. Activate the timing chain tensioner by using a flat bladed pry tool to gently pry tensioner arm towards the tensioner slightly. Then release the tensioner arm. Verify the tensioner is activated (extends).

37. Install camshaft position sensor and connect electrical connector.

38. Install the timing chain cover, crankshaft vibration damper, and cylinder head covers.

39. Install upper intake manifold.

➡After installation of a reset tensioner, engine noise will occur after initial start-up. This noise will normally disappear within 5–10 seconds.

40. Fill cooling system.

41. Connect negative battery cable.

CRANKSHAFT SPROCKET

1. Before servicing the vehicle, refer to the precautions in the beginning of this section.
2. Remove primary timing chain.

✲✲ WARNING

Use care not to turn crankshaft while removing crankshaft sprocket, as damage to valves and or pistons could occur.

3. Remove crankshaft sprocket by first installing the crankshaft damper bolt. Apply grease or equivalent to damper bolt head and position Special Tools 5048-1(3), 5048-6 (2), and 8539 (1) on sprocket and crankshaft nose. Remove sprocket using care not to rotate the crankshaft.

To install:

4. Install crankshaft sprocket using Special Tools 6780-1(1) and 8179 (2) until sprocket bottoms against crankshaft step flange. Use care not to rotate crankshaft.
5. Verify that crankshaft sprocket is installed to proper depth by measuring from sprocket outer face to end of crankshaft. Measurement should read: 39.05 ± 0.50 mm (1.5374 ± 0.020 in.).
6. Install primary timing chain.

5.7L Engine

1. Before servicing the vehicle, refer to the precautions in the beginning of this section.
2. Disconnect battery negative cable.

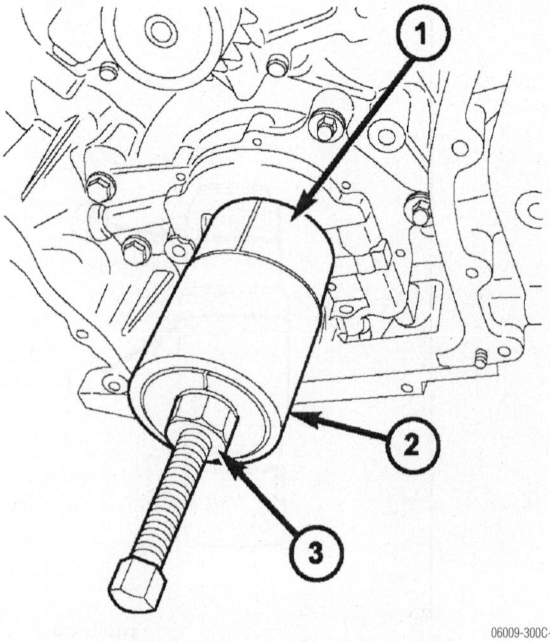

Crankshaft sprocket removal—2.7L engine

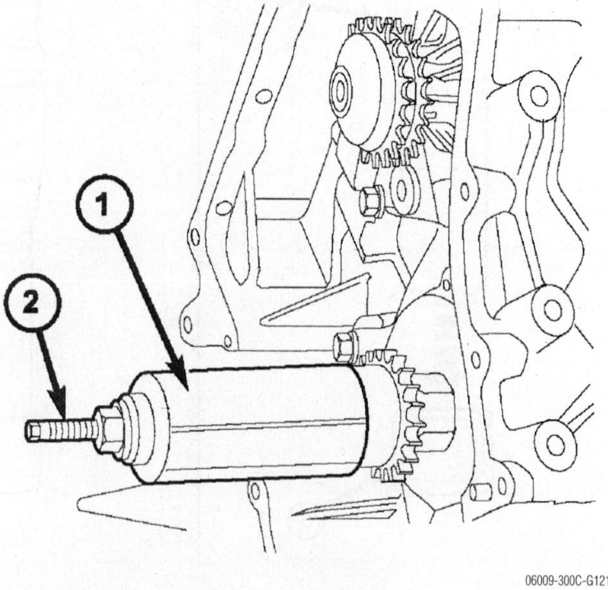

Crankshaft sprocket installation—2.7L engine

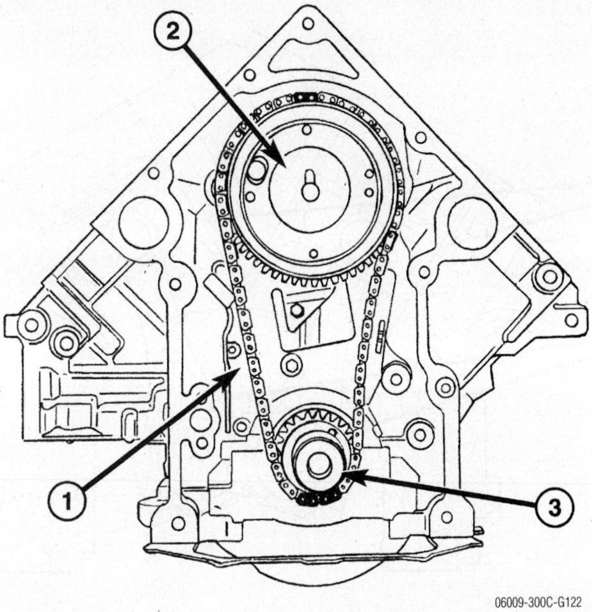

Timing mark alignment—5.7L engine

3. Drain cooling system.
4. Remove timing chain cover.
5. Re-install the vibration damper bolt finger tight. Using a suitable socket and breaker bar, rotate the crankshaft to align timing chain sprockets and keyways as shown.

⁂ WARNING

The camshaft pin and the slot in the cam sprocket must be positioned at 12:00 (2). The crankshaft keyway must be positioned at 2:00 (3). The crankshaft sprocket must be installed so that the dots and or paint marking is at 6:00.

6. Retract tensioner shoe until hole in shoe lines up with hole in bracket.
7. Slide a suitable pin into the holes.
8. Remove camshaft sprocket attaching bolt and remove timing chain with crankshaft and camshaft sprockets.
9. If tensioner assembly is to be replaced, remove the tensioner to block bolts and remove tensioner assembly.

To install:

10. If tensioner assembly is being replaced, install tensioner and mounting bolts. Torque bolts to 28 Nm (250 inch lbs.).
11. Retract tensioner if required.

⁂ WARNING

The timing chain must be installed with the single plated link aligned with the dot and or paint marking on the camshaft sprocket. The crankshaft sprocket is aligned with the dot and or paint marking on the sprocket between two plated timing chain links.

⁂ WARNING

The camshaft pin and the slot in the cam sprocket must be clocked at 12:00. The crankshaft keyway must be clocked at 2:00. The crankshaft sprocket must be installed so that the dots and or paint marking is at 6:00.

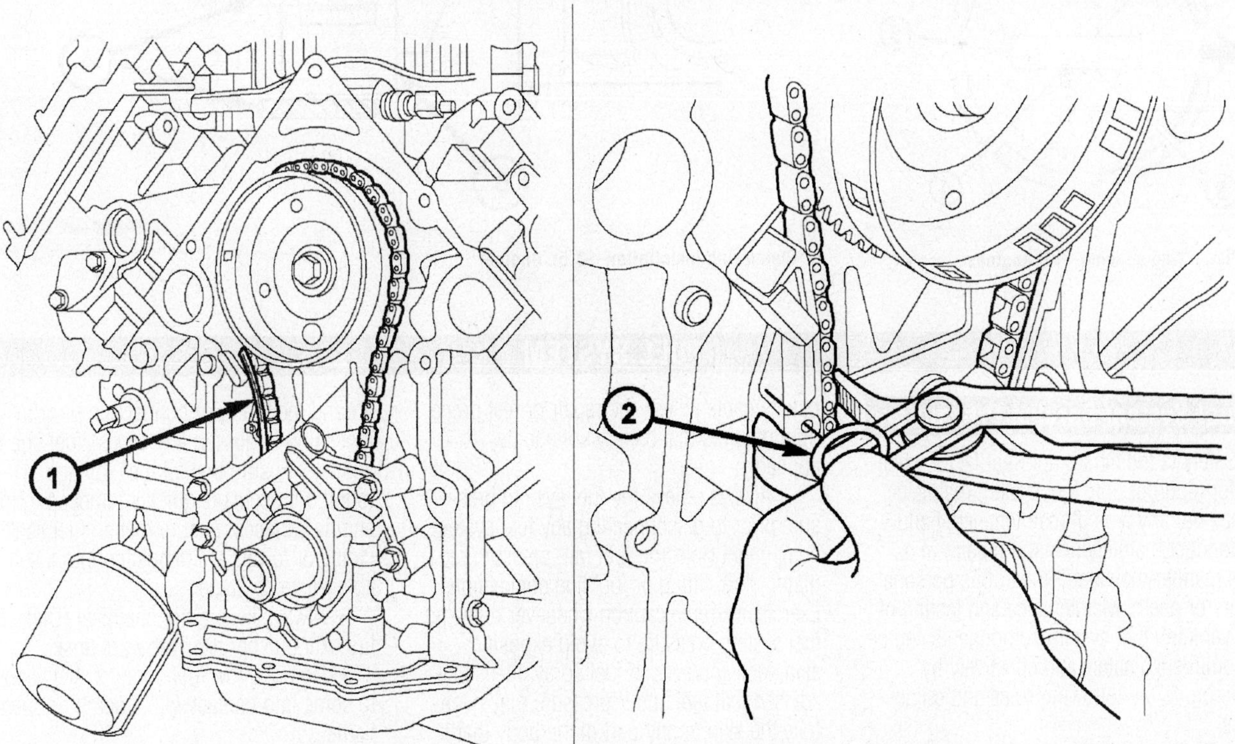

Retract tensioner shoe (1) until hole in shoe lines up with hole in bracket. Slide a suitable pin (2) into the holes—5.7L engine

12. Place both camshaft sprocket and crankshaft sprocket on the bench with timing marks on exact imaginary center line through both camshaft and crankshaft bores.
13. Place timing chain around both sprockets.
14. Lift sprockets and chain. Keep sprockets tight against the chain.
15. Slide both sprockets evenly over their respective shafts and check alignment of timing marks.
16. Install the camshaft bolt. Tighten the bolt to 122 Nm (90 ft. lbs.) torque.
17. Remove tensioner pin. Again, verify alignment of timing marks.
18. Install the oil pump.
19. Install the oil pan and pick up.
20. Install the timing chain cover
21. Refill engine oil.
22. Fill cooling system.
23. Connect battery negative cable.
24. Start engine and check for oil and coolant leaks.

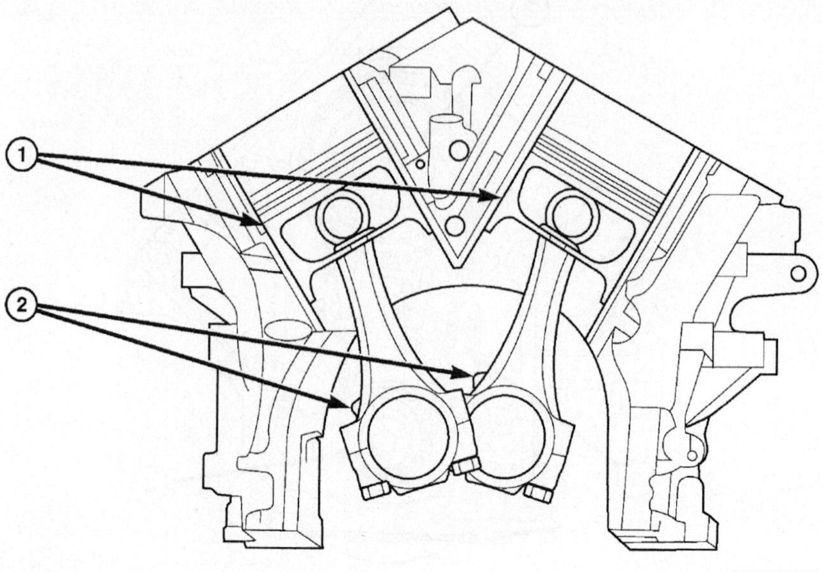

Piston installation—2.7L and 3.5L engines

Piston and Ring

POSITIONING

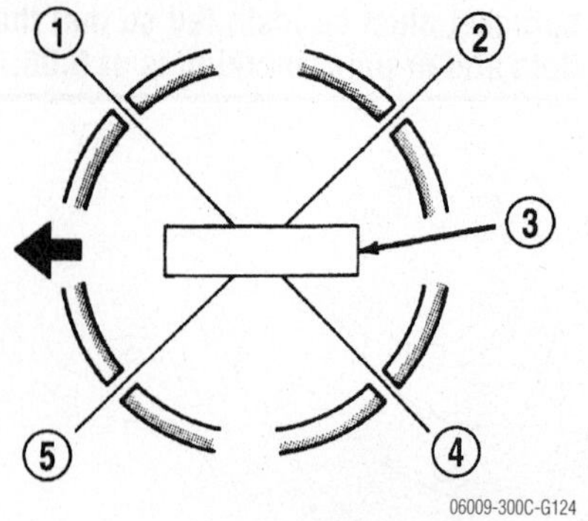

Piston ring spacing—all engines

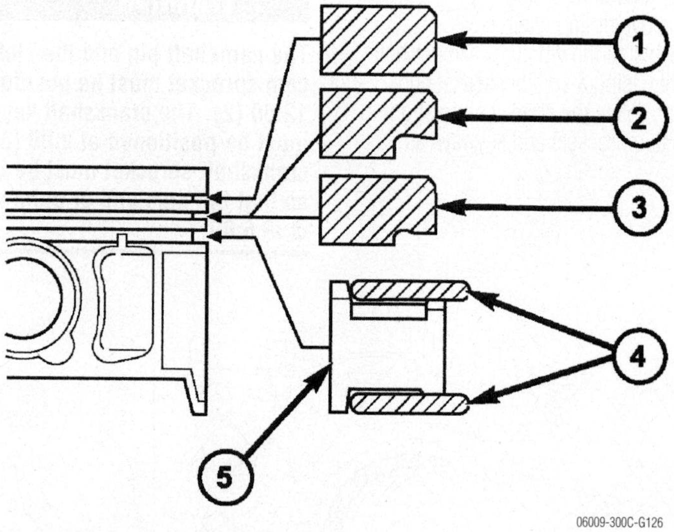

Piston ring installation—3.5L engine

FUEL SYSTEM

Fuel System Service Precautions

Safety is the most important factor when performing not only fuel system maintenance but any type of maintenance. Failure to conduct maintenance and repairs in a safe manner may result in serious personal injury or death. Maintenance and testing of the vehicle's fuel system components can be accomplished safely and effectively by adhering to the following rules and guidelines.

• To avoid the possibility of fire and personal injury, always disconnect the negative battery cable unless the repair or test procedure requires that battery voltage be applied.

• Always relieve the fuel system pressure prior to disconnecting any fuel system component (injector, fuel rail, pressure regulator, etc.), fitting or fuel line connection. Exercise extreme caution whenever relieving fuel system pressure to avoid exposing skin, face and eyes to fuel spray. Please be advised that fuel under pressure may penetrate the skin or any part of the body that it contacts.

• Always place a shop towel or cloth around the fitting or connection prior to loosening to absorb any excess fuel due to spillage. Ensure that all fuel spillage (should it occur) is quickly removed from engine surfaces. Ensure that all fuel soaked cloths or towels are deposited into a suitable waste container.

• Always keep a dry chemical (Class B) fire extinguisher near the work area.

• Do not allow fuel spray or fuel vapors to come into contact with a spark or open flame.

• Always use a back-up wrench when loosening and tightening fuel line connection fittings. This will prevent unnecessary stress and torsion to fuel line piping.

• Always replace worn fuel fitting O-rings with new. Do not substitute fuel hose or equivalent where fuel pipe is installed.

Before servicing the vehicle, make sure to also refer to the precautions in the beginning of this section as well.

Relieving Fuel System Pressure

1. Before servicing the vehicle, refer to the precautions in the beginning of this section.
2. Remove fuel pump relay from Power Distribution Center (PDC). For location of relay, refer to label on underside of PDC cover.
3. Start and run engine until it stalls.
4. Attempt restarting engine until it will no longer run.
5. Turn ignition key to OFF position.
6. Return fuel pump relay to PDC.
7. One or more Diagnostic Trouble Codes (DTC's) may have been stored in PCM memory due to fuel pump relay removal. A scan tool must be used to erase a DTC.

Fuel Module

REMOVAL & INSTALLATION

1. Before servicing the vehicle, refer to the precautions in the beginning of this section.
2. Release the fuel pressure.
3. Disconnect negative battery cable.

➡The fuel level of the vehicle must be below ⅝ of a tank before you remove the module lock-rings. If it is above that you can spill fuel in the vehicle.

4. Drain partial fuel from fuel tank through the filler tube. Use a hard nylon tube, with a 30° cut on the end, to push the check valve open to drain fuel from tank.
5. Remove the rear lower seat cushion.
6. Push seat back and up to remove seat cushion.
7. Fold back the foam pad covering access cover for modules.
8. Disconnect the electrical connector from left side module.
9. Mark the module orientation.
10. Use special tool 9340 to remove left side module lockring.
11. Drain fuel from left side of fuel tank. Lift module up enough to push hose into tank and drain. Do not spill fuel in interior of vehicle.

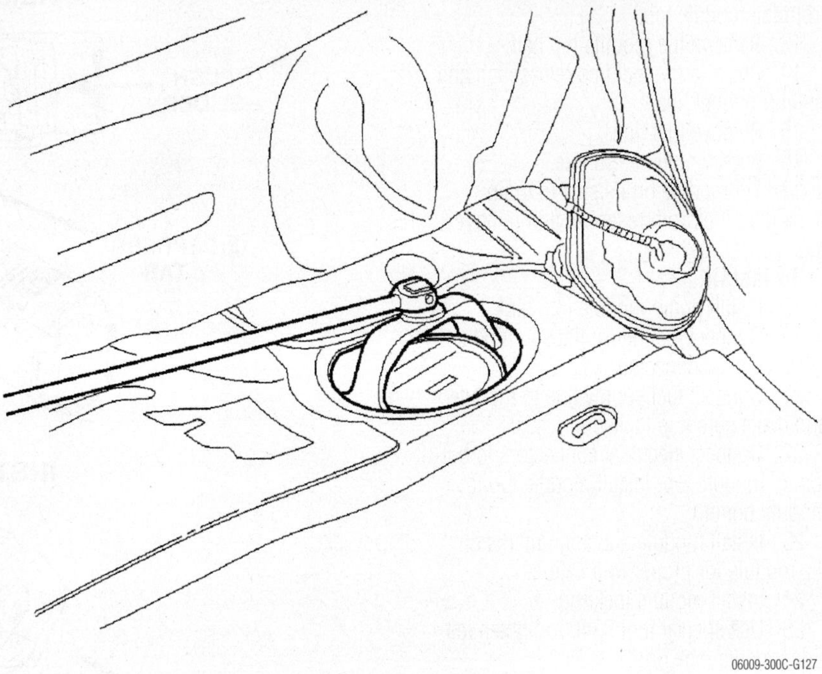

Use special tool 9340 to remove left side module lock ring

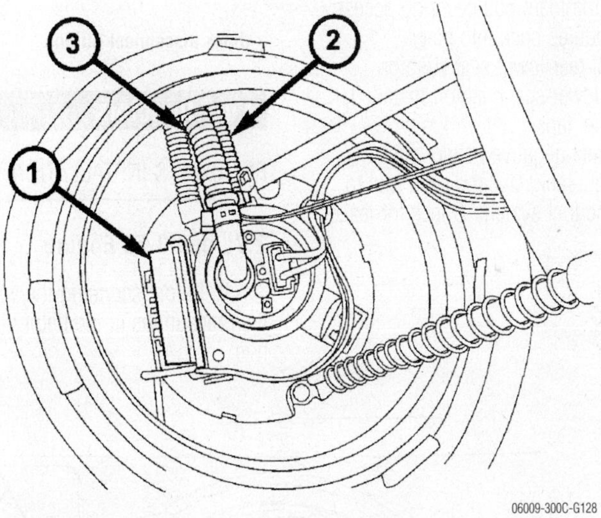

Fuel level sending card (1), fuel return lines, fuel supply line (3)

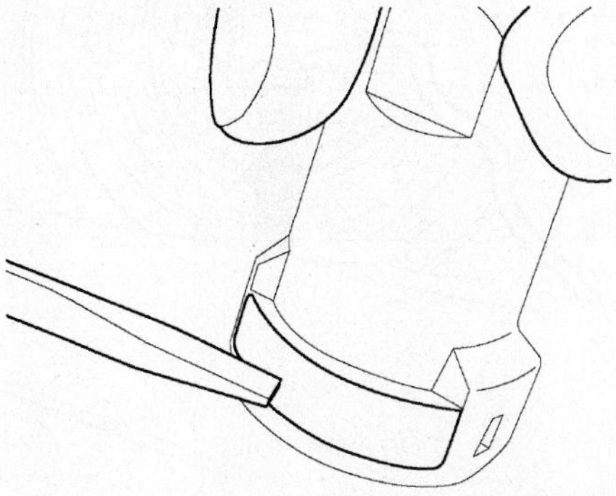

Fuel line release tab

12. Disconnect the electrical connectors from the module top.
13. Remove the module top half.
14. Press in the fuel line release tab and pull up on fuel line.
15. Remove fuel line.
16. Remove fuel return line.
17. Tip module on its side to drain remaining fuel from reservoir and remove module from vehicle.

To install:

18. Install pump module into tank.
19. Connect fuel return lines to module.
20. Lines connected.
21. Connect fuel supply line to module and make sure it is locked in place.
22. Connect electrical connectors to bottom of module top, Install module top to module bottom.
23. Install module and align marks on the module for proper orientation.
24. Install module lockring.
25. Use special tool 9340 to tighten left side module lock ring.
26. Connect the electrical connector to left side module.
27. Install plastic access cover.
28. Fold the foam pad covering access cover for modules back into place.
29. Install rear lower seat cushion
30. Rear lower seat cushion installed.
31. Fill fuel tank.
32. Connect negative battery cable.
33. Fill fuel tank. Use the scan tool to pressurize the fuel system. Check for leaks.

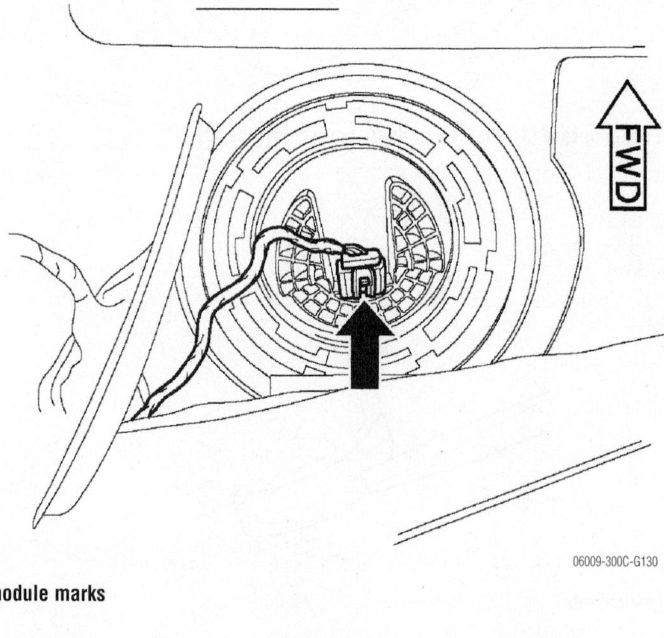

Align module marks

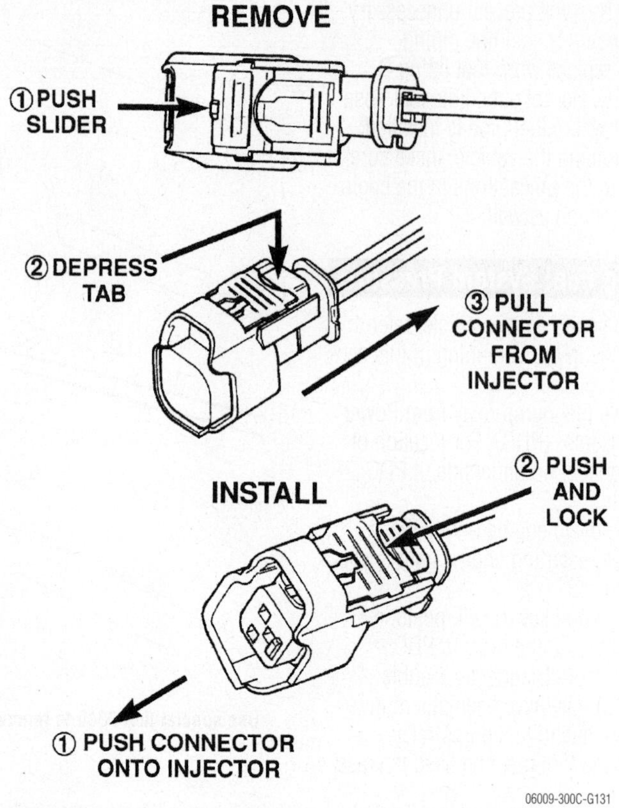

Quick disconnect fittings

Fuel Injectors

REMOVAL & INSTALLATION

2.7L and 3.5L Engine

1. Before servicing the vehicle, refer to the precautions in the beginning of this section.

✲✲ CAUTION

The fuel system is under constant pressure even with engine off. Before servicing fuel injector(s), fuel system pressure must be released.

2. Release fuel system pressure.
3. Disconnect negative cable to battery.
4. Remove intake manifold plenum.
5. Lift Plenum up off of engine. Cover intake manifold to prevent foreign material from entering engine.
6. Disconnect fuel supply tube quick connect fittings at the rear of intake manifold.

➡If the injector connectors are not tagged with their cylinder number, tag them to identify the correct cylinder.

7. Disconnect electrical connectors at fuel injectors. Push red colored slider away from injector (1). While pushing slider, depress tab (2) and remove connector (3) from injector. The factory fuel injection wiring harness is numerically tagged (INJ 1, INJ 2, etc.) for injector position identification. If harness is not tagged, note wiring location before removal.
8. Remove fuel rail-mounting bolts.
9. Lift fuel rail straight up off of the cylinder head.

10. Remove retaining clips from fuel injectors at fuel rail.
11. Remove fuel injectors.
12. Repeat for remaining injectors.
13. Check injector O-ring for damage. If O-ring is damaged, it must be replaced. Replace the injector clip if it is damaged.

To install:

14. Lightly lubricate the fuel injector O-rings with a couple drops of clean engine oil.
15. Install retaining clips on fuel injectors.
16. Push injectors into fuel injector rail until clips are in the correct position.
17. Position fuel rail over cylinder head, and push rail into place. Tighten fuel rail mounting bolts to 11 Nm (100 inch lbs.) torque.
18. Connect fuel supply tube quick connect fittings at the rear of intake manifold.
19. Connect electrical connectors to fuel injectors.
20. Install intake manifold plenum.
21. Connect negative cable to battery.
22. Use a scan tool to pressurize the fuel system. Check for leaks.

5.7L Engine

1. Before servicing the vehicle, refer to the precautions in the beginning of this section.

**** CAUTION**

The fuel system is under constant pressure even with engine off. Before servicing fuel rail, fuel system pressure must be released.

**** WARNING**

The left and right fuel rails are replaced as an assembly. Do not attempt to separate rail halves at connector tube. Due to design of tube, it does not use any clamps. Never attempt to install a clamping device of any kind to tube. When removing fuel rail assembly for any reason, be careful not to bend or kink tube.

2. Remove fuel tank filler tube cap.
3. Perform fuel system pressure release procedure.
4. Remove negative battery cable at battery.
5. Remove flex tube (air cleaner housing to engine).
6. Remove air resonator box at throttle body.
7. Disconnect all spark plug cables from all spark plugs and ignition coils. Do not remove cables from cable routing tray. Note original cable positions while removing.
8. Remove spark plug cable tray from engine by releasing 4 retaining clips. Remove tray and cables from engine as an assembly.

➡Before removing or disconnecting any spark plug cables, note their original position. Remove cables one-at-a-time. To prevent ignition crossfire, spark plug cables must be placed in cable tray (routing loom) into their original position.

An individual ignition coil is used at each cylinder. The coil mounts to the top of the valve cover with two bolts. The bottom of the coil is equipped with a rubber boot to seal the spark plug to the coil. Inside each rubber boot is a spring. The spring is used for a mechanical contact between the coil and the top of the spark plug.

9. Depending on which coil is being removed, the throttle body air intake tube or intake box may need to be removed to gain access to coil.
10. Unlock electrical connector by moving slide lock first. Press on release lock while pulling electrical connector from coil.
11. Disconnect secondary high-voltage cable from coil with a twisting action.
12. Clean area at base of coil with compressed air before removal.
13. Remove two mounting bolts (note that mounting bolts are retained to coil).
14. Carefully pull up coil from cylinder head opening with a slight twisting action.
15. Remove coil from vehicle.
16. Before installing spark plug cables to either the spark plugs or coils, or before installing a coil to a spark plug, apply dielectric grease to inside of boots.
17. Disconnect fuel line latch clip and fuel line at fuel rail. A special tool will be necessary for fuel line disconnection.
18. Disconnect electrical connectors at all 8 fuel injectors. Push red colored slider away from injector (1). While pushing slider, depress tab (2) and remove connector (3) from injector. The factory fuel injection wiring harness is numerically tagged (INJ 1, INJ 2, etc.) for injector position identification. If harness is not tagged, note wiring location before removal.
19. Disconnect electrical connectors at all throttle body sensors.
20. Remove four fuel rail mounting bolts and hold-down clamps.
21. Gently rock and pull left side of fuel rail until fuel injectors just start to clear machined holes in intake manifold. Gently rock and pull right side of rail until injectors just start to clear intake manifold head holes. Repeat this procedure (left/right) until all injectors have cleared machined holes.
22. Remove fuel rail (with injectors attached) from engine.
23. Remove clip(s) retaining injector(s) to fuel rail.
24. Remove injector(s) from fuel rail.

To install:

25. Clean out fuel injector machined bores in intake manifold.
26. Apply a small amount of engine oil to each fuel injector O-ring. This will help in fuel rail installation.
27. Position fuel rail/fuel injector assem-

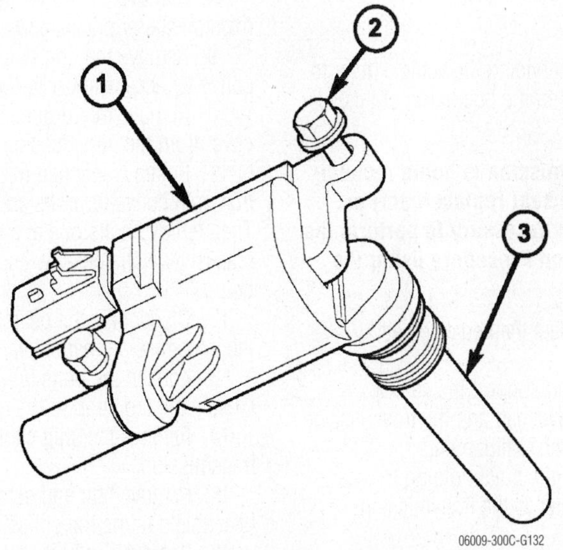

Ignition coil (1), bolt (2), spark plug boot (3)—5.7L engine

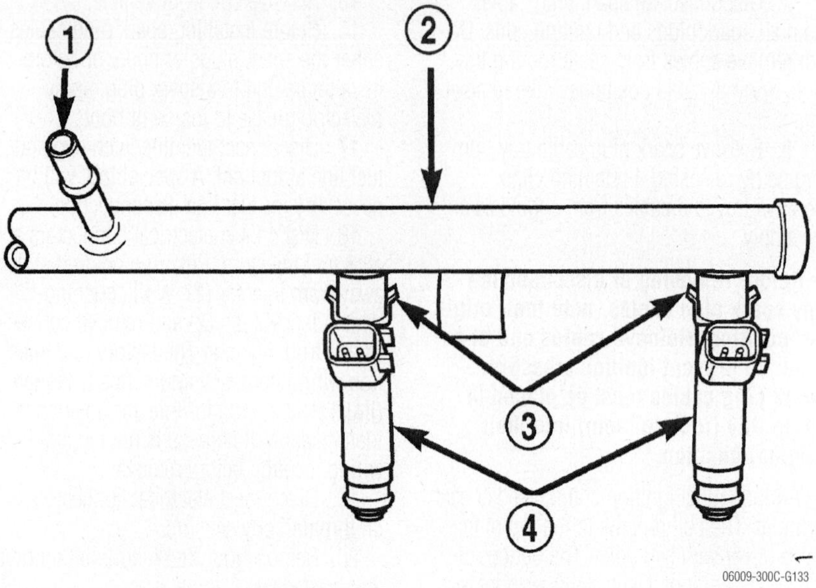

Fuel rail (2) with injectors (4) attached by clips (3)—5.7L engine

bly to machined injector openings in intake manifold.

28. Guide each injector into intake manifold. Be careful not to tear injector O-rings.
29. Push right side of fuel rail down until fuel injectors have bottomed on shoulders. Push left fuel rail down until injectors have bottomed on shoulders.
30. Install 4 fuel rail hold-down clamps and 4 mounting bolts and. Torque to 100 inch lbs. (11 Nm).
31. Position spark plug cable tray (7) and cable assembly to intake manifold. Snap 4 cable tray retaining clips into intake manifold.
32. Install all cables to spark plugs and ignition coils.
33. Connect electrical connector to throttle body.
34. Using compressed air, blow out any dirt or contaminants from around top of spark plug.
35. Before installing spark plug cables to either the spark plugs or coils, or before installing a coil to a spark plug, apply dielectric grease to inside of boots.
36. Position ignition coil into cylinder head opening and push boot onto spark plug. Twist coil into position.
37. Install two coil mounting bolts and tighten to 12 Nm (105 inch lbs.).
38. Connect electrical connector to coil and lock connector.
39. Install spark plug cables to coils. To prevent ignition crossfire, spark plug cables MUST be placed in cable tray (routing loom) into their original position. See the Specifications section of this chapter, following the Tune-up Chart, for correct wiring.
40. If necessary, install throttle body air tube.
41. Connect electrical connector to throttle body.
42. Connect electrical connectors at all fuel injectors. Push connector onto injector and then push and lock red colored slider. Verify connector is locked to injector by lightly tugging on connector.
43. Connect fuel line latch clip and fuel line to fuel rail.
44. Install air resonator to throttle body (2 bolts).
45. Install flexible air duct to air box.
46. Connect battery cable to battery.
47. Start engine and check for leaks.

DRIVE TRAIN

Transmission

REMOVAL & INSTALLATION

NAG1

1. Before servicing the vehicle, refer to the precautions in the beginning of this section.

➡If the transmission is being reconditioned (clutch/seal replacement) or replaced, it is necessary to perform the TCM Adaptation Procedure using the scan tool.

2. Disconnect the negative battery cable.
3. Raise and support the vehicle.
4. Mark driveshaft and the transmission flange for assembly alignment.
5. Remove the bolts holding the rear driveshaft coupler to the transmission flange.
6. Slide the driveshaft rearward until the coupler clears the driveshaft pilot on the transmission output shaft.
7. Remove the bolts holding the starter motor to the transmission.
8. Remove the starter from the transmission starter pocket and safely relocate.
9. Remove the bolt holding the torque converter access cover to the transmission.
10. Remove the torque converter access cover from the transmission.
11. Rotate crankshaft in clockwise direction until converter bolts are accessible. Then remove bolts one at a time. Rotate crankshaft with socket wrench on dampener bolt.
12. Disconnect the gearshift cable) from the transmission manual valve lever.
13. Loosen the bolts holding the shift cable retaining strap to the transmission.
14. Remove the shift cable from the transmission.
15. Remove bolt and screw holding the heat shield to the transmission.
16. Remove the heat shield from the transmission.
17. Disconnect 13-pin plug connector. Turn bayonet lock of guide bushing counterclockwise.
18. Remove the 13-pin connector from the transmission.
19. Disconnect transmission fluid cooler lines at transmission fittings and clips.
20. Disconnect the transmission vent hose from the transmission.
21. Support rear of engine with safety stand or jack.
22. Raise transmission slightly with service jack to relieve load on crossmember and supports.
23. Remove bolts securing rear support and cushion to transmission crossmember.
24. Remove the bolts holding the engine oil pan to the transmission.
25. Remove all remaining bolts holding the engine to the transmission.
26. Carefully work transmission and torque converter assembly rearward off engine block dowels.
27. Hold torque converter in place during transmission removal.

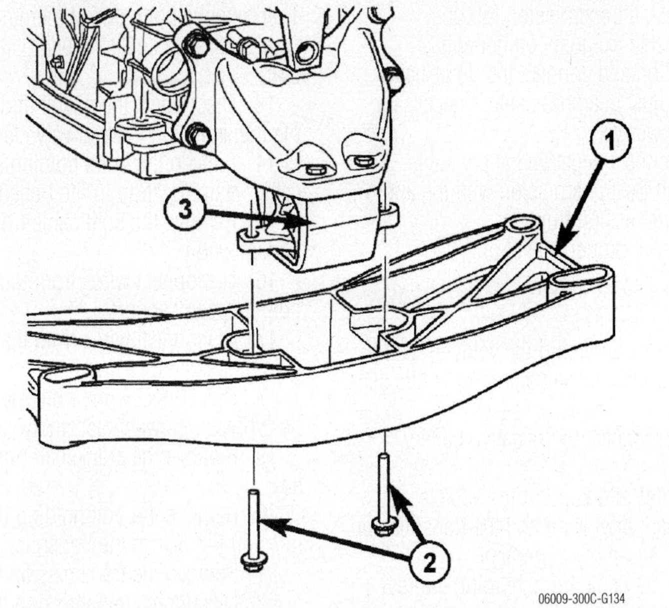

Rear crossmember—NAG1

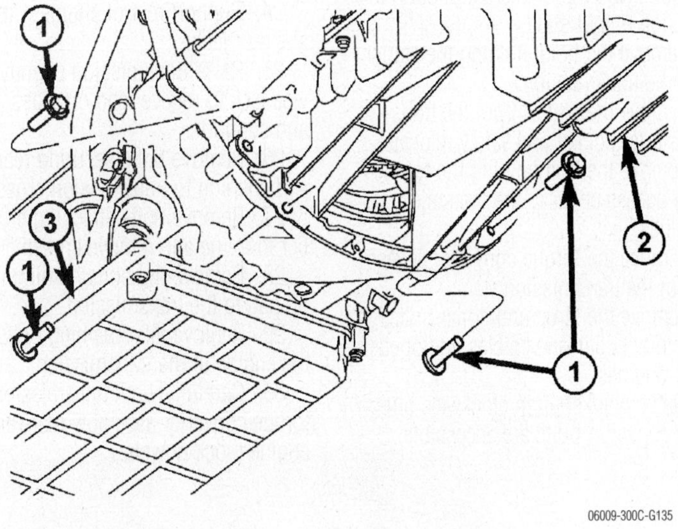

Transmission-to-oil pan bolts—NAG1

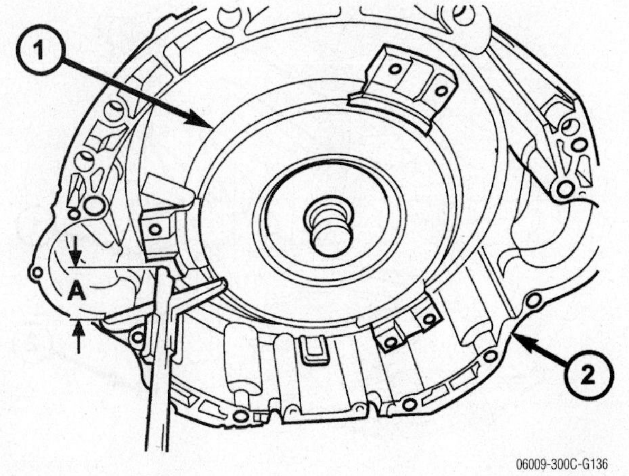

Check torque converter seating

28. Lower transmission and remove assembly from under the vehicle.

29. To remove torque converter, carefully slide torque converter out of the transmission.

To install:

30. Check torque converter hub and hub drive flats for sharp edges burrs, scratches, or nicks. Polish the hub and flats with 320/400 grit paper and crocus cloth if necessary. The hub must be smooth to avoid damaging pump seal at installation.

31. If a replacement transmission is being installed, transfer any components necessary, such as the manual shift lever and shift cable bracket, from the original transmission onto the replacement transmission.

32. Lubricate oil pump seal lip with transmission fluid.

33. Place torque converter in position in transmission.

**** WARNING**

Do not damage oil pump seal or converter hub while inserting torque converter into the front of the transmission.

34. Align torque converter to oil pump seal opening.

35. Insert torque converter hub into oil pump.

36. While pushing torque converter inward, rotate converter until converter is fully seated in the oil pump gears.

37. Check converter seating with a scale and straight-edge at "A". Surface of converter lugs should be at least 19 mm (¾ in.) to rear of straight-edge when converter is fully seated.

38. If necessary, temporarily secure converter with C-clamp attached to the converter housing.

39. Check condition of converter driveplate. Replace the plate if cracked, distorted or damaged. Also be sure transmission dowel pins are seated in engine block and protrude far enough to hold transmission in alignment.

40. Apply a light coating of Mopar® High Temp Grease, or equivalent, to the torque converter hub pocket in the rear pocket of the engine's crankshaft.

41. Raise transmission and align the torque converter with the drive plate and the transmission converter housing with the engine block.

42. Move transmission forward. Then raise, lower, or tilt transmission to align the converter housing with the engine block dowels.

43. Carefully work transmission forward and over engine block dowels until converter hub is seated in crankshaft. Verify that no wires, or the transmission vent hose, have become trapped between the engine block and the transmission.

44. Install two bolts to attach the transmission to the engine.

45. Install remaining torque converter housing to engine bolts. Tighten to 39 Nm (29 ft. lbs.).

46. Install rear transmission crossmember. Tighten crossmember to frame bolts to 68 Nm (50 ft. lbs.).

47. Install rear support to transmission. Tighten bolts to 47 Nm (35 ft. lbs.).

48. Lower transmission onto crossmember and install bolts attaching transmission mount to crossmember. Tighten clevis bracket to crossmember bolts to 47 Nm (35 ft. lbs.).

49. Remove engine support fixture.

50. Install the transmission to engine oil pan bolts. Tighten to 39 Nm (29 ft. lbs.).

51. Connect the gearshift cable to the transmission manual shift lever.

52. Check O-ring on plug connector, and replace if necessary.

53. Install the plug connector into the guide bushing. Turn bayonet lock of guide bushing clockwise to connect plug connector.

54. Position the heat shield onto the transmission housing and install the screw and bolt to hold the shield in place.

⁂ WARNING

It is essential that correct length bolts be used to attach the converter to the driveplate. Bolts that are too long will damage the clutch surface inside the converter.

55. Install all torque converter-to-driveplate bolts by hand.

56. Verify that the torque converter is pulled flush to the driveplate. Tighten bolts to 42 Nm (31 ft. lbs.).

57. Install the torque converter bolt access cover onto the transmission. Install the access cover bolt and tighten to 11 Nm (96 inch lbs.).

58. Install starter motor.

59. Connect the cooler line fittings and cooler lines to the transmission.

➡Inspect the fill tube grommet to determine if the grommet is new or used. If a new fill tube grommet is in the transmission case, pierce the center of the grommet with a ballpoint pen, or similar instrument, to prepare the grommet for the fill tube installation.

60. Install transmission fill tube.

61. Install exhaust components.

62. Align and connect the driveshaft.

63. Adjust gearshift cable if necessary.

64. Lower vehicle.

65. Connect negative battery cable.

66. Fill the transmission with the appropriate transmission fluid.

67. Verify proper operation.

42RLE

1. Before servicing the vehicle, refer to the precautions in the beginning of this section.

2. Disconnect the negative battery cable.

3. Raise and support the vehicle.

4. Mark driveshaft and the transmission flange for assembly alignment.

5. Remove the bolts holding the rear driveshaft coupler to the transmission flange.

6. Slide the driveshaft rearward until the coupler clears the driveshaft pilot on the transmission output shaft.

7. Remove the bolts holding the starter motor to the transmission.

8. Remove the starter from the transmission starter pocket and safely relocate.

9. Remove the bolt holding the torque converter access cover to the transmission, 3.5L engines.

10. Remove the torque converter access cover from the transmission.

11. Remove the structural collar bolts and structural collar on vehicles equipped with 2.7L engines.

12. Rotate crankshaft in clockwise direction until converter bolts are accessible. Then remove bolts one at a time. Rotate crankshaft with socket wrench on dampener bolt.

13. Disconnect the gearshift cable from the transmission manual valve lever.

14. Loosen the bolts holding the shift cable retaining strap to the transmission.

15. Remove the shift cable from the transmission.

16. Disconnect wires from the input and output speed sensors.

17. Disconnect wires from the transmission range sensor.

18. Disconnect wires from the solenoid/pressure switch assembly.

19. Remove the crankshaft position sensor.

20. Remove the bolt holding the transmission fill tube to the transmission.

21. Remove the transmission fill tube.

22. Disconnect transmission fluid cooler lines at transmission fittings and clips.

23. Disconnect the transmission vent hose from the transmission.

24. Support rear of engine with safety stand or jack.

25. Raise transmission slightly with service jack to relieve load on crossmember and supports.

26. Remove bolts securing rear support and cushion to transmission crossmember.

27. Remove bolts attaching crossmember to frame and remove crossmember.

28. Remove the bolts holding the engine oil pan to the transmission.

29. Remove all remaining bolts holding the engine to the transmission.

30. Carefully work transmission and torque converter assembly rearward off engine block dowels.

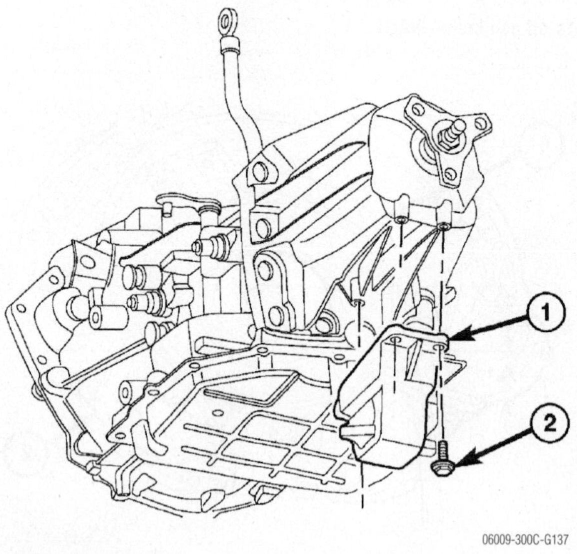

Rear crossmember cushion—42RLE

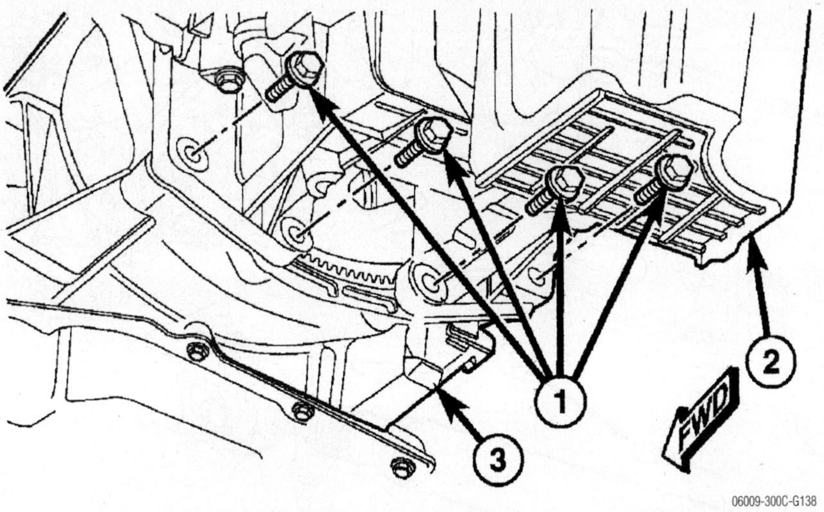

Transmission-to-oil pan bolts—42RLE

31. Hold torque converter in place during transmission removal.
32. Lower transmission and remove assembly from under the vehicle.
33. To remove torque converter, carefully slide torque converter out of the transmission.

To install:

34. Check torque converter hub and hub drive flats for sharp edges burrs, scratches, or nicks. Polish the hub and flats with 320/400 grit paper and crocus cloth if necessary. The hub must be smooth to avoid damaging pump seal at installation.
35. If a replacement transmission is being installed, transfer any components necessary, such as the manual shift lever and shift cable bracket, from the original transmission onto the replacement transmission.
36. Lubricate oil pump seal lip with transmission fluid.
37. Place torque converter in position in transmission.

⁂ WARNING

Do not damage oil pump seal or converter hub while inserting torque converter into the front of the transmission.

38. Align torque converter to oil pump seal opening.
39. Insert torque converter hub into oil pump.
40. While pushing torque converter inward, rotate converter until converter is fully seated in the oil pump gears.
41. Check converter seating with a scale and straight-edge at "A". Surface of converter lugs should be at least 19 mm (¾ in.) to rear of straight-edge when converter is fully seated.
42. If necessary, temporarily secure converter with C-clamp attached to the converter housing.
43. Check condition of converter driveplate. Replace the plate if cracked, distorted or damaged. Also be sure transmission dowel pins are seated in engine block and protrude far enough to hold transmission in alignment.
44. Apply a light coating of Mopar® High Temp Grease to the torque converter hub pocket in the rear pocket of the engine's crankshaft.
45. Raise transmission and align the torque converter with the drive plate and the transmission converter housing with the engine block.
46. Move transmission forward. Then raise, lower, or tilt transmission to align the converter housing with the engine block dowels.
47. Carefully work transmission forward and over engine block dowels until converter hub is seated in crankshaft. Verify that no wires, or the transmission vent hose, have become trapped between the engine block and the transmission.
48. Install two bolts to attach the transmission to the engine.
49. Install remaining torque converter housing to engine bolts. Tighten to 39 Nm (29 ft. lbs.).
50. Install rear transmission crossmember. Tighten crossmember to frame bolts to 68 Nm (50 ft. lbs.).
51. Install rear support to transmission. Tighten bolts to 47 Nm (35 ft. lbs.).
52. Lower transmission onto crossmember and install bolts attaching transmission mount to crossmember. Tighten clevis bracket to crossmember bolts to 47 Nm (39 ft. lbs.).
53. Remove engine support fixture.
54. Install the transmission to engine oil pan bolts. Tighten to 39 Nm (29 ft. lbs.).
55. Connect input and output speed sensor wires and the transmission range sensor.
56. Connect wires to the solenoid/pressure switch assembly.
57. Install the crankshaft position sensor.
58. Connect the gearshift cable to the transmission manual shift lever.

⁂ WARNING

It is essential that correct length bolts be used to attach the converter to the driveplate. Bolts that are too long will damage the clutch surface inside the converter.

59. Install all torque converter-to-driveplate bolts by hand.
60. Verify that the torque converter is pulled flush to the driveplate. Tighten bolts to 42 Nm (31 ft. lbs.).
61. Install the torque converter bolt access cover onto the transmission, for vehicles equipped with 3.5L engines. Install the access cover bolt and tighten to 11 Nm (96 inch lbs.).
62. Install the structural collar and bolts for vehicles equipped with a 2.7L engine.
63. Install starter motor.
64. Connect the cooler line fittings and cooler lines to the transmission.

➡Inspect the fill tube grommet to determine if the grommet is new or used. If a new fill tube grommet is in the transmission case, pierce the center of the grommet with a ballpoint pen, or similar instrument, to prepare the grommet for the fill tube installation.

65. Install the transmission fill tube.
66. Install exhaust components.
67. Align and connect the driveshaft.
68. Adjust gearshift cable if necessary.
69. Lower vehicle.
70. Connect negative battery cable.
71. Fill transmission with appropriate transmission fluid.
72. Verify proper operation.

Driveshaft

REMOVAL & INSTALLATION

Front

1. Before servicing the vehicle, refer to the precautions in the beginning of this section.

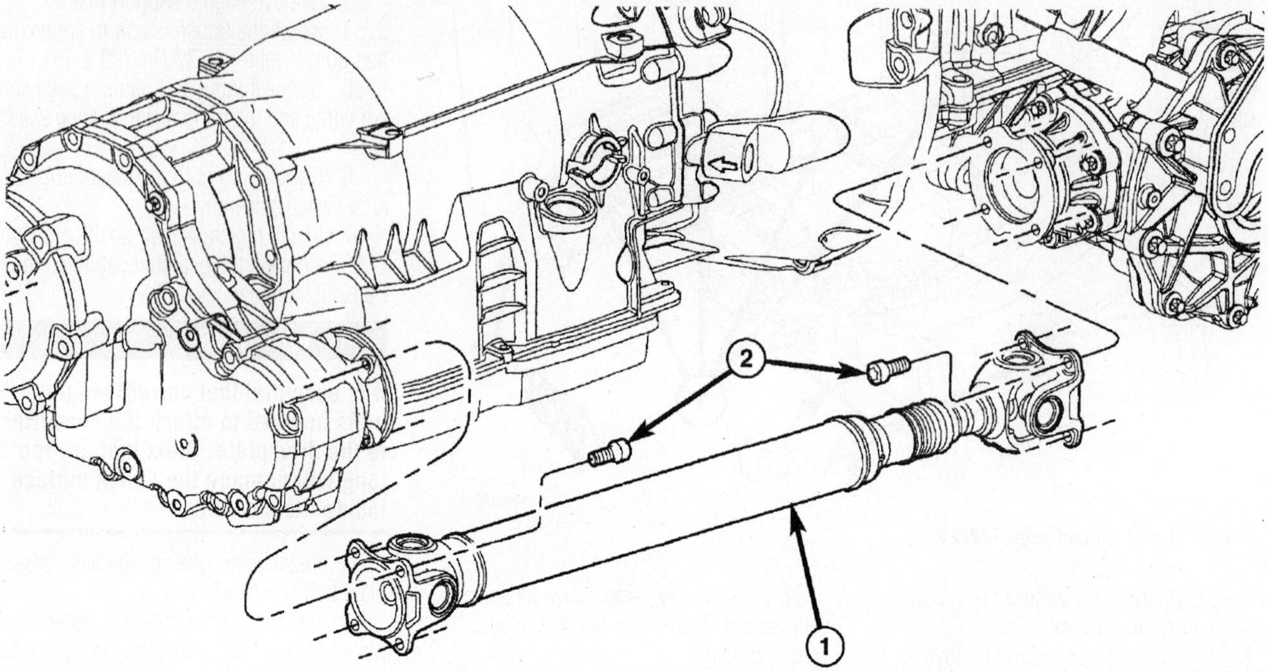

Front driveshaft

2. Place gearshift lever in NEUTRAL and raise vehicle on hoist.
3. Remove driveshaft heat shield.
4. Apply alignment index marks on the driveshaft and front axle flanges.
5. Remove four front driveshaft-to-axle flange bolts.
6. Apply alignment index marks on the driveshaft and transfer case flanges.
7. Remove four front driveshaft-to-transfer case flange bolts.
8. Remove driveshaft assembly.

To install:

9. Install driveshaft into position.
10. Starting at transfer case end, align index marks. Install driveshaft-to-transfer case flange bolts.
11. Align index marks at front axle end of shaft, and loose-install driveshaft-to-front axle flange bolts.
12. Torque driveshaft-to-transfer case and axle flange bolts to 30 Nm (22 ft. lbs.).
13. Install driveshaft heat shield.
14. Lower vehicle.

Rear

1. Before servicing the vehicle, refer to the precautions in the beginning of this section.

⁂ WARNING

Driveshaft removal is a 2-person operation. Never allow driveshaft to hang from the center bearing, or while only connected to the transmission or rear axle flanges. A helper is required. If a driveshaft section is hung unsupported, damage may occur to the shaft, coupler, and/or center bearing from over-angulations. This may result in driveline vibrations and/or component failure.

2. With vehicle in neutral, position on hoist.
3. Apply alignment index marks on the transmission and axle flanges and rubber couplers.
4. Remove crossmember.
5. Remove rear exhaust system.
6. Remove heat shield.
7. Remove driveshaft front coupler-to-flange bolts.
8. Remove driveshaft rear coupler-to-flange bolts.
9. Remove center bearing mounting bolts.

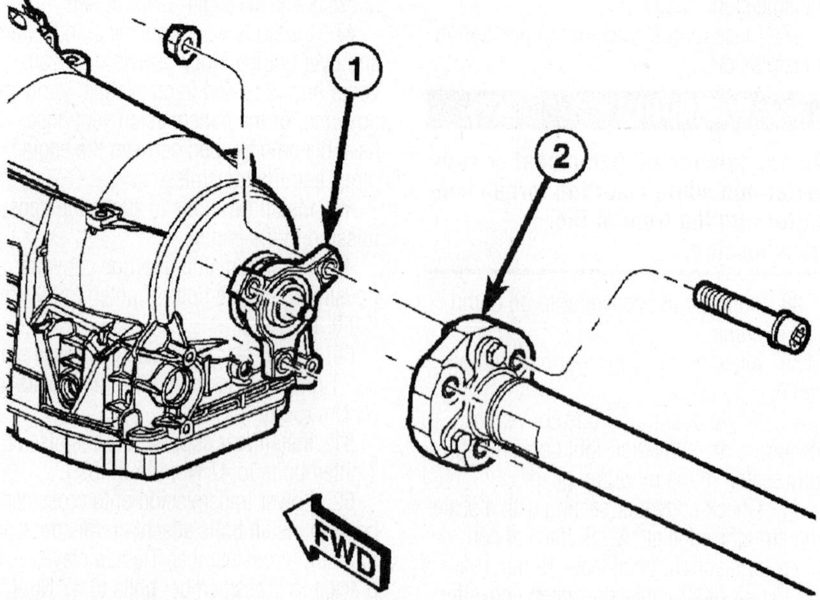

Driveshaft (2)-to-transmission flange (1) attachment

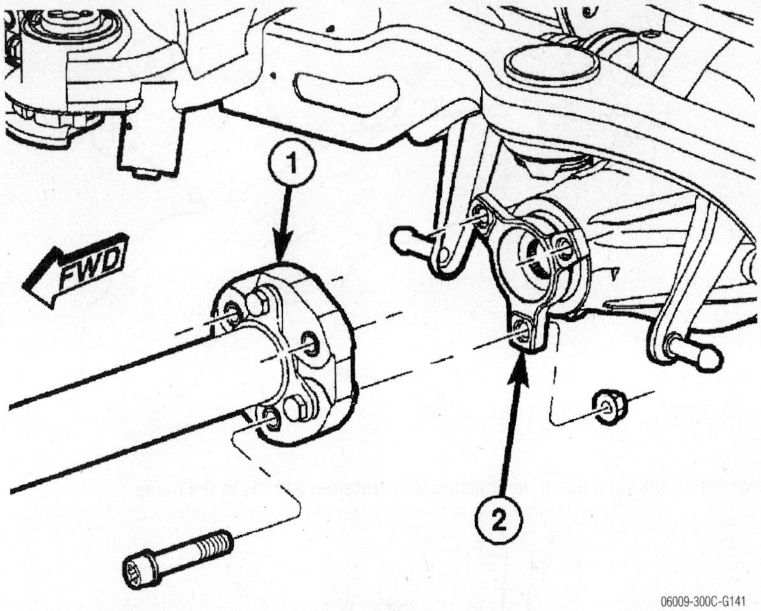

Driveshaft (1)-to-rear axle flange (2) attachment

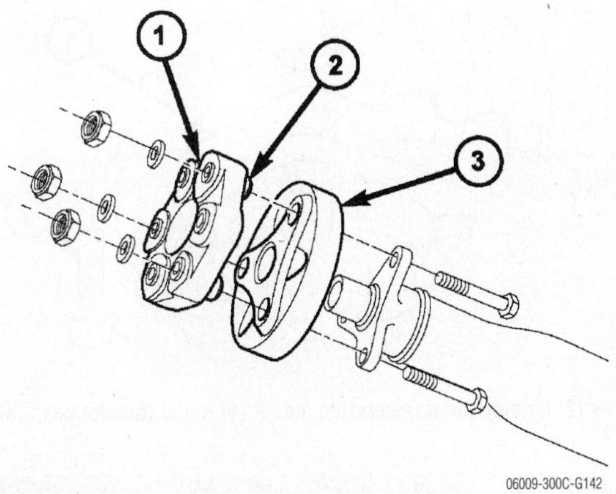

Coupler/damper. (1) coupler, (2) protruding sleeve, (3) damper

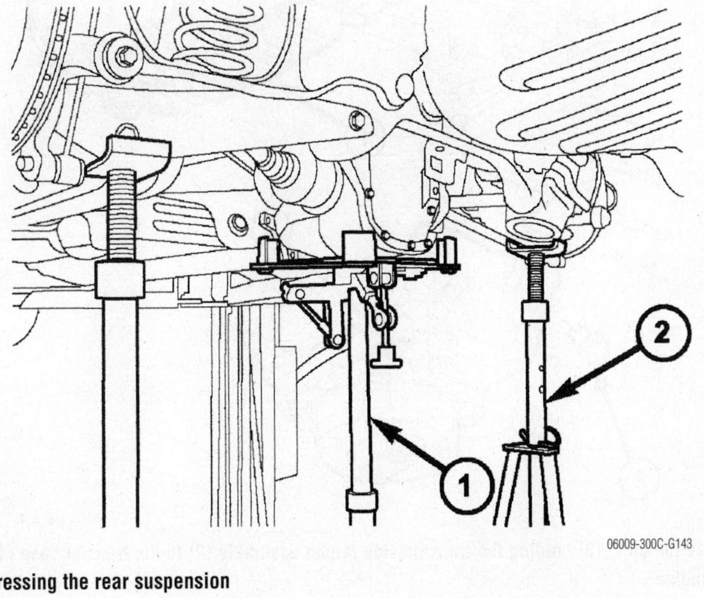

Compressing the rear suspension

10. With the aid of a helper, remove driveshaft assembly.

11. Remove three coupler-to-driveshaft bolt/nuts.

12. Separate coupler and damper (if equipped) from driveshaft. Note orientation and direction of components. It is imperative that they are properly reinstalled.

To install:

13. If coupler and/or damper (V6 Models) were removed, align index marks and reinstall. Make sure protruding sleeve is properly seated into driveshaft or damper counterbores.

14. Install three bolts with washers and nuts and torque to 58 Nm (43 ft. lbs.) torque.

15. Obtain helper and install driveshaft into position at axle. Align index marks placed upon removal. Install driveshaft rear coupler-to-axle flange bolt/nuts by hand. Do not torque at this time.

16. Install driveshaft into position at transmission flange. Align index marks placed upon removal. Install driveshaft front coupler-to-transmission flange bolt/nuts by hand. Do not torque at this time.

17. Loosely install center bearing-to-body bolts. Do not torque at this time.

18. Torque driveshaft front coupler-to-transmission flange bolt/nuts to 58 Nm (43 ft. lbs.)

19. Torque driveshaft rear coupler-to-axle flange bolt/nuts to 58 Nm (43 ft. lbs.)

➡Note: It is necessary to compress rear suspension to ride height before securing center bearing to body. Failure to compress suspension may result in objectionable noise and premature bearing wear.

20. Compress rear suspension with suitable jackstands.

21. Torque center bearing-to-body bolts to 27 Nm (20 ft. lbs.).

22. Install heat shield.

23. Install rear exhaust system.

24. Install crossmember.

Transfer Case Assembly

REMOVAL & INSTALLATION

1. Before servicing the vehicle, refer to the precautions in the beginning of this section.
2. Raise support vehicle.
3. Disconnect the front and rear driveshafts from the transfer case.
4. Support transmission with jackstand.
5. Remove the bolts holding the transmission crossmember to the frame.

6. Remove the bolts holding the transmission crossmember to the transmission mount.
7. Remove the bolts holding the transmission mount to the transfer case, 5.7L engine, and remove the transmission mount.
8. Remove the bolts holding the transmission mount assembly to the transfer case, 3.5L engine, and remove the transmission mount assembly.
9. Remove the transfer case damper.
10. Remove the transfer case drain plug and drain transfer case lubricant.
11. Support transfer case with transmission jack.
12. Secure transfer case to jack with chains.
13. Lower the transmission and transfer case slightly, to improve access to the upper transfer case to transmission fasteners.
14. Remove nuts and bolts attaching transfer case to transmission.
15. Pull transfer case and jack rearward to disengage transfer case.
16. Remove transfer case from under vehicle.

To install:

17. Support transfer case on a transmission jack.
18. Secure transfer case to jack with chains.
19. Position the transfer case under the vehicle and align the input shaft with the transmission output shaft.
20. Move the transfer case and jack forward to engage transfer case to the transmission.
21. Install the nuts and bolts to attach transfer case to transmission. Tighten the 120mm bolts to 25 Nm (18 ft. lbs.), the 30mm bolts to 27 Nm (20 ft. lbs.), and the nuts to 20 Nm (15 ft. lbs.)
22. Raise the transmission and transfer case into their normal position.
23. Fill the transfer case with the appropriate fluid.
24. Install the transfer case damper.
25. Install the transmission mount assembly to the transfer case and install the bolts to hold the transmission mount assembly to the transfer case, 3.5L engine. Tighten the bolts to 55 Nm (41 ft. lbs.).
26. Install the transmission mount to the transfer case and install the bolts to hold the transmission mount to the transfer case, 5.7L engine. Tighten the bolts to 68 Nm (50 ft. lbs.).
27. Install the bolts to hold the transmission crossmember to the transmission

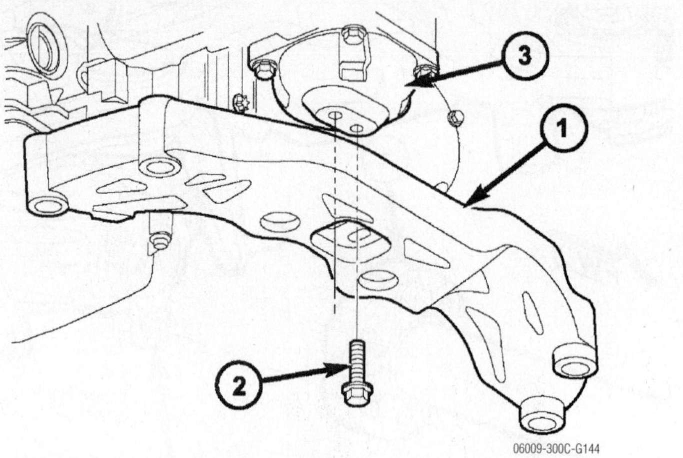

Remove the bolts (2) holding the transmission crossmember (1) to the frame

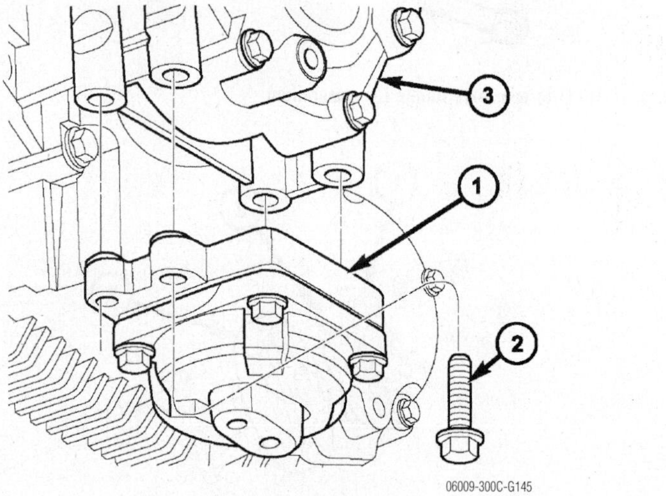

Remove the bolts (2) holding the transmission mount (1) to the transfer case (3)—5.7L engine

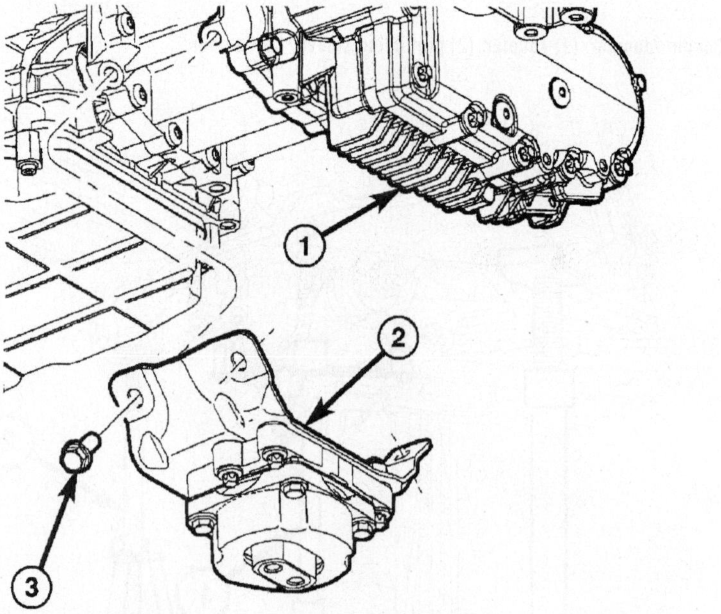

Remove the bolts (3) holding the transmission mount assembly (2) to the transfer case (1)—3.5L engine

mount. Tighten the bolts to 65 Nm (48 ft. lbs.).

28. Install the bolts to hold the transmission crossmember to the frame. Tighten the bolts to 68 Nm (50 ft. lbs.).

29. Install the front and rear driveshafts to the transfer case.

Front Halfshaft

REMOVAL & INSTALLATION

1. Before servicing the vehicle, refer to the precautions in the beginning of this section.
2. Raise and support vehicle.
3. Remove wheel mounting nuts, then tire and wheel assembly.
4. While holding link ball joint stem from rotating, remove nut fastening stabilizer link to shock clevis bracket. Slide link ball joint stem from clevis bracket.
5. Remove nut and pinch bolt fastening clevis bracket to bottom of shock assembly.
6. Remove nut and bolt attaching shock clevis bracket to lower control arm.
7. Pull lower end of clevis bracket outward away from lower control arm bushing, then slide it off shock assembly. It may be necessary to use an appropriate prying tool to spread clamp area of clevis bracket allowing removal from shock assembly.
8. While a helper applies brakes to keep hub from rotating, remove hub nut from axle halfshaft.

➡In some cases, it may be necessary to retract caliper piston in its bore a small amount in order to provide sufficient clearance between shoes and rotor to easily remove caliper from knuckle. This can usually be accomplished before mounting bolts are removed, by grasping rear of caliper and pulling outward working with guide pins, thus retracting piston. Never push on piston directly as it may get damaged.

9. Remove two bolts securing disc brake caliper and adapter to knuckle.
10. Remove disc brake caliper and adapter from knuckle as an assembly. Hang assembly out of way using wire or a bungee cord. Use care not to overextend brake hose when doing this.
11. Remove any clips retaining brake rotor to wheel studs.
12. Slide brake rotor off hub and bearing.

**** WARNING**

In following step, use care not to damage ball joint seal boot while sliding the puller into place past seal boot.

13. Separate upper ball joint stud from knuckle.
14. Remove tool.
15. Remove nut from end of upper ball joint stud.
16. Remove clip fastening wheel speed sensor to knuckle.
17. Disengage right halfshaft from axle and remove from vehicle.
18. Remove left halfshaft from vehicle.

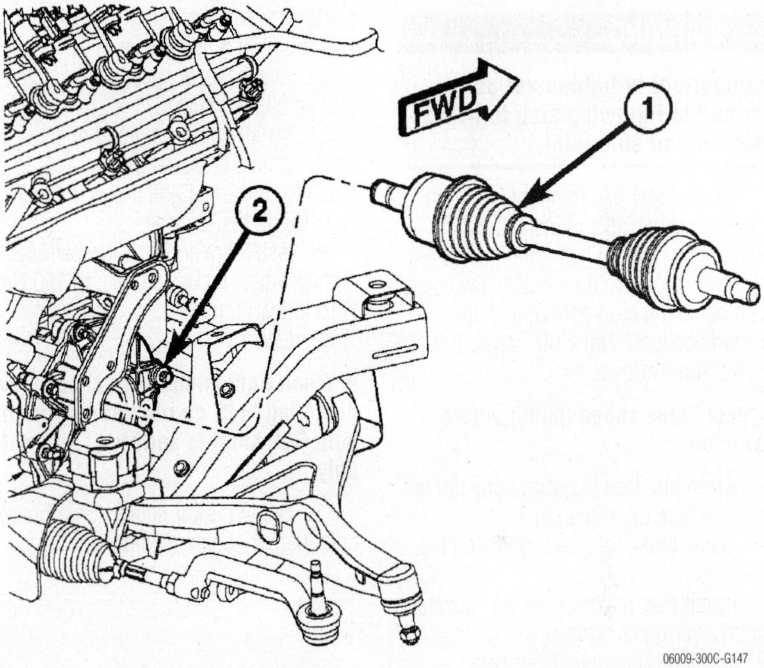

06009-300C-G147

Right front halfshaft removal/installation

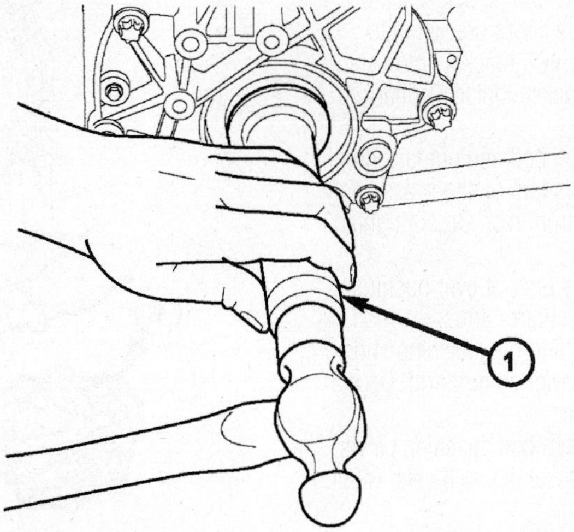

06009-300C-G148

Axle seal installation

To install:

19. Install left halfshaft assembly.
20. Using Tool C-4193-A, install new axle seal.
21. Install right halfshaft assembly.

➡Before installing halfshafts into hub/bearing assemblies, ensure isolation washer is present on end of halfshaft. Inspect washer making sure it is not worn or damaged. Washer is bi-directional and can be installed in either direction on shaft.

22. Install halfshaft isolation washer.
23. Install halfshaft into hub/bearing assembly.

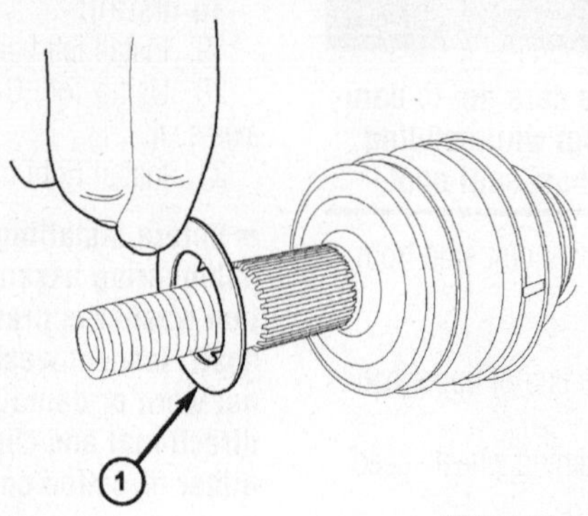

Halfshaft isolation washer

24. Loosely install halfshaft hub nut, and do not tighten at this time.

⁂ WARNING

It is important to tighten nut as described in following step to avoid damaging ball stud joint.

25. Place upper ball joint stud through hole in top of knuckle and install nut. Tighten nut by holding ball joint stud with a hex wrench while turning nut with a wrench. Tighten nut using crow foot wrench on torque wrench to 47 Nm + 90° turn (35 ft. lbs. + 90° turn) torque.

➡Inspect brake shoes (pads) before installation.

26. Clean hub face to remove any dirt or corrosion where rotor mounts.
27. Install brake rotor over studs on hub and bearing.
28. Install disc brake caliper and adapter assembly over brake rotor.
29. Install mounting bolts securing caliper adapter to knuckle. Tighten bolts to 169 Nm (125 ft. lbs.) torque.
30. Pull lower end of shock assembly outward, then slide clevis bracket onto lower end. Slide clevis bracket onto shock assembly until bracket contacts collar on shock housing.
31. Install pinch bolt and nut fastening clevis bracket to bottom of shock assembly. Install pinch bolt from rear. Do not tighten at this time.
32. Slide clevis bracket over bushing mounted in lower control arm.
33. Install bolt and nut attaching shock clevis bracket to lower control arm. Do not tighten at this time.
34. Tighten pinch bolt attaching clevis bracket to shock assembly to 61 Nm (45 ft. lbs.) torque.
35. Slide stabilizer link ball joint stem into clevis bracket. Install nut fastening link to clevis bracket. Tighten nut by holding ball joint stud while turning nut. Tighten nut using crow foot wrench on torque wrench to 146 Nm (108 ft. lbs.) torque.
36. Attach wheel speed sensor cable routing clip at knuckle.
37. Install hub nut on end of axle halfshaft. While a helper applies brakes to keep hub from turning, tighten hub nut to 213 Nm (157 ft. lbs.) torque.
38. Install tire and wheel assembly. Tighten wheel mounting nuts to 150 Nm (110 ft. lbs.) torque.
39. Lower vehicle.

➡When tightening lower shock clevis mounting bolt, do not attempt rotating bolt. Bolt shaft is serrated. Turn nut only.

40. Tighten lower shock clevis bracket bolt nut to 174 Nm (128 ft. lbs.) torque.

Intermediate Shaft

REMOVAL & INSTALLATION

➡The intermediate shaft assembly is serviced only as an assembly.

1. Before servicing the vehicle, refer to the precautions in the beginning of this section.
2. Remove four intermediate shaft assembly-to-oil pan bolts.
3. Remove intermediate shaft assembly from engine oil pan.

To install:

4. If previously removed, install intermediate shaft assembly.
5. Install four intermediate shaft-to-oil pan bolts and torque to 25 Nm (18 ft. lbs.).

Rear Halfshaft

REMOVAL & INSTALLATION

1. Before servicing the vehicle, refer to the precautions in the beginning of this section.

➡This procedure requires the compression of the rear suspension to ride height. A drive-on hoist should be used. If a drive-on hoist is not used, screw-style under-hoist jackstands are required to compress the rear suspension, facilitating rear halfshaft removal.

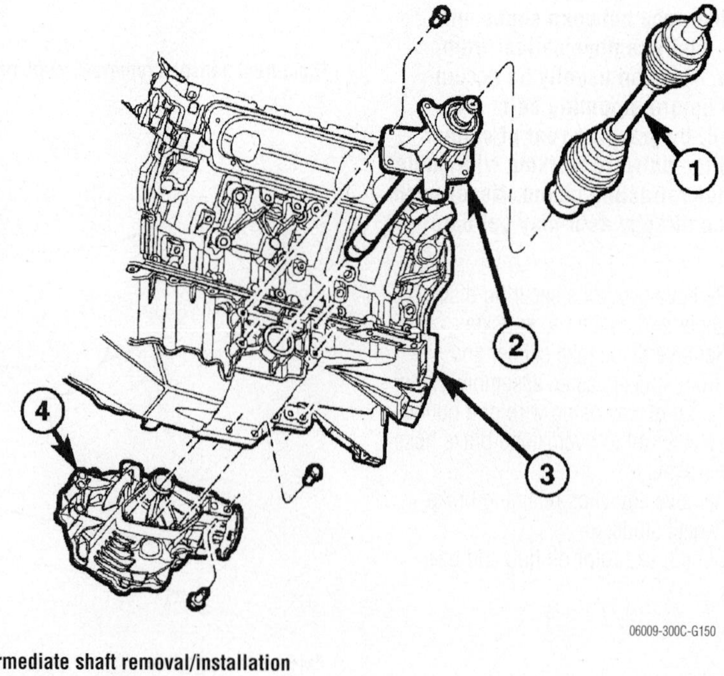

Intermediate shaft removal/installation

➡Halfshaft inner and outer boots are not serviceable separately. Boot replacement requires entire shaft assembly replacement.

⁂ WARNING

Unequal-length halfshafts are used. The left halfshaft is shorter than the right, and it is necessary to identify and tag halfshafts upon removal to ensure proper installation.

⁂ WARNING

Never grasp halfshaft assembly by the inner or outer boots. Doing so may cause the boot to pucker or crease, reducing the service life of the boot and joint. Avoid over angulating or stroking the CV-joints when handling the halfshaft.

2. With vehicle in neutral, position and raise vehicle on hoist.
3. Using 14mm hex, remove axle drain plug and drain rear axle fluid into container suitable for fluid reuse.
4. Install drain plug and torque to:
   - 198 Axle: 60 Nm (44 ft. lbs.)
   - 210 Axle: 50 Nm (37 ft. lbs.)
5. Remove rear exhaust system on V8-equipped models.
6. Remove wheel/tire assembly from sides that shaft is to be removed.
7. Remove wheel hub nut and discard.
8. Apply alignment index marks to the driveshaft rubber coupler and axle flange.
9. Remove three driveshaft coupler-to-axle flange bolt/nuts.
10. Using suitable screwdriver, partially disengage halfshaft(s) from axle assembly.
11. If a drive-on hoist is used, position transmission jack to rear axle assembly. If a drive-on hoist is not used, compress rear suspension using screw-style under-hoist jackstands, then position transmission jack to rear axle assembly.
12. Remove rear axle forward mount isolator bolt/nut.

⁂ WARNING

Access to rear axle-to-crossmember bolts is best achieved by use of short socket and a flexible-head ratchet.

13. Remove two rear axle-to-crossmember bolts.
14. Carefully lower rear axle. While lowering axle, separate driveshaft from axle and support with suitable rope or wire.

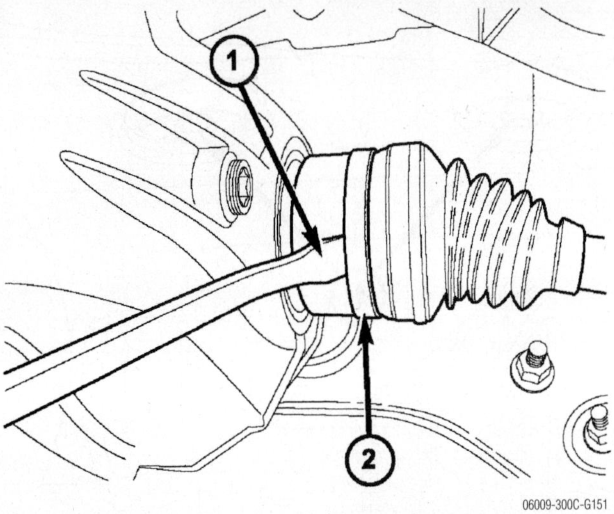

Using suitable screwdriver (1), partially disengage halfshaft(s) (2) from rear axle assembly

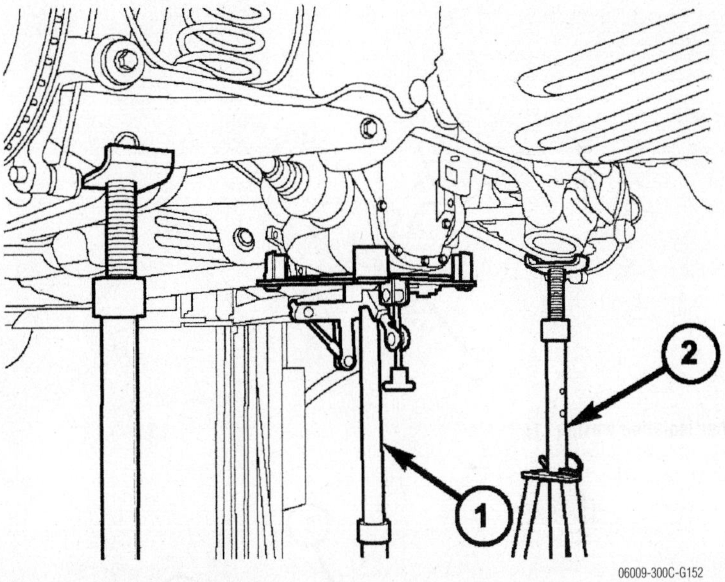

If a drive-on hoist is used, position transmission jack (1) to rear axle assembly. If a drive-on hoist is not used, compress rear suspension using screw-style under-hoist jackstands (2), then position transmission jack to rear axle assembly

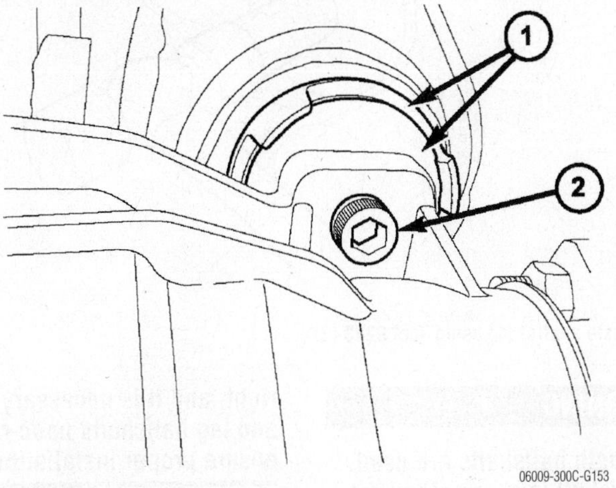

Rear axle forward mount isolator (1) bolt/nut (2)

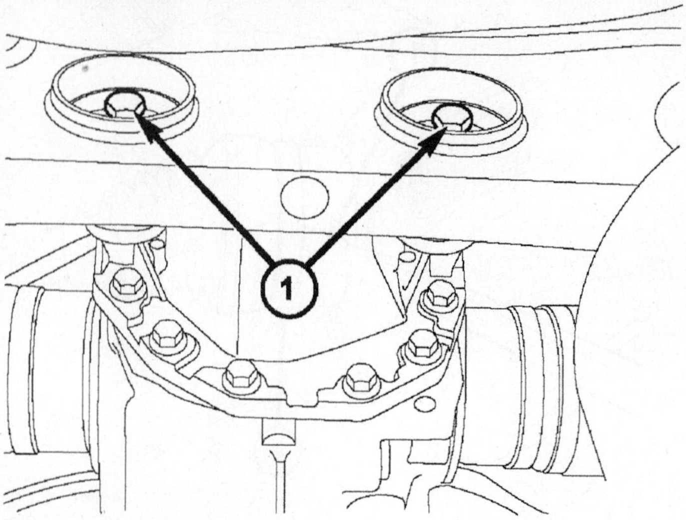

06009-300C-G154

Rear axle-to-crossmember bolts (1)

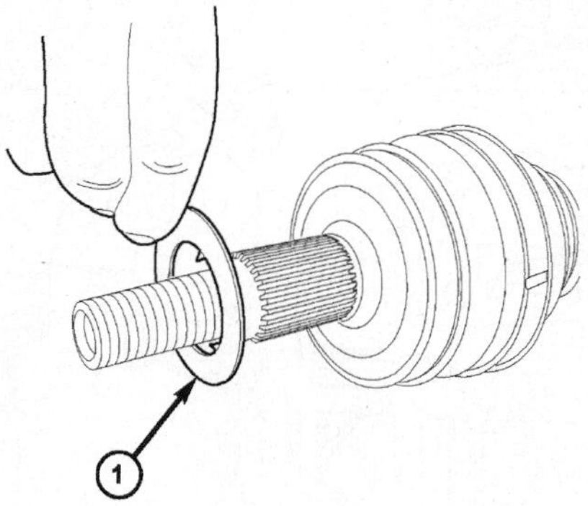

06009-300C-G155

Rubber isolation washer (1)

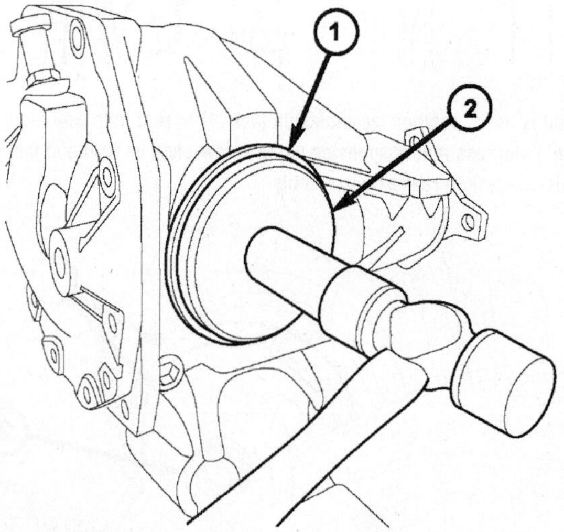

06009-300C-G156

Install new axle seal(s) (1) using Tool 9223 (2)

⁂ WARNING

Unequal-length halfshafts are used. The left halfshaft is shorter than the right, and it is necessary to identify and tag halfshafts upon removal to ensure proper installation.

15. Lower axle just enough to remove halfshafts one at a time. Shift axle assembly in one direction, compressing one halfshaft while removing the other (1). Use caution to protect axle seal and journal.

⁂ WARNING

Never grasp halfshaft assembly by the inner or outer boots. Doing so may cause the boot to pucker or crease, reducing the service life of the boot and joint. Avoid over angulating or stroking the CV-joints when handling the halfshaft.

⁂ WARNING

Use care while handling/storing halfshaft assembly. Damage to the slinger can result from improper handling. If slinger gets bent or damaged, straighten slinger to avoid contact with axle seal assembly.

16. Remove halfshaft from hub. Remove and inspect rubber isolation washer. Discard washer if rubber surfaces are worn away. Repeat on other side if necessary.
17. Remove axle seals using suitable screwdriver.

To install:

➡Halfshaft inner and outer boots are not serviceable separately. Boot replacement requires entire shaft assembly replacement.

⁂ WARNING

Unequal-length halfshafts are used. The left halfshaft is shorter than the right, and it is necessary to identify and tag halfshafts upon removal to ensure proper installation.

⁂ WARNING

Never grasp halfshaft assembly by the inner or outer boots. Doing so may cause the boot to pucker or crease, reducing the service life of the boot and joint. Avoid over angulating or stroking the CV-joints when handling the halfshaft.

18. Install new axle seal(s) using Tool 9223, or equivalent.
19. Install halfshaft isolation washer. Washer is bi-directional, and can be installed in either direction.
20. Install halfshaft to wheel hub/knuckle assembly. Install new hub nut by hand.

➡Inspect slinger(s) for handling damage. Straighten as necessary to avoid contact with axle seal.

➡Use care when installing halfshaft to axle assembly. The halfshaft installation angle should be minimized to avoid damage to seal upon installation.

21. Lubricate halfshaft inner joint bearing journal with Mopar® Gear and Axle Lubricant (75W-140). Using new circlip(s), install halfshaft to rear axle assembly. Use care not to damage axle seals. Verify proper installation by pulling outward on joint by hand.
22. Raise rear axle assembly into position. Align driveshaft index marks and start driveshaft coupler-to-axle bolt/nuts by hand.
23. Install two rear axle-to-crossmember bolts and torque to 220 Nm (162 ft. lbs.).
24. Install rear axle front mount isolator as shown and torque bolt/nut to 65 Nm (48 ft. lbs.).
25. Again verify halfshaft inner joints are fully engaged to axle assembly.
26. Remove transmission jack.
27. If used, remove screw-type under hoist jackstands.
28. Torque driveshaft coupler-to-axle flange bolt/nuts to 58 Nm (43 ft. lbs.).
29. Using a 14mm hex, remove rear axle fill plug. Fill axle with Mopar® 75W-140 Synthetic Gear & Axle Lubricant in the following quantities:
   - 198 Axle: 1.4L (1.5 qts.)
   - 210 Axle: 1.6L (1.7 qts.)
30. Install and torque fill plug to:
   - 198 Axle: 60 Nm (44 ft. lbs.)
   - 210 Axle: 50 Nm (37 ft. lbs.)
31. V8 Models: Install exhaust system. Tighten band clamps to 61 Nm (45 ft. lbs.).
32. Lower vehicle. Torque halfshaft hub nut to 213 Nm (157 ft. lbs.). Install wheel center cap.
33. Install wheel/tire assembly and torque lug nuts to 150 Nm (110 ft. lbs.).

Front Halfshaft CV-Joints

OVERHAUL

Inner

1. Before servicing the vehicle, refer to the precautions in the beginning of this section.
2. Remove large boot clamp which retains inner tripod joint sealing boot to tripod joint housing and discard.
3. Remove small clamp which retains inner tripod joint sealing boot to interconnecting shaft and discard.
4. Remove the sealing boot from the tripod housing and slide it down the interconnecting shaft.

⁂ WARNING

When removing the tripod joint housing from the spider assembly, hold the bearings in place on the spider trunions to prevent the bearings from falling away.

5. Slide the tripod joint housing off the spider assembly and the interconnecting shaft.
6. Remove snapring which retains spider assembly to interconnecting shaft. Do not hit the outer tripod bearings in an attempt to remove spider assembly from interconnecting shaft.
7. Remove the spider assembly from interconnecting shaft. If spider assembly will not come off interconnecting shaft by hand, it can be removed by tapping spider assembly with a brass drift.
8. Slide sealing boot off interconnecting shaft.
9. Thoroughly clean and inspect spider assembly, tripod joint housing, and interconnecting shaft for any signs of excessive wear. If any parts show signs of excessive wear, the driveshaft assembly will require replacement. Component parts of these driveshaft assemblies are not serviceable.

To assemble:

10. Slide inner tripod joint seal boot retaining clamp, onto interconnecting shaft. Then, slide the replacement inner tripod joint sealing boot onto the interconnecting shaft. Inner tripod joint seal boot MUST be positioned on interconnecting shaft, so the raised bead on the inside of the seal boot is in groove on interconnecting shaft.
11. Install spider assembly onto interconnecting shaft. Spider assembly must be installed on interconnecting shaft far enough to fully install spider retaining snapring.
12. If spider assembly will not fully install on interconnecting shaft by hand, it can be installed by tapping the spider body with a brass drift. Do not hit the outer tripod bearings in an attempt to install spider assembly on interconnecting shaft.
13. Install the spider assembly to interconnecting shaft retaining snapring into groove on end of interconnecting shaft. Be sure the snapring is fully seated into groove on interconnecting shaft.

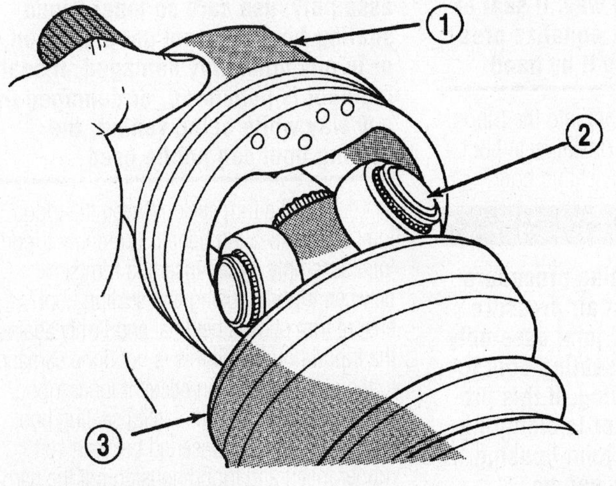

06009-300C-G157

Front halfshaft inner joint tripod housing (1), spider (2), boot (3)

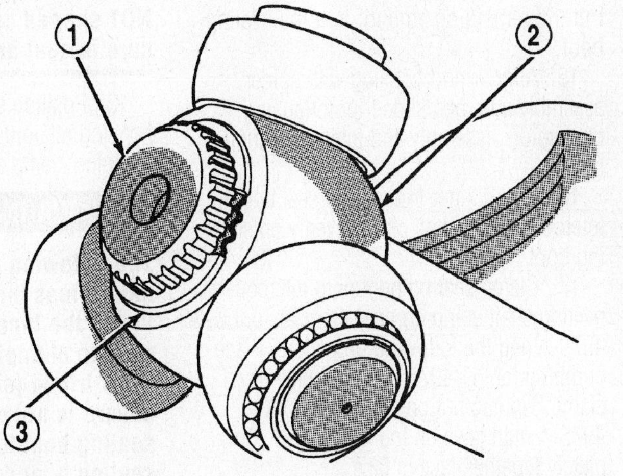

06009-300C-G158

Front shaft inner joint (1) shaft, (2) spider, (3) snapring

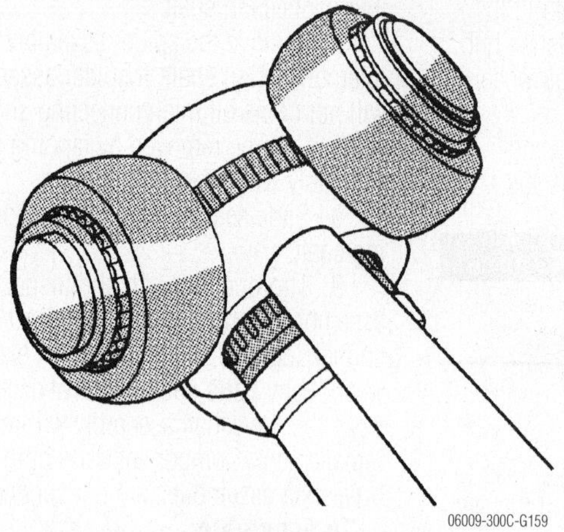

Drive the spider from the shaft with a brass drift

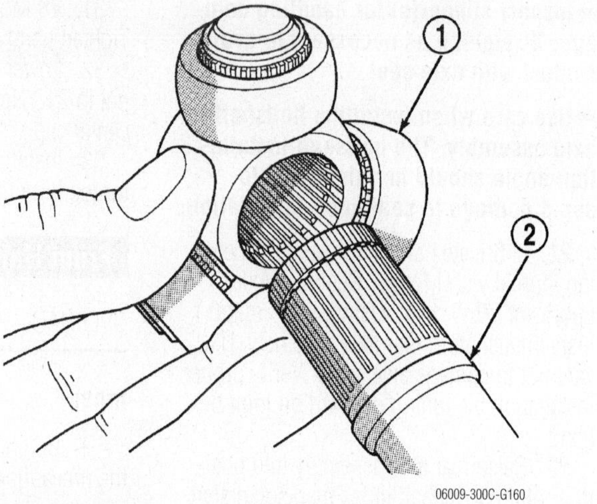

Install spider assembly (1) onto interconnecting shaft (2)

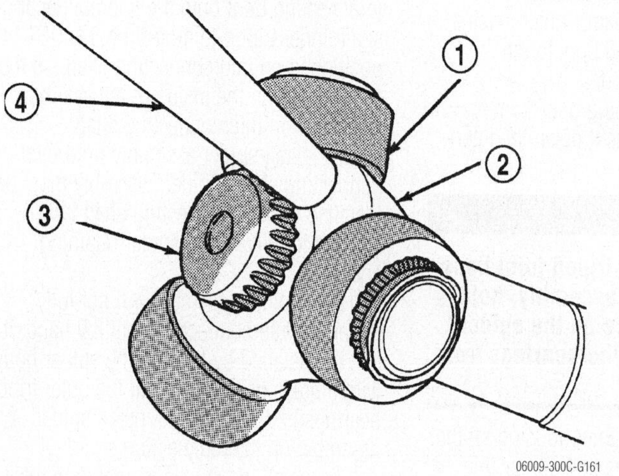

Drive the spider (2) into place with a brass drift (4)

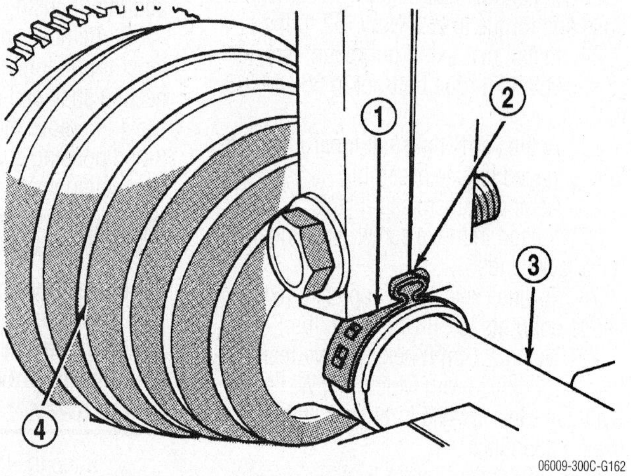

Clamp type band tool. (1) clamp, (2) crimp completely closed, (3) shaft, boot (4)

14. Distribute ½ the amount of grease provided in the seal boot service package (DO NOT USE ANY OTHER TYPE OF GREASE) into tripod housing. Put the remaining amount into the sealing boot.

15. Align tripod housing with spider assembly and then slide tripod housing over spider assembly and interconnecting shaft.

16. Install inner tripod joint seal boot to interconnecting shaft clamp evenly on sealing boot.

17. Clamp sealing boot onto interconnecting shaft using crimper, Special Tool C-4975-A and the following procedure. Place crimping tool C-4975-A over bridge of clamp. Tighten nut on crimping tool C-4975-A until jaws on tool are closed completely together, face to face.

**** WARNING**

Seal must not be dimpled, stretched or out of shape in any way. If seal is NOT shaped correctly, equalize pressure in seal and shape it by hand.

18. Position sealing boot into the tripod housing retaining groove. Install seal boot retaining clamp evenly on sealing boot.

**** WARNING**

The following positioning procedure determines the correct air pressure inside the inner tripod joint assembly prior to clamping the sealing boot to inner tripod joint housing. If this procedure is not done prior to clamping sealing boot to tripod joint housing sealing boot durability can be adversely affected.

**** WARNING**

When venting the inner tripod joint assembly, use care so inner tripod sealing boot does not get punctured, or in any other way damaged. If sealing boot is punctured, or damaged in any way while being vented, the sealing boot can not be used.

19. Insert a trim stick between the tripod joint and the sealing boot to vent inner tripod joint assembly. When inserting trim stick between tripod housing and sealing boot ensure trim stick is held flat and firmly against the tripod housing. If this is not done damage to the sealing boot can occur. If inner tripod joint has a Hytrel (hard plastic) sealing boot, be sure trim stick is inserted between soft rubber insert and tripod housing not the hard plastic sealing boot and soft rubber insert.

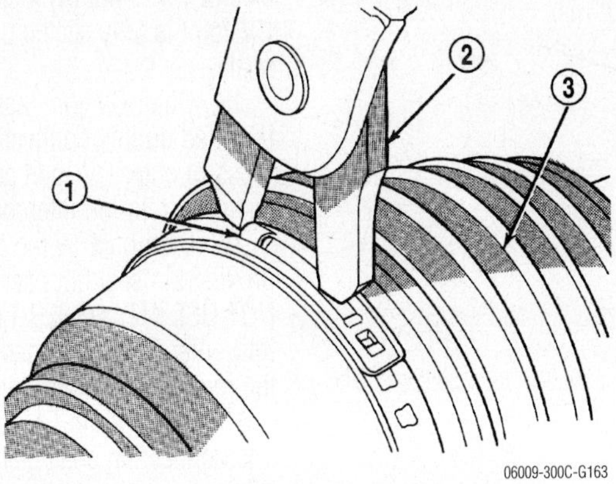

Latching type boot clamp tool

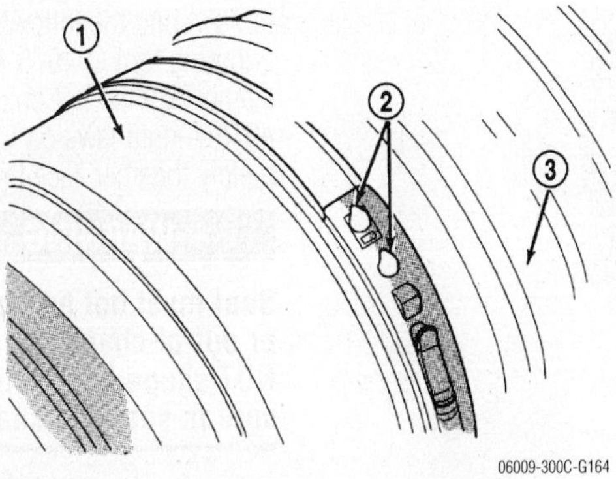

Latching type clamp installed

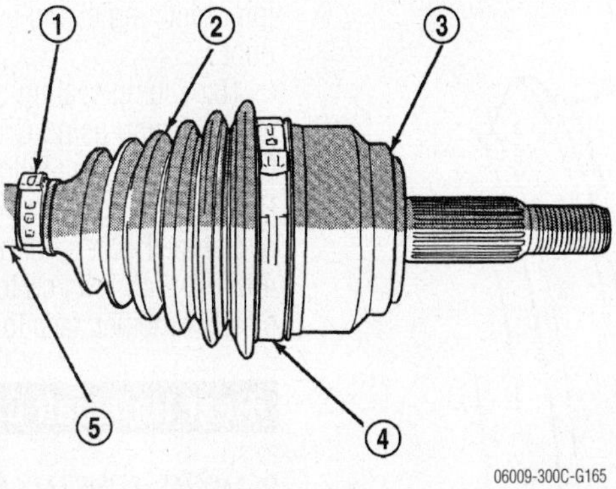

Outer CV-joint. (1) small clamp, (2) boot, (3) joint housing, (4) large clamp, (5) shaft

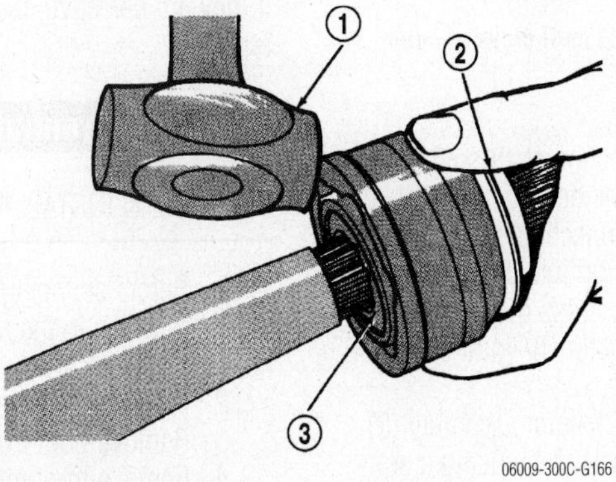

Outer CV-joint removal. (1) soft-faced hammer, (2) joint, (3) circlip

20. With trim stick inserted between sealing boot and tripod joint housing, position the interconnecting shaft so it is at the center of its travel in the tripod joint housing. Remove the trim stick from between the sealing boot and the tripod joint housing. This procedure will equalize the air pressure in the tripod joint, preventing premature sealing boot failure.
21. Position trilobal boot to interface with the tripod housing. The lobes of the boot must be properly aligned with the recesses of the tripod housing.
22. Clamp sealing boot (3) onto tripod housing using clamp locking tool, Snap-On YA3050 (2) or an equivalent.
23. Place prongs of clamp locking tool in the holes of the clamp (1).
24. Squeeze tool together until top band of clamp is latched behind the two tabs on lower band of clamp (2).

Outer

1. Before servicing the vehicle, refer to the precautions in the beginning of this section.
2. Remove large boot clamp retaining CV-joint sealing boot to CV-joint housing and discard.
3. Remove small clamp that retains outer CV-joint sealing boot to interconnecting shaft and discard.
4. Remove sealing boot from outer CV-joint housing and slide it down interconnecting shaft.
5. Wipe away grease to expose outer CV-joint and interconnecting shaft.
6. Remove outer CV-joint from interconnecting shaft using the following procedure:
   a. Support interconnecting shaft in a vise equipped with protective caps on jaws of vise to prevent damage to interconnecting shaft.
   b. Then, using a soft-faced hammer, sharply hit the end of the CV-joint housing to dislodge housing from internal circlip on interconnecting shaft.
   c. Then slide outer CV-joint off end of interconnecting shaft, joint may have to be tapped off shaft using a soft-faced hammer.
7. Remove large circlip from the interconnecting shaft before attempting to remove outer CV-joint sealing boot.
8. Slide failed sealing boot off interconnecting shaft.
9. Thoroughly clean and inspect outer CV-joint assembly and interconnecting joint for any signs of excessive wear. If any parts

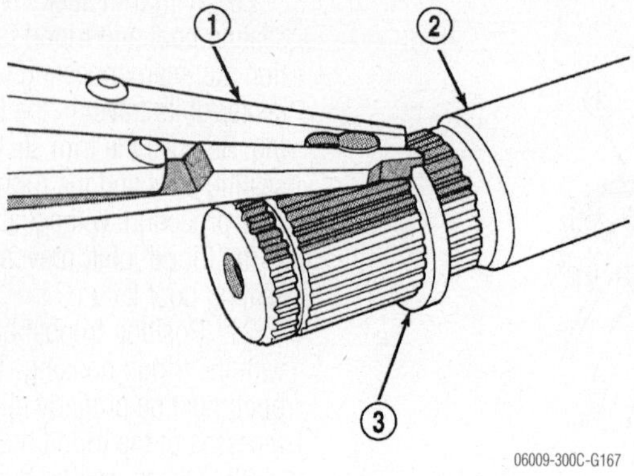

Large circlip removal

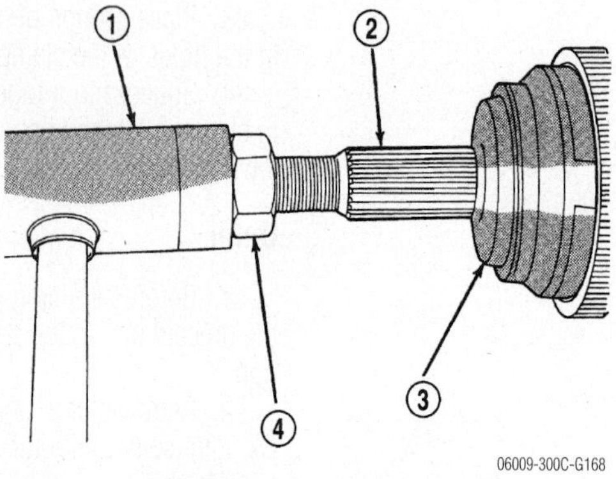

Outer CV-joint installation

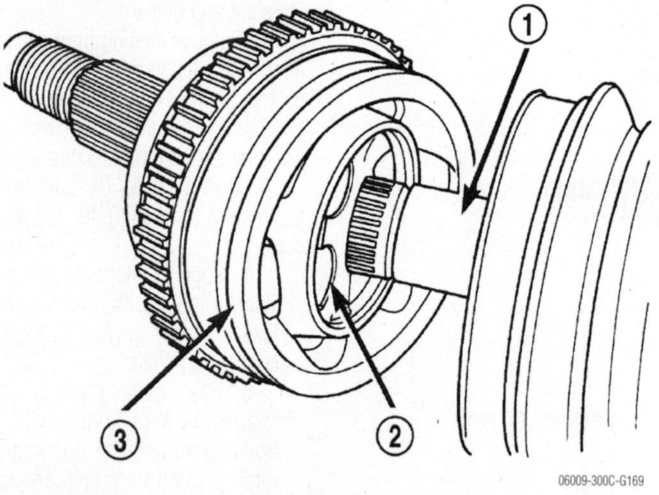

Outer CV-joint assembly (3) must be installed on interconnecting shaft (1) until cross of outer CV-joint assembly (2) is seated against circlip on interconnecting shaft

show signs of excessive wear, the driveshaft assembly will require replacement. Component parts of these driveshaft assemblies are not serviceable.

To assemble

10. Slide new sealing boot to interconnecting shaft retaining clamp onto interconnecting shaft. Slide the outer CV-joint assembly sealing boot onto the interconnecting shaft. Seal boot MUST be positioned on interconnecting shaft so the raised bead on the inside of the seal boot is in groove on interconnecting shaft.

11. Align splines on interconnecting shaft with splines on cross of outer CV-joint assembly and start outer CV-joint onto interconnecting shaft.

12. Install outer CV-joint assembly (3) onto interconnecting shaft by using a soft-faced hammer (1) and tapping end of stub axle (2), with nut (4) installed, until outer CV-joint is fully seated on interconnecting shaft.

13. Outer CV-joint assembly must be installed on interconnecting shaft until cross of outer CV-joint assembly is seated against circlip on interconnecting shaft.

14. Distribute ½ the amount of grease provided in seal boot service package (DO NOT USE ANY OTHER TYPE OF GREASE) into outer CV-joint assembly housing. Put the remaining amount into the sealing boot.

15. Install outer CV-joint sealing boot to interconnecting shaft clamp evenly on sealing boot.

16. Clamp sealing boot onto interconnecting shaft using crimper, Special Tool C-4975-A and the following procedure. Place crimping tool C-4975-A over bridge of clamp. Tighten nut on crimping tool C-4975-A until jaws on tool are closed completely together, face to face.

⁂ WARNING

Seal must not be dimpled, stretched, or out-of-shape in any way. If seal is NOT shaped correctly, equalize pressure in seal and shape it by hand.

17. Position outer CV-joint sealing boot into its retaining groove on outer CV-joint housing. Install sealing boot to outer CV-joint retaining clamp evenly on sealing boot.

18. Clamp sealing boot onto outer CV-joint housing using Crimper, Special Tool C-4975-A and the following procedure. Place crimping tool C-4975-A over bridge of clamp. Tighten nut on crimping tool C-4975-A until jaws on tool are closed completely together, face to face.

Rear Halfshaft CV-Joints

REMOVAL & INSTALLATION

These joints are not serviced separately. If they are defective, the entire shaft must be replaced.

Front Differential

REMOVAL & INSTALLATION

1. Before servicing the vehicle, refer to the precautions in the beginning of this section.
2. Raise and support vehicle.
3. Remove both front halfshafts.
4. Remove front intermediate shaft assembly.

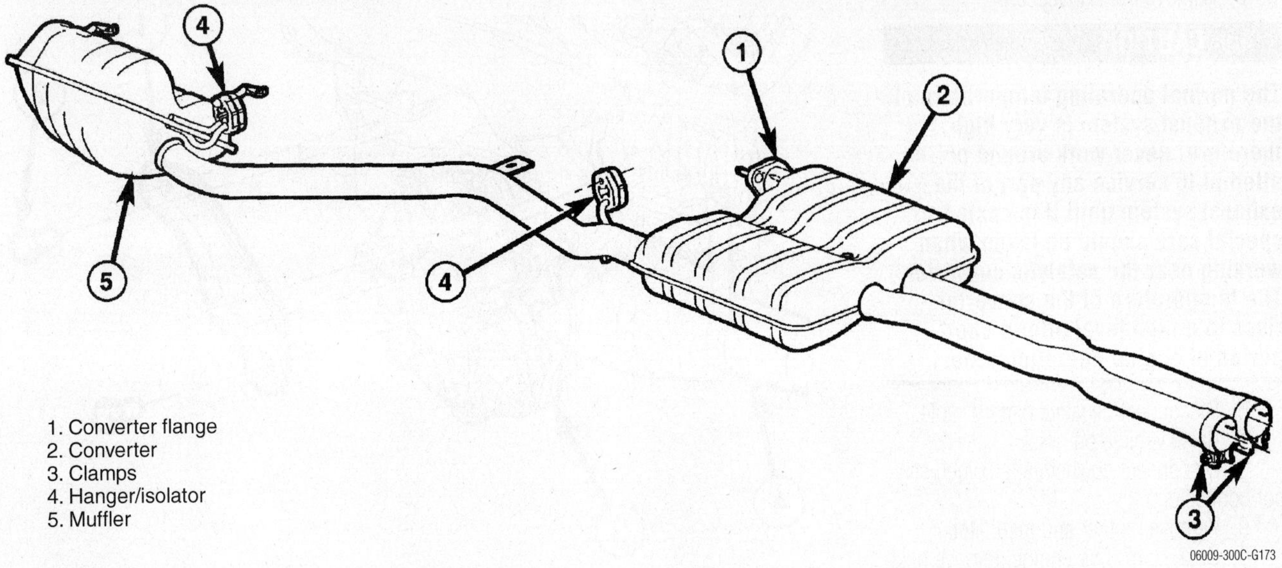

Exhaust system

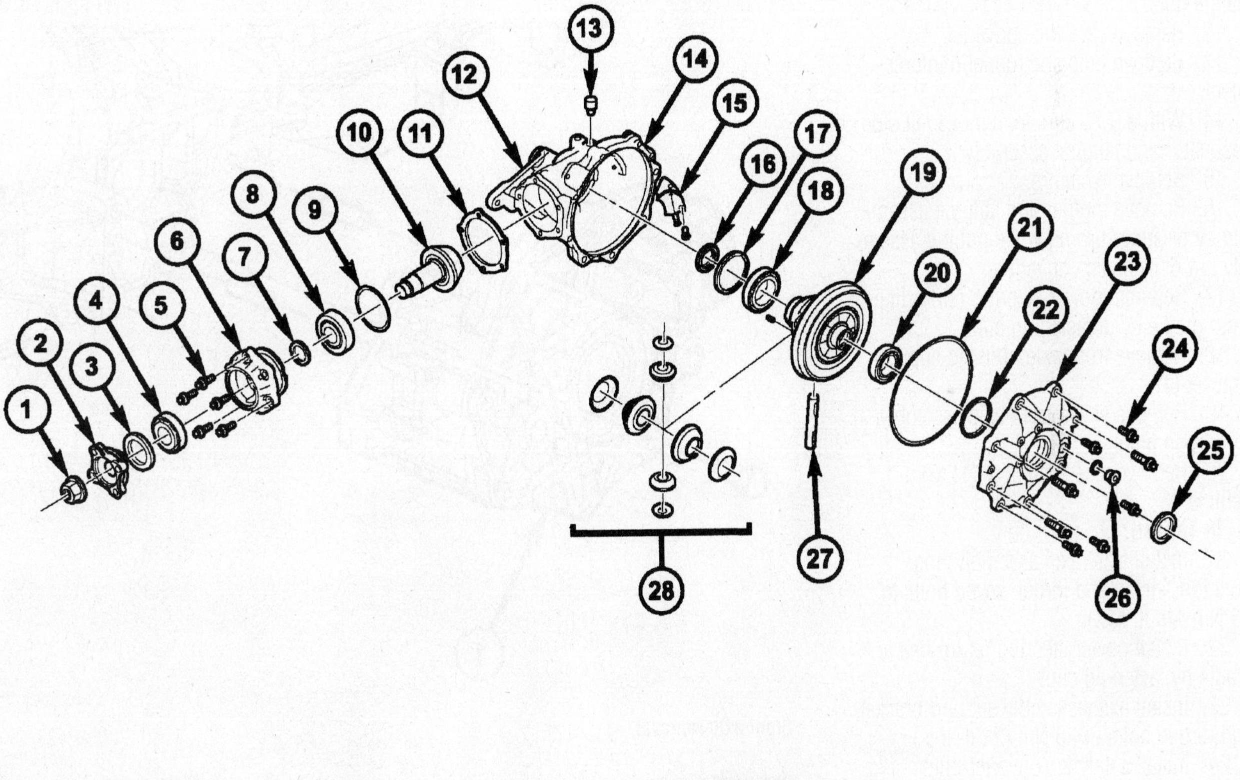

Front axle exploded view

5. Remove front driveshaft.

**** CAUTION**

The normal operating temperature of the exhaust system is very high. therefore, never work around or attempt to service any part of the exhaust system until it is cooled. special care should be taken when working near the catalytic converter. The temperature of the converter rises to a high level after a short period of engine operation time.

6. Disconnect negative battery cable.
7. Raise vehicle on hoist.
8. Disconnect downstream oxygen sensor connectors.
9. Remove muffler and resonator.
10. On 2.7L or 3.5L engine, remove nuts and cross-brace.
11. Remove catalytic converter to ball flange nuts.
12. Remove catalytic converter.
13. Remove nuts and tunnel reinforcement.
14. With a 5.7L engine, remove left side resonator and tailpipe assembly.
15. Remove isolators.
16. Remove muffler and tailpipe assembly by twisting/turning while pulling assembly out of catalytic converters.
17. Separate power steering return line from cradle by unfastening clips.
18. Remove front axle housing support bracket (1).
19. Remove seven front axle assembly-to-oil pan bolts.
20. Remove axle assembly (1) from vehicle.

To install:

21. Install front axle assembly) into position. Install and torque seven bolts to 65 Nm (48 ft. lbs.).
22. Install power steering return line to cradle by fastening clips.
23. Install axle assembly support bracket and torque bolts to 28 Nm (21 ft. lbs.).
24. Install catalytic converter onto exhaust manifold ball flange. Only finger tighten nuts at this time.
25. Install muffler and resonator assembly.
26. Position front exhaust pipe module clamps on exhaust pipe assembly.
27. Install exhaust pipe assembly into catalytic converters.

➡**Isolators will have an approximately 10° forward angle when installed.**

28. Install isolators.

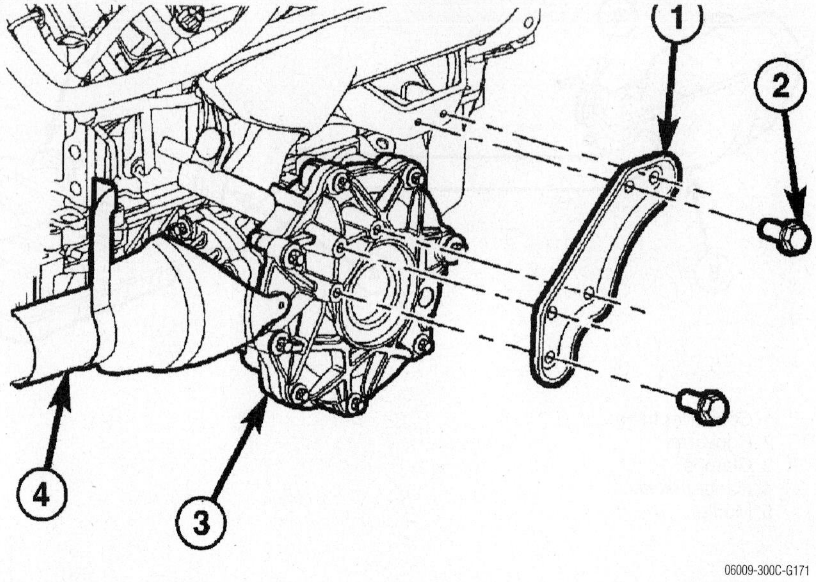

Front axle support bracket

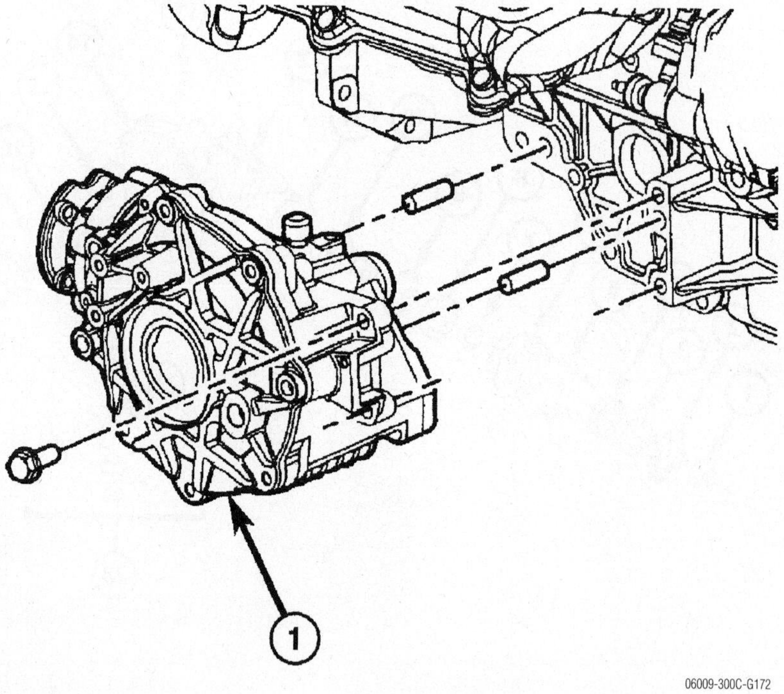

Front axle removal

29. With a 5.7L engine, install left side rear resonator and tailpipe assembly.

➡**Check for proper alignment and clearance to underbody and engine compartment components before tightening clamps.**

30. Tighten band clamps to 61 Nm (45 ft. lbs.).
31. Install tunnel reinforcement and nuts. Tighten nuts to 25 Nm (215 inch lbs.)
32. Check clearance between muffler and resonator assembly and fuel tank. Clearance is 14mm (0.55 in.) for V8 engine and 16mm (0.62 in.) for V6 engine.
33. Check clearance at rear tunnel reinforcement. Clearance should be 15–20mm (0.59–0.78 in.).
34. The tailpipe should be centered in the rear fascia opening.
35. Adjust clearance as necessary.
36. Lower vehicle.
37. Start the engine and inspect for exhaust leaks. Repair exhaust leaks as necessary.
38. Install front driveshaft.

39. Install intermediate shaft.
40. Install both front halfshafts.
41. Fill axle with 600ml (0.64 qts.) of Mopar® 75W-90 Synthetic Gear Lubricant. Fluid level should be at or near bottom of fill hole. Torque fill plug to 30 Nm (22 ft. lbs.).
42. Lower vehicle.

Rear Differential

REMOVAL & INSTALLATION

198 RII Axle

1. Before servicing the vehicle, refer to the precautions in the beginning of this section.

➡This procedure requires the compression of the rear suspension to ride height. A drive-on hoist should be used. If a drive-on hoist is not used, screw-style under-hoist jackstands are required to compress the rear suspension, facilitating rear halfshaft removal.

⁂ WARNING

Never grasp halfshaft assembly by the inner or outer boots. Doing so may cause the boot to pucker or crease, reducing the service life of the boot and joint. Avoid over angling or stroking the CV-joints when handling the halfshaft.

2. With vehicle in neutral, position and raise vehicle on hoist.
3. Using 14mm hex, remove axle drain plug and drain rear axle fluid into container suitable for fluid reuse.
4. Install drain plug and torque to 60 Nm (44 ft. lbs.) torque.
5. Remove rear exhaust system on dual-outlet exhaust models, otherwise, lower exhaust system at rear hanger(s) to provide adequate clearance.
6. Remove the driveshaft.
7. Remove the halfshafts.
8. Remove two rear axle-to-crossmember bolts.
9. Carefully lower rear axle. While lowering axle, separate driveshaft from axle and support with suitable rope or wire.
10. Remove axle assembly from vehicle and transfer to bench.
11. Using suitable screwdriver, remove axle seals and discard.

To install:

12. Install new axle seal(s).

➡Use care when installing halfshaft to axle assembly. The halfshaft installation angle should be minimized to avoid damage to seal upon installation.

13. Install the halfshafts.
14. Raise rear axle assembly into position. Align driveshaft index marks and start driveshaft coupler-to-axle bolt/nuts by hand.
15. Install two rear axle-to-crossmember bolts and torque to 220 Nm (162 ft. lbs.).
16. Install rear axle front mount isolator and torque bolt/nut to 65 Nm (48 ft. lbs.).
17. Again verify halfshaft inner joints are fully engaged to axle assembly.
18. Remove transmission jack.
19. If used, remove screw-type under-hoist jackstands.
20. Torque driveshaft coupler-to-axle flange bolt/nuts to 58 Nm (43 ft. lbs.).
21. Using a 14mm hex, remove rear axle fill plug. Fill axle with 1.4L (1.5 qts.) of Mopar® 75W-140 Synthetic Gear & Axle Lubricant. Install fill plug and torque to 60 Nm (44 ft. lbs.).

210RII

1. Before servicing the vehicle, refer to the precautions in the beginning of this section.

➡This procedure requires the compression of the rear suspension to ride height. A drive-on hoist should be used. If a drive-on hoist is not used, screw-style under-hoist jackstands are required to compress the rear suspension, facilitating rear halfshaft removal.

⁂ WARNING

Never grasp halfshaft assembly by the inner or outer boots. Doing so may cause the boot to pucker or crease, reducing the service life of the boot and joint. Avoid over angling or stroking the CV-joints when handling the halfshaft.

2. With vehicle in neutral, position and raise vehicle on hoist.
3. Using 14mm hex, remove axle drain plug and drain rear axle fluid into container suitable for fluid reuse.
4. Install drain plug and torque to 50 Nm (37 ft. lbs.) torque.
5. Remove rear exhaust system on dual-outlet exhaust models, otherwise, lower exhaust system at rear hanger.
6. Remove the rear driveshaft.
7. Remove the halfshafts.
8. Remove two rear axle-to-crossmember bolts.
9. Carefully lower rear axle. While lowering axle, separate driveshaft from axle and support with suitable rope or wire.
10. Remove axle assembly from vehicle and transfer to bench.
11. Using suitable screwdriver, remove axle seals and discard.

To install:

12. Install new axle seal(s) using Tool 9223.

➡Use care when installing halfshaft to axle assembly. The halfshaft installation angle should be minimized to avoid damage to seal upon installation.

13. Install the halfshafts.
14. Raise rear axle assembly into position. Align driveshaft index marks and start driveshaft coupler-to-axle bolt/nuts by hand.
15. Install two rear axle-to-crossmember bolts and torque to 220 Nm (162 ft. lbs.).
16. Install rear axle front mount isolator and torque bolt/nut to 65 Nm (48 ft. lbs.).
17. Again verify halfshaft inner joints are fully engaged to axle assembly.
18. Remove transmission jack.
19. If used, remove screw-type under-hoist jackstands.
20. Torque driveshaft coupler-to-axle flange bolt/nuts to 58 Nm (43 ft. lbs.).
21. Using a 14mm hex, remove rear axle fill plug. Fill axle with 1.6L (1.7 qts.) of Mopar® 75W-140 Synthetic Gear & Axle Lubricant. Install fill plug and torque to 50 Nm (37 ft. lbs.).

Front Differential Pinion Seal

REMOVAL & INSTALLATION

1. Before servicing the vehicle, refer to the precautions in the beginning of this section.
2. Remove driveshaft and heat shield.
3. Using Tool 6958 (2), remove pinion flange nut.
4. Using Puller 1026 (1), remove pinion flange.
5. Remove pinion flange seal with suitable screwdriver and discard.

To install:

6. Using a seal driver, install new pinion flange seal.
7. Lightly tap pinion flange onto the shaft, just enough to start flange nut by hand.
8. Install pinion flange nut and torque to 475 Nm (350 ft. lbs.).

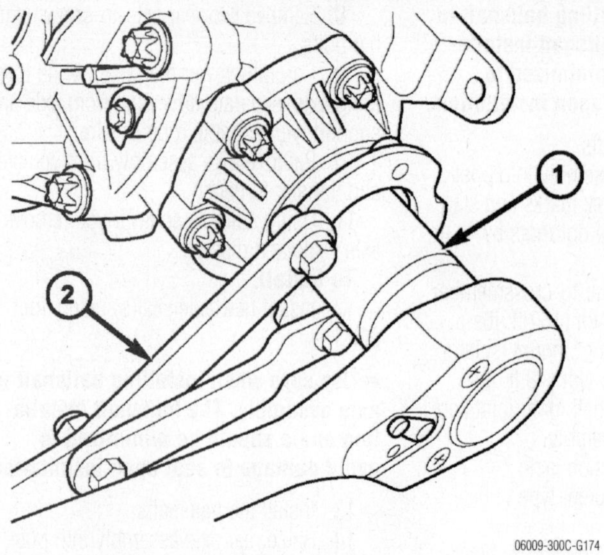

Remove the pinion flange nut—Front drive axle

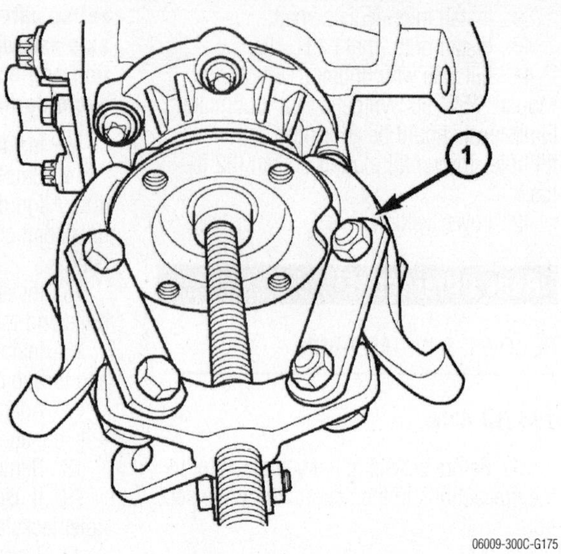

Remove the pinion flange—Front drive axle

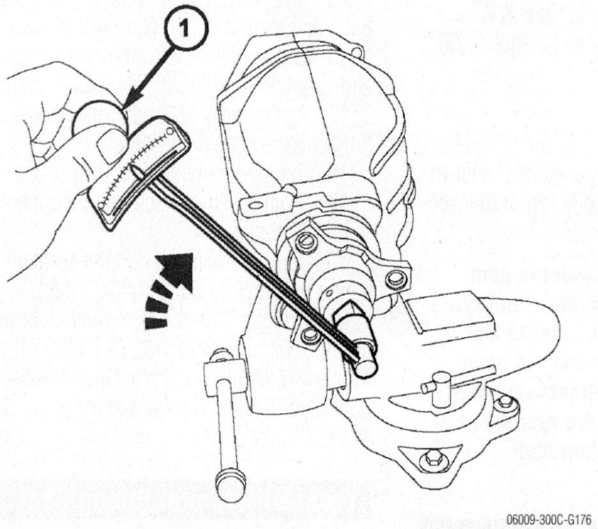

Measure axle assembly rotating torque

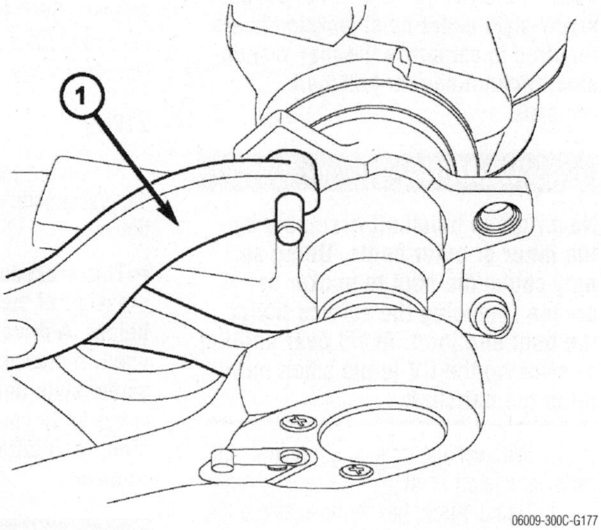

Remove flange nut

9. Install driveshaft and heat shield.

Rear Differential Pinion Seal

REMOVAL & INSTALLATION

1. Before servicing the vehicle, refer to the precautions in the beginning of this section.
2. Remove rear axle assembly from vehicle.
3. Measure axle assembly rotating torque and record measurement for reuse on assembly.
4. Using flange holder C-3281(1) and 41mm socket, remove pinion flange nut and discard.

➡Due to axle imbalance concerns, it is necessary to make sure pinion flange-to-shaft orientation is maintained. If alignment marks are not visible, apply appropriate marks before removing pinion flange.

5. Using puller 1026 (1), remove pinion flange (2) from pinion shaft.
6. Using suitable tool, remove pinion seal and discard.

To install:

7. Apply light coating of gear lubricant to the lip of the pinion seal.
8. Using a seal driver, install pinion seal until tool bottoms on carrier.
9. Install pinion flange into position. Align index marks to maintain assembly balance.
10. Lightly tap on pinion flange until adequate pinion shaft threads are exposed.
11. Install new pinion flange nut. Using flange holder tool C-3281and 41mm socket, torque nut to 136 Nm (100 ft. lbs.).
12. Measure assembly turning torque. Axle assembly rotating torque must be should be equal to the reading recorded upon seal/flange removal.
13. If rotating torque is low, increase pinion flange nut torque in 7 Nm (60 inch lbs.) increments. Repeat until proper rotating torque is received.
14. Stake pinion flange nut (1) as shown.
15. Install rear axle assembly.

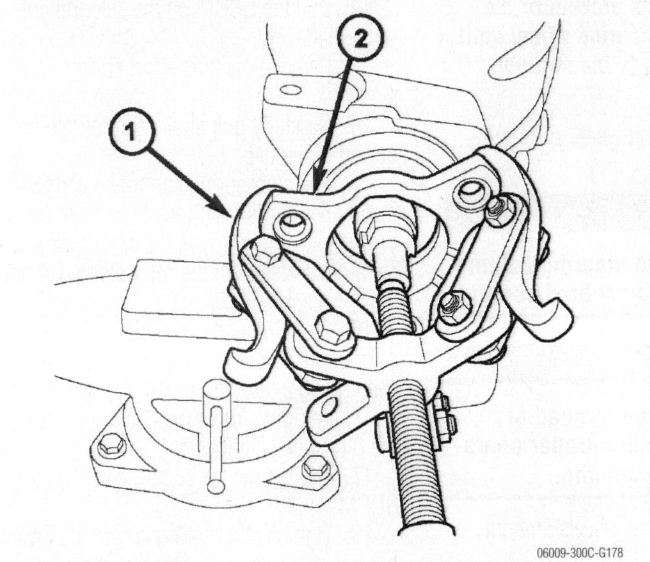

Remove pinion flange

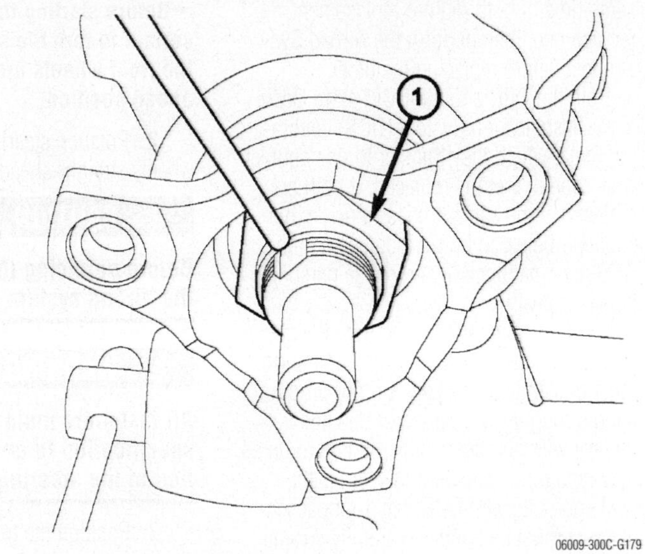

Stake the pinion flange nut

STEERING

➡Suspension components with rubber/urethane bushings should be tightened with the vehicle at normal ride height. It is important to have the springs supporting the weight of the vehicle when the fasteners are torqued. If springs are not at their normal ride position, vehicle ride comfort could be affected and premature bushing wear may occur.

Air Bag

⁂ CAUTION

Some vehicles are equipped with an air bag system. The system must be disarmed before performing service on, or around, system components, the steering column, instrument panel components, wiring and sensors. Failure to follow the safety precautions and the disarming procedure could result in accidental air bag deployment, possible injury and unnecessary system repairs.

PRECAUTIONS

Disconnect and isolate the battery negative cable before beginning any airbag system component diagnosis, testing, removal, or installation procedures. Allow system capacitor to discharge for two minutes before beginning any component service. This will disable the airbag system. Failure to disable the airbag system may result in accidental airbag deployment, personal injury, or death.

Do not place an intact undeployed airbag face down on a solid surface. The airbag will propel into the air if accidentally deployed and may result in personal injury or death.

When carrying or handling an undeployed airbag, the trim side (face) of the airbag should be pointing towards the body to minimize possibility of injury if accidental deployment occurs. Failure to do this may result in personal injury or death.

Replace airbag system components with Mopar® replacement parts. Substitute parts may appear interchangeable, but internal differences may result in inferior occupant protection. Failure to do so may result in occupant personal injury or death.

Wear safety glasses, rubber gloves, and long sleeved clothing when cleaning powder residue from vehicle after an airbag deployment. Powder residue emitted from a deployed airbag can cause skin irritation. Flush affected area with cool water if irritation is experienced. If nasal or throat irritation is experienced, exit the vehicle for fresh air until the irritation ceases. If irritation continues, see a physician.

Do not use a replacement airbag that is not in the original packaging. This may result in improper deployment, personal injury, or death.

The factory installed fasteners, screws and bolts used to fasten airbag components have a special coating and are specifically designed for the airbag system. Do not use substitute fasteners. Use only original equipment fasteners listed in the parts catalog when fastener replacement is required.

During, and following, any child restraint anchor service, due to impact event or vehicle repair, carefully inspect all mounting hardware, tether straps, and anchors for proper installation, operation, or damage. If a child restraint anchor is found damaged in any way, the anchor must be replaced. Failure to do this may result in personal injury or death.

Deployed and non-deployed airbags may or may not have live pyrotechnic material within the airbag inflator.

Do not dispose of driver/passenger/curtain airbags or seat belt tensioners unless you are sure of complete deployment. Refer to the Hazardous Substance Control System for proper disposal.

Dispose of deployed airbags and tensioners consistent with state, provincial, local, and federal regulations.

After any airbag component testing or service, do not connect the battery negative cable. Personal injury or death may result if the system test is not performed first.

If the vehicle is equipped with the Occupant Classification System (OCS), do not connect the battery negative cable before performing the OCS Verification Test using the scan tool and the appropriate diagnostic information. Personal injury or death may result if the system test is not performed properly.

Never replace both the Occupant Restraint Controller (ORC) and the Occupant Classification Module (OCM) at the same time. If both require replacement, replace one, then perform the Airbag System test before replacing the other.

Both the ORC and the OCM store Occupant Classification System (OCS) calibration data, which they transfer to one another when one of them is replaced. If both are replaced at the same time, an irreversible fault will be set in both modules and the OCS may malfunction and cause personal injury or death.

If equipped with OCS, the Seat Weight Sensor is a sensitive, calibrated unit and must be handled carefully. Do not drop or handle roughly. If dropped or damaged, replace with another sensor. Failure to do so may result in occupant injury or death.

If equipped with OCS, the front passenger seat must be handled carefully as well. When removing the seat, be careful when setting on floor not to drop. If dropped, the sensor may be inoperative, could result in occupant injury, or possibly death.

If equipped with OCS, when the passenger front seat is on the floor, no one should sit in the front passenger seat. This uneven force may damage the sensing ability of the seat weight sensors. If sat on and damaged, the sensor may be inoperative, could result in occupant injury, or possibly death.

DISARMING THE SYSTEM

1. Before servicing the vehicle, refer to the precautions in the beginning of this section.

Disconnect and isolate the negative battery cable. Wait 2 minutes for the system capacitor to discharge before performing any service.

ARMING THE SYSTEM

1. Before servicing the vehicle, refer to the precautions in the beginning of this section.

To arm the system, connect the negative battery cable.

CLOCKSPRING CENTERING

1. Before servicing the vehicle, refer to the precautions in the beginning of this section.

If the rotating tape (wire coil) in the clockspring is not positioned properly with the steering wheel and the front wheels, the clockspring may fail. The following procedure MUST BE USED to center the clockspring if it is not known to be properly positioned, or if the front wheels were moved from the straight ahead position.

➡Before starting this procedure, be certain to turn the steering wheel until the front wheels are in the straight-ahead position.

2. Position steering wheel and front wheels straight-ahead.

✻✻ CAUTION

Before servicing the steering column the airbag system must be disarmed.

✻✻ WARNING

All fasteners must be torqued to specification to ensure proper operation of the steering column.

3. Position the front wheels straight-ahead.
4. Fully extend or pull out adjustable steering column.
5. Disconnect the negative (ground) cable from the battery.
6. From behind the steering wheel, remove the two screw covers and screws to the driver airbag.

✻✻ WARNING

Do not pull on the horn switch feed pigtail wire to disengage the connector from the driver's airbag or to disconnect the horn switch to steering wheel wire harness connection. Improper pulling on this pigtail wire or connection can result in damage to the horn switch membrane or feed circuit.

7. Carefully pull airbag rearward just far enough to disconnect the two airbag squib connectors and the horn connector.
8. Separate driver airbag from steering column.

➡The driver's airbag trim cover is available for service separately. If the horn switch is faulty or the trim cover is damaged, the airbag cushion assembly may be transferred to a new trim cover. If the airbag is defective, the entire driver airbag and driver airbag trim cover must be replaced as an assembly.

9. Remove the steering wheel retaining bolt, then slide the steering wheel off the shaft.
10. Remove at least one clockspring screw. This will help keep the clockspring from un-centering itself.
11. Back out the set screw through the access hole in the bottom of the Steering Column Control Module (SCCM).
12. Pull the SCCM off the steering column shaft.
13. Remove the three clockspring screws.
14. Carefully pull straight up on clockspring to remove.
15. The clockspring can rotate approximately 5¾ turns from lock to lock. To be properly centered, rotate the clockspring rotor clockwise until the rotor stops. Do not apply excessive force.
16. From the end of travel, rotate the rotor counterclockwise two turns and then keep going a little more until the wires end up on the right side of the column shaft (at the 3 o'clock position).
17. Tighten the set screw securing the SCCM.
18. Tighten clockspring screws that were backed out to keep the clockspring from unwinding.
19. Align the spline on the steering wheel hub to shaft and install the steering wheel.
20. Install a NEW retaining bolt. Torque the bolt to 70 Nm (52 ft. lbs.).
21. Install the driver's airbag.
22. Connect the battery negative cable. If equipped with Electronic Stability Program (ESP), the steering angle sensor must be calibrated.
23. Install the steering column control module.

✻✻ CAUTION

Do not connect the battery negative cable. Personal injury or death may result if the following system test is not performed properly:

a. With the battery negative cable disconnected, connect the scan tool to the Data Link Connector (DLC).
b. Turn the ignition key to the ON position, then exit vehicle with the scan tool.
c. After checking that no one is inside the vehicle, connect the battery negative remote terminal.
d. Read and record the ACTIVE Diagnostic Trouble Code (DTC) data.
e. Read and record any STORED DTCs.
f. Refer to the proper diagnostic information if any DTCs are found in Step 4 and Step 5.
g. If the airbag warning lamp either fails to light, or goes ON and stays ON, there is a system malfunction.

24. Position driver airbag in steering wheel near mounting location.

25. Connect the two airbag squib connectors and the horn connector.

26. Position the driver's airbag mounting screws into the driver airbag. From behind the steering wheel install the two screws to the driver airbag. Torque the two screws to 10 Nm (89 inch lbs.).

Power Steering Gear

REMOVAL & INSTALLATION

2-Wheel Drive

1. Before servicing the vehicle, refer to the precautions in the beginning of this section.
2. Disconnect and isolate battery negative cable from battery post.
3. Siphon power steering fluid from pump reservoir.
4. Raise and support vehicle.
5. Remove wheel mounting nuts, then both front tire and wheel assemblies.

✲✲ WARNING

When loosening jam nut and rotating inner tie rod, use care not to twist bellows at inner tie rod. Remove clamp at inner tie rod and make sure bellows moves freely before rotating inner tie rod.

6. Loosen tie rod jam nut (3) at each outer tie rod (5).
7. Remove outer tie rod nut (1) at each knuckle (2).
8. Using Remover (2), Special Tool 9630, separate outer tie rod (1) from each knuckle.
9. Remove steering coupling (2) pinch bolt (1) at steering gear (3).
10. Unthread pressure hose tube nut from steering gear. Remove pressure hose from steering gear.
11. Unthread return hose tube nut from steering gear. Remove return hose from steering gear.
12. Remove steering gear mounting bolts.
13. If necessary, remove outer tie rods from gear. Count number of revolutions off for each tie rod for reference upon installation to replacement gear.

To install:

14. If necessary, install outer tie rods from original gear to replacement inner tie rods. Install each outer tie rod same amount of threads as it was installed on original gear. This will get toe setting close, saving some time when toe is set later in this procedure.

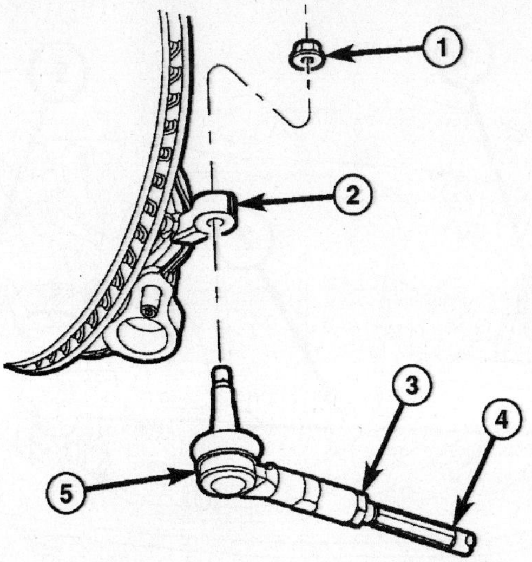

06009-300C-G180

Outer tie rod end and knuckle—2-wheel drive

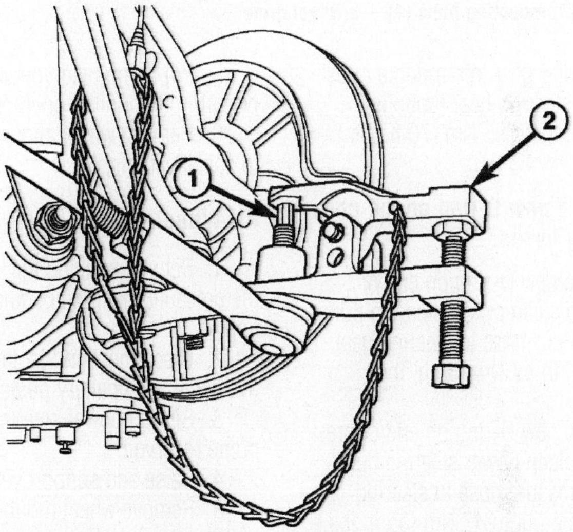

06009-300C-G181

Removing outer tie rod end from the knuckle with tool 9630

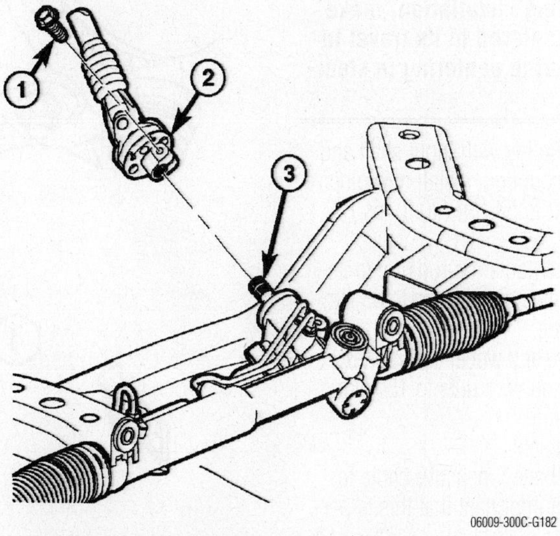

06009-300C-G182

Remove the steering coupling from the gear—2-wheel drive

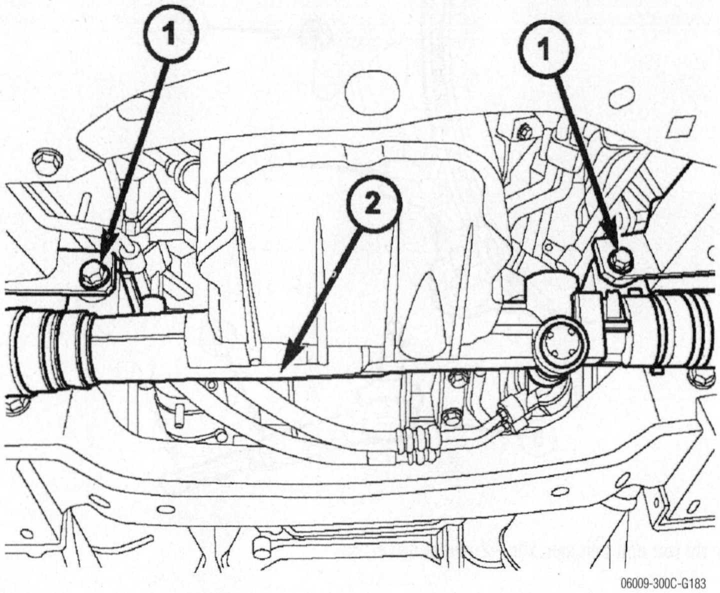

Steering gear (2) mounting bolts (1) —2-wheel drive

15. Lift steering gear into mounted position and install steering gear mounting bolts. Tighten bolts to 95 Nm (70 ft. lbs.) torque.

➡**Always use a new O-ring on the ends of the steering hoses.**

16. Lubricate new O-ring on end of return hose with clean power steering fluid.
17. Install return hose to steering gear. Tighten tube nut to 47 Nm (35 ft. lbs.) torque.
18. Lubricate new O-ring on end of pressure hose with clean power steering fluid.
19. Install pressure hose to steering gear. Tighten tube nut to 47 Nm (35 ft. lbs.) torque.

⁂ WARNING

Prior to coupling installation, make sure gear is centered in its travel to match clockspring centering in steering column.

20. Align coupling with input shaft and install steering coupling. Install new pinch bolt. Tighten bolt to 54 Nm (40 ft. lbs.) torque.
21. Install each outer tie rod end to its knuckle. Install nuts and tighten to 85 Nm (63 ft. lbs.) torque.
22. Install tire and wheel assemblies. Tighten wheel mounting nuts to 150 Nm (110 ft. lbs.) torque.
23. Lower vehicle.
24. Connect battery negative cable to battery post. It is important that this is performed properly.
25. Fill pump reservoir with fluid and perform pump initial operation procedure.
26. Perform wheel alignment setting toe to specifications.

All Wheel Drive

1. Before servicing the vehicle, refer to the precautions in the beginning of this section.
2. Disconnect and isolate battery negative cable from battery post.
3. Siphon power steering fluid from pump reservoir.
4. Raise and support vehicle.
5. Remove wheel mounting nuts, then both front tire and wheel assemblies.

⁂ WARNING

When loosening jam nut and rotating inner tie rod, use care not to twist bellows at inner tie rod.

6. Remove clamp (2) at inner tie rod (3) and make sure bellows moves freely before rotating inner tie rod.
7. Loosen tie rod jam nut (1) at each outer tie rod (4).
8. Remove outer tie rod end nut at each knuckle.
9. Using Remover, Special Tool 9630, separate outer tie rod from each knuckle.
10. Remove steering coupling pinch bolt at steering gear.
11. Unthread return hose tube nut from steering gear. Remove return hose from steering gear.
12. Unthread pressure hose tube nut from steering gear. Remove pressure hose from steering gear.
13. Remove mounting screws, then heat shield above each inner tie rod bellows.
14. Remove steering gear upper mounting bolt and nut.
15. Remove steering gear lower mounting bolts.
16. Remove steering gear.
17. If necessary, remove outer tie rods from gear. Count number of revolutions off for each tie rod for reference upon installation to replacement gear.

To install:

18. If necessary, install outer tie rods from original gear to replacement inner tie rods. Install each outer tie rod same amount of threads as it was installed on original

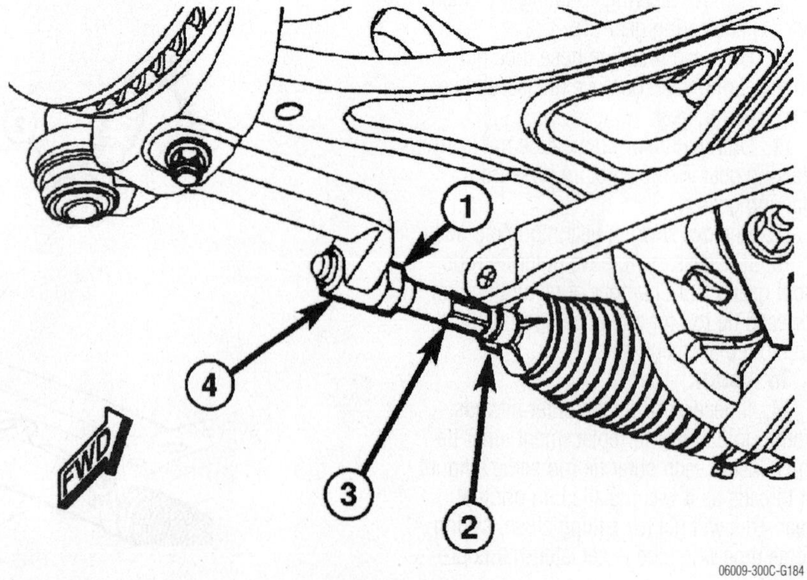

Inner-to-outer tie rod connection—all wheel drive

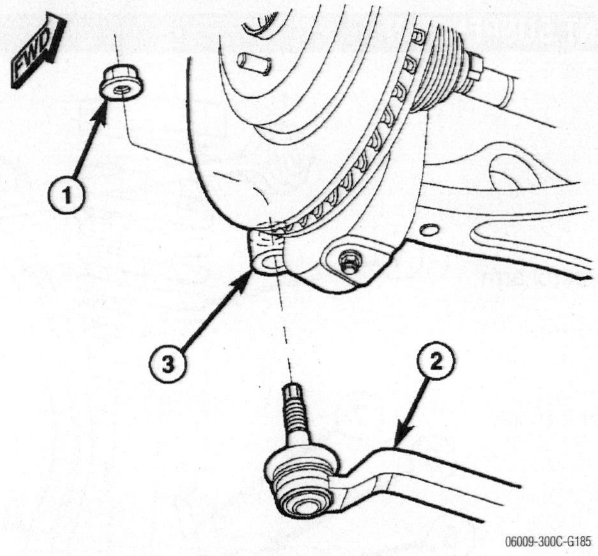

Outer tie rod end (2) to knuckle (3) connection—all wheel drive

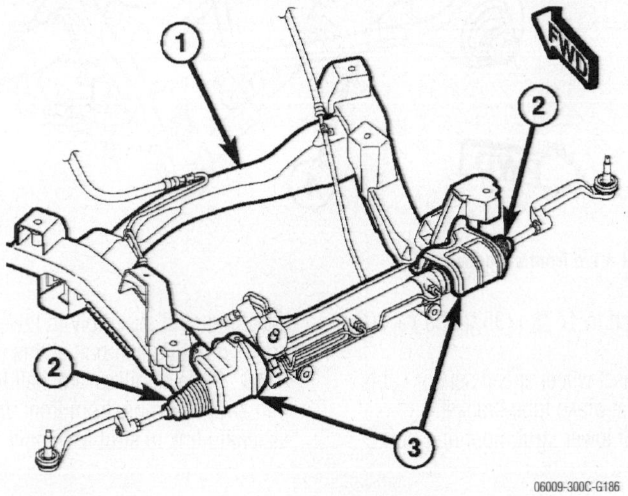

Steering gear (3) heat shields, (2) boots, (1) crossmember—all wheel drive

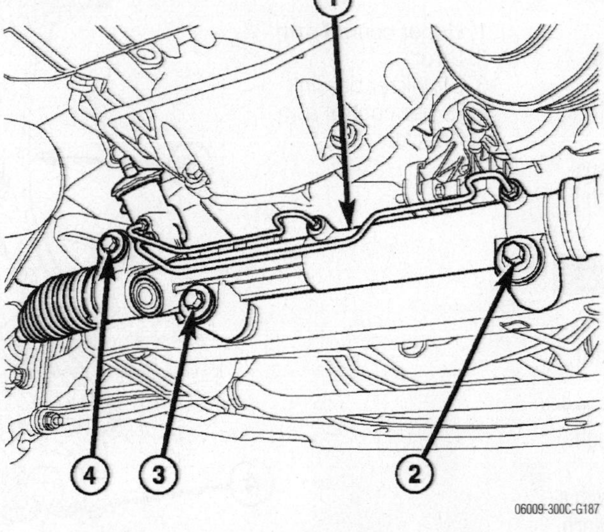

Steering gear mounting (1) gear, (2, 3) lower mounting bolts, (4) upper mounting bolt—all wheel drive

gear. This will get toe setting close, saving some time when toe is set later in this procedure.

19. Lift steering gear into mounted position and install steering gear lower mounting bolts. Install upper mounting bolt and nut. Tighten all bolts to 102 Nm (75 ft. lbs.) torque.

20. Install heat shield above each inner tie rod bellows. Tighten screws to 10 Nm (89 inch lbs.) torque.

➡Always use a NEW O-ring on the ends of the steering hoses.

21. Lubricate NEW O-ring on end of pressure hose with clean power steering fluid.

22. Install pressure hose to steering gear. Tighten tube nut to 47 Nm (35 ft. lbs.) torque.

23. Lubricate NEW O-ring on end of return hose with clean power steering fluid.

24. Install return hose to steering gear. Tighten tube nut to 47 Nm (35 ft. lbs.) torque.

25. Align splines and install steering coupling to steering gear shaft. Install NEW pinch bolt. Tighten bolt to 54 Nm (40 ft. lbs.) torque.

26. Install each outer tie rod end to its knuckle. Install nut and tighten to 85 Nm (63 ft. lbs.) torque.

27. Install tire and wheel assemblies. Tighten wheel mounting nuts to 150 Nm (110 ft. lbs.) torque.

28. Lower vehicle.

FRONT SUSPENSION

MacPherson Strut

REMOVAL & INSTALLATION

2-Wheel Drive

1. Before servicing the vehicle, refer to the precautions in the beginning of this section.
2. If equipped, remove front strut tower cap from top of strut assembly.
3. Remove three nuts fastening strut assembly to strut tower.
4. Raise and support vehicle.
5. Remove wheel mounting nuts, then tire and wheel assembly.
6. Remove nut (2) fastening stabilizer link (3) to strut assembly (1). Slide link ball joint stem from strut assembly.
7. Remove bolt securing strut assembly to lower control arm.
8. Disconnect wheel speed sensor cable routing clip at brake tube bracket.
9. Loosen nut attaching upper ball joint stud to knuckle. Back nut off until nut is even with end of stud. Keeping nut on at this location will help keep end of stud from distorting while using Puller in next step.

⁂ WARNING

In following step, use care not to damage ball joint seal boot while sliding Puller, Special Tool 9360, into place past seal boot.

10. Using Puller (2), Special Tool 9360, separate upper ball joint stud (1) from knuckle.
11. Remove nut from end of upper ball joint stud.
12. Tip top of knuckle outward using care not to overextend bake flex hose.
13. Remove strut assembly from vehicle.

To install:

14. Place strut assembly into front suspension using reverse direction in which it was removed.

⁂ WARNING

It is important to tighten nut as described in following step to avoid damaging ball stud joint.

15. Place upper ball joint stud through hole in top of knuckle and install nut. Tighten nut by holding ball joint stud with a hex wrench while turning nut with a wrench. Tighten nut using crow foot wrench on torque wrench to 47 Nm (35 ft. lbs.) + 90° turn torque.
16. Connect wheel speed sensor cable routing clip at brake tube bracket.
17. Install lower strut mounting bolt attaching strut assembly to lower control arm. Do not tighten bolt at this time.
18. Slide stabilizer link ball joint stem into strut assembly from front. Install nut fastening link to strut assembly. Tighten nut

1. Strut
2. Upper control arm
3. Kuckle
4. Tension strut
5. Lower control arm
6. Stabilizer bar
7. Stabilizer bar link

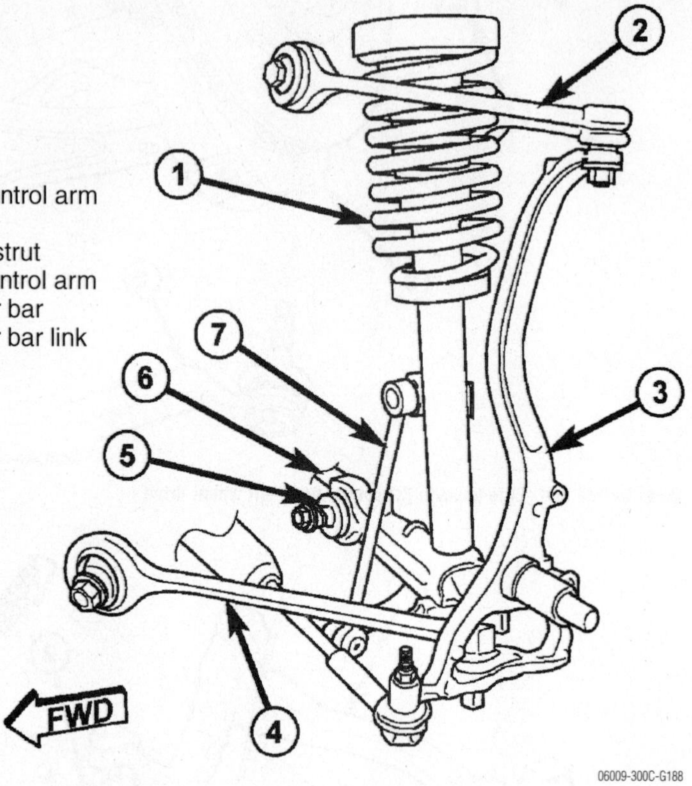

Rear wheel drive front suspension

1. Upper control arm
2. Strut
3. Stabilizer bar link
4. Lower control arm
5. Knuckle

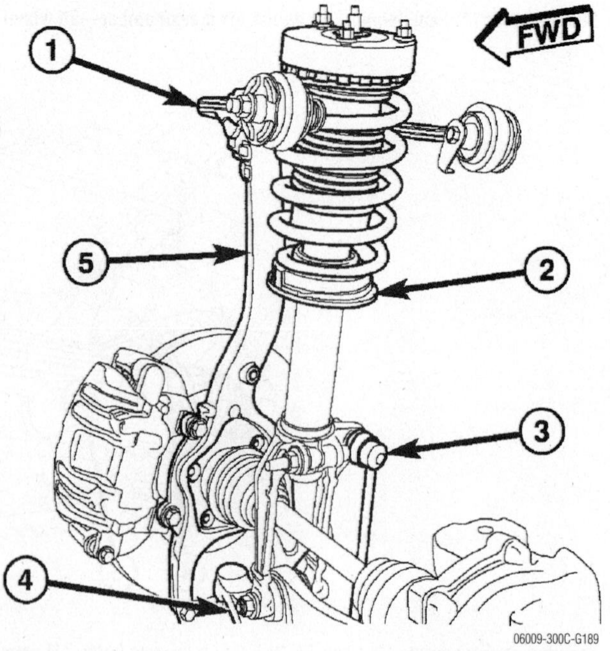

All wheel drive front suspension

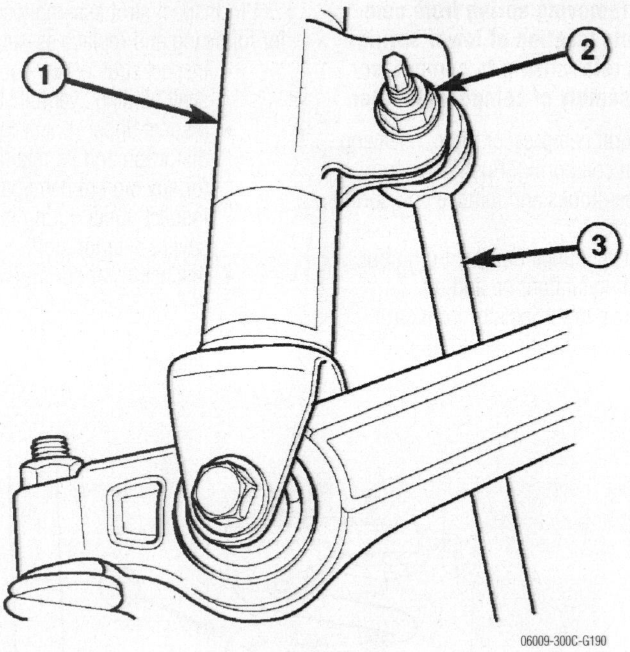

Stabilizer bar link connection—rear wheel drive

by holding ball joint stud while turning nut. Tighten nut using crow foot wrench on torque wrench to 128 Nm (95 ft. lbs.) torque.

19. Install tire and wheel assembly. Tighten wheel mounting nuts to 150 Nm (110 ft. lbs.) torque.

20. Lower vehicle.

21. Install three nuts fastening strut assembly to strut tower. Tighten nuts to 27 Nm (20 ft. lbs.) torque.

22. If equipped, align strut tower cap with strut mounting nuts and snap into place.

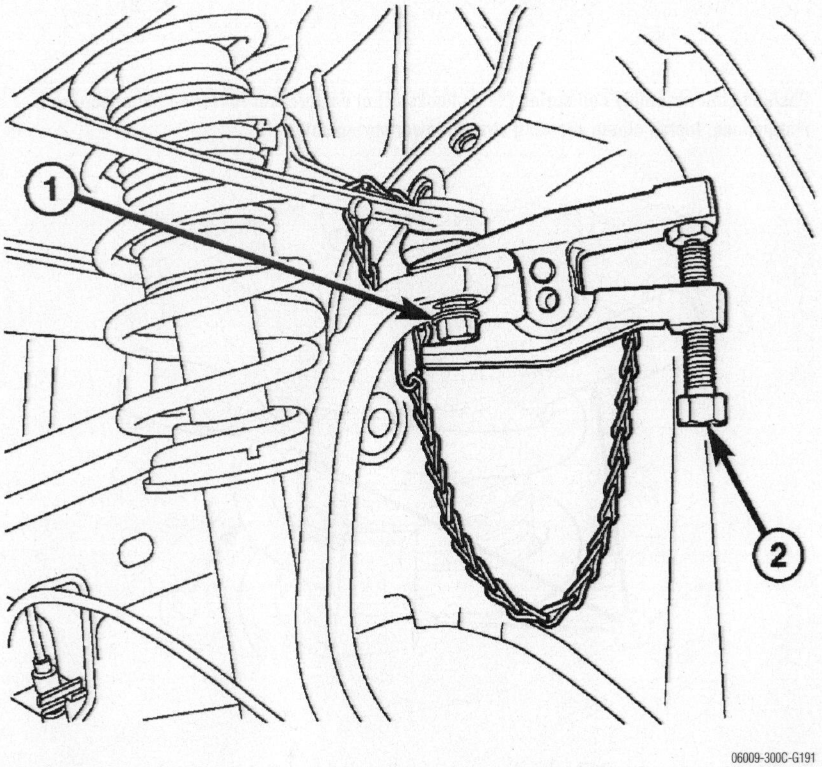

Separating the upper ball stud from the knuckle—rear wheel drive

All Wheel Drive

1. Before servicing the vehicle, refer to the precautions in the beginning of this section.

2. Raise and support vehicle.

3. Remove wheel mounting nuts, then tire and wheel assembly.

4. While holding link ball joint stem from rotating, remove nut fastening stabilizer link to strut clevis bracket. Slide link ball joint stem from clevis bracket.

5. Remove nut and pinch bolt fastening clevis bracket to bottom of strut assembly.

6. Remove nut and bolt attaching strut clevis bracket to lower control arm.

7. Pull lower end of clevis bracket outward away from lower control arm bushing, then slide it off strut assembly. It may be necessary to use an appropriate prying tool to spread clamp area of clevis bracket allowing removal from strut assembly.

8. Lower vehicle just enough to access upper strut assembly mounting nuts.

9. If equipped, remove strut tower cap from top of strut assembly.

10. Remove three nuts fastening strut assembly to strut tower.

11. Remove strut assembly from vehicle.

To install:

12. Guide strut assembly up into strut tower and into mounting holes.

13. Install three nuts fastening strut assembly to strut tower. Tighten nuts to 27 Nm (20 ft. lbs.) torque.

14. If equipped, align strut tower cap with strut mounting nuts and snap into place.

15. Raise and support vehicle.

16. Pull lower end of strut assembly outward, then slide clevis bracket onto lower end. Slide clevis bracket onto strut assembly until bracket contacts collar on strut housing.

17. Install pinch bolt and nut fastening clevis bracket to bottom of strut assembly. Install pinch bolt from rear. Do not tighten at this time.

18. Slide clevis bracket over bushing mounted in lower control arm.

19. Install bolt and nut attaching strut clevis bracket to lower control arm. Do not tighten at this time.

20. Tighten pinch bolt attaching clevis bracket to strut assembly to 61 Nm (45 ft. lbs.) torque.

21. Slide stabilizer link ball joint stem into clevis bracket. Install nut fastening link to clevis bracket. Tighten nut by holding ball joint stud while turning nut. Tighten nut using crow foot wrench on torque wrench to 128 Nm (95 ft. lbs.) torque.

22. Install tire and wheel assembly. Tighten wheel mounting nuts to 150 Nm (110 ft. lbs.) torque.

23. Lower vehicle.

➡When tightening lower strut clevis mounting bolt, do not attempt rotating bolt. Bolt shaft is serrated. Turn nut only.

24. Tighten lower strut clevis bracket bolt nut to 174 Nm (128 ft. lbs.) torque.

Coil Spring

REMOVAL & INSTALLATION

1. Before servicing the vehicle, refer to the precautions in the beginning of this section.

The strut assembly must be removed from vehicle for it to be disassembled and assembled. For strut assembly disassembly and assembly, use of strut Spring Compressor, Pentastar Service Equipment (PSE) tool W-7200, or equivalent, is recommended to compress coil spring. Follow manufacturer's instructions closely.

⁂ CAUTION

Do not remove strut shaft nut before coil spring is compressed. Coil spring is held under pressure and must be compressed, removing spring tension from upper and lower mounts, before strut removal.

2. Position strut assembly coil spring on hooks of compressor following manufacturer's instructions. Install clamp securing strut to lower spring coil.

3. Position compressor upper hooks on upper coil spring following manufacturer's instructions. To ease installation, rotate strut as necessary positioning strut in compressor so that upper spring coil ends (step in upper mount) at straight outward position from compressor.

4. Compress coil spring until all spring tension is removed from upper mount.

5. Position Wrench (2), Special Tool 9362, on strut shaft (1) retaining nut. Next, insert 8 mm socket though wrench onto hex located on end of strut shaft. While holding strut shaft from turning, remove nut from strut shaft using wrench.

6. Remove clamp from bottom of coil spring and remove strut and lower isolator out through bottom of coil spring.

7. Remove upper mount from strut shaft and coil spring.

➡Prior to removing spring from compressor, note location of lower spring coil end in relationship to compressor to ease assembly of components later.

8. Back off compressor drive, releasing tension from coil spring. Push back compressor upper hooks and remove coil spring from compressor.

9. Remove jounce bumper from strut shaft by pulling straight up and off.

10. Remove lower isolator from strut body by pulling straight up and off strut shaft.

11. Inspect strut assembly components for following and replace as necessary:

- Inspect strut for any condition of shaft binding over full stroke of shaft.
- Inspect upper mount for cracks and distortion and its retaining studs for any sign of damage.
- Inspect upper spring isolator for severe deterioration.
- Inspect lower spring isolator for severe deterioration.
- Inspect dust shield for tears and deterioration.

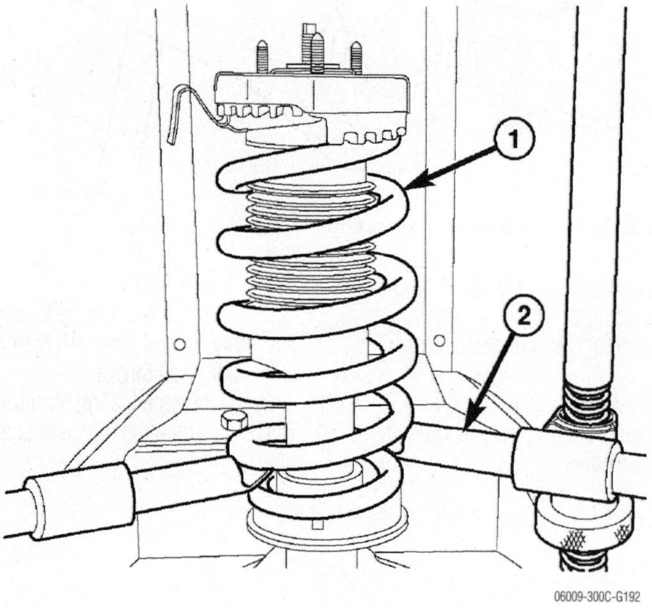

Position strut assembly coil spring (1) on hooks (2) of compressor following manufacturer's instructions. Install clamp securing strut to lower spring coil

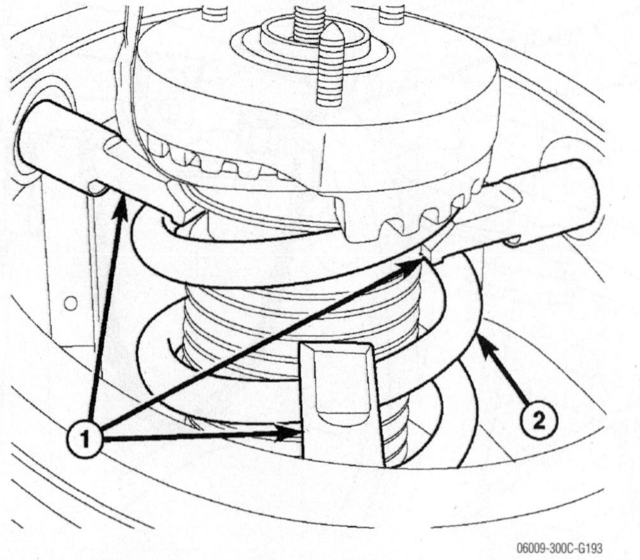

Position compressor upper hooks (1) on upper coil spring (2) following manufacturer's instructions

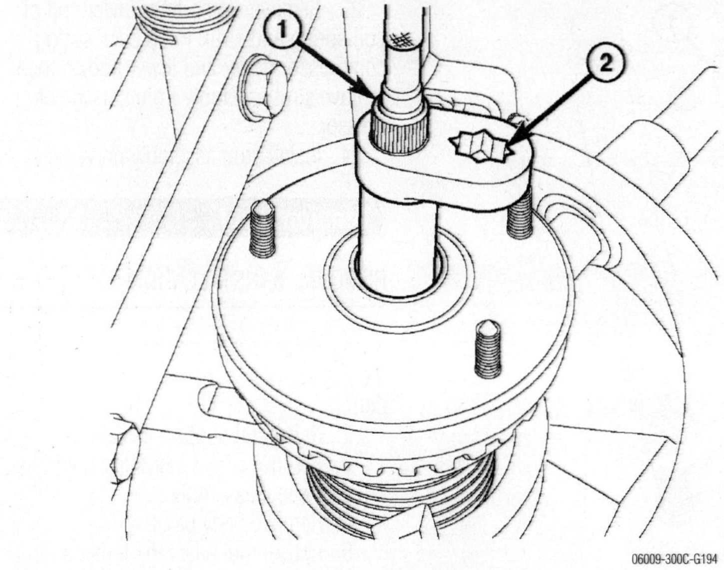

Position Wrench (2), Special Tool 9362, on strut shaft (1) retaining nut

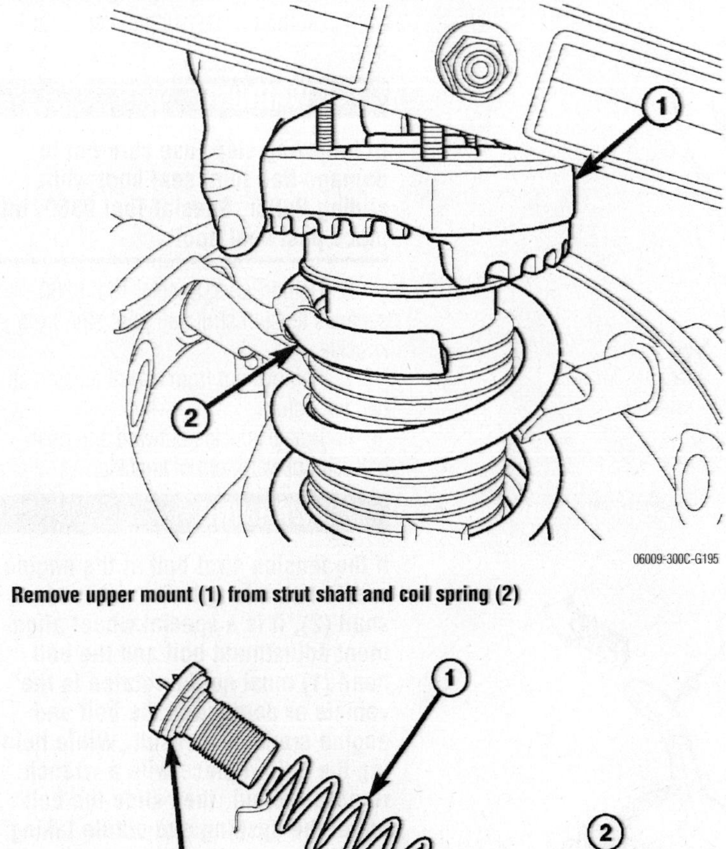

Remove upper mount (1) from strut shaft and coil spring (2)

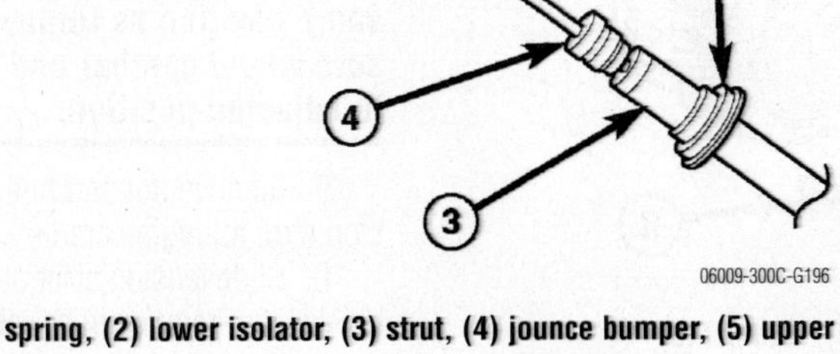

Strut parts: (1) spring, (2) lower isolator, (3) strut, (4) jounce bumper, (5) upper mount

- Inspect coil spring for cracks in the coating and corrosion.
- Inspect jounce bumper for cracks and signs of deterioration.

To assemble

⁂ WARNING

Use care not to damage coil spring coating during spring assembly. Damage to coating will jeopardize its corrosion protection.

➡Left and right springs must not be interchanged.

12. Place coil spring (part number tag end upward) in compressor lower hooks following manufacturer's instructions. To ease strut reassembly, rotate coil spring around until upper coil ends at straight outward position from compressor. Proper orientation of spring to upper mount (once installed) is necessary.
13. Position compressor upper hooks over coil spring following manufacturer's instructions.
14. Compress coil spring far enough to allow strut installation.
15. If separated, install upper mount onto coil spring. Match step in upper isolator to end of spring coil.
16. Install lower spring isolator on strut body.
17. Install jounce bumper on strut shaft, small end first.
18. Install strut through bottom of coil spring until lower spring isolator (on strut) contacts lower end of coil spring. Match step built into isolator to lower coil end. Once in this position, stabilizer bar bracket, or clevis key on AWD models, should point straight inward toward compressor body. If not, rotate isolator on strut body until alignment is achieved when isolator is correctly positioned with lower spring coil.
19. Install clamp to hold strut and coil spring together.
20. Install retaining nut on strut shaft as far as possible by hand. Make sure nut is installed far enough for 8 mm socket to grasp hex on end of shaft for tightening.
21. Install Wrench (on end of a torque wrench), Special Tool 9362, on strut shaft retaining nut. Next, insert 8 mm socket though wrench onto hex located on end of strut shaft. While holding strut shaft from turning, tighten nut using wrench to 90 Nm (66 ft. lbs.) torque.
22. Slowly release tension from coil spring by backing off compressor drive fully. As tension is relieved, make sure strut components are properly in place.

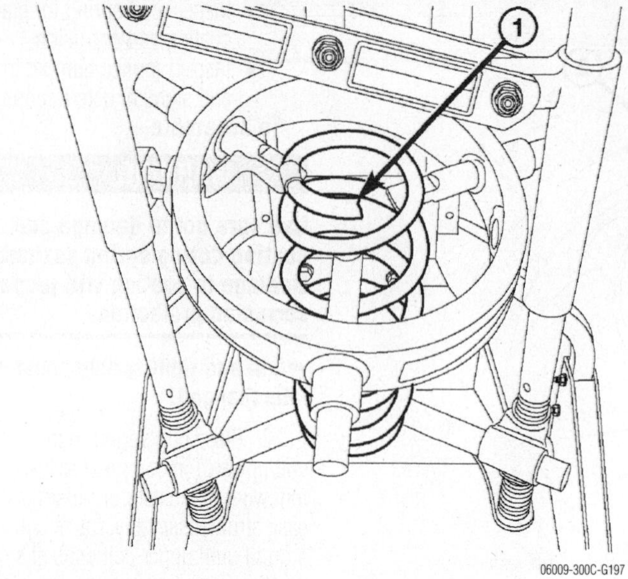

Coil spring upper end (1) positioning

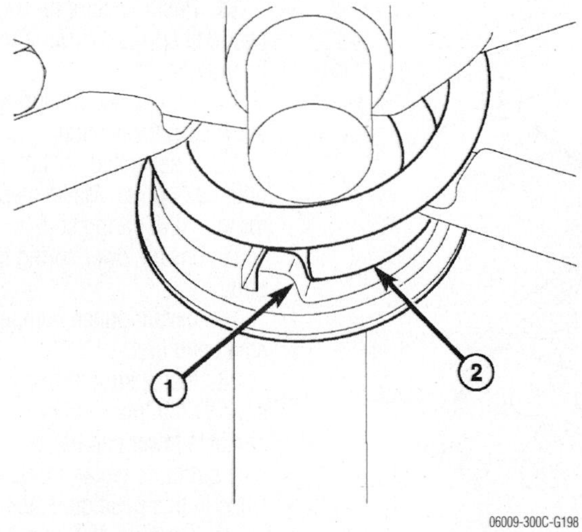

Match step built into isolator (1) to lower coil end (2)

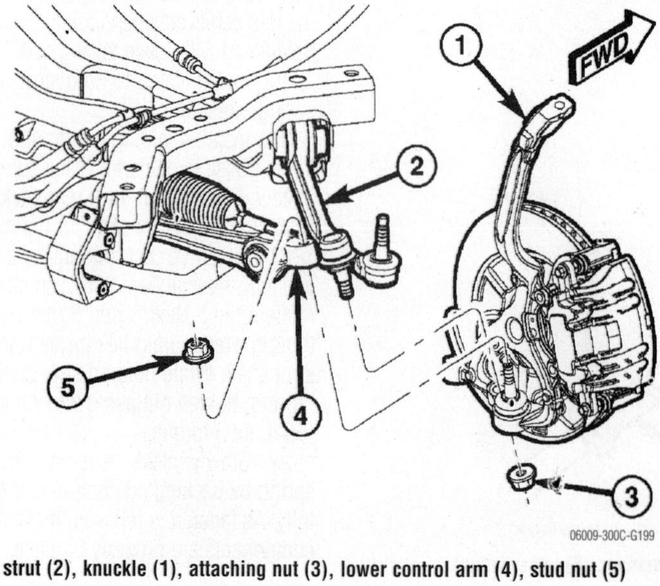

Tension strut (2), knuckle (1), attaching nut (3), lower control arm (4), stud nut (5)

23. Remove clamp from lower end of coil spring and strut. Push back spring compressor upper and lower hooks, then remove strut assembly from spring compressor.

24. Install strut assembly on vehicle.

Tension Strut

REMOVAL & INSTALLATION

1. Before servicing the vehicle, refer to the precautions in the beginning of this section.
2. Raise and support vehicle.
3. Remove wheel mounting nuts, then tire and wheel assembly.
4. Remove belly pan.
5. Loosen nut attaching tension strut ball joint stud to knuckle. Back nut off until nut is even with end of stud. Keeping nut on at this location will help keep end of stud from distorting while using puller in next step.

**** WARNING**

In following step, use care not to damage ball joint seal boot while sliding Puller, Special Tool 9360, into place past seai boot.

6. Using Puller, Special Tool 9360, separate tension strut ball joint stud from knuckle.
7. Remove nut from end of tension strut ball joint stud.
8. Rotate knuckle outward and push ball joint upward, out of knuckle.

**** WARNING**

If the tension strut bolt at the engine cradle has a lengthwise grooved shaft (2), it is a special wheel alignment adjustment bolt and the bolt head (1) must not be rotated in the vehicle or damage to the bolt and engine cradle will result. While holding the bolt in place with a wrench, remove the nut, then slide the bolt out of the bushing and cradle taking note of bolt positioning in engine cradle for reassembly purposes. The bolt needs to be installed in the same position as removed to make sure wheel camber and caster return to adjusted position.

9. Remove nut and bolt securing tension strut to engine cradle.
10. Slide tension strut out of cradle bracket and remove from vehicle.

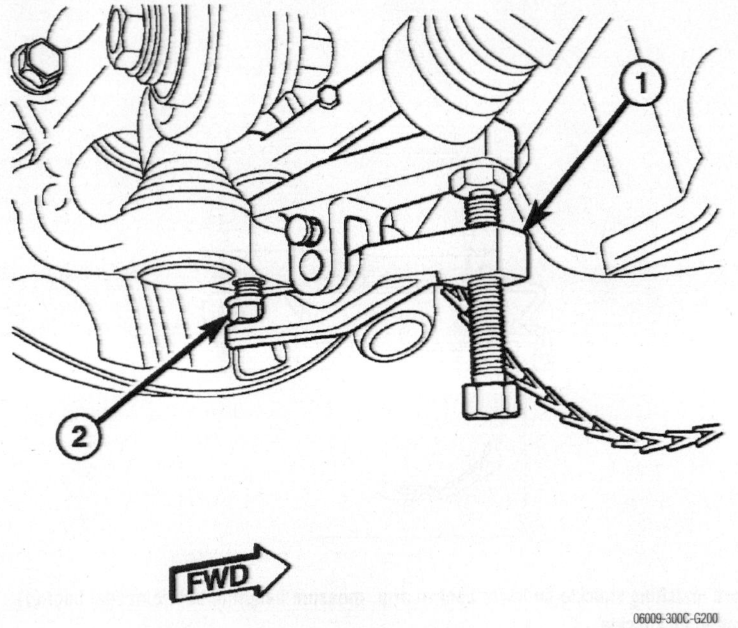

Using Puller (1), Special Tool 9360, separate tension strut ball joint stud (2) from knuckle

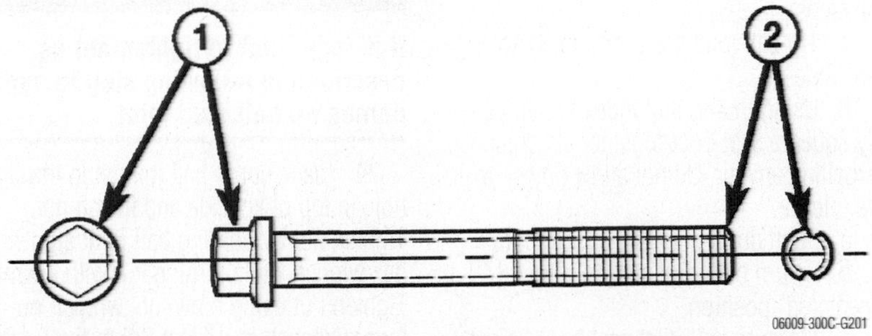

Tension strut bolt

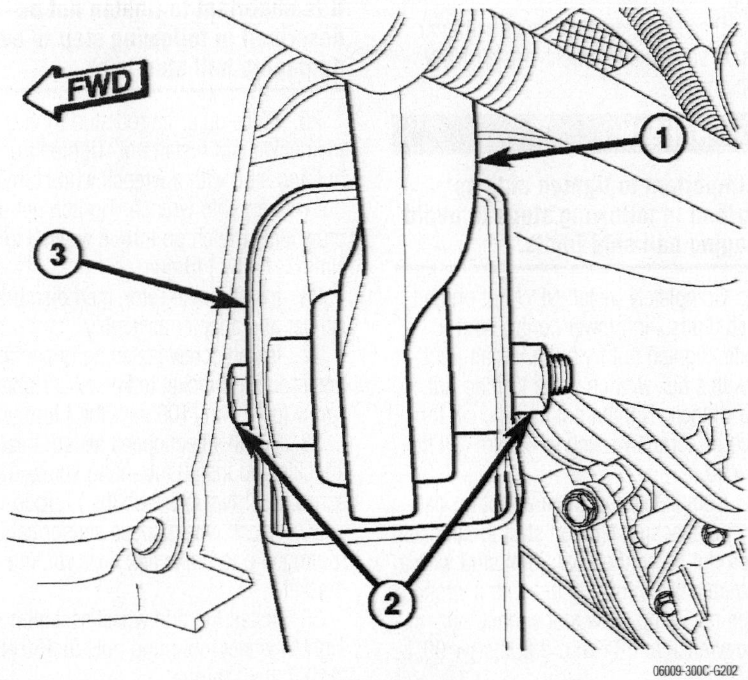

Remove nut and bolt (2) securing tension strut (1) to engine cradle (3)

To install:

➡If installing a tension strut engine cradle bolt that is a wheel alignment adjustment bolt (lengthwise grooved shaft), make sure to install it in the same position which it was in upon removal.

11. Slide bushing end of tension strut into cradle bracket.
12. Install mounting bolt from front through cradle and bushing.
13. Install nut, but do not tighten at this time.
14. Insert tension strut ball joint stud downward, into knuckle hole.
15. Completely install NEW nut on ball joint stud attaching tension strut to knuckle. Tighten nut by holding ball joint stud with a hex wrench while turning nut with a wrench. Tighten nut using crow foot wrench on torque wrench to 68 Nm (50 ft. lbs.) + 90° turn.
16. Install tire and wheel assembly. Tighten wheel mounting nuts to 150 Nm (110 ft. lbs.).
17. Lower vehicle.
18. Position vehicle on an alignment rack/drive-on lift.
19. Perform wheel alignment.

WARNING

If the tension strut engine cradle bolt is a wheel alignment adjustment bolt (lengthwise grooved shaft), be sure to only tighten the nut. Do not rotate the bolt head or damage to the bushing will occur.

20. Once alignment is found to be within specifications, using a crowfoot wrench, tighten tension strut cradle bolt nut to 176 Nm (130 ft. lbs.) torque while holding the bolt stationary.
21. Install belly pan.

Steering Knuckle

REMOVAL & INSTALLATION

2-Wheel Drive

1. Before servicing the vehicle, refer to the precautions in the beginning of this section.
2. Raise and support vehicle.
3. Remove wheel mounting nuts, then tire and wheel assembly.
4. Remove screw fastening wheel speed sensor to knuckle. Pull sensor head out of knuckle.

5. Remove wheel speed sensor cable routing clip from brake flex hose routing bracket.
6. Remove screw fastening brake flex hose routing bracket to knuckle.
7. Access and remove front brake rotor.
8. Remove nut from outer tie rod end stud.
9. Using a separator, separate tie rod stud from knuckle.
10. Loosen nut attaching upper ball joint stud to knuckle. Back nut off until nut is even with end of stud. Keeping nut on at this location will help keep end of stud from distorting in next step.

** WARNING

In following step, use care not to damage ball joint seal boot while sliding the separator into place past seal boot.

11. Separate upper ball joint stud from knuckle.
12. Remove nut from end of upper ball joint stud.
13. Loosen nut attaching tension strut ball joint stud to knuckle. Back nut off until nut is even with end of stud. Keeping nut on at this location will help keep end of stud from distorting while using Puller in next step.

** WARNING

In following step, use care not to damage ball joint seal boot.

14. Separate tension strut ball joint stud from knuckle.
15. Remove nut from end of tension strut ball joint stud.
16. Loosen nut attaching ball joint stud to lower control arm. Back nut off until nut is even with end of stud. Keeping nut on at this location will help keep end of stud from distorting while using Puller in next step.

** WARNING

In following step, use care not to damage ball joint seal boot.

17. Separate ball joint stud from lower control arm.
18. Remove nut from end of ball joint stud attaching lower control arm to knuckle.
19. Remove knuckle from vehicle.

To install:

20. Before installing knuckle on lower control arm, measure height of ball joint seal boot mounted on knuckle. If seal boot height is above 25.5 mm, any air inside seal boot must be expelled. To do so, follow these steps.

a. Tip ball joint stud completely to one side.
b. Using thumb and index finger, gently squeeze seal boot together at center expelling any air. Do not allow grease to be release.
c. Push down very top of seal boot.
d. Return ball joint stud to original "centered" position.
e. Measure ball joint seal boot height (1) making sure it is within specification.
f. Wipe any grease from ball joint stud.

21. Place knuckle over lower ball joint studs on vehicle and loosely install NEW nuts by hand.

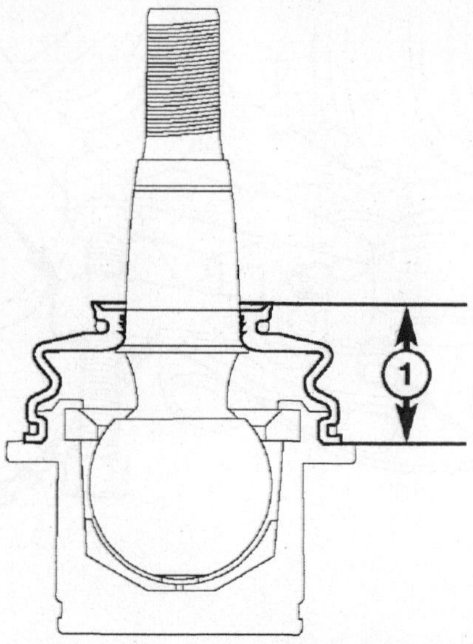

Before installing knuckle on lower control arm, measure height of ball joint seal boot (1) mounted on knuckle

** WARNING

It is important to tighten nuts as described in following steps to avoid damaging ball stud joints.

22. Completely install NEW nut on ball joint stud attaching lower control arm to knuckle. Tighten nut by holding ball joint stud with a hex wrench while turning nut with a wrench. Tighten nut using crow foot wrench on torque wrench to 68 Nm (50 ft. lbs.) + 90°.
23. Completely install NEW nut on ball joint stud attaching tension strut to knuckle. Tighten nut by holding ball joint stud with a hex wrench while turning nut with a wrench. Tighten nut using crow foot wrench on torque wrench to 68 Nm (50 ft. lbs.) + 90°.

** WARNING

It is important to tighten nut as described in following step to avoid damaging ball stud joint.

24. Place upper ball joint stud through hole in top of knuckle and install nut. Tighten nut by holding ball joint stud with a hex wrench while turning nut with a wrench. Tighten nut using crow foot wrench on torque wrench to 47 Nm (35 ft. lbs.) + 90°.

** WARNING

It is important to tighten nut as described in following step to avoid damaging ball stud joint.

25. Place outer tie rod stud through hole in knuckle and install nut. Tighten nut by holding stud with a wrench while turning nut with another wrench. Tighten nut using crow foot wrench on torque wrench to 85 Nm (63 ft. lbs.) torque.
26. Install brake rotor, then disc brake caliper and adapter assembly.
27. Install screw fastening brake flex hose routing bracket to knuckle. Tighten screw to 12 Nm (106 inch lbs.) torque.
28. Install wheel speed sensor head into knuckle and install mounting screw. Tighten screw to 11 Nm (95 inch lbs.) torque.
29. Attach wheel speed sensor cable and routing clip to brake flex hose routing bracket.
30. Install tire and wheel assembly. Tighten wheel mounting nuts to 150 Nm (110 ft. lbs.) torque.

31. Lower vehicle.
32. Pump brake pedal several times to ensure vehicle has a firm brake pedal before moving vehicle.
33. Check and adjust brake fluid level in reservoir as necessary.
34. Perform wheel alignment.

All Wheel Drive

1. Before servicing the vehicle, refer to the precautions in the beginning of this section.
2. Raise and support vehicle.
3. Remove wheel mounting nuts, then tire and wheel assembly.
4. While a helper applies brakes to keep hub from rotating, remove hub nut from axle halfshaft.
5. Remove clip fastening wheel speed sensor to knuckle.
6. Remove screw fastening wheel speed sensor to knuckle. Pull sensor head out of knuckle.
7. Access and remove front brake rotor.
8. Remove nut from outer tie rod end stud.
9. Separate tie rod stud from knuckle.
10. Loosen nut attaching lower control arm ball joint stud to knuckle. Back nut off until nut is even with end of stud. Keeping nut on at this location will help keep end of stud from distorting while using puller in next step.

**** WARNING**

In following step, use care not to damage ball joint seal boot.

11. Separate ball joint stud from knuckle.
12. Remove tool.
13. Remove nut from end of ball joint stud.
14. Loosen nut attaching upper ball joint stud to knuckle. Back nut off until nut is even with end of stud. Keeping nut on at this location will help keep end of stud from distorting while using Puller in next step.

**** WARNING**

In following step, use care not to damage ball joint seal boot.

15. Separate upper ball joint stud from knuckle.
16. Remove tool.
17. Remove nut from end of upper ball joint stud.
18. Slide knuckle off halfshaft and remove from vehicle.

To install:

19. Place knuckle over lower ball joint stud and guide hub and bearing onto axle halfshaft.
20. Start NEW nut on lower ball joint stud. Do not tighten at this time.

**** WARNING**

It is important to tighten nut as described in following step to avoid damaging ball stud joint.

21. Place upper ball joint stud through hole in top of knuckle and install nut. Tighten nut by holding ball joint stud with a hex wrench while turning nut with a wrench. Tighten nut using crow foot wrench on torque wrench to 47 Nm (35 ft. lbs.) +90°.

**** WARNING**

It is important to tighten nut as described in following step to avoid damaging ball stud joint.

22. Tighten lower ball joint nut by holding ball joint stud with a hex wrench while turning nut with a wrench. Tighten nut using crow foot wrench on torque wrench to 68 Nm (50 ft. lbs.) +90°.

**** WARNING**

It is important to tighten nut as described in following step to avoid damaging ball stud joint.

23. Place outer tie rod stud through hole in knuckle and install nut. Tighten nut by holding stud with a wrench while turning nut with another wrench. Tighten nut using crow foot wrench on torque wrench to 85 Nm (63 ft. lbs.).
24. Install brake rotor, then disc brake caliper and adapter assembly.
25. Install hub nut on end of axle halfshaft. While a helper applies brakes to keep hub from turning, tighten hub nut to 213 Nm (157 ft. lbs.) torque.
26. Install wheel speed sensor head into knuckle and install mounting screw. Tighten screw to 11 Nm (95 inch lbs.) torque.
27. Attach wheel speed sensor cable routing clip at knuckle.
28. Install tire and wheel assembly. Tighten wheel mounting nuts to 150 Nm (110 ft. lbs.) torque.
29. Lower vehicle.
30. Pump brake pedal several times to ensure vehicle has a firm brake pedal before moving vehicle.
31. Check and adjust brake fluid level in reservoir as necessary.
32. Perform wheel alignment.

Upper Ball Joint

REMOVAL & INSTALLATION

The upper ball joint is not replaceable. If defective, the control arm must be replaced.

Lower Ball Joint

REMOVAL & INSTALLATION

1. Before servicing the vehicle, refer to the precautions in the beginning of this section.

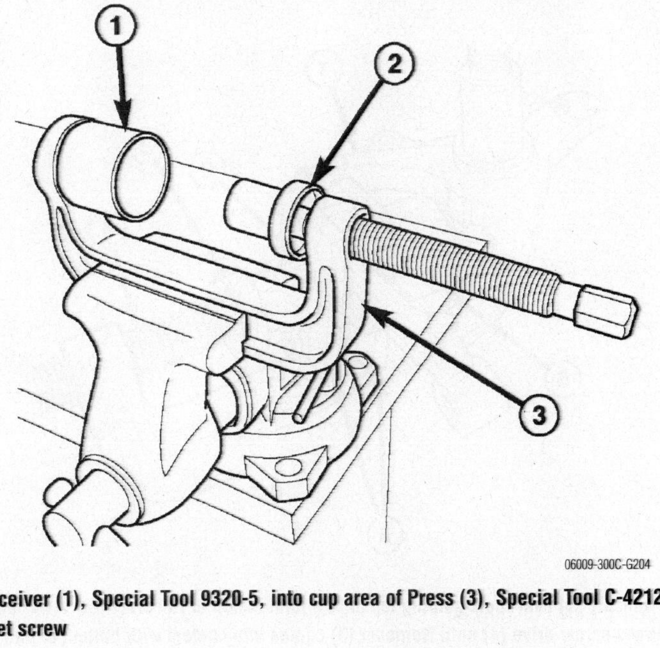

Place Receiver (1), Special Tool 9320-5, into cup area of Press (3), Special Tool C-4212F, and tighten set screw

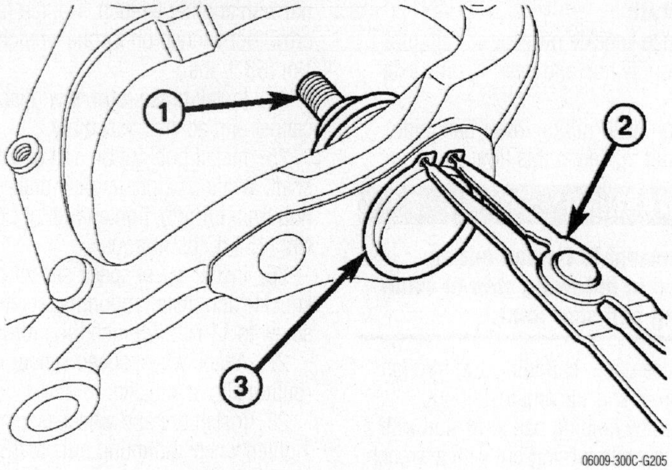

Ball joint snapring removal/installation

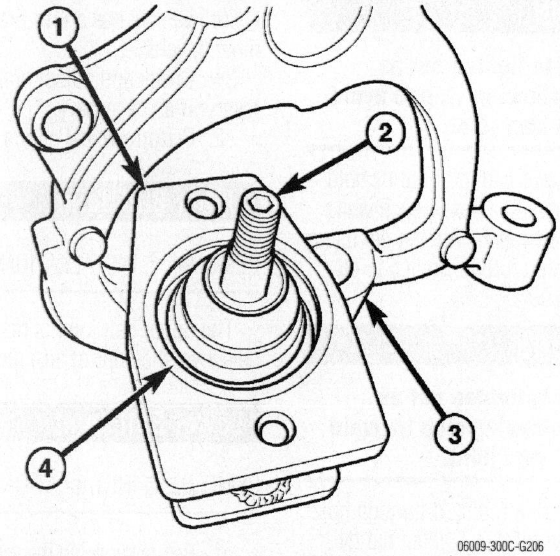

Install halves of Support Clamp (1), Special Tool 9320-1, over ball joint (2) and around knuckle surface (3)

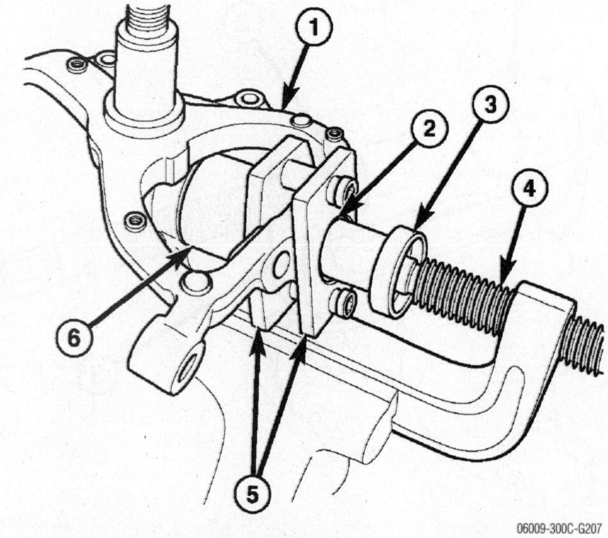

Position knuckle (1) over tools guiding top of ball joint inside of Receiver (6), then hand tighten Press screw-drive (4) until Remover (3) comes into contact with bottom of ball joint (2)

2. Remove the knuckle.

➡To perform this procedure it works best to mount Press, Special Tool C-4212F, in a vise and hold the component in your hands.

3. Place Receiver, Special Tool 9320-5, into cup area of Press, Special Tool C-4212F, as shown and tighten set screw.
4. Place Remover, Special Tool 9320-3, onto end of screw-drive of Press, Special Tool C-4212F, as shown.
5. Using a pair of snapring pliers (2), remove snapring (3) from bottom of ball joint (1).
6. Install halves of Support Clamp, Special Tool 9320-1, over ball joint and around knuckle surface as shown. Install and snug Support Clamp screws from underside.
7. Position knuckle over tools guiding top of ball joint inside of receiver, then hand tighten press screw-drive until remover comes into contact with bottom of ball joint.

➡When positioning knuckle over tools, make sure Receiver, Special Tool 9320-5, sets into recessed area of Support Clamp, Special Tool 9320-1.

8. Tighten Press screw-drive forcing ball joint out of knuckle and into receiver.
9. Loosen screw-drive and remove knuckle from press. Remove ball joint from receiver.
10. Remove Support Clamp from knuckle.

To install:

11. Place Installer, Special Tool 9320-4, into cup area of press, Special Tool C-4212F, as shown and tighten set screw.
12. Place Remover, Special Tool 9320-3, onto end of screw-drive of Press, Special Tool C-4212F, as shown.
13. Start NEW ball joint into bore of knuckle.
14. Position knuckle (1) over tools guiding top of ball joint inside of Installer (5) until outside flange of ball joint (4) comes into contact with Installer, then hand tighten Press screw-drive (3) until Remover (2) comes into contact with bottom of knuckle (1).
15. Using hand tools, tighten screw-drive, pressing ball joint into knuckle until flange comes to a stop against the knuckle.
16. Loosen screw-drive and remove knuckle from Press.
17. Install snapring into groove on bottom of ball joint.
18. Inspect ball joint for proper fit. Make sure seal boot is uniform and wire rings are in place.

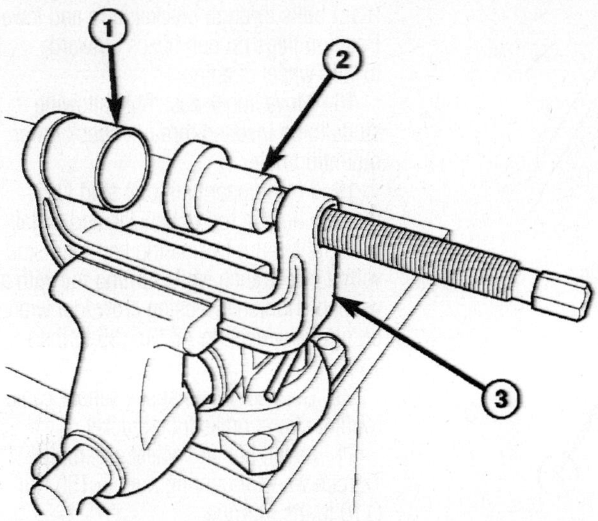

Installer assembled for ball joint installation. (1) installer, (2) remover, (3) press

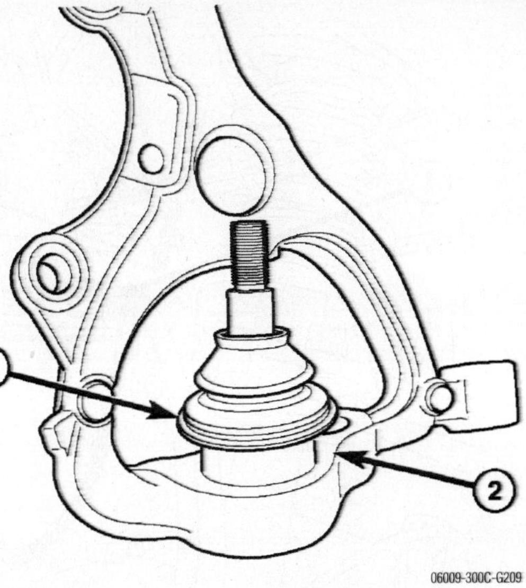

Ball joint in position

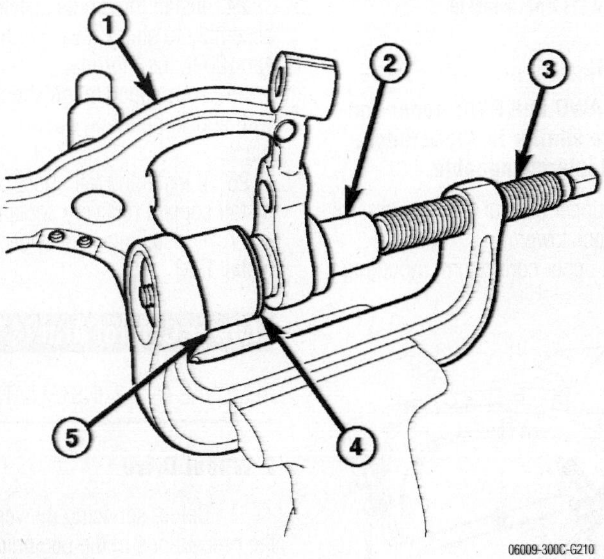

Pressing the ball joint into the knuckle

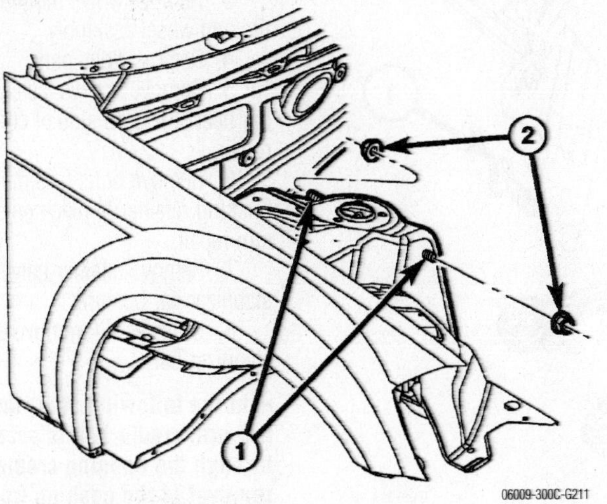

Upper control arm mounting bolts (1) and nuts (2)

Upper Control Arm

REMOVAL & INSTALLATION

1. Before servicing the vehicle, refer to the precautions in the beginning of this section.
2. If removing left upper control arm, remove and reposition coolant recovery container.
3. If removing right upper control arm, remove IPM from mount and reposition.
4. If equipped, remove front shock tower cap from top of shock assembly.
5. Remove three nuts fastening shock assembly to shock tower.
6. Remove nuts from upper control arm mounting bolts.
7. Raise and support vehicle.
8. Remove wheel mounting nuts, then tire and wheel assembly.
9. Disconnect wheel speed sensor cable routing clip at brake tube bracket.
10. Loosen nut attaching upper ball joint stud to knuckle. Back nut off until nut is even with end of stud. Keeping nut on at this location will help keep end of stud from distorting while using puller in next step.

WARNING

In following step, use care not to damage ball joint seal boot.

11. Using Puller, Special Tool 9360, separate upper ball joint stud from knuckle.
12. Remove nut from end of upper ball joint stud.
13. Pull shock assembly downward until

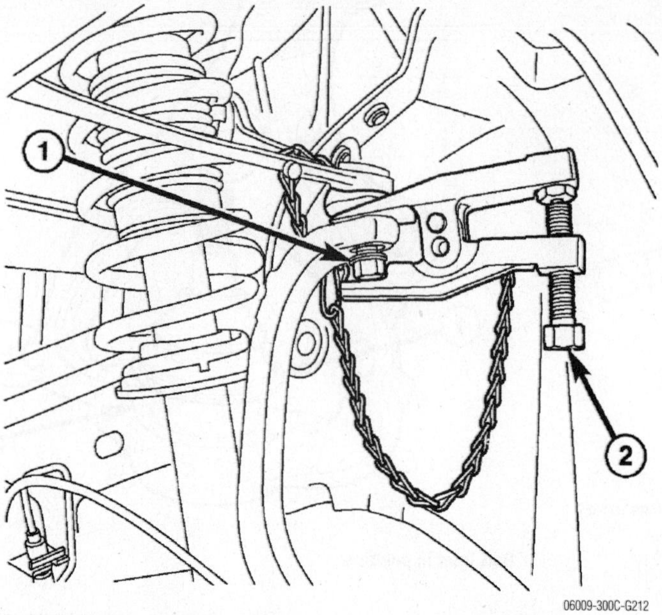

Using Puller (2), Special Tool 9360, separate upper ball joint stud (1) from knuckle

studs clear shock tower, then pull it outward allowing access to upper control arm mounting bolts.

14. Remove upper control arm mounting (flag) bolts.

15. Remove upper control arm (2) from bracket (3) in shock tower (4).

To install:

➡Although AWD and RWD upper control arms are similar in appearance, they are not interchangeable.

16. Slide upper control arm into bracket located in shock tower.

17. Install upper control arm mounting (flag) bolts through bracket, arm and tower. Position flags on bolt heads outward, toward wheel opening.

18. Move shock assembly allowing studs to be inserted through shock tower mounting holes.

19. Place upper ball joint stud (1) through hole in top of knuckle and install nut. Tighten nut by holding ball joint stud with a hex wrench while turning nut with a wrench. Tighten nut using crow foot wrench on torque wrench to 47 Nm (35 ft. lbs.) +90°.

20. Connect wheel speed sensor cable routing clip at brake tube bracket.

21. Install tire and wheel assembly. Tighten wheel mounting nuts to 150 Nm (110 ft. lbs.) torque.

22. Lower vehicle to curb position.

23. Install nuts on upper control arm body mounting bolts. Tighten nuts to 75 Nm (55 ft. lbs.).

24. Install three nuts fastening shock assembly to shock tower. Tighten nuts to 27 Nm (20 ft. lbs.) torque.

25. If equipped, align shock tower cap with shock mounting nuts and snap into place.

26. If installing left upper control arm, install coolant recovery container.

27. If installing right upper control arm, install IPM.

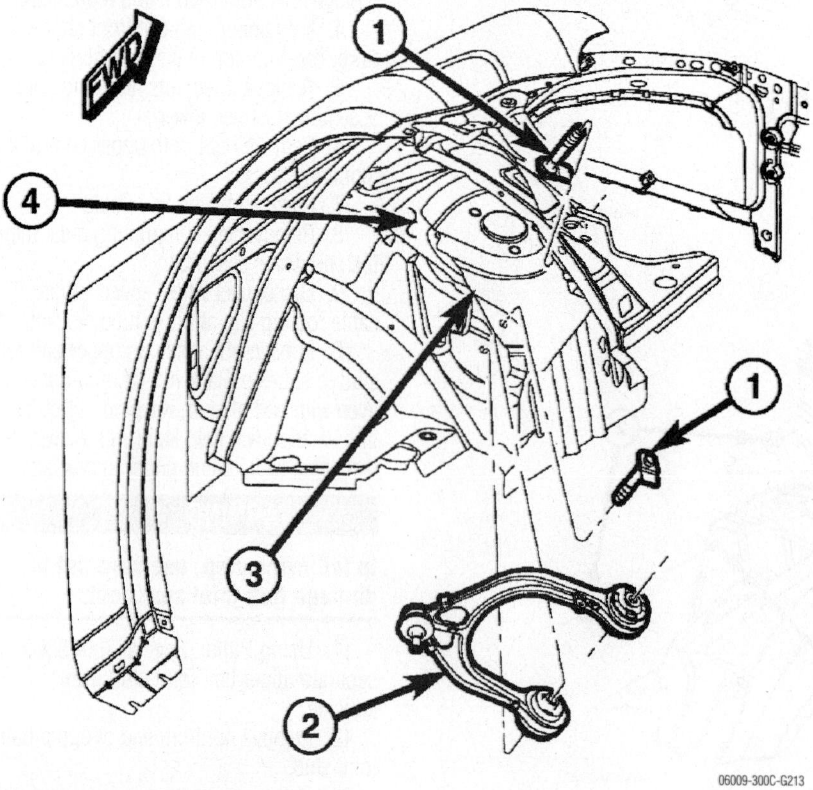

Upper control arm removal

Lower Control Arm

REMOVAL AND & INSTALLATION

2-Wheel Drive

1. Before servicing the vehicle, refer to the precautions in the beginning of this section.
2. Raise and support vehicle.
3. Remove wheel mounting nuts, then tire and wheel assembly.
4. Remove belly pan.
5. Remove screws fastening stabilizer bar heat shield on side of control arm repair.
6. Remove bolts fastening stabilizer bar bushing retainer in place on side of control arm repair.
7. Remove retainer halves from around stabilizer bar bushing.
8. Utilizing slit, remove bushing from stabilizer bar.

➡In the following step, the lower control arm cradle bolt is accessed through the opening created by removal of the bushing from the stabilizer bar.

⁂ WARNING

If the lower control arm bolt at the engine cradle has a lengthwise grooved shaft, it is a special wheel alignment adjustment bolt and the bolt head must not be rotated in the vehicle or damage to the bolt and engine cradle will result. While holding the bolt in place with a wrench, remove the nut, then slide the bolt out of the bushing and cradle taking note of bolt positioning in engine cradle for reassembly purposes. The bolt needs to be installed in the same position as removed to make sure wheel camber and caster return to adjusted position.

9. Remove bolt and nut securing lower control arm to engine cradle. If bolt has a lengthwise grooved shaft (see above note), remove bolt and nut by holding the bolt in place with a wrench, removing nut, then sliding bolt out of bushing and cradle while taking note of bolt positioning in lower control arm bushing for reassembly purposes.
10. Remove bolt securing shock assembly to lower control arm.
11. Remove screw fastening wheel speed sensor to knuckle. Pull sensor head out of knuckle.
12. Remove wheel speed sensor cable routing clip from brake flex hose routing bracket.
13. Loosen nut attaching ball joint stud to lower control arm. Back nut off until nut is even with end of stud. Keeping nut on at this location will help keep end of stud from distorting while using Puller in next step.

➡In following step, use care not to damage ball joint seal boot.

14. Using Puller, Special Tool 9360, separate ball joint stud from lower control arm.
15. Remove nut from end of ball joint stud attaching lower control arm to knuckle.
16. Pry knuckle downward and slide ball joint stud out of lower control arm. Position knuckle outward, away from lower control arm.
17. Slide lower control arm out of engine cradle and remove from vehicle.

To install:

➡If installing a lower control arm engine cradle bolt that is a wheel alignment adjustment bolt (lengthwise grooved shaft), make sure to install it in the same position which it was in upon removal.

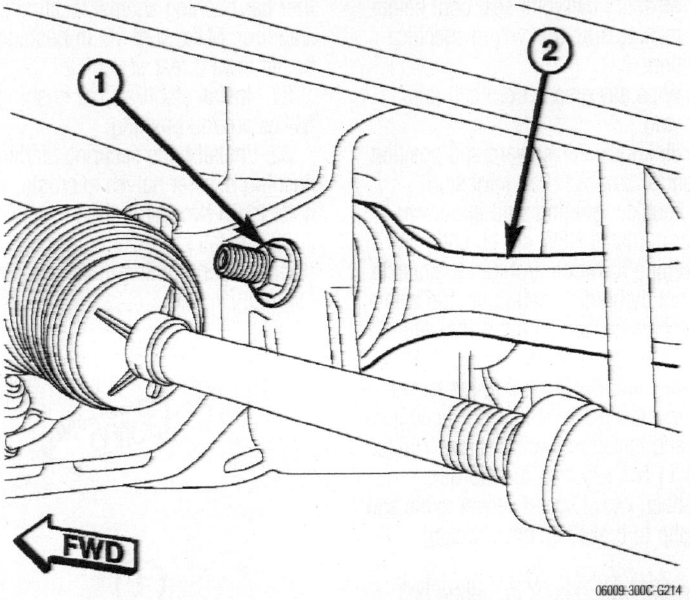

Remove bolt and nut (1) securing lower control arm (2) to engine cradle—2-wheel drive

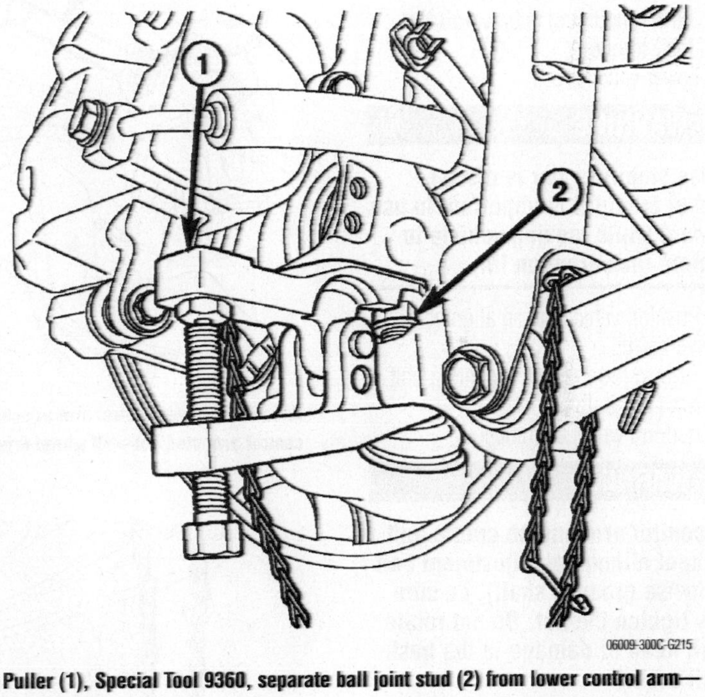

Using Puller (1), Special Tool 9360, separate ball joint stud (2) from lower control arm—2-wheel drive

18. Slide lower control arm into position in engine cradle and install mounting bolt from rear.
19. Install nut on lower control arm cradle bolt, but do not tighten at this time.

⁂ WARNING

Before installing knuckle to lower control arm, measure height of ball joint seal boot mounted on knuckle. If seal boot height is above 25.5 mm, any air inside seal boot must be expelled. To do so, follow these steps:

- Tip ball joint stud completely to one side.
- Using thumb and index finger, gently squeeze seal boot together at center expelling any air. Do not allow grease to be release.
- Push down very top of seal boot.
- Return ball joint stud to original "centered" position.

- Measure ball joint seal boot height making sure it is within specification.
- Wipe any grease from ball joint stud.

20. Pull knuckle downward and position lower control arm over ball joint stud. Release knuckle, guiding stud into lower control arm. Install NEW nut on ball joint stud attaching lower control arm to knuckle. Tighten nut by holding ball joint stud with a hex wrench while turning nut with a wrench. Tighten nut using crow foot wrench on torque wrench to 68 Nm (50 ft. lbs.) + 90°.
21. Install wheel speed sensor head into knuckle and install mounting screw. Tighten screw to 11 Nm (95 inch lbs.) torque.
22. Attach wheel speed sensor cable and routing clip to brake flex hose routing bracket.
23. Install lower shock mounting bolt attaching shock assembly to lower control arm. Do not tighten bolt at this time.
24. Install tire and wheel assembly. Tighten wheel mounting nuts to 150 Nm (110 ft. lbs.) torque.
25. Lower vehicle.

✲✲ WARNING

Because stabilizer bar is disconnected at cradle it is important to use extra care while moving vehicle to alignment rack/drive-on lift.

26. Position vehicle on an alignment rack/drive-on lift.
27. Tighten lower shock mounting bolt to 174 Nm (128 ft. lbs.).
28. Perform wheel alignment.

✲✲ WARNING

If the control arm engine cradle bolt is a wheel alignment adjustment bolt (lengthwise grooved shaft), be sure to only tighten the nut. Do not rotate the bolt head or damage to the bushing will occur.

29. Once camber is found to be within specifications, using a crowfoot wrench, tighten lower control arm cradle bolt nut to 176 Nm (130 ft. lbs.) torque while holding the bolt stationary.

✲✲ WARNING

Because of stabilizer bushing outer shape, it is very important to install bushings in position discussed in following step.

30. Utilizing slit in bushing, install stabilizer bar bushing against locating collar on stabilizer. Make sure slit in bushing is positioned toward rear of vehicle.
31. Install stabilizer bar bushing retainer halves around bushing.
32. Install bolts securing stabilizer bar bushing retainer halves to cradle. Tighten bolts to 60 Nm (44 ft. lbs.) torque.
33. Install stabilizer bar heat shield over stabilizer bar bushing retainer. Install mounting screws.
34. Install belly pan.

All Wheel Drive

1. Before servicing the vehicle, refer to the precautions in the beginning of this section.
2. Raise and support vehicle.
3. Remove wheel mounting nuts, then tire and wheel assembly.
4. While a helper applies brakes to

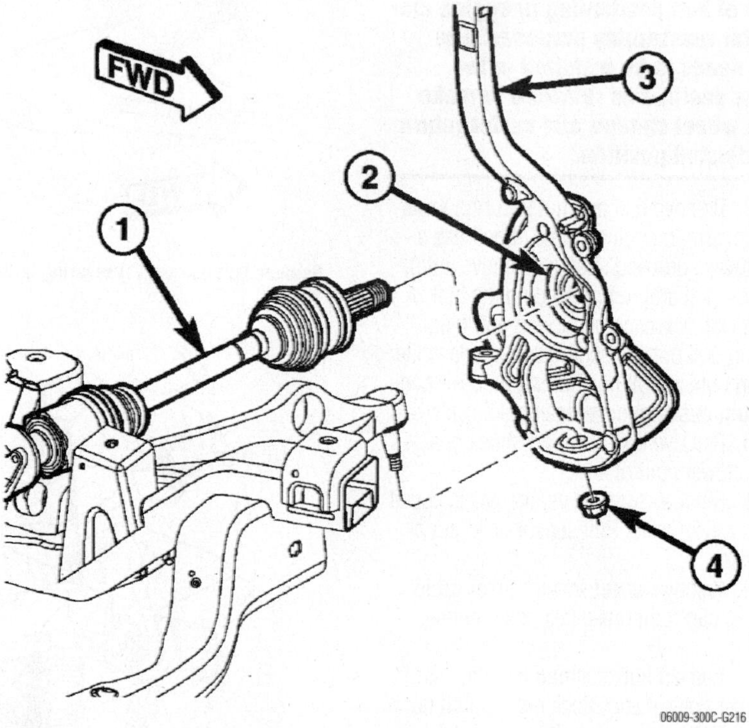

Knuckle-to-lower control arm attachment. (1) halfshaft, (2) hub/bearing, (3) knuckle, (4) lower control arm stud nut—all wheel drive

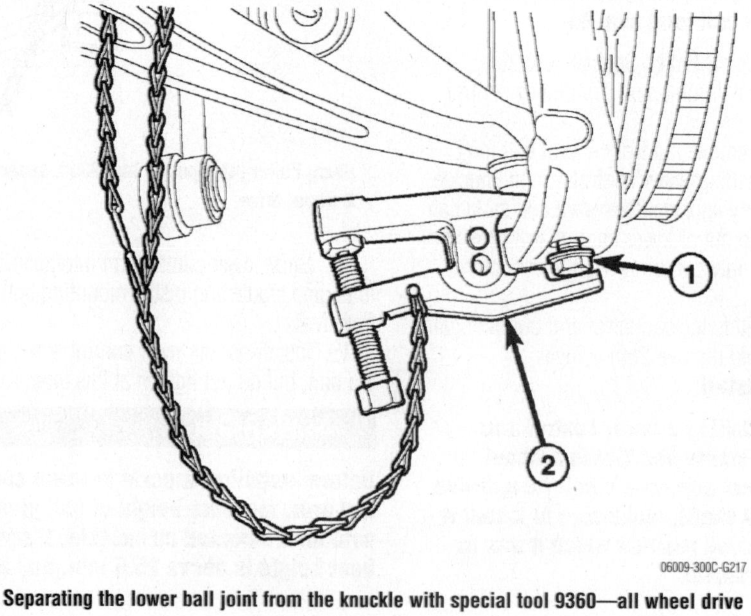

Separating the lower ball joint from the knuckle with special tool 9360—all wheel drive

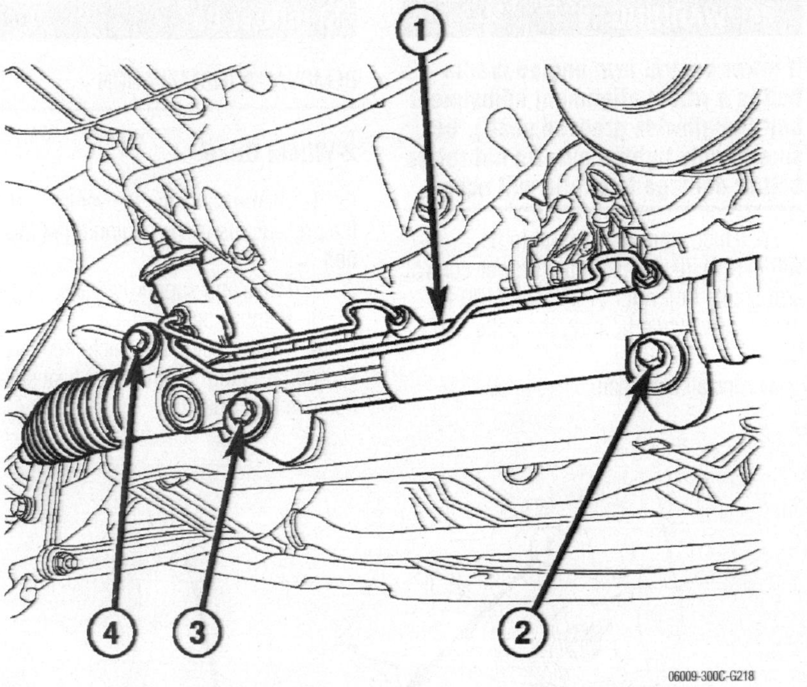

Steering gear mounting bolt identification—all wheel drive

keep hub from rotating, remove hub nut from the axle halfshaft.

5. Remove belly pan.

6. Loosen nut attaching lower control arm ball joint stud to knuckle. Back nut off until nut is even with end of stud. Keeping nut on at this location will help keep end of stud from distorting while using Puller in next step.

⁂ WARNING

In following step, use care not to damage ball joint seal boot.

7. Using Special Tool 9360, separate ball joint stud from knuckle.

8. Remove tool.

9. Remove nut from end of ball joint stud.

10. Back off nut from bolt attaching shock clevis bracket to lower control arm until it is flush with end of bolt.

11. Using a brass drift punch, tap the bolt out of the clevis bracket until bolt serrations clear bracket.

12. Remove nut and bolt for clevis bracket and control arm.

13. Remove mounting screws, then heat shields above both inner tie rod bellows.

⁂ WARNING

While steering gear bolts are removed, it important to avoid putting downward force on steering gear.

14. See the accompanying illustration, and remove steering gear mounting bolts as follows:

- Left side arm: Remove bolts (3) and (4). Loosen, but do not remove, bolt (2). It is important leave bolt (2) installed to avoid dropping gear too far, putting excessive force on steering coupling.
- Right side arm: Remove bolts (2) and (3). Loosen, but do not remove, bolt (4). It is important to leave bolt (4) installed to avoid dropping gear too far, putting excessive force on steering coupling.

➡If the lower control arm bolt at the engine cradle has a lengthwise grooved shaft, it is a special wheel alignment adjustment bolt and the bolt head must not be rotated in the vehicle or damage to the bolt and engine cradle will result. While holding the bolt in place with a wrench, remove the nut, then slide the bolt out of the bushing and cradle taking note of bolt positioning in engine cradle for reassembly purposes. The bolt needs to be installed in the same position as removed to make sure wheel camber and caster return to adjusted position.

15. Remove bolt (3) and nut securing forward end of lower control arm (1) to engine cradle (2). If bolt has a lengthwise grooved shaft (see above note), remove bolt and nut by holding bolt stationary with a wrench, removing nut, then sliding bolt out of bushing and cradle while taking note of bolt positioning in engine cradle for reassembly purposes.

16. Remove bolt and nut (4) securing rearward end of lower control arm (1) to engine cradle (2). If bolt has a lengthwise grooved shaft (see above note), remove bolt and nut by holding bolt stationary with a wrench, removing nut, then sliding bolt out of bushing and cradle while taking note of bolt positioning in engine cradle for reassembly purposes.

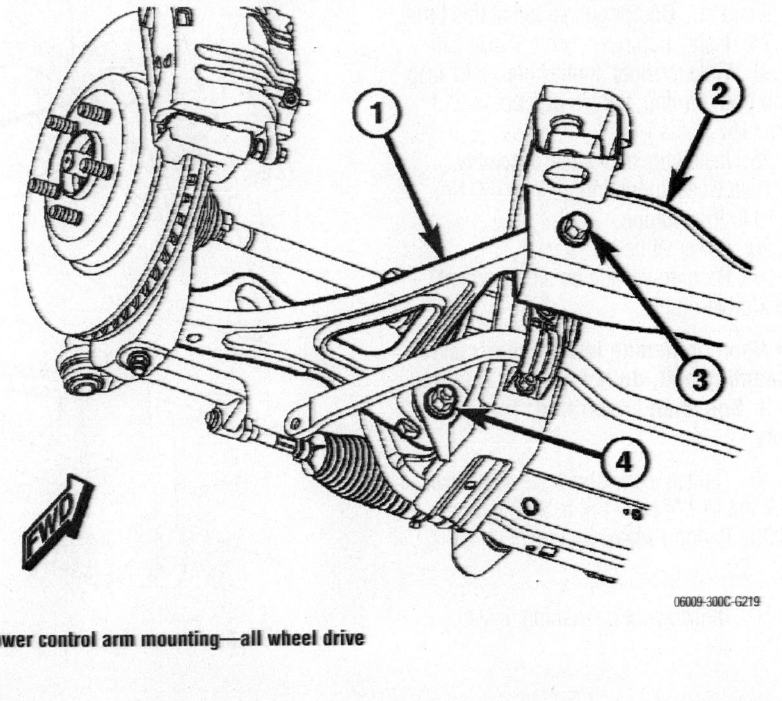

Lower control arm mounting—all wheel drive

17. Slide lower control arm (1) from engine cradle (2) and knuckle, and remove from vehicle.

To install:

➡If installing a lower control arm engine cradle bolt that is a wheel alignment adjustment bolt (identifying lengthwise grooved shaft), make sure to install it in the same position which it was in upon removal.

18. Slide lower control arm into position in engine cradle and place ball joint stem into mounting hole in knuckle.

➡When installing lower control arm engine cradle bolts, it important to note that the forward bolt is installed front-to-rear and the rearward bolt is installed rear-to-front.

19. Install lower control arm mounting bolts and nuts. Do not tighten bolts at this time.
20. Raise steering gear to mounted position and install lower mounting bolts. Install upper mounting bolt and nut. Tighten all mounting bolts to 122 Nm (90 ft. lbs.) torque.
21. Install heat shields above both inner tie rod bellows. Tighten mounting screws to 7 Nm (62 inch lbs.).
22. Pull downward on control arm and guide ball joint stud into knuckle. Install NEW nut on ball joint stud. Tighten nut by holding ball joint stud with a hex wrench while turning nut with a wrench. Tighten nut using crow foot wrench on torque wrench to 122 Nm (90 ft. lbs.) torque.
23. Align shock clevis bracket with lower control arm bushing and install mounting bolt and nut. Do not tighten nut at this time.
24. Install hub nut on end of axle halfshaft. While a helper applies brakes to keep hub from turning, tighten hub nut to 213 Nm (157 ft. lbs.).
25. Install tire and wheel assembly. Tighten wheel mounting nuts to 150 Nm (110 ft. lbs.) torque.
26. Lower vehicle.
27. Position vehicle on an alignment rack/drive-on lift.

➡When tightening lower shock clevis mounting bolt, do not attempt rotating bolt. Bolt shaft is serrated. Turn nut only.

28. Tighten lower shock clevis bracket bolt nut to 174 Nm (128 ft. lbs.) torque.
29. Perform wheel alignment.

⁂ WARNING

If lower control arm engine cradle bolt is a wheel alignment adjustment bolt (lengthwise grooved shaft), be sure to only tighten nut. Do not rotate bolt or damage to cradle will occur.

30. Once camber is found to be within specifications, tighten lower control arm cradle bolt nuts to 176 Nm (130 ft. lbs.) torque while holding the bolts stationary.
31. Install belly pan.

Stabilizer Bar

REMOVAL & INSTALLATION

2-Wheel Drive

1. Before servicing the vehicle, refer to the precautions in the beginning of this section.
2. Raise and support vehicle.
3. Remove belly pan.
4. On each side of vehicle, remove screws fastening stabilizer bar heat shield. Remove heat shield.

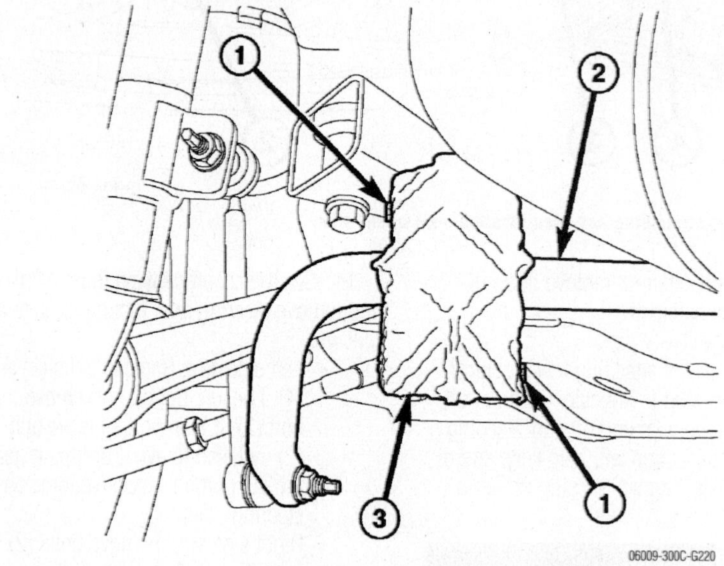

Stabilizer bat heat shield (3), stabilizer bar (2), heat shield screws (1)—2-wheel drive

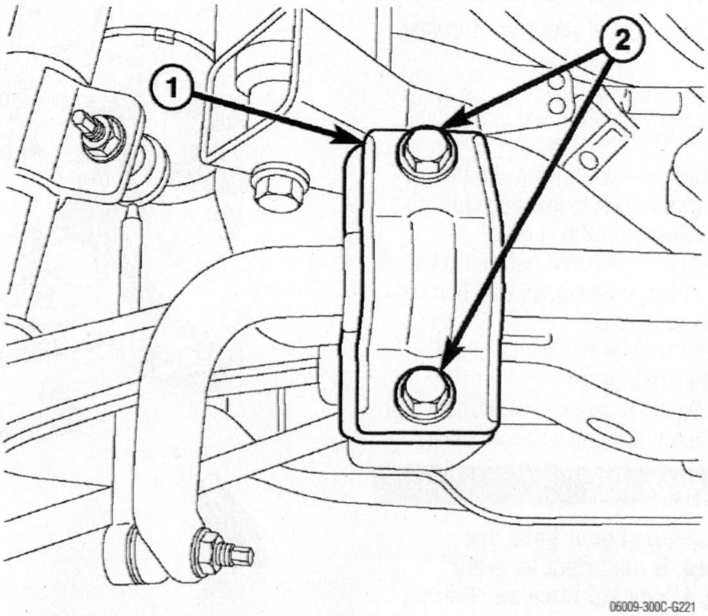

Isolator retainer (1) and bolts (2)—2-wheel drive

5. On each side of vehicle, remove bolts fastening stabilizer bar isolator retainer in place.

6. On each side of vehicle, remove retainer halves from around stabilizer bar isolator.

7. Utilizing slit, remove each isolator from stabilizer bar.

8. On each side of vehicle, remove nut (5) fastening stabilizer link (3) to stabilizer bar (4). Slide link ball joint stem (1) from bar, then remove bar from vehicle.

To install:

➡When attaching stabilizer link to stabilizer bar, make sure link ball joint stem is pointed inboard toward engine cradle.

9. On each side of vehicle, raise stabilizer bar to stabilizer link and slide link ball joint stem through mounting hole in bar. Loosely install nut at this time.

WARNING

Because of stabilizer isolator outer shape, it is very important to install isolators in position discussed in following step.

10. Utilizing slit in isolator, install each stabilizer bar isolator on bar resting against locating collar as shown. Make sure slit in isolator is positioned toward rear of vehicle.

11. On each side of vehicle, install stabilizer bar isolator retainer halves around isolator.

12. On each side of vehicle, install bolts securing stabilizer bar isolator retainer halves to cradle. Tighten bolts to 60 Nm (44 ft. lbs.) torque.

13. On each side of vehicle, install stabilizer bar heat shield over stabilizer bar isolator retainer. Install mounting screws.

14. While holding stem from rotating at hex or flat tighten stabilizer link nuts at each end of stabilizer bar to 128 Nm (95 ft. lbs.) torque.

15. Install belly pan.

16. Lower vehicle.

All Wheel Drive

1. Before servicing the vehicle, refer to the precautions in the beginning of this section.

2. Raise and support vehicle.

3. Remove belly pan.

4. On each side of vehicle, remove nut fastening stabilizer link to stabilizer bar. Slide link ball joint stem from bar.

5. On each side of vehicle, remove bolts fastening stabilizer bar isolator retainer in place.

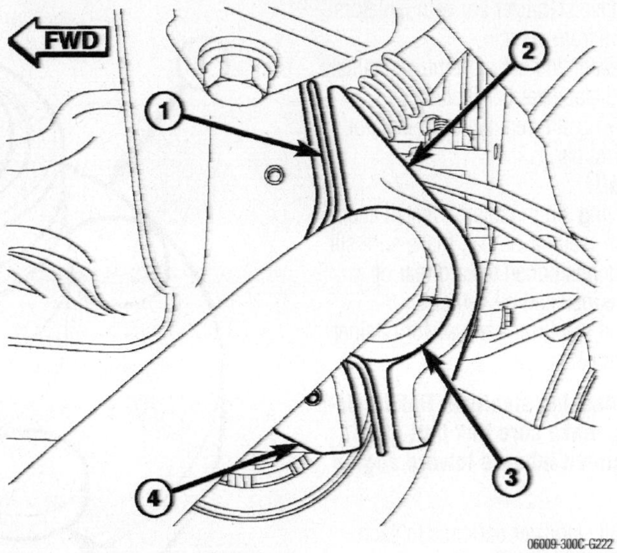

Correct retainer (1, 2) positioning—2-wheel drive

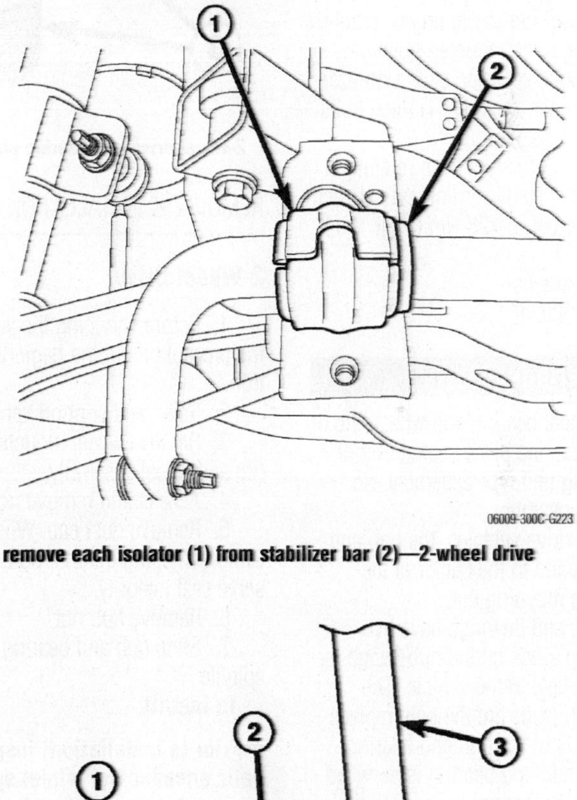

Utilizing slit, remove each isolator (1) from stabilizer bar (2)—2-wheel drive

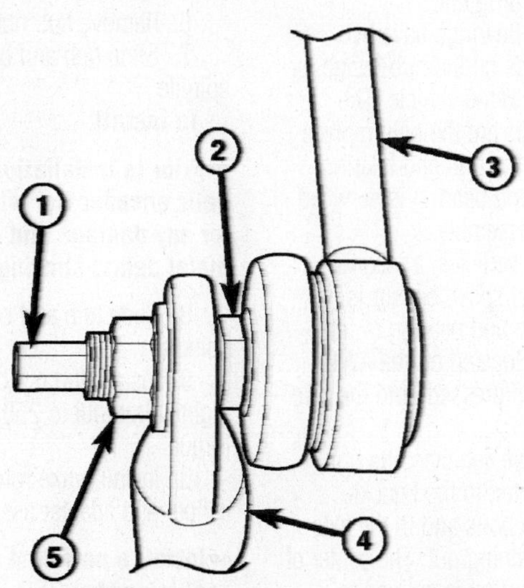

Stabilizer link-to-bar attachment—2-wheel drive

6. Remove stabilizer bar with isolators and retainers from vehicle.

7. On each side of bar, remove retainers from around stabilizer bar isolators.

8. Utilizing slit, remove each isolator from stabilizer bar.

To install:

9. Utilizing slit in isolator, install each stabilizer bar isolator on bar. Make sure slit in isolator is positioned toward rear of vehicle once installed on vehicle.

10. Install stabilizer bar isolator retainer over each isolator.

➡When attaching stabilizer link to stabilizer bar, make sure link ball joint stem is pointed inboard toward engine cradle.

11. Install stabilizer bar ends to each stabilizer link. Slide link ball joint stem through mounting hole in bar. Loosely install nuts at this time.

12. Raise stabilizer bar to engine cradle, placing bushings into mounted position.

13. Install bolts securing each stabilizer bar isolator retainer to cradle. Tighten bolts to 60 Nm (44 ft. lbs.) torque.

14. While holding stem from rotating at hex or flat tighten stabilizer link nuts at each end of stabilizer bar to 128 Nm (95 ft. lbs.) torque.

15. Install belly pan.

16. Lower vehicle.

Wheel Bearings

The front wheel bearing and wheel hub of this vehicle are a one piece sealed unit or hub and bearing unit type assembly. No adjustment is possible.

On 2-wheel-drive vehicles, the hub and bearing is mounted to the center of the spindle using a retaining nut.

All front hub and bearings have five wheel mounting studs on the hub flange for mounting the wheel to the vehicle. The wheel mounting studs are the only replaceable components of the hub and bearing. Otherwise, the hub and bearing is serviced only as a complete assembly.

On 2-wheel drive vehicles, a magnetic encoder for the wheel speed sensor is pressed onto the hub and bearing. A magnetic encoder is not located on the AWD hub and bearing. It is pressed onto the axle halfshaft outer joint.

On All-Wheel-Drive vehicles, the hub and bearing is mounted to the knuckle using four mounting bolts and to the axle halfshaft using a retaining nut. The center of the AWD hub and bearing is splined to match the outer joint of the axle halfshaft.

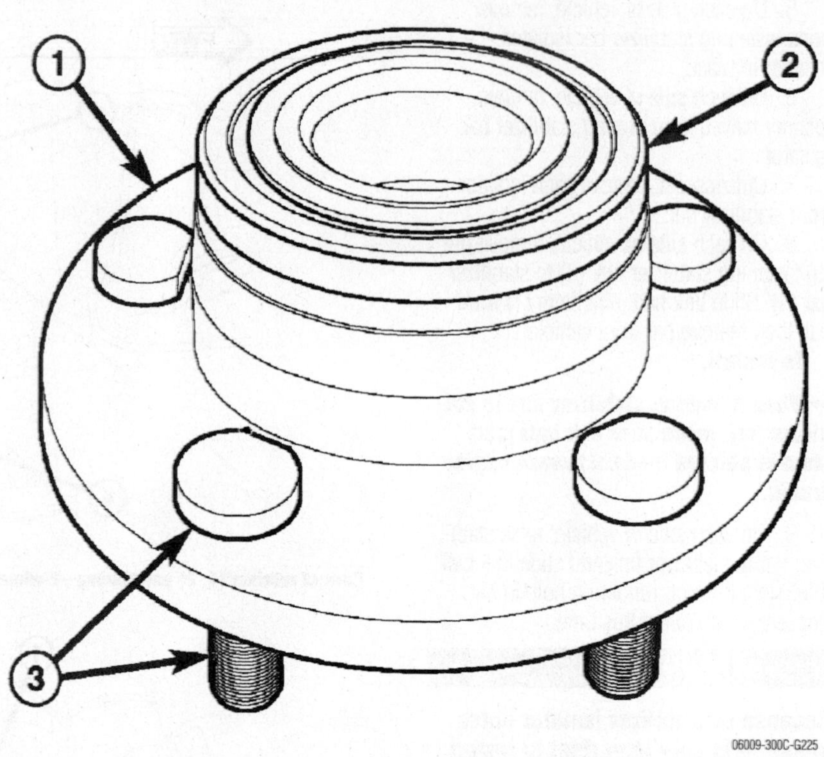

2-wheel drive hub/bearing assembly (1), magnetic encoder (2), stud (3)

REMOVAL & INSTALLATION

2-Wheel Drive

1. Before servicing the vehicle, refer to the precautions in the beginning of this section.

2. Raise and support vehicle.

3. Remove wheel mounting nuts, then tire and wheel assembly.

4. Access and remove front brake rotor.

5. Remove dust cap. When doing this, avoid damaging internal bore of hub to preserve seal integrity.

6. Remove hub nut.

7. Slide hub and bearing off knuckle spindle.

To install:

➡Prior to installation, inspect magnetic encoder (for wheel speed sensor) for any damage and make sure any metal debris sticking to it is removed.

8. Slide hub and bearing onto knuckle spindle.

9. Install hub nut on end of spindle. Tighten hub nut to 250 Nm (184 ft. lbs.) torque.

10. Install brake rotor, then disc brake caliper and adapter assembly.

➡Install a new dust cap to preserve seal integrity.

11. Install NEW dust cap on hub and bearing.

12. Install tire and wheel assembly. Tighten wheel mounting nuts to 150 Nm (110 ft. lbs.) torque.

13. Lower vehicle.

14. Pump brake pedal several times to ensure vehicle has a firm brake pedal before moving vehicle.

15. Check and adjust brake fluid level in reservoir as necessary.

16. Road test vehicle and make several stops to wear off any foreign material on brakes and to seat brake pads.

All Wheel Drive

1. Before servicing the vehicle, refer to the precautions in the beginning of this section.

2. Raise and support vehicle.

3. Remove wheel mounting nuts, then tire and wheel assembly.

4. While a helper applies brakes to keep hub from rotating, remove hub nut from the axle halfshaft.

5. Access and remove front brake rotor.

6. Remove four bolts fastening hub and bearing to knuckle.

7. Slide hub and bearing off axle halfshaft and knuckle.

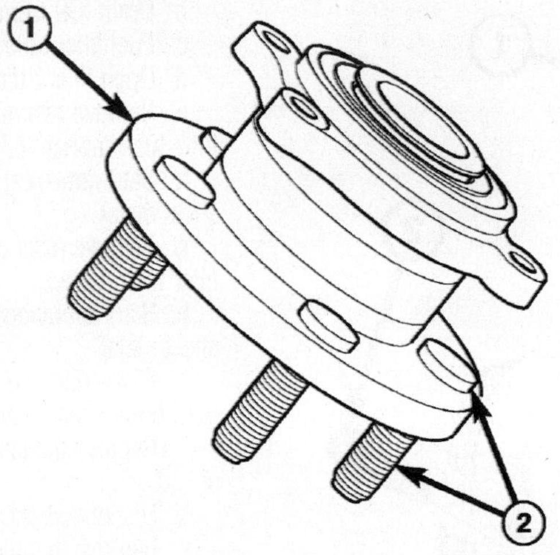

All wheel drive front hub/bearing (1), studs (2)

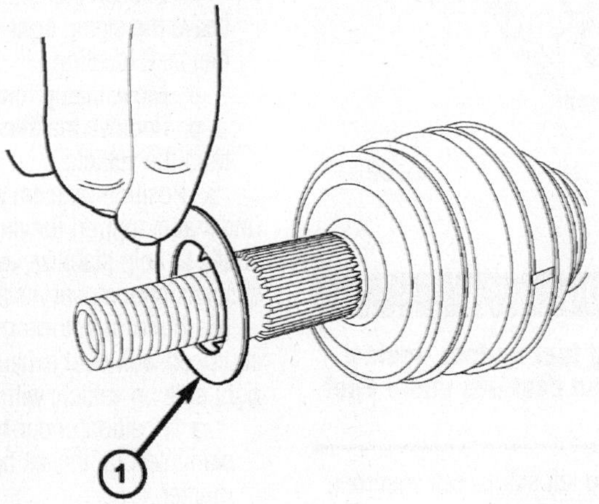

Isolation washer—all wheel drive

To install:

➡Before installing hub and bearing on end of axle halfshaft, ensure isolation washer is present on end of halfshaft. Inspect washer making sure it is not worn or damaged. Washer can be installed in either direction on shaft.

8. Slide hub and bearing onto axle halfshaft. Position hub and bearing onto knuckle, lining up mounting bolt holes.
9. Install four bolts fastening hub and bearing in place. Tighten mounting bolts to 68 Nm (50 ft. lbs.) torque.
10. Install brake rotor, then disc brake caliper and adapter assembly.
11. Install hub nut on end of axle halfshaft. While a helper applies brakes to keep hub from turning, tighten hub nut to 213 Nm (157 ft. lbs.).
12. Install tire and wheel assembly. Tighten wheel mounting nuts to 150 Nm (110 ft. lbs.) torque.
13. Lower vehicle.
14. Pump brake pedal several times to ensure vehicle has a firm brake pedal before moving vehicle.
15. Check and adjust brake fluid level in reservoir as necessary.
16. Road test vehicle and make several stops to wear off any foreign material on brakes and to seat brake pads.

REAR SUSPENSION

Shock Absorber

REMOVAL & INSTALLATION

Standard

1. Before servicing the vehicle, refer to the precautions in the beginning of this section.
2. Raise and support vehicle.
3. Remove wheel mounting nuts, then tire and wheel assembly.
4. Position under-hoist utility jack or jackstand under outer spring link adding just enough support to keep suspension from going into full-rebound when shock absorber mounting bolts are removed.
5. Remove shock absorber lower mounting bolt and nut.
6. Remove shock absorber upper mounting bolts.

1. Camber link
2. Spring link
3. Toe link
4. Knuckle
5. Compression link
6. Tension link

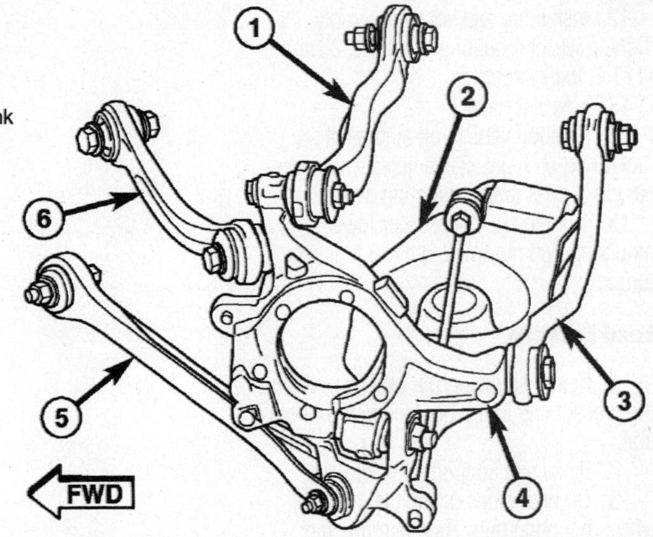

Rear suspension linkage

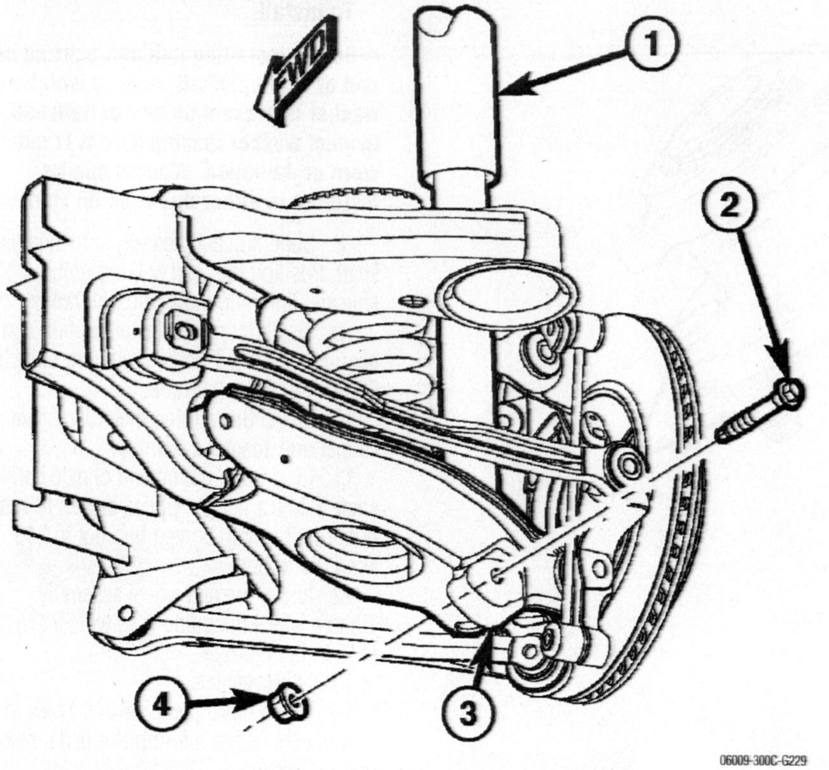

Shock absorber (1) lower bolt (2) and nut (4)

7. Remove shock absorber.

To install:

8. Insert lower end of shock absorber into well of spring link.
9. Raise upper end of shock absorber up into mounted position on body and install upper mounting screws. Tighten upper mounting screws to 52 Nm (38 ft. lbs.) torque.
10. Install lower shock mounting bolt and nut. Do not tighten at this time.
11. Remove under-hoist utility jack or jackstand from under spring link.
12. Install tire and wheel assembly. Tighten wheel mounting nuts to 150 Nm (110 ft. lbs.) torque.
13. Lower vehicle.
14. Position vehicle on alignment rack/drive-on hoist. Raise vehicle as necessary to access lower mounting bolt.
15. Tighten shock absorber lower mounting bolt nut to 72 Nm (53 ft. lbs.) torque.

Load Leveling

1. Before servicing the vehicle, refer to the precautions in the beginning of this section.
2. Raise and support vehicle.
3. On both sides of vehicle, remove wheel mounting nuts, then rear tire and wheel assembly.

CAUTION

Before opening fuel system, review all warnings and cautions under Fuel System.

4. If servicing left side shock absorber, remove fuel filler tube:
   a. Disconnect negative battery cable.
   b. Drain fuel from tank.
   c. Fuel filler tube assembly.
   d. Open filler tube door.
   e. Remove retaining wire from inside filler tube rubber.
   f. Start removing rubber from body sheet metal.
   g. Remove retaining wire from inside filler tube rubber.
   h. Start removing rubber from body sheet metal.
   i. Filler tube and rubber removed from body sheet metal.
   j. Remove the left inner splash shield.
   k. Disconnect the filler tube vent line.
   l. Remove the filler tube mounting bolt.
   m. Remove the under body splash shield.
   n. Loosen the filler tube hose clamp. Leave the clamp tight on the hose and fuel tank location.
   o. Move clamp toward the fuel tank.
   p. Remove the filler tube assembly from the vehicle.
5. Position an extra pair of jackstands under and support forward end of engine cradle to help stabilize vehicle during rear suspension removal/installation.
6. Perform following if vehicle is equipped with dual exhaust or are servicing right side on vehicle with single exhaust:
   a. Position under-hoist utility jack or stand several inches below exhaust at muffler.
   b. Disconnect exhaust isolators at muffler and resonators hangers.

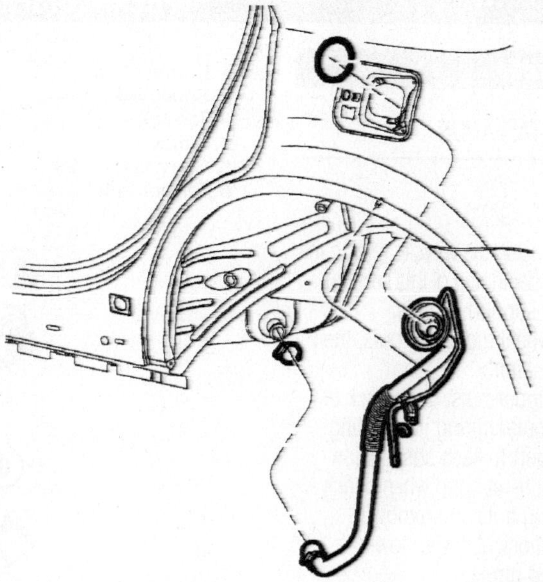

Fuel filler tube

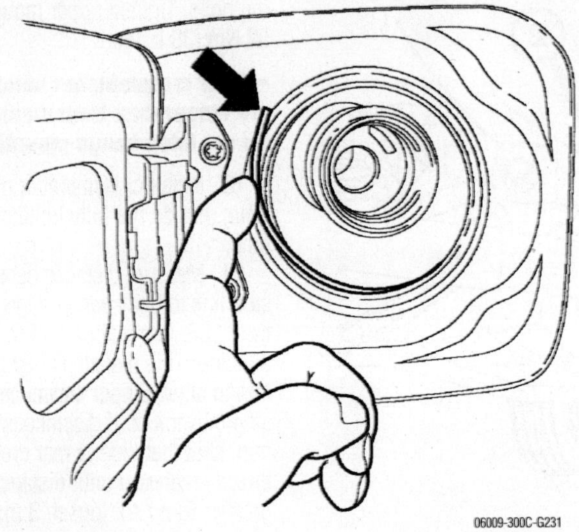

Fuel filler tube retaining wire

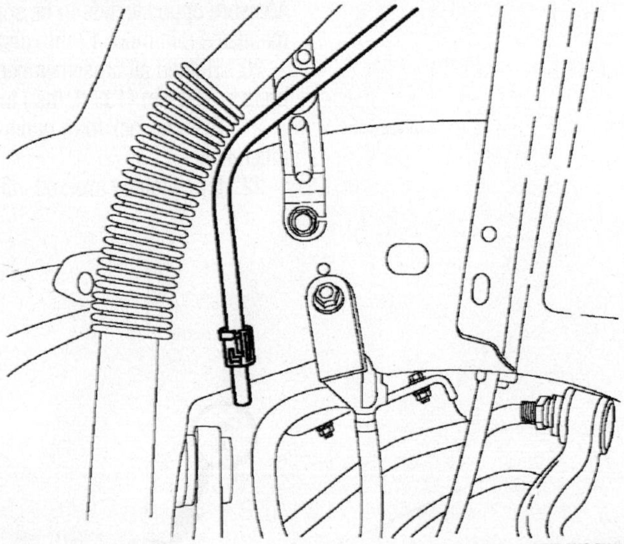

Filler tube vent line

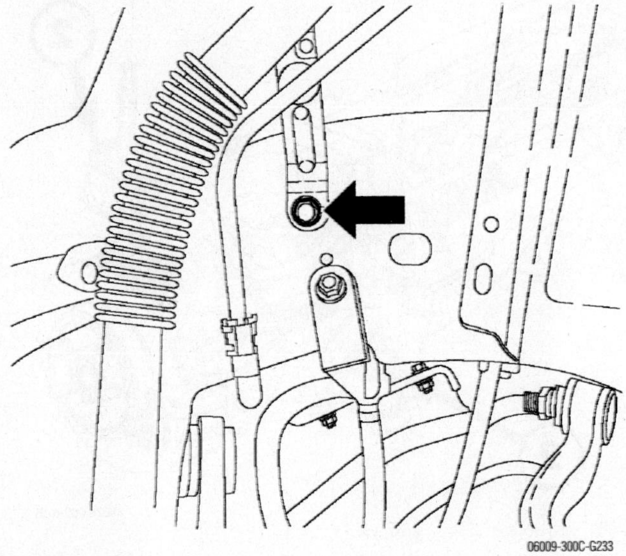

Filler tube mounting bolt

c. Lower exhaust down to rest upon top of jack or stand placed below muffler.

7. Position under-hoist utility jack or transmission jack (3) under center of rear axle differential (1). Raise jack head to contact differential and secure in place. When securing crossmember to jack, be sure not to secure stabilizer bar.

8. Remove shock absorber upper mounting screws.

9. Remove shock absorber lower mounting bolt and nut.

➡When removing crossmember mounting bolts in following step, be sure to not to misplace spacers between crossmember mounts and body.

⁂ WARNING

If equipped with AWD, when removing crossmember mounting bolts it is important NOT to loosen or remove crossmember mounting bolts on opposite side of vehicle. Doing so will require rear wheel alignment following reinstallation to ensure proper thrust angle.

10. Remove both front and rear crossmember mounting bolts on side of vehicle being serviced.

⁂ WARNING

To avoid damaging other components of vehicle, do not lower crossmember any further than necessary to remove shock absorber.

11. Slowly lower jack allowing repair-side of crossmember to drop. Do not lower jack at a fast rate. Lower jack just enough to allow top of shock absorber to clear body flange.

12. Remove shock absorber by tipping top outward and lifting lower end out of pocket in spring link.

To install:

13. Install shock absorber by setting lower end into pocket in spring link, then tipping top inward until aligned with upper mounting holes.

14. Install lower shock mounting bolt and nut. Do not tighten at this time.

15. Make sure spacers on top of crossmember mount bushings on side of repair are in position.

16. Carefully raise jack, guiding coil spring and upper end of shock absorber into mounted positions.

17. Install shock absorber upper mount-

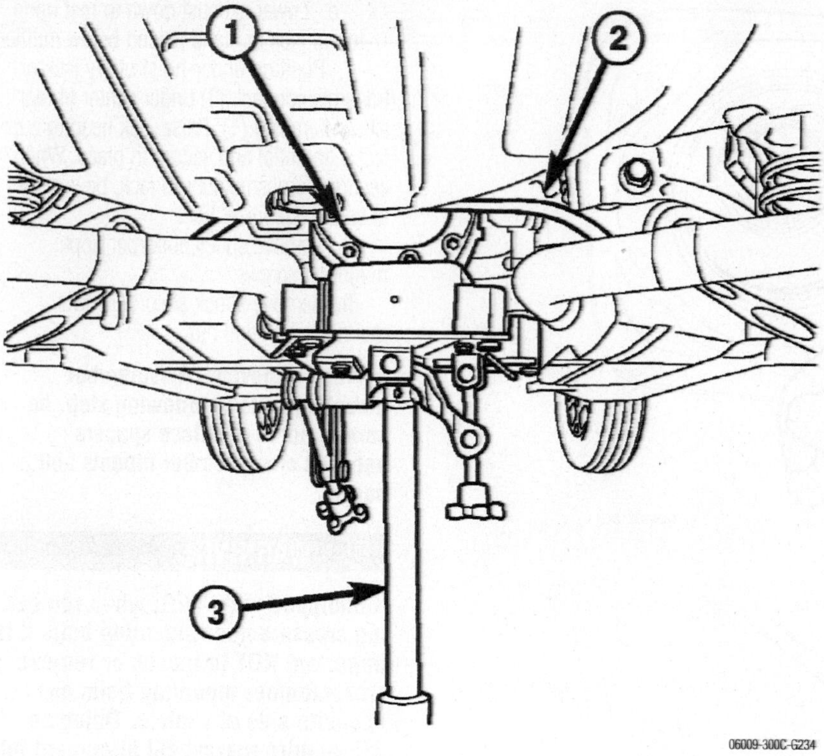

Utility hoist under the exhaust system

ing bolts. Tighten upper mounting bolts to 52 Nm (38 ft. lbs.).

➡**Rear crossmember mounting bolts are longer than front mounting bolts. Do not interchange mounting bolts.**

18. Install crossmember mounting bolts. Snug, but do not fully tighten bolts at this time.

19. Measure distance between from tension link to body weld flange directly in front of it, just outboard of front mount bushing. This distance must be at least 12 mm to allow proper clearance for suspension movement. If distance is less than 12 mm, shift that side of rear crossmember directly rearward until distance is 12 mm or greater. To do so, loosen 3 mounting bolts slightly, leaving one on opposite side shifted, snugged to pivot off of. Shift crossmember rearward and snug loosened bolts. Measure opposite side to be sure it also maintains minimum 12 mm distance.

20. Tighten all crossmember mounting bolts to 180 Nm (133 ft. lbs.) torque.

21. Remove jack from under rear axle differential.

22. If previously lowered, raise rear

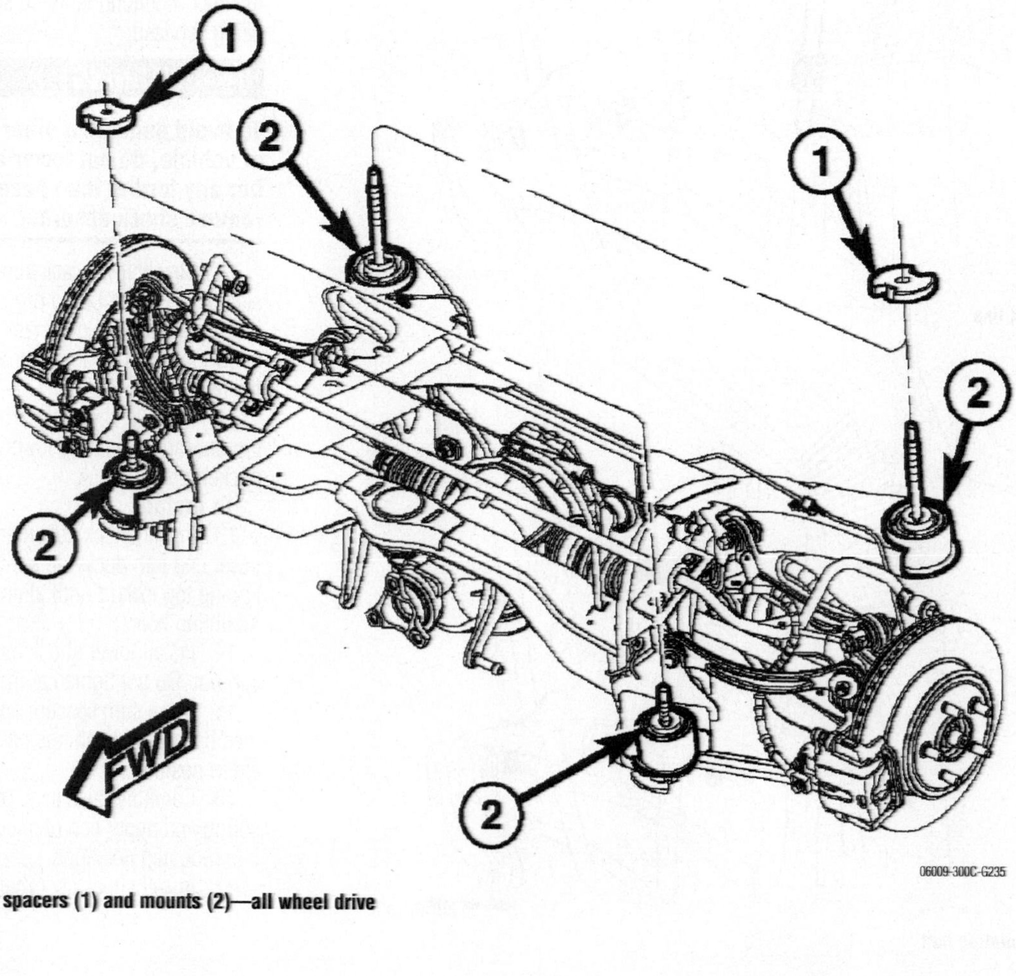

Crossmember spacers (1) and mounts (2)—all wheel drive

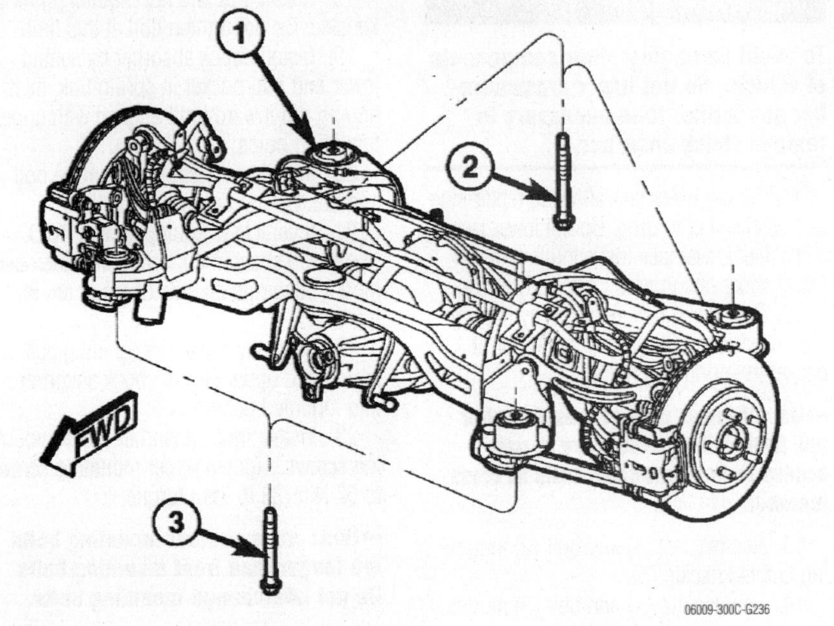

Crossmember (1) mounting bolts (2, 3)—all wheel drive

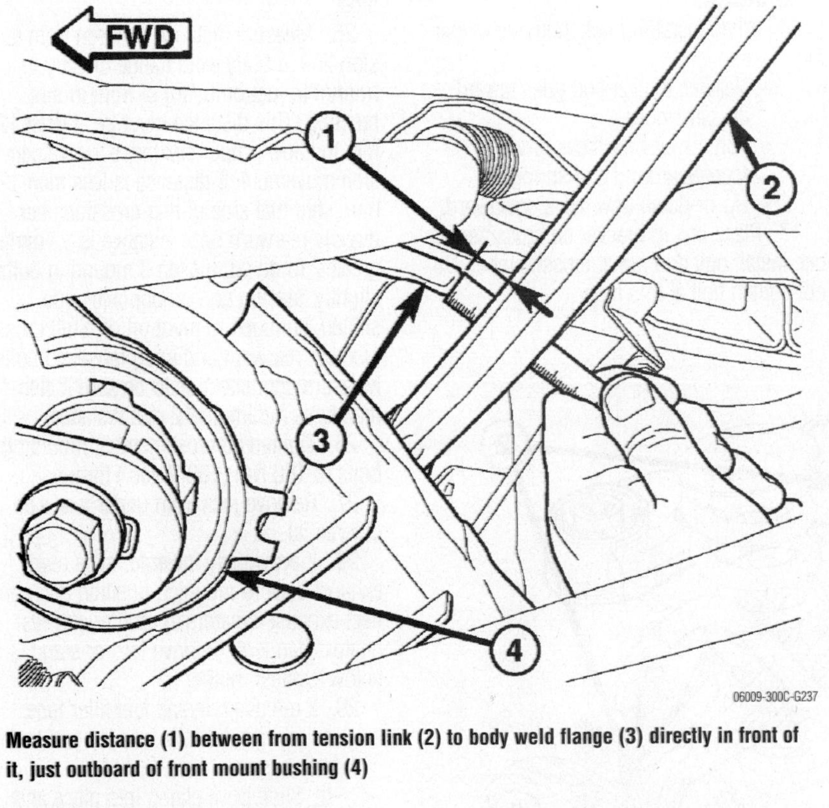

Measure distance (1) between from tension link (2) to body weld flange (3) directly in front of it, just outboard of front mount bushing (4)

exhaust back to mounted position and connect exhaust isolators at muffler and resonators hangers. Remove jack or stand below exhaust muffler.

23. If removed, install fuel filler tube:
    a. Insert filler tube into the fuel tank rubber hose.
    b. Slide hose clamp into place and tighten.
    c. Hose clamp in place and tighten.
    d. Install filler tube mounting bolt.
    e. Install Filler tube vent line.
    f. Install the underbody splash shield.
    g. Install the left inner splash shield.
24. Install tire and wheel assemblies. Tighten wheel mounting nuts to 150 Nm (110 ft. lbs.) torque.
25. Lower vehicle.
26. Position vehicle on alignment rack/drive-on lift. Raise lift as necessary to access lower mounting bolt.
27. Tighten shock absorber lower mounting bolt nut to 72 Nm (53 ft. lbs.) torque.

Camber Link

REMOVAL & INSTALLATION

1. Before servicing the vehicle, refer to the precautions in the beginning of this section.
2. Raise and support vehicle.
3. On both sides of vehicle, remove wheel mounting nuts, then rear tire and wheel assembly.

CAUTION

Before opening fuel system, review all warnings and cautions under Fuel System.

4. If servicing left side shock absorber, remove fuel filler tube:
    a. Disconnect negative battery cable.
    b. Drain fuel from tank.
    c. Fuel filler tube assembly.
    d. Open filler tube door.
    e. Remove retaining wire from inside filler tube rubber.
    f. Start removing rubber from body sheet metal.
    g. Remove retaining wire from inside filler tube rubber.
    h. Start removing rubber from body sheet metal.
    i. Filler tube and rubber removed from body sheet metal.
    j. Remove the left inner splash shield.
    k. Disconnect the filler tube vent line.
    l. Remove the filler tube mounting bolt.
    m. Remove the under body splash shield.
    n. Loosen the filler tube hose clamp. Leave the clamp tight on the hose and fuel tank location.
    o. Move clamp toward the fuel tank.
    p. Remove the filler tube assembly from the vehicle.
5. Position an extra pair of jackstands under and support forward end of engine cradle to help stabilize vehicle during rear suspension removal/installation.
6. Perform following if vehicle is equipped with dual exhaust or are servicing right side on vehicle with single exhaust:
    a. Position under-hoist utility jack or stand several inches below exhaust at muffler.

b. Disconnect exhaust isolators at muffler and resonators hangers.

c. Lower exhaust down to rest upon top of jack or stand placed below muffler.

7. Position under-hoist utility jack or transmission jack under center of rear axle differential. Raise jack head to contact differential and secure in place. When securing crossmember to jack, be sure not to secure stabilizer bar.

8. Remove shock absorber upper mounting screws.

9. Remove shock absorber lower mounting bolt and nut.

➡If equipped with AWD, when removing crossmember mounting bolts in following step, be sure to not to misplace spacers between crossmember mounts and body.

⁂ WARNING

When removing crossmember mounting bolts it is important NOT to loosen or remove crossmember mounting bolts on opposite side of vehicle. Doing so will require rear wheel alignment following reinstallation to ensure proper thrust angle.

10. Remove both front and rear crossmember mounting bolts on repair-side of vehicle.

⁂ WARNING

To avoid damaging other components of vehicle, do not lower crossmember any further than necessary to remove shock absorber.

11. Slowly lower jack allowing repair-side of crossmember to drop. Do not lower jack at a fast rate. Lower jack just enough to allow top of shock absorber to clear body flange.

12. Remove shock absorber by tipping top outward and lifting lower end out of pocket in spring link.

➡Do not lower side of crossmember any further than necessary to gain access to link mounting bolts at crossmember.

13. Remove nut (4) and bolt (2) mounting link to knuckle (3).

14. Remove nut (4) and bolt (2) mounting link to crossmember (1).

15. Remove link.

To install:

16. When installing link, note the following:

- Heavier, thicker end goes toward crossmember.
- Fore-or-aft bow faces forward (curves around coil spring).
- Up-or-down bow faces downward.

17. Place link in bracket on crossmember. Install bolt and nut at crossmember. Do not tighten bolt at this time.

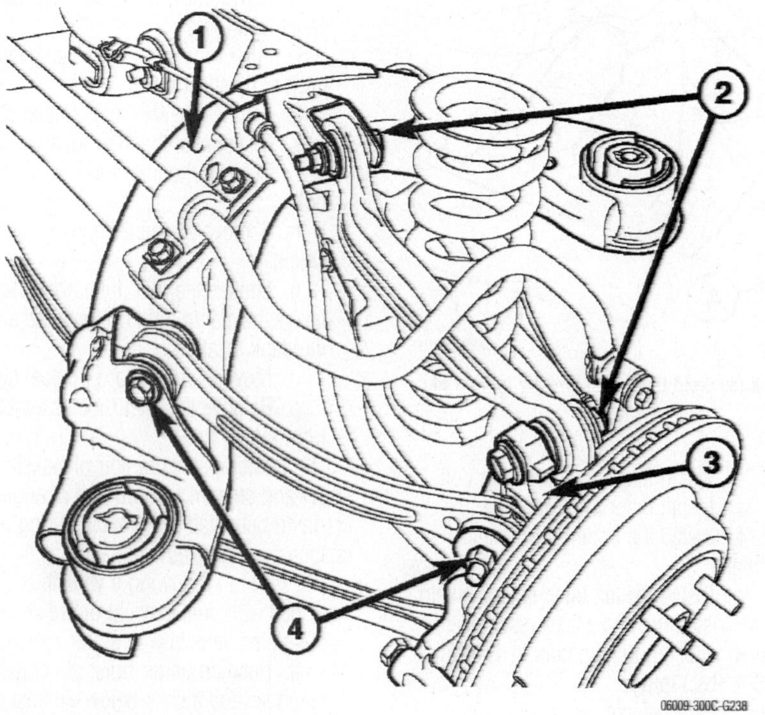

Camber link attachment points

18. Install bolt and nut mounting link to knuckle. Do not tighten bolt at this time.

19. Install shock absorber by setting lower end into pocket in spring link, then tipping top inward until aligned with upper mounting holes.

20. Install lower shock mounting bolt and nut. Do not tighten at this time.

21. If vehicle is equipped with AWD, make sure spacers on top of crossmember mount bushings on side of repair are in position.

22. Carefully raise jack, guiding coil spring and upper end of shock absorber into mounted positions.

23. Install shock absorber upper mounting screws. Tighten upper mounting screws to 52 Nm (38 ft. lbs.) torque.

➡Rear crossmember mounting bolts are longer than front mounting bolts. Do not interchange mounting bolts.

24. Install crossmember mounting bolts. Snug, but do not fully tighten bolts at this time.

25. Measure distance between from tension link to body weld flange directly in front of it, just outboard of front mount bushing. This distance must be at least 12 mm to allow proper clearance for suspension movement. If distance is less than 12 mm, shift that side of rear crossmember directly rearward until distance is 12 mm or greater. To do so, loosen 3 mounting bolts slightly, leaving one on opposite side shifted, snugged to pivot off of. Shift crossmember rearward and snug loosened bolts. Measure opposite side to be sure it also maintains minimum 12 mm distance.

26. Tighten all crossmember mounting bolts to 180 Nm (133 ft. lbs.) torque.

27. Remove jack from under rear axle differential.

28. If previously lowered, raise rear exhaust back to mounted position and connect exhaust isolators at muffler and resonators hangers. Remove jack or stand below exhaust muffler.

29. If removed, install fuel filler tube:

a. Insert filler tube into the fuel tank rubber hose.

b. Slide hose clamp into place and tighten.

c. Hose clamp in place and tighten.

d. Install filler tube mounting bolt.

e. Install Filler tube vent line.

f. Install the underbody splash shield.

g. Install the left inner splash shield.

30. Lower vehicle until front tires contact floor but rear is still suspended. Place jackstands under each rear suspension spring

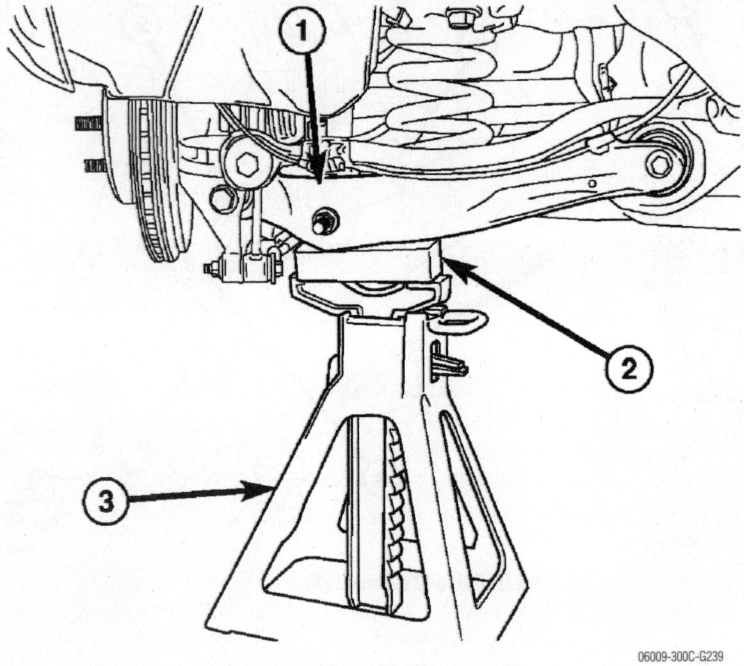

Jackstand (3) and wood block (2) placement under link (1)

link. Place an appropriate wooden block between stand and link to avoid damaging spring link, then lower vehicle until full vehicle weight is supported by suspension.

31. Tighten camber link fasteners to:
    - Bolt at crossmember: 85 Nm (63 ft. lbs.) torque.
    - Bolt nut at knuckle: 98 Nm (72 ft. lbs.) torque.
32. Tighten shock absorber lower mounting bolt nut to 72 Nm (53 ft. lbs.) torque.
33. Raise vehicle and remove jackstands.
34. Install tire and wheel assemblies. Tighten wheel mounting nuts to 150 Nm (110 ft. lbs.) torque.
35. Lower vehicle.
36. Perform wheel alignment.

Compression Link

REMOVAL & INSTALLATION

1. Before servicing the vehicle, refer to the precautions in the beginning of this section.
2. Raise and support vehicle.
3. Remove wheel mounting nuts, then rear tire and wheel assembly.
4. Remove bolt and nut mounting link at knuckle.
5. Remove bolt and nut mounting link at crossmember.
6. Remove link.

To install:

➡Although the compression link is different end-to-end, there is no top and bottom.

7. Position link and install bolt and nut mounting link at crossmember. Do not tighten bolt at this time.
8. Install bolt and nut mounting link at knuckle. Do not tighten bolt at this time.
9. Install tire and wheel assembly. Tighten wheel mounting nuts to 150 Nm (110 ft. lbs.).
10. Lower vehicle.
11. Position vehicle on alignment rack/drive-on lift. Raise vehicle as necessary to access link fasteners.
12. Tighten compression link fasteners to:
    - Bolt at crossmember: 85 Nm (63 ft. lbs.) torque.
    - Bolt at knuckle: 81 Nm (60 ft. lbs.) torque.
13. Perform wheel alignment.

Spring Link

REMOVAL & INSTALLATION

1. Before servicing the vehicle, refer to the precautions in the beginning of this section.
2. Raise and support vehicle.
3. Remove rear spring.
4. Remove spring link-to-knuckle nut and bolt.

⁂ WARNING

It important to use Guide, Special Tool 9361-2, when tapping sleeve in knuckle to help keep Tap, Special Tool 9361-1, straight during use or damage to Tap may occur.

5. Place Guide 9361-2 against sleeve in knuckle to keep Tap 9361-1 straight. Using Tap with an appropriate handle, cut

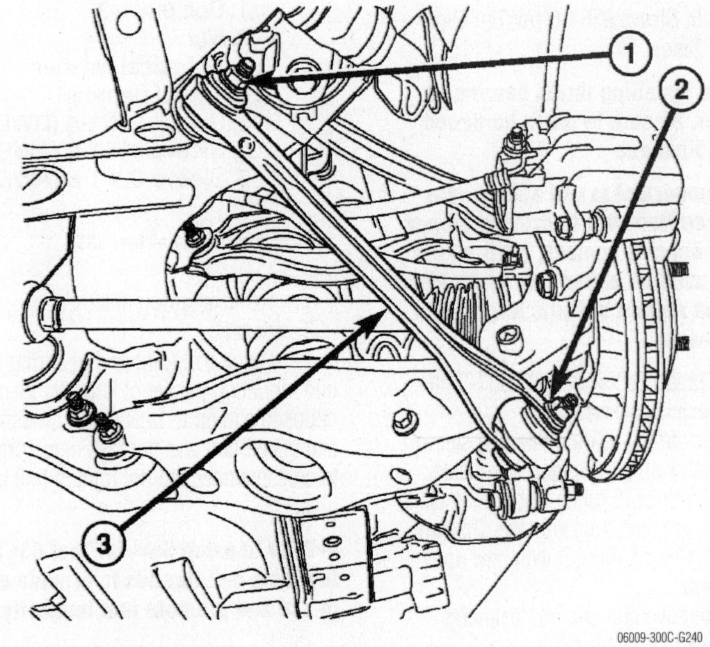

Compression link (3) and fasteners (1, 2)

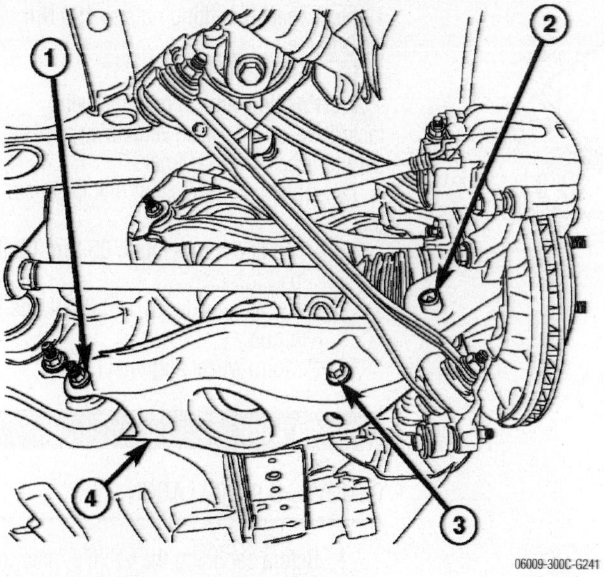

Spring link (4) and fasteners (1, 2, 3)

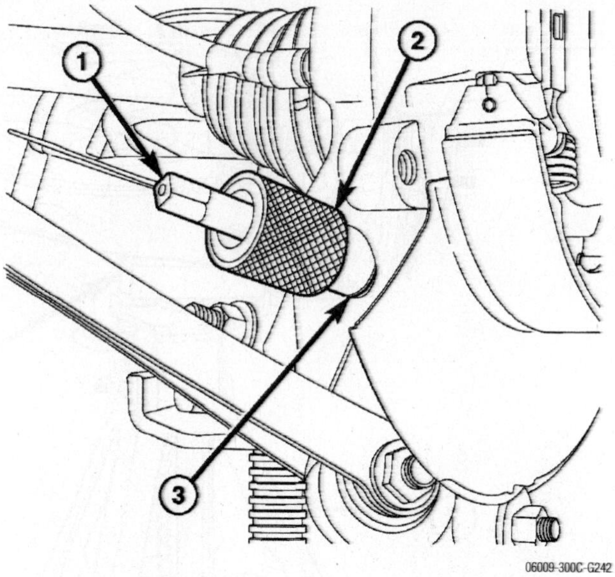

Tap (1), Guide (2), Sleeve (3)

threads approximately halfway through bushing (or about six complete threads). It is important to back tap out, clean out burrs and lubricate Tap often during process.

➡Prior to using Special Tool 9361, lubricate bolt threads to provide ease of use and promote tool longevity.

6. Assemble Remover, Special Tool 9361, as shown.
- 1. Bolt 9361-3
- 2. Nut
- 3. Spherical Washer
- 4. Thrust Bearing
- 5. Sleeve 9361-4 (RWD)
- 5. Sleeve 9361-5 (AWD—Left Side)
- 5. Sleeve 9361-6 (AWD—Right Side)

➡When installing thrust bearing on remover, be sure to place hardened side against nut.

➡It is important to use appropriate Sleeve on Remover to provide proper Tool-to-Knuckle contact. RWD sleeve can be used on either side while AWD knuckles require specific left or right side Sleeves.

7. Thread Remover Bolt 9361-3 into tapped knuckle sleeve.

8. Rotate nut down, matching sleeve angled end with angled face of knuckle.

9. Continue to rotate Nut until knuckle sleeve is removed from knuckle. Discard knuckle sleeve; replace it with new upon installation.

10. Remove bolt and nut fastening spring link to crossmember.

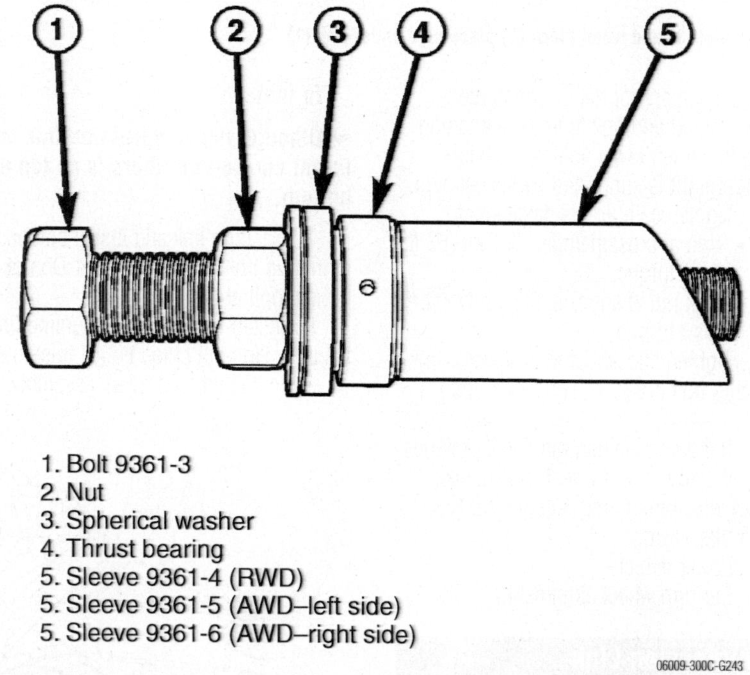

Correct assembly of tool 9361

11. Remove spring link.

To install:

12. Guide ball joint end of spring link into mounting pocket of knuckle, then swing opposite end up to bushing in crossmember and install bolt and nut fastening spring link to crossmember. Do not tighten bolt at this time.

➡Prior to using Special Tool 9361, lubricate bolt threads to provide ease of use and promote tool longevity.

13. Place new knuckle sleeve onto Installer Bolt 9361-7, and slide it up to bolt's head.

14. Slide Bolt 9361-7 with sleeve through knuckle and spring link ball joint starting from knuckle forward end.

15. Install thrust bearing and nut on end of bolt.

16. While holding bolt head stationary, rotate nut (using hand tools) installing sleeve in knuckle. Install sleeve until nut stops turning. Do not overtighten nut.

17. Remove special tool.

18. Install spring link-to-knuckle bolt

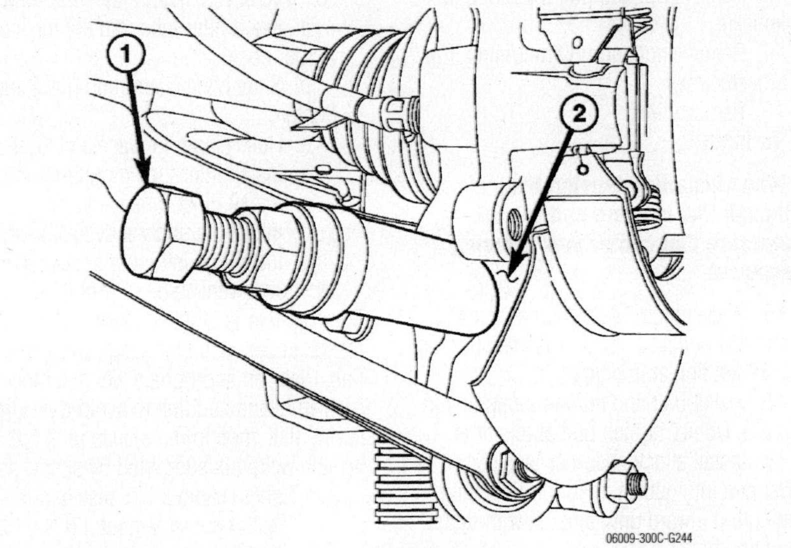

Tool 9361 installed

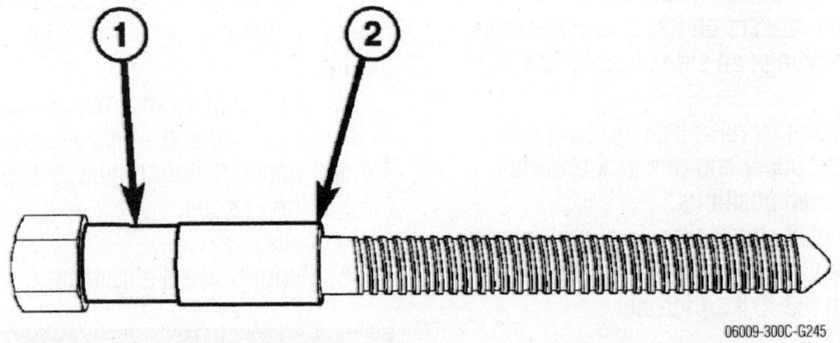

Place new knuckle sleeve (2) onto Installer Bolt 9361-7 (1), and slide it up to bolt's head

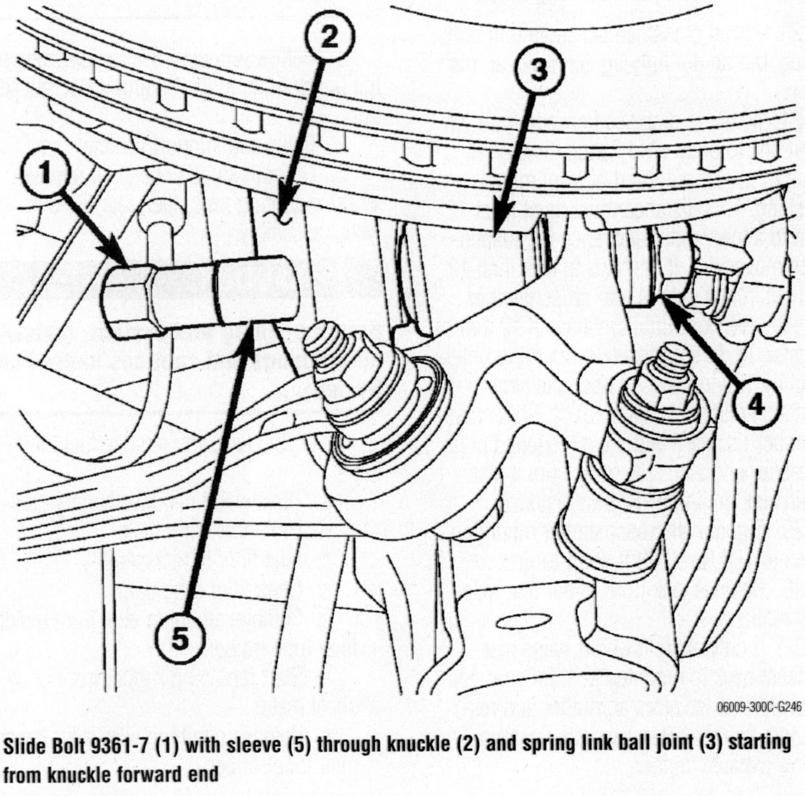

Slide Bolt 9361-7 (1) with sleeve (5) through knuckle (2) and spring link ball joint (3) starting from knuckle forward end

front-to-rear through knuckle and link, then install nut. While holding bolt head stationary, tighten nut to 138 Nm (102 ft. lbs.) torque.

19. Install rear spring.
20. Lower vehicle.
21. Position vehicle on alignment rack/drive-on lift. Raise vehicle as necessary to access mounting bolt.
22. Tighten spring link bolt at crossmember to 108 Nm (80 ft. lbs.) torque.
23. Perform wheel alignment.

Tension Link

REMOVAL AND INSTALLATION

1. Before servicing the vehicle, refer to the precautions in the beginning of this section.
2. Raise and support vehicle.
3. On both sides of vehicle, remove wheel mounting nuts, then rear tire and wheel assembly.

**** CAUTION**

Before opening fuel system, review all warnings and cautions under Fuel System.

4. If servicing left side shock absorber, remove fuel filler tube:
   a. Disconnect negative battery cable.
   b. Drain fuel from tank.
   c. Fuel filler tube assembly.
   d. Open filler tube door.
   e. Remove retaining wire from inside filler tube rubber.
   f. Start removing rubber from body sheet metal.
   g. Remove retaining wire from inside filler tube rubber.
   h. Start removing rubber from body sheet metal.
   i. Filler tube and rubber removed from body sheet metal.
   j. Remove the left inner splash shield.
   k. Disconnect the filler tube vent line.
   l. Remove the filler tube mounting bolt.
   m. Remove the under body splash shield.
   n. Loosen the filler tube hose clamp. Leave the clamp tight on the hose and fuel tank location.
   o. Move clamp toward the fuel tank.
   p. Remove the filler tube assembly from the vehicle.
5. Position an extra pair of jackstands under and support forward end of engine

cradle to help stabilize vehicle during rear suspension removal/installation.

6. Perform following if vehicle is equipped with dual exhaust or are servicing right side on vehicle with single exhaust:
   a. Position under-hoist utility jack or stand several inches below exhaust at muffler.
   b. Disconnect exhaust isolators at muffler and resonators hangers.
   c. Lower exhaust down to rest upon top of jack or stand placed below muffler.
7. Position under-hoist utility jack or transmission jack under center of rear axle differential. Raise jack head to contact differential and secure in place. When securing crossmember to jack, be sure not to secure stabilizer bar.
8. Remove shock absorber upper mounting screws.
9. Remove shock absorber lower mounting bolt and nut.

➡If equipped with AWD, when removing crossmember mounting bolts in following step, be sure to not to misplace spacers between crossmember mounts and body.

⁂ WARNING

When removing crossmember mounting bolts it is important NOT to loosen or remove crossmember mounting bolts on opposite side of vehicle. Doing so will require rear wheel alignment following reinstallation to ensure proper thrust angle.

10. Remove both front and rear crossmember mounting bolts on repair-side of vehicle.

⁂ WARNING

To avoid damaging other components of vehicle, do not lower crossmember any further than necessary to remove shock absorber.

11. Slowly lower jack allowing repair-side of crossmember to drop. Do not lower jack at a fast rate. Lower jack just enough to allow top of shock absorber to clear body flange.
12. Remove shock absorber by tipping top outward and lifting lower end out of pocket in spring link.

➡Do not lower side of crossmember any further than necessary to gain access to link mounting bolts at crossmember.

13. Remove nut and bolt mounting link to knuckle.
14. Remove nut and bolt mounting link to crossmember.
15. Remove link.

To install:

➡When installing tension link, although link is same end-to-end, make sure that center bow is facing downward.

16. Place link in bracket on crossmember. Install bolt and nut at crossmember. Do not tighten bolt at this time.
17. Install bolt and nut mounting link to knuckle. Do not tighten bolt at this time.
18. Install shock absorber by setting lower end into pocket in spring link, then tipping top inward until aligned with upper mounting holes.
19. Install lower shock mounting bolt and nut. Do not tighten at this time.
20. If vehicle is equipped with AWD, make sure spacers on top of crossmember mount bushings on side of repair are in position.
21. Carefully raise jack, guiding coil spring and upper end of shock absorber into mounted positions.
22. Install shock absorber upper mounting screws. Tighten upper mounting screws to 52 Nm (38 ft. lbs.) torque.

➡Rear crossmember mounting bolts are longer than front mounting bolts. Do not interchange mounting bolts.

23. Install crossmember mounting bolts. Snug, but do not fully tighten bolts at this time.
24. Measure distance between from tension link to body weld flange directly in front of it, just outboard of front mount bushing. This distance must be at least 12 mm to allow proper clearance for suspension movement. If distance is less than 12 mm, shift that side of rear crossmember directly rearward until distance is 12 mm or greater. To do so, loosen 3 mounting bolts slightly, leaving one on opposite side shifted, snugged to pivot off of. Shift crossmember rearward and snug loosened bolts. Measure opposite side to be sure it also maintains minimum 12 mm distance.
25. Tighten all crossmember mounting bolts to 180 Nm (133 ft. lbs.) torque.
26. Remove jack from under rear axle differential.
27. If previously lowered, raise rear exhaust back to mounted position and connect exhaust isolators at muffler and resonators hangers. Remove jack or stand below exhaust muffler.
28. If removed, install fuel filler tube:
   a. Insert filler tube into the fuel tank rubber hose.
   b. Slide hose clamp into place and tighten.
   c. Hose clamp in place and tighten.
   d. Install filler tube mounting bolt.
   e. Install Filler tube vent line.
   f. Install the underbody splash shield.
   g. Install the left inner splash shield.
29. Lower vehicle until front tires contact floor but rear is still suspended. Place jackstands under each rear suspension spring link. Place an appropriate wooden block between stand and link to avoid damaging spring link, then lower vehicle until full vehicle weight is supported by suspension.
30. Tighten camber link fasteners to:
   - Bolt at crossmember: 85 Nm (63 ft. lbs.) torque.
   - Bolt nut at knuckle: 98 Nm (72 ft. lbs.) torque.
31. Tighten shock absorber lower mounting bolt nut to 72 Nm (53 ft. lbs.) torque.
32. Raise vehicle and remove jackstands.
33. Install tire and wheel assemblies. Tighten wheel mounting nuts to 150 Nm (110 ft. lbs.) torque.
34. Lower vehicle.
35. Perform wheel alignment.

Coil Spring

REMOVAL & INSTALLATION

1. Before servicing the vehicle, refer to the precautions in the beginning of this section.
2. Raise and support vehicle.
3. On both sides of vehicle, remove wheel mounting nuts, then rear tire and wheel assembly.

⁂ CAUTION

Before opening fuel system, review all warnings and cautions under Fuel System.

4. On the left side, remove fuel filler tube:
   a. Disconnect negative battery cable.
   b. Drain fuel from tank.
   c. Fuel filler tube assembly.
   d. Open filler tube door.
   e. Remove retaining wire from inside filler tube rubber.
   f. Start removing rubber from body sheet metal.
   g. Remove retaining wire from inside filler tube rubber.

h. Start removing rubber from body sheet metal.

i. Filler tube and rubber removed from body sheet metal.

j. Remove the left inner splash shield.

k. Disconnect the filler tube vent line.

l. Remove the filler tube mounting bolt.

m. Remove the under body splash shield.

n. Loosen the filler tube hose clamp. Leave the clamp tight on the hose and fuel tank location.

o. Move clamp toward the fuel tank.

p. Remove the filler tube assembly from the vehicle.

5. Position an extra pair of jackstands under and support forward end of engine cradle to help stabilize vehicle during rear suspension removal/installation.

6. Perform following if vehicle is equipped with dual exhaust or are servicing right side on vehicle with single exhaust:

a. Position under-hoist utility jack or stand several inches below exhaust at muffler.

b. Disconnect exhaust isolators at muffler and resonators hangers.

c. Lower exhaust down to rest upon top of jack or stand placed below muffler.

7. Position under-hoist utility jack or transmission jack (3) under center of rear axle differential (1). Raise jack head to contact differential and secure in place. When securing crossmember to jack, be sure not to secure stabilizer bar.

8. Remove shock absorber upper mounting screws.

9. Remove shock absorber lower mounting bolt and nut.

➡When removing crossmember mounting bolts in following step, be sure to not to misplace spacers between crossmember mounts and body.

⁂ WARNING

If equipped with AWD, when removing crossmember mounting bolts it is important NOT to loosen or remove crossmember mounting bolts on opposite side of vehicle. Doing so will require rear wheel alignment following reinstallation to ensure proper thrust angle.

10. Remove both front and rear crossmember mounting bolts on side of vehicle being serviced.

⁂ WARNING

To avoid damaging other components of vehicle, do not lower crossmember any further than necessary to remove shock absorber.

11. Slowly lower jack allowing repair-side of crossmember to drop. Do not lower jack at a fast rate. Lower jack just enough to allow top of shock absorber to clear body flange.

12. Remove shock absorber by tipping top outward and lifting lower end out of pocket in spring link.

13. Disconnect brake hose at bracket mounted to body to allow to avoid overextending hose, damaging it, during following step.

14. Slowly lower jack until crossmember is low enough to remove coil spring. Do not lower jack any further than necessary to remove spring.

15. Remove coil spring and isolators (1, 2 and 5).

To install:

16. Install upper and lower isolators on coil spring.

➡Before installing coil spring, make sure isolators are completely installed on ends of spring.

17. Install coil spring with isolators into spring pocket of spring link fitting lower isolator to shape of pocket, then align top of spring with body mount.

18. Install shock absorber by setting lower end into pocket in spring link, then tipping top inward until aligned with upper mounting holes.

19. Install lower shock mounting bolt and nut. Do not tighten at this time.

20. Make sure spacers on top of crossmember mount bushings on side of repair are in position.

21. Carefully raise jack, guiding coil spring and upper end of shock absorber into mounted positions.

22. Install shock absorber upper mounting bolts. Tighten upper mounting bolts to 52 Nm (38 ft. lbs.).

➡Rear crossmember mounting bolts are longer than front mounting bolts. Do not interchange mounting bolts.

23. Install crossmember mounting bolts. Snug, but do not fully tighten bolts at this time.

24. Measure distance between from tension link to body weld flange directly in front of it, just outboard of front mount bushing. This distance must be at least 12 mm to allow proper clearance for suspension movement. If distance is less than 12 mm, shift that side of rear crossmember directly rearward until distance is 12 mm or greater. To do so, loosen 3 mounting bolts slightly, leaving one on opposite side shifted, snugged to pivot off of. Shift crossmember rearward and snug loosened bolts. Measure opposite side to be sure it also maintains minimum 12 mm distance.

25. Tighten all crossmember mounting bolts to 180 Nm (133 ft. lbs.) torque.

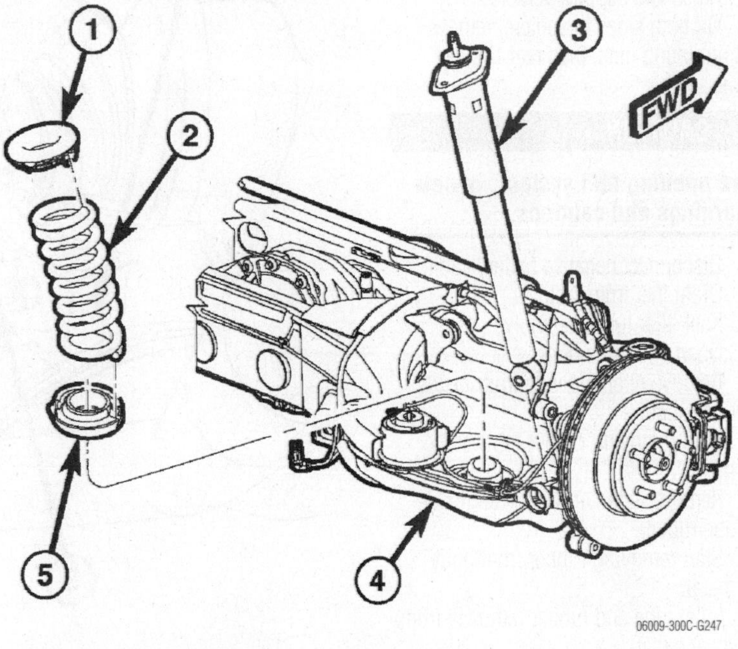

Rear coil spring removal/installation

26. Remove jack from under rear axle differential.

27. If previously lowered, raise rear exhaust back to mounted position and connect exhaust isolators at muffler and resonators hangers. Remove jack or stand below exhaust muffler.

28. If removed, install fuel filler tube:

a. Insert filler tube into the fuel tank rubber hose.

b. Slide hose clamp into place and tighten.

c. Hose clamp in place and tighten.

d. Install filler tube mounting bolt.

e. Install filler tube vent line.

f. Install the underbody splash shield.

g. Install the left inner splash shield.

29. Install tire and wheel assemblies. Tighten wheel mounting nuts to 150 Nm (110 ft. lbs.) torque.

30. Lower vehicle.

31. Position vehicle on alignment rack/drive-on lift. Raise lift as necessary to access lower mounting bolt.

32. Tighten shock absorber lower mounting bolt nut to 72 Nm (53 ft. lbs.) torque.

Toe Link

REMOVAL & INSTALLATION

Left Side

1. Before servicing the vehicle, refer to the precautions in the beginning of this section.

2. Raise and support vehicle.

3. On both sides of vehicle, remove wheel mounting nuts, then rear tire and wheel assembly.

CAUTION

Before opening fuel system, review all warnings and cautions.

4. Disconnect negative battery cable.

5. Drain fuel from tank.

6. Fuel filler tube assembly.

7. Open filler tube door.

8. Remove retaining wire from inside filler tube rubber.

9. Start removing rubber from body sheet metal.

10. Remove retaining wire from inside filler tube rubber.

11. Start removing rubber from body sheet metal.

12. Filler tube and rubber removed from body sheet metal.

13. Remove the left inner splash shield.

14. Disconnect the filler tube vent line.

15. Remove the filler tube mounting bolt.

16. Remove the under body splash shield.

17. Loosen the filler tube hose clamp. Leave the clamp tight on the hose and fuel tank location.

18. Move clamp toward the fuel tank.

19. Remove the filler tube assembly from the vehicle.

20. Position an extra pair of jackstands under and support forward end of engine cradle to help stabilize vehicle during rear suspension removal/installation.

21. Perform following if vehicle is equipped with dual exhaust:

a. Position under-hoist utility jack or stand several inches below exhaust at muffler.

b. Disconnect exhaust isolators at muffler and resonators hangers.

c. Lower exhaust down to rest upon top of jack or stand placed below muffler.

22. Position under-hoist utility jack or transmission jack under center of rear axle differential. Raise jack head to contact differential and secure in place.

23. Remove shock absorber lower mounting bolt and nut.

WARNING

When removing crossmember mounting bolts it is important NOT to loosen or remove crossmember mounting bolts on opposite side of vehicle. Doing so will require rear wheel alignment following reinstallation to ensure proper thrust angle.

24. Remove both front and rear crossmember mounting bolts on repair-side of vehicle. If equipped with AWD, be sure to not to misplace spacers between crossmember mounts and body.

WARNING

To avoid damaging other components of vehicle, do not lower crossmember any further than necessary to remove shock absorber.

25. Slowly lower jack allowing repair-side of crossmember to drop. Do not lower jack at a fast rate. Lower jack just enough to allow toe link mounting bolt at crossmember.

26. If equipped, remove wheel speed sensor cable from toe link.

27. While holding toe adjustment cam bolt from rotating, remove nut (3) securing toe link at crossmember (1).

28. Slide cam bolt rearward out of crossmember and link.

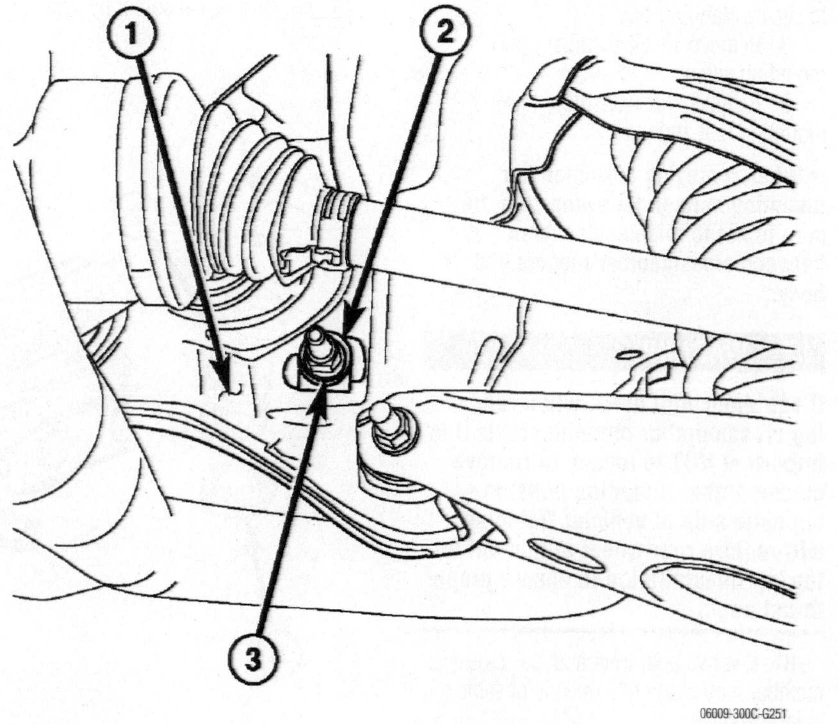

Toe adjustment cam bolt

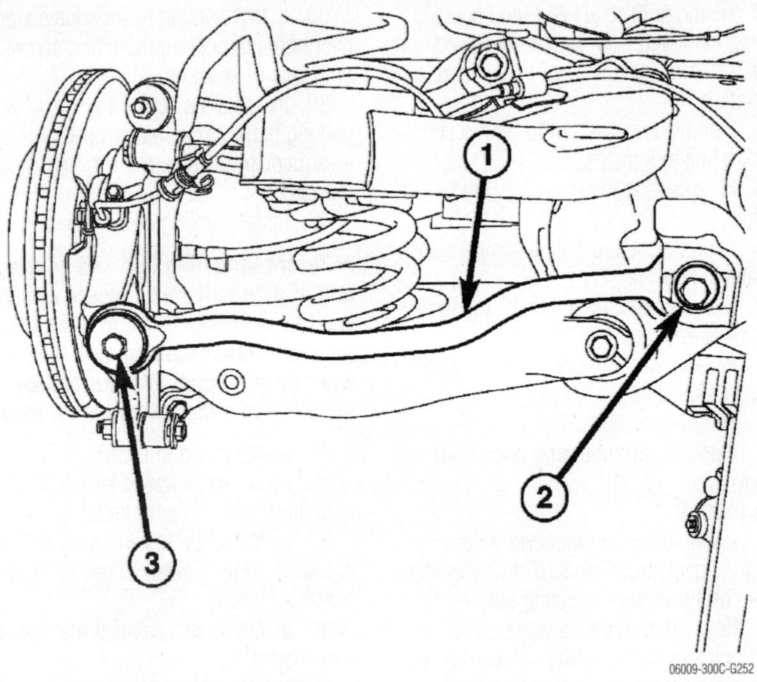

Cam bolt (2), toe link (1) and mounting bolt (3)

29. Remove mounting bolt and nut at knuckle.

30. Remove toe link.

To install:

31. Slide crossmember end of toe link into box bracket on crossmember. Slide cam bolt through bracket and link from rear.

32. Install bolt and nut securing link to knuckle. Do not tighten bolt at this time.

33. While holding toe adjustment cam bolt from rotating (cam facing upward), Install cam washer and nut securing toe link at crossmember. Do not tighten nut at this time.

34. If equipped, attach wheel speed sensor cable to toe link.

35. Carefully raise jack, guiding coil spring and lower end of shock absorber into mounted positions.

36. When lower shock mounting bolt holes line up install bolt and nut. Do not tighten at this time.

➡Rear crossmember mounting bolts are longer than front mounting bolts. Do not interchange mounting bolts.

37. Continue to raise crossmember in not already in mounted position, then install crossmember mounting bolts. Snug, but do not fully tighten bolts at this time.

38. Measure distance between from tension link to body weld flange directly in front of it, just outboard of front mount bushing. This distance must be at least 12 mm to allow proper clearance for suspension movement. If distance is less than 12 mm, shift that side of rear crossmember directly rearward until distance is 12 mm or greater. To do so, loosen 3 mounting bolts slightly, leaving one on opposite side of shift snugged to pivot off of. Shift crossmember rearward and snug loosened bolts. Measure opposite side to be sure it also maintains minimum 12 mm distance.

39. Tighten all crossmember mounting bolts to 180 Nm (133 ft. lbs.) torque.

40. Remove jack from under rear axle differential.

41. If previously lowered, raise rear exhaust back to mounted position and connect exhaust isolators at muffler and resonators hangers. Remove jack or stand below exhaust muffler.

42. Insert filler tube into the fuel tank rubber hose.

43. Slide hose clamp into place and tighten.

44. Hose clamp in place and tighten.

45. Install filler tube mounting bolt.

46. Install filler tube vent line.

47. Install the underbody splash shield.

48. Install the left inner splash shield.

49. Install tire and wheel assemblies. Tighten wheel mounting nuts to 150 Nm (110 ft. lbs.) torque.

50. Lower vehicle.

51. Position vehicle on alignment rack/drive-on lift. Raise vehicle as necessary to access mounting bolts.

52. Tighten shock absorber lower mounting bolt nut to 72 Nm (53 ft. lbs.) torque.

53. Tighten toe link fasteners to:
- Nut at crossmember: 108 Nm (80 ft. lbs.) torque (This nut may be tightened after rear wheel alignment toe is set. Do not tighten from bolt head end.).
- Bolt at knuckle: 81 Nm (60 ft. lbs.).

54. Perform wheel alignment.

Right Side

1. Before servicing the vehicle, refer to the precautions in the beginning of this section.

2. Raise and support vehicle.

3. Remove wheel mounting nuts, then rear tire and wheel assembly.

4. If equipped, remove wheel speed sensor cable from toe link.

5. While holding toe adjustment cam bolt from rotating, remove nut securing toe link at crossmember.

6. Slide cam bolt rearward out of crossmember and link.

7. Remove mounting bolt and nut at knuckle.

8. Remove toe link.

To install:

9. Slide crossmember end of toe link into box bracket on crossmember. Slide cam bolt through bracket and link from rear of vehicle.

10. Install bolt and nut securing link to knuckle. Do not tighten bolt at this time.

11. While holding toe adjustment cam bolt from rotating (cam facing upward), Install cam washer and nut securing toe link at crossmember. Do not tighten nut at this time.

12. If equipped, attach wheel speed sensor cable to toe link.

13. Raise rear exhaust back to mounted position and connect exhaust isolators at muffler and resonators hangers. Remove jack or stand below exhaust muffler.

14. Install tire and wheel assembly. Tighten wheel mounting nuts to 150 Nm (110 ft. lbs.) torque.

15. Lower vehicle.

16. Position vehicle on alignment rack/drive-on lift. Raise vehicle as necessary to access mounting bolts.

17. Tighten toe link fasteners to:
- Nut at crossmember: 108 Nm (80 ft. lbs.). (This nut may be tightened after rear wheel alignment toe is set. Do not tighten from bolt head end.).
- Bolt at knuckle: 81 Nm (60 ft. lbs.) torque.

18. Perform wheel alignment.

Knuckle

REMOVAL & INSTALLATION

1. Before servicing the vehicle, refer to the precautions in the beginning of this section.
2. Raise and support vehicle.
3. Unclip wheel speed sensor cable at rear brake rotor shield.
4. Remove screw fastening sensor head to rear knuckle.
5. Remove wheel speed sensor head from knuckle.
6. Remove rear hub and bearing.
7. Remove parking brake shoes.
8. If not removed, remove parking brake shoe actuator lever from end of cable.
9. Remove shoe support from knuckle.
10. Remove parking brake cable screw at knuckle and pull cable out of knuckle.
11. Position under-hoist utility jack or jackstand under spring link. Raise jack head to contact spring link at shock mount secure in place.
12. Remove spring link-to-knuckle nut and bolt.

⁂ WARNING

It important to use Guide, Special Tool 9361-2, when tapping sleeve in knuckle to help keep Tap, Special Tool 9361-1, straight during use or damage to Tap may occur.

13. Place Guide 9361-2 against sleeve in knuckle to keep Tap 9361-1 straight. Using Tap with an appropriate handle, cut threads approximately halfway through bushing (or about six complete threads). It is important to back tap out, clean out burrs and lubricate Tap often during process.

➡Prior to using Special Tool 9361, lubricate Bolt (1) threads to provide ease of use and promote tool longevity.

➡When installing thrust bearing on Remover, be sure to place hardened side against nut.

➡It is important to use appropriate Sleeve on Remover to provide proper Tool-to-Knuckle contact. RWD sleeve can be used on either side while AWD knuckles require specific left or right side Sleeves.

14. Thread Remover Bolt 9361-3 into tapped knuckle sleeve.
15. Rotate Nut down, matching Sleeve angled end with angled face of knuckle.
16. Continue to rotate Nut until knuckle sleeve is removed from knuckle. Discard knuckle sleeve; replace it with new upon installation.
17. Remove bolt and nut fastening compression link to knuckle.
18. Remove bolt fastening toe link to knuckle.
19. Remove nut and bolt fastening stabilizer link to knuckle.
20. Remove nut and bolt fastening tension link to knuckle.
21. Remove nut and bolt fastening camber link to knuckle.
22. Remove knuckle.
23. Remove hub mounting bolts from knuckle.

To install:

24. Install four hub mounting bolts through knuckle from inboard side allowing ends to protrude from opposite side.
25. Position knuckle on vehicle and install bolt and nut fastening camber link to knuckle. Do not tighten bolt at this time.
26. Install bolt and nut fastening tension link to knuckle. Do not tighten bolt at this time.
27. Install bolt and nut fastening stabilizer link to knuckle. Do not tighten bolt at this time.
28. Install bolt fastening toe link to knuckle. Do not tighten bolt at this time.
29. Install bolt and nut fastening compression link to knuckle. Do not tighten bolt at this time.

➡Prior to using Special Tool 9361, lubricate bolt threads to provide ease of use and promote tool longevity.

30. Place NEW knuckle sleeve onto Installer Bolt 9361-7, and slide it up to Bolt's head.
31. Slide Bolt 9361-7 with sleeve through knuckle and spring link ball joint starting from knuckle forward end.
32. Install thrust bearing and nut on end of Bolt.
33. While holding Bolt head stationary, rotate Nut (using hand tools) installing sleeve in knuckle. Install sleeve until Nut stops turning. Do not overtighten Nut.
34. Remove special tool.
35. Install spring link-to-knuckle bolt front-to-rear through knuckle and link, then install nut. While holding bolt head stationary, tighten nut to 138 Nm (102 ft. lbs.) torque.
36. Remove under-hoist utility jack or jackstand from under spring link.
37. Insert end of cable through rear knuckle and install mounting screw. Tighten screw to 8 Nm (71 inch lbs.).
38. Install parking brake shoe support over hub and bearing mounting screws and onto face of knuckle.
39. Install shoe actuator lever on end of parking brake cable. Make sure actuator lever is positioned with word "UP" facing outward.
40. Install parking brake shoes.

➡Before installing hub and bearing on end of axle halfshaft, ensure isolation washer is present on end of halfshaft. Inspect washer making sure it is not worn or damaged. Washer can be installed in either direction on shaft.

41. Install hub and bearing.
42. Insert wheel speed sensor head into mounting hole in rear of knuckle.
43. Install screw fastening sensor head to rear knuckle. Tighten Screw to 11 Nm (97 inch lbs.) torque.
44. Install sensor cable at rear brake rotor shield.
45. Lower vehicle.
46. Adjust parking brake shoes as necessary.
47. Position vehicle on alignment rack/drive-on hoist. Raise vehicle as necessary to access mounting bolts.
48. Tighten fasteners at knuckle (vehicle at curb height) as follows:
    - Camber Link: 98 Nm (72 ft. lbs.).
    - Compression Link: 81 Nm (60 ft. lbs.).
    - Stabilizer Link: 61 Nm (45 ft. lbs.).
    - Tension Link: 98 Nm (72 ft. lbs.).
    - Toe Link: 81 Nm (60 ft. lbs.).
49. Perform wheel alignment.

Stabilizer Bar

REMOVAL & INSTALLATION

1. Before servicing the vehicle, refer to the precautions in the beginning of this section.
2. Disconnect and isolate battery negative cable from battery post.
3. Raise and support vehicle.
4. On each side of vehicle rear, remove wheel mounting nuts, then tire and wheel assembly.
5. Remove rear exhaust system.
6. Apply alignment index marks to the driveshaft rubber coupler and axle flange.
7. Remove three driveshaft coupler-to-axle flange bolts and nuts.
8. Support driveshaft using a bungee cord. Attach ends of cord to fuel tank straps as shown.

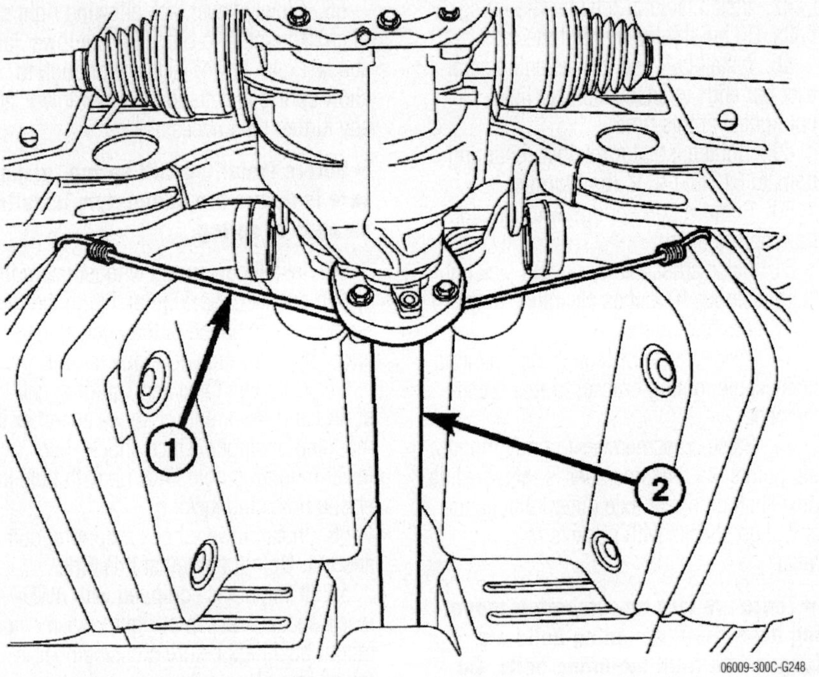

Support driveshaft (2) using a bungee cord (1)

➡Due to short travel and low spring tension, it is not necessary to lock-out parking brake lever to service parking brake components.

9. Disconnect front parking brake cable at connector to right rear parking brake cable.
10. Remove front parking brake cable from equalizer.
11. On each rear disc brake:
    a. While holding guide pins from turning, remove disc brake caliper guide pin bolts.
    b. Remove brake caliper from brake adapter and pads.
    c. Guide brake caliper up through suspension, following brake hose path. Support caliper above rear suspension using with bungee cord or wire to keep caliper from overextending brake hose when crossmember is lowered.

➡To remove wheel speed sensor connector from body wiring harness connector, move retaining clip and pull sensor connector outward.

12. Remove wheel speed sensor connectors from body wiring harness connector located in luggage compartment floor pan.
13. Unclip left wheel speed sensor cable from routing clip near body connector.
14. On each side of vehicle, remove shock absorber lower mounting bolt and nut.
15. Carefully mark location of rear crossmember on body at all four mount (bushing) locations using a marker or crayon. Do not use a scratch awl to mark location.
16. Position an extra pair of jackstands under and support forward end of engine cradle to help stabilize vehicle during rear suspension removal/installation.
17. Position under-hoist utility jack or transmission jack under center of rear axle differential. Raise jack head to contact differential and secure in place. When securing crossmember to jack, be sure not to secure stabilizer bar.

✲✲ CAUTION

Before opening fuel system, review all warnings and cautions under Fuel System.

18. Disconnect negative battery cable.
19. Drain fuel from tank.
20. Fuel filler tube assembly.
21. Open filler tube door.
22. Remove retaining wire from inside filler tube rubber.
23. Start removing rubber from body sheet metal.
24. Remove retaining wire from inside filler tube rubber.
25. Start removing rubber from body sheet metal.
26. Filler tube and rubber removed from body sheet metal.
27. Remove the left inner splash shield.
28. Disconnect the filler tube vent line.
29. Remove the filler tube mounting bolt.
30. Remove the under body splash shield.
31. Loosen the filler tube hose clamp. Leave the clamp tight on the hose and fuel tank location.
32. Move clamp toward the fuel tank.
33. Remove the filler tube assembly from the vehicle.

➡If equipped with AWD, when removing crossmember mounting bolts in following step, be sure to not to misplace spacers between crossmember mounts and body.

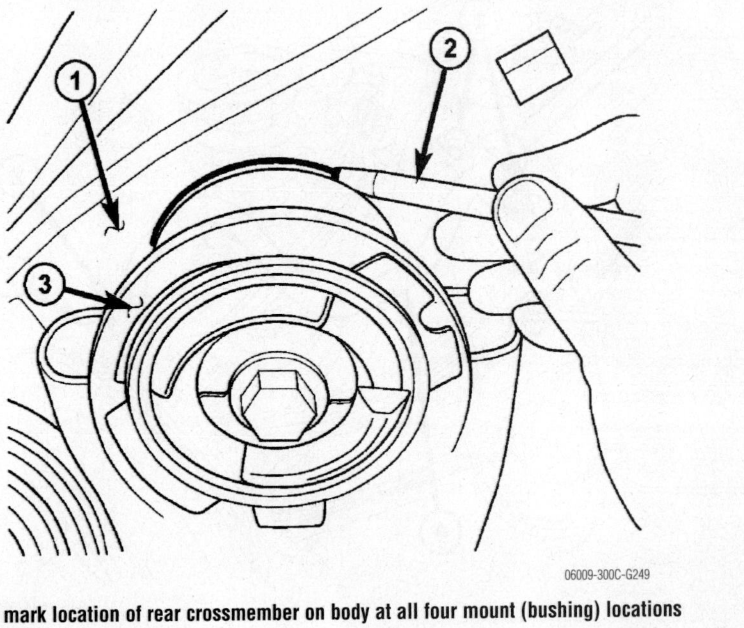

Carefully mark location of rear crossmember on body at all four mount (bushing) locations using a marker

34. Remove both front and both rear mounting bolts fastening crossmember in place.

35. Slowly lower crossmember using jack. Do not lower jack at a fast rate. Lower just enough to allow driveshaft removal from rear axle differential. Do not lower jack any further than necessary. Slide driveshaft out of rear axle differential and allow bungee cord previously installed to support.

36. Slowly lower crossmember several inches.

37. Remove screw fastening front parking brake cable routing bracket to rear crossmember.

38. Continue to lower jack until crossmember is at a comfortable working level to access stabilizer bar fasteners.

39. On each end, remove bolt (3) and nut fastening stabilizer bar (4) to stabilizer link (2).

40. Remove bolts (1) fastening each stabilizer bar isolator retainer to crossmember.

41. Remove stabilizer bar (4) with isolators and retainers.

42. Remove retainers from isolators.

43. Remove isolators from stabilizer bar utilizing slits in bushings.

To install:

44. Install isolators on stabilizer bar utilizing slits in bushings. Install each isolator so its slit faces forward and flat side is positioned toward crossmember once installed.

45. Install retainers on isolators.

46. Install stabilizer bar with isolators and retainers on crossmember.

47. Install isolator retainer mounting bolts. Do not tighten at this time.

48. Install bolt and nut fastening stabilizer bar ends to each stabilizer links. Do not tighten at this time.

49. Tighten isolator retainer mounting bolts to 61 Nm (45 ft. lbs.) torque.

50. Remove coil springs with isolators from spring links.

51. Raise crossmember using jack until there is about 10 inches clearance to the to body mounting points.

52. Install screw fastening front parking brake cable routing bracket to rear crossmember.

53. Raise crossmember to body mounting points. As crossmember is raised, slide driveshaft onto rear axle differential flange and align shocks with pockets in spring links.

➡There are four crossmember mounting bolts. Rear mounting bolts are longer than front mounting bolts. Do not interchange mounting bolts.

54. Continue to raise crossmember with jack until crossmember mounting bolts can be installed. Install left side crossmember mounting bolts, but not the right side bolts. It is not necessary to tighten bolts at this point.

⁂ WARNING

To avoid damaging other components of vehicle, do not lower crossmember any further than necessary to install coil spring.

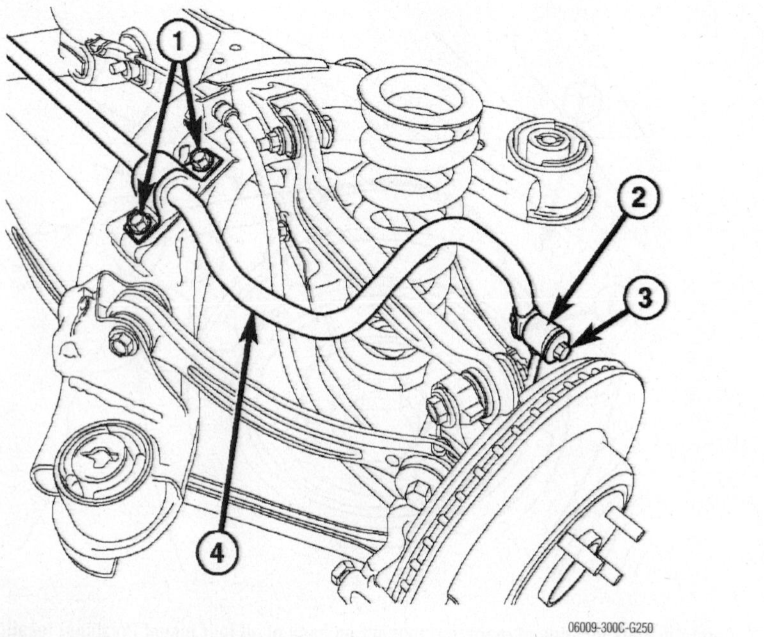

Rear stabilizer bar and fasteners

55. Slowly lower jack allowing right side of crossmember to drop. Do not lower jack at a fast rate. Lower jack just enough to allow spring Installation. Do not lower jack any further than necessary.

➡Before installing coil spring, make sure isolators are completely installed on ends of spring.

56. Install coil spring with isolators into spring pocket of spring link fitting the lower isolator to the shape of the pocket, then align top of spring with body mount.

57. Carefully raise jack, guiding coil spring and lower end of shock absorber into mounted positions. Once shock absorber lower mounting hole lines up with hole in spring link, stop jacking.

58. Install lower shock mounting bolt and nut. Do not tighten at this time.

59. If vehicle is equipped with AWD, insert spacers on top of right crossmember mount bushings before crossmember is raised into place.

➡There are four crossmember mounting bolts. Rear mounting bolts are longer than front mounting bolts. Do not interchange mounting bolts.

60. Raise right side of crossmember into mounted position. Install right side crossmember mounting bolts. Snug, but do not fully tighten bolts at this time.

61. Remove both front and rear crossmember mounting bolts on left side of vehicle.

⁂ WARNING

To avoid damaging other components of vehicle, do not lower crossmember any further than necessary to install coil spring.

62. Slowly lower jack allowing left side of crossmember to drop. Do not lower jack at a fast rate. Lower jack just enough to allow spring installation. Do not lower jack any further than necessary.

➡Before installing coil spring, make sure isolators are completely installed on ends of spring.

63. Install coil spring with isolators into spring pocket of spring link fitting the lower isolator to the shape of the pocket, then align top of spring with body mount.

64. Carefully raise jack, guiding coil spring and lower end of shock absorber into mounted positions. Once shock absorber lower mounting hole lines up with hole in spring link, stop jacking.

65. Install lower shock mounting bolt and nut. Do not tighten at this time.

66. If vehicle is equipped with AWD, insert spacers on top of left crossmember mount bushings before crossmember is raised into place.

➡**There are four crossmember mounting bolts. Rear mounting bolts are longer than front mounting bolts. Do not interchange mounting bolts.**

67. Raise left side of crossmember into mounted position. Install left side crossmember mounting bolts. Snug, but do not fully tighten bolts at this time.

68. Shift crossmember as necessary to line up mounts with location marks drawn on body before removal.

69. Once mounts are lined up with location marks, on both sides of vehicle, measure distance between the tension link and weld flange on body directly in front of it, just outboard of the front mount bushing. This distance must be at least 12 mm to allow proper clearance for suspension movement. If distance is less than 12 mm on either side of vehicle, shift that side of rear crossmember directly rearward until distance is 12 mm or greater. To do so, loosen 3 mounting bolts slightly, leaving one on opposite side of shift snugged to pivot off of. Shift crossmember rearward and snug loosened bolts. Re-measure opposite side to be sure it still maintains minimum 12 mm distance.

70. Tighten all four crossmember mounting bolts to 180 Nm (133 ft. lbs.) torque.

71. Remove jack from under rear axle differential.

72. Remove bungee cord supporting driveshaft.

73. Align driveshaft index marks placed upon removal. Install driveshaft rear coupler-to-axle flange bolts and nuts by hand. Tighten driveshaft rear coupler-to-axle flange bolts to 81 Nm (60 ft. lbs.) torque.

74. Insert filler tube into the fuel tank rubber hose.

75. Slide hose clamp into place and tighten.

76. Hose clamp in place and tighten.

77. Install filler tube mounting bolt.

78. Install filler tube vent line.

79. Install the underbody splash shield.

80. Install the left inner splash shield.

81. Clip left rear wheel speed sensor cable to routing clip near body connector.

82. Match left rear wheel speed sensor connector to right sensor connector to make one connector.

83. Insert speed sensor connectors into body wiring harness connector located in luggage compartment floor pan. When installing connector, make sure retaining clip on body connector is properly in place and sensor connector cannot be pulled out.

84. On each rear disc brake:

a. Push caliper guide pins into caliper adapter to clear caliper mounting bosses when installing.

b. Guide caliper and brake hose down through rear suspension, then slide caliper over brake pads and onto caliper adapter.

✱✱ WARNING

Extreme caution should be taken not to cross-thread caliper guide pin bolts when they are installed. Align caliper mounting holes with guide pins, then install guide pin bolts. While holding guide pins from turning, tighten bolts to 60 Nm (44 ft. lbs.).

c. Make sure brake hose is properly routed and will not come in contact with suspension components.

85. Route parking brake cable above rear crossmember, then slide cable through equalizer above rear differential.

➡**Due to short travel and low spring tension, it is not necessary to lock-out parking brake lever to service parking brake components.**

86. Connect front parking brake cable at connector to right rear parking brake cable.

87. Install rear exhaust system.

88. Install tire and wheel assemblies. Tighten wheel mounting nuts to 150 Nm (110 ft. lbs.) torque.

89. Lower vehicle until rear wheels are just above floor level.

90. Apply parking brake lever. Release lever, then reapply.

91. Check to make sure rear wheels will not rotate with lever applied.

92. Lower vehicle.

93. Connect battery negative cable to battery post. It is important that this is performed properly.

94. Pump brake pedal several times to ensure vehicle has a firm brake pedal before moving vehicle.

95. Position vehicle on alignment rack/drive-on hoist. Raise vehicle as necessary to access mounting bolts.

96. Tighten shock absorber lower mounting bolt nuts to 72 Nm (53 ft. lbs.) torque.

97. Tighten stabilizer link fasteners to 61 Nm (45 ft. lbs.).

98. Perform wheel alignment.

Wheel Bearings

The rear wheel bearing and wheel hub of this vehicle are a one piece sealed unit or hub and bearing unit type assembly. The hub and bearing (1) is mounted to the center of the knuckle using 4 bolts. It has five wheel mounting studs (2) on the hub flange.

The center of the hub and bearing is splined to match the axle halfshaft.

The wheel mounting studs used to mount the tire and wheel to the vehicle are the only replaceable components of the hub and bearing. Otherwise, the hub and bearing is serviced only as a complete assembly.

No adjustments are possible.

REMOVAL & INSTALLATION

1. Before servicing the vehicle, refer to the precautions in the beginning of this section.

2. Raise and support vehicle.

3. Remove wheel mounting nuts, then tire and wheel assembly.

4. While a helper applies brakes to keep hub from rotating, remove hub nut from the halfshaft.

➡**In some cases, it may be necessary to retract caliper piston in its bore a small amount in order to provide sufficient clearance between shoes and rotor to easily remove caliper from knuckle. This can usually be accomplished before guide pin bolts are removed, by grasping rear of caliper and pulling outward working with guide pins, thus retracting piston. Never push on piston directly as it may get damaged.**

5. Remove two bolts securing disc brake caliper adapter to knuckle.

6. Remove disc brake caliper and adapter from knuckle as an assembly. Hang assembly out of way using wire or a bungee cord. Use care not to overextend brake hose when doing this.

7. Remove any clips retaining brake rotor to wheel mounting studs.

8. Slide brake rotor off hub and bearing.

9. Loosen each hub and bearing mounting bolt a turn or two at a time while pulling outward on hub and bearing to avoid bolt contact with halfshaft outer joint. Once removed from threads in hub and bearing (but not knuckle), allow bolts to stay in and protrude through knuckle and brake support plate to keep brake support plate in place when hub and bearing is removed.

10. Slide hub and bearing off knuckle and halfshaft.

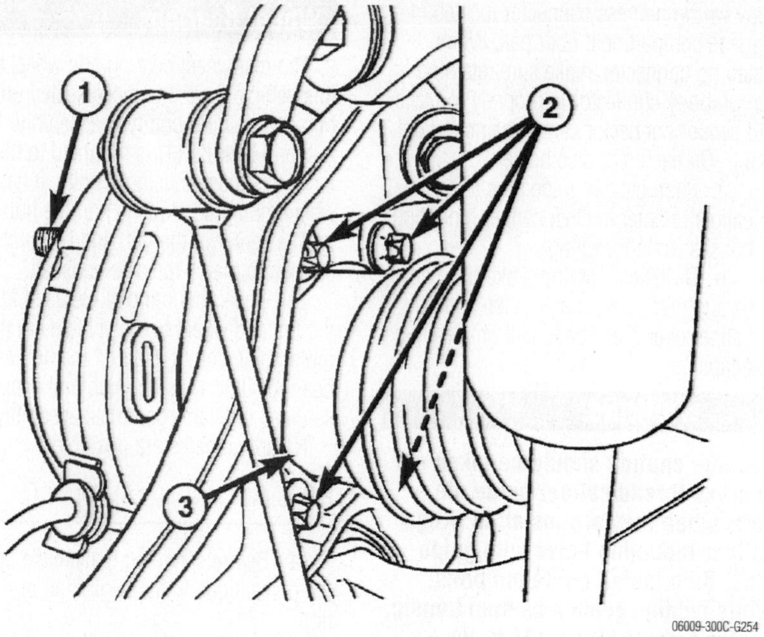

Rear hub/bearing mounting bolts (2)

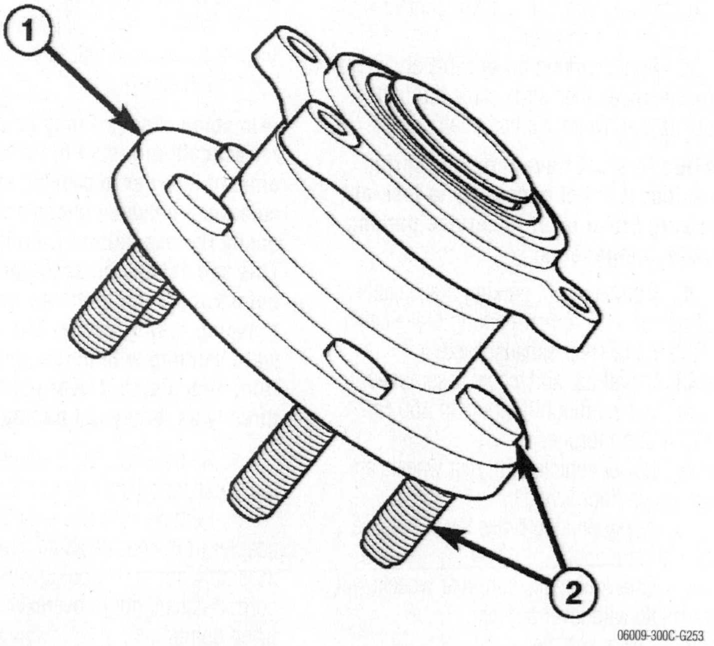

Rear hub/bearing assembly

To install:

➡Before installing hub and bearing on end of axle halfshaft, ensure isolation washer is present on end of halfshaft. Inspect washer making sure it is not worn or damaged. Washer can be installed in either direction on shaft.

11. Position hub and bearing bolts though rear of knuckle and parking brake support just enough to hold support in place as hub and bearing is installed.
12. Slide hub and bearing onto halfshaft. Place hub and bearing through brake support, onto knuckle, lining up mounting bolt holes with bolts.
13. Install four bolts fastening hub and bearing in place. Tighten mounting bolts to 68 Nm (50 ft. lbs.) torque.

➡Inspect disc brake pads and parking brake shoes before brake rotor installation.

14. Install brake rotor over wheel mounting studs and onto hub.
15. Install disc brake caliper and adapter assembly over brake rotor.
16. Install mounting bolts securing caliper adapter to knuckle. Tighten bolts to 95 Nm (70 ft. lbs.) torque.
17. Install hub nut on end of halfshaft. While a helper applies brakes to keep hub from turning, tighten hub nut to 213 Nm (157 ft. lbs.) torque.
18. Verify proper adjustment of the parking brake shoes and adjust as necessary.
19. Install tire and wheel assembly. Tighten wheel mounting nuts to 150 Nm (110 ft. lbs.) torque.
20. Lower vehicle.
21. Pump brake pedal several times to ensure vehicle has a firm brake pedal before moving vehicle.
22. Check and adjust brake fluid level in reservoir as necessary.
23. Road test vehicle and make several stops to wear off any foreign material on brakes and to seat brake pads.

FRONT DISC BRAKES

⁂ CAUTION

Dust and dirt accumulating on brake parts during normal use may contain asbestos fibers from production or aftermarket brake linings. Breathing excessive concentrations of asbestos fibers can cause serious bodily harm. Exercise care when servicing brake parts. Do not sand or grind brake lining unless equipment used is designed to contain the dust residue. Do not clean brake parts with compressed air or by dry brushing. Cleaning should be done by dampening the brake components with a fine mist of water, then wiping the brake components clean with a dampened cloth. Dispose of cloth and all residue containing asbestos fibers in an impermeable container with the appropriate label. Follow practices prescribed by the occupational safety and health administration (OSHA) and the environmental protection agency (EPA) for the handling, processing, and disposing of dust or debris that may contain asbestos fibers.

Hydraulic System

BLEEDING

Base System

1. Before servicing the vehicle, refer to the precautions in the beginning of this section.

➡To bleed the base brake system manually, an assistant's help is required. To ensure all air is bled from the ICU or junction block in a timely manner, it is recommended to raise the rear of the vehicle approximately 10–12 inches as measured at the rear bumper.

2. Raise and support vehicle placing rear of vehicle approximately 10–12 inches above level. It will be necessary to add extra support stands under vehicle to support this angle.
3. Remove rubber duct caps from all 4 bleeder screws.
4. Attach a clear hose to the bleeder screw at one wheel and feed the other end of the hose into a clear jar containing fresh brake fluid.
5. Have an assistant pump the brake pedal three or four times and hold it down before the bleeder screw is opened.

⁂ WARNING

Open the bleeder screw at least one full turn when instructed. Some air may be trapped in the brake lines or valves far upstream, as far as ten feet or more from the bleeder screw. If the bleeder screw is not opened sufficiently, fluid flow is restricted causing a slow, weak fluid discharge. This will NOT get all the air out. Therefore, it is essential to open the bleeder screw at least one full turn to allow a fast, large volume discharge of brake fluid.

6. While the pedal is being held down, open the bleeder screw at least 1 full turn. When the bleeder screw opens the brake pedal will drop all the way to the floor. Continue to hold the pedal all the way down.
7. Once the brake pedal has dropped, close the bleeder screw. The pedal can then be released.
8. Repeat steps 1 through 5 until all trapped air is removed from that wheel circuit (usually four or five times). This should pass a sufficient amount of fluid to expel all the trapped air from the brakes hydraulic system. Be sure to monitor brake fluid level in master cylinder fluid reservoir making sure it stays at a proper level. This will ensure air does not reenter brake hydraulic system through master cylinder.

➡Monitor the brake fluid level in the fluid reservoir periodically to make sure it does not go too low. This will ensure that air does not reenter the brake hydraulic system.

9. Bleed the remaining wheel circuits in the same manner until all air is removed from the brake hydraulic system.
10. Check brake pedal travel. If pedal travel is excessive or has not improved, some air may still be trapped in the hydraulic system. Re-bleed the brake system as necessary.
11. If equipped with antilock brakes, the hydraulic control unit may need to be bled, then re-bleed base brakes.
12. Reinstall all 4 bleeder screw dust caps.
13. Test drive vehicle to ensure brakes are operating properly and pedal feel is correct.

ABS Units

1. Before servicing the vehicle, refer to the precautions in the beginning of this section.

The base brake's hydraulic system must be bled anytime air enters the hydraulic system. The ABS must always be bled anytime it is suspected that the HCU has ingested air.

Brake systems with ABS must be bled as two independent braking systems. The non-ABS portion of the brake system with ABS is to be bled the same as any non-ABS system.

The ABS portion of the brake system must be bled separately. Use the following procedure to properly bleed the brake hydraulic system including the ABS.

➡During the brake bleeding procedure, be sure the brake fluid level remains close to the FULL level in the master cylinder fluid reservoir. Check the fluid level periodically during the bleeding procedure and add Mopar® DOT 3 brake fluid as required.

When bleeding the ABS system, the following bleeding sequence must be followed to insure complete and adequate bleeding.

2. Make sure all hydraulic fluid lines are installed and properly torqued.
3. Connect the scan tool to the diagnostics connector. The diagnostic connector is located under the lower steering column cover to the left of the steering column.
4. Using the scan tool, check to make sure the ABM does not have any fault codes stored. If it does, clear them.

⁂ CAUTION

When bleeding the brake system wear safety glasses. A clear bleed tube must be attached to the bleeder screws and submerged in a clear container filled part way with clean brake fluid. Direct the flow of brake fluid away from yourself and the painted surfaces of the vehicle. Brake fluid at high pressure may come out of the bleeder screws when opened.

5. Bleed the base brake system.
6. Using a scan tool, select ECU VIEW, followed by ABS MISCELLANEOUS FUNCTIONS to access bleeding. Follow the instructions displayed. When finished, disconnect the scan tool and proceed.
7. Bleed the base brake system a second time. Check brake fluid level in the reservoir periodically to prevent emptying, causing air to enter the hydraulic system.
8. Fill the master cylinder fluid reservoir to the MAX level.
9. Test drive the vehicle to be sure the brakes are operating correctly and that the brake pedal does not feel spongy.

Brake Caliper

REMOVAL & INSTALLATION

Single Piston

1. Before servicing the vehicle, refer to the precautions in the beginning of this section.
2. Disconnect and isolate battery negative cable from battery post.
3. Using a brake pedal holding tool, depress brake pedal past its first inch of travel and hold it in this position. Holding pedal in this position will isolate master cylinder from hydraulic brake system and will not allow brake fluid to drain out of brake fluid reservoir while brake lines are open.
4. Raise and support vehicle.
5. Remove wheel mounting nuts, then tire and wheel assembly.
6. Remove banjo bolt connecting flexible brake hose to caliper. There are two sealing washers (one on each side of hose fitting) that will come off when bolt is removed. Discard these washers; install NEW washers on installation.

7. While holding guide pins from turning, remove caliper guide pin bolts.
8. Remove brake caliper from brake adapter and pads.

To install:

⁂ WARNING

Always inspect brake pads before installing disc brake caliper and replace as necessary.

9. Completely retract caliper piston back into bore of caliper. Use hand pressure or a C-clamp may be used to retract piston, first placing a wood block over piston before installing C-clamp to avoid damaging piston.

⁂ WARNING

Use care when installing caliper onto disc brake adapter to avoid damaging boots on caliper guide pins.

10. Push caliper guide pins into caliper adapter to clear caliper mounting bosses when installing.
11. Slide caliper over brake pads and onto caliper adapter.

⁂ WARNING

Extreme caution should be taken not to cross-thread caliper guide pin bolts when they are installed.

12. Align caliper mounting holes with guide pins, then install guide pin bolts. While holding guide pins from turning, tighten bolts to 60 Nm (44 ft. lbs.) torque.
13. Install banjo bolt attaching brake hose to caliper. Install NEW washers on each side of hose fitting as banjo bolt is placed through fitting. Thread banjo bolt into caliper and tighten to 43 Nm (32 ft. lbs.) torque.
14. Install tire and wheel assembly. Tighten wheel mounting nuts to 150 Nm (110 ft. lbs.) torque.
15. Lower vehicle.
16. Remove brake pedal holding tool.
17. Connect battery negative cable to battery post. It is important that this is performed properly.
18. Bleed base brake hydraulic system as necessary.
19. Road test vehicle making several stops to wear off any foreign material on brakes and to seat brake shoes.

Dual Piston

2-WHEEL DRIVE

1. Before servicing the vehicle, refer to the precautions in the beginning of this section.
2. Disconnect and isolate battery negative cable from battery post.
3. Using a brake pedal holding tool, depress brake pedal past its first inch of travel and hold it in this position. Holding pedal in this position will isolate master cylinder from hydraulic brake system and will not allow brake fluid to drain out of brake fluid reservoir while brake lines are open.
4. Raise and support vehicle.
5. Remove wheel mounting nuts, then tire and wheel assembly.
6. Remove banjo bolt connecting flexible brake hose to caliper. There are two sealing washers (one on each side of hose fitting) that will come off when bolt is removed. Discard these washers; install NEW washers on installation.
7. While holding guide pins from turning, remove caliper guide pin bolts.
8. Remove brake caliper from brake adapter and pads.

To install:

⁂ WARNING

Always inspect brake pads before installing disc brake caliper and replace as necessary.

9. Completely retract caliper pistons back into bores of caliper. Use hand pressure or a C-clamp may also be used to retract pistons, first placing a wood block over piston before installing C-clamp to avoid damaging piston.

⁂ WARNING

Use care when installing caliper onto disc brake adapter to avoid damaging boots on caliper guide pins.

10. Push caliper guide pins into caliper adapter to clear caliper mounting bosses when installing.
11. Slide caliper over brake pads and onto caliper adapter.

⁂ WARNING

Extreme caution should be taken not to cross-thread caliper guide pin bolts when they are installed.

12. Align caliper mounting holes with guide pins, then install guide pin bolts. While holding guide pins from turning, tighten bolts to 60 Nm (44 ft. lbs.) torque.
13. Install banjo bolt attaching brake hose to caliper. Install NEW washers on each side of hose fitting as banjo bolt is placed through fitting. Thread banjo bolt into caliper and tighten to 43 Nm (32 ft. lbs.) torque.
14. Install tire and wheel assembly.

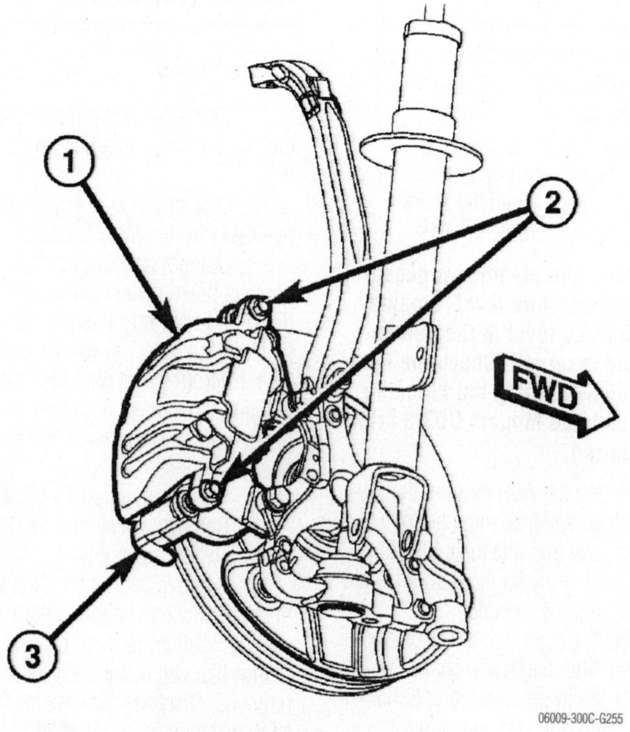

Single piston caliper (1), mounting pins (2), adapter (3)

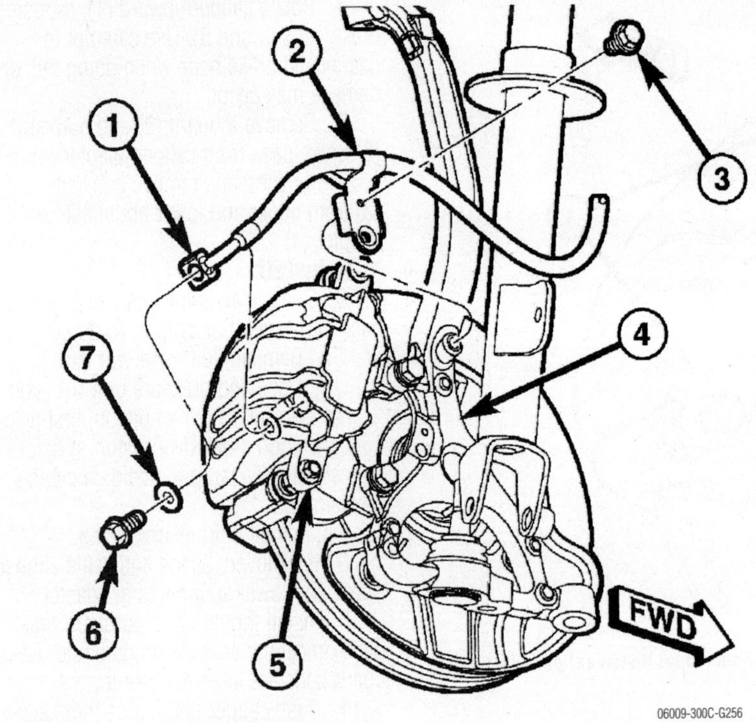

Dual piston caliper mounting. (1) banjo fitting, (2) hose, (3) bolt, (4) adaptor, (5) caliper pin, (6) banjo bolt, (7) washer

Tighten wheel mounting nuts to 150 Nm (110 ft. lbs.).

15. Lower vehicle.
16. Remove brake pedal holding tool.
17. Connect battery negative cable to battery post. It is important that this is performed properly.
18. Bleed base brake hydraulic system as necessary.
19. Road test vehicle making several stops to wear off any foreign material on brakes and to seat brake shoes.

ALL WHEEL DRIVE

1. Before servicing the vehicle, refer to the precautions in the beginning of this section.
2. Disconnect and isolate battery negative cable from battery post.
3. Using a brake pedal holding tool, depress brake pedal past its first inch of travel and hold it in this position. Holding pedal in this position will isolate master cylinder from hydraulic brake system and will not allow brake fluid to drain out of brake fluid reservoir while brake lines are open.
4. Raise and support vehicle.
5. Remove wheel mounting nuts, then tire and wheel assembly.
6. Remove banjo bolt (3) connecting flexible brake hose (1) to caliper (4). There are two sealing washers (2) (one on each side of hose fitting) that will come off when bolt is removed. Discard these washers; install NEW washers on installation.
7. While holding guide pins from turning, remove caliper guide pin bolts.
8. Remove brake caliper from brake adapter and pads.

To install:

※※ WARNING

Always inspect brake pads before installing disc brake caliper and replace as necessary.

9. Completely retract caliper pistons back into bores of caliper. Use hand pressure or a C-clamp may also be used to retract pistons, first placing a wood block over piston before installing C-clamp to avoid damaging piston.

※※ WARNING

Use care when installing caliper onto disc brake adapter to avoid damaging boots on caliper guide pins.

10. Push caliper guide pins into caliper adapter to clear caliper mounting bosses when installing.
11. Slide caliper over brake pads and onto caliper adapter.

※※ WARNING

Extreme caution should be taken not to cross-thread caliper guide pin bolts when they are installed.

12. Align caliper mounting holes with guide pins, then install guide pin bolts. While holding guide pins from turning, tighten bolts to 60 Nm (44 ft. lbs.) torque.

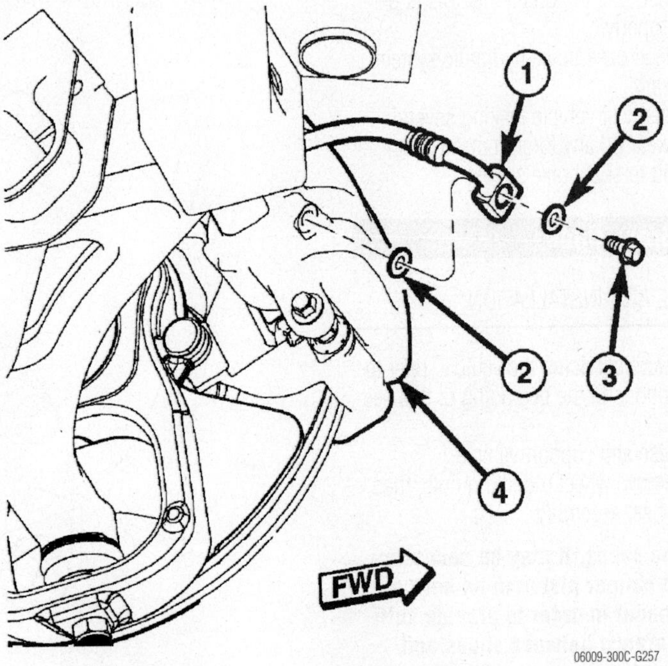

Brake line-to-caliper attachment

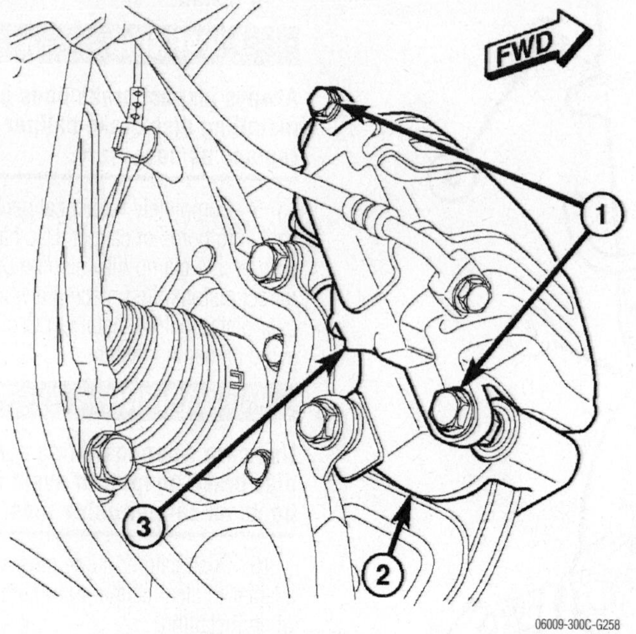

Caliper attachment. (1) caliper pins, (2) anchor plate, (3) caliper—dual piston caliper with AWD

13. Install banjo bolt attaching brake hose to caliper. Install NEW washers on each side of hose fitting as banjo bolt is placed through fitting. Thread banjo bolt into caliper and tighten to 43 Nm (32 ft. lbs.) torque.
14. Install tire and wheel assembly. Tighten wheel mounting nuts to 150 Nm (110 ft. lbs.) torque.
15. Lower vehicle.
16. Remove brake pedal holding tool.
17. Connect battery negative cable to battery post. It is important that this is performed properly.
18. Bleed base brake hydraulic system as necessary.
19. Road test vehicle making several stops to wear off any foreign material on brakes and to seat brake shoes.

Disc Brake Pads

REMOVAL AND INSTALLATION

1. Before servicing the vehicle, refer to the precautions in the beginning of this section.
2. Raise and support vehicle.
3. Remove wheel mounting nuts, then tire and wheel assembly.

➡In some cases, it may be necessary to retract caliper piston in its bore a small amount in order to provide sufficient clearance between shoes and rotor to easily remove caliper from knuckle. This can usually be accomplished before guide pin bolts are removed by grasping rear of caliper and pulling outward working with guide pins, thus retracting piston. Never push on piston directly as it may get damaged.

4. Remove Lower caliper guide pin bolt. To do so, hold the guide pin (3) stationary while turning bolt (1).

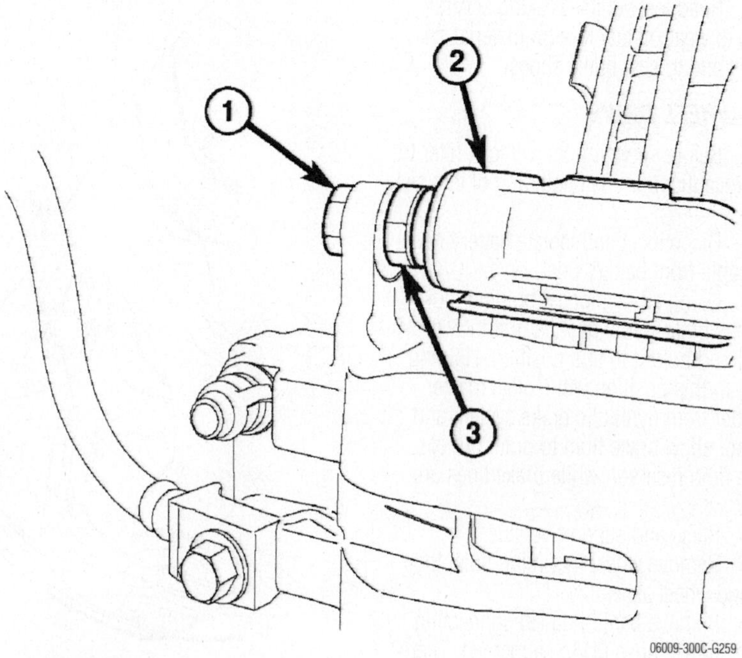

Front caliper guide pin

5. Rotate caliper upward (1), exposing brake pads (2 and 5). Use care not to overextend brake hose when doing this or damage may occur.
6. Remove inboard (2) and outboard (5) brake pads from caliper adapter (4).
7. If necessary, remove anti-rattle clips (3) from upper and lower abutments of adapter (4).

To install:

8. Completely retract caliper piston(s) back into bore(s) of caliper. To do so:
   a. Remove fluid reservoir cap.
   b. Use hand pressure or a C-clamp may be used to retract piston, first placing a wood block over piston(s) before installing C-clamp to avoid damaging piston(s).
   c. Install fluid reservoir cap.
9. If removed, attach anti-rattle clips to upper and lower abutments of adapter.
10. Install inboard and outboard brake pads on caliper adapter. Inboard and outboard pads are interchangeable.
11. Push caliper guide pins into caliper adapter to clear caliper mounting bosses when installing.
12. Rotate caliper downward, aligning upper mounting boss with lower guide pin.
13. Install upper caliper guide pin bolt. While holding guide pin stationary tighten bolt to 60 Nm (44 ft. lbs.) torque.
14. Install tire and wheel assembly. Tighten wheel mounting nuts to 150 Nm (110 ft. lbs.) torque.

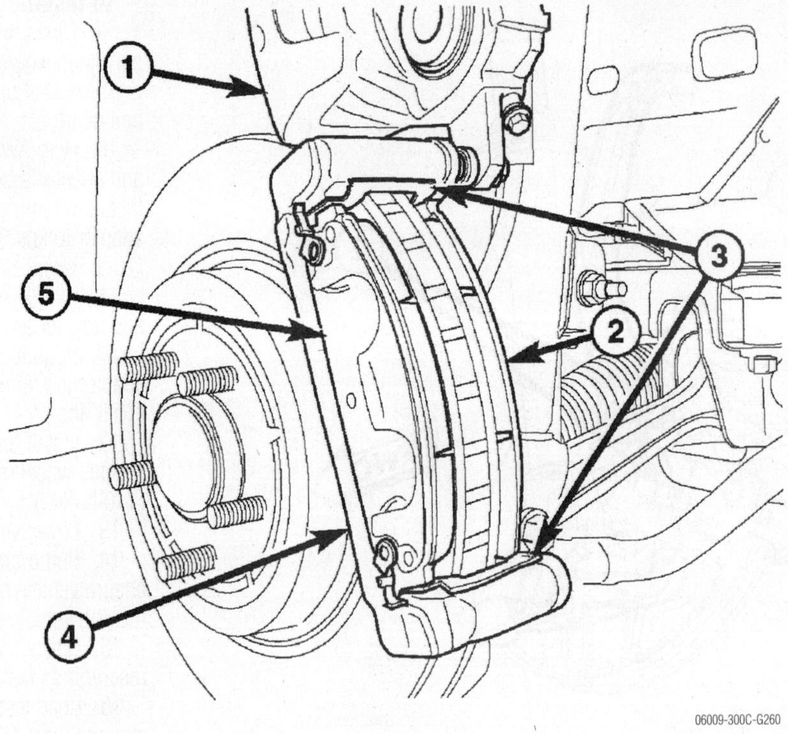

Front pad removal

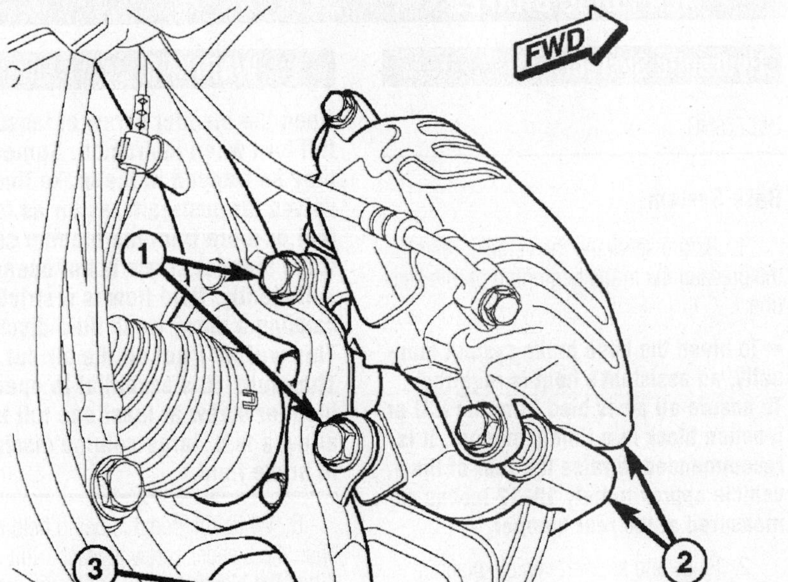

Caliper/adapter (2) mounting bolts (1) to knuckle (3)—all wheel drive

15. Lower vehicle.
16. Pump brake pedal several times to set pads to caliper and brake rotor.
17. Check and adjust brake fluid level in reservoir.
18. Road test vehicle making several stops to wear off any foreign material on brakes and to seat brake shoes.

Rotor

REMOVAL & INSTALLATION

1. Before servicing the vehicle, refer to the precautions in the beginning of this section.
2. Raise and support vehicle.
3. Remove wheel mounting nuts, then tire and wheel assembly.

➡In some cases, it may be necessary to retract caliper piston in its bore a small amount in order to provide sufficient clearance between shoes and rotor to easily remove caliper from knuckle. This can usually be accomplished before mounting bolts are removed, by grasping rear of caliper and pulling outward working with guide pins, thus retracting piston. Never push on piston directly as it may get damaged.

4. With AWD: Remove two bolts securing disc brake caliper and adapter to knuckle. Remove disc brake caliper and adapter from knuckle as an assembly. Hang assembly out of way using wire or a bungee cord. Use care not to overextend brake hose when doing this.
5. With 2-WD: Remove two bolts securing disc brake caliper and adapter to knuckle. Remove disc brake caliper and adapter from knuckle as an assembly. Hang assembly out of way using wire or a bungee cord. Use care not to overextend brake hose when doing this.
6. Remove any clips retaining brake rotor to wheel studs.
7. Slide brake rotor off hub and bearing.

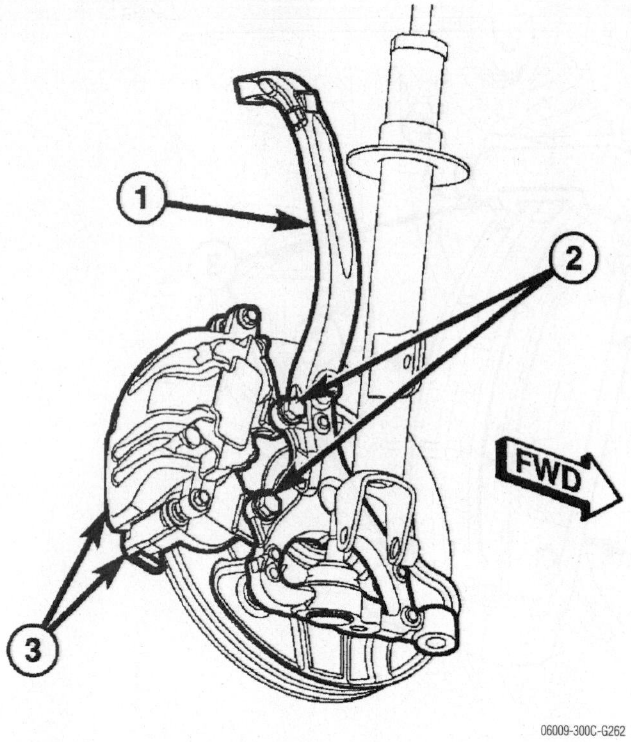

Caliper/adapter (3) mounting bolts (2) to knuckle (1)—2-wheel drive

To install:

8. Clean hub face to remove any dirt or corrosion where rotor mounts.
9. Install brake rotor over studs on hub and bearing.
10. With AWD: Install disc brake caliper and adapter assembly over brake rotor. Install mounting bolts securing caliper adapter to knuckle. Tighten bolts to 95 Nm (70 ft. lbs.).
11. With 2-WD: Install disc brake caliper and adapter assembly over brake rotor. Install mounting bolts securing caliper adapter to knuckle. Tighten bolts to 95 Nm (70 ft. lbs.).
12. Install tire and wheel assembly. Tighten wheel mounting nuts to 150 Nm (110 ft. lbs.).
13. Lower vehicle.
14. Pump brake pedal several times to ensure vehicle has a firm brake pedal before moving vehicle.
15. Check and adjust brake fluid level in reservoir as necessary.
16. Road test vehicle and make several stops to wear off any foreign material on brakes and to seat brake shoes.

REAR DISC BRAKES

CAUTION

Dust and dirt accumulating on brake parts during normal use may contain asbestos fibers from production or aftermarket brake linings. Breathing excessive concentrations of asbestos fibers can cause serious bodily harm. Exercise care when servicing brake parts. Do not sand or grind brake lining unless equipment used is designed to contain the dust residue. Do not clean brake parts with compressed air or by dry brushing. Cleaning should be done by dampening the brake components with a fine mist of water, then wiping the brake components clean with a dampened cloth. Dispose of cloth and all residue containing asbestos fibers in an impermeable container with the appropriate label. Follow practices prescribed by the occupational safety and health administration (OSHA) and the environmental protection agency (EPA) for the handling, processing, and disposing of dust or debris that may contain asbestos fibers.

Hydraulic System

BLEEDING

Base System

1. Before servicing the vehicle, refer to the precautions in the beginning of this section.

➡To bleed the base brake system manually, an assistant's help is required. To ensure all air is bled from the ICU or junction block in a timely manner, it is recommended to raise the rear of the vehicle approximately 10–12 inches as measured at the rear bumper.

2. Raise and support vehicle placing rear of vehicle approximately 10–12 inches above level. It will be necessary to add extra support stands under vehicle to support this angle.
3. Remove rubber duct caps from all 4 bleeder screws.
4. Attach a clear hose to the bleeder screw at one wheel and feed the other end of the hose into a clear jar containing fresh brake fluid.
5. Have an assistant pump the brake pedal three or four times and hold it down before the bleeder screw is opened.

WARNING

Open the bleeder screw at least one full turn when instructed. Some air may be trapped in the brake lines or valves far upstream, as far as ten feet or more from the bleeder screw. If the bleeder screw is not opened sufficiently, fluid flow is restricted causing a slow, weak fluid discharge. This will NOT get all the air out. Therefore, it is essential to open the bleeder screw at least one full turn to allow a fast, large volume discharge of brake fluid.

6. While the pedal is being held down, open the bleeder screw at least 1 full turn. When the bleeder screw opens the brake pedal will drop all the way to the floor. Continue to hold the pedal all the way down.
7. Once the brake pedal has dropped, close the bleeder screw. The pedal can then be released.
8. Repeat steps 1 through 5 until all trapped air is removed from that wheel circuit (usually four or five times). This should pass a sufficient amount of fluid to expel all the trapped air from the brakes hydraulic system. Be sure to monitor brake fluid level in master cylinder fluid reservoir making

sure it stays at a proper level. This will ensure air does not reenter brake hydraulic system through master cylinder.

➡**Monitor the brake fluid level in the fluid reservoir periodically to make sure it does not go too low. This will ensure that air does not reenter the brake hydraulic system.**

9. Bleed the remaining wheel circuits in the same manner until all air is removed from the brake hydraulic system.

10. Check brake pedal travel. If pedal travel is excessive or has not improved, some air may still be trapped in the hydraulic system. Re-bleed the brake system as necessary.

11. If equipped with antilock brakes, the hydraulic control unit may need to be bled, then re-bleed base brakes.

12. Reinstall all 4 bleeder screw dust caps.

13. Test drive vehicle to ensure brakes are operating properly and pedal feel is correct.

ABS Units

1. Before servicing the vehicle, refer to the precautions in the beginning of this section.

The base brake's hydraulic system must be bled anytime air enters the hydraulic system. The ABS must always be bled anytime it is suspected that the HCU has ingested air.

Brake systems with ABS must be bled as two independent braking systems. The non-ABS portion of the brake system with ABS is to be bled the same as any non-ABS system.

The ABS portion of the brake system must be bled separately. Use the following procedure to properly bleed the brake hydraulic system including the ABS.

➡**During the brake bleeding procedure, be sure the brake fluid level remains close to the FULL level in the master cylinder fluid reservoir. Check the fluid level periodically during the bleeding procedure and add Mopar® DOT 3 brake fluid as required.**

When bleeding the ABS system, the following bleeding sequence must be followed to insure complete and adequate bleeding.

2. Make sure all hydraulic fluid lines are installed and properly torqued.

3. Connect the scan tool to the diagnostics connector. The diagnostic connector is located under the lower steering column cover to the left of the steering column.

4. Using the scan tool, check to make sure the ABM does not have any fault codes stored. If it does, clear them.

✲✲ CAUTION

When bleeding the brake system wear safety glasses. A clear bleed tube must be attached to the bleeder screws and submerged in a clear container filled part way with clean brake fluid. Direct the flow of brake fluid away from yourself and the painted surfaces of the vehicle. Brake fluid at high pressure may come out of the bleeder screws when opened.

5. Bleed the base brake system.

6. Using a scan tool, select ECU VIEW, followed by ABS MISCELLANEOUS FUNCTIONS to access bleeding. Follow the instructions displayed. When finished, disconnect the scan tool and proceed.

7. Bleed the base brake system a second time. Check brake fluid level in the reservoir periodically to prevent emptying, causing air to enter the hydraulic system.

8. Fill the master cylinder fluid reservoir to the MAX level.

9. Test drive the vehicle to be sure the brakes are operating correctly and that the brake pedal does not feel spongy.

Brake Caliper

REMOVAL & INSTALLATION

1. Before servicing the vehicle, refer to the precautions in the beginning of this section.

2. Disconnect and isolate battery negative cable from battery post.

3. Using a brake pedal holding tool, depress brake pedal past its first inch of travel and hold it in this position. Holding pedal in this position will isolate master cylinder from hydraulic brake system and will not allow brake fluid to drain out of brake fluid reservoir while brake lines are open.

4. Raise and support vehicle.

5. Remove wheel mounting nuts, then tire and wheel assembly.

6. Remove banjo bolt connecting flexible brake hose to caliper. There are two sealing washers (one on each side of hose fitting) that will come off when bolt is removed. Discard these washers; install NEW washers on installation.

7. While holding guide pins from turning, remove caliper guide pin bolts.

8. Remove brake caliper from brake adapter and pads.

To install:

9. Completely retract caliper piston back into bore of caliper. Use hand pressure

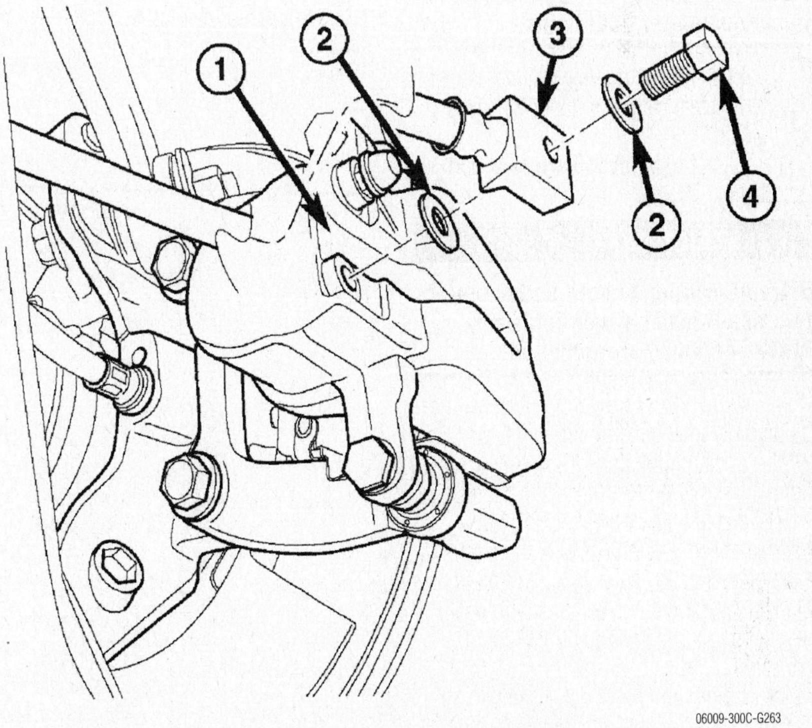

Rear brake line (3) connection. (1) caliper, (2) washers, (4) banjo bolt

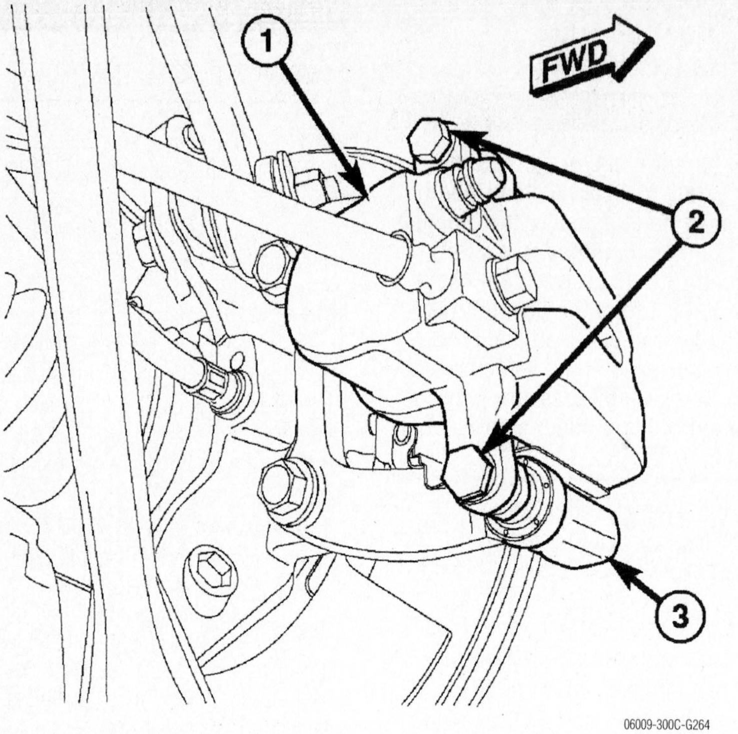

Rear caliper mounting. (1) caliper, (2) mounting pins, (3) adapter

or a C-clamp may be used to retract piston, first placing a wood block over piston before installing C-clamp to avoid damaging piston.

⁂ WARNING

Use care when installing caliper onto disc brake adapter to avoid damaging boots on caliper guide pins.

10. Push caliper guide pins into caliper adapter to clear caliper mounting bosses when installing.
11. Slide caliper over brake pads and onto caliper adapter.

⁂ WARNING

Extreme caution should be taken not to cross-thread caliper guide pin bolts when they are installed.

12. Align caliper mounting holes with guide pins, then install guide pin bolts. While holding guide pins from turning, tighten bolts to 60 Nm (44 ft. lbs.) torque.
13. Install banjo bolt attaching brake hose to caliper. Install NEW washers on each side of hose fitting as banjo bolt is placed through fitting. Thread banjo bolt into caliper and tighten to 43 Nm (32 ft. lbs.).
14. Install tire and wheel assembly. Tighten wheel mounting nuts to 150 Nm (110 ft. lbs.).
15. Lower vehicle.
16. Remove brake pedal holding tool.
17. Connect battery negative cable to battery post. It is important that this is performed properly.
18. Bleed base brake hydraulic system as necessary.
19. Road test vehicle making several stops to wear off any foreign material on brakes and to seat brake shoes.

Disc Brake Pads

REMOVAL AND INSTALLATION

1. Before servicing the vehicle, refer to the precautions in the beginning of this section.
2. Raise and support vehicle.
3. Remove wheel mounting nuts, then tire and wheel assembly.

➡In some cases, it may be necessary to retract caliper piston in its bore a small amount in order to provide sufficient clearance between shoes and rotor to easily remove caliper from knuckle. This can usually be accomplished before guide pin bolts are removed by grasping rear of caliper and pulling outward working with guide pins, thus retracting piston. Never push on piston directly as it may get damaged.

4. Remove upper caliper guide pin bolt. To do so, hold the guide pin stationary while turning bolt.
5. Rotate caliper downward, exposing brake pads. Use care not to overextend brake hose when doing this or damage may occur.
6. Remove inboard and outboard brake pads from caliper adapter.
7. If necessary, remove anti-rattle clips from upper and lower abutments of adapter.

To install:

8. Completely retract caliper piston back into bore of caliper. To do so:

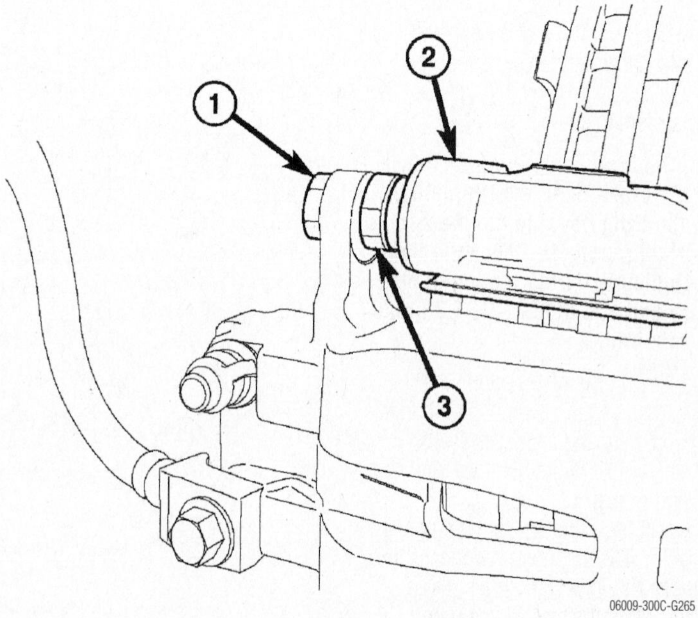

Upper guide pin bolt (1) adapter (2), guide pin (3)—rear brakes

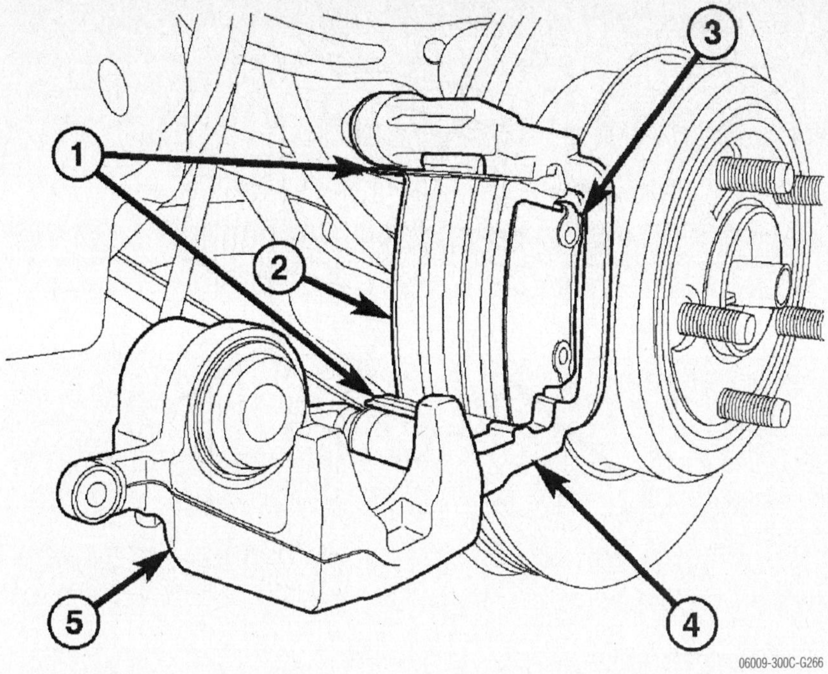

Rear brake pad removal/installation. (1) anti-rattle clips, (2, 3) brake pads, (4) adapter, (5) caliper

a. Remove fluid reservoir cap.

b. Use hand pressure or a C-clamp may be used to retract piston, first placing a wood block over piston before installing C-clamp to avoid damaging piston.

c. Install fluid reservoir cap.

9. If removed, attach anti-rattle clips to upper and lower abutments of adapter.
10. Install NEW inboard and outboard brake pads on caliper adapter. NEW Inboard and outboard pads are interchangeable.
11. Push caliper guide pins into caliper adapter to clear caliper mounting bosses when installing.
12. Rotate caliper upward, aligning upper mounting boss with upper guide pin.
13. Install Upper caliper guide pin bolt. While holding guide pin stationary tighten bolt to 60 Nm (44 ft. lbs.) torque.
14. Install tire and wheel assembly. Tighten wheel mounting nuts to 150 Nm (110 ft. lbs.) torque.
15. Lower vehicle.
16. Pump brake pedal several times to set pads to caliper and brake rotor.
17. Check and adjust brake fluid level in reservoir.
18. Road test vehicle making several stops to wear off any foreign material on brakes and to seat brake shoes.

Rotor

REMOVAL & INSTALLATION

1. Before servicing the vehicle, refer to the precautions in the beginning of this section.
2. Raise and support vehicle.
3. Remove wheel mounting nuts, then tire and wheel assembly.

➡In some cases, it may be necessary to retract caliper piston in its bore a small amount in order to provide sufficient clearance between shoes and rotor to easily remove caliper from knuckle. This can usually be accomplished before guide pin bolts are removed, by grasping rear of caliper and pulling outward working with guide pins, thus retracting piston. Never push on piston directly as it may get damaged.

4. Remove two bolts securing disc brake caliper adapter to knuckle.
5. Remove disc brake caliper and adapter from knuckle as an assembly. Hang assembly out of way using wire or a bungee cord. Use care not to overextend brake hose when doing this.
6. Remove any clips retaining brake rotor to wheel mounting studs.
7. Slide brake rotor off hub and bearing.

To install:

➡Inspect disc brake shoes and parking brake shoes before installation.

8. Clean hub face to remove any dirt or corrosion where brake rotor mounts.
9. Install brake rotor over wheel mounting studs and onto hub.

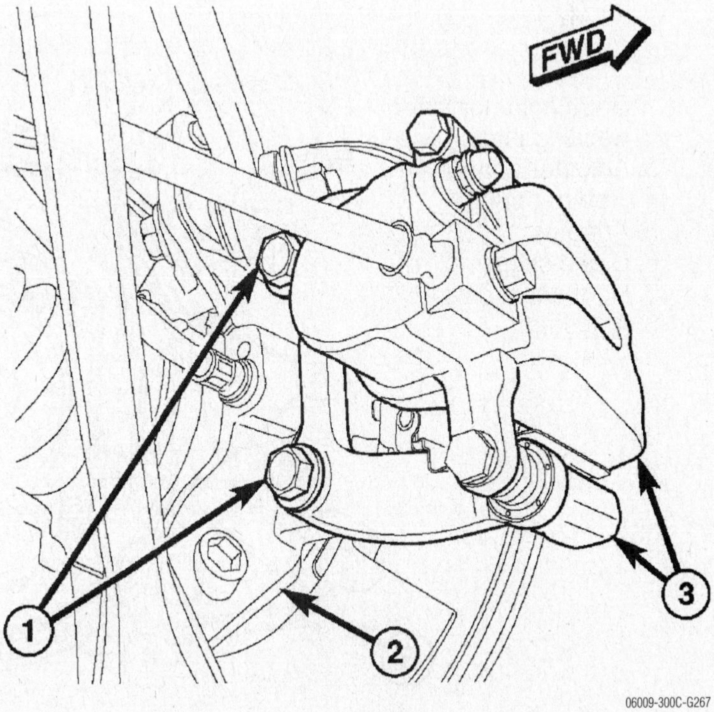

Rear caliper/adapter (3) to knuckle (2) mounting bolts (1)

10. Install disc brake caliper and adapter assembly over brake rotor.
11. Install mounting bolts securing caliper adapter to knuckle. Tighten bolts to 115 Nm (85 ft. lbs.) torque.
12. Install tire and wheel assembly. Tighten wheel mounting nuts to 150 Nm (110 ft. lbs.) torque.
13. Lower vehicle.
14. Pump brake pedal several times to ensure vehicle has a firm brake pedal before moving vehicle.
15. Check and adjust brake fluid level in reservoir as necessary.
16. Road test vehicle and make several stops to wear off any foreign material on brakes and to seat brake shoes.

Parking Brake Shoes

REMOVAL & INSTALLATION

1. Before servicing the vehicle, refer to the precautions in the beginning of this section.

➡The following procedure may be used to remove shoes on either side of the vehicle.

2. Raise and support vehicle.
3. Remove rear hub and bearing.
4. Completely back off parking brake shoe adjustment.
5. Remove parking brake shoe adjuster spring.
6. Remove shoe adjuster.

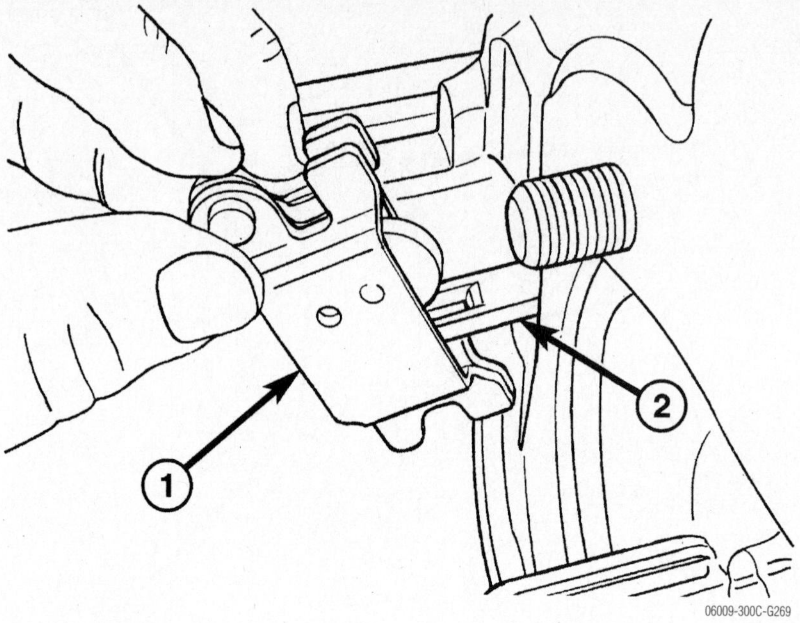

Removing actuator lever

7. Remove upper brake shoe hold-down clip and pin.
8. Remove upper shoe from return spring and shoe actuator lever.
9. Remove return spring from lower shoe.
10. Remove shoe actuator lever from end of cable.
11. Remove lower brake shoe hold-down clip and pin.
12. Remove lower shoe.
13. Inspect springs, adjuster, lever and aluminum shoe anchor pin for wear or damage. Replace as necessary.

To install:

➡The following procedure may be used to install shoes on either side of the vehicle.

➡Inspect springs, adjuster, lever and aluminum shoe anchor pin for wear or damage prior to installation. Replace as necessary.

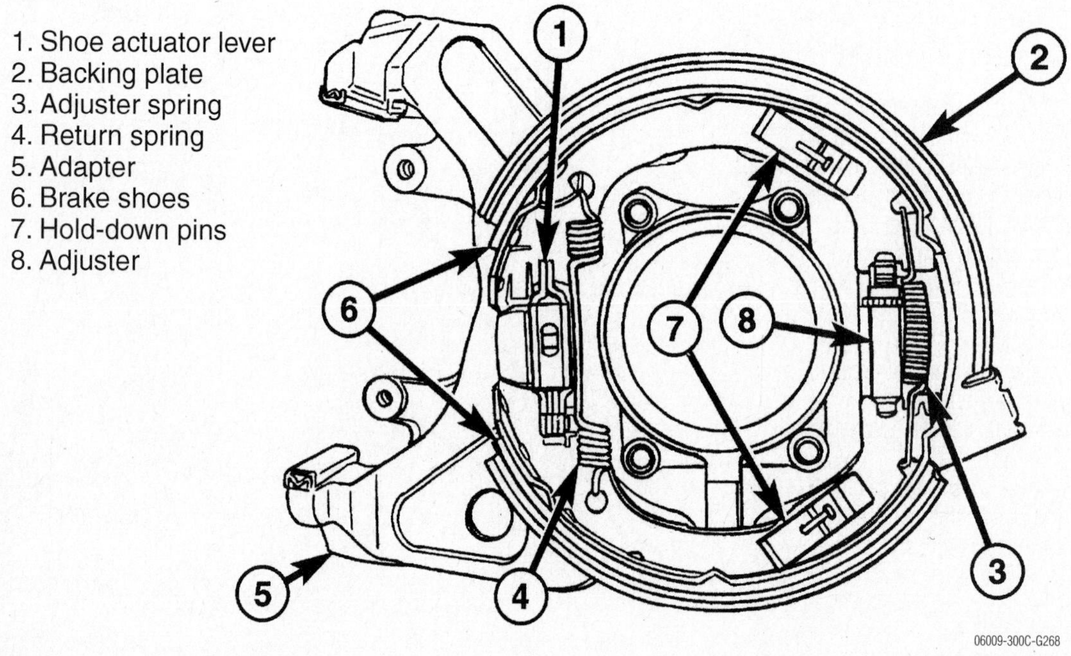

Parking brake

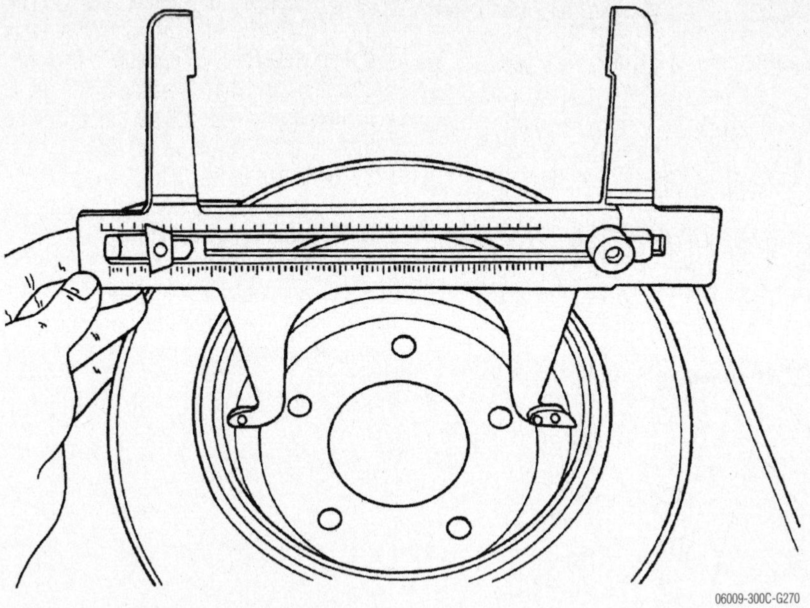

Measuring inside diameter of parking brake drum

14. Install lower brake shoe hold-down pin through rear of support.
15. Install lower shoe against support plate.
16. Install lower brake shoe hold-down clip.
17. Install shoe actuator lever on end of parking brake cable. Make sure actuator lever is positioned with word "UP" facing outward.
18. Install return spring to lower shoe.
19. Install upper shoe against support plate and onto shoe actuator lever.
20. Install upper brake shoe hold-down pin through rear of support and upper shoe.
21. Install upper brake shoe hold-down clip.
22. Attach return spring to upper shoe.
23. Install shoe adjuster. Place end of adjuster with star wheel upward.
24. Install parking brake shoe adjuster spring.
25. Using Brake Shoe Gauge, Special Tool C-3919, or equivalent, measure inside diameter of parking brake drum portion of rotor.
26. Place Gauge over parking brake shoes at widest point.
27. Using adjuster star wheel, adjust parking brake shoes until linings on both park brake shoes just touch jaws on gauge. This will give a good preliminary adjustment of parking brake shoes, before a final adjustment is made at end of this procedure.
28. Install hub and bearing with wheel speed sensor as well as all components necessary to access it.
29. Lower vehicle.
30. Perform final adjustment of parking brake shoes.

ADJUSTMENT

1. Before servicing the vehicle, refer to the precautions in the beginning of this section.
2. Place parking brake lever in "full released" position.
3. Raise and support vehicle.
4. Remove plug in parking brake shoe support to access adjuster star-wheel.

➡Through the access hole, rotate the adjuster star wheel in the following direction to expand the shoes outward against the drum. Left brake: rotate star-wheel toward rear of vehicle. Right brake: rotate starwheel toward front of vehicle.

5. Using an appropriate tool, turn adjuster starwheel until wheel will not rotate.

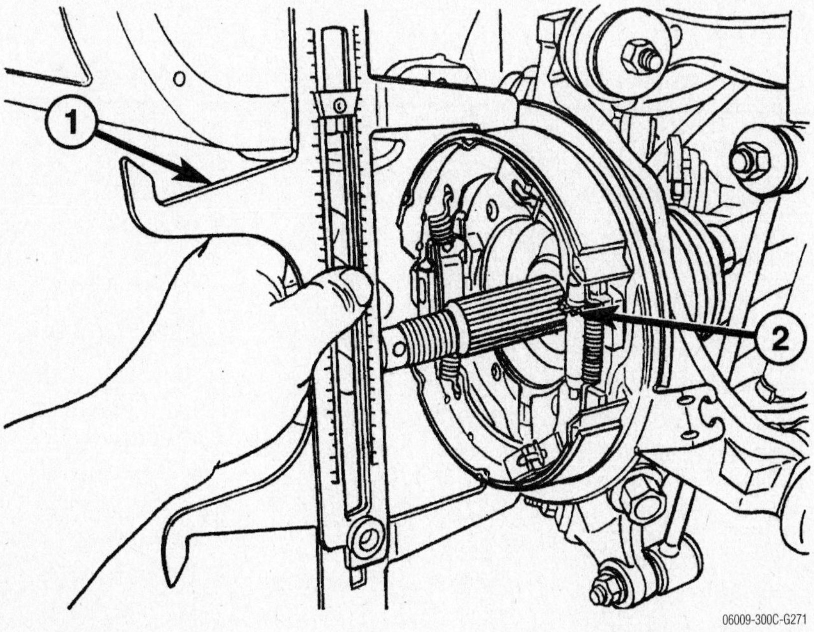

Measuring brake shoes at the widest point

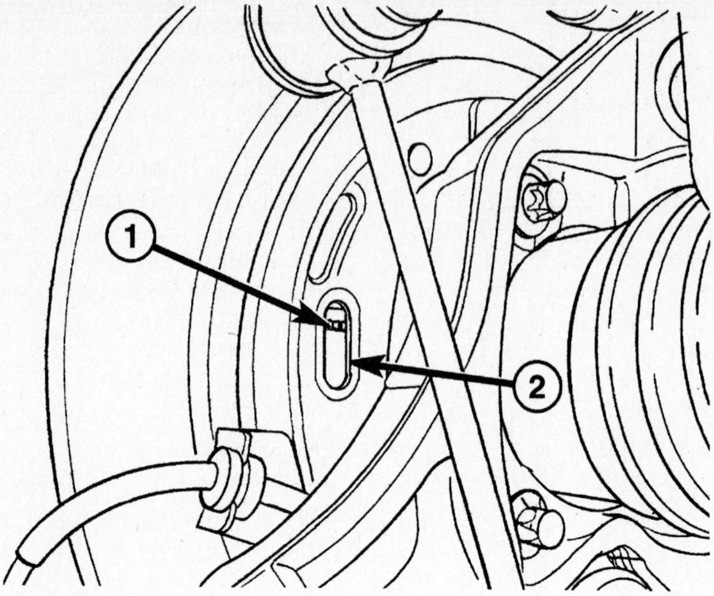

Adjuster access hole

6. Back off adjuster six detents (teeth).
7. Rotate wheel, checking for light drag. If drag is too heavy, continue to back off adjuster one detent at a time until light drag is present. Do not back off star-wheel more than 17 detents from wheel lock.
8. Install access plug.
9. Adjust opposite wheel parking brake shoes using same method.
10. Lower vehicle.
11. Apply and release parking brake lever once to ensure proper operation of parking brakes.

CHRYSLER AND DODGE

300M • Concorde • Intrepid • LHS

2

SPECIFICATIONS AND MAINTENANCE CHARTS

ENGINE AND VEHICLE IDENTIFICATION

Engine						
Code ①	Liters (cc)	Cu. In.	Cyl.	Fuel Sys.	Engine Type	Eng. Mfg.
G	3.5 (3518)	215	V6	MFI	SOHC	Chrysler
R	2.7 (2736)	167	V6	MFI	DOHC	Chrysler
U	2.7 (2736) ③	167	V6	MFI	DOHC	Chrysler
V	3.5 (3518) ④	215	V6	MFI	SOHC	Chrysler
K	3.5 (3518)	215	V6	MFI	SOHC	Chrysler
M	3.5 (3518)	215	V6	MFI	SOHC	Chrysler

Model Year	
Code ②	Year
2	2002
3	2003
4	2004
5	2005

MFI: Multi-point Fuel Injection

DOHC: Double Overhead Camshafts

SOHC: Single Overhead Camshaft

① 8th position of the Vehicle Identification Number (VIN)

② 10th position of VIN

③ 120 amp alternator

④ Magnum

06009-300M-C01

GENERAL ENGINE SPECIFICATIONS

All measurements are given in inches.

Year	Model	Engine Displacement Liters	Engine Series VIN	Net Horsepower @ rpm	Net Torque @ rpm (ft. lbs.)	Bore x Stroke (in.)	Com-pression Ratio	Oil Pressure @ rpm
2002	300M	3.5	G	253@6400	255@3950	3.78x3.19	9.6:1	25-105@3000
	Concorde	2.7	R	200@5800	190@4850	3.39x3.09	9.6:1	25-105@3000
		2.7	U	200@5800	190@4850	3.39x3.09	9.6:1	25-105@3000
		3.5	K	232@6200	241@4000	3.78X3.19	9.6:1	25-105@3000
		3.5	M	232@6200	241@4000	3.78X3.19	9.6:1	25-105@3000
		3.5	V	253@6400	255@3950	3.78x3.19	9.6:1	25-105@3000
	Intrepid	2.7	R	200@5800	190@4850	3.39x3.09	9.6:1	25-105@3000
		2.7	U	200@5800	190@4850	3.39x3.09	9.6:1	25-105@3000
		3.5	K	232@6200	241@4000	3.78X3.19	9.6:1	25-105@3000
		3.5	M	232@6200	241@4000	3.78X3.19	9.6:1	25-105@3000
		3.5	V	253@6400	255@3950	3.78x3.19	9.6:1	25-105@3000
	Concord Limited	3.5	G	253@6400	255@3950	3.78x3.19	9.6:1	25-105@3000
2003	300M	3.5	G	253@6400	255@3950	3.78x3.19	9.6:1	25-105@3000
	Concorde	2.7	R	200@5800	190@4850	3.39x3.09	9.6:1	25-105@3000
		2.7	U	200@5800	190@4850	3.39x3.09	9.6:1	25-105@3000
		3.5	K	232@6200	241@4000	3.78X3.19	9.6:1	25-105@3000
		3.5	M	232@6200	241@4000	3.78X3.19	9.6:1	25-105@3000
		3.5	V	253@6400	255@3950	3.78x3.19	9.6:1	25-105@3000
	Intrepid	2.7	R	200@5800	190@4850	3.39x3.09	9.6:1	25-105@3000
		2.7	U	200@5800	190@4850	3.39x3.09	9.6:1	25-105@3000
		3.5	K	232@6200	241@4000	3.78X3.19	9.6:1	25-105@3000
		3.5	M	232@6200	241@4000	3.78X3.19	9.6:1	25-105@3000
		3.5	V	253@6400	255@3950	3.78x3.19	9.6:1	25-105@3000
	Concord Limited	3.5	G	253@6400	255@3950	3.78x3.19	9.6:1	25-105@3000
2004	300M	3.5	G	253@6400	255@3950	3.78x3.19	9.6:1	25-105@3000
	Concorde	2.7	R	200@5800	190@4850	3.39x3.09	9.6:1	25-105@3000
		3.5	K	232@6200	241@4000	3.78X3.19	9.6:1	25-105@3000
		3.5	M	232@6200	241@4000	3.78X3.19	9.6:1	25-105@3000
		3.5	V	253@6400	255@3950	3.78x3.19	9.6:1	25-105@3000
	Intrepid	2.7	R	200@5800	190@4850	3.39x3.09	9.6:1	25-105@3000
		3.5	K	232@6200	241@4000	3.78X3.19	9.6:1	25-105@3000
		3.5	M	232@6200	241@4000	3.78X3.19	9.6:1	25-105@3000
		3.5	V	253@6400	255@3950	3.78x3.19	9.6:1	25-105@3000
	Concord Limited	3.5	G	253@6400	255@3950	3.78x3.19	9.6:1	25-105@3000

06009-300M-C02

ENGINE TUNE-UP SPECIFICATIONS

Year	Engine Displacement Liters	Engine VIN	Spark Plug Gap (in.)	Ignition Timing (deg.)	Fuel Pump (psi)	Idle Speed (rpm)	Valve Clearance Intake	Valve Clearance Exhaust
2002	2.7	R	0.048-0.053	①	58	①	HYD	HYD
	2.7	U	0.048-0.053	①	58	①	HYD	HYD
	3.5	K	0.048-0.053	①	58	①	HYD	HYD
	3.5	M	0.048-0.053	①	58	①	HYD	HYD
	3.5	V	0.048-0.053	①	58	①	HYD	HYD
	3.5	G	0.048-0.053	①	58	①	HYD	HYD
2003	2.7	R	0.048-0.053	①	58	①	HYD	HYD
	2.7	U	0.048-0.053	①	58	①	HYD	HYD
	3.5	K	0.048-0.053	①	58	①	HYD	HYD
	3.5	M	0.048-0.053	①	58	①	HYD	HYD
	3.5	V	0.048-0.053	①	58	①	HYD	HYD
	3.5	G	0.048-0.053	①	58	①	HYD	HYD
2004	2.7	R	0.048-0.053	①	58	①	HYD	HYD
	3.5	K	0.048-0.053	①	58	①	HYD	HYD
	3.5	M	0.048-0.053	①	58	①	HYD	HYD
	3.5	V	0.048-0.053	①	58	①	HYD	HYD
	3.5	G	0.048-0.053	①	58	①	HYD	HYD

NOTE: The Vehicle Emission Control Information label often reflects specification changes made during production.
The label figures must be used if they differ from those in this chart.

HYD: Hydraulic

① Controlled by the Powertrain Control Module (PCM)

06009-300M-C03

3.5L (VIN G) Engines
Firing Order: 1–2–3–4–5–6
Distributorless ignition system (One coil per cylinder)

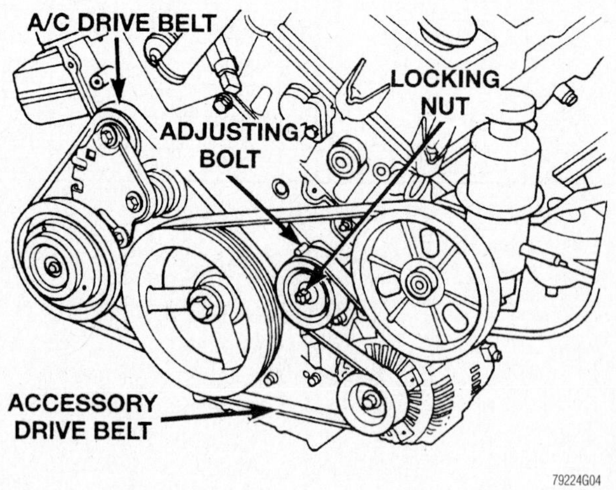

Accessory drive belt routing—2.7L engine

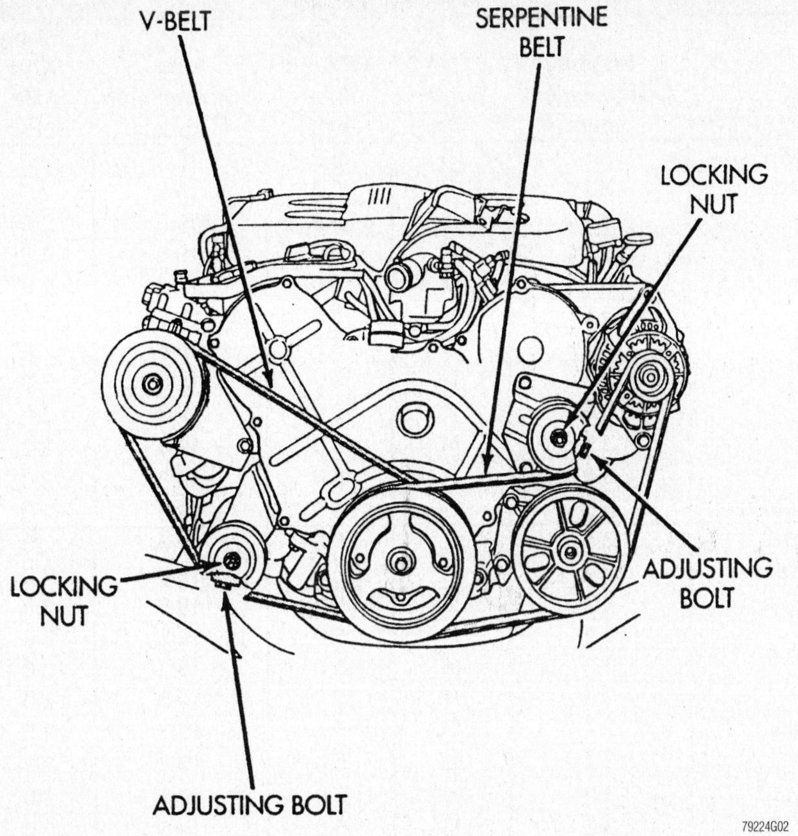

Accessory drive belt routing—3.5L engine

CAPACITIES

Year	Model	Engine Displacement Liters	Engine VIN	Engine Oil with Filter (qts.)	Auto. Transmission (pts.) ①	Front Drive Axle (pts.)	Fuel Tank (gal.)	Cooling System (qts.)
2002	300M	3.5	G	5.0	19.8	2.0	18.0	9.4
	Concorde	2.7	R	5.0	19.8	2.0	18.0	9.4
		2.7	U	5.0	19.8	2.0	18.0	9.4
		3.5	K	5.0	19.8	2.0	18.0	9.4
		3.5	M	5.0	19.8	2.0	18.0	9.4
		3.5	V	5.0	19.8	2.0	18.0	9.4
	Intrepid	2.7	R	5.0	19.8	2.0	18.0	9.4
		2.7	U	5.0	19.8	2.0	18.0	9.4
		3.5	K	5.0	19.8	2.0	18.0	9.4
		3.5	M	5.0	19.8	2.0	18.0	9.4
		3.5	V	5.0	19.8	2.0	18.0	9.4
	Concorde Limited	3.5	G	5.0	19.8	2.0	18.0	9.4
2003	300M	3.5	G	5.0	19.8	2.0	18.0	9.4
	Concorde	2.7	R	5.0	19.8	2.0	18.0	9.4
		2.7	U	5.0	19.8	2.0	18.0	9.4
		3.5	K	5.0	19.8	2.0	18.0	9.4
		3.5	M	5.0	19.8	2.0	18.0	9.4
		3.5	V	5.0	19.8	2.0	18.0	9.4
	Intrepid	2.7	R	5.0	19.8	2.0	18.0	9.4
		2.7	U	5.0	19.8	2.0	18.0	9.4
		3.5	K	5.0	19.8	2.0	18.0	9.4
		3.5	M	5.0	19.8	2.0	18.0	9.4
		3.5	V	5.0	19.8	2.0	18.0	9.4
	Concorde Limited	3.5	G	5.0	19.8	2.0	18.0	9.4
2004	300M	3.5	G	5.0	19.8	2.0	18.0	9.4
	Concorde	2.7	R	5.0	19.8	2.0	18.0	9.4
		3.5	K	5.0	19.8	2.0	18.0	9.4
		3.5	M	5.0	19.8	2.0	18.0	9.4
		3.5	V	5.0	19.8	2.0	18.0	9.4
	Intrepid	2.7	R	5.0	19.8	2.0	18.0	9.4
		3.5	K	5.0	19.8	2.0	18.0	9.4
		3.5	M	5.0	19.8	2.0	18.0	9.4
		3.5	V	5.0	19.8	2.0	18.0	9.4
	Concorde Limited	3.5	G	5.0	19.8	2.0	18.0	9.4

NOTE: All capacities are approximate. Add fluid gradually and ensure a proper fluid level is obtained.

① Overhaul fill capacity with torque converter empty

Estimated service fill: 9 pts.

06009-300M-C04

VALVE SPECIFICATIONS

Year	Engine Displacement Liters	Engine VIN	Seat Angle (deg.)	Face Angle (deg.)	Spring Test Pressure (lbs. @ in.)	Spring Installed Height (in.)	Stem-to-Guide Clearance (in.) Intake	Stem-to-Guide Clearance (in.) Exhaust	Stem Diameter (in.) Intake	Stem Diameter (in.) Exhaust
2002	3.5	G	45-45.5	44.5-45	①	1.496	0.0009-0.0026	0.0020-0.0037	0.2730-0.2737	0.2719-0.2726
	2.7	R	44.5-45	45-45.5	②	1.496	0.0009-0.0026	0.0020-0.0033	0.2337-0.2344	0.2326-0.2333
	2.7	U	44.5-45	45-45.5	②	1.496	0.0009-0.0026	0.0020-0.0033	0.2337-0.2344	0.2326-0.2333
	3.5	K	45-45.5	44.5-45	①	1.496	0.0009-0.0026	0.0020-0.0037	0.2730-0.2737	0.2719-0.2726
	3.5	M	45-45.5	44.5-45	①	1.496	0.0009-0.0026	0.0020-0.0037	0.2730-0.2737	0.2719-0.2726
	3.5	V	45-45.5	44.5-45	①	1.496	0.0009-0.0026	0.0020-0.0037	0.2730-0.2737	0.2719-0.2726
2003	3.5	G	45-45.5	44.5-45	①	1.496	0.0009-0.0026	0.0020-0.0037	0.2730-0.2737	0.2719-0.2726
	2.7	R	44.5-45	45-45.5	②	1.496	0.0009-0.0026	0.0020-0.0033	0.2337-0.2344	0.2326-0.2333
	2.7	U	44.5-45	45-45.5	②	1.496	0.0009-0.0026	0.0020-0.0033	0.2337-0.2344	0.2326-0.2333
	3.5	K	45-45.5	44.5-45	①	1.496	0.0009-0.0026	0.0020-0.0037	0.2730-0.2737	0.2719-0.2726
	3.5	M	45-45.5	44.5-45	①	1.496	0.0009-0.0026	0.0020-0.0037	0.2730-0.2737	0.2719-0.2726
	3.5	V	45-45.5	44.5-45	①	1.496	0.0009-0.0026	0.0020-0.0037	0.2730-0.2737	0.2719-0.2726
2004	3.5	G	45-45.5	44.5-45	①	1.496	0.0009-0.0026	0.0020-0.0037	0.2730-0.2737	0.2719-0.2726
	2.7	R	44.5-45	45-45.5	②	1.496	0.0009-0.0026	0.0020-0.0033	0.2337-0.2344	0.2326-0.2333
	3.5	K	45-45.5	44.5-45	①	1.496	0.0009-0.0026	0.0020-0.0037	0.2730-0.2737	0.2719-0.2726
	3.5	M	45-45.5	44.5-45	①	1.496	0.0009-0.0026	0.0020-0.0037	0.2730-0.2737	0.2719-0.2726
	3.5	V	45-45.5	44.5-45	①	1.496	0.0009-0.0026	0.0020-0.0037	0.2730-0.2737	0.2719-0.2726

① Intake: 69.5-80.5 lbs.@1.496 in. valve closed
Intake: 188.0-204.0 lbs.@1.1594 in. valve opened
Exhaust: 71-79 lbs.@1.4961 in. valve closed
Exhaust: 130-144 lbs.@1.239 in. valve opened

② 56-64 lbs.@ 1.496 in. valve closed
Intake: 147.9-162.1 lbs.@1.1417 in. valve opened
Exhaust: 138.0-150.8 lbs.@1.811 in. valve opened

06009-300M-C05

CRANKSHAFT AND CONNECTING ROD SPECIFICATIONS

All measurements are given in inches.

	Engine		Crankshaft				Connecting Rod		
Year	Displacement Liters	Engine VIN	Main Brg. Journal Dia.	Main Brg. Oil Clearance	Shaft End-play	Thrust on No.	Journal Diameter	Oil Clearance	Side Clearance
2002	2.7	R	2.4997-2.5004	0.0014-0.0021	0.017 max.	3	2.1067-2.1060	0.0010-0.0026	0.0052-0.0150
	2.7	U	2.4997-2.5004	0.0014-0.0021	0.017 max.	3	2.1067-2.1060	0.0010-0.0026	0.0052-0.0150
	3.5	K	2.5190-2.5200	0.0004-0.0022	0.0040-0.0120	2	2.2830-2.2840	0.0008-0.0034	0.0050-0.0150
	3.5	M	2.5190-2.5200	0.0004-0.0022	0.0040-0.0120	2	2.2830-2.2840	0.0008-0.0034	0.0050-0.0150
	3.5	V	2.5190-2.5200	0.0004-0.0022	0.0040-0.0120	2	2.2830-2.2840	0.0008-0.0034	0.0050-0.0150
	3.5	G	2.5190-2.5200	0.0004-0.0022	0.0040-0.0120	2	2.2830-2.2840	0.0008-0.0034	0.0050-0.0150
2003	2.7	R	2.4997-2.5004	0.0014-0.0021	0.017 max.	3	2.1067-2.1060	0.0010-0.0026	0.0052-0.0150
	2.7	U	2.4997-2.5004	0.0014-0.0021	0.017 max.	3	2.1067-2.1060	0.0010-0.0026	0.0052-0.0150
	3.5	K	2.5190-2.5200	0.0004-0.0022	0.0040-0.0120	2	2.2830-2.2840	0.0008-0.0034	0.0050-0.0150
	3.5	M	2.5190-2.5200	0.0004-0.0022	0.0040-0.0120	2	2.2830-2.2840	0.0008-0.0034	0.0050-0.0150
	3.5	V	2.5190-2.5200	0.0004-0.0022	0.0040-0.0120	2	2.2830-2.2840	0.0008-0.0034	0.0050-0.0150
	3.5	G	2.5190-2.5200	0.0004-0.0022	0.0040-0.0120	2	2.2830-2.2840	0.0008-0.0034	0.0050-0.0150
2004	2.7	R	2.4997-2.5004	0.0014-0.0021	0.017 max.	3	2.1067-2.1060	0.0010-0.0026	0.0052-0.0150
	3.5	K	2.5190-2.5200	0.0004-0.0022	0.0040-0.0120	2	2.2830-2.2840	0.0008-0.0034	0.0050-0.0150
	3.5	M	2.5190-2.5200	0.0004-0.0022	0.0040-0.0120	2	2.2830-2.2840	0.0008-0.0034	0.0050-0.0150
	3.5	V	2.5190-2.5200	0.0004-0.0022	0.0040-0.0120	2	2.2830-2.2840	0.0008-0.0034	0.0050-0.0150
	3.5	G	2.5190-2.5200	0.0004-0.0022	0.0040-0.0120	2	2.2830-2.2840	0.0008-0.0034	0.0050-0.0150

06009-300M-C06

PISTON AND RING SPECIFICATIONS

All measurements are given in inches.

	Engine			Ring Gap			Ring Side Clearance		
Year	Displacement Liters	Engine VIN	Piston Clearance	Top Compression	Bottom Compression	Oil Control	Top Compression	Bottom Compression	Oil Control
2002	2.7	R	0.0001-0.0016	0.008-0.014	0.0146-0.0249	0.010-0.030	0.0016-0.0031	0.0016-0.0031	0.0025-0.0082
	2.7	U	0.0003-0.0018	0.008-0.014	0.0146-0.0249	0.010-0.030	0.0016-0.0031	0.0016-0.0031	0.0025-0.0082
	3.5	K	0.0003-0.0018	0.008-0.014	0.0087-0.0193	0.010-0.030	0.0016-0.0031	0.0016-0.0031	0.0015-0.0073
	3.5	M	0.0003-0.0018	0.008-0.014	0.0087-0.0193	0.010-0.030	0.0016-0.0031	0.0016-0.0031	0.0015-0.0073
	3.5	V	0.0003-0.0018	0.008-0.014	0.0087-0.0193	0.010-0.030	0.0016-0.0031	0.0016-0.0031	0.0015-0.0073
	3.5	G	0.0003-0.0018	0.008-0.014	0.0087-0.0193	0.010-0.030	0.0016-0.0031	0.0016-0.0031	0.0015-0.0073
2003	2.7	R	0.0001-0.0016	0.008-0.014	0.0146-0.0249	0.010-0.030	0.0016-0.0031	0.0016-0.0031	0.0025-0.0082
	2.7	U	0.0003-0.0018	0.008-0.014	0.0146-0.0249	0.010-0.030	0.0016-0.0031	0.0016-0.0031	0.0025-0.0082
	3.5	K	0.0003-0.0018	0.008-0.014	0.0087-0.0193	0.010-0.030	0.0016-0.0031	0.0016-0.0031	0.0015-0.0073
	3.5	M	0.0003-0.0018	0.008-0.014	0.0087-0.0193	0.010-0.030	0.0016-0.0031	0.0016-0.0031	0.0015-0.0073
	3.5	V	0.0003-0.0018	0.008-0.014	0.0087-0.0193	0.010-0.030	0.0016-0.0031	0.0016-0.0031	0.0015-0.0073
	3.5	G	0.0003-0.0018	0.008-0.014	0.0087-0.0193	0.010-0.030	0.0016-0.0031	0.0016-0.0031	0.0015-0.0073
2004	2.7	R	0.0001-0.0016	0.008-0.014	0.0146-0.0249	0.010-0.030	0.0016-0.0031	0.0016-0.0031	0.0025-0.0082
	3.5	K	0.0003-0.0018	0.008-0.014	0.0087-0.0193	0.010-0.030	0.0016-0.0031	0.0016-0.0031	0.0015-0.0073
	3.5	M	0.0003-0.0018	0.008-0.014	0.0087-0.0193	0.010-0.030	0.0016-0.0031	0.0016-0.0031	0.0015-0.0073
	3.5	V	0.0003-0.0018	0.008-0.014	0.0087-0.0193	0.010-0.030	0.0016-0.0031	0.0016-0.0031	0.0015-0.0073
	3.5	G	0.0003-0.0018	0.008-0.014	0.0087-0.0193	0.010-0.030	0.0016-0.0031	0.0016-0.0031	0.0015-0.0073

06009-300M-C07

TORQUE SPECIFICATIONS

All readings in ft. lbs.

Year	Engine Displacement Liters	Engine VIN	Cylinder Head Bolts	Main Bearing Bolts	Rod Bearing Bolts	Crankshaft Damper Bolts	Flywheel Bolts	Manifold Intake	Manifold Exhaust	Spark Plugs	Oil Pan Drain Plug
2002	3.5	G	①	②	③	70	70	109④	200④	20	20
	3.5	K	①	②	③	70	70	109④	200④	20	20
	3.5	M	①	②	③	70	70	109④	200④	20	20
	2.7	R	⑤	②	⑥	125	70	111④	200④	15	20
	2.7	U	⑤	②	⑥	125	70	112④	200④	15	20
	3.5	V	①	②	③	70	70	113④	200④	20	20
2003	3.5	G	①	②	③	70	70	109④	200④	20	20
	3.5	K	①	②	③	70	70	109④	200④	20	20
	3.5	M	①	②	③	70	70	109④	200④	20	20
	2.7	R	⑤	②	⑥	125	70	111④	200④	15	20
	2.7	U	⑤	②	⑥	125	70	112④	200④	15	20
	3.5	V	①	②	③	70	70	113④	200④	20	20
2004	3.5	G	①	②	③	70	70	109④	200④	20	20
	3.5	K	①	②	③	70	70	109④	200④	20	20
	3.5	M	①	②	③	70	70	109④	200④	20	20
	2.7	R	⑤	②	⑥	125	70	111④	200④	15	20
	3.5	V	①	②	③	70	70	113④	200④	20	20

① Step 1: 45 ft. lbs.
Step 2: 65 ft. lbs.
Step 3: 65 ft. lbs.
Step 4: Plus 1/4 turn
Final torque should be over 90 ft. lbs.

② Main cap inside bolts: 15 ft. lbs. plus 1/4 turn
Main cap outside bolts: 20 ft. lbs. plus 1/4 turn
Main cap tie bolts: 250 inch lbs.

③ Step 1: 40 ft. lbs.
Step 2: Plus 1/4 turn

④ Specification is in Inch lbs.

⑤ Step 1: 35 ft. lbs.
Step 2: 55 ft. lbs.
Step 3: 55 ft. lbs.
Step 4: Plus 90 degrees
Step 5: M8 bolts (3 front) 250 inch lbs.

⑥ Step 1: 20 ft. lbs.
Step 2: Plus 1/4 turn

06009-300M-C08

WHEEL ALIGNMENT

Year	Model		Caster Range (+/-Deg.)	Caster Preferred Setting (Deg.)	Camber Range (+/-Deg.)	Camber Preferred Setting (Deg.)	Toe-in (in.)
2002	300M	F	+1.00	+3.00	+0.60	0	0.10 +/- 0.20
		R	—	—	+0.50	-0.20	0.10 +/- 0.30
	Concorde	F	+1.00	+3.00	+0.56	0	0.09 +/- 0.10
		R	—	—	+0.50	-0.20	0.09 +/- 0.15
	Concorde Limited	F	+1.00	+3.00	+0.60	0	0.09 +/- 0.10
		R	—	—	+0.50	-0.20	0.05 +/- 0.15
	Intrepid	F	+1.00	+3.00	+0.56	0	0.05 +/- 0.10
		R	—	—	+0.50	-0.20	0.05 +/- 0.15
2003	300M	F	+1.00	+3.00	+0.60	0	0.10 +/- 0.20
		R	—	—	+0.50	-0.20	0.10 +/- 0.30
	Concorde	F	+1.00	+3.00	+0.56	0	0.09 +/- 0.10
		R	—	—	+0.50	-0.20	0.09 +/- 0.15
	Concorde Limited	F	+1.00	+3.00	+0.60	0	0.09 +/- 0.10
		R	—	—	+0.50	-0.20	0.05 +/- 0.15
	Intrepid	F	+1.00	+3.00	+0.56	0	0.05 +/- 0.10
		R	—	—	+0.50	-0.20	0.05 +/- 0.15
2004	300M	F	+1.00	+3.00	+0.60	0	0.10 +/- 0.20
		R	—	—	+0.50	-0.20	0.10 +/- 0.30
	Concorde	F	+1.00	+3.00	+0.56	0	0.09 +/- 0.10
		R	—	—	+0.50	-0.20	0.09 +/- 0.15
	Concorde Limited	F	+1.00	+3.00	+0.60	0	0.09 +/- 0.10
		R	—	—	+0.50	-0.20	0.05 +/- 0.15
	Intrepid	F	+1.00	+3.00	+0.56	0	0.05 +/- 0.10
		R	—	—	+0.50	-0.20	0.05 +/- 0.15

06009-300M-C09

TIRE, WHEEL AND BALL JOINT SPECIFICATIONS

Year	Model	OEM Tires Standard	OEM Tires Optional	Tire Pressures (psi) Front	Tire Pressures (psi) Rear	Wheel Size	Ball Joint Inspection	Lug Nuts (ft. lbs.)
2002	300M	P225/55/R17	None	30	30	7-JJ	①	100
	Concorde	P205/70R15	P225/60R16	32	32	6-JJ	①	100
	Concorde Limited	P225/55/R17	None	30	30	7-JJ	①	100
	Intrepid	P225/55/R17	None	30	30	7-JJ	①	100
2003	300M	P225/55/R17	None	30	30	7-JJ	①	100
	Concorde	P205/70R15	P225/60R16	32	32	6-JJ	①	100
	Concorde Limited	P225/55/R17	None	30	30	7-JJ	①	100
	Intrepid	P225/55/R17	None	30	30	7-JJ	①	100
2004	300M	P225/55/R17	None	30	30	7-JJ	①	100
	Concorde	P205/70R15	P225/60R16	32	32	6-JJ	①	100
	Concorde Limited	P225/55/R17	None	30	30	7-JJ	①	100
	Intrepid	P225/55/R17	None	30	30	7-JJ	①	100

OEM: Original Equipment Manufacturer

PSI: Pounds Per Square Inch

① Do not lift car. Grasp the grease fitting and attempt to move or rotate. Replace if any movement is found.

06009-300M-C10

BRAKE SPECIFICATIONS

All measurements in inches unless noted

			Brake Disc				Brake Caliper
Year	Model		Original Thickness	Minimum Thickness	Maximum Runout	Minimum Lining Thickness	Mounting Bolts (ft. lbs.)
2002	300M	F	1.024	0.960	0.0035	0.250	17
		R	0.468	0.409	0.0035	0.280	17
	Concorde	F	1.024	0.960	0.0035	0.250	17
		R	0.468	0.409	0.0035	0.280	17
	Intrepid	F	1.024	0.960	0.0035	0.250	17
		R	0.468	0.409	0.0035	0.280	17
	Concord Ltd	F	1.024	0.960	0.0035	0.250	17
		R	0.468	0.409	0.0035	0.280	17
2003	300M	F	1.024	0.960	0.0035	0.250	17
		R	0.468	0.409	0.0035	0.280	17
	Concorde	F	1.024	0.960	0.0035	0.250	17
		R	0.468	0.409	0.0035	0.280	17
	Intrepid	F	1.024	0.960	0.0035	0.250	17
		R	0.468	0.409	0.0035	0.280	17
	Concord Ltd	F	1.024	0.960	0.0035	0.250	17
		R	0.468	0.409	0.0035	0.280	17
2004	300M	F	1.024	0.960	0.0035	0.250	17
		R	0.468	0.409	0.0035	0.280	17
	Concorde	F	1.024	0.960	0.0035	0.250	17
		R	0.468	0.409	0.0035	0.280	17
	Intrepid	F	1.024	0.960	0.0035	0.250	17
		R	0.468	0.409	0.0035	0.280	17
	Concord Ltd	F	1.024	0.960	0.0035	0.250	17
		R	0.468	0.409	0.0035	0.280	17

F: Front
R: Rear

06009-300M-C11

SCHEDULED MAINTENANCE INTERVALS

2002 Chrysler—300M, Concorde, LHS & Dodge—Intrepid

TO BE SERVICED	TYPE OF SERVICE	VEHICLE MILEAGE INTERVAL (x1000)												
		7.5	15	22.5	30	37.5	45	52.5	60	67.5	75	82.5	90	97.5
Engine oil & filter	R	✓	✓	✓	✓	✓	✓	✓	✓	✓	✓	✓	✓	✓
Exhaust system	S/I	✓	✓	✓	✓	✓	✓	✓	✓	✓	✓	✓	✓	✓
Brake hoses	S/I	✓	✓	✓	✓	✓	✓	✓	✓	✓	✓	✓	✓	✓
CV joints & front suspension components	S/I	✓	✓	✓	✓	✓	✓	✓	✓	✓	✓	✓	✓	✓
Rotate tires	S/I	✓	✓	✓	✓	✓	✓	✓	✓	✓	✓	✓	✓	✓
Coolant level, hoses & clamps	S/I	✓	✓	✓	✓	✓	✓	✓	✓	✓	✓	✓	✓	✓
Accessory drive belts	S/I					✓			✓		✓		✓	
Brake linings	S/I			✓				✓	✓				✓	
Spark plugs	R				✓				✓				✓	
Air filter element	R					✓			✓				✓	
Lubricate steering linkage & tie rod ends	S/I				✓				✓				✓	
Engine coolant	R						✓				✓			
PCV valve	S/I								✓				✓	
Ignition cables	R								✓					
Camshaft timing belt	R								✓					

R: Replace S/I: Service or Inspect

FREQUENT OPERATION MAINTENANCE (SEVERE SERVICE)

If a vehicle is operated under any of the following conditions it is considered severe service:

- Extremely dusty areas

- 50% or more of the vehicle operation is in 32°C (90°F) or higher temperatures, or constant operation in temperatures below 0°C (32°F)

Prolonged idling (vehicle operation in stop and go traffic)

- Frequent short running periods (engine does not warm to normal operating temperatures)

- Police, taxi, delivery usage or trailer towing usage

CV joints & front suspension components: check every 3000 miles

Oil & oil filter change: change every 3000 miles

Rotate tires: every 3000 miles

Brake linings: check every 9000 miles

Air filter element: change every 15,000 miles

Automatic transaxle fluid: change every 15,000 miles

Differential fluid: change every 15,000 miles

Tie rod ends & steering linkage: lubricate every 15,000 miles

PCV valve: check every 30,000 miles

06009-300M-C12

SCHEDULED MAINTENANCE INTERVALS

2003-04 Chrysler—300M, Concorde, LHS & Dodge—Intrepid

		VEHICLE MILEAGE INTERVAL (x1000)												
TO BE SERVICED	TYPE OF SERVICE	6	12	18	24	30	36	42	48	54	60	66	72	78
Engine oil & filter	R	✓	✓	✓	✓	✓	✓	✓	✓	✓	✓	✓	✓	✓
Exhaust system	S/I	✓	✓	✓	✓	✓	✓	✓	✓	✓	✓	✓	✓	✓
Brake hoses	S/I	✓	✓	✓	✓	✓	✓	✓	✓	✓	✓	✓	✓	✓
CV joints & front suspension components	S/I	✓	✓	✓	✓	✓	✓	✓	✓	✓	✓	✓	✓	✓
Rotate tires	S/I	✓	✓	✓	✓	✓	✓	✓	✓	✓	✓	✓	✓	✓
Coolant level, hoses & clamps	S/I	✓	✓	✓	✓	✓	✓	✓	✓	✓	✓	✓	✓	✓
Accessory drive belts	S/I					✓			✓		✓		✓	
Brake linings	S/I			✓				✓	✓				✓	
Spark plugs	R				✓				✓				✓	
Air filter element	R					✓			✓				✓	
Lubricate steering linkage & tie rod ends	S/I				✓				✓				✓	
Engine coolant	R						✓				✓			
PCV valve	S/I								✓				✓	
Ignition cables	R								✓					
Camshaft timing belt	R								✓					

R: Replace S/I: Service or Inspect

FREQUENT OPERATION MAINTENANCE (SEVERE SERVICE)

If a vehicle is operated under any of the following conditions it is considered severe service:

- Extremely dusty areas

- 50% or more of the vehicle operation is in 32°C (90°F) or higher temperatures, or constant operation in temperatures below 0°C (32°F)

Prolonged idling (vehicle operation in stop and go traffic)

- Frequent short running periods (engine does not warm to normal operating temperatures)

- Police, taxi, delivery usage or trailer towing usage

CV joints & front suspension components: check every 3000 miles

Oil & oil filter change: change every 3000 miles

Rotate tires: every 3000 miles

Brake linings: check every 9000 miles

Air filter element: change every 15,000 miles

Automatic transaxle fluid: change every 15,000 miles

Differential fluid: change every 15,000 miles

Tie rod ends & steering linkage: lubricate every 15,000 miles

PCV valve: check every 30,000 miles

06009-300M-C13

PRECAUTIONS

Before servicing any vehicle, please be sure to read all of the following precautions, which deal with personal safety, prevention of component damage, and important points to take into consideration when servicing a motor vehicle:

- Never open, service or drain the radiator or cooling system when the engine is hot; serious burns can occur from the steam and hot coolant.
- Observe all applicable safety precautions when working around fuel. Whenever servicing the fuel system, always work in a well-ventilated area. Do not allow fuel spray or vapors to come in contact with a spark, open flame or excessive heat (a hot drop light, for example). Keep a dry chemical fire extinguisher near the work area. Always keep fuel in a container specifically designed for fuel storage; also, always properly seal fuel containers to avoid the possibility of fire or explosion. Refer to the additional fuel system precautions later in this section.
- Fuel injection systems often remain pressurized, even after the engine has been turned **OFF**. The fuel system pressure must be relieved before disconnecting any fuel lines. Failure to do so may result in fire and/or personal injury.
- Brake fluid often contains polyglycol ethers and polyglycols. Avoid contact with the eyes and wash your hands thoroughly after handling brake fluid. If you do get brake fluid in your eyes, flush your eyes with clean, running water for 15 minutes. If eye irritation persists, or if you have taken brake fluid internally, IMMEDIATELY seek medical assistance.
- The EPA warns that prolonged contact with used engine oil may cause a number of skin disorders, including cancer! You should make every effort to minimize your exposure to used engine oil. Protective gloves should be worn when changing oil. Wash your hands and any other exposed skin areas as soon as possible after exposure to used engine oil. Soap and water, or waterless hand cleaner should be used.
- All new vehicles are now equipped with an air bag system. The system must be disabled before performing service on or around system components, steering column, instrument panel components, wiring and sensors. Failure to follow safety and disabling procedures could result in accidental air bag deployment, possible personal injury and unnecessary system repairs.
- Always wear safety goggles when working with, or around, the air bag system. When carrying a non-deployed air bag, be sure the bag and trim cover are pointed away from your body. When placing a non-deployed air bag on a work surface, always face the bag and trim cover upward, away from the surface. This will reduce the motion of the module if it is accidentally deployed. Refer to the additional air bag system precautions later in this section.
- Clean, high quality brake fluid from a sealed container is essential to the safe and proper operation of the brake system. You should always buy the correct type of brake fluid for your vehicle. If the brake fluid becomes contaminated, completely flush the system with new fluid. Never reuse any brake fluid. Any brake fluid that is removed from the system should be discarded. Also, do not allow any brake fluid to come in contact with a painted surface; it will damage the paint.
- Never operate the engine without the proper amount and type of engine oil; doing so WILL result in severe engine damage.
- Timing belt maintenance is extremely important! Many models utilize an interference-type, non-freewheeling engine. If the timing belt breaks, the valves in the cylinder head may strike the pistons, causing potentially serious (also time-consuming and expensive) engine damage.
- Disconnecting the negative battery cable on some vehicles may interfere with the functions of the on-board computer system(s) and may require the computer to undergo a relearning process once the negative battery cable is reconnected.
- When servicing drum brakes, only disassemble and assemble one side at a time, leaving the remaining side intact for reference.
- Only an MVAC-trained, EPA-certified automotive technician should service the air conditioning system or its components.

ENGINE REPAIR

Alternator

REMOVAL

2.7L Engine

1. Before servicing the vehicle, refer to the precautions in the beginning of this section.
2. Remove or disconnect the following:
   - Negative battery cable
   - Lower plastic splash shield
   - Transmission cooler and position it aside
   - Lower radiator crossmember support
3. Loosen but do not remove:
   - Adjusting "T" bolt and the pivot bolt
   - Drive belt adjusting bolt
4. Remove or disconnect the following:
   - Drive belt
   - Alternator field circuit plug
   - Alternator B+ terminal nut and wire
   - Pivot bolt; be careful not to lose the spacer
   - Alternator

3.5L Engines

1. Before servicing the vehicle, refer to the precautions in the beginning of this section.
2. Remove or disconnect the following:
   - Negative battery cable
   - Lower mounting bolt and pivot bolt
   - Drive belt tension bolt
   - Drive belt
   - Bracket, lower mounting bolt and pivot bolt
   - Alternator
   - Alternator field circuit plug
   - Alternator B+ terminal nut and wire

INSTALLATION

2.7L Engine

1. Install or connect the following:
   - Alternator
   - Pivot bolt with the spacer and leave loose
   - Alternator field circuit plug
   - Alternator B+ wire and terminal nut; then, torque the nut to 89 inch lbs. (10 Nm)
   - Drive belt
2. Using a belt tension gauge, torque the adjusting "T" bolt until the tension gauge reads 120 lbs. (534 N) for a used belt or 180–200 lbs. (792–880 N) for a new belt.

- Lower radiator crossmember support
- Transmission cooler
- Lower plastic splash shield
- Negative battery cable

3. Torque the bolts to the following specifications:
    a. Pivot bolt to 40 ft. lbs. (54 Nm).
    b. 8mm mounting bolt to 30 ft. lbs. (41 Nm).
    c. 10mm mounting bolt to 40 ft. lbs. (54 Nm).

3.5L Engines

1. Install or connect the following:
    - Alternator B+ wire and terminal nut; then, torque the nut to 89 inch lbs. (10 Nm)
    - Alternator field circuit plug
    - Alternator
    - Drive belt
    - Pivot bolt and lower mounting bolt
2. Using a belt tension gauge, torque the "V" drive belt until the tension gauge reads 120 lbs. (534 N) for a used belt or 140–160 lbs. (623–711 N) for a new belt.
3. Using a belt tension gauge, torque the Poly "V" belt until the tension gauge reads 120 lbs. (534 N) for a used belt or 180–200 lbs. (792–880 N) for a new belt.
4. Install the bracket.
5. Torque the bolts to the following specifications:
    a. Pivot bolt to 40 ft. lbs. (54 Nm).
    b. 8mm mounting bolt to 30 ft. lbs. (41 Nm).
    c. 10mm mounting bolt to 40 ft. lbs. (54 Nm).
6. Connect the negative battery cable.

Ignition Timing

ADJUSTMENT

All models utilize a Distributorless Ignition System (DIS). It is a fixed ignition timing system, which means that basic ignition timing cannot be adjusted. All spark advance is permanently set by the Powertrain Control Module (PCM).

Engine Assembly

REMOVAL & INSTALLATION

2.7L Engine

1. Before servicing the vehicle, refer to the precautions in the beginning of this section.

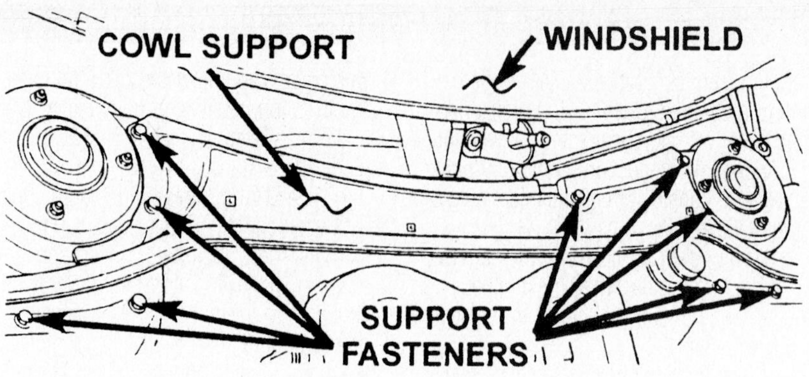

Remove the cowl support for clearance to remove the engine—2.7L engine

2. Relieve the fuel system pressure.
3. Drain the cooling system.
4. Drain the engine oil.
5. Remove or disconnect the following:
    - Negative battery cable at the remote terminal near the right strut tower
    - Hood
    - Wiper arms, cowl covers and cowl support
    - Air intake duct and air cleaner assembly
    - Upper radiator crossmember
    - Hood release cable from the latch
    - Cooling fan assembly
    - Upper and lower radiator hoses
    - A/C condenser-to-radiator fasteners
    - Radiator
    - Accessory drive belts
    - Power steering pump and position it aside without disconnecting the lines

➡Remove the alternator with the engine.

- A/C compressor and position it aside without disconnecting the lines
- V-band clamps at the exhaust manifolds
- Speed control and throttle cables from the throttle body
- Heater hoses and the coolant hoses at the recovery tank
- All vacuum lines, electrical connectors and ground straps from the engine
- Catalytic converter down pipes-to-rear mount fasteners
- Structural collar mounting bolts and collar

6. Matchmark the flexplate-to-torque converter and remove the torque converter bolts.
    - Both transaxle cooler line-to-engine brackets

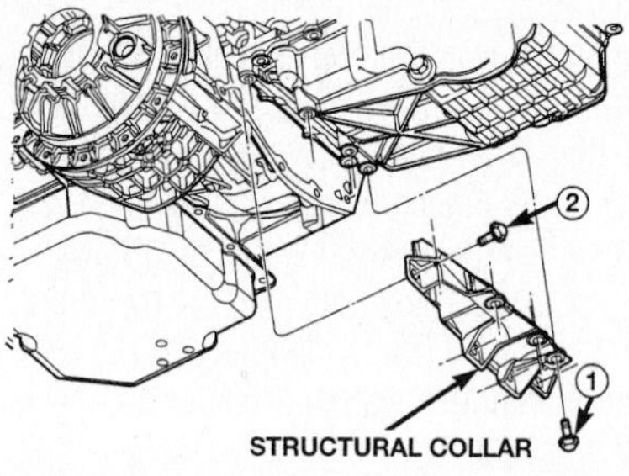

Removing the structural collar—2.7L engine

- Starter and Crankshaft Position (CKP) sensor
- 2 lower transaxle-to-engine bolts
- Exhaust Gas Recirculation (EGR) valve assembly
- Fuel line and engine harness from the throttle body bracket

7. Remove the following brackets from the double-ended transaxle-to-engine bolts:
    - Wiring harness bracket
    - Transaxle shift cable bracket
    - Throttle body support bracket
    - Upper transaxle-to-engine bolts
    - Both (right and left) engine mount isolators-to-engine mount bracket fasteners

⁂ WARNING

Be careful not to damage the intake manifold or cylinder head covers when lifting the engine.

8. Attach an engine lifting fixture to the engine.
9. Support the transaxle with a jack and remove the engine.

To install:

10. Install the engine.
11. Align the engine mounts and install the fasteners, but do not tighten them at this time.

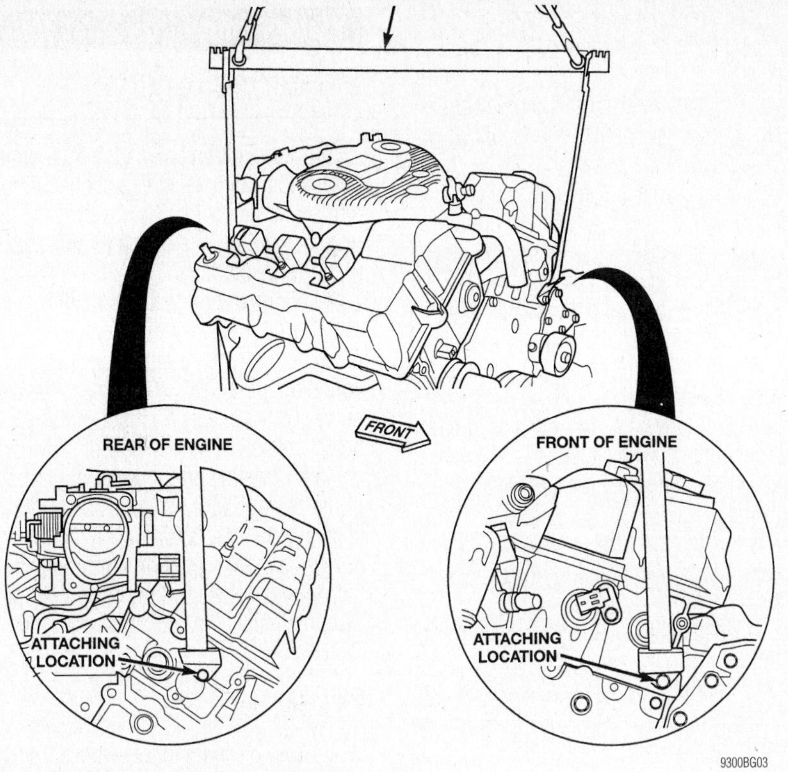

Engine support fixture attaching points—2.7L engine

⁂ WARNING

Do not tighten the transaxle-to-engine bolts until all of the bolts have been hand-started and the engine is flush against the transaxle.

12. Install the transaxle-to-engine bolts. Torque the bolts to 75 ft. lbs. (102 Nm).
13. Remove the engine support fixture.
14. Torque the engine mount isolator-to-bracket fasteners to 45 ft. lbs. (61 Nm).
15. Align the matchmarks and install the flexplate-to-torque converter bolts. Torque the bolts to 55 ft. lbs. (75 Nm).
16. Install the starter and the CKP sensor.
17. Install the structural collar using the following sub-steps:
    a. Torque the vertical collar-to-oil pan bolts to 10 inch lbs. (1.1 Nm).
    b. Torque the collar-to-transaxle bolts to 40 ft. lbs. (55 Nm).
    c. Torque the center vertical bolt to 40 ft. lbs. (55 Nm); then, the remaining vertical bolts to the same torque.
18. Complete the installation by reversing the removal procedure. Keep the following in mind:
    - Use new V-band clamps and torque them to 100 inch lbs. (11 Nm)
    - Torque the A/C compressor and power steering pump fasteners to 21 ft. lbs. (28 Nm)
    - Negative battery cable
19. Fill the cooling system to the proper level.
20. Fill the engine with clean oil.
21. Start the vehicle, check for leaks and repair if necessary.

3.5L Engines

1. Before servicing the vehicle, refer to the precautions in the beginning of this section.
2. Drain the engine oil.
3. Drain the engine coolant.
4. Properly relieve the fuel system pressure.
5. Remove or disconnect the following:
    - Negative battery cable
    - Hood
    - Wiper arms
    - Cowl covers and supports
    - Air cleaner and air inlet duct
    - Upper radiator support and hood release cable
    - Fan module

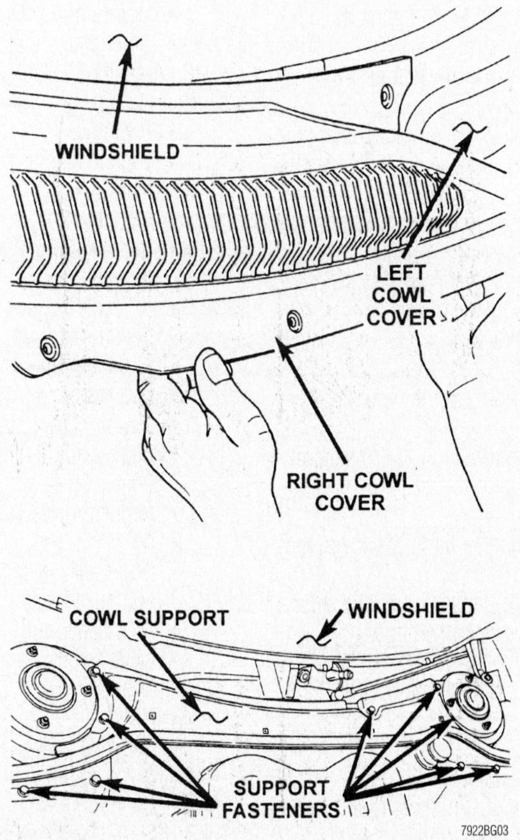

Be careful not to damage the windshield when removing the cowl covers and supports 3.5L models

- Accessory drive belts
- Upper and lower radiator hoses
- Engine oil and transmission cooler lines at the radiator
- Alternator, if necessary.
- A/C compressor mounting bolts and position it aside with the lines connected
- Power steering pump mounting bolts and position the pump aside
- V-band clamp at the right exhaust manifold
- Right side catalytic converter down pipe bracket fasteners
- Fuel lines
- Throttle and cruise control cables
- All vacuum hoses
- Ground straps at both cylinder heads

⁂ WARNING

The intake manifold is a composite design and should be removed before lifting the engine or it may be damaged and require replacement.

- Upper intake manifold and cover the openings
- Heater hoses
- Throttle body support brackets
- Upper transaxle-to-engine mounting bolts
- All electrical and vacuum hose connections
- Structural collar and matchmark the flexplate-to-torque converter position
- Flexplate-to-torque converter mounting bolts
- Left exhaust manifold V-band clamp and left catalytic converter support brackets
- Starter
- Left and right engine mounting bolts
- Crankshaft Position (CKP) sensor
- Lower engine-to-transaxle mounting bolts

6. Attach a suitable lifting device to the engine.
7. Support the transaxle using a floor jack with a small block of wood in between.
8. Slowly lift the engine from the vehicle.

To install:

9. Install the engine.
10. Align the engine mounts and install the fasteners, but do not tighten them until all of the mounting bolts have been installed.
11. Install the engine-to-transaxle mounting bolts and torque to 75 ft. lbs. (102 Nm).
12. Remove the engine lifting device.
13. Torque the engine mount nuts/bolts 45 ft. lbs. (61 Nm).
14. Align the flexplate-to-torque converter matchmarks and torque the bolts to 55 ft. lbs. (75 Nm).
15. Install or connect the following:
    - CKP sensor and the starter
    - Left-side exhaust manifold V-band clamp and torque to 100 inch lbs. (11 Nm)
    - Left-side catalytic converter mounting bracket fasteners
16. Install the structural collar using the following procedure:
    a. Install the vertical collar-to-oil pan bolts and tighten, temporarily to 10 inch lbs. (1.1 Nm).
    b. Install the collar-to-transaxle bolts and torque to 40 ft. lbs. (55 Nm).
    c. Starting with the center vertical bolt and working outward, torque the bolts to 40 ft. lbs. (55 Nm).
17. Install or connect the following:
    - Throttle body support bracket
    - Heater hoses, all ground straps and vacuum hoses
    - Upper intake manifold
    - All engine wiring harnesses
18. Adjust the throttle and cruise control cables.
    - Fuel lines
    - V-band clamp on the right exhaust manifold and torque to 100 inch lbs. (11 Nm)
    - Right-side catalytic converter mounting bracket fasteners
    - A/C compressor and torque the bolts to 21 ft. lbs. (28 Nm)
    - Alternator and its wiring harness
    - Radiator
    - Power steering pump
    - Engine oil and transaxle cooler lines
    - Radiator hoses and accessory drive belt
    - Hood release cable and upper radiator support
    - Air cleaner and inlet hose
    - Cowl covers, supports and wiper arms
    - Hood
    - Negative battery cable
19. Fill the cooling system to the proper level.
20. Fill the engine with clean oil.
21. Start the vehicle, check for leaks and repair if necessary.

Water Pump

REMOVAL & INSTALLATION

The water pump has a die cast aluminum body and a stamped steel impeller. It bolts directly to the chain case cover using an O-ring for sealing. It is driven by the back side of the serpentine belt.

It is normal for a small amount of coolant to drip from the weep hole located on the water pump body (small black spot). If this condition exists, DO NOT replace the water pump. Only replace the water pump if a heavy deposit or steady flow of brown/green coolant is visible on the water pump body from the weep hole, which would indicate shaft seal failure. Before replacing the water pump, be sure to perform a thorough inspection. A defective pump will not be able to circulate heated coolant through the long heater hose.

2.7L Engine

1. Before servicing the vehicle, refer to the precautions in the beginning of this section.
2. Drain the cooling system.
3. Remove or disconnect the following:
    - Negative battery cable
    - Upper radiator crossmember
    - Fan module
    - Accessory drive belts

➡The water pump is driven by the primary timing chain.

- Crankshaft damper, timing chain cover, timing chain and all guides
- Water pump mounting bolts
- Water pump

4. Clean the water pump mounting surfaces.

To install:

5. Install or connect the following:
    - Water pump and gasket and torque the bolts to 105 inch lbs. (12 Nm)
    - Guides, timing chain and timing chain cover
    - Crankshaft damper and torque the center bolt to 125 ft. lbs. (170 Nm)
    - Accessory drive belts
    - Fan module and upper radiator crossmember
    - Negative battery cable
6. Fill the cooling system to the proper level.
7. Start the vehicle, check for leaks and repair if necessary.

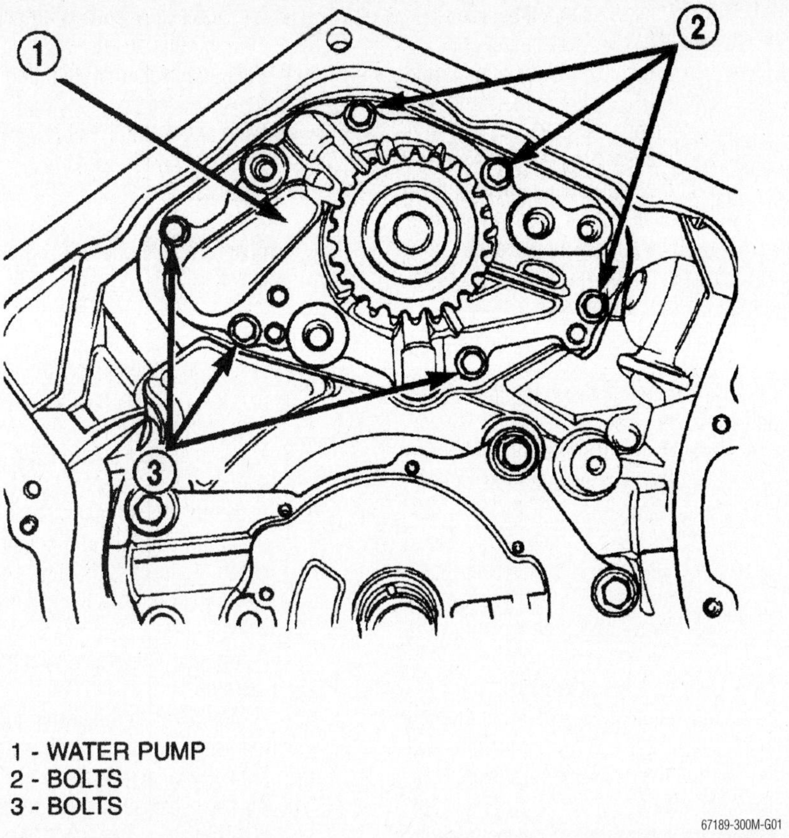

Water pump mounting—2.7L engine

3.5L Engines

1. Before servicing the vehicle, refer to the precautions in the beginning of this section.
2. Drain the cooling system.

⁂ WARNING

Do not use pliers to open the plastic drain.

3. Remove or disconnect the following:
   - Negative battery cable
   - Coolant recovery cap and open the thermostat bleed valve
   - Drive belts
   - Timing belt

➡It is good practice to turn the crankshaft until the No. 1 cylinder is at Top Dead Center (TDC) of its compression stroke (firing position).

   - Water pump mounting bolts, noting the position of the longer bolt
   - Water pump. Discard the O-ring seal.

4. Clean the gasket sealing surfaces, being careful not to scratch the aluminum surfaces.

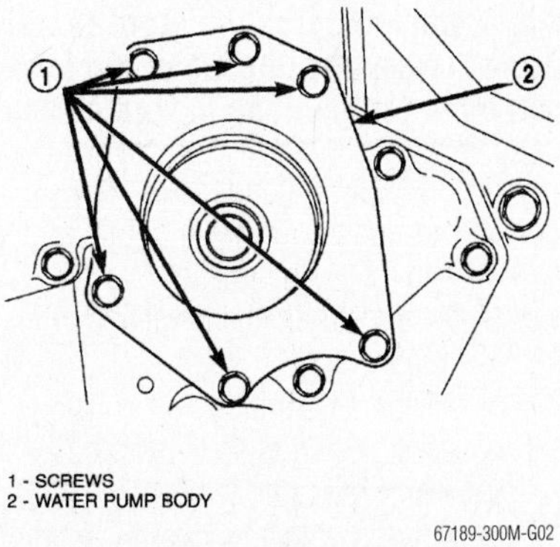

Water pump bolt locations, make sure to note the position of the longer bolt—3.5L engine

To install:

5. Install or connect the following:
   - New O-ring and wet with clean coolant prior to installation. Be sure to keep the new O-ring free of any oil or grease.
   - Water pump with a new O-ring. Torque the water pump-to-engine bolts to 105 inch lbs. (12 Nm).

⁂ WARNING

Rotate the pump and check for freedom of movement.

   - Timing belt
   - Drive belts

6. Fill the cooling system by performing the following procedure:
   a. Close the radiator drain.
   b. Open the thermostat bleed valve. Install a 1/4 in. (6mm) clear hose about 48 in. (1.2m) long to the end of the bleed valve and the other end into a clean container. The intent is to keep coolant off of the drive belt(s).
   c. Slowly, refill the coolant recovery bottle until a steady stream of coolant flows out of the thermostat bleed valve. Gently squeeze the upper radiator hose until all of the air is removed from the system.
   d. Close the bleed valve and continue to fill the coolant recovery bottle to the proper level. Install the cap on the bottle and remove the hose from the bleed valve.
7. Reconnect the negative battery cable. Start the engine and allow it to reach normal operating temperatures.
8. Check the cooling system for leaks and correct coolant level. Be sure that the thermostat bleed valve is closed once the cooling system has been bled of any trapped air.

Heater Core

REMOVAL & INSTALLATION

1. Disconnect the negative battery cable from the remote battery post located near the right strut tower.

⁂ CAUTION

After disconnecting the negative battery cable, wait 2 minutes for the driver's/passenger's air bag system capacitor to discharge before attempting to do any work around the steering column or instrument

2. Remove the instrument panel by removing or disconnecting the following:
   - Shifter knob Allen screw
   - Both instrument panel end covers

➡On the LHS or 300M, remove the center bezel prior to the shifter bezel

3. Using a trim stick tool, gently pry upward on the shifter bezel; then, disconnect both wiring connectors and a light bulb socket connector; then, remove the socket.
4. Remove the 2 lower instrument panel cover screws (outside end) and disconnect the deck lid release switch wiring connector.

Pull the lower instrument panel cover rearward releasing the clips. Remove the cable-to-brake release handle. Pull the cover rearward and remove it.

5. If equipped for 6 passengers, remove the lower floor bin.

6. If equipped for 5 passengers, remove or disconnect the following items:

- Center bezel using the trim stick
- Heater/air conditioning housing assembly control switch and the traction control switch wiring harnesses
- Bezel from the vehicle
- Two left side console side cover screws, pull the cover outward and remove from the vehicle
- Lower glove box door; the 2 screws are located at the right console side cover, pull outward and remove it
- 2 bracket screws and the 2 inside the console bin; then, remove the console
- 2 center lower instrument panel nuts
- 4 steel reinforcement-to-lower instrument panel cover bottom bolts
- 16-way DLC from the reinforcement
- left floor duct/silencer pad screw and the pad
- Steering column shrouds
- Shift interlock cable at the ignition switch
- Column wiring
- The under column duct section
- Left panel air conditioning outlet duct
- The 4 brake pedal support bracket-to-column mounting bolts
- Lower the steering column
- Heater/air conditioning housing and the Air bag Control Module (ACM) harness connectors then, the 2 ground eyelets from the floor tunnel near the bulkhead

7. Using a trim stick, carefully, pry out on the left and right A-pillar trim moldings and slide them rearward to remove them.

8. Using a trim stick, carefully pry upward on the instrument panel top cover, and remove it.

9. From both sides, remove the scuff plate screw. Using a trim stick, carefully, pry out the scuff plates and remove them.

10. Remove or disconnect the following:

- 3 right side cowl panel screws and the panel
- 3 left side cowl panel screws and the panel
- Remove the right side under dash silencer and pad
- 2 right side radio antenna and amplifier harness connectors
- Left side junction block and Body Control Module (BCM) harness connectors
- 8 instrument panel-to-chassis screws
- Pull the instrument panel rearward
- Instrument panel from the vehicle with the help of an assistant

11. Discharge and recover the air conditioning system refrigerant.

12. Drain the cooling system into a clean container for reuse.

13. Place shop cloths in the interior to catch any excess coolant.

14. Remove or disconnect the following:

- Air cleaner hose and the air distribution duct
- Heater hoses-to-dash panel spring type fasteners and the heater hoses from the heater core. Plug the openings to prevent coolant spillage.
- Refrigerant lines-to-expansion valve nut and separate the refrigerant lines from the expansion valve. Plug the openings to prevent contamination.

⁂ WARNING

The air conditioning lubricant used in this system absorbs moisture readily. Do not leave any portion of the system open for any long period of time.

- 3 heater/air conditioning housing assembly-to-chassis nuts from the engine compartment
- 2 defroster duct screws and the duct
- Heater/air conditioning housing assembly-to-dash panel nuts (2) and screws (2)
- 4 rear seat heat duct nuts and the duct
- Rear seat heat duct elbow push-pin fastener
- Electrical connector from the heater/air conditioning housing assembly
- Pull the heater/air conditioning housing assembly rearward and remove it from the vehicle
- 2 heater core-to-heater/air conditioning housing assembly screws
- Heater core from the heater/air conditioning housing assembly

To install:

15. Install or connect the following:

- Heater core to the heater/air conditioning housing assembly
- 2 heater core-to-heater/air conditioning housing assembly screws
- Push the heater/air conditioning housing assembly forward (carefully) and install it to the vehicle
- Electrical connector to the heater/air conditioning housing assembly
- Rear seat heat duct elbow push pin fastener
- Duct and the 4 rear seat heat duct nuts
- Heater/air conditioning housing assembly-to-dash panel nuts (2) and screws (2)
- 2 defroster duct and the duct screws
- In the engine compartment, the 3 heater/air conditioning housing assembly-to-chassis nuts
- Refrigerant lines to the expansion valve
- Refrigerant lines-to-expansion valve nut
- Heater hoses to the heater core and the heater hoses-to-dash panel spring type fasteners
- Air cleaner hose and the air distribution duct

16. Refill the cooling system.

17. Evacuate, charge and leak-test the air conditioning system refrigerant.

18. Install the instrument panel by installing or connecting the following:

- Instrument panel to the vehicle
- Push the instrument panel forward
- 8 instrument panel-to-chassis screws
- Left side junction block and Body Control Module (BCM) harness connectors
- 2 right side radio antenna and amplifier harness connectors
- Right side under dash silencer and pad
- Left side cowl panel and the 3 panel screws
- Right side cowl panel and the 3 panel screws
- Install the scuff plate and the screw (both sides)
- Instrument panel top cover
- Left and right A-pillar trim moldings
- Heater/air conditioning housing and the Air bag Control Module (ACM) harness connectors then, the 2 ground eyelets to the floor tunnel near the bulkhead
- Raise the steering column
- The 4 brake pedal support bracket-to-column mounting bolts

- Left panel air conditioning outlet duct
- The under column duct section
- Column wiring
- Shift interlock cable at the ignition switch
- Steering column shrouds
- Left floor duct/silencer pad and the pad screw
- 4 steel reinforcement-to-lower instrument panel cover bottom bolts
- 6-way Diagnostic Link Connector (DLC) to the reinforcement
- 2 center lower instrument panel nuts
- The 2 bracket screws and the 2 inside the console bin; then, the console (5 passenger models only)
- 2 screws are located at the right console side cover (5 passenger models only)

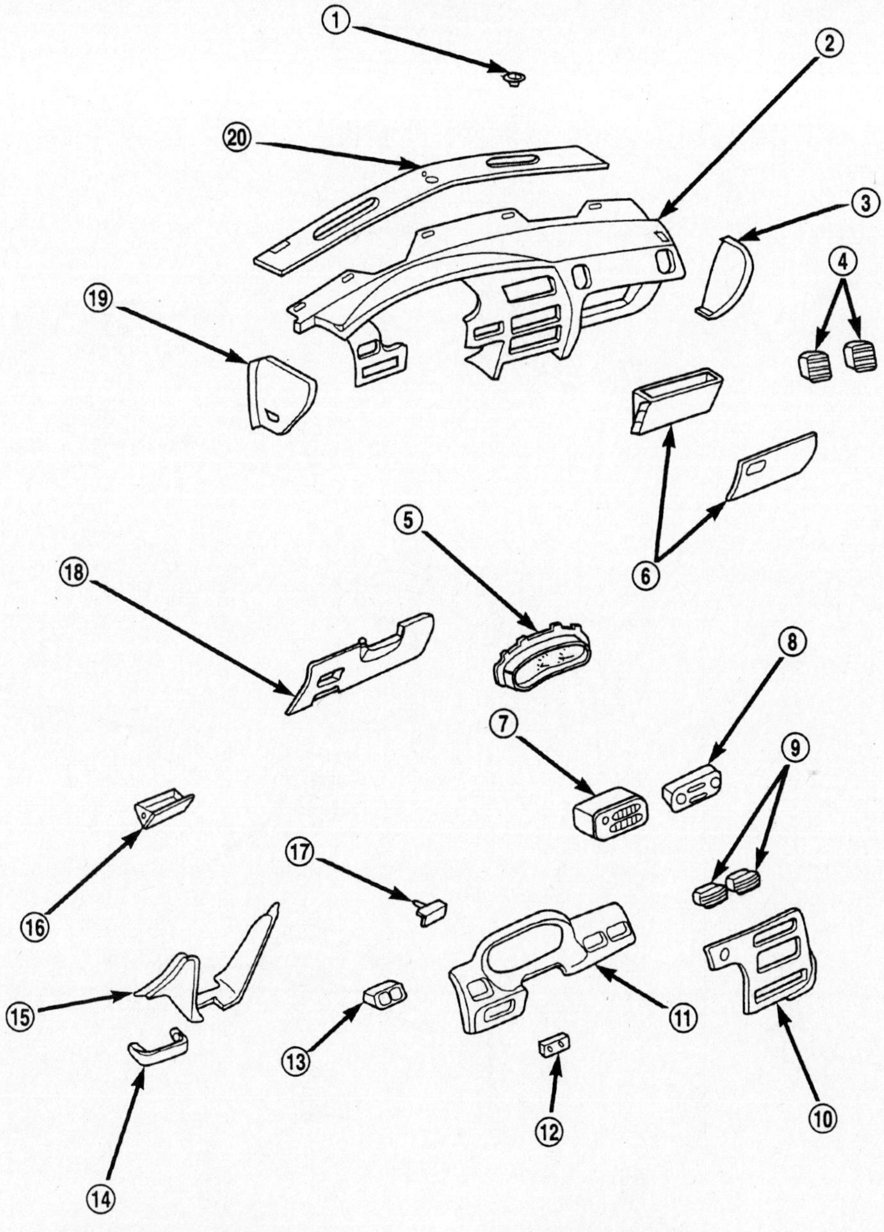

1 – SPEAKER, INSTRUMENT PANEL CENTER
2 – INSTRUMENT PANEL ASSEMBLY
3 – BEZEL, INSTRUMENT PANEL END CAP
4 – LOUVER, AIR OUTLET
5 – HOUSING INSTRUMENT CLUSTER
6 – GLOVE BOX ASSEMBLY
7 – RADIO
8 – CONTROL ASSEMBLY, INSTRUMENT PANEL
9 – LOUVER, AIR OUTLET
10 – BEZEL, INSTRUMENT PANEL TRIM-CENTER
11 – BEZEL INSTRUMENT CLUSTER
12 – SWITCH HEADLAMP
13 – LOUVER, AIR OUTLET
14 – COVER, INSTRUMENT PANEL CENTER SUPPORT/BIN (6 PASS. ONLY)
15 – 6 PASS. ONLY
16 – ASH RECEIVER
17 – LEVER, PARKING BRAKE
18 – COVER, LOWER INSTRUMENT PANEL
19 – BEZEL, INSTRUMENT PANEL END CAP
20 – COVER, UPPER INSTRUMENT PANEL

93111G81

Exploded view of the instrument panel—Intrepid

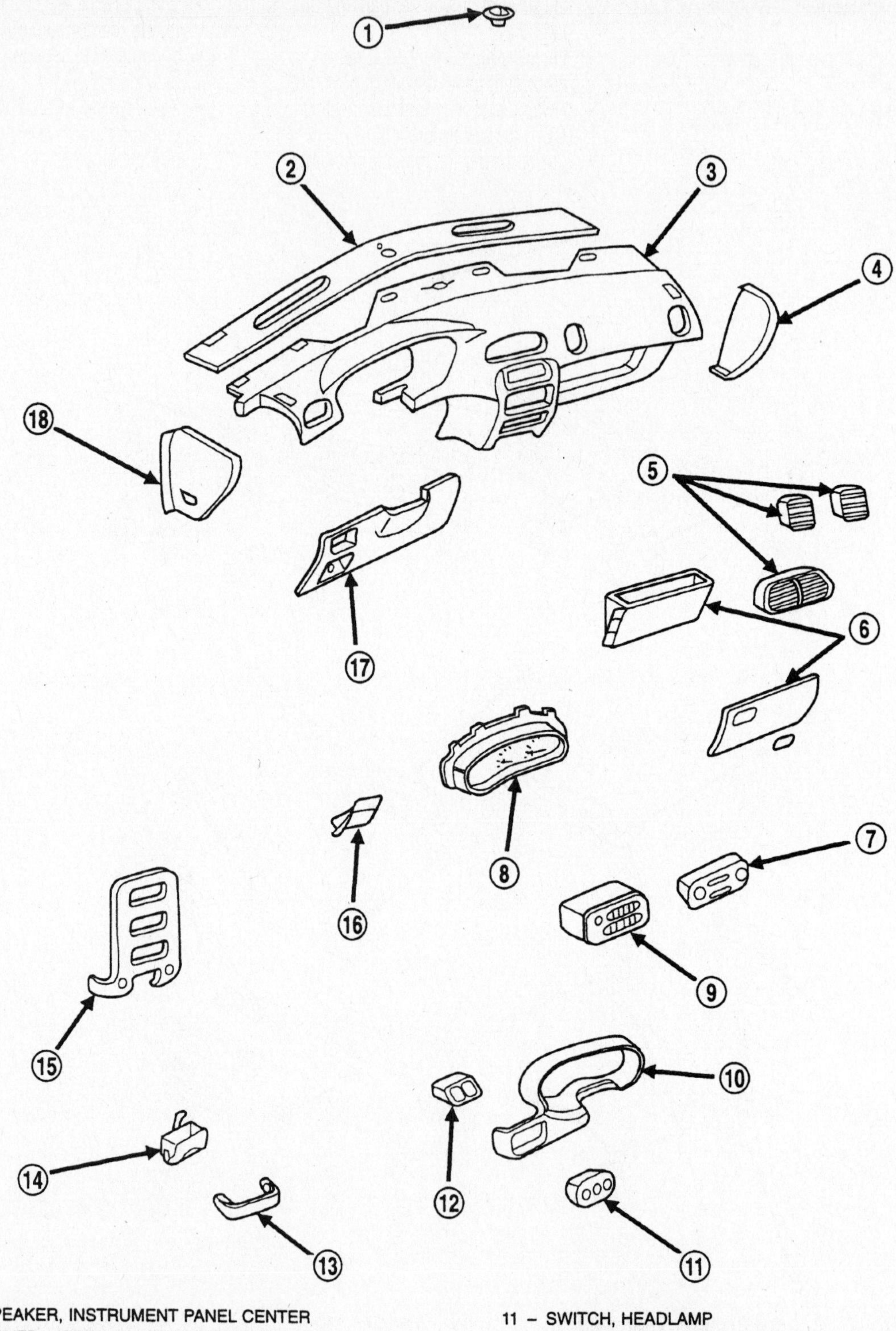

1 – SPEAKER, INSTRUMENT PANEL CENTER
2 – COVER, UPPER INSTRUMENT PANEL
3 – INSTRUMENT PANEL ASSEMBLY
4 – BEZEL, INSTRUMENT PANEL END CAP
5 – LOUVER, AIR OUTLET
6 – GLOVE BOX ASSEMBLY
7 – CONTROL ASSEMBLY, INSTRUMENT PANEL
8 – HOUSING, INSTRUMENT CLUSTER
9 – RADIO
10 – BEZEL, INSTRUMENT CLUSTER
11 – SWITCH, HEADLAMP
12 – LOUVER, AIR OUTLET
13 – COVER, INSTRUMENT PANEL CENTER SUPPORT/BIN (6 PASS. ONLY)
14 – ASH RECEIVER
15 – BEZEL, INSTRUMENT PANEL TRIM-CENTER
16 – LEVER, PARKING BRAKE
17 – COVER, LOWER INSTRUMENT PANEL
18 – BEZEL, INSTRUMENT PANEL END CAP

93111G82

Exploded view of the instrument panel—Concorde

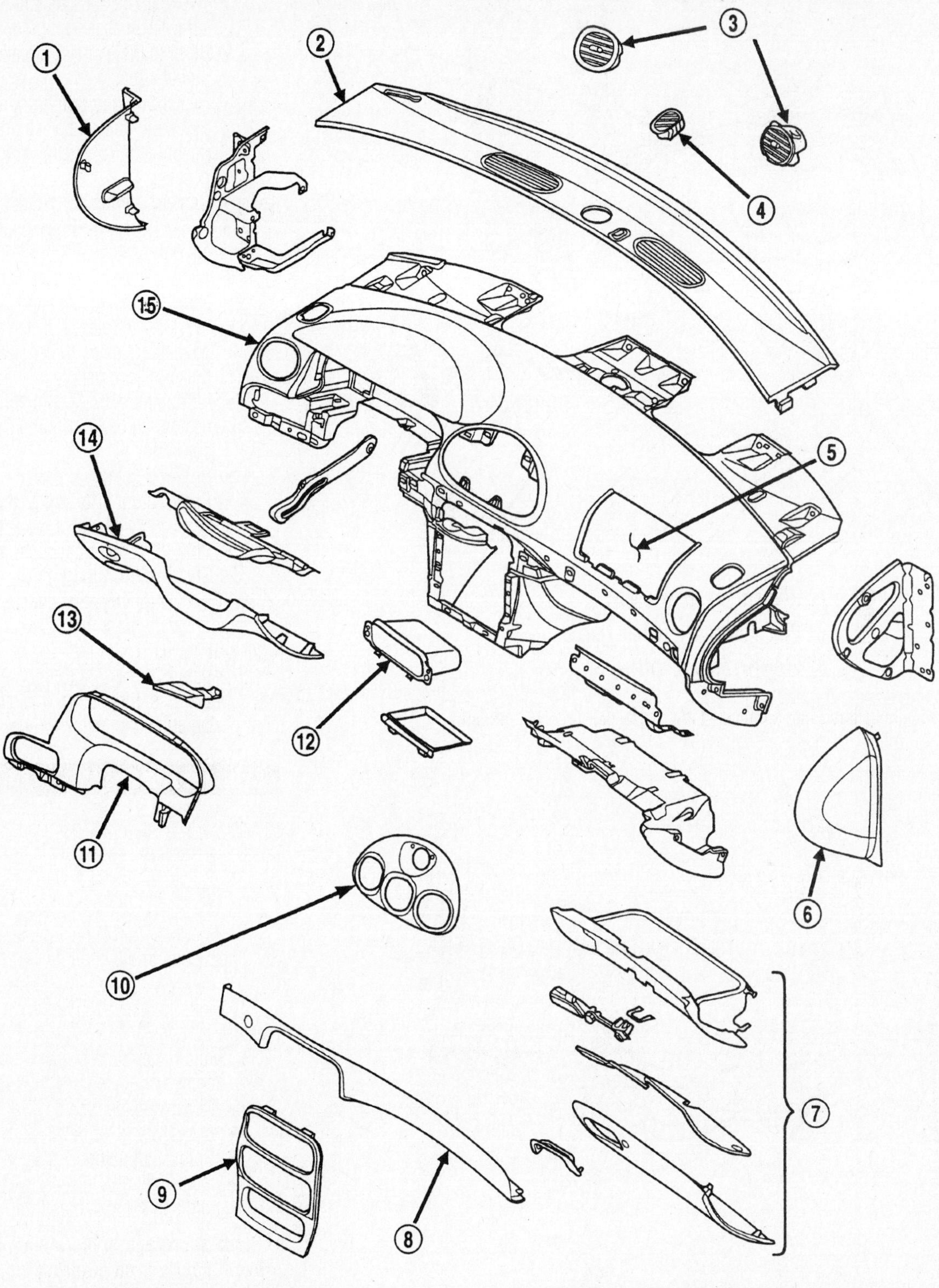

1 – BEZEL, INSTRUMENT PANEL END CAP
2 – COVER, UPPER INSTRUMENT PANEL
3 – LOUVER, AIR OUTLET
4 – LOUVER/DEMISTER SIDE WINDOW
5 – MODULE, PASSENGER SIDE AIRBAG
6 – BEZEL, INSTRUMENT PANEL END CAP
7 – GLOVE BOX ASSEMBLY, INSTRUMENT PANEL
8 – BEZEL, INSTRUMENT PANEL UPPER RIGHT TRIM
9 – BEZEL, INSTRUMENT PANEL TRIM-CENTER
10 – BEZEL, INSTRUMENT PANEL AIR DISTRIBUTION OUTLET
11 – BEZEL, INSTRUMENT CLUSTER
12 – STORAGE COMPARTMENT CUBBY BOX
13 – LEVER, PARKING BRAKE
14 – COVER, LOWER INSTRUMENT PANEL-LEFT SIDE
15 – INSTRUMENT PANEL ASSEMBLY

93111G83

Exploded view of the instrument panel—LHS and 300M

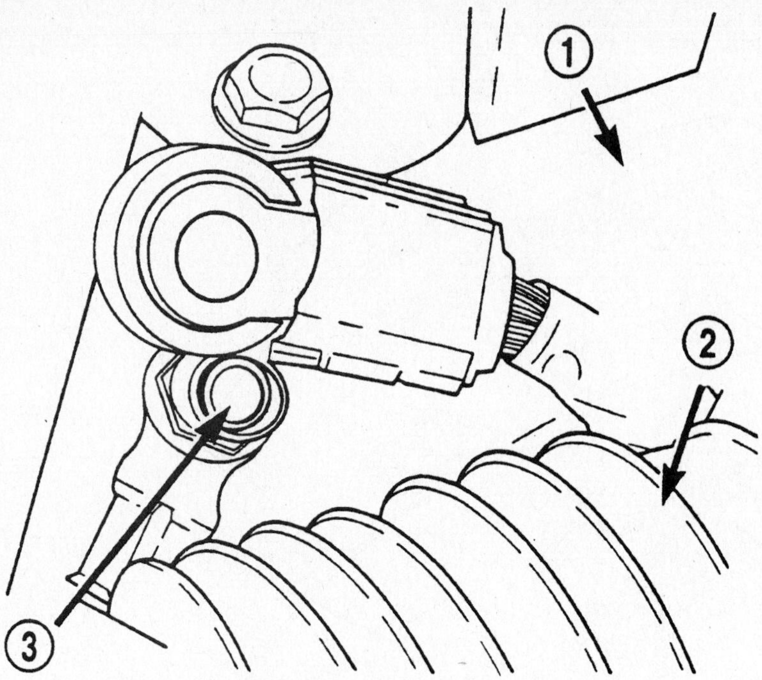

1. Right strut tower
2. Air cleaner inlet tube
3. Remote terminal

View of the remote battery cable terminal—LHS, 300M, Concorde and Intrepid

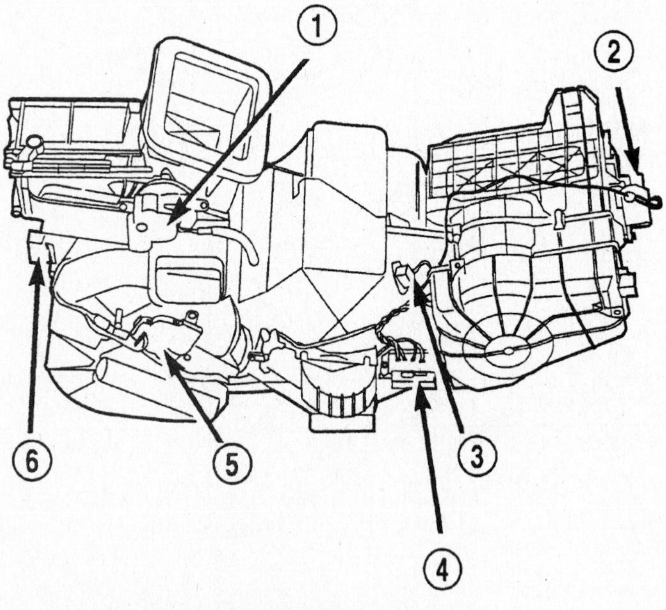

1 – MODE DOOR ACTUATOR
2 – RECIRCULATION DOOR ACTUATOR
3 – EVAPORATOR TEMPERATURE SENSOR
4 – POWER MODULE OR BLOWER RESISTOR
5 – BLEND DOOR ACTUATOR
6 – HVAC PLENUM CONNECTOR

View of the heater/air conditioning housing assembly—LHS, 300M, Concorde and Intrepid

- Two left side console side cover screws (5 passenger models only)
- Bezel to the vehicle (5 passenger models only).
- Heater/air conditioning housing assembly control switch and the traction control switch wiring harnesses.
- Center bezel using the trim stick.
- Lower floor bin (6 passenger models only).
- 2 lower instrument panel cover screws (outside end) and connect the deck lid release switch wiring connector. Install the cable-to-brake release handle
- Light bulb socket. Both wiring connectors and a light bulb socket connector
- Shifter bezel
- Center bezel prior to the shifter bezel (LHS or 300M models)
- Both instrument panel end covers
- Shifter knob Allen screw

19. Connect the negative battery cable to the remote battery post located near the right strut tower.

20. Operate the engine to normal operating temperatures; then, check the climate control operation and check for leaks.

Cylinder Head

REMOVAL & INSTALLATION

2.7L Engine

1. Before servicing the vehicle, refer to the precautions in the beginning of this section.
2. Properly relieve the fuel system pressure.
3. Drain the cooling system.
4. Remove or disconnect the following:
   - Negative battery cable
   - Accessory drive belts
   - Crankshaft damper
   - Intake plenum, lower intake manifold and exhaust manifold

➡Place shop rags in the openings to prevent debris from entering the engine.

   - Valve and timing chain covers
   - Coolant connections for the cylinder heads

5. Rotate the crankshaft until the crankshaft timing mark aligns with the timing mark on the oil pump.
   - Primary timing chain
   - Camshaft bearing caps, gradually,

in the reverse order of the tightening sequence
- Camshafts

⁂ WARNING

Be sure the head bolts 9–11 are removed before attempting to remove the cylinder head, the head and/or block may be damaged.

- Cylinder head bolts in reverse order of installation starting with bolts 11–9, then 8–1
- Cylinder head(s)

To install:

6. Thoroughly clean and dry the mating surfaces of the head and block. Check the cylinder head for cracks, damage or engine coolant leakage. Remove scale, sealing compound and carbon. Clean the oil passages thoroughly.
7. Place a new head gasket on the cylinder block over the locating dowels.
8. Inspect the cylinder head bolts for necking (stretching) by holding a straightedge against the threads of each bolt. If all of the threads are not contacting the straightedge, the bolt should be replaced. New head bolts are recommended.

⁂ WARNING

Due to the cylinder head bolt torque method used, it is imperative that the bolt threads be inspected for necking (stretching) prior to installation. If the threads are necked down, the bolt should be replaced. Failure to do so may result in parts failure or damage.

9. Lubricate the bolt threads with clean engine oil, then install them.
10. Torque the head bolts in the sequence shown in the illustration, utilizing the following steps and tightening values:
    a. Step 1: bolts 1 through 8: 35 ft. lbs. (48 Nm).
    b. Step 2: bolts 1 through 8: 55 ft. lbs. (75 Nm).
    c. Step 3: bolts 1 through 8: 55 ft. lbs. (75 Nm).
    d. Step 4: bolts 1 through 8: plus 90 degree turn using a torque angle meter.
    e. Step 5: bolts 9 through 11: 21 ft. lbs. (28 Nm).
11. Install or connect the following:
    - Camshafts, timing chain and sprockets
    - Water connections to the cylinder head

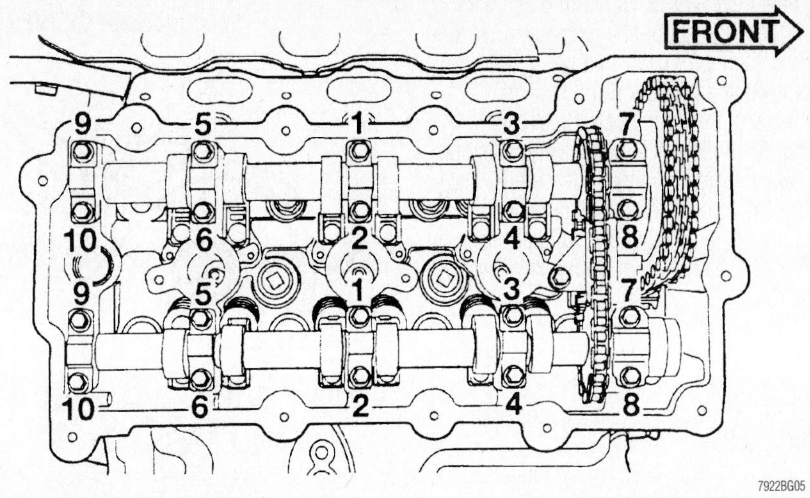

Camshaft bolt tightening sequence—2.7L engine

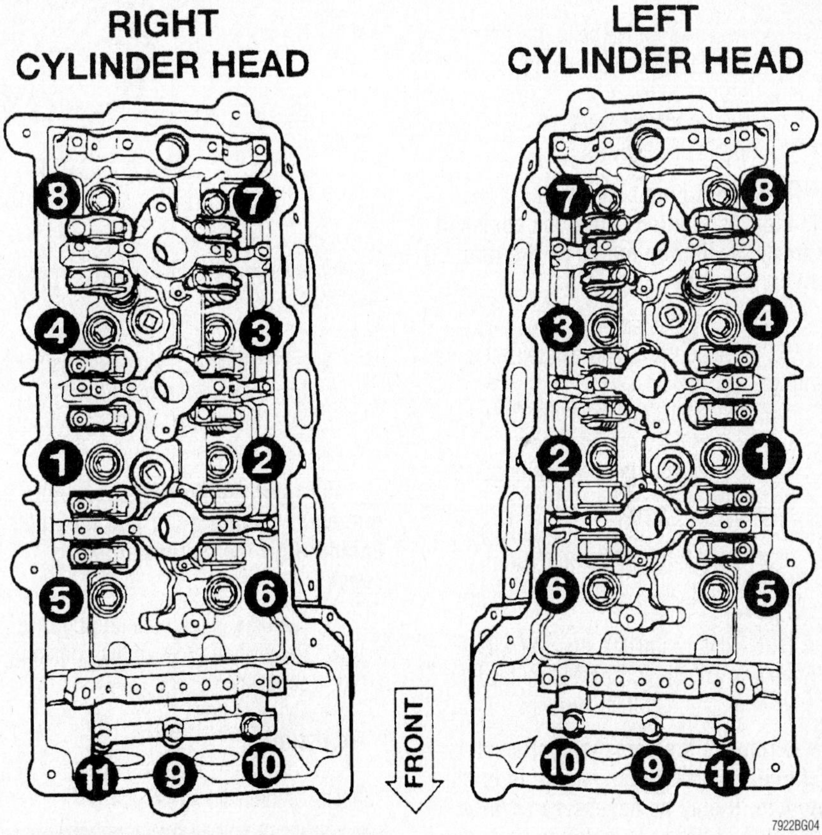

Cylinder head bolt tightening sequence—2.7L engine

- Valve and timing chain covers
- Crankshaft damper and torque the center bolt to 125 ft. lbs. (170 Nm)
- Lower intake manifold and intake plenum
- Exhaust manifolds
- Accessory drive belts
- Negative battery cable

12. Fill the cooling system to the proper level.
13. Start the vehicle, check for leaks and repair if necessary.

3.5L Engines

1. Before servicing the vehicle, refer to the precautions in the beginning of this section.

This engine uses aluminum alloy cylinder heads. Use care when working with light alloy components. The heads are common to either cylinder bank, but in practice a cylinder head should be returned to the side of the engine from which it was removed. Removal of a cylinder head involves removal of the timing belt. Great care is required to install the belt, paying attention to all valve timing marks. Please note that camshaft removal on this engine does require the removal of the cylinder head.

2. Properly relieve the fuel system pressure.
3. Drain the cooling system.
4. Remove or disconnect the following:
   - Negative battery cable
   - Radiator and cooling fan assemblies
   - Air cleaner assembly and intake manifold plenum

➡Cover the lower intake manifold during service.

   - Accessory drive belts
   - Crankshaft damper using the proper puller
   - Engine valve covers
   - Timing belt covers

➡Mark the timing belt running direction for installation. Align the camshaft sprockets with the marks on the rear covers.

   - Timing belt and tensioner

5. Pre-load the timing belt tensioner as follows:
   a. Place tensioner in a vise the same way it is mounted on the engine.
   b. Slowly compress the plunger into the tensioner body.
   c. Once the plunger is compressed, install a pin through the body and plunger to retain it in place until the tensioner is installed.
6. Hold the camshaft sprocket with a 36mm box wrench, loosen and remove the sprocket retaining bolt and washer.

➡To remove the camshaft sprocket retainer bolt while the engine is in the vehicle, it may be necessary to raise that side of the engine due to the length of the retainer bolt. The right bolt is 8.370 in. (212.6mm) long, while the left bolt is 10.0 in. (253mm) long. These bolts are not interchangeable and their original location during removal should be noted.

7. Remove or disconnect the following:
   - Camshaft sprocket from the camshaft

➡The camshaft sprockets are not interchangeable.

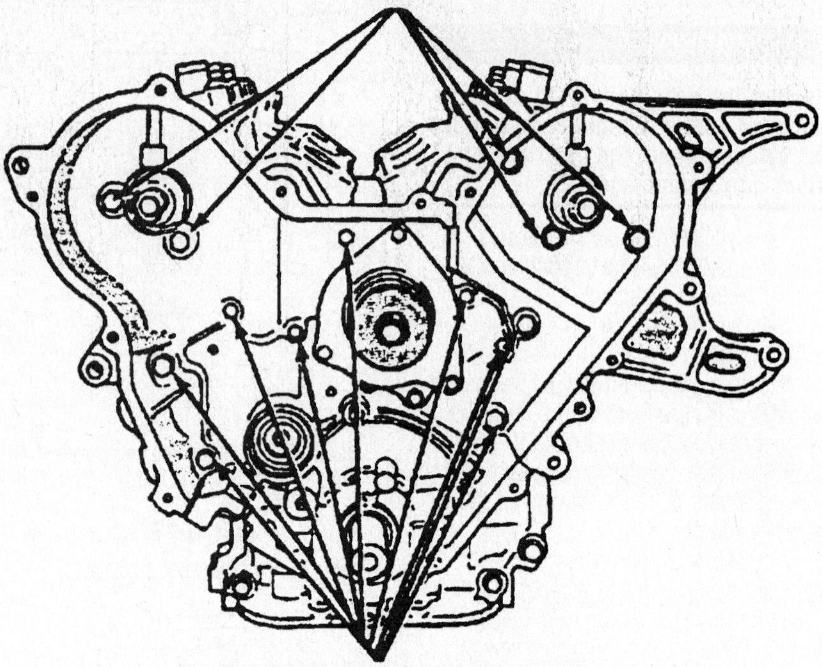

Remove the rear timing belt cover-to-cylinder head bolts, noting the bolt locations—3.5L engines

   - Intake manifold assembly using the recommended procedure
   - Rear timing belt cover-to-cylinder head fasteners

➡If the right timing belt cover is to be removed, there are O-rings located behind it for the water pump passages.

   - Cylinder head mounting bolts in the reverse order of the tightening sequence
   - Cylinder head

To install:

8. Thoroughly clean and dry the mating surfaces of the head and block.

⁂ WARNING

When cleaning the cylinder head and block mating surfaces, do not use a metal scraper because the soft aluminum surfaces could be cut or damaged. Instead, use a scraper made of wood or plastic.

9. Check the cylinder head for cracks, damage or engine coolant leakage. Check the head for flatness. End-to-end, the head should be within 0.002 in. (0.051mm) normally with 0.008 in. (0.203mm) the maximum allowed out of true. The resurface limit is 0.008 in. (0.203mm) maximum, the combined total dimension of stock removal from the cylinder head, if any, and block top surface.

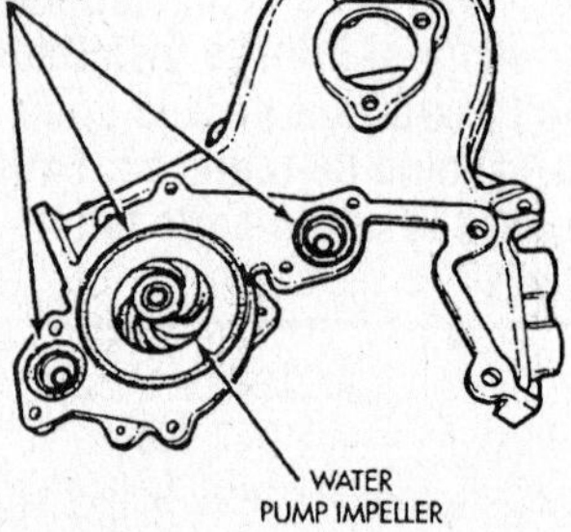

Right side belt cover, water pump and O-rings—3.5L engines

10. Place a new head gasket on the cylinder block locating dowels, being sure the gasket is on the correct side.
11. Inspect the cylinder head bolts for necking (stretching) by holding a straightedge against the threads of each bolt. If all of the threads are not contacting the scale, the bolt should be replaced.

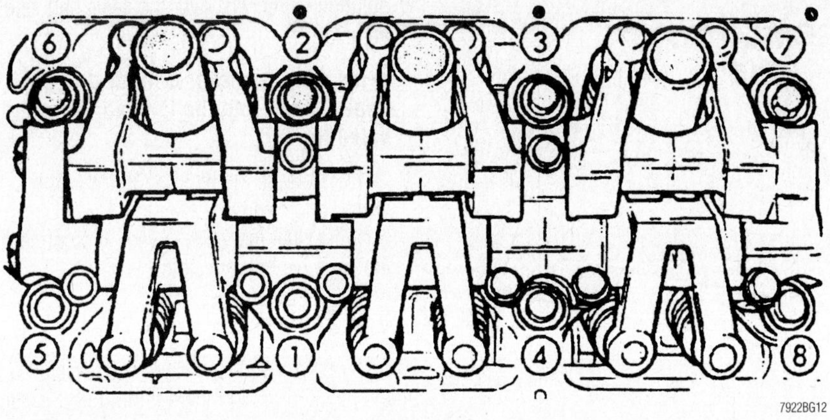

Cylinder head bolt tightening sequence—3.5L engines

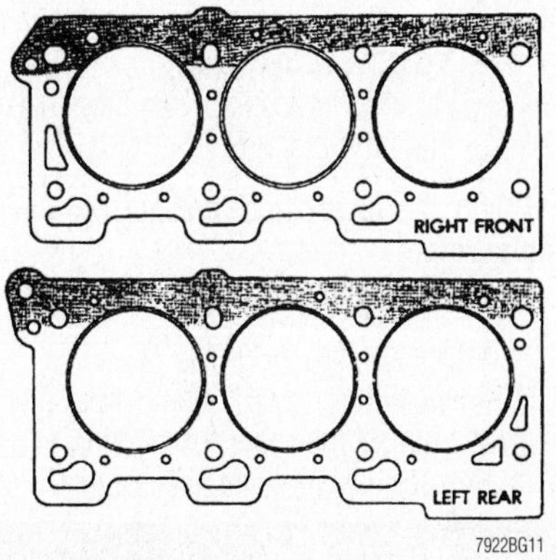

Correct positioning for the head gaskets—3.5L engines

⁂ WARNING

Due to the cylinder head bolt torque method used, it is imperative that the threads of the bolts be inspected for necking prior to installation. If the threads are necked down, the bolt should be replaced. Failure to do so may result in parts failure or damage. New bolts are always recommended.

12. Install the cylinder head into position on the engine block and over the dowels. Install the cylinder head bolts, lubricating the threads with clean engine oil prior to installation.
13. Torque the cylinder head bolts using the proper sequence as follows:
    a. Step 1: torque in sequence to 45 ft. lbs. (61 Nm).
    b. Step 2: torque in sequence to 65 ft. lbs. (88 Nm).
    c. Step 3: torque in sequence to 65 ft. lbs. (88 Nm).
    d. Step 4: torque in sequence an additional ¼ turn by using a torque angle meter.

➡Inspect the bolt torque after tightening. The torque should be over 90 ft. lbs. (122 Nm). If not, replace the cylinder head bolt.

14. Install the rear timing belt cover bolts and torque as follows:
    a. M6 bolts: 105 inch lbs. (12 Nm).
    b. M8 bolts: 21 ft. lbs. (28 Nm).
    c. M10 bolts: 40 ft. lbs. (54 Nm).
15. Install the intake manifold assembly and torque the bolts following the proper sequence to 21 ft. lbs. (28 Nm).

➡The following procedure can only be used when the camshaft sprockets have been loosened or removed from the shafts.

16. When the camshaft sprockets are loosened or removed, the camshafts must be timed to the engine. Install the Camshaft Alignment tools 6642-A, to the rear of the cylinder heads.
17. Install both camshaft sprockets to the appropriate shafts. The left camshaft sprocket has the Distributorless Ignition System (DIS) pickup as part of the sprocket.
18. Apply thread locking compound to the threads of the camshaft sprocket retainer bolts and install to the appropriate shafts. The right bolt is 8.380 in. (21.3cm) long, while the left bolt is 10.0 in. (25.4cm) long. These bolts are not interchangeable. Do not tighten the bolts at this time. The camshaft marks should be positioned between the marks on the cover.
19. Place the crankshaft sprocket to the Top Dead Center (TDC) mark on the oil pump housing. Install the timing belt starting at the crankshaft sprocket and working in a counterclockwise direction.
20. After the belt is installed around the last sprocket keep tension on the belt until it is past the tensioner pulley.
21. Holding the tensioner pulley against the belt, install the tensioner housing and torque to 21 ft. lbs. (28 Nm).
22. When the tensioner is in place pull the retainer pin to allow the tensioner to extend to the pulley bracket.
23. Hold the right camshaft sprocket hex with a 36mm box wrench and torque the right camshaft sprocket bolt to 75 ft. lbs. (102 Nm). Turn the sprocket bolt an additional 90 degrees.
24. Hold the left camshaft sprocket hex with a 36mm box wrench and torque the left camshaft sprocket bolt to 85 ft. lbs. (115 Nm). Turn the sprocket bolt an additional 90 degrees.
25. Install or connect the following:
    - Camshaft alignment tools from the back of the cylinder heads
    - Cam covers and new O-rings. Torque the fasteners to 20 ft. lbs. (27 Nm). Repeat this procedure on the other camshaft.
    - Timing belt covers and crankshaft damper and torque the crankshaft damper bolt to 85 ft. lbs. (115 Nm)
    - Valve covers and torque the bolts to 105 inch lbs. (12 Nm)
    - Spark plug tube nut and O-ring and torque to 60 inch lbs. (7 Nm)
    - Spark plug and torque to 20 ft. lbs. (28 Nm)
    - A/C compressor and torque the mounting bracket bolts to 30 ft. lbs. (41 Nm)
    - Spark plug wires
    - Accessory drive belts and adjust to the proper tension
    - Intake manifold plenum using the recommended procedure
    - Air cleaner assembly
    - Radiator and cooling fan assemblies
    - Negative battery cable
26. Check to be sure that all hoses, wiring connectors, cables, fluid and vacuum lines are reconnected.
27. Change the engine oil and oil filter.
28. Fill and bleed the cooling system.
29. Run the vehicle with the radiator cap off so coolant can be added as required until the thermostat opens. Watch for leaks and for unusual engine noises. Refill the radiator completely as required.
30. Once the vehicle has cooled, recheck the coolant and oil level.

Rocker Arm/Shafts

REMOVAL & INSTALLATION

2.7L Engine

1. Before servicing the vehicle, refer to the precautions in the beginning of this section.

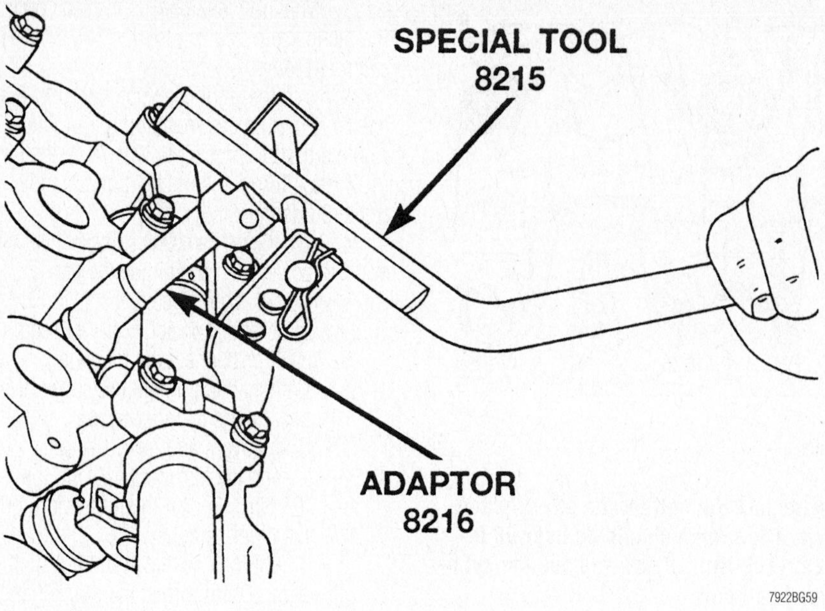

Only depress the valve spring enough to remove the rocker arm—2.7L engine

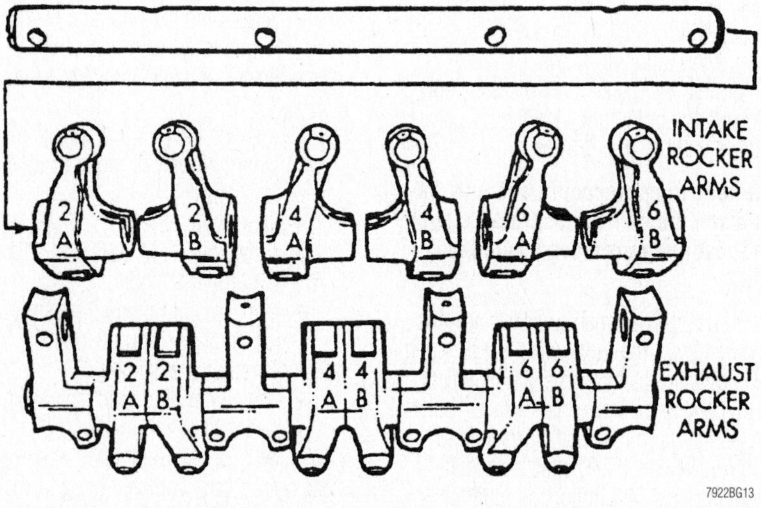

Left bank rocker arm and shaft identification—3.5L engines

2. Disconnect the negative battery cable.
3. Remove the valve covers.
4. Position the camshaft so that the base circle (heel) is facing the rocker arm being serviced.

WARNING

Depress the valve spring only enough to remove the rocker arm or damage to the spring may result.

5. Using Valve Spring tool 8215 and Adapter 8216, depress the valve spring enough to release the tension on the rocker arm.

➡If the rocker arms are to be reused, identify their positions for reassembly in their original positions.

6. Repeat this procedure for each rocker arm being removed.

To install:

7. Lubricate the rocker arms with clean engine oil, prior to installation.
8. Position the camshaft so that the base circle (heel) is facing the rocker arm being installed.

WARNING

Depress the valve spring only enough to install the rocker arm or damage to the spring may result.

9. Using Valve Spring tool 8215 and Adapter 8216, depress the valve spring enough to install the rocker arm.
10. Install the rocker arm in the original position (if reused) over the valve and lash adjuster.

➡Inspect the rocker arm for proper engagement into the lash adjuster and valve tip.

11. Release the tension on the valve spring and remove the tools.
12. Install the valve covers and connect the negative battery cable.

3.5L Engines

1. Before servicing the vehicle, refer to the precautions in the beginning of this section.
2. Relieve the fuel system pressure.
3. Remove or disconnect the following:
   - Negative battery cable
   - Air cleaner assembly and the intake manifold plenum

➡Cover the lower intake manifold during service.

   - Cylinder head covers
   - Rocker arm assembly
4. Inspect the rocker arms for wear or damage. Inspect the roller for scuffing or wear. Replace assembly as necessary.

WARNING

Do not remove the lash adjusters from the rocker arm assembly. The rocker arm and the adjuster are serviced as an assembly.

5. Identify the rocker arm assemblies and rocker arms and disassemble the shaft as follows:
   a. Thread a nut, washer and spacer onto a 4mm screw.
   b. Insert and tighten the 4mm screw into the dowel pin on the shaft.
   c. Loosen the nut on the screw. This will pull the dowel pin from the shaft support.

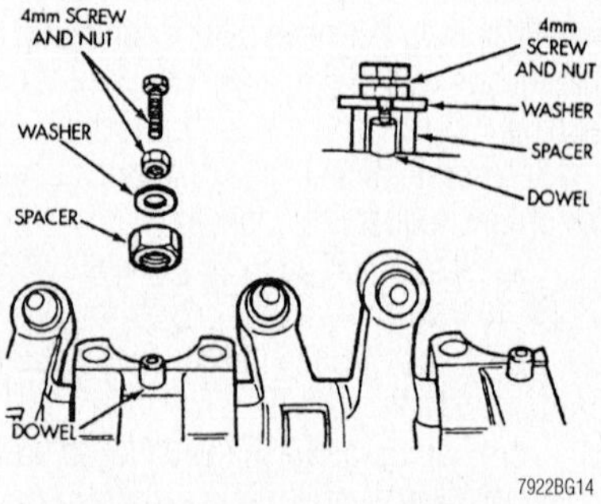

Remove the dowel pin using a 4mm screw, nut, spacer and washer installed into the pin—3.5L engines

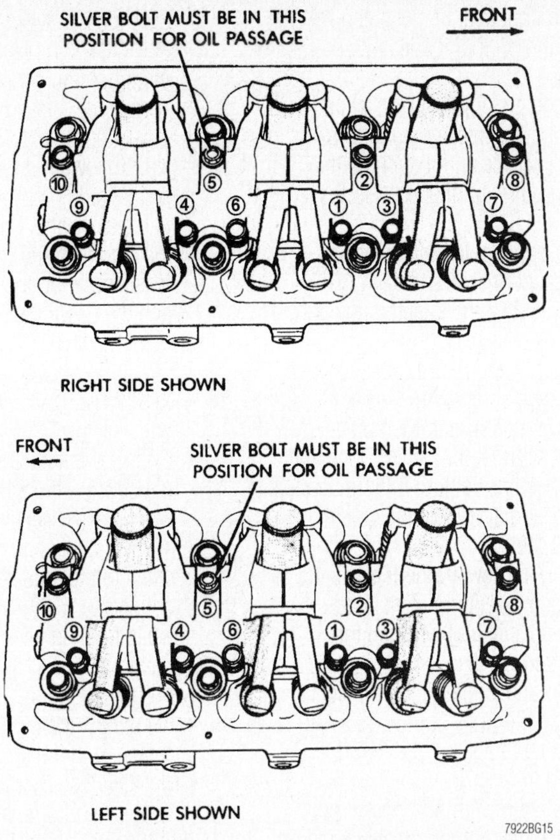

Proper torque sequence for the rocker arm and shaft assemblies—3.5L engines

d. Remove the rocker arms and pedestals, keeping them in order.

e. Check the oil holes for restrictions with a small wire and clean as required.

To install:

6. Assemble the rocker shaft as follows:

a. Install the rocker arms and pedestals onto the shaft, keeping them in the original order.

b. Press the dowel pins into the pedestals until they bottom out in the pedestals.

7. Position the camshaft so that the timing mark on the right camshaft timing belt sprocket aligns with the timing mark on the rear timing belt cover and the timing mark on the left sprocket is 45 degrees from the mark on the rear timing belt cover. There will be no load on the shaft during installation. Install the rocker shafts so the identification marks are facing toward the front of the engine.

8. Install the oil feed bolt in the correct location on the rocker shaft retainer. Torque the bolts in proper sequence to 23 ft. lbs. (31 Nm).

9. Install or connect the following:

- Valve covers and torque the bolts to 105 inch lbs. (12 Nm)
- Intake manifold plenum
- Air cleaner assembly
- Negative battery cable

Intake Manifold

REMOVAL & INSTALLATION

2.7L Engine

1. Before servicing the vehicle, refer to the precautions in the beginning of this section.

2. Disconnect the negative battery cable.

3. Properly relieve the fuel system pressure.

4. Drain the cooling system.

5. Disconnect the air tube from the air cleaner and the throttle body.

6. Hold the throttle lever in the wide-open position and remove the throttle cable and the speed control cable from the lever. Compress the locking tabs on the cables and remove them from the mounting brackets.

7. Unplug the electrical connections from the following:

- Solenoid on the Exhaust Gas Recirculation (EGR) valve transducer
- Manifold Absolute Pressure (MAP) sensor
- Throttle Position (TP) sensor
- Idle Air Control (IAC) motor
- Vacuum hose from the Positive Crankcase Ventilation (PCV) valve as well as the power brake booster at the intake manifold nipple
- Vacuum line at the fuel pressure regulator
- Purge hose from the throttle body
- Electrical connector from the TP sensor and the IAC motor
- EGR tube-to-intake manifold plenum screws
- Intake manifold plenum (upper part of the manifold)

➡Cover the lower part of the intake manifold to prevent foreign material from entering the engine.

- Fuel supply and return tubes from the fuel rail at the rear of the intake manifold

8. Disconnect the fuel/return tubes by pushing the quick-connect fitting toward the fuel tube while depressing the built-in disconnect tool with Quick-Connect Fitting tool 6751. To disconnect the fitting from the fuel rail, slightly twist the fitting while maintaining downward pressure on tool 6751. Wrap shop towels around the fuel hoses to absorb any fuel spillage.

9. Plug the fuel line openings to prevent system contamination.

10. Remove or disconnect the following:

- Fuel clamp screw and tubes from the bracket
- Electrical harness from the injectors and turn toward the center of the engine
- Fuel rail mounting bolts and lift the fuel rail with the injectors attached

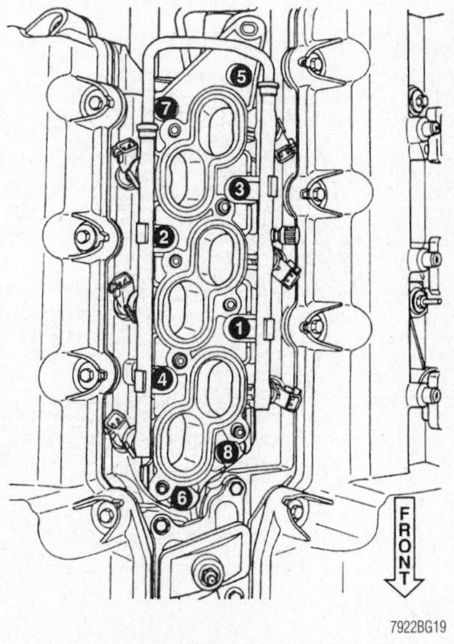

Intake manifold loosening and tightening sequence—2.7L engine

straight up and off the engine. Cover the injector openings.
- Intake manifold bolts and the manifold
- Intake manifold seal retainer screws and intake manifold gasket

11. Clean all mating surfaces.
12. Inspect the manifold for damage, cracks or clogged passages. Repair, clean or replace the manifold as required.

To install:

13. Verify that all intake manifold and cylinder head sealing surfaces are clean. Place a drop of sealant onto each of the 4 corners of the intake manifold gasket, where the cylinder head meets the engine block.

**** CAUTION**

The intake manifold gasket is made of very thin metal and can cause cuts if handled carelessly.

14. Install the intake manifold and 8 mounting bolts. Snug down evenly to just 10 inch lbs. (1.1 Nm).
15. Torque the lower intake manifold bolts in the proper sequence to 105 inch lbs. (12 Nm).
16. Install the fuel injectors by performing the following procedure:

a. Apply a light coat of clean engine oil to the O-ring on the nozzle end of each injector.

b. Insert the fuel injector nozzles into the openings in the intake manifold. Seat the injectors in place and install the fuel rail mounting bolts, torque to 16 ft. lbs. (22 Nm).

17. Install or connect the following:
    - Electrical connectors to each fuel injector. Rotate the injectors toward the cylinder head covers.
    - Fuel supply and return tubes to the fuel rail. Be sure that the black plastic release ring to the quick-connect fitting is in the OUT position. Place special tool 6751 under the largest diameter of the quick-connect fitting.
18. Pull tool 6751 toward the fuel rail until the quick-connect fitting clicks into place. Place the special tool between the shoulder of the built-in disconnect tool and top of the quick-connect fitting, then inspect the security of the fitting by applying a slight downward force against the fitting. It should be locked in place.
    - Intake plenum with new gasket onto the intake manifold. Loosely install the mounting bolts.
    - EGR tube to the manifold with a new gasket in place. Loosely install the mounting screws.
    - Torque the intake manifold plenum mounting bolts to 105 inch lbs. (12 Nm) following the outlined sequence
    - Left and right support brackets to the manifold and torque the lower fasteners to 50 inch lbs. (6 Nm) and the upper fasteners to 105 inch lbs. (12 Nm)
    - EGR tube mounting bolts
    - PCV valve hose and power brake booster hose
    - Electrical connectors to the EGR transducer solenoid, IAC motor, MAP and TP sensors
    - Throttle cable and speed control cable to the mounting bracket and connect to the throttle body lever while holding lever in the wide-open position
    - Purge hose to the throttle body
    - Reconnect the air tube to the air cleaner and the throttle body
    - Negative battery cable
19. Change the engine oil and oil filter.
20. Fill the cooling system. Run the vehicle with the radiator cap removed until the thermostat opens, adding coolant as required. Watch for fuel and coolant leaks and for correct engine operation.
21. Once the vehicle has cooled, recheck the coolant level and add, if necessary.

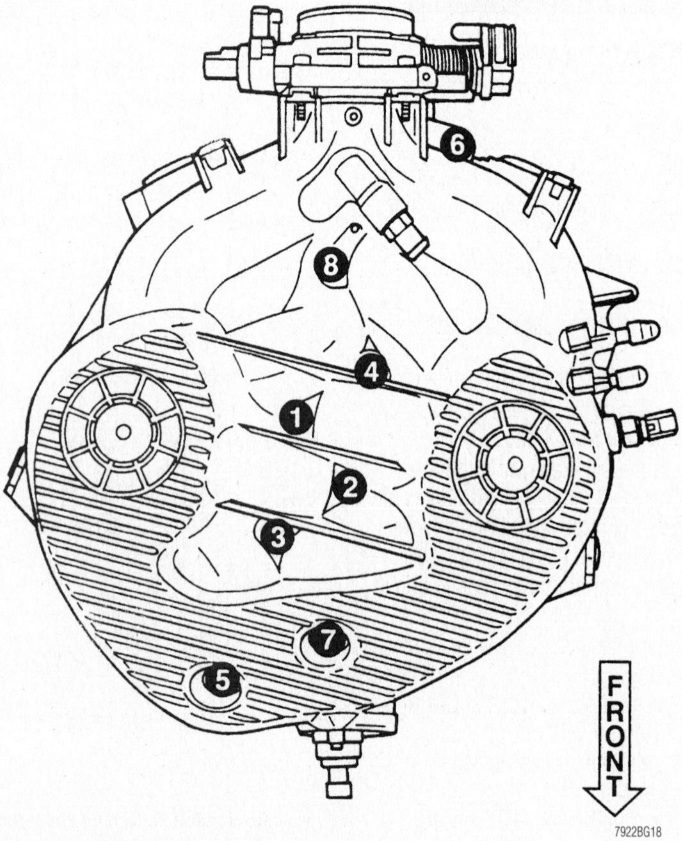

Intake plenum tightening sequence—2.7L engine

3.5L Engines

1. Before servicing the vehicle, refer to the precautions in the beginning of this section.
2. Properly relieve the fuel system pressure.
3. Drain the cooling system.
4. Remove or disconnect the following:
    - Negative battery cable
    - Engine cover from the top of the intake manifold
    - Accelerator and the speed control cable from the throttle lever
    - Idle Air Control (IAC) motor
    - Intake Air Temperature (IAT) sensor
    - Manifold Absolute Pressure (MAP) sensor
    - Ground screw from the intake manifold
    - Electrical connector from the Throttle Position (TP) sensor
    - Vacuum hoses from the manifold tuning valve, Positive Crankcase Ventilation (PCV) make-up air hose, IAC motor supply hose and

the purge hose from the throttle bodies
- Brake booster hose, PCV hose and the remaining vacuum hoses from the intake manifold
- Exhaust Gas Recirculation (EGR) tube-to-intake manifold plenum bolts
- Plenum support bracket mounting bolts on each side of the plenum
- Intake plenum mounting bolts

➡The intake manifold plenum (upper half of the intake manifold assembly) uses 2 different length bolts. Take note of their position and be sure they are installed in the same location during installation.

5. Remove the intake manifold plenum from the intake manifold.

➡Discard the old gasket. Cover the intake manifold openings with tape to keep debris from entering the engine.

6. Remove or disconnect the following:
   - Upper radiator hose from the thermostat housing
   - Heater hose from the rear of the intake manifold
   - Lower intake manifold bolts and manifold

➡Clean all gasket mating surfaces and inspect for distortion with a good straightedge.

To install:

➡Verify that all intake manifold and cylinder head sealing surfaces are clean.

7. Install or connect the following:
   - Intake manifold gasket, then the lower manifold. Torque the bolts in the proper sequence to 21 ft. lbs. (28 Nm).

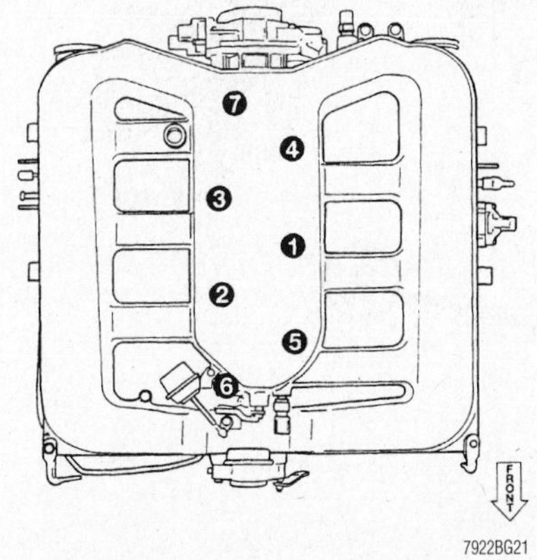

Upper intake manifold loosening and tightening sequence—3.5L engines

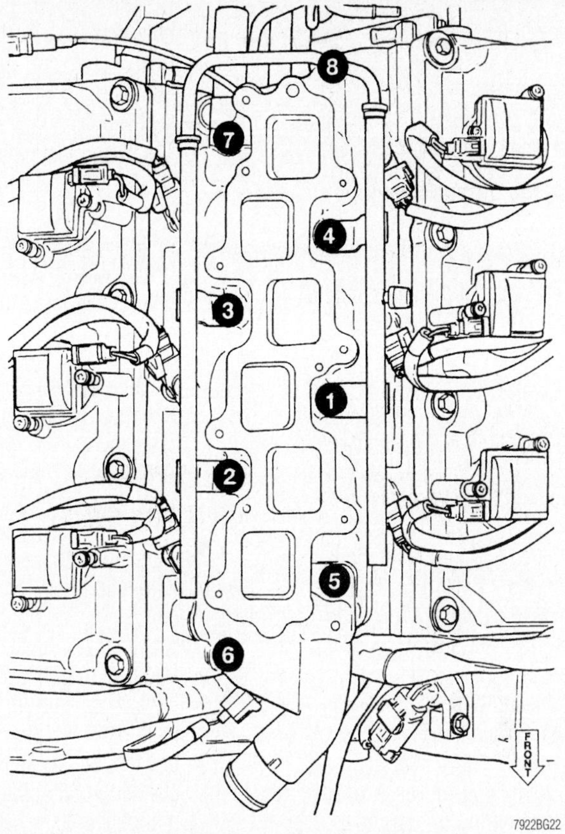

Lower intake manifold loosening and tightening sequence—3.5L engines

- Upper radiator hose to the thermostat housing
- Heater hose to the rear of the intake manifold

➡Ensure the ignition cables are routed out of the way of the intake plenum.

- Intake manifold plenum with a new gasket in place. Torque the mounting bolts, working from the center outward, to 21 ft. lbs. (28 Nm)

➡Do not overtighten bolts when working with light alloys.

- Support bracket bolts
- Electrical connectors to the MAP sensor, TP sensor, IAC motor and IAT sensor
- Vacuum hose to the manifold tuning valve
- EGR tube. Torque the EGR tube-to-intake manifold plenum screws to 17 ft. lbs. (22 Nm).

➡Be sure that the insulation on the EGR tube aligns with and contacts the insulation on the vacuum harness at the rear of the engine. Rotate the throttle lever to the wide-open position and reconnect the speed control and throttle cables.

- PCV valve hose
- Air cleaner plenum and plenum hose
- Ground wire to the intake manifold plenum
- Brake booster hose to intake manifold plenum fitting
- Throttle body purge tubes
- Intake manifold plenum cover
- Negative battery cable

8. Fill and bleed the cooling system.
9. Change the engine oil and filter.
10. Test run the engine, check for fuel and coolant leaks and verify correct engine operation.

Exhaust Manifold

REMOVAL & INSTALLATION

2.7L Engine

RIGHT MANIFOLD

1. Before servicing the vehicle, refer to the precautions in the beginning of this section.
2. Remove or disconnect the following:
   - Negative battery cable
   - Air intake plenum and the air filter housing

- Battery cable housing tube-to-transaxle bolt
- Exhaust Gas Recirculation (EGR) valve and tube
- Oxygen (O_2S) sensor
- V-band clamp from the manifold

➡**Do not reuse the V-band clamps.**

- Heat shield
- Exhaust manifold

3. Remove all traces of the old manifold gasket and clean both gasket mating surfaces.

To install:

4. Install or connect the following:
   - Exhaust manifold and new gasket. Torque the bolts to 17 ft. lbs. (23 Nm) working from the center outward.
   - Heat shields and torque the bolts to 105 inch lbs. (12 Nm)
   - New V-band clamp and torque to 100 inch lbs. (11.3 Nm)
   - O_2S sensor
   - EGR valve and tube using new gaskets, then torque to 95 inch lbs. (11 Nm)
   - Battery cable tube-to-transaxle and torque the bolt to 75 ft. lbs. (101 Nm)
   - Air inlet plenum and air filter housing
   - Negative battery cable

LEFT MANIFOLD

1. Before servicing the vehicle, refer to the precautions in the beginning of this section.
2. Remove or disconnect the following:
   - Negative battery cable
   - Exhaust system
   - V-band clamps and left catalytic converter

➡**Do not reuse the V-band clamps.**

3. Loosen and rotate the transaxle dipstick tube out of the way.
4. Remove or disconnect the following:
   - Engine wiring harness support bracket from the cylinder head
   - Oxygen (O_2S) sensor
   - Engine oil dipstick tube and manifold heat shield
   - Exhaust manifold bolts and manifold

To install:

5. Remove all traces of the old manifold gasket and clean both gasket mating surfaces.
6. Install or connect the following:
   - Exhaust manifold and new gasket and torque the bolts to 17 ft. lbs. (23 Nm) working from the center outward
   - Heat shields and torque the bolts to 105 inch lbs. (12 Nm)
   - O_2S sensor
   - Transaxle dipstick tube
   - O-ring for the engine oil dipstick tube and tube
   - Catalytic converter with a new V-band clamp and torque to 89 inch lbs. (10 Nm)
   - Engine wiring harness support bracket to the cylinder head
   - Exhaust system
   - Negative battery cable

3.5L Engines

RIGHT MANIFOLD

1. Before servicing the vehicle, refer to the precautions in the beginning of this section.
2. Remove or disconnect the following:
   - Negative battery cable
   - Exhaust pipes from the exhaust manifold
   - Loosen the converter pipe support attaching bolt at the transaxle mount
   - A/C belt, loosen only
   - Air cleaner housing and air inlet tube
   - V-Band clamp from the exhaust manifold connector

➡**Do not reuse the V-band clamps.**

   - A/C compressor bolts and position aside
   - Engine oil dipstick
   - A/C compressor bracket
   - Heated Oxygen Sensor (HO_2S) electrical connector and sensor
   - Heat shield-to-exhaust manifold screws
   - Exhaust manifold bolts and manifold

3. Inspect the manifold for damage or cracks. Check for distortion against a straight-edge or thickness gauge. Replace manifold if required.
4. Remove all traces of the old manifold gasket and clean both gasket mating surfaces.

To install:

5. Install or connect the following:
   - New manifold gasket and exhaust manifold to the cylinder head and torque to 17 ft. lbs. (23 Nm)
   - Heat shield and torque the retaining screws to 11 ft. lbs. (15 Nm)

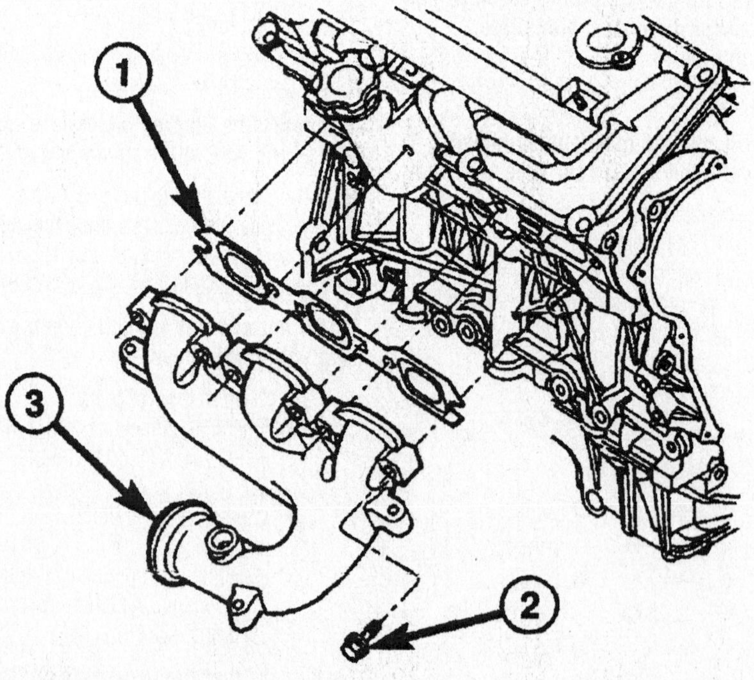

Exploded view of the right side exhaust manifold mounting—3.5L engine

- HO_2S and electrical connector
- A/C compressor mounting bracket
- New V-band clamp and tighten to 100 inch lbs. (11 Nm)
- Engine oil dipstick tube
- A/C compressor and drive belt
- New V-band clamp and tighten to 100 inch lbs. (11 Nm)
- Exhaust pipe to the exhaust manifold and torque the nuts to 21 ft. lbs. (28 Nm)
- Nut attaching converter pipe support to the transaxle mount. Torque to 35 ft. lbs. (47 Nm).
- Air cleaner housing and air inlet tube
- Negative battery cable

6. Operate the vehicle and inspect for exhaust leaks.

LEFT MANIFOLD

1. Before servicing the vehicle, refer to the precautions in the beginning of this section.
2. Remove or disconnect the following:
   - Negative battery cable
   - Exhaust pipes or system, as necessary from the exhaust manifold
   - Loosen the converter pipe support attaching bolt at the transaxle mount
   - V-Band clamp from the exhaust manifold connector

➡Do not reuse the V-band clamps.

   - Electrical connector harness bracket
   - Heated Oxygen Sensor (HO_2S) electrical connector and sensor
   - Heat shield-to-exhaust manifold screws
   - Exhaust manifold bolts and manifold

3. Inspect the manifold for damage or cracks. Check for distortion against a straight-edge or thickness gauge. Replace manifold if required.
4. Remove all traces of the old manifold gasket and clean both gasket mating surfaces.

To install:

5. Install or connect the following:
   - New manifold gasket and exhaust manifold to the cylinder head and torque to 17 ft. lbs. (23 Nm)
   - Heat shield and torque the retaining screws to 11 ft. lbs. (15 Nm)
   - HO_2S and electrical connector
   - Electrical connector bracket to brace
   - New V-band clamp and tighten to 100 inch lbs. (11 Nm)
   - Nut attaching converter pipe support to 35 ft. lbs. (47 Nm)
   - Exhaust pipes/system to the manifold
   - Negative battery cable
6. Operate the vehicle and inspect for exhaust leaks.

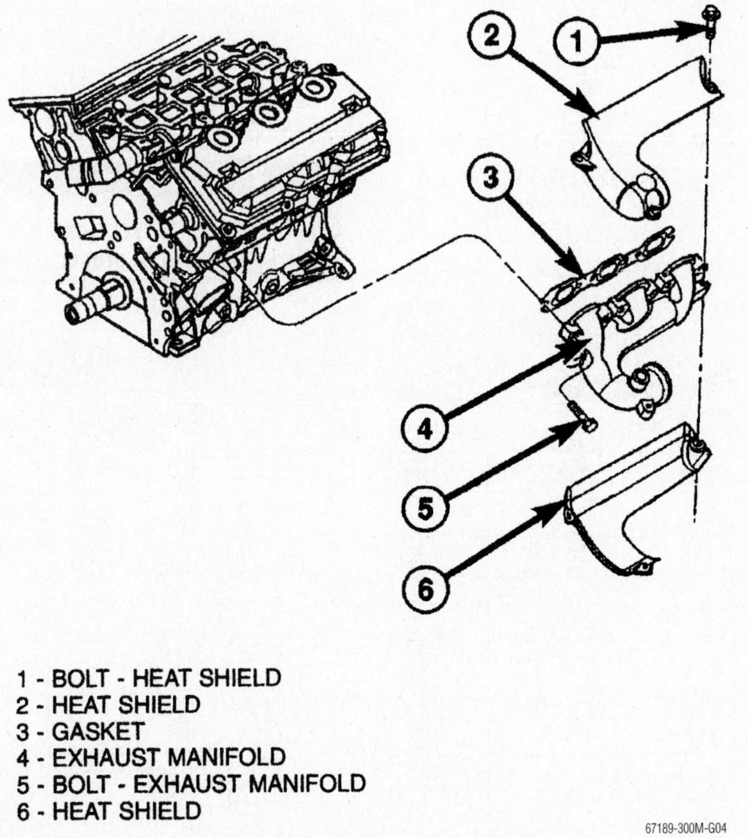

Exploded view of the left side exhaust manifold mounting—3.5L engine

Front Crankshaft Seal

➡The front crankshaft seal procedures are for timing belt equipped engines only. For engines that utilize timing chains, please refer to the applicable procedure later in this section.

REMOVAL & INSTALLATION

3.5L Engines

Note that the timing belt must be removed from the vehicle to perform this service. Use care to be sure all valve timing marks are carefully aligned both before removing the belt and after belt installation and all service has been completed. It may be good practice to set the engine to Top Dead Center (TDC) No. 1 cylinder compression stroke (firing position) and aligning all timing marks before removing the timing belt. This serves as a reference for all work that follows.

1. Before servicing the vehicle, refer to the precautions at the beginning of this section.
2. Properly relieve the fuel system pressure.
3. Drain the cooling system.
4. Remove or disconnect the following:
   - Negative battery cable
   - Radiator and cooling fan module assembly
   - Accessory drive belts
   - Crankshaft damper bolt
   - Timing belt front cover

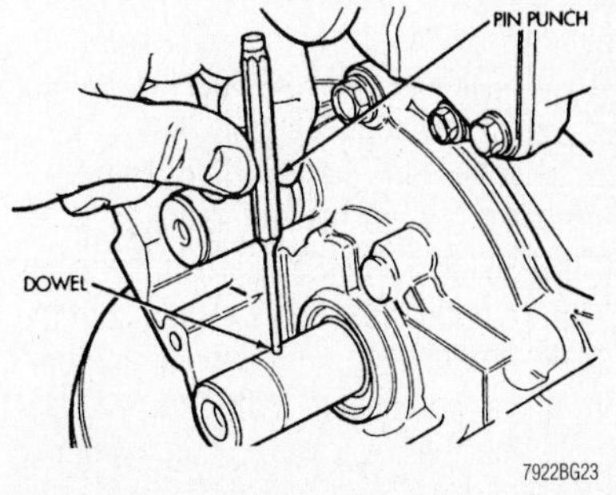

Removing the timing belt sprocket dowel pin from the crankshaft—3.5L engines

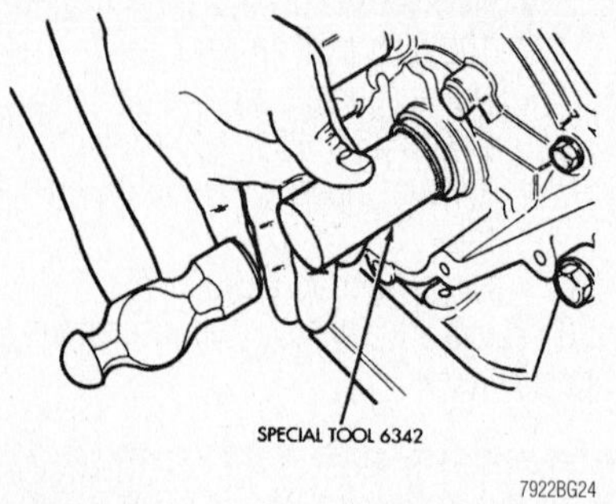

Installing the crankshaft oil seal—3.5L engines

➡The sealer on the timing belt front cover may be reusable and should not be removed. Use silicone rubber adhesive sealant to replace any missing sealer.

- Timing belt and tensioner. Refer to the appropriate section for the removal and installation procedure.
- Crankshaft timing belt sprocket

5. Locate the small dowel pin in the crankshaft. With a small punch, carefully tap out the dowel from the end of the crankshaft.
6. Remove the crankshaft seal using tool 6341A, taking care not to nick the shaft seal surface or seal bore during removal.

To install:

7. Inspect the crankshaft seal lip surface for varnish and dirt. Polish the area using 400 grit sandpaper to remove varnish as necessary.
8. Install or connect the following:
   - Crankshaft seal using seal installer tool 6342
   - Rear lower timing belt cover
   - Dowel into the crankshaft so that it protrudes 0.047 in. (1.2mm)
   - Timing belt sprocket at the crankshaft using tool C-4685C1, thrust bearing, washer and 12mm bolt or an equivalent setup to pull the sprocket onto crankshaft. Do not hammer on the sprocket.
9. Verify that all valve timing marks are aligned.
   - Timing belt and tensioner using the recommended procedure
10. Rotate the crankshaft 2 complete turns and recheck the timing marks on the camshafts and crankshaft. The marks must align with their respective locations. If the marks do not align, repeat the timing belt installation procedure. When correct valve timing has been verified, install the timing belt covers.
    - Crankshaft damper. Hold the crankshaft damper, using tool L-3281, and torque the bolt to 85 ft. lbs. (115 Nm).
    - Accessory drive belts and adjust to the proper tension
    - Radiator and cooling fan assemblies
    - Negative battery cable
11. Fill and bleed the cooling system.
12. Start the vehicle, check for leaks and repair if necessary.

Camshaft and Valve Lifters

REMOVAL & INSTALLATION

2.7L Engine

1. Before servicing the vehicle, refer to the precautions in the beginning of this section.

⁂ WARNING

When the timing chain is removed and the cylinder heads are installed, DO NOT turn the crankshaft or camshaft without first locating the proper crankshaft position. Failure to do so will result in piston-to-valve contact.

2. Remove or disconnect the following:
   - Primary timing chain
   - Second chain tensioner mounting bolts
3. Slowly loosen the camshaft bearing cap retaining bolts in the reverse order of the tightening sequence.
   - Bearing caps
   - Camshafts, secondary chain and tensioner as an assembly
   - Tensioner and chain from the crankshaft

To install:

4. Assemble the chain on the camshafts. Ensure the plated links are facing toward the front. Align the plated links to the dots on the camshaft sprockets.

➡There are 2 different styles of camshaft (secondary) chain tensioners. The Early Build tensioners will separate into sub-components, the Later Build tensioners will not. Compress the camshaft (secondary) chain as follows:

5. For early build vehicles:
   a. Separate the tensioner cylinder from the tensioner housing.
   b. Carefully drain the oil from the housing using care not to remove the internal tensioner components.
   c. Assemble the tensioner housing.
   d. Using hand pressure, compress and lock the tensioner using a fabricated lockpin.
6. For late build vehicles:
   a. Place the tensioner in a soft-jawed vise.
   b. Slowly compress the tensioner until the fabricated lockpin can be installed.
   c. Remove the compressed and locked tensioner from the vise.
7. Insert the compressed and locked camshaft chain tensioner in between the camshafts and chain.
8. Position the camshafts so that the plated links and dots are facing upward.
9. Install or connect the following:
   - Camshafts

⁂ WARNING

Ensure that the rocker arms are correctly seated and in proper positions.

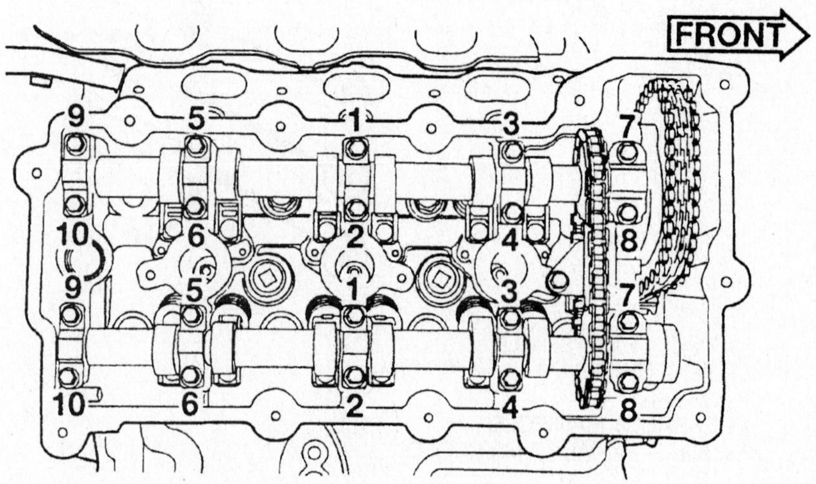

Camshaft bearing cap tightening sequence—2.7L engine

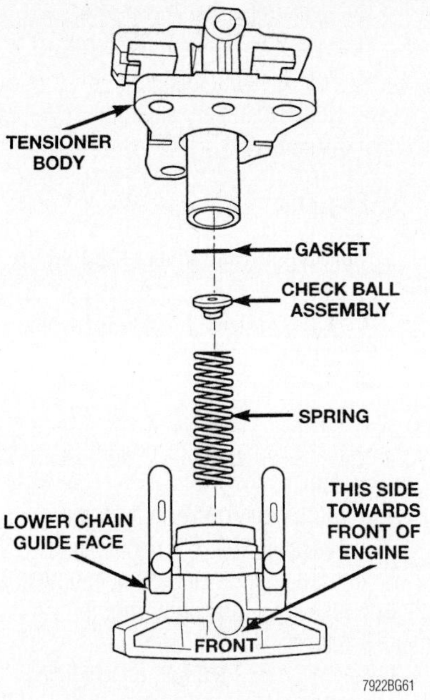

Exploded view of the camshaft (secondary) chain tensioner, early build—2.7L engine

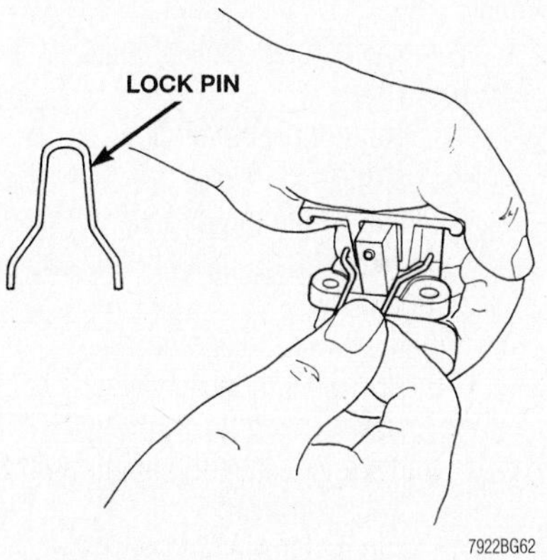

Fabricate a lockpin, as shown, to keep the tensioner compressed—2.7L engine

- Camshaft bearing caps and torque the bolts gradually, in sequence, to 105 inch lbs. (12 Nm)
- Secondary chain tensioner and torque the bolts to 105 inch lbs. (12 Nm)
- Primary timing chain
- Negative battery cable

10. Remove the lockpin from the secondary chain tensioner.

3.5L Engines

Camshafts are serviced from the rear of the cylinder head. Although the engine does not need to be removed for camshaft service, the cylinder head must be removed from the vehicle. Note too, that the camshaft sprockets have a D-shaped hole that allows it to rotate several degrees in each direction on its shaft.

1. Before servicing the vehicle, refer to the precautions in the beginning of this section.
2. Properly relieve the fuel system pressure.
3. Drain the cooling system.
4. Remove or disconnect the following:
   - Negative battery cable
   - Radiator/cooling fan assemblies and the accessory drive belts
   - Crankshaft damper and the timing belt covers

➡Mark the timing belt rotation direction for installation. Align the timing belt sprockets with marks on the rear timing belt covers before removing the timing belt.

   - Timing belt tensioner and timing belt. Refer to the appropriate section for the removal and installation procedure.
   - Camshaft timing belt sprockets
   - Intake manifold assembly using the recommended procedure
   - Exhaust manifold

➡Be sure to clean the gasket mating surfaces between the exhaust manifold and the cylinder head.

5. The rear timing belt cover must be removed to remove the cylinder heads. Remove the rear timing belt cover-to-cylinder head bolts. Remove the rear timing belt covers.

➡The right-hand side timing belt cover has O-rings located behind it for the water pump passages.

6. Remove the cylinder head bolts and cylinder head.

➡Mark the rocker arm assembly to note component locations before disassembly.

7. Remove the rocker arm and shaft assemblies.
8. Remove the rear camshaft cover and O-ring.

⁂ WARNING

Carefully, remove the camshaft from the rear of the head taking care not to nick or scratch the journals.

9. Inspect camshaft journals for wear or damage. If wear is present, inspect the cylinder head for damage. Inspect the head oil holes for clogging. Replace the camshaft as required.
10. Measure the height of the cam using a micrometer. Measure in 2 places: the unworn area and in the wear zone. Subtract the figures to get cam wear. The standard

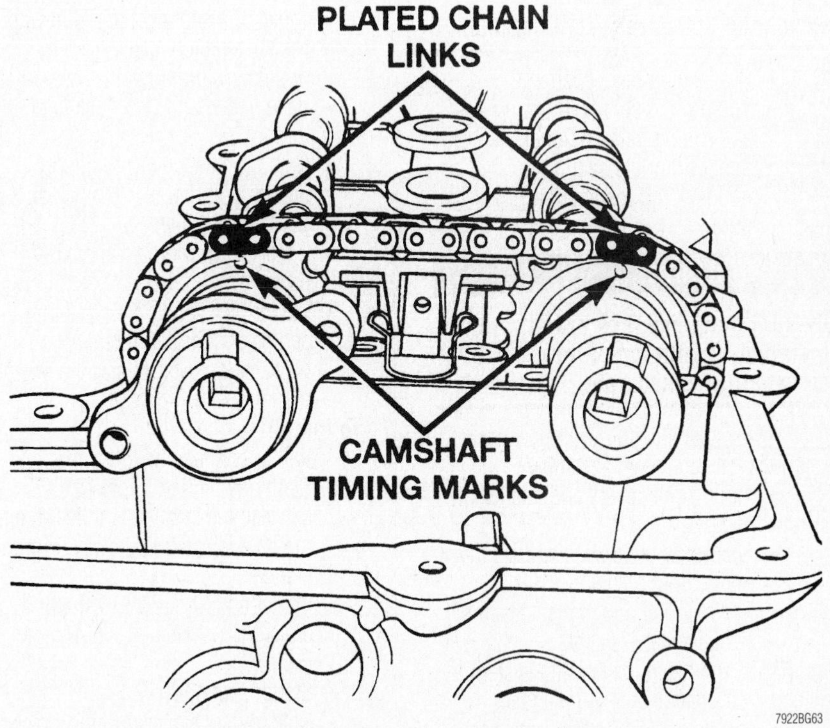

Proper camshaft (secondary) chain alignment—2.7L engine

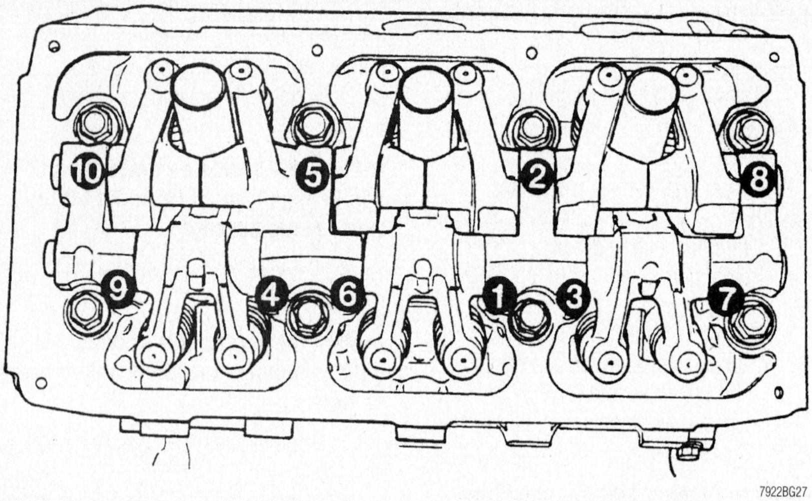

Rocker arm/shaft tightening sequence—3.5L engines

specification is 0.001 in. (0.0254mm) with the wear limit being 0.010 in. (0.254mm). Replace the camshaft if it is worn beyond this specification.

To install:

11. Lubricate the camshaft journals and lobes with clean engine oil.
12. Install or connect the following:
    - Camshaft
    - Camshaft cover and O-ring and torque the bolts to 21 ft. lbs. (28 Nm)
    - Rocker arm assemblies
    - Cylinder head assembly

**** WARNING**

New head bolts are recommended.

- Rear timing belt covers
- Timing belt sprocket, timing belt and timing belt tensioner. Refer to the appropriate section for the removal and installation procedure.

**** WARNING**

Be sure that once they are all installed, the timing of the camshaft(s) is accurate.

- Exhaust manifold with new gasket
- Intake manifold assembly
- Timing belt covers and crankshaft damper
- Accessory drive belts and set them to the proper tension
- Radiator and cooling fan assembly
- Negative battery cable

13. Fill and bleed the cooling system. An oil and filter change is recommended.
14. Start the vehicle, check for leaks and repair if necessary.

Valve Lash

ADJUSTMENT

These engines use hydraulic roller lifters to take up the free-play in the valve train system, therefore no lash adjustments are necessary.

Starter Motor

REMOVAL & INSTALLATION

2.7L Engine

1. Remove or disconnect the following:
    - Negative battery cable from the remote ground post
    - Battery feed and posi-lock connectors
    - Nuts from the catalyst support bracket and mount
    - Starter heat shield
    - 3 starter-to-engine/transaxle bolts
    - Starter by rotating it toward the engine and sliding it rearward between the catalyst and the engine mount

To install:

2. Install or connect the following:
    - Starter by sliding it forward between the catalyst and the engine mount and rotating it away from the engine
    - Starter and torque the 3 starter-to-engine/transaxle bolts to 40 ft. lbs. (54 Nm)
    - Starter heat shield
    - Nuts to the catalyst support bracket and mount
    - Positive battery cable and torque the nut to 89 inch lbs. (10 Nm)
    - Posi-lock connectors
    - Negative battery cable to the remote ground post

3.5L Engines

1. Remove or disconnect the following:
    - Negative battery cable from the remote ground post
    - Starter-to-engine/transaxle nut and bolts
    - Positive battery feed wire from the starter
    - Starter and position it to gain access to the posi-lock connector
2. Place a support under the engine and slightly relieve the pressure from the left engine mount.
3. Remove the 3 left engine mount-to-engine bolts.
4. Slightly, raise the engine to provide more room.
5. Remove the starter by sliding it rearward between the catalyst and the engine mount.
6. Remove the posi-lock connector.

To install:

7. Install or connect the following:
    - Posi-lock connector
    - Starter by sliding it forward between the catalyst and the engine mount
8. Slightly, lower the engine.
    - 3 left engine mount-to-engine bolts
    - Positive battery feed wire to the starter and torque the nut to 89 inch lbs. (10 Nm)
    - Starter and torque the starter-to-engine/transaxle bolts to 40 ft. lbs. (54 Nm)
    - Negative battery cable to the remote ground post

Oil Pan

REMOVAL & INSTALLATION

1. Before servicing the vehicle, refer to the precautions in the beginning of this section.
2. Drain the engine oil and remove the oil filter.
3. Remove or disconnect the following:
    - Negative battery cable
    - Dipstick and housing
    - Structural collar from the rear of the oil pan and transmission housing
    - Engine oil cooler lines from the oil pan, if equipped

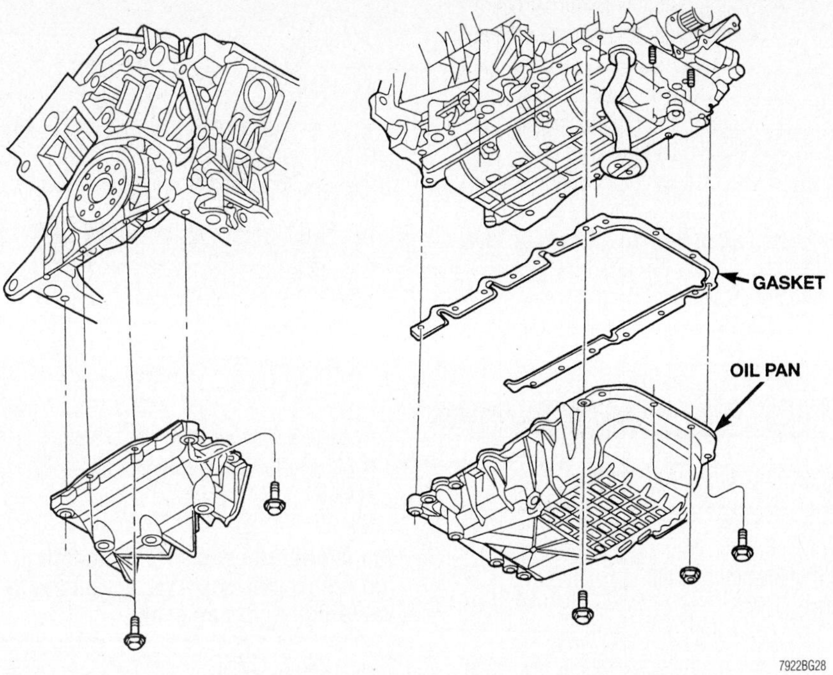

Exploded view of the oil pan removal and installation—2.7L and 3.5L engines

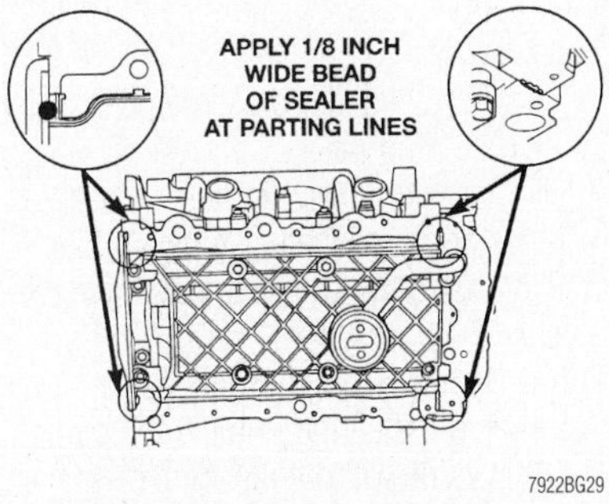

To ensure a proper seal, apply sealer as shown—2.7L and 3.5L engines

- Transmission oil cooler line clips, if necessary
- Oil pan mounting bolts, oil pan and gasket

4. Clean the oil pan and all gasket surfaces.

To install:

5. Apply a 1/8 in. (3mm) bead of sealer at the parting line of the oil pump body and the rear seal retainer.
6. Install or connect the following:
   - Oil pan and torque the M8 nuts/bolts to 21 ft. lbs. (28 Nm) and the M6 nuts/bolts to 105 inch lbs. (12 Nm)
7. Install the structural collar using the following procedure:
   a. Install the vertical collar to the oil pan mounting bolts and tighten, temporarily to 10 inch lbs. (1.1 Nm).
   b. Install the collar-to-transaxle bolts and torque to 40 ft. lbs. (55 Nm).
   c. Starting with the center vertical bolt and working outward. Torque the bolts to 40 ft. lbs. (55 Nm).
8. Install the dipstick and housing and connect the negative battery cable.
9. Fill the engine with the proper amount of clean SAE 5W-30 or SAE 10W-30 engine oil only. Do not mix the two grades of oil.
10. Start the engine, check for leaks and repair if necessary.

Oil Pump

REMOVAL & INSTALLATION

2.7L Engine

1. Before servicing the vehicle, refer to the precautions in the beginning of this section.
2. Drain the engine oil.
3. Remove or disconnect the following:
   - Crankshaft damper
   - Timing chain cover
   - Timing chain
   - Crankshaft sprocket
   - Oil pan
   - Oil pickup tube and O-ring
   - Oil pump

To install:

4. Fill the oil pump rotor cavity with clean engine oil.

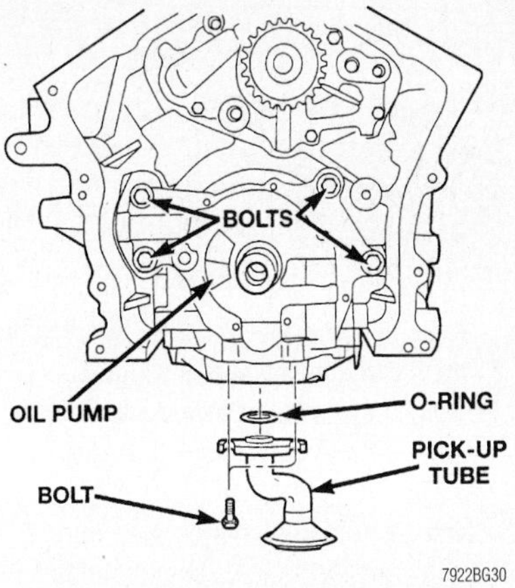

Oil pump mounting bolt locations—2.7L engine

5. Carefully, install the oil pump over the crankshaft and into position.
6. Install the oil pump mounting bolts and torque to 21 ft. lbs. (28 Nm).
7. Lubricate the new pickup tube O-ring with clean engine oil and torque the pickup tube mounting bolts to 21 ft. lbs. (28 Nm).
8. Install or connect the following:
   - Oil pan
   - Crankshaft sprocket
   - Timing chain
   - Timing chain cover
   - Crankshaft damper
9. Fill the engine with clean engine oil.
10. Start the vehicle, check for leaks and repair if necessary.

3.5L Engines

The timing belt must be removed to access the oil pump located behind the crankshaft drive sprocket. It is good practice to turn the crankshaft to Top Dead Center (TDC) No. 1 cylinder compression stroke (firing position) before starting disassembly. This should align all timing marks and be a good point of reference for all work to follow.

1. Before servicing the vehicle, refer to the precautions in the beginning of this section.
2. Drain the engine oil.
3. Drain the cooling system.
4. Remove or disconnect the following:
   - Negative battery cable
   - Accessory drive belts
   - Oil filter
   - Oil pan
   - Oil pump pickup tube
   - Windage tray/oil pan gasket
   - Crankshaft damper using a suitable puller tool
   - Timing belt covers

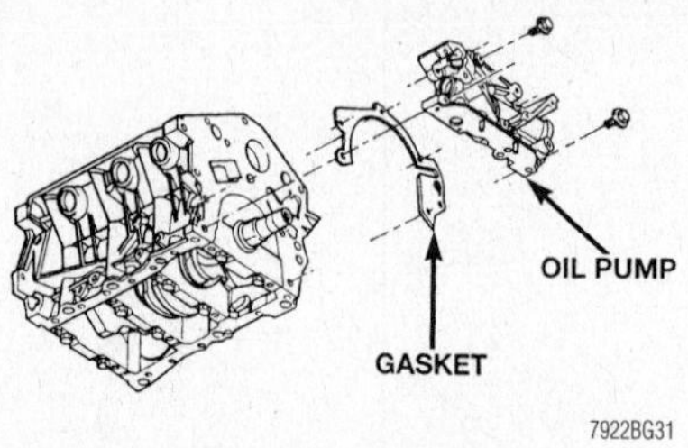

Prime the oil pump before installation, because a dry pump will wear prematurely and cause low oil pressure—3.5L engines

5. Place matchmarks on the timing belt to aid installation. Align the matchmarks on the camshaft sprockets to marks on the rear timing belt covers before removing the timing.
6. Remove or disconnect the following:
   - Timing belt. Refer to the appropriate section for the removal and installation procedure.
   - Crankshaft sprocket using a suitable puller tool
   - Oil pump-to-engine screws and pump
   - Oil pump cover screws and cover
   - Oil pump rotors
7. Wash all parts in solvent and inspect carefully for damage or wear.

To install:

8. Clean all parts well. There should be no traces of old gasket/sealer on any components.
9. Assemble the oil pump with new parts as required.
10. Install the oil pump cover. Torque the fasteners to 108 inch lbs. (12 Nm).
11. Prime the oil pump prior to installation by filling the rotor cavity with clean engine oil.
12. Install the oil pump and tighten the oil pump-to-engine screws as follows:
    a. M8 screws: 21 ft. lbs. (28 Nm).
    b. M10 screws: 40 ft. lbs. (55 Nm).
13. Tighten the oil pan drain plug and install a new oil filter.
14. Install or connect the following:
    - Oil pump pickup tube
    - Windage tray/oil pan gasket
    - Oil pan and torque fasteners to 108 inch lbs. (12 Nm). Pay attention to sealing the oil pan gasket and its integral windage tray.
    - Crankshaft sprocket using tool C-4685C1, thrust bearing, washer and 12mm bolt to draw the sprocket onto the crankshaft
    - Timing belt. Refer to the appropriate section for the removal and installation procedure.
    - Timing belt covers, thrust bearing, washer plate and vibration damper using tool L-4524
    - Accessory drive belts
    - Radiator and radiator hoses
    - Negative battery cable
15. Fill and bleed the cooling system.
16. Fill the engine with the correct amount of clean SAE 5W-30 or SAE 10W-30 engine oil only. Do not mix the two grades of oil.
17. Start the engine, check for leaks and proper oil pressure.

Rear Main Seal

REMOVAL & INSTALLATION

1. Before servicing the vehicle, refer to the precautions in the beginning of this section.
2. Drain the transaxle fluid.
3. Remove the negative battery cable.
4. Remove the transaxle, inspection cover and flywheel/flexplate.
5. Using a small prytool, carefully pry out the rear oil seal. Be careful not to nick or damage the crankshaft flange seal surface or the retainer bore.

To install:

6. Place the Seal Pilot tool C-4681 on the crankshaft.
7. Lightly coat the oil seal outside diameter with Loctite® Stud N' Bearing Mount® or the equivalent.
8. Apply a light coating of engine oil to the entire circumference of the oil seal lip.
9. Place the seal over the special tool and tap the seal in place with a plastic mallet.
10. Install the flexplate/flywheel and transaxle.
11. Connect the negative battery cable.
12. Fill the transaxle with the proper fluid.
13. Start the vehicle, check for leaks and repair if necessary.

Timing Chain, Sprockets, Front Cover and Seal

REMOVAL & INSTALLATION

2.7L Engine

⁂ WARNING

When aligning the timing marks, rotate the crankshaft, not the camshafts. DO NOT rotate the camshafts or crankshaft with the timing chain removed without locating the crankshaft position, piston and/or valve damage may occur.

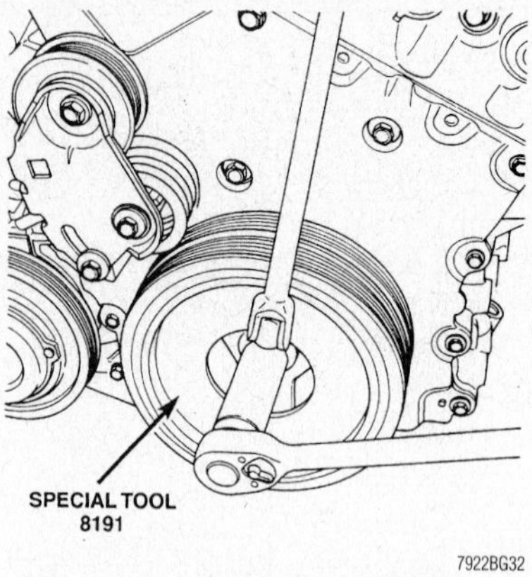

Removing the crankshaft center bolt using the Crankshaft Damper Holder tool—2.7L engine

1. Before servicing the vehicle, refer to the precautions in the beginning of this section.
2. Remove or disconnect the following:
   - Upper intake manifold and valve covers
   - Upper radiator crossmember
   - Fan module
   - Accessory drive belts
3. Using Crankshaft Damper Holder tool 8191, hold the crankshaft and remove the center bolt.
   - Damper using a 3-jaw puller
   - Power steering pump and position it aside without disconnecting the hydraulic lines
   - Accessory drive belt tensioner pulley
   - Timing chain cover bolts
4. Clean and inspect the sealing surfaces.
5. Align the crankshaft sprocket timing mark with the oil pump housing mark.

➡The mark on the oil pump housing is 60 degrees at Top Dead Center (TDC).

6. Remove or disconnect the following:
   - Primary timing chain tensioner from the right cylinder head
   - Camshaft Position (CKP) sensor and timing chain access plug from the left cylinder head

➡The camshafts will rotate clockwise, when the camshaft sprocket bolts are removed.

   - Right camshaft sprocket mounting bolts, camshaft damper and sprocket

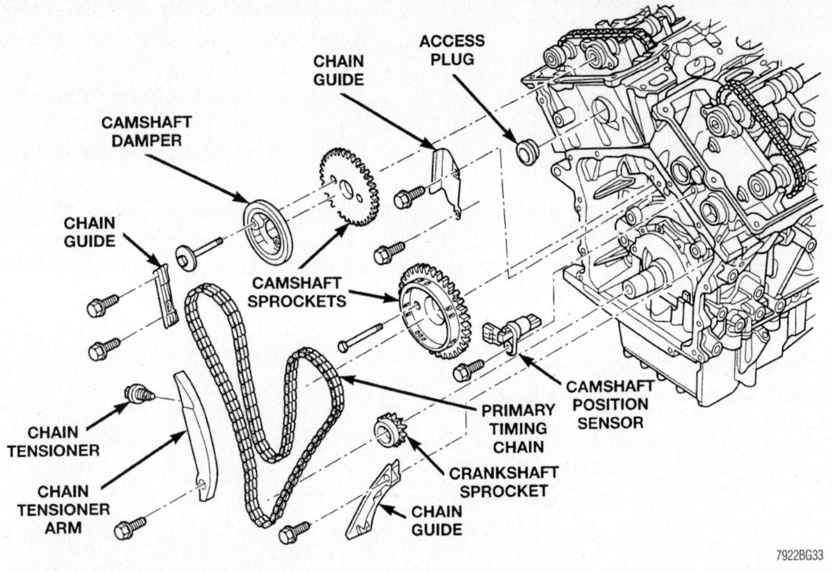

Exploded view of the timing chain drive assembly—2.7L engine

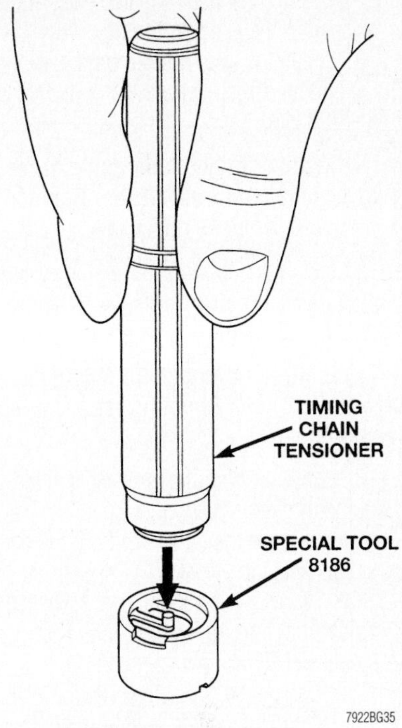

Using the Tensioner Resetting Special tool, to purge the oil from the tensioner—2.7L engine

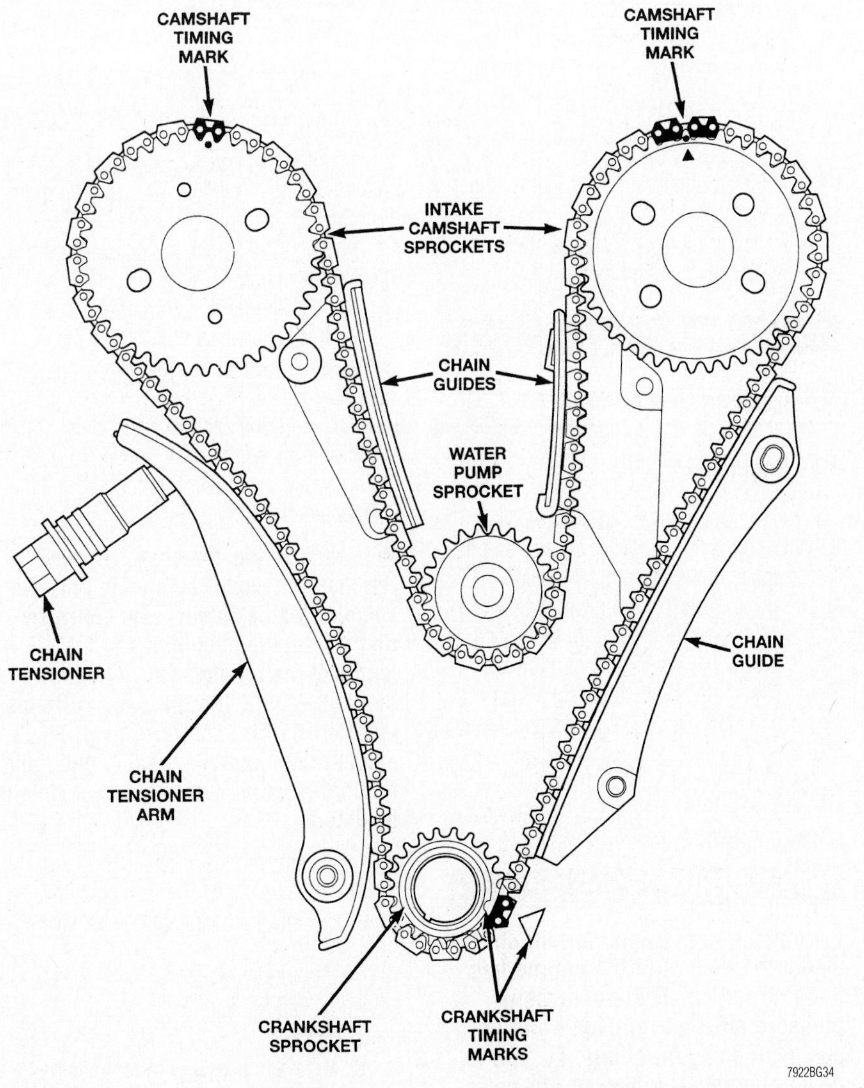

After the timing chain is installed, the timing marks should be aligned—2.7L engine

- Left camshaft sprocket bolts and sprocket
- Lower timing chain guide, tensioner arm and primary timing chain

To install:

➡Lubricate the timing chain and guides with clean engine oil before installation.

7. Install the timing chain by performing the following procedure:
   a. Verify that the crankshaft sprocket timing mark is aligned with the mark on the oil pump housing.
   b. Place the left side primary timing chain sprocket onto the chain, while aligning the timing mark on the sprocket to the 2 plated links on the chain.
   c. Lower the chain with the left sprocket through the left cylinder head opening.
   d. Loosely position the left camshaft sprocket over the camshaft hub.
   e. Align the plated link to the crankshaft sprocket timing mark.
   f. Position the timing chain around the water pump drive sprocket.
   g. Align the right camshaft sprocket timing mark to the plated link on the timing chain and loosely position the sprocket over the camshaft hub.

h. Verify that all the plated links are aligned to their proper timing marks.

8. Install the left lower timing chain guide and tensioner. Torque the bolts to 21 ft. lbs. (28 Nm).

➡**Inspect the timing chain guide access plug O-rings before installing. Replace damaged O-rings as necessary.**

9. Install the timing chain guide access plug to the left cylinder head and torque to 15 ft. lbs. (20 Nm).

➡**To reset the timing chain tensioner, oil will first need to be purged from the tensioner.**

10. Purge oil from the timing chain tensioner using the following procedure:

a. Remove the tensioner from the tensioner housing.

b. Place the check ball end of the tensioner into the shallow end of the Tensioner Resetting Special tool 8186.

c. Using hand pressure, slowly depress the tensioner until oil is purged from the cylinder.

d. Reinstall the tensioner into the tensioner housing.

11. Reset the timing chain tensioner using the following procedure:

a. Position the cylinder plunger into the deeper side of the Tensioner Resetting special tool 8186.

b. Apply a downward force until the tensioner is reset.

**** WARNING**

Ensure that the tensioner is properly reset. The tensioner body must be bottomed against the top edge of the Tensioner Resetting Special tool 8186. Failure to properly perform the resetting procedure may cause tensioner jamming.

12. Install the chain tensioner into the right cylinder head.

13. At the right cylinder head, insert a 3/8 in. square drive extension with a breaker bar into the intake camshaft drive hub. Rotate the camshaft until the camshaft hub aligns with the camshaft sprocket and damper attaching holes. Install the sprocket attaching bolts and torque to 21 ft. lbs. (28 Nm).

14. Turn the left camshaft by inserting a 3/8 in. square drive extension with a breaker bar into the intake camshaft drive hub. Rotate the camshaft until the camshaft hub aligns with the camshaft sprocket and damper attaching holes. Install the sprocket attaching bolts and torque to 21 ft. lbs. (28 Nm).

15. If necessary, rotate the engine slightly clockwise to remove any slack in the timing chain.

16. To arm the timing chain tensioner: Use a flat-bladed prytool to gently pry the tensioner arm towards the tensioner slightly. Then, release the tensioner arm. Verify the tensioner extends.

17. Inspect and replace the timing chain cover gasket and oil seal.

18. Apply a 1/8 in. (3mm) bead of sealer at the parting line of the oil pan and engine block.

➡**When installing the timing cover, guide the seal over the crankshaft to prevent damage to the seal's lip.**

19. Install or connect the following:

- Timing cover and gasket and torque the M10 bolts to 40 ft. lbs. (54 Nm) and the M6 bolts to 105 inch lbs. (12 Nm)
- Crankshaft damper
- Accessory drive belt tensioner pulley
- Power steering pump
- Crankshaft damper
- Crankshaft center bolt and torque the bolt to 125 ft. lbs. (170 Nm) using tool 8191
- Accessory drive belts
- Fan module and electrical wiring harness
- Upper radiator crossmember
- Negative battery cable

Timing Belt

REMOVAL & INSTALLATION

3.5L Engines

Use care when servicing a timing belt. Valve timing is absolutely critical to engine performance. If the valve timing marks on all drive sprockets are not properly aligned, engine damage will result. If only the belt and tensioner are being serviced, do not loosen the camshaft drive sprockets unless they are to be replaced. The sprockets have oversized openings and can be rotated several degrees in each direction on their shafts. This means the sprockets must be re-timed, requiring some special tools.

**** CAUTION**

Fuel injection systems remain under pressure, even after the engine has been turned off. The fuel system pressure must be relieved before disconnecting any fuel lines. Failure to do so may result in fire and/or personal injury.

1. Before servicing the vehicle, refer to the precautions in the beginning of this section.

2. Disconnect the negative battery cable.

3. Rotate the engine to Top Dead Center (TDC) on the compression stroke for cylinder No. 1.

4. Release the fuel system pressure using the recommended procedure.

5. Place a pan under the radiator and drain the coolant.

6. Remove or disconnect the following:

- Radiator and cooling fan assemblies
- Accessory drive belts
- Upper radiator hose
- Crankshaft damper with a quality puller tool gripping the inside of the pulley
- Stamped steel cover. Do not remove the sealer on the cover; it may be reusable.
- Left side cast cover
- Lower belt cover, located behind the crankshaft damper, if necessary

7. If the timing belt is to be reused, mark the timing belt with the running direction for installation.

8. Align the camshaft sprockets with the marks on the rear covers.

- Timing belt and tensioner

9. If it is necessary to service the camshaft sprockets, use the following procedure:

a. Hold the camshaft sprocket with a 36mm box end wrench, loosen and remove the sprocket retaining bolt and washer.

➡**To remove the camshaft sprocket retainer bolt while the engine is in the vehicle, it may be necessary to raise that side of the engine due to the length of the retainer bolt. The right bolt is 8 3/8 in. (213mm) long, while the left bolt is 10 in. (254mm) long. These bolts are not interchangeable and their original location during removal should be noted.**

b. Remove the camshaft sprocket from the camshaft. The camshaft sprockets are not interchangeable from side-to-side.

c. Remove the crankshaft sprocket using Puller L-4407A.

To install:

10. If it was necessary to remove the camshaft sprockets, use the following procedure:

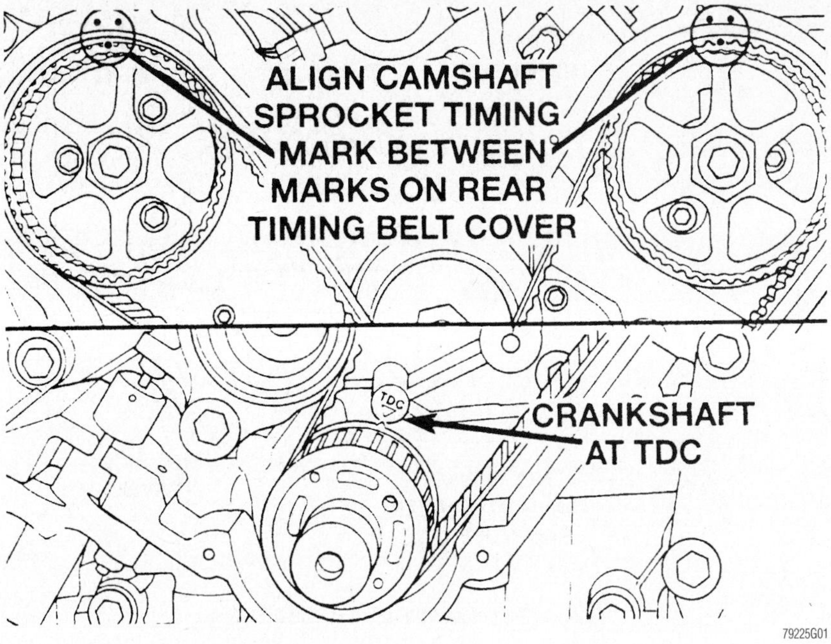

Timing belt alignment marks—3.5L engines

➡This procedure can only be used when the camshaft sprockets have been loosened or removed from the camshafts. Each sprocket has a D shaped hole that allows it to be rotated several degrees in each direction on its shaft. This design must be timed with the engine to ensure proper performance.

a. Install the crankshaft sprocket, using tool C-4685-C1, thrust bearing, washer and 12mm bolt.

b. When the camshaft sprockets are loosened or removed, the camshafts must be timed to the engine. Install the Camshaft Alignment tools 6642-A, to the rear of the cylinder heads. These tools lock the camshafts in the proper position.

11. Preload the belt tensioner as follows:

a. Place the tensioner in a vise the same way it is mounted on the engine.

b. Slowly compress the plunger into the tensioner body.

c. When the plunger is compressed into the tensioner body, install a pin through the body and plunger to retain the plunger in place until the tensioner is installed.

12. Install both camshaft sprockets to the appropriate shafts. The left camshaft sprocket has the Distributorless Ignition System (DIS) pick-up as part of the sprocket.

➡The right bolt is 8⅜ in. (213mm) long, while the left bolt is 10 in. (254mm) long. These bolts are not interchangeable.

13. Apply Loctite®271, to the threads of the camshaft sprocket retainer bolts and install to the appropriate shafts. Do not tighten the bolts at this time.

14. Align the camshaft sprockets between the marks on the rear belt covers.

15. Align the crankshaft sprocket with the TDC mark on the oil pump cover.

16. Install the timing belt, starting at the crankshaft sprocket and going in a counter-clockwise direction. After the belt is installed on the right sprocket, keep tension on the belt until it is past the tensioner pulley.

17. Holding the tensioner pulley against the belt, install the timing belt tensioner into the housing and tighten to 21 ft. lbs. (28 Nm).

18. When the tensioner is in place, pull the retainer pin out to allow tensioner to extend to the pulley bracket.

➡Be sure that the timing marks on the cam sprockets are still between the marks on the rear cover.

19. Remove the spark plug in the No.1 cylinder and install a dial indicator to check for TDC of the piston. Rotate the crankshaft until the piston is exactly at TDC.

20. Hold the camshaft sprocket hex with a 36mm wrench and tighten the right camshaft sprocket bolt to 75 ft. lbs. (102 Nm) plus an additional 90 degree turn. Tighten the left camshaft sprocket bolt to 85 ft. lbs. (115 Nm) plus an additional 90 degree turn.

21. Remove the dial indicator. Install the spark plug and tighten to 20 ft. lbs. (28 Nm).

22. Remove the camshaft alignment tools from the back of the cylinder heads and install the cam covers with new O-rings.

23. Tighten the fasteners to 20 ft. lbs. (27 Nm). Repeat this procedure on the other camshaft.

24. Rotate the crankshaft sprocket two revolutions and check for proper alignment of the timing marks on the camshaft and the crankshaft. If the timing marks do not align, repeat the procedure.

25. Before installing, inspect the sealer on the stamped steel cover. If some sealer is missing, use MOPAR Silicone Rubber Adhesive sealant to replace the missing sealer.

26. Install or connect the following:

- Lower belt cover behind the crankshaft damper, if necessary
- Stamped steel cover and the left side cast cover. Tighten the 6mm bolts to 105 inch lbs. (12 Nm), the 8mm bolts to 250 inch lbs. (28 Nm) and the 10mm bolts to 40 ft. lbs. (54 Nm).
- Crankshaft damper using special tool L-4524, a 5.9 in. long bolt, thrust bearing and washer. Tighten the center bolt to 85 ft. lbs. (115 Nm).
- Upper radiator hose
- Accessory drive belts and adjust them to the proper tension
- Radiator and cooling fan assemblies

27. Refill and bleed the cooling system.

28. Connect the negative battery cable.

29. With the radiator cap off so coolant can be added, run the engine. Watch for leaks and listen for unusual engine noises.

Piston and Rings

POSITIONING

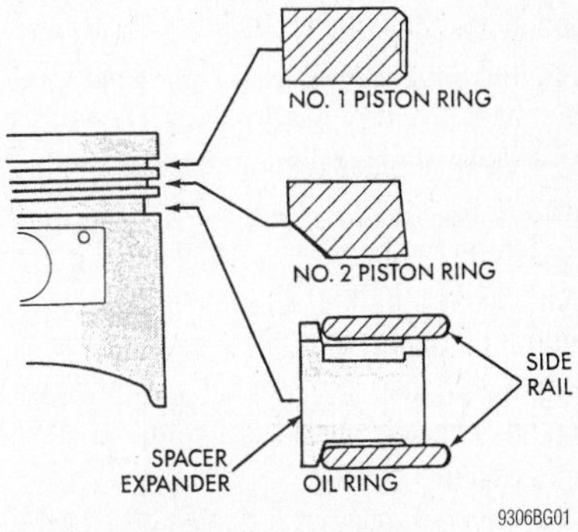

Cross-sectional view of the piston rings—2.7L Engines

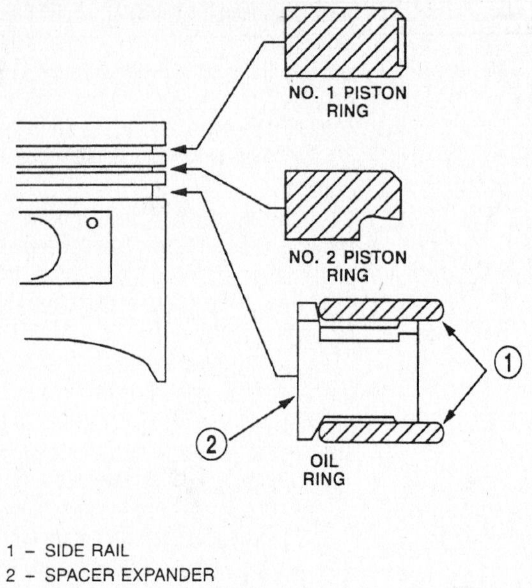

Cross-sectional view of the piston rings—3.5L Engine

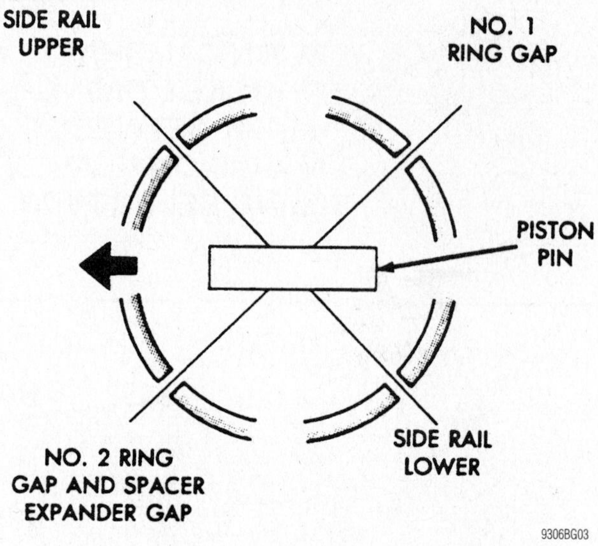

Piston ring gap positions—2.7L and 3.5L Engines

FUEL SYSTEM

Fuel System Service Precautions

Safety is the most important factor when performing not only fuel system maintenance but any type of maintenance. Failure to conduct maintenance and repairs in a safe manner may result in serious personal injury or death. Maintenance and testing of the vehicle's fuel system components can be accomplished safely and effectively by adhering to the following rules and guidelines:

- To avoid the possibility of fire and personal injury, always disconnect the negative battery cable unless the repair or test procedure requires that battery voltage be applied.
- Always relieve the fuel system pressure prior to disconnecting any fuel system component (injector, fuel rail, pressure regulator, etc.), fitting or fuel line connection. Exercise extreme caution whenever relieving fuel system pressure, to avoid exposing skin, face and eyes to fuel spray. Please be advised that fuel under pressure may penetrate the skin or any part of the body that it contacts.
- Always place a shop towel or cloth around the fitting or connection prior to loosening to absorb any excess fuel due to spillage. Ensure that all fuel spillage (should it occur) is quickly removed from engine surfaces. Ensure that all fuel soaked cloths or towels are deposited into a suitable waste container.
- Always keep a dry chemical (Class B) fire extinguisher near the work area.
- Do not allow fuel spray or fuel vapors to come into contact with a spark or open flame.
- Always use a back-up wrench when loosening and tightening fuel line connection fittings. This will prevent unnecessary stress and torsion to fuel line piping.
- Always replace worn fuel fitting O-rings with new. Do not substitute fuel hose, where fuel pipe is installed.

Before servicing the vehicle, also make sure to refer to the precautions in the beginning of this section as well.

Fuel System Pressure

RELIEVING

1. Before servicing the vehicle, refer to the precautions in the beginning of this section.
2. At the Power Distribution Center (PDC), remove the Fuel Pump Relay.
3. Operate the engine until the engine stalls; then, continue restarting the engine until it will no longer run.
4. Turn the ignition switch **OFF**.

⁂ CAUTION

The previous steps must be performed to relieve the high pressure fuel from the fuel rail. The following steps must be performed to remove excess fuel from the fuel rail. Do not use the following steps to relieve high pressure for excessive fuel will be forced into a cylinder chamber.

5. Disconnect the electrical connector from any injector.
6. Connect 1 end of a jumper wire (with an alligator clip) to either injector terminal and the other end to the positive side of the battery.
7. Connect 1 end of a second jumper wire to the other injector terminal.

⁂ WARNING

Applying power to an injector for more than a few seconds will damage the injector.

8. Momentarily, touch the other end of the jumper wire to a ground for no more than a few seconds.
9. At the Power Distribution Center (PDC), install the fuel pump relay.

➡When the fuel pump relay is removed, 1 or more Diagnostic Trouble Codes (DTC's) may be stored in the Powertrain Control Module (PCM) memory. A DRB scan tool must by used to clear the DTC's.

Fuel Filter

The fuel filter mounts to the frame rail in front of the fuel tank. The inlet and outlet ends of the filter are marked for installation purposes. Install the fuel filter making certain that it is properly orientated.

REMOVAL & INSTALLATION

➡The fuel filter is part of the fuel pressure regulator mounted on the fuel pump module.

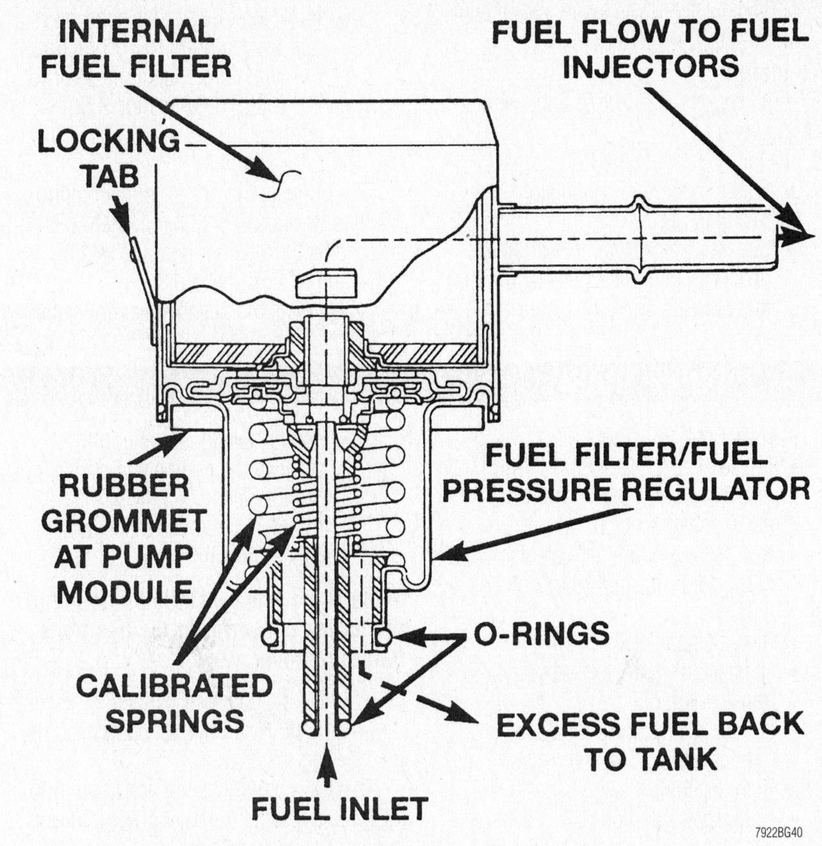

Cut away view of the fuel filter/pressure regulator

1. Before servicing the vehicle, refer to the precautions in the beginning of this section.
2. Properly relieve the fuel system pressure.
3. Lower the fuel tank.
4. Remove or disconnect the following:
   - Negative battery cable
   - Purge and vent lines
   - Fuel line from the pressure regulator
   - Filter/regulator by pushing in the locking tab, turning the regulator to unlock it and pulling the it straight up

To install:

5. Push the fuel filter/regulator into the fuel pump module and turn to lock it into position.
6. Connect the fuel lines and install the tank.
7. Start the engine, check for leaks and repair if necessary.

Fuel Pump

REMOVAL & INSTALLATION

The in-tank fuel pump module contains the fuel pump and pressure regulator which adjusts fuel system pressure. Fuel pump voltage is supplied through the fuel pump relay.

The fuel pump is serviced as part of the fuel pump module. The fuel pump module is installed in the top of the fuel tank and contains the electric fuel pump, fuel pump reservoir, inlet strainer fuel gauge sending unit, fuel supply and return line connections and the pressure regulator. The inlet strainer, fuel pressure regulator and level sensor are the only serviceable items. If the fuel pump requires service, replace the fuel pump module.

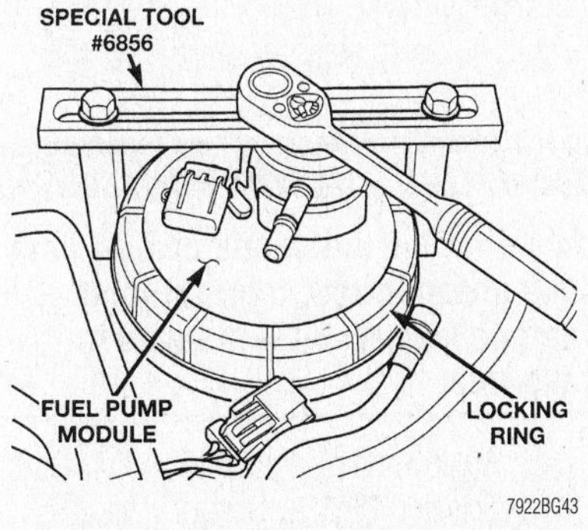

Using special tool 6856, remove the fuel pump module locknut

1. Before servicing the vehicle, refer to the precautions in the beginning of this section.
2. Properly relieve the fuel system pressure.
3. Remove or disconnect the following:
   - Negative battery cable
   - Fuel tank
4. Clean the top of the tank to remove any loose dirt.
   - Fuel lines from the fuel pump module by squeezing the quick-connect fitting with thumb and forefinger
   - Fuel pump module electrical connector from the top of the fuel pump module
5. Using special tool 6856, remove the fuel pump locknut by turning it counterclockwise.

⁂ CAUTION

The fuel reservoir of the fuel pump module does not empty out when the tank is drained. The fuel in the reservoir may spill out when the module is removed.

6. Remove the fuel pump and O-ring from the tank and discard the O-ring.

To install:

7. Thoroughly clean all parts. Wipe the seal area of the tank clean. Place a new O-ring on the ledge between the tank threads and the pump module opening.
8. Position the fuel pump module in the tank. Be sure the alignment tab on the underside of the pump module flange sits in the corresponding notch in the fuel tank.
9. While holding the fuel pump module in place install the locking ring and torque to 40 inch lbs. (5 Nm) using special tool 6856 or a spanner-type tool.
10. Install or connect the following:
    - Fuel tank
    - Fuel pump module electrical connector
    - Negative battery cable
11. Fill the fuel tank with fuel. Install the fuel filler cap. Turn the ignition switch to the **ON** position to pressurize the system. Check the fuel system for leaks.

Fuel Injector

REMOVAL & INSTALLATION

1. Before servicing the vehicle, refer to the precautions in the beginning of this section.
2. Relieve the fuel system pressure.

3. Remove or disconnect the following:
   - Negative battery cable
   - Intake manifold plenum and cover opening with a clean cloth

4. Place a shop rag under the fuel rail's quick-connect fitting; then, squeeze the quick-connect fitting's retainer tabs together and pull the fitting assembly off of the fuel tube nipple.
   - Fuel injector electrical connectors
   - Fuel rail-to-engine bolts and fuel rail
   - Fuel injector-to-fuel rail retainer clips
   - Fuel injectors

To install:

5. Lubricate the injector O-rings with clean engine oil.
6. Install or connect the following:
   - Fuel injector and secure with retaining clips
   - Fuel rail onto cylinder head and press rail into place. Make sure that the injectors are fully seated.
   - Fuel rail-to-cylinder head bolts and torque bolts to 100 inch lbs. (11 Nm) for 2.7L engines, or to 21 ft. lbs. (28 Nm) for 3.5L engine.
   - Intake plenum
7. Lubricate the quick-connect fitting's O-rings with clean engine oil; then, push the connector together until the retainer seats and a click is heard.
8. Connect the negative battery cable.

DRIVE TRAIN

Transaxle Assembly

REMOVAL & INSTALLATION

The 42LE four speed transaxle uses fully-adaptive controls. Adaptive controls are those which perform their functions based on real-time feedback sensor information. The transaxle is conventional in the use of hydraulically applied clutches to shift a planetary gear train. However, it uses electronics to control virtually all other functions. The following components are serviceable in the vehicle: valve body assembly, solenoid pack, manual valve lever position sensor, input and output speed sensors, transfer chain and sprockets, short (right side) stub shaft seal and the long (left side) stub shaft and ball bearing. Note that the factory recommends that before attempting any repair on the 42LE four-speed automatic transaxle, always check for proper shift linkage adjustment. Also, check for diagnostic trouble codes with the Chrysler DRB scan tool.

Use MOPAR Type 7176 Automatic Transmission Fluid only. Do not substitute transaxle fluid. If the differential sump requires fluid, use 80W-90 petroleum based Hypoid gear lubricant.

1. Before servicing the vehicle, refer to the precautions in the beginning of this section.
2. Remove or disconnect the following:
   - Negative battery cable
   - Wiper blades
   - Right and left wiper module covers
   - Steel cowl/strut support

⁂ CAUTION

Be careful of sharp edges cowl/wiper area.

   - Engine air inlet tube
   - Transaxle harness connectors at cowl area
   - Upper bell housing studs
   - Loosen clamps at intersection of rear exhaust system to front catalytic converter pipe
   - Rear exhaust from left catalytic converter pipe and right extension pipe
   - Rear exhaust system
   - Nuts securing the exhaust pipes to transaxle mount
   - Loosen clamp at right extension at right catalytic converter
   - Right extension
   - Crankshaft Position (CKP) sensor connector and remove sensor. The sensor is located on the upper right side of the transaxle bell housing.
   - Dipstick tube
   - Gear selector cable from the transaxle
   - Transmission range sensor connector
   - Input and output speed sensor connector
   - Transaxle cooler lines from the transaxle and plug them
   - Lower control arm pinch bolts
   - Lower control arms from steering knuckles by prying them out
3. Use a prybar to separate the inner tripod joints from the transaxle
4. Pull the bottom of the knuckles and halfshafts outward to allow clearance during transaxle removal. The halfshafts do not have to be completely removed from the vehicle.

⁂ WARNING

Do not let the halfshafts or CV-joints hang unsupported. Internal joint damage may result if allowed to hang free.

5. Remove or disconnect the following:
   - Heated Oxygen Sensor (HO_2S) wiring and left catalytic converter pipe
   - Starter bolts and position it between the engine and transaxle. Do not disconnect the wiring or completely remove the starter.
   - Engine oil pan cooler

➡Matchmark the installed position of the torque converter to the flex plate before removal.

   - Torque converter bolts and discard them. New bolts must be used during installation.
6. Use a transmission jack to slightly raise the transaxle, removing weight from the rear transaxle mount.

➡The crossmember bridge bolts are different lengths for each side, so note their locations during removal.

   - Rear crossmember bridge bolts
   - Rear mount adapter plate mounting bolts
   - Rear crossmember bridge, mount and adapter plate as an assembly

➡Lower the rear of the transaxle to access to the bell housing bolts.

   - Side bell housing bolts
   - Dipstick tube from the transaxle and plug to prevent leakage
   - Solenoid/pressure switch assembly connector from the top of the transaxle
   - Transaxle by lowering it from the vehicle.

To install:

7. To prevent damaging the structural collar, hand-tighten all fasteners before torquing them to specifications.
8. If the torque converter is being replaced, apply a light coat of oil to the crankshaft pilot hole.
9. Check the drive plate for cracks before reinstalling the transaxle and replace the plate as necessary.
10. Use new torque converter-to-drive plate bolts, as different year models use different bolts.

11. Installation is the reverse of the removal procedure, noting the following torque specifications:

a. Converter-to-driveplate bolts: 65 ft. lbs. (88 Nm)

b. Driveplate-to-crankshaft: 70 ft. lbs. (95 Nm)

c. Engine-to-transaxle case bolts to 75 ft. lbs. (101 Nm)

d. Input and output speed sensor-to-case bolts: 20 ft. lbs. (27 Nm).

12. Fill the transaxle with fluid. Start the engine and allow it to idle for 2 minutes. Apply the parking brake and move the selector through each gear position, ending in **N**. Recheck the fluid level and add if necessary. Be sure the vehicle is level when refilling the transaxle. Use Mopar Type 9602 Automatic Transmission Fluid (ATF+4) only. Do not substitute transaxle fluid.

13. Check the transaxle or proper operation. Adjust the shift linkage, if necessary. Be sure the reverse lamps come on when in reverse.

Halfshaft

REMOVAL & INSTALLATION

**** WARNING**

Allowing the CV-joint assemblies to dangle unsupported, or pulling or pushing the ends, can damage boots or CV-joints. Always support both ends of the halfshaft to prevent damage or disengagement of the Tri-pot joint.

1. Before servicing the vehicle, refer to the precautions in the beginning of this section.
2. Remove or disconnect the following:
   - Negative battery cable
   - Front wheels
   - Front caliper assembly from steering knuckle
   - Front brake rotor from the hub
   - Speed sensor cable routing bracket from strut assembly
   - Hub and bearing-to-stub axle retainer nut
3. Install a puller tool onto the hub and bearing assembly and secure it into place using the wheel lug nuts.
4. Protect wheel stud threads by installing a wheel lug nut onto a wheel stud. Use a flat-bladed prying tool to prevent the hub from turning. Using the puller tool, force the halfshaft outer stub axle from the hub and bearing assembly.
5. Dislodge the inner Tri-pot joint from the stub shaft retaining snapring on the transaxle. To do this, insert a prybar between the transaxle case and the inner Tri-pot joint and pry on Tri-pot joint.

➡Do not try to remove the inner Tri-pot joint from the transaxle stub shaft at this time. Only disengage the inner Tri-pot joint from the retainer snapring.

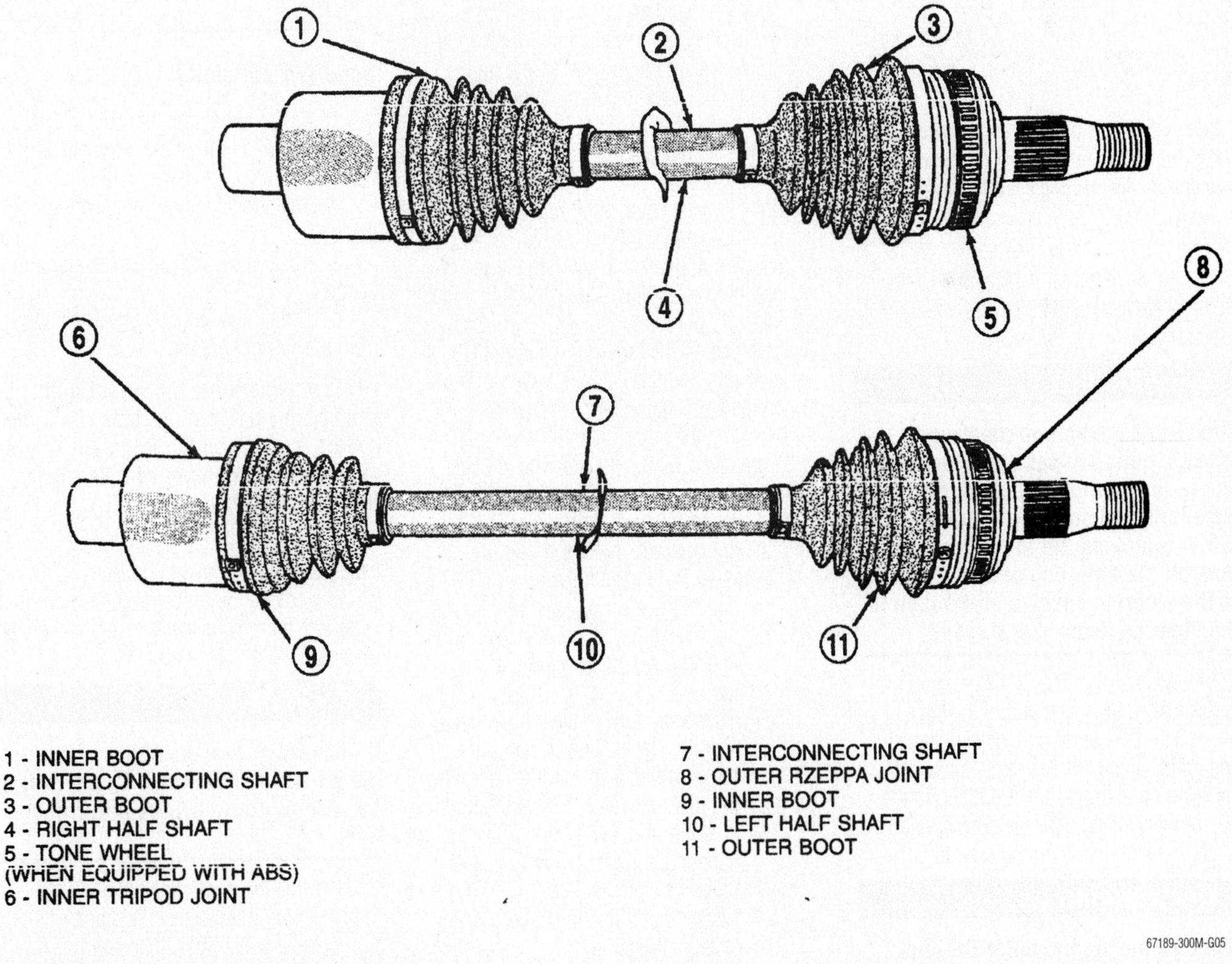

Front halfshaft assemblies—2004 shown, others similar

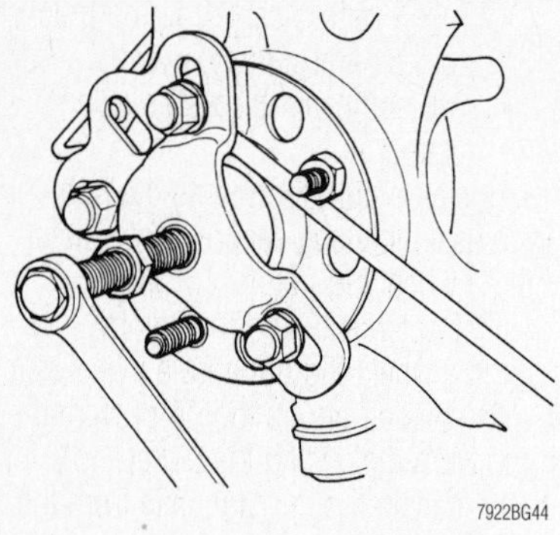

Removing the stub axle from the front hub/bearing assembly—halfshaft service

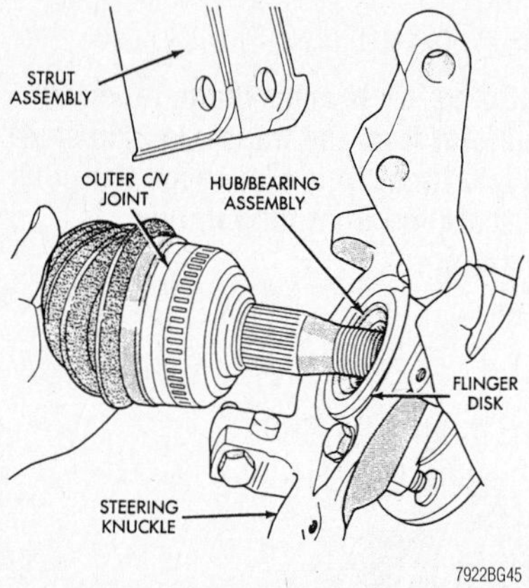

Be careful not to damage the threads for the axle nut when removing the outer CV-joint from the steering knuckle—halfshaft service

6. Remove the strut assembly-to-steering knuckle attaching bolts from the strut assembly.

⁂ WARNING

The strut assembly-to-steering knuckle bolts are serrated (toothed) where they go through the strut assembly and steering knuckle. When removing the bolts, turn the nuts off the bolt; do not turn the bolts in the steering knuckle or damage to the steering knuckle will result.

7. Separate the top of the steering knuckle from the lower end of the strut.

8. Hold the outer joint assembly with one hand. Grasp the steering knuckle with the other hand and rotate it out and to the rear of the vehicle, until the outer CV-joint clears the hub and bearing assembly.

⁂ WARNING

When removing the outer CV-joint from the hub and bearing assembly, do not allow the flange disc on the hub and bearing assembly to become damaged. If this happens, dirt and water can enter the bearing, which will cause premature bearing failure.

9. Remove the halfshaft inner joint from the transaxle stub shaft by grasping the inner Tri-pot joint and the interconnecting shaft and pulling both pieces at the same time. Take care not to pull on the interconnecting shaft to remove or separation of the spider assembly will occur.

To install:

10. Install the Tri-pot joint side of the halfshaft by performing the following procedure:

a. Replace the inner Tri-pot joint retaining circlip and O-ring seal on the transaxle stub shaft. These components are not reusable and must be replaced whenever the halfshaft is removed.

b. Apply an even coat of grease on the splines of the inner Tri-pot joint, where the O-ring seats against the Tri-pot joint.

c. Install the halfshaft through the hole in the splash shield. Grasp the inner joint in 1 hand and interconnecting shaft in the other. Align the inner Tri-pot joint spline with the stub shaft spline on the transaxle. Use a rocking motion with the inner Tri-pot joint to get it past the circlip on the transaxle stub shaft.

d. Continue pushing the Tri-pot joint onto transaxle stub shaft until it stops moving. The O-ring on the stub shaft should not be visible when the inner Tri-pot joint is fully installed. Check that the inner Tri-pot joint is locked in position by grasping the inner joint and pulling. If locked in position, the joint will not move on the stub shaft.

11. Hold the outer CV-joint assembly with one hand. Grasp the steering knuckle with the other and rotate it out and to the rear of the vehicle. Install the outer CV-joint into the hub and bearing assembly.

12. Install or connect the following:

- Top of the steering knuckle into the strut assembly. Align the steering knuckle-to-strut assembly mounting holes.
- Strut assembly-to-steering knuckle attaching bolts. Install the nuts to the attaching bolts and while holding the bolt heads, torque the nuts to 125 ft. lbs. (170 Nm). Turn the nuts on the bolts. DO NOT turn the bolts.
- New hub and bearing assembly-to-stub shaft retainer nut. Tighten but do not torque the nut at this time.
- Speed sensor cable routing bracket and screw
- Brake rotor and caliper assembly; then, tighten the caliper guide pin bolts to 30 ft. lbs. (41 Nm)
- Front wheels and lug nuts
- Negative battery cable

13. Pump the brakes until a firm pedal is obtained.

14. Apply the brakes and torque the new stub shaft-to-hub and bearing assembly retainer nut to 120 ft. lbs. (163 Nm).

⁂ WARNING

When tightening the stub shaft retaining nut, be careful not to exceed the maximum torque specification of 120 ft. lbs. (163 Nm). If this specification is exceeded, failure of the halfshaft could result.

15. Road test the vehicle to check for noise or vibration.

CV-Joints

OVERHAUL

Inner (Tri-pot) Joint

1. Before servicing the vehicle, refer to the precautions in the beginning of this section.

2. Disconnect the negative battery cable.

3. Remove the halfshaft and retaining clamps.

4. Slide the boot down the shaft away from the tri-pot housing.

➡When separating the spider joint from the tri-pot joint housing, hold the rollers in place on the trunnions to prevent the rollers and needle bearings from falling away.

5. Carefully, slide the shaft/spider assembly from the tri-pot housing.

6. Remove the spider assembly-to-shaft snapring; then, slide the spider assembly off the shaft.

⁂ WARNING

If necessary, tap the spider assembly off the shaft using a brass drift; be careful not to hit the outer bearings.

7. Slide the boot off the shaft.

8. Thoroughly, inspect all parts for signs of excessive wear; if necessary, replace the halfshaft.

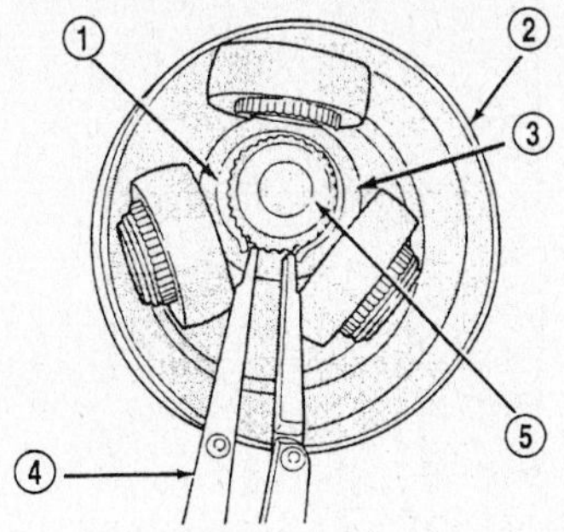

View of the halfshaft inner tri-pot joint and snapring

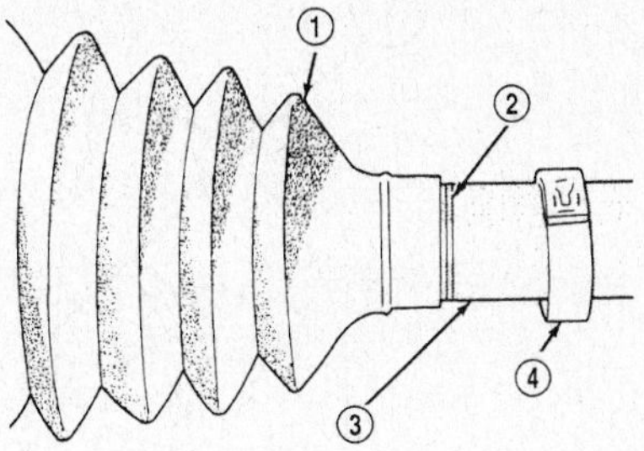

View of the halfshaft boot, shaft and boot clamp

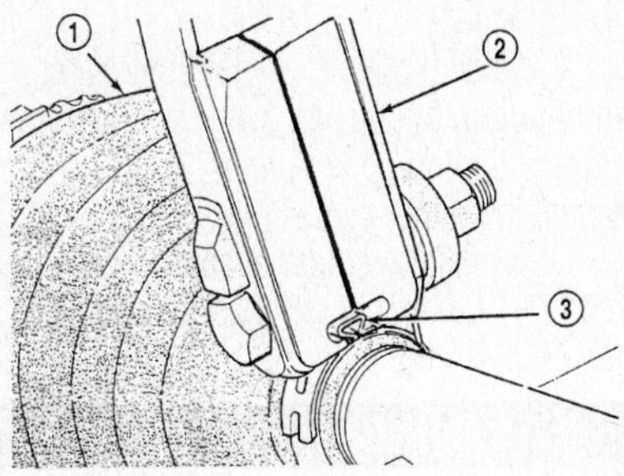

Securing the halfshaft boot clamp

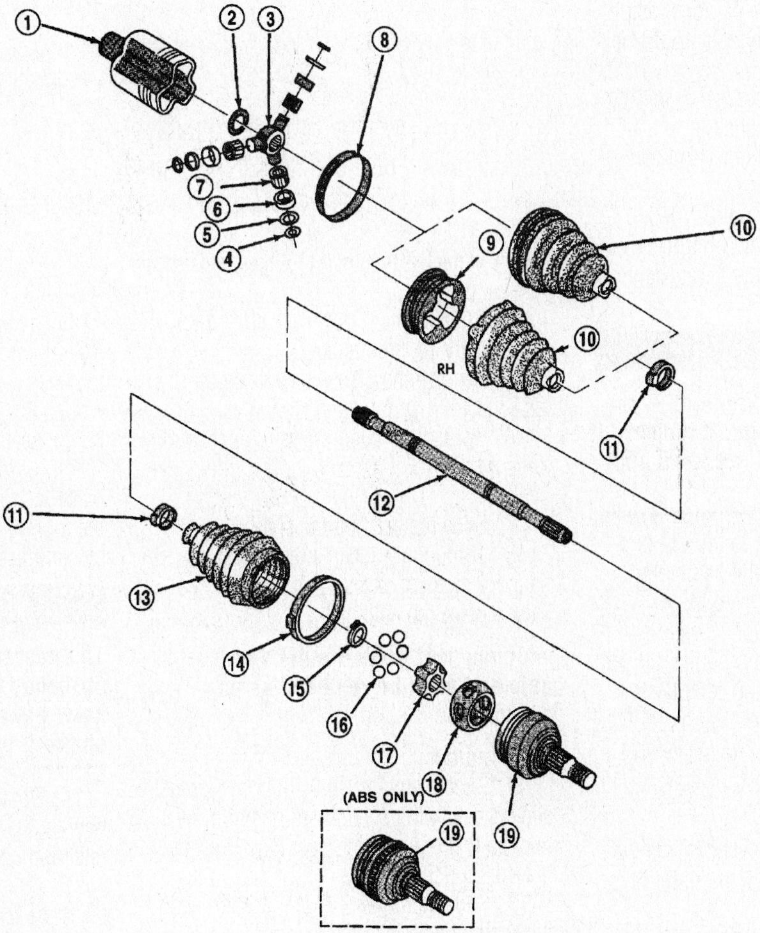

Exploded view of the halfshaft assemblies

➡**Component parts are not serviceable and must be replaced as an assembly.**

To install:

9. Slide the inner tri-pot boot clamp and boot onto the shaft; then, position the boot so that only the thinnest (sight) groove is visible on the shaft.
10. Install the spider assembly onto the shaft just far enough so that the snapring can be installed.

⁂ WARNING

If necessary, tap the spider assembly onto the shaft using a brass drift; be careful not to hit the outer bearings.

11. Install the snapring onto the shaft; make sure that the snapring is fully seated in the groove.
12. If installing a new boot, distribute ½ of the grease in the service package inside the tri-pot housing and the other ½ inside the boot.
13. Carefully, slide the spider assembly and shaft into the tri-pot housing.
14. Position the inner boot clamp evenly on the sealing boot.
15. Using the Crimper tool C-4975, place the tool over the clamp bridge, tighten the tool nut until the jaws are completely closed (face-to-face).

⁂ WARNING

The seal must not be dimpled, stretched or out of shape. If necessary, equalize the seal pressure and shape it by hand.

16. Position the boot onto the tri-pot housing retaining groove and install the retaining clamp evenly on the boot.
17. Using the Crimper tool C-4975, place the tool over the clamp bridge, tighten the tool nut until the jaws are completely closed (face-to-face).
18. Install the halfshaft into the vehicle.

Outer CV-Joint

1. Before servicing the vehicle, refer to the precautions in the beginning of this section.
2. Disconnect the negative battery cable.
3. Remove the halfshaft and the retaining clamps.

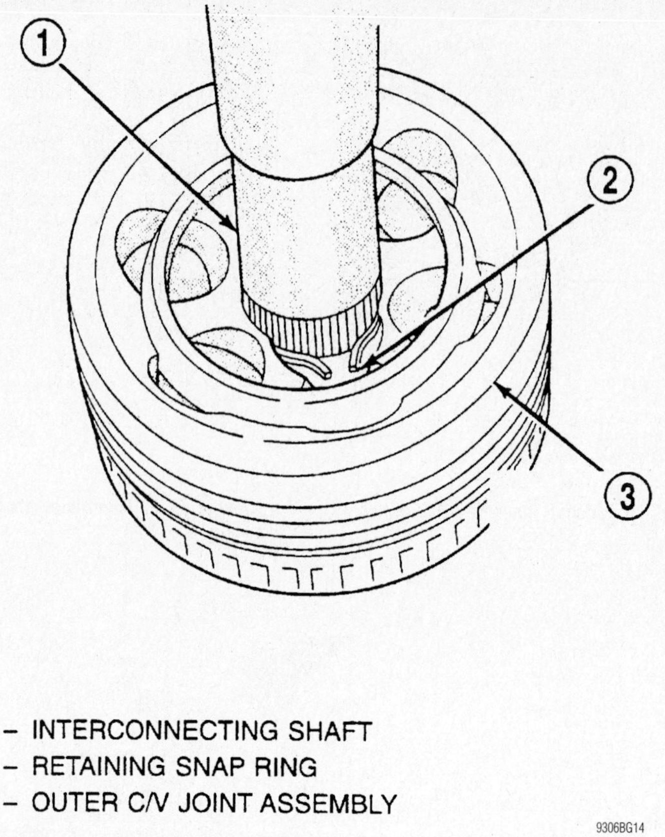

View of the halfshaft outer CV-joint and snapring

4. Slide the boot down the shaft away from the CV-joint housing.
5. Remove the grease to expose the CV-joint-to-shaft retaining ring.
6. Spread the snapring ears apart and slide the CV-joint assembly off of the shaft.
7. Slide the boot off the shaft.
8. Thoroughly, clean and inspect all parts for signs of excessive wear; if necessary, replace the halfshaft.

➡**Component parts are not serviceable and must be replaced as an assembly.**

To install:

9. Slide the outer CV-joint boot clamp and boot onto the shaft; then, position the boot so that only the thinnest (sight) groove is visible on the shaft.
10. Slide the outer CV-joint assembly on the shaft, spread the snapring ears, position the CV-joint and verify that the snapring is fully seated in the shaft groove.
11. If installing a new boot, distribute ½ of the grease in the service package into the CV-joint housing and the other ½ inside the boot.
12. Position the outer boot clamp evenly on the sealing boot.
13. Using the Crimper tool C-4975, place the tool over the clamp bridge, tighten the tool nut until the jaws are completely closed (face-to-face).

⁂ WARNING

The seal must not be dimpled, stretched or out of shape. If necessary, equalize the seal pressure and shape it by hand.

14. Position the boot onto the CV-joint housing retaining groove and install the retaining clamp evenly on the boot.
15. Using the Crimper tool C-4975, place the tool over the clamp bridge, tighten the tool nut until the jaws are completely closed (face-to-face).
16. Install the halfshaft into the vehicle.

STEERING AND SUSPENSION

Air Bag

⁂ CAUTION

Some vehicles are equipped with an air bag system. The system must be disabled before performing service on or around system components, steering column, instrument panel components, wiring and sensors. Failure to follow safety and disabling procedures could result in accidental air bag deployment, possible personal injury and unnecessary system repairs.

PRECAUTIONS

Several precautions must be observed when handling the inflator module to avoid accidental deployment and possible personal injury.

- Never carry the inflator module by the wires or connector on the underside of the module.
- When carrying a live inflator module, hold securely with both hands, and ensure that the bag and trim cover are pointed away.
- Place the inflator module on a bench or other surface with the bag and trim cover facing up.
- With the inflator module on the bench, never place anything on or close to the module which may be thrown in the event of an accidental deployment.

Before servicing the vehicle, also make sure to refer to the precautions in the beginning of this section as well.

DISARMING

⁂ CAUTION

The Air Bag system must be disarmed before repair and/or removal of any component in its immediate area including the air bag itself. Failure to do so may cause accidental deployment of the air bag, resulting in unnecessary system repairs and/or personal injury.

1. Disconnect the negative battery cable and isolate the cable using an appropriate insulator (wrap with quality electrical tape).
2. Allow the system capacitor to discharge for 2 minutes before starting any repair on any air bag system or related components. This will disable the air bag system.

⁂ CAUTION

Always wear safety goggles when working with or around the air bag system. When carrying a live air bag, be sure the bag and trim cover are pointed away from the body. In the unlikely event of an accidental deployment, the bag will, then deploy with minimal chance of injury. When placing a live air bag on a bench or other surface, always face the bag and trim cover up, away from the surface. This will reduce the motion of the module if it is accidentally deployed.

Power Rack and Pinion Steering Gear

REMOVAL & INSTALLATION

1. Before servicing the vehicle, refer to the precautions in the beginning of this section.
2. Turn the front wheels to the straight-ahead position.
3. Remove or disconnect the following:
   - Negative battery cable
   - Wiper arms
   - Wiper module cover and cowl cover
   - Reinforcement from the strut towers and wiper module
   - Throttle body's intake air duct and resonator
4. Clamp the steering wheel in the center position with some type of holding device.
   - Intermediate shaft from the steering column coupler
   - Both tie rods from the steering gear and place them on the transaxle bell housing
   - Siphon the power steering fluid from the reservoir
   - Power steering gear fluid lines
5. If equipped with speed proportional steering, disconnect the wiring from the solenoid control valve on the end of the steering gear under the master cylinder.
   - Master cylinder from the booster and position it to the side without disconnecting the fluid lines
   - Booster vacuum line

➡It may be helpful to loosen the 2 bolts attaching the right mounting bracket to the steering gear in order the clear the air conditioning lines.

   - 4 steering gear mounting bolts (2 on each side)
6. Move the steering gear and intermediate shaft into the engine compartment to gain access to the roll pin.

➡Special tools are commercially available for removing roll pins.

   - Roll pin and separate the intermediate shaft from the steering gear
   - Right front wheel
   - Tie rod end from the steering arm on the right strut

➡If equipped with a 2.7L engine, turn the left front tire towards the left as far as possible to provide clearance.

7. Slide the right end of the steering gear out through the tie rod hole while raising the left end of the steering gear upward.

To install:

8. Position the tie rod attaching points at the center of the steering gear travel.
9. Install or connect the following:
   - Steering gear

➡If equipped with a 2.7L engine, turn the left front tire straight-ahead.

   - Right tie rod on the steering arm and torque the nut to 27 ft. lbs. (37 Nm)
   - Right front wheel. Torque the lug nuts in sequence to half of the specified torque, then tighten the nuts in sequence again to 100 ft. lbs. (135 Nm).
   - Intermediate shaft to the steering gear using the roll pin. Be sure the pin is centered in the joint.
   - 4 steering gear mounting bolts and torque the bolts to 43 ft. lbs. (58 Nm)

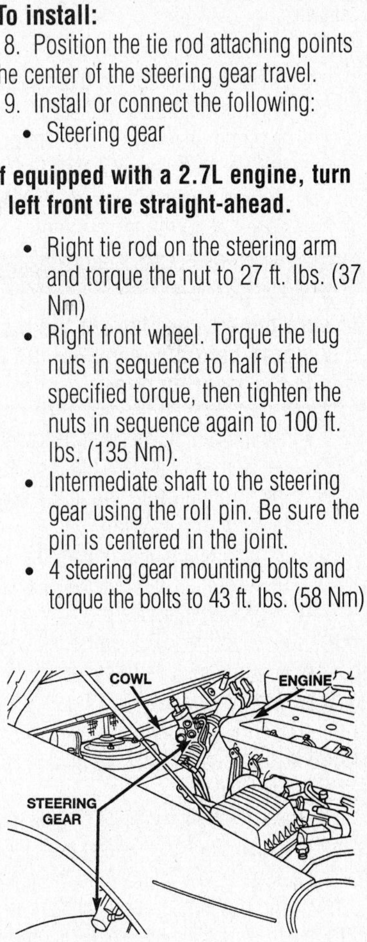

The steering gear is removed or installed through the right side of the vehicle

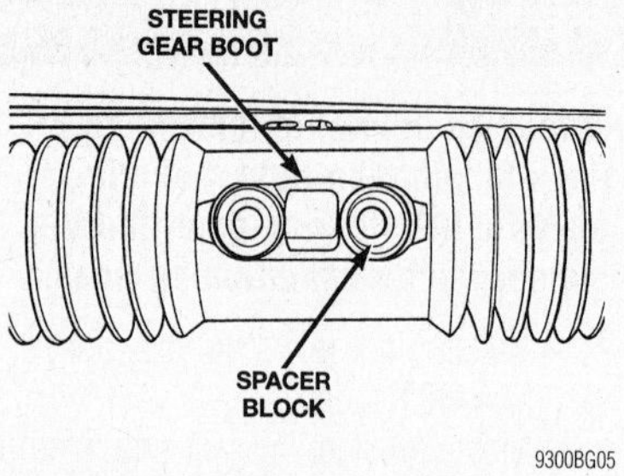

Center the spacer block in the steering gear before installing the tie rods

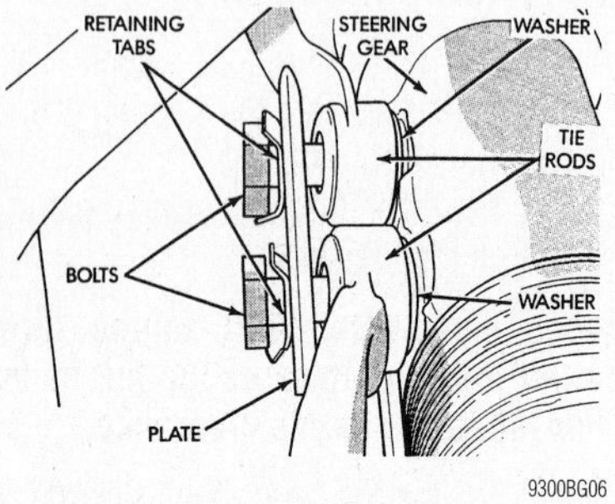

Be sure to position the washers between the tie rods and the steering gear as shown

- Bracket bolts; tighten them (if loosened) to 27 ft. lbs. (37 Nm)
- Steering gear fluid lines; torque the nuts to 35 ft. lbs. (47 Nm)
- Solenoid valve, if equipped with speed proportional steering

⁂ WARNING

Be sure the spacer block in the rack is centered before connecting the tie rods to the steering gear.

- Tie rods to the steering gear with the washers between the rods and the steering gear. Torque the bolts to 60 ft. lbs. (82 Nm).
- Remaining components in the reverse order of the removal steps
- Negative battery cable

10. Add Mopar® power steering fluid to the reservoir and let it settle for at least 2 minutes. Start the engine for a few seconds and turn it **OFF**. Add fluid as necessary. Repeat this procedure until the fluid level remains constant.

11. Raise the front wheels off the floor and start the engine. Turn the steering from lock-to-lock several times. Do not hold the wheel in the locked position for more than 2 seconds at a time. Refill the reservoir as needed.

12. Lower the vehicle and repeat the procedure. If the fluid is extremely foamy, allow the vehicle to stand with the engine **OFF** for a few minutes, then repeat the procedure.

Strut

REMOVAL & INSTALLATION

Front

1. Before servicing the vehicle, refer to the precautions in the beginning of this section.

➡Service of the coil spring requires the use of a coil spring compressor tool. It is required that 5 coils be captured within the jaws of the compressor tool.

➡Do not support the vehicle by placing supports under the suspension arms. The suspension arms must hang freely.

2. Remove or disconnect the following:
   - Negative battery cable
   - Front wheel(s)
   - Stabilizer bar attaching link at the strut assembly

3. Loosen, but do not remove the outer tie rod end-to-strut assembly steering arm attaching nut. Then, remove the outer tie rod end from the steering arm using puller MB-990635/C-3894A, or equivalent.
   - Speed sensor wiring harness mounting bracket from the strut
   - Brake caliper assembly. Support the caliper assembly from the vehicle frame with a strong piece of wire. Do not allow the assembly to hang by the brake hose.
   - Front brake rotor

⁂ WARNING

The strut assembly-to-steering knuckle bolts are serrated where they go through the strut and steering knuckle. Do not turn the bolts during removal. If the bolts are turned, damage to the steering knuckle will result.

4. The strut assembly-to-steering knuckle bolts must not be turned during strut removal. Hold the bolt head with a wrench and turn the nuts off the bolts.

5. Remove the 3 or 4 strut assembly upper mount-to-shock tower mounting nuts and washers. Remove the strut from the vehicle.

6. Disassemble the strut by performing the following procedure:
   a. Securely mount the strut assembly into a vise. Using paint, mark the strut unit, lower spring isolator, spring and upper strut mount for indexing of the parts at assembly.
   b. Position the spring compressor tool onto the strut. Compress the coil spring until all load is off the upper strut mount assembly.
   c. Install Strut Rod Socket tool L-4558A on the strut shaft nut and a 10mm socket on the end of the strut shaft to prevent it from turning. Remove the strut shaft nut.
   d. Remove the upper mount assembly, jounce bumper and seat bearing and dust shield as an assembly.
   e. Remove the coil spring and compressor as an assembly from the strut. Remove the lower spring isolator from the strut assembly lower spring seat.
   f. Inspect all components for abnormal wear, oil leakage or failure. Replace parts as required.

To install:

7. Assemble the strut by performing the following procedure:
   a. Inspect the strut assembly for signs of leakage. Actual leakage will be a stream of fluid running down the side and dripping off the lower end of the strut. A slight amount of seepage between the strut rod and strut shaft seal is not unusual and does not affect performance of the strut assembly.
   b. Install the lower spring isolator on the strut unit. Install the compressed coil spring onto the strut assembly aligning the paint marks made during removal.
   c. Install the strut bearing into the bearing seat. The bearing must be installed into the seat with the notches on the bearings facing down.
   d. Lower the seat bearing and dust shield onto the strut and spring assembly. Align the paint marks made during removal.
   e. Install the jounce bumper and upper mount on the strut shaft, aligning the paint marks.
   f. Install the strut mount-to-shaft retainer nut. Inspect all alignment marks made during removal and align as required. While holding the strut shaft from turning with a 10mm socket, torque the strut shaft nut to 70 ft. lbs. (94 Nm).
   g. Equally loosen the spring compressor tool until all tension is released. Remove the spring compressor tool.

8. Install the front strut into the strut tower. Torque the 3 or 4 upper nuts to 25–28 ft. lbs. (33–37 Nm).

9. Position the steering knuckle neck

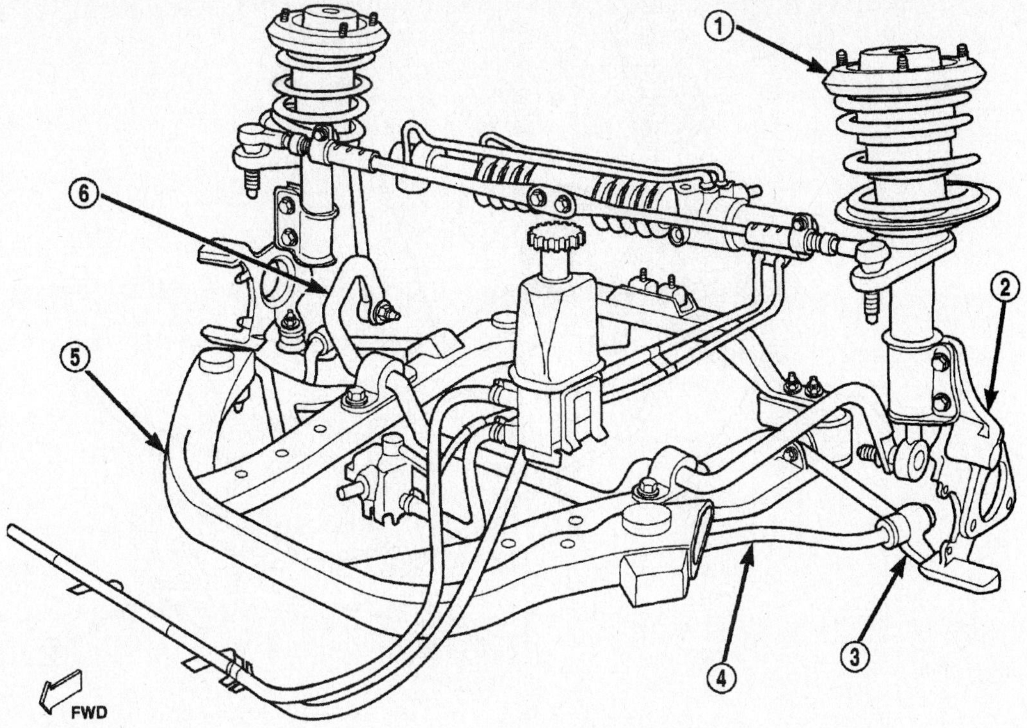

View of the front suspension components—2004 vehicle shown, others similar

into the strut assembly. Install the strut assembly-to-steering knuckle bolts. Install the nuts onto the attaching bolts and torque to 150 ft. lbs. (203 Nm). Do not turn the serrated bolt heads during installation. Turn only the nuts.

WARNING

The strut assembly-to-steering knuckle bolts are serrated (toothed) where they go through the strut and steering knuckle. Do not turn the bolts during removal. If bolts are turned, damage to the steering knuckle will result.

10. Install or connect the following:
    - Brake rotor and caliper assembly to the adapter and torque the caliper bolts to 14 ft. lbs. (19 Nm)
    - Front speed sensor cable routing bracket onto the front strut, if equipped
    - Outer tie rod on the steering arm and torque the attaching nut to 27 ft. lbs. (37 Nm)
    - Stabilizer link assembly onto the strut assembly and torque the attaching nut to 70 ft. lbs. (95 Nm)
    - Front wheel and lug nuts. Torque the lug nuts, in sequence, to 100 ft. lbs. (135 Nm).
    - Negative battery cable

Rear

1. Before servicing the vehicle, refer to the precautions in the beginning of this section.

The rear strut assemblies support the weight of the vehicle using coil springs positioned around the struts. The coil springs are contained between the upper mount of the strut assembly and a lower spring seat on the body of the strut assembly. The strut is attached to the spindle by a split collar on the rear spindle with a pinch bolt to hold the spindle to the strut.

2. Remove or disconnect the following:
    - Rear wheel
    - Caliper assembly and rotor from the hub, if equipped with rear disc brakes
    - Brake flex hose from the support bracket and wheel cylinder, if equipped with rear drum brakes. Plug the brake flex hose to prevent system contamination. Do not allow the rear caliper to hang by the brake hose. Support the caliper off of the frame with a strong piece of wire.
    - Speed sensor cable routing bracket and tube, if equipped with Anti-lock Brake System (ABS)
    - Lateral links to the rear spindle assembly bolts
    - Rear strut assembly-to-stabilizer bar attaching link at the stabilizer bar

➡Hold the hex on the attaching link stud while breaking the nut loose. The attaching link does not have to be removed from the strut.

3. Remove the rear spindle-to-strut assembly pinch bolt. Install a center punch in the hole on the spindle and tap the punch into the hole until jammed. This will spread the spindle casting allowing it to be removed from the strut.
4. Using a hammer, tap on the top surface of the spindle, driving the spindle down and off the end of the strut assembly. Let the spindle and assembled components hang from the trailing arm while the strut is being serviced.
5. From inside the trunk of the vehicle,

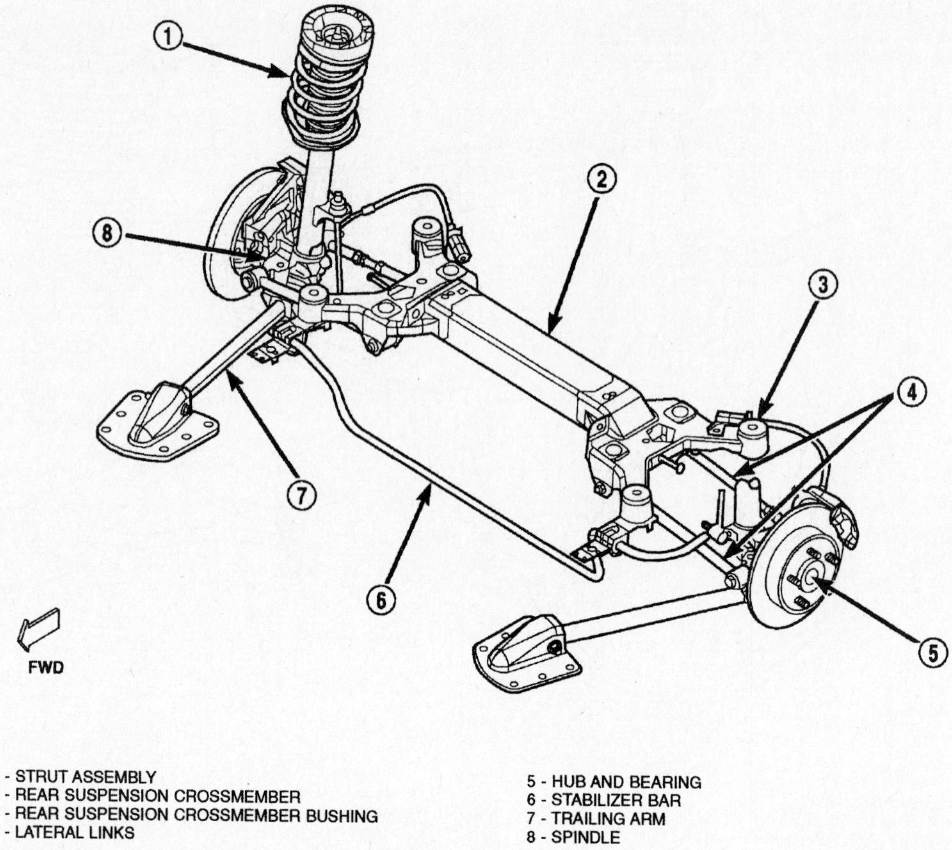

View of the rear suspension components—2004 vehicle shown, others similar

remove the 3 upper strut mounting bolts and remove the strut from the vehicle.

6. Disassemble the strut by performing the following procedure:

➡**Service of the coil spring requires the use of a coil spring compressor tool. It is required that 5 coils be captured within the jaws of the compressor tool.**

a. Securely mount the strut assembly into a vise. Using paint, mark the strut assembly, lower spring isolator, spring and upper strut mount for indexing of the parts at reassembly.

b. Position a spring compressor tool onto the coil spring. Compress the coil spring until all load is off of the upper strut mount assembly.

c. Install the strut rod socket tool L-4558 on the strut shaft nut and an 8mm Allen wrench on the end of the strut shaft to prevent it from turning. Remove the strut shaft nut.

d. Remove the upper strut mount assembly off of the strut shaft. Remove the coil spring and compressor tool as an assembly from the strut.

e. Remove the plate, dust shield and jounce bumper off of the strut unit.

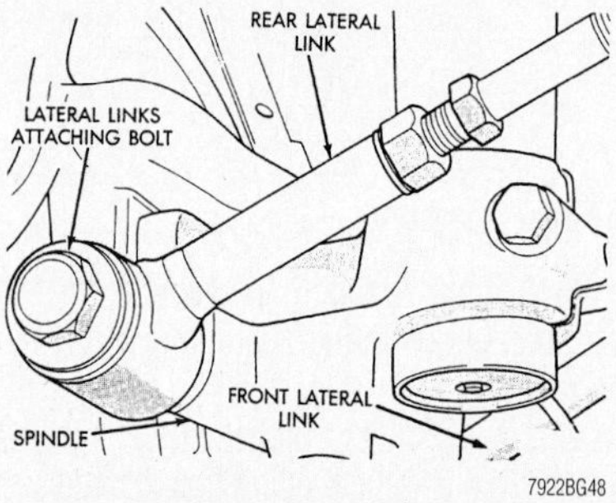

Separate the lateral links from the spindle—rear strut service

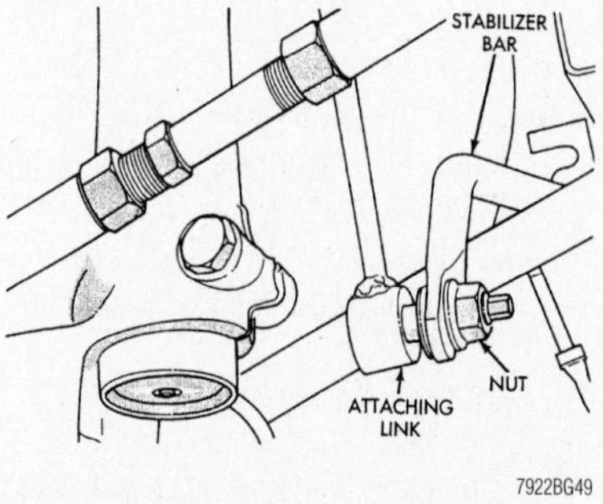

Remove the nut from the stabilizer-to-strut attaching link stud at the bar—rear strut service

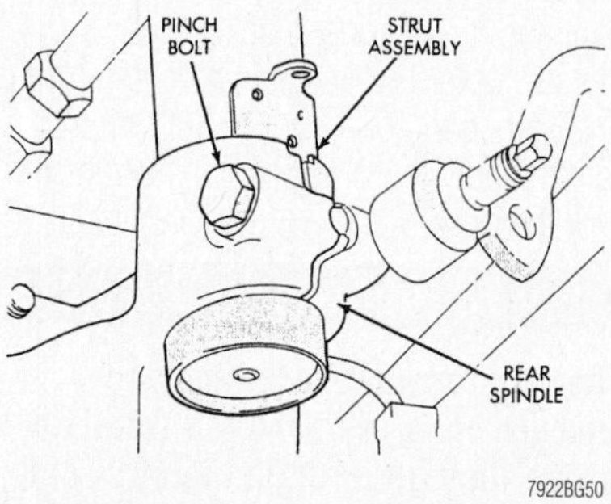

Loosen, then remove the rear spindle-to-strut pinch bolt—rear strut service

f. Inspect all components for abnormal wear, oil leakage or failure. Replace parts as required.

To install:

7. Assemble the strut by performing the following procedure:

a. Install the lower spring isolator on the strut unit. If it is the original isolator, align the paint marks.

b. Install the jounce bumper into the dust shield. Install the plate on top of the dust shield and into the jounce bumper.

c. Install the dust shield, jounce

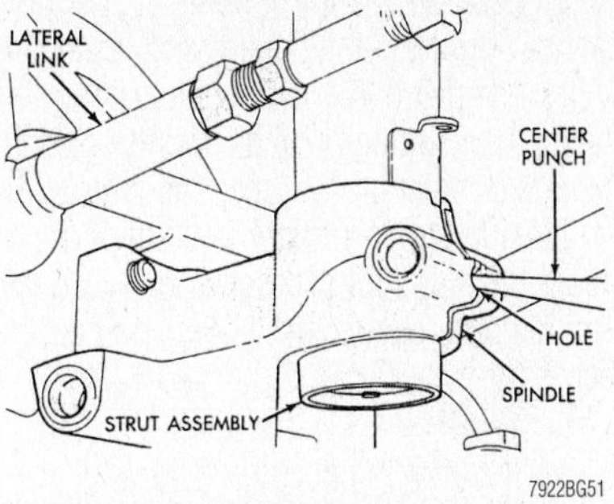

7922BG51

Insert a center punch into the hole on the spindle and tap until the casting is spread—rear strut service

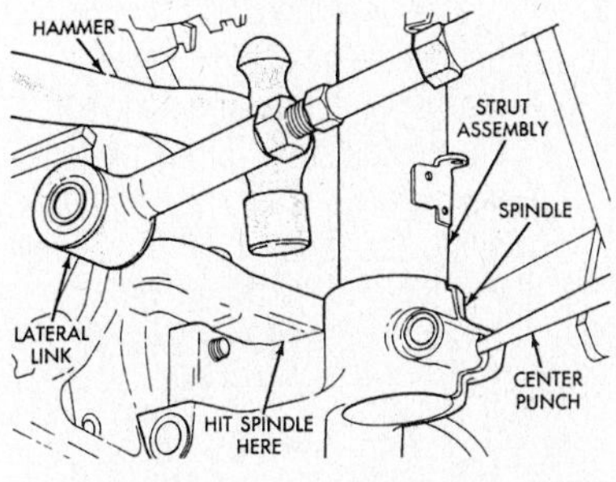

7922BG52

Tap with a hammer on the surface of the spindle driving it down and off the end of the strut—rear strut service

bumper and the top plate onto the strut unit as an assembly.

d. Install the coil spring and compressor tool onto the strut unit and align the paint marks on the spring to that of the strut unit.

e. Install the upper strut mount assembly onto the strut shaft. Align the paint marks and install the strut shaft retaining nut.

f. Using the strut rod socket tool L-4558 and the 8mm Allen wrench to prevent the strut shaft from turning, torque the strut shaft nut to 70 ft. lbs. (95 Nm).

g. Equally loosen the spring compressor tool until all tension is released. Remove the spring compressor tool.

8. Position the strut in the vehicle. Torque the 3 upper mounting nuts to 20 ft. lbs. (28 Nm).

9. Install the spindle assembly onto the bottom of the strut. Push or tap the spindle assembly onto the strut, until the notch in the spindle is tightly seated against the locating tap on the strut assembly. Remove the center punch from the hole in the spindle.

10. Install or connect the following:

- Strut-to-spindle pinch bolt and torque to 40 ft. lbs. (55 Nm)
- Lateral link-to-spindle attaching bolt, hand-tight only

⁂ WARNING

Do not tighten the lateral link-to-spindle bolt at this time, or the bushings will be contorted and fail. You must tighten the bolt at curb height.

- Stabilizer bar attaching link and torque the stabilizer link-to-stabilizer bar attaching nut to 70 ft. lbs. (95 Nm), while holding the stabilizer link stud at the hex with a wrench
- Rear speed sensor cable routing tube and bracket, if equipped with ABS
- Rotor and caliper assembly (if equipped with rear disc brakes) and torque the bolts to 16 ft. lbs. (22 Nm)
- Rear brake flex hose to the wheel cylinder and support plate, if equipped with rear drum brakes
- Rear wheel(s) and torque the lug nuts, in sequence, to 100 ft. lbs. (135 Nm)
- With the weight of the vehicle on the tires, tighten the lateral link-to-spindle attaching bolt to 105 ft. lbs. (140 Nm)

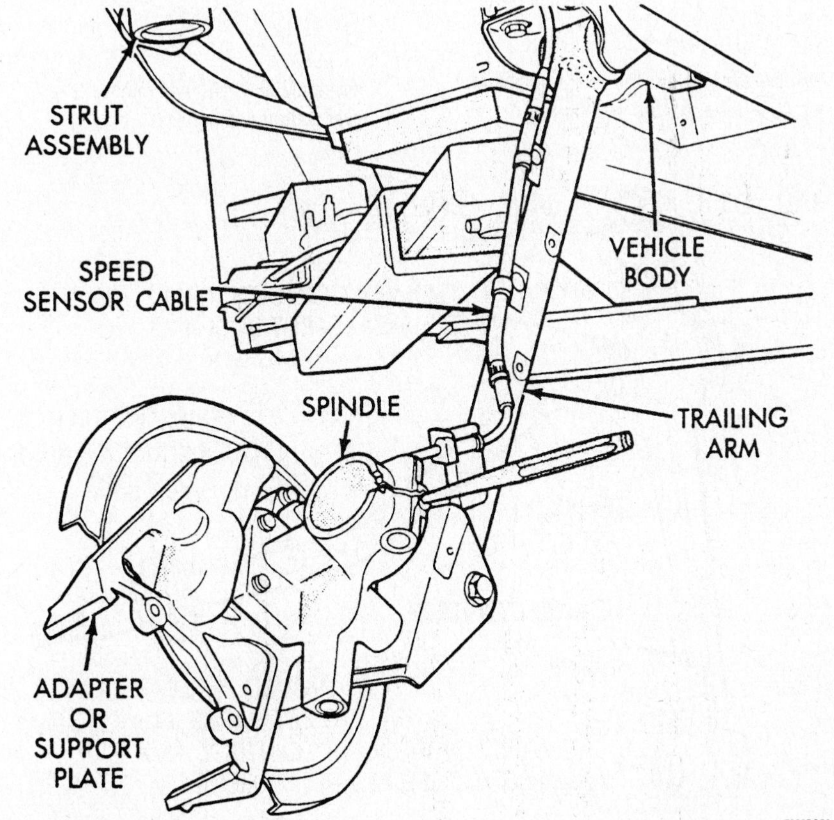

7922BG53

Allow the components to hang from the trailing arm as shown—rear strut service

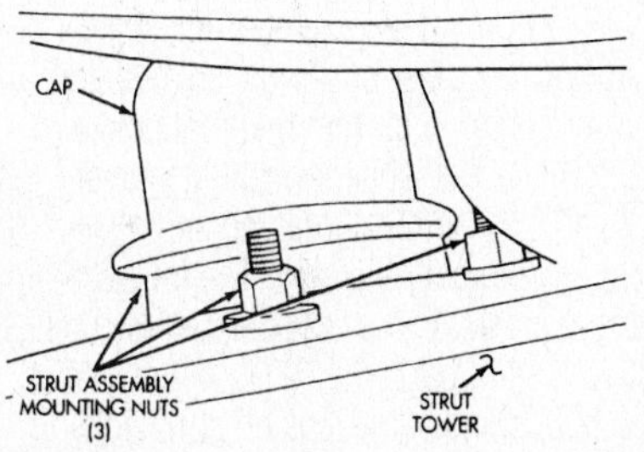

7922BG54

Access the 3 upper strut mounting nuts through the trunk—rear strut service

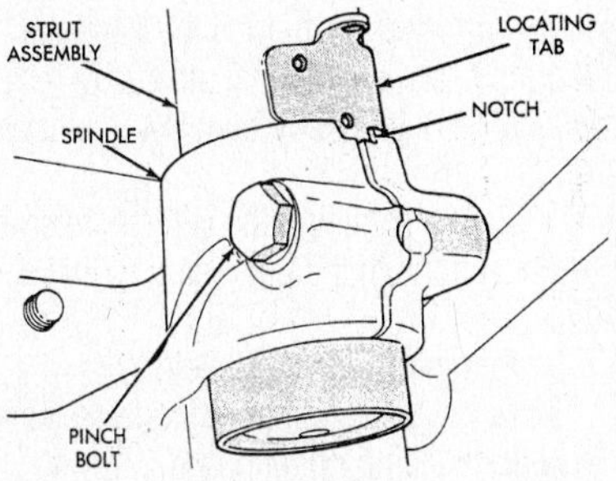

7922BG55

During reassembly, tap the spindle onto the strut until the notch in the spindle is tightly seated against the tab—rear strut service

11. Bleed the brake system, if equipped with rear drum brakes.
12. Have the rear wheel toe set to specifications.

Coil Springs

REMOVAL & INSTALLATION

Front

Refer to the front strut removal and installation procedure for coil spring service information.

Rear

Refer to the rear strut removal and installation procedure for coil spring service information.

Lower Ball Joint

➡The lower ball joints on these vehicles are not serviced separately. The lower ball joints operate with no free-play. If defective, the entire lower control arm must be replaced.

Lower Control Arm

REMOVAL & INSTALLATION

The front lower control arm is a steel forging with 2 rubber bushings isolating the lower control arm from the front cradle assembly. The isolator bushings consist of a metal encased pivot bushing and a solid rubber tension strut bushing. The lower control arm is bolted to the cradle assembly using a pivot bolt through the center of the rubber pivot bushing at the tension strut isolator bushing. The ball joint is built into the lower control arm and is non-serviceable. If the ball joint becomes worn, the entire lower control arm must be replaced. The ball joint seal, however, is replaceable as well as the lower control arm inner bushing. If the lower control arm is damaged, do not attempt to repair or straighten a broken or bent lower control arm.

1. Before servicing the vehicle, refer to the precautions in the beginning of this section.
2. Remove the front wheel(s).
3. Remove the ball joint stud-to-steering knuckle clamp nut and bolt.
4. Carefully insert a prybar between the lower control arm and the steering knuckle and separate ball joint from knuckle. Be sure the ball joint seal does not get damaged during separation.

⁂ WARNING

Pulling the steering knuckle out from the vehicle after releasing from the ball joint can separate the inner CV-joint. Do not separate the inner CV-joint or it can be damaged.

5. Remove the tension strut-to-cradle attaching nut and washer from the end of the tension strut. When removing the nut, keep the strut from turning by holding the tension strut at the flats using an open end wrench. Discard the tension strut-to-cradle retainer nut. A new nut must be used during installation.

➡A new tension strut-to-cradle attaching nut must be used when installing the tension strut.

6. Loosen and remove the lower control arm pivot bushing-to-cradle assembly pivot bolt.
7. Separate the lower control arm and tension strut from the cradle as an assembly by first removing the pivot bushing from the cradle, then sliding the tension strut out of the isolator bushing. Inspect the control arm and tension strut for distortion, check the rubber bushings for excessive wear or deterioration and replace these components, if necessary.

To install:

8. Install or connect the following:
   - Tension strut and isolator bushing into the cradle first, then install lower control arm pivot bushing into bracket on the cradle
   - Lower control arm-to-cradle bracket attaching bolt. Do not tighten the bolt at this time.
   - Washer and new nut on the end of the tension strut. Torque the tension strut-to-cradle bracket retainer nut to 110 ft. lbs. (150 Nm), while

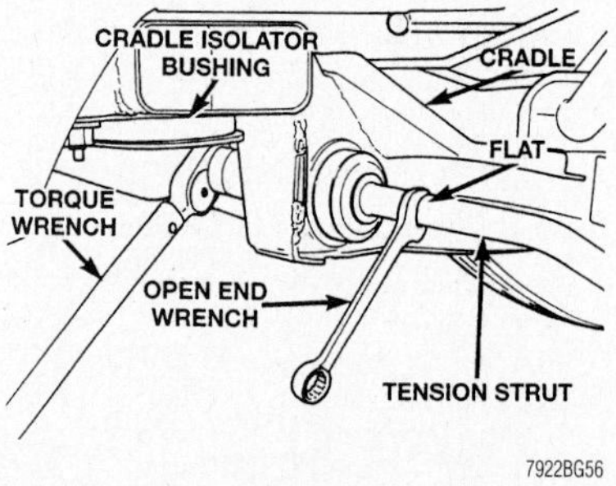

Remove the tension strut-to-cradle nut and washer—lower control arm service

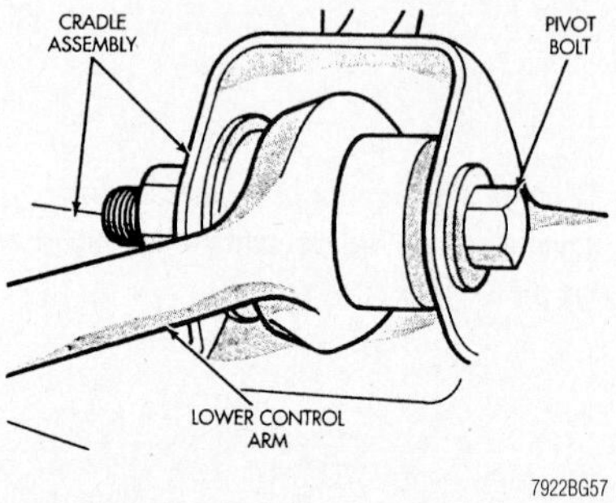

Loosen the pivot bolt and remove it—lower control arm service

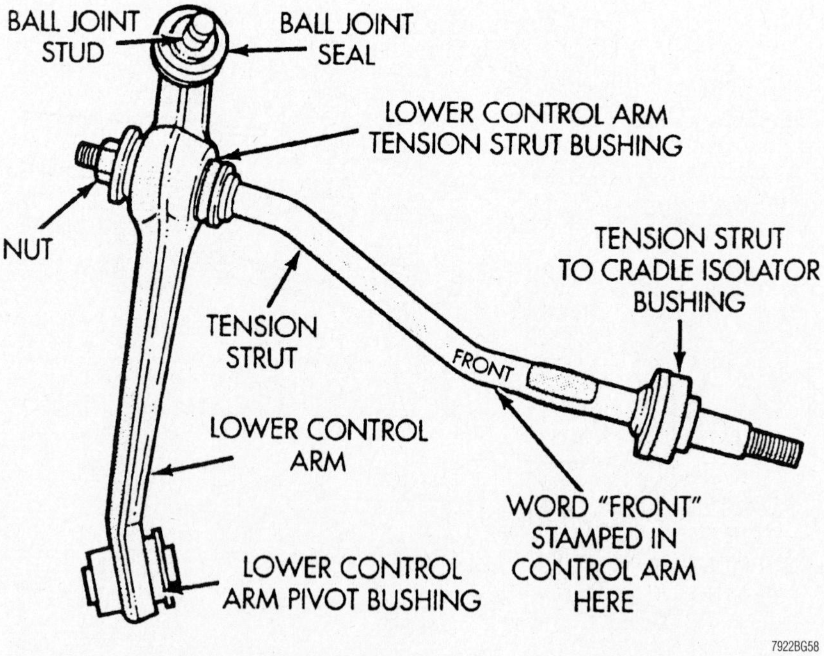

Inspect the control arm and tension strut for distortion—lower control arm service

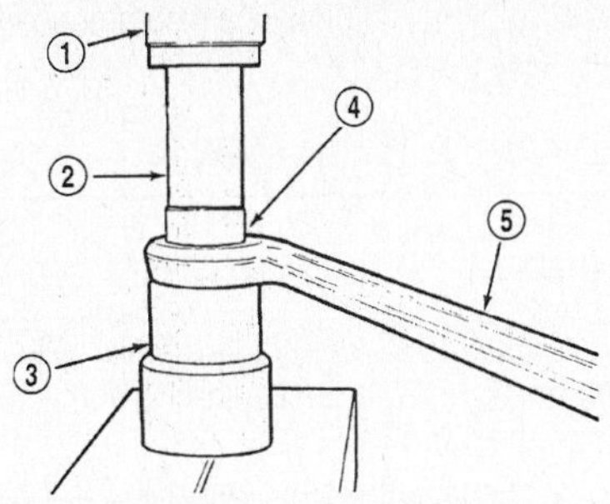

Removing the lower control arm pivot bushing

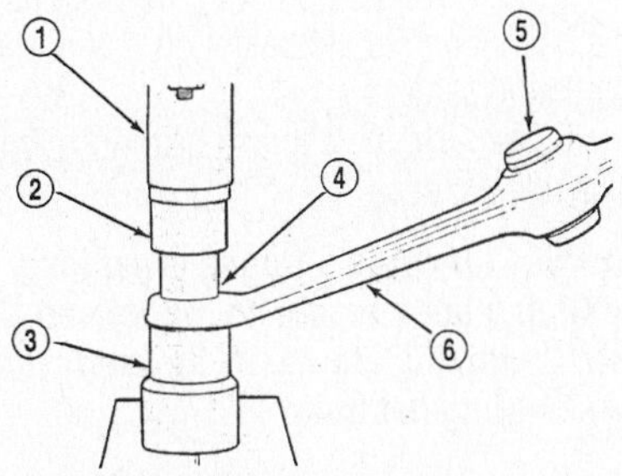

Installing the lower control arm pivot bushing

holding the tension strut flat with an open end wrench.

9. Inspect the ball joint seal and replace it if damaged. Install the lower ball joint stud into the steering knuckle and install the clamp bolt and nut. Torque the bolt to 40 ft. lbs. (55 Nm).

10. Install the front wheel and lug nuts. Torque the lug nuts, in a star pattern, to 100 ft. lbs. (135 Nm). Lower the vehicle so the suspension is supporting the weight of the vehicle.

11. Torque the lower control arm pivot bushing-to-cradle bracket attaching bolt to 105 ft. lbs. (142 Nm).

CONTROL ARM BUSHING REPLACEMENT

1. Remove the lower control arm/tension strut assembly from the vehicle.

2. Remove the nut and separate the tensions strut from the lower control arm.

3. Using an arbor press, position the lower control arm with the large end of the pivot bushing inside the Receiver tool MB-990799.

4. Position Remover tool 6644-2 on top of the pivot bushing.

5. Using the arbor press, press the pivot bushing out of the lower control arm.

To install:

6. Turn the lower control arm over and reposition it on the Receiver tool MB-990799.

7. Using a new pivot bushing, position it in the lower control arm so that it is square with the bushing hole.

8. Position Installer tool 6644-1 on top of the pivot bushing with the pivot bushing setting in the recessed area of the installer tool.

9. Using the arbor press, press the pivot bushing into the lower control arm until the installer tool squarely bottoms against the surface of the lower control arm.

10. Install the tension strut into the lower control arm's tension strut bushing with the word FRONT, stamped on the tension strut, facing away from the control arm.

11. Position an open end wrench on the flat of the tension strut to keep it from turning. Torque the tension strut-to-lower control arm nut to 95 ft. lbs. (130 Nm).

TENSION STRUT BUSHING REPLACEMENT

1. Remove the lower control arm/tension strut assembly from the vehicle.

2. Remove the nut and separate the tensions strut from the lower control arm.

3. Using an arbor press, position the lower control arm with the tension strut bushing inside the Receiver tool MB-990799.

4. Position Remover tool 6644-4 on top of the tension strut bushing.

5. Using the arbor press, press the tension strut bushing out of the lower control arm.

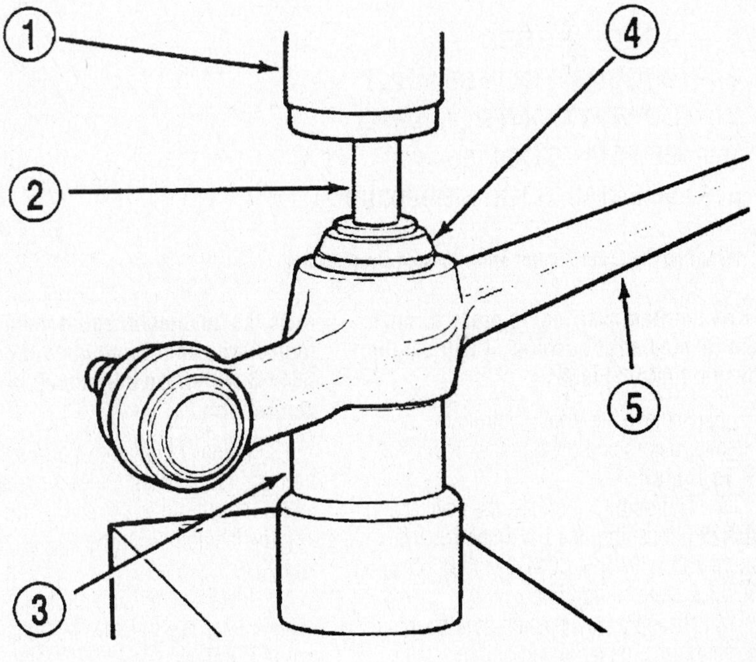

Removing the lower control arm's tension strut bushing

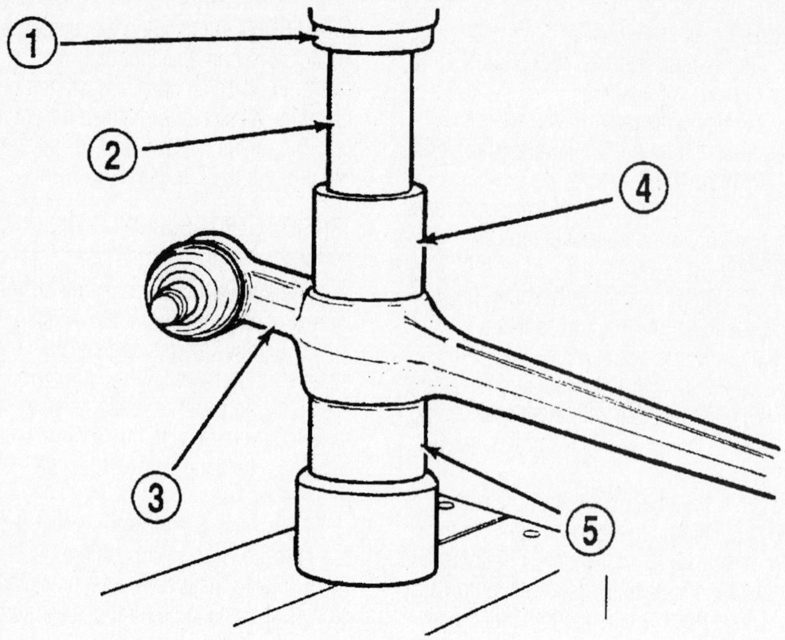

1 – ARBOR PRESS
2 – SPECIAL TOOL 6644-2
3 – LOWER CONTROL ARM
4 – SPECIAL TOOL 6644-3
5 – SPECIAL TOOL MB990799

Installing the lower control arm's tension strut bushing

➡As the Remover tool is press through the tension strut bushing, it will cut the bushing into 2 pieces.

6. Remove the lower control arm, both bushing pieces and the Remover tool.

To install:

7. Thoroughly, lubricate the new tension strut bushing, the lower control arm and the Installer tool 6644-3 with Rubber Bushing Installation Lube.
8. Press the new tension strut bushing by hand, into the large end of the Installer tool 6644-3 as far as it will go.
9. Position the lower control arm onto on the arbor press so that the tension strut hole is centered on the Receiver tool MB-990799.
10. Position Installer tool 6644-3, with the bushing installed, inside the tension strut bushing hole in the lower control arm.
11. Position Installer tool 6644-2 on top of the tension strut bushing.
12. Using the arbor press, press the tension strut bushing into the lower control arm.

➡As the bushing is being installed, a pop will be heard and Installer tool 6644-3 will move slightly up off the control arm.

13. Remove the control arm assembly from the arbor press and remove Installer tool 6644-3 from the tension strut bushing; the tension strut bushing is now installed.
14. Install the tension strut into the lower control arm's tension strut bushing with the word FRONT, stamped on the tension strut, facing away from the control arm.
15. Position an open end wrench on the flat of the tension strut to keep it from turning. Torque the tension strut-to-lower control arm nut to 95 ft. lbs. (130 Nm).

Wheel Bearings

ADJUSTMENT

These front wheel drive vehicles are equipped with permanently sealed front and rear wheel bearings. There is no periodic lubrication or maintenance recommended for these units.

REMOVAL & INSTALLATION

Front

1. Before servicing the vehicle, refer to the precautions in the beginning of this section.
2. Remove the front wheel.
3. Remove the front caliper assembly from the steering knuckle by removing the 2 guide pin bolts, then rotating the top of the caliper away from the knuckle and lifting the caliper off the machined abutment on the steering knuckle.
4. Suspend the caliper out of the way with a piece of wire.
5. Remove the front brake rotor from the hub by pulling it straight off of the wheel mounting stud.
6. Remove the hub and bearing-to-stub axle retainer nut.

➡This hub nut is a torque prevailing retaining nut and can not be reused. A NEW retaining nut MUST be used when assembling the hub.

7. Remove the 3 attaching bolts that mount the hub and bearing assembly to the steering knuckle assembly.

➡If the metal seal on the hub and bearing assembly is seized to the steering knuckle and becomes dislodged on the hub and bearing during removal, the hub and bearing must be replaced. If the flinger disc becomes damaged during the removal procedure, the hub and bearing assembly must be replaced.

8. Remove the hub and bearing assembly from the steering knuckle by sliding it straight out of the knuckle and off the ends of the stub shaft.
9. Gently pry the assembly out with a prybar or tap it out with a soft-faced hammer, if necessary. Be very careful not to damage the hub and bearing assembly.

To install:

10. Clean the hub and bearing mounting surfaces of dirt and be sure there are no nicks present.
11. Install or connect the following:
   - Hub and bearing squarely onto the stub shaft and the steering knuckle
   - Bearing assembly mounting bolts and tighten equally until the bearing assembly is seated squarely against the front of the steering

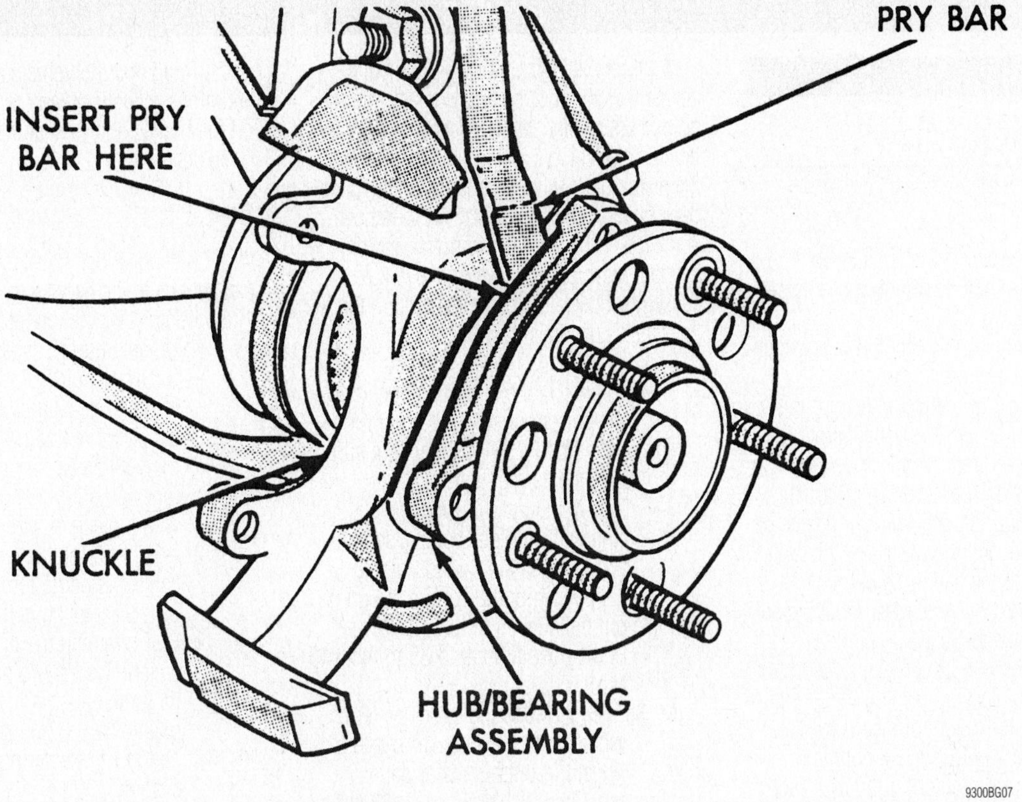

Carefully pry the front bearing/hub assembly away from the steering knuckle

knuckle. Torque the mounting bolts to 80 ft. lbs. (110 Nm).

- New hub and bearing assembly-to-stub shaft retainer nut. A NEW retaining nut MUST be used when assembling the hub. Tighten but do not torque the nut at this time.
- Brake rotor and the caliper assembly. Torque the brake caliper guide bolts to 16 ft. lbs. (22 Nm).
- Wheel and lug nuts. Torque the lug nuts in a star pattern to 100 ft. lbs. (135 Nm).

12. Pump the brakes until a firm pedal is obtained.
13. With the weight on the vehicle on its wheels, apply the brakes to keep the vehicle from moving. Torque the hub and bearing assembly-to-stub shaft retaining nut to 120 ft. lbs. (163 Nm).

WARNING

When tightening the hub and bearing assembly to stub shaft retaining nut, do not exceed the maximum torque of 120 ft. lbs. (163 Nm). If the maximum torque is exceeded this may result in a failure of the halfshaft.

14. Inspect the toe setting on the vehicle and adjust, if necessary.

Rear

1. Before servicing the vehicle, refer to the precautions in the beginning of this section.
2. Remove or disconnect the following:
   - Rear wheel
   - Brake caliper and rotor, if equipped with rear disc brakes
   - Brake drum, if equipped with drum brakes
   - Bearing dust cap
   - Cotter pin, nut retainer, nut, washer and bearing/hub assembly from the spindle

To install:

3. Install or connect the following:
   - Bearing/hub assembly, the bearing/hub assembly washer and torque the nut to 124 ft. lbs. (168 Nm)
   - Nut retainer and a new cotter pin
   - Dust cap
   - Brake drum or rotor and caliper assembly
   - Rear wheel and torque the lug nuts, in a star pattern, to 100 ft. lbs. (135 Nm)
4. Road test the vehicle to verify no excessive noise from the rear wheel bearing area.

BRAKES

Caliper

REMOVAL & INSTALLATION

Front

1. Before servicing the vehicle, refer to the precautions in the beginning of this section.
2. Remove or disconnect the following:
   - Wheel and tire assemblies
   - 2 caliper guide pin bolts and the caliper assembly. If the caliper is not being removed from the vehicle as during brake pad renewal, simply hang the caliper with a piece of wire to take the weight off the brake hose. If the caliper is being removed for rebuild or replacement, continue to Step 5.
   - Bolt retaining the brake hose to the caliper. Be sure to plug the end of the brake hose or cover it with a plastic bag to prevent contamination from entering the hydraulic system.
   - Caliper from the vehicle

To install:

3. Install or connect the following:
   - Brake hose to the caliper, if removed, using new sealing washers
4. If new linings are being installed, the caliper pistons must be pushed back into their bore to accommodate the thickness of the new lining. Special tools are available for pushing the piston back although a large C-clamp can often be used. It is good practice to remove some (⅓–½) brake fluid from the master cylinder reservoir. This prevents overflow caused by brake fluid being forced through the lines as the piston is pushed back.
   - Brake pads into the caliper
   - Caliper to the steering knuckle in the correct position. Lightly lubricate the machined areas that support the caliper with high-temperature grease.
   - Caliper guide pin bolts. Use care not to cross the threads of the caliper pin bolts. Torque the guide pin bolts to 15 ft. lbs. (20 Nm).
   - Brake hose fittings to 35 ft. lbs. (48 Nm)
5. Be sure to bleed the brake system.
   - Wheels and lug nuts. Torque the wheel lug nuts in a star pattern sequence to 100 ft. lbs. (135 Nm).
6. Before attempting to move the vehicle, pump the brake pedal to seat the pads against the rotors. Make sure the vehicle has a firm brake pedal. Check the level of the brake fluid and add DOT 3 brake fluid, if necessary.
7. Road test the vehicle and make several stops to wear off any foreign material on the brakes and to seat the brake linings, if replaced.

Rear

1. Before servicing the vehicle, refer to the precautions in the beginning of this section.
2. Remove or disconnect the following:
   - Wheel and tire assemblies
   - 2 caliper guide pin bolts and remove the caliper assembly. If the caliper is not being removed from the vehicle as during brake pad renewal, simply hang the caliper with a piece of wire to take the weight off the brake hose. If the caliper is being removed for rebuild or replacement, continue to Step 5.
   - Bolt retaining the brake hose to the caliper. Be sure to plug the end of the brake hose or cover it with a plastic bag to prevent contamination from entering the hydraulic system.
   - Caliper from the vehicle

To install:

3. Install or connect the following:
   - Brake hose to the caliper, if removed, using new sealing washers

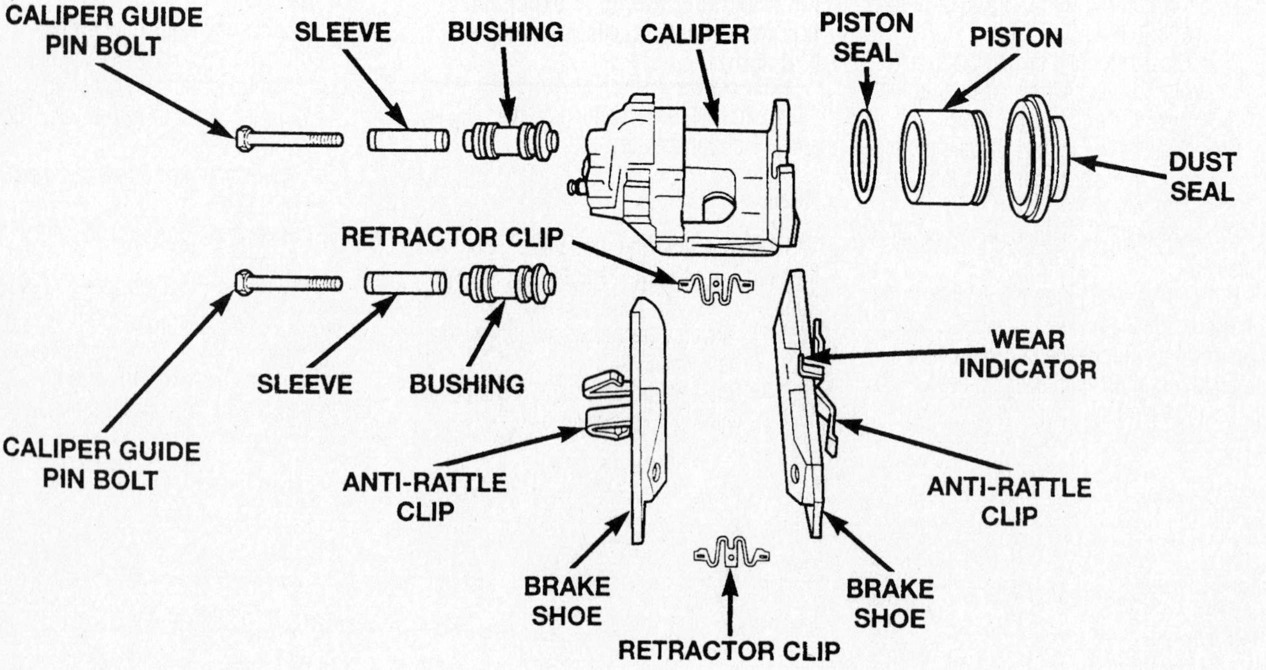

Front brake caliper and related components

4. If new linings are being installed, the caliper pistons must be pushed back into their bore to accommodate the thickness of the new lining. Special tools are available for this, although a large C-clamp can often be used. It is good practice to remove some brake fluid from the master cylinder reservoir. This prevents overflow caused by brake fluid being forced through the lines as the piston is pushed back.
- Brake pads into the caliper
- Brake caliper to the caliper adapter. Lightly lubricate machined areas that support the caliper with high-temperature grease.
- Caliper guide pin bolts. Use care not to cross the threads of the caliper pin bolts. Torque the guide pin bolts to 17 ft. lbs. (22 Nm).
- Brake hose fittings to 35 ft. lbs. (48 Nm)

5. Be sure to bleed the brake system.
- Wheels and lug nuts. Torque the wheel lug nuts in a star pattern sequence to 100 ft. lbs. (135 Nm).

6. Before attempting to move the vehicle, pump the brake pedal to seat the pads against the rotors. Make sure the vehicle has a firm brake pedal.

7. Check the level of the brake fluid and add DOT 3 brake fluid, if necessary.

8. Road test the vehicle and make several stops to wear off any foreign material on the brakes and to seat the brake linings, if replaced.

Disc Brake Pads

REMOVAL & INSTALLATION

1. Before servicing the vehicle, refer to the precautions in the beginning of this section.

2. Remove or disconnect the following:
- Some of the fluid from the master cylinder
- Wheels
- 2 caliper guide pin bolts
- Caliper assembly by swinging the top part of the caliper away from the brake rotor edge, then lift the caliper assembly up

➡Prevent strain or other damage to the brake hose by supporting the caliper assembly with a strong piece of wire hanging from the strut.

- Outboard brake pad by prying the brake pad retaining clip over raised area on the caliper. Then slide the pad down and off the caliper.

3. Before removing the inboard brake pad, use a large C-clamp to press the piston back into the caliper. This will prevent possible damage to the caliper piston. It is good practice to remove some (1/3–1/2) of the brake fluid from the reservoir. This is because as the caliper piston is pushed back into the caliper, brake fluid will be pushed back through the lines, back into the master cylinder and fluid reservoir, possibly causing the reservoir to overflow.
- Inboard brake pad by pulling away from piston until the retainer clip is free from the cavity in the piston.

To install:

4. Lubricate both the caliper mating surface and the machined abutment surfaces with multi-purpose lubricant.

5. Before brake pad installation, be sure to lightly coat the outer backing plate surface of the new brake pads with a disc brake pad anti-squeal lubricant, usually a gel-like material that deadens any high-frequency vibration that can be the source of disc brake squeal.

6. Install or connect the following:
- Brake pads into the caliper assembly making sure both pads are seated securely onto the caliper
- Caliper assembly back into position over the brake rotor
- Caliper guide pin bolts and torque to 17 ft. lbs. (23 Nm)
- Wheels and lug nuts. Torque the lug nuts, in a star pattern sequence, to 100 ft. lbs. (135 Nm).

7. Top off the master cylinder to the appropriate level, using DOT 3 type brake fluid only.

8. Before moving the vehicle, pump the brakes until a firm pedal is obtained. Road test the vehicle to make sure the brake operation is normal.

Brake Drums

REMOVAL & INSTALLATION

1. Before servicing the vehicle, refer to the precautions in the beginning of this section.

2. Remove or disconnect the following:
- Rear wheels
- Brake drum from the rear hub and bearing assembly

3. If the drum is difficult to remove, increase the clearance between the brake shoes and the drum as follows:

a. Remove the rubber plug from the top of the brake support plate.

b. Rotate the automatic shoe adjuster screw with an upward motion using a medium size flat tipped tool.

c. Remove the brake drum from the rear hub and bearing.

4. Inspect the brake shoe linings and drums for wear, contamination and scoring.

To install:

5. Install the rear brake drum onto the rear hub and bearing assembly.

6. Adjust the brake shoes as follows:

a. Rotate the automatic shoe adjuster in a downward motion using a medium size flat tip tool. Turn the adjuster until there is a slight drag felt while turning the drum.

7. Install or connect the following:
- Rubber plug back into the top part of the brake support plate
- Rear wheels and lug nuts. Torque the lug nuts in a star pattern sequence to 100 ft. lbs. (135 Nm).

Brake Shoes

REMOVAL & INSTALLATION

1. Before servicing the vehicle, refer to the precautions in the beginning of this section.

2. Remove or disconnect the following:
- Rear wheels and brake drums
- Dust cap from the rear hub and bearing assembly
- Cotter pin, nut retainer and wave washer. Discard the old cotter pin.
- Rear hub and bearing assembly retainer nut and washer
- Rear hub and bearing assembly from the spindle
- Automatic adjuster spring from the adjuster lever

3. Rotate the automatic adjuster starwheel enough so both shoes move out far enough to be free of the wheel cylinder boots.
- Parking brake cable from the actuating lever. Disconnect parking brake cable one side at a time.
- Both lower brake shoe to anchor springs
- 2 brake shoe hold-down springs from the brake shoes
- Brake shoes, upper shoe-to-shoe return spring, automatic adjuster and automatic adjuster lever from the backing plate as an assembly
- Brake shoes from the automatic adjuster mechanism
- Brake shoe automatic adjuster lever from the leading brake shoe

To install:

4. Thoroughly clean and dry the backing plate. To prepare the backing plate, lubricate the bosses, anchor pin and parking brake actuating lever pivot surface lightly with lithium based grease.

5. Remove, clean and dry all parts still on the old shoes. Lubricate the starwheel shaft threads with anti-seize lubricant.

6. Assemble both brake shoes, the top shoe to shoe return spring, automatic adjuster and automatic adjuster lever before mounting on vehicle. Make sure the ends of the automatic adjusters are positioned above the extruded pins in the webbing of the brake shoes prior to installation.

7. Install or connect the following:

- Brake shoe assembly onto the brake support plate and secure with the hold-down springs
- Lower anchor springs and reconnect the parking brake cable to the park brake lever of the trailing brake shoe

8. Rotate the serrated adjuster nut to remove the free-play from the adjuster assembly.

- Automatic adjuster lever spring on the lead brake shoe assembly and the automatic adjuster lever
- Rear hub and bearing assembly
- Washer and retainer nut and torque to 124 ft. lbs. (168 Nm)
- Wave washer, nut retainer and a new cotter pin onto the spindle. Install dust cap.

9. Adjust brake shoes so not to interfere with brake drum installation. Install the rear brake drum.

➡After installing the brake drums, pump the brake pedal several times to partially adjust the brake shoes. To verify proper operation of the self-adjusting parking brake, be sure that both rear brakes are not dragging when the parking brake pedal is released.

- Rear wheels and lug nuts. Torque the lug nuts, in a star pattern sequence, to 100 ft. lbs. (135 Nm).

10. Road test the vehicle. The automatic adjusters will continue brake adjustment during the road test of the vehicle.

CHRYSLER, DODGE AND PLYMOUTH

3

Caravan • Town & Country • Voyager

SPECIFICATION AND MAINTENANCE CHARTS

ENGINE AND VEHICLE IDENTIFICATION

Engine						
Code ①	Liters (cc)	Cu. In.	Cyl.	Fuel Sys.	Engine Type	Eng. Mfg.
B	2.4 (2429)	148	4	SMFI	DOHC	Chrysler
L	3.8 (3785)	231	6	SMFI	OHV	Chrysler
R	3.3 (3300)	201	6	SMFI	OHV	Chrysler

Model Year	
Code ②	Year
2	2002
3	2003
4	2004
5	2005
6	2006

SMFI: Sequential Multi-port Fuel Injection

DOHC: Double Overhead Camshaft

SOHC: Single Overhead Camshaft

OHV: Overhead Valve

① 8th position of VIN

② 10th position of VIN

06009-MINI-C01

GENERAL ENGINE SPECIFICATIONS

Year	Model	Engine Displacement Liters (VIN)	Net Horsepower @ rpm	Net Torque @ rpm (ft. lbs.)	Bore x Stroke (in.)	Com-pression Ratio	Oil Pressure @ rpm
2002	Caravan	2.4 (B)	150@5200	167@4000	3.44x3.98	9.4:1	25-80@3000
		3.3 (R)	158@4800	203@3600	3.66x3.19	8.9:1	30-80@3000
		3.8 (L)	180@4400	240@3300	3.78x3.43	9.0:1	30-80@3000
	Town & Country	3.3 (R)	158@4800	203@3600	3.66x3.19	8.9:1	30-80@3000
		3.8 (L)	180@4400	240@3300	3.78x3.43	9.0:1	30-80@3000
	Voyager	2.4 (B)	150@5200	167@4000	3.44x3.98	9.4:1	25-80@3000
		3.3 (R)	158@4800	203@3600	3.66x3.19	8.9:1	30-80@3000
		3.8 (L)	180@4400	240@3300	3.78x3.43	9.0:1	30-80@3000
2003	Caravan	2.4 (B)	150@5200	167@4000	3.44x3.98	9.4:1	25-80@3000
		3.3 (R)	158@4800	203@3600	3.66x3.19	8.9:1	30-80@3000
	Town & Country	3.3 (R)	158@4800	203@3600	3.66x3.19	8.9:1	30-80@3000
		3.8 (L)	180@4400	240@3300	3.78x3.43	9.0:1	30-80@3000
	Voyager	2.4 (B)	150@5200	167@4000	3.44x3.98	9.4:1	25-80@3000
		3.3 (R)	158@4800	203@3600	3.66x3.19	8.9:1	30-80@3000
		3.8 (L)	180@4400	240@3300	3.78x3.43	9.0:1	30-80@3000
2004	Caravan	2.4 (B)	150@5200	167@4000	3.44x3.98	9.4:1	25-80@3000
		3.3 (R)	158@4800	203@3600	3.66x3.19	8.9:1	30-80@3000
	Town & Country	3.3 (R)	158@4800	203@3600	3.66x3.19	8.9:1	30-80@3000
		3.8 (L)	180@4400	240@3300	3.78x3.43	9.0:1	30-80@3000
2005-06	Caravan	2.4 (B)	150@5200	167@4000	3.44x3.98	9.4:1	25-80@3000
		3.3 (R)	158@4800	203@3600	3.66x3.19	8.9:1	30-80@3000
	Town & Country	3.3 (R)	158@4800	203@3600	3.66x3.19	8.9:1	30-80@3000
		3.8 (L)	180@4400	240@3300	3.78x3.43	9.0:1	30-80@3000

SMFI: Sequential Multi-port Fuel Injection

06009-MINI-C02

ENGINE TUNE-UP SPECIFICATIONS

Year	Engine Displacement Liters (VIN)	Spark Plug Gap (in.)	Ignition Timing (deg.)	Fuel Pump (psi)	Idle Speed (rpm)	Valve Clearance	
						In.	Ex.
2002	2.4 (B)	0.048-0.053	①	49	②	HYD	HYD
	3.3 (R)	0.048-0.053	①	55	②	HYD	HYD
	3.8 (L)	0.048-0.053	①	49	②	HYD	HYD
2003	2.4 (B)	0.048-0.053	①	49	②	HYD	HYD
	3.3 (R)	0.048-0.053	①	55	②	HYD	HYD
	3.8 (L)	0.048-0.053	①	49	②	HYD	HYD
2004	2.4 (B)	0.048-0.053	①	49	②	HYD	HYD
	3.3 (R)	0.048-0.053	①	55	②	HYD	HYD
	3.8 (L)	0.048-0.053	①	49	②	HYD	HYD
2005-06	2.4 (B)	0.048-0.053	①	49	②	HYD	HYD
	3.3 (R)	0.048-0.053	①	55	②	HYD	HYD
	3.8 (L)	0.048-0.053	①	49	②	HYD	HYD

NOTE: The Vehicle Emission Control Information label often reflects specification changes made during production. The label figures must be used if they differ from those in this chart.

HYD: Hydraulic

① Ignition timing is regulated by the Powertrain Control Module (PCM), and cannot be adjusted.

② Idle speed is controled by the Powertrain Control Module (PCM), and cannot be adjusted.

06009-MINI-C03

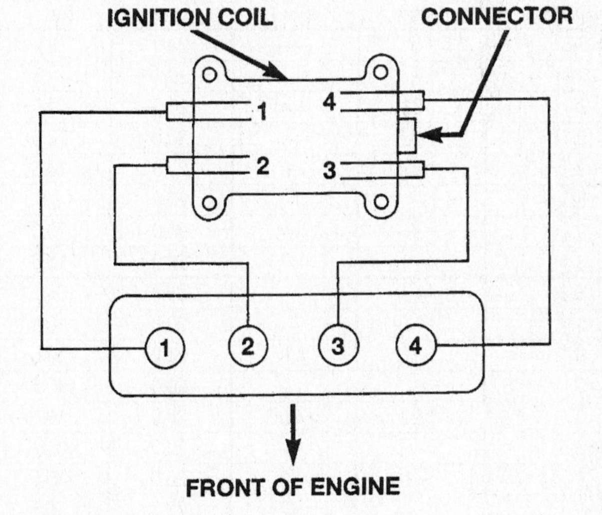

79243G20

2.4L engine
Firing order: 1–3–4–2
Distributorless ignition system

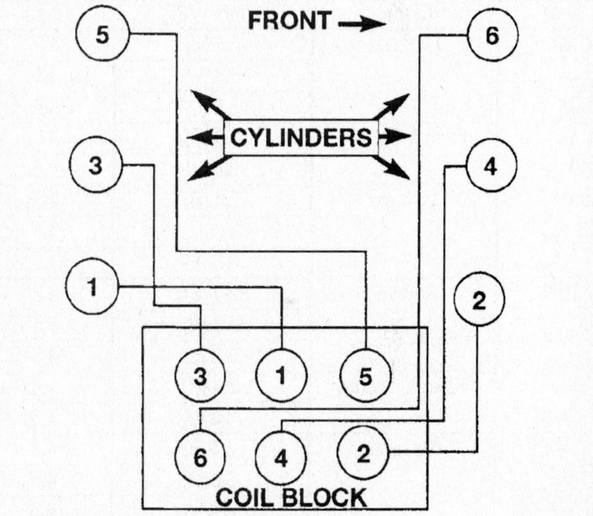

79243G22

3.3L and 3.8L engines
Firing order: 1–2–3–4–5–6
Distributorless ignition system

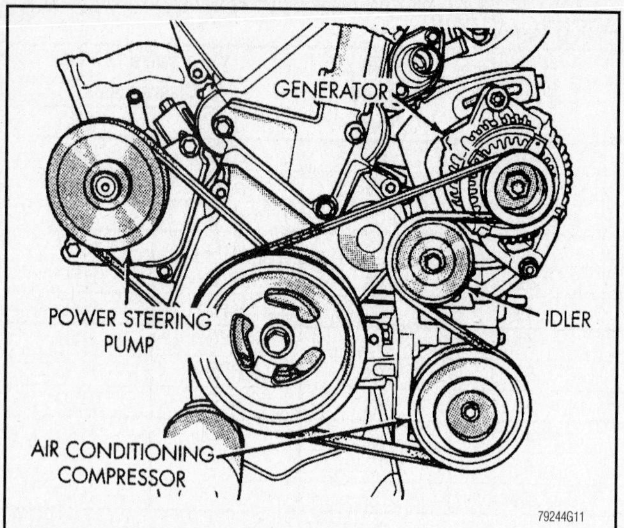

Accessory V-belt routing—2.4L engine

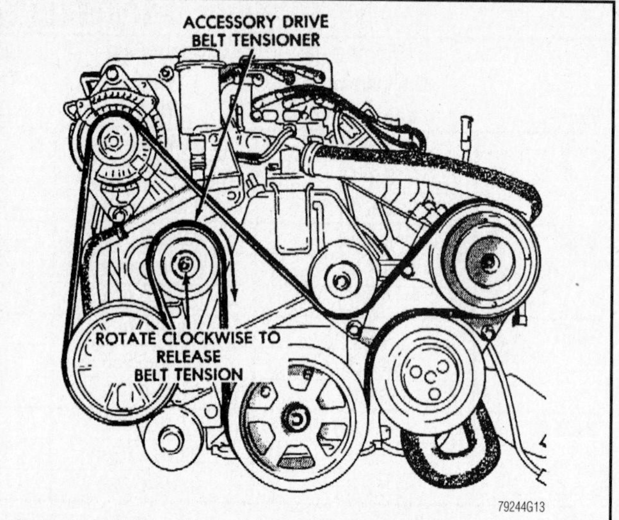

Accessory V-belt routing—3.3L and 3.8L engines

CAPACITIES

Year	Model	Engine Displacement Liters (VIN)	Engine Oil with Filter (qts.)	Automatic Transaxle (qts.)	Transfer Case (qts.)	Rear Drive Axle (pts.)	Fuel Tank (gal.)	Cooling System (qts.)
2002	Caravan	2.4 (B)	4.5	8.5	—	—	20.0	9.5
		3.3 (R)	4.5	9.1	—	—	20.0	10.5
		3.8 (L)	4.5	9.1	1.22	4.0	20.0	10.5
	Town & Country	3.3 (R)	4.5	9.1	—	—	20.0	10.5
		3.8 (L)	4.5	9.1	1.22	4.0	20.0	10.5
	Voyager	2.4 (B)	4.5	8.5	—	—	20.0	9.5
		3.3 (R)	4.5	9.1	—	—	20.0	10.5
		3.8 (L)	4.5	9.1	—	—	20.0	10.5
2003	Caravan	2.4 (B)	4.5	8.5	—	—	20.0	9.5
		3.3 (R)	4.5	9.1	—	—	20.0	10.5
		3.8 (L)	4.5	9.1	1.22	4.0	20.0	10.5
	Town & Country	3.3 (R)	4.5	9.1	—	—	20.0	10.5
		3.8 (L)	4.5	9.1	1.22	4.0	20.0	10.5
	Voyager	2.4 (B)	4.5	8.5	—	—	20.0	9.5
		3.3 (R)	4.5	9.1	—	—	20.0	10.5
		3.8 (L)	4.5	9.1	—	—	20.0	10.5
2004	Caravan	2.4 (B)	4.5	8.5	—	—	20.0	9.5
		3.3 (R)	4.5	9.1	—	—	20.0	10.5
	Town & Country	3.3 (R)	4.5	9.1	—	—	20.0	10.5
		3.8 (L)	4.5	9.1	1.22	4.0	20.0	10.5
2005-06	Caravan	2.4 (B)	4.5	8.5	—	—	20.0	9.5
		3.3 (R)	4.5	9.1	—	—	20.0	10.5
	Town & Country	3.3 (R)	4.5	9.1	—	—	20.0	10.5
		3.8 (L)	4.5	9.1	1.22	4.0	20.0	10.5

NOTE: All capacities are approximate. Add fluid gradually and check to be sure a proper fluid level is obtained.

06009-MINI-C04

VALVE SPECIFICATIONS

Year	Engine Displacement Liters (VIN)	Seat Angle (deg.)	Face Angle (deg.)	Spring Test Pressure (lbs. @ in.)	Spring Installed Height (in.)	Stem-to-Guide Clearance (in.)		Stem Diameter (in.)	
						Intake	Exhaust	Intake	Exhaust
2002	2.4 (B)	45	44.5-45.0	129-143@1.17	1.50	0.0018-0.0025	0.0029-0.0037	0.2340	0.2330
	3.3 (R)	45.0-45.5	①	207-229@1.169	1.62-1.68	0.0010-0.0030	0.0020-0.0060	0.3120-0.3130	0.3112-0.3119
	3.8 (L)	45.0-45.5	①	207-229@1.169	1.62-1.68	0.0010-0.0030	0.0020-0.0060	0.3120-0.3130	0.3112-0.3119
2003	2.4 (B)	45	44.5-45.0	129-143@1.17	1.50	0.0018-0.0025	0.0029-0.0037	0.2340	0.2330
	3.3 (R)	45.0-45.5	①	207-229@1.169	1.62-1.68	0.0010-0.0030	0.0020-0.0060	0.3120-0.3130	0.3112-0.3119
	3.8 (L)	45.0-45.5	①	207-229@1.169	1.62-1.68	0.0010-0.0030	0.0020-0.0060	0.3120-0.3130	0.3112-0.3119
2004	2.4 (B)	45	44.5-45.0	129-143@1.17	1.50	0.0018-0.0025	0.0029-0.0037	0.2340	0.2330
	3.3 (R)	45.0-45.5	①	207-229@1.169	1.62-1.68	0.0010-0.0030	0.0020-0.0060	0.3120-0.3130	0.3112-0.3119
	3.8 (L)	45.0-45.5	①	207-229@1.169	1.62-1.68	0.0010-0.0030	0.0020-0.0060	0.3120-0.3130	0.3112-0.3119
2005-06	2.4 (B)	45	44.5-45.0	129-143@1.17	1.50	0.0018-0.0025	0.0029-0.0037	0.2340	0.2330
	3.3 (R)	45.0-45.5	①	207-229@1.169	1.62-1.68	0.0010-0.0030	0.0020-0.0060	0.3120-0.3130	0.3112-0.3119
	3.8 (L)	45.0-45.5	①	207-229@1.169	1.62-1.68	0.0010-0.0030	0.0020-0.0060	0.3120-0.3130	0.3112-0.3119

① Intake valve: 44.5 degrees
Exhaust valve: 45 degrees

06009-MINI-C05

CRANKSHAFT AND CONNECTING ROD SPECIFICATIONS

All measurements are given in inches.

Year	Engine Displacement Liters (VIN)	Crankshaft				Connecting Rod		
		Main Brg. Journal Dia.	Main Brg. Oil Clearance	Shaft End-play	Thrust on No.	Journal Diameter	Oil Clearance	Side Clearance
2002	2.4 (B)	2.3610-2.3625	0.0007-0.0023	0.0035-0.0094	2	1.9670-1.9685	0.0009-0.0027	0.0051-0.0150
	3.3 (R)	2.5202-2.5195	0.0023-0.0043	0.0036-0.0095	2	2.1240-2.1250	0.0008-0.0026	0.0050-0.0150
	3.8 (L)	2.5202-2.5195	0.0023-0.0043	0.0036-0.0095	2	2.1240-2.1250	0.0008-0.0026	0.0050-0.0150
2003	2.4 (B)	2.3610-2.3625	0.0007-0.0023	0.0035-0.0094	2	1.9670-1.9685	0.0009-0.0027	0.0051-0.0150
	3.3 (R)	2.5202-2.5195	0.0023-0.0043	0.0036-0.0095	2	2.1240-2.1250	0.0008-0.0026	0.0050-0.0150
	3.8 (L)	2.5202-2.5195	0.0023-0.0043	0.0036-0.0095	2	2.1240-2.1250	0.0008-0.0026	0.0050-0.0150
2004	2.4 (B)	2.3610-2.3625	0.0007-0.0023	0.0035-0.0094	2	1.9670-1.9685	0.0009-0.0027	0.0051-0.0150
	3.3 (R)	2.5202-2.5195	0.0023-0.0043	0.0036-0.0095	2	2.1240-2.1250	0.0008-0.0026	0.0050-0.0150
	3.8 (L)	2.5202-2.5195	0.0023-0.0043	0.0036-0.0095	2	2.1240-2.1250	0.0008-0.0026	0.0050-0.0150
2005-06	2.4 (B)	2.3610-2.3625	0.0007-0.0023	0.0035-0.0094	2	1.9670-1.9685	0.0009-0.0027	0.0051-0.0150
	3.3 (R)	2.5202-2.5195	0.0023-0.0043	0.0036-0.0095	2	2.1240-2.1250	0.0008-0.0026	0.0050-0.0150
	3.8 (L)	2.5202-2.5195	0.0023-0.0043	0.0036-0.0095	2	2.1240-2.1250	0.0008-0.0026	0.0050-0.0150

06009-MINI-C06

PISTON AND RING SPECIFICATIONS

All measurements are given in inches.

Year	Engine Displacement Liters (VIN)	Piston Clearance	Ring Gap Top Compression	Ring Gap Bottom Compression	Ring Gap Oil Control	Ring Side Clearance Top Compression	Ring Side Clearance Bottom Compression	Ring Side Clearance Oil Control
2002	2.4 (B)	0.0009-0.0022	0.0098-0.0200	0.0090-0.0180	0.0098-0.0250	0.0011-0.0031	0.0011-0.0031	0.0004-0.0070
	3.3 (R)	0.0010-0.0022	0.0118-0.0217	0.0118-0.0217	0.0098-0.0394	0.0012-0.0037	0.0012-0.0037	0.0005-0.0089
	3.8 (L)	0.0010-0.0022	0.0118-0.0217	0.0118-0.0217	0.0098-0.0394	0.0012-0.0037	0.0012-0.0037	0.0005-0.0089
2003	2.4 (B)	0.0009-0.0022	0.0098-0.0200	0.0090-0.0180	0.0098-0.0250	0.0011-0.0031	0.0011-0.0031	0.0004-0.0070
	3.3 (R)	0.0010-0.0022	0.0118-0.0217	0.0118-0.0217	0.0098-0.0394	0.0012-0.0037	0.0012-0.0037	0.0005-0.0089
	3.8 (L)	0.0010-0.0022	0.0118-0.0217	0.0118-0.0217	0.0098-0.0394	0.0012-0.0037	0.0012-0.0037	0.0005-0.0089
2004	2.4 (B)	0.0009-0.0022	0.0098-0.0200	0.0090-0.0180	0.0098-0.0250	0.0011-0.0031	0.0011-0.0031	0.0004-0.0070
	3.3 (R)	0.0010-0.0022	0.0118-0.0217	0.0118-0.0217	0.0098-0.0394	0.0012-0.0037	0.0012-0.0037	0.0005-0.0089
	3.8 (L)	0.0010-0.0022	0.0118-0.0217	0.0118-0.0217	0.0098-0.0394	0.0012-0.0037	0.0012-0.0037	0.0005-0.0089
2005-06	2.4 (B)	0.0009-0.0022	0.0098-0.0200	0.0090-0.0180	0.0098-0.0250	0.0011-0.0031	0.0011-0.0031	0.0004-0.0070
	3.3 (R)	0.0010-0.0022	0.0118-0.0217	0.0118-0.0217	0.0098-0.0394	0.0012-0.0037	0.0012-0.0037	0.0005-0.0089
	3.8 (L)	0.0010-0.0022	0.0118-0.0217	0.0118-0.0217	0.0098-0.0394	0.0012-0.0037	0.0012-0.0037	0.0005-0.0089

① Oil control ring side rails must be free to rotate after assembly

06009-MINI-C07

TORQUE SPECIFICATIONS

All readings in ft. lbs.

Year	Engine Displacement Liters (cc)	Cylinder Head Bolts	Main Bearing Bolts	Rod Bearing Bolts	Crankshaft Damper Bolts	Flywheel Bolts	Manifold Intake	Manifold Exhaust	Spark Plugs	Oil Pan Drain Plug
2002	2.4 (B)	①	②	③	100	70	④	15	20	20
	3.3 (R)	⑤	⑥	⑦	40	70	⑧	⑨	20	20
	3.8 (L)	⑤	⑥	⑦	40	70	⑧	⑨	20	20
2003	2.4 (B)	①	②	③	100	70	④	15	20	20
	3.3 (R)	⑤	⑥	⑦	40	70	⑧	⑨	20	20
	3.8 (L)	⑤	⑥	⑦	40	70	⑧	⑨	20	20
2004	2.4 (B)	①	②	③	100	70	④	15	20	20
	3.3 (R)	⑤	⑥	⑦	40	70	⑧	⑨	20	20
	3.8 (L)	⑤	⑥	⑦	40	70	⑧	⑨	20	20
2005-06	2.4 (B)	①	②	③	100	70	④	15	20	20
	3.3 (R)	⑤	⑥	⑦	40	70	⑧	⑨	20	20
	3.8 (L)	⑤	⑥	⑦	40	70	⑧	⑨	20	20

① Step 1: 25 ft. lbs.
Step 2: 50 ft. lbs.
Step 3: 50 ft. lbs.
Step 4: Plus 1/4 turn

② M8 bolts: 21 ft. lbs.
M11 bolts: 30 ft. lbs. plus 90 degrees

③ Step 1: 20 ft. lbs.
Step 2: Plus 90 degrees

④ 2.4L: Upper and lower 250 inch lbs

⑤ Step 1: 45 ft. lbs.
Step 2: 65 ft. lbs.
Step 3: 65 ft. lbs.
Step 4: Plus 90 degrees

⑥ Step 1: 30 ft. lbs.
Step 2: Plus 90 degrees

⑦ Step 1: 40 ft. lbs.
Step 2: Plus 90 degrees

⑧ Step 1: 10 inch lbs.
Step 2: 17 ft. lbs.
Step 3: 17 ft. lbs.

⑨ Refer to procedure for torque sequence and specifications

06009-MINI-C08

WHEEL ALIGNMENT

Year	Model		Caster Range (+/-Deg.)	Caster Preferred Setting (Deg.)	Camber Range (+/-Deg.)	Camber Preferred Setting (Deg.)	Toe-in (in.)
2002	All	F	1.00	+1.40	0.30	+0.05	0.10+/-0.20
		R	—	—	0.25	0	0+/-0.30
		F	1.00	+1.40	0.40	+0.15	0.10+/-0.20
		R	—	—	0.25	0	0+/-0.30
2003	All	F	1.00	+1.40	0.30	+0.05	0.10+/-0.20
		R	—	—	0.25	0	0+/-0.30
		F	1.00	+1.40	0.40	+0.15	0.10+/-0.20
		R	—	—	0.25	0	0+/-0.30
2004	All	F	1.00	+1.40	0.30	+0.05	0.10+/-0.20
		R	—	—	0.25	0	0+/-0.30
		F	1.00	+1.40	0.40	+0.15	0.10+/-0.20
		R	—	—	0.25	0	0+/-0.30
2005-06	All	F	1.00	+1.40	0.30	+0.05	0.10+/-0.20
		R	—	—	0.25	0	0+/-0.30
		F	1.00	+1.40	0.40	+0.15	0.10+/-0.20
		R	—	—	0.25	0	0+/-0.30

06009-MINI-C09

TIRE, WHEEL AND BALL JOINT SPECIFICATIONS

Year	Model	OEM Tires Standard	OEM Tires Optional	Tire Pressures (psi) Front	Tire Pressures (psi) Rear	Wheel Size	Ball Joint Inspection	Lug Nuts
2002	Town & Country base, LX	P215/70R15	None	①	①	6-J	0.030 in. ①	100
	Except Town & Country base, LX	P215/65R16	None	①	①	6-J	0.030 in. ①	100
	Caravan/Voyager, base	P205/75SR14	None	①	①	6-J	0.030 in. ①	100
	Caravan/Voyager SE, LE	P215/65SR15	P215/65R16	①	①	6-J	0.030 in. ①	100
	Caravan/Voyager ES	P215/65R16	None	①	①	6-J	0.030 in. ①	100
2003	Town & Country base, LX	P215/70R15	None	①	①	6-J	0.030 in. ①	100
	Except Town & Country base, LX	P215/65R16	None	①	①	6.5-J	0.030 in. ①	100
	Caravan/Voyager, base	P205/75SR14	None	①	①	6-J	0.030 in. ①	100
	Caravan/Voyager SE, LE	P215/65SR15	P215/65R16	①	①	6-J	0.030 in. ①	100
	Caravan/Voyager ES	P215/65R16	None	①	①	6-J	0.030 in. ①	100
2004	Town & Country base, LX	P215/70R15	None	①	①	6-J	0.030 in. ①	100
	Except Town & Country base, LX	P215/65R16	None	①	①	6.5-J	0.030 in. ①	100
	Caravan base	P205/75SR14	None	①	①	6-J	0.030 in. ①	100
	Caravan SE, LE	P215/65SR15	P215/65R16	①	①	6-J	0.030 in. ①	100
	Caravan ES	P215/65R16	None	①	①	6-J	0.030 in. ①	100
2005-06	Town & Country base, LX	P215/70R15	None	①	①	6-J	0.030 in. ①	100
	Except Town & Country base, LX	P215/65R16	None	①	①	6.5-J	0.030 in. ①	100
	Caravan base	P205/75SR14	None	①	①	6-J	0.030 in. ①	100
	Caravan SE, LE	P215/65SR15	P215/65R16	①	①	6-J	0.030 in. ①	100
	Caravan ES	P215/65R16	None	①	①	6-J	0.030 in. ①	100

OEM: Original Equipment Manufacturer
PSI: Pounds Per Square Inch
① Refer to label on drivers side door pillar
② Both upper and lower

06009-MINI-C10

BRAKE SPECIFICATIONS
All measurements in inches unless noted

Year	Model		Brake Disc Original Thickness	Brake Disc Minimum Thickness	Brake Disc Maximum Run-out	Brake Drum Diameter Original Inside Diameter	Brake Drum Diameter Max. Wear Limit	Brake Drum Diameter Maximum Machine Diameter	Min. Lining Thick-ness	Caliper Guide Pin Bolts (ft. lbs.)
2002	Caravan	F	0.939-0.949	0.881	0.005	—	—	—	0.313	35
		R	0.482-0.502	0.443	0.005	9.84	9.93	9.90	①	35
	Town & Country	F	0.939-0.949	0.881	0.005	—	—	—	0.313	35
		R	0.482-0.502	0.443	0.005	9.84	9.93	9.90	①	35
	Voyager	F	0.939-0.949	0.881	0.005	—	—	—	0.313	35
		R	—	—	—	9.84	9.93	9.90	0.031	35
2003	Caravan	F	0.939-0.949	0.881	0.005	—	—	—	0.313	35
		R	0.482-0.502	0.443	0.005	9.84	9.93	9.90	①	35
	Town & Country	F	0.939-0.949	0.881	0.005	—	—	—	0.313	35
		R	0.482-0.502	0.443	0.005	9.84	9.93	9.90	①	35
2004	Caravan	F	0.939-0.949	0.881	0.005	—	—	—	0.313	35
		R	0.482-0.502	0.443	0.005	9.84	9.93	9.90	①	35
	Town & Country	F	0.939-0.949	0.881	0.005	—	—	—	0.313	35
		R	0.482-0.502	0.443	0.005	9.84	9.93	9.90	①	35
2005-06	Caravan	F	0.939-0.949	0.881	0.005	—	—	—	0.313	26
		R	0.482-0.502	0.443	0.005	9.84	9.93	9.90	①	26
	Town & Country	F	0.939-0.949	0.881	0.005	—	—	—	0.313	26
		R	0.482-0.502	0.443	0.005	9.84	9.93	9.90	①	26

F: Front
R: Rear
① Drum brakes: 0.031 in
Disc brakes: 0.281 in.

06009-MINI-C11

SCHEDULED MAINTENANCE INTERVALS

CHRYSLER—TOWN & COUNTRY, DODGE—CARAVAN & PLYMOUTH—VOYAGER

		VEHICLE MILEAGE INTERVAL (x1000)												
TO BE SERVICED	TYPE OF SERVICE	7.5	15	22.5	30	37.5	45	52.5	60	67.5	75	82.5	90	97.5
Engine oil & filter	R	✓	✓	✓	✓	✓	✓	✓	✓	✓	✓	✓	✓	✓
Driveshaft boots	S/I	✓	✓	✓	✓	✓	✓	✓	✓	✓	✓	✓	✓	✓
Exhaust system	S/I	✓	✓	✓	✓	✓	✓	✓	✓	✓	✓	✓	✓	✓
Engine coolant level, hoses & clamps	S/I	✓	✓	✓	✓	✓	✓	✓	✓	✓	✓	✓	✓	✓
Rotate tires	S/I	✓	✓	✓	✓	✓	✓	✓	✓	✓	✓	✓	✓	✓
Drive belts	S/I		✓		✓		✓		✓		✓		✓	
Brake hoses & linings	S/I			✓			✓			✓			✓	
Automatic transaxle fluid & filter	R				✓				✓				✓	
Air filter	R				✓				✓				✓	
Spark plugs ①	R				✓				✓				✓	
Serpentine belts (3.0L & 3.3L)	S/I								✓		✓		✓	
Lubricate tie rod ends	S/I				✓				✓				✓	
PCV valve	S/I				✓				✓				✓	
Engine coolant	R								✓			✓		
Timing belt (3.0L)	R								✓					
Distributor cap & rotor	R							✓						
Ignition cables (3.0L)	R								✓					
Ignition timing	S/I								✓					

R: Replace S/I: Service or Inspect

① Platinum tip spark plugs & ignition cables (3.3L & 3.8L): replace every 100,000 miles.

FREQUENT OPERATION MAINTENANCE (SEVERE SERVICE)

If a vehicle is operated under any of the following conditions it is considered severe service:

- Extremely dusty areas.

- 50% or more of the vehicle operation is in 32°C (90°F) or higher temperatures, or constant operation in temperatures below 0°C (32°F).

- Prolonged idling (vehicle operation in stop and go traffic.

- Frequent short running periods (engine does not warm to normal operating temperatures).

- Police, taxi, delivery usage or trailer towing usage.

Oil & oil filter change: change every 3000 miles.

Automatic transaxle fluid & filter: change every 15,000 miles.

Brake hoses & linings: check every 9000 miles.

CV-joints & front suspension ball joints: check every 3000 miles.

Tie rod ends & steering linkage: check every 15,000 miles.

Air filter: change every 15,000 miles.

06009-MINI-C12

PRECAUTIONS

Before servicing any vehicle, please be sure to read all of the following precautions, which deal with personal safety, prevention of component damage, and important points to take into consideration when servicing a motor vehicle:

- Never open, service or drain the radiator or cooling system when the engine is hot; serious burns can occur from the steam and hot coolant.
- Observe all applicable safety precautions when working around fuel. Whenever servicing the fuel system, always work in a well-ventilated area. Do not allow fuel spray or vapors to come in contact with a spark, open flame, or excessive heat (a hot drop light, for example). Keep a dry chemical fire extinguisher near the work area. Always keep fuel in a container specifically designed for fuel storage; also, always properly seal fuel containers to avoid the possibility of fire or explosion. Refer to the additional fuel system precautions later in this section.
- Fuel injection systems often remain pressurized, even after the engine has been turned **OFF**. The fuel system pressure must be relieved before disconnecting any fuel lines. Failure to do so may result in fire and/or personal injury.
- Brake fluid often contains polyglycol ethers and polyglycols. Avoid contact with the eyes and wash your hands thoroughly after handling brake fluid. If you do get brake fluid in your eyes, flush your eyes with clean, running water for 15 minutes. If eye irritation persists, or if you have taken brake fluid internally, IMMEDIATELY seek medical assistance.
- The EPA warns that prolonged contact with used engine oil may cause a number of skin disorders, including cancer! You should make every effort to minimize your exposure to used engine oil. Protective gloves should be worn when changing oil. Wash your hands and any other exposed skin areas as soon as possible after exposure to used engine oil. Soap and water, or waterless hand cleaner should be used.
- All new vehicles are now equipped with an air bag system, often referred to as a Supplemental Restraint System (SRS) or Supplemental Inflatable Restraint (SIR) system. The system must be disabled before performing service on or around system components, steering column, instrument panel components, wiring and sensors. Failure to follow safety and disabling procedures could result in accidental air bag deployment, possible personal injury and unnecessary system repairs.
- Always wear safety goggles when working with, or around, the air bag system. When carrying a non-deployed air bag, be sure the bag and trim cover are pointed away from your body. When placing a non-deployed air bag on a work surface, always face the bag and trim cover upward, away from the surface. This will reduce the motion of the module if it is accidentally deployed. Refer to the additional air bag system precautions later in this section.
- Clean, high quality brake fluid from a sealed container is essential to the safe and proper operation of the brake system. You should always buy the correct type of brake fluid for your vehicle. If the brake fluid becomes contaminated, completely flush the system with new fluid. Never reuse any brake fluid. Any brake fluid that is removed from the system should be discarded. Also, do not allow any brake fluid to come in contact with a painted surface; it will damage the paint.
- Never operate the engine without the proper amount and type of engine oil; doing so WILL result in severe engine damage.
- Timing belt maintenance is extremely important! Many models utilize an interference-type, non-freewheeling engine. If the timing belt breaks, the valves in the cylinder head may strike the pistons, causing potentially serious (also time-consuming and expensive) engine damage. Refer to the maintenance interval charts in this manual for the recommended replacement interval for the timing belt, and to the timing belt repair procedure in this section for belt replacement and inspection.
- Disconnecting the negative battery cable on some vehicles may interfere with the functions of the on-board computer system(s) and may require the computer to undergo a relearning process once the negative battery cable is reconnected.
- When servicing drum brakes, only disassemble and assemble one side at a time, leaving the remaining side intact for reference.
- Only an MVAC-trained, EPA-certified automotive technician should service the air conditioning system or its components.

ENGINE REPAIR

➡Disconnecting the negative battery cable on some vehicles may interfere with the functions of the on board computer system. The computer may undergo a relearning process once the negative battery cable is reconnected.

Alternator

REMOVAL

2.4L Engine

1. Before servicing the vehicle, refer to the precautions in the beginning of this section.
2. Remove or disconnect the following:
   - Negative battery cable
   - Air inlet temperature sensor connector
   - Air cleaner assembly
   - Evaporative emissions (EVAP) purge solenoid from the bracket and position aside
   - Alternator harness connections
   - Accessory drive belt
   - Alternator

3.3L and 3.8L Engines

1. Before servicing the vehicle, refer to the precautions in the beginning of this section.
2. Remove or disconnect the following:
   - Negative battery cable
   - Alternator harness connectors
   - Right front lower splash shield
   - Accessory drive belt
   - Lower oil dipstick tube bolt and the wiring harness from the tube
   - 3 mounting bolts
   - Dipstick tube
   - Alternator

INSTALLATION

2.4L Engine

1. Install or connect the following:
   - Alternator
   - Accessory drive belt. Tighten the alternator mounting fasteners to 20 ft. lbs. (28 Nm).
   - Alternator harness connections
   - EVAP purge solenoid
   - Air cleaner assembly
   - Air inlet temperature sensor connector
   - Negative battery cable

3.3L and 3.8L Engines

1. Install or connect the following:
   - Alternator in position and the 3 mounting bolts. Tighten to 40 ft. lbs. (54 Nm).
   - Dipstick tube, lubricate the O–ring with clean engine oil prior to installation, then tighten the upper dipstick tube bolt.
   - Alternator harness connectors
   - Lower oil dipstick tube bolt
   - Accessory drive belt
   - Right front lower splash shield
   - Wiring harness to the dipstick tube
   - Negative battery cable

Ignition Timing

ADJUSTMENT

The base ignition timing cannot be adjusted. The Powertrain Control Module (PCM) regulates the ignition timing automatically.

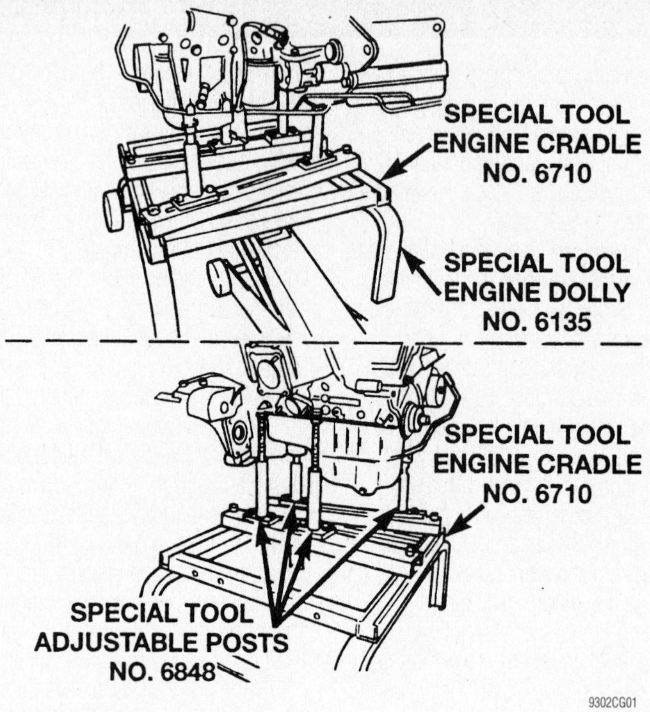

Chrysler powertrain support tools

Engine Assembly

REMOVAL & INSTALLATION

2.4L Engine

1. Before servicing the vehicle, refer to the precautions in the beginning of this section.
2. Drain the cooling system.
3. Drain the engine oil.
4. Relieve the fuel system pressure.
5. Recover the A/C refrigerant.
6. Remove or disconnect the following:
   - Negative battery cable
   - Air cleaner and hoses
   - Fuel line
   - Vacuum hoses
   - Radiator fans
   - Radiator hoses

➡When the transaxle lines are removed from the fittings at the transaxle damage to the inner wall of the hose will occur. To prevent leakage, cut the cooler hoses off flush at the transaxle fitting and use a service cooler hose splice kit upon installation.

   - Transaxle cooler lines, cut the lines flush at the transaxle fittings
   - Transaxle shift linkage
   - Throttle body linkage
   - Engine wiring harness
   - Heater hoses
   - Front wheels

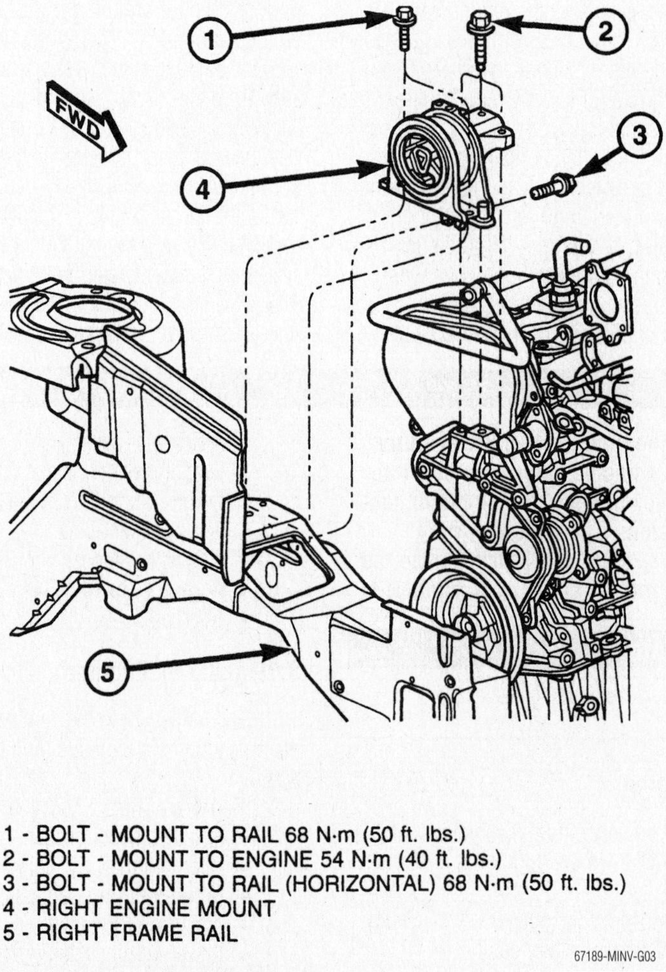

Right mount-to-rail and engine installation—2.4L engine

- Right inner splash shield
- Drive belts
- Axle halfshafts
- Crossmember cradle plate
- Exhaust front pipe
- Front motor mount
- Structural collar
- Rear motor mount
- Power steering pump. Pinch off the supply hose at the pump, disconnect the hose but leave the pressure line attached and then set the pump assembly aside.
- A/C compressor lines
- Body ground straps

7. Raise the vehicle enough to position dolly 6135, cradle tool 6710 with posts 6848 under vehicle.

8. Loosen the cradle posts to allow movement for proper positioning. Position the two rear posts (right side of engine) into the engine bedplate holes. Position the two front posts (left side of engine) on the oil pan rails. Lower the vehicle and position the cradle mounts until the engine is resting on the mounts and then tighten the mounts to the cradle frame to prevent mounts from moving when removing/installing the powertrain assembly.

9. Attach safety straps around the engine-to-cradle assembly. Lower the vehicle so the weight of the powertrain assembly, not the vehicle are on the cradle.

10. Remove the engine and transmission bolts.

11. Raise the vehicle away from the powertrain.

To install:

12. Lower the vehicle over the powertrain.

13. Align the engine and transmission mounts to their attaching points. Install and tighten the right mount-to-rail and engine and left mount-to-frame brackets as shown in the two accompanying illustrations.

14. Remove the safety straps and the dolly.

15. Install the rear mount bracket and through bolt.

16. Install or connect the following:
- Torque converter
- Axle halfshafts

17. Install the bending struts and structural collar, as follows:

a. Step 1: Position the collar between the transaxle and oil pan and install bolts hand tight.

b. Step 2: Position the collar to oil pan bolts and install bolts hand tight.

c. Step 3: Tighten collar to transaxle bolts to 75 ft. lbs. (101 Nm).

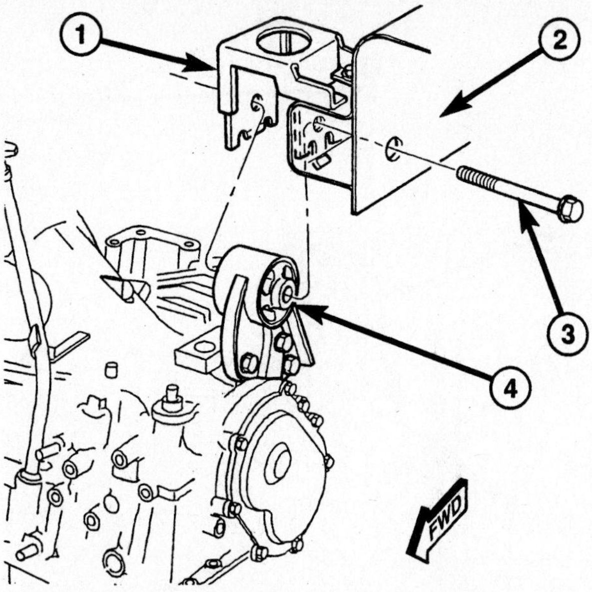

Left mount-to-frame bracket installation—2.4L engine

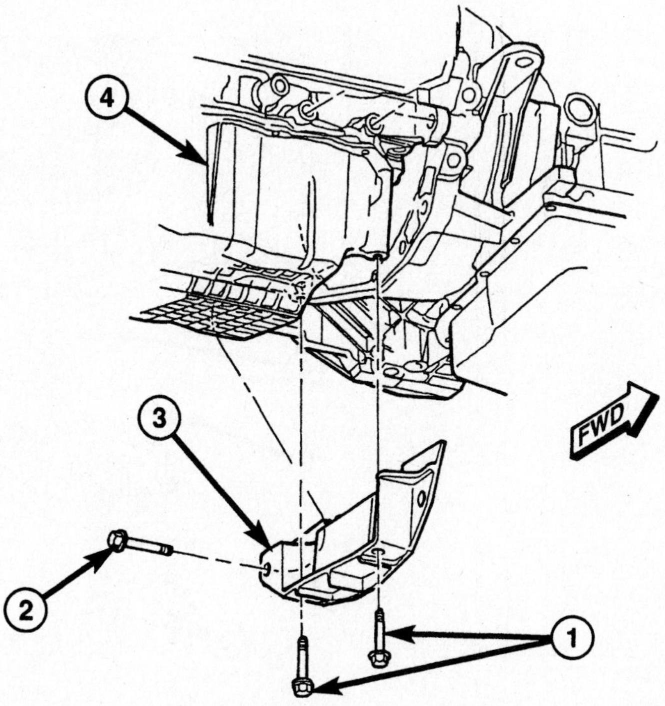

Structural collar and bending strut torque sequence—2.4L engine

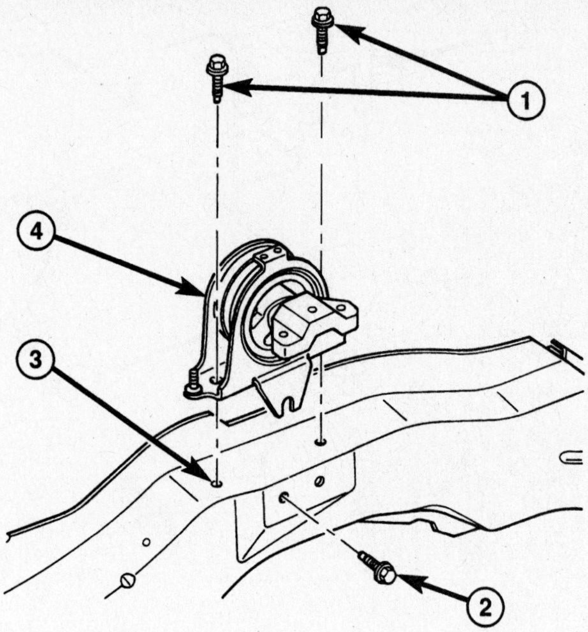

1 - BOLT
2 - BOLT
3 - FRAME RAIL
4 - RIGHT MOUNT - 2.4L ENGINE

67189-MINV-G06

Exploded view of the right side engine hydro–type mount—2.4L engine

d. Step 4: Tighten collar to oil pan bolts to 40 ft. lbs. (54 Nm).

e. Step 5: Install the engine front mount assembly.

18. Install or connect the following:
 - Exhaust front pipe
 - Crossmember cradle plate
 - Power steering pump and lines
 - A/C compressor lines
 - Drive belts
 - Right inner splash shield
 - Axle halfshafts
 - Front wheels
 - Transaxle cooler lines using a splice service kit
 - Transaxle shift linkage
 - Heater hoses
 - Body ground straps
 - Engine wiring harness
 - Vacuum lines
 - Throttle body linkage
 - Engine cooling fans
 - Fuel line
 - Radiator hoses
 - Air cleaner and hoses
 - Negative battery cable
19. Fill the cooling system.
20. Fill the engine with clean oil.

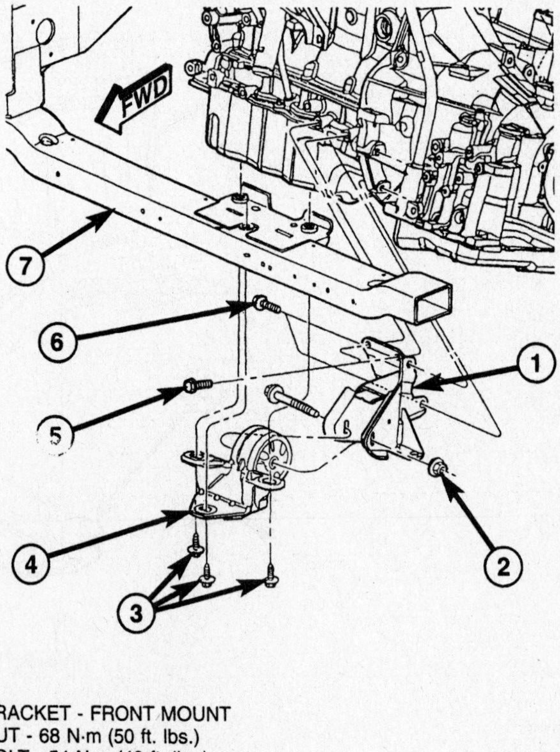

1 - BRACKET - FRONT MOUNT
2 - NUT - 68 N·m (50 ft. lbs.)
3 - BOLT - 54 N·m (40 ft. lbs.)
4 - MOUNT - FRONT INSULATOR
5 - BOLT - 68 N·m (50 ft. lbs.)
6 - BOLT - 68 N·m (50 ft. lbs.)
7 - FRONT CROSSMEMBER

67189-MINV-G07

Exploded view of the front mount and bracket assembly—2.4L engine

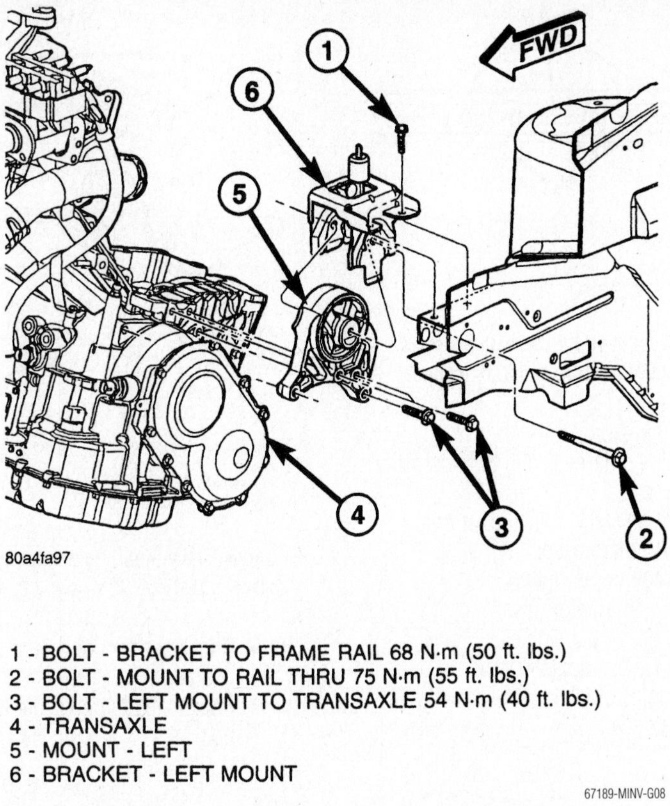

Exploded view of the left mount and bracket assembly—2.4L engine

21. Recharge the A/C system.
22. Start the engine and check for leaks.

3.3L and 3.8L Engines

1. Before servicing the vehicle, refer to the precautions in the beginning of this section.
2. Drain the cooling system.
3. Drain the engine oil.
4. Relieve the fuel system pressure.
5. Recover the A/C refrigerant.
6. Remove or disconnect the following:
   - Battery and tray
   - Air cleaner and hoses
   - Fuel line from the rail
   - Wiper module
7. Block off the heater hoses to the rear system by using pinch off pliers, if equipped.
   - Heater hoses
   - Radiator upper support crossmember
   - Engine cooling fan
   - Throttle body linkage
   - Manifold Absolute Pressure (MAP) sensor connector
   - Idle Air Control (IAC) valve connector
   - Throttle Position (TPS) sensor connector
   - Exhaust Gas Recirculation (EGR) transducer connector
   - Vacuum lines from the throttle body, brake booster and speed control
   - Wiring harness clip from the right side mount
   - Power steering reservoir and set aside without disconnecting the lines
   - Ground strap from the rear of the cylinder head
   - Engine Coolant Temperature (ECT) sensor connector
   - Ignition coil connectors
   - Fuel injector connectors and clip from the bracket
   - Camshaft (CMP) and Crankshaft (CKP) position sensor connectors
   - A/C compressor electrical connectors and hoses. Cap the lines to avoid system contamination.
   - Radiator upper hose
   - Electrical connector at the transaxle dipstick tube
   - Transaxle dipstick tube and seal the opening to avoid system contamination

➡When the transaxle lines are removed from the fittings at the transaxle damage to the inner wall of the hose will occur. To prevent leakage, cut the cooler hoses off flush at the transaxle fitting and use a service cooler hose splice kit upon installation.

   - Transaxle cooler lines, cut the lines flush at the transaxle fittings
   - Transaxle shift linkage and electrical connectors
   - Axle halfshafts
   - Crossmember cradle plate
   - Power Transfer Unit (PTO), if equipped
   - Exhaust front pipe from the manifold
   - Front motor mount and bracket
   - Rear motor mount bracket
   - Engine-to-transaxle struts
   - Transaxle case cover
   - Torque converter
   - Power steering pressure hose clip bolt
   - Knock sensor connector, if equipped
   - Engine block heater connector, if equipped
   - Drive belt splash shield
   - Accessory drive belt
   - Lower radiator hose
   - A/C compressor
   - Alternator
   - Water pump pulley bolts and position the pulley between the pump and the housing
   - Oil pressure switch connector
   - Wiring harness clip from the dipstick tube
8. Install adapter tools on the right side of the engine block as illustrated
   - Power steering pump and set aside
9. Raise the vehicle enough to position dolly 6135, cradle tool 6710 with posts 6848 and adapter tool 6909 under vehicle.
10. Loosen the cradle posts to allow movement for proper positioning. Lower the vehicle and position the cradle/post mounts until the engine is resting on the posts and then tighten the mounts to the cradle frame to prevent mounts from moving when removing/installing the powertrain assembly.
11. Attach safety straps around the engine-to-cradle assembly. Lower the vehicle so the weight of the powertrain assembly, not the vehicle are on the cradle.
12. Remove or disconnect the following:
   - Engine right side mount-to-engine bolts
   - Left mount through bolt
13. Raise the vehicle away from the powertrain.

To install:

14. Lower the vehicle over the powertrain.

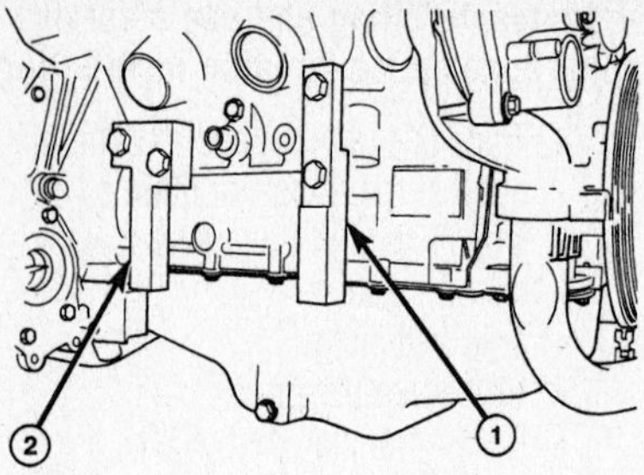

Adapter tools 6912 and 8444 mounted on the block—3.3L and 3.8L engines

15. Align the engine and transmission mounts to their attaching points. Install and tighten the right mount and left transmission mounts as shown in the two accompanying illustrations.

16. Remove the safety straps and the dolly and adapters.

17. Install or connect the following:

- Power steering pump and pressure line

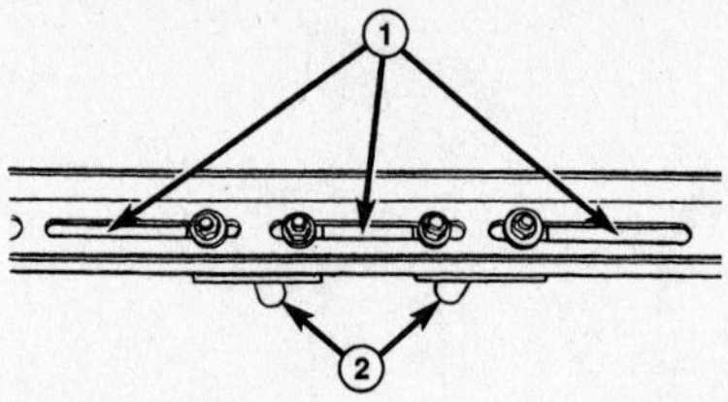

Position tool 6848 for use with adapters 8444 and 6912—3.3L and 3.8L engines

- Alternator
- Wiring harness clip to the dipstick tube
- Oil pressure switch connector
- A/C compressor
- Water pump pulley
- Lower radiator hose
- Accessory drive belt
- Drive belt splash shield
- Engine block heater connector, if equipped

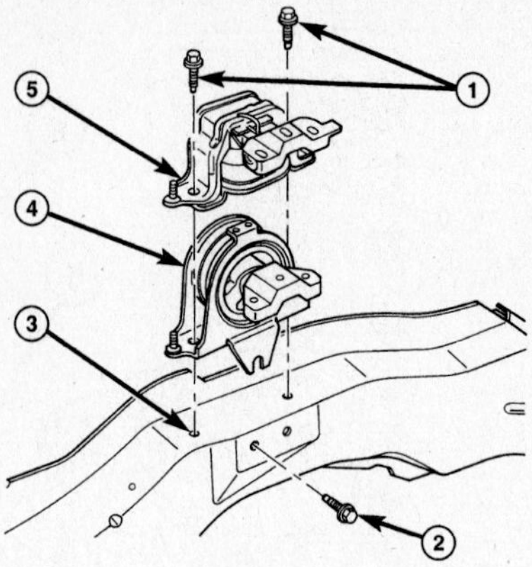

Exploded view of the right side hrydo–type engine mounting—3.3L and 3.8L engines

- Knock sensor connector, if equipped
- Power steering pressure hose clip bolt

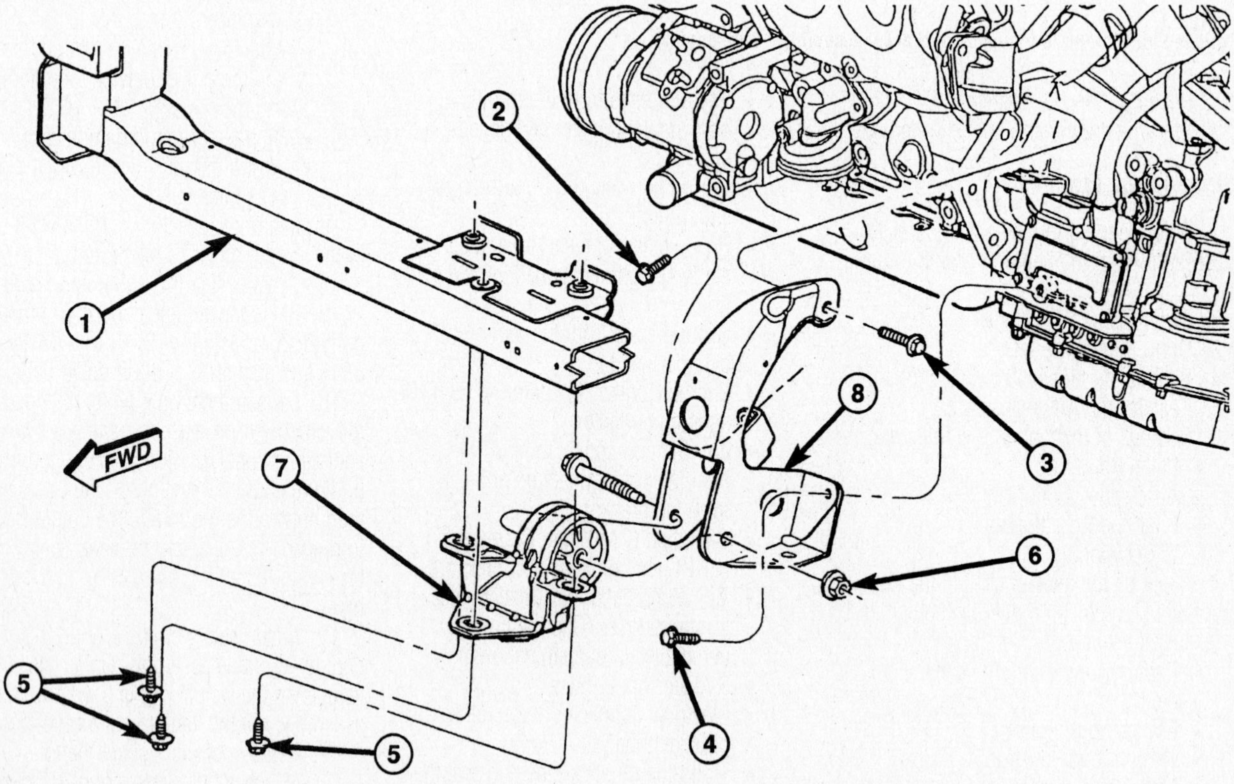

Exploded view of the front engine mounting—3.3L and 3.8L engines

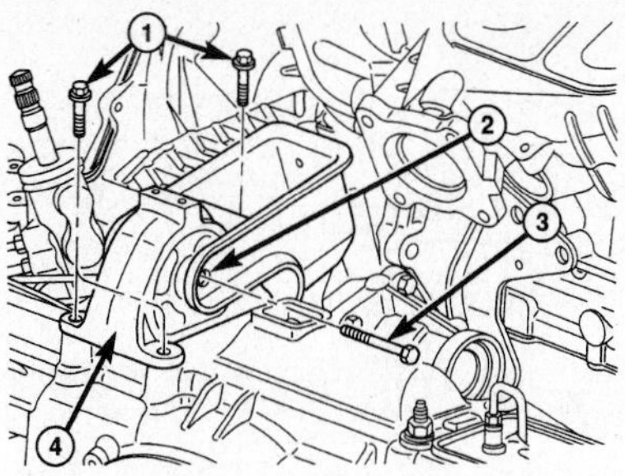

1 - BOLT 54 N·m (40 ft. lbs.)
2 - REAR MOUNT BRACKET
3 - THRU-BOLT 54 N·m (40 ft. lbs.)
4 - REAR MOUNT

67189-MINV-G22

Exploded view of the rear engine mounting—3.3L and 3.8L engines

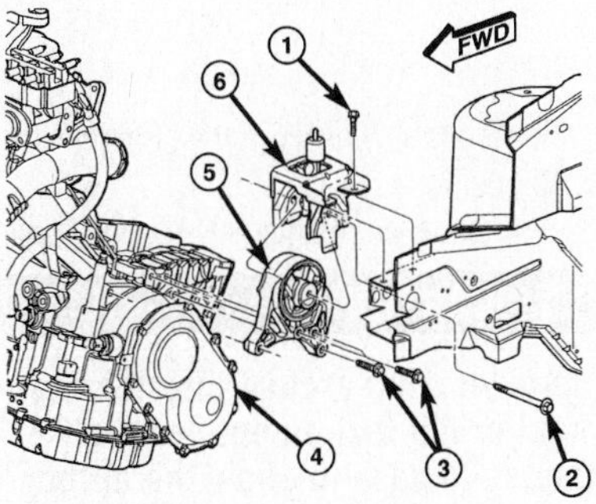

1 - BOLT - BRACKET TO FRAME RAIL 68 N·m (50 ft. lbs.)
2 - BOLT - MOUNT TO RAIL THRU 75 N·m (55 ft. lbs.)
3 - BOLT - LEFT MOUNT TO TRANSAXLE 54 N·m (40 ft. lbs.)
4 - TRANSAXLE
5 - MOUNT - LEFT
6 - BRACKET - LEFT MOUNT

67189-MINV-G23

Exploded view of the left engine mounting—3.3L and 3.8L engines

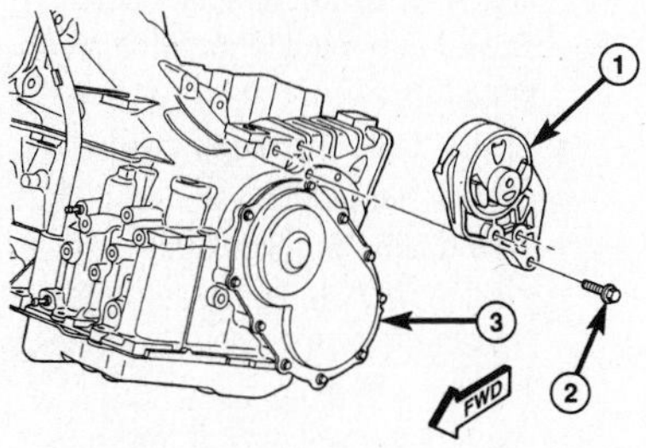

1 - LEFT MOUNT ASEMBLY
2 - BOLT - 54 N·m (40 ft. lbs.)
3 - TRANSAXLE - 31TH

67189-MINV-G24

Exploded view of the left engine mounting on models equipped with a 31TH transaxle—3.3L and 3.8L engines

- Torque converter
- Transaxle case cover
- Engine-to-transaxle struts
- Rear motor mount bracket
- Front motor mount and bracket
- PTO, if equipped

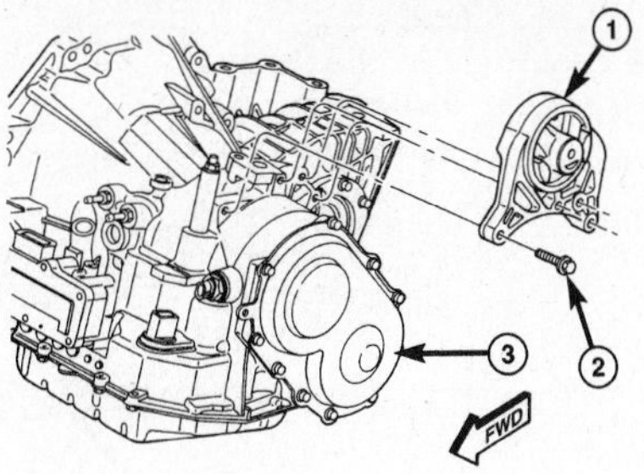

1 - LEFT MOUNT ASSEMBLY
2 - BOLT - 54 N·m (40 ft. lbs.)
3 - TRANSAXLE - 41TE

67189-MINV-G25

Exploded view of the left engine mounting on models equipped with a 41TE transaxle—3.3L and 3.8L engines

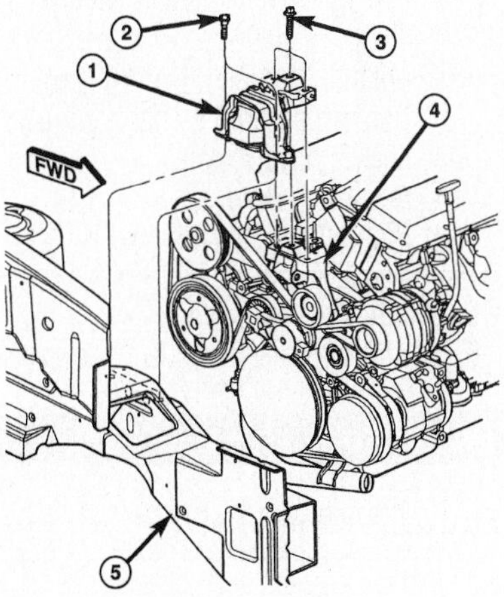

1 - RIGHT ENGINE MOUNT
2 - BOLT - MOUNT TO FRAME RAIL
3 - BOLT - MOUNT TO ENGINE
4 - ENGINE MOUNT BRACKET
5 - RIGHT FRAME RAIL

67189-MINV-G26

Exploded view of the right engine mounting—3.3L and 3.8L engines

- Axle halfshafts
- Exhaust front pipe to the manifold
- Crossmember cradle plate
- Transaxle shift linkage and electrical connectors
- Transaxle cooler lines using a splice kit
- Transaxle dipstick tube and
- Electrical connector at the transaxle dipstick tube
- A/C compressor electrical connectors and hoses
- CMP and CKP position sensor connectors
- Fuel injector connectors and clip to the bracket
- ECT sensor connector
- Ignition coil connectors
- Ground strap to the rear of the cylinder head
- Power steering reservoir
- Wiring harness clip to the right side mount
- Vacuum lines to the throttle body, brake booster and speed control
- EGR transducer connector
- TPS sensor connector
- IAC valve connector
- MAP sensor connector
- Throttle body linkage
- Engine cooling fan
- Radiator upper hose
- Heater hoses
- Radiator upper support crossmember
- Wiper module
- Fuel line to the rail
- Air cleaner and hoses
- Battery and tray

18. Fill the cooling system.
19. Fill the engine with clean oil.
20. Recharge the A/C system.
21. Start the engine and check for leaks.

Water Pump

REMOVAL & INSTALLATION

2.4L Engine

1. Before servicing the vehicle, refer to the precautions in the beginning of this section.
2. Drain the cooling system.
3. Remove or disconnect the following:
   - Negative battery cable
   - Right inner splash shield
   - Accessory drive belts
   - Right motor mount and bracket, support the engine from below before removing the mount
   - Front cover
   - Timing belt.
   - Timing belt idler pulley
   - Camshaft sprockets
   - Timing belt rear cover
   - Alternator and bracket
   - Water pump

To install:

4. Install or connect the following:
   - Water pump with a new O–ring gasket. Tighten the bolts to 105 inch lbs. (12 Nm). Rotate the pump by hand to make sure it moves freely.
   - Timing belt rear cover
   - Camshaft sprockets. Tighten the bolts to 75 ft. lbs. (101 Nm).
   - Timing belt idler pulley. Tighten the bolt to 45 ft. lbs. (61 Nm).
   - Timing belt

- Front cover
- Alternator and bracket
- Right motor mount
- Accessory drive belts
- Right inner splash shield
- Negative battery cable
5. Fill the cooling system.
6. Start the engine and check for leaks.

3.3L and 3.8L Engines

1. Before servicing the vehicle, refer to the precautions in the beginning of this section.
2. Drain the cooling system.
3. Remove or disconnect the following:
   - Negative battery cable
   - Right inner splash shield
   - Accessory drive belt
   - Water pump pulley bolts

➡To remove the water pump it must be positioned between the pump housing and the drive hub. Refer to the accompanying illustration for further clarification.

- Water pump with the pulley loosely positioned between the hub and pump body

To install:

4. Install or connect the following:
   - New seal into the water pump housing groove
   - Water pump with the pulley loosely positioned between the hub and pump body. Tighten the bolts to 105 inch lbs. (12 Nm).
   - Water pump pulley and tighten the bolts to 250 inch lbs. (28 Nm). Rotate the pump by hand to make sure it moves freely.
   - Accessory drive belt
   - Right inner splash shield
   - Negative battery cable
5. Fill the cooling system.
6. Start the engine and check for leaks.

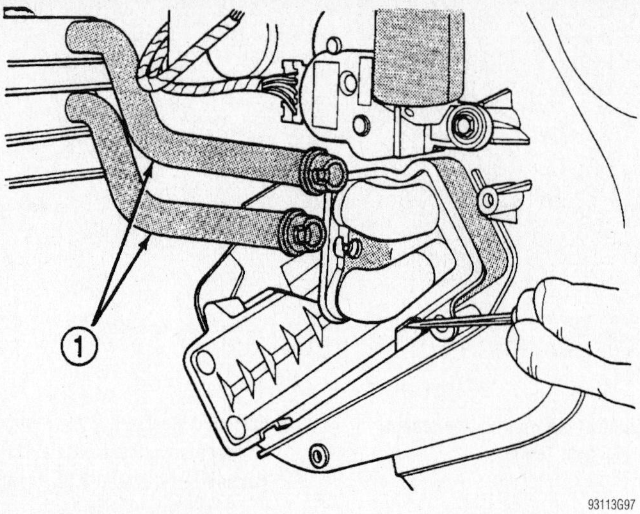

View of the front heater core assembly

Heater Core

REMOVAL & INSTALLATION

Front System

1. Before servicing the vehicle, refer to the precautions in the beginning of this section.
2. Align the wheels in the straight-ahead position.
3. Disconnect the negative battery cable.

⁂ CAUTION

Before working around the steering wheel or the instrument panel all 2 minutes to pass to allow the air bag module to discharge.

4. Drain the cooling system into a clean container for reuse.
   - Lower silencer boot, at the base of the steering shaft
   - Brake lamp switch
   - Power brake booster push rod from the pin on the brake pedal arm
   - 3 screws from the heater core shield at the left side of the HVAC housing
   - Heater core shield rearwards so the 2 tabs that position the front of the shield can be disengaged
   - Heater core shield. Position some towels under the heater core hoses.
   - Heater core plate screw
5. Push both core tubes at the same time towards the dash far enough to disengage the fittings from the heater core supply and return ports.
   - 2 screws that attach the core mounting plate to the distribution housing
6. Lift the accelerator pedal and slide the heater core past it.
7. Depress the brake pedal and remove the heater core from the heater/air conditioning housing assembly.

To install:

8. Depress the brake pedal and install the heater core into the heater/air conditioning housing assembly.

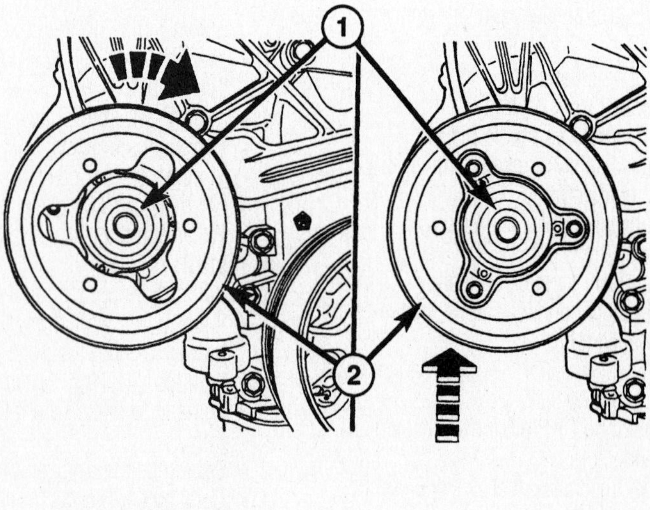

The water pump must be positioned between the pump housing and the drive hub so it can be removed—3.3L and 3.8L engines

9. Lift the accelerator pedal and slide the heater core past it.

10. Install or connect the following:

- 2 screws that attach the core plate to the distribution housing

➡The heater core tubes each have a slot that must be indexed to a location tab within each of the ports. Adjust the tube position as necessary so the sealing plate fits flush against the core supply and return ports which will indicated the ports are properly indexed.

- Heater core tubes and the sealing plate at the same time to the core supply and return ports.
- Heater core sealing plate screw.
- Heater core shield onto the distribution housing making sure the two location tabs and fully engaged
- 3 screws that attach the core shield to the left end of the HVAC housing
- Power brake booster push rod to the pin on the brake pedal arm
- Brake lamp switch
- Lower silencer boot, at the base of the steering shaft

11. Refill the cooling system.

12. Connect the negative battery cable.

13. Run the engine to normal operating temperatures; then, check the climate control operation and check for leaks.

Rear Auxiliary System

1. Disconnect the negative battery cable.

2. Partially drain the cooling system into a clean container for reuse.

3. Remove the right rear quarter trim panel by removing or disconnecting the following:

- 1st row seat
- 2nd row seat, if equipped
- Sliding door sill trim panel and the quarter trim bolster
- C-pillar and D-pillar trim panels
- 1st row seat belt anchor and the 2nd row seat anchor, if equipped
- Quarter trim-to-quarter panel screws from the bolster area
- Quarter trim's rear edge-to-bracket screws
- Quarter trim-to-quarter panel hidden clips, located rearward of the sliding door
- Quarter trim from the quarter panel
- Electrical connector from the accessory power outlet, if equipped
- Pass the 2nd row seat belt through the slot in the trim panel, if equipped
- Pass the 1st row seat belt through the slot in the trim panel
- Quarter trim panel from the vehicle

4. Remove the rear heater distribution duct from the right quarter inner panel.

5. Remove the screw attaching the back of the rear heater/AC unit to the to the right D panel.

6. Remove the screw that attaches the front of the rear heater/AC unit to the right quarter inner panel.

7. Remove the rear auxiliary heater hoses clamps from the rear heater core.

8. Remove the rear heater hoses from the rear heater core.

9. Remove the 4 latch tabs that attach the core in the heater/AC housing.

10. Carefully lift the rear heater core and tubes up and out of the rear auxiliary housing.

To install:

11. Carefully, install the rear heater core into the housing. Press firmly and evenly to engage the 4 latch tabs.

12. Install the rear heater hoses to the rear heater core.

13. Install the rear auxiliary heater hoses clamps to the rear heater core.

14. Install the screw that attaches the front of the rear heater/AC unit to the right quarter inner panel.

15. Install the screw attaching the back of the rear heater/AC unit to the to the right D panel.

16. Install the rear heater distribution duct to the right quarter inner panel.

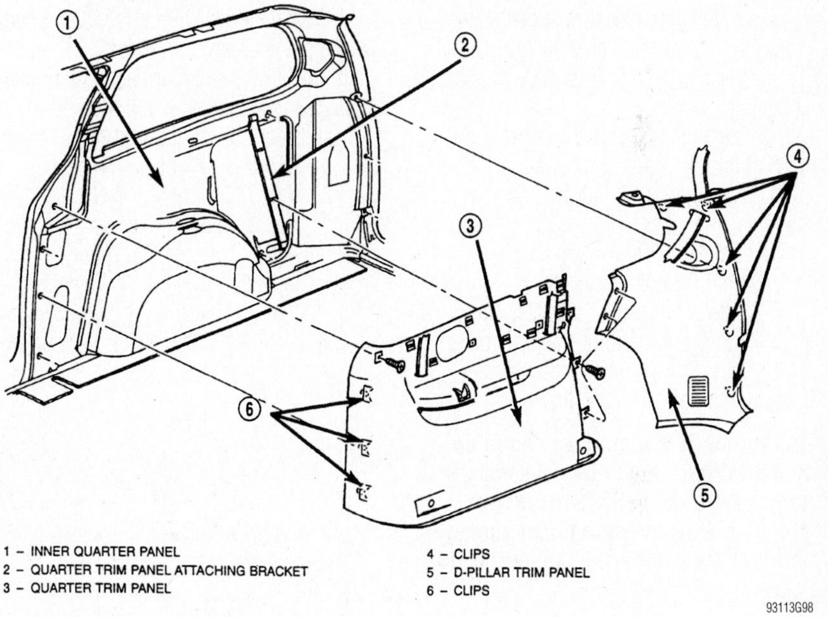

View of the right quarter and D-trim panels

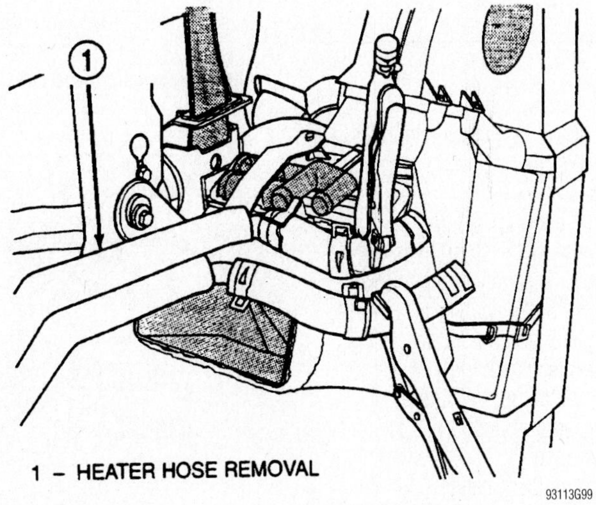

View of the rear heater core

17. Install the right rear quarter trim panel by performing the following procedure:
    a. Install the quarter trim panel to the vehicle.
    b. Pass the 1st row seat belt through the slot in the trim panel.
    c. If equipped, pass the 2nd row seat belt through the slot in the trim panel.
    d. If equipped, connect the electrical connector to the accessory power outlet.
    e. Install the quarter trim to the quarter panel.
    f. Rearward of the sliding door, connect the quarter trim-to-quarter panel hidden clips.
    g. Install the quarter trim's rear edge-to-bracket screws.
    h. At the bolster area, install the quarter trim-to-quarter panel screws.
    i. Install the 1st row seat belt anchor and the 2nd row seat anchor, if equipped.
    j. Install the C-pillar and D-pillar trim panels.
    k. Install the quarter trim bolster and the sliding door sill trim panel.
    l. If equipped, install the 2nd row seat.
    m. Install the 1st row seat.
18. Refill the cooling system.
19. Connect the negative battery cable.
20. Run the engine to normal operating temperatures; then, check the climate control operation and check for leaks.

➡If the rear heater core was emptied and not refilled, it will be necessary to thermal cycle the vehicle twice. Run the engine until the thermostat opens; then, turn the engine OFF and allow it to cool.

Cylinder Head

REMOVAL & INSTALLATION

2.4L Engine

1. Before servicing the vehicle, refer to the precautions in the beginning of this section.
2. Drain the cooling system.
3. Relieve the fuel system pressure.
4. Remove or disconnect the following:
    - Negative battery cable
    - Air cleaner and hoses
    - Upper intake manifold
    - Heater tube support bracket
    - Upper radiator hose
    - Heater hose
    - Accessory drive belts
    - Exhaust front pipe
    - Power steering pump reservoir and line support bracket
    - Ignition coil and spark plug wires
    - Camshaft Position (CMP) sensor connector
    - Fuel injector harness connectors
    - Timing belt
    - Camshaft sprockets
    - Timing belt idler pulley
    - Rear timing cover
    - Valve cover

➡Keep all valvetrain components in order for assembly.

    - Camshafts and cam followers
    - Cylinder head bolts in the reverse of the tightening sequence using several passes
    - Cylinder head and gasket

To install:

5. Examine the cylinder head bolts and replace any that have stretched.
6. Install the cylinder head and tighten the bolts in sequence, as follows:
    a. Step 1: 25 ft. lbs. (34 Nm).
    b. Step 2: 50 ft. lbs. (68 Nm).
    c. Step 3: 50 ft. lbs. (68 Nm).
    d. Step 4: Plus 90 degrees.
7. Install or connect the following:
    - Camshafts and cam followers
8. Install the valve cover as follows:
    a. Install new head cover gaskets and spark plug well seals.
    b. Apply Mopar RTV GEN II at the camshaft cap corners and the top edges of the ½ round seal.
    c. Valve cover and use new bolts making sure the single stud used to attach the upper intake manifold support bracket is located in the # 8 as shown in the illustration. Tighten the bolts in the sequence illustrated in 3 steps. First step to 40 inch lbs. (4.5 Nm). Second step to 80 inch lbs. (9 Nm) and the third step to 105 inch lbs. (12 Nm).
9. Install or connect the following:
    - Rear timing cover
    - Timing belt idler pulley. Tighten the bolt to 45 ft. lbs. (61 Nm).
    - Camshaft sprockets. Tighten the bolts to 75 ft. lbs. (101 Nm).
    - Timing belt
    - Front cover
    - CMP sensor connector
    - Fuel injector harness connectors
    - Ignition coil and spark plug wires
    - Power steering pump reservoir and line support bracket
    - Exhaust front pipe
    - Accessory drive belts
    - Upper radiator hose
    - Heater hose
    - Intake manifold and the EGR tube with new gaskets
    - Air cleaner and hoses
    - Negative battery cable

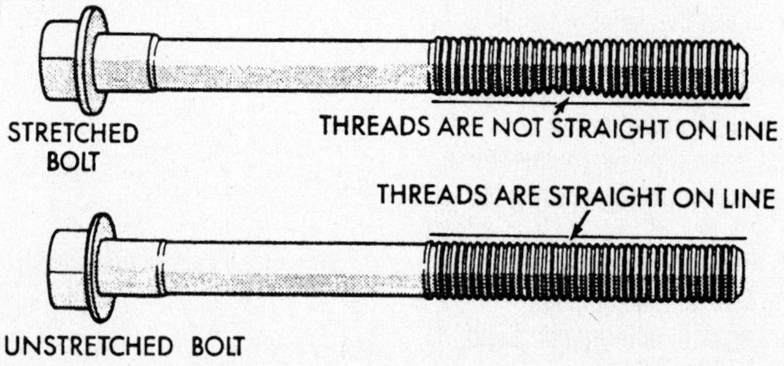

Check the cylinder head bolts for stretching—2.4L engine

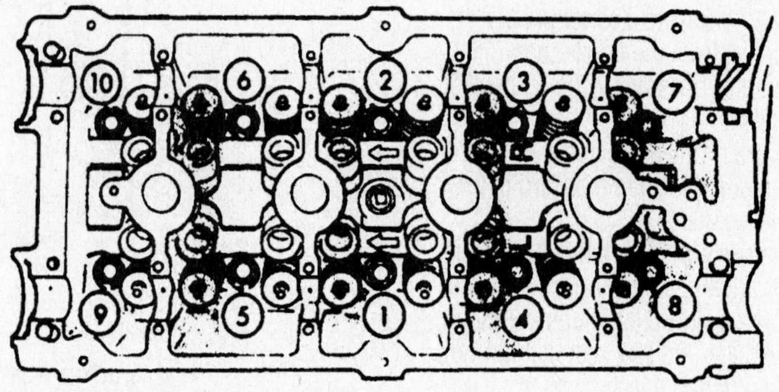

Cylinder head torque sequence—2.4L engine

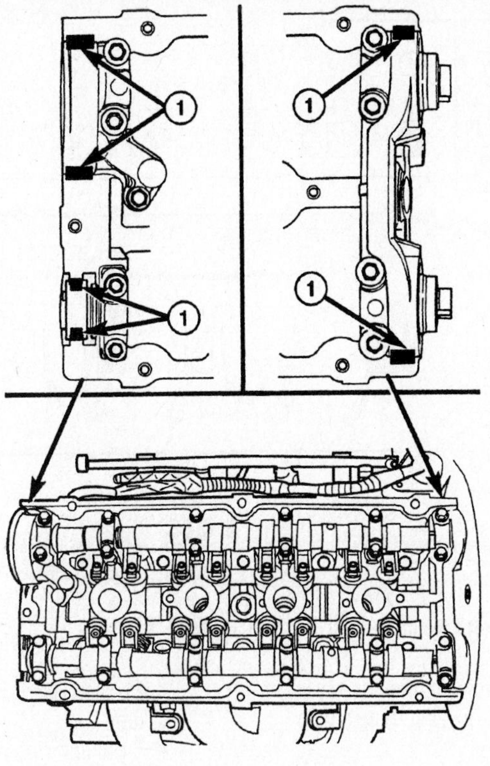

1 – SEALER LOCATION

67189-MINV-G12

Apply sealant at these locations when installing the valve cover—2.4L engines

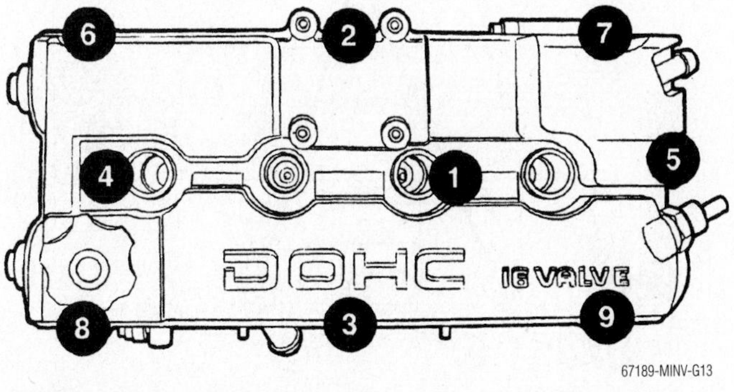

67189-MINV-G13

Valve cover torque sequence. The stud should be located at the # 8 position—2.4L engines

10. Fill the cooling system.
11. Start the engine, check for leaks and repair if necessary.

3.3L and 3.8L Engines

1. Before servicing the vehicle, refer to the precautions in the beginning of this section.
2. Drain the cooling system.
3. Relieve the fuel system pressure.
4. Remove or disconnect the following:
   - Negative battery cable
   - Wiper module if removing the right cylinder head
   - Upper and lower intake manifolds

✲✲ CAUTION

The intake manifold gaskets are made of a very thin metal and can cause injury if not properly handled.

   - Valve covers.
   - Spark plugs
   - Dipstick tube
   - Exhaust manifolds

➡Keep all valvetrain components in order for assembly

   - Rocker arm and shaft assemblies
   - Cylinder head bolts, heads and gaskets
5. Clean all gasket mating surfaces.

To install:

6. Examine the cylinder head bolts and replace any that have stretched.
7. Install the cylinder head with a new gasket. The left bank gasket has an L on it and is located at the front of the engine. The right gasket has an R stamped on it and is located at the rear of the engine. Tighten the bolts in sequence, as follows:
   a. Step 1: Tighten bolts 1–8 to 45 ft. lbs. (61 Nm).
   b. Step 2: Tighten bolts 1–8 to 65 ft. lbs. (88 Nm).
   c. Step 3: Tighten bolts 1–8 to 65 ft. lbs. (88 Nm).
   d. Step 4: Bolts 1–8 plus 90 degrees.
   e. Step 5: Tighten bolt 9 to 25 ft. lbs. (33 Nm).
8. Check that the torque on bolts 1–8 has exceeded 90 ft. lbs. (122 Nm). If not, replace the bolt.
9. Install or connect the following:
   - Pushrods
   - Rocker arm and shaft assemblies
   - Valve covers
   - Exhaust manifolds
   - Dipstick tube with a new O–ring
   - Spark plugs
   - Intake manifolds
   - Wiper module, if removed
   - Negative battery cable

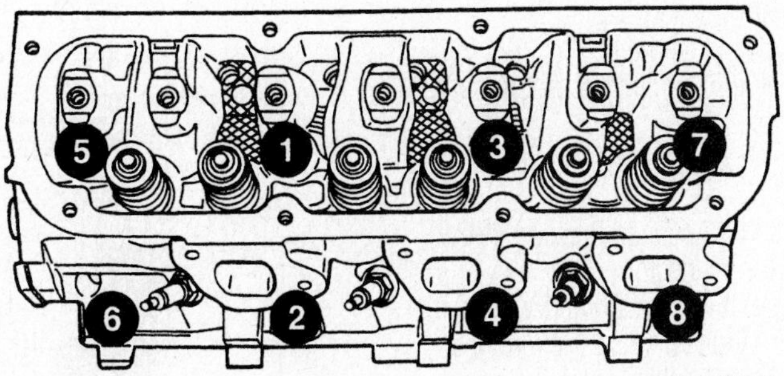

67189-MINV-G14

Cylinder head torque sequence—3.3L and 3.8L engines

10. Fill the cooling system.
11. Start the engine and check for leaks.

Rocker Arms/Shafts

REMOVAL & INSTALLATION

3.3L and 3.8L Engines

1. Before servicing the vehicle, refer to the precautions in the beginning of this section.
2. Remove or disconnect the following:
   - Negative battery cable
   - Wiper module
   - Upper intake manifold
   - Valve covers
   - Rocker arm and shaft assemblies

➡Keep all valvetrain components in order for assembly.

To install:

➡Do not rotate the crankshaft during or immediately after rocker arm installation. Wait 20 minutes for the hydraulic lash adjusters to bleed down.

3. Install or connect the following:
   - Rocker arm and shaft assemblies. Tighten the bolts to several passes to 200 inch lbs. (23 Nm).
   - Valve covers
   - Upper intake manifold
   - Wiper module
   - Negative battery cable
4. Start the engine and check for proper operation.

Intake Manifold

REMOVAL & INSTALLATION

2.4L Engine

1. Before servicing the vehicle, refer to the precautions in the beginning of this section.
2. Drain the cooling system.
3. Relieve the fuel system pressure.
4. Remove or disconnect the following:
   - Negative battery cable
   - Air cleaner and tube
   - Throttle Position (TP) sensor connector
   - Idle Air Control (IAC) valve connector
   - Manifold Absolute Pressure (MAP) sensor connector
   - Evaporative Emissions (EVAP) canister purge solenoid vacuum line
   - Positive Crankcase Ventilation (PCV) valve and hose
   - Brake booster vacuum line
   - Cruise control vacuum reservoir line
   - Accelerator cable
   - Cruise control cable
   - Exhaust Gas Recirculation (EGR) tube
   - Front and rear intake manifold support brackets
   - Engine oil dipstick tube

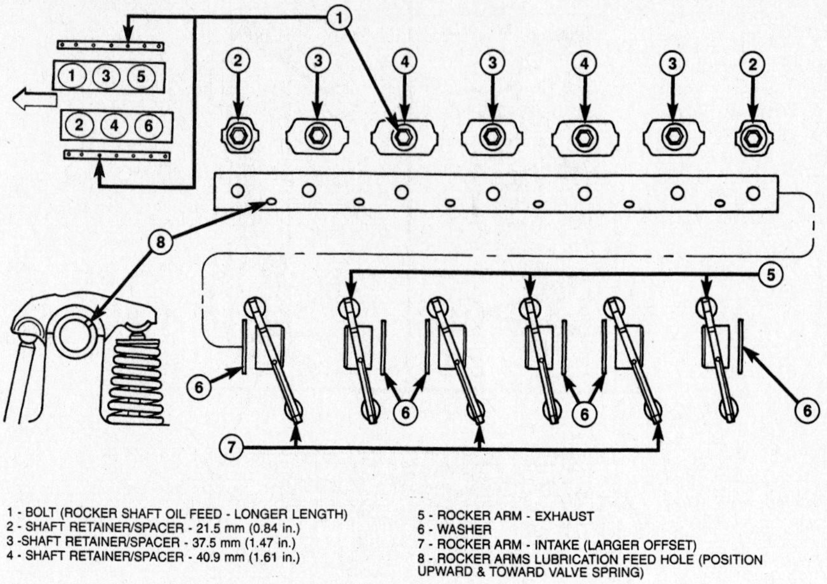

Rocker arm and shaft assembly—3.3L and 3.8L engines

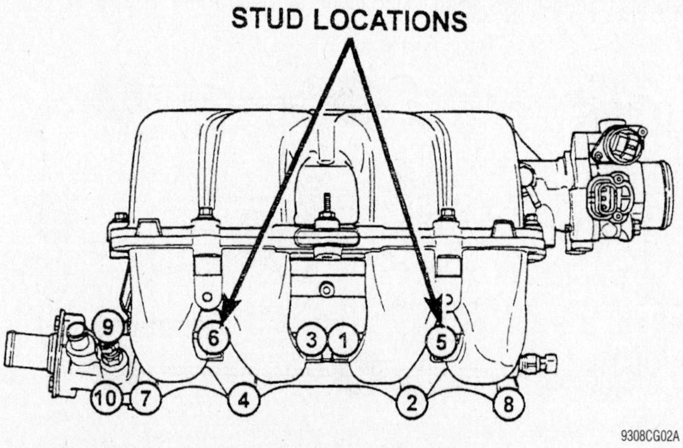

Lower intake manifold torque sequence—2002–04 2.4L engine

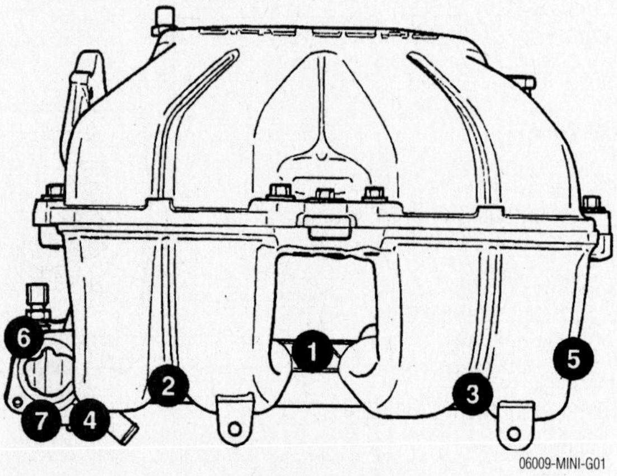

Lower intake manifold torque sequence—2005 2.4L engine

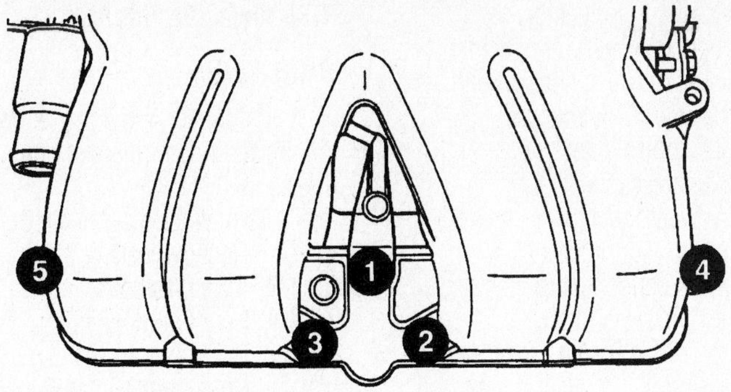

06009-MINI-G02

Upper intake manifold torque sequence—2005 2.4L engine

- Upper intake manifold
- Fuel line
- Upper radiator hose
- Heater hose
- Engine Coolant Temperature (ECT) sensor connector
- Lower manifold support upper bolts and lower bolt
- Fuel injector harness connectors
- Power steering reservoir bolts and position aside without disconnecting the lines
- Lower intake manifold bolts and the manifold

To install:

5. Install or connect the following:
   - Lower intake manifold using a new gasket. Tighten the bolts in several steps to 250 inch lbs. (28 Nm).
   - Fuel injector harness connectors
   - Intake manifold Y-bracket. Tighten the block bolts to 40 ft. lbs. (54 Nm) and the intake manifold bolts to 250 inch (28 Nm).
   - Power steering reservoir
   - Fuel line
   - ECT sensor connector
   - Heater hose
   - Upper radiator hose
6. Apply a 0.060 inch (1.5mm) bead of gasket maker to the perimeter of the lower intake manifold runner openings.
   - Upper intake manifold. Tighten the bolts in several passes to 250 inch lbs. (28 Nm).
   - Engine oil dipstick tube
   - Upper bolt on the manifold-to-front support bracket to 250 inch lbs. (28 Nm)
   - EGR tube
   - Cruise control cable
   - Accelerator cable
   - Cruise control vacuum reservoir line
   - Brake booster vacuum line
   - PCV valve and hose
   - EVAP canister purge solenoid vacuum line
   - MAP sensor connector
   - IAC valve connector
   - TP sensor connector
   - Air cleaner and tube
   - Negative battery cable
7. Fill the cooling system.
8. Start the engine, check for leaks and repair if necessary.

3.3L and 3.8L Engines

1. Before servicing the vehicle, refer to the precautions in the beginning of this section.
2. Drain the cooling system.
3. Relieve the fuel system pressure.
4. Remove or disconnect the following:
   - Negative battery cable
   - Air cleaner and tube
   - Idle Air Control (IAC) valve connector
   - Accelerator cable
   - Cruise control cable
   - Make up air hose support clip from the throttle cable bracket
   - Automatic Idle Speed (IAS) and Throttle Position (TP) sensor connectors
   - Manifold Absolute Pressure (MAP) sensor connector
   - Exhaust Gas Recirculation (EGR) transducer connector, if equipped
   - EGR tube, if equipped

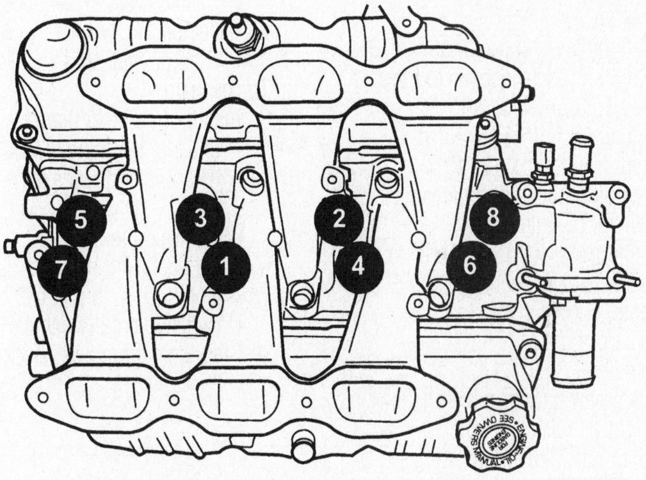

67189-MINV-G29

Lower intake manifold torque sequence—3.3L and 3.8L engines

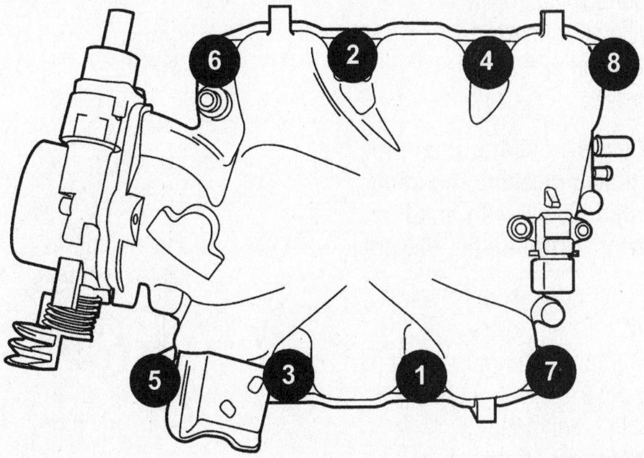

67189-MINV-G28

Upper intake manifold torque sequence—3.3L and 3.8L engines

- Vapor purge vacuum hose from the throttle body
- Positive Crankcase Ventilation (PCV) hose
- Power steering reservoir bolts and loosen the nut, lift the reservoir up to disengage the lower mount from the stud and position aside with the lines still attached.
- Brake booster and leak detection pump hoses
- Upper intake manifold
- Fuel line
- Ignition coil and bracket
- Heater supply hose and Engine Coolant temperature (ECT) sensor
- Fuel injector harness connectors
- Fuel supply manifold
- Upper radiator hose
- Lower intake manifold

To install:

5. Install a ¼ inch bead of gasket maker onto each of the 4 manifold-to-cylinder head gasket corners.
6. Install the lower intake manifold gasket and tighten the seal retainer screws to 105 inch lbs. (12 Nm)
7. Install the lower intake manifold and tighten the bolts in sequence, as follows:
   a. Step 1: 10 inch lbs. (1 Nm).
   b. Step 2: 17 ft. lbs. (22 Nm).
   c. Step 3: 17 ft. lbs. (22 Nm).
8. Install or connect the following:
   - Fuel supply manifold
   - Fuel injector harness connectors
   - ECT sensor connector
   - Heater supply hose
   - Radiator hose
   - Fuel line
9. Place a new gasket in the channel and press lightly in place.
   - Upper intake manifold. Apply Mopar thread and lock adhesive (medium strength) to each upper intake manifold bolt. Tighten the bolts to 105 inch lbs. (12 Nm) in the sequence illustrated.
   - MAP sensor connector
   - Brake booster and leak detection pump hoses

➡**The special screws used to retain the EGR tube and power steering reservoir must be tightened slowly with hand tools only to avoid stripping the threads.**

- Power steering reservoir
- PCV hose
- Vapor purge vacuum hose from the throttle body
- EGR tube, if equipped
- EGR transducer connector, if equipped
- MAP sensor connector
- IAS and TP sensor connectors
- Make up air hose support clip to the throttle cable bracket
- Cruise control cable
- Accelerator cable
- IAC valve connector
- Air cleaner and tube
- Negative battery cable

10. Fill the cooling system.
11. Start the engine, check for leaks and repair if necessary.

Exhaust Manifold

REMOVAL & INSTALLATION

2.4L Engine

1. Before servicing the vehicle, refer to the precautions in the beginning of this section.
2. Remove or disconnect the following:
   - Negative battery cable
   - Exhaust front pipe
   - Heated Oxygen Sensor (HO_2S) connector
   - Exhaust manifold

To install:

3. Install or connect the following:
   - Exhaust manifold. Torque the fasteners in sequence to 15 ft. lbs. (20 Nm).
   - HO_2S connector
   - Exhaust front pipe. Torque the fasteners to 27 ft. lbs. (37 Nm).
   - Negative battery cable
4. Start the engine and check for leaks.

3.3L and 3.8L Engines

RIGHT SIDE

1. Before servicing the vehicle, refer to the precautions in the beginning of this section.
2. Remove or disconnect the following:
   - Negative battery cable
   - Wiper module
   - Spark plug wires
   - Crossover pipe
   - Upstream Oxygen Sensor (O_2S) connectors
   - Heat shield
   - Accessory drive belt
   - Power steering pump support strut lower bolt
   - Downstream O_2S connectors
   - Catalytic converter pipe from the manifold
   - Power steering support strut upper bolt and strut
   - Exhaust manifold bolt
   - Exhaust manifold and gasket

To install:

3. Position the exhaust manifold on the cylinder head and install the bolts to the center runner (cyl # 3), and temporarily tighten to 25 inch lbs. (2.8 Nm).

➡**Examine the crossover pipe bolts for damage caused due to heat or corrosion and replace using new OEM bolts if found to be defective.**

4. Install a new gasket, attach the crossover pipe to the manifold and tighten the bolts to 30 ft. lbs. (41 Nm).
5. Install or connect the following:

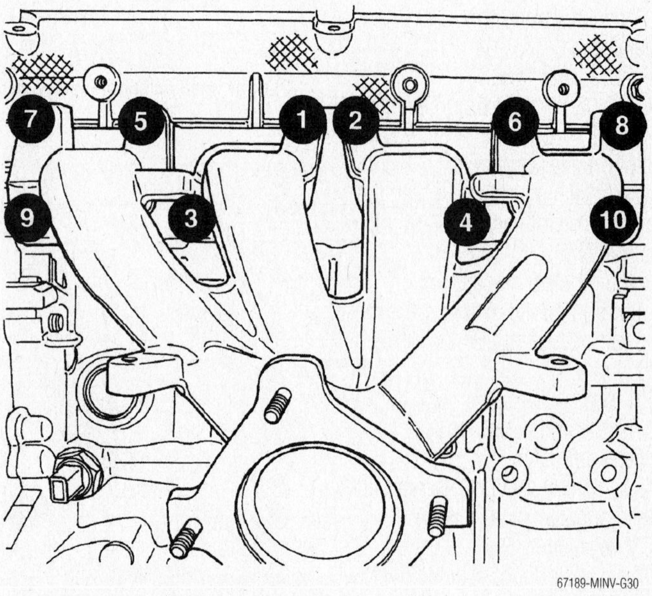

Exhaust manifold torque sequence—2.4L engine

- Remaining manifold bolts. Torque the fasteners to 17 ft. lbs. (23 Nm).
- Power steering support strut and upper bolt
- Heat shield
- Upstream O_2S connectors
- Catalytic converter pipe to the manifold
- Downstream O_2S connectors
- Power steering pump support strut lower bolt
- Accessory drive belt
- Spark plug wires
- Wiper module
- Negative battery cable

6. Start the engine and check for leaks.

LEFT SIDE

1. Before servicing the vehicle, refer to the precautions in the beginning of this section.
2. Remove or disconnect the following:
   - Negative battery cable
   - Crossover pipe
   - Spark plug wires
   - Heat shield
   - Exhaust manifold bolt
   - Exhaust manifold and gasket

To install:

3. Position the exhaust manifold on the cylinder head and install the bolts to the center runner (cyl # 4), and temporarily tighten to 25 inch lbs. (2.8 Nm).

➡Examine the crossover pipe bolts for damage caused due to heat or corrosion and replace using new OEM bolts if found to be defective.

4. Install a new gasket, attach the crossover pipe to the manifold and tighten the bolts to 30 ft. lbs. (41 Nm).
5. Install or connect the following:
   - Remaining manifold bolts. Torque the fasteners to 17 ft. lbs. (23 Nm).
   - Heat shield
   - Negative battery cable
6. Start the engine and check for leaks.

Front Crankshaft Seal

REMOVAL & INSTALLATION

2.4L Engine

1. Before servicing the vehicle, refer to the precautions in the beginning of this section.
2. Remove or disconnect the following:
   - Negative battery cable
   - Accessory drive belts
   - Crankshaft pulley
   - Front cover
   - Timing belt
   - Crankshaft timing sprocket using puller 6793 and insert C–4685–C2,

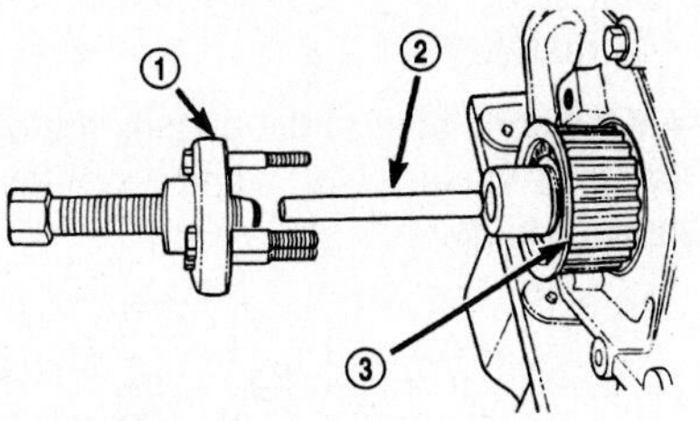

1 - SPECIAL TOOL 6793
2 - SPECIAL TOOL C-4685-C2
3 - CRANKSHAFT SPROCKET

67189-MINV-G31

Remove the crankshaft timing sprocket using puller 6793 and insert C–4685–C2—2.4L engine

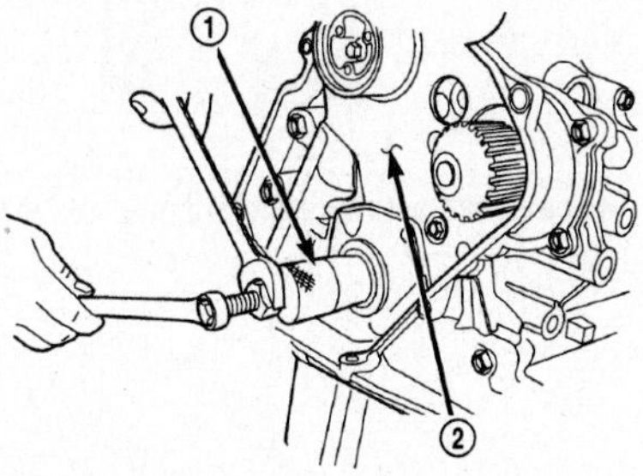

1 - SPECIAL TOOL 6771
2 - REAR TIMING BELT COVER

67189-MINV-G32

Remove the front crankshaft seal using tool 6771—2.4L engine

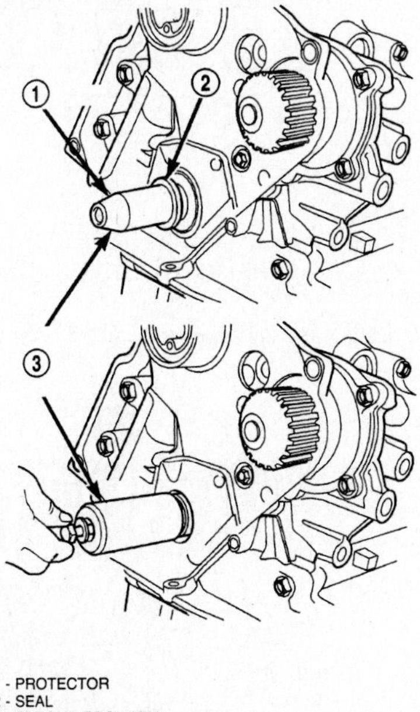

1 - PROTECTOR
2 - SEAL
3 - SPECIAL TOOL 6780

67189-MINV-G33

Install the front crankshaft seal using tool 6780—2.4L engine

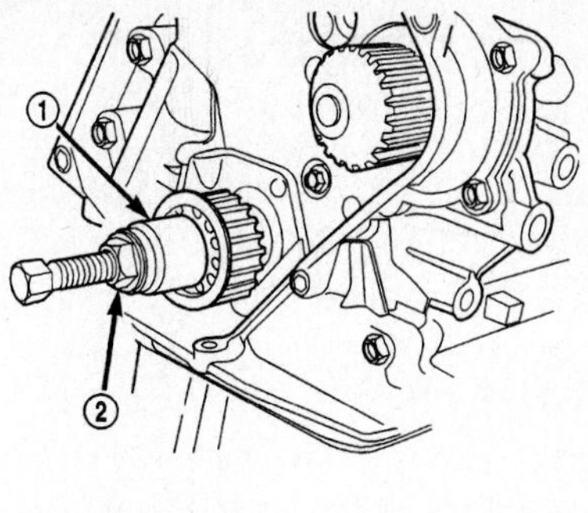

1 - SPECIAL TOOL 6792
2 - TIGHTEN NUT TO INSTALL

67189-MINV-G34

Install the crankshaft timing sprocket using tool 6792—2.4L engine

being careful not to damage the seal bore

   - Front crankshaft seal using tool 6771

To install:

3. Install or connect the following:
   - Front crankshaft seal using tool 6780. Place the seal in position, with the seal spring towards the engine and install so that it is flush with the cover.
   - Crankshaft timing sprocket using tool 6792
   - Timing belt
   - Front cover
   - Crankshaft pulley. Torque the bolt to 105 ft. lbs. (142 Nm).
   - Accessory drive belts
   - Negative battery cable
4. Start the engine and check for leaks.

3.3 and 3.8L Engines

1. Before servicing the vehicle, refer to the precautions in the beginning of this section.
2. Remove or disconnect the following:
   - Negative battery cable
   - Right wheel and splash shield
   - Accessory drive belts
   - Crankshaft pulley

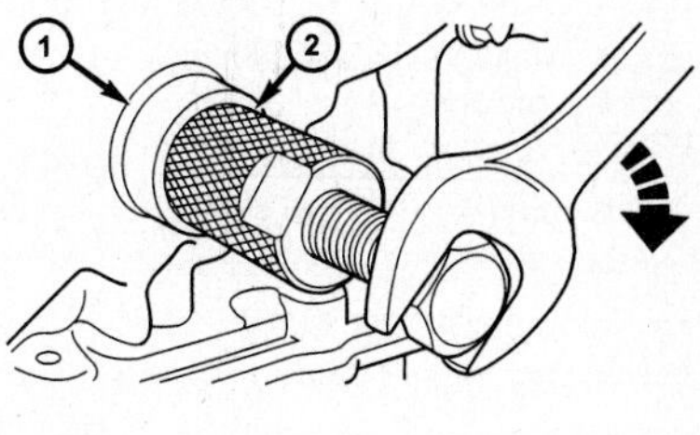

1 - SEAL
2 - SPECIAL TOOL 6341A

67189-MINV-G38

Removing the front seal using tool 6341A—3.3L and 3.8L engines

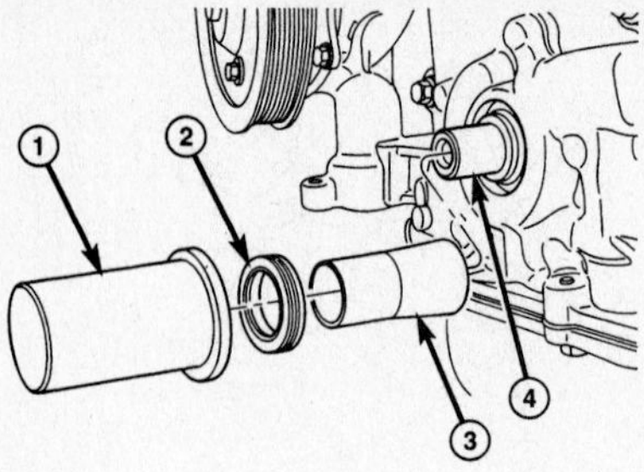

1 - SPECIAL TOOL C-4992-1
2 - SEAL
3 - SPECIAL TOOL C-4992-2
4 - CRANKSHAFT

67189-MINV-G35

Removing the front seal using tools C–4992–2 and C–42992–1—3.3L and 3.8L engines

3. Place tool 6341A on the crankshaft nose and screw the tool into the seal until it is engaged being careful not to damage the crankshaft seal surface cover.
4. Turn the forcing screw on the tool to remove the seal

To install:

5. Position guide C–4992–2 on the crankshaft nose.
6. Position a new seal over the guide with the seal spring facing the engine cover.
7. Install the seal using tool C–42992–1 until the seal is flush with the cover and remove the tools.
8. Install or connect the following:
   - Crankshaft pulley and tighten the bolt to 40 ft. lbs. (54 Nm)
   - Accessory drive belts
   - Right wheel and splash shield
   - Negative battery cable

Camshaft and Valve Lifters

REMOVAL & INSTALLATION

2.4L Engine

1. Before servicing the vehicle, refer to the precautions in the beginning of this section.
2. Remove or disconnect the following:
   - Negative battery cable
   - Valve cover
   - Camshaft Position (CMP) sensor and camshaft target magnet
   - Timing belt
   - Camshaft sprockets
   - Rear timing cover

➡The bearing caps are marked for location, Remove the outside caps first. The caps are not interchangeable the intake cam # 6 thrust bearing face spacing is wider.

3. Make sure to identify the caps before removal, then loosen the caps in the sequence illustrated one at a time and remove the camshafts and cam followers. Always make sure to remove the outside bearing caps first.

To install:

➡Make sure none of the pistons are at Top Dead Center (TDC) when installing the camshafts.

4. Install or connect the following:
   - Lubricate then install the cam followers in their original positions
   - Lubricate then install the camshafts
5. Install right and left camshaft bearing caps 2–5 and right # 6. Tighten the M6 fasteners to 105 inch lbs. (12 Nm) in the sequence illustrated.
6. Apply a 0.060 (1–1.5mm) bead of gasket maker to the # 1 and # 6 bearing

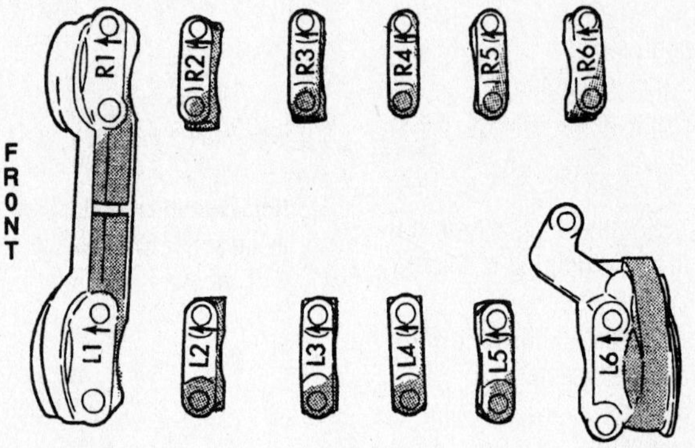

67189-MINV-G15

The camshaft bearing caps are marked for location—2.4L engines

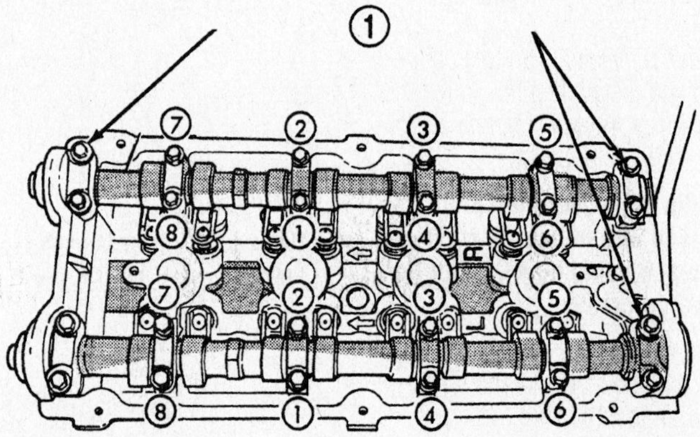

1 - REMOVE OUTSIDE BEARING CAPS FIRST

67189-MINV-G16

Camshaft bearing caps removal sequence—2.4L engines

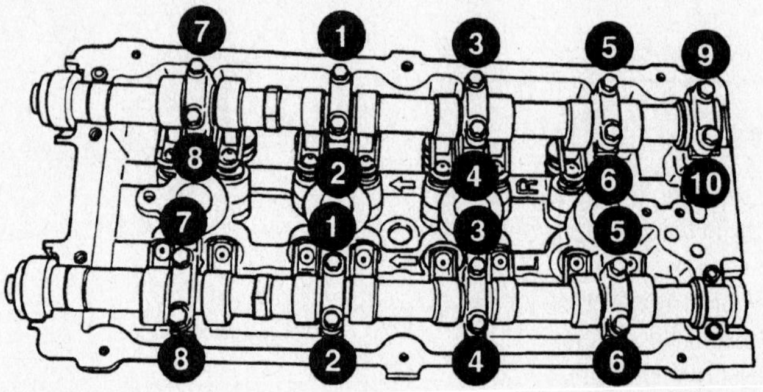

67189-MINV-G17

Apply a 0.060 (1–1.5mm) bead of gasket maker to the # 1 and # 6 bearing caps—2.4L engines

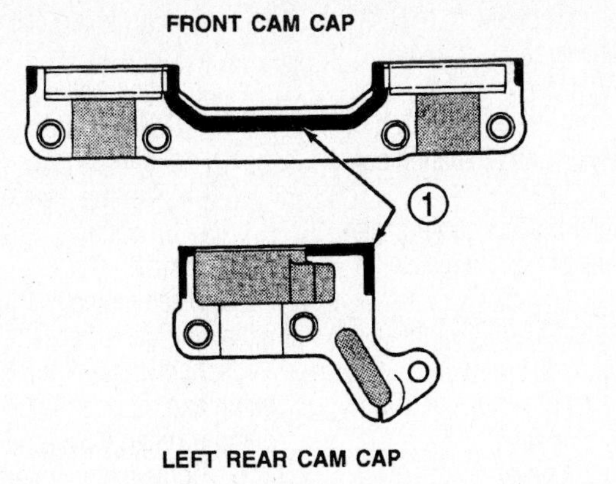

Camshaft bearing caps tightening sequence—2.4L engines

caps. Install the bearing caps in their original positions and tighten the M8 fasteners to 250 inch lbs. (28 Nm)

- Camshaft oil seals
- Camshaft target magnet and CMP sensor
- Valve cover
- Rear timing cover
- Timing belt idler pulley
- Camshaft sprockets. Tighten the bolts to 75 ft. lbs. (101 Nm).
- Timing belt
- Front cover
- Negative battery cable

7. Start the engine and check for proper operation.

3.3L and 3.8L Engines

➡Keep all valvetrain components in order for assembly.

1. Before servicing the vehicle, refer to the precautions in the beginning of this section.
2. Remove the engine from the vehicle and mount it on a stand.
3. Remove or disconnect the following:
   - Valve covers
   - Rocker arm and shaft assemblies
   - Pushrods
   - Intake manifold
   - Cylinder heads
   - Yoke retainer
   - Aligning yokes
   - Hydraulic lifters
   - Oil pan
   - Oil pump pickup tube
   - Crankshaft pulley
   - Front cover
   - Timing chain and sprockets
   - Camshaft thrust plate
   - Camshaft

To install:

4. Install or connect the following:
   - Camshaft
   - Camshaft thrust plate. Torque the bolts to 105 inch lbs. (12 Nm).
   - Timing chain and sprockets. Torque the camshaft sprocket bolt to 40 ft. lbs. (54 Nm).
   - Front cover. Torque the bolts to 20 ft. lbs. (27 Nm).
   - Crankshaft pulley. Torque the bolt to 40 ft. lbs. (54 Nm).
   - Oil pump pickup tube
   - Oil pan
   - Hydraulic lifters
   - Aligning yokes
   - Yoke retainer. Torque the bolts to 105 inch lbs. (12 Nm).
   - Cylinder heads
   - Intake manifold
   - Pushrods
   - Rocker arm and shaft assemblies
   - Valve covers
5. Install the engine to the vehicle.

Valve Lash

ADJUSTMENT

All engines covered in this section are equipped with hydraulic lash adjusters. No adjustment is necessary.

Starter Motor

REMOVAL & INSTALLATION

1. Before servicing the vehicle, refer to the precautions in the beginning of this section.
2. Remove or disconnect the following:
   - Negative battery cable
   - Starter harness connectors
   - Starter motor

To install:

3. Install or connect the following:
   - Starter motor. Torque the bolts to 35 ft. lbs. (47 Nm).
   - Starter harness connectors. Torque the battery cable nut to 100 inch lbs. (11 Nm).
   - Negative battery cable

Oil Pan

REMOVAL & INSTALLATION

2.4L Engine

1. Before servicing the vehicle, refer to the precautions in the beginning of this section.

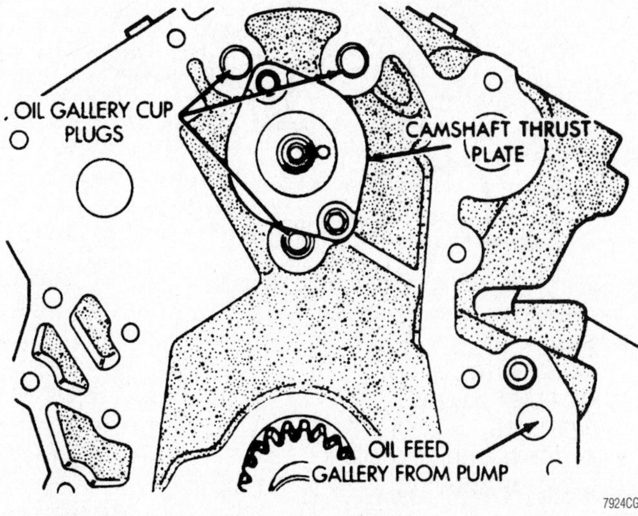

Remove the thrust plate and withdraw the camshaft from the engine—3.3L and 3.8L engines

2. Drain the engine oil.
3. Remove or disconnect the following:
   - Negative battery cable
   - Structural collar
   - A/C compressor bracket to oil pan bolt
   - Oil pan
4. Clean the oil pan and all gasket mating surfaces.

To install:

5. Install or connect the following:
   - Oil pan gasket to the block after applying engine RTV at the oil pump parting line
   - Oil pan. Torque the bolts to 105 inch lbs. (12 Nm).
   - Structural collar
   - Negative battery cable
6. Fill the crankcase to the correct level.
7. Start the engine, check for leaks and repair if necessary.

3.3L and 3.8L Engines

1. Before servicing the vehicle, refer to the precautions in the beginning of this section.
2. Drain the engine oil.
3. Remove or disconnect the following:
   - Negative battery cable
   - Engine oil dipstick
   - Drive belt splash shield
   - Strut to transaxle attaching bolt and loosen the strut to engine attaching bolts
   - Transaxle case cover
   - Oil pan fasteners
   - Oil pan and gasket

To install:

4. Clean the oil pan and all mating surfaces.
5. Apply a ⅛ inch bead of gasket material at the parting line of the chain case cover and the real seal retainer.
6. Install or connect the following:
   - New gasket on the oil pan
   - Oil pan. Torque the bolts to 105 inch lbs. (12 Nm).
   - Transaxle case cover
   - All bending brace bolts
   - Drive belt splash shield
   - Engine oil dipstick
   - Negative battery cable
7. Fill the crankcase to the correct level.
8. Start the engine, check for leaks and repair if necessary.

Oil Pump

REMOVAL & INSTALLATION

2.4L Engine

1. Before servicing the vehicle, refer to the precautions in the beginning of this section.
2. Drain the engine oil.
3. Remove or disconnect the following:
   - Negative battery cable
   - Timing belt.
   - Rear timing belt cover
   - Oil pan
   - Crankshaft sprocket using puller 6793 and insert C–4685–C2
   - Crankshaft key
   - Oil pump pickup tube
   - Oil pump

To install:

4. Clean all mating surfaces of any remaining gasket material.
5. Apply gasket maker material to the oil pump and replace the O-ring in the oil pump discharge passage.
6. Prime the pump with oil prior to installation.
7. Install or connect the following:
   - Oil pump, align the flats on the rotor with the flats on the crankshaft. Torque the bolts to 21 ft. lbs. (28 Nm).

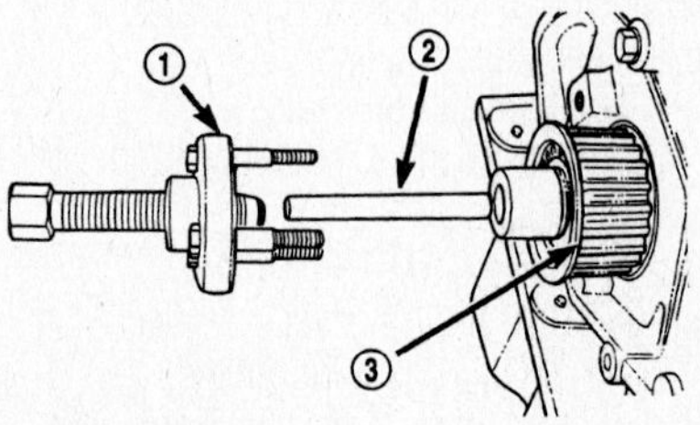

Remove the crankshaft timing sprocket using puller 6793 and insert C–4685–C2—2.4L engine

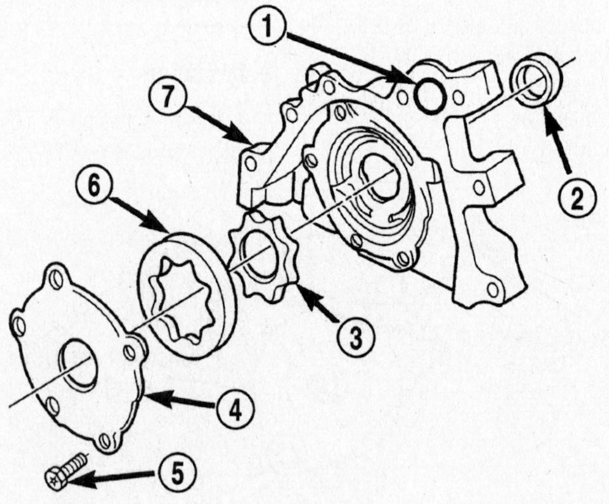

Exploded view of the oil pump—2.4L engine

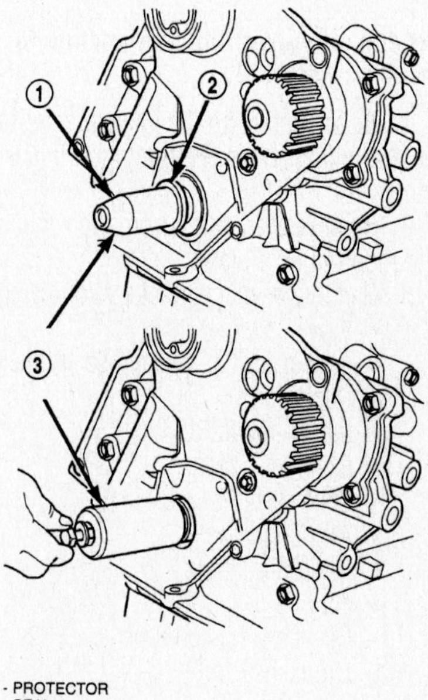

Install the front crankshaft seal using tool 6780—2.4L engine

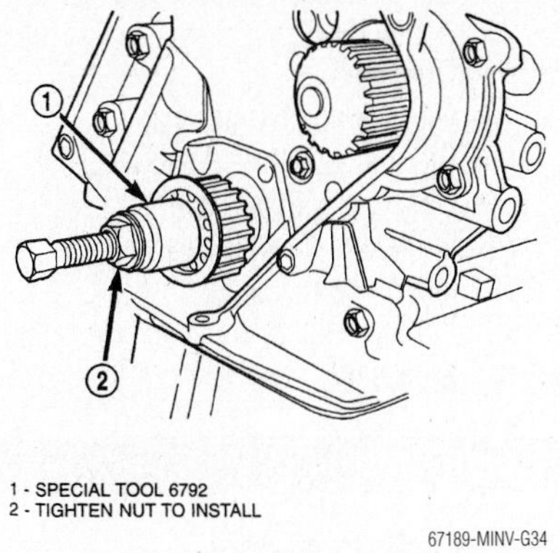

Install the crankshaft timing sprocket using tool 6792—2.4L engine

➡The front crankshaft seal must be out of the pump to align it or the pump may be damaged.

- Front crankshaft seal using tool 6780
- Crankshaft key
- Crankshaft sprocket using tool 6792
- Oil pump pickup tube. Torque the bolt to 21 ft. lbs. (28 Nm).
- Oil pan
- Rear timing belt cover
- Timing belt
- Negative battery cable

8. Fill the engine with clean oil.
9. Start the engine, check for leaks and repair if necessary.

3.3L and 3.8L Engines

1. Before servicing the vehicle, refer to the precautions in the beginning of this section.
2. Drain the cooling system.
3. Drain the engine oil.
4. Remove or disconnect the following:
   - Negative battery cable
   - Oil pan
   - Timing chain cover
   - Oil pump from the case cover

To install:

5. Install or connect the following:
   - Oil pump to the front cover. Torque the cover screws to 105 inch lbs. (12 Nm).
   - Front cover. Torque the bolts to 20 ft. lbs. 927 Nm).
   - Oil pan
   - Negative battery cable
6. Fill the engine with clean oil.
7. Fill the cooling system.
8. Start the engine, check for leaks and repair if necessary.

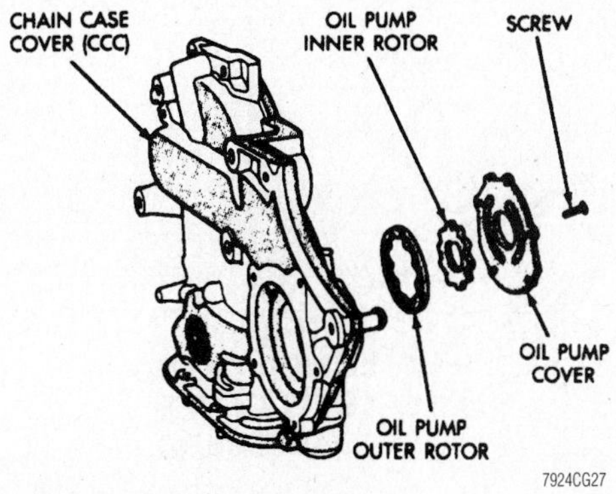

Exploded view of the oil pump assembly—3.3L and 3.8L engines

Rear Main Seal

REMOVAL & INSTALLATION

1. Before servicing the vehicle, refer to the precautions at the beginning of this section.
2. Remove or disconnect the following:
   - Negative battery cable
   - Transaxle

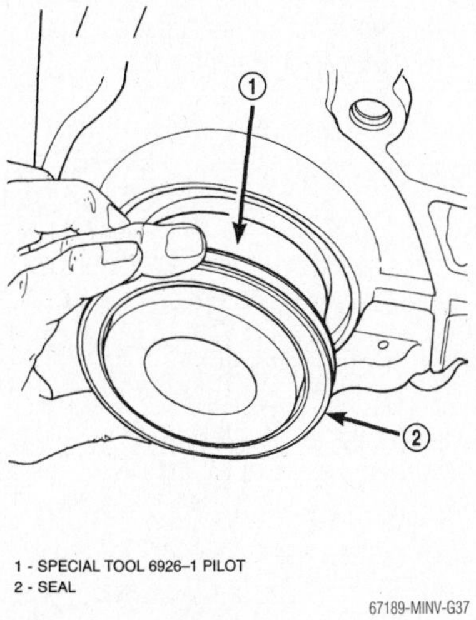

Install the rear main seal flush with the cylinder block surface

   - Flexplate
   - Rear main seal

To install:

3. Place seal guide tool 6926–1 on the crankshaft and place the seal over the tool. Make sure the lip of the seal is facing towards the crankcase.
4. Install or connect the following:
   - Rear main seal flush with the cylinder block surface using the tools illustrated.
   - Flexplate
   - Transaxle
   - Negative battery cable
5. Run the engine and check for leaks.

Timing Chain, Sprockets, Front Cover and Seal

REMOVAL & INSTALLATION

3.3L and 3.8L Engines

1. Before servicing the vehicle, refer to the precautions in the beginning of this section.
2. Drain the cooling system.
3. Drain the engine oil.
4. Remove the timing chain cover as follows:
   - Negative battery cable
   - Right wheel and splash shield
   - Oil pan and pick up tube
   - Drive belt
   - A/C compressor and set aside with the lines still attached
   - Crankshaft damper
   - Lower radiator hose
   - Camshaft Position (CMP), if necessary
   - Heater hose from the timing cover or water pump inlet tub if equipped with an engine oil cooler
   - Right side engine mount
   - Idler pulley from the bracket
   - Engine mount bracket
   - Water pump
   - Power steering support strut-to-cover bolt
   - Timing chain cover bolts and the cover
5. Rotate the engine so that the timing marks are aligned.
6. Remove or disconnect the following:
   - Camshaft sprocket attaching bolt
   - Timing chain and camshaft sprocket
   - Crankshaft sprocket with special tools 8539, 5048–6 and 5048–1

To install:

7. Rotate the engine so the timing arrow is at the 12 o'clock position.

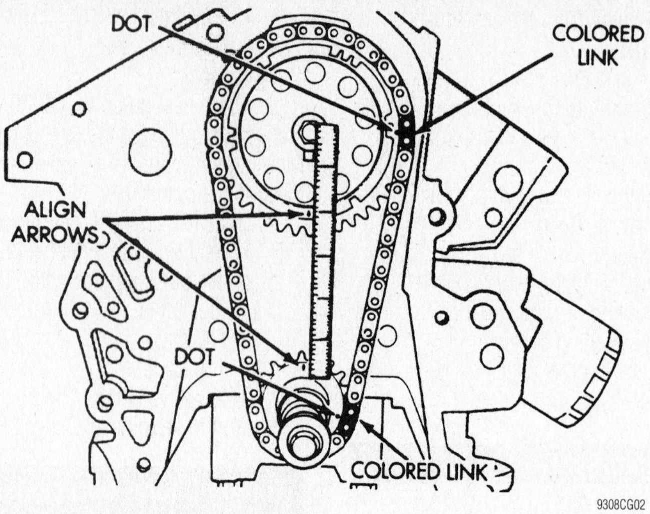

Timing mark alignment—3.3L and 3.8L engines

8. Lubricate the chain and sprockets with clean oil.

9. Hold the camshaft sprocket and chain and place the chain around the sprocket aligning the plated link with the dot on the sprocket. Position the timing arrow at the 6 o'clock position.

10. Place the chain around the crankshaft sprocket with the plated link lined up with the dot on the sprocket and install the camshaft sprocket.

11. Use a straight edge to check the timing alignment marks.

12. Align the timing chain colored links with the dots on the timing sprockets.

13. Tighten the camshaft sprocket bolt to 40 ft. lbs. (54 Nm).

14. Rotate the crankshaft 2 revolutions and verify the proper timing chain alignment, if not remove the components and reinstall as described above.

15. Clean the timing cover mating surfaces.

16. Install or the front cover as follows:
- New gasket on the cover making sure the lower edge of the gasket is flush to 0.020 inch (0.5mm) past the lower edge of the cover. Rotate the crankshaft so the oil pump drive flats are in a vertical position.
- Position the oil pump inner rotor so the mating flats are in the same position as the crankshaft drive flats or damage may occur
- Timing chain cover and bolts. Tighten the M8 bolts to 20 ft. lbs. (27 Nm) and the M10 bolts to 40 ft. lbs. (54 Nm).
- Crankshaft front oil seal
- Water pump
- Power steering support strut-to-cover bolt
- Crankshaft damper
- Engine mount bracket. Tighten the M10 bolts to 40 ft. lbs. (54 Nm) and the M8 bolts to 21 ft. lbs. (28 Nm).
- Idler pulley on the bracket
- Right side engine mount
- CMP sensor, if removed
- Heater hose to the timing cover or water pump inlet tub if equipped with an engine oil cooler
- Lower radiator hose
- A/C compressor
- Drive belt
- Oil pick up tube and pan
- Right splash shield and wheel
- Negative battery cable

17. Fill the engine with clean oil.

18. Fill the cooling system.

19. Start the engine, check for leaks and repair if necessary.

Timing Belt

REMOVAL & INSTALLATION

2.4L Engine

➡You may need DRB scan tool to perform the crankshaft and camshaft relearn alignment procedure.

1. Remove or disconnect the following:
- Negative battery cable
- Right front wheel and inner splash shield
- Accessory drive belts
- Crankshaft damper
- A/C compressor/alternator tensioner and pulley assembly
- Lower front timing cover

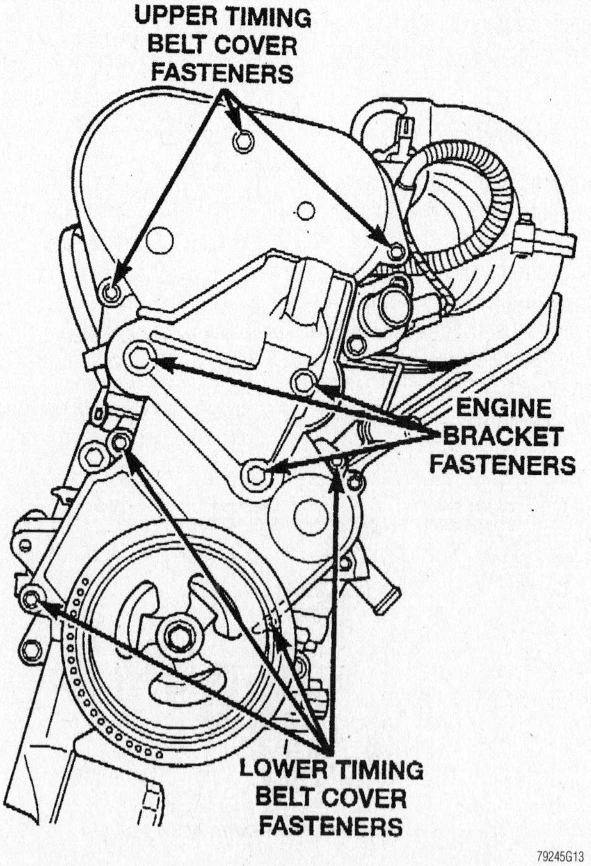

Timing cover and engine mounting bracket bolt locations—2.4L engine

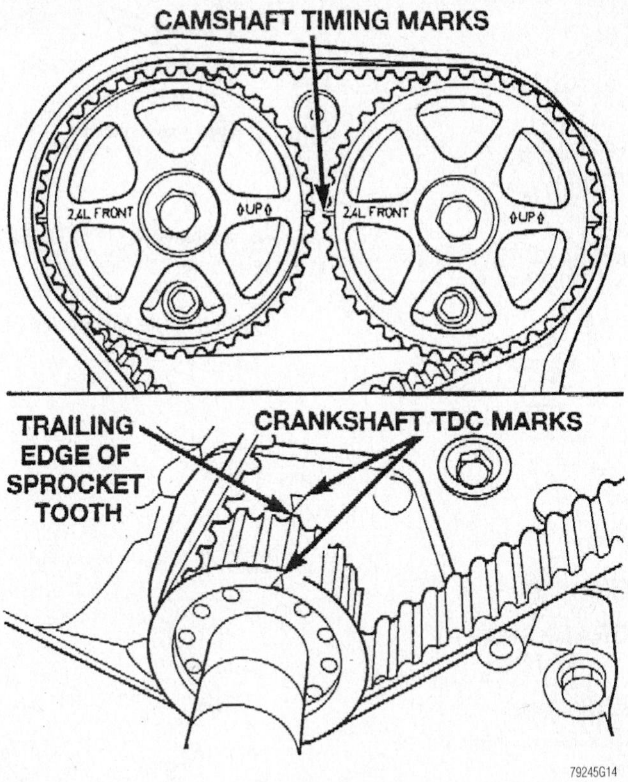

Camshaft and crankshaft alignment marks—2.4L engine

- Upper front timing cover
- Right engine mount
- Engine mount bracket

➡**This is an interference engine. Do not rotate the crankshaft or the camshafts after the timing belt has been removed. Damage to the valve components may occur. Before removing the timing belt, always align the timing marks.**

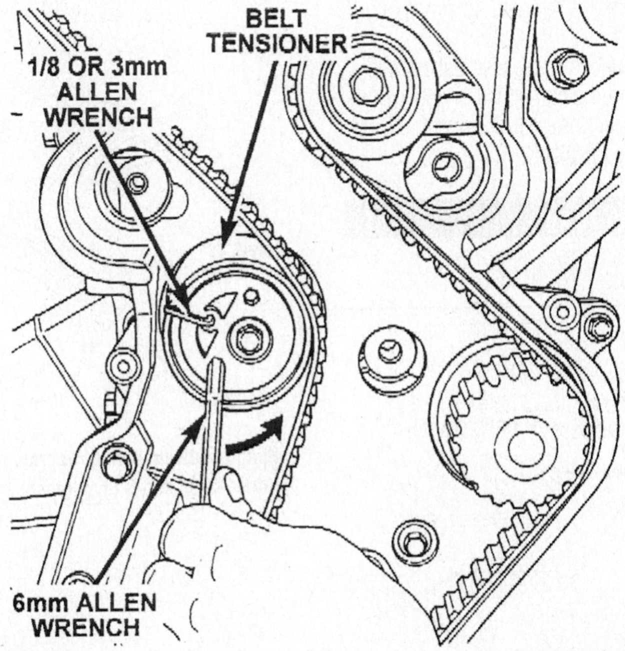

To lock the timing belt tensioner, be sure to fully insert the smaller Allen wrench into the tensioner as shown—2.4L engine

2. Rotate the crankshaft until the Top Dead Center (TDC) mark on the oil pump housing aligns with the TDC mark on the camshaft sprocket (located on the trailing edge of the tooth).
3. Loosen the timing belt tensioner lock bolt.
4. Insert a 6mm Allen wrench into the hexagon opening located on the top plate of the belt tensioner pulley. Rotate the top plate clockwise until there is enough slack to remove the belt.
5. If necessary, remove the camshaft timing belt sprockets.
6. If necessary, remove the crankshaft timing belt sprocket using removal tool No. 6793, or equivalent.
7. Place the tensioner into a soft-jawed vise to compress the tensioner.
8. After compressing the tensioner, insert a pin (a 5⁄64 in. Allen wrench will also work) into the plunger side hole to retain the plunger until installation.

To install:

9. If necessary, use tool No. 6792, or equivalent, to install the crankshaft timing belt sprocket onto the crankshaft.
10. If necessary, install the camshaft sprockets onto the camshafts. Install and tighten the camshaft sprocket bolts to 75 ft. lbs. (101 Nm).
11. Set the crankshaft sprocket to Top Dead Center (TDC) by aligning the notch on the sprocket with the arrow on the oil pump housing.
12. Set the camshafts timing marks so the exhaust camshaft is ½ notch below intake camshaft sprocket. Make sure the arrows on both camshaft sprockets are facing up
13. Install the timing belt starting at the crankshaft, then around the water pump sprocket, idler pulley, camshaft sprockets and around the tensioner pulley.
14. Move the exhaust camshaft sprocket counterclockwise to take up the belt slack.
15. Insert a 6mm Allen wrench into the hexagon opening located on the top plate of the belt tensioner pulley. Rotate the top plate counterclockwise until there is no slack on the belt. The tensioner setting notch will start to move clockwise. Watch the notch and continue rotating the top plate counterclockwise until the setting notch is aligned with the spring tang. Using the Allen wrench to prevent the top plate from moving, tighten the tensioner lock bolt to 220 inch lbs. (25 Nm). The setting notch and spring tang should remain aligned after the lock nut is tightened.
16. Rotate the crankshaft 2 revolutions and recheck the timing marks.

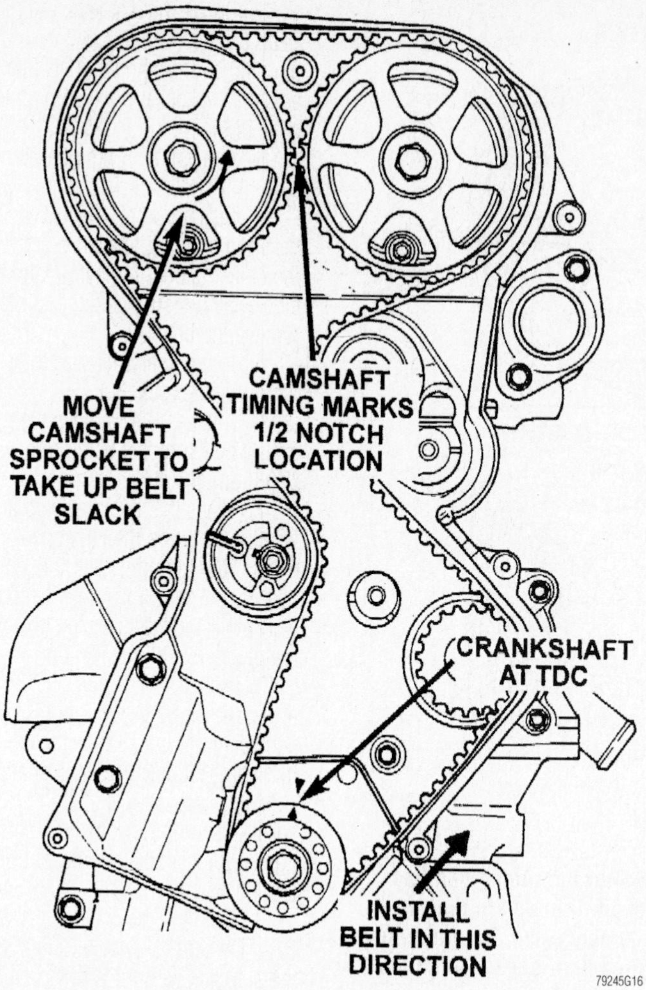

Installation of the timing belt, notice the camshaft alignment— 2.4L (VIN B) engine

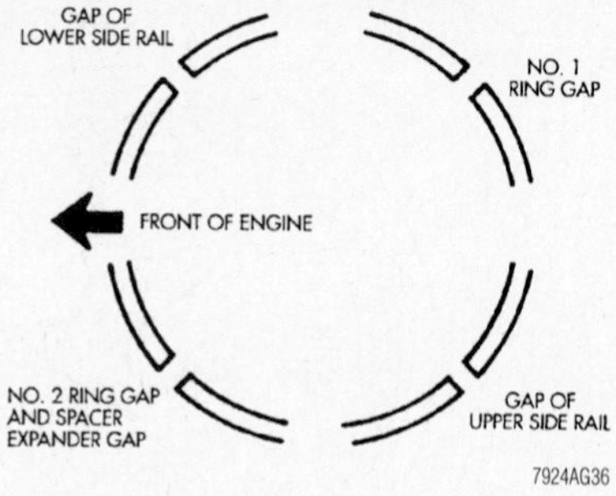

Piston ring end-gap spacing—2.4L engine

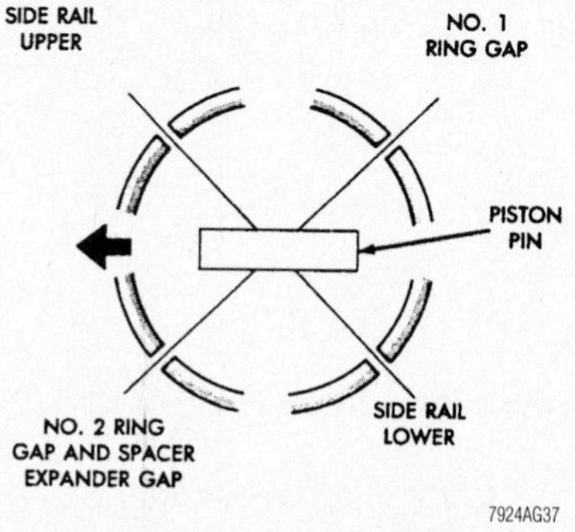

Piston ring end-gap spacing—3.3L and 3.8L engines

17. Check the spring tang is within the tolerance window. If not within the window, reinsert the Allen wrench and into the hexagon opening located on the top plate of the belt tensioner pulley. Rotate the top plate counterclockwise until there is no slack on the belt. The tensioner setting notch will start to move clockwise. Watch the notch and continue rotating the top plate counterclockwise until the setting notch is aligned with the spring tang. Using the Allen wrench to prevent the top plate from moving, tighten the tensioner lock bolt to 220 inch lbs. (25 Nm). The setting notch and spring tang should remain aligned after the lock nut is tightened.
18. Install the engine mount bracket.
19. Install the front timing belt covers.
20. Install the A/C compressor/alternator tensioner and pulley assembly.
21. Install the right engine mount.
22. Install the crankshaft damper
23. Install the accessory drive belts and adjust to the proper tension.
24. Install the right inner splash shield and wheel.
25. Reconnect the negative battery cable.
26. Perform the crankshaft and camshaft relearn alignment procedure using the DRB scan tool, or equivalent.

Piston and Ring

POSITIONING

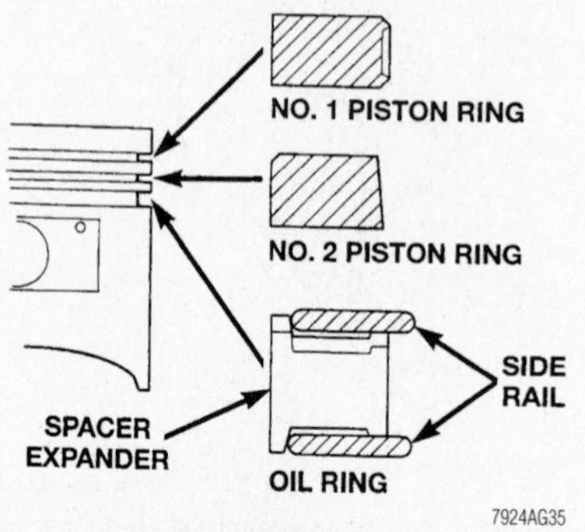

Piston ring positioning—2.4L engine

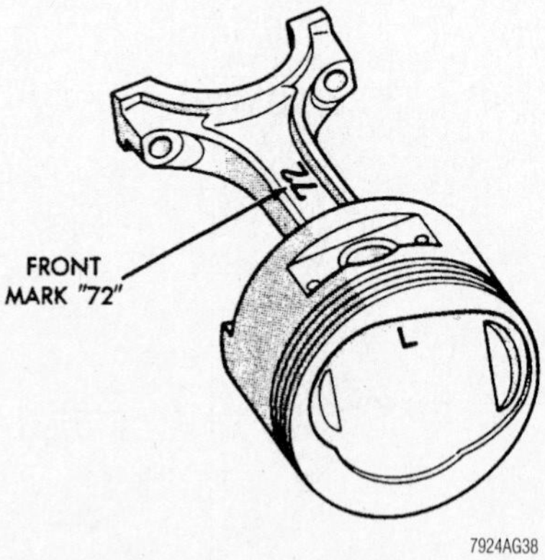

Piston and connecting rod front mark locations—3.3L and 3.8L engines

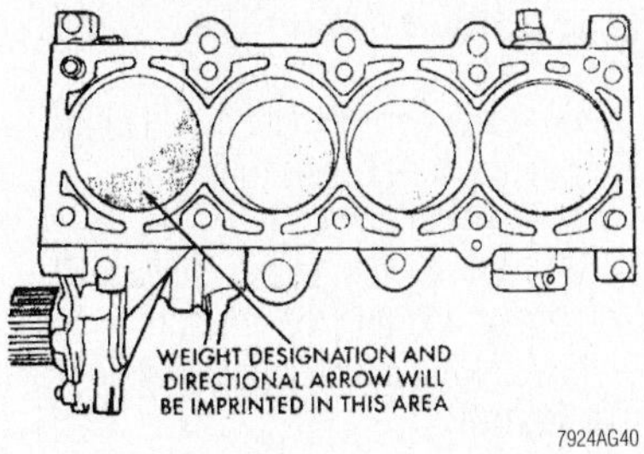

Piston-to-engine positioning mark locations—2.4L engine

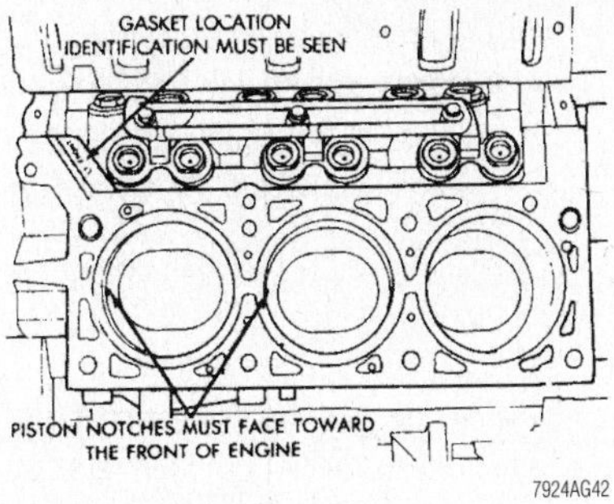

Piston-to-engine positioning mark locations—3.3L and 3.8L engines

FUEL SYSTEM

Fuel System Service Precautions

Safety is the most important factor when performing not only fuel system maintenance but any type of maintenance. Failure to conduct maintenance and repairs in a safe manner may result in serious personal injury or death. Maintenance and testing of the vehicle's fuel system components can be accomplished safely and effectively by adhering to the following rules and guidelines.

- To avoid the possibility of fire and personal injury, always disconnect the negative battery cable unless the repair or test procedure requires that battery voltage be applied.
- Always relieve the fuel system pressure prior to disconnecting any fuel system component (injector, fuel rail, pressure regulator, etc.), fitting or fuel line connection. Exercise extreme caution whenever relieving fuel system pressure to avoid exposing skin, face and eyes to fuel spray. Please be advised that fuel under pressure may penetrate the skin or any part of the body that it contacts.
- Always place a shop towel or cloth around the fitting or connection prior to loosening to absorb any excess fuel due to spillage. Ensure that all fuel spillage (should it occur) is quickly removed from engine surfaces. Ensure that all fuel soaked cloths or towels are deposited into a suitable waste container.
- Always keep a dry chemical (Class B) fire extinguisher near the work area.
- Do not allow fuel spray or fuel vapors to come into contact with a spark or open flame.
- Always use a back-up wrench when loosening and tightening fuel line connection fittings. This will prevent unnecessary stress and torsion to fuel line piping.
- Always replace worn fuel fitting O-rings with new ones. Do not substitute fuel hose or equivalent, where fuel pipe is installed.

Fuel System Pressure

RELIEVING

1. Before servicing the vehicle, refer to the preceding fuel system precautions, as well as the precautions in the beginning of this section.
2. Remove the fuel pump relay from the Power Distribution Center (PDC).
3. Start the vehicle until it stalls.
4. Try and start the vehicle several times to verify the system is fully relieved.
5. Reinstall the relay once all repair procedures have been completed.

Fuel Filter

REMOVAL & INSTALLATION

➡The fuel filter mounts to the top of the fuel tank.

1. Before servicing the vehicle, refer to the precautions in the beginning of this section.
2. Relieve the fuel system pressure.
3. Remove or disconnect the following:
   - Negative battery cable
4. Raise the vehicle and support the fuel tank with a transmission jack.
   - Fuel line from the front of the tank
   - Ground strap
   - Inboard side of the fuel tank straps and front T strap fastener and lower the tank about 6 inches
   - Fuel lines from the fuel pump module
   - Fuel filter
5. Installation is the reverse of removal. Tighten the fuel filter bolts to 40 inch lbs. (4.5 Nm), the main tank straps to 40 ft. lbs. (54 Nm) and the T strap to 250 inch lbs. (28 Nm).
6. Start the engine and check for leaks.

Fuel Pump

REMOVAL & INSTALLATION

1. Before servicing the vehicle, refer to the precautions in the beginning of this section.
2. Relieve the fuel system pressure.
3. Drain the fuel tank.
4. Remove or disconnect the following:
   - Negative battery cable
   - Fuel tank straps. Support the fuel tank before loosening the strap bolts.
   - Fuel lines
   - Fuel pump module harness connector
5. Lower the tank for access and remove the fuel pump module locking ring and the fuel pump module.

To install:

6. Install or connect the following:
   - Fuel pump module. Tighten the locking ring to 40 ft. lbs. (54 Nm).
   - Fuel pump module harness connector
   - Fuel lines
   - Fuel tank straps. Tighten the main tank straps to 40 ft. lbs. (54 Nm) and the T strap to 250 inch lbs. (28 Nm).
   - Negative battery cable
7. Start the engine and check for leaks.

Fuel Injector

REMOVAL & INSTALLATION

2.4L Engine

1. Before servicing the vehicle, refer to the precautions in the beginning of this section.
2. Relieve the fuel system pressure.
3. Remove or disconnect the following:
   - Negative battery cable
   - Wiring connectors for the injector harness
   - Wiring harness from the brackets
   - Fuel injector harness connectors
   - Harness from the vehicle
   - Quick connect fuel hose fittings from the chassis tube
   - Fuel rail bolts
   - Fuel rail with injectors attached
4. Rotate and pull the injectors to separate them from the fuel supply manifold.

To install:

5. Install or connect the following:
   - Injectors with new O-ring seals
   - Fuel rail with injectors attached. Tighten the bolts to 16 ft. lbs. (22 Nm).
   - Quick connect fuel hose fittings to the chassis tube
   - Harness from the vehicle
   - Fuel injector harness connectors
   - Wiring harness from the brackets
   - Wiring connectors to the injector harness
   - Negative battery cable
6. Start the engine, check for leaks and repair if necessary.

3.3L and 3.8L Engines

1. Before servicing the vehicle, refer to the precautions in the beginning of this section.
2. Relieve the fuel system pressure.
3. Remove or disconnect the following:
   - Negative battery cable
   - Upper intake manifold
   - Quick connect fuel hose fittings from the chassis tube
   - Fuel rail bolts
   - Fuel rail with injectors attached
4. Rotate and pull the injectors to separate them from the fuel supply manifold.

To install:

5. Install or connect the following:
   - Injectors with new O-ring seals
   - Fuel rail with injectors attached. Tighten the bolts to 16 ft. lbs. (22 Nm).
   - Quick connect fuel hose fittings to the chassis tube
   - Upper intake manifold
   - Negative battery cable
6. Start the engine, check for leaks and repair if necessary.

DRIVE TRAIN

Manual Transaxle Assembly

REMOVAL & INSTALLATION

1. Before servicing the vehicle, refer to the precautions in the beginning of this section.
2. Drain the transaxle fluid.
3. Remove or disconnect the following:
   - Negative battery cable
   - Gearshift cables from the shift levers cover assembly
   - Gearshift cable from the mounting bracket
   - 3 right engine mount bracket-to-transaxle bolts
   - Front wheels
   - Axle shafts
   - Front harness retainer and move the harness aside
   - Clutch quick connect fitting from the slave cylinder tube using tool 6638A by depressing towards the case and turning counterclockwise 60 degrees while lifting the anti-rotation tab out of the case slot using a suitable prytool
   - Engine left mount bracket
   - Starter
   - Back-up lamp switch connector
   - Structural collar
   - Modular clutch-to-drive plate bolts
4. Support the engine with a screw jack and wood block.
   - Upper mount through bolt from the left frame rail
5. Lower the engine/transaxle assembly.
   - 4 upper mount-to-transaxle bolts and the mount
6. Attach a transmission jack to the transaxle.
   - Transaxle

To install:

7. Install or connect the following:
   - Transaxle and tighten the bolts to 70 ft. lbs. (95 Nm)
8. Raise the engine/transaxle assembly.
   - Upper mount through bolt to the left frame rail and tighten to 55 ft. lbs. (75 Nm)
   - Modular clutch-to-drive plate bolts and tighten to 65 ft. lbs. (88 Nm)
   - Structural collar
   - Back-up lamp switch connector
   - Starter
   - Front mount bracket and tighten the bolts and nut to 50 ft. lbs. (68 Nm)
   - Slave cylinder making sure to note the location of the different size lugs. While pressing inward, rotate the slave cylinder 60 degrees clockwise until the nylon locating tab rests inside the case cutout and the hydraulic tube is vertical. Attach the quick connect fitting, a click should be heard when it is properly attached.
   - Front harness
   - Axle shafts
   - Front wheels
   - 3 right engine mount bracket-to-transaxle bolts
   - Gearshift cable
   - Gearshift cables to the shift levers cover assembly
   - Negative battery cable
9. Fill the transaxle with the correct type and amount of fluid
10. Start the engine and check for proper operation.

Automatic Transaxle Assembly

REMOVAL & INSTALLATION

31TH Transaxle

1. Before servicing the vehicle, refer to the precautions in the beginning of this section.
2. Attach a support fixture to the engine lifting eyes.
3. Remove or disconnect the following:
   - Negative battery cable
   - Dipstick tube and plug the opening
   - Torque converter clutch harness connector
   - Kickdown cable
   - Shift cable
   - Gear position switch connector
   - Back-up lamp switch connector

- 2 upper transaxle bolts
- 3 rear mount bracket-to-case bolts
- Front wheels
- Axle halfshafts
- Starter motor
- Transaxle fluid cooler lines. Cut the lines flush with the fittings. A service kit will be used upon installation.
- Structural collar
- Vehicle Speed (VSS) sensor connector
- Rear mount shield
- Rear mount through bolt

4. Support the engine with a screw jack and wood block.
   - Cradle plate
   - Torque converter bolts
   - Left wheel splash shield
   - Left upper mount through bolt
5. Lower the engine/transaxle assembly and attach a transmission jack to the transaxle.
   - Remaining transaxle bolts
   - Transaxle

To install:

6. Install or connect the following:
   - Transaxle. Torque the flange bolts to 70 ft. lbs. (95 Nm).
7. Raise the transaxle/engine assembly into position using a screw jack and a wood block.
   - Left mount bolt. Torque the through bolt to 55 ft. lbs. (75 Nm).
   - Left wheel splash shield
   - Torque converter bolts and tighten to 65 ft. lbs. (88 Nm)
   - Cradle plate
   - Rear mount bracket-to-transaxle and hand tighten the bolts. Align the rear mount bracket to the mount and install the through bolt. Hand tighten the bolt.
   - VSS sensor connector
   - Structural collar
   - Transaxle fluid cooler lines. Follow the instruction with service kit to install.
   - Starter motor
   - Front mount and bracket. Torque the large bracket bolts to 80 ft. lbs. (108 Nm), the small bracket bolts to 40 ft. lbs. (54 Nm) and the through bolt to 45 ft. lbs. (61 Nm).
   - Axle halfshafts
   - Rear mount bracket vertical bolts to 75 ft. lbs. (102 Nm)
   - Back-up lamp switch connector
   - Gear position switch connector
   - Shift cable
   - Kickdown cable
   - Torque converter clutch harness connector
   - Dipstick tube
   - Tighten the rear mount bract horizontal bolt to 75 ft. lbs. (102 Nm)
   - Tighten the rear mount through bolt to 40 ft. lbs. (54 Nm)
   - Negative battery cable
8. Fill the transaxle with the correct type and amount of fluid
9. Start the engine and check for proper operation.

41TE Transaxle

1. Before servicing the vehicle, refer to the precautions in the beginning of this section.
2. Attach a support fixture to the engine lifting eyes.
3. Remove or disconnect the following:
   - Battery cables
   - Battery shield
   - Coolant recovery bottle
   - Dipstick tube and plug the opening
   - Torque converter clutch harness connector
   - Transaxle fluid cooler lines. Cut the lines flush with the fittings. A service kit will be used upon installation.
   - Input and output shaft sensor connectors
   - Transmission Range Sensor (TRS) connector
   - Solenoid/pressure switch connector
   - Shift cable
   - Crankshaft Position (CKP) sensor from the bell housing, if equipped
   - Position the leak detection pump harness and hoses aside
   - Rear mount bracket-to-case bolts
   - Upper transaxle bolts
   - Transaxle pan
   - Front wheels
   - Axle halfshafts
   - Power Transfer Unit (PTO), if equipped
   - Rear mount bracket-to-case lower bolt
   - Front mount/bracket assembly
   - Starter motor
   - Lateral bending brace
   - Inspection cover
   - Torque converter bolts
4. Support the engine with a screw jack and wood block.
   - Left wheel splash shield
   - Left upper mount through bolt
5. Lower the engine/transaxle assembly and attach a transmission jack to the transaxle.
   - Upper mount bracket from the transaxle
   - Remaining transaxle bolts
   - Transaxle

To install:

6. Install or connect the following:
   - Transaxle. Torque the flange bolts to 70 ft. lbs. (95 Nm).
   - Upper mount assembly and tighten the bolts to 40 ft. lbs. (54 Nm)
7. Raise the transaxle/engine assembly into position using a screw jack and a wood block.
   - Left upper mount through bolt and tighten to 55 ft. lbs. (75 Nm)
   - Left wheel splash shield
   - Torque converter bolts and tighten to 65 ft. lbs. (88 Nm)
   - Inspection cover
   - Lateral bending brace
   - Starter motor
   - Front mount/bracket assembly
   - Rear mount bracket-to-case lower bolt. Align the assembly, hand tighten the bolt first, then tighten to 75 ft. lbs. (102 Nm) on the assembly is aligned.
   - PTO, if equipped
   - Axle halfshafts
   - Front wheels
   - Transaxle pan
   - Rear mount bracket-to-case vertical bolts to 75 ft. lbs. (102 Nm)
   - Upper transaxle bolts to 70 ft. lbs. (95 Nm)
   - CKP sensor to the bell housing, if equipped
   - Shift cable
   - Attach the leak detection pump harness and hoses
   - Solenoid/pressure switch connector
   - TRS connector
   - Input and output shaft sensor connectors
   - Transaxle fluid cooler lines. Attach using the service kit supplied following the kit instructions.
   - Torque converter clutch harness connector
   - Dipstick tube
   - Coolant recovery bottle
   - Battery shield
   - Battery cables
8. Fill the transaxle with the correct type and amount of fluid
9. Start the engine and check for proper operation.

40TE Transaxle

1. Before servicing the vehicle, refer to the precautions in the beginning of this section.

2. Attach a support fixture to the engine lifting eyes.

3. Remove or disconnect the following:
- Battery cables
- Battery shield
- Coolant recovery bottle
- Dipstick tube and plug the opening
- Transaxle fluid cooler lines. Cut the lines flush with the fittings. A service kit will be used upon installation.
- Input and output shaft sensor connectors
- Transmission Range Sensor (TRS) connector
- Solenoid/pressure switch connector
- Shift cable
- Crankshaft Position (CKP) sensor from the bell housing, if equipped
- Position the leak detection pump harness and hoses aside
- Rear mount bracket-to-case bolts
- Upper transaxle bolts
- Transaxle pan
- Front wheels
- Axle halfshafts
- Power Transfer Unit (PTO), if equipped
- Rear mount bracket-to-case lower bolt
- Front mount/bracket assembly
- Starter motor
- Lateral bending brace
- Inspection cover
- Torque converter bolts

4. Support the engine with a screw jack and wood block.
- Left wheel splash shield
- Left upper mount through bolt

5. Lower the engine/transaxle assembly and attach a transmission jack to the transaxle.
- Upper mount bracket from the transaxle
- Remaining transaxle bolts
- Transaxle

To install:

6. Install or connect the following:
- Transaxle. Torque the flange bolts to 70 ft. lbs. (95 Nm).
- Upper mount assembly and tighten the bolts to 40 ft. lbs. (54 Nm)

7. Raise the transaxle/engine assembly into position using a screw jack and a wood block.
- Left upper mount through bolt and tighten to 55 ft. lbs. (75 Nm)
- Left wheel splash shield
- Torque converter bolts and tighten to 65 ft. lbs. (88 Nm)
- Inspection cover
- Lateral bending brace
- Starter motor
- Front mount/bracket assembly
- Rear mount bracket-to-case lower bolt. Align the assembly, hand tighten the bolt first, then tighten to 75 ft. lbs. (102 Nm) on the assembly is aligned.
- PTO, if equipped
- Axle halfshafts
- Front wheels
- Transaxle pan
- Rear mount bracket-to-case vertical bolts to 75 ft. lbs. (102 Nm)
- Upper transaxle bolts to 70 ft. lbs. (95 Nm)
- CKP sensor to the bell housing, if equipped
- Shift cable
- Attach the leak detection pump harness and hoses
- Solenoid/pressure switch connector
- TRS connector
- Input and output shaft sensor connectors
- Transaxle fluid cooler lines. Attach using the service kit supplied following the kit instructions.
- Dipstick tube
- Coolant recovery bottle
- Battery shield
- Battery cables

8. Fill the transaxle with the correct type and amount of fluid

9. Start the engine and check for proper operation.

Modular Clutch Assembly

REMOVAL & INSTALLATION

1. Before servicing the vehicle, refer to the precautions in the beginning of this section.

2. Remove or disconnect the following:
- Transaxle
- Clutch assembly from the input shaft

3. Installation is the reverse of removal

Power Transfer Unit

REMOVAL & INSTALLATION

1. Before servicing the vehicle, refer to the precautions in the beginning of this section.

2. Remove or disconnect the following:
- Right front wheel
- Right axle halfshaft
- Propeller shaft

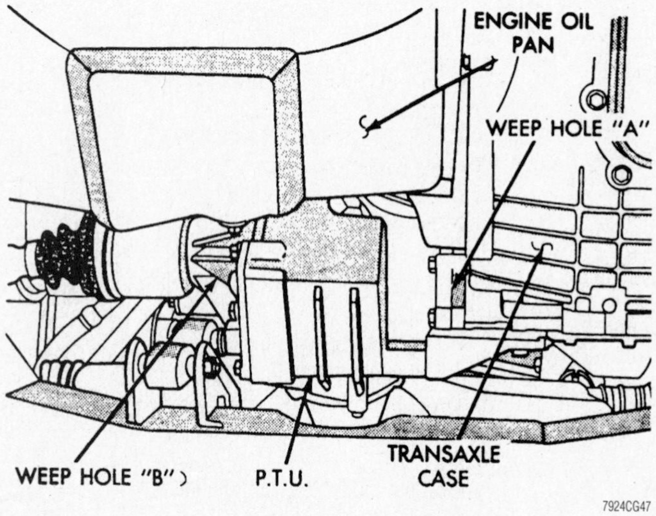

Power transfer unit and related components

- Cradle plate
- Power transfer unit brackets
- Power transfer unit

To install:

3. Install or connect the following:
   - Power transfer unit and brackets. Tighten the bolts to 37 ft. lbs. (50 Nm).
   - Rear driveshaft
   - Cradle plate. Tighten the bolts to 123 ft. lbs. (166 Nm).
   - Right axle halfshaft
   - Right front wheel

Halfshaft

REMOVAL & INSTALLATION

Front

1. Before servicing the vehicle, refer to the precautions in the beginning of this section.
2. Remove or disconnect the following:
   - Front wheel
   - Split pin
   - Nut lock
   - Spring washer
   - Hub nut
   - Brake caliper and rotor
   - Outer tie rod end
   - Wheel speed sensor harness, if equipped
   - Lower ball joint
3. Separate the outer CV-joint stub shaft from the steering knuckle.
4. Pry the inner tri-pot joint out of the transaxle and remove the axle halfshaft.

To install:

5. Install the axle halfshaft so that the inner joint circlip seats in the transaxle side gear.
6. Guide the outer CV-joint stub shaft through the steering knuckle hub.
7. Install or connect the following:
   - Lower ball joint. Torque the nut to 65 ft. lbs. (88 Nm) plus a 90 degree turn.
   - Wheel speed sensor harness, if equipped
   - Outer tie rod end. Torque the nut to 55 ft. lbs. (75 Nm).
   - Brake caliper and rotor. Torque the caliper bolts to 35 ft. lbs. (47 Nm).
   - Hub nut. Torque the nut to 180 ft. lbs. (245 Nm).
   - Spring washer
   - Nut lock
   - Split pin
   - Front wheel
8. Check the wheel alignment and adjust as necessary.

Rear

1. Before servicing the vehicle, refer to the precautions in the beginning of this section.
2. Remove or disconnect the following:
   - Rear wheel
   - Cotter pin, nut lock and washer
   - Half shaft nut and washer
   - Rear halfshafts from the output flanges
   - Axle halfshaft

To install:

3. Guide the outer CV-joint stub shaft through the rear wheel hub.
4. Install or connect the following:
   - Inner joint to the differential. Torque the flange bolts to 45 ft. lbs. (61 Nm).
   - Half shaft nut and washer and tighten the nut to 180 ft. lbs. (244 Nm)
   - Waver washer, nut lock and new cotter pin
   - Rear wheel

CV-Joints

OVERHAUL

Outer CV-Joint

The outer CV-joint and boot are serviced with the axle halfshaft as an assembly.

Inner Tri-pot Joint

1. Before servicing the vehicle, refer to the precautions in the beginning of this section.
2. Remove or disconnect the following:
   - Negative battery cable
   - Axle halfshaft from the vehicle
   - Inner tri-pot joint boot clamps
   - Tri-pot joint housing
   - Snapring
   - Tri-pot joint

To install:

➡Use new snaprings, clips, and boot clamps for assembly.

3. Install or connect the following:
   - Tri-pot joint
   - Snapring
   - Tri-pot joint housing
4. Fill the tri-pot joint housing and boot with grease and tighten the boot clamps.
5. Install the axle halfshaft.
6. Connect the negative battery cable

STEERING AND SUSPENSION

Air Bag

⁂ CAUTION

Some vehicles are equipped with an air bag system. The system must be disarmed before performing service on, or around, system components, the steering column, instrument panel components, wiring and sensors. Failure to follow the safety precautions and the disarming procedure could result in accidental air bag deployment, possible injury and unnecessary system repairs.

PRECAUTIONS

Several precautions must be observed when handling the inflator module to avoid accidental deployment and possible personal injury.

- Never carry the inflator module by the wires or connector on the underside of the module.
- When carrying a live inflator module, hold securely with both hands, and ensure that the bag and trim cover are pointed away from your body. In the unlikely event of an accidental deployment, the bag will then deploy with minimal chance of injury.
- Place the inflator module on a bench or other surface with the bag and trim cover facing up. This will reduce the motion of the module if accidentally deployed.
- With the inflator module on the bench, never place anything on or close to the module which may be thrown in the event of an accidental deployment.

DISARMING

1. Disconnect and isolate the negative battery cable from the battery.
2. Allow the SIR system capacitor to discharge for at least two (2) minutes, before performing any repairs.
3. When repairs are completed, connect the negative battery cable.

Power Rack and Pinion Steering Gear

REMOVAL & INSTALLATION

1. Before servicing the vehicle, refer to the precautions in the beginning of this section.
2. Lock the steering wheel to the straight ahead position.
3. Drain the power steering fluid from the reservoir.
4. Remove or disconnect the following:
   - Negative battery cable
   - Front wheels
   - Steering column shaft coupler
   - Front emissions vapor canister, if equipped
   - Single hose at the power steering cooler and let the fluid drain into a suitable container
   - Power steering cooler-to-cradle bolts, if equipped
   - Leak detection pump-to-cradle crossmember bolts, if equipped
   - Outer tie rod ends

➡The bolts retaining the cradle crossmember are different sizes, makes sure to note the locations of the bolts and their sizes prior to removal.

   - Lower control arm rear bushing bolts
   - Crossmember reinforcement
   - Power transfer unit, if equipped with AWD
   - Power steering pressure and return lines
   - Steering gear mounting fasteners
   - Intermediate shaft coupler
   - Power steering gear

To install:

5. Install or connect the following:
   - Power steering gear
   - Intermediate shaft coupler. Start the roll pin into the coupler before installing the coupler, start the roll pin and then user a hammer to tap it into the coupler. Install the coupler on the shaft of the gear. Install removal/installer tool 6831A through the center of the roll pin, using the knurled nut to secure it. Hold the threaded rod stationary while turning the nut to pull the pin into the coupler.
   - Steering gear mounting fasteners. Torque the 14mm fasteners to 135 ft. lbs. (183 Nm) and the 12mm fastener to 70 ft. lbs. (95 Nm).
   - Power steering pressure and return lines. Torque the line to 25 ft. lbs. (31 Nm).
   - Power transfer unit, if equipped with AWD
   - Crossmember reinforcement. Tighten the M14 bolts to 113 ft. lbs. (153 Nm) and the M12 bolts to 78 ft. lbs. (106 Nm).
   - Lower control arm rear bushing bolts. Torque the bolts to 45 ft. lbs. (61 Nm).
   - Outer tie rod ends to the steering knuckle. Torque the nut to 55 ft. lbs. (75 Nm).
   - Leak detection pump-to-cradle crossmember bolts, if equipped
   - Power steering cooler-to-cradle bolts, if equipped
   - Single hose at the power steering cooler and fasten the clamp
   - Front emissions vapor canister, if equipped
   - Front wheels
   - Steering column shaft coupler. Tighten the pinch bolt to 21 ft. lbs. (28 Nm).
   - Negative battery cable
6. Fill and bleed the power steering reservoir
7. Inspect the power steering system for leaks and repair if necessary.
8. Check the wheel alignment and adjust as necessary.

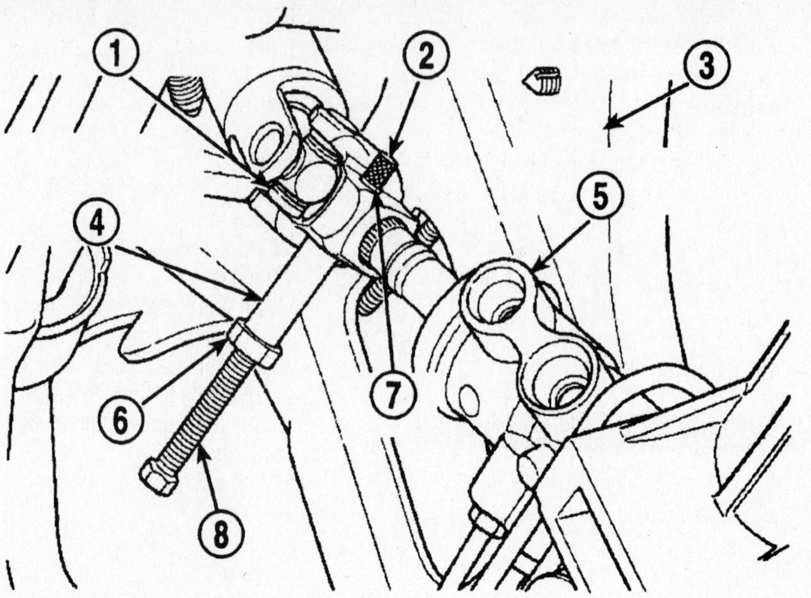

1 - INTERMEDIATE COUPLER
2 - KNURLED NUT
3 - SUSPENSION CRADLE
4 - REMOVER SPECIAL TOOL 6831A
5 - STEERING GEAR
6 - NUT
7 - ROLL PIN
8 - THREADED ROD

67189-MINV-G02

Installing the intermediate shaft coupler to the gear

Strut

REMOVAL & INSTALLATION

1. Before servicing the vehicle, refer to the precautions in the beginning of this section.
2. Remove or disconnect the following:
   - Front wheel
   - Brake hose bracket
   - Wheel speed sensor harness bracket, if equipped
   - Stabilizer bar link
   - Steering knuckle bracket bolts
   - Upper strut mount nuts
   - Strut assembly

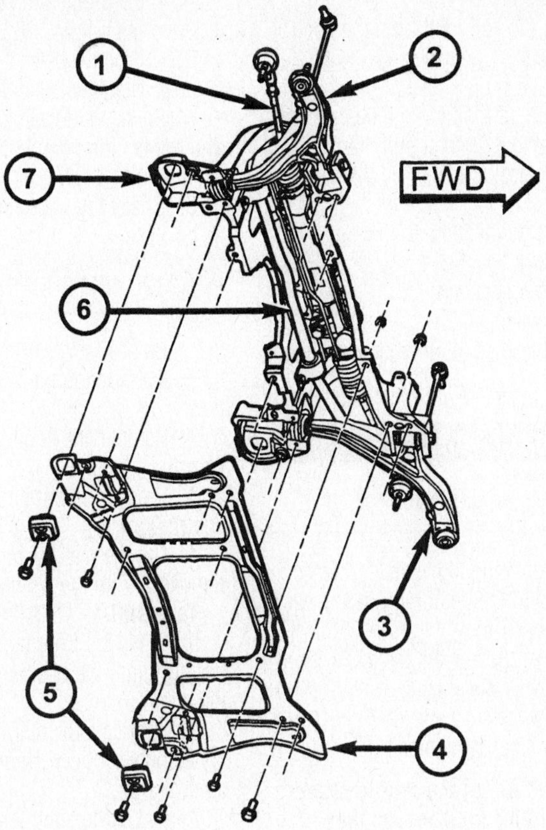

1 - STEERING GEAR
2 - RIGHT LOWER CONTROL ARM
3 - LEFT LOWER CONTROL ARM
4 - CRADLE CROSSMEMBER REINFORCEMENT
5 - REAR CRADLE CROSSMEMBER ISOLATOR BUSHING
6 - STABILIZER BAR
7 - CRADLE CROSSMEMBER

67189-MINV-G01

Cradle crossmember reinforcement mounting

To install:

3. Install or connect the following:
   - Strut assembly. Torque the upper strut mount nuts to 21 ft. lbs. (28 Nm).
   - Steering knuckle bracket bolts. Torque the bolts to 60 ft. lbs. (81 Nm) plus 90 degrees.
   - Stabilizer bar link. Torque the nut to 65 ft. lbs. (88 Nm).
   - Wheel speed sensor harness bracket, if equipped. Torque the bolt to 10 ft. lbs. (13 Nm).
   - Brake hose bracket. Tighten the bolt to 10 ft. lbs. (13 Nm).
   - Front wheel
4. Check the wheel alignment and adjust as necessary.

Shock Absorber

REMOVAL & INSTALLATION

Rear

1. Before servicing the vehicle, refer to the precautions in the beginning of this section.
2. Support the rear axle with a jackstand.
3. Remove the top and bottom shock absorber bolts.
4. Remove the shock absorber.

To install:

5. Install the shock absorber. Torque the mounting bolts to 65 ft. lbs. (88 Nm).
6. Remove the jackstand.

Coil Spring

REMOVAL & INSTALLATION

Front

1. Before servicing the vehicle, refer to the precautions in the beginning of this section.
2. Remove the strut assembly from the vehicle.
3. Compress the coil spring and remove the piston rod nut.
4. Remove or disconnect the following:
   - Upper strut mount
   - Pivot bearing
   - Spring upper seat
   - Coil spring
   - Jounce bumper and dust shield

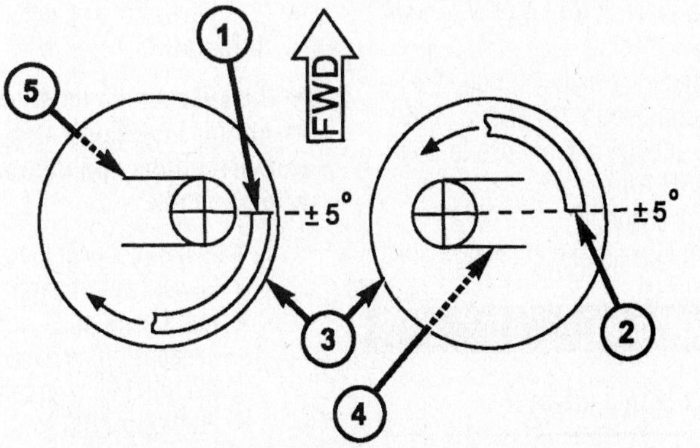

1 - END OF LEFT COIL SPRING AT STRUT LOWER SEAT
2 - END OF RIGHT COIL SPRING AT STRUT LOWER SEAT
3 - LOWER SEATS OF STRUTS
4 - RIGHT STRUT CLEVIS BRACKET
5 - LEFT STRUT CLEVIS BRACKET

06009-MINI-G03

Position the coil spring as shown during assembly

- Lower spring isolator

To install:

5. Install or connect the following:
   - Lower spring isolator
   - Jounce bumper and dust shield
   - Coil spring
   - Spring upper seat
   - Pivot bearing
   - Upper strut mount. Torque the piston rod nut to 75 ft. lbs. (100 Nm).
6. Remove the spring compressor and install the strut assembly to the vehicle.

Leaf Springs

REMOVAL & INSTALLATION

1. Before servicing the vehicle, refer to the precautions in the beginning of this section.
2. Remove the axle halfshaft, if equipped.
3. Support the vehicle at the frame rail with jackstands.
4. Support the axle with a floor jack.
5. Remove or disconnect the following:
   - Shock absorber
   - Axle plate from the axle and spring
6. Slowly lower the axle so that the leaf spring hangs free.
7. Remove or disconnect the following:
   - Front spring mount
   - Rear spring shackle plate
   - Leaf spring

To install:

8. Install or connect the following:
   - Leaf spring
   - Rear spring shackle plate. Torque the nuts to 45 ft. lbs. (61 Nm).
   - Axle plate. Torque the bolts to 75 ft. lbs. (101 Nm) on AWD models or 70 ft. lbs. (95 Nm) on FWD models.
   - Front spring mount. Torque the through bolt to 115 ft. lbs. (156 Nm) and the mount bolts to 45 ft. lbs. (61 Nm).
   - Shock absorber
   - Axle halfshaft, if equipped

Lower Ball Joint

REMOVAL & INSTALLATION

1. Before servicing the vehicle, refer to the precautions in the beginning of this section.
2. Remove or disconnect the following:
   - Front wheel
   - Lower control arm
   - Ball joint seal boot
3. Remove the ball joint from the control arm with a press.

To install:

4. Install the ball joint to the control arm with a press so that the ball joint flange contacts the control arm with no visible gaps.
5. Install or connect the following:
   - Ball joint seal boot
   - Lower control arm
   - Front wheel
6. Check the wheel alignment and adjust as necessary.

Lower Control Arm

REMOVAL & INSTALLATION

1. Before servicing the vehicle, refer to the precautions in the beginning of this section.
2. Remove or disconnect the following:
   - Negative battery cable
   - Front wheels
   - Power steering cooler

➡The bolts retaining the cradle crossmember are different sizes, makes sure to note the locations of the bolts and their sizes prior to removal.

   - Cradle plate
   - Lower ball joints
   - Rear control arm bushing retainers
3. Matchmark the suspension cradle and the frame rail.
4. Loosen the left suspension cradle bolts and lower the cradle to allow the pivot bolt to be removed.
5. Remove or disconnect the following:
   - Front pivot bolts
   - Lower control arms

To install:

➡The suspension must be at curb height for final tightening of the control arm pivot bolts and the rear bushing retainer bolts.

6. Install or connect the following:
   - Lower control arms
   - Front pivot bolts. Torque the M14 suspension cradle bolts to 113 ft. lbs. (153 Nm) and the M12 bolts to 78 ft. lbs. (106 Nm).
   - Rear control arm bushing retainers. Torque the bolts to 45 ft. lbs. (61 Nm).
   - Lower ball joints. Tighten the pinch bolts 105 ft. lbs. (145 Nm).
   - Cradle plate. Tighten the bolts to 120 ft. lbs. (163 Nm).
   - Power steering cooler and tighten the bolts to 100 inch lbs. (11 Nm)
   - Front wheels
   - Negative battery cable
7. Raise the suspension to curb height and tighten the pivot bolts to 135 ft. lbs. (183 Nm).
8. Check the wheel alignment and adjust as necessary.

CONTROL ARM BUSHING REPLACEMENT

1. Before servicing the vehicle, refer to the precautions in the beginning of this section.
2. Remove the control arm from the vehicle.
3. Press the front bushing out of the control arm.
4. Cut the rear bushing lengthwise and remove it from the control arm.

To install:

5. Press the front bushing into the control arm until the bushing flange contacts the control arm.
6. Lubricate the rear bushing with silicone spray lubricant and push it onto the control arm.
7. Install the control arm to the vehicle.

Wheel Bearings

ADJUSTMENT

The front and rear wheel bearings are designed for the life of the vehicle and require no type of adjustment or periodic maintenance. The bearing is a sealed unit with the wheel hub and can only be removed and/or replaced as an assembly.

REMOVAL & INSTALLATION

Front

1. Before servicing the vehicle, refer to the precautions in the beginning of this section.
2. Remove or disconnect the following:
   - Front wheel
   - Brake caliper and rotor
   - Hub retaining bolts
   - Hub and bearing assembly

To install:

3. Install or connect the following:
   - Hub and bearing assembly. Torque the bolts to 45 ft. lbs. (65 Nm).
   - Caliper and adapter to the brake rotor and align the assembly to the steering knuckle. Torque the mounting bolts to 125 ft. lbs. (169 Nm).

- Hub retaining nut. Tighten the nut to 180 ft. lbs. (244 Nm).
- Front wheel

4. Check and adjust the front end alignment, if necessary.

Rear

FRONT WHEEL DRIVE

1. Before servicing the vehicle, refer to the precautions in the beginning of this section.
2. Remove or disconnect the following:
   - Rear wheel
   - Brake drum or caliper
   - Wheel speed sensor, if equipped
   - Hub and bearing assembly from the rear axle

To install:

3. Install or connect the following:
   - Hub and bearing assembly. Tighten the bolts to 95 ft. lbs. (129 Nm).
   - Wheel speed sensor, if equipped. Tighten the bolt to 105 inch lbs. (12 Nm).
   - Brake drum or caliper
   - Rear wheel

ALL WHEEL DRIVE

1. Before servicing the vehicle, refer to the precautions in the beginning of this section.
2. Set the parking brake.
3. Remove or disconnect the following:
   - Rear wheel
   - Brake caliper and rotor
   - Mounting bolts from the driveshaft inner joint to the output shaft
   - Wheel speed sensor and release the parking brake
   - Axle halfshaft
   - Hub and bearing assembly

To install:

4. Install or connect the following:
   - Hub and bearing assembly. Torque the bolts to 95 ft. lbs. (129 Nm).
   - Axle halfshaft
   - Wheel speed sensor and set the parking brake. Torque the fastener to 105 inch lbs. (12 Nm).
   - Driveshaft inner joint to output shaft mounting bolts. Torque the bolts to 45 ft. lbs. (61 Nm).
   - Torque the outer CV Joint hub nut to 180 ft. lbs. (244 Nm).
   - Brake caliper and rotor
   - Rear wheel

BRAKES

Brake Caliper

REMOVAL & INSTALLATION

1. Before servicing the vehicle, refer to the precautions in the beginning of this section.
2. Raise and safely support the front of the vehicle. Remove the front wheels.
3. If the caliper is only being removed from the bracket (as for a brake pad change), move to Step 3. If the caliper is being removed from the vehicle (as for replacement or an overhaul and reseal), remove the brake hose attaching bolt from the caliper. Remove the hose from the caliper and discard the washers. New seal washers will be required at assembly. Plug the brake hose to prevent fluid leakage.
4. Remove the caliper guide pin bolts that secure the caliper to the steering knuckle.
5. Remove the caliper by slowly sliding it away from the steering knuckle. Slide the opposite end of the brake caliper out from under the machined abutment on the steering knuckle.
6. Using a strong piece of wire, support the brake caliper assembly off the strut unit. Do NOT allow the caliper to hang from the brake fluid flex hose or damage to the hose will result.

To install:

7. Clean both steering knuckle abutment surfaces of any dirt, grease or corrosion. Then lubricate the abutment surfaces with a liberal amount of MOPAR® Multipurpose Lubricant or equivalent.
8. Properly position the brake caliper over the brake pads and disc rotor. Be careful not to allow the caliper seals or guide pin bushings to get damaged by the steering knuckle bosses. Install the caliper guide pin bolts and torque to: 35 ft. lbs. (47 Nm) on 2002–04 models or 26 ft. lbs. (35 Nm) on 2005 models. Be careful not to cross thread the guide pin bolts.
9. If removed, attach the brake hose to the caliper using new washers. Tighten the banjo bolt to 35 ft. lbs. (47 Nm).
10. Bleed the brake system.
11. Install the front wheels and lug nuts. Torque the lug nuts, in a star pattern sequence, to ½ torque specifications. Then repeat the tightening sequence to the full torque specification of 100 ft. lbs. (135 Nm). Lower the vehicle.
12. Pump the brake pedal several times to insure that the brake pedal is firm. Road-test the vehicle.

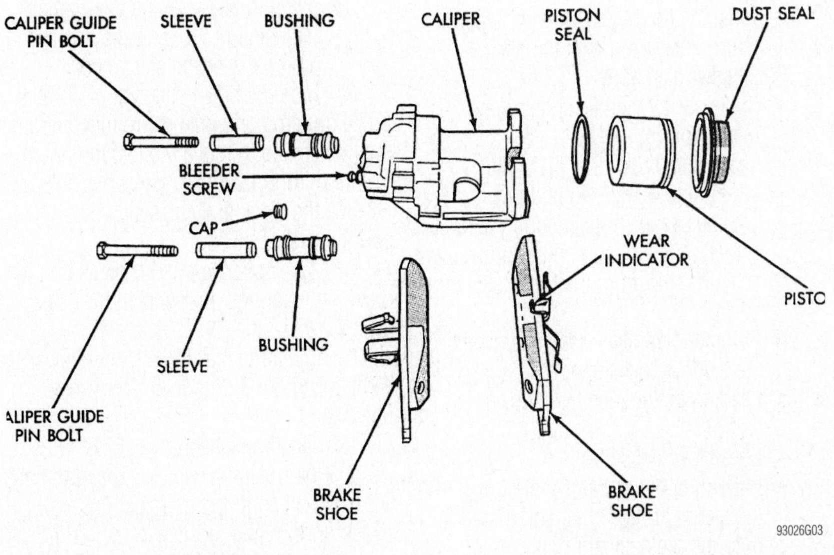

Front brake caliper assembly

Disc Brake Pads

REMOVAL & INSTALLATION

1. Before servicing the vehicle, refer to the precautions in the beginning of this section.
2. Remove brake fluid from the master cylinder brake fluid reservoir until the reservoir is approximately ½ full. Discard the removed fluid.
3. Raise and safely support the front of the vehicle. Remove the front wheels.
4. Remove or disconnect the following:
   - Front brake caliper guide pin bolts

- Brake caliper by slowly sliding it up and off the adapter and brake rotor. Support the caliper out of the way with a strong piece of wire. Do not let the caliper hang by the brake hose or damage to the brake hose will result.

5. If necessary, compress the caliper piston into the bore using a C-clamp. Insert a suitable piece of wood between the C-clamp and caliper piston to protect the piston.

- Outboard disc brake pad from the caliper by prying the brake pad retaining clip over the raised area on the caliper. Slide the brake pad down and off the caliper.
- Inboard disc brake pad from the caliper by pulling the brake pad away from the caliper piston until the retaining clip on the pad is free from the caliper piston cavity

To install:

6. Be sure the caliper piston has been completely retracted into the piston bore of the caliper assembly. This is required when installing the brake caliper equipped with new brake pads.

7. If equipped, remove the protective paper from the noise suppression gaskets on the new disc brake pads.

8. Install or connect the following:

- New inboard disc brake pad into the caliper piston by pressing the pad firmly into the cavity of the caliper piston. Be sure the new inboard brake pad is seated squarely against the face of the brake caliper piston.
- Outboard disc brake pad by sliding it onto the caliper assembly
- Brake caliper assembly over the brake rotor and onto the steering knuckle adapter
- Caliper guide pin bolts and torque to: 35 ft. lbs. (47 Nm) on 2002–04 models or 26 ft. lbs. (35 Nm) on 2005 models
- Front. Apply the brake pedal several times until a firm pedal is obtained.

9. Check the fluid level in the master cylinder and add fluid as necessary. Road-test the vehicle.

Brake Drums

REMOVAL & INSTALLATION

1. Before servicing the vehicle, refer to the precautions in the beginning of this section.
2. Raise and safely support the vehicle.
3. Remove or disconnect the following:
   - Rear wheels
   - Brake drum from the hub assembly by pulling the drum straight off the wheel studs
4. Inspect the brake drum for thickness and runout. Replace or machine as necessary.

To install:

5. Install the brake drum to the hub assembly.
6. Adjust the brake shoes.
7. Install the rear wheel.

Brake Shoes

REMOVAL & INSTALLATION

1. Before servicing the vehicle, refer to the precautions in the beginning of this section.
2. Raise and safely support the vehicle.
3. Remove or disconnect the following:
   - Rear wheels and the brake drums
4. Be sure the parking brake pedal is in the released position. Create slack in the rear parking brake cables by grasping an exposed section of the front parking brake cable, pulling it down and rearward. Maintain the slack in the brake cable by clamping a pair of locking pliers onto the parking brake cable just rearward of **the rear** body outrigger bracket.
   - Adjustment lever spring from the automatic adjustment lever and front brake shoe (leading brake shoe)
   - Automatic adjustment lever from the front brake shoe (leading brake shoe)
   - Brake shoe-to-brake shoe lower return spring
   - Tension clip that secures the upper return spring to the automatic adjuster assembly
   - Brake shoe-to-brake shoe upper return spring
   - Rear brake shoe (trailing brake shoe) hold-down clip and pin
   - Trailing brake shoe, parking brake actuating lever and parking brake actuator strut from the brake support plate
   - Automatic adjuster assembly from the leading brake shoe
   - Leading brake shoe hold-down clip and pin
   - Leading brake shoe
   - Parking brake actuator plate from the leading brake shoe and install onto the replacement brake shoe

To install:

5. Thoroughly clean and dry the backing plate. To prepare the backing plate, lubricate the 8 brake shoe contact areas and brake shoe anchor, using suitable grease.
6. Install or connect the following:
   - Leading brake shoe into position on the brake shoe support plate. Secure the leading brake shoe by installing the brake shoe hold-down clip and pin.
   - Parking brake actuating strut onto the leading brake shoe and then install the parking brake actuating lever onto the strut
7. Lubricate the shaft threads of the automatic adjuster screw assembly with anti-seize lubricant.
   - Automatic adjuster screw assembly onto the leading brake shoe
   - Trailing brake shoe onto the parking brake actuating lever and parking brake actuating strut
   - Trailing brake shoe into position on the brake support plate and install the brake shoe hold-down clip and pin
   - Brake shoe-to-brake shoe upper return spring
   - Tension clip that secures the upper return spring to the automatic adjuster assembly. Be sure the tension clip is positioned on the threaded area of the adjuster assembly or the function of the automatic adjuster will be affected.
   - Brake shoe-to-brake shoe lower return spring
   - Automatic adjustment lever onto the leading brake shoe
   - Actuating spring onto the automatic adjustment lever and leading brake shoe. Make sure the automatic adjustment lever makes positive contact with the star wheel.
8. Once the brake shoes and all other brake system components are fully and correctly installed, remove the locking pliers from the front parking brake cable. This will remove the slack and correctly adjust the parking brake cables.
9. Make sure there is no grease on the brake shoe linings, then install the brake drums.
10. Adjust the rear brakes, then lower the vehicle and check the brakes for proper operation.
11. Install the wheels.
12. Road-test the vehicle. The automatic adjuster will continue to adjust the brake shoes during the road-test.

CHRYSLER

4

Crossfire

SPECIFICATIONS AND MAINTENANCE CHARTS

ENGINE AND VEHICLE IDENTIFICATION

Engine						
Code	Liters (cc)	Cu. In.	Cyl.	Fuel Sys.	Engine Type	Eng. Mfg.
L	3.2 (3199)	195	6	MFI	SOHC	DaimlerChrysler
N	3.2 (3199)	195	6	MFI	SOHC	DaimlerChrysler

Model Year	
Code ①	Year
5	2005
6	2006

MFI: Multi-point Fuel Injection

SOHC: Single Overhead Camshaft

① 10th digit of the Vehicle Identification Number (VIN)

06009-CSFR-C01

GENERAL ENGINE SPECIFICATIONS

Year	Model	Engine Displacement Liters (VIN)	Net Horsepower @ rpm	Net Torque @ rpm (ft. lbs.)	Bore x Stroke (in.)	Com-pression Ratio	Oil Pressure @ rpm
2005	Crossfire	3.2 (L)	215@5700	230@3000	3.54x3.31	10.0:1	45-105@3000
		3.2 (N)	330@6100	310@3500	3.54x3.31	9.0:1	45-105@3000

06009-CSFR-C02

ENGINE TUNE-UP SPECIFICATIONS

Year	Engine Displacement Liters (VIN)	Spark Plug Gap (in.)	Ignition Timing (deg.)		Fuel Pump (psi)	Idle Speed (rpm)		Valve Clearance	
			MT	AT		MT	AT	In.	Ex.
2005	3.2 (L)	0.040	①	①	54-60	①	①	HYD	HYD
	3.2 (N)	0.040	①	①	54-60	①	①	HYD	HYD

NOTE: The Vehicle Emission Control Information label reflects specification changes made during production and must be used if difernent from this chart.

NOTE: Fuel pump pressure specifications with the fuel pressure regulator vacuum hose attached.

HYD: Hydraulic

① The basic setting is controlled by the PCM and is not adjustable

06009-CSFR-C03

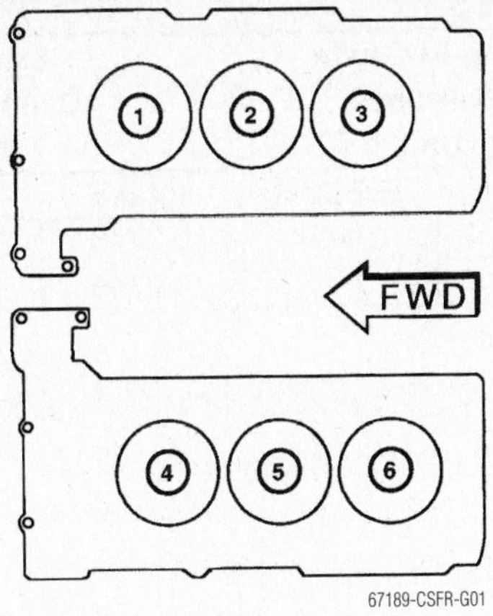

3.2L Engine
Firing Order 1-4-3-6-2-5
Distributorless ignition system (one coil on each cylinder)

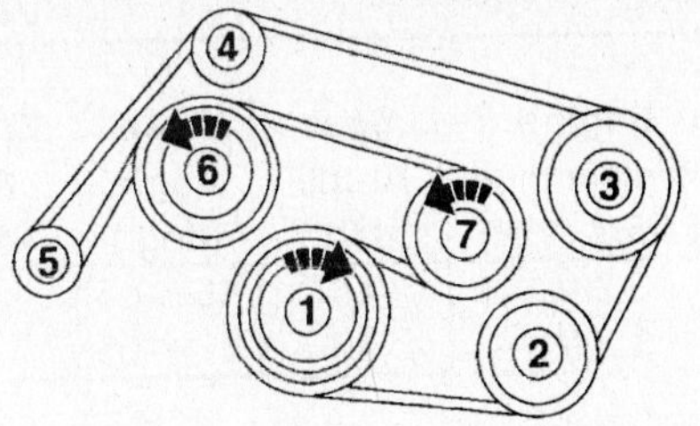

1. Crankshaft
2. A/C compressor
3. Power steering pump
4. Idler pulley
5. Alternator
6. Coolant pump
7. Automatic belt tensioner

67189-CSFR-G02

Accessory drive belt routing–3.2L Engine

CAPACITIES

Year	Model	Engine Displacement Liters (VIN)	Engine Oil with Filter	Transmission (pts.) 6-Spd	Transmission (pts.) Auto	Drive Axle Front (pts.)	Drive Axle Rear (pts.)	Fuel Tank (gal.)	Cooling System (qts.)
2005	Crossfire	3.2 (L)	8.5	3.8	17.0	—	2.8	15.8	11.8
		3.2 (N)	8.5	3.8	17.0	—	2.8	15.8	15.3

06009-CSFR-C04

CRANKSHAFT AND CONNECTING ROD SPECIFICATIONS

All measurements are given in inches.

Year	Engine Displacement Liters (VIN)	Crankshaft Main Brg. Journal Dia.	Crankshaft Main Brg. Oil Clearance	Crankshaft Shaft End-play	Crankshaft Thrust on No.	Connecting Rod Journal Diameter	Connecting Rod Oil Clearance	Connecting Rod Side Clearance
2005	3.2 (L)	NA	0.0010-0.0020	NA	3	NA	0.0010-0.0020	NA
	3.2 (N)	NA	0.0010-0.0020	NA	3	NA	0.0010-0.0020	NA

NA: Information not available

06009-CSFR-C05

VALVE SPECIFICATIONS

Year	Engine Displacement Liters (VIN)	Seat Angle (deg.)	Face Angle (deg.)	Spring Test Pressure (lbs. @ in.)	Spring Installed Height (in.)	Stem-to-Guide Clearance (in.)		Stem Diameter (in.)	
						Intake	Exhaust	Intake	Exhaust
2005	3.2 (L)	45	45+/-1.5	NA	NA	NA	NA	0.2746	0.2745
	3.2 (N)	45	45+/-1.5	NA	NA	NA	NA	0.2746	0.2745

NA: Information not available

06009-CSFR-C06

PISTON AND RING SPECIFICATIONS

All measurements are given in inches

Year	Engine Displacement Liters (VIN)	Piston Clearance	Ring Gap			Ring Side Clearance		
			Top Compression	Bottom Compression	Oil Control	Top Compression	Bottom Compression	Oil Control
2005	3.2 (L)	0.0010-0.0020	0.0078-0.0138	0.0078-0.0158	NA	0.0004-0.0023	0.0003-0.0011	NA
	3.2 (N)	0.0010-0.0020	0.0078-0.0138	0.0078-0.0158	NA	0.0004-0.0023	0.0003-0.0011	NA

NA: Information not available

06009-CSFR-C07

TORQUE SPECIFICATIONS

All readings in ft. lbs.

Year	Engine Displacement Liters (VIN)	Cylinder Head Bolts	Main Bearing Bolts	Rod Bearing Bolts	Crankshaft Damper Bolts	Flywheel Bolts	Manifold		Spark Plugs	Oil Pan Drain Plug
							Intake	Exhaust		
2005	3.2 (L)	①	②	③	④	⑤	15	26	21	22
	3.2 (N)	①	②	③	④	⑤	15	26	21	22

① Step 1: 15 ft. lbs.
Step 2: 37 ft. lbs.
Step 3: 65 degrees
Step 4: 65 degrees

② M8 Bolt
Step 1: 15 ft. lbs.
Step 2: plus 90 degrees
M10 bolt
Step 1: 22 ft. lbs.
Step 2: plus 90 degres

③ Step 1: 44 inch lbs.
Step 2: 18 ft. lbs.
Step 3: 90 degrees

④ Step 1: 148 ft. lbs.
Step 2: 90 degrees

⑤ Step 1: 33 ft. lbs.
Step 2: 90 degrees

06009-CSFR-C08

WHEEL ALIGNMENT

Year	Model		Caster Range (+/-Deg.)	Caster Preferred Setting (Deg.)	Camber Range (+/-Deg.)	Camber Preferred Setting (Deg.)	Toe-in (in.)
2005	Crossfire	F	—	5.20	0.33	-1.22	0.16 +/- 0.12
		R	—	—	—	-1.13	—

06009-CSFR-C09

TIRE, WHEEL AND BALL JOINT SPECIFICATIONS

Year	Model	OEM Tires Standard	OEM Tires Optional	Tire Pressures (psi) Front	Tire Pressures (psi) Rear	Wheel Size (in.)	Wheel Lug Nut Torque (ft. lbs.)
2005	Crossfire	Front-225/40ZR18	—	32	—	18	81
		Rear-255/35ZR19	—	—	33	19	81

OEM: Original Equipment Manufacturer

PSI: Pounds Per Square Inch

06009-CSFR-C10

BRAKE SPECIFICATIONS

All measurements in inches unless noted

Year	Model		Brake Disc Original Thickness	Brake Disc Minimum Thickness	Brake Disc Maximum Runout	Minimum Lining Thickness	Brake Caliper Bracket Bolts (ft. lbs.)	Brake Caliper Mounting Bolts (ft. lbs.)
2005	Crossfire	F	1.100	0.990	0.001	0.080	85	18
		R	0.350	0.290	0.001	0.080	—	41
	Crossfire SRT	F	1.250	1.160	0.001	0.080	85	18
		R	0.870	0.790	0.001	0.080	—	41

06009-CSFR-C11

SCHEDULED MAINTENANCE INTERVALS

Chrysler—Crossfire

TO BE SERVICED	TYPE OF SERVICE	VEHICLE MILEAGE INTERVAL (x1000)										
		5	10	20	30	40	50	60	70	80	90	100
Engine oil & filter ①	R	✓	✓	✓	✓	✓	✓	✓	✓	✓	✓	✓
Automatic shiftlock operation	S/I		✓	✓	✓	✓	✓	✓	✓	✓	✓	✓
Cooling system	S/I			✓		✓		✓		✓		✓
Passenger compartment air filter	R			✓		✓		✓		✓		✓
Automatic transmission fluid, filter & final drive	S/I			✓		✓		✓		✓		✓
Battery electrolyte level	S/I			✓		✓		✓		✓		✓
Brake system (brake pads & fluid level)	S/I		✓	✓	✓	✓	✓	✓	✓	✓	✓	✓
Drive axle shaft boots	S/I			✓		✓		✓		✓		✓
Engine (check for leaks)	S/I			✓		✓		✓		✓		✓
Exhaust system	S/I			✓		✓		✓		✓		✓
Idle speed	S/I											
Manual transmission fluid	S/I											
OBD System check for codes	S/I			✓		✓		✓		✓		✓
V-belts	S/I			✓		✓				✓		✓
Air cleaner element	R					✓				✓		
Spark plugs	R					✓				✓		
Power steering fluid level	S/I					✓				✓		
Automatic transmission fluid	R					✓				✓		
Timing belt	R											
Brake Fluid ②	R					✓				✓		
Front axle dust seals on ball joints & tie rod ends	S/I					✓				✓		✓
Poly-ribbed belt	S/I					✓				✓		
Rotate tires	S/I		✓	✓	✓	✓	✓	✓	✓	✓	✓	✓

R: Replace S/I: Service or Inspect

① Reset service interval display, if equipped.

② Replace every 2 years regardless of mileage.

FREQUENT OPERATION MAINTENANCE (SEVERE SERVICE)

If a vehicle is operated under any of the following conditions it is considered severe service:

- Extremely dusty areas.

- 50% or more of the vehicle operation is in 32°C (90°F) or higher temperatures, or constant operation in temperatures below 0°C (32°F).

- Prolonged idling (vehicle operation in stop and go traffic).

- Frequent short running periods (engine does not warm to normal operating temperatures).

- Police, taxi, delivery usage or trailer towing usage.

Oil & oil filter: change every 5000 miles.

06009-CSFR-C12

PRECAUTIONS

Before servicing any vehicle, please be sure to read all of the following precautions, which deal with personal safety, prevention of component damage, and important points to take into consideration when servicing a motor vehicle:

- Never open, service or drain the radiator or cooling system when the engine is hot; serious burns can occur from the steam and hot coolant.
- Observe all applicable safety precautions when working around fuel. Whenever servicing the fuel system, always work in a well-ventilated area. Do not allow fuel spray or vapors to come in contact with a spark, open flame, or excessive heat (a hot drop light, for example). Keep a dry chemical fire extinguisher near the work area. Always keep fuel in a container specifically designed for fuel storage; also, always properly seal fuel containers to avoid the possibility of fire or explosion. Refer to the additional fuel system precautions later in this section.
- Fuel injection systems often remain pressurized, even after the engine has been turned **OFF**. The fuel system pressure must be relieved before disconnecting any fuel lines. Failure to do so may result in fire and/or personal injury.
- Brake fluid often contains polyglycol ethers and polyglycols. Avoid contact with the eyes and wash your hands thoroughly after handling brake fluid. If you do get brake fluid in your eyes, flush your eyes with clean, running water for 15 minutes. If eye irritation persists, or if you have taken brake fluid internally, IMMEDIATELY seek medical assistance.
- The EPA warns that prolonged contact with used engine oil may cause a number of skin disorders, including cancer! You should make every effort to minimize your exposure to used engine oil. Protective gloves should be worn when changing oil. Wash your hands and any other exposed skin areas as soon as possible after exposure to used engine oil. Soap and water, or waterless hand cleaner should be used.
- All new vehicles are now equipped with an air bag system, often referred to as a Supplemental Restraint System (SRS) or Supplemental Inflatable Restraint (SIR) system. The system must be disabled before performing service on or around system components, steering column, instrument panel components, wiring and sensors. Failure to follow safety and disabling procedures could result in accidental air bag deployment, possible personal injury and unnecessary system repairs.
- Always wear safety goggles when working with, or around, the air bag system. When carrying a non-deployed air bag, be sure the bag and trim cover are pointed away from your body. When placing a non-deployed air bag on a work surface, always face the bag and trim cover upward, away from the surface. This will reduce the motion of the module if it is accidentally deployed. Refer to the additional air bag system precautions later in this section.
- Clean, high quality brake fluid from a sealed container is essential to the safe and proper operation of the brake system. You should always buy the correct type of brake fluid for your vehicle. If the brake fluid becomes contaminated, completely flush the system with new fluid. Never reuse any brake fluid. Any brake fluid that is removed from the system should be discarded. Also, do not allow any brake fluid to come in contact with a painted surface; it will damage the paint.
- Never operate the engine without the proper amount and type of engine oil; doing so WILL result in severe engine damage.
- Timing belt maintenance is extremely important! Many models utilize an interference-type, non-freewheeling engine. If the timing belt breaks, the valves in the cylinder head may strike the pistons, causing potentially serious (also time-consuming and expensive) engine damage. Refer to the maintenance interval charts in the front of this manual for the recommended replacement interval for the timing belt, and to the timing belt section for belt replacement and inspection.
- Disconnecting the negative battery cable on some vehicles may interfere with the functions of the on-board computer system(s) and may require the computer to undergo a relearning process once the negative battery cable is reconnected.
- When servicing drum brakes, only disassemble and assemble one side at a time, leaving the remaining side intact for reference.
- Only an MVAC-trained, EPA-certified automotive technician should service the air conditioning system or its components.

ENGINE REPAIR

➡Disconnecting the negative battery cable on some vehicles may interfere with the functions of the on board computer system. The computer may undergo a relearning process once the negative battery cable is reconnected.

Distributor

➡The 3.2L engine does not use a distributor.

Alternator

REMOVAL

1. Before servicing the vehicle, refer to the precautions in the beginning of this section.
2. Remove or disconnect the following:
   - Negative battery cable
   - Right side air inlet tube
   - Accessory drive belt
   - Alternator harness connectors
   - Alternator

INSTALLATION

1. Before servicing the vehicle, refer to the precautions in the beginning of this section.
2. Install the alternator and tighten the bolts to 31 ft. lbs. (42 Nm).
3. Install or connect the following:
   - Alternator harness connectors and tighten the retaining nut to 11 ft. lbs. (15Nm)
   - Accessory drive belt
   - Right side air inlet tube
   - Negative battery cable

Ignition Timing

ADJUSTMENT

The ignition timing is controlled by the Powertrain Control Module (PCM) and is not adjustable.

Engine Assembly

REMOVAL & INSTALLATION

1. Before servicing the vehicle, refer to the precautions in the beginning of this section.

2. Drain the cooling system.
3. Recover the A/C refrigerant, if equipped.
4. Drain the engine oil.
5. Relieve the fuel system pressure.
6. Remove or disconnect the following:
   - Negative battery cable
   - Air cleaner housing assembly
   - Radiator hoses
   - Radiator and fan shroud
   - Accessory drive belt
   - Vacuum hoses from brake booster, intake manifold and purge valve
   - Power steering fluid from pump
   - Ground lead at power steering pump
   - Power steering pressure and return lines at pump
   - Oxygen (O_2) sensors
   - Mass Air Flow (MAF) sensor
   - Heater hose
   - Fuel supply line
   - Engine wiring harness
   - Lower radiator hose and coolant bypass hose from water pump
   - Upper radiator hose from thermostat housing
7. Raise and support vehicle on jack stands.
8. Remove or disconnect the following:
   - Engine undercover
   - Front exhaust pipes
   - Ground cable at transmission
   - Driveshaft
   - Starter
9. Place a transmission jack under the transmission.
10. Remove or disconnect the following:
   - Transmission assembly
   - A/C compressor and place aside with lines attached
   - Front engine mount bolts

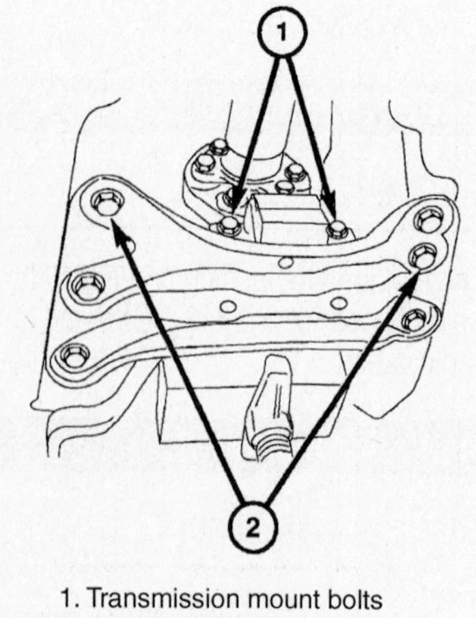

Rear crossmember/transmission mount

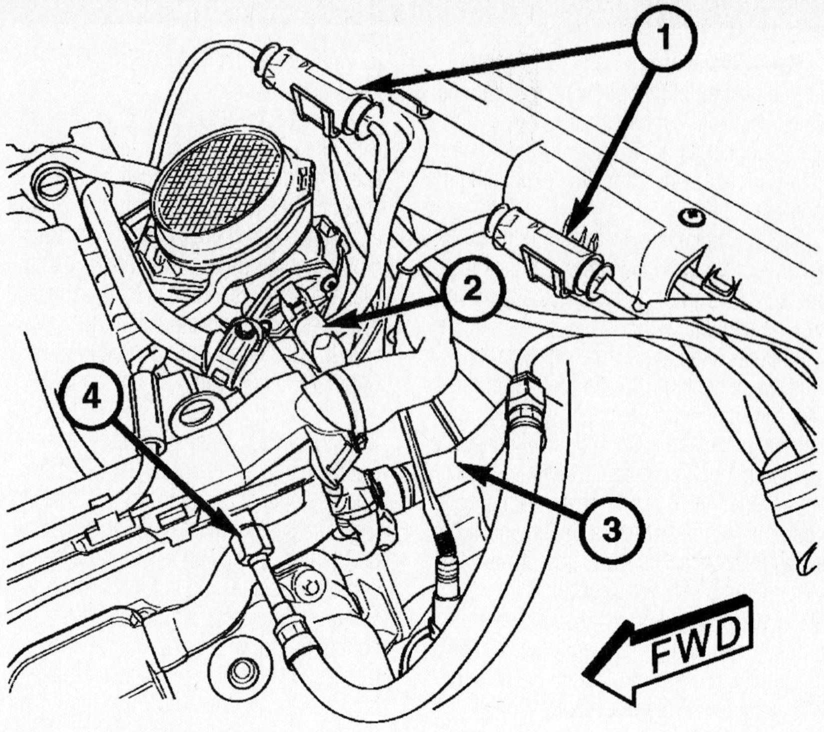

Oxygen sensor locations

11. Lower the vehicle.
12. Attach a suitable engine lifting device to the engine lifting eyes.
13. Remove engine assembly from the vehicle.

To install:

14. Lower the engine into the vehicle.
15. Install or connect the following:
   - Front engine mount. Tighten the bolt to 41 ft. lbs. (55 Nm).
   - Transmission
   - A/C compressor. Tighten the bolts to 17 ft. lbs. (23 Nm).
   - Transmission ground cable. Tighten to 30 ft. lbs. (40 Nm).
   - Starter
   - Driveshaft
   - Front exhaust pipes
   - Engine undercover
   - Lower radiator hose and coolant bypass hose
   - Heater hoses to the rear of the intake manifold
   - Upper radiator hose
   - Engine wiring harness
   - Fuel supply line. Tighten fitting to 28 ft. lbs. (38 Nm).
   - O_2 sensors, MAF sensor and heater hose
   - Power steering lines. Tighten fittings to 33 ft. lbs. (45 Nm).
   - Power steering pump ground. Tighten bolt to 18 ft. lbs. (25 Nm).
   - Vacuum hoses
   - Accessory drive belt
   - Radiator and fan
   - Air cleaner housing
   - Negative battery cable
16. Fill the power steering pump to the correct level.
17. Fill the crankcase to the correct level.
18. Fill the cooling system.
19. Start the engine and check for leaks.

Water Pump

REMOVAL & INSTALLATION

Except SRT Models

1. Before servicing the vehicle, refer to the precautions in the beginning of this section.
2. Drain the cooling system.
3. Remove or disconnect the following:
   - Negative battery cable
   - Air intake assembly
   - Radiator fan
   - Accessory drive belt
   - Belt tensioner
   - Lower radiator hoses
   - Alternator
   - Pump pulley and idler pulley
   - Water pump (16 bolts)
4. Installation is the reverse of removal. Tighten the bolts to 28 ft. lbs. (35 Nm).

To install:

5. Install or connect the following:
   - Water pump with a new gasket. Tighten bolts to 26 ft. lbs. (35 Nm).
   - Alternator
   - Belt tensioner
   - Pump pulley and idler pulley. Tighten bolts to 89 inch lbs. (10 Nm).
   - Lower radiator hoses
   - Accessory drive belt
   - Radiator fan
   - Air intake assembly
   - Negative battery cable.
6. Fill cooling system.
7. Start the engine and check for leaks.

SRT Models

1. Before servicing the vehicle, refer to the precautions in the beginning of this section.
2. Drain the cooling system.
3. Remove or disconnect the following:
   - Negative battery cable
   - Radiator fan
   - Air pump tube assembly
4. Release the tension on the accessory drive belt and remove the belt from the pulleys.
5. Remove or disconnect the following:
   - Supercharger idler pulley
   - Accessory drive belt
   - Belt tensioner pulley and belt tensioner
   - Lower radiator hoses
   - Alternator
   - Pump pulley and idler pulley
   - Oil cooler hose clamp
   - Water pump
   - Oil cooler hose from water pump

To install:

6. Install or connect the following:
   - Water pump with a new gasket. Tighten bolts to 26 ft. lbs. (35 Nm).
   - Oil cooler hose to the water pump and hose clamp
   - Pump pulley and idler pulley. Tighten bolts to 89 inch lbs. (10 Nm).
   - Lower radiator hoses
   - Alternator
   - Supercharger idler pulley
   - Belt tensioner pulley and belt tensioner
   - Accessory drive belt
   - Air pump tube assembly
   - Radiator fan
   - Negative battery cable
7. Fill the cooling system.
8. Start the engine and check for leaks.

Heater Core

REMOVAL & INSTALLATION

1. Disconnect the negative battery cable.

**** CAUTION**

After disconnecting the negative battery cable, wait 2 minutes for the driver's/passenger's air bag system capacitor to discharge before attempting to do any work around the steering column or instrument panel.

2. Drain the cooling system.
3. Recover the A/C refrigerant, if equipped.
4. Remove or disconnect the following:
   - Heater hoses
   - A pillar trim
   - Fuse block covers
   - Left and right air vents
   - Center console-to-instrument top panel screws
   - Glove box
   - Defroster vents
   - A pillar sheet metal clips
   - Instrument panel top section
   - Steering column cover
   - Lower instrument panel
   - Instrument cluster cover
   - Instrument cluster
   - Steering wheel
   - Sentry Key Remote Entry Module (SKREEM)
   - Wiring connectors at transmission tunnel
   - Steering column bolts-to-instrument panel support
   - Position steering column aside
   - Electrical harness connector on passenger airbag module
   - Left and right heater ducts
   - Heater core attaching nuts
   - Instrument panel support bolts
   - Vacuum reservoir lines
   - Instrument panel support
   - HVAC housing electrical connectors
   - HVAC housing
5. Separate heater core from HVAC housing

To install:

6. Install or connect the following:
   - Heater core to HVAC housing
   - HVAC housing
   - HVAC housing electrical connectors
   - Instrument panel support

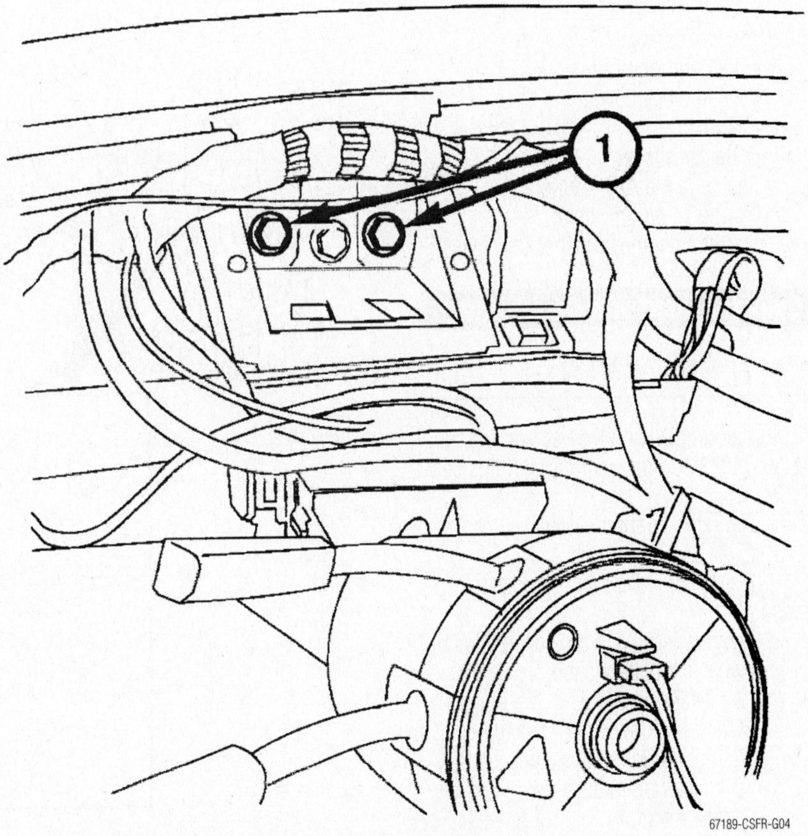

Removing instrument panel support bolts from bulkhead—Crossfire

- Instrument panel support bolts
- Steering column bolts-to-instrument panel support
- Wiring connectors at transmission tunnel
- Sentry Key Remote Entry Module (SKREM)
- Instrument cluster
- Instrument cluster cover
- Steering wheel
- Left and right heater ducts
- Vacuum reservoir lines
- Instrument panel support bolts
- HVAC housing attaching nuts
- Position steering column into place
- Steering column bolts-to-instrument panel support
- Wiring connectors at transmission tunnel
- Sentry Key Remote Entry Module (SKREM)
- Steering wheel
- Instrument cluster
- Instrument cluster cover
- Left and right heater ducts
- Lower instrument panel cover
- Lower instrument panel screws
- Steering column cover
- Instrument panel top section
- A pillar sheet metal clips
- Defroster vents
- Glove box
- Center console-to-instrument top panel screws
- Left and right air vents
- Fuse block covers
- A pillar trim
- Heater hoses

7. Recharge the A/C refrigerant.
8. Fill the cooling system.
9. Start the engine and check for leaks.

Cylinder Head and Camshafts

REMOVAL & INSTALLATION

1. Before servicing the vehicle, refer to the precautions in the beginning of this section.
2. Drain the cooling system.
3. Drain the engine oil
4. Properly relieve the fuel system pressure.
5. Remove or disconnect the following:
   - Negative battery cable
   - Air intake assembly
   - Mass Air Flow (MAF) sensor
   - Radiator and radiator fan
   - Camshaft position sensor
   - Intake manifold
   - Cylinder head cover
   - Exhaust pipes from exhaust manifold
6. Rotate the crankshaft so that the crankshaft timing mark aligns with the 40 degree After Top Dead Center (ATDC) mark

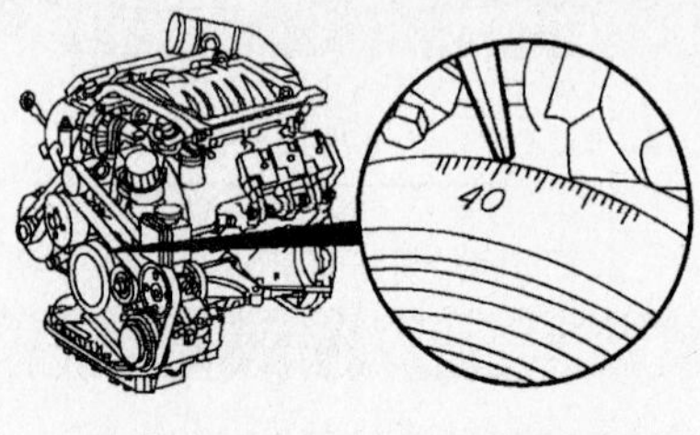

67189-CSFR-G05

Crankshaft timing mark set to 40 degrees ATDC—3.2L engine

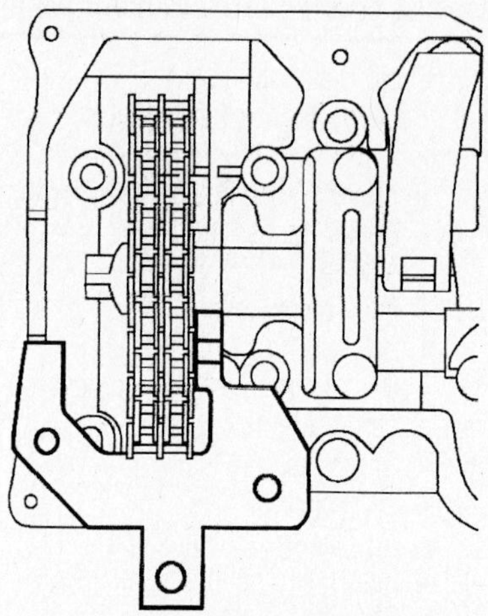

67189-CSFR-G06

Camshaft locking plate—3.2L engine

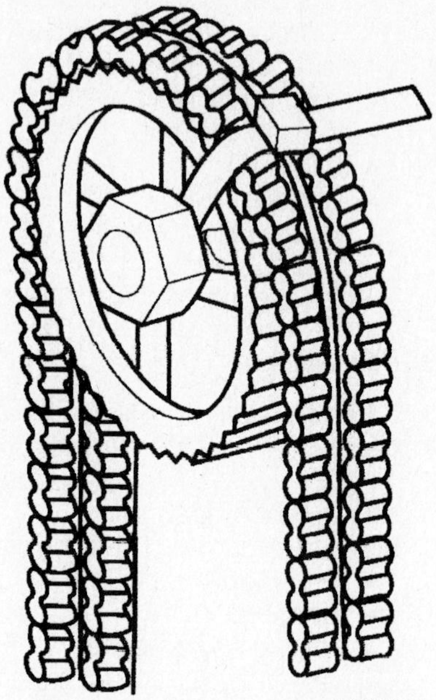

67189-CSFR-G08

Securing timing chain to camshaft sprocket—3.2L engine

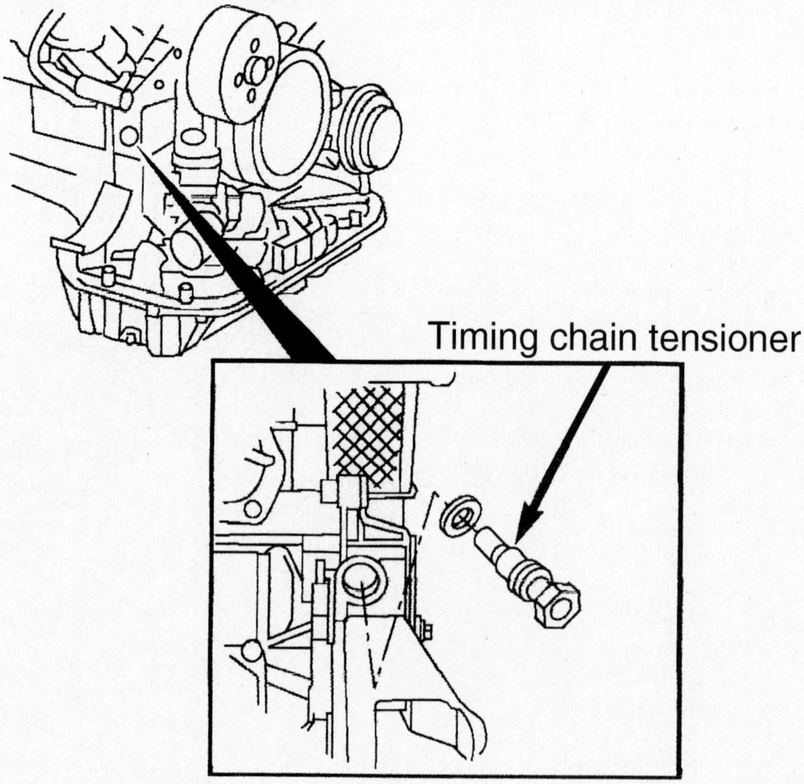

67189-CSFR-G07

Removing timing chain tensioner—3.2L engine

on the front cover as shown. Grooves in the camshafts must be toward the inside of the wedge.

7. Lock the camshafts in place with Camshaft Locking Plates 9104 and 9105.

8. Remove the timing chain tensioner as shown.

9. Use a cable tie and secure the timing chain the camshaft sprocket.

10. Remove the camshaft sprockets.

11. Remove the camshaft locking plate tools.

12. Loosen and remove the camshaft bearing cap bolts in the sequence shown.

13. Remove the rocker arms and the camshafts.

14. Remove the cylinder head bolts in the sequence shown.

15. Remove the cylinder head.

➡The cylinder head is retained by sixteen bolts. Eight of the bolts are smaller and are at the front of the head.

To install:

16. Check the cylinder head bolts for signs of stretching. If length exceeds 5.67 inches (144 mm), replace bolts as necessary.

17. Lubricate the threads of the bolts with clean engine oil.

18. Install the cylinder heads. Use new gaskets and tighten the bolts, in sequence, as follows:

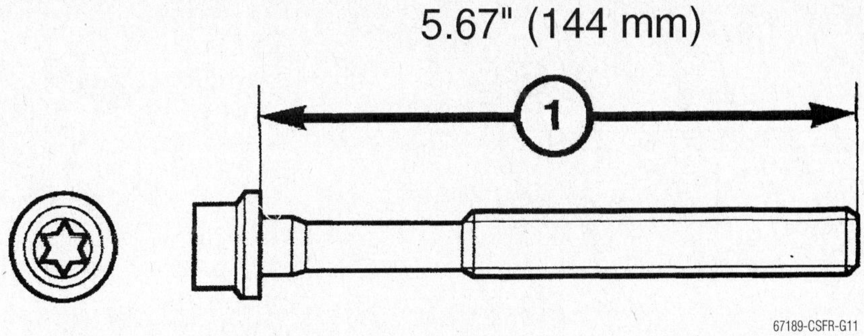

Examine the head bolts for signs of stretching—3.2L engine

a. Step 1: Bolts 1–8 to 7 ft. lbs. (10 Nm)

b. Step 2: Bolts 1–8 to 37 ft. lbs. (50 Nm)

c. Step 3: Bolts 1–8 plus 65 degrees clockwise

d. Step 4: Bolts 1–8 plus 65 degrees clockwise

e. Step 5: Bolts 9 and 10 to 18 ft. lbs. (25 Nm)

19. Install the rocker arms and the camshafts.

20. Install the camshaft bearing cap bolts in the sequence shown. Tighten the bolts to 11 ft. lbs. (15 Nm), plus an additional 90 degrees.

21. Lock the camshafts in place with Camshaft Locking Plates 9104 and 9105.

22. Install the camshaft sprockets. Tighten the bolts to 37 ft. lbs. (50 Nm), plus an additional 90 degrees.

23. Remove cable tie from timing chain.

24. Install the timing chain tensioner. Tighten the bolt to 59 ft. lbs. (80 Nm).

25. Remove camshaft locking plates

26. Ensure crankshaft is set to 40 degrees ATDC

27. Install or connect the following:

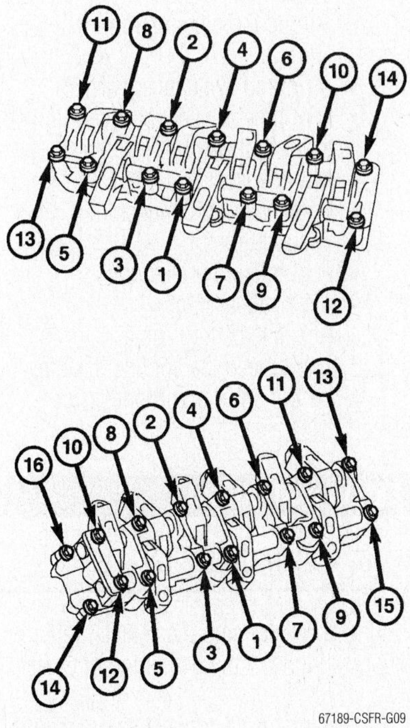

Camshaft bearing cap bolt removal sequence—3.2L engine

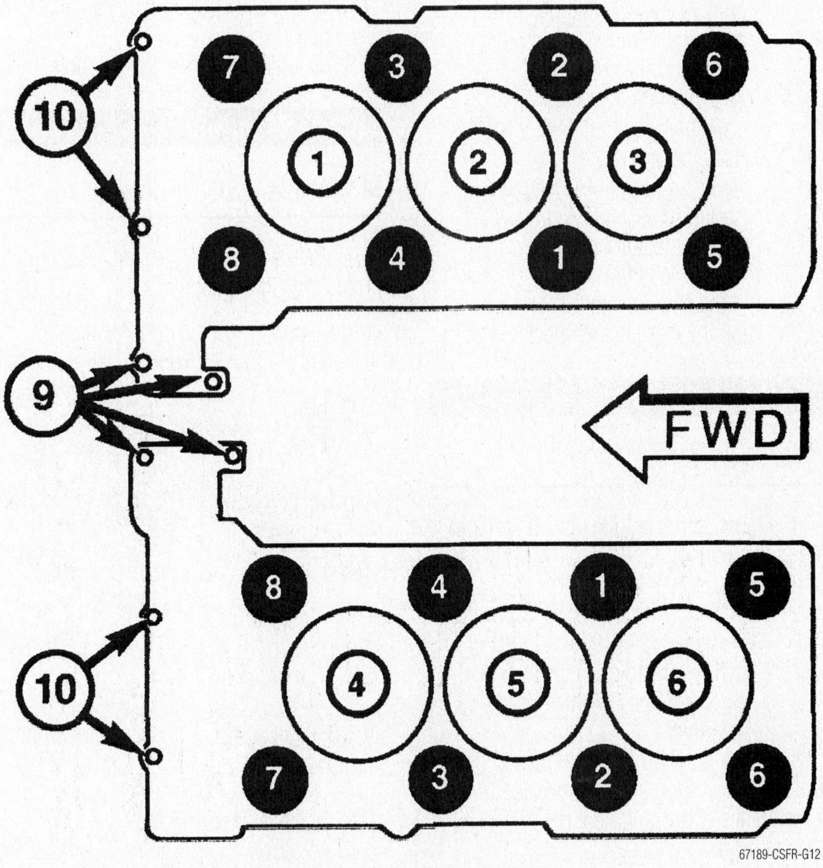

Cylinder head bolt loosening and torque sequence—3.2L engine

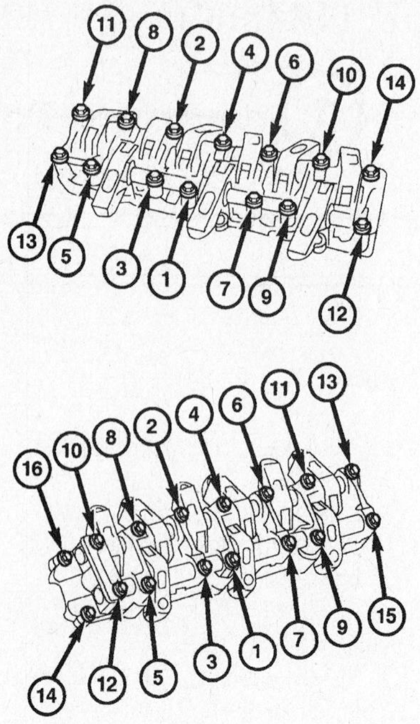

67189-CSFR-G13

Camshaft bearing cap bolt torque sequence—3.2L engine

- Exhaust pipes to manifold. Tighten the bolts to 15 ft. lbs. (20 Nm).
- Cylinder head covers. Tighten the bolts to 89 inch lbs. (10 Nm).
- Intake manifold.
- Ignition coils
- Accessory drive belt
- Camshaft position sensor
- Radiator and radiator fan
- Mass Air Flow (MAF) sensor
- Air intake assembly
- Negative battery cable

28. Fill and bleed the cooling system.
29. Fill the engine with oil.
30. Start the engine and check for leaks.

Rocker Arms/Shafts

REMOVAL & INSTALLATION

1. Before servicing the vehicle, refer to the precautions in the beginning of this section.
2. Remove or disconnect the following:
   - Negative battery cable
   - Cylinder head and camshaft bearing caps. See Cylinder Head and Camshaft.
3. Use a .63 inch (16 mm) drift to drive out the rocker shaft from the bearing bridge. If resistance is present, heat the bridge to ease removal.
4. Remove the rocker arms after marking them for reinstallation reference.

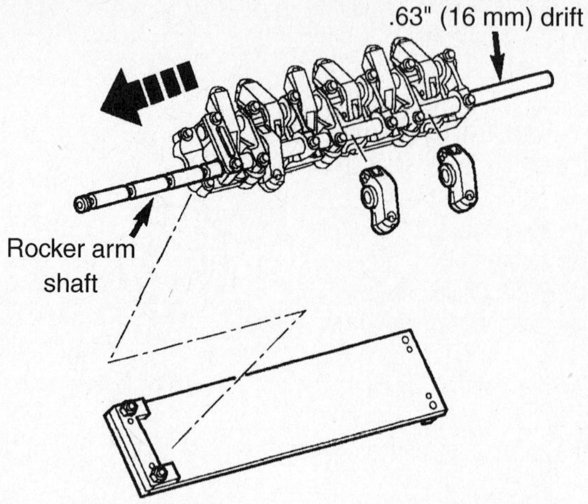

67189-CSFR-G14

Removing rocker arm shaft—3.2L engine

To install:

5. Place rocker arms in their correct locations.
6. Tap the rocker arm shaft into the bearing bridge using a mallet. Ensure the rocker arms are straight.
7. Insert two camshaft bearing bridge bolts to secure the shaft. The oil supply holes in the shaft point downward toward the cylinder head.
8. Install the camshaft bearing caps and cylinder head. See Cylinder Head and Camshaft.
9. Connect negative battery cable.

Intake Manifold

REMOVAL & INSTALLATION

1. Before servicing the vehicle, refer to the precautions in the beginning of this section.

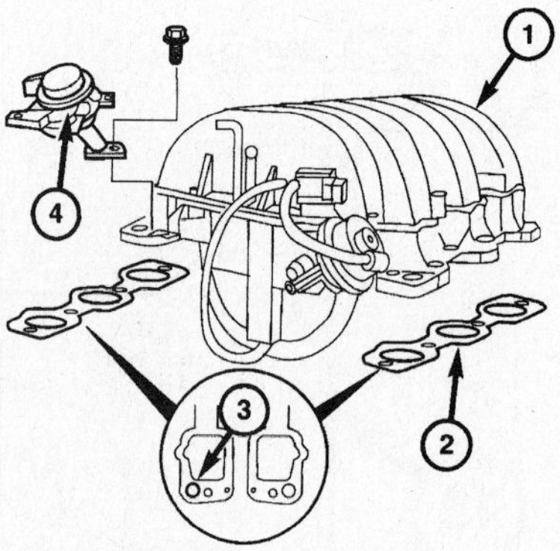

1. Intake manifold
2. Gasket
3. Locating hole
4. Air pump switchover valve

67189-CSFR-15

Exploded view of intake manifold—3.2L engine

2. Drain the cooling system.
3. Properly relieve the fuel system pressure.
4. Remove or disconnect the following:
   - Negative battery cable
   - Air intake assembly
   - Mass Air Flow (MAF) sensor
   - Fuel rail and injectors
   - Vacuum lines
   - Engine wiring connectors
   - EGR pipe at EGR valve
   - Air pump switchover valves
   - Intake manifold

To install:

5. Install or connect the following:
   - Intake manifold using new gaskets. Torque the bolts to 15 ft. lbs. (20 Nm).
   - Air pump switchover valves
   - EGR pipe at EGR valve and tighten to 30 ft. lbs. (40 Nm)
   - MAF sensor
   - Engine wiring connectors
   - Vacuum lines
   - Fuel rail and injectors
   - Air intake assembly
   - Negative battery cable
6. Fill and bleed the cooling system.
7. Start the engine and check for leaks.

Exhaust Manifold

REMOVAL & INSTALLATION

1. Before servicing the vehicle, refer to the precautions in the beginning of this section.
2. Remove or disconnect the following:
   - Air intake assembly
   - Engine undercover

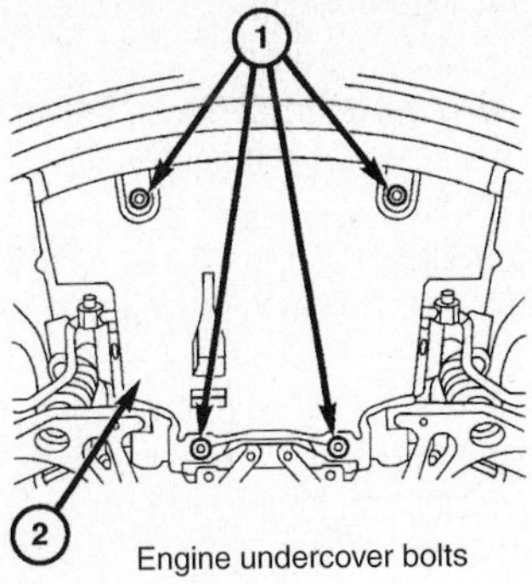

Engine undercover bolts

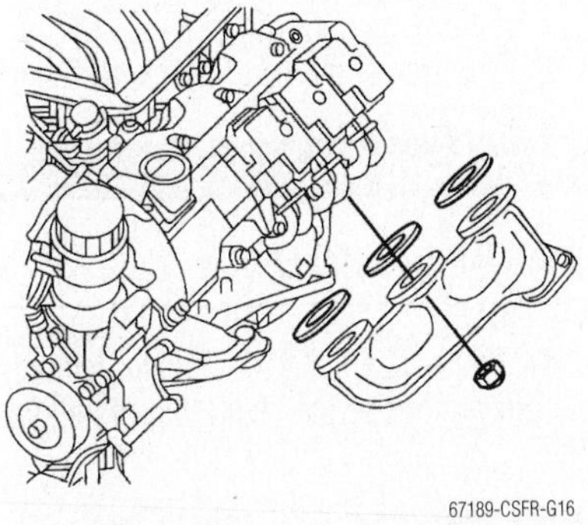

Exhaust manifold left side, right side similar—3.2L engine

- Catalytic converters
- Exhaust manifolds

To install:

3. Install or connect the following:
   - Exhaust manifolds, using new gaskets. Tighten the bolts to 26 ft. lbs. (35 Nm).
   - Catalytic converters
   - Engine undercover
   - Air intake assembly
4. Start the engine and check for leaks.

Valve Lash

ADJUSTMENT

The 3.2L engine uses hydraulic lifters. No maintenance or periodic adjustment is required.

Starter Motor

REMOVAL & INSTALLATION

1. Before servicing the vehicle, refer to the precautions in the beginning of this section.
2. Raise and support the vehicle.
3. Remove or disconnect the following:
   - Negative battery cable
   - Engine undercover
   - Right side oxygen (O_2) sensor
   - Right side exhaust pipe
   - Starter harness connections
   - Starter

To install:

4. Install or connect the following:
   - Starter. Tighten the bolts to 31 ft. lbs. (42 Nm).
   - Starter harness connections
   - Right side exhaust pipe
   - Right side O_2sensor
   - Engine undercover
   - Negative battery cable

Oil Pan

REMOVAL & INSTALLATION

Lower Oil Pan

1. Before servicing the vehicle, refer to the precautions in the beginning of this section.
2. Drain the engine oil.
3. Remove or disconnect the following:
   - Negative battery cable
   - Engine undercover
   - Transmission oil cooler line retainer bolts and place line aside
   - Oil pan
   - Oil pan gasket

To install:

4. Install a bead of sealant around oil pan perimeter.
5. Install or connect the following:
   - Oil pan. Tighten the bolts to 10 ft. lbs. (14 Nm).
   - Transmission oil cooler lines
   - Oil drain plug. Tighten to 22 ft. lbs. (30 Nm).
   - Engine undercover
   - Negative battery cable
6. Fill the engine with oil.
7. Start the engine and check for leaks.

Upper Oil Pan

1. Before servicing the vehicle, refer to the precautions in the beginning of this section.
2. Drain the engine oil.
3. Remove or disconnect the following:
   - Negative battery cable
   - Engine undercover
   - Lower oil pan
   - Front exhaust pipes
   - Oil level/temperature sensor harness connector
   - Front engine mount bolts
   - Oil dipstick tube assembly
4. Install an engine support bracket and attach a suitable lifting device.

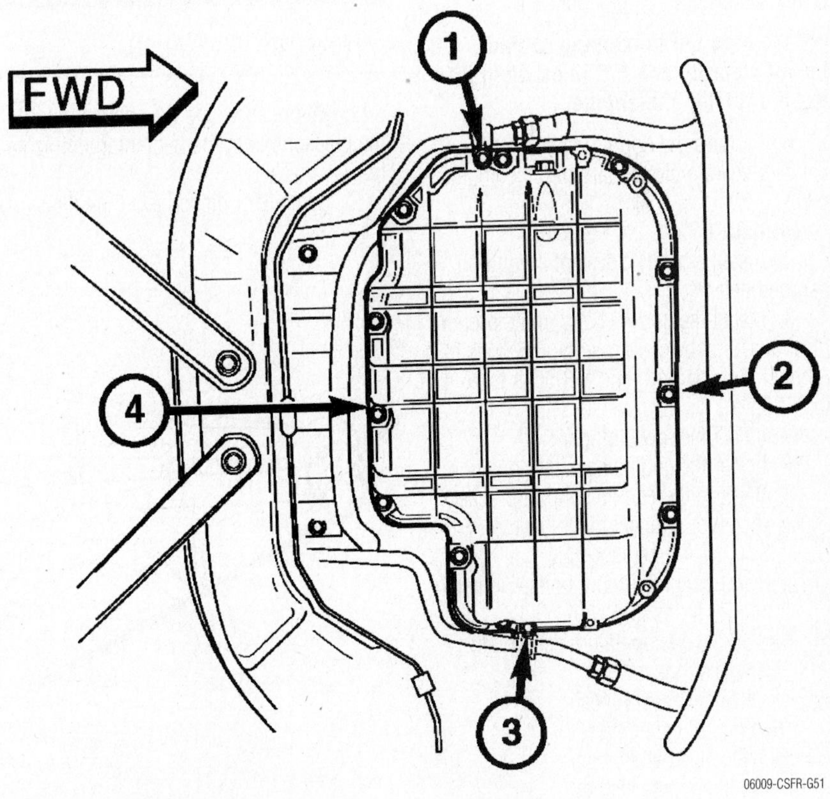

Lower oil pan mounting bolt locations.

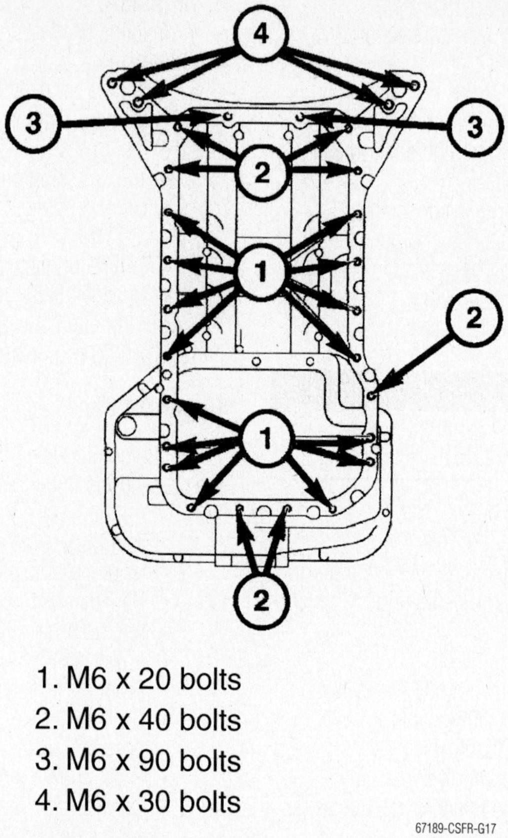

Upper oil pan mounting bolt locations—3.2L engine

5. Raise the engine slightly to provide clearance to upper oil pan bolts.

➡Take care not to damage coolant, power steering and A/C lines when lifting the raising the engine.

6. Remove the upper oil pan bolts.
7. Separate the oil pan from the engine block.

To install:

8. Install a bead of sealant around oil pan perimeter.
9. Install the correct bolts in the proper locations as shown. Torque the M6 bolts to 89 inch lbs. (10 Nm), and M8 bolts to 15 ft. lbs. (20 Nm).
10. Install the lower oil pan.
11. Lower the engine back into position.
12. Remove the engine support bracket.
13. Install or connect the following:
    - Dipstick tube assembly
    - Front engine mount bolts. Tighten to 26 ft. lbs. (35 Nm).
    - Oil level/temperature sensor harness connector
    - Front exhaust pipes
    - Engine undercover
    - Negative battery cable
14. Fill the engine with oil.
15. Start the engine and check for leaks.

Oil Pump

REMOVAL & INSTALLATION

1. Before servicing the vehicle, refer to the precautions in the beginning of this section.
2. Remove or disconnect the following:
    - Lower oil pan
    - Oil pump bolts
    - Release the oil pump timing chain tensioner and remove the oil pump

To install:

3. Fill the oil pump with clean engine oil.
4. Position the pump driven sprocket into the drive chain.
5. Install and torque the bolts to 15 ft. lbs. (20 Nm).
6. Install the lower oil pan.
7. Fill the engine with oil.
8. Start the engine and check for leaks.

Rear Main Seal

REMOVAL & INSTALLATION

1. Before servicing the vehicle, refer to the precautions in the beginning of this section.

➡The rear main seal cannot be replaced separately. The crankshaft end cover and oil seal are replaced as a set.

2. Remove or disconnect the following:
    - Transmission
    - Flywheel
    - Crankshaft end cover
3. Clean the engine block and oil pan sealing surfaces.

To install:

4. Install a bead of sealant as shown on the NEW end cover.
5. Install the end cover. Torque the bolts to 89 inch lbs. (10 Nm).

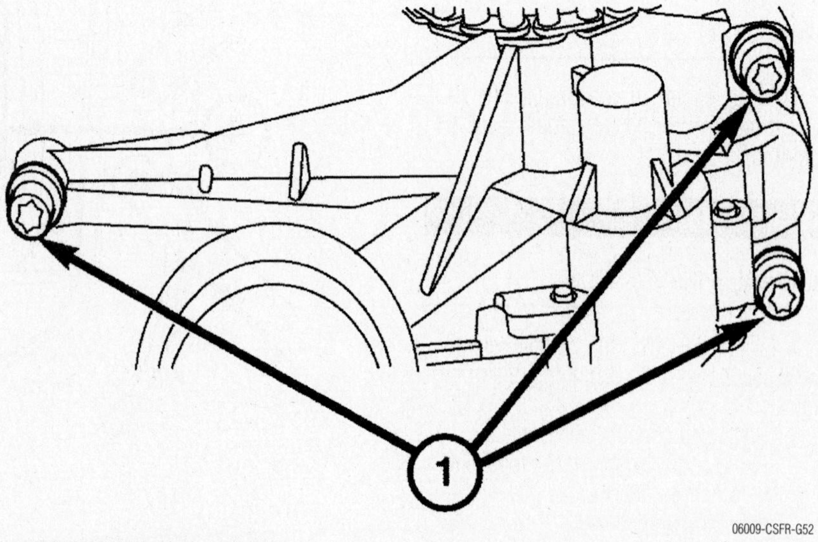

Oil pump mounting bolt locations.

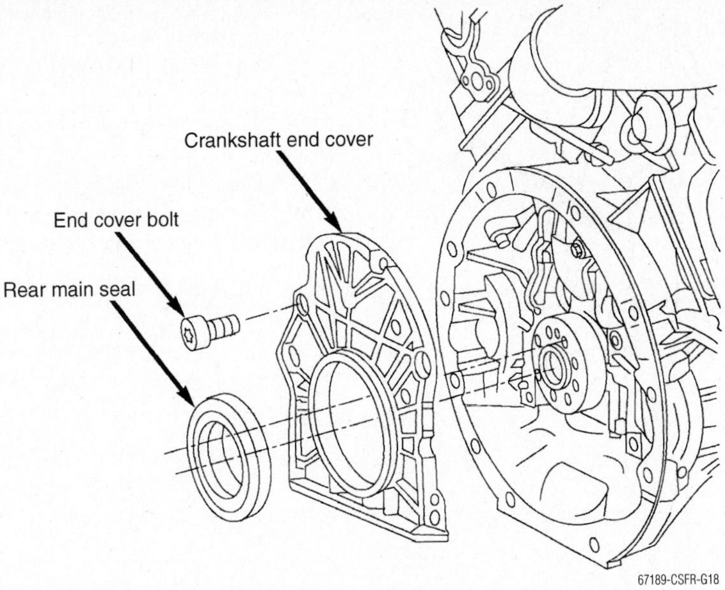

Crankshaft end cover assembly—3.2L engine

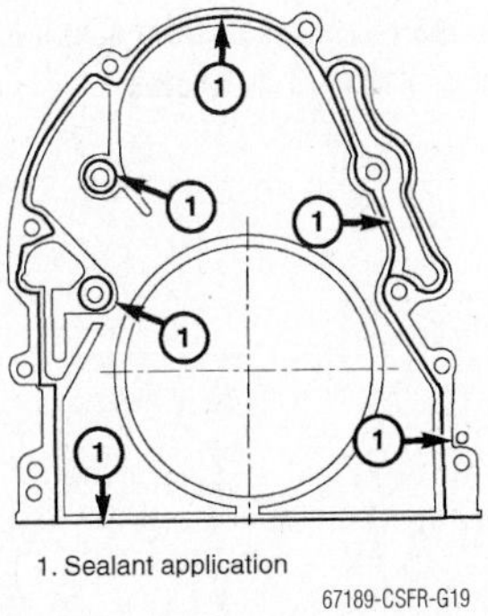

Crankshaft end cover sealant application—3.2L engine

6. Use Special Tool 9100 and install the real seal into the end cover.
7. Remove the seal installer.
8. Install the following:
   - Flywheel
   - Transmission

Timing Chain, Balance Shafts, Front Cover and Seal

REMOVAL & INSTALLATION

1. Before servicing the vehicle, refer to the precautions in the beginning of this section.

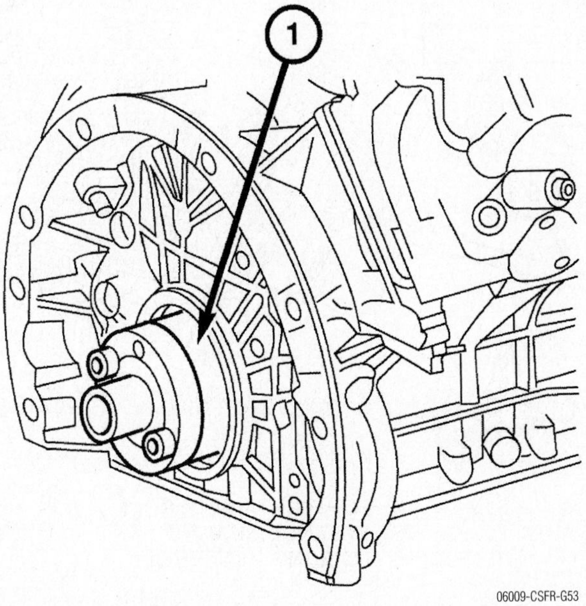

Using special tool 9110 to install the seal.

2. Drain the cooling system.
3. Drain the power steering pump fluid.
4. Remove or disconnect the following:
   - Negative battery cable
   - Radiator fan
   - Accessory drive belt
   - Idler pulley
   - Accessory drive belt tensioner
   - Vibration damper
   - Coolant and heater hoses from water pump
   - Power steering pump
   - AIR pump
   - Cylinder head covers
   - Front cylinder head to timing chain cover retaining bolts
   - Alternator
   - Lower oil pan
   - Upper oil pan
   - Starter
5. Rotate crankshaft to 40 degrees After Top Dead Center (ATDC).
6. Lock the flywheel in place using Special Tool 9102 by inserting the tool into the starter opening.
7. Remove or disconnect the following:
   - Timing chain tensioner
   - Timing chain cover
   - Oil pump drive chain and tensioner
   - Left and right camshaft sprocket bolts
   - Timing chain and camshaft sprockets
   - Timing chain camshaft sprocket
   - Rear main seal
8. Hold the balance shaft rear counter weight using a drift, and remove the rear retaining bolt.
9. Remove the balance shaft counterweight.
10. Remove the balance shaft locking plate at the front of the engine block.
11. Remove the balance shaft.
12. Pry out the front crankshaft seal using a suitable pry tool.

To install:

13. Position the front crankshaft seal into Special Tool 9103 as shown.

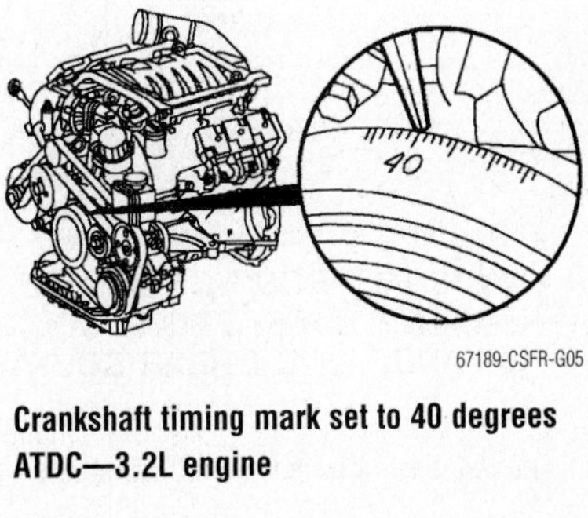

Crankshaft timing mark set to 40 degrees ATDC—3.2L engine

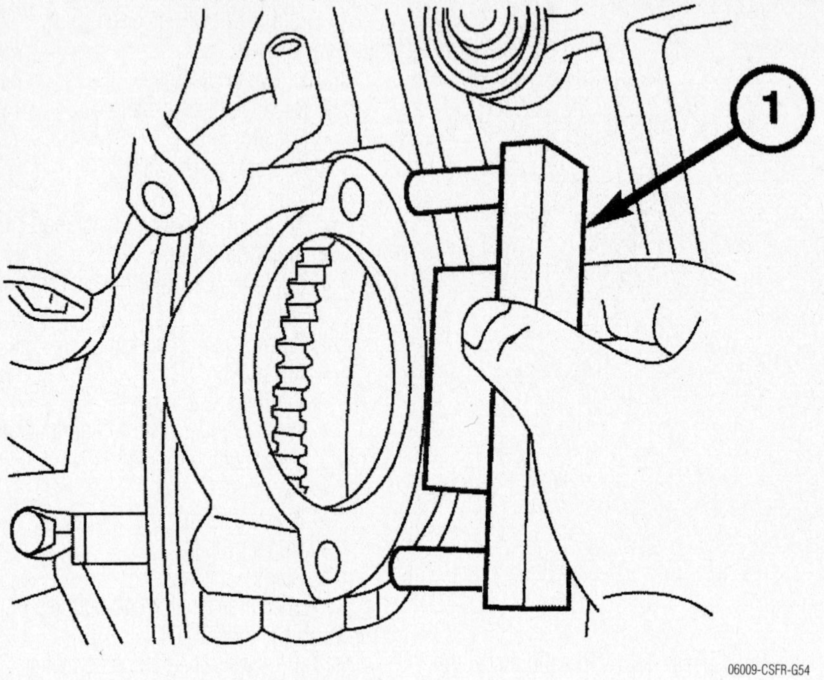

Locking the flywheel in place with tool 9102.

14. Align the slot of the installation tool into the crankshaft keyway.

15. Tap the tool in until the crankshaft damper bolt can be inserted.

16. Tighten the damper bolt until the seal is installed.

17. Remove the installer tool.

18. Place the balance shaft into the bore in the block from the front of the engine.

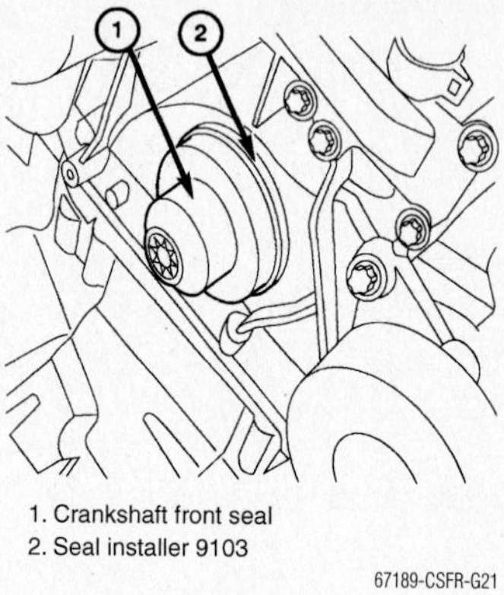

Positioning crankshaft front seal into Seal Installer 9103—3.2L engine

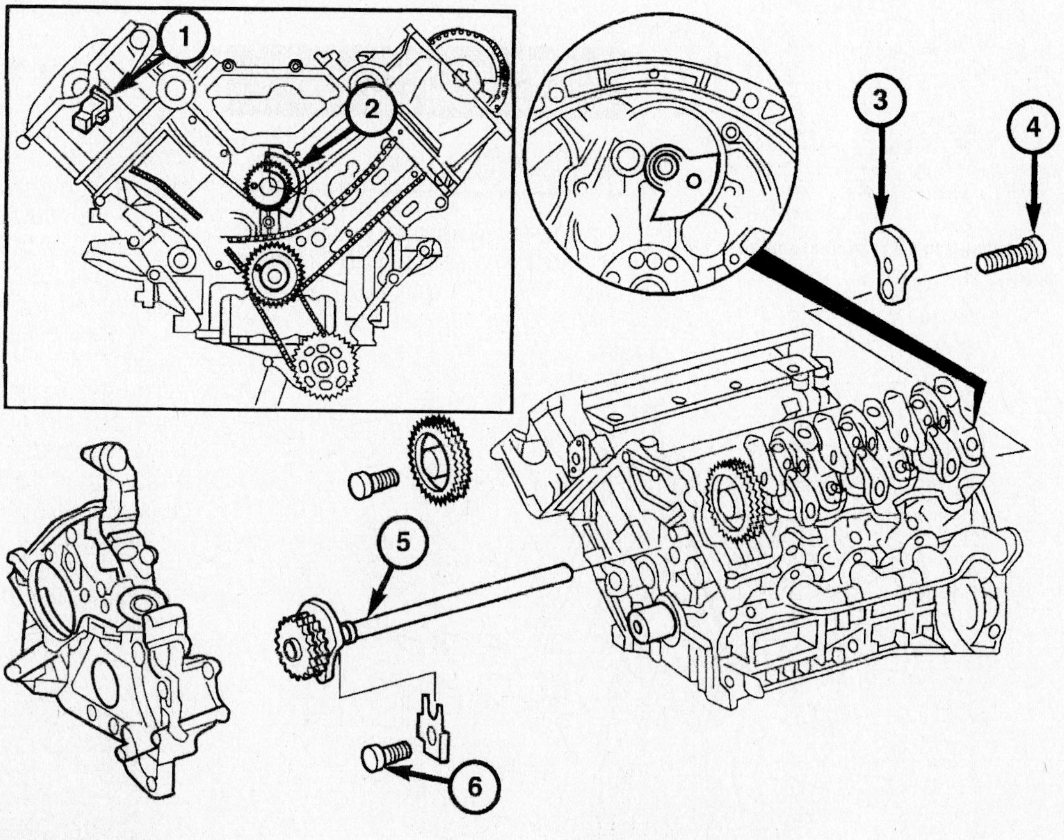

Exploded view of balance shaft assembly—3.2L engine

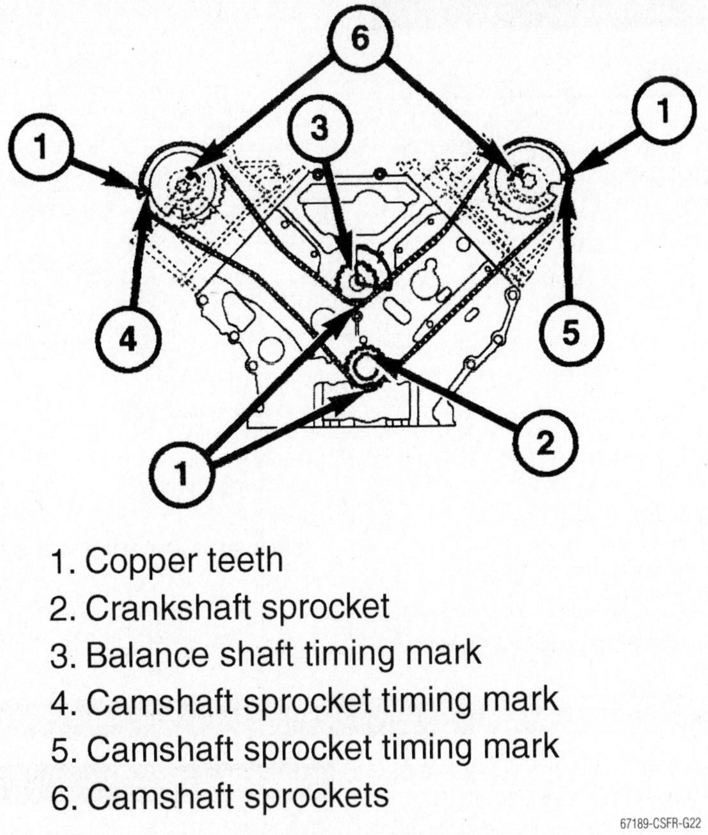

Positioning timing chain on camshaft sprockets—3.2L engine

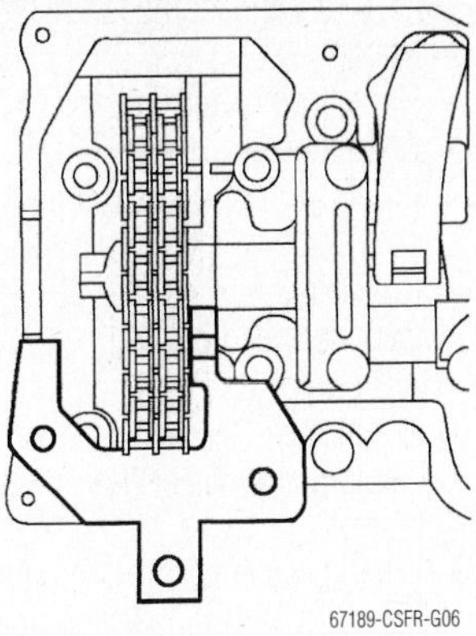

Camshaft locking plate—3.2L engine

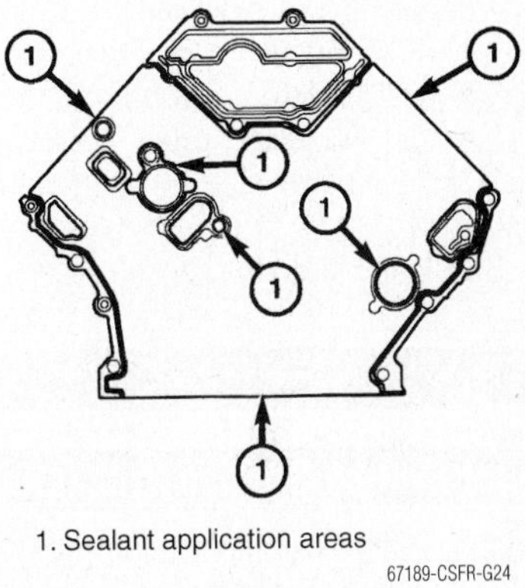

Applying sealant to front cover—3.2L engine

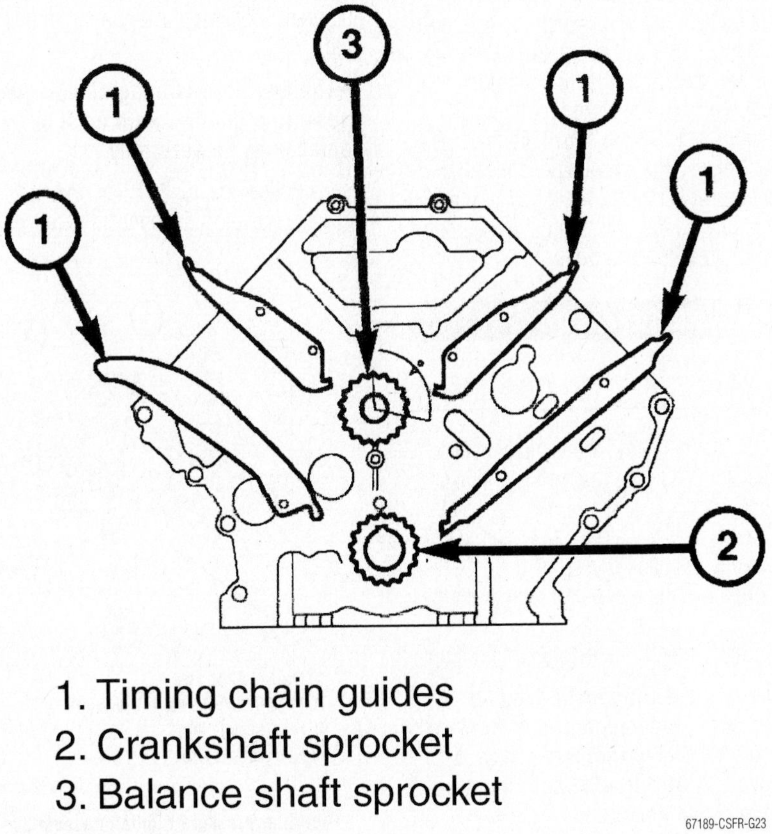

Timing chain guide locations—3.2L engine

19. Install the rear counter weight onto the shaft and torque the retaining bolt to 15 ft. lbs. (20 Nm), plus an additional 90 degrees.
20. Install the balance shaft locking plate and bolt and torque to 15 ft. lbs. (20 Nm).
21. Install the rear crankshaft seal/end cover assembly.
22. Install the crankshaft sprocket.
23. Ensure that the crankshaft is still positioned at 40 degrees ATDC.
24. Align the balance shaft sprocket with the timing chain (3) and check that the copper teeth (1) of the timing chain are aligned on the camshaft sprockets (6).
25. Route the timing chain within the timing chain guides as shown.
26. Lock the camshafts in place with Camshaft Locking Plates 9104 and 9105.
27. Install the oil pump drive chain and tensioner.
28. Apply sealant to timing chain cover as shown.
29. Install or connect the following:
    - Timing chain cover. Torque the bolts to 15 ft. lbs. (20 Nm).
    - Timing chain tensioner. Torque the bolt to 59 ft. lbs. (80 Nm).
30. Remove Special Tool 9102 to unlock the flywheel.

31. Install or connect the following:
    - Starter
    - Upper and lower oil pans
    - Alternator
    - Front cylinder head to timing chain cover bolts. Tighten to 15 ft. lbs. (20 Nm).
    - Cylinder head covers
    - AIR pump
    - Power steering pump
    - Coolant and heater hoses to water pump
    - Vibration damper. Tighten to 148 ft. lbs. (200 Nm) plus 90 degrees clockwise.
    - Idler pulley
    - Accessory belt tensioner
    - Accessory drive belt
    - Radiator fan
    - Negative battery cable
32. Fill the power steering pump with fluid.
33. Fill and bleed the cooling system.

Piston and Ring

POSITIONING

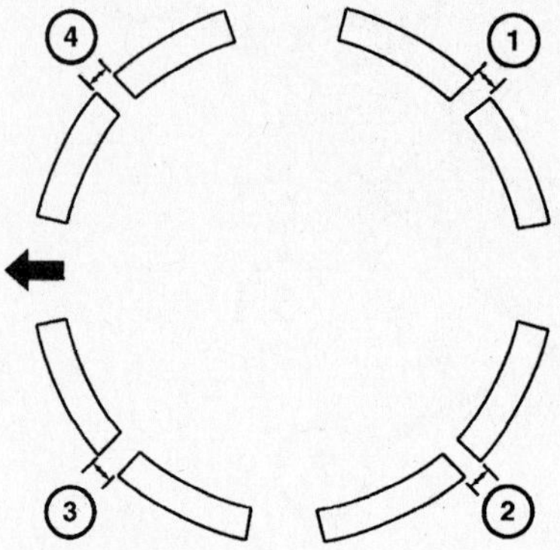

1. Upper ring
2. Upper side rail
3. Middle ring and spacer expander
4. Lower side rail

67189-CSFR-G25

Piston ring end-gap spacing—3.2L engine

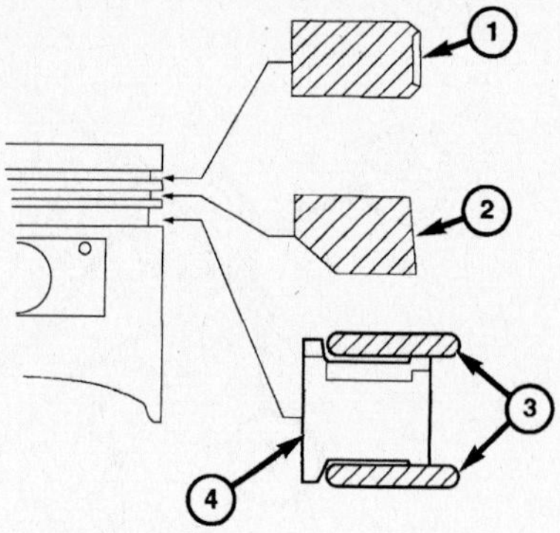

1. Compression ring
2. Sealing ring
3. Oil ring rail
4. Oil ring expander

67189-CSFR-G26

Ring to piston positioning—3.2L engine

FUEL SYSTEM

Fuel System Service Precautions

Safety is the most important factor when performing not only fuel system maintenance but any type of maintenance. Failure to conduct maintenance and repairs in a safe manner may result in serious personal injury or death. Maintenance and testing of the vehicle's fuel system components can be accomplished safely and effectively by adhering to the following rules and guidelines.

- To avoid the possibility of fire and personal injury, always disconnect the negative battery cable unless the repair or test procedure requires that battery voltage be applied.
- Always relieve the fuel system pressure prior to detaching any fuel system component (injector, fuel rail, pressure regulator, etc.), fitting or fuel line connection. Exercise extreme caution whenever relieving fuel system pressure to avoid exposing skin, face and eyes to fuel spray. Please be advised that fuel under pressure may penetrate the skin or any part of the body that it contacts.
- Always place a shop towel or cloth around the fitting or connection prior to loosening to absorb any excess fuel due to spillage. Ensure that all fuel spillage (should it occur) is quickly removed from engine surfaces. Ensure that all fuel soaked cloths or towels are deposited into a suitable waste container.
- Always keep a dry chemical (Class B) fire extinguisher near the work area.
- Do not allow fuel spray or fuel vapors to come into contact with a spark or open flame.
- Always use a back-up wrench when loosening and tightening fuel line connection fittings. This will prevent unnecessary stress and torsion to fuel line piping.
- Always replace worn fuel fitting O-rings with new. Do not substitute fuel hose or equivalent, where fuel pipe is installed.

Before servicing the vehicle, also make sure to refer to the precautions in the beginning of this section as well.

Fuel System Pressure

RELIEVING

1. Before servicing the vehicle, refer to the precautions in the beginning of this section.
2. Remove the fuel tank filler cap to release any fuel tank pressure.
3. Remove the fuel pump relay from the PDC.
4. Start and run the engine until it stops.

➡**One or more Diagnostic Trouble Codes (DTC) may be stored in PCM memory when the fuel pump fuse is removed. A DRB III scan tool must be used to erase the code.**

Fuel Filter

REMOVAL & INSTALLATION

1. Before servicing the vehicle, refer to the precautions in the beginning of this section.

➡**The fuel filter is integral with the fuel pressure regulator. Both must be replaced as an assembly.**

2. Relieve the fuel system pressure.
3. Remove or disconnect the following:

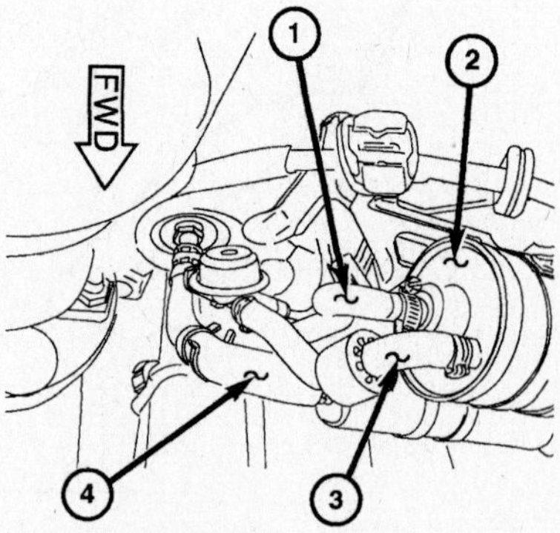

1. Degaussing line
2. Fuel filter/pressure regulator
3. Fuel supply line
4. Fuel return line

67189-CSFR-G27

Fuel filter/pressure regulator assembly—3.2L engine

- Fuel pump splash shield
- Degassing line
- Fuel supply line
- Fuel delivery line
- Fuel return line

4. Pull the filter/regulator out of the mounting clamp.

To install:

5. Install the filter/regulator into the mounting clamp.
6. Install or connect the following in order:
   - Fuel return line
   - Fuel delivery line
   - Fuel supply line
   - Degassing line
   - Splash shield
7. Start the engine and check for leaks.

Fuel Pump

REMOVAL & INSTALLATION

1. Before servicing the vehicle, refer to the precautions in the beginning of this section.
2. Relieve the fuel system pressure.
3. Remove or disconnect the following:
   - Negative battery cable
   - Splash shield
   - Fuel suction hose
   - Fuel delivery hose
   - Fuel pump electrical connectors
   - Fuel pump retaining clamp
   - Fuel pump

To install:

4. Install or connect the following:
   - Fuel pump
   - Fuel pump retaining clamp
   - Electrical connectors
   - Fuel delivery hose
   - Fuel suction hose
   - Splash shield
   - Negative battery cable
5. Start the engine and check for leaks.

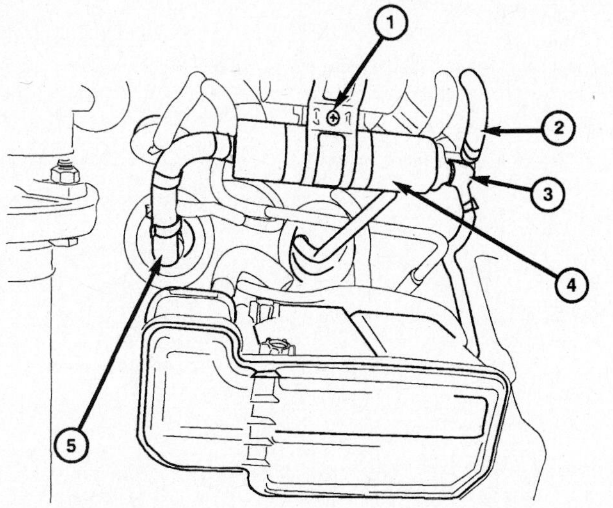

1. Retaining clamp
2. Fuel delivery hose
3. Harness connectors
4. Fuel pump
5. Suction hose

67189-CSFR-G28

Fuel pump removal—3.2L engine

Fuel Rail And Injectors

REMOVAL & INSTALLATION

1. Before servicing the vehicle, refer to the precautions in the beginning of this section.
2. Relieve the fuel system pressure.
3. Remove or disconnect the following:
   - Negative battery cable
   - Wiring harness cover

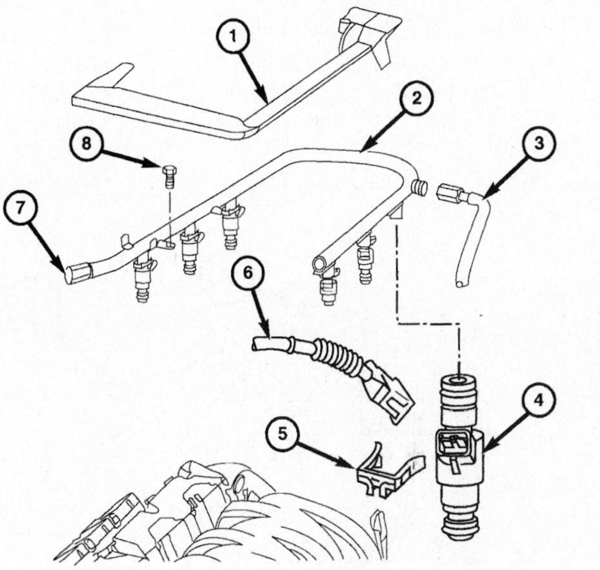

1 - FUEL RAIL TRIM PANEL
2 - FUEL RAIL
3 - FUEL LINE
4 - FUEL INJECTOR
5 - FUEL INJECTOR RETAINING CLIP
6 - FUEL INJECTOR HARNESS CONNECTOR
7 - FUEL RAIL SERVICE VALVE
8 - FUEL RAIL BOLT

67189-CSFR-G29

Fuel rail and injectors—3.2L engine

4. Open fuel rail service valve.
5. Remove or disconnect:
   - Fuel feed line
   - Fuel injector retaining clamps
   - Fuel supply manifold with injectors attached
   - Fuel injectors

To install:

6. Install or connect the following:
   - Fuel injectors, using new O-rings
   - Fuel supply manifold with injectors attached. Tighten the bolts to 80 inch lbs. (9 Nm).
   - Injector retaining clamps
   - Fuel feed line. Tighten the fuel line to 28 ft. lbs. (38 Nm).
7. Close fuel rail service valve
8. Install or connect:
   - Fuel rail trim cover
   - Air cleaner housing
   - Negative battery cable
9. Start the engine and check for leaks.

Throttle Body

REMOVAL & INSTALLATION

Except SRT Models

1. Before servicing the vehicle, refer to the precautions in the beginning of this section.
2. Remove or disconnect the following:
   - Negative battery cable
   - Throttle body intake tube
   - Electrical connector
   - PCV hose from the throttle body
   - Throttle body retaining bolts and PCV hose bracket
   - Throttle body with gasket

To install:

3. Install or connect the following:
   - PCV hose to the throttle body
   - Throttle body with new gasket. Tighten bolts to 15 ft. lbs. (20 Nm).
   - Electrical connector
   - Throttle body intake tube
   - Negative battery cable

SRT Models

1. Before servicing the vehicle, refer to the precautions in the beginning of this section.
2. Remove or disconnect the following:
   - Negative battery cable
   - Upper air intake assembly
   - Throttle body intake tube by lifting up on the inlet tube lever
   - Electrical connector
   - PCV hose from throttle body

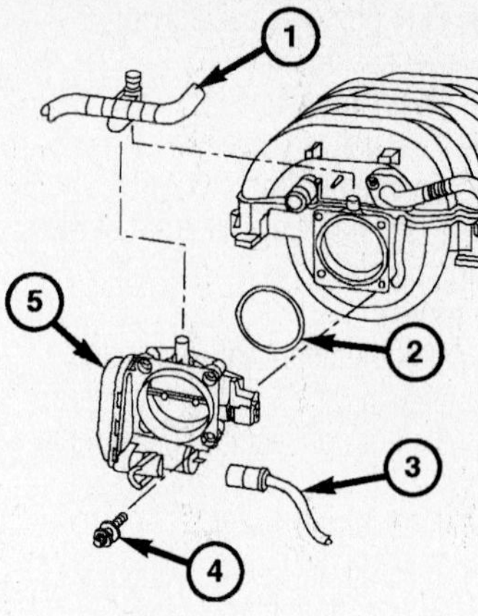

1. PCV hose
2. O ring
3. Harness connector
4. Bolts (4)
5. Throttle body

67189-CSFR-G30

Throttle body—3.2L engine

   - Throttle body retaining bolts and PCV hose bracket
   - Throttle body and gasket

To install:

3. Install or connect the following:
   - PCV hose to the throttle body
   - Throttle body and gasket. Tighten to 15 ft. lbs. (20 Nm).
   - Electrical connector
   - Throttle body intake tube as follows:

   a. Install the upper air intake assembly.

   b. Install the throttle body intake tube by firmly inserting tube into the throttle body.

   c. Ensure the intake tube lever is attached.

   - Negative battery cable

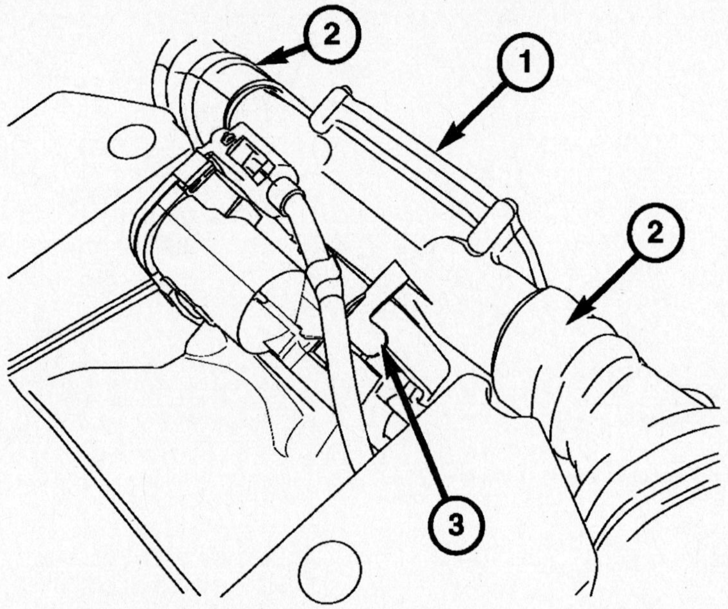

06009-CSFR-G55

Throttle body intake tube and components

DRIVE TRAIN

Transmission Assembly

REMOVAL & INSTALLATION

Automatic

1. Before servicing the vehicle, refer to the precautions in the beginning of this section.
2. Drain the transmission fluid.
3. Remove or disconnect the following:
   - Negative battery cable
   - Exhaust system
   - Center exhaust heat shield
   - Transmission support bracket
   - Transmission mount and crossmember. Support the transmission with a jack.
   - Driveshaft
   - Front engine undercover
   - Transmission ground bolt
   - Shift rod
   - Transmission oil cooler fittings
   - Torque converter access cover
   - Torque converter bolts
   - Oxygen sensor connectors
   - Heat shield
   - Controller harness connector
4. Support the transmission with a jack under the transmission oil pan rail.
5. Remove or disconnect:
   - Starter
   - Engine to transmission retaining bolts
   - Transmission

To install:

6. Ensure the torque converter is fully seated in the transmission.
7. Install or connect the following:
   - Transmission. Torque the mounting bolts to 28 ft. lbs. (38 Nm).
   - Torque converter. Torque the mounting bolts to 37 ft. lbs. (50 Nm).
   - Torque converter access cover
   - Starter. Torque the mounting bolts to 31 ft. lbs. (42 Nm).
   - Controller harness connector
   - Heat shield. Torque the mounting bolts to 15 ft. lbs. (20 Nm).
   - Oxygen sensor connector
   - Transmission cooler fittings. Torque the mounting bolts to 25 ft. lbs. (34 Nm).
   - Shift rod
   - Driveshaft. Torque the mounting bolts to 44 ft. lbs. (60 Nm).
   - Transmission support bracket. Torque the mounting bolts to 15 ft. lbs. (20 Nm).
   - Transmission mount and crossmember. Torque the mounting bolts to 33 ft. lbs. (45 Nm).
   - Center heat shield
   - Exhaust system
   - Engine undercover
   - Negative battery cable
8. Fill the transmission to the correct level.
9. Start engine and check for proper transmission operation.

Manual

1. Before servicing the vehicle, refer to the precautions in the beginning of this section.
2. Remove or disconnect the following:
   - Negative battery cable
   - Exhaust system
   - Center exhaust heat shield
   - Driveshaft
   - Transmission support bracket
   - Transmission mount and crossmember. Support the transmission with a jack stand.
   - Reverse light switch connector
   - Clutch slave cylinder pressure line
   - Reverse lockout cable
   - Shift rod
   - Ground cable
   - Transmission

To install:

3. Raise the transmission into place.
4. Install or connect the following:
   - Ground cable. Tighten the bolt to 30 ft. lbs. (40 Nm).
   - Transmission to engine block retaining bolts. Tighten the bolts to 30 ft. lbs. (40 Nm).
   - Transmission mount bolts to the transmission. Tighten to 22 ft. lbs. (30 Nm).
   - Rear crossmember retaining bolts. Tighten to 30 ft. lbs. (40 Nm).
   - Shift rod
   - Reverse lockout cable
   - Reverse light connector
   - Clutch slave cylinder pressure line
   - Driveshaft. Torque the mounting bolts to 44 ft. lbs. (60 Nm).
   - Center exhaust heat shield
   - Exhaust system
   - Negative battery cable
5. Fill transmission to the correct level.
6. Check for proper clutch operation.

Clutch

REMOVAL & INSTALLATION

1. Before servicing the vehicle, refer to the precautions in the beginning of this section.
2. Remove or disconnect the following:
   - Negative battery cable
   - Transmission
3. Matchmark the pressure plate to the flywheel.
   - Pressure plate. Loosen the bolts evenly in 1 to 1 ½ turn steps.
   - Clutch disc

To install:

4. If the pressure plate is going to be reused, an adjustment is required on the

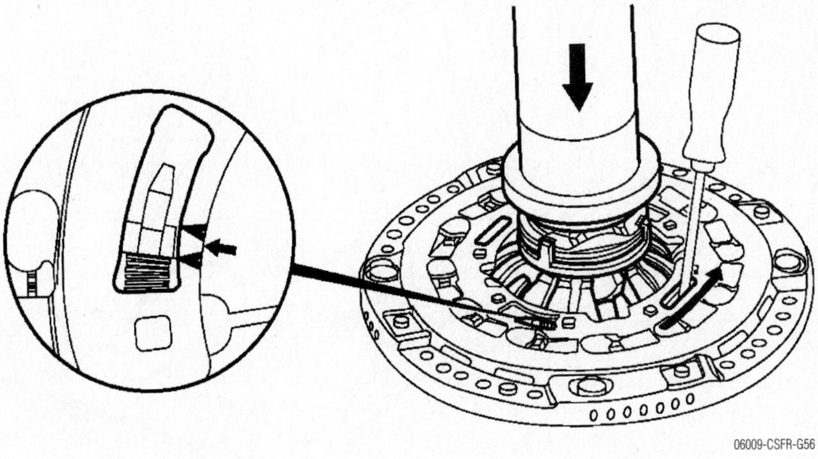

Adjusting the pressure plate.

plate. Release the diaphragm pressure and set to the adjusting ring to the mark shown.

5. Position the clutch disc and the pressure plate on the flywheel and loosely install the pressure plate bolts.

6. Align the clutch disc to the flywheel using a clutch alignment tool.

7. Tighten the pressure plate bolts evenly in 1 to 1 ½ turns to 18 ft. lbs. (25 Nm).

8. Remove the clutch alignment tool.

9. Install the transmission.

10. Connect the negative battery cable.

Hydraulic Clutch System

BLEEDING

1. Before servicing the vehicle, refer to the precautions in the beginning of this section.

2. Connect a pressure bleeder to the brake master cylinder.

3. Open the bleed screw on the clutch slave cylinder.

4. Allow the fluid to flow until there are no bubbles and the fluid is clear.

5. Close the bleed screw and disconnect the pressure bleeder.

6. Check the clutch for proper operation.

Halfshaft

REMOVAL & INSTALLATION

1. Before servicing the vehicle, refer to the precautions in the beginning of this section.

2. Raise and support the rear of the vehicle.

3. Remove or disconnect the following:
   - Rear wheels
   - Rear knuckle assembly

4. Using a pry bar between the differential housing and halfshaft, pry against the housing until the halfshaft retaining ring is disengaged from the differential side gear.

5. Carefully pull the halfshaft out of the differential housing.

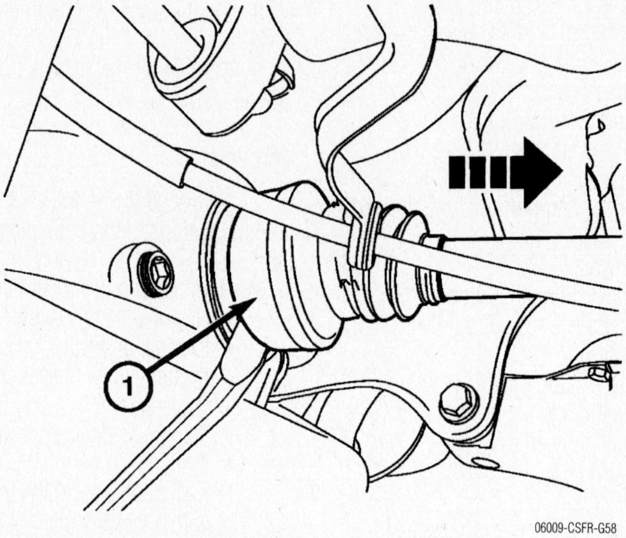

Prying the halfshaft from the differential.

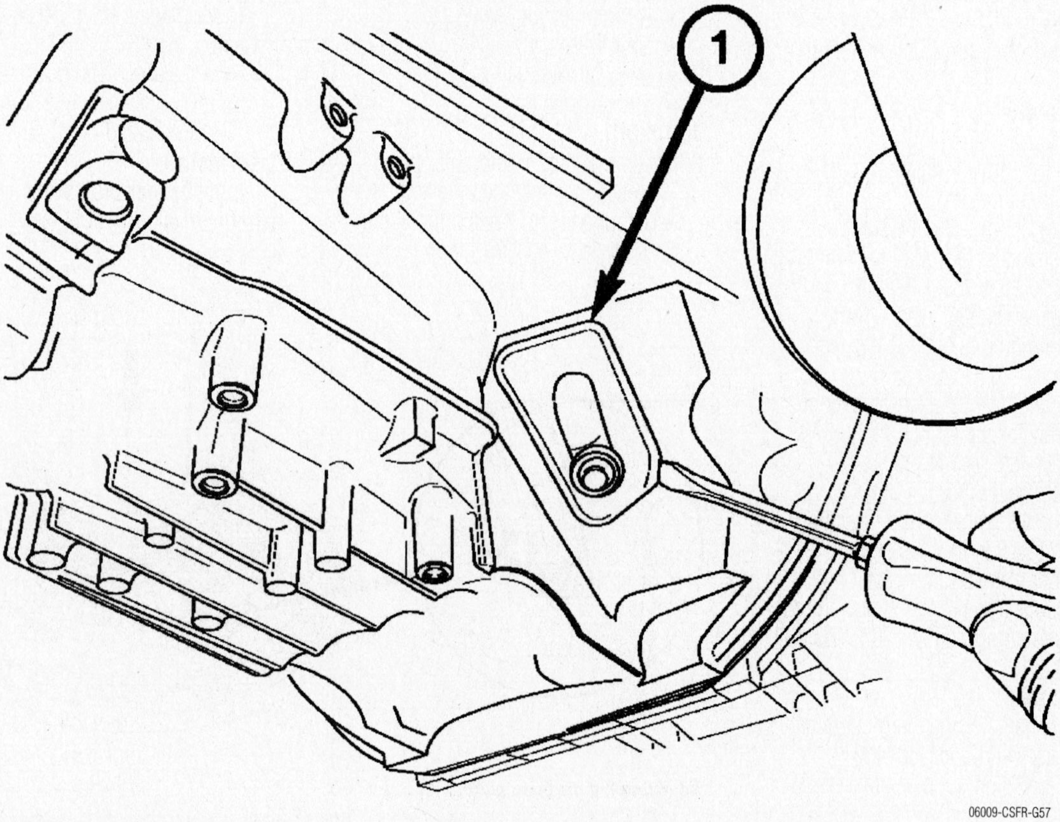

Clutch bleed screw access plug.

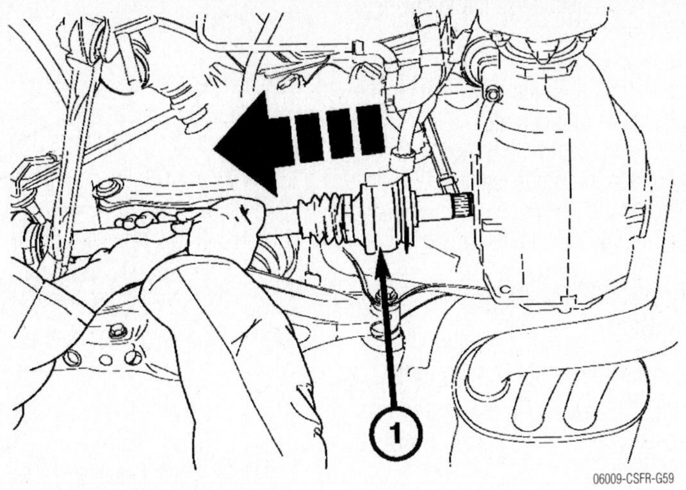

Removing the halfshaft from the vehicle.

To install:

6. Insert the halfshaft outer end through the wheel hub.
7. Install the halfshaft into the differential side gear until the snap ring is engaged.
8. Install the rear knuckle.
9. Install the rear wheels.

CV-Joints

OVERHAUL

Outer CV-Joint

1. Before servicing the vehicle, refer to the precautions in the beginning of this section.
2. Remove or disconnect the following:
   - Axle halfshaft from the vehicle
   - Inner CV-joint boot and clamps
   - Outer CV-joint and boot

To install:

3. Pack the outer CV-joint with grease supplied in repair kit.
4. Install or connect the following:
   - CV-joint into inner seal
5. Crimp the retaining boot clamps.
6. Install the axle halfshaft.

Inner CV Joint

1. Before servicing the vehicle, refer to the precautions in the beginning of this section.
2. Remove or disconnect the following:
   - Axle halfshaft from the vehicle
   - CV joint boot clamps
3. Clamp the halfshaft in a vise.
4. Pry the joint cover from the bearing retainer.
5. Remove the boot from the inner seal.
6. Pry the inner seal from the bearing retainer.
7. Slide the boot, inner seal and bearing retainer off the halfshaft.
8. Remove the circlip from the axle shaft.
9. Using a press, remove the inner joint bearing from the halfshaft.

To install:

➡Use new circlips and boot clamps for assembly.

10. Install or connect the following:
   - Inner boot
   - Inner joint bearing
   - Inner seal
   - Circlip
11. Fill the bearing housing with grease.
12. Apply a bead of sealant to the outer sealing surface of the joint ring.

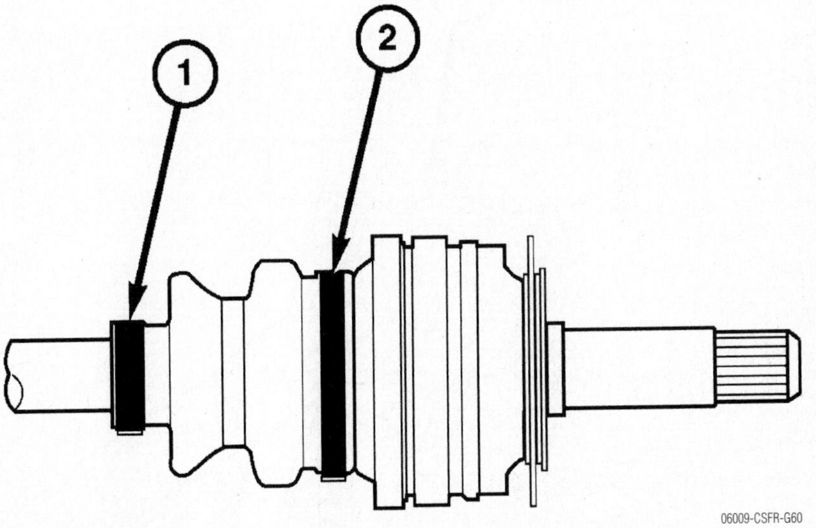

CV Joint boot clamps

Removing the boot from inner seal.

13. Using a mallet, tap on a NEW joint cover.
14. Install the boot into the sealing groove of the inner seal.
15. Install the retaining clamps and crimp in place.
16. Install the axle halfshaft.

Pinion Seal

REMOVAL & INSTALLATION

1. Before servicing the vehicle, refer to the precautions in the beginning of this section.
2. Drain the differential of fluid.
3. Remove or disconnect the following:
   - Wheels
   - Muffler
   - Driveshaft from the rear differential
   - Halfshafts at differential
4. Check and record the bearing preload with an inch lb. torque wrench.
5. Hold the pinion flange using special tool C-3281 or equivalent and remove the collared nut.
6. Pry out the pinion flange seal.

To install:

➡Use a new pinion nut for assembly.

7. Install the new pinion seal using special tool 9321 drift and a mallet. Tighten the pinion nut to 133 ft. lbs. (180 Nm).

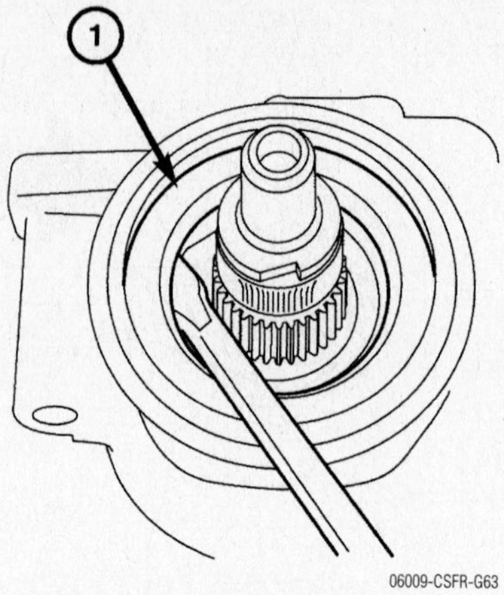

Prying out pinion flange seal.

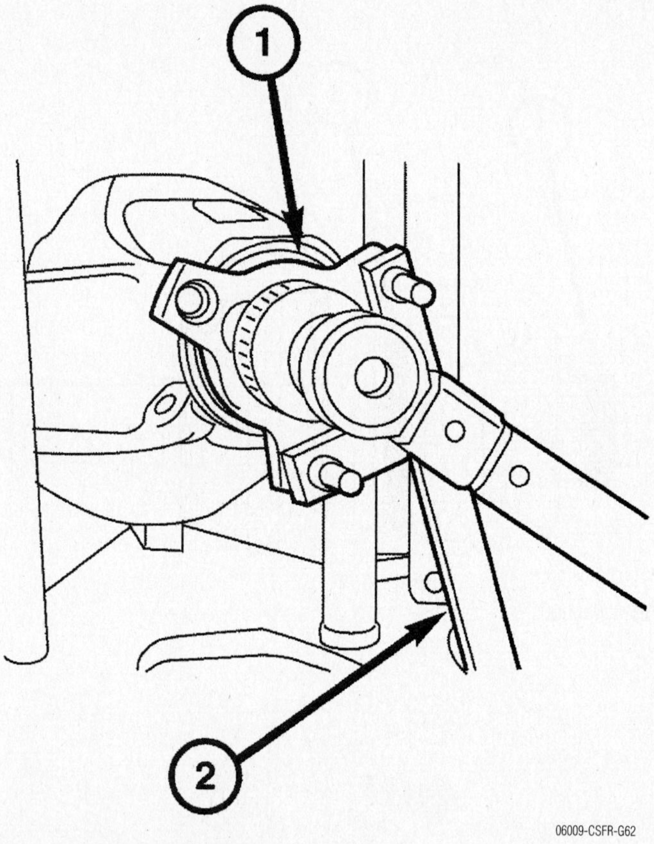

Holding pinion flange to remove pinion nut.

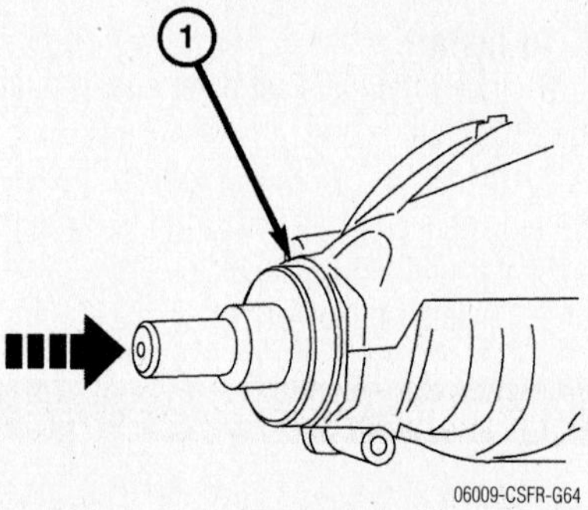

Installing pinion seal using tool 9321.

8. Check the bearing preload. The bearing preload should be equal to the reading taken earlier.
9. If the preload torque is low, tighten the pinion nut in 44 inch lb. (5 Nm) increments until the torque value is reached.
10. If the pinion bearing preload torque cannot be attained at maximum pinion nut torque, replace the collapsible spacer.
11. Install or connect the following:
    - Halfshafts
    - Driveshaft
    - Muffler
    - Wheels
12. Fill the axle assembly to correct level.

STEERING AND SUSPENSION

Air Bag

⁂ CAUTION

Some vehicles are equipped with an air bag system. The system must be disarmed before performing service on, or around, system components, the steering column, instrument panel components, wiring and sensors. Failure to follow the safety precautions and the disarming procedure could result in accidental air bag deployment, possible injury and unnecessary system repairs.

PRECAUTIONS

Several precautions must be observed when handling the inflator module to avoid accidental deployment and possible personal injury.

- Never carry the inflator module by the wires or connector on the underside of the module.
- When carrying a live inflator module, hold securely with both hands, and ensure that the bag and trim cover are pointed away.
- Place the inflator module on a bench or other surface with the bag and trim cover facing up.
- With the inflator module on the bench, never place anything on or close to the module which may be thrown in the event of an accidental deployment.

Before servicing the vehicle, also make sure to refer to the precautions in the beginning of this section as well.

DISARMING

1. Disconnect and isolate the negative battery cable. Wait 2 minutes for the system capacitor to discharge before performing any service.
2. When repairs are completed, connect the negative battery cable.

Recirculating Ball Power Steering Gear

REMOVAL & INSTALLATION

1. Before servicing the vehicle, refer to the precautions in the beginning of this section.
2. Remove ignition key and allow the steering wheel to lock.
3. Draw off fluid from the power steering reservoir.
4. Remove or disconnect the following:
   - Negative battery cable
   - Drag link and tie rod using special tool 9168 or suitable equivalent
   - Left front cross brace
   - Power steering pressure and return lines
   - Pinch bolt from the steering coupling
   - Steering shaft

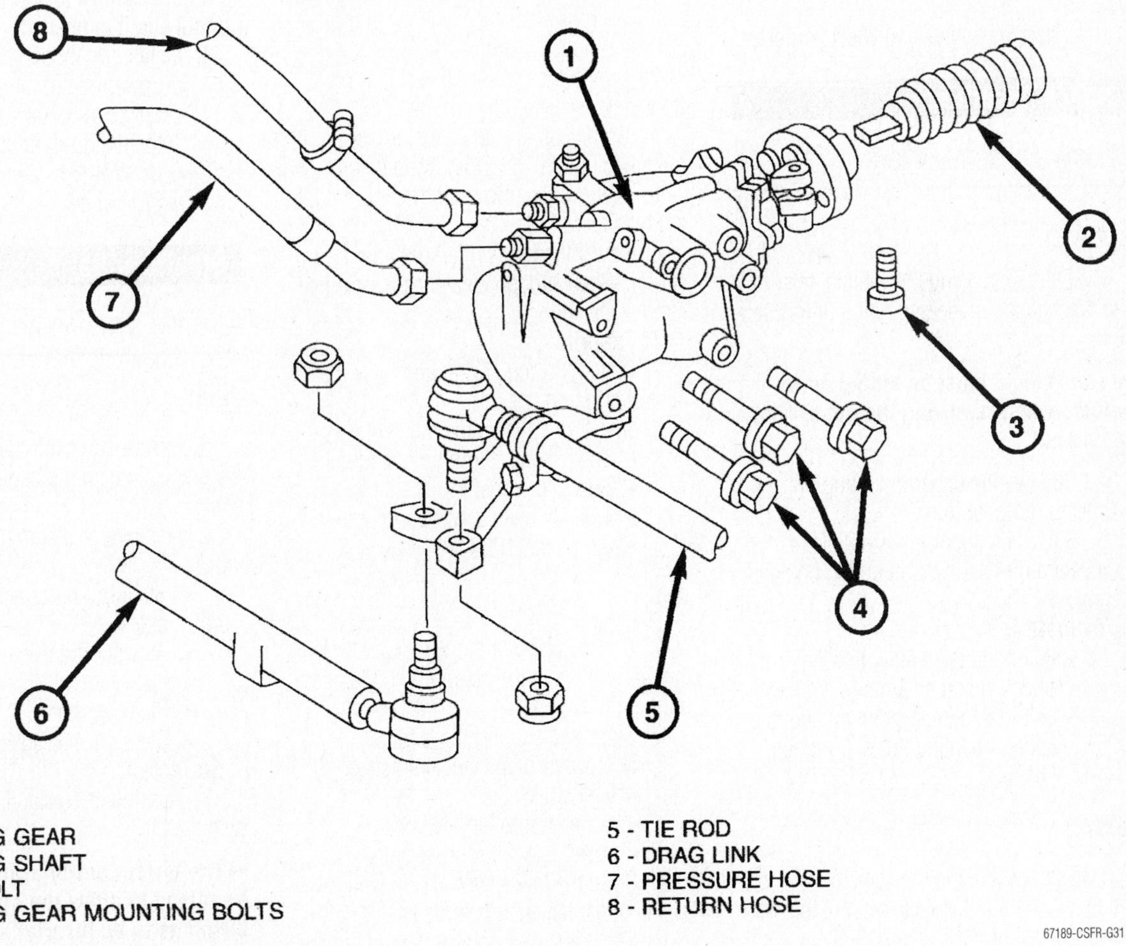

Exploded view of power steering gear—Crossfire

5. Lower engine mounting bolts on both sides.
6. Place a jack under the left side of the engine and raise the engine about 2 inches.
7. Remove 3 bolts and remove the steering gear.

To install:

8. Align the marks on the steering shaft with the housing cover to ensure gear is in the center position.
9. Install or connect the following:
   - Gear coupling onto the steering shaft
   - Steering gear. Tighten the bolts to 44 ft. lbs. (60 Nm).
   - Engine mount bolts. Tighten the bolts to 18 ft. lbs. (25 Nm).
   - Intermediate shaft. Tighten the pinch bolt to 15 ft. lbs. (20 Nm).
   - Power steering pressure line and tighten to 22 ft. lbs. (30 Nm)
   - Power steering return line and tighten to 30 ft. lbs. (40 Nm)
   - Drag link/tie rod. Tighten the nut to 44 ft. lbs. (60 Nm).
   - Left cross brace
   - Negative battery cable
10. Fill the power steering fluid reservoir.
11. Start the engine and check for leaks.

Shock Absorber

REMOVAL & INSTALLATION

Front

1. Before servicing the vehicle, refer to the precautions in the beginning of this section.

➡The vehicle must be sitting on its wheels when removing the upper shock mount nut.

2. Remove the upper mounting nut, washer and rubber nut.
3. Raise and support the vehicle.
4. Remove the lower mounting bolt and remove the shock.

To install:

5. Install or connect the following:
   - Shock absorber. Tighten the lower bolt to 41 ft. lbs. (55 Nm) and the upper mounting nut to 13 ft. lbs. (18 Nm).

Rear

1. Before servicing the vehicle, refer to the precautions in the beginning of this section.
2. Remove or disconnect the following:
   - Rear wheel
   - Upper mounting bolt
   - Lower control arm cover
   - Lower mounting bolts
   - Shock absorber

To install:

3. Install or connect the following:
   - Shock absorber. Tighten the upper bolt nut to 13 ft. lbs. (18 Nm) and the lower bolt to 41 ft. lbs. (55 Nm).
   - Lower control arm cover
   - Rear wheel

Coil Spring

REMOVAL & INSTALLATION

Front

1. Before servicing the vehicle, refer to the precautions in the beginning of this section.

➡The vehicle must be sitting on its wheels when removing the upper shock mount nut.

2. Remove or disconnect the following:
   - Front wheel
   - Shock absorber
3. Support the lower control arm on a floor jack.
4. Compress the spring using special tools 9151 and 9152 or equivalent.
5. Remove the lower control arm mounting nuts and bolts.
6. Lower the jack while holding the spring. When enough clearance is obtained, remove the spring.

To install:

7. Install the coil spring and raise the control arm into position.
8. Install or connect the following:
   - Lower control arm bolts and nuts
   - Shock absorber
   - Tighten lower control arm nuts to 88 ft. lbs. (120 Nm)

Rear

1. Before servicing the vehicle, refer to the precautions in the beginning of this section.
2. Raise and support the vehicle.
3. Remove or disconnect the following:
   - Rear wheel
   - Lower control arm cover
   - Lower shock absorber bolts
4. Raise the lower control arm with a jack until the halfshaft is horizontal.
5. Using a spring compressor, compress the rear spring.

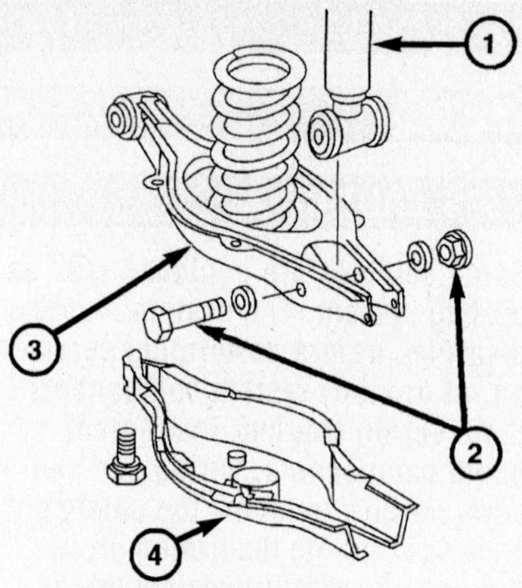

1. Shock absorber
2. Lower shock bolt
3. Lower control arm
4. Control arm cover

67189-CSFR-G32

Rear coil spring mounting—Crossfire

6. Remove the rear lower control arm attaching bolts, and then lower the jack allowing the control arm to swing away from the mounting tabs.
7. Remove the rear spring.

To install:

8. Install the coil spring and raise the control arm into position.
9. Install or connect the following:
   - Lower control arm bolts and nuts.
   - Lower shock absorber bolt. Torque the bolt to 41 ft. lbs. (55 Nm).
   - Control arm cover
   - Rear wheel

Stabilizer Bar

REMOVAL & INSTALLATION

Front

1. Before servicing the vehicle, refer to the precautions in the beginning of this section.
2. Remove or disconnect the following:
   - Stabilizer bar brackets on both sides
   - Retainer bracket and mounting plate
   - Stabilizer bar
   - Stabilizer bar bushings

To install:

3. Assemble the stabilizer bushings onto the bar.

➡The lower control arms may have to be raised to allow the stabilizer bar bushings to be inserted into their mating surfaces.

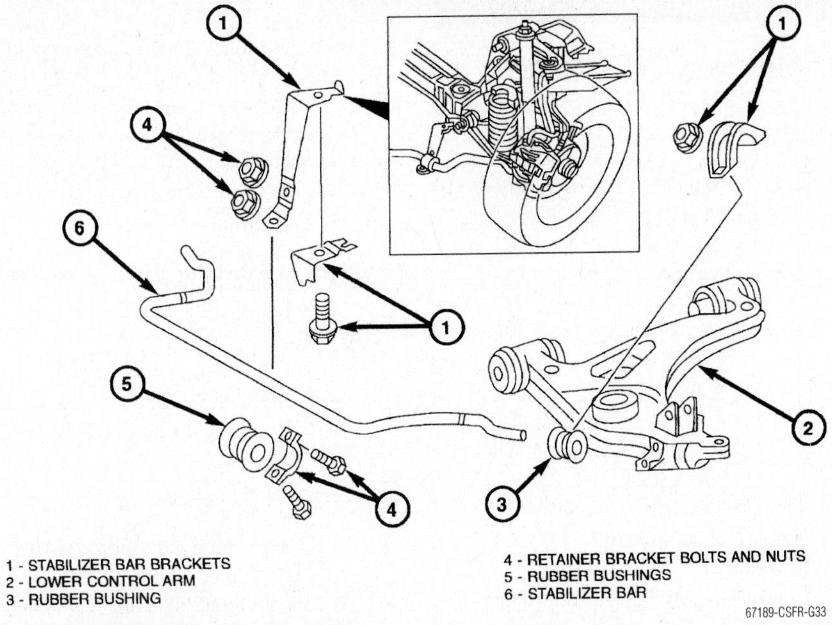

Front stabilizer bar mounting—Crossfire

4. Install the stabilizer bar by loosely installing the bolts and nuts to all 4 brackets.
5. Align the stabilizer bar so it is centered in the vehicle.
6. Due to the preload of the bracket, install the hexagon bolt first.
7. Tighten both stabilizer bolts to the frame.
8. Tighten the hexagon socket bolts to the retainer.
9. Tighten both bar bracket nuts to the lower control arm. Torque the bolts to 15 ft. lbs. (20 Nm).
10. Install the bar retainer and mounting plate.
11. Tighten the retainer bolts to 44 ft. lbs. 60 Nm).
12. Tighten the lower retainer bracket nut to 15 ft. lbs. (20 Nm).
13. Tighten the upper retainer nuts to 29 ft. lbs. (40 Nm).

Rear

1. Before servicing the vehicle, refer to the precautions in the beginning of this section.
2. Raise and support the vehicle.
3. Remove or disconnect the following:
    - Rear wheels
    - Wheel speed sensors
    - Brake hoses from caliper
    - Coil springs
    - Shock absorbers
    - Parking brake cables from equalizer

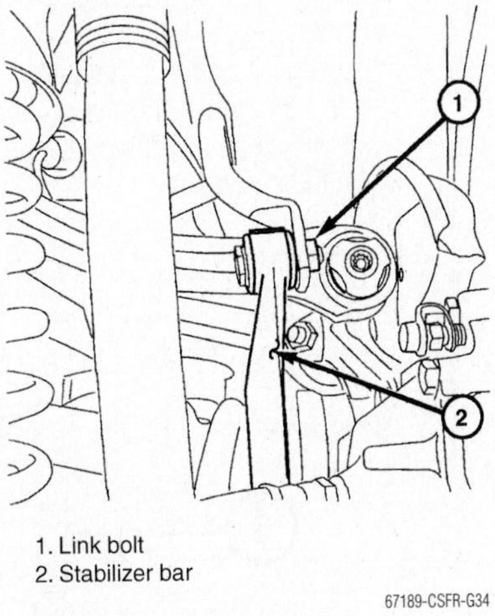

Rear stabilizer bar link bolt—Crossfire

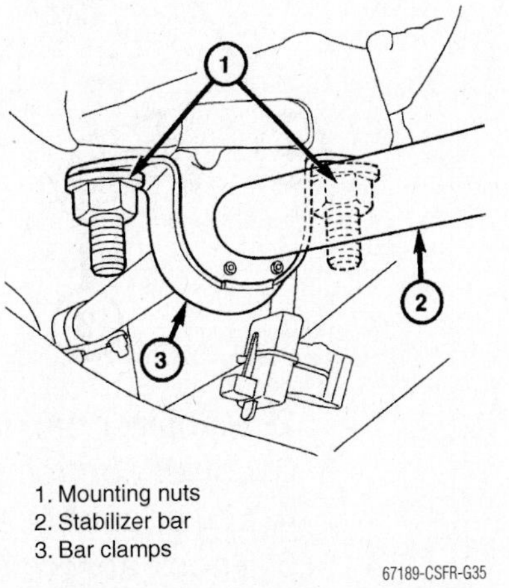

Rear stabilizer bar clamp bolts—Crossfire

    - Rear driveshaft bolts
    - Stabilizer bar links
4. Raise and support the rear axle carrier.
5. Remove 4 bolts attaching the rear axle carrier to the frame.
6. Lower the rear axle carrier enough to allow the stabilizer bar to be removed.
7. Remove the stabilizer bar clamps and remove the stabilizer bar.

To install:

8. Position the stabilizer bar into place.
9. Raise the rear axle carrier and install the 4 carrier to frame bolts. Torque the bolts to 66 ft. lb. 90 Nm).
10. Install or connect the following:
    - Rear springs
    - Shock absorbers
    - Stabilizer bar clamps
    - Stabilizer bar links. Torque the bolts to 22 ft. lbs. (30 Nm).
    - Tighten the bar clamp nuts 15 ft. lbs. (20 Nm).
    - Rear driveshaft bolts.
    - Parking brake cables to equalizer
    - Brake hoses to calipers
    - Wheel seed sensors
    - Rear wheels

Upper Ball Joint and Upper Control Arm

REMOVAL & INSTALLATION

➡If replacing the ball joint on the right side, the air cleaner housing must be removed for access. The shock absorbers must remain installed.

➡To remove the upper control arm the shock absorber must remain installed.

1. Before servicing the vehicle, refer to the precautions in the beginning of this section.
2. Remove the front wheel.
3. Wire tie the steering knuckle to the shock absorber.
4. Remove the ball joint nut from the upper control arm.
5. Using special tool 9168 or an equivalent puller, press the ball joint out of the steering knuckle.
6. From inside the engine compartment, remove the upper control arm nut and bolt, and remove the control arm.
7. Installation is the reverse of the removal procedure. Tighten the upper control arm nut to 48 ft. lbs. (65 Nm). Tighten the ball joint nut to 33 ft. lbs. (45 Nm).

Lower Ball Joint

REMOVAL & INSTALLATION

1. Before servicing the vehicle, refer to the precautions in the beginning of this section.
2. Raise and support the vehicle.
3. Remove or disconnect the following:
   - Front wheels
   - Brake caliper
   - Brake rotor
   - Dust shield
   - Lower shock mount bolt
   - Coil spring
   - Ball joint nuts
   - Outer ball joint stud from steering knuckle using special tool 9168 or equivalent puller
   - Upper ball joint stud from lower control arm using special tool 9168 or equivalent puller

To install:

4. Install or connect the following:
   - Ball joint. Tighten ball joint nuts to 77 ft. lbs. (105 Nm).
   - Coil spring
   - Lower shock bolt
   - Dust shield
   - Brake rotor and caliper
   - Front wheel

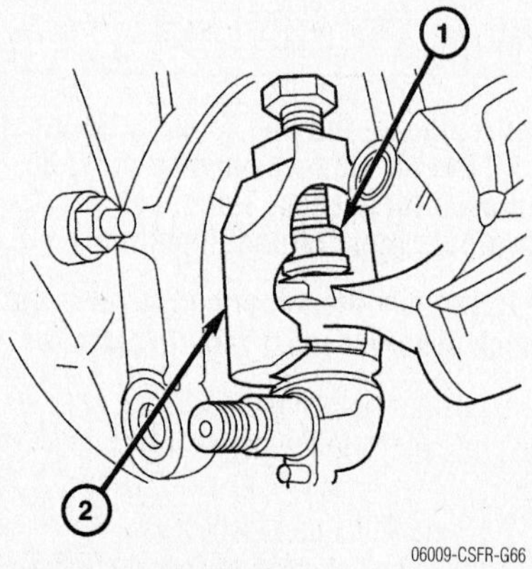

Upper ball joint removal

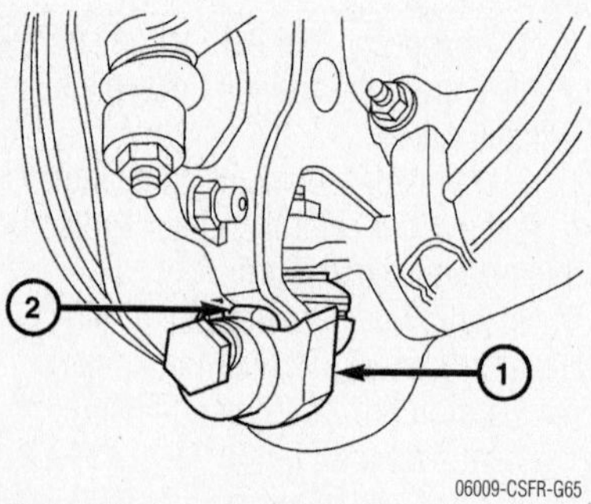

Lower ball joint removal.

Lower Control Arm

REMOVAL & INSTALLATION

Front

1. Before servicing the vehicle, refer to the precautions in the beginning of this section.
2. Remove or disconnect the following:
   - Front wheel
   - Coil spring
   - Lower shock bolt
   - Stabilizer bar clamp
   - Lower ball joint nuts
   - Press ball joint out of steering knuckle using special tool 9168 or equivalent puller
   - Lower control arm bolts
   - Lower control arm

To install:

3. Install or connect the following:
   - Lower control arm. Tighten the bolts to the frame to 88 ft. lbs. (120 Nm).
   - Lower control arm to steering knuckle lower ball joint. Tighten the nut to 33 ft. lbs. (45 Nm).
   - Stabilizer bar clamp. Tighten the nuts to 15 ft. lbs. (20 Nm).
   - Lower shock bolt. Tighten the nut to 41 ft. lbs. (55 Nm).
   - Coil spring
   - Front wheel

Rear

1. Before servicing the vehicle, refer to the precautions in the beginning of this section.
2. Remove or disconnect the following:
   - Rear wheel
   - Lower control arm cover
   - Coil spring
   - Shock absorber
   - Stabilizer bar link
   - Lower control arm

To install:

3. Install or connect the following:
   - Lower control arm to wheel carrier bolt. Tighten to 52 ft. lbs. (70 Nm).
   - Shock absorber
   - Coil spring
   - Lower control arm to frame bolt. Tighten to 52 ft. lbs. (70 Nm).
   - Stabilizer bar link. Tighten the nut to 15 ft. lbs. (20 Nm).
   - Lower control arm cover
   - Rear wheel

Front Wheel Bearings

ADJUSTMENT

1. Before servicing the vehicle, refer to the precautions in the beginning of this section.

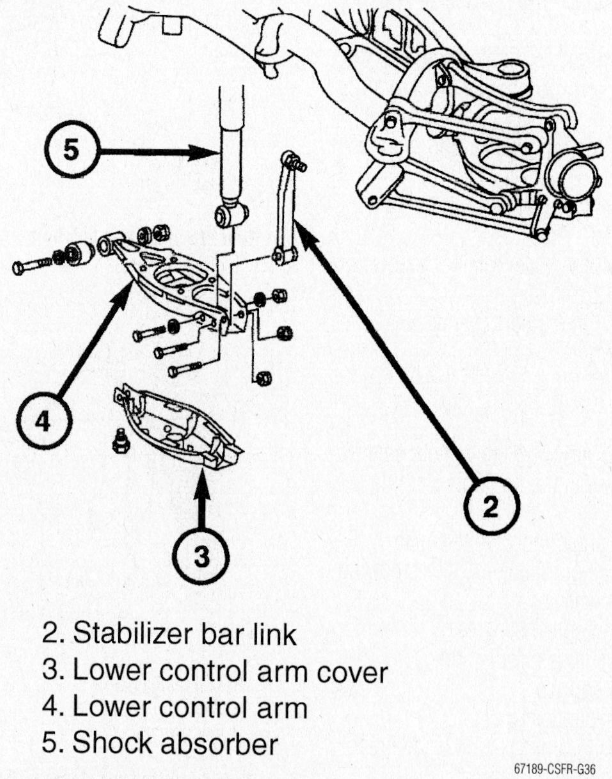

Exploded view of rear lower control arm—Crossfire

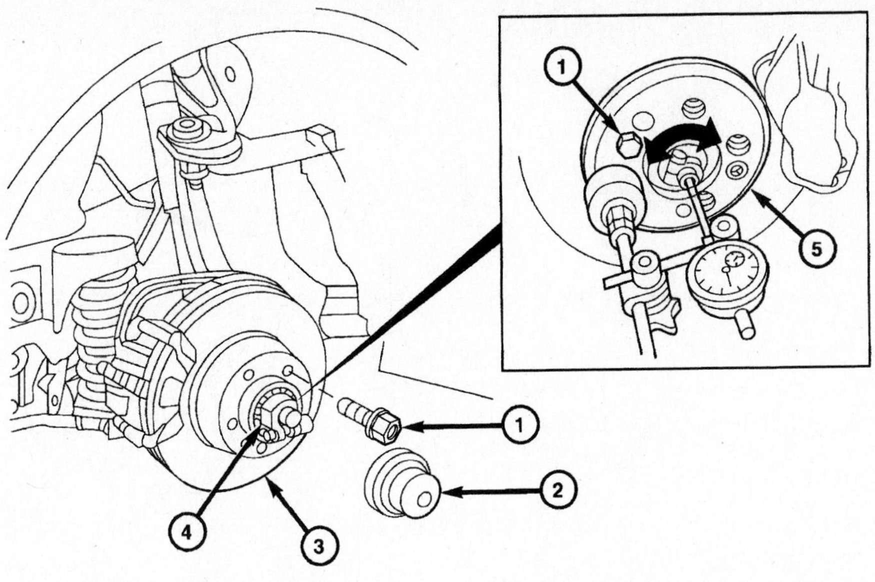

Adjusting front wheel bearing—Crossfire

2. With the wheel removed, install a wheel bolt on the opposite side from the brake rotor retaining bolt
3. Press the brake pads back into the caliper so the rotor turns free.
4. Remove the bearing grease cap.
5. Loosen the wheel bearing adjusting nut until some end play is present.
6. Attach a dial indicator to the hub.

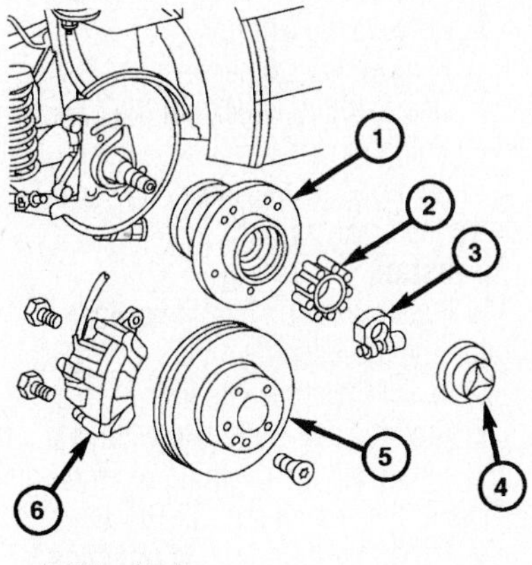

Exploded view of front wheel bearing assembly—Crossfire

7. Turn the hub nut in stages while pushing and pulling the rotor firmly back and forth. Adjust the end play to .0004-.0008 inches (.01-.02 mm).
8. Tighten the hub nut to 97 inch lbs. (11 Nm).
9. Remove dial indicator and install grease cap.
10. Remove the wheel bolt from rotor and install the wheel.

REMOVAL & INSTALLATION

1. Before servicing the vehicle, refer to the precautions in the beginning of this section.
2. Remove or disconnect the following:
   - Front wheel
   - Brake caliper
   - Brake rotor
   - Grease cap
   - Adjusting nut
   - Bearing
   - Wheel hub

To install:

3. Pack the wheel hub with grease and install the hub, bearing and seal ring onto the spindle.
4. Install or connect the following:
   - Brake rotor
   - Hub nut
   - Adjust bearing end play
   - Brake caliper
   - Grease cap
   - Front wheel

Steering Knuckle

REMOVAL & INSTALLATION

1. Before servicing the vehicle, refer to the precautions in the beginning of this section.
2. Raise and support the vehicle.
3. Remove or disconnect the following:
   - Front wheel
   - Brake caliper
   - Brake rotor
   - Wheel hub
   - Dust shield
   - Wheel speed sensor
   - Tie rod end
   - Upper and lower ball joints from knuckle
   - Steering knuckle

To install:

4. Install or connect the following:
   - Steering knuckle to upper ball joint stud. Tighten the nut to 33 ft. lbs. (45 Nm).
   - Steering knuckle to lower ball joint stud. Tighten the nut to 77 ft. lbs. (105 Nm).
   - Tie rod end. Tighten the nut to 37 ft. lbs. (50 Nm).
   - Wheel speed sensor. Tighten bolt to 16 ft. lbs. (22 Nm).
   - Dust shield. Tighten bolts to 16 ft. lbs. (22 Nm).
   - Wheel hub
   - Brake rotor
   - Brake caliper
   - Wheel

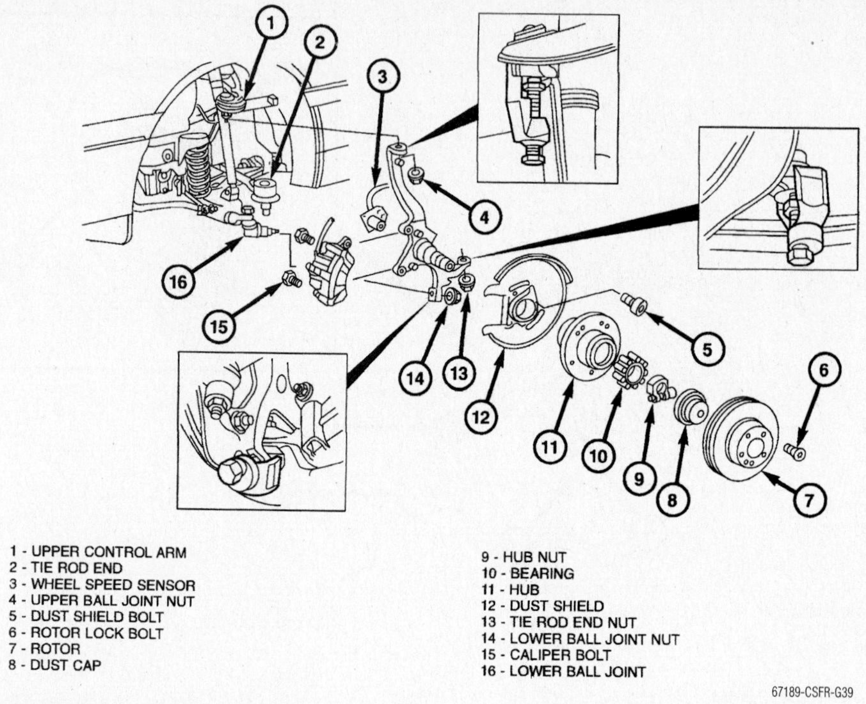

Exploded view of front steering knuckle assembly—Crossfire

BRAKES

Brake Caliper

REMOVAL & INSTALLATION

Front

1. Raise and support the front end on jackstands.
2. Drain a small amount of fluid from the master cylinder reservoir.
3. Remove the wheels.
4. Remove the brake pad wear indicator.
5. Press the caliper piston back into the bore with a suitable prytool. Use a large C-clamp to drive the piston into the bore of additional force is required.
6. Remove the caliper support spring.
7. Remove the caliper mounting pins.
8. Loosen the bolt that secures the front brake hose fitting bolt in the caliper.
9. Slide the caliper off the rotor and out from its mount.
10. Remove the front brake hose fitting bolt completely, then remove the caliper with the pads installed as an assembly. Take care not to drip fluid onto the pad surfaces.
11. Cover the open end of the front brake hose fitting to prevent dirt entry.

To install:

12. Lubricate the caliper slide pins and bushings silicone grease.
13. Install the caliper over the rotor and seat it in its original position until flush.
14. Install the slide pins by hand, and then tighten them to 18 ft. lbs. (25 Nm).
15. Connect the brake hose.
16. Install the brake pad wear indicator. Tighten the bolt to 6 ft. lbs. (8 Nm).
17. Install the caliper support spring.
18. Install the wheels.
19. Lower the vehicle.
20. Pump the brakes several times to seat the pads.

Rear

1. Before servicing the vehicle, refer to the precautions in the beginning of this section.
2. Remove or disconnect the following:
   - Rear wheel
   - Caliper mounting bolts
   - Brake hose
   - Caliper
3. Installation is the reverse of removal. Torque the caliper mounting bolts to 41 ft. lbs. (55 Nm).

Disc Brake Pads

REMOVAL & INSTALLATION

Front

1. Raise and support the front end on jackstands.
2. Remove the wheels.
3. Remove the caliper.
4. Remove the brake pads.
5. Installation is the reverse of the removal procedure.

Rear

1. Raise and support the rear end on jackstands.
2. Remove the wheels.
3. Remove the caliper.
4. Use a drift to knock out the brake pad retaining pin.
5. Remove the anti-rattle clips and remove the brake pads.

To install:

6. Install the pads into the caliper through the top opening.
7. Install the anti-rattle lip.
8. Hold the anti-rattle clip with your thumb and install the retaining pin through the hole in the side of the caliper.
9. Tap the retaining pin in until it is threaded through both pads and over the anti-rattle clip.
10. Install the caliper.
11. Install the wheels.

DODGE

Dakota

5

SPECIFICATIONS AND MAINTENANCE CHARTS

ENGINE AND VEHICLE IDENTIFICATION

Engine						
Code ①	Liters (cc)	Cu. In.	Cyl.	Fuel Sys.	Engine Type	Eng. Mfg.
P	2.5 (2507)	153	4	SMFI	OHV	Chrysler
K	3.7 (3701)	226	6	MFI	SOHC	Chrysler
X	3.9 (3916)	238	6	SMFI	OHV	Chrysler
N	4.7 (4701)	287	8	SMFI	SOHC	Chrysler
J	4.7 (4701)	287	8	SMFI	SOHC	Chrysler
Z	5.9 (5899)	360	8	SMFI	OHV	Chrysler

Model Year	
Code ②	Year
2	2002
3	2003
4	2004
5	2005
6	2006

OHV: Overhead Valve
SMFI: Sequential Multi-port Fuel Injection

① 8th position of VIN

② 10th position of VIN

06009-DAKO-C01

GENERAL ENGINE SPECIFICATIONS

Year	Engine Displacement Liters	Engine VIN	Net Horsepower @ rpm	Net Torque @ rpm (ft. lbs.)	Bore x Stroke (in.)	Com-pression Ratio	Oil Pressure @ rpm
2002	2.5	P	120@5200	145@3400	3.88x3.19	9.2:1	25-80@3000
	3.9	X	175@4800	220@3200	3.91x3.31	9.1:1	30-80@3000
	4.7	N	235@4800	295@3200	3.66x3.40	9.3:1	25@3000
	5.9	Z	230@4000	330@3250	4.00x3.58	9.1:1	30-80@3000
2003	3.9	X	175@4800	220@3200	3.91x3.31	9.1:1	30-80@3000
	4.7	N	235@4800	295@3200	3.66x3.40	9.3:1	25@3000
	5.9	Z	230@4000	330@3250	4.00x3.58	9.1:1	30-80@3000
2004	3.7	K	210@5200	225@4200	3.66x3.40	9.2:1	25-110@3000
	4.7	N	235@4800	295@3200	3.66x3.40	9.0:1	25-110@3000
2005	3.7	K	210@5200	225@4200	3.66x3.40	9.2:1	25-110@3000
	4.7 HO	J	270@5100	330@3600	3.66x3.40	9.7:1	25-110@3000
	4.7	N	235@4800	295@3200	3.66x3.40	9.0:1	25-110@3000

06009-DAKO-C02

GASOLINE ENGINE TUNE-UP SPECIFICATIONS

Year	Engine Dispacement Liters	Engine VIN	Spark Plug Gap (in.)	Ignition Timing (deg.)	Fuel Pump (psi)	Idle Speed (rpm)	Valve Clearance Intake	Valve Clearance Exhaust
2002	2.5	P	0.035	①	44.2-54.2	②	HYD	HYD
	3.9	X	0.040	①	44.2-54.2	②	HYD	HYD
	4.7	N	0.040	①	47-51	②	HYD	HYD
	5.9	Z	0.040	①	44.2-54.2	②	HYD	HYD
2003	3.9	X	0.040	①	44.2-54.2	②	HYD	HYD
	4.7	N	0.040	①	47-51	②	HYD	HYD
	5.9	Z	0.040	①	44.2-54.2	②	HYD	HYD
2004	3.7	K	0.042	①	44-54	②	HYD	HYD
	4.7	N	0.040	①	47-51	②	HYD	HYD
2005	3.7	K	0.042	①	44-54	②	HYD	HYD
	4.7 HO	J	0.040	①	56-60	②	HYD	HYD
	4.7	N	0.040	①	47-51	②	HYD	HYD

NOTE: The Vehicle Emission Control Information (VECI) label often reflects specification changes made during production.
The label figures must be used if they differ from those in this chart.
HYD: Hydraulic

① Ignition timing is controlled by the PCM and is not adjustable.

② Idle speed is controlled by the PCM and is not adjustable

06009-DAKO-C03

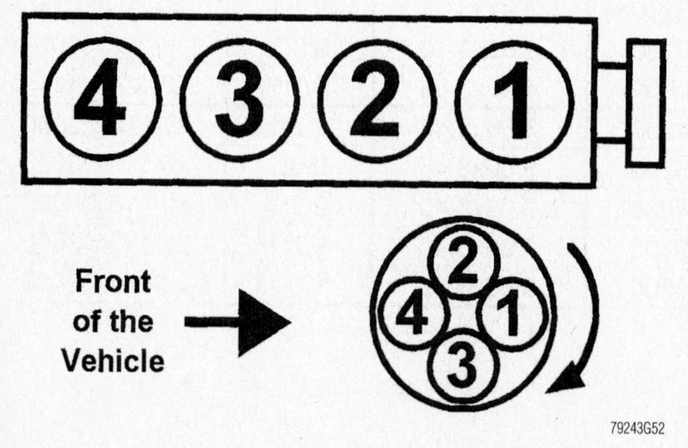

2.5L engine
Firing order: 1-3-4-2
Distributor rotation: clockwise

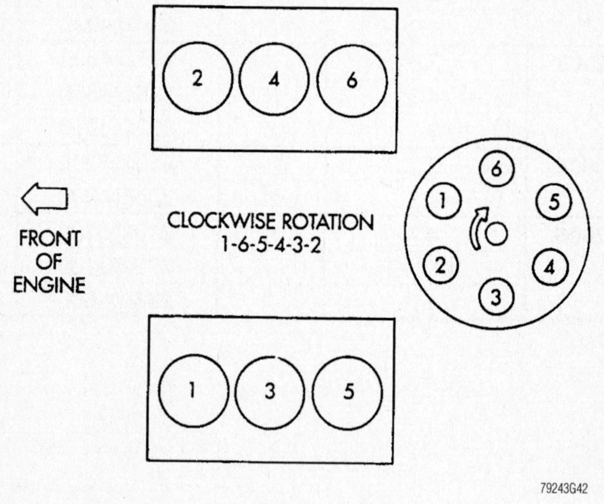

3.9L engine
Firing order: 1-6-5-4-3-2

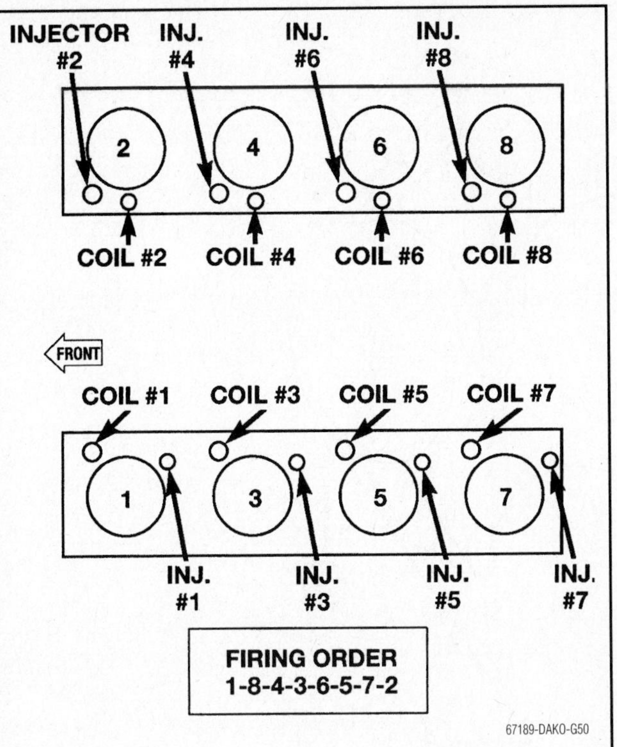

4.7L engine
Distributorless ignition

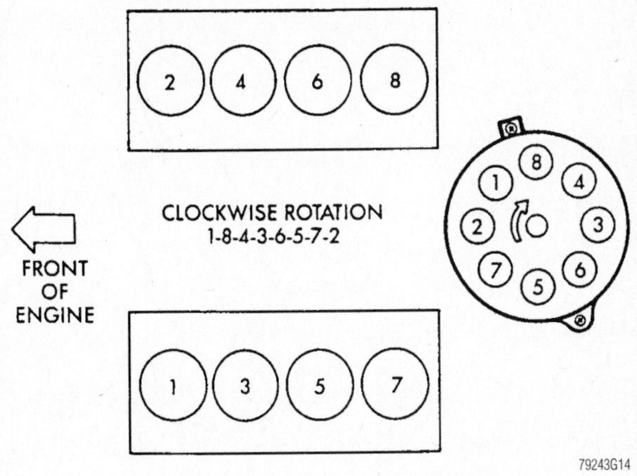

5.9L engine
Firing order: 1-8-4-3-6-5-7-2
Distributor rotation: clockwise

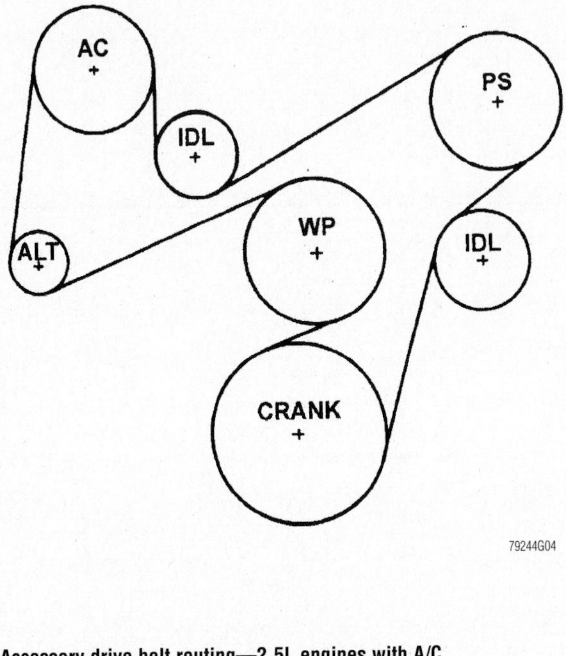

Accessory drive belt routing—2.5L engines with A/C

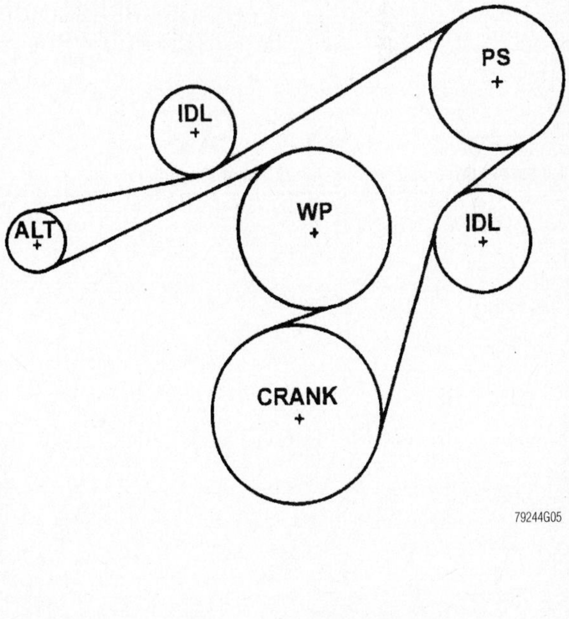

Accessory drive belt routing—2.5L engines without A/C

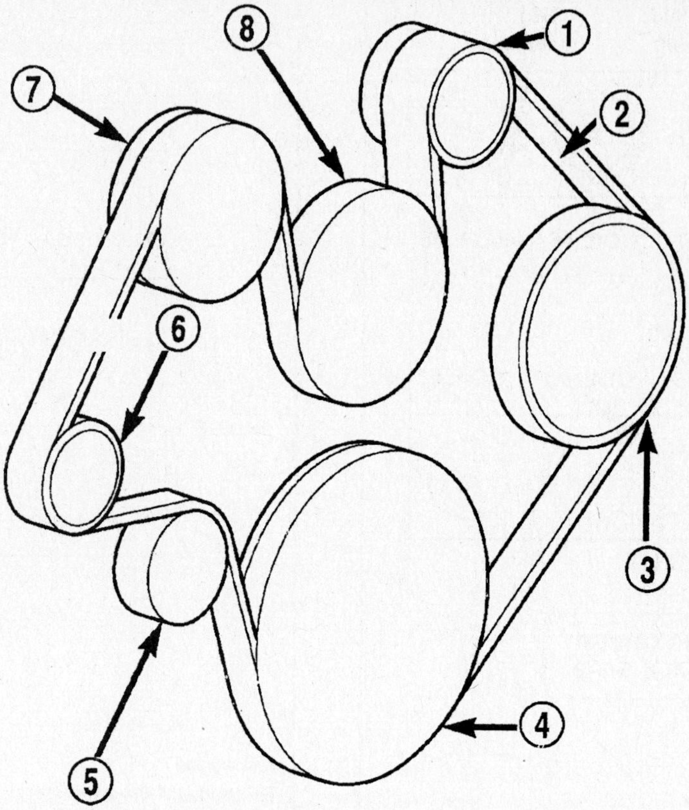

1 - GENERATOR PULLEY
2 - ACCESSORY DRIVE BELT
3 - POWER STEERING PUMP PULLEY
4 - CRANKSHAFT PULLEY
5 - IDLER PULLEY
6 - TENSIONER
7 - A/C COMPRESSOR PULLEY
8 - WATER PUMP PULLEY

67189-DAKO-G51

Accessory drive belt routing—3.7L and 4.7L engines

CAPACITIES

Year	Engine Displ. Liters	Engine VIN	Oil with Filter (qts.)	Transmission (pts.) Manual	Transmission (pts.) Auto.	Transfer Case (pts.)	Drive Axle Front (pts.)	Drive Axle Rear (pts.)	Fuel Tank (gal.)	Cooling System (qts.)
2002	2.5	P	4.5	4.8	—	—	—	①	②	9.8
	3.9	X	4.0	4.8	③	④	3.5	①	②	14.0
	4.7	N	6.0	4.8	③	④	3.5	①	②	13.0
	5.9	Z	5.0	—	③	④	3.5	①	②	14.6
2003	3.9	X	4.0	4.8	③	④	3.5	①	②	14.0
	4.7	N	6.0	4.8	③	④	3.5	①	②	13.0
	5.9	Z	5.0	—	③	④	3.5	①	②	14.6
2004	3.7	K	5.0	⑤	⑥	④	3.5	①	②	13.0
	4.7	N	6.0	⑤	⑥	④	3.5	①	②	13.0
2005	3.7	K	5.0	4.65	⑥	④	3.5	①	25.0	16.2
	4.7 HO	J	6.0	4.65	⑥	④	3.5	①	25.0	16.2
	4.7	N	6.0	4.65	⑥	④	3.5	①	25.0	16.2

NOTE: All capacities are approximate. Add fluid gradually and check to be sure a proper fluid level is obtained.

① The following values include 0.25 pt. of friction modifier for LSD axles.
8.25 in. axle: 4.4 pts.
9.25 in. axle: 4.9 pts.

② 2-door: 22 gal.
4-door: 24 gal.

③ Drain and refill: 8.0 pts.
42RE overhaul: 20 pts.
46RE overhaul: 20 pts.
45RFE overhaul: 28.0 pts.

④ NV233: 2.5 pts.
NV244: 2.85 pts.

⑤ 2wd: 4.8 pts.
4wd: 4.2 pts.

⑥ 42RLE: 8.0 pts.
545RFE 2wd: 11.0
545RFE 4wd: 13 pts.

06009-DAKO-C04

VALVE SPECIFICATIONS

Year	Engine Displ. Liters	Engine VIN	Seat Angle (deg.)	Face Angle (deg.)	Spring Test Pressure (lbs. @ in.)	Spring Installed Height (in.)	Stem-to-Guide Clearance (in.)		Stem Diameter (in.)	
							Intake	Exhaust	Intake	Exhaust
2002	2.5	P	44.5	45.0	184-196@ 1.216	1.64	0.001-0.003	0.001-0.003	0.311-0.312	0.311-0.312
	3.9	X	44.25-44.75	43.25-43.75	200@1.21	1.64	0.001-0.003	0.001-0.003	0.311-0.312	0.311-0.312
	4.7	N	44.5-45	45-45.5	176.7-193.3 @1.1670	1.601	0.0008-0.0028	0.0019-0.0039	0.2729-0.2739	0.2717-0.2728
	5.9	Z	44.25-44.75	43.25-43.75	200@1.212	1.640	0.0010-0.0030	0.0020-0.0040	0.3720-0.3730	0.3710-0.3720
2003	3.9	X	44.25-44.75	43.25-43.75	200@1.21	1.64	0.001-0.003	0.001-0.003	0.311-0.312	0.311-0.312
	4.7	N	44.5-45	45-45.5	176.7-193.3 @1.1670	1.601	0.0008-0.0028	0.0019-0.0039	0.2729-0.2739	0.2717-0.2728
	5.9	Z	44.25-44.75	43.25-43.75	200@1.212	1.640	0.0010-0.0030	0.0020-0.0040	0.3720-0.3730	0.3710-0.3720
2004	3.7	K	44.5-45	45-45.5	221-234@ 1.107	1.579	0.0008-0.0028	0.0019-0.0039	0.2729-0.2739	0.2717-0.2728
	4.7	N	44.5-45	45-45.5	176.7-193.3 @1.1670	1.601	0.0008-0.0028	0.0019-0.0039	0.2729-0.2739	0.2717-0.2728
2005	3.7	K	44.5-45	45-45.5	①	1.579	0.0008-0.0028	0.0019-0.0039	0.2729-0.2739	0.2717-0.2728
	4.7 HO	J	44.5-45	45-45.5	②	1.579	0.0008-0.0028	0.0019-0.0039	0.2729-0.2739	0.2717-0.2728
	4.7	N	44.5-45	45-45.5	174.4-195.5 @1.1370	1.579	0.0008-0.0028	0.0019-0.0039	0.2729-0.2739	0.2717-0.2728

① Intake: 213-234@1.107
Exhaust: 215-219@1.067

② Intake: 221.2-233.8@1.107
Exhaust: 216.9-237.1@1.146

06009-DAKO-C05

CRANKSHAFT AND CONNECTING ROD SPECIFICATIONS

All measurements are given in inches.

	Engine		Crankshaft				Connecting Rod		
Year	Engine Displ. Liters	Engine VIN	Main Brg. Journal Dia.	Main Brg. Oil Clearance	Shaft End-play	Thrust on No.	Journal Diameter	Oil Clearance	Side Clearance
2002	2.5	P	2.4996-2.5001	0.0010-0.0025	0.0015-0.0065	2	2.2080-2.2085	0.0015-0.0020	0.010-0.0190
	3.9	X	2.4995-2.5005	0.0005-0.0015	0.0020-0.0070	2	2.1240-2.1250	0.0005-0.0022	0.0060-0.0140
	4.7	N	2.4996-2.5005	0.0008-0.0021	0.0021-0.0112	2	2.0076-2.0082	0.0006-0.0022	0.0040-0.0138
	5.9	Z	2.8095-2.8105	①	0.0020-0.0070	2	2.1240-2.1250	0.0005-0.0022	0.0060-0.0140
2003	3.9	X	2.4995-2.5005	0.0005-0.0015	0.0020-0.0070	2	2.1240-2.1250	0.0005-0.0022	0.0060-0.0140
	4.7	N	2.4996-2.5005	0.0008-0.0021	0.0021-0.0112	2	2.0076-2.0082	0.0006-0.0022	0.0040-0.0138
	5.9	Z	2.8095-2.8105	①	0.0020-0.0070	2	2.1240-2.1250	0.0005-0.0022	0.0060-0.0140
2004	3.7	K	2.4996-2.5005	0.0020-0.0034	0.0021-0.0112	2	2.2794-2.2797	0.0004-0.0019	0.0040-0.0138
	4.7	N	2.4996-2.5005	0.0008-0.0021	0.0021-0.0112	2	2.0076-2.0082	0.0006-0.0022	0.0040-0.0138
2005	3.7	K	2.4996-2.5005	0.0008-0.0018	0.0021-0.0112	2	2.2792-2.2798	0.0002-0.0011	0.0040-0.0138
	4.7 HO	J	2.4996-2.5005	0.0008-0.0021	0.0021-0.0112	2	2.0076-2.0082	0.0006-0.0022	0.0040-0.0138
	4.7	N	2.4996-2.5005	0.0008-0.0021	0.0021-0.0112	2	2.0076-2.0082	0.0006-0.0022	0.0040-0.0138

① No. 1: 0.0005-0.0015
Nos. 2-5: 0.0005-0.0020

06009-DAKO-C06

PISTON AND RING SPECIFICATIONS

All measurements are given in inches.

Year	Engine Displ. Liters	Engine VIN	Piston Clearance	Ring Gap Top Comp.	Ring Gap Bottom Comp.	Ring Gap Oil Control	Ring Side Clearance Top Comp.	Ring Side Clearance Bottom Comp.	Ring Side Clearance Oil Control
2002	2.5	P	0.0013-0.0021	0.0090-0.0240	0.0190-0.0380	0.0100-0.0600	0.0017-0.0033	0.0017-0.0033	0.0024-0.0083
	3.9	X	0.0005-0.0015	0.0100-0.0200	0.0100-0.0200	0.0020-0.0080	0.0015-0.0030	0.0015-0.0030	0.1515-0.1565
	4.7	N	0.0014	0.0146-0.0249	0.0146-0.0249	0.0099-0.0300	0.0020-0.0037	0.0016-0.0031	0.0175-0.0185
	5.9	Z	0.0005-0.0015	0.0120-0.0220	0.0220-0.0310	0.0150-0.0550	0.0016-0.0033	0.0016-0.0033	0.0020-0.0080
2003	3.9	X	0.0005-0.0015	0.0100-0.0200	0.0100-0.0200	0.0020-0.0080	0.0015-0.0030	0.0015-0.0030	0.1515-0.1565
	4.7	N	0.0014	0.0146-0.0249	0.0146-0.0249	0.0099-0.0300	0.0020-0.0037	0.0016-0.0031	0.0175-0.0185
	5.9	Z	0.0005-0.0015	0.0120-0.0220	0.0220-0.0310	0.0150-0.0550	0.0016-0.0033	0.0016-0.0033	0.0020-0.0080
2004	3.7	K	0.0014	0.0079-0.0142	0.0146-0.0249	0.0100-0.0300	0.0020-0.0037	0.0016-0.0031	0.0007-0.0091
	4.7	N	0.0014	0.0146-0.0249	0.0146-0.0249	0.0099-0.0300	0.0020-0.0037	0.0016-0.0031	0.0175-0.0185
2005	3.7	K	0.0014	0.0079-0.0142	0.0146-0.0249	0.0100-0.0300	0.0020-0.0037	0.0016-0.0031	0.0007-0.0091
	4.7 HO	J	0.0014	0.0146-0.0249	0.0146-0.0249	0.0099-0.0300	0.0020-0.0037	0.0016-0.0031	0.0175-0.0185
	4.7	N	0.0014	0.0146-0.0249	0.0146-0.0249	0.0099-0.0300	0.0020-0.0037	0.0016-0.0031	0.0175-0.0185

06009-DAKO-C07

TORQUE SPECIFICATIONS

All readings in ft. lbs.

Year	Engine Displ. Liters	Engine VIN	Cylinder Head Bolts	Main Bearing Bolts	Rod Bearing Bolts	Crankshaft Damper Bolts	Flywheel Bolts	Manifold Intake	Manifold Exhaust	Spark Plugs	Oil Pan Drain Plug
2002	2.5	P	①	80	33	80	105	②	②	27	25
	3.9	X	③	85	45	18	55	④	25	30	25
	4.7	N	⑤	⑥	⑦	130	45	⑧	18	20	25
	5.9	Z	③	85	45	18	55	④	25	30	25
2003	3.9	X	③	85	45	18	55	④	25	30	25
	4.7	N	⑤	⑥	⑦	130	45	⑧	18	20	25
	5.9	Z	③	85	45	18	55	④	25	30	25
2004	3.7	K	⑤	⑩	⑦	130	70	9	18	20	25
	4.7	N	⑨	⑥	⑦	130	45	④	18	20	25
2005	3.7	K	⑤	⑩	⑦	130	70	9	18	20	25
	4.7 HO	J	⑤	⑥	⑦	130	45	9	18	20	25
	4.7	N	⑤	⑥	⑦	130	45	④	18	20	25

① See illustration in text section
- Bolts 1-10 and 12-14: 110 ft. lbs.
- Bolt 11: 100 ft. lbs.

② Exhaust manifold bolt 1: 30 ft. lbs.
- Intake/exhaust manifold bolts 2-5: 23 ft. lbs.
- Exhaust manifold nuts 6 & 7: 23 ft. lbs.

③ Step 1: 50 ft. lbs.
- Step 2: 105 ft. lbs.

④ See illustration in text section
- Step 1: 1-4 to 72 inch lbs. in 12 inch lb. Increments
- Step 2: bolts 5-12: 72 inch lbs.
- Step 3: Check that all bolts are at 72 inch lbs.
- Step 4: All bolts, in sequence, to 12 ft. lbs.
- Step 5: Check that all bolts are at 12 ft. lbs.

⑤ See text

⑥ Bed plate bolt sequence. Refer to illustration
- Step 1: Bolts A-L to 40 ft. lbs.
- Step 2: Bolts 1-10 25 inch lbs.
- Step 3: Bolts 1-10 plus 90 degrees
- Step 4: Bolts A1-A6 20 ft. lbs.

⑦ 20 ft. lbs. plus 90 degrees

⑧ 105 inch lbs.

⑨ M8 bolts: 19 ft. lbs.
- M11 bolts: 60 ft. lbs.

⑩ Bed plate bolt sequence. Refer to illustration
- Step 1: Hand tighten bolts 1D,1G and 1F until the bedplate contacts the block.
- Step 2: Tighten bolts 1A - 1J to 54 N·m (40 ft. lbs.)
- Step 3: Tighten bolts 1 - 8 to 7 N·m (5 ft. lbs.)
- Step 4: Turn bolts 1 - 8 an additional 90°.
- Step 5: Tighten bolts A - E 27 N·m (20 ft. lbs.).

06009-DAKO-C08

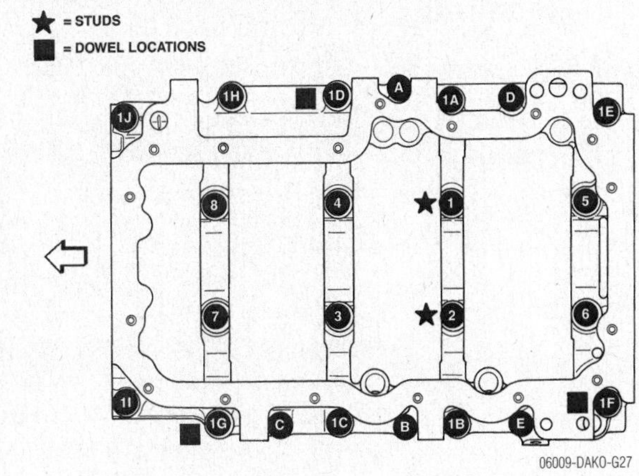

3.7L engine bedplate bolt torque sequence

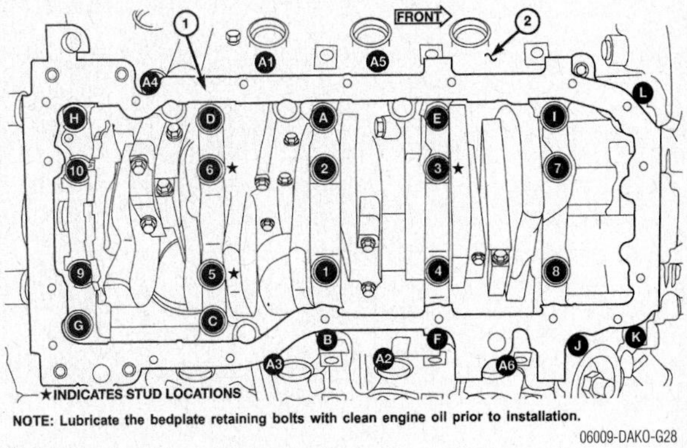

4.7L engine bedplate bolt torque sequence

WHEEL ALIGNMENT

Year	Model	Wheel Base (in.)	Caster Range (+/-Deg.)	Caster Preferred Setting (Deg.)	Camber Range (+/-Deg.)	Camber Preferred Setting (Deg.)	Toe-in (in.)
2002	2WD	111.9	0.50	+2.99	0.50	-0.25	0.10+/-0.06
	2WD	130.9	0.50	+3.13	0.50	-0.25	0.10+/-0.06
	4WD	111.9	0.50	+3.16	0.50	-0.25	0.10+/-0.06
	4WD	130.9	0.50	+3.27	0.50	-0.25	0.10+/-0.06
	RT	111.9	0.50	+3.67	0.50	-0.34	0.10+/-0.06
	RT	130.9	0.50	+3.81	0.50	-0.34	0.10+/-0.06
2003	2WD	111.9	0.50	+2.99	0.50	-0.25	0.10+/-0.06
	2WD	130.9	0.50	+3.13	0.50	-0.25	0.10+/-0.06
	4WD	111.9	0.50	+3.16	0.50	-0.25	0.10+/-0.06
	4WD	130.9	0.50	+3.27	0.50	-0.25	0.10+/-0.06
	RT	111.9	0.50	+3.67	0.50	-0.34	0.10+/-0.06
	RT	130.9	0.50	+3.81	0.50	-0.34	0.10+/-0.06
2004	2WD	111.9	0.50	+2.99	0.50	0	0.10+/-0.06
	2WD	130.9	0.50	+3.13	0.50	0	0.10+/-0.06
	4WD	111.9	0.50	+3.16	0.50	0	0.10+/-0.06
	4WD	130.9	0.50	+3.27	0.50	0	0.10+/-0.06
	RT	111.9	0.50	+3.67	0.50	-0.09	0.10+/-0.06
	RT	130.9	0.50	+3.81	0.50	-0.09	0.10+/-0.06
2005	All	131	0.50	+3.50	0.50	+0.25	0.20+/-0.50

06009-DAKO-C11

TIRE, WHEEL AND BALL JOINT SPECIFICATIONS

Year	Model	OEM Tires Standard	OEM Tires Optional	Tire Pressures (psi) Front	Tire Pressures (psi) Rear	Wheel Size	Ball Joint Inspection	Lug Nut Torque ft. lbs.
2002	2wd	P215/75R15	P235/75R15	①	①	6.5-JJ	0.060 in. ②	100
	4wd	P235/75R15 XL	P255/65R16	①	①	NA		
			P265/70R16	①	①			
	RT	P255/55R17	none	①	①	NA		
2003	2wd	P215/75R15	P235/75R15	①	①	6.5-JJ	0.060 in. ②	100
	4wd	P235/75R15 XL	P255/65R16	①	①	NA		
			P265/70R16	①	①			
	RT	P255/55R17	none	①	①	NA		
2004	2wd	P245/70R16	P255/65R16	①	①	std: 7	0.060 in. ②	100
	4wd	P245/70R16	P265/70R16	①	①	opt: 8		
2005	2wd/4wd ST Club Cab	P245/70R16	none	①	①	NA	0.020 in.	135
	2wd/4wd SLT Club Cab	P245/70R16	P255/65R16	①	①	NA		
	Laramie 2wd Club Cab	P255/65R17	P265/65R17	①	①	NA		
	Laramie 4wd Club Cab	P265/70R16	P265/65R17	①	①	NA		
	2wd/4wd ST Quad Cab	P245/70R16	none	①	①	NA		
	2wd SLT Quad Cab	P245/70R16	P255/65R16	①	①	NA		
	Laramie 2wd Quad Cab	P255/65R16	P265/65R17	①	①	NA		
	Laramie 4wd Quad Cab	P245/70R16	P265/70R16	①	①	NA		
			P255/65R16	①	①	NA		
			P265/75R17	①	①	NA		

OEM: Original Equipment Manufacturer

PSI: Pounds Per Square Inch

STD: Standard

OPT: Optional

① See the tire placard on the vehicle

② Both upper and lower

06009-DAKO-C10

BRAKE SPECIFICATIONS

All measurements in inches unless noted

Year		Brake Disc			Brake Drum			Minimum Lining Thickness		Brake Caliper	
		Original Thickness	Minimum Thickness	Maximum Run-out	Original Inside Diameter	Max. Wear Limit	Maximum Machine Diameter	Front	Rear	Bracket Bolts (ft. lbs.)	Mounting Bolts (ft. lbs.)
2002		0.944	0.890	0.004	①	②	②	③	④	47	22
2003	F	1.000	0.965	0.004	11.00	②	②	③	④	⑤	⑥
	R	0.600	0.585	0.0014							
2004	F	1.000	0.965	0.004	11.00	②	②	③	④	⑤	⑥
	R	0.600	0.585	0.0014							
2005	F	1.102	1.039	0.0010	11.50	②	②	③	④	130	26

F: Front

R: Rear

NA: Not Available

① Available with both 9 in. and 11 in. rear brakes

② Maximum allowable drum diameter, either from wear or machining, is stamped on the drum.

③ Riveted brake pads: 0.0625 in.
Bonded brake pads: 0.1875 in.

④ Riveted brake shoes: 0.031 in.
Bonded brake shoes: 0.0625 in.

⑤ Front adapter bolts: 147.5 ft. lbs.
Support plate: 50 ft. lbs.

⑥ Front: 22 ft. lbs.
Rear: 19 ft. lbs.

06009-DAKO-C09

SCHEDULED MAINTENANCE INTERVALS

2002-2003 Dodge Dakota

TO BE SERVICED	TYPE OF SERVICE	VEHICLE MILEAGE INTERVAL (x1000)													
		7.5	15	22.5	30	37.5	45	52.5	60	67.5	75	82.5	90	97.5	100
Engine oil & filter	R	✓	✓	✓	✓	✓	✓	✓	✓	✓	✓	✓	✓	✓	
Ball joints	L			✓			✓			✓			✓		
Front wheel bearings	S/I			✓			✓			✓			✓		
Brake linings	S/I			✓			✓			✓			✓		
Air cleaner element	R				✓				✓				✓		
Spark plugs	R				✓				✓				✓		
Transfer case fluid	R					✓					✓				
Automatic transmission fluid	R					✓					✓				
Automatic transmission bands	Adj					✓					✓				
Engine coolant ①	R						✓						✓		
Spark plug cables	R								✓						
PCV valve ②	S/I								✓						
Drive belt tensioner (3.9L & 5.9L) ②	S/I								✓						
Drive belt tension (2.5L)	Adj								✓						

R: Replace S/I: Service or Inspect L: Lubricate Adj: Adjust

① Change every 36 months, regardless of mileage

② Replace if necessary.

FREQUENT OPERATION MAINTENANCE (SEVERE SERVICE)

If a vehicle is operated under any of the following conditions it is considered severe service:

- Extremely dusty areas.

- 50% or more of the vehicle operation is in 32°C (90°F) or higher temperatures, or constant operation in temperatures below 0°C (32°F).

- Prolonged idling (vehicle operation in stop and go traffic.

- Frequent short running periods (engine does not warm to normal operating temperatures).

- Police, taxi, delivery usage or trailer towing usage.

Oil & oil filter change: change every 3000 miles.

Air filter/air pump air filter: change every 24,000 miles.

Engine coolant level, hoses & clamps: check every 6,000 miles.

Exhaust system: check every 6000 miles.

Drive belts: check every 18,000 miles; replace every 24,000 miles.

Crankcase inlet air filter (6 & 8 cyl.): clean every 24,000 miles.

Oxygen sensor: replace every 82,500 miles.

Automatic transmission fluid, filter & bands: change & adjust every 12,000 miles.

Steering linkage: lubricate every 6000 miles.

Rear axle fluid: change every 12,000 miles.

06009-DAKO-C12

SCHEDULED MAINTENANCE INTERVALS

2004 Dodge Dakota

TO BE SERVICED	TYPE OF SERVICE	VEHICLE MILEAGE INTERVAL (x1000)													
		6	12	18	24	30	36	42	48	54	60	66	72	78	84
Engine coolant ①	R						✓						✓		
Accessory drive belt ②	S/I										✓				
Engine oil & filter	R	✓	✓	✓	✓	✓	✓	✓	✓	✓	✓	✓	✓	✓	
PCV valve ②	S/I										✓				
Ball joints	L			✓			✓			✓			✓		
Front wheel bearings	S/I			✓			✓			✓			✓		
Brake linings	S/I			✓			✓			✓			✓		
Air cleaner element	R					✓					✓				
Spark plugs	R					✓					✓				
Auto. Trans. fluid ③	R	every 100,000 miles													
Transfer case fluid ④	I										✓				

R: Replace S/I: Service or Inspect L: Lubricate Adj: Adjust

① Change every 36 months, regardless of mileage

② Replace if necessary.

③ w/3.7L engine: change fluid, filter and adjust bands

w/4.7L engine: change fluid, main filter and cooler return filter (if equipped)

④ Replace every 120,000 miles

FREQUENT OPERATION MAINTENANCE (SEVERE SERVICE)

If a vehicle is operated under any of the following conditions it is considered severe service:

- Extremely dusty areas.

- 50% or more of the vehicle operation is in 32°C (90°F) or higher temperatures, or constant operation in temperatures below 0°C (32°F).

- Prolonged idling (vehicle operation in stop and go traffic.

- Frequent short running periods (engine does not warm to normal operating temperatures).

- Police, taxi, delivery usage or trailer towing usage.

Oil & oil filter change: change every 3000 miles.

Air filter/air pump air filter: change every 24,000 miles.

Engine coolant level, hoses & clamps: check every 6,000 miles.

Exhaust system: check every 6000 miles.

Drive belts: check every 18,000 miles; replace every 24,000 miles.

Crankcase inlet air filter (6 & 8 cyl.): clean every 24,000 miles.

Oxygen sensor: replace every 82,500 miles.

Automatic transmission fluid, filter & bands: change & adjust every 12,000 miles.

Steering linkage: lubricate every 6000 miles.

Rear axle fluid: change every 12,000 miles.

06009-DAKO-C13

SCHEDULED MAINTENANCE INTERVALS

2005 Dodge Dakota

TO BE SERVICED	TYPE OF SERVICE	VEHICLE MILEAGE INTERVAL (x1000)													
		6	12	18	24	30	36	42	48	54	60	66	72	78	84
Engine coolant	R	Replace every 60 months, regardless of milage													
Accessory drive belt ①	S/I										✓				
Engine oil & filter	R	✓	✓	✓	✓	✓	✓	✓	✓	✓	✓	✓	✓	✓	✓
Tires	Rotate	✓	✓	✓	✓	✓	✓	✓	✓	✓	✓	✓	✓	✓	✓
PCV valve ①	S/I										✓				
Brake linings	S/I			✓			✓			✓			✓		
Air cleaner element	S/I					✓					✓				
Air cleaner element	R										✓				
Spark plugs	R					✓					✓				
Transfer case fluid level ②	I					✓					✓				

R: Replace S/I: Service or Inspect L: Lubricate Adj: Adjust

① Replace if necessary.

② Replace every 120,000 miles

FREQUENT OPERATION MAINTENANCE (SEVERE SERVICE)

If a vehicle is operated under any of the following conditions it is considered severe service:

- Extremely dusty areas.

- 50% or more of the vehicle operation is in 32°C (90°F) or higher temperatures, or constant operation in temperatures below 0°C (32°F).

- Prolonged idling (vehicle operation in stop and go traffic.

- Frequent short running periods (engine does not warm to normal operating temperatures).

- Police, taxi, delivery usage or trailer towing usage.

Oil & oil filter change: change every 3000 miles.

Air filter/air pump air filter: change every 24,000 miles.

Engine coolant level, hoses & clamps: check every 6,000 miles.

Exhaust system: check every 6000 miles.

Drive belts: check every 18,000 miles; replace every 24,000 miles.

Crankcase inlet air filter (6 & 8 cyl.): clean every 24,000 miles.

Oxygen sensor: replace every 82,500 miles.

Automatic transmission fluid, filter & bands: change & adjust every 12,000 miles.

Steering linkage: lubricate every 6000 miles.

Rear axle fluid: change every 12,000 miles.

06009-DAKO-C14

PRECAUTIONS

Before servicing any vehicle, please be sure to read all of the following precautions, which deal with personal safety, prevention of component damage, and important points to take into consideration when servicing a motor vehicle:

- Never open, service or drain the radiator or cooling system when the engine is hot; serious burns can occur from the steam and hot coolant.
- Observe all applicable safety precautions when working around fuel. Whenever servicing the fuel system, always work in a well-ventilated area. Do not allow fuel spray or vapors to come in contact with a spark, open flame, or excessive heat (a hot drop light, for example). Keep a dry chemical fire extinguisher near the work area. Always keep fuel in a container specifically designed for fuel storage; also, always properly seal fuel containers to avoid the possibility of fire or explosion. Refer to the additional fuel system precautions later in this section.
- Fuel injection systems often remain pressurized, even after the engine has been turned **OFF**. The fuel system pressure must be relieved before disconnecting any fuel lines. Failure to do so may result in fire and/or personal injury.
- Brake fluid often contains polyglycol ethers and polyglycols. Avoid contact with the eyes and wash your hands thoroughly after handling brake fluid. If you do get brake fluid in your eyes, flush your eyes with clean, running water for 15 minutes. If eye irritation persists, or if you have taken brake fluid internally, IMMEDIATELY seek medical assistance.
- The EPA warns that prolonged contact with used engine oil may cause a number of skin disorders, including cancer! You should make every effort to minimize your exposure to used engine oil. Protective gloves should be worn when changing oil. Wash your hands and any other exposed skin areas as soon as possible after exposure to used engine oil. Soap and water, or waterless hand cleaner should be used.
- All new vehicles are now equipped with an air bag system, often referred to as a Supplemental Restraint System (SRS) or Supplemental Inflatable Restraint (SIR) system. The system must be disabled before performing service on or around system components, steering column, instrument panel components, wiring and sensors. Failure to follow safety and disabling procedures could result in accidental air bag deployment, possible personal injury and unnecessary system repairs.
- Always wear safety goggles when working with, or around, the air bag system. When carrying a non-deployed air bag, be sure the bag and trim cover are pointed away from your body. When placing a non-deployed air bag on a work surface, always face the bag and trim cover upward, away from the surface. This will reduce the motion of the module if it is accidentally deployed. Refer to the additional air bag system precautions later in this section.
- Clean, high quality brake fluid from a sealed container is essential to the safe and proper operation of the brake system. You should always buy the correct type of brake fluid for your vehicle. If the brake fluid becomes contaminated, completely flush the system with new fluid. Never reuse any brake fluid. Any brake fluid that is removed from the system should be discarded. Also, do not allow any brake fluid to come in contact with a painted surface; it will damage the paint.
- Never operate the engine without the proper amount and type of engine oil; doing so WILL result in severe engine damage.
- Timing belt maintenance is extremely important! Many models utilize an interference-type, non-freewheeling engine. If the timing belt breaks, the valves in the cylinder head may strike the pistons, causing potentially serious (also time-consuming and expensive) engine damage. Refer to the maintenance interval charts in the front of this manual for the recommended replacement interval for the timing belt, and to the timing belt section for belt replacement and inspection.
- Disconnecting the negative battery cable on some vehicles may interfere with the functions of the on-board computer system(s) and may require the computer to undergo a relearning process once the negative battery cable is reconnected.
- When servicing drum brakes, only disassemble and assemble one side at a time, leaving the remaining side intact for reference.
- Only an MVAC-trained, EPA-certified automotive technician should service the air conditioning system or its components.

ENGINE REPAIR

➡Disconnecting the negative battery cable on some vehicles may interfere with the functions of the on board computer system. The computer may undergo a relearning process once the negative battery cable is reconnected.

Distributor

REMOVAL

2.5L Engine

1. Before servicing the vehicle, refer to the precautions in the beginning of this section.
2. Remove or disconnect the following:
   - Negative battery cable
   - Distributor cap
   - Camshaft Position (CMP) sensor connector
3. Matchmark the distributor housing and the rotor.
   - Distributor

3.9L and 5.9L Engines

1. Before servicing the vehicle, refer to the precautions in the beginning of this section.
2. Remove or disconnect the following:
   - Negative battery cable
   - Air cleaner tube
   - Distributor cap
   - Camshaft Position (CMP) sensor connector
3. Matchmark the distributor housing and the rotor.
4. Matchmark the distributor housing and the intake manifold.
5. Remove the distributor.

INSTALLATION

Timing Not Disturbed

2.5L ENGINE

1. Before servicing the vehicle, refer to the precautions in the beginning of this section.

➡The rotor will rotate clockwise as the gears engage.

2. Position the rotor slightly counterclockwise of the matchmark made during removal.
3. Install the distributor. Ensure that the

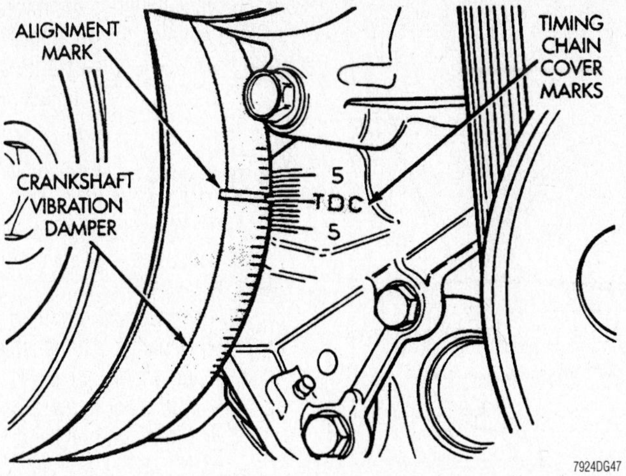

Crankshaft pulley and timing chain cover marks aligned at Top Dead Center (TDC)

rotor moves into alignment with the matchmark.

4. Align the locating fork with the clamp bolt hole. Install the clamp and bolt. Tighten the bolt to 17 ft. lbs. (23 Nm).

5. Install or connect the following:
   - CMP sensor connector
   - Distributor cap
   - Air cleaner tube
   - Negative battery cable

3.9L AND 5.9L ENGINES

1. Before servicing the vehicle, refer to the precautions in the beginning of this section.

➡The rotor will rotate clockwise as the gears engage.

2. Position the rotor slightly counterclockwise of the matchmark made during removal.

3. Install the distributor.

4. Align the distributor housing and intake manifold matchmarks and check that the distributor housing matchmark and rotor are also aligned.

5. Install or connect the following:
   - Distributor housing clamp and bolt. Tighten the bolt to 17 ft. lbs. (23 Nm).
   - CMP sensor connector
   - Distributor cap
   - Air cleaner tube
   - Negative battery cable

Timing Disturbed

2.5L ENGINE

1. Before servicing the vehicle, refer to the precautions in the beginning of this section.

2. Set the engine at Top Dead Center (TDC) of the No. 1 cylinder compression stroke.

3. Position the slot in the oil pump drive gear as shown.

4. Locate the alignment holes in the plastic ring and align the correct hole with the mating hole in the distributor housing as shown. Install a locking pin.

➡The distributor will rotate clockwise as the gears engage.

5. Position the base mounting slot at the 1 o'clock position and install the distributor.

6. Check that the centerline of the mounting slot aligns with the centerline of the clamp bolt hole.

7. Install the clamp and bolt. Tighten the bolt to 17 ft. lbs. (23 Nm).

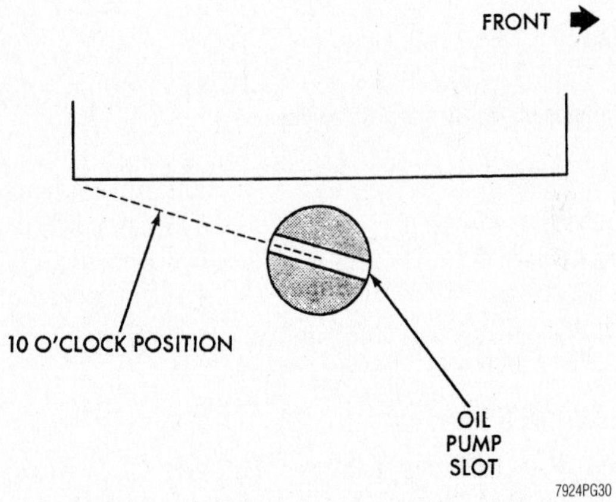

Slot in the oil pump gear at 10 o'clock position—2.5L engine

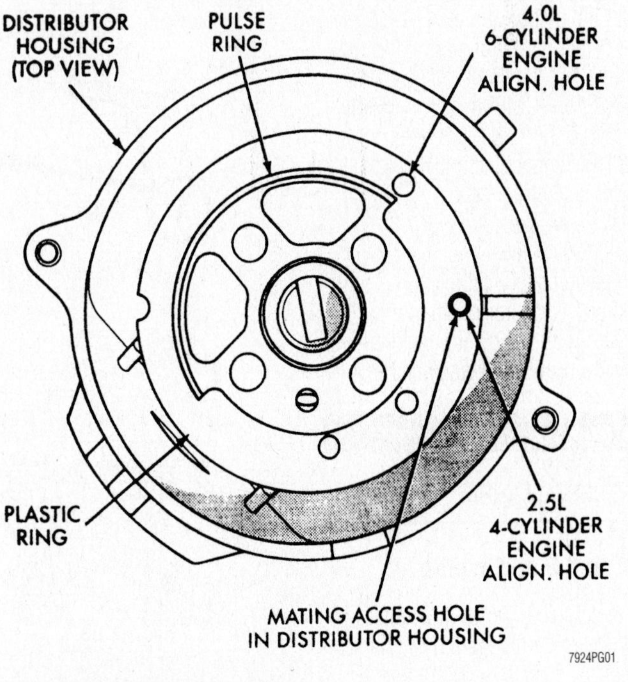

Distributor pin alignment holes—2.5L engine

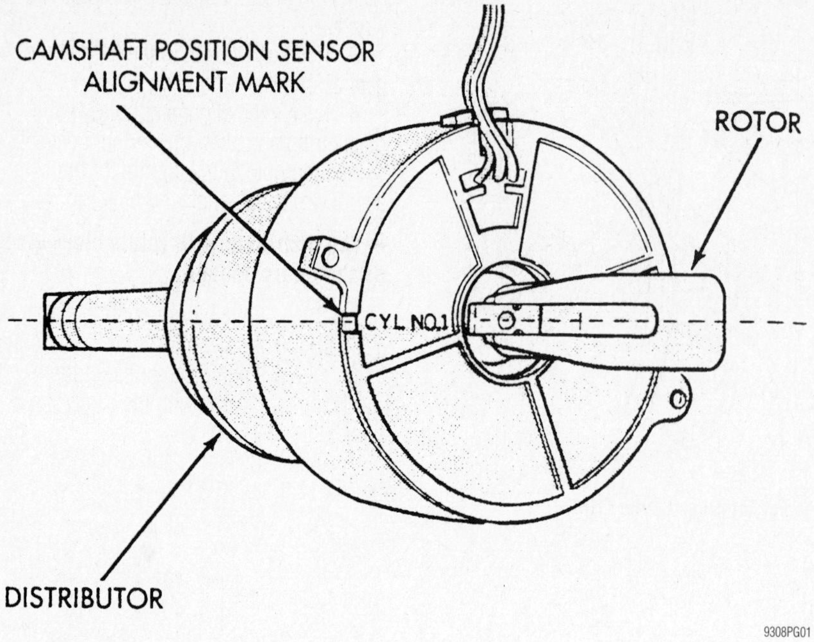

Distributor rotor alignment—3.9L and 5.9L engines

8. Remove the locking pin.
9. Install or connect the following:
   - CMP sensor connector
   - Distributor cap
   - Air cleaner tube
   - Negative battery cable

3.9L AND 5.9L ENGINES

1. Before servicing the vehicle, refer to the precautions in the beginning of this section.
2. Set the engine at Top Dead Center (TDC) of the No. 1 cylinder compression stroke.
3. Install the distributor, clamp and bolt.
4. Rotate the distributor so that the rotor aligns with the No. 1 cylinder mark on the Camshaft Position (CMP) sensor. Tighten the clamp bolt to 17 ft. lbs. (23 Nm).
5. Install or connect the following:
   - CMP sensor connector
   - Distributor cap
   - Air cleaner tube
   - Negative battery cable
6. The final distributor position must be set with a scan tool. Follow the instructions supplied by the scan tool manufacturer.

Alternator

REMOVAL & INSTALLATION

2.5L, 3.9L and 5.9L Engines

1. Before servicing the vehicle, refer to the precautions in the beginning of this section.
2. Remove or disconnect the following:
   - Negative battery cable
   - Accessory drive belt
   - Alternator harness connectors
   - Alternator
3. Installation is the reverse of removal. Observe the following torques:
   a. 2.5L engine: 41 ft. lbs. (56 Nm).
   b. 3.9L, 5.9L engines: 30 ft. lbs. (41 Nm).

3.7L Engine

1. Before servicing the vehicle, refer to the precautions in the beginning of this section.
2. Remove or disconnect the following:
   - Negative battery cable
   - Accessory drive belt
   - Alternator harness connectors
   - Mounting bolts and alternator

➡There are 1 vertical and 2 horizontal bolts.

To install:

3. Before servicing the vehicle, refer to the precautions in the beginning of this section.
4. Install the alternator and tighten the bolts to the following specifications:
   - Vertical bolt and long horizontal bolt: 41 ft. lbs. (56 Nm)
   - Short horizontal bolt: 55 ft. lbs. (74 Nm).
5. Install or connect the following:
   - Alternator harness connectors

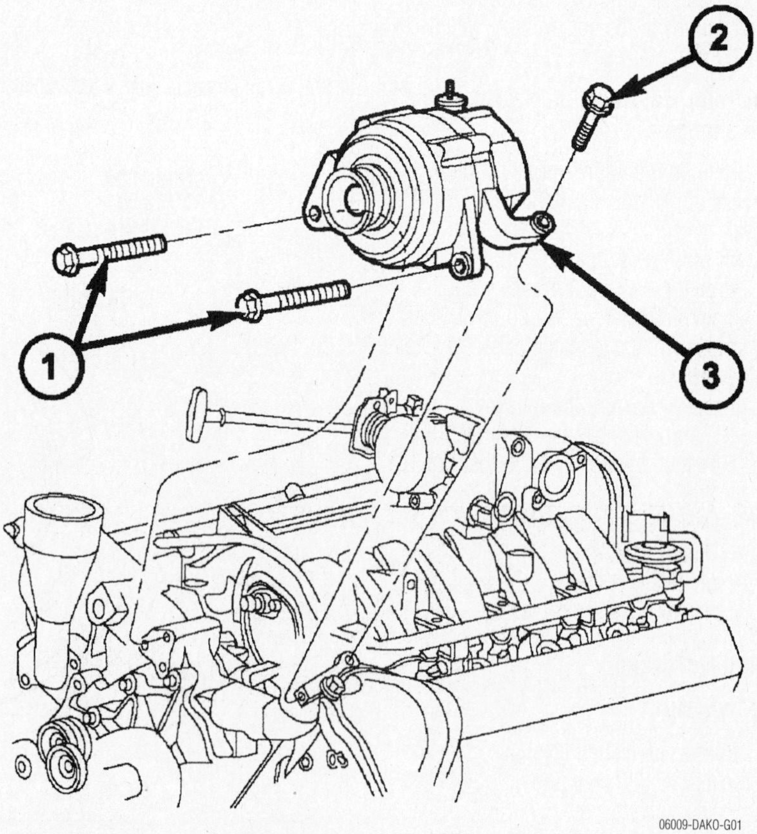

Alternator mounting—3.7L and 4.7L engines

- Accessory drive belt
- Negative battery cable

4.7L Engine

1. Before servicing the vehicle, refer to the precautions in the beginning of this section.
2. Remove or disconnect the following:
   - Negative battery cable
   - Accessory drive belt
   - Alternator harness connectors
   - Alternator
3. Installation is the reverse of removal. Observe the following torques:

Ignition Timing

ADJUSTMENT

The ignition timing is controlled by the Powertrain Control Module (PCM) and is not adjustable.

Engine Assembly

REMOVAL & INSTALLATION

2.5L, 3.9L & 5.9L Engines

1. Before servicing the vehicle, refer to the precautions in the beginning of this section.
2. Drain the cooling system.
3. Drain the engine oil.
4. Relieve the fuel system pressure.
5. Remove or disconnect the following:
   - Negative battery cable
   - Hood
   - Upper crossmember and top core support
   - Radiator hoses
   - Cooling fan and shroud
   - Radiator
   - Accelerator cable
   - Cruise control cable, if equipped
   - Transmission cable, if equipped
   - Heater hoses
   - Intake manifold vacuum lines
   - Accessory drive belt
   - Power steering pump
   - A/C compressor, if equipped
   - Engine control sensor harness connectors
   - Engine block heater, if equipped
   - Fuel line
   - Exhaust front pipe
   - Starter motor
   - Torque converter, if equipped
   - Transmission oil cooler lines, if equipped
   - Engine mounts
   - Transmission flange bolts. Support the transmission.
   - Engine

WARNING

Do not lift the engine by the intake manifold. Damage to the manifold may result.

To install:

6. Install or connect the following:
   - Engine. Tighten the engine mount bolts to 70 ft. lbs. (95 Nm).
   - Transmission flange bolts. Tighten the bolts to 40–45 ft. lbs. (54–61 Nm).
   - Transmission oil cooler lines, if equipped
   - Torque converter, if equipped. Tighten the bolts to 23 ft. lbs. (31 Nm).
   - Starter motor
   - Exhaust front pipe
   - Fuel line
   - Engine block heater, if equipped
   - Engine control sensor harness connectors
   - A/C compressor, if equipped
   - Power steering pump
   - Accessory drive belt
   - Intake manifold vacuum lines
   - Heater hoses
   - Transmission cable, if equipped
   - Cruise control cable, if equipped
   - Accelerator cable
   - Radiator
   - Cooling fan and shroud
   - Radiator hoses
   - Upper crossmember and top core support
   - Hood
   - Negative battery cable
7. Fill the crankcase to the correct level.
8. Fill the cooling system.
9. Start the engine and check for leaks.

3.7L Engine

1. Discharge the A/C system.
2. Drain the cooling system.
3. Release the fuel rail pressure.
4. Remove the air cleaner assembly.
5. Disconnect the battery.
6. Remove the upper fan shroud.
7. Remove the accessory drive belt.
8. Remove the viscous fan.
9. Remove the A/C compressor and position out of the way.
10. Remove the generator and secure away from engine.

➡Do not remove the phenolic pulley from the P/S pump. It is not required for P/S pump removal.

11. Remove the power steering pump with lines attached and secure away from engine.
12. Disconnect the heater hoses from the engine.
13. Disconnect the heater hoses from heater core and remove the hose assembly.
14. Remove the upper radiator hose from engine.
15. Remove the lower radiator hose from engine.
16. Disconnect the transmission oil cooler lines at the radiator.
17. Remove the radiator core support bracket.
18. Remove the radiator assembly, A/C Condenser and transmission oil cooler.
19. Disconnect throttle and speed control cables.
20. Disconnect the engine to body ground straps at the left side of cowl.
21. Disconnect the engine wiring harness at the following points:
    - Intake air temperature (IAT) sensor
    - Fuel Injectors
    - Throttle Position (TPS) Switch
    - Idle Air Control (IAC) Motor
    - Engine Oil Pressure Switch
    - Engine Coolant Temperature (ECT) Sensor
    - Manifold Absolute Pressure MAP) Sensor
    - Camshaft Position (CMP) Sensor
    - Coil Over Plugs
    - Crankshaft Position Sensor
22. Remove the coil over plugs.
23. Remove fuel rail and secure away from engine.

➡It is not necessary to release the quick connect fitting from the fuel supply line for engine removal.

24. Remove the PCV hose.
25. Remove the breather hoses.
26. Remove the vacuum hose for the power brake booster.
27. Disconnect the knock sensors.
28. Remove the engine oil dipstick tube.
29. Remove the intake manifold.
30. Install the engine lifting fixture, special tool 8247, using original fasteners from the removed intake manifold, and fuel rail.
31. Raise the vehicle on hoist.
32. Remove the exhaust crossover pipe from exhaust manifolds.
33. On 4wd vehicles, disconnect the axle vent tube from the left side engine mount.
34. Remove the through bolt retaining

nut and bolt from both the left and right side engine mounts.

35. On 4wd vehicles, remove the locknut from the left and right side engine mount brackets.

36. Disconnect two ground straps from the lower left hand side and one ground strap from the lower right hand side of the engine.

37. Disconnect the crankshaft position sensor.

➡The following step applies to 4wd vehicles equipped with automatic transmission only.

38. On 4wd vehicles, remove the axle isolator bracket from the engine, transmission and the axle.

39. Remove the structural cover.

40. Remove the starter.

41. Remove the torque converter bolts (automatic transmission only).

42. Remove transmission to engine mounting bolts.

43. Disconnect the engine block heater power cable from the block heater, if equipped.

44. Lower the vehicle.

45. Remove throttle body resonator assembly and air inlet hose.

46. Disconnect the throttle and speed control cables.

47. Disconnect the tube from both the left and right side crankcase breathers. Remove the breathers.

48. Remove the generator.

49. Disconnect the two heater hoses from the timing chain cover and heater core.

50. Unclip and remove the heater hoses and tubes from the intake manifold.

51. Disconnect the engine harness at the following points:

- Intake air temperature (IAT) sensor
- Fuel Injectors
- Throttle Position (TPS) Switch
- Idle Air Control (IAC) Motor
- Engine Oil Pressure Switch
- Engine Coolant Temperature (ECT) Sensor
- Manifold absolute pressure (MAP) Sensor
- Camshaft Position (CMP) Sensor
- Coil Over Plugs

52. Disconnect the vacuum lines at the throttle body and intake manifold.

53. Remove the power steering pump and position out of the way.

54. Disconnect the body ground strap at the right side cowl.

55. Disconnect the body ground strap at the left side cowl.

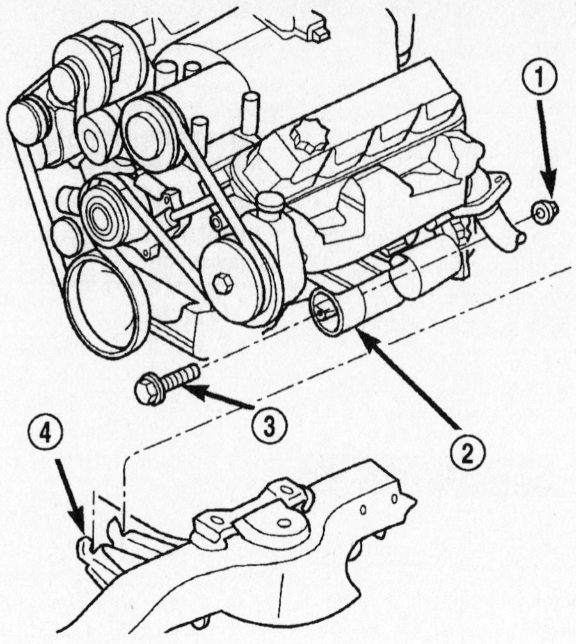

1 - LOCKNUT AND WASHER
2 - ENGINE MOUNT/INSULATOR
3 - THROUGH BOLT
4 - FRAME

67189-DAKO-G49

Engine mount through-bolt removal—3.7L engine w/2wd

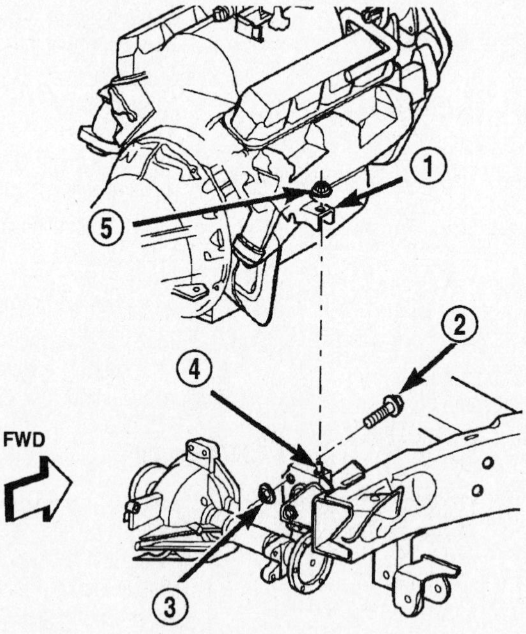

1 - ENGINE MOUNT BRACKET (2)
2 - THROUGH BOLT (2)
3 - LOCKNUT AND WASHER (2)
4 - ENGINE ISOLATOR TO ENGINE MOUNT BRACKET STUD (2)
5 - LOCKNUT (2)

67189-DAKO-G47

Engine mount through-bolt removal—3.7L engine w/4wd

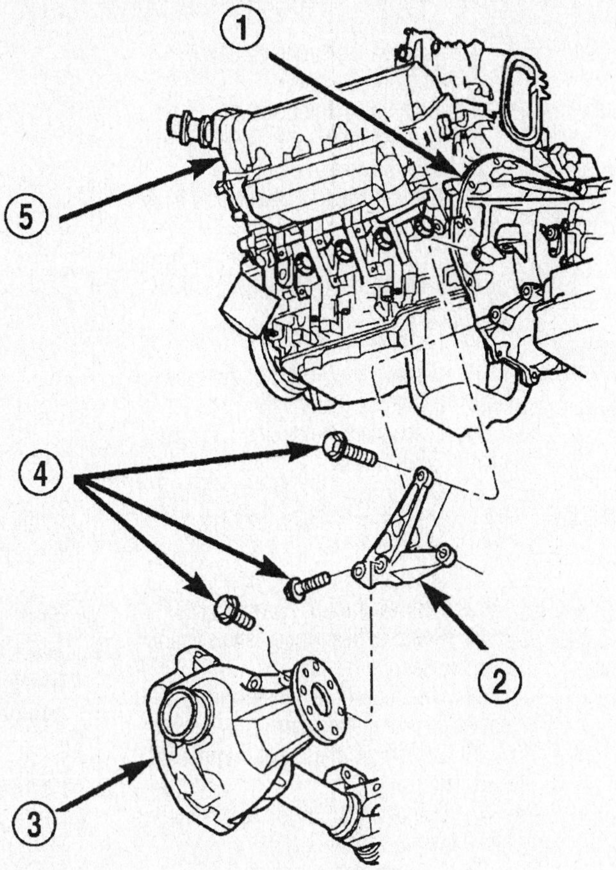

Axle isolator bracket removal—3.7L engine w/4wd

➡It will be necessary to support the transmission in order to remove the engine.

56. Position a suitable jack under the transmission.

57. Remove the engine from the vehicle.

To install:

58. Position the engine in the vehicle. Position both the left and right side engine mount brackets and install the through bolts and nuts. On 2wd vehicles, tighten nuts to 95 Nm (70 ft. lbs.); on 4wd vehicles, to 102 Nm (75 ft. lbs.).

59. On 4wd vehicles, install the locknuts onto the engine mount brackets. Tighten the locknuts to 41 Nm (30 ft. lbs.).

60. Remove the jack from under the transmission.

61. Remove Engine Lifting Fixture Tool 8347.

62. Remove Special Tools 8400 Lifting Studs.

63. Position the generator wiring behind the oil dipstick tube, then install the oil dipstick tube upper mounting bolt.

64. Connect both left and right side body ground straps.

65. Install the power steering pump.

66. Connect the fuel supply line quick connect fitting.

67. Connect the vacuum lines at the throttle body and intake manifold.

68. Connect the engine harness at the following points:

- Intake Air Temperature (IAT) Sensor
- Idle Air Control (IAC) Motor
- Fuel Injectors
- Throttle Position (TPS) Switch
- Engine Oil Pressure Switch
- Engine Coolant Temperature (ECT) Sensor
- Manifold Absolute Pressure (MAP) Sensor
- Camshaft Position (CMP) Sensor
- Coil Over Plugs

69. Position and install the heater hoses and tubes onto intake manifold.

70. Install the heater hoses onto the heater core and the engine front cover.

71. Install the generator.

72. Install the A/C condenser, radiator and transmission oil cooler.

73. Connect the radiator upper and lower hoses.

74. Connect the transmission oil cooler lines to the radiator.

75. Install the accessory drive belt, fan assembly and shroud.

76. Install A/C compressor.

77. Install both breathers. Connect the tube to both crankcase breathers.

78. Connect the throttle and speed control cables.

79. Install the throttle body resonator assembly and air inlet hose. Tighten the clamps to 4 Nm (35 inch lbs.).

80. Raise the vehicle.
81. Install the transmission to engine mounting bolts. Tighten the bolts to 41 Nm (30 ft. lbs.).
82. Install the torque converter bolts (automatic transmission only).
83. Connect the crankshaft position sensor.
84. On 4wd vehicles, position and install the axle isolator bracket onto the axle, transmission and engine block.
85. Install the starter.

⁂ WARNING

The structural cover requires a specific torque sequence. Failure to follow this sequence may cause severe damage to the cover.

86. Install the structural cover.
87. Install the exhaust crossover pipe.
88. Install the engine block heater power cable, if equipped.
89. On 4wd vehicles, connect the axle vent tube to the left side engine mount.
90. Lower the vehicle.
91. Check and fill engine oil.
92. Recharge the A/C system.
93. Refill the engine cooling system.
94. Install the battery tray and battery.
95. Connect the battery positive and negative cables.
96. Start the engine and check for leaks.

4.7L Engine

1. Disconnect the battery negative and positive cables.
2. Remove the battery and the battery tray.
3. Raise the vehicle on hoist.
4. Remove exhaust crossover pipe from the exhaust manifolds.
5. On 4wd vehicles, disconnect the axle vent tube from left side engine mount.
6. Remove the through bolt retaining nut and bolt from both the left and right side engine mounts.
7. On 4wd vehicles, remove the locknut from left and right side engine mount brackets.
8. Disconnect two ground straps from the lower left hand side and one ground strap from the lower right hand side of the engine.
9. Disconnect the crankshaft position sensor.

➡The following step applies to 4X4 vehicles equipped with automatic transmission only.

10. On 4wd vehicles, remove the axle isolator bracket from the engine, transmission and the axle.
11. Remove the structural cover.
12. Remove the starter.
13. Drain the cooling system.
14. Remove the torque converter bolts (automatic transmission only).
15. Remove the transmission to engine mounting bolts.
16. Disconnect the engine block heater power cable from the block heater, if equipped.
17. Lower the vehicle.
18. Remove the throttle body resonator assembly and air inlet hose.
19. Disconnect the throttle and speed control cables.
20. Disconnect the tube from both the left and right side crankcase breathers. Remove breathers.
21. Discharge the A/C system.
22. Remove the A/C compressor.
23. Remove the shroud, fan assembly and accessory drive belt.
24. Disconnect the transmission oil cooler lines at the radiator.
25. Disconnect the radiator upper and lower hoses.
26. Remove the radiator, A/C condenser and transmission oil cooler.
27. Remove the generator.
28. Disconnect the two heater hoses from the timing chain cover and heater core.
29. Unclip and remove the heater hoses and tubes from the intake manifold.
30. Disconnect the engine harness at the following points :
    - Intake air temperature (IAT) sensor
    - Fuel Injectors
    - Throttle Position (TPS) Switch
    - Idle Air Control (IAC) Motor
    - Engine Oil Pressure Switch
    - Engine Coolant Temperature (ECT) Sensor
    - Manifold absolute pressure (MAP) Sensor
    - Camshaft Position (CMP) Sensor
    - Coil Over Plugs
31. Disconnect the vacuum lines at the throttle body and intake manifold.
32. Release the fuel rail pressure then disconnect the fuel supply quick connect fitting at the fuel rail.
33. Remove the power steering pump and position out of the way.
34. Install Special Tools 8400 Lifting Studs, into the cylinder heads.
35. Install Engine Lifting Fixture Special Tool 8347 following these steps:
    - Holding the lifting fixture at a slight angle, slide the large bore in the front plate over the hex portion of the lifting stud.
    - Position the two remaining fixture arms onto the two Special Tools 8400 Lifting Studs, in the cylinder heads.
    - Pull forward and upward on the lifting fixture so that the lifting stud rest in the slotted area below the large bore.
    - Secure the lifting fixture to the three studs using three 7⁄16—14 N/C locknuts.
    - Make sure the lifting loop in the lifting fixture is in the last hole (closest to the throttle body) to minimize the angle of engine during removal.
36. Disconnect the body ground strap at the right side cowl.
37. Disconnect the body ground strap at the left side cowl.

➡It will be necessary to support the transmission in order to remove the engine.

38. Position a suitable jack under the transmission.
39. Remove the engine from the vehicle.

To install:

40. Position the engine in the vehicle. Position both the left and right side engine mount brackets and install the through bolts and nuts. On 2wd vehicles, tighten nuts to 95 Nm (70 ft. lbs.); On 4wd vehicles, tighten to 102 Nm (75 ft. lbs.).
41. On 4wd vehicles, install locknuts onto the engine mount brackets. Tighten the locknuts to 41 Nm (30 ft. lbs.).
42. Remove the jack from under the transmission.
43. Remove Engine Lifting Fixture Special Tool 8347.
44. Remove Special Tools 8400 Lifting Studs.
45. Position the generator wiring behind the oil dipstick tube, then install the oil dipstick tube upper mounting bolt.
46. Connect both left and right side body ground straps.
47. Install the power steering pump.
48. Connect the fuel supply line quick connect fitting.
49. Connect the vacuum lines at the throttle body and intake manifold.
50. Connect the engine harness at the following points:
    - Intake Air Temperature (IAT) Sensor
    - Idle Air Control (IAC) Motor
    - Fuel Injectors
    - Throttle Position (TPS) Switch

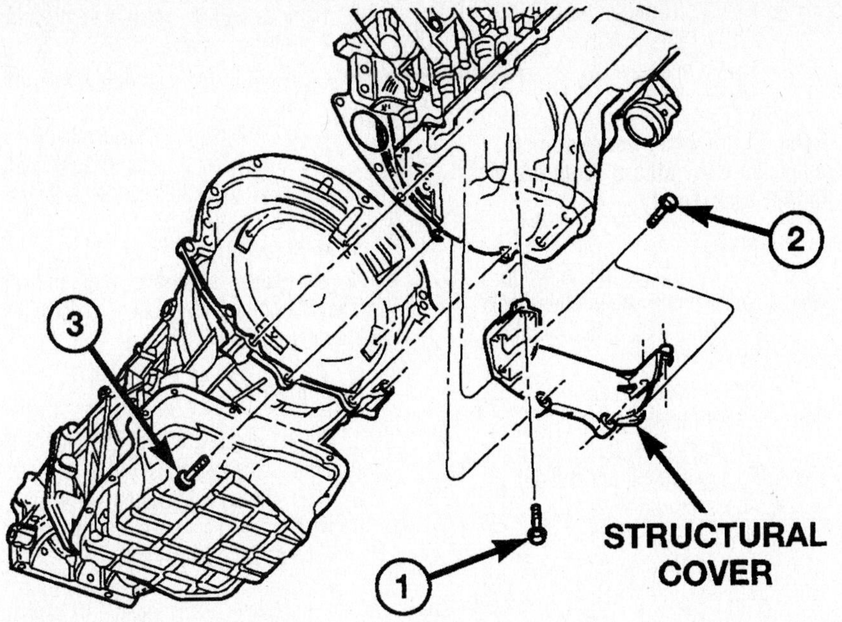

Structural cover torque sequence—3.7L and 4.7L engines

- Engine Oil Pressure Switch
- Engine Coolant Temperature (ECT) Sensor
- Manifold Absolute Pressure (MAP) Sensor
- Camshaft Position (CMP) Sensor
- Coil Over Plugs

51. Position and install the heater hoses and tubes onto intake manifold.
52. Install the heater hoses onto the heater core and the engine front cover.
53. Install the generator.
54. Install the A/C condenser, radiator and transmission oil cooler.
55. Connect the radiator upper and lower hoses.
56. Connect the transmission oil cooler lines to the radiator.
57. Install the accessory drive belt, fan assembly and shroud.
58. Install the A/C compressor.
59. Install both breathers. Connect the tube to both crankcase breathers.
60. Connect the throttle and speed control cables.
61. Install the throttle body resonator assembly and air inlet hose. Tighten clamps 4 Nm (35 inch lbs.).
62. Raise the vehicle.
63. Install transmission to engine mounting bolts. Tighten the bolts to 41 Nm (30 ft. lbs.).

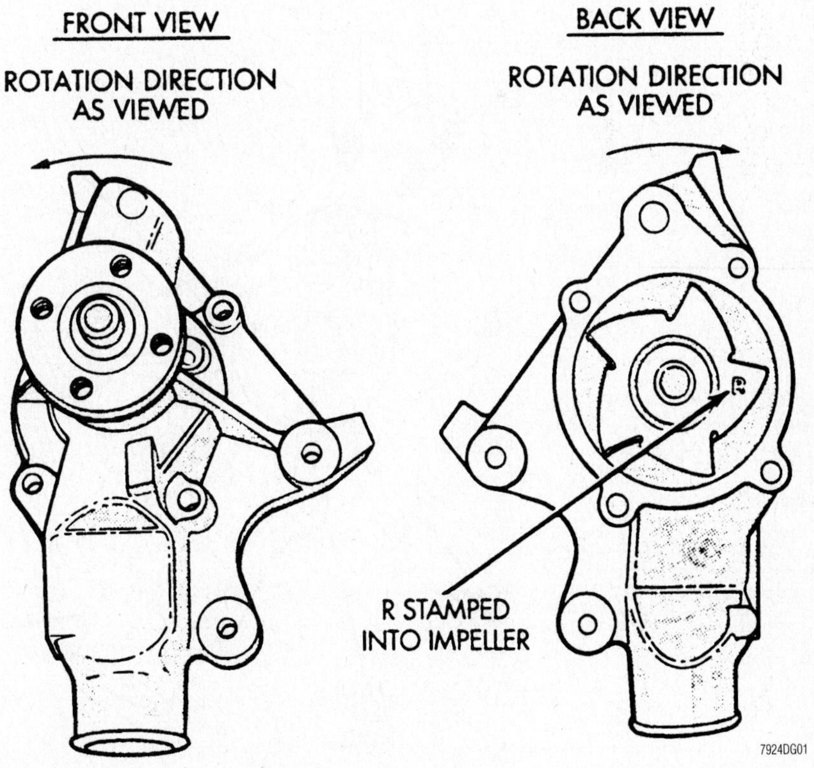

Reverse rotation water pump—2.5L engine

64. Install the torque converter bolts (automatic transmission only).
65. Connect crankshaft position sensor.
66. On 4wd vehicles, position and install the axle isolator bracket onto the axle, transmission and engine block. Tighten the bolts to specification.
67. Install the starter.

⁂ WARNING

The structural cover requires a specific torque sequence. Failure to follow this sequence may cause severe damage to the cover.

68. Install the structural cover.
69. Install the exhaust crossover pipe.
70. Install the engine block heater power cable, if equipped.
71. On 4wd vehicles, connect the axle vent tube to left side engine mount.
72. Lower the vehicle.
73. Check and fill the engine oil.
74. Recharge the A/C system.
75. Refill the engine cooling system.
76. Install the battery tray and battery.
77. Connect the battery positive and negative cables.
78. Start the engine and check for leaks.

Water Pump

REMOVAL & INSTALLATION

2.5L Engine

➡The 2.5L engine uses a reverse rotation water pump. The letter R is stamped on the impeller to identify. Engines from previous years may be equipped with forward rotation water pumps. Installation of the wrong water pump will cause engine over heating.

1. Before servicing the vehicle, refer to the precautions in the beginning of this section.
2. Drain the cooling system.
3. Remove or disconnect the following:
   - Negative battery cable
   - Accessory drive belt
   - Engine cooling fan and pulley

➡Do not store the fan clutch assembly horizontally, silicone may leak into the bearing grease and cause contamination.

   - Power steering pump
   - Lower radiator hose
   - Heater hose
   - Water pump

➡One of the water pump bolts is longer than the others. Note the location for reassembly.

To install:

4. Install or connect the following:
   - Water pump using a new gasket. Tighten the bolts to 17 ft. lbs. (23 Nm).
   - Heater hose
   - Lower radiator hose
   - Power steering pump
   - Engine cooling fan and pulley
   - Accessory drive belt
   - Negative battery cable
5. Fill the cooling system.
6. Run the engine and check for leaks.

3.7L Engine

1. Before servicing the vehicle, refer to the precautions in the beginning of this section.
2. Drain the cooling system.
3. Remove or disconnect the following:
   - Negative battery cable
   - Fan and clutch assembly from the pump
   - Fan shroud and fan assembly. If you're reusing the fan clutch, keep it upright to avoid silicone fluid loss!
   - Lower hose
   - Water pump (8 bolts)
4. Installation is the reverse of removal. Tighten the bolts, in sequence, to 40 ft. lbs. (54 Nm).

4.7L Engine

1. Before servicing the vehicle, refer to the precautions in the beginning of this section.
2. Drain the cooling system.
3. Remove or disconnect the following:
   - Negative battery cable
   - Fan and fan drive assembly from the pump. Don't attempt to remove it from the vehicle, yet.

➡If a new pump is being installed, don't separate the fan from the drive.

   - Shroud and fan

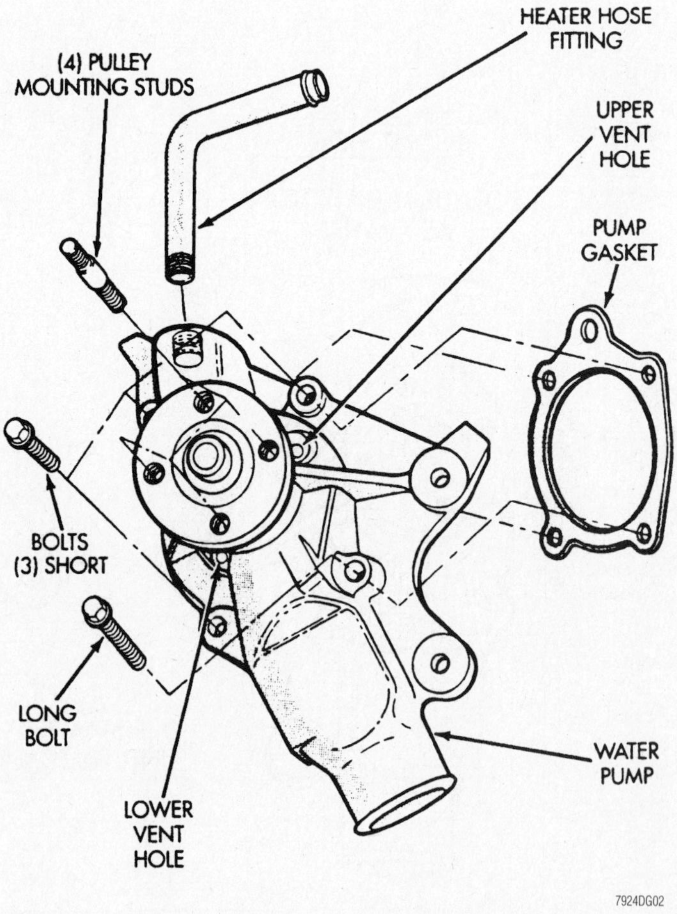

Water pump assembly—2.5L engine

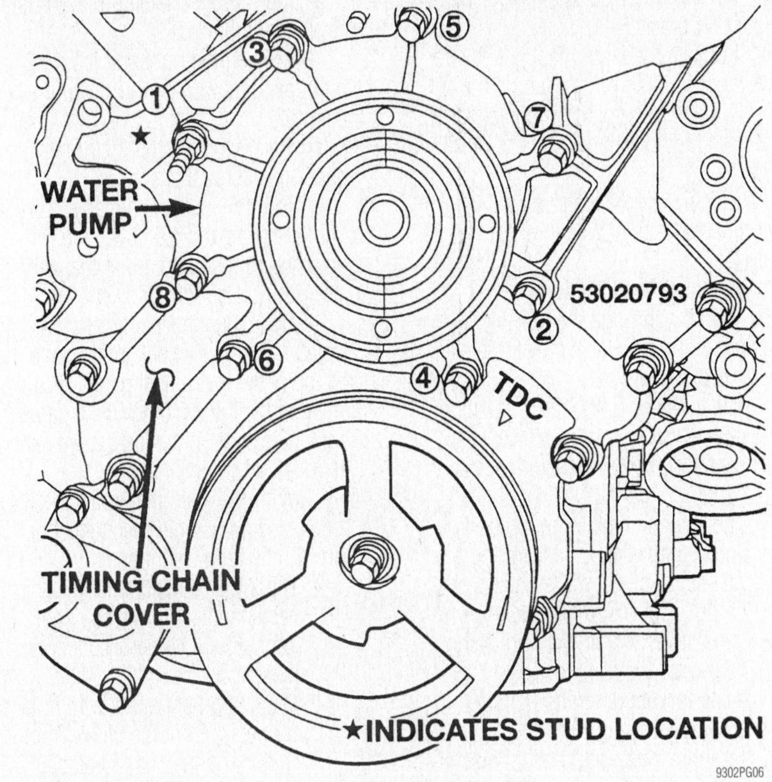

Water pump torque sequence—3.7L and 4.7L engines

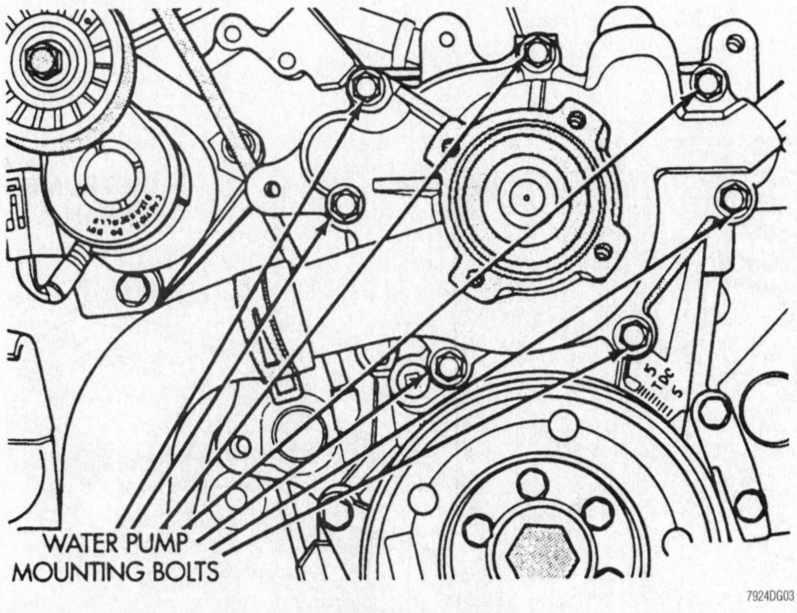

Water pump mounting bolt locations—3.9L and 5.9L engines

⁂ WARNING

Keep the fan upright to avoid fluid loss from the drive.

- Accessory drive belt
- Lower radiator hose
- Water pump

4. Installation is the reverse of removal. Tighten the bolts in sequence to 40 ft. lbs. (54 Nm).

3.9L and 5.9L Engines

1. Before servicing the vehicle, refer to the precautions in the beginning of this section.
2. Drain the cooling system.
3. Remove or disconnect the following:
   - Negative battery cable
   - Engine cooling fan and shroud

➡Do not store the fan clutch assembly horizontally, silicone may leak into the bearing grease and cause contamination.

   - Accessory drive belt
   - Water pump pulley
   - Lower radiator hose
   - Heater hose and tube
   - Bypass hose
   - Water pump

To install:

4. Install or connect the following:
   - Water pump, using a new gasket. Tighten the bolts to 30 ft. lbs. (40 Nm).
   - Bypass hose
   - Heater hose and tube. Use a new O-ring seal.
   - Lower radiator hose
   - Water pump pulley. Tighten the bolts to 20 ft. lbs. (27 Nm).
   - Accessory drive belt
   - Engine cooling fan and shroud
   - Negative battery cable
5. Fill the cooling system.
6. Start the engine and check for leaks.

Heater Core

REMOVAL & INSTALLATION

2002–03

1. Disconnect the negative battery cable.

⁂ CAUTION

After disconnecting the negative battery cable, wait 2 minutes for the driver's/passenger's air bag system capacitor to discharge before attempting to do any work around the steering column or instrument

2. Remove the instrument panel by performing the following procedure:
   a. Remove the trim from the right and left door sills.
   b. Remove the trim from the right and left cowl side inner panels.
   c. Remove the steering column opening cover from the instrument panel.
   d. Remove the 2 hood latch release handle-to-instrument panel lower reinforcement screws and lower the release handle to the floor.

e. Disconnect the driver's side air bag module wire harness connector.

f. If equipped, disconnect the overdrive lockout switch harness connector.

g. Do not disassemble the steering column but remove the assembly from the vehicle.

h. From under the driver's side of the instrument panel, disconnect or remove the following items:

- The screw from the center of the headlight/dash-to-instrument panel bulkhead wire harness connector and disconnect the connector.
- The 2 body wire harness connectors from the 2 instrument panel wire harness connector that secure the outboard side of the instrument panel bulkhead connector.
- The 3 wire harness connectors from the 3 junction block connector receptacles located closest to the dash panel; 1 from the body wire harness and 2 from the headlight/dash wire harness.
- The plastic park brake release linkage rod-to-rear parking brake release handle lever. Disengage the linkage rod from the handle lever.
- The stoplight switch connector.
- The vacuum harness connector from the left side of the heater/air conditioning housing assembly.

i. Remove the instrument panel center support bracket.

j. Remove the instrument panel wire harness ground screw located on the left side of the air bag control module (ACM) mount on the floor panel transmission tunnel.

k. Disconnect the instrument panel wiring harness-to-ACM connector receptacle.

l. Remove the glove box.

m. Working through the glove box opening, disconnect or remove the following items:

- The 2 halves of the radio antenna coaxial cable connector near the center of the lower instrument panel.
- The antenna half of the radio antenna coaxial cable from the retainer clip near the outboard side of the lower instrument panel.
- The blower motor wire harness connector located near the heater/air conditioning housing assembly support brace.

n. From the passenger's side, disconnect or remove the following items:

- The 2 instrument panel wire harness connectors from the infinity speaker amplifier connector receptacles on the right cowl.
- The instrument panel wire harness radio ground eyelet-to-stud nut on the right cowl.

o. Loosen the right and left instrument panel cowl side roll-down bracket screws.

p. Remove the 5 top instrument panel-to-dash panel screws.

q. Pull the instrument panel rearward until the right and left cowl side roll-down bracket screws are in the roll-down slot position of both brackets.

r. Roll down the instrument panel and install a temporary hook in the center hole on top of the instrument panel; secure the other end to the top of the dash panel. The hook is to support the instrument panel in its rolled down position.

s. With the instrument panel in the rolled-down position, disconnect or remove the following items:

- The 2 instrument panel wire harness connector from the door jumper wire harness connectors located on the right bracket.
- The instrument panel wire harness connector from the blower motor resistor connector receptacle.
- The temperature control cable flag retainer from the top of the heater/air conditioning housing assembly. Pull the cable core adjuster clip off the blend-air door lever.

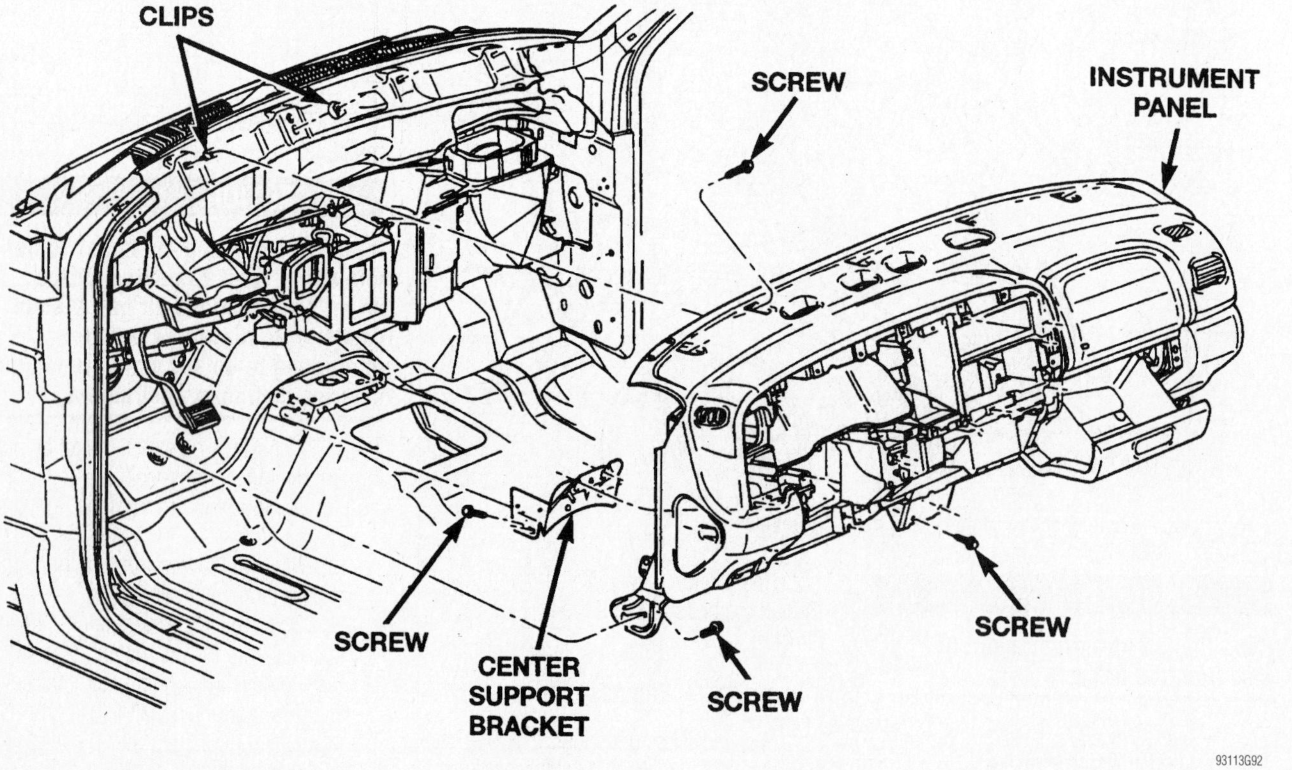

View of the instrument panel assembly

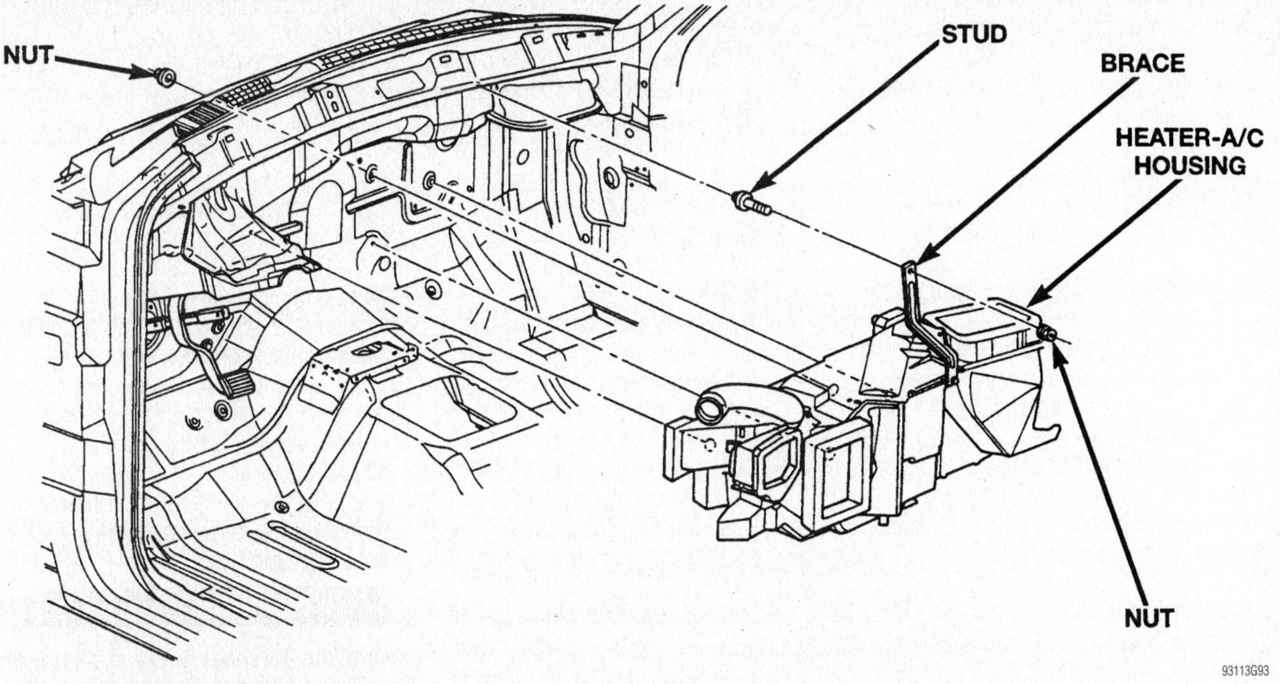

View of the heater/air conditioning assembly

- The demister duct flexible hose from the adapter on the top of the heater/air conditioning housing assembly.

t. With aid of an assistant, lift the instrument panel from the vehicle.

3. If equipped with air conditioning, perform the following procedure:

a. Discharge and recover the air conditioning system refrigerant.

b. Disconnect the refrigerant line from the evaporator inlet and outlet tubes. Plug the openings to prevent contamination.

4. Drain the cooling system into a clean container for reuse.

5. Disconnect the heater hoses from the heater core. Plug the openings.

6. Remove the 4 heater/air conditioning housing assembly-to-chassis nuts.

7. Remove the heater/air conditioning housing assembly-to-mounting brace nut; the nut is located on the passenger side of the vehicle.

8. Pull the heater/air conditioning housing assembly rearward for the studs and drain tube to clear the dash panel hole.

9. Remove the heater/air conditioning housing assembly from the vehicle.

10. Remove the heater housing cover screws and the cover.

11. Remove the heater core from the heater/air conditioning housing assembly.

To install:

12. Install the heater core to the heater/air conditioning housing assembly.

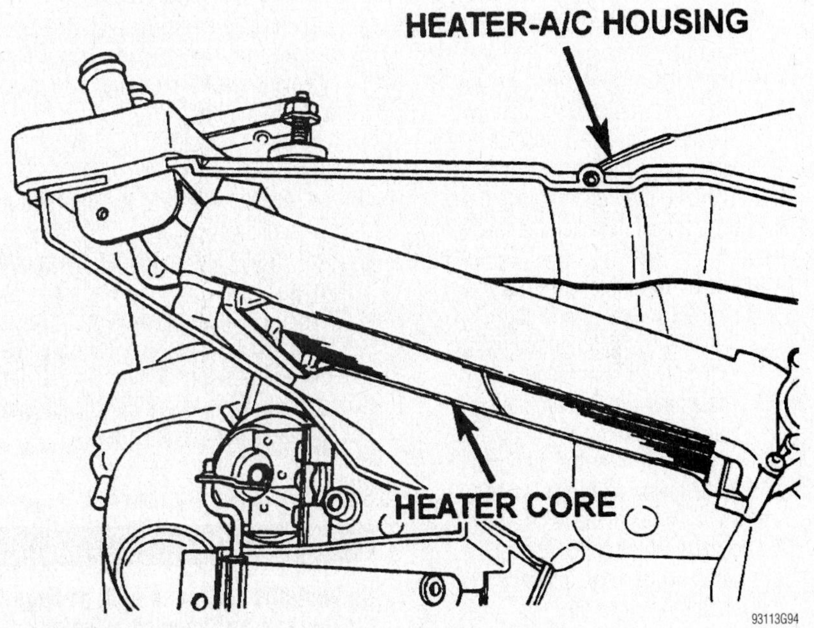

View of the heater core

13. Install the heater housing cover and the cover screws.

14. Install the heater/air conditioning housing assembly.

15. Install the heater/air conditioning housing assembly-to-mounting brace nut; the nut is located on the passenger side of the vehicle.

16. Install the 4 heater/air conditioning housing assembly-to-chassis nuts.

17. Connect the heater hoses to the heater core.

18. If equipped with air conditioning, perform the following procedure:

a. Using new gaskets, connect the refrigerant line to the evaporator inlet and outlet tubes.

b. Evacuate and charge the air conditioning system.

19. Install the instrument panel by performing the following procedures:

a. With aid of an assistant, install the instrument panel into the vehicle.

b. With the instrument panel in the rolled-down position, connect or install the following items:

- The demister duct flexible hose to the adapter on the top of the heater/air conditioning housing assembly.
- The temperature control cable flag retainer to the top of the heater/air conditioning housing assembly.
- The instrument panel wire harness connector to the blower motor resistor connector receptacle.
- The 2 instrument panel wire harness connector to the door jumper wire harness connectors located on the right bracket.

c. Move the instrument panel forward until the right and left cowl side roll-down bracket screws are in the roll-down slot position of both brackets.

d. Install the 5 top instrument panel-to-dash panel screws.

e. Install the right and left instrument panel cowl side roll-down bracket screws.

f. At the passenger's side, connect or install the following items:

- The instrument panel wire harness radio ground eyelet-to-stud nut on the right cowl.
- The 2 instrument panel wire harness connectors to the infinity speaker amplifier connector receptacles on the right cowl.

g. Working through the glove box opening, connect or install the following items:

- The blower motor wire harness connector located near the heater/air conditioning housing assembly support brace.
- The antenna half of the radio antenna coaxial cable to the retainer clip near the outboard side of the lower instrument panel.
- The 2 halves of the radio antenna coaxial cable connector near the center of the lower instrument panel.

h. Install the glove box.

i. Connect the instrument panel wiring harness-to-ACM connector receptacle.

j. Install the instrument panel wire harness ground screw located on the left side of the air bag control module (ACM) mount on the floor panel transmission tunnel.

k. Install the instrument panel center support bracket.

l. Under the driver's side of the instrument panel, connect or install the following items:

- The vacuum harness connector to the left side of the heater/air conditioning housing assembly.
- The stoplight switch connector.
- Engage the linkage rod to the handle lever. Install the plastic park brake release linkage rod-to-rear parking brake release handle lever.
- The 3 wire harness connectors to the 3 junction block connector receptacles located closest to the dash panel; 1 at the body wire harness and 2 at the headlight/dash wire harness.
- The 2 body wire harness connectors to the 2 instrument panel wire harness connector that secure the outboard side of the instrument panel bulkhead connector.
- Connect the connector. Install the screw at the center of the headlight/dash-to-instrument panel bulkhead wire harness connector.

m. Install the steering column assembly.

n. If equipped, connect the overdrive lockout switch harness connector.

o. Connect the driver's side air bag module wire harness connector.

p. Install the 2 hood latch release handle-to-instrument panel lower reinforcement screws and lower the release handle to the floor.

q. Install the steering column opening cover to the instrument panel.

r. Install the trim to the right and left cowl side inner panels.

s. Install the trim to the right and left door sills.

20. Refill the cooling system.
21. Connect the negative battery cable.
22. Run the engine to normal operating temperatures; then, check the climate control operation and check for leaks.

2004–05

⁂ CAUTION

On vehicles equipped with airbags, disable the airbag system before attempting any steering wheel, steering column, or instrument panel component diagnosis or service. Disconnect and isolate the battery negative (ground) cable, then wait two minutes for the airbag system capacitor to discharge before performing further diagnosis or service. This is the only sure way to disable the airbag system. Failure to take the proper precautions could result in an accidental airbag deployment and possible personal injury.

1. Disconnect and isolate the negative battery cable.
2. Recover refrigerant from the refrigerant system.
3. Drain the engine cooling system.
4. Disconnect the liquid line from the evaporator inlet tube. Install plugs in, or tape over the opened liquid line fitting and the evaporator inlet tube.
5. Disconnect the accumulator from the evaporator outlet tube. Install plugs in, or tape over the opened refrigerant line fitting and the evaporator outlet tube.
6. Disconnect the heater hoses from the heater core tubes. Install plugs in, or tape over the opened heater core tubes.
7. Remove the four nuts that secure the HVAC housing to the dash panel in the engine compartment.

⁂ WARNING

To avoid personal injury or death, on vehicles equipped with airbags, disable the supplemental restraint system before attempting any steering wheel, steering column, airbag, occupant classification system, seat belt tensioner, impact sensor, or instrument panel component diagnosis or service. Disconnect and isolate the battery negative (ground) cable, then wait two minutes for the system capacitor to discharge before performing further diagnosis or service. This is the only sure way to disable the supplemental restraint system. Failure to take the proper precautions could result in accidental airbag deployment.

⁂ WARNING

To avoid personal injury or death, never strike or drop the occupant restraint controller, as it can damage the impact sensor or affect its calibration. The occupant restraint controller contains the impact sensor, which enables the system to deploy the supplemental restraints. If an airbag control module is accidentally dropped during service, the module must be scrapped and replaced with a new unit. Failure to observe this warning could result in accidental, incomplete, or improper supplemental restraint deployment.

⁂ CAUTION

On vehicles equipped with the Occupant Classification System (OCS), never replace both the Occupant Restraint Controller (ORC) and the Occupant Classification Module (OCM) at the same time. If both require replacement, replace one. Then perform the supplemental restraint verification test before replacing the other. Both the ORC and the OCM store OCS calibration data, which they transfer to one another when one of them is replaced. If both are replaced at the same time, an irreversible fault will be set in both modules.

8. Disconnect and isolate battery negative cable.
9. Remove the floor console.
10. Remove the driver side door sill trim cover.
11. Remove the steering column opening cover.
12. Position the front wheels straight ahead.
13. Remove the steering column tilt lever knob.
14. Remove the column shrouds.
15. Disconnect the brake switch electrical connector.
16. Disconnect the wiring harness to the column.
17. Remove the shift cable from the column shift lever actuator.
18. Release the shift cable from the column bracket and remove it from the bracket.
19. Remove the SKIM module in order to disconnect the electrical connector.
20. Remove the upper steering shaft coupler bolt from the column.
21. Separate the shaft from the coupler.
22. Remove the four steering column mounting nuts.
23. Remove the steering column assembly from the vehicle.
24. Remove the bolts from the pedal support.
25. Remove the defroster grille.
26. Remove the five upper fence line bolts.
27. Disconnect the HVAC electrical connector.
28. Loosen the screws and disconnect the electrical connectors.
29. Remove the bolt and the ground from the left a-pillar.
30. Disconnect the ABS module electrical connectors.
31. Separate the center wire harness from the center support.
32. Remove the two center support bolts.
33. Remove the two left support bolts.
34. Remove the passenger door sill trim cover.
35. Remove the passenger side end cap.
36. Disconnect the antenna cable.
37. Remove the bolt and nut and disconnect the grounds.
38. Disconnect the antenna amplifier connector, if equipped.
39. Remove the right side support bolts and with the help of another person, remove the instrument panel assembly from the vehicle.
40. Remove the two bolts that secure the HVAC housing to the dash panel in the passenger compartment.
41. Remove the HVAC housing from the vehicle.
42. Remove the foam seal from the heater core tubes.
43. Disassemble the HVAC housing to access the heater core.
44. Lift the heater core out of the HVAC housing.

To install:

45. Install the heater core into the HVAC housing.
46. Install the foam seal onto the heater core tubes.
47. Position the HVAC housing to the dash panel. Be certain that the evaporator condensate drain tube and the housing mounting studs are inserted into their correct mounting holes.
48. Install the two bolts that secure the HVAC housing to the dash panel in the passenger compartment. Tighten the bolts to 7 Nm (60 inch lbs.).
49. Install the instrument into the vehicle and guide onto the a-pillar studs.
50. Install the right side bolts.
51. Install the fenceline bolts.
52. Install the bolts into the pedal support bracket.
53. Install the left side support bolts.
54. Install the center support bolts.
55. Tighten the left side support bolts to 12 Nm (105 inch lbs.).
56. Tighten the right side support bolts to 12 Nm (105 inch lbs.).
57. Tighten the center support bolts to 12 Nm (105 inch lbs.).
58. Tighten the pedal support bracket bolts to 14 Nm (10 ft. lbs.).
59. Tighten the fenceline bolts to 10 Nm (85 inch lbs.).
60. Connect the antenna amplifier connector, if equipped.
61. Install the grounds and install the bolt and nut.
62. Connect the antenna cable.
63. Install the passenger side end cap.
64. Install the passenger door sill trim cover.
65. Position the center wire harness back and connect the ABS module connectors and grounds.
66. Install the ground and install the bolts.

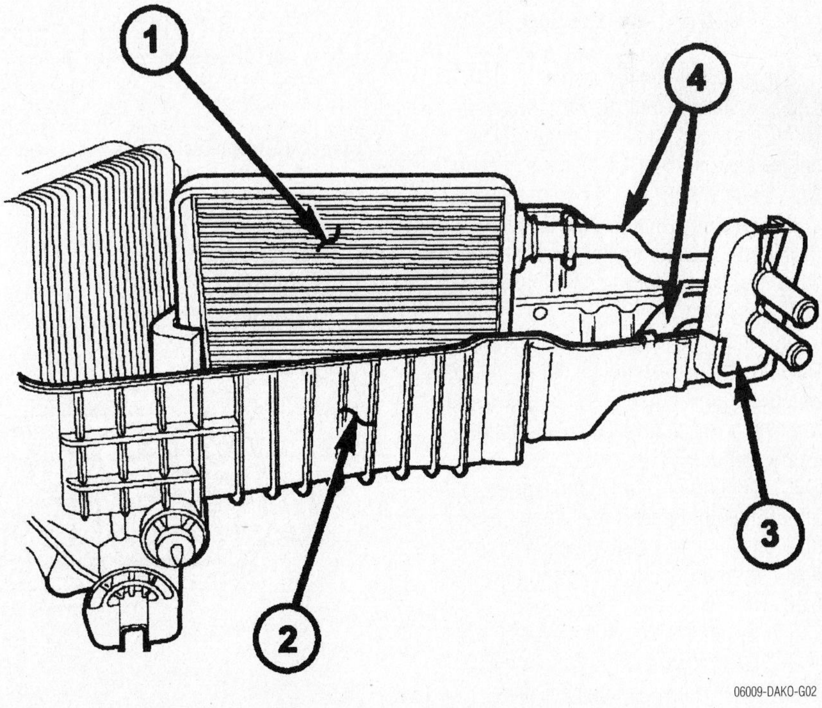

Heater core installation—2004–05 models

67. Connect the electrical connectors and tighten the screws to 5 Nm (40 inch lbs.).
68. Connect the HVAC electrical connector.
69. Install the defroster grille.

➡Torque the upper left nut first then the lower right nut. Then torque the lower left nut then the upper right nut.

70. Position the steering column on the dash panel support and loosely install the mounting nuts.
71. Firmly slide the steering column upward against the studs in dash panel and hand tighten the nuts.
72. Center steering column in dash opening and tighten mounting nuts to 28 Nm (21 ft. lbs.).
73. Install the steering shaft coupler on the steering shaft and loosely install a new bolt.

➡A new bolt must be used for reinstallation.

74. Tighten the coupler bolt to 57 Nm (42 ft. lbs.).
75. Install the shifter cable.
76. Connect the wiring harness to the column.
77. Install the SKIM module.
78. Install the shrouds.
79. Install the lilt lever knob.
80. Connect the brake switch electrical connector.
81. Install the steering column opening cover.
82. Install the driver side door sill trim cover.
83. Install the floor console.
84. Install the four nuts that secure the HVAC housing to the dash panel in the engine compartment. Tighten the nuts to 7 Nm (60 inch lbs.).
85. Unplug or remove the tape from the heater core tubes. Connect the heater hoses to the heater core tubes and fill the engine cooling system.
86. Unplug or remove the tape from the accumulator refrigerant line fitting and the evaporator outlet tube. Install a new O-ring seal and connect the accumulator to the evaporator outlet tube.
87. Unplug or remove the tape from the liquid line fitting and the evaporator inlet tube. Install a new O-ring seal and connect the liquid line to the evaporator inlet tube.
88. Reconnect the negative battery cable.
89. Evacuate the refrigerant system.
90. Charge the refrigerant system.
91. Refill the cooling system.

Cylinder Head

REMOVAL & INSTALLATION

2.5L Engine

1. Before servicing the vehicle, refer to the precautions in the beginning of this section.
2. Drain the cooling system.
3. Remove or disconnect the following:
   - Negative battery cable
   - Crankcase Ventilation (CCV) hoses
   - Air cleaner assembly
   - Valve cover

➡Keep valve train components in order for reassembly.

   - Rocker arms
   - Pushrods
   - Accessory drive belt
   - A/C compressor and bracket, if equipped
   - Power steering pump and bracket, if equipped
   - Fuel line
   - Combination manifold
   - Thermostat housing coolant hoses
   - Spark plugs
   - Engine Coolant Temperature (ECT) sensor connector
   - Cylinder head

To install:

⁂ WARNING

Cylinder head bolts may only be reused one time. If reusing a cylinder head bolt, place a paint mark on the bolt after installation. If a cylinder head bolt has a paint mark, discard it and use a new bolt.

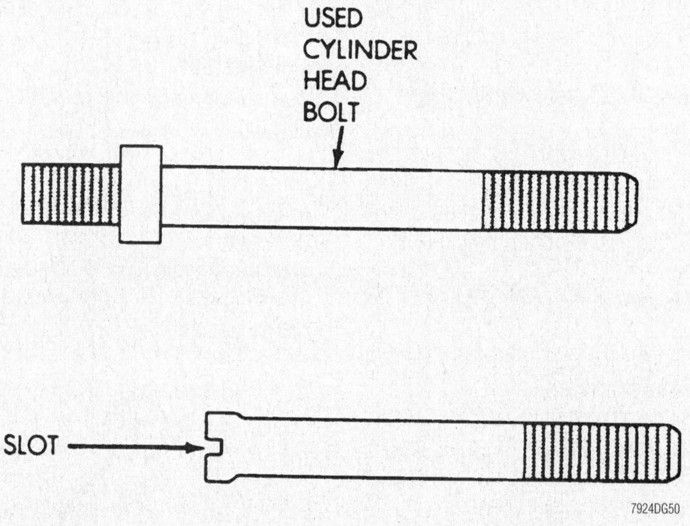

Fabricate 2 alignment dowels out of used cylinder head bolts—2.5L engine

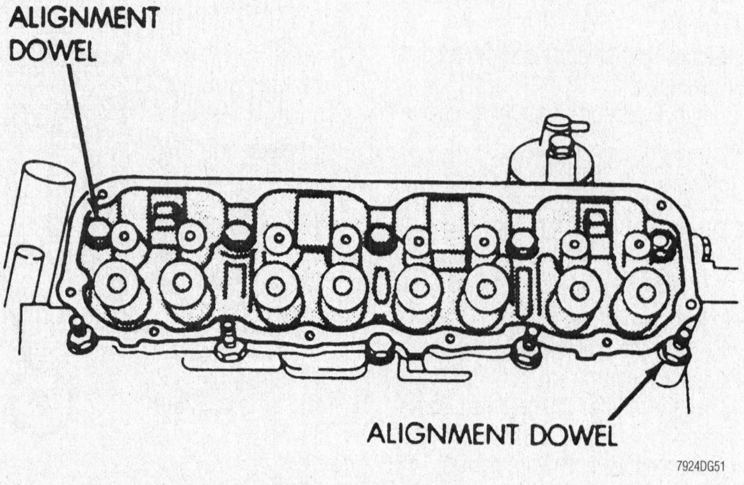

Alignment dowel locations—2.5L engine

4. Fabricate two alignment dowels from old cylinder head bolts. Cut the hex head off of the bolts, and cut a slot in each dowel to ease removal.
5. Install or connect the following:
   - One dowel in bolt hole No. 8, and one dowel in bolt hole No. 10.
   - Cylinder head and gasket.
   - Cylinder head bolts except for No 8 and No 10. Coat the threads of bolt No. 7 with Loctite® 592 sealant.
6. Remove the alignment dowels and install the No. 8 and No. 10 head bolts.

WARNING

During the final tightening sequence, bolt No. 7 will be tightened to a lower torque value than the rest of the bolts. Do not overtighten bolt No. 7.

7. Tighten the cylinder head bolts, in sequence, as follows:
   a. Step 1: 22 ft. lbs. (30 Nm).
   b. Step 2: 45 ft. lbs. (61 Nm).
   c. Step 3: 45 ft. lbs. (61 Nm).
   d. Step 4: Bolts 1–6 to 110 ft. lbs. (149 Nm).
   e. Step 5: Bolt 7 to 100 ft. lbs. (136 Nm).
   f. Step 6: Bolts 8–10 to 110 ft. lbs. (149 Nm).
   g. Step 7: Repeat steps 4, 5 and 6.
8. Install or connect the following:
   - ECT sensor connector
   - Spark plugs
   - Thermostat housing coolant hoses
   - Combination manifold
   - Fuel line
   - Power steering pump and bracket, if equipped
   - A/C compressor and bracket, if equipped
   - Accessory drive belt
   - Pushrods and rocker arms in their original positions
   - Valve cover
   - Air cleaner assembly
   - CCV hoses
   - Negative battery cable
9. Fill the cooling system.
10. Start the engine and check for leaks.

3.7L Engine

LEFT SIDE

1. Before servicing the vehicle, refer to the precautions in the beginning of this section.
2. Drain the cooling system.
3. Properly relieve the fuel system pressure.

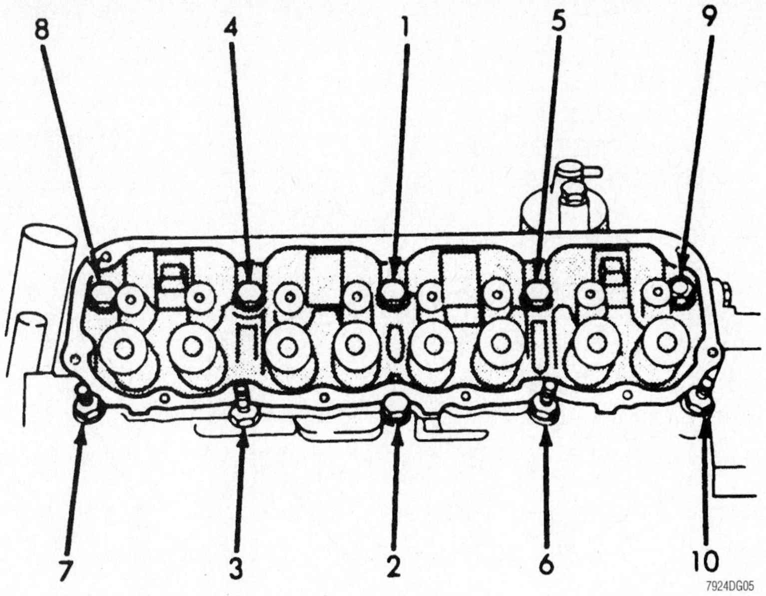

Cylinder head torque sequence—2.5L engine

4. Remove or disconnect the following:
   - Negative battery cable
   - Exhaust Y-pipe
   - Intake manifold
   - Cylinder head cover
   - Engine cooling fan and shroud
   - Accessory drive belt
   - Power steering pump
5. Rotate the crankshaft so that the crankshaft timing mark aligns with the Top Dead Center (TDC) mark on the front cover, and the **V6** marks on the camshaft sprockets are at 12 o'clock as shown.
   - Crankshaft damper
   - Front cover
6. Lock the secondary timing chain to the idler sprocket with Timing Chain Locking tool 8429.
7. Matchmark the secondary timing chain one link on each side of the V6 mark to the camshaft sprocket.
   - Left secondary timing chain tensioner
   - Cylinder head access plug
   - Secondary timing chain guide
   - Camshaft sprocket
   - Cylinder head

➡The cylinder head is retained by twelve bolts. Four of the bolts are smaller and are at the front of the head.

To install:

8. Check the cylinder head bolts for signs of stretching and replace as necessary.
9. Lubricate the threads of the 11mm bolts with clean engine oil.
10. Coat the threads of the 8mm bolts with Mopar® Lock and Seal Adhesive.
11. Install the cylinder heads. Use new gaskets and tighten the bolts, in sequence, as follows:
   a. Step 1: Bolts 1–8 to 20 ft. lbs. (27 Nm)
   b. Step 2: Bolts 1–10 verify torque without loosening
   c. Step 3: Bolts 9–12 to 10 ft. lbs. (14 Nm)
   d. Step 4: Bolts 1–8 plus ¼ (90 degree) turn
   e. Step 5: Bolts 9–12 to 19 ft. lbs. (26 Nm)
12. Install or connect the following:
   - Camshaft sprocket. Align the secondary chain matchmarks and tighten the bolt to 90 ft. lbs. (122 Nm).
   - Secondary timing chain guide
   - Cylinder head access plug
   - Secondary timing chain tensioner. Refer to the timing chain procedure in this section.
13. Remove the Timing Chain Locking tool.
14. Install or connect the following:
   - Front cover
   - Crankshaft damper. Torque the bolt to 130 ft. lbs. (175 Nm).
   - Power steering pump
   - Accessory drive belt
   - Engine cooling fan and shroud
   - Cover
   - Intake manifold
   - Exhaust Y-pipe
   - Negative battery cable

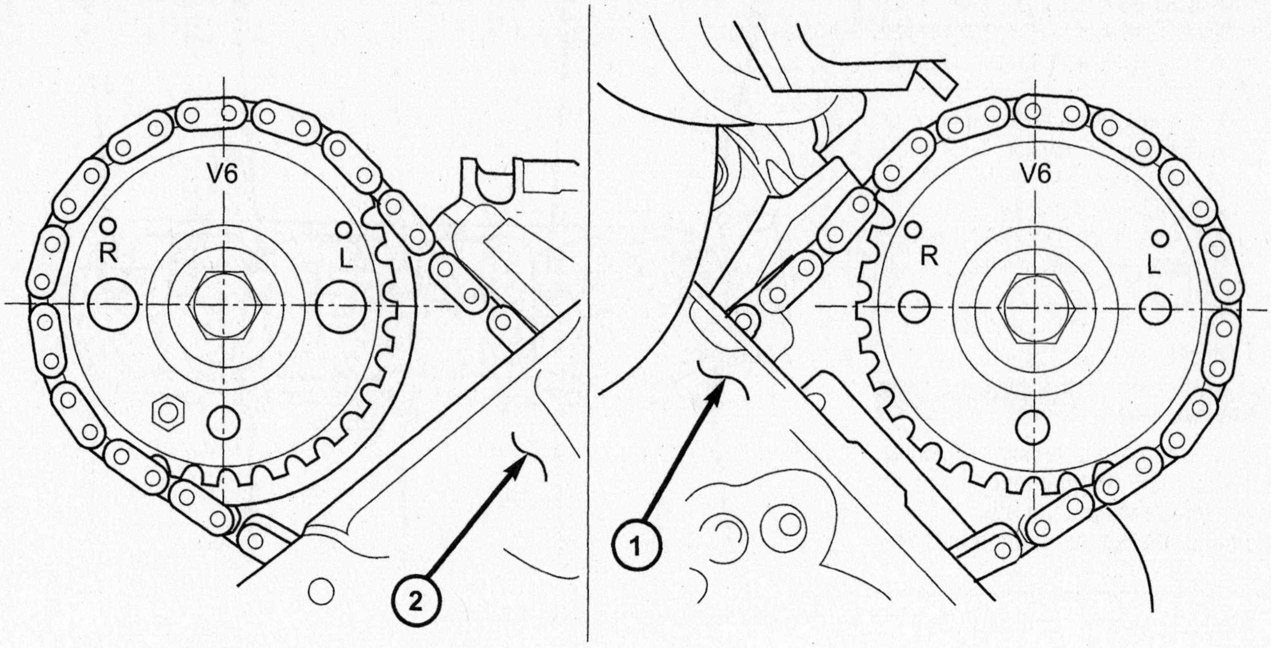

1 - LEFT CYLINDER HEAD
2 - RIGHT CYLINDER HEAD

Camshaft sprocket timing marks—3.7L

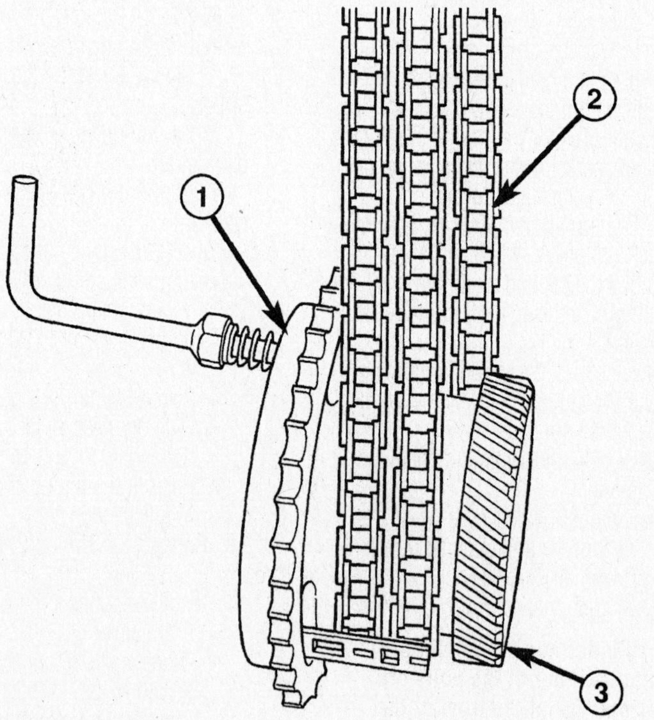

1 - SPECIAL TOOL 8429
2 - CAMSHAFT CHAIN
3 - CRANKSHAFT TIMING GEAR

Camshaft locking tool—3.7L

15. Fill and bleed the cooling system.
16. Start the engine, check for leaks and repair if necessary.

RIGHT SIDE

1. Before servicing the vehicle, refer to the precautions in the beginning of this section.
2. Drain the cooling system.
3. Properly relieve the fuel system pressure.
4. Remove or disconnect the following:
   - Negative battery cable
   - Exhaust Y-pipe
   - Intake manifold
   - Valve cover
   - Engine cooling fan and shroud
   - Accessory drive belt
   - Oil fill housing
   - Power steering pump
5. Rotate the crankshaft so that the crankshaft timing mark aligns with the Top Dead Center (TDC) mark on the front cover, and the **V6** marks on the camshaft sprockets are at 12 o'clock as shown.
6. Remove or disconnect the following:
   - Crankshaft damper
   - Front cover
7. Lock the secondary timing chains to the idler sprocket with Timing Chain Locking tool 8429.

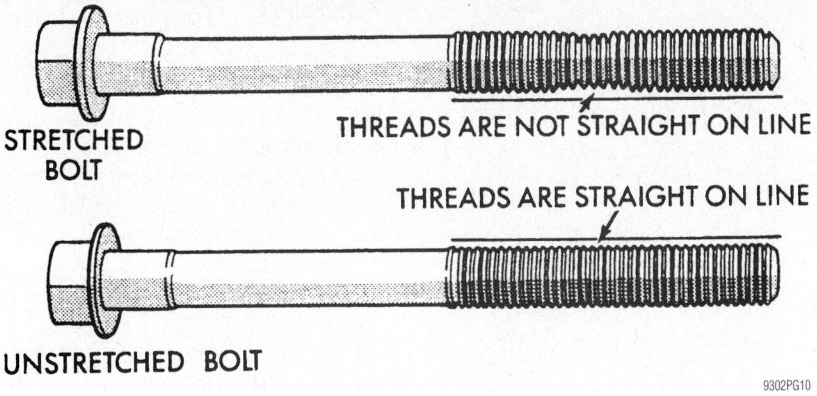

Examine the head bolts for signs of stretching—3.7L engine

8. Matchmark the secondary timing chains to the camshaft sprockets.

9. Remove or disconnect the following:
   - Secondary timing chain tensioners
   - Cylinder head access plugs
   - Secondary timing chain guides
   - Camshaft sprockets
   - Cylinder heads

➡**Each cylinder head is retained by eight 11mm bolts and four 8mm bolts.**

To install:

10. Check the cylinder head bolts for signs of stretching and replace as necessary.

11. Lubricate the threads of the 11mm bolts with clean engine oil.

12. Coat the threads of the 8mm bolts with Mopar® Lock and Seal Adhesive.

13. Install the cylinder heads. Use new gaskets and tighten the bolts, in sequence, as follows:
   a. Step 1: Bolts 1–8 to 20 ft. lbs. (27 Nm)
   b. Step 2: Bolts 1–10 verify torque without loosening
   c. Step 3: Bolts 9–12 to 10 ft. lbs. (14 Nm)
   d. Step 4: Bolts 1–8 plus ¼ (90 degree) turn
   e. Step 5: Bolts 9–12 to 19 ft. lbs. (26 Nm)

14. Install or connect the following:
   - Camshaft sprockets. Align the secondary chain matchmarks and tighten the bolts to 90 ft. lbs. (122 Nm).
   - Secondary timing chain guides
   - Cylinder head access plugs
   - Secondary timing chain tensioners. Refer to the timing chain procedure in this section.

15. Remove the Timing Chain Locking tool.

16. Install or connect the following:
   - Front cover
   - Crankshaft damper. Torque the bolt to 130 ft. lbs. (175 Nm).
   - Rocker arms
   - Power steering pump
   - Oil fill housing
   - Accessory drive belt
   - Engine cooling fan and shroud
   - Valve covers
   - Intake manifold
   - Exhaust Y-pipe
   - Negative battery cable

17. Fill and bleed the cooling system.

18. Start the engine, check for leaks and repair if necessary.

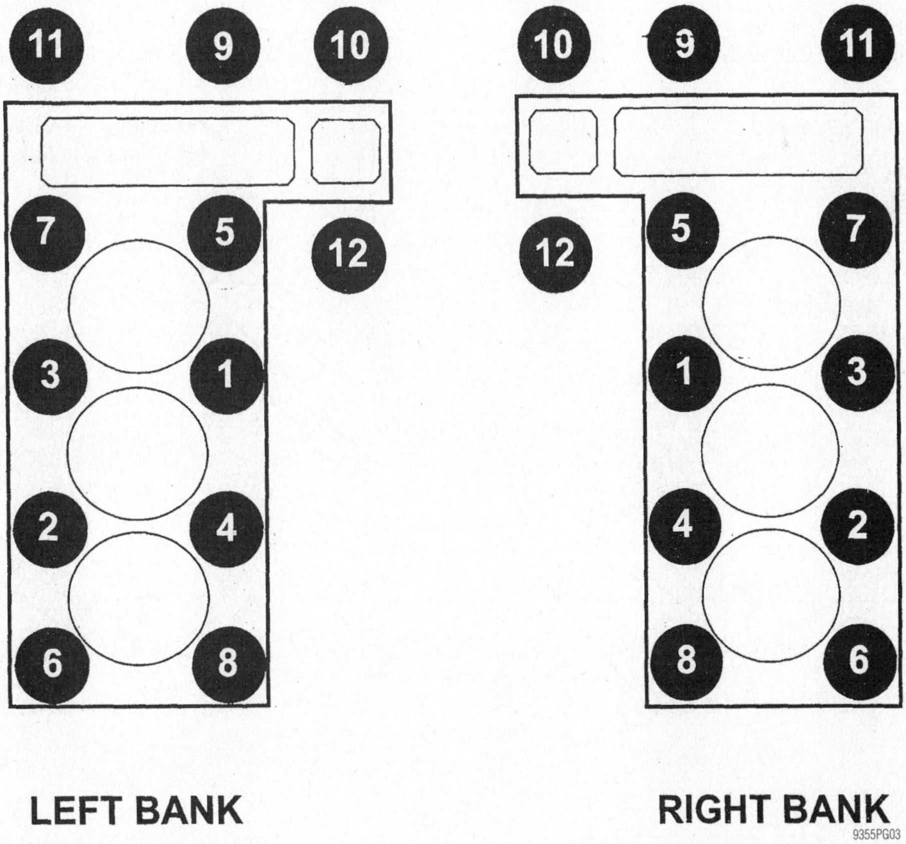

Cylinder head bolt torque sequence—3.7L

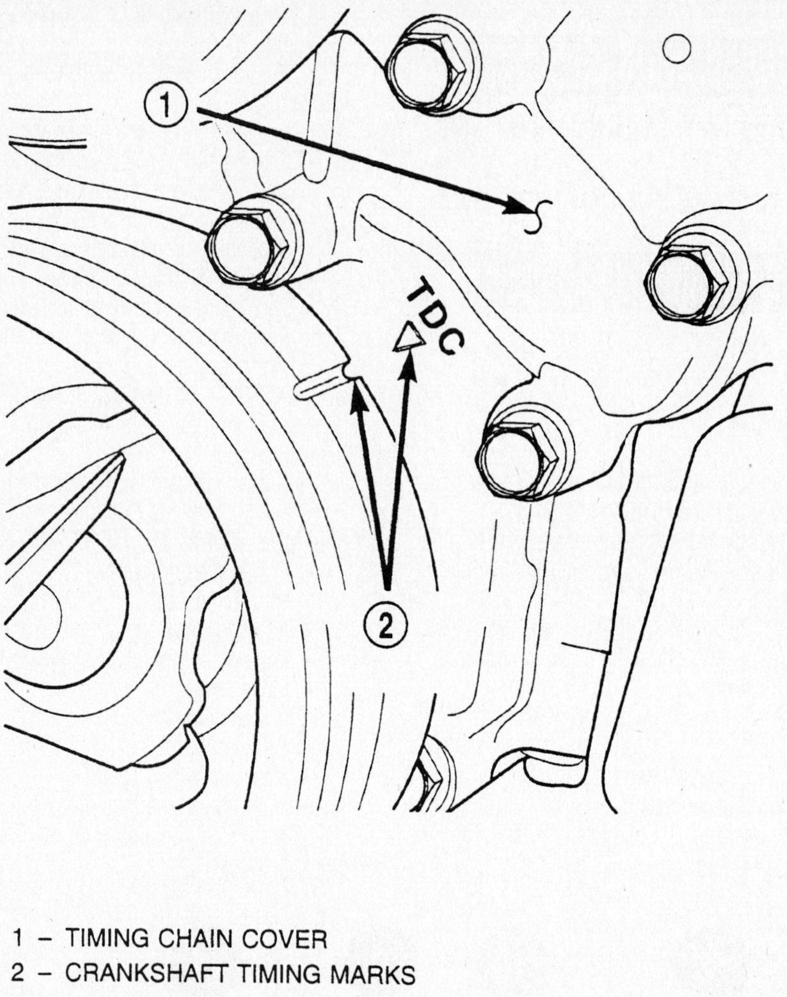

Crankshaft timing marks—3.7L engine

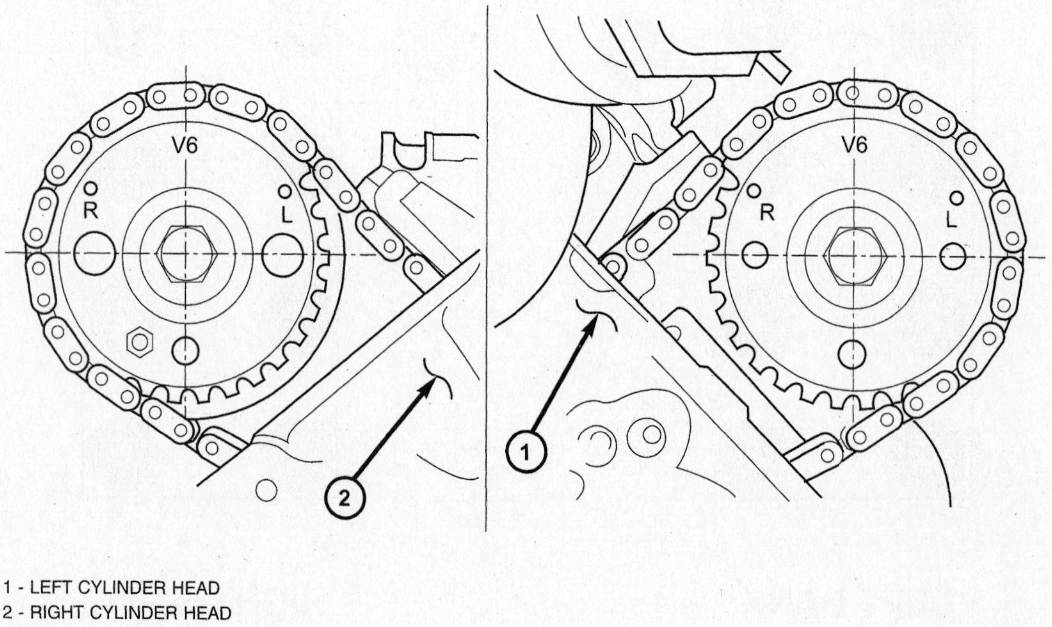

Camshaft sprocket timing marks—3.7L

3.9L Engine

1. Before servicing the vehicle, refer to the precautions in the beginning of this section.
2. Relieve the fuel pressure.
3. Drain the cooling system.
4. Remove or disconnect the following:
   - Negative battery cable
   - Accessory drive belt
   - Alternator
   - A/C compressor, if equipped
   - Alternator and A/C compressor bracket
   - Air injection pump, if equipped
   - Closed Crankcase Ventilation (CCV) system
   - Air cleaner and hose
   - Fuel line
   - Accelerator linkage
   - Cruise control cable, if equipped
   - Transmission cable, if equipped
   - Spark plug wires
   - Distributor
   - Ignition coil harness connectors
   - Engine Coolant Temperature (ECT) sensor connector
   - Heater hoses
   - Bypass hose
   - Intake manifold vacuum lines
   - Fuel injector harness connectors
   - Valve covers
   - Intake manifold
   - Exhaust front pipe
   - Exhaust manifolds

➡Keep all valvetrain components in order for assembly.

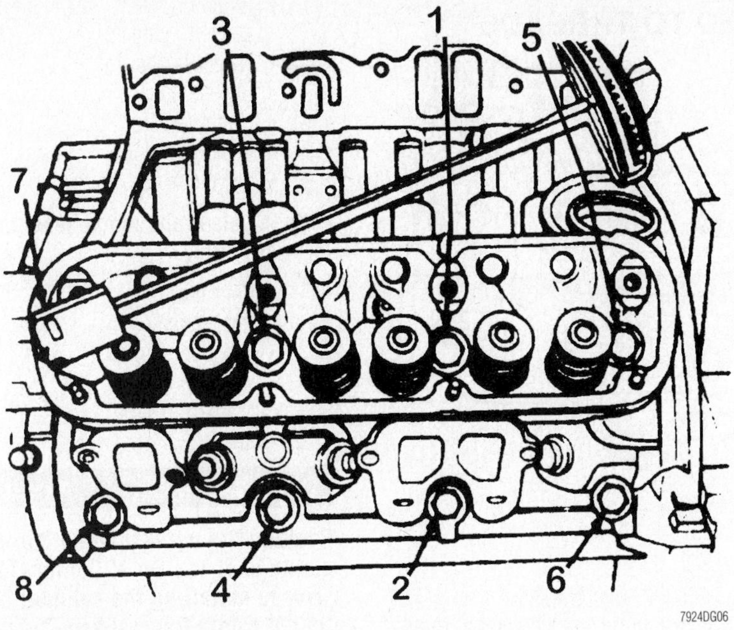

Cylinder head torque sequence—3.9L engine

- Rocker arms
- Pushrods
- Cylinder heads

To install:

WARNING

Position the crankshaft so that no piston is at Top Dead Center (TDC) prior to installing the cylinder heads. Do not rotate the crankshaft during or immediately after rocker arm installation. Wait 5 minutes for the hydraulic lash adjusters to bleed down.

5. Install the cylinder heads with new gaskets. Tighten the bolts in sequence as follows:
   a. Step 1: 50 ft. lbs. (68 Nm).
   b. Step 2: 105 ft. lbs. (143 Nm).
   c. Step 3: 105 ft. lbs. (143 Nm).
6. Install or connect the following:
   - Pushrods in their original locations
   - Rocker arms in their original locations. Tighten the bolts to 21 ft. lbs. (28 Nm).
   - Exhaust manifolds
   - Exhaust front pipe
   - Intake manifold
   - Valve covers
   - Fuel injector harness connectors
   - Intake manifold vacuum lines
   - Bypass hose
   - Heater hoses
   - ECT sensor connector
   - Ignition coil harness connectors
   - Distributor
   - Spark plug wires
   - Transmission cable, if equipped
   - Cruise control cable, if equipped
   - Accelerator linkage
   - Fuel line
   - Air cleaner and hose
   - CCV system
   - Air injection pump, if equipped
   - Alternator and A/C compressor bracket
   - Alternator
   - A/C compressor
   - Accessory drive belt
   - Negative battery cable
7. Fill the cooling system.
8. Start the engine and check for leaks.

4.7L Engine

LEFT SIDE

1. Before servicing the vehicle, refer to the precautions in the beginning of this section.
2. Drain the cooling system.
3. Remove or disconnect the following:
   - Negative battery cable
   - Exhaust pipe
   - Intake manifold
   - Cylinder head cover
   - Fan shroud and fan
   - Accessory drive belt
   - Power steering pump
4. Rotate the crankshaft until the damper mark is aligned with the TDC mark. Verify that the V8 mark on the camshaft sprocket is at the 12 o'clock position.
5. Remove or disconnect the following:
   - Vibration damper
   - Timing chain cover
6. Lock the secondary timing chains to the idler sprocket with tool 8515, or equivalent.
7. Mark the secondary timing chain, on link on either side of the V8 mark on the cam sprocket.
8. Remove the left side secondary chain tensioner.
9. Remove the cylinder head access plug.
10. Remove the chain guide.
11. Remove the camshaft sprocket.

➡There are 4 smaller bolts at the front of the head. Don't overlook these.

12. Remove the head bolts and head.

WARNING

Don't lay the head on its sealing surface. Due to the design of the head gasket, any distortion to the head sealing surface will result in leaks.

13. Installation is the reverse of removal. Observe the following:
   - Check the head bolts. If any necking is observed, replace the bolt.
   - The 4 small bolts must be coated with sealer.
   - The head bolts are tightened in the following sequence:

   Step 1: Bolts 1-10 to 15 ft. lbs. (20 Nm)
   Step 2: Bolts 1-10 to 35 ft. lbs. (47 Nm)
   Step 3: Bolts 11-14 to 18 ft. lbs. (25 Nm)
   Step 4: Bolts 1-10 90 degrees
   Step 5: Bolts 11-14 to 22 ft. lbs.

RIGHT SIDE

1. Before servicing the vehicle, refer to the precautions in the beginning of this section.
2. Drain the cooling system.
3. Remove or disconnect the following:
   - Negative battery cable
   - Exhaust pipe
   - Intake manifold
   - Cylinder head cover
   - Fan shroud and fan
   - Oil filler housing
   - Accessory drive belt
4. Rotate the crankshaft until the damper mark is aligned with the TDC mark. Verify that the V8 mark on the camshaft sprocket is at the 12 o'clock position.
5. Remove or disconnect the following:
   - Vibration damper
   - Timing chain cover
6. Lock the secondary timing chains to the idler sprocket with tool 8515, or equivalent.

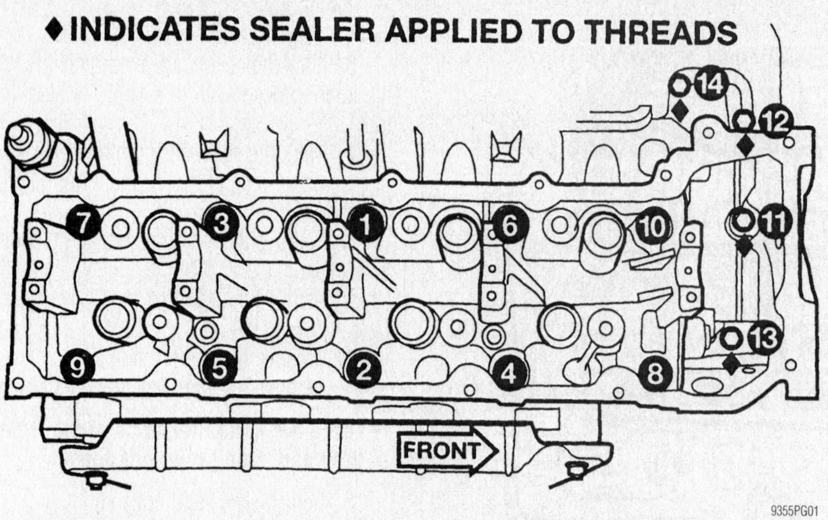

Cylinder head tightening sequence—4.7L

7. Mark the secondary timing chain, on link on either side of the V8 mark on the cam sprocket.

8. Remove the left side secondary chain tensioner.

9. Remove the cylinder head access plug.

10. Remove the chain guide.

11. Remove the camshaft sprocket.

⁂ WARNING

Do not pry on the target wheel for any reason!

➡There are 4 smaller bolts at the front of the head. Don't overlook these.

12. Remove the head bolts and head.

⁂ WARNING

Don't lay the head on its sealing surface. Due to the design of the head gasket, any distortion to the head sealing surface will result in leaks.

13. Installation is the reverse of removal. Observe the following:

- Check the head bolts. If any necking is observed, replace the bolt.
- The 4 small bolts must be coated with sealer.
- The head bolts are tightened in the following sequence:

Step 1: Bolts 1-10 to 15 ft. lbs. (20 Nm)
Step 2: Bolts 1-10 to 35 ft. lbs. (47 Nm)
Step 3: Bolts 11-14 to 18 ft. lbs. (25 Nm)
Step 4: Bolts 1-10 90 degrees
Step 5: Bolts 11-14 to 22 ft. lbs.

5.9L Engines

1. Before servicing the vehicle, refer to the precautions in the beginning of this section.

2. Drain the cooling system.

3. Remove or disconnect the following:

- Negative battery cable
- Accessory drive belt
- Alternator
- A/C compressor, if equipped
- Air injection pump, if equipped
- Air cleaner assembly
- Closed Crankcase Ventilation (CCV) system
- Evaporative emissions control system
- Fuel line
- Accelerator linkage
- Cruise control cable
- Transmission cable
- Distributor cap and wires
- Ignition coil wiring
- Engine Coolant Temperature (ECT) sensor connector
- Heater hoses
- Bypass hose
- Upper radiator hose
- Intake manifold
- Valve covers

➡Keep valvetrain components in order for reassembly.

- Rocker arms
- Pushrods
- Exhaust manifolds
- Spark plugs
- Cylinder heads

To install:

⁂ WARNING

Position the crankshaft so that no piston is at Top Dead Center (TDC) prior to installing the cylinder heads. Do not rotate the crankshaft during or immediately after rocker arm installation. Wait 5 minutes for the hydraulic lash adjusters to bleed down.

4. Install the cylinder heads. Use new gaskets and tighten the bolts in sequence as follows:

a. Step 1: 50 ft. lbs. (68 Nm).
b. Step 2: 105 ft. lbs. (143 Nm).
c. Step 3: 105 ft. lbs. (143 Nm).

5. Install or connect the following:

- Spark plugs
- Exhaust manifolds
- Pushrods and rocker arms in their original positions

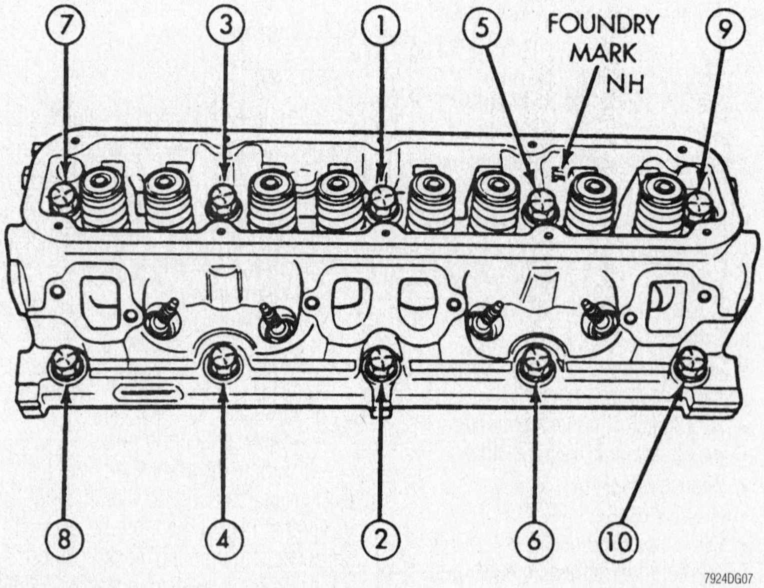

Cylinder head torque sequence—5.9L engines

- Valve covers
- Intake manifold
- Upper radiator hose
- Bypass hose
- Heater hoses
- ECT sensor connector
- Ignition coil wiring
- Distributor cap and wires
- Transmission cable
- Cruise control cable
- Accelerator linkage
- Fuel line
- Evaporative emissions control system
- CCV system
- Air cleaner assembly
- A/C compressor, if equipped
- Air injection pump, if equipped
- Alternator
- Accessory drive belt
- Negative battery cable

6. Fill the cooling system.
7. Start the engine and check for leaks.

Rocker Arms/Shafts

REMOVAL & INSTALLATION

2.5L Engine

1. Before servicing the vehicle, refer to the precautions in the beginning of this section.
2. Remove or disconnect the following:
   - Negative battery cable
   - Accelerator cable
   - Transmission cable, if equipped
   - Cruise control cable, if equipped
   - Valve cover
   - Rocker arm bolts, loosen them evenly to avoid damaging the alignment bridges
   - Rocker arms

➡Keep valvetrain components in order for reassembly.

To install:

3. Install or connect the following:
   - Rocker arms, pivots and bridges in their original positions. Tighten the bolts for each bridged pair one turn at a time to 21 ft. lbs. (28 Nm).
   - Valve cover
   - Cruise control cable, if equipped
   - Transmission cable, if equipped
   - Accelerator cable
   - Negative battery cable

3.9L Engine

1. Before servicing the vehicle, refer to the precautions in the beginning of this section.
2. Remove or disconnect the following:
   - Negative battery cable
   - Valve covers
   - Rocker arms

➡Keep all valvetrain components in order for assembly.

To install:

3. Rotate the crankshaft so that the **V6** mark on the crankshaft damper aligns with the timing mark on the front cover. The **V6** mark is located 147 degrees **AFTER** Top Dead Center (TDC).
4. Install the rocker arms in their original positions and tighten the bolts to 21 ft. lbs. (28 Nm).

⁂ CAUTION

Do not rotate the crankshaft during or immediately after rocker arm installation. Wait 5 minutes for the hydraulic lash adjusters to bleed down.

5. Install or connect the following:
   - Valve covers
   - Negative battery cable

3.7L and 4.7L Engines

1. Before servicing the vehicle, refer to the precautions in the beginning of this section.
2. Remove or disconnect the following:
   - Negative battery cable
   - Valve covers
3. Rotate the crankshaft so that the piston of the cylinder to be serviced is at Bottom Dead Center (BDC) and both valves are closed.
4. Use special tool 8516 to depress the valve and remove the rocker arm.
5. Repeat for each rocker arm to be serviced.

➡Keep valvetrain components in order for reassembly.

To install:

6. Rotate the crankshaft so that the piston of the cylinder to be serviced is at BDC.
7. Compress the valve spring and install each rocker arm in its original position.
8. Repeat for each rocker arm to be installed.
9. Install or connect the following:
   - Cylinder head cover
   - Negative battery cable

5.9L Engines

1. Before servicing the vehicle, refer to the precautions in the beginning of this section.
2. Remove or disconnect the following:
   - Negative battery cable

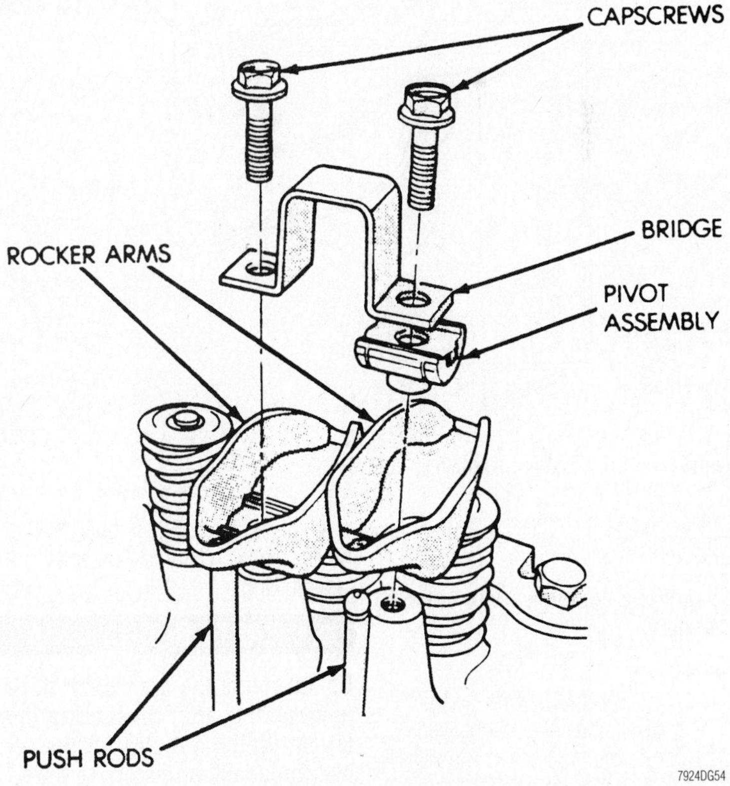

Exploded view of the rocker arm mounting—2.5L engine

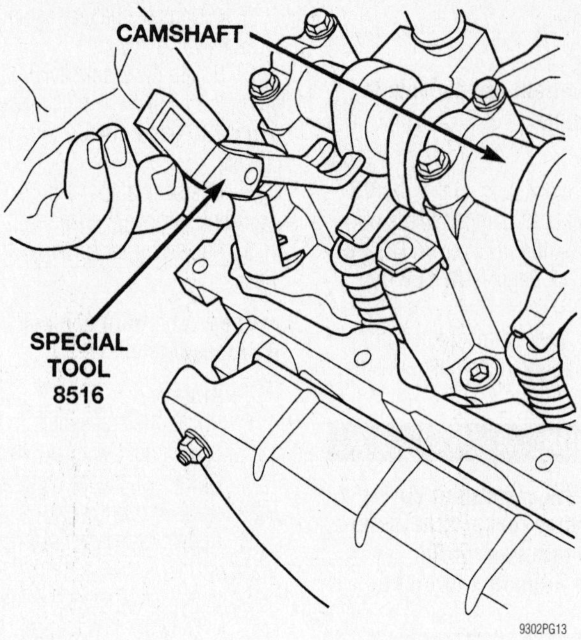

Rocker arm service—3.7L and 4.7L engines

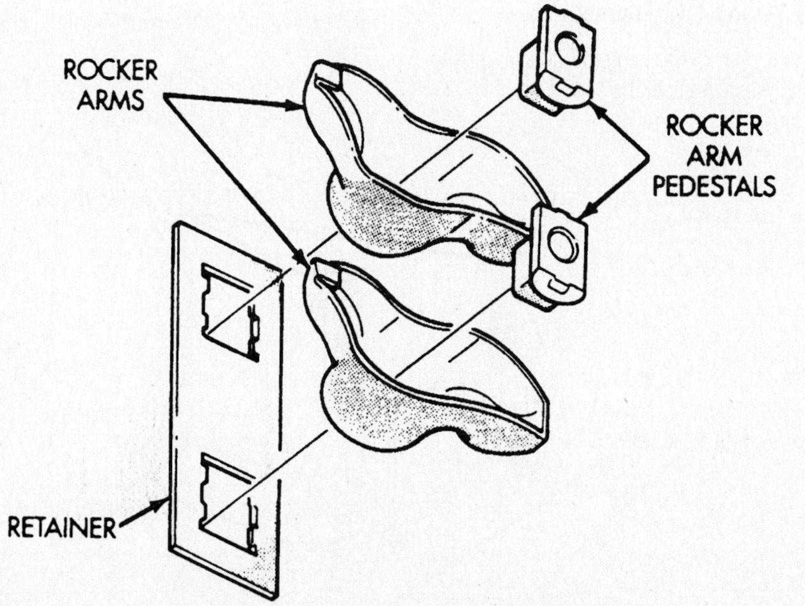

Exploded view of the rocker arm assembly—5.9L engine

- Valve covers
- Rocker arms

➡Keep valvetrain components in order for reassembly.

To install:

3. Rotate the crankshaft so that the **V8** mark on the crankshaft damper aligns with the timing mark on the front cover. The **V8** mark is located 147 degrees **AFTER** Top Dead Center (TDC).

4. Install the rocker arms in their original positions and tighten the bolts to 21 ft. lbs. (28 Nm).

⁂ CAUTION

Do not rotate the crankshaft during or immediately after rocker arm installation. Wait 5 minutes for the hydraulic lash adjusters to bleed down.

5. Install or connect the following:
   - Valve covers
   - Negative battery cable

Intake Manifold

REMOVAL & INSTALLATION

2.5L Engine

1. Before servicing the vehicle, refer to the precautions in the beginning of this section.
2. Drain the cooling system.
3. Relieve the fuel system pressure.
4. Remove or disconnect the following:
   - Negative battery cable
   - Air intake hose and resonator
   - Accessory drive belt
   - Power steering pump and brackets
   - Fuel line
   - Accelerator cable
   - Cruise control cable, if equipped
   - Transmission cable, if equipped
   - Throttle Position (TP) sensor connector
   - Intake Air Temperature (IAT) sensor connector
   - Idle Air Control (IAC) valve connector
   - Engine Coolant Temperature (ECT) sensor connector
   - Heated Oxygen (HO_2S) sensor connector
   - Fuel injector harness connectors
   - Manifold Absolute Pressure (MAP) sensor vacuum line
   - Closed Crankcase Ventilation (CCV) hose
   - Intake manifold vacuum hoses
   - Molded vacuum hose harness
5. Remove bolts 2–5. Loosen bolt No. 1 and nuts 6–7.
6. Remove the intake manifold.

To install:

7. Install the intake manifold with a new gasket. Tighten the fasteners in sequence as follows:
   a. Step 1: Tighten bolt No. 1 to 30 ft. lbs. (41 Nm)
   b. Step 2: Tighten bolts 2–5 to 23 ft. lbs. (31 Nm)
   c. Step 3: Tighten nuts 6–7 to 17 ft. lbs. (23 Nm)
8. Install or connect the following:
   - Molded vacuum hose harness
   - Intake manifold vacuum hoses
   - CCV hose
   - MAP sensor vacuum line
   - Fuel injector harness connectors
   - HO_2S sensor connector

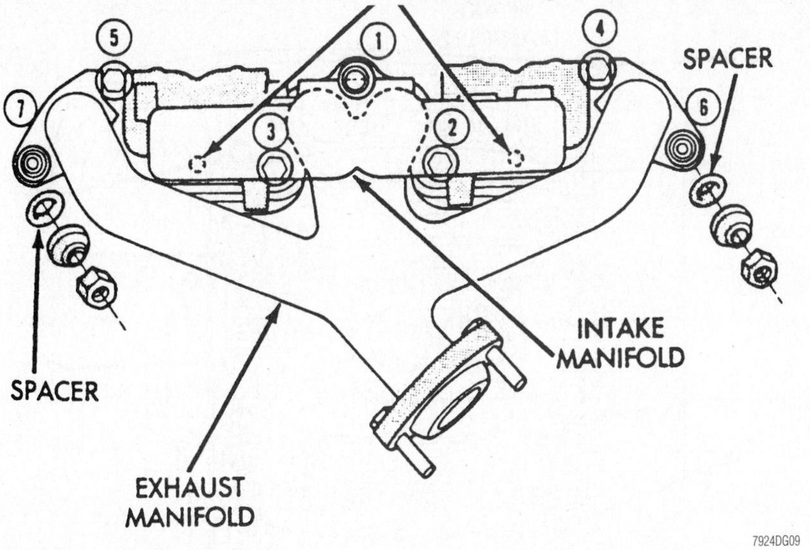

Manifold torque sequence—2.5L engine

- ECT sensor connector
- IAC valve connector
- IAT sensor connector
- TP sensor connector
- Transmission cable, if equipped
- Cruise control cable, if equipped
- Accelerator cable
- Fuel line
- Power steering pump and brackets
- Accessory drive belt
- Air intake hose and resonator
- Negative battery cable

9. Fill the cooling system.
10. Start the engine and check for leaks.

3.7L Engine

1. Before servicing the vehicle, refer to the precautions in the beginning of this section.
2. Drain the cooling system.
3. Remove or disconnect the following:
   - Negative battery cable
   - Air cleaner assembly
   - Accelerator cable
   - Cruise control cable
   - Manifold Absolute Pressure (MAP) sensor connector
   - Intake Air Temperature (IAT) sensor connector
   - Throttle Position (TP) sensor connector
   - Idle Air Control (IAC) valve connector
   - Engine Coolant Temperature (ECT) sensor
   - Positive Crankcase Ventilation (PCV) valve and hose
   - Canister purge vacuum line
   - Brake booster vacuum line
   - Cruise control servo hose
   - Accessory drive belt
   - Alternator
   - A/C compressor
   - Engine ground straps
   - Ignition coil towers
   - Oil dipstick tube
   - Fuel line
   - Fuel supply manifold
   - Throttle body and mounting bracket
   - Cowl seal
   - Right engine lifting stud
   - Intake manifold. Remove the fasteners in reverse of the tightening sequence.

To install:

4. Install or connect the following:
   - Intake manifold using new gaskets. Tighten the bolts, in sequence, to 105 inch lbs. (12 Nm).
   - Right engine lifting stud
   - Cowl seal
   - Throttle body and mounting bracket
   - Fuel supply manifold
   - Fuel line
   - Oil dipstick tube
   - Ignition coil towers
   - Engine ground straps
   - A/C compressor
   - Alternator
   - Accessory drive belt
   - Cruise control servo hose
   - Brake booster vacuum line
   - Canister purge vacuum line
   - PCV valve and hose
   - ECT sensor
   - IAC valve connector
   - TP sensor connector
   - IAT sensor connector
   - MAP sensor connector
   - Cruise control cable
   - Accelerator cable
   - Air cleaner assembly
   - Negative battery cable
5. Fill the cooling system.
6. Start the engine and check for leaks.

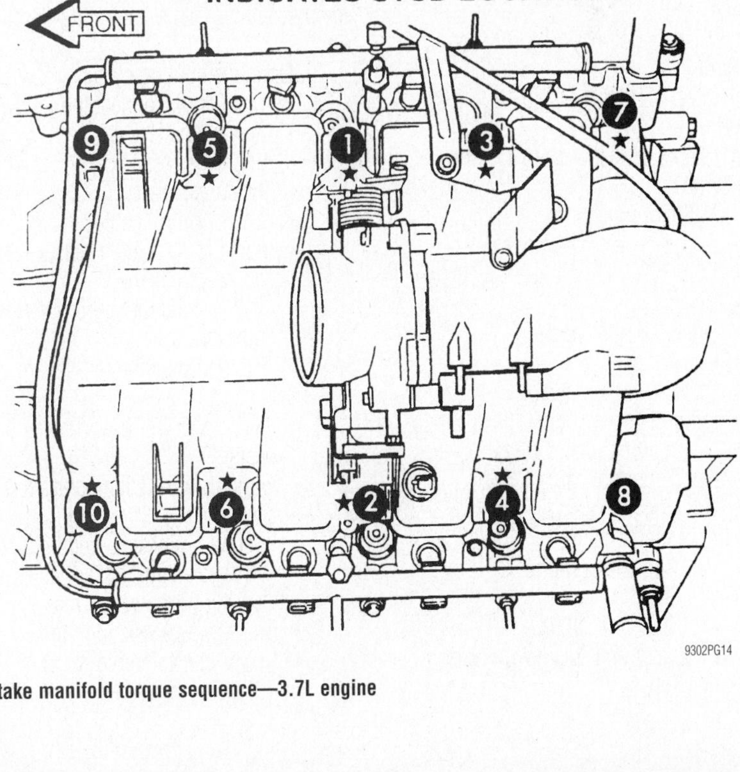

Intake manifold torque sequence—3.7L engine

3.9L Engine

1. Before servicing the vehicle, refer to the precautions in the beginning of this section.
2. Drain the cooling system.
3. Remove or disconnect the following:
   - Negative battery cable
   - Accessory drive belt
   - Alternator
   - A/C compressor
   - Alternator and A/C compressor bracket
   - Air cleaner assembly
   - Fuel line
   - Fuel supply manifold
   - Accelerator cable
   - Transmission cable
   - Cruise control cable
   - Distributor cap and wires
   - Ignition coil wiring
   - Engine Coolant Temperature (ECT) sensor connector
   - Heater hose
   - Upper radiator hose
   - Bypass hose
   - Closed Crankcase Ventilation (CCV) system
   - Evaporative emissions system
   - Intake manifold

To install:

4. Install the intake manifold. Use a new gasket and tighten the bolts in sequence as follows:
   a. Step 1: Bolts 1–2 to 72 inch lbs. (8 Nm) using 12 inch lb. (1.4 Nm) increments.
   b. Step 2: Bolts 3–12 to 72 inch lbs. (8 Nm).
   c. Step 3: Bolts 1–12 to 72 inch lbs. (8 Nm).
   d. Step 4: Bolts 1–12 to 12 ft. lbs. (16 Nm).
   e. Step 5: Bolts 1–12 to 12 ft. lbs. (16 Nm).
5. Install or connect the following:
   - Evaporative emissions system
   - CCV system
   - Bypass hose
   - Upper radiator hose
   - Heater hose
   - ECT sensor connector
   - Ignition coil wiring
   - Distributor cap and wires
   - Cruise control cable
   - Transmission cable
   - Accelerator cable
   - Fuel supply manifold
   - Fuel line
   - Air cleaner assembly
   - Alternator and A/C compressor bracket

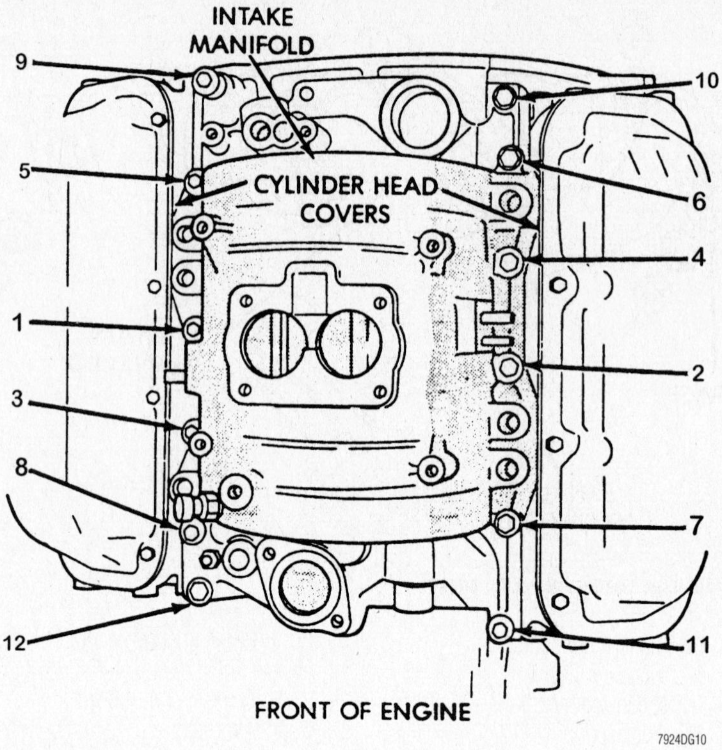

Intake manifold torque sequence—3.9L engine

   - A/C compressor
   - Alternator
   - Accessory drive belt
   - Negative battery cable
6. Fill the cooling system.
7. Start the engine and check for leaks.

4.7L Engine

1. Before servicing the vehicle, refer to the precautions in the beginning of this section.
2. Drain the cooling system.
3. Remove or disconnect the following:
   - Negative battery cable
   - Air cleaner assembly
   - Accelerator cable
   - Cruise control cable
   - Manifold Absolute Pressure (MAP) sensor connector
   - Intake Air Temperature (IAT) sensor connector
   - Throttle Position (TP) sensor connector
   - Idle Air Control (IAC) valve connector
   - Engine Coolant Temperature (ECT) sensor
   - Positive Crankcase Ventilation (PCV) valve and hose
   - Canister purge vacuum line
   - Brake booster vacuum line
   - Cruise control servo hose
   - Accessory drive belt
   - Alternator
   - A/C compressor
   - Engine ground straps
   - Ignition coil towers
   - Oil dipstick tube
   - Fuel line
   - Fuel supply manifold
   - Throttle body and mounting bracket
   - Cowl seal
   - Right engine lifting stud
   - Intake manifold. Remove the fasteners in reverse of the tightening sequence.

To install:

4. Install or connect the following:
   - Intake manifold using new gaskets. Tighten the bolts, in sequence, to 105 inch lbs. (12 Nm).
   - Right engine lifting stud
   - Cowl seal
   - Throttle body and mounting bracket
   - Fuel supply manifold
   - Fuel line
   - Oil dipstick tube
   - Ignition coil towers
   - Engine ground straps
   - A/C compressor
   - Alternator
   - Accessory drive belt
   - Cruise control servo hose
   - Brake booster vacuum line
   - Canister purge vacuum line
   - PCV valve and hose

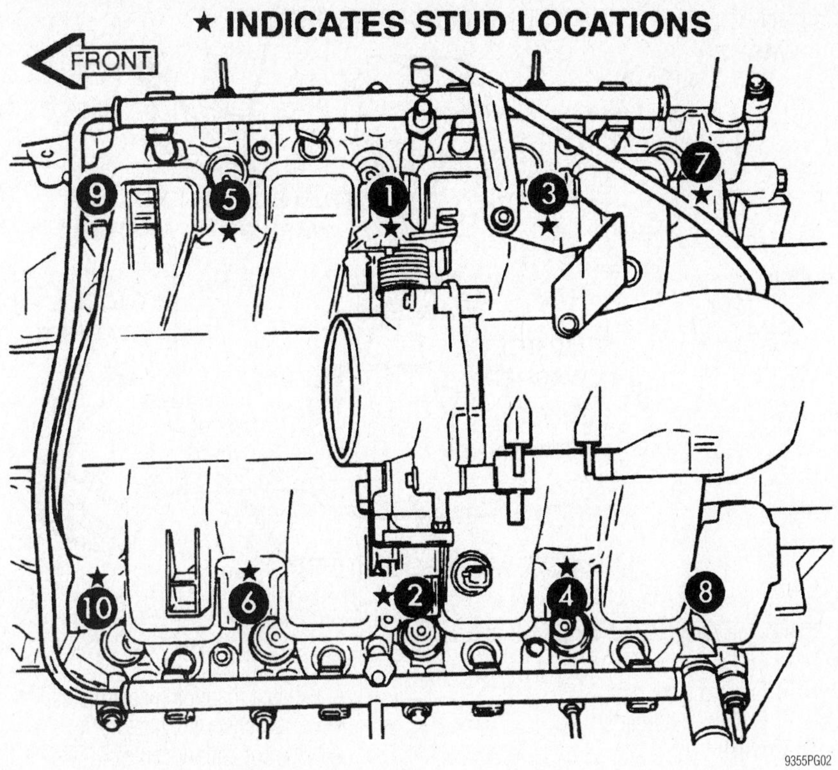

Intake manifold torque sequence—4.7L

- ECT sensor
- IAC valve connector
- TP sensor connector
- IAT sensor connector
- MAP sensor connector
- Cruise control cable
- Accelerator cable
- Air cleaner assembly
- Negative battery cable

5. Fill the cooling system.
6. Start the engine and check for leaks.

5.9L Engines

1. Before servicing the vehicle, refer to the precautions in the beginning of this section.
2. Drain the cooling system.
3. Remove or disconnect the following:
   - Negative battery cable
   - Accessory drive belt
   - Alternator
   - A/C compressor
   - Alternator and A/C compressor bracket
   - Air cleaner assembly
   - Fuel line
   - Fuel supply manifold
   - Accelerator cable
   - Transmission cable
   - Cruise control cable
   - Distributor cap and wires
   - Ignition coil wiring
   - Engine Coolant Temperature (ECT) sensor connector
   - Heater hose
   - Upper radiator hose
   - Bypass hose
   - Closed Crankcase Ventilation (CCV) system
   - Evaporative emissions system
   - Intake manifold

To install:

4. Install the intake manifold. Use a new gasket and tighten the bolts in sequence as follows:
   a. Step 1: Bolts 1–4 to 72 inch lbs. (8 Nm) using 12 inch lb. (1.4 Nm) increments.
   b. Step 2: Bolts 5–12 to 72 inch lbs. (8 Nm).
   c. Step 3: Bolts 1–12 to 72 inch lbs. (8 Nm).
   d. Step 4: Bolts 1–12 to 12 ft. lbs. (16 Nm).
   e. Step 5: Bolts 1–12 to 12 ft. lbs. (16 Nm).
5. Install or connect the following:
   - Evaporative emissions system
   - CCV system
   - Bypass hose
   - Upper radiator hose
   - Heater hose
   - ECT sensor connector
   - Ignition coil wiring
   - Distributor cap and wires
   - Cruise control cable
   - Transmission cable
   - Accelerator cable
   - Fuel supply manifold
   - Fuel line
   - Air cleaner assembly
   - Alternator and A/C compressor bracket
   - A/C compressor
   - Alternator
   - Accessory drive belt
   - Negative battery cable
6. Fill the cooling system.
7. Start the engine and check for leaks.

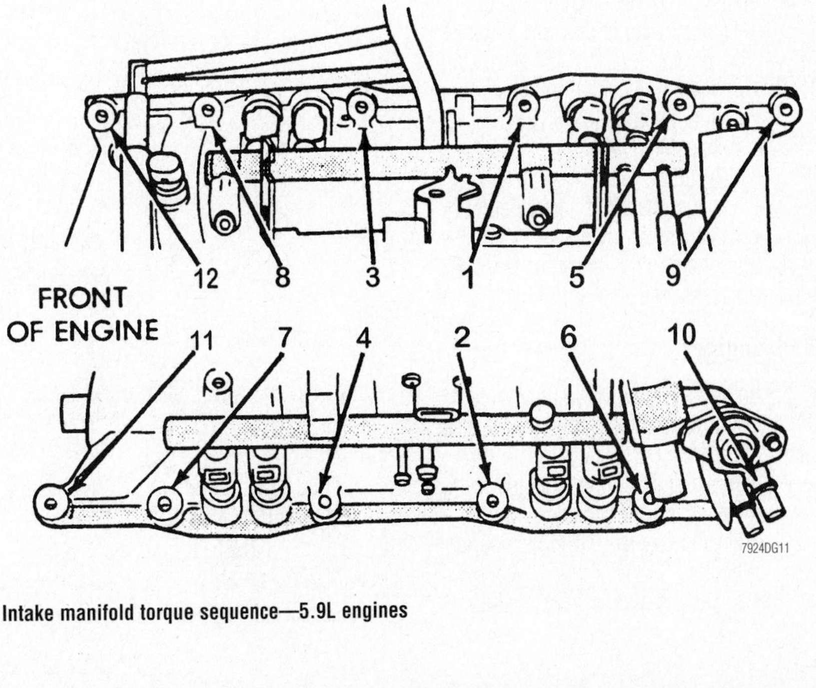

Intake manifold torque sequence—5.9L engines

Exhaust Manifold

REMOVAL & INSTALLATION

2.5L Engine

1. Before servicing the vehicle, refer to the precautions in the beginning of this section.
2. Remove or disconnect the following:
   - Negative battery cable
   - Exhaust front pipe
   - Intake manifold
   - Exhaust manifold

To install:

3. Install or connect the following:
   - Exhaust manifold
   - Intake manifold
   - Exhaust front pipe
   - Negative battery cable
4. Start the engine and check for leaks.

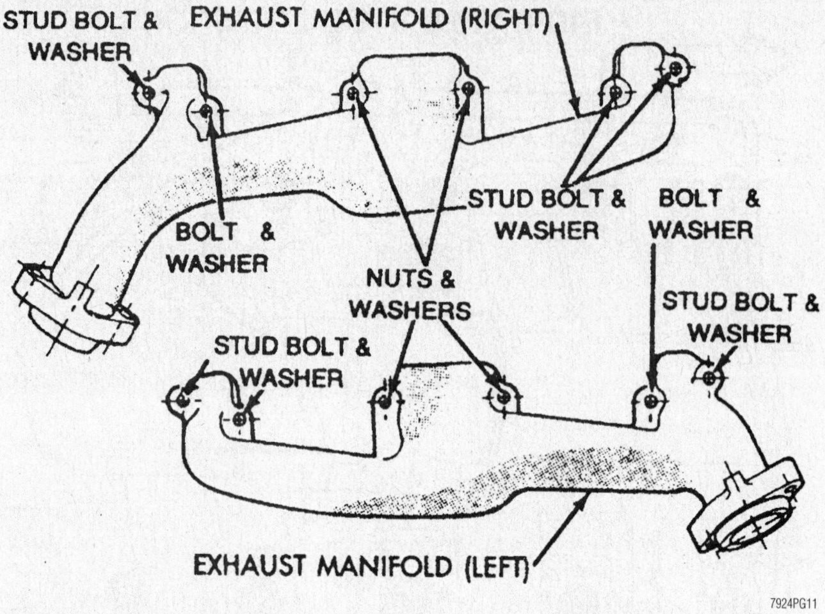

Exhaust manifold fastener locations—3.7L engines

3.7L Engines

1. Before servicing the vehicle, refer to the precautions in the beginning of this section.
2. Remove or disconnect the following:
   - Negative battery cable
   - Exhaust manifold heat shields
   - Exhaust Gas Recirculation (EGR) tube
   - Exhaust Y-pipe
   - Exhaust manifolds

To install:

➡If the exhaust manifold studs came out with the nuts when removing the exhaust manifolds, replace them with new studs.

3. Install or connect the following:
   - Exhaust manifolds. Torque the fasteners to 20 ft. lbs. (27 Nm), starting with the center nuts and work out to the ends.
   - Exhaust Y-pipe
   - EGR tube
   - Exhaust manifold heat shields
   - Negative battery cable
4. Start the engine, check for leaks and repair if necessary.

3.9L Engine

1. Before servicing the vehicle, refer to the precautions in the beginning of this section.
2. Remove or disconnect the following:
   - Negative battery cable
   - Heated Oxygen (HO_2S) sensor connectors
   - Exhaust manifold heat shields
   - Exhaust front pipe
   - Exhaust manifolds

To install:

➡If the exhaust manifold studs came out with the nuts when removing the exhaust manifolds, replace them with new studs.

3. Install or connect the following:
   - Exhaust manifolds. Tighten the fasteners to 25 ft. lbs. (34 Nm), starting with the center fasteners and working out to the ends.
   - Exhaust front pipe
   - Exhaust manifold heat shields
   - HO_2S sensor connectors
   - Negative battery cable
4. Start the engine and check for leaks.

4.7L Engine

1. Before servicing the vehicle, refer to the precautions in the beginning of this section.
2. Drain the cooling system.
3. Remove or disconnect the following:
   - Battery
   - Power distribution center
   - Battery tray

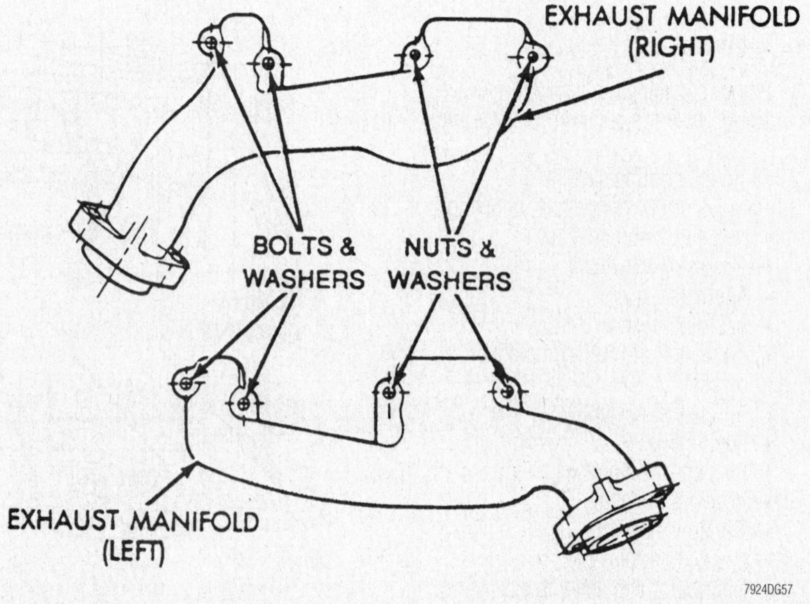

Stud and bolt locations—3.9L engine

- Windshield washer fluid bottle
- Air cleaner assembly
- Accessory drive belt
- A/C compressor
- A/C accumulator bracket
- Heater hoses
- Exhaust manifold heat shields
- Exhaust Y-pipe
- Starter motor
- Exhaust manifolds

To install:

4. Install or connect the following:
   - Exhaust manifolds, using new gaskets. Tighten the bolts to 18 ft. lbs. (25 Nm), starting with the inner bolts and work out to the ends.
   - Starter motor
   - Exhaust Y-pipe
   - Exhaust manifold heat shields
   - Heater hoses
   - A/C accumulator bracket
   - A/C compressor
   - Accessory drive belt
   - Air cleaner assembly
   - Windshield washer fluid bottle
   - Battery tray
   - Power distribution center
   - Battery
5. Fill the cooling system.
6. Start the engine and check for leaks.

5.9L Engines

1. Before servicing the vehicle, refer to the precautions in the beginning of this section.
2. Remove or disconnect the following:
   - Negative battery cable
   - Exhaust manifold heat shields
   - Exhaust Gas Recirculation (EGR) tube
   - Exhaust Y-pipe
   - Exhaust manifolds

To install:

➡If the exhaust manifold studs came out with the nuts when removing the exhaust manifolds, replace them with new studs.

3. Install or connect the following:
   - Exhaust manifolds. Tighten the fasteners to 20 ft. lbs. (27 Nm), starting with the center nuts and work out to the ends.
   - Exhaust Y-pipe
   - EGR tube
   - Exhaust manifold heat shields
   - Negative battery cable
4. Fill the cooling system.
5. Start the engine and check for leaks.

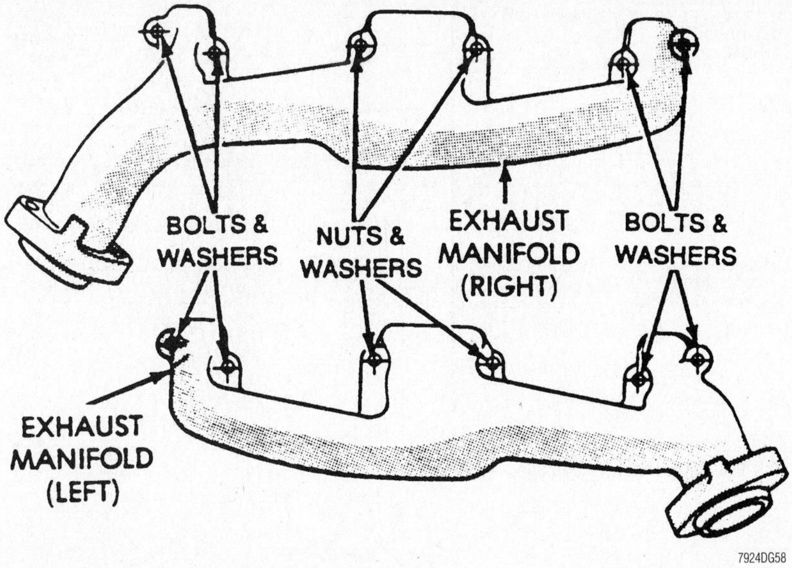

Exhaust manifold fastener locations—5.9L engines

Camshaft and Valve Lifters

REMOVAL & INSTALLATION

2.5L Engine

1. Before servicing the vehicle, refer to the precautions in the beginning of this section.
2. Drain the cooling system.
3. Recover the A/C refrigerant, if equipped with air conditioning.
4. Remove or disconnect the following:
   - Negative battery cable
   - Grille, if necessary
   - Radiator
   - A/C condenser, if equipped
   - Distributor
   - Valve cover

➡Keep all valvetrain components in order for assembly.

   - Rocker arms and pushrods
   - Hydraulic valve tappets
   - Accessory drive belt
   - Crankshaft damper
   - Front cover
   - Timing chain and gears
   - Camshaft

To install:

➡If the camshaft sprocket appears to have been rubbing against the cover, check the oil pressure relief holes in the rear cam journal for debris.

5. Lubricate the camshaft with clean engine oil.
6. Install or connect the following:
   - Camshaft
   - Timing chain and gears
   - Front cover
   - Crankshaft damper
   - Accessory drive belt
   - Hydraulic valve tappets
   - Rocker arms and pushrods
   - Valve cover
   - Distributor
   - A/C condenser, if equipped
   - Radiator
   - Grille, if removed
   - Negative battery cable
7. Fill the cooling system.
8. Recharge the A/C system, if equipped.
9. Start the engine and check for leaks.

3.7L Engines

1. Before servicing the vehicle, refer to the precautions in the beginning of this section.
2. Remove or disconnect the following:
   - Negative battery cable
   - Cylinder head covers
   - Rocker arms
   - Hydraulic lash adjusters

➡Keep all valvetrain components in order for assembly.

3. Set the engine at Top Dead Center (TDC) of the compression stroke for the No. 1 cylinder.
4. Install Timing Chain Wedge 8350 to retain the chain tensioners.
5. Matchmark the timing chains to the camshaft sprockets.
6. Install Camshaft Holding Tool 6958 and Adapter Pins 8346 to the left camshaft sprocket.

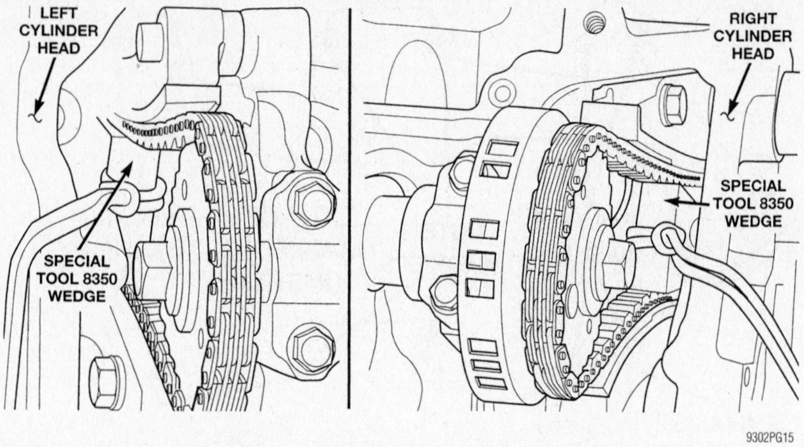

Chain Tensioner Retaining Wedges—3.7L engine

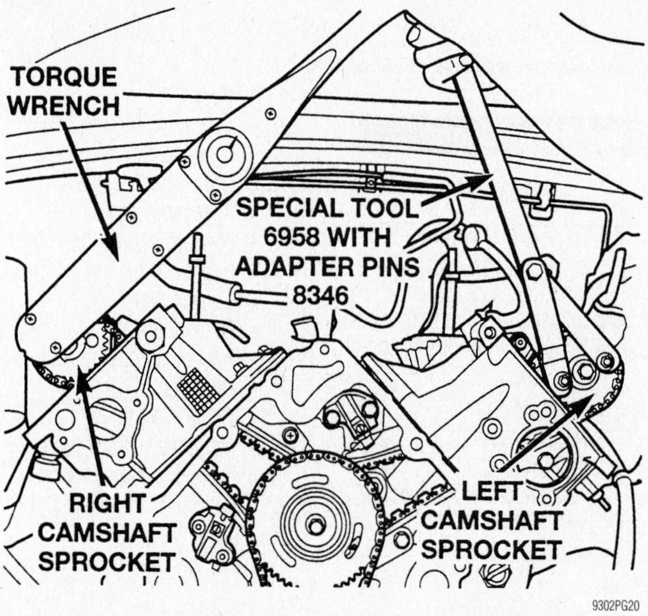

Hold the left camshaft sprocket with a spanner wrench while removing or installing the camshaft sprocket bolts—3.7L engine

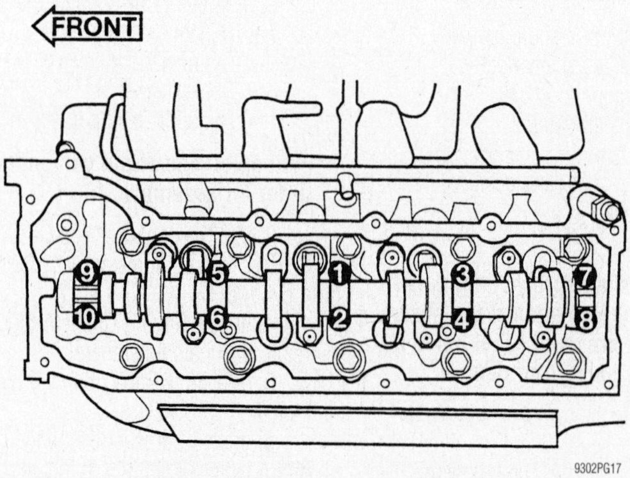

Camshaft bearing cap bolt tightening sequence—3.7L engine

7. Remove or disconnect the following:
   - Right camshaft timing sprocket and target wheel
   - Left camshaft sprocket
   - Camshaft bearing caps, by reversing the tightening sequence
   - Camshafts

To install:

8. Install or connect the following:
   - Camshafts. Tighten the bearing cap bolts in ½ turn increments, in sequence, to 100 inch lbs. (11 Nm).
   - Target wheel to the right camshaft
   - Camshaft timing sprockets and chains, by aligning the matchmarks
9. Remove the tensioner wedges and tighten the camshaft sprocket bolts to 90 ft. lbs. (122 Nm).
10. Install or connect the following:
    - Hydraulic lash adjusters in their original locations
    - Rocker arms in their original locations
    - Cylinder head covers
    - Negative battery cable

4.7L Engine

1. Before servicing the vehicle, refer to the precautions in the beginning of this section.
2. Remove or disconnect the following:
   - Negative battery cable
   - Cylinder head covers
   - Rocker arms
   - Hydraulic lash adjusters

➡Keep all valvetrain components in order for assembly.

3. Set the engine at Top Dead Center (TDC) of the compression stroke for the No. 1 cylinder.
4. Install Timing Chain Wedge 8350 to retain the chain tensioners.
5. Matchmark the timing chains to the camshaft sprockets.
6. Install Camshaft Holding Tool 6958 and Adapter Pins 8346 to the left camshaft sprocket.
7. Remove or disconnect the following:
   - Right camshaft timing sprocket and target wheel
   - Left camshaft sprocket
   - Camshaft bearing caps, by reversing the tightening sequence
   - Camshafts

To install:

8. Install or connect the following:
   - Camshafts. Tighten the bearing cap bolts in ½ turn increments, in sequence, to 100 inch lbs. (11 Nm).

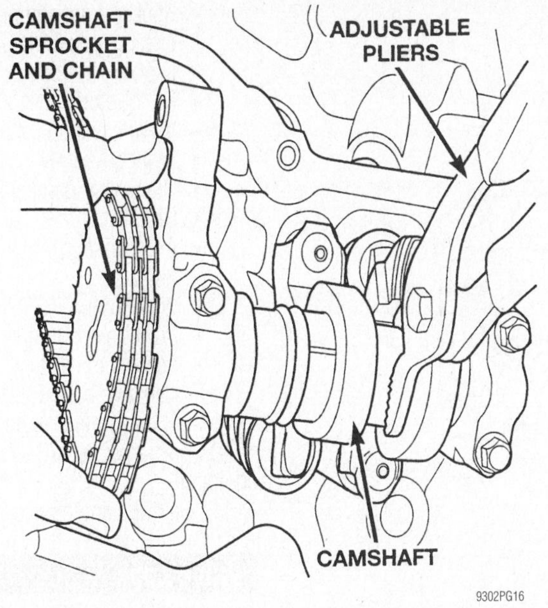

Turn the camshaft with pliers, if needed, to align the dowel in the sprocket—3.7L engine

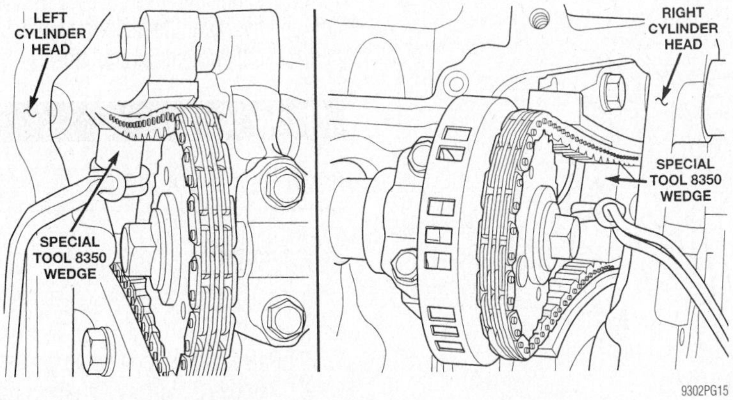

Chain Tensioner Retaining Wedges—4.7L engine

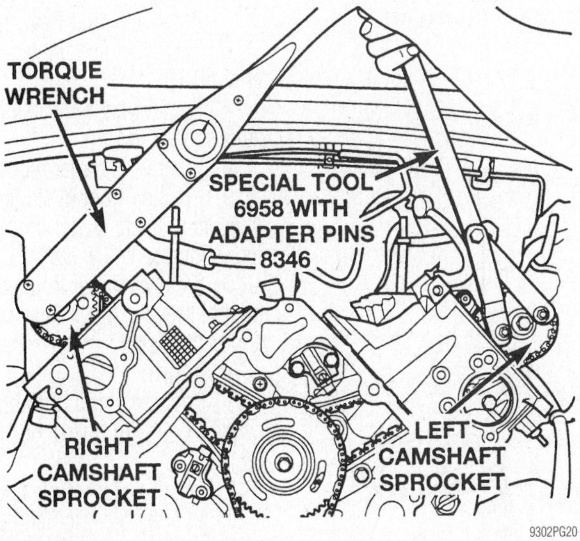

Hold the left camshaft sprocket with a spanner wrench while removing or installing the camshaft sprocket bolts—4.7L engine

- Target wheel to the right camshaft
- Camshaft timing sprockets and chains, by aligning the matchmarks

9. Remove the tensioner wedges and tighten the camshaft sprocket bolts to 90 ft. lbs. (122 Nm).
10. Install or connect the following:
    - Hydraulic lash adjusters in their original locations
    - Rocker arms in their original locations
    - Cylinder head covers
    - Negative battery cable

3.9L and 5.9L Engines

1. Before servicing the vehicle, refer to the precautions in the beginning of this section.
2. Drain the cooling system.
3. Recover the A/C refrigerant, if equipped with air conditioning.
4. Set the crankshaft to Top Dead Center (TDC) of the compression stroke for the No. 1 cylinder.
5. Remove or disconnect the following:
    - Negative battery cable
    - Accessory drive belt
    - Power steering pump
    - Water pump
    - Radiator
    - A/C condenser
    - Grille
    - Crankshaft damper
    - Front cover
    - Valve covers
    - Distributor
    - Intake manifold

➡Keep all valvetrain components in order for assembly.

- Rocker arms and pushrods
- Hydraulic lifters
- Timing chain and sprockets
- Camshaft thrust plate and chain oil tab
- Camshaft

To install:

6. Install or connect the following:
    - Camshaft
    - Camshaft Holding Tool C-3509
    - Camshaft thrust plate and chain oil tab. Tighten the bolts to 18 ft. lbs. (24 Nm).
    - Timing chain and sprockets. Tighten the camshaft sprocket bolt to 50 ft. lbs. (68 Nm).
7. Remove the camshaft holding tool.
8. Install or connect the following:
    - Hydraulic lifters in their original positions

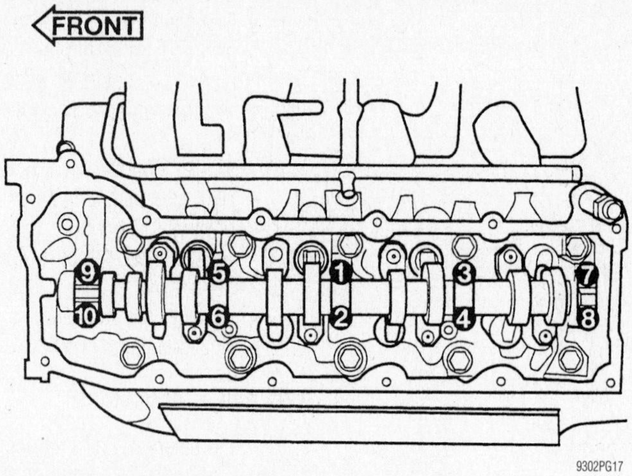

Camshaft bearing cap bolt tightening sequence—4.7L engine

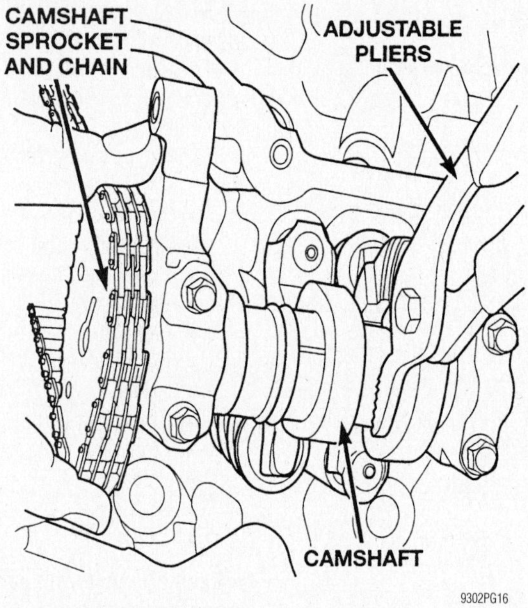

Turn the camshaft with pliers, if needed, to align the dowel in the sprocket—4.7L engine

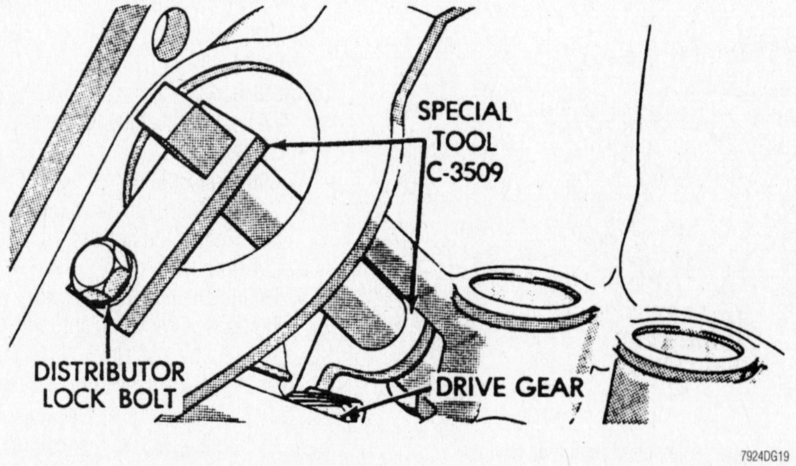

Camshaft holding tool C-3509—3.9L and 5.9L engines

- Rocker arms and pushrods in their original positions
- Intake manifold
- Distributor
- Valve covers
- Front cover
- Crankshaft damper
- Grille
- A/C condenser
- Radiator
- Water pump
- Power steering pump
- Accessory drive belt
- Negative battery cable

9. Fill the cooling system.
10. Recharge the A/C system, if equipped.
11. Start the engine and check for leaks.

Valve Lash

ADJUSTMENT

All gasoline engines covered in this section use hydraulic lifters. No maintenance or periodic adjustment is required.

Starter Motor

REMOVAL & INSTALLATION

3.7L Engine

1. Disconnect and isolate negative battery cable.
2. Raise and support vehicle.
3. Remove 2 starter heat shield bolts at side of starter.
4. Remove starter heat shield nut at front of starter.
5. Remove starter heat shield.
6. Remove solenoid wire from solenoid terminal.
7. Remove battery cable from stud on starter solenoid.
8. Remove 2 starter mounting bolts.
9. Position front of starter to face rear of vehicle. Rotate starter until solenoid position is located below starter.
10. Remove starter from vehicle by passing it between exhaust pipe and transmission bellhousing.

To install:

11. Position starter into bellhousing and install 2 bolts. Torque to 68 Nm (50 ft. lbs.).
12. Install battery cable and nut to stud on starter solenoid. Tighten nut to 13.6 Nm (120 inch lbs.).
13. Install solenoid wire connector to solenoid terminal.

14. Position starter heat shield and install nut at front of starter.
15. Install 2 starter heat shield bolts at side of starter.
16. Lower vehicle.
17. Connect negative battery cable.

4.7L Engine

WITH MANUAL TRANSMISSION

1. Disconnect and isolate negative battery cable.
2. Raise and support vehicle.
3. Remove nut securing starter motor to stud on transmission housing.
4. While supporting starter motor, remove bolt securing starter motor to transmission housing.
5. If equipped with automatic transmission, slide transmission cooler tube bracket forward on tubes far enough for starter motor to be removed from lower mounting stud.
6. Lower starter motor from front transmission housing far enough to access and remove nut securing battery cable eyelet to starter solenoid stud.

Always support starter motor during this process. Do not let starter motor hang from wire harness.

7. Remove solenoid wire solenoid terminal stud.
8. Disconnect battery cable solenoid wire from receptacle on starter solenoid.
9. Remove starter motor from transmission housing.

To install:

10. Position starter motor to transmission housing.
11. Connect battery cable solenoid terminal wire harness connector to connector receptacle on starter solenoid. Always support the starter motor during this process. Do not let the starter motor hang from the wire harness.
12. Install battery cable eyelet terminal onto solenoid B(+) terminal stud.
13. Install and tighten nut securing battery cable eyelet terminal to starter solenoid B (+) terminal stud. Tighten nut to 13.6 Nm (120 inch lbs.).
14. Position starter motor over stud on transmission housing.
15. If equipped with automatic transmission, slide automatic transmission cooler tube bracket rearward on tubes and into position over starter motor flange.
16. Loosely install the washers, bolt, and nut to starter. Tighten bolt and nut to 67.8 Nm (50 ft. lbs.).
17. Lower vehicle.
18. Connect negative battery cable.

WITH AUTOMATIC TRANSMISSION

1. Disconnect and isolate negative battery cable.
2. Raise and support vehicle.
3. Remove bolt and washer (rearward facing) securing starter motor to the transmission housing.
4. While supporting starter motor, remove bolt and washer (rearward facing) securing starter motor to the transmission housing.
5. Lower starter motor from front of transmission housing far enough to access and remove nut securing battery positive cable eyelet terminal to the starter solenoid B (+) terminal stud. Always support starter motor during this process. Do not let starter motor hang from wire harness.
6. Remove battery cable eyelet terminal from solenoid B (+) terminal stud.
7. Disconnect battery cable solenoid terminal wire harness connector from receptacle on starter solenoid.
8. Remove starter motor from transmission housing.

To install:

9. Position starter motor to transmission housing.
10. Connect battery cable solenoid terminal wire harness connector to connector receptacle on starter solenoid. Always support the starter motor during this process. Do not let the starter motor hang from the wire harness.
11. Install battery cable eyelet terminal onto solenoid B (+) terminal stud.
12. Install and tighten nut securing battery cable eyelet terminal to starter solenoid B (+) terminal stud. Tighten nut to 13.6 Nm (120 inch lbs.).
13. Position starter motor to transmission housing and loosely install two bolts/washers.
14. Tighten bolts to 67.8 Nm (50 ft. lbs.).
15. Lower vehicle.
16. Connect negative battery cable.

2.5L, 3.9L, and 5.9L Engines

1. Before servicing the vehicle, refer to the precautions in the beginning of this section.
2. Remove or disconnect the following:
   - Negative battery cable
   - Starter mounting bolts
   - Starter solenoid harness connections
   - Starter

To install:

3. Connect the starter solenoid wiring connectors.
4. Install the starter and tighten the bolts to the following specifications:
   a. 2.5L engine: 33 ft. lbs. (45 Nm).
   b. 4.7L engine: 40 ft. lbs. (54 Nm).
   c. 3.9L, 5.9L engines: 50 ft. lbs. (68 Nm).
5. Install the negative battery cable and check for proper operation.

Oil Pan

REMOVAL & INSTALLATION

2.5L Engine

1. Before servicing the vehicle, refer to the precautions in the beginning of this section.
2. Drain the engine oil.
3. Remove or disconnect the following:
   - Negative battery cable
   - Exhaust front pipe
   - Starter motor
   - Bell housing access cover
   - Oil level sensor connector, if equipped
   - Left and right motor mounts
4. Place a jack under the crankshaft damper and raise the engine for clearance.
5. Remove the oil pan.

To install:

6. Fabricate 4 alignment dowels from 1½ inch x ¼ inch bolts. Cut the heads off the bolts and cut a slot into the top of the dowel to allow installation/removal with a screwdriver.
7. Install or connect the following:
   - Dowels
   - Oil pan, using a new gasket. Tighten the ¼ inch bolts to 85 inch lbs. (9.5 Nm) and the 5⁄16 inch bolts to 11 ft. lbs. (15 Nm).
8. Replace the alignment dowels with ¼ inch bolts and tighten them to 85 inch lbs. (9.5 Nm).
9. Install or connect the following:
   - Left and right motor mounts
   - Oil level sensor connector, if equipped
   - Bell housing access cover
   - Starter motor
   - Exhaust front pipe
   - Negative battery cable
10. Fill the crankcase.
11. Start the engine and check for leaks.

3.7L Engine

1. Disconnect the negative battery cable.

2. Install an engine support fixture. Do not raise engine at this time.
3. Loosen both left and right side engine mount through bolts. Do not remove bolts.
4. Remove the structural dust cover, if equipped.
5. Drain engine oil.
6. Remove the front crossmember.

➡Raise the engine just enough to provide clearance for oil pan removal. Check for proper clearance at fan shroud to fan and cowl to intake manifold.

7. Raise engine to provide clearance to remove oil pan.

➡Do not pry on oil pan or oil pan gasket. Gasket is integral to engine windage tray and does not come out with oil pan.

8. Remove the oil pan mounting bolts and oil pan.
9. Unbolt oil pump pickup tube and remove tube.
10. Inspect the integral windage tray and gasket and replace as needed.

To install:

11. Clean the oil pan gasket mating surface of the bedplate and oil pan.
12. Inspect integrated oil pan gasket, and replace as necessary.
13. Position the integrated oil pan gasket/windage tray assembly.
14. Install the oil pickup tube
15. If removed, install stud at position No. 9.
16. Install the mounting bolt and nuts. Tighten nuts to 28 Nm (20 ft. lbs.).
17. Position the oil pan and install the mounting bolts. Tighten the mounting bolts to 15 Nm (11 ft. lbs.) in the sequence shown.
18. Lower the engine into mounts.
19. Install both the left and right side engine mount through bolts. Tighten the nuts to 68 Nm (50 ft. lbs.).
20. Remove the lifting device.
21. Install structural dust cover, if equipped.
22. Install the front crossmember.
23. Fill engine oil.
24. Reconnect the negative battery cable.
25. Start engine and check for leaks.

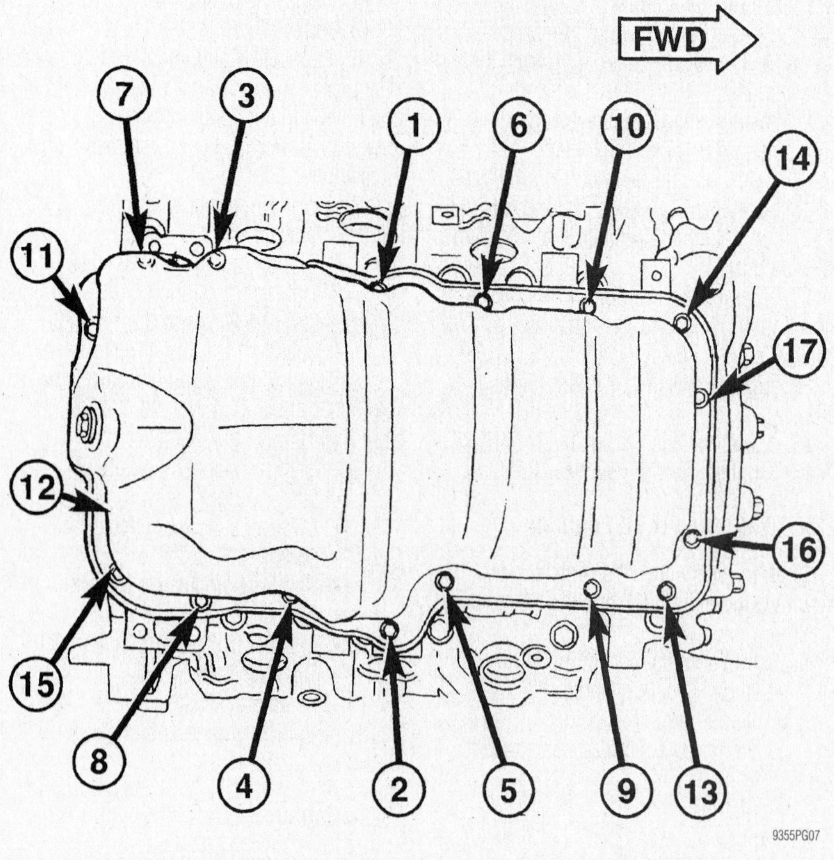

Oil pan bolt torque sequence—3.7L Engine

3.9L Engine

2-WHEEL DRIVE MODELS

1. Before servicing the vehicle, refer to the precautions in the beginning of this section.
2. Drain the engine oil.
3. Remove or disconnect the following:
   - Negative battery cable
   - Distributor cap
   - Oil dipstick
   - Exhaust front pipe
   - Flywheel access panel, if equipped
   - Left and right motor mount through bolts
   - Oil pan. Raise the engine as necessary for clearance.

To install:

4. Fabricate 4 alignment dowels from 1½ inch x ¼ inch bolts. Cut the heads off the bolts and cut a slot into the top of the dowel to allow installation/removal with a screwdriver.
5. Install or connect the following:
   - Alignment dowels
   - Oil pan. Replace the dowels with bolts and tighten all bolts to 17 ft. lbs. (23 Nm).
   - Left and right motor mount through bolts
   - Flywheel access panel, if equipped
   - Exhaust front pipe
   - Oil dipstick
   - Distributor cap
   - Negative battery cable
6. Fill the crankcase to the correct level.
7. Start the engine and check for leaks.

4-WHEEL DRIVE MODELS

1. Before servicing the vehicle, refer to the precautions in the beginning of this section.
2. Drain the engine oil.
3. Remove or disconnect the following:
   - Negative battery cable
   - Engine oil dipstick
   - Front axle. Support the engine
   - Exhaust front pipe
   - Flywheel access panel, if equipped
   - Oil pan

To install:

4. Fabricate 4 alignment dowels from 1½ inch x ¼ inch bolts. Cut the heads off the bolts and cut a slot into the top of the dowel to allow installation/removal with a screwdriver.
5. Install or connect the following:
   - Alignment dowels
   - Oil pan. Replace the dowels with bolts and tighten all bolts to 17 ft. lbs. (23 Nm).
   - Flywheel access panel, if equipped

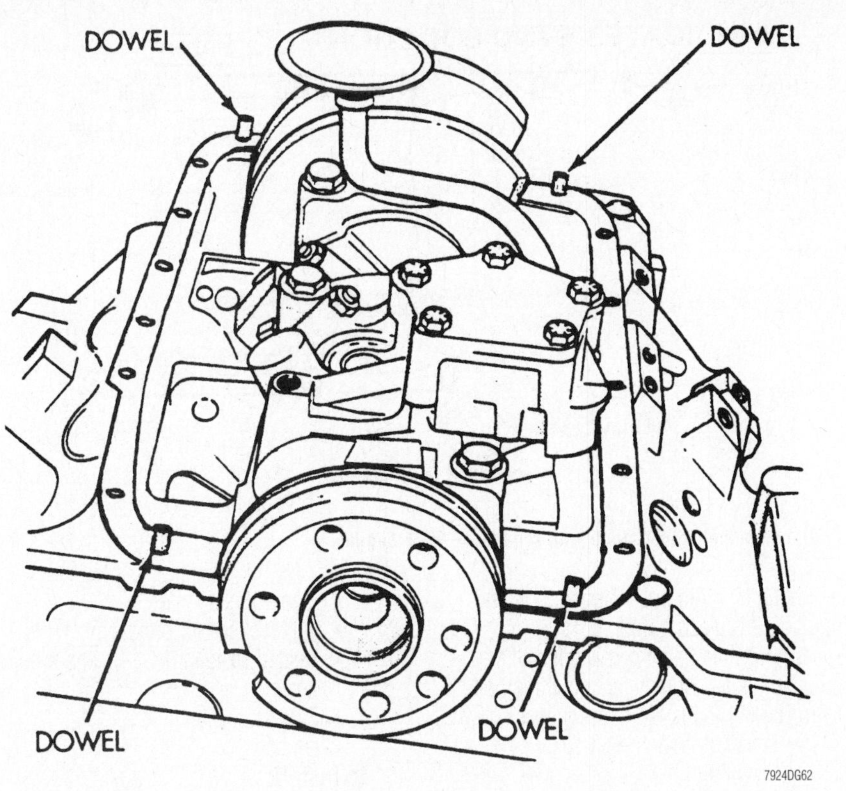

Oil pan alignment dowel placement—3.9L engine shown

- Exhaust front pipe
- Front axle
- Engine oil dipstick
- Negative battery cable

6. Fill the crankcase to the correct level.
7. Start the engine and check for leaks.

4.7L Engine

2-WHEEL DRIVE

1. Drain the cooling system.
2. Remove the upper fan shroud.
3. Remove the throttle body resonator and air inlet hose.
4. Remove the intake manifold.
5. Raise vehicle on hoist.
6. Disconnect exhaust pipe at exhaust manifolds.
7. Remove the structural dust cover using sequence shown.
8. Drain engine oil and remove oil filter.
9. Position suitable jack under engine.
10. Remove both left and right side engine mount through bolts.
11. Raise engine to provide clearance to remove oil pan.
12. Place blocks of wood between

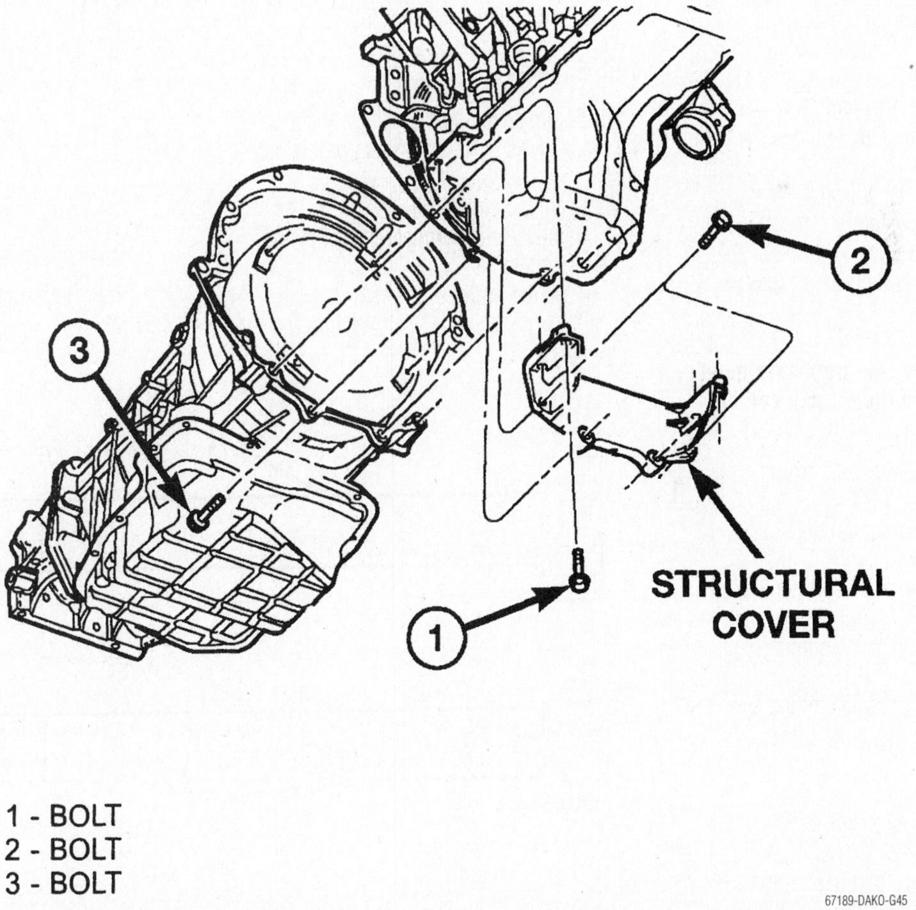

Structural cover torque sequence—3.7L and 4.7L engines

engine brackets and lower mounts to provide stability to engine.

➡Do not pry on oil pan or oil pan gasket. Gasket is mounted to engine and does not come out with oil pan.

13. Remove the oil pan mounting bolts and oil pan.
14. Unbolt oil pump pickup tube and remove tube and oil pan gasket from engine.

To install:

15. Clean the oil pan gasket mating surface of the bedplate and oil pan.
16. Position the oil pan gasket and pickup tube with new o-ring. Install the mounting bolt and nuts. Tighten bolt and nuts to 28 Nm (20 ft. lbs.).
17. Position the oil pan and install the mounting bolts. Tighten the mounting bolts to 15 Nm (11 ft. lbs.) in the sequence shown.
18. Raise the engine and remove the blocks of wood.
19. Lower engine and install both the left and right side engine mount through bolts. Tighten the nuts to 68 Nm (50 ft. lbs.).
20. Remove jack and install oil filter.
21. Install structural dust cover.
22. Install exhaust pipe onto exhaust manifolds.
23. Lower vehicle.
24. Install intake manifold.
25. Install throttle body resonator and air inlet hose.
26. Install upper fan shroud.
27. Fill cooling system.
28. Fill engine oil.
29. Start engine and check for leaks.

4-WHEEL DRIVE

➡On 4wd vehicles, the front axle must be removed before the oil pan can be removed.

1. Remove the front axle from vehicle.
2. Remove the structural dust cover using sequence shown.
3. Drain the engine oil and remove oil filter.
4. Remove the oil pan mounting bolts and oil pan.
5. Unbolt oil pump pickup tube and remove tube and oil pan gasket from engine.

To install:

6. Clean the oil pan gasket mating surface of the bedplate and oil pan.
7. Position the oil pan gasket and pickup tube with new o-ring. Install the mounting bolt and nuts. Tighten bolt and nuts to 28 Nm (20 ft. lbs.).
8. Position the oil pan and install the mounting bolts. Tighten the mounting bolts to 15 Nm (11 ft. lbs.) in the sequence shown.
9. Install structural dust cover.
10. Install oil filter.
11. Install front axle.
12. Lower vehicle.
13. Fill engine oil.
14. Start engine check for leaks.

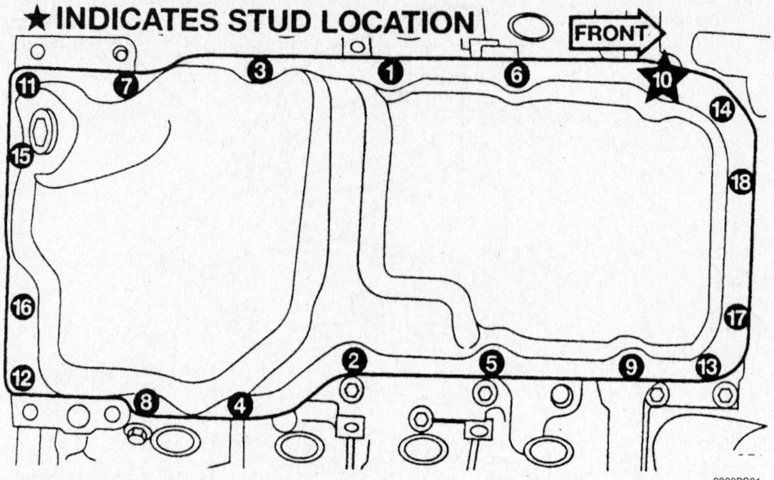

Oil pan mounting bolt tightening sequence—4.7L engine

5.9L Engines

1. Before servicing the vehicle, refer to the precautions in the beginning of this section.
2. Drain the engine oil.
3. Remove or disconnect the following:
   - Oil filter
   - Starter motor
   - Cooler lines
   - Oil level sensor connector
   - Heated Oxygen (HO_2S) sensor connector
   - Exhaust Y-pipe
   - Oil pan

To install:

4. Install or connect the following:
   - Oil pan, using a new gasket. Tighten the bolts to 18 ft. lbs. (24 Nm).
   - Exhaust Y-pipe
   - Heated Oxygen (HO_2S) sensor connector
   - Oil level sensor connector
   - Cooler lines
   - Starter motor
   - Oil filter
5. Fill the crankcase.
6. Start the engine and check for leaks.

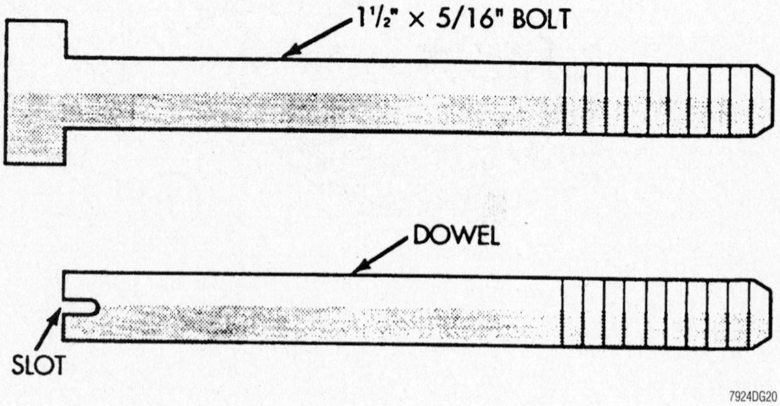

Oil pan alignment dowels—3.9L, 5.9L engines

Oil Pump

REMOVAL & INSTALLATION

2.5L Engine

1. Before servicing the vehicle, refer to the precautions in the beginning of this section.
2. Drain the engine oil.
3. Remove or disconnect the following:
   - Negative battery cable
   - Oil pan
   - Oil pump and pickup tube

➡If the oil pump is not to be serviced, do not disturb the position of the oil inlet tube and strainer assembly in the pump body. If the tube is moved within the pump body, a replacement tube and strainer assembly must be installed to assure an airtight seal.

To install:

4. Install or connect the following:
   - Oil pump. Tighten the mounting bolts to 17 ft. lbs. (23 Nm).
   - Oil pan
   - Negative battery cable
5. Fill the crankcase to the correct level.
6. Start the engine and check for leaks.

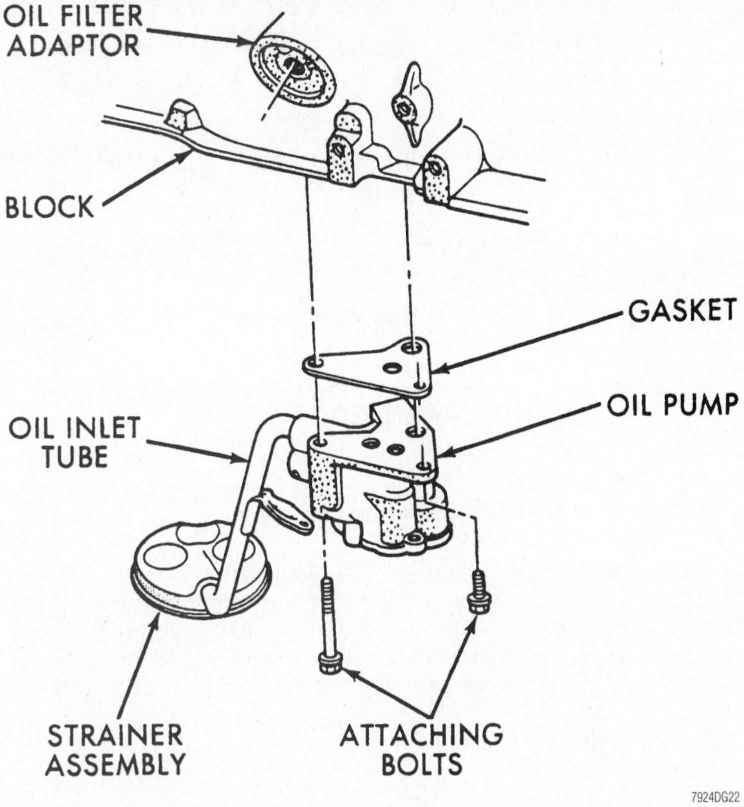

Exploded view of the oil pump assembly—2.5L engine

3.7L Engine

1. Before servicing the vehicle, refer to the precautions in the beginning of this section.
2. Remove or disconnect the following:
   - Oil Pan
   - Timing chain cover
   - Timing chains and tensioners
   - Oil pump
3. Installation is the reverse of removal. Torque the pump bolts, in sequence, to 21 ft. lbs. (28 Nm),

4.7L Engine

1. Before servicing the vehicle, refer to the precautions in the beginning of this section.
2. Drain the engine oil.
3. Remove or disconnect the following:
   - Negative battery cable
   - Oil pan
   - Oil pump pick-up tube
   - Timing chains and tensioners
   - Oil pump

To install:

4. Install or connect the following:
   - Oil pump. Tighten the bolts to 21 ft. lbs. (28 Nm).
   - Timing chains and tensioners

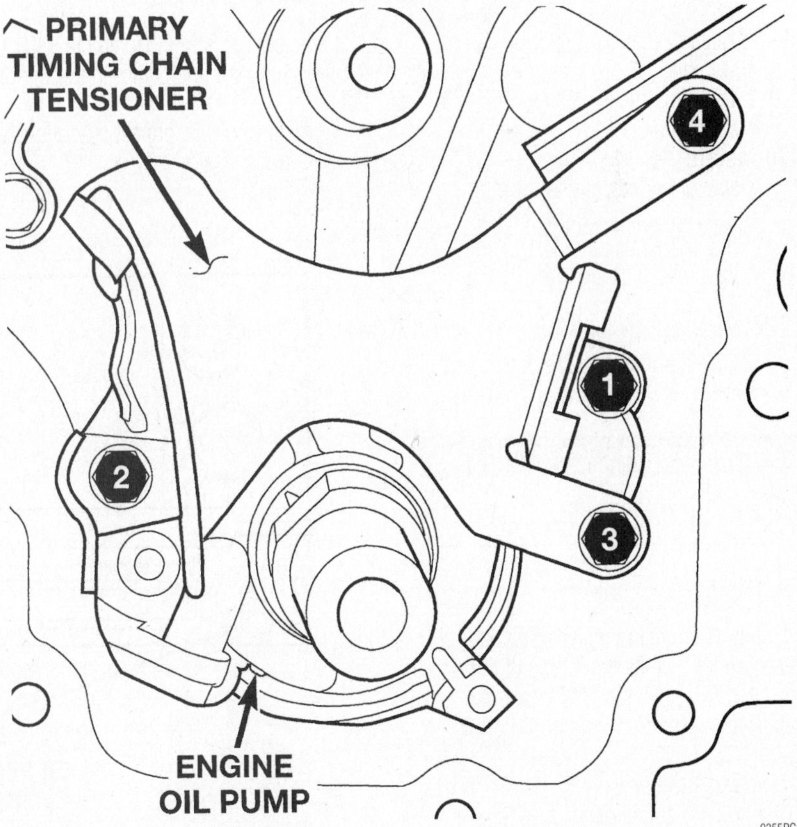

Oil pump bolt torque sequence—3.7L

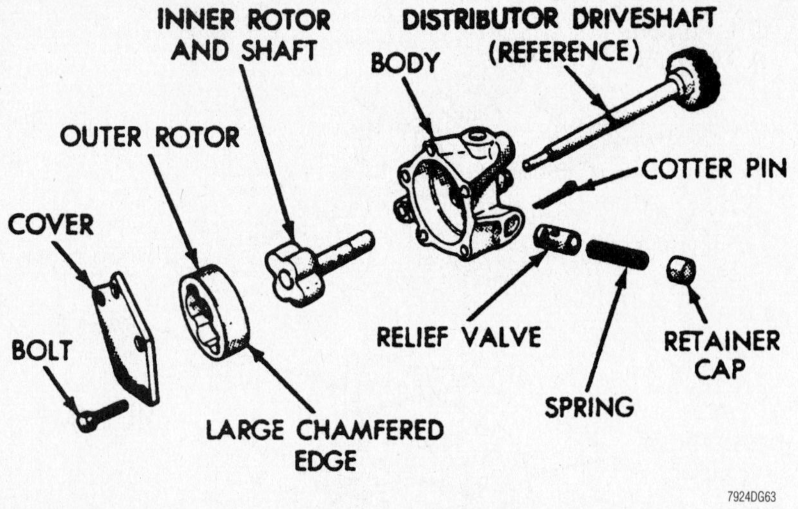

Exploded view of the oil pump assembly—3.9L and 5.9L engines

- Oil pump pick-up tube
- Oil pan
- Negative battery cable

5. Fill the crankcase to the correct level.
6. Start the engine and check for leaks.

3.9L and 5.9L Engines

1. Before servicing the vehicle, refer to the precautions in the beginning of this section.
2. Drain the engine oil.
3. Remove or disconnect the following:
   - Negative battery cable
   - Oil pan
   - Oil pump pick-up tube
   - Oil pump

To install:

4. Install or connect the following:
   - Oil pump. Tighten the bolts to 30 ft. lbs. (41 Nm).
   - Oil pump pick-up tube
   - Oil pan
   - Negative battery cable
5. Fill the crankcase to the correct level.
6. Start the engine and check for leaks.

Rear Main Seal

REMOVAL & INSTALLATION

2.5L Engine

1. Before servicing the vehicle, refer to the precautions in the beginning of this section.
2. Remove or disconnect the following:
   - Transmission
   - Clutch, if equipped
   - Flywheel
   - Rear main seal

To install:

3. Install the rear main seal so that it is flush with the cylinder block.
4. Install or connect the following:
   - Flywheel. Use new bolts and tighten to 50 ft. lbs. (68 Nm) plus 60 degrees
   - Clutch, if equipped
   - Transmission
5. Start the engine and check for leaks.

3.7L and 4.7L Engine

1. Before servicing the vehicle, refer to the precautions in the beginning of this section.
2. Remove or disconnect the following:
   - Transmission
   - Flexplate
3. Thread Oil Seal Remover 8506 into the rear main seal as far as possible and remove the rear main seal.

To install:

4. Install or connect the following:
   - Seal Guide 8349-2 onto the crankshaft
   - Rear main seal on the seal guide
   - Rear main seal, using the Crankshaft Rear Oil Seal Installer 8349 and Driver Handle C-4171; tap it into place until the installer is flush with the cylinder block
   - Flexplate. Tighten the bolts to 45 ft. lbs. (60 Nm).
   - Transmission
5. Start the engine and check for leaks.

3.9L and 5.9L Engines

1. Before servicing the vehicle, refer to the precautions in the beginning of this section.
2. Drain the engine oil.
3. Remove or disconnect the following:
   - Oil pan
   - Oil pump
   - Rear main bearing cap
4. Loosen the other main bearing cap bolts for clearance and remove the rear main seal halves.

To install:

5. Install or connect the following:
   - New upper seal half to the cylinder block
   - New lower seal half to the bearing cap
6. Apply sealant to the rear main bearing cap.

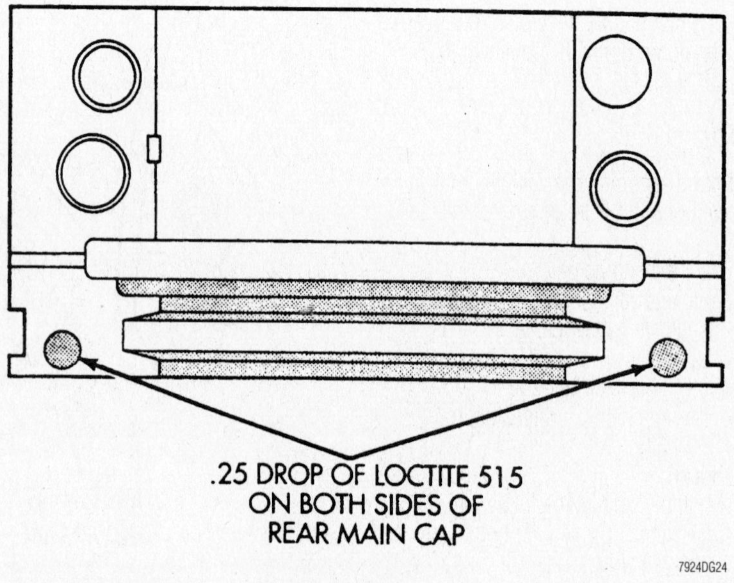

Sealant application locations —3.9L and 5.9L engines

7. Install or connect the following:
    - Rear main bearing cap. Tighten **all** main bearing cap bolts to 85 ft. lbs. (115 Nm).
    - Oil pump and oil pan
8. Fill the engine.
9. Start the engine and check for leaks.

Timing Chain, Sprockets, Front Cover and Seal

REMOVAL & INSTALLATION

2.5L Engine

1. Before servicing the vehicle, refer to the precautions in the beginning of this section.
2. Remove or disconnect the following:
    - Negative battery cable
    - Accessory drive belt
    - Cooling fan and shroud
    - Crankshaft damper
    - Front crankshaft seal
    - Accessory brackets
    - Front cover
    - Oil slinger
3. Rotate the crankshaft so that the timing marks are aligned.
4. Remove the timing chain and sprockets.

To install:

5. Turn the timing chain tensioner lever to the unlock (down) position. Pull the tensioner block toward the tensioner lever to compress the spring. Hold the block and turn the tensioner lever to the lock (up) position.
6. Install the timing chain and sprockets with the timing marks aligned. Tighten the camshaft sprocket bolt to 80 ft. lbs. (108 Nm) for 2.5L engines or to 50 ft. lbs. (68 Nm) for 4.0L engines.
7. Release the timing chain tensioner.
8. Install or connect the following:
    - Oil slinger
    - New front crankshaft seal to the front cover
    - Front cover, using a new gasket
    - Timing Case Cover Alignment and Seal Installation Tool 6139 in the crankshaft opening to center the front cover
9. Tighten the front cover bolts as follows:
    a. Step 1: Cover-to-block ¼ inch bolts to 60 inch lbs. (7 Nm).
    b. Step 2: Cover-to-block 5⁄16 inch bolts to 16 ft. lbs. (22 Nm).
    c. Step 3: Oil pan-to-cover ¼ inch bolts to 85 inch lbs. (9.5 Nm).
    d. Step 4: Oil pan-to-cover 5⁄16 inch bolts to 11 ft. lbs. (15 Nm).
10. Install or connect the following:

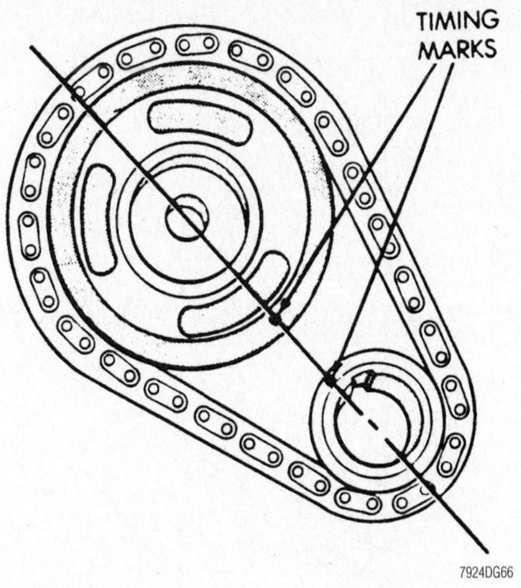

Timing mark alignment—2.5L engine

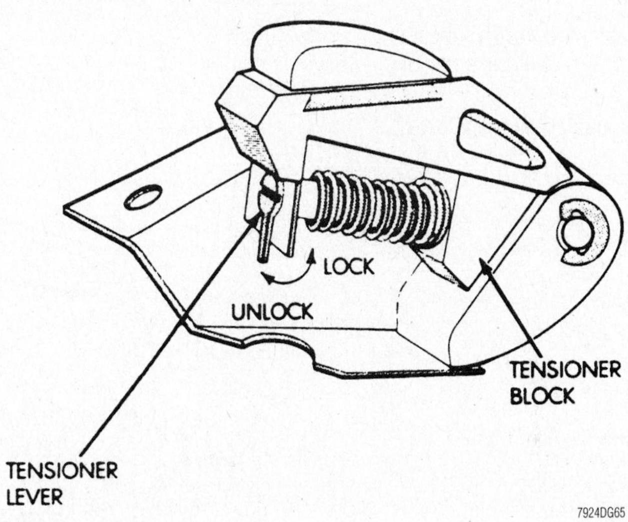

Timing chain tensioner—2.5L engines

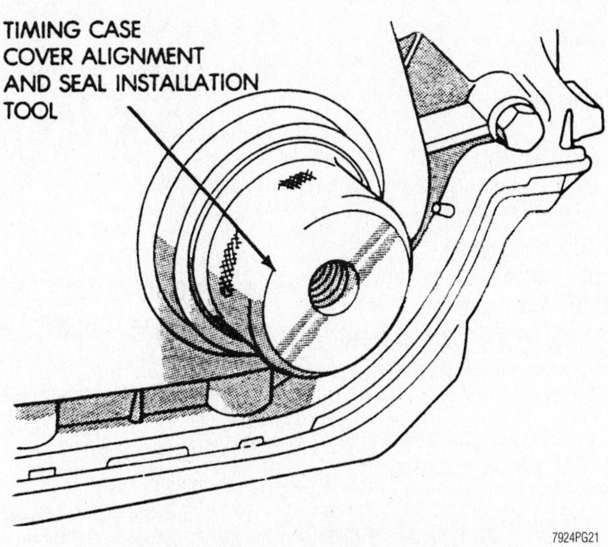

Timing Case Cover Alignment and Seal Installation Tool 6139—2.5L engine

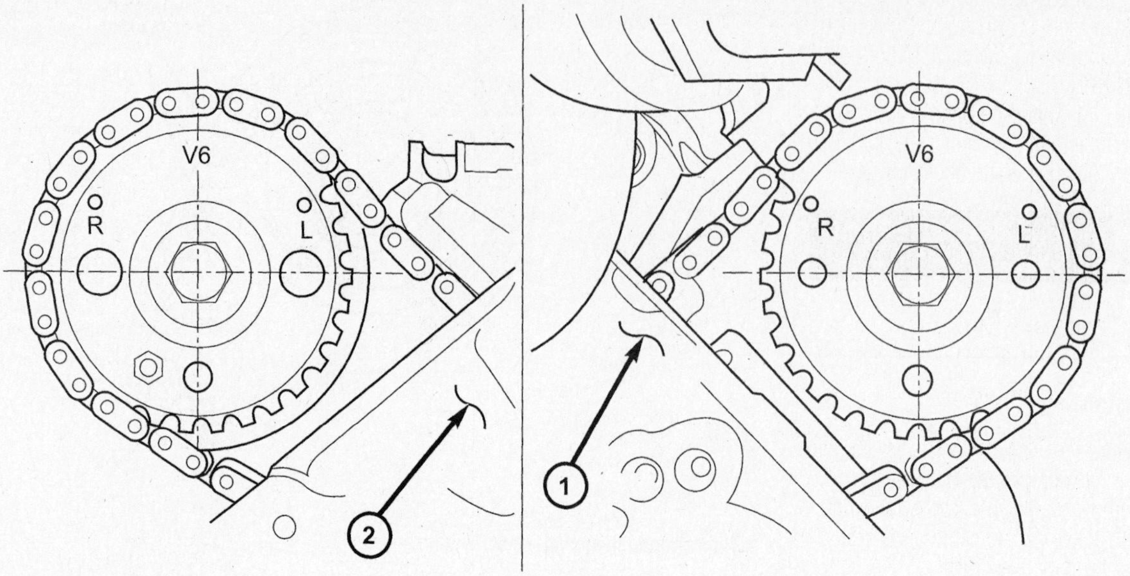

1 - LEFT CYLINDER HEAD
2 - RIGHT CYLINDER HEAD

9355PG09

Camshaft sprocket timing marks—3.7L

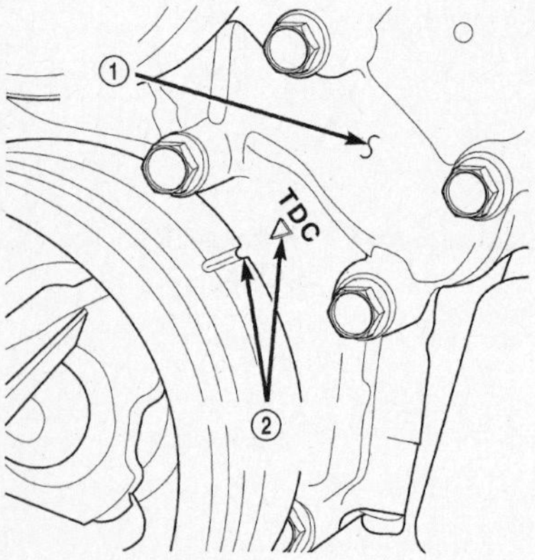

1 - TIMING CHAIN COVER
2 - CRANKSHAFT TIMING MARKS

9355PG10

Crankshaft timing marks—3.7L

- Accessory brackets
- Crankshaft damper. Tighten the bolt to 80 ft. lbs. (108 Nm).
- Cooling fan and shroud
- Accessory drive belt
- Negative battery cable

11. Start the engine and check for leaks.

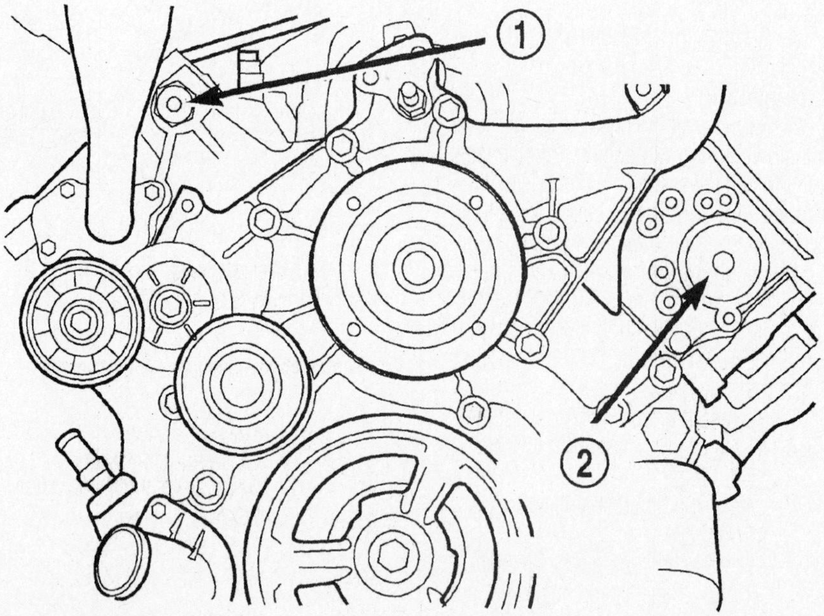

1 - RIGHT CYLINDER HEAD ACCESS PLUG
2 - LEFT CYLINDER HEAD ACCESS PLUG

9355PG11

Cylinder head access plugs—3.7L

3.7L Engines

1. Before servicing the vehicle, refer to the precautions in the beginning of this section.
2. Drain the cooling system.
3. Remove or disconnect the following:
   - Negative battery cable
   - Valve covers
   - Radiator fan
4. Rotate the crankshaft so that the crankshaft timing mark aligns with the Top Dead Center (TDC) mark on the front cover, and the **V6** marks on the camshaft sprockets are at 12 o'clock.
   - Power steering pump
   - Access plugs from the cylinder heads

- Oil fill housing
- Crankshaft damper

5. Compress the primary timing chain tensioner and install a lockpin.
6. Remove the secondary timing chain tensioners.
7. Hold the left camshaft with adjustable pliers and remove the sprocket and chain. Rotate the **left** camshaft 15 degrees **clockwise** to the neutral position.
8. Hold the right camshaft with adjustable pliers and remove the camshaft sprocket. Rotate the **right** camshaft 45 degrees **counterclockwise** to the neutral position.
9. Remove the primary timing chain and sprockets.

To install:

10. Use a small prytool to hold the ratchet pawl and compress the secondary timing chain tensioners in a vise and install locking pins.

➡**The black bolts fasten the guide to the engine block and the silver bolts fasten the guide to the cylinder head.**

11. Install or connect the following:
    - Secondary timing chain guides. Tighten the bolts to 21 ft. lbs. (28 Nm).

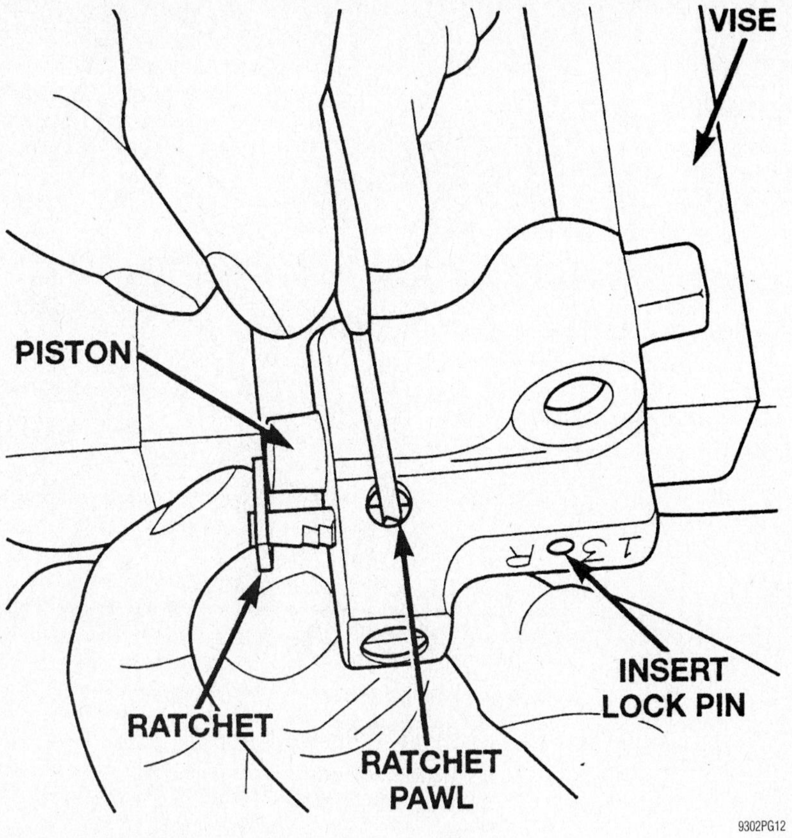

Secondary timing chain tensioner preparation—3.7L engine

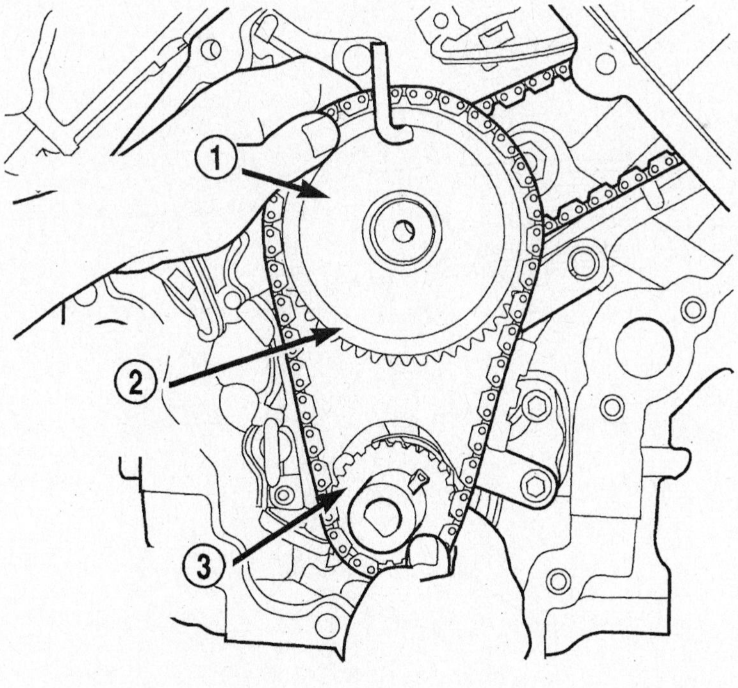

1 - SPECIAL TOOL 8429
2 - PRIMARY CHAIN IDLER SPROCKET
3 - CRANKSHAFT SPROCKET

Installing the idler gear and timing chains—3.7L

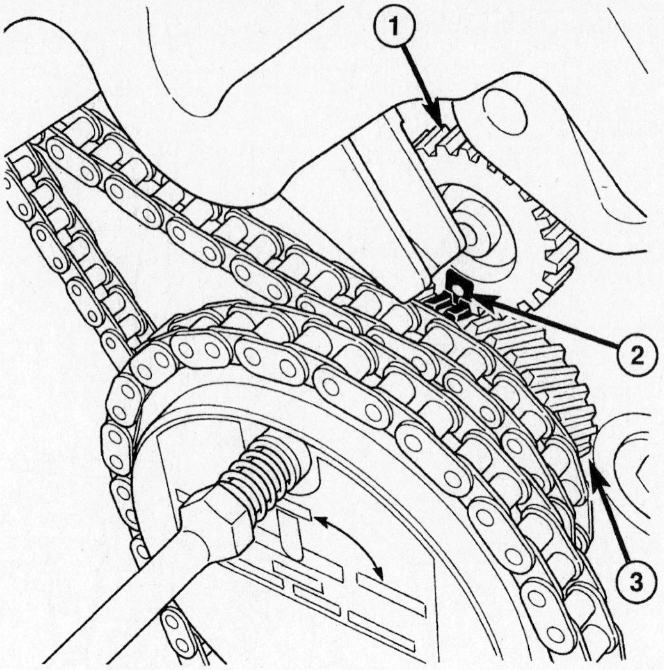

Counterbalance shaft timing marks—3.7L

Tightening the left side camshaft sprocket—3.7L

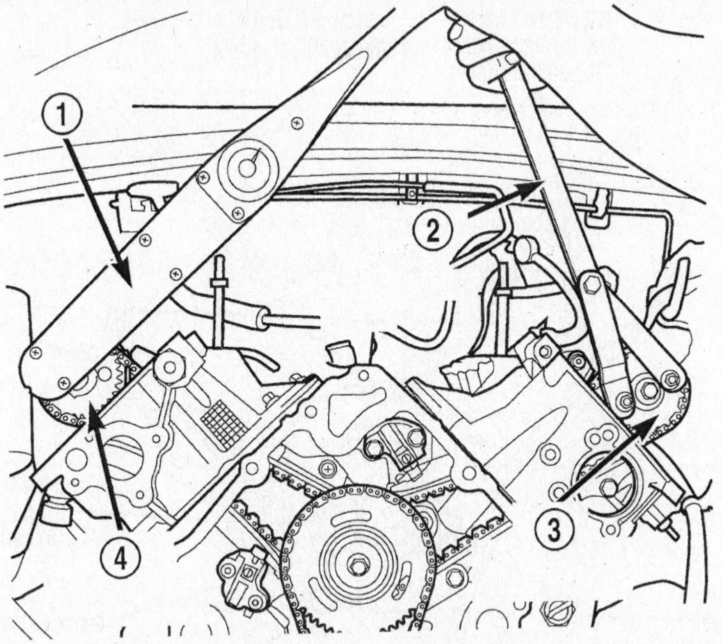

Tightening the right side camshaft sprocket—3.7L

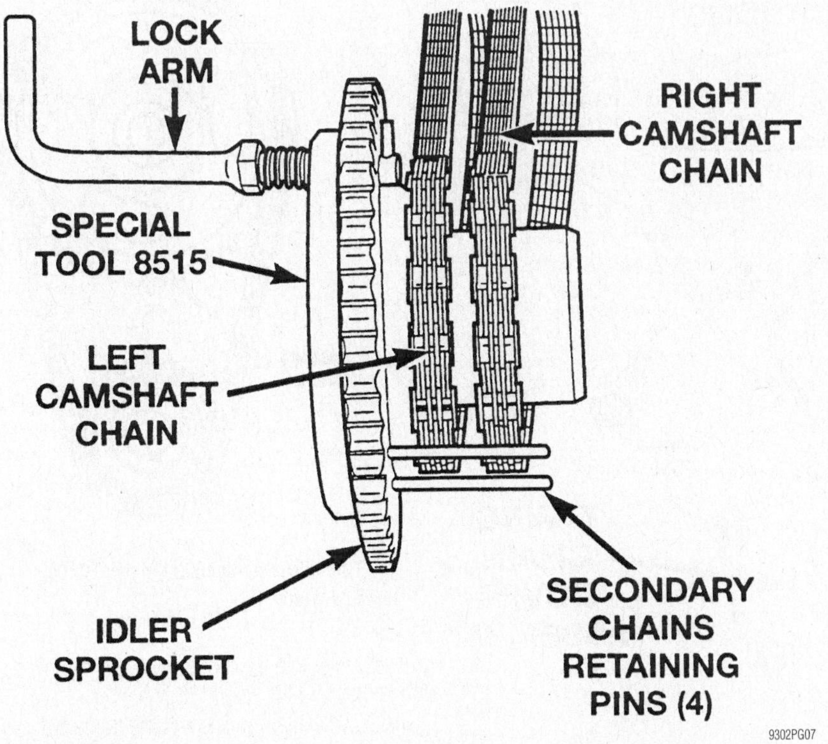

Use the Timing Chain Locking tool to lock the timing chains on the idler gear—3.7L engine

- Secondary timing chains to the idler sprocket so that the double plated links on each chain are visible through the slots in the primary idler sprocket

12. Lock the secondary timing chains to the idler sprocket with Timing Chain Locking tool as shown.

13. Align the primary chain double plated links with the idler sprocket timing mark and the single plated link with the crankshaft sprocket timing mark.

14. Install the primary chain and sprockets. Tighten the idler sprocket bolt to 25 ft. lbs. (34 Nm).

15. Align the secondary chain single plated links with the timing marks on the secondary sprockets. Align the dot at the **L** mark on the left sprocket with the plated link on the left chain and the dot at the **R** mark on the right sprocket with the plated link on the right chain.

16. Rotate the camshafts back from the neutral position and install the camshaft sprockets.

17. Remove the secondary chain locking tool.

18. Remove the primary and secondary timing chain tensioner locking pins.

19. Hold the camshaft sprockets with a

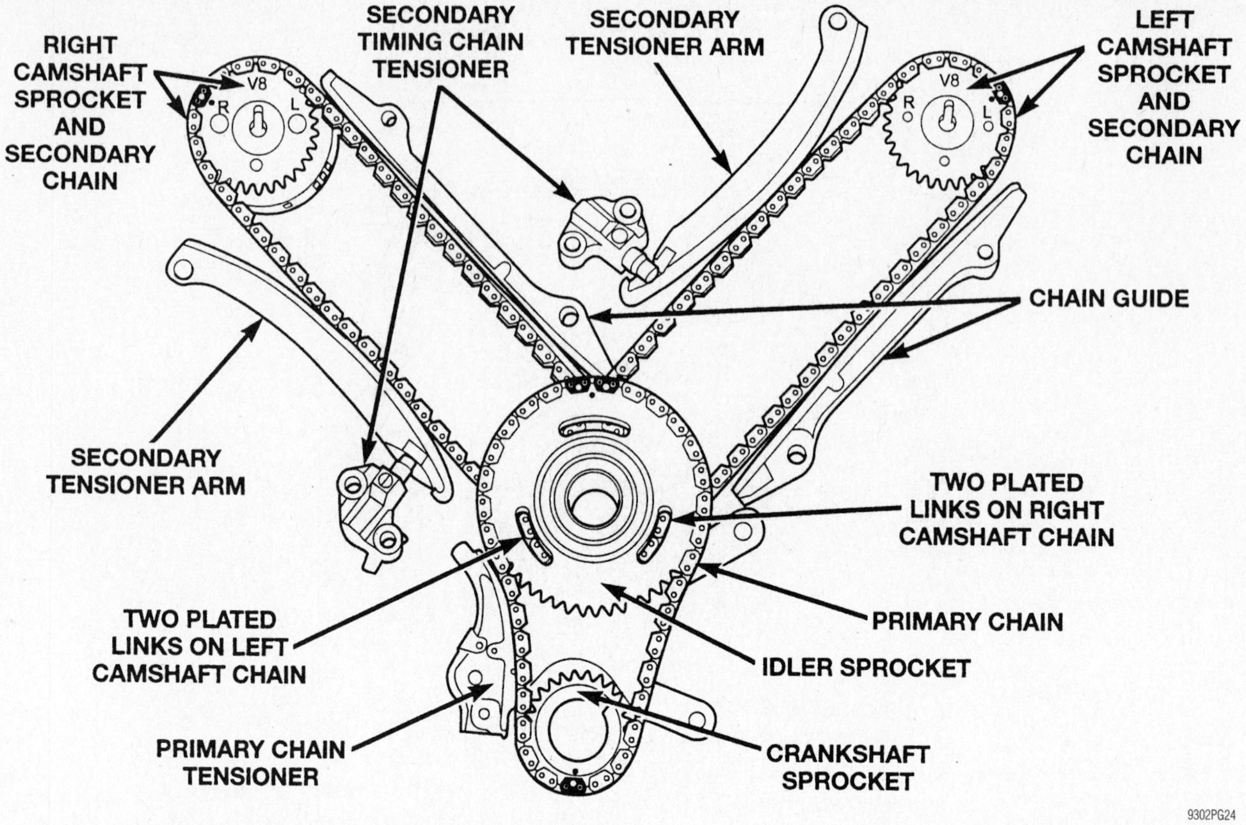

Timing chain system and alignment marks—3.7L engine

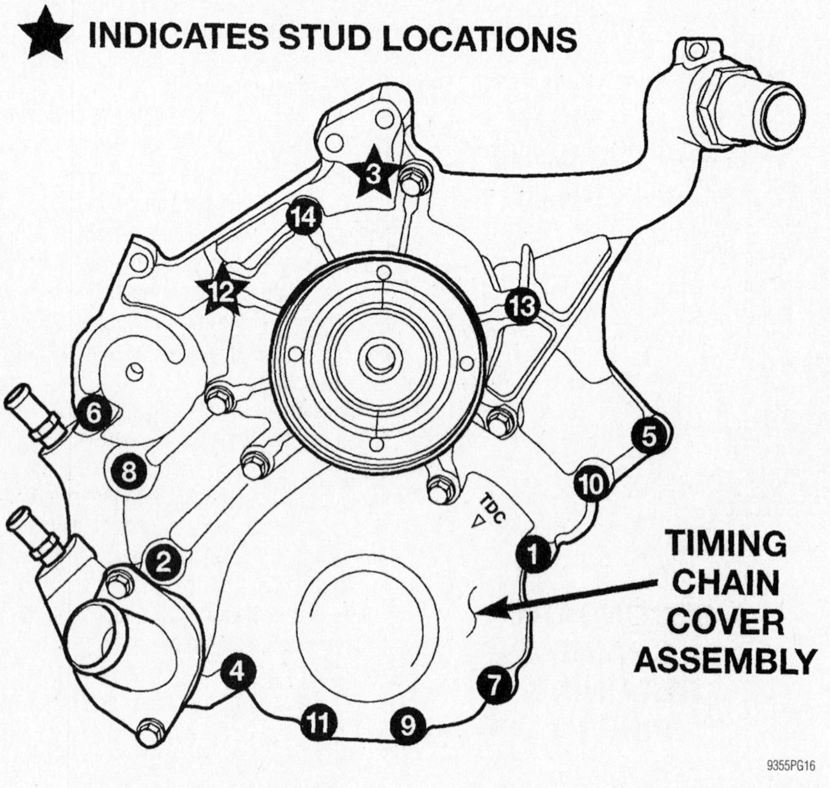

Timing cover bolt torque sequence—3.7L

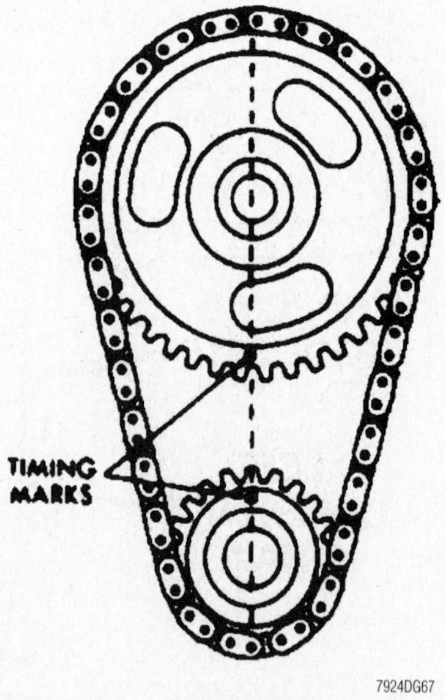

Timing chain alignment marks—3.9L, 5.9L engines

spanner wrench and tighten the retaining bolts to 90 ft. lbs. (122 Nm).

20. Install or connect the following:
   - Front cover. Tighten the bolts, in sequence, to 40 ft. lbs. (54 Nm).
   - Front crankshaft seal

- Cylinder head access plugs
- A/C compressor
- Alternator
- Accessory drive belt tensioner. Tighten the bolt to 40 ft. lbs. (54 Nm).
- Oil fill housing
- Crankshaft damper. Tighten the bolt to 130 ft. lbs. (175 Nm).
- Power steering pump
- Lower radiator hose
- Heater hoses
- Accessory drive belt
- Engine cooling fan and shroud
- Camshaft Position (CMP) sensor
- Valve covers
- Negative battery cable

21. Fill and bleed the cooling system.
22. Start the engine, check for leaks and repair if necessary.

3.9L Engine

1. Before servicing the vehicle, refer to the precautions in the beginning of this section.
2. Drain the cooling system.
3. Remove or disconnect the following:
   - Negative battery cable
   - Accessory drive belt
   - Radiator
   - Cooling fan
   - Water pump
   - Crankshaft pulley
   - Front crankshaft seal
   - Front cover
   - Timing chain and gears

To install:

4. Install or connect the following:
   - Timing chain and gears. Align the timing marks and tighten the camshaft sprocket bolt to 35 ft. lbs. (47 Nm).
   - Front cover. Tighten the bolts to 30 ft. lbs. (41 Nm).
   - Front crankshaft seal
   - Crankshaft pulley. Tighten the bolt to 135 ft. lbs. (183 Nm).

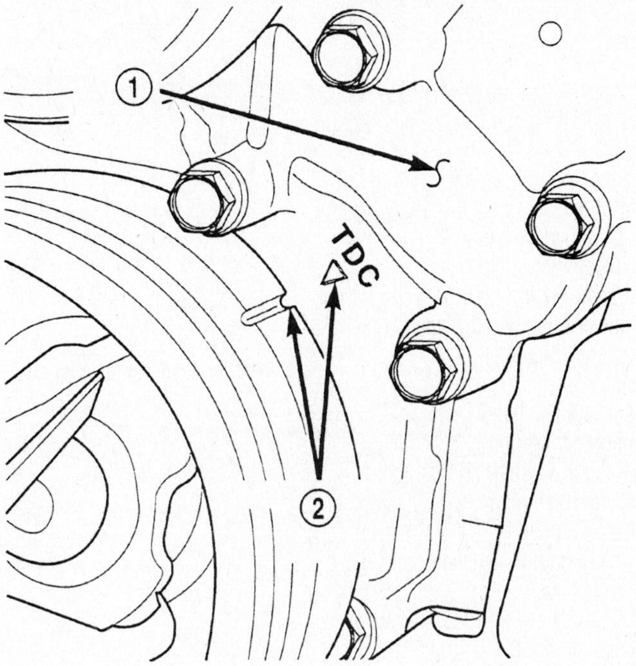

Crankshaft timing marks—4.7L engine

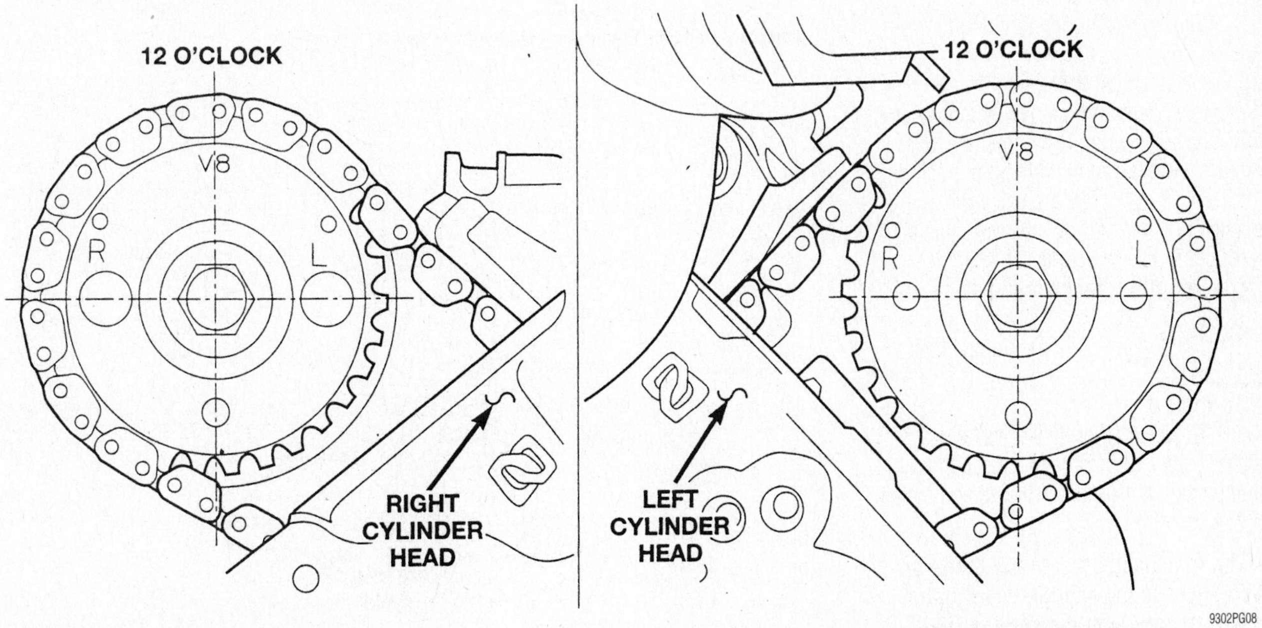

Camshaft positioning—4.7L engine

- Water pump
- Cooling fan
- Radiator
- Accessory drive belt
- Negative battery cable

5. Fill the cooling system.
6. Start the engine and check for leaks.

4.7L Engine

1. Before servicing the vehicle, refer to the precautions in the beginning of this section.
2. Drain the cooling system.
3. Remove or disconnect the following:
   - Negative battery cable
   - Valve covers
   - Camshaft Position (CMP) sensor
   - Engine cooling fan and shroud
   - Accessory drive belt
   - Heater hoses
   - Lower radiator hose
   - Power steering pump
4. Rotate the crankshaft so that the crankshaft timing mark aligns with the Top Dead Center (TDC) mark on the front cover, and the **V8** marks on the camshaft sprockets are at 12 o'clock.
5. Remove or disconnect the following:
   - Crankshaft damper
   - Oil fill housing
   - Accessory drive belt tensioner
   - Alternator
   - A/C compressor
   - Front cover
   - Front crankshaft seal
   - Cylinder head access plugs
   - Secondary timing chain guides
6. Compress the primary timing chain tensioner and install a lockpin.
7. Remove the secondary timing chain tensioners.
8. Hold the left camshaft with adjustable pliers and remove the sprocket and chain. Rotate the **left** camshaft 15 degrees **clockwise** to the neutral position.
9. Hold the right camshaft with adjustable pliers and remove the camshaft sprocket. Rotate the **right** camshaft 45 degrees **counterclockwise** to the neutral position.
10. Remove the primary timing chain and sprockets.

To install:

11. Use a small prytool to hold the ratchet pawl and compress the secondary timing chain tensioners in a vise and install locking pins.

➡The black bolts fasten the guide to the engine block and the silver bolts fasten the guide to the cylinder head.

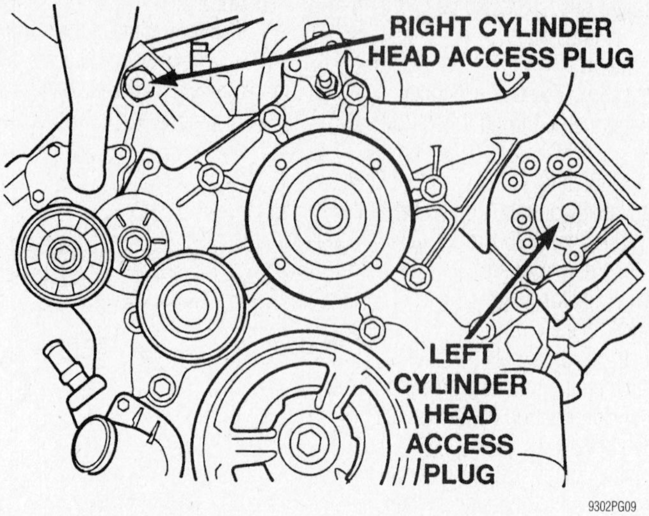

Cylinder head access plug locations—4.7L engine

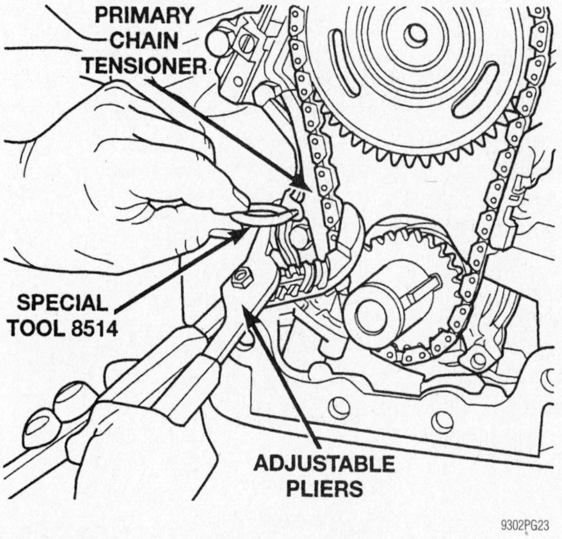

Compress and lock the primary chain tensioner—4.7L engine

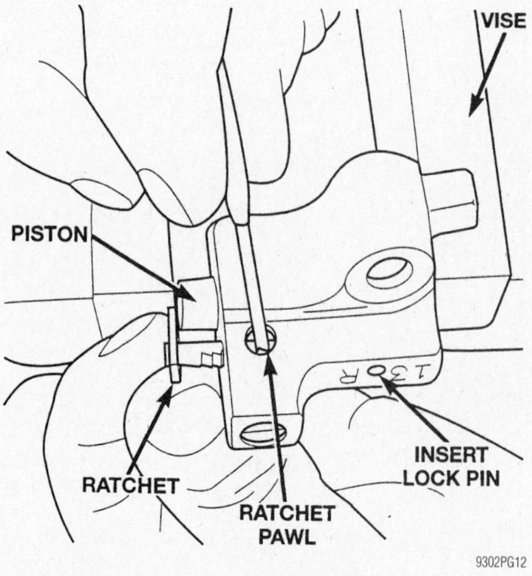

Secondary timing chain tensioner preparation—4.7L engine

12. Install or connect the following:
- Secondary timing chain guides. Tighten the bolts to 21 ft. lbs. (28 Nm).
- Secondary timing chains to the idler sprocket so that the double plated links on each chain are visible through the slots in the primary idler sprocket

13. Lock the secondary timing chains to the idler sprocket with Timing Chain Locking tool 8515 as shown.

14. Align the primary chain double plated links with the idler sprocket timing mark and the single plated link with the crankshaft sprocket timing mark.

15. Install the primary chain and sprockets. Tighten the idler sprocket bolt to 25 ft. lbs. (34 Nm).

16. Align the secondary chain single plated links with the timing marks on the secondary sprockets. Align the dot at the **L** mark on the left sprocket with the plated link on the left chain and the dot at the **R** mark on the right sprocket with the plated link on the right chain.

17. Rotate the camshafts back from the neutral position and install the camshaft sprockets.

18. Remove the secondary chain locking tool.

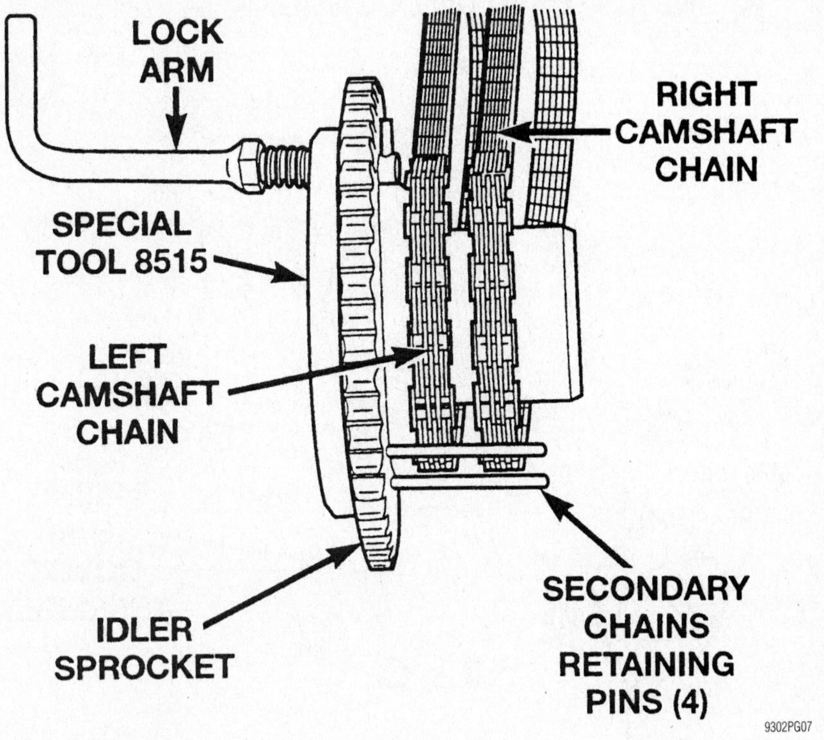

Use the Timing Chain Locking tool to lock the timing chains on the idler gear—4.7L engine

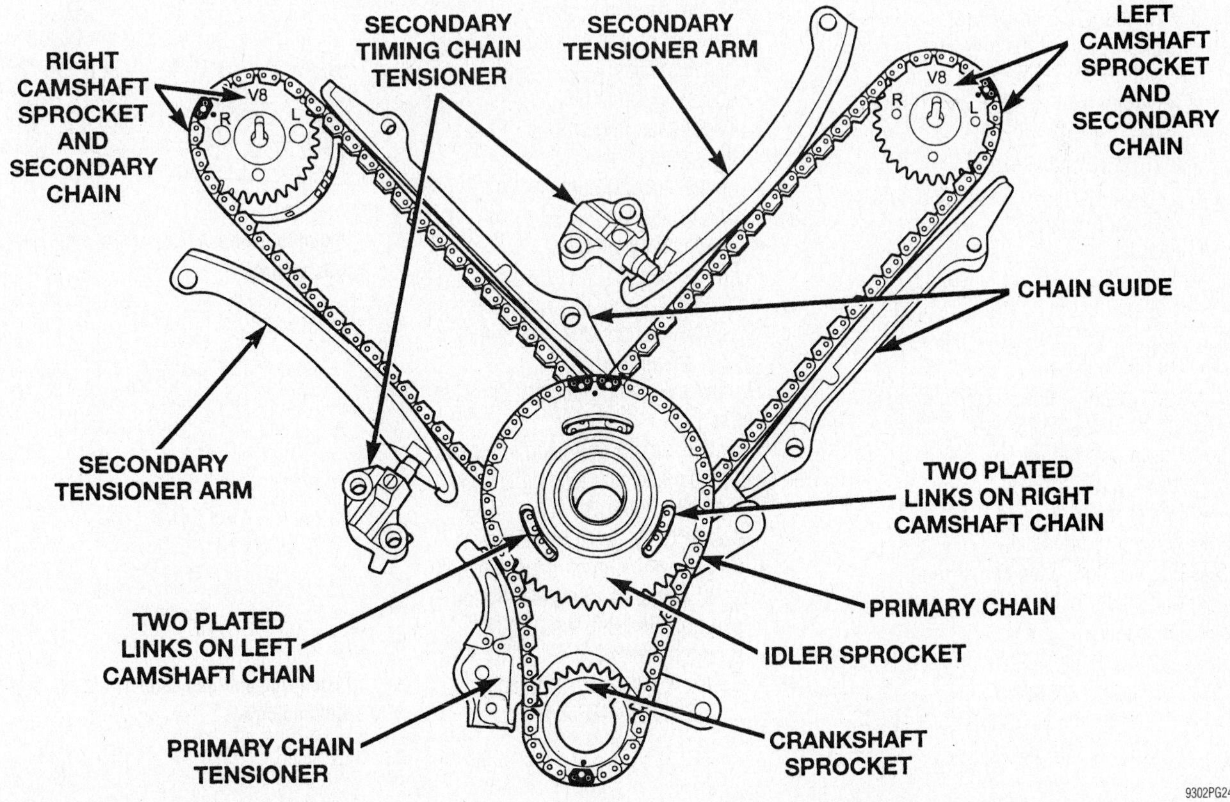

Timing chain system and alignment marks—4.7L engine

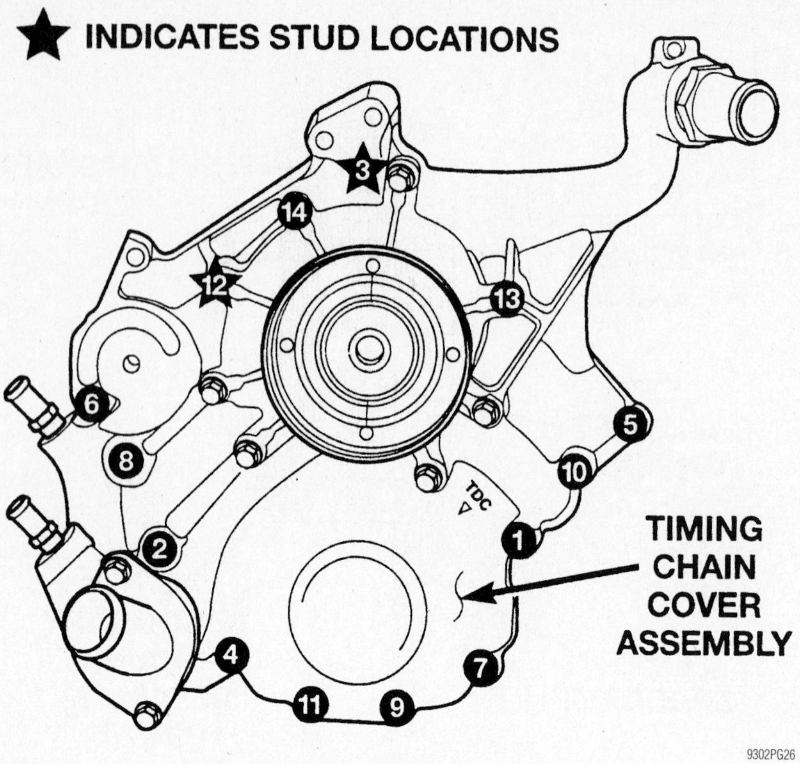

Timing chain cover bolt torque sequence—4.7L engine

19. Remove the primary and secondary timing chain tensioner locking pins.
20. Hold the camshaft sprockets with a spanner wrench and tighten the retaining bolts to 90 ft. lbs. (122 Nm).
21. Install or connect the following:
    - Front cover. Tighten the bolts, in sequence, to 40 ft. lbs. (54 Nm).
    - Front crankshaft seal
    - Cylinder head access plugs
    - A/C compressor
    - Alternator
    - Accessory drive belt tensioner. Tighten the bolt to 40 ft. lbs. (54 Nm).
    - Oil fill housing
    - Crankshaft damper. Tighten the bolt to 130 ft. lbs. (175 Nm).
    - Power steering pump
    - Lower radiator hose
    - Heater hoses
    - Accessory drive belt
    - Engine cooling fan and shroud
    - Camshaft Position (CMP) sensor
    - Valve covers
    - Negative battery cable
22. Fill the cooling system.
23. Start the engine and check for leaks.

5.9L Engines

1. Before servicing the vehicle, refer to the precautions in the beginning of this section.
2. Drain the cooling system.
3. Remove or disconnect the following:
    - Negative battery cable
    - Accessory drive belt
    - Cooling fan and shroud
    - Water pump
    - Power steering pump
    - Crankshaft damper
    - Front crankshaft seal
    - Front cover
4. Rotate the crankshaft so that the camshaft sprocket and crankshaft sprocket timing marks are aligned.
5. Remove the timing chain and sprockets.

To install:

6. Install the timing chain and sprockets with the timing marks aligned. Tighten the camshaft sprocket bolt to 50 ft. lbs. (68 Nm).
7. Install or connect the following:
    - Front cover. Tighten the cover bolts to 30 ft. lbs. (41 Nm) and the oil pan bolts to 18 ft. lbs. (24 Nm).
    - Front crankshaft seal
    - Crankshaft damper. Tighten the bolt to 135 ft. lbs. (183 Nm).
    - Power steering pump
    - Water pump
    - Cooling fan and shroud
    - Accessory drive belt
    - Negative battery cable
8. Fill the cooling system.
9. Start the engine and check for leaks.

Piston and Ring

POSITIONING

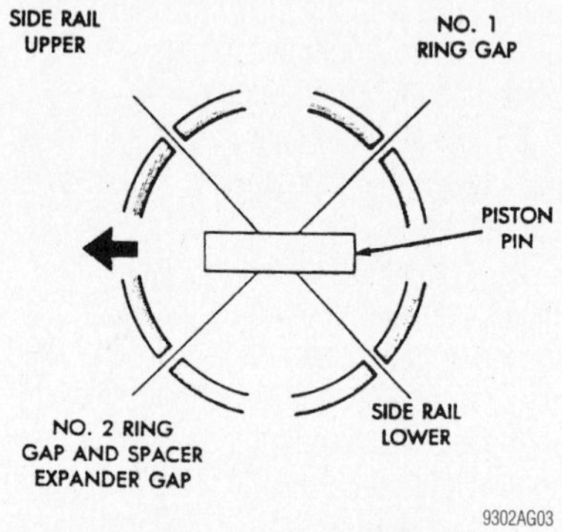

Piston ring end-gap spacing. Position raised "F" on piston towards front of engine—3.7L and 4.7L engines

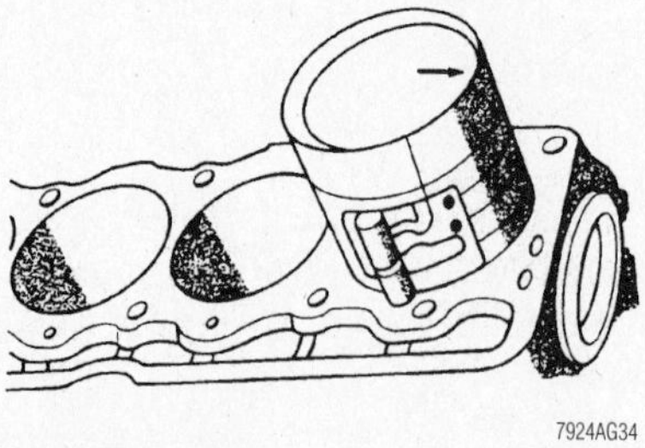

Piston to engine positioning—2.5L, 3.9L, 5.9L engines

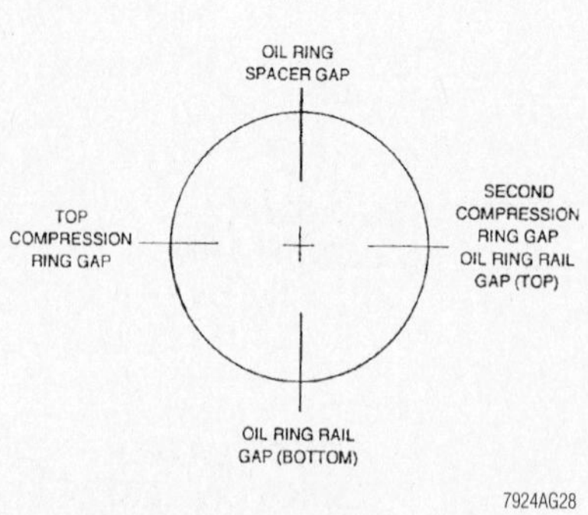

Piston ring end-gap spacing—2.5L, 3.9L, 5.9L engines

GASOLINE FUEL SYSTEM

Fuel System Service Precautions

Safety is the most important factor when performing not only fuel system maintenance but any type of maintenance. Failure to conduct maintenance and repairs in a safe manner may result in serious personal injury or death. Maintenance and testing of the vehicle's fuel system components can be accomplished safely and effectively by adhering to the following rules and guidelines.

- To avoid the possibility of fire and personal injury, always disconnect the negative battery cable unless the repair or test procedure requires that battery voltage be applied.
- Always relieve the fuel system pressure prior to detaching any fuel system component (injector, fuel rail, pressure regulator, etc.), fitting or fuel line connection. Exercise extreme caution whenever relieving fuel system pressure to avoid exposing skin, face and eyes to fuel spray. Please be advised that fuel under pressure may penetrate the skin or any part of the body that it contacts.
- Always place a shop towel or cloth around the fitting or connection prior to loosening to absorb any excess fuel due to spillage. Ensure that all fuel spillage (should it occur) is quickly removed from engine surfaces. Ensure that all fuel soaked cloths or towels are deposited into a suitable waste container.
- Always keep a dry chemical (Class B) fire extinguisher near the work area.
- Do not allow fuel spray or fuel vapors to come into contact with a spark or open flame.
- Always use a back-up wrench when loosening and tightening fuel line connection fittings. This will prevent unnecessary stress and torsion to fuel line piping.
- Always replace worn fuel fitting O-rings with new. Do not substitute fuel hose or equivalent, where fuel pipe is installed.

Before servicing the vehicle, also make sure to refer to the precautions in the beginning of this section as well.

Fuel System Pressure

RELIEVING

2002

1. Remove the fuel filler cap.
2. Remove the fuel pump relay from the power distribution center.
3. Run the engine until it stalls.
4. Turn the key to **OFF**.

2003–05

1. Before servicing the vehicle, refer to the precautions in the beginning of this section.
2. Disconnect the negative battery cable.
3. Remove the fuel tank filler cap to release any fuel tank pressure.
4. Remove the fuel pump relay from the PDC.
5. Start and run the engine until it stops.
6. Unplug the connector from any injector and connect a jumper wire from either injector terminal to the positive battery terminal. Connect another jumper wire to the other terminal and momentarily touch the other end to the negative battery terminal.

WARNING

Just touch the jumper to the battery. Powering the injector for more than a few seconds will permanently damage it.

7. Place a rag below the quick-disconnect coupling at the fuel rail and disconnect it.

Fuel Filter

REMOVAL & INSTALLATION

These engines are equipped with a fuel filter/pressure regulator. This unit is not a regularly maintained unit and is serviced only when a DTC indicates a fault.

Fuel Pump

REMOVAL & INSTALLATION

2002–04

1. Before servicing the vehicle, refer to the precautions in the beginning of this section.
2. Relieve the fuel system pressure.
3. Remove or disconnect the following:
   - Negative battery cable
   - Fuel pump module harness connector
   - Fuel line
   - Fuel tank
   - Fuel pump module locknut
   - Fuel pump module

To install:

4. Install or connect the following:
   - Fuel pump module
   - Fuel pump module locknut
   - Fuel tank
   - Fuel line
   - Fuel pump module harness connector
   - Negative battery cable
5. Start the engine and check for leaks.

2005

1. Release fuel system pressure.
2. Raise vehicle.
3. Remove left rear tire/wheel.
4. Remove plastic fender liner in front of left-rear tire/wheel.

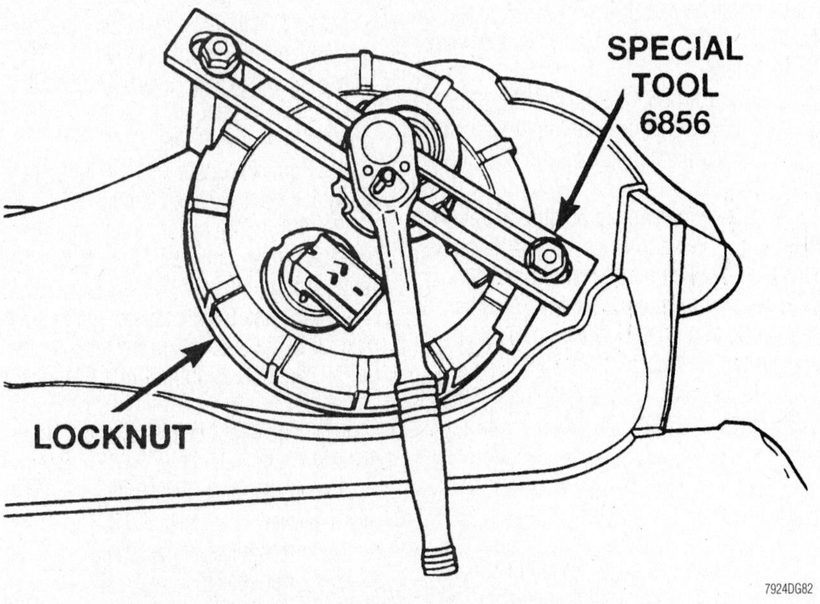

Fuel pump module locknut removal—2002–04 models

5. Thoroughly clean area around fuel fill fitting and rubber fuel fill hose at fuel tank.

6. Loosen clamp and disconnect rubber fuel fill hose at fuel tank fitting. Using an approved gas holding tank, drain fuel tank through this fitting.

7. Disconnect vent line from tank.

8. If equipped, remove fuel tank skid plate.

9. Disconnect NVLD, ORVR and EVAP lines at front of tank.

10. Disconnect electrical connector at NVLD pump.

11. Support tank with a hydraulic jack.

12. Remove two fuel tank strap nuts and remove both tank support straps.

13. Carefully lower tank a few inches and disconnect fuel pump module electrical connector at top of tank. To disconnect electrical connector: Push upward on red colored tab to unlock. Push on black colored tab while removing connector.

14. Disconnect fuel line at fuel pump module fitting by pressing on tabs at side of quick-connect fitting.

15. Continue to lower tank for removal.

16. Note rotational position of module before attempting removal. An indexing arrow is located on top of module for this purpose.

17. Position Special Tool 9340 into notches on outside edge of lockring.

18. Install ½ inch drive breaker bar to tool 9340.

19. Rotate breaker bar counter-clockwise to remove lockring.

20. Remove lockring. The module will spring up slightly when lockring is removed.

21. Remove module from fuel tank. Be careful not to bend float arm while removing.

To install:

22. Using a new seal (gasket), position fuel pump module into opening in fuel tank.

23. Position lockring over top of fuel pump module.

24. Rotate module until embossed alignment arrow points to center alignment mark. This step must be performed to prevent float from contacting side of fuel tank. Also be sure fuel fitting on top of pump module is pointed to drivers side of vehicle.

25. Install Special Tool 9340 to lockring.

26. Install breaker into Special Tool 9340.

27. Tighten lockring (clockwise) until all seven notches have engaged.

28. Install fuel tank.

29. Position fuel tank to hydraulic jack.

30. Raise tank until positioned near body.

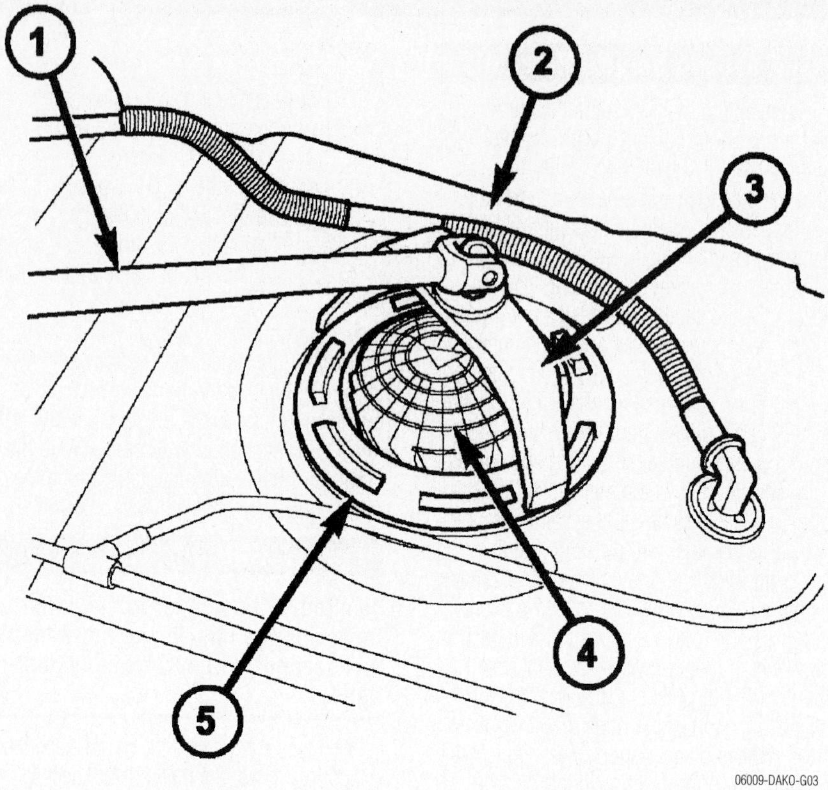

Lockring removal—2005 models

31. Connect fuel pump module electrical connector at top of tank.

32. Connect NVLD pump electrical connector to NVLD pump.

33. Connect fuel line quick-connect fitting to pump module.

34. Continue raising tank until positioned snug to body.

35. Install and position both tank support straps. Install two fuel tank strap nuts and tighten to 30 ft. lbs. (41 Nm).

36. Connect EVAP, ORVR and NVLD lines at front of tank.

37. Connect vent line.

38. Connect rubber fill hose to fuel tank fitting and tighten hose clamps.

39. The vapor/vacuum lines and hoses must be firmly connected. Also check the vapor/vacuum lines at the NVLD pump, NVLD filter and EVAP canister purge solenoid for damage or leaks. If a leak is present, a Diagnostic Trouble Code (DTC) may be set.

40. If equipped, install fuel tank skid plate.

41. Install plastic liner in front of left-rear tire/wheel.

42. Install left rear tire/wheel.

43. Lower vehicle.

44. Fill fuel tank with fuel.

45. Start engine and check for fuel leaks near top of module.

Fuel Injector

REMOVAL & INSTALLATION

2.5L Engine

1. Before servicing the vehicle, refer to the precautions in the beginning of this section.
2. Relieve fuel system pressure.
3. Remove or disconnect the following:
   - Negative battery cable
   - Air intake tube
   - Fuel injector connectors
   - Fuel line
   - Accelerator cable
   - Cruise control cable, if equipped
   - Transmission cable, if equipped
   - Cable routing bracket
   - Fuel supply manifold with injectors attached
   - Fuel injectors

To install:

4. Install or connect the following:
   - Fuel injectors, using new O-rings
   - Fuel supply manifold with injectors.

Tighten the bolts to 75–125 inch lbs. (8–14 Nm).

- Cable routing bracket
- Transmission cable, if equipped
- Cruise control cable, if equipped
- Accelerator cable
- Fuel line
- Fuel injector connectors
- Air intake tube
- Negative battery cable

5. Start the engine and check for leaks.

3.7L Engine

⁂ CAUTION

The fuel system is under constant pressure even with engine off. Before servicing fuel rail, fuel system pressure must be released.

⁂ CAUTION

The left and right fuel rails are replaced as an assembly. Do not attempt to separate rail halves at connector tube. Due to design of tube, it does not use any clamps. Never attempt to install a clamping device of any kind to tube. When removing fuel rail assembly for any reason, be careful not to bend or kink tube.

1. Remove fuel tank filler tube cap.
2. Perform Fuel System Pressure Release Procedure.
3. Remove negative battery cable at battery.
4. Remove air duct at throttle body air box.
5. Remove air box at throttle body.
6. Disconnect fuel line latch clip and fuel line at fuel rail. A special tool will be necessary for fuel line disconnection.
7. Remove necessary vacuum lines at throttle body.
8. Disconnect electrical connectors at all 6 fuel injectors. To remove connector, push red colored slider away from injector. While pushing slider, depress tab and remove connector from injector. The factory fuel injection wiring harness is numerically tagged (INJ 1, INJ 2, etc.) for injector position identification. If harness is not tagged, note wiring location before removal.
9. Disconnect electrical connectors at throttle body sensors.
10. Remove 6 ignition coils.
11. Remove 4 fuel rail mounting bolts.
12. Gently rock and pull left side of fuel rail until fuel injectors just start to clear machined holes in cylinder head. Gently rock and pull right side of rail until injectors just start to clear cylinder head holes. Repeat this procedure (left/right) until all injectors have cleared cylinder head holes.
13. Remove fuel rail (with injectors attached) from engine.
14. Disconnect clip(s) that retain fuel injector(s) to fuel rail.

To install:

15. Apply a small amount of clean engine oil to each fuel injector o-ring. This will help in fuel rail installation.
16. Install injector(s) and injector clip(s) to fuel rail.
17. Apply a small amount of engine oil to each fuel injector o-ring. This will help in fuel rail installation.
18. Position fuel rail/fuel injector assembly to machined injector openings in cylinder head.
19. Guide each injector into cylinder head. Be careful not to tear injector o-rings.
20. Push right side of fuel rail down until fuel injectors have bottomed on cylinder head shoulder. Push left fuel rail down until injectors have bottomed on cylinder head shoulder.
21. Install 4 fuel rail mounting bolts and tighten to 27 Nm (20 ft. lbs.).
22. Install ignition coils.
23. Connect electrical connectors to throttle body.
24. Connect electrical connectors to MAP and IAT sensors.
25. Connect electrical connectors at all fuel injectors. To install connector, push connector onto injector and then push and lock red colored slider. Verify connector is locked to injector by lightly tugging on connector.

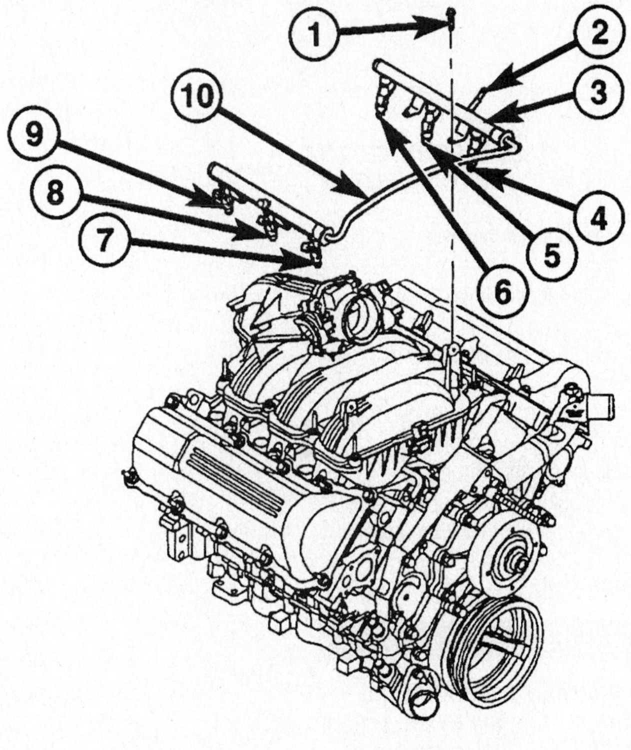

1 - MOUNTING BOLTS (4)
2 - QUICK-CONNECT FITTING
3 - FUEL RAIL
4 - INJ. #1
5 - INJ. #3
6 - INJ. #5
7 - INJ. #2
8 - INJ. #4
9 - INJ. #6
10 - CONNECTOR TUBE

67189-DAKO-G42

Fuel rail components—3.7L engine

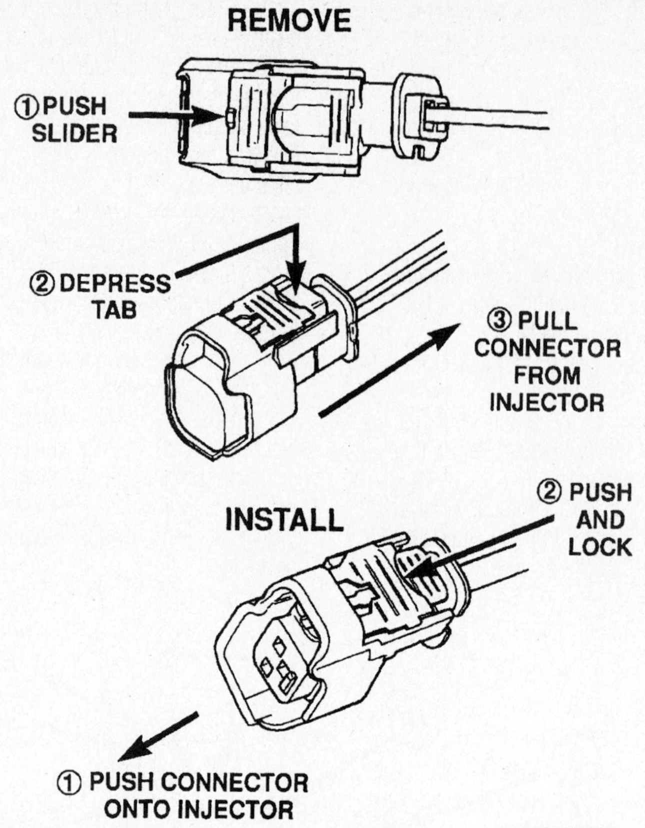

Fuel injector connector removal

26. Connect vacuum lines to throttle body.
27. Connect fuel line latch clip and fuel line to fuel rail.
28. Connect wiring to rear of generator.
29. Install air box to throttle body.
30. Install air duct to air box.
31. Connect battery cable to battery.
32. Start engine and check for leaks.

4.7L Engine

⁂ CAUTION

The fuel system is under constant pressure even with engine off. Before servicing fuel rail, fuel system pressure must be released.

⁂ WARNING

The left and right fuel rails are replaced as an assembly. Do not attempt to separate rail halves at connector tube. Due to design of tube, it does not use any clamps. Never attempt to install a clamping device of any kind to tube. When removing fuel rail assembly for any reason, be careful not to bend or kink tube.

1. Remove fuel tank filler tube cap.
2. Perform Fuel System Pressure Release Procedure.
3. Remove negative battery cable at battery.
4. Remove air duct at throttle body air box.
5. Remove air box at throttle body.
6. Remove wiring at rear of generator.
7. Disconnect fuel line latch clip and fuel line at fuel rail. A special tool will be necessary for fuel line disconnection.
8. Remove vacuum lines at throttle body.
9. Disconnect electrical connectors at all 8 fuel injectors. To remove, push red colored slider away from injector. While pushing slider, depress tab and remove connector from injector. The factory fuel injection wiring harness is numerically tagged (INJ 1, INJ 2, etc.) for injector position identification. If harness is not tagged, note wiring location before removal.
10. Disconnect electrical connectors at throttle body.
11. Disconnect electrical connectors at MAP and IAT sensors.
12. Remove first three ignition coils on each bank (cylinders #1, 3, 5, 2, 4 and 6).
13. Remove 4 fuel rail mounting bolts.
14. Gently rock and pull left side of fuel rail until fuel injectors just start to clear

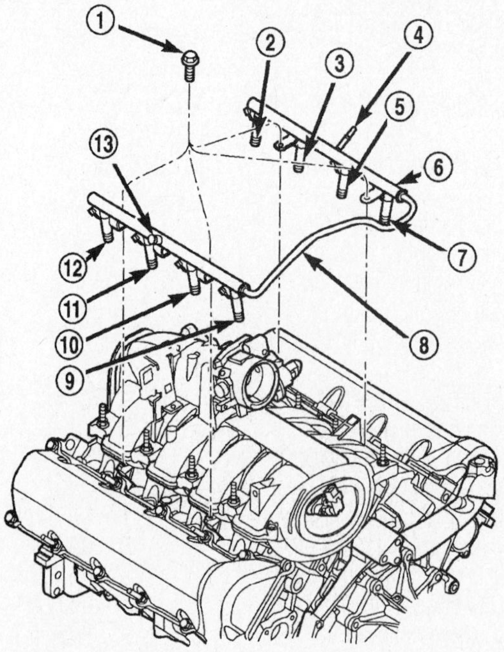

Fuel rail components—4.7L engine

machined holes in cylinder head. Gently rock and pull right side of rail until injectors just start to clear cylinder head holes. Repeat this procedure (left/right) until all injectors have cleared cylinder head holes.

15. Remove fuel rail (with injectors attached) from engine.

16. Disconnect clip(s) that retain fuel injector(s) to fuel rail.

To install:

17. Apply a small amount of clean engine oil to each fuel injector O-ring. This will help in fuel rail installation.

18. Install injector(s) and injector clip(s) to fuel rail.

19. Apply a small amount of engine oil to each fuel injector o-ring. This will help in fuel rail installation.

20. Position fuel rail/fuel injector assembly to machined injector openings in cylinder head.

21. Guide each injector into cylinder head. Be careful not to tear injector o-rings.

22. Push right side of fuel rail down until fuel injectors have bottomed on cylinder head shoulder. Push left fuel rail down until injectors have bottomed on cylinder head shoulder.

23. Install 4 fuel rail mounting bolts and tighten to 27 Nm (20 ft. lbs.).

24. Install ignition coils.

25. Connect electrical connectors to throttle body.

26. Connect electrical connectors to MAP and IAT sensors.

27. Connect electrical connectors at all fuel injectors. To install connector, push connector onto injector and then push and lock red colored slider. Verify connector is locked to injector by lightly tugging on connector.

28. Connect vacuum lines to throttle body.

29. Connect fuel line latch clip and fuel line to fuel rail.

30. Connect wiring to rear of generator.

31. Install air box to throttle body.

32. Install air duct to air box.

33. Connect battery cable to battery.

34. Start engine and check for leaks.

3.9L and 5.9L Engines

1. Before servicing the vehicle, refer to the precautions in the beginning of this section.

2. Relieve fuel system pressure.

3. Remove or disconnect the following:

- Negative battery cable
- Air intake tube
- Throttle body
- A/C compressor bracket
- Fuel injector connectors
- Fuel line
- Fuel supply manifold with injectors
- Fuel injectors

To install:

4. Install or connect the following:

- Fuel injectors with new O-ring seals
- Fuel supply manifold with injectors. Tighten the bolts to 17 ft. lbs. (23 Nm).
- Fuel line
- Fuel injector connectors
- A/C compressor bracket
- Throttle body
- Air intake tube
- Negative battery cable

5. Start the engine and check for leaks.

DRIVE TRAIN

Transmission Assembly

REMOVAL & INSTALLATION

Manual

2002–03

1. Before servicing the vehicle, refer to the precautions in the beginning of this section.

2. Remove or disconnect the following:

- Negative battery cable
- Shift lever and tower assembly
- Crankshaft Position (CKP) sensor
- Skidplate, if equipped
- Rear driveshaft
- Front driveshaft, if equipped
- Transfer case shift linkage, if equipped
- Transmission mount and crossmember. Support the transmission.
- Exhaust front pipe
- Clutch slave cylinder
- Starter motor
- Vehicle Speed (VSS) sensor connector
- Reverse light switch connector
- Transmission flange bolts
- Transmission

To install:

3. Install or connect the following:

- Transmission. Tighten the flange bolts to 40–45 ft. lbs. (54–61 Nm).
- Reverse light switch connector
- Vehicle Speed (VSS) sensor connector
- Starter motor
- Clutch slave cylinder
- Exhaust front pipe
- Transmission mount and crossmember. Tighten the fasteners to 50 ft. lbs. (68 Nm).
- Transfer case shift linkage, if equipped
- Front driveshaft, if equipped
- Rear driveshaft
- Skidplate, if equipped
- CKP sensor
- Shift lever and tower assembly
- Negative battery cable

2004–05—2-WHEEL DRIVE

1. Disconnect battery negative cable.
2. Shift transmission into Neutral.
3. Remove floor console.
4. Remove shift lever boot.
5. Remove the shift lever extension from the shift tower and lever assembly.
6. Raise vehicle.
7. Remove skid plate, if equipped (4WD).
8. If transmission will be disassembled for repair, remove drain plug and drain lubricant from transmission.
9. Mark propeller shafts and companion flange for assembly reference.
10. Disconnect and remove propeller shafts.
11. Disconnect and remove exhaust system Y-pipe. Then disconnect and lower remaining exhaust pipes for clearance as necessary.
12. Disconnect backup light switch wires.
13. Remove bolts/nuts attaching transmission to rear mount.
14. Support transmission with a transmission jack. Secure transmission to jack with safety chains.
15. Remove rear crossmember.
16. Remove bolts attaching clutch slave cylinder to clutch housing. Then move cylinder aside for working clearance.
17. Remove starter.
18. Remove transmission dust shield.
19. Remove transmission harness wires from clips on transmission shift cover.
20. Lower transmission slightly.
21. Remove the bolts attaching the shift tower and lever assembly to the transmis-

sion housing. Then remove the shift tower and lever assembly.

22. Remove bolts attaching transmission to engine.

23. Slide transmission and jack rearward until input shaft clears clutch disc.

24. Lower transmission jack and remove transmission from under vehicle.

To install:

➡If a new transmission is being installed, be sure to use all components supplied with the new transmission. For example, if a new shift tower is supplied with the new transmission, do not re-use the original shift tower.

25. Apply light coat of Mopar® high temperature bearing grease to contact surfaces of following components:
- input shaft splines.
- release bearing slide surface of front retainer.
- release bearing bore.
- release fork.
- release fork ball stud.
- propeller shaft slip yoke.

26. Apply sealer to threads of drain plug, then install plug in case.

27. Mount transmission on jack and position transmission under vehicle. Secure transmission to jack with safety chains.

28. Raise transmission until input shaft is centered in clutch disc hub.

29. Move transmission forward and start input shaft in clutch disc.

30. Work transmission forward until seated against engine. Do not allow transmission to remain unsupported after input shaft has entered clutch disc.

31. Install and tighten transmission to engine bolts to 108 Nm (80 ft. lbs.).

32. Position transmission harness wires in clips on shift cover.

33. Install slave cylinder and shield, if equipped.

34. Install transmission mount on transmission or rear crossmember.

2004–05—4-WHEEL DRIVE

1. Disconnect battery negative cable.
2. Shift transmission into Neutral.
3. Remove floor console.
4. Remove shift lever boot.
5. Remove the shift lever extension from the shift tower and lever assembly.
6. Raise vehicle.
7. Remove skid plate, if equipped (4WD).
8. If transmission will be disassembled for repair, remove drain plug and drain lubricant from transmission.
9. Mark propeller shafts and companion flange for assembly reference.
10. Disconnect and remove propeller shafts.
11. Disconnect and remove exhaust system Y-pipe. Then disconnect and lower remaining exhaust pipes for clearance as necessary.
12. Disconnect backup light switch wires.
13. Support engine with adjustable safety stand.
14. Disconnect transfer case shift linkage at transfer case range lever.
15. Remove transfer case shift lever from transmission.
16. Remove bolts/nuts attaching transmission to rear support.
17. Remove crossmember bolts/nuts and remove crossmember.
18. Support transfer case with transmission jack. Secure transfer case to jack with safety chains.
19. Remove transfer case attaching nuts.
20. Move transfer case rearward until input gear clears transmission output shaft.
21. Lower transfer case assembly and move it from under vehicle.
22. Support transmission with transmission jack. Secure transmission to jack with safety chains.
23. Remove transmission harness from retaining clips on transmission shift cover.
24. Remove clutch slave cylinder splash shield, if equipped.
25. Remove clutch slave cylinder attaching nuts. Move cylinder aside for working clearance.
26. Remove starter.
27. Remove transmission splash shield.
28. Lower transmission slightly
29. Remove bolts attaching shift tower and lever assembly to rear case. Then remove shift tower and lever as an assembly.
30. Remove bolts attaching transmission to engine.
31. Move transmission rearward until input shaft clears clutch disc.
32. Lower transmission and remove it from under vehicle.

To install:

➡If a new transmission is being installed, be sure to use all components supplied with the new transmission. For example, if a new shift tower is supplied with the new transmission, do not re-use the original shift tower.

33. Apply light coat of Mopar® high temperature bearing grease to contact surfaces of following components:
- input shaft splines.
- release bearing slide surface of front retainer.
- release bearing bore.
- release fork.
- release fork ball stud.
- propeller shaft slip yoke.

34. Apply sealer to threads of drain plug, then install plug in case.

35. Mount transmission on jack and position transmission under vehicle. Secure transmission to jack with safety chains.

36. Raise transmission until input shaft is centered in clutch disc hub.

37. Move transmission forward and start input shaft in clutch disc.

38. Work transmission forward until seated against engine. Do not allow transmission to remain unsupported after input shaft has entered clutch disc.

39. Install and tighten transmission to engine bolts to 108 Nm (80 ft. lbs.).

40. Position transmission harness wires in clips on shift cover.

41. Install slave cylinder and shield, if equipped.

42. Install transmission mount on transmission or rear crossmember.

43. Install transfer case shift lever on transmission.

44. Install rear crossmember.

➡Ensure wiring harness is clear before installing crossmember.

45. Remove transmission jack and engine support fixture.

46. Install transfer case on transmission jack. Secure transfer case to jack with safety chains.

47. Raise jack and align transfer case input gear with transmission output shaft.

48. Move transfer case forward and seat it on transmission.

49. Install and tighten transfer case attaching nuts. Tighten nuts to 41–47 Nm (30–35 ft. lbs.) if case has 3/8 studs, or 30–41 Nm (22–30 ft. lbs.) if case has 5/16 studs.

50. Connect backup light switch wires.

51. Install transmission dust cover.

52. Install starter.

53. Install transfer case shift lever to side of transfer case.

54. Connect transfer case shift lever to range lever on transfer case.

55. Align and connect propeller shafts.

56. Fill transmission with required lubricant. Check lubricant level in transfer case and add lubricant if necessary.

57. Install transfer case skid plate, if equipped, and crossmember. Tighten attaching bolts/nuts to 41 Nm (30 ft. lbs.).

58. Install exhaust system components.

59. Lower vehicle.
60. Install shift tower and lever assembly. Tighten shift tower bolts to 7–10 Nm (5–7 ft. lbs.).
61. Install shift lever boot.
62. Install floor console.
63. Connect battery negative cable.

Automatic

2002–03

1. Before servicing the vehicle, refer to the precautions in the beginning of this section.
2. Remove or disconnect the following:
   - Negative battery cable
   - Rear driveshaft
   - Crankshaft Position (CKP) sensor
   - Exhaust front pipe
   - Transmission braces, if equipped
   - Starter motor
   - Transmission oil cooler lines
   - Torque converter access cover
   - Torque converter
   - Transmission oil dipstick tube
   - Vehicle Speed (VSS) sensor connector
   - Park/Neutral switch connector
   - Shift cable
   - Throttle valve cable
   - Transmission mount and crossmember. Support the transmission.
   - Front driveshaft and transfer case, if equipped
   - Transmission flange bolts
   - Transmission

To install:

3. Install or connect the following:
   - Transmission. Tighten the flange bolts to 65 ft. lbs. (87 Nm).
   - Front driveshaft and transfer case, if equipped
   - Transmission mount and crossmember
   - Throttle valve cable
   - Shift cable
   - Park/Neutral switch connector
   - VSS sensor connector
   - Transmission oil dipstick tube
   - Torque converter. Tighten the bolts to 23 ft. lbs. (31 Nm) for 10.75 inch converters and to 35 ft. lbs. (47 Nm) for 12.2 inch converters.
   - Torque converter access cover
   - Transmission oil cooler lines
   - Starter motor
   - Transmission braces, if equipped. Tighten the bolts to 30 ft. lbs. (41 Nm).
   - Exhaust front pipe
   - CKP sensor
   - Rear driveshaft
   - Negative battery cable

2004–05 545RFE

1. Disconnect the negative battery cable.
2. Raise and support the vehicle
3. Remove any necessary skid plates.
4. Mark propeller shaft and axle companion flanges for assembly alignment.
5. Remove the rear propeller shaft.
6. Remove the front propeller shaft, if necessary.
7. Remove the engine to transmission collar.
8. Remove the exhaust support bracket from the rear of the transmission.
9. Disconnect and lower or remove any necessary exhaust components.
10. Remove the starter motor.
11. Rotate crankshaft in clockwise direction until converter bolts are accessible. Then remove bolts one at a time. Rotate crankshaft with socket wrench on dampener bolt.
12. Disengage the output speed sensor connector from the output speed sensor.
13. Disengage the input speed sensor connector from the input speed sensor.
14. Disengage the transmission solenoid/TRS assembly connector from the transmission solenoid/TRS assembly.
15. Disengage the line pressure sensor connector from the line pressure sensor.
16. Disconnect gearshift cable from transmission manual valve lever.
17. Disconnect transmission fluid cooler lines at transmission fittings and clips.
18. Disconnect the transmission vent hose from the transmission.
19. Support rear of engine with safety stand or jack.
20. Raise transmission slightly with service jack to relieve load on crossmember and supports.
21. Remove bolts securing rear support and cushion to transmission and crossmember.
22. Remove bolts attaching crossmember to frame and remove crossmember.
23. Remove transfer case.
24. Remove all remaining converter housing bolts.
25. Carefully work transmission and torque converter assembly rearward off engine block dowels.
26. Hold torque converter in place during transmission removal.
27. Lower transmission and remove assembly from under the vehicle.
28. To remove torque converter, carefully slide torque converter out of the transmission.

To install:

29. Check torque converter hub and hub drive flats for sharp edges burrs, scratches, or nicks. Polish the hub and flats with 320/400 grit paper and crocus cloth if necessary. Verify that the converter hub O-ring is properly installed and is free of any debris. The hub must be smooth to avoid damaging pump seal at installation.
30. If a replacement transmission is being installed, transfer any components necessary, such as the manual shift lever and shift cable bracket, from the original transmission onto the replacement transmission.
31. Lubricate oil pump seal lip with transmission fluid.
32. Align converter and oil pump.
33. Carefully insert converter in oil pump. Then rotate converter back and forth until fully seated in pump gears.
34. Check converter seating with steel scale and straightedge. Surface of converter lugs should be at least 13mm (½ in.) to rear of straightedge when converter is fully seated.
35. Temporarily secure converter with C-clamp.
36. Position transmission on jack and secure it with chains.
37. Check condition of converter driveplate. Replace the plate if cracked, distorted or damaged. Also be sure transmission dowel pins are seated in engine block and protrude far enough to hold transmission in alignment.
38. Apply a light coating of high temperature grease to the torque converter hub pocket in the rear pocket of the engine's crankshaft.
39. Raise transmission and align the torque converter with the drive plate and transmission converter housing with the engine block.
40. Move transmission forward. Then raise, lower or tilt transmission to align the converter housing with engine block dowels.
41. Carefully work transmission forward and over engine block dowels until converter hub is seated in crankshaft. Verify that no wires, or the transmission vent hose, have become trapped between the engine block and the transmission.
42. Install two bolts to attach the transmission to the engine.
43. Install remaining torque converter housing to engine bolts. Tighten to 68 Nm (50 ft. lbs.).
44. Install transfer case, if equipped. Tighten transfer case nuts to 35 Nm (26 ft. lbs.).

45. Install rear transmission crossmember. Tighten crossmember to frame bolts to 68 Nm (50 ft. lbs.).

46. Install rear support to transmission. Tighten bolts to 47 Nm (35 ft. lbs.).

47. Lower transmission onto crossmember and install bolts attaching transmission mount to crossmember. Tighten clevis bracket to crossmember bolts to 47 Nm (35 ft. lbs.). Tighten the clevis bracket to rear support bolt to 68 Nm (50 ft. lbs.).

48. Remove engine support fixture.

49. Connect gearshift cable to transmission.

50. Connect wires to solenoid and pressure switch assembly connector, input and output speed sensors, and line pressure sensor. Be sure transmission harnesses are properly routed.

51. Install torque converter-to-driveplate bolts. Tighten bolts to 31 Nm (270 inch lbs.).

52. Install starter motor and cooler line bracket.

53. Connect cooler lines to transmission.

54. Install transmission fill tube.

55. Install exhaust components.

56. Install the engine collar onto the transmission and the engine. Tighten the bolts to 54 Nm (40 ft. lbs.).

57. Align and connect propeller shaft(s).

58. Adjust gearshift cable if necessary.

59. Install any skid plates removed previously.

60. Lower vehicle.

61. Fill transmission with Mopar® ATF +4, Automatic Transmission Fluid.

2004–05 42RLE

1. Disconnect the negative battery cable.

2. Raise and support the vehicle

3. Remove any necessary skid plates.

4. Mark the propeller shaft and axle companion flanges for assembly alignment.

5. Remove the rear propeller shaft.

6. Remove the front propeller shaft, if necessary.

7. Disconnect the wires from the input and output speed sensors.

8. Disconnect the wires from the transmission range sensor and the solenoid/pressure switch assembly.

9. Remove the bolts holding the exhaust crossover pipe to the pre-catalytic converter pipe flanges.

10. Remove the bolts holding the exhaust crossover pipe to the catalytic converter flange.

11. Disconnect gearshift cable from transmission manual valve lever.

12. Disengage the shift cable from the cable support bracket.

13. Remove the starter motor.

14. Remove the engine to transmission collar.

15. Rotate the crankshaft in clockwise direction until the converter bolts are accessible. Then remove the bolts one at a time. Rotate the crankshaft with a socket wrench on dampener bolt.

16. Disconnect the transmission vent hose from the transmission.

17. Remove the transfer case.

18. Support the rear of engine with a safety stand or jack.

19. Raise the transmission slightly with a service jack to relieve the load on the crossmember and supports.

20. Remove the bolts securing the rear support and cushion to the transmission and crossmember.

21. Remove the bolts attaching the crossmember to frame and remove crossmember.

22. Disconnect the transmission fluid cooler lines at the transmission fittings and clips.

23. Remove all remaining converter housing bolts.

24. Carefully work the transmission and torque converter assembly rearward off the engine block dowels.

25. Hold the torque converter in place during transmission removal.

26. Lower the transmission and remove the assembly from under the vehicle.

27. To remove the torque converter, carefully slide the torque converter out of the transmission.

To install:

28. Check the torque converter hub and hub drive flats for sharp edges burrs, scratches, or nicks. Polish the hub and flats with 320/400 grit paper and crocus cloth if necessary. Verify that the converter hub O-ring is properly installed and is free of any debris. The hub must be smooth to avoid damaging pump seal at installation.

29. If a replacement transmission is being installed, transfer any components necessary, such as the manual shift lever and shift cable bracket, from the original transmission onto the replacement transmission.

30. Lubricate the oil pump seal lip with transmission fluid.

31. Align the converter and oil pump.

32. Carefully insert the converter in the oil pump. Then rotate the converter back and forth until fully seated in the pump gears.

33. Check the converter seating with a steel scale and straightedge. The surface of the converter lugs should be at least 13mm (½ in.) to the rear of straightedge when the converter is fully seated.

34. Temporarily secure the converter with a C-clamp.

35. Position the transmission on the jack and secure it with chains.

36. Check the condition of the converter driveplate. Replace the plate if cracked, distorted or damaged. Also, be sure the transmission dowel pins are seated in the engine block and protrude far enough to hold the transmission in alignment.

37. Apply a light coating of high temperature grease to the torque converter hub pocket in the rear pocket of the engine's crankshaft.

38. Raise the transmission and align the torque converter with the drive plate and transmission converter housing with the engine block.

39. Move the transmission forward. Then raise, lower or tilt transmission to align the converter housing with engine block dowels.

40. Carefully work the transmission forward and over the engine block dowels until the converter hub is seated in the crankshaft. Verify that no wires, or the transmission vent hose, have become trapped between the engine block and the transmission.

41. Install the two bolts to attach the transmission to the engine.

42. Install the remaining torque converter housing to engine bolts. Tighten to 68 Nm (50 ft. lbs.).

43. Install the transfer case, if equipped. Tighten the transfer case nuts to 35 Nm (26 ft. lbs.).

44. Install the rear transmission crossmember. Tighten the crossmember to frame bolts to 68 Nm (50 ft. lbs.).

45. Install the rear support to the transmission. Tighten the bolts to 47 Nm (35 ft. lbs.).

46. Lower the transmission onto the crossmember and install the bolts attaching the transmission mount to the crossmember. Tighten clevis bracket to crossmember bolts to 47 Nm (35 ft. lbs.). Tighten the clevis bracket to rear support bolt to 68 Nm (50 ft. lbs.).

47. Remove the engine support fixture.

48. Connect the gearshift cable to the support bracket and transmission manual lever.

49. Connect the input and output speed sensor wires.

50. Connect the wires to the transmis-

sion range sensor and the solenoid/pressure switch assembly.

51. Install the torque converter-to-driveplate bolts. Tighten the bolts to 88 Nm (65 inch lbs.).
52. Install the starter motor and cooler line bracket.
53. Connect the cooler lines to transmission.
54. Install the transmission fill tube.
55. Install the exhaust components.
56. Align and connect the propeller shaft(s).
57. Adjust the gearshift cable if necessary.
58. Install any skid plates removed previously.
59. Lower the vehicle.
60. Fill the transmission with the required amount of Mopar® ATF +4, Automatic Transmission Fluid.

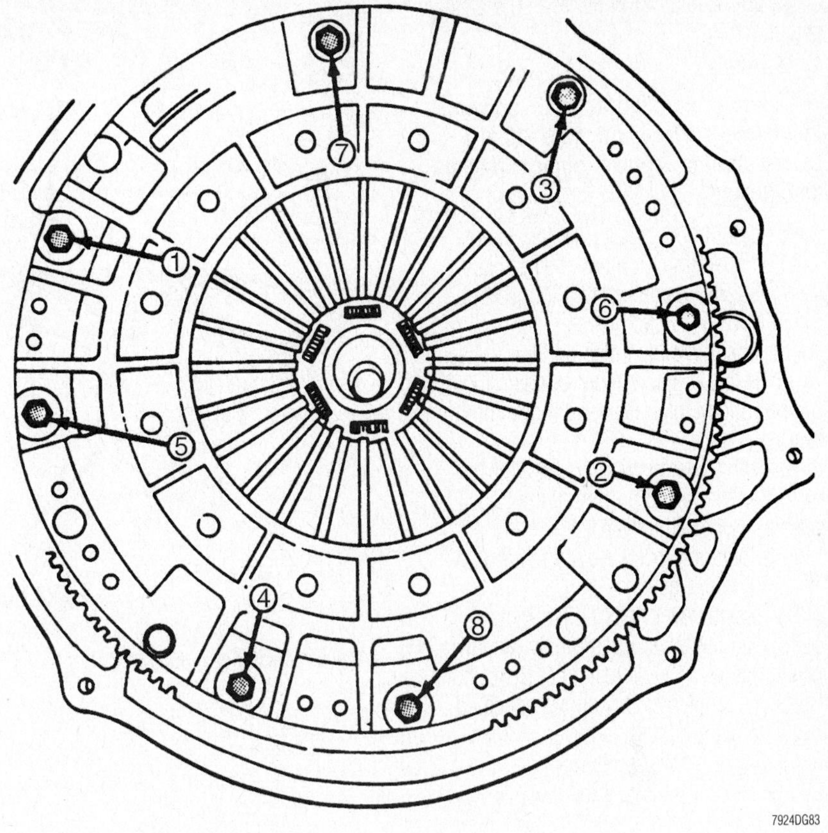

Pressure plate torque sequence

Clutch

REMOVAL & INSTALLATION

2002–03

1. Before servicing the vehicle, refer to the precautions in the beginning of this section.
2. Remove or disconnect the following:
   - Negative battery cable
   - Transfer case, if equipped
   - Transmission
   - Pressure plate. Loosen the bolts evenly in ½ turn steps.
   - Clutch disc

To install:

3. Install or connect the following:
   - Clutch disc and pressure plate. Tighten the pressure plate bolts evenly in ½ turns to 21 ft. lbs. (28 Nm).
   - Transmission
   - Transfer case, if equipped
   - Negative battery cable

2004–05

1. Raise vehicle.
2. Remove transmission and clutch housing as assembly.
3. If pressure plate is being removed for access to another component, mark position of pressure plate cover on flywheel with small punch marks.
4. Loosen pressure plate cover bolts evenly and in rotation to relieve spring tension. Loosen bolts a few threads at a time to avoid warping cover.

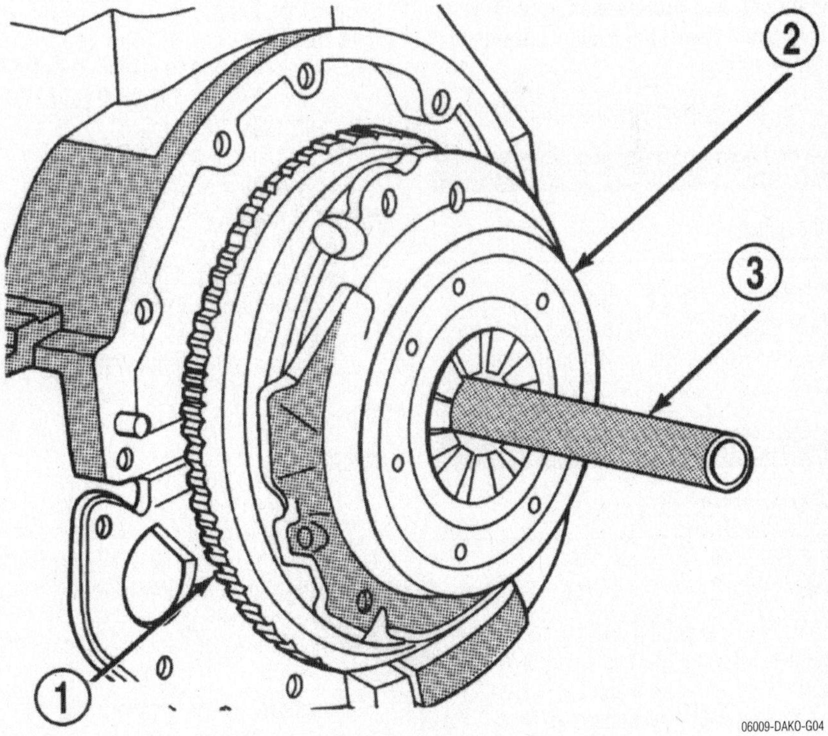

Clutch installation. (1) flywheel, (2) pressure plate, (3) alignment tool

5. Remove cover bolts, pressure plate and clutch disc.

To install:

➡Clean flywheel surface with solvent. Scuff sand surface with 120/180 grit emery cloth to remove minor scratches and glazing.

6. Check new clutch disc for runout and free operation on input shaft splines.
7. Lubricate crankshaft pilot bearing with a NLGI—2 rated grease.
8. Position clutch disc on the flywheel.
9. Insert alignment tool or spare input shaft through clutch disc and into pilot bearing.
10. Verify that disc hub is positioned correctly. The raised portion of the hub faces away from the flywheel.
11. Position pressure plate cover over disc and on flywheel.
12. Install cover bolts finger tight.
13. Tighten cover bolts evenly (and in rotation) a few threads at a time. Cover bolts must be tightened evenly and to specified torque to avoid distorting cover.
14. Tighten cover bolts to:
    - 5⁄16 in. bolts to 23 Nm (17 ft. lbs.).
    - 3⁄8 in. bolts to 41 Nm (30 ft. lbs.).
15. Apply light coat of high temperature bearing grease to splines of transmission input shaft and to release bearing slide surface of front bearing retainer.

➡Do not over-lubricate shaft splines. This could result in grease contamination of disc.

16. Install transmission as assembly.

Hydraulic Clutch System

BLEEDING

The system is self-bleeding. Press the clutch pedal repeatedly to release air from the fluid. The air will be vented from the reservoir.

Transfer Case Assembly

REMOVAL & INSTALLATION

2002–03

1. Before servicing the vehicle, refer to the precautions in the beginning of this section.
2. Shift the transfer case into **N**.
3. Remove or disconnect the following:
    - Front and rear driveshafts
    - Transmission mount and crossmember. Support the transmission.
    - Vehicle Speed (VSS) sensor connector
    - Shift linkage
    - Vent hose
    - Vacuum hose
    - Indicator switch connector
    - Transfer case attaching nuts
    - Transfer case

To install:

4. Install or connect the following:
    - Transfer case. Tighten the nuts to 26 ft. lbs. (35 Nm).
    - Indicator switch connector
    - Vacuum hose
    - Vent hose
    - Shift linkage
    - VSS sensor connector
    - Transmission mount and crossmember
    - Front and rear driveshafts

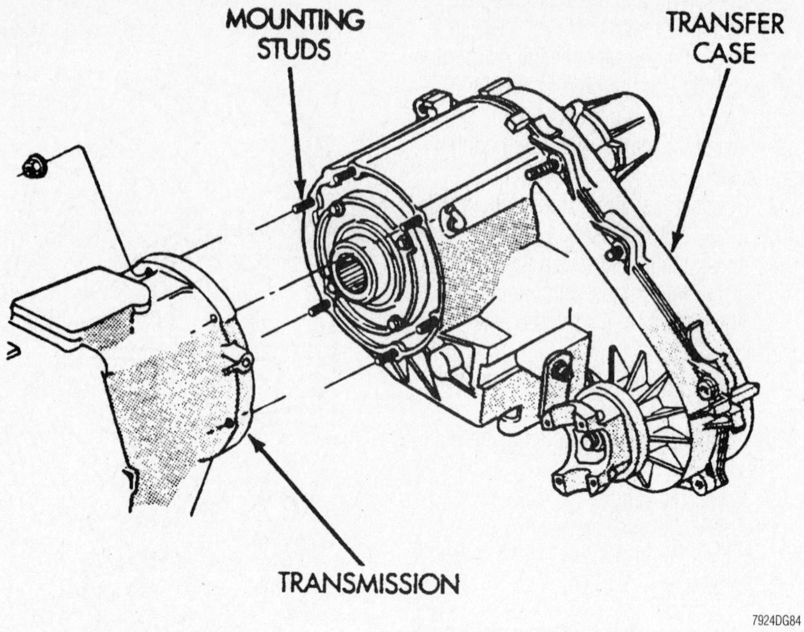

Typical transfer case mounting

2004–05

NV233

1. Shift the transfer case into 2WD.
2. Raise the vehicle.
3. Drain the transfer case lubricant.
4. Mark the front and rear propeller shafts for alignment reference.
5. Support the transmission with jack stand.
6. Remove the rear crossmember and skid plate, if equipped.
7. Disconnect the front and rear propeller shafts at transfer case.
8. Disconnect the transfer case shift motor and mode sensor wire connectors.
9. Disconnect the transfer case vent hose.
10. Support the transfer case with a transmission jack.
11. Secure the transfer case to the jack with chains.
12. Remove the nuts attaching transfer case to the transmission.
13. Pull the transfer case and jack rearward to disengage the transfer case.
14. Remove the transfer case from under the vehicle.

To install:

15. Mount the transfer case on a transmission jack.
16. Secure the transfer case to the jack with chains.
17. Position transfer case under vehicle.
18. Align the transfer case and transmission shafts and install the transfer case onto the transmission.
19. Install and tighten the transfer case attaching nuts to 27–34 Nm (20–25 ft. lbs.).
20. Connect the vent hose.
21. Connect the shift motor and mode sensor wiring connectors. Secure the wire harness to clips on the transfer case.
22. Align and connect the propeller shafts.
23. Fill the transfer case with the correct fluid.
24. Install the rear crossmember and skid plate, if equipped. Tighten the crossmember bolts to 41 Nm (30 ft. lbs.).
25. Remove the transmission jack and support stand.

26. Lower the vehicle and verify transfer case shift operation.

NV244

1. Shift transfer case into AWD.
2. Raise vehicle.
3. Drain transfer case lubricant.
4. Mark front and rear propeller shafts for alignment reference.
5. Disconnect front and rear propeller shafts at transfer case.
6. Support transmission with jack stand.
7. Remove rear crossmember and skid plate, if equipped.
8. Disconnect transfer case shift motor and mode sensor wire connectors.
9. Disconnect transfer case vent hose.
10. Support transfer case with transmission jack.
11. Secure transfer case to jack with chains.
12. Remove nuts attaching transfer case to transmission.
13. Pull transfer case and jack rearward to disengage transfer case from the transmission adapter housing and output shaft.
14. Remove transfer case from under vehicle.

To install:

15. Mount transfer case on a transmission jack.
16. Secure transfer case to jack with chains.
17. Position transfer case under vehicle.
18. Align transfer case and transmission shafts and install transfer case onto the transmission.
19. Install and tighten transfer case attaching nuts to 27–34 Nm (20–25 ft. lbs.). Connect the vent hose.
20. Connect the shift motor and mode sensor wiring connectors. Secure wire harness to clips on transfer case.
21. Align and connect the propeller shafts.
22. Fill transfer case with correct fluid.
23. Install rear crossmember and skid plate, if equipped. Tighten crossmember bolts to 41 Nm (30 ft. lbs.).
24. Remove transmission jack and support stand.
25. Lower vehicle and verify transfer case shift operation.

Halfshaft

REMOVAL & INSTALLATION

2002–03

1. Before servicing the vehicle, refer to the precautions in the beginning of this section.
2. Remove or disconnect the following:
   - Skid plate, if equipped
   - Front wheel
   - Split pin
   - Nut lock
   - Spring washer
   - Hub nut
   - Brake caliper and rotor
   - Wheel speed sensor, if equipped
   - Wheel bearing and hub assembly
3. Pry the inner tripod joint out of the differential and remove the axle halfshaft.

To install:

4. Install the axle halfshaft so that the snapring is felt to seat in the joint housing groove.
5. Install or connect the following:
   - Wheel bearing and hub assembly
   - Wheel speed sensor, if equipped
   - Brake caliper and rotor
   - Hub nut. Tighten the nut to 180 ft. lbs. (244 Nm).
   - Spring washer
   - Nut lock
   - Split pin
   - Front wheel
   - Skid plate, if equipped

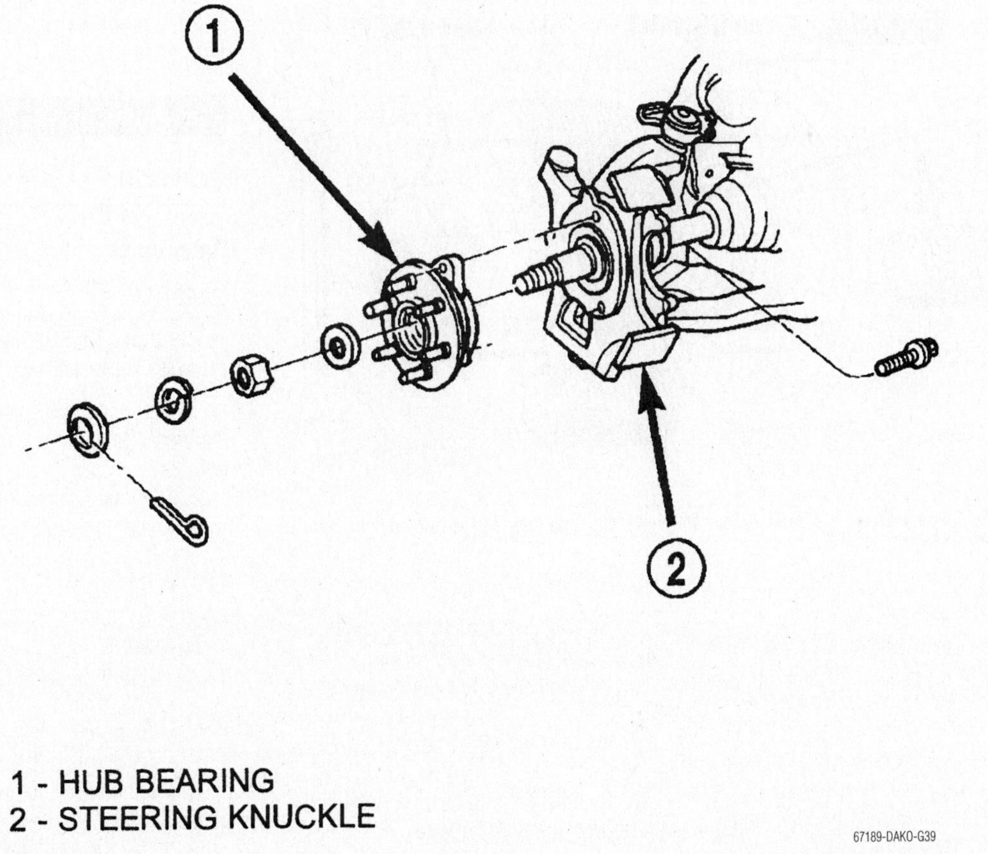

Hub/bearing assembly

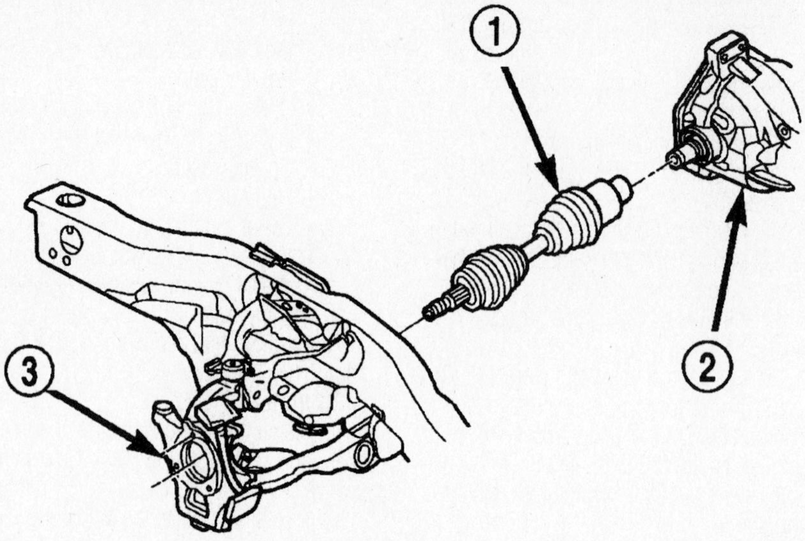

Halfshaft

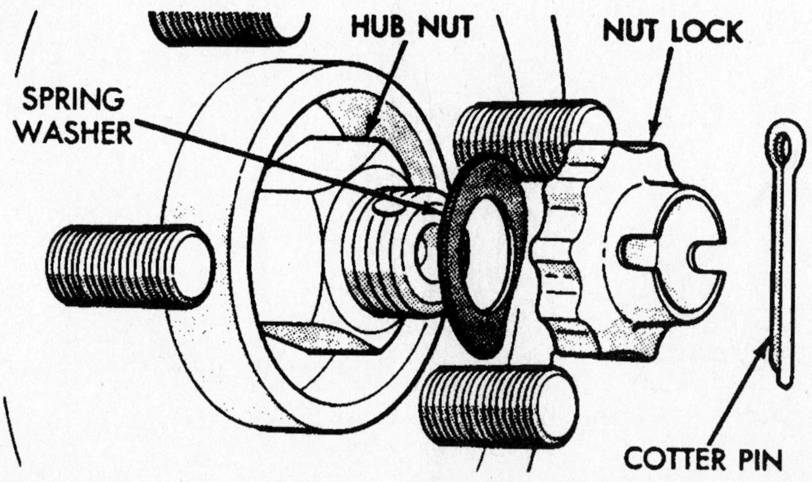

To separate the halfshaft from the hub, remove the cotter pin, nut lock and spring washer from the axle shaft

2004

1. Raise the vehicle.
2. Remove the wheel and tire assembly.
3. Remove the skid plate, if equipped.
4. Remove the cotter pin, nut lock, and spring washer from the stub shaft.
5. Remove the hub nut and washer from the stub shaft.
6. Remove the brake caliper and rotor.
7. Remove the ABS wheel speed sensor if equipped.
8. Remove the hub bearing bolts and hub bearing from the knuckle.
9. Support the half shaft at the CV joint housings.
10. Position two pry bars behind the inner CV housing and disengage the CV joint from the axle.
11. Remove the half shaft from the vehicle.

To install:

12. Apply a light coating of wheel bearing grease on the axle splines.
13. Insert the half shaft stub through the steering knuckle and onto the axle. Verify the shaft snapring engages with the groove on the inside of the joint housing.
14. Clean the hub bearing bore and hub bearing mating surface of all foreign materials. Apply a light coating of grease to all mating surfaces.
15. Install the hub bearing onto the axle half shaft and steering knuckle.
16. Install the hub bearing bolts and tighten to 120 ft. lbs. (163 Nm).
17. Install the ABS wheel speed sensor, if equipped.
18. Install brake rotor and caliper.
19. Apply the brakes and tighten hub nut to 244 Nm (180 ft. lbs.) for 2002–04; 251 Nm (185 ft. lbs.) for 2005.
20. Install the spring washer, nut lock and cotter pin.
21. Install the skid plate, if equipped.
22. Install the wheel and tire assembly.

CV-Joints

OVERHAUL

Outer Joint

2002–04

➡The outer joint is not serviceable and must be replaced as a unit.

1. Before servicing the vehicle, refer to the precautions in the beginning of this section.
2. Remove or disconnect the following:
   - Axle halfshaft from the vehicle
   - CV-joint boot and clamps
   - Snapring
   - CV-joint

To install:

3. Install or connect the following:
   - CV-joint
   - Snapring
   - CV-joint boot and clamps
4. Fill the joint housing and boot with grease and tighten the boot clamps.
5. Install the axle halfshaft.

2005

1. Place shaft in vise with soft jaws and support CV-joint.

CAUTION

Do not damage CV-joint housing or halfshaft.

2. Remove clamps with a cut-off wheel or grinder.
3. Slide the boot down the shaft.
4. Remove lubricant to expose the CV-joint snapring.
5. Spread snapring and slide the joint off the shaft.
6. Slide boot off the shaft and discard old boot.
7. Mark alignment marks on the inner race/hub, bearing cage and housing with dabs of paint.
8. Clamp CV-joint in a vertical position in a soft jawed vise.
9. Press down one side of the bearing cage to gain access to the ball at the opposite side.

➡If joint is tight, use a hammer and brass drift to loosen the bearing hub. Do not contact the bearing cage with the drift.

10. Remove ball from the bearing cage.
11. Repeat step above until all six balls are removed from the bearing cage.
12. Lift cage and inner race upward and out from the housing.
13. Turn inner race 90° in the cage and rotate the inner race/hub out of the cage.

To Install:

14. Apply a light coat of grease to the CV-joint components before assembling them.
15. Align inner race, cage and housing according to the alignment reference marks.
16. Insert inner race into the cage and rotate race into the cage.
17. Rotate inner race/hub in the cage.
18. Insert cage into the housing.
19. Rotate cage 90° into the housing.
20. Apply lubricant included with replacement boot/joint to the ball races. Spread lubricant equally among all the races.
21. Tilt inner race/hub and cage and install the balls.
22. Place new clamps onto new boot and slide boot onto the shaft to its original position.
23. Apply the rest of lubricant to the CV-joint and boot.
24. Push the joint onto the shaft until the

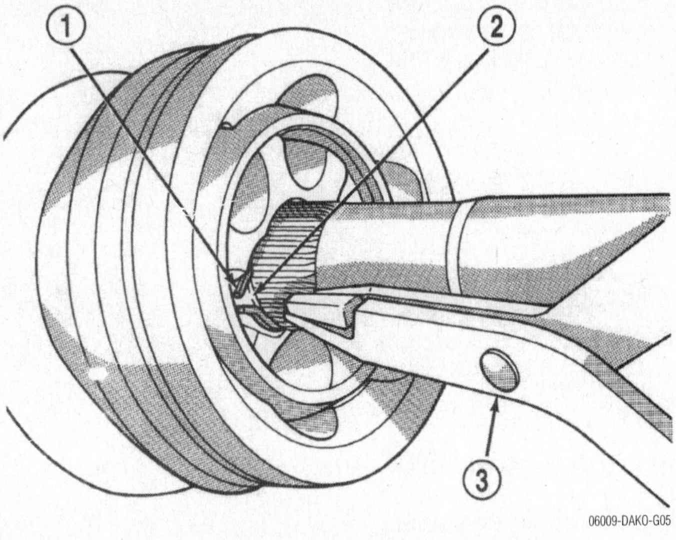

Removing the snapring (1) from the shaft (2) with snapring pliers (3)

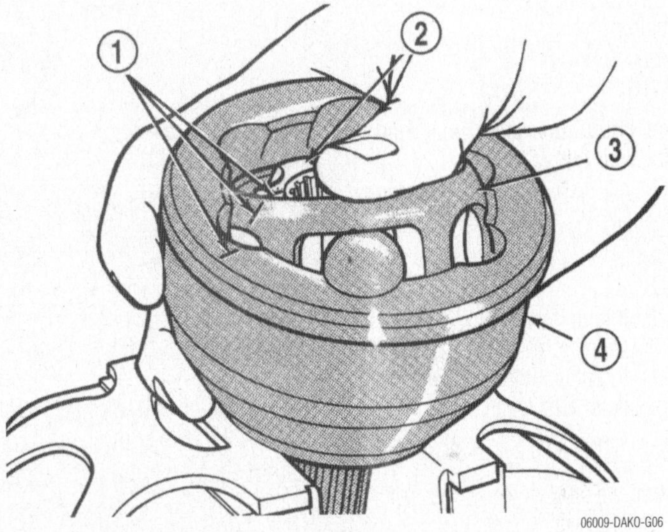

Make alignment marks (1) on the inner race/hub (2) and cage (3)

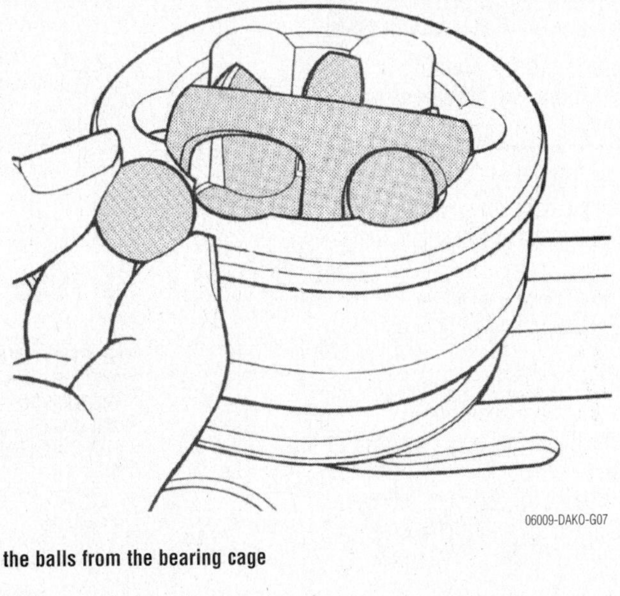

Removing the balls from the bearing cage

snapring seats in the groove. Pull on the joint to verify the span ring has engaged.

25. Position boot on the joint in its original position. Ensure boot is not twisted and remove any excess air.

26. Secure both boot clamps with Clamp Installer C-4975A, or equivalent. Place tool on clamp bridge and tighten tool until the jaws of the tool are closed.

Inner Joint

2002–03

1. Before servicing the vehicle, refer to the precautions in the beginning of this section.
2. Remove or disconnect the following:
   - Axle halfshaft from the vehicle
   - Inner tripod joint boot clamps
   - Tripod joint housing
   - Snapring
   - Circlip
   - Tripod joint

To install:

➡Use new snaprings, clips, and boot clamps for assembly.

3. Install or connect the following:
   - Tripod joint
   - Circlip
   - Snapring
   - Tripod joint housing
4. Fill the tripod joint housing and boot with grease and tighten the boot clamps.
5. Install the axle halfshaft.

2004–05

1. Clamp the shaft in a vise with soft jaws and support the CV joint.
2. Remove the clamps with a cut-off wheel or grinder.

⁂ WARNING

Do not damage the CV housing or half shaft with the cut-off wheel or grinder.

3. Remove the housing from the half shaft and slide the boot down shaft.
4. Remove the housing bushing from the housing.
5. Remove the tripod snapring.
6. Remove the tripod and boot from the halfshaft.
7. Clean and inspect the CV components for excessive wear and damage. Replace the tripod as a unit only if necessary.

To install:

8. Slide a new boot down the halfshaft.
9. Install the tripod and tripod snapring on the halfshaft.

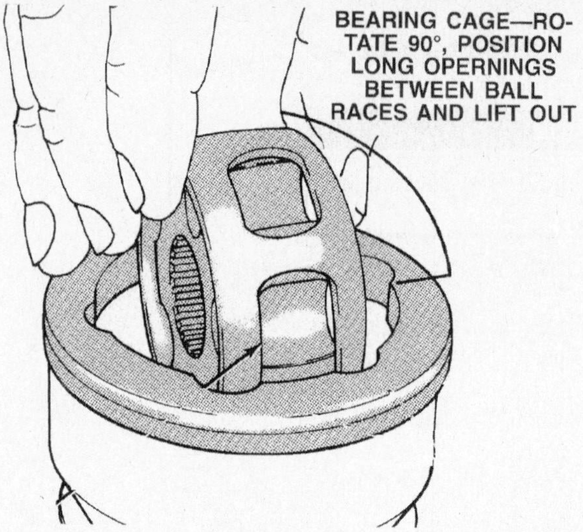

Removing the cage and inner race from the housing

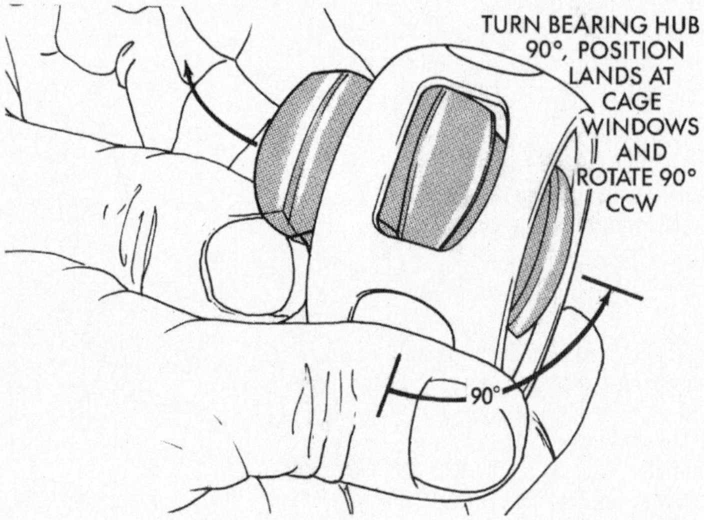

Removing the inner race/hub from the cage

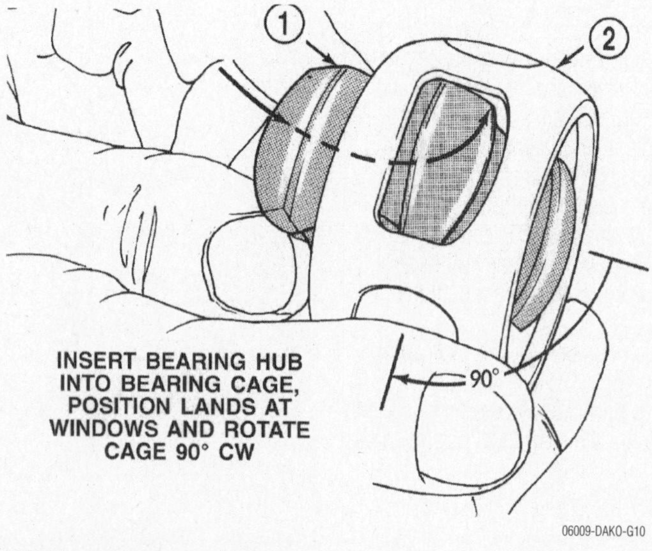

Assembling the inner race cage and housing

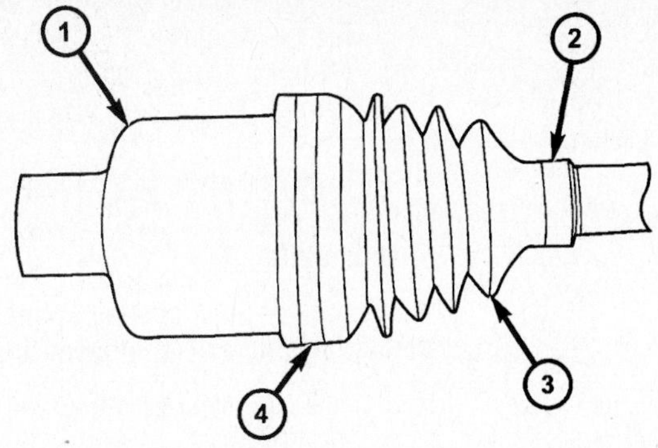

1 - C/V HOUSING
2 - CLAMP
3 - BOOT
4 - CLAMP

67189-DAKO-G33

Boot clamp location

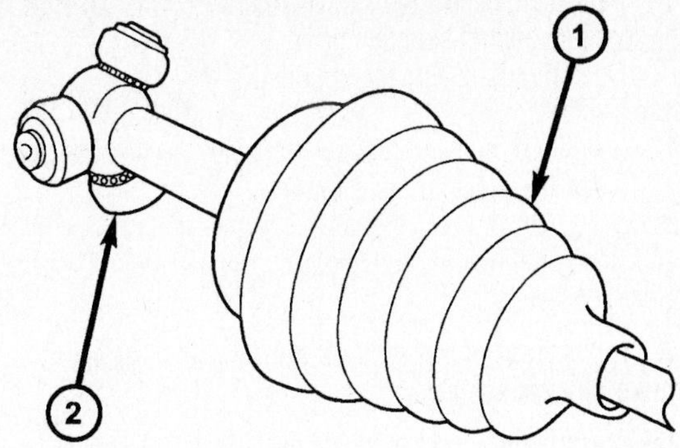

1 - BOOT
2 - TRIPOD

67189-DAKO-G36

Tripod joint

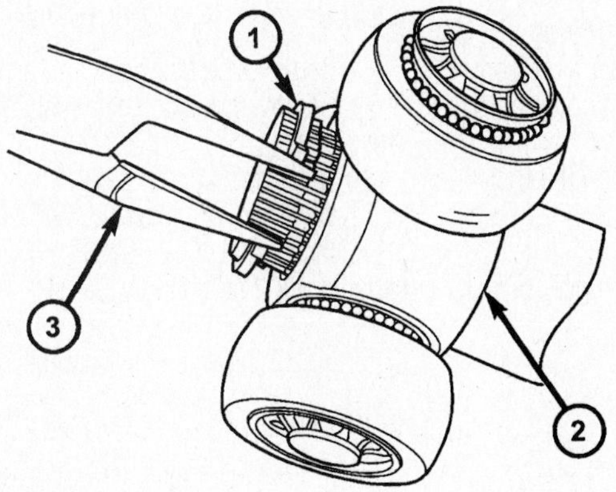

1 - SNAP RING
2 - TRIPOD
3 - PLIERS

67189-DAKO-G34

Removing the snapring

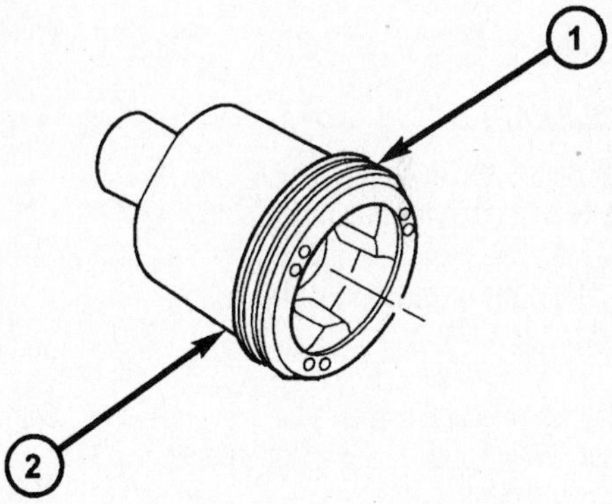

1 - HOUSING
2 - BUSHING

67189-DAKO-G37

Housing bushing

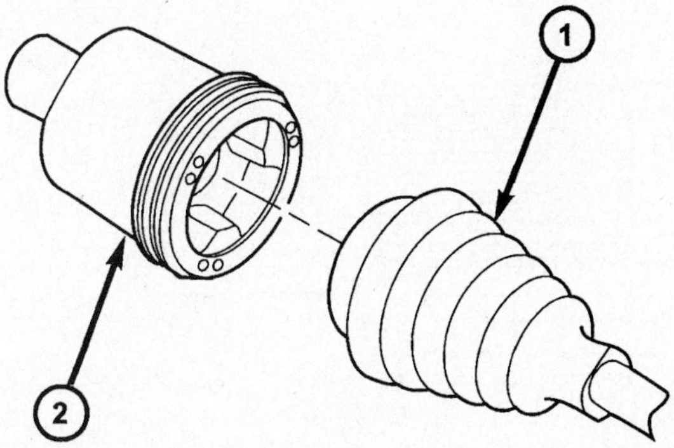

1 - BOOT
2 - HOUSING

67189-DAKO-G35

CV joint housing

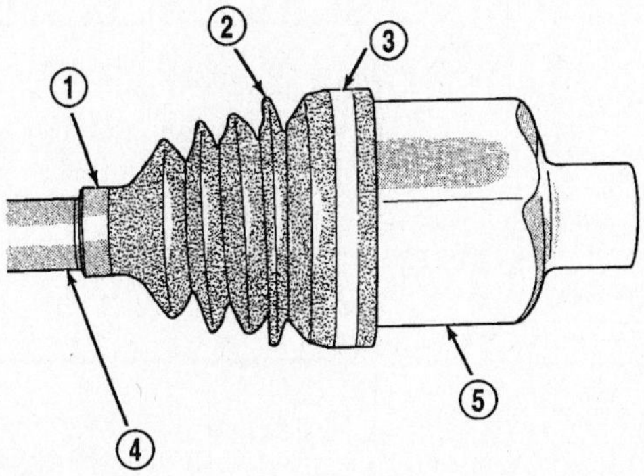

1 - CLAMP
2 - BOOT
3 - CLAMP
4 - SHAFT
5 - HOUSING

67189-DAKO-G38

Inner CV joint boot

10. Pack the grease supplied with the joint/boot into the housing and boot.
11. Coat the tripod with the supplied grease.
12. Install new bushing onto the housing.
13. Insert the tripod and shaft in the housing.
14. Position the boot on the joint in its original position.

➡Verify the boot is not twisted and remove any excess air.

15. Secure both boot clamps with Clamp Installer C-4975A, or equivalent. Place the tool on the clamp bridge and tighten the tool until the jaws of the tool are closed.

Front Axle Shaft, Bearing and Seal

REMOVAL & INSTALLATION

2002–03

AXLE SHAFT

1. Before servicing the vehicle, refer to the precautions in the beginning of this section.
2. Remove or disconnect the following:
    - Front wheel
    - Brake caliper and rotor
    - Wheel speed sensor, if equipped
    - Axle hub nut
    - Wheel bearing and hub assembly
    - Axle shaft

To install:

3. Install or connect the following:
    - Axle shaft
    - Wheel bearing and hub assembly
    - Axle hub nut. Tighten the nut to 175 ft. lbs. (237 Nm).
    - Wheel speed sensor, if equipped
    - Brake caliper and rotor
    - Front wheel

SEAL

1. Before servicing the vehicle, refer to the precautions in the beginning of this section.
2. Remove or disconnect the following:
    - Front axle shafts
    - Differential cover
    - Differential and ring gear assembly
    - Axle seals

To install:

3. Press the axle seals into the differential housing with Turnbuckle 6797 and Disc set 8110.
4. Install or connect the following:
    - Differential and ring gear assembly. Tighten the bearing cap bolts to 45 ft. lbs. (61 Nm).
    - Differential cover. Tighten the bolts to 30 ft. lbs. (41 Nm).
    - Front axle shafts
5. Fill the axle assembly with gear oil and check for leaks.

2004–05

AXLE SHAFTS

1. Place the transmission in neutral.
2. Raise and support the vehicle.
3. Remove the half shaft from vehicle.
4. Remove skid plate, if equipped.
5. Clean the axle seal area.
6. Remove the snapring from the axle shaft.
7. Remove the axle with remover 8420A and slide hammer C-3752, or equivalent.

To install:

➡Use care to prevent shaft splines from damaging axle shaft seal lip.

8. Lubricate the bearing bore and seal lip with gear lubricant.
9. Insert the axle shaft through seal, bearing, and engage it into the side gear splines. Push firmly on the axle shaft to engage the snapring.
10. Check the differential fluid level and add fluid if necessary.
11. Install the skid plate, if necessary.
12. Install the half shaft.

SEALS

1. Remove the half shaft and axle shaft.
2. Remove the axle shaft seal with a small pybar.

To install:

3. Wipe the axle shaft tube bore clean.
4. Install a new axle shaft seal with installer 8402 and handle C-4171, or equivalent.
5. Install the axle shaft and half shaft.

BEARINGS

1. Remove the half shaft from vehicle.
2. Remove the skid plate, if equipped.
3. Clean the axle seal area.

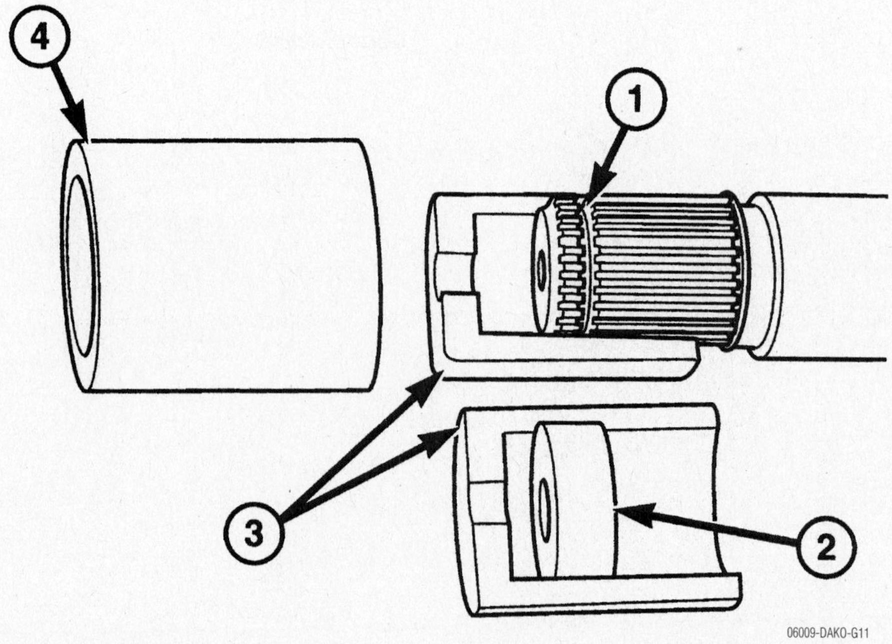

Assembling the axle shaft removal tool 8420A

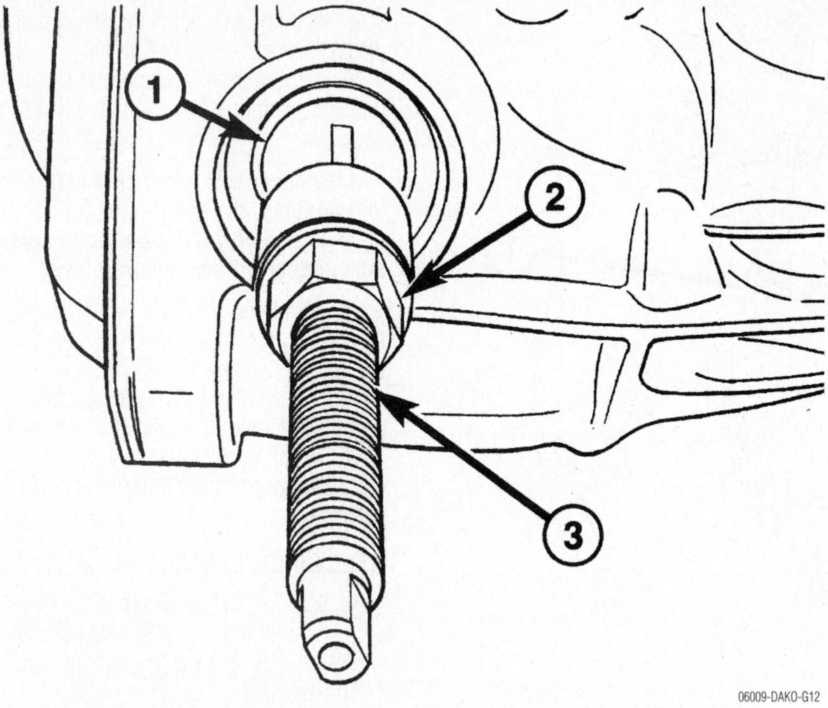

Removing the axle shaft bearing (1) using remover C-4660-A (3) and nut (2)

4. Remove axle shaft O-ring.
5. Remove the axle shaft.
6. Remove the axle shaft seal.
7. Install the axle shaft bearing remover C-4660-A, or equivalent, in the bearing. Then tighten the nut to spread the remover in the bearing.
8. Install the bearing remove cup, bearing and nut. Then tighten the nut to draw the bearing out.
9. Inspect the axle shaft tube bore for roughness and burrs. Remove as necessary.

To install:

10. Wipe the axle shaft tube bore clean.
11. Install the axle shaft bearing with installer 5063 and handle C-4171, or equivalent.
12. Install a new axle shaft seal with installer 8402 and handle C-4171, or equivalent.
13. Install the axle shaft and half shaft.

Rear Axle Shaft, Bearing and Seal

REMOVAL & INSTALLATION

2002–03

C-CLIP TYPE

1. Before servicing the vehicle, refer to the precautions in the beginning of this section.
2. Remove or disconnect the following:
   - Rear wheel
   - Brake drum
   - Differential cover
   - Differential gear shaft retainer
   - Differential gear shaft
   - C-clip
   - Axle shaft
   - Axle seal
   - Axle bearing

To install:

3. Install or connect the following:
   - Axle bearing
   - Axle seal
   - Axle shaft
   - C-clip
   - Differential gear shaft. Use Loctite® and tighten the retainer to 14 ft. lbs. (19 Nm).
   - Differential cover. Tighten the bolts to 30 ft. lbs. (41 Nm).
   - Brake drum
   - Rear wheel
4. Fill the axle assembly with gear oil and check for leaks.

NON C-CLIP TYPE

1. Before servicing the vehicle, refer to the precautions in the beginning of this section.
2. Remove or disconnect the following:
   - Rear wheel
   - Brake caliper and rotor, if equipped
   - Brake drum, if equipped
   - Axle retainer nuts
   - Axle shaft, seal and bearing assembly
3. Split the bearing retainer with a chisel and remove the retainer ring.
4. Press the bearing off the axle shaft.
5. Remove the axle seal and retaining plate.

To install:

6. Install the retaining plate and axle seal onto the axle shaft.
7. Pack the wheel bearing with axle grease and press the bearing on to the axle shaft.
8. Press the retaining ring onto the axle shaft.
9. Install or connect the following:
   - Axle shaft, seal and bearing assembly. Tighten the nuts to 45 ft. lbs. (61 Nm).
   - Brake caliper and rotor, if equipped
   - Brake drum, if equipped
   - Rear wheel
10. Fill the axle assembly with gear oil and check for leaks.

2004–05 8¼ inch Axle

AXLE SHAFTS

1. Place the transmission in neutral and raise and support the vehicle.
2. Remove the brake drum/caliper and rotor.
3. Remove the differential cover and drain the lubricant.
4. Rotate the differential case to access the pinion shaft lock screw. Remove the lock screw and pinion shaft from the differential case.
5. Push the axle shaft inward then remove axle shaft C-lock.
6. Remove the axle shaft being careful not to damage the shaft bearing and seal.
7. Inspect the axle shaft seal for leakage or damage.
8. Inspect the axle shaft bearing contact surface for signs of brinelling, galling and pitting.

To install:

9. Lubricate the bearing bore and seal lip with gear lubricant. Insert the axle shaft through the seal, bearing and engage it into side gear splines.

➡Use care to prevent shaft splines from damaging axle shaft seal lip.

10. Insert the C-lock in end of axle shaft. Push the axle shaft outward to seat the C-lock in side gear.
11. Insert the pinion shaft into differential case and through thrust washers and differential pinions.
12. Align hole in shaft with the hole in

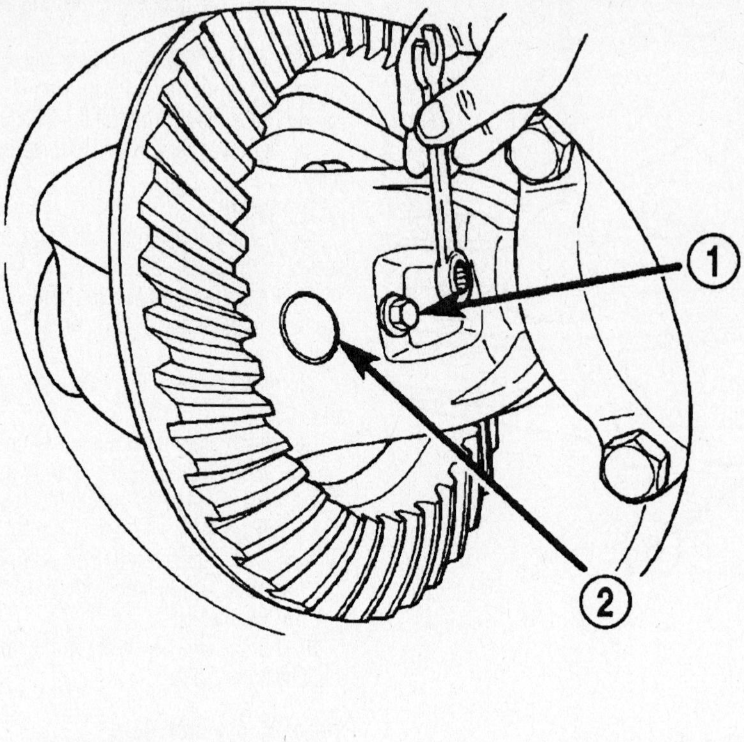

Pinion mate shaft lock screw

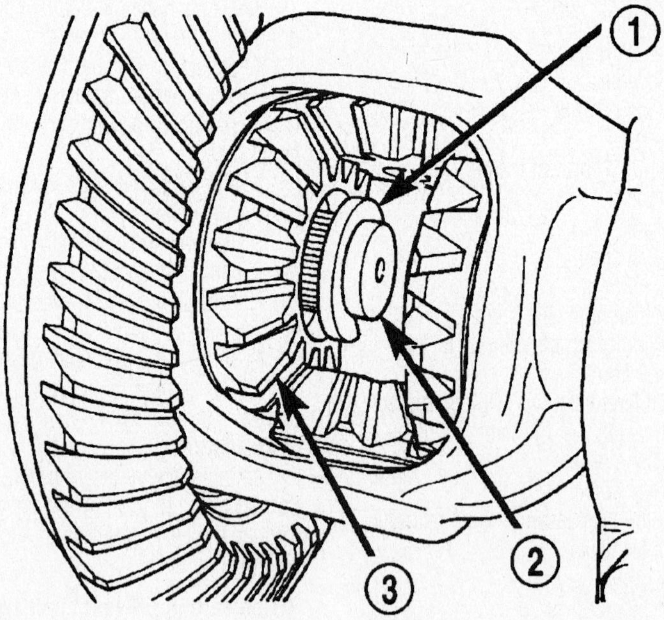

Axle shaft C-lock

the differential case and install the lock screw with Loctite® on the threads. Tighten the lock screw to 11 Nm (8 ft. lbs.).

13. Apply a bead of Mopar red silicone rubber sealant or equivalent to the housing cover.

➡If the cover is not installed within 3 to 5 minutes, the cover must be cleaned and new RTV applied or adhesion quality will be compromised.

14. Install the cover and tighten the bolts in a criss-cross pattern to 41 Nm (30 ft. lbs.).

15. Fill the differential with gear lubricant to the bottom of the fill plug hole.

16. Install the fill hole plug.

17. Install the brake drum/rotor and caliper.

AXLE SHAFT SEALS

1. Remove the axle shaft.
2. Remove the axle shaft seal from the end of the axle tube with a small pry bar.

To install:

3. Wipe the axle tube bore clean. Remove any old sealer or burrs from the tube.
4. Install a new axle seal with Installer C-4076-B and Handle C-4735-1, or equivalent. When the tool contacts the axle tube, the seal is installed to the correct depth.
5. Coat the lip of the seal with axle lubricant for protection prior to installing the axle shaft.
6. Install the axle shaft.
7. Apply a bead of Mopar red Silicone Rubber Sealant or equivalent to the housing cover.

➡If the cover is not installed within 3 to 5 minutes, the cover must be cleaned and new RTV applied or adhesion quality will be compromised.

8. Install the cover and tighten the bolts in a criss-cross pattern to 41 Nm (30 ft. lbs.).
9. Fill the differential with gear lubricant to the bottom of the fill plug hole and install the fill plug.

AXLE BEARINGS

1. Remove the axle shaft.
2. Remove the axle shaft seal from the axle tube with a small pry bar.

➡The seal and bearing can be removed at the same time with the bearing removal tool.

3. Remove the axle shaft bearing with Bearing Removal Tool Set 6310 and Adapter Foot 6310-9, or equivalent.

To install:

4. Wipe the axle tube bore clean. Remove any old sealer or burrs from the tube.
5. Install the axle shaft bearing with Installer C-4198 and Handle C-4171, or equivalent.

➡**Install the bearing with part number against the installer.**

6. Install a new axle seal with Installer C-4076-B and Handle C-4735-1, or equivalent. When the tool contacts the axle tube, the seal is installed to the correct depth.
7. Coat the lip of the seal with axle lubricant and install the axle shaft.
8. Apply a bead of Mopar red Silicone Rubber Sealant or equivalent to the housing cover.

➡**If the cover is not installed within 3 to 5 minutes, the cover must be cleaned and new RTV applied or adhesion quality will be compromised.**

9. Install the cover and tighten the bolts in a criss-cross pattern to 41 Nm (30 ft. lbs.).
10. Fill the differential with gear lubricant to bottom of the fill plug hole and install fill plug.

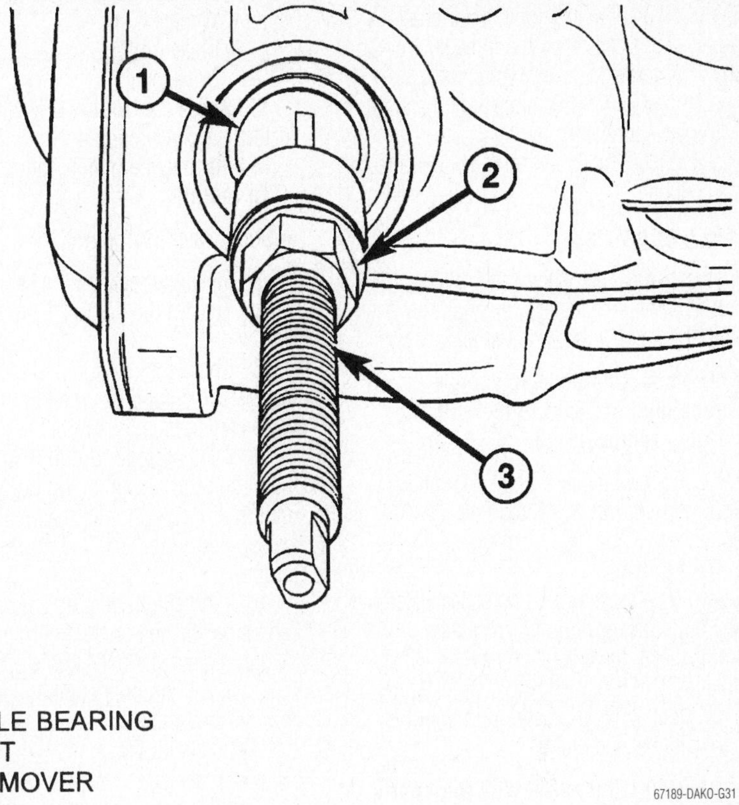

Bearing remover

2004–05 9¼ inch Axle

AXLE SHAFTS

1. Place the transmission in neutral.
2. Remove the brake caliper, adapter and rotor.
3. Remove the differential housing cover and drain lubricant.
4. Rotate the differential case so the pinion mate shaft lock screw is accessible. Remove the lock screw and pinion mate shaft from the differential case.
5. Push the axle shaft inward and remove the axle shaft C-lock from the axle shaft.
6. Remove the axle shaft. Use care to prevent damage to the axle shaft bearing and seal in the axle tube.

To install:

7. Lubricate the bearing bore and seal lip with gear lubricant. Insert the axle shaft through seal, bearing, and engage it into the side gear splines.

➡**Use care to prevent the shaft splines from damaging the axle shaft seal.**

8. Insert the C-lock in end of the axle shaft then push the axle shaft outward to seat the C-lock in the side gear.
9. Insert the pinion shaft into the differential case and through the thrust washers and differential pinions.

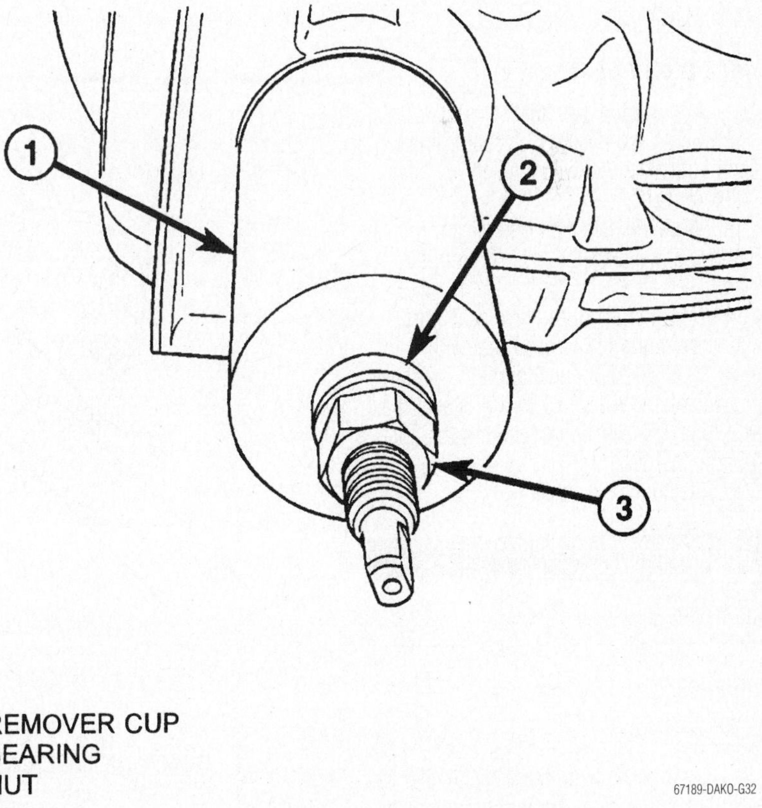

Bearing cup remover

10. Align the hole in shaft with the hole in the differential case and install the lock screw with Loctite® on the threads. Tighten the lock screw to 11 Nm (8 ft. lbs.).

11. Install the differential cover and fill with gear lubricant.

12. Install the brake rotor, caliper adapter and caliper.

AXLE BEARINGS

1. Remove the axle shaft.
2. Remove the axle shaft seal from the end of the axle tube with a small pry bar.

➡The seal and bearing can be removed at the same time with the bearing removal tool.

3. Remove the axle shaft bearing with Bearing Remover 6310 and Foot 6310-9, or equivalent.

To install:

4. Wipe the axle tube bore clean. Remove any old sealer or burrs from the tube.
5. Install the axle shaft bearing with Installer C-4198 and Handle C-4171, or equivalent. Drive the bearing in until tool contacts the axle tube.

➡Bearing is installed with the bearing part number against the installer.

6. Coat the lip of the new axle seal with axle lubricant and install with Installer C-4076-B and Handle C-4735-1, or equivalent.
7. Install the axle shaft.

AXLE SHAFT SEALS

1. Remove the axle shaft.
2. Remove the axle shaft seal from the end of the axle tube with a small pry bar.

To install:

3. Wipe the axle tube bore clean. Remove any old sealer or burrs from the tube.
4. Coat the lip of the new seal with axle lubricant and install a seal with Installer C-4076-B and Handle C-4735-1, or equivalent. When the tool contacts the axle tube, the seal is installed to the correct depth.
5. Install the axle shaft.

Pinion Seal

REMOVAL & INSTALLATION

2002–03

C-CLIP TYPE

1. Before servicing the vehicle, refer to the precautions in the beginning of this section.
2. Remove or disconnect the following:
   - Wheels
   - Brake drums
   - Driveshaft
3. Check the bearing preload with an inch lb. torque wrench.
4. Remove the pinion flange and seal.

To install:

➡Use a new pinion nut for assembly.

5. Install the new pinion seal and flange. Tighten the nut to 210 ft. lbs. (285 Nm).
6. Check the bearing preload. The bearing preload should be equal to the reading taken earlier, plus 5 inch lbs.
7. If the preload torque is low, tighten the pinion nut in 5 inch lb. increments until the torque value is reached. Do not exceed 350 ft. lbs. (474 Nm) pinion nut torque.
8. If the pinion bearing preload torque cannot be attained at maximum pinion nut torque, replace the collapsible spacer.
9. Install or connect the following:
   - Driveshaft
   - Brake drums
   - Wheels
10. Fill the axle assembly with gear oil and check for leaks.

NON C-CLIP TYPE—FRONT

1. Before servicing the vehicle, refer to the precautions in the beginning of this section.
2. Remove or disconnect the following:
   - Wheels
   - Brake rotors
   - Driveshaft
3. Check the bearing preload with an inch lb. torque wrench.
4. Remove the pinion flange and seal.

To install:

➡Use a new pinion nut for assembly.

5. Install the new pinion seal and flange. Tighten the nut to 160 ft. lbs. (217 Nm).
6. Check the bearing preload. The bearing preload should be equal to the reading taken earlier, plus 5 inch lbs.
7. If the preload torque is low, tighten the pinion nut in 5 inch lb. increments until the torque value is reached. Do not exceed 260 ft. lbs. (353 Nm) pinion nut torque.
8. If the pinion bearing preload torque can not be attained at maximum pinion nut torque, replace the collapsible spacer.
9. Install or connect the following:
   - Driveshaft
   - Brake rotors

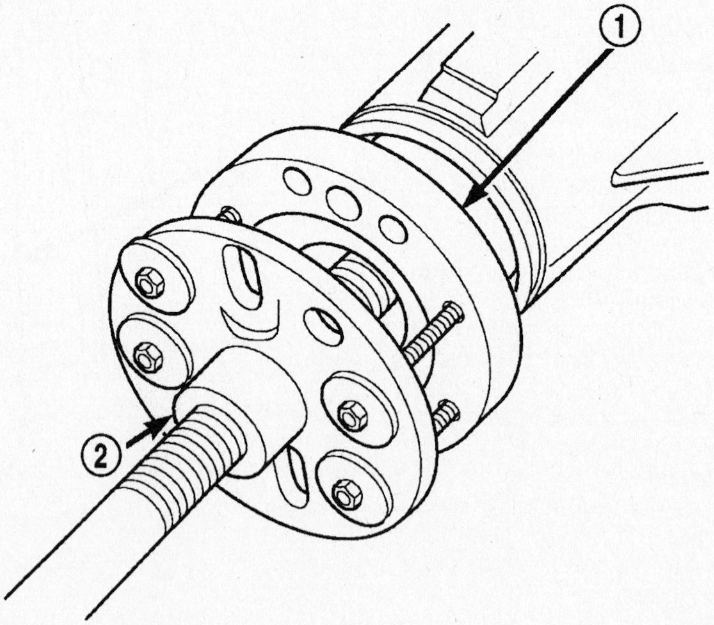

1 - COMPANION FLANGE
2 - PULLER

67189-DAKO-G28

Removing the companion flange

- Wheels

10. Fill the axle assembly with gear oil and check for leaks.

NON C-CLIP TYPE—REAR

1. Before servicing the vehicle, refer to the precautions in the beginning of this section.
2. Remove or disconnect the following:
   - Wheels
   - Brake rotors or drums
   - Driveshaft
3. Check the bearing preload with an inch lb. torque wrench.
4. Remove the pinion flange and seal.

To install:

➡Use a new pinion nut for assembly.

5. Install the new pinion seal and flange. Tighten the nut to 160 ft. lbs. (217 Nm).
6. Check the bearing preload. The bearing preload should be equal to the reading taken earlier, plus 5 inch lbs.
7. If the preload torque is low, tighten the pinion nut in 5 inch lb. increments until the torque value is reached. Do not exceed 260 ft. lbs. (353 Nm) pinion nut torque.
8. If the pinion bearing preload torque can not be attained at maximum pinion nut torque, remove one or more pinion preload shims.
9. Install or connect the following:
   - Driveshaft
   - Brake rotors or drums
   - Wheels
10. Fill the axle assembly with gear oil and check for leaks.

2004–05

FRONT AXLE

1. Raise and support the vehicle.
2. Remove skid plate, if equipped.
3. Remove both half shafts.
4. Mark the propeller shaft and pinion companion flange for installation reference.
5. Remove the front propeller shaft.
6. Rotate the pinion gear three or four times and verify pinion rotates smoothly.
7. Record pinion rotating torque with an inch pound torque wrench, for installation reference.
8. Position Holder 6719, or equivalent, against the companion flange and install a four bolts and washers into the threaded holes and tighten the bolts.
9. Remove the pinion nut.
10. Remove the companion flange with Remover C-452, or equivalent.
11. Remove pinion seal with a pry tool or a slide hammer mounted screw.

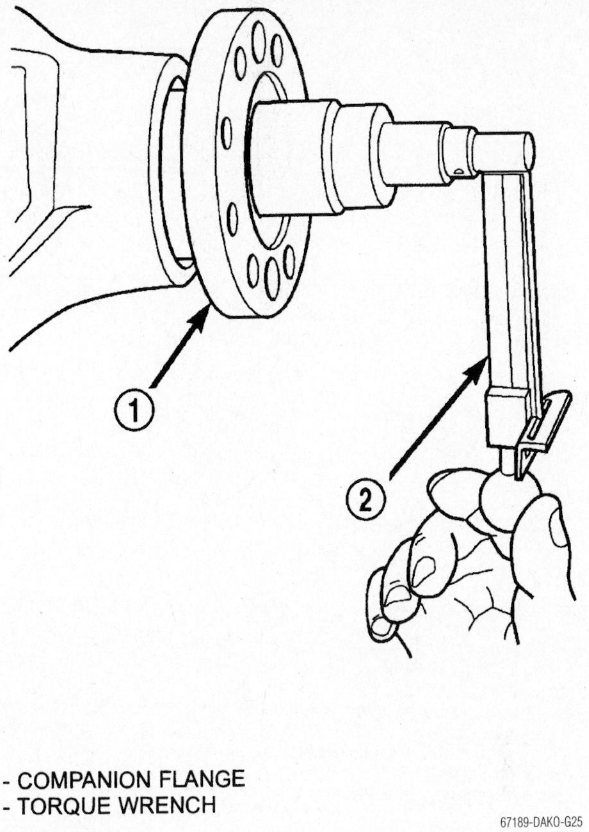

Measuring pinion rotating torque

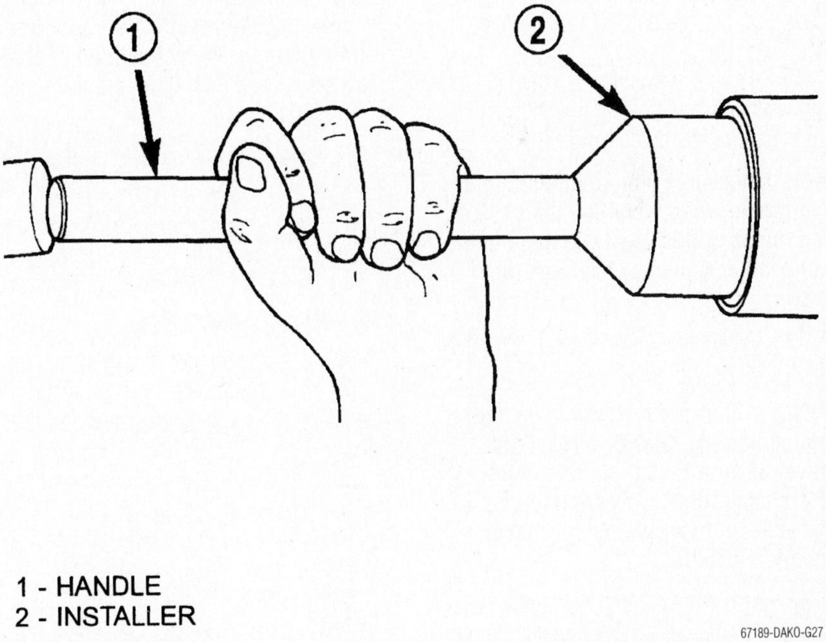

Pinion seal installer

To install:

12. Apply a light coating of gear lubricant on the lip of pinion seal.
13. install seal with Installer C-3972-A and Handle C-4171, or equivalent,
14. Install the companion flange onto the pinion with Installer C-3718 and Holder 6719A, or equivalent.
15. Position holder against the companion flange and install four bolts and washers

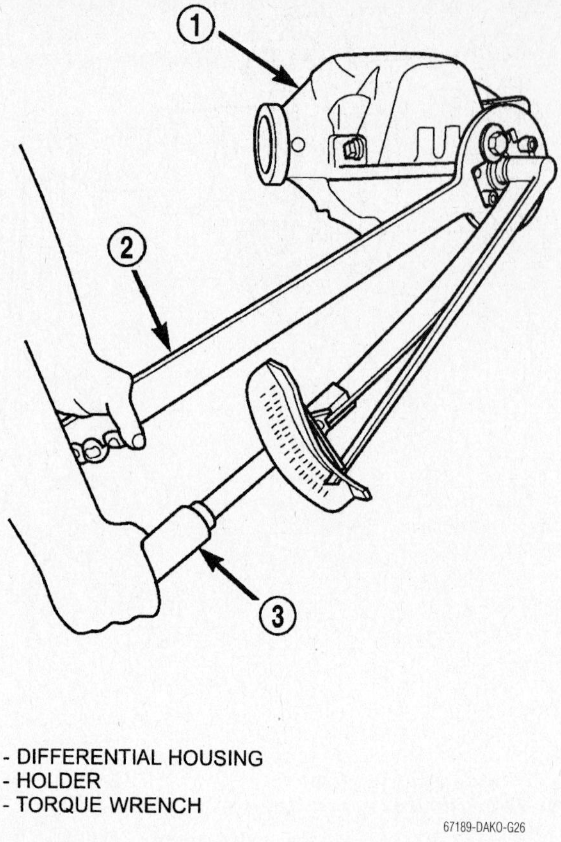

Tightening the pinion nut

into the threaded holes. Tighten the bolt and washer so that the holder is held to the flange.

16. Install a new pinion nut onto the pinion shaft and tighten the pinion nut until there is zero bearing end-play.

➡Do not exceed the minimum tightening torque when installing the companion flange at this point. Damage to the collapsible spacer or bearings may result.

17. Tighten the nut to 271 Nm (200 ft. lbs.).

➡Never loosen pinion nut to decrease pinion bearing rotating torque and never exceed specified preload torque. If preload torque or rotating torque is exceeded a new collapsible spacer must be installed.

18. Record the pinion rotating torque using a torque wrench. The rotating torque should be equal to the reading recorded during removal plus an additional 0.56 Nm (5 inch lbs.).
19. If the rotating torque is low, tighten the pinion nut in 6.8 Nm (5 ft. lbs.) increments until the proper rotating torque is achieved.

➡If the maximum tightening torque is reached prior to reaching the required rotating torque, the collapsible spacer may have been damaged. Replace the collapsible spacer.

20. Install propeller shaft with reference marks aligned.
21. Add gear lubricant to differential housing if necessary.
22. Install half shafts.

$8^1/_4$ INCH REAR AXLE

1. Raise and support the vehicle.
2. Mark the universal joint, companion flange and pinion shaft for installation reference.
3. Remove companion flange bolts and secure the shaft in an upright position to prevent damage to the rear universal joint.
4. Remove the wheel and tire assemblies.
5. Remove brake drums to prevent any drag.
6. Rotate companion flange three or four times and verify flange rotates smoothly.
7. Measure rotating torque of the pinion with an inch pound torque wrench and record the reading for installation reference.
8. Install bolts into two of the threaded holes in the companion flange 180° apart.
9. Position Holder 6719, or equivalent, against the companion flange and install a bolt and washer into one of the remaining threaded holes. Tighten the bolts so the Holder 6719, or equivalent, is held to the flange.
10. Remove the pinion nut and washer.
11. Remove companion flange with Remover C-452, or equivalent.
12. Remove pinion seal with a pry tool or slidehammer mounted screw.

To install:

➡The outer perimeter of the seal is pre-coated with a special sealant.

13. Apply a light coating of gear lubricant on the lip of pinion seal.
14. Install new pinion seal with Installer C-4076-B and Handle C-4735-1, or equivalent.
15. Install companion flange on the end of the shaft with the reference marks aligned.
16. Install bolts into two of the threaded holes in the companion flange 180° apart.
17. Position Holder 6719, or equivalent, against the companion flange and install a bolt and washer into one of the remaining threaded holes. Tighten the bolts so Holder 6719 is held to the flange.
18. Install companion flange on pinion shaft with Installer C-3718 and Holder 6719, or equivalent.
19. Install the pinion washer and a new pinion nut. The convex side of the washer must face outward.

➡Do not exceed the minimum tightening torque when installing the companion flange retaining nut at this point. Damage to collapsible spacer or bearings may result.

20. Hold companion flange with Holder 6719 and tighten the pinion nut to 285 Nm (210 ft. lbs.). Rotate pinion several revolutions to ensure the bearing rollers are seated.
21. Rotate pinion with an inch pound torque wrench. Rotating torque should be equal to the reading recorded during removal plus an additional 0.56 Nm (5 inch lbs.).

➡Never loosen pinion nut to decrease pinion bearing rotating torque and never exceed specified preload torque. If rotating torque is exceeded, a new collapsible spacer must be installed.

22. If rotating torque is low use Holder 6719 to hold the companion flange and tighten pinion nut in 6.8 Nm (5 ft. lbs.)

increments until proper rotating torque is achieved.

➡**The seal replacement is unacceptable if final pinion nut torque is less than 285 Nm (210 ft. lbs.).**

➡**The bearing rotating torque should be constant during a complete revolution of the pinion. If the rotating torque varies, this indicates a binding condition.**

23. Install propeller shaft with the installation reference marks aligned.
24. Tighten companion flange bolts to 108 Nm (80 ft. lbs.).
25. Install the brake drums.
26. Check the differential housing lubricant level.
27. Install wheel and tire assemblies and lower the vehicle.

9 1/4 INCH REAR AXLE

REMOVAL

1. Raise and support the vehicle.
2. Remove the wheel and tire assemblies.
3. Mark the universal joint, companion flange and pinion shaft for installation reference.
4. Remove the propeller shaft from the companion flange.
5. Remove the brake drums to prevent any drag.
6. Rotate the companion flange three or four times and record the pinion rotating torque with an inch pound torque wrench.
7. Install two bolts into the companion flange threaded holes, 180° apart. Position Holder 6719A, or equivalent, against the companion flange and install and tighten two bolts and washers into the remaining holes.
8. Hold the companion flange with Holder 6719A, or equivalent, and remove pinion nut and washer.
9. Remove the companion flange with Remover C-452, or equivalent.
10. Remove the pinion seal with pry tool or slidehammer mounted screw.

To install:

11. Apply a light coating of gear lubricant on the lip of pinion seal.
12. Install a new pinion seal with Installer C-3860-A and Handle C-4171, or equivalent.
13. Install the companion flange on the end of the shaft with the reference marks aligned.
14. Install two bolts into the threaded holes in the companion flange, 180° apart.
15. Position Holder 6719, or equivalent, against the companion flange and install a bolt and washer into one of the remaining threaded holes. Tighten the bolts so holder is held to the flange.
16. Install the companion flange on the pinion shaft with Installer C-3718 and Holder 6719, or equivalent.
17. Install the pinion washer and a new pinion nut. The convex side of the washer must face outward.

➡**Never exceed the minimum tightening torque 285 Nm (210 ft. lbs.) when installing the companion flange retaining nut at this point. Damage to the collapsible spacer or bearings may result.**

18. Hold the companion flange with Holder 6719 and tighten the pinion nut with a torque set to 285 Nm (210 ft. lbs.). Rotate the pinion several revolutions to ensure the bearing rollers are seated.
19. Rotate the pinion with an inch pound torque wrench. Rotating torque should be equal to the reading recorded during removal plus an additional 0.56 Nm (5 inch lbs.).

➡**Never loosen the pinion nut to decrease pinion bearing rotating torque and never exceed the specified preload torque. If the rotating torque is exceeded, a new collapsible spacer must be installed.**

20. If the rotating torque is low, use Holder 6719 to hold the companion flange and tighten the pinion nut in 6.8 Nm (5 ft. lbs.) increments until the proper rotating torque is achieved.

➡**The bearing rotating torque should be constant during a complete revolution of the pinion. If the rotating torque varies, this indicates a binding condition.**

➡**The seal replacement is unacceptable if the final pinion nut torque is less than 285 Nm (210 ft. lbs.).**

21. Install the propeller shaft with the installation reference marks aligned.
22. Tighten the companion flange bolts to 108 Nm (80 ft. lbs.).
23. Install the brake drums, wheel and tire assemblies and lower the vehicle.
24. Check the differential lubricant level.

STEERING AND SUSPENSION

Air Bag

CAUTION

Some vehicles are equipped with an air bag system. The system must be disarmed before performing service on, or around, system components, the steering column, instrument panel components, wiring and sensors. Failure to follow the safety precautions and the disarming procedure could result in accidental air bag deployment, possible injury and unnecessary system repairs.

PRECAUTIONS

Several precautions must be observed when handling the inflator module to avoid accidental deployment and possible personal injury.

- Never carry the inflator module by the wires or connector on the underside of the module.
- When carrying a live inflator module, hold securely with both hands, and ensure that the bag and trim cover are pointed away.
- Place the inflator module on a bench or other surface with the bag and trim cover facing up.
- With the inflator module on the bench, never place anything on or close to the module which may be thrown in the event of an accidental deployment.

Before servicing the vehicle, also make sure to refer to the precautions in the beginning of this section as well.

DISARMING

1. Disconnect and isolate the negative battery cable. Wait 2 minutes for the system capacitor to discharge before performing any service.
2. When repairs are completed, connect the negative battery cable.

Rack and Pinion Steering Gear

REMOVAL & INSTALLATION

2002–03 Models

1. Before servicing the vehicle, refer to the precautions in the beginning of this section.
2. Remove or disconnect the following:
   - Front wheels

- Outer tie rod ends
- Steering shaft coupler
- Power steering hoses
- Steering gear

To install:

3. Install or connect the following:
   - Steering gear. Tighten the bolts to 190 ft. lbs. (258 Nm).
   - Power steering hoses. Tighten the fittings to 25 ft. lbs. (35 Nm).
   - Steering shaft coupler. Tighten the bolt to 36 ft. lbs. (49 Nm).
   - Outer tie rod ends. Tighten the nuts to 65 ft. lbs. (88 Nm).
   - Front wheels

2004 Models

2-WHEEL DRIVE

➡The steering column on vehicles with an automatic transmission may not be equipped with an internal locking shaft that allows the ignition key cylinder to be locked with the key. Alternative methods of locking the steering wheel for service will have to be used.

1. Siphon out as much power steering fluid as possible from the pump.
2. Lock the steering wheel.
3. Raise and support the vehicle.
4. Remove the front tires.
5. Remove the nuts from the tie rod ends.
6. Separate tie rod ends from the knuckles with Puller C-3894-A, or equivalent.
7. Remove the power steering lines from the gear.
8. Remove the lower coupler bolt and slide the coupler off the gear.
9. Remove the mounting bolts from the gear to the front crossmember and remove the gear.

To install:

➡Before installing gear inspect bushings and replace if worn or damaged.

10. Install gear on front crossmember and tighten mounting bolts to 271 Nm (200 ft. lbs.).
11. Slide shaft coupler onto gear. Install new bolt and tighten to 49 Nm (36 ft. lbs.).
12. Clean tie rod end studs and knuckle tapers.
13. Install tie rod ends into the steering knuckles and tighten the nuts to 81 Nm (60 ft. lbs.).
14. Install power steering lines to steering gear and tighten the hose to 31 Nm (23 ft. lbs.).
15. Install the front tires.

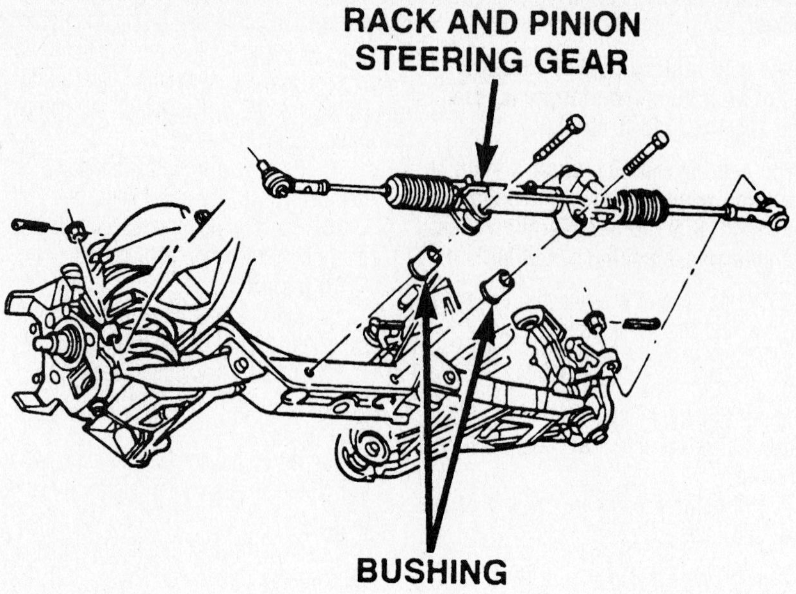

Rack and pinion steering gear mounting used on the 2002–03 2WD Dakota models

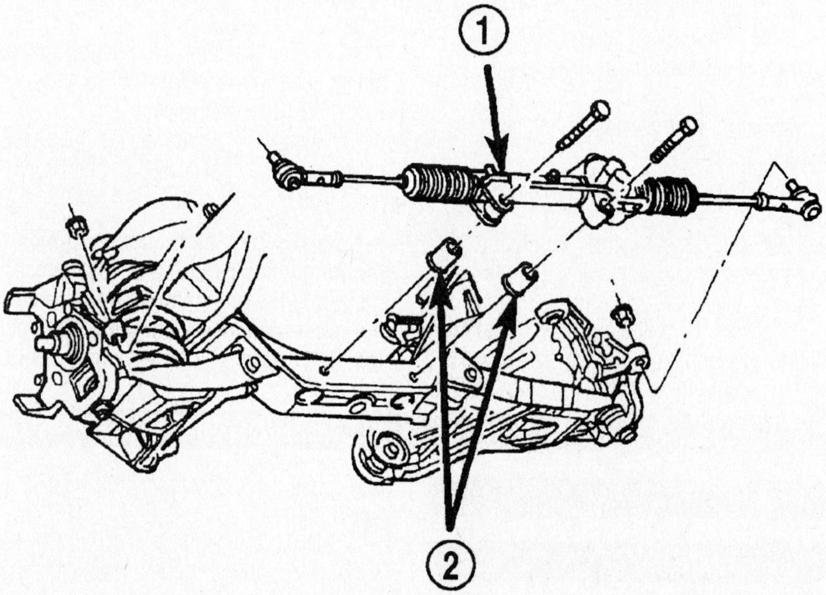

Steering gear mounting—2004 2 wheel drive

16. Remove support and lower vehicle.
17. Unlock the steering wheel.
18. Fill system with fluid.
19. Adjust the toe position.

4-WHEEL DRIVE

➡The steering column on vehicles with an automatic transmission may not be equipped with an internal locking shaft that allows the ignition key cylinder to be locked with the key. Alternative methods of locking the steering wheel for service will have to be used.

1. Siphon out as much power steering fluid as possible from the pump.
2. Lock the steering wheel.

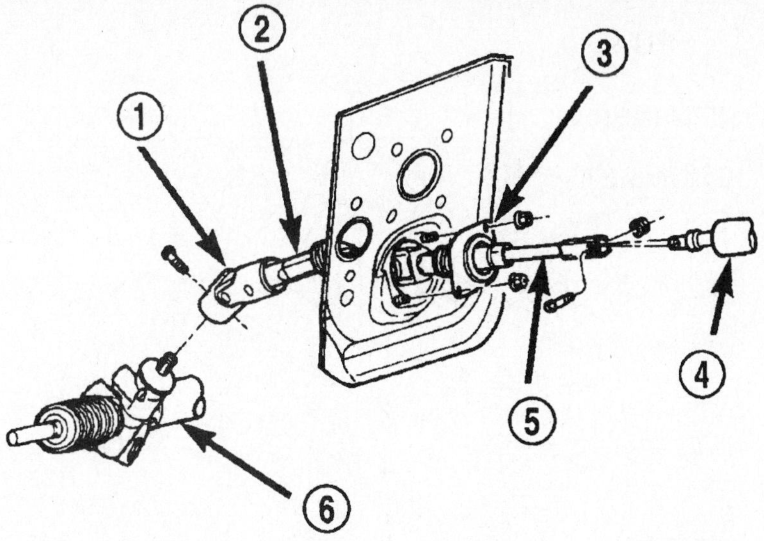

1 - COUPLER
2 - LOWER SHAFT
3 - TOE PLATE
4 - STEERING COLUMN
5 - UPPER SHAFT
6 - RACK AND PINION STEERING GEAR

67189-DAKO-G24

Steering gear coupler

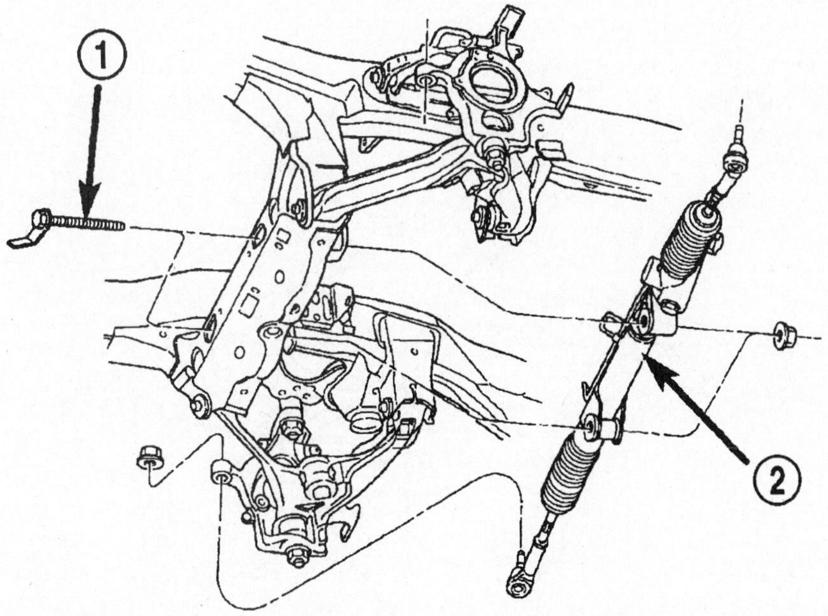

1 - MOUNTING BOLT
2 - RACK AND PINION STEERING GEAR

67189-DAKO-G22

Steering gear mounting—2004 4 wheel drive

3. Raise and support the vehicle.
4. Remove the front tires.
5. Remove the nuts from the tie rod ends.
6. Separate tie rod ends from the knuckles with Puller C-3894-A, or equivalent.
7. Remove the splash shield from under the front end to gain access to the gear.
8. Remove the skid plate.
9. Remove the power steering lines from the gear.
10. Remove the lower coupler bolt and slide the coupler off the gear.
11. Remove the mounting bolts from the gear to the front crossmember. Slide the gear to the right side of the vehicle. Then tilt the left end of the gear down and remove the gear.
12. Adjust the toe position.

To install:

➡Before installing the gear inspect bushings and replace if worn or damaged.

13. Install the gear on the front crossmember and tighten the mounting bolts to 271 Nm (200 ft. lbs.).
14. Slide the shaft coupler onto gear. Install a new bolt and tighten to 49 Nm (36 ft. lbs.).
15. Clean the tie rod end studs and knuckle tapers.
16. Install the tie rod ends into the steering knuckles and tighten the nuts to 81 Nm (60 ft. lbs.).
17. Install the power steering lines to steering gear and tighten the hose to 31 Nm (23 ft. lbs.).
18. Install the splash shield.
19. Install the skid plate.
20. Install the front tires.
21. Remove the support and lower the vehicle.
22. Unlock the steering wheel.
23. Fill the system with fluid.

2005 Models

1. Siphon out as much power steering fluid as possible from the pump.
2. Lock the steering wheel.
3. Raise and support the vehicle.
4. Remove the front tires.
5. Remove the nuts from the tie rod ends.
6. Separate tie rod ends from the knuckles.
7. Remove the steering gear pinch bolt.
8. Remove the lower steering coupling from the steering gear.

9. Turn the steering gear to the full right position.

➡**Protect the end of hoses to prevent contamination to the system and damage to the O-rings.**

10. Remove the power steering lines from the gear.
11. Remove the steering gear mounting bolts and nuts.
12. Tip the gear forward to allow clearance and move to the right then tip the gear downward on the left side to remove from the vehicle.

To install:

➡**Before installing gear inspect bushings and replace if worn or damaged.**

13. Install gear to the vehicle and tighten mounting nuts and bolts to 258 Nm (190 ft. lbs.).
14. Install power steering lines to steering gear and tighten the pressure hose to 31 Nm (23 ft. lbs.) and tighten the return hose to 37 Nm (27 ft. lbs.).
15. Slide the shaft coupler onto gear. Install new bolt and tighten to 49 Nm (36 ft. lbs.).
16. Clean tie rod end studs and knuckle tapers.
17. Install tie rod ends into the steering knuckles and tighten the nuts to 75 Nm (55 ft. lbs.).
18. Install the front tires.
19. Remove the support and lower the vehicle.
20. Unlock the steering wheel.
21. Fill system with fluid.
22. Adjust the toe position.

Front Shock Absorber

REMOVAL & INSTALLATION

2002–03

1. Before servicing the vehicle, refer to the precautions in the beginning of this section.
2. Remove or disconnect the following:
3. Install or connect the following:
   - Front wheel
   - Upper mount nut
   - Lower mount bolt
   - Shock absorber

To install:

4. Install or connect the following:
   - Shock absorber. Tighten the lower bolt to 100 ft. lbs. (136 Nm) and the upper nut to 30 ft. lbs. (41 Nm).
   - Front wheel

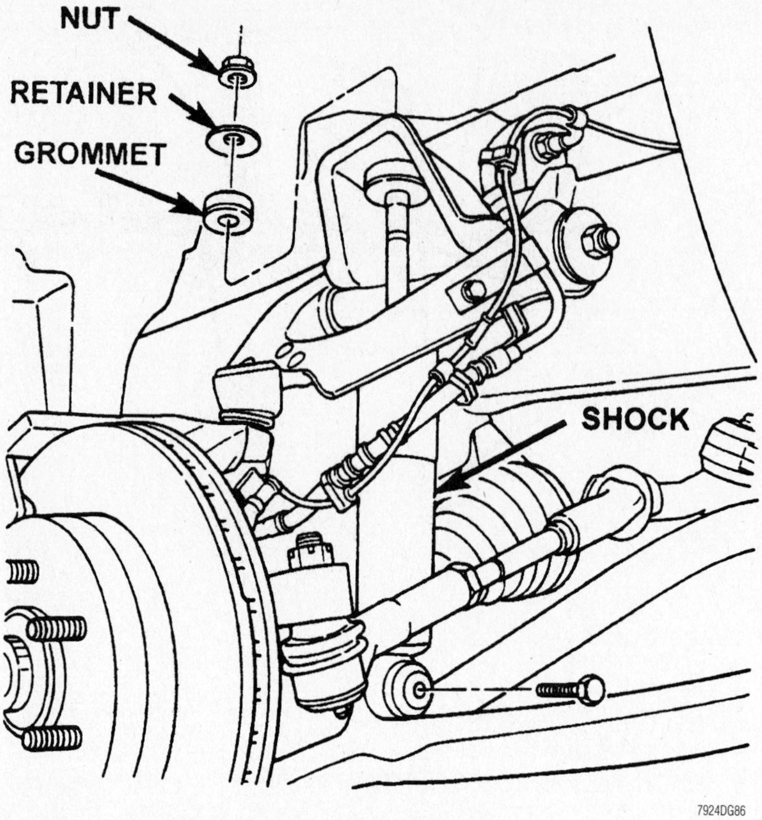

Front shock absorber mounting—2002–03 4WD Dakota models

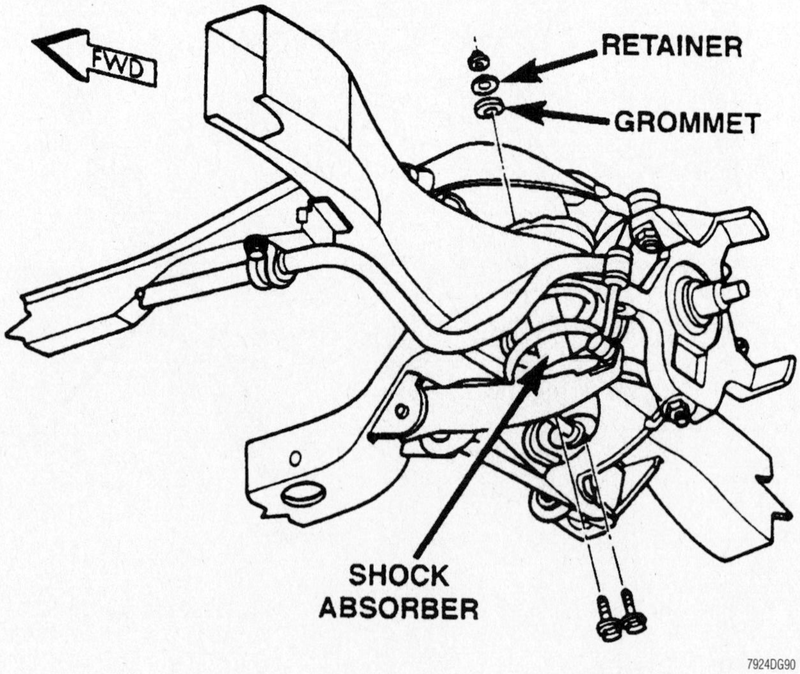

Front shock absorber mounting—2002–03 2WD Dakota models

2004 2-Wheel Drive

1. Remove the upper shock nut, retainer and grommet from the shock absorber stud.
2. Raise and support the vehicle.
3. Remove the lower mounting bolts and remove shock absorber through the lower suspension arm.

To install:

➡The upper shock nut must be replaced or use Mopar Lock 'N Seal or Loctite® 242 on existing nut.

4. Install the lower retainer (lower retainer is stamped with an L) and grommet on the shock absorber stud and extend the shock. Insert the shock absorber through the lower suspension arm and upper mounting hole.
5. Install the lower mounting bolts and tighten to 28 Nm (21 ft. lbs.).
6. Remove the support and lower the vehicle.
7. Install the upper grommet and retainer (upper retainer is stamped with a U) on the shock absorber stud. Install a new nut or use Mopar Lock 'N Seal or Loctite® 242 on the existing nut and tighten to 26 Nm (19 ft. lbs.).

2004 4-WHEEL DRIVE

1. Raise and support vehicle.
2. Remove the upper shock absorber nut, retainer and grommet.
3. Remove the lower bolt and remove the shock absorber.

To install:

➡Upper shock nut must be replaced or use Mopar Lock 'N Seal or Loctite® 242 on existing nut.

4. Install the lower retainer (lower retainer is stamped with an L) and grommet on the shock absorber stud. Insert the shock absorber through the frame bracket hole.
5. Install the lower bolt and tighten the bolt to 108 Nm (80 ft. lbs.).
6. Install the upper grommet, retainer (upper retainer is stamped with a U) and new nut or use Mopar Lock 'N Seal or Loctite® 242 on existing nut, on the shock absorber stud. Tighten nut to 26 Nm (19 ft. lbs.).

2005

The unit is a MacPherson strut design.

1. Raise and support the vehicle.
2. Remove the tire and wheel assembly.
3. Support the lower control arm outboard end.
4. Remove the upper shock nuts.

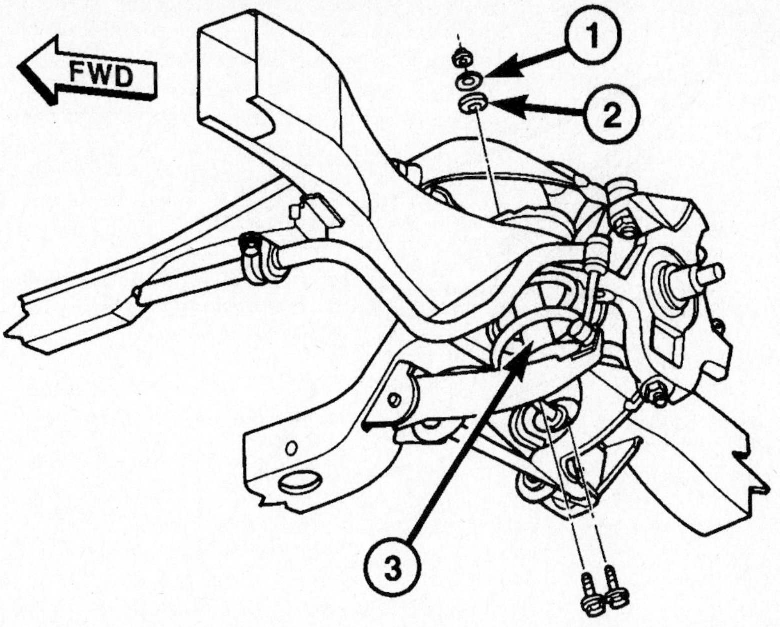

1 - RETAINER
2 - GROMMET
3 - SHOCK ABSORBER

67189-DAKO-G21

Front shock absorber mounting—2004 2 wheel drive

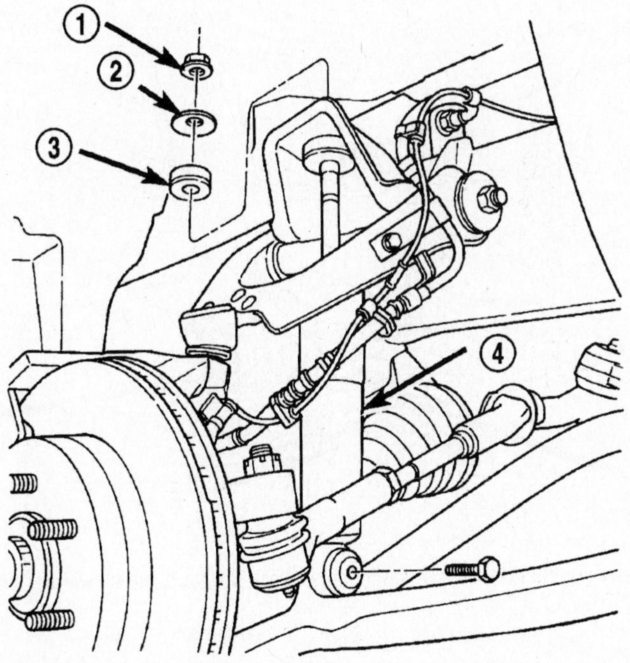

1 - NUT
2 - RETAINER
3 - GROMMET
4 - SHOCK

67189-DAKO-G20

Front shock absorber mounting—2004 4 wheel drive

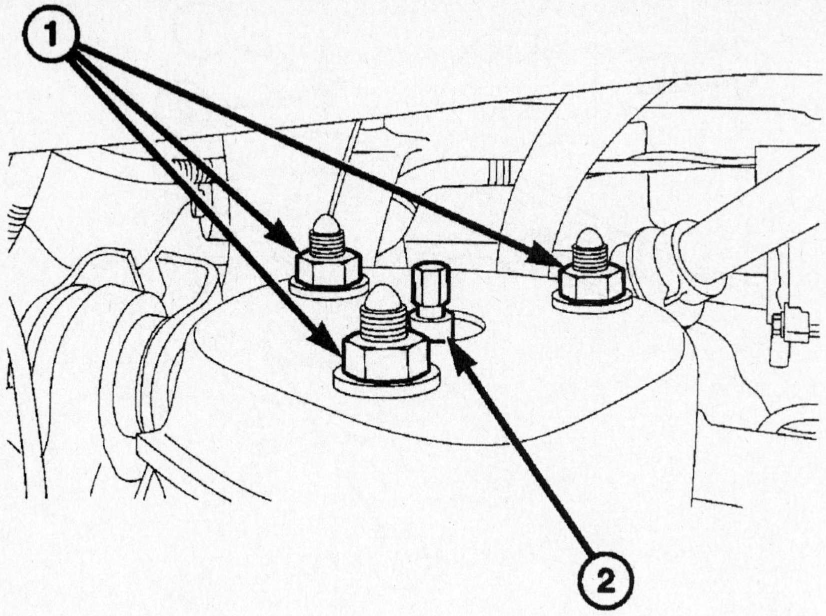

Remove the upper shock nuts (1)—2005 models

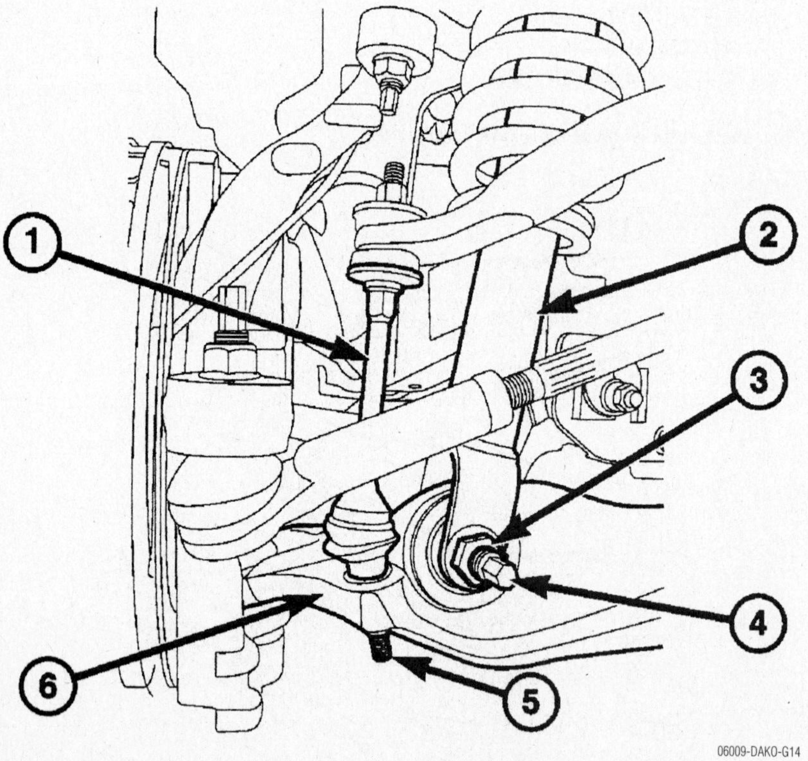

Stabilizer link (1), strut (2), nut (3), lower strut bolt (4), link lower nut (5), lower control arm (6)

5. Remove the stabilizer link lower nut and then separate the stabilizer link from the lower control arm to gain access to the lower shock nut.
6. Remove the lower shock bolt and nut.
7. Remove the shock.

To install:

➡All suspension components should be tightened with the weight of the vehicle on them (curb height).

8. Install the upper part of the shock into the frame bracket.
9. Install the upper nuts. Tighten to 102 Nm (75 ft. lbs.).
10. Install the lower part of the shock into the lower control arm shock bushing.
11. Position shock module clevis to lower control arm. Install bolt so head of bolt is facing rear of vehicle and hand start nut. Tighten the bolt and nut to 81 Nm (60 ft. lbs.).
12. Install the stabilizer link lower nut to the lower control arm. Tighten to 125 ft. lbs. (169 Nm).
13. Remove the support from the lower control arm outboard end.
14. Install the tire and wheel assembly.
15. Remove the support and lower the vehicle.

Rear Shock Absorbers

REMOVAL & INSTALLATION

2002–03

1. Before servicing the vehicle, refer to the precautions in the beginning of this section.
2. Support the axle.
3. Remove or disconnect the following:
   - Upper bolt
   - Lower bolt
   - Shock absorber

To install:

4. Install the bolts through the brackets and shock and tighten them as follows:
   - Tighten the lower bolt and nut to 60 ft. lbs. (81 Nm) and the upper bracket nuts to 20 ft. lbs. (27 Nm)

2004–05 Models

1. Raise the vehicle and support rear axle.
2. Remove the shock absorber lower nut and bolt from the axle bracket.
3. Remove the shock absorber upper nut and bolt from the frame bracket and remove the shock absorber.

To install:

4. Install the shock absorber and upper mounting bolt and nut. Tighten the nut to 95 Nm (70 ft. lbs.) for 2004 models; 102 Nm (75 ft. lbs.) for 2005 models.
5. Install the shock absorber into the axle bracket. Install the bolt and nut and tighten the nut to 95 Nm (70 ft. lbs.) for 2004 models; 102 Nm (75 ft. lbs.) for 2005 models.
6. Remove the axle support and lower the vehicle.

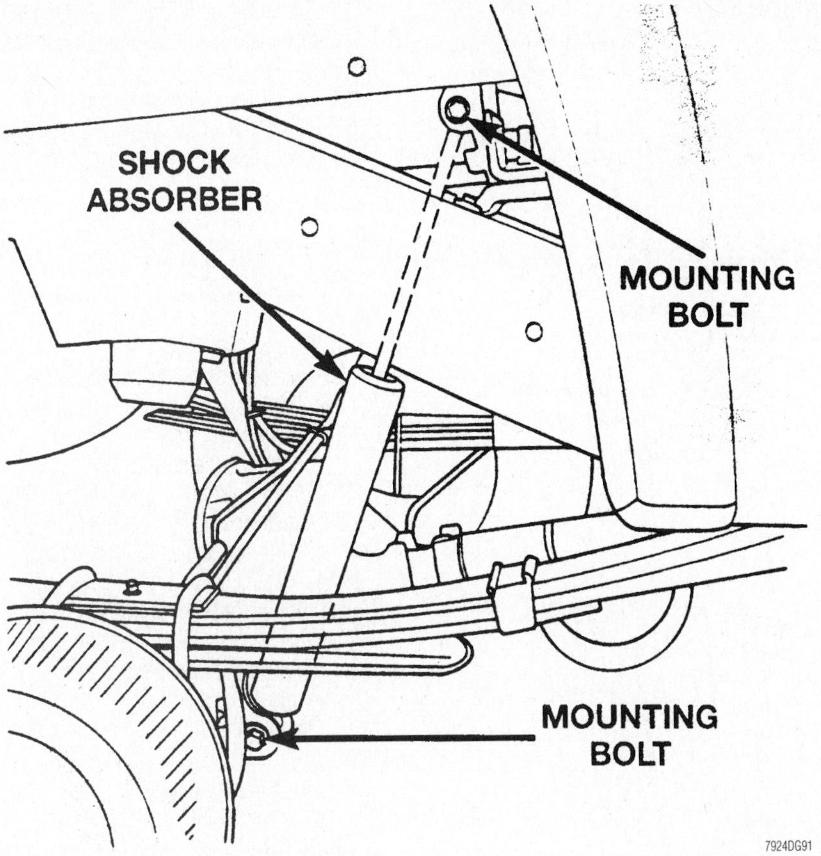

Rear shock absorber mounting—2002–03

Coil Spring

REMOVAL & INSTALLATION

2002–03

1. Before servicing the vehicle, refer to the precautions in the beginning of this section.
2. Support the lower control arm on a floor jack.
3. Remove or disconnect the following:
   - Front wheel
   - Shock absorber
   - Stabilizer bar link
   - Lower ball joint.
4. Lower the jack and remove the coil spring.

To install:

5. Install the coil spring and raise the control arm into position.
6. Install or connect the following:
   - Lower ball joint. Tighten the nut to 135 ft. lbs. (183 Nm).
   - Stabilizer bar link
   - Shock absorber
   - Front wheel

2004

1. Raise and support the vehicle.
2. Remove wheel and tire assembly.
3. Remove the stabilizer bar link from the lower suspension arm.
4. Remove the shock absorber.
5. Install a spring compressor up through the lower suspension arm, coil spring and upper shock mounting hole.
6. Tighten the tool lower nut to compress the coil spring.
7. Remove the lower ball joint nut and separate the ball joint from the knuckle.
8. Loosen the spring compressor lower nut to relieve spring tension.
9. Remove the tool and pull down on the lower suspension arm to remove the spring.

To install:

➡The ramped or open end of the coil spring is the bottom of the spring.

10. Tape the isolator pad to the top of the coil spring. Position the spring in the lower suspension arm pocket. Be sure that the coil spring is seated in the pocket.
11. Install a spring compressor up through the lower suspension arm, coil spring upper shock mounting hole.
12. Tighten the tool nut to compress the coil spring.
13. Install the lower ball joint into the knuckle and tighten the nut to 127 Nm (94 ft. lbs.). Install cotter pin.

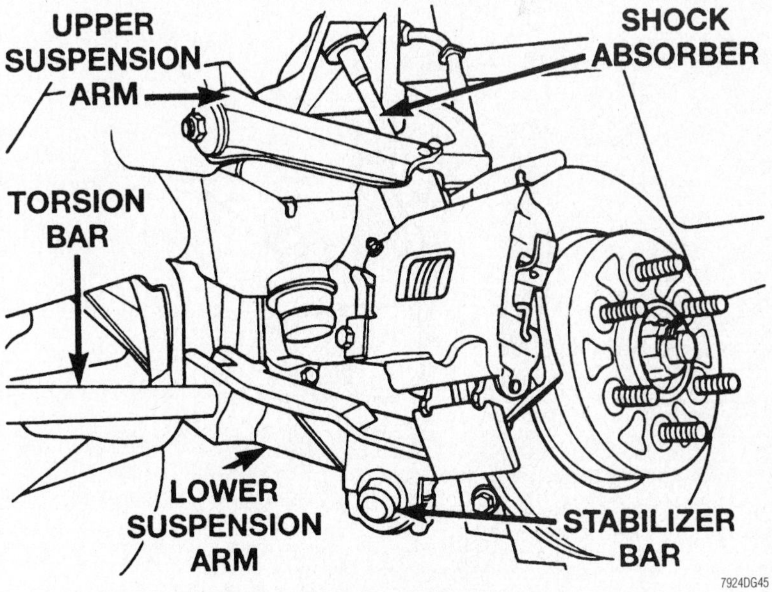

Front suspension components—2002–03 4WD models

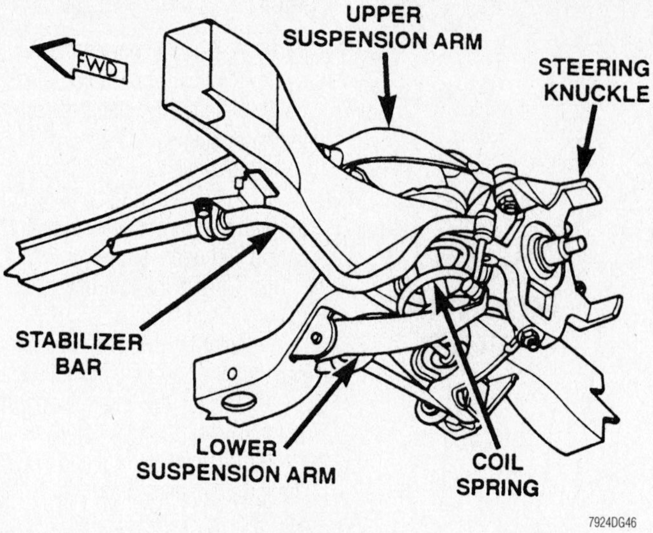

Front suspension components—2002–03 2WD models

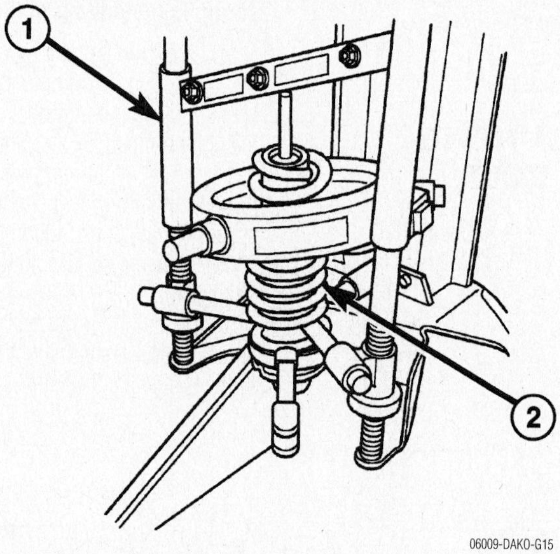

Strut assembly (2) mounted in the tool (1)

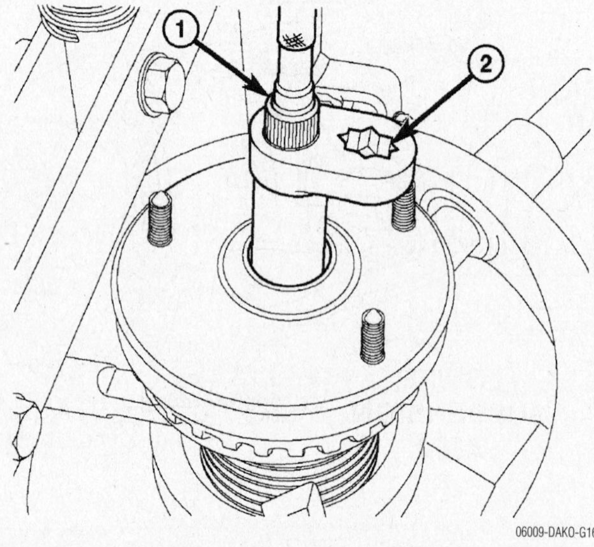

Position Wrench, Special Tool 9362 (2), or equivalent, on shock shaft retaining nut (1)

14. Remove the spring compressor tool.
15. Install the stabilizer bar link to the lower suspension arm and tighten nut to 47 Nm (35 ft. lbs.).
16. Install the shock absorber.
17. Install the wheel and tire assembly.
18. Remove support and lower the vehicle.

2005

The coil spring is part of the strut assembly.

1. Remove the shock.
2. Install the shock assembly in the Branick 7200T spring removal/installation tool or equivalent.
3. Compress the spring.
4. Position Wrench, Special Tool 9362, or equivalent, on shock shaft retaining nut. Next, insert 8 mm socket though Wrench onto hex located on end of shock shaft. While holding shock shaft from turning, remove nut from shock shaft using Wrench.
5. Remove the upper shock nut.
6. Remove the shock.
7. Remove the shock upper mounting plate.
8. Remove and inspect the upper and lower spring isolators.

To install:

9. Install the lower isolator.
10. Install the upper isolator.
11. Position the shock into the coil spring.
12. Install the upper shock mounting plate.
13. Install Wrench (on end of a torque wrench), Special Tool 9362, on shock shaft retaining nut. Next, insert 8 mm socket though Wrench onto hex located on end of shock shaft. While holding shock shaft from turning, tighten nut using Wrench to 90 Nm (66 ft. lbs.) torque.
14. Install the shock upper mounting nut.
15. Decompress the spring.
16. Remove the shock assembly from the spring compressor tool.
17. Install the shock assembly.

Leaf Spring

REMOVAL & INSTALLATION

2002–03

1. Before servicing the vehicle, refer to the precautions in the beginning of this section.
2. Support the vehicle at the frame rails.
3. Support the rear axle with a jack.

4. Remove or disconnect the following:
   - Rear wheel
   - Stabilizer bar link
   - Axle U-bolts
   - Spring bracket
   - Leaf spring

To install:

➡The weight of the vehicle must be supported by the springs when the spring eye and stabilizer bar fasteners are tightened.

5. Install or connect the following:
   - Leaf spring
   - Spring bracket
   - Axle U-bolts. Tighten the nuts to 52 ft. lbs. (70 Nm).
   - Stabilizer bar link
   - Rear wheel
6. Tighten the front spring eye bolt and nut to 115 ft. lbs. (156 Nm). Tighten the rear spring eye bolt and nut to 80 ft. lbs. (108 Nm). Tighten the stabilizer bar nuts 55 ft. lbs. (74 Nm).

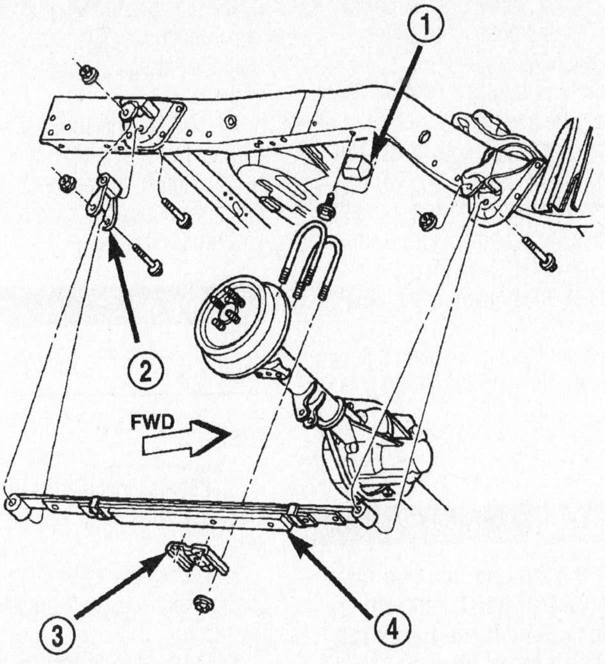

Leaf spring—2 wheel drive

2004

CAUTION

The rear of the vehicle must be lifted only with a jack or hoist. The lift must be placed under the frame rail crossmember located aft of the rear axle. Use care to avoid bending the side rail flange.

1. Raise the vehicle at the frame.
2. Use a hydraulic jack to relieve the axle weight.
3. Remove the wheel and tire assemblies.
4. Remove the nuts, the U-bolts and spring plate from the axle and.
5. Remove the nut and bolt from the spring front eye.
6. Remove the nut and bolt that attaches the spring shackle to the rear frame bracket.
7. Remove the spring from the vehicle.
8. Remove the shackle from the spring.

To install:

9. Install the spring shackle on the spring finger tight.
10. Position the spring on the rear axle pad. Make sure the spring center bolt is inserted in the pad locating hole.
11. Align front spring eye with the bolt hole in the front frame bracket. Install the spring eye bolt and nut and tighten the spring eye nut finger-tight.
12. Align spring shackle eye with the

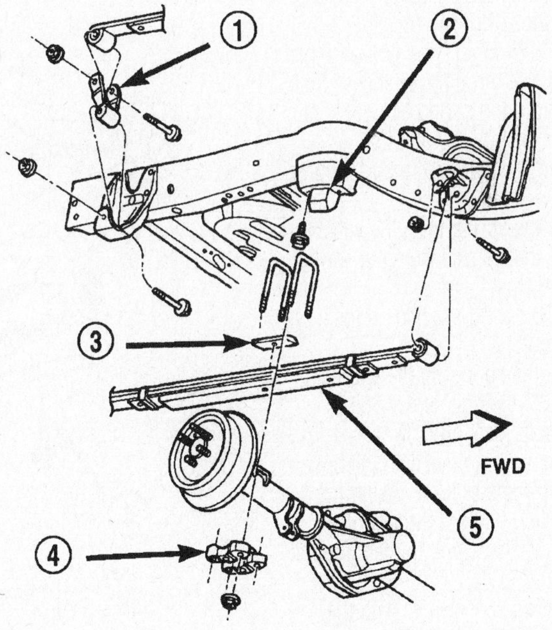

Leaf spring—4 wheel drive

bolt hole in the rear frame bracket. Install the bolt and nut and tighten the spring shackle eye nut finger-tight.

13. Install the spring seat (4x4 only), U-bolts, spring plate and nuts.
14. Tighten the U-bolt nuts to 149 Nm (110 ft. lbs.).
15. Install the wheel and tire assemblies.
16. Remove the support stands from under the frame rails. Lower the vehicle until the springs are supporting the weight of the vehicle.
17. Tighten the spring eye pivot bolt nut and all shackle nuts to 163 Nm (120 ft. lbs.).

2005

CAUTION

The rear of the vehicle must be lifted only with a jack or hoist. The lift must be placed under the frame rail crossmember located aft of the rear axle. Use care to avoid bending the side rail flange.

1. Raise the vehicle at the frame.
2. Use a hydraulic jack to relieve the axle weight.
3. Remove the wheel and tire assemblies.
4. Remove the nuts, the U-bolts and spring plate from the axle.
5. Loosen and remove the bolt and then remove the flag nut through the access hole in the bracket from the spring front eye.
6. Remove the nut and bolt that attaches the spring shackle to the rear frame bracket.
7. Remove the spring from the vehicle.
8. Remove the shackle from the spring.

To install:

9. Install the spring shackle on the spring finger tight.
10. Position the spring on the rear axle pad. Make sure the spring center bolt is inserted in the pad locating hole.
11. Align front spring eye with the bolt hole in the front frame bracket. Install the spring eye bolt and flag nut through the access hole in the frame and tighten the bolt finger-tight.
12. Align spring shackle eye with the bolt hole in the rear frame bracket. Install the bolt and nut and tighten the spring shackle eye nut finger-tight.
13. Install the U-bolts, spring plate and nuts.
14. Tighten the U-bolt nuts to 149 Nm (110 ft. lbs.).
15. Install the wheel and tire assemblies.
16. Remove the support stands from under the frame rails. Lower the vehicle until the springs are supporting the weight of the vehicle.
17. Tighten the spring eye pivot bolt and flag nut to 163 Nm (120 ft. lbs.).
18. Tighten the upper shackle bolt and nut and the lower shackle bolt and nut to 163 Nm (120 ft. lbs.).

Torsion bar

REMOVAL & INSTALLATION

2002–03

1. Before servicing the vehicle, refer to the precautions in the beginning of this section.
2. Loosen the adjustment bolt to remove spring load. Note the number of turns for installation.
3. Remove or disconnect the following:
   - Adjustment bolt, swivel and bearing
   - Torsion bar and anchor
4. Separate the torsion bar and anchor.

To install:

5. Assemble the torsion bar and anchor.
6. Install or connect the following:
   - Torsion bar and anchor
   - Adjustment bolt, swivel and bearing
7. Tighten the adjustment bolt the recorded number of turns.

2004

➡The left and right side torsion bars are NOT interchangeable. The bars are identified and stamped R or L, for right or left. The bars do not have a front or rear end and can be installed with either end facing forward.

1. Raise and support the vehicle with the front suspension hanging.
2. Turn the adjustment bolt counterclockwise to release spring load.

➡Count and record the number of turns for installation reference.

3. Remove the adjustment bolt from swivel.
4. Remove torsion bar and anchor. Remove anchor from torsion bar.
5. Remove all foreign material from torsion bar mounting in anchor and suspension arm.
6. Inspect adjustment bolt, bearing and swivel for damage.

To install:

7. Insert torsion bar ends into anchor and suspension arm.

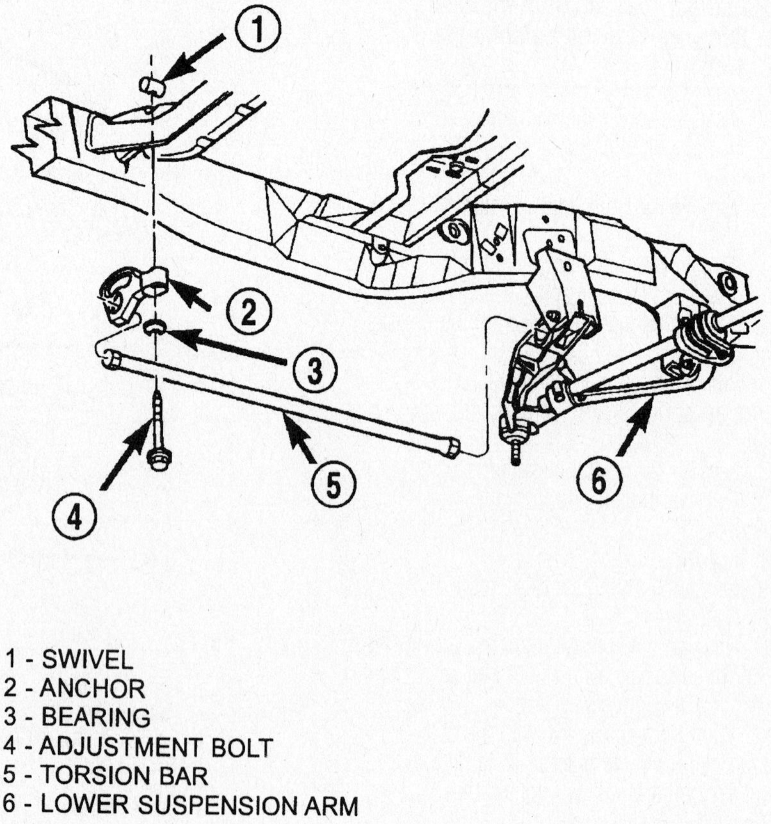

Torsion bar

8. Position anchor and bearing in frame crossmember. Install adjustment bolt through bearing, anchor and into the swivel.
9. Turn adjustment bolt clockwise the recorded amount of turns.
10. Lower vehicle and adjust the front suspension height.

Upper Ball Joint

REMOVAL & INSTALLATION

The Dakota models utilize an upper control arm with an integral ball joint. If the ball joint is damaged or worn, the upper control arm must be replaced.

Lower Ball Joint

REMOVAL & INSTALLATION

2002–04

These models utilize a lower control arm with an integral ball joint. If the ball joint is damaged or worn, the upper control arm must be replaced.

2005

1. Remove the tire and wheel assembly.
2. Remove the brake caliper and rotor.
3. Remove the outer tie rod retaining nut from the knuckle.
4. Separate the tie rod from the steering knuckle using special tool C-3894-A.
5. Remove the upper ball joint nut, then separate the upper ball joint from the knuckle using special tool 8677, or equivalent.
6. Remove the lower ball joint nut, then separate the lower ball joint from the steering knuckle using special tool 8677.
7. Remove the steering knuckle.
8. Move the halfshaft to the side and support the halfshaft out of the way (4WD) only.
9. Remove the snapring from the ball joint flange.

➡Extreme pressure lubrication must be used on the threaded portions of the tool. This will increase the longevity of the tool and insure proper operation during the removal and installation process.

10. Press the ball joint (3) from the lower control arm (4) using special tools C-4212-F (Press) (1), 8445-3 (Driver) (2) and 9604 (Receiver) (5).

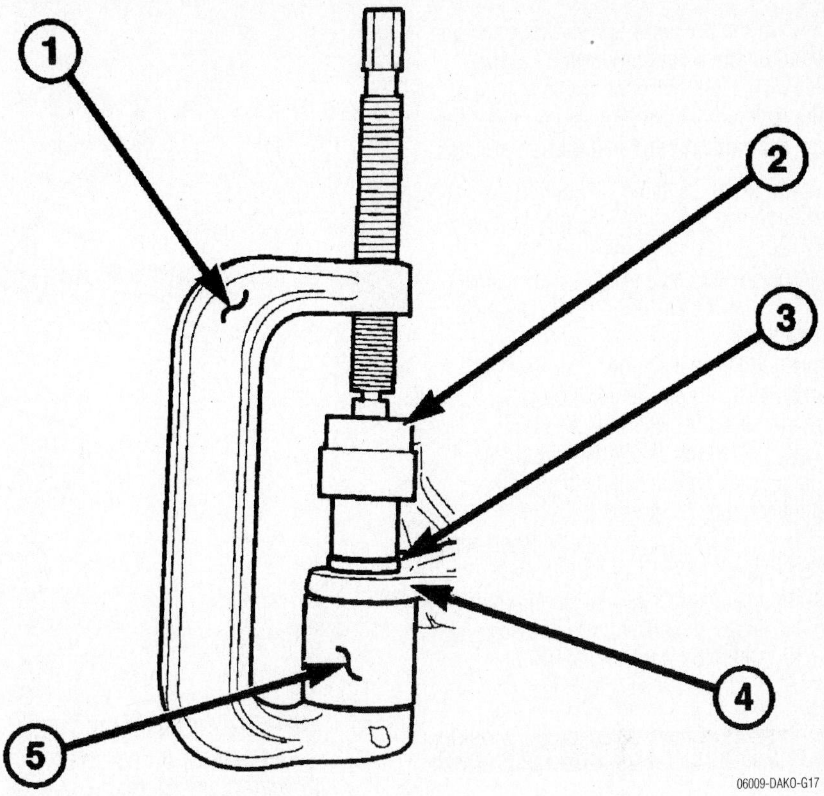

06009-DAKO-G17

Removing the lower ball joint from the control arm

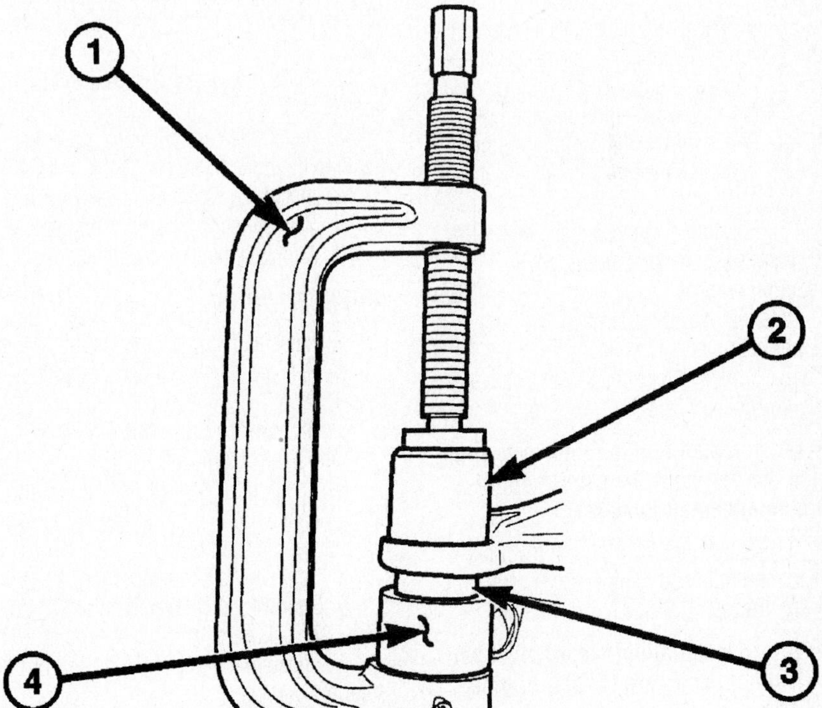

06009-DAKO-G18

Installing the lower ball joint

To install:

➡Extreme pressure lubrication must be used on the threaded portions of the tool. This will increase the longevity of the tool and insure proper operation during the removal and installation process.

11. Install the ball joint (3) into the control arm and press in using special tools C-4212-F (press) (1), 8441-4 (Receiver) (2) and 9654-1 (Driver) (4).
12. Install the snapring around the ball joint flange.
13. Remove the support for the halfshaft and install into position 4X4 only.
14. Install the steering knuckle.
15. Install the tie rod end into the steering knuckle, then install the retaining nut and tighten to 75 Nm (55 ft. lbs.).
16. Install and tighten the halfshaft nut (if equipped) to 251 Nm (185 ft. lbs.).
17. Install the brake caliper and rotor.
18. Install the tire and wheel assembly.
19. Check the vehicle ride height.
20. Perform a wheel alignment.

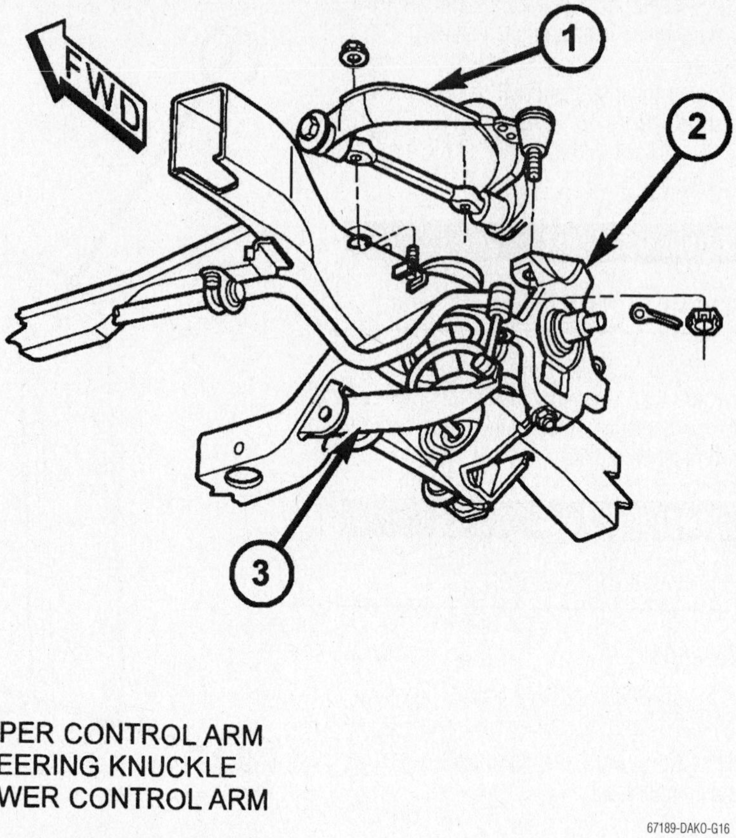

Upper control arm—2002–04 2 wheel drive

Upper Control Arm

REMOVAL & INSTALLATION

2002–04

2-WHEEL DRIVE

1. Before servicing the vehicle, refer to the precautions in the beginning of this section.
2. Raise and support the vehicle.
3. Remove wheel and tire assembly.
4. Remove the caliper mounting bolts.
5. Remove brake hose bracket from the arm.
6. Position a hydraulic jack under the arm and raise the jack to unload the rebound bumper.
7. Remove cotter pin and nut from upper ball joint.
8. Separate upper ball joint from steering knuckle.

➡When installing the remover to separate the ball joint, be careful not to damage the ball joint seal.

9. Remove the control arm pivot bar mounting nuts and remove the control arm.

To install:

➡Before installation, insure pivot bar adjustment bolts are in their original location.

10. Position the control arm pivot bar on adjustment bolts. Install nuts and tighten to 210 Nm (155 ft. lbs.).
11. Position steering knuckle on upper ball joint. Tighten the upper ball joint nut to 81 Nm (60 ft. lbs.) and install a new cotter pin.
12. Install the brake hose bracket to the arm.
13. Install the caliper mounting bolts. Tighten the bolts to 30 Nm (22 ft. lbs.).
14. Install the wheel and tire assembly.
15. Remove support and lower vehicle.
16. Align front end to specifications.

4-WHEEL DRIVE

1. Raise and support the vehicle.
2. Remove the wheel and tire assembly.
3. Remove the brake hose brackets from the control arm.
4. Position a hydraulic jack under the lower suspension arm and raise the jack to unload the rebound bumper.
5. Remove the shock absorber.
6. Remove the cotter pin and nut from the upper ball joint.
7. Separate the upper ball joint from the steering knuckle.

➡When installing the remover to separate the ball joint, be careful not to damage the ball joint seal.

To install:

8. Position the control arm pivot bar on the mounting bracket. Install the bolts and tighten (temporarily) to 136 Nm (100 ft. lbs.).
9. Insert the ball joint in the steering knuckle and tighten the ball joint nut to 81 Nm (60 ft. lbs.) then install a new cotter pin.
10. Install the shock absorber.

➡The upper shock nut must be replaced or use Mopar Lock 'N Seal or Loctite® 242 on the existing nut.

11. Remove the hydraulic jack.
12. Attach the brake hose brackets to the control arm.
13. Tighten the upper control arm pivot bolts to 204 Nm (150 ft. lbs.).
14. Install the wheel and tire assembly.
15. Remove the support and lower the vehicle.
16. Align the front suspension.

2005

1. Raise and support vehicle.
2. Remove wheel and tire assembly.
3. Remove the nut from upper ball joint.
4. Separate upper ball joint from the steering knuckle with a ball joint tool.

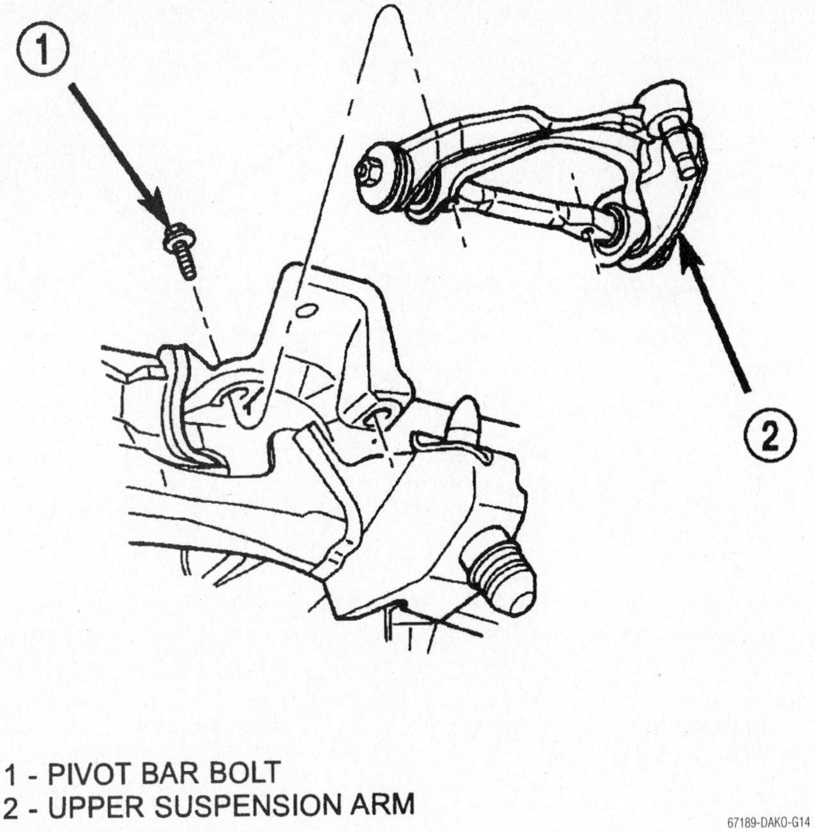

Upper control arm—2002–04 4 wheel drive

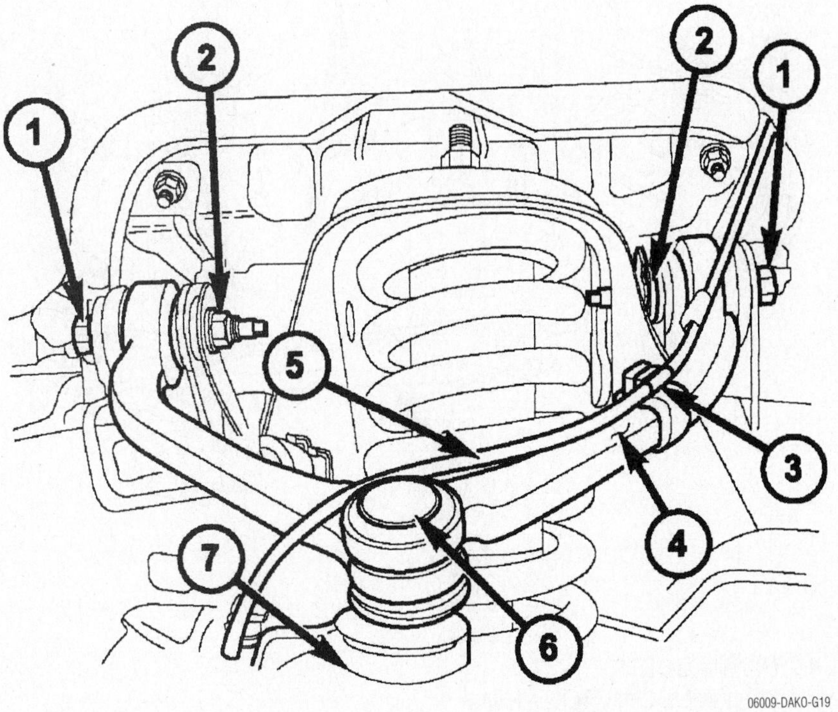

Upper control arm mounting, 2005 models. (1) bolts, (2) nuts, (3) brackets, (4) control arm, (5) ABS wheel speed wire, (6) ball joint, (7) knuckle

✱✱ WARNING

When installing the tool to separate the ball joint, be careful not to damage the ball joint seal.

5. Remove the wheel speed sensor wire from the retaining brackets to the upper control arm.
6. Remove the control arm pivot bolts and nuts and remove control arm.

To install:

➡All suspension components should be tightened with the weight of the vehicle on them (curb height).

7. Position the control arm into the frame brackets. Install bolts and nuts. Tighten to 102 Nm (75 ft. lbs.).
8. Reposition the wheel speed wire into the retaining brackets.
9. Insert ball joint in steering knuckle and tighten ball joint nut to 75 Nm (55 ft. lbs.).
10. Install the wheel and tire assembly.
11. Remove the support and lower vehicle.
12. Perform a wheel alignment.

CONTROL ARM BUSHING REPLACEMENT

The control arm bushings are serviced with the control arm as an assembly.

Lower Control Arm

REMOVAL & INSTALLATION

2002–04

2 WHEEL DRIVE

1. Before servicing the vehicle, refer to the precautions in the beginning of this section.
2. Remove or disconnect the following:
   - Front wheel
   - Shock absorber
   - Brake caliper and rotor
   - Stabilizer bar link
   - Coil spring
   - Inner mounting bolts
   - Lower control arm

To install:

3. Install or connect the following:
   - Lower control arm. Tighten the front bolt to 130 ft. lbs. (175 Nm) and the rear bolt to 80 ft. lbs. (108 Nm).
   - Coil spring. Tighten the lower ball joint nut to 94 ft. lbs. (127 Nm).
   - Stabilizer bar link

- Brake caliper and rotor
- Shock absorber
- Front wheel

4 WHEEL DRIVE

1. Before servicing the vehicle, refer to the precautions in the beginning of this section.
2. Remove or disconnect the following:
   - Front wheel
   - Outer tie rod end
   - Halfshaft
   - Torsion bar
   - Shock absorber
   - Stabilizer bar
   - Lower ball joint
   - Pivot bolts
   - Lower control arm

To install:

3. Install or connect the following:
   - Lower control arm. Tighten the front pivot bolt to 80 ft. lbs. (108 Nm) and the rear bolt to 140 ft. lbs. (190 Nm).
   - Lower ball joint. Tighten the nut to 135 ft. lbs. (183 Nm).
   - Stabilizer bar
   - Shock absorber
   - Torsion bar
   - Halfshaft
   - Outer tie rod end
   - Front wheel

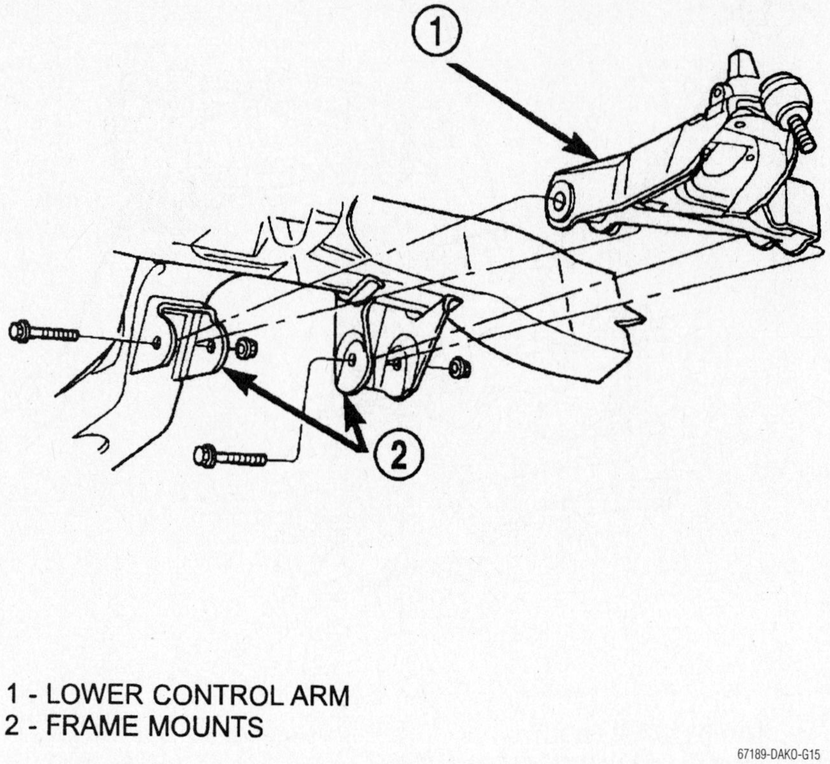

Lower control arm—2002–04 2 wheel drive

2005

1. Raise and support the vehicle.
2. Remove the wheel and tire assembly.
3. Remove the disc brake caliper assembly.
4. Remove the disc brake rotor.
5. Disconnect the wheel speed sensor at the wheel well.
6. Remove tie rod end jam nut.
7. Disconnect the tie rod from the knuckle using special tool C-3894-A, or equivalent.
8. Remove the front halfshaft nut (4WD).
9. Remove the upper ball joint nut. Separate the upper ball joint from the steering knuckle with a remover.
10. Remove the lower ball joint nut. Separate the lower ball joint from the steering knuckle with a remover.
11. Remove the steering knuckle.
12. Remove the stabilizer bar link
13. Remove the shock absorber lower bolt and nut.
14. Remove the lower control arm bolts, nuts and washers.
15. Remove the lower control arm from the vehicle.

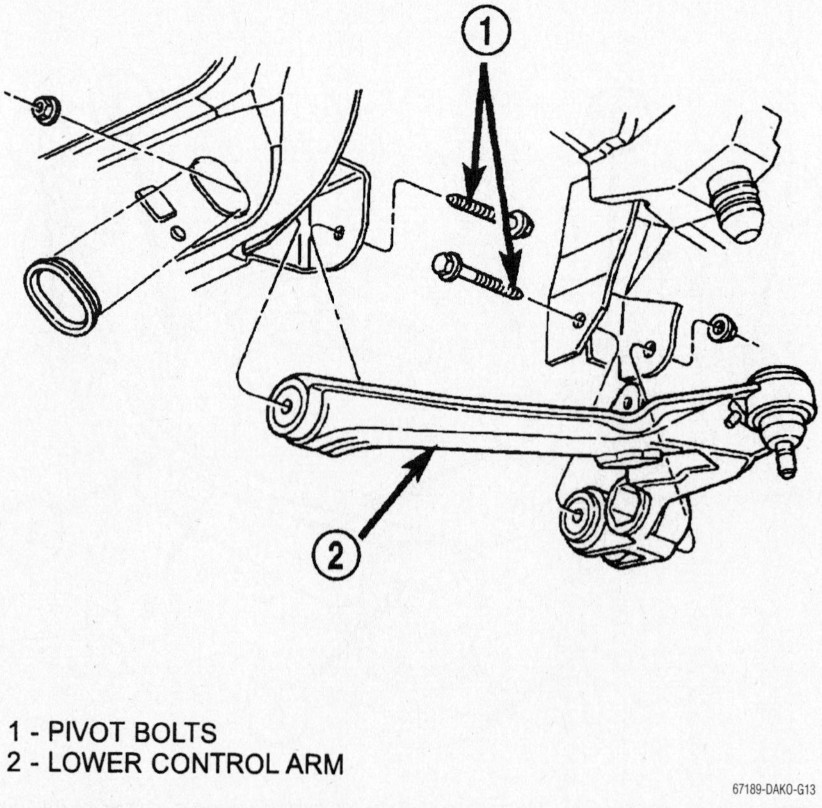

Lower control arm—2002–04 4 wheel drive

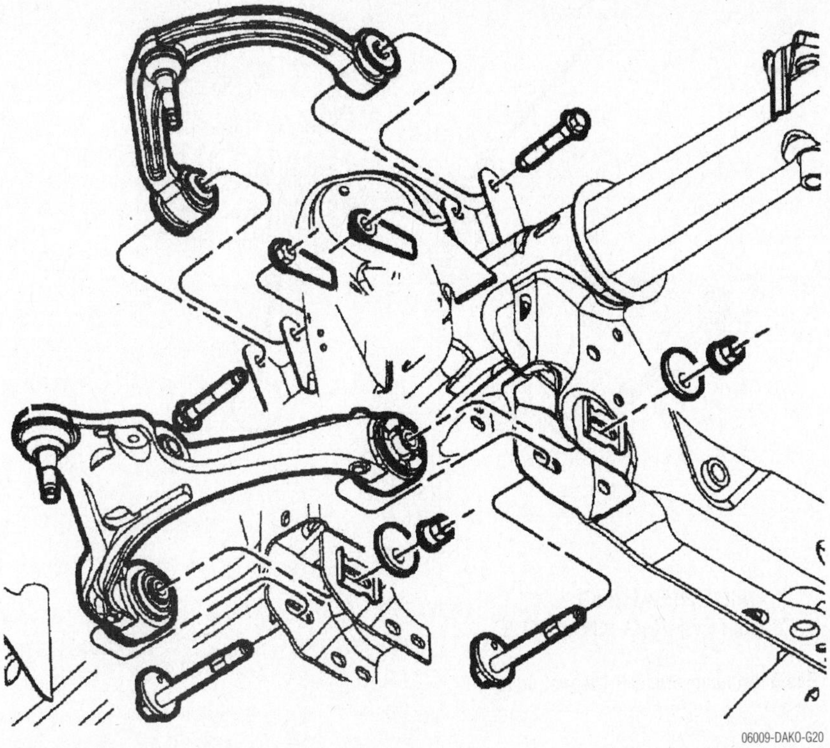

Lower and upper control arm installation—2005 models

To install:

➡**All suspension components should be tightened with the weight of the vehicle on them (curb height).**

16. Position the lower control arm at the frame rail brackets. Install the pivot bolts washers and nuts. Tighten the nuts finger-tight.

⁂ WARNING

The ball joint stud taper must be CLEAN and DRY before installing the knuckle. Clean the stud taper with mineral spirits to remove dirt and grease.

17. Install the steering knuckle.
18. Insert the lower ball joint into the steering knuckle. Install and tighten the retaining nut to 81 Nm (60 ft. lbs.).
19. Install shock absorber lower bolt and nut. Tighten to 81 Nm (60 ft. lbs.).
20. Install the front halfshaft nut (4WD).
21. Insert the upper ball joint into the steering knuckle. Install and tighten the retaining nut to 75 Nm (55 ft. lbs.).
22. Install the stabilizer bar link and tighten to 169 Nm (125 ft. lbs.).
23. Tighten the lower control arm pivot nut and bolts to 244 Nm (180 ft. lbs.).
24. Insert the outer tie rod end into the steering knuckle. Install and tighten the retaining nut to 75 Nm (55 ft. lbs.).
25. Install the disc brake rotor.
26. Install the disc brake caliper and adaptor assembly and tighten to 135 Nm (100 ft. lbs.).
27. Install the wheel and tire assembly.
28. Remove the support and lower the vehicle.
29. Perform a wheel alignment.

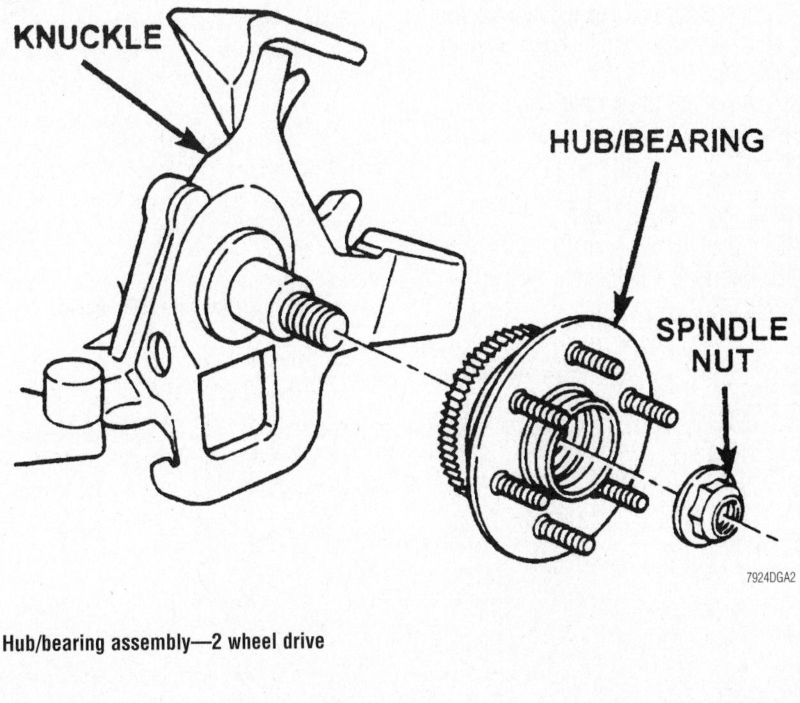

Hub/bearing assembly—2 wheel drive

CONTROL ARM BUSHING REPLACEMENT

The control arm bushings are serviced with the control arm as an assembly.

Wheel Bearings

ADJUSTMENT

The Dakota models utilize a hub/bearing assembly which is not adjustable.

REMOVAL & INSTALLATION

2002–04

2 WHEEL DRIVE

1. Before servicing the vehicle, refer to the precautions in the beginning of this section.
2. Remove or disconnect the following:
   - Front wheel
   - Brake caliper and rotor
   - Spindle nut
   - Hub and bearing assembly

To install:

3. Install or connect the following:
   - Hub and bearing assembly
   - Spindle nut. Tighten the nut to 185 ft. lbs. (251 Nm).
   - Brake caliper and rotor
   - Front wheel

4 WHEEL DRIVE

1. Before servicing the vehicle, refer to the precautions in the beginning of this section.

2. Remove or disconnect the following:
   - Front wheel
   - Brake caliper and rotor
   - Hub retainer nut
   - Hub and bearing assembly

To install:

3. Install or connect the following:
   - Hub and bearing assembly. Tighten the bolts to 123 ft. lbs. (166 Nm).
   - Hub retainer nut. Tighten the nut to 180 ft. lbs. (244 Nm).
   - Brake caliper and rotor
   - Front wheel

2005

1. Raise and support the vehicle.
2. Remove the wheel and tire assembly.
3. Remove the brake caliper and rotor.
4. Remove the ABS wheel speed sensor if equipped.
5. Remove the halfshaft nut (4WD).

➡Do not strike the knuckle with a hammer to remove the tie rod end or the ball joint. Damage to the steering knuckle will occur.

6. Pull down on the steering knuckle to separate the halfshaft from the hub/bearing (4WD).
7. Remove the three hub/bearing mounting bolts from the steering knuckle.
8. Slide the hub/bearing out of the steering knuckle.
9. Remove the brake dust shield.

To install:

10. Install the brake dust shield.
11. Install the hub/bearing into the steering knuckle and tighten the bolts to 163 Nm (120 ft. lbs.).
12. Install the brake rotor and caliper.
13. Install the ABS wheel speed sensor if equipped.
14. Install the halfshaft nut and tighten to 251 Nm (185 ft. lbs.) (4WD).
15. Install the wheel and tire assembly.
16. Remove the support and lower vehicle.

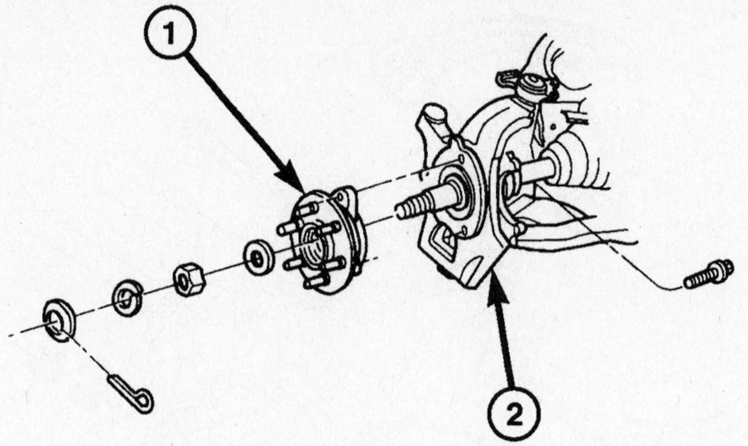

Hub/bearing assembly—4 wheel drive

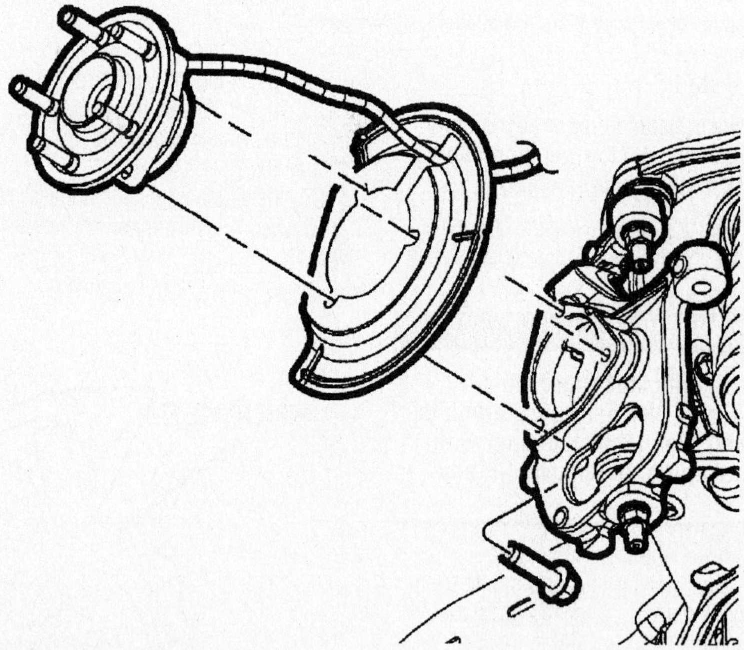

Front hub installation—2005 models

Front Brake Caliper

REMOVAL AND INSTALLATION

2002–04

1. Raise and support the front end on jackstands.
2. Remove the wheels.
3. Disconnect the rubber brake hose from the tubing at the frame mount. If the pistons are to be removed from the caliper, leave the brake hose connected to the caliper. Check the rubber hose for cracks or chafed spots.
4. Plug the brake line to prevent loss of fluid.
5. Remove the caliper slide pins.
6. Remove the caliper and brake pads from the rotor adapter.

To install:

7. Position the outboard shoe in the caliper. The shoe should not rattle in the caliper. If it does, or if any movement is obvious, bend the shoe tabs over the caliper to tighten the fit.
8. Slide the caliper into position on the adapter and over the rotor.
9. Align the caliper and start the pins in by hand.
10. Tighten the pins to 22 ft. lbs. (30 Nm).
11. Connect the brake hose to the caliper. Use new washers to attach the hose fitting if the original washers are scored, worn or damaged.
12. Fill and bleed the brake system.
13. Install the wheels.
14. Lower the vehicle.

2005

CAUTION

Never allow the disc brake caliper to hang from the brake hose. Damage to the brake hose with result. Provide a suitable support to hang the caliper securely.

1. Raise and support the vehicle.
2. Remove the tire and wheel assembly.
3. Compress the disc brake caliper.
4. Remove the banjo bolt and discard the copper washers.
5. Remove the caliper slide pin bolts.
6. Remove the disc brake caliper from the caliper adapter.
7. Remove the caliper slide pins from the adapter.

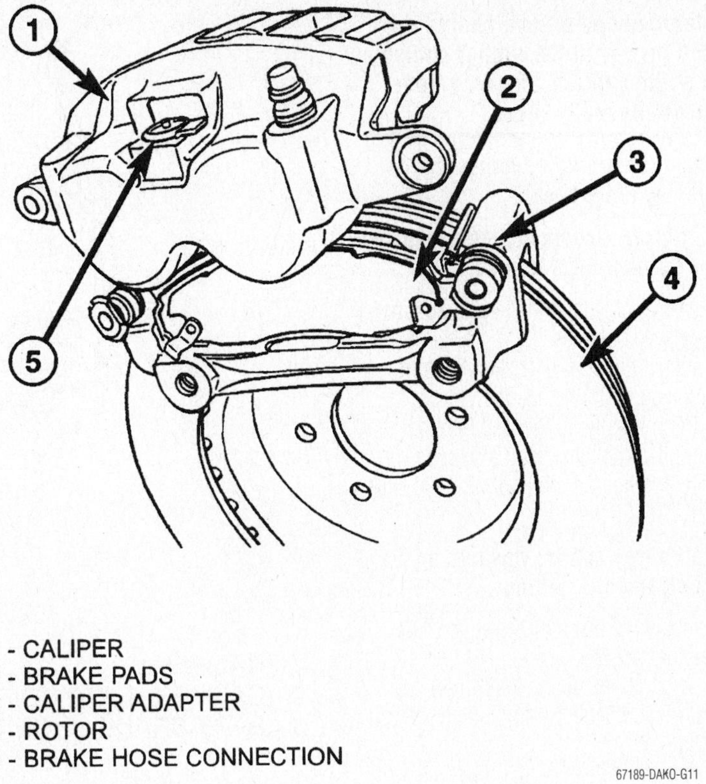

2002–04 front caliper removed

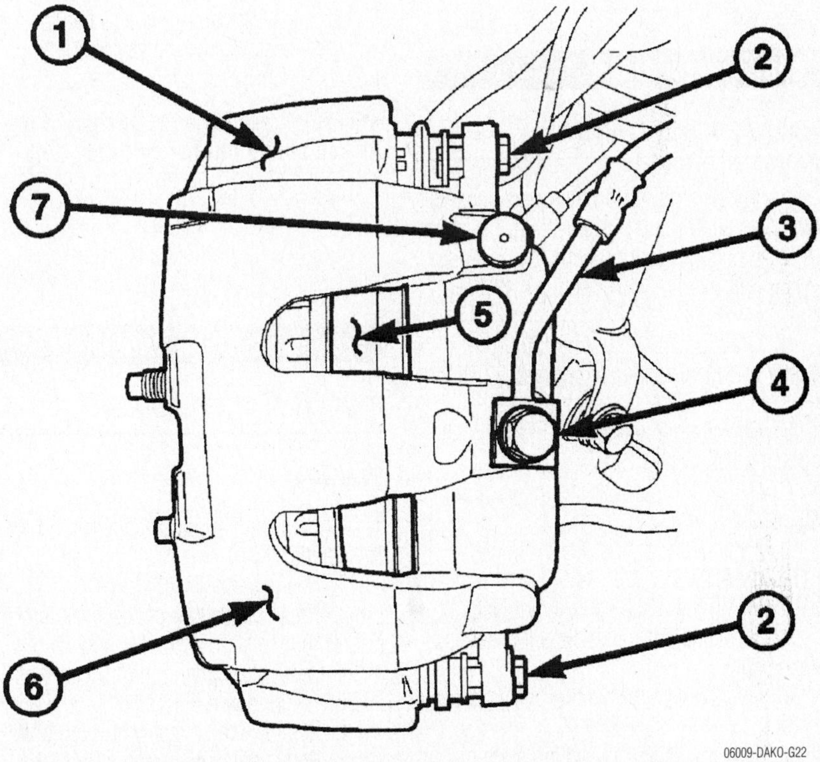

(1) caliper mounting adaptor, (2) caliper mounting bolts, (3) brake line, (4) banjo bolt, (5) pad, (6) caliper—2005 models

To install:

⁂ WARNING

Petroleum based grease should not be used on any of the rubber components of the caliper, Use only Non-Petroleum based grease.

➡**Clean slide pin bores thoroughly to remove any old grease.**

➡**Use grease packets included with kit or Dow Corning-807T grease.**

8. Thoroughly coat the new slide pins on all working surfaces.
9. Install the boot onto the slide pin and then insert into the adapter.
10. Push the pin all the way into the adapter and carefully expel the trapped air by gently pushing on the boot near the slide pin head.

➡**Install a new copper washers on the banjo bolt when installing**

11. Install the disc brake caliper to the brake caliper adapter.
12. Install the banjo bolt with new copper washers to the caliper. Tighten to 28 Nm (250 inch lbs.)
13. Install the caliper slide pin bolts. Tighten to 32 Nm (24 ft. lbs.).
14. Bleed the base brake system.
15. Install the tire and wheel assembly.
16. Lower the vehicle.

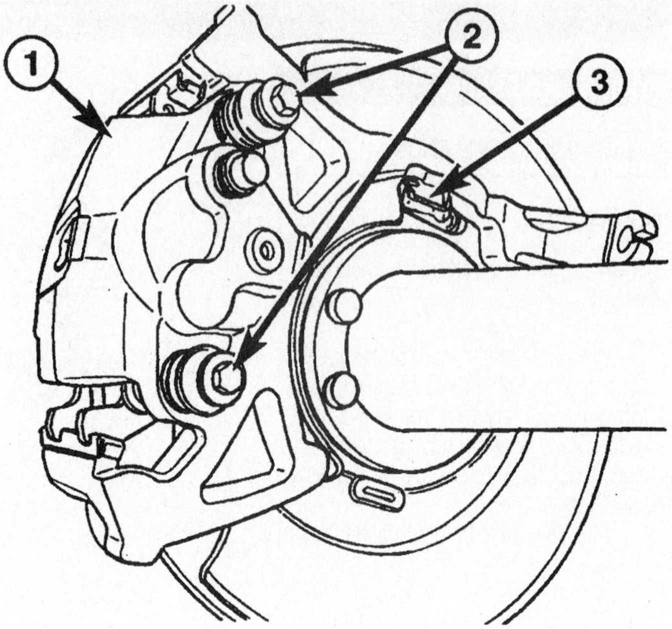

1 - CALIPER
2 - CALIPER SLIDE BOLTS
3 - PARK BRAKE SHOE LEVER

67189-DAKO-G10

Rear caliper installed

Rear Brake Caliper

REMOVAL & INSTALLATION

1. Install prop rod on the brake pedal to keep pressure on the brake system.
2. Raise and support vehicle.
3. Remove the wheel and tire assembly.
4. Drain small amount of fluid from master cylinder brake reservoir with suction gun.
5. Remove the brake hose banjo bolt if replacing caliper.
6. Remove the caliper mounting slide pin bolts.
7. Remove the caliper from vehicle.

To install:

8. Install the brake pads if removed.
9. Lubricate anti-rattle clips for the disc brake pads.
10. Install the caliper to the brake caliper adapter.
11. Coat the caliper mounting slide pin bolts with silicone grease. Then install and tighten the bolts to 25 Nm (18 ft. lbs.).
12. Install the brake hose banjo bolt if removed.
13. Install the brake hose to the caliper with new seal washers and tighten fitting bolt to 31 Nm (23 ft. lbs.).

➡**Verify brake hose is not twisted or kinked before tightening fitting bolt.**

14. Remove the prop rod from the vehicle.
15. Bleed the base brake system.
16. Install the wheel and tire assemblies.

Front Disc Brake Pads

REMOVAL AND INSTALLATION

2002–04

1. Raise and support the front end on jackstands.
2. Remove the wheels.
3. Press the caliper piston back into the bore with a suitable prytool. Use a large C-clamp to drive the piston into the bore of additional force is required.
4. Remove the caliper mounting bolts.
5. Rotate the caliper rearward off the rotor and out from its mount.
6. Set the caliper on a crate or sturdy box, then remove the inboard and outboard brake pads. The inboard pad has a spring clip that holds it in the caliper. Tilt this pad out at the top to unseat the clip. The outboard pad has a retaining spring that secures it in the caliper. Unseat 1 spring end and rotate the pad out of the caliper.
7. Secure the caliper to a chassis or suspension component with a sturdy wire. Do not let it hang from the hose.

To install:

8. Clean the caliper and steering knuckle sliding surfaces with a wire brush. Then, apply a coat of Mopar® multi-mileage grease or equivalent.
9. Clean the caliper slide pins with brake cleaner or brake fluid. Then apply a light coating of silicone grease to the pins.

➡**If there is minor rust or corrosion on the pins, first polish them with a crocus cloth. If they are severely rusted, replace them.**

10. Install the inboard pad and its spring clip.
11. Install the outboard brake pad.
12. Install the caliper over the rotor and seat it in its original position until flush.
13. Final tighten the caliper slide pins to 22 ft. lbs. (30 Nm).

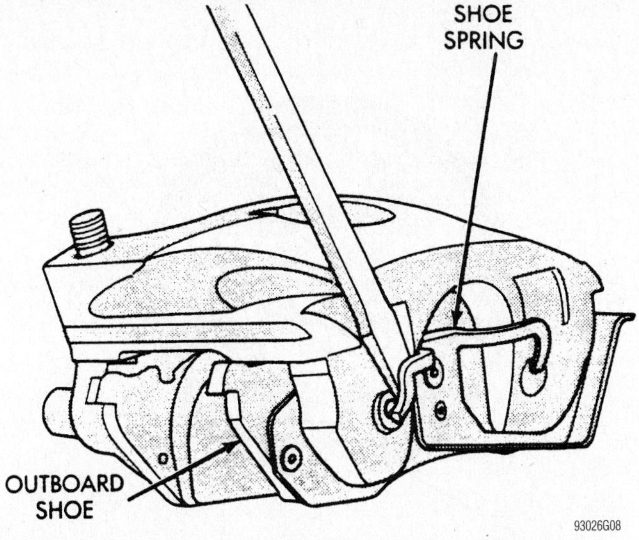

Prying the disc brake from the 4WD front brake caliper assembly—2002–04 models

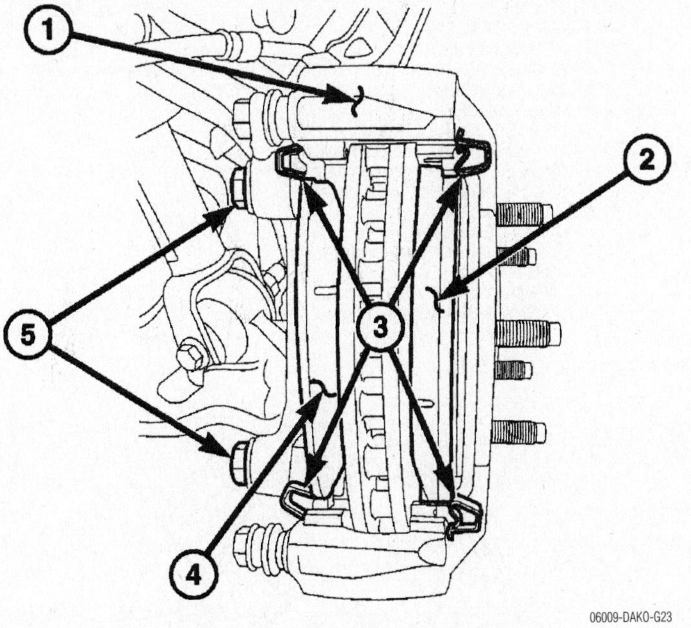

(1) caliper adapter, (2) outboard pad, (3) anti-rattle clips, (4) inboard pad, (5) caliper mounting bolts—2005 models

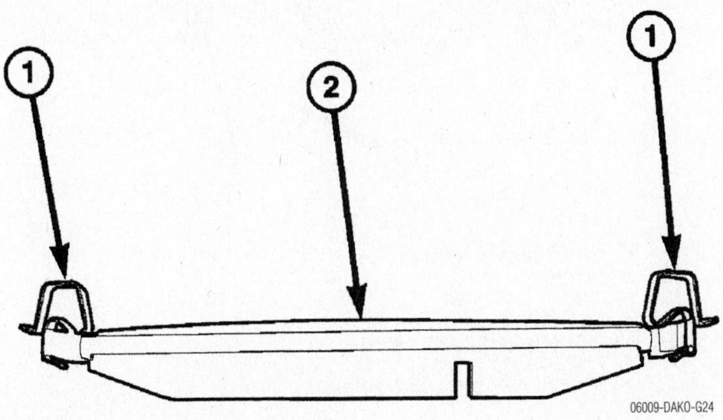

Anti-rattle clips installed—2005 models

14. Install the wheels.
15. Lower the vehicle.
16. Pump the brakes several times to seat the pads.

2005

1. Raise and support vehicle.
2. Remove the wheel and tire assemblies.
3. Compress the caliper.
4. Remove the caliper slide pin bolts.
5. Remove the caliper from the caliper adapter.

➡Do not allow brake hose to support caliper assembly.

6. Support and hang the caliper.
7. Remove the inboard brake pad from the caliper adapter.
8. Remove the outboard brake pad from the caliper adapter.
9. Remove the anti-rattle clips from the pad.

To install:

10. Bottom pistons in caliper bore with C-clamp. Place an old brake shoe between a C-clamp and caliper piston.
11. Clean caliper mounting adapter.
12. Install new anti-rattle clips to the brake pads.
13. Install inboard brake pad in adapter.
14. Install outboard brake pad in adapter.
15. Install the caliper over rotor. Then, push the caliper onto the adapter.
16. Install caliper slide pin bolts.
17. Install wheel and tire assemblies and lower vehicle.
18. Apply brakes several times to seat caliper pistons and brake shoes and obtain firm pedal.
19. Top off master cylinder fluid level.

Rear Disc Brake Pads

REMOVAL & INSTALLATION

1. Raise and support vehicle.
2. Remove the wheel and tire assemblies.
3. Compress the caliper.
4. Remove the caliper.
5. Remove the caliper by tilting the top up and off the caliper adapter.

➡Do not allow brake hose to support caliper assembly.

6. Support and hang the caliper.
7. Remove the inboard brake pad from the caliper adapter.
8. Remove the outboard brake pad from the caliper adapter.

➡To install:

9. Bottom the piston in the caliper bore with C-clamp. Place an old brake shoe between a C-clamp and caliper piston.
10. Clean the caliper mounting surface.
11. Install anti-rattle clips.

➡When servicing the rear brake pads, replace the anti-rattle clips on the brake adapter. Anti-rattle clips are provided with the shoe & lining kit.

12. Install inboard brake pad in adapter.
13. Install outboard brake pad in adapter.
14. Tilt the top of the caliper over rotor and secure to the mounting holes on the brake caliper adapter.
15. Install the caliper.
16. Install wheel and tire assemblies and lower vehicle.
17. Apply brakes several times to seat caliper pistons and brake shoes and obtain firm pedal.
18. Top off master cylinder fluid level.

Brake Drums

REMOVAL AND INSTALLATION

1. Remove the axle shaft nuts, washers and cones. If the cones do not readily release, rap the axle shaft sharply in the center.
2. Remove the axle shaft.
3. Remove the outer hub nut.
4. Straighten the lockwasher tab and remove it along with the inner nut and bearing.
5. Carefully remove the drum.

To install:

6. Position the drum on the axle housing.
7. Install the bearing and inner nut. While rotating the wheel and tire, tighten the adjusting nut until a slight drag is felt.
8. Back off the adjusting nut ⅙ turn so that the wheel rotates freely without excessive end-play.
9. Install the lockrings and nut. Place a new gasket on the hub and install the axle shaft, cones, lockwashers and nuts.
10. Install the wheel and tire.
11. Road-test the vehicle.

Brake Shoes

REMOVAL AND INSTALLATION

2002–04

1. Raise and support the vehicle.
2. Remove the wheel and tire assembly.

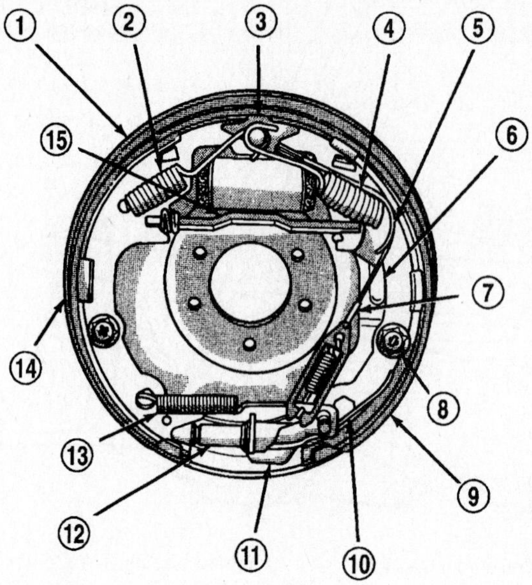

Brake shoes assembled—2002–04

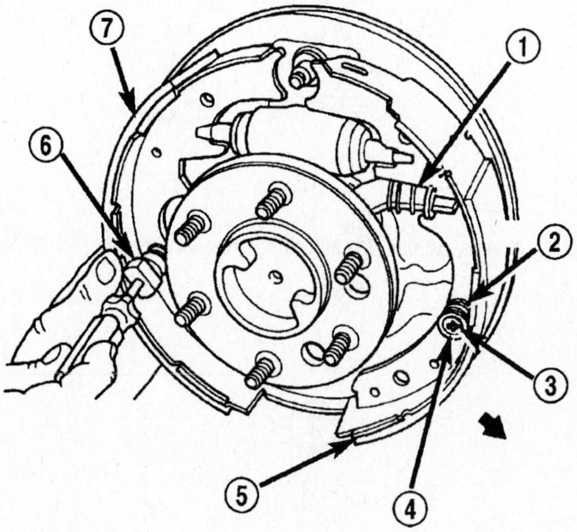

Shoe retainers, springs and pins—2002–04

3. Remove the clip nuts securing the brake drum to wheel studs.

4. Remove the drum.

5. Vacuum the brake components to remove brake lining dust, or use a liquid spray brake cleaner.

⁂ CAUTION

Never clean the brake shoes or parts with compressed air!

6. Remove the shoe return springs with brake spring pliers tool.

7. Remove the adjuster cable. Slide the cable eye off anchor pin. Then unhook and remove the cable from the adjuster lever.

8. Remove the cable guide from the secondary shoe and anchor plate from anchor pin.

9. Remove the adjuster lever. Disengage the lever from spring by sliding the lever forward to clear the pivot and work the lever out from under the spring.

10. Remove the adjuster lever spring from the pivot.

11. Disengage and remove the shoe spring from the brake shoes.

12. Disengage and remove the adjuster screw assembly from the brake shoes.

13. Remove the brake shoe retainers, springs.

14. Remove the secondary brake shoe from the support plate.

15. Remove the strut and spring.

16. Remove the parking brake lever retaining clip from the secondary shoe and remove the lever.

17. Remove the primary shoe from the support plate.

18. Disengage the parking brake lever from the parking brake cable.

To install:

19. Clean and inspect individual brake components.

20. Lubricate anchor pin and brake shoe contact pads on support plate with high temperature grease.

21. Lubricate adjuster screw socket, nut, button and screw thread surfaces with grease.

22. Install the parking brake cable to the parking brake lever.

23. Install parking brake lever to the secondary shoe and install retaining clip.

24. Install primary shoe on support plate. Secure shoe with new spring retainers and pin.

25. Install spring on parking brake strut and engage strut in primary.

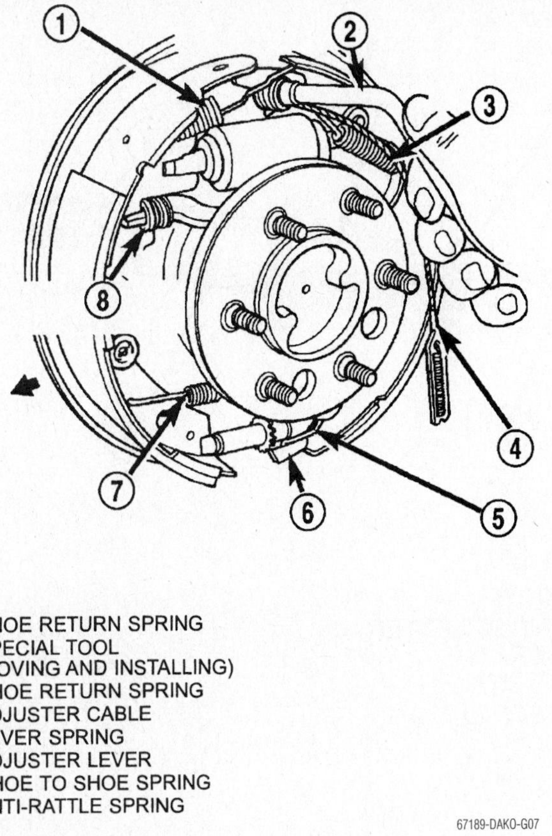

Shoe return springs—2002–04

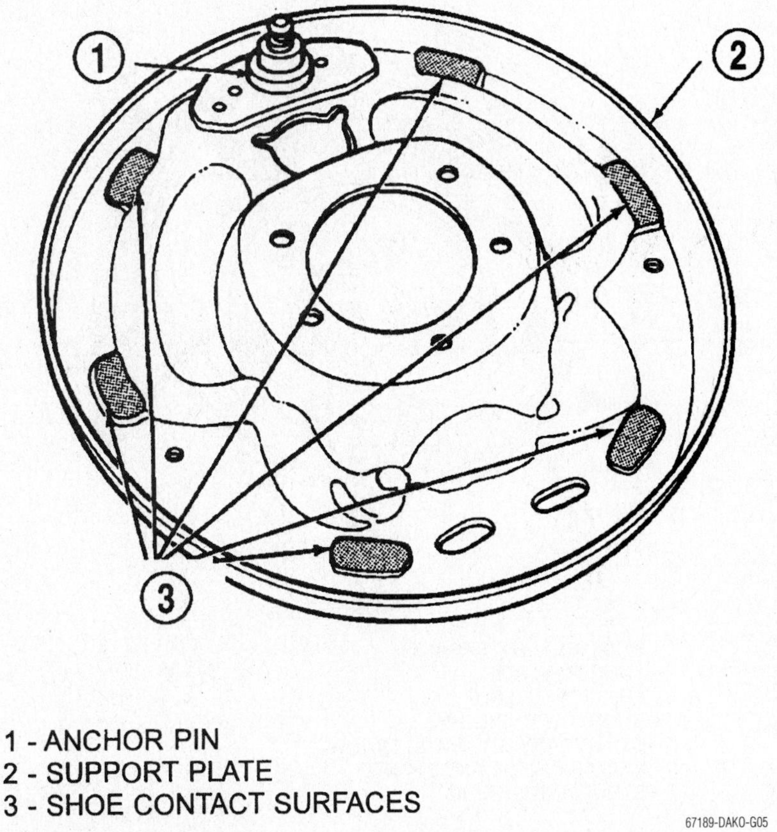

Shoe contact surfaces—2002–04

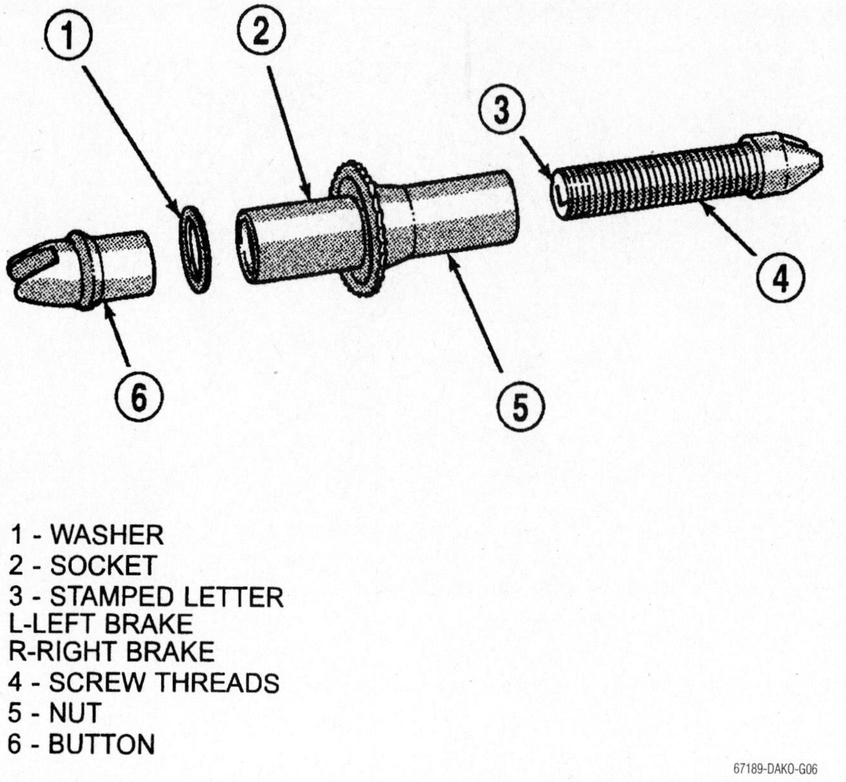

Adjuster screw—2002-04

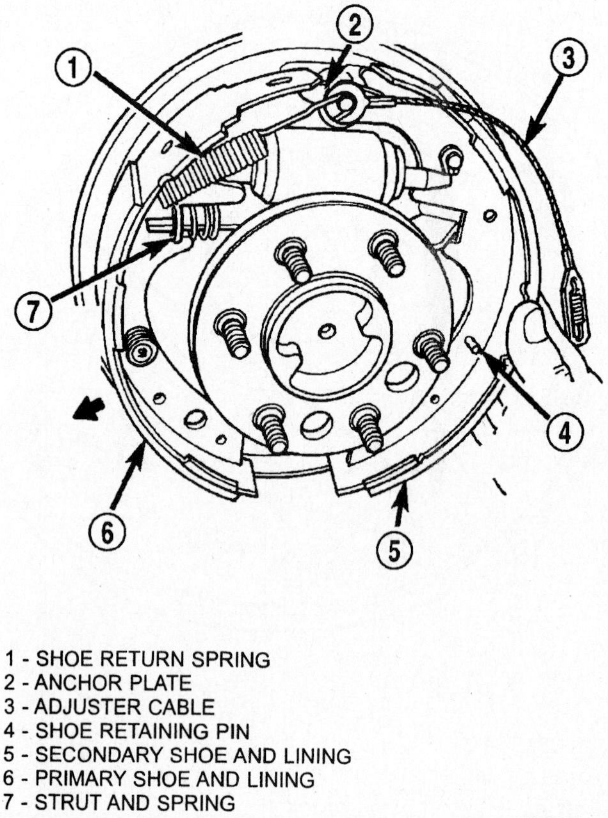

Brake shoe installation—2002-04

26. Install secondary shoe on support plate. Insert strut in shoe and guide shoe onto anchor pin. Temporarily secure shoe with retaining pin.
27. Install anchor plate and adjuster cable eyelet on support plate anchor pin.
28. Install cable guide in secondary shoe and position cable in guide.
29. Assemble adjuster screw. Then install and adjuster screw between the brake shoes.

➡Be sure the adjuster screws are installed on the correct brake unit. The adjuster screws are marked L (left) and R (right) for identification.

30. Install adjuster lever and spring and connect adjuster cable to lever.
31. Install secondary shoe retainers and spring.
32. Install shoe spring. Connect spring to secondary shoe first. Then to primary shoe.
33. Verify adjuster operation. Pull adjuster cable upward, cable should lift lever and rotate star wheel. Be sure adjuster lever properly engages star wheel teeth.
34. Adjust brake shoes to drum with brake gauge.
35. Install wheel and tire assembly.

2005

1. Raise and support vehicle.
2. Remove wheel and tire assembly.
3. Remove clip nuts securing brake drum to wheel studs.
4. Remove drum. If drum is difficult to remove, remove rear plug from access hole in support plate. Back-off self adjusting by inserting a thin screwdriver into access hole and push lever away from adjuster star wheel. Then insert an adjuster tool into brake adjusting hole rotate adjuster star wheel to retract brake shoes.
5. Vacuum brake components to remove brake lining dust.
6. Remove shoe return spring with brake spring pliers tool.
7. Remove adjuster spring and lever. Disengage lever from spring by sliding lever forward to clear pivot and work lever out from under spring.
8. Disengage and remove shoe return spring from brake shoes.
9. Remove brake shoe hold down clips.
10. Remove rear brake shoe from support plate.
11. Remove front brake shoe from support plate.

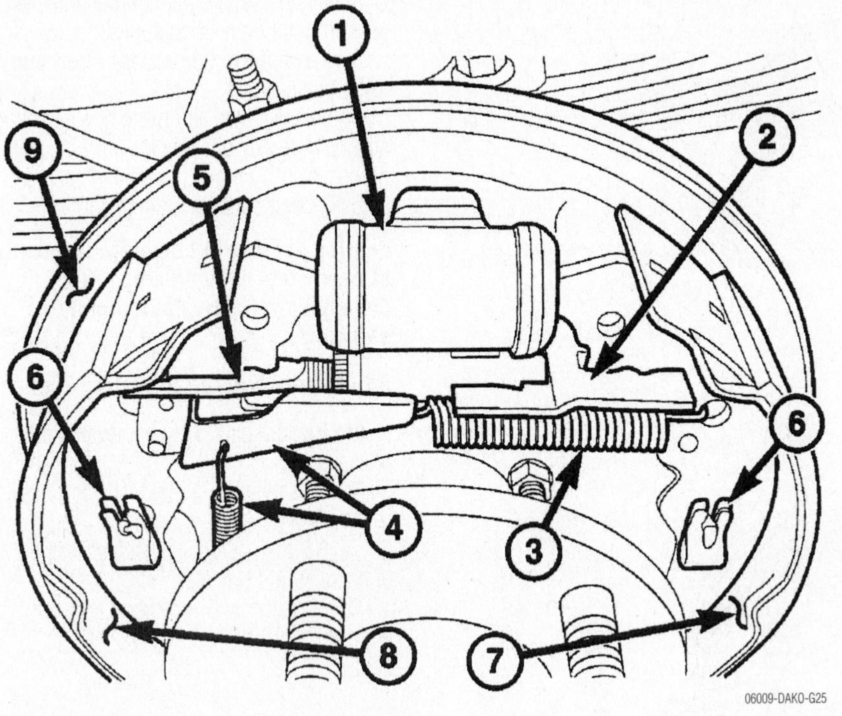

Rear brake parts, 2005 models. (1) wheel cylinder, (2) parking brake lever, (3) return spring, (4 & 5) adjuster spring and lever, (6) hold-down clips, (7 & 8) brake shoes, (9) backing plate

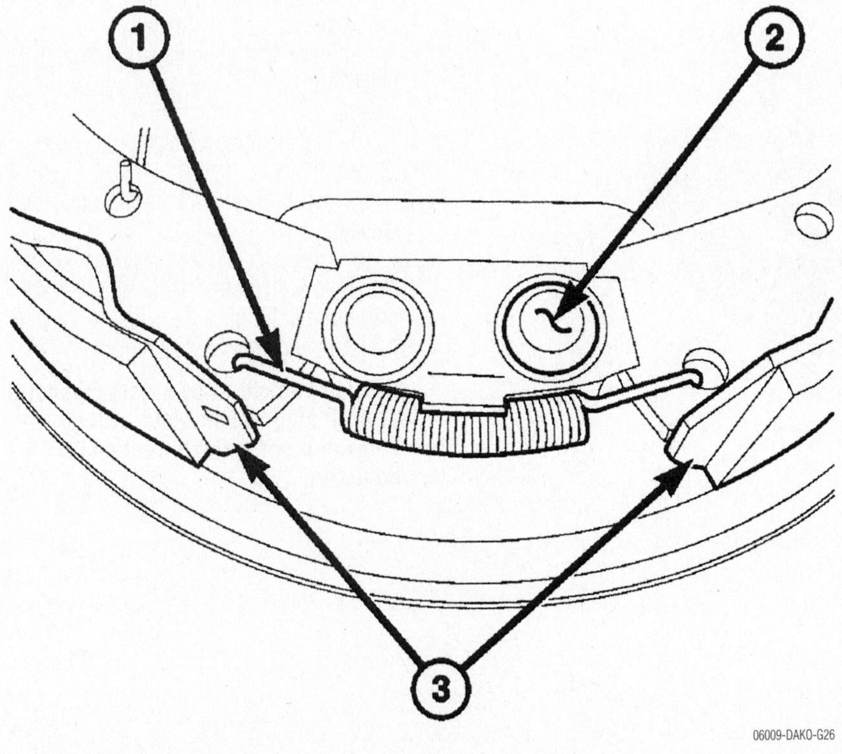

Lower return spring (1), cam (2) and brake shoes, (3)—2005 models

12. Remove park brake lever from the brake shoe.

To install:

13. Clean and inspect individual brake components.
14. Lubricate where the brake shoe contacts the support plate with high temperature grease or Lubriplate®.
15. Lubricate adjuster screw socket, nut, button and screw thread surfaces with grease or Lubriplate®.
16. Install parking brake lever to the rear shoe and install the hold down clip.
17. Install the adjuster strut onto the shoes and park brake lever.
18. Install the front shoe on support plate, and install the hold down clip.
19. Install the adjuster spring and lever in the slot in the adjuster strut.
20. Install the lower return spring to the shoes.
21. Verify adjuster operation. Pull both shoes outward to move the adjuster lever to rotate the star wheel. Be sure adjuster lever properly engages star wheel teeth.
22. Adjust brake shoes to drum with brake gauge.
23. Install wheel and tire assembly.

ADJUSTMENTS

The rear drum brakes are equipped with a self-adjusting mechanism. Under normal circumstances, the only time adjustment is required is when the shoes are replaced, removed for access to other parts, or when one or both drums are replaced. Adjustment can be made with a standard brake gauge or with adjusting tool. Adjustment is performed with the complete brake assembly installed on the backing plate.

Adjustment with a Brake Gauge

1. Be sure parking brakes are fully released.
2. Raise rear of vehicle and remove wheels and brake drums.
3. Verify that left and right automatic adjuster levers and cables are properly connected.
4. Insert brake gauge in drum. Expand gauge until gauge inner legs contact drum braking surface. Then lock gauge in position.
5. Reverse gauge and install it on brake shoes. Position gauge legs at shoe centers as shown. If gauge does not fit (too loose/too tight), adjust shoes.
6. Pull shoe adjuster lever away from adjuster screw star wheel.

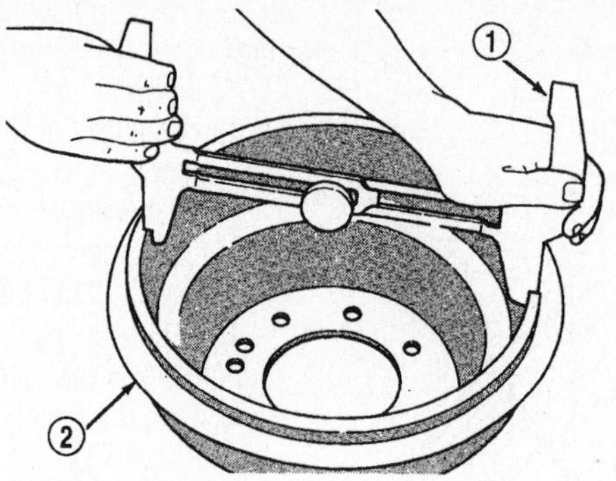

1 - BRAKE GAUGE
2 - BRAKE DRUM

67189-DAKO-G03

Adjusting gauge on the drum

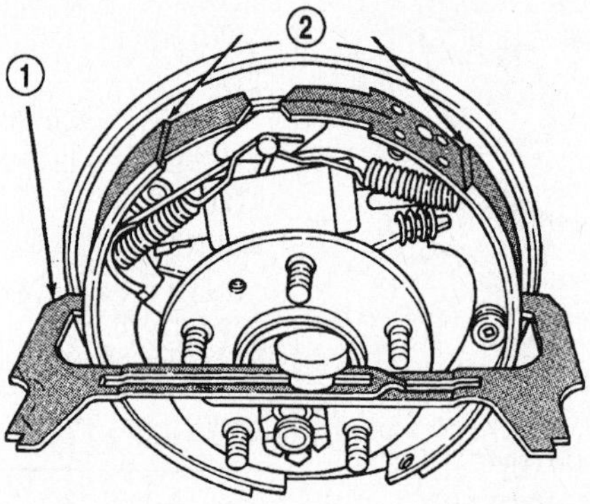

1 - BRAKE GAUGE
2 - BRAKE SHOES

67189-DAKO-G01

Adjustment with a brake adjusting gauge

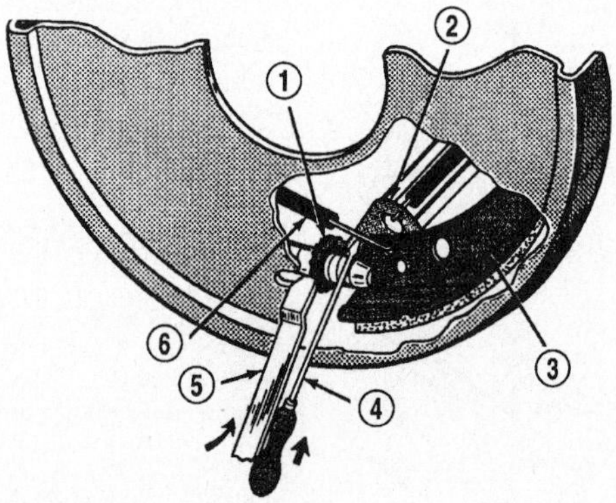

1 - STAR WHEEL
2 - LEVER
3 - BRAKE SHOE WEB
4 - SCREWDRIVER
5 - ADJUSTING TOOL
6 - ADJUSTER SPRING

67189-DAKO-G02

Adjustment with a brake adjusting tool

7. Turn adjuster screw star wheel (by hand) to expand or retract brake shoes. Continue adjustment until gauge outside legs are light drag-fit on shoes.
8. Install brake drums and wheels and lower vehicle.
9. Drive vehicle and make one forward stop followed by one reverse stop. Repeat procedure 8-10 times to operate automatic adjusters and equalize adjustment.

➡Bring vehicle to complete standstill at each stop. Incomplete, rolling stops will not activate automatic adjusters.

Adjustment with an Adjusting Tool

1. Be sure parking brake lever is fully released.
2. Raise vehicle so rear wheels can be rotated freely.
3. Remove plug from each access hole in brake support plates.
4. Loosen parking brake cable adjustment nut until there is slack in front cable.
5. Insert adjusting tool through support plate access hole and engage tool in teeth of adjusting screw star wheel.
6. Rotate adjuster screw star wheel (move tool handle upward) until slight drag can be felt when wheel is rotated.
7. Push and hold adjuster lever away from star wheel with thin screwdriver.
8. Back off adjuster screw star wheel until brake drag is eliminated.
9. Repeat adjustment at opposite wheel. Be sure adjustment is equal at both wheels.
10. Install support plate access hole plugs.
11. Adjust parking brake cable and lower vehicle.
12. Drive vehicle and make one forward stop followed by one reverse stop. Repeat procedure 8-10 times to operate automatic adjusters and equalize adjustment.

➡Bring vehicle to complete standstill at each stop. Incomplete, rolling stops will not activate automatic adjusters.

DODGE

6

Durango

SPECIFICATIONS AND MAINTENANCE CHARTS

ENGINE AND VEHICLE IDENTIFICATION

Engine						
Code ①	Liters (cc)	Cu. In.	Cyl.	Fuel Sys.	Engine Type	Eng. Mfg.
N	4.7 (4701)	287	8	SMFI	SOHC	Chrysler
K	3.7 (3701)	226	6	MFI	SOHC	Chrysler
D	5.7 (5653)	345	8	SMPI	OHV	Chrysler
Z	5.9 (5899)	360	8	SMFI	OHV	Chrysler

Model Year	
Code ②	Year
2	2002
3	2003
4	2004
5	2005
6	2006

SOHC: Single overhead camshaft

OHV: Overhead Valve

SMFI: Sequential Multi-port Fuel Injection

① 8th position of VIN

② 10th position of VIN

06009-DURA-C01

GENERAL ENGINE SPECIFICATIONS

Year	Model	Engine Displ. Liters	Engine VIN	Net Horsepower @ rpm	Net Torque @ rpm (ft. lbs.)	Bore x Stroke (in.)	Com-pression Ratio	Oil Pressure @ rpm
2002	Durango	4.7	N	235@4800	295@3200	3.66x3.40	9.3:1	25-110@3000
		5.9	Z	230@4000	330@3250	4.00x3.58	9.1:1	30-80@3000
2003	Durango	4.7	N	235@4800	295@3200	3.66x3.40	9.0:1	25-110@3000
		5.9	Z	230@4000	330@3250	4.00x3.58	9.1:1	30-80@3000
2004	Durango	3.7	K	210@5200	225@4200	3.66x3.40	9.1:1	25-110@3000
		4.7	N	235@4800	295@3200	3.66x3.40	9.0:1	25-110@3000
		5.7	D	345@5400	375@4200	3.91x3.58	9.6:1	25-110@3000
2005	Durango	3.7	K	210@5200	225@4200	3.66x3.40	9.1:1	25-110@3000
		4.7	N	235@4800	295@3200	3.66x3.40	9.0:1	25-110@3000
		5.7	D	345@5400	375@4200	3.91x3.58	9.6:1	25-110@3000

06009-DURA-C02

GASOLINE ENGINE TUNE-UP SPECIFICATIONS

Year	Engine Displacement Liters	Engine VIN	Spark Plug Gap (in.)	Ignition Timing (deg.)	Fuel Pump (psi)	Idle Speed (rpm)	Valve Clearance	
							Intake	Exhaust
2002	4.7	N	0.040	①	47-51	②	HYD	HYD
	5.9	Z	0.040	①	44.2-54.2	②	HYD	HYD
2003	4.7	N	0.040	①	47-51	②	HYD	HYD
	5.9	Z	0.040	①	44.2-54.2	②	HYD	HYD
2004	3.7	K	0.042	①	44-54	②	HYD	HYD
	4.7	N	0.040	①	47-51	②	HYD	HYD
	5.7	D	0.045	①	49.0-49.4	②	HYD	HYD
2005	3.7	K	0.042	①	56-60	②	HYD	HYD
	4.7	N	0.040	①	56-60	②	HYD	HYD
	5.7	D	0.045	①	56-60	②	HYD	HYD

NOTE: The Vehicle Emission Control Information (VECI) label often reflects specification changes made during production.
The label figures must be used if they differ from those in this chart.

HYD: Hydraulic

① Ignition timing is controlled by the PCM and is not adjustable.

② Idle speed is controlled by the PCM and is not adjustable

06009-DURA-C03

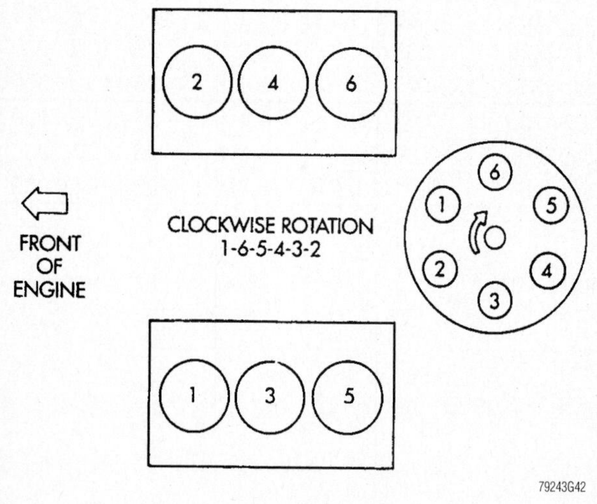

3.9L engine

4.7L engine
Firing order: 1-8-4-3-6-5-7-2
One coil per cylinder

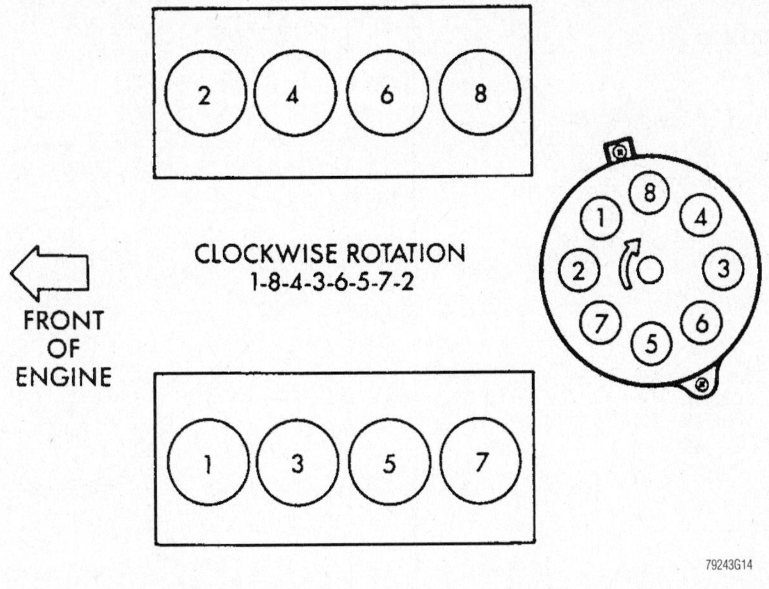

5.9L engine

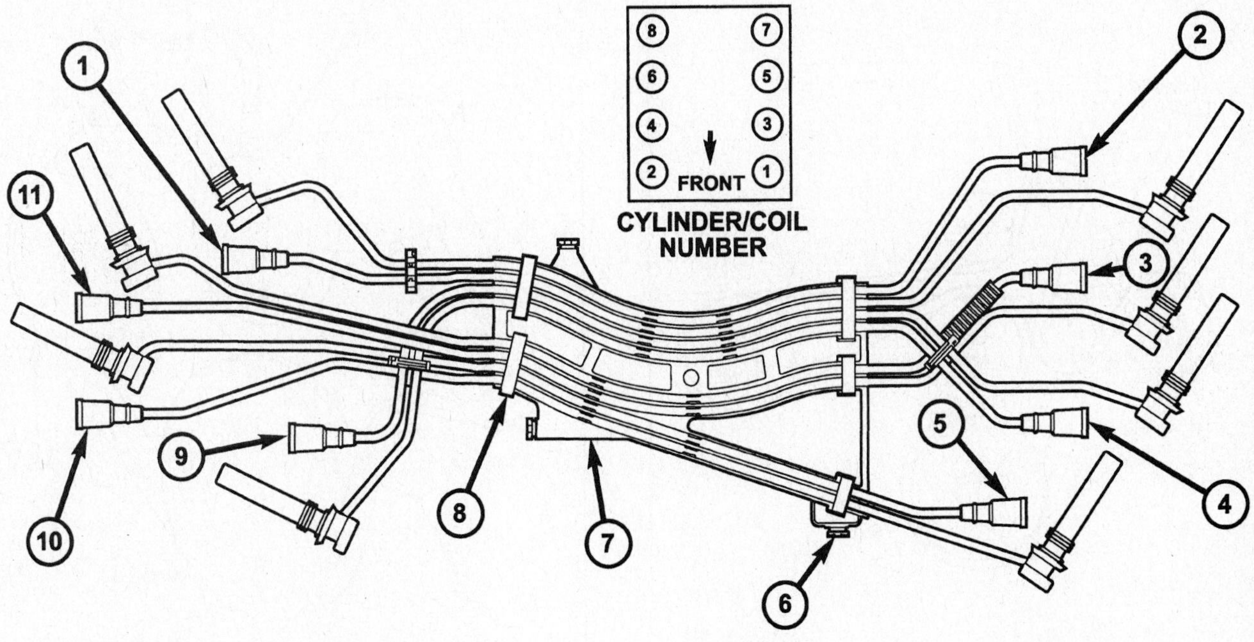

1 - #8 COIL-TO- #5 SPARK PLUG (MARKED 5/8)
2 - #5 COIL-TO- #8 SPARK PLUG (MARKED 5/8)
3 - #7 COIL-TO- #4 SPARK PLUG (MARKED 4/7)
4 - #3 COIL-TO- #2 SPARK PLUG (MARKED 2/3)
5 - #1 COIL-TO- #6 SPARK PLUG (MARKED 1/6)
6 - CLIPS (TRAY-TO-MANIFOLD RETENTION)
7 - CABLE TRAY
8 - CLIPS (SPARK PLUG CABLE-TO-TRAY- RETENTION)
9 - #2 COIL-TO- #3 SPARK PLUG (MARKED 2/3)
10 - #6 COIL-TO- #1 SPARK PLUG (MARKED 1/6)
11 - #4 COIL-TO- #7 SPARK PLUG (MARKED 4/7)

67189-DURA-G05

5.7L engine firing order

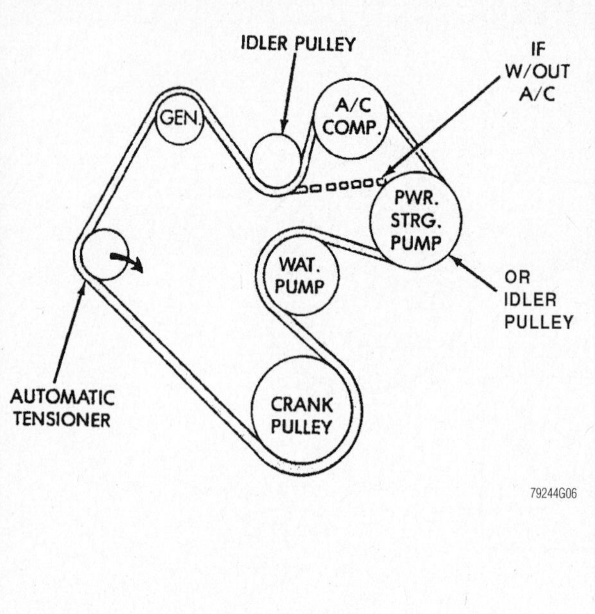

Accessory drive belt routing—3.9L and 5.9L LDC engines

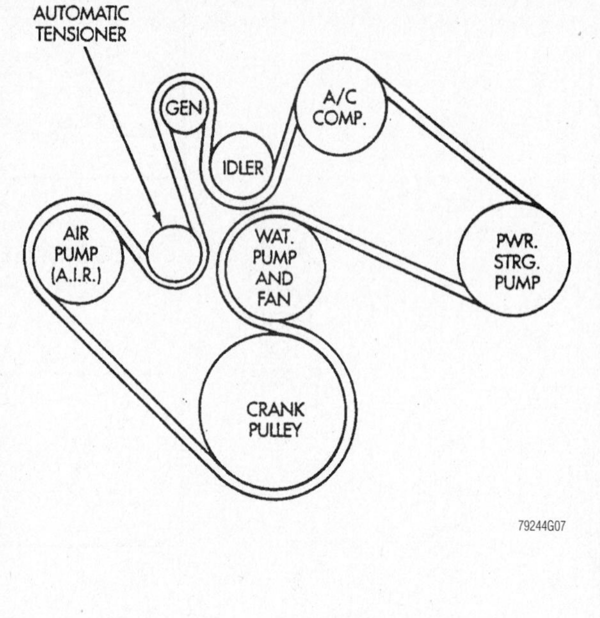

Accessory drive belt routing—5.9L HDC engines with A/C

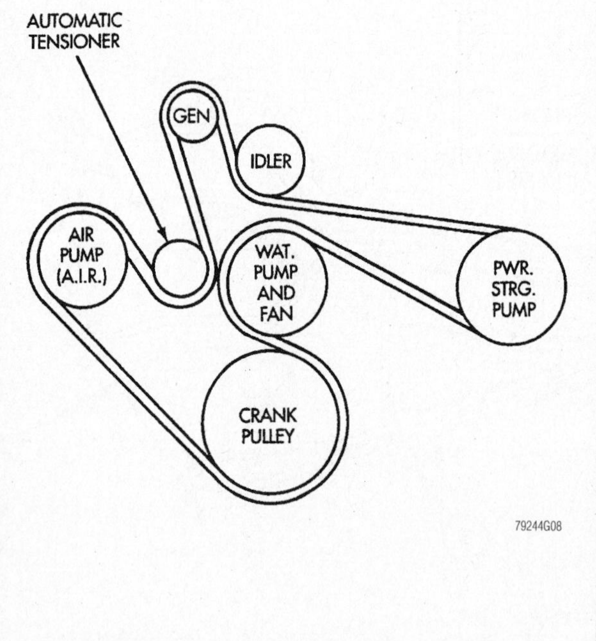

Accessory drive belt routing—5.9L HDC engines without A/C

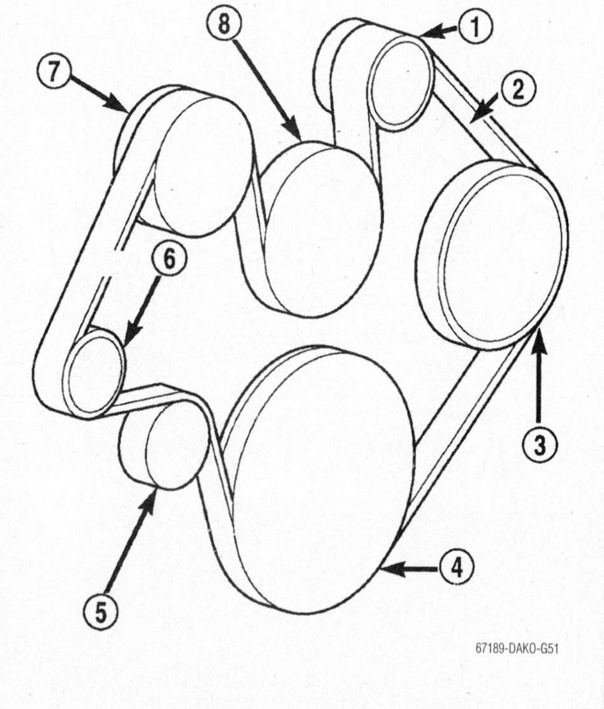

Accessory drive belt routing—3.7L and 4.7L engines

CAPACITIES

Year	Model	Engine Displ. Liters	Engine VIN	Engine Oil with Filter (qts.)	Transmission (pts.)	Transfer Case (pts.)	Drive Axle Front (pts.)	Drive Axle Rear (pts.)	Fuel Tank (gal.)	Cooling System (qts.)
2002	Durango	4.7	N	6.0	①	②	3.5	③	25.0	17.0
		5.9	Z	5.0	①	②	3.5	③	25.0	14.3
2003	Durango	4.7	N	6.0	①	②	3.5	③	25.0	④
		5.9	Z	5.0	①	②	3.5	③	25.0	⑤
2004	Durango	3.7	K	5.0	⑥	⑦	3.4	③	27.0	16.2
		4.7	N	6.0	⑥	⑦	3.4	③	27.0	16.2
		5.7	D	7.0	⑥	⑦	3.4	③	27.0	16.2
2005	Durango	3.7	K	5.0	⑥	⑦	3.4	③	27.0	16.2
		4.7	N	6.0	⑥	⑦	3.4	③	27.0	16.2
		5.7	D	7.0	⑥	⑦	3.4	③	27.0	16.2

NOTE: All capacities are approximate. Add fluid gradually and check to be sure a proper fluid level is obtained.

① Fluid drain/filter service: 46RE 8.0 pts.
Fluid drain/filter service: 45RFE 2wd 11 pts; 4wd 13 pts.
Overhaul, 46RE: 20 pts.
Overhaul, 45RFE: 28 pts.

② NV133 and NV233: 2.5 pts.
NV244: 2.85 pts.

③ The following values include 0.25 pt. of friction modifier for LSD axles.
8.25 axle: 4.4 pts.
9.25 axle: 4.9 pts.

④ Without rear heat: 13.3 qts.
With rear heat: 14.1 qts.

⑤ Without rear heat: 16.3 qts.
With rear heat: 17.2 qts.

⑥ 42RLE Fluid drain/filter service: 8pts.; Overhaul: 17.6 pts.
545RFE Fluid drain/filter service, 2wd: 11 pts.; 4wd: 13 pts.; Overhaul: 28 pts.

⑦ NV144: 1.8 pts.; NV244 GEN II: 3.4 pts.

06009-DURA-C04

VALVE SPECIFICATIONS

Year	Engine Displ. Liters	Engine VIN	Seat Angle (deg.)	Face Angle (deg.)	Spring Test Pressure (lbs. @ in.)	Spring Installed Height (in.)	Stem-to-Guide Clearance (in.) Intake	Stem-to-Guide Clearance (in.) Exhaust	Stem Diameter (in.) Intake	Stem Diameter (in.) Exhaust
2002	4.7	N	44.5-45	45-45.5	176.7-193.3 @1.1670	1.601	0.0008-0.0028	0.0019-0.0039	0.2729-0.2739	0.2717-0.2728
	5.9	Z	44.25-44.75	43.25-43.75	200@1.212	1.640	0.0010-0.0030	0.0020-0.0040	0.3720-0.3730	0.3710-0.3720
2003	4.7	N	44.5-45	45-45.5	176.7-193.3 @1.1670	1.601	0.0008-0.0028	0.0019-0.0039	0.2729-0.2739	0.2717-0.2728
	5.9	Z	44.25-44.75	43.25-43.75	200@1.212	1.640	0.0010-0.0030	0.0020-0.0040	0.3720-0.3730	0.3710-0.3720
2004	3.7	K	44.5-45	45-45.5	221-242@ 1.107	1.619	0.0008-0.0028	0.0019-0.0039	0.2729-0.2739	0.2717-0.2728
	4.7	N	44.5-45	45-45.5	176.7-193.3 @1.1670	1.601	0.0008-0.0028	0.0019-0.0039	0.2729-0.2739	0.2717-0.2728
	5.7	D	44.5-45	45-45.5	95@1.81	1.81	0.0008-0.0025	0.0019-0.0037	0.3120-0.3130	0.3110-0.3120
2005	3.7	K	44.5-45	45-45.5	221-242@ 1.107	1.619	0.0008-0.0028	0.0019-0.0039	0.2729-0.2739	0.2717-0.2728
	4.7	N	44.5-45	45-45.5	176.7-193.3 @1.1670	1.601	0.0008-0.0028	0.0019-0.0039	0.2729-0.2739	0.2717-0.2728
	5.7	D	44.5-45	45-45.5	95@1.81	1.81	0.0008-0.0025	0.0019-0.0037	0.3120-0.3130	0.3110-0.3120

06009-DURA-C05

CRANKSHAFT AND CONNECTING ROD SPECIFICATIONS

All measurements are given in inches.

			Crankshaft				Connecting Rod		
Year	Engine Displ. Liters	Engine VIN	Main Brg. Journal Dia.	Main Brg. Oil Clearance	Shaft End-play	Thrust on No.	Journal Diameter	Oil Clearance	Side Clearance
2002	4.7	N	2.4996-2.5005	0.0008-0.0021	0.0021-0.0112	2	2.0076-2.0082	0.0006-0.0022	0.0040-0.0138
	5.9	Z	2.8095-2.8105	①	0.0020-0.0070	2	2.1240-2.1250	0.0005-0.0022	0.0060-0.0140
2003	4.7	N	2.4996-2.5005	0.0008-0.0021	0.0021-0.0112	2	2.0076-2.0082	0.0006-0.0022	0.0040-0.0138
	5.9	Z	2.8095-2.8105	①	0.0020-0.0070	2	2.1240-2.1250	0.0005-0.0022	0.0060-0.0140
2004	3.7	K	2.4996-2.5005	0.0020-0.0034	0.0021-0.0112	2	2.2794-2.2797	0.0004-0.0019	0.0040-0.0138
	4.7	N	2.4996-2.5005	0.0008-0.0021	0.0021-0.0112	2	2.0076-2.0082	0.0006-0.0022	0.0040-0.0138
	5.7	D	2.5585-2.5595	0.0009-0.0020	0.0020-0.0110	3	2.1250-2.1260	0.0007-0.0023	0.0030-0.0137
2005	3.7	K	2.4996-2.5005	0.0020-0.0034	0.0021-0.0112	2	2.2794-2.2797	0.0004-0.0019	0.0040-0.0138
	4.7	N	2.4996-2.5005	0.0008-0.0021	0.0021-0.0112	2	2.0076-2.0082	0.0006-0.0022	0.0040-0.0138
	5.7	D	2.5585-2.5595	0.0009-0.0020	0.0020-0.0110	3	2.1250-2.1260	0.0007-0.0023	0.0030-0.0137

① No. 1: 0.0005-0.0015
Nos. 2-5: 0.0005-0.0020

06009-DURA-C06

PISTON AND RING SPECIFICATIONS

All measurements are given in inches.

	Engine			Ring Gap			Ring Side Clearance		
Year	Displ. Liters	Engine VIN	Piston Clearance	Top Comp.	Bottom Comp.	Oil Control	Top Comp.	Bottom Comp.	Oil Control
2002	4.7	N	0.0014	0.0146-0.0249	0.0146-0.0249	0.0099-0.0300	0.0020-0.0037	0.0016-0.0031	0.0175-0.0185
	5.9	Z	0.0005-0.0015	0.0120-0.0220	0.0220-0.0310	0.0150-0.0550	0.0016-0.0033	0.0016-0.0033	0.0020-0.0080
2003	4.7	N	0.0014	0.0146-0.0249	0.0146-0.0249	0.0099-0.0300	0.0020-0.0037	0.0016-0.0031	0.0175-0.0185
	5.9	Z	0.0005-0.0015	0.0120-0.0220	0.0220-0.0310	0.0150-0.0550	0.0016-0.0033	0.0016-0.0033	0.0020-0.0080
2004	3.7	K	0.0014	0.0146-0.0249	0.0146-0.0249	0.0100-0.0300	0.0020-0.0037	0.0016-0.0031	0.0007-0.0091
	4.7	N	0.0014	0.0146-0.0249	0.0146-0.0249	0.0099-0.0300	0.0020-0.0037	0.0016-0.0031	0.0175-0.0185
	5.7	D	0.0008-0.0019	0.0090-0.0149	0.0137-0.0236	0.0059-0.0259	0.0007-0.0026	0.0007-0.0022	0.0007-0.0091
2005	3.7	K	0.0014	0.0146-0.0249	0.0146-0.0249	0.0100-0.0300	0.0020-0.0037	0.0016-0.0031	0.0007-0.0091
	4.7	N	0.0014	0.0146-0.0249	0.0146-0.0249	0.0099-0.0300	0.0020-0.0037	0.0016-0.0031	0.0175-0.0185
	5.7	D	0.0008-0.0019	0.0090-0.0149	0.0137-0.0236	0.0059-0.0259	0.0007-0.0026	0.0007-0.0022	0.0007-0.0091

06009-DURA-C07

TORQUE SPECIFICATIONS

All readings in ft. lbs.

Year	Engine Displ. Liters	Engine VIN	Cylinder Head Bolts	Main Bearing Bolts	Rod Bearing Bolts	Crankshaft Damper Bolts	Flexplate Bolts	Manifold Intake	Manifold Exhaust	Spark Plugs	Oil Pan Drain Plug
2002	4.7	N	①	②	③	130	45	④	18	20	25
	5.9	Z	⑤	85	45	18	55	⑥	25	30	25
2003	4.7	N	①	②	③	130	45	④	18	20	25
	5.9	Z	⑤	85	45	18	55	⑥	25	30	25
2004	3.7	K	①	⑦	③	130	45	④	18	27	25
	4.7	N	①	②	③	130	45	④	18	20	25
	5.7	D	⑧	⑩	⑨	90	70	①	18	18	25
2005	3.7	K	①	⑦	③	130	70	④	18	20	25
	4.7	N	①	②	③	130	45	④	18	20	25
	5.7	D	⑧	⑩	⑨	90	70	④	18	13	25

① See text

② Bed plate bolt sequence. Refer to illustration
Step 1: Bolts A-L to 40 ft. lbs.
Step 2: Bolts 1-10 25 inch lbs.
Step 3: Bolts 1-10 plus 90 degrees
Step 4: Bolts A1-A6 20 ft. lbs.

③ 20 ft. lbs. plus 90 degrees

④ 105 inch lbs.

⑤ Step 1: 50 ft. lbs.
Step 2: 105 ft. lbs.

⑥ See illustration in text section
Step 1: 1-4 to 72 inch lbs. in 12 inch lb. Increments
Step 2: bolts 5-12: 72 inch lbs.
Step 3: Check that all bolts are at 72 inch lbs.
Step 4: All bolts, in sequence, to 12 ft. lbs.
Step 5: Check that all bolts are at 12 ft. lbs.

⑦ See the illustration
Step 1: Hand tighten bolts 1D, 1G and 1F until bedplate contacts the block
Step 2: tighten bolts 1A-1J to 40 ft. lbs.
Step 3: Tighten bolts 1-8 to 60 inch lbs.
Step 4: Tighten bolts 1-8 an additional 90 degrees
Step 5: Tighten bolts A-E to 20 ft. lbs.

⑧ M8
Step 1: 15 ft. lbs.
Step 2: 25 ft. lbs.
M12
Step 1: 25 ft. lbs.
Step 2: 40 ft. lbs.
Step 3: plus 90 degree turn

⑨ 15 ft. lbs. Plus 90 degrees

⑩ See the illustration
Step 1: Bolts 1-10 to 20 ft. lbs.
Step 2: bolts 1-10 an additional 90 degrees
Step 3: Bolts (cross bolts) A-J to 21 ft. lbs.

06009-DURA-C08

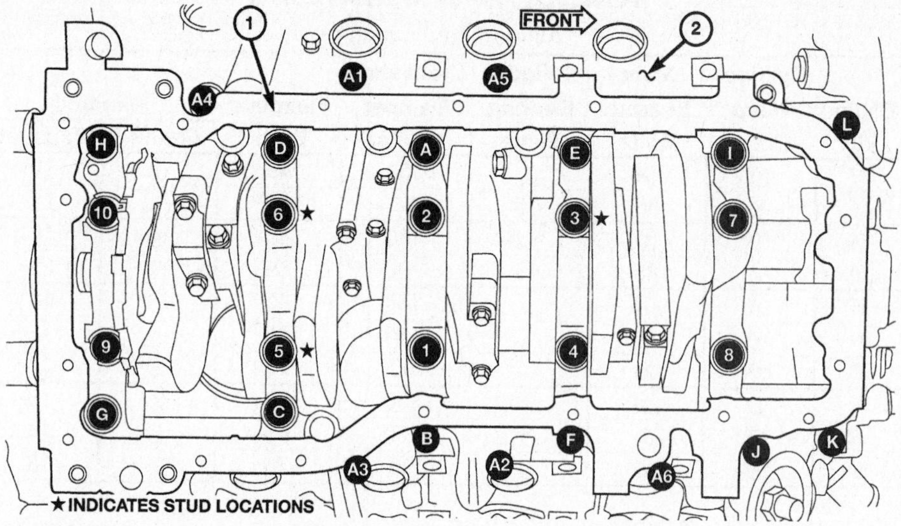

NOTE: Lubricate the bedplate retaining bolts with clean engine oil prior to installation.

06009-DURA-G01

4.7L crankshaft installation torque sequence

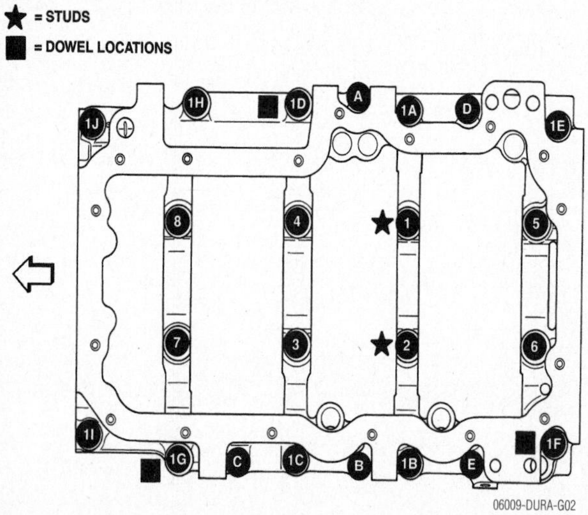

3.7L crankshaft installation torque sequence

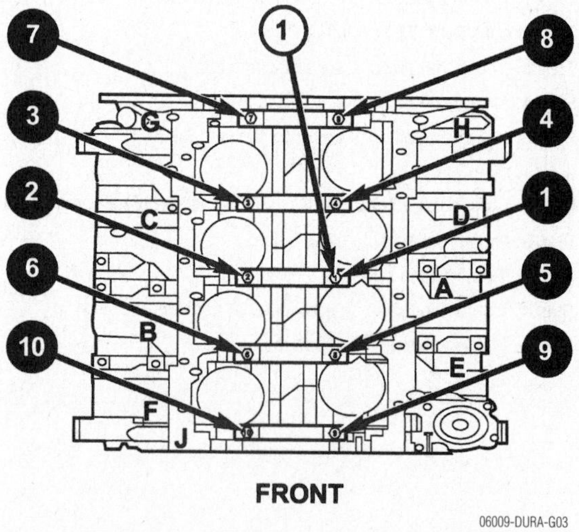

5.7L crankshaft installation torque sequence

WHEEL ALIGNMENT

		Caster		Camber		
Year	Model	Range (+/-Deg.)	Preferred Setting (Deg.)	Range (+/-Deg.)	Preferred Setting (Deg.)	Toe-in (in.)
2002	2WD	0.50	+3.10	0.50	-0.25	0.10+/-0.06
	4WD	0.50	+3.30	0.50	-0.25	0.10+/-0.06
2003	2WD	0.50	+3.10	0.50	-0.25	0.10+/-0.06
	4WD	0.50	+3.30	0.50	0	0.10+/-0.06
2004	Front	0.50	+3.50	0.50	0	0.05+/-0.05
	Rear	—	—	0.35	-0.10	0.30+/-0.35
2005	Front	0.50	+3.50	0.50	0	0.05+/-0.05
	Rear	—	—	0.35	-0.10	0.30+/-0.35

06009-DURA-C09

TIRE, WHEEL AND BALL JOINT SPECIFICATIONS

		OEM Tires		Tire Pressures (psi)		Wheel	Ball Joint	Lug Nut Torque
Year	Model	Standard	Optional	Front	Rear	Size	Inspection	ft. lbs.
2002	Durango	P235/75R15XL	P255/65R16 P265/80R16 P275/60R17	①	①	NA	0.060 in. ②	100
2003	Durango	P235/75R15XL	P255/65R16 P265/80R16 P275/60R17	①	①	NA	0.060 in. ②	100
2004	Durango	P245/70R17	P265/65R17	①	①	std: 7 opt: 8	0.020 in. ②	100
2005	Durango	P245/70R17	P265/65R17 P275/60R17	①	①	std: 7 opt: 8	0.020 in. ②	135

NA: Information not available

OEM: Original Equipment Manufacturer

PSI: Pounds Per Square Inch

STD: Standard

OPT: Optional

① See placard on vehicle

② Both upper and lower

06009-DURA-C10

BRAKE SPECIFICATIONS

All measurements in inches unless noted

		Brake Disc			Brake Drum			Minimum Lining Thickness		Brake Caliper	
Year	Model	Original Thickness	Minimum Thickness	Maximum Run-out	Original Inside Diameter	Max. Wear Limit	Maximum Machine Diameter	Front	Rear	Bracket Bolts (ft. lbs.)	Mounting Bolts (ft. lbs.)
2002	Durango	0.900	0.890	0.004	11.00	①	①	②	③	④	22
2003	Durango	F 1.100	F 1.047	F 0.0009	—	—	—	NA	NA	④	⑤
		R 0.866	R 0.811	R 0.0009							
2004	Durango	F 1.100	F 1.039	F 0.0009	—	—	—	NA	NA	⑥	F 24
		R 0.866	R 0.811	R 0.0009							R 11
2005	Durango	F 1.100	F 1.039	F 0.0009	—	—	—	NA	NA	⑥	F 24
		R 0.866	R 0.811	R 0.0009							R 11

NA: Not Available

① Maximum allowable drum diameter, either from wear or machining, is stamped on the drum.

② Riveted brake pads: 0.0625 in.
Bonded brake pads: 0.1875 in.

③ Riveted brake shoes: 0.031 in.
Bonded brake shoes: 0.0625 in.

④ Support plate bolts/nuts: 47 ft. lbs.
Adapter plate bolts: 130 ft. lbs.

⑤ Front and rear: 23.5 ft. lbs.

⑥ Adapter plate: front 130; rear 100
Support plate: 47

06009-DURA-C11

SCHEDULED MAINTENANCE INTERVALS

2002-03 DODGE DURANGO

		VEHICLE MILEAGE INTERVAL (x1000)													
TO BE SERVICED	TYPE OF SERVICE	7.5	15	22.5	30	37.5	45	52.5	60	67.5	75	82.5	90	97.5	100
Engine oil & filter	R	✓	✓	✓	✓	✓	✓	✓	✓	✓	✓	✓	✓	✓	✓
Front wheel bearings	S/I			✓			✓			✓			✓		
Brake linings	S/I			✓			✓			✓			✓		
Air cleaner element	R				✓				✓				✓		
Spark plugs	R				✓				✓				✓		
Transfer case fluid	R					✓					✓				
Engine coolant ①	R						✓						✓		
Spark plug wires (5.9L)	R								✓						
PCV valve	R								✓						
Drive belt tensioner ②	S/I								✓						
Automatic transmission fluid	R														✓
Automatic transmission bands	Adj.														✓

R: Replace S/I: Service or Inspect

① Replace every 36 months, regardless of mileage.

② Replace if necessary

FREQUENT OPERATION MAINTENANCE (SEVERE SERVICE)

If a vehicle is operated under any of the following conditions it is considered severe service:

- Extremely dusty areas.

- 50% or more of the vehicle operation is in 32°C (90°F) or higher temperatures, or constant operation in temperatures below 0°C (32°F).

- Prolonged idling (vehicle operation in stop and go traffic.

- Frequent short running periods (engine does not warm to normal operating temperatures).

- Police, taxi, delivery usage or trailer towing usage.

Oil & oil filter change: change every 3000 miles.

Air filter: change every 24,000 miles.

Engine coolant level, hoses & clamps: check every 6,000 miles.

Exhaust system: check every 6000 miles.

Drive belts: check every 18,000 miles; replace every 24,000 miles.

Oxygen sensor: replace every 82,500 miles.

Automatic transmission fluid, filter & bands: change & adjust every 12,000 miles.

Rear axle fluid: change every 12,000 miles.

06009-DURA-C12

SCHEDULED MAINTENANCE INTERVALS

2004-05 DODGE DURANGO

TO BE SERVICED	TYPE OF SERVICE	VEHICLE MILEAGE INTERVAL (x1000)																
		6	12	18	24	30	36	42	48	54	60	66	72	78	84	90	96	102
Engine oil & filter	R	✓	✓	✓	✓	✓	✓	✓	✓	✓	✓	✓	✓	✓	✓	✓	✓	✓
Tires	Rotate	✓	✓	✓	✓	✓	✓	✓	✓	✓	✓	✓	✓	✓	✓	✓	✓	✓
Brake linings	S/I			✓			✓			✓			✓					
Air cleaner element	R					✓					✓					✓		
Spark plugs	R					✓					✓					✓		
Transfer case fluid	I					✓					✓							
Engine coolant	R	Replace every 60 months regardless of mileage																
Spark plug wires (5.7L)	R								✓									
PCV valve	S/I										✓							
Drive belt tensioner	S/I										✓							
Automatic transmission fluid ①	R																	✓
Transfer case fluid	R	Drain the fluid every 120,000 miles																

R: Replace S/I: Service or Inspect

① On 4.7L and 5.7L, change the filter, if equipped

FREQUENT OPERATION MAINTENANCE (SEVERE SERVICE)

If a vehicle is operated under any of the following conditions it is considered severe service:

- **Extremely dusty areas.**
- **Day or night time temperatures below 0°C (32°F).**
- **Prolonged idling (vehicle operation in stop and go traffic.)**
- **Frequent short running periods (engine does not warm to normal operating temperatures).**
- **Police, taxi, delivery usage or trailer towing usage.**
- **Off road or desert operation**
- **50% or more of your driving is done in temperatures above 90 degrees F (32 deg. C)**

Oil & oil filter change: change every 3000 miles.

Air filter: change every 15,000 miles and change as necessary.

Drive belts: check and replace as necessary every 60,000 miles.

Automatic transmission fluid, filter every 30,000 miles 4.7L and 5.7L.

Automatic transmission fluid, filter every 60,000 miles 3.7L.

Front and rear axle fluid: change every 15,000 miles.

Inspect PCV valve every 30,000 miles

06009-DURA-C13

PRECAUTIONS

Before servicing any vehicle, please be sure to read all of the following precautions, which deal with personal safety, prevention of component damage, and important points to take into consideration when servicing a motor vehicle:

- Never open, service or drain the radiator or cooling system when the engine is hot; serious burns can occur from the steam and hot coolant.
- Observe all applicable safety precautions when working around fuel. Whenever servicing the fuel system, always work in a well-ventilated area. Do not allow fuel spray or vapors to come in contact with a spark, open flame, or excessive heat (a hot drop light, for example). Keep a dry chemical fire extinguisher near the work area. Always keep fuel in a container specifically designed for fuel storage; also, always properly seal fuel containers to avoid the possibility of fire or explosion. Refer to the additional fuel system precautions later in this section.
- Fuel injection systems often remain pressurized, even after the engine has been turned **OFF**. The fuel system pressure must be relieved before disconnecting any fuel lines. Failure to do so may result in fire and/or personal injury.
- Brake fluid often contains polyglycol ethers and polyglycols. Avoid contact with the eyes and wash your hands thoroughly after handling brake fluid. If you do get brake fluid in your eyes, flush your eyes with clean, running water for 15 minutes. If eye irritation persists, or if you have taken brake fluid internally, IMMEDIATELY seek medical assistance.
- The EPA warns that prolonged contact with used engine oil may cause a number of skin disorders, including cancer! You should make every effort to minimize your exposure to used engine oil. Protective gloves should be worn when changing oil. Wash your hands and any other exposed skin areas as soon as possible after exposure to used engine oil. Soap and water, or waterless hand cleaner should be used.
- All new vehicles are now equipped with an air bag system, often referred to as a Supplemental Restraint System (SRS) or Supplemental Inflatable Restraint (SIR) system. The system must be disabled before performing service on or around system components, steering column, instrument panel components, wiring and sensors. Failure to follow safety and disabling procedures could result in accidental air bag deployment, possible personal injury and unnecessary system repairs.
- Always wear safety goggles when working with, or around, the air bag system. When carrying a non-deployed air bag, be sure the bag and trim cover are pointed away from your body. When placing a non-deployed air bag on a work surface, always face the bag and trim cover upward, away from the surface. This will reduce the motion of the module if it is accidentally deployed. Refer to the additional air bag system precautions later in this section.
- Clean, high quality brake fluid from a sealed container is essential to the safe and proper operation of the brake system. You should always buy the correct type of brake fluid for your vehicle. If the brake fluid becomes contaminated, completely flush the system with new fluid. Never reuse any brake fluid. Any brake fluid that is removed from the system should be discarded. Also, do not allow any brake fluid to come in contact with a painted surface; it will damage the paint.
- Never operate the engine without the proper amount and type of engine oil; doing so WILL result in severe engine damage.
- Timing belt maintenance is extremely important! Many models utilize an interference-type, non-freewheeling engine. If the timing belt breaks, the valves in the cylinder head may strike the pistons, causing potentially serious (also time-consuming and expensive) engine damage. Refer to the maintenance interval charts in the front of this manual for the recommended replacement interval for the timing belt, and to the timing belt section for belt replacement and inspection.
- Disconnecting the negative battery cable on some vehicles may interfere with the functions of the on-board computer system(s) and may require the computer to undergo a relearning process once the negative battery cable is reconnected.
- When servicing drum brakes, only disassemble and assemble one side at a time, leaving the remaining side intact for reference.

ENGINE REPAIR

➡Disconnecting the negative battery cable on some vehicles may interfere with the functions of the on board computer system. The computer may undergo a relearning process once the negative battery cable is reconnected.

Distributor

REMOVAL

5.9L Engines

1. Before servicing the vehicle, refer to the precautions in the beginning of this section.
2. Remove or disconnect the following:
   - Negative battery cable

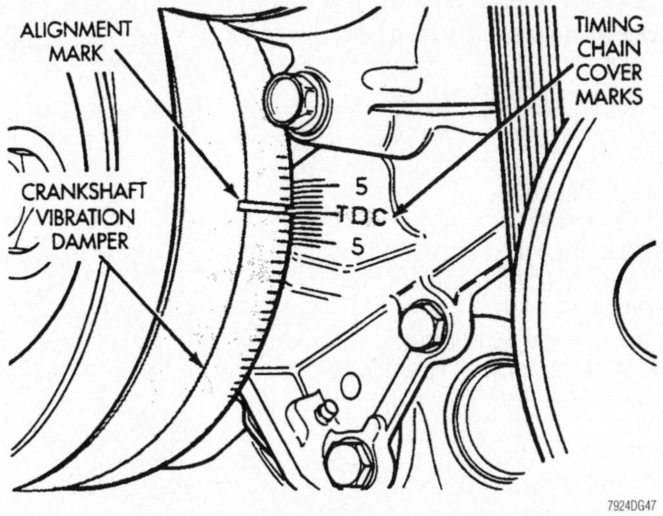

Crankshaft pulley and timing chain cover marks aligned at Top Dead Center (TDC)

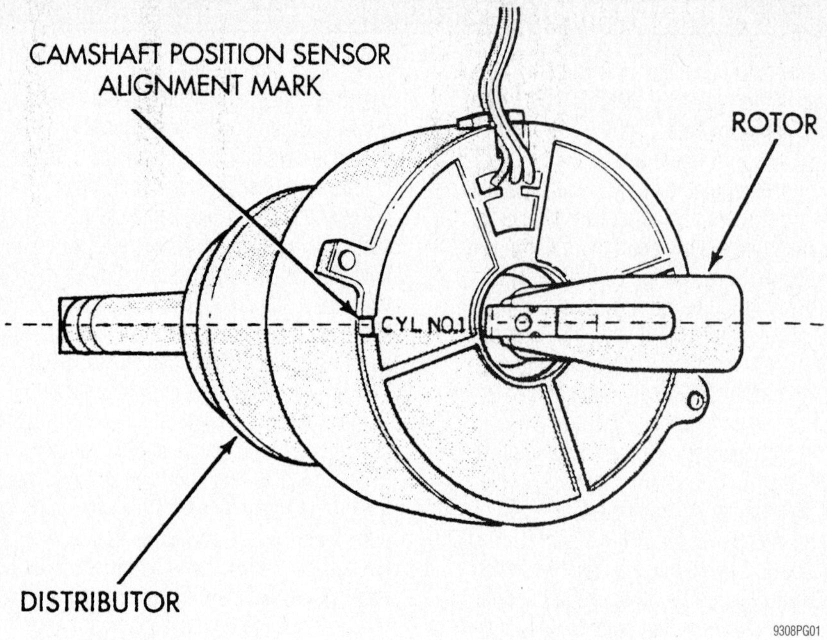

Distributor rotor alignment—5.9L engines

- Air cleaner tube
- Distributor cap
- Camshaft Position (CMP) sensor connector

3. Matchmark the distributor housing and the rotor.
4. Matchmark the distributor housing and the intake manifold.
5. Remove the distributor.

INSTALLATION

Timing Not Disturbed

5.9L ENGINES

1. Before servicing the vehicle, refer to the precautions in the beginning of this section.

➡The rotor will rotate clockwise as the gears engage.

2. Position the rotor slightly counterclockwise of the matchmark made during removal.
3. Install the distributor.
4. Align the distributor housing and intake manifold matchmarks and check that the distributor housing matchmark and rotor are also aligned.
5. Install or connect the following:
   - Distributor housing clamp and bolt. Tighten the bolt to 17 ft. lbs. (23 Nm).
   - CMP sensor connector
   - Distributor cap
   - Air cleaner tube
   - Negative battery cable

Timing Disturbed

5.9L ENGINES

1. Before servicing the vehicle, refer to the precautions in the beginning of this section.
2. Set the engine at Top Dead Center (TDC) of the No. 1 cylinder compression stroke.
3. Install the distributor, clamp and bolt.
4. Rotate the distributor so that the rotor aligns with the No. 1 cylinder mark on the Camshaft Position (CMP) sensor. Tighten the clamp bolt to 17 ft. lbs. (23 Nm).
5. Install or connect the following:
   - CMP sensor connector
   - Distributor cap
   - Air cleaner tube
   - Negative battery cable
6. The final distributor position must be set with a scan tool. Follow the instructions supplied by the scan tool manufacturer.

Alternator

REMOVAL & INSTALLATION

3.7L and 4.7L Engines

1. Before servicing the vehicle, refer to the precautions in the beginning of this section.
2. Remove or disconnect the following:
   - Negative battery cable
   - Accessory drive belt
   - Alternator harness connectors
   - Mounting bolts and alternator

➡There are 1 vertical and 2 horizontal bolts.

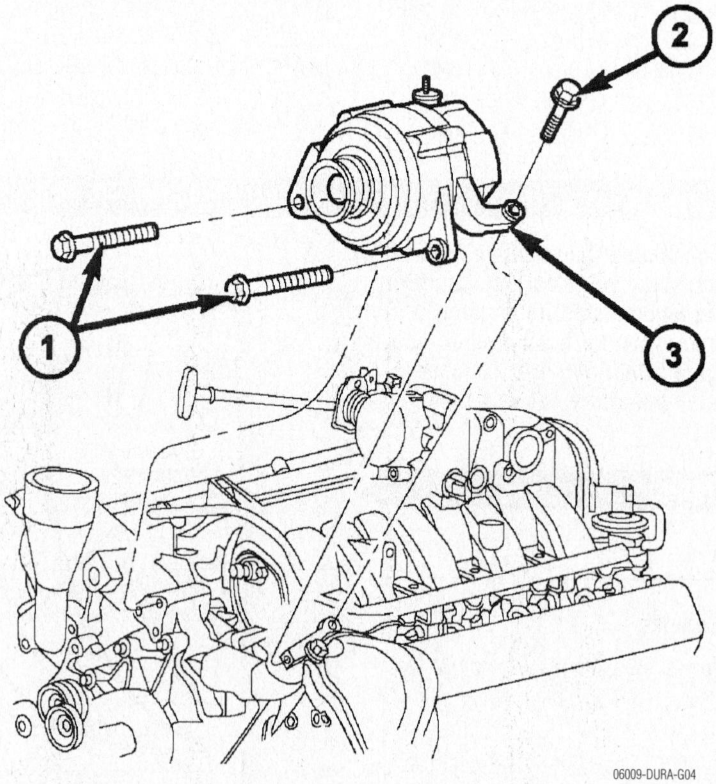

Alternator mounting, 3.7L and 4.7L engines. (1) horizontal bolts, (2) vertical bolt, (3) alternator

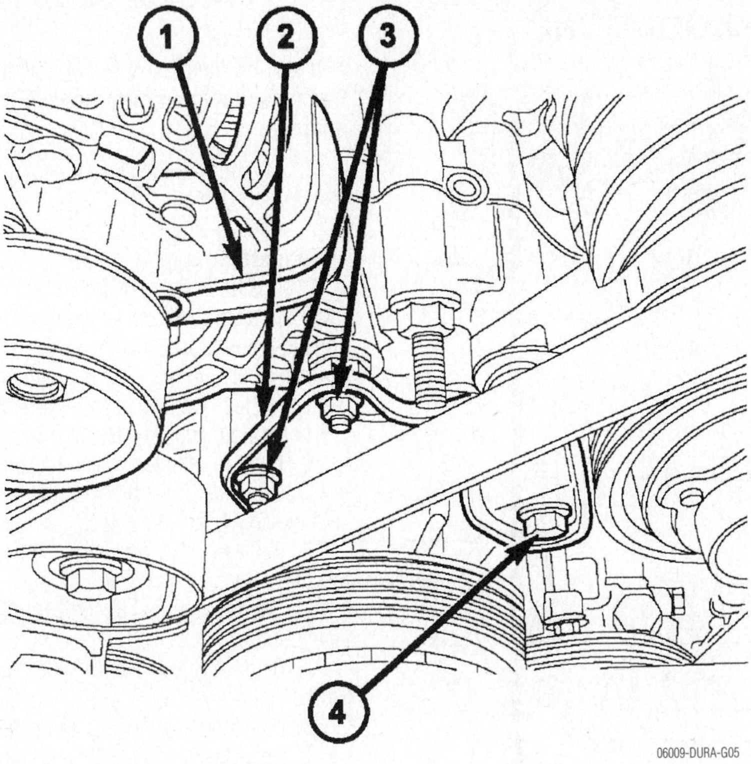

Alternator mounting—5.7L engine

To install:

3. Before servicing the vehicle, refer to the precautions in the beginning of this section.
4. Install the alternator and tighten the bolts to the following specifications:
   - Short horizontal bolt to 55 ft. lbs. (74 Nm)
   - Vertical bolt and long horizontal bolt to 40 ft. lbs. (55 Nm)
5. Install or connect the following:
   - Alternator harness connectors
   - Accessory drive belt
   - Negative battery cable

5.7L Engines

1. Before servicing the vehicle, refer to the precautions in the beginning of this section.
2. Remove or disconnect the following:
   - Negative battery cable
   - Accessory drive belt
   - Alternator harness connectors
   - Support bracket nuts and bolt
   - Mounting bolts and alternator
3. Installation is the reverse of removal. Torque the bolts to 30 ft. lbs. (41Nm).

5.9L Engines

1. Before servicing the vehicle, refer to the precautions in the beginning of this section.
2. Remove or disconnect the following:
   - Negative battery cable
   - Accessory drive belt
   - Alternator harness connectors
   - Alternator
3. Installation is the reverse of removal. Tighten the bolts to 30 ft. lbs. (41 Nm).

Ignition Timing

ADJUSTMENT

The ignition timing is controlled by the Powertrain Control Module (PCM) and is not adjustable.

Engine Assembly

REMOVAL & INSTALLATION

3.7L Engine

1. Discharge the A/C system.
2. Drain the cooling system.
3. Release the fuel system pressure.
4. Remove the air cleaner assembly.
5. Disconnect the battery.
6. Remove the accessory drive belt.
7. Remove the viscous fan.
8. Remove the A/C compressor.
9. Remove the generator and secure away from engine.

➡Do not remove the phenolic pulley from the P/S pump. It is not required for P/S pump removal.

10. Remove the power steering pump with lines attached and secure away from engine.
11. Disconnect the heater hoses from the engine.
12. Disconnect the heater hoses from heater core and remove hose assembly.
13. Remove the upper radiator hose from engine.
14. Remove the lower radiator hose from engine.
15. Disconnect the transmission oil cooler lines at the radiator.
16. Disconnect the power steering cooler lines.
17. Remove the radiator assembly.
18. Disconnect the throttle and speed control cables.
19. Disconnect the engine to body ground straps at the left side of cowl.
20. Disconnect the engine wiring harness at the following points:
   - Intake air temperature (IAT) sensor
   - Fuel Injectors
   - Throttle Position (TPS) Switch
   - Idle Air Control (IAC) Motor
   - Engine Oil Pressure Switch
   - Engine Coolant Temperature (ECT) Sensor
   - Manifold Absolute Pressure MAP) Sensor
   - Camshaft Position (CMP) Sensor
   - Coil Over Plugs
   - Crankshaft Position Sensor
21. Remove the coil over plugs.
22. Remove fuel rail and secure away from engine.

➡It is not necessary to release the quick connect fitting from the fuel supply line for engine removal.

23. Remove the PCV hose.
24. Remove the breather hoses.
25. Remove the vacuum hose for the power brake booster.
26. Disconnect the knock sensors.
27. Remove the engine oil dipstick tube.
28. Remove intake manifold.
29. Install an engine lifting fixture, Tool 8247, using the original fasteners from the removed intake manifold, and fuel rail.
30. Raise the vehicle on hoist.
31. Disconnect the exhaust pipes from exhaust manifolds.
32. On 4wd vehicles disconnect axle vent tube from left side engine mount.
33. Remove the through bolt retaining

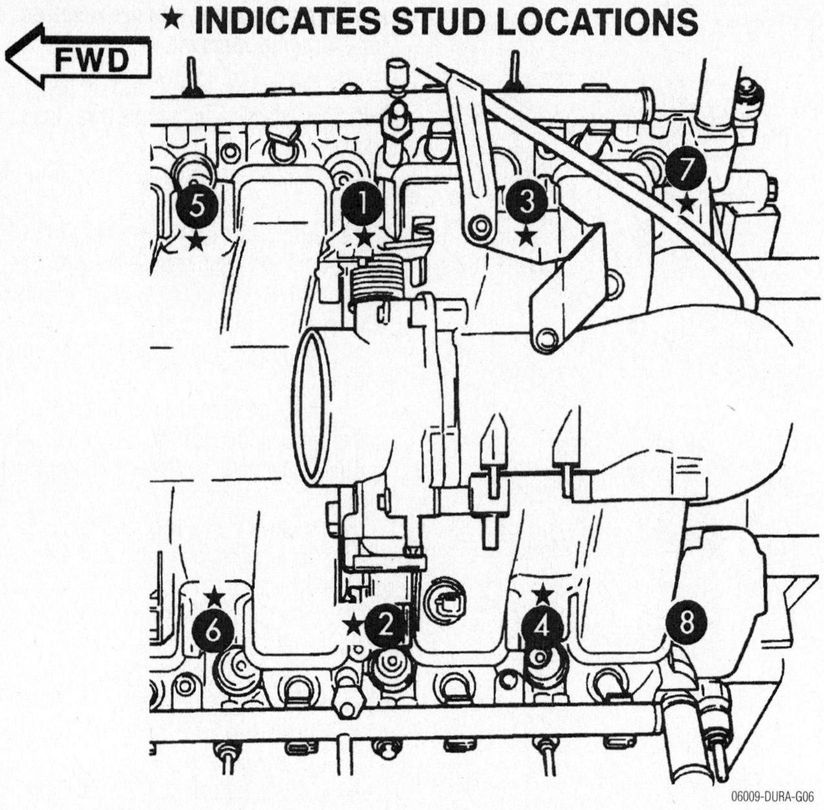

Engine lifting fixture mounting locations—3.7L engines

nut and bolt from both the left and right side engine mounts.

34. On 4wd vehicles Remove locknut from left and right side engine mount brackets.

35. Disconnect two ground straps from the lower left hand side and one ground strap from the lower right hand side of the engine.

36. Disconnect the crankshaft position sensor.

➡**The following step applies to 4wd vehicles equipped with automatic transmission only.**

37. On 4wd vehicles, remove the axle isolator bracket from the engine, transmission and the axle.

38. Remove the structural cover.

39. Remove the starter.

40. Remove the torque converter bolts (automatic transmission only).

41. Remove the transmission to engine mounting bolts.

42. Disconnect the engine block heater power cable from the block heater, if equipped.

43. Lower the vehicle.

44. Remove throttle body resonator assembly and air inlet hose.

45. Disconnect throttle and speed control cables.

46. Disconnect the tube from both the left and right side crankcase breathers. Remove the breathers.

47. Remove the generator.

48. Disconnect the two heater hoses from the timing chain cover and heater core.

49. Unclip and remove the heater hoses and tubes from the intake manifold.

50. Disconnect engine harness at the following points:

- Intake air temperature (IAT) sensor
- Fuel Injectors
- Throttle Position (TPS) Switch
- Idle Air Control (IAC) Motor
- Engine Oil Pressure Switch
- Engine Coolant Temperature (ECT) Sensor
- Manifold absolute pressure (MAP) Sensor
- Camshaft Position (CMP) Sensor
- Coil Over Plugs

51. Disconnect the vacuum lines at the throttle body and intake manifold.

52. Remove power steering pump and position out of the way.

53. Disconnect body ground strap at the right side cowl.

54. Disconnect body ground strap at the left side cowl.

➡**It will be necessary to support the transmission in order to remove the engine.**

55. Position a suitable jack under the transmission.

56. Remove engine from the vehicle.

To install:

57. Position engine in the vehicle.

58. Position both the left and right side engine mount brackets and install the through bolts and nuts. Tighten nuts to 95 Nm (70 ft. lbs.) on 2wd vehicles; 102 Nm (75 ft. lbs.) on 4wd vehicles.

59. On 4wd vehicles, install locknuts onto the engine mount brackets. Tighten locknuts to 41 Nm (30 ft. lbs.)

60. Remove jack from under the transmission.

61. Remove Engine Lifting Fixture Special Tool.

62. Remove Special Tools 8400 Lifting Studs.

63. Position generator wiring behind the oil dipstick tube, then install the oil dipstick tube upper mounting bolt.

64. Connect both left and right side body ground straps.

65. Install power steering pump.

66. Connect fuel supply line quick connect fitting.

67. Connect the vacuum lines at the throttle body and intake manifold.

68. Connect engine harness at the following points:

- Intake Air Temperature (IAT) Sensor
- Idle Air Control (IAC) Motor
- Fuel Injectors
- Throttle Position (TPS) Switch
- Engine Oil Pressure Switch
- Engine Coolant Temperature (ECT) Sensor
- Manifold Absolute Pressure (MAP) Sensor
- Camshaft Position (CMP) Sensor
- Coil Over Plugs

69. Position and install heater hoses and tubes (1) onto intake manifold.

70. Install the heater hoses onto the heater core and the engine front cover.

71. Install generator.

72. Install radiator assembly.

73. Connect radiator upper and lower hoses.

74. Connect power steering cooler lines.

75. Connect the transmission oil cooler lines to the radiator.

76. Install accessory drive belt.

77. Install A/C compressor.

78. Install both breathers. Connect tube to both crankcase breathers.

79. Connect throttle and speed control cables.

80. Install throttle body resonator assembly and air inlet hose. Tighten clamps 4 Nm (35 in. lbs.).

81. Raise vehicle.

82. Install transmission to engine mounting bolts. Tighten the bolts to 41 Nm (30 ft. lbs.).

83. Install torque converter bolts (Automatic Transmission Only).

84. Connect crankshaft position sensor.

85. On 4wd vehicles, position and install the axle isolator bracket onto the axle, transmission and engine block. Tighten bolts to specification.

86. Install starter.

CAUTION

The structural cover requires a specific torque sequence. Failure to follow this sequence may cause severe damage to the cover.

87. Install structural cover.

88. Install exhaust crossover pipe.

89. Install engine block heater power cable, if equipped.

90. On 4wd vehicles, connect axle vent tube to left side engine mount.

91. Lower vehicle.

92. Check and fill engine oil.

93. Recharge the A/C system.

94. Refill the engine cooling system.

95. Install the battery tray and battery.

96. Connect the battery positive and negative cables.

97. Start the engine and check for leaks.

4.7L Engine

1. Before servicing the vehicle, refer to the precautions in the beginning of this section.

2. Drain the cooling system and engine oil.

3. Remove or disconnect the following:
   - Negative battery cable
   - Battery and tray
   - Exhaust crossover pipe
   - On 4wd, the axle vent tube
   - Left and right engine mount through bolts
   - On 4wd, the left and right engine mount bracket locknuts
   - Ground straps
   - CKP sensor
   - On 4wd, the axle isolator bracket
   - Structural cover
   - Starter
   - Torque converter bolts
   - Transmission-to-engine bolts
   - Engine block heater
   - Resonator and air inlet
   - Throttle and speed control cables
   - Crankcase breathers
   - A/C compressor
   - Shroud and fan assemblies
   - Transmission cooler lines
   - Radiator hoses
   - Radiator
   - Alternator
   - Heater hoses
   - Engine harness
   - Vacuum lines
   - Fuel system pressure
   - Fuel line at the rail
   - Power steering pump

4. Install lifting eyes and take up the weight of the engine with a crane.

5. Support the transmission with a jack.

6. Remove the engine.

7. Installation is the reverse of removal. Observe the following:
   - Left and right engine mount through bolts: 2wd 70 ft. lbs. (95 Nm); 4wd 75 ft. lbs. (102 Nm)
   - On 4wd, the bracket locknuts: 30 ft. lbs. (41 Nm)
   - Transmission-to-engine bolts: 30 ft. lbs. (41 Nm)

➡The structural cover has a specific torque sequence.

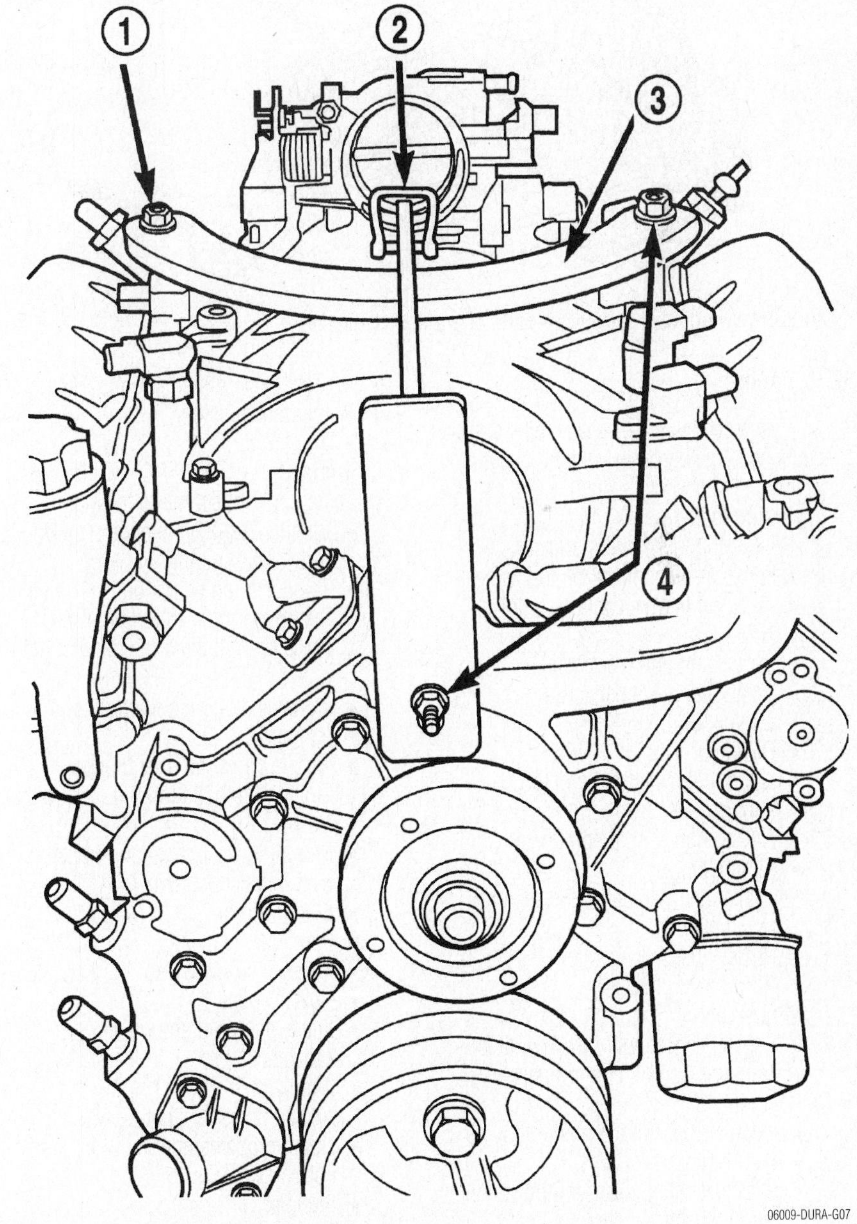

Engine lifting fixture mounting locations—4.7L engines

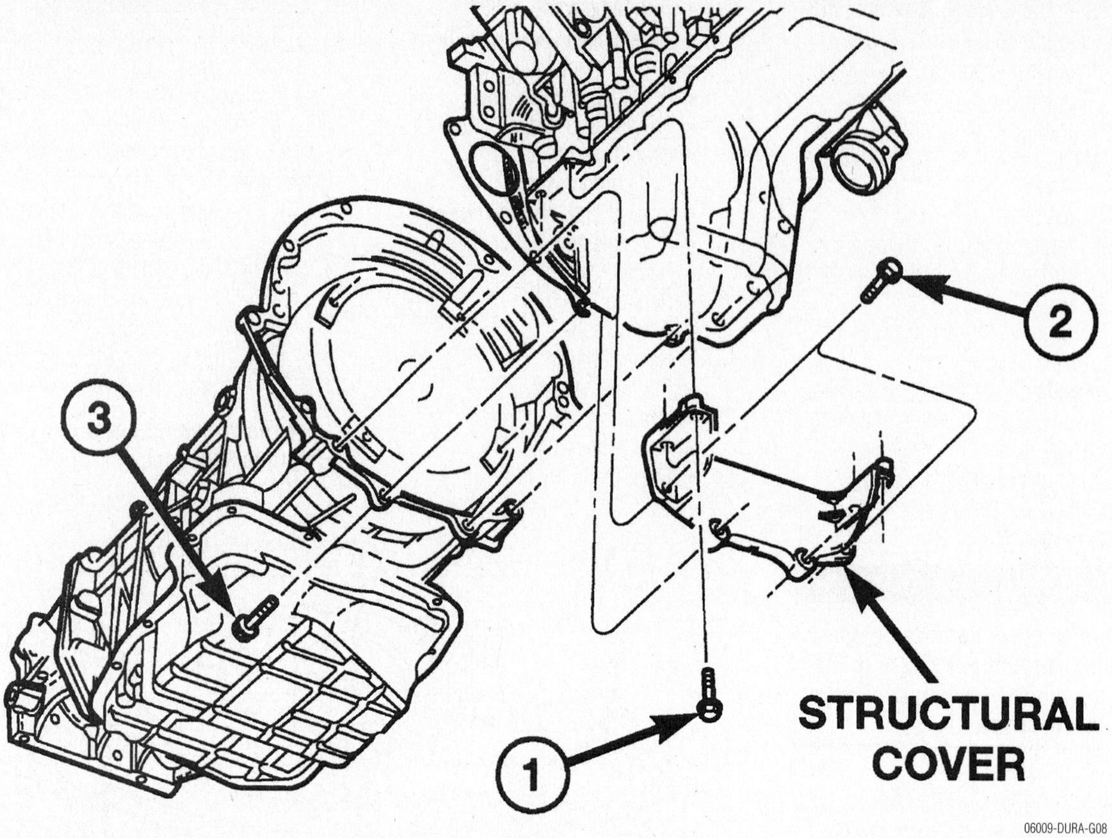

Structural cover torque sequence—3.7L, 4.7L and 5.7L engines

5.7L Engine

1. Before servicing the vehicle, refer to the precautions in the beginning of this section.
2. Drain the cooling system.
3. Drain the engine oil.
4. Relieve the fuel system pressure.
5. Remove or disconnect the following:
   - Negative battery cable
   - Hood
   - Air cleaner and resonator
   - Accessory drive belt
   - Engine fan
   - Radiator
   - Upper crossmember and top core support
   - A/C compressor, if equipped
   - Alternator
   - Intake manifold and IAFM as an assembly
   - Heater hoses
   - Power steering pump
   - Fuel line
   - Engine front mount thru-bolt nuts
   - Transmission oil cooler lines, if equipped
   - Exhaust pipes at the manifolds
   - Starter motor
   - Structural dust cover and transmission inspection cover
   - Torque converter-to-flexplate bolts
   - Transmission flange bolts. Support the transmission.
   - Engine

To install:

6. Install or connect the following:
   - Engine. Tighten the engine mount thru-bolt finger tight.
   - Transmission flange bolts. Tighten the bolts to 40–45 ft. lbs. (54–61 Nm). Then, tighten the mount bolt nuts to 70 ft. lbs. (95 Nm)
   - Transmission oil cooler lines, if equipped
   - Torque converter, if equipped. Tighten the bolts to 23 ft. lbs. (31 Nm).
   - Structural dust cover and transmission inspection cover
   - Starter motor
   - Fuel line
   - Power steering pump
   - Heater hoses
   - Intake manifold and IAFM as an assembly
   - Alternator
   - A/C compressor, if equipped
   - Upper crossmember and top core support
   - Radiator
   - Engine fan
   - Accessory drive belt
   - Air cleaner and resonator
   - Hood
   - Negative battery cable
7. Fill the crankcase to the correct level.
8. Fill the cooling system.
9. Start the engine and check for leaks.

5.9L Engines

1. Before servicing the vehicle, refer to the precautions in the beginning of this section.
2. Drain the cooling system.
3. Drain the engine oil.
4. Relieve the fuel system pressure.
5. Remove or disconnect the following:
   - Negative battery cable
   - Hood
   - Upper crossmember and top core support
   - Radiator hoses
   - Cooling fan and shroud
   - Radiator
   - Accelerator cable
   - Cruise control cable, if equipped
   - Transmission cable, if equipped
   - Heater hoses
   - Intake manifold vacuum lines
   - Accessory drive belt
   - Power steering pump
   - A/C compressor, if equipped

- Engine control sensor harness connectors
- Engine block heater, if equipped
- Fuel line
- Exhaust front pipe
- Starter motor
- Torque converter, if equipped
- Transmission oil cooler lines, if equipped
- Engine mounts
- Transmission flange bolts. Support the transmission.
- Engine

✲✲ WARNING

Do not lift the engine by the intake manifold. Damage to the manifold may result.

To install:

6. Install or connect the following:
   - Engine. Tighten the engine mount bolts to 70 ft. lbs. (95 Nm).
   - Transmission flange bolts. Tighten the bolts to 40–45 ft. lbs. (54–61 Nm).
   - Transmission oil cooler lines, if equipped
   - Torque converter, if equipped. Tighten the bolts to 23 ft. lbs. (31 Nm).
   - Starter motor
   - Exhaust front pipe
   - Fuel line
   - Engine block heater, if equipped
   - Engine control sensor harness connectors
   - A/C compressor, if equipped
   - Power steering pump
   - Accessory drive belt
   - Intake manifold vacuum lines
   - Heater hoses
   - Transmission cable, if equipped
   - Cruise control cable, if equipped
   - Accelerator cable
   - Radiator
   - Cooling fan and shroud
   - Radiator hoses
   - Upper crossmember and top core support
   - Hood
   - Negative battery cable
7. Fill the crankcase to the correct level.
8. Fill the cooling system.
9. Start the engine and check for leaks.

Water Pump

REMOVAL & INSTALLATION

3.7L Engine

1. Before servicing the vehicle, refer to the precautions in the beginning of this section.
2. Drain the cooling system.
3. Remove or disconnect the following:
   - Negative battery cable
   - Fan and clutch assembly from the pump
   - Fan shroud and fan assembly. If you're reusing the fan clutch, keep it upright to avoid silicone fluid loss!
   - Lower hose
   - Water pump (8 bolts)
4. Installation is the reverse of removal. Tighten the bolts, in sequence, to 40 ft. lbs. (54 Nm).

4.7L Engine

1. Before servicing the vehicle, refer to the precautions in the beginning of this section.
2. Drain the cooling system.
3. Remove or disconnect the following:
   - Negative battery cable
   - Fan and fan drive assembly from the pump. Don't attempt to remove it from the vehicle, yet.

➡If a new pump is being installed, don't separate the fan from the drive.

   - Shroud and fan

✲✲ WARNING

Keep the fan upright to avoid fluid loss from the drive.

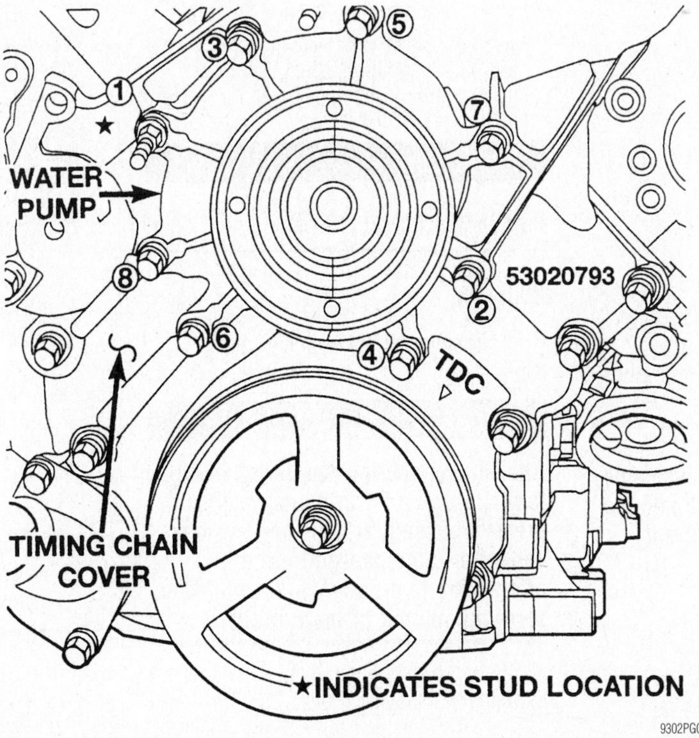

Water pump torque sequence—3.7L and 4.7L engine

- Accessory drive belt
- Lower radiator hose
- Water pump

4. Installation is the reverse of removal. Tighten the bolts in sequence to 40 ft. lbs. (54 Nm).

5.7L Engines

1. Before servicing the vehicle, refer to the precautions in the beginning of this section.
2. Drain the cooling system.
3. Remove or disconnect the following:
   - Negative battery cable
   - Accessory drive belt
   - Engine cooling fan
   - Coolant recovery bottle
   - Washer bottle
   - Fan shroud
   - A/C compressor and alternator brace
   - Idler pulleys
   - Belt tensioner
   - Radiator hoses
   - Heater hoses
   - Water pump

To install:

4. Install or connect the following:
   - Water pump. Tighten the bolts to 18 ft. lbs. (24 Nm).
   - Heater hoses
   - Radiator hoses
   - Idler pulleys
   - A/C compressor and alternator brace
   - Fan shroud
   - Washer bottle
   - Coolant recovery bottle
   - Accessory drive belt
   - Negative battery cable
   - Negative battery cable
5. Fill the cooling system.
6. Start the engine and check for leaks.

5.9L Engines

1. Before servicing the vehicle, refer to the precautions in the beginning of this section.
2. Drain the cooling system.
3. Remove or disconnect the following:
   - Negative battery cable
   - Engine cooling fan and shroud

➡Do not store the fan clutch assembly horizontally, silicone may leak into the bearing grease and cause contamination.

   - Accessory drive belt
   - Water pump pulley
   - Lower radiator hose
   - Heater hose and tube
   - Bypass hose
   - Water pump

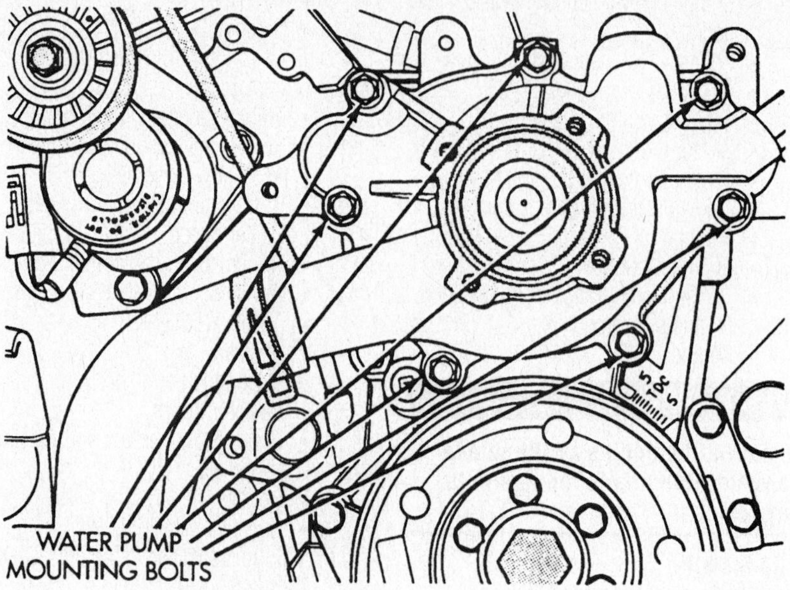

Water pump mounting bolt locations—5.9L engines

To install:

4. Install or connect the following:
   - Water pump, using a new gasket. Tighten the bolts to 30 ft. lbs. (40 Nm).
   - Bypass hose
   - Heater hose and tube. Use a new O-ring seal.
   - Lower radiator hose
   - Water pump pulley. Tighten the bolts to 20 ft. lbs. (27 Nm).
   - Accessory drive belt
   - Engine cooling fan and shroud
   - Negative battery cable
5. Fill the cooling system.
6. Start the engine and check for leaks.

Heater Core

REMOVAL & INSTALLATION

2002–03

1. Disconnect the negative battery cable.

⁂ CAUTION

After disconnecting the negative battery cable, wait 2 minutes for the driver's/passenger's air bag system capacitor to discharge before attempting to do any work around the steering column or instrument

2. Remove the instrument panel by performing the following procedure:
   a. Remove the trim from the right and left door sills.
   b. Remove the trim from the right and left cowl side inner panels.
   c. Remove the steering column opening cover from the instrument panel.
   d. Remove the 2 hood latch release handle-to-instrument panel lower reinforcement screws and lower the release handle to the floor.
   e. Disconnect the driver's side air bag module wire harness connector.
   f. If equipped, disconnect the overdrive lockout switch harness connector.
   g. Do not disassemble the steering column but remove the assembly from the vehicle.
   h. From under the driver's side of the instrument panel, disconnect or remove the following items:
      - The screw from the center of the headlight/dash-to-instrument panel bulkhead wire harness connector and disconnect the connector.
      - The 2 body wire harness connectors from the 2 instrument panel wire harness connector that secure the outboard side of the instrument panel bulkhead connector.
      - The 3 wire harness connectors from the 3 junction block connector receptacles located closest to the dash panel; 1 from the body wire harness and 2 from the headlight/dash wire harness.
      - The plastic park brake release linkage rod-to-rear parking brake release handle lever. Disengage the linkage rod from the handle lever.
      - The stoplight switch connector.

- The vacuum harness connector from the left side of the heater/air conditioning housing assembly.

i. Remove the instrument panel center support bracket.

j. Remove the instrument panel wire harness ground screw located on the left side of the air bag control module (ACM) mount on the floor panel transmission tunnel.

k. Disconnect the instrument panel wiring harness-to-ACM connector receptacle.

l. Remove the glove box.

m. Working through the glove box opening, disconnect or remove the following items:

- The 2 halves of the radio antenna coaxial cable connector near the center of the lower instrument panel.
- The antenna half of the radio antenna coaxial cable from the retainer clip near the outboard side of the lower instrument panel.
- The blower motor wire harness connector located near the heater/air conditioning housing assembly support brace.

n. From the passenger's side, disconnect or remove the following items:

- The 2 instrument panel wire harness connectors from the infinity speaker amplifier connector receptacles on the right cowl.
- The instrument panel wire harness radio ground eyelet-to-stud nut on the right cowl.

o. Loosen the right and left instrument panel cowl side roll-down bracket screws.

p. Remove the 5 top instrument panel-to-dash panel screws.

q. Pull the instrument panel rearward until the right and left cowl side roll-down bracket screws are in the roll-down slot position of both brackets.

r. Roll down the instrument panel and install a temporary hook in the center hole on top of the instrument panel; secure the other end to the top of the dash panel. The hook is to support the instrument panel in its rolled down position.

s. With the instrument panel in the rolled-down position, disconnect or remove the following items:

- The 2 instrument panel wire harness connector from the door jumper wire harness connectors located on the right bracket.
- The instrument panel wire harness connector from the blower motor resistor connector receptacle.
- The temperature control cable flag retainer from the top of the heater/air conditioning housing assembly. Pull the cable core adjuster clip off the blend-air door lever.
- The demister duct flexible hose from the adapter on the top of the heater/air conditioning housing assembly.

t. With aid of an assistant, lift the instrument panel from the vehicle.

3. If equipped with air conditioning, perform the following procedure:

a. Discharge and recover the air conditioning system refrigerant.

b. Disconnect the refrigerant line from the evaporator inlet and outlet tubes. Plug the openings to prevent contamination.

4. Drain the cooling system into a clean container for reuse.

5. Disconnect the heater hoses from the heater core. Plug the openings.

6. Remove the 4 heater/air conditioning housing assembly-to-chassis nuts.

7. Remove the heater/air conditioning housing assembly-to-mounting brace nut; the nut is located on the passenger side of the vehicle.

8. Pull the heater/air conditioning housing assembly rearward for the studs and drain tube to clear the dash panel hole.

9. Remove the heater/air conditioning housing assembly from the vehicle.

10. Remove the heater housing cover screws and the cover.

11. Remove the heater core from the heater/air conditioning housing assembly.

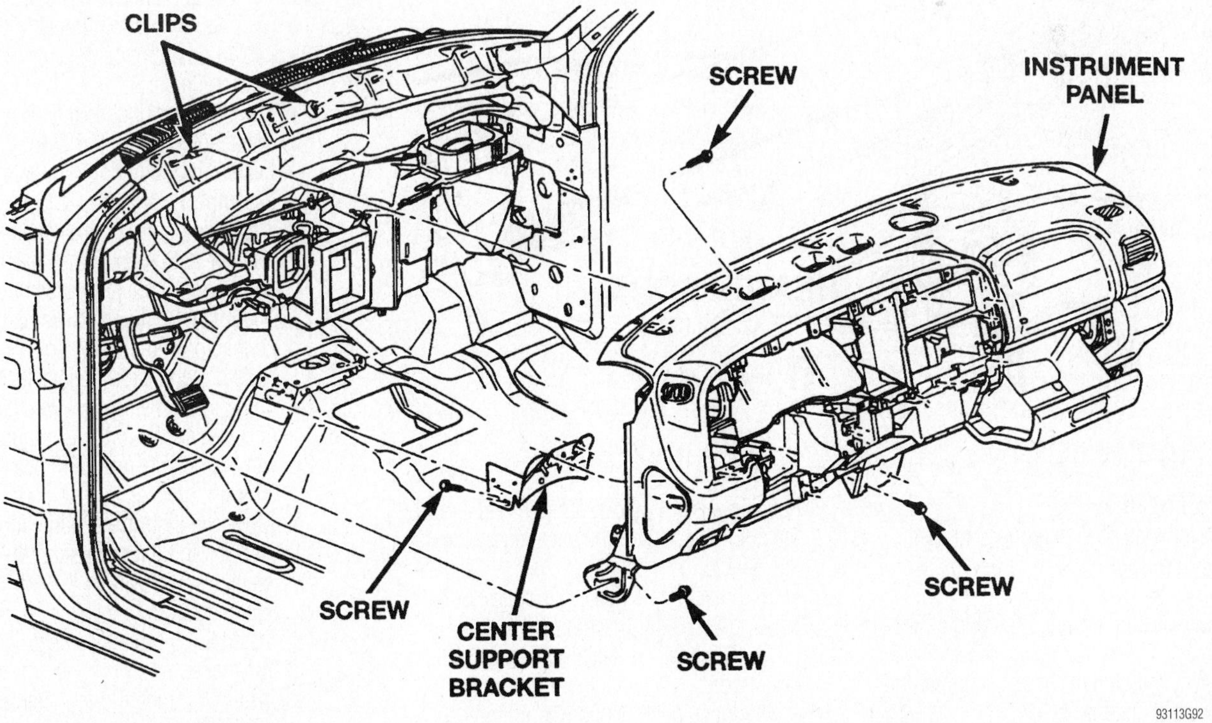

View of the instrument panel assembly

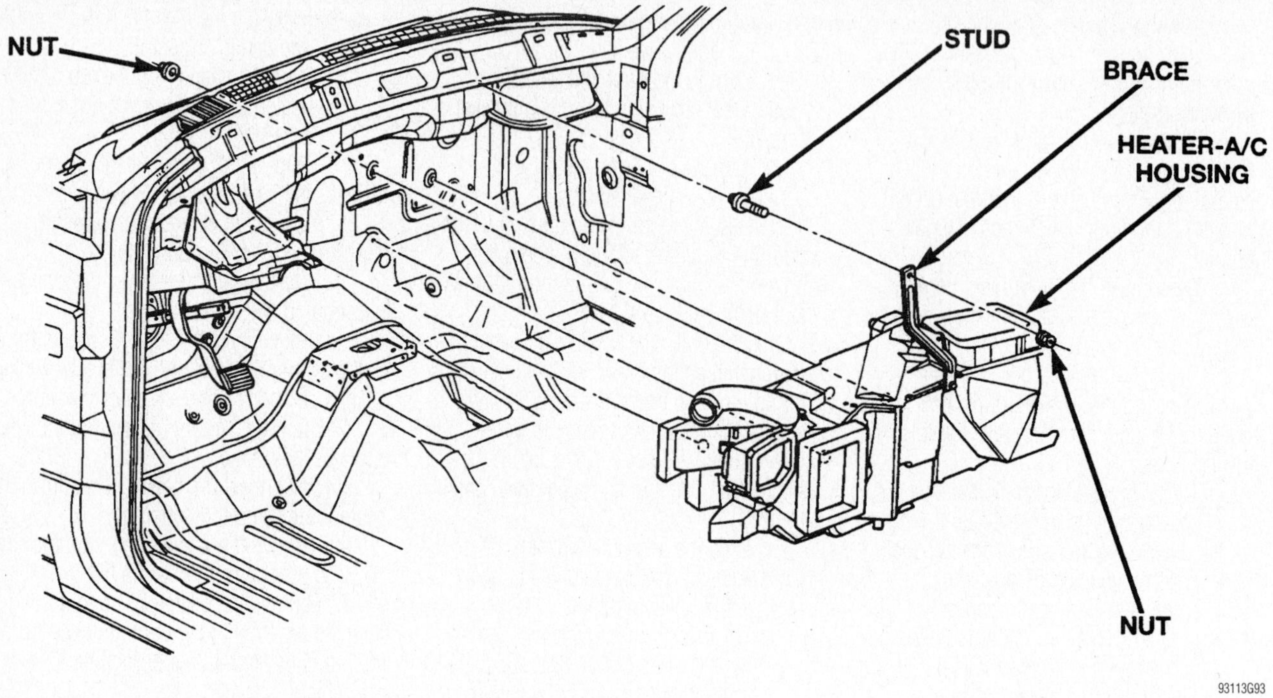

View of the heater/air conditioning assembly

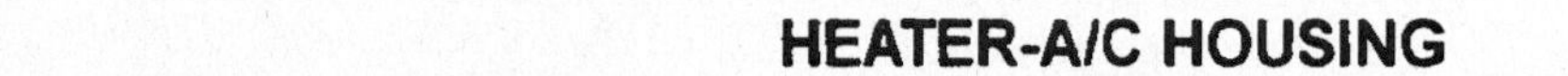

View of the heater core

To install:

12. Install the heater core to the heater/air conditioning housing assembly.

13. Install the heater housing cover and the cover screws.

14. Install the heater/air conditioning housing assembly.

15. Install the heater/air conditioning housing assembly-to-mounting brace nut; the nut is located on the passenger side of the vehicle.

16. Install the 4 heater/air conditioning housing assembly-to-chassis nuts.

17. Connect the heater hoses to the heater core.

18. If equipped with air conditioning, perform the following procedure:

a. Using new gaskets, connect the refrigerant line to the evaporator inlet and outlet tubes.

b. Evacuate and charge the air conditioning system.

19. Install the instrument panel by performing the following procedures:

a. With aid of an assistant, install the instrument panel into the vehicle.

b. With the instrument panel in the rolled-down position, connect or install the following items:

- The demister duct flexible hose to the adapter on the top of the heater/air conditioning housing assembly.
- The temperature control cable flag retainer to the top of the heater/air conditioning housing assembly.
- The instrument panel wire harness connector to the blower motor resistor connector receptacle.
- The 2 instrument panel wire harness connector to the door jumper wire harness connectors located on the right bracket.

c. Move the instrument panel forward until the right and left cowl side roll-down bracket screws are in the roll-down slot position of both brackets.

d. Install the 5 top instrument panel-to-dash panel screws.

e. Install the right and left instrument panel cowl side roll-down bracket screws.

f. At the passenger's side, connect or install the following items:

- The instrument panel wire harness radio ground eyelet-to-stud nut on the right cowl.
- The 2 instrument panel wire harness connectors to the infinity speaker amplifier connector receptacles on the right cowl.

g. Working through the glove box opening, connect or install the following items:

- The blower motor wire harness connector located near the heater/air conditioning housing assembly support brace.
- The antenna half of the radio antenna coaxial cable to the retainer clip near the outboard side of the lower instrument panel.
- The 2 halves of the radio antenna coaxial cable connector near the center of the lower instrument panel.

h. Install the glove box.

i. Connect the instrument panel wiring harness-to-ACM connector receptacle.

j. Install the instrument panel wire harness ground screw located on the left side of the air bag control module (ACM) mount on the floor panel transmission tunnel.

k. Install the instrument panel center support bracket.

l. Under the driver's side of the instrument panel, connect or install the following items:

- The vacuum harness connector to the left side of the heater/air conditioning housing assembly.
- The stoplight switch connector.
- Engage the linkage rod to the handle lever. Install the plastic park brake release linkage rod-to-rear parking brake release handle lever.
- The 3 wire harness connectors to the 3 junction block connector receptacles located closest to the dash panel; 1 at the body wire harness and 2 at the headlight/dash wire harness.
- The 2 body wire harness connectors to the 2 instrument panel wire harness connector that secure the outboard side of the instrument panel bulkhead connector.
- Connect the connector. Install the screw at the center of the headlight/dash-to-instrument panel bulkhead wire harness connector.

m. Install the steering column assembly.

n. If equipped, connect the overdrive lockout switch harness connector.

o. Connect the driver's side air bag module wire harness connector.

p. Install the 2 hood latch release handle-to-instrument panel lower reinforcement screws and lower the release handle to the floor.

q. Install the steering column opening cover to the instrument panel.

r. Install the trim to the right and left cowl side inner panels.

s. Install the trim to the right and left door sills.

20. Refill the cooling system.

21. Connect the negative battery cable.

22. Run the engine to normal operating temperatures; then, check the climate control operation and check for leaks.

2004–05

1. Drain the engine cooling system.
2. Raise and support the vehicle.
3. Remove the right front wheelhouse splash shield.
4. Remove the heater hoses from the heater core tubes in the engine compartment.
5. Lower the vehicle.

✲✲ WARNING

To avoid personal injury or death, on vehicles equipped with airbags, disable The supplemental restraint system before attempting any steering wheel, steering column, airbag, occupant classification system, seat belt tensioner, impact sensor, or Instrument panel component diagnosis or service. Disconnect and isolate the battery negative (ground) cable, then wait two minutes for the system capacitor to discharge before performing further diagnosis or service. This is the only sure way to disable the supplemental restraint system. Failure to take the proper precautions could result in accidental airbag deployment.

6. Before proceeding with the following repair procedure, review all warnings and cautions at the beginning of this chapter.
7. Disconnect and isolate battery negative cable.
8. Using a trim stick C-4755 or equivalent, remove the left door sill trim cover.
9. Remove the screw and remove the left cowl trim cover.
10. Remove the left instrument panel end cap.
11. Remove the two screws and remove the steering column opening cover.
12. Remove the screws and position aside the hood release handle.
13. Remove the four screws and remove the steering column opening reinforcement.
14. Remove the steering column tilt lever.
15. Remove the upper and lower column shrouds.
16. Disconnect the wiring harness connectors to the column.
17. Remove the shift cable from the column shift lever actuator.
18. Release the shift cable from the column bracket and remove it from the bracket.
19. Remove the SKIM module in order to disconnect the electrical connector.
20. Remove the upper steering shaft coupler bolt and slide the shaft down.
21. Remove the brake light switch and discard.
22. Remove the four steering column mounting nuts.
23. Lower the column from the mounting studs.
24. Remove the steering column assembly from the vehicle.
25. Remove the pedal support bracket bolts.
26. Disengage the release rod from the arm on the pedal assembly.
27. Disconnect the electrical connectors from the fuse block.
28. Remove the bolt and remove the ground wire.
29. Open the trim covers in the drivers side a-pillar grab handle and remove the bolts.
30. Remove the a-pillar trim panel.
31. Remove the floor console.
32. Disconnect the two body wire harness connectors from the Occupant Restraint Controller (ORC) connector receptacles located on the forward facing side of the module. To disconnect the wire harness connectors from the ORC, depress the release tab and lift the lever arm on each connector.
33. Position the carpet aside and remove the center harness screws. Pull instrument panel wiring harness from under the carpet.
34. Remove the center support bolts.
35. Remove the driver's seat.
36. Remove the drivers side floor duct.
37. Remove the drivers side support bolts.
38. Remove the right instrument panel end cap.
39. Remove the passenger side support bolts.
40. Using a trim stick C-4755 or equivalent, remove the right door sill trim cover.

41. Remove the screw and remove the right cowl trim cover.
42. Remove the bolts and remove the amplifier.
43. Disconnect the amplifier and antenna electrical connectors.
44. Remove the harness bolt and the ground wire bolt.
45. Remove the passenger side rear floor duct.
46. Open the trim covers in the passenger side a-pillar grab handle and remove the bolts.
47. Remove the a-pillar trim panel.
48. Using a trim stick C-4755 or equivalent, remove the instrument panel defroster grille.
49. Disconnect the sensor electrical connector.
50. Remove the four fenceline bolts.
51. Lift the instrument panel assembly off the side support pins and remove assembly through the driver's door.
52. Remove the bolt that secures the HVAC housing bracket to the dash panel.
53. Remove the two screws that secure the HVAC housing bracket to the top of the HVAC housing.
54. Remove the HVAC housing bracket from the vehicle.
55. Remove the screw that secures the heater core tube retaining bracket to the top of the HVAC housing.
56. Remove the heater core tube retaining bracket from the HVAC housing.
57. Remove the screw that secures the heater core tubes to the heater core.
58. Remove the heater core tubes from the heater core and the dash panel. Remove the O-ring seals from the heater core tube fittings and discard.
59. Remove the two screws that secure the heater core retaining bracket to the top of the HVAC housing.
60. Remove the heater core retaining bracket from the top of the HVAC housing.
61. Carefully lift the heater core out of the HVAC housing.

To install:

62. Carefully install the heater core and the heater core retaining bracket to the top of the HVAC housing. Make sure that the heater core insulator is properly positioned.
63. Install the two screws that secure the heater core and retaining bracket to the HVAC housing. Tighten the screws to 2.2 Nm (20 inch lbs.).
64. Lubricate new rubber O-ring seals with clean engine coolant and install them onto the heater core tube fittings. Use only the specified O-ring as it is made of a special material for the engine cooling system.
65. Install the heater core tubes through the dash panel and to the heater core.
66. Install the screw that secures the heater core tubes to the heater core. Tighten the screw securely.
67. Install the heater core tube retaining bracket to the top of the HVAC housing.
68. Install the screw that secures the heater core tube retaining bracket to the HVAC housing. Tighten the screw to 2.2 Nm (20 inch lbs.).
69. Install the HVAC housing bracket to the top of HVAC housing and to the dash panel.
70. Install the two screws that secure the HVAC housing bracket to the HVAC housing. Tighten the screws to 2.2 Nm (20 inch lbs.).
71. Install the bolt that secures the HVAC housing bracket to the dash panel. Tighten the bolt to 3 Nm (26 inch lbs.).
72. Position the instrument panel assembly into the vehicle through the driver's side door and install onto the side support pins.
73. Install the four fenceline bolts and tighten to 8 Nm (70 inch lbs.).
74. Connect the sensor electrical connector.
75. Install the instrument panel defroster grille and seat fully.
76. Position the passenger side a-pillar trim panel into place and seat fully.
77. Install the bolts and tighten to 6 Nm (55 inch lbs.).
78. Install the passenger side rear floor duct.
79. Install the screws for the ground wire and passenger side wire harness.
80. Tighten the ground wire and harness bolts to 14 Nm (10 ft. lbs.).
81. Connect the amplifier and antenna electrical connectors.
82. Install the amplifier and install the bolts.
83. Tighten the bolts to 6 Nm (50 inch lbs.).
84. Install the right cowl trim cover and install the screw.
85. Install the right door sill trim cover.
86. Install the passenger side support bolts and tighten to 27 Nm (20 ft. lbs.).
87. Install the right instrument panel end cap.
88. Install the drivers side support bolts and tighten to 27 Nm (20 ft. lbs.).
89. Install the drivers side floor duct.
90. Install the driver's seat.
91. Install the center support bolts and tighten to 11 Nm (95 inch lbs.).
92. Position the center instrument panel wiring harness under the carpet and install the bolts.

CAUTION

The lever arms of the wire harness connectors for the ORC MUST be in the unlatched position before they are inserted into their connector receptacles on the ORC or they may become damaged.

93. Reconnect the two body wire harness connectors to the ORC connector receptacles located on the forward facing side of the module. Be certain that the latches on both connectors are each fully engaged.
94. Install the floor console.
95. Position the driver's side a-pillar trim into place and seat the retaining clips fully.
96. Install the bolts and tighten to 6 Nm (55 inch lbs.).
97. Install the ground wire and install the bolt.
98. Tighten the ground wire bolts to 14 Nm (10 ft. lbs.)
99. Connect the electrical connectors at the fuse block.
100. Connect the brake release rod to the arm on the pedal assembly.
101. Install the pedal support bolts and tighten to 14 Nm (10 ft. lbs.).

CAUTION

All fasteners must be torqued to specification to ensure proper operation of the steering column.

102. Position the steering column on the dash panel support and loosely install the mounting nuts.
103. Firmly slide the steering column upward against the studs in dash panel and hand tighten the nuts.
104. Install the steering shaft coupler on the steering shaft and loosely install a new bolt.
105. Center steering column in dash opening and tighten mounting nuts to 28 Nm (21 ft. lbs.).

➡Torque the upper left nut first then the lower right nut. Then torque the lower left nut then the upper right nut.

➡A new bolt must be used for reinstallation.

106. Tighten the coupler bolt to 38 Nm (28 ft. lbs.).
107. Install a new brake light switch.
108. Install the shifter cable.
109. Connect the wiring harness to the column.
110. Install the SKIM module.

111. Install the upper and lower column shrouds and install the screws.
112. Install the column tilt lever.
113. Install the steering column opening reinforcement and install the four screws.
114. Install the hood release handle and install the screws.
115. Install the steering column opening cover and install the two screws.
116. Install the left instrument panel end cap.
117. Install the left cowl trim cover and install the screw.
118. Install the left door sill trim cover and seat fully.
119. Do not reconnect the battery negative cable at this time. The supplemental restraint system verification test procedure should be performed following service of any supplemental restraint system component.
120. Raise the vehicle and install the heater hoses to the heater core tubes in the engine compartment.
121. Install the right front wheelhouse splash shield.
122. Lower the vehicle.
123. Refill the engine cooling system.

Cylinder Head

REMOVAL & INSTALLATION

3.7L Engine

LEFT SIDE

1. Before servicing the vehicle, refer to the precautions in the beginning of this section.
2. Drain the cooling system.
3. Properly relieve the fuel system pressure.
4. Remove or disconnect the following:
   - Negative battery cable
   - Exhaust Y-pipe
   - Intake manifold
   - Cylinder head cover
   - Engine cooling fan and shroud
   - Accessory drive belt
   - Power steering pump
5. Rotate the crankshaft so that the crankshaft timing mark aligns with the Top Dead Center (TDC) mark on the front cover, and the **V6** marks on the camshaft sprockets are at 12 o'clock as shown.
   - Crankshaft damper
   - Front cover
6. Lock the secondary timing chain to the idler sprocket with Timing Chain Locking tool 8429.
7. Matchmark the secondary timing chain one link on each side of the V6 mark to the camshaft sprocket.
   - Left secondary timing chain tensioner
   - Cylinder head access plug
   - Secondary timing chain guide
   - Camshaft sprocket
   - Cylinder head

➡The cylinder head is retained by twelve bolts. Four of the bolts are smaller and are at the front of the head.

To install:

8. Check the cylinder head bolts for signs of stretching and replace as necessary.
9. Lubricate the threads of the 11mm bolts with clean engine oil.
10. Coat the threads of the 8mm bolts with Mopar® Lock and Seal Adhesive, or equivalent.
11. Install the cylinder heads. Use new gaskets and tighten the bolts, in sequence, as follows:
    a. Step 1: Bolts 1–8 to 20 ft. lbs. (27 Nm)

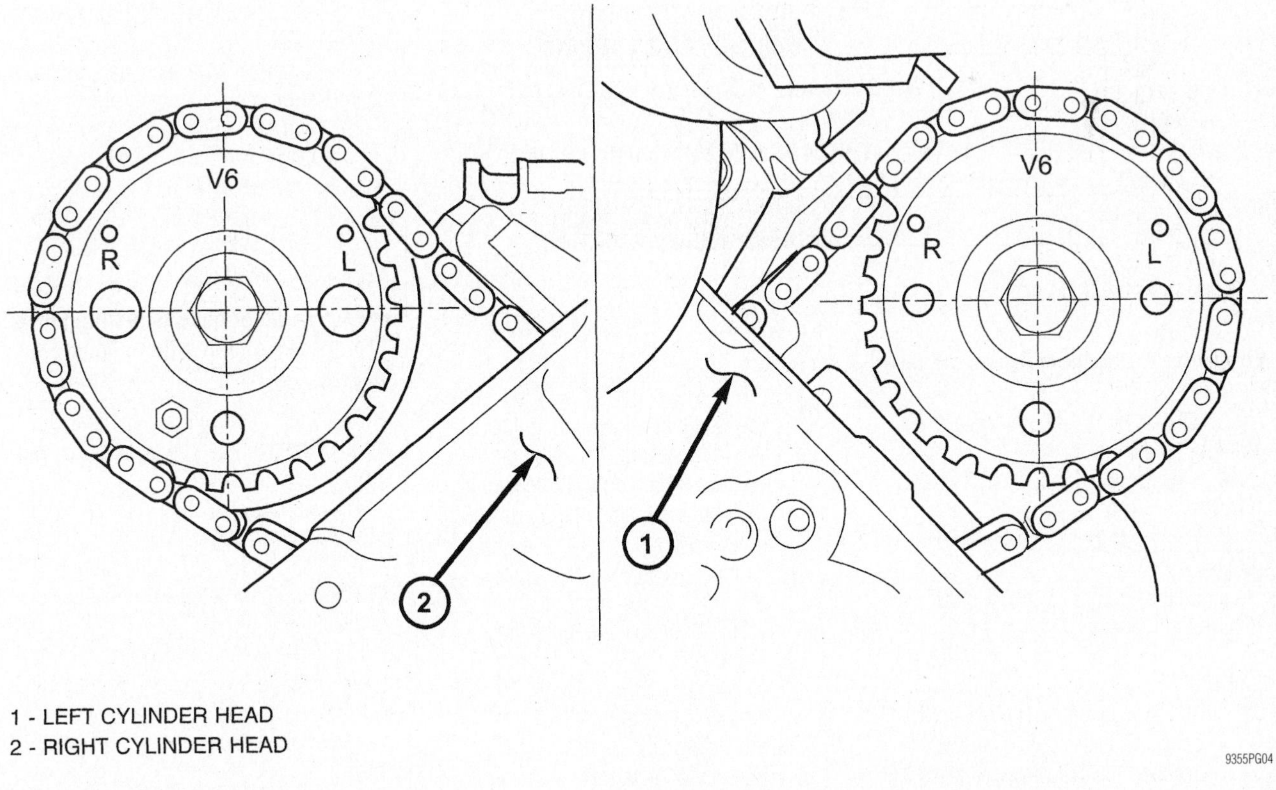

Camshaft sprocket timing marks—3.7L

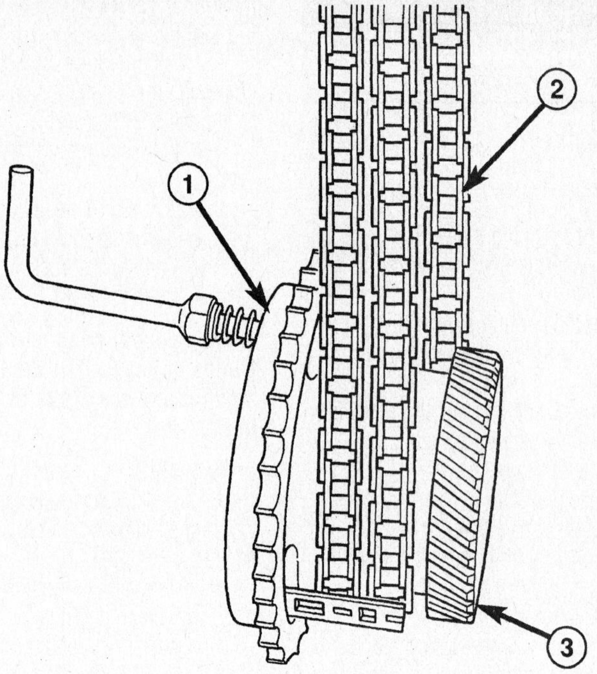

Camshaft locking tool—3.7L

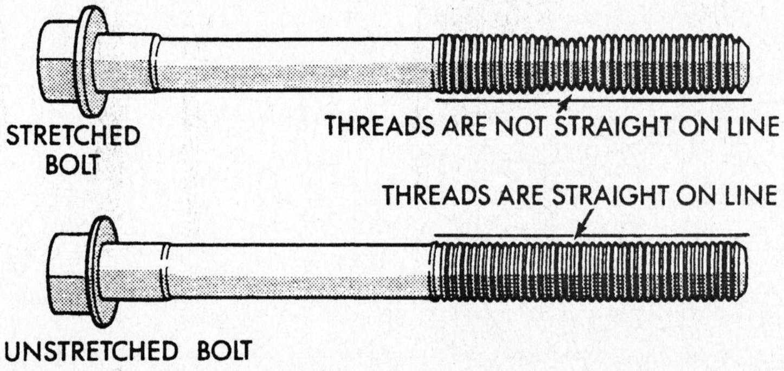

Examine the head bolts for signs of stretching—3.7L engine

b. Step 2: Bolts 1–10 verify torque without loosening

c. Step 3: Bolts 9–12 to 10 ft. lbs. (14 Nm)

d. Step 4: Bolts 1–8 plus ¼ (90 degree) turn

e. Step 5: Bolts 1–8 plus ¼ (90 degree) turn again

f. Step 6: Bolts 9–12 to 19 ft. lbs. (26 Nm)

12. Install or connect the following:
    - Camshaft sprocket. Align the secondary chain matchmarks and tighten the bolt to 90 ft. lbs. (122 Nm).
    - Secondary timing chain guide
    - Cylinder head access plug
    - Secondary timing chain tensioner. Refer to the timing chain procedure in this section.
13. Remove the Timing Chain Locking tool.
14. Install or connect the following:
    - Front cover
    - Crankshaft damper. Torque the bolt to 130 ft. lbs. (175 Nm).
    - Power steering pump
    - Accessory drive belt
    - Engine cooling fan and shroud
    - Cover
    - Intake manifold
    - Exhaust Y-pipe
    - Negative battery cable
15. Fill and bleed the cooling system.
16. Start the engine, check for leaks and repair if necessary.

RIGHT SIDE

1. Before servicing the vehicle, refer to the precautions in the beginning of this section.
2. Drain the cooling system.
3. Properly relieve the fuel system pressure.
4. Remove or disconnect the following:
    - Negative battery cable
    - Exhaust Y-pipe
    - Intake manifold
    - Valve cover
    - Engine cooling fan and shroud
    - Accessory drive belt
    - Oil fill housing
    - Power steering pump
5. Rotate the crankshaft so that the crankshaft timing mark aligns with the Top Dead Center (TDC) mark on the front cover, and the **V6** marks on the camshaft sprockets are at 12 o'clock as shown.
6. Remove or disconnect the following:
    - Crankshaft damper
    - Front cover
7. Lock the secondary timing chains to the idler sprocket with Timing Chain Locking tool 8429.
8. Matchmark the secondary timing chains to the camshaft sprockets.
9. Remove or disconnect the following:
    - Secondary timing chain tensioners
    - Cylinder head access plugs
    - Secondary timing chain guides
    - Camshaft sprockets
    - Cylinder heads

➡Each cylinder head is retained by 8 11mm bolts and four 8mm bolts.

To install:

10. Check the cylinder head bolts for signs of stretching and replace as necessary.
11. Lubricate the threads of the 11mm bolts with clean engine oil.
12. Coat the threads of the 8mm bolts with Mopar® Lock and Seal Adhesive, or equivalent.
13. Install the cylinder heads. Use new gaskets and tighten the bolts, in sequence, as follows:

    a. Step 1: Bolts 1–8 to 20 ft. lbs. (27 Nm)

    b. Step 2: Bolts 1–10 verify torque without loosening

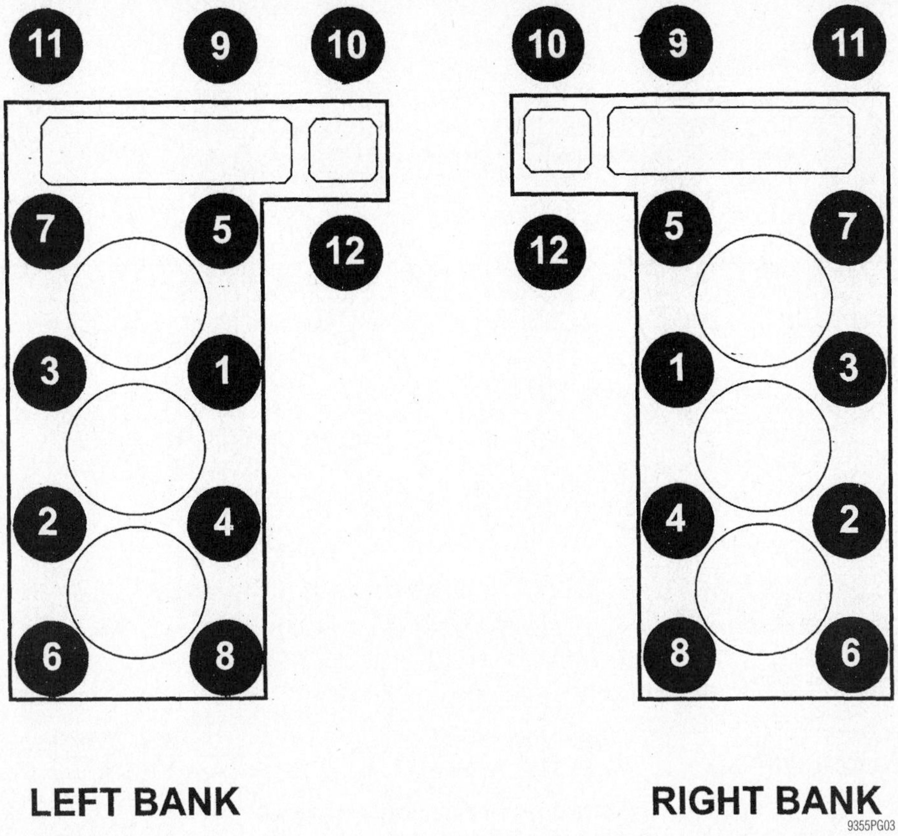

Cylinder head bolt torque sequence—3.7L

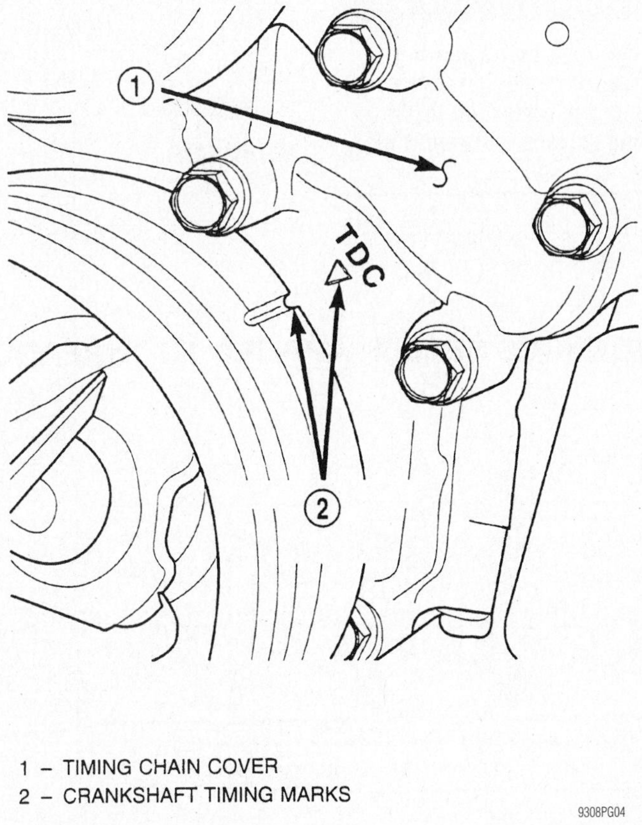

Crankshaft timing marks—3.7L engine

c. Step 3: Bolts 9–12 to 10 ft. lbs. (14 Nm)

d. Step 4: Bolts 1–8 plus ¼ (90 degree) turn

e. Step 5: Bolts 9–12 to 19 ft. lbs. (26 Nm)

14. Install or connect the following:
    - Camshaft sprockets. Align the secondary chain matchmarks and tighten the bolts to 90 ft. lbs. (122 Nm).
    - Secondary timing chain guides
    - Cylinder head access plugs
    - Secondary timing chain tensioners. Refer to the timing chain procedure in this section.
15. Remove the Timing Chain Locking tool.
16. Install or connect the following:
    - Front cover
    - Crankshaft damper. Torque the bolt to 130 ft. lbs. (175 Nm).
    - Rocker arms
    - Power steering pump
    - Oil fill housing
    - Accessory drive belt
    - Engine cooling fan and shroud
    - Valve covers
    - Intake manifold
    - Exhaust Y-pipe

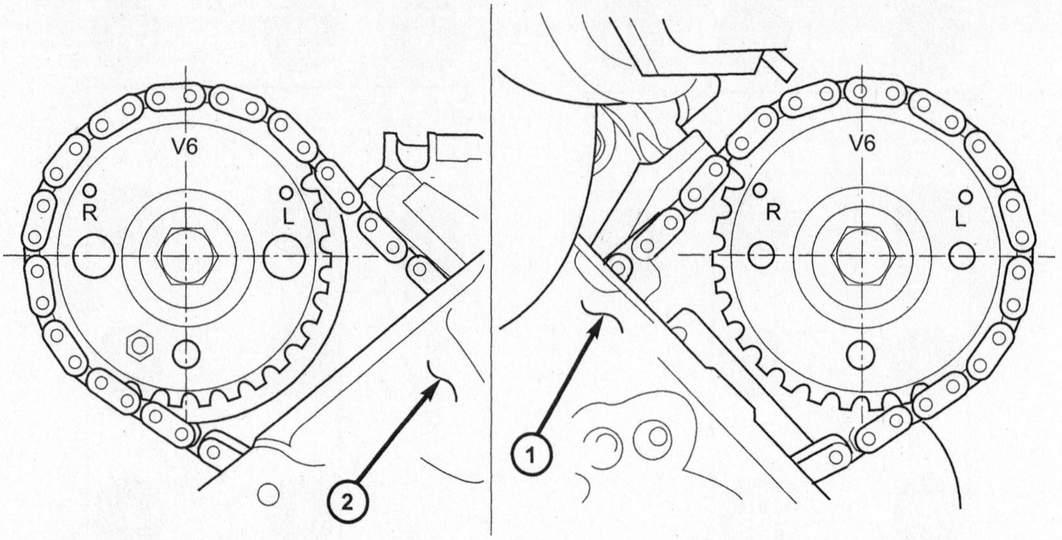

1 - LEFT CYLINDER HEAD
2 - RIGHT CYLINDER HEAD

9355PG04

Camshaft sprocket timing marks—3.7L

- Negative battery cable

17. Fill and bleed the cooling system.
18. Start the engine, check for leaks and repair if necessary.

4.7L Engine

LEFT SIDE

1. Before servicing the vehicle, refer to the precautions in the beginning of this section.
2. Drain the cooling system.
3. Remove or disconnect the following:
   - Negative battery cable
   - Exhaust pipe
   - Intake manifold
   - Cylinder head cover
   - Fan shroud and fan
   - Accessory drive belt
   - Power steering pump
4. Rotate the crankshaft until the damper mark is aligned with the TDC mark. Verify that the V8 mark on the camshaft sprocket is at the 12 o'clock position.
5. Remove or disconnect the following:
   - Vibration damper
   - Timing chain cover
6. Lock the secondary timing chains to the idler sprocket with tool 8515, or equivalent.
7. Mark the secondary timing chain, on link on either side of the V8 mark on the cam sprocket.
8. Remove the left side secondary chain tensioner.
9. Remove the cylinder head access plug.
10. Remove the chain guide.
11. Remove the camshaft sprocket.

➡There are 4 smaller bolts at the front of the head. Don't overlook these.

12. Remove the head bolts and head.

⁂ WARNING

Don't lay the head on its sealing surface. Due to the design of the head gasket, any distortion to the head sealing surface will result in leaks.

13. Installation is the reverse of removal. Observe the following:
   - Check the head bolts. If any necking is observed, replace the bolt.
   - The 4 small bolts must be coated with sealer.
   - The head bolts are tightened in the following sequence:

Step 1: Bolts 1-10 to 15 ft. lbs. (20 Nm)
Step 2: Bolts 1-10 to 35 ft. lbs. (47 Nm)
Step 3: Bolts 11-14 to 18 ft. lbs. (25 Nm)
Step 4: Bolts 1-10 90 degrees
Step 5: Bolts 11-14 to 22 ft. lbs.

RIGHT SIDE

1. Before servicing the vehicle, refer to the precautions in the beginning of this section.

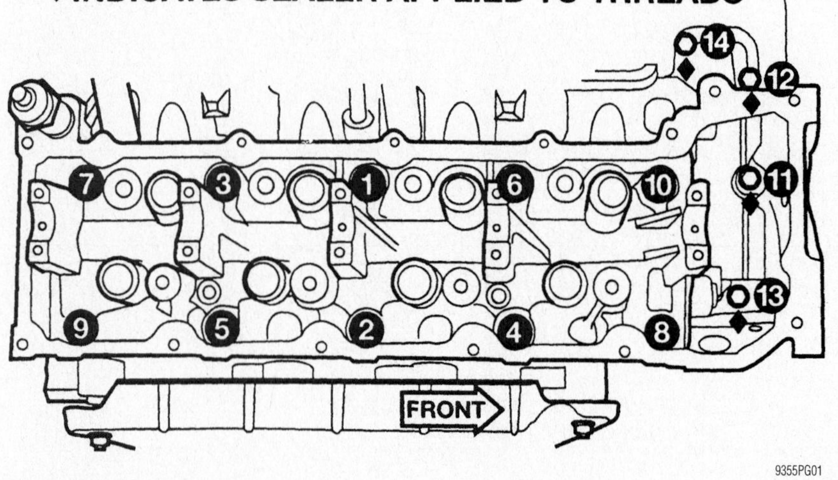

9355PG01

Cylinder head tightening sequence—4.7L

2. Drain the cooling system.
3. Remove or disconnect the following:
- Negative battery cable
- Exhaust pipe
- Intake manifold
- Cylinder head cover
- Fan shroud and fan
- Oil filler housing
- Accessory drive belt

4. Rotate the crankshaft until the damper mark is aligned with the TDC mark. Verify that the V8 mark on the camshaft sprocket is at the 12 o'clock position.
5. Remove or disconnect the following:
- Vibration damper
- Timing chain cover

6. Lock the secondary timing chains to the idler sprocket with tool 8515, or equivalent.
7. Mark the secondary timing chain, on link on either side of the V8 mark on the cam sprocket.
8. Remove the left side secondary chain tensioner.
9. Remove the cylinder head access plug.
10. Remove the chain guide.
11. Remove the camshaft sprocket.

⁂ WARNING

Do not pry on the target wheel for any reason!

➡There are 4 smaller bolts at the front of the head. Don't overlook these.

12. Remove the head bolts and head.

⁂ WARNING

Don't lay the head on its sealing surface. Due to the design of the head gasket, any distortion to the head sealing surface will result in leaks.

13. Installation is the reverse of removal. Observe the following:
- Check the head bolts. If any necking is observed, replace the bolt.
- The 4 small bolts must be coated with sealer.
- The head bolts are tightened in the following sequence:

Step 1: Bolts 1-10 to 15 ft. lbs. (20 Nm)
Step 2: Bolts 1-10 to 35 ft. lbs. (47 Nm)
Step 3: Bolts 11-14 to 18 ft. lbs. (25 Nm)
Step 4: Bolts 1-10 90 degrees
Step 5: Bolts 11-14 to 22 ft. lbs.

5.7L Engine

1. Before servicing the vehicle, refer to the precautions in the beginning of this section.
2. Drain the cooling system.
3. Properly relieve the fuel system pressure.
4. Remove or disconnect the following:
- Negative battery cable
- Air cleaner resonator and ducts
- Alternator
- Closed crankcase ventilation system
- EVAP control system
- Heater hoses
- Cylinder head covers
- Intake manifold
- Rocker arms and pushrods
- Cylinder heads

To install:

➡The head gaskets are not interchangeable. They are marked "L" and "R".

5. Install the cylinder heads. Use new gaskets and tighten the bolts, in sequence, as follows:
   a. Step 1: 12 mm bolts–25 ft. lbs. (34 Nm); 8mm bolts–15 ft. lbs. (20 Nm)
   b. Step 2: 12mm bolts–40 ft. lbs. (54 Nm); 8mm bolts retorque–15 ft. lbs. (20 Nm)
   c. Step 3: 12mm bolts–plus 90 degrees; 8mm bolts–25 ft. lbs. (34 Nm)
6. Install or connect the following:
- Rocker arms and pushrods
- Intake manifold
- Heater hoses
- Alternator
- Cylinder head covers. Torque the studs and bolts to 70 inch lbs.
- Air cleaner resonator and ducts
- Negative battery cable

5.9L Engines

1. Before servicing the vehicle, refer to the precautions in the beginning of this section.
2. Drain the cooling system.
3. Remove or disconnect the following:
- Negative battery cable
- Accessory drive belt
- Alternator
- A/C compressor, if equipped
- Air injection pump, if equipped
- Air cleaner assembly
- Closed Crankcase Ventilation (CCV) system
- Evaporative emissions control system
- Fuel line
- Accelerator linkage
- Cruise control cable

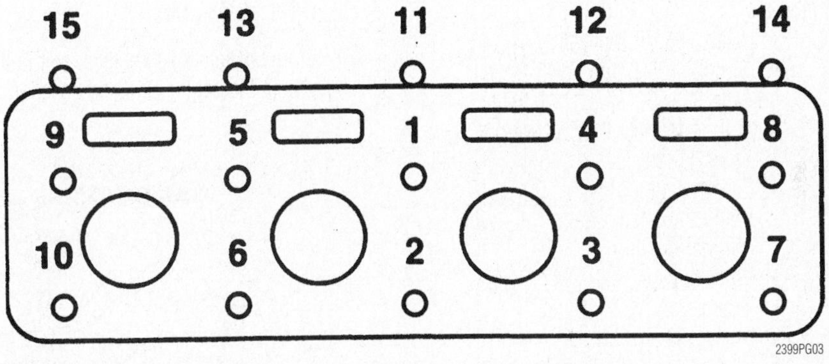

Cylinder head torque sequence—5.7L engine

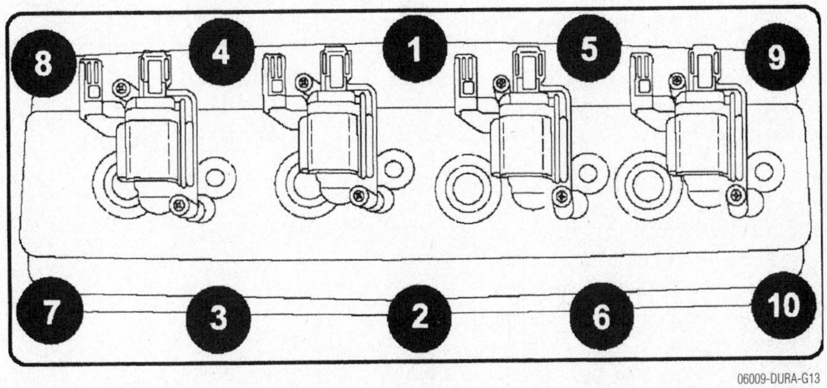

Cylinder head cover torque sequence—5.7L engine

- Transmission cable
- Distributor cap and wires
- Ignition coil wiring
- Engine Coolant Temperature (ECT) sensor connector
- Heater hoses
- Bypass hose
- Upper radiator hose
- Intake manifold
- Valve covers

➡**Keep valvetrain components in order for reassembly.**

- Rocker arms
- Pushrods
- Exhaust manifolds
- Spark plugs
- Cylinder heads

To install:

**** WARNING**

Position the crankshaft so that no piston is at Top Dead Center (TDC) prior to installing the cylinder heads. Do not rotate the crankshaft during or immediately after rocker arm installation. Wait 5 minutes for the hydraulic lash adjusters to bleed down.

4. Install the cylinder heads. Use new gaskets and tighten the bolts in sequence as follows:
   a. Step 1: 50 ft. lbs. (68 Nm).
   b. Step 2: 105 ft. lbs. (143 Nm).
   c. Step 3: 105 ft. lbs. (143 Nm).
5. Install or connect the following:
   - Spark plugs
   - Exhaust manifolds
   - Pushrods and rocker arms in their original positions
   - Valve covers
   - Intake manifold
   - Upper radiator hose
   - Bypass hose
   - Heater hoses
   - ECT sensor connector
   - Ignition coil wiring
   - Distributor cap and wires
   - Transmission cable
   - Cruise control cable
   - Accelerator linkage
   - Fuel line
   - Evaporative emissions control system
   - CCV system
   - Air cleaner assembly
   - A/C compressor, if equipped
   - Air injection pump, if equipped
   - Alternator
   - Accessory drive belt
   - Negative battery cable

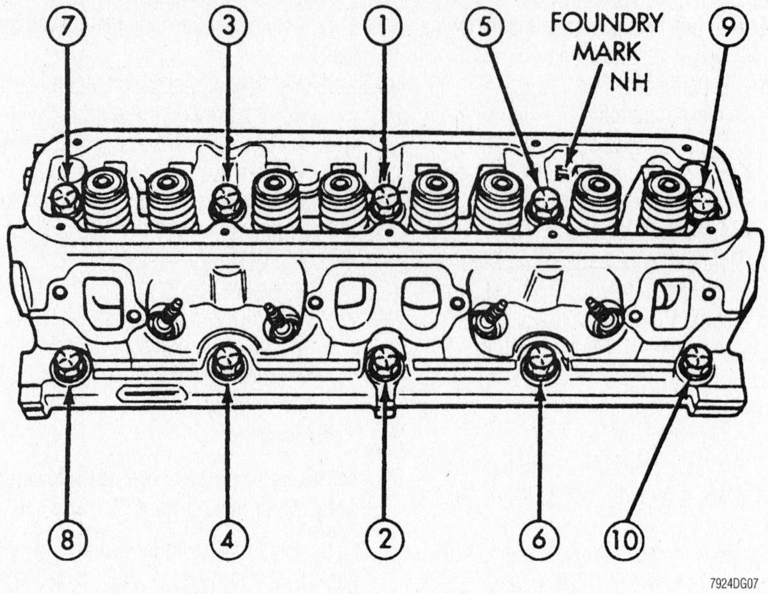

Cylinder head torque sequence—5.9L engines

6. Fill the cooling system.
7. Start the engine and check for leaks.

Rocker Arms/Shafts

REMOVAL & INSTALLATION

3.7L and 4.7L Engine

1. Before servicing the vehicle, refer to the precautions in the beginning of this section.
2. Remove or disconnect the following:
   - Negative battery cable
   - Valve covers
3. Rotate the crankshaft so that the piston of the cylinder to be serviced is at Bottom Dead Center (BDC) and both valves are closed.
4. Use special tool 8516 to depress the valve and remove the rocker arm.
5. Repeat for each rocker arm to be serviced.

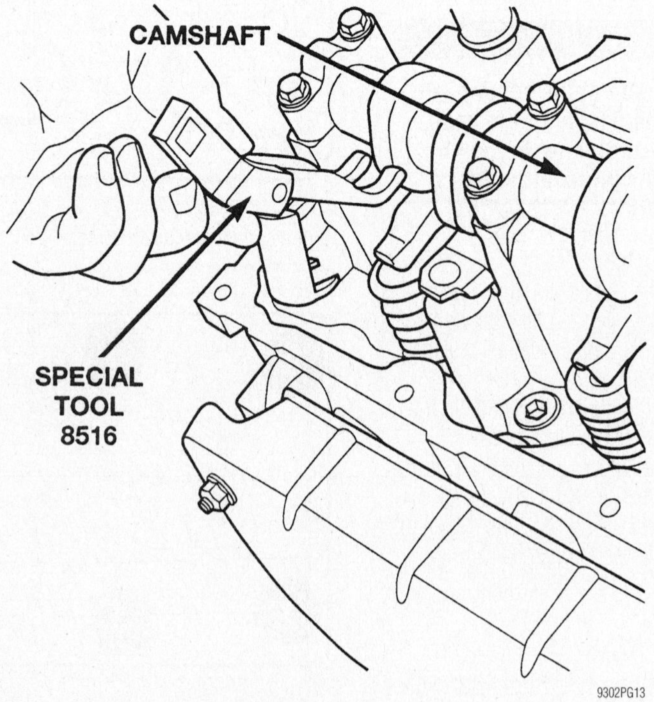

Rocker arm service—3.7L and 4.7L engine

➡**Keep valvetrain components in order for reassembly.**

To install:

6. Rotate the crankshaft so that the piston of the cylinder to be serviced is at BDC.
7. Compress the valve spring and install each rocker arm in its original position.
8. Repeat for each rocker arm to be installed.
9. Install or connect the following:
   - Cylinder head cover
   - Negative battery cable

5.7L Engines

1. Before servicing the vehicle, refer to the precautions in the beginning of this section.
2. Remove or disconnect the following:
   - Negative battery cable
   - Cylinder head cover
3. Loosen the rocker shaft bolts as follows: center, center left, center right, left, right.

➡**The rocker shaft assemblies are not interchangeable. The intake side is marked "I". The shorter pushrods are used on the intake side.**

To install:

4. Install or connect the following:
   - Pushrods
   - Rocker shaft assemblies. Torque the shafts to 16 ft. lbs. (22 Nm) in this sequence: center, center right, center left, right, left.

⁂ WARNING

Do not rotate the crankshaft during or immediately after rocker shaft installation. Allow the roller tappets to leakdown for about 5 minutes.

   - Cylinder head covers. Torque the fasteners to 70 inch lbs.

5.9L Engines

1. Before servicing the vehicle, refer to the precautions in the beginning of this section.
2. Remove or disconnect the following:
   - Negative battery cable
   - Valve covers
   - Rocker arms

➡**Keep valvetrain components in order for reassembly.**

To install:

3. Rotate the crankshaft so that the **V8** mark on the crankshaft damper aligns with the timing mark on the front cover. The **V8**

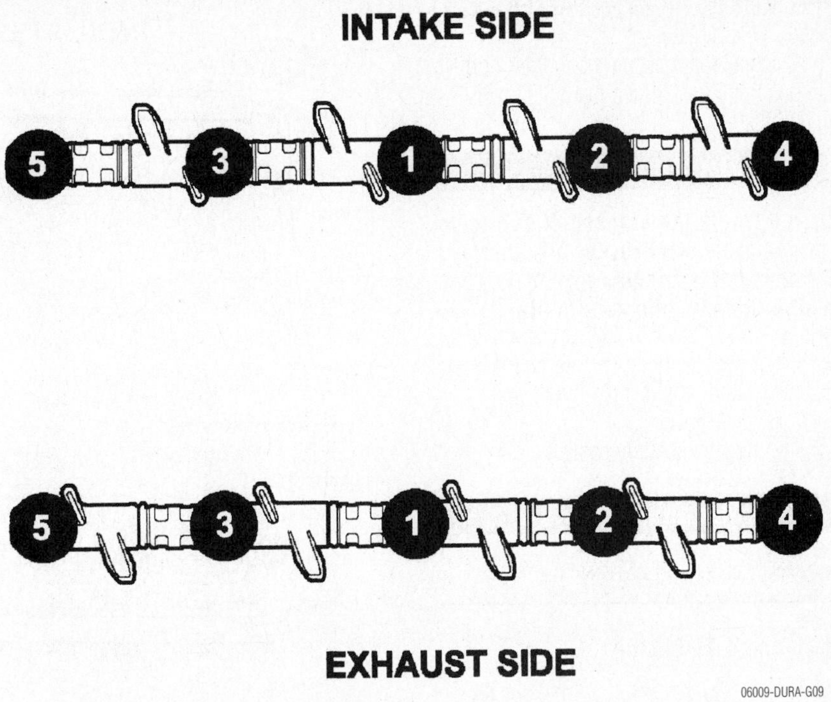

Rocker shaft identification—5.7L engine

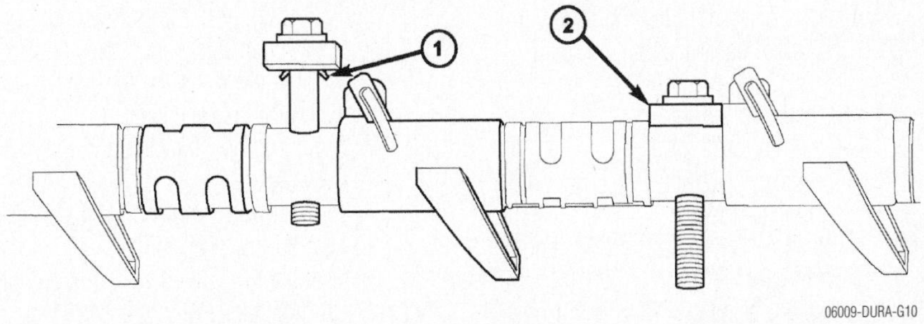

Make sure that the bolt retainers (1) fit properly as shown (2), when torquing the bolts—5.7L engines

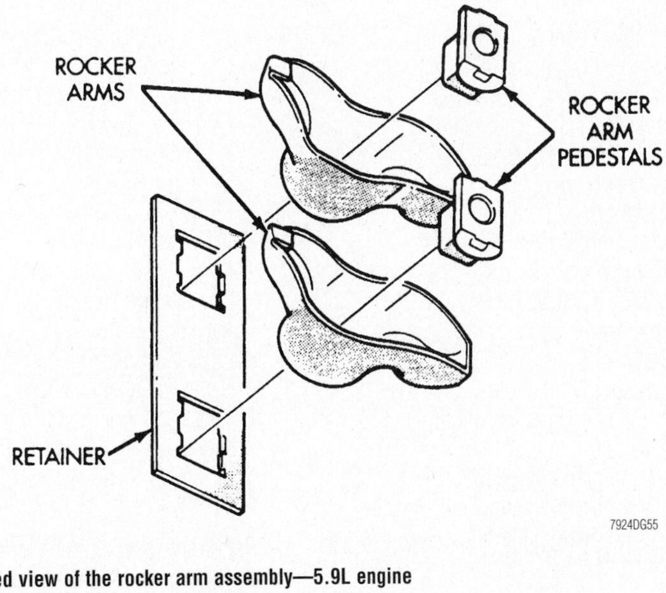

Exploded view of the rocker arm assembly—5.9L engine

mark is located 147 degrees **AFTER** Top Dead Center (TDC).

4. Install the rocker arms in their original positions and tighten the bolts to 21 ft. lbs. (28 Nm).

CAUTION

Do not rotate the crankshaft during or immediately after rocker arm installation. Wait 5 minutes for the hydraulic lash adjusters to bleed down.

5. Install or connect the following:
   - Valve covers
   - Negative battery cable

Intake Manifold

REMOVAL & INSTALLATION

3.7L and 4.7L Engine

1. Before servicing the vehicle, refer to the precautions in the beginning of this section.
2. Drain the cooling system.
3. Remove or disconnect the following:
   - Negative battery cable
   - Air cleaner assembly
   - Accelerator cable
   - Cruise control cable
   - Manifold Absolute Pressure (MAP) sensor connector
   - Intake Air Temperature (IAT) sensor connector
   - Throttle Position (TP) sensor connector
   - Idle Air Control (IAC) valve connector
   - Engine Coolant Temperature (ECT) sensor
   - Positive Crankcase Ventilation (PCV) valve and hose
   - Canister purge vacuum line
   - Brake booster vacuum line
   - Cruise control servo hose
   - Accessory drive belt
   - Alternator
   - A/C compressor
   - Engine ground straps
   - Ignition coil towers
   - Oil dipstick tube
   - Fuel line
   - Fuel supply manifold
   - Throttle body and mounting bracket
   - Cowl seal
   - Right engine lifting stud
   - Intake manifold. Remove the fasteners in reverse of the tightening sequence.

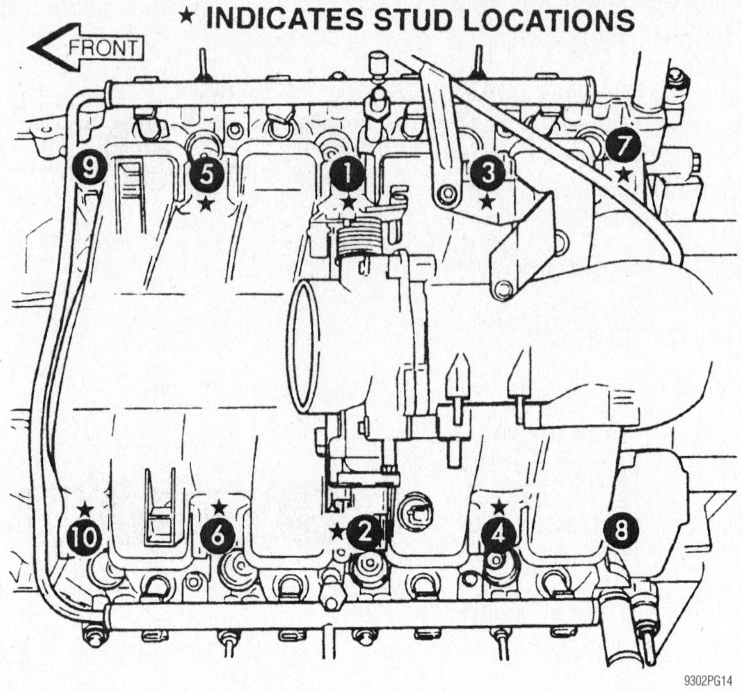

Intake manifold torque sequence—3.7L and 4.7L engine

To install:

4. Install or connect the following:
   - Intake manifold using new gaskets. Tighten the bolts, in sequence, to 105 inch lbs. (12 Nm).
   - Right engine lifting stud
   - Cowl seal
   - Throttle body and mounting bracket
   - Fuel supply manifold
   - Fuel line
   - Oil dipstick tube
   - Ignition coil towers
   - Engine ground straps
   - A/C compressor
   - Alternator
   - Accessory drive belt
   - Cruise control servo hose
   - Brake booster vacuum line
   - Canister purge vacuum line
   - PCV valve and hose
   - ECT sensor
   - IAC valve connector
   - TP sensor connector
   - IAT sensor connector
   - MAP sensor connector
   - Cruise control cable
   - Accelerator cable
   - Air cleaner assembly
   - Negative battery cable
5. Fill the cooling system.
6. Start the engine and check for leaks.

5.7L Engines

1. Before servicing the vehicle, refer to the precautions in the beginning of this section.
2. Drain the cooling system.
3. Relieve the fuel system pressure.
4. Remove or disconnect the following:
   - Negative battery cable
   - Air cleaner assembly
   - Accessory drive belt
   - MAP connector
   - IAT connector
   - TPS connector
   - CTS connector
   - Brake booster hose
   - PCV hose
   - Alternator
   - A/C compressor
   - Intake manifold bolts, in a criss-cross pattern, from the outside to the center
   - Intake manifold/IAFM

To install:

5. Position new intake manifold seals.
6. Install the intake manifold. Tighten the bolts in sequence from the center outwards, to 105 inch lbs. (12 Nm).
7. Install or connect the following:
   - Electrical connectors
   - Alternator
   - A/C compressor
   - Brake booster hose
   - PCV hose
   - Accessory drive belt
   - Negative battery cable
   - Air cleaner assembly

5.9L Engines

1. Before servicing the vehicle, refer to the precautions in the beginning of this section.

2. Drain the cooling system.
3. Remove or disconnect the following:
   - Negative battery cable
   - Accessory drive belt
   - Alternator
   - A/C compressor
   - Alternator and A/C compressor bracket
   - Air cleaner assembly
   - Fuel line
   - Fuel supply manifold
   - Accelerator cable
   - Transmission cable
   - Cruise control cable
   - Distributor cap and wires
   - Ignition coil wiring
   - Engine Coolant Temperature (ECT) sensor connector
   - Heater hose
   - Upper radiator hose
   - Bypass hose
   - Closed Crankcase Ventilation (CCV) system
   - Evaporative emissions system
   - Intake manifold

To install:

4. Install the intake manifold. Use a new gasket and tighten the bolts in sequence as follows:
   a. Step 1: Bolts 1–4 to 72 inch lbs. (8 Nm) using 12 inch lb. (1.4 Nm) increments.
   b. Step 2: Bolts 5–12 to 72 inch lbs. (8 Nm).
   c. Step 3: Bolts 1–12 to 72 inch lbs. (8 Nm).
   d. Step 4: Bolts 1–12 to 12 ft. lbs. (16 Nm).
   e. Step 5: Bolts 1–12 to 12 ft. lbs. (16 Nm).
5. Install or connect the following:
   - Evaporative emissions system
   - CCV system
   - Bypass hose
   - Upper radiator hose
   - Heater hose
   - ECT sensor connector
   - Ignition coil wiring
   - Distributor cap and wires
   - Cruise control cable
   - Transmission cable
   - Accelerator cable
   - Fuel supply manifold
   - Fuel line
   - Air cleaner assembly
   - Alternator and A/C compressor bracket
   - A/C compressor
   - Alternator
   - Accessory drive belt
   - Negative battery cable

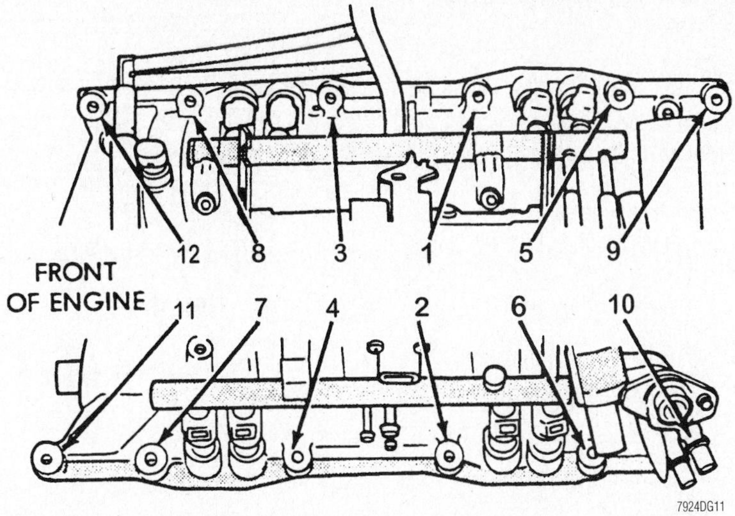

Intake manifold torque sequence—5.9L engines

6. Fill the cooling system.
7. Start the engine and check for leaks.

Exhaust Manifold

REMOVAL & INSTALLATION

3.7L Engines

1. Before servicing the vehicle, refer to the precautions in the beginning of this section.
2. Remove or disconnect the following:
   - Negative battery cable
   - Exhaust manifold heat shields
   - Exhaust Gas Recirculation (EGR) tube
   - Exhaust Y-pipe
   - Exhaust manifolds

To install:

➡If the exhaust manifold studs came out with the nuts when removing the exhaust manifolds, replace them with new studs.

3. Install or connect the following:
   - Exhaust manifolds. Torque the fasteners to 20 ft. lbs. (27 Nm), starting with the center nuts and work out to the ends.
   - Exhaust Y-pipe
   - EGR tube
   - Exhaust manifold heat shields
   - Negative battery cable

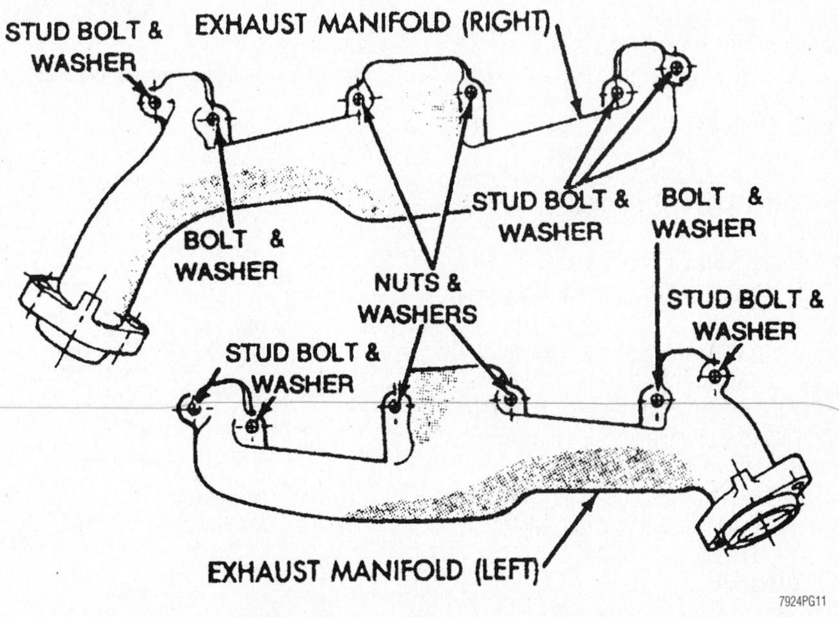

Exhaust manifold fastener locations—3.7L engines

4. Start the engine, check for leaks and repair if necessary.

4.7L Engine

1. Before servicing the vehicle, refer to the precautions in the beginning of this section.
2. Drain the cooling system.
3. Remove or disconnect the following:
   - Battery
   - Power distribution center
   - Battery tray
   - Windshield washer fluid bottle
   - Air cleaner assembly
   - Accessory drive belt
   - A/C compressor
   - A/C accumulator bracket
   - Heater hoses
   - Exhaust manifold heat shields
   - Exhaust Y-pipe
   - Starter motor
   - Exhaust manifolds

To install:

4. Install or connect the following:
   - Exhaust manifolds, using new gaskets. Tighten the bolts to 18 ft. lbs. (25 Nm), starting with the inner bolts and work out to the ends.
   - Starter motor
   - Exhaust Y-pipe
   - Exhaust manifold heat shields
   - Heater hoses
   - A/C accumulator bracket
   - A/C compressor
   - Accessory drive belt
   - Air cleaner assembly
   - Windshield washer fluid bottle
   - Battery tray
   - Power distribution center
   - Battery
5. Fill the cooling system.
6. Start the engine and check for leaks.

5.7L Engines

1. Before servicing the vehicle, refer to the precautions in the beginning of this section.
2. Drain the cooling system.
3. Remove or disconnect the following:
   - Battery
   - Exhaust pipe-to-manifold bolts
4. Install an engine crane. Remove the right and left mount through bolts. Raise the engine just enough to provide clearance for manifold removal.
5. Remove or disconnect the following:
   - Exhaust manifold heat shields
   - Exhaust manifolds

To install:

6. Install or connect the following:
   - Exhaust manifolds, using new gas-

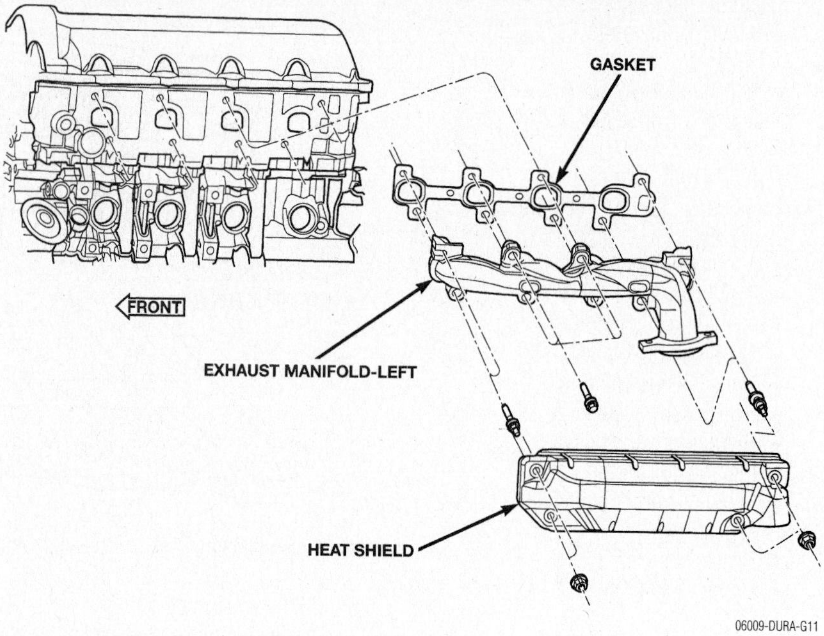

Left side exhaust manifold installation; right side similar—4.7L engine

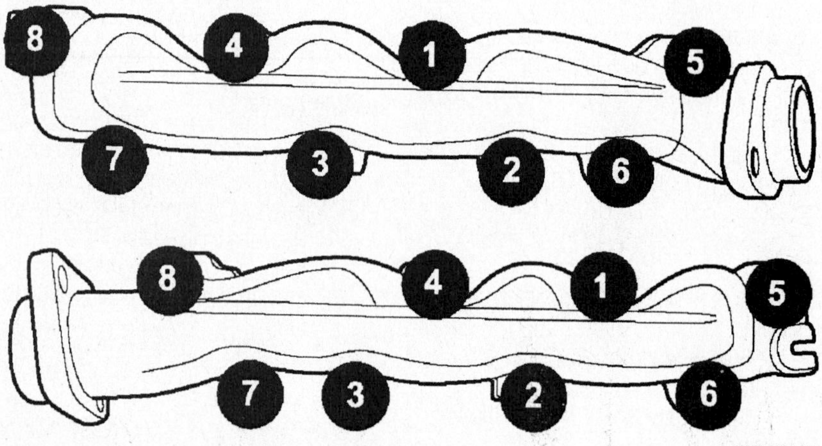

Exhaust manifold bolt torque sequences—5.7L engine

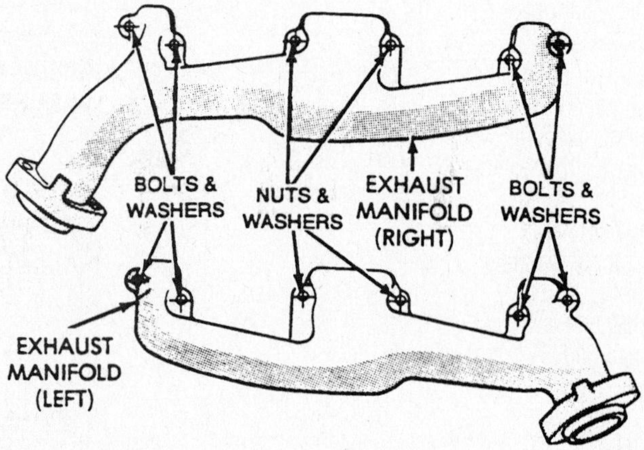

Exhaust manifold fastener locations—5.9L engines

kets. Tighten the bolts to 18 ft. lbs. (25 Nm), starting with the inner bolts and work out to the ends.
- Exhaust manifold heat shields

7. Install the right and left mount through bolts. Remove the crane.
8. Install or connect the following:
    - Exhaust pipe-to-manifold bolts
    - Battery

5.9L Engines

1. Before servicing the vehicle, refer to the precautions in the beginning of this section.
2. Remove or disconnect the following:
    - Negative battery cable
    - Exhaust manifold heat shields
    - Exhaust Gas Recirculation (EGR) tube
    - Exhaust Y-pipe
    - Exhaust manifolds

To install:

➡If the exhaust manifold studs came out with the nuts when removing the exhaust manifolds, replace them with new studs.

3. Install or connect the following:
    - Exhaust manifolds. Tighten the fasteners to 20 ft. lbs. (27 Nm), starting with the center nuts and work out to the ends.
    - Exhaust Y-pipe
    - EGR tube
    - Exhaust manifold heat shields
    - Negative battery cable
4. Fill the cooling system.
5. Start the engine and check for leaks.

Camshaft and Valve Lifters

REMOVAL & INSTALLATION

3.7L and 4.7L Engines

1. Before servicing the vehicle, refer to the precautions in the beginning of this section.
2. Remove or disconnect the following:
    - Negative battery cable
    - Cylinder head covers
    - Rocker arms
    - Hydraulic lash adjusters

➡Keep all valvetrain components in order for assembly.

3. Set the engine at Top Dead Center (TDC) of the compression stroke for the No. 1 cylinder.

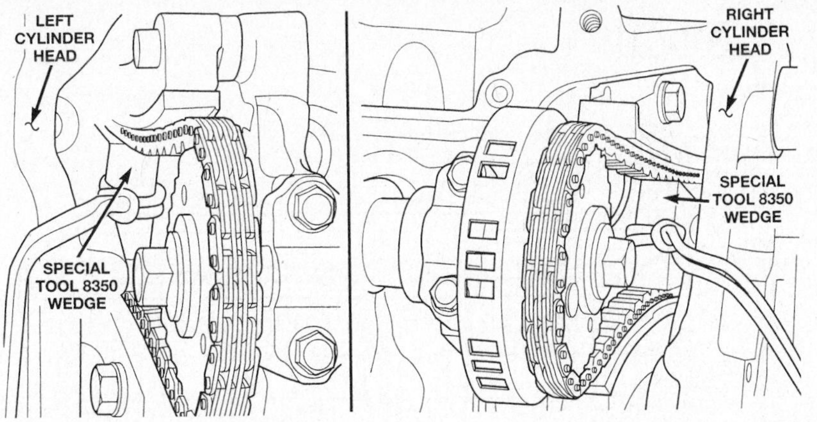

Chain Tensioner Retaining Wedges—4.7L engine

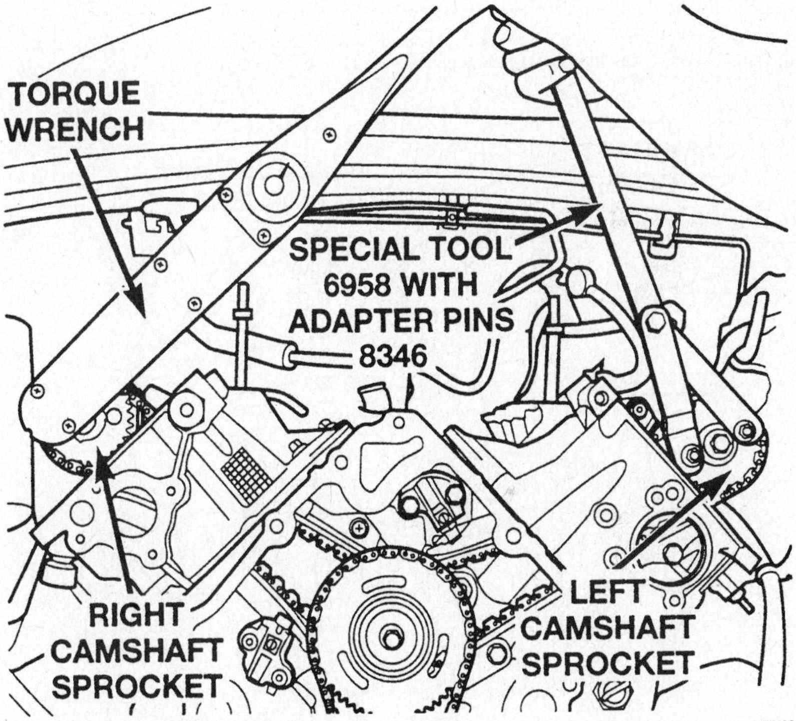

Hold the left camshaft sprocket with a spanner wrench while removing or installing the camshaft sprocket bolts—3.7L and 4.7L engine

4. Install Timing Chain Wedge 8350 to retain the chain tensioners.
5. Matchmark the timing chains to the camshaft sprockets.
6. Install Camshaft Holding Tool 6958 and Adapter Pins 8346 to the left camshaft sprocket.
7. Remove or disconnect the following:
    - Right camshaft timing sprocket and target wheel
    - Left camshaft sprocket
    - Camshaft bearing caps, by reversing the tightening sequence
    - Camshafts

To install:

8. Install or connect the following:
    - Camshafts. Tighten the bearing cap bolts in ½ turn increments, in sequence, to 100 inch lbs. (11 Nm).
    - Target wheel to the right camshaft
    - Camshaft timing sprockets and chains, by aligning the matchmarks
9. Remove the tensioner wedges and tighten the camshaft sprocket bolts to 90 ft. lbs. (122 Nm).
10. Install or connect the following:
    - Hydraulic lash adjusters in their original locations

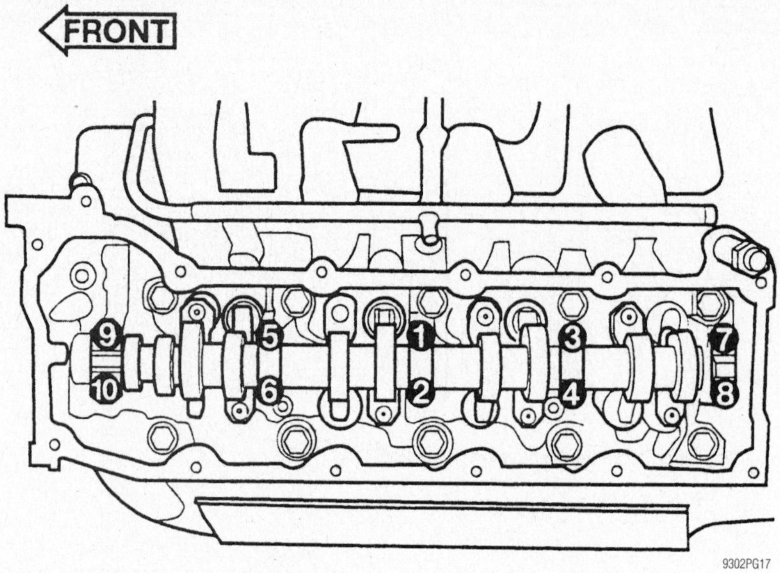

Camshaft bearing cap bolt tightening sequence—3.7L and 4.7L engine

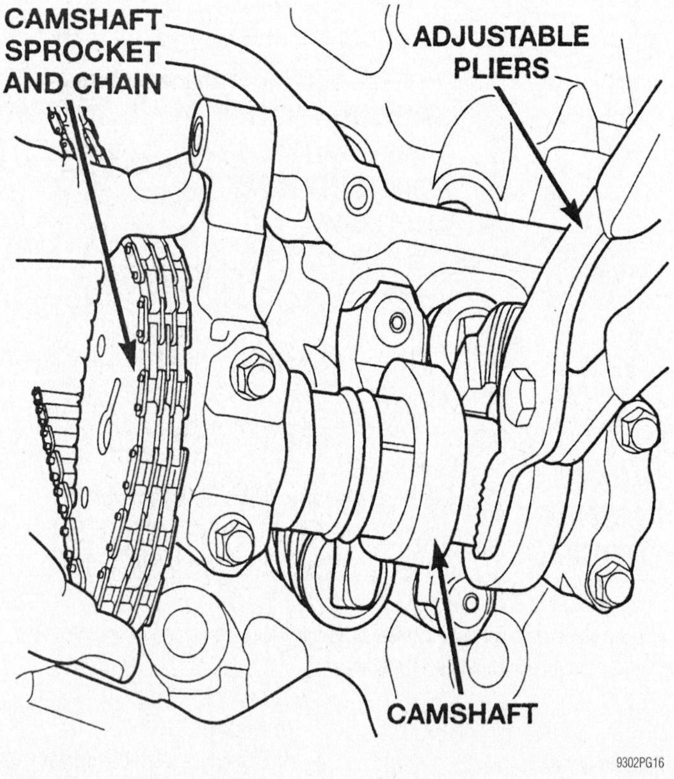

Turn the camshaft with pliers, if needed, to align the dowel in the sprocket—4.7L engine

- Rocker arms in their original locations
- Cylinder head covers
- Negative battery cable

5.7L Engines

1. Before servicing the vehicle, refer to the precautions in the beginning of this section.
2. Drain the cooling system.
3. Recover the A/C refrigerant, if equipped with air conditioning.
4. Set the crankshaft to Top Dead Center (TDC) of the compression stroke for the No. 1 cylinder.
5. Remove or disconnect the following:
    - Negative battery cable
    - Camshaft rear cam bearing core plug
    - Air cleaner
    - Accessory drive belt
    - Alternator
    - A/C compressor
    - Radiator
    - Intake manifold
    - Cylinder head covers
    - Cylinder heads
    - Oil pan
    - Front cover
    - Oil pickup tube
    - Oil pump
    - Timing chain and sprockets
    - Camshaft thrust plate
    - Hydraulic lifters
    - Camshaft

To install:

6. Install or connect the following:
    - Camshaft
    - Camshaft thrust plate. Tighten the bolts to 21 ft. lbs. (28 Nm).
    - Timing chain and sprockets
    - Oil pump
    - Oil pickup tube

➡Lifters must be replaced in their original positions.

- Hydraulic lifters
- Cylinder heads
- Pushrods
- Rocker arms
- Front cover
- Oil pan
- Cylinder head covers
- Intake manifold
- A/C compressor
- Alternator
- Accessory drive belt
- Radiator
- Air cleaner
- Camshaft rear cam bearing core plug
- Negative battery cable

7. Fill the cooling system.
8. Recharge the A/C system, if equipped.
9. Start the engine and check for leaks.

5.9L Engines

1. Before servicing the vehicle, refer to the precautions in the beginning of this section.
2. Drain the cooling system.
3. Recover the A/C refrigerant, if equipped with air conditioning.
4. Set the crankshaft to Top Dead Center (TDC) of the compression stroke for the No. 1 cylinder.
5. Remove or disconnect the following:
    - Negative battery cable

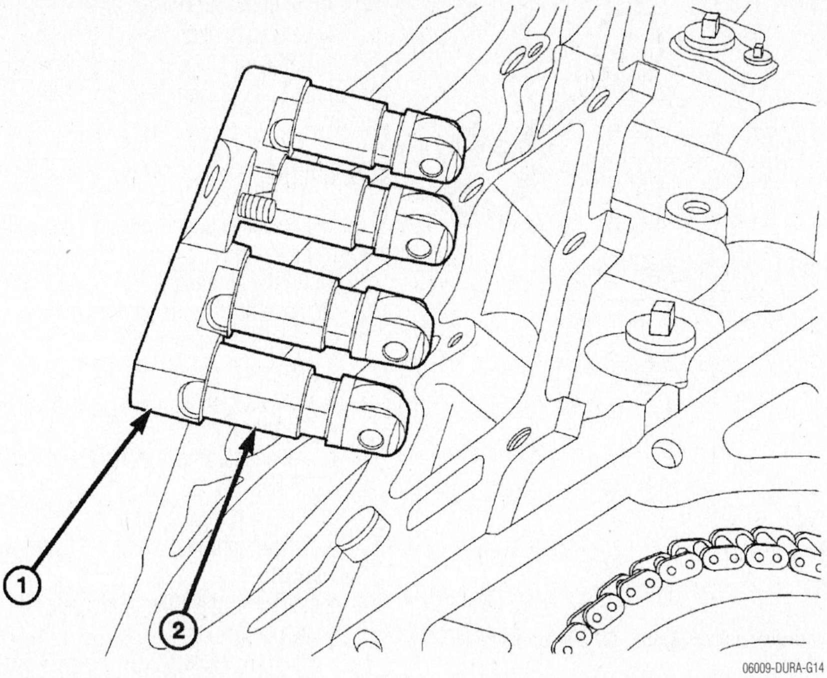

Lifter group—5.7L engine

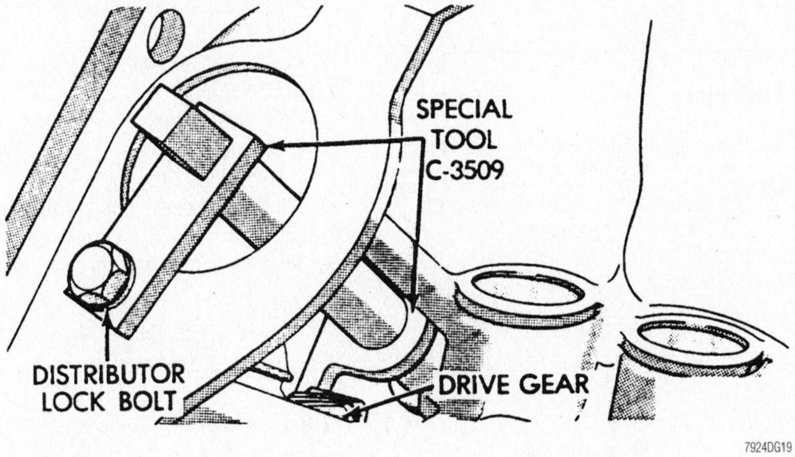

Camshaft holding tool C-3509—5.9L engines

- Accessory drive belt
- Power steering pump
- Water pump
- Radiator
- A/C condenser
- Grille
- Crankshaft damper
- Front cover
- Valve covers
- Distributor
- Intake manifold

➡Keep all valvetrain components in order for assembly.

- Rocker arms and pushrods
- Hydraulic lifters
- Timing chain and sprockets
- Camshaft thrust plate and chain oil tab
- Camshaft

To install:

6. Install or connect the following:
   - Camshaft
   - Camshaft Holding Tool C-3509
   - Camshaft thrust plate and chain oil tab. Tighten the bolts to 18 ft. lbs. (24 Nm).
   - Timing chain and sprockets. Tighten the camshaft sprocket bolt to 50 ft. lbs. (68 Nm).
7. Remove the camshaft holding tool.
8. Install or connect the following:
   - Hydraulic lifters in their original positions
   - Rocker arms and pushrods in their original positions
   - Intake manifold
   - Distributor
   - Valve covers
   - Front cover
   - Crankshaft damper
   - Grille
   - A/C condenser
   - Radiator
   - Water pump
   - Power steering pump
   - Accessory drive belt
   - Negative battery cable
9. Fill the cooling system.
10. Recharge the A/C system, if equipped.
11. Start the engine and check for leaks.

Valve Lash

ADJUSTMENT

All gasoline engines covered in this section use hydraulic lifters. No maintenance or periodic adjustment is required.

Starter Motor

REMOVAL & INSTALLATION

3.7L and 4.7L Engines

1. Before servicing the vehicle, refer to the precautions in the beginning of this section.

➡If equipped with 4WD and certain transmissions, a support bracket is used between front axle and side of transmission. Remove 2 support bracket bolts at transmission. Pry support bracket slightly to gain access to lower starter mounting bolt.

2. Remove or disconnect the following:
   - Negative battery cable
   - Starter mounting bolts

➡The left side exhaust pipe and front driveshaft must be disconnected.

   - Starter solenoid harness connections
   - Starter

To install:

3. Connect the starter solenoid wiring connectors.
4. Install the starter and torque the bolts to 40 ft. lbs. (54 Nm).
5. Install the negative battery cable and check for proper operation.

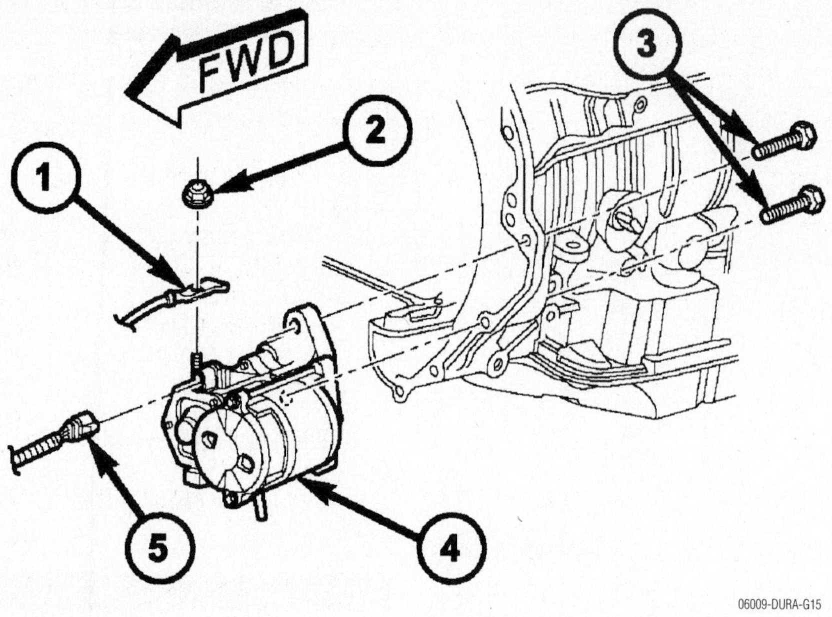

Starter mounting, 3.7L and 4.7L engines. (1) battery positive cable, (2) nut, (3) starter mounting bolts, (4) starter, (5) wiring harness

5.9L Engines

1. Before servicing the vehicle, refer to the precautions in the beginning of this section.
2. Remove or disconnect the following:
   - Negative battery cable
   - Starter mounting bolts
   - Starter solenoid harness connections
   - Starter

To install:

3. Connect the starter solenoid wiring connectors.
4. Install the starter and tighten the bolts to 50 ft. lbs. (68 Nm).
5. Install the negative battery cable and check for proper operation.

5.7L Engines

1. Before servicing the vehicle, refer to the precautions in the beginning of this section.
2. Remove or disconnect the following:
   - Negative battery cable

➡Depending on drivetrain configuration, a support bracket may be used.

   - Starter mounting bolts
   - Starter solenoid harness connections
   - Starter

To install:

3. Connect the starter solenoid wiring connectors.
4. Install the starter and torque the bolts to 50 ft. lbs. (68 Nm).
5. Install the negative battery cable and check for proper operation.

Oil Pan

REMOVAL & INSTALLATION

3.7L Engine

1. Before servicing the vehicle, refer to the precautions in the beginning of this section.
2. Remove or disconnect the following:
   - Engine from the vehicle
   - Oil pan
   - Oil pump pickup tube
   - Oil pan gasket
3. Installation is the reverse of removal. Torque the bolts, in sequence, to 11 ft. lbs. (15 Nm).

4.7L Engine

1. Before servicing the vehicle, refer to the precautions in the beginning of this section.
2. Drain the engine oil.
3. Remove or disconnect the following:
   - Negative battery cable
   - Structural cover
   - Exhaust Y-pipe
   - Starter motor
   - Transmission oil cooler lines
   - Oil pan
   - Oil pump pickup tube
   - Oil pan gasket

To install:

4. Install or connect the following:
   - Oil pan gasket
   - Oil pump pickup tube, using a new O-ring. Tighten the tube bolts to 20 ft. lbs. (28 Nm); tighten the O-ring end bolt first.
   - Oil pan. Tighten the bolts, in sequence, to 11 ft. lbs. (15 Nm).
   - Transmission oil cooler lines
   - Starter motor
   - Exhaust Y-pipe
   - Structural cover
   - Negative battery cable
5. Fill the crankcase to the proper level with engine oil.
6. Start the engine and check for leaks.

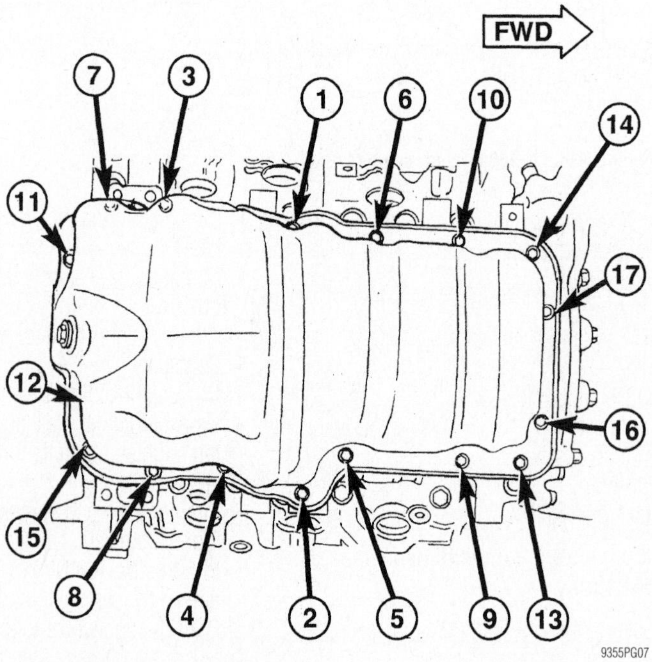

Oil pan bolt torque sequence—3.7L Engine

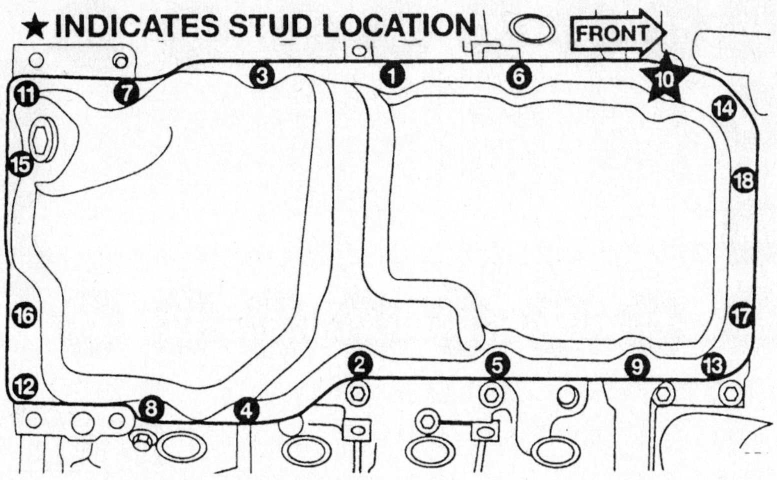

Oil pan mounting bolt tightening sequence—4.7L engine

5.7L Engine

2-WHEEL DRIVE

1. Disconnect the negative battery cable.
2. Loosen both left and right side engine mount through bolts. Do not remove bolts.
3. Install an engine support fixture tool. Do not raise engine at this time.
4. Remove the structural dust cover.
5. Remove fan and fan shroud.
6. Drain engine oil.
7. Remove the front crossmember
8. Raise engine to provide clearance to remove oil pan.

➡Do not pry on oil pan or oil pan gasket. Gasket is integral to engine windage tray and does not come out with oil pan.

➡If more clearance is needed to remove oil pan, the transmission mount can be removed, and the transmission raised to gain clearance.

➡The double ended oil pan studs must be installed in the same location that they were removed from.

9. Remove the oil pan mounting bolts using the sequence provided.
10. Unbolt oil pump pickup tube and remove tube.

➡When the oil pan is removed, a new integral windage tray and gasket assembly must be installed. The old gasket cannot be reused.

11. Discard the integral windage tray and gasket and replace.

To install:

12. Clean the oil pan gasket mating surface of the block and oil pan.

➡Mopar® Engine RTV must be applied to the 4 T-joints (1, 2), (area where front cover, rear retainer, block, and oil pan gasket meet). The bead of RTV should cover the bottom of the gasket. This area is approximately 4.5 mm x 25 mm in each of the 4 T-joint locations.

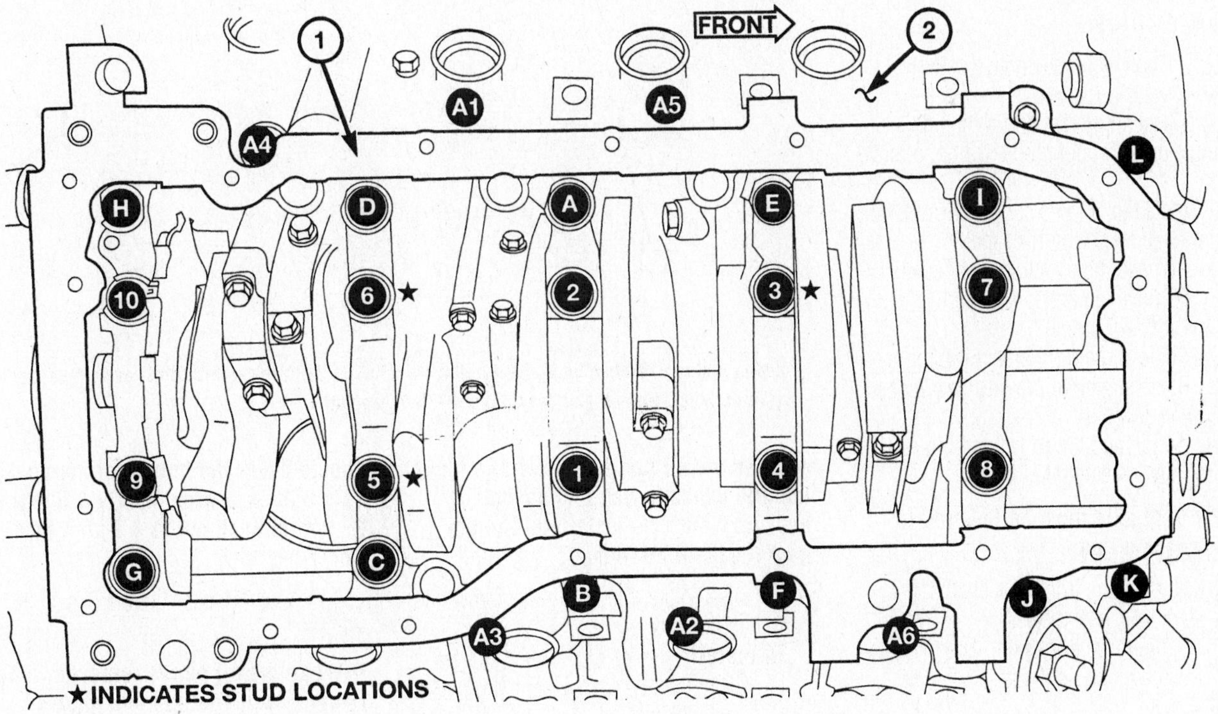

1 - BEDPLATE
2 - CYLINDER BLOCK

9355PG03A

Bedplate bolt tightening sequence—4.7L engine

13. Apply Mopar® Engine RTV at the 4 T-joints.

➡When the oil pan is removed, a new integral windage tray and gasket assembly must be installed. The old gasket cannot be reused.

14. Install a new integral windage tray and gasket.
15. Reinstall the oil pump pickup tube with new O-ring. Tighten tube to pump fasteners to 28 Nm (250 inch lbs.).

➡The double ended oil pan studs must be installed in the same location that they were removed from.

16. Position the oil pan and install the mounting bolts and studs. Tighten the mounting bolts to 12 Nm (105 inch lbs.).
17. Install the structural cover.
18. Lower the engine into mounts.
19. Install both the left and right side engine mount through bolts. Tighten the nuts to 68 Nm (50 ft. lbs.).
20. Remove the lifting fixture.
21. Install the front crossmember. Torque to 75 ft. lbs. (102 Nm).
22. Install the fan shroud and fan.
23. Fill engine oil.
24. Reconnect the negative battery cable.
25. Start engine and check for leaks.

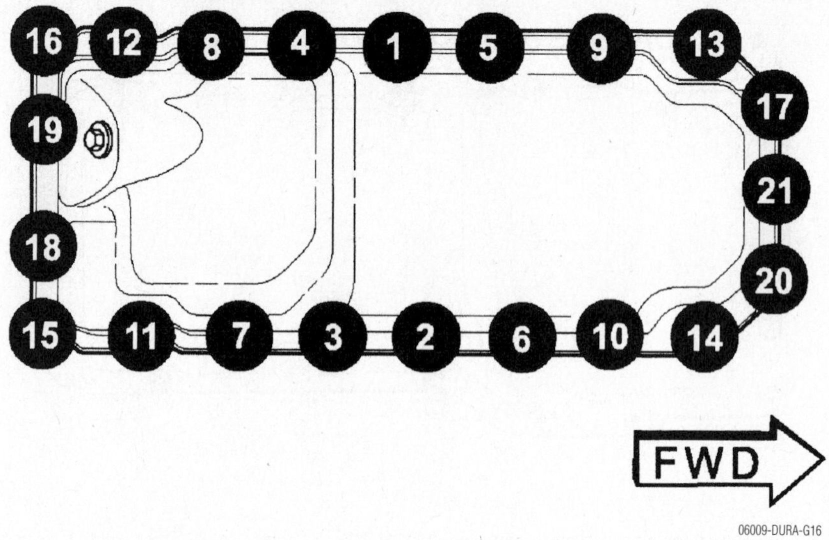

Oil pan bolt torque sequence—5.7L engine

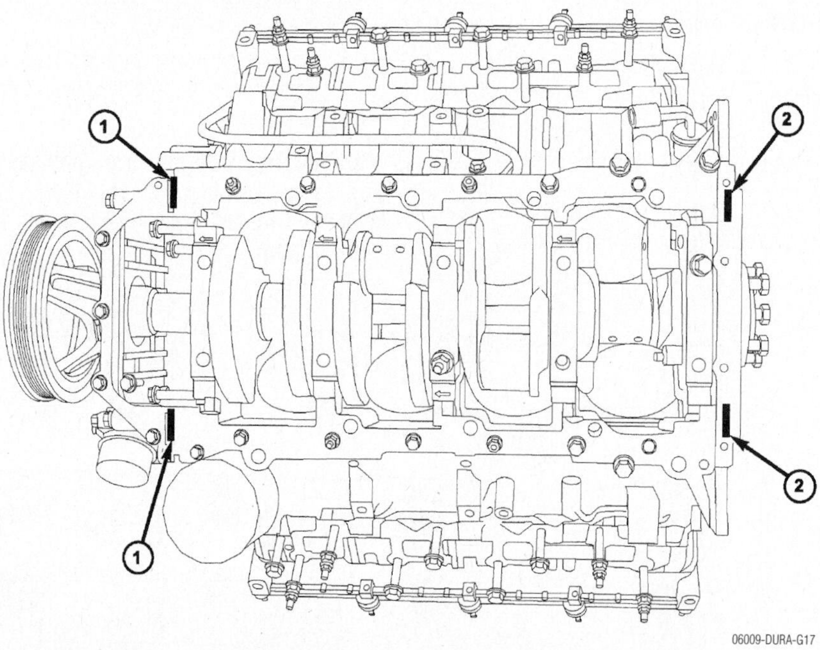

Mopar Engine RTV must be applied to the 4 T-joints (1,2), (area where front cover, rear retainer, block, and oil pan gasket meet)—5.7L engine

4-WHEEL DRIVE

1. Disconnect the negative battery cable.
2. Loosen both left and right side engine mount through bolts. Do not remove bolts.
3. Install an engine support fixture tool. Do not raise engine at this time.
4. Remove the structural dust cover.
5. Remove fan and fan shroud.
6. Drain engine oil.
7. Remove the front crossmember.
8. Raise engine to provide clearance to remove oil pan.
9. Unbolt and lower the steering rack, without disconnecting the lines.

➡The front axle must be lowered to remove the oil pan.

10. Remove the front driveshaft at the axle. Mark for reassembly.
11. Support the front axle.
12. Remove the right and left axle to mount, bolts.
13. Lower axle.
14. Remove the oil pan mounting bolts and oil pan.
15. Unbolt oil pump pickup tube and remove tube.

➡When the oil pan is removed, a new integral windage tray and gasket assembly must be installed. The old gasket cannot be reused.

16. Discard the integral windage tray and gasket and replace.

To install:

17. Clean the oil pan gasket mating surface of the block and oil pan.

➡Mopar® Engine RTV must be applied to the 4 T-joints (1, 2), (area where front cover, rear retainer, block, and oil pan gasket meet). The bead of RTV should cover the bottom of the gasket. This area is approximately 4.5 mm x 25 mm in each of the 4 T-joint locations.

18. Apply Mopar® Engine RTV at the 4 T-joints.

➡When the oil pan is removed, a new integral windage tray and gasket assembly must be installed. The old gasket cannot be reused.

19. Install a new integral windage tray and gasket.
20. Reinstall the oil pump pickup tube

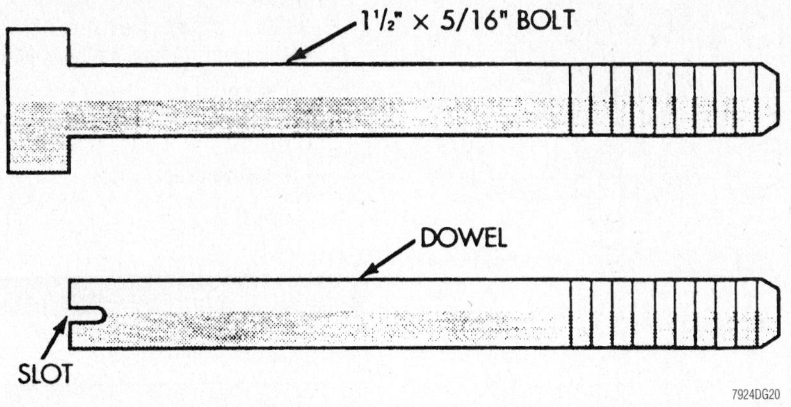

Oil pan alignment dowels—5.9L engines

with new O-ring. Tighten tube to pump fasteners to 28 Nm (250 inch lbs.).

➡**The double ended oil pan studs must be installed in the same location that they were removed from.**

21. Position the oil pan and install the mounting bolts and studs.. Tighten the mounting bolts to 12 Nm (105 inch lbs.).
22. Install the structural cover.
23. Lower the engine into mounts.
24. Install both the left and right side engine mount through bolts. Tighten the nuts to 68 Nm (50 ft. lbs.).
25. Reinstall the front axle.
26. Install the steering rack.
27. Install the rear transmission mount.
28. Remove lifting fixture.
29. Install the front crossmember. Torque to 75 ft. lbs. (102 Nm).
30. Install the fan shroud and fan.
31. Fill engine oil.
32. Reconnect the negative battery cable.
33. Start engine and check for leaks.

5.9L Engines

1. Before servicing the vehicle, refer to the precautions in the beginning of this section.
2. Drain the engine oil.
3. Remove or disconnect the following:
   - Oil filter
   - Starter motor
   - Cooler lines
   - Oil level sensor connector
   - Heated Oxygen (HO_2S) sensor connector
   - Exhaust Y-pipe
   - Oil pan

To install:

4. Install or connect the following:
   - Oil pan, using a new gasket. Tighten the bolts to 18 ft. lbs. (24 Nm).
   - Exhaust Y-pipe
   - Heated Oxygen (HO_2S) sensor connector
   - Oil level sensor connector
   - Cooler lines
   - Starter motor
   - Oil filter
5. Fill the crankcase.
6. Start the engine and check for leaks.

Oil Pump

REMOVAL & INSTALLATION

3.7L Engine

1. Before servicing the vehicle, refer to the precautions in the beginning of this section.
2. Remove or disconnect the following:
   - Oil Pan
   - Timing chain cover
   - Timing chains and tensioners
   - Oil pump
3. Installation is the reverse of removal. Torque the pump bolts, in sequence, to 21 ft. lbs. (28 Nm),

4.7L Engine

1. Before servicing the vehicle, refer to the precautions in the beginning of this section.
2. Drain the engine oil.
3. Remove or disconnect the following:
   - Negative battery cable
   - Oil pan
   - Oil pump pick-up tube
   - Timing chains and tensioners
   - Oil pump

To install:

4. Install or connect the following:
   - Oil pump. Tighten the bolts to 21 ft. lbs. (28 Nm).
   - Timing chains and tensioners
   - Oil pump pick-up tube
   - Oil pan
   - Negative battery cable
5. Fill the crankcase to the correct level.
6. Start the engine and check for leaks.

5.7L Engine

1. Before servicing the vehicle, refer to the precautions in the beginning of this section.

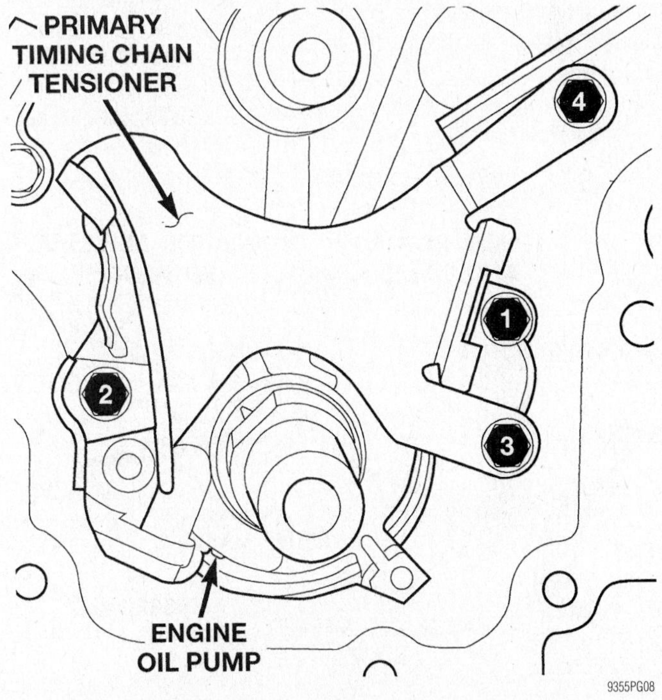

Oil pump bolt torque sequence—3.7L and 4.7L engines

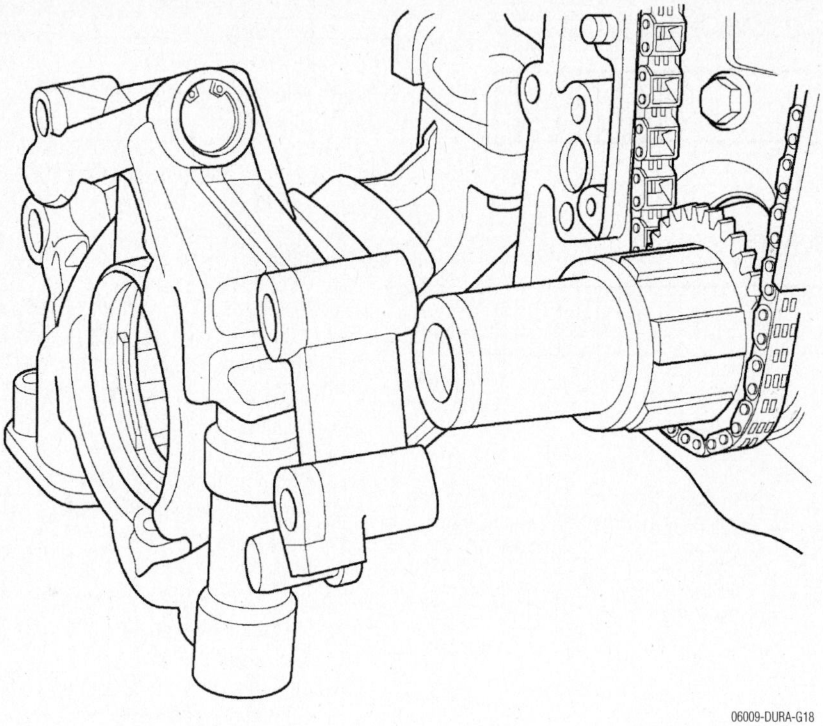

Oil pump installation—5.7L engine

2. Drain the engine oil.
3. Remove or disconnect the following:
   - Negative battery cable
   - Oil pan
   - Oil pump pick-up tube
   - Timing chains and tensioners
   - Oil pump

To install:

4. Install or connect the following:
   - Oil pump. Tighten the bolts to 21 ft. lbs. (28 Nm).
   - Timing chains and tensioners
   - Oil pump pick-up tube
   - Oil pan
   - Negative battery cable
5. Fill the crankcase to the correct level.
6. Start the engine and check for leaks.

5.9L Engines

1. Before servicing the vehicle, refer to the precautions in the beginning of this section.
2. Drain the engine oil.
3. Remove or disconnect the following:
   - Negative battery cable
   - Oil pan
   - Oil pump pick-up tube
   - Oil pump

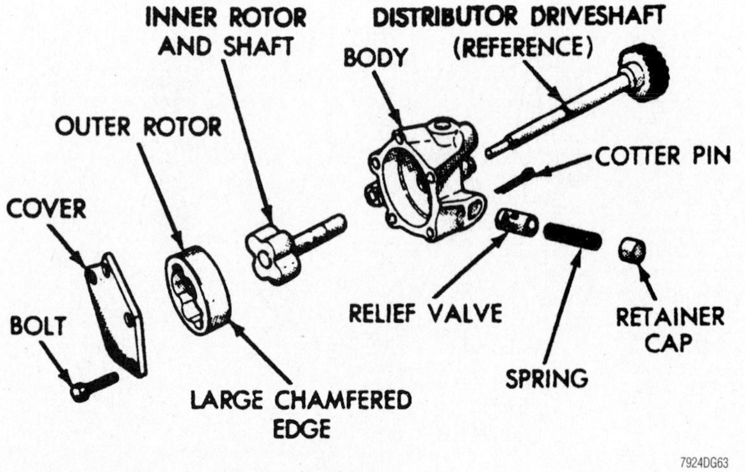

Exploded view of the oil pump assembly—5.9L engines

To install:

4. Install or connect the following:
   - Oil pump. Tighten the bolts to 30 ft. lbs. (41 Nm).
   - Oil pump pick-up tube
   - Oil pan
   - Negative battery cable
5. Fill the crankcase to the correct level.
6. Start the engine and check for leaks.

Rear Main Seal

REMOVAL & INSTALLATION

3.7L and 4.7L Engines

1. Before servicing the vehicle, refer to the precautions in the beginning of this section.
2. Remove or disconnect the following:
   - Transmission
   - Flexplate
3. Thread Oil Seal Remover 8506 into the rear main seal as far as possible and remove the rear main seal.

To install:

4. Install or connect the following:
   - Seal Guide 8349-2 onto the crankshaft
   - Rear main seal on the seal guide
   - Rear main seal, using the Crankshaft Rear Oil Seal Installer 8349 and Driver Handle C-4171; tap it into place until the installer is flush with the cylinder block
   - Flexplate. Tighten the bolts to 45 ft. lbs. (60 Nm) for 2004 3.7L and all 4.7L; 70 ft. lbs. (95 Nm) for 5.7L and 2005 3.7L engine.
   - Transmission
5. Start the engine and check for leaks.

5.7L Engine

1. Disconnect negative cable from battery.
2. Remove the transmission.
3. Remove the flexplate.
4. Remove the oil pan.
5. Remove the rear oil seal retainer mounting bolts.
6. Carefully remove the retainer from the engine block.
7. Using a seal puller, remove the crankshaft rear oil seal.

To install:

➡**The rear seal must be installed dry for proper operation. Do not lubricate the seal lip or outer edge.**

8. Position the plastic seal guide onto the crankshaft rear face. Then position the crankshaft rear oil seal onto the guide.

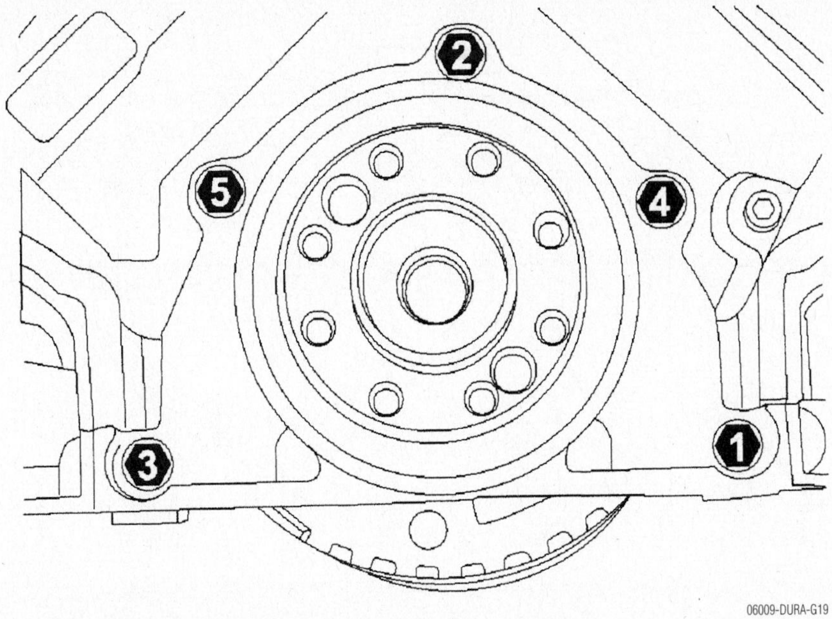

Rear seal retainer torque sequence—5.7L engine

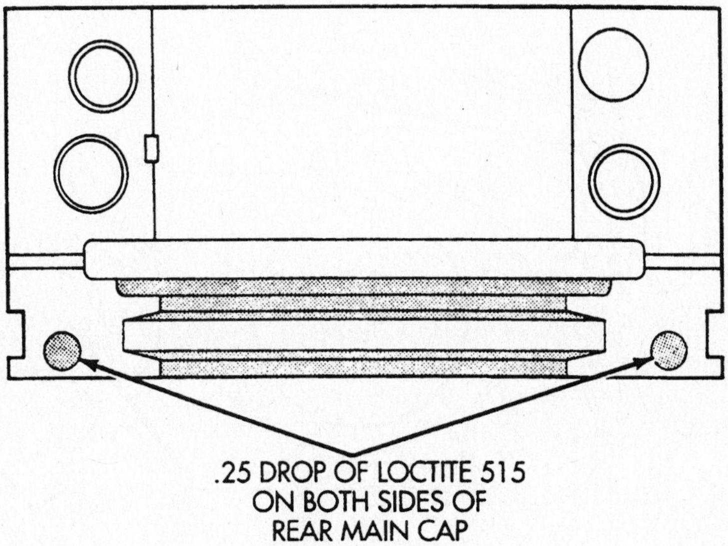

Sealant application locations —5.9L engines

9. Using a seal driver and hammer, tap the seal into place. Continue to tap on the driver handle until the seal installer seats against the cylinder block crankshaft bore.
10. Thoroughly clean all gasket residue from the engine block.
11. Use extreme care and clean all gasket residue from the retainer.
12. Position the gasket onto the retainer.
13. Position the retainer onto the engine block.
14. Install the retainer mounting bolts. Tighten the bolts to 15 Nm (132 inch lbs.) using the procedure shown.
15. Install the oil pan.
16. Install the flexplate.
17. Install the transmission.
18. Check and verify engine oil level.
19. Start engine and check for leaks.

5.9L Engines

1. Before servicing the vehicle, refer to the precautions in the beginning of this section.
2. Drain the engine oil.
3. Remove or disconnect the following:
   - Oil pan
   - Oil pump
   - Rear main bearing cap
4. Loosen the other main bearing cap bolts for clearance and remove the rear main seal halfs.

To install:

5. Install or connect the following:
   - New upper seal half to the cylinder block
   - New lower seal half to the bearing cap
6. Apply sealant to the rear main bearing cap.
7. Install or connect the following:
   - Rear main bearing cap. Tighten **all** main bearing cap bolts to 85 ft. lbs. (115 Nm).
   - Oil pump and oil pan
8. Fill the engine.
9. Start the engine and check for leaks.

Timing Chain, Sprockets, Front Cover and Seal

REMOVAL & INSTALLATION

3.7L Engines

1. Before servicing the vehicle, refer to the precautions in the beginning of this section.
2. Drain the cooling system.
3. Remove or disconnect the following:
   - Negative battery cable
   - Valve covers
   - Radiator fan
4. Rotate the crankshaft so that the crankshaft timing mark aligns with the Top Dead Center (TDC) mark on the front cover, and the **V6** marks on the camshaft sprockets are at 12 o'clock.
   - Power steering pump
   - Access plugs from the cylinder heads
   - Oil fill housing
   - Crankshaft damper
5. Compress the primary timing chain tensioner and install a lockpin.
6. Remove the secondary timing chain tensioners.
7. Hold the left camshaft with adjustable pliers and remove the sprocket and chain. Rotate the **left** camshaft 15 degrees **clockwise** to the neutral position.
8. Hold the right camshaft with adjustable pliers and remove the camshaft sprocket. Rotate the **right** camshaft 45 degrees **counterclockwise** to the neutral position.
9. Remove the primary timing chain and sprockets.

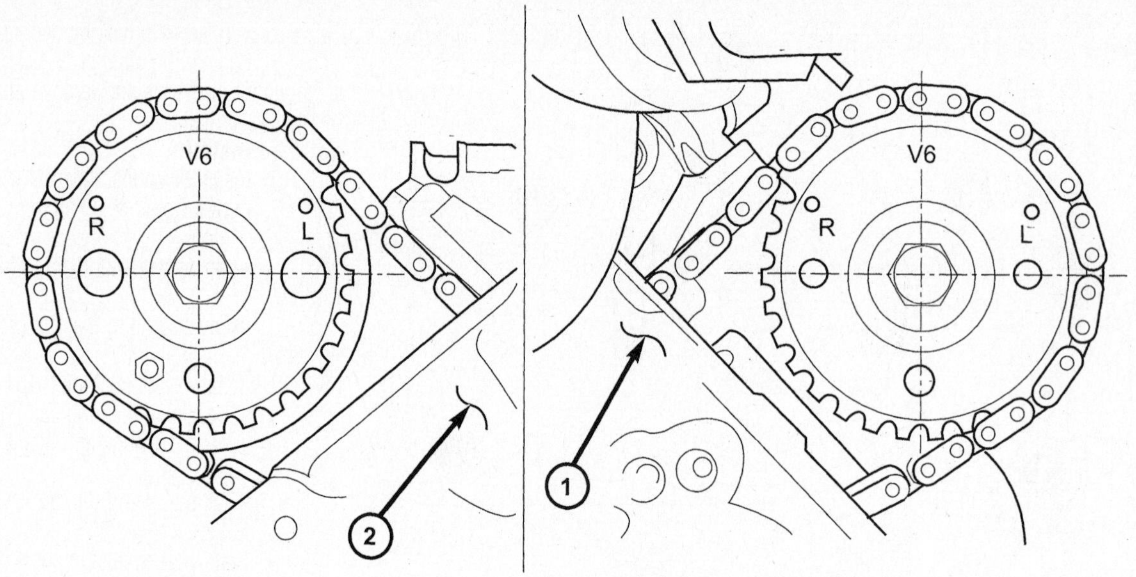

1 - LEFT CYLINDER HEAD
2 - RIGHT CYLINDER HEAD

9355PG09

Camshaft sprocket timing marks—3.7L

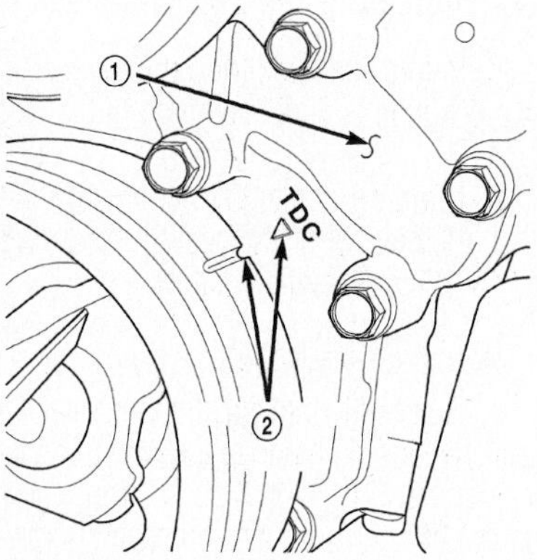

1 - TIMING CHAIN COVER
2 - CRANKSHAFT TIMING MARKS

9355PG10

Crankshaft timing marks—3.7L

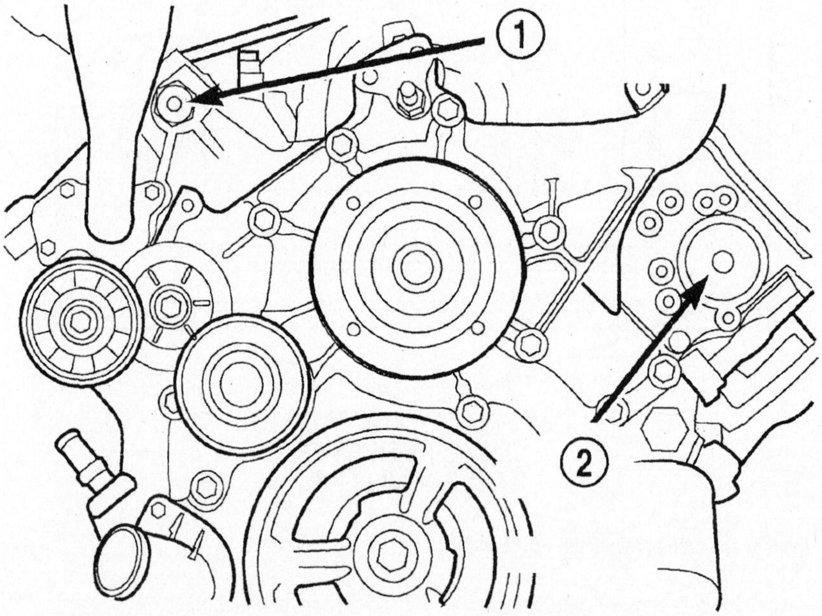

1 - RIGHT CYLINDER HEAD ACCESS PLUG
2 - LEFT CYLINDER HEAD ACCESS PLUG

9355PG11

Cylinder head access plugs—3.7L

To install:

10. Use a small prytool to hold the ratchet pawl and compress the secondary timing chain tensioners in a vise and install locking pins.

➡The black bolts fasten the guide to the engine block and the silver bolts fasten the guide to the cylinder head.

11. Install or connect the following:
    - Secondary timing chain guides. Tighten the bolts to 21 ft. lbs. (28 Nm).
    - Secondary timing chains to the idler sprocket so that the double plated links on each chain are visible through the slots in the primary idler sprocket
12. Lock the secondary timing chains to the idler sprocket with Timing Chain Locking tool as shown.
13. Align the primary chain double plated links with the idler sprocket timing mark and the single plated link with the crankshaft sprocket timing mark.
14. Install the primary chain and sprock-

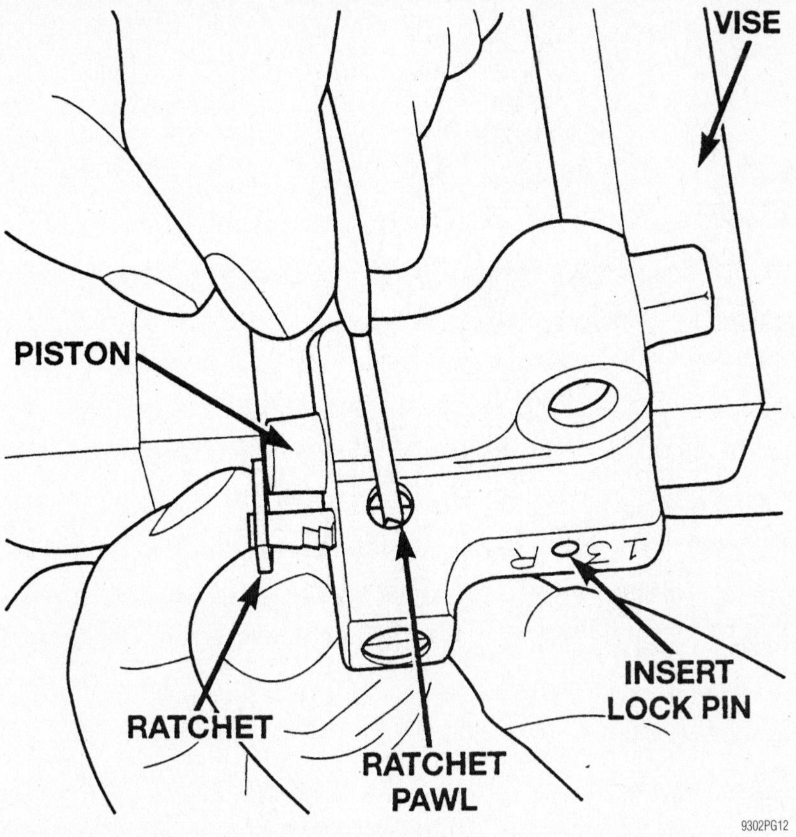

Secondary timing chain tensioner preparation—3.7L engine

ets. Tighten the idler sprocket bolt to 25 ft. lbs. (34 Nm).

15. Align the secondary chain single plated links with the timing marks on the secondary sprockets. Align the dot at the **L** mark on the left sprocket with the plated link on the left chain and the dot at the **R** mark on the right sprocket with the plated link on the right chain.

16. Rotate the camshafts back from the neutral position and install the camshaft sprockets.

17. Remove the secondary chain locking tool.

18. Remove the primary and secondary timing chain tensioner locking pins.

19. Hold the camshaft sprockets with a spanner wrench and tighten the retaining bolts to 90 ft. lbs. (122 Nm).

20. Install or connect the following:

- Front cover. Tighten the bolts, in sequence, to 40 ft. lbs. (54 Nm).
- Front crankshaft seal
- Cylinder head access plugs
- A/C compressor
- Alternator
- Accessory drive belt tensioner. Tighten the bolt to 40 ft. lbs. (54 Nm).
- Oil fill housing

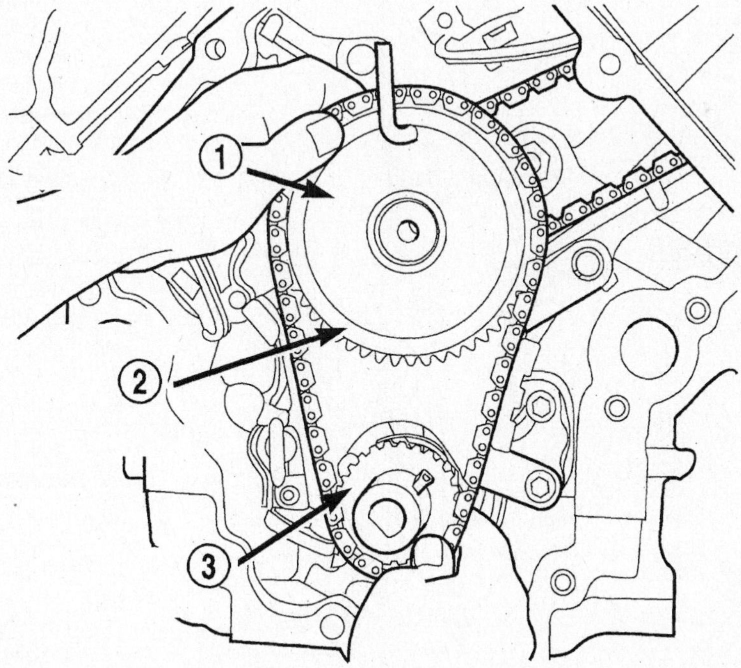

1 - SPECIAL TOOL 8429
2 - PRIMARY CHAIN IDLER SPROCKET
3 - CRANKSHAFT SPROCKET

9355PG12

Installing the idler gear and timing chain—3.7L

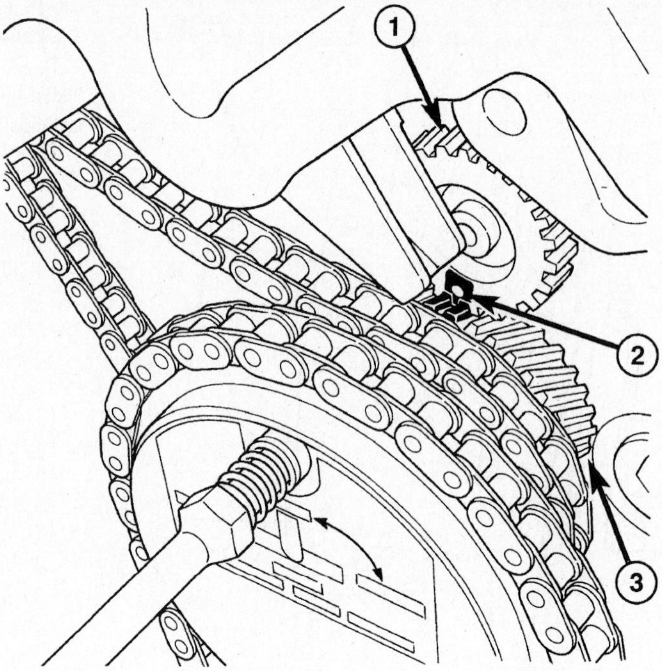

1 - COUNTERBALANCE SHAFT
2 - TIMING MARKS
3 - IDLER SPROCKET

9355PG13

Counterbalance shaft timing marks—3.7L

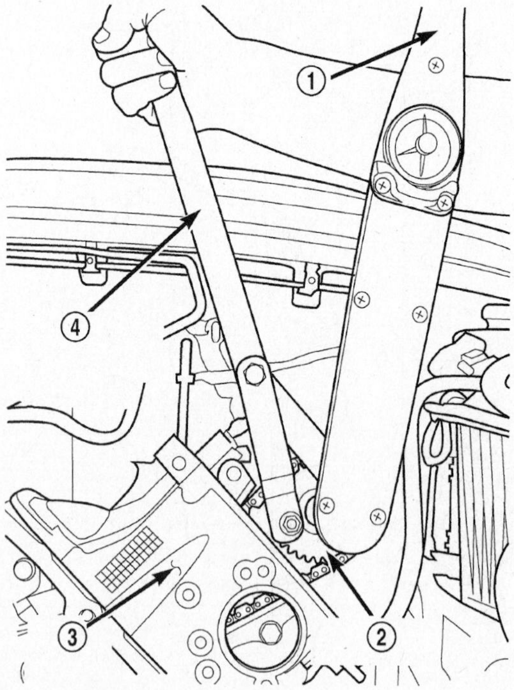

1 - TORQUE WRENCH
2 - CAMSHAFT SPROCKET
3 - LEFT CYLINDER HEAD
4 - SPECIAL TOOL 6958 SPANNER WITH ADAPTER PINS 8346

9355PG14

Tightening the left side camshaft sprocket—3.7L

- Crankshaft damper. Tighten the bolt to 130 ft. lbs. (175 Nm).
- Power steering pump
- Lower radiator hose
- Heater hoses
- Accessory drive belt
- Engine cooling fan and shroud
- Camshaft Position (CMP) sensor
- Valve covers
- Negative battery cable

21. Fill and bleed the cooling system.
22. Start the engine, check for leaks and repair if necessary.

4.7L Engine

1. Before servicing the vehicle, refer to the precautions in the beginning of this section.
2. Drain the cooling system.
3. Remove or disconnect the following:
   - Negative battery cable
   - Valve covers
   - Camshaft Position (CMP) sensor
   - Engine cooling fan and shroud
   - Accessory drive belt
   - Heater hoses
   - Lower radiator hose
   - Power steering pump
4. Rotate the crankshaft so that the

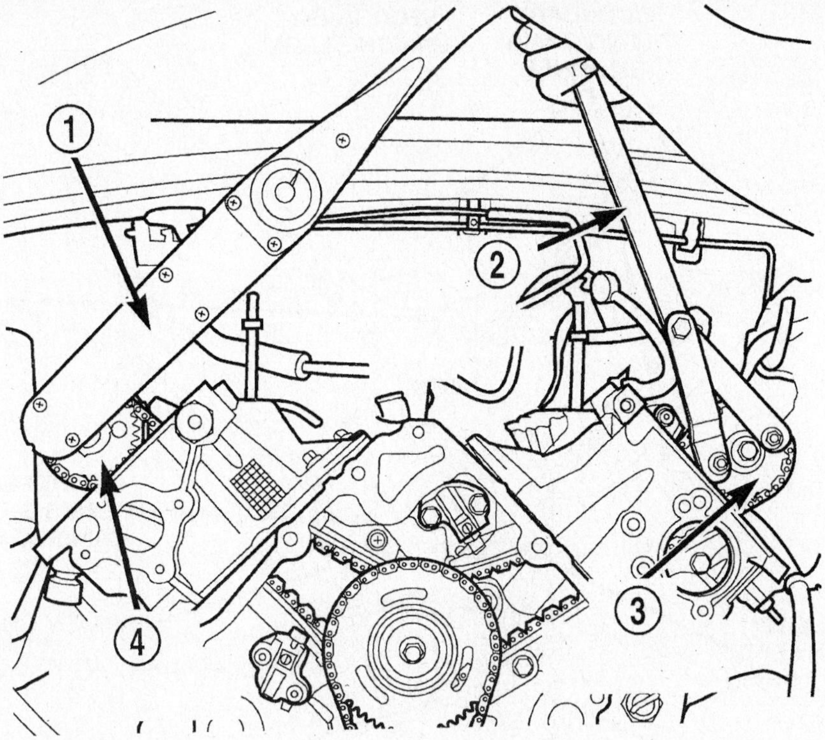

1 - TORQUE WRENCH
2 - SPECIAL TOOL 6958 WITH ADAPTER PINS 8346
3 - LEFT CAMSHAFT SPROCKET
4 - RIGHT CAMSHAFT SPROCKET

9355PG15

Tightening the right side camshaft sprocket—3.7L

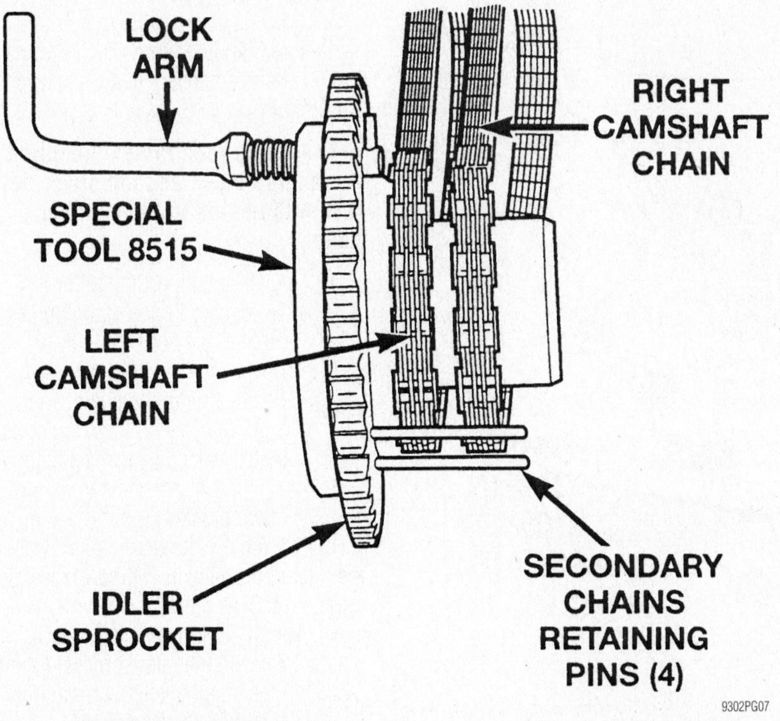

Use the Timing Chain Locking tool to lock the timing chains on the idler gear—3.7L engine

crankshaft timing mark aligns with the Top Dead Center (TDC) mark on the front cover, and the **V8** marks on the camshaft sprockets are at 12 o'clock.

5. Remove or disconnect the following:
   - Crankshaft damper
   - Oil fill housing
   - Accessory drive belt tensioner
   - Alternator
   - A/C compressor
   - Front cover
   - Front crankshaft seal
   - Cylinder head access plugs
   - Secondary timing chain guides
6. Compress the primary timing chain tensioner and install a lockpin.
7. Remove the secondary timing chain tensioners.
8. Hold the left camshaft with adjustable pliers and remove the sprocket and chain. Rotate the **left** camshaft 15 degrees **clockwise** to the neutral position.
9. Hold the right camshaft with adjustable pliers and remove the camshaft sprocket. Rotate the **right** camshaft 45

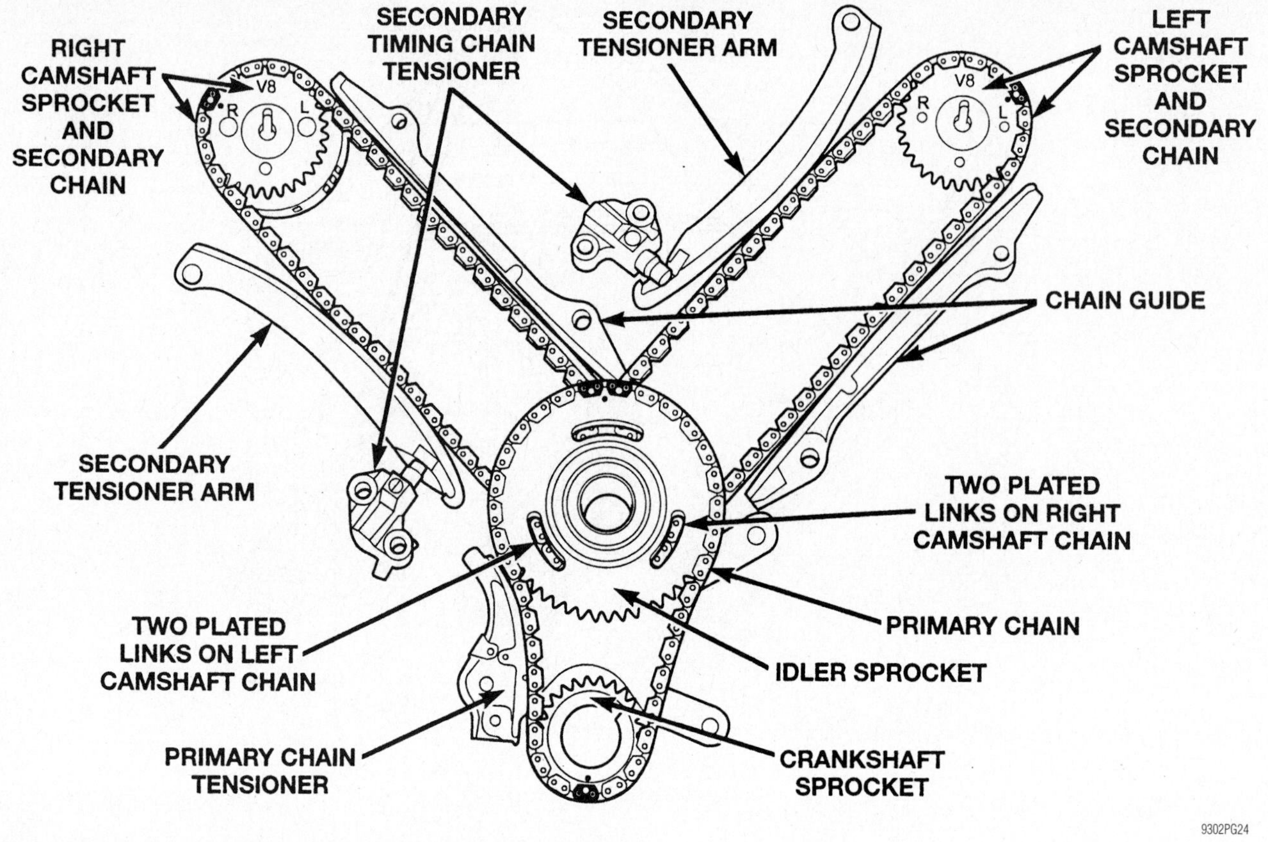

Timing chain system and alignment marks—4.7L engine shown. V6 similar

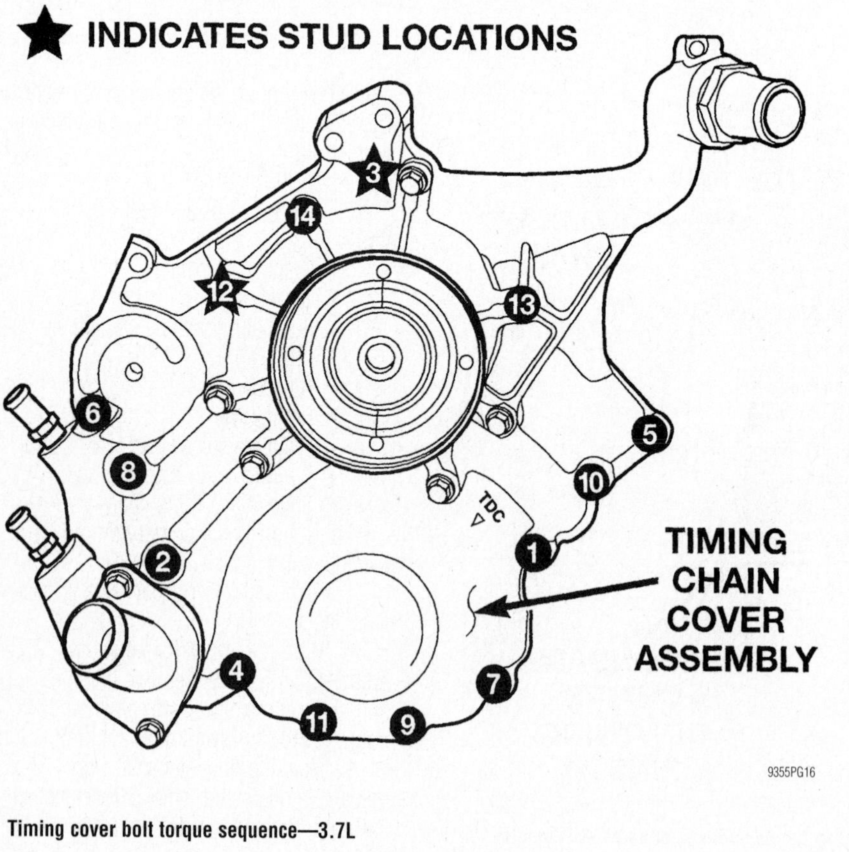

Timing cover bolt torque sequence—3.7L

degrees **counterclockwise** to the neutral position.

10. Remove the primary timing chain and sprockets.

To install:

11. Use a small prytool to hold the ratchet pawl and compress the secondary timing chain tensioners in a vise and install locking pins.

➡The black bolts fasten the guide to the engine block and the silver bolts fasten the guide to the cylinder head.

12. Install or connect the following:
    - Secondary timing chain guides. Tighten the bolts to 21 ft. lbs. (28 Nm).
    - Secondary timing chains to the idler sprocket so that the double plated links on each chain are visible through the slots in the primary idler sprocket

13. Lock the secondary timing chains to the idler sprocket with Timing Chain Locking tool 8515 as shown.

14. Align the primary chain double plated links with the idler sprocket timing mark and the single plated link with the crankshaft sprocket timing mark.

15. Install the primary chain and sprock-

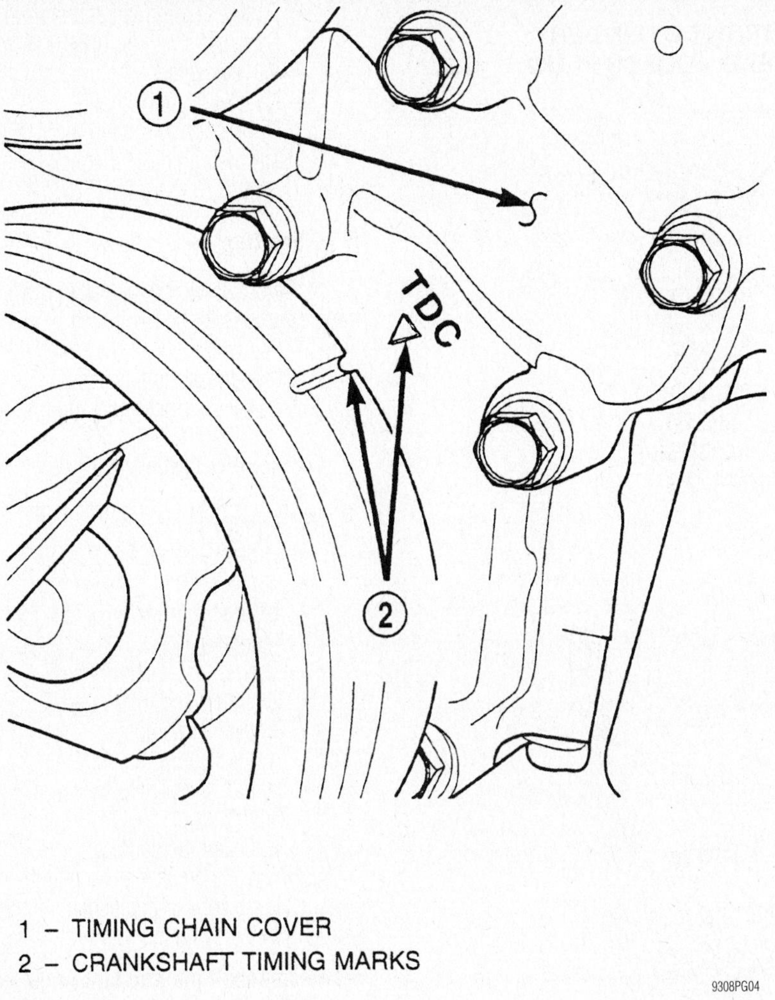

Crankshaft timing marks—4.7L engine

ets. Tighten the idler sprocket bolt to 25 ft. lbs. (34 Nm).

16. Align the secondary chain single plated links with the timing marks on the secondary sprockets. Align the dot at the **L** mark on the left sprocket with the plated link on the left chain and the dot at the **R** mark on the right sprocket with the plated link on the right chain.

17. Rotate the camshafts back from the neutral position and install the camshaft sprockets.

18. Remove the secondary chain locking tool.

19. Remove the primary and secondary timing chain tensioner locking pins.

20. Hold the camshaft sprockets with a spanner wrench and tighten the retaining bolts to 90 ft. lbs. (122 Nm).

21. Install or connect the following:

- Front cover. Tighten the bolts, in sequence, to 40 ft. lbs. (54 Nm).
- Front crankshaft seal
- Cylinder head access plugs
- A/C compressor
- Alternator
- Accessory drive belt tensioner. Tighten the bolt to 40 ft. lbs. (54 Nm).
- Oil fill housing
- Crankshaft damper. Tighten the bolt to 130 ft. lbs. (175 Nm).
- Power steering pump
- Lower radiator hose
- Heater hoses
- Accessory drive belt

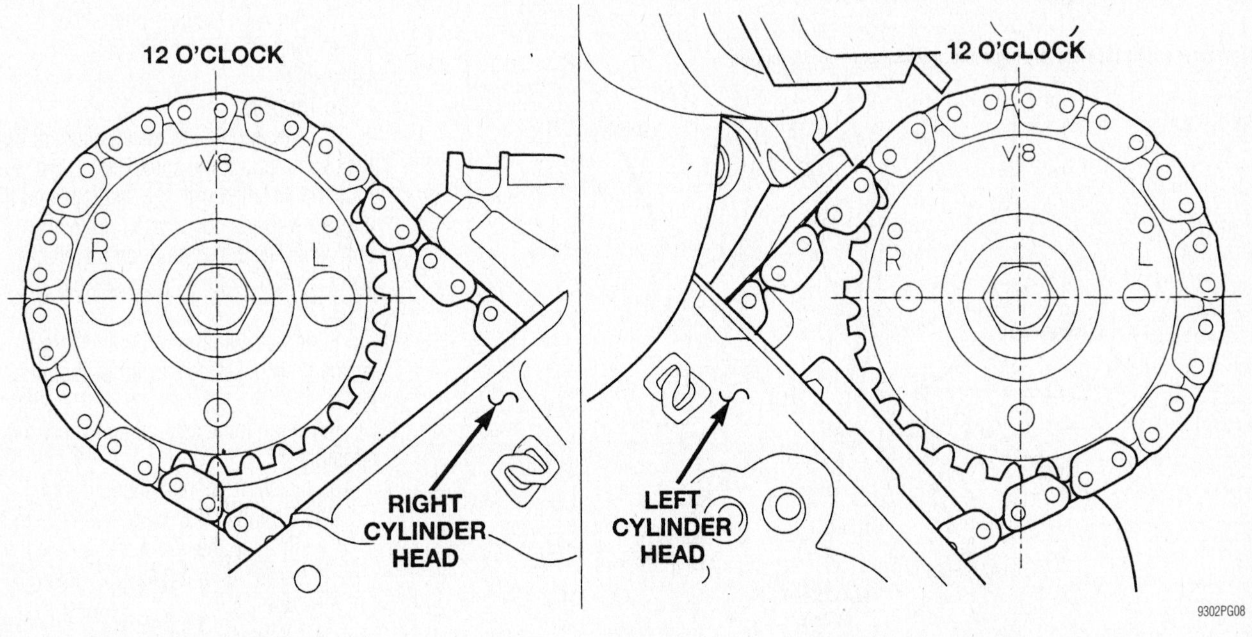

Camshaft positioning—4.7L engine

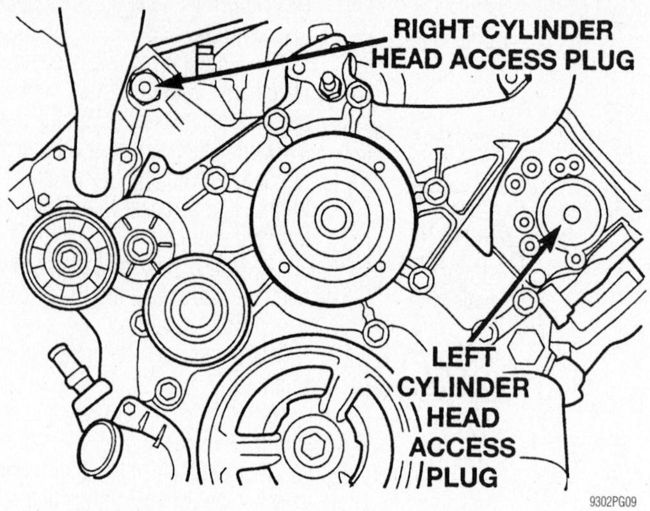

Cylinder head access plug locations—4.7L engine

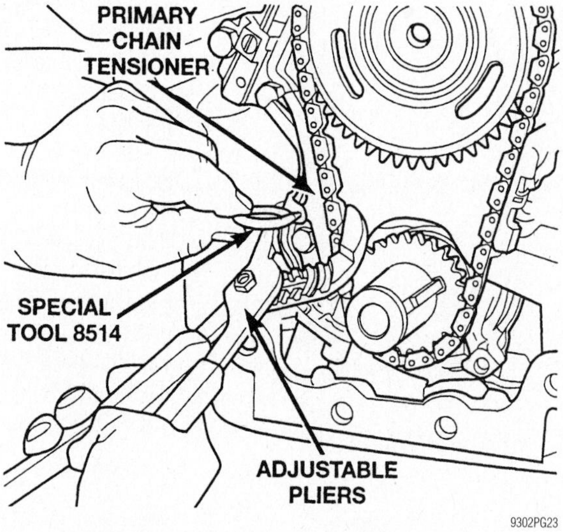

Compress and lock the primary chain tensioner—4.7L engine

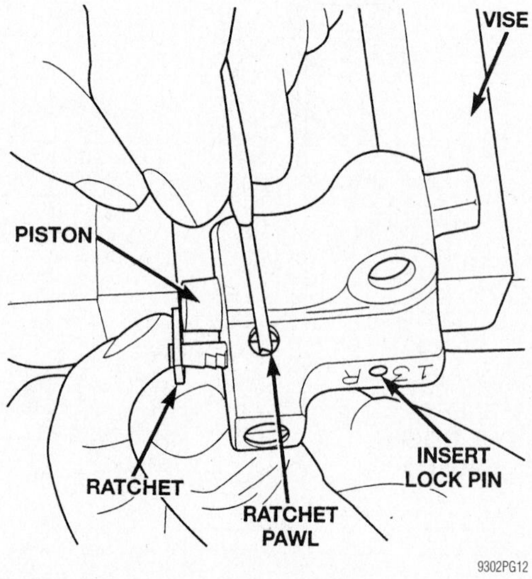

Secondary timing chain tensioner preparation—4.7L engine

- Engine cooling fan and shroud
- Camshaft Position (CMP) sensor
- Valve covers
- Negative battery cable

22. Fill the cooling system.
23. Start the engine and check for leaks.

5.7L Engines

1. Before servicing the vehicle, refer to the precautions in the beginning of this section.
2. Drain the cooling system.
3. Remove or disconnect the following:
   - Negative battery cable
   - Drive belt
   - Radiator fan
   - Coolant and washer bottles
   - Fan shroud
   - A/C compressor
   - Alternator
   - Radiator and heater hoses
   - Tensioner and idler pulleys
   - Crankshaft damper
   - Power steering pump
   - Oil pan and pickup tube
   - Timing cover
   - Re-install the damper
4. Rotate the crankshaft so that the camshaft sprocket and crankshaft sprocket timing marks are aligned.

➡The camshaft pin and slot in the cam sprocket must be a 12 o'clock, the crankshaft keyway must be at 2 o'clock, and the dots or paint on the crank sprocket must be at 6 o'clock.

5. Pin back the tensioner shoe.
6. Remove the timing chain and sprockets.

To install:

7. With the timing marks aligned, wrap the chain around the sprockets. The chain must be installed with the single plated link aligned with the dot or paint on the cam sprocket. The dot or paint on the crank sprocket should be aligned between the 2 plated links.
8. Install the assembly and torque the cam sprocket bolt to 90 ft. lbs. (122 Nm).
9. Unpin the tensioner and verify the alignment.
10. Install or connect the following:
    - Timing cover. Torque all fasteners to 21 ft. lbs. (28 Nm). Torque the large lifting lug to 40 ft. lbs. (55 Nm)
    - Oil pan and pickup tube
    - Power steering pump

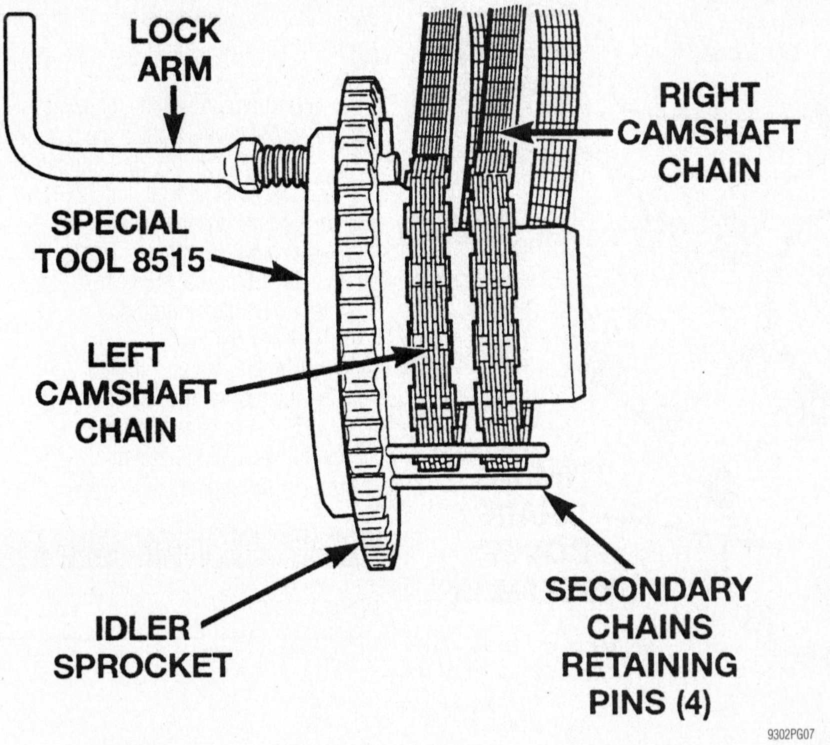

Use the Timing Chain Locking tool to lock the timing chains on the idler gear—4.7L engine

- Crankshaft damper
- Tensioner and idler pulleys
- Radiator and heater hoses
- Alternator
- A/C compressor
- Fan shroud
- Coolant and washer bottles
- Radiator fan
- Drive belt
- Negative battery cable

5.9L Engines

1. Before servicing the vehicle, refer to the precautions in the beginning of this section.
2. Drain the cooling system.
3. Remove or disconnect the following:
   - Negative battery cable
   - Accessory drive belt
   - Cooling fan and shroud
   - Water pump
   - Power steering pump
   - Crankshaft damper
   - Front crankshaft seal
   - Front cover
4. Rotate the crankshaft so that the camshaft sprocket and crankshaft sprocket timing marks are aligned.
5. Remove the timing chain and sprockets.

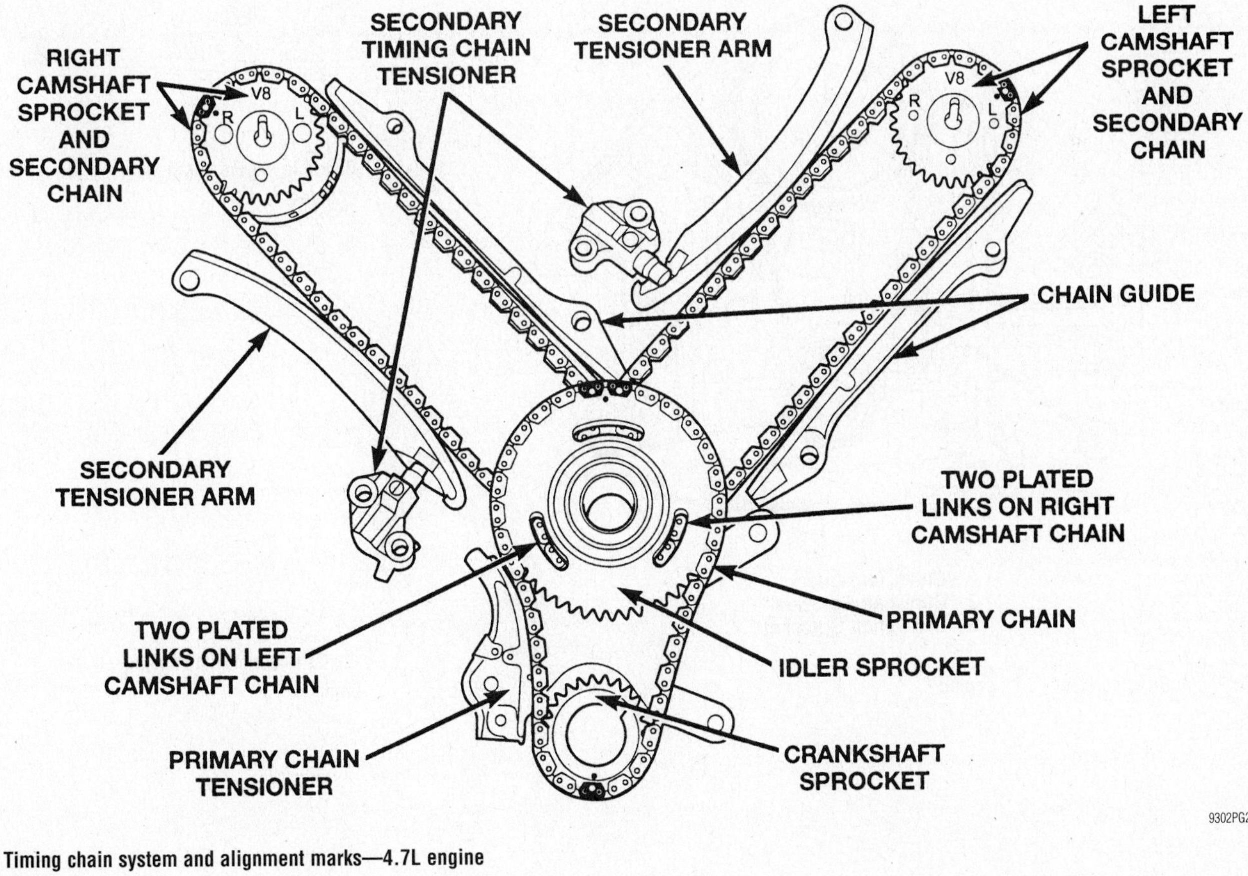

Timing chain system and alignment marks—4.7L engine

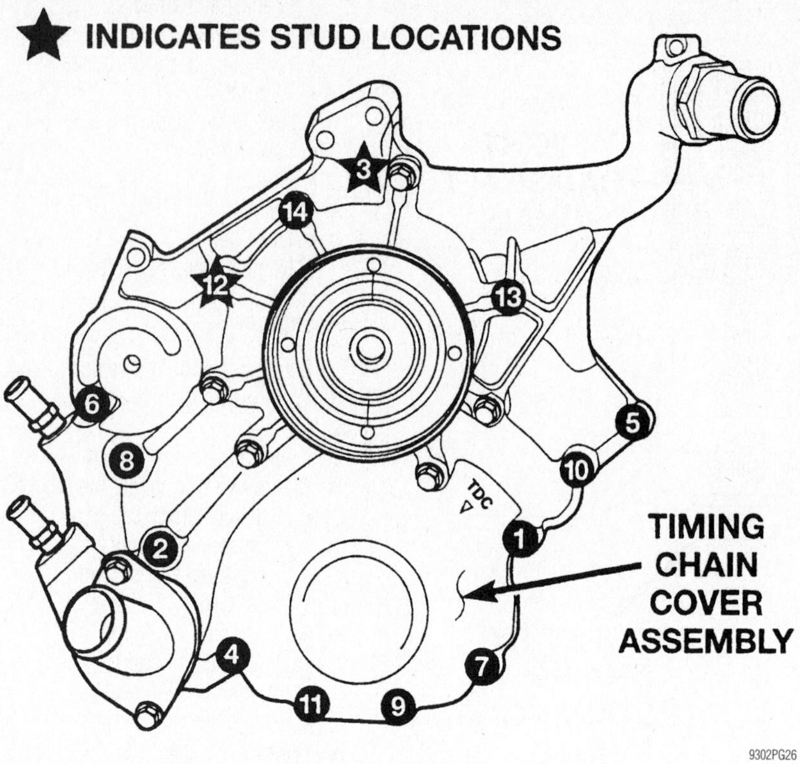

Timing chain cover bolt torque sequence—4.7L engine

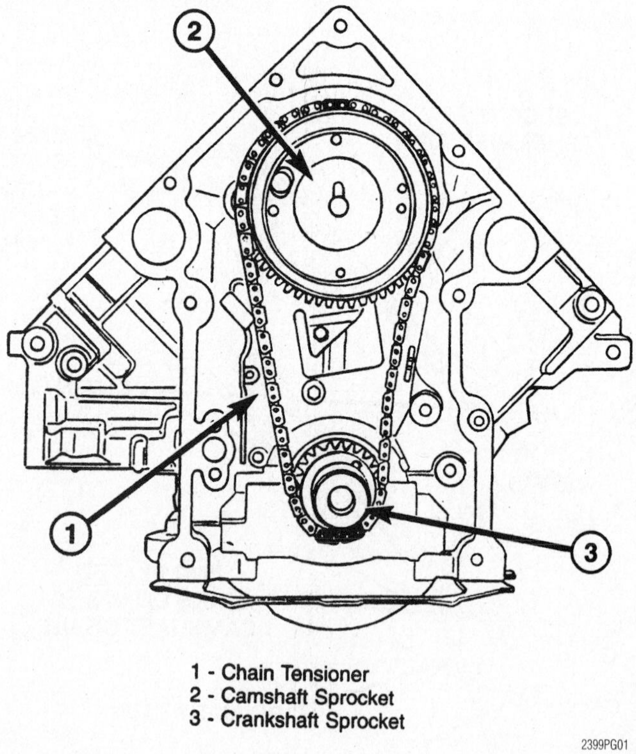

Timing mark alignment—5.7L engine

To install:

6. Install the timing chain and sprockets with the timing marks aligned. Tighten the camshaft sprocket bolt to 50 ft. lbs. (68 Nm).
7. Install or connect the following:
   - Front cover. Tighten the cover bolts to 30 ft. lbs. (41 Nm) and the oil pan bolts to 18 ft. lbs. (24 Nm).
   - Front crankshaft seal
   - Crankshaft damper. Tighten the bolt to 135 ft. lbs. (183 Nm).
   - Power steering pump
   - Water pump
   - Cooling fan and shroud
   - Accessory drive belt
   - Negative battery cable
8. Fill the cooling system.
9. Start the engine and check for leaks.

Piston and Ring

POSITIONING

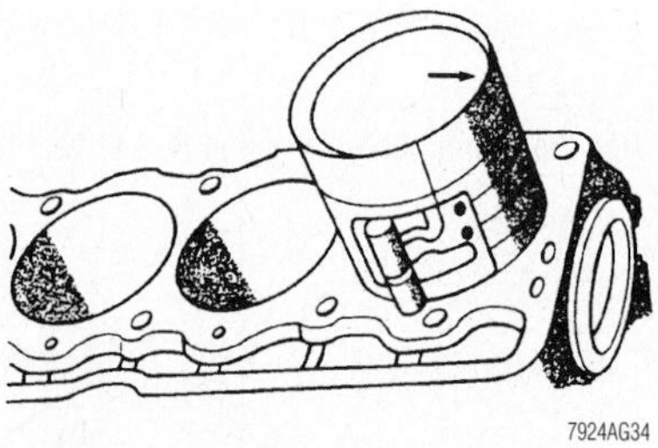

Piston to engine positioning—5.9L engines

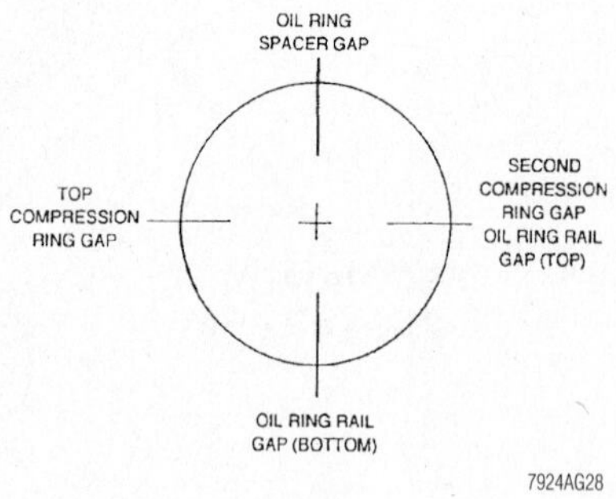

Piston ring end-gap spacing—5.9L engines

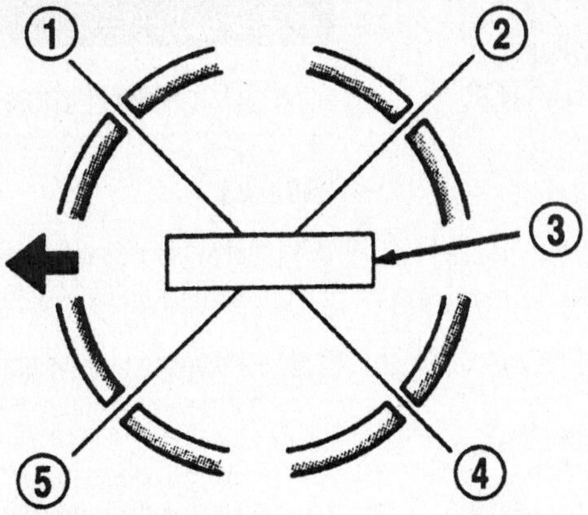

Piston ring end-gap spacing—5.7L engines

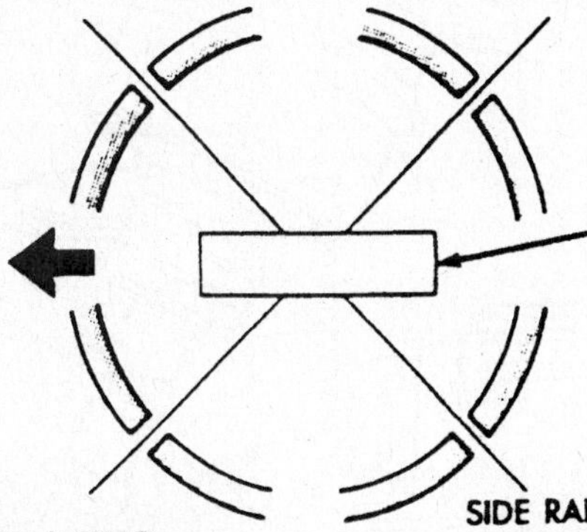

Piston ring end-gap spacing. Position raised "F" on piston towards front of engine—3.7L and 4.7L engine

FUEL SYSTEM

Fuel System Service Precautions

Safety is the most important factor when performing not only fuel system maintenance but any type of maintenance. Failure to conduct maintenance and repairs in a safe manner may result in serious personal injury or death. Maintenance and testing of the vehicle's fuel system components can be accomplished safely and effectively by adhering to the following rules and guidelines.

- To avoid the possibility of fire and personal injury, always disconnect the negative battery cable unless the repair or test procedure requires that battery voltage be applied.
- Always relieve the fuel system pressure prior to detaching any fuel system component (injector, fuel rail, pressure regulator, etc.), fitting or fuel line connection. Exercise extreme caution whenever relieving fuel system pressure to avoid exposing skin, face and eyes to fuel spray. Please be advised that fuel under pressure may penetrate the skin or any part of the body that it contacts.
- Always place a shop towel or cloth around the fitting or connection prior to loosening to absorb any excess fuel due to spillage. Ensure that all fuel spillage (should it occur) is quickly removed from engine surfaces. Ensure that all fuel soaked cloths or towels are deposited into a suitable waste container.
- Always keep a dry chemical (Class B) fire extinguisher near the work area.
- Do not allow fuel spray or fuel vapors to come into contact with a spark or open flame.
- Always use a back-up wrench when loosening and tightening fuel line connection fittings. This will prevent unnecessary stress and torsion to fuel line piping.
- Always replace worn fuel fitting O-rings with new. Do not substitute fuel hose or equivalent, where fuel pipe is installed.

Before servicing the vehicle, also make sure to refer to the precautions in the beginning of this section as well.

Fuel System Pressure

RELIEVING

2002

1. Remove the fuel filler cap.
2. Remove the fuel pump relay from the power distribution center.
3. Run the engine until it stalls.
4. Turn the key to **OFF**.

2003–05

1. Before servicing the vehicle, refer to the precautions in the beginning of this section.
2. Disconnect the negative battery cable.
3. Remove the fuel tank filler cap to release any fuel tank pressure.
4. Remove the fuel pump relay from the PDC.
5. Start and run the engine until it stops.
6. Unplug the connector from any injector and connect a jumper wire from either injector terminal to the positive battery terminal. Connect another jumper wire to the other terminal and momentarily touch the other end to the negative battery terminal.

**** WARNING**

Just touch the jumper to the battery. Powering the injector for more than a few seconds will permanently damage it.

7. Place a rag below the quick-disconnect coupling at the fuel rail and disconnect it.

Fuel Filter

REMOVAL & INSTALLATION

2002–03

These engines are equipped with a fuel filter/pressure regulator. This unit is not a regularly maintained unit and is serviced only when a DTC indicates a fault.

1. Before servicing the vehicle, refer to the precautions in the beginning of this section.
2. Relieve the fuel system pressure.
3. Remove or disconnect the following:
   - Negative battery cable
   - Fuel tank
4. Pull the filter/regulator out of the rubber grommet. Cut the hose clamp and remove the fuel line.

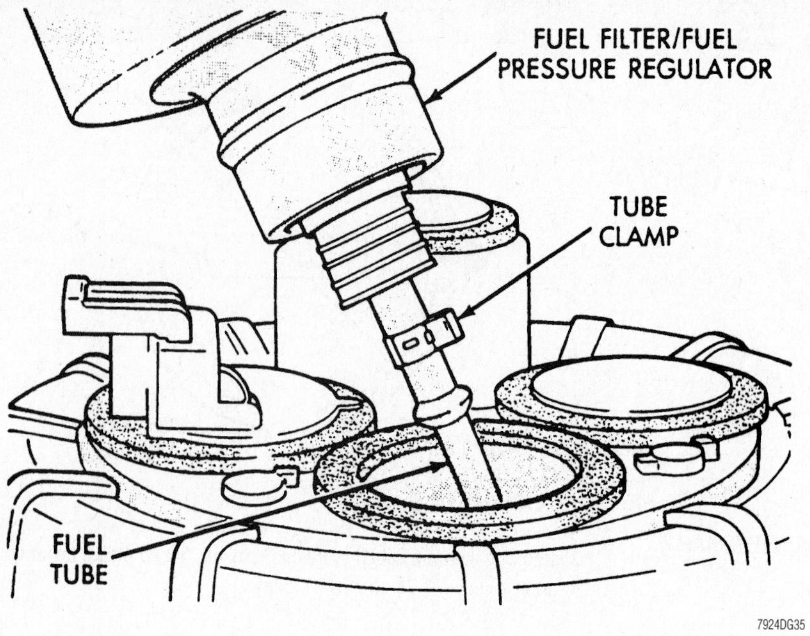

Pull and twist the filter/regulator to remove it from the top of the fuel pump module

To install:

5. Install the filter/regulator with a new clamp and push it into the rubber grommet.
6. Install or connect the following:
   - Fuel tank
   - Negative battery cable
7. Start the engine and check for leaks.

2004–05

1. Drain and remove fuel tank.
2. Note rotational position of module before attempting removal. An indexing arrow is located on top of module for this purpose.
3. Position Special Tool 9340 into notches on outside edge of lockring.
4. Install ½ inch drive breaker bar to tool 9340.
5. Rotate breaker bar counter-clockwise to remove lockring.
6. Remove lockring. The module will spring up slightly when lockring is removed.
7. Remove module from fuel tank. Be careful not to bend float arm while removing.

To install:

CAUTION

Whenever the fuel pump module is serviced, the rubber seal (gasket) must be replaced.

8. Using a new seal (gasket), position fuel pump module into opening in fuel tank.
9. Position lockring over top of fuel pump module.
10. Rotate module until embossed alignment arrow points to center alignment mark. This step must be performed to prevent float from contacting side of fuel tank. Also be sure fuel fitting on top of pump module is pointed to driver's side of vehicle.
11. Install Special Tool 9340 to lockring.
12. Install ½ inch drive breaker (1) into Special Tool 9340.
13. Tighten lockring (clockwise) until all seven notches have engaged.
14. Install fuel tank.

Fuel Pump

REMOVAL & INSTALLATION

2002–03

1. Before servicing the vehicle, refer to the precautions in the beginning of this section.
2. Relieve the fuel system pressure.
3. Remove or disconnect the following:
   - Negative battery cable
   - Fuel pump module harness connector
   - Fuel line
   - Fuel tank
   - Fuel pump module locknut
   - Fuel pump module

To install:

4. Install or connect the following:
   - Fuel pump module
   - Fuel pump module locknut
   - Fuel tank
   - Fuel line
   - Fuel pump module harness connector
   - Negative battery cable
5. Start the engine and check for leaks.

2004–05

1. Drain and remove fuel tank.
2. Note rotational position of module before attempting removal. An indexing arrow is located on top of module for this purpose.

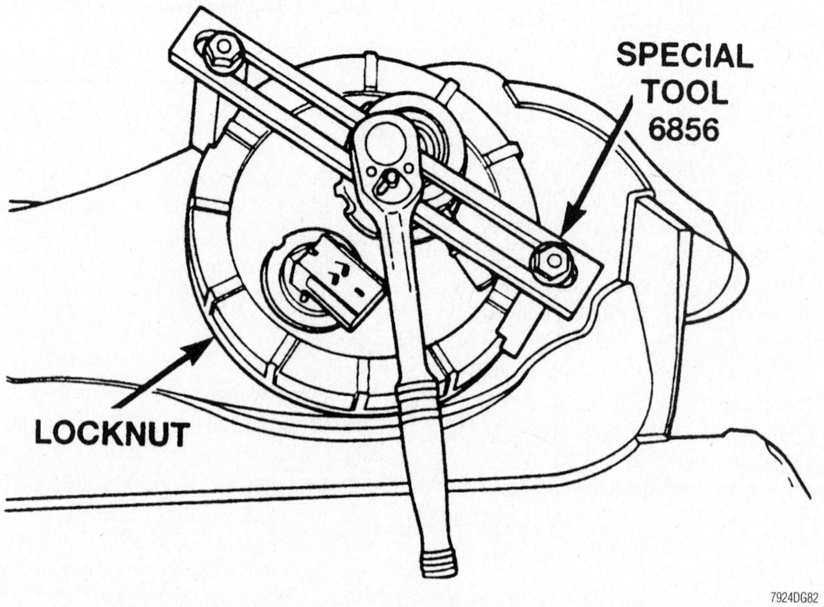

Fuel pump module locknut removal—2002–03 models

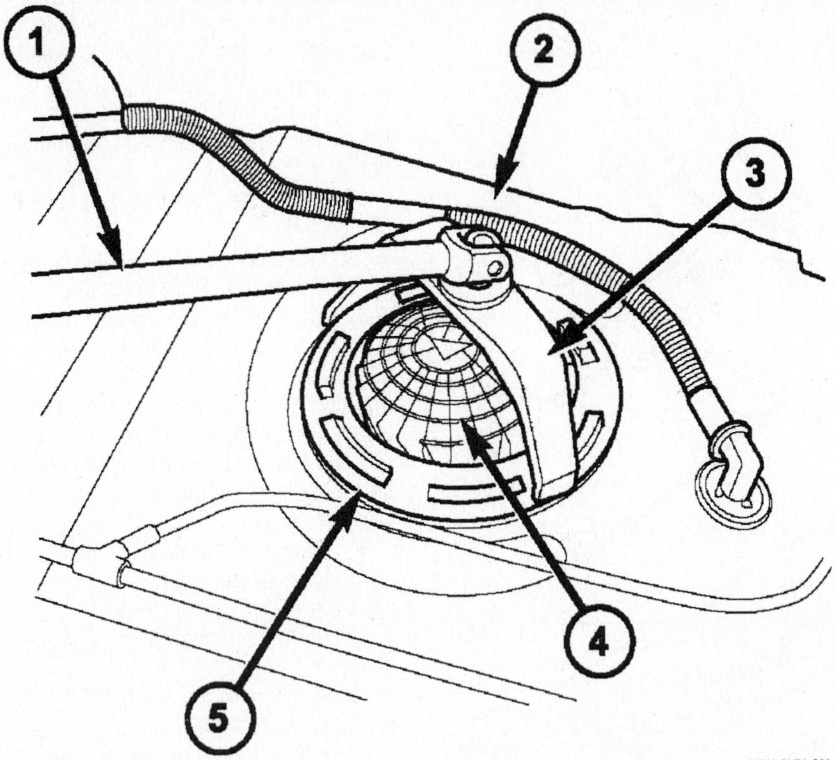

Module lockring removal. (1) breaker bar, (2) fuel return line, (3) lockring tool, (4) module, (5) lockring—2004–05 models

3. Position Special Tool 9340 into notches on outside edge of lockring.
4. Install ½ inch drive breaker bar to tool 9340.
5. Rotate breaker bar counter-clockwise to remove lockring.
6. Remove lockring. The module will spring up slightly when lockring is removed.
7. Remove module from fuel tank. Be careful not to bend float arm while removing.

To install:

⁂ CAUTION

Whenever the fuel pump module is serviced, the rubber seal (gasket) must be replaced.

8. Using a new seal (gasket), position fuel pump module into opening in fuel tank.
9. Position lockring over top of fuel pump module.
10. Rotate module until embossed alignment arrow points to center alignment mark. This step must be performed to prevent float from contacting side of fuel tank. Also be sure fuel fitting on top of pump module is pointed to driver's side of vehicle.
11. Install Special Tool 9340 to lockring.
12. Install ½ inch drive breaker (1) into Special Tool 9340.
13. Tighten lockring (clockwise) until all seven notches have engaged.
14. Install fuel tank.

Fuel Injector

REMOVAL & INSTALLATION

3.7L Engine

⁂ CAUTION

The left and right fuel rails are replaced as an assembly. Do not attempt to separate rail halves at connector tubes. Due to design of tubes, it does not use any clamps. Never attempt to install a clamping device of any kind to tubes. When removing fuel rail assembly for any reason, be careful not to bend or kink tubes.

1. Remove fuel tank filler tube cap.
2. Perform Fuel System Pressure Release Procedure.
3. Remove negative battery cable at battery.
4. Remove air duct at throttle body air box.
5. Remove air box at throttle body.
6. Remove air resonator mounting bracket at front of throttle body (2 bolts).
7. Disconnect fuel line latch clip and fuel line at fuel rail. A special tool will be necessary for fuel line disconnection.
8. Remove necessary vacuum lines at throttle body.
9. Disconnect electrical connectors at all 6 fuel injectors. Push red colored slider away from injector. While pushing slider, depress tab and remove connector from injector. The factory fuel injection wiring harness is numerically tagged (INJ 1, INJ 2, etc.) for injector position identification. If harness is not tagged, note wiring location before removal.
10. Disconnect electrical connectors at all throttle body sensors.
11. Remove 6 ignition coils.
12. Remove four fuel rail mounting bolts.
13. Gently rock and pull left side of fuel rail until fuel injectors just start to clear machined holes in cylinder head. Gently rock and pull right side of rail until injectors just start to clear cylinder head holes. Repeat this procedure (left/right) until all injectors have cleared cylinder head holes.
14. Remove fuel rail (with injectors attached) from engine.
15. Disconnect clip(s) that retain fuel injector(s) to fuel rail.

To install:

16. Install fuel injector(s) into fuel rail assembly and install retaining clip(s).
17. If same injector(s) is being reinstalled, install new o-ring(s).
18. Apply a small amount of clean engine oil to each injector o-ring. This will aid in installation.
19. Clean out fuel injector machined bores in intake manifold.
20. Apply a small amount of engine oil to each fuel injector o-ring. This will help in fuel rail installation.
21. Position fuel rail/fuel injector assembly to machined injector openings in cylinder head.
22. Guide each injector into cylinder head. Be careful not to tear injector o-rings.
23. Push right side of fuel rail down until fuel injectors have bottomed on cylinder head shoulder. Push left fuel rail down until injectors have bottomed on cylinder head shoulder.
24. Install 4 fuel rail mounting bolts and tighten.

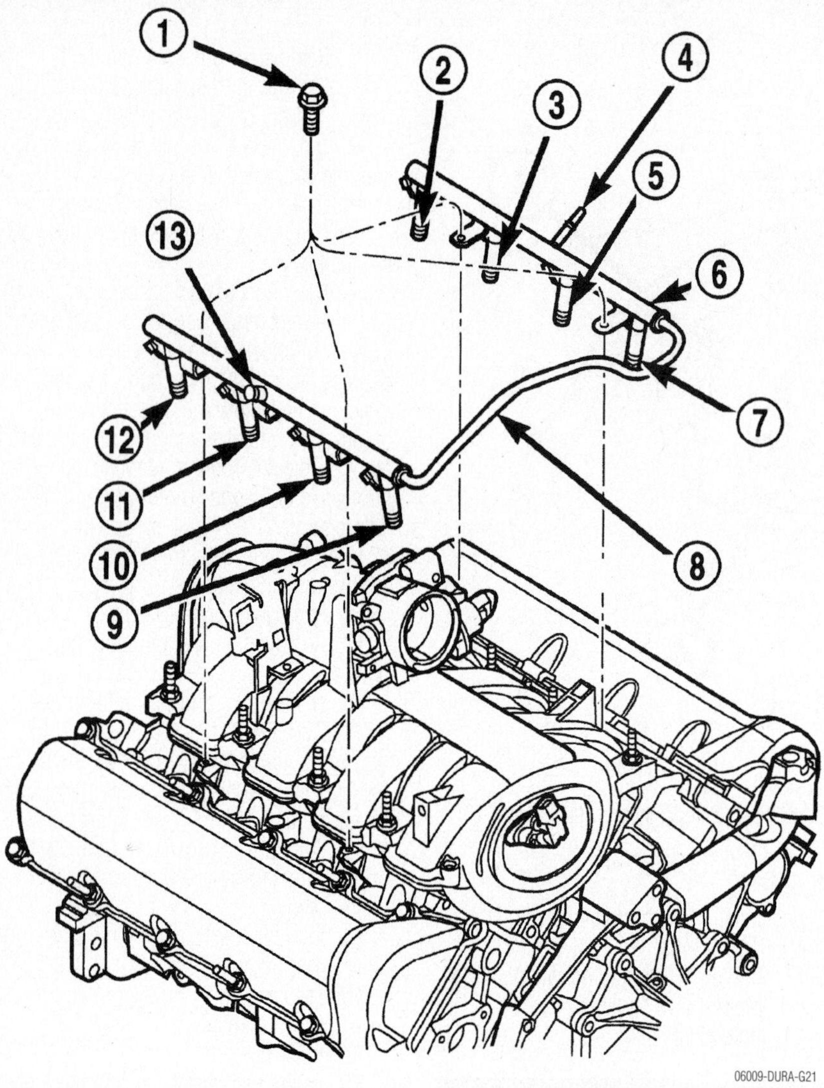

Fuel rail removal. 4.7L shown, 3.7L and 5.7L similar. (1) retaining bolts (2, 3, 5, 7, 9, 10, 11, 12) injectors. (6) rail, (4 and 13) fuel fee and return line connections, (8) crossover line

25. Install 6 ignition coils.
26. Connect electrical connectors to throttle body.
27. Connect electrical connectors at all fuel injectors. Refer to graphic. Push connector onto injector and then push and lock red colored slider. Verify connector is locked to injector by lightly tugging on connector.
28. Connect necessary vacuum lines to throttle body.
29. Install air resonator mounting bracket near front of throttle body (2 bolts).
30. Connect fuel line latch clip and fuel line to fuel rail.
31. Install air box to throttle body.
32. Install air duct to air box.
33. Connect battery cable to battery.
34. Start engine and check for leaks.

4.7L Engine

⁂ CAUTION

The left and right fuel rails are replaced as an assembly. Do not attempt to separate rail halves at connector tubes. Due to design of tubes, it does not use any clamps. Never attempt to install a clamping device of any kind to tubes. When removing fuel rail assembly for any reason, be careful not to bend or kink tubes.

1. Remove fuel tank filler tube cap.
2. Perform Fuel System Pressure Release Procedure.
3. Remove negative battery cable at battery.
4. Remove air duct at throttle body air box.
5. Remove air box at throttle body.
6. Remove air resonator mounting bracket at front of throttle body (2 bolts).
7. Disconnect fuel line latch clip and fuel line at fuel rail. A special tool will be necessary for fuel line disconnection.
8. Remove necessary vacuum lines at throttle body.
9. Disconnect electrical connectors at all 8 fuel injectors. Push red colored slider away from injector. While pushing slider, depress tab and remove connector from injector. The factory fuel injection wiring harness is numerically tagged (INJ 1, INJ 2, etc.) for injector position identification. If harness is not tagged, note wiring location before removal.
10. Disconnect electrical connectors at all throttle body sensors.
11. Remove 8 ignition coils.
12. Remove four fuel rail mounting bolts.
13. Gently rock and pull left side of fuel rail until fuel injectors just start to clear machined holes in cylinder head. Gently rock and pull right side of rail until injectors just start to clear cylinder head holes. Repeat this procedure (left/right) until all injectors have cleared cylinder head holes.
14. Remove fuel rail (with injectors attached) from engine.
15. Disconnect clip(s) that retain fuel injector(s) to fuel rail.

To install:

16. Install fuel injector(s) into fuel rail assembly and install retaining clip(s).
17. If same injector(s) is being reinstalled, install new o-ring(s).
18. Apply a small amount of clean engine oil to each injector o-ring. This will aid in installation.
19. Clean out fuel injector machined bores in intake manifold.
20. Apply a small amount of engine oil to each fuel injector o-ring. This will help in fuel rail installation.
21. Position fuel rail/fuel injector assembly to machined injector openings in cylinder head.
22. Guide each injector into cylinder head. Be careful not to tear injector o-rings.
23. Push right side of fuel rail down until fuel injectors have bottomed on cylinder head shoulder. Push left fuel rail down until injectors have bottomed on cylinder head shoulder.
24. Install 4 fuel rail mounting bolts and tighten.

25. Install 8 ignition coils.

26. Connect electrical connectors to throttle body.

27. Connect electrical connectors at all fuel injectors. Refer to graphic. Push connector onto injector and then push and lock red colored slider. Verify connector is locked to injector by lightly tugging on connector.

28. Connect necessary vacuum lines to throttle body.

29. Install air resonator mounting bracket near front of throttle body (2 bolts).

30. Connect fuel line latch clip and fuel line to fuel rail.

31. Install air box to throttle body.

32. Install air duct to air box.

33. Connect battery cable to battery.

34. Start engine and check for leaks.

5.7L Engine

⁂ CAUTION

The left and right fuel rails are replaced as an assembly. Do not attempt to separate rail halves at connector tube. Due to design of tube, it does not use any clamps. Never attempt to install a clamping device of any kind to tube. When removing fuel rail assembly for any reason, be careful not to bend or kink tube.

1. Remove fuel tank filler tube cap.

2. Perform Fuel System Pressure Release Procedure.

3. Remove negative battery cable at battery.

4. Remove flex tube (air cleaner housing to engine).

5. Remove air resonator box at throttle body.

6. Disconnect all spark plug cables from all spark plugs and ignition coils. Do not remove cables from cable routing tray. Note original cable positions while removing.

7. Remove spark plug cable tray from engine by releasing 4 retaining clips. Remove tray and cables from engine as an assembly.

8. Disconnect electrical connectors at all 8 ignition coils.

9. Disconnect fuel line latch clip and fuel line at fuel rail. A special tool will be necessary for fuel line disconnection.

10. Disconnect electrical connectors at all 8 fuel injectors. Refer to graphic. Push red colored slider away from injector. While pushing slider, depress tab and remove connector from injector. The factory fuel injection wiring harness is numerically tagged (INJ 1, INJ 2, etc.) for injector position identification. If harness is not tagged, note wiring location before removal.

11. Disconnect electrical connectors at all throttle body sensors.

12. Remove four fuel rail mounting bolts (2) and hold-down clamps.

13. Gently rock and pull left side of fuel rail until fuel injectors just start to clear machined holes in intake manifold. Gently rock and pull right side of rail until injectors just start to clear intake manifold head holes. Repeat this procedure (left/right) until all injectors have cleared machined holes.

14. Remove fuel rail (with injectors attached) from engine.

15. Disconnect clip(s) that retain fuel injector(s) to fuel rail (2).

To install:

16. Install fuel injector(s) into fuel rail assembly and install retaining clip(s).

17. If same injector(s) is being reinstalled, install new o-ring(s).

18. Clean out fuel injector machined bores in intake manifold.

19. Apply a small amount of engine oil to each fuel injector o-ring. This will help in fuel rail installation.

20. Position fuel rail/fuel injector assembly to machined injector openings in intake manifold.

21. Guide each injector into intake manifold. Be careful not to tear injector o-rings.

22. Push right side of fuel rail down until fuel injectors have bottomed on shoulders. Push left fuel rail down until injectors have bottomed on shoulders.

23. Install 4 fuel rail hold-down clamps and 4 mounting bolts.

24. Position spark plug cable tray and cable assembly to intake manifold. Snap 4 cable tray retaining clips into intake manifold.

25. Install all cables to spark plugs and ignition coils.

26. Connect electrical connector to throttle body.

27. Install electrical connectors to all 8 ignition coils.

28. Connect electrical connector to throttle body.

29. Connect electrical connectors at all fuel injectors. Refer to graphic. Push connector onto injector and then push and lock red colored slider. Verify connector is locked to injector by lightly tugging on connector.

30. Connect fuel line latch clip and fuel line to fuel rail.

31. Install air resonator to throttle body (2 bolts).

32. Install flexible air duct to air box.

33. Connect battery cable to battery.

34. Start engine and check for leaks.

5.9L Engines

1. Before servicing the vehicle, refer to the precautions in the beginning of this section.
2. Relieve fuel system pressure.
3. Remove or disconnect the following:
   - Negative battery cable
   - Air intake tube
   - Throttle body
   - A/C compressor bracket
   - Fuel injector connectors
   - Fuel line
   - Fuel supply manifold with injectors
   - Fuel injectors

To install:

4. Install or connect the following:
   - Fuel injectors with new O-ring seals
   - Fuel supply manifold with injectors. Tighten the bolts to 17 ft. lbs. (23 Nm).
   - Fuel line
   - Fuel injector connectors
   - A/C compressor bracket
   - Throttle body
   - Air intake tube
   - Negative battery cable
5. Start the engine and check for leaks.

DRIVE TRAIN

Transmission Assembly

REMOVAL & INSTALLATION

2002–03

1. Before servicing the vehicle, refer to the precautions in the beginning of this section.
2. Remove or disconnect the following:
   - Negative battery cable
   - Rear driveshaft
   - Crankshaft Position (CKP) sensor
   - Exhaust front pipe
   - Transmission braces, if equipped
   - Starter motor
   - Transmission oil cooler lines
   - Torque converter access cover
   - Torque converter
   - Transmission oil dipstick tube
   - Vehicle Speed (VSS) sensor connector
   - Park/Neutral switch connector
   - Shift cable
   - Throttle valve cable
   - Transmission mount and crossmember. Support the transmission.
   - Front driveshaft and transfer case, if equipped
   - Transmission flange bolts
   - Transmission

To install:

3. Install or connect the following:
   - Transmission. Tighten the flange bolts to 65 ft. lbs. (87 Nm).
   - Front driveshaft and transfer case, if equipped
   - Transmission mount and crossmember
   - Throttle valve cable
   - Shift cable
   - Park/Neutral switch connector
   - VSS sensor connector
   - Transmission oil dipstick tube
   - Torque converter. Tighten the bolts to 23 ft. lbs. (31 Nm) for 10.75 inch converters and to 35 ft. lbs. (47 Nm) for 12.2 inch converters.
   - Torque converter access cover
   - Transmission oil cooler lines
   - Starter motor
   - Transmission braces, if equipped. Tighten the bolts to 30 ft. lbs. (41 Nm).
   - Exhaust front pipe
   - CKP sensor
   - Rear driveshaft
   - Negative battery cable

2004–05

42RLE

1. Disconnect the negative battery cable.
2. Raise and support the vehicle
3. Remove any necessary skid plates.
4. Mark propeller shaft and axle companion flanges for assembly alignment.
5. Remove the rear propeller shaft.
6. Remove the front propeller shaft, if necessary.
7. Disconnect wires from the input and output speed sensors.
8. Disconnect wires from the transmission range sensor.
9. Disconnect wires from the solenoid/pressure switch assembly.
10. Remove the bolts holding the exhaust crossover pipe to the pre-catalytic converter pipe flanges.
11. Remove the bolts holding the exhaust crossover pipe to the catalytic converter flange.
12. Disconnect gearshift cable from transmission manual valve lever.
13. Disengage the shift cable from the cable support bracket.
14. Remove the starter motor.
15. Remove the engine to transmission collar.
16. Rotate crankshaft in clockwise direction until converter bolts are accessible. Then remove bolts one at a time. Rotate crankshaft with socket wrench on dampener bolt.
17. Disconnect the transmission vent hose from the transmission.
18. Remove transfer case.
19. Support rear of engine with safety stand or jack.
20. Raise transmission slightly with service jack to relieve load on crossmember and supports.
21. Remove bolts securing rear support and cushion to transmission and crossmember.
22. Remove bolts attaching crossmember to frame and remove crossmember.
23. Disconnect transmission fluid cooler lines at transmission fittings and clips.
24. Remove all remaining converter housing bolts.
25. Carefully work transmission and torque converter assembly rearward off engine block dowels.
26. Hold torque converter in place during transmission removal.
27. Lower transmission and remove assembly from under the vehicle.
28. To remove torque converter, carefully slide torque converter out of the transmission.

To install:

➡Check torque converter hub and hub drive flats for sharp edges burrs, scratches, or nicks. Polish the hub and flats with 320/400 grit paper and crocus cloth if necessary. The hub must be smooth to avoid damaging pump seal at installation.

29. If a replacement transmission is being installed, transfer any components necessary, such as the manual shift lever and shift cable bracket, from the original transmission onto the replacement transmission.
30. Lubricate oil pump seal lip with transmission fluid.
31. Align converter and oil pump.
32. Carefully insert converter in oil pump. Then rotate converter back and forth until fully seated in pump gears.
33. Check converter seating with steel scale and straightedge. Surface of converter lugs should be at least 13mm to rear of straightedge when converter is fully seated.
34. Temporarily secure converter with C-clamp.
35. Position transmission on jack and secure it with chains.
36. Check condition of converter driveplate. Replace the plate if cracked, distorted or damaged. Also be sure transmission dowel pins are seated in engine block and protrude far enough to hold transmission in alignment.
37. Apply a light coating of high temperature grease to the torque converter hub pocket in the rear pocket of the engine's crankshaft.
38. Raise transmission and align the torque converter with the drive plate and transmission converter housing with the engine block.
39. Move transmission forward. Then raise, lower or tilt transmission to align the converter housing with engine block dowels.
40. Carefully work transmission forward and over engine block dowels until converter hub is seated in crankshaft. Verify that no wires, or the transmission vent hose, have become trapped between the engine block and the transmission.
41. Install two bolts to attach the transmission to the engine.

42. Install remaining torque converter housing to engine bolts. Tighten to 68 Nm (50 ft. lbs.).

43. Install transfer case, if equipped. Tighten transfer case nuts to 35 Nm (26 ft. lbs.).

44. Install rear transmission crossmember. Tighten crossmember to frame bolts to 68 Nm (50 ft. lbs.).

45. Install rear support to transmission. Tighten bolts to 47 Nm (35 ft. lbs.).

46. Lower transmission onto crossmember and install bolts attaching transmission mount to crossmember. Tighten clevis bracket to crossmember bolts to 47 Nm (35 ft. lbs.). Tighten the clevis bracket to rear support bolt to 68 Nm (50 ft. lbs.).

47. Connect gearshift cable to support bracket and transmission manual lever.

48. Connect input and output speed sensor wires and the transmission range sensor.

49. Connect wires to the solenoid/pressure switch assembly.

⁂ CAUTION

It is essential that correct length bolts be used to attach the converter to the driveplate. Bolts that are too long will damage the clutch surface inside the converter.

50. Install torque converter-to-driveplate bolts. Tighten bolts to 88 Nm (65 ft. lbs.).

51. Install starter motor and cooler line bracket.

52. Connect cooler lines to transmission.

53. Install transmission fill tube.

54. Install exhaust components.

55. Align and connect propeller shaft(s).

56. Adjust gearshift cable if necessary.

57. Install any skid plates removed previously.

58. Lower vehicle.

59. Fill transmission with Mopar® ATF +4, Automatic Transmission Fluid or equivalent.

545RFE

1. Disconnect the negative battery cable.

2. Raise and support the vehicle

3. Remove any necessary skid plates.

4. Mark propeller shaft and axle companion flanges for assembly alignment.

5. Remove the rear propeller shaft.

6. Remove the front propeller shaft, if necessary.

7. Remove the engine to transmission collar.

8. Remove the exhaust support bracket from the rear of the transmission.

9. Disconnect and lower or remove any necessary exhaust components.

10. Remove the starter motor.

11. Rotate crankshaft in clockwise direction until converter bolts are accessible. Then remove bolts one at a time. Rotate crankshaft with socket wrench on dampener bolt.

12. Disengage the output speed sensor connector from the output speed sensor.

13. Disengage the input speed sensor connector from the input speed sensor.

14. Disengage the transmission solenoid/TRS assembly connector from the transmission solenoid/TRS assembly.

15. Disengage the line pressure sensor connector from the line pressure sensor.

16. Disconnect gearshift cable from transmission manual valve lever.

17. Remove the gearshift cable from the shift cable support bracket.

18. Disconnect transmission fluid cooler lines at transmission fittings and clips.

19. Support rear of engine with safety stand or jack.

20. Raise transmission slightly with service jack to relieve load on crossmember and supports.

21. Remove bolts securing rear support and cushion to transmission and crossmember.

22. Remove bolts attaching crossmember to frame and remove crossmember.

23. Remove transfer case.

24. Remove all remaining converter housing bolts.

25. Carefully work transmission and torque converter assembly rearward off engine block dowels.

26. Hold torque converter in place during transmission removal.

27. Lower transmission and remove assembly from under the vehicle.

28. To remove torque converter, carefully slide torque converter out of the transmission.

To install:

➡Check torque converter hub and hub drive flats for sharp edges burrs, scratches, or nicks. Polish the hub and flats with 320/400 grit paper and crocus cloth if necessary. Verify that the converter hub o-ring is properly installed and is free of any debris. The hub must be smooth to avoid damaging pump seal at installation.

29. If a replacement transmission is being installed, transfer any components necessary, such as the manual shift lever and shift cable bracket, from the original transmission onto the replacement transmission.

30. Lubricate oil pump seal lip with transmission fluid.

31. Align converter and oil pump.

32. Carefully insert converter in oil pump. Then rotate converter back and forth until fully seated in pump gears.

33. Check converter seating with steel scale and straightedge. Surface of converter lugs should be at least 13mm to rear of straightedge when converter is fully seated.

34. Temporarily secure converter with C-clamp.

35. Position transmission on jack and secure it with chains.

36. Check condition of converter driveplate. Replace the plate if cracked, distorted or damaged. Also be sure transmission dowel pins are seated in engine block and protrude far enough to hold transmission in alignment.

37. Apply a light coating of high temperature grease to the torque converter hub pocket in the rear pocket of the engine's crankshaft.

38. Raise transmission and align the torque converter with the drive plate and transmission converter housing with the engine block.

39. Move transmission forward. Then raise, lower or tilt transmission to align the converter housing with engine block dowels.

40. Carefully work transmission forward and over engine block dowels until converter hub is seated in crankshaft. Verify that no wires, or the transmission vent hose, have become trapped between the engine block and the transmission.

41. Install two bolts to attach the transmission to the engine.

42. Install remaining torque converter housing to engine bolts. Tighten to 68 Nm (50 ft. lbs.).

43. Install transfer case, if equipped. Tighten transfer case nuts to 35 Nm (26 ft. lbs.).

44. Install rear transmission crossmember. Tighten crossmember to frame bolts to 68 Nm (50 ft. lbs.).

45. Install rear support to transmission. Tighten bolts to 47 Nm (35 ft. lbs.).

46. Lower transmission onto crossmember and install bolts attaching transmission mount to crossmember. Tighten clevis bracket to crossmember bolts to 47 Nm (35 ft. lbs.). Tighten the clevis bracket to rear support bolt to 68 Nm (50 ft. lbs.).

47. Remove engine support fixture.

48. Connect gearshift cable to transmission.

49. Connect the wiring harness connector to the solenoid and pressure switch assembly connector. Be sure transmission harnesses are properly routed.

50. Connect the wiring harness connector to the input speed sensor.

51. Connect the wiring harness connector to the output speed sensor.

52. Connect the wiring harness connector to the line pressure sensor.

CAUTION

It is essential that correct length bolts be used to attach the converter to the driveplate. Bolts that are too long will damage the clutch surface inside the converter.

53. Install torque converter-to-driveplate bolts. Tighten bolts to 31 Nm (270 inch lbs.).

54. Install starter motor and cooler line bracket.

55. Connect cooler lines to transmission.

56. Install transmission fill tube.

57. Install exhaust components.

58. Install the structural dust cover onto the transmission and the engine.

59. Align and install the front propeller shaft, if necessary.

60. Align and install the rear propeller shaft.

61. Adjust gearshift cable if necessary.

62. Install any skid plates removed previously.

63. Lower vehicle.

64. Fill transmission with Mopar® ATF +4, Automatic Transmission Fluid.

Transfer Case Assembly

REMOVAL & INSTALLATION

2002–03

1. Before servicing the vehicle, refer to the precautions in the beginning of this section.
2. Shift the transfer case into **N**.
3. Remove or disconnect the following:
   - Front and rear driveshafts
   - Transmission mount and crossmember. Support the transmission.
   - Vehicle Speed (VSS) sensor connector
   - Shift linkage
   - Vent hose
   - Vacuum hose
   - Indicator switch connector
   - Transfer case attaching nuts
   - Transfer case

To install:

4. Install or connect the following:
   - Transfer case. Tighten the nuts to 26 ft. lbs. (35 Nm).
   - Indicator switch connector
   - Vacuum hose
   - Vent hose
   - Shift linkage
   - VSS sensor connector
   - Transmission mount and crossmember
   - Front and rear driveshafts

2004–05

NV144

1. Shift transfer case into AWD.
2. Raise vehicle.
3. Drain transfer case lubricant.
4. Mark front and rear propeller shafts for alignment reference.
5. Support transmission with jack stand.
6. Remove the transfer case skid plate, if equipped.
7. Disconnect front and rear propeller shafts at transfer case.
8. Disconnect transfer case shift motor and mode sensor wire connectors.
9. Disconnect transfer case vent hose.
10. Support transfer case with transmission jack.
11. Secure transfer case to jack with chains.
12. Remove nuts attaching transfer case to transmission.
13. Pull transfer case and jack rearward to disengage transfer case.
14. Remove transfer case from under vehicle.

To install:

15. Mount transfer case on a transmission jack.
16. Secure transfer case to jack with chains.
17. Position transfer case under vehicle.
18. Align transfer case and transmission shafts and install transfer case onto the transmission.
19. Install and tighten transfer case attaching nuts to 27–34 Nm (20–25 ft. lbs.) torque.
20. Connect the vent hose.
21. Connect the shift motor and mode sensor assembly wiring connector. Secure wire harness to clips on transfer case.
22. Align and connect the propeller shafts.
23. Fill transfer case with correct fluid.
24. Install skid plate, if equipped.
25. Remove transmission jack and support stand.
26. Lower vehicle and verify transfer case shift operation.

NV244

1. Shift transfer case into AWD.
2. Raise vehicle.
3. Drain transfer case lubricant.
4. Mark front and rear propeller shafts for alignment reference.

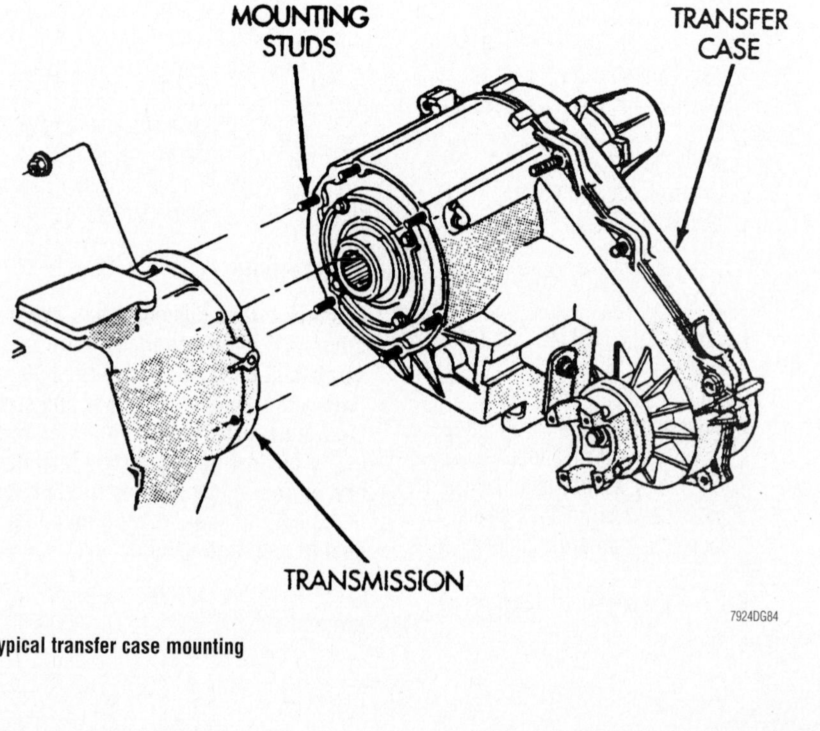

Typical transfer case mounting

5. Support transmission with jack stand.
6. Remove the transfer case skid plate, if equipped.
7. Disconnect front and rear propeller shafts at transfer case.
8. Disconnect transfer case shift motor and mode sensor wire connectors.
9. Disconnect transfer case vent hose.
10. Support transfer case with transmission jack.
11. Secure transfer case to jack with chains.
12. Remove nuts attaching transfer case to transmission.
13. Pull transfer case and jack rearward to disengage transfer case.
14. Remove transfer case from under vehicle.

To install:

15. Mount transfer case on a transmission jack.
16. Secure transfer case to jack with chains.
17. Position transfer case under vehicle.
18. Align transfer case and transmission shafts and install transfer case onto the transmission.
19. Install and tighten transfer case attaching nuts to 27–34 Nm (20–25 ft. lbs.) torque.
20. Connect the vent hose.
21. Connect the shift motor and mode sensor assembly wiring connector. Secure wire harness to clips on transfer case.
22. Align and connect the propeller shafts.
23. Fill transfer case with correct fluid.
24. Install skid plate, if equipped.
25. Remove transmission jack and support stand.
26. Lower vehicle and verify transfer case shift operation.

Halfshaft

REMOVAL & INSTALLATION

2002–03

1. Before servicing the vehicle, refer to the precautions in the beginning of this section.
2. Remove or disconnect the following:
   - Skid plate, if equipped
   - Front wheel
   - Split pin
   - Nut lock
   - Spring washer
   - Hub nut
   - Brake caliper and rotor
   - Wheel speed sensor, if equipped
   - Wheel bearing and hub assembly
3. Pry the inner tripod joint out of the differential and remove the axle halfshaft.

To install:

4. Install the axle halfshaft so that the snapring is felt to seat in the joint housing groove.
5. Install or connect the following:
   - Wheel bearing and hub assembly
   - Wheel speed sensor, if equipped
   - Brake caliper and rotor
   - Hub nut. Tighten the nut to 180 ft. lbs. (244 Nm).
   - Spring washer
   - Nut lock
   - Split pin
   - Front wheel
   - Skid plate, if equipped

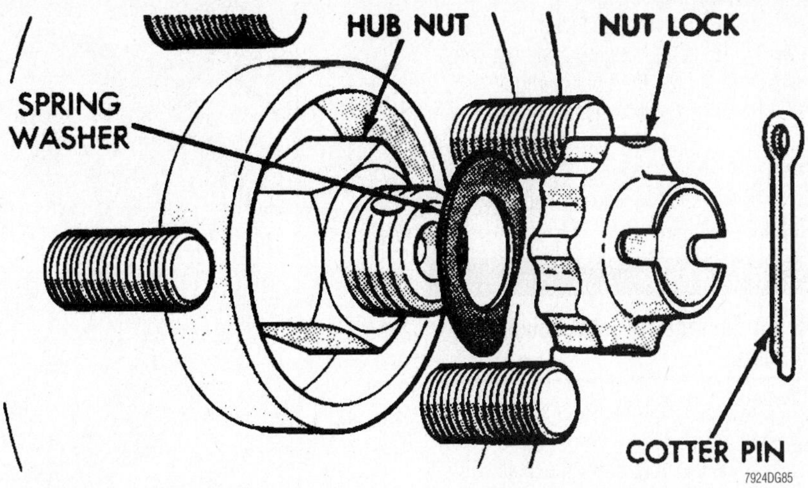

To separate the halfshaft from the hub, remove the cotter pin, nut lock and spring washer from the axle shaft

2004–05

1. With vehicle in neutral, position vehicle on hoist.
2. Remove skid plate, if equipped.
3. Remove hub nut from the halfshaft.
4. Remove brake caliper and rotor.
5. Remove wheel speed sensor if equipped.
6. Remove hub bearing bolts from the knuckle.
7. Remove hub bearing and brake shield from knuckle.
8. Support halfshaft at the CV-joint housings.
9. Position two pry bars behind the inner C/V housing and disengage the CV-joint from the axle.
10. Remove halfshaft through the knuckle.

To install:

11. Apply a light coating of wheel bearing grease on the axle splines.
12. Insert halfshaft through the steering knuckle and onto the axle. Verify shaft snapring engages with the groove on the inside of the joint housing.
13. Clean hub bearing bore and hub bearing mating surface. Lightly coat mating surfaces with grease.
14. Install hub bearing onto the axle halfshaft and into steering knuckle. Tighten hub bearing bolts to 163 Nm (120 ft. lbs.).
15. Install wheel speed sensor, if equipped.
16. Install brake rotor and caliper adapter with caliper.
17. Install halfshaft nut. Apply brakes and tighten shaft nut to 251 Nm (185 ft. lbs.).
18. Install skid plate, if equipped.

CV-Joints

OVERHAUL

Outer CV-Joint

2002–04

1. Before servicing the vehicle, refer to the precautions in the beginning of this section.
2. Remove or disconnect the following:
   - Axle halfshaft from the vehicle
   - CV-joint boot and clamps
   - Snapring
   - CV-joint

To install:

3. Install or connect the following:
   - CV-joint

- Snapring
- CV-joint boot and clamps

4. Fill the joint housing and boot with grease and tighten the boot clamps.
5. Install the axle halfshaft.

2005

1. Place shaft in vise with soft jaws and support CV-joint.

CAUTION

Do not damage CV-joint housing or halfshaft.

2. Remove clamps with a cut-off wheel or grinder.
3. Slide the boot down the shaft.
4. Remove lubricant to expose the CV-joint snapring.
5. Spread snapring and slide the joint off the shaft.
6. Slide boot off the shaft and discard old boot.
7. Mark alignment marks on the inner race/hub, bearing cage and housing with dabs of paint.
8. Clamp CV-joint in a vertical position in a soft jawed vise.
9. Press down one side of the bearing cage to gain access to the ball at the opposite side.

➡If joint is tight, use a hammer and brass drift to loosen the bearing hub. Do not contact the bearing cage with the drift.

10. Remove ball from the bearing cage.
11. Repeat step above until all six balls are removed from the bearing cage.
12. Lift cage and inner race upward and out from the housing.
13. Turn inner race 90° in the cage and rotate the inner race/hub out of the cage.

To Install:

14. Apply a light coat of grease to the CV-joint components before assembling them.
15. Align inner race, cage and housing according to the alignment reference marks.
16. Insert inner race into the cage and rotate race into the cage.
17. Rotate inner race/hub in the cage.
18. Insert cage into the housing.
19. Rotate cage 90° into the housing.
20. Apply lubricant included with replacement boot/joint to the ball races. Spread lubricant equally among all the races.
21. Tilt inner race/hub and cage and install the balls.
22. Place new clamps onto new boot and

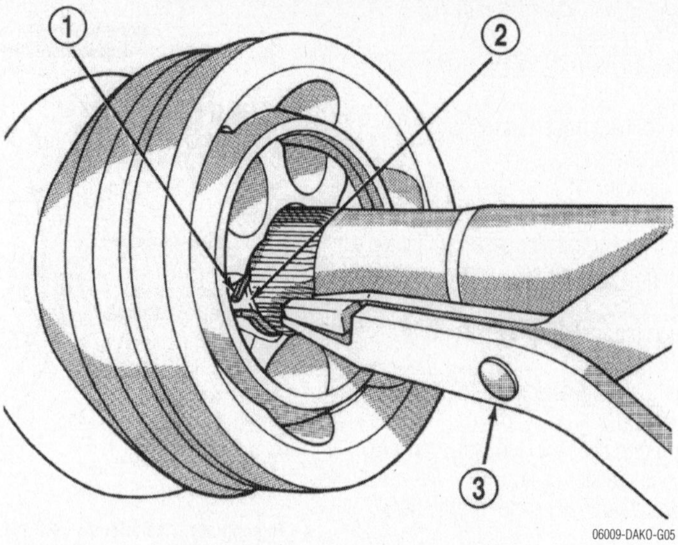

Removing the snapring (1) from the shaft (2) with snapring pliers (3)

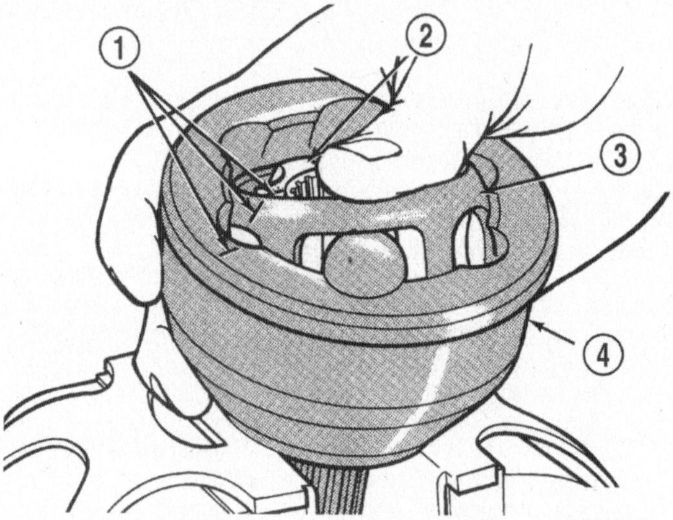

Make alignment marks (1) on the inner race/hub (2) and cage (3)

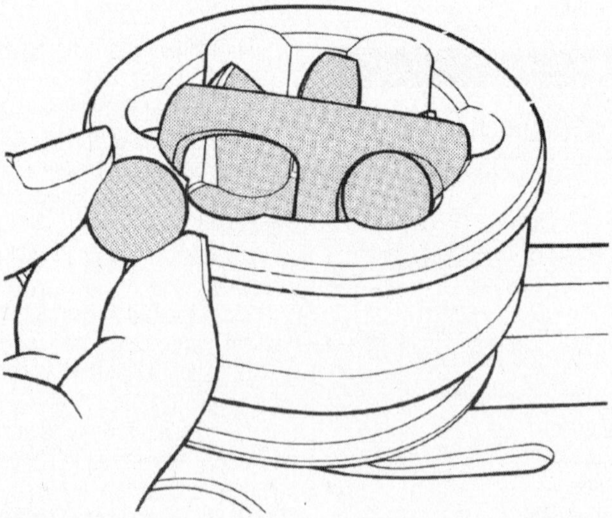

Removing the balls from the bearing cage

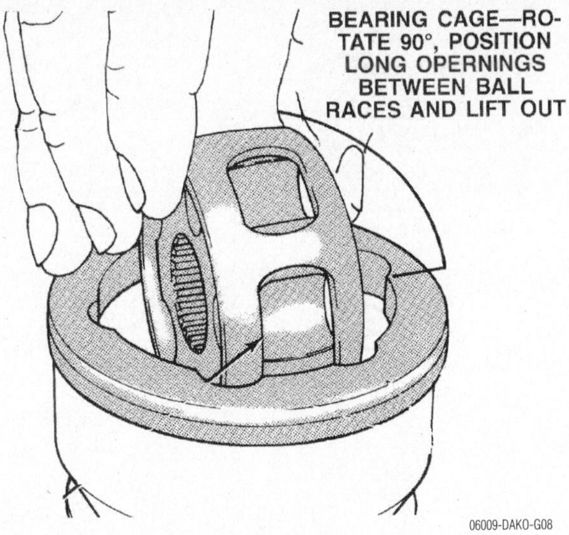

Removing the cage and inner race from the housing

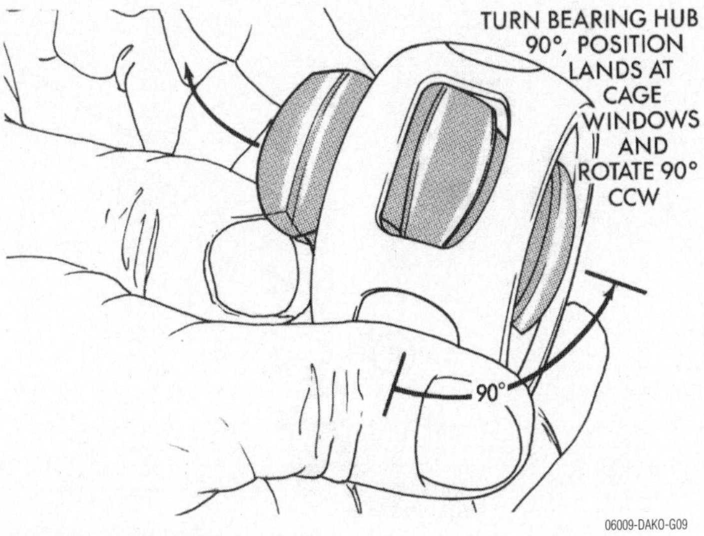

Removing the inner race/hub from the cage

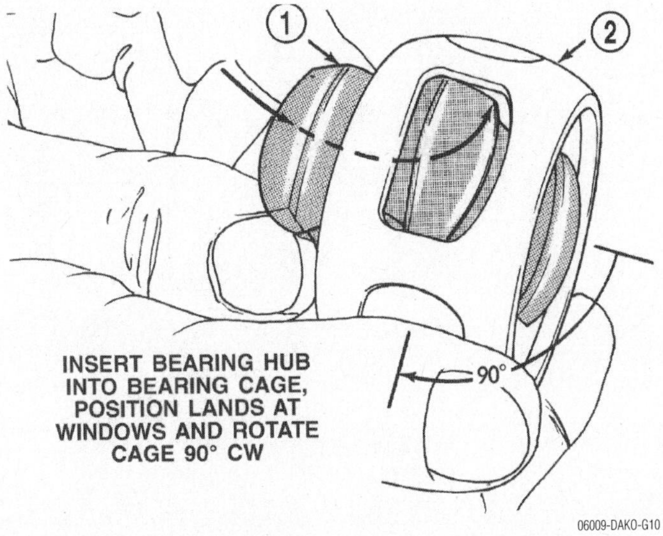

Assembling the inner race cage and housing

slide boot onto the shaft to its original position.

23. Apply the rest of lubricant to the CV-joint and boot.
24. Push the joint onto the shaft until the snapring seats in the groove. Pull on the joint to verify the span ring has engaged.
25. Position boot on the joint in its original position. Ensure boot is not twisted and remove any excess air.
26. Secure both boot clamps with Clamp Installer C-4975A, or equivalent. Place tool on clamp bridge and tighten tool until the jaws of the tool are closed.

Inner Tripod Joint

2002–03

1. Before servicing the vehicle, refer to the precautions in the beginning of this section.
2. Remove or disconnect the following:
    - Axle halfshaft from the vehicle
    - Inner tripod joint boot clamps
    - Tripod joint housing
    - Snapring
    - Circlip
    - Tripod joint

To install:

➡Use new snaprings, clips, and boot clamps for assembly.

3. Install or connect the following:
    - Tripod joint
    - Circlip
    - Snapring
    - Tripod joint housing
4. Fill the tripod joint housing and boot with grease and tighten the boot clamps.
5. Install the axle halfshaft.

2004–05

1. Clamp the shaft in a vise with soft jaws and support the CV joint.
2. Remove the clamps with a cut-off wheel or grinder.

⁂ WARNING

Do not damage the CV housing or half shaft with the cut-off wheel or grinder.

3. Remove the housing from the half shaft and slide the boot down shaft.
4. Remove the housing bushing from the housing.
5. Remove the tripod snapring.
6. Remove the tripod and boot from the halfshaft.
7. Clean and inspect the CV components for excessive wear and damage. Replace the tripod as a unit only if necessary.

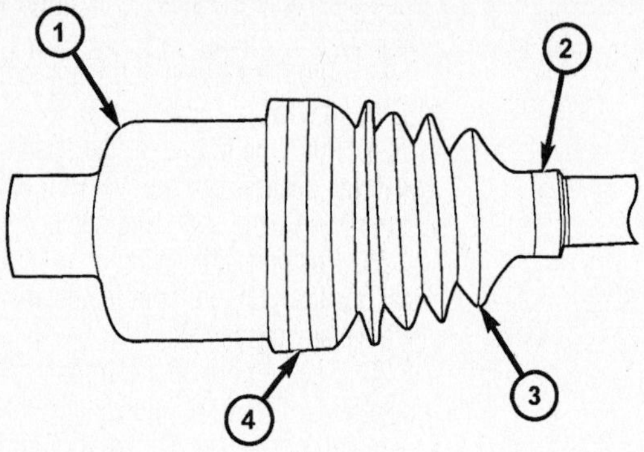

1 - C/V HOUSING
2 - CLAMP
3 - BOOT
4 - CLAMP

67189-DAKO-G33

Boot clamp location

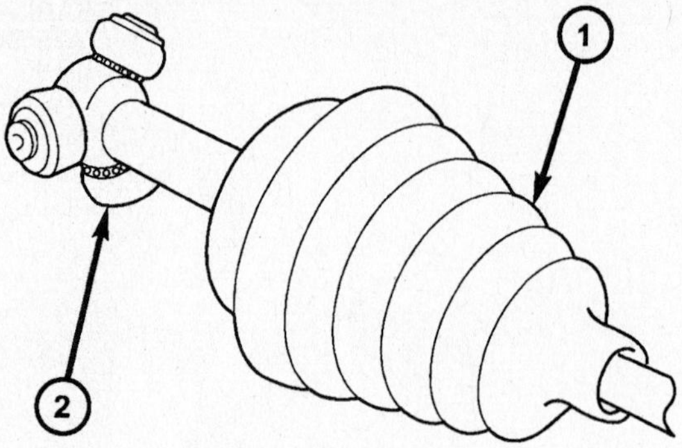

1 - BOOT
2 - TRIPOD

67189-DAKO-G36

Tripod joint

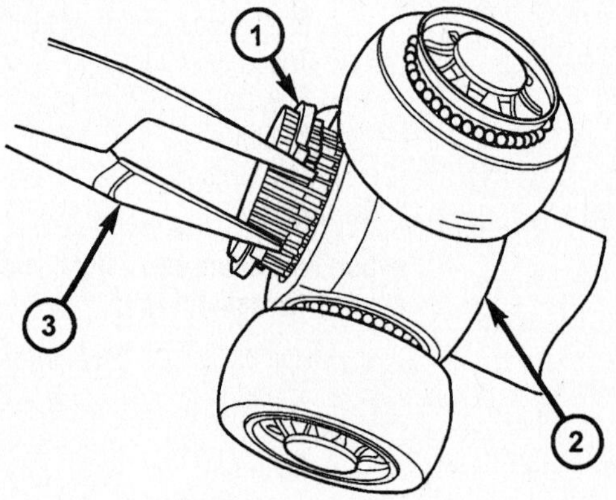

1 - SNAP RING
2 - TRIPOD
3 - PLIERS

67189-DAKO-G34

Removing the snapring

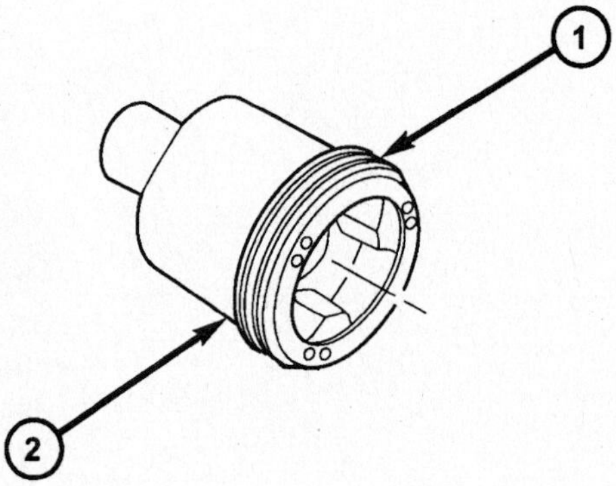

1 - HOUSING
2 - BUSHING

67189-DAKO-G37

Housing bushing

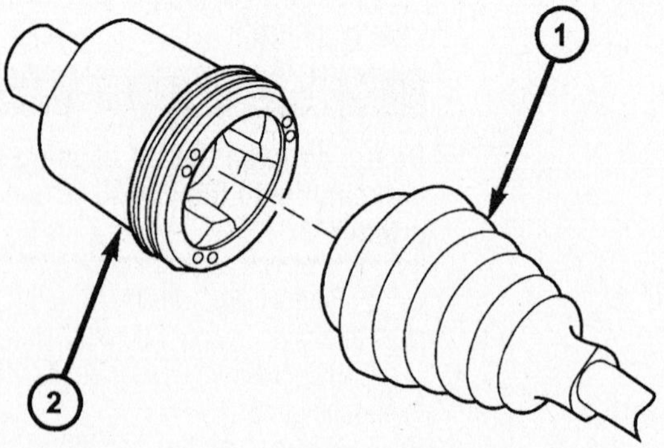

1 - BOOT
2 - HOUSING

67189-DAKO-G35

CV joint housing

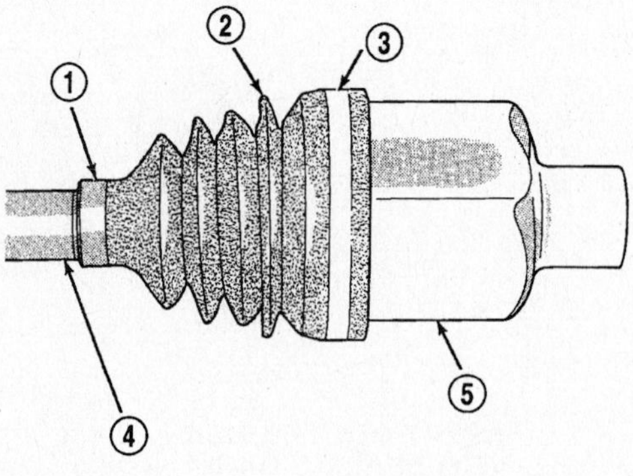

1 - CLAMP
2 - BOOT
3 - CLAMP
4 - SHAFT
5 - HOUSING

67189-DAKO-G38

Inner CV joint boot

To install:

8. Slide a new boot down the halfshaft.
9. Install the tripod and tripod snapring on the halfshaft.
10. Pack the grease supplied with the joint/boot into the housing and boot.
11. Coat the tripod with the supplied grease.
12. Install new bushing onto the housing.
13. Insert the tripod and shaft in the housing.
14. Position the boot on the joint in its original position.

➡Verify the boot is not twisted and remove any excess air.

15. Secure both boot clamps with Clamp Installer C-4975A, or equivalent. Place the tool on the clamp bridge and tighten the tool until the jaws of the tool are closed.

Front Axle Shaft, Bearing and Seal

REMOVAL & INSTALLATION

2002–03

AXLE SHAFT

1. Before servicing the vehicle, refer to the precautions in the beginning of this section.
2. Remove or disconnect the following:
   - Front wheel
   - Brake caliper and rotor
   - Wheel speed sensor, if equipped
   - Axle hub nut
   - Wheel bearing and hub assembly
   - Axle shaft

To install:

3. Install or connect the following:
   - Axle shaft
   - Wheel bearing and hub assembly
   - Axle hub nut. Tighten the nut to 175 ft. lbs. (237 Nm).
   - Wheel speed sensor, if equipped
   - Brake caliper and rotor
   - Front wheel

SEAL

1. Before servicing the vehicle, refer to the precautions in the beginning of this section.
2. Remove or disconnect the following:
   - Front axle shafts
   - Differential cover
   - Differential and ring gear assembly
   - Axle seals

To install:

3. Press the axle seals into the differential housing with Turnbuckle 6797 and Disc set 8110.
4. Install or connect the following:
   - Differential and ring gear assembly. Tighten the bearing cap bolts to 45 ft. lbs. (61 Nm).
   - Differential cover. Tighten the bolts to 30 ft. lbs. (41 Nm).
   - Front axle shafts
5. Fill the axle assembly with gear oil and check for leaks.

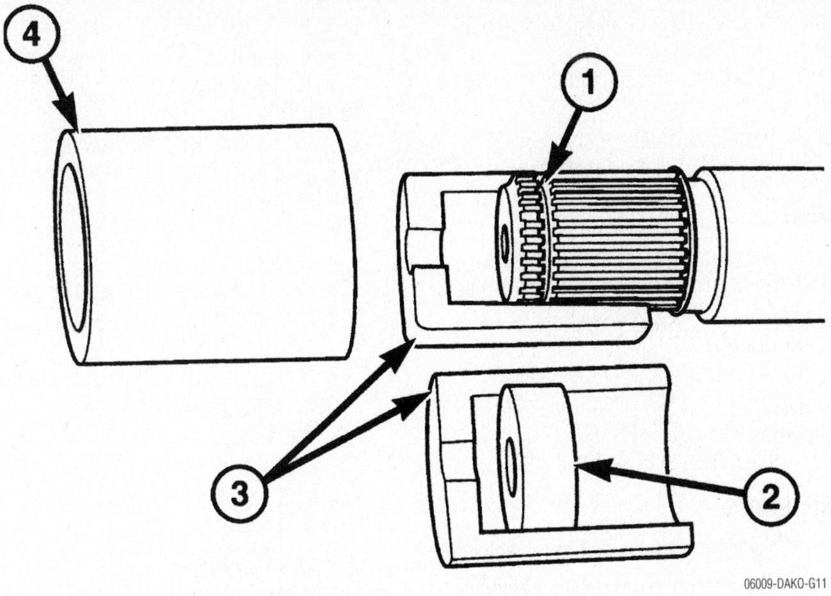

Assembling the axle shaft removal tool 8420A

2004–05

AXLE SHAFT

1. Remove half shaft from vehicle.
2. Clean axle seal area.
3. Remove snap ring from the axle shaft.
4. Remove axle shaft with Remover 8420A and slide hammer.

To install:

5. Lubricate bearing bore and seal lip with gear lubricant.

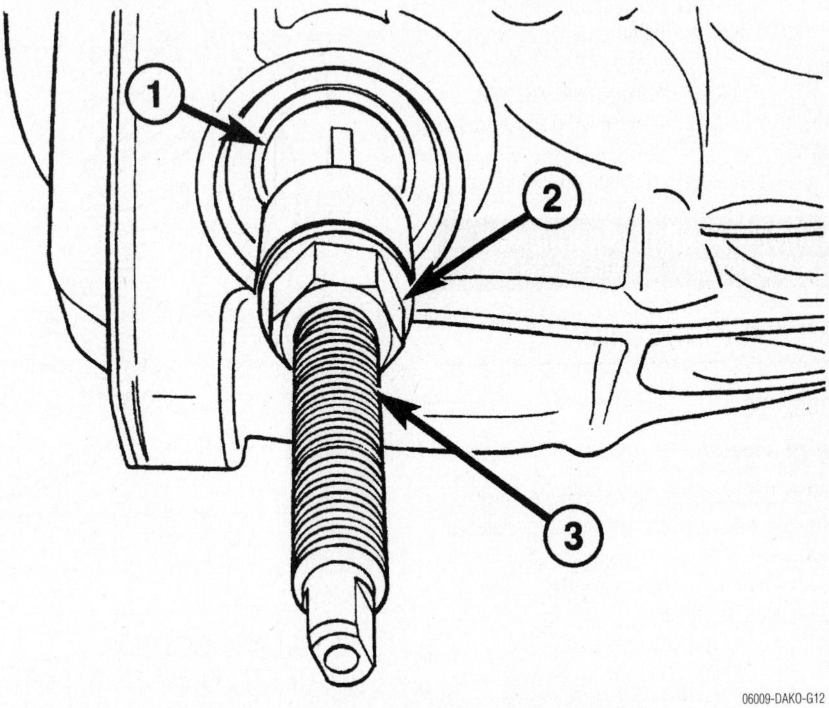

Removing the axle shaft bearing (1) using remover C-4660-A (3) and nut (2)

6. Install axle shaft and engage shaft into side gear. Push firmly on axle shaft to engage snap-ring.
7. Check the differential fluid level and add fluid if necessary.
8. Install half shaft.
9. Install skid plate, if necessary.

SEAL

1. Remove half shaft and axle shaft.
2. Remove axle shaft seal with a small pry bar.

To install:

3. Wipe axle shaft tube bore clean.
4. Install new axle shaft seal with Installer 8402 and Handle C-4171.
5. Install axle shaft and half shaft.

BEARINGS

1. Remove the half shaft from vehicle.
2. Remove the skid plate, if equipped.
3. Clean the axle seal area.
4. Remove axle shaft O-ring.
5. Remove the axle shaft.
6. Remove the axle shaft seal.
7. Install the axle shaft bearing remover C-4660-A, or equivalent, in the bearing. Then tighten the nut to spread the remover in the bearing.
8. Install the bearing remove cup, bearing and nut. Then tighten the nut to draw the bearing out.
9. Inspect the axle shaft tube bore for roughness and burrs. Remove as necessary.

To install:

10. Wipe the axle shaft tube bore clean.
11. Install the axle shaft bearing with installer 5063 and handle C-4171, or equivalent.
12. Install a new axle shaft seal with installer 8402 and handle C-4171, or equivalent.
13. Install the axle shaft and half shaft.

Rear Axle Shaft, Bearing and Seal

REMOVAL & INSTALLATION

2002–03

C-CLIP TYPE

1. Before servicing the vehicle, refer to the precautions in the beginning of this section.
2. Remove or disconnect the following:
   - Rear wheel
   - Brake drum
   - Differential cover
   - Differential gear shaft retainer
   - Differential gear shaft
   - C-clip
   - Axle shaft
   - Axle seal
   - Axle bearing

To install:

3. Install or connect the following:
   - Axle bearing
   - Axle seal
   - Axle shaft
   - C-clip
   - Differential gear shaft. Use Loctite® and tighten the retainer to 14 ft. lbs. (19 Nm).
   - Differential cover. Tighten the bolts to 30 ft. lbs. (41 Nm).
   - Brake drum
   - Rear wheel
4. Fill the axle assembly with gear oil and check for leaks.

NON C-CLIP TYPE

1. Before servicing the vehicle, refer to the precautions in the beginning of this section.
2. Remove or disconnect the following:
   - Rear wheel
   - Brake caliper and rotor, if equipped
   - Brake drum, if equipped
   - Axle retainer nuts
   - Axle shaft, seal and bearing assembly
3. Split the bearing retainer with a chisel and remove the retainer ring.
4. Press the bearing off the axle shaft.
5. Remove the axle seal and retaining plate.

To install:

6. Install the retaining plate and axle seal onto the axle shaft.
7. Pack the wheel bearing with axle grease and press the bearing on to the axle shaft.
8. Press the retaining ring onto the axle shaft.
9. Install or connect the following:
   - Axle shaft, seal and bearing assembly. Tighten the nuts to 45 ft. lbs. (61 Nm).
   - Brake caliper and rotor, if equipped
   - Brake drum, if equipped
   - Rear wheel
10. Fill the axle assembly with gear oil and check for leaks.

2004–05 8¼ inch axle

AXLE SHAFT

1. With vehicle in neutral, position it on a hoist.
2. Remove brake caliper adapter with caliper and remove rotor.

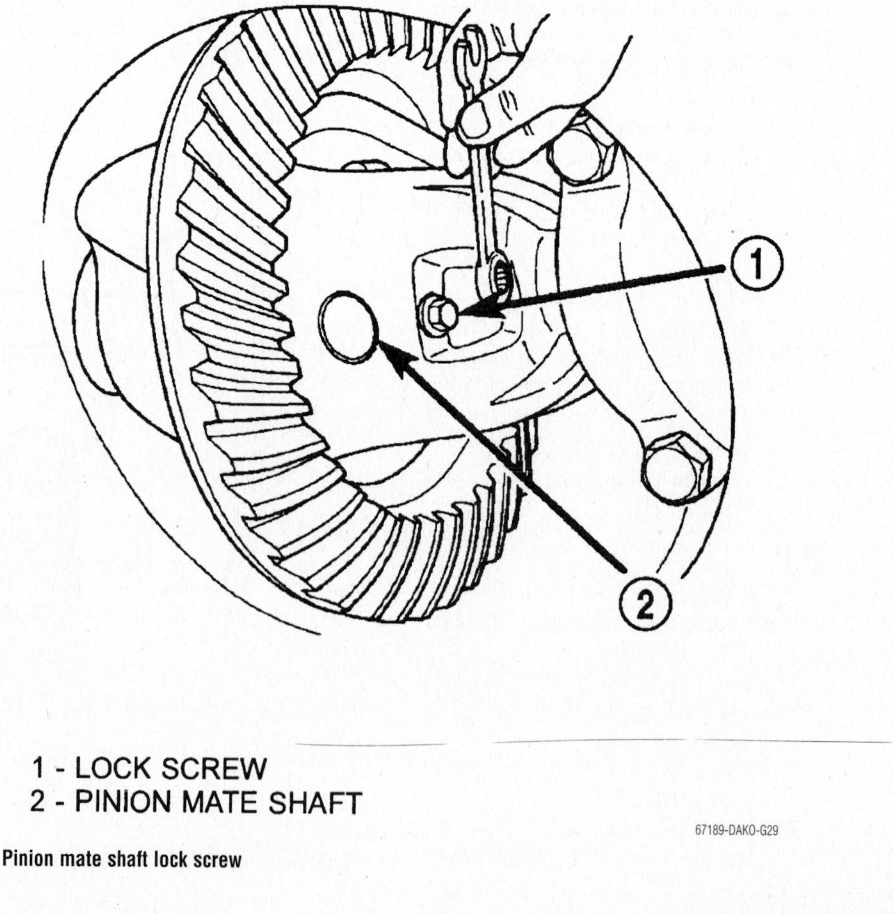

Pinion mate shaft lock screw

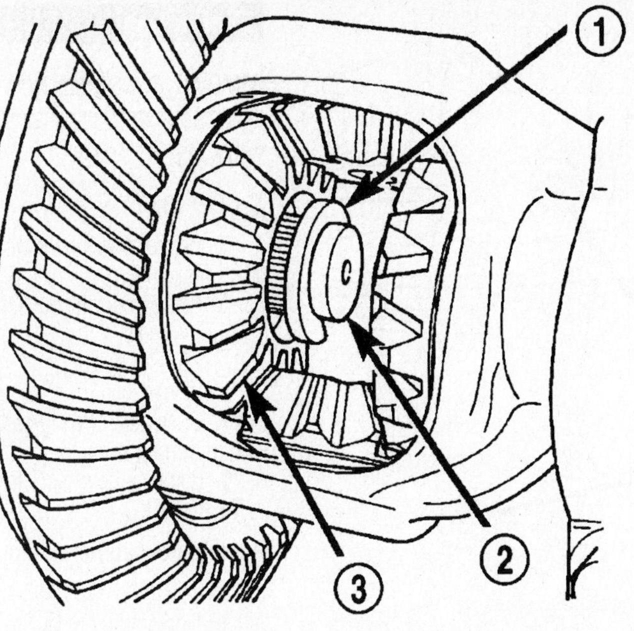

Axle shaft C-lock

3. Remove differential housing cover and drain lubricant.
4. Rotate differential case, to access pinion mate shaft lock screw. Remove screw and pinion mate shaft from differential case.
5. Push axle shaft inward and remove axle shaft C-lock.
6. Remove axle shaft.

To install:

7. Lubricate bearing bore and seal lip with gear lubricant.
8. Insert axle shaft through seal and engage into side gear splines.
9. Insert C-lock in end of axle shaft then push axle shaft outward to seat C-lock in side gear.
10. Insert pinion shaft into differential case and through thrust washers and differential pinions.
11. Align hole in shaft with hole in differential case and install lock screw with Loctite® on the threads. Tighten lock screw to 25 Nm (220 inch lbs.).
12. Install differential cover and fill with gear lubricant.
13. Install brake rotor and caliper adapter with caliper.

BEARING AND SEAL

1. Remove axle shaft.
2. Remove axle seal with pry bar.

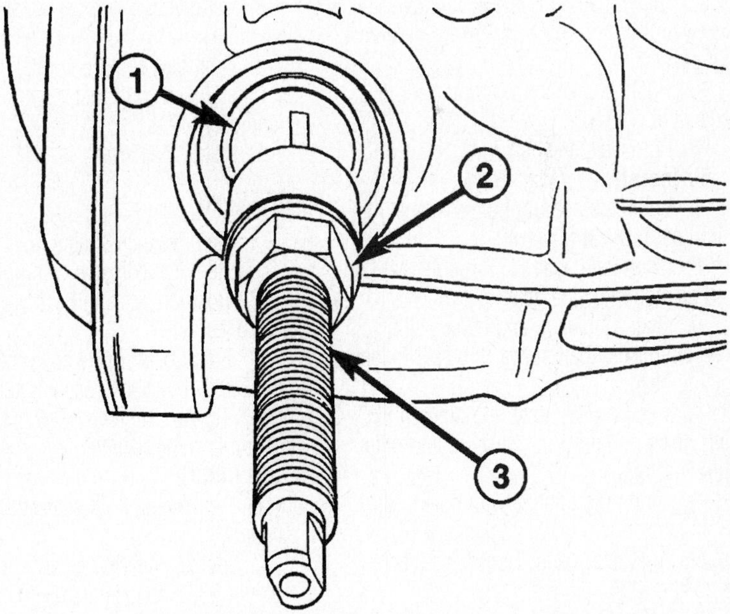

Bearing remover

3. Position bearing Receiver 9338 on axle tube.
4. Insert bearing Remover 6310 with Foot 6310-9 through receiver and bearing.
5. Tighten Remove 6310 nut to pull bearing into the receiver.

To install:

6. Remove any old sealer/burrs from axle tube.
7. Install axle shaft bearing with Installer 9337 and Handle. Drive bearing in until tool contacts the axle tube.

➡Bearing is installed with the bearing part number against the installer.

8. Coat new axle seal lip with axle lubricant and install with Installer 9337 and Handle C-4171.
9. Install axle shaft.

2004–05 9¼ inch axle

AXLE SHAFT

1. With vehicle in neutral, position it on a hoist.
2. Remove brake caliper adapter with caliper and remove rotor.
3. Remove differential housing cover and drain lubricant.

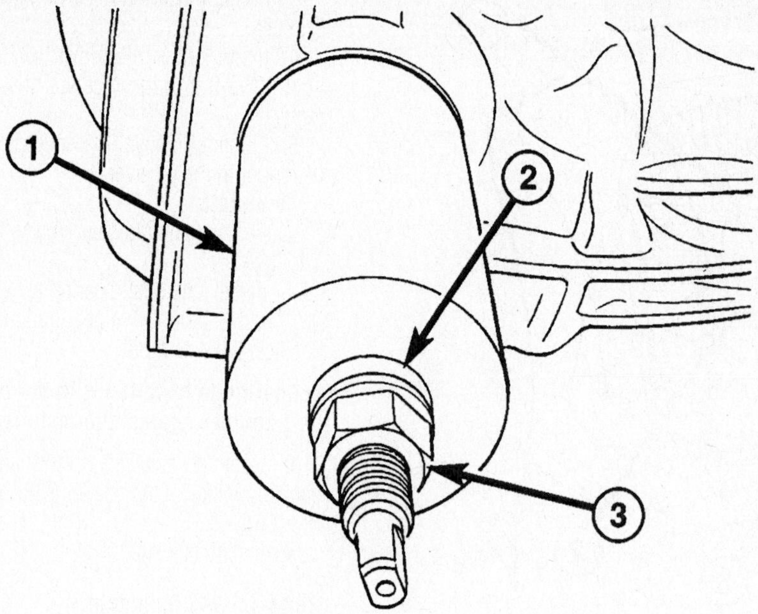

1 - REMOVER CUP
2 - BEARING
3 - NUT

Bearing cup remover

4. Rotate differential case, to access pinion mate shaft lock screw. Remove screw and pinion mate shaft from differential case.
5. Push axle shaft inward and remove axle shaft C-lock.
6. Remove axle shaft.

To install:

7. Lubricate bearing bore and seal lip with gear lubricant.
8. Insert axle shaft (1) through seal (2) and engage into side gear splines.
9. Insert C-lock in end of axle shaft then push axle shaft outward to seat C-lock in side gear.
10. Insert pinion shaft into differential case and through thrust washers and differential pinions.
11. Align hole in shaft with hole in differential case and install lock screw (4) with Loctite® on the threads. Tighten lock screw to 25 Nm (220 inch lbs.).
12. Install differential cover and fill with gear lubricant.
13. Install brake rotor and caliper adapter with caliper.

SEAL

1. Remove axle shaft.
2. Remove axle shaft seal from end of the axle tube with a pry bar.

To install:

3. Remove any old sealer/burrs from axle tube.
4. Coat new seal lip with axle lubricant and install seal with Installer 9337 and Handle C-4171.
5. Install axle shaft.

BEARING

1. Remove axle shaft.
2. Remove axle seal with pry bar.
3. Position bearing Receiver 9338 on axle tube.
4. Insert bearing Remover 6310 with Foot 6310-9 through receiver and bearing.
5. Tighten Remove 6310 nut to pull bearing into the receiver.

To install:

6. Remove any old sealer/burrs from axle tube.
7. Install axle shaft bearing with Installer 9337 and Handle C-417
8. Drive bearing in until tool contacts the axle tube.

➡Bearing is installed with the bearing part number against the installer.

9. Coat new axle seal lip with axle lubricant and install with Installer 9337 and Handle C-4171.
10. Install axle shaft.

Pinion Seal

REMOVAL & INSTALLATION

2002–03

C-CLIP TYPE

1. Before servicing the vehicle, refer to the precautions in the beginning of this section.
2. Remove or disconnect the following:
   - Wheels
   - Brake drums
   - Driveshaft
3. Check the bearing preload with an inch lb. torque wrench.
4. Remove the pinion flange and seal.

To install:

➡Use a new pinion nut for assembly.

5. Install the new pinion seal and flange. Tighten the nut to 210 ft. lbs. (285 Nm).
6. Check the bearing preload. The bearing preload should be equal to the reading taken earlier, plus 5 inch lbs.
7. If the preload torque is low, tighten the pinion nut in 5 inch lb. increments until the torque value is reached. Do not exceed 350 ft. lbs. (474 Nm) pinion nut torque.
8. If the pinion bearing preload torque cannot be attained at maximum pinion nut torque, replace the collapsible spacer.
9. Install or connect the following:
   - Driveshaft
   - Brake drums
   - Wheels
10. Fill the axle assembly with gear oil and check for leaks.

Non C-Clip Type

FRONT

1. Before servicing the vehicle, refer to the precautions in the beginning of this section.
2. Remove or disconnect the following:
   - Wheels
   - Brake rotors
   - Driveshaft
3. Check the bearing preload with an inch lb. torque wrench.
4. Remove the pinion flange and seal.

To install:

➡Use a new pinion nut for assembly.

5. Install the new pinion seal and flange. Tighten the nut to 160 ft. lbs. (217 Nm).
6. Check the bearing preload. The bearing preload should be equal to the reading taken earlier, plus 5 inch lbs.

7. If the preload torque is low, tighten the pinion nut in 5 inch lb. increments until the torque value is reached. Do not exceed 260 ft. lbs. (353 Nm) pinion nut torque.

8. If the pinion bearing preload torque can not be attained at maximum pinion nut torque, replace the collapsible spacer.

9. Install or connect the following:
- Driveshaft
- Brake rotors
- Wheels

10. Fill the axle assembly with gear oil and check for leaks.

NON C-CLIP TYPE—REAR

1. Before servicing the vehicle, refer to the precautions in the beginning of this section.

2. Remove or disconnect the following:
- Wheels
- Brake rotors or drums
- Driveshaft

3. Check the bearing preload with an inch lb. torque wrench.

4. Remove the pinion flange and seal.

To install:

➡Use a new pinion nut for assembly.

5. Install the new pinion seal and flange. Tighten the nut to 160 ft. lbs. (217 Nm).

6. Check the bearing preload. The bearing preload should be equal to the reading taken earlier, plus 5 inch lbs.

7. If the preload torque is low, tighten the pinion nut in 5 inch lb. increments until the torque value is reached. Do not exceed 260 ft. lbs. (353 Nm) pinion nut torque.

8. If the pinion bearing preload torque can not be attained at maximum pinion nut torque, remove one or more pinion preload shims.

9. Install or connect the following:
- Driveshaft
- Brake rotors or drums
- Wheels

10. Fill the axle assembly with gear oil and check for leaks.

2004–05 Front Axle

1. Remove both half shafts.

2. Mark propeller shaft and pinion flange for installation reference.

3. Remove front propeller shaft.

4. Rotate pinion gear three to four times, to verify pinion rotates smoothly.

5. Record pinion flange rotating torque with an inch pound torque wrench for installation reference.

6. Hold flange with Holder 6719 and four bolts and washers.

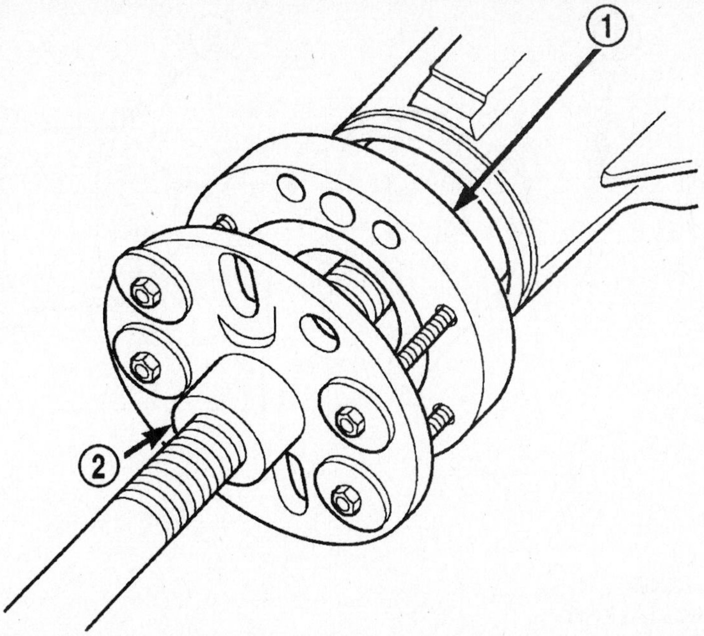

Removing the companion flange

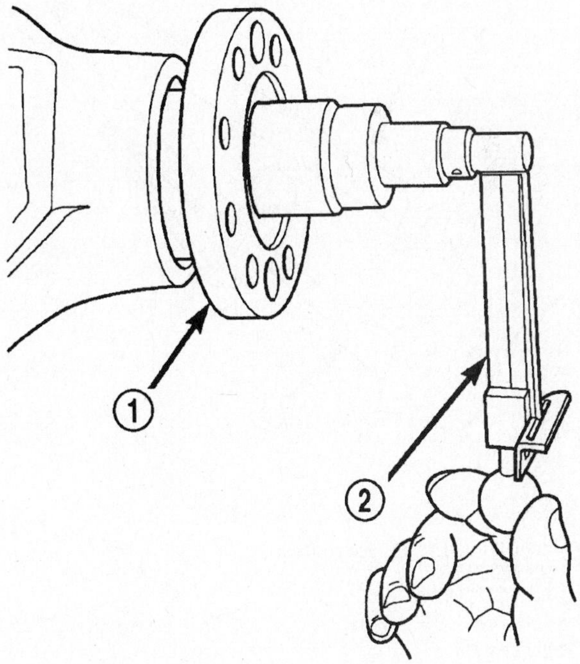

Measuring pinion rotating torque

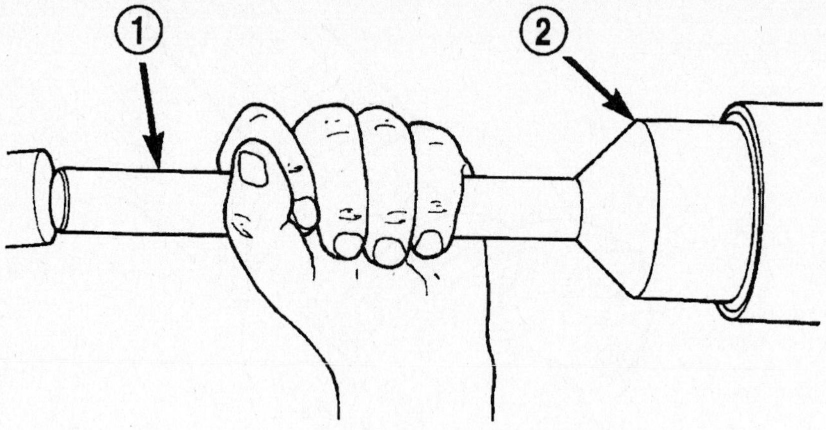

Pinion seal installer

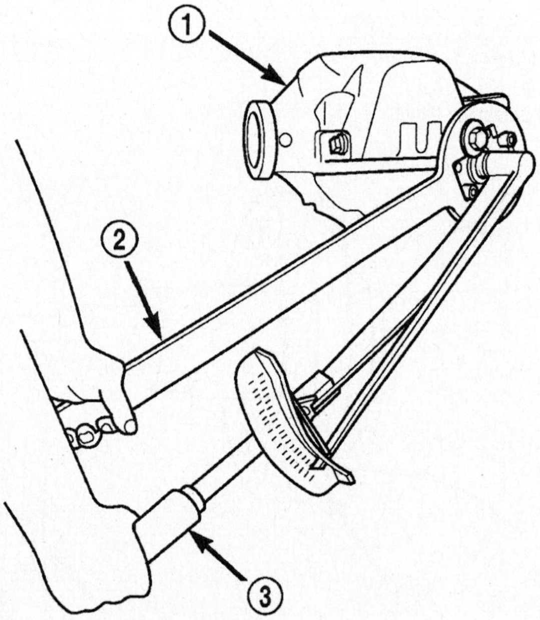

Tightening the pinion nut

7. Remove pinion nut.
8. Remove flange with Remover C-452.
9. Remove pinion seal with a pry tool.

To install:

10. Apply a light coating of gear lubricant on the lip of pinion seal.
11. Install seal with Installer C-3972-A and Handle C-4171.
12. Install pinion flange onto the pinion with Installer C-3718 and holder.
13. Hold pinion flange with Holder 6719A.
14. Install new pinion nut and tighten nut until there is zero bearing end-play.

**** CAUTION**

Do not exceed the minimum tightening torque when installing the companion flange at this point. Damage to the collapsible spacer or bearings may result.

15. Tighten pinion nut to 271 Nm (200 ft. lbs.).

**** CAUTION**

Never loosen pinion nut to decrease pinion bearing rotating torque and never exceed specified preload torque. If preload torque or rotating torque is exceeded a new collapsible spacer must be installed.

16. Record pinion flange rotating torque, with a torque wrench. Rotating torque should be equal to the reading recorded during removal plus an additional 0.56 Nm (5 inch lbs.).
17. If rotating torque is low, tighten pinion nut in 6.8 Nm (5 ft. lbs.) increments until rotating torque is achieved.

**** CAUTION**

If maximum tightening torque is reached prior to reaching the required rotating torque, the collapsible spacer may have been damaged. Replace the collapsible spacer.

18. Install propeller shaft with reference marks aligned.
19. Install half shafts.

2004–05 Rear Axle

8 ¼ INCH AXLE

1. With vehicle in neutral, position vehicle on hoist.
2. Mark a reference line across the axle flange and propeller shaft flange.
3. Remove propeller shaft.
4. Remove brake calipers and rotors to prevent any drag.
5. Rotate flange three or four times and verify flange rotates smoothly.
6. Measure torque to rotating pinion flange with an inch pound torque wrench. Record reading for installation reference.
7. Install bolts into two of the threaded holes in the flange 180° apart.
8. Position Holder 6719 against the flange and install a bolt and washer into one of the remaining threaded holes. Tighten the

bolts so the Holder 6719 is held to the flange.

9. Remove pinion nut and washer.
10. Remove flange with Remover C-452.
11. Remove pinion seal with a pry tool or slide-hammer mounted screw.

To install:

12. Apply a light coating of gear lubricant on the lip of pinion seal.
13. Install new pinion seal with Installer C-4076-B and Handle C-4735.
14. Install flange on the end of the shaft with the reference marks aligned.
15. Install bolts into two of the threaded holes in the flange 180° apart.
16. Position Holder 6719 against flange. Install a bolt and washer into one of the remaining threaded holes. Tighten bolts so Holder 6719 is held to the flange.
17. Install flange on pinion shaft with Installer C-3718 and Holder 6719.
18. Install pinion washer and a new pinion nut. The convex side of the washer must face outward.

⁂ CAUTION

Do not exceed the minimum tightening torque when installing the companion flange retaining nut at this point. Failure to follow these instructions can damage the collapsible spacer or bearings.

19. Hold flange with Holder 6719 and tighten pinion nut to 285 Nm (210 ft. lbs.). Rotate pinion several revolutions to ensure bearing rollers are seated.
20. Rotate pinion flange with an inch pound torque wrench. Rotating torque should be equal to the reading recorded during removal plus an additional 0.56 Nm (5 inch lbs.).

⁂ CAUTION

Never loosen pinion nut to decrease pinion bearing rotating torque and never exceed specified preload torque. If rotating torque is exceeded, a new collapsible spacer must be installed. Failure to follow these instructions can damage the collapsible spacer or bearings.

21. If rotating torque is low use Holder 6719, to hold flange and tighten pinion nut in 6.8 Nm (5 ft. lbs.) increments until proper rotating torque is achieved.

➡The seal replacement is unacceptable if final pinion nut torque is less than 285 Nm (210 ft. lbs.).

➡The bearing rotating torque should be constant during a complete revolution of the pinion. If the rotating torque varies, this indicates a binding condition.

22. Install propeller shaft.
23. Install rear brake rotors components.

9 ¼ INCH AXLE

1. With vehicle in neutral, position it on a hoist.
2. Mark an installation reference line across the pinion flange and driveshaft flange.
3. Remove driveshaft.
4. Remove brake calipers and rotors to prevent any drag.
5. Rotate pinion flange three or four times and verify flange rotates smoothly.
6. Measure rotating torque of the pinion with an inch pound torque wrench and record reading.
7. Install two bolts into the pinion flange threaded holes, 180° apart. Position Holder 6719A, or equivalent, against the flange and install and tighten two bolts and washers into the remaining holes.
8. Hold the flange with Holder 6719A, or equivalent, and remove pinion nut and washer.
9. Remove companion flange with Remover C-452, or equivalent.
10. Remove pinion seal with pry tool or slide-hammer mounted screw.

To install:

11. Apply a light coating of gear lubricant on the lip of pinion seal.
12. Install new pinion seal with Installer C-4076-B and Handle C-4735, or equivalent.
13. Install pinion flange on pinion shaft with Installer C-3718, or equivalent.
14. Install two bolts into the threaded holes in the flange, 180° apart.
15. Position Holder 6719 against the flange and install a bolt and washer into one of the remaining threaded holes. Tighten the bolts so holder is held to the flange.
16. Install pinion washer and a new pinion nut. The convex side of the washer must face outward.

⁂ CAUTION

Never exceed the minimum tightening torque 285 Nm (210 ft. lbs.) when installing the companion flange retaining nut at this point. Failure to follow these instructions will result in damage to collapsible spacer or bearings.

17. Hold companion flange with Holder 6719, or equivalent, and tighten pinion nut with a torque set to 285 Nm (210 ft. lbs.). Rotate pinion several revolutions to ensure the bearing rollers are seated.
18. Rotate pinion with an inch pound torque wrench. Rotating torque should be equal to the reading recorded during removal plus an additional 0.56 Nm (5 inch lbs.).

⁂ CAUTION

Never loosen pinion nut to decrease pinion bearing rotating torque and never exceed specified preload torque. If rotating torque is exceeded, a new collapsible spacer must be installed. Failure to follow these instructions will result in damage to the collapsible spacer

19. If rotating torque is low, use Holder 6719, or equivalent, to hold the companion flange and tighten pinion nut in 6.8 Nm (5 ft. lbs.) increments until proper rotating torque is achieved.

➡The bearing rotating torque should be constant during a complete revolution of the pinion. If the rotating torque varies, this indicates a binding condition.

➡The seal replacement is unacceptable if the final pinion nut torque is less than 285 Nm (210 ft. lbs.).

20. Install driveshaft.

STEERING AND SUSPENSION

Air Bag

⁂ CAUTION

Some vehicles are equipped with an air bag system. The system must be disarmed before performing service on, or around, system components, the steering column, instrument panel components, wiring and sensors. Failure to follow the safety precautions and the disarming procedure could result in accidental air bag deployment, possible injury and unnecessary system repairs.

PRECAUTIONS

Several precautions must be observed when handling the inflator module to avoid accidental deployment and possible personal injury.

- Never carry the inflator module by the wires or connector on the underside of the module.
- When carrying a live inflator module, hold securely with both hands, and ensure that the bag and trim cover are pointed away.
- Place the inflator module on a bench or other surface with the bag and trim cover facing up.
- With the inflator module on the bench, never place anything on or close to the module which may be thrown in the event of an accidental deployment.

Before servicing the vehicle, also make sure to refer to the precautions in the beginning of this section as well.

DISARMING

1. Disconnect and isolate the negative battery cable. Wait 2 minutes for the system capacitor to discharge before performing any service.
2. When repairs are completed, connect the negative battery cable.

Rack and Pinion Steering Gear

REMOVAL & INSTALLATION

2002–03

1. Before servicing the vehicle, refer to the precautions in the beginning of this section.
2. Remove or disconnect the following:
   - Front wheels
   - Outer tie rod ends
   - Steering shaft coupler
   - Power steering hoses
   - Steering gear

To install:

3. Install or connect the following:
   - Steering gear. Tighten the bolts to 190 ft. lbs. (258 Nm).
   - Power steering hoses. Tighten the fittings to 25 ft. lbs. (35 Nm).
   - Steering shaft coupler. Tighten the bolt to 36 ft. lbs. (49 Nm).
   - Outer tie rod ends. Tighten the nuts to 65 ft. lbs. (88 Nm).
   - Front wheels

2004–05

1. Siphon out as much power steering fluid as possible from the pump.
2. Lock the steering wheel.
3. Raise and support the vehicle.
4. Remove the front tires.
5. Remove the nuts from the tie rod ends.
6. Separate tie rod ends from the knuckles.
7. Remove the steering gear pinch bolt.
8. Remove the lower steering coupling from the steering gear.
9. Turn the steering gear to the full right position.
10. Remove the exhaust Y-pipe (3.7L & 4.7L engines only).
11. Remove the power steering lines from the gear.
12. Remove the front crossmember.
13. Remove the steering gear mounting bolts, washers and nuts.
14. Tip the gear forward to allow clearance and move to the right then tip the gear downward on the left side to remove from the vehicle.

To install:

➡Before installing gear inspect the bushings and replace if worn or damaged, also use new gear mounting bolts and nuts.

15. Install gear to the vehicle and tighten mounting nuts and bolts to 258 Nm (190 ft. lbs.).
16. Install power steering lines to steering gear and tighten the pressure hose to 38 Nm (28 ft. lbs.) and tighten the return hose to 65 Nm (48 ft. lbs.).
17. Slide the shaft coupler onto gear. Install new bolt and tighten to 49 Nm (36 ft. lbs.).
18. Clean tie rod end studs and knuckle tapers.
19. Install tie rod ends into the steering knuckles and tighten the nuts to 75 Nm (55 ft. lbs.).
20. Install the Y-pipe (3.7L & 4.7L engines only).
21. Install the front crossmember.
22. Install the front tires.
23. Remove the support and lower the vehicle.
24. Unlock the steering wheel.
25. Fill system with fluid.
26. Adjust the toe position.

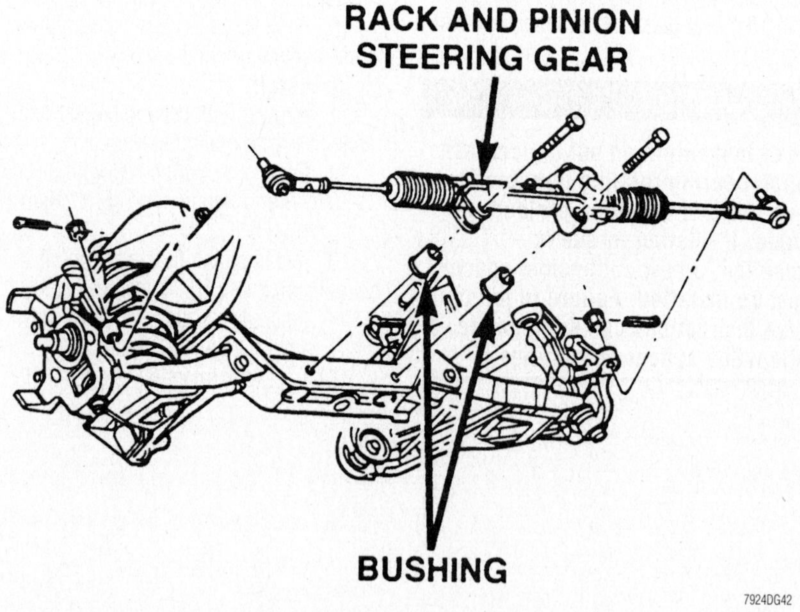

Rack and pinion steering gear mounting used on the 2WD 2002–03 models

Front Shock Absorber

REMOVAL & INSTALLATION

2002–03

1. Before servicing the vehicle, refer to the precautions in the beginning of this section.
2. Remove or disconnect the following:
3. Install or connect the following:
   - Front wheel
   - Upper mount nut
   - Lower mount bolt
   - Shock absorber

To install:

4. Install or connect the following:
   - Shock absorber. Tighten the lower bolt to 100 ft. lbs. (136 Nm) and the upper nut to 30 ft. lbs. (41 Nm).
   - Front wheel

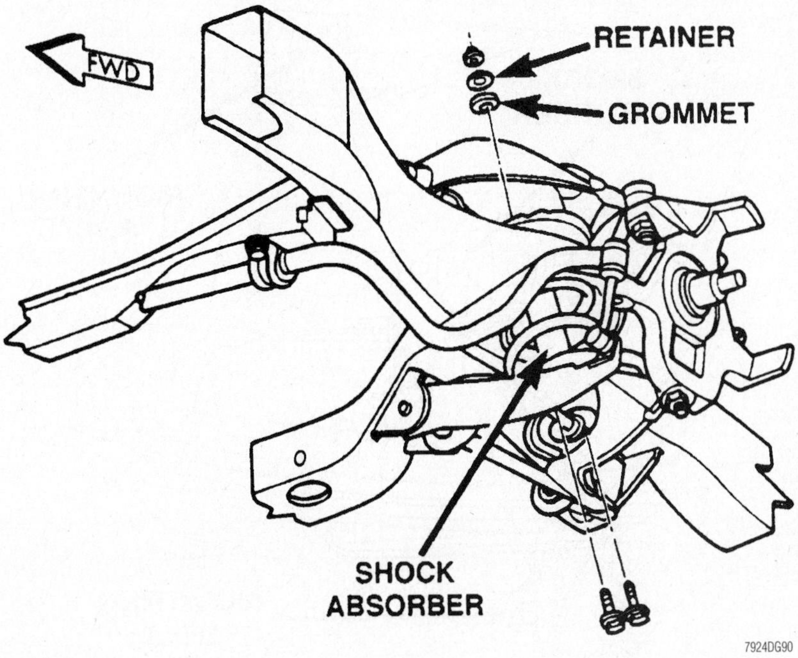

Front shock absorber mounting—2002–03 2WD models

2004–05

1. Raise and support the vehicle.
2. Remove the tire and wheel assembly.
3. Support the lower control arm outboard end.
4. Remove the upper shock bolt and nut.
5. Remove the stabilizer link lower nut and then separate the stabilizer link from the lower control arm to gain access to the lower shock bolt.
6. Remove the lower shock bolt.
7. Remove the shock.

To install:

8. Install the upper part of the shock into the frame bracket.
9. Install the nut and bolt. Tighten to 102 Nm (75 ft. lbs.).
10. Install the lower part of the shock into the lower control arm and Tighten the bolt to 81 Nm (60 ft. lbs.).
11. Install the stabilizer link lower nut to the lower control arm.
12. Remove the support from the lower control arm outboard end.
13. Install the tire and wheel assembly.
14. Remove the support and lower the vehicle.

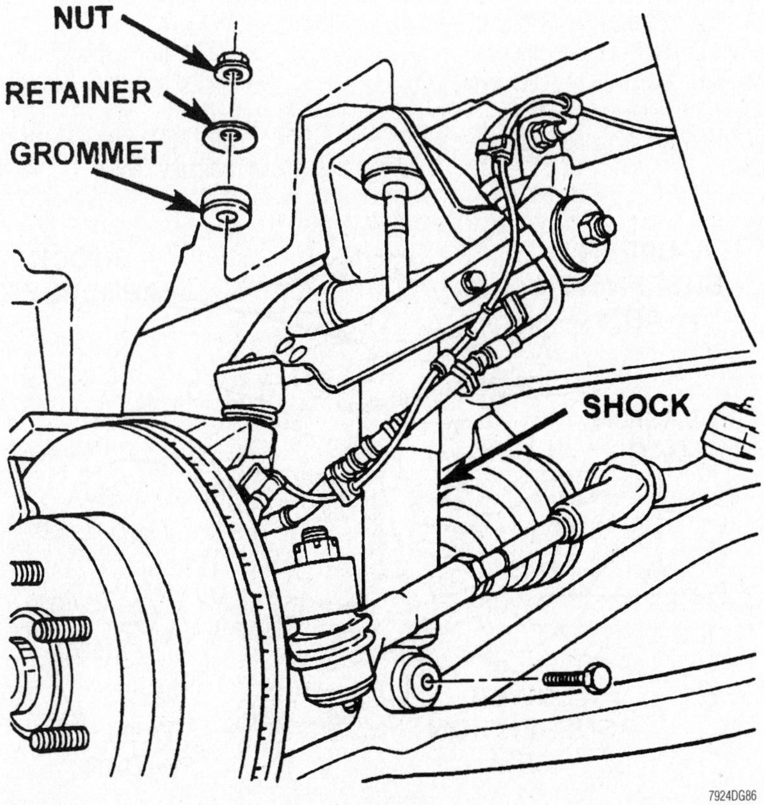

Front shock absorber mounting—2002–03 4WD models

Rear Shock Absorber

REMOVAL & INSTALLATION

2002–03

1. Before servicing the vehicle, refer to the precautions in the beginning of this section.

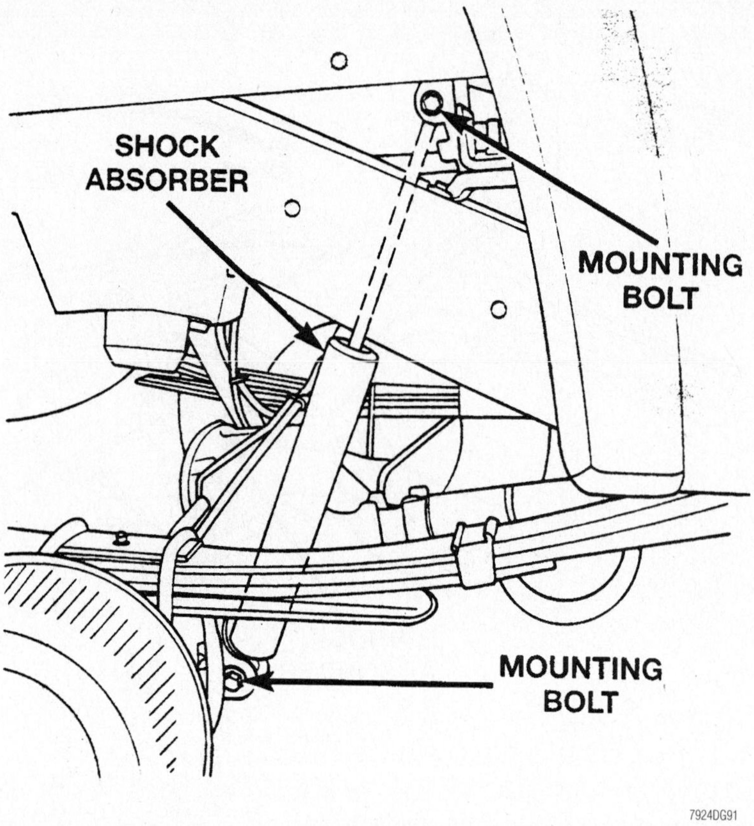

Rear shock absorber mounting—2002–03 models

2. Support the axle.
3. Remove or disconnect the following:
   - Upper bolt
   - Lower bolt
   - Shock absorber

To install:

4. Install the bolts through the brackets and shock and tighten them as follows: Tighten the lower bolt and nut to 60 ft. lbs. (81 Nm) and the upper bracket nuts to 20 ft. lbs. (27 Nm)

2004–05

1. Raise vehicle and support the axle.
2. Lower the spare tire.

➡This step must be done if replacing the left side shock.

3. Remove the upper shock bolt and flag nut.
4. Remove the lower shock bolt and nut.
5. Remove the rear shock absorber from the vehicle.

To install:

6. Position the shock absorber in the brackets.
7. Install the bolts through the brackets and the shock. Install the flag nut on the top bolt and nut on lower bolt.
8. Tighten the upper and lower bolt/nuts to 102 Nm (75 ft. lbs.)
9. Raise the spare tire back in place if lowered for the left shock.
10. Remove the support and lower the vehicle.

Coil Spring

REMOVAL & INSTALLATION

Front

1. Before servicing the vehicle, refer to the precautions in the beginning of this section.
2. Support the lower control arm on a floor jack.
3. Remove or disconnect the following:
   - Front wheel
   - Shock absorber
   - Stabilizer bar link
   - Lower ball joint.
4. Lower the jack and remove the coil spring.

To install:

5. Install the coil spring and raise the control arm into position.
6. Install or connect the following:
   - Lower ball joint. Tighten the nut to 135 ft. lbs. (183 Nm).
   - Stabilizer bar link
   - Shock absorber
   - Front wheel

Rear

1. Raise and support the vehicle.
2. Support the axle with a suitable holding fixture.
3. Remove the lower shock bolt.
4. Remove the bell crank bolt (4) from the rear axle.
5. Lower the jack to remove the spring and isolator from the vehicle.

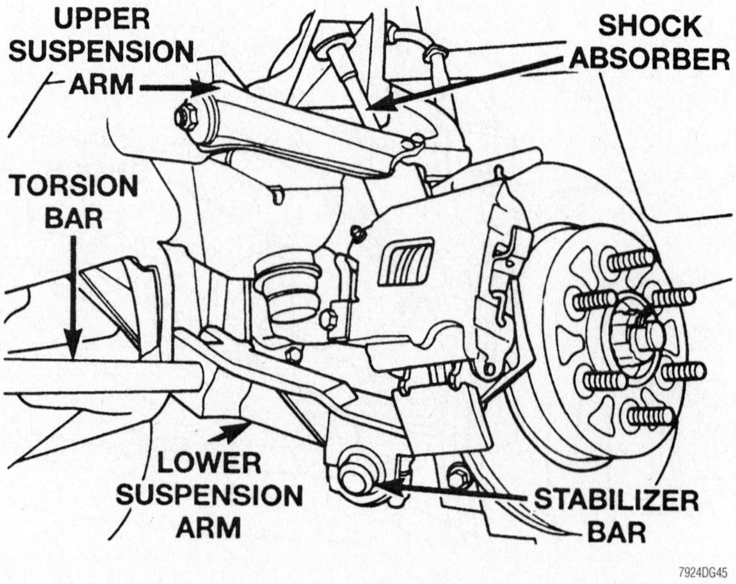

Coil spring type front suspension components—4WD models

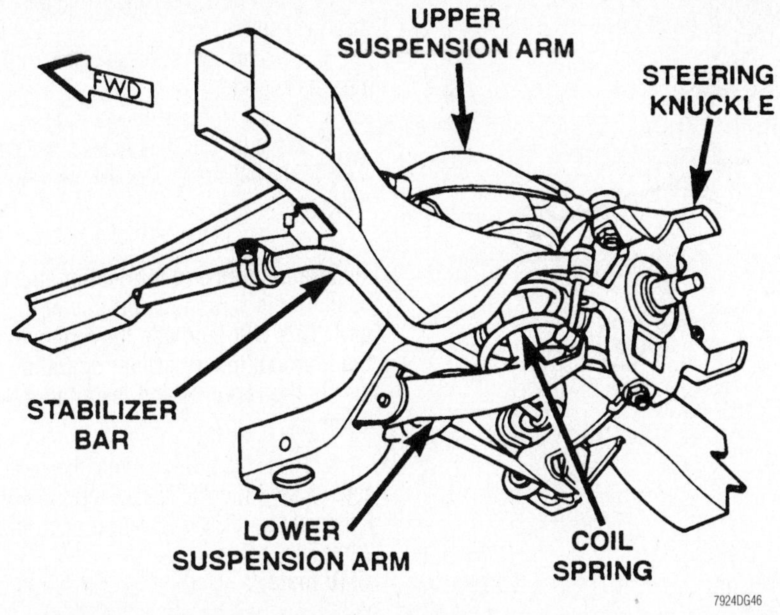

Coil spring type front suspension components—2WD models

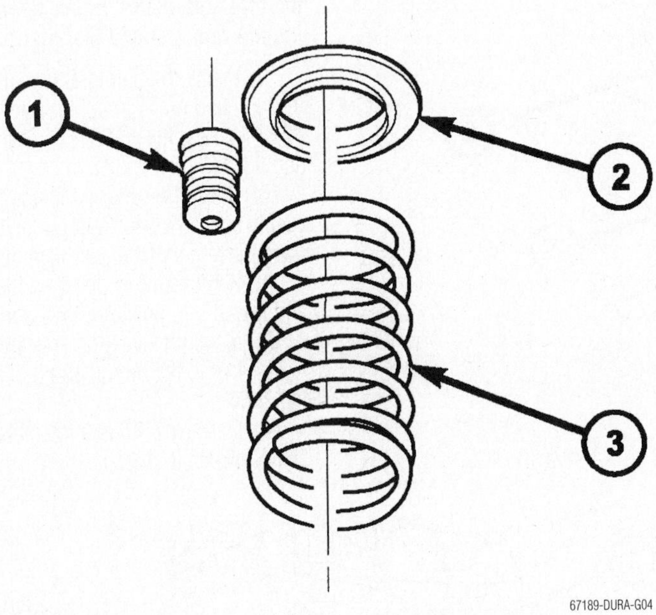

2004–05 rear coil spring removal

To install:

➡**All torques should be made with the full vehicle weight on the ground being supported by the tires.**

6. Position spring (3) to the vehicle on top of the isolator (2).
7. Align the springs (3) to the spring pockets.
8. Raise the rear axle into place.
9. Install the bell crank bolt (4) to the rear axle. Tighten to 251 Nm (185 ft. lbs.).
10. Install the lower shock bolts to the rear axle. Tighten to 102 Nm (75 ft. lbs.).
11. Remove the holding fixture for the rear axle.

Torsion Bar

REMOVAL & INSTALLATION

Except 2002–03

1. Before servicing the vehicle, refer to the precautions in the beginning of this section.
2. Loosen the adjustment bolt to remove spring load. Note the number of turns for installation.
3. Remove or disconnect the following:
   - Adjustment bolt, swivel and bearing
   - Torsion bar and anchor
4. Separate the torsion bar and anchor.

To install:

5. Assemble the torsion bar and anchor.
6. Install or connect the following:
   - Torsion bar and anchor
   - Adjustment bolt, swivel and bearing
7. Tighten the adjustment bolt the recorded number of turns.

2004–05

➡**The left and right side torsion bars are NOT interchangeable. The bars are identified and stamped R or L, for right or left. The bars do not have a front or rear end and can be installed with either end facing forward.**

1. Raise and support the vehicle with the front suspension hanging.
2. Remove the transfer case skid plate.

➡**Count and record the number of turns for installation reference.**

3. Mark the adjustment bolt setting.
4. Install Special Tool 8686 (1) to the anchor arm (2) and the cross member (3).
5. Increase the tension on the anchor arm tool 8686 (1) until the load is removed from the adjustment bolt and the adjuster nut.
6. Turn the adjustment bolt counter-clockwise to remove the bolt and the adjuster nut.
7. Remove the Special Tool 8686 (1), allowing the torsion bar to unload.

To install:

➡**The left and right side torsion bars are NOT interchangeable. The bars are identified and stamped R or L, for right or left. The bars do not have a front or rear end and can be installed with either end facing forward.**

8. Insert torsion bar ends into anchor and suspension arm.
9. Position the anchor (2) in the cross-member frame (3).
10. Install Special Tool 8686 (1) to the anchor (2) and the crossmember (3).
11. Increase the tension on the anchor in order to load the torsion bar.
12. Install the adjustment bolt and the adjuster nut.
13. Turn adjustment bolt clockwise the recorded amount of turns.
14. Remove tool 8686 (1) from the torsion bar crossmember (3).
15. Install the transfer case skid plate.

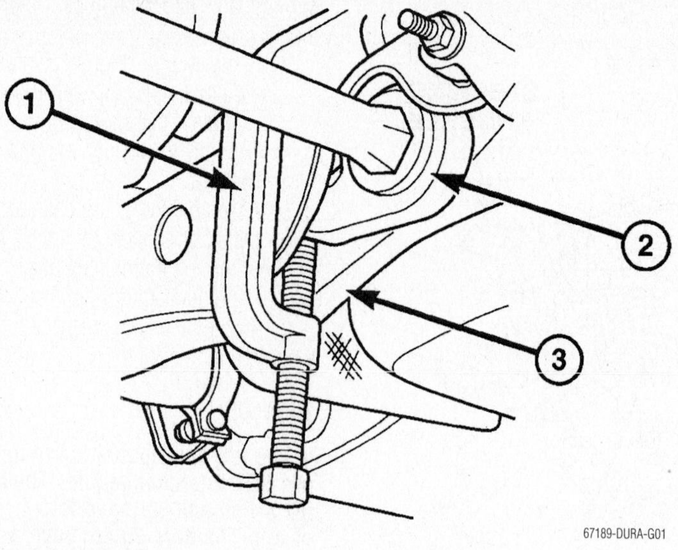

Unloading the torsion bar

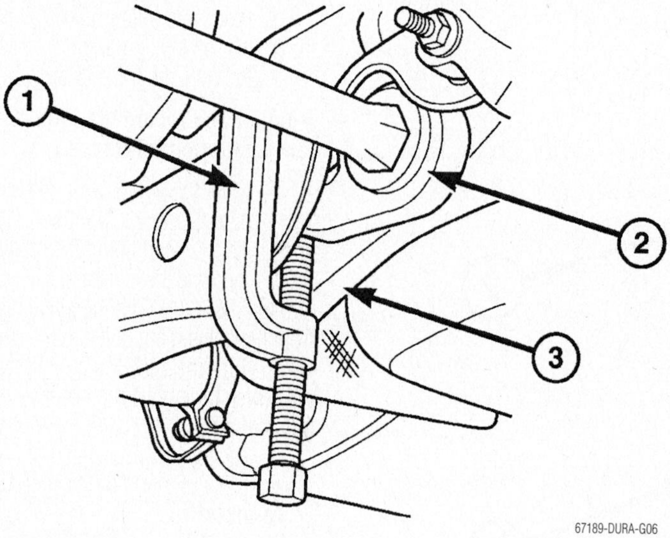

Torsion bar removal

16. Lower vehicle and adjust the front suspension height.
17. Perform a wheel alignment.

Upper Ball Joint

REMOVAL & INSTALLATION

Durango models utilize an upper control arm with an integral ball joint. If the ball joint is damaged or worn, the upper control arm must be replaced.

Lower Ball Joint

REMOVAL & INSTALLATION

1. Remove the tire and wheel assembly.
2. Remove the brake caliper and rotor.
3. Disconnect the tie rod from the steering knuckle.
4. Separate the lower ball joint from the steering knuckle.
5. Remove the steering knuckle.
6. Move the halfshaft to the side and support the halfshaft out of the way (4wd only).
7. Chisel out the ball joint stakes.

➡Extreme pressure lubrication must be used on the threaded portions of the tool. This will increase the longevity of the tool and insure proper operation during the removal and installation process.

8. Press the ball joint from the lower control arm (3) using special tools C-4212-F (press) (1), 9331-1 (driver) (2) and 9331-2 (receiver) (4).

To install:

➡Extreme pressure lubrication must be used on the threaded portions of the tool. This will increase the longevity of the tool and insure proper operation during the removal and installation process.

9. Install the ball joint (3) into the control arm and press in using special tools C-4212-F (press) (1), 9331-1 (receiver) (2) and 9331-3 (driver) (4).
10. Stake the ball joint flange in four evenly spaced places around the ball joint flange, using a chisel and hammer.
11. Remove the support for the halfshaft and install into position (4wd only).
12. Install the steering knuckle.
13. Install the tie rod end into the steering knuckle.
14. Install and tighten the halfshaft nut to 251 Nm (185 ft. lbs.).

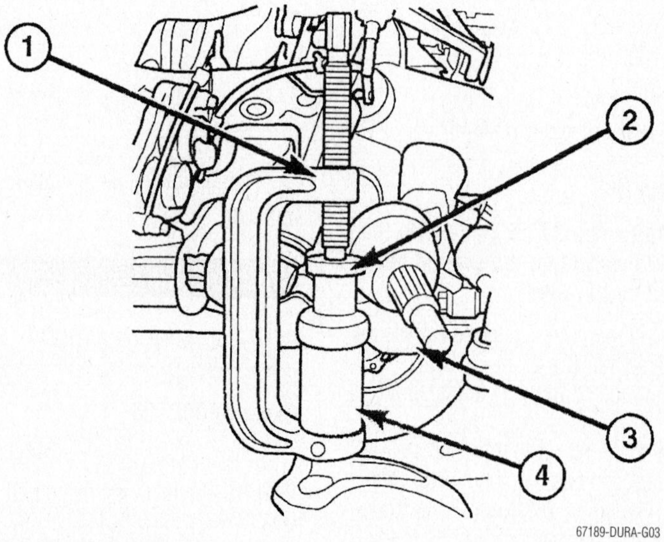

Removing the lower ball joint

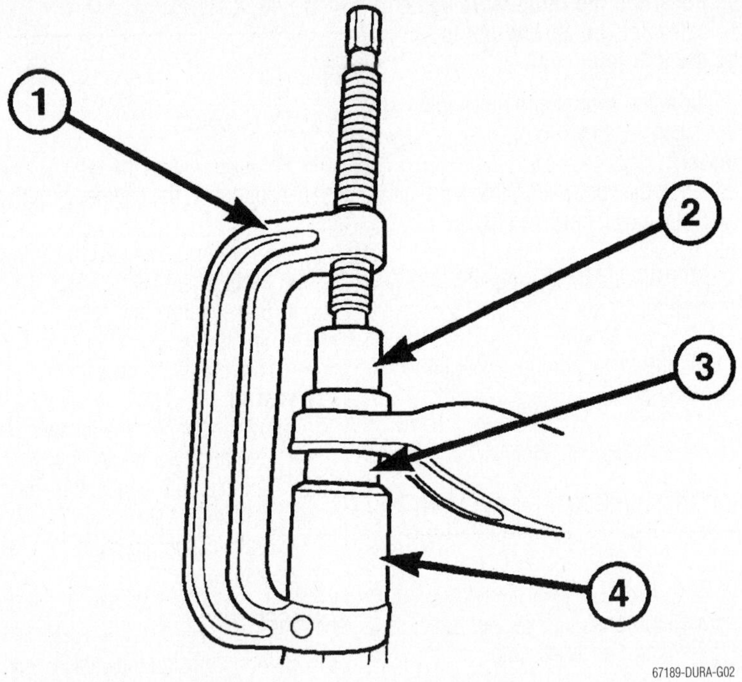

Installing the lower ball joint

15. Install the brake caliper and rotor.
16. Install the tire and wheel assembly.
17. Check the vehicle ride height.
18. Perform a wheel alignment.

Upper Control Arm

REMOVAL & INSTALLATION

2002–03

1. Before servicing the vehicle, refer to the precautions in the beginning of this section.
2. Support the lower control arm.
3. Remove or disconnect the following:
   - Front wheel
   - Brake hose brackets
   - Upper ball joint
   - Pivot mounting nuts
   - Upper control arm

To install:

4. Install or connect the following:
   - Upper control arm. Tighten the pivot nuts to 155 ft. lbs. (210 Nm).
   - Upper ball joint. Tighten the nut to 60 ft. lbs. (81 Nm).
   - Brake hose brackets
   - Front wheel
5. Check the wheel alignment and adjust as necessary.

2004–05

1. Raise and support vehicle.
2. Remove wheel and tire assembly.
3. Remove the nut from upper ball joint.
4. Separate upper ball joint from the steering knuckle.

➡When installing the tool to separate the ball joint, be careful not to damage the ball joint seal.

5. Remove the control arm pivot bolts and remove control arm.

To install:

6. Position the control arm into the frame brackets. Install bolts and tighten to 102 Nm (75 ft. lbs.).
7. Insert ball joint in steering knuckle and tighten ball joint nut to 75 Nm (55 ft. lbs.).
8. Install the wheel and tire assembly.
9. Remove the support and lower vehicle.
10. Perform a wheel alignment.

CONTROL ARM BUSHING REPLACEMENT

The control arm bushings are serviced with the control arm as an assembly.

Lower Control Arm

REMOVAL & INSTALLATION

2002–03

2-WHEEL DRIVE

1. Before servicing the vehicle, refer to the precautions in the beginning of this section.
2. Remove or disconnect the following:
   - Front wheel
   - Shock absorber
   - Brake caliper and rotor
   - Stabilizer bar link
   - Coil spring
   - Inner mounting bolts
   - Lower control arm

To install:

3. Install or connect the following:
   - Lower control arm. Tighten the front bolt to 130 ft. lbs. (175 Nm) and the rear bolt to 80 ft. lbs. (108 Nm).
   - Coil spring. Tighten the lower ball joint nut to 94 ft. lbs. (127 Nm).
   - Stabilizer bar link
   - Brake caliper and rotor
   - Shock absorber
   - Front wheel

4-WHEEL DRIVE

1. Before servicing the vehicle, refer to the precautions in the beginning of this section.
2. Remove or disconnect the following:
   - Front wheel
   - Front driveshaft
   - Torsion bar
   - Shock absorber
   - Stabilizer bar
   - Lower ball joint
   - Pivot bolts
   - Lower control arm

To install:

3. Install or connect the following:
   - Lower control arm. Tighten the front pivot bolt to 80 ft. lbs. (108 Nm) and the rear bolt to 140 ft. lbs. (190 Nm).
   - Lower ball joint. Tighten the nut to 135 ft. lbs. (183 Nm).
   - Stabilizer bar
   - Shock absorber
   - Torsion bar
   - Front driveshaft
   - Front wheel

2004

1. Raise and support the vehicle.
2. Remove the wheel and tire assembly.
3. Remove the disc brake caliper assembly.
4. Remove the disc brake rotor.
5. Disconnect the wheel speed sensor at the wheel well.
6. Disconnect the tie rod from the knuckle.
7. Remove the front halfshaft nut (4wd).
8. Remove the upper ball joint nut. Separate the upper ball joint from the steering knuckle.

9. Remove the lower ball joint nut. Separate the lower ball joint from the steering knuckle.
10. Remove the steering knuckle.
11. Unload the torsion bar using special tool 8686 (1).
12. Remove the torsion bar from the vehicle.
13. Remove the stabilizer bar link.
14. Remove the shock absorber lower bolt.
15. Remove the front and rear pivot bolts.
16. Remove the lower control arm from the vehicle.

To install:

17. Position the lower control arm at the frame rail brackets. Install the pivot bolts and nuts. Tighten the nuts finger-tight.

➡The ball joint stud taper must be CLEAN and DRY before installing the knuckle.

Clean the stud taper with mineral spirits to remove dirt and grease.

18. Install the steering knuckle.
19. Insert the lower ball joint into the steering knuckle. Install and tighten the retaining nut to 81 Nm (60 ft. lbs.).
20. Install the torsion bar.
21. Install shock absorber lower bolt and tighten to 81 Nm (60 ft. lbs.).
22. Install the front halfshaft nut (4wd).
23. Insert the upper ball joint into the steering knuckle. Install and tighten the retaining nut to 75 Nm (55 ft. lbs.).
24. Install the stabilizer bar link and tighten to 169 Nm (125 ft. lbs.).
25. Tighten the lower control arm pivot nut and bolts to 244 Nm (180 ft. lbs.).
26. Insert the outer tie rod end into the steering knuckle (1). Install and tighten the retaining nut to 75 Nm (55 ft. lbs.).
27. Install the disc brake rotor.
28. Install the disc brake caliper and adaptor assembly and tighten to 135 Nm (100 ft. lbs.).
29. Install the wheel and tire assembly.
30. Remove the support and lower the vehicle.
31. Adjust the front suspension height and perform a wheel alignment.

2005

1. Raise and support vehicle.
2. Remove wheel and tire assembly.
3. Remove the nut from upper ball joint.
4. Separate upper ball joint from the steering knuckle with a ball joint tool.

➡When installing the remover to separate the ball joint, be careful not to damage the ball joint seal.

5. Remove the control arm pivot bolts and remove control arm.

To install:

6. Position the control arm into the frame brackets. Install bolts and tighten to 102 Nm (75 ft. lbs.).
7. Insert ball joint in steering knuckle and tighten ball joint nut to 75 Nm (55 ft. lbs.).
8. Install the wheel and tire assembly.
9. Remove the support and lower vehicle.
10. Perform a wheel alignment.

CONTROL ARM BUSHING REPLACEMENT

2002–03

The control arm bushings are serviced with the control arm as an assembly.

2004–05

1. Remove lower control.
2. Secure the control arm in a vise.
3. Press the bushings from the control arm.
4. Installation is the reverse of removal.

Front Wheel Bearings

ADJUSTMENT

Durango models utilize a hub/bearing assembly which is not adjustable.

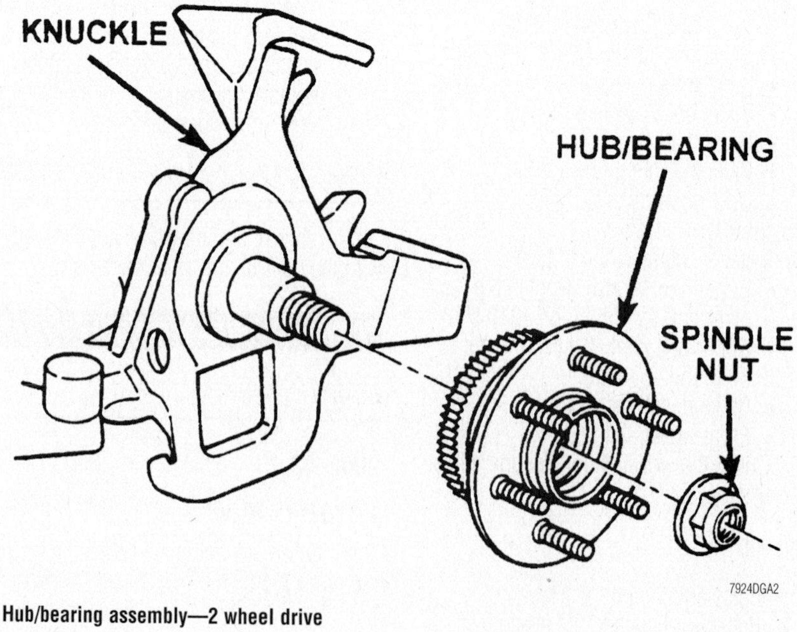

Hub/bearing assembly—2 wheel drive

REMOVAL & INSTALLATION

2002–03

2-WHEEL DRIVE

1. Before servicing the vehicle, refer to the precautions in the beginning of this section.
2. Remove or disconnect the following:
   - Front wheel
   - Brake caliper and rotor
   - Spindle nut
   - Hub and bearing assembly

To install:

3. Install or connect the following:
   - Hub and bearing assembly
   - Spindle nut. Tighten the nut to 185 ft. lbs. (251 Nm).
   - Brake caliper and rotor
   - Front wheel

4-WHEEL DRIVE

1. Before servicing the vehicle, refer to the precautions in the beginning of this section.
2. Remove or disconnect the following:
   - Front wheel
   - Brake caliper and rotor
   - Hub retainer nut
   - Hub and bearing assembly

To install:

3. Install or connect the following:
   - Hub and bearing assembly. Tighten the bolts to 123 ft. lbs. (166 Nm).
   - Hub retainer nut. Tighten the nut to 173 ft. lbs. (235 Nm).
   - Brake caliper and rotor
   - Front wheel

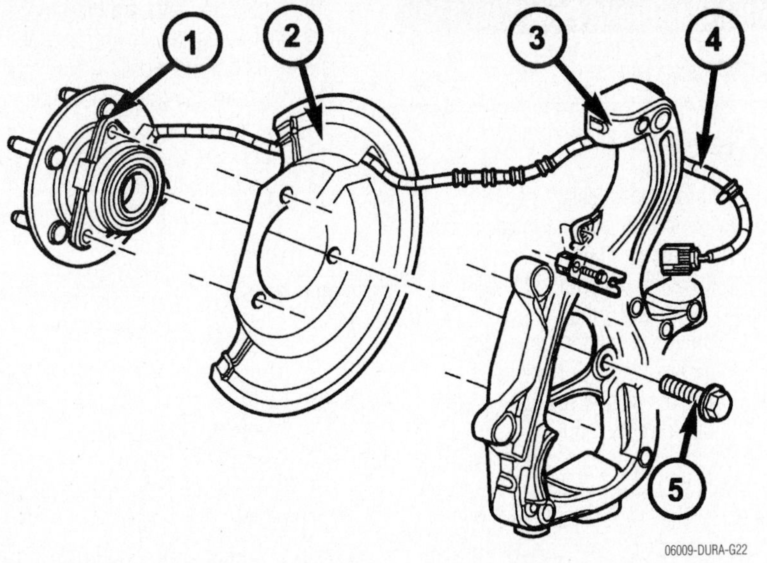

Hub/bearing installation, 2004–05 models. (1) hub/bearing, (2) dust shield, (3) knuckle, (4) ABS harness, (5) hub bolt

2004–05

1. Raise and support the vehicle.
2. Remove the wheel and tire assembly.
3. Remove the brake caliper and rotor.
4. Remove the ABS wheel speed sensor if equipped,
5. Remove the halfshaft nut, (4wd only).

➡Do not strike the knuckle with a hammer to remove the tie rod end or the ball joint. Damage to the steering knuckle will occur.

6. Remove the tie rod end nut and separate the tie rod from the knuckle.
7. Remove the upper ball joint nut and separate the upper ball joint from the knuckle.
8. Pull down on the steering knuckle to separate the halfshaft from the hub/bearing, (4wd).
9. Remove the three hub/bearing mounting bolts from the steering knuckle.
10. Slide the hub/bearing out of the steering knuckle.
11. Remove the brake dust shield.

To install:

12. Install the brake dust shield.
13. Install the hub/bearing into the steering knuckle and tighten the bolts to 163 Nm (120 ft. lbs.).
14. Install the brake rotor and caliper.
15. Install the ABS wheel speed sensor if equipped.
16. Install the upper ball joint nut to the steering knuckle and tighten to 75 Nm (55 ft. lbs.).
17. Install the tie rod end nut to the steering knuckle and tighten to 81 Nm (60 ft. lbs.).
18. Install the halfshaft nut and tighten to 251 Nm (185 ft. lbs.) (4wd only).
19. Install the wheel and tire assembly.
20. Remove the support and lower vehicle.

BRAKES

Front Brake Caliper

REMOVAL AND INSTALLATION

2002–03

1. Remove the wheels.
2. Disconnect the rubber brake hose from the tubing at the frame mount. If the pistons are to be removed from the caliper, leave the brake hose connected to the caliper. Check the rubber hose for cracks or chafed spots.
3. Plug the brake line to prevent loss of fluid.
4. Remove the caliper slide pins.
5. Remove the caliper and brake pads from the rotor adapter.

To install:

6. Position the outboard shoe in the caliper. The shoe should not rattle in the caliper. If it does, or if any movement is obvious, bend the shoe tabs over the caliper to tighten the fit.
7. Slide the caliper into position on the adapter and over the rotor.
8. Align the caliper and start the pins in by hand.
9. Tighten the pins to 22 ft. lbs. (30 Nm) on 2002–02; 23.5 ft. lbs. on 2003 models.
10. Connect the brake hose to the caliper. Use new washers to attach the hose fitting if the original washers are scored, worn or damaged. Torque the bolt to 21 ft. lbs. (28 Nm).
11. Fill and bleed the brake system.
12. Install the wheels.
13. Lower the vehicle.

2004–05

1. Install prop rod on the brake pedal to keep pressure on the brake system.
2. Raise and support the vehicle.
3. Remove the tire and wheel assembly.
4. Compress the disc brake caliper.
5. Remove the banjo bolt and discard the copper washers.
6. Remove the caliper slide pin bolts.
7. Remove the disc brake caliper from the caliper adapter.

To install:

➡Install a new copper washers on the banjo bolt when installing

8. Install the disc brake caliper on the brake caliper adapter.

⁂ CAUTION

Verify brake hose is not twisted or kinked before tightening fitting bolt.

9. Install the banjo bolt with new copper washers on the caliper. Tighten to 28 Nm (250 inch lbs.)
10. Install the caliper slide pin bolts. Tighten to 32 Nm (24 ft. lbs.)
11. Remove the prop rod.
12. Bleed the base brake system.
13. Install the tire and wheel assembly.

Rear Brake Caliper

REMOVAL & INSTALLATION

2002–03

1. Remove the wheels.
2. Disconnect the rubber brake hose from the tubing at the frame mount. If the pistons are to be removed from the caliper, leave the brake hose connected to the caliper. Check the rubber hose for cracks or chafed spots.
3. Plug the brake line to prevent loss of fluid.

4. Remove the caliper slide pins.
5. Remove the caliper and brake pads from the rotor adapter.
6. Remove the anti-rattle springs.

To install:

7. Position the outboard shoe in the caliper. The shoe should not rattle in the caliper. If it does, or if any movement is obvious, bend the shoe tabs over the caliper to tighten the fit.
8. Slide the caliper into position on the adapter and over the rotor.
9. Align the caliper and start the pins in by hand.
10. Tighten the pins to 22 ft. lbs. (30 Nm) on 2002; 23.5 ft. lbs. on 2003 models.
11. Connect the brake hose to the caliper. Use new washers to attach the hose fitting if the original washers are scored, worn or damaged.
12. Fill and bleed the brake system.
13. Install the wheels.
14. Lower the vehicle.

2004–05

1. Install prop rod on the brake pedal to keep pressure on the brake system.
2. Raise and support vehicle.
3. Remove the wheel and tire assembly.
4. Drain small amount of fluid from master cylinder brake reservoir with suction gun.
5. Remove the brake hose banjo bolt if replacing caliper.
6. Remove the caliper mounting slide pin bolts.
7. Remove the caliper from vehicle.

To install:

8. Install caliper (4) to the caliper adapter.
9. Coat the caliper mounting slide pin bolts with silicone grease. Then install and tighten the bolts to 15 Nm (11 ft. lbs.).
10. Install the brake hose banjo bolt and new copper seal washers if caliper was removed.
11. Install the brake hose to the caliper with and tighten fitting bolt to 28 Nm (250 inch lbs.).

**** CAUTION**

Verify brake hose is not twisted or kinked before tightening fitting bolt.

12. Remove the prop rod from the vehicle.
13. Bleed the base brake system.
14. Install the wheel and tire assemblies.
15. Remove the supports and lower the vehicle.
16. Verify a firm pedal before moving the vehicle.

Front Disc Brake Pads

REMOVAL AND INSTALLATION

2002–03

1. Remove the wheels.
2. Press the caliper piston back into the bore with a suitable prytool. Use a large C-clamp to drive the piston into the bore of additional force is required.
3. Remove the caliper mounting bolts with a ⅜ in. hex wrench or socket.
4. Rotate the caliper rearward off the rotor and out from its mount.
5. Set the caliper on a crate or sturdy box, then remove the inboard and outboard brake pads. The inboard pad has a spring clip that holds it in the caliper. Tilt this pad out at the top to unseat the clip. The outboard pad has a retaining spring that secures it in the caliper. Unseat 1 spring end and rotate the pad out of the caliper.
6. Secure the caliper to a chassis or suspension component with a sturdy wire. Do not let it hang from the hose.

To install:

7. Clean the caliper and steering knuckle sliding surfaces with a wire brush. Then, apply a coat of Mopar® multi-mileage grease or equivalent.
8. Clean the caliper slide pins with brake cleaner or brake fluid. Then apply a light coating of silicone grease to the pins.

➡If there is minor rust or corrosion on the pins, first polish them with a crocus cloth. If they are severely rusted, replace them.

9. Install the inboard pad and its spring clip.
10. Install the outboard brake pad.
11. Install the caliper over the rotor and seat it in its original position until flush.
12. Final tighten the caliper slide pins to 22 ft. lbs. (30 Nm).
13. Install the wheels.
14. Lower the vehicle.
15. Pump the brakes several times to seat the pads.

2004–05

1. Raise and support vehicle.
2. Remove the wheel and tire assemblies.
3. Compress the caliper.
4. Remove the caliper.
5. Remove the caliper by tilting the top up and off the caliper adapter.

➡Do not allow brake hose to support caliper assembly.

6. Support and hang the caliper.
7. Remove the inboard brake shoe from the caliper adapter.
8. Remove the outboard brake shoe from the caliper adapter.

➡Anti-rattle springs are not interchangeable.

9. Remove the top anti-rattle springs from the caliper adapter.
10. Remove the bottom anti-rattle springs from the caliper adapter.

To install:

11. Bottom pistons in caliper bore with C-clamp. Place an old brake shoe between a C-clamp and caliper piston.
12. Clean caliper mounting adapter and anti-rattle springs.
13. Lubricate anti-rattle springs with brake grease.

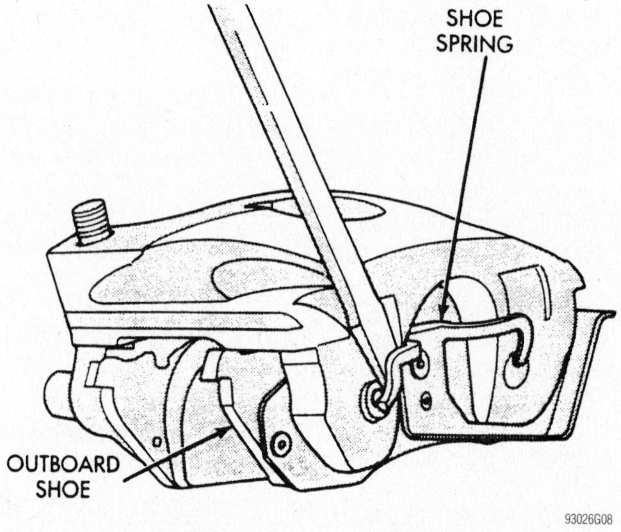

Prying the disc brake from the 4WD front brake caliper assembly—2002–03 models

➡Anti-rattle springs are not interchangeable.

14. Install the bottom anti-rattle springs.
15. Install the top anti-rattle springs.
16. Install inboard brake shoe in adapter.
17. Install outboard brake shoe in adapter.
18. Tilt the top of the caliper over rotor and under adapter. Then push the bottom of the caliper down onto the adapter.
19. Install caliper.
20. Install wheel and tire assemblies and lower vehicle.
21. Apply brakes several times to seat caliper pistons and brake shoes and obtain firm pedal.
22. Top off master cylinder fluid level.

Rear Disc Brake Pads

REMOVAL & INSTALLATION

2002–03

1. Remove the wheels.
2. Press the caliper piston back into the bore with a suitable prytool. Use a large C-clamp to drive the piston into the bore of additional force is required.
3. Remove the caliper mounting bolts with a ⅜ in. hex wrench or socket.
4. Rotate the caliper rearward off the rotor and out from its mount.
5. Set the caliper on a crate or sturdy box, then remove the inboard and outboard brake pads. The inboard pad has a spring clip that holds it in the caliper. Tilt this pad out at the top to unseat the clip. The outboard pad has a retaining spring that secures it in the caliper. Unseat 1 spring end and rotate the pad out of the caliper.
6. Secure the caliper to a chassis or suspension component with a sturdy wire. Do not let it hang from the hose.

To install:

7. Clean the caliper, anti-rattle springs and steering knuckle sliding surfaces with a wire brush. Then, apply a coat of Mopar® multi-mileage grease or equivalent.
8. Clean the caliper slide pins with brake cleaner or brake fluid. Then apply a light coating of silicone grease to the pins.

➡If there is minor rust or corrosion on the pins, first polish them with a crocus cloth. If they are severely rusted, replace them.

9. Install the inboard pad and its spring clip.
10. Install the outboard brake pad.
11. Install the caliper over the rotor and seat it in its original position until flush.
12. Final tighten the caliper slide pins to 22 ft. lbs. (30 Nm).
13. Install the wheels.
14. Lower the vehicle.
15. Pump the brakes several times to seat the pads.

2004–05

1. Raise and support vehicle.
2. Remove the wheel and tire assemblies.
3. Compress the caliper.
4. Remove the caliper.
5. Remove the caliper by tilting the top up and off the caliper adapter.

➡Do not allow brake hose to support caliper assembly.

6. Support and hang the caliper.
7. Remove the inboard brake shoe from the caliper adapter.
8. Remove the outboard brake shoe from the caliper adapter.

➡Anti-rattle springs are not interchangeable.

9. Remove the top anti-rattle springs from the caliper adapter.
10. Remove the bottom anti-rattle springs from the caliper adapter.

To install:

11. Bottom pistons in caliper bore with C-clamp. Place an old brake shoe between a C-clamp and caliper piston.
12. Clean caliper mounting adapter and anti-rattle springs.
13. Lubricate anti-rattle springs with brake grease.

➡Anti-rattle springs are not interchangeable.

14. Install the bottom anti-rattle springs.
15. Install the top anti-rattle springs.
16. Install inboard brake shoe in adapter.
17. Install outboard brake shoe in adapter.
18. Tilt the top of the caliper over rotor and under adapter. Then push the bottom of the caliper down onto the adapter.
19. Install caliper.
20. Install wheel and tire assemblies and lower vehicle.
21. Apply brakes several times to seat caliper pistons and brake shoes and obtain firm pedal.
22. Top off master cylinder fluid level.

Brake Drums

REMOVAL AND INSTALLATION

Chrysler Servo Type With Single Anchor

1. Raise and safely support the truck.
2. Remove the plug from the brake adjustment access hole.
3. Insert a thin bladed screwdriver through the adjusting hole and hold the adjusting lever away from the starwheel.
4. Release the brake by prying down against the starwheel with a brake spoon.
5. Remove the rear wheel and clips from the wheel studs. Remove the brake drum.
6. Installation is the reverse of removal. Adjust the brakes.

Bendix Duo-Servo Type

1. Raise and safely support the vehicle.
2. Remove the rear wheel and tire.

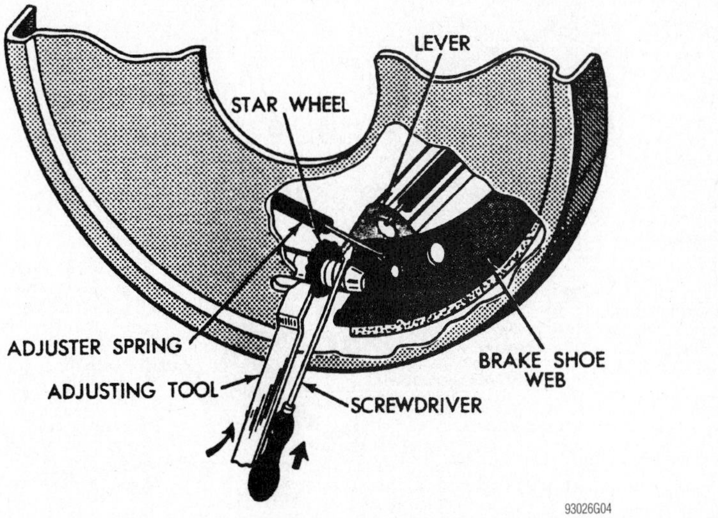

Use a lever releasing tool to depress the adjuster lever while turning the starwheel with a prytool—Bendix brakes

3. Remove the axle shaft nuts, washers and cones. If the cones do not readily release, rap the axle shaft sharply in the center.

4. Remove the axle shaft.

5. Remove the outer hub nut.

6. Straighten the lockwasher tab and remove it along with the inner nut and bearing.

7. Carefully remove the drum.

To install:

8. Position the drum on the axle housing.

9. Install the bearing and inner nut. While rotating the wheel and tire, tighten the adjusting nut until a slight drag is felt.

10. Back off the adjusting nut ⅙ turn so that the wheel rotates freely without excessive end-play.

11. Install the lockrings and nut. Place a new gasket on the hub and install the axle shaft, cones, lockwashers and nuts.

12. Install the wheel and tire.

13. Road-test the vehicle.

Brake Shoes

REMOVAL AND INSTALLATION

Servo Type With Single Anchor

1. Raise and support the vehicle.

2. Remove the rear wheel, drum retaining clips and the brake drum.

3. Remove the brake shoe return springs, noting how the secondary spring overlaps the primary spring.

4. Remove the brake shoe retainer, springs and nails.

5. Disconnect the automatic adjuster cable from the anchor and unhook it from the lever. Remove the cable, cable guide, and anchor plate.

6. Remove the spring and lever from the shoe web.

7. Spread the anchor ends of the primary and secondary shoes and remove the parking brake spring and strut.

8. Disconnect the parking brake cable and remove the brake assembly.

9. Remove the primary and secondary brake shoe assemblies and the star adjuster as an assembly. Block the wheel cylinders to retain the pistons.

To install:

10. Measure the drum as described in this section.

11. Apply a thin coat of lubricant to the support platforms.

12. Attach the parking brake lever to the rear of the secondary shoe.

13. Place the primary and secondary shoes in their relative positions on a workbench.

14. Lubricate the adjuster screw threads. Install it between the primary and secondary shoes with the star wheel next to the secondary shoe. The star wheels are stamped with an **L** (left) and **R** (right).

15. Overlap the ends of the primary and second brake shoes and install the adjusting spring and lever at the anchor end.

16. Hold the shoes in position and install the parking brake cable into the lever.

17. Install the parking brake strut and spring between the parking brake lever and primary shoe.

18. Place the brake shoes on the support and install the retainer nails and springs.

19. Install the anchor pin plate.

20. Install the eye of the adjusting cable over the anchor pin and install the return spring between the anchor pin and primary shoe.

21. Install the cable guide in the secondary shoe and install the secondary return spring. Be sure that the primary spring overlaps the secondary spring.

22. Position the adjusting cable in the groove of the cable guide and engage the hook of the cable in the adjusting lever.

23. Install the brake drum and retaining clips. Install the wheel and tire.

24. Adjust the brakes and road-test the truck.

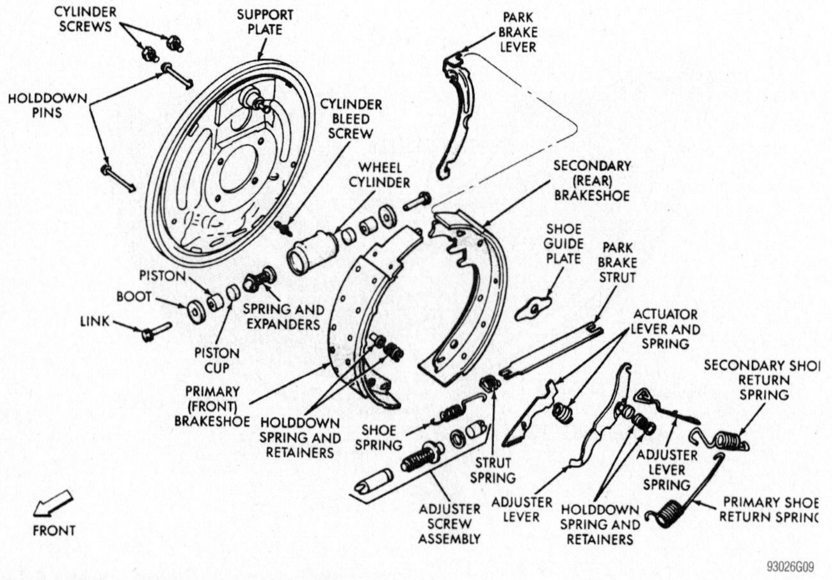

Exploded view of the rear brake components

Bendix Duo-Servo Type

1. Unhook and remove the adjusting lever return spring.

2. Remove the lever from the lever pivot pin.

3. Unhook the adjuster lever from the adjuster cable.

4. Unhook the upper shoe-to-shoe spring.

5. Unhook and remove the shoe hold-down springs.

6. Disconnect the parking brake cable from the parking brake lever.

7. Remove the shoes with the lower shoe-to-shoe spring and star wheel as an assembly.

To install:

8. The pivot screw and adjusting nut on the left side have left-hand threads and right-hand threads on the right side.

9. Lubricate and assemble the star wheel assembly. Lubricate the guide pads on the support plates.

10. Assemble the star wheel, lower shoe-to-shoe spring, and the primary and secondary shoes. Position this assembly on the support plate.

11. Install and hook the hold-down springs.

12. Install the upper shoe-to-shoe spring.

13. Install the cable and retaining clip.

14. Position the adjuster lever return spring on the pivot (green springs on left brakes and red springs on right brakes).

15. Install the adjuster lever. Route the adjuster cable and connect it to the adjuster.

16. Install the brake drum and adjust the brakes.

JEEP

7

Grand Cherokee • Liberty • Wrangler

SPECIFICATIONS AND MAINTENANCE CHARTS

ENGINE AND VEHICLE IDENTIFICATION

Engine						
Code ①	Liters (cc)	Cu. In.	Cyl.	Fuel Sys.	Engine Type	Eng. Mfg.
N	4.7 (4701)	287	8	MFI	SOHC	Chrysler
P	2.5 (2458)	150	4	MFI	OHV	Chrysler
S	4.0 (3966)	242	6	MFI	OHV	Chrysler
K	3.7 (3701)	226	6	MFI	SOHC	Chrysler
1	2.4 (2429)	148	4	MFI	DOHC	Chrysler
2	5.7 (5653)	345	8	MFI	OHV	Chrysler
5	2.8 (2776)	169	4	Turbo Diesel	DOHC	VM Motori

Model Year	
Code ②	Year
2	2002
3	2003
4	2004
5	2005
6	2006

MFI: Multi-port Fuel Injection

OHV: Over Head Valve

SOHC: Single Overhead Camshaft

DOHC: Dual Overhead Camshafts

① 8th position of VIN

② 10th position of VIN

06009-JEEP-C01

GENERAL ENGINE SPECIFICATIONS

Year	Model	Engine Displ. Liters	Engine VIN	Net Horsepower @ rpm	Net Torque @ rpm (ft. lbs.)	Bore x Stroke (in.)	Comp. Ratio	Oil Pressure @ rpm
2002	Grand Cherokee	4.0	S	195@4600	230@3000	3.88x3.44	8.8:1	37@1600
		4.7	N	235@4800	295@3200	3.66x3.40	9.3:1	25@3000
	Wrangler	2.5	P	120@5400	139@3500	3.88x3.19	9.1:1	37@1600
		4.0	S	181@4600	222@2800	3.88x3.44	8.8:1	37@1600
	Liberty	2.4	1	150@5200	165@4000	3.44x3.98	9.5:1	25-80@3000
		3.7	K	210@5200	225@4200	3.66x3.40	9.1:1	25-110@3000
2003	Grand Cherokee	4.0	S	195@4600	230@3000	3.88x3.44	8.8:1	37@1600
		4.7	N	235@4800	295@3200	3.66x3.40	9.3:1	25@3000
	Wrangler	2.4	1	150@5200	165@4000	3.44x3.98	9.5:1	25-80@3000
		4	S	181@4600	222@2800	3.88x3.44	8.8:1	37@1600
	Liberty	2.4	1	150@5200	165@4000	3.44x3.98	9.5:1	25-80@3000
		3.7	K	210@5200	225@4200	3.66x3.40	9.1:1	25-110@3000
2004	Grand Cherokee	4.0	S	195@4600	230@3000	3.88x3.44	8.8:1	37@1600
		4.7	N	235@4800	295@3200	3.66x3.40	9.3:1	25@3000
	Wrangler	2.4	1	150@5200	165@4000	3.44x3.98	9.5:1	25-80@3000
		4.0	S	181@4600	222@2800	3.88x3.44	8.8:1	37@1600
	Liberty	2.4	1	150@5200	165@4000	3.44x3.98	9.5:1	25-80@3000
		3.7	K	210@5200	225@4200	3.66x3.40	9.1:1	25-110@3000
2005	Grand Cherokee	3.7	K	210@5200	225@4200	3.66x3.40	9.1:1	25-110@3000
		4.7	N	235@4800	295@3200	3.66x3.40	9.3:1	25@3000
		5.7	2	345@5400	375@4200	3.91x3.58	9.6:1	25-110@3000
	Wrangler	2.4	1	150@5200	165@4000	3.44x3.98	9.5:1	25-80@3000
		4.0	S	181@4600	222@2800	3.88x3.44	8.8:1	37@1600
	Liberty	2.4	1	150@5200	165@4000	3.44x3.98	9.5:1	25-80@3000
		3.7	K	210@5200	225@4200	3.66x3.40	9.1:1	25-110@3000
		2.8	5	163@3800	295@1800	3.70x3.94	17.5:1	29@3800

06009-JEEP-C02

GASOLINE ENGINE TUNE-UP SPECIFICATIONS

Year	Engine Displ. Liters	Engine VIN	Spark Plug Gap (in.)	Ignition Timing (deg.)	Fuel Pump (psi)	Idle Speed (rpm)	Valve Clearance Intake	Valve Clearance Exhaust
2002	2.4	1	0.050	①	44-54	①	HYD	HYD
	2.5	P	0.035	①	47-51	①	HYD	HYD
	3.7	K	0.042	①	44-54	①	HYD	HYD
	4.0	S	0.035	①	47-51	①	HYD	HYD
	4.7	N	0.040	①	47-51	①	HYD	HYD
2003	2.4	1	0.050	①	44-54	①	HYD	HYD
	3.7	K	0.042	①	44-54	①	HYD	HYD
	4.0	S	0.035	①	47-51	①	HYD	HYD
	4.7	N	0.040	①	47-51	①	HYD	HYD
2004	2.4	1	0.050	①	44-54	①	HYD	HYD
	3.7	K	0.042	①	44-54	①	HYD	HYD
	4.0	S	0.040	①	47-51	①	HYD	HYD
	4.7	N	0.040	①	47-51	①	HYD	HYD
2005	2.4	1	0.050	①	②	①	HYD	HYD
	3.7	K	0.042	①	②	①	HYD	HYD
	4.0	S	0.040	①	56-60	①	HYD	HYD
	4.7	N	0.040	①	56-60	①	HYD	HYD
	5.7	2	0.045	①	56-60	①	HYD	HYD

Note: The information on the Vehicle Emission Control labe must be used, if different from the figures in this chart.

HYD: Hydraulic

① Ignition timing and idle speed are controlled by the PCM. No adjustment is necessary.

② Liberty:

If equipped with JTEC PCM (3 connectors): 44-54

If equipped with NGC PCM (4 connectors): 53-63

Wrangler and Grand Cherokee 3.7L : 56-60

06009-JEEP-C03

DIESEL ENGINE TUNE-UP SPECIFICATIONS

Year	Engine Displacement Liters	Engine VIN	Valve Clearance Intake (in.)	Valve Clearance Exhaust (in.)	Intake Valve Opens (deg.)	Injection Pump Setting (deg.)	Injection Nozzle Pressure psi (new)	Idle Speed (rpm)	Cranking Compression Pressure (psi)
2005	2.8	5	HYD.	HYD.	15.6 A	①	23,000 max.	NA	②

NOTE: The Vehicle Emission Control Information (VECI) label figures must be used if they differ from those in this chart

NA: Information not available

A: After TDC

① PCM controlled

② Maximum difference among cylinders: 72.5 psi

06009-JEEP-C04

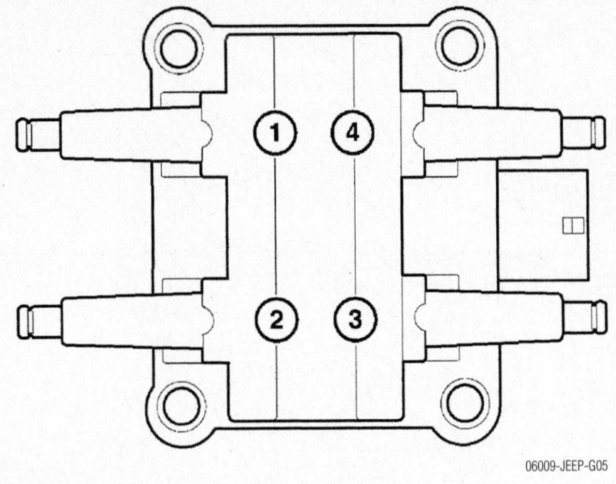

2.4L Engine
Firing order: 1–3–4–2
Coil pack

2.5L Engine
Firing order: 1–3–4–2
Distributor rotation: clockwise

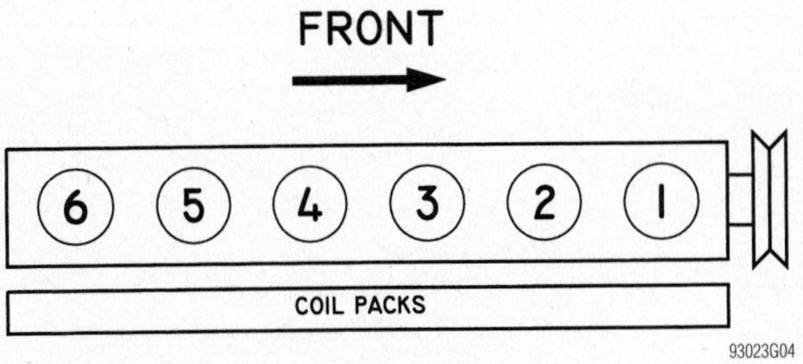

4.0L Engine
Firing order: 1–5–3–6–2–4
Distributorless ignition

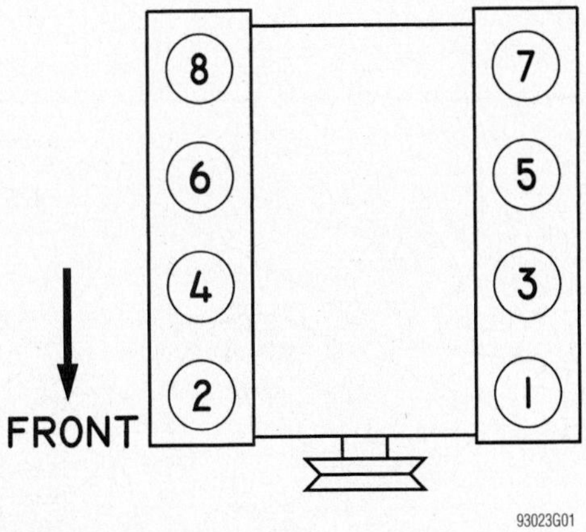

4.7L Engine
Firing order: 1–8–4–3–6–5–7–2
Distributorless ignition

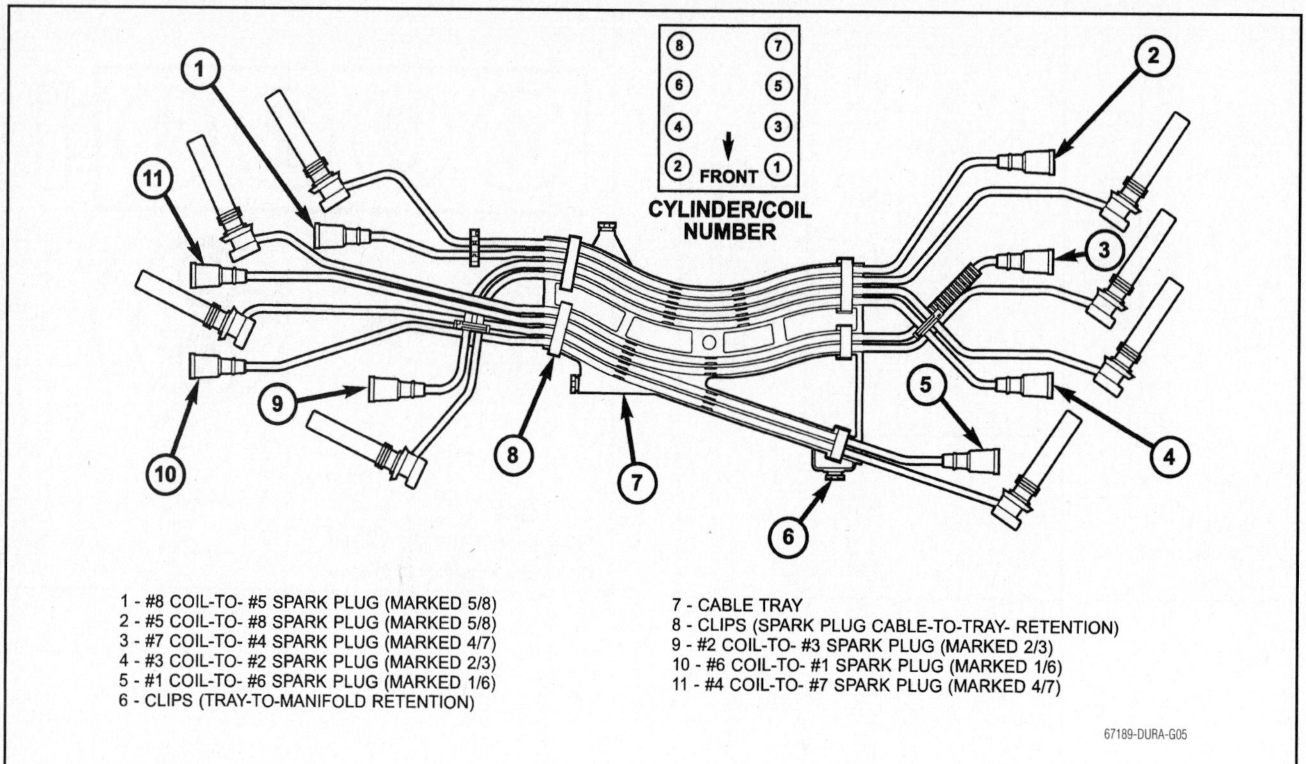

5.7L engine firing order
Distributorless ignition

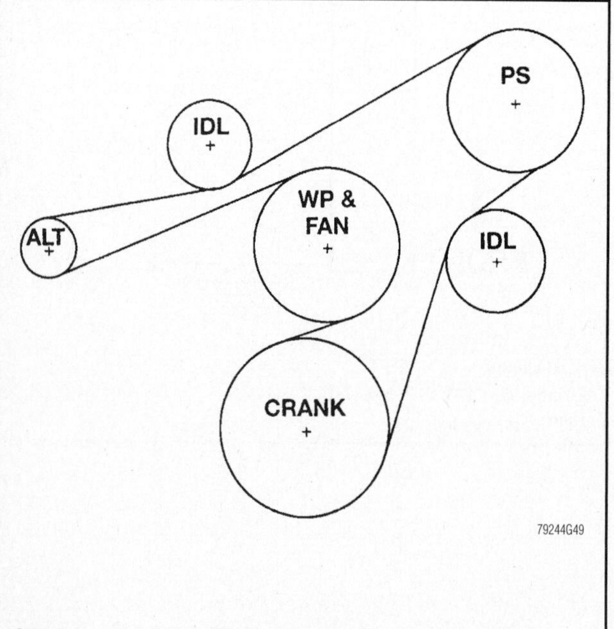

Accessory drive belt routing—2.5L and 2002–04 4.0L engines, without A/C

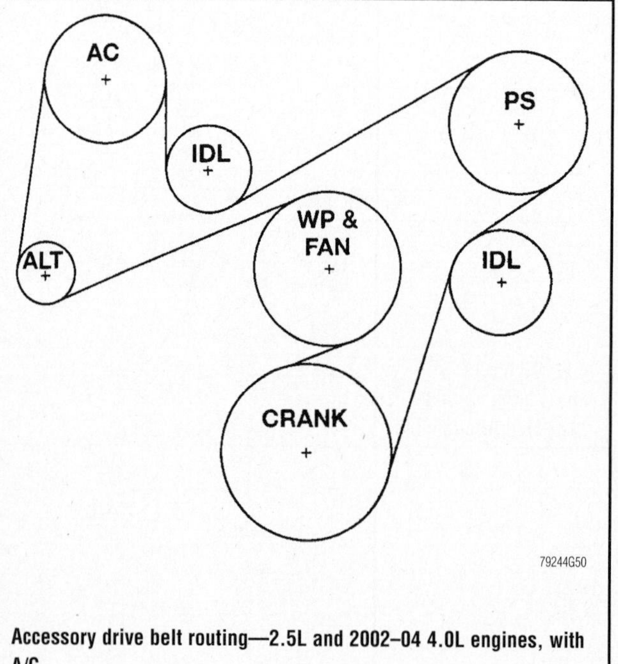

Accessory drive belt routing—2.5L and 2002–04 4.0L engines, with A/C

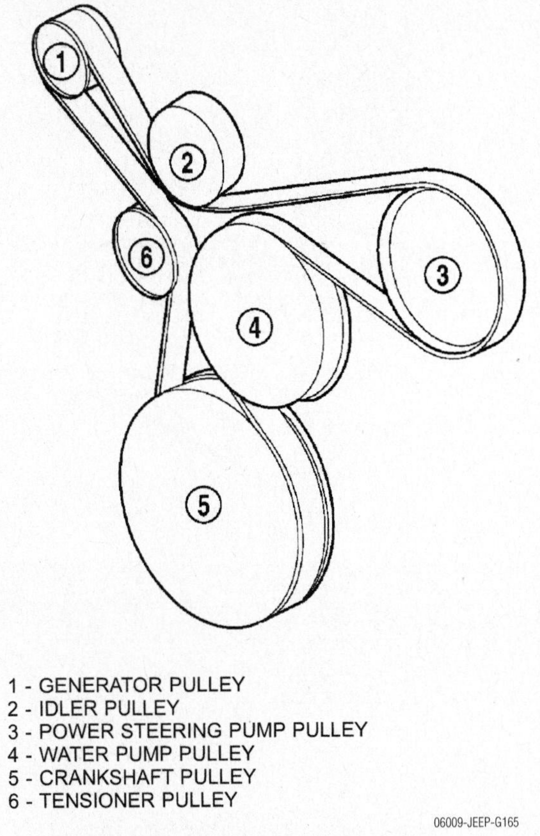

Accessory drive belt routing—2005 4.0L engine without A/C

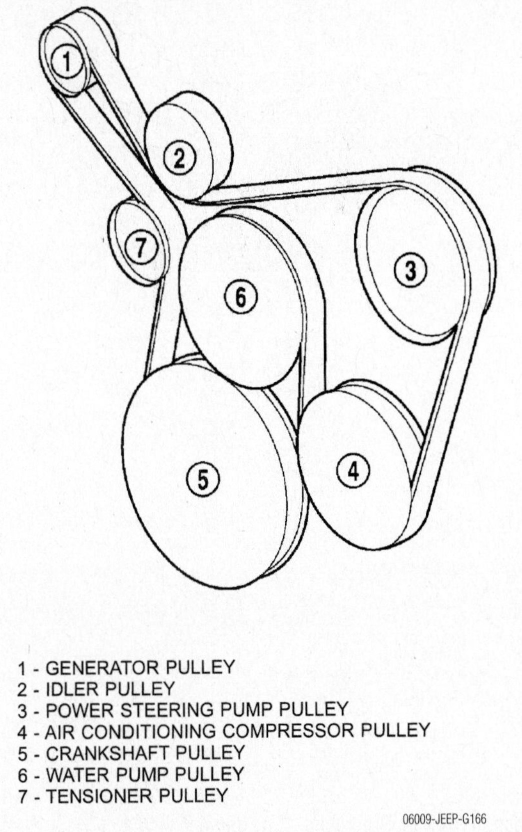

Accessory drive belt routing—2005 4.0L engine with A/C

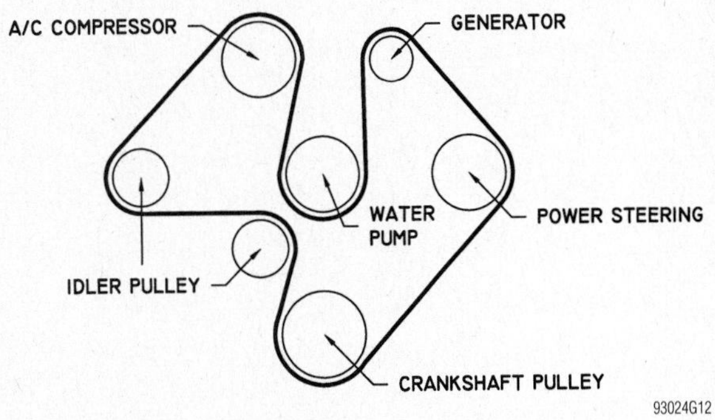

Accessory drive belt routing—3.7L and 4.7L engines

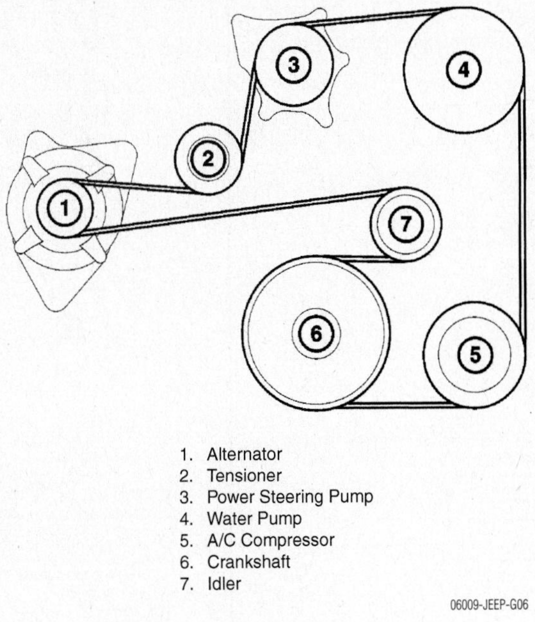

Accessory drive belt routing—5.7L engines

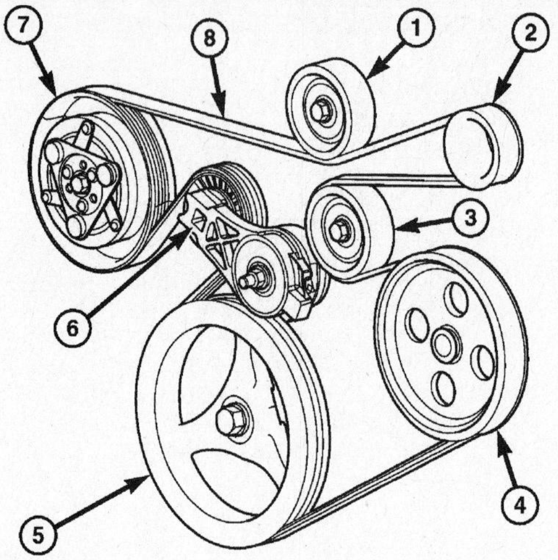

1 - IDLER PULLEY
2 - GENERATOR PULLEY
3 - IDLER PULLEY
4 - POWER STEERING PUMP PULLEY
5 -CRANKSHAFT PULLEY
6 - TENSIONER
7 - A/C COMPRESSOR PULLEY
8 - ACCESSORY DRIVE BELT

06009-JEEP-G07

Accessory drive belt routing—2.4L engines with A/C

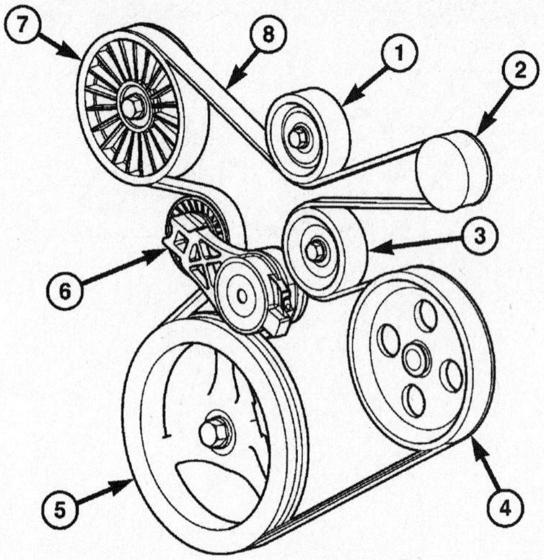

1 - IDLER PULLEY
2 - GENERATOR PULLEY
3 - IDLER PULLEY
4 - POWER STEERING PUMP PULLEY
5 -CRANKSHAFT PULLEY
6 - TENSIONER
7 - NON A/C IDLER PULLEY
8 - ACCESSORY DRIVE BELT

06009-JEEP-G08

Accessory drive belt routing—2.4L engines without A/C

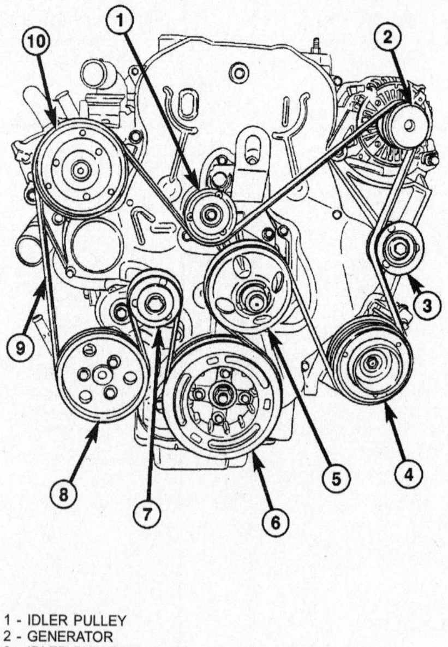

1 - IDLER PULLEY
2 - GENERATOR
3 - IDLER PULLEY
4 - A/C COMPRESSOR
5 - COOLING FAN SUPPORT
6 - VIBRATION DAMPER
7 - BELT TENSIONER
8 - POWER STEERING PUMP
9 - ACCESSORY DRIVE BELT
10 - VISCOUS HEATER

06009-JEEP-G09

Accessory drive belt routing—2.8L diesel engine

CAPACITIES

Year	Model	Engine Displ. Liters	Engine VIN	Engine Oil with Filter	Transmission (pts.) Man.	Transmission (pts.) Auto.**	Transfer Case (pts.)	Drive Axle Front (pts.)	Drive Axle Rear* (pts.)	Fuel Tank (gal.)	Cooling System (qts.)
2002	Grand Cherokee	4.0	S	6.0	—	①	②	3.1	③	20.5	12.0
		4.7	N	6.0	—	①	②	2.5	③	20.5	13.0
	Wrangler	2.5	P	4.0	6.6	17.5	④	3.7	3.5	⑤	9.0
		4.0	S	6.0	6.6	17.5	④	3.7	3.5	⑤	10.5
	Liberty	2.4	1	5.0	4.8	—	⑥	2.6	⑦	18.5	10.1
		3.7	K	5.0	4.8	10.0	⑥	2.6	⑦	18.5	13.0
2003	Grand Cherokee	4.0	S	6.0	—	⑧	⑨	3.1	③	20.0	15.0
		4.7	N	6.0	—	⑧	⑨	2.5	③	20.0	14.5
	Wrangler	2.4	1	4.0	⑩	17.5	⑪	⑫	⑬	19.0	9.0
		4.0	S	6.0	⑩	17.5	⑪	⑫	⑬	19.0	10.5
	Liberty	2.4	1	5.0	4.8	—	⑭	2.6	4.4	18.5	10.1
		3.7	K	5.0	4.8	⑮	⑭	2.6	4.4	18.5	14.0
2004	Grand Cherokee	4.0	S	6.0	—	⑧	⑨	2.5	③	20.0	15.0
		4.7	N	6.0	—	⑧	⑨	2.5	③	20.0	14.5
	Wrangler	2.4	1	4.0	⑯	17.5	⑪	⑫	⑬	19.0	9.0
		4.0	S	6.0	⑯	17.5	⑪	⑫	⑬	19.0	10.5
	Liberty	2.4	1	5.0	4.8	—	⑭	2.6	4.4	19.5	10.1
		3.7	K	5.0	4.8	⑮	⑭	2.6	4.4	19.5	14.0
2005	Grand Cherokee	3.7	K	5.0	—	⑰	⑱	3.6	4.4	20.0	9.0
		4.7	N	6.0	—	⑰	⑱	3.6	4.4	20.0	14.5
		5.7	2	7.0	—	⑰	⑱	3.6	4.4	20.0	14.5
	Wrangler	2.4	1	4.0	3.17	17.5	⑪	⑫	⑬	19.0	9.0
		4.0	S	6.0	3.17	17.5	⑪	⑫	⑬	19.0	10.5
	Liberty	2.4	1	5.0	4.8	—	⑭	2.6	4.4	19.5	10.1
		3.7	K	5.0	4.8	⑮	⑭	2.6	4.4	19.5	14.0
		2.8	5	6.4	3.17	⑮	⑲	2.6	4.4	19.5	11.8

* When equipped with Trac Lok, add 4 oz. of limited slip additive

**Overhaul

① 42RE: 19-22 pts.
44RE: 19-22 pts.
45RFE: 28.0 pts.
46RE: 19.5-28.0 pts.

② 242 NVG: 3.0 pts.
247 NVG: 2.5 pts.
249 NVG: 2.5 pts.

③ 194 RBI: 3.5 pts.
198 RBI: 3.75 pts.
216 RBA: 4.75 pts.
226 RBA: 4.75 pts.

④ Command-Trac:
Automatic: 2.2 pts.
Manual: 3.3 pts.

⑤ Standard: 15.0 gals.
Optional: 19.6 gals.

⑥ NV231: 2.2 pts.
NV242: 2.85 pts.

⑦ Model 35: 3.7 pts.
8 1/4 inch: 4.4 pts.

⑧ 42RE: 19-20 pts.
545RFE: 28.0 pts.
NAG1: 16.3 pts.

⑨ NV147: 3.4 pts.
NV242: 2.85 pts.
NV247: 3.4 pts.

⑩ NV1500: 4.8 pts
NV3550: 4.2 pts

⑪ NV231: 2.2 pts.
NV241: 4.2 pts.

⑫ Model 30: 2.5 pts.
Model 44: 4.0 pts.

⑬ Model 35: 3.5 pts.
Model 44: 4.0 pts.

⑭ NV231: 2.95 pts.
NV241: 3.4 pts.

⑮ 45RFE: 28.0
42RLE: 17.6

⑯ NV231: 2.2 pts.
NV241: 4.2 pts.

⑰ 545RFE 2WD: 11.0
545RFE 4WD: 13.0
NAG1: 16.3

⑱ NV140: 1.4
NV245: 3.8

⑲ NV231: 2.95
NV241 GEN II: 4.2
NV242: 3.4

06009-JEEP-C05

VALVE SPECIFICATIONS

Year	Engine Displ. Liters	Engine VIN	Seat Angle (deg.)	Face Angle (deg.)	Spring Test Pressure (lbs. @ in.)	Spring Installed Height (in.)	Stem-to-Guide Clearance (in.)		Stem Diameter (in.)	
							Intake	Exhaust	Intake	Exhaust
2002	2.4	1	44.5-45	44.5-45	136@1.172	1.496	0.0018-0.0025	0.0029-0.0037	0.2337-0.2344	0.2326-0.2333
	2.5	P	44.5	45	184-196@1.216	1.640	0.0010-0.0030	0.0010-0.0030	0.3110-0.3120	0.3110-0.3120
	3.7	K	44.5-45	45-45.5	221-242@1.107	1.619	0.0008-0.0028	0.0019-0.0039	0.2729-0.2739	0.2717-0.2728
	4.0	S	44.5	45	184-196@1.216	1.640	0.0010-0.0030	0.0010-0.0030	0.3110-0.3120	0.3110-0.3120
	4.7	N	44.5-45	45-45.5	176.2-192.4@1.1532	1.602	0.0011-0.0017	0.0029	0.2728-0.2739	0.2717-0.2728
2003	2.4	1	44.5-45	44.5-45	136@1.172	1.496	0.0018-0.0025	0.0029-0.0037	0.2337-0.2344	0.2326-0.2333
	3.7	K	44.5-45	45-45.5	221-242@1.107	1.619	0.0008-0.0028	0.0019-0.0039	0.2729-0.2739	0.2717-0.2728
	4.0	S	44.5	45	184-196@1.216	1.640	0.0010-0.0030	0.0010-0.0030	0.3110-0.3120	0.3110-0.3120
	4.7	N	44.5-45	45-45.5	176.2-192.4@1.1532	1.602	0.0011-0.0017	0.0029	0.2728-0.2739	0.2717-0.2728
2004	2.4	1	44.5-45	44.5-45	136@1.172	1.496	0.0018-0.0025	0.0029-0.0037	0.2337-0.2344	0.2326-0.2333
	3.7	K	44.5-45	45-45.5	221-242@1.107	1.619	0.0008-0.0028	0.0019-0.0039	0.2729-0.2739	0.2717-0.2728
	4.0	S	44.5	46.5	202-218@1.216	1.640	0.0010-0.0030	0.0010-0.0030	0.3110-0.3120	0.3110-0.3120
	4.7	N	44.5-45	45-45.5	176.2-192.4@1.1532	1.602	0.0011-0.0017	0.0029	0.2728-0.2739	0.2717-0.2728
2005	2.4	1	44.5-45	44.5-45	136@1.172	1.496	0.0018-0.0025	0.0029-0.0037	0.2337-0.2344	0.2326-0.2333
	2.8	5	NA	45.45-55.55	401@1.496	1.496	0.0012-0.0023	0.0012-0.0023	0.2343-0.2350	0.2334-0.2343
	3.7	K	44.5-45	45-45.5	221-242@1.107	1.619	0.0008-0.0028	0.0019-0.0039	0.2729-0.2739	0.2717-0.2728
	4.0	S	44.5	46.5	202-218@1.216	1.640	0.0010-0.0030	0.0010-0.0030	0.3110-0.3120	0.3110-0.3120
	4.7	N	44.5-45	45-45.5	176.2-192.4@1.1532	1.602	0.0011-0.0017	0.0029	0.2728-0.2739	0.2717-0.2728
	5.7	2	44.5-45	45-45.5	95@1.81	1.81	0.0008-0.0025	0.0019-0.0037	0.3120-0.3130	0.3110-0.3120

NA: Information not available

06009-JEEP-C06

CRANKSHAFT AND CONNECTING ROD SPECIFICATIONS

All measurements are given in inches.

	Engine		Crankshaft				Connecting Rod		
Year	Displ. Liters	Engine VIN	Main Brg. Journal Dia.	Main Brg. Oil Clearance	Shaft End-play	Thrust on No.	Journal Diameter	Oil Clearance	Side Clearance
2002	2.4	1	2.3620-2.3625	0.0007-0.0024	0.0035-0.0094	NA	1.9680-1.9685	0.0009-0.0027	0.0050-0.0150
	2.5	P	2.4996-2.5001	0.0010-0.0025	0.0015-0.0065	2	2.2080-2.2085	0.0015-0.0020	0.0100-0.0190
	3.7	K	2.4996-2.5005	0.0020-0.0034	0.0021-0.0112	2	2.2794-2.2797	0.0004-0.0019	0.0040-0.0138
	4.0	S	2.4996-2.5001 ①	0.0010-0.0025	0.0015-0.0065	2	2.0934-2.0955	0.0015-0.0020	0.0100-0.0190
	4.7	N	2.4996-2.5005	0.0002-0.0013	0.0021-0.0112	2	2.0076-2.0082	0.0004-0.0019	0.0040-0.0138
2003	2.4	1	2.3620-2.3625	0.0007-0.0024	0.0035-0.0094	NA	1.9680-1.9685	0.0009-0.0027	0.0050-0.0150
	3.7	K	2.4996-2.5005	0.0020-0.0034	0.0021-0.0112	2	2.2794-2.2797	0.0004-0.0019	0.0040-0.0138
	4.0	S	2.4996-2.5001 ①	0.0010-0.0025	0.0015-0.0065	2	2.0934-2.0955	0.0015-0.0020	0.0100-0.0190
	4.7	N	2.4996-2.5005	0.0002-0.0013	0.0021-0.0112	2	2.0076-2.0082	0.0004-0.0019	0.0040-0.0138
2004	2.4	1	2.3620-2.3625	0.0007-0.0024	0.0035-0.0094	NA	1.9680-1.9685	0.0009-0.0027	0.0050-0.0150
	3.7	K	2.4996-2.5005	0.0020-0.0034	0.0021-0.0112	2	2.2794-2.2797	0.0004-0.0019	0.0040-0.0138
	4.0	S	2.4996-2.5001 ①	0.0010-0.0025	0.0015-0.0065	2	2.0934-2.0955	0.0015-0.0020	0.0100-0.0190
	4.7	N	2.4996-2.5005	0.0002-0.0013	0.0021-0.0112	2	2.0076-2.0082	0.0004-0.0019	0.0040-0.0138
2005	2.4	1	2.3620-2.3625	0.0007-0.0024	0.0035-0.0094	NA	1.9680-1.9685	0.0009-0.0027	0.0050-0.0150
	2.8	5	②	③	0.0031-0.0110	3	2.1236-2.1242	0.0008-0.0029	NA
	3.7	K	2.4996-2.5005	0.0020-0.0034	0.0021-0.0112	2	2.2794-2.2797	0.0004-0.0019	0.0040-0.0138
	4.0	S	2.4996-2.5001 ①	0.0010-0.0025	0.0015-0.0065	2	2.0934-2.0955	0.0015-0.0020	0.0100-0.0190
	4.7	N	2.4996-2.5005	0.0002-0.0013	0.0021-0.0112	2	2.0076-2.0082	0.0004-0.0019	0.0040-0.0138
	5.7	2	2.5585-2.5595	0.0009-0.0020	0.0020-0.0110	3	2.1250-2.1260	0.0007-0.0023	0.0030-0.0137

NA: Information not available

① No 7: 2.4980-2.4995

② No.1: 2.4797-2.4805
No. 2: 2.4805-2.4811
No. 3: 3.5425-3.5433

③ No.1: 0.0010 max.
No.2: 0.0002-0.0003
No. 3: 0.0010-0.0030

06009-JEEP-C07

PISTON AND RING SPECIFICATIONS

All measurements are given in inches.

Year	Engine Displ. Liters	Engine VIN	Piston Clearance	Ring Gap: Top Compression	Ring Gap: Bottom Compression	Ring Gap: Oil Control	Ring Side Clearance: Top Compression	Ring Side Clearance: Bottom Compression	Ring Side Clearance: Oil Control
2002	2.4	1	0.0009-0.0022	0.0098-0.0200	0.0090-0.0180	0.0098-0.0250	0.0011-0.0031	0.0011-0.0031	0.0004-0.0070
	2.5	P	0.0013-0.0021	0.0090-0.0240	0.0190-0.0380	0.0100-0.0600	0.0017-0.0033	0.0017-0.0033	0.0024-0.0083
	3.7	K	0.0014	0.0146-0.0249	0.0146-0.0249	0.0100-0.0300	0.0020-0.0037	0.0016-0.0031	0.0007-0.0091
	4.0	S	0.0008-0.0015	0.0090-0.0240	0.0190-0.0380	0.0100-0.0600	0.0017-0.0033	0.0017-0.0033	0.0024-0.0083
	4.7	N	0.0008-0.0020	0.0146-0.0249	0.0146-0.0249	0.0100-0.0500	0.0020-0.0041	0.0016-0.0032	0.0007-0.0091
2003	2.4	1	0.0009-0.0022	0.0098-0.0200	0.0090-0.0180	0.0098-0.0250	0.0011-0.0031	0.0011-0.0031	0.0004-0.0070
	3.7	K	0.0014	0.0146-0.0249	0.0146-0.0249	0.0100-0.0300	0.0020-0.0037	0.0016-0.0031	0.0007-0.0091
	4.0	S	0.0008-0.0015	0.0090-0.0240	0.0190-0.0380	0.0100-0.0600	0.0017-0.0033	0.0017-0.0033	0.0024-0.0083
	4.7	N	0.0008-0.0020	0.0146-0.0249	0.0146-0.0249	0.0100-0.0500	0.0020-0.0041	0.0016-0.0032	0.0007-0.0091
2004	2.4	1	0.0009-0.0022	0.0098-0.0200	0.0090-0.0180	0.0098-0.0250	0.0011-0.0031	0.0011-0.0031	0.0004-0.0070
	3.7	K	0.0014	0.0146-0.0249	0.0146-0.0249	0.0100-0.0300	0.0020-0.0037	0.0016-0.0031	0.0007-0.0091
	4.0	S	0.0008-0.0015	0.0090-0.0240	0.0190-0.0380	0.0100-0.0600	0.0017-0.0033	0.0017-0.0033	0.0024-0.0083
	4.7	N	0.0008-0.0020	0.0146-0.0249	0.0146-0.0249	0.0100-0.0500	0.0020-0.0041	0.0016-0.0032	0.0007-0.0091
2005	2.4	1	0.0009-0.0022	0.0098-0.0200	0.0090-0.0180	0.0098-0.0250	0.0011-0.0031	0.0011-0.0031	0.0004-0.0070
	2.8	5	0.0003-0.0008	0.0012-0.0018	0.0012-0.0019	0.0009-0.0019	0.0031-0.0054	0.0028-0.0043	0.0016-0.0031
	3.7	K	0.0014	0.0146-0.0249	0.0146-0.0249	0.0100-0.0300	0.0020-0.0037	0.0016-0.0031	0.0007-0.0091
	4.0	S	0.0008-0.0015	0.0090-0.0240	0.0190-0.0380	0.0100-0.0600	0.0017-0.0033	0.0017-0.0033	0.0024-0.0083
	4.7	N	0.0008-0.0020	0.0146-0.0249	0.0146-0.0249	0.0100-0.0500	0.0020-0.0041	0.0016-0.0032	0.0007-0.0091
	5.7	2	0.0008-0.0019	0.0090-0.0149	0.0137-0.0236	0.0059-0.0259	0.0007-0.0026	0.0007-0.0022	0.0007-0.0091

06009-JEEP-C08

TORQUE SPECIFICATIONS

All readings in ft. lbs.

Year	Engine Displ. Liters	Engine VIN	Cylinder Head Bolts	Main Bearing Bolts	Rod Bearing Bolts	Crankshaft Damper Bolts	Flywheel Bolts	Manifold Intake	Manifold Exhaust	Spark Plugs	Oil Pan Drain Plug
2002	2.4	1	①	②	③	100	④	21 ⑤	17	20	20
	2.5	P	⑥	80	33	80	105	⑦	⑦	27	25
	3.7	K	⑧	⑨	⑩	130	45	9	18	20	25
	4.0	S	⑪	80	33	80	105	⑫	⑫	27	25
	4.7	N	⑬	⑭	15 ⑮	130	45	9	18	27	25
2003	2.4	1	①	②	③	100	④	21	17	20	20
	3.7	K	⑧	⑨	⑩	130	45	9	18	20	25
	4.0	S	⑪	80	33	80	105	⑫	⑫	27	25
	4.7	N	⑬	⑭	15 ⑮	130	45	9	18	27	25
2004	2.4	1	①	⑯	⑰	100	④	21 ⑤	17	20	20
	3.7	K	⑧	⑨	⑩	130	45	9	18	20	25
	4.0	S	⑪	80	33	80	105	⑱	⑫	27	25
	4.7	N	⑬	⑭	15 ⑮	130	45	9	18	27	25
2005	2.4	1	①	⑲	⑩	100	④	21 ⑤	17	13	20
	2.8	5	⑬	32 ⑳	㉑	203	32 ㉒	⑬	26	—	NA
	3.7	K	⑧	⑨	⑩	130	70	9	18	20	25
	4.0	S	⑪	80	33	80	105	⑱	⑫	22	25
	4.7	N	⑬	㉓	⑩	130	45	9	18	20	25
	5.7	2	⑬	㉔	㉕	130	55	9	18	13	20

NA: Information not available

① Step 1: 25 ft. lbs.
Step 2: 50 ft. lbs.
Step 3: 50 ft. lbs.
Step 4: Plus 1/4 turn

② M8 bolts: 21 ft. lbs.
M11 bolts: 30 ft. lbs. plus 1/4 turn

③ 40 ft. lbs. plus 1/4 turn

④ Flexplate: 70 ft. lbs.
Flywheel: 60 ft. lbs.

⑤ Lower manifold

⑥ Step 1: 22 ft. lbs.
Step 2: 45 ft. lbs.
Step 3: Bolts 1-6 to 110 ft. lbs.
Step 4: Bolt 7 to 100 ft. lbs.
Step 5: Bolts 8-10 to 110 ft. lbs.

⑦ Bolt 1: 30 ft. lbs.
Bolts 2-7: 23 ft. lbs.

⑧ See procedure

⑨ Bed plate bolt sequence. Refer to illustration
Step 1: Hand tighten bolts 1D,1G and 1F until the bedplate contacts the block.
Step 2: Tighten bolts 1A - 1J to 54 N·m (40 ft. lbs.)
Step 3: Tighten bolts 1 - 8 to 7 N·m (5 ft. lbs.)
Step 4: Turn bolts 1 - 8 an additional 90°.
Step 5: Tighten bolts A - E 27 N·m (20 ft. lbs.).

⑩ 20 ft. lbs. + 90 degrees

⑪ Step 1: 22 ft. lbs.
Step 2: 45 ft. lbs.
Step 3: Bolts 1-10 to 110 ft. lbs.
Step 4: Bolt 11 to 100 ft. lbs.
Step 5: Bolts 12-14 to 110 ft. lbs.

⑫ Bolts 1-5 and 8-11: 24 ft. lbs.
Bolts 6-7: 23 ft. lbs.

⑬ See text

⑭ See text
Step 1: Bolts 1-10 to 25 inch lbs
Step 2: Bolts 1-10 plus 90 degrees
Step 3: Bolts A-K to 40 ft. lbs.
Step 4: Bolts A1-A5 to 20 ft. lbs.

⑮ Plus 110 degrees

⑯ M8 bolts: 21 ft. lbs.
M11 bolts: 55 ft. lbs.

⑱ 1-5 and 8-11: 24 ft. lbs.
6 and 7: 11 ft. lbs.

⑲ See accompanying illustration
Step 1: Hold crankshaft forward with a wood wedge
Step 2: Bolts 1-10 30 ft. lbs.
Step 3: Remove wedge
Step 4: Bolts 1-10 30 ft. lbs.
Step 5: Bolts 1-10: Plus 90 degree turn
Step 6: Install bedplate
Step 7: Bolts 11-20: 20 ft. lbs.

⑳ Main bearing supports

㉑ Step 1: 88 inch lbs.
Step 2: 22 ft. lbs.
Step 3: Plus 60 degrees
Step 4: 65 ft. lbs.

㉒ Flexplate to flexplate hub

㉓ Bed plate bolt sequence. Refer to illustration
Step 1: Bolts A-L to 40 ft. lbs.
Step 2: Bolts 1-10 25 inch lbs.
Step 3: Bolts 1-10 plus 90 degrees
Step 4: Bolts A1-A6 20 ft. lbs.

㉔ See the illustration
Step 1: Bolts 1-10 to 20 ft. lbs.
Step 2: bolts 1-10 an additional 90 degrees
Step 3: Bolts (cross bolts) A-J to 21 ft. lbs.

㉕ 15 ft. lbs., plus 90 degrees

06009-JEEP-C09

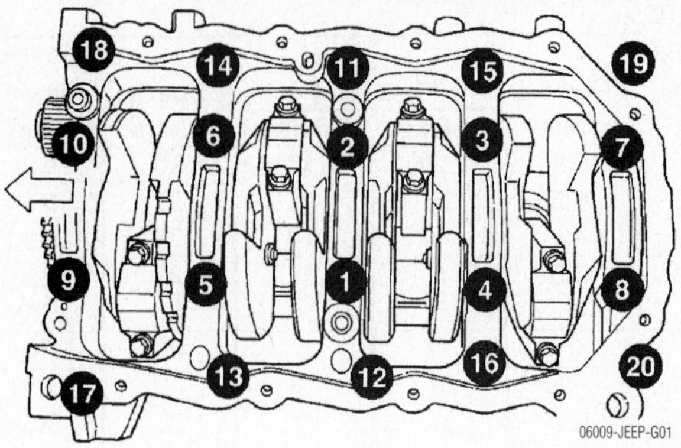

Bedplate torque sequence—2.4L engines

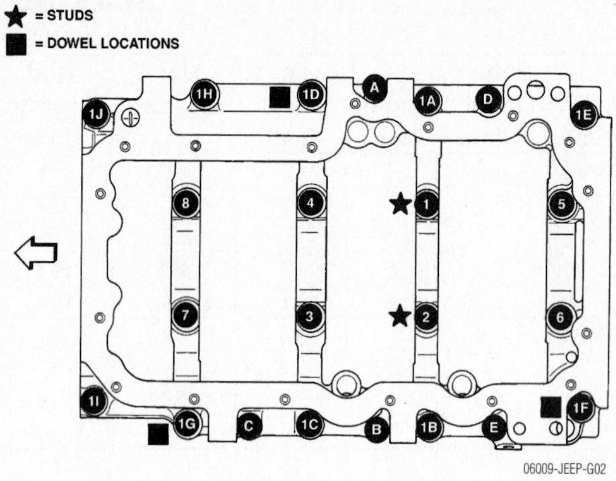

Bedplate torque sequence—3.7L engines

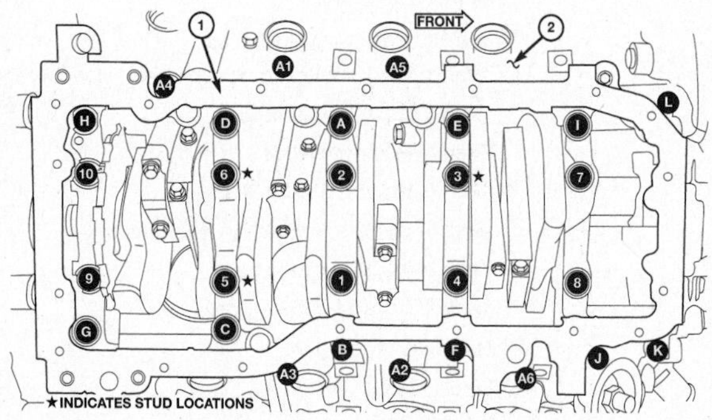

NOTE: Lubricate the bedplate retaining bolts with clean engine oil prior to installation.

06009-JEEP-G03

Bedplate torque sequence—4.7L engines

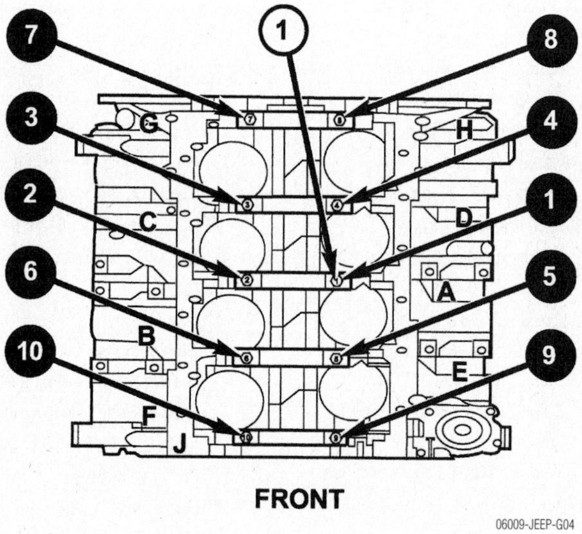

Main bearing cap torque sequence—5.7L engines

WHEEL ALIGNMENT

Year	Model		Caster Range (+/-Deg.)	Caster Preferred Setting (Deg.)	Camber Range (+/-Deg.)	Camber Preferred Setting (Deg.)	Toe-in (deg.)
2002	Liberty	F	0.60	+3.5	0.375	0	0.20+/-0.125
		R	—	—	0.375	-0.25	0.25+/-0.41
	Grand Cherokee	F	0.75	+6.75	0.37	-0.37	0.20+/-0.20
		R	—	—	0.25	-0.25	0.25+/-0.25
	Wrangler	F	1.00	+7.00	0.63	-0.25	0.15+/-0.07
		R	—	—	0.25	-0.25	0.25+/-0.25
2003	Liberty	F	0.50	+3.9	0.50	0	0.20+/-0.125
		R	—	—	0.375	-0.25	0.25+/-0.41
	Grand Cherokee	F	0.75	+6.75	0.37	-0.37	0.20+/-0.20
		R	—	—	0.25	-0.25	0.25+/-0.25
	Wrangler	F	1.00	+7.00	0.63	-0.25	0.15+/-0.07
		R	—	—	0.25	-0.25	0.25+/-0.25
2004	Liberty	F	0.50	+3.9	0.375	-0.375	0.20+/-0.125
		R	—	—	0.375	-0.25	0.25+/-0.41
	Grand Cherokee	F	0.75	+6.75	0.37	-0.37	0.20+/-0.20
	Standard	R	—	—	0.25	-0.25	0.25+/-0.25
	Grand Cherokee	F	0.80	+6.50	0.37	-0.75	0.20+/-0.20
	Up-Country	R	—	—	0.25	-0.25	0.25+/-0.25
	Wrangler	F	1.00	+7.00	0.63	-0.25	0.15+/-0.07
		R	—	—	0.25	-0.25	0.25+/-0.25
2005	Liberty	F	0.50	+3.9	0.375	-0.375	0.20+/-0.125
		R	—	—	0.375	-0.25	0.25+/-0.41
	Grand Cherokee	F	0.50	+4.00	0.50	-0.25	0.25+/-0.25
		R	—	—	0.25	-0.25	0.25+/-0.25
	Wrangler	F	1.00	+7.00	0.63	-0.25	0.15+/-0.07
		R	—	—	0.25	-0.25	0.25+/-0.25

06009-JEEP-C10

TIRE, WHEEL AND BALL JOINT SPECIFICATIONS

Year	Model	OEM Tires Standard	OEM Tires Optional	Tire Pressures (psi) Front	Tire Pressures (psi) Rear	Wheel Size	Ball Joint Inspection	Lug Nut Torque (ft. lbs.)
2002	Liberty	P215/75R16	P235/70R16 P225/70R15	①	①	①	②	85-115
	Grand Cherokee	P215/75R15	P225/75R15 P235/75R15 P245/70R15 P225/70R16	33	33	7-J	②	85-115
	Wrangler SE	P205/75R15	P215/75R15 P225/75R15	33	33	6JJ/7JJ 7JJ	②	85-115
	Wrangler Sport	P215/75R15	P225/75R15 30x9.5R15LT	33 29	33 29	6JJ/7JJ 8JJ	②	85-115
	Wrangler Sahara	P225/70R15	30x9.5R15LT	Std: 33 Opt: 29	Std: 33 Opt: 29	7JJ/8JJ	②	85-115
2003	Liberty	P215/75R16	P235/70R16 P235/65R17	①	①	①	②	85-115
	Grand Cherokee	P215/75R15	P225/75R15 P235/75R15 P245/70R15 P225/70R16	33	33	7-J	②	85-115
	Wrangler	P215/75R15	P225/75R15 LT245/75R16 30x9.5R15LT	①	①	①	②	85-115
2004	Liberty	P215/75R16	P235/70R16 P235/65R17	①	①	①	②	85-115
	Grand Cherokee	P225/75R16	P245/70R16 P235/65R17 235/65HR17	①	①	①	②	85-115
	Wrangler SE	P215/75R15	P225/75R15	①	①	std: 6 opt.: 7	②	85-115
	Wrangler Sport	P215/75R15	P225/75R15 30x9.5R15LT	①	①	6JJ/7JJ 8JJ	②	85-115
	Wrangler Sahara	30x9.5R15LT	none	①	①	8	②	85-115
	Wrangler X	P215/75R15	P225/75R15	①	①	7	②	85-115
	Wrangler Rubicon	LT245/75R16	none	①	①	8	②	85-115
2005	Liberty	P215/75R16	P225/75R15 P235/70R16 P235/65R17	①	①	①	②	85-115
	Grand Cherokee	P235/65R17	P245/65R17	①	①	①	②	85-115
	Wrangler SE	P215/75R15	P225/75R15	①	①	std: 6 opt.: 7	②	85-115
	Wrangler Sport	P215/75R15	P225/75R15 30x9.5R15LT	①	①	6JJ/7JJ 8JJ	②	85-115
	Wrangler X	P215/75R15	P225/75R15	①	①	7	②	85-115
	Wrangler Rubicon	LT245/75R16	none	①	①	8	②	85-115

OEM: Original Equipment Manufacturer
STD: Standard
OPT: Optional
① See placard on vehicle
② Replace if any measurable movement is found.

06009-JEEP-C11

BRAKE SPECIFICATIONS

All measurements in inches unless noted

			Brake Disc			Brake Drum			Minimum Lining Thickness		Caliper Mounting Bolts
Year	Model		Original Thickness	Minimum Thickness	Maximum Run-out	Original Inside Diameter	Max. Wear Limit	Maximum Machine Diameter	Front	Rear	(ft. lbs.)
2002	Liberty		NA	0.8937	0.005	①	—	②	0.030	③	11
	Grand Cherokee	F	NA	0.965	0.003	—	—	—	0.030	—	21-30
		R	NA	0.335	0.003	—	—	—	—	0.030	21-30
	Wrangler		0.94	0.89	0.005	9.00	—	9.06	0.030	③	11
2003	Liberty	F	1.102	1.0236	0.004	—	—	—	NA	NA	11
		R	0.472	0.4331	0.004	—	—	—	NA	NA	18
	Grand Cherokee	F	NA	0.965	0.003	—	—	—	0.030	—	53
		R	NA	0.335	0.003	—	—	—	—	0.030	26
	Wrangler	F	0.940	0.894	0.005	—	—	—	0.030	③	11
		R	0.470	0.433	0.004	9.00	—	9.06	—	0.030	18
2004	Liberty	F	1.102	1.0236	0.004	—	—	—	NA	NA	11
		R	0.472	0.4331	0.004	—	—	—	NA	NA	18
	Grand Cherokee	F	NA	0.965	0.003	—	—	—	0.030	—	53
		R	NA	0.335	0.003	—	—	—	—	0.030	26
	Wrangler	F	0.940	0.0937	0.005	—	—	—	0.030	—	11
		R	0.472	0.4330	0.004	9.00	—	9.06	—	③	18
2005	Liberty	F	1.102	1.0236	0.004	—	—	—	NA	NA	11
		R	0.472	0.4331	0.004	—	—	—	NA	NA	18
	Grand Cherokee	F	NA	1.122	0.0008	—	—	—	0.030	—	18
		R	NA	0.492	0.0008	—	—	—	—	0.030	18
	Wrangler	F	0.940	0.837	0.005	—	—	—	0.030	—	11
		R	0.472	0.4330	0.004	9.00	—	9.06	—	③	18

NA: Information not available

① Standard: 9.00 in.
Optional 10.00 in.

② Standard: 9.06 in.
Optional 10.06 in.

③ Riveted brake shoes: 0.030 in.
Bonded brake shoes: 0.060 in.

06009-JEEP-C12

SCHEDULED MAINTENANCE INTERVALS
2002-03 Wrangler, Grand Cherokee, Liberty

TO BE SERVICED	TYPE OF SERVICE	VEHICLE MILEAGE INTERVAL (x1000)												
		7.5	15	22.5	30	37.5	45	52.5	60	67.5	75	82.5	90	97.5
Engine oil & filter	R	✓	✓	✓	✓	✓	✓	✓	✓	✓	✓	✓	✓	✓
Brake hoses & linings	S/I	✓	✓	✓	✓	✓	✓	✓	✓	✓	✓	✓	✓	✓
Engine coolant level, hoses & clamps	S/I	✓	✓	✓	✓	✓	✓	✓	✓	✓	✓	✓	✓	✓
Exhaust system	S/I	✓	✓	✓	✓	✓	✓	✓	✓	✓	✓	✓	✓	✓
Lubricate steering linkage (4x2)	S/I	✓	✓	✓	✓	✓	✓	✓	✓	✓	✓	✓	✓	✓
Lubricate steering linkage (4x4)	S/I	✓		✓		✓		✓		✓		✓		✓
Air filter	R				✓				✓				✓	
Automatic transmission fluid & filter	R				✓				✓				✓	
Spark plugs	R				✓				✓				✓	
Transfer case fluid	R				✓				✓				✓	
Drive belts	S/I				✓				✓				✓	
Front & rear axle oil	R				✓				✓				✓	
Prop shaft universal joints	S/I				✓				✓				✓	
Rotate tires	S/I				✓				✓				✓	
Engine coolant	R						✓				✓			
Manual transmission fluid	R					✓					✓			
Distributor cap & rotor	R								✓					
Fuel filter	R								✓					
Ignition cables	R								✓					

R: Replace S/I: Service or Inspect

FREQUENT OPERATION MAINTENANCE (SEVERE SERVICE)

If a vehicle is operated under any of the following conditions it is considered severe service:

- Extremely dusty areas.

- 50% or more of the vehicle operation is in 32°C (90°F) or higher temperatures, or constant operation in temperatures below 0°C (32°F).

- Prolonged idling (vehicle operation in stop and go traffic.)

- Frequent short running periods (engine does not warm to normal operating temperatures).

- Police, taxi, delivery usage or trailer towing usage.

Oil & oil filter change: change every 3000 miles.

Automatic transmission fluid, filter & bands: change & adjust every 12,000 miles.

Brake hoses & linings: check every 12,000 miles.

Liubricate steering linkage: check every 3000 miles.

Manual transmission fluid: change every 18,000 miles.

Prop shaft universal joints: lubricate every 3000 miles.

Front & rear axle oil: change every 12,000 miles.

06009-JEEP-C13

SCHEDULED MAINTENANCE INTERVALS
2004 Liberty

TO BE SERVICED	TYPE OF SERVICE	VEHICLE MILEAGE INTERVAL (x1000)												
		6	12	18	24	30	36	42	48	54	60	66	72	78
Engine oil & filter	R	✓	✓	✓	✓	✓	✓	✓	✓	✓	✓	✓	✓	✓
Brake hoses & linings	I			✓			✓			✓			✓	
Air filter	R					✓					✓			
Spark plugs	R				✓				✓				✓	
Transfer case fluid level	I					✓					✓			
Accessory drive belt ①	I/R												✓	
PCV valve	R										✓			
Ignition cables (2.4L)	R										✓			
Engine coolant	R	Replace every 5 years, regardless of mileage												
Transfer case fluid	R	Drain, flush and refill every 120,000 miles												
Timing belt (2.4L)	R	Replace every 120,000 miles												

R: Replace S/I: Service or Inspect I/R: Inspect and replace as necessary

① Replace at 72,000 miles if not replaced at 60,000 during inspection

FREQUENT OPERATION MAINTENANCE (SEVERE SERVICE)

If a vehicle is operated under any of the following conditions it is considered severe service:

- Extremely dusty areas.

- 50% or more of the vehicle operation is in 32°C (90°F) or higher temperatures, or constant operation in temperatures below 0°C (32°F).

- Prolonged idling (vehicle operation in stop and go traffic.)

- Frequent short running periods (engine does not warm to normal operating temperatures).

- Police, taxi, delivery usage or trailer towing usage.

Oil & oil filter change: change every 3000 miles.

Air filter element: inspect and replace as necessary every 12,000 miles

PCV valve: replace every 30,000 miles

Automatic transmission fluid, filter: change every 12,000 miles.

Brake hoses & linings: check every 12,000 miles.

Accessory drive belt: inspect and replace as necessary every 45,000 miles

Transfer case fluid: replace every 60,000 miles

Engine coolant: replace every 99,000 miles

Front & rear axle oil: change every 12,000 miles.

06009-JEEP-C14

SCHEDULED MAINTENANCE INTERVALS
2005 Liberty w/Gasoline Engines

TO BE SERVICED	TYPE OF SERVICE	VEHICLE MILEAGE INTERVAL (x1000)												
		3	6	9	12	15	18	21	24	27	30	33	36	39
Engine oil & filter	R	✓	✓	✓	✓	✓	✓	✓	✓	✓	✓	✓	✓	✓
Tires	Rotate		✓		✓		✓		✓		✓		✓	
Brake hoses & linings	I				✓				✓				✓	
Drive axle fluid	R				✓				✓				✓	
Air filter	I/R					✓					✓			
Spark plugs	R										✓			
Transfer case fluid level	I										✓			
PCV valve	I/R										✓			
Transfer case fluid	R	Drain, flush and refill every 60,000 miles												
Accessory drive belt	I/R	Every 45,000 miles												
Engine coolant	R	Every 60,000 miles												
Automatic trans. Fluid & filter	R	Every 60,000 miles												
Timing belt (2.4L)	R	Replace every 90,000 miles												
Ignition cables (2.4L)	R	Every 60,000 miles												

R: Replace S/I: Service or Inspect I/R: Inspect and replace as necessary

The above schedule is to be used if you drive under any of the following conditions:

Driving in temperatures under 32 degrees F
Stop and go traffic
Extensive engine idling
Driving in dusty conditions
Frequent trips under 10 miles
More than 50 % of your driving is in hot weather (90 deg. F) above 50 miles per hour
Trailer towing
Taxi, police or delivery service
Off-road driving

If none of these conditions is met, double the maintenance intervals

06009-JEEP-C15

SCHEDULED MAINTENANCE INTERVALS
2005 Liberty w/2.8L Diesel Engine

TO BE SERVICED	TYPE OF SERVICE	VEHICLE MILEAGE INTERVAL (x1000)												
		6.25	12.5	18.75	25	31.25	37.5	43.75	50	56.25	62.5	68.75	75	81.25
Engine oil & filter	R	✓	✓	✓	✓	✓	✓	✓	✓	✓	✓	✓	✓	✓
Brake linings	I		✓		✓		✓		✓		✓		✓	
Fuel/water separator	R				✓				✓				✓	
Air filter	I/R	✓	✓	✓	✓	✓	✓	✓	✓	✓	✓	✓	✓	✓
Air filter	R		✓		✓		✓		✓		✓		✓	
Drive axle fluid	R		✓		✓		✓		✓		✓		✓	
Automatic trans. fluid & filter	R										✓			
Transfer case fluid	R										✓			
Accessory drive belt	R						✓						✓	
Engine coolant	R	Every 100,000 miles												
Timing belt and idler pulleys	R	Every 100,000 miles												
Timing belt tensioner	R	Every 100,000 miles												

R: Replace S/I: Service or Inspect I/R: Inspect and replace as necessary

The above schedule is to be used if you drive under any of the following conditions:

Driving in temperatures under 32 degrees F
Stop and go traffic
Extensive engine idling
Driving in dusty conditions
Frequent trips under 10 miles
More than 50 % of your driving is in hot weather (90 deg. F) above 50 miles per hour
Trailer towing
Taxi, police or delivery service
Off-road driving

If none of these conditions is met, double the maintenance intervals

06009-JEEP-C16

SCHEDULED MAINTENANCE INTERVALS
2004 Wrangler

		VEHICLE MILEAGE INTERVAL (x1000)												
TO BE SERVICED	**TYPE OF SERVICE**	**6**	**12**	**18**	**24**	**30**	**36**	**42**	**48**	**54**	**60**	**66**	**72**	**78**
Engine oil & filter	R	✓	✓	✓	✓	✓	✓	✓	✓	✓	✓	✓	✓	✓
Brake hoses & linings	S/I			✓			✓			✓			✓	
Engine coolant level, hoses & clamps	S/I	✓	✓	✓	✓	✓	✓	✓	✓	✓	✓	✓	✓	✓
Lubricate steering linkage and outer tie rod ends	S/I	✓	✓	✓	✓	✓	✓	✓	✓	✓	✓	✓	✓	✓
Lubricate steering and suspension ball joints	S/I		✓		✓		✓		✓		✓		✓	
Air filter	R					✓					✓			
Spark plugs	R					✓					✓			
Engine coolant	R						✓				✓			
Ignition cables (2.4L)	R										✓			
Transfer case fluid level	I					✓					✓			
PCV valve	I/R										✓			
Accessory drive belt ①	S/I										✓			
Timing belt	R	Replace every 120,000 miles												
Transfer case fluid	R	Replace every 120,000 miles												

R: Replace S/I: Service or Inspect

① Replace every 72,000 miles

FREQUENT OPERATION MAINTENANCE (SEVERE SERVICE)

If a vehicle is operated under any of the following conditions it is considered severe service:

- Extremely dusty areas.

- 50% or more of the vehicle operation is in 32°C (90°F) or higher temperatures, or constant operation in temperatures below 0°C (32°F).

- Prolonged idling (vehicle operation in stop and go traffic.)

- Frequent short running periods (engine does not warm to normal operating temperatures).

- Police, taxi, delivery usage or trailer towing usage.

Oil & oil filter change: change every 3000 miles.

Air filter element: inspect and replace as necessary every 12,000 miles

PCV valve: replace every 30,000 miles

Automatic transmission fluid, filter: change every 12,000 miles.

Brake hoses & linings: check every 12,000 miles.

Accessory drive belt: inspect and replace as necessary every 45,000 miles

Transfer case fluid: replace every 60,000 miles

Engine coolant: replace every 99,000 miles

Front & rear axle oil: change every 12,000 miles.

06009-JEEP-C17

SCHEDULED MAINTENANCE INTERVALS
2005 Wrangler

TO BE SERVICED	TYPE OF SERVICE	VEHICLE MILEAGE INTERVAL (x1000)												
		3	6	9	12	15	18	21	24	27	30	33	36	39
Engine oil & filter	R	✓	✓	✓	✓	✓	✓	✓	✓	✓	✓	✓	✓	✓
Tires	Rotate		✓		✓		✓		✓		✓		✓	
Brake hoses & linings	S/I				✓				✓				✓	
Drive axle fluid	R				✓				✓				✓	
Engine coolant level, hoses & clamps	S/I	✓	✓	✓	✓	✓	✓	✓	✓	✓	✓	✓	✓	✓
Outer tie rod ends	L	✓	✓	✓	✓	✓	✓	✓	✓	✓	✓	✓	✓	✓
Ball joints	L		✓		✓		✓		✓		✓		✓	
Air filter	I/R					✓					✓			
Spark plugs	R										✓			
Transfer case fluid level	I										✓			
PCV valve	I/R										✓			
Transfer case fluid	R	Every 60,000 miles												
Automatic trans. Fluid & filter	R	Every 60,000 miles												
Accessory drive belt	I/R	Every 45,000 miles												
Timing belt (2.4L)	R	Every 90,000 miles												
Engine coolant	R	Every 60,000 miles												
Ignition cables (2.4L)	I/R	Every 60,000 miles												

R: Replace S/I: Service or Inspect I/R: Inspect and replace if necessary L: Lubricate

The above schedule is to be used if you drive under any of the following conditions:

Driving in temperatures under 32 degrees F
Stop and go traffic
Extensive engine idling
Driving in dusty conditions
Frequent trips under 10 miles
More than 50 % of your driving is in hot weather (90 deg. F) above 50 miles per hour
Trailer towing
Taxi, police or delivery service
Off-road driving

If none of these conditions is met, double the maintenance intervals

06009-JEEP-C18

SCHEDULED MAINTENANCE INTERVALS
2004 Grand Cherokee

TO BE SERVICED	TYPE OF SERVICE	VEHICLE MILEAGE INTERVAL (x1000)												
		6	12	18	24	30	36	42	48	54	60	66	72	78
Accessory drive belt ①	S/I													✓
Engine oil & filter	R	✓	✓	✓	✓	✓	✓	✓	✓	✓	✓	✓	✓	✓
Brake hoses & linings	S/I			✓			✓			✓			✓	
Lubricate steering and suspension ball joints	S/I		✓		✓		✓		✓		✓		✓	
Brake caliper pins	C/L			✓			✓			✓			✓	
Air filter	R					✓					✓			
Spark plugs	R					✓					✓			
Transfer case fluid level ②	I					✓					✓			
Transfer case fluid ③	R					✓					✓			
Engine coolant	R										✓			
PCV valve ④	I/R										✓			
Automatic transmission	⑤	every 100,000 miles												
Transfer case fluid ②	R	Replace every 120,000 miles												

R: Replace S/I: Service or Inspect C/L: Clean and lubricate

① Replace as necessary....4.0L only

② Selec-Trac only

③ Quadra-Trac only

④ 4.7L only

⑤ 4.0L only: Drain and refill the fluid, adjust the bands and change the filter
4.7L Only: Drain and refill the automatic transmission fluid, replace main sump filter, and spin-on cooler return filter (if equipped)

FREQUENT OPERATION MAINTENANCE (SEVERE SERVICE)

If a vehicle is operated under any of the following conditions it is considered severe service:

- Extremely dusty areas.

- 50% or more of the vehicle operation is in 32°C (90°F) or higher temperatures, or constant operation in temperatures below 0°C (32°F).

- Prolonged idling (vehicle operation in stop and go traffic.)

- Frequent short running periods (engine does not warm to normal operating temperatures).

- Police, taxi, delivery usage or trailer towing usage.

Oil & oil filter change: change every 3000 miles.

Air filter element: inspect and replace as necessary every 12,000 miles

PCV valve: replace every 30,000 miles

Automatic transmission fluid, filter: change every 12,000 miles.

Brake hoses & linings: check every 12,000 miles.

Accessory drive belt: inspect and replace as necessary every 45,000 miles

Front & rear axle oil: change every 12,000 miles.

06009-JEEP-C19

SCHEDULED MAINTENANCE INTERVALS
2005 Grand Cherokee

		VEHICLE MILEAGE INTERVAL (x1000)												
TO BE SERVICED	TYPE OF SERVICE	3	6	9	12	15	18	21	24	27	30	33	36	39
Engine oil & filter	R	✓	✓	✓	✓	✓	✓	✓	✓	✓	✓	✓	✓	✓
Tires	Rotate		✓		✓		✓		✓		✓		✓	
Brake hoses & linings	S/I				✓				✓				✓	
Lubricate steering and suspension ball joints	C/L		✓		✓		✓		✓		✓		✓	
Brake caliper pins	C/L				✓				✓				✓	
Air filter	I/R		✓		✓		✓		✓		✓		✓	
Spark plugs	R										✓			
Drive axle lubricant	R				✓				✓				✓	
Transfer case fluid	R										✓			
PCV valve	I/R										✓			
Accessory drive belt	S/I	Every 60,000 miles												
Spark plug cables (5.7L)	I/R	Every 60,000 miles												
Automatic trans. fluid and filter (4.7L and 5.7L)	R	Every 60,000 miles												
Engine coolant	R	Every 60,000 miles												

R: Replace S/I: Service or Inspect C/L: Clean and lubricate I/R: Inspect and rerplace if necessary

The above schedule is to be used if you drive under any of the following conditions:

Driving in temperatures under 32 degrees F
Stop and go traffic
Extensive engine idling
Driving in dusty conditions
Frequent trips under 10 miles
More than 50 % of your driving is in hot weather (90 deg. F) above 50 miles per hour
Trailer towing
Taxi, police or delivery service
Off-road driving

If none of these conditions is met, double the maintenance intervals

06009-JEEP-C20

PRECAUTIONS

Before servicing any vehicle, please be sure to read all of the following precautions, which deal with personal safety, prevention of component damage, and important points to take into consideration when servicing a motor vehicle:

- Never open, service or drain the radiator or cooling system when the engine is hot; serious burns can occur from the steam and hot coolant.
- Observe all applicable safety precautions when working around fuel. Whenever servicing the fuel system, always work in a well-ventilated area. Do not allow fuel spray or vapors to come in contact with a spark, open flame, or excessive heat (a hot drop light, for example). Keep a dry chemical fire extinguisher near the work area. Always keep fuel in a container specifically designed for fuel storage; also, always properly seal fuel containers to avoid the possibility of fire or explosion. Refer to the additional fuel system precautions later in this section.
- Fuel injection systems often remain pressurized, even after the engine has been turned **OFF**. The fuel system pressure must be relieved before disconnecting any fuel lines. Failure to do so may result in fire and/or personal injury.
- Brake fluid often contains polyglycol ethers and polyglycols. Avoid contact with the eyes and wash your hands thoroughly after handling brake fluid. If you do get brake fluid in your eyes, flush your eyes with clean, running water for 15 minutes. If eye irritation persists, or if you have taken brake fluid internally, IMMEDIATELY seek medical assistance.
- The EPA warns that prolonged contact with used engine oil may cause a number of skin disorders, including cancer. You should make every effort to minimize your exposure to used engine oil. Protective gloves should be worn when changing oil. Wash your hands and any other exposed skin areas as soon as possible after exposure to used engine oil. Soap and water, or waterless hand cleaner should be used.
- All new vehicles are now equipped with an air bag system, often referred to as a Supplemental Restraint System (SRS) or Supplemental Inflatable Restraint (SIR) system. The system must be disabled before performing service on or around system components, steering column, instrument panel components, wiring and sensors. Failure to follow safety and disabling procedures could result in accidental air bag deployment, possible personal injury and unnecessary system repairs.
- Always wear safety goggles when working with, or around, the air bag system. When carrying a non-deployed air bag, be sure the bag and trim cover are pointed away from your body. When placing a non-deployed air bag on a work surface, always face the bag and trim cover upward, away from the surface. This will reduce the motion of the module if it is accidentally deployed. Refer to the additional air bag system precautions later in this section.
- Clean, high quality brake fluid from a sealed container is essential to the safe and proper operation of the brake system. You should always buy the correct type of brake fluid for your vehicle. If the brake fluid becomes contaminated, completely flush the system with new fluid. Never reuse any brake fluid. Any brake fluid that is removed from the system should be discarded. Also, do not allow any brake fluid to come in contact with a painted surface; it will damage the paint.
- Never operate the engine without the proper amount and type of engine oil; doing so WILL result in severe engine damage.
- Timing belt maintenance is extremely important. Many models utilize an interference-type, non-freewheeling engine. If the timing belt breaks, the valves in the cylinder head may strike the pistons, causing potentially serious (also time-consuming and expensive) engine damage. Refer to the maintenance interval charts for the recommended replacement interval for the timing belt, and to the timing belt section for belt replacement and inspection.
- Disconnecting the negative battery cable on some vehicles may interfere with the functions of the on-board computer system(s) and may require the computer to undergo a relearning process once the negative battery cable is reconnected.
- When servicing drum brakes, only disassemble and assemble one side at a time, leaving the remaining side intact for reference.
- Only an MVAC-trained, EPA-certified automotive technician should service the air conditioning system or its components.

GASOLINE ENGINE REPAIR

➡Disconnecting the negative battery cable on some vehicles may interfere with the functions of the on-board computer system. The computer may undergo a relearning process once the negative battery cable is reconnected.

Distributor

REMOVAL

2.5L Engines

1. Before servicing the vehicle, refer to the precautions in the beginning of this section.
2. Remove or disconnect the following:
   - Negative battery cable
   - Distributor cap
   - Camshaft Position (CMP) sensor connector
   - Distributor wiring harness from the main engine harness
   - Cylinder No. 1 spark plug
3. Matchmark the distributor housing and the rotor.
4. Remove the distributor.

To install:

Timing Not Disturbed

5. Before servicing the vehicle, refer to the precautions in the beginning of this section.

➡The rotor will rotate clockwise as the gears engage.

6. Position the rotor slightly counterclockwise of the matchmark made during removal.
7. Install the distributor. Ensure that the rotor moves into alignment with the matchmark.
8. Align the locating fork with the clamp bolt hole. Install the clamp and bolt. Tighten the bolt to 17 ft. lbs. (23 Nm).
9. Install or connect the following:
   - Cylinder No. 1 spark plug
   - Distributor wiring harness to the main engine harness
   - CMP sensor connector
   - Distributor cap
   - Air cleaner tube
   - Negative battery cable

Timing Disturbed

10. Before servicing the vehicle, refer to the precautions in the beginning of this section.
11. Set the engine at Top Dead Center

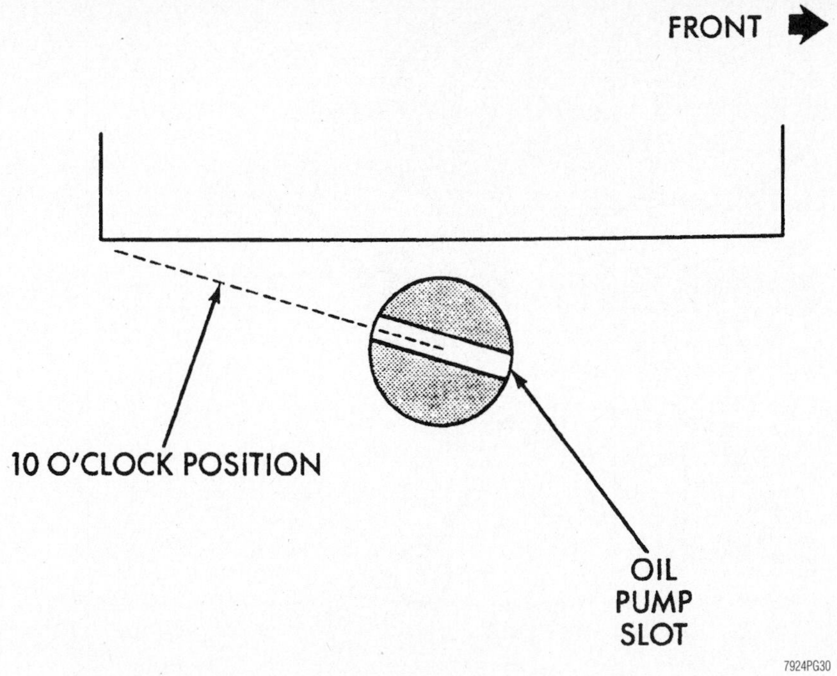

Slot in the oil pump gear at 10 o'clock position—2.5L engine

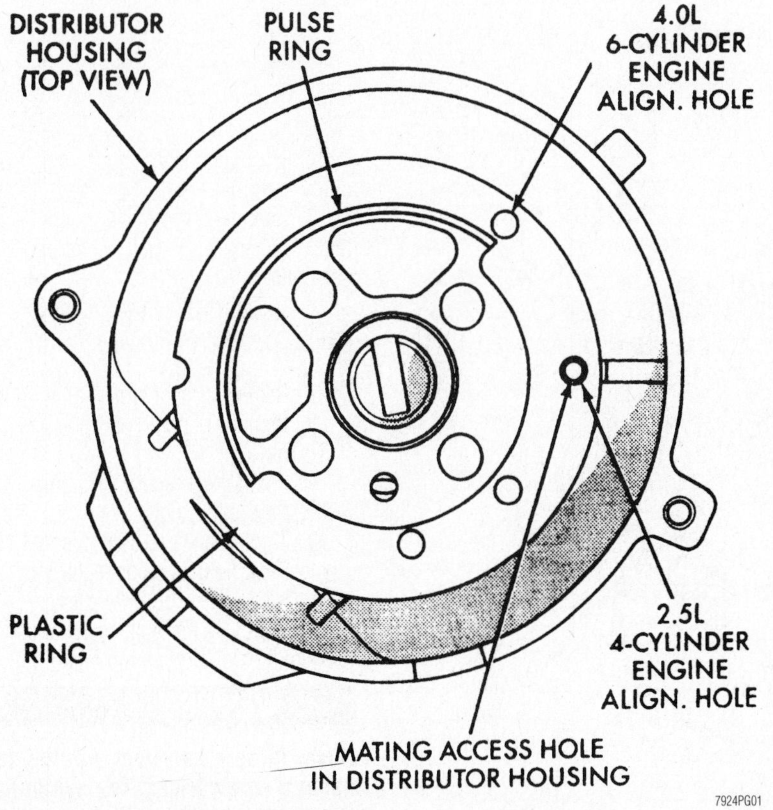

Distributor pin alignment holes—2.5L engines

(TDC) of the No. 1 cylinder compression stroke.

12. Position the slot in the oil pump drive gear as shown.

13. Locate the alignment holes in the plastic ring and align the correct hole with the mating hole in the distributor housing as shown. Install a locking pin.

➡**The distributor will rotate clockwise as the gears engage.**

14. Position the base mounting slot at the 1 o'clock position and install the distributor.

15. Check that the centerline of the mounting slot aligns with the centerline of the clamp bolt hole.

16. Install the clamp and bolt. Tighten the bolt to 17 ft. lbs. (23 Nm).

17. Remove the locking pin.

18. Install or connect the following:
- CMP sensor connector
- Distributor cap
- Air cleaner tube
- Negative battery cable

Alternator

REMOVAL & INSTALLATION

2.4L, 2.5L, 3.7L, 4.0L and 4.7L Engines

1. Before servicing the vehicle, refer to the precautions in the beginning of this section.

2. Remove or disconnect the following:
- Negative battery cable
- Accessory drive belt
- Alternator harness connectors
- Mounting bolts and alternator

➡**The 3.7L and 4.7L engines have 1 vertical and 2 horizontal bolts.**

To install:

3. Before servicing the vehicle, refer to the precautions in the beginning of this section.

4. Install the alternator and tighten the bolts to the following specifications:
- 2.4L, 2.5L and 4.0L engines: 41 ft. lbs. (55 Nm).
- 3.7L engine: Tighten the horizontal bolts to 42 ft. lbs. (57 Nm), then the vertical bolt to 29 ft. lbs. (40 Nm)
- 4.7L engine: Vertical bolt and long horizontal bolt to 41 ft. lbs. (56 Nm), short horizontal bolt to 55 ft. lbs. (74 Nm)

5. Install or connect the following:
- Alternator harness connectors
- Accessory drive belt
- Negative battery cable

5.7L Engine

1. Before servicing the vehicle, refer to the precautions in the beginning of this section.

2. Disconnect negative battery cable at battery.

3. Remove generator drive belt.

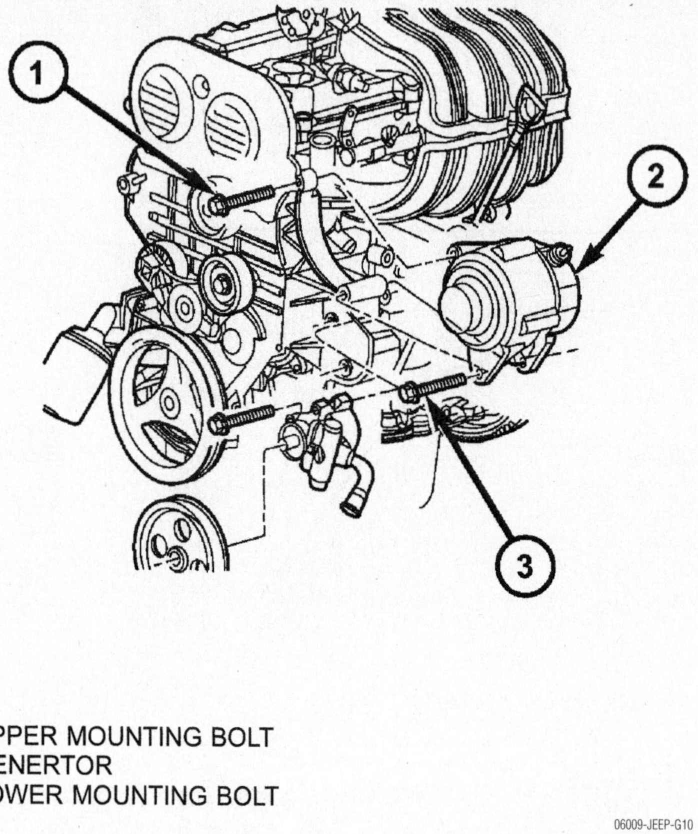

2.4L engine alternator mounting

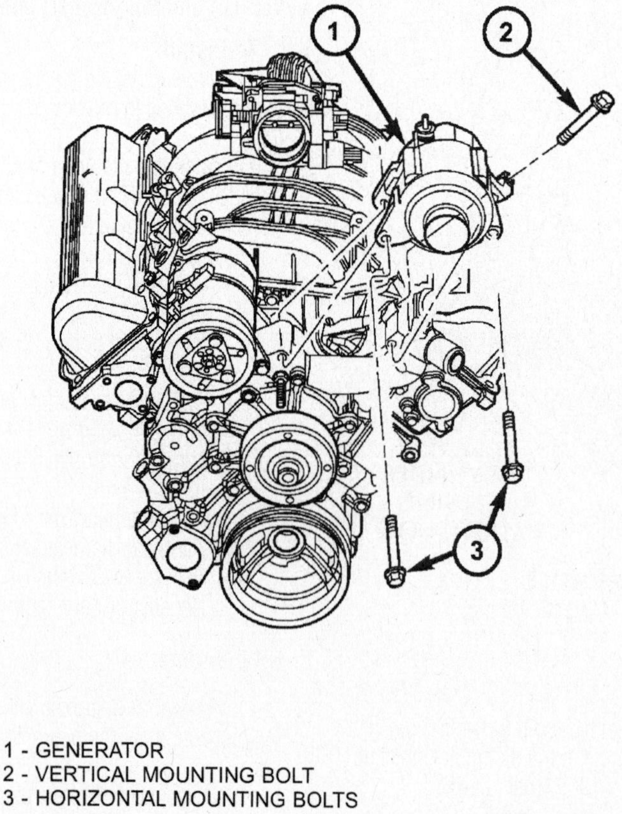

3.7L engine alternator mounting

4. Unsnap plastic insulator cap from B+ output terminal.
5. Remove B+ terminal mounting nut at rear of generator. Disconnect terminal from generator.
6. Disconnect field wire connector at rear of generator by pushing on connector tab.
7. Remove 2 generator mounting bolts.
8. Remove generator from vehicle.

To install:

9. Position generator to engine and install 2 mounting bolts.
10. Tighten bolts to 30 ft. lbs. (41 Nm).
11. Snap field wire connector into rear of generator.
12. Install B+ terminal eyelet to generator output stud. Tighten mounting nut.

**** WARNING**

Never force a belt over a pulley rim using a screwdriver. The synthetic fiber of the belt can be damaged.

➡When installing a serpentine accessory drive belt, the belt MUST be routed correctly. The water pump may be rotating in the wrong direction if the belt is installed incorrectly, causing the engine to overheat.

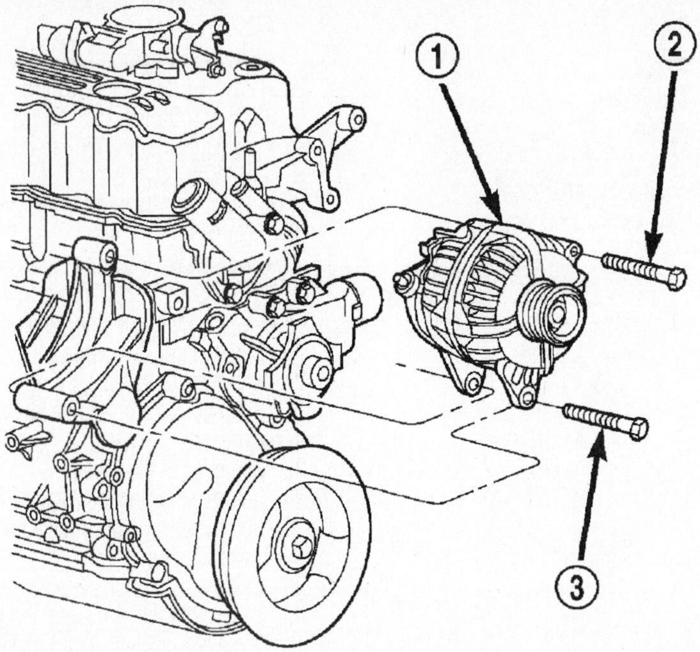

1 - GENERATOR
2 - UPPER BOLT
3 - LOWER BOLT

06009-JEEP-G152

4.0L engine alternator mounting

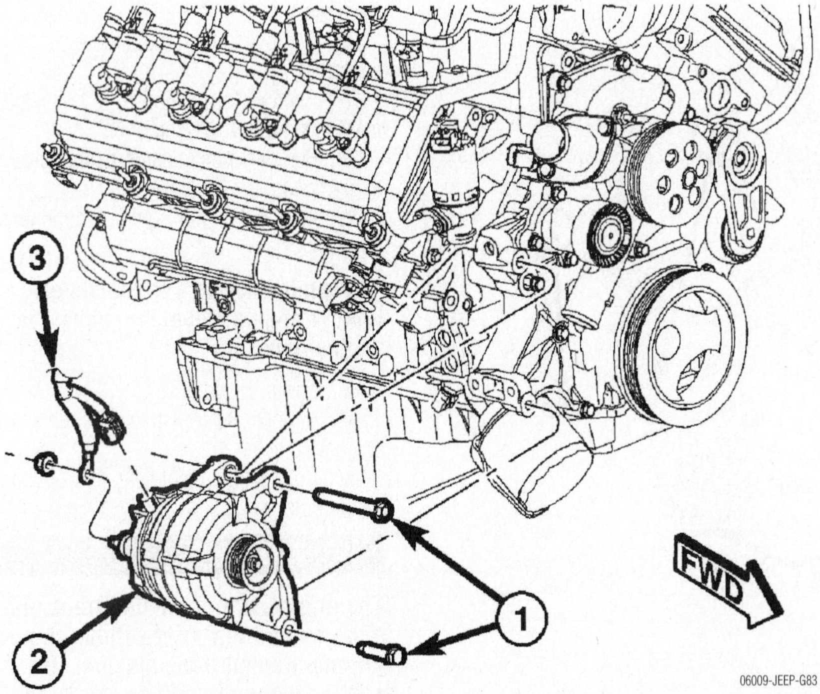

06009-JEEP-G83

Alternator mounting—5.7L engine

13. Install generator drive belt.
14. Install negative battery cable to battery.

Ignition Timing

ADJUSTMENT

Ignition timing is controlled by the Powertrain Control Module (PCM). No adjustment is possible.

Engine Assembly

REMOVAL & INSTALLATION

Liberty

2.4L ENGINE

1. Before servicing the vehicle, refer to the precautions in the beginning of this section.
2. Disconnect the battery negative cable.
3. Remove hood. Mark hood hinge location for reinstallation.
4. Remove air cleaner assembly.
5. Remove radiator core support bracket.
6. Remove fan shroud with electric fan assembly.
7. Remove drive belt.

➡It is not necessary to discharge the A/C system to remove the engine.

8. Remove A/C compressor and secure away from engine with lines attached.
9. Remove generator and secure away from engine.

➡Do not remove the phenolic pulley from the P/S pump. It is not required for P/S pump removal.

10. Remove power steering pump with lines attached and secure away from engine.
11. Drain cooling system.
12. Remove coolant bottle.
13. Disconnect the heater hoses from the engine.
14. Disconnect heater hoses from heater core and remove hose assembly.
15. Disconnect throttle and speed control cables.
16. Remove upper radiator hose from engine.
17. Remove lower radiator hose from engine.
18. Disconnect the engine to body ground straps at the left side of cowl.
19. Disconnect the engine wiring harness at the following points:

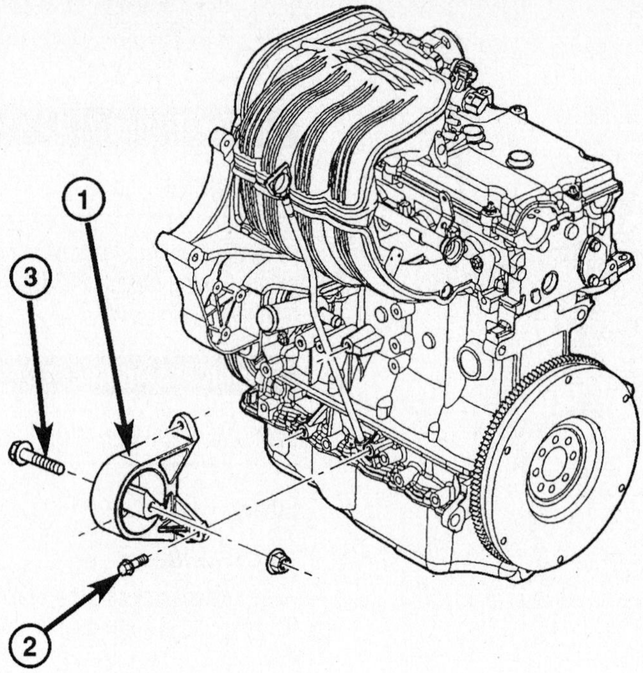

1 - ENGINE MOUNT
2 - ENGINE MOUNT BOLT (3)
3 - ENGINE MOUNT THROUGH BOLT

67189-JEEP-G01

Left side engine mount—2.4L engines

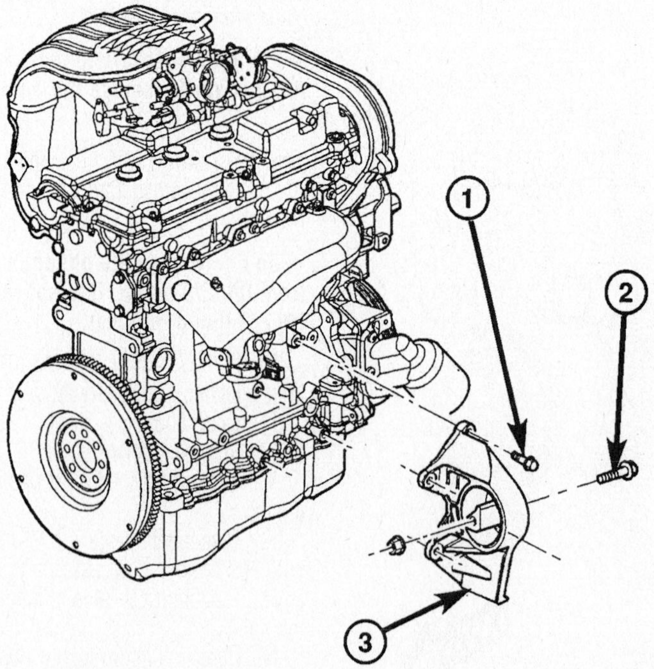

1 - ENGINE MOUNT BOLT (4)
2 - ENGINE MOUNT THROUGH BOLT
3 - ENGINE MOUNT

67189-JEEP-G02

Right side engine mount—2.4L engines

- Intake air temperature (IAT) sensor
- Fuel Injectors
- Throttle Position (TPS) Switch
- Idle Air Control (IAC) Motor
- Engine Oil Pressure Switch
- Engine Coolant Temperature (ECT) Sensor
- Manifold Absolute Pressure MAP) Sensor
- Camshaft Position (CMP) Sensor
- Coil Over Plugs
- Crankshaft Position Sensor

20. Remove coil over plugs.
21. Release fuel rail pressure.
22. Remove fuel rail and secure away from engine.
23. Remove the PCV hose.
24. Remove the breather hoses.
25. Remove the vacuum hose for the power brake booster.
26. Disconnect knock sensors.
27. Secure the left and right engine wiring harnesses away from engine.
28. Raise vehicle.
29. Disconnect oxygen sensor wiring.
30. Disconnect crankshaft position sensor.
31. Disconnect the engine block heater power cable, if equipped.
32. Disconnect the front driveshaft at the front differential and secure out of way.
33. Remove the starter.
34. Remove the ground straps from the engine
35. Disconnect the exhaust pipes at the manifold.
36. Remove the structural cover, if equipped.
37. Remove torque converter bolts, and mark location for reassembly.
38. Remove transmission bellhousing to engine bolts.
39. Loosen left and right engine mount thru bolts.

➡It is not necessary to completely remove engine mount thru bolts, for engine removal.

40. Lower the vehicle.
41. Support the transmission with a suitable jack.
42. Connect a suitable engine hoist to the engine.

✱✱ WARNING

The engine with a manual transmission, can be removed without removing the manual transmission. Use caution when attempting this procedure as the clearance is tight.

43. Remove engine from vehicle.

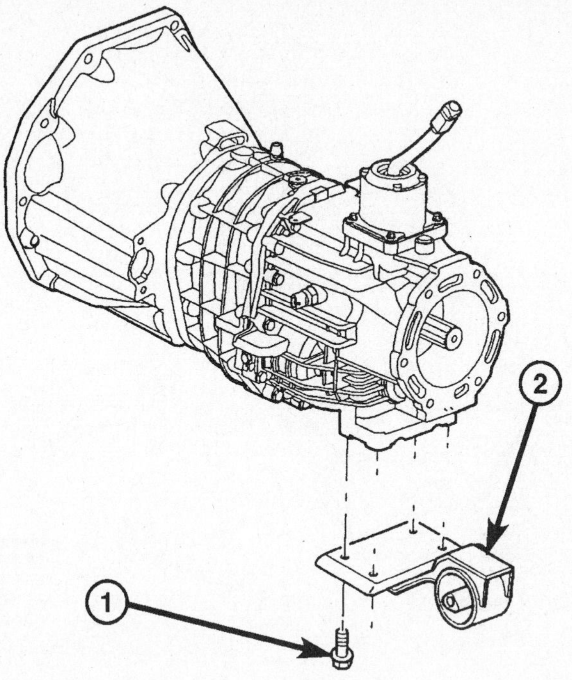

Transmission mount—2.4L engines

To install:

44. Position the engine in the vehicle.

45. Install both left and right side engine mounts into the frame mounts.

46. Raise the vehicle.

47. Install the transmission bellhousing to engine mounting bolts. Tighten the bolts to 41 Nm (30 ft. lbs.).

48. Tighten the engine mount thru bolts.

49. Install the torque converter bolts.

50. Connect the ground straps on the left and right side of the engine.

51. Install the starter.

52. Connect the crankshaft position sensor.

53. Install the engine block heater power cable, if equipped.

WARNING

The structural collar requires a specific torque sequence. Failure to follow this sequence may cause severe damage to the collar.

54. Install the structural collar.

Perform the following steps for installing structural collar.

a. Step 1: Position collar between transmission and oil pan. Install collar to transmission bolts, hand start only.

b. Step 2: Install collar to oil pan bolts, hand snug only.

c. Step 3: Tighten collar to transmission bolts.

d. Step 4: Tighten collar to oil pan bolts.

55. Install the exhaust pipe.

56. Connect the oxygen sensors.

57. Lower vehicle.

58. Connect the knock sensors.

59. Connect the engine to body ground straps.

60. Install the power brake booster vacuum hose.

61. Install the breather hoses.

62. Install the PCV hose.

63. Install the fuel rail.

64. Install the coil over plugs.

65. Reconnect the engine wiring harness at the following points:

- Intake air temperature (IAT) sensor
- Fuel Injectors
- Throttle Position (TPS) Switch
- Idle Air Control (IAC) Motor
- Engine Oil Pressure Switch
- Engine Coolant Temperature (ECT) Sensor
- Manifold Absolute Pressure MAP) Sensor
- Camshaft Position (CMP) Sensor
- Coil Over Plugs
- Crankshaft Position Sensor

66. Connect lower radiator hose.

67. Connect upper radiator hose.

68. Connect throttle and speed control cables.

69. Install the heater hose assembly.

70. Install coolant recovery bottle.

71. Install the power steering pump.

72. Install the generator.

73. Install the A/C compressor.

74. Install the drive belt.

75. Install the fan shroud with the electric fan assembly.

76. Install the radiator core support bracket.

77. Install the air cleaner assembly.

78. Refill the engine cooling system.

79. Install the hood.

80. Check and fill engine oil.

81. Connect the battery negative cable.

82. Start the engine and check for leaks.

3.7L ENGINE

1. Before servicing the vehicle, refer to the precautions in the beginning of this section.

2. Properly relieve the fuel system pressure.

3. Drain the cooling system.

4. Drain the engine oil.

5. Remove or disconnect the following:

- Negative battery cable
- Hood
- Air cleaner assembly
- Radiator
- Electric and mechanical fan assemblies
- A/C compressor, if equipped, and secure it out of the way with the lines attached. DO NOT DISCHARGE!
- Power steering pump, with the lines attached
- Alternator
- Coolant bottle
- Heater hoses
- Accelerator and speed control cables
- Lower and upper radiator hoses
- Engine ground straps
- Intake Air Temperature (IAT) sensor
- Fuel injection wiring connectors
- Throttle Position (TP) sensor
- Idle Air Control (IAC) motor
- Oil pressure sender connector
- Engine Coolant Temperature (ECT) sensor
- Manifold Absolute Pressure (MAP) sensor

- Camshaft position sensor
- Ignition coil wiring connector
- Crankshaft Position (CKP) sensor
- Coil pack
- Fuel rail
- PCV hose
- Vacuum hoses from the intake manifold
- Knock sensor connectors
- Oil dipstick tube
- Intake manifold
- Heated Oxygen (HO_2S) sensor connector
- Block heater connector
- Front driveshaft at the differential
- Starter
- Structural cover
- With a manual transmission, remove the transmission
- Torque converter bolts and matchmark the converter
- Automatic transmission-to-engine bolts
- Exhaust front pipes
- Left and right engine mounts

6. Place a support stand under the transmission.
7. Install an engine lift plate
8. Lift the engine out of the vehicle.

To install:

9. If equipped with a manual transmission, install the transmission
10. Lower the engine and install the mounts. Don't tighten the bolts yet.
11. If equipped with an automatic transmission, perform the following steps:
    a. Align the torque converter housing to the engine.
    b. Torque the bolts to 30 ft. lbs. (41 Nm).
    - Install the torque converter to flexplate bolts. Torque the bolts to 50 ft. lbs. (68 Nm).
12. Install or connect the following:
    - Torque the through bolts to 45 ft. lbs. (61 Nm).
    - Engine ground strap
    - Starter motor
    - CKP sensor
    - Block heater cable
    - Structural cover.

⁂ WARNING

The structural cover must be held tightly against the engine and bellhousing during tightening. The torque for all bolts is 40 ft. lbs. (54 Nm); the bolts must be tightened in the order shown.

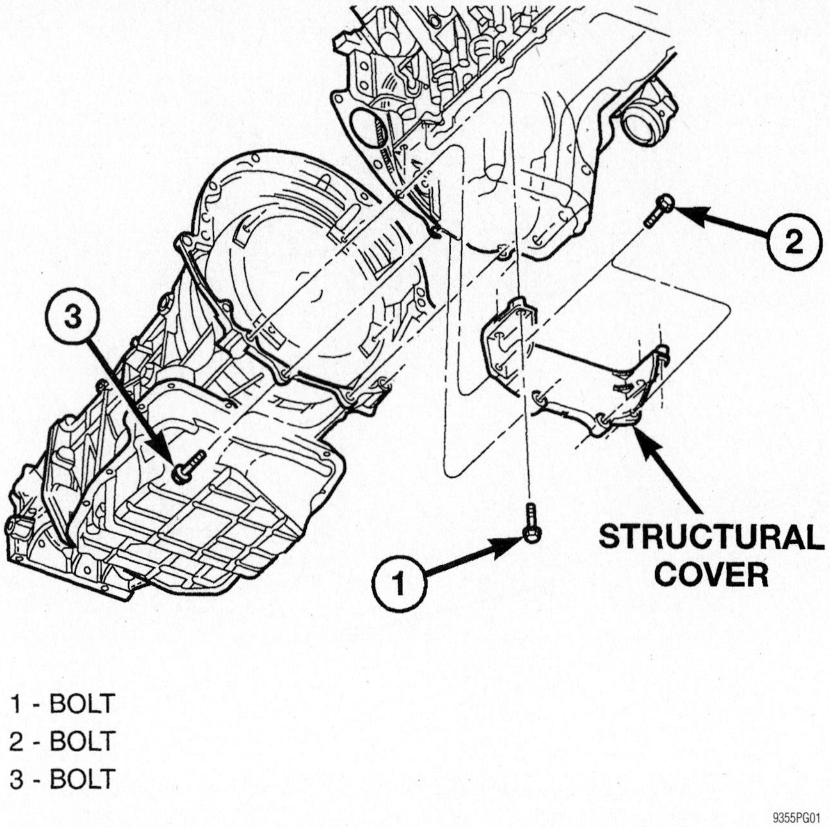

Tighten the structural cover bolts in this order—3.7L

- Exhaust pipes. New flange clamps MUST be used!
- HO_2S sensor connectors
- KS sensors
- Intake Manifold
- Dipstick tube
- Vacuum hoses to the intake manifold
- PCV and breather hoses
- Fuel rail
- Ignition coil
- IAT sensor
- Fuel injector connectors
- TP sensor
- IAC motor
- Oil pressure sender
- ECT sensor electrical connector
- MAP sensor
- CMP sensor
- Radiator hoses
- Cruise control cable, if equipped
- Throttle cable
- Heater hoses
- Coolant bottle
- Power steering pump
- Alternator
- A/C compressor
- Radiator
- Fan assemblies
- Air cleaner assembly
- Negative battery cable

13. Fill and bleed the power steering system.
14. Fill the engine with clean oil.
15. Start the engine and check for leaks, repair if necessary.

Wrangler

2.4L ENGINE

1. Before servicing the vehicle, refer to the precautions in the beginning of this section.
2. Disconnect the battery negative cable.
3. Remove hood. Mark hood hinge location for reinstallation.
4. Remove air cleaner assembly.
5. Remove radiator core support bracket.
6. Remove fan shroud with electric fan assembly.
7. Remove drive belt.

➡It is not necessary to discharge the A/C system to remove the engine.

8. Remove A/C compressor and secure away from engine with lines attached.
9. Remove generator and secure away from engine.

➡Do not remove the phenolic pulley from the power steering pump. It is not required for power steering pump removal.

10. Remove power steering pump with lines attached and secure away from engine.
11. Drain cooling system.
12. Remove coolant bottle.
13. Disconnect the heater hoses from the engine.
14. Disconnect heater hoses from heater core and remove hose assembly.
15. Disconnect throttle and speed control cables.
16. Remove upper radiator hose from engine.
17. Remove lower radiator hose from engine.
18. Disconnect the engine to body ground straps at the left side of cowl.
19. Disconnect the engine wiring harness at the following points:
    - Intake air temperature (IAT) sensor
    - Fuel Injectors
    - Throttle Position (TPS) Switch
    - Idle Air Control (IAC) Motor
    - Engine Oil Pressure Switch
    - Engine Coolant Temperature (ECT) Sensor
    - Manifold Absolute Pressure MAP) Sensor
    - Camshaft Position (CMP) Sensor
    - Coil Over Plugs
    - Crankshaft Position Sensor
20. Remove coil over plugs.
21. Release fuel rail pressure.
22. Remove fuel rail and secure away from engine.
23. Remove the PCV hose.
24. Remove the breather hoses.
25. Remove the vacuum hose for the power brake booster.
26. Disconnect knock sensors.
27. Secure the left and right engine wiring harnesses away from engine.
28. Raise vehicle.
29. Disconnect oxygen sensor wiring.
30. Disconnect crankshaft position sensor.
31. Disconnect the engine block heater power cable, if equipped.
32. Disconnect the front driveshaft at the front differential and secure out of way.
33. Remove the starter.
34. Remove the ground straps from the engine
35. Disconnect the exhaust pipes at the manifold.
36. Remove the structural cover, if equipped.
37. Remove torque converter bolts, and mark location for reassembly.
38. Remove transmission bellhousing to engine bolts.
39. Loosen left and right engine mount through bolts.

➡It is not necessary to completely remove engine mount through bolts, for engine removal.

40. Lower the vehicle.
41. Support the transmission with a suitable jack.
42. Connect a suitable engine hoist to the engine.

WARNING

The 2.4L engine with manual transmission can be removed without removing the manual transmission. Use caution when attempting this procedure as the clearance is tight.

43. Remove engine from vehicle.

To install:

44. Position the engine in the vehicle.

WARNING

Use caution when installing 2.4L engine into vehicle equipped with manual transmission, as clearance is tight.

45. Install both left and right side engine mounts into the frame mounts.
46. Raise the vehicle.
47. Install the transmission bellhousing to engine mounting bolts. Tighten the bolts to 41 Nm (30 ft. lbs.).
48. Tighten the engine mount through bolts.
49. Install the torque converter bolts.
50. Connect the ground straps on the left and right side of the engine.
51. Install the starter.
52. Connect the crankshaft position sensor.
53. Install the engine block heater power cable, if equipped.

WARNING

The structural cover requires a specific torque sequence. Failure to follow this sequence may cause severe damage to the cover.

54. Install the structural cover.

Perform the following steps for installing structural collar.

a. Step 1: Position collar between transmission and oil pan. Install collar to transmission bolts, hand start only.
b. Step 2: Install collar to oil pan bolts, hand snug only.
c. Step 3: Tighten collar to transmission bolts.
d. Step 4: Tighten collar to oil pan bolts.

55. Install the exhaust pipe.
56. Connect the oxygen sensors.
57. Lower vehicle.
58. Connect the knock sensors.
59. Connect the engine to body ground straps.
60. Install the power brake booster vacuum hose.
61. Install the breather hoses.
62. Install the PCV hose.
63. Install the fuel rail.
64. Install the coil over plugs.
65. Reconnect the engine wiring harness at the following points:
    - Intake air temperature (IAT) sensor
    - Fuel Injectors
    - Throttle Position (TPS) Switch
    - Idle Air Control (IAC) Motor
    - Engine Oil Pressure Switch
    - Engine Coolant Temperature (ECT) Sensor
    - Manifold Absolute Pressure MAP) Sensor
    - Camshaft Position (CMP) Sensor
    - Coil Over Plugs
    - Crankshaft Position Sensor
66. Connect lower radiator hose.
67. Connect upper radiator hose.
68. Connect throttle and speed control cables.
69. Install the heater hose assembly.
70. Install coolant recovery bottle.
71. Install the power steering pump.
72. Install the generator.
73. Install the A/C compressor.
74. Install the drive belt.
75. Install the fan shroud with the electric fan assembly.
76. Install the radiator core support bracket.
77. Install the air cleaner assembly.
78. Refill the engine cooling system.
79. Install the hood.
80. Check and fill engine oil.
81. Connect the battery negative cable.
82. Start the engine and check for leaks.

2.5L ENGINE

1. Before servicing the vehicle, refer to the precautions in the beginning of this section.
2. Properly relieve the fuel system pressure.
3. Drain the cooling system.
4. Drain the engine oil.
5. Evacuate the A/C system.

6. Drain the power steering fluid, if equipped.

7. Remove or disconnect the following:

- Negative battery cable
- Air cleaner assembly
- Upper radiator hose
- Accessory drive belt and fan drive assembly
- Transmission cooler lines, if equipped
- Lower radiator hose
- Radiator and fan shroud
- Starter electrical connectors
- Alternator wiring connectors
- A/C compressor, if equipped
- Ignition coil wiring connector
- Oil pressure sender connector
- Wiring harness ground at the dipstick tube
- Heater hoses from the thermostat housing
- Water pump inlet tube
- Closed Crankcase Ventilation (CCV) hoses from the cylinder head and intake manifold
- Accelerator, transmission line pressure and speed control cables, if equipped
- Engine ground strap
- Power steering pressure switch, if equipped
- Engine Coolant Temperature (ECT) sensor
- Fuel injection wiring connectors
- Intake Air Temperature (IAT) sensor
- Idle Air Control (IAC) motor
- Throttle Position (TP) sensor
- Manifold Absolute Pressure (MAP) sensor
- Crankshaft Position (CKP) sensor
- Heated Oxygen (HO_2S) sensor connector
- Vacuum hoses from the intake manifold
- Fuel line and bracket
- Power steering hoses from the steering gear, if equipped
- Oil filter
- Starter
- Engine support through bolts
- Exhaust front pipe
- Flywheel cover
- Torque converter, if equipped with automatic transmission
- Transmission flange bolts
- Left and right motor mounts
- Engine shock damper bracket

8. Place a support stand under the transmission.

9. Lift the engine out of the vehicle.

To install:

10. If equipped with a manual transmission, perform the following steps:

a. Insert the transmission shaft into the clutch spline.

b. Align the flywheel housing to the engine.

c. Install the flywheel housing bolts. Torque the bolts to 28 ft. lbs. (38 Nm).

11. If equipped with an automatic transmission, perform the following steps:

a. Align the torque converter housing to the engine.

b. Torque the bolts to 28 ft. lbs. (38 Nm).

- Install the torque converter to flexplate bolts. Torque the bolts to 50 ft. lbs. (68 Nm).

12. Install or connect the following:

- Engine onto the engine brackets. When properly aligned, torque the through bolts to 60 ft. lbs. (81 Nm).
- Inspection cover. Torque the bolts to 138 inch lbs. (16 Nm).
- Exhaust pipe. Torque the bolts to 23 ft. lbs. (31 Nm).
- Starter motor. Tighten the bolts to 33 ft. lbs. (45 Nm).
- Oil filter
- Power steering hoses
- Fuel supply line to the fuel rail
- Fuel supply rail to the intake manifold
- Vacuum hoses to the intake manifold
- Power steering pressure switch electrical connector
- ECT sensor electrical connector
- Fuel injector connectors
- IAT sensor
- IAC motor
- TP sensor
- MAP sensor
- CKP sensor
- HO_2S sensor connector
- Engine ground strap
- Heater hoses and water pump inlet tube
- Cruise control cable, if equipped
- Throttle cable
- Fan shroud and radiator. Torque the bolts to 75 inch lbs. (8 Nm).
- Transmission cooler lines, if equipped. Torque the bolts to 10 ft. lbs. (15 Nm).
- Fan drive assembly. Torque the bolts to 20 ft. lbs. (27 Nm).
- Drive belt
- Radiator hoses
- Ignition coil
- Distributor
- A/C compressor
- A/C high pressure switch
- Alternator
- Oil pressure sender
- Harness ground at the oil dipstick tube bracket
- A/C hoses
- Air cleaner assembly
- Negative battery cable

13. Fill and bleed the power steering system.

14. Fill the engine with clean oil.

15. Fill and bleed the cooling system.

16. Recharge the A/C system.

17. Start the engine and check for leaks, repair if necessary.

4.0L ENGINE

1. Before servicing the vehicle, refer to the precautions in the beginning of this section.

2. Disconnect the battery negative cable.

3. Mark the hinge locations on the hood panel for alignment reference during installation. Remove the engine compartment lamp. Remove the hood.

4. Drain engine coolant. Drain the coolant into a clean container for reuse.

5. Remove the upper radiator hose and coolant recovery hose.

6. Remove the lower radiator hose.

7. Remove upper radiator support retaining bolts and remove radiator support.

8. The thermal viscous fan drive/fan blade assembly is attached (threaded) to water pump hub shaft. Remove fan blade/viscous fan drive assembly from water pump by turning mounting nut counterclockwise as viewed from front. Threads on viscous fan drive are RIGHT HAND. Using a suitable fan wrench loosen the fan drive.

9. Remove accessory drive belt.

10. Some models with certain engines may require the removal of the fan shroud to remove the viscous fan drive. The fan shroud and fan blade/viscous fan drive should be removed from the vehicle as one assembly.

11. After removing fan blade/viscous fan drive assembly, do not place thermal viscous fan drive in horizontal position. If stored horizontally, silicone fluid in viscous fan drive could drain into its bearing assembly and contaminate lubricant.

12. Remove the fan shroud.

13. Disconnect the transmission fluid cooler lines (automatic transmission).

14. Discharge the A/C system.

15. Remove the service valves and cap the compressor ports.

⁂ CAUTION

Plastic tanks, while stronger than brass, are subject to damage by impact, such as wrenches, mishandling, etc.

⁂ CAUTION

Do not remove the cylinder block drain plugs or loosen the radiator drain cock with the system hot and pressurized. Serious burns from the coolant can occur.

➡Do not waste reusable coolant. If the solution is clean, drain the coolant into a clean container for reuse.

⁂ CAUTION

Constant tension hose clamps are used on most cooling system hoses When removing or installing, use only tools designed for servicing this type of clamp, such as special clamp tool number 6094. Always wear safety glasses when servicing constant tension clamps.

⁂ WARNING

A number or letter is stamped into the tongue of constant tension clamps. If replacement is necessary, use only an original equipment clamp with matching number or letter.

⁂ CAUTION

When removing the radiator or A/C condenser for any reason, note the location of all radiator-to-body and radiator-to-A/C condenser rubber air seals. These are used at the top, bottom and sides of the radiator and A/C condenser. To prevent overheating, these seals must be installed to their original positions.

16. Disconnect negative battery cable at battery.
17. Observe the previous warning. Remove the radiator cap.
18. Remove the condenser lower seal from the lower core support.
19. Drain cooling system. Drain coolant into a clean container for reuse.
20. Remove radiator upper and lower hose clamps. Remove radiator hoses.
21. Disconnect coolant reserve/overflow tank hose from radiator.
22. Remove the reservoir tank and power steering reservoir and position out of the way.
23. Remove the four fan shroud mounting bolts. Position the fan shroud back over the fan blades.
24. If equipped, disconnect and plug automatic transmission fluid cooler lines.
25. Remove six radiator mounting bolts. Position the front axle vent hose to the side.
26. The lower part of radiator is equipped with two alignment dowel pins. They are located on the bottom of radiator tank and fit into rubber grommets. These rubber grommets are pressed into the radiator lower crossmember.
27. Lift radiator straight up and out of vehicle taking care not to damage radiator fins.
28. When removing radiator, note position of the rubber seals located on the top and bottom of radiator (on certain models only). To prevent possible overheating, these seals must be installed to their original positions.
29. Disconnect the heater hoses at the engine thermostat housing and water pump.
30. Disconnect the accelerator cable, transmission line pressure cable and speed control cable (if equipped) from the throttle body.
31. Remove cables from the bracket and secure out of the way.
32. Disconnect the body ground at the engine.
33. Disconnect the following connectors and secure their harness out of the way.
    - Power steering pressure switch
    - Coolant temperature sensor
    - Six
34. Fuel injector connectors
    - Intake air temperature sensor
    - Throttle position sensor
    - Map sensor
    - Crankshaft position sensor
    - Oxygen sensor
    - Camshaft position sensor
    - Generator connector and B+ terminal wire
35. Disconnect the coil rail electrical connections and the oil pressure switch connector.
36. Perform the fuel pressure release procedure.
37. Disconnect the fuel supply line at the injector rail.
38. Remove the fuel line bracket from the intake manifold.
39. Remove the air cleaner assembly.
40. Disconnect the hoses from the fittings at the steering gear.
41. Drain the pump reservoir.
42. Cap the fittings on the hoses and steering gear to prevent foreign objects from entering the system.
43. Raise and support the vehicle.
44. Disconnect the wires from the engine starter motor solenoid.
45. Remove the engine starter motor.
46. Disconnect the oxygen sensor from the exhaust pipe.
47. Disconnect the exhaust pipe from the manifold.
48. Remove the exhaust pipe support.
49. Remove the exhaust hanger bracket retaining bolt.
50. Remove locknut and transmission bending brace bar.
51. Remove engine-to-bending brace retaining bolt, bending brace bar and cross bar.
52. Remove the engine flywheel/converter housing access cover.
53. Mark the converter and drive plate location.
54. Remove the converter-to-drive plate bolts.
55. Remove the upper engine flywheel/converter housing bolts and loosen the bottom bolts.
56. Remove the engine mount cushion-to-engine compartment bracket bolts.
57. Lower the vehicle.
58. Attach a lifting device to the engine.
59. Raise the engine off the front supports.
60. Place a support or floor jack under the converter (or engine flywheel) housing.
61. Remove the remaining converter (or engine flywheel) housing bolts.
62. Lift the engine out of the engine compartment.

To install:

⁂ CAUTION

When installing the engine into a vehicle equipped with an automatic transmission, be careful not to damage the trigger wheel on the engine flywheel.

63. Attach a lifting device to the engine and lower the engine into the engine compartment. For easier installation, it may be necessary to remove the engine mount bracket as an aid in alignment of the engine to the transmission.
64. Align the transmission torque converter housing with the engine.
65. Loosely install the converter housing lower bolts and install the next higher bolt and nut on each side.
66. Tighten all 4 bolts finger tight.
67. If the engine support bracket was removed, position the bracket onto the

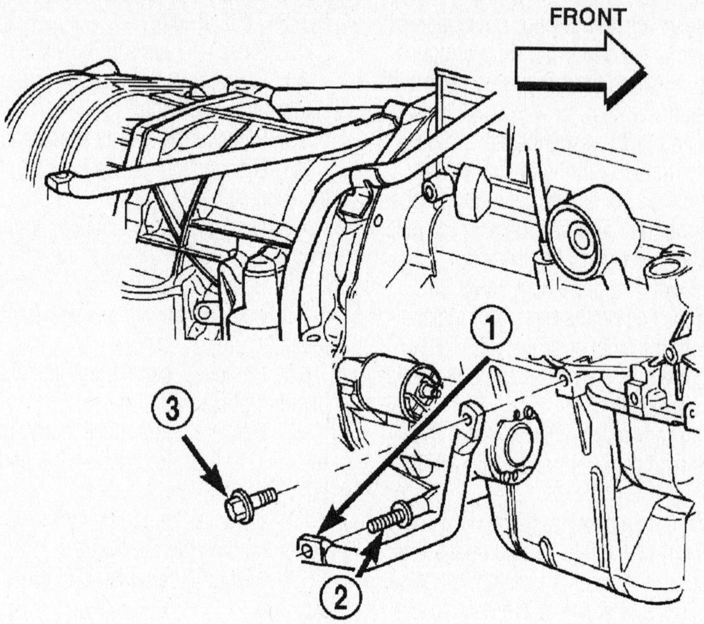

1 - ENGINE-TO-TRANSMISSION BENDING BRACE
2 - CROSS BRACE
3 - ENGINE-TO-TRANSMISSION BENDING BRACE RETAINING BOLT

06009-JEEP-G153

Engine-to-transmission bending braces—4.0L Wrangler

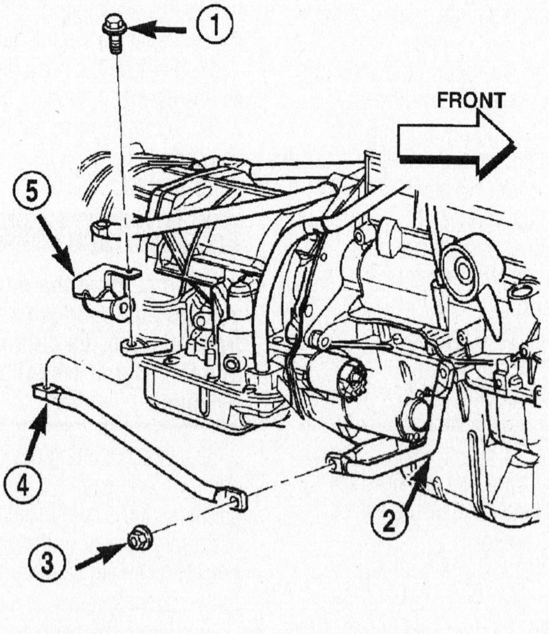

1 - TRANSMISSION BENDING BRACE RETAINING BOLT
2 - ENGINE-TO-TRANSMISSION BENDING BRACE
3 - LOCKNUT
4 - TRANSMISSION BRACE
5 - EXHAUST HANGER

06009-JEEP-G154

Transmission bending braces and exhaust hanger—4.0L Wrangler

block and install the attaching bolts. Tighten the bolts to 50 Nm (37 ft. lbs.) torque.

68. Place the insulator on the support bracket. Install the insulator retaining bolts and nuts. Tighten the bolts and nuts to 40 Nm (30 ft. lbs) torque.
69. Install the through bolt and the retaining nut. Tighten the through bolt nut to 48 Nm (35 ft. lbs.) torque.
70. Remove the engine support.
71. Lower the vehicle.
72. Connect negative cable to battery.
73. Lower the engine and engine mount brackets onto the engine compartment cushions. Install the bolts and finger tighten the nuts.
74. Remove the engine lifting device.
75. Raise and support the vehicle.
76. Install the remaining engine flywheel/converter housing bolts. Tighten all bolts to 38 Nm (28 ft. lbs.) torque.
77. Install the converter-to-drive plate bolts.
78. Ensure the installation reference marks are aligned.
79. Install the engine flywheel/converter housing access cover.
80. Install the exhaust pipe support and tighten the screw.

➡Do not tighten the retaining hardware until all bending braces are in place.

81. Position the cross brace into the engine-to-transmission brace, then position the engine-to-transmission brace and install retaining bolt.
82. Position the transmission bending brace onto through brace and install new locknut.
83. Position exhaust hanger and transmission brace, install retaining bolt.
84. Tighten engine-to-transmission brace retaining bolt to 40 Nm (30 ft. lbs.).
85. Tighten transmission brace retaining bolts to 40 Nm (30 ft. lbs.), then tighten transmission brace retaining lock nuts to 108 Nm (80 ft. lbs.).
86. Tighten the engine mount-to-bracket bolts.
87. Connect the vehicle speed sensor wire connections and tighten the screws.
88. Connect the exhaust pipe to the manifold.
89. Install the engine starter motor.
90. Connect the wires to the engine starter motor solenoid.
91. Lower the vehicle.
92. Connect vacuum hoses and wire connectors disconnected during engine removal.
93. Remove protective caps from the power steering hoses.
94. Connect the hoses to the fittings at the steering gear. Tighten the nut to 52 Nm (38 ft. lbs.) torque.

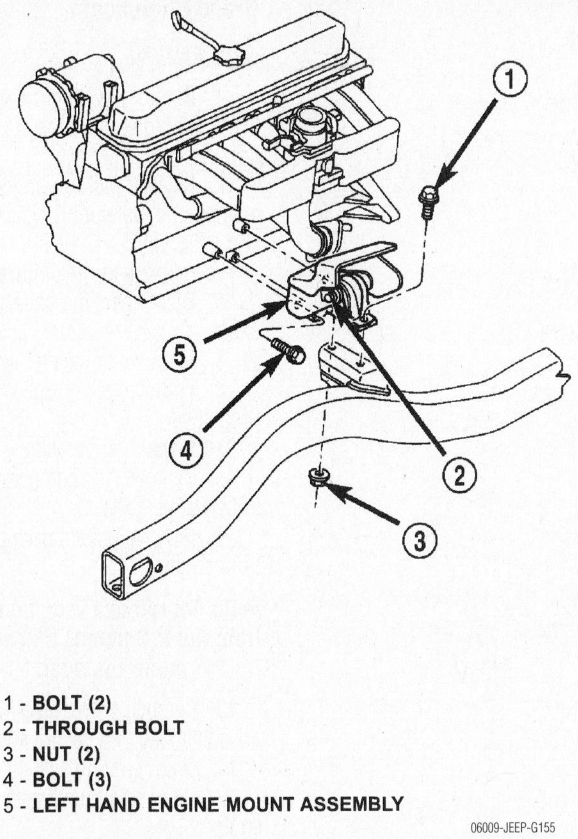

Left engine mount—4.0L Wrangler

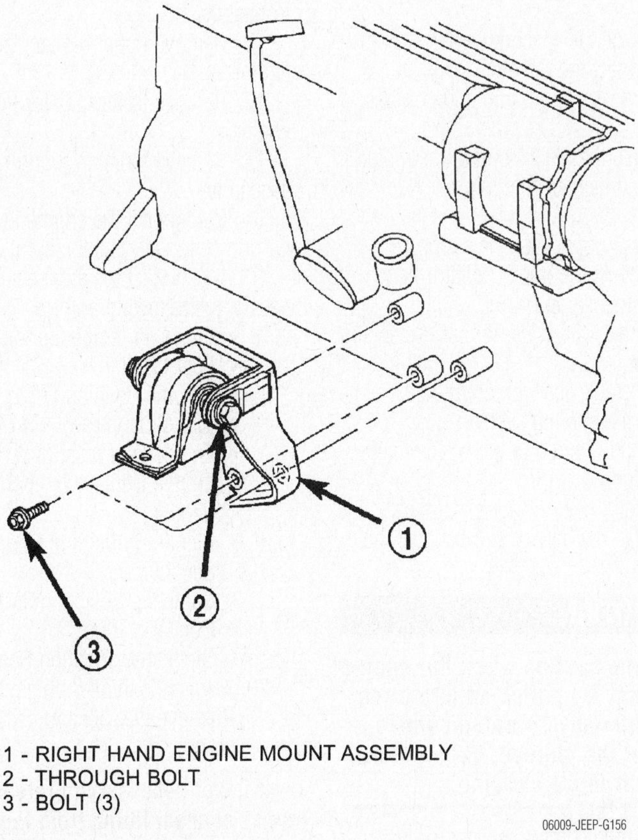

Right engine mount—4.0L Wrangler

95. Fill the pump reservoir with fluid.
96. Inspect the fuel supply line O-ring(s) and replace if necessary. Connect fuel supply line to injector rail and verify connection by pulling outward on the line.
97. Install the fuel line bracket to the intake manifold.
98. Connect the coil rail electrical connectors and oil pressure switch connector.
99. Connect the following electrical connectors:
    - Power steering pressure switch
    - Coolant temperature sensor
    - Six fuel injector connectors
    - Intake air temperature sensor
    - Throttle position sensor
    - Map sensor
    - Crankshaft position sensor
    - Oxygen sensor
    - Camshaft position sensor
    - Generator connector and B+ terminal wire
100. Connect all previously removed vacuum hoses.
101. Connect the body ground strap.
102. Connect the body ground strap.
103. Install the throttle, transmission line pressure, and speed control cables to their mounting bracket and connect them to the throttle body.
104. Connect the heater hoses at the engine thermostat housing and water pump.
105. Assemble fan blade to viscous fan drive. Tighten mounting bolts to 27 Nm (20 ft. lbs.) torque.
106. Thread the fan and fan drive onto the water pump pulley.

**** WARNING**

When installing a serpentine accessory drive belt, the belt MUST be routed correctly. If not, the engine may overheat due to the water pump rotating in the wrong direction.

107. Install accessory drive belt.
108. Place the fan shroud in position over the fan.

**** WARNING**

Before installing the radiator or A/C condenser, be sure the radiator-to-body and radiator-to-A/C condenser rubber air seals are properly fastened to their original positions. These are used at the top, bottom and sides of the radiator and A/C condenser. To prevent overheating, these seals must be installed to their original positions.

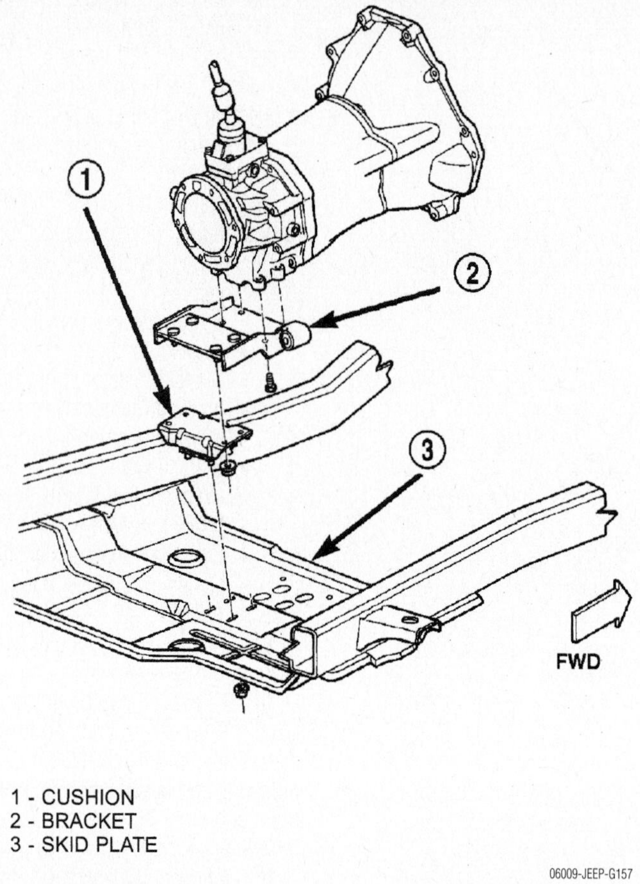

Rear mount—4.0L Wrangler

109. Guide the two radiator alignment dowels into the rubber grommets located in lower radiator crossmember. Install and tighten the six mounting bolts to 8 Nm (72 inch lbs.) torque.

110. Close radiator draincock.

111. Position fan shroud and power steering reservoir tank (if equipped). Install and tighten four mounting bolts to 8 Nm (72 inch lbs.) torque.

112. If equipped, remove plugs and connect automatic transmission fluid cooler lines and constant tension clamps.

➡The tangs on the hose clamps must be positioned straight down.

113. Connect radiator hoses and install hose clamps.

114. Position and install the condenser lower seal.

115. Connect battery negative cable.

116. Fill cooling system with correct coolant.

117. Connect coolant recovery bottle hose.

118. Install radiator cap.

119. Connect the service valves to the A/C compressor ports, if equipped with A/C.

120. Charge the air conditioner system.

121. Connect the radiator hoses and automatic transmission fluid cooler pipes, if equipped.

122. Install the fan shroud to the radiator or radiator/condenser (if equipped with A/C).

123. Install upper radiator support.

124. Connect the upper radiator hose and tighten the clamp.

125. Connect the lower radiator hose and tighten the clamp.

126. Fill crankcase with engine oil.

127. Fill the cooling system.

128. Align the hood to the scribe marks. Install the hood.

129. Install the air cleaner assembly.

130. Install the battery and connect the battery cable.

⁂ CAUTION

Use extreme caution when the engine is operating. Do not stand in a direct line with the fan. Do not put your hands near the pulleys, belts or fan. Do not wear loose clothing.

131. Start the engine, inspect for leaks and correct the fluid levels, as necessary.

Grand Cherokee

3.7L ENGINE

1. Before servicing the vehicle, refer to the precautions in the beginning of this section.
2. Release fuel rail pressure then disconnect the fuel supply quick connect fitting at the fuel rail.
3. Remove the strut tower support.
4. Disconnect the battery negative cable.
5. Remove air cleaner assembly.
6. Remove fan shroud with viscous fan assembly.
7. Remove drive belt.
8. Remove A/C compressor and secure away from engine.
9. Remove generator and secure away from engine.

➡Do not remove the phenolic pulley from the P/S pump. It is not required for P/S pump removal.

10. Remove power steering pump with lines attached and secure away from engine.
11. Drain cooling system.
12. Disconnect the heater hoses from the engine.
13. Disconnect heater hoses from heater core and remove hose assembly.
14. Disconnect throttle and speed control cables.
15. Remove upper radiator hose from engine.
16. Remove lower radiator hose from engine.
17. Remove radiator/cooling module assembly.
18. Disconnect the engine to body ground straps at the left side of cowl.
19. Disconnect the engine wiring harness at the following points:
    - Intake air temperature (IAT) sensor
    - Fuel Injectors
    - Throttle Position (TPS) Switch
    - Idle Air Control (IAC) Motor
    - Engine Oil Pressure Switch
    - Engine Coolant Temperature (ECT) Sensor
    - Manifold Absolute Pressure MAP) Sensor
    - Camshaft Position (CMP) Sensor
    - Coil Over Plugs
    - Crankshaft Position Sensor
20. Remove coil over plugs.
21. Remove fuel rail and secure away from engine.

➡It is not necessary to release the quick connect fitting from the fuel supply line for engine removal.

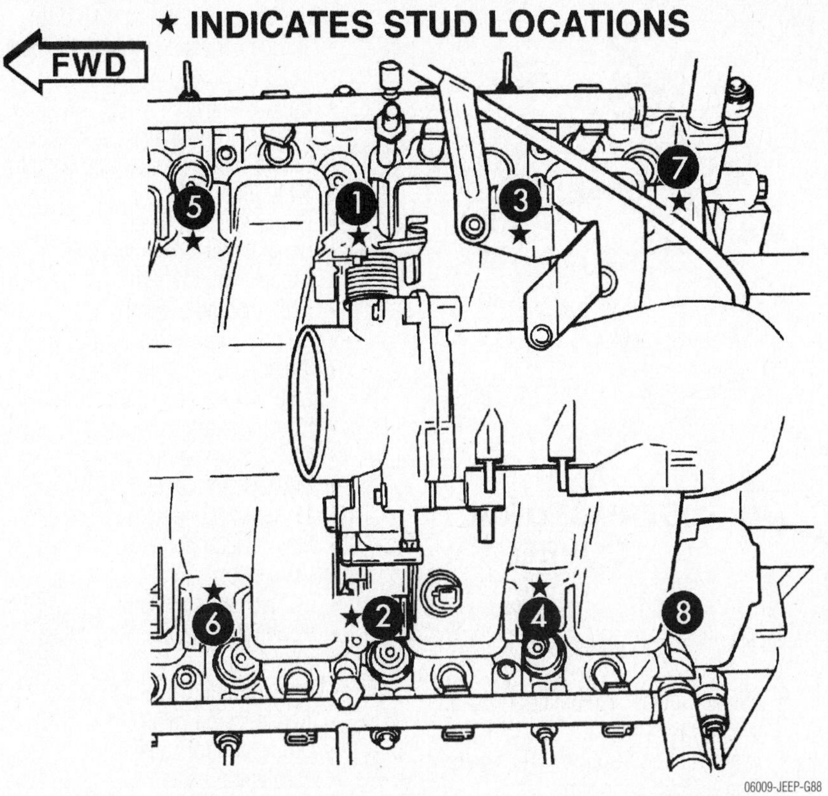

Engine lifting fixture stud locations—3.7L engine

22. Remove the PCV hose.
23. Remove the breather hoses.
24. Remove the vacuum hose for the power brake booster.
25. Disconnect knock sensors.
26. Remove engine oil dipstick tube.
27. Remove intake manifold.
28. Install engine lifting fixture, special tool 8247, using original fasteners from the removed intake manifold, and fuel rail.

➡Recheck bolt torque for engine lift plate before removing engine.

29. Disconnect oxygen sensor wiring.
30. Disconnect crankshaft position sensor.
31. Disconnect the engine block heater power cable, if equipped.
32. Disconnect the front driveshaft at the front differential and secure out of way.

➡It is necessary to disconnect the front driveshaft for access to the starter and left side exhaust flange.

33. Remove the starter.
34. Remove the ground straps from the left and right side of the block.
35. Remove the structural cover.
36. Disconnect the right and left exhaust pipes at the manifolds and from the crossover, and remove from the vehicle.

➡The exhaust clamps at the manifolds cannot be reused. New clamps must be used or leaks may occur.

37. Remove torque converter bolts, and mark location for reassembly.
38. Remove transmission bellhousing to engine bolts.
39. Remove left and right engine mount thru bolts.
40. Lower the vehicle.
41. Support the transmission with a suitable jack.
42. Connect a suitable engine hoist to the engine lift plate.
43. Remove engine from vehicle.

To install

44. Position the engine in the vehicle.
45. Install the transmission bellhousing to engine mounting bolts. Tighten the bolts to 41 Nm (30 ft. lbs.)
46. Install the engine mount through bolts.
47. Install the torque converter bolts.
48. Install the starter.
49. Connect the crankshaft position sensor.
50. Install the engine block heater power cable, if equipped.

⁂ WARNING

The structural cover requires a specific torque sequence. Failure to follow this sequence may cause severe damage to the cover.

51. Install the structural cover.

➡New clamps must be used on exhaust manifold flanges. Failure to use new clamps may result in exhaust leaks.

52. Install the left and right exhaust pipes.
53. Connect the left and right oxygen sensors.
54. Remove the engine lift plate.
55. Connect the knock sensors.
56. Connect the engine to body ground straps at the left side of the cowl.
57. Install the intake manifold using the sequence provided.
58. Install the engine oil dipstick tube.
59. Install the power brake booster vacuum hose.
60. Install the breather hoses.
61. Install the PCV hose.
62. Install the fuel rail.
63. Install the coil over plugs.
64. Connect the engine wiring harness at the following points:
- Intake air temperature (IAT) sensor
- Fuel Injectors
- Throttle Position (TPS) Switch
- Idle Air Control (IAC) Motor
- Engine Oil Pressure Switch
- Engine Coolant Temperature (ECT) Sensor
- Manifold Absolute Pressure MAP) Sensor
- Camshaft Position (CMP) Sensor
- Coil Over Plugs
- Crankshaft Position Sensor

65. Connect the ground straps on the right side of the engine.
66. Connect the ground straps on the left side of the engine.
67. Reinstall the radiator/cooling module assembly.
68. Connect lower radiator hose.
69. Connect upper radiator hose.
70. Connect throttle and speed control cables.
71. Install the heater hose assembly.
72. Install coolant recovery bottle.
73. Install the power steering pump.
74. Install the generator.
75. Install the A/C compressor. See the accompanying illustration for the proper torque sequence.
76. Install the drive belt.

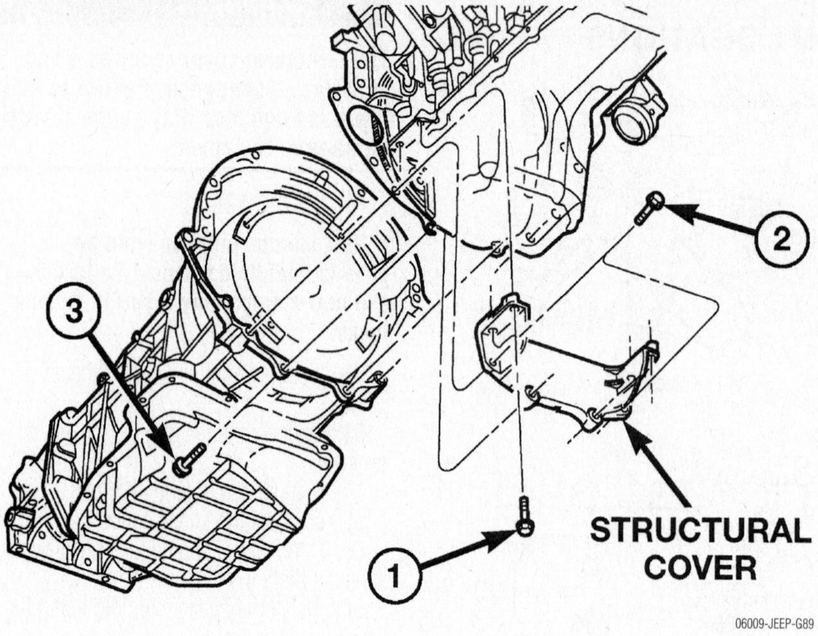

Structural cover torque sequence—Grand Cherokee 3.7L engine

77. Install the fan shroud with the viscous fan assembly.
78. Install the radiator core support bracket.
79. Recharge the A/C system.
80. Install the strut tower support.
81. Install the air cleaner assembly.
82. Refill the engine cooling system.
83. Check and fill engine oil.
84. Connect the battery negative cable.
85. Start the engine and check for leaks.

4.0L ENGINE

1. Before servicing the vehicle, refer to the precautions in the beginning of this section.
2. Drain the cooling system.
3. Drain the engine oil.
4. Drain the power steering fluid.
5. Properly relieve the fuel system pressure.
6. Recover the A/C refrigerant.
7. Remove or disconnect the following:
   - Negative battery cable
   - Hood
   - Radiator hoses
   - Upper radiator support
   - Cooling fan and shroud
   - Radiator
   - A/C condenser
   - Heater hoses
   - Accelerator cable
   - Cruise control cable
   - Transmission cable
   - Body ground cable
   - Power steering pressure switch connector
   - Engine Coolant Temperature (ECT) sensor connector
   - Intake Air Temperature (IAT) sensor
   - Fuel injector connectors
   - Throttle Position (TP) sensor connector
   - Manifold Absolute Pressure (MAP) sensor connector
   - Crankshaft Position (CKP) sensor connector
   - Heated Oxygen (HO_2S) sensor connector
   - Camshaft Position (CMP) sensor connector
   - Ignition coil wiring connector
   - Oil pressure sender connector
   - Fuel line and bracket
   - Air cleaner assembly
   - Brake booster vacuum line
   - Power steering hoses
   - Starter motor
   - Exhaust front pipe
   - Flywheel cover
   - Torque converter
   - Transmission flange bolts
   - Left and right motor mounts
8. Place a support stand under the transmission.
9. Lift the engine out of the vehicle.

To install:

10. Lower the engine in to the vehicle and position to the transmission.
11. Install or connect the following:
    - Left and right motor mounts. Tighten the through bolts to 51 ft. lbs. (69 Nm).
    - Transmission flange bolts. Tighten the bolts to 30 ft. lbs. (41 Nm).
    - Torque converter. Tighten the bolts to 23 ft. lbs. (31 Nm).
    - Flywheel cover
    - Exhaust front pipe
    - Starter motor. Tighten the bolts to 33 ft. lbs. (45 Nm).
    - Power steering hoses
    - Brake booster vacuum line
    - Air cleaner assembly
    - Fuel line and bracket
    - Oil pressure sender connector
    - Ignition coil wiring connector
    - CMP sensor connector
    - HO_2S sensor connector
    - CKP sensor connector
    - MAP sensor connector
    - TP sensor connector
    - Fuel injector connectors
    - IAT sensor
    - ECT sensor connector
    - Power steering pressure switch connector
    - Body ground cable
    - Transmission cable
    - Cruise control cable
    - Accelerator cable
    - Heater hoses
    - A/C condenser
    - Radiator
    - Cooling fan and shroud
    - Upper radiator support
    - Radiator hoses
    - Hood
    - Negative battery cable
12. Fill and bleed the cooling system.
13. Fill and bleed the cooing system.
14. Fill the engine with clean oil
15. Recharge the A/C system.
16. Start the engine, check for leaks and repair if necessary.

4.7L ENGINE

1. Before servicing the vehicle, refer to the precautions in the beginning of this section.
2. Drain the cooling system.
3. Drain the engine oil.
4. Drain the power steering system.
5. Properly relieve the fuel system pressure.
6. Recover the A/C refrigerant.
7. Remove or disconnect the following:
   - Negative battery cable
   - Front fascia
   - Exhaust crossover pipe
   - Engine ground straps
   - Crankshaft Position (CKP) sensor
   - Structural collar
   - Starter
   - Left and right inner fender liners

- Headlamp mounting module
- Air intake resonator
- Accelerator cable
- Cruise control cable
- Crankcase breather tubes
- Accessory drive belt
- A/C compressor
- Cooling fan assemblies
- Radiator hoses
- Transmission oil cooler lines
- Radiator
- A/C condenser
- Alternator
- Heater hoses
- Throttle Position (TP) sensor connector
- Intake Air Temperature (IAT) sensor connector
- Fuel injector harness connectors
- Engine Coolant Temperature (ECT) sensor connector
- Idle Air Control (IAC) valve connector
- Manifold Absolute Pressure (MAP) sensor connector
- Ignition coils
- Fuel line
- Power steering pump
- Oil fill tube
- Oil dipstick tube
- Heated Oxygen (HO_2S) sensor connectors
- Engine oil filter
- Exhaust crossover pipe
- Structural cover
- Rubber splash shield
- Starter motor
- Crankshaft Position (CKP) sensor connector
- Camshaft Position (CMP) sensor connector
- Torque converter
- Engine ground straps
- Left and right motor mounts
- Transmission flange bolts

8. Install Engine Lifting Fixture 8347 as shown.
9. Place a support stand under the transmission.
10. Lift the engine out of the vehicle.

To install:

11. Lower the engine in to the vehicle and position to the transmission.
12. Remove the engine lifting fixture.
13. Install or connect the following:
    - Transmission flange bolts. Tighten the bolts to 50 ft. lbs. (68 Nm).
    - Left and right motor mounts. Tighten the bolts to 45 ft. lbs. (61 Nm).
    - Engine ground straps
    - Torque converter. Tighten the bolts to 23 ft. lbs. (31 Nm).

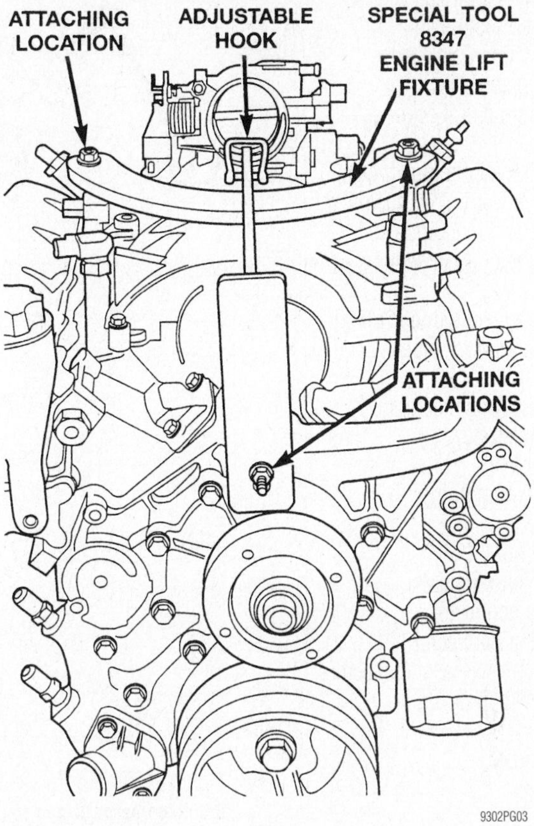

Engine Lifting Fixture—4.7L engine

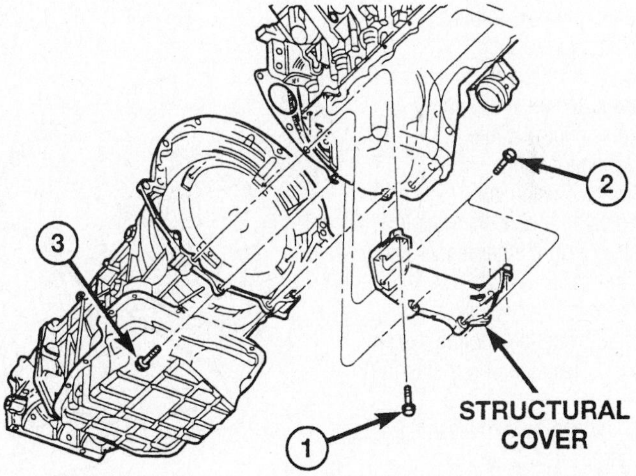

SEQUENCE	ITEM	TORQUE
1	BOLT (Qty 4)	54 N·m (40 ft. lbs.)
2	BOLT (Qty 2)	54 N·m (40 ft. lbs.)
3	BOLT (Qty 2)	54 N·m (40 ft. lbs.)

9308PG02

Structural cover torque sequence—4.7L engine

- CMP sensor connector
- CKP sensor connector
- Starter motor
- Rubber splash shield

14. Install the structural cover as follows:
    a. Install all the bolts finger-tight.
    b. Hold the cover tightly against the transmission and the engine.
    c. Tighten the bolts in sequence to 40 ft. lbs. (54 Nm).
15. Install or connect the following:
    - Exhaust crossover pipe
    - Engine oil filter
    - HO_2S sensor connectors
    - Oil dipstick tube
    - Oil fill tube
    - Power steering pump
    - Fuel line
    - Ignition coils
    - MAP sensor connector
    - IAC valve connector
    - ECT sensor connector
    - Fuel injector harness connectors
    - IAT sensor connector
    - TP sensor connector
    - Heater hoses
    - Alternator
    - A/C condenser
    - Radiator
    - Transmission oil cooler lines
    - Radiator hoses
    - Cooling fan assemblies
    - A/C compressor
    - Accessory drive belt
    - Crankcase breather tubes
    - Cruise control cable
    - Accelerator cable
    - Air intake resonator
    - Headlamp mounting module
    - Left and right inner fender liners
    - Front fascia
    - Negative battery cable
16. Fill and bleed the cooling system.
17. Fill and bleed the power steering system.
    - Fill the engine with clean oil.
18. Recharge the A/C system.
19. Start the engine, check for leaks and repair if necessary.

5.7L ENGINE

1. Before servicing the vehicle, refer to the precautions in the beginning of this section.
2. Remove the strut tower support.
3. Remove the engine cover.
4. Perform the fuel system pressure release procedure.
5. Disconnect the battery negative cable.
6. Remove the air cleaner resonator and duct work as an assembly.

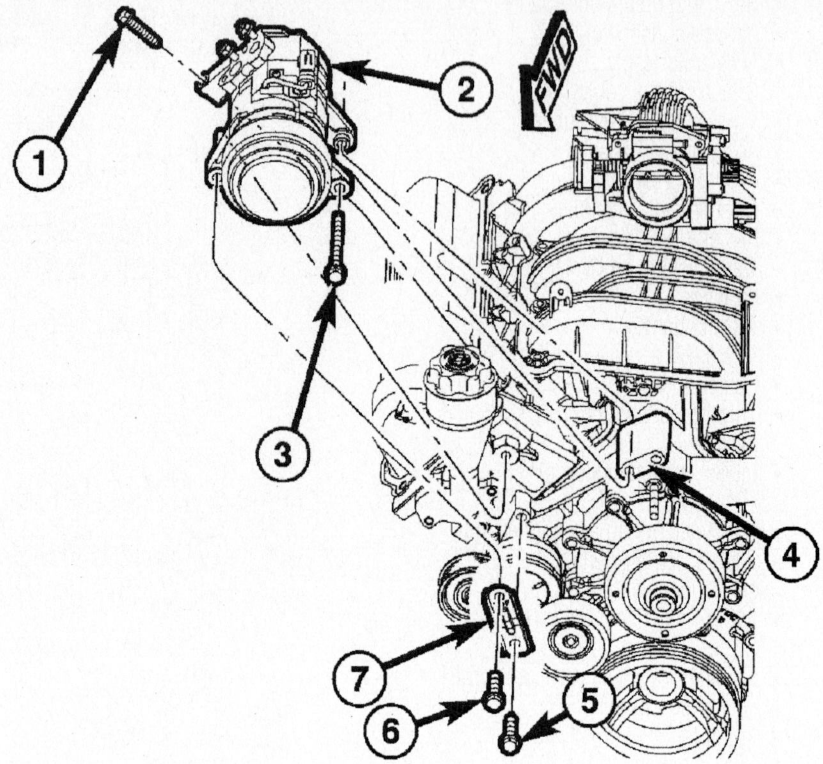

A/C compressor torque sequence—3.7L and 4.7L engines. Hand tighten all bolts, then, torque in the following sequence: bolt 3: 30 ft. lbs. (41 Nm); bolt 6: 41 ft. lbs. (55 Nm); bolt 1: 41 ft. lbs. (55 Nm); bolt 5: 41 ft. lbs. (55 Nm)

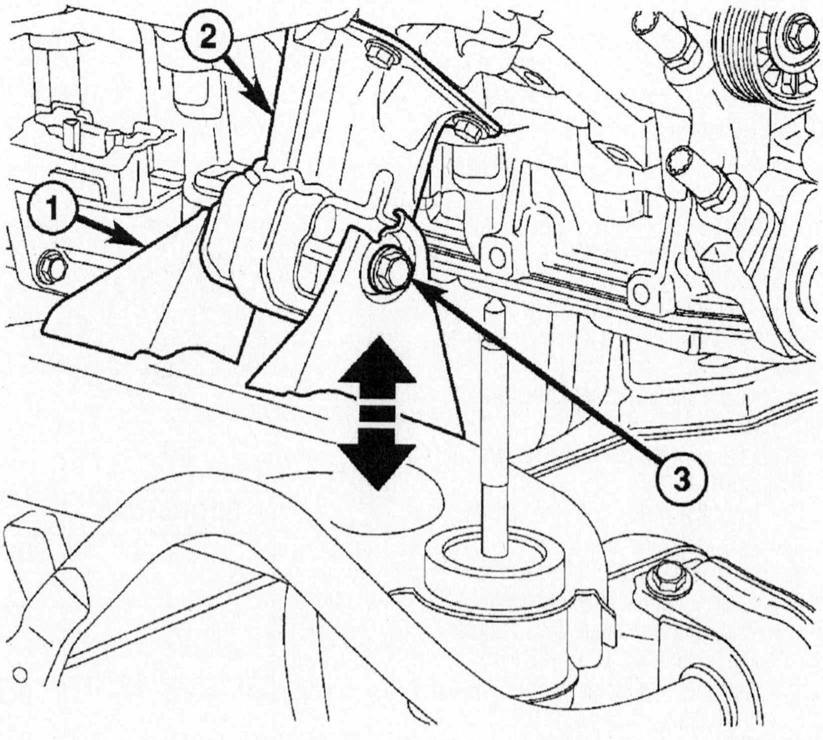

Front engine mount—5.7L engine

7. Drain cooling system.
8. Remove the accessory drive belt.
9. Remove radiator fan shroud.
10. Remove the A/C compressor with the lines attached. Secure compressor out of the way.
11. Remove generator assembly.
12. Remove the intake manifold and IAFM as an assembly.
13. Remove the ground wires from the rear of each cylinder head.
14. Disconnect the heater hoses.

➡It is not necessary to disconnect P/S hoses from pump, for P/S pump removal.

15. Remove the power steering pump and set aside.
16. Disconnect the fuel supply line.
17. Raise and support the vehicle on a hoist and drain the engine oil.
18. Remove engine front mount to frame bolts and nuts.
19. Disconnect the transmission oil cooler lines from their retainers at the oil pan bolts.
20. Disconnect exhaust pipe at manifolds.
21. Disconnect the starter wires. Remove starter motor.
22. Remove the structural dust cover.
23. Remove drive plate to converter bolts.
24. Remove the oil pan to transmission bolts.
25. Remove transmission bell housing to engine block bolts.
26. Lower the vehicle.
27. Install engine lift fixture, special tool 8984 and 8984-UPD.
28. Separate engine from transmission, remove engine from vehicle, and install engine assembly on a repair stand.

To install

29. Install engine lift fixture Special tool 8984 and 8984-UPD.
30. Position the engine in the engine compartment.
31. Lower engine into compartment and align engine with transmission.
32. Mate engine and transmission and install two transmission to engine block mounting bolts finger tight.
33. Lower engine assembly until the engine mounts rests in frame perches.
34. Install remaining transmission to engine block mounting bolts and tighten to 50 ft. lbs. (68 Nm), and the oil pan to transmission bolts and torque to 39 ft. lbs. (54 Nm).
35. Install and tighten engine mount to frame bolts and nuts to 70 ft. lbs. (95 Nm).

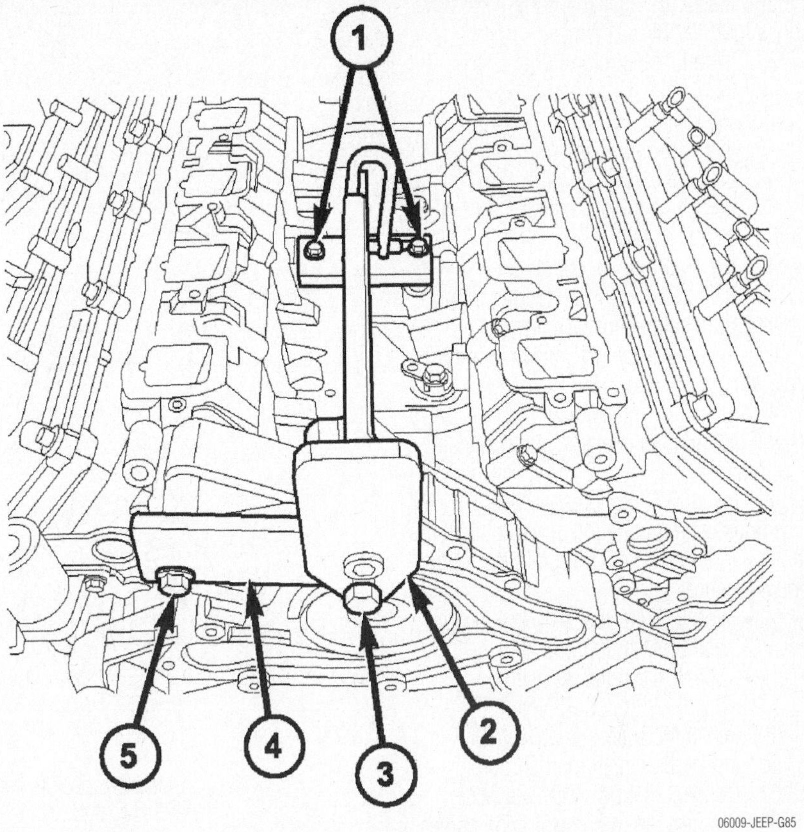

Install engine lift fixture, special tool 8984 (2) and 8984-UPD (4) with bolt positions (1, 3, and 5) —5.7L engine

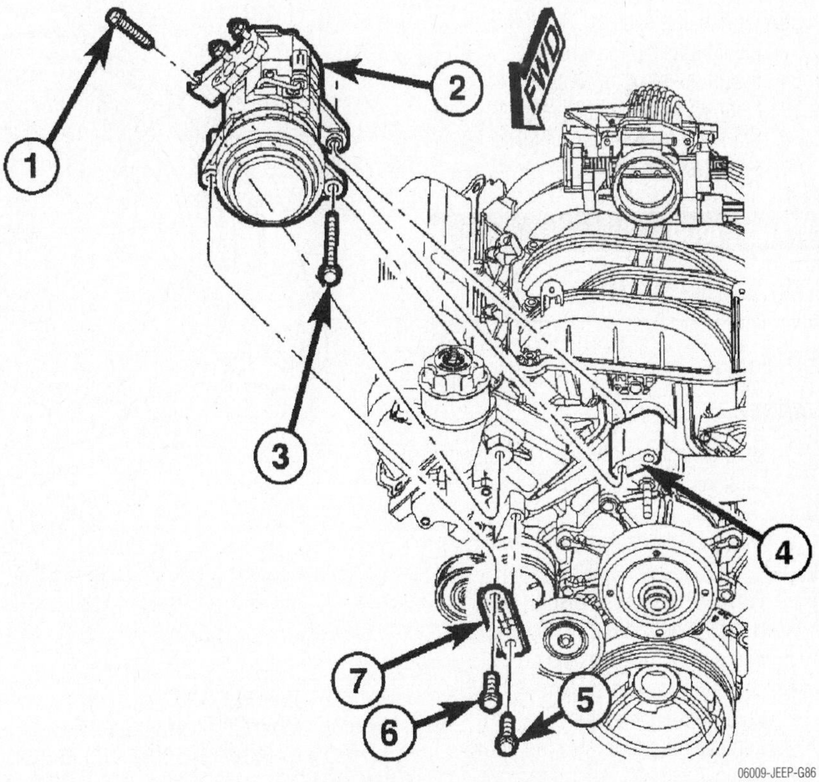

A/C compressor installation torque sequence—5.7L engine

36. Install drive plate to torque converter bolts to 23 ft. lbs. (31 Nm).
37. Install the structural dust cover.
38. Install the starter and connect the starter wires.
39. Install exhaust pipe to manifold.
40. Lower the vehicle.
41. Remove engine lift fixture, special tool 8984 and 8984-UPD.
42. Connect the fuel supply line.
43. Reinstall the power steering pump.
44. Connect the heater hoses.
45. Reconnect the ground wires to the rear of each cylinder head.
46. Install the intake manifold.
47. Install the generator, and wire connections.
48. Install A/C compressor. See the accompanying illustration and torque all the bolts hand tight. Then, torque in the following sequence:
    - bolt 3: 30 ft. lbs. (41 Nm)
    - bolt 6: 41 ft. lbs. (55 Nm)
    - bolt 1: 41 ft. lbs. (55 Nm)
    - bolt 5: 41 ft. lbs. (55 Nm)
49. Install the accessory drive belt.
50. Install the radiator fan shroud.
51. Connect the radiator lower hose.
52. Connect the transmission oil cooler lines to the radiator.
53. Connect the radiator upper hose.
54. Install the air cleaner resonator and duct work.
55. Add engine oil to crankcase.
56. Fill cooling system.
57. Install the engine cover.
58. Install the strut tower support.
59. Connect battery negative cable.
60. Start engine and inspect for leaks.
61. Road test vehicle.

Water Pump

REMOVAL & INSTALLATION

2.4L Engine

LIBERTY

1. Before servicing the vehicle, refer to the precautions in the beginning of this section.
2. Disconnect negative cable from battery.
3. Raise vehicle on a hoist.
4. Remove the accessory drive belts.
5. Remove the belt tensioner.
6. Drain the cooling system.
7. Remove the generator.
8. Remove the power steering pump.
9. Remove the A/C compressor.
10. Remove the accessory drive bracket.

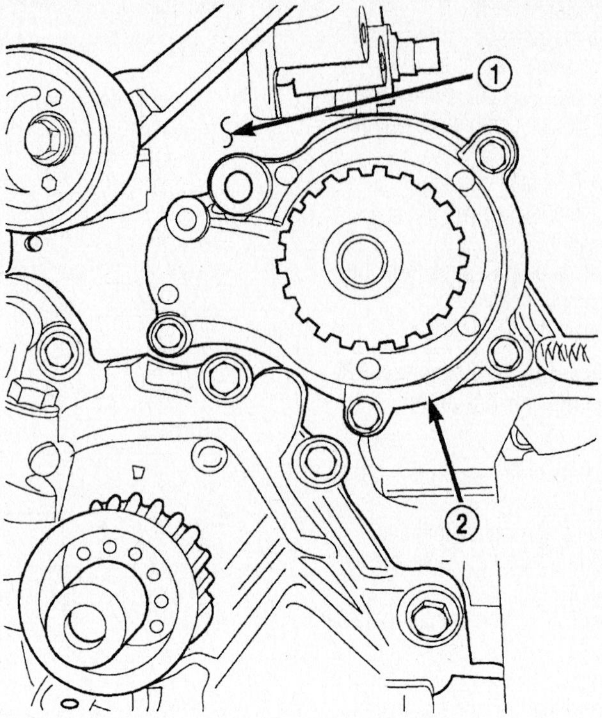

Water pump mounting—2.4L engines

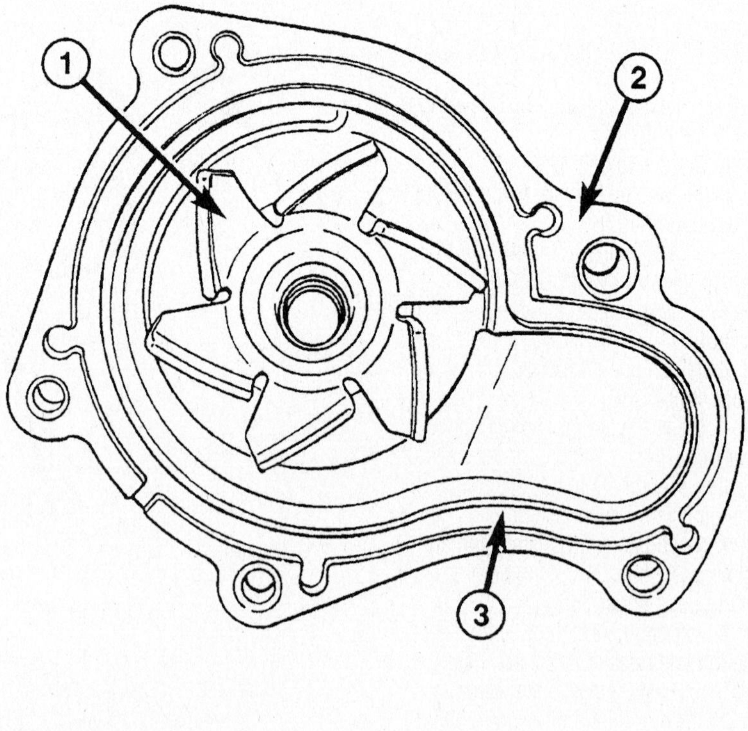

Water pump body—2.4L engines

11. Remove the timing belt.
12. Remove timing belt idler pulley.
13. Hold camshaft sprocket with Special tool C-4687 and adaptor C-4687-1, or equivalent, while removing bolt. Remove both cam sprockets.
14. Remove the timing belt rear cover.
15. Remove water pump to engine attaching screws.
16. Replace water pump body assembly if it has any of these defects:
    a. Cracks or damage on the body.
    b. Coolant leaks from the shaft seal, evident by wet coolant traces on the pump body.
    c. Loose or rough turning bearing.
    d. Impeller rubs either the pump body or the engine block.
    e. Impeller loose or damaged.
    f. Sprocket or sprocket flange loose or damaged.

To install:

17. Install new O-ring gasket in water pump body O-ring locating groove.

➡Make sure O-ring is properly seated in water pump groove before tightening screws. An improperly located O-ring may be damaged and cause a coolant leak.

18. Assemble pump body to block and tighten screws to 12 Nm (105 inch lbs.). Pressurize cooling system to 103.4 Kpa (15 psi) with pressure tester and check water pump shaft seal and O-ring for leaks.
19. Rotate pump by hand to check for freedom of movement.
20. Install the timing belt rear cover.
21. Install camshaft sprockets and target ring. Torque bolts to 101 Nm (75 ft. lbs.) while holding camshaft sprocket.
22. Install timing belt idler pulley and torque mounting bolt to 61 Nm (45 ft. lbs.).
23. Install the timing belt.
24. Install the accessory drive mounting bracket.
25. Install the power steering pump.
26. Install the generator.
27. Install the A/C compressor.
28. Install the belt tensioner.
29. Install the accessory drive belts.
30. Fill the cooling system.
31. Lower vehicle and connect battery cable.

WRANGLER

1. Before servicing the vehicle, refer to the precautions in the beginning of this section.
2. Drain cooling system.
3. Remove timing belt.
4. Remove camshaft sprockets and rear timing belt cover.
5. Remove screws attaching water pump to engine. Remove pump.

To install:

6. Apply Mopar® Dielectric Grease, or equivalent, to new O-ring before installation.
7. Install O-ring gasket in water pump body groove.

⁂ WARNING

Make sure O-ring gasket is properly seated in water pump groove before tightening screws. An improperly located O-ring may cause damage to the O-ring, resulting in a coolant leak.

8. Assemble pump body to block and tighten screws to 12 Nm (105 inch lbs.).
9. Rotate pump by hand to check for freedom of movement.
10. Fill cooling system. Pressurize cooling system to 103 Kpa (15 psi) with pressure tester and check water pump shaft seal and O-ring for leaks.
11. Install rear timing belt cover and camshaft sprockets.
12. Install timing belt.

2.5L and 4.0L Engines

➡The 2.5L and 4.0L engines covered use a reverse rotation water pump. The letter R is stamped on the impeller to identify. Engines from previous years may be equipped with forward rotation water pumps. Installation of the wrong water pump will cause engine over heating.

1. Before servicing the vehicle, refer to the precautions in the beginning of this section.
2. Drain the cooling system.
3. Remove or disconnect the following:
    - Negative battery cable
    - Electric cooling fan connector
    - Accessory drive belt
    - Engine cooling fan and pulley

➡Some 4.0L engines are equipped with a fan clutch that threads directly on to the water pump shaft. This fan clutch is equipped with right-hand threads.

➡Do not store the fan clutch assembly horizontally, silicone may leak into the bearing grease and cause contamination.

- Water pump pulley
- Power steering pump

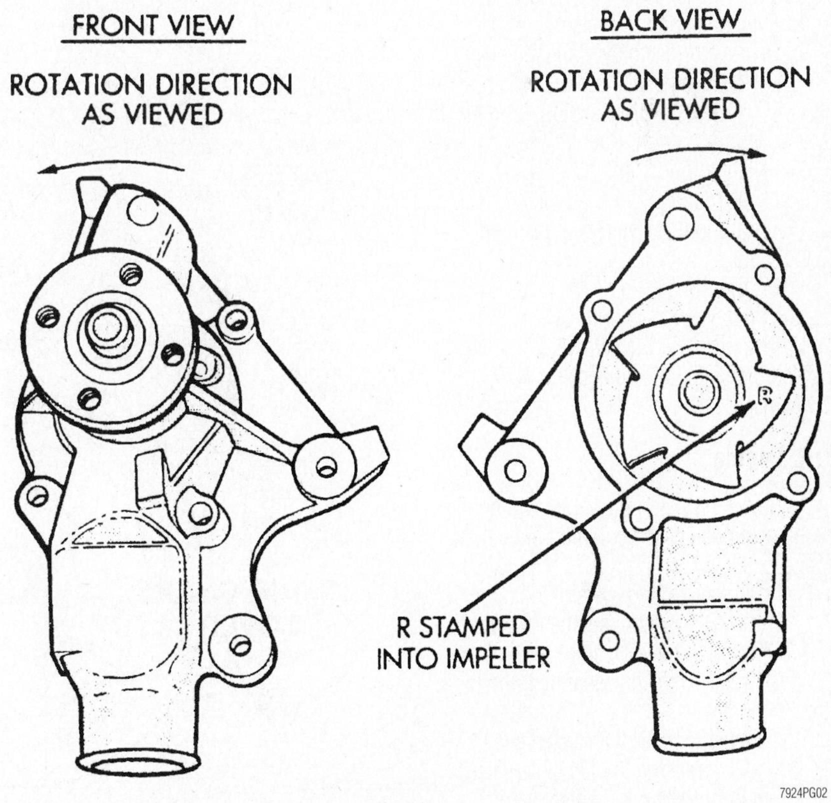

Reverse rotation water pump—2.5L and 4.0L engines

- Lower radiator hose from the water pump
- Heater hose
- Water pump and discard the gasket

➡One of the water pump bolts is longer than the others. Note the location for reassembly.

To install:

4. Clean the mating surfaces of all gasket material.
5. Install or connect the following:
   - Water pump using a new gasket. Torque the bolts to 17 ft. lbs. (23 Nm).
   - Heater hose
   - Lower radiator hose
   - Water pump pulley. Torque the bolts to 20 ft. lbs. (27 Nm).
   - Power steering pump
   - Engine cooling fan and shroud. Torque the bolts to 31 inch lbs. (3 Nm).
   - Accessory drive belt
   - Electric cooling fan connector
   - Negative battery cable
6. Fill and bleed the cooling system.
7. Start the engine, check for leaks and repair if necessary.

3.7L Engine

1. Before servicing the vehicle, refer to the precautions in the beginning of this section.
2. Drain the cooling system.
3. Remove or disconnect the following:
   - Negative battery cable
   - Fan and clutch assembly from the pump
   - Fan shroud and fan assembly. If you're reusing the fan clutch, keep it upright to avoid silicone fluid loss!
   - Lower hose
   - Water pump (8 bolts)
4. Installation is the reverse of removal. Tighten the bolts, in sequence, to 40 ft. lbs. (54 Nm).

4.7L Engine

1. Before servicing the vehicle, refer to the precautions in the beginning of this section.
2. Drain the cooling system.
3. Remove or disconnect the following:
   - Negative battery cable
   - Fan and clutch assembly from the pump
   - Fan shroud and fan assembly. If you're reusing the fan clutch, keep it upright to avoid silicone fluid loss!

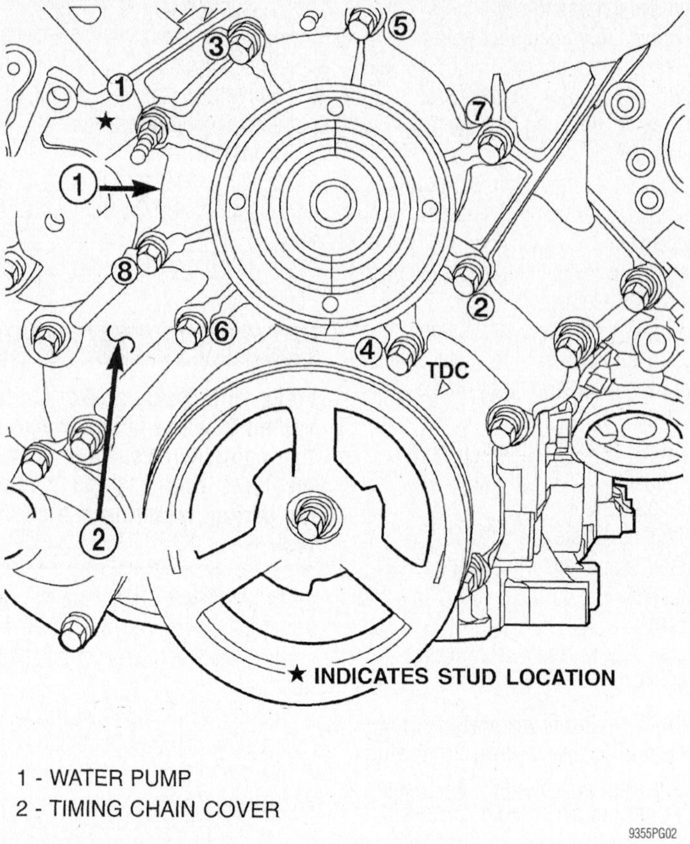

Water pump tightening sequence—3.7L

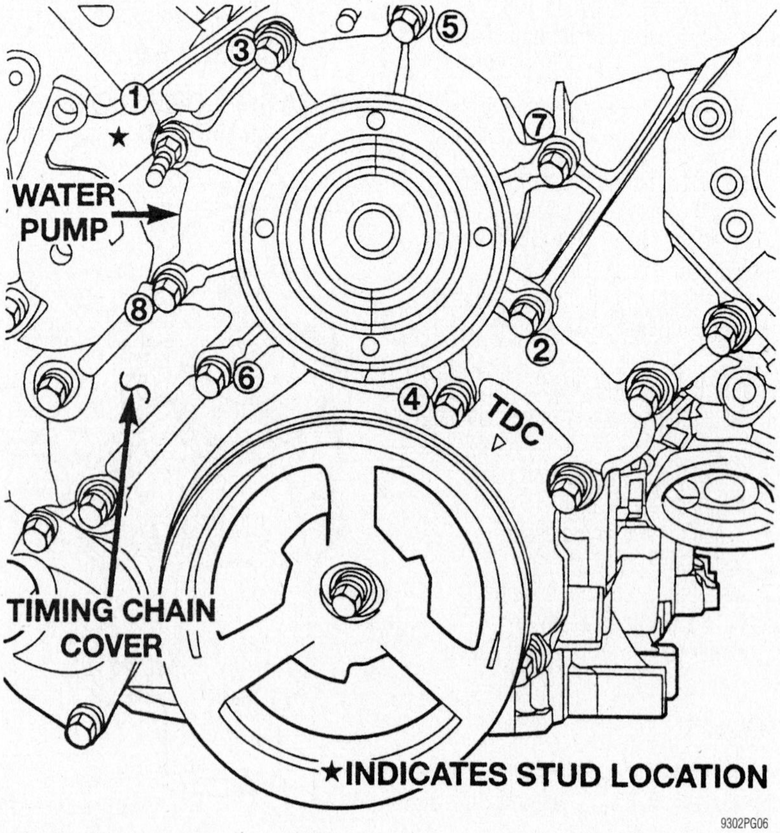

Water pump torque sequence—4.7L engine

- Drive belt
- Lower radiator hose from the water pump
- Water pump

4. Clean the mating surfaces of all gasket material.

To install:

5. Install or connect the following:
   - Water pump using a new gasket. Torque the bolts to 40 ft. lbs. (54 Nm).
   - Lower radiator hose
   - Drive belt
   - Negative battery cable
6. Fill and bleed the cooling system.
7. Start the engine, check for leaks and repair if necessary.

5.7L Engine

1. Before servicing the vehicle, refer to the precautions in the beginning of this section.
2. Disconnect negative battery cable.
3. Drain coolant.
4. Remove serpentine belt.
5. Remove fan clutch assembly.
6. Remove coolant fill bottle.
7. Disconnect washer bottle wiring and hose.
8. Remove fan shroud assembly.
9. Remove A/C compressor and generator brace.
10. Remove idler pulleys.
11. Remove belt tensioner assembly.
12. Remove upper and lower radiator hoses.
13. Remove heater hoses.
14. Remove water pump mounting bolts and remove pump.

To install:

15. Install water pump and mounting bolts. Tighten mounting bolts to 24 Nm (18 ft. lbs.).
16. Install heater hoses.
17. Install upper and lower radiator hoses.
18. Install belt tensioner assembly.
19. Install idler pulleys.
20. Install A/C compressor and alternator brace. Tighten bolt and nuts to 28 Nm (21 ft. lbs.).
21. Install fan shroud assembly.
22. Connect washer bottle wiring and hose.
23. Install coolant fill bottle.
24. Install fan clutch assembly.
25. Install serpentine belt.
26. Connect negative battery cable.
27. Fill coolant.
28. Pressure test coolant system

Heater Core

REMOVAL & INSTALLATION

Wrangler

1. Before servicing the vehicle, refer to the precautions in the beginning of this section.
2. Disconnect and remove the negative battery.

⁂ CAUTION

After disconnecting the negative battery cable, wait 2 minutes for the driver's/passenger's air bag system capacitor to discharge before attempting to do any work around the steering column or instrument.

3. Drain the cooling system into a clean container for reuse.
4. Remove the instrument panel by performing the following procedure:
   a. Turn the steering wheel in the straight-ahead position.
   b. Remove the knee blocker from the instrument panel.
   c. Remove the steering column; do not remove the air bag module, the steering wheel or switches from the steering column.
   d. From under the driver's side of the instrument panel, disconnect the following items:
      - Instrument panel wiring harness connector from the 100-way wiring harness connector at the left side of the inner panel.
      - Side window demister hose at the heater/air conditioning housing demister/defroster duct on the driver's side.
   e. Remove the glove box.
   f. Reaching through the glove box opening, disconnect the following items:
      - Two halves of the heater/air conditioning system vacuum harness connector.
      - Instrument panel wiring harness connector from the heater/air conditioning system wiring harness connector.
      - Instrument panel wiring harness connector from the passenger's side air bag module wiring harness connector.
      - Side window demister hose at the heater/air conditioning housing demister/defroster duct (passenger's side).
      - Two halves of the radio antenna coaxial cable connector.
      - Two instrument panel wiring harness connectors from the passenger air bag ON/OFF switch wiring harness connector.
      - Passenger's side air bag ON/OFF switch wiring harness from the retainer clip on the plenum bracket that supports the heater/air conditioning housing just inboard of the fuse block module.
      - Two lower passenger's side air bag module bracket-to-dash panel nuts.
   g. Remove the upper cover from the instrument panel.
   h. Remove the 3 instrument panel-to-door hinge pillar screws.
   i. Remove the 4 upper instrument panel-to-dash nuts.
   j. Using an assistant, remove the instrument panel from the vehicle.
5. If equipped with air conditioning, discharge and recover the air conditioning system refrigerant.
6. Disconnect the refrigerant lines from the evaporator. Plug the refrigerant openings to prevent evaporation.
7. Disconnect the heater hoses from the heater core tubes.
8. Disconnect the heater/air conditioning system vacuum supply line connector from the T-fitting near the heater core tubes.
9. In the engine compartment, remove the 5 heater/air conditioning housing-to-chassis nuts. If necessary, loosen the battery hold-downs and reposition the battery for access.
10. Remove the cowl plenum drain tube from the heater/air conditioning housing stud; it's located behind the cylinder head on the cowl.
11. From the bottom of the heater/air conditioning housing, remove the floor duct.
12. On the passenger side, remove the heater/air conditioning housing-to-plenum bracket screw.
13. Pull the heater/air conditioning housing down far enough to clear the defrost/demist and fresh air ducts, then, rearward far enough to clear the mounting studs and the evaporator drain tube to clear the dash panel holes.
14. Remove the heater/air conditioning housing assembly from the vehicle.
15. Remove the heater/air conditioning housing upper case.
16. Lift the heater core from the lower half of the heater/air conditioning housing.

To install:

17. Assemble the heater core into the lower half of the heater/air conditioning housing.
18. Install the heater/air conditioning housing upper case.
19. Install the heater/air conditioning housing assembly to the vehicle.
20. On the passenger's side, install the heater/air conditioning housing-to-plenum bracket screw.
21. At the bottom of the heater/air conditioning housing, install the floor duct.
22. Install the cowl plenum drain tube to the heater/air conditioning housing stud; it's located behind the cylinder head on the cowl.
23. In the engine compartment, install the 5 heater/air conditioning housing-to-chassis nuts.
24. Connect the heater/air conditioning system vacuum supply line connector to the T-fitting near the heater core tubes.
25. Connect the heater hoses to the heater core tubes.
26. Connect the refrigerant lines to the evaporator.
27. If equipped with air conditioning, evacuate and charge the air conditioning system refrigerant.
28. Install the instrument panel by performing the following procedure:
    a. Using an assistant, install the instrument panel to the vehicle.
    b. Install the 4 upper instrument panel-to-dash nuts.
    c. Install the 3 instrument panel-to-door hinge pillar screws.
    d. Install the upper cover to the instrument panel.
    e. Reaching through the glove box opening, connect the following items.
    - Two lower passenger's side air bag module bracket-to-dash panel nuts.
    - Passenger's side air bag ON/OFF switch wiring harness to the retainer clip on the plenum bracket that supports the heater/air conditioning housing just inboard of the fuse block module.
    - Two instrument panel wiring harness connectors to the passenger air bag ON/OFF switch wiring harness connector.
    - Two halves of the radio antenna coaxial cable connector.
    - Side window demister hose at the heater/air conditioning housing demister/defroster duct (passenger's side).
    - Instrument panel wiring harness connector to the passenger's side

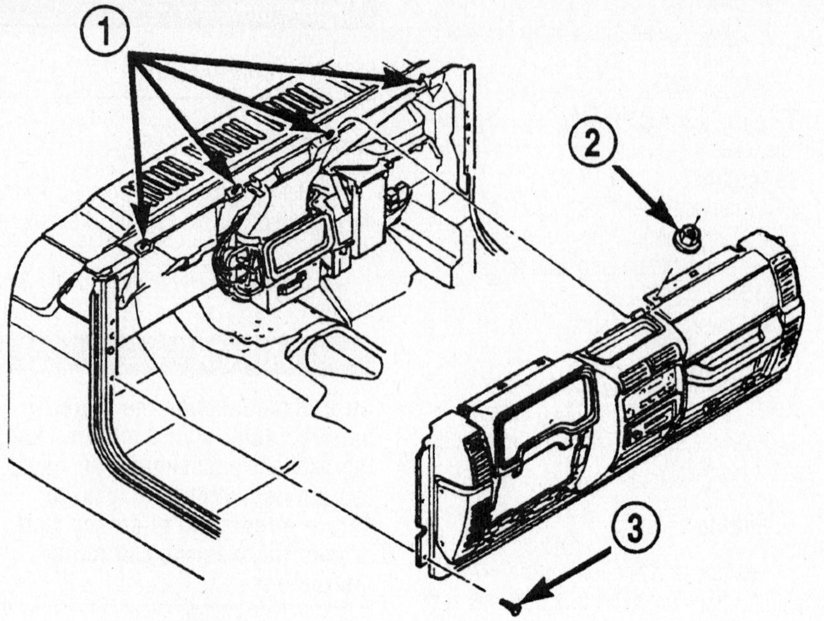

View of the instrument panel and fasteners—Wrangler

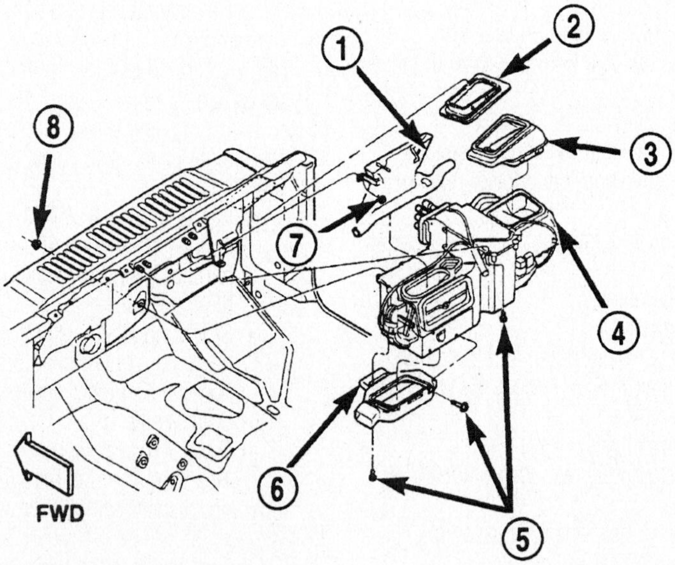

Exploded view of the heater core assembly—Wrangler

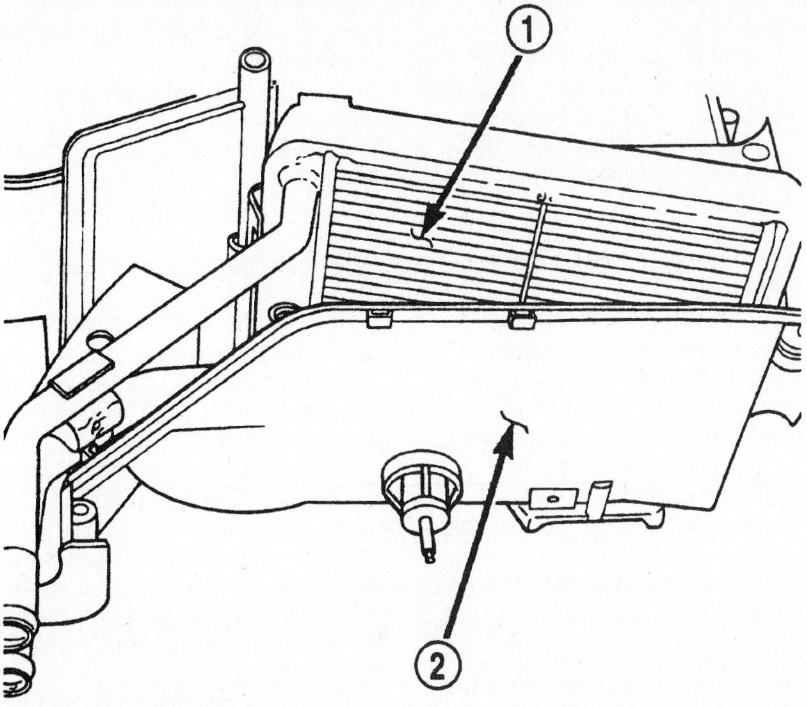

1 – HEATER CORE
2 – LOWER HEATER-A/C HOUSING

93113GA8

View of the heater core—Wrangler

air bag module wiring harness connector.

- Instrument panel wiring harness connector to the heater/air conditioning system wiring harness connector.
- Two halves of the heater/air conditioning system vacuum harness connector.

f. Install the glove box.

g. Under the driver's side of the instrument panel, connect the following items:

- Side window demister hose at the heater/air conditioning housing demister/defroster duct on the driver's side.
- Instrument panel wiring harness connector to the 100-way wiring harness connector at the left side of the inner panel.

h. Install the steering column.

i. Install the knee blocker to the instrument panel.

29. Connect and remove the negative battery.

30. Refill the cooling system.

31. Run the engine to normal operating temperatures; then, check the climate control operation and check for leaks.

2002–04 Grand Cherokee

1. Before servicing the vehicle, refer to the precautions in the beginning of this section.

2. Disconnect and remove the negative battery.

CAUTION

After disconnecting the negative battery cable, wait 2 minutes for the driver's/passenger's air bag system capacitor to discharge before attempting to do any work around the steering column or instrument.

3. Drain the cooling system into a clean container for reuse.

4. Remove the instrument panel by performing the following procedure:

a. Turn the steering wheel in the straight-ahead position.

b. Remove the A-pillar trim from both sides of the vehicle.

c. Remove the top cover from the instrument panel.

d. Near the windshield line, remove the 4 instrument panel-to-chassis nuts.

e. Remove the scuff plates from both front door sills.

f. Remove the trim panels from both sides of the inner cowl.

g. Remove the floor console.

h. Remove the fuse cover from the junction box.

i. Remove the instrument panel cluster bezel.

j. Remove the steering column opening cover from the instrument panel.

k. Remove the steering column bracket from the instrument panel column support bracket.

l. Remove the lower steering column shroud cover-to-multifunction switch screw; then, unsnap both halves of the shroud cover from the steering column.

m. Disconnect the instrument panel wiring harness connectors from the following steering column components:

- Both lower clockspring connector receptacles
- Left multifunction switch receptacle
- Right multifunction switch receptacle
- Both ignition switch receptacles
- Shifter interlock solenoid receptacle
- Sentry Key Immobilizer Module (SKIM) receptacle, if equipped

n. Turn the ignition switch to ON position; then, release and remove the shifter interlock cable connector from the ignition lock housing receptacle.

o. Turn the ignition switch to OFF position; this will prevent the steering wheel from turning and the loss of the clockspring centering following steering column removal.

p. Remove the 4 steering column-to-instrument panel steering column bracket nuts.

q. Remove the steering column from the instrument panel.

r. Disconnect both side body wiring harness bulkhead connectors, the Ignition Off Draw (IOD) wiring harness connector and the fused wiring harness connector from the junction block connector receptacles.

s. Disconnect the instrument panel wiring harness-to-floor console component connectors:

- Air bag control module connector receptacle
- Parking brake switch terminal
- Transmission shifter connector receptacle

t. Remove the 2 instrument panel wiring harness-to-floor console ground terminals located behind the air bag control module.

u. Disconnect the instrument panel wiring harness-to-floor console retainers.

v. Remove the instrument panel-to-floor console bracket screws and the bracket.

w. Remove the driver's side floor duct-to-heater/air conditioning housing assembly screw and remove the duct.

x. If equipped with a manual heating-air conditioning system, disconnect the vacuum harness connector from behind the driver's side floor duct.

y. Remove the instrument panel steering column support bracket-to-driver's side of the heater/air conditioning housing assembly screw

z. Remove the instrument panel steering column support bracket-to-intermediate bracket screw.

aa. Remove the instrument panel steering column support bracket-to-driver's side cowl plenum panel nut.

bb. Remove the 2 instrument panel-to-driver's side cowl side inner panel screws.

cc. Remove the instrument panel end cap.

dd. Remove the lower right center bezel from the instrument panel.

ee. At the passenger's side cowl side inner panel, disconnect the instrument panel wiring harness bulkhead connector from the lower cavity of the inline connector.

ff. Near the right side cowl inner panel, located under the end of the instrument panel, disconnect both halves of the radio antenna coaxial cable connector.

gg. Disconnect the 2 instrument panel-to-heater/air conditioning assembly wiring harness connectors.

hh. At the passenger's side, remove the 2 instrument panel structural duct-to-heater/air conditioning housing assembly screws.

ii. At the passenger's side cowl side inner panel, remove the 2 instrument panel-to-passenger's side cowl side inner panel screws.

jj. With the help of an assistant, lift the instrument panel from the vehicle.

5. Discharge and recover the air conditioning system refrigerant.

6. Disconnect the air conditioning system lines at the evaporator. Plug the openings to prevent contamination.

7. Disconnect the heater hoses from the heater core. Plug the openings to prevent coolant loss.

8. If equipped with a manual temperature control system, unplug the heater/air conditioning system vacuum supply line connector from the T-fitting located near the heater core tubes.

9. From the passenger's side inner fender shield, remove the coolant reservoir/overflow bottle.

10. From the passenger's side in the engine compartment dash panel, remove the PCM; DO NOT unplug it, just move it aside.

11. In the engine compartment, remove the heater/air conditioning housing-to-chassis nuts.

12. At the center of the dashboard, remove the rear floor ducts from the floor heat duct outlets.

13. Disconnect the heater/air conditioning housing wire harness connectors.

14. In the passenger compartment, remove the heater/air conditioning housing-to-chassis nuts.

15. Place covers inside the vehicle to catch any spilt coolant.

16. Remove the heater/air conditioning housing assembly from the vehicle.

17. Remove the foam gasket from around the heater core tubes.

➡Note the position of the irregular shaped gasket so that it may be reinstalled in its correct position.

18. Remove the heater core retainers and screws.

19. If necessary, remove the mode door actuator for clearance to remove the core.

20. Remove the heater core from the heater/air conditioning housing assembly.

To install:

21. Install the heater core tom the heater/air conditioning housing assembly.

22. If removed, install the mode door actuator.

23. Install the heater core retainers and screws.

24. Install the foam gasket around the heater core tubes.

25. Install the heater/air conditioning housing assembly to the vehicle.

26. In the passenger compartment, install the heater/air conditioning housing-to-chassis nuts.

27. Connect the heater/air conditioning housing wire harness connectors.

28. At the center of the dashboard, install the rear floor ducts to the floor heat duct outlets.

29. In the engine compartment, install the heater/air conditioning housing-to-chassis nuts.

30. At the passenger's side in the engine compartment dash panel, install the PCM.

31. At the passenger's side inner fender shield, install the coolant reservoir/overflow bottle.

32. If equipped with a manual temperature

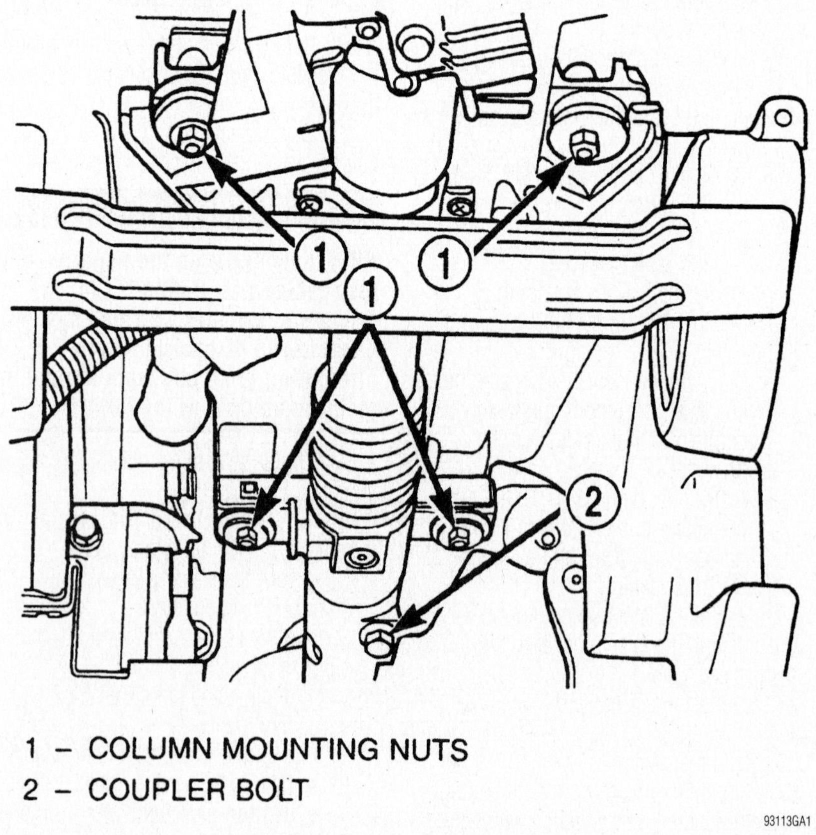

View of the steering column mounting nuts—2002–04 Grand Cherokee

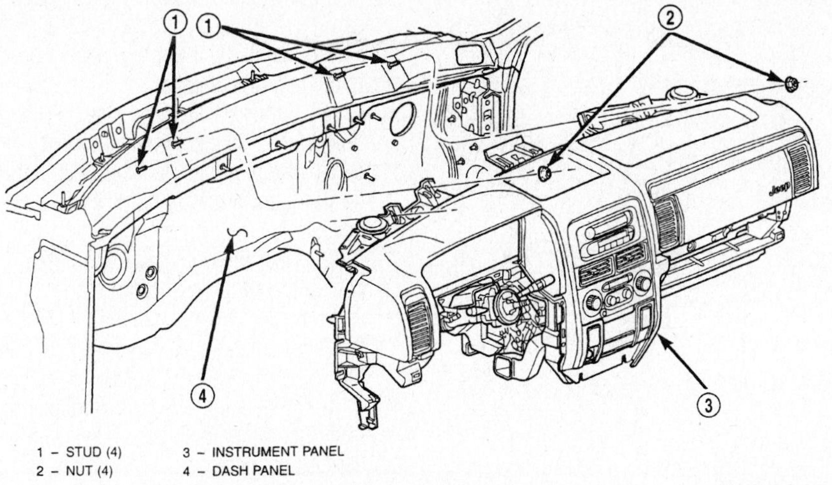

View of the instrument panel assembly—2002–04 Grand Cherokee

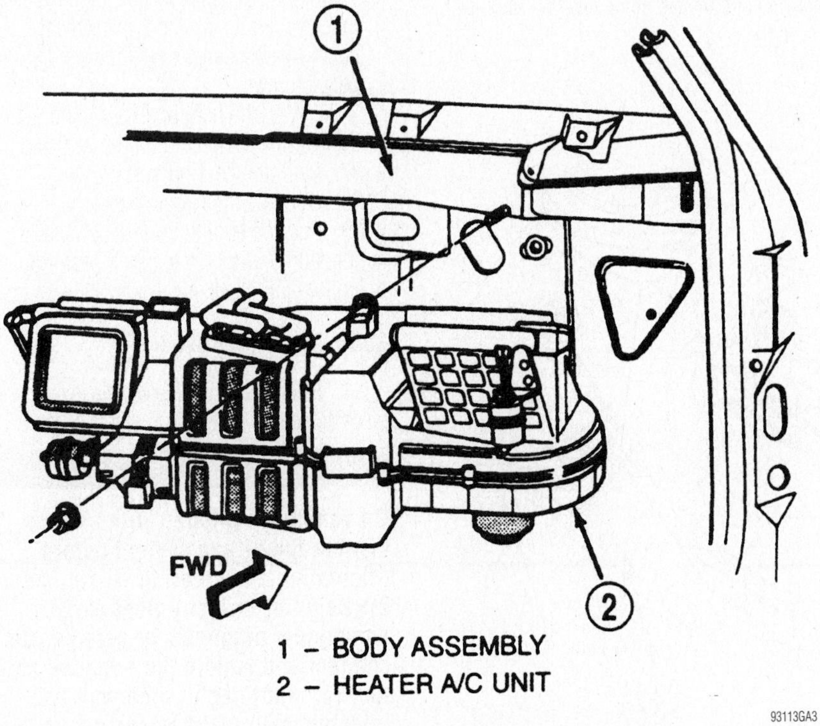

View of the heater/air conditioning housing assembly—2002–04 Grand Cherokee

control system, plug the heater/air conditioning system vacuum supply line connector to the T-fitting located near the heater core tubes.

33. Connect the heater hoses to the heater core.

34. Connect the air conditioning system lines to the evaporator.

35. Evacuate and charge the air conditioning system refrigerant.

36. Remove the instrument panel by performing the following procedure:

a. Turn the steering wheel in the straight-ahead position.

b. Install the A-pillar trim to both sides of the vehicle.

c. Install the top cover to the instrument panel.

d. Near the windshield line, install the 4 instrument panel-to-chassis nuts.

e. Install the scuff plates to both front door sills.

f. Install the trim panels to both sides of the inner cowl.

g. Install the floor console.

h. Install the fuse cover to the junction box.

i. Install the instrument panel cluster bezel.

j. Install the steering column opening cover to the instrument panel.

k. Install the steering column bracket to the instrument panel column support bracket.

l. Install the lower steering column shroud cover-to-multifunction switch screw; then, snap both halves of the shroud cover to the steering column.

m. Connect the instrument panel wiring harness connectors to the following steering column components:

- Sentry Key Immobilizer Module (SKIM) receptacle, if equipped
- Shifter interlock solenoid receptacle
- Both ignition switch receptacles
- Right multifunction switch receptacle
- Left multifunction switch receptacle
- Both lower clockspring connector receptacles

n. Turn the ignition switch to ON position; then, release and install the shifter interlock cable connector to the ignition lock housing receptacle.

o. Install the 4 steering column-to-instrument panel steering column bracket nuts.

p. Install the steering column to the instrument panel.

q. Connect both side body wiring harness bulkhead connectors, the Ignition Off Draw (IOD) wiring harness connector and the fused wiring harness connector to the junction block connector receptacles.

r. Disconnect the instrument panel wiring harness-to-floor console component connectors:

- Transmission shifter connector receptacle
- Parking brake switch terminal
- Air bag control module connector receptacle

s. Install the 2 instrument panel wiring harness-to-floor console ground terminals located behind the air bag control module.

t. Connect the instrument panel wiring harness-to-floor console retainers.

u. Install the instrument panel-to-floor console bracket and the screws.

v. Install the driver's side floor duct and the duct-to-heater/air conditioning housing assembly screw.

w. If equipped with a manual heating-air conditioning system, connect the vacuum harness connector behind the driver's side floor duct.

x. Install the instrument panel steer-

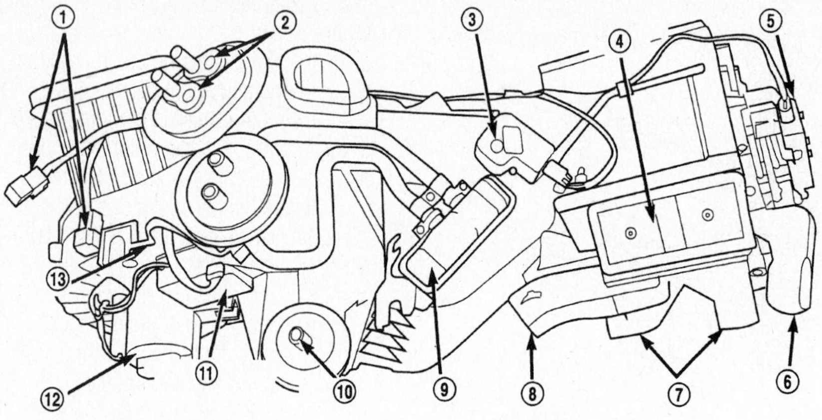

View of the heater core, the heater/air conditioning housing and related components—2002–04 Grand Cherokee

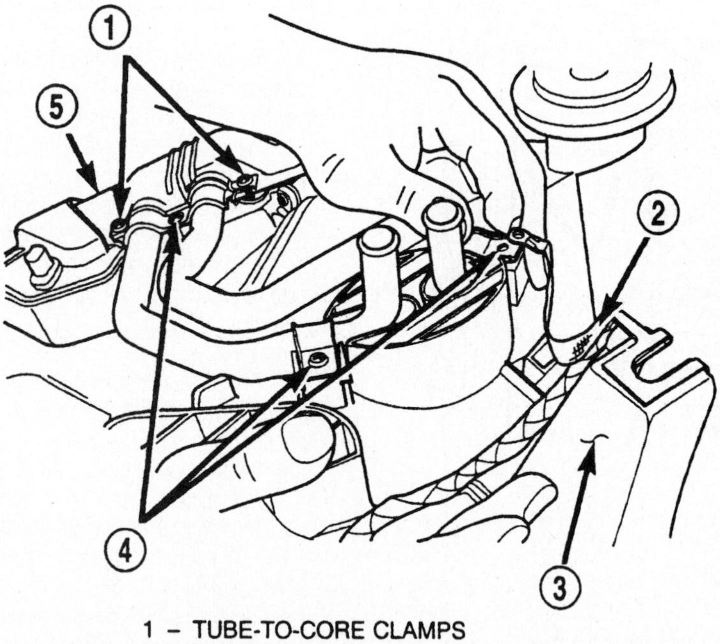

View of the heater core screws, gasket and retainers—2002–04 Grand Cherokee

ing column support bracket-to-driver's side of the heater/air conditioning housing assembly screw.

y. Install the instrument panel steering column support bracket-to-intermediate bracket screw.

z. Install the instrument panel steering column support bracket-to-driver's side cowl plenum panel nut.

aa. Install the 2 instrument panel-to-driver's side cowl side inner panel screws.

bb. Install the instrument panel end cap.

cc. Install the lower right center bezel to the instrument panel.

dd. At the passenger's side cowl side inner panel, connect the instrument panel wiring harness bulkhead connector to the lower cavity of the inline connector.

ee. Near the right cowl side inner panel located under the end of the instrument panel, connect both halves of the radio antenna coaxial cable connector.

ff. Connect the 2 instrument panel-to-heater/air conditioning assembly wiring harness connectors.

gg. At the passenger's side, install the 2 instrument panel structural duct-to-heater/air conditioning housing assembly screws.

hh. At the passenger's side cowl side inner panel, install the 2 instrument panel-to-passenger's side cowl side inner panel screws.

ii. With the help of an assistant, lift the instrument panel into the vehicle.

37. Refill the cooling system.
38. Connect the negative battery.
39. Run the engine to normal operating temperatures. Check the climate control operation and check for leaks.

2005 Grand Cherokee

1. Before servicing the vehicle, refer to the precautions in the beginning of this section.

✲✲ CAUTION

On vehicles equipped with airbags, disable the airbag system before attempting any steering wheel, steering column, or instrument panel component diagnosis or service. Disconnect and isolate the negative battery (ground) cable, then wait two minutes for the airbag system capacitor to discharge before performing further diagnosis or service. This is the only sure way to disable the airbag system. Failure to take the proper precautions could result in accidental airbag deployment and possible personal injury or death.

2. Drain the engine cooling system.
3. Disconnect and isolate the negative battery cable.
4. If required, disconnect the heater hoses from the heater core tubes in the engine compartment.

5. Disconnect and isolate battery negative cable.

6. Remove the bolts and remove the driver's side hush panel.

7. Using a trim stick, separate the upper clips and rotate the cover down and release the lower hinges at the bottom and remove the opening cover.

CAUTION

Before servicing the steering Column the airbag system must be disarmed. Failure to do so may result in accidental deployment of the airbag and possible personal injury.

8. Remove the three bolts in the lower shroud.

➡Use care not to break off the tangs on the shrouds.

9. Then unsnap the lower shroud from the upper shroud.

10. Disconnect the steering column electrical connectors.

11. Remove the pinch bolt.

12. Remove the column support bolts.

13. Remove the column cross bolt and slide the column downward off the bracket and remove the steering column.

14. Remove the nuts attaching the instrument panel to the pedal support bracket.

15. Using a trim stick, remove the ignition cylinder bezel.

16. Remove the screws and remove the ignition cylinder

17. Disconnect the electrical connector.

18. Using a trim stick, remove the left side cover.

19. Remove the left side door sill trim.

20. Remove the driver's side cowl trim panel.

21. Disconnect the electrical connectors.

22. Disconnect the electrical connectors.

23. Remove the bolts and disconnect the ground wires within the steering column opening to the left of the steering column.

24. Disconnect the white adjustable pedal electrical connector.

25. Remove the bolts and remove the left a-pillar trim.

26. Remove the two left side bolts and one screw at the a-pillar support.

27. Using a trim stick, remove the shifter bezel ring.

28. Using a trim stick, remove the shifter bezel.

29. Remove the center console back cover and disconnect the electrical connector.

30. Remove the screws and remove the console.

31. Disconnect the electrical connectors under the console.

32. Open the glove box and push the stop tabs down to drop the glove box out of the instrument panel.

33. Rotate the box down and release the door hinges at the bottom and remove the glove box.

34. Remove the passenger side hush panel.

35. Remove the right side door sill trim.

36. Remove the nut, separate the right cowl trim panel and remove.

37. Remove the right side end cap.

38. Disconnect the electrical connectors.

39. Remove the bolts and remove the right a-pillar trim panel.

40. Remove the two right side bolts and one screw at the a-pillar support.

41. Using a trim stick, remove the defroster grills.

42. Remove the two fence line nuts.

43. Using a trim stick, remove the radio bezel.

44. Using a trim stick, remove the center bezel and disconnect the electrical connectors.

45. Remove the nut from the HVAC support stud behind the center bezel.

46. Remove the screw to the HVAC from under the glove box opening.

47. Remove the HVAC bracket bolt from the glove box opening.

48. Remove the left and right center support brackets.

49. Remove the center floor duct.

50. Remove the radio.

51. Remove the two HVAC bolts from the radio opening.

52. Remove the instrument panel assembly.

53. Remove the five screws that secure the heater core and tube cover to the HVAC housing.

54. Remove the heater core and tube cover from the HVAC housing.

55. If equipped with dual zone heating-A/C, remove the blend door actuator from the passenger side of the HVAC air distribution housing.

56. Remove the screw that secure the heater core tubes and retaining bracket to the HVAC housing.

➡Take proper precautions to protect the carpeting from engine coolant. Have absorbent toweling readily available to clean up any spills.

57. Remove the bolt that secures the heater tubes to the heater core.

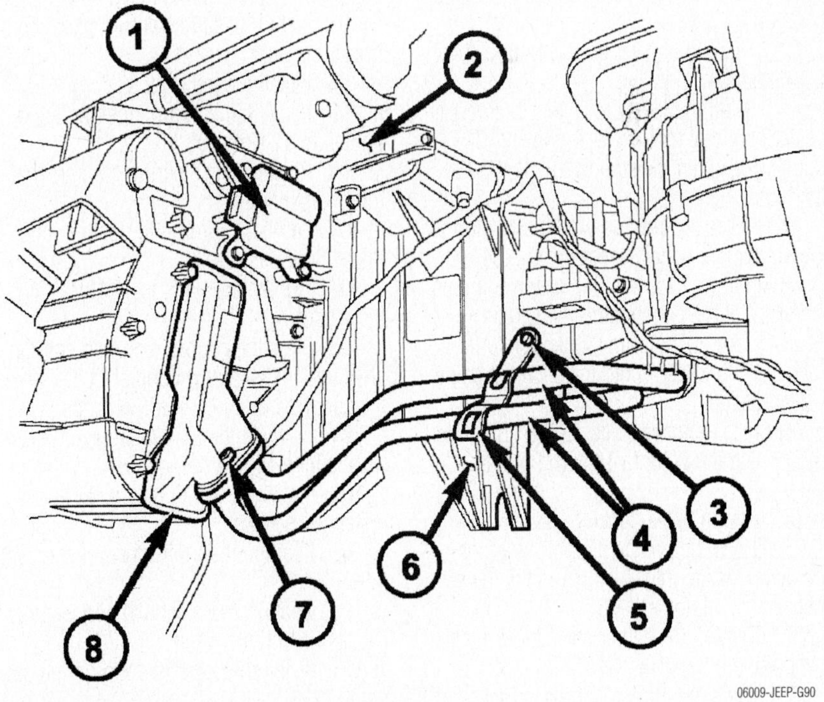

Heater core plumbing and related parts—2005 Grand Cherokee. (1) blend door actuator, (2) HVAC air distribution housing, (3) heater core tube screw, (4) heater core tubes, (5) retaining bracket, (6) HVAC housing, (7) core tube bolt, (8) heater core

58. Disconnect the heater core tubes from the heater core and remove and discard the O-ring seals.

59. Install plugs in, or tape over the opened heater core ports.

60. Carefully pull the heater core out of the HVAC air distribution housing.

61. If required, remove the heater core tubes from the vehicle.

To install:

62. Carefully install the heater core into the passenger's side of the HVAC air distribution housing.

63. Remove the tape or plugs from the heater core ports.

64. If removed, position the heater core tubes into the vehicle.

65. Lubricate new rubber O-ring seals with clean engine coolant and install them onto the heater core tubes. Use only the specified O-ring as they are made of a special material for the engine cooling system.

66. Connect the heater core tubes to the heater core.

67. Install the bolt that secures the heater core tubes to the heater core. Tighten the bolt securely.

68. Install the screw that secures the heater core tube retaining bracket to the HVAC housing. Tighten the screw to 2.2 Nm (20 inch lbs.).

69. If equipped with dual zone heating-A/C, install the blend door actuator onto the passenger side of the HVAC air distribution housing.

70. Install the heater core and tube cover onto the HVAC housing.

71. Install the five screws that secure the heater core and tube cover to the HVAC housing. Tighten the screws to 2.2 Nm (20 inch lbs.).

72. Before proceeding with installation of the instrument panel assembly, remove the load clips from each a-pillar and install onto the sides of the panel using the screw.

73. Place the instrument panel into the vehicle and position onto the a-pillars using the load clips.

74. Install the two bolts on the right side and tighten the bolts to 11 Nm (95 inch lbs.).

75. Install the two fence line nuts.

76. Guide the peddle support studs on the driver's side, into the slots in the instrument panel reinforcement.

77. Position the center stack brackets in front of the duct work.

78. Rotate the instrument panel back and guide the HVAC stud.

79. Continue to rotate the instrument panel back and guide the tube brackets located in the radio opening, into the HVAC.

80. Install the screw at the bottom of the column opening next to the center stack up from the bottom attaching the HVAC.

81. Install the bolt at the HVAC blower motor housing.

82. Install the screw at the bottom of the glove box opening to the blower motor housing support

83. Snap the lower glove box hinges over the pivots and rotate up.

84. Push the stop tabs down and close the glove box.

85. Connect the right side wiring.

86. Install the right hush panel trim.

87. Install the right cowl trim and position the door seal back.

88. Position the right side door sill trim and seat fully.

89. Install the right a-pillar trim, the bolts and position the door seal back.

90. Install the two bolts on the left side and tighten to 11 Nm (95 inch lbs.).

91. Position the right side end cap and seat fully.

92. Install the two nuts on the studs above the column tighten the nuts to 12 Nm (105 inch lbs.).

93. Connect the two left bulk connectors.

94. Connect the two fuse block connectors.

95. Connect the white connector to adjustable brake pedal.

96. Install the grounds within the steering column opening and install the bolts.

97. Install the left cowl trim panel and install the nut

98. Install the left door sill.

99. Install the column cross bolt and tighten to 12 Nm (105 inch lbs.).

100. Position the steering column shaft into the slip joint.

101. Lift the column into position and install the bolts.

102. Tighten the bolts to 12 Nm (105 inch lbs.).

103. Slide the coupler onto the column shaft and install a new coupler bolt. Tighten the coupler bolt to 49 Nm (36 ft. lbs.).

104. Connect the shifter cable and electrical connectors to the ignition.

105. Install the ignition cylinder and install the screws.

106. Position the ignition bezel and seat fully.

107. Connect the electrical connector for the multifunction switch

108. Install the upper shroud on the column.

109. Install the lower shroud to the column.

110. Snap the two halves back into place.

111. Install the three mounting bolts to secure the shroud in place.

112. Snap the steering column opening cover lower hinges onto the pins and roll cover up.

113. Seat the upper retaining clips fully.

114. Install the hush panel.

115. Connect the column electrical connectors.

116. Install the column covers.

117. Install the left a-pillar trim, the bolts and position the door seal back.

118. Position the left side end cap and seat fully.

119. Install the center support nut onto the stud.

120. Install the two HVAC bolts within the radio opening.

121. Install the radio.

122. Install the radio bezel.

123. Install the left and right center support brackets and nuts and tighten to 12 Nm (105 inch lbs.).

124. Position the center wire harness and connect the electrical connectors.

125. Install the center bezel.

126. Position the defroster vents and seat fully.

127. Connect the center harness electrical connectors (ground, module, shifter, brake lever).

128. Install the center floor duct.

129. Position the console into the vehicle and install the screws.

130. Connect the electrical connector and install the rear cover onto the console.

131. Position the shifter bezel and seat fully.

132. Position the shifter bezel trim ring and seat fully.

133. If disconnected, connect the heater hoses to the heater core tubes in the engine compartment.

134. Connect the negative battery cable.

135. If the heater core is being replaced, flush the cooling system.

136. Refill the engine cooling system.

Liberty

1. Before servicing the vehicle, refer to the precautions in the beginning of this section.
2. Disconnect and remove the negative battery.

CAUTION

After disconnecting the negative battery cable, wait 2 minutes for the driver's/passenger's air bag system capacitor to discharge before attempting to do any work around the steering column or instrument.

3. Drain the cooling system into a clean container for reuse.

4. Remove the instrument panel by performing the following procedure:

a. Turn the steering wheel in the straight-ahead position.

b. Remove the A-pillar trim from both sides of the vehicle.

c. Remove the top cover from the instrument panel.

d. Remove the speakers.

e. Remove the floor console.

f. Remove the radio.

g. Remove the center support bracket.

h. Remove the trim panels from both sides of the inner cowl.

i. Remove the fuse cover from the junction box.

j. Remove the instrument panel cluster bezel.

k. Remove the steering column opening cover from the instrument panel.

l. Remove the steering column bracket from the instrument panel column support bracket.

m. Remove the lower steering column shroud cover-to-multifunction switch screw; then, unsnap both halves of the shroud cover from the steering column.

n. Disconnect the instrument panel wiring harness connectors from the following steering column components:

- Both lower clockspring connector receptacles
- Left multifunction switch receptacle
- Right multifunction switch receptacle
- Both ignition switch receptacles
- Shifter interlock solenoid receptacle
- Sentry Key Immobilizer Module (SKIM) receptacle, if equipped

o. Turn the ignition switch to ON position; then, release and remove the shifter interlock cable connector from the ignition lock housing receptacle.

p. Turn the ignition switch to OFF position; this will prevent the steering wheel from turning and the loss of the clockspring centering following steering column removal.

q. Remove the 4 steering column-to-instrument panel steering column bracket nuts.

r. Remove the steering column from the instrument panel.

5. Remove the driver's side cowl trim cover.

6. Disconnect the green and light blue wire harness bulk connectors at the junction block.

7. Disconnect the electrical connector at the inner side of the pedal support bracket.

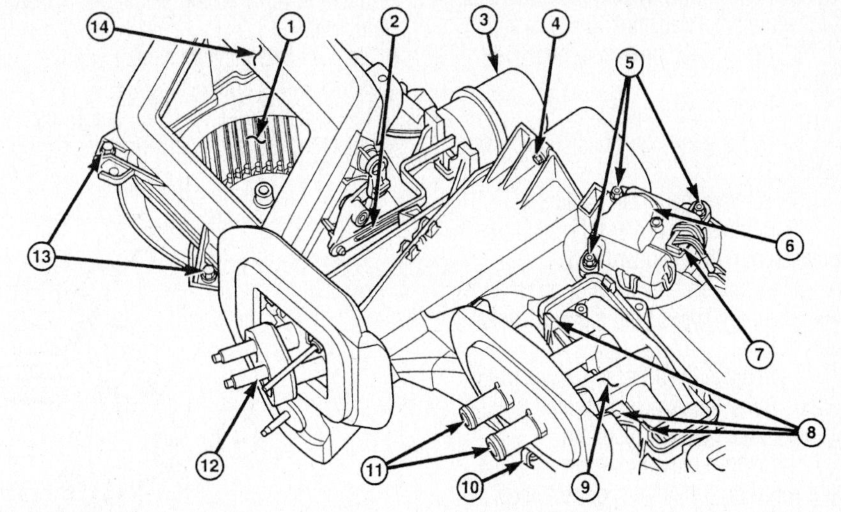

1 - BLOWER MOTOR AND CAGE
2 - RECIRCULATION DOOR ACTUATOR LINKAGE
3 - RECIRCULATION DOOR VACUUM ACTUATOR
4 - CASE RETAINER SCREW
5 - BLEND DOOR ACTUATOR MOUNTING SCREWS
6 - ELECTRIC BLEND DOOR ACTUATOR
7 - ELECTRICAL CONNECTOR FOR BLEND DOOR ACTUATOR
8 - HEATER CORE RETAINER TABS (4) AND SCREWS (2)
9 - HEATER CORE
10 - HVAC CASE RETAINER CLIP
11 - HEATER CORE INPUT AND OUTPUT CONNECTIONS
12 - EVAPORATOR CONNECTION FLANGE
13 - HVAC CASE RETAINER SCREWS
14 - HVAC HOUSING

9355PG99

HVAC case components—Liberty

8. Remove the 2 bolts from the front and the 2 from the side of the pedal support bracket.

9. Remove the glove box.

10. Remove the 2 HVAC mounting bolts behind the center trim.

11. Remove the passenger trim bezel.

12. Remove the HVAC mount bolt above the glove box.

13. Remove the HVAC bolt at the lower outside glove box opening.

14. Remove the passenger trim cover, disconnect the blower resistor, remove the rolldown brackets at the right cowl side panel.

15. Disconnect the vacuum check valve and the vacuum reservoir.

16. Disconnect the blower connectors.

17. Remove the 4 top bolts connecting the instrument panel to the cowl.

18. Roll the instrument panel rearward and disconnect the wiring.

19. Remove the panel.

20. Discharge and recover the air conditioning system refrigerant.

21. Disconnect the air conditioning system lines at the evaporator. Plug the openings to prevent contamination.

22. Disconnect the heater hoses from the heater core. Plug the openings to prevent coolant loss.

23. If equipped with a manual temperature control system, unplug the heater/air conditioning system vacuum supply line connector from the T-fitting located near the heater core tubes.

24. Remove all remaining fasteners and connections and remove the HVAC unit.

25. Disconnect all remaining hoses and wires.

26. Remove the blower motor.

27. Pop out the grommet on the vacuum supply line and slide hole.

28. Remove the foam gasket from around the heater core tubes.

29. Pry off the 4 snap clips that hold the halves of the unit together and separate the unit halves.

30. Installation is the reverse of removal.

31. Refill the cooling system.

32. Connect the negative battery.

33. Evacuate, charge and leak test the system.

34. Run the engine to normal operating temperatures. Check the climate control operation and check for leaks.

Cylinder Head

REMOVAL & INSTALLATION

2.4L Engine

1. Before servicing the vehicle, refer to the precautions in the beginning of this section.

2. Perform fuel system pressure release procedure before attempting any repairs.
3. Disconnect the battery negative cable.
4. Drain the cooling system.
5. Remove air filter housing and inlet tube.
6. Remove the intake manifold.
7. Remove the heater tube support bracket from the cylinder head.
8. Disconnect the radiator upper and heater supply hoses from the water outlet connections.
9. Remove the accessory drive belts.
10. Raise the vehicle and remove the exhaust pipe from the manifold.
11. Remove the power steering pump and set aside. Do not disconnect lines.
12. Remove the accessory drive bracket
13. Remove the ignition coil and wires from the engine.
14. Disconnect the cam sensor and fuel injector wiring connectors.
15. Remove the timing belt and camshaft sprockets.
16. Remove the timing belt idler pulley and rear timing belt cover.
17. Remove the cylinder head cover.
18. Remove camshafts.

➡Identify the rocker arm positions to ensure correct installation in original position, if reused.

19. Remove the rocker arms.
20. Remove the cylinder head bolts in reverse of the tightening sequence.
21. Remove the cylinder head from the engine block.
22. Inspect and clean the cylinder head.

To install:

➡The cylinder head bolts should be examined before reuse. If the threads are necked down, the bolts must be replaced.

23. Before installing the bolts, the threads should be coated with engine oil.
24. Position the cylinder head gasket on the engine block.
25. Install the cylinder head on the engine block.
26. Tighten the cylinder head bolts in the sequence shown. Tighten in 4 steps:
   - First: 34 Nm (25 ft. lbs.)
   - Second: 68 Nm (50 ft. lbs.)
   - Third: 68 Nm (50 ft. lbs.)
   - Fourth: an additional ¼ turn
27. Install the rocker arms.
28. Install the camshafts.
29. Install the cylinder head cover.
30. Install the timing belt rear cover and timing belt idler pulley.
31. Install the timing belt and camshaft sprockets.
32. Connect the cam sensor and fuel injectors wiring connectors.
33. Install the ignition coil and wires. Connect the ignition coil wiring connector.
34. Install the accessory drive bracket.
35. Install the power steering pump to the cylinder head.
36. Raise the vehicle and install the exhaust pipe to the manifold.
37. Install the accessory drive belts.

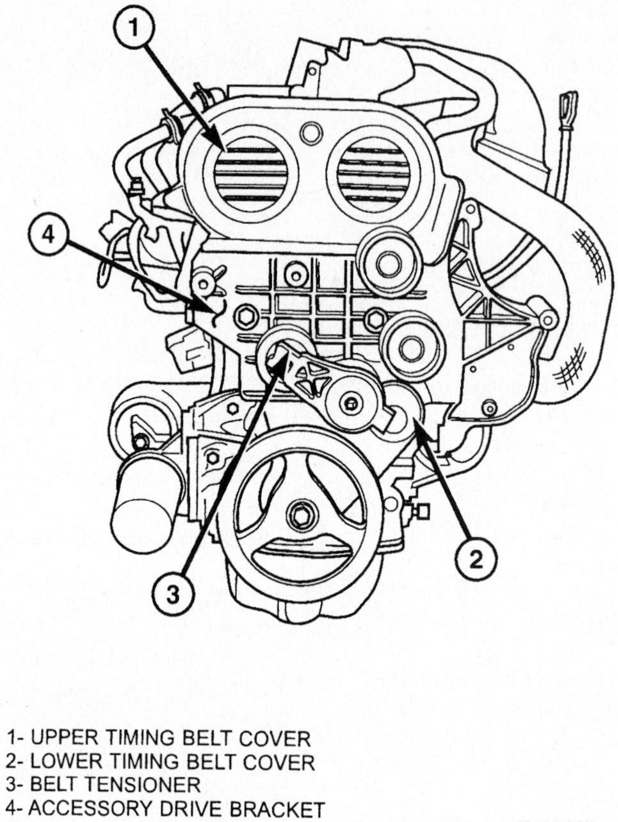

Accessory drive belt bracket—2.4L engines

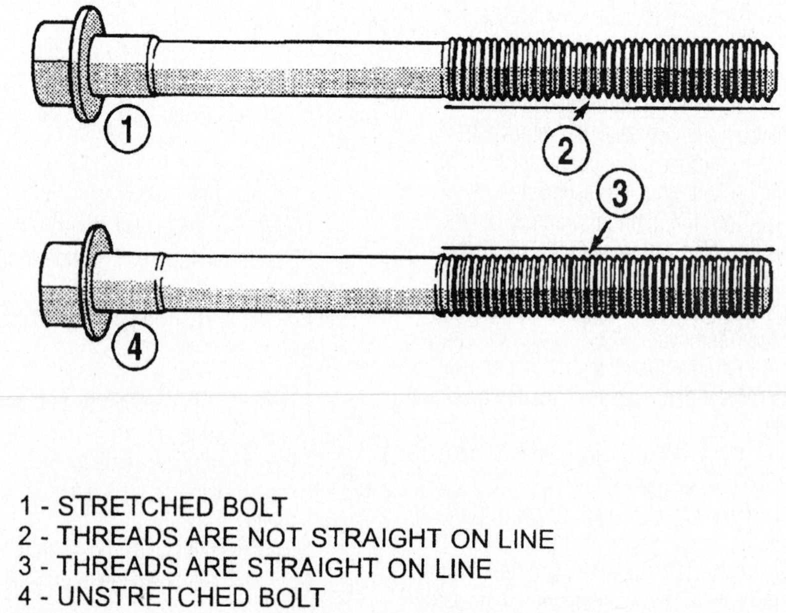

Checking the head bolts for stretching—2.4L engines

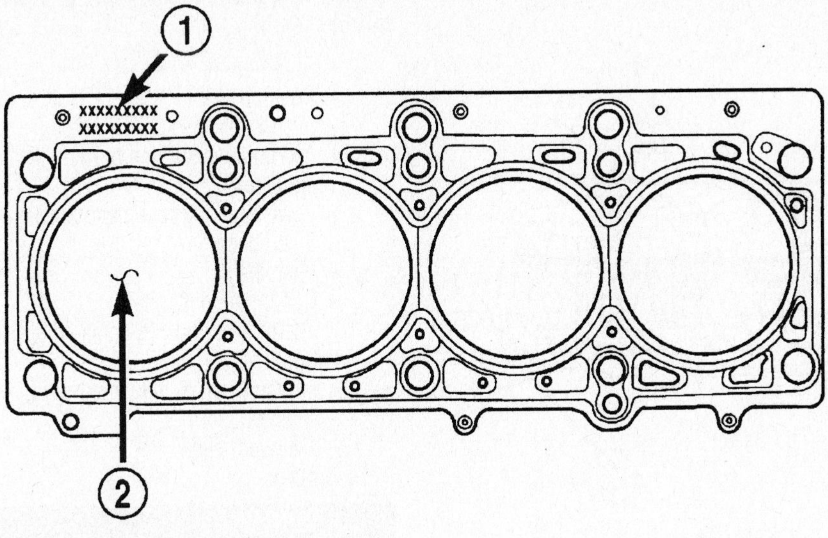

1 - PART NUMBER FACES UP
2 - NO. 1 CYLINDER

67189-JEEP-G08

Cylinder head gasket positioning—2.4L engines

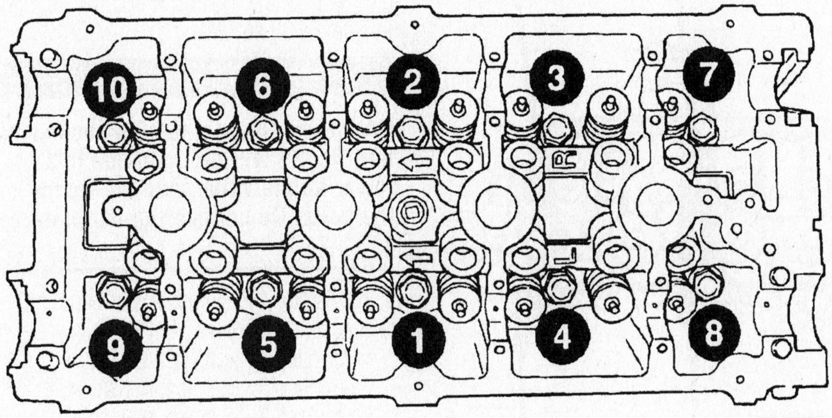

67189-JEEP-G09

Cylinder head tightening sequence—2.4L engines

38. Install the heater tube support bracket to the cylinder head.
39. Install the intake manifold.
40. Connect all vacuum lines, electrical wiring, ground straps and fuel line.
41. Fill the cooling system.
42. Connect the battery negative cable.

2.5L Engine

1. Before servicing the vehicle, refer to the precautions in the beginning of this section.
2. Properly relieve the fuel system pressure.
3. Drain the cooling system.
4. Remove or disconnect the following:
   - Negative battery cable
   - Crankcase Ventilation (CCV) hoses
   - Air cleaner assembly
   - Valve cover

➡Keep valvetrain components in order for reassembly.

   - Rocker arms
   - Pushrods
   - Accessory drive belt
   - A/C compressor and bracket, if equipped
   - Power steering pump and bracket, if equipped
   - Fuel line
   - Combination manifold
   - Thermostat housing coolant hoses
   - Spark plugs
   - Engine Coolant Temperature (ECT) sensor connector
   - Cylinder head

To install:

⁂ WARNING

Cylinder head bolts may only be reused one time. If reusing a cylinder head bolt, place a paint mark on the bolt after installation. If a cylinder head bolt has a paint mark, discard it and use a new bolt.

5. Fabricate two alignment dowels from old cylinder head bolts. Cut the hex head off of the bolts, and cut a slot in each dowel to ease removal.
6. Install or connect the following:
   - One dowel in bolt hole No. 8, and one dowel in bolt hole No. 10.
   - Cylinder head and gasket.
   - Cylinder head bolts except for No 8 and No 10. Coat the threads of bolt No. 7 with Loctite® 592 sealant.
7. Remove the alignment dowels and install the No. 8 and No. 10 head bolts.

⁂ WARNING

During the final tightening sequence, bolt No. 7 will be tightened to a lower torque value than the rest of the bolts. Do not overtighten bolt No. 7.

8. Tighten the cylinder head bolts, in sequence, as follows:
   a. Step 1: 22 ft. lbs. (30 Nm)
   b. Step 2: 45 ft. lbs. (61 Nm)
   c. Step 3: 45 ft. lbs. (61 Nm)
   d. Step 4: Bolts 1–6 to 110 ft. lbs. (149 Nm)
   e. Step 5: Bolt 7 to 100 ft. lbs. (136 Nm)
   f. Step 6: Bolts 8–10 to 110 ft. lbs. (149 Nm)
   g. Step 7: Repeat steps 4, 5 and 6
9. Install or connect the following:
   - ECT sensor connector
   - Spark plugs
   - Thermostat housing coolant hoses
   - Combination manifold
   - Fuel line
   - Power steering pump and bracket, if equipped
   - A/C compressor and bracket, if equipped
   - Accessory drive belt
   - Pushrods and rocker arms in their original positions

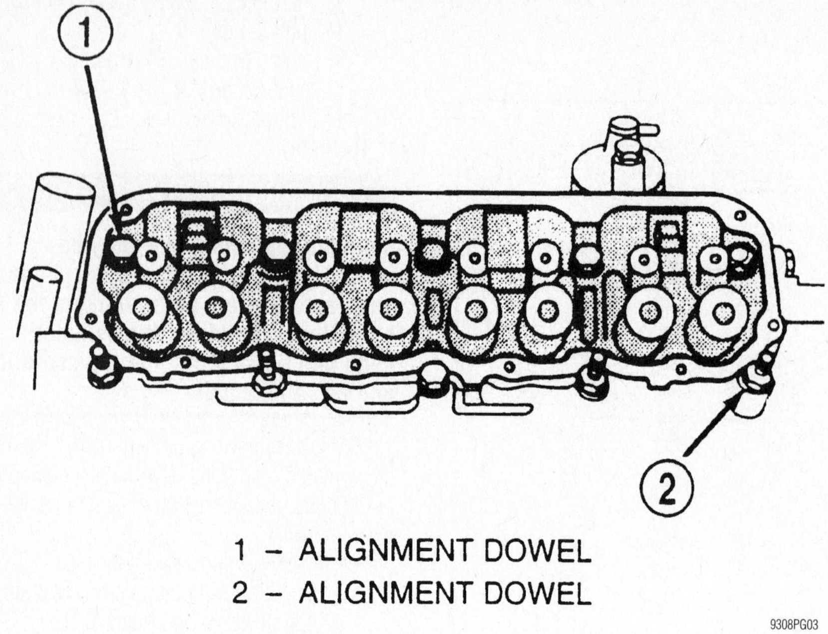

Alignment dowel locations—2.5L engine

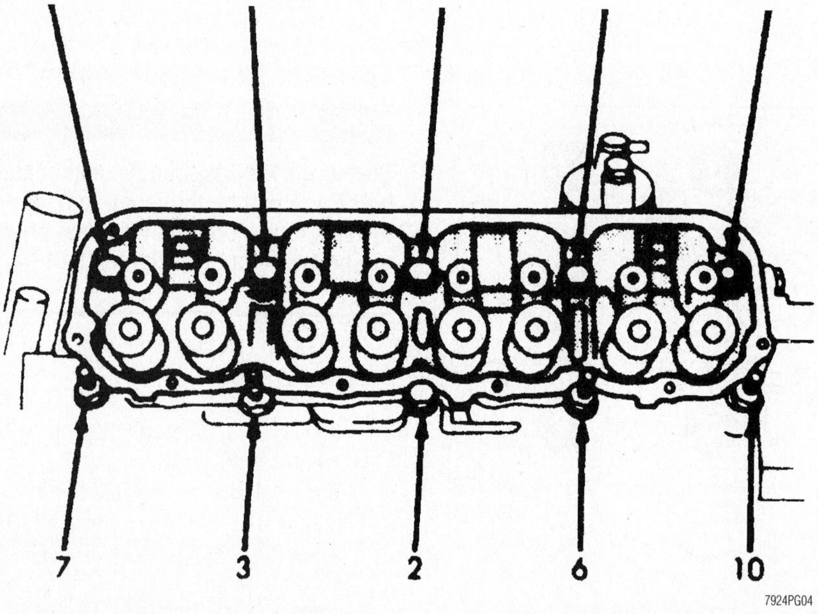

Cylinder head torque sequence—2.5L engine

- Valve cover
- Air cleaner assembly
- CCV hoses
- Negative battery cable

10. Fill and bleed the cooling system.

11. Start the engine, check for leaks and repair if necessary.

4.0L Engines

GRAND CHEROKEE

1. Before servicing the vehicle, refer to the precautions in the beginning of this section.

2. Drain the cooling system.

3. Properly relieve the fuel system pressure.

4. Remove or disconnect the following:

- Negative battery cable
- Crankcase Ventilation (CCV) hoses
- Air cleaner assembly
- Accelerator cable
- Cruise control cable, if equipped
- Transmission cable, if equipped
- Control cable bracket
- Valve cover

➡Keep valvetrain components in order for reassembly.

- Rocker arms
- Pushrods
- Accessory drive belt
- A/C compressor and bracket, if equipped
- Power steering pump and bracket, if equipped
- Fuel line
- Combination manifold
- Thermostat housing coolant hoses
- Spark plugs
- Engine Coolant Temperature (ECT) sensor connector
- Cylinder head

To install:

⁂ WARNING

Cylinder head bolts may only be reused one time. If reusing a cylinder head bolt, place a paint mark on the bolt after installation. If a cylinder head bolt has a paint mark, discard it and use a new bolt.

5. Install the cylinder head with a new gasket. Coat the threads of bolt No. 11 with Loctite®F 592 sealant.

⁂ CAUTION

During the final tightening sequence, bolt No. 11 will be tightened to a lower torque value than the rest of the bolts. Do not overtighten bolt No. 11.

6. Tighten the cylinder head bolts, in sequence, as follows:
   a. Step 1: 22 ft. lbs. (30 Nm)
   b. Step 2: 45 ft. lbs. (61 Nm)
   c. Step 3: 45 ft. lbs. (61 Nm)
   d. Step 4: Bolts 1–10 to 110 ft. lbs. (149 Nm)
   e. Step 5: Bolt 11 to 100 ft. lbs. (136 Nm)
   f. Step 6: Bolts 12–14 to 110 ft. lbs. (149 Nm)
   g. Step 7: Repeat steps 4, 5 and 6
7. Install or connect the following:
   - ECT sensor connector
   - Spark plugs
   - Thermostat housing coolant hoses
   - Combination manifold
   - Fuel line
   - Power steering pump and bracket, if equipped
   - A/C compressor and bracket, if equipped. Torque the bolts to 30 ft. lbs. (40 Nm).
   - Accessory drive belt

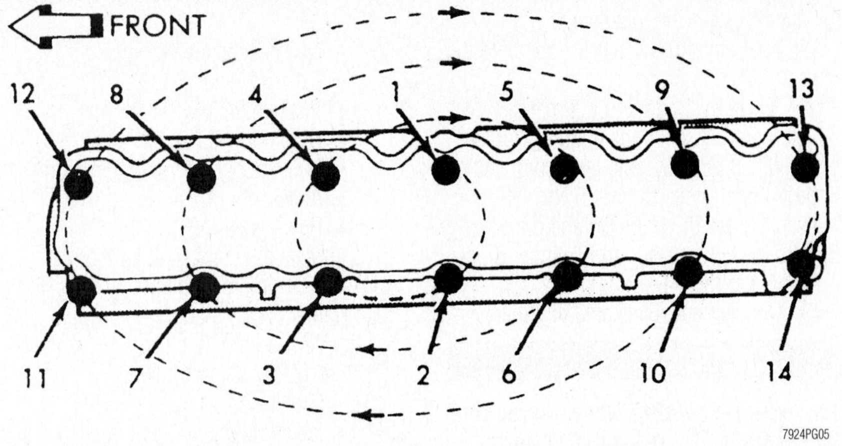

Cylinder head torque sequence—4.0L engine

- Pushrods and rocker arms in their original positions
- Valve cover
- Control cable bracket
- Transmission cable, if equipped
- Cruise control cable, if equipped
- Accelerator cable
- Air cleaner assembly
- CCV hoses
- Negative battery cable

8. Fill and bleed the cooling system.
9. Start the engine, check for leaks and repair if necessary.

WRANGLER

1. Before servicing the vehicle, refer to the precautions in the beginning of this section.
2. Disconnect the battery negative cable.

CAUTION

Do not remove the cylinder block drain plugs or loosen the radiator drain cock with the system hot and pressurized because serious burns from the coolant can occur.

3. Drain the coolant and disconnect the hoses at the engine thermostat housing and the water pump inlet. DO NOT waste reusable coolant. If the solution is clean and is being drained only to service the engine or cooling system, drain the coolant into a clean container for reuse.
4. Remove the air cleaner assembly.
5. Remove the cylinder head cover.
6. Remove the capscrews, bridge and pivot assemblies and rocker arms.
7. Remove the push rods. Retain the pushrods, bridges, pivots and rocker arms in the same order as removed.
8. Remove the accessory drive belt.
9. Remove the A/C compressor mounting bolts and secure the compressor to the side.
10. Remove the power steering pump and bracket from the intake manifold and water pump. Set the pump and bracket aside. DO NOT disconnect the hoses.
11. Perform the Fuel System Pressure Release procedure.
12. Disconnect the fuel supply line at the fuel rail.
13. Remove the intake and exhaust manifolds from the engine cylinder head.
14. Disconnect the coil rail electrical connectors and remove the coil rail.
15. Remove spark plugs.
16. Disconnect the temperature sending unit wire connector.
17. Remove the engine cylinder head bolts.

➡Bolt No.14 cannot be removed until the head is moved forward. Pull bolt No.14 out as far as it will go and then suspend the bolt in this position (tape around the bolt).

18. Remove the engine cylinder head and gasket.
19. If this was the first time the bolts were removed, put a paint dab on the top of the bolt. If the bolts have a paint dab on the top of the bolt or it isn't known if they were used before, discard the bolts.
20. Stuff clean lint free shop towels into the cylinder bores.

To install:

The engine cylinder head gasket may be a composition gasket. This gasket is to be installed DRY. Do not use a gasket sealing compound on the gasket. If the engine cylinder head is to be replaced and the original valves used, measure the valve stem diameter. Only standard size valves can be used with a service replacement engine cylinder head unless the replacement head valve stem guide bores are reamed to accommodate oversize valve stems. Remove all carbon buildup and reface the valves.

21. Remove the shop towels from the cylinder bores. Coat the bores with clean engine oil.
22. Position the engine cylinder head gasket (with the numbers facing up) using the alignment dowels in the cylinder block, to position the gasket.

WARNING

Engine cylinder head bolts should be reused only once. Replace the head bolts if they were used before or if they have a paint dab on the top of the bolt.

23. With bolt No.14 held in place (tape around bolt), install the engine cylinder head over the same dowels used to locate the gasket. Remove the tape from bolt No.14.
24. Coat the threads of stud bolt No.11 with Loctite® 592 sealant, or equivalent.
25. Tighten the engine cylinder head bolts in sequence according to the following procedure.

WARNING

During the final tightening sequence, bolt No.11 will be tightened to a lower torque than the rest of the bolts. DO NOT overtighten bolt No.11.

- Bolts 1 through 14 to 30 Nm (22 ft. lbs.).
- Bolts 1 through 14 to 61 Nm (45 ft. lbs.).
- Check all bolts to verify they are set to 61 Nm (45 ft. lbs.).
- Bolts 1 through 10 to 149 Nm (110 ft. lbs.).
- Bolt 11 to 135 Nm (100 ft. lbs.).
- Bolts 12 through 14 to 149 Nm (110 ft. lbs.).
- Check all bolts in sequence to verify the correct torque.
- If not already done, clean and mark each bolt with a dab of paint after tightening. Should you encounter bolts which were painted in an earlier service operation, replace them.

26. Install the spark plugs.
27. Connect the temperature sending unit wire connector.
28. Install the ignition coil rail and coil rail electrical connectors.

29. Install the intake and exhaust manifolds.

30. Install the fuel line and the vacuum advance hose.

31. Attach the power steering pump and bracket.

32. Install the push rods, rocker arms, pivots and bridges in the order they were removed.

33. Install the engine cylinder head cover.

34. Attach the air conditioner compressor mounting bracket to the engine cylinder head and block. Tighten the bolts to 40 Nm (30 ft. lbs.) torque.

35. Attach the air conditioning compressor to the bracket. Tighten the bolts to 27 Nm (20 ft. lbs.) torque.

⁂ WARNING

The serpentine drive belt must be routed correctly. Incorrect routing can cause the water pump to turn in the opposite direction causing the engine to overheat.

36. Install the serpentine drive belt.

37. Install the air cleaner and ducting.

38. Connect the hoses to the engine thermostat housing and fill the cooling system to the specified level.

39. The automatic transmission throttle linkage and cable must be adjusted after completing the engine cylinder head installation.

40. Install the temperature sending unit and connect the wire connector.

41. Position the A/C compressor into the engine compartment.

42. Install the bolts and nut that secure the A/C compressor to the engine. Tighten the bolts and nut fastening the compressor to the cylinder block to 45–65 Nm (35–50 ft. lbs.). Tighten the bolts fastening the rear brace to the compressor and cylinder block to 40–55 Nm (30–40 ft. lbs.).

43. Connect negative cable to battery.

⁂ CAUTION

Use extreme caution when the engine is operating. Do not stand in direct line with the fan. Do not put hands near the pulleys, belts or fan. Do not wear loose clothing.

44. Operate the engine with the radiator cap off. Inspect for leaks and continue operating the engine until the engine thermostat opens. Add coolant, if required.

3.7L Engine

LIBERTY—LEFT SIDE

1. Before servicing the vehicle, refer to the precautions in the beginning of this section.

2. Drain the cooling system.

3. Properly relieve the fuel system pressure.

4. Remove or disconnect the following:

- Negative battery cable
- Exhaust Y-pipe
- Intake manifold
- Cylinder head cover
- Engine cooling fan and shroud
- Accessory drive belt
- Power steering pump

5. Rotate the crankshaft so that the crankshaft timing mark aligns with the Top Dead Center (TDC) mark on the front cover, and the **V6** marks on the camshaft sprockets are at 12 o'clock as shown.

- Crankshaft damper
- Front cover

6. Lock the secondary timing chain to the idler sprocket with Timing Chain Locking tool 8429.

7. Matchmark the secondary timing chain one link on each side of the V6 mark to the camshaft sprocket.

- Left secondary timing chain tensioner
- Cylinder head access plug
- Secondary timing chain guide
- Camshaft sprocket
- Cylinder head

➡The cylinder head is retained by twelve bolts. Four of the bolts are smaller and are at the front of the head.

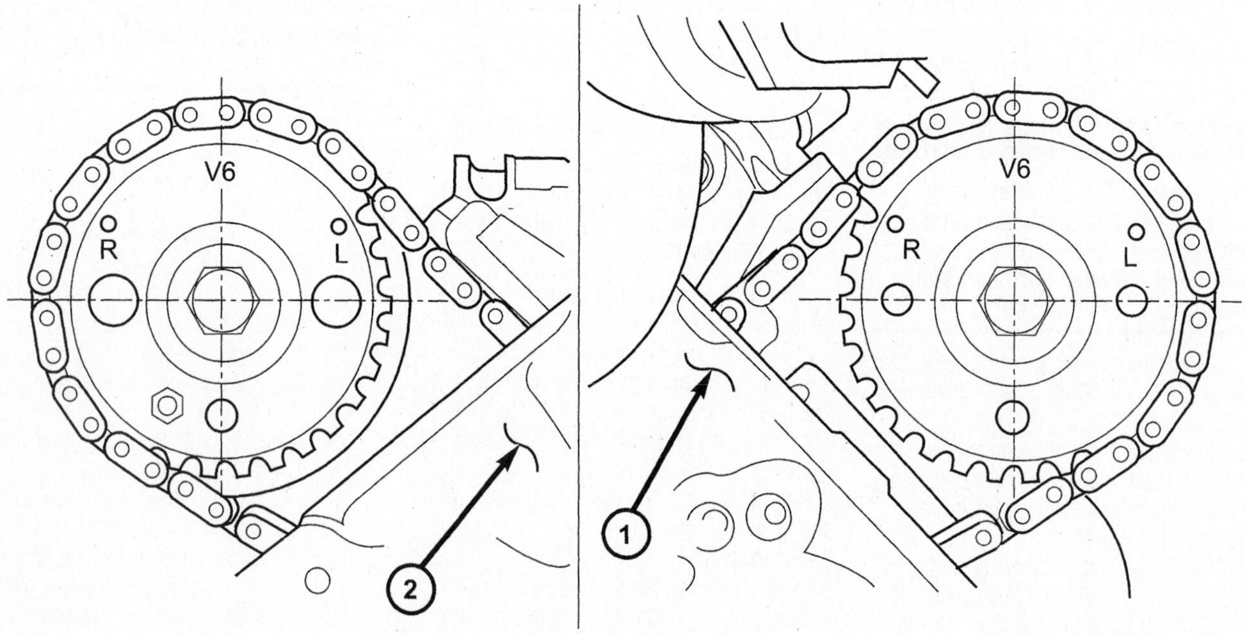

1 - LEFT CYLINDER HEAD
2 - RIGHT CYLINDER HEAD

9355PG04

Camshaft sprocket timing marks—3.7L

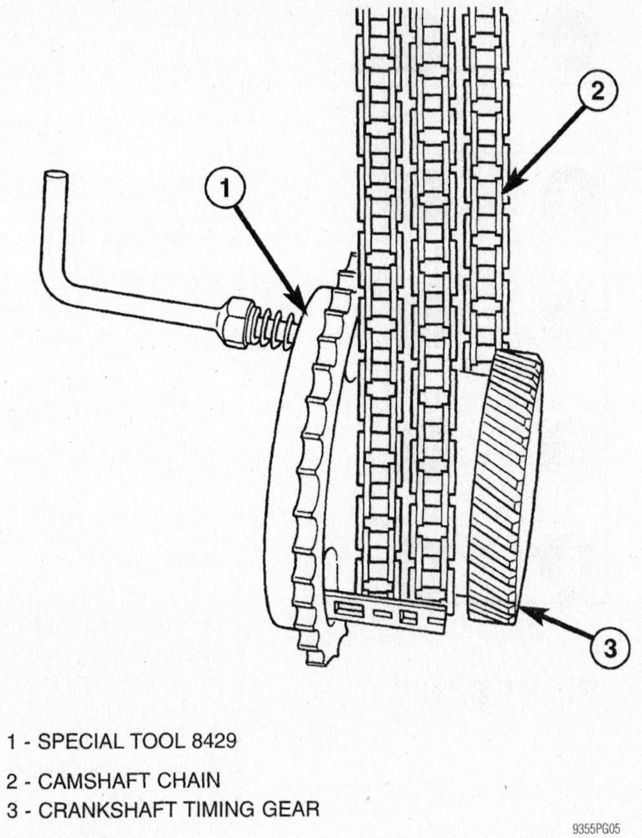

Camshaft locking tool—3.7L

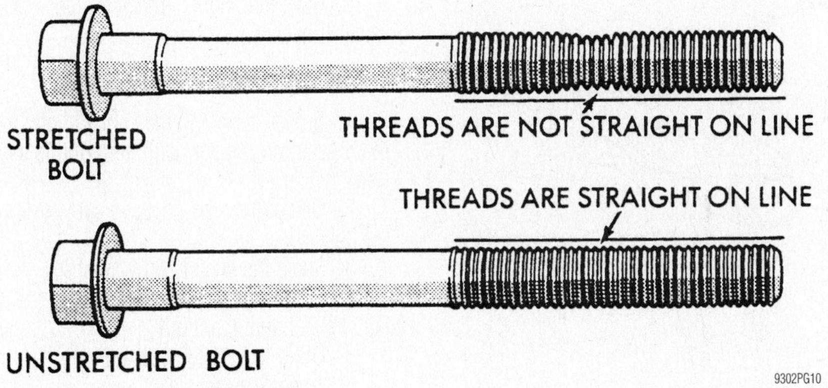

Examine the head bolts for signs of stretching—3.7L engine

To install:

8. Check the cylinder head bolts for signs of stretching and replace as necessary.
9. Lubricate the threads of the 11mm bolts with clean engine oil.
10. Coat the threads of the 8mm bolts with Mopar® Lock and Seal Adhesive.
11. Install the cylinder heads. Use new gaskets and tighten the bolts, in sequence, as follows:
    a. Step 1: Bolts 1–8 to 20 ft. lbs. (27 Nm)
    b. Step 2: Bolts 1–10 verify torque without loosening
    c. Step 3: Bolts 9–12 to 10 ft. lbs. (14 Nm)
    d. Step 4: Bolts 1–8 plus ¼ (90 degree) turn
    e. Step 5: Bolts 9–12 to 19 ft. lbs. (26 Nm)
12. Install or connect the following:
    - Camshaft sprocket. Align the secondary chain matchmarks and tighten the bolt to 90 ft. lbs. (122 Nm).
    - Secondary timing chain guide
    - Cylinder head access plug
    - Secondary timing chain tensioner. Refer to the timing chain procedure in this section.
13. Remove the Timing Chain Locking tool.
14. Install or connect the following:
    - Front cover
    - Crankshaft damper. Torque the bolt to 130 ft. lbs. (175 Nm).
    - Power steering pump
    - Accessory drive belt
    - Engine cooling fan and shroud
    - Cover
    - Intake manifold
    - Exhaust Y-pipe
    - Negative battery cable
15. Fill and bleed the cooling system.
16. Start the engine, check for leaks and repair if necessary.

LIBERTY—RIGHT SIDE

1. Before servicing the vehicle, refer to the precautions in the beginning of this section.
2. Drain the cooling system.
3. Properly relieve the fuel system pressure.
4. Remove or disconnect the following:
    - Negative battery cable
    - Exhaust Y-pipe
    - Intake manifold
    - Valve cover
    - Engine cooling fan and shroud
    - Accessory drive belt
    - Oil fill housing
    - Power steering pump
5. Rotate the crankshaft so that the crankshaft timing mark aligns with the Top Dead Center (TDC) mark on the front cover, and the **V6** marks on the camshaft sprockets are at 12 o'clock as shown.
6. Remove or disconnect the following:
    - Crankshaft damper
    - Front cover
7. Lock the secondary timing chains to the idler sprocket with Timing Chain Locking tool 8429.
8. Matchmark the secondary timing chains to the camshaft sprockets.
9. Remove or disconnect the following:
    - Secondary timing chain tensioners
    - Cylinder head access plugs
    - Secondary timing chain guides
    - Camshaft sprockets
    - Cylinder heads

➡Each cylinder head is retained by 8 11mm bolts and four 8mm bolts.

To install:

10. Check the cylinder head bolts for signs of stretching and replace as necessary.

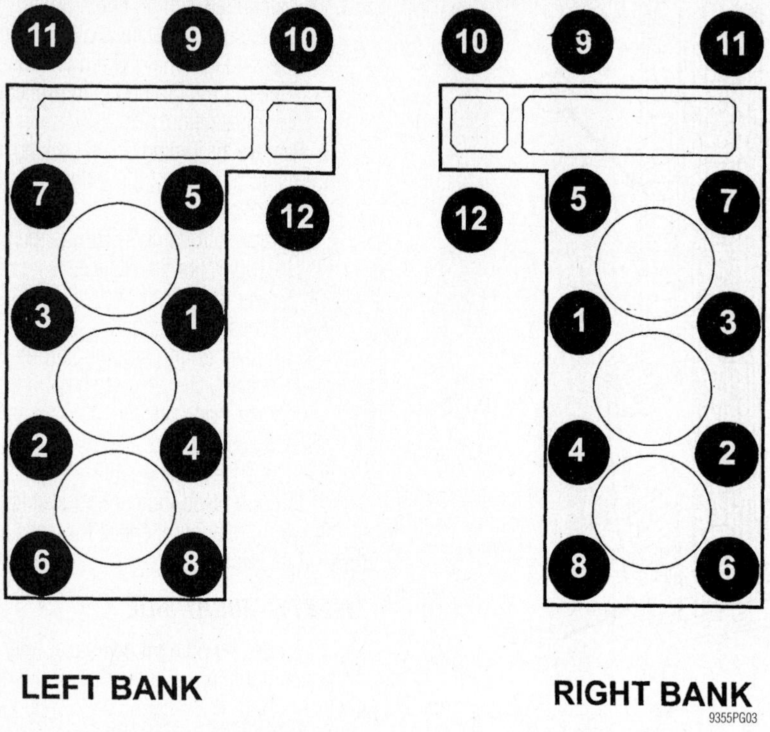

Cylinder head bolt torque sequence—3.7L

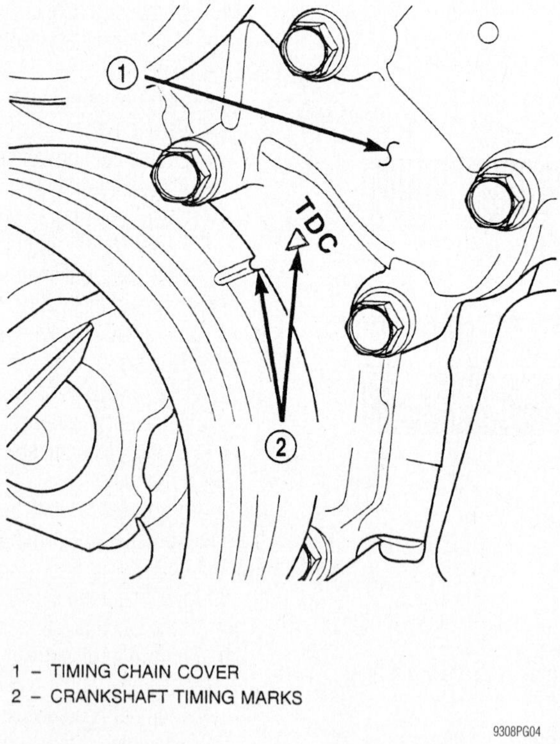

Crankshaft timing marks—3.7L engine

11. Lubricate the threads of the 11mm bolts with clean engine oil.
12. Coat the threads of the 8mm bolts with Mopar® Lock and Seal Adhesive.
13. Install the cylinder heads. Use new gaskets and tighten the bolts, in sequence, as follows:
    a. Step 1: Bolts 1–8 to 20 ft. lbs. (27 Nm)
    b. Step 2: Bolts 1–10 verify torque without loosening
    c. Step 3: Bolts 9–12 to 10 ft. lbs. (14 Nm)
    d. Step 4: Bolts 1–8 plus 1/4 (90 degree) turn
    e. Step 5: Bolts 9–12 to 19 ft. lbs. (26 Nm)
14. Install or connect the following:
    - Camshaft sprockets. Align the secondary chain matchmarks and tighten the bolts to 90 ft. lbs. (122 Nm).
    - Secondary timing chain guides
    - Cylinder head access plugs
    - Secondary timing chain tensioners. Refer to the timing chain procedure in this section.
15. Remove the Timing Chain Locking tool.
16. Install or connect the following:
    - Front cover
    - Crankshaft damper. Torque the bolt to 130 ft. lbs. (175 Nm).
    - Rocker arms
    - Power steering pump
    - Oil fill housing
    - Accessory drive belt
    - Engine cooling fan and shroud
    - Valve covers
    - Intake manifold
    - Exhaust Y-pipe
    - Negative battery cable
17. Fill and bleed the cooling system.
18. Start the engine, check for leaks and repair if necessary.

GRAND CHEROKEE—LEFT SIDE

1. Before servicing the vehicle, refer to the precautions in the beginning of this section.
2. Disconnect the negative cable from the battery.
3. Raise the vehicle on a hoist.
4. Disconnect the exhaust pipe at the left side exhaust manifold.
5. Drain the engine coolant.
6. Lower the vehicle.
7. Remove the intake manifold.
8. Siphon the master cylinder.
9. Remove the primary brake line at the master cylinder.
10. Remove the primary brake line from the HCU.
11. Remove the secondary brake line at the master cylinder.
12. Remove the secondary brake line at the HCU.
13. Remove the 4 chassis lines at the HCU.
14. Disconnect the HCU electrical connector.
15. Remove the 3 mounting nuts at the HCU bracket.

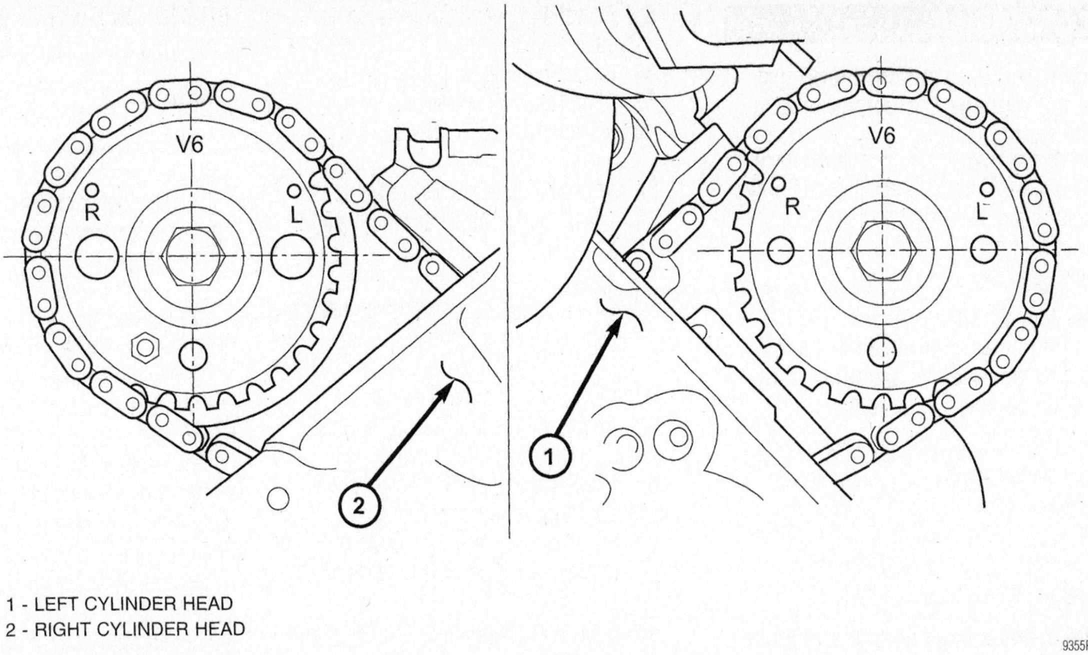

Camshaft positioning—3.7L engine

16. Remove the HCU with the bracket from the vehicle.
17. Disconnect the brake fluid level sensor electrical connector from the fluid reservoir.
18. Remove the 2 master cylinder mounting nuts.
19. Remove the master cylinder.
20. Disconnect vacuum hose at booster check valve.
21. Remove the hush panel.
22. Remove the steering column opening cover and hush panel.
23. Remove retainer clip that holds booster push rod on pedal pin. Then slide push rod off pin.
24. Disconnect the brake lamp switch wire connector.
25. Remove the brake lamp switch.
26. Remove four nuts that attach booster to dash panel.
27. In the engine compartment, disconnect the wire harness routing clips and move the harness to the side and downward.
28. Remove the booster from the vehicle.
29. Remove the cylinder head cover
30. Remove the fan shroud and fan blade assembly
31. Remove accessory drive belt.
32. Remove the power steering pump and set aside.
33. Rotate the crankshaft until the damper timing mark is aligned with TDC indicator mark.
34. Verify the V6 mark on the camshaft sprocket is at the 12 o'clock position, with the No. 1 cylinder at TDC on the exhaust stroke. Rotate the crankshaft one turn if necessary.
35. Remove the crankshaft damper.
36. Remove the timing chain cover.
37. Lock the secondary timing chains to the idler sprocket using Special Tool 8429 Timing Chain Holding Fixture.

➡Mark the secondary timing chain prior to removal to aid in installation.

38. Mark the secondary timing chain, one link on each side of the V6 mark on the camshaft drive gear.
39. Remove the left side secondary chain tensioner.
40. Remove the cylinder head access plug.
41. Remove the left side secondary chain guide.
42. Remove the retaining bolt and the camshaft drive gear.

⁂ WARNING

Do not allow the engine to rotate. Severe damage to the valve train can occur.

➡Do not overlook the four smaller bolts at the front of the cylinder head. Do not attempt to remove the cylinder head without removing these four bolts.

➡The cylinder head is attached to the cylinder block with twelve bolts.

43. Remove the cylinder head retaining bolts.
44. Remove the cylinder head and gasket. Discard the gasket.

⁂ WARNING

Do not lay the cylinder head on its gasket sealing surface, due to the design of the cylinder head gasket any distortion to the cylinder head sealing surface may prevent the gasket from properly sealing resulting in leaks.

To install:

➡The cylinder head bolts are tightened using a torque plus angle procedure. The bolts must be examined before reuse. If the threads are necked down the bolts should be replaced. Necking can be checked by holding a straight edge against the threads. If all the threads do not contact the scale, the bolt should be replaced.

⁂ WARNING

When cleaning cylinder head and cylinder block surfaces, DO NOT use a metal scraper because the surfaces could be cut or ground. Use only a wooden or plastic scraper.

45. Clean the cylinder head and cylinder block mating surfaces.
46. Position the new cylinder head gasket on the locating dowels.

⁂ WARNING

When installing cylinder head, use care not damage the tensioner arm or the guide arm.

47. Position the cylinder head onto the cylinder block. Make sure the cylinder head seats fully over the locating dowels.

➡The four smaller cylinder head mounting bolts require sealant to be added to them before installing. Failure to do so may cause leaks.

48. Lubricate the cylinder head bolt threads with clean engine oil and install the eight M11 bolts.
49. Coat the four M8 cylinder head bolts with Mopar® Lock and Seal Adhesive, or equivalent, then install the bolts.

➡The cylinder head bolts are tightened using an angle torque procedure, however, the bolts are not a torque-to-yield design.

50. Tighten the bolts in sequence using the following steps and torque values:
    - Step 1: Tighten bolts 1-8 to 27 Nm (20 ft. lbs.).
    - Step 2: Verify that bolts 1-8, all reached 27 Nm (20 ft. lbs.), by repeating step 1 without loosening the bolts.
    - Step 3: Tighten bolts 9-12 to 14 Nm (10 ft. lbs.).
    - Step 4: Tighten bolts 1-8, 90 degrees.
    - Step 5: Tighten bolts 1-8, 90 degrees, again.
    - Step 6: Tighten bolts 9-12 to 26 Nm (19 ft. lbs.)
51. Position the secondary chain onto the camshaft drive gear, making sure one marked chain link is on either side of the V6 mark on the gear then using Special Tool 8428 Camshaft Wrench, position the gear onto the camshaft.

⁂ WARNING

Remove excess oil from camshaft sprocket retaining bolt before reinstalling bolt. Failure to do so may cause over-torquing of bolt resulting in bolt failure.

52. Install the camshaft drive gear retaining bolt.
53. Install the left side secondary chain guide.
54. Install the cylinder head access plug.
55. Re-set and install the left side secondary chain tensioner.
56. Install the timing chain cover.
57. Install the crankshaft damper. Tighten damper bolt 175 Nm (130 ft. lbs.).
58. Install the power steering pump.
59. Install the fan blade assembly and fan shroud.
60. Install the cylinder head cover.
61. Check condition of grommet that secures check valve in booster. Replace grommet if cut, torn, or loose.
62. Install new booster dash seal.
63. Align and position booster in the mounting holes of the dash panel.
64. Inside the vehicle, Lubricate pedal pin Mopar multi-mileage grease.
65. Install booster attaching nuts on studs. Tighten attaching nuts to 39 Nm (29 ft. lbs.).
66. Slide booster push rod on pedal pin. Then secure rod to pin with retainer clip.
67. Install a new brake lamp switch and reconnect the electrical connector.
68. Install the steering column opening cover.
69. Install the hush panel.
70. In engine compartment, attach vacuum hose to booster check valve.
71. Install the master cylinder to the power brake booster with new gasket and nuts.
72. Reroute the wire harness back into position and install the routing clip.
73. Reconnect the brake fluid level electrical connector.
74. Install the HCU with bracket to the vehicle and tighten the 2 mounting nuts.
75. Reconnect the HCU electrical connector.
76. Install the 4 chassis lines at the HCU.
77. Install the secondary brake line at the HCU.
78. Install the secondary brake line at the master cylinder.
79. Install the primary brake line at the HCU.
80. Install the primary brake line at the master cylinder.
81. Fill and bleed brake system.
82. Install the intake manifold.
83. Refill the cooling system.
84. Raise the vehicle.
85. Install the exhaust pipe onto the left exhaust manifold.
86. Lower the vehicle.
87. Connect the negative cable to the battery.
88. Start the engine and check for leaks.

GRAND CHEROKEE—RIGHT SIDE

1. Before servicing the vehicle, refer to the precautions in the beginning of this section.
2. Disconnect battery negative cable.
3. Raise the vehicle on a hoist.
4. Disconnect the exhaust pipe at the right side exhaust manifold.
5. Drain the engine coolant.
6. Lower the vehicle.
7. Remove the intake manifold.
8. Remove the cylinder head cover.
9. Remove the fan shroud.
10. Remove oil fill housing from cylinder head.
11. Remove accessory drive belt.
12. Rotate the crankshaft until the damper timing mark is aligned with TDC indicator mark.
13. Verify the V6 mark on the camshaft sprocket is at the 12 o'clock position. Rotate the crankshaft one turn if necessary.
14. Remove the crankshaft damper.
15. Remove the timing chain cover.
16. Lock the secondary timing chains to the idler sprocket using Special Tool 8429 Timing Chain Holding Fixture.

➡Mark the secondary timing chain prior to removal to aid in installation.

17. Mark the secondary timing chain, one link on each side of the V6 mark on the camshaft drive gear.
18. Remove the right side secondary chain tensioner.
19. Remove the cylinder head access plug.
20. Remove the right side secondary chain guide.

⁂ WARNING

The nut on the right side camshaft sprocket should not be removed for any reason, as the sprocket and camshaft sensor target wheel is serviced as an assembly. If the nut was removed, torque nut to 5 Nm (44 inch lbs.).

21. Remove the retaining bolt and the camshaft drive gear.

⁂ WARNING

Do not allow the engine to rotate. Severe damage to the valve train can occur.

➡Do not overlook the four smaller bolts at the front of the cylinder head. Do not attempt to remove the cylinder head without removing these four bolts.

WARNING

Do not hold or pry on the camshaft target wheel for any reason. A damaged target wheel can result in a vehicle no start condition.

➡The cylinder head is attached to the cylinder block with twelve bolts.

22. Remove the cylinder head retaining bolts.
23. Remove the cylinder head and gasket. Discard the gasket.

WARNING

Do not lay the cylinder head on its gasket sealing surface, do to the design of the cylinder head gasket any distortion to the cylinder head sealing surface may prevent the gasket from properly sealing resulting in leaks.

To install:

➡The cylinder head bolts are tightened using a torque plus angle procedure. The bolts must be examined before reuse. If the threads are necked down the bolts should be replaced. Necking can be checked by holding a straight edge against the threads. If all the threads do not contact the scale, the bolt should be replaced.

WARNING

When cleaning cylinder head and cylinder block surfaces, DO NOT use a metal scraper because the surfaces could be cut or ground. Use only a wooden or plastic scraper.

24. Clean the cylinder head and cylinder block mating surfaces.
25. Position the new cylinder head gasket on the locating dowels.

WARNING

When installing cylinder head, use care not damage the tensioner arm or the guide arm.

26. Position the cylinder head onto the cylinder block. Make sure the cylinder head seats fully over the locating dowels.

➡The four M8 cylinder head mounting bolts require sealant to be added to them before installing. Failure to do so may cause leaks.

27. Lubricate the cylinder head bolt threads with clean engine oil and install the eight M10 bolts.
28. Coat the four M8 cylinder head bolts with Mopar Lock and Seal Adhesive, or equivalent, then install the bolts.

➡The cylinder head bolts are tightened using an angle torque procedure, however, the bolts are not a torque-to-yield design.

29. Tighten the bolts in sequence using the following steps and torque values:
   - Step 1: Tighten bolts 1-8, to 27 Nm (20 ft. lbs.).
   - Step 2: Verify that bolts 1-8, all reached 27 Nm (20 ft. lbs.), by repeating step 1 without loosening the bolts.
   - Step 3: Tighten bolts 9 through 12 to 14 Nm (10 ft. lbs.).
   - Step 4: Tighten bolts 1-8, 90°.
   - Step 5: Tighten bolts 1-8, 90°, again.
   - Step 6: Tighten bolts 9-12, to 26 Nm (19 ft. lbs.)

WARNING

The nut on the right side camshaft sprocket should not be removed for any reason, as the sprocket and camshaft sensor target wheel is serviced as an assembly. If the nut was removed, torque nut to 5 Nm (60 in. lbs.).

30. Position the secondary chain onto the camshaft drive gear, making sure one marked chain link is on either side of the V6 mark on the gear then using Special Tool 8428 Camshaft Wrench, position the gear onto the camshaft.

WARNING

Remove excess oil from camshaft sprocket retaining bolt before reinstalling bolt. Failure to do so may cause over-torquing of bolt resulting in bolt failure.

31. Install the camshaft drive gear retaining bolt.

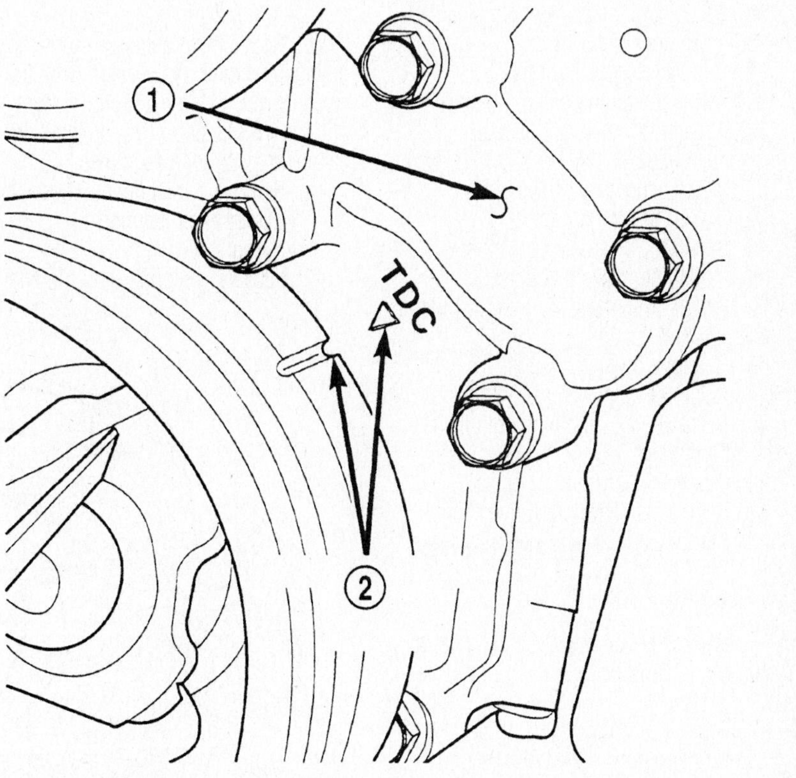

1 – TIMING CHAIN COVER
2 – CRANKSHAFT TIMING MARKS

Crankshaft timing marks—4.7L engine

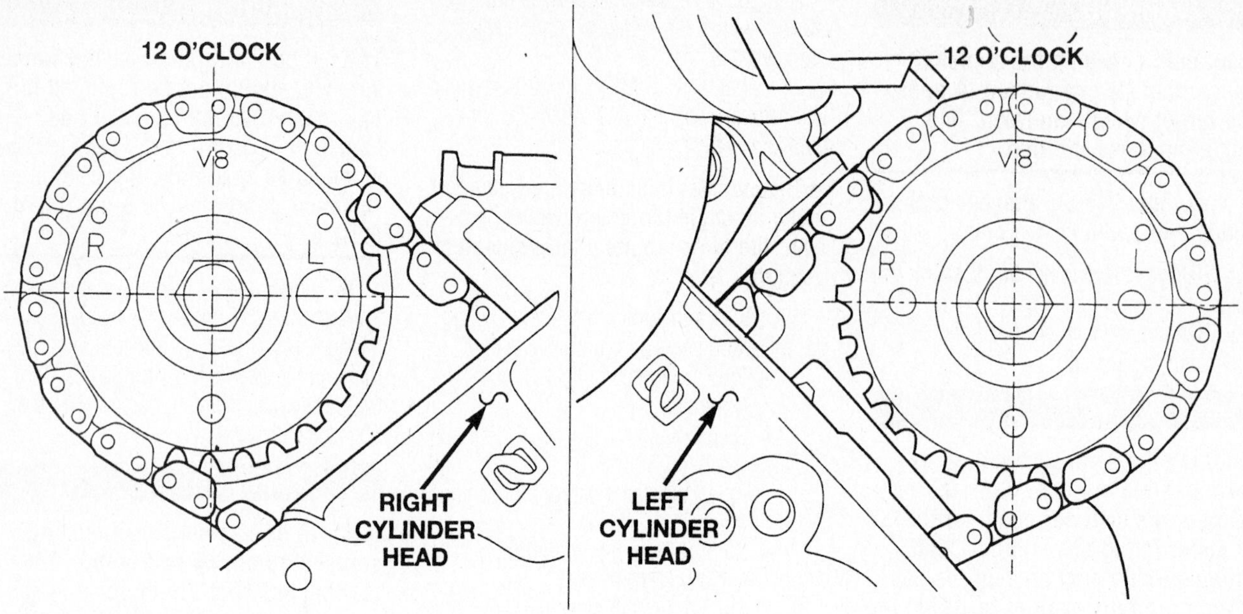

Camshaft positioning—4.7L engine

32. Install the right side secondary chain guide.
33. Install the cylinder head access plug.
34. Re-set and install the right side secondary chain tensioner.
35. Remove Special Tool 8429.
36. Install the timing chain cover.
37. Install the crankshaft damper. Tighten damper bolt 175 Nm (130 Ft. Lbs.).
38. Install accessory drive belt.
39. Install the fan shroud.
40. Install the cylinder head cover.
41. Install the intake manifold.
42. Install oil fill housing onto cylinder head.
43. Refill the cooling system.
44. Raise the vehicle.
45. Install the exhaust pipe onto the right exhaust manifold.
46. Lower the vehicle.
47. Reconnect battery negative cable.
48. Start the engine and check for leaks.

4.7L Engine

1. Before servicing the vehicle, refer to the precautions in the beginning of this section.
2. Drain the cooling system.
3. Properly relieve the fuel system pressure.
4. Remove or disconnect the following:
   - Negative battery cable
   - Exhaust Y-pipe
   - Intake manifold
   - Valve covers
   - Engine cooling fan and shroud
   - Accessory drive belt
   - Oil fill housing
   - Power steering pump
   - Rocker arms
5. Rotate the crankshaft so that the crankshaft timing mark aligns with the Top Dead Center (TDC) mark on the front cover, and the **V8** marks on the camshaft sprockets are at 12 o'clock as shown.
6. Remove or disconnect the following:
   - Crankshaft damper
   - Front cover
7. Lock the secondary timing chains to the idler sprocket with Timing Chain Locking tool 8515.
8. Matchmark the secondary timing chains to the camshaft sprockets.
9. Remove or disconnect the following:
   - Secondary timing chain tensioners
   - Cylinder head access plugs
   - Secondary timing chain guides
   - Camshaft sprockets
   - Cylinder heads

➡Each cylinder head is retained by ten 11mm bolts and four 8mm bolts.

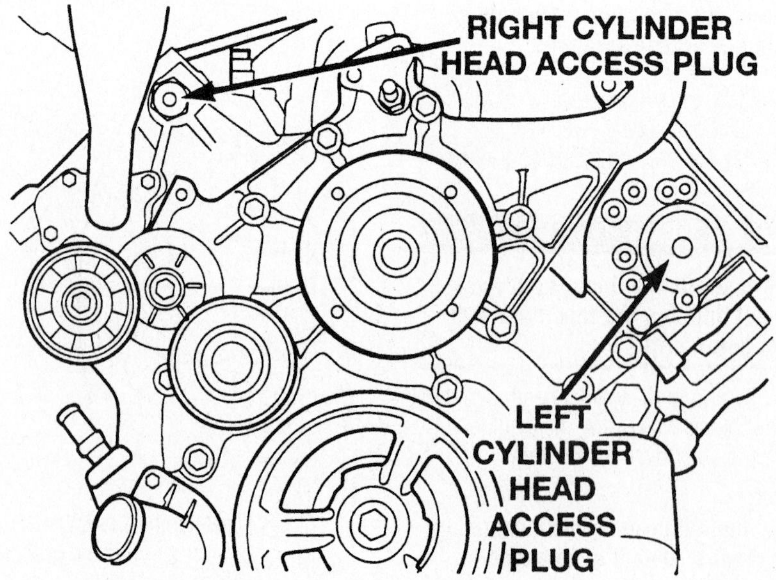

Cylinder head access plug locations—4.7L engine

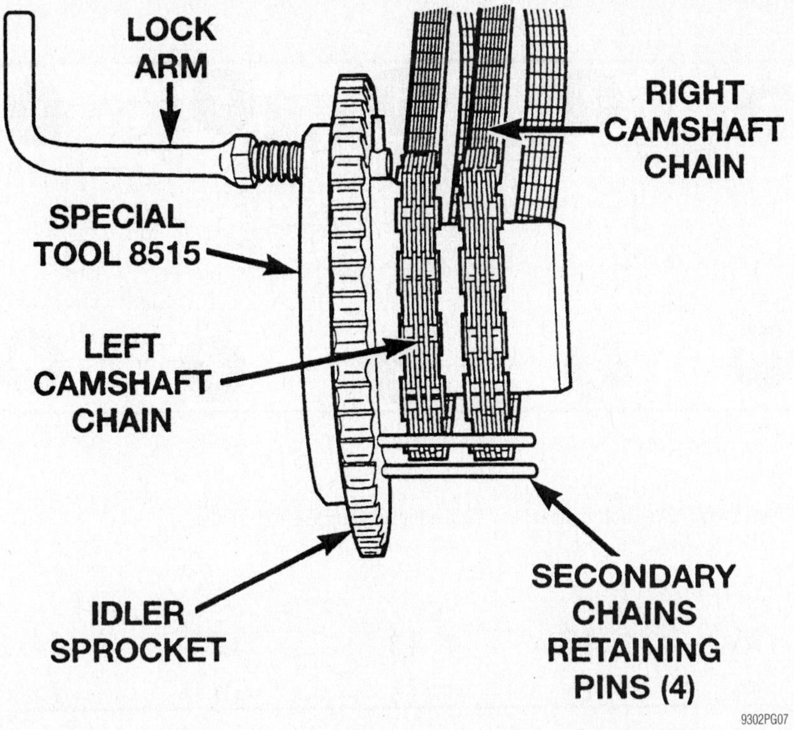

Use the special tool to lock the timing chains on the idler gear—4.7L engine

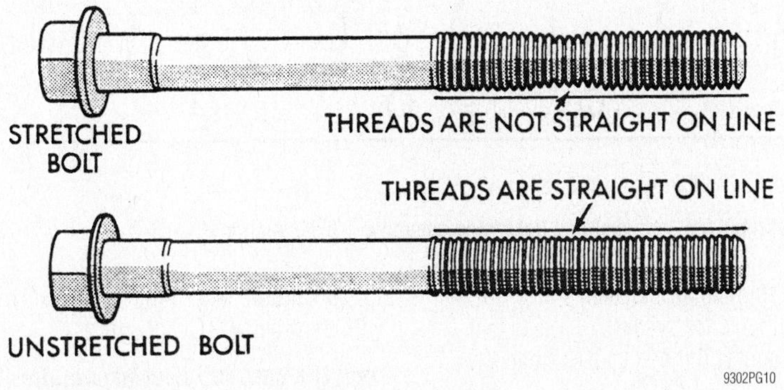

Examine the head bolts for signs of stretching—4.7L engine

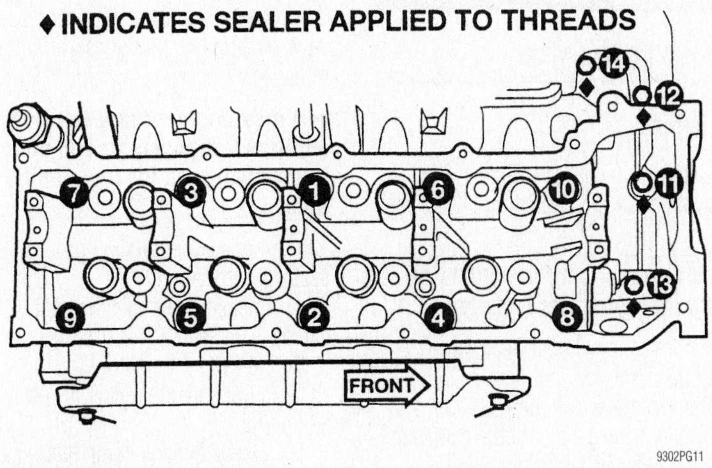

Cylinder head torque sequence—4.7L engine

To install:

10. Check the cylinder head bolts for signs of stretching and replace as necessary.

11. Lubricate the threads of the 11mm bolts with clean engine oil.

12. Coat the threads of the 8mm bolts with Mopar® Lock and Seal Adhesive.

13. Install the cylinder heads. Use new gaskets and tighten the bolts, in sequence, as follows:

a. Step 1: Bolts 1–10 to 15 ft. lbs. (20 Nm)

b. Step 2: Bolts 1–10 to 35 ft. lbs. (47 Nm)

c. Step 3: Bolts 11–14 to 18 ft. lbs. (25 Nm)

d. Step 4: Bolts 1–10 plus ¼ (90 degree) turn

e. Step 5: Bolts 11–14 to 19 ft. lbs. (26 Nm)

14. Install or connect the following:

- Camshaft sprockets. Align the secondary chain matchmarks and tighten the bolts to 90 ft. lbs. (122 Nm).
- Secondary timing chain guides
- Cylinder head access plugs
- Secondary timing chain tensioners. Refer to the timing chain procedure in this section.

15. Remove the Timing Chain Locking tool 8515.

16. Install or connect the following:

- Front cover
- Crankshaft damper. Torque the bolt to 130 ft. lbs. (175 Nm).
- Rocker arms
- Power steering pump
- Oil fill housing
- Accessory drive belt
- Engine cooling fan and shroud
- Valve covers
- Intake manifold
- Exhaust Y-pipe
- Negative battery cable

17. Fill and bleed the cooling system.

18. Start the engine, check for leaks and repair if necessary.

5.7L Engine

1. Before servicing the vehicle, refer to the precautions in the beginning of this section.

2. Perform the Fuel System Pressure Release procedure. Disconnect the fuel supply line.

3. Disconnect the battery negative cable.

4. Drain cooling system.

5. Remove the air cleaner resonator and duct work.

6. Remove closed crankcase ventilation system.
7. Disconnect the exhaust at the exhaust manifolds.
8. Disconnect the evaporation control system.
9. Disconnect heater hoses.
10. Remove the power steering pump.
11. Remove cylinder head covers, using the sequence provided, and gaskets.
12. Remove intake manifold and throttle body as an assembly.
13. Remove rocker arm assemblies and push rods. Identify to ensure installation in original locations.
14. Remove the head bolts from each cylinder head, using the sequence provided, and remove cylinder heads. Discard the cylinder head gasket.

To install:

15. Clean all surfaces of cylinder block and cylinder heads.
16. Clean cylinder block front and rear gasket surfaces using a suitable solvent.

➡The head gaskets are not interchangeable between left and right sides. They are marked "L" and "R" to indicate left and right sides. The head gaskets are also marked "TOP" to indicate which side goes up.

17. Position new cylinder head gaskets onto the cylinder block.
18. Position cylinder heads onto head gaskets and cylinder block.
19. Tighten the cylinder head bolts in three steps using the sequence provided:
   - Step 1: Tighten M12 cylinder head bolts, in sequence, to 34 Nm (25 ft. lbs.) and M8 bolts to 20 Nm (15 ft. lbs.) torque.
   - Step 2: Tighten M12 cylinder head bolts, in sequence, to 54 Nm (40 ft. lbs.) and verify M8 bolts to 20 Nm (15 ft. lbs.) torque.
   - Step 3: Turn M12 cylinder head bolts, in sequence, 90 degrees and tighten M8 bolts to 34 Nm (25 ft. lbs.) torque.

20. Install pushrods and rocker arm assemblies in their original position.
21. Install the intake manifold and throttle body assembly.
22. If required, adjust spark plugs to specifications. Install the plugs.
23. Connect the heater hoses.
24. Install the fuel supply line.
25. Install the power steering pump.
26. Install the drive belt.
27. Install cylinder head covers.
28. Connect the evaporation control system.

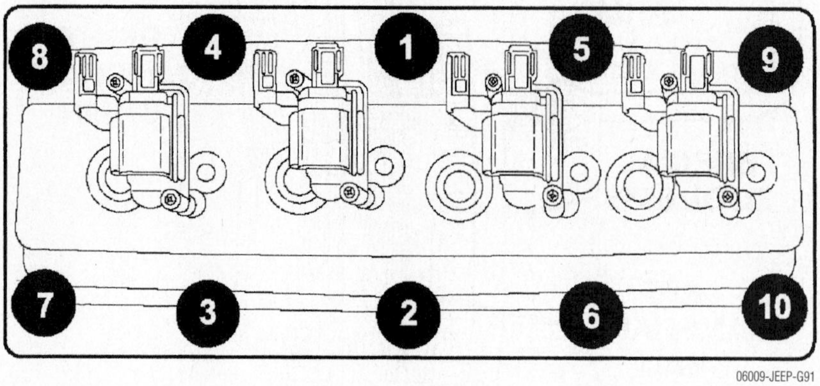

Cylinder head cover bolt removal and tightening sequence—5.7L engine

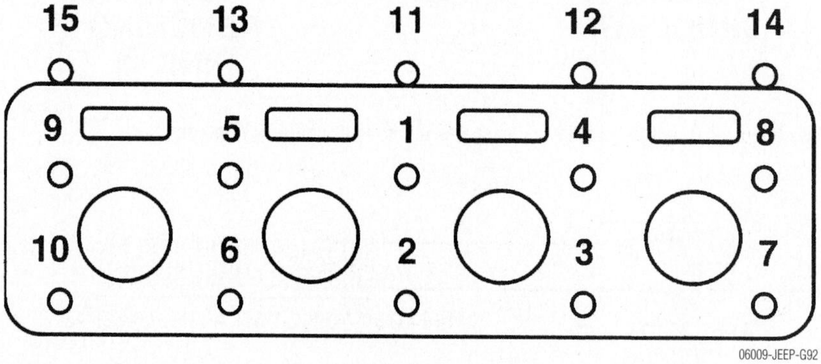

Cylinder head bolt removal and tightening sequence—5.7L engine

29. Install the air cleaner.
30. Fill cooling system.
31. Connect the negative cable to the battery.
32. Start engine check for leaks.

Rocker Arms

REMOVAL & INSTALLATION

2.4L Engines

1. Before servicing the vehicle, refer to the precautions in the beginning of this section.

➡This procedure is for in-vehicle service with the camshafts installed.

2. Remove the cylinder head cover.
3. Remove the fuel rail.
4. Remove the spark plugs.
5. Rotate the engine until the camshaft lobe, on the follower being removed, is position on its base circle (heel). Also, the piston should be a minimum of 6.3 mm (0.25 in) below TDC position.

➡If the cam follower assemblies are to be reused, always mark the position for reassembly in their original positions.

6. Using Special Tools 8215 and 8436, or their equivalents, slowly depress the valve assembly until the rocker arm can be removed.

➡It may be necessary to remove additional brackets or components to allow clearance for tool handle movement.

7. Repeat the removal procedure for each rocker arm.
8. Inspect the rocker arm for wear or damage. Replace as necessary.

To install:

9. Lubricate the rocker arm with clean engine oil.
10. Using Special Tools 8215 and 8436, slowly depress the valve assembly until the

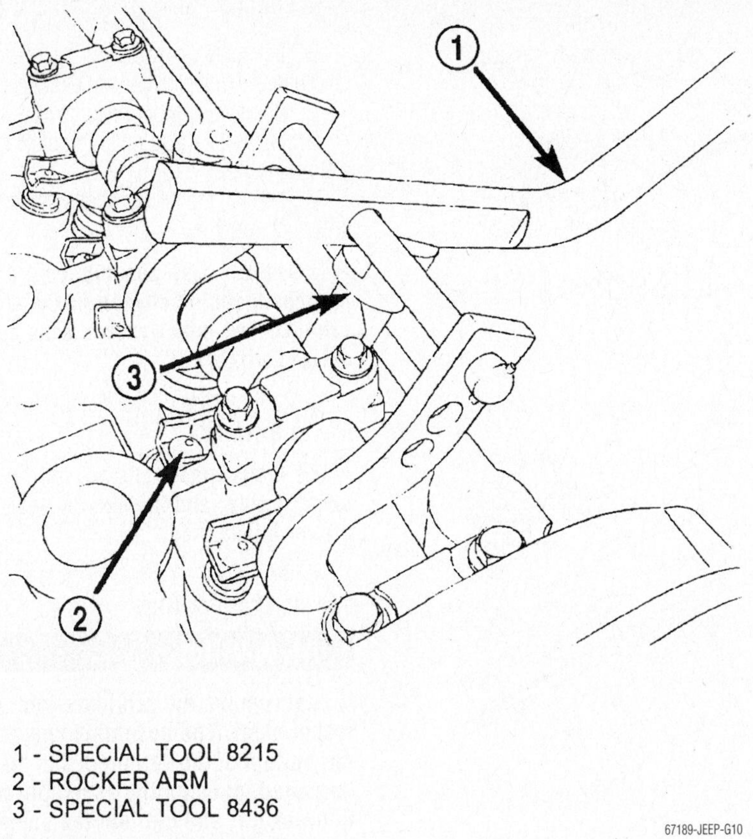

Rocker arm removal—2.4L engines

rocker arm can be installed on the hydraulic lifter and valve stem.

11. Repeat the installation procedure for each rocker arm.
12. Install the spark plugs.
13. Install the fuel rail.
14. Install the cylinder head cover.

2.5L and 4.0L Engines

1. Before servicing the vehicle, refer to the precautions in the beginning of this section.
2. Remove or disconnect the following:
   - Negative battery cable
   - Valve cover
   - Rocker arm bolts, loosen them evenly to avoid damaging the alignment bridges
   - Rocker arms

➡Keep valvetrain components in order for reassembly.

To install:

3. Install or connect the following:
   - Rocker arms, pivots and bridges in their original positions. Torque the bolts for each bridged pair one turn at a time to 21 ft. lbs. (28 Nm).
   - Valve cover
   - Negative battery cable

3.7L and 4.7L Engines

1. Before servicing the vehicle, refer to the precautions in the beginning of this section.
2. Remove or disconnect the following:
   - Negative battery cable
   - Valve covers
3. Rotate the crankshaft so that the piston of the cylinder to be serviced is at Top Dead Center (TDC) and both valves are closed.
4. Use special tool 8516 to depress the valve and remove the rocker arm.
5. Repeat for each rocker arm to be serviced.

➡Keep valvetrain components in order for reassembly.

To install:

6. Rotate the crankshaft so that the piston of the cylinder to be serviced is at BDC.
7. Compress the valve spring and install each rocker arm in its original position.
8. Repeat for each rocker arm to be installed.
9. Install or connect the following:
   - Cylinder head cover
   - Negative battery cable

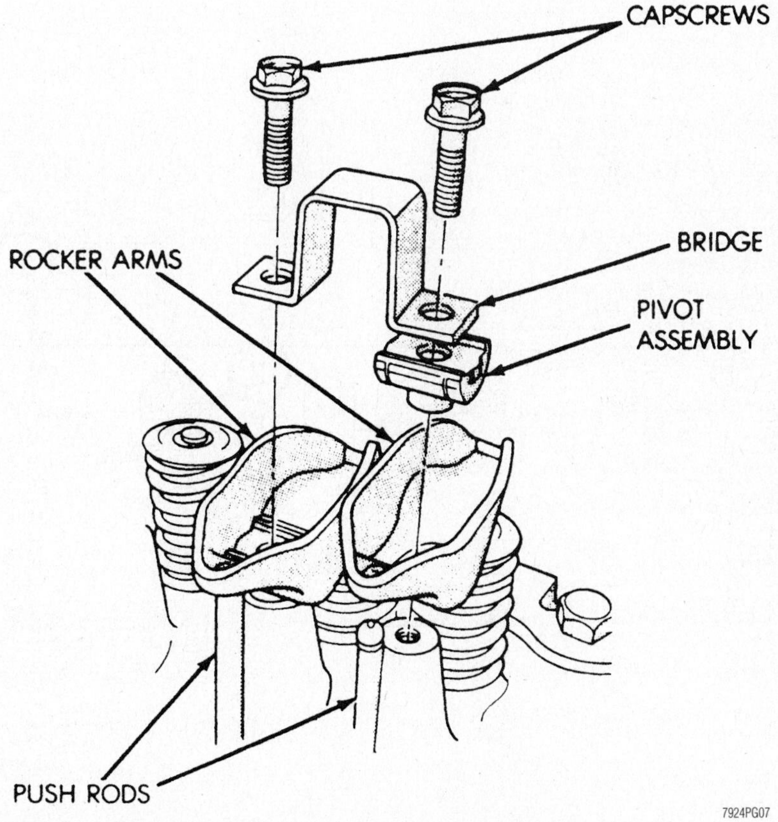

Exploded view of the rocker arm assembly—2.5L and 4.0L engines

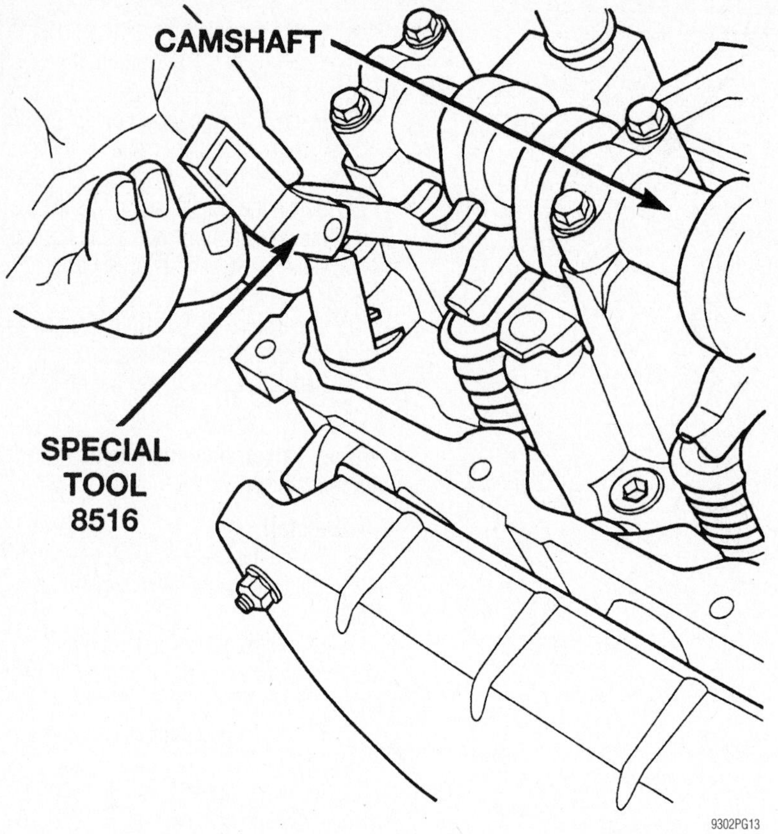

Rocker arm service—3.7L and 4.7L engines

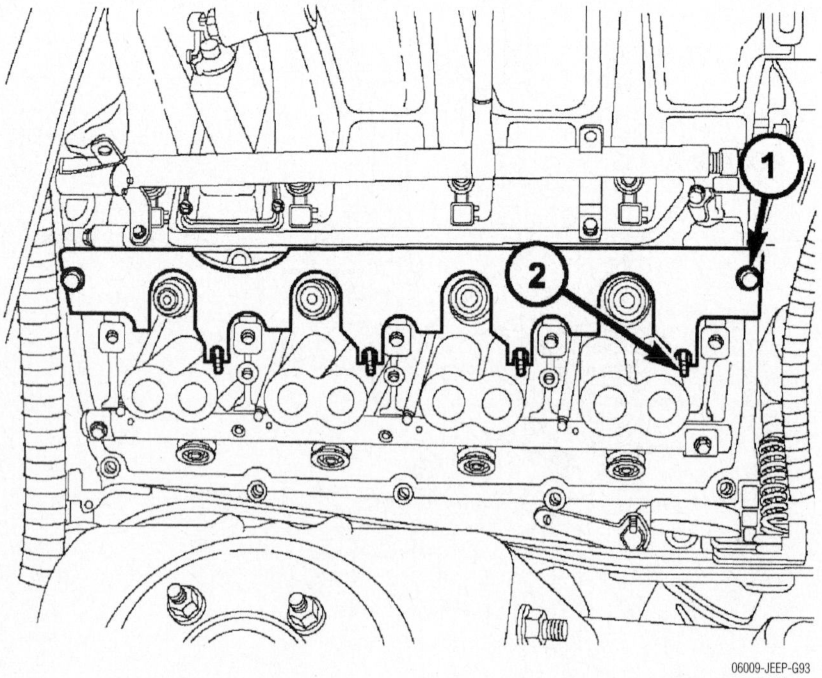

Pushrod retaining plate—5.7L engine

5.7L Engine

1. Before servicing the vehicle, refer to the precautions in the beginning of this section.
2. Remove cylinder head cover.
3. Install pushrod retaining plate special tool 9070.
4. Loosen the rocker shafts using the sequence provided.

➡The rocker shaft assemblies are not interchangeable between intake and exhaust. The intake rocker arms are marked with an "I".

5. Remove the rocker shafts. Note location for reassembly.

➡The longer push rods are for the exhaust side, and the shorter push rods are for intake side.

6. Remove the pushrods. Note pushrod location for reassembly.

✱✱ WARNING

Do not remove the retainers from the rocker shaft. The assembly tangs at the bottom of the retainers can be damaged, causing the assembly tangs to break off, and get into the engine.

To install:

➡The longer push rods are for the exhaust side, and the shorter push rods are for intake side.

7. Install the push rods in the same order as removed.
8. Install the pushrod retaining plate special tool 9070.

➡Verify that pushrod is installed into rocker arm and tappet correctly while installing rocker shaft assembly. Recheck after rocker shaft has been torqued to specification. The rocker shaft assemblies are not interchangeable between intake and exhaust. The intake rocker arms are marked with the letter "I".

9. Install rocker shaft assemblies in the same order as removed.
10. Tighten the rocker shaft bolts to 22 Nm (16 ft. lbs.) torque, using the sequence provided.

✱✱ WARNING

DO NOT rotate or crank the engine during or immediately after rocker arm installation. Allow the hydraulic roller tappets adequate time to bleed down (about 5 minutes).

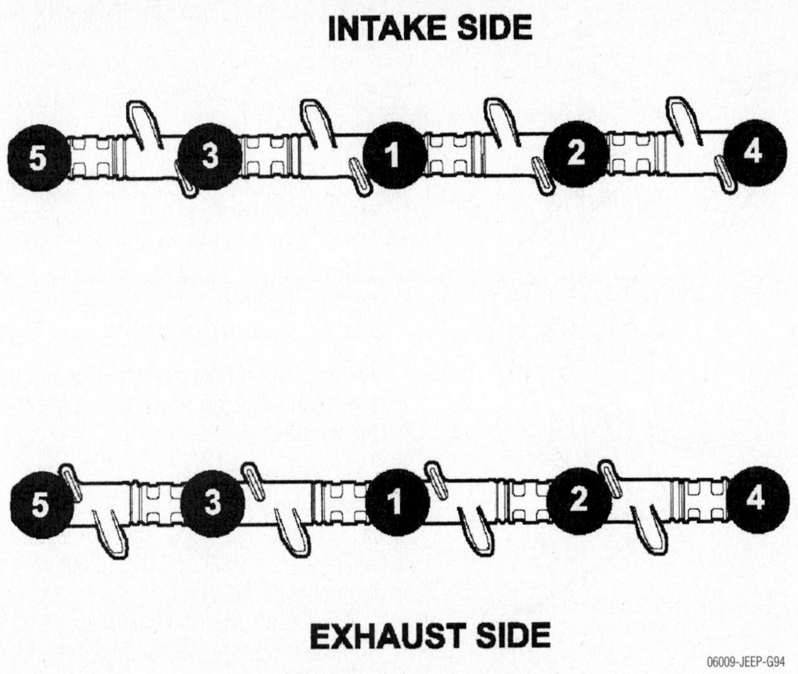

Rocker arm shaft removal and tightening sequence—5.7L engine

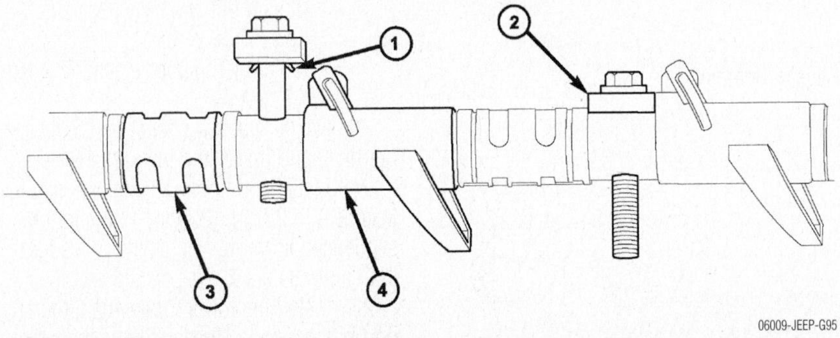

Do not remove the retainers from the rocker shaft. The assembly tangs (1) at the bottom of the retainers (2) can be damaged, causing the assembly tangs to break off, and get into the engine. Ensure that retainers (2) and rocker arms (4) are not overlapped when torquing bolts—5.7L engine

11. Remove pushrod retaining plate special tool 9070.
12. Install cylinder head cover.

Intake Manifold

REMOVAL & INSTALLATION

2.4L Engine w/2-Piece Aluminum Manifold

UPPER

The intake manifold is a two piece aluminum casting that attaches to the cylinder head with fasteners.

1. Before servicing the vehicle, refer to the precautions in the beginning of this section.
2. Disconnect the negative cable from the battery.
3. Disconnect the connector from the inlet air temperature sensor.
4. Disconnect the air intake tube at the throttle body and remove the upper air cleaner housing.
5. Disconnect the connector from the throttle position sensor (TPS).
6. Disconnect the connector from the idle air control (IAC) motor.
7. Disconnect the connector from the MAP sensor.
8. Remove the vacuum lines for the purge solenoid and the PCV valve at the intake manifold.
9. Remove the vacuum lines for the power brake booster, LDP, EGR transducer, and speed control vacuum reservoir (if equipped) at the upper intake manifold fittings.
10. Disconnect the throttle, speed control (if equipped), and transaxle control (if equipped) and cables from the throttle lever and bracket.
11. Perform the fuel system pressure release procedure before attempting any repairs.
12. Disconnect the fuel line.
13. Disconnect the coolant temperature sensor.

➡Cover the intake manifold openings to prevent foreign material from entering the engine. (16) Inspect the manifold injector wire harness connector.

14. Disconnect the fuel injector harness.
15. Remove the intake manifold to cylinder head fasteners.
16. Remove the manifold from engine.

To install:

17. Clean the manifold sealing surfaces.
18. Apply a 1.5 mm (0.060 in.) bead Mopar® Gasket Maker, or equivalent, to the perimeter of the lower intake manifold runner openings.
19. Install the upper intake manifold and tighten the fasteners to 28 Nm (250 inch lbs.) in the sequence shown in. Repeat this procedures until all fasteners are at the specified torque.
20. Install the engine oil dipstick.
21. Install the upper bolt in the intake manifold to front support bracket.
22. Install the EGR tube.
23. Install the throttle cables in the bracket.
24. Connect the throttle, speed control, (if equipped), cables to the throttle lever.
25. Connect the vacuum lines for the power brake booster, LDP, EGR transducer, and speed control vacuum reservoir (if equipped) at the upper intake manifold fittings.
26. Connect the vacuum lines for the purge solenoid and PCV valve.
27. Connect the electrical connectors for the MAP sensor, throttle position sensor (TPS), and idle air control (IAC) motor.
28. Install the air cleaner upper housing and air intake tube to throttle body.
29. Connect the inlet air temperature sensor connector.
30. Connect the negative cable to the battery.

LOWER

1. Before servicing the vehicle, refer to the precautions in the beginning of this section.

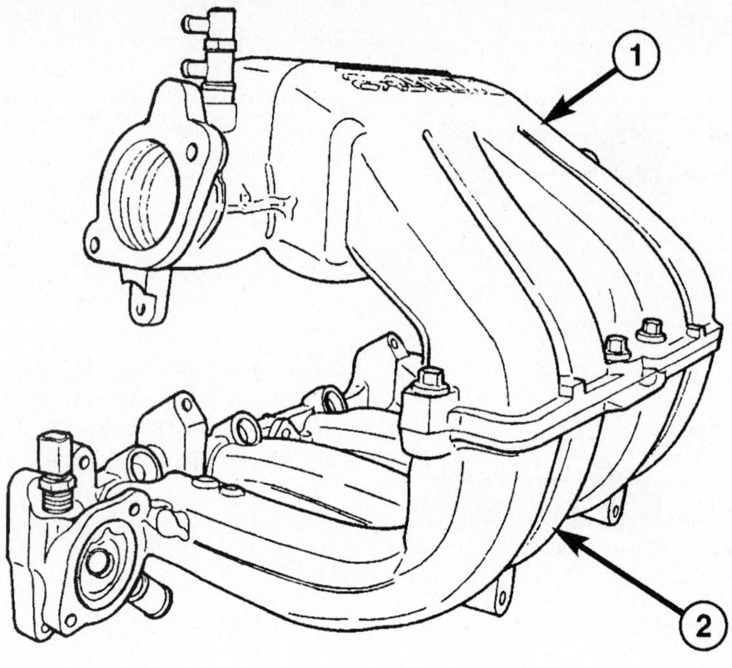

1 - UPPER INTAKE MANIFOLD
2 - LOWER INTAKE MANIFOLD

67189-JEEP-G11

Upper and lower intake manifolds—2.4L engines w/2-piece manifold

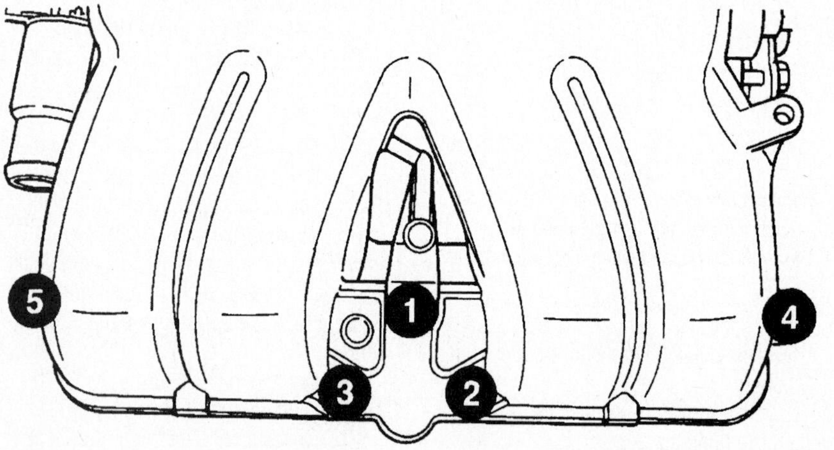

67189-JEEP-G12

Upper intake manifold torque sequence—2.4L engines w/2-piece manifold

2. Disconnect the negative cable from the battery.
3. Disconnect the connector from the inlet air temperature sensor.
4. Disconnect the air intake tube at the throttle body and remove the upper air cleaner housing.
5. Disconnect the connector from the throttle position sensor (TPS).
6. Disconnect the connector from the idle air control (IAC) motor.
7. Disconnect the connector from the MAP sensor.
8. Remove the vacuum lines for the purge solenoid and PCV valve at the intake manifold.
9. Remove the vacuum lines for the power brake booster, LDP, EGR transducer, and speed control vacuum reservoir (if equipped) at the intake manifold fittings.
10. Disconnect the throttle, speed control (if equipped), and transaxle control (if equipped) and cables from the throttle lever and bracket.
11. Perform the fuel system pressure release procedure before attempting any repairs.
12. Disconnect the fuel line.
13. Disconnect the coolant temperature sensor/fuel injector wire harness connector.
14. Disconnect the fuel injector harness.
15. Remove the intake manifold to cylinder head fasteners.
16. Remove the manifold from engine.

➡Cover the intake manifold openings to prevent foreign material from entering engine.

17. Inspect the manifold.
18. Check the manifold surfaces for flatness with straight edge. Surface must be flat within 0.15 mm per 300 mm (0.006 in. per foot) of manifold length.
19. Inspect the manifold for cracks or distortion. Replace the manifold if necessary.

To install:

If the following items were removed, install and torque to specifications:

- Fuel rail bolts: 22 Nm (200 inch lbs.)
- Coolant outlet connector bolts: 28 Nm (250 inch lbs.)
- Coolant temperature sensor: 7 Nm (60 inch lbs.)

20. Position a new gasket on the cylinder head and install the lower manifold.
21. Install and tighten the intake manifold fasteners to 28 Nm (250 inch lbs.) in the sequence shown in. Repeat the procedure until all bolts are at the specified torque.
22. Install the lower intake manifold support bracket bolts. Torque the bolts to the intake to 28 Nm (250 inch lbs.). Torque the bolts to the engine block to 54 Nm (40 ft. lbs.)
23. Connect the fuel line.
24. Connect the coolant temperature sensor/fuel injector wiring harness electrical connector.
25. Install the radiator upper and heater supply hoses.
26. Install the upper intake manifold.
27. Fill the cooling system.

2.4L Engine w/1-Piece Composite Manifold

The intake manifold is a one piece composite module that attaches to the cylinder head with fasteners.

1. Before servicing the vehicle, refer to the precautions in the beginning of this section.
2. Disconnect negative cable from battery.

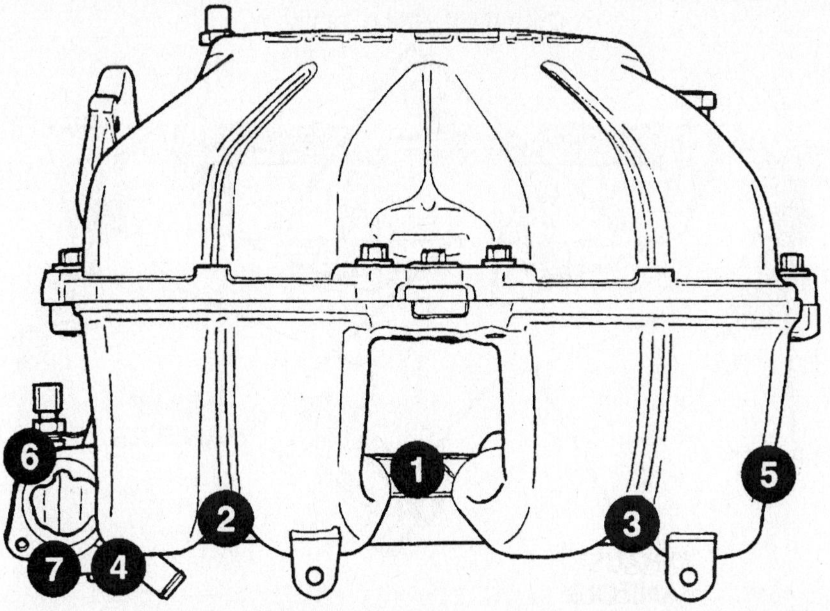

Lower intake manifold torque sequence—2.4L engines w/2-piece manifold

3. Disconnect connector from inlet air temperature sensor.
4. Disconnect air intake tube at throttle body and remove upper air cleaner housing.
5. Disconnect connector from throttle position sensor (TPS).
6. Disconnect connector from idle air control (IAC) motor.
7. Disconnect connector from MAP sensor.
8. Remove vacuum lines for purge solenoid and PCV valve at intake manifold.
9. Remove vacuum lines for power brake booster, LDP, EGR transducer, and speed control vacuum reservoir (if equipped) at intake manifold fittings.
10. Disconnect throttle, speed control (if equipped), and transaxle control (if equipped) and cables from throttle lever and bracket.
11. Perform fuel system pressure release procedure before attempting any repairs.
12. Disconnect fuel line.
13. Disconnect coolant temperature sensor/fuel injector wire harness connector.
14. Disconnect fuel injector harness.
15. Remove intake manifold to cylinder head fasteners.
16. Remove the manifold from engine.

➡Cover intake manifold openings to prevent foreign material from entering engine.

To install:

17. Check manifold surfaces for flatness with straightedge. Surface must be flat within 0.15 mm per 300 mm (0.006 in. per foot) of manifold length.
18. Inspect manifold for cracks or distortion. Replace manifold if necessary.
19. Clean manifold sealing surfaces.
20. Install new manifold to cylinder head seals.
21. Install manifold to head.
22. Install and tighten intake manifold fasteners to 28 Nm (250 inch lbs.) in the sequence shown. Repeat procedure until all bolts are at specified torque.
23. Install throttle cables in bracket.
24. Connect throttle, speed control, (if equipped), cables to throttle lever.
25. Connect vacuum lines for power brake booster, LDP, EGR transducer, and speed control vacuum reservoir (if equipped) at upper intake manifold fittings.
26. Connect vacuum lines for purge solenoid and PCV valve.
27. Connect electrical connectors for MAP sensor, throttle position sensor (TPS), and idle air control (IAC) motor.
28. Connect the fuel line.
29. Connect coolant temperature sensor/fuel injector wiring harness electrical connector.
30. Install the air cleaner housing and air intake tube to throttle body.
31. Connect inlet air temperature sensor connector.
32. Connect negative cable to battery.

2.5L Engine

1. Before servicing the vehicle, refer to the precautions in the beginning of this section.
2. Properly relieve the fuel system pressure.
3. Drain the cooling system.
4. Remove or disconnect the following:
   - Negative battery cable
   - Air intake hose
   - Accessory drive belt
   - Power steering pump and brackets, if equipped
   - Fuel line
   - Accelerator cable
   - Cruise control cable, if equipped
   - Transmission cable, if equipped
   - Throttle Position (TP) sensor connector
   - Idle Air Control (IAC) valve connector
   - Engine Coolant Temperature (ECT) sensor connector

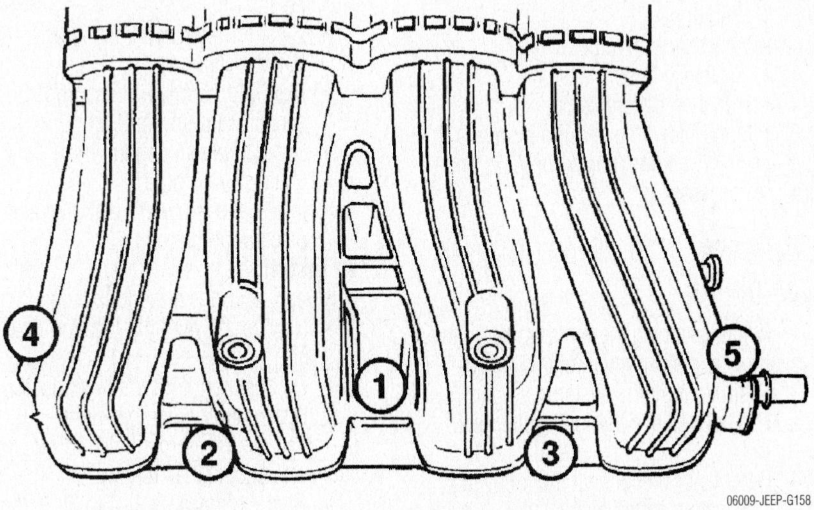

Intake manifold torque sequence—2.4L engines w/1-piece composite manifold

- Intake Air Temperature (IAT) sensor connector
- Heated Oxygen (HO_2S) sensor connector
- Fuel injector connectors
- Closed Crankcase Ventilation (CCV) system
- Manifold Absolute Pressure (MAP) sensor vacuum line
- Brake booster vacuum line
- Exhaust front pipe
- Intake manifold and discard the gaskets
- Exhaust manifold and discard the gaskets

To install:

5. Install or connect the following:
   - New gasket over the locating dowels
   - Exhaust manifold to the studs and tighten the nuts finger-tight
6. Install the intake manifold. Torque the fasteners in sequence as follows:
   a. Step 1: Bolt 1 to 30 ft. lbs. (41 Nm).
   b. Step 2: Fasteners 2–7 to 23 ft. lbs. (31 Nm).
7. Install or connect the following:
   - Exhaust front pipe
   - Brake booster vacuum line
   - MAP sensor vacuum line
   - CCV system
   - Fuel injector connectors
   - HO_2S sensor connector
   - IAT sensor connector
   - ECT sensor connector
   - IAC valve connector
   - TP sensor connector
   - Transmission cable, if equipped
   - Cruise control cable, if equipped
   - Accelerator cable
   - Fuel line
   - Power steering pump and brackets, if equipped
   - Accessory drive belt
   - Air intake hose
   - Negative battery cable
8. Fill and bleed the cooling system.
9. Start the engine, check for leaks and repair if necessary..

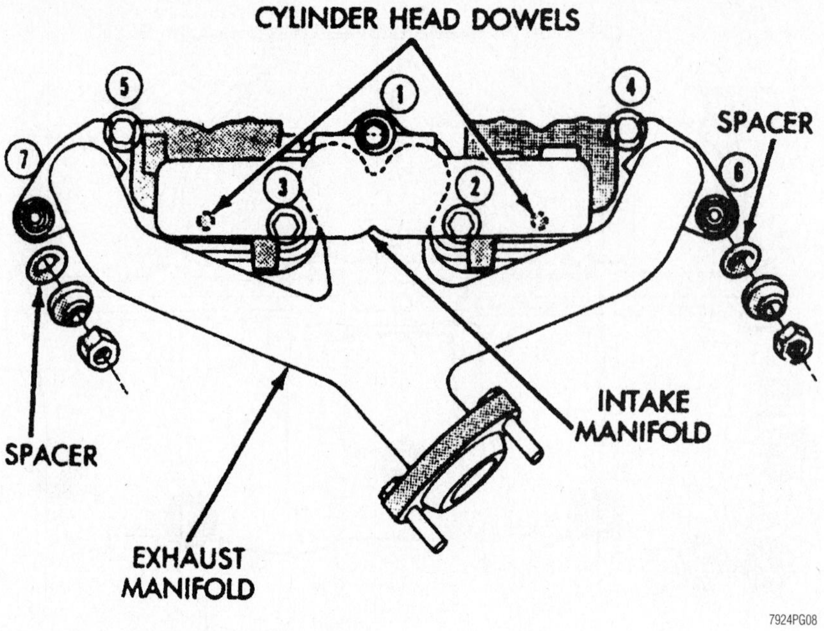

Manifold torque sequence—2.5L engine

4.0L Engine

2002–04

1. Before servicing the vehicle, refer to the precautions in the beginning of this section.
2. Drain the cooling system.
3. Properly relieve the fuel system pressure.
4. Remove or disconnect the following:
   - Negative battery cable
   - Air cleaner assembly
   - Accessory drive belt
   - Power steering pump and brackets, if equipped
   - Fuel line
   - Fuel injector connectors
   - Fuel supply manifold and injectors
   - Accelerator cable
   - Cruise control cable, if equipped
   - Transmission cable, if equipped
   - Throttle Position (TP) sensor connector
   - Idle Air Control (IAC) valve connector
   - Engine Coolant Temperature (ECT) sensor connector
   - Intake Air Temperature (IAT) sensor connector
   - Heated Oxygen (HO_2S) sensor connector
   - Closed Crankcase Ventilation (CCV) system
   - Manifold Absolute Pressure (MAP) sensor vacuum line
   - Brake booster vacuum line
   - Exhaust front pipe
   - Intake and exhaust manifolds and discard the gaskets

To install:

5. Install or connect the following:
   - New gasket over the locating dowels
   - Exhaust manifold to the studs and tighten the nuts finger-tight
6. Install the intake manifold. Torque the fasteners in sequence as follows:
   a. Step 1: Fasteners 1–5 to 30 ft. lbs. (41 Nm).
   b. Step 2: Fasteners 6 and 7 to 23 ft. lbs. (31 Nm).
   c. Step 3: Fasteners 8–11 to 24 ft. lbs. (33 Nm)
7. Install or connect the following:
   - Exhaust front pipe
   - Brake booster vacuum line
   - MAP sensor vacuum line
   - CCV system
   - HO_2S sensor connector
   - IAT sensor connector
   - ECT sensor connector
   - IAC valve connector
   - TP sensor connector
   - Transmission cable, if equipped
   - Cruise control cable, if equipped
   - Accelerator cable
   - Fuel supply manifold and injectors
   - Fuel injector connectors
   - Fuel line
   - Power steering pump and brackets, if equipped
   - Accessory drive belt
   - Air cleaner assembly
   - Negative battery cable
8. Fill and bleed the cooling system.
9. Start the engine, check for leaks and repair if necessary.

2005

1. Before servicing the vehicle, refer to the precautions in the beginning of this section.
2. Disconnect the battery negative cable.
3. Remove air cleaner inlet hose from the resonator assembly.
4. Remove the air cleaner assembly.

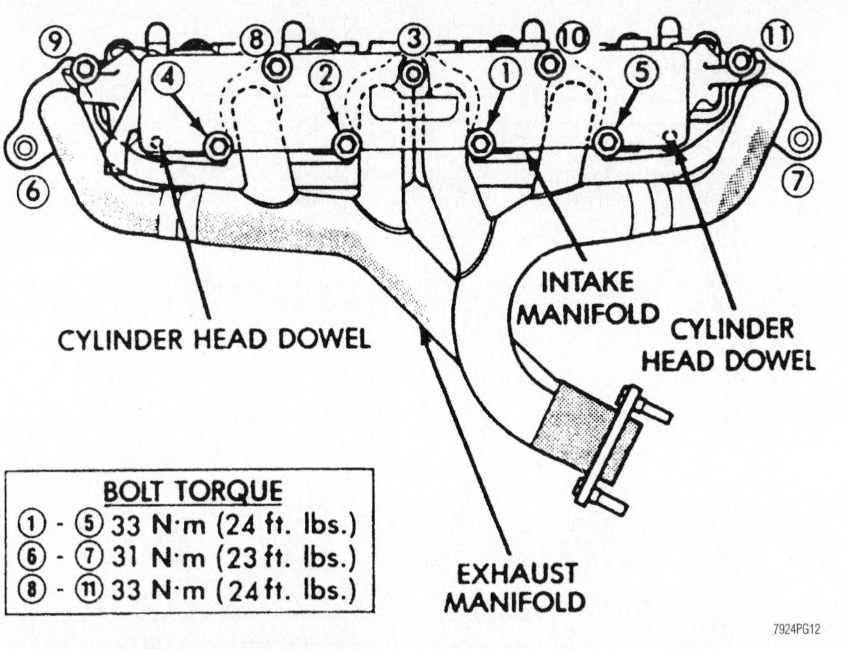

Manifold torque sequence—2002–04 4.0L engine

5. Remove the throttle cable, vehicle speed control cable (if equipped) and the transmission line pressure cable.

6. Disconnect the following electrical connections and secure their harness out of the way:
   - Throttle Position Sensor
   - Idle Air Control Motor
   - Coolant Temperature Sensor (at thermostat housing)
   - Intake Air Temperature Sensor
   - Oxygen Sensor
   - Crank Position Sensor
   - Six

7. Fuel Injector Connectors
   - Manifold Absolute Pressure (MAP) Sensor.

8. Disconnect HVAC, and Brake Booster vacuum supply hoses at the intake manifold.

9. Perform the fuel pressure release procedure.

10. Disconnect and remove the fuel system supply line from the fuel rail assembly.

11. Remove the accessory drive belt.

12. Remove the power steering pump from the intake manifold and set aside.

13. Raise the vehicle.

14. Disconnect the exhaust pipes from the engine exhaust manifolds.

15. Lower the vehicle.

16. Remove the intake manifold and exhaust manifold bolts and manifolds.

To install:

➡If the manifold is being replaced, ensure all the fittings, etc. are transferred to the replacement manifold.

17. Install a new engine exhaust/intake manifold gasket over the alignment dowels on the cylinder head.

18. Position the engine exhaust manifolds to the cylinder head. Install fastener Number 3 and finger tighten at this time.

19. Install intake manifold on the cylinder head dowels.

20. Install washers and fasteners 1, 2, 4, 5, 8, 9, 10 and 11.

21. Install washers and fasteners 6 and 7.

22. Tighten the fasteners in sequence and to the specified torque.
   - Fasteners 1 through 5: 33 Nm (24 ft. lbs.).
   - Fasteners 6 and 7: Tighten to 31 Nm (23 ft. lbs.).
   - Fasteners 8 through 11: 33 Nm (24 ft. lbs.).

23. Install the power steering pump to the intake manifold.

24. Install the accessory drive belt.

25. Install the fuel system supply line to the fuel rail assembly.

26. Connect all electrical connections on the intake manifold.

27. Connect the vacuum hoses previously removed.

28. Install throttle cable, vehicle speed control cable (if equipped).

29. Install the transmission line pressure cable (if equipped).

30. Install air cleaner assembly.

31. Connect air inlet hose to the resonator assembly.

32. Raise the vehicle.

33. Connect the exhaust pipes to the engine exhaust manifolds. Tighten the bolts to 31 Nm (23 ft. lbs.)

34. Lower the vehicle.

35. Connect the battery negative cable.

36. Start the engine and check for leaks.

3.7L and 4.7L Engines

1. Before servicing the vehicle, refer to the precautions in the beginning of this section.

2. Drain the cooling system.

3. Properly relieve the fuel system pressure.

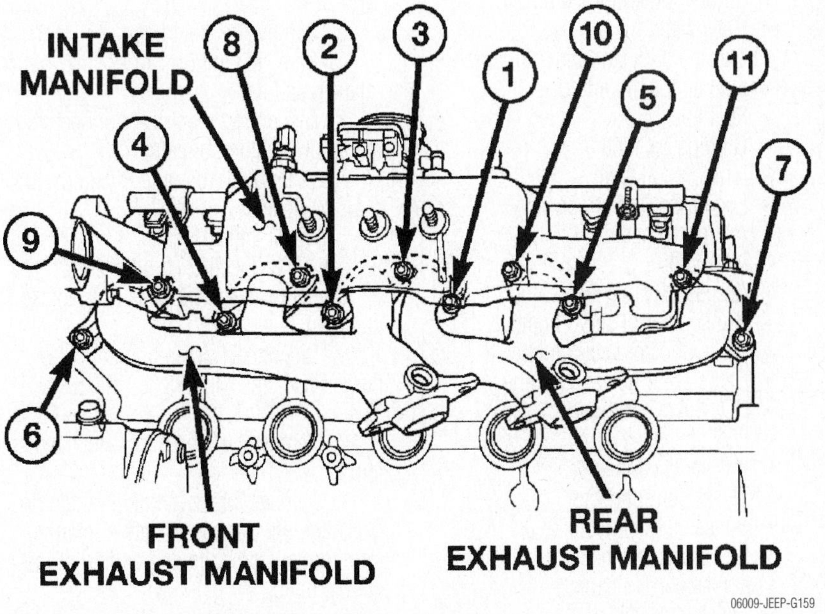

Manifold torque sequence—2005 4.0L engine

4. Remove or disconnect the following:
   - Negative battery cable
   - Air cleaner assembly
   - Accelerator cable
   - Cruise control cable
   - Manifold Absolute Pressure (MAP) sensor connector
   - Intake Air Temperature (IAT) sensor connector
   - Throttle Position (TP) sensor connector
   - Idle Air Control (IAC) valve connector
   - Engine Coolant Temperature (ECT) sensor
   - Positive Crankcase Ventilation (PCV) valve and hose
   - Canister purge vacuum line
   - Brake booster vacuum line
   - Cruise control servo hose
   - Accessory drive belt
   - Alternator
   - A/C compressor
   - Engine ground straps
   - Ignition coil towers
   - Oil dipstick tube
   - Fuel line
   - Fuel supply manifold
   - Throttle body and mounting bracket
   - Cowl seal
   - Right engine lifting stud
   - Intake manifold. Remove the fasteners in reverse of the tightening sequence.

To install:

5. Install or connect the following:
   - Intake manifold using new gaskets. Torque the bolts, in sequence, to 105 inch lbs. (12 Nm).
   - Right engine lifting stud
   - Cowl seal
   - Throttle body and mounting bracket
   - Fuel supply manifold
   - Fuel line
   - Oil dipstick tube
   - Ignition coil towers
   - Engine ground straps
   - A/C compressor
   - Alternator
   - Accessory drive belt
   - Cruise control servo hose
   - Brake booster vacuum line
   - Canister purge vacuum line
   - PCV valve and hose
   - ECT sensor
   - IAC valve connector
   - TP sensor connector
   - IAT sensor connector
   - MAP sensor connector
   - Cruise control cable
   - Accelerator cable
   - Air cleaner assembly

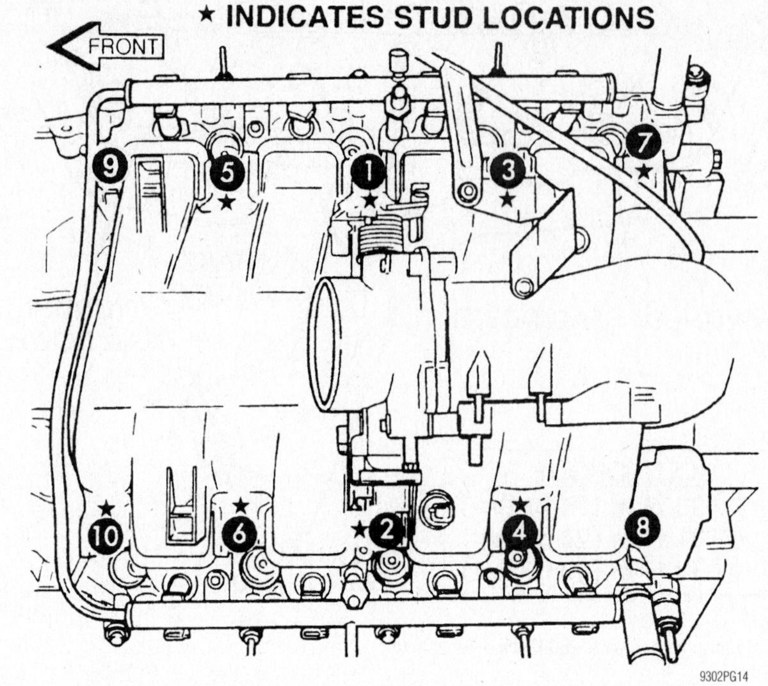

Intake manifold torque sequence—3.7 & 4.7L engines

   - Negative battery cable

6. Fill and bleed the cooling system.
7. Start the engine, check for leaks and repair if necessary.

5.7L Engine

1. Before servicing the vehicle, refer to the precautions in the beginning of this section.
2. Remove engine cover.
3. Bleed fuel system.
4. Disconnect negative cable from battery.
5. Remove air inlet hose.
6. Remove ignition wires from on top of intake manifold.
7. Disconnect electrical connectors for the following components:
   - Manifold Absolute Pressure (MAP) Sensor
   - Fuel Injectors
   - ETC (Electric Throttle Control)
8. Remove wire harness from intake manifold.
9. Disconnect brake booster hose, purge hose, and MUA hose (Make Up Air Hose).
10. Remove EGR tube from intake manifold.
11. Remove intake manifold retaining fasteners in a crisscross pattern starting from the outside bolts and ending at the middle bolts.
12. Remove intake manifold as an assembly.

➡There is no approved repair procedure for the intake manifold. If severe damage is found during inspection, the intake manifold must be replaced.

➡Before installing the intake manifold thoroughly clean the mating surfaces. Use a suitable cleaning solvent, then air dry.

13. Inspect the intake sealing surface for cracks, nicks and distortion.
14. Inspect the intake manifold vacuum hose fittings for looseness or blockage.

To install:

15. Position intake manifold.
16. Install intake manifold retaining bolts, and tighten from the middle bolts towards the outside in a criss-cross pattern. Torque fasteners to 12 Nm (105 inch lbs.).
17. Install EGR tube.
18. Install wire harness on intake manifold.
19. Connect electrical connectors for the following components:
    - Manifold Absolute Pressure (MAP) Sensor
    - Fuel Injectors
    - ETC (Electronic Throttle Control)
20. Install ignition wires.
21. Connect Brake booster hose, purge hose, and MUA hose (Make Up Air hose).
22. Install air inlet hose.
23. Connect negative cable to battery.
24. Install engine cover.

Exhaust Manifold

REMOVAL & INSTALLATION

2.4L Engine

1. Before servicing the vehicle, refer to the precautions in the beginning of this section.
2. Raise the vehicle and disconnect the exhaust pipe from the exhaust manifold.
3. Lower the vehicle.
4. Disconnect the upstream oxygen sensor connector at the rear of the exhaust manifold.
5. Remove the air cleaner bracket.
6. Remove the heat shield.
7. Remove the bolts attaching the manifold to the cylinder head.
8. Remove the exhaust manifold.
9. Inspect the manifold.
10. Discard the gasket (if equipped) and clean all surfaces of the manifold and cylinder head.
11. Inspect the manifold gasket surfaces for flatness with a straight edge. Surface must be flat within 0.15 mm per 300 mm (0.006 in. per foot) of manifold length.
12. Inspect the manifolds for cracks or distortion. Replace the manifold as necessary.

To install:

13. Clean the manifold mating surfaces.
14. Install the exhaust manifold with a new gasket. Tighten the attaching nuts to 175 inch lbs. (20 Nm).
15. Attach the exhaust pipe to the exhaust manifold and tighten the fasteners to 27 ft. lbs. (37 Nm)
16. Install and connect the oxygen sensor.
17. Install the heat shield.
18. Install the air cleaner bracket.

4.7L Engine

1. Before servicing the vehicle, refer to the precautions in the beginning of this section.
2. Drain the cooling system.
3. Remove or disconnect the following:
   - Battery
   - Power distribution center
   - Battery tray
   - Windshield washer fluid bottle
   - Air cleaner assembly
   - Accessory drive belt
   - A/C compressor
   - A/C accumulator bracket
   - Heater hoses
   - Exhaust manifold heat shields
   - Exhaust Y-pipe
   - Starter motor
   - Exhaust manifolds

To install:

4. Install or connect the following:
   - Exhaust manifolds, using new gaskets. Torque the bolts to 18 ft. lbs. (25 Nm), starting with the inner bolts and work out to the ends.
   - Starter motor
   - Exhaust Y-pipe
   - Exhaust manifold heat shields
   - Heater hoses
   - A/C accumulator bracket
   - A/C compressor
   - Accessory drive belt
   - Air cleaner assembly
   - Windshield washer fluid bottle
   - Battery tray
   - Power distribution center
   - Battery
5. Fill and bleed the cooling system.
6. Start the engine, check for leaks and repair if necessary.

3.7L Engines

LIBERTY

1. Before servicing the vehicle, refer to the precautions in the beginning of this section.
2. Remove or disconnect the following:
   - Negative battery cable
   - Exhaust manifold heat shields
   - Exhaust Gas Recirculation (EGR) tube
   - Exhaust Y-pipe
   - Exhaust manifolds

To install:

➡If the exhaust manifold studs came out with the nuts when removing the exhaust manifolds, replace them with new studs.

3. Install or connect the following:
   - Exhaust manifolds. Torque the fasteners to 20 ft. lbs. (27 Nm), starting with the center nuts and work out to the ends.
   - Exhaust Y-pipe
   - EGR tube
   - Exhaust manifold heat shields
   - Negative battery cable
4. Start the engine, check for leaks and repair if necessary.

GRAND CHEROKEE—RIGHT SIDE

1. Before servicing the vehicle, refer to the precautions in the beginning of this section.
2. Disconnect the negative cable from the battery.
3. Raise and support the vehicle.
4. Remove the bolts and nuts attaching the exhaust pipe to the engine exhaust manifold.
5. Lower the vehicle.
6. Remove the exhaust heat shield.
7. Remove bolts, nuts and washers attaching manifold to cylinder head.
8. Remove manifold and gasket from the cylinder head.

To install:

**** WARNING**

If the studs came out with the nuts when removing the engine exhaust manifold, install new studs. Apply sealer on the coarse thread ends. Water leaks may develop at the studs if this precaution is not taken.

9. Position the engine exhaust manifold and gasket on the two studs located on the cylinder head. Install conical washers and nuts on these studs.
10. Install remaining conical washers. Starting at the center arm and working outward, tighten the bolts and nuts to 25 Nm (18 ft. lbs.) torque.
11. Install the exhaust heat shields.
12. Raise and support the vehicle.

**** WARNING**

Over tightening heat shield fasteners, may cause shield to distort and/or crack.

13. Assemble exhaust pipe to manifold and secure with bolts, nuts and retainers. Tighten the bolts and nuts to 34 Nm (25 ft. lbs.) torque.

GRAND CHEROKEE—LEFT SIDE

1. Before servicing the vehicle, refer to the precautions in the beginning of this section.
2. Disconnect the negative cable from the battery.
3. Raise and support the vehicle.
4. Remove the bolts and nuts attaching the exhaust pipe to the engine exhaust manifold.
5. Lower the vehicle.
6. Remove the exhaust heat shields.
7. Remove bolts, nuts and washers attaching manifold to cylinder head.
8. Remove manifold and gasket from the cylinder head.

To install:

**** WARNING**

If the studs came out with the nuts when removing the engine exhaust manifold, install new studs. Apply sealer on the coarse thread ends. Water leaks may develop at the studs if this precaution is not taken.

9. Position the engine exhaust manifold and gasket on the two studs located on the cylinder head. Install conical washers and nuts on these studs.
10. Install remaining conical washers. Starting at the center arm and working outward, tighten the bolts and nuts to 25 Nm (18 ft. lbs.) torque.
11. Install the exhaust heat shields.
12. Raise and support the vehicle.

⁂ WARNING

Over tightening heat shield fasteners, may cause shield to distort and/or crack.

13. Assemble exhaust pipe to manifold and secure with bolts, nuts and retainers. Tighten the bolts and nuts to 34 Nm (25 ft. lbs.) torque.

2.5L Engine

1. Before servicing the vehicle, refer to the precautions in the beginning of this section.
2. Properly relieve the fuel system pressure.
3. Drain the cooling system.
4. Remove or disconnect the following:
   - Negative battery cable
   - Air intake hose
   - Accessory drive belt
   - Power steering pump and brackets, if equipped
   - Fuel line
   - Accelerator cable
   - Cruise control cable, if equipped
   - Transmission cable, if equipped
   - Throttle Position (TP) sensor connector
   - Idle Air Control (IAC) valve connector
   - Engine Coolant Temperature (ECT) sensor connector
   - Intake Air Temperature (IAT) sensor connector
   - Heated Oxygen (HO_2S) sensor connector
   - Fuel injector connectors
   - Closed Crankcase Ventilation (CCV) system
   - Manifold Absolute Pressure (MAP) sensor vacuum line
   - Brake booster vacuum line
   - Exhaust front pipe
   - Intake manifold and discard the gaskets
   - Exhaust manifold and discard the gaskets

To install:

5. Install or connect the following:
   - New gasket over the locating dowels
   - Exhaust manifold to the studs and tighten the nuts finger-tight
6. Install the intake manifold. Torque the fasteners in sequence as follows:
   a. Step 1: Bolt 1 to 30 ft. lbs. (41 Nm).
   b. Step 2: Fasteners 2–7 to 23 ft. lbs. (31 Nm).
7. Install or connect the following:
   - Exhaust front pipe
   - Brake booster vacuum line
   - MAP sensor vacuum line
   - CCV system
   - Fuel injector connectors
   - HO_2S sensor connector
   - IAT sensor connector
   - ECT sensor connector
   - IAC valve connector
   - TP sensor connector
   - Transmission cable, if equipped
   - Cruise control cable, if equipped
   - Accelerator cable
   - Fuel line
   - Power steering pump and brackets, if equipped
   - Accessory drive belt
   - Air intake hose
   - Negative battery cable
8. Fill and bleed the cooling system.
9. Start the engine, check for leaks and repair if necessary.

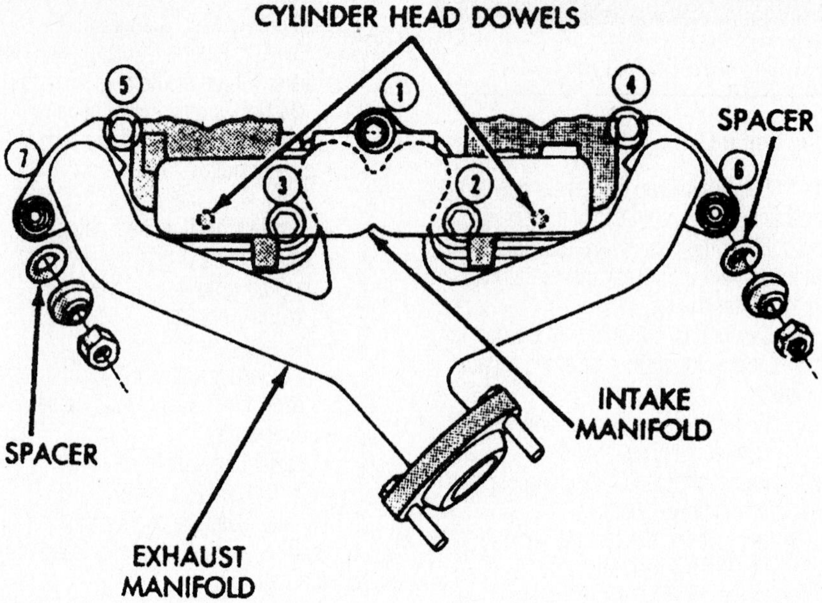

Manifold torque sequence—2.5L engine

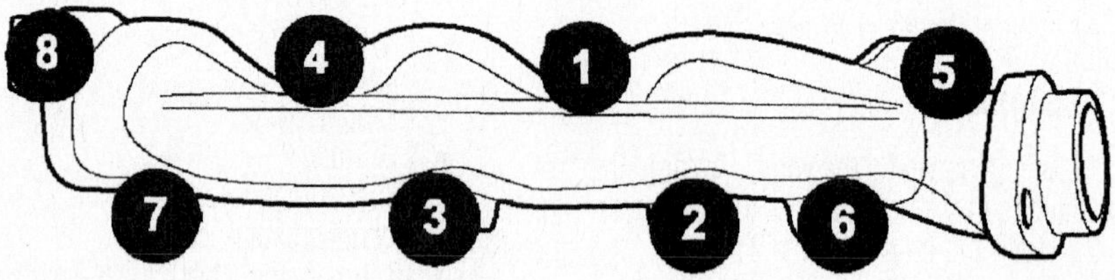

Left side exhaust manifold—5.7L engine

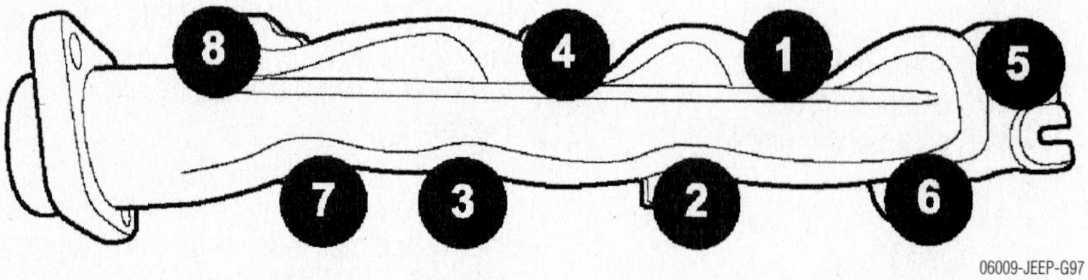

Right side exhaust manifold—5.7L engine

4.0L Engine

See the procedure under Intake Manifold.

5.7L Engine

1. Before servicing the vehicle, refer to the precautions in the beginning of this section.
2. Disconnect negative battery cable.
3. Raise vehicle.
4. Remove exhaust pipe to manifold bolts.
5. Remove engine mount to frame fasteners.
6. Using suitable jack, raise engine enough to remove manifolds.

➡Do not damage engine harness while raising the engine.

7. Remove the engine mount.
8. Remove heat shield.
9. Remove manifold bolts using sequence provided.
10. Remove manifold and gasket.

To install:

11. Inspect manifold for cracks.
12. Inspect mating surfaces of manifold for flatness with a straightedge. Gasket surfaces must be flat within 0.2 mm per 300 mm (0.008 inch per foot).
13. Install manifold gasket and manifold.
14. Install manifold bolts and tighten to 25 Nm (18 ft. lbs.).
15. Install heat shield and tighten nuts to 8 Nm (70 inch lbs.).
16. Install front engine mount. Torque to 70 ft. lbs. (95 Nm).
17. Lower engine.

✲✲ WARNING

Do not damage engine harness while lowering the engine.

18. Install and tighten right and left side engine mount to frame fasteners. Torque to 70 ft. lbs. (95 Nm).
19. Install exhaust flange to pipe bolts.
20. Lower vehicle.
21. Connect negative battery cable.

Camshaft and Valve Lifters

REMOVAL & INSTALLATION

2.4L Engine

1. Before servicing the vehicle, refer to the precautions in the beginning of this section.
2. Remove the cylinder head cover.
3. Remove the camshaft position sensor and camshaft target magnet.
4. Remove the timing belt.
5. Remove the camshaft sprockets and timing belt rear cover.
6. Bearing caps are identified for location. Remove the outside bearing caps first.
7. Loosen the camshaft bearing cap attaching fasteners in the sequence shown, one camshaft at a time.

➡Camshafts are not interchangeable. The intake cam number 6 thrust bearing face spacing is wider.

8. Identify the camshafts before removing from the head. The camshafts are not interchangeable.
9. Remove the camshafts from the cylinder head.

➡If removing the rocker arms, identify for reinstallation in the original position.

10. Inspect the camshaft bearing journals for damage and binding. If the journals are binding, check the cylinder head for damage. Also the check cylinder head oil holes for clogging.
11. Check the cam lobe and bearing surfaces for abnormal wear and damage. Replace the camshaft if defective.

➡If the camshaft is replaced due to lobe wear or damage, always replace the rocker arms.

12. Measure the lobe actual wear (unworn area minus wear zone = actual wear) and replace the camshaft if out of limit. Standard value is 0.0254 mm (0.001 in.), wear limit is 0.254 mm (0.010 in.).

To install:

➡Ensure that NONE of the pistons are at top dead center when installing the camshafts.

13. Lubricate all camshaft bearing journals, rocker arms and camshaft lobes.
14. Install all rocker arms in original positions, if reused.
15. Position the camshafts on cylinder head bearing journals. Install the right and left camshaft bearing caps No. 2—5 and right No. 6. Tighten M6 fasteners to 12 Nm (105 inch lbs.) in the sequence shown.
16. Apply Mopar® Gasket Maker, or equivalent, to the No. 1 and No. 6 bearing caps. Install the bearing caps and tighten the M8 fasteners to 28 Nm (250 inch lbs.).
17. Install the camshaft oil seals.
18. Install the camshaft target magnet and camshaft position sensor.
19. Install the cylinder head cover.
20. Install the timing belt rear cover and camshaft sprocket.
21. Install the timing belt.

2.5L Engine

1. Before servicing the vehicle, refer to the precautions in the beginning of this section.
2. Drain the cooling system.
3. Recover the A/C refrigerant, if equipped with air conditioning.
4. Remove or disconnect the following:
   - Negative battery cable
   - Grille, if necessary
   - Radiator
   - A/C condenser, if equipped

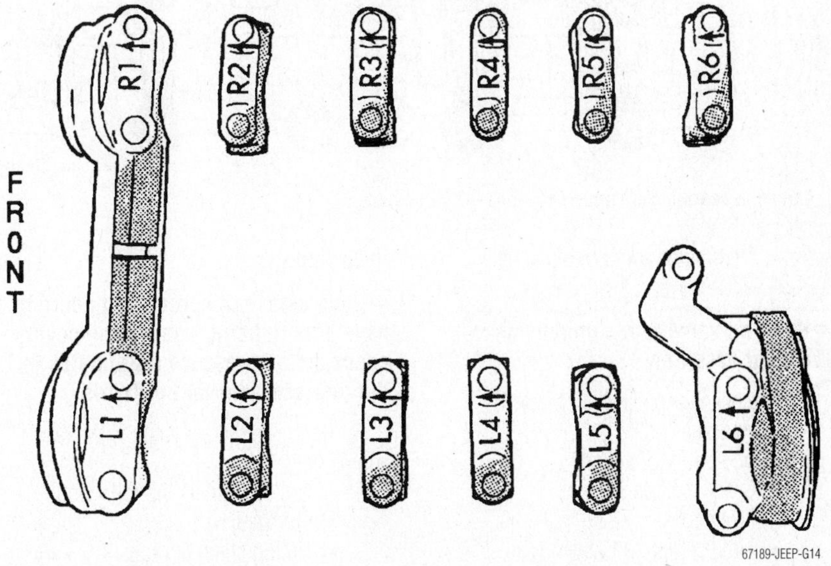

Camshaft bearing cap identification—2.4L engines

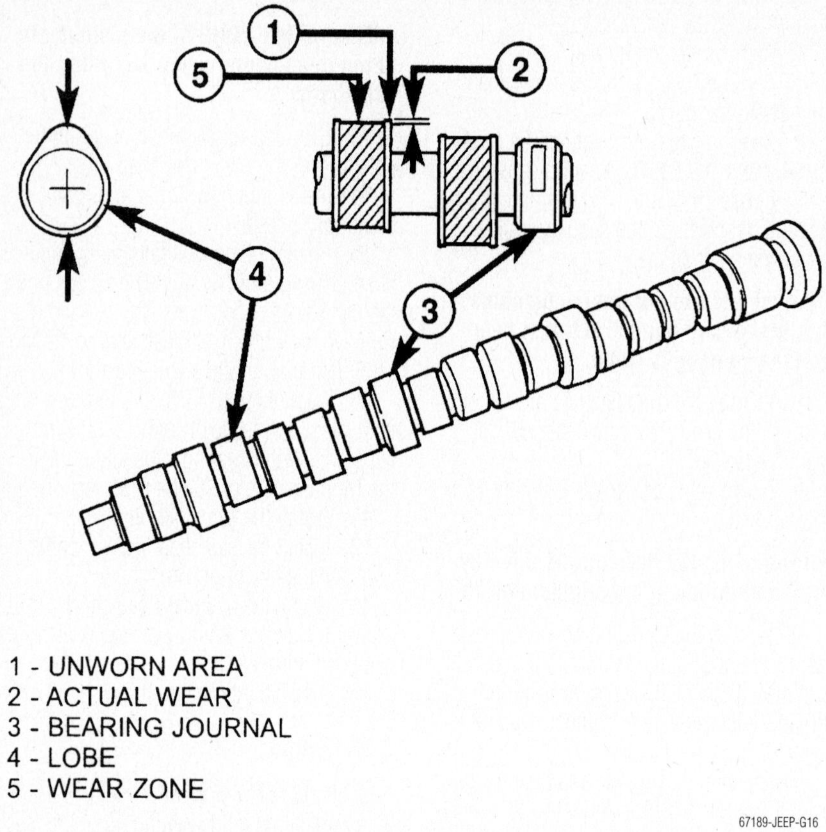

Checking camshaft wear—2.4L engines

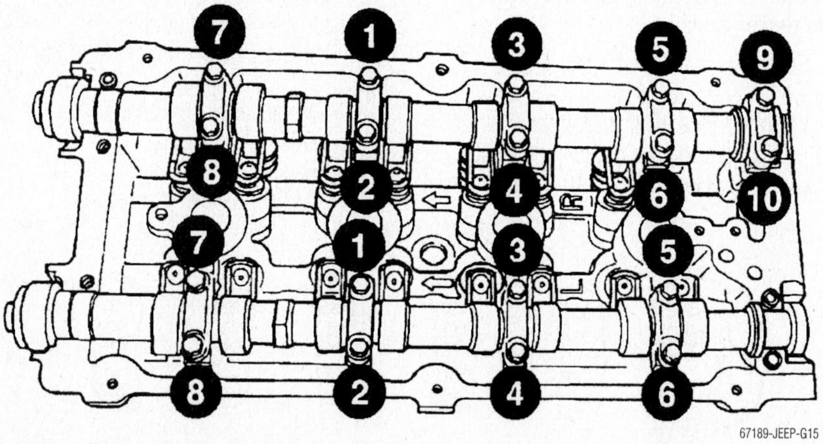

Camshaft bearing cap torque sequence—2.4L engines

- Distributor and ignition wires
- Valve cover

➡Keep all valvetrain components in order for assembly.

- Rocker arms and pushrods
- Hydraulic valve tappets
- Accessory drive belt
- Crankshaft damper
- Front cover
- Timing chain and gears
- Camshaft

To install:

➡If the camshaft sprocket appears to have been rubbing against the cover, check the oil pressure relief holes in the rear cam journal for debris.

5. Lubricate the camshaft with clean engine oil.
6. Install or connect the following:
   - Camshaft
   - Timing chain and gears
   - Front cover
   - Crankshaft damper
   - Accessory drive belt
   - Hydraulic valve tappets
   - Rocker arms and pushrods
   - Valve cover
   - Distributor
   - A/C condenser, if equipped
   - Radiator
   - Grille, if removed
   - Negative battery cable
7. Fill and bleed the cooling system.
8. Recharge the A/C system, if equipped.
9. Start the engine, check for leaks and repair if necessary.

4.0L Engine

1. Before servicing the vehicle, refer to the precautions in the beginning of this section.
2. Drain the cooling system.
3. Recover the A/C refrigerant, if equipped with air conditioning.
4. Remove or disconnect the following:
   - Negative battery cable
   - Grille, if necessary
   - Radiator
   - A/C condenser, if equipped
   - Distributor or camshaft sensor housing
   - Valve cover

➡Keep all valvetrain components in order for assembly.

- Rocker arms and pushrods
- Cylinder head
- Hydraulic valve tappets
- Accessory drive belt
- Crankshaft damper
- Front cover
- Timing chain and gears
- Thrust plate
- Camshaft

To install:

5. Lubricate the camshaft with clean engine oil.
6. Install or connect the following:
   - Camshaft
   - Thrust plate. Torque the bolts to 18 ft. lbs. (24 Nm).
   - Timing chain and gears
   - Front cover
   - Crankshaft damper
   - Accessory drive belt
   - Hydraulic valve tappets
   - Cylinder head
   - Rocker arms and pushrods
   - Valve cover
   - Distributor or camshaft sensor housing
   - A/C condenser, if equipped

- Radiator
- Grille, if removed
- Negative battery cable

7. Fill and bleed the cooling system.
8. Recharge the A/C system, if equipped.
9. Start the engine, check for leaks and repair if necessary.

3.7L Engine

LIBERTY

1. Before servicing the vehicle, refer to the precautions in the beginning of this section.
2. Remove or disconnect the following:
   - Negative battery cable
   - Cylinder head covers
   - Rocker arms
   - Hydraulic lash adjusters

➡Keep all valvetrain components in order for assembly.

3. Set the engine at Top Dead Center (TDC) of the compression stroke for the No. 1 cylinder.
4. Install Timing Chain Wedge 8379 to retain the chain tensioners.
5. Matchmark the timing chains to the camshaft sprockets.
6. Install Camshaft Holding Tool 8428 to the left camshaft sprocket.
7. Remove or disconnect the following:
   - Right camshaft timing sprocket and target wheel
   - Left camshaft sprocket
   - Camshaft bearing caps, by reversing the tightening sequence
   - Camshafts

To install:

8. Install or connect the following:
   - Camshafts. Torque the bearing cap bolts in ½ turn increments, in sequence, to 100 inch lbs. (11 Nm).
   - Target wheel to the right camshaft
   - Camshaft timing sprockets and chains, by aligning the matchmarks
9. Remove the tensioner wedges and tighten the camshaft sprocket bolts to 90 ft. lbs. (122 Nm).
10. Install or connect the following:
    - Hydraulic lash adjusters in their original locations
    - Rocker arms in their original locations
    - Cylinder head covers
    - Negative battery cable

GRAND CHEROKEE—LEFT SIDE

1. Before servicing the vehicle, refer to the precautions in the beginning of this section.

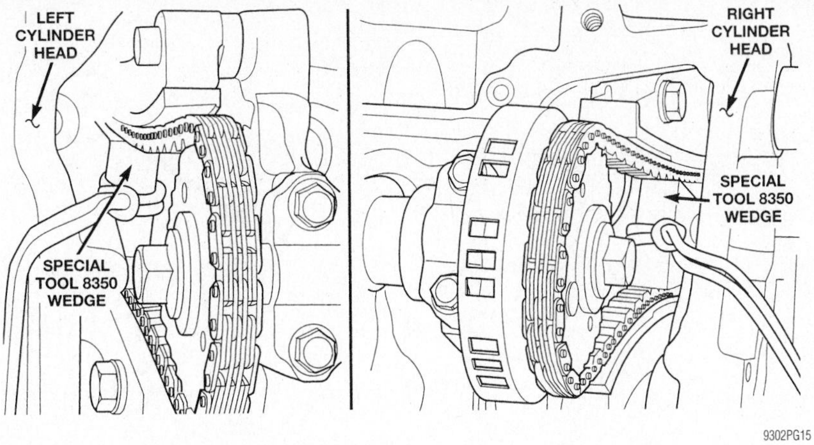

Chain Tensioner Retaining Wedges—3.7L Engine

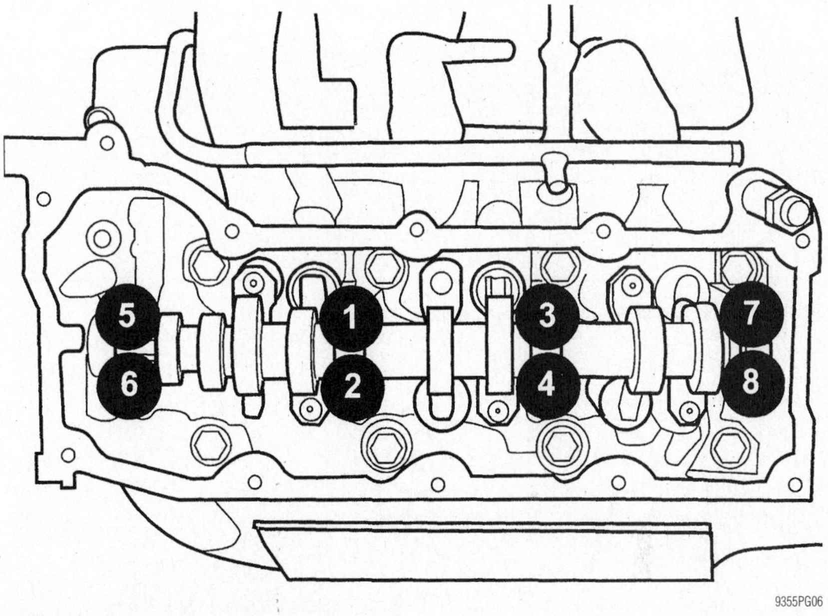

Camshaft bearing cap bolt tightening sequence—3.7L

⁂ WARNING

When the timing chain is removed and the cylinder heads are still installed, DO NOT forcefully rotate the camshafts or crankshaft independently of each other. Severe valve and/or piston damage can occur.

⁂ WARNING

When removing the cam sprocket, timing chains or camshaft, Failure to use Special Tool 8379 will result in hydraulic tensioner ratchet over extension, requiring timing chain cover removal to reset the tensioner ratchet.

2. Remove cylinder head cover.
3. Set engine to TDC cylinder No. 1, camshaft sprocket V6 marks at the 12 o'clock position.
4. Mark one link on the secondary timing chain on both sides of the V6 mark on the camshaft sprocket to aid in installation.

⁂ WARNING

Do not hold or pry on the camshaft target wheel (located on the right side camshaft sprocket) for any reason, severe damage will occur to the target wheel resulting in a vehicle no start condition.

5. Loosen but do not remove the camshaft sprocket retaining bolt. Leave the bolt snug against the sprocket.

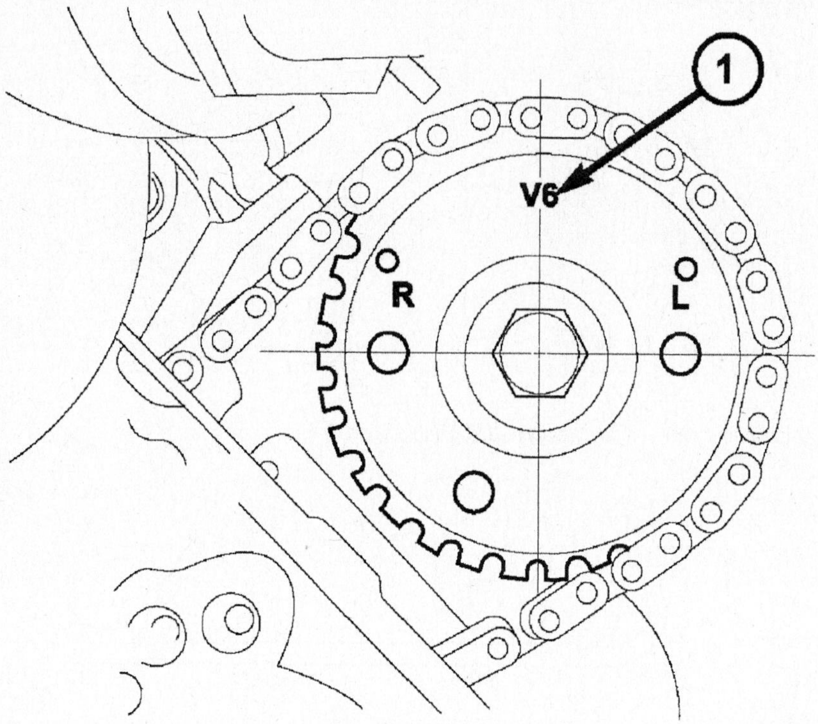

V6 mark at the 12 o'clock position—3.7L engine

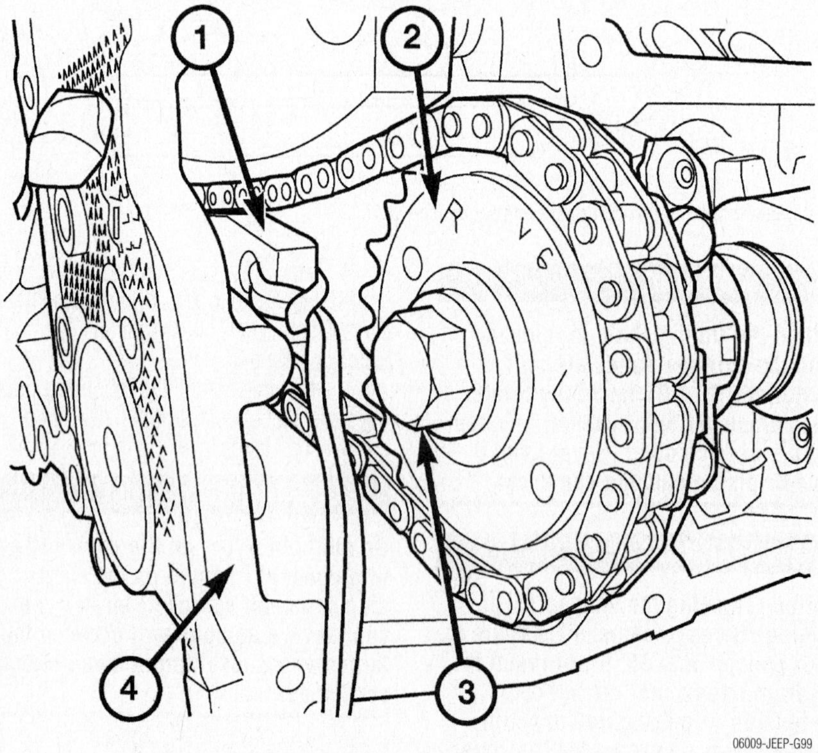

Position Special Tool 8379 timing chain wedge between the timing chain strands—3.7L engine

✲✲ WARNING

The timing chain tensioners must be secured prior to removing the camshaft sprockets. Failure to secure tensioners will allow the tensioners to extend, requiring timing chain cover removal in order to reset tensioners.

✲✲ WARNING

Do not force wedge past the narrowest point between the chain strands. Damage to the tensioners may occur.

6. Position Special Tool 8379 timing chain wedge between the timing chain strands, tap the tool to securely wedge the timing chain against the tensioner arm and guide.
7. Hold the camshaft with Special Tool 8428 Camshaft Wrench, or equivalent, while removing the camshaft sprocket bolt and sprocket.
8. Using Special Tool 8428 Camshaft Wrench, gently allow the camshaft to rotate 5° clockwise until the camshaft is in the neutral position (no valve load).
9. Starting at the outside working inward, loosen the camshaft bearing cap retaining bolts ½ turn at a time. Repeat until all load is off the bearing caps.

✲✲ WARNING

Do not stamp or strike the camshaft bearing caps. Severe damage will occur to the bearing caps.

➡When the camshaft is removed the rocker arms may slide downward, mark the rocker arms before removing camshaft.

10. Remove the camshaft bearing caps and the camshaft.

To install:

11. Lubricate camshaft journals with clean engine oil.

➡Position the left side camshaft so that the camshaft sprocket dowel is near the 1 o'clock position. This will place the camshaft at the neutral position easing the installation of the camshaft bearing caps.

12. Position the camshaft into the cylinder head.
13. Install the camshaft bearing caps, hand tighten the retaining bolts.

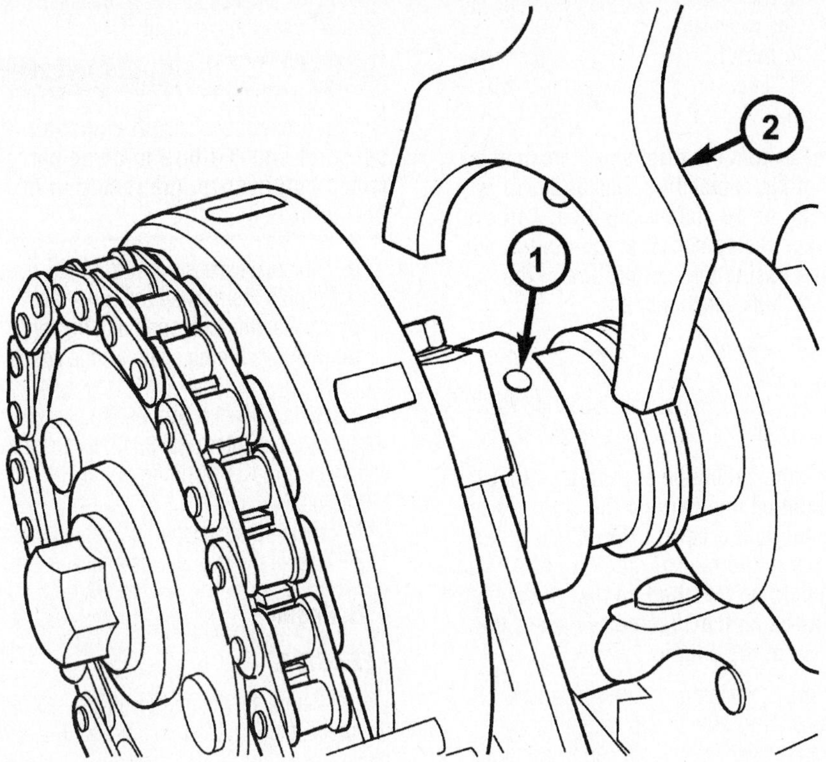

Hold the camshaft with Special Tool 8428 Camshaft Wrench, or equivalent—3.7L engine

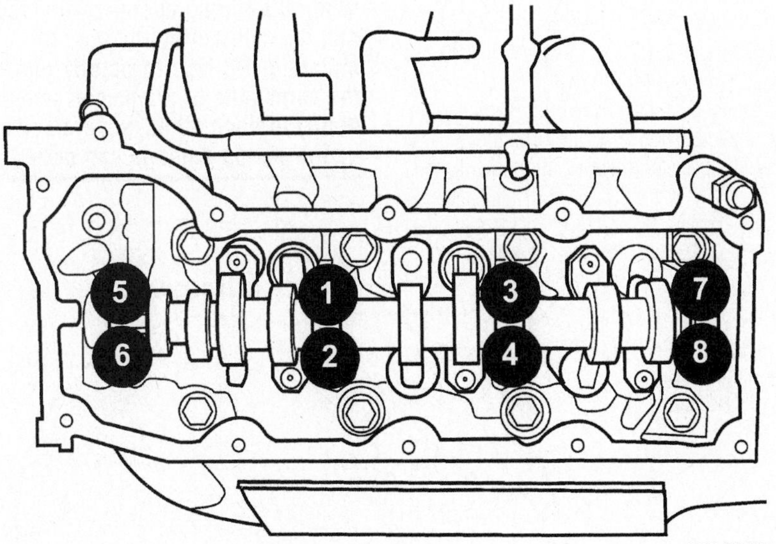

Camshaft bearing cap torque sequence—3.7L engine

➡Caps should be installed so that the stamped numbers on the caps are in numerical order, (1 thru 4) from the front to the rear of the engine. All caps should be installed so that the stamped arrows on the caps point toward the front of the engine.

14. Working in ½ turn increments, tighten the bearing cap retaining bolts starting with the middle cap working outward.

15. Torque the camshaft bearing cap retaining bolts to 11 Nm (100 inch lbs.).

16. Position the camshaft drive gear into the timing chain aligning the V6 mark between the two marked chain links (two links marked during removal).

17. Using Tool 8428 Camshaft Wrench, rotate the camshaft until the camshaft sprocket dowel is aligned with the slot in the camshaft sprocket. Install the sprocket onto the camshaft.

✻✻ WARNING

Remove excess oil from camshaft sprocket bolt. Failure to do so can cause bolt over-torque resulting in bolt failure.

18. Remove excess oil from bolt, then install the camshaft sprocket retaining bolt and hand tighten.

19. Remove Special Tool 8379 timing chain wedge.

20. Using Special Tool 6958 spanner wrench with adapter pins 8346, torque the camshaft sprocket retaining bolt to 122 Nm (90 ft. lbs.).

21. Install the cylinder head cover.

GRAND CHEROKEE—RIGHT SIDE

1. Before servicing the vehicle, refer to the precautions in the beginning of this section.

✻✻ WARNING

When the timing chain is removed and the cylinder heads are still installed, do not forcefully rotate the camshafts or crankshaft independently of each other. Severe valve and/or piston damage can occur.

✻✻ WARNING

When removing the cam sprocket, timing chains or camshaft, Failure to use special tool 8379 will result in hydraulic tensioner ratchet over extension, Requiring timing chain cover removal to re-set the tensioner ratchet.

2. Remove the cylinder head cover.

3. Set engine to TDC cylinder No. 1, camshaft sprocket V6 marks at the 12 o'clock position.

4. Mark one link on the secondary timing chain on both sides of the V6 mark on the camshaft sprocket to aid in installation.

✻✻ WARNING

Do not hold or pry on the camshaft target wheel for any reason, Severe damage will occur to the target wheel. A damaged target wheel could cause a vehicle no start condition.

5. Loosen but do not remove the camshaft sprocket retaining bolt. Leave bolt snug against sprocket.

➡The timing chain tensioners must be secured prior to removing the camshaft sprockets. Failure to secure tensioners will allow the tensioners to extend, requiring timing chain cover removal in order to reset tensioners.

⁂ WARNING

Do not force wedge past the narrowest point between the chain strands. Damage to the tensioners may occur.

6. Position Special Tool 8379 timing chain wedge between the timing chain strands. Tap the tool to securely wedge the timing chain against the tensioner arm and guide.
7. Remove the camshaft position sensor.
8. Hold the camshaft with Special Tool 8428 Camshaft Wrench, while removing the camshaft sprocket bolt and sprocket.
9. Starting at the outside working inward, loosen the camshaft bearing cap retaining bolts ½ turn at a time. Repeat until all load is off the bearing caps.

⁂ WARNING

Do not stamp or strike the camshaft bearing caps. Severe damage will occur to the bearing caps.

➡When the camshaft is removed the rocker arms may slide downward, mark the rocker arms before removing camshaft.

10. Remove the camshaft bearing caps and the camshaft.

To install:

11. Lubricate camshaft journals with clean engine oil.

➡Position the right side camshaft so that the camshaft sprocket dowel is near the 10 o'clock position. This will place the camshaft at the neutral position easing the installation of the camshaft bearing caps.

12. Position the camshaft into the cylinder head.
13. Install the camshaft bearing caps, hand tighten the retaining bolts.

➡Caps should be installed so that the stamped numbers on the caps are in numerical order, (1 thru 4) from the front to the rear of the engine. All caps should be installed so that the stamped arrows on the caps point toward the front of the engine.

14. Working in ½ turn increments, tighten the bearing cap retaining bolts starting with the middle cap working outward.
15. Torque the camshaft bearing cap retaining bolts to 11 Nm (100 inch lbs.).
16. Position the camshaft drive gear into the timing chain aligning the V6 mark between the two marked chain links (two links marked during removal).
17. Using Special Tool 8428 Camshaft Wrench, rotate the camshaft until the camshaft sprocket dowel is aligned with the slot in the camshaft sprocket. Install the sprocket onto the camshaft.

⁂ WARNING

Remove excess oil from camshaft sprocket bolt. Failure to do so can cause bolt over-torque resulting in bolt failure.

18. Remove excess oil from camshaft sprocket bolt, then install the camshaft sprocket retaining bolt and hand tighten.
19. Remove timing chain wedge special tool 8379.
20. Using Special Tool 6958 spanner wrench with adapter pins 8346, torque the camshaft sprocket retaining bolt to 122 Nm (90 ft. lbs.).
21. Install the camshaft position sensor.
22. Install the cylinder head cover.

4.7L Engine

LEFT SIDE

1. Before servicing the vehicle, refer to the precautions in the beginning of this section.
2. Remove cylinder head cover.

⁂ WARNING

When the timing chain is removed and the cylinder heads are still installed, DO NOT forcefully rotate the camshafts or crankshaft independently of each other. Severe valve and/or piston damage can occur.

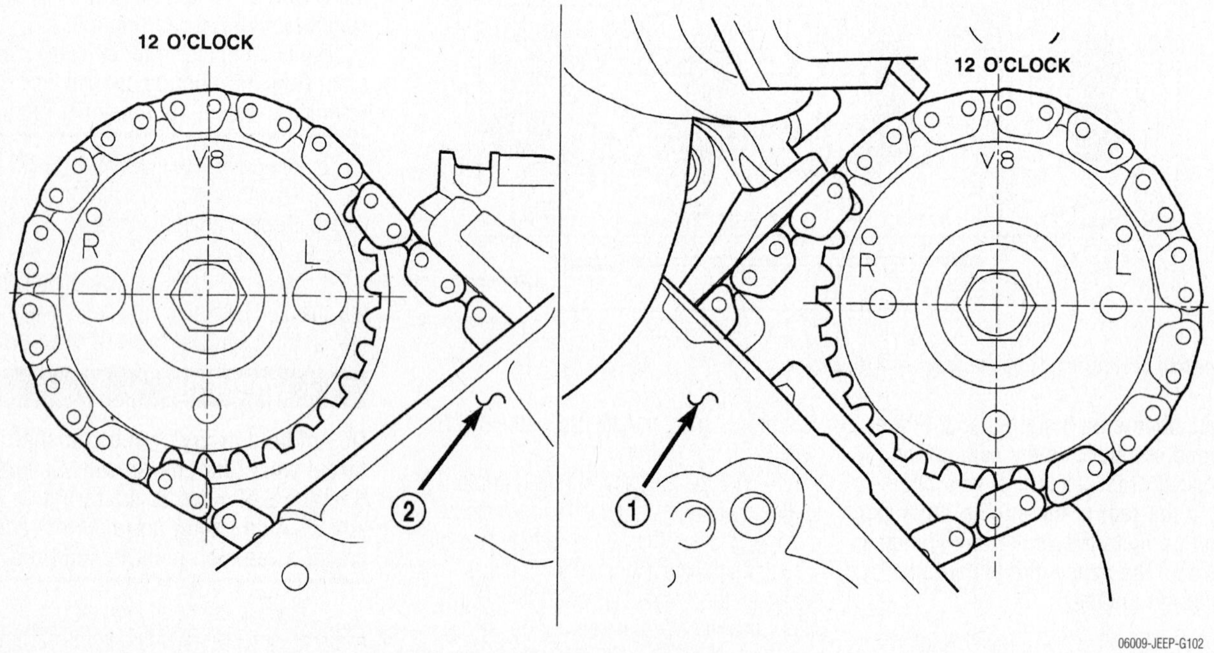

Set engine to TDC cylinder no.1, camshaft sprocket V8 marks at the 12 o'clock position—4.7L engine

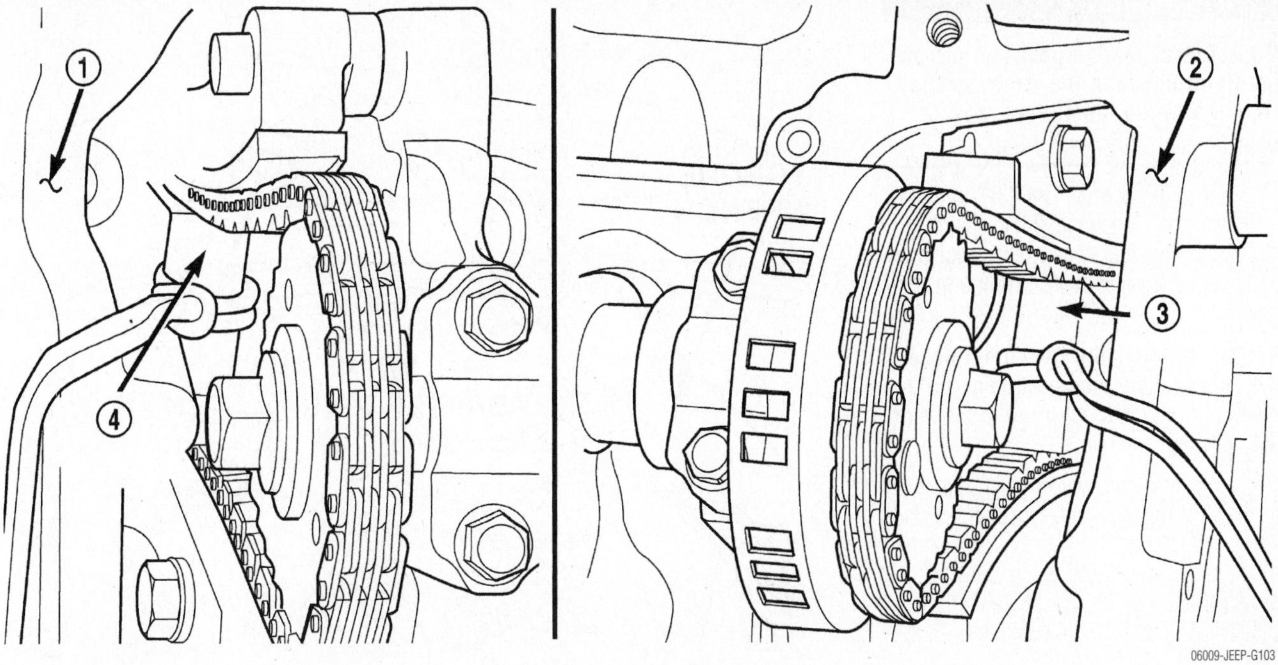

The timing chain tensioners must be secured prior to removing the camshaft sprockets—4.7L engine

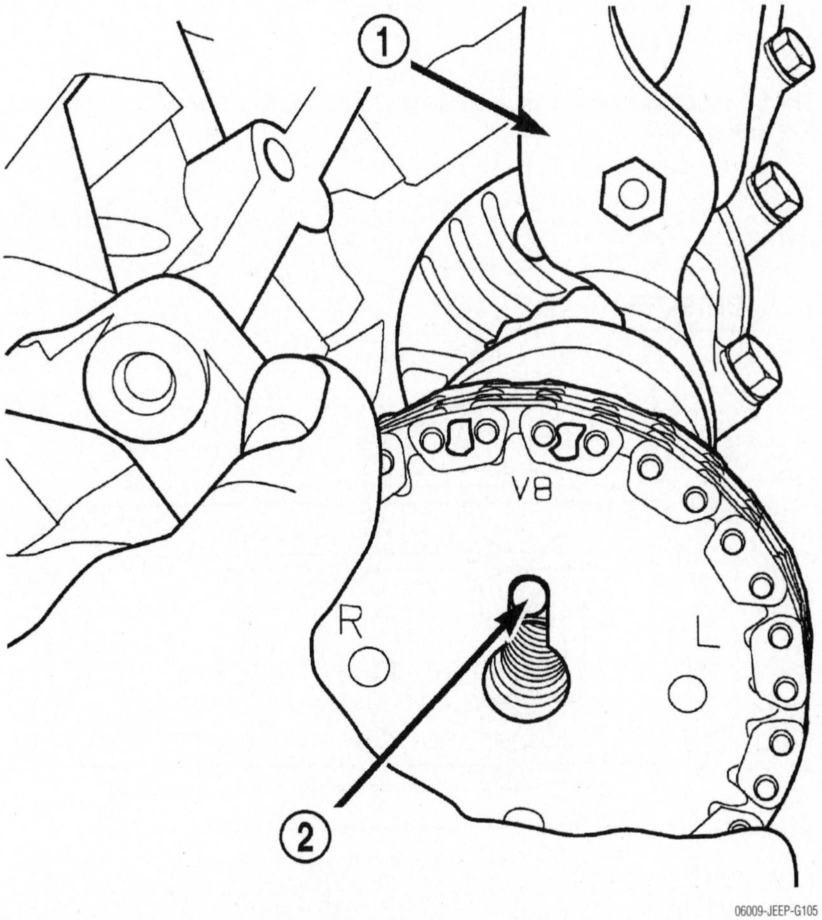

Hold the camshaft with adjustable pliers (1). (2) is the camshaft dowel—4.7L engine

**** WARNING**

When removing the cam sprocket, timing chains or camshaft, failure to use Special Tool 8350 will result in hydraulic tensioner ratchet over extension, requiring timing chain cover removal to reset the tensioner ratchet.

3. Set engine to TDC cylinder no.1, camshaft sprocket V8 marks at the 12 o'clock position.
4. Mark one link on the secondary timing chain on both sides of the V8 mark on the camshaft sprocket to aid in installation.

**** WARNING**

Do not hold or pry on the camshaft target wheel (located on the right side camshaft sprocket) for any reason, severe damage will occur to the target wheel resulting in a vehicle no start condition.

5. Loosen but do not remove the camshaft sprocket retaining bolt. Leave the bolt snug against the sprocket.

➡The timing chain tensioners must be secured prior to removing the camshaft sprockets. Failure to secure tensioners will allow the tensioners to extend, requiring timing chain cover removal in order to reset tensioners.

**** WARNING**

Do not force wedge past the narrowest point between the chain strands. Damage to the tensioners may occur.

6. Position Special Tool 8350 timing chain wedge between the timing chain strands, tap the tool to securely wedge the timing chain against the tensioner arm and guide.

➡When gripping the camshaft, place the pliers on the tube portion of the camshaft only. Do not grip the lobes or the sprocket areas.

7. Hold the camshaft with adjustable pliers while removing the camshaft sprocket bolt and sprocket.
8. Using the pliers, gently allow the camshaft to rotate 15° clockwise until the camshaft is in the neutral position (no valve load).
9. Starting at the outside working inward, loosen the camshaft bearing cap retaining bolts ½ turn at a time. Repeat until all load is off the bearing caps.

**** WARNING**

Do not stamp or strike the camshaft bearing caps. Severe damage will occur to the bearing caps.

➡When the camshaft is removed the rocker arms may slide downward, mark the rocker arms before removing camshaft.

10. Remove the camshaft bearing caps and the camshaft.

To install:

11. Lubricate camshaft journals with clean engine oil.

➡Position the left side camshaft so that the camshaft sprocket dowel is near the 1 o'clock position. This will place the camshaft at the neutral position easing the installation of the camshaft bearing caps.

12. Position the camshaft into the cylinder head.
13. Install the camshaft bearing caps, hand tighten the retaining bolts.
14. Working in ½ turn increments, tighten the bearing cap retaining bolts starting with the middle cap working outward.
15. Torque the camshaft bearing cap retaining bolts to 11 Nm (100 inch lbs.).

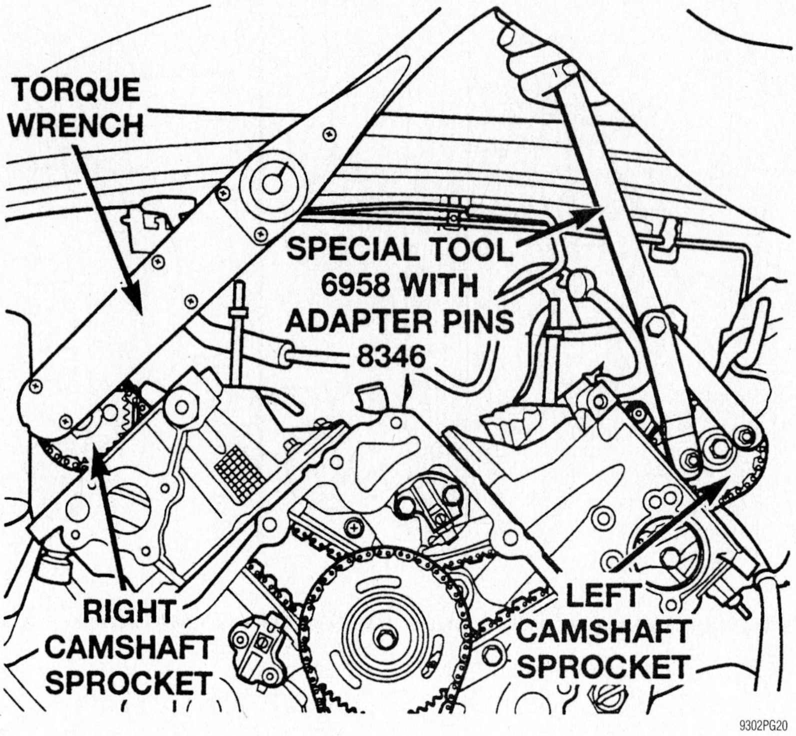

Hold the left camshaft sprocket with a spanner wrench while removing he camshaft sprocket—4.7L engine

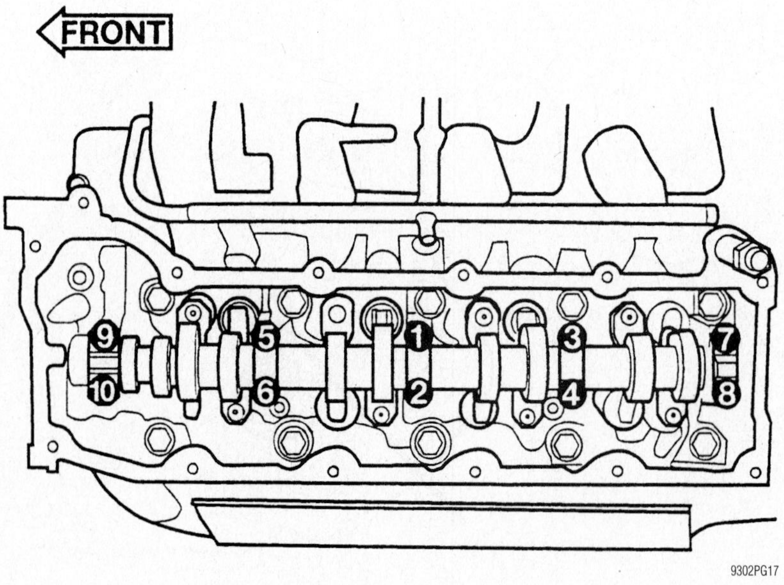

Camshaft bearing cap bolt tightening sequence—4.7L engine

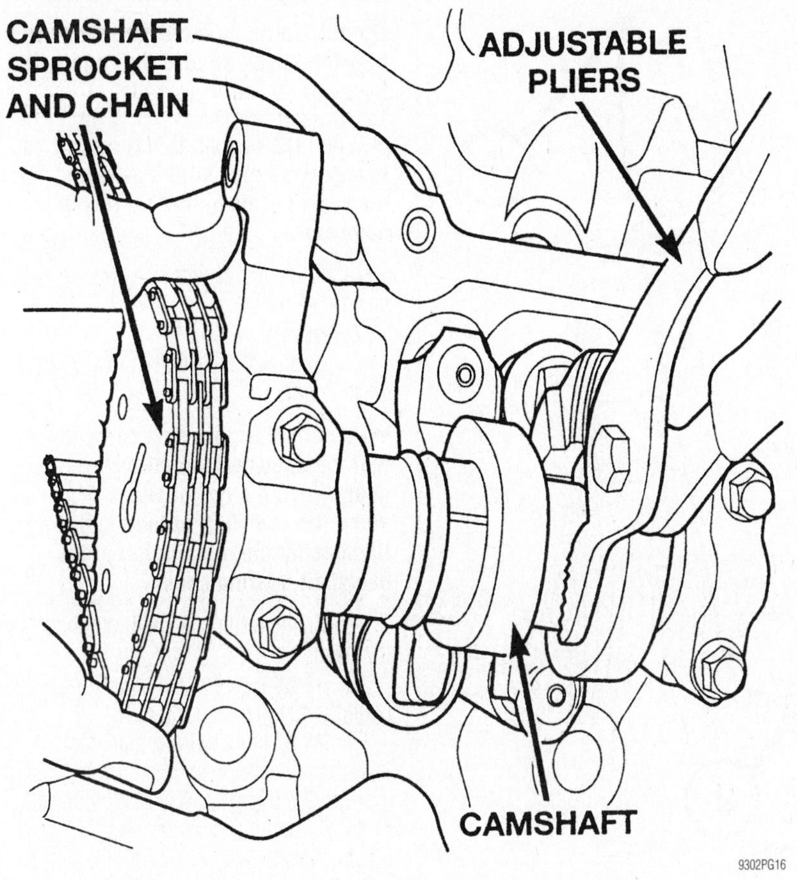

Turn the camshaft with pliers, if needed, to align the dowel in the sprocket—4.7L engine

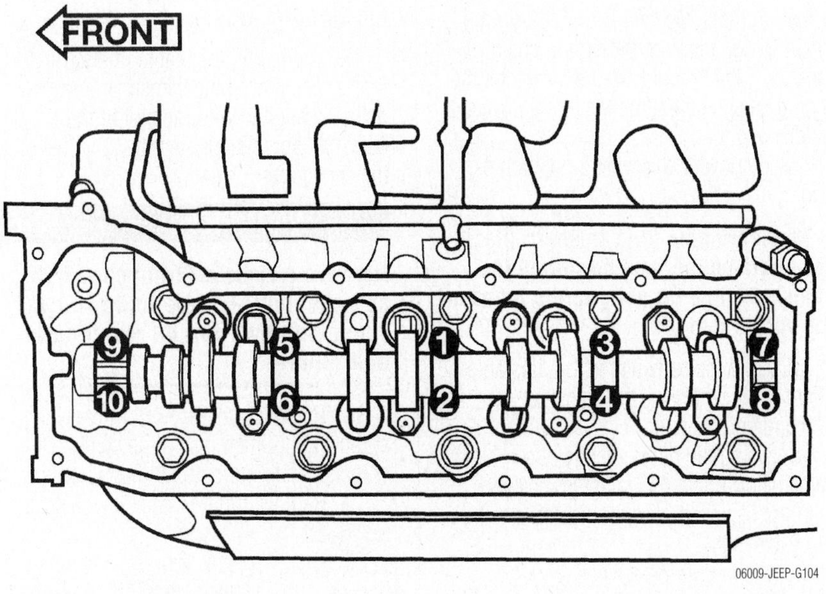

Camshaft tightening sequence—4.7L engine

16. Position the camshaft drive gear into the timing chain aligning the V8 mark between the two marked chain links (two links marked during removal).

➡When gripping the camshaft, place the pliers on the tube portion of the camshaft only. Do not grip the lobes or the sprocket areas.

17. Using the adjustable pliers, rotate the camshaft until the camshaft sprocket dowel is aligned with the slot in the camshaft sprocket. Install the sprocket onto the camshaft.

WARNING

Remove excess oil from camshaft sprocket bolt. Failure to do so can cause bolt over-torque resulting in bolt failure.

18. Remove excess oil from bolt, then install the camshaft sprocket retaining bolt and hand tighten.
19. Remove Special Tool 8350 timing chain wedge.
20. Using Special Tool 6958 spanner wrench with adapter pins 8346, torque the camshaft sprocket retaining bolt to 122 Nm (90 ft. lbs.).
21. Install the cylinder head cover.

RIGHT SIDE

1. Before servicing the vehicle, refer to the precautions in the beginning of this section.

WARNING

When the timing chain is removed and the cylinder heads are still installed, do not forcefully rotate the camshafts or crankshaft independently of each other. Severe valve and/or piston damage can occur.

WARNING

When removing the cam sprocket, timing chains or camshaft, failure to use special tool 8350 will result in hydraulic tensioner ratchet over extension, requiring timing chain cover removal to reset the tensioner ratchet.

2. Set engine to TDC cylinder no.1, camshaft sprocket V8 marks at the 12 o'clock position.
3. Mark one link on the secondary timing chain on both sides of the V8 mark

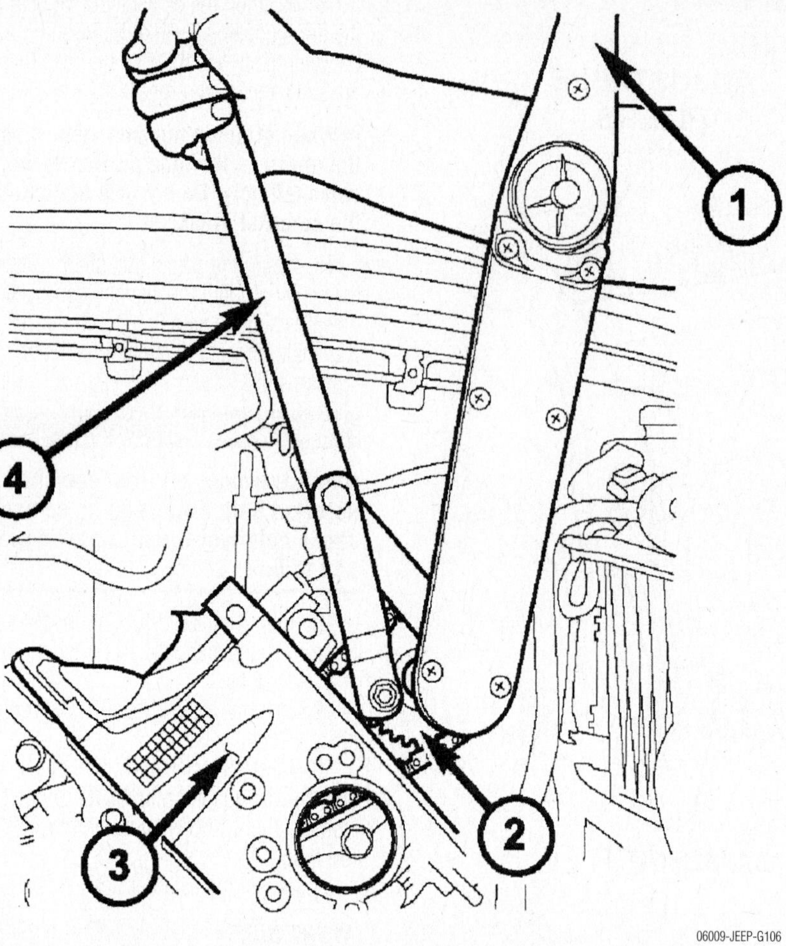

Using Special Tool 6958 spanner wrench (4) with adapter pins 8346, torque the camshaft sprocket retaining bolt (2). (1) is the torque wrench, (3) is the head—4.7L engine

on the camshaft sprocket to aid in installation.

⁂ WARNING

Do not hold or pry on the camshaft target wheel for any reason, severe damage will occur to the target wheel. A damaged target wheel could cause a vehicle no start condition.

4. Loosen but do not remove the camshaft sprocket retaining bolt. Leave bolt snug against sprocket.

➡The timing chain tensioners must be secured prior to removing the camshaft sprockets. Failure to secure tensioners will allow the tensioners to extend, requiring timing chain cover removal in order to reset tensioners.

⁂ WARNING

Do not force wedge past the narrowest point between the chain strands. Damage to the tensioners may occur.

5. Position Special Tool 8350 timing chain wedge between the timing chain strands. Tap the tool to securely wedge the timing chain against the tensioner arm and guide.

6. Remove the camshaft position sensor.

➡When gripping the camshaft, place the pliers on the tube portion of the camshaft only. Do not grip the lobes or the sprocket areas.

7. Hold the camshaft with adjustable pliers while removing the camshaft sprocket bolt and sprocket.

8. Using the pliers, gently allow the camshaft to rotate 45° counterclockwise until the camshaft is in the neutral position (no valve load).

9. Starting at the outside working inward, loosen the camshaft bearing cap retaining bolts ½ turn at a time. Repeat until all load is off the bearing caps.

⁂ WARNING

Do not stamp or strike the camshaft bearing caps. Severe damage will occur to the bearing caps.

➡When the camshaft is removed the rocker arms may slide downward, mark the rocker arms before removing camshaft.

10. Remove the camshaft bearing caps and the camshaft.

To install:

11. Lubricate camshaft journals with clean engine oil.

➡Position the right side camshaft so that the camshaft sprocket dowel is near the 10 o'clock position. This will place the camshaft at the neutral position easing the installation of the camshaft bearing caps.

12. Position the camshaft into the cylinder head.

13. Install the camshaft bearing caps, hand tighten the retaining bolts.

14. Working in ½ turn increments, tighten the bearing cap retaining bolts starting with the middle cap working outward.

15. Torque the camshaft bearing cap retaining bolts to 11 Nm (100 inch lbs.).

16. Position the camshaft drive gear into the timing chain aligning the V8 mark between the two marked chain links (two links marked during removal).

➡When gripping the camshaft, place the pliers on the tube portion of the camshaft only. Do not grip the lobes or the sprocket areas.

17. Using the adjustable pliers, rotate the camshaft until the camshaft sprocket dowel is aligned with the slot in the camshaft sprocket. Install the sprocket onto the camshaft.

⁂ WARNING

Remove excess oil from camshaft sprocket bolt. Failure to do so can cause bolt over-torque resulting in bolt failure.

18. Remove excess oil from camshaft sprocket bolt, then install the camshaft sprocket retaining bolt and hand tighten.

19. Remove timing chain wedge special tool 8350.

20. Using Special Tool 6958 spanner wrench with adapter pins 8346, torque the camshaft sprocket retaining bolt to 122 Nm (90 ft. lbs.).

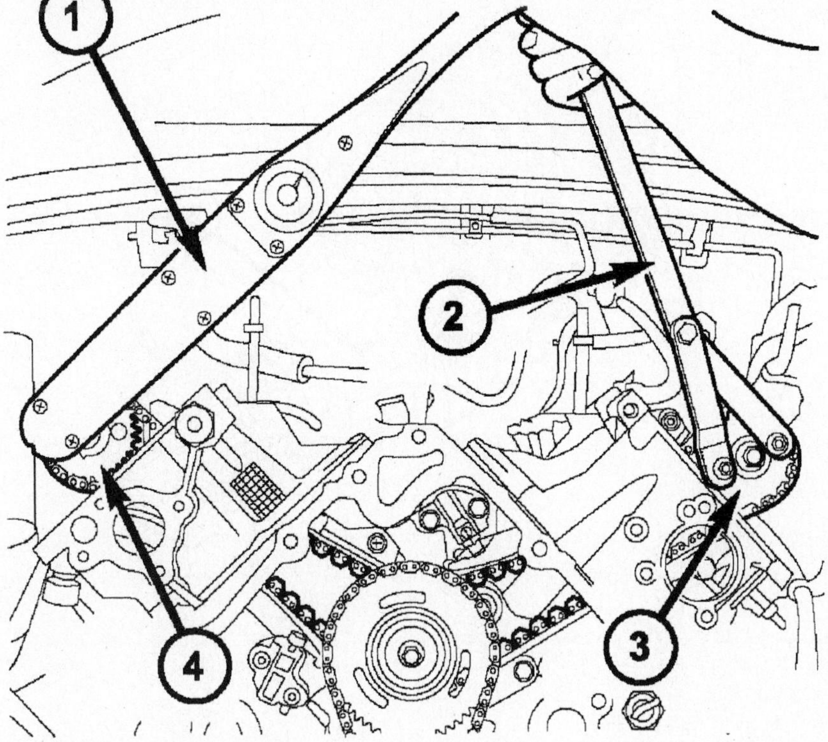

06009-JEEP-G107

Tightening the right side camshaft sprocket retaining bolt—4.7L engine

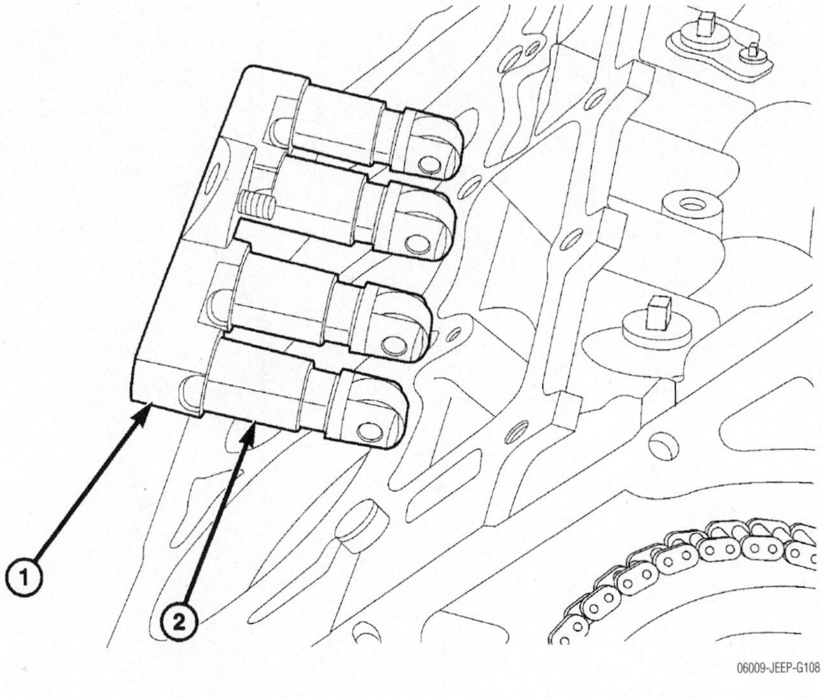

06009-JEEP-G108

Lifter set—5.7L engine

21. Install the camshaft position sensor.
22. Install the cylinder head cover.

5.7L Engine

1. Before servicing the vehicle, refer to the precautions in the beginning of this section.
2. Remove the battery negative cable.
3. Remove the air cleaner assembly.
4. Drain coolant.
5. Remove the accessory drive belt.
6. Remove the generator.
7. Remove the A/C compressor, and set aside.
8. Remove the radiator.
9. Remove intake manifold.
10. Remove cylinder head covers.
11. Remove both left and right cylinder heads.
12. Remove the oil pan.
13. Remove timing case cover.
14. Remove the oil pick up tube.
15. Remove the oil pump.
16. Remove timing chain.
17. Remove camshaft tensioner/thrust plate assembly.

➡Identify lifters to ensure installation in original location.

18. Remove the tappets and retainer assembly.
19. Install a long bolt into front of camshaft to aid in removal of the camshaft. Remove camshaft, being careful not to damage cam bearings with the cam lobes.

To install:

➡The 5.7L LX engine uses a unique camshaft for use with the Multi Displacement System. When installing a new camshaft, the replacement camshaft must be compatible with the Multi Displacement System.

20. Lubricate camshaft lobes and camshaft bearing journals and insert the camshaft.
21. Install camshaft tensioner plate assembly. Tighten bolts to 28 Nm (250 inch lbs.) torque.
22. Install timing chain and sprockets.
23. Measure camshaft end play. If not within limits install a new thrust plate.
24. Install the oil pump.
25. Install the oil pick-up tube.
26. Each tappet reused must be installed in the same position from which it was removed.

➡When camshaft is replaced, all of the tappets must be replaced.

⁂ WARNING

The 5.7L LX engine uses both standard roller tappets and deactivating roller tappets, for use with the Multi Displacement System. The deactivating roller tappets must be used in cylinders 1, 4, 6, 7. The deactivating tappets can be identified by the two holes in the side of the tappet body, for the latching pins.

27. Install tappets and retaining yoke assembly.
28. Install both left and right cylinder heads.
29. Install pushrods.
30. Install rocker arms.
31. Install timing case cover.
32. Install the oil pan.
33. Install cylinder head covers.
34. Install intake manifold.
35. Install the A/C compressor.
36. Install the generator.
37. Install the accessory drive belt.
38. Install the radiator.
39. Install the air cleaner assembly.
40. Install the battery negative cable.
41. Refill coolant.
42. Refill engine oil.
43. Start engine and check for leaks.

Valve Lash

ADJUSTMENT

These engines are equipped with hydraulic valve lifters. No valve clearance adjustments are necessary.

Starter Motor

REMOVAL & INSTALLATION

1. Before servicing the vehicle, refer to the precautions in the beginning of this section.
2. Remove or disconnect the following:
   - Negative battery cable
   - Starter mounting bolts

➡On the 3.7L Liberty engine, the left side exhaust pipe and front driveshaft must be disconnected.

   - Starter solenoid harness connections
   - Starter

To install:

3. Connect the starter solenoid wiring connectors.
4. Install the starter and torque the bolts to the following specifications:

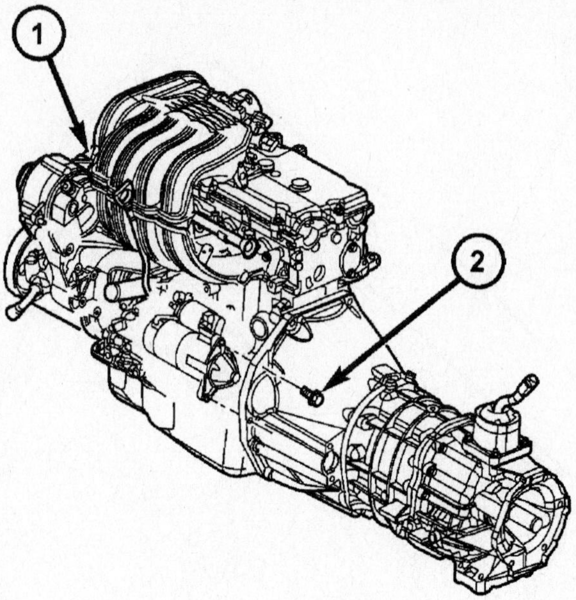

Starter mounting—2.4L engine

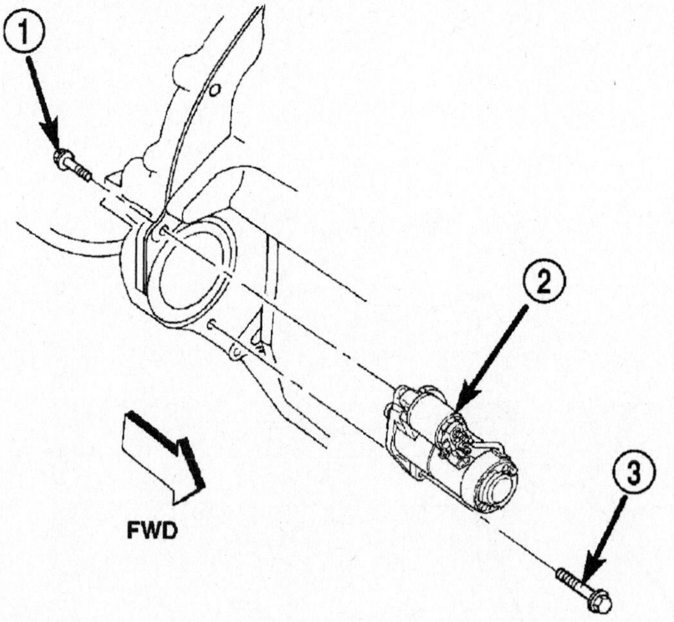

Starter mounting—4.0L engine w/manual transmission

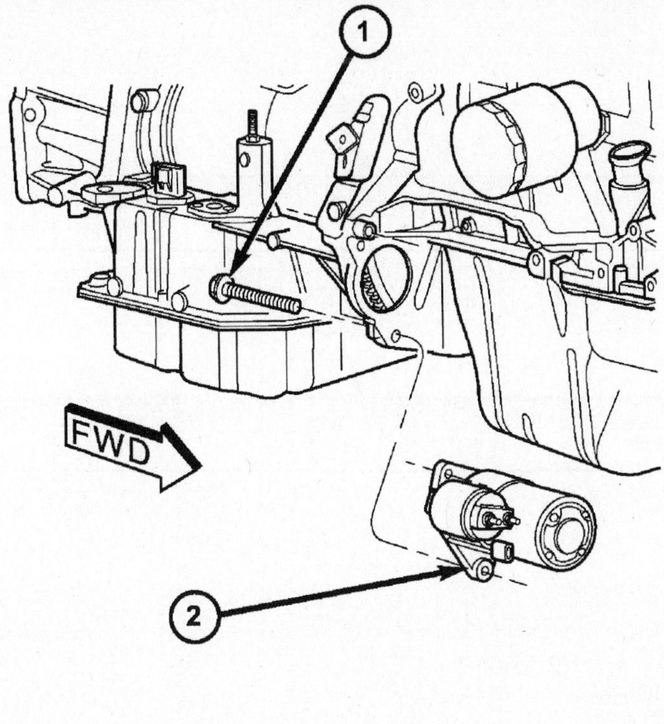

Starter mounting—4.0L engine w/42RLE automatic transmission

- 2.5L engine: 33 ft. lbs. (45 Nm).
- 4.0L engine: Upper bolt to 40 ft. lbs. (54 Nm) for 2002–04; 35 ft. lbs. (48 Nm) for 2005. Torque the lower bolt to 30 ft. lbs. (41 Nm).
- 2.4L, 3.7L and 4.7L engine: 40 ft. lbs. (54 Nm).
- 5.7L engine: 50 ft. lbs. (68 Nm).
- On the 3.7L Liberty engine, the left side exhaust pipe and front driveshaft

5. Install the negative battery cable and check for proper operation.

Oil Pan

REMOVAL & INSTALLATION

2.4L Engine

1. Before servicing the vehicle, refer to the precautions in the beginning of this section.
2. Remove air cleaner assembly.
3. Raise the vehicle on a hoist and drain the engine oil.
4. Loosen the engine mount through bolts.
5. Disconnect the exhaust pipe at the manifold.
6. Remove the structural collar, if equipped.
7. Remove the front axle mounting bolts, and lower the axle as far possible, if equipped.
8. Position Special Tool 8534, or equivalent, on the fender lip and align the slots in the brackets with the fender mounting holes.
9. Secure the brackets to the fender using four M6 X 1.0 X 25 mm flanged cap screws.
10. Tighten the thumbscrews to secure the sleeves to the support tube.
11. Secure the support tube in an upright position.
12. Assemble the flat washer, thrust bearing, hook and T handle.
13. Using the M10 X 1.5 X 40 mm capscrew supplied with the support fixture, secure the chain to the front cover and the hook.
14. Support the engine as needed.
15. Remove the oil pan attaching bolts.
16. Remove the oil pan.
17. Clean the oil pan and all gasket surfaces.

To install:

18. Install the oil pan gasket to the block.

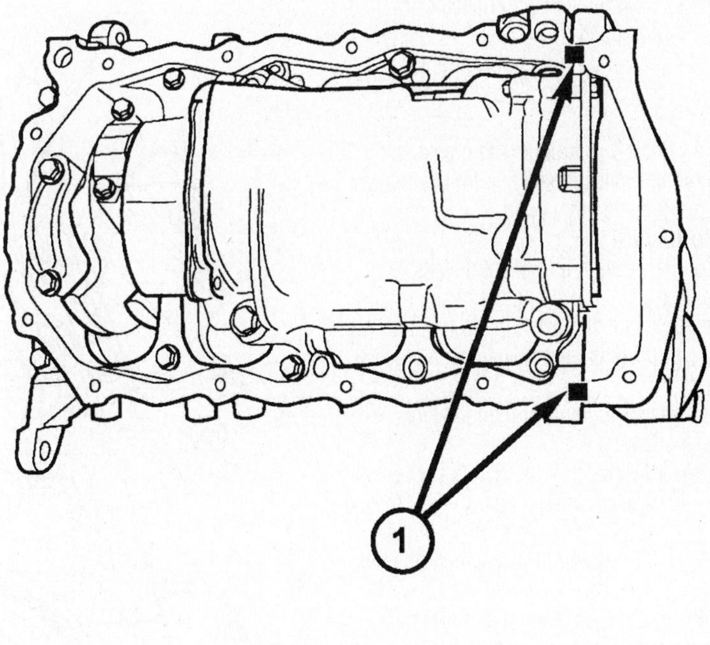

Oil pan gasket positioning—2.4L engines

19. Apply a 3mm (1⁄8 inch) bead of Mopar® Engine RTV, or equivalent, at the oil pump to engine block parting line.

20. Install the pan and tighten the screws to 12 Nm (105 inch lbs.).

21. Lower the engine, and remove Special Tool 8534.

22. Tighten the engine mount through bolts.

23. Raise the front axle into position, and reinstall the front axle mounting bolts, if equipped.

24. Reconnect the exhaust pipe to the manifold.

25. Install the structural collar, if equipped.

26. Lower the vehicle.

27. Fill the engine crankcase with the proper oil to the correct level.

28. Reinstall the air cleaner assembly.

2.5L and 4.0L Engines

1. Before servicing the vehicle, refer to the precautions in the beginning of this section.
2. Drain the engine oil.
3. Remove or disconnect the following:
   - Negative battery cable
   - Exhaust front pipe
   - Starter motor
   - Bell housing access cover
   - Oil level sensor connector, if equipped
   - Left and right motor mounts
   - Transmission oil cooler lines, if equipped
4. Place a jack under the crankshaft damper and raise the engine for clearance.
5. Remove the oil pan.

To install:

6. Fabricate 4 alignment dowels from 1½ in. x ¼ in. bolts. Cut the heads off the bolts and cut a slot into the top of the dowel to allow installation/removal with a screwdriver.
7. Install or connect the following:
   - Dowels
   - RTV sealer at the points shown
   - Oil pan, using a new gasket. Torque the ¼ inch bolts to 85 inch lbs. (9.5 Nm) and the 5⁄16 inch bolts to 11 ft. lbs. (15 Nm).
8. Replace the alignment dowels with ¼ inch bolts and torque them to 85 inch lbs. (9.5 Nm).
9. Install or connect the following:
   - Left and right motor mounts
   - Oil level sensor connector, if equipped
   - Bell housing access cover
   - Starter motor

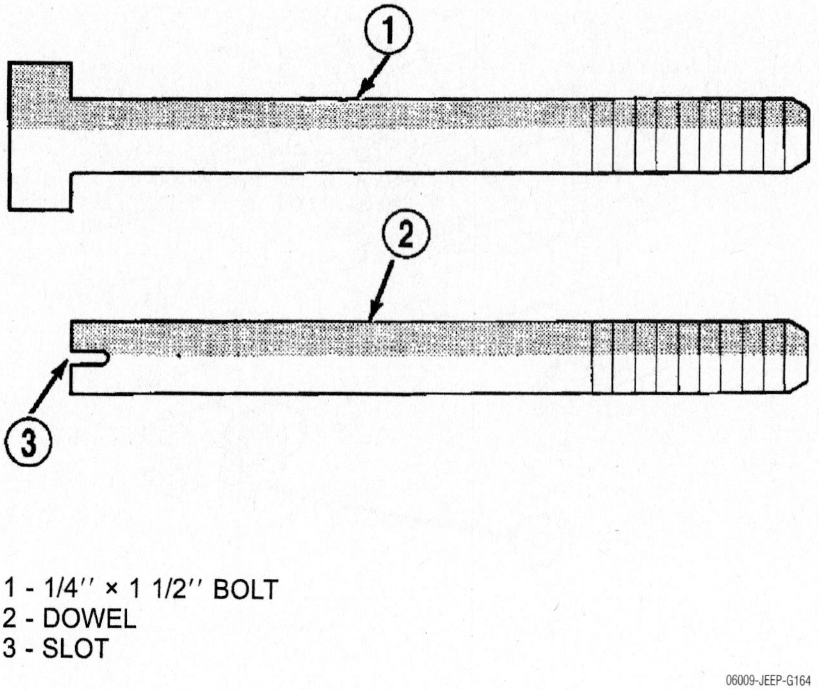

Fabrication of alignment dowels—4.0L engine

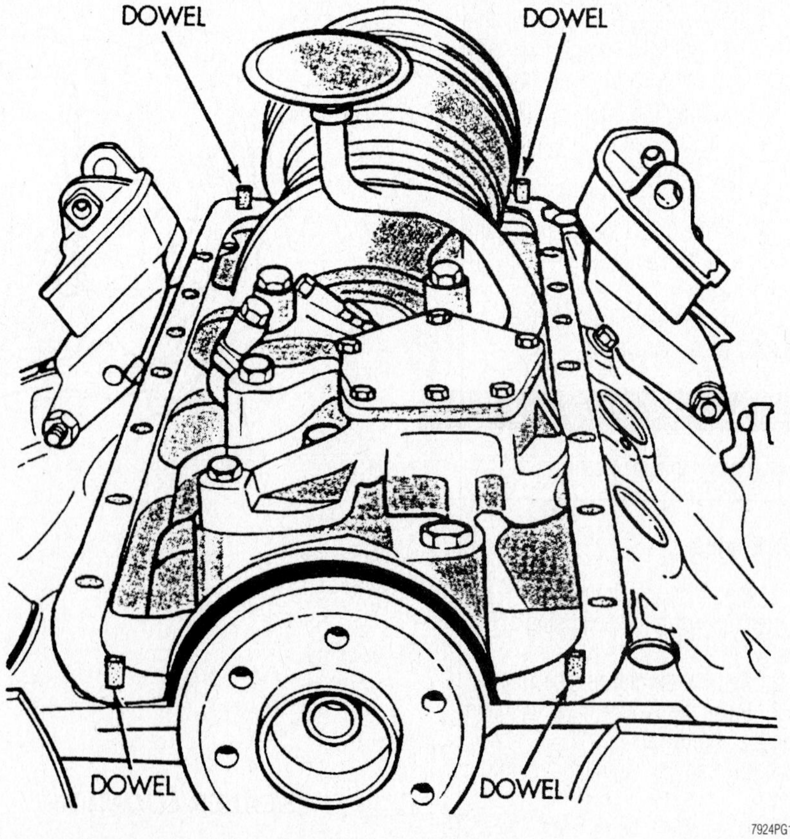

Oil pan alignment dowel placement—2.5L and 4.0L engines

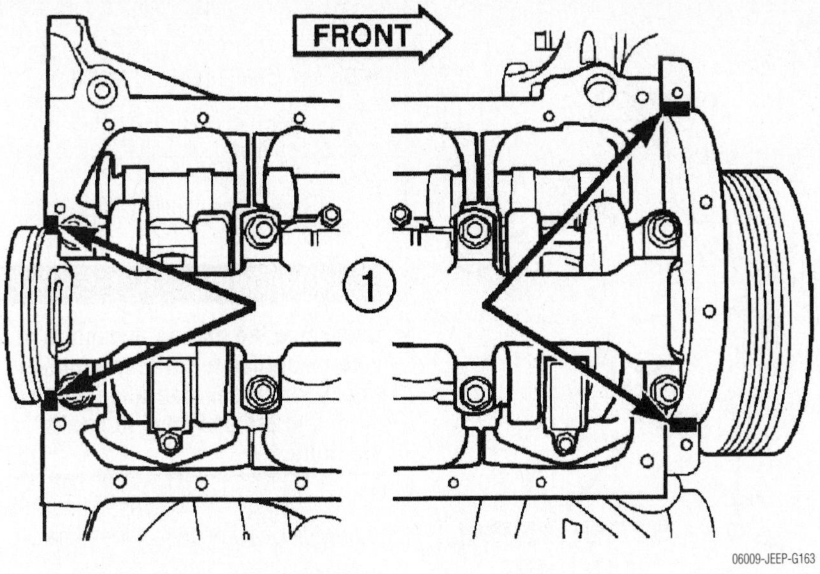

RTV sealer application points—4.0L engine

- Exhaust front pipe
- Negative battery cable

10. Fill the engine with clean oil.
11. Start the engine, check for leaks and repair if necessary.

3.7L Engine

LIBERTY

1. Before servicing the vehicle, refer to the precautions in the beginning of this section.
2. Remove or disconnect the following:
   - Engine from the vehicle
   - Oil pan
   - Oil pump pickup tube
   - Oil pan gasket
3. Installation is the reverse of removal. Torque the bolts, in sequence, to 11 ft. lbs. (15 Nm).

GRAND CHEROKEE

1. Before servicing the vehicle, refer to the precautions in the beginning of this section.
2. Disconnect negative battery cable.
3. Remove the radiator fan.
4. Remove the intake manifold.
5. Install engine lifting Tool 8534, or equivalent.

➡Do not raise engine at this time.

6. Remove the structural cover using sequence shown.
7. Remove both left and right side engine mount through bolts.
8. Raise engine to provide clearance to remove oil pan.
9. Drain engine oil and remove oil filter.

➡Do not pry on oil pan or oil pan gasket. Gasket is mounted to engine and does not come out with oil pan.

10. Remove the oil pan mounting bolts and oil pan.
11. Unbolt oil pump pickup tube and remove tube and oil pan gasket from engine.

To install:

12. Clean the oil pan gasket mating surface of the bedplate and oil pan.
13. Inspect integrated oil pan gasket, and replace as necessary.
14. Position the integrated oil pan gasket/windage tray assembly.
15. Install the oil pickup tube
16. If removed, install stud at position No. 9.
17. Install the mounting bolt and nuts. Tighten nuts to 28 Nm (20 ft. lbs.).
18. Position the oil pan and install the mounting bolts. Tighten the mounting bolts to 15 Nm (11 ft. lbs.) in the sequence shown.
19. Install structural dust cover
20. Lower the engine into mounts.
21. Remove the lifting tool.
22. Install both the left and right side engine mount through bolts. Tighten the nuts to 68 Nm (50 ft. lbs.).
23. Install the intake manifold.
24. Fill engine oil.
25. Reconnect the negative battery cable.
26. Start engine and check for leaks.

4.7L Engine

2002–04 MODELS

1. Before servicing the vehicle, refer to the precautions in the beginning of this section.

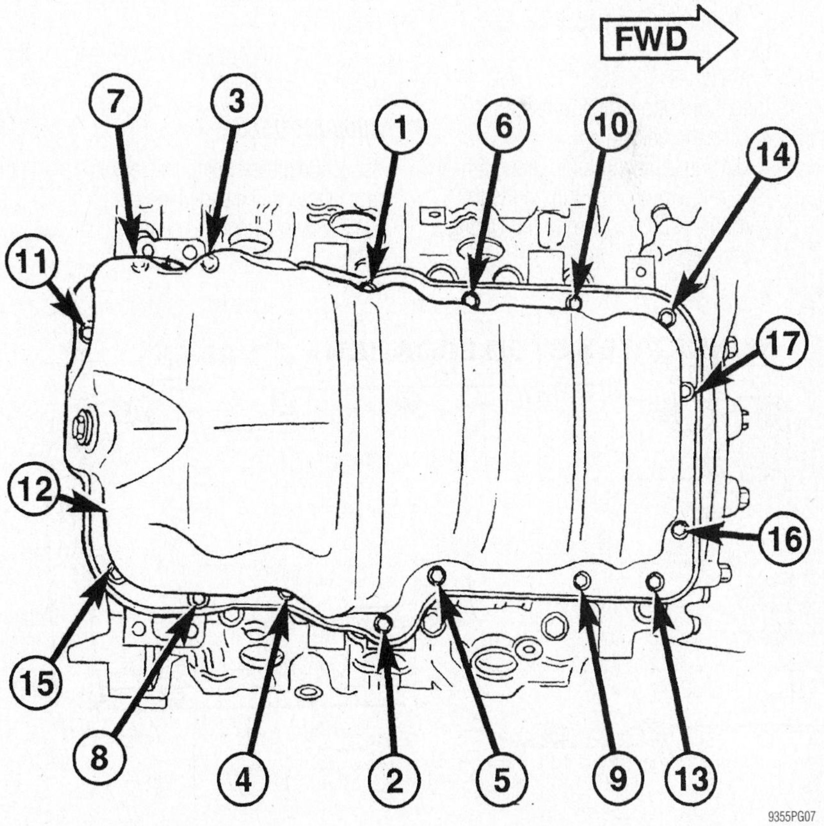

Oil pan bolt torque sequence—3.7L

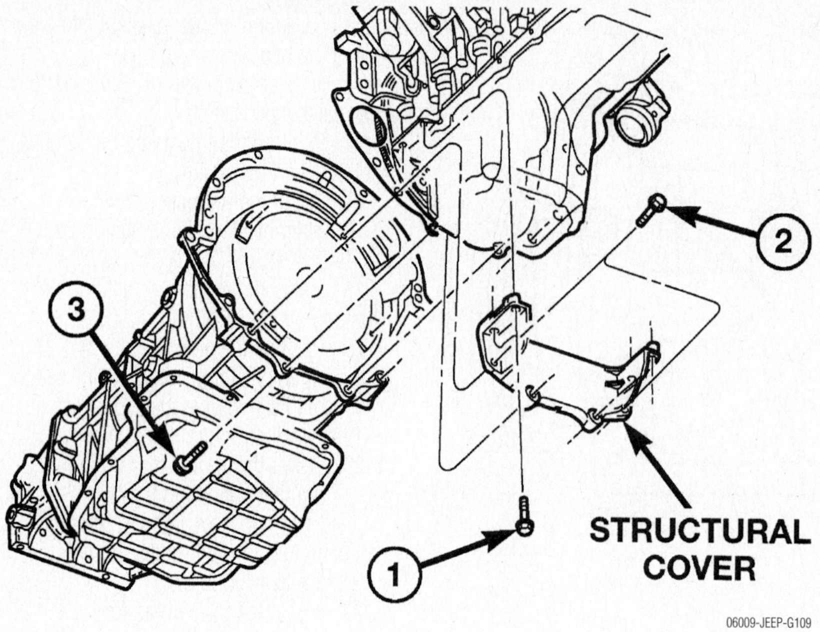

Structural cover removal and installation sequence—3.7L and 4.7L engines

2. Drain the engine oil.
3. Remove or disconnect the following:
   - Negative battery cable
   - Structural cover
   - Exhaust Y-pipe
   - Starter motor
   - Transmission oil cooler lines
   - Oil pan
   - Oil pump pickup tube
   - Oil pan gasket

To install:

4. Install or connect the following:
   - Oil pan gasket
   - Oil pump pickup tube, using a new O-ring. Torque the tube bolts to 20 ft. lbs. (28 Nm); torque the O-ring end bolt first.
   - Oil pan. Torque the bolts, in sequence, to 11 ft. lbs. (15 Nm).
   - Transmission oil cooler lines
   - Starter motor
   - Exhaust Y-pipe
   - Structural cover
   - Negative battery cable
5. Fill the engine to the proper level with clean oil.
6. Start the engine, check for leaks and repair if necessary.

2005 MODELS

1. Before servicing the vehicle, refer to the precautions in the beginning of this section.
2. Disconnect the negative battery cable.
3. Install engine support fixture special tool 8534, or equivalent.

➡Do not raise engine at this time.

4. Loosen both left and right side engine mount through bolts. Do not remove bolts.
5. Remove the structural dust cover.
6. Drain engine oil.

⁂ WARNING

Only raise the engine enough to provide clearance for oil pan removal. Check for proper clearance at fan shroud to fan and cowl to intake manifold.

7. Raise engine using special tool 8534 to provide clearance to remove oil pan.

➡Do not pry on oil pan or oil pan gasket. Gasket is integral to engine windage tray and does not come out with oil pan.

8. Remove the oil pan mounting bolts and oil pan.
9. Unbolt oil pump pickup tube and remove tube.
10. Inspect the integral windage tray and gasket and replace as needed.

To install:

11. Clean the oil pan gasket mating surface of the bedplate and oil pan.
12. Position the oil pan gasket and pickup tube with new O-ring. Install the mounting bolt and nuts. Tighten bolt and nuts to 28 Nm (20 ft. lbs.).
13. Position the oil pan and install the mounting bolts. Tighten the mounting bolts to 15 Nm (11 ft. lbs.) in the sequence shown.
14. Lower the engine into mounts using special tool 8534.
15. Install both the left and right side engine mount through bolts. Tighten the nuts to 68 Nm (50 ft. lbs.).
16. Remove special tool 8534.
17. Install structural dust cover.
18. Fill engine oil.
19. Reconnect the negative battery cable.
20. Start engine and check for leaks.

5.7L Engine

1. Before servicing the vehicle, refer to the precautions in the beginning of this section.
2. Disconnect the negative battery cable.
3. Remove the engine cover.
4. Remove the intake manifold.

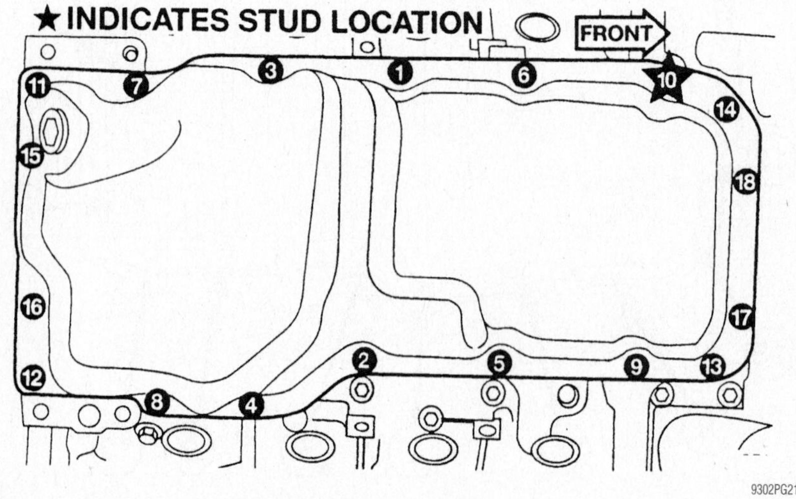

Oil pan mounting bolt tightening sequence—4.7L engine

5. Raise vehicle.

6. Remove both left and right side engine mount to frame bolts.

7. Drain engine oil and remove the oil filter.

8. Remove the engine oil dipstick and tube from the oil pan.

9. Lower the vehicle.

10. Install engine support fixture special tool 8534, or equivalent.

➡Do not use the third leg.

11. Raise engine using special tool 8534 to provide clearance to remove oil pan.

12. Raise the vehicle.

13. Remove the front axle.

14. Remove the structural dust cover.

➡Do not pry on oil pan or oil pan gasket. Gasket is integral to engine windage tray and does not come out with oil pan.

➡The horizontal M10 fasteners are 5 mm longer in length, and must be reinstalled in original locations.

15. Remove the M10 fasteners (vertical and horizontal) from the rear of the oil pan to the transmission and engine.

16. Remove the oil pan mounting bolts using the sequence provided, and oil pan.

➡When the oil pan is removed, a new oil pan gasket/windage tray assembly must be installed. The old gasket cannot be reused.

17. Discard the integral windage tray and gasket and replace.

To install:

18. Clean the oil pan gasket mating surface of the block and oil pan.

➡Mopar® Engine RTV, or equivalent, must be applied to the 4 T-joints, (area where front cover, rear retainer, block, and oil pan gasket meet). The bead of RTV should cover the bottom of the gasket. This area is approximately 4.5 mm x 25 mm in each of the 4 T-joint locations.

19. Apply Mopar T Engine RTV at the 4 T- joints.

➡When the oil pan is removed, a new oil pan gasket/windage tray assembly must be installed. The old gasket cannot be reused.

20. Install a new oil pan gasket/windage tray assembly.

21. If removed, reinstall the oil pump pickup tube with new O-ring. Tighten tube to pump fasteners to 28 Nm (250 inch lbs.).

➡The horizontal M10 fasteners are 5 mm longer in length, and must be reinstalled in original locations.

➡The horizontal M10 fasteners are 5 mm longer in length, and must be reinstalled in original locations.

22. Align the rear of the oil pan with the rear face of the engine block, and install the M10 and M6 oil pan fasteners finger tight. Using the following torque sequence, torque the M6 mounting bolts to 5 Nm (44 inch lbs.).

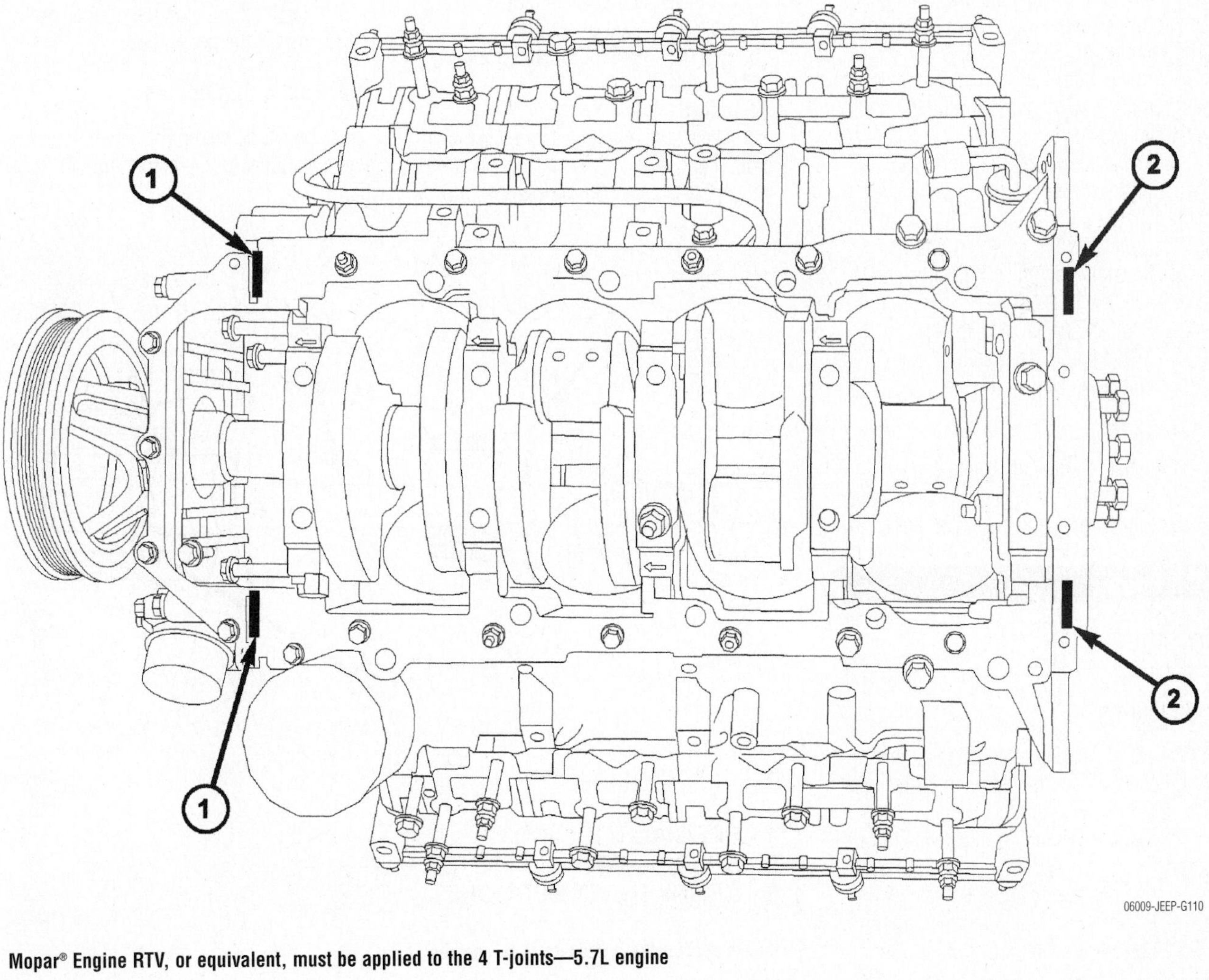

Mopar® Engine RTV, or equivalent, must be applied to the 4 T-joints—5.7L engine

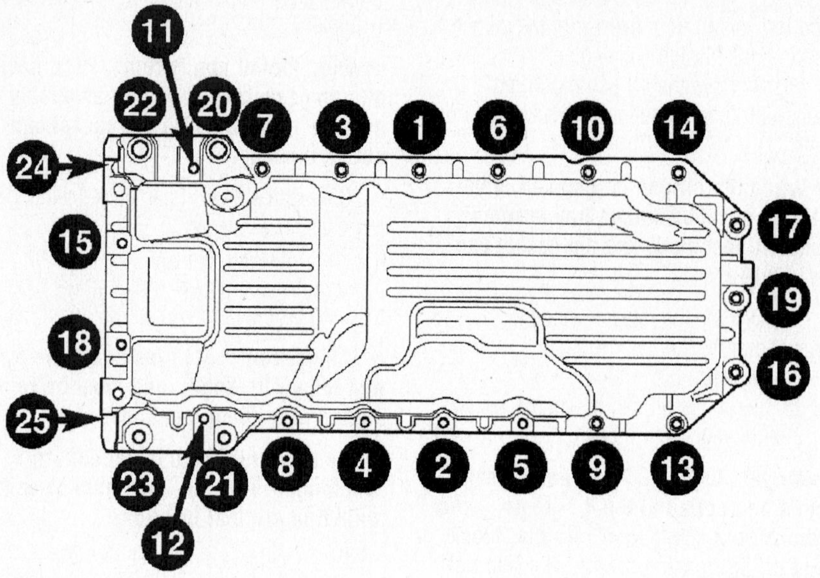

Oil pan bolt torque sequence—5.7L engine

23. Using the following torque sequence, torque the M10 oil pan fasteners to 54 Nm (39 ft. lbs.).
24. Using the following torque sequence, torque the M6 oil pan fasteners to 12 Nm (106 inch lbs.).
25. Install both the left and right side oil pan to transmission bolts. Torque the bolts to 54 Nm (39 ft. lbs.).
26. Lower the engine into mounts.
27. Install both the left and right side engine mount bolts and nuts. Torque the studs and nuts to 54 Nm (39 ft. lbs.).
28. Install the engine oil dipstick and tube.
29. Remove special tool 8534.
30. Install the intake manifold.
31. Install the front axle.
32. Install the engine cover.
33. Fill engine oil.
34. Install oil filter, if removed.
35. Reconnect the negative battery cable.
36. Start engine and check for leaks.

Oil Pump

REMOVAL & INSTALLATION

2.4L Engine

1. Before servicing the vehicle, refer to the precautions in the beginning of this section.
2. Disconnect the negative cable from battery.
3. Remove the timing belt.
4. Remove the timing belt rear cover.
5. Remove the oil pan.
6. Remove the crankshaft sprocket using Special Tools 6793 and C-4685-C2, or equivalent.
7. Remove the crankshaft key.
8. Remove the oil pick-up tube.
9. Remove the oil pump and front crankshaft seal.

To install:

10. Make sure all surfaces are clean and free of oil and dirt.
11. Apply Mopar® Gasket Maker, or equivalent, to the oil pump as shown. Install the O-ring into the oil pump body discharge passage.
12. Prime the oil pump with engine oil before installation.
13. Align the oil pump rotor flats with the flats on the crankshaft. Install the oil pump to the block.

➡To align, the front crankshaft seal MUST be out of the pump, or damage may result.

14. Install a new front crankshaft seal.
15. Install the crankshaft key.

➡The crankshaft sprocket is set to a predetermined depth from the factory for correct timing belt tracking. If removed, use of Special Tool 6792, or equivalent, is required to set the sprocket to original installation depth. An incorrectly installed sprocket will result in timing belt and engine damage.

16. Install the crankshaft sprocket using Special Tool 6792.
17. Install the oil pump pick-up tube.
18. Install the oil pan.
19. Install the timing belt rear cover.
20. Install the timing belt.

2.5L and 4.0L Engines

1. Before servicing the vehicle, refer to the precautions in the beginning of this section.

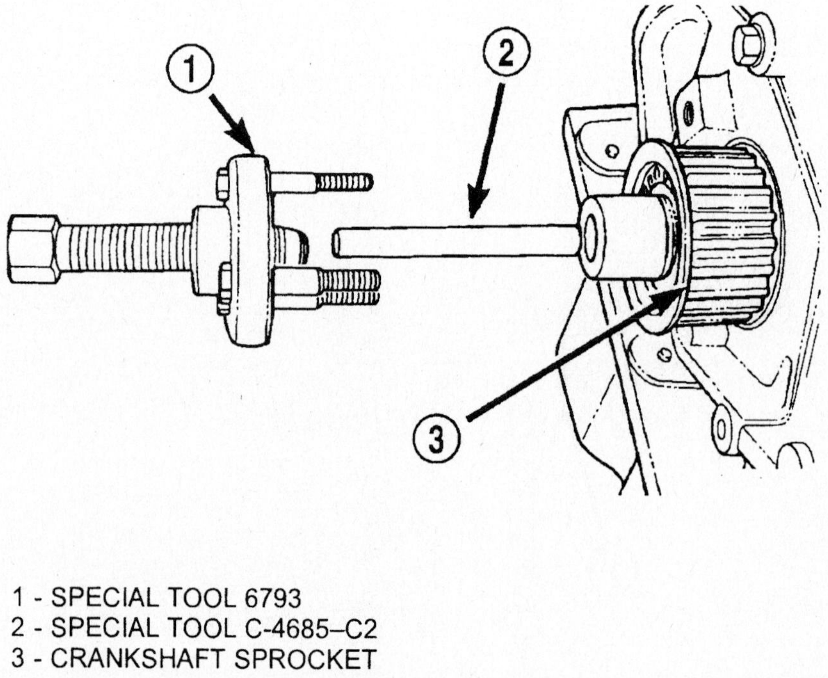

Crankshaft sprocket removal—2.4L engines

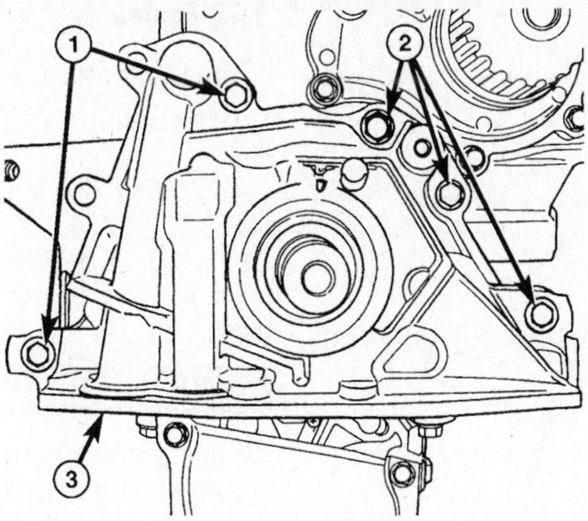

Oil pump—2.4L engines

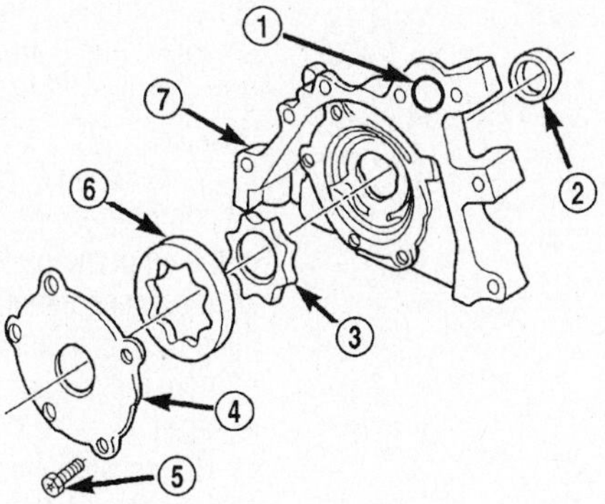

Oil pump parts—2.4L engines

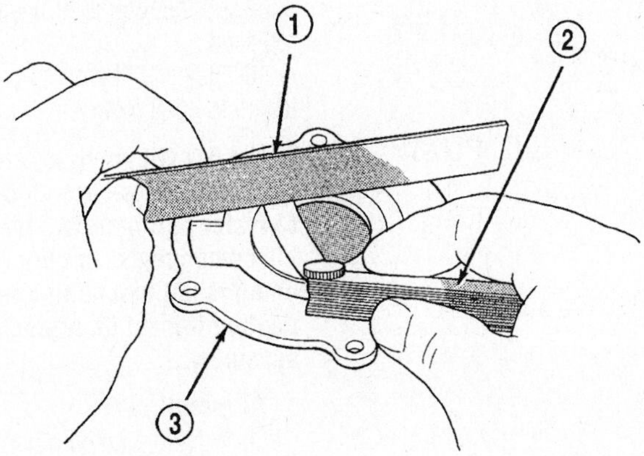

Checking the oil pump body for flatness—2.4L engines

2. Drain the engine oil.
3. Remove or disconnect the following:
   - Negative battery cable
   - Oil pan
   - Oil pump and pickup tube

➡If the oil pump is not to be serviced, do not disturb the position of the oil inlet tube and strainer assembly in the pump body. If the tube is moved within the pump body, a replacement tube and strainer assembly must be installed to assure an airtight seal.

To install:

4. Install or connect the following:
   - Oil pump. Torque the mounting bolts to 17 ft. lbs. (23 Nm).
   - Oil pan
   - Negative battery cable
5. Fill the engine with the proper type and quantity of oil.
6. Start the engine, check for leaks and repair if necessary.

3.7L Engine

1. Before servicing the vehicle, refer to the precautions in the beginning of this section.
2. Remove or disconnect the following:
   - Oil Pan
   - Timing chain cover
   - Timing chains and tensioners
   - Oil pump
3. Installation is the reverse of removal. Torque the pump bolts, in sequence, to 21 ft. lbs. (28 Nm),

4.7L Engine

1. Before servicing the vehicle, refer to the precautions in the beginning of this section.
2. Drain the engine oil.
3. Remove or disconnect the following:
   - Valve covers
   - Front cover
   - Timing chains and sprockets
   - Oil pan and pickup tube
   - Oil pump and primary timing chain tensioner

To install:

4. Install or connect the following:
   - Oil pump and primary timing chain tensioner. Torque the bolts in sequence to 21 ft. lbs. (28 Nm).
   - Oil pan and pickup tube
   - Timing chains and sprockets
   - Front cover
   - Valve covers
5. Fill the engine with clean oil.
6. Start the engine, check for leaks and repair if necessary.

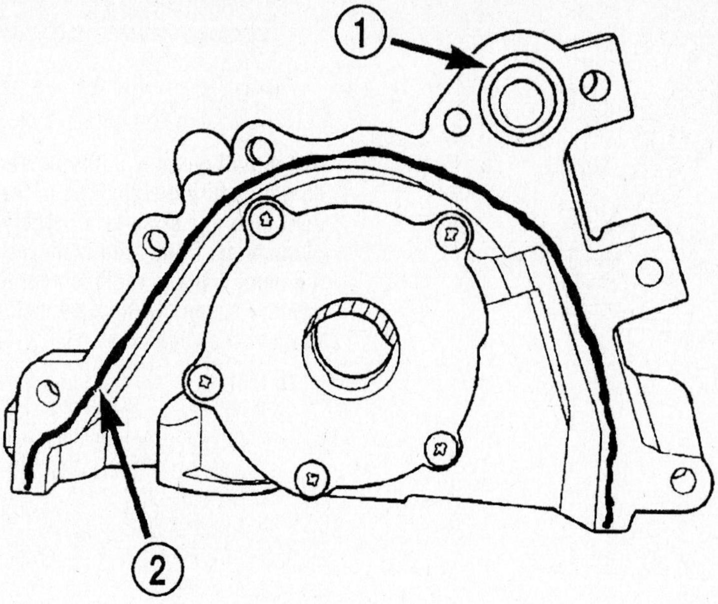

Oil pump sealing—2.4L engines

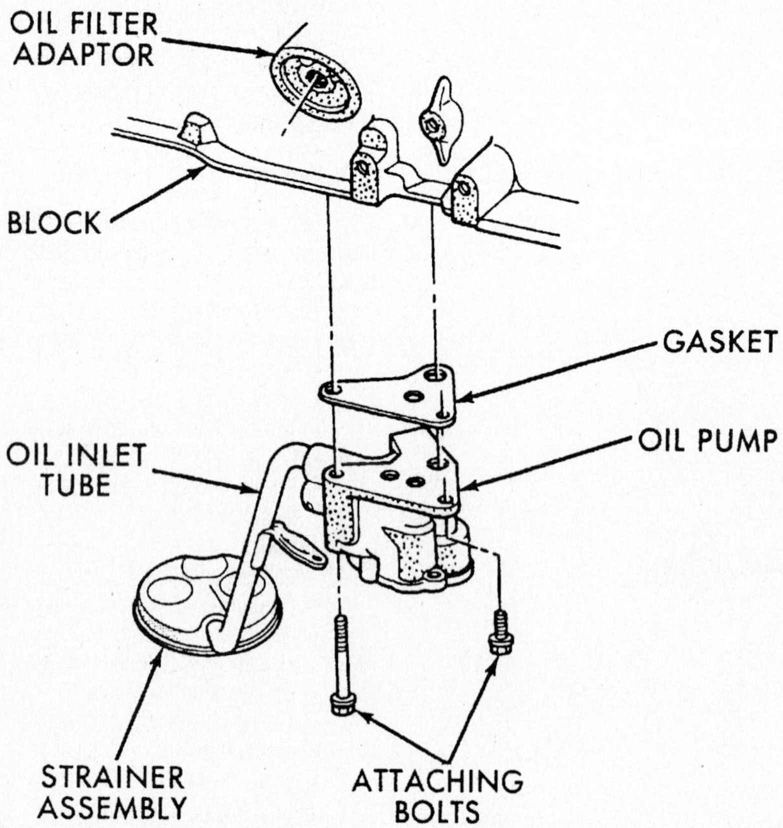

Exploded view of the oil pump assembly—2.5L and 4.0L engine

5.7L Engine

1. Before servicing the vehicle, refer to the precautions in the beginning of this section.
2. Remove the oil pan and pick-up tube.
3. Remove the timing chain cover.
4. Remove the four bolts, and the oil pump.

⁂ WARNING

Oil pump pressure relief valve and spring should not be removed from the oil pump. If these components are disassembled and or removed from the pump the entire oil pump assembly must be replaced.

5. Remove the pump cover.
6. Clean all parts thoroughly. Mating surface of the oil pump housing should be smooth. If the pump cover is scratched or grooved the oil pump assembly should be replaced.
7. Slide outer rotor into the body of the oil pump. Press the outer rotor to one side of the oil pump body and measure clearance between the outer rotor and the body. If the measurement is 0.235mm (0.009 in.) or more the oil pump assembly must be replaced.
8. Install the inner rotor in the into the oil pump body. Measure the clearance between the inner and outer rotors. If the clearance between the rotors is 0.150 mm (0.006 in.) or more the oil pump assembly must be replaced.
9. Place a straightedge across the body of the oil pump (between the bolt holes), if a feeler gauge of 0.095 mm (0.0038 in.) or greater can be inserted between the straightedge and the rotors, the pump must be replaced.
10. Reinstall the pump cover. Torque fasteners to 15 Nm (132 inch lbs.).

➡The 5.7 Oil pump is released as an assembly. There are no DaimlerChrysler part numbers for sub-assembly components. In the event the oil pump is not functioning or out of specification it must be replaced as an assembly.

To install:

11. Position the oil pump onto the crankshaft and install the 4 oil pump retaining bolts.
12. Tighten the oil pump retaining bolts to 28 Nm (250 inch lbs.).
13. Install the timing chain cover.
14. Install the pick-up tube and oil pan.

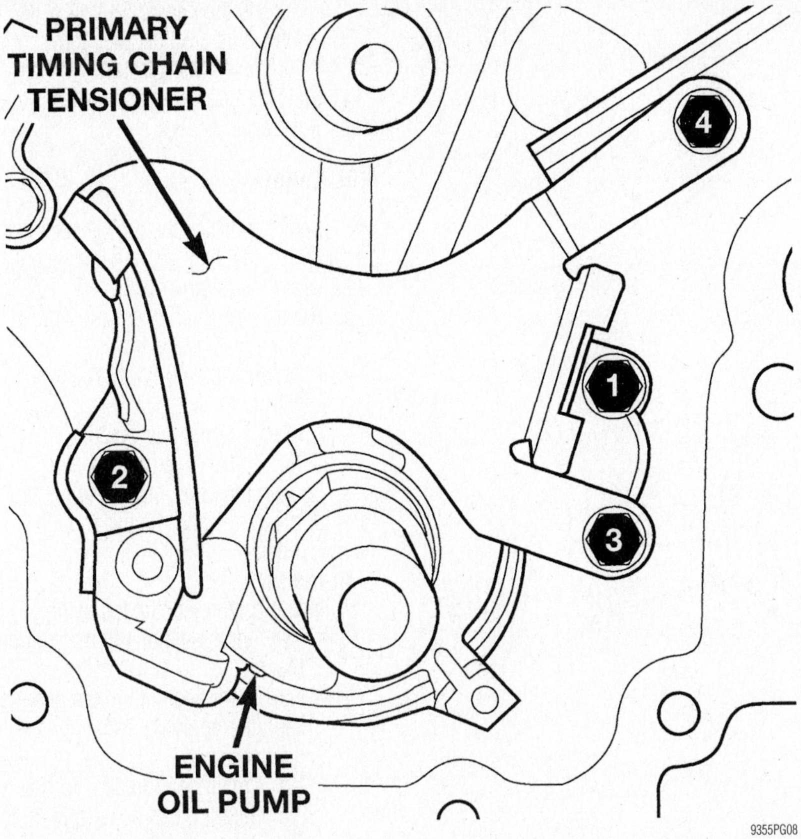

Oil pump bolt torque sequence—3.7L

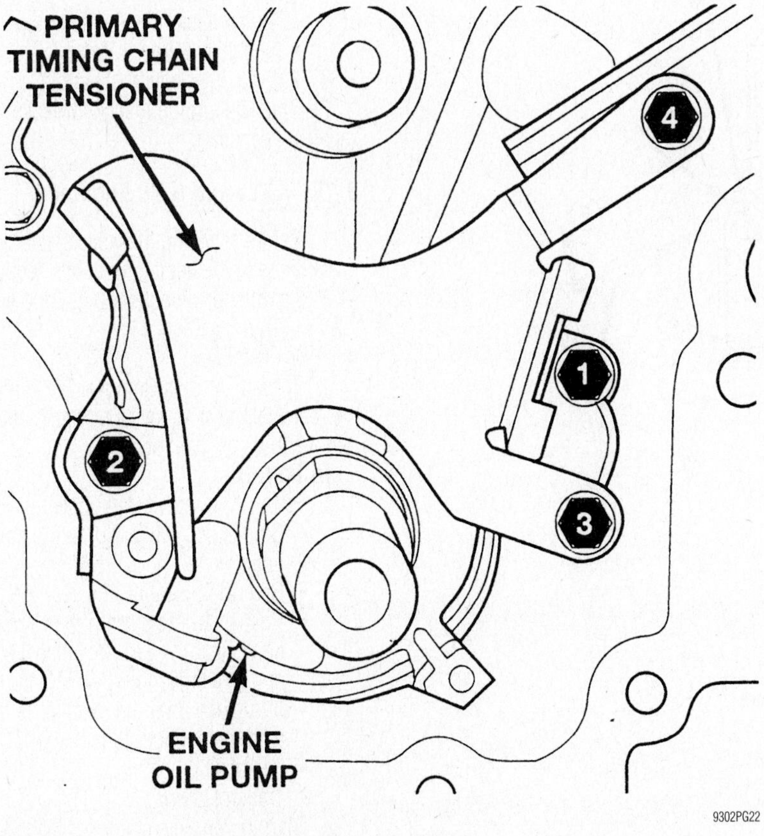

Oil pump and chain tensioner torque sequence—4.7L engine

Rear Main Seal

REMOVAL & INSTALLATION

2.4L Engine

1. Before servicing the vehicle, refer to the precautions in the beginning of this section.
2. Remove the transmission.
3. Remove the flex plate.
4. Insert a 3⁄16 flat bladed screwdriver between the dust lip and the metal case of the crankshaft seal. Angle the screwdriver through the dust lip against metal case of the seal. Pry out seal.

➡Do not permit the screwdriver blade to contact the crankshaft seal surface. Contact of the screwdriver blade against the crankshaft edge (chamfer) is permitted.

To install:

➡If burrs or scratches are present on the crankshaft edge (chamfer), cleanup with 400 grit sand paper to prevent seal damage during the installation of the new seal.

5. Lubricate the crankshaft flange with engine oil.
6. Place Special Tool 6926-1 Seal Guide, or equivalent, on the crankshaft.
7. Position the seal over the guide tool. The guide tool should remain on the crankshaft during installation of the seal. Ensure that the lip of the seal is facing towards the crankcase during installation.

➡If the seal is driven into the block past flush, this may cause an oil leak.

8. Drive the seal into the block using Special Tool 6926-2 and handle C-4171, or equivalent, until the tool bottoms out against the block.
9. Install the flex plate. Apply Mopar® Lock & Seal Adhesive to the bolt threads and tighten the bolts to 95 Nm (70 ft. lbs.).
10. Install the transmission.

2.5L Engine

1. Before servicing the vehicle, refer to the precautions in the beginning of this section.
2. Remove or disconnect the following:
   - Negative battery cable
   - Flywheel/converter drive plate and discard the bolt
   - Rear main seal

To install:

3. Install or connect the following:
   - Rear main seal so that it is flush with the cylinder block
   - Flywheel/ converter drive plate.

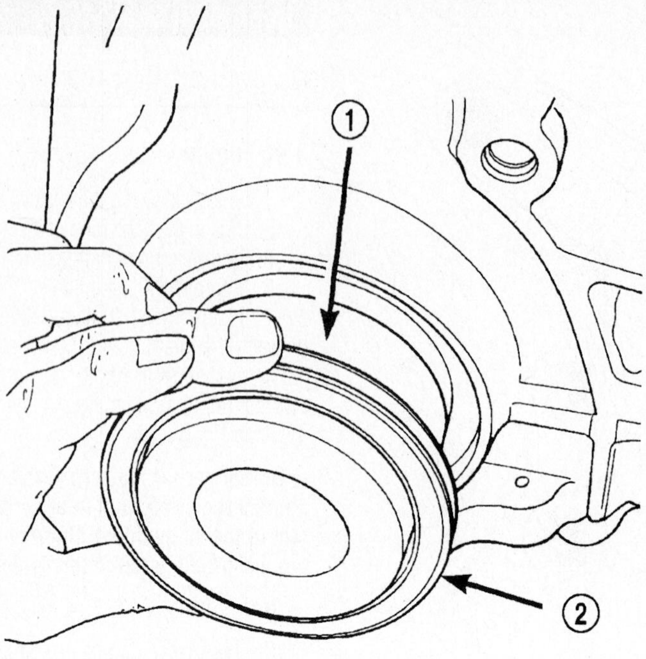

1 - SPECIAL TOOL 6926–1 PILOT
2 - SEAL

67189-JEEP-G23

Rear crankshaft seal—2.4L engines

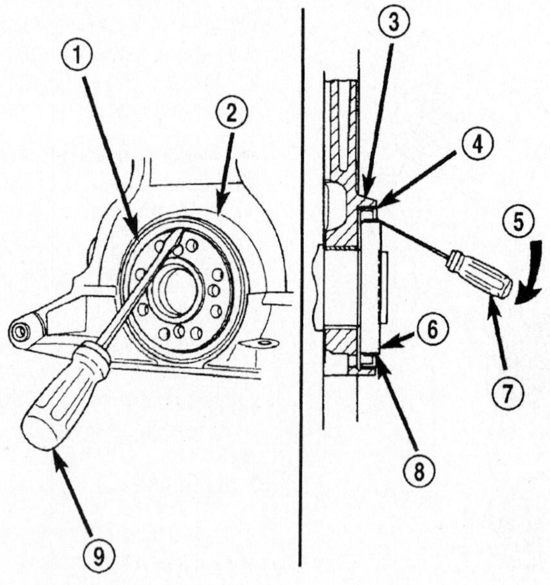

1 - REAR CRANKSHAFT SEAL
2 - ENGINE BLOCK
3 - ENGINE BLOCK
4 - REAR CRANKSHAFT SEAL METAL CASE
5 - PRY IN THIS DIRECTION
6 - CRANKSHAFT
7 - SCREWDRIVER
8 - REAR CRANKSHAFT SEAL DUST LIP
9 - SCREWDRIVER

67189-JEEP-G24

Rear crankshaft seal removal—2.4L engines

Torque the new to 50 ft. lbs. (68 Nm) plus a 60 degrees turn
- Negative battery cable

4. Start the engine, check for leaks and repair if necessary.

4.0L Engine

1. Before servicing the vehicle, refer to the precautions in the beginning of this section.
2. Drain the engine oil.
3. Remove or disconnect the following:
   - Negative battery cable
   - Transmission inspection cover
   - Oil pan
   - Main bearing cap brace
   - No. 7 main bearing cap
4. Loosen the other main bearing cap bolts for clearance and remove the rear main seal halfs.

To install:

5. Install or connect the following:
   - New upper seal half to the cylinder block
   - New lower seal half to the bearing cap after applying sealant to the bearing cap
   - No. 7 main bearing cap. Torque **all** main bearing cap bolts to 80 ft. lbs. (108 Nm).
   - Main bearing cap brace. Tighten the nuts to 35 ft. lbs. (47 Nm).
   - Oil pan
   - Transmission inspection cover
   - Negative battery cable
6. Fill the engine with clean oil.
7. Start the engine, check for leaks and repair if necessary.

3.7L, 4.7L, and 5.7L Engines

1. Before servicing the vehicle, refer to the precautions in the beginning of this section.
2. Remove or disconnect the following:
   - Transmission
   - Flexplate
3. Thread Oil Seal Remover 8506 into the rear main seal as far as possible and remove the rear main seal.

To install:

4. Install or connect the following:
   - Seal Guide 8349-2 onto the crankshaft
   - Rear main seal on the seal guide
   - Rear main seal, using the Crankshaft Rear Oil Seal Installer 8349 and Driver Handle C-4171; tap it into place until the installer is flush with the cylinder block
   - Flexplate.
   - Transmission
5. Start the engine, check for leaks and repair if necessary.

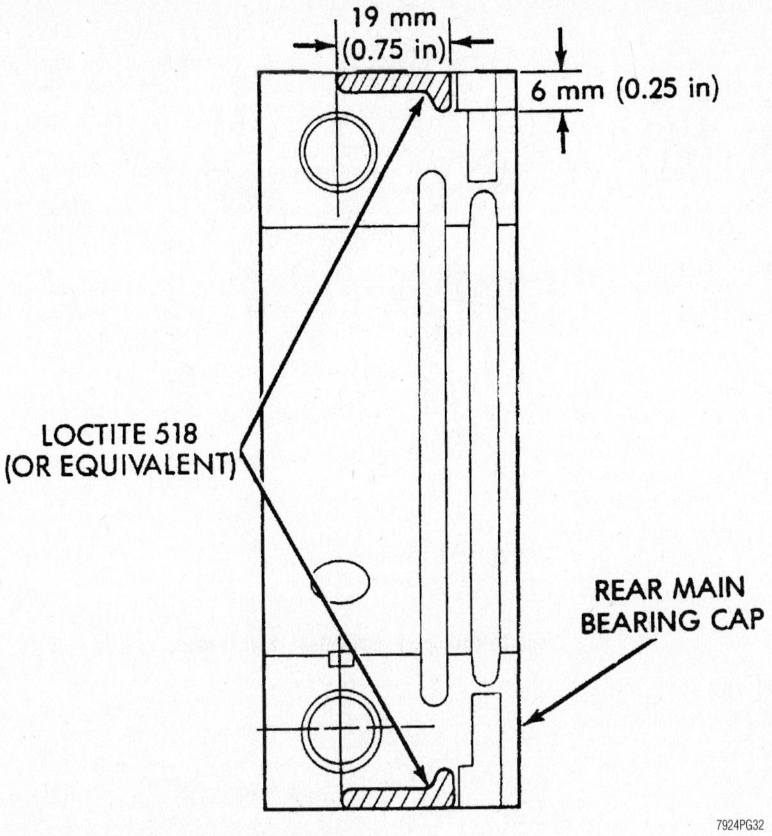

Sealant application locations—4.0L engine

Timing Belt Covers

REMOVAL & INSTALLATION

2.4L Engine

FRONT COVER

1. Before servicing the vehicle, refer to the precautions in the beginning of this section.
2. Remove the crankshaft vibration damper.
3. Remove the generator drive belt tensioner assembly.
4. Remove the timing belt front cover bolts, and remove the covers.

To install:

5. Install the timing belt front covers. Tighten the fasteners to 7 Nm (60 inch lbs.).
6. Install the generator drive belt tensioner.
7. Install the crankshaft vibration damper.

REAR COVER

1. Before servicing the vehicle, refer to the precautions in the beginning of this section.
2. Remove the front covers.
3. Remove the timing belt.
4. Hold the camshaft sprocket with Special Tool 6847, or equivalent, while removing the center bolt.
5. Remove the timing belt idler pulley.
6. Remove the rear cover fasteners and remove the cover from engine.

To install:

7. Install the timing belt rear cover and bolts.

➡Do not use an impact wrench for tightening the camshaft sprocket bolt. Damage to the timing locating pin can occur. Hand tighten using a wrench ONLY.

8. Install the camshaft sprockets. Hold the sprockets with Special Tool 6848, or equivalent, and tighten the center bolt to 101 Nm (75 ft. lbs.).
9. Install the timing belt idler pulley and tighten the mounting bolt to 61 Nm (45 ft. lbs.).
10. Install the timing belt.
11. Install the front covers.

Front Crankshaft Seal

The following applies only to the 2.4L engine, which has a timing belt.

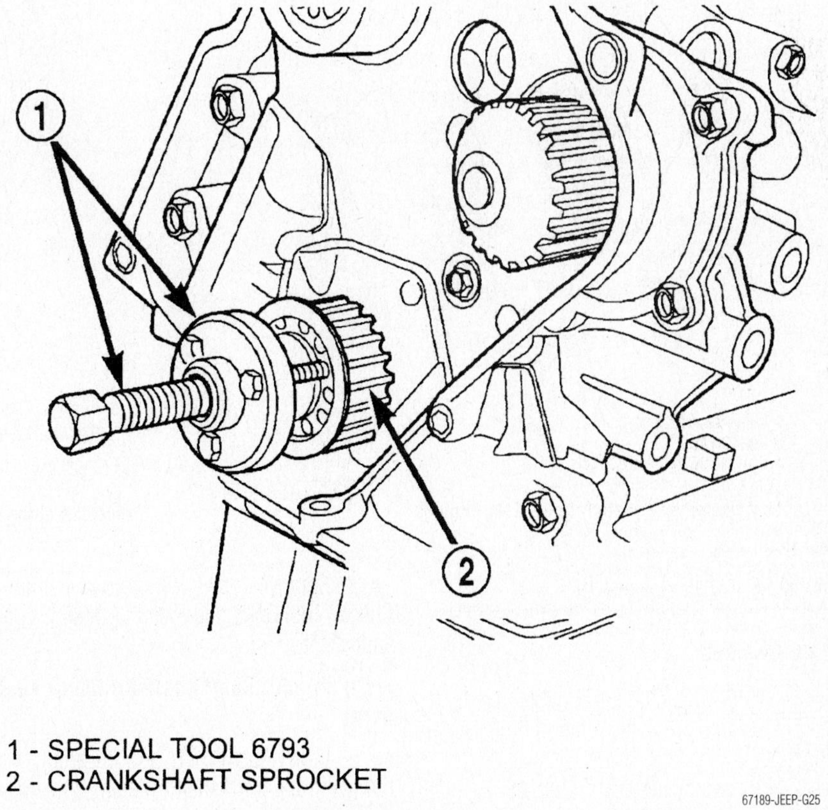

Crankshaft sprocket removal—2.4L engines

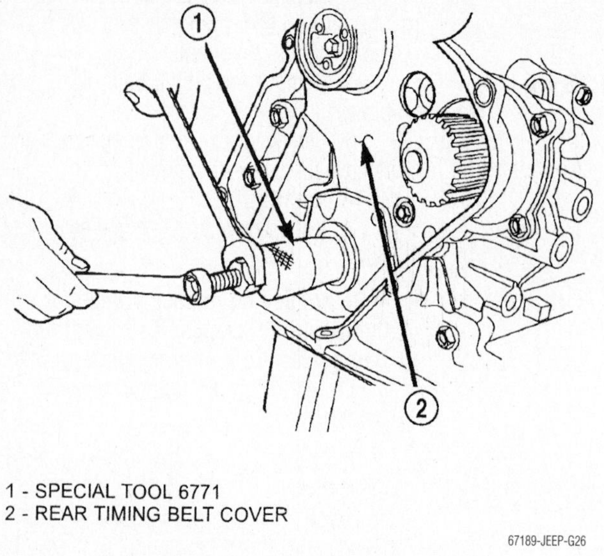

Front crankshaft seal removal—2.4L engines

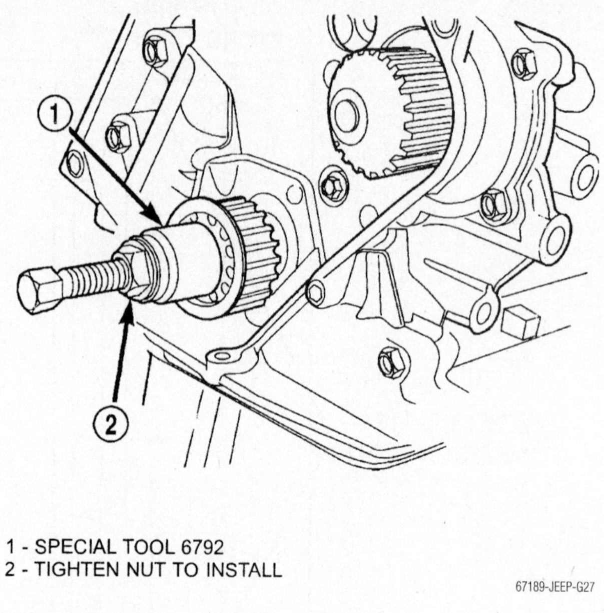

Crankshaft sprocket installation—2.4L engines

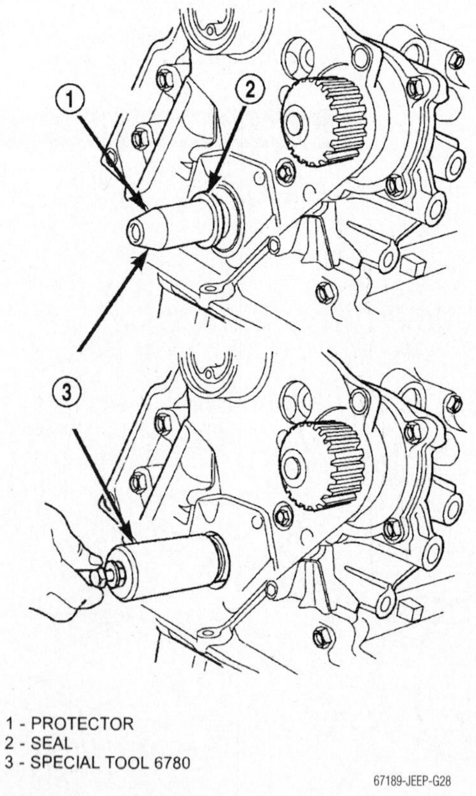

Front crankshaft seal installation—2.4L engines

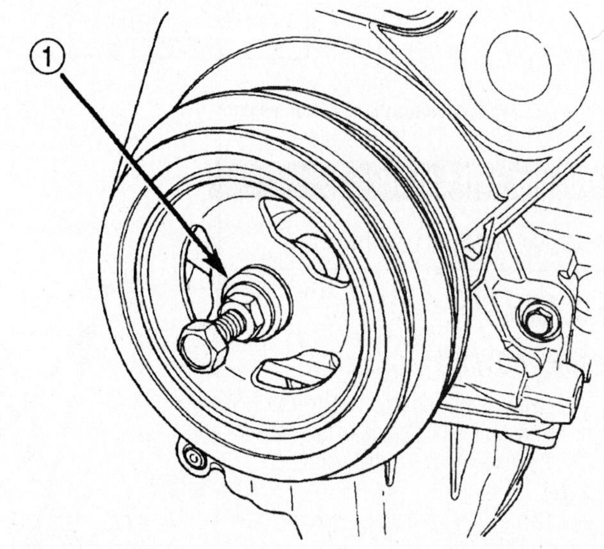

Crankshaft damper installation—2.4L engines

REMOVAL & INSTALLATION

2.4L Engines

1. Before servicing the vehicle, refer to the precautions in the beginning of this section.
2. Remove the crankshaft vibration damper.
3. Remove timing belt.
4. Remove crankshaft sprocket using Special Tool 6793 and insert C-4685-C2, or equivalent.

➡Do not nick shaft seal surface or seal bore.

5. Using Tool 6771, or equivalent, remove front crankshaft oil seal. Be careful not to damage the seal surface of cover.

To install:

6. Install new seal by using Special Tool 6780, or equivalent.
7. Place seal into opening with seal spring towards the inside of engine. Install seal until flush with cover.
8. Install crankshaft sprocket using Special Tool 6792, or equivalent.
9. Install timing belt.
10. Install crankshaft vibration damper.

Timing Belt and Crankshaft Sprocket

REMOVAL & INSTALLATION

2.4L Engines

1. Before servicing the vehicle, refer to the precautions in the beginning of this section.
2. Remove the air cleaner upper cover, housing, and clean air tube.
3. Raise the vehicle on hoist.
4. Remove the accessory drive belts.
5. Remove the crankshaft vibration damper.
6. Remove the air conditioner/generator belt tensioner and pulley assembly.
7. Remove the timing belt lower front cover bolts and remove cover.
8. Lower the vehicle.
9. Remove the bolts attaching timing belt upper front cover and remove cover.

➡When aligning the crankshaft and camshaft timing marks always rotate engine by the crankshaft. Camshaft should not be rotated after the timing belt is removed. Damage to the valve components may occur. Always align the timing marks before removing the timing belt.

10. Before removal of the timing belt, rotate the crankshaft until the TDC mark on the oil pump housing aligns with the TDC mark on the crankshaft sprocket (trailing edge of the sprocket tooth).

➡The crankshaft sprocket TDC mark is located on the trailing edge of the sprocket tooth. Failure to align the trailing edge of sprocket tooth to TDC mark on oil pump housing will cause the camshaft timing marks to be misaligned.

11. Install 6 mm Allen wrench into belt tensioner. Before rotating the tensioner, insert the long end of a ⅛ inch or 3 mm Allen wrench into the pin hole on the front of the tensioner. While rotating the tensioner counterclockwise, push in lightly on the ⅛ inch or 3 mm Allen wrench, until it slides into the locking hole.
12. Remove timing belt.
13. Remove crankshaft sprocket using Special Tools 6793 and insert C-4685-C2, or equivalent.

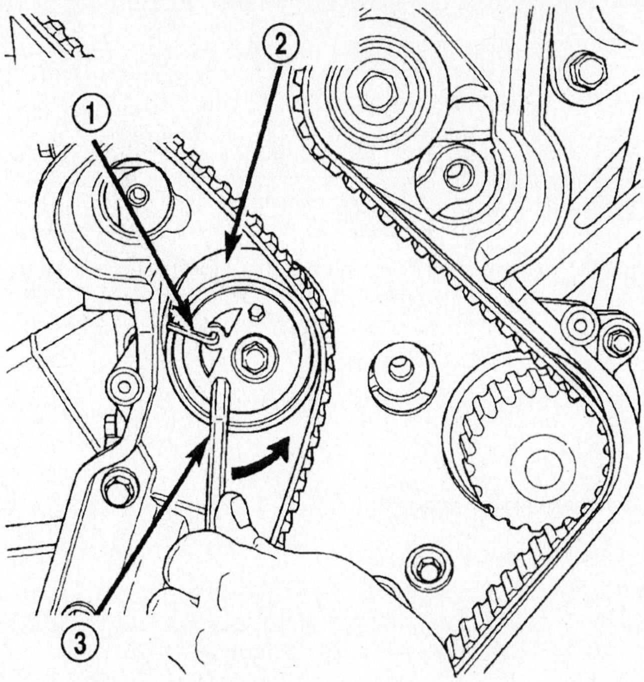

1 - 1/8 OR 3mm ALLEN WRENCH
2 - BELT TENSIONER
3 - 6mm ALLEN WRENCH

67189-JEEP-G30

Locking the tensioner—2.4L engines

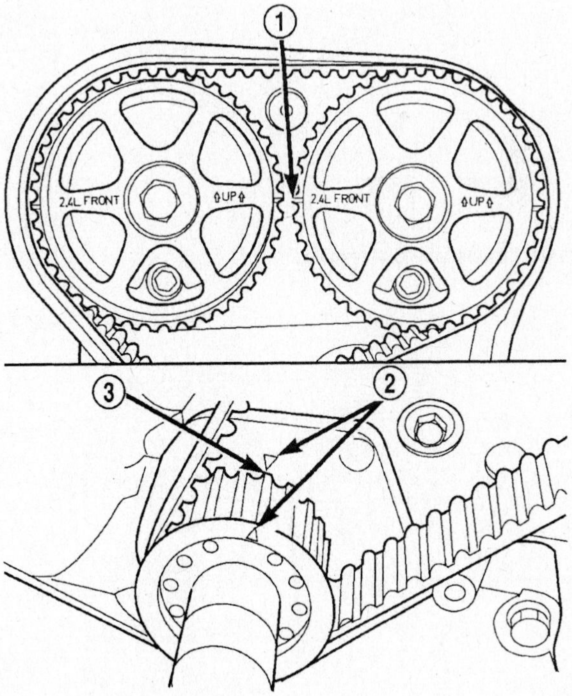

1 - CAMSHAFT TIMING MARKS
2 - CRANKSHAFT TDC MARKS
3 - TRAILING EDGE OF SPROCKET TOOTH

67189-JEEP-G31

Timing mark alignment—2.4L engines

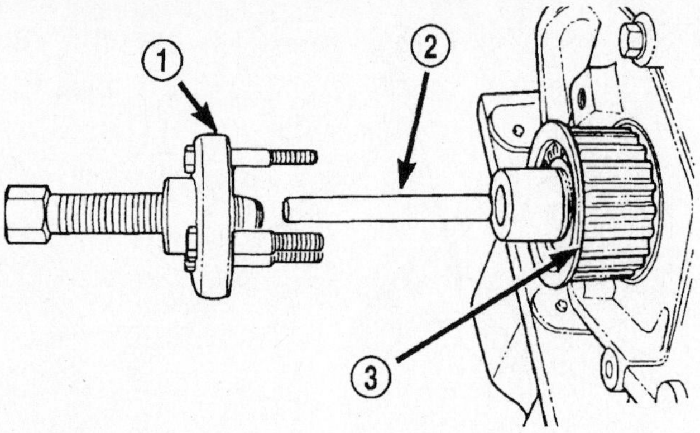

1 - SPECIAL TOOL 6793
2 - SPECIAL TOOL C-4685–C2
3 - CRANKSHAFT SPROCKET

67189-JEEP-G32

Crankshaft sprocket removal—2.4L engines

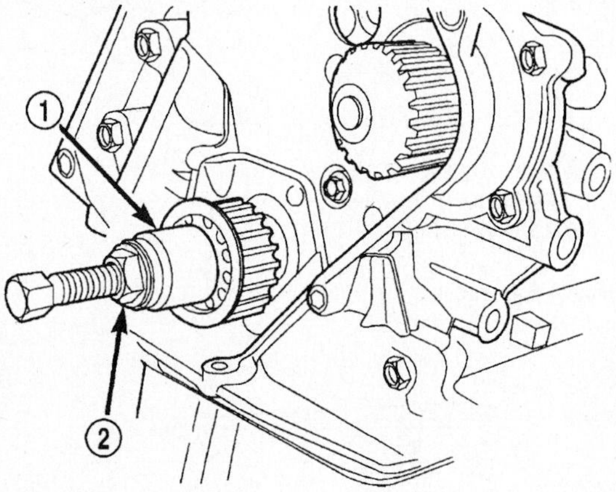

1 - SPECIAL TOOL 6792
2 - TIGHTEN NUT TO INSTALL

67189-JEEP-G33

Crankshaft sprocket installation—2.4L engines

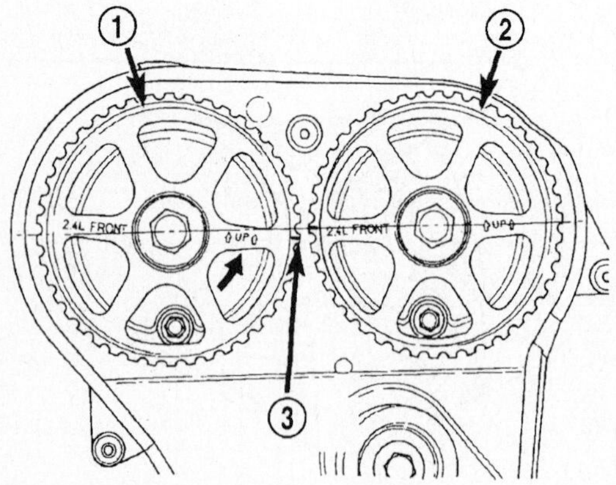

1 - CAMSHAFT SPROCKET-EXHAUST
2 - CAMSHAFT SPROCKET-INTAKE
3 - 1/2 NOTCH LOCATION

67189-JEEP-G34

Camshaft sprocket alignment—2.4L engines

To install:

➡The crankshaft sprocket is set to a predetermined depth from the factory for correct timing belt tracking. If removed, use of Special Tool 6792 is required to set the sprocket to original installation depth. An incorrectly installed sprocket will result in timing belt and engine damage.

14. Install crankshaft sprocket using Special Tool 6792, or equivalent.
15. Set crankshaft sprocket to TDC by aligning the sprocket with the arrow on the oil pump housing.
16. Set camshafts timing marks so that the exhaust camshaft sprocket is a ½ notch below the intake camshaft sprocket.

✲✲ CAUTION

Ensure that the arrows on both camshaft sprockets are facing up.

17. Install timing belt. Starting at the crankshaft, go around the water pump sprocket, idler pulley, camshaft sprockets and then around the tensioner.
18. Move the exhaust camshaft sprocket counterclockwise to align marks and take up belt slack.
19. Insert a 6 mm Allen wrench into the hexagon opening located on the top plate of the belt tensioner pulley. Rotate the top plate COUNTERCLOCKWISE. The tensioner pulley will move against the belt and the tensioner setting notch will eventually start to move clockwise. Watching the movement of the setting notch, continue rotating the top plate counterclockwise until the setting notch is aligned with the spring tang. Using the Allen wrench to prevent the top plate from moving, torque the tensioner lock nut to 30 Nm (22 ft. lbs.). Setting notch and spring tang should remain aligned after lock nut is torqued.
20. Remove Allen wrench and torque wrench.

➡Repositioning the crankshaft to the TDC position must be done only during the CLOCKWISE rotation movement. If TDC is missed, rotate a further two revolutions until TDC is achieved. DO NOT rotate crankshaft counterclockwise as this will make verification of proper tensioner setting impossible.

21. Once the timing belt has been installed and tensioner adjusted, rotate the crankshaft CLOCKWISE two complete revolutions manually for seating of the belt, until the crankshaft is repositioned at the TDC position. Verify that the camshaft and

crankshaft timing marks are in proper position.

22. Check if the spring tang is within the tolerance window. If the spring tang is within the tolerance window, the installation process is complete and nothing further is required. If the spring tang is not within the tolerance window, repeat Steps 5 through 7.

23. Install timing belt front covers and bolts.

24. Install air conditioning/generator belt tensioner and pulley.

25. Install crankshaft vibration damper.

26. Install accessory drive belts.

27. Install drive belt splash shield.

28. Install air cleaner housing, upper cover, and clean air tube.

Timing Belt Tensioner and Pulley

REMOVAL AND INSTALLATION

2.4L Engine

1. Before servicing the vehicle, refer to the precautions in the beginning of this section.
2. Remove the timing belt.
3. Remove the timing belt idler pulley.
4. Hold the camshaft sprocket with Special Tool 6847, or equivalent, while removing the bolt. Remove both cam sprockets.
5. Remove the rear timing belt cover fasteners and remove the cover from engine.
6. Remove the lower bolt attaching the

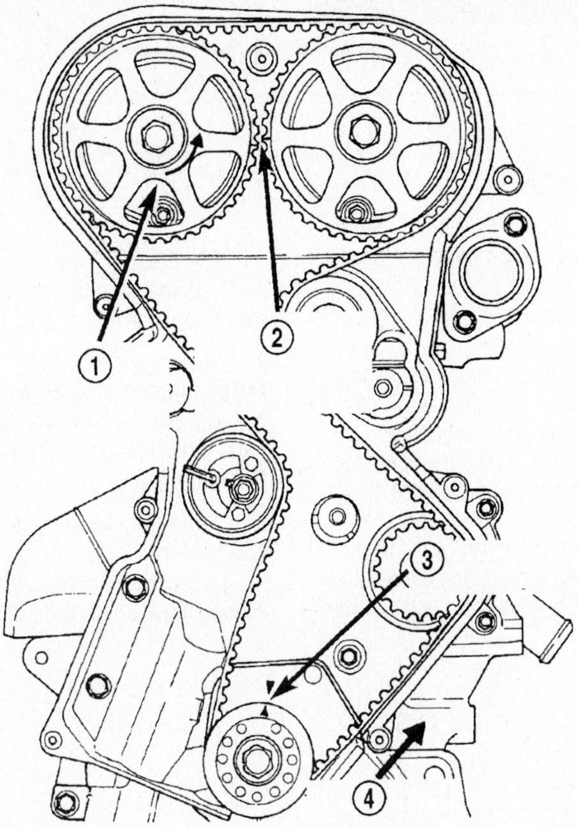

1 - ROTATE CAMSHAFT SPROCKET TO TAKE UP BELT SLACK
2 - CAMSHAFT TIMING MARKS 1/2 NOTCH LOCATION
3 - CRANKSHAFT AT TDC
4 - INSTALL BELT IN THIS DIRECTION

67189-JEEP-G35

Timing belt installation—2.4L engines

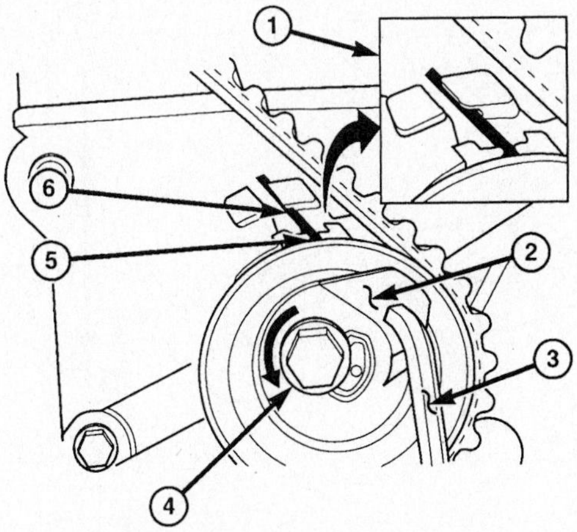

1 - ALIGN SETTING NOTCH WITH SPRING TANG
2 - TOP PLATE
3 - 6mm ALLEN WRENCH
4 - LOCK NUT
5 - SETTING NOTCH
6 - SPRING TANG

67189-JEEP-G36

Timing belt tension adjustment—2.4L engines

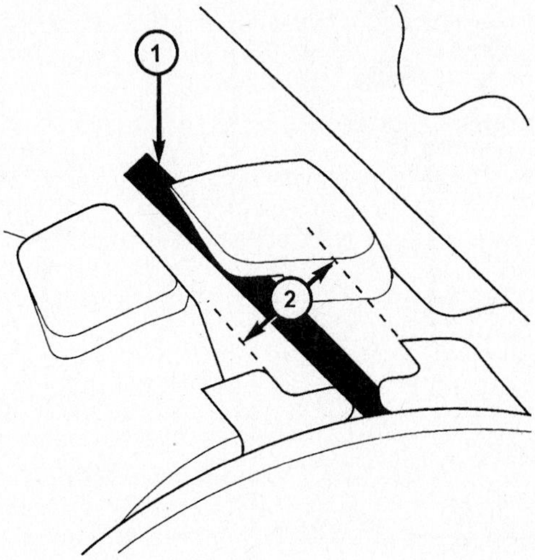

1 - SPRING TANG
2 - TOLERANCE WINDOW

67189-JEEP-G37

Timing belt tension verification—2.4L engines

timing belt tensioner assembly to engine and remove the tensioner as an assembly.

To install:

7. Align the timing belt tensioner assembly to engine and install the lower mounting bolt but do not tighten. To properly align the tensioner assembly, install one of the engine bracket mounting bolts (M10) 5 to 7 turns into the tensioner's upper mounting location.
8. Torque the tensioner's lower mounting bolt to 61 Nm (45 ft. lbs.). Remove the upper bolt used for tensioner alignment.
9. Install the rear timing belt cover and fasteners.
10. Install the timing belt idler pulley and torque the mounting bolt to 61 Nm (45 ft. lbs.).
11. Install the camshaft sprockets. Use Special Tool 6847 to hold the sprockets. Torque the bolts to 101 Nm (75 ft. lbs.).
12. Install the timing belt.

Balance Shafts and Carrier

REMOVAL AND INSTALLATION

2.4L Engine

BALANCE SHAFTS

1. Before servicing the vehicle, refer to the precautions in the beginning of this section.
2. Drain engine oil.
3. Remove the oil pan.
4. Remove chain cover, guide and tensioner.
5. Remove gear cover retaining stud (double ended to also retain chain guide). Remove cover and balance shaft gears.
6. Remove balance shaft gear, chain sprocket retaining screws, and crankshaft chain sprocket. Remove chain and sprocket assembly. Using two wide pry bars, work the sprocket back and forth until it is off the shaft.
7. Remove carrier gear cover and balance shafts.
8. Remove four carrier to crankcase attaching bolts to separate carrier from engine bedplate.

BALANCE SHAFT CARRIER

1. Before servicing the vehicle, refer to the precautions in the beginning of this section.

➡The following components will remain intact during carrier removal. Gear cover, gears, balance shafts and the rear cover.

2. Remove chain cover and driven balance shaft chain sprocket screw.
3. Loosen tensioner pivot and adjusting screws, move driven balance shaft inboard

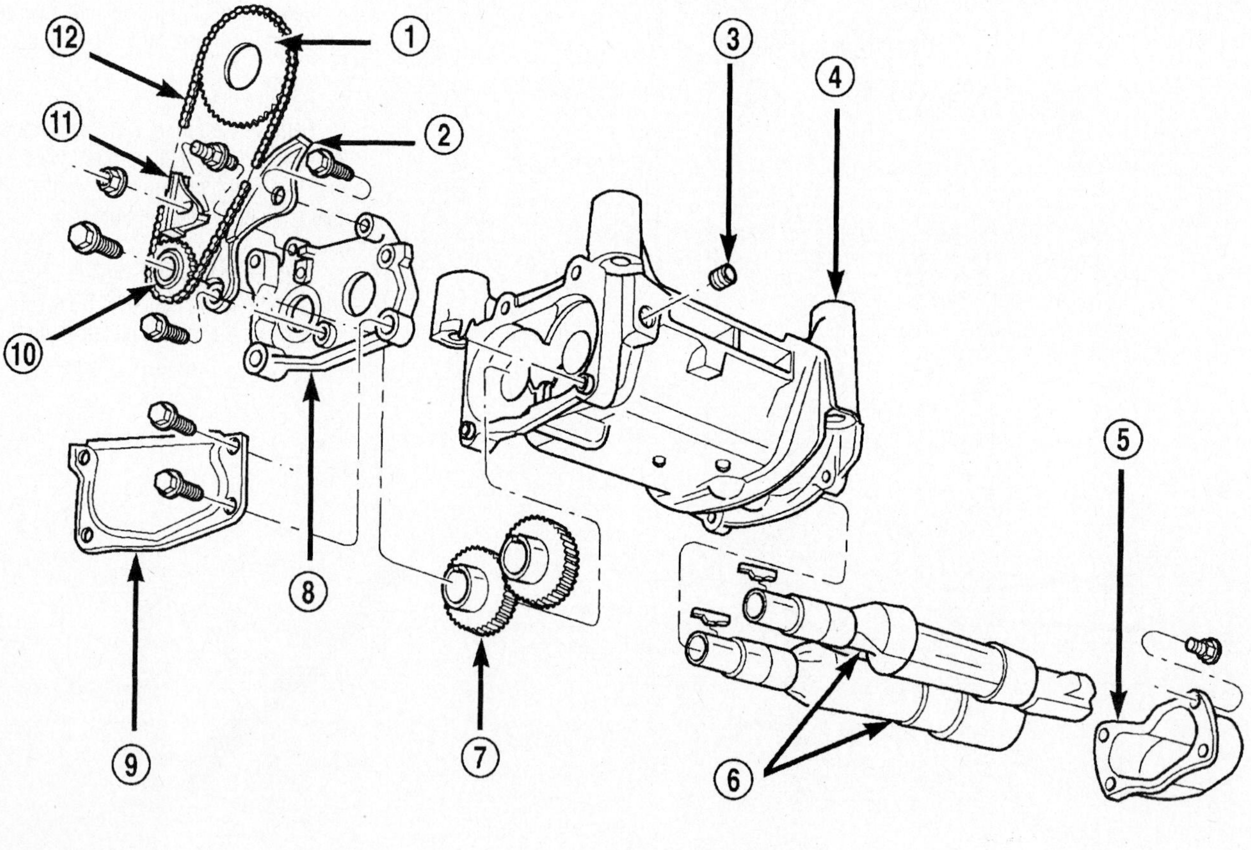

1 - SPROCKET
2 - TENSIONER
3 - PLUG
4 - CARRIER
5 - REAR COVER
6 - BALANCE SHAFTS
7 - GEARS
8 - GEAR COVER
9 - CHAIN COVER
10 - SPROCKET
11 - GUIDE
12 - CHAIN

67189-JEEP-G38

Balance shafts and carrier—2.4L engines

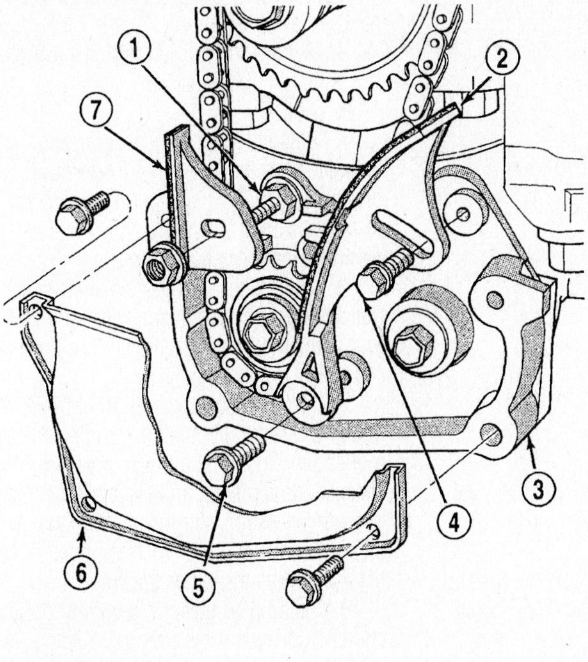

1 - STUD
2 - TENSIONER (ADJUSTER)
3 - GEAR COVER
4 - ADJUST SCREW
5 - PIVOT SCREW
6 - CHAIN COVER (CUTAWAY)
7 - GUIDE

67189-JEEP-G39

Balance shaft chain cover, guide and tensioner—2.4L engines

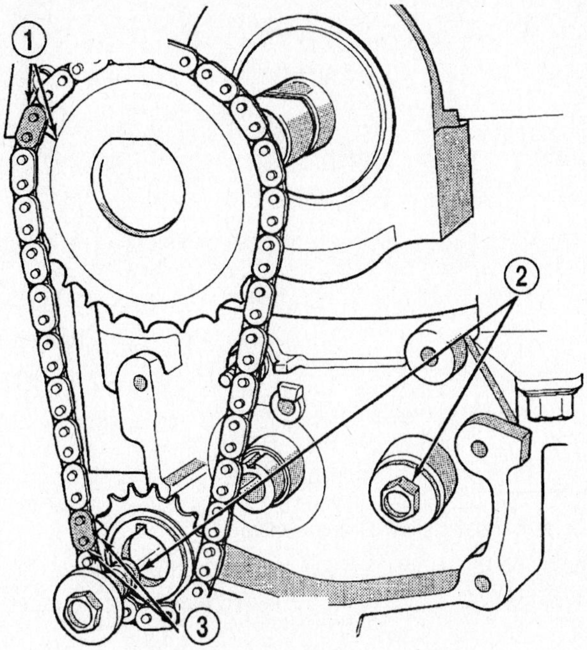

1 - NICKEL PLATED LINK AND MARK
2 - GEAR/SPROCKET SCREWS
3 - NICKEL PLATED LINK AND DOT

67189-JEEP-G41

Balance shaft chain and sprockets—2.4L engines

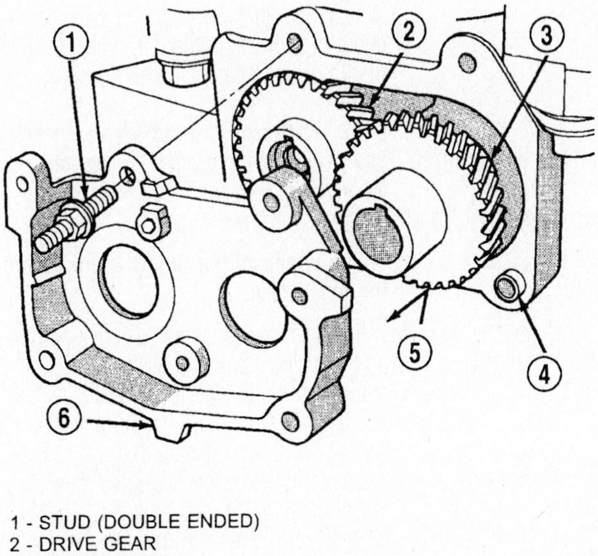

1 - STUD (DOUBLE ENDED)
2 - DRIVE GEAR
3 - DRIVEN GEAR
4 - CARRIER DOWEL
5 - GEAR(S)
6 - GEAR COVER

67189-JEEP-G40

Balance shaft gear cover and gears—2.4L engines

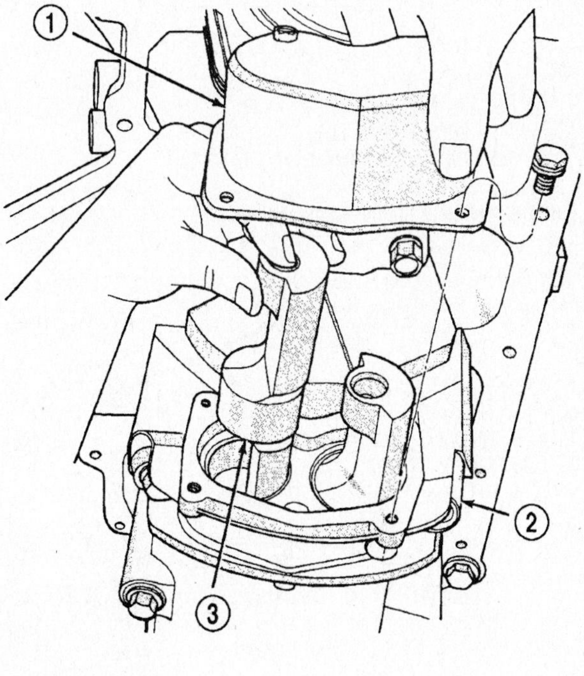

1 - REAR COVER
2 - CARRIER
3 - BALANCE SHAFT

67189-JEEP-G42

Balance shaft removal—2.4L engines

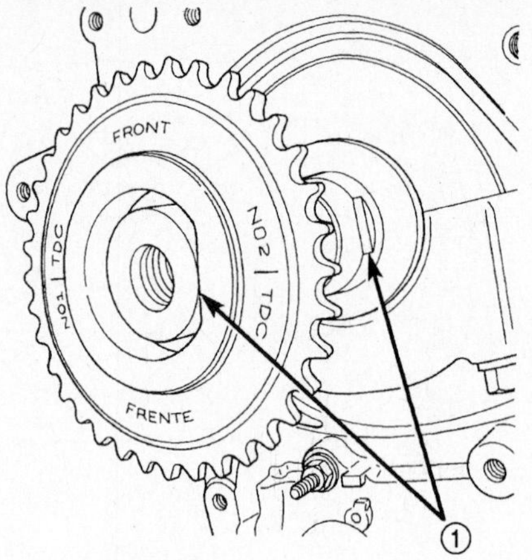

1 - ALIGN FLATS

67189-JEEP-G43

Balance shaft installation—2.4L engines

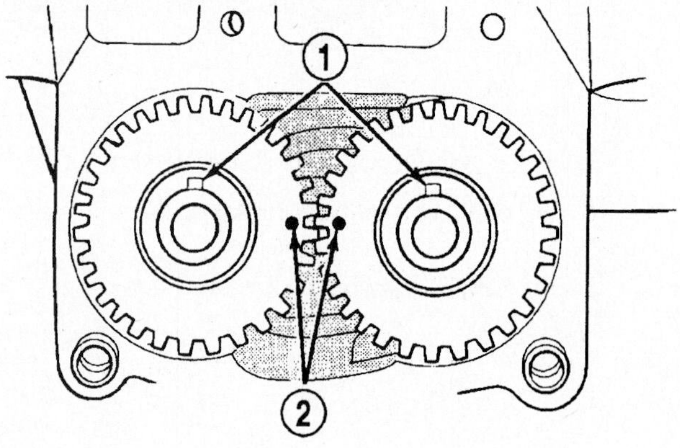

1 - KEY WAYS UP
2 - GEAR ALIGNMENT DOTS

67189-JEEP-G44

Balance shaft gear timing—2.4L engines

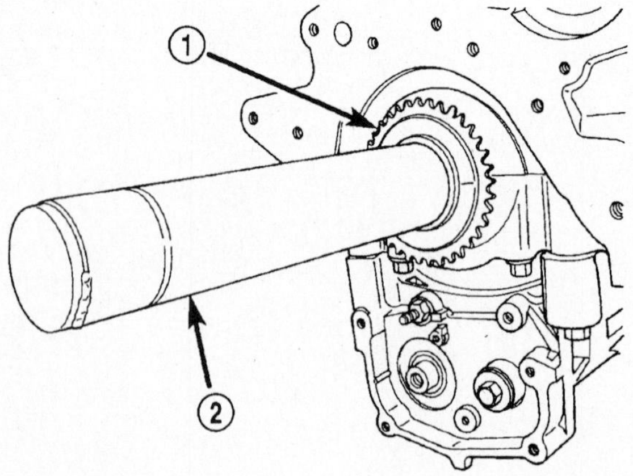

1 - SPROCKET
2 - SPECIAL TOOL 6052

67189-JEEP-G45

Balance shaft drive sprocket installation—2.4L engines

through driven chain sprocket. Sprocket will hang in lower chain loop.

4. Remove carrier to crankcase attaching bolts to remove carrier.

To install:

5. Balance shaft and carrier assembly installation is the reverse of the removal procedure. During installation crankshaft-to-balance shaft timing must be established. Timing is performed as follows:

6. With balance shafts installed in carrier position carrier on crankcase and install four attaching bolts and tighten to 54 Nm (40 ft. lbs.).

7. Turn balance shafts until both shaft key ways are up, parallel to vertical centerli e of engine. Install short hub drive gear on sprocket driven shaft and long hub gear on gear driven shaft. After installation gear and balance shaft keyways must be up with gear timing marks meshed as shown in.

8. Install gear cover and tighten double ended stud/washer fastener to 12 Nm (105 inch lbs.).

9. Align flat on balance shaft drive sprocket to the flat on crankshaft.

10. Install balance shaft drive sprocket on crankshaft using Special Tool 6052, or equivalent.

11. Turn crankshaft until number 1 cylinder is at top dead center (TDC). The timing marks on the chain sprocket should line up with the parting line on the left side of number one main bearing cap.

12. Place chain over crankshaft sprocket so that the plated link of the chain is over the number 1 cylinder timing mark on the balance shaft crankshaft sprocket.

13. Place balance shaft sprocket into the timing chain and align the timing mark on the sprocket (dot) with the (lower) plated link on the chain.

➡The lower plated link is 8 links from the upper link.

14. With balance shaft keyways pointing up (12 o'clock) slide the balance shaft sprocket onto the nose of the balance shaft. The balance shaft may have to be pushed in slightly to allow for clearance.

15. If the sprockets are timed correctly, install the balance shaft bolts and tighten to 28 Nm (250 in. lbs.). A wood block placed between crankcase and crankshaft counterbalance will prevent crankshaft and gear rotation.

CHAIN TENSIONING

1. Before servicing the vehicle, refer to the precautions in the beginning of this section.

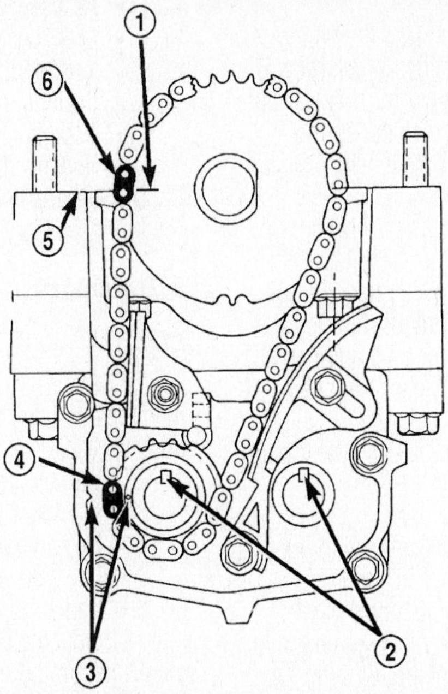

1 - MARK ON SPROCKET
2 - KEYWAYS UP
3 - ALIGN MARKS
4 - PLATED LINK
5 - PARTING LINE (BEDPLATE TO BLOCK)
6 - PLATED LINK

67189-JEEP-G46

Balance shaft timing—2.4L engines

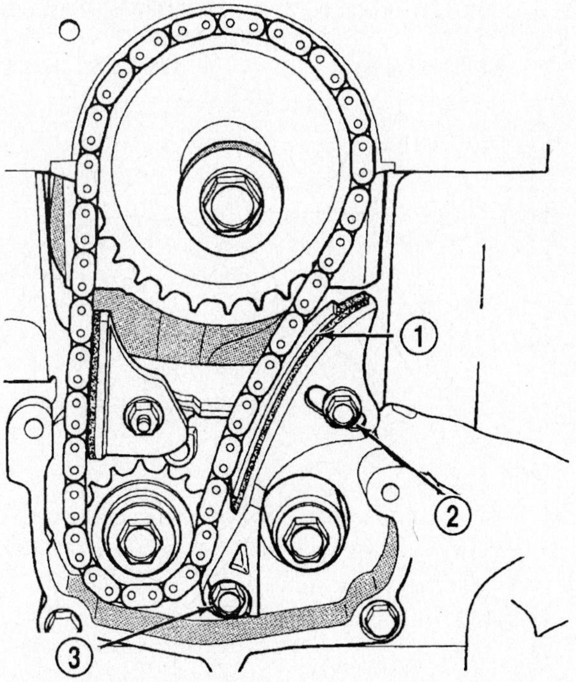

1 - 1MM (0.039 IN.) SHIM
2 - TENSIONER (ADJUSTER) BOLT
3 - PIVOT BOLT

67189-JEEP-G47

Balance shaft chain tension adjustment—2.4L engines

2. Install chain tensioner loosely assembled.
3. Position guide on double ended stud making sure tab on the guide fits into slot on the gear cover. Install and tighten nut/washer assembly to 12 Nm (105 inch lbs.).
4. Place a shim 1 mm (0.039 in.) thick x 70 mm (2.75 in.) long or between tensioner and chain. Push tensioner and shim up against the chain. Apply firm pressure 2.5–3 Kg (5.5–6.6 lbs.) directly behind the adjustment slot to take up all slack. Chain must have shoe radius contact as shown.
5. With the load applied, tighten top tensioner bolt first, then bottom pivot bolt. Tighten bolts to 12 Nm (105 in. lbs.). Remove shim.
6. Install carrier covers and tighten screws to 12 Nm (105 inch lbs.).
7. Install pick-up tube and oil pan.
8. Fill engine crankcase with proper oil to correct level.

Timing Chain, Sprockets, and Front Cover

REMOVAL & INSTALLATION

2.5L and 4.0L Engines

1. Before servicing the vehicle, refer to the precautions in the beginning of this section.
2. Remove or disconnect the following:
   - Negative battery cable
   - Accessory drive belt
   - Cooling fan and shroud
   - Crankshaft damper
   - Front crankshaft seal
   - Accessory brackets
   - Front cover

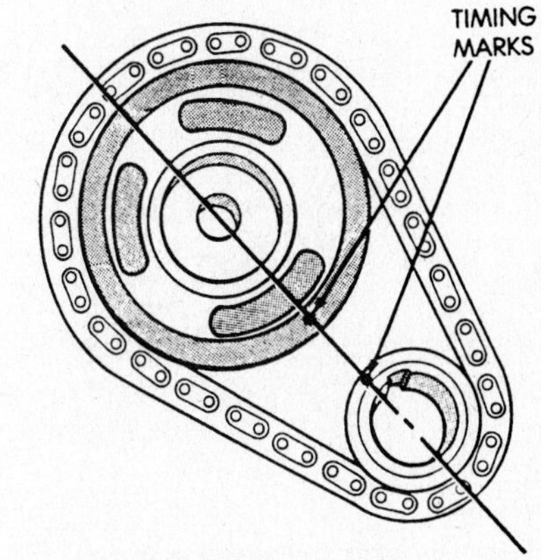

7924PG20

Timing chain alignment marks—2.5L and 4.0L engines

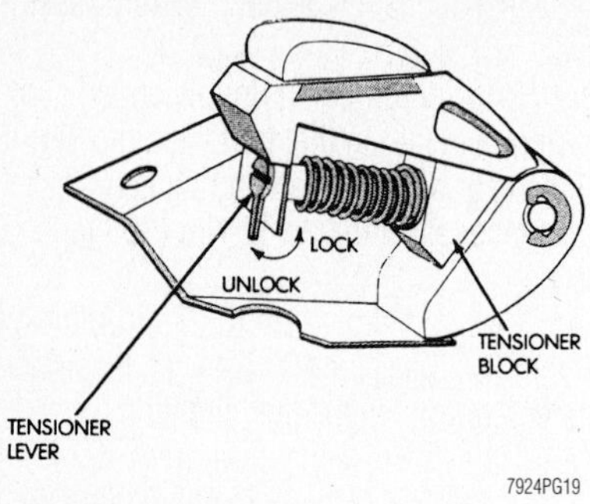

Timing chain tensioner—2.5L engines

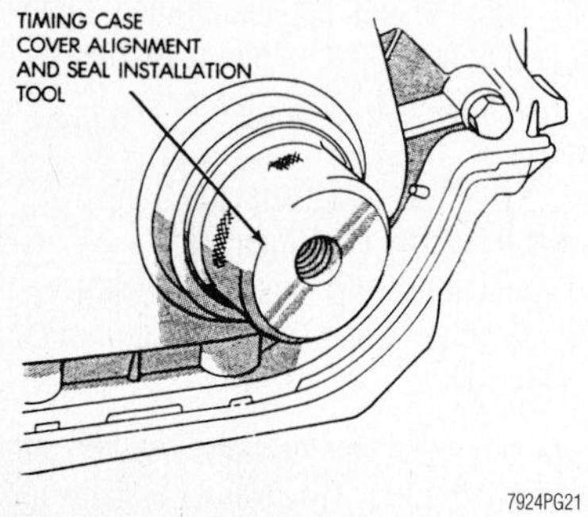

Timing Case Cover Alignment and Seal Installation Tool 6139—2.5L and 4.0L engines

- Oil slinger

3. Rotate the crankshaft so that the timing marks are aligned.
4. Remove the timing chain and sprockets.

To install:

5. For 2.5L engines, turn the timing chain tensioner lever to the unlock (down) position. Pull the tensioner block toward the tensioner lever to compress the spring. Hold the block and turn the tensioner lever to the lock (up) position.
6. Install or connect the following:
   - Timing chain and sprockets with the timing marks aligned. Torque the camshaft sprocket bolt to 80 ft. lbs. (108 Nm) for 2.5L engines or to 50 ft. lbs. (68 Nm) for 4.0L engines.
7. For 2.5L engines, release the timing chain tensioner.
   - Oil slinger
   - New front crankshaft seal to the front cover
   - Front cover, using a new gasket
   - Timing Case Cover Alignment and Seal Installation Tool 6139 in the crankshaft opening to center the front cover
8. Torque the front cover bolts as follows:
   a. Step 1: Cover-to-block ¼ inch bolts to 60 inch lbs. (7 Nm).
   b. Step 2: Cover-to-block 5⁄16 inch bolts to 16 ft. lbs. (22 Nm).
   c. Step 3: Oil pan-to-cover ¼ inch bolts to 85 inch lbs. (9.5 Nm).
   d. Step 4: Oil pan-to-cover 5⁄16 inch bolts to 11 ft. lbs. (15 Nm).
9. Install or connect the following:
   - Accessory brackets
   - Crankshaft damper. Torque the bolt to 80 ft. lbs. (108 Nm).
   - Cooling fan and shroud
   - Accessory drive belt
   - Negative battery cable
10. Start the engine, check for leaks and repair if necessary.

3.7L Engines

LIBERTY

1. Before servicing the vehicle, refer to the precautions in the beginning of this section.
2. Drain the cooling system.
3. Remove or disconnect the following:
   - Negative battery cable
   - Valve covers
   - Radiator fan
4. Rotate the crankshaft so that the crankshaft timing mark aligns with the Top Dead Center (TDC) mark on the front cover, and the **V6** marks on the camshaft sprockets are at 12 o'clock.
   - Power steering pump
   - Access plugs from the cylinder heads
   - Oil fill housing
   - Crankshaft damper
5. Compress the primary timing chain tensioner and install a lockpin.
6. Remove the secondary timing chain tensioners.
7. Hold the left camshaft with

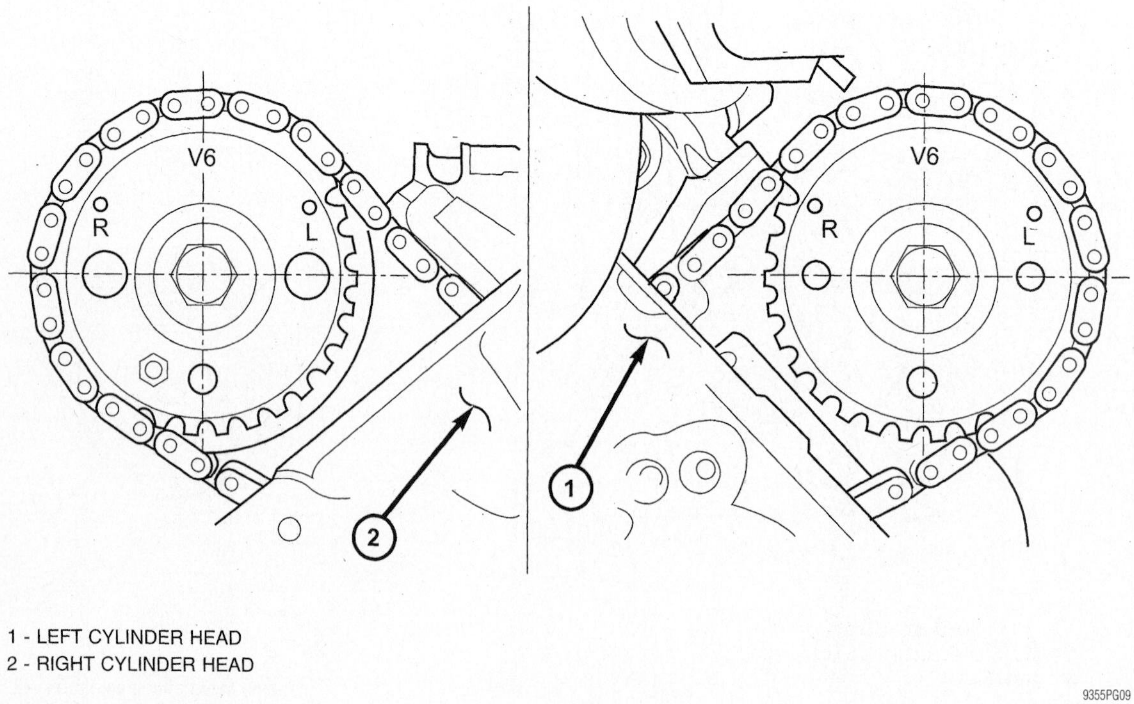

Camshaft sprocket timing marks—3.7L

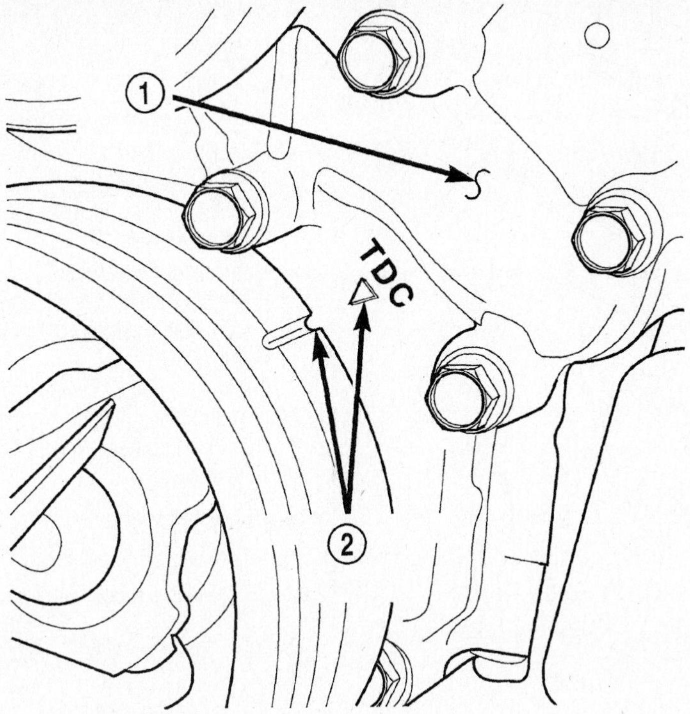

Crankshaft timing marks—3.7L

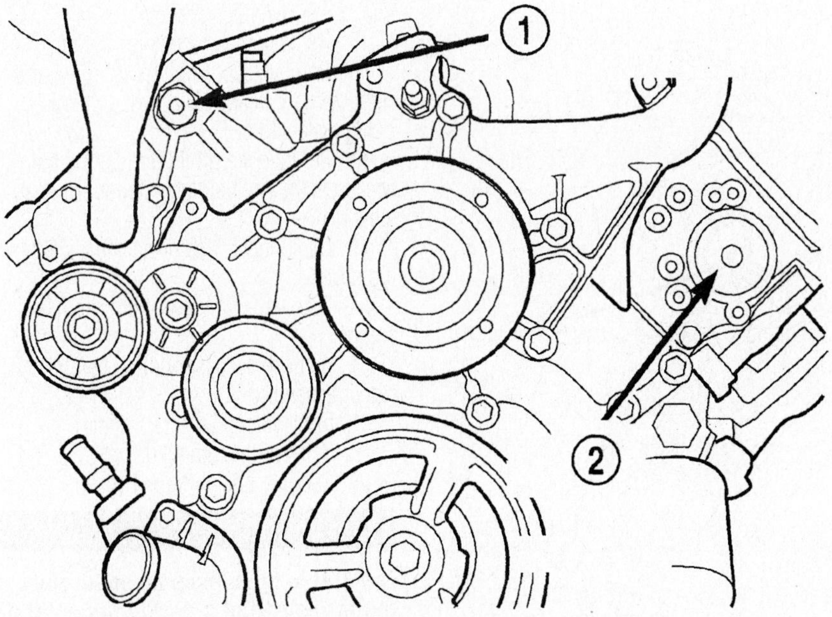

Cylinder head access plugs—3.7L

adjustable pliers and remove the sprocket and chain. Rotate the **left** camshaft 15 degrees **clockwise** to the neutral position.

8. Hold the right camshaft with adjustable pliers and remove the camshaft sprocket. Rotate the **right** camshaft 45 degrees **counterclockwise** to the neutral position.

9. Remove the primary timing chain and sprockets.

To install:

10. Use a small prytool to hold the ratchet pawl and compress the secondary timing chain tensioners in a vise and install locking pins.

➡The black bolts fasten the guide to the engine block and the silver bolts fasten the guide to the cylinder head.

11. Install or connect the following:
    - Secondary timing chain guides. Tighten the bolts to 21 ft. lbs. (28 Nm).
    - Secondary timing chains to the idler sprocket so that the double plated links on each chain are visible through the slots in the primary idler sprocket

12. Lock the secondary timing chains to the idler sprocket with Timing Chain Locking tool as shown.

13. Align the primary chain double plated links with the idler sprocket timing mark and the single plated link with the crankshaft sprocket timing mark.

14. Install the primary chain and sprockets. Tighten the idler sprocket bolt to 25 ft. lbs. (34 Nm).

15. Align the secondary chain single plated links with the timing marks on the secondary sprockets. Align the dot at the **L** mark on the left sprocket with the plated link on the left chain and the dot at the **R** mark on the right sprocket with the plated link on the right chain.

16. Rotate the camshafts back from the neutral position and install the camshaft sprockets.

17. Remove the secondary chain locking tool.

18. Remove the primary and secondary timing chain tensioner locking pins.

19. Hold the camshaft sprockets with a spanner wrench and tighten the retaining bolts to 90 ft. lbs. (122 Nm).

20. Install or connect the following:
    - Front cover. Tighten the bolts, in sequence, to 40 ft. lbs. (54 Nm).
    - Front crankshaft seal
    - Cylinder head access plugs
    - A/C compressor
    - Alternator

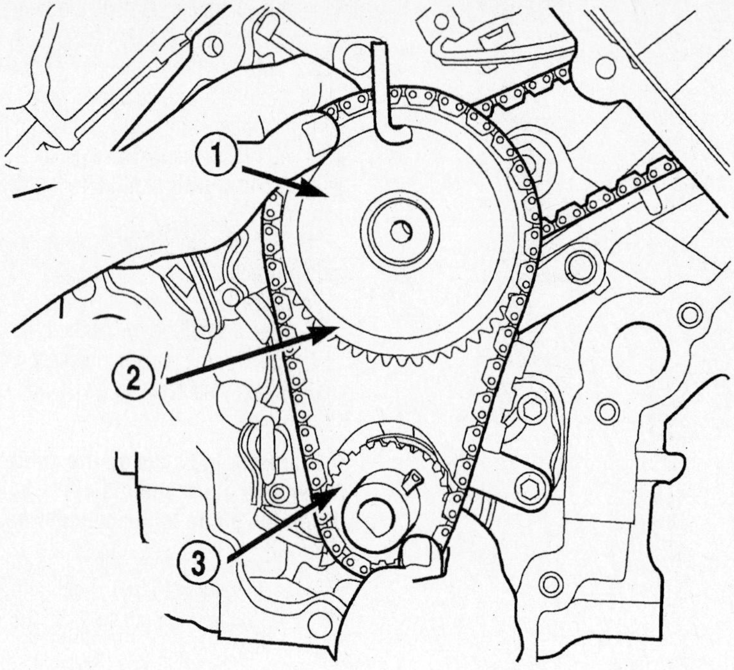

Installing the idler gear and timing chains—3.7L

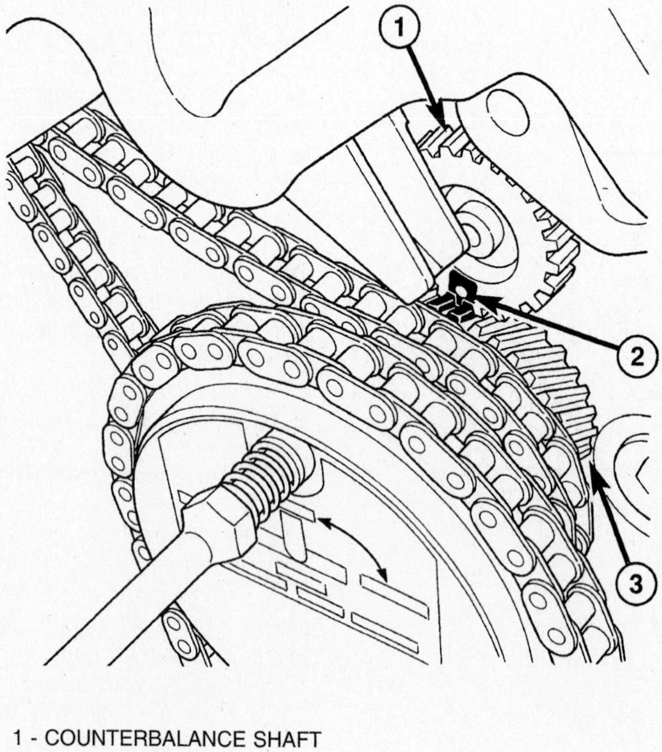

Counterbalance shaft timing marks—3.7L

- Accessory drive belt tensioner. Tighten the bolt to 40 ft. lbs. (54 Nm).
- Oil fill housing
- Crankshaft damper. Tighten the bolt to 130 ft. lbs. (175 Nm).
- Power steering pump
- Lower radiator hose
- Heater hoses
- Accessory drive belt
- Engine cooling fan and shroud
- Camshaft Position (CMP) sensor
- Valve covers
- Negative battery cable

21. Fill and bleed the cooling system.
22. Start the engine, check for leaks and repair if necessary.

GRAND CHEROKEE

1. Before servicing the vehicle, refer to the precautions in the beginning of this section.
2. Disconnect the battery negative cable.
3. Drain cooling system.
4. Remove electric cooling fan and fan shroud assembly.
5. Remove radiator fan.
6. Rotate engine until timing mark on crankshaft damper aligns with TDC mark on timing chain cover.
7. Make sure the camshaft sprocket "V6" marks are at the 12 o'clock position (No. 1 TDC exhaust stroke).
8. Remove power steering pump.
9. Remove access plugs from left and right cylinder heads for access to chain guide fasteners.
10. Remove the oil fill housing to gain access to the right side tensioner arm fastener.
11. Disconnect both heater hoses at timing cover.
12. Disconnect lower radiator hose at engine.
13. Remove accessory drive belt tensioner assembly.
14. Remove crankshaft damper.
15. Remove the generator.
16. Remove A/C compressor.

⁂ WARNING

The 3.7L engine uses an anaerobic sealer instead of a gasket to seal the front cover to the engine block, from the factory. For service, Mopar° Grey Engine RTV sealant must be substituted.

➡It is not necessary to remove the water pump for timing cover removal.

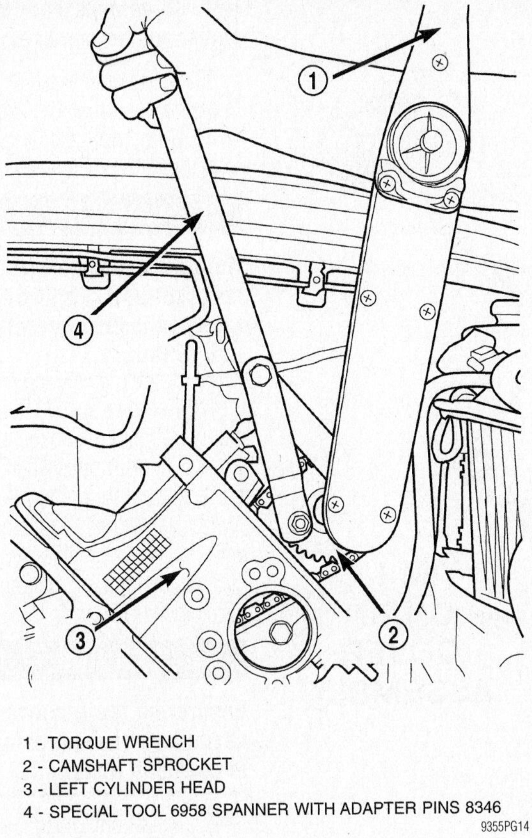

Tightening the left side camshaft sprocket—3.7L

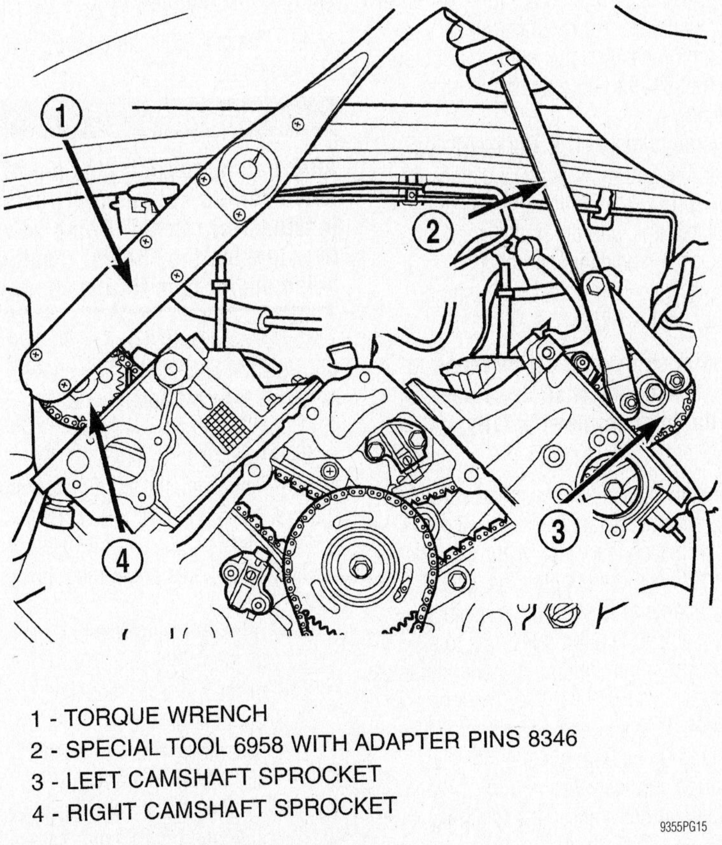

Tightening the right side camshaft sprocket—3.7L

17. Remove the bolts holding the timing cover to engine block.
18. Remove the timing cover.
19. Collapse and pin primary chain tensioner.

✲✲ WARNING

Plate behind left secondary chain tensioner could fall into oil pan. Therefore, cover pan opening.

20. Remove secondary chain tensioners.
21. Remove camshaft position sensor.

✲✲ WARNING

Care should be taken not to damage camshaft target wheel. Do not hold target wheel while loosening or tightening camshaft sprocket. Do not place the target wheel near a magnetic source of any kind. A damaged or magnetized target wheel could cause a vehicle no start condition.

✲✲ WARNING

Do not forcefully rotate the camshafts or crankshaft independently of each other. Damaging intake valve to piston contact will occur. Ensure negative battery cable is disconnected to guard against accidental starter engagement.

22. Remove left and right camshaft sprocket bolts.
23. While holding the left camshaft steel tube with Special Tool 8428 Camshaft Wrench, remove the left camshaft sprocket. Slowly rotate the camshaft approximately 5 degrees clockwise to a neutral position.
24. While holding the right camshaft steel tube with Special Tool 8428 Camshaft Wrench, remove the right camshaft sprocket.
25. Remove idler sprocket assembly bolt.
26. Slide the idler sprocket assembly and crank sprocket forward simultaneously to remove the primary and secondary chains.
27. Remove both pivoting tensioner arms and chain guides.
28. Remove primary chain tensioner.

To install:

29. Using a vise, lightly compress the secondary chain tensioner piston until the piston step is flush with the tensioner body. Using a pin or suitable tool, release ratchet pawl by pulling pawl back against spring force through access hole on side of ten-

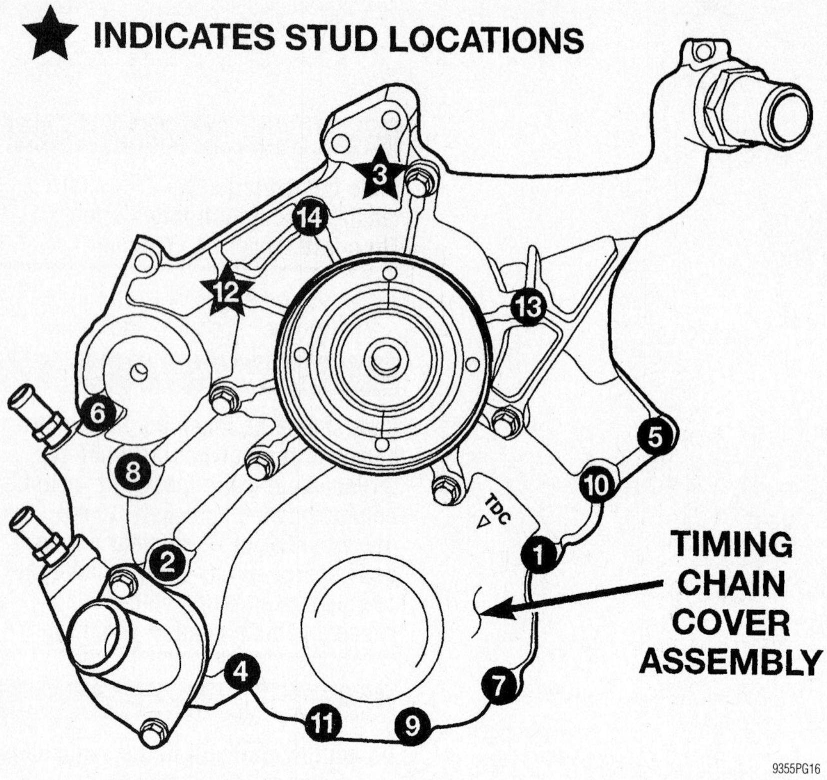

Timing cover bolt torque sequence—3.7L

sioner. While continuing to hold pawl back, push ratchet device to approximately 2 mm from the tensioner body. Install Special Tool 8514 lock pin into hole on front of tensioner. Slowly open vise to transfer piston spring force to lock pin.

30. Position primary chain tensioner over oil pump and insert bolts into lower two holes on tensioner bracket. Tighten bolts to 28 Nm (250 inch lbs.).

31. Install right side chain tensioner arm. Install Torx® bolt. Tighten bolt to 28 Nm (250 inch lbs.).

✱✱ WARNING

The silver bolts retain the guides to the cylinder heads and the black bolts retain the guides to the engine block.

32. Install the left side chain guide. Tighten the bolts to 28 Nm (250 inch lbs.).

33. Install left side chain tensioner arm, and Torx® bolt. Tighten bolt to 28 Nm (250 inch lbs.).

34. Install the right side chain guide. Tighten the bolts to 28 Nm (250 inch lbs.).

35. Install both secondary chains onto the idler sprocket. Align two plated links on the secondary chains to be visible through the two lower openings on the idler sprocket (4 o'clock and 8 o'clock). Once the secondary timing chains are installed, position special tool 8429 to hold chains in place for installation.

36. Align primary chain double plated links with the timing mark at 12 o'clock on the idler sprocket. Align the primary chain single plated link with the timing mark at 6 o'clock on the crankshaft sprocket.

37. Lubricate idler shaft and bushings with clean engine oil.

➡The idler sprocket must be timed to the counterbalance shaft drive gear before the idler sprocket is fully seated.

38. Install all chains, crankshaft sprocket, and idler sprocket as an assembly. After guiding both secondary chains through the block and cylinder head openings, affix chains with an elastic strap or equivalent. This will maintain tension on chains to aid in installation. Align the timing mark on the idler sprocket gear to the timing mark on the counterbalance shaft drive gear, then seat idler sprocket fully. Before installing idler sprocket bolt, lubricate washer with oil, and tighten idler sprocket assembly retaining bolt to 34 Nm (25 ft. lbs.).

➡It will be necessary to slightly rotate camshafts for sprocket installation.

39. Align left camshaft sprocket "L" dot to plated link on chain.

40. Align right camshaft sprocket "R" dot to plated link on chain.

✱✱ WARNING

Remove excess oil from the camshaft sprocket bolt. Failure to do so can result in overtorque of bolt resulting in bolt failure.

41. Remove Special Tool 8429, then attach both sprockets to camshafts. Remove excess oil from bolts, then install sprocket bolts, but do not tighten at this time.

42. Verify that all plated links are aligned with the marks on all sprockets and the "V6" marks on camshaft sprockets are at the 12 o'clock position.

✱✱ WARNING

Ensure the plate between the left secondary chain tensioner and block is correctly installed.

43. Install both secondary chain tensioners. Tighten bolts to 28 Nm (250 inch lbs.).

➡Left and right secondary chain tensioners are not common.

44. Remove all locking pins from tensioners.

✱✱ WARNING

After pulling locking pins out of each tensioner, do not manually extend the tensioner(s) ratchet. Doing so will over tension the chains, resulting in noise and/or high timing chain loads.

45. Using Special Tool 6958, Spanner with Adaptor Pins 8346, tighten left camshaft sprocket bolts to 122 Nm (90 ft. lbs.).

46. Using Special Tool 6958, Spanner with Adaptor Pins 8346, tighten right camshaft sprocket bolts to 122 Nm (90 ft. lbs.).

47. Rotate engine two full revolutions. Verify timing marks are at the follow locations:

- primary chain idler sprocket dot is at 12 o'clock
- primary chain crankshaft sprocket dot is at 6 o'clock
- secondary chain camshaft sprockets "V6" marks are at 12 o'clock
- counterbalance shaft drive gear dot is aligned to the idler sprocket gear dot

48. Lubricate all three chains with engine oil.

49. After installing all chains, it is recommended that the idler gear end play be checked. The end play must be within 0.10–0.25 mm (0.004–0.010 in.). If not within specification, the idler gear must be replaced.

**** WARNING**

Do not use oil based liquids to clean timing cover or block surfaces. Use only rubbing alcohol, along with plastic or wooden scrapers. Use no wire brushes or abrasive wheels or metal scrapers, or damage to surfaces could result.

50. Clean timing chain cover and block surface using rubbing alcohol.

**** WARNING**

The 3.7L uses a special anaerobic sealer instead of a gasket to seal the timing cover to the engine block, from the factory. For service repairs, Mopar® Engine RTV must be used as a substitute.

51. Inspect the water passage o-rings for any damage, and replace as necessary.

52. Apply Mopar® Engine RTV sealer to front cover as shown using a 3 to 4mm thick bead.

53. Install cover. Tighten fasteners in sequence as shown in to 58 Nm (43 ft. lbs.).

54. Install crankshaft damper.

55. Install the A/C compressor.

56. Install the generator.

57. Install accessory drive belt tensioner.

58. Install radiator upper and lower hoses.

59. Install both heater hoses.

60. Install radiator fan.

61. Install cylinder head covers.

➡Before installing threaded plug in right cylinder head, the plug must be coated with sealant to prevent leaks.

62. Coat the large threaded access plug with Mopar® Thread Sealant with Teflon, then install into the right cylinder head and tighten to 81 Nm (60 ft. lbs.).

63. Install the oil fill housing.

64. Install access plug in left cylinder head.

65. Install power steering pump.

66. Fill cooling system.

67. Connect negative cable to battery.

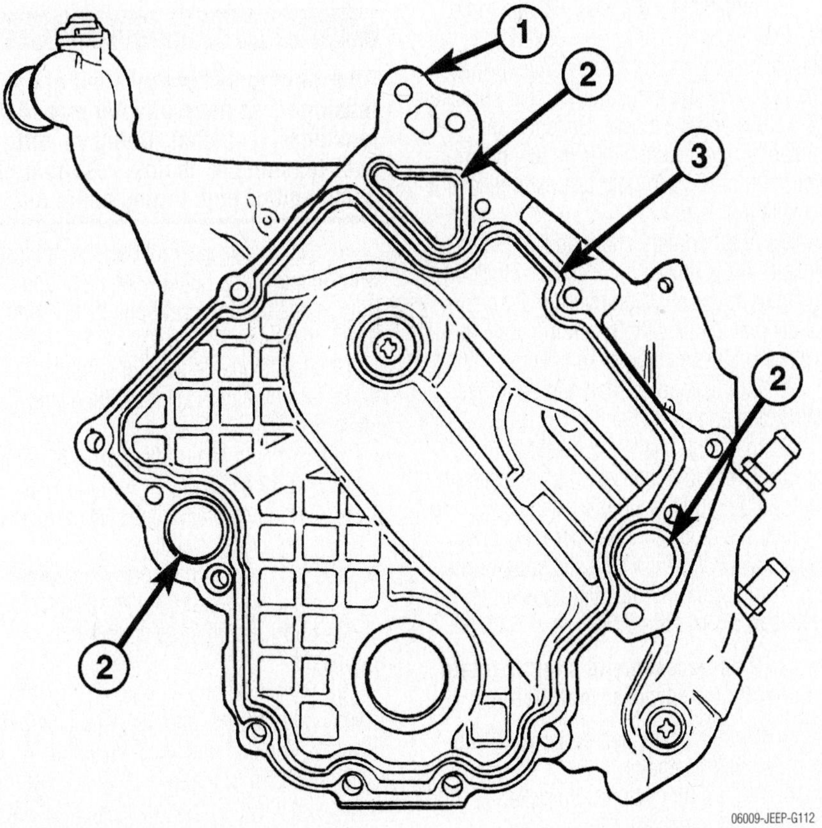

Front cover sealant application—3.7L engine

4.7L Engine

1. Before servicing the vehicle, refer to the precautions in the beginning of this section.

2. Disconnect the battery negative cable.

3. Drain cooling system.

4. Disconnect both heater hoses at timing cover.

5. Disconnect lower radiator hose at engine.

6. Remove crankshaft damper.

7. Remove accessory drive belt tensioner assembly.

8. Rotate engine until timing mark on crankshaft damper aligns with TDC mark on timing chain cover cylinder exhaust stroke) and the camshaft sprocket "V8" marks are at the 12 o'clock position.

9. Remove power steering pump.

10. Remove access plugs from left and right cylinder heads for access to chain guide fasteners.

11. Remove the oil fill housing to gain access to the right side tensioner arm fastener.

12. Remove the generator and A/C compressor.

**** WARNING**

The 4.7L engine uses an RTV sealer instead of a gasket to seal the front cover to the engine block, from the factory. For service, Mopar® Grey Engine RTV sealant must be substituted.

➡It is not necessary to remove the water pump for timing cover removal.

13. Remove the bolts holding the timing cover to engine block.

14. Remove cover.

15. Disconnect negative cable from battery.

16. Drain cooling system.

17. Remove right and left cylinder head covers.

18. Remove radiator fan.

19. Collapse and pin primary chain tensioner.

**** WARNING**

Plate behind left secondary chain tensioner could fall into oil pan. Therefore, cover pan opening.

20. Remove secondary chain tensioners.

21. Remove camshaft position sensor from right cylinder head.

**** WARNING**

Care should be taken not to damage camshaft target wheel. Do not hold target wheel while loosening or tightening camshaft sprocket. Do not place the target wheel near a magnetic source of any kind. A damaged or magnetized target wheel could cause a vehicle no start condition.

**** WARNING**

Do not forcefully rotate the camshafts or crankshaft independently of each other. Damaging intake valve to piston contact will occur. Ensure negative battery cable is disconnected to guard against accidental starter engagement.

22. Remove left and right camshaft sprocket bolts.
23. While holding the left camshaft steel tube with adjustable pliers, remove the left camshaft sprocket. Slowly rotate the camshaft approximately 15 degrees clockwise to a neutral position.
24. While holding the right camshaft steel tube with adjustable pliers, remove the right camshaft sprocket. Slowly rotate the camshaft approximately 45 degrees counterclockwise to a neutral position.
25. Remove idler sprocket assembly bolt.
26. Slide the idler sprocket assembly and crank sprocket forward simultaneously to remove the primary and secondary chains.
27. Remove both pivoting tensioner arms and chain guides.
28. Remove chain tensioner.

To install:

29. Using a vise, lightly compress the secondary chain tensioner piston until the piston step is flush with the tensioner body. Using a pin or suitable tool, release ratchet pawl by pulling pawl back against spring force through access hole on side of tensioner. While continuing to hold pawl back, Push ratchet device to approximately 2 mm from the tensioner body. Install Special Tool 8514 lock pin into hole on front of tensioner. Slowly open vise to transfer piston spring force to lock pin.
30. Position primary chain tensioner over oil pump and insert bolts into lower two holes on tensioner bracket. Tighten bolts to 28 Nm (250 inch lbs.).

**** WARNING**

Over-tightening the tensioner arm Torx® bolt can cause severe damage to the cylinder head. Tighten bolt to specified torque only.

31. Install right side chain tensioner arm. Apply Mopar® Lock N, Seal to Torx® bolt, tighten bolt to 17 Nm (150 inch lbs.).

➡The silver bolts retain the guides to the cylinder heads and the black bolts retain the guides to the engine block.

32. Install the left side chain guide. Tighten the bolts to 28 Nm (250 inch lbs.).

**** WARNING**

Over-tightening the tensioner arm can cause severe damage to the cylinder head. Tighten bolt to specified torque only.

33. Install left side chain tensioner arm. Apply Mopar® Lock N, Seal to bolt, tighten bolt to 17 Nm (150 inch lbs.).
34. Install the right side chain guide. Tighten the bolts to 28 Nm (250 inch lbs.).
35. Install both secondary chains onto the idler sprocket. Align two plated links on the secondary chains to be visible through the two lower openings on the idler sprocket (4 o'clock and 8 o'clock). Once the secondary timing chains are installed, position special tool 8515 to hold chains in place for installation.
36. Align primary chain double plated links with the timing mark at 12 o'clock on the idler sprocket. Align the primary chain single plated link with the timing mark at 6 o'clock on the crankshaft sprocket.
37. Lubricate idler shaft and bushings with clean engine oil.
38. Install all chains, crankshaft sprocket, and idler sprocket as an assembly. After guiding both secondary chains through the block and cylinder head openings, affix chains with a elastic strap or the equivalent, This will maintain tension on chains to aid in installation.

➡It will be necessary to slightly rotate camshafts for sprocket installation.

39. Align left camshaft sprocket "L" dot to plated link on chain.
40. Align right camshaft sprocket "R" dot to plated link on chain.

**** WARNING**

Remove excess oil from the camshaft sprocket bolt. Failure to do so can result in overtorque of bolt resulting in bolt failure.

41. Remove Special Tool 8515, then attach both sprockets to camshafts. Remove excess oil from bolts, then Install sprocket bolts, but do not tighten at this time.
42. Verify that all plated links are aligned with the marks on all sprockets and the "V8" marks on camshaft sprockets are at the 12 o'clock position.

**** WARNING**

Ensure the plate between the left secondary chain tensioner and block is correctly installed.

43. Install both secondary chain tensioners. Tighten bolts to 28 Nm (250 inch lbs.).

➡Left and right secondary chain tensioners are not common.

44. Before installing idler sprocket bolt, lubricate washer with oil, and tighten idler sprocket assembly retaining bolt to 34 Nm (25 ft. lbs.).
45. Remove all locking pins from tensioners.

**** WARNING**

After pulling locking pins out of each tensioner, do not manually extend the tensioner(s) ratchet. Doing so will over tension the chains, resulting in noise and/or high timing chain loads.

46. Using Special Tool 6958, Spanner with Adaptor Pins 8346, tighten left and right. camshaft sprocket bolts to 122 Nm (90 ft. lbs.).
47. Rotate engine two full revolutions. Verify timing marks are at the following locations:
    - primary chain idler sprocket dot is at 12 o'clock
    - primary chain crankshaft sprocket dot is at 6 o'clock
    - secondary chain camshaft sprockets "V8" marks are at 12 o'clock
48. Lubricate all three chains with engine oil.
49. After installing all chains, it is recommended that the idler gear end play be checked. The end play must be within 0.10–0.25 mm (0.004–0.010 in.). If not within specification, the idler gear and idler shaft must be replaced.

⁂ WARNING

Do not use oil based liquids to clean timing cover or block surfaces. Use only rubbing alcohol, along with plastic or wooden scrapers. Use no wire brushes or abrasive wheels or metal scrapers, or damage to surfaces could result.

50. Clean timing chain cover and block surface using rubbing alcohol.

⁂ WARNING

The 4.7L can use a special RTV sealer instead of a carrier gasket to seal the timing cover to the engine block, from the factory. For service repairs, Mopar® Grey Engine RTV must be used as a substitute, if RTV is present. If the front cover being used has no provisions for the water passage O-rings, then Mopar® Grey Engine RTV must be applied around the water passages.

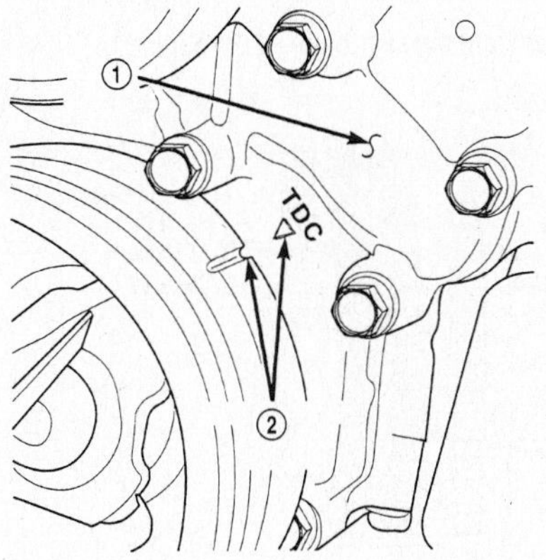

1 – TIMING CHAIN COVER
2 – CRANKSHAFT TIMING MARKS

9308PG04

Crankshaft timing marks—4.7L engine

51. Inspect the water passage o-rings, if equipped for any damage, and replace as necessary.
52. Apply Mopar® Grey Engine RTV sealer to the front cover following the path above, using a 3 to 4mm thick bead.
53. Install cover. Tighten flange head fasteners in sequence as shown in to 58 Nm (43 ft. lbs.).
54. Install the A/C compressor and generator.
55. Install crankshaft damper.
56. Install accessory drive belt tensioner assembly. Tighten fastener to 54 Nm (40 ft. lbs.).
57. Install lower radiator hose.
58. Install both heater hoses.
59. Install cylinder head covers.

➡Before installing threaded plug in right cylinder head, the plug must be coated with sealant to prevent leaks.

60. Coat the large threaded access plug with Mopar® Thread Sealant with Teflon, then install into the right cylinder head and tighten to 81 Nm (60 ft. lbs.).
61. Install the oil fill housing.
62. Install access plug in left cylinder head.
63. Install power steering pump.
64. Install radiator fan.
65. Fill cooling system.
66. Connect negative cable to battery.

5.7L Engine

1. Before servicing the vehicle, refer to the precautions in the beginning of this section.
2. Disconnect the battery negative cable.
3. Remove the engine cover.
4. Remove air cleaner assembly.
5. Drain cooling system.
6. Remove accessory drive belt.
7. Remove fan and fan drive assembly.
8. Remove coolant bottle and washer bottle.
9. Remove fan shroud.

➡It is not necessary to disconnect A/C lines or discharge refrigerant.

10. Remove A/C compressor and set aside.
11. Remove the generator.
12. Remove upper radiator hose.
13. Disconnect both heater hoses at timing cover.
14. Disconnect lower radiator hose at engine.
15. Remove accessory drive belt tensioner and both idler pulleys.
16. Remove crankshaft damper.

➡Do not remove the hoses from the power steering pump.

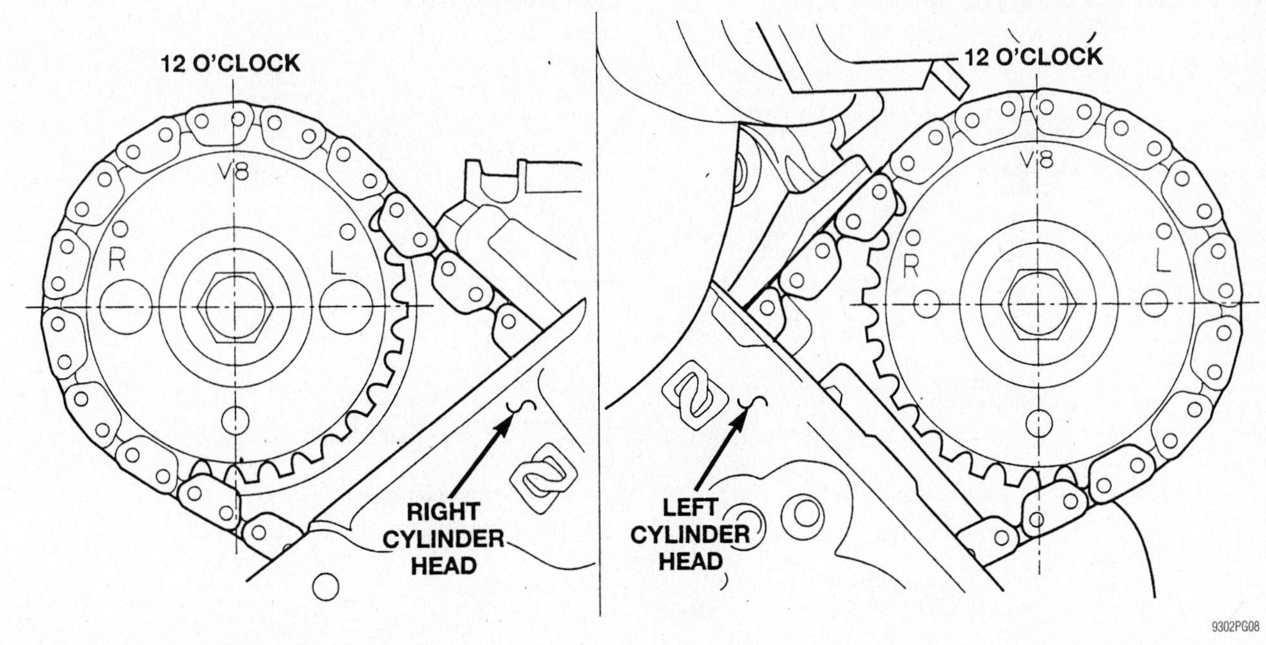

9302PG08

Camshaft positioning—4.7L engine

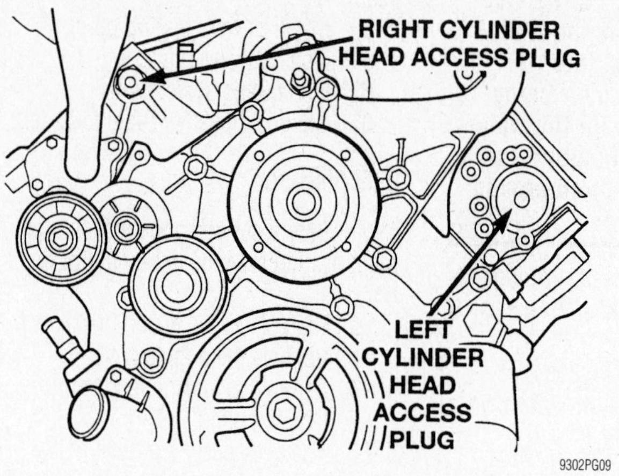

Cylinder head access plug locations—4.7L engine

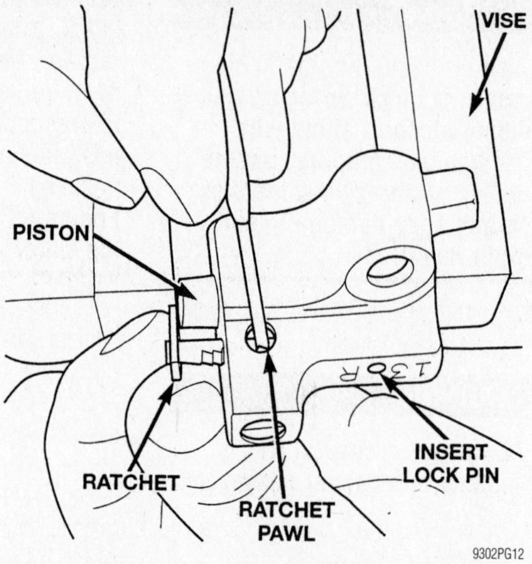

Secondary timing chain tensioner preparation—4.7L engine

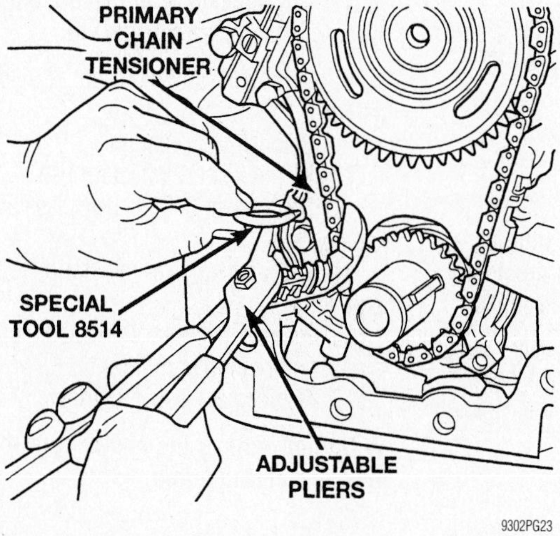

Compress and lock the primary chain tensioner—4.7L engine

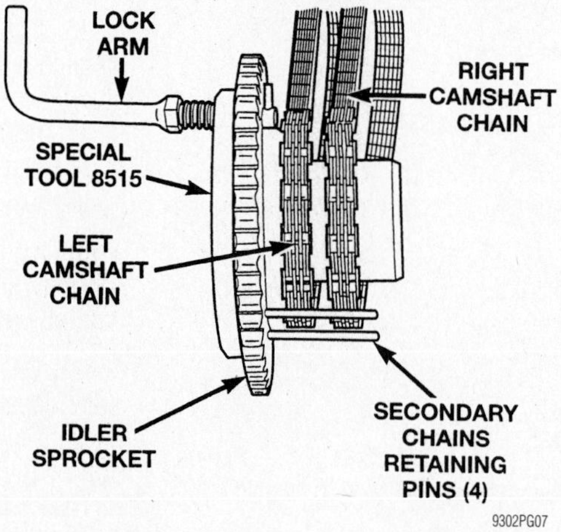

Use the Timing Chain Locking tool to lock the timing chains on the idler gear—4.7L engine

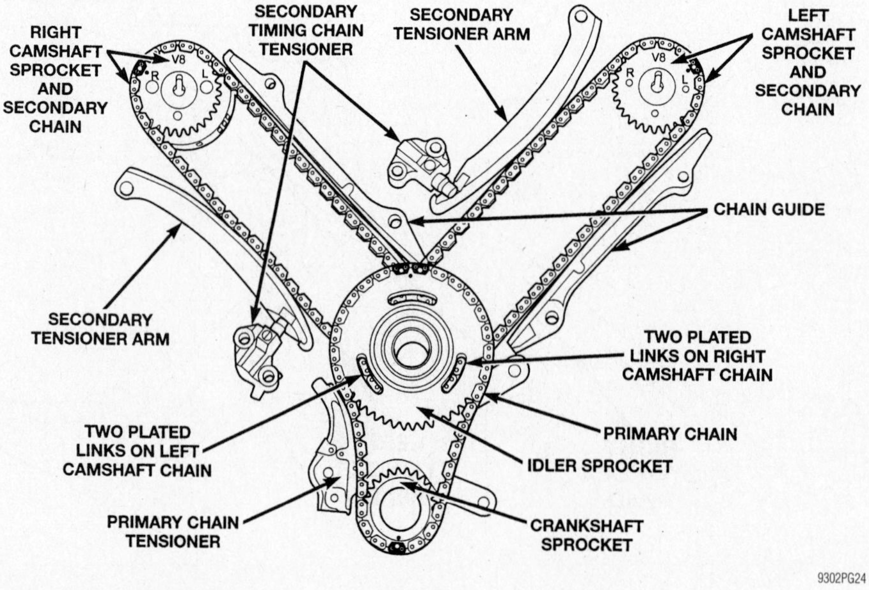

Timing chain system and alignment marks—4.7L engine

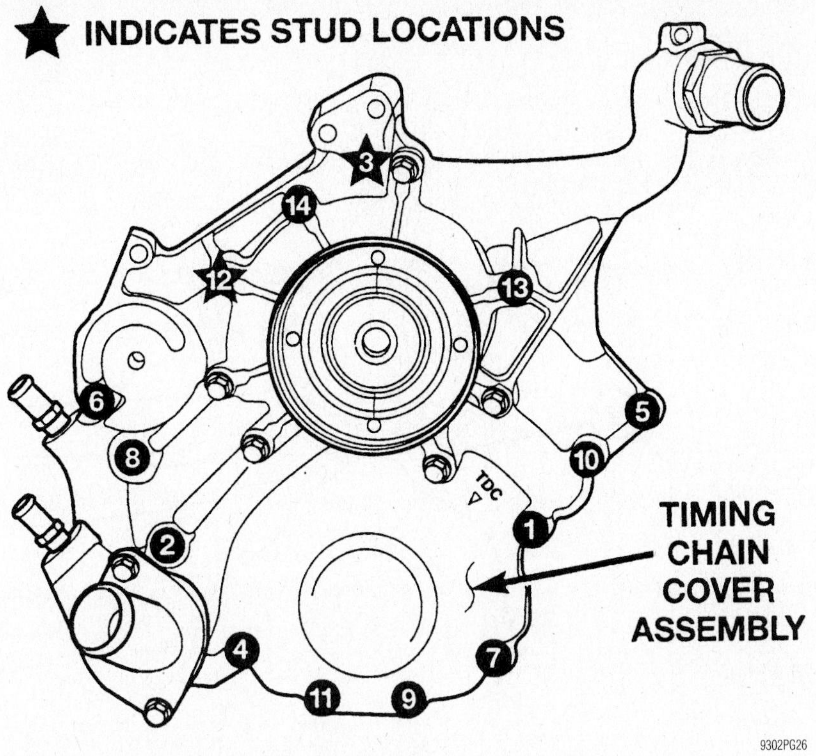

Timing chain cover bolt torque sequence—4.7L engine

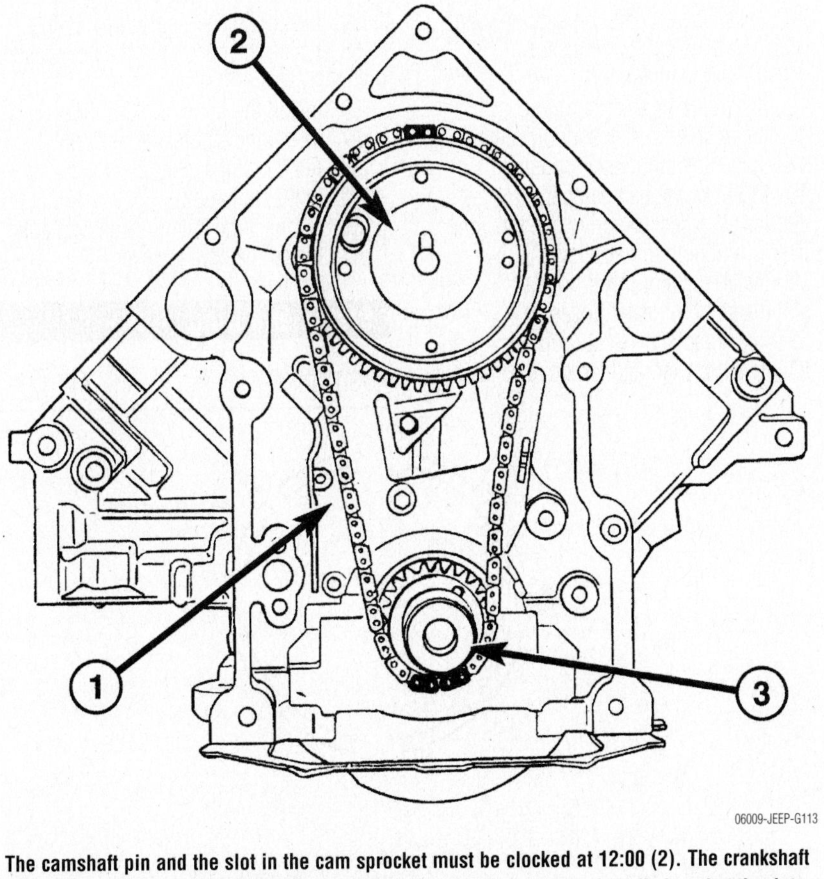

The camshaft pin and the slot in the cam sprocket must be clocked at 12:00 (2). The crankshaft keyway must be clocked at 2:00 (3). The crankshaft sprocket must be installed so that the dots and or paint marking is at 6:00—5.7L Engine

17. Remove power steering pump and set aside.
18. Remove the dipstick support bolt.
19. Drain the engine oil.
20. Remove the oil pan and pick up tube.

➡It is not necessary to remove water pump for timing cover removal.

21. Remove timing cover bolts and remove cover.
22. Verify that timing cover slide bushings are located in timing cover.
23. Reinstall the vibration damper bolt finger tight. Using a suitable socket and breaker bar, rotate the crankshaft to align timing chain sprockets and keyways as shown.

**** WARNING**

The camshaft pin and the slot in the cam sprocket must be clocked at 12:00 (2). The crankshaft keyway must be clocked at 2:00 (3). The crankshaft sprocket must be installed so that the dots and or paint marking is at 6:00.

24. Remove oil pump.
25. Retract tensioner shoe until hole in shoe lines up with hole in bracket.
26. Slide a suitable pin into the holes.
27. Remove camshaft sprocket attaching bolt and remove timing chain with crankshaft and camshaft sprockets.
28. If tensioner assembly is to be replaced, remove the tensioner to block bolts and remove tensioner assembly.

To install:

29. If tensioner assembly is being replaced, install tensioner and mounting bolts. Torque bolts to 28 Nm (250 inch lbs.).
30. Retract tensioner if required.

**** WARNING**

The timing chain must be installed with the single plated link aligned with the dot and or paint marking on the camshaft sprocket. The crankshaft sprocket is aligned with the dot and or paint marking on the sprocket between two plated timing chain links.

**** WARNING**

The camshaft pin and the slot in the cam sprocket must be clocked at 12:00. The crankshaft keyway must be clocked at 2:00. The crankshaft sprocket must be installed so that the dots and or paint marking is at 6:00.

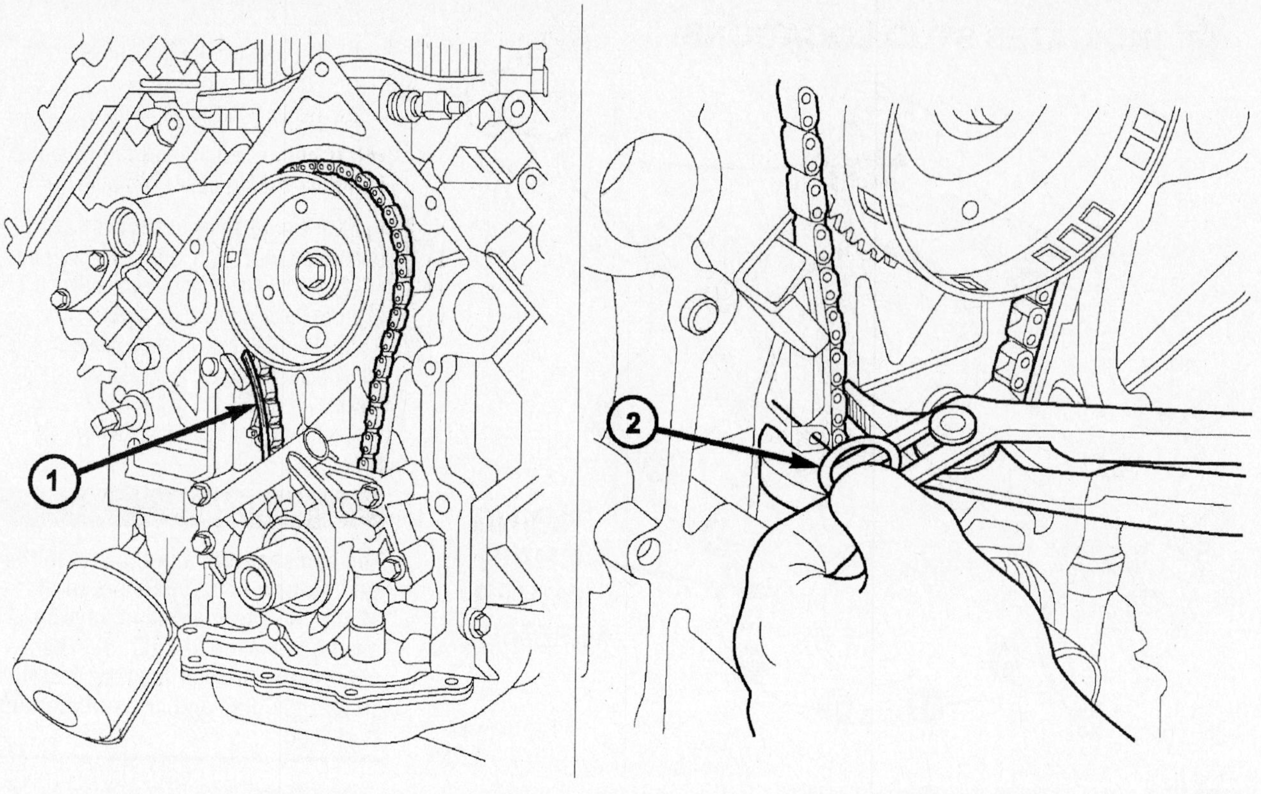

Retract tensioner shoe (1) until hole in shoe lines up with hole in bracket. Slide a suitable pin (2) into the holes—5.7L Engine

31. Place both camshaft sprocket and crankshaft sprocket on the bench with timing marks on exact imaginary center line through both camshaft and crankshaft bores.
32. Place timing chain around both sprockets.
33. Lift sprockets and chain (keep sprockets tight against the chain in position as shown).
34. Slide both sprockets evenly over their respective shafts and check alignment of timing marks.
35. Install the camshaft bolt. Tighten the bolt to 122 Nm (90 ft. lbs.) torque.
36. Remove tensioner pin. Again, verify alignment of timing marks.
37. Install the oil pump.
38. Install the oil pan and pick up.
39. Clean timing chain cover and block surface.

➡Always install a new gasket on timing cover.

40. Verify that the slide bushings are installed in timing cover.
41. Install cover and new gasket. Tighten fasteners to 28 Nm (250 inch lbs.).

➡The large lifting stud is torqued to 55 Nm (40 ft. lbs.).

42. Install the oil pan and pick up tube.
43. Install the A/C compressor.
44. Install the generator.
45. Install power steering pump.
46. Install the dipstick support bolt.
47. Install the thermostat housing.
48. Install crankshaft damper.
49. Install accessory drive belt tensioner assembly and both idler pulleys.
50. Install radiator lower hose.
51. Install both heater hoses.
52. Install radiator fan shroud.
53. Install the fan and fan drive assembly
54. Install the accessory drive belt.
55. Install the coolant bottle and washer bottle.
56. Install the upper radiator hose.
57. Install the air cleaner assembly.
58. Fill cooling system.
59. Refill engine oil.
60. Connect the battery negative cable.
61. Install the engine cover.

Piston and Ring

POSITIONING

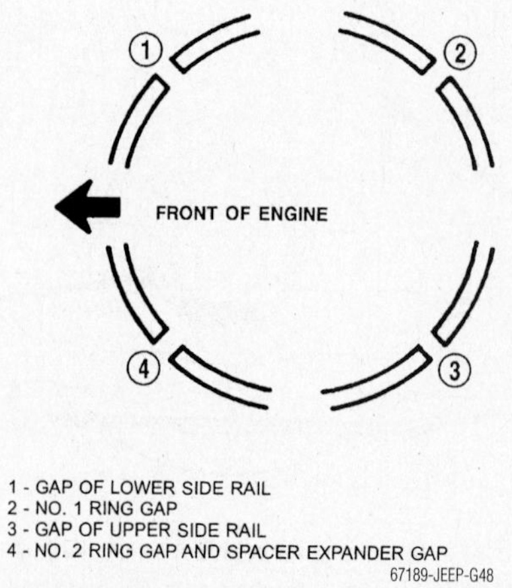

Piston ring end gap spacing—2.4L engines

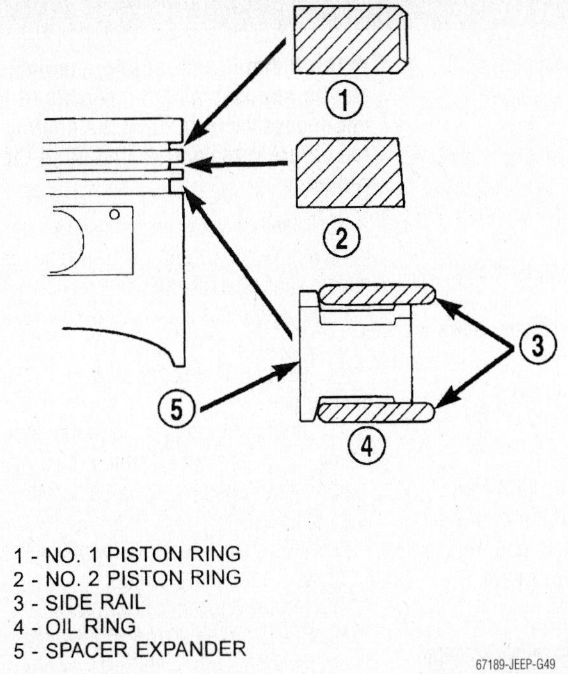

Piston ring installation—2.4L engines

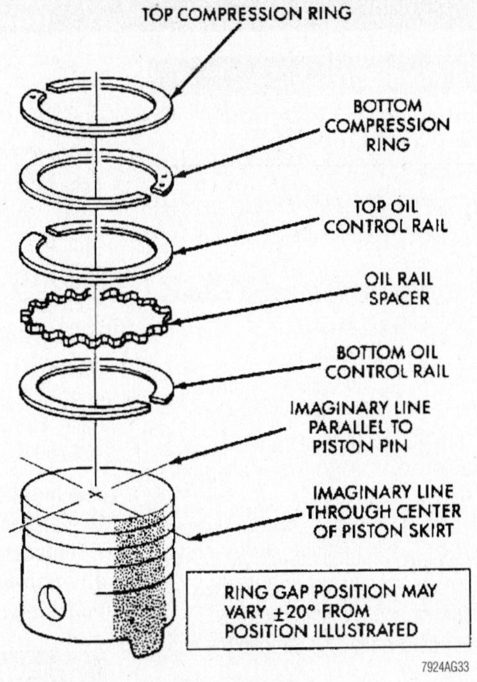

Piston ring end-gap spacing—4.0L engine

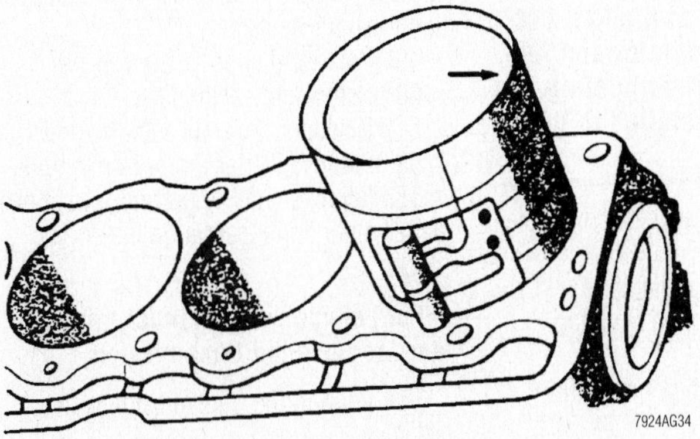

Piston to engine positioning—2.5L, 4.0L engines

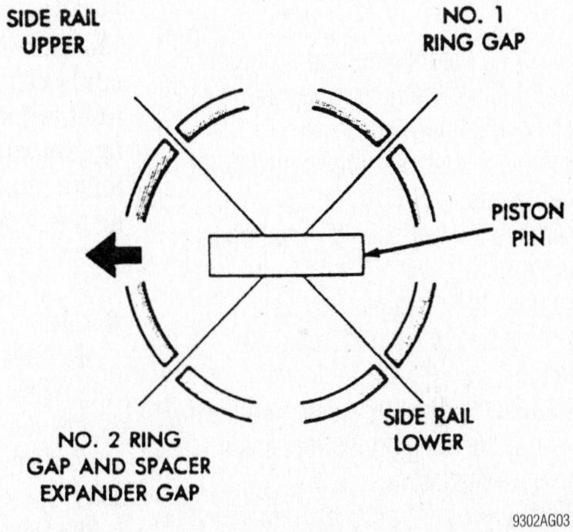

Piston ring end-gap spacing. Position raised "F" on piston towards front of engine—3.7L and 4.7L engines

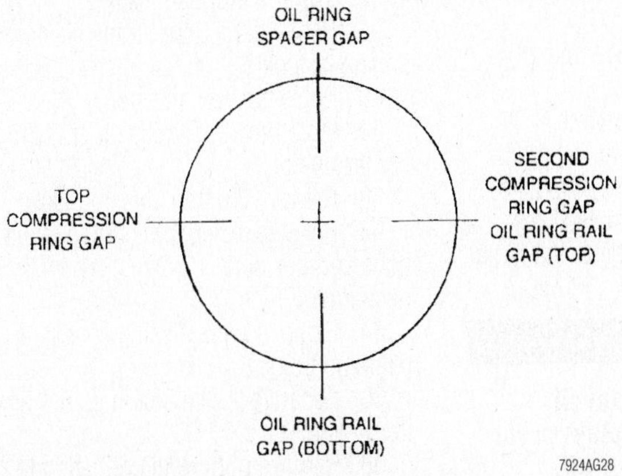

Piston ring end-gap spacing—2.5L engines

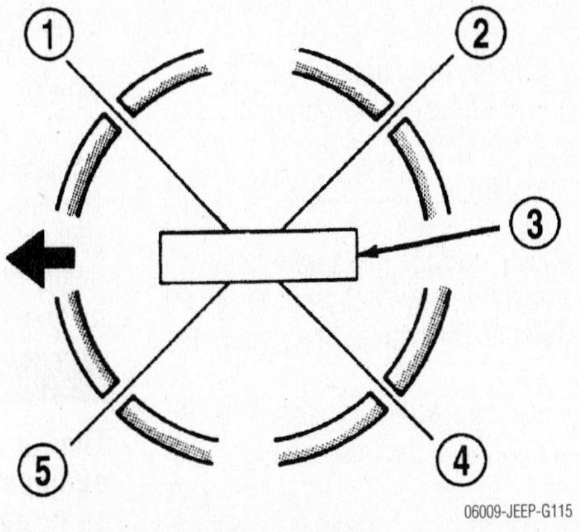

Piston ring end-gap spacing—5.7L engine

DIESEL ENGINE SERVICE

Engine Assembly

REMOVAL & INSTALLATION

1. Before servicing the vehicle, refer to the precautions in the beginning of this section.
2. Disconnect negative battery cable.
3. Drain cooling system.
4. Remove engine cover.
5. Remove engine cover mounting bracket retaining bolts and remove the bracket from the top of the engine.
6. Remove air cleaner assembly from the engine bay.
7. Recover refrigerant from A/C system.
8. Disconnect high side refrigerant line from the upper radiator support bracket.
9. Remove upper radiator support bracket retaining bolts and remove the support bracket.
10. Release the upper grille clips.
11. Roll the grille forward and disengage the two grille hooks under the headlamp units.
12. Lift the grille forward and up off of the location tabs at the bottom and remove.
13. Remove the turn signal lamps.
14. Remove the fascia screws next to the turn signal lamps.
15. Remove the two screws at the bottom of the fascia.
16. Remove the three plastic rivets securing the fascia to each wheelhouse splash shield.
17. Release the three tabs securing the fascia to the grille opening reinforcement.
18. Remove the fascia.
19. Disconnect the electrical connectors.
20. Disconnect the rubber side flap push pin connectors.
21. Remove the seven bolts and remove the grille opening reinforcement.
22. Disconnect the headlamp units electrical connectors.
23. Remove the headlamp units.
24. Remove high side refrigerant line retaining nut and remove the line from the condenser assembly. Position the line out of the way.
25. Remove low side refrigerant line retaining nut and remove line from the condenser assembly. Position the line out of the way.
26. Remove high side refrigerant line from A/C compressor and remove high side line.
27. Remove cooling fan and fan drive viscous clutch assembly.
28. Remove fan shroud retaining bolts and remove fan assembly and shroud together.
29. Disconnect charge air cooler outlet hose from charge air cooler.
30. Remove the charge air inlet hose from the charge air cooler.
31. Disconnect engine coolant hoses from engine assembly.
32. Disconnect coolant reservoir hoses and remove reservoir.
33. Remove power steering cooler retaining bolts and unclip air deflectors from both sides of the radiator assembly.

➡A number or letter is stamped into the tongue of constant tension clamps. If replacement is necessary, use only an original equipment clamp with matching number or letter.

⁂ WARNING

When removing the radiator or A/C condenser for any reason, note the location of all radiator-to-body and radiator-to-A/C condenser rubber air seals. These are used at the top, bottom and sides of the radiator and A/C condenser. To prevent overheating, these seals must be installed to their original positions.

34. Disconnect the negative battery cable at battery.
35. Drain coolant from radiator.
36. Remove the cooling fan from the engine, if equipped.
37. Remove the two radiator mounting bolts.
38. Disconnect the connector for the electric fan.
39. Disconnect the power steering cooler line from cooler.
40. Disconnect the radiator upper and lower hoses.
41. Disconnect the overflow hose from radiator.
42. The lower part of radiator is equipped with two alignment dowel pins. They are located on the bottom of radiator tank and fit into rubber grommets. These rubber grommets are pressed into the radiator lower crossmember.

⁂ CAUTION

The air conditioning system (if equipped) is under a constant pressure, even with the engine off.

➡The radiator and radiator cooling fan can be removed as an assembly. It is not necessary to remove the cooling fan before removing or installing the radiator.

43. Gently lift up and remove radiator from vehicle. Be careful not to scrape the radiator fins against any other component. Also be careful not to disturb the air conditioning condenser (if equipped).
44. Remove accessory drive belt from the engine.
45. Accessing bolts through the pump pulley, remove power steering pump retaining bolts and position pump aside with lines still attached.
46. Disconnect heater core inlet and outlet hoses from the heater core.
47. Remove generator from engine.
48. Trace engine wiring and disconnect electrical connectors and tie straps one at a time until all wiring is disconnected from the engine assembly. When all the engine electrical harness is disconnected, separate the main engine wiring harness connectors at the right rear inner fender well then position the harness aside.
49. Raise and support the vehicle.
50. Remove the skid plate.
51. Disconnect the oil sending unit.
52. Remove starter motor from engine.
53. Remove chassis ground wire above starter mounting location on the engine block.

➡For vehicles with manual transmission, remove the transmission.

54. Support the transmission with a jackstand.
55. Remove the flex plate access cover.
56. Remove the flex plate fasteners.
57. Remove transmission cross member fasteners.
58. Lower the transmission.
59. Remove the upper transmission to engine fasteners.
60. Raise the transmission.
61. Install the transmission cross member fasteners.
62. Remove the transmission jack.
63. Remove exhaust inlet pipe retaining bolts and disconnect exhaust pipe from turbocharger.
64. Remove transmission to engine retaining bolts.
65. Separate transmission cooler line(s) from retainer.
66. Remove both right and left side engine mount retaining nut.

67. Lower vehicle.
68. Remove left side engine mount retaining nut.
69. Remove exhaust manifold rear heat shield.
70. Disconnect the fuel supply and return lines.
71. Remove crankshaft sensor heat shield.
72. Disconnect crankshaft position sensor, located on the right rear of the engine.
73. Remove the oil separator from the cylinder head cover/intake manifold.
74. Connect a suitable lifting device to engine assembly.
75. Place a floor jack under the transmission to support the transmission.
76. With engine and transmission supported by a lifting device separate the engine from the transmission.
77. Lift the engine assembly out of the engine bay.

To install:

78. Install engine assembly and align with the transmission.

➡For vehicle equipped with a manual transmission, Install the transmission.

79. Remove engine lifting device.
80. Raise and support the vehicle.
81. Install accessible engine to transmission housing bolts. Tighten bolts to 68 Nm (50 ft. lbs.).
82. Install the left upper engine mount bolt. Tighten bolt to 110 Nm (82 lbs. ft.).
83. Tighten remaining left mount to engine bolts to 60 Nm (44 lbs. ft.) starting with the left upper bolt and continuing clockwise.
84. Tighten engine left mount through bolts to 110 Nm (82 lbs. ft.).
85. Position the right engine mount and hand tighten the engine mount bracket bolts. Tighten bolts to 60 Nm (44 lbs. ft.).
86. Tighten the right upper engine mount fastener to 110 Nm (82 ft. lbs.). Tighten the right engine mount bracket bolts to 60 Nm (44 lbs. ft.), starting with the right upper then left upper bolt, continuing to the right lower and finishing with the left lower bolt.
87. Tighten the right engine mount lower fastener to 110 Nm (82 ft. lbs.).
88. Support the transmission and remove the rear transmission crossmember.
89. Lower the transmission and install the upper transmission to engine fasteners. Tighten bolts to 68 Nm (50 ft. lbs.).
90. Raise the transmission and install the transmission crossmember.
91. Install the flex plate to torque converter fasteners, torque to 50 ft. lbs. (68 Nm) and install the shield.

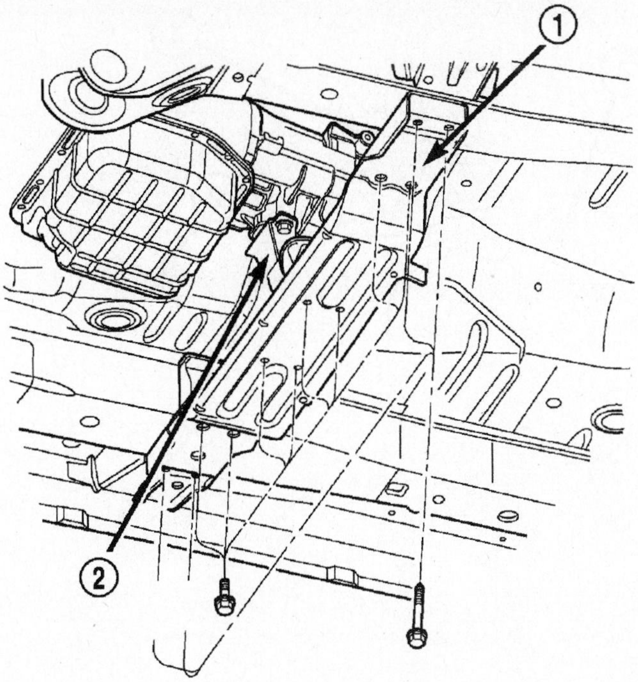

1 - CROSSMEMBER
2 - REAR TRANSMISSION MOUNT

06009-JEEP-G12

Rear transmission crossmember—2.8L diesel

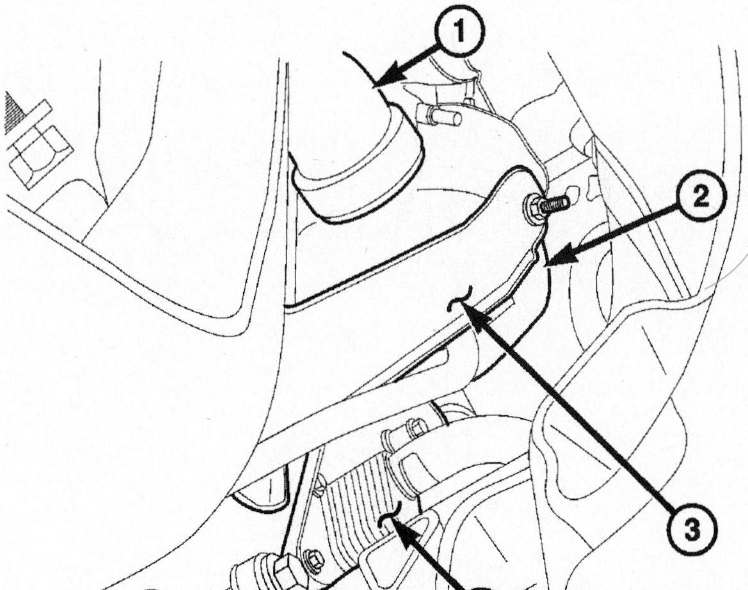

1 - EXHAUST PIPE
2 - TURBOCHARGER
3 - BRACKET
4 - ENGINE OIL COOLER

06009-JEEP-G13

Exhaust connect at the turbocharger—2.8L diesel

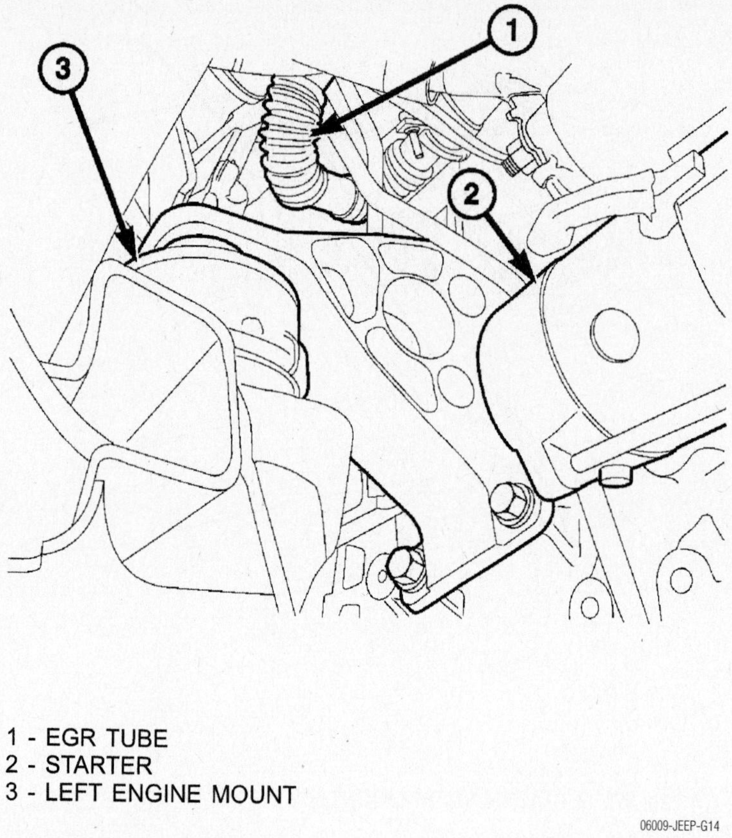

Left engine mount—2.8L diesel

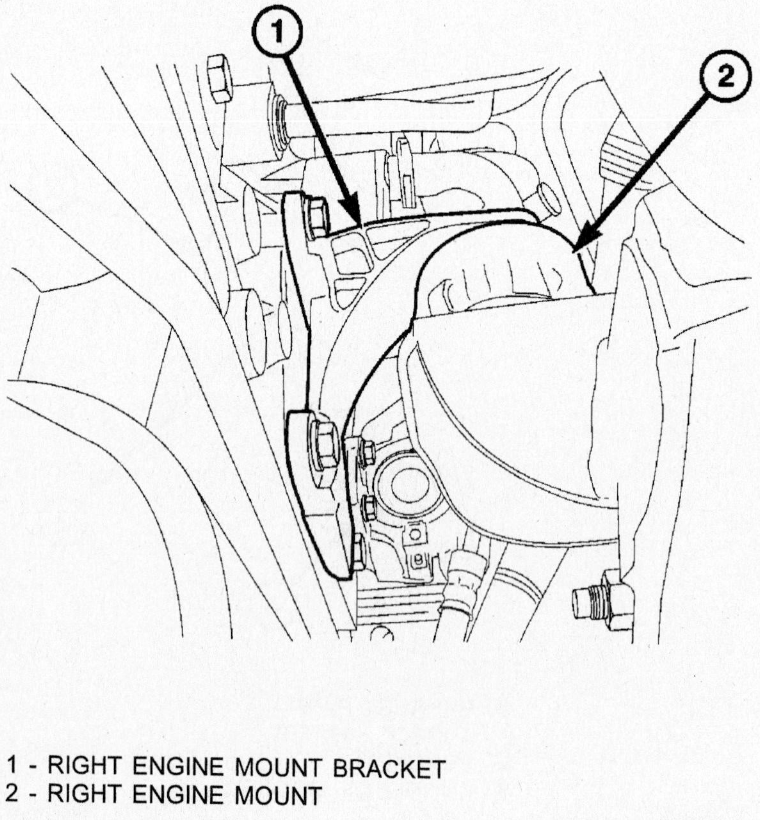

Right engine mount—2.8L diesel

92. Install the engine ground wire above the starter.
93. Install the starter.
94. Connect the transmission cooler lines to the oil pan retainers.
95. Connect the exhaust.
96. Lower the vehicle.
97. Install the oil separator.
98. Connect the crankshaft sensor and install the heat shield.
99. Connect the heater hose inlet and out let hoses.
100. Install the power steering pump.
101. Install the generator.
102. Connect the fuel lines.
103. Position and connect the main engine wiring harness.
104. Install the accessory drive belt.

**** WARNING**

Before installing the radiator or A/C condenser, be sure the radiator-to-body and radiator-to-A/C condenser rubber air seals are properly fastened to their original positions. These are used at the top, bottom and sides of the radiator and A/C condenser. To prevent overheating, these seals must be installed to their original positions.

105. Gently lower the radiator and fan shroud into the vehicle. Guide the two radiator alignment dowels into the rubber grommets located in lower radiator crossmember.
106. Connect the radiator upper and lower hoses and hose clamps to radiator.

**** WARNING**

The tangs on the hose clamps must be positioned straight down.

107. Install coolant reserve/overflow tank hose at radiator.
108. Install both radiator mounting bolts.
109. Reconnect the electric cooling fan.
110. Reinstall the cooling fan to the engine.
111. Rotate the fan blades (by hand) and check for interference at fan shroud.
112. Refill cooling system.
113. Connect battery cable at battery.
114. Start and warm engine. Check for leaks.
115. Install the cooling module.
116. Connect the power steering cooler and fasten the air deflectors.
117. Install the coolant reservoir.
118. Connect the engine coolant hoses.
119. Connect the charge air inlet hose.

120. Connect the charge air outlet hose.
121. Install the cooling fan assembly and shroud.
122. Install high side refrigerant line.
123. Install the low side refrigerant line.
124. Install the headlamp units.
125. Connect the headlamp unit electrical connectors.
126. Install the grille opening reinforcement and install the seven bolts.
127. Tighten the bolts to 10 Nm (85 inch lbs.).
128. Connect the rubber side flap and install the push pin connectors.
129. Connect the electrical connectors.
130. Position the fascia onto the vehicle and align the center pin into the hole located near the center of the grille opening reinforcement.
131. Secure the three tabs onto the grille reinforcement fully.
132. Install the two screws next to the turn signal lamps.
133. Install the two lower screws.
134. Install new rivets attaching the wheelhouse splash shield to the fascia.
135. Install the turn signal lamps.
136. Install the grille onto the locating tabs at the bottom.
137. Push the grille back and snap the hooks onto the grille opening reinforcement.
138. Install the upper radiator support bracket.
139. Install the air cleaner assembly.
140. Install the engine cover bracket and engine cover.
141. Evacuate and recharge A/C.
142. Fill cooling system.
143. Fill crankcase with the correct viscosity oil, to the proper level.
144. Start engine, allow to warm, turn engine off and inspect for leaks.

Water Pump

REMOVAL & INSTALLATION

1. Before servicing the vehicle, refer to the precautions in the beginning of this section.
2. Disconnect negative battery cable.
3. Drain cooling system.
4. Remove timing belt inner and outer covers.
5. Remove water pump retaining bolts and pump.

To install:

6. Clean mating surfaces of water pump housing and engine block as necessary.
7. Place new O-ring in groove in water pump housing. Install water pump and retaining bolts. Torque bolts to 18 ft. lbs. (244 Nm).
8. Install both inner and outer timing belt covers.
9. Refill cooling system.

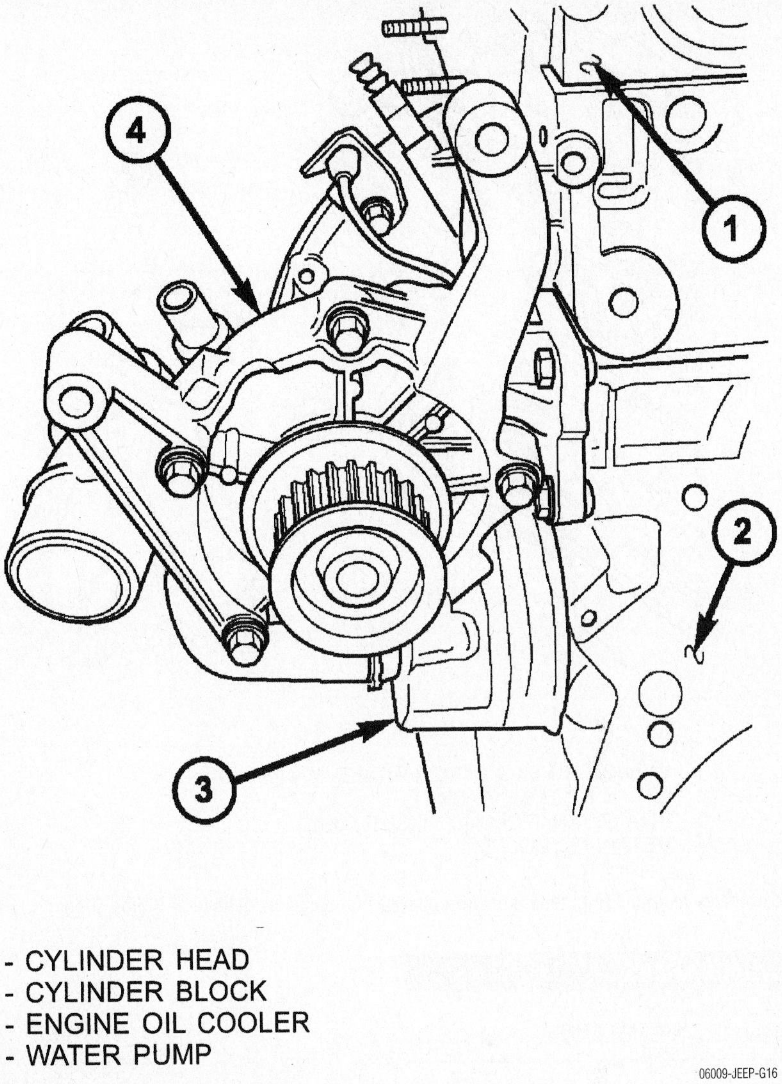

Water pump mounting—2.8L diesel

Heater Core

REMOVAL & INSTALLATION

See the procedure under, Gasoline Engine Service.

Glow Plugs

REMOVAL & INSTALLATION

1. Before servicing the vehicle, refer to the precautions in the beginning of this section.
2. Disconnect negative battery cable.
3. Remove generator.
4. Disconnect glow plug electrical connectors.

➡The intake manifold inlet tube must be removed to remove the cylinder # 3 glow plug.

5. Remove glow plugs from cylinder head.

To install:

6. Install glow plugs all the way into cylinder head, hand tight, until the thread stops.
7. Tighten glow plugs to 12.5 Nm (110 inch lbs.).
8. Connect glow plug electrical connectors.
9. Install intake manifold air inlet tube.
10. Install generator.
11. Connect negative battery cable.

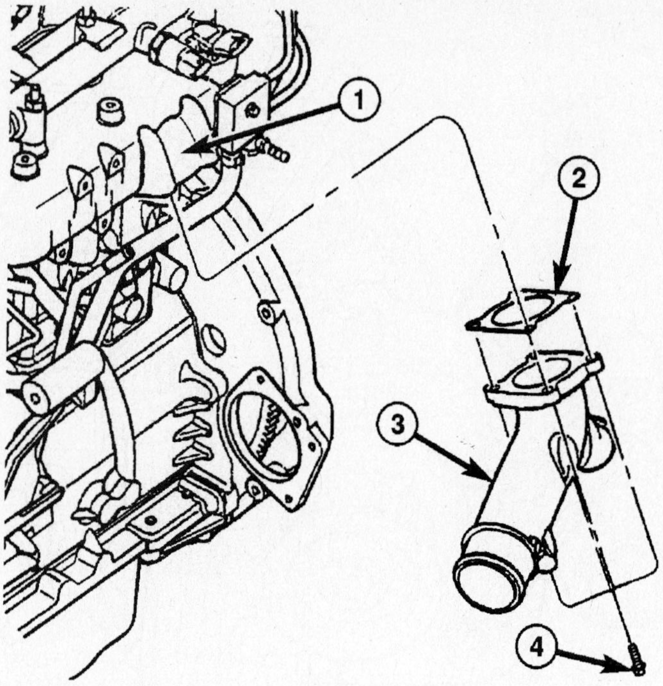

1 - CYLINDER HEAD COVER/INTAKE MANIFOLD
2 - AIR INLET TUBE GASKET
3 - INTAKE MANIFOLD AIR INLET TUBE
4 - RETAINING BOLTS

06009-JEEP-G17

The intake manifold inlet tube must be removed to remove the cylinder # 3 glow plug

Cylinder Head

REMOVAL & INSTALLATION

1. Before servicing the vehicle, refer to the precautions in the beginning of this section.
2. Disconnect negative battery cable.
3. Remove engine cover and bracket.
4. Drain cooling system.
5. Evacuate A/C.
6. Remove radiator core support.
7. Disconnect negative battery cable.

➡The thermal viscous fan drive/fan blade assembly is attached (threaded) to fan support.

8. Remove fan blade/viscous fan drive assembly from water pump using special tool 6958 spanner wrench, or equivalent, by turning mounting nut counterclockwise as viewed from front. Threads on viscous fan drive are RIGHT HAND.
9. Do not attempt to remove fan/fan drive viscous clutch assembly from vehicle at this time.
10. Do not unbolt fan blade assembly from fan drive viscous clutch at this time.
11. Remove fan shroud to radiator bolts.
12. Remove fan shroud and fan blade/fan drive viscous clutch assembly as a complete unit from vehicle.
13. After removing fan blade/fan drive viscous clutch assembly, do not place viscous clutch in horizontal position. If stored horizontally, silicone fluid in the fan drive viscous clutch could drain into its bearing assembly and contaminate lubricant.
14. Remove four bolts securing fan blade assembly to fan drive viscous clutch.
15. Remove cooling fan support from engine block.
16. Remove accessory drive belt.
17. Remove accessory drive belt tensioner and both idler pulleys. Idler pulley bolts are L.H. thread.
18. Remove power steering pump pulley.
19. Remove front engine lift bracket.
20. Remove generator and support bracketing.
21. Remove viscous heater.
22. Rotate the engine to 90 degrees ATDC, or the 3 O'clock position at the crankshaft hub.
23. Remove vibration damper.
24. Disconnect main engine wiring harness connectors from right inner wheel housing.

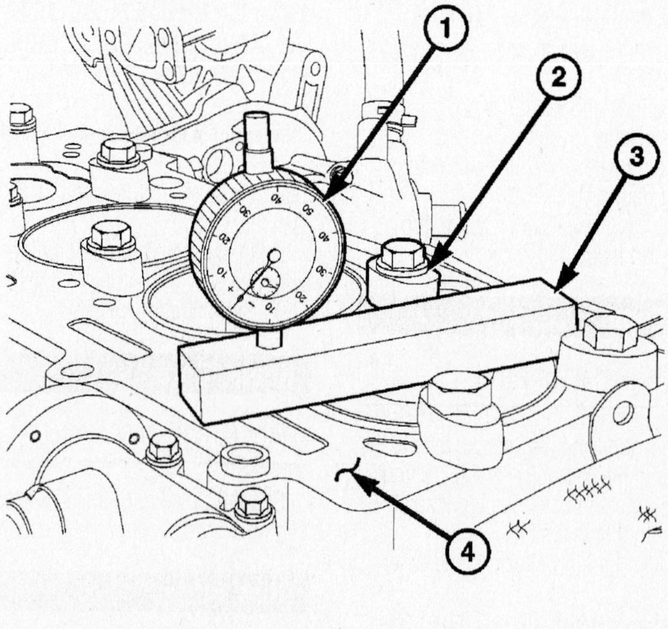

1 - VM.1010 DIAL INDICATOR
2 - VM.1076 CYLINDER LINER RETAINERS
3 - VM.1013 STRAIGHT EDGE
4 - CYLINDER BLOCK SURFACE

06009-JEEP-G18

Cylinder protrusion set up—2.8L diesel

25. Disconnect main engine wiring harness ancillary components and set harness aside.
26. Remove air cleaner housing.
27. Remove exhaust manifold heat shield.
28. Remove turbocharger heat shield retaining bolt and position shield aside.
29. Remove turbocharger oil supply line from turbocharger.
30. Raise and support vehicle.
31. Disconnect exhaust stabilizer bracket at lower exhaust manifold.
32. Disconnect exhaust system bracket at transmission crossmember.
33. Lower the vehicle and remove the exhaust manifold retaining nuts.
34. Slide the exhaust manifold and turbocharger off of exhaust manifold studs.
35. Remove coolant hoses at thermostat housing.
36. Disconnect fuel return hose from fuel injectors and set aside.
37. Remove fuel injector pressure lines.
38. Disconnect fuel pump high pressure line at fuel rail.
39. Disconnect oil level indicator tube from intake manifold.
40. Disconnect brake booster line bracket from intake manifold and position aside.
41. Remove fuel injectors.

⁂ WARNING

Before removing the cylinder head cover/intake manifold or timing belt the crankshaft hub must placed at 90° ATDC or the 3 O'clock position. Failure to do so could result in valve and/or piston damage during reassembly.

42. Remove timing belt outer cover.
43. Remove timing belt.
44. Remove both camshaft gears.
45. Remove timing belt inner cover.
46. Remove cylinder head cover/intake manifold.
47. Remove rocker arm and lifter assemblies from cylinder head. Be sure to keep in same order as removed.
48. Remove cylinder head cover/intake manifold gasket from cylinder head.
49. Disconnect glow plug and engine coolant temperature electrical connectors.
50. Remove turbocharger outlet to charge air cooler hose.
51. Remove cylinder head bolts.
52. Remove cylinder head assembly from engine block.

Measure Dimension (mm)	0.460-0.609
Cylinder Head Gasket Thickness (mm)	1.32 No Holes or Notches
Piston Clearance (mm)	0.71-0.86
Measure Dimension (mm)	0.610-0.709
Cylinder Head Gasket Thickness (mm)	1.42 1 Hole or Notch
Piston Clearance (mm)	0.711-0.81
Measure Dimension (mm)	0.710-0.810
Cylinder Head Gasket Thickness (mm)	1.52 2 Holes or Notches
Piston Clearance (mm)	0.71-0.81

06009-JEEP-G19

Head gasket measurement chart—2.8L diesel

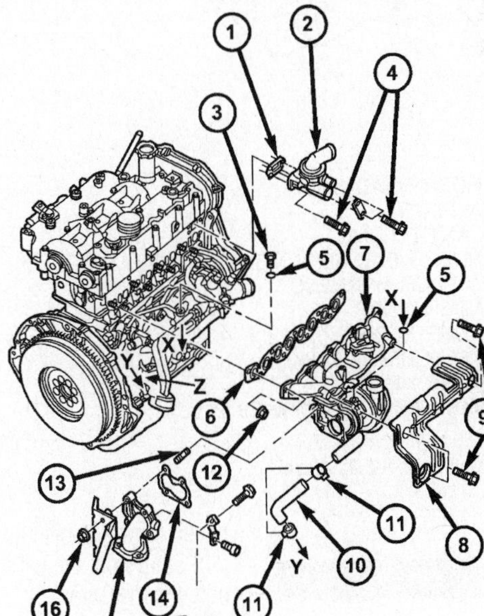

1 - THERMOSTAT HOUSING GASKET
2 - THERMOSTAT HOUSING
3 - TURBOCHARGER OIL SUPPLY LINE BANJO BOLT
4 - THERMOSTAT HOUSING RETAINING BOLTS
5 - BRASS WASHER
6 - EXHAUST MANIFOLD GASKET
7 - EXHAUST MANIFOLD
8 - EXHAUST MANIFOLD HEATSHIELD
9 - EXHAUST MANIFOLD HEATSHIELD RETAINING BOLTS
10 - OIL RETURN HOSE
11 - HOSE CLAMPS
12 - EXHAUST MANIFOLD RETAINING NUTS
13 - TURBOCHARGER DOWNPIPE STUDS
14 - TURBOCHARGER DOWN PIPE GASKET
15 - TURBOCHARGER DOWNPIPE
16 - TURBOCHARGER DOWNPIPE RETAINING NUT

06009-JEEP-G20

Remove these parts for cylinder head removal—2.8L diesel

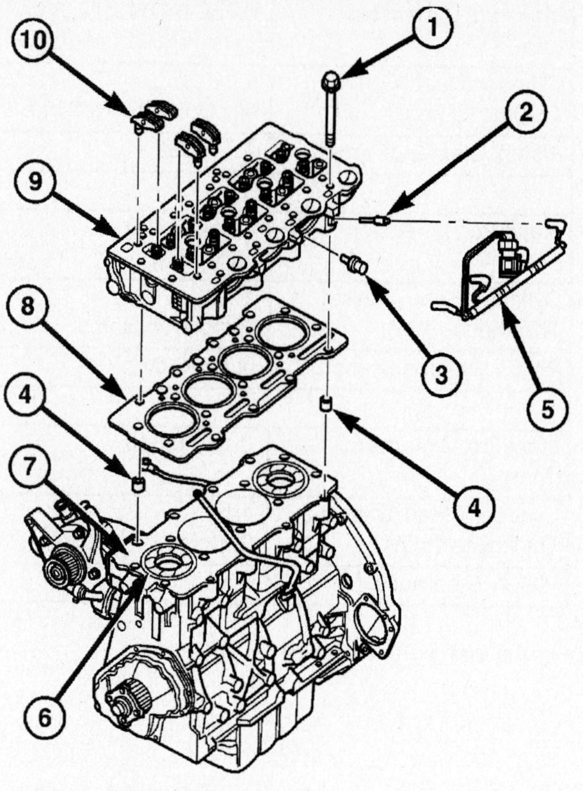

Cylinder head and related parts—2.8L diesel

To install:

⁂ WARNING

Piston protrusion must be measured to determine cylinder head gasket thickness if one or more cylinder liners have been replaced

➡If cylinder liner(s) have not been removed, the same thickness head gasket that was removed can be used.

53. Bring the piston of cylinder number 1 exactly to top dead center.
54. Lay a straight edge across the cylinder sleeve and zero a dial indicator on the cylinder block mating surface.
55. Lay a straight edge across the cylinder sleeve, setup the dial indicator on the piston crown (above the center of the piston pin) 5mm from the edge of the piston and note the measurement.
56. Repeat the procedure with the rest of the cylinders.
57. Establish the thickness of the steel gasket by averaging the four piston protrusion readings.
58. Clean and inspect gasket mating surfaces.
59. Position correct head gasket on engine block.
60. Place cylinder head on engine block.

⁂ WARNING

New cylinder head bolts must be used. Do Not lubricate new cylinder head bolts. They already are coated with an anti-scuff treatment.

61. Tighten cylinder head bolts following procedure below.
    a. Tighten all cylinder head bolts to 30 Nm (22 ft. lbs.) starting from the center bolts, following the cylinder head scheme and the following sequence: 3-2-1-10-9-8-7-6-5-4-11-12-13-14-15-16-17-18.
    b. Without loosening any bolts, starting from the center bolts, tighten each bolt an additional 75° in the following sequence: 10-9-8-7-6-5-4-3-2-1.
    c. Tighten the lateral cylinder head bolts an additional 50° in the following sequence: 11-12-13-14-15-16-17-18.
    d. Finally tighten all bolts an additional 75° in the following sequence: 10-9-8-7-6-5-4-3-2-1-11-12-13-14-15-16-17-18.
62. Slide exhaust manifold and turbocharger on exhaust manifold studs.
63. Install exhaust manifold retaining nuts. Tighten the nuts to 36 Nm. (27 ft. lbs) in a cross sequence beginning in the middle and working outward, then perform the tightening sequence to the exhaust manifold nuts again.
64. Install exhaust manifold heat shield. Torque bolts to 24.5Nm (217 inch lbs).
65. Install turbocharger outlet to charge air cooler pipe.
66. Install upper radiator hose.
67. Connect glow plug and coolant temperature sensor electrical connectors.
68. Install new cylinder head cover/intake manifold gasket.
69. Install rocker arm and lifter assemblies. Be sure to put rocker arm and lifter assemblies in same location as removed.

➡Care must be taken not to knock the rocker arms off of the valves when installing the cylinder head cover/intake manifold.

70. Install cylinder head cover/intake manifold.
71. Install timing belt inner cover.

➡Make sure the crankshaft is rotated to the 90 degree ATCD, or the 3 O'clock position.

72. Install both camshaft gears and tighten bolts finger tight.
73. Install timing belt.
74. Torque cam gear retaining bolts to 108 Nm.
75. Install timing belt outer cover.
76. Remove both camshaft locking pins at this time.
77. Install fuel injectors.
78. Install brake booster line bracket to intake manifold.
79. Install oil level indicator tube to intake manifold.

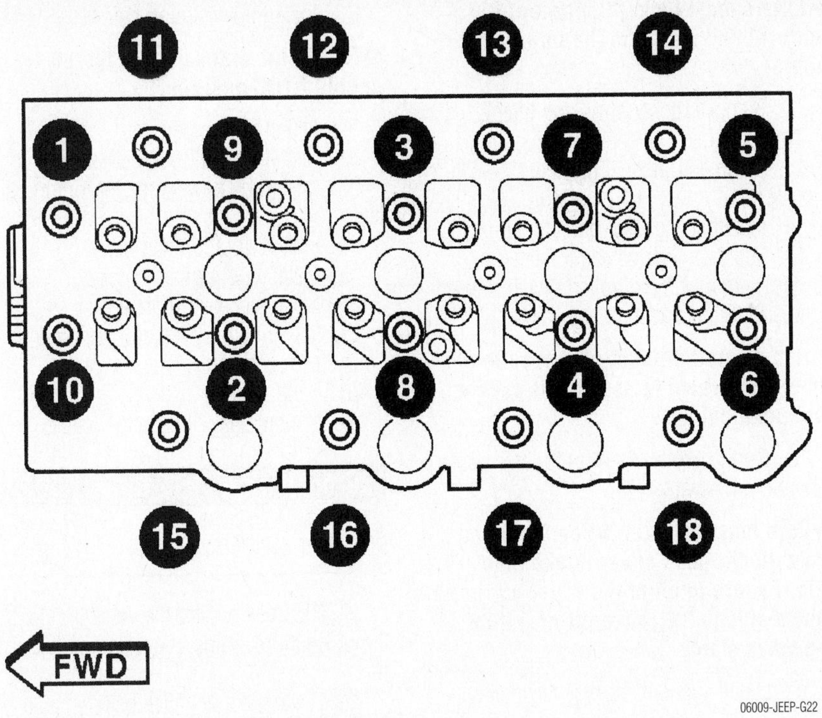

Cylinder head torque sequence—2.8L diesel

80. Connect fuel pump high pressure line to fuel rail.
81. Install fuel injector pressure lines.
82. Connect fuel return hose to fuel injectors.
83. Connect engine coolant hoses to thermostat housing.
84. Raise vehicle.
85. Reconnect exhaust system bracket at transmission crossmember.
86. Connect exhaust stabilizer bracket to lower exhaust manifold.
87. Install turbocharger oil supply line to turbocharger.
88. Lower vehicle.
89. Install turbocharger heat shield.
90. Install exhaust manifold heat shield.
91. Install air cleaner housing.
92. Install main engine wiring harness and connect all ancillary electrical components.
93. Install viscous heater.
94. Install vibration damper.
95. Install generator.
96. Assemble fan blade to viscous fan drive. Tighten mounting bolts to 23.7 Nm (210 inch lbs.) torque.

➡The viscous fan and fan shroud must be installed as an assembly.

97. Gently lay fan and viscous drive into fan shroud.
98. Install the fan shroud to radiator mounting bolt. Tighten bolts to 9 Nm (80 inch lbs.) torque.
99. Thread the fan and viscous drive onto the fan support and tighten nut.
100. Install cooling fan support to engine block. Torque bolts to 149 Nm (110 ft. lbs.).
101. Install fan drive viscous clutch and fan assembly.
102. Install front engine lift bracket.
103. Install power steering pump pulley.
104. Install accessory drive belt tensioner and both idler pulleys. Idler pulley retaining bolts are LH Thread.
105. Install accessory drive belt.
106. Install cooling fan and fan drive viscous clutch assembly.
107. Install upper radiator core support.
108. Refill cooling system.
109. Install engine cover and bracket.
110. Connect negative battery cable.
111. Start engine and allow to warm. Turn engine off and inspect for leaks.

Rocker Arms and Lifters

REMOVAL & INSTALLATION

1. Before servicing the vehicle, refer to the precautions in the beginning of this section.

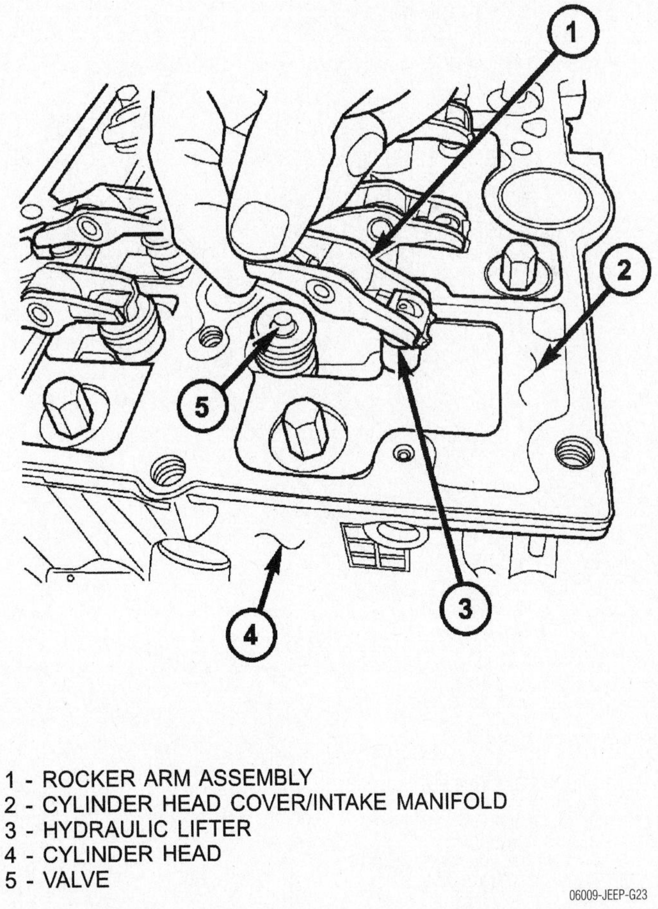

Rocker arms—2.8L diesel

⁂ WARNING

Before removing the cylinder head cover/intake manifold the engine must rotated to 90° after TDC to assure proper alignment of the engine timing components. Failure to do so could result in valve and/or piston damage during reassembly.

2. Disconnect negative battery cable.
3. Drain cooling system.
4. Remove cooling fan and fan drive viscous clutch assembly.
5. Remove accessory drive belt.
6. Remove the vibration damper.
7. Remove cooling fan support.
8. Remove the power steering pump pulley.
9. Rotate the crankshaft to 90 degrees ATDC.
10. Remove outer timing belt cover.
11. Remove timing belt.
12. Remove inner timing belt cover.
13. Remove cylinder head cover/intake manifold.

➡Lifters must be kept in order of removal and stored in the up right position.

14. Remove rocker arms and lifters.

To install:

15. Clean and inspect gasket sealing surfaces.
16. Install new gasket on cylinder head.
17. Lubricate lifter ball end of lifter(s), valve(s), and rocker arm roller(s) with Mopar® Engine Oil Supplement, or equivalent.

➡Add a small amount of grease to the tip of each valve to assist with rocker arm positioning.

18. Connect rocker arm(s) to lifter and position on valve(s).

➡Care must be taken when installing the cylinder head cover/intake manifold. Failure to align the cover correctly may dislodge the rocker arm(s) from the valve stem.

19. Install cylinder head cover/intake manifold.
20. Install inner timing belt cover.

➡Rotate the crankshaft to the 90 degree ATDC position.

21. Install timing belt.
22. Install outer timing belt cover.
23. Install the power steering pump pulley.
24. Install vibration damper.
25. Install cooling fan support.
26. Install accessory drive belt.
27. Install cooling fan and fan drive viscous clutch assembly.
28. Refill cooling system.
29. Connect negative battery cable.

Turbocharger

REMOVAL & INSTALLATION

1. Before servicing the vehicle, refer to the precautions in the beginning of this section.
2. Disconnect negative battery cable.
3. Remove engine cover.

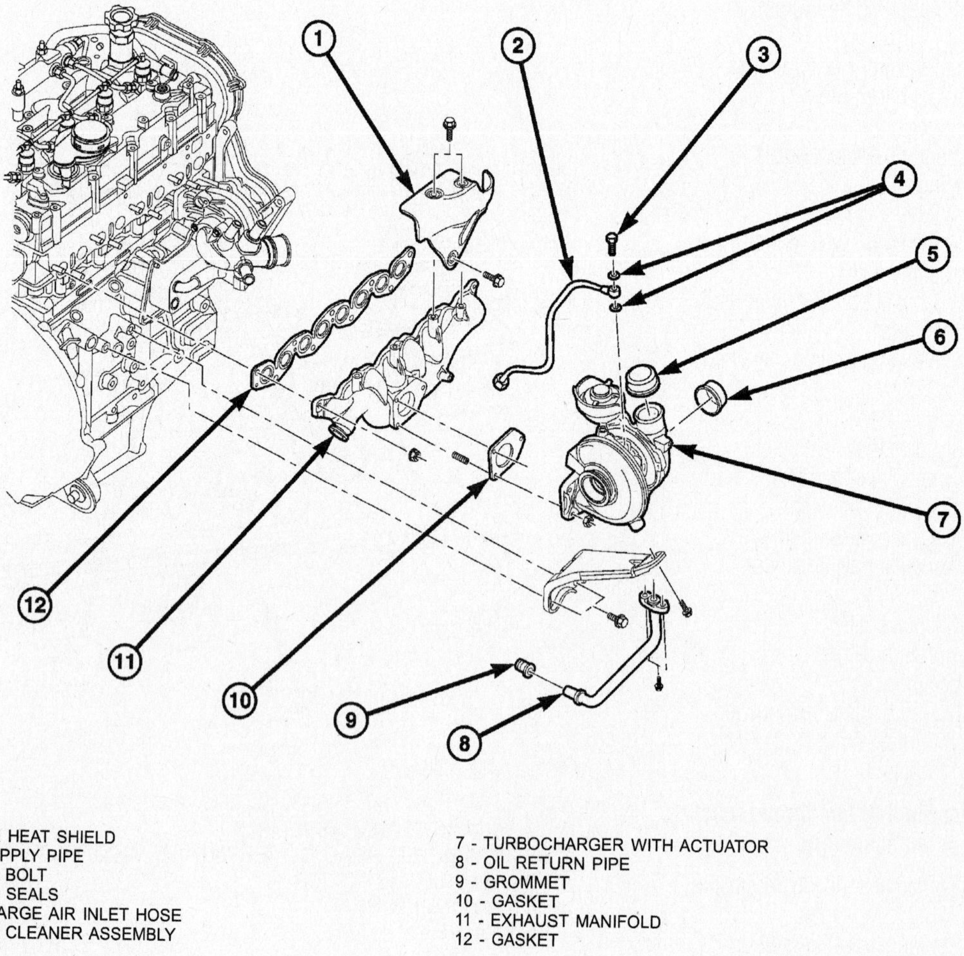

Turbocharger and oil lines—2.8L diesel

4. Disconnect the MAF and Inlet air pressure sensors wiring harness connectors, disconnect the air outlet duct from the turbocharger, and remove air cleaner assembly.
5. Remove charge air cooler outlet hose from turbocharger.
6. Drain cooling system.
7. Remove coolant reservoir.
8. Disconnect the turbocharger actuator vacuum hose and position aside.
9. Remove turbocharger upper heat shield.
10. Raise and support the vehicle.
11. Remove the lower splash shield.
12. Disconnect the front exhaust pipe from the turbocharger.
13. Remove the turbocharger support bracket.
14. Disconnect turbocharger oil return line at turbocharger.
15. Lower the vehicle.
16. Remove the turbocharger oil supply line.
17. Remove turbocharger to exhaust manifold retaining nuts and separate turbocharger from exhaust manifold.

To install:

18. Install a new exhaust manifold gasket and position the exhaust manifold into place. Hand tighten fasteners.

➡After tightening the exhaust manifold to specification using a diagonal-cross pattern, retrace the pattern checking the correct torque value again.

19. Tighten the exhaust manifold to cylinder head fasteners in a diagonal-cross pattern to 36 Nm (26 ft. lbs.).
20. Connect turbocharger to exhaust manifold with new gasket. Torque retaining nuts to 32 Nm.
21. Connect oil supply line at turbocharger. Torque banjo fitting to 24 Nm (18 ft. lbs.).
22. Raise and support the vehicle.
23. Connect the oil return pipe to the turbocharger with new gasket and tighten the fasteners to 10 Nm (95 inch lbs.).
24. Install the turbocharger support bracket and tighten fasteners to 24 Nm (18 ft. lbs.).
25. Connect the front exhaust pipe to the turbocharger.
26. Install the lower skid plate.
27. Lower the vehicle.
28. Install the upper turbocharger heat shield. Tighten the fasteners to 24 Nm (18 ft. lbs.).
29. Connect the turbocharger actuator vacuum hose to the actuator.
30. Install the coolant reservoir.
31. Install the charge air outlet hose
32. Install the air cleaner assembly and connect the electrical connectors.
33. Install engine cover.
34. Refill cooling system.
35. Connect negative battery cable.

Cooling Fan

REMOVAL & INSTALLATION

1. Before servicing the vehicle, refer to the precautions in the beginning of this section.
2. Disconnect negative battery cable.

➡The thermal viscous fan drive/fan blade assembly is attached (threaded) to fan support.

3. Remove fan blade/viscous fan drive assembly from water pump using special tool 6958 spanner wrench, or equivalent, by turning mounting nut counterclockwise as viewed from front. Threads on viscous fan drive are RIGHT HAND.
4. Do not attempt to remove fan/fan drive viscous clutch assembly from vehicle at this time.
5. Do not unbolt fan blade assembly from fan drive viscous clutch at this time.
6. Remove fan shroud to radiator bolts.
7. Remove fan shroud and fan blade/fan drive viscous clutch assembly as a complete unit from vehicle.
8. After removing fan blade/fan drive viscous clutch assembly, do not place viscous clutch in horizontal position. If stored horizontally, silicone fluid in the fan drive viscous clutch could drain into its bearing assembly and contaminate lubricant.
9. Remove four bolts securing fan blade assembly to fan drive viscous clutch.
10. Remove cooling fan support from engine block.

To install:

11. Assemble fan blade to viscous fan drive. Tighten mounting bolts to 23 Nm (210 inch lbs.) torque.

➡The viscous fan and fan shroud must be installed as an assembly.

12. Gently lay fan and viscous drive into fan shroud.
13. Install the fan shroud to radiator mounting bolt. Tighten bolts to 9 Nm (80 inch lbs.) torque.
14. Thread the fan and viscous drive onto the fan support and tighten nut using special tool 6958 spanner wrench.

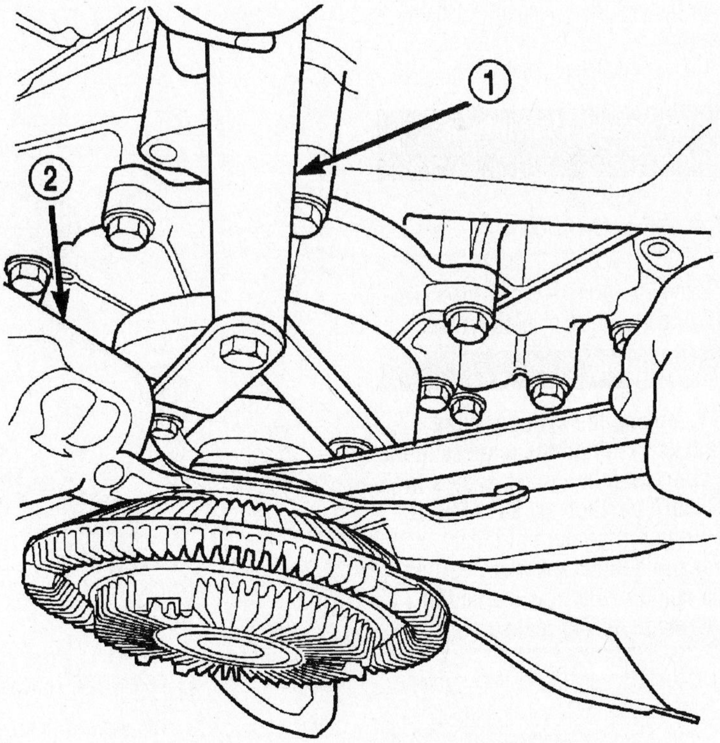

Removing the fan clutch—2.8L diesel

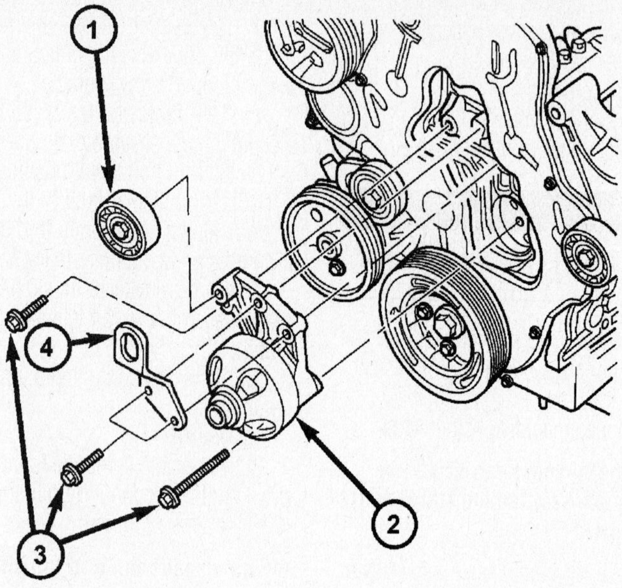

1 - IDLER PULLEY
2 - COOLING FAN SUPPORT
3 - RETAINING BOLTS
4 - ENGINE LIFT HOOK

06009-JEEP-G26

Cooling fan support—2.8L diesel

15. Install cooling fan support to engine block. Torque bolts to 149 Nm (110 ft. lbs.).
16. Install fan drive viscous clutch and fan assembly.
17. Connect negative battery cable.

Cylinder Head Cover/Intake Manifold

REMOVAL & INSTALLATION

1. Before servicing the vehicle, refer to the precautions in the beginning of this section.

WARNING

Before removing the cylinder head cover/intake manifold the witness mark on the crankshaft hub must rotated to the 3 O'clock position or, 90° ATDC to assure proper alignment of the camshafts and the crankshaft. Failure to do so could result in valve and/or piston damage during reassembly.

2. Disconnect negative battery cable.
3. Remove engine cover and bracket.
4. Remove the air cleaner housing.
5. Drain cooling system.
6. Evacuate the air conditioning.
7. Disconnect the coolant recovery hose at the radiator.
8. Remove the charge air inlet hose at the coolant module.
9. Remove the upper radiator hose at the radiator.
10. Separate the A/C hoses from the cooling fan shroud.
11. Remove cooling fan and fan drive viscous drive assembly, along with the fan shroud.
12. Remove accessory drive belt.
13. Remove cooling fan support.
14. Disconnect the EGR cooler pipe behind the inner timing cover.
15. Remove generator and generator rear support bracket.
16. Remove the accessory drive belt idler pulleys (LH thread).
17. Remove the drive belt tensioner.
18. Remove the power steering pump pulley.
19. Remove the vibration damper.
20. Rotate the crankshaft hub to 90° after TDC (3 O'clock position).
21. Remove heater hose pipe fasteners.

➡It may be necessary to rotate the camshaft gear bolt slightly to gain proper camshaft alignment pin seating against the intake manifold. Alignment pins must seat flush against the intake manifold.

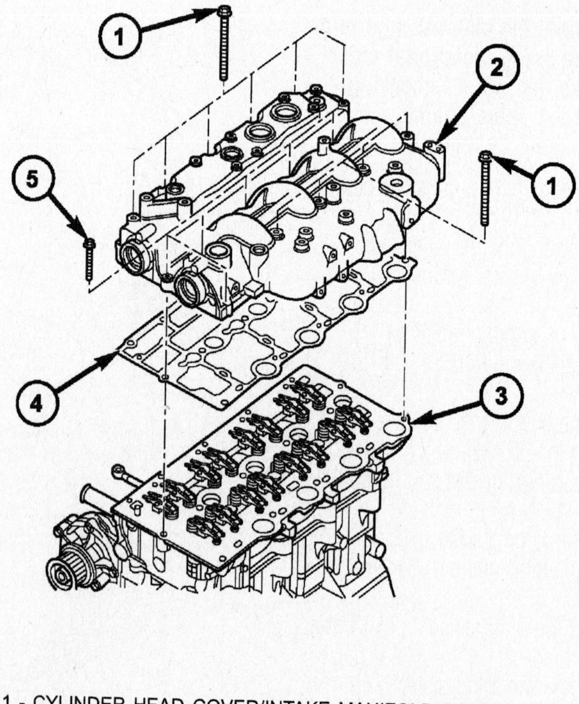

1 - CYLINDER HEAD COVER/INTAKE MANIFOLD BOLTS (LONG)
2 - CYLINDER HEAD COVER/INTAKE MANIFOLD
3 - CYLINDER HEAD
4 - CYLINDER HEAD COVER/INTAKE MANIFOLD GASKET
5 - CYLINDER HEAD COVER/INTAKE MANIFOLD BOLTS (SHORT)

06009-JEEP-G27

Cylinder head cover/intake manifold—2.8L diesel engine

22. Install the intake camshaft locking pin VM.1052 and exhaust camshaft locking pin VM.1053.
23. Remove outer timing belt cover.
24. Remove timing belt.
25. Install special tool VM.1085 and remove camshaft gears.
26. Remove timing belt idler pulleys (LH thread).
27. Remove inner timing belt cover.
28. Disconnect coolant temperature sensor, camshaft position sensor, boost pressure/intake air temperature sensor, fuel injectors, fuel temperature sensor, fuel rail solenoid, and EGR air flow electrical connectors.
29. Disconnect the main engine harness connectors at the right inner fender well and position the harness over the left side of the engine and aside.
30. Disconnect the brake booster vacuum pipe from the EGR air control valve.
31. Separate the main engine harness from the bracket on the EGR air control valve.
32. Remove the EGR tube from the underside of the EGR air control valve.
33. Separate the block heater wiring harness from the oil level indicator tube.
34. Remove the oil level indicator tube from the EGR air control valve.
35. Remove fuel injectors.
36. Remove fuel rail.
37. Disconnect oil separator outlet hose at separator.
38. Remove the oil separator from the intake manifold/cylinder head cover.
39. Disconnect the return fuel junction block from the intake manifold/cylinder head cover.
40. Remove cylinder head cover/intake manifold retaining bolts.
41. Lift cylinder head cover/intake manifold from cylinder head.

➡When removing rocker arm and lifter assemblies, always keep lifters in an upright position and in the order that they were removed from the cylinder head.

42. Remove rocker arm and lifter assemblies from cylinder head.
43. Remove cylinder head cover/intake manifold gasket from cylinder head.

To install:

44. Clean and inspect sealing surfaces.
45. Install new gasket on cylinder head.

➡Add a small amount of grease to the top of each valve to assist with rocker arm position retention.

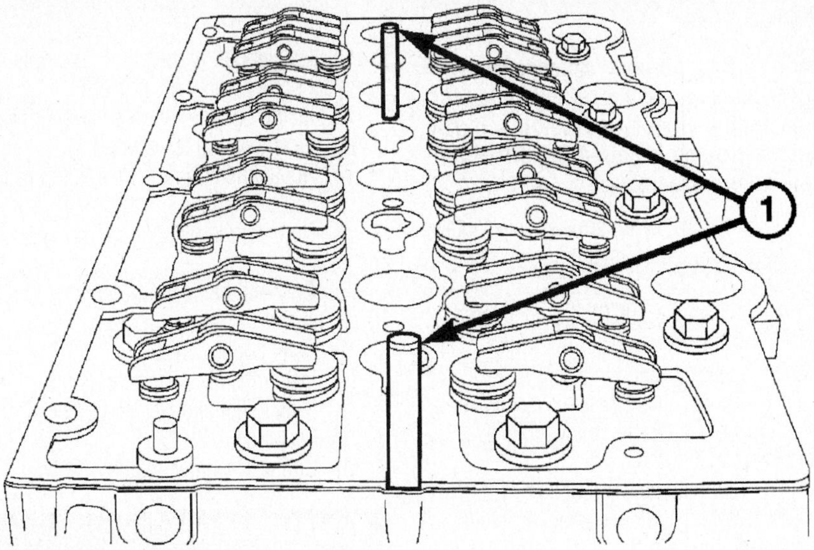

1 - #9673 ALIGNMENT STUDS

06009-JEEP-G28

Cylinder head cover/intake manifold alignment studs—2.8L diesel engine

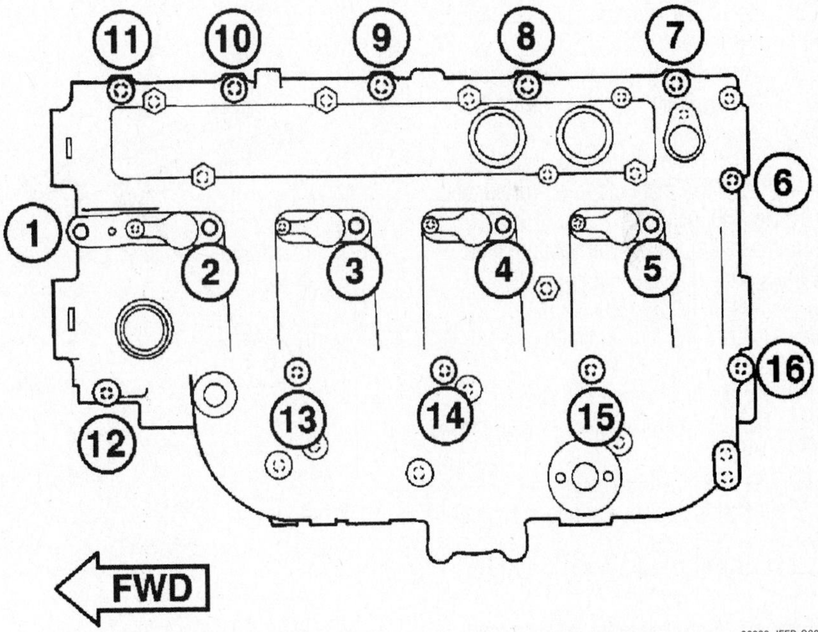

06009-JEEP-G29

Cylinder head cover/intake manifold bolt torque sequence—2.8L diesel engine

46. Install rocker arm and lifter assemblies in cylinder head. Be sure to put rocker arm and lifter assemblies in same location as removed.
47. Install special tool #9673 intake manifold/cylinder head cover alignment studs.

⁂ WARNING

Care must be taken when installing the cylinder head cover/intake manifold. Do not dislodge the rocker arm(s) off of the valve stem(s) when installing the cover.

48. Install cylinder head cover/intake manifold.

➡**Be sure to lubricate cylinder head cover/intake manifold retaining bolts with engine oil before assembly. If new bolts are being installed, DO NOT lubricate before assembly.**

49. Install two cylinder head cover/intake manifold retaining bolts and tighten finger tight.
50. Remove the #9673 alignment studs and install remaining retaining bolts. Tighten retaining bolts finger tight.
51. Torque cylinder head cover/intake manifold retaining bolts following procedure below.
    a. Alternate between bolts #11 and #16 to seat cylinder head cover/intake manifold on cylinder head. Torque bolts to 7 Nm.
    b. Torque all cylinder head cover/intake manifold retaining bolts to 25 Nm in numerical order starting with #1 and ending with #16.
52. Connect the fuel return junction block to the intake manifold/cylinder head cover.
53. Install the oil separator and outlet hose.
54. Install fuel injectors and fuel injector supply lines.
55. Connect EGR tube at EGR air control valve. Torque clamp to 10 Nm. (88 inch lbs.).
56. Install oil level indicator tube retaining bolt at intake manifold inlet. Torque bolt to 10 Nm (88 inch lbs.).
57. Install brake booster vacuum tube retaining bolt at intake manifold inlet. Torque bolt to 10 Nm (88 inch lbs.).
58. Install inner timing belt cover.
59. Install timing belt.
60. Install outer timing belt cover.
61. Remove both camshaft locking pins and install plugs.
62. Install vibration damper.
63. Install the power steering pump pulley.
64. Install the accessory drive belt tensioner
65. Install the accessory drive belt pulleys (LH thread).
66. Install generator.
67. Install cooling fan support.
68. Install accessory drive belt.
69. Position heater hose pipe to intake manifold/cylinder head cover, and install fasteners.
70. Position main engine wiring harness over the engine and connect the harness connectors at the right inner fender well.
71. Connect camshaft position sensor, boost pressure/intake air temperature sensor, EGR solenoid, and fuel rail and pressure sensor electrical connectors.
72. Install cooling fan, fan drive viscous clutch assembly and fan shroud.
73. Route and connect the A/C hoses to fan shroud.
74. Connect the radiator and charge air hoses.
75. Connect the coolant recovery container hose to radiator.
76. Refill cooling system.
77. Install the air cleaner housing.
78. Install engine cover and bracket.
79. Recharge A/C.
80. Connect negative battery cable.

Exhaust Manifold

REMOVAL & INSTALLATION

See the instructions under, Turbocharger Removal & Installation.

Camshaft

REMOVAL & INSTALLATION

1. Before servicing the vehicle, refer to the precautions in the beginning of this section.
2. Disconnect negative battery cable.
3. Remove engine cover and bracket.
4. Drain cooling system.
5. Remove cooling fan and fan drive viscous clutch assembly.
6. Remove accessory drive belt.
7. Remove cooling fan support.
8. Rotate the engine to 90 degrees ATDC.
9. Remove vibration damper.
10. Remove generator.

WARNING

Before removing the cylinder head cover/intake manifold or timing belt the engine must rotated to the 90°

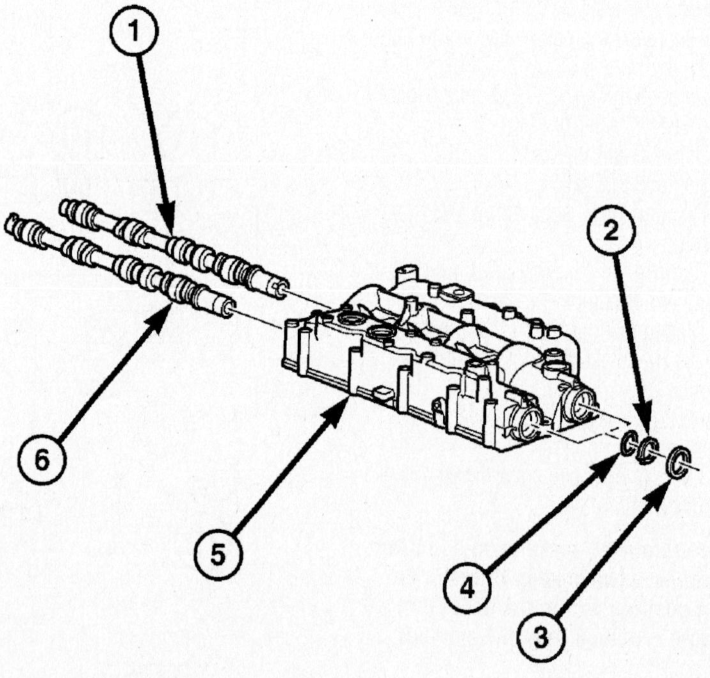

1 - INTAKE CAMSHAFT
2 - SNAPRING
3 - CAMSHAFT OIL SEAL
4 - THRUST WASHER
5 - CYLINDER HEAD COVER/INTAKE MANIFOLD
6 - EXHAUST MANIFOLD

06009-JEEP-G30

Camshaft assemblies—2.8L diesel engine

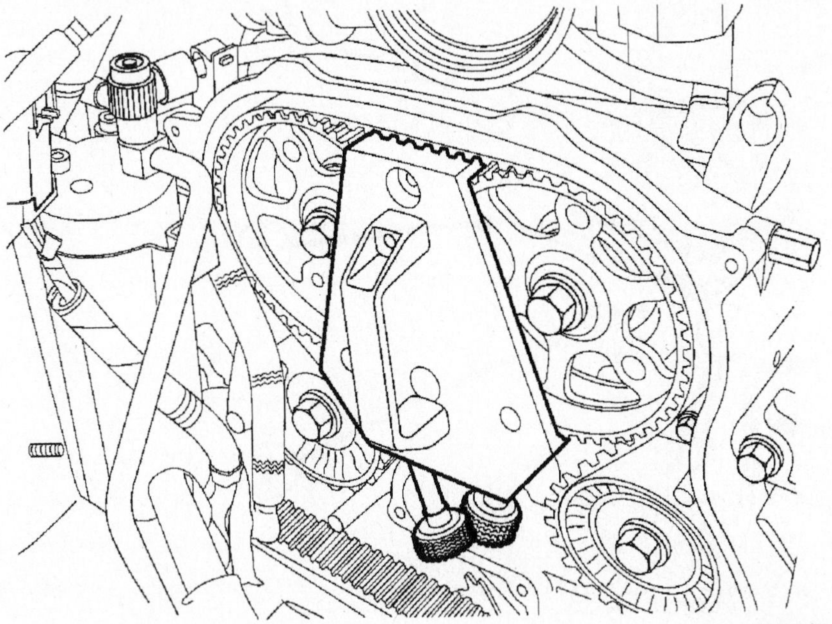

Camshaft locking tool—2.8L diesel engine

ATDC, or the 3 O'clock position. Failure to do so could result in valve and/or piston damage during reassembly.

11. Remove timing belt outer cover.
12. Remove timing belt.
13. Remove timing belt inner cover.
14. Remove cylinder head cover/intake manifold.
15. With cylinder head cover/intake manifold on work bench, remove plugs at rear of cylinder head cover/intake manifold.
16. Remove camshaft oil seals.
17. Remove snap ring and thrust washer from camshaft.
18. Slide camshaft through access hole at rear of cylinder head cover/intake manifold.

To install:

19. Lubricate camshafts with Mopar® Engine Oil Supplement, or equivalent.
20. Carefully install camshafts into access holes in rear of cylinder head cover/intake manifold.
21. Install thrust washer, snap ring, and camshaft oil seal.

➡If the camshaft endplay is not within specification, measure thickness of the camshaft spacer. Camshaft spacer thickness should be 2.8mm +/- 0.02mm.

22. Measure the camshaft end play with a dial indicator. The end play should be between 0.10mm+/-0.55 mm.
23. Install access hole plugs and gaskets at rear of cylinder head cover/intake manifold. Torque plugs to 80 Nm (59 ft. lbs.).
24. Install cylinder head cover/intake manifold on engine block.
25. Align the camshafts and install camshaft alignment pins, VM1052 and VM1053 into the camshaft cover.
26. Install timing belt inner cover.

➡Crankshaft must be rotated to the 90 degree ATDC, or the 3 O'clock position.

27. Install timing.
28. Install timing belt outer cover.
29. Install generator.
30. Install vibration damper.
31. Install cooling fan support.
32. Install accessory drive belt.
33. Install cooling fan and fan drive viscous clutch assembly.
34. Refill cooling system.
35. Connect negative battery cable.

Oil Pan

REMOVAL & INSTALLATION

1. Before servicing the vehicle, refer to the precautions in the beginning of this section.
2. Disconnect negative battery cable.
3. Remove engine cover.

➡When installing engine support fixture, care must be taken not to damage the hood ajar switch mounted to the right inner fender.

4. Install engine support fixture, special tool 8534, or equivalent.
5. Raise vehicle on hoist.
6. Remove both front wheel and tire assemblies.
7. Remove front skid plate (if equipped).
8. Drain engine oil.
9. Disconnect the front drive shaft from the front drive axle,
10. Loosen both engine mount through bolts.
11. Lower vehicle.
12. Raise engine using the support fixture, until the viscous fan almost touches the fan shroud.
13. Raise vehicle on hoist.
14. Support the front cradle assembly with a suitable lifting devise.
15. Mark the front cradle to under body position to assure proper alignment during assembly.
16. Remove both inner rail cradle alignment bolts in the front wheel housing.
17. Loosen both power steering gear retaining bolts, leave the retaining nuts on the bolts.
18. Loosen power steering lines from cradle.
19. Remove both lower front strut bolts.
20. Loosen both front cradle mounting bolts.
21. Loosen both rear cradle mounting bolts.
22. Disconnect the oil sending unit harness connector.
23. Remove all oil pan retaining bolts, separate the oil level indicator tube from the pan and lower oil pan.
24. Lower front cradle using the suitable lifting devise until enough clearance is obtained to remove the oil pan.

To install:

25. Clean oil pan and sealing surfaces. Inspect oil pan and engine block.
26. Install a new oil filter on the oil pan.

➡If installing a new oil pan, exchange the oil sending unit and install a new oil filter.

27. Install oil pan, gasket, oil pan retaining bolts, hand tight.
28. Push the oil pan against adaptor plate and tighten transmission to oil pan bolts first.

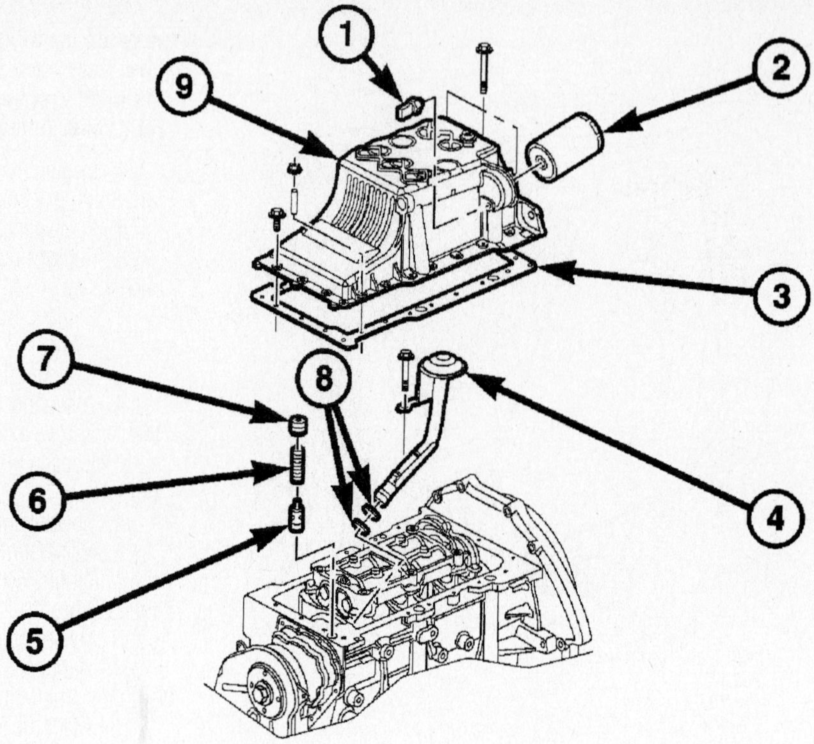

Oil pan and related parts—2.8L diesel engine

29. Tighten oil pan bolts to 14.7 Nm (130 inch lbs.), tighten oil pan nuts to 24.5 Nm (217 inch lbs.), beginning with the center bolts and then tighten the remaining fasteners in a clockwise rotation. Retighten the center two bolts again after all bolts are tightened to specification.
30. Raise the front cradle using a suitable lifting devise and align the cradle with the underbody marks made during the removal procedure.
31. Torque the cradle mounting bolts to 122 Nm (90 ft. lbs.).
32. Torque the clevis bolts to 136 Nm (100 ft. lbs.).
33. Torque the inner rail mounting bolts to 47 Nm (35 ft. lbs.).
34. Torque the steering gear mounting bolts to 162 Nm (120 ft. lbs.).
35. Remove the cradle support devise and lower the vehicle.
36. Lower the engine using the support fixture, until the engine mount through bolts are seated in the cradle.
37. Raise the vehicle.
38. Torque engine mount through bolts to 88 Nm (65 ft. lbs.).
39. Connect the oil sending unit harness connector.
40. Install front axle skid plate (if equipped).
41. Install both front wheel and tire assemblies. Tighten to 115–155 Nm (85–115 ft. lbs.).

➡When removing engine support fixture, care must be taken not to damage the hood ajar switch mounted to the right inner fender well (if equipped).

42. Lower the vehicle and remove the engine support fixture.
43. Refill engine to proper level with the correct viscosity engine oil.
44. Connect negative battery cable.
45. Start engine and inspect for leaks.
46. Install engine cover.
47. Perform complete front wheel alignment.

Oil Pump

REMOVAL & INSTALLATION

1. Before servicing the vehicle, refer to the precautions in the beginning of this section.
2. Disconnect negative battery cable.
3. Remove cooling fan and fan drive viscous clutch assembly.

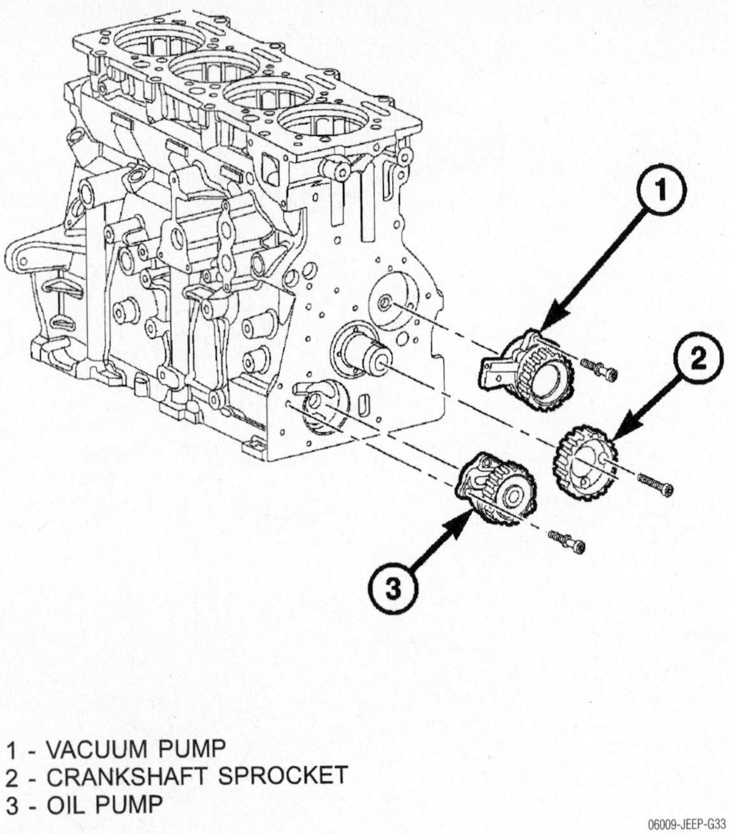

Oil pump—2.8L diesel engine

4. Remove accessory drive belt.
5. Remove cooling fan support.
6. Remove vibration damper.
7. Remove the power steering pump pulley.
8. Remove timing belt outer cover.

➡The crankshaft must be rotated to 90 degrees ATDC.

9. Remove timing belt.
10. Remove timing belt inner cover.
11. Remove front engine cover.
12. Remove crankshaft sprocket.
13. Remove oil pump retaining bolts and remove pump from engine block.

To install:

14. Lubricate oil pump rotor with engine oil.
15. Install oil pump in bore in engine block.
16. Install oil pump retaining bolts. Torque bolts to 10.8 Nm. (96 inch lbs.).
17. Install crankshaft sprocket. Torque bolts to 10.8 Nm. (96 inch lbs.).
18. Install front engine cover and seal.
19. Install timing belt inner cover.
20. Install the lower timing belt gear and hand tighten fastener.
21. Connect special tool 6958, or equivalent, to the lower timing gear using the vibration damper bolts and tighten the timing gear fastener to 275 Nm (203 ft. lbs).
22. Remove special tool 6958.
23. Install timing belt.
24. Install timing belt outer cover.
25. Install vibration damper.
26. Install power steering pump pulley.
27. Install cooling fan support.
28. Install accessory drive belt.
29. Install cooling fan and fan drive viscous clutch assembly.
30. Connect negative battery cable.

Rear Main Seal

REMOVAL & INSTALLATION

1. Before servicing the vehicle, refer to the precautions in the beginning of this section.

➡This can be done with either the engine or transmission removed from vehicle.

2. Remove flexplate/flywheel assembly.
3. Paint mark or scribe a witness mark on the engine block to be used for alignment purposes during assembly.
4. Pry out old crankshaft oil seal.

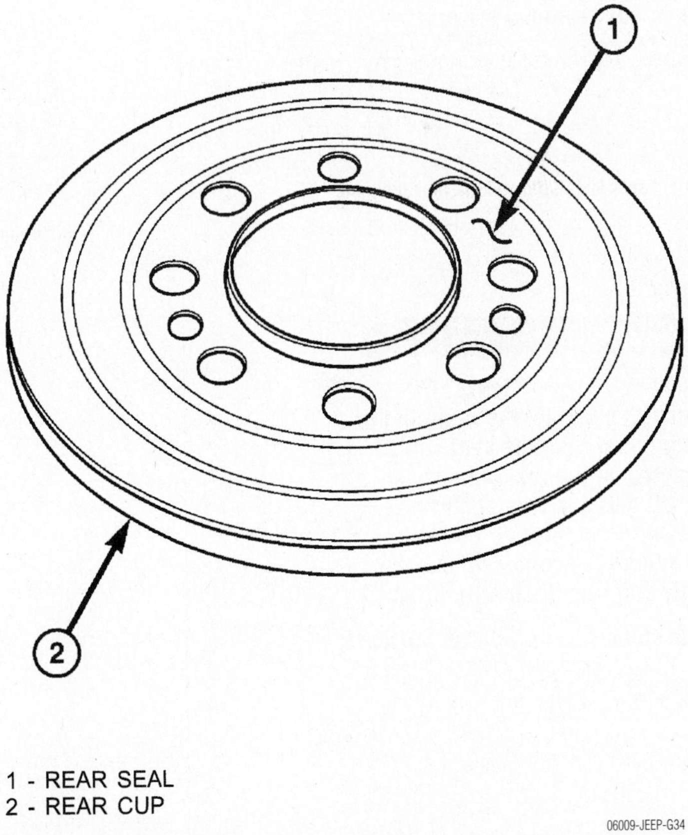

Rear main seal—2.8L diesel engine

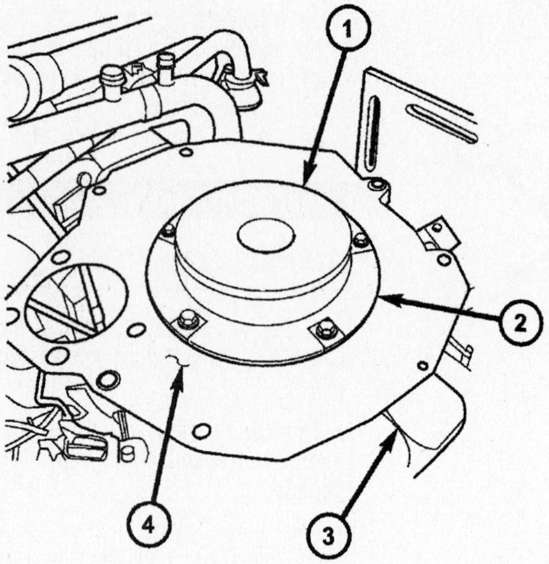

1 - SPECIAL TOOL VM.1050
2 - REAR MAIN BEARING SUPPORT
3 - OIL PAN
4 - ENGINE TO TRANSMISSION ADAPTER PLATE

06009-JEEP-G35

Rear main seal installation—2.8L diesel engine

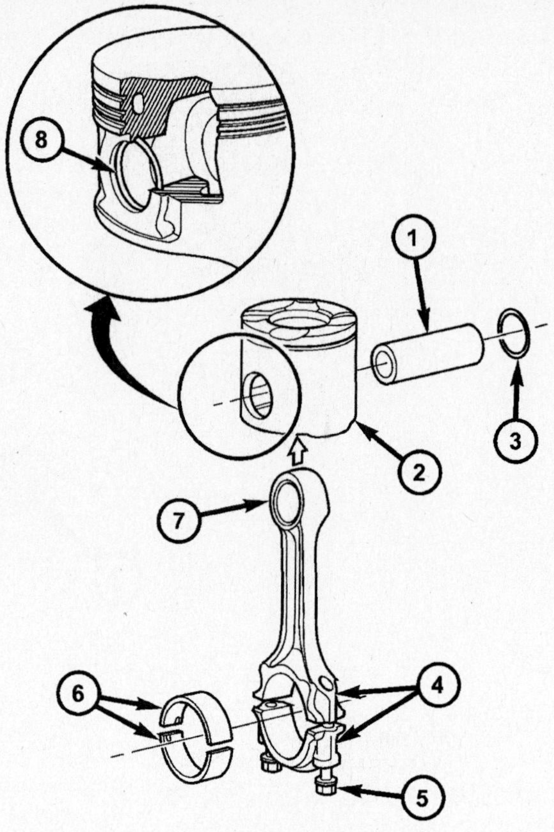

1 - PISTON PIN
2 - PISTON
3 - SNAP RING
4 - CONNECTING ROD ALIGNMENT NUMBERS
5 - CONNECTING ROD BOLT
6 - CONNECTING ROD BEARING
7 - CONNECTING ROD
8 - SNAP RING

06009-JEEP-G36

Piston and connecting rod assembly—2.8L diesel engine

To install:

⁂ WARNING

To prevent potential oil leaks, DO NOT touch or separate the rear crankshaft inner seal from the seal cup. When replacing the rear crankshaft carrier, DO NOT touch or separate the seal cup from the rear oil seal. Always handle the components as an assembly and from their outer diameter.

5. Using special tool VM.1050, install rear crankshaft oil seal in rear main bearing support.
6. Install engine or transmission in vehicle.

Piston and Ring Positioning

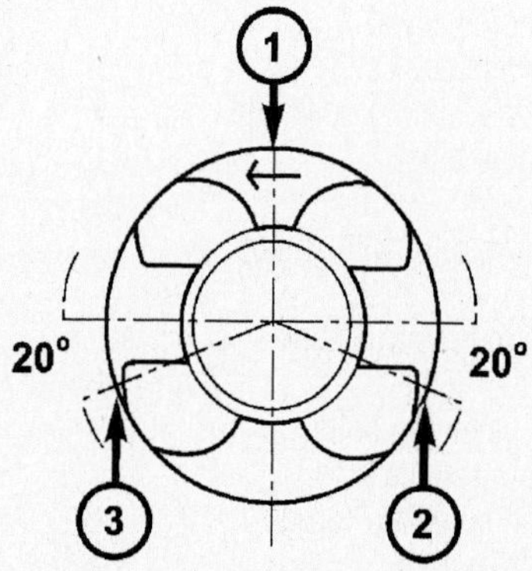

1 - SECOND COMPRESSION RING GAP POSITION
2 - OIL CONTROL RING GAP POSITION
3 - TOP COMPRESSION RING GAP POSITION

06009-JEEP-G37

Ring gap locations—2.8L diesel engine

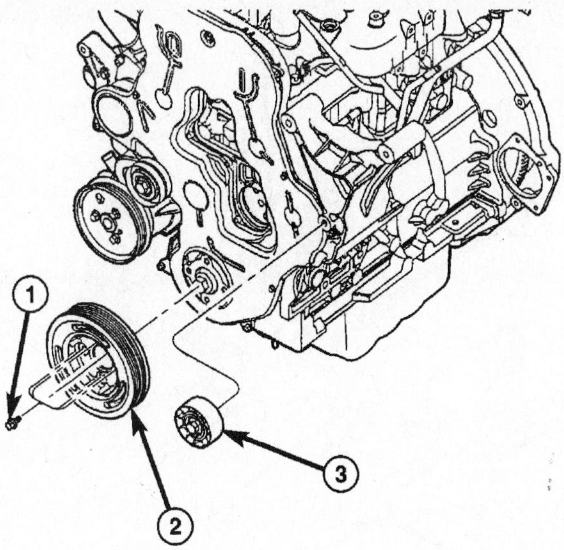

1 - VIBRATION DAMPER/CRANKSHAFT PULLEY RETAINING BOLTS
2 - VIBRATION DAMPER/CRANKSHAFT PULLEY
3 - IDLER PULLEY

06009-JEEP-G38

Vibration damper—2.8L diesel engine

Vibration Damper

REMOVAL & INSTALLATION

1. Before servicing the vehicle, refer to the precautions in the beginning of this section.
2. Disconnect negative battery cable.
3. Remove viscous cooling fan and shroud.
4. Remove accessory drive belt.
5. Remove vibration damper retaining bolts and damper.

To install:

6. Install vibration damper and retaining bolts. Torque bolts to 32 Nm. (24 ft. lbs.).
7. Install accessory drive belt).
8. Install viscous fan and fan shroud.
9. Connect negative battery cable.

Timing Belt Outer Cover

REMOVAL & INSTALLATION

1. Before servicing the vehicle, refer to the precautions in the beginning of this section.
2. Disconnect negative battery cable.
3. Remove engine cover.
4. Remove cooling fan and fan drive viscous clutch assembly.
5. Remove accessory drive belt.
6. Remove cooling fan support.
7. Remove the power steering pump pulley.
8. Remove vibration damper.
9. Remove timing belt outer cover retaining bolts and remove cover.

To install:

10. Install timing belt outer cover seal and cover. Torque 3mm bolts to 6 Nm (53 inch lbs.).
11. Install vibration damper.
12. Install the power steering pump pulley.
13. Install cooling fan support.
14. Install accessory drive belt.
15. Install cooling fan and fan drive viscous clutch assembly.
16. Install engine cover.
17. Connect negative battery cable.

Timing Belt

REMOVAL & INSTALLATION

1. Before servicing the vehicle, refer to the precautions in the beginning of this section.

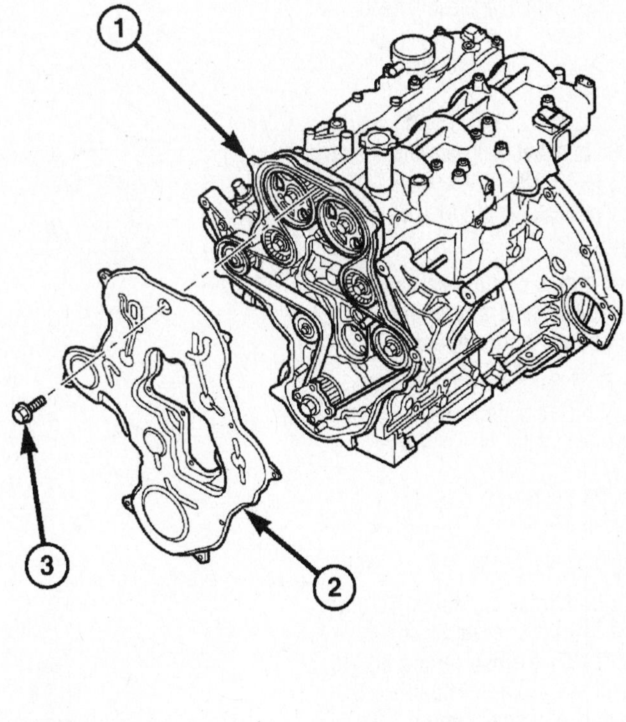

1 - TIMING BELT
2 - TIMING BELT OUTER COVER
3 - TIMING BELT OUTER COVER RETAINING BOLTS

06009-JEEP-G39

Timing belt outer cover—2.8L diesel engine

WARNING

Before removing the timing belt, the engine must be placed at 90° after TDC. Failure to do so may result in valve and/or piston damage during assembly.

2. Disconnect negative battery cable.
3. Remove engine cover.
4. Remove cooling fan and fan drive viscous clutch assembly.
5. Remove accessory drive belt.
6. Remove cooling fan support.
7. Remove vibration damper.
8. Bring piston #1 to TDC, turn crankshaft until the witness line on the crankshaft hub is at the 12 o'clock position.
9. Looking at the engine from the belt side, rotate the crankshaft 90° clockwise so the witness mark is now at the 3 o'clock position.
10. Remove the alternator.
11. Remove the intake and exhaust camshaft plugs from the camshaft cover, to introduce the camshaft timing pins VM.1052 Intake, and VM.1053 Exhaust (if the engine is timed correctly, the pins can be installed flush with the intake manifold/cylinder head cover).
12. Remove timing belt outer cover.
13. Loosen timing belt tensioner and remove timing belt.
14. Remove the intake and exhaust camshaft alignment pins.
15. Loosen camshaft gears using special tool VM 1085 to retain the gears when loosening the bolts.
16. Remove special tool VM 1085.

To install:

➡There are marks on both camshaft gears. These ARE NOT alignment marks and should be disregarded.

17. With both camshaft alignment pins still installed and the engine rotated at 90° after TDC, verify that the camshaft gears are loose.
18. Align the timing mark on the high pressure injection pump gear with the timing mark on the inner cover.

➡DO NOT remove the timing belt from the package until it's ready to be installed. DO NOT expose timing belt to oil, grease or water contamination. DO NOT crimp belt at a sharp angle. DO NOT clean belt, pulleys or tensioner with solvent. Check that pulleys and bearings are not seized or damage before installing belt.

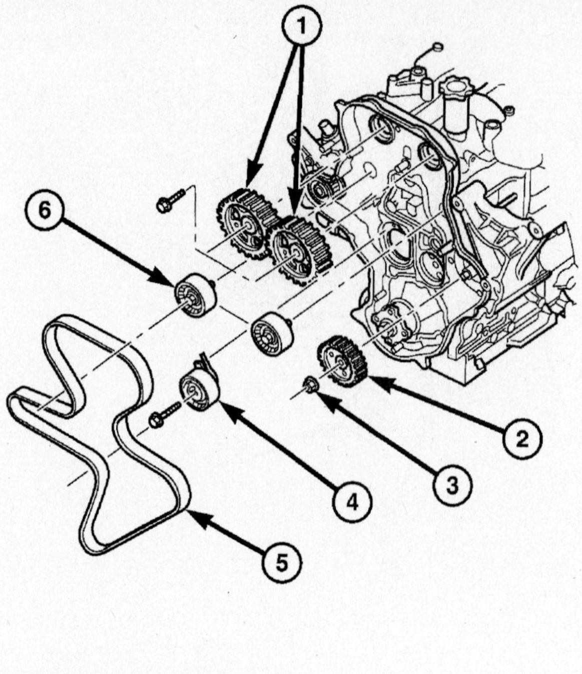

Timing belt and sprockets—2.8L diesel engine

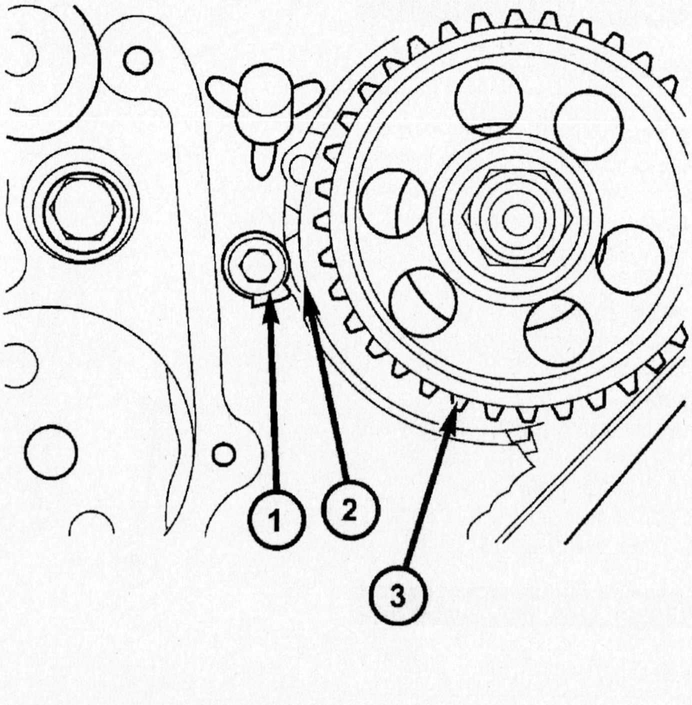

Injection pump gear timing marks—2.8L diesel engine

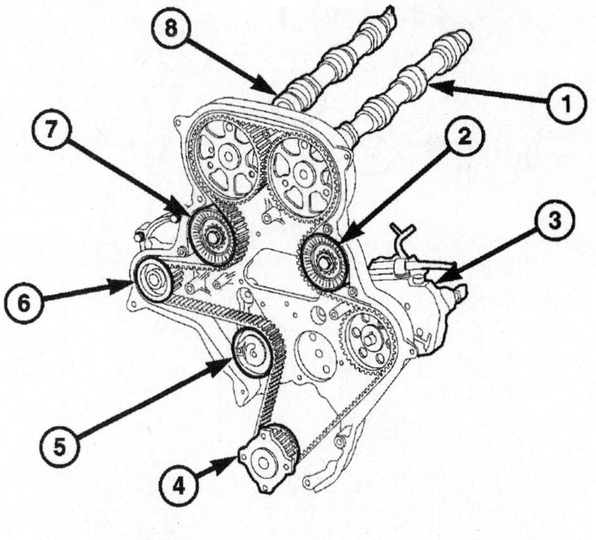

1 - INTAKE CAMSHAFT
2 - IDLER PULLEY
3 - HIGH PRESSURE PUMP
4 - CRANKSHAFT GEAR
5 - TIMING BELT TENSIONER
6 - WATER PUMP
7 - IDLER PULLEY
8 - EXHAUST CAMSHAFT

06009-JEEP-G43

Timing belt routing—2.8L diesel engine

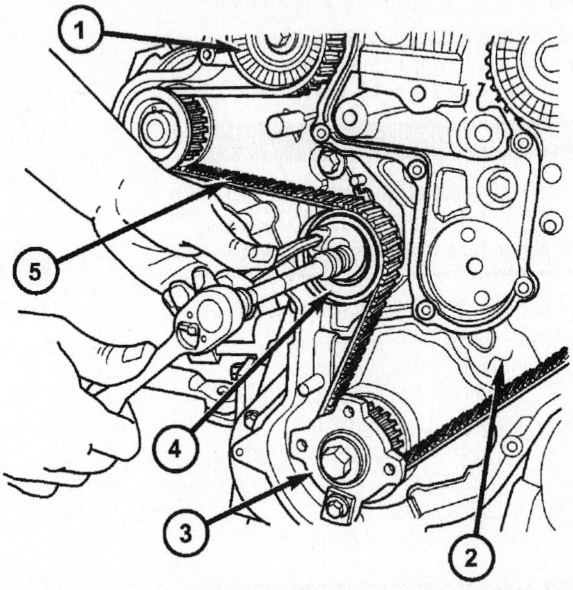

1 - TIMING BELT IDLER PULLEY
2 - ENGINE FRONT COVER
3 - CRANKSHAFT HUB
4 - TIMING BELT TENSIONER
5 - TIMING BELT

06009-JEEP-G44

Timing belt tensioner adjustment—2.8L diesel engine

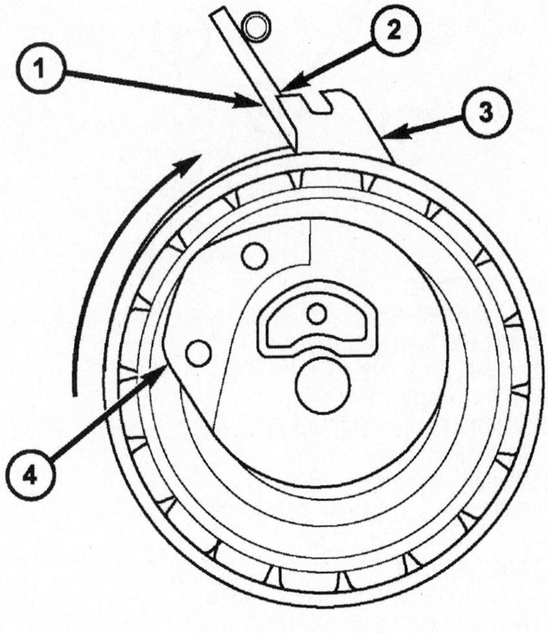

1 - TENSIONER SPRING
2 - 1MM ALIGNMENT POINTER OVERLAP
3 - TENSIONER ALIGNMENT POINTER
4 - TENSIONER ASSEMBLY

06009-JEEP-G45

Timing belt tensioner adjustment—2.8L diesel engine

19. Install timing belt on crankshaft hub.
20. Route the belt around high pressure injection pump, idler pulley, intake camshaft gear, exhaust camshaft gear, idler pulley, and water pump gear.
21. Adjust the timing belt tensioner (turn it clockwise) using special tool VM.9660, lining up the center notch with the aluminum cover dowel pin. Tighten the retaining bolt to 28 Nm (20 ft. lbs.).
22. Install special tool VM.1085 between camshaft gears and tighten thumb screws to engage and retain the camshaft gears.
23. Tighten the camshaft gear bolts to 108 Nm (80 ft. lbs.) while holding the gears with special tool VM 1085.
24. Remove camshaft gear locking tool.
25. Remove intake and exhaust alignment pins.

⁂ WARNING

If intake manifold/cylinder head cover was removed wait 30 minutes before rotating crankshaft.

26. Rotate the engine clockwise 2 revolutions (looking at engine from the belt side).
27. With the crankshaft hub witness mark aligned to 90 degrees ATDC, or the 3

o'clock position, check that the intake and exhaust camshaft alignment pins (VM 1052 and VM 1053) can be installed into the camshaft alignment access holes.

28. Install timing belt outer cover.

⁂ WARNING

If the camshaft alignment pins can not be installed at this time, repeat the procedure from the beginning.

29. Install the camshaft access plugs.
30. Install vibration damper.
31. Install the alternator.
32. Install cooling fan support.
33. Install accessory drive belt.
34. Install cooling fan and fan drive viscous clutch assembly.
35. Install engine cover.
36. Connect negative battery cable.

Timing Belt Inner Cover

REMOVAL & INSTALLATION

1. Before servicing the vehicle, refer to the precautions in the beginning of this section.
2. Disconnect negative battery cable.
3. Remove engine cover.
4. Remove cooling fan and fan drive viscous clutch assembly.
5. Remove accessory drive belt.
6. Remove cooling fan support.
7. Remove vibration damper.
8. Rotate the engine to 90 degrees ATDC.
9. Remove timing belt outer cover.
10. Remove timing belt.
11. Remove timing belt idler pulleys.
12. Using special tool VM.1085, remove camshaft sprockets.
13. Remove timing belt tensioner.
14. Remove timing belt inner cover retaining bolts and remove cover.

To install:

➡Rotate crankshaft hub to 90 degrees ATDC, or the 3 o'clock position.

15. Install timing belt inner cover to engine front cover seal.
16. Install timing belt inner cover to cylinder head cover gaskets.
17. Install timing belt inner cover and retaining bolts. Torque 10mm bolts to 47 Nm and 8mm bolts to 11 Nm (96 inch lbs.).
18. Install timing belt idler pulleys.
19. Install camshaft sprockets. Hand tighten bolts.
20. Install timing belt and tensioner.
21. Install special tool VM.1085 between

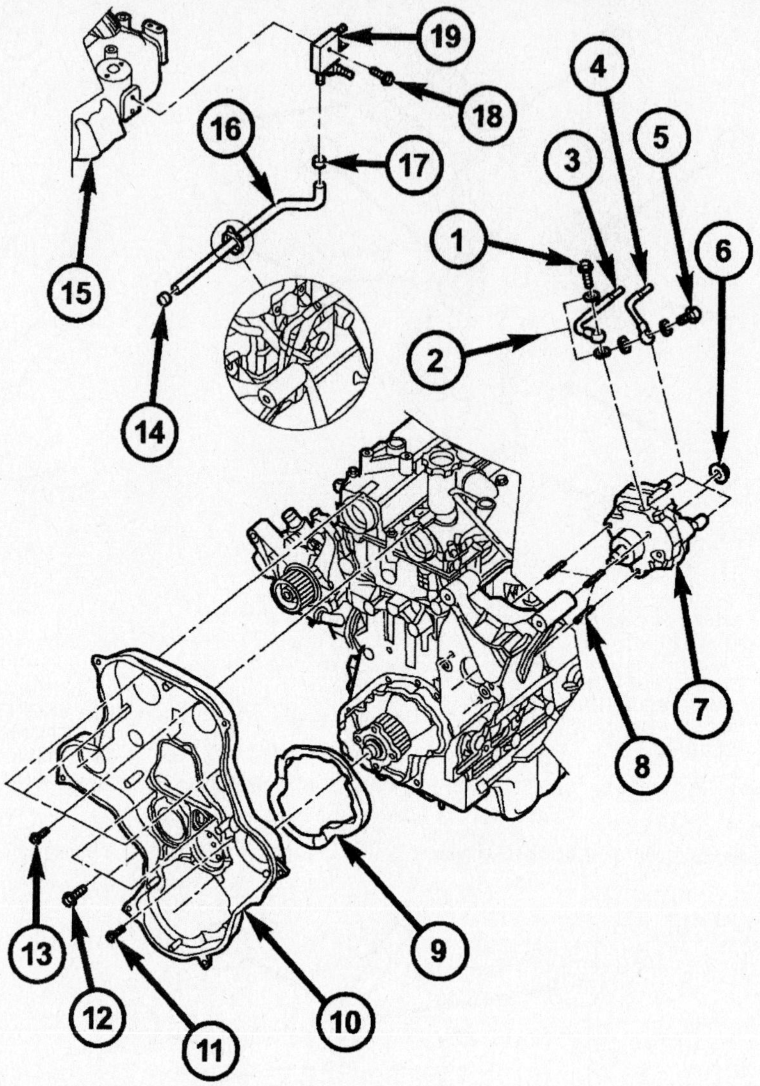

1 - BANJO FITTING
2 - BRASS WASHERS
3 - FUEL INLET LINE
4 - FUEL OUTLET LINE
5 - BANJO FITTING
6 - INJECTION PUMP RETAINING NUT
7 - INJECTION PUMP
8 - MOUNTING STUDS
9 - INNER TIMING COVER SEAL
10 - INNER TIMING COVER
11 - RETAINING BOLT
12 - RETAINING BOLT
13 - RETAINING BOLT
14 - HOSE CLAMP
15 - CYLINDER HEAD COVER/INTAKE MANIFOLD
16 - FUEL RETURN LINE TO INJECTION PUMP
17 - HOSE CLAMP
18 - FUEL RETURN JUNCTION BLOCK RETAINING BOLT
19 - FUEL RETURN JUNCTION BLOCK

06009-JEEP-G40

Timing belt inner cover—2.8L diesel engine

camshaft gears and tighten thumb screws to engage and retain the camshaft gears.

22. Tighten the camshaft gear bolts to 108 Nm (80 ft. lbs.) while holding the gears with special tool VM 1085.
23. Install timing belt outer cover.
24. Install vibration damper.
25. Install cooling fan support.
26. Install accessory drive belt.
27. Install cooling fan and fan drive viscous clutch assembly.
28. Install engine cover.
29. Connect negative battery cable.

Engine Front Cover

REMOVAL & INSTALLATION

1. Before servicing the vehicle, refer to the precautions in the beginning of this section.
2. Disconnect negative battery cable.
3. Remove cooling fan and fan drive viscous clutch assembly.
4. Remove accessory drive belt.
5. Remove cooling fan support.
6. Remove vibration damper.
7. Remove timing belt outer cover.

⁂ WARNING

Before removing the cylinder head cover/intake manifold or timing belt the engine must aligned at 90° ATDC. Failure to do so could result in valve and/or piston damage during reassembly.

8. Remove timing belt.
9. Remove timing belt inner cover.

➡Crankshaft hub has left hand thread.

10. Remove crankshaft hub.
11. Remove front engine cover.

To install:

12. Clean engine block and front engine cover sealing surfaces.
13. Apply a continuous 3mm bead of silicone sealer to cover, install within 10 minutes. Torque bolts to 12 Nm (104 inch lbs.).
14. Install crankshaft hub. Using special tool 6958, torque bolt to 304 Nm. (203 ft. lbs.).
15. Install timing belt inner cover.
16. Install timing belt.
17. Install timing belt outer cover.
18. Install vibration damper.
19. Install cooling fan support.
20. Install accessory drive belt.
21. Install cooling fan and fan drive viscous clutch assembly.
22. Connect negative battery cable.

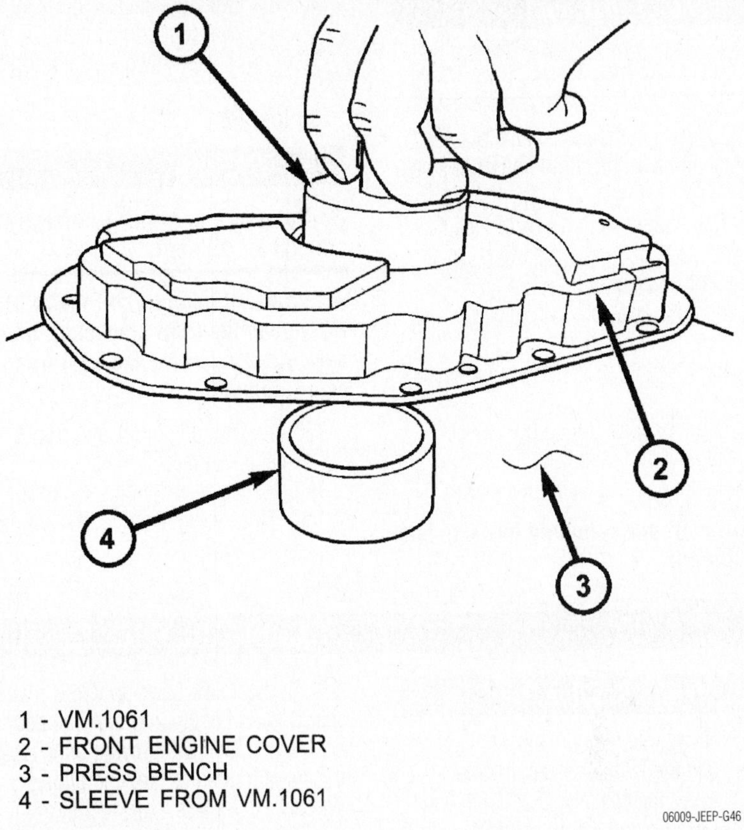

Special tool placement—2.8L diesel

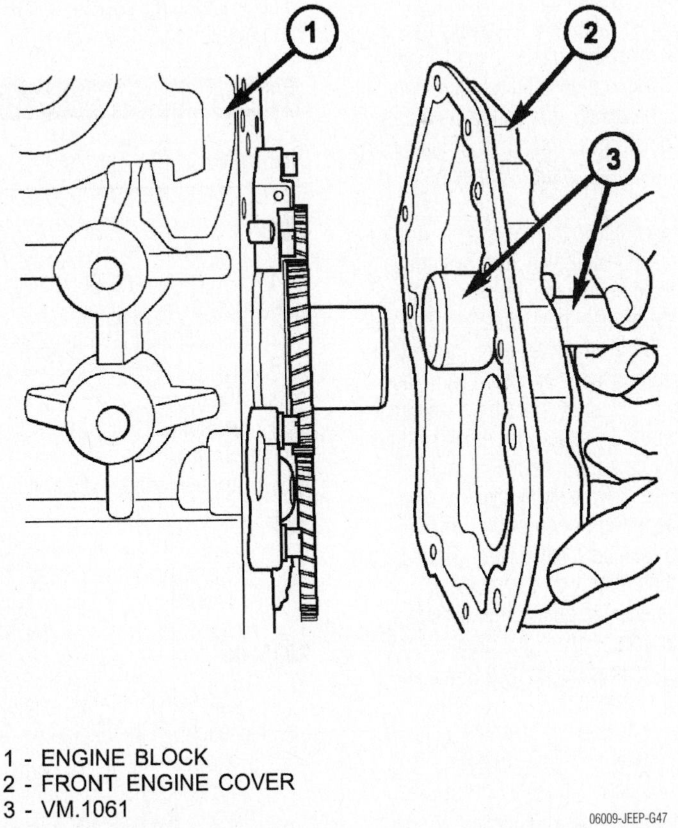

Engine front cover installation—2.8L diesel

Crankshaft Front Seal

REMOVAL & INSTALLATION

1. Before servicing the vehicle, refer to the precautions in the beginning of this section.
2. Disconnect negative battery cable.
3. Remove cooling fan and fan drive viscous clutch assembly.
4. Remove accessory drive belt.
5. Remove cooling fan support.
6. Remove vibration damper/crankshaft pulley.
7. Remove outer timing belt cover.
8. Remove timing belt.
9. Remove timing belt inner cover.

➡Crankshaft hub retaining bolt has left hand thread.

10. Remove crankshaft hub.
11. Remove front engine cover.
12. With cover on work bench, pry out old seal.

To install:

⁂ WARNING

Do Not use a hammer to install the crankshaft oil seal.

➡To prevent potential oil leaks, DO NOT touch the front crankshaft inner seal. Always handle the seal from the outer diameter.

13. Clean engine block and front engine cover sealing surfaces.
14. Install crankshaft oil seal on VM.1061.
15. Place sleeve for VM.1061 on press bench as shown.
16. Position VM.1061 and press in new seal into front engine cover.
17. Install front engine cover on engine using VM 101 as a guide, care must be taken not to damage the seal.
18. Install crankshaft hub and retaining bolt. Torque bolt to 304 Nm:
19. Install timing belt inner cover.

➡Crankshaft must be rotated to 90 degrees ATDC.

20. Install timing belt.
21. Install timing belt outer cover.
22. Install vibration damper.
23. Install cooling fan support.
24. Install accessory drive belt.
25. Install cooling fan and fan drive viscous clutch assembly.
26. Connect negative battery cable.

GASOLINE FUEL SYSTEM

Fuel System Service Precautions

Safety is the most important factor when performing not only fuel system maintenance but any type of maintenance. Failure to conduct maintenance and repairs in a safe manner may result in serious personal injury or death. Maintenance and testing of the vehicle's fuel system components can be accomplished safely and effectively by adhering to the following rules and guidelines.

- To avoid the possibility of fire and personal injury, always disconnect the negative battery cable unless the repair or test procedure requires that battery voltage be applied.
- Always relieve the fuel system pressure prior to disconnecting any fuel system component (injector, fuel rail, pressure regulator, etc.), fitting or fuel line connection. Exercise extreme caution whenever relieving fuel system pressure to avoid exposing skin, face and eyes to fuel spray. Please be advised that fuel under pressure may penetrate the skin or any part of the body that it contacts.
- Always place a shop towel or cloth around the fitting or connection prior to loosening to absorb any excess fuel due to spillage. Ensure that all fuel spillage (should it occur) is quickly removed from engine surfaces. Ensure that all fuel soaked cloths or towels are deposited into a suitable waste container.
- Always keep a dry chemical (Class B) fire extinguisher near the work area.
- Do not allow fuel spray or fuel vapors to come into contact with a spark or open flame.
- Always use a back-up wrench when loosening and tightening fuel line connection fittings. This will prevent unnecessary stress and torsion to fuel line piping.
- Always replace worn fuel fitting O-rings with new. Do not substitute fuel hose or equivalent where fuel pipe is installed.

Before servicing the vehicle, make sure to also refer to the precautions in the beginning of this section as well.

Fuel System Pressure

RELIEVING

2002

1. Before servicing the vehicle, refer to the precautions in the beginning of this section.
2. Remove the fuel pump relay.
3. Start the engine and allow it to run until it stalls.
4. Attempt restarting the engine until it no longer runs.
5. Turn the ignition key to the **OFF** position.
6. Disconnect the negative battery cable.
7. After repairs are complete, replace the relay and connect the negative battery cable.

2003–05

1. Before servicing the vehicle, refer to the precautions in the beginning of this section.
2. Disconnect the negative battery cable.
3. Remove the fuel tank filler cap to release any fuel tank pressure.
4. Remove the fuel pump relay from the PDC.
5. Start and run the engine until it stops.
6. Unplug the connector from any injector and connect a jumper wire from either injector terminal to the positive battery terminal. Connect another jumper wire to the other terminal and momentarily touch the other end to the negative battery terminal.

⁂ WARNING

Just touch the jumper to the battery. Powering the injector for more than a few seconds will permanently damage it.

7. Place a rag below the quick-disconnect coupling at the fuel rail and disconnect it.

Fuel Filter

REMOVAL & INSTALLATION

Wrangler

The combination Fuel Filter/Fuel Pressure Regulator is located on the fuel pump module. The fuel pump module is located on top of fuel tank.

1. Before servicing the vehicle, refer to the precautions in the beginning of this section.
2. Remove fuel tank.
3. Clean area around filter/regulator.
4. Disconnect fuel line at filter/regulator.
5. Remove retainer clamp from top of filter/regulator. Clamp snaps to tabs on pump module. Discard old clamp.

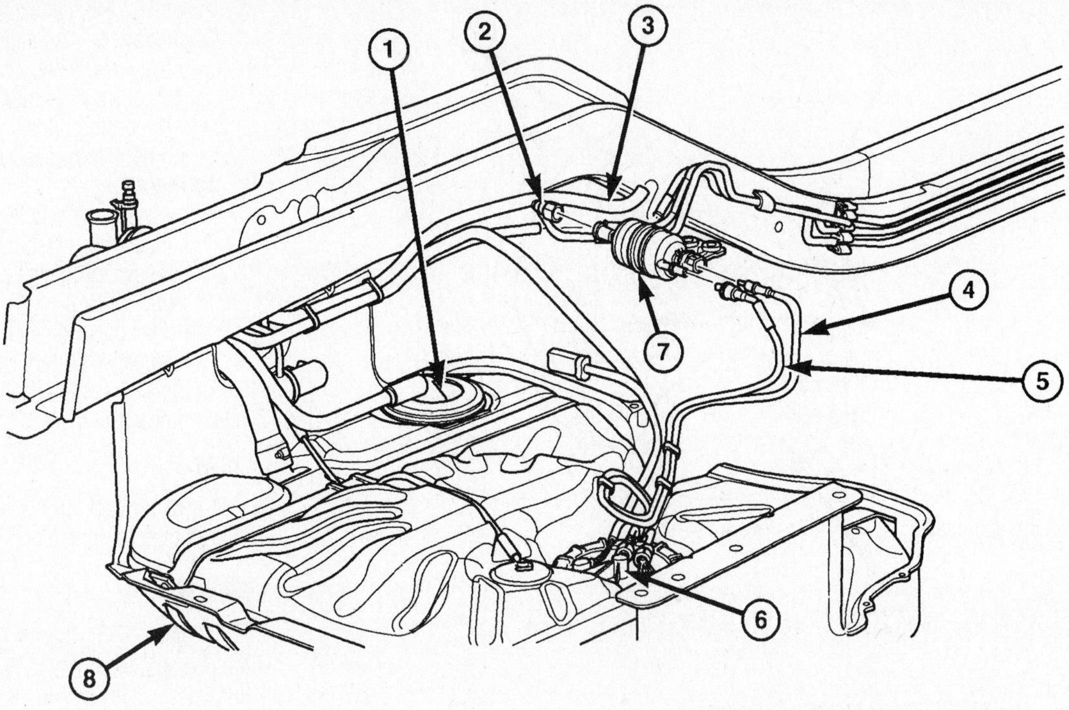

1 - ORVR CONTROL VALVE
2 - FUEL SUPPLY LINE (TO FUEL RAIL)
3 - EVAP LINE
4 - FUEL RETURN LINE (MALE)
5 - FUEL PRESSURE LINE (FEMALE)
6 - FUEL PUMP MODULE ASSEMBLY
7 - FUEL FILTER/PRESSURE REGULATOR
8 - FUEL TANK

67189-JEEP-G55

Fuel filter/pressure regulator—Grand Cherokee

6. Pry filter/regulator from top of pump module with 2 screwdrivers. Unit is snapped into module.
7. Discard gasket below filter/regulator.

To install:

8. Clean recessed area in pump module where filter/regulator is to be installed.
9. Obtain new filter/regulator (two new O-rings should already be installed).
10. Apply a small amount of clean engine oil to O-rings. Do not install o-rings separately into fuel pump module. They will be damaged when installing filter/regulator.
11. Install new gasket to top of fuel pump module.
12. Press new filter/regulator into pump module until it snaps into position (a positive click must be heard or felt).
13. Install the module.
14. Install fuel tank.

Grand Cherokee

2002–04

The fuel filter/pressure regulator is mounted to the body above the rear axle and near the front of the fuel tank.

1. Before servicing the vehicle, refer to the precautions in the beginning of this section.

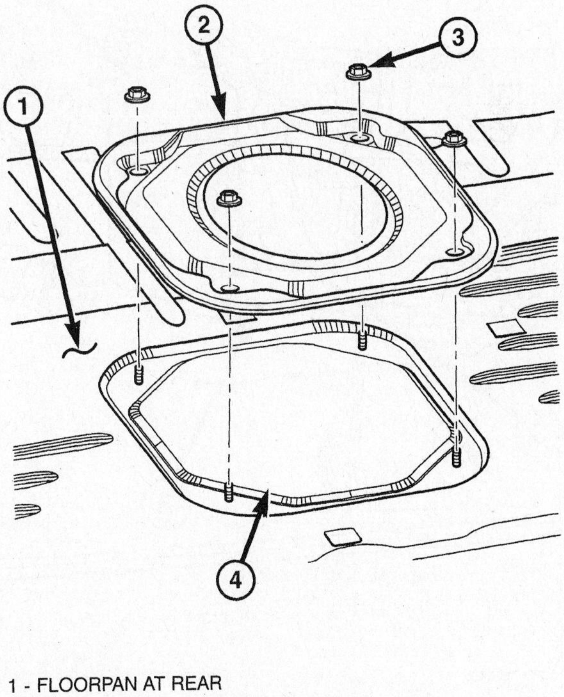

1 - FLOORPAN AT REAR
2 - FUEL PUMP MODULE ACCESS PLATE
3 - NUTS (4)
4 - OPENING TO PUMP MODULE

9355PG17

Access plate—Liberty

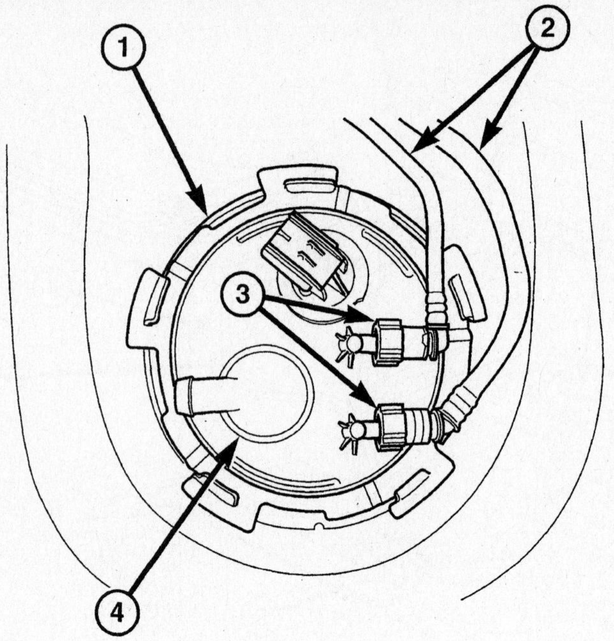

1 - FUEL PUMP MODULE LOCKRING
2 - FUEL LINES TO FUEL FILTER (2)
3 - QUICK-CONNECT FITTINGS (2)
4 - ROLLOVER VALVE

9355PG18

Fuel lines at the pump module—Liberty

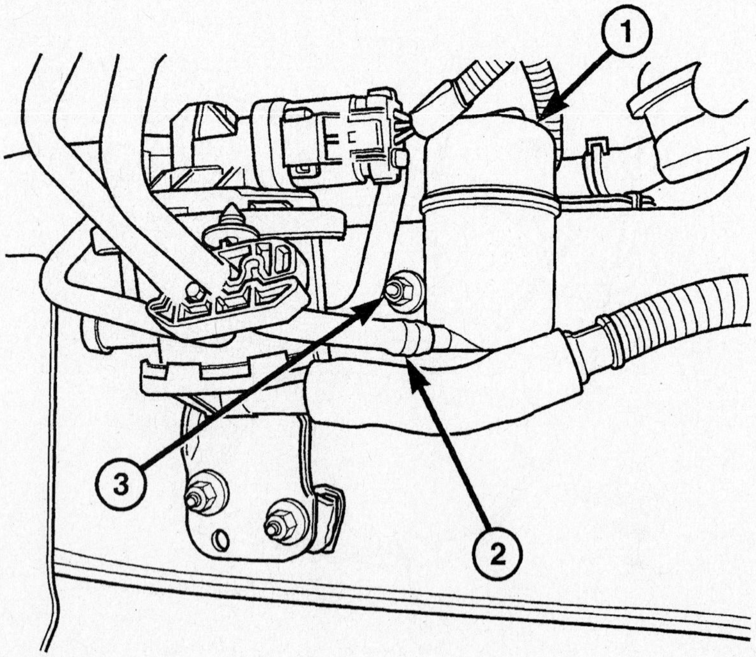

1 - FUEL FILTER
2 - 3RD FUEL LINE TO ENGINE
3 - FILTER MOUNTING NUT

9355PG19

Fuel filter location—Liberty

2. Relieve the fuel system pressure.
3. Remove or disconnect the following:
   - Negative battery cable
   - Fuel supply, return and pressure lines
   - Mounting bolts
   - Fuel filter/pressure regulator

To install:

4. Install or connect the following:
   - Fuel filter/pressure regulator to the body. Torque the bolts to 30 inch lbs. (3 Nm).
   - Fuel pressure, return and supply lines
   - Negative battery cable
5. Start the vehicle, check for leaks and repair if necessary.

2005

A separate fuel filter is not used. The fuel filter is built into the fuel pump module.

Liberty

1. Before servicing the vehicle, refer to the precautions in the beginning of this section.

The fuel filter is attached to the front of the fuel tank.

2. Before servicing the vehicle, refer to the precautions in the beginning of this section.
3. Relieve the fuel system pressure.
4. Remove or disconnect the following:
   - Negative battery cable
   - 2 rear cargo hold-down clamps by drilling the rivets
   - 4 fuel pump module access plate nuts

➡Once the nuts are removed, the plate must be removed by applying a heat gun to melt the adhesive. Take care to avoid bending the plate. Once removed, you can disconnect the 2 top hoses from the filter. The bottom hose must be removed from under the vehicle. The disconnect point on this hose is about a foot in front of the filter.

   - Ground strap
   - Mounting nut and filter.
5. Installation is the reverse of removal.

Fuel Pump

REMOVAL & INSTALLATION

Wrangler and Grand Cherokee

2002–04

1. Before servicing the vehicle, refer to the precautions in the beginning of this section.

2. Relieve the fuel system pressure.
3. Remove or disconnect the following:
   - Negative battery cable
   - Fuel tank
   - Fuel lines
   - Fuel pump module locknut
   - Fuel pump module

To install:

4. Install or connect the following:
   - Fuel pump module with a new gasket. Rotate the module until the arrow is pointed toward the front of the vehicle, on all models except Grand Cherokee. The arrow must pointed toward the rear of the vehicle.
   - Locknut. Torque the locknut to 55 ft. lbs. (74 Nm).
   - Fuel lines
   - Fuel tank
   - Negative battery cable
5. Start the engine, check for leaks and repair if necessary.

2005 GRAND CHEROKEE

1. Before servicing the vehicle, refer to the precautions in the beginning of this section.

CAUTION

The fuel system is under a constant pressure (even with the engine off). Before servicing the fuel pump module, the fuel system pressure must be released.

2. Drain the tank.
3. Release fuel system pressure.
4. Raise vehicle.
5. Thoroughly clean area around fuel fill fitting and rubber fuel fill hose at rear of fuel tank.
6. Loosen clamp at tank fitting and disconnect rubber fuel fill hose at fuel tank fitting. Using an pproved gas holding tank, drain fuel tank through this fitting.
7. Loosen clamp and disconnect rubber fill hose at tank fitting.
8. At rear of tank, disconnect fuel pump module electrical jumper connector from body connector.
9. At rear of tank, disconnect EVAP lines and from lines and.
10. At front of tank, disconnect fuel and EVAP lines and from lines and.
11. Support tank with a hydraulic jack.
12. Remove bolts and at right side of fuel tank.
13. Remove bolts at left side of fuel tank.
14. Lower tank for removal.
15. Note rotational position of module before attempting removal. An indexing arrow is located on top of module for this purpose.
16. Position Special Tool 9340, or equivalent, into notches on outside edge of lockring.
17. Install ½ inch drive breaker bar to tool 9340.
18. Rotate breaker bar counter-clockwise to remove lockring.
19. Remove lockring. The module will spring up slightly when lockring is removed.
20. Remove module from fuel tank. Be careful not to bend float arm while removing.

To install:

➡Whenever the fuel pump module is serviced, the rubber seal (gasket) must be replaced.

21. Using a new seal (gasket), position fuel pump module into opening in fuel tank.
22. Position lockring over top of fuel pump module.
23. Rotate module until embossed alignment arrow points to center alignment mark. This step must be performed to prevent float from contacting side of fuel tank. Also be sure fuel fitting on top of pump module is pointed to drivers side of vehicle.
24. Install Special Tool 9340 to lockring.
25. Install ½ drive breaker into Special Tool 9340.
26. Tighten lockring (clockwise) until all seven notches have engaged.
27. Position fuel tank to hydraulic jack.
28. Raise tank until positioned to body.
29. Install and tighten bolts to 50 ft. lbs. (68 Nm).
30. Remove hydraulic jack.
31. Connect EVAP, ORVR, fuel and NVLD lines at front and rear of tank.
32. Connect fuel pump module electrical jumper connector to body connector.
33. Connect rubber fill hose to tank fitting and tighten clamp.
34. Lower vehicle.
35. Fill fuel tank with fuel.
36. Start engine and check for fuel leaks near top of module.

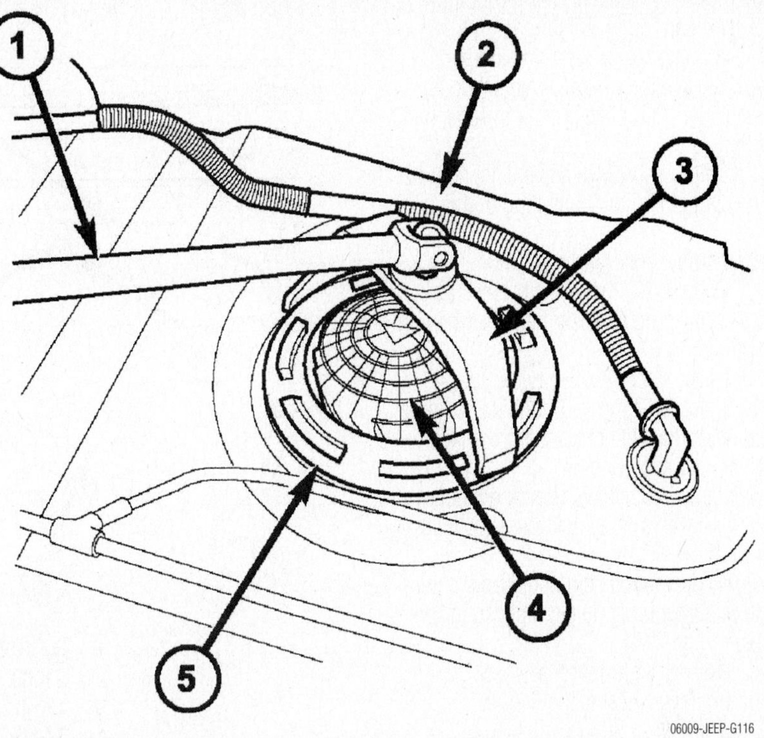

Installing the fuel module (1) breaker bar, (2) tank, (3) tool 9340, (4) module, (5) locking ring—2005 Grand Cherokee and Wrangler

2005 WRANGLER

CAUTION

The fuel system is under a constant pressure (even with the engine off). Before servicing the fuel pump module, the fuel system pressure must be released.

This vehicle is equipped with an ORVR (On-Board Refueling Vapor Recovery) system. Because of this, the fuel tank may be drained the conventional way through the filler cap opening. On this model, the fuel tank is mounted to the vehicle skid plate. The skid plate is mounted to vehicle body. The tank and skid plate are removed as one assembly.

1. Before servicing the vehicle, refer to the precautions in the beginning of this section.
2. Remove fuel filler cap.
3. Perform the fuel system pressure release procedure as described elsewhere in this group.
4. Disconnect negative battery cable.
5. Using an approved portable gasoline siphon/storage tank, drain fuel from tank through filler cap opening.
6. Remove 8 screws retaining plastic fuel filler bezel to body. Remove plastic fuel filler bezel.
7. To prevent contaminants from entering tank, temporarily install fuel cap to fill hoses.
8. Remove right/rear tire/wheel.
9. Remove wheelhouse liner at right/rear wheel.
10. Remove vertical support bracket to gain access to ORVR vapor line.
11. A vapor line connects the fuel tank to the EVAP canister and NVLD pump. This connection is made near the right/rear corner of the fuel tank. Carefully disconnect this vapor line. Be very careful not to bend or kink the vapor lines. If vapor lines leak, a Diagnostic Trouble Code (DTC) will be set.
12. Cut plastic tie wrap securing rear axle vent hose to fuel fill hose.
13. Disconnect fuel tank electrical connector at left/front of fuel tank.
14. Disconnect 2 vapor lines at left/front of fuel tank.
15. Disconnect quick-connect fitting from fuel supply line at front of fuel tank.
16. The fuel tank and skid plate are removed as an assembly. Centrally position a transmission jack (or equivalent lifting device) under skid plate/fuel tank assembly. Secure tank assembly to jack.
17. Remove three skid plate-to-body nuts at front of tank. Remove one of the nuts through access hole on skid plate.
18. Remove four skid plate-to-body nuts at rear of tank. Do not loosen tank strap nuts.
19. Lower the tank assembly.
20. Disconnect fuel filler hose at tank. Before disconnecting, mark and note the hose rotational position in relation to tank fitting.
21. To separate tank from skid plate, remove two protective caps at tank strap studs and remove tank strap nuts.

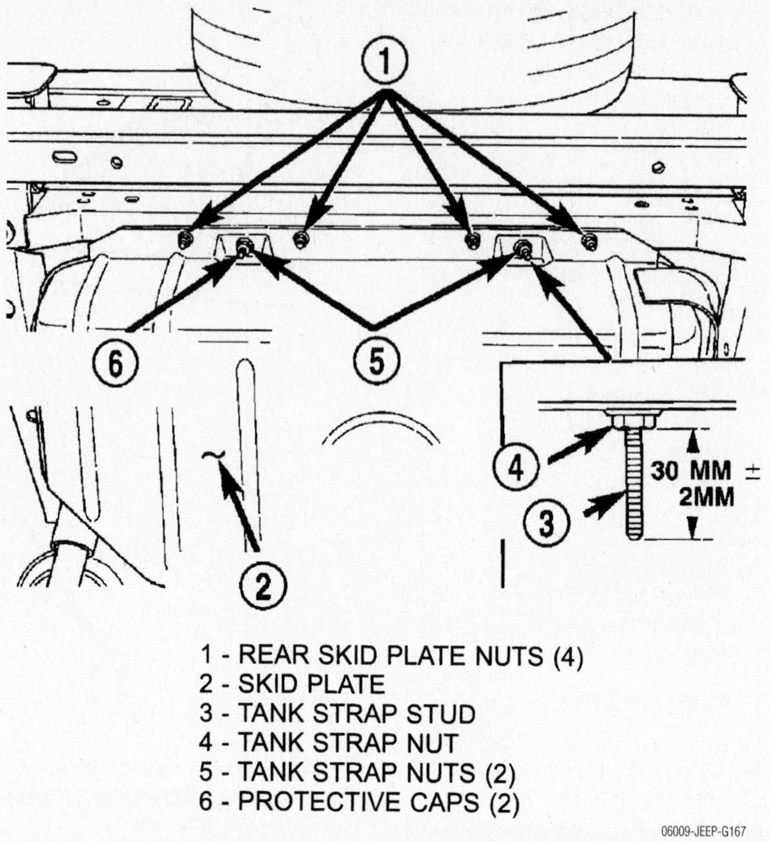

Fuel tank rear mounting nuts—Wrangler

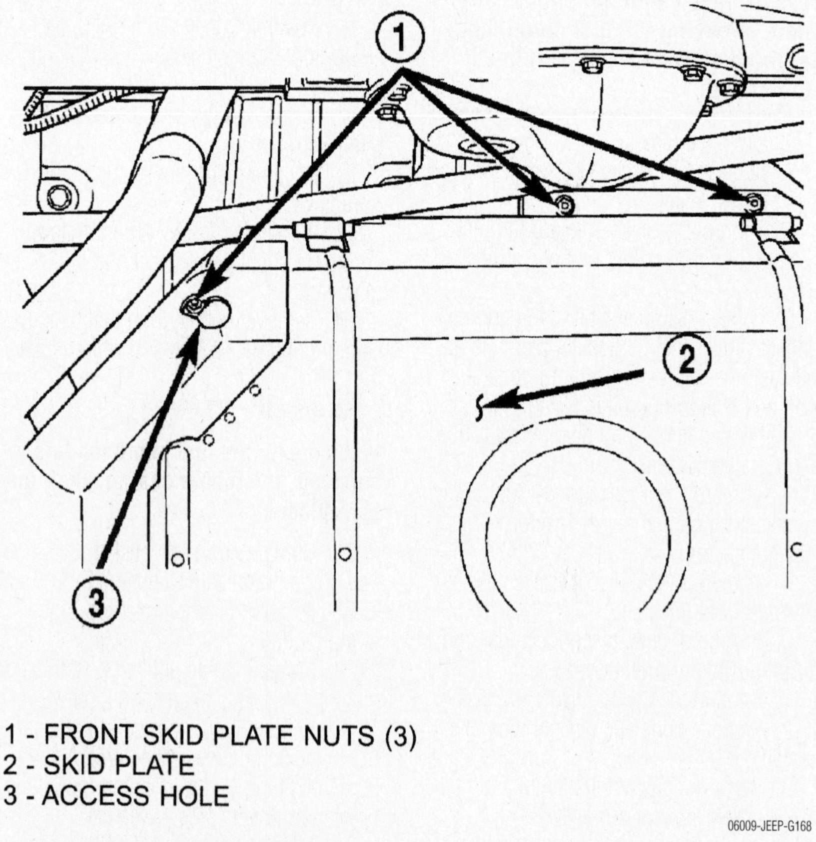

Fuel tank front mounting nuts—Wrangler

22. Remove both straps and remove tank from skid plate.

23. Note rotational position of module before attempting removal. An indexing arrow is located on top of module for this purpose.

24. Position Special Tool 9340 into notches on outside edge of lockring.

25. Install ½ inch drive breaker bar to tool 9340.

26. Rotate breaker bar counter-clockwise to remove lockring.

27. Remove lockring. The module will spring up slightly when lockring is removed.

28. Remove module from fuel tank. Be careful not to bend float arm while removing.

To install:

29. Using a new seal (gasket), position fuel pump module into opening in fuel tank.

30. Position lockring over top of fuel pump module.

31. Rotate module until embossed alignment arrow points to center alignment mark. This step must be performed to prevent float from contacting side of fuel tank. Also be sure fuel fitting on top of pump module is pointed to front of vehicle.

32. Install Special Tool 9340 to lockring.

33. Install ½ inch drive breaker into Special Tool 9340.

34. Tighten lockring (clockwise) until all seven notches have engaged.

35. Place fuel tank into skid plate. Wrap straps around tank with strap studs inserted through holes in skid plate. Tighten strap nuts to attain 30 mm (+/-2 mm) between bottom of nut to end of strap stud. Do not over tighten nuts.

36. Install two protective caps to tank strap studs.

37. Connect fuel fill hose at tank. Tighten hose clamp.

38. Raise skid plate/fuel tank assembly into position on body while carefully guiding plastic vapor lines and fill hose.

39. Install four skid plate mounting nuts at rear of tank. Tighten to 16 Nm (141 inch lbs.) torque.

40. Install three skid plate mounting nuts at front of tank. Tighten to 16 Nm (141 inch lbs.) torque.

41. Remove tank jacking device.

42. Carefully connect the vapor line.

➡Be very careful not to bend or kink the vapor lines. If lines leak, a Diagnostic Trouble Code (DTC) will be set.

43. Install EVAP canister bracket.

44. Install wheelhouse liner at right/rear wheel.

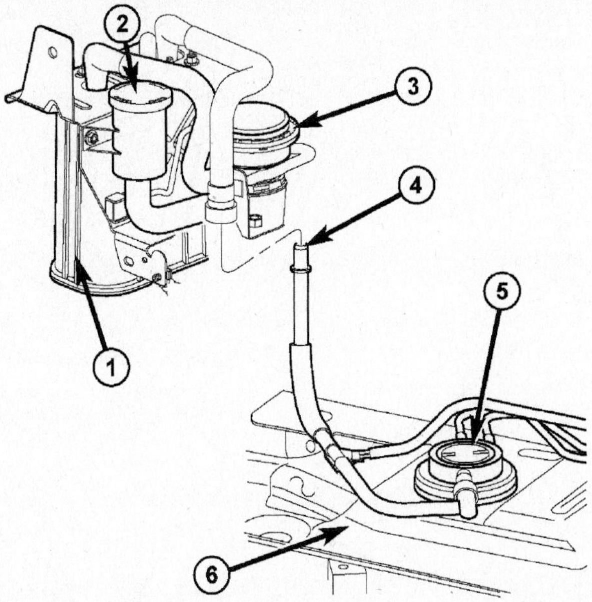

1 - EVAP CANISTER
2 - NVLD FILTER
3 - NVLD PUMP
4 - VAPOR LINES
5 - FLOW VALVE
6 - TOP OF FUEL TANK)

06009-JEEP-G169

ORVR vapor line—Wrangler

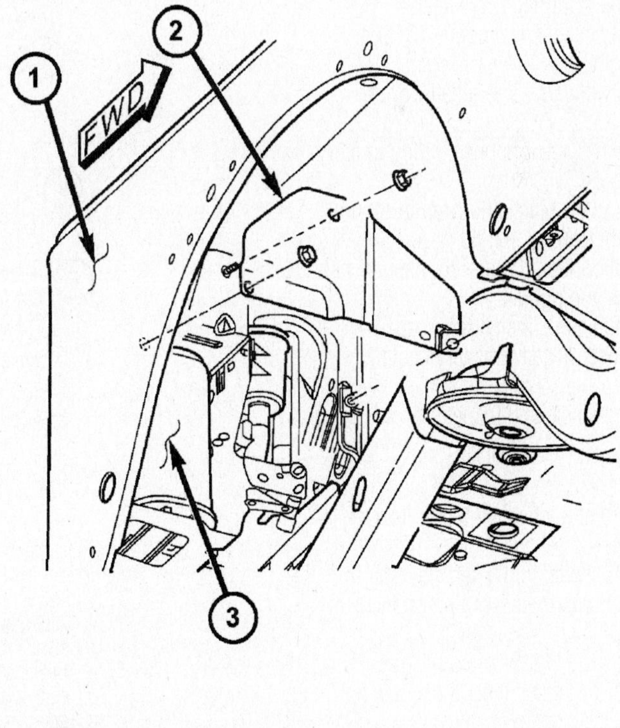

1 - RIGHT / REAR FENDER (WHEELHOUSE)
2 - BRACKET
3 - EVAP CANISTER

06009-JEEP-G170

EVAP canister/NVLD pump location—Wrangler

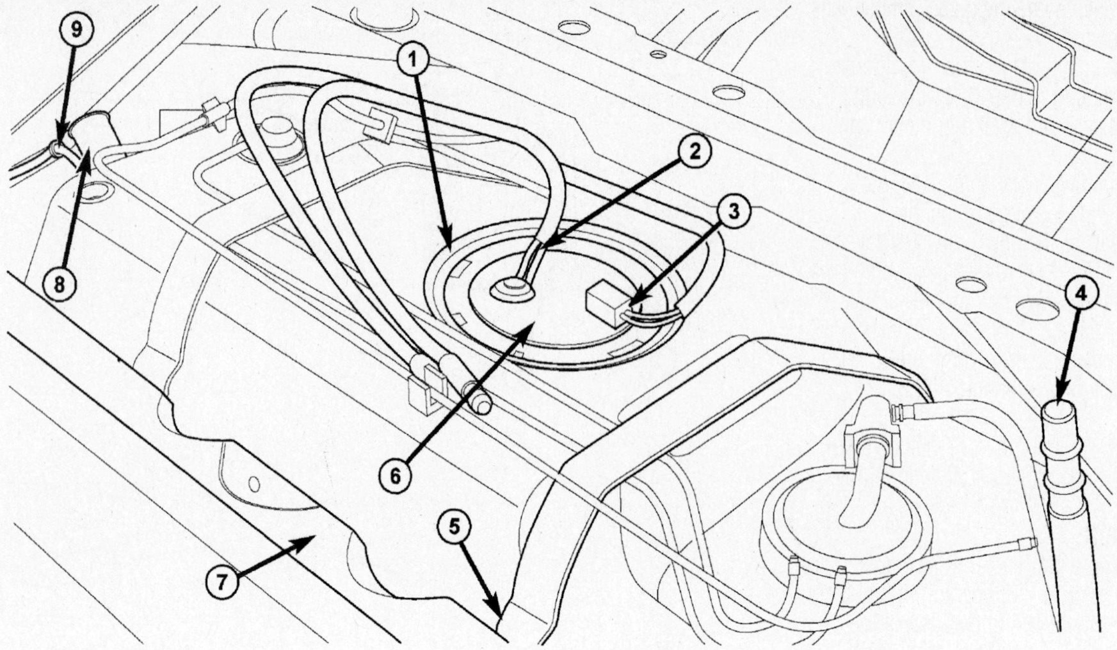

Top of fuel tank—Wrangler

45. Install right/rear tire/wheel.
46. Connect electrical connector at of fuel tank.
47. Connect vapor line at left/front of fuel tank.
48. Connect quick-connect fitting to fuel supply fitting.
49. Use a new plastic tie wrap to secure rear axle vent hose to fuel fill hose.
50. Position fuel fill bezel to body. Install eight screws and tighten.
51. Fill fuel tank. Install filler cap.
52. Connect negative battery cable to battery.
53. Start vehicle and inspect for leaks.

Liberty

The fuel filter is attached to the front of the fuel tank.

1. Before servicing the vehicle, refer to the precautions in the beginning of this section.
2. Relieve the fuel system pressure.
3. Remove or disconnect the following:
   - Negative battery cable
   - 2 rear cargo hold-down clamps by drilling the rivets
   - 4 fuel pump module access plate nuts

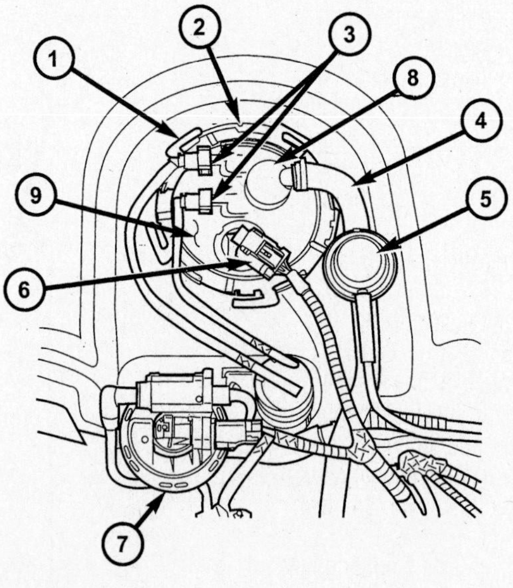

Top view of the fuel pump module—Liberty

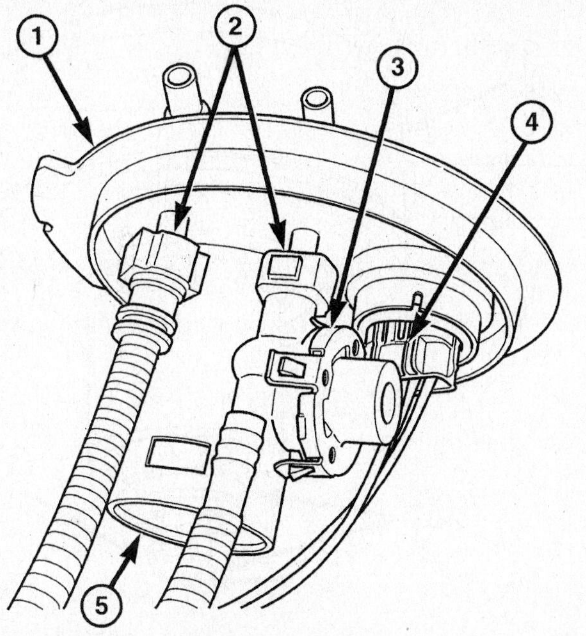

Fuel pressure regulator—Liberty

➡Once the nuts are removed, the plate must be removed by applying a heat gun to melt the adhesive. Take care to avoid bending the plate.

- Fuel lines at the module
- Electrical connector by first sliding the red tab, then pushing the gray tab
- ORVR hose
- Module lockring
- Module from the tank

⁂ WARNING

Lift the module out slowly and carefully until you can secure the rubber gasket. If you're not careful, the gasket will fall into the tank.

- Electrical connector from the upper module section
- Fule pressure regulator
- Fuel return line
- Upper module section

4. Drain the tank through the module opening.
5. Remove or disconnect the following:
   - Lower module section by pushing gently on the release tab while sliding the lock tab upward
6. Installation is the reverse of removal.

Fuel Rail

REMOVAL AND INSTALLATION

2.4L Engine

1. Before servicing the vehicle, refer to the precautions in the beginning of this section.

⁂ CAUTION

The fuel system is under constant pressure even with engine off. Before servicing fuel rail, fuel system pressure must be released.

The fuel rail can be removed without removing the intake manifold if the following procedures are followed.

2. Remove fuel tank filler tube cap.
3. Perform Fuel System Pressure Release Procedure.
4. Remove negative battery cable at battery.
5. Remove air duct at throttle body.
6. Disconnect fuel line latch clip and fuel line at fuel rail. A special tool will be necessary for fuel line disconnection.
7. Remove necessary vacuum lines at throttle body.

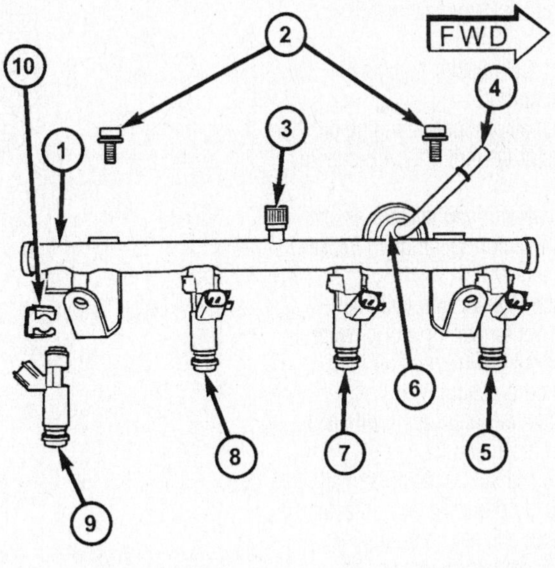

Fuel rail—2.4L engines

8. Drain engine coolant and remove thermostat and thermostat housing.

9. Remove PCV hose and valve at valve cover.

10. Remove 3 upper intake manifold mounting bolts, but only loosen 2 lower bolts about 2 turns.

11. Disconnect 2 main engine harness connectors at rear of intake manifold.

12. Disconnect 2 injection wiring harness clips at harness mounting bracket.

13. Disconnect electrical connectors at all 4 fuel injectors. Push red colored slider away from injector. While pushing slider, depress tab and remove connector from injector. The factory fuel injection wiring harness is numerically tagged (INJ 1, INJ 2, etc.) for injector position identification. If harness is not tagged, note wiring location before removal.

14. Remove 2 injection rail mounting bolts.

15. Gently rock and pull fuel rail until fuel injectors just start to clear machined holes in intake manifold.

16. Remove fuel rail (with injectors attached) from intake manifold.

17. If fuel injectors are to be removed, refer to Fuel Injector Removal/Installation.

To install:

18. If fuel injectors are to be installed, refer to Fuel Injector Removal/Installation.

19. Clean out fuel injector machined bores in intake manifold.

20. Apply a small amount of engine oil to each fuel injector O-ring. This will help in fuel rail installation.

21. Position fuel rail/fuel injector assembly to machined injector openings in intake manifold.

22. Guide each injector into cylinder head. Be careful not to tear injector O-rings.

23. Push fuel rail down until fuel injectors have bottomed on shoulders.

24. Install 2 fuel rail mounting bolts and tighten to 28 Nm (250 inch lbs.).

25. Connect electrical connectors at all fuel injectors. Push connector onto injector and then push and lock red colored slider. Verify connector is locked to injector by lightly tugging on connector.

26. Snap 2 injection wiring harness clips into brackets.

27. Connect 2 main engine harness connectors at rear of intake manifold.

28. Tighten 5 intake manifold mounting bolts.

29. Install PCV valve and hose.

30. Install thermostat and radiator hose. Fill with coolant.

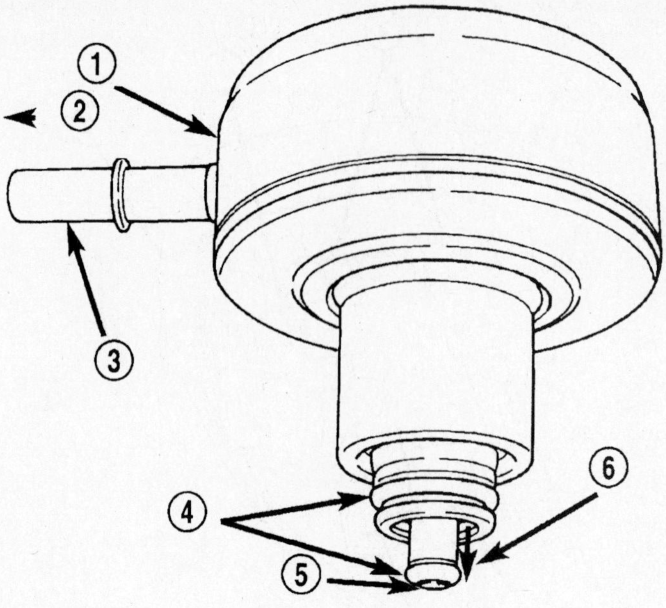

1 - FUEL FILTER/FUEL PRESSURE REGULATOR
2 - TO FUEL INJECTORS
3 - FUEL SUPPLY TUBE
4 - O-RINGS
5 - FUEL INLET FROM PUMP
6 - FUEL RETURN TO TANK

67189-JEEP-G53

Fuel pressure regulator—2.4L engines

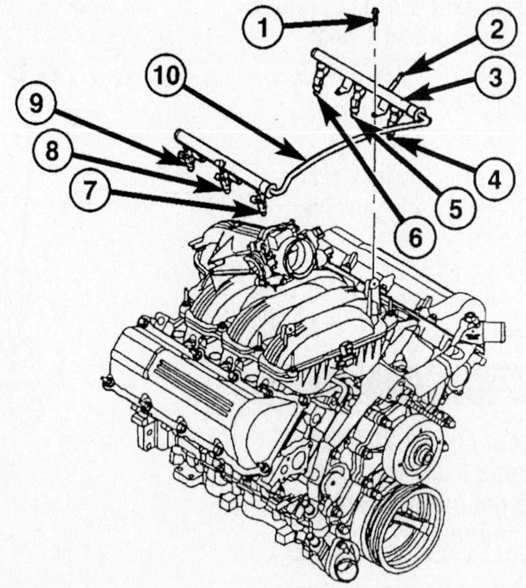

1 - MOUNTING BOLTS (4)
2 - QUICK-CONNECT FITTING
3 - FUEL RAIL
4 - INJ. #1
5 - INJ. #3
6 - INJ. #5
7 - INJ. #2
8 - INJ. #4
9 - INJ. #6
10 - CONNECTOR TUBE

67189-JEEP-G52

Fuel rail—3.7L engines

31. Connect necessary vacuum lines to throttle body.
32. Connect fuel line latch clip and fuel line to fuel rail.
33. Install air duct to throttle body.
34. Connect battery cable to battery.
35. Start engine and check for leaks.

3.7L Engine

1. Before servicing the vehicle, refer to the precautions in the beginning of this section.

CAUTION

the fuel system is under constant pressure even with engine off. Before servicing fuel rail, fuel system pressure must be released.

➡The left and right fuel rails are replaced as an assembly. Do not attempt to separate rail halves at connector tube. Due to design of tube, it does not use any clamps. Never attempt to install a clamping device of any kind to tube. When removing fuel rail assembly for any reason, be careful not to bend or kink tube.

2. Remove fuel tank filler tube cap.
3. Perform Fuel System Pressure Release Procedure.
4. Remove negative battery cable at battery.
5. Remove air duct at throttle body air box.
6. Remove air box at throttle body.
7. Disconnect fuel line latch clip and fuel line at fuel rail. A special tool will be necessary for fuel line disconnection.
8. Remove necessary vacuum lines at throttle body.
9. Disconnect electrical connectors at all 6 fuel injectors. Push red colored slider away from injector. While pushing slider, depress tab and remove connector from injector. The factory fuel injection wiring harness is numerically tagged (INJ 1, INJ 2, etc.) for injector position identification. If harness is not tagged, note wiring location before removal.
10. Disconnect electrical connectors at throttle body sensors.
11. Remove 6 ignition coils.
12. Remove 4 fuel rail mounting bolts.
13. Gently rock and pull left side of fuel rail until fuel injectors just start to clear machined holes in cylinder head. Gently rock and pull right side of rail until injectors just start to clear cylinder head holes. Repeat this procedure (left/right) until all injectors have cleared cylinder head holes.
14. Remove fuel rail (with injectors attached) from engine.
15. If fuel injectors are to be removed, refer to Fuel Injector Removal/Installation.

To install:

16. If fuel injectors are to be installed, refer to Fuel Injector Removal/Installation.
17. Clean out fuel injector machined bores in intake manifold.
18. Apply a small amount of engine oil to each fuel injector O-ring. This will help in fuel rail installation.
19. Position fuel rail/fuel injector assembly to machined injector openings in cylinder head.
20. Guide each injector into cylinder head. Be careful not to tear injector O-rings.
21. Push right side of fuel rail down until fuel injectors have bottomed on cylinder h ad shoulder. Push left fuel rail down until injectors have bottomed on cylinder head shoulder.
22. Install 4 fuel rail mounting bolts and tighten to 11 Nm (100 inch lbs.)
23. Install 6 ignition coils.
24. Connect electrical connectors to throttle body.
25. Connect electrical connectors at all fuel injectors. Push connector onto injector and then push and lock red colored slider.

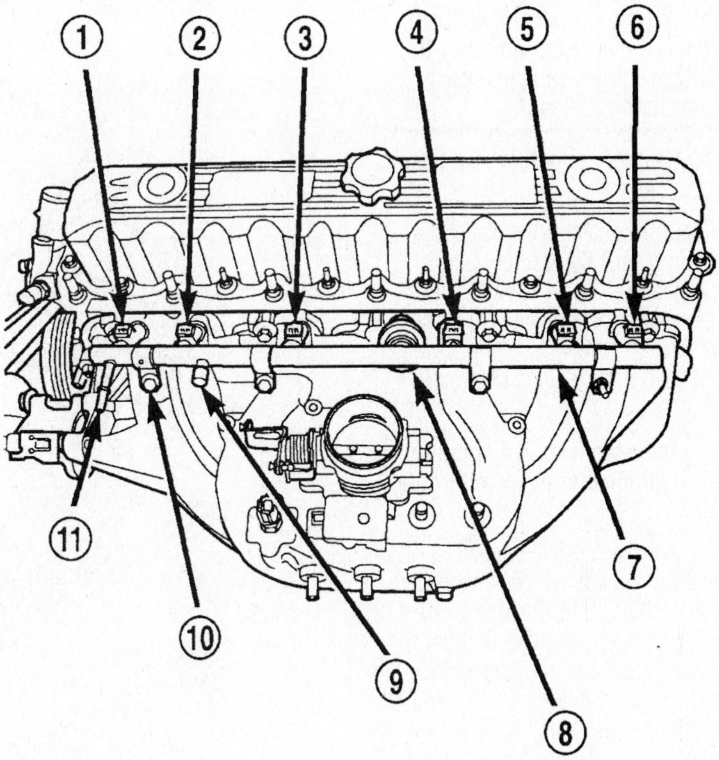

1 - INJ. #1
2 - INJ. #2
3 - INJ. #3
4 - INJ. #4
5 - INJ. #5
6 - INJ. #6
7 - FUEL INJECTOR RAIL
8 - FUEL DAMPER
9 - PRESSURE TEST PORT CAP
10 - MOUNTING BOLTS (4)
11 - QUICK-CONNECT FITTING

67189-JEEP-G54

Fuel rail—4.0L engines

Verify connector is locked to injector by lightly tugging on connector.

26. Connect necessary vacuum lines to throttle body.
27. Connect fuel line latch clip and fuel line to fuel rail.
28. Install air box to throttle body.
29. Install air duct to air box.
30. Connect battery cable to battery.
31. Start engine and check for leaks.

2.5L and 4.0L Engines

1. Before servicing the vehicle, refer to the precautions in the beginning of this section.

**** CAUTION**

The fuel system is under constant fuel pressure even with engine off. This pressure must be released before servicing fuel rail.

➡The fuel damper is not serviced separately

2. Remove fuel tank filler tube cap.
3. Perform Fuel System Pressure Release Procedure.
4. Disconnect negative battery cable from battery.
5. Remove air tube at top of throttle body.

➡Some engine/vehicles may require removal of air cleaner ducts at throttle body.

6. Disconnect electrical connectors at all 6 fuel injectors. Push red colored slider away from injector. While pushing slider, depress tab and remove connector from injector. The factory fuel injection wiring harness is numerically tagged (INJ 1, INJ 2, etc.) for injector position identification. If harness is not tagged, note wiring location before removal.
7. Disconnect fuel supply line latch clip and fuel line at fuel rail.
8. Disconnect throttle cable at throttle body.
9. Disconnect speed control cable at throttle body (if equipped).
10. Disconnect automatic transmission cable at throttle body (if equipped).
11. Remove cable routing bracket at intake manifold.
12. If equipped, remove wiring harnesses at injection rail studs by removing nuts.
13. Clean dirt/debris from each fuel injector at intake manifold.
14. Remove fuel rail mounting nuts/bolts.
15. Remove fuel rail by gently rocking until all the fuel injectors are out of intake manifold.

To install:

16. Clean each injector bore at intake manifold.
17. Apply a small amount of clean engine oil to each injector O-ring. This will aid in installation.
18. Position tips of all fuel injectors into the corresponding injector bore in intake manifold. Seat injectors into manifold.
19. Install and tighten fuel rail mounting bolts to 11 +/- 3 Nm (100 +/- 25 inch lbs.) torque.
20. If equipped, connect wiring harnesses to injection rail studs.
21. Connect electrical connectors at all fuel injectors. Push connector onto injector and then push and lock red colored slider. Verify connector is locked to injector by lightly tugging on connector.
22. Connect fuel line and fuel line latch clip to fuel rail.

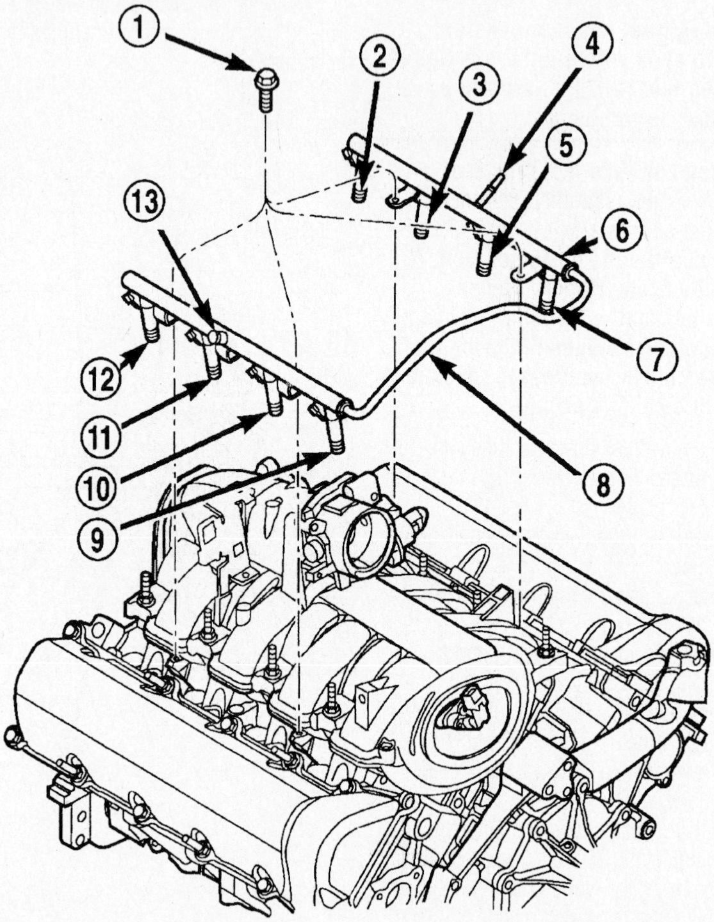

1 - MOUNTING BOLTS (4)
2 - INJ.#7
3 - INJ.#5
4 - QUICK-CONNECT FITTING
5 - INJ.#3
6 - FUEL INJECTOR RAIL
7 - INJ.#1
8 - CONNECTOR TUBE
9 - INJ.#2
10 - INJ.#4
11 - INJ.#6
12 - INJ.#8
13 - PRESSURE TEST PORT CAP

67189-JEEP-G56

Fuel rail—4.7L engines

23. Install protective cap to pressure test port fitting (if equipped).
24. Install cable routing bracket to intake manifold.
25. Connect throttle cable at throttle body.
26. Connect speed control cable at throttle body (if equipped).
27. Connect automatic transmission cable at throttle body (if equipped).
28. Install air tube (or duct) at top of throttle body.
29. Install fuel tank cap.
30. Connect negative battery cable to battery.
31. Start engine and check for fuel leaks.

4.7L Engine

1. Before servicing the vehicle, refer to the precautions in the beginning of this section.
2. Remove fuel tank filler tube cap.
3. Perform Fuel System Pressure Release Procedure.
4. Remove negative battery cable at battery.
5. Remove air duct at throttle body air box.
6. Remove air box at throttle body.
7. Remove wiring at rear of generator.
8. Disconnect fuel line latch clip and fuel line at fuel rail. A special tool will be necessary for fuel line disconnection.
9. Remove vacuum lines at throttle body.
10. Disconnect electrical connectors at all 8 fuel injectors. Push red colored slider away from injector. While pushing slider, depress tab and remove connector from injector. The factory fuel injection wiring harness is numerically tagged (INJ 1, INJ 2, etc.) for injector position identification. If harness is not tagged, note wiring location before removal.
11. Disconnect electrical connectors at throttle body.
12. Disconnect electrical connectors at MAP and IAT sensors.
13. Remove first three ignition coils on each bank (cylinders #1, 3, 5, 2, 4 and 6).
14. Remove 4 fuel rail mounting bolts.
15. Gently rock and pull left side of fuel rail until fuel injectors just start to clear machined holes in cylinder head. Gently rock and pull right side of rail until injectors just start to clear cylinder head holes. Repeat this procedure (left/right) until all injectors have cleared cylinder head holes.
16. Remove fuel rail (with injectors attached) from engine.
17. If fuel injectors are to be removed, refer to Fuel Injector Removal/Installation.

To install:

18. If fuel injectors are to be installed, refer to Fuel Injector Removal/Installation.
19. Apply a small amount of engine oil to each fuel injector o-ring. This will help in fuel rail installation.
20. Position fuel rail/fuel injector assembly to machined injector openings in cylinder head.
21. Guide each injector into cylinder head. Be careful not to tear injector o-rings.
22. Push right side of fuel rail down until fuel injectors have bottomed on cylinder head shoulder. Push left fuel rail down until injectors have bottomed on cylinder head shoulder.
23. Install 4 fuel rail mounting bolts and tighten to 27 Nm (20 ft. lbs.).
24. Install ignition coils.
25. Connect electrical connectors to throttle body.
26. Connect electrical connectors to MAP and IAT sensors.
27. Connect electrical connectors at all fuel injectors. Push connector onto injector and then push and lock red colored slider. Verify connector is locked to injector by lightly tugging on connector.
28. Connect vacuum lines to throttle body.
29. Connect fuel line latch clip and fuel line to fuel rail.
30. Connect wiring to rear of generator.
31. Install air box to throttle body.
32. Install air duct to air box.
33. Connect battery cable to battery.
34. Start engine and check for leaks.

5.7L Engine

1. Before servicing the vehicle, refer to the precautions in the beginning of this section.

⁂ CAUTION

the fuel system is under constant pressure even with engine off. Before servicing fuel rail, fuel system pressure must be released.

⁂ WARNING

The left and right fuel rails are replaced as an assembly. Do not attempt to separate rail halves at connector tube. Due to design of tube, it does not use any clamps. Never attempt to install a clamping device of any kind to tube. When removing fuel rail assembly for any reason, be careful not to bend or kink tube.

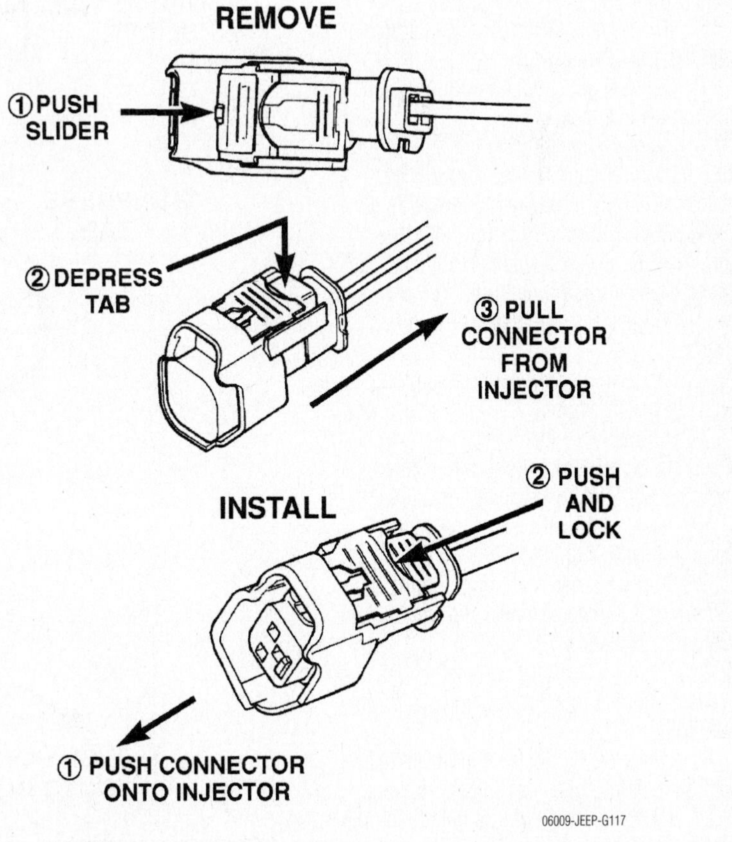

Quick-disconnect fittings

2. Remove fuel tank filler tube cap.
3. Perform fuel system pressure release procedure.
4. Remove negative battery cable at battery.
5. Remove flex tube (air cleaner housing to engine).
6. Remove air resonator box at throttle body.
7. Disconnect all spark plug cables from all spark plugs and ignition coils. Do not remove cables from cable routing tray. Note original cable positions while removing.
8. Remove spark plug cable tray from engine by releasing 4 retaining clips. Remove tray and cables from engine as an assembly.
9. Disconnect electrical connectors at all 8 ignition coils. Refer to Ignition Coil Removal/Installation.
10. Disconnect fuel line latch clip and fuel line at fuel rail. A special tool will be necessary for fuel line disconnection.
11. Disconnect electrical connectors at all 8 fuel injectors. Push red colored slider away from injector. While pushing slider, depress tab and remove connector from injector. The factory fuel injection wiring harness is numerically tagged (INJ 1, INJ 2, etc.) for injector position identification. If harness is not tagged, note wiring location before removal.
12. Disconnect electrical connectors at all throttle body sensors.
13. Remove four fuel rail mounting bolts and hold-down clamps.
14. Gently rock and pull left side of fuel rail until fuel injectors just start to clear machined holes in intake manifold. Gently rock and pull right side of rail until injectors just start to clear intake manifold head holes. Repeat this procedure (left/right) until all injectors have cleared machined holes.
15. Remove fuel rail (with injectors attached) from engine.

To install:

16. If fuel injectors are to be installed, refer to Fuel Injector Removal/Installation.
17. Clean out fuel injector machined bores in intake manifold.
18. Apply a small amount of engine oil to each fuel injector o-ring. This will help in fuel rail installation.
19. Position fuel rail/fuel injector assembly to machined injector openings in intake manifold.
20. Guide each injector into intake manifold. Be careful not to tear injector o-rings.
21. Push right side of fuel rail down until fuel injectors have bottomed on shoulders. Push left fuel rail down until injectors have bottomed on shoulders.
22. Install 4 fuel rail hold-down clamps and 4 mounting bolts. Torque to 100 inch lbs. (11 Nm).
23. Position spark plug cable tray and cable assembly to intake manifold. Snap 4 cable tray retaining clips into intake manifold.
24. Install all cables to spark plugs and ignition coils.
25. Connect electrical connector to throttle body.
26. Install electrical connectors to all 8 ignition coils. Refer to Ignition Coil Removal/Installation.
27. Connect electrical connector to throttle body.
28. Connect electrical connectors at all fuel injectors. Push connector onto injector and then push and lock red colored slider. Verify connector is locked to injector by lightly tugging on connector.
29. Connect fuel line latch clip and fuel line to fuel rail.
30. Install air resonator to throttle body (2 bolts).
31. Install flexible air duct to air box.
32. Connect battery cable to battery.
33. Start engine and check for leaks.

Fuel Injector

REMOVAL & INSTALLATION

1. Before servicing the vehicle, refer to the precautions in the beginning of this section.
2. Relieve fuel system pressure.
3. Remove or disconnect the following:
   - Negative battery cable
   - Fuel rail
   - Fuel injector retaining clips
   - Fuel injectors

To install:

4. Install or connect the following:
   - Fuel injectors, using new O-rings coated with a small amount of engine oil
   - Fuel injector clips to the fuel rail
   - Fuel rail
   - Negative battery cable
5. Start the vehicle and check for leaks, repair if necessary.

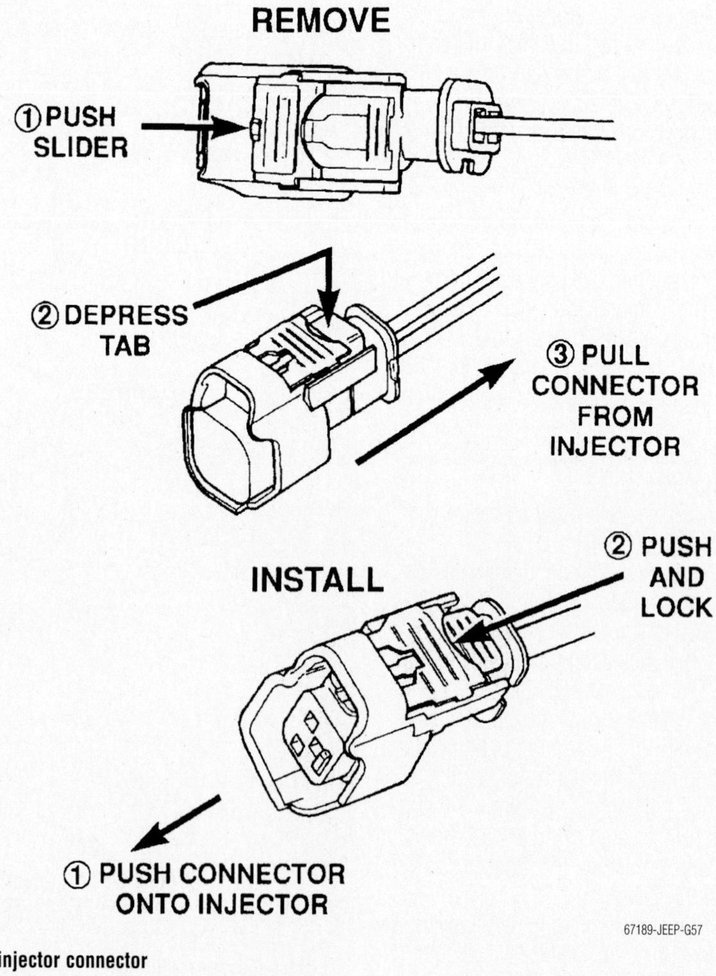

Fuel injector connector

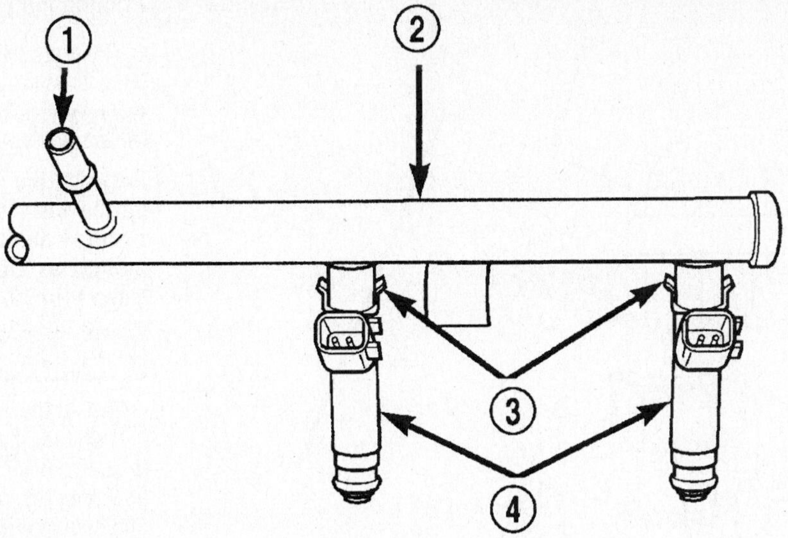

Typical fuel injector mounting—4.7L engine shown

DIESEL FUEL SYSTEM

Fuel System Pressure

CAUTION

High-pressure fuel lines deliver fuel under extreme pressure from the Injection pump to the injectors. This may be as high as 1600bar (23,200psi). Use extreme caution when inspecting for high pressure fuel leaks. Fuel under this amount of pressure can penetrate skin causing personal injury or death. Inspect high-pressure fuel leaks with a sheet of cardboard. Wear safety goggles and adequate protective clothing when servicing fuel system.

PRECAUTIONS

High Pressure Fuel Leaks

High-pressure fuel leaks can cause starting problems and poor engine performance. Carefully place a piece of cardboard over the high pressure fuel lines or suspected area. Move your body and hands away from the area. Start the engine and run till warm. TURN THE ENGINE OFF. Inspect the piece of card board for marks. If a high-pressure line connection is leaking, replace damaged, restricted or leaking high-pressure fuel lines with the correct replacement line.

CAUTION

The high-pressure fuel lines cannot contact each other or other components. Do not attempt to weld high-pressure fuel lines or to repair lines that are damaged. Only use the recommended lines when replacement of high-pressure fuel line is necessary.

Air in the Fuel System

Air will enter the fuel system whenever fuel supply lines, separator filters, injection pump, high-pressure lines or injectors are removed or disconnected. Air trapped in the fuel system can result in hard starting, a rough running engine, engine misfire, low power, excessive smoke and fuel knock. Inspect the fuel system from the fuel tank to the injectors for loose connections. Leaking fuel is an indicator of loose connections or defective seals. Air can also enter the fuel system between the fuel tank and the fuel filter. Inspect the fuel tank and fuel lines for damage that might allow air into the system. Trapped air or leaking lines may also be identified by placing a clear piece of fuel line between the fuel supply hose to the filter and the fuel supply hose to the high pressure pump. If a steady stream of large air bubbles are visible in the clear hose while the engine is cranking or running the air intrusion is occurring before the high pressure pump. With the scan tool connected to the vehicle, select Engine and then select Sensor Display. Page down to view Fuel Pressure Set Point and Actual Fuel Pressure. Start the engine and observe the Fuel Pressure Set Point and the Actual Fuel Pressure. If the Actual Fuel Pressure Oscillates above and below the Fuel Pressure Set Point in a regular cycle, inspect the fuel system for air intrusion. If the Actual Fuel Pressure gradually drops below the Fuel Pressure Set Point then spikes well above the Fuel Pressure Set Point, replace the fuel rail pressure solenoid.

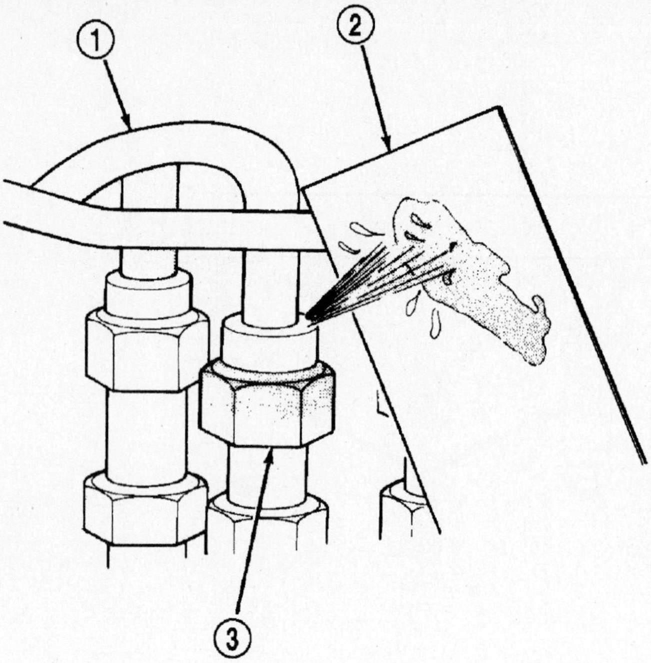

1 - HIGH-PRESSURE LINE
2 - CARDBOARD
3 - FITTING

06009-JEEP-G48

Pressure leak testing with cardboard

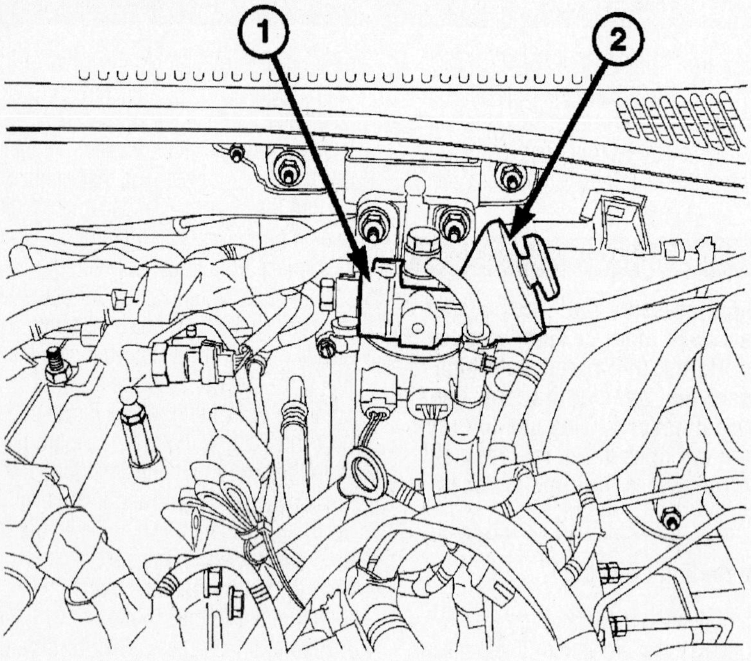

1 - FUEL FILTER AND HOUSING ASSEMBLY
2 - PRIMER BUTTON

06009-JEEP-G49

Fuel system primer

Priming the Fuel System

1. Before servicing the vehicle, refer to the precautions in the beginning of this section.

**** WARNING**

Cranking the engine for an extended period with out a fuel supply may result in damage to the high pressure fuel pump. DO NOT force the plunger when priming the fuel system. Damage to the plunger or fuel filter/water separate will result.

➡Should the vehicle run out of fuel, the fuel system must be re-primed in order to start. Factors such as temperature, how long the vehicle has been sitting, and engine cranking speed will influence how quickly the engine starts. If the water in fuel light is illuminated in the instrument panel, refer to the water in fuel sensor for draining procedure.

The fuel system must be primed if the fuel system has been serviced. This is done using the fuel primer button located at the top of the fuel filter/water separator.

2. With service completed on the fuel system.
3. Depress the fuel primer 20 consecutive times then open the bleeder screw on top of the housing to dispel trapped air.
4. Close the bleeder screw and continue the step above until the primer button becomes slightly harder to depress.
5. Turn the ignition to START and crank the engine a maximum of ten seconds.

➡If the engine does not start with in ten seconds, repeat the priming procedure. The engine will typically start within ten seconds; the engine may idle, idle rough, or stall, purging any trapped air from the lines and filter.

6. Tighten bleeder screw to 11 Nm (96 inch lbs.).

Fuel Filter/Water Separator

DRAINING WATER

1. Before servicing the vehicle, refer to the precautions in the beginning of this section.

**** WARNING**

Store fuel in approved and properly marked containers. Wear safety goggles and adequate protective clothing when servicing fuel system.

2. Disconnect the negative battery cable.
3. Disconnect the water in fuel (WIF) wiring harness connector located under the fuel filter.
4. Connect a drain hose to the WIF sensor.
5. Place the other end of the drain hose in a approved and properly marked container.
6. Open the bleed screw on top of the fuel filter housing.
7. Loosen the WIF sensor on the bottom of the fuel filter to begin draining.
8. Allow the filter to drain into the container until fuel is visible.
9. Tighten the WIF sensor, remove the drain hose and clean any spillage.
10. Tighten the bleeder screw on top of the fuel filter housing.
11. Connect the WIF wiring harness connector.
12. Connect the negative battery cable.

REMOVAL & INSTALLATION

Fuel Filter/Water Separator Assembly Removal

1. Before servicing the vehicle, refer to the precautions in the beginning of this section.

⁂ WARNING

Store fuel in approved and properly marked containers. Wear safety goggles and adequate protective clothing when servicing fuel system.

2. Disconnect negative battery cable.
3. Drain fuel filter/water separator assembly.

➡DO NOT loosen the banjo bolt fittings. Loosen the fuel hose clamps and remove hoses from the fuel lines.

4. Remove two fuel lines from the water separator. Loosen retaining clamps and slide the hose from barbed fittings.
5. Disconnect the fuel heater and temperature sensor harness connectors.
6. Remove the fuel filter/water separator assembly from the engine cowl.

Fuel Filter Removal

1. Before servicing the vehicle, refer to the precautions in the beginning of this section.

➡Capture all fuel in approved and appropriately marked containers. Wear safety goggles and adequate protective clothing when servicing the fuel system.

2. Disconnect the negative battery cable.
3. Disconnect the water in fuel (WIF) sensor wiring harness connector.
4. Drain the fuel filter/water separator.
5. Unscrew the filter assembly from the head assembly by rotating the housing counterclockwise.
6. Separate the WIF sensor and seal from the housing by rotating counterclockwise.

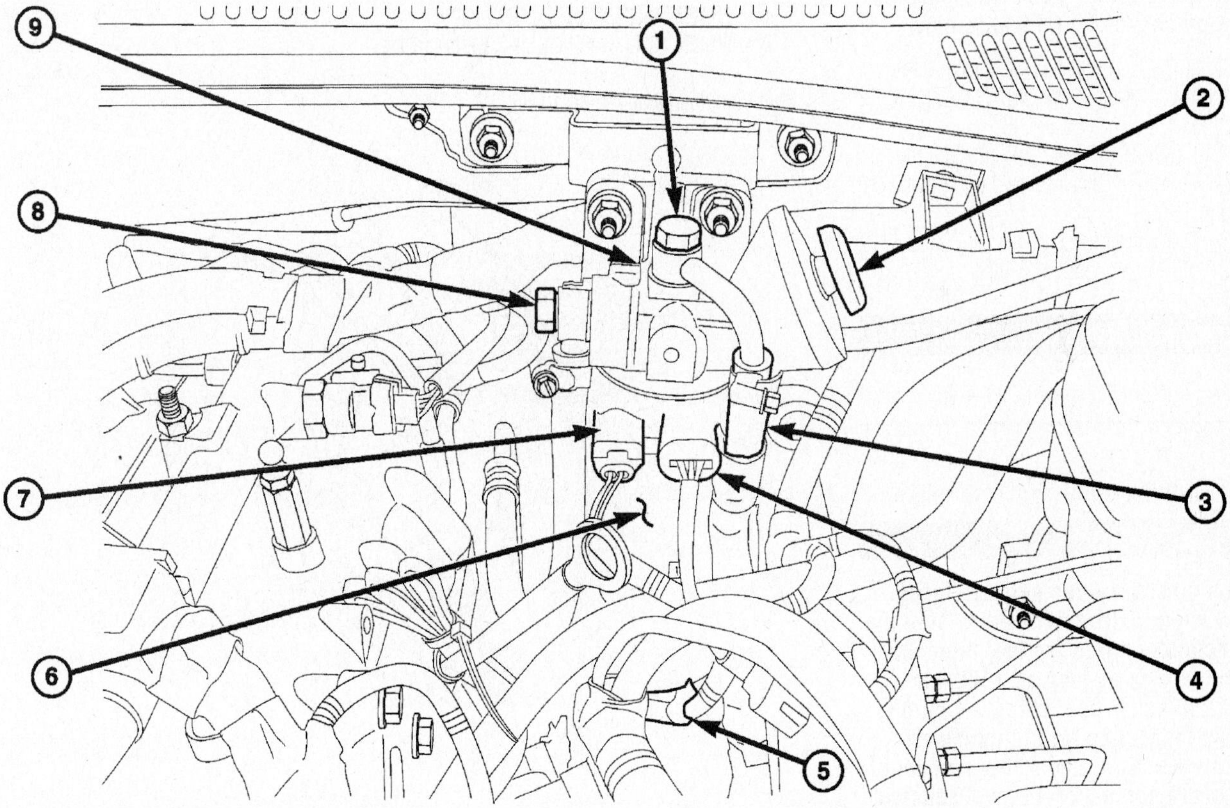

1 - FUEL FILTER ASSEMBLY
2 - PRIMER BUTTON
3 - FUEL SUPPLY TO HIGH PRESSURE PUMP
4 - FUEL HEATER
5 - WIF SENSOR
6 - FUEL FILTER
7 - FUEL TEMPERATURE SENSOR
8 - FUEL SUPPLY FROM TANK
9 - AIR BLEED SCREW

06009-JEEP-G50

Fuel filter/water separator

Fuel Filter Installation

1. Before servicing the vehicle, refer to the precautions in the beginning of this section.
2. Lubricate the fuel filter seal with clean diesel fuel.
3. Install the water in fuel (WIF) sensor hand tight.
4. Screw filter assembly onto the fuel filter/water separator head. Tighten filter to 18 Nm (118 in. lbs.).
5. Connect the WIF wiring harness connector.
6. Prime the fuel system.
7. Start the engine and allow to warm.
8. Turn engine off and inspect for leaks.

Fuel Filter/Water Separator Assembly Installation

1. Before servicing the vehicle, refer to the precautions in the beginning of this section.
2. Carefully position fuel water separator over the mounting studs, install fasteners and tighten to 24.5 Nm (18 ft. lbs.).

➡Make sure the fuel supply hose to the high pressure pump is mounted to the left side of the filter assembly.

3. Install the fuel lines to the housing assembly. Tighten hose clamps to 7 Nm (62 inch lbs.).
4. Connect the water in fuel, fuel temperature sensor and the fuel heater electrical connectors.
5. Prime fuel system.
6. Connect negative battery cable.

Injection Pump

REMOVAL & INSTALLATION

1. Before servicing the vehicle, refer to the precautions in the beginning of this section.

**** CAUTION**

High pressure lines deliver diesel fuel under extreme pressure from the injection pump to the fuel injectors. This may be as high as 1600 bar (23,200 psi.). Use extreme caution when inspecting for high pressure fuel leaks. Fuel under this amount of pressure can penetrate skin causing personal injury or death. Inspect for high pressure fuel leaks with a sheet of cardboard. Wear safety goggles and adequate protective clothing when servicing fuel system.

2. Disconnect negative battery cable.
3. Remove engine cover and bracket.

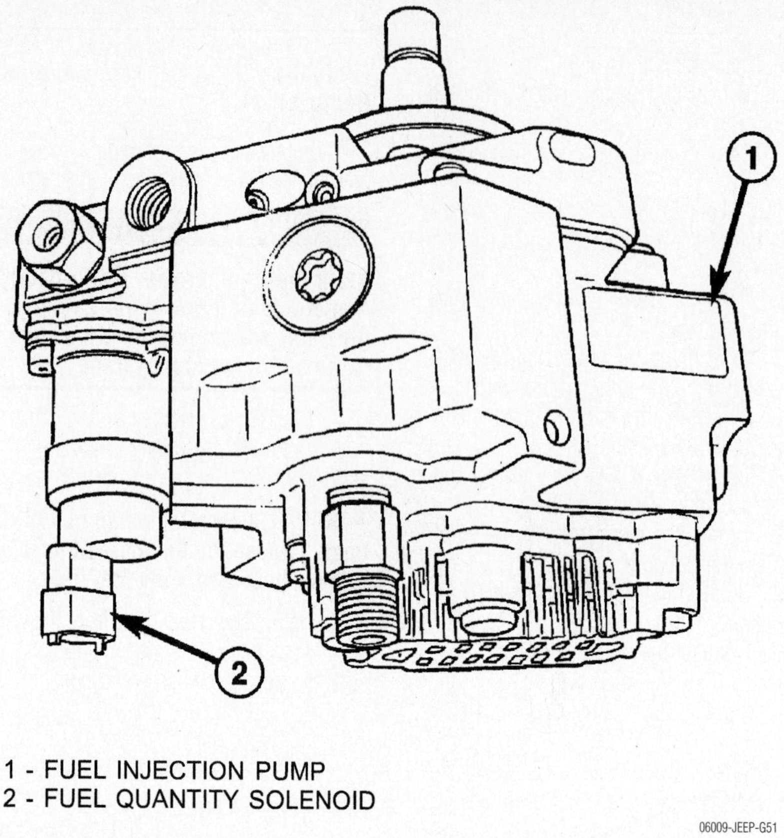

Fuel injection pump—2.8L diesel

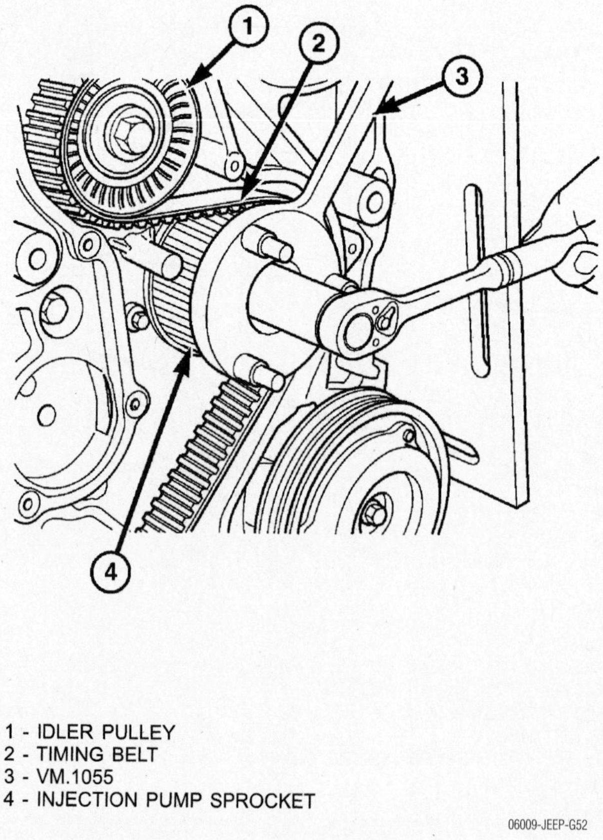

Injection pump sprocket retaining nut removal—2.8L diesel

4. Evacuate A/C system.
5. Remove accessory drive belt.
6. Remove cooling fan and fan shroud.
7. Remove charge air cooler outlet hose to intake manifold.
8. Remove fan support assembly.
9. Remove timing belt outer cover.
10. Using special tool VM.1055, remove high pressure injection pump sprocket retaining nut.

➡The use of special tool VM.1067 will allow you to remove the high pressure injection pump without removing the timing belt from the engine. This will allow you to remove and install the high pressure injection pump without altering injection pump timing.

11. Install feet from VM.1067 in injection pump sprocket as shown.
12. Install inner flange of special tool VM.1067 on injection pump sprocket as shown. Secure flange to feet in injection pump sprocket with Allen bolts supplied with tool.
13. Screw the high pressure injection pump sprocket holding plate assembly into flange of VM.1067. Using left hand threaded bolt supplied, secure holding plate assembly to timing belt inner cover.
14. Disconnect A/C lines at compressor.
15. Remove charge air inlet hose.
16. Remove the EGR airflow control valve from the intake manifold.
17. Remove high pressure injection pump to fuel rail high pressure line.
18. Disconnect high pressure injection pump quantity control valve electrical connector.
19. Disconnect fuel supply and return lines at high pressure injection pump.
20. Remove alternator to intake manifold bracket.

WARNING

Care must be taken not to bend the brake vacuum tube when removing high pressure pump.

21. Remove high pressure injection pump retaining nuts and remove pump.

To install:

22. Loosen bolt in center of injection pump holding plate and slide high pressure injection pump through the accessory bracket into the injection pump sprocket.
23. Install high pressure injection pump retaining nuts. Torque nuts to 27 Nm (20 ft. lbs.).
24. Unscrew injection pump holding plate (part of VM.1067) from inner timing belt cover and remove.

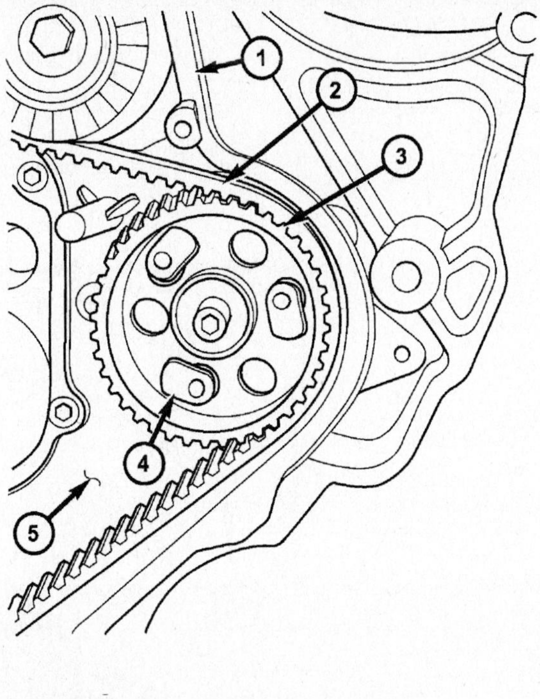

Special tool feet installation—2.8L diesel

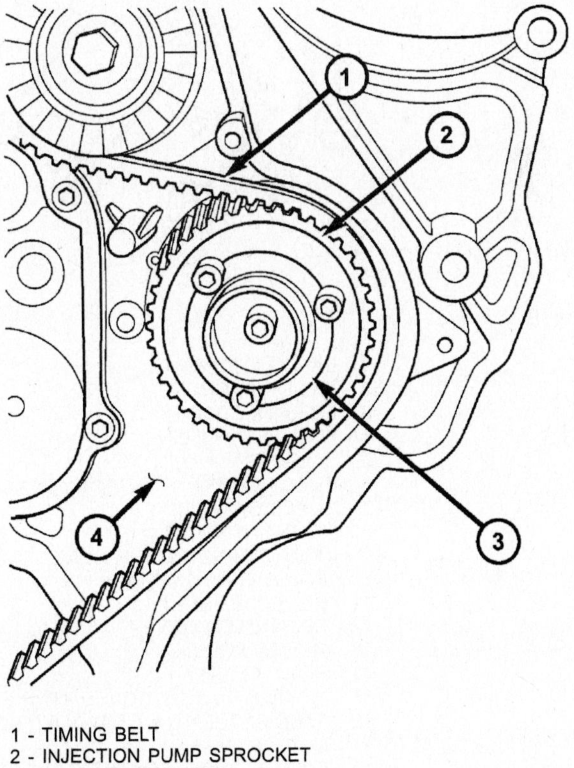

Special tool installation—2.8L diesel

25. Install high pressure injection pump sprocket retaining nut to hold sprocket in place.

26. Remove flange and feet (both part of VM.1067) from high pressure injection pump sprocket.

27. Using special tool VM.1055, torque high pressure injection pump sprocket retaining nut to 88 Nm (65 ft. lbs.).

28. Connect fuel quantity control valve electrical connector.

29. Connect fuel supply and return lines at high pressure injection pump.

30. Install the EGR airflow control valve.

31. Install outer timing belt cover.

32. Install fan support assembly.

33. Install accessory drive belt.

34. Install cooling fan and shroud assembly.

35. Install charge air cooler outlet hose.

36. Install engine cover and bracket.

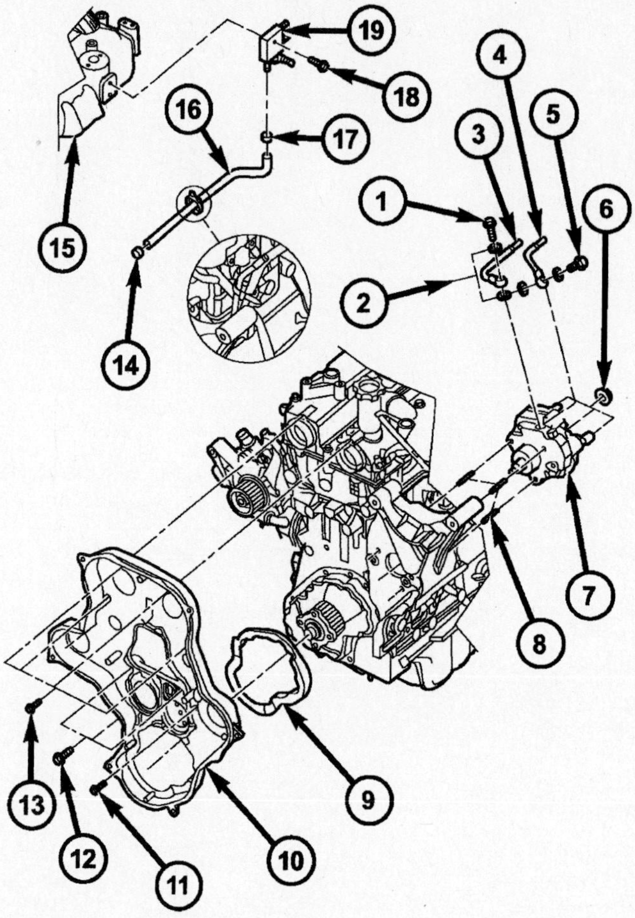

1 - BANJO FITTING
2 - BRASS WASHERS
3 - FUEL INLET LINE
4 - FUEL OUTLET LINE
5 - BANJO FITTING
6 - INJECTION PUMP RETAINING NUT
7 - INJECTION PUMP
8 - MOUNTING STUDS
9 - INNER TIMING COVER SEAL
10 - INNER TIMING COVER
11 - RETAINING BOLT
12 - RETAINING BOLT
13 - RETAINING BOLT
14 - HOSE CLAMP
15 - CYLINDER HEAD COVER/INTAKE MANIFOLD
16 - FUEL RETURN LINE TO INJECTION PUMP
17 - HOSE CLAMP
18 - FUEL RETURN JUNCTION BLOCK RETAINING BOLT
19 - FUEL RETURN JUNCTION BLOCK

06009-JEEP-G55

Injection pump removal—2.8L diesel

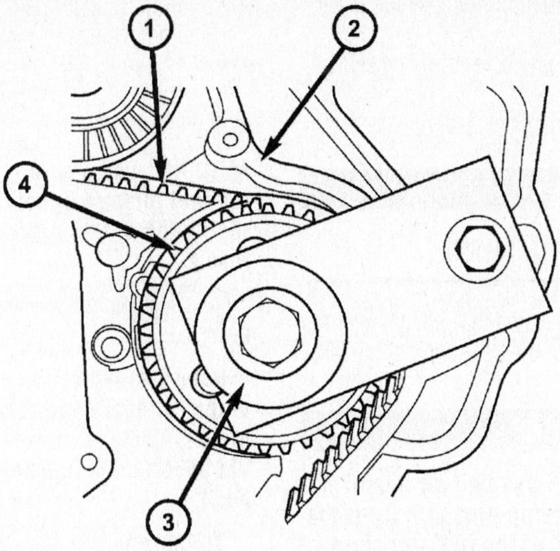

1 - TIMING BELT
2 - INNER TIMING BELT COVER
3 - INJECTION PUMP SPROCKET HOLDING PLATE ASSEMBLY PART OF VM.1067
4 - INJECTION PUMP SPROCKET

06009-JEEP-G56

Injection pump gear removal—2.8L diesel

37. Connect negative battery cable.
38. Evacuate and recharge A/C system.

Fuel Rail

REMOVAL & INSTALLATION

1. Before servicing the vehicle, refer to the precautions in the beginning of this section.

CAUTION

High pressure lines deliver diesel fuel under extreme pressure from the injection pump to the fuel injectors. This may be as high as 1600 bar (23,200 psi.). Use extreme caution when inspecting for high pressure fuel leaks. Fuel under this amount of pressure can penetrate skin causing personal injury or death. Inspect for high pressure fuel leaks with a sheet of cardboard. Wear safety goggles and adequate protective clothing when servicing fuel system.

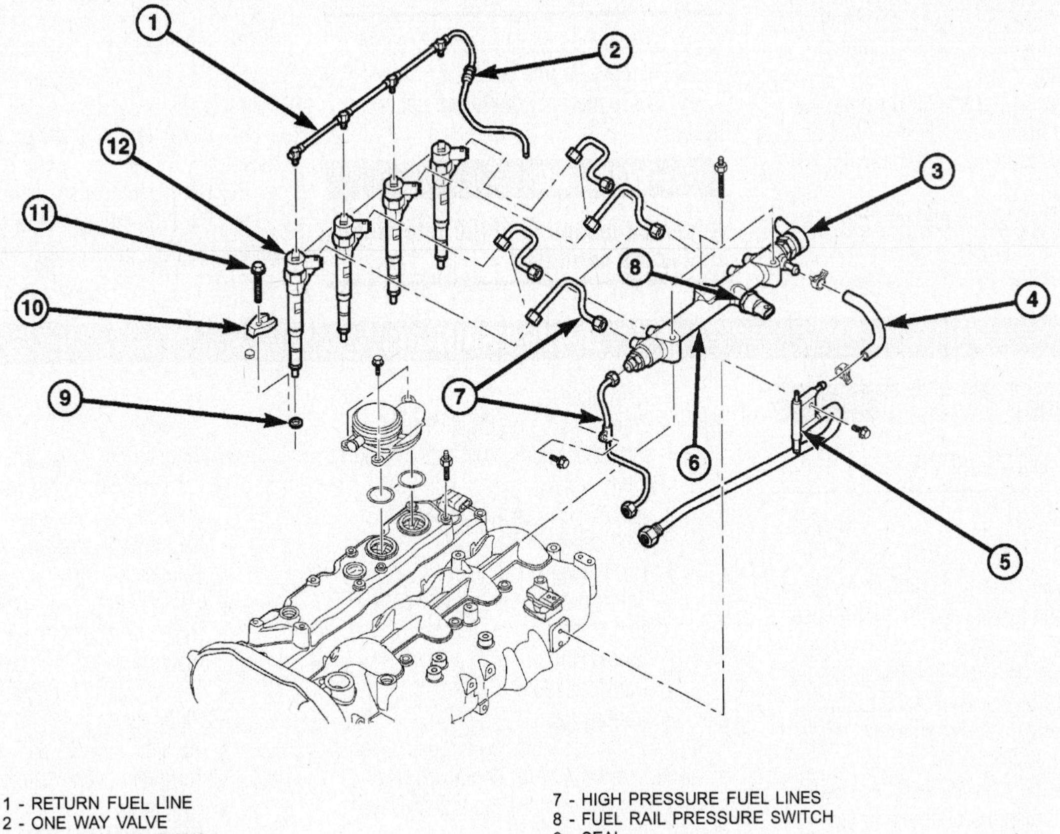

1 - RETURN FUEL LINE
2 - ONE WAY VALVE
3 - FUEL RAIL SOLENOID
4 - RETURN FUEL HOSE
5 - RETURN FUEL JUNCTION
6 - FUEL RAIL
7 - HIGH PRESSURE FUEL LINES
8 - FUEL RAIL PRESSURE SWITCH
9 - SEAL
10 - INJECTOR HOLD DOWN
11 - BOLT
12 - FUEL INJECTOR

06009-JEEP-G57

Diesel fuel system components

2. Disconnect negative battery cable.
3. Remove engine cover and bracket assembly.
4. Disconnect fuel pressure sensor electrical connector.

➡If fuel rail is being replaced it is necessary to replace the fuel rail solenoid.

5. Disconnect the fuel rail solenoid electrical connector.
6. Disconnect fuel rail return line at fuel rail.
7. Disconnect fuel high pressure line from injection pump to fuel rail at fuel rail.
8. Disconnect fuel high pressure line from fuel rail to fuel injector at fuel rail.
9. Remove fuel rail retaining bolts and remove rail from cylinder head cover/intake manifold

To install:

➡The fuel rail solenoid must be replaced upon removal. When replacing fuel lines or solenoid it is necessary to counterhold and correctly torque the fitting. Refer to the proper procedure.

10. Replace the fuel rail solenoid.
11. Install the fuel pressure sensor.
12. Install fuel rail on cylinder head cover/intake manifold. Torque retaining bolts to 27 Nm.
13. Connect injector high pressure fuel lines at fuel rail.
14. Connect fuel rail high pressure fuel line at fuel rail.
15. Connect fuel rail fuel return line at fuel rail.
16. Connect fuel pressure sensor electrical connector.
17. Install engine cover and bracket assembly.
18. Connect negative battery cable.

Fuel Injectors

REMOVAL & INSTALLATION

1. Before servicing the vehicle, refer to the precautions in the beginning of this section.

**** CAUTION**

High pressure lines deliver diesel fuel under extreme pressure from the injection pump to the fuel injectors. This may be as high as 1600 bar (23,200 psi.). Use extreme caution when inspecting for high pressure fuel leaks. Fuel under this amount of pressure can penetrate skin causing personal injury or death. Inspect for high pressure fuel leaks with a sheet of cardboard. Wear safety goggles and adequate protective clothing when servicing fuel system.

2. Disconnect negative battery cable.
3. Remove engine cover.
4. Disconnect injector electrical connector.

**** WARNING**

Repeated mounting of the retaining ring is not permitted.

5. Remove fuel return line from injector.
6. Remove fuel injector high pressure line.
7. Remove fuel injector retainer and retaining bolt.

➡DO NOT use a wire brush to clean the fuel injector or nozzle. Possible restriction of the injector needle may result.

8. Remove fuel injector from cylinder head.

➡If the fuel injectors will not come out of the cylinder head, assemble and install injector extractor special tool VM 9075A on to injector and cylinder head.

To install:

➡Be sure a new copper washer/seal is installed on end of injector before installing in cylinder head. Apply anti-seize compound to injector body.

9. Install fuel injector in cylinder head with new seal and O-ring.
10. Install fuel injector retainer and bolt. Torque bolt to 32 Nm.
11. Install fuel injector high pressure line.
12. Install fuel return line to injector.
13. Connect fuel injector electrical connector.
14. Install engine cover and bracket assembly.
15. Connect negative battery cable.
16. Start the engine, allow to warm, turn off the ignition, inspect for leaks.

DRIVE TRAIN

Automatic Transmission

REMOVAL & INSTALLATION

2002–03

1. Before servicing the vehicle, refer to the precautions in the beginning of this section.
2. Drain the transmission fluid.
3. Remove or disconnect the following:
   - Negative battery cable
   - Exhaust front pipe
   - Transmission braces, if equipped
   - Transmission oil cooler lines
   - Starter motor
   - Crankshaft Position (CKP) sensor
   - Torque converter access cover
   - Transmission oil pan
   - Skid plate
   - Transmission oil dipstick tube
   - Propellar shafts after matchmarking them
   - Park/Neutral switch connector
   - Shift cable
   - Throttle valve cable
   - Shift rod from the transfer case shift lever
   - Crossmember
   - Transfer case vent hose
   - Transfer case
4. Slide the transmission/torque converter assembly rearward off the engine dowels
5. Lower and remove the transmission assembly

To install:

6. Install or connect the following:
   - Transmission. Torque the flange bolts to 55 ft. lbs. (75 Nm).
   - Front driveshaft and transfer case, if equipped
   - Transfer case vent hose
   - Transmission mount and crossmember
   - Throttle valve cable
   - Shift cable
   - Park/Neutral switch connector
   - Propellar shafts using the matchmarks made during removal
   - Transmission oil dipstick tube
   - Skid plate
   - Transmission oil pan
   - Torque converter access cover
   - CKP sensor
   - Starter motor
   - Transmission oil cooler lines
   - Transmission braces, if equipped. Torque the bolts to 30 ft. lbs. (41 Nm).

- Exhaust front pipe
- Negative battery cable

7. Fill the transmission with the proper fluid.
8. Start the vehicle, check for leaks and repair if necessary.

2004–05 Liberty

1. Before servicing the vehicle, refer to the precautions in the beginning of this section.
2. Disconnect the negative battery cable.
3. Raise and support the vehicle
4. Remove any necessary skid plates.
5. Mark propeller shaft and axle companion flanges for assembly alignment.
6. Remove the rear propeller shaft.
7. Remove the front propeller shaft, if necessary.
8. Disconnect wires from the input and output speed sensors.
9. Disconnect wires from the transmission range sensor and the solenoid/pressure switch assembly.
10. Remove the bolts holding the exhaust crossover pipe to the pre-catalytic converter pipe flanges.
11. Remove the bolts holding the exhaust crossover pipe to the catalytic converter flange.
12. Disconnect gearshift cable from transmission manual valve lever.
13. Disengage the shift cable from the cable support bracket.
14. Remove the starter motor.
15. Remove the engine to transmission collar.
16. Rotate crankshaft in clockwise direction until converter bolts are accessible. Then remove bolts one at a time. Rotate crankshaft with socket wrench on dampener bolt.
17. Disconnect the transmission vent hose from the transmission.
18. Remove transfer case.
19. Support rear of engine with safety stand or jack.
20. Raise transmission slightly with service jack to relieve load on crossmember and supports.
21. Remove bolts securing rear support and cushion to transmission and crossmember.
22. Remove bolts attaching crossmember to frame and remove crossmember.
23. Disconnect transmission fluid cooler lines at transmission fittings and clips.
24. Remove all remaining converter housing bolts.

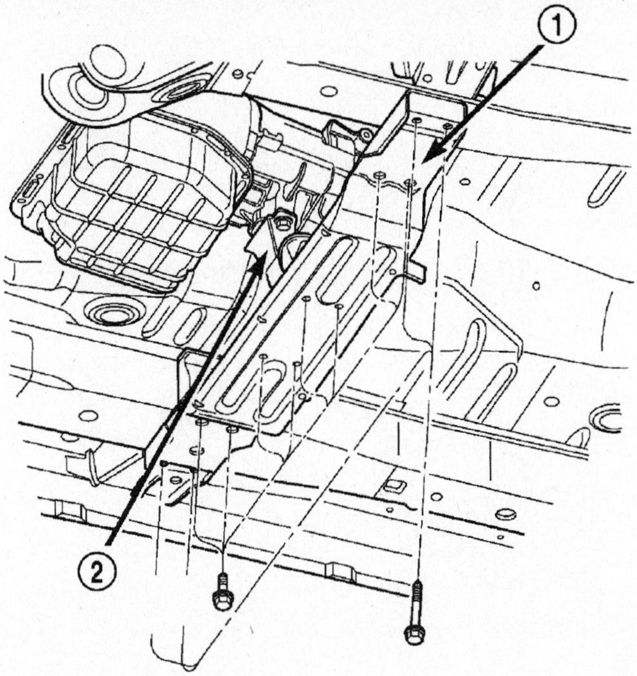

Rear transmission crossmember—2004–05 Liberty

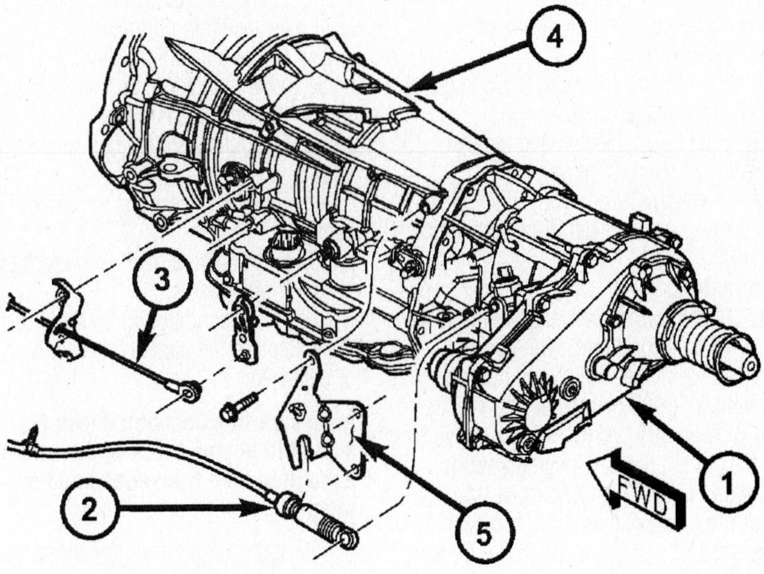

Transmission shift cables—2004–05 Liberty

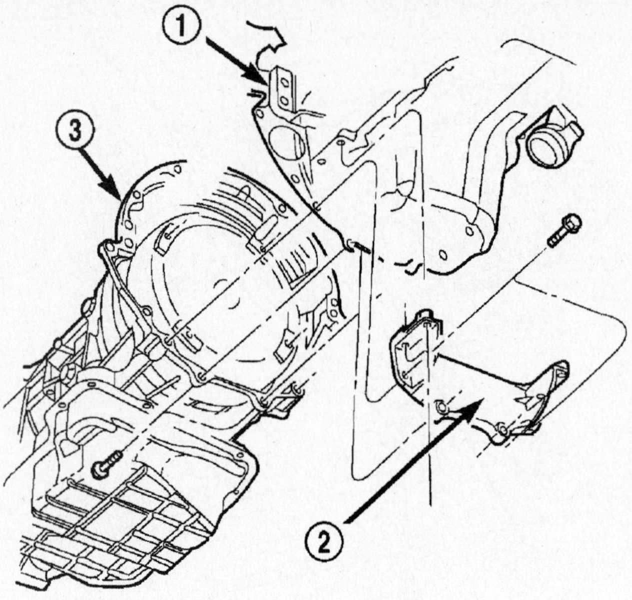

Transmission collar—2004–05 Liberty

25. Carefully work transmission and torque converter assembly rearward off engine block dowels.

26. Hold torque converter in place during transmission removal.

27. Lower transmission and remove assembly from under the vehicle.

28. To remove torque converter, carefully slide torque converter out of the transmission.

To install:

29. Check torque converter hub and hub drive flats for sharp edges burrs, scratches, or nicks. Polish the hub and flats with 320/400 grit paper and crocus cloth if necessary. Verify that the converter hub O-ring is properly installed and is free of any debris. The hub must be smooth to avoid damaging pump seal at installation.

30. If a replacement transmission is being installed, transfer any components necessary, such as the manual shift lever and shift cable bracket, from the original transmission onto the replacement transmission.

31. Lubricate oil pump seal lip with transmission fluid.

32. Align converter and oil pump.

33. Carefully insert converter in oil pump. Then rotate converter back and forth until fully seated in pump gears.

34. Check converter seating with steel scale and straightedge. Surface of converter lugs should be at least 13mm (½ in.) to rear of straightedge when converter is fully seated.

35. Temporarily secure converter with C-clamp.

36. Position transmission on jack and secure it with chains

37. Check condition of converter driveplate. Replace the plate if cracked, distorted or damaged.

➡Be sure transmission dowel pins are seated in engine block and protrude far enough to hold transmission in alignment.

38. Apply a light coating of Mopar® High Temp Grease to the torque converter hub pocket in the rear pocket of the engine's crankshaft.

39. Raise transmission and align the torque converter with the drive plate and transmission converter housing with the engine block.

40. Move transmission forward. Then raise, lower or tilt transmission to align the converter housing with engine block dowels.

41. Carefully work transmission forward and over engine block dowels until converter hub is seated in crankshaft. Verify that no wires, or the transmission vent hose, have become trapped between the engine block and the transmission.

42. Install two bolts to attach the transmission to the engine.

43. Install remaining torque converter housing to engine bolts. Tighten to 68 Nm (50 ft. lbs.).

44. Install transfer case, if equipped. Tighten transfer case nuts to 35 Nm (26 ft. lbs.).

45. Install rear transmission crossmember. Tighten crossmember to frame bolts to 68 Nm (50 ft. lbs.).

46. Install rear support to transmission. Tighten bolts to 47 Nm (35 ft. lbs.).

47. Lower transmission onto crossmember and install bolts attaching transmission mount to crossmember. Tighten clevis bracket to crossmember bolts to 47 Nm (35 ft. lbs.). Tighten the clevis bracket to rear support bolt to 68 Nm (50 ft. lbs.).

48. Remove engine support fixture.

49. Connect gearshift cable to support bracket and transmission manual lever.

50. Connect input and output speed sensor wires.

51. Connect wires to the transmission range sensor and the solenoid/pressure switch assembly.

52. Install torque converter-to-driveplate bolts. Tighten bolts to 88 Nm (65 inch lbs.).

53. Install starter motor and cooler line bracket.

54. Connect cooler lines to transmission.

55. Install transmission fill tube.

56. Install exhaust components.

57. Align and connect propeller shaft(s).

58. Adjust gearshift cable if necessary.

59. Install any skid plates removed previously.

2004 Grand Cherokee w/42RE

➡The transmission and torque converter must be removed as an assembly to avoid component damage. The converter driveplate, pump bushing, or oil seal can be damaged if the converter is left attached to the driveplate during removal. Be sure to remove the transmission and converter as an assembly.

1. Before servicing the vehicle, refer to the precautions in the beginning of this section.

2. Disconnect battery negative cable.

3. Disconnect and lower or remove necessary exhaust components.

4. Disconnect fluid cooler lines at transmission.

5. Remove starter motor.

6. Disconnect and remove crankshaft position sensor. Retain sensor attaching bolts.

➡The crankshaft position sensor will be damaged if the transmission is removed, or installed, while the sensor is still bolted to the engine block, or transmission (4.0L only). To avoid damage, be sure to remove the sensor before removing the transmission.

7. Remove the bolts holding the bell housing brace to the transmission.

8. Remove nut holding the bell housing brace to the engine to transmission bending brace.

9. Remove the bell housing brace from the transmission.

10. Remove the bolt holding the torque converter cover to the transmission.

11. Remove the torque converter cover from the transmission.

12. If transmission is being removed for overhaul, remove transmission oil pan, drain fluid and reinstall pan.

13. Remove fill tube bracket bolts and pull tube out of transmission. Retain fill tube seal. On 4 x 4 models, it will also be necessary to remove bolt attaching transfer case vent tube to converter housing.

14. Rotate crankshaft in clockwise direction until converter bolts are accessible. Then remove bolts one at a time. Rotate crankshaft with socket wrench on dampener bolt.

15. Mark propeller shaft and axle yokes for assembly alignment. Then disconnect and remove propeller shaft. On 4 x 4 models, remove both propeller shafts.

16. Disconnect wires from park/neutral position switch and transmission solenoid.

17. Disconnect gearshift cable from transmission manual valve lever.

18. Disconnect throttle valve cable from transmission bracket and throttle valve lever.

19. Disconnect transfer case shift cable from the transfer case shift lever.

20. Remove the clip securing the transfer case shift cable into the cable support bracket.

21. Disconnect transmission fluid cooler lines at transmission fittings and clips.

22. Support rear of engine with safety stand or jack.

23. Raise transmission slightly with service jack to relieve load on crossmember and supports.

24. Remove bolts securing rear support and cushion to transmission and crossmember.

25. Remove bolts attaching crossmember to frame and remove crossmember.

26. Remove transfer case.

To install:

27. Check torque converter hub and hub drive notches for sharp edges burrs, scratches, or nicks. Polish the hub and notches with 320/400 grit paper and crocus cloth if necessary. The hub must be smooth to avoid damaging pump seal during installation.

28. Lubricate oil pump seal lip with transmission fluid.

29. Align converter and oil pump.

30. Carefully insert converter in oil pump. Then rotate converter back and forth until fully seated in pump gears.

31. Check converter seating with steel scale and straightedge. Surface of converter lugs should be ½ in. to rear of straightedge when converter is fully seated.

32. Temporarily secure converter with C-clamp.

33. Position transmission on jack and secure it with chains.

34. Check condition of converter driveplate. Replace the plate if cracked, distorted or damaged. Also be sure transmission dowel pins are seated in engine block and protrude far enough to hold transmission in alignment.

35. Apply a light coating of Mopar® High Temp grease to the torque converter hub pocket in the rear of the crankshaft.

36. Raise transmission and align converter with drive plate and converter housing with engine block.

37. Move transmission forward. Then raise, lower or tilt transmission to align converter housing with engine block dowels.

38. Carefully work transmission forward and over engine block dowels until converter hub is seated in crankshaft.

39. Install two bolts to attach converter housing to engine.

40. Install the upper transmission bending braces to the torque converter housing and the overdrive unit. Tighten the bolts to 41 Nm (30 ft. lbs.).

41. Install remaining torque converter housing to engine bolts. Tighten to 68 Nm (50 ft. lbs.).

42. Install rear transmission crossmember. Tighten crossmember to frame bolts to 68 Nm (50ft. lbs.).

43. Install rear support to transmission. Tighten bolts to 47 Nm (35 ft. lbs.).

44. Lower transmission onto crossmember and install bolts attaching transmission mount to crossmember. Tighten clevis bracket to crossmember bolts to 47 Nm (35 ft. lbs.). Tighten the clevis bracket to rear support bolt to 68 Nm (50 ft. lbs.).

45. Remove engine support fixture.

46. Install crankshaft position sensor.

47. Install new plastic retainer grommet on any shift cable that was disconnected. Grommets should not be reused. Use pry tool to remove rod from grommet and cut away old grommet. Use pliers to snap new grommet into cable and to snap grommet onto lever.

48. Connect gearshift and throttle valve cable to transmission.

49. Connect wires to park/neutral position switch and transmission solenoid connector. Be sure transmission harnesses are properly routed.

➡It is essential that correct length bolts be used to attach the converter to the driveplate. Bolts that are too long will damage the clutch surface inside the converter.

50. Install all torque converter-to-driveplate bolts by hand.

51. Verify that the torque converter is pulled flush to the driveplate. Tighten bolts to 31 Nm (270 inch lbs.).

52. Install converter housing access cover. Tighten bolt to 23 Nm (200 inch lbs.).

53. Install the bell housing brace to the torque converter cover and the engine to transmission bending brace. Tighten the bolts and nut to 41 Nm (30 ft. lbs.).

54. Install starter motor and cooler line bracket.

55. Connect cooler lines to transmission.

56. Install transmission fill tube. Install new seal on tube before installation.

57. Install exhaust components.

58. Install transfer case. Tighten transfer case nuts to 35 Nm (26 ft. lbs.).

59. Install the transfer case shift cable to the cable support bracket and the transfer case shift lever.

60. Align and connect propeller shaft(s).

61. Adjust gearshift linkage and throttle valve cable if necessary.

62. Lower vehicle.

63. Fill transmission with Mopar® ATF +4, type 9602, fluid.

2004–05 Grand Cherokee w/545RFE

➡The transmission and torque converter must be removed as an assembly to avoid component damage. The converter driveplate, converter hub O-ring, or oil seal can be damaged if the converter is left attached to the driveplate during removal. Be sure to remove the transmission and converter as an assembly.

1. Before servicing the vehicle, refer to the precautions in the beginning of this section.
2. Disconnect the negative battery cable.
3. Raise and support the vehicle
4. Mark propeller shaft and axle yokes for assembly alignment.
5. Remove the rear propeller shaft
6. Remove the front propeller shaft.
7. Remove the engine to transmission collar.
8. Remove the exhaust support bracket from the rear of the transmission.
9. Disconnect and lower or remove any necessary exhaust components.
10. Remove the starter motor.
11. Rotate crankshaft in clockwise direction until converter bolts are accessible. Then remove bolts one at a time. Rotate crankshaft with socket wrench on dampener bolt.
12. Disconnect wires from solenoid and pressure switch assembly, input and output speed sensors, and line pressure sensor.
13. Disconnect gearshift cable from transmission manual valve lever.
14. Disconnect transfer case shift cable from the transfer case shift lever.
15. Remove the clip securing the transfer case shift cable into the cable support bracket.
16. Disconnect transmission fluid cooler lines at transmission fittings and clips.
17. Disconnect the transmission vent hose from the transmission.
18. Support rear of engine with safety stand or jack.
19. Raise transmission slightly with service jack to relieve load on crossmember and supports.
20. Remove bolts securing rear support and cushion to transmission and crossmember.
21. Remove bolts attaching crossmember to frame and remove crossmember.
22. Remove transfer case.
23. Remove all remaining converter housing bolts.
24. Carefully work transmission and torque converter assembly rearward off engine block dowels.
25. Hold torque converter in place during transmission removal.
26. Lower transmission and remove assembly from under the vehicle.
27. To remove torque converter, carefully slide torque converter out of the transmission.

To install:

28. Check torque converter hub and hub drive flats for sharp edges burrs, scratches, or nicks. Polish the hub and flats with 320/400 grit paper and crocus cloth if nece sary. Verify that the converter hub O-ring is properly installed and is free of any debris. The hub must be smooth to avoid damaging pump seal at installation.
29. If a replacement transmission is being installed, transfer any components necessary, such as the manual shift lever and shift cable bracket, from the original transmission onto the replacement transmission.
30. Lubricate oil pump seal lip with transmission fluid.
31. Align converter and oil pump.
32. Carefully insert converter in oil pump. Then rotate converter back and forth until fully seated in pump gears.
33. Check converter seating with steel scale and straightedge. Surface of converter lugs should be at least 13 mm (½ in.) to rear of straightedge when converter is fully seated.
34. Temporarily secure converter with C-clamp.
35. Position transmission on jack and secure it with chains.
36. Check condition of converter driveplate. Replace the plate if cracked, distorted or damaged.

➡Be sure transmission dowel pins are seated in engine block and protrude far enough to hold transmission in alignment.

37. Apply a light coating of Mopar® High Temp Grease to the torque converter hub pocket in the rear pocket of the engine's crankshaft.
38. Raise transmission and align the torque converter with the drive plate and the transmission converter housing with the engine block.
39. Move transmission forward. Then raise, lower, or tilt transmission to align the converter housing with the engine block dowels.
40. Carefully work transmission forward and over engine block dowels until converter hub is seated in crankshaft. Verify that no wires, or the transmission vent hose, have become trapped between the engine block and the transmission.
41. Install two bolts to attach the transmission to the engine.
42. Install remaining torque converter housing to engine bolts. Tighten to 68 Nm (50 ft. lbs.).
43. Install rear transmission crossmember. Tighten crossmember to frame bolts to 68 Nm (50 ft. lbs.).
44. Install rear support to transmission. Tighten bolts to 47 Nm (35 ft. lbs.).
45. Lower transmission onto crossmember and install bolts attaching transmission mount to crossmember. Tighten clevis bracket to crossmember bolts to 47 Nm (35 ft. lbs.). Tighten the clevis bracket to rear support bolt to 68 Nm (50 ft. lbs.).
46. Install both the left and right side oil pan to transmission bolts, 5.7L only. Torque the bolts to 54 Nm (39 ft. lbs.).
47. Remove engine support fixture.
48. Install new plastic retainer grommet on any shift cable that was disconnected. Grommets should not be reused. Use pry tool to remove rod from grommet and cut away old grommet. Use pliers to snap new grommet into cable and to snap grommet onto lever.
49. Connect gearshift cable to transmission.
50. Connect wires to solenoid and pressure switch assembly connector, input and output speed sensors, and line pressure sensor. Be sure transmission harnesses are properly routed.

➡It is essential that correct length bolts be used to attach the converter to the driveplate. Bolts that are too long will damage the clutch surface inside the converter.

51. Install all torque converter-to-driveplate bolts by hand.
52. Verify that the torque converter is pulled flush to the driveplate. Tighten bolts to 31 Nm (270 inch lbs.).
53. Install starter motor and cooler line bracket.
54. Connect cooler lines to transmission.
55. Install transmission fill tube.
56. Install exhaust components.
57. Install transfer case. Tighten transfer case nuts to 35 Nm (26 ft. lbs.).
58. Install the transfer case shift cable to the cable support bracket and the transfer case shift lever.
59. Install the structural dust cover onto the transmission and the engine.

60. Align and connect propeller shaft(s).
61. Adjust gearshift cable if necessary.
62. Lower vehicle.
63. Fill transmission with Mopar® ATF +4, Automatic Transmission fluid.

2005 Grand Cherokee w/NAG1

➡If the transmission is being reconditioned (clutch/seal replacement) or replaced, it is necessary to perform the TCM Adaptation Procedure using the scan tool.

1. Before servicing the vehicle, refer to the precautions in the beginning of this section.
2. Disconnect the negative battery cable.
3. Raise and support the vehicle.
4. Remove the driveshafts.
5. Remove the bolts holding the starter motor to the transmission.
6. Remove the starter from the transmission starter pocket and safely relocate.
7. Remove the bolts holding the structural cover to the transmission and engine.
8. Remove the structural cover from the vehicle.
9. Rotate crankshaft in clockwise direction until converter bolts are accessible. Then remove bolts one at a time. Rotate crankshaft with socket wrench on dampener bolt.
10. Disconnect the gearshift cable from the transmission manual valve lever.
11. Remove the shift cable from the gearshift cable bracket.
12. Disconnect 13-pin plug connector. Turn bayonet lock of guide bushing counterclockwise.
13. Remove the 13-pin connector from the transmission.
14. Disconnect transmission fluid cooler lines at transmission.
15. Disconnect the transmission vent hose from the transmission.
16. Remove the bolts holding the transmission fill tube to the transmission.
17. Support rear of engine with safety stand or jack.
18. Raise transmission slightly with service jack to relieve load on crossmember and supports.
19. Remove bolts securing rear support and cushion to transmission crossmember.
20. Remove bolts attaching crossmember to frame and remove crossmember.
21. Remove all remaining bolts holding the engine to the transmission. Note the location of any wiring harness clips.
22. Carefully work transmission and torque converter assembly rearward off engine block dowels.
23. Hold torque converter in place during transmission removal.
24. Lower transmission and remove assembly from under the vehicle.
25. To remove torque converter, carefully slide torque converter out of the transmission.

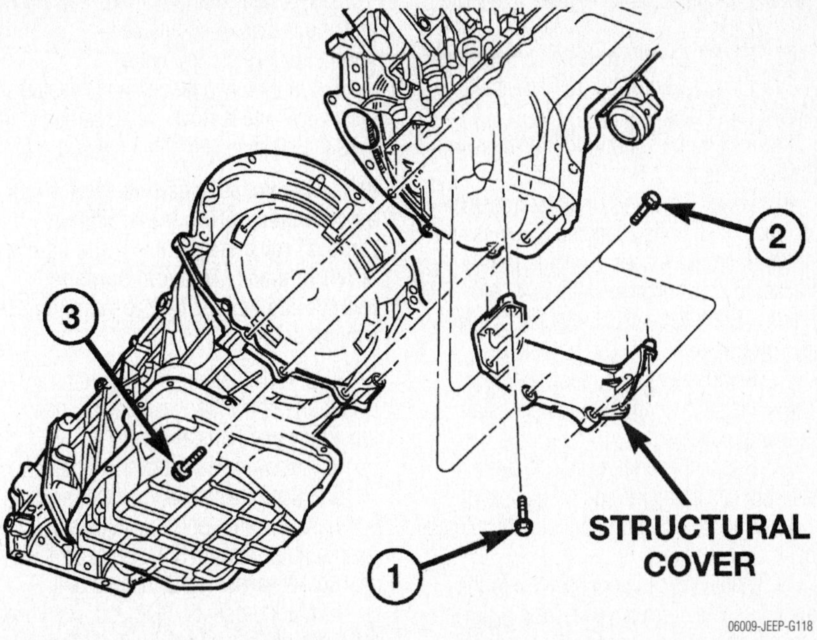

Structural cover—2005 Grand Cherokee w/NAG1

To install:

26. Check torque converter hub and hub drive flats for sharp edges burrs, scratches, or nicks. Polish the hub and flats with 320/400 grit paper and crocus cloth if necessary. The hub must be smooth to avoid damaging pump seal at installation.
27. If a replacement transmission is being installed, transfer any components necessary, such as the manual shift lever and shift cable bracket, from the original transmission onto the replacement transmission.
28. Lubricate oil pump seal lip with transmission fluid.
29. Place torque converter in position on transmission.

⁂ WARNING

Do not damage oil pump seal or converter hub while inserting torque converter into the front of the transmission.

30. Align torque converter to oil pump seal opening.
31. Insert torque converter hub into oil pump.
32. While pushing torque converter inward, rotate converter until converter is fully seated in the oil pump gears.
33. Check converter seating with a scale and straightedge. Surface of converter lugs should be at least 19 mm (¾ in.) to rear of straightedge when converter is fully seated.
34. If necessary, temporarily secure converter with C-clamp attached to the converter housing.
35. Check condition of converter driveplate. Replace the plate if cracked, distorted or damaged. Also be sure transmission dowel pins are seated in engine block and protrude far enough to hold transmission in alignment.
36. Apply a light coating of Mopar® High Temp Grease to the torque converter hub pocket in the rear pocket of the engine's crankshaft.
37. Raise transmission and align the torque converter with the drive plate and the transmission converter housing with the engine block.
38. Move transmission forward. Then raise, lower, or tilt transmission to align the converter housing with the engine block dowels.
39. Carefully work transmission forward and over engine block dowels until converter hub is seated in crankshaft. Verify that no wires, or the transmission vent hose, have become trapped between the engine block and the transmission.
40. Install two bolts to attach the transmission to the engine.
41. Install remaining torque converter

housing to engine bolts. Tighten to 39 Nm (29 ft. lbs.).

42. Install rear transmission crossmember. Tighten crossmember to frame bolts to 68 Nm (50 ft. lbs.).

43. Install rear support to transmission. Tighten bolts to 47 Nm (35 ft. lbs.).

44. Lower transmission onto crossmember and install bolts attaching transmission mount to crossmember. Tighten clevis bracket to crossmember bolts to 47 Nm (39 ft. lbs.). Tighten the clevis bracket to rear support bolt to 68 Nm (50 ft. lbs.).

45. Remove engine support fixture.

46. Install the engine to transmission structural cover.

47. Connect gearshift cable to the gearshift cable bracket and transmission.

48. Check O-ring on plug connector, and replace if necessary.

49. Install the plug connector into the guide bushing. Turn bayonet lock of guide bushing clockwise to connect plug connector.

WARNING

It is essential that correct length bolts be used to attach the converter to the driveplate. Bolts that are too long will damage the clutch surface inside the converter.

50. Install all torque converter-to-driveplate bolts by hand.

51. Verify that the torque converter is pulled flush to the driveplate. Tighten bolts to 42 Nm (30 ft. lbs.).

52. Install starter motor.

53. Install transmission fill tube.

54. Connect cooler lines to transmission.

55. Install exhaust components.

56. Install transfer case, if necessary. Tighten transfer case nuts to 35 Nm (26 ft. lbs.).

57. Align and connect driveshafts.

58. Adjust gearshift cable if necessary.

59. Lower vehicle.

60. Connect negative battery cable.

61. Fill transmission with the appropriate fluid.

62. Verify proper operation.

2004–05 Wrangler

1. Before servicing the vehicle, refer to the precautions in the beginning of this section.

2. Disconnect battery negative cable.

3. Raise and support vehicle.

4. Disconnect and lower or remove necessary exhaust components.

5. Remove engine-to-transmission bending braces or engine collar.

6. Remove starter motor.

7. On 4.0L engine equipped vehicles, disconnect and remove crankshaft position sensor. Retain sensor attaching bolt.

➡The crankshaft position sensor can be damaged during transmission removal (or installation) if the sensor is left in place. To avoid damage, remove the sensor before removing the transmission.

8. If transmission is being removed for overhaul, remove transmission oil pan, drain fluid and reinstall pan.

9. Remove torque converter access cover.

10. Rotate crankshaft in clockwise direction until converter bolts are accessible. Then remove bolts one at a time. Rotate crankshaft with socket wrench on dampener bolt.

11. Mark propeller shaft and axle yokes for assembly alignment. Then disconnect and remove propeller shafts.

12. Disconnect wires from the input and output speed sensors.

13. Disconnect wires from the transmission range sensor and the solenoid/pressure switch assembly.

14. Disconnect gearshift cable from transmission manual valve lever.

15. Disconnect shift rod from transfer case shift lever or remove shift lever from transfer case.

16. Support rear of engine with safety stand or jack.

17. Raise transmission slightly with service jack to relieve load on skid plate and transmission support.

18. Remove bolts securing rear support and cushion to transmission and skid plate. Raise transmission slightly, slide exhaust hanger arm from bracket and remove rear support.

19. Remove bolts attaching skid plate to frame and remove skid plate.

20. Disconnect transfer case vent hose.

21. Remove transfer case.

22. Remove fill tube bracket bolts and pull tube out of transmission. Retain fill tube seal. Remove the bolt attaching transfer case vent tube to converter housing.

23. Disconnect fluid cooler lines at transmission.

24. Remove all converter housing bolts.

25. Carefully work transmission and torque converter assembly rearward off engine block dowels.

26. Hold torque converter in place during transmission removal.

27. Lower transmission and remove assembly from under the vehicle.

28. To remove torque converter, carefully slide torque converter out of the transmission.

To install:

29. Check torque converter hub and hub drive notches for sharp edges burrs, scratches, or nicks. Polish the hub and notches with 320/400 grit paper and crocus cloth if necessary. The hub must be smooth to avoid damaging pump seal at installation.

30. Lubricate converter drive hub and oil pump seal lip with transmission fluid.

31. Align converter and oil pump.

32. Carefully insert converter in oil pump. Then rotate converter back and forth until fully seated in pump gears.

33. Check converter seating with steel scale and straightedge. Surface of converter lugs should be ½ in. to rear of straightedge when converter is fully seated.

34. Temporarily secure converter with C-clamp.

35. Lightly grease crankshaft flange hole.

36. Position transmission on jack and secure it with safety chains.

37. Check condition of converter driveplate. Replace the plate if cracked, distorted or damaged.

➡Be sure transmission dowel pins are seated in engine block and protrude far enough to hold transmission in alignment.

38. Raise transmission and align converter with drive plate and converter housing with engine block.

39. Move transmission forward. Then raise, lower or tilt transmission to align converter housing with engine block dowels.

40. Carefully work transmission forward and over engine block dowels until converter hub is seated in crankshaft.

41. Install and tighten bolts that attach transmission converter housing to engine block to 40 ft. lbs. (54 Nm).

➡Be sure the converter housing is fully seated on the engine block dowels before tightening any bolts.

42. Install torque converter attaching bolts. Tighten bolts to 88 Nm (65 ft. lbs.).

43. On 4.0L engine equipped vehicles, install the crankshaft position sensor.

44. Install transmission fill tube and seal. Install new fill tube seal in transmission before installation.

45. Connect transmission cooler lines to transmission.

46. Install transfer case onto transmission.

47. Install skid plate. Torque to 33 ft. lbs.

(45 Nm). Attach transmission rear support to skid plate. Torque to 26 ft. lbs. (33 Nm).

48. Remove engine support fixture.
49. Remove transmission jack.
50. Connect input and output speed sensor wires.
51. Connect wires to the transmission range sensor and the solenoid/pressure switch assembly.
52. Install converter housing access cover.
53. Install exhaust pipes and support brackets, if removed.
54. Install starter motor and cooler line bracket.
55. Install new plastic retainer grommet on any shift linkage rod or lever that was disconnected. Grommets should not be reused. Use pry tool to remove rod from grommet and cut away old grommet. Use pliers to snap new grommet into lever and to snap rod into grommet at assembly.
56. Connect gearshift cable.
57. Connect transfer case shift linkage.
58. Adjust gearshift linkage, if necessary.
59. Align and connect propeller shaft(s).
60. Fill transfer case to bottom edge of fill plug hole.
61. Lower vehicle and connect battery negative cable.
62. Fill transmission to correct level with Mopar® ATF +4

Manual Transmission

REMOVAL & INSTALLATION

2002–03

1. Before servicing the vehicle, refer to the precautions in the beginning of this section.
2. Place the transmission in first or third gear.
3. Drain the fluid from the transmission.
4. Remove or disconnect the following:
    - Negative battery cable
    - Exhaust front pipe
    - Skid plate
    - Clutch slave cylinder
    - Rear propeller shaft
    - Front propeller shaft, if equipped
    - Wire harness from the transmission and transfer case, if equipped
    - Transfer case shift linkage, if equipped
    - Transfer case, if equipped
    - Crankcase Position (CKP) sensor
    - Rear cushion and transmission bracket after supporting the engine and transmission
    - Rear crossmember
    - Transmission shift lever
    - Clutch housing brace rod
    - Transmission

To install:

5. Install or connect the following:
    - Transmission. Tighten the 3⁄8 inch bolts to 27 ft. lbs. (37 Nm), the 7⁄16 inch bolts to 43 ft. lbs. (58 Nm), and the 12mm bolts to 55 ft. lbs. (75 Nm).
    - Shift lever
    - Rear crossmember. Torque the bolts to 31 ft. lbs. (41 Nm).
    - Rear cushion and bracket. Torque the bolts to 33 ft. lbs. (45 Nm).
    - CKP sensor
    - Transfer case, if equipped. Torque the bolts to 26 ft. lbs. (35 Nm).
    - Transfer case shift linkage, if equipped
    - Wire harness to the transmission
    - Front driveshaft and transfer case, if equipped. Torque the bolts to 12 ft. lbs. (19 Nm).
    - Rear driveshaft. Torque the bolts to 12 ft. lbs. (19 Nm).
    - Clutch slave cylinder
    - Skid plate. Torque the bolts to 31 ft. lbs. (42 Nm).
    - Exhaust front pipe
    - Negative battery cable
6. Fill the transmission with the proper fluid.
7. Start the vehicle, check for leaks and repair if necessary.

2004 Liberty w/NV1500

1. Before servicing the vehicle, refer to the precautions in the beginning of this section.
2. Shift transmission into neutral.
3. Raise and support the vehicle.
4. Remove skid plate if equipped.
5. Remove wiring connectors from the transmission.
6. Remove propeller shaft/shafts.
7. Remove transfer case shift cable, vent hose and transfer case, if equipped.
8. Remove slave cylinder from clutch housing.
9. Remove starter.
10. Support engine with jack stand. Position wood block between jack and oil pan to avoid damaging pan.
11. Support transmission with a transmission jack.
12. Remove crossover pipe from manifold extensions.
13. Remove exhaust hanger from the transmission crossmember.
14. Remove transmission dust shield.
15. Remove transmission mount and crossmember.
16. Lower transmission jack enough to remove shift tower bolts.
17. Lower transmission jack and remove transmission from under vehicle.
18. Pull transmission jack rearward until input shaft clears clutch.
19. Remove clutch release bearing, release fork and retainer clip.
20. Remove clutch housing from transmission.

To install:

21. Install clutch housing on transmission and tighten housing bolts to 46 Nm (34 ft. lbs.).
22. Lubricate contact surfaces of release fork pivot ball stud and release fork with high temp grease.
23. Install release bearing, fork and retainer clip.
24. Position and secure transmission on transmission jack.
25. Lightly lubricate pilot bearing and transmission input shaft splines with Mopar high temp grease.
26. Raise transmission and align transmission input shaft and clutch disc splines. Then slide transmission into place.
27. Install clutch housing-to-engine bolts and tighten to 58 Nm (43 ft. lbs.).

➡Be sure the housing is properly seated on engine block before tightening bolts.

28. Install shift tower and bolts. Tighten bolts to 7–10 Nm (5–7 ft. lbs.).
29. Install rear crossmember and tighten crossmember bolts to 41 Nm (31 ft. lbs.).
30. Install transmission mount bolts and to 54 Nm (40 ft. lbs.).
31. Install exhaust bracket to crossmember.
32. Install crossover pipe to manifold extensions.
33. Remove support stands from engine and transmission.
34. Install transfer case, shift cable and vent hose if equipped.
35. Install wire connectors to transmission/transfer case.
36. Install propeller shaft/shafts.
37. Install slave cylinder in clutch housing.
38. Install skid plate if equipped.
39. Fill transmission and transfer case if equipped, with recommended lubricants.

2004 Liberty w/NV3550

1. Before servicing the vehicle, refer to the precautions in the beginning of this section.
2. Shift transmission into neutral.
3. Raise and support the vehicle.
4. Remove skid plate if equipped.
5. Remove wiring connectors from the transmission.
6. Remove propeller shaft/shafts.
7. Remove transfer case shift cable and vent hose from transfer case, if equipped.
8. Remove slave cylinder from clutch housing.
9. Support engine with jack stand. Position wood block between jack and oil pan to avoid damaging pan.
10. Support transmission with a transmission jack.
11. Remove exhaust hanger from the transmission crossmember.
12. Remove transmission mount and crossmember.
13. Lower transmission jack enough to remove shift tower bolts.
14. Remove clutch housing-to-engine bolts.
15. Pull transmission jack rearward until input shaft clears clutch.
16. Remove clutch release bearing, release fork and retainer clip.
17. Remove clutch housing from transmission.

To install:

18. Install clutch housing on transmission and tighten housing bolts to 46 Nm (34 ft. lbs.).
19. Lubricate contact surfaces of release fork pivot ball stud and release fork with high temp grease.
20. Install release bearing, fork and retainer clip.
21. Position and secure transmission on transmission jack.
22. Lightly lubricate the transmission input shaft splines with Mopar high temp grease.
23. Raise transmission and align transmission input shaft and clutch disc splines. Then slide transmission into place.
24. Install clutch housing-to-engine bolts and tighten to 58 Nm (43 ft. lbs.).

➡Be sure the housing is properly seated on engine block before tightening bolts.

25. Install shift tower and bolts. Tighten bolts to 7–10 Nm (5–7 ft. lbs.).
26. Install rear crossmember and tighten crossmember bolts to 41 Nm (31 ft. lbs.).
27. Install transmission mount bolts and to 54 Nm (40 ft. lbs.).
28. Install exhaust bracket to crossmember.
29. Remove support stands from engine and transmission.
30. Install transfer case, shift cable and vent hose if equipped.
31. Install wire connectors to transmission/transfer case.
32. Install propeller shaft(s).
33. Install slave cylinder in clutch housing.
34. Fill transmission and transfer case if equipped, with recommended lubricants.
35. Install skid plate if equipped.

2005 Liberty w/NSG370

1. Before servicing the vehicle, refer to the precautions in the beginning of this section.
2. Disconnect negative battery cable.
3. With vehicle in neutral, position vehicle on hoist.
4. Remove drain plug and drain fluid.
5. Mark installation reference marks on propeller shaft/shafts and remove shafts.
6. Remove transfer case shift cable bracket nuts, cable, wiring connector, and vent hose, if equipped.
7. Remove transfer case, if equipped.
8. Support transmission with jack.
9. Remove transmission mount and crossmember.
10. On 2.4L engine remove catalyst assembly.
11. Remove clutch slave cylinder nuts and remove cylinder.
12. Remove backup lamp switch wiring connector.
13. Remove shift lever tower bolts and remove shift lever housing.
14. Remove starter bolts and remove starter.
15. Remove transmission bolt and remove transmission.

To install:

16. Install transmission on engine.
17. On 3.7L engine tighten bolts to 41 Nm (30 ft. lbs.). Tighten bolts to 67 Nm (50 ft. lbs.). Tighten bolts to 54 Nm (40 ft. lbs.).
18. On 2.4L engine tighten bolts to 75 Nm (55 ft. lbs.). Install lower dust shield and tighten bolts to 34 Nm (25 ft. lbs.)
19. On 2.8L diesel engine tighten bolts to 41 Nm (30 ft. lbs.). Tighten bolts to 67 Nm (50 ft. lbs.). Tighten bolts to 54 Nm (40 ft. lbs.).
20. Clean shift tower and mating surface then apply Mopar Gasket Maker to shift tower.
21. Install shift tower and tighten bolts to 10 Nm (7 ft. lbs.).
22. Install back-up lamp wiring connector.
23. Install clutch slave cylinder and mounting nuts.
24. Install catalyst assembly on 2.4L engine.
25. Install transmission crossmember and tighten bolts to 47 Nm (35 ft. lbs.). Install transmission mount bolts and tighten to 47 Nm (35 ft. lbs.).
26. Install transfer case, if equipped.
27. Install transfer case shift cable on 2.4L. Install transfer case shift linkage on 3.7L.
28. Install transfer case wiring connector and vent hose.
29. Install propeller shaft/shafts with reference marks aligned.
30. Remove fill plug and fill transmission with lubricant.

2004 Wrangler w/NV1500

1. Before servicing the vehicle, refer to the precautions in the beginning of this section.
2. With vehicle in neutral, position vehicle on hoist.
3. Support engine with jack stand. Position wood block between jack and oil pan to avoid damaging pan.
4. Remove skid plate/crossmember.
5. Support transmission with a transmission jack.
6. Remove transmission mount from transmission and exhaust.
7. Remove propeller shafts.
8. Remove transfer case shift linkage and vent hose.
9. Remove wiring connectors from transmission and transfer case.
10. Remove transfer case.
11. Remove slave cylinder from clutch housing.
12. Remove starter.
13. Remove transmission dust shield.
14. Lower transmission jack enough to remove shift tower bolts.
15. Lower transmission jack and remove transmission from under vehicle.
16. Pull transmission jack rearward until input shaft clears clutch.
17. Remove clutch release bearing, release fork and retainer clip.
18. Remove clutch housing from transmission.

To install:

19. Install clutch housing on transmission and tighten housing bolts to 46 Nm (34 ft. lbs.).

20. Lubricate contact surfaces of release fork pivot ball stud and release fork with high temp grease.

21. Install release bearing, fork and retainer clip.

22. Position and secure transmission on transmission jack.

23. Lightly lubricate pilot bearing and transmission input shaft splines with Mopar high temp grease.

24. Raise transmission and align transmission input shaft and clutch disc splines. Then slide transmission into place.

25. Install clutch housing-to-engine bolts and tighten to 75 Nm (55 ft. lbs.).

➡Be sure the housing is properly seated on engine block before tightening bolts.

26. Install shift tower and bolts. Tighten bolts to 11 Nm (8 ft. lbs.).

27. Install transmission mount and tighten bolts to 54 Nm (40 ft. lbs.).

28. Install transfer case, shift linkage and vent hose.

29. Install wire connectors to transmission and transfer case.

30. Install skid plate/crossmember and tighten bolts to 41 Nm (31 ft. lbs.).

31. Remove support stands from engine and transmission.

2004 Wrangler w/NV3550

1. Before servicing the vehicle, refer to the precautions in the beginning of this section.

2. Shift transmission into first or third gear.

3. Remove floor console and shift boot as necessary to access the bottom of the shift lever at the shift tower attachment.

4. Remove shift tower bolts and remove shift tower and shift lever assembly.

5. Raise and support vehicle on suitable safety stands.

6. Support engine with adjustable jack stand. Position wood block between jack and oil pan to avoid damaging pan.

7. Remove skid plate, if equipped.

8. Remove crossmember.

9. Disconnect necessary exhaust system components.

10. Remove slave cylinder from clutch housing.

11. Remove propeller shafts.

12. Unclip wire harnesses from transmission and transfer case.

13. Disconnect transfer case shift linkage at transfer case.

14. Remove nuts attaching transfer case to transmission.

15. Remove transfer case.

16. Remove crankshaft position sensor.

➡The crankshaft position sensor must be removed prior to transmission removal. Failure to heed caution may result in damage.

17. Support engine with adjustable jack stand. Position wood block between jack and oil pan to avoid damaging pan.

18. Support transmission with transmission jack.

19. Secure transmission to jack with safety chains.

20. Disconnect rear cushion and bracket from transmission.

21. Remove rear crossmember.

22. Remove clutch housing-to-engine bolts.

23. Pull transmission jack rearward until input shaft clears clutch. Then slide transmission out from under vehicle.

24. Remove clutch release bearing, release fork and retainer clip.

25. Remove clutch housing from transmission.

To install:

26. Install clutch housing on transmission and tighten housing bolts to 46 Nm (34 ft. lbs.).

27. Lubricate contact surfaces of release fork pivot ball stud and release fork with high temp grease.

28. Install release bearing, fork and retainer clip.

29. Position and secure transmission on transmission jack.

30. Lightly lubricate pilot bearing and transmission input shaft splines with Mopar high temp grease.

31. Raise transmission and align transmission input shaft and clutch disc splines. Then slide transmission into place.

➡Be sure the housing is properly seated on engine block before tightening bolts.

32. Install and tighten clutch housing-to-engine bolts to:

- 3/8 in. diameter bolts to 37 Nm (27 ft. lbs.)
- 7/16 in. diameter bolts to 58 Nm (43 ft. lbs.)
- M12 bolts to 75 Nm (55 ft. lbs.)

33. Be sure transmission is in first or third gear.

34. Install crossmember and tighten crossmember-to-frame bolts to 41 Nm (31 ft. lbs.).

35. Install fasteners to hold rear cushion and bracket to transmission. Then tighten

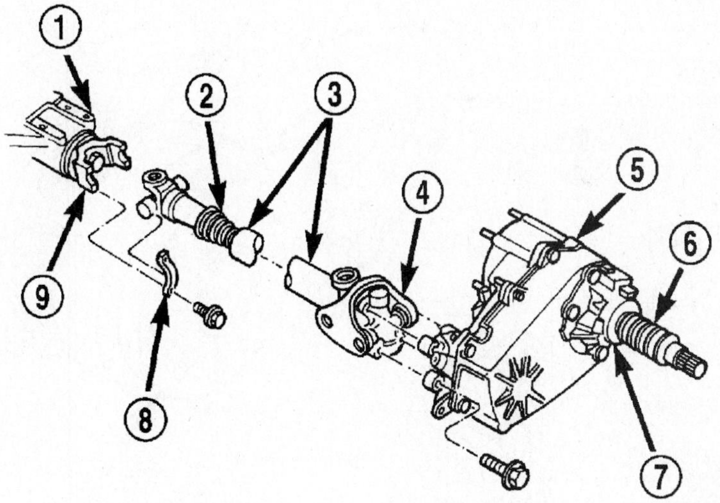

Front driveshaft—Wrangler

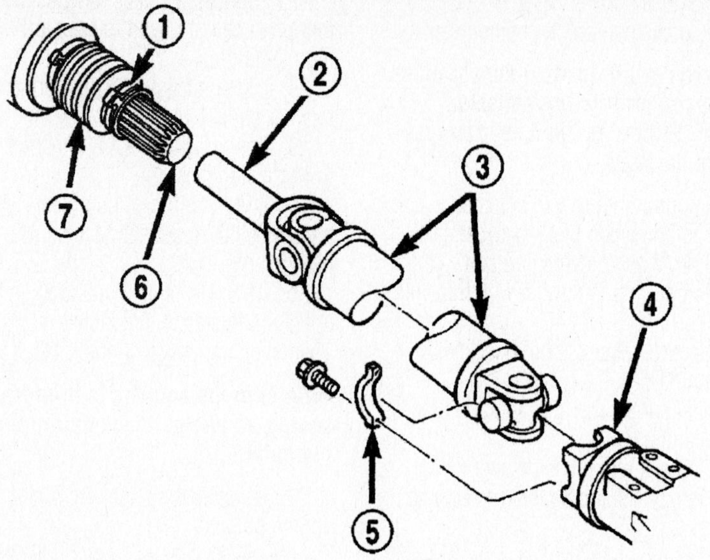

1 - CLAMP
2 - YOKE
3 - PROPELLER SHAFT
4 - AXLE YOKE
5 - CLAMP
6 - OUTPUT SHAFT
7 - BOOT

06009-JEEP-G175

Rear driveshaft—Wrangler, exc. Rubicon

1 - REAR PROPELLER SHAFT
2 - TRANSFER CASE FLANGE

06009-JEEP-G176

Rear driveshaft—Rubicon

transmission-to- rear support bolts/nuts to 54 Nm (40 ft. lbs.).

36. Remove support stands from engine and transmission.
37. Install and connect crankshaft position sensor.
38. Install transfer case.
39. Install propeller shafts.
40. Install slave cylinder in clutch housing.
41. Install skid plate, if equipped and tighten bolts to 42 Nm (31 ft. lbs.). Tighten stud nuts to 17 Nm (150 inch lbs.).
42. Fill transmission and transfer case if equipped, with recommended lubricants.
43. Install nuts on two M6 x 1.0 bolts and thread the bolts into the threaded holes at the base of the shift lever.
44. Tighten the nuts equally until the shift lever will slide over the shift tower stub shaft.
45. Install the floor console and shift boot.

2005 Wrangler w/NSG370

1. Before servicing the vehicle, refer to the precautions in the beginning of this section.
2. Disconnect negative battery cable.
3. With vehicle in neutral, position vehicle on hoist.
4. Remove drain plug and drain fluid.
5. Mark installation reference marks on driveshafts and remove shafts.
6. Remove transfer case shift linkage, wiring connector, and vent hose, if equipped.
7. Remove transfer case, if equipped.
8. Support transmission with jack.
9. Remove transmission mount and crossmember.
10. Remove catalyst assembly.
11. Remove clutch slave cylinder nuts and remove cylinder.
12. Remove backup lamp switch wiring connector.
13. Remove shift lever tower bolts and remove shift lever tower.
14. Remove starter bolts and remove starter.
15. Remove transmission bolt and remove transmission.

To install:

16. Install transmission on engine.
17. On 2.4L engine tighten bolts to 75 Nm (55 ft. lbs.). Install lower dust shield and tighten bolts to 34 Nm (25 ft. lbs.).
18. On 4.0L engine tighten bolts to 37 Nm (27 ft. lbs.). Install side and lower dust shield and tighten bolts to 58 Nm (43 ft. lbs.).

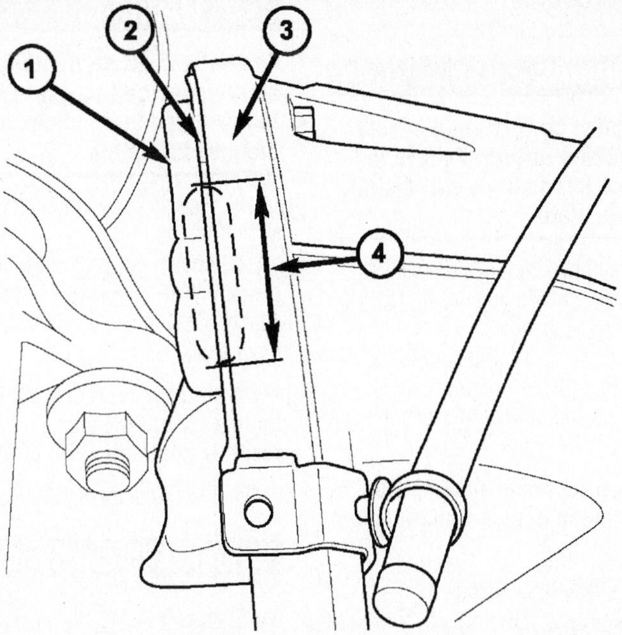

1 - BACK OF ENGINE
2 - ADAPTER PLATE
3- BELLHOUSING - MANUAL TRANSMISSION
4 - ADAPTER PLATE OPENING - APPLY SEALER HERE

06009-JEEP-G172

Adapter opening detail—2005 Wrangler

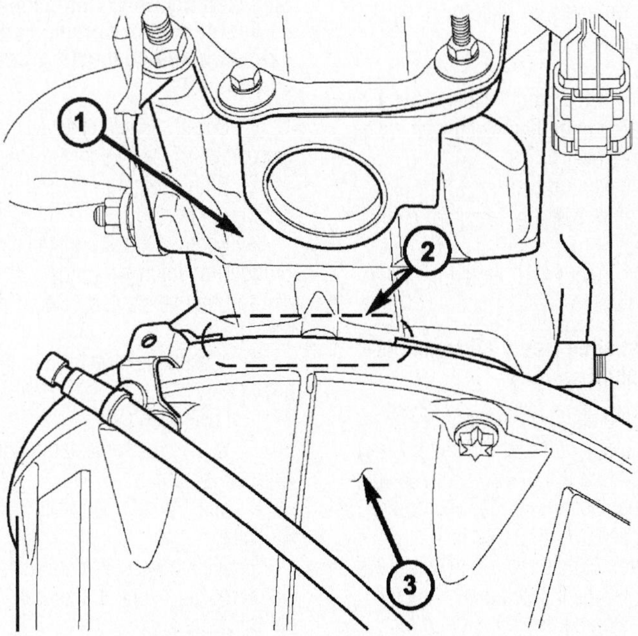

1 - BACK OF ENGINE
2 - LOCATION OF NARROW OPENING DUE TO SLOT IN ADAPTER PLATE
3 - TOP OF BELLHOUSING - MANUAL TRANSMISSION

06009-JEEP-G173

Location of adapter opening—2005 Wrangler

19. Clean shift tower and mating surface then apply Mopar® Gasket Maker, or equivalent, to shift tower.
20. Install shift tower and tighten bolts to 10 Nm (7 ft. lbs.).
21. Install backup lamp wiring connector.
22. Install clutch slave cylinder and mounting nuts.
23. Install catalyst assembly on 2.4L engine.
24. Install transmission crossmember and tighten bolts to 47 Nm (35 ft. lbs.). Install transmission mount bolts and tighten to 47 Nm (35 ft. lbs.).
25. Install transfer case, if equipped.
26. Install transfer case shift linkage.
27. Install transfer case wiring connector and vent hose.
28. Install driveshafts with reference marks aligned. Torque as follows:
    - Front driveshaft to axle: 14 ft. lbs. (19 Nm)
    - Front driveshaft to transfer case: 20 ft. lbs. (27 Nm)
    - Rear driveshaft to axle: 14 ft. lbs. (19 Nm)
    - Rear driveshaft to transfer case (Rubicon): 85 ft. lbs. (115 Nm)
29. Remove fill plug and fill transmission with lubricant.
30. Locate narrow adapter plate opening towards the rear of the engine and top of the transmission bellhousing from the driver's side of the engine compartment.
31. Thoroughly clean around the adapter opening with throttle body cleaner and allow cleaner to dry.
32. With your finger, apply a sufficient amount of RTV over the spacer opening. Verify that the adapter opening is completely sealed off by the RTV.

Clutch

REMOVAL & INSTALLATION

2002–05 Wrangler

1. Before servicing the vehicle, refer to the precautions in the beginning of this section.
2. Remove or disconnect the following:
    - Negative battery cable
    - Transfer case, if equipped
    - Transmission
    - Pressure plate. Loosen the bolts evenly in ½ turn steps.
    - Clutch disc

To install:

3. Install or connect the following:
    - Clutch disc and pressure plate.

Tighten the pressure plate bolts evenly in ½ turns to 23 ft. lbs. (31 Nm) for 2.4L and 2.5L engines; 37 ft. lbs. (50 Nm) for 4.0L engines.
- Transmission
- Transfer case, if equipped
- Negative battery cable

2005 Liberty

2.4L AND 3.7L ENGINES

1. Before servicing the vehicle, refer to the precautions in the beginning of this section.
2. Remove the transmission.
3. Mark position of pressure plate on flywheel with paint or a scriber for assembly reference, if clutch is not being replaced.
4. Loosen pressure plate bolts evenly and in rotation to relieve spring tension and avoid warping the plate.
5. Remove pressure plate bolts and pressure plate and disc.

To install:

6. Lightly scuff sand flywheel face with 180 grit emery cloth, then clean with a wax and grease remover.
7. Lubricate pilot bearing with Mopar high temperature bearing grease or equivalent.
8. Check runout and operation of a new clutch disc.

➡Disc must slide freely on transmission input shaft splines.

9. With the disc on the input shaft, check face runout with dial indicator. Check runout at disc hub 6 mm (¼ in.) from outer edge of facing. Obtain another clutch disc if runout exceed 0.5 mm 0.020 in.).
10. Position clutch disc on flywheel with side marked flywheel against the flywheel.

➡If not marked, the flat side of disc hub goes towards the flywheel on the 3.7L engine and towards the transmission on 2.4L engine.

11. Insert clutch alignment tool through the clutch disc and into the pilot bearing.
12. Position clutch pressure plate over disc and on the flywheel.
13. Install pressure plate bolts finger tight.

WARNING

Use only the factory bolts to mount the pressure plate. The bolts must be the correct size. If bolts are too short, there isn't enough thread engagement, if too long bolts interfere with the Dual Mass Flywheel.

14. Tighten pressure plate bolts evenly and in rotation a few threads at a time.

WARNING

The bolts must be tightened evenly and to specified torque. Failure to follow these instructions will distort the pressure plate.

15. Tighten pressure plate bolts to 31 Nm (23 ft. lbs.) on 2.4L engines and 50 Nm (37ft. lbs.) on 3.7L engines.
16. Apply light coat of Mopar high temperature bearing grease or equivalent to clutch disc hub and splines of transmission input shaft.

➡Do not over lubricate shaft splines. This will result in grease contamination of disc.

17. Install the transmission.

2.8L DIESEL ENGINE

1. Before servicing the vehicle, refer to the precautions in the beginning of this section.
2. Remove transmission.
3. Mark position of pressure plate on flywheel with paint or a scriber for assembly reference, if clutch is not being replaced.
4. Loosen pressure plate bolts evenly and in rotation to relieve spring tension and avoid warping the plate.
5. Remove pressure plate bolts and pressure plate and disc.

To install:

6. Lightly scuff sand flywheel face with 180 grit emery cloth, then clean with a wax and grease remover.
7. Lubricate pilot bearing with Mopar high temperature bearing grease or equivalent.
8. Check runout and operation of the new clutch disc.

➡Disc must slide freely on transmission input shaft splines.

9. With the disc on the input shaft, check face runout with dial indicator. Check runout at disc hub 6 mm (¼ in.) from outer edge of facing. Obtain another clutch disc if runout exceeds 0.5 mm 0.020 in.).
10. Position clutch disc on flywheel with side marked flywheel side against the flywheel. If not marked, the flat side of disc hub goes towards the flywheel.
11. Insert clutch alignment tool through the clutch disc and into the pilot bearing.
12. Position clutch pressure plate over disc and on the flywheel.
13. Install pressure plate bolts finger tight.
14. Tighten pressure plate bolts evenly and in rotation a few threads at a time.

WARNING

The bolts must be tightened evenly and to specified torque. Failure to follow these instructions will distort the pressure plate.

15. Tighten pressure plate bolts to 54 Nm (40 ft. lbs.).
16. Apply light coat of Mopar high temperature bearing grease, or equivalent, to clutch disc hub and splines of transmission input shaft.

➡Do not over lubricate shaft splines. Failure to follow these instructions will result in contamination of disc.

17. Install the transmission.

Hydraulic Clutch System

BLEEDING

➡The clutch master cylinder, slave cylinder and fluid line are serviced only as an assembly. Bleeding is not possible.

REMOVAL & INSTALLATION

➡Clutch master cylinder, slave cylinder and connecting line are serviced as an assembly only. Components cannot be overhauled or serviced separately. Cylinders and connecting line are sealed units.

1. Before servicing the vehicle, refer to the precautions in the beginning of this section.
2. With vehicle in neutral, position vehicle on hoist
3. Remove fasteners attaching slave cylinder to clutch housing.
4. Remove slave cylinder from clutch housing.
5. Disengage clutch fluid line from body clips.
6. Lower vehicle.
7. Verify clutch master cylinder reservoir cap is tight.
8. Remove clutch master cylinder attaching nuts.
9. Disengage captured bushing on clutch master cylinder actuator from pivot pin on pedal arm.
10. Slide actuator off pivot pin.
11. Disconnect clutch interlock safety switch wires.
12. Remove clutch hydraulic linkage through engine compartment.

To install:

13. Tighten clutch master cylinder cap.
14. Position connecting line and slave

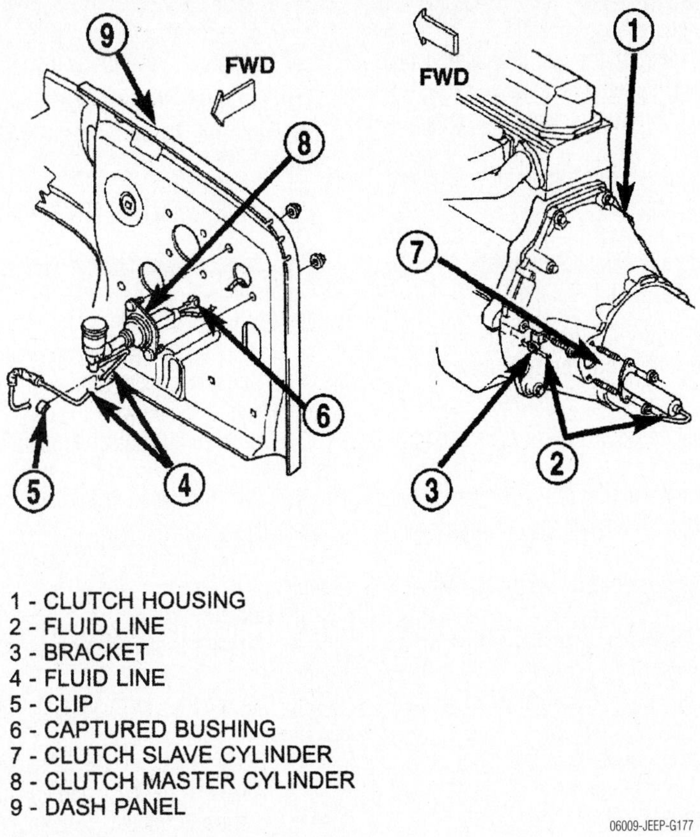

Clutch hydraulic system—Wrangler

cylinder downward past engine and next to clutch housing.

15. Position clutch master cylinder on dash panel.
16. Install clutch master cylinder actuator on clutch pedal pivot pin.
17. Install clutch master cylinder and tighten nuts to 38 Nm (28 ft. lbs.).
18. Insert slave cylinder push rod through clutch housing opening and into release lever. Be sure cap on end of rod is securely engaged in lever. Check this before installing cylinder attaching nuts.
19. Install slave cylinder and tighten nuts to 23 Nm (17 ft. lbs.).
20. Secure clutch fluid line in body and transmission clips.
21. Connect clutch interlock safety switch wires.

Transfer Case Assembly

REMOVAL & INSTALLATION

2002–03

1. Before servicing the vehicle, refer to the precautions in the beginning of this section.
2. Shift the transfer case into **N**.
3. Drain the transfer case fluid.
4. Remove or disconnect the following:
   - Negative battery cable
   - Front and rear driveshafts
   - Transmission mount and crossmember. Support the transmission with a jackstand.
   - Vehicle Speed (VSS) sensor connector
   - Indicator switch connector
   - Vacuum hose
   - Shift linkage
   - Vent hose
   - Transfer case attaching nuts
   - Transfer case

To install:

5. Install or connect the following:
   - Transfer case. Tighten the nuts to 26 ft. lbs. (35 Nm).
   - Indicator switch connector
   - Vacuum hose
   - Vent hose
   - Shift linkage
   - VSS sensor connector
   - Transmission mount and crossmember. Torque the bolts to 30 ft. lbs. (41 Nm).
   - Front and rear driveshafts
   - Negative battery cable
6. Fill the transfer case with the proper fluid.
7. Start the vehicle, check for leaks and repair if necessary.

2004 Grand Cherokee w/NV147

1. Before servicing the vehicle, refer to the precautions in the beginning of this section.
2. Raise vehicle.
3. Remove transfer case drain plug and drain transfer case lubricant.
4. Mark front and rear propeller shaft yokes for alignment reference.
5. Support transmission with jack stand.
6. Remove rear crossmember and skid plate, if equipped.
7. Disconnect front propeller shaft from transfer case at companion flange. Remove rear propeller shaft from vehicle.

➡Do not allow driveshafts to hang at attached end. Damage to joint can result.

8. Disconnect transfer case vent hose.
9. Support transfer case with transmission jack.
10. Secure transfer case to jack with chains.
11. Remove nuts attaching transfer case to transmission.
12. Pull transfer case and jack rearward to disengage transfer case.
13. Remove transfer case from under vehicle.

To install:

14. Mount transfer case on a transmission jack.
15. Secure transfer case to jack with chains.
16. Position transfer case under vehicle.
17. Align transfer case and transmission shafts and install transfer case on transmission.
18. Install and tighten transfer case attaching nuts to 35 Nm (26 ft. lbs.) torque.
19. Connect front propeller shaft and install rear propeller shaft.
20. Fill transfer case with correct fluid. Check transmission fluid level. Correct as necessary.
21. Install rear crossmember and skid plate, if equipped. Tighten crossmember bolts to 41 Nm (30 ft. lbs.) torque.
22. Remove transmission jack and support stand.
23. Lower vehicle and verify transfer case shift operation.

2004 Grand Cherokee w/NV242

1. Before servicing the vehicle, refer to the precautions in the beginning of this section.

2. Shift transfer case into NEUTRAL.
3. Raise vehicle.
4. Remove transfer case drain plug and drain transfer case lubricant.
5. Mark front and rear propeller shaft yokes for alignment reference.
6. Support transmission with jack stand.
7. Remove rear crossmember and skid plate, if equipped.
8. Disconnect front/rear propeller shafts at transfer case.
9. Disconnect transfer case cable from range lever.
10. Disconnect transfer case vent hose and transfer case position sensor.
11. Support transfer case with transmission jack.
12. Secure transfer case to jack with chains.
13. Remove nuts attaching transfer case to transmission.
14. Pull transfer case and jack rearward to disengage transfer case.
15. Remove transfer case from under vehicle.

To install:

16. Mount transfer case on a transmission jack.
17. Secure transfer case to jack with chains.
18. Position transfer case under vehicle.
19. Align transfer case and transmission shafts and install transfer case on transmission.
20. Install and tighten transfer case attaching nuts to 35 Nm (26 ft. lbs.) torque.
21. Align and connect propeller shafts.
22. Fill transfer case with correct fluid. Check transmission fluid level. Correct as necessary.
23. Install rear crossmember and skid plate, if equipped. Tighten crossmember bolts to 41 Nm (30 ft. lbs.) torque.
24. Remove transmission jack and support stand.
25. Connect shift rod to transfer case range lever.
26. Connect transfer case vent hose and transfer case position sensor.
27. Adjust transfer case shift cable.
28. Lower vehicle and verify transfer case shift operation.

2004 Grand Cherokee w/NV247

1. Before servicing the vehicle, refer to the precautions in the beginning of this section.
2. Shift transfer case into NEUTRAL.
3. Raise vehicle.
4. Remove transfer case drain plug and drain transfer case lubricant.
5. Mark front and rear propeller shaft yokes for alignment reference.
6. Support transmission with jack stand.
7. Remove rear crossmember and skid plate, if equipped.
8. Disconnect front propeller shaft from transfer case at companion flange. Remove rear propeller shaft from vehicle.

➡Do not allow driveshafts to hang at attached end. Damage to joint can result.

9. Disconnect transfer case cable from range lever.
10. Disconnect transfer case vent hose.
11. Support transfer case with transmission jack.
12. Secure transfer case to jack with chains.
13. Remove nuts attaching transfer case to transmission.
14. Pull transfer case and jack rearward to disengage transfer case.
15. Remove transfer case from under vehicle.

To install:

16. Mount transfer case on a transmission jack.
17. Secure transfer case to jack with chains.
18. Position transfer case under vehicle.
19. Align transfer case and transmission shafts and install transfer case on transmission.
20. Install and tighten transfer case attaching nuts to 35 Nm (26 ft. lbs.).
21. Connect front propeller shaft and install rear propeller shaft.
22. Fill transfer case with correct fluid. Check transmission fluid level. Correct as necessary.
23. Install rear crossmember and skid plate, if equipped. Tighten crossmember bolts to 41 Nm (30 ft. lbs.).
24. Remove transmission jack and support stand.
25. Verify transfer case is in NEUTRAL. Connect shift cable to transfer case range lever.
26. Lower vehicle and verify transfer case shift operation.
27. Adjust the transfer case shift cable, if necessary.

2005 Grand Cherokee w/NV140

1. Before servicing the vehicle, refer to the precautions in the beginning of this section.
2. Raise vehicle.
3. Remove transfer case drain plug and drain the transfer case lubricant.

➡Do not allow driveshafts to hang at attached end. Damage to joint can result.

4. Remove the front and rear propeller shafts.
5. Support transmission with jack stand.
6. Remove rear crossmember and skid plate, if equipped.
7. Disconnect transfer case vent hose.
8. Support transfer case with transmission jack and secure with chains.
9. Remove nuts attaching transfer case to transmission.
10. Pull transfer case and jack rearward to disengage transfer case.
11. Remove transfer case from under vehicle.

To install:

12. Mount transfer case on a transmission jack.
13. Secure transfer case to jack with chains.
14. Position transfer case under vehicle.
15. Align transfer case and transmission shafts and install transfer case onto the transmission.
16. Install and tighten transfer case attaching nuts to 35 Nm (26 ft. lbs.) torque.
17. Connect front and rear driveshafts. Torque the front driveshaft-to-transfer case bolts to 24 ft. lbs. (32 Nm). Torque the rear driveshaft-to-transfer case bolts to 80 ft. lbs. (108 Nm).
18. Fill transfer case with correct fluid. Check transmission fluid level. Correct as necessary.
19. Install the transfer case fill plug. Tighten the plug to 20–34 Nm (15–25 ft. lbs.).
20. Install rear crossmember and skid plate, if equipped. Tighten crossmember bolts to 41 Nm (30 ft. lbs.) torque.
21. Remove transmission jack and support stand.
22. Lower vehicle and verify transfer case shift operation.

2005 Grand Cherokee w/NV245

1. Before servicing the vehicle, refer to the precautions in the beginning of this section.
2. Shift transfer case into NEUTRAL.
3. Raise vehicle.
4. Remove transfer case drain plug and drain transfer case lubricant.
5. Support transmission with jack stand.

6. Remove rear crossmember and skid plate, if equipped.
7. Disconnect front propeller shaft from transfer case at companion flange. Remove rear propeller shaft from vehicle.

➡Do not allow driveshafts to hang at attached end. Damage to joint can result.

8. Disconnect the transfer case shift motor and mode sensor connector.
9. Disconnect transfer case vent hose.
10. Support transfer case with transmission jack.
11. Secure transfer case to jack with chains.
12. Remove nuts attaching transfer case to transmission.
13. Pull transfer case and jack rearward to disengage transfer case.
14. Remove transfer case from under vehicle.

To install:

15. Mount transfer case on a transmission jack.
16. Secure transfer case to jack with chains.
17. Position transfer case under vehicle.
18. Align transfer case and transmission shafts and install transfer case onto transmission.
19. Install and tighten transfer case attaching nuts to 35 Nm (26 ft. lbs.) torque.
20. Connect the transfer case vent hose to the transfer case.
21. Install rear crossmember and skid plate, if equipped.
22. Remove transmission jack and support stand.
23. Connect front and rear driveshafts. Torque the front driveshaft-to-transfer case bolts to 24 ft. lbs. (32 Nm). Torque the rear driveshaft-to-transfer case bolts to 80 ft. lbs. (108 Nm).
24. Fill transfer case with correct fluid. Check transmission fluid level. Correct as necessary.
25. Install the transfer case fill plug. Tighten the plug to 20–34 Nm (15–25 ft. lbs.).
26. Connect the shift motor and mode sensor wiring connector.
27. Lower vehicle and verify transfer case shift operation.

2004–05 Wrangler w/NV231

1. Before servicing the vehicle, refer to the precautions in the beginning of this section.
2. Shift transfer case into NEUTRAL.
3. Raise vehicle.
4. Drain transfer case lubricant.
5. Mark front and rear propeller shaft yokes for alignment reference.
6. Support transmission with jack stand.
7. Remove rear crossmember, or skid plate.
8. Disconnect front/rear propeller shafts at transfer case.
9. Disconnect vehicle speed sensor wires.
10. Disconnect transfer case linkage rod from range lever.
11. Disconnect transfer case vent hose and indicator switch harness, if necessary.
12. Support transfer case with transmission jack.
13. Secure transfer case to jack with chains.
14. Remove nuts attaching transfer case to transmission.
15. Pull transfer case and jack rearward to disengage transfer case.
16. Remove transfer case from under vehicle.

To install:

17. Mount transfer case on a transmission jack.
18. Secure transfer case to jack with chains.
19. Position transfer case under vehicle.
20. Align transfer case and transmission shafts and install transfer case on transmission.
21. Install and tighten transfer case attaching nuts to 35 Nm (26 ft. lbs.) torque.
22. Connect vehicle speed sensor wires, and vent hose.
23. Connect indicator switch harness to transfer case switch, if necessary. Secure wire harness to clips on transfer case.
24. Align and connect propeller shafts.
25. Fill transfer case with correct fluid. Check transmission fluid level. Correct as necessary.
26. Install rear crossmember, or skid plate. Tighten crossmember bolts to 41 Nm (30 ft. lbs.).
27. Remove transmission jack and support stand.
28. Connect shift rod to transfer case range lever.
29. Adjust transfer case shift linkage.
30. Lower vehicle and verify transfer case shift operation.

2004 Wrangler w/NV242

1. Before servicing the vehicle, refer to the precautions in the beginning of this section.
2. Raise and support vehicle.
3. Remove skid plate, if equipped.
4. Position drain oil container under transfer case.
5. Remove transfer case drain plug and drain lubricant into container.
6. Disconnect vent hose and vacuum harness at transfer case switch.
7. Disconnect shift rod from grommet in transfer case shift lever, or from floor shift arm whichever provides easy access. Use pliers to press rod out of lever grommet.
8. Support transmission with jack stand.
9. Remove rear crossmember.
10. Mark front and rear propeller shafts for assembly reference.
11. Remove front and rear propeller shafts.
12. Support transfer case with suitable jack. Secure transfer case to jack with safety chains.
13. Remove nuts attaching transfer case to transmission.
14. Move transfer case assembly rearward until free of transmission output shaft.
15. Lower jack and move transfer case from under vehicle.

To install:

16. Align and seat transfer case on transmission. Be sure transfer case input gear splines are aligned with transmission output shaft. Align splines by rotating transfer case rear output shaft yoke if necessary.

➡Do not install any transfer case attaching nuts until the transfer case is completely seated against the transmission.

17. Install and tighten transfer case attaching nuts. Tighten nuts to 30–41 Nm (20–30 ft. lbs.).
18. Install rear crossmember.
19. Remove jack stand from under transmission.
20. Align and connect propeller shafts.
21. Connect vacuum harness and vent hose.
22. Connect shift rod to transfer case lever or floor shift arm. Use pliers to press rod back into lever grommet.
23. Adjust shift linkage, if necessary.
24. Fill transfer case with recommended transmission fluid and install fill plug.
25. Install skid plate, if equipped.
26. Lower vehicle

2005 Wrangler w/NV241

1. Before servicing the vehicle, refer to the precautions in the beginning of this section.

2. Raise and support vehicle.
3. Remove skid plate, if equipped.
4. Position drain oil container under transfer case.
5. Remove transfer case drain plug and drain lubricant into container.
6. Disconnect vent hose and vacuum harness at transfer case switch.
7. Disconnect shift rod from grommet in transfer case shift lever, or from floor shift arm whichever provides easy access. Use channel lock style pliers to press rod out of lever grommet.
8. Support transmission with jackstand.
9. Remove rear crossmember.
10. Mark front and rear propeller shafts for assembly reference.
11. Remove front and rear propeller shafts.
12. Support transfer case with suitable jack. Secure transfer case to jack with safety chains.
13. Remove nuts attaching transfer case to transmission.
14. Move transfer case assembly rearward until free of transmission output shaft.
15. Lower jack and move transfer case from under vehicle.

To install:

16. Align and seat transfer case on transmission. Be sure transfer case input gear splines are aligned with transmission output shaft. Align splines by rotating transfer case rear output shaft yoke if necessary. Do not install any transfer case attaching nuts until the transfer case is completely seated against the transmission.
17. Install and tighten transfer case attaching nuts. Tighten nuts to 30–41 Nm (20–30 ft. lbs.).
18. Install rear crossmember.
19. Remove jackstand from under transmission.
20. Align and connect propeller shafts. Torque as follows:
    - Front driveshaft to axle: 14 ft. lbs. (19 Nm)
    - Front driveshaft to transfer case: 20 ft. lbs. (27 Nm)
    - Rear driveshaft to axle: 14 ft. lbs. (19 Nm)
    - Rear driveshaft to transfer case (Rubicon): 85 ft. lbs. (115 Nm)
21. Connect vacuum harness and vent hose.
22. Connect shift rod to transfer case lever or floor shift arm. Use channel lock style pliers to press rod back into lever grommet.
23. Adjust shift linkage, if necessary.
24. Fill transfer case with recommended transmission fluid and install fill plug.
25. Install skid plate, if equipped.
26. Lower vehicle

2004 Liberty w/NV231

1. Before servicing the vehicle, refer to the precautions in the beginning of this section.
2. Shift transfer case into NEUTRAL.
3. Raise vehicle.
4. Remove skid plate.
5. Drain transfer case lubricant.
6. Mark front and rear propeller shaft yokes for alignment reference.
7. Disconnect front/rear propeller shafts at transfer case.
8. Disconnect transfer case position sensor connector.
9. Disconnect transfer case shift cable at the range lever.
10. Disconnect the transfer case shift cable from the shift cable bracket.
11. Disconnect transfer case vent hose.
12. Support transfer case with transmission jack.
13. Secure transfer case to jack with chains.
14. Remove nuts attaching transfer case to transmission.
15. Pull transfer case and jack rearward to disengage transfer case.
16. Remove transfer case from under vehicle.

To install:

17. Mount transfer case on a transmission jack.
18. Secure transfer case to jack with chains.
19. Position transfer case under vehicle.
20. Align transfer case and transmission shafts and install transfer case on transmission.
21. Install and tighten transfer case attaching nuts to 35 Nm (26 ft. lbs.) torque.
22. Connect vent hose.
23. Connect transfer case position sensor connector to sensor.
24. Align and connect propeller shafts.
25. Fill transfer case with correct fluid. Check transmission fluid level. Correct as necessary.
26. Install skid plate.
27. Remove transmission jack and support stand.
28. Connect shift cable to transfer case range lever.
29. Lower vehicle and verify transfer case shift operation.

2004 Liberty w/NV242

1. Before servicing the vehicle, refer to the precautions in the beginning of this section.
2. Shift transfer case into NEUTRAL.
3. Raise vehicle.
4. Remove skid plate.
5. Drain transfer case lubricant.
6. Mark front and rear propeller shaft yokes for alignment reference.
7. Disconnect front/rear propeller shafts at transfer case.
8. Disconnect transfer case position sensor connector.
9. Disconnect transfer case shift cable at the range lever.
10. Disconnect the transfer case shift cable from the shift cable bracket.
11. Disconnect transfer case vent hose.
12. Support transfer case with transmission jack.
13. Secure transfer case to jack with chains.
14. Remove nuts attaching transfer case to transmission.
15. Pull transfer case and jack rearward to disengage transfer case.
16. Remove transfer case from under vehicle

To install:

17. Mount transfer case on a transmission jack.
18. Secure transfer case to jack with chains.
19. Position transfer case under vehicle.
20. Align transfer case and transmission shafts and install transfer case on transmission.
21. Install and tighten transfer case attaching nuts to 35 Nm (26 ft. lbs.) torque.
22. Connect vent hose.
23. Connect transfer case position sensor connector to sensor.
24. Align and connect propeller shafts.
25. Fill transfer case with correct fluid. Check transmission fluid level. Correct as necessary.
26. Install skid plate
27. Remove transmission jack and support stand.
28. Connect shift cable to transfer case range lever.
29. Lower vehicle and verify transfer case shift operation.

2005 Liberty w/NV231

1. Before servicing the vehicle, refer to the precautions in the beginning of this section.
2. Shift transfer case into NEUTRAL.
3. Raise the vehicle.
4. Remove skid plate.
5. Drain transfer case lubricant.
6. Mark front and rear propeller shaft yokes for alignment reference.

7. Disconnect front/rear propeller shafts at transfer case.
8. Disconnect transfer case position sensor connector.
9. Disconnect transfer case shift cable at the range lever.
10. Disconnect the transfer case shift cable from the shift cable bracket.
11. Disconnect transfer case vent hose.
12. Support transfer case with transmission jack.
13. Secure transfer case to jack with chains.
14. Remove the nuts attaching transfer case to transmission.
15. Pull transfer case and jack rearward to disengage transfer case.
16. Remove transfer case from under vehicle.

To install:

17. Mount transfer case on a transmission jack.
18. Secure transfer case to jack with chains.
19. Position transfer case under vehicle.
20. Align transfer case and transmission shafts and install transfer case on transmission.
21. Install and tighten transfer case attaching nuts to 35 Nm (26 ft. lbs.) torque.
22. Connect vent hose.
23. Connect transfer case position sensor connector to sensor.
24. Align and connect propeller shafts. Torque the rear companion flange nuts to 85 ft. lbs. (115 Nm), the front companion flange nuts to 22 ft. lbs. (30 Nm).
25. Fill transfer case with correct fluid. Check transmission fluid level. Correct as necessary.
26. Install skid plate.
27. Remove transmission jack and support stand.
28. Connect shift cable to transfer case range lever.
29. Lower vehicle and verify transfer case shift operation.

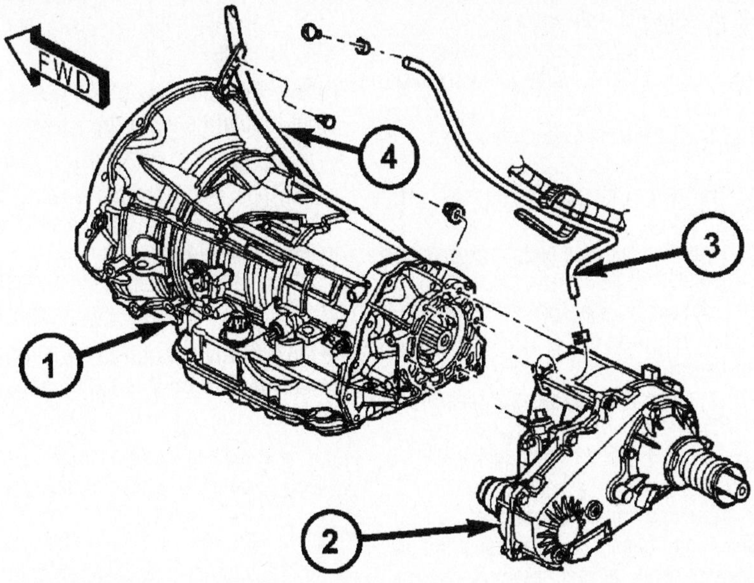

Typical transfer case mounting—2004–05 Liberty

2005 Liberty w/NV241 GENII

1. Before servicing the vehicle, refer to the precautions in the beginning of this section.
2. Shift transfer case into NEUTRAL.
3. Raise vehicle.
4. Remove skid plate.
5. Drain transfer case lubricant.
6. Mark front and rear propeller shaft yokes for alignment reference.
7. Remove the front/rear propeller shafts at transfer case.
8. Disconnect transfer case position sensor connector from the position sensor.
9. Disconnect transfer case shift cable at the range lever.
10. Disconnect the transfer case shift cable from the shift cable bracket.
11. Disconnect transfer case vent hose.
12. Support transfer case with transmission jack.
13. Secure transfer case to jack with chains.
14. Remove nuts attaching transfer case to transmission.
15. Pull transfer case and jack rearward to disengage transfer case.
16. Remove transfer case from under vehicle.

To install:

17. Mount transfer case on a transmission jack.
18. Secure transfer case to jack with chains.
19. Position transfer case under vehicle.
20. Align transfer case and transmission shafts and install transfer case on transmission.
21. Install and tighten transfer case attaching nuts to 35 Nm (26 ft. lbs.) torque.
22. Connect vent hose.
23. Connect transfer case position sensor connector to sensor.
24. Align and connect propeller shafts. Torque the rear companion flange nuts to 85 ft. lbs. (115 Nm), the front companion flange nuts to 22 ft. lbs. (30 Nm).
25. Connect shift cable to transfer case range lever.
26. Fill transfer case with correct fluid. Check transmission fluid level. Correct as necessary.
27. Install skid plate.
28. Remove transmission jack and support stand.
29. Lower vehicle and verify transfer case shift operation.

2005 Liberty w/NV242

1. Before servicing the vehicle, refer to the precautions in the beginning of this section.
2. Shift transfer case into NEUTRAL.
3. Raise vehicle.
4. Remove skid plate.
5. Drain transfer case lubricant.
6. Mark front and rear propeller shaft yokes for alignment reference.
7. Disconnect front/rear propeller shafts at transfer case.
8. Disconnect transfer case position sensor connector.
9. Disconnect transfer case shift cable at the range lever.

10. Disconnect the transfer case shift cable from the shift cable bracket.
11. Disconnect transfer case vent hose.
12. Support transfer case with transmission jack.
13. Secure transfer case to jack with chains.
14. Remove nuts attaching transfer case to transmission.
15. Pull transfer case and jack rearward to disengage transfer case.
16. Remove transfer case from under vehicle.

To install:

17. Mount transfer case on a transmission jack.
18. Secure transfer case to jack with chains.
19. Position transfer case under vehicle.
20. Align transfer case and transmission shafts and install transfer case on transmission.
21. Install and tighten transfer case attaching nuts to 35 Nm (26 ft. lbs.) torque.
22. Connect vent hose.
23. Connect transfer case position sensor connector to sensor.
24. Align and connect propeller shafts. Torque the rear companion flange nuts to 85 ft. lbs. (115 Nm), the front companion flange nuts to 22 ft. lbs. (30 Nm).
25. Fill transfer case with correct fluid. Check transmission fluid level. Correct as necessary.
26. Install skid plate.
27. Remove transmission jack and support stand.
28. Connect shift cable to transfer case range lever.
29. Lower vehicle and verify transfer case shift operation.

CV-Joints

CV-joints are serviced only as an assembly, as are halfshafts and CV-joints.

Halfshafts

REMOVAL & INSTALLATION

Liberty

1. Before servicing the vehicle, refer to the precautions in the beginning of this section.
2. Remove or disconnect the following:
   - Wheel
   - Hub nut
   - Stabilizer link
   - Lower clevis bolt
   - Ball joint from the lower arm
1. Pull on the hub and push the halfshaft from the knuckle

➡The right side has a splined axle shaft that will stay in the axle.

To install:

2. Apply a light coating of wheel bearing grease on the splines of the inner joint, and in the hub bearing bore.
3. Install or connect the following:
   - Halfshaft on the axle shaft splines

➡Push firmly enough to engage the snapring. Pull on it to verify engagement.

   - Halfshaft into the knuckle
   - Lower ball joint and pinch bolt
   - Lower clevis bolt
   - Stabilizer link
   - Hub nut. Torque to 100 ft. lbs. (136 Nm).
   - Wheel

Grand Cherokee

1. Before servicing the vehicle, refer to the precautions in the beginning of this section.
2. With vehicle in neutral, position vehicle on hoist.
3. Remove halfshaft hub/bearing nut.
4. Remove wheel speed sensor from hub/bearing.
5. Remove brake calipers bolts and remove calipers from caliper adapters.
6. Remove lower stabilizer link bolt from control arm.
7. Remove outer tie rod end nuts and separate tie rods from knuckles with the proper tool.
8. Remove upper ball joint nuts and separate ball joints from knuckles with the proper tool.
9. Remove shock clevis bolt and nut from lower control arm
10. Lean the knuckle out and push halfshaft out of the hub/bearing.
11. Pry halfshafts from axle/axle tube with pry bar.

To install:

12. Install halfshaft on the axle and through the hub/bearing. Verify halfshaft has engaged.
13. Install shock clevis on lower control arm and tighten nut to 60 ft. lbs. (81 Nm).
14. Install upper control arm on knuckle. Torque to 55 ft. lbs. (75 Nm).
15. Install tie rod end on knuckle. Torque to 70 ft. lbs. (95 Nm).

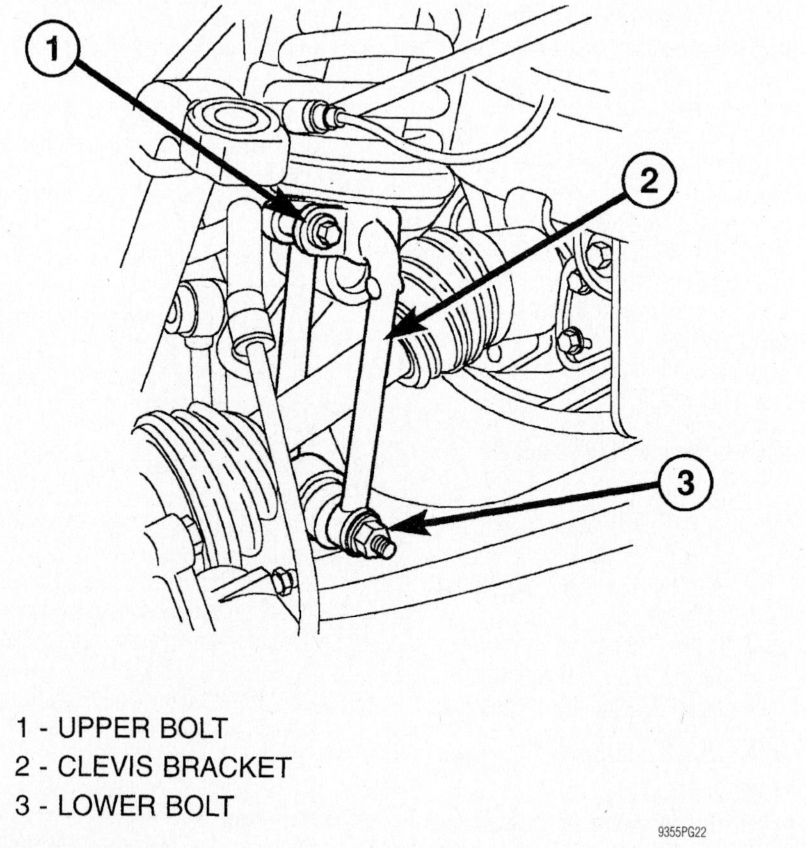

Clevis bracket—Liberty

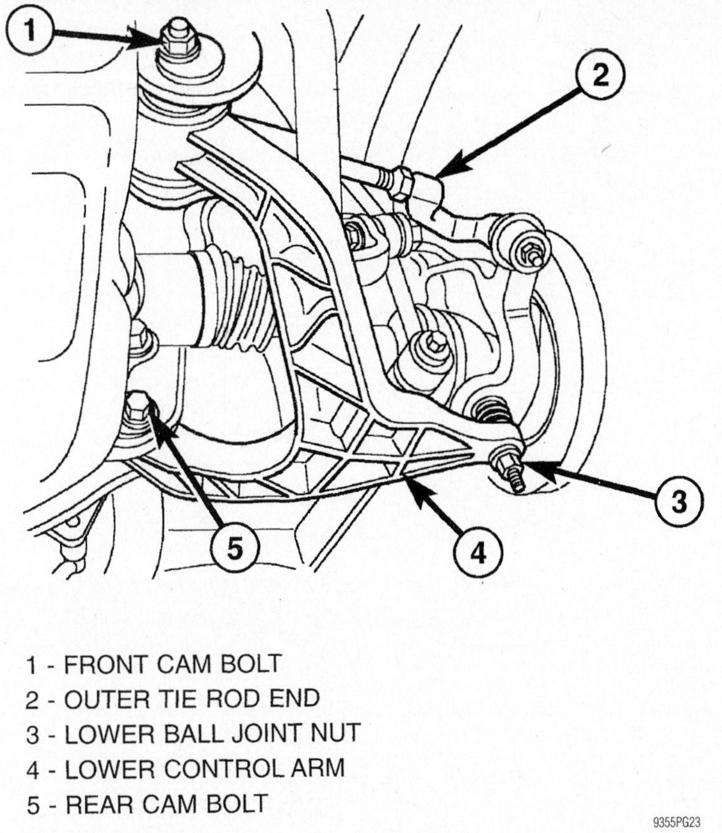

Lower control arm—Liberty

16. Install stabilizer link on lower control arm. Torque to 85 ft. lbs. (115 Nm).
17. Install caliper on caliper adapter.
18. Install wheel speed sensor on the hub/bearing.
19. Install halfshaft hub/bearing nut and tighten to 135 Nm (100 ft. lbs.).

Front Axle Shaft, Bearing and Seal

REMOVAL & INSTALLATION

Liberty

RIGHT SIDE SHAFT ONLY

1. Before servicing the vehicle, refer to the precautions in the beginning of this section.
2. Remove or disconnect the following:
   - Negative battery cable
   - Right front wheel
   - Halfshaft
   - Snapring from the axle
   - Axle shaft using remover 8420A and a slide hammer

To install:

3. Coat the bearing bore and seal lip with gear oil.
4. Install or connect the following:
   - Axle shaft and snapring
   - Halfshaft
   - Front wheel
   - Negative battery cable

RIGHT SIDE SEAL AND BEARING ONLY

1. Before servicing the vehicle, refer to the precautions in the beginning of this section.
2. Remove or disconnect the following:
   - Negative battery cable
   - Right front wheel
   - Halfshaft
   - Snapring from the axle
   - Axle shaft using remover 8420A and a slide hammer
   - Seal, using Remover 7794A and a slide hammer
   - Bearing, using Remover 7794A and a slide hammer

To install:

3. Coat the bearing bore and seal lip with gear oil.
4. Install or connect the following:
   - New seal, using Installer 8806 and handle C4171, or their equivalents
   - Axle shaft and snapring
   - Halfshaft

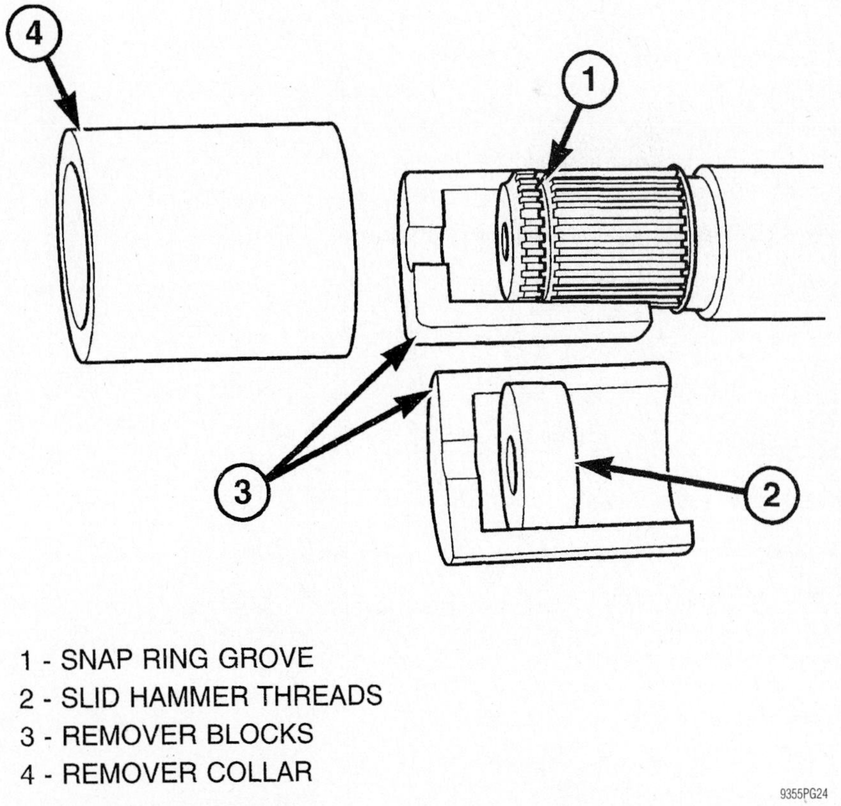

Axle shaft remover tool—Liberty

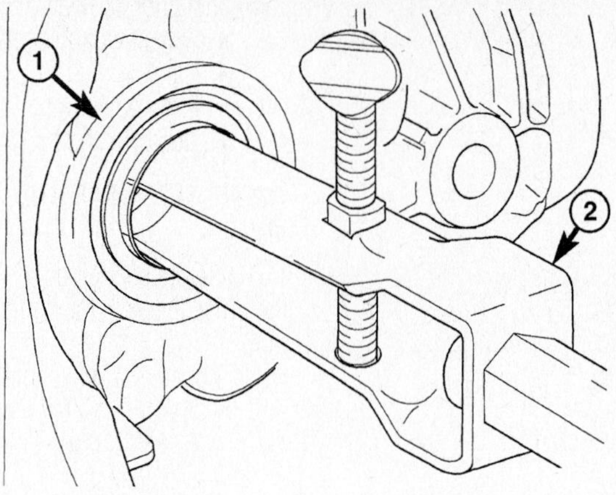

Seal removal—Liberty

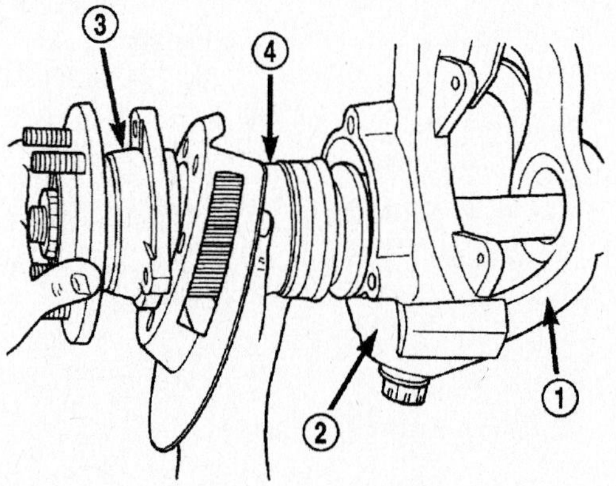

Front axle shaft removal—2005 Wrangler w/181FBI axle

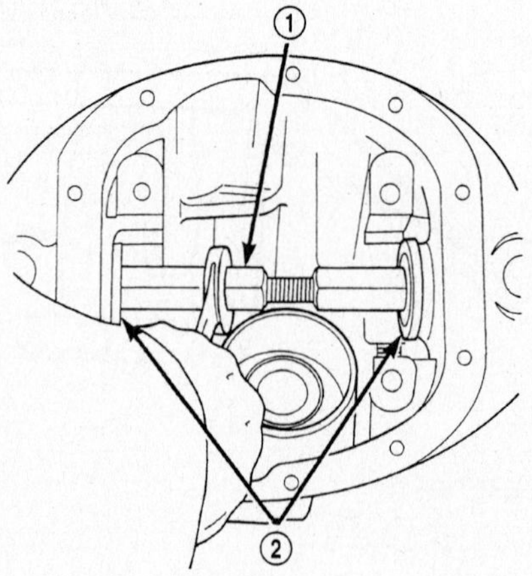

Axle seal installation

- Front wheel
- Negative battery cable

2002–04 Grand Cherokee; 2002–05 Wrangler

FRONT AXLE SHAFT

1. Before servicing the vehicle, refer to the precautions in the beginning of this section.
2. Remove or disconnect the following:
   - Negative battery cable
   - Front wheel
   - Brake caliper and rotor
   - Wheel speed sensor, if equipped
   - Axle hub nut
   - Wheel bearing and hub assembly
   - Axle shaft

To install:

3. Install or connect the following:
   - Axle shaft
   - Wheel bearing and hub assembly
   - Axle hub nut. Tighten the nut to 175 ft. lbs. (237 Nm).
   - Wheel speed sensor, if equipped
   - Brake caliper and rotor
   - Front wheel
   - Negative battery cable

FRONT SHAFT SEAL

1. Before servicing the vehicle, refer to the precautions in the beginning of this section.
2. Remove or disconnect the following:
   - Front axle shafts
   - Differential cover
   - Differential and ring gear assembly
   - Axle seals

To install:

3. Press the axle seals into the differential housing with Turnbuckle 6797 and Disc set 8110.
4. Install or connect the following:
   - Differential and ring gear assembly. Tighten the bearing cap bolts to 45 ft. lbs. (61 Nm).
   - Differential cover. Tighten the bolts to 30 ft. lbs. (41 Nm).
   - Front axle shafts
5. Fill the axle assembly with gear oil and check for leaks.

2005 Grand Cherokee

AXLE SHAFT

1. Before servicing the vehicle, refer to the precautions in the beginning of this section.
2. Remove halfshaft from vehicle.
3. Clean axle seal area.
4. Remove snapring and O-ring from the axle shaft.
5. Remove axle (1) with Remover 8420A (2) (3) (4) and slide hammer.

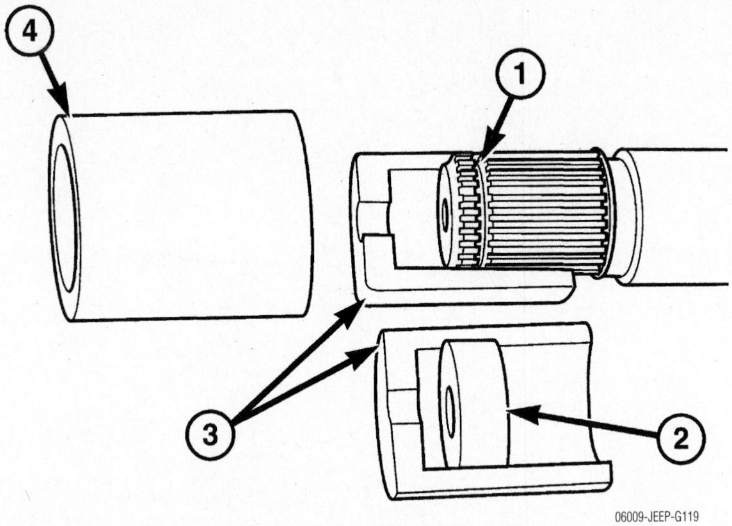

Remove axle (1) with Remover 8420A (2) (3) (4) and slide hammer

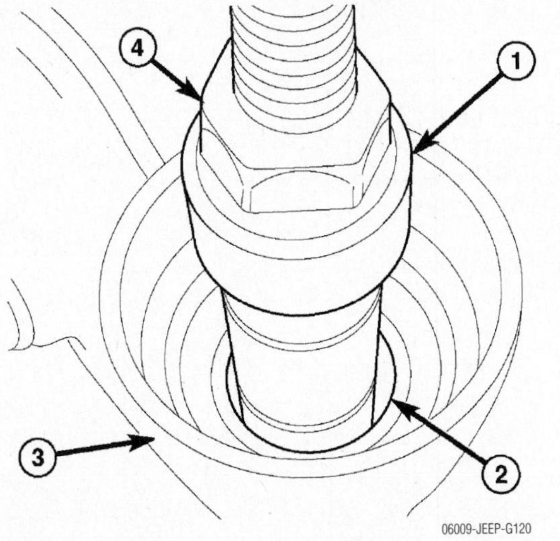

Install axle bearing remover C-4660-A (1) into the axle bearing (2) in the axle tube (3) and tighten remover nut (4)

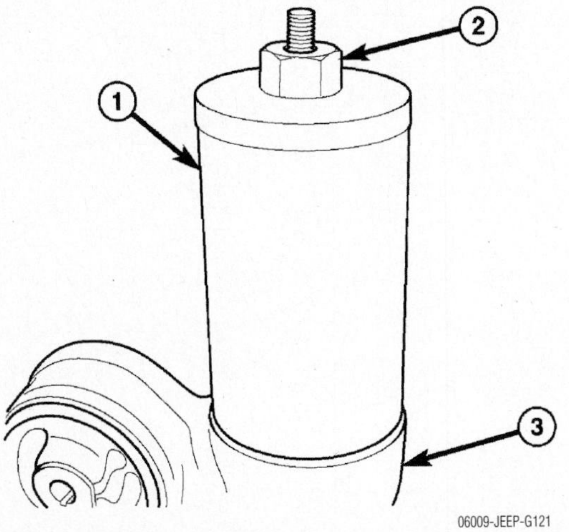

Install bearing Receiver 9537 (1) and nut (2) on remover and tube (3). Tighten nut (2) and draw bearing into receiver (1)

To install:

6. Lubricate bearing bore and seal lip with gear lubricant.
7. Install axle shaft and engage shaft into side gear. Push firmly on axle shaft to engage snapring.
8. Check the differential fluid level and add fluid if necessary.
9. Install halfshaft.
10. Install skid plate, if necessary.

AXLE SHAFT SEAL

1. Before servicing the vehicle, refer to the precautions in the beginning of this section.
2. Remove axle shaft.
3. Remove axle shaft tube seal with a seal puller and slide hammer.

To install:

4. Install axle seal with a seal driver.
5. Install axle shaft.

AXLE SHAFT BEARING

1. Before servicing the vehicle, refer to the precautions in the beginning of this section.
2. Remove axle shaft.
3. Remove axle shaft.
4. Install axle bearing remover C-4660-A (1) into the axle bearing (2) in the axle tube (3) and tighten remover nut (4).
5. Install bearing Receiver 9537 (1) and nut (2) on remover and tube (3). Tighten nut (2) and draw bearing into receiver (1).

To install:

6. Install axle bushing with a driver.
7. Install axle shaft and seal.

Rear Axle Shaft, Bearing and Seal

REMOVAL & INSTALLATION

C-Clip Type Rear Axle Shaft

2002–05 EXC. 2005 LIBERTY

1. Before servicing the vehicle, refer to the precautions in the beginning of this section.
2. Remove or disconnect the following:
   - Negative battery cable
   - Rear wheel
   - Brake drum
   - Differential cover
   - Differential gear shaft retainer
   - Differential gear shaft
   - C-clip
   - Axle shaft
   - Axle seal
   - Axle bearing

To install:

3. Install or connect the following:
   - Axle bearing

- Axle seal
- Axle shaft
- C-clip
- Differential gear shaft. Use Loctite® and tighten the retainer to 14 ft. lbs. (19 Nm).
- Differential cover. Tighten the bolts to 30 ft. lbs. (41 Nm).
- Brake drum
- Rear wheel

4. Fill the axle assembly with gear oil and check for leaks.

2005 LIBERTY

1. Before servicing the vehicle, refer to the precautions in the beginning of this section.
2. With vehicle in neutral, position vehicle on hoist.
3. Remove differential housing cover and drain lubricant.
4. Rotate differential case so pinion mate shaft lock screw is accessible. Remove lock screw and pinion mate shaft from differential case.
5. Push axle shaft inward and remove axle shaft C-lock from the axle shaft.
6. Remove axle shaft from side gear and axle tube.
7. Remove axle shaft seal from axle tube with a seal pick.

To install:

8. Remove any old sealer/burrs from axle tube.
9. Coat new seal lip with axle lubricant and install seal with a seal driver.
10. Lubricate bearing bore and seal lip with gear lubricant.
11. Install axle shaft through seal, bearing and engage into side gear splines.
12. Install C-lock in axle shaft end, then push axle shaft outward to seat C-lock in side gear.
13. Install pinion mate shaft into differential case and through thrust washers and differential pinions.
14. Align hole in shaft with hole in the differential case and install lock screw with Loctite on the threads. Tighten lock screw to 11 Nm (8 ft. lbs.).
15. Install the differential cover.
16. Install rear brake components.

Non C-Clip Type Rear Axle Shaft

⁂ CAUTION

Never reuse axle retaining bolts and nuts. Used nuts can loosen. Failure to follow these instructions may result in personal injury.

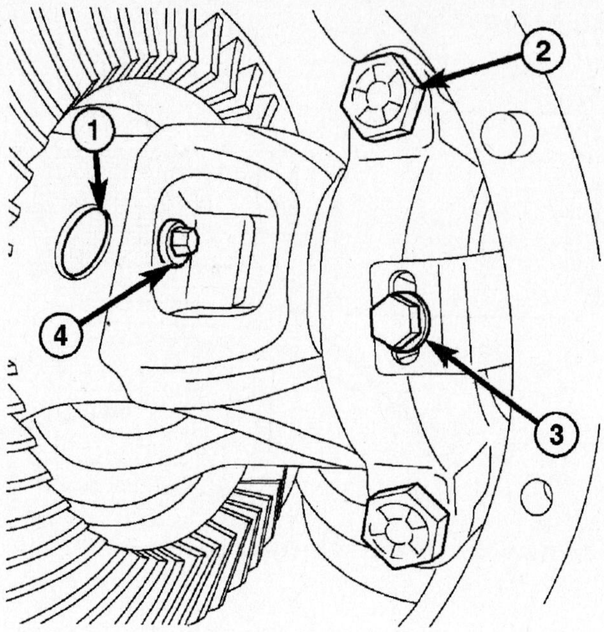

Typical rear axle shaft lock screw

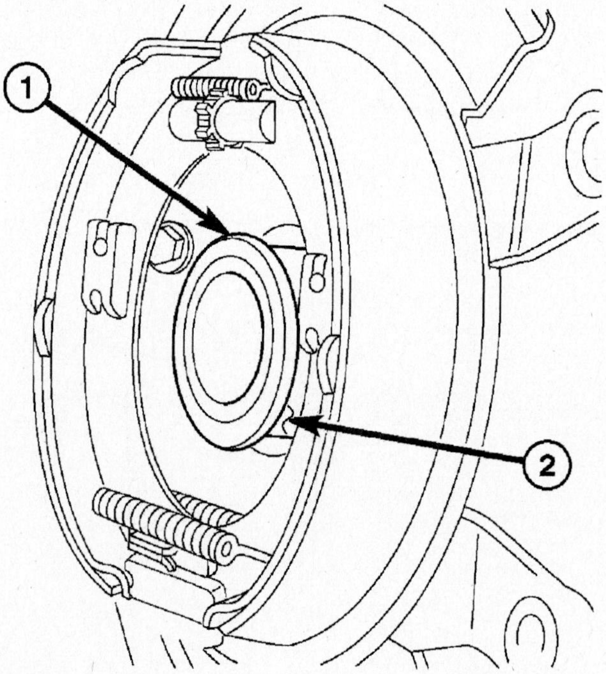

Typical rear axle shaft seal

2002–05 EXC. 2005 GRAND CHEROKEE

1. Before servicing the vehicle, refer to the precautions in the beginning of this section.
2. Remove or disconnect the following:
   - Rear wheel
   - Brake caliper and rotor, if equipped
   - Brake drum, if equipped
   - Axle retainer nuts
   - Axle shaft, seal and bearing assembly
3. Split the bearing retainer with a chisel and remove the retainer ring.
4. Press the bearing off the axle shaft.
5. Remove the axle seal and retaining plate.

To install:

6. Install the retaining plate and axle seal onto the axle shaft.
7. Pack the wheel bearing with axle grease and press the bearing on to the axle shaft.
8. Press the retaining ring onto the axle shaft.
9. Install or connect the following:
   - Axle shaft, seal and bearing assembly. Tighten the nuts to 45 ft. lbs. (61 Nm).
   - Brake caliper and rotor, if equipped
   - Brake drum, if equipped
   - Rear wheel
10. Fill the axle assembly with gear oil and check for leaks.

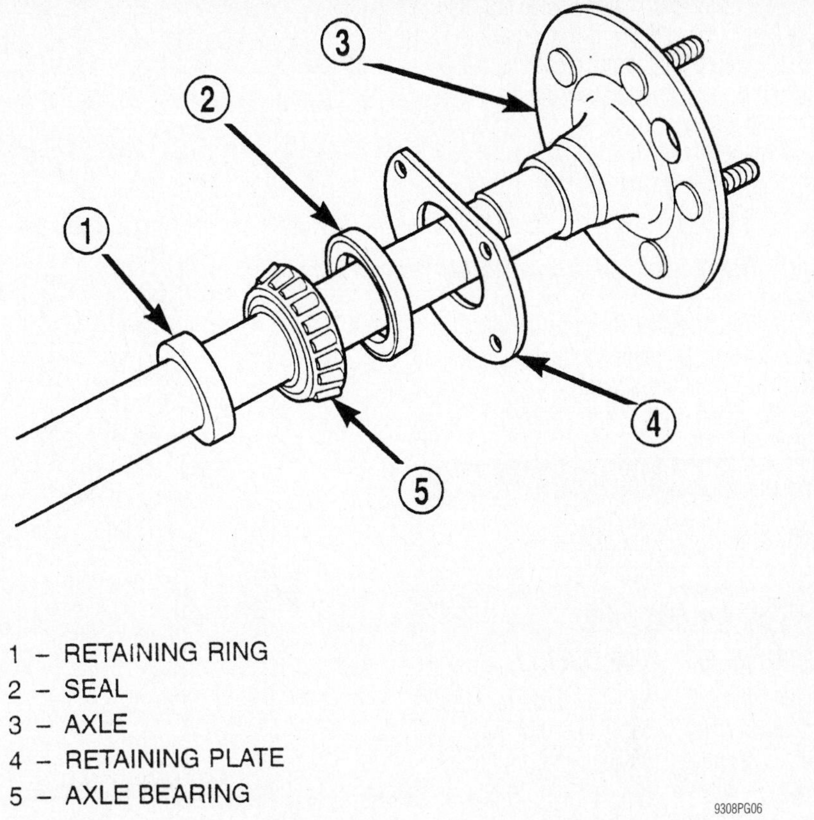

Rear axle seal and bearing components—2002–05 exc. 2005 Grand Cherokee

2005 GRAND CHEROKEE

1. Before servicing the vehicle, refer to the precautions in the beginning of this section.
2. With vehicle in neutral, position on hoist.
3. Remove calipers and rotors.
4. Tap axle end plug (1) loose from the axle flange (2) with a hammer and punch. Pull plug out of axle flange.
5. Remove speed sensors from axle tube flange.
6. Remove axle flange nuts from axle.
7. Pull axle shaft and backing plate out of axle tube until axle bearing is exposed.
8. Remove O-ring from the axle bearing.
9. Slide axle shaft from axle tube and backing plate.
10. Tap axle shaft out of the bearing and axle flange through the plug hole with a hammer and brass drift.

To install:

11. Tap axle shaft into through axle bearing into axle flange.

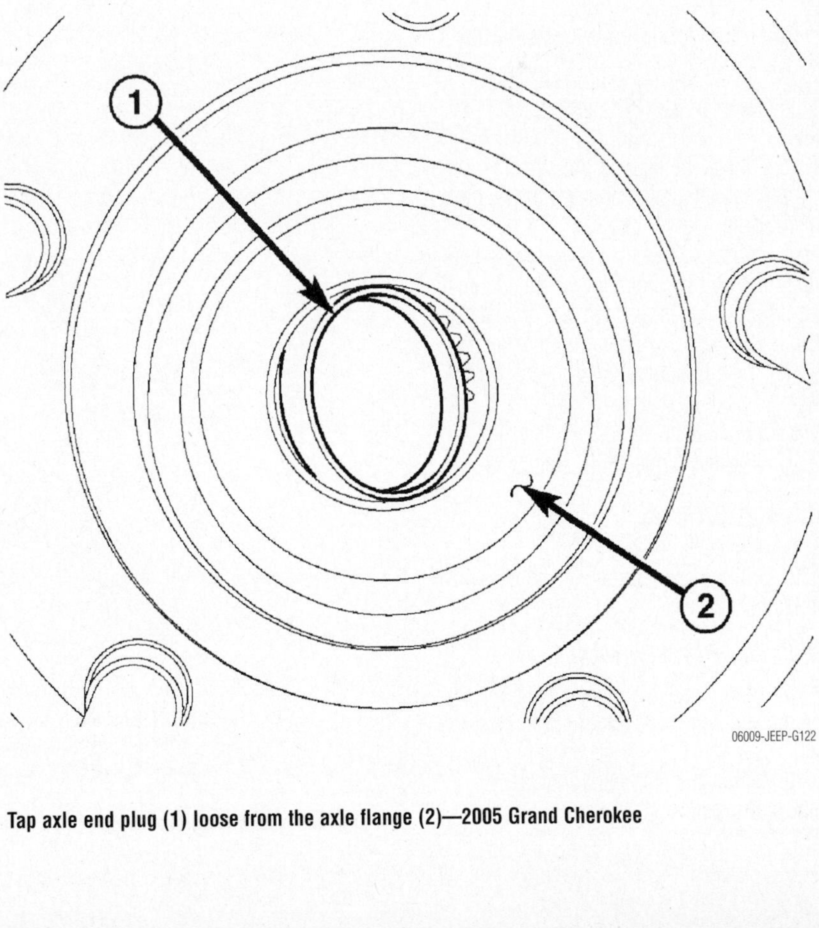

Tap axle end plug (1) loose from the axle flange (2)—2005 Grand Cherokee

12. Install axle shaft into axle tube and backing plate with new o-ring on axle.
13. Slip O-ring through backing plate, then push axle through backing plate until bearing is exposed.
14. Install O-ring axle bearing.
15. Push axle into axle tube.
16. Install axle flange nuts and tighten to 119 Nm (88 ft. lbs.).
17. Install speed sensors in axle tube flange.
18. Coat new axle flange plug with Mopar® Stud N' Bearing Mount Adhesive, or equivalent, and install plug.
19. Install calipers and rotors.

Pinion Seal

REMOVAL & INSTALLATION

C-Clip Type Axle

2002–05 EXC. 2005 LIBERTY

1. Before servicing the vehicle, refer to the precautions in the beginning of this section.
2. Remove or disconnect the following:
   - Wheels
   - Brake drums
   - Driveshaft
3. Check the bearing preload with an inch lb. torque wrench.
4. Remove the pinion flange and seal.

To install:

➡Use a new pinion nut for assembly.

5. Install the new pinion seal and flange. Tighten the nut to 200 ft. lbs. (271 Nm).
6. Check the bearing preload. The bearing preload should be equal to the reading taken earlier, plus 5 inch lbs.
7. If the preload torque is low, tighten the pinion nut in 5 inch lb. increments until the torque value is reached. Do not exceed 350 ft. lbs. (474 Nm) pinion nut torque.
8. If the pinion bearing preload torque cannot be attained at maximum pinion nut torque, replace the collapsible spacer.
9. Install or connect the following:
   - Driveshaft
   - Brake drums
   - Wheels
10. Fill the axle assembly with gear oil and check for leaks.

2005 LIBERTY REAR AXLE

1. Before servicing the vehicle, refer to the precautions in the beginning of this section.
2. With vehicle in neutral, position vehicle on hoist.

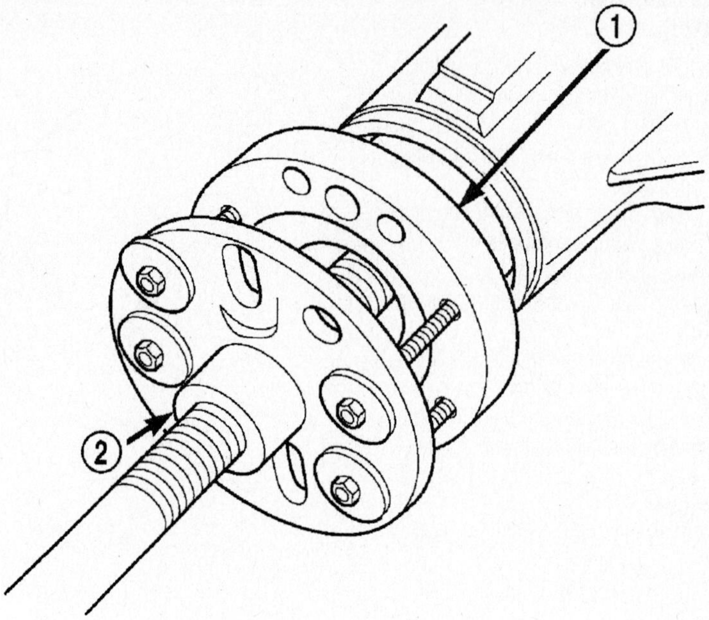

Companion shaft puller——C-Clip Type Axle C-Clip Type Axle

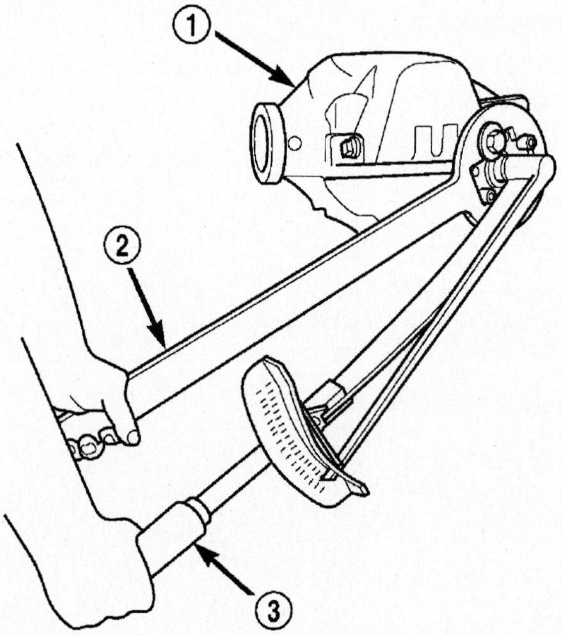

Tightening the pinion nut—C-Clip Type Axle

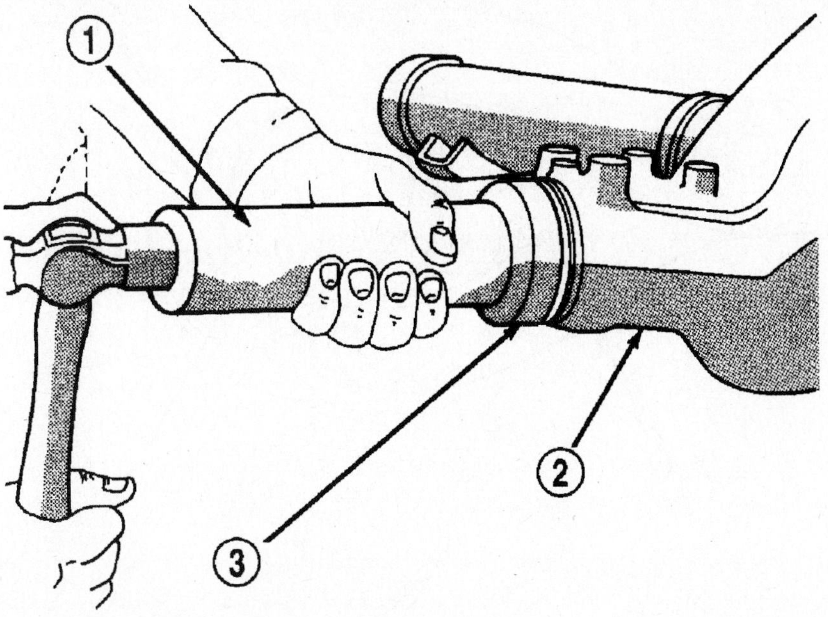

Pinion seal installer—C-Clip Type Axle

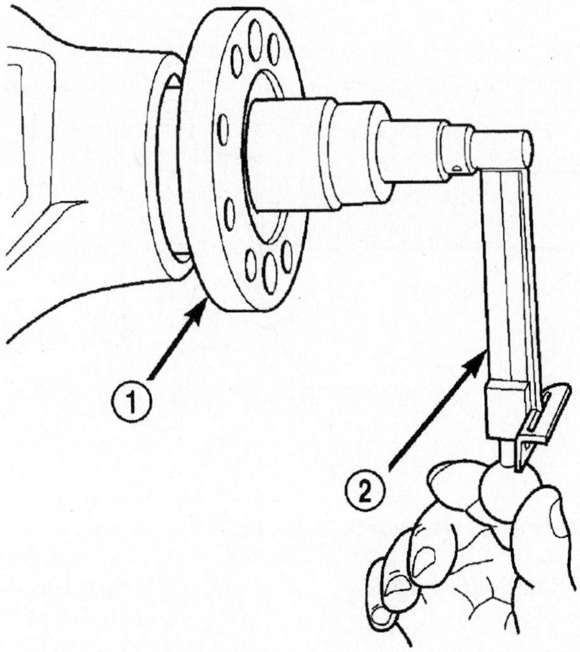

Measuring pinion rotating torque—C-Clip Type Axle

3. Mark a reference line across the axle companion flange and propeller shaft flange yoke.
4. Remove companion flange bolts and remove propeller shaft
5. Remove wheel and tire assemblies.
6. Remove brake calipers and rotors to prevent any drag.
7. Rotate companion flange three or four times and verify flange rotates smoothly.
8. Measure rotating torque of the pinion with a inch pound torque wrench and record reading for installation reference.
9. Install bolts into two of the threaded holes in the companion flange 180° apart.
10. Position Holder 6719, or equivalent, against the companion flange and install a bolt and washer into one of the remaining threaded holes. Tighten the bolts so the Holder 6719 is held to the flange.
11. Remove the pinion nut and washer.
12. Remove companion flange with Remover C-452, or equivalent.
13. Remove pinion seal with a seal pick.

To install:

14. Apply a light coating of gear lubricant on the lip of pinion seal.
15. Install new pinion seal with Installer C-4076-B and Handle C-4735, or equivalent.
16. Install companion flange on the end of the shaft with the reference marks aligned.
17. Install bolts into two of the threaded holes in the companion flange 180° apart.
18. Position Holder 6719 against the companion flange and install a bolt and washer into one of the remaining threaded holes. Tighten the bolts so Holder 6719 is held to the flange.
19. Install companion flange on pinion shaft with Installer C-3718 and Holder 6719.
20. Install the pinion washer and a new pinion nut. The convex side of the washer must face outward.

➡Do not exceed the minimum tightening torque when installing the companion flange retaining nut at this point. Failure to heed caution may result in damage.

21. Hold companion flange with Holder 6719 and with a torque wrench tighten pinion nut to 285 Nm (210 ft. lbs.). Rotate pinion several revolutions to ensure the bearing rollers are seated.
22. Rotate pinion with an inch pound torque wrench. Rotating torque should be equal to the reading recorded during removal plus an additional 0.56 Nm (5 inch lbs.).

⁂ WARNING

Never loosen pinion nut to decrease pinion bearing rotating torque and never exceed specified preload torque. Failure to follow these instructions will result in damage.

23. If rotating torque is low use Holder 6719 to hold the companion flange and tighten pinion nut in
6.8 Nm (60 inch lbs.) increments until proper rotating torque is achieved.

➡**NOTE: The seal replacement is unacceptable if final pinion nut torque is less than 285 Nm (210 ft. lbs.). The bearing rotating torque should be constant during a complete revolution of the pinion. If the rotating torque varies, this indicates a binding condition.**

24. Install propeller shaft with installation reference marks aligned.
25. Tighten companion flange bolts to 115 Nm (85 ft. lbs.).
26. Install rear brake components.

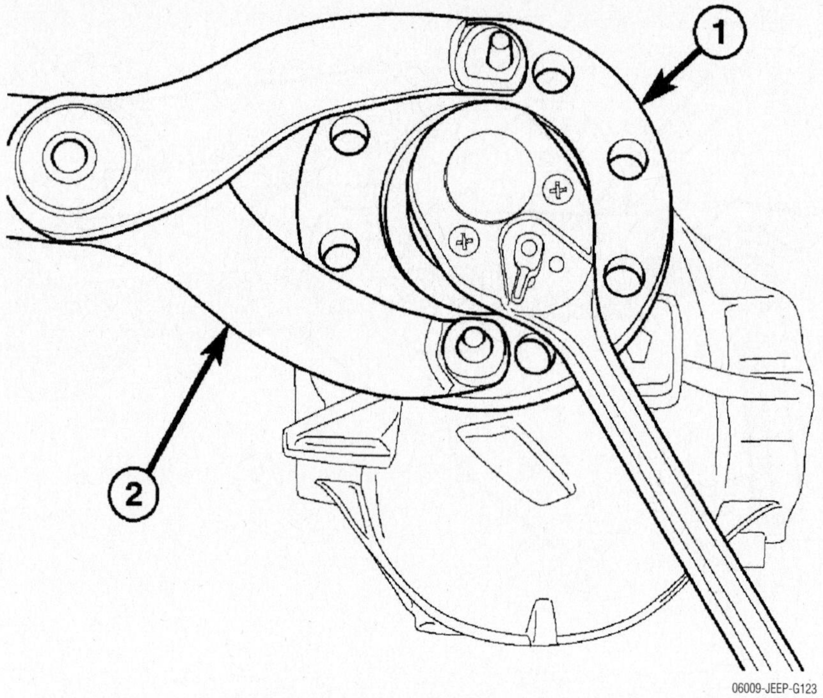

Hold pinion flange (1) with Wrench C-3281(2) and remove pinion nut—2005 Grand Cherokee

Non C-Clip Type

2002–04 FRONT AXLE

1. Before servicing the vehicle, refer to the precautions in the beginning of this section.
2. Remove or disconnect the following:
   - Wheels
   - Brake rotors
   - Driveshaft
3. Check the bearing preload with an inch lb. torque wrench.
4. Remove the pinion flange and seal.

To install:

➡**Use a new pinion nut for assembly.**

5. Install the new pinion seal and flange. Tighten the nut to 160 ft. lbs. (217 Nm).
6. Check the bearing preload. The bearing preload should be equal to the reading taken earlier, plus 5 inch lbs.
7. If the preload torque is low, tighten the pinion nut in 5 inch lb. increments until the torque value is reached. Do not exceed 260 ft. lbs. (353 Nm) pinion nut torque.
8. If the pinion bearing preload torque can not be attained at maximum pinion nut torque, replace the collapsible spacer.
9. Install or connect the following:
   - Driveshaft
   - Brake rotors
   - Wheels
10. Fill the axle assembly with gear oil and check for leaks.

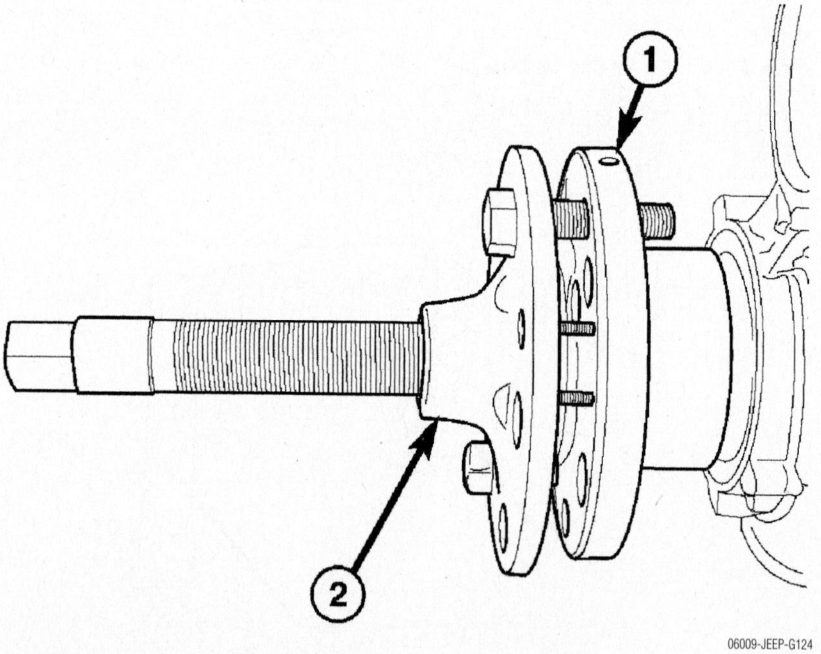

Remove pinion flange (1) with Puller C-452 (2)—2005 Grand Cherokee

2005 GRAND CHEROKEE FRONT AXLE

1. Before servicing the vehicle, refer to the precautions in the beginning of this section.
2. Remove wheels.
3. Push rear brake pads and release hand brake.
4. Remove propeller shaft.
5. Rotate pinion with inch pound torque and record torque to rotate.
6. Mark installation position of collared nut with respect to drive pinion.
7. Bend pinion nut lock back with a punch and hammer.
8. Hold pinion flange with Wrench C-3281 and remove pinion nut.
9. Remove pinion flange with Puller C-452.

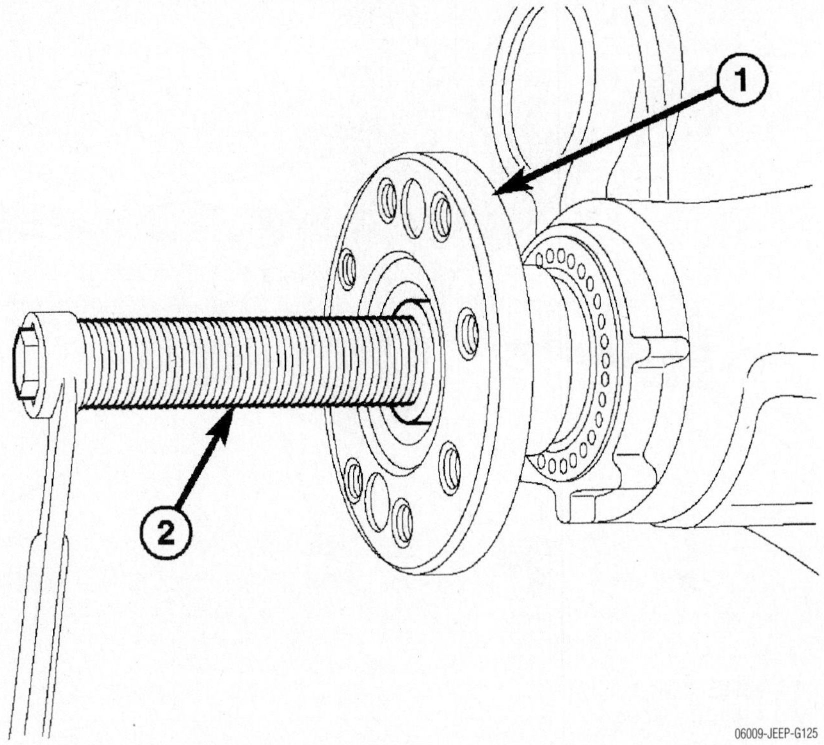

Install pinion flange (1) on pinion with Installer 9616 (2)—2005 Grand Cherokee

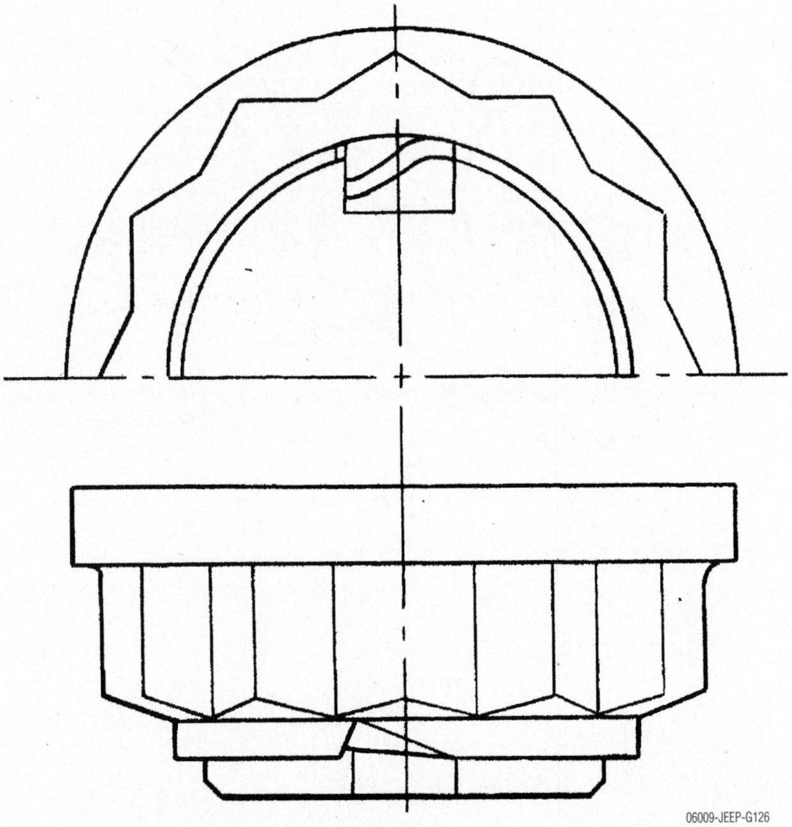

Bend nut collar so it touches the wall of the slot in the pinion shaft—2005 Grand Cherokee

10. Remove pinion seal with a seal pick.

To install:

11. Install pinion seal with Installer C-3972A and Handle C-4171, or equivalent.
12. Install pinion flange on pinion with Installer 9616. Tap flange on pinion, then thread installer center bolt on pinion shaft and draw flange onto the pinion.
13. Install new pinion collared nut and carefully tighten nut in stages holding flange with Wrench C-3281. Check torque to rotate after each stage, until previously value of torque to rotate is exceeded by 0.5 Nm. (4.4 inch lbs.).
14. Cut the pinion nut collar.
15. Bend nut collar so it touches the wall of the slot in the pinion shaft.
16. Connect propeller shaft to pinion flange.
17. Install wheel and tires.
18. Operate brake pedal several times until brake pads contact brake discs (brake pressure built up).

2005 LIBERTY FRONT AXLE

1. Before servicing the vehicle, refer to the precautions in the beginning of this section.
2. With vehicle in neutral, position vehicle on hoist.
3. Remove brake calipers and rotors.
4. Remove propeller shaft.
5. Rotate pinion gear a minimum of ten times and verify the pinion rotates smoothly.
6. Record torque to rotate the pinion gear with an inch pound torque wrench.
7. Using a short piece of pipe and Spanner Wrench 6958 to hold the pinion companion flange and remove the pinion nut.
8. Remove pinion companion flange with Remover C-452 and Spanner Wrench 6958, or equivalent.
9. Pry pinion seal out with a seal pick.

To install:

10. Apply a light coating of gear lubricant on the lip of pinion seal. Install seal with a seal driver.
11. Install pinion companion flange on the pinion gear with Installer W-162-D, Cup 8109 and Wrench 6958, or equivalent.

✻✻ WARNING

Do not exceed the minimum tightening torque 216 Nm (160 ft. lbs.) while installing pinion nut at this point. Failure to follow these instructions will result in damage.

12. Install a new nut on the pinion gear. Tighten the nut only enough to remove the shaft end play.

✲✲ WARNING

Never loosen pinion nut to decrease pinion rotating torque and never exceed specified preload torque. Failure to follow these instructions will result in damage.

13. Rotate pinion a minimum of ten time and verify pinion rotates smoothly. Rotate the pinion shaft with an inch pound torque wrench. Rotating torque should be equal to the reading recorded during removal plus 0.56 Nm (5 inch lbs.).

14. If rotating torque is low, use Spanner Wrench 6958 to hold the pinion companion flange and tighten the pinion nut in 6.8 Nm (60 inch lbs.) increments until proper rotating torque is achieved.

✲✲ WARNING

If maximum tightening torque is reached prior to reaching the required rotating torque, the collapsible spacer may have been damaged. Failure to follow these instructions will result in damage.

15. Install propeller shaft.

16. Fill differential with gear lubricant.

2005 WRANGLER W/181FBI FRONT AXLE

1. Before servicing the vehicle, refer to the precautions in the beginning of this section.
2. With vehicle in neutral, position vehicle on hoist.
3. Remove brake rotors and calipers.
4. Remove propeller shaft from the yoke.
5. Rotate pinion gear three or four times.
6. Record rotating torque of the pinion gear with an inch pound torque wrench, for installation reference.
7. Hold pinion yoke with Wrench 6958 and remove pinion nut and washer, or equivalent.
8. Remove pinion yoke with Remover C-452 and Wrench C-3281, or equivalent.
9. Remove seal with a pry tool or a slide hammer mounted screw.

To install:

10. Apply a light coating of gear lubricant on the lip of pinion seal. Install seal with an appropriate installer.
11. Install yoke on pinion gear with Installer W-162-D, Cup 8109 and Holder 6958, or equivalent.

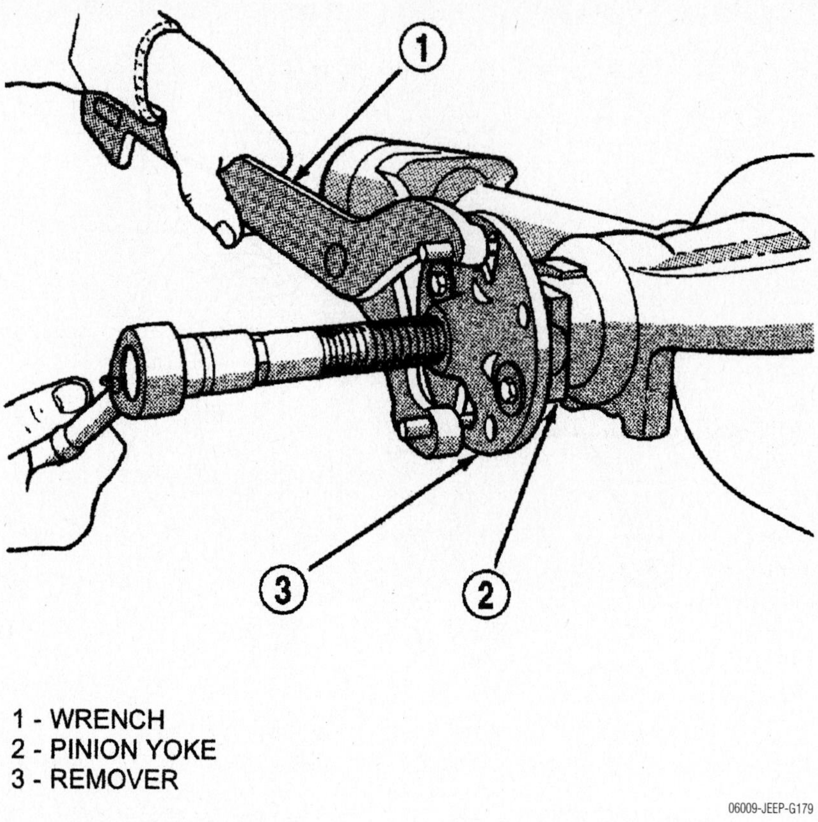

Pinion yoke remover—181FBI axle

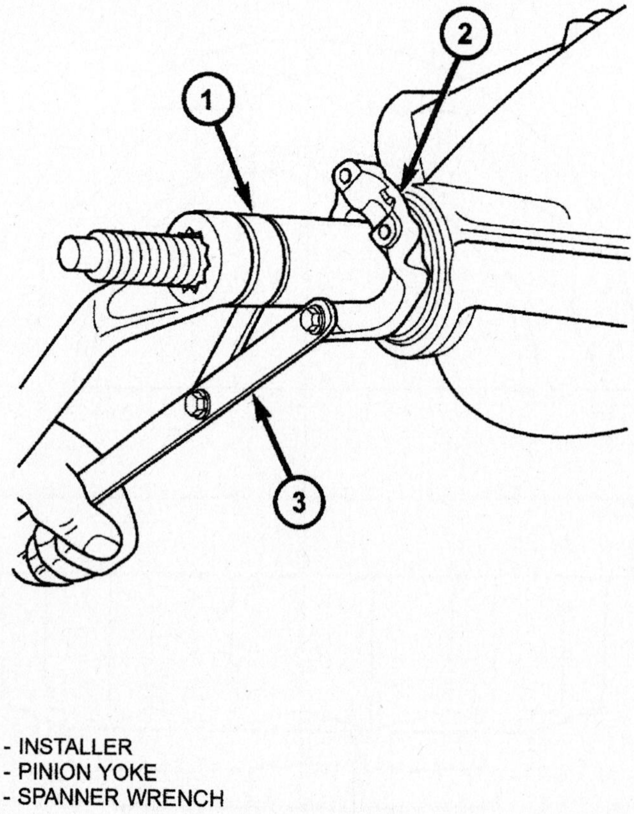

Pinion yoke installer—181FBI axle

WARNING

Do not exceed the minimum tightening torque when installing the pinion yoke retaining nut at this point. Failure to heed caution may result in damage

12. Install pinion washer and a new nut on the pinion gear shaft.

➡Tighten nut only enough to remove the shaft end play.

13. Tighten pinion nut to 217 Nm (160 ft. lbs.).
14. Rotate pinion shaft using an inch pound torque wrench. Rotating torque should be equal to the reading recorded during removal, plus an additional 0.56 Nm (5 inch lbs.).
15. If rotating torque is low, use Spanner 6958, or equivalent, to hold the pinion yoke and tighten pinion shaft nut in 6.8 Nm (60 inch lbs.) increments until proper rotating torque is achieved.
16. Install propeller shaft with reference marks aligned.
17. Install brake rotors and calipers.

2005 WRANGLER W/261FBI FRONT AXLE

2005 WRANGLER W/216RBI REAR AXLE

1. Before servicing the vehicle, refer to the precautions in the beginning of this section.
2. Mark the propeller shaft and pinion yoke for installation reference.
3. Remove propeller shaft from the yoke.
4. Rotate pinion gear three or four times and verify it rotates smoothly.
5. Remove pinion yoke nut and washer with Remover C-452 and Flange Wrench C-3281, or equivalent.
6. Remove pinion shaft seal with a pry tool or slide hammer mounted screw.

To install:

7. Apply a light coating of gear lubricant on the lip of pinion seal. Install seal with an appropriate installer.
8. Install yoke on the pinion gear with Installer W-162-D and Spanner Wrench 6958, or equivalent.
9. Install a new nut on the pinion gear and tighten the nut to 217–271 Nm (160–200 ft. lbs.).
10. Installation propeller shaft with reference marks aligned.
11. Check and add gear lubricant to axle if necessary.

2002–04 REAR AXLE

1. Before servicing the vehicle, refer to the precautions in the beginning of this section.
2. Remove or disconnect the following:
   - Wheels
   - Brake rotors or drums
   - Driveshaft
3. Check the bearing preload with an inch lb. torque wrench.
4. Remove the pinion flange and seal.

To install:

➡Use a new pinion nut for assembly.

5. Install the new pinion seal and flange. Tighten the nut to 160 ft. lbs. (217 Nm).
6. Check the bearing preload. The bearing preload should be equal to the reading taken earlier, plus 5 inch lbs.
7. If the preload torque is low, tighten the pinion nut in 5 inch lb. increments until the torque value is reached. Do not exceed 260 ft. lbs. (353 Nm) pinion nut torque.
8. If the pinion bearing preload torque can not be attained at maximum pinion nut torque, remove one or more pinion preload shims.
9. Install or connect the following:
   - Driveshaft
   - Brake rotors or drums
   - Wheels
10. Fill the axle assembly with gear oil and check for leaks.

2005 GRAND CHEROKEE REAR AXLE

1. Before servicing the vehicle, refer to the precautions in the beginning of this section.
2. With vehicle in neutral, position vehicle on hoist.
3. Mark a reference line across the axle flange and propeller shaft flange.
4. Remove propeller shaft
5. Remove brake calipers and rotors to prevent any drag.
6. Rotate flange three or four times and verify flange rotates smoothly.
7. Measure torque to rotating pinion flange with an inch pound torque wrench. Record the reading for installation reference.
8. Hold pinion flange with Wrench C-3281 and remove pinion nut and washer.
9. Remove flange with two jaw puller.
10. Remove pinion seal with a seal pick or slide-hammer mounted screw.

To install:

11. Apply a light coating of gear lubricant on the lip of pinion seal.
12. Install new pinion seal with Installer C-3972A and Handle C-4171.
13. Install flange on the end of the shaft with the reference marks aligned.
14. Install flange on pinion shaft with Installer C-3718 and Wrench C-3281.
15. Install pinion washer and a new pinion nut. The convex side of the washer must face outward.

WARNING

Do not exceed the minimum tightening torque when installing the companion flange retaining nut at this point. Failure to follow these instructions can damage the collapsible spacer or bearings.

16. Hold flange with Wrench C-3281 and tighten pinion nut to 285 Nm (210 ft. lbs.). Rotate pinion several revolutions to ensure bearing rollers are seated.
17. Rotate pinion flange with an inch pound torque wrench. Rotating torque should be equal to the reading recorded during removal plus an additional 0.56 Nm (5 inch lbs.).

WARNING

Never loosen pinion nut to decrease pinion bearing rotating torque and never exceed specified preload torque. If rotating torque is exceeded, a new collapsible spacer must be installed. Failure to follow these instructions can damage the collapsible spacer or bearings.

18. If rotating torque is low use hold the flange and tighten pinion nut in 6.8 Nm (60 inch lbs.) increments until proper rotating torque is achieved.

➡The seal replacement is unacceptable if final pinion nut torque is less than 285 Nm (210 ft. lbs.).

➡The bearing rotating torque should be constant during a complete revolution of the pinion. If the rotating torque varies, this indicates a binding condition.

19. Install propeller shaft.
20. Install rear brake components.

STEERING AND SUSPENSION

➡Suspension components with rubber/urethane bushings should be tightened with the vehicle at normal ride height. It is important to have the springs supporting the weight of the vehicle when the fasteners are torqued. If springs are not at their normal ride position, vehicle ride comfort could be affected and premature bushing wear may occur.

Air Bag

⁂ CAUTION

Some vehicles are equipped with an air bag system. The system must be disarmed before performing service on, or around, system components, the steering column, instrument panel components, wiring and sensors. Failure to follow the safety precautions and the disarming procedure could result in accidental air bag deployment, possible injury and unnecessary system repairs.

PRECAUTIONS

Several precautions must be observed when handling the inflator module to avoid accidental deployment and possible personal injury.

- Never carry the inflator module by the wires or connector on the underside of the module.
- When carrying a live inflator module, hold securely with both hands, and ensure that the bag and trim cover are pointed away.
- Place the inflator module on a bench or other surface with the bag and trim cover facing up.
- With the inflator module on the bench, never place anything on or close to the module, which may be thrown in the event of an accidental deployment. Before servicing the vehicle, also make sure to refer to the precautions in the beginning of this section as well.

DISARMING

Disconnect and isolate the negative battery cable. Wait 2 minutes for the system capacitor to discharge before performing any service.

ARMING

To arm the system, connect the negative battery cable.

Recirculating Ball Power Steering Gear

REMOVAL & INSTALLATION

2002–04

1. Before servicing the vehicle, refer to the precautions in the beginning of this section.
2. Place the front wheels in the straight ahead position.
3. Drain the power steering system.
4. Remove or disconnect the following:
   - Negative battery cable
   - Air cleaner housing
   - Power steering pressure and return lines
   - Column coupler shaft bolt
   - Left front wheel assembly
   - Pitman arm
   - Windshield washer reservoir
   - Steering gear

To install:

5. Install or connect the following:
   - Steering gear. Tighten the bolts to 80 ft. lbs. (108 Nm).
   - Pitman arm. Tighten the nut to 185 ft. lbs. (251 Nm).
   - Windshield washer reservoir
   - Power steering pressure and return lines. Torque the fasteners to 14 ft. lbs. (20 Nm).
   - Intermediate shaft. Tighten the pinch bolt to 36 ft. lbs. (49 Nm).
   - Air cleaner assembly
   - Left front wheel
   - Negative battery cable
6. Fill and bleed the power steering fluid reservoir.
7. Start the engine, check for leaks and repair if necessary.

2005 Wrangler

1. Before servicing the vehicle, refer to the precautions in the beginning of this section.
2. Place the front wheels in the straight ahead position with the steering wheel centered and locked.
3. Siphon out as much power steering fluid as possible.
4. Remove the bumper shield.
5. Remove power steering hoses/tubes from steering gear.
6. Remove the column coupler shaft from the gear.
7. Remove pitman arm from gear.
8. Remove the steering gear retaining bolts and remove the gear.

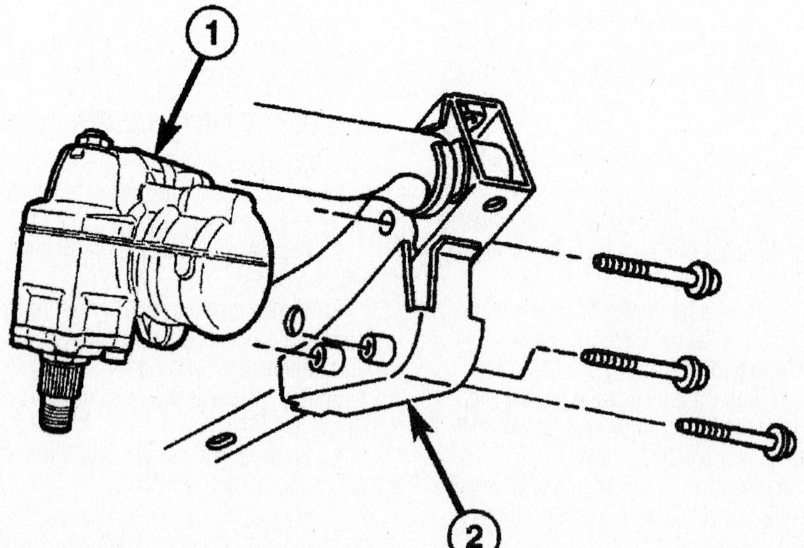

Steering gear mounting—Wrangler

To install:

9. Install steering gear on the frame rail and tighten bolts to 95 Nm (70 ft. lbs.)
10. Align and install the pitman arm and tighten nut to 251 Nm (185 ft. lbs.).
11. Align the column coupler shaft to steering gear. Install a new coupler pinch bolt and tighten to 49 Nm (36 ft. lbs.).
12. Install power steering hoses/tubes to steering gear and tighten to 28 Nm (21 ft. lbs.).
13. Install the bumper shield.
14. Fill power steering system to proper level.

Rack and Pinion Power Steering Gear

REMOVAL & INSTALLATION

Liberty 2-Wheel Drive w/2.4L or 3.7L Engine

1. Before servicing the vehicle, refer to the precautions in the beginning of this section.
2. Siphon the power steering fluid from the power steering reservoir.

➡The steering column on vehicles with an automatic transmission may not be equipped with an internal locking shaft that allows the ignition key cylinder to be locked with the key. Alternative methods of locking the steering wheel for service will have to be used.

3. Lock the steering wheel to prevent spinning of the clockspring.
4. Raise and support the vehicle.
5. Remove the skid plate from under the front end to gain access to the gear.
6. Remove the tire and wheel assembly.

➡Mark the alignment adjusting cams and tie rod end jam nuts on the steering gear for easier installation.

7. Remove the tie rod end nuts.
8. Separate tie rod ends from the knuckles.
9. Remove the lower intermediate shaft coupler bolt and slide the coupler off the gear.
10. Remove power steering pressure hose bracket.
11. Remove the power steering lines from the gear.
12. Remove the mounting bolts from the gear to the front cradle.
13. Remove the steering gear from the vehicle.

To install:

14. Transfer the outer tie rod ends to the new steering gear (if needed).
15. Install the steering gear to the vehicle.
16. Install the gear mounting bolts to the front cradle. Tighten the gear mounting bolts to 162 Nm (120 ft. lbs.)
17. Install the power steering lines to the gear.
18. Install the power steering pressure hose bracket.
19. Install the lower coupler bolt and slide the coupler on to the gear.
20. Install the tie rod end to the knuckle and tighten the nuts.
21. Install the tire and wheel assembly.
22. Install the skid plate.
23. Lower the vehicle.
24. Unlock the steering wheel.
25. Fill the power steering fluid.
26. Reset the toe and center the steering wheel.

Liberty 4-Wheel Drive w/2.4L or 3.7L Engine

1. Before servicing the vehicle, refer to the precautions in the beginning of this section.
2. Siphon the power steering fluid from the power steering reservoir.

➡The steering column on vehicles with an automatic transmission may not be equipped with an internal locking shaft that allows the ignition key cylinder to be locked with the key. Alternative methods of locking the steering wheel for service will have to be used.

3. Lock the steering wheel to prevent spinning of the clockspring.
4. Raise and support the vehicle.
5. Remove the skid plate from under the front end to gain access to the gear.
6. Remove the front tire and wheel assemblies.

➡Mark the alignment adjusting cams for easier installation.

7. Remove the lower control arms.
8. Remove the front axle.
9. Remove the tie rod end nuts.
10. Separate tie rod ends from the knuckles.
11. Remove the intermediate shaft lower coupler bolt and slide the coupler off the gear).
12. Remove power steering pressure hose bracket.
13. Remove the power steering lines from the gear.
14. Remove the mounting bolts from the gear to the front cradle.
15. Remove the steering gear from the vehicle.

To install:

16. Transfer the tie rod ends to the new steering gear (if needed).
17. Install the steering gear to the vehicle.
18. Install the gear mounting bolts to the front cradle. Tighten the gear mounting bolts to 162 Nm (120 ft. lbs.)
19. Install the power steering lines to the gear.
20. Install the power steering pressure hose bracket.

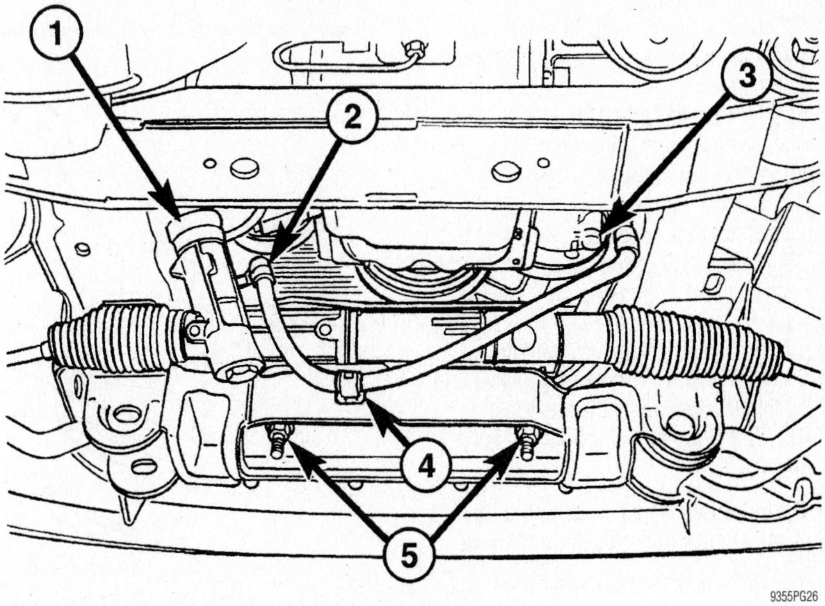

Rack mounting points—Liberty w/2.4L or 3.7L Engine

21. Install the lower coupler bolt and slide the coupler on to the gear.
22. Install the tie rod end to the knuckle and tighten the nuts.
23. Install the front axle.
24. Install the lower control arms.
25. Install the tire and wheel assembly.
26. Install the skid plate
27. Lower the vehicle.
28. Unlock the steering wheel.
29. Fill the power steering fluid.
30. Reset the toe and center the steering wheel.

Liberty w/2.8L Diesel

1. Before servicing the vehicle, refer to the precautions in the beginning of this section.
2. Siphon the power steering fluid from the power steering reservoir.

➡The steering column on vehicles with an automatic transmission may not be equipped with an internal locking shaft that allows the ignition key cylinder to be locked with the key. Alternative methods of locking the steering wheel for service will have to be used.

3. Lock the steering wheel to prevent spinning of the clockspring.
4. Raise and support the vehicle.
5. Remove the skid plate from under the front end to gain access to the gear.
6. Remove the tire and wheel assembly.

➡Mark the alignment adjusting cams for easier installation.

7. Remove the lower control arms.
8. Remove differential housing drain plug and drain fluid.
9. Mark the front propeller shaft and pinion flange. Remove propeller shaft from pinion flange.
10. Remove halfshaft hub nuts.
11. Remove the stabilizer bar links from the lower control arms.
12. Remove tie rod end nuts and separate ends from the knuckles.
13. Remove the lower ball joint nuts and separate ball joints from the lower control arms.
14. Remove shock clevis lower bolts.
15. Pull out on the steering knuckles and push the halfshaft out of the knuckles.
16. With a pry bar remove the halfshafts from the axle.

➡Right halfshaft has a splined axle that may come out with the halfshaft.

17. Remove differential vent hose from cover and remove drain trough nut.

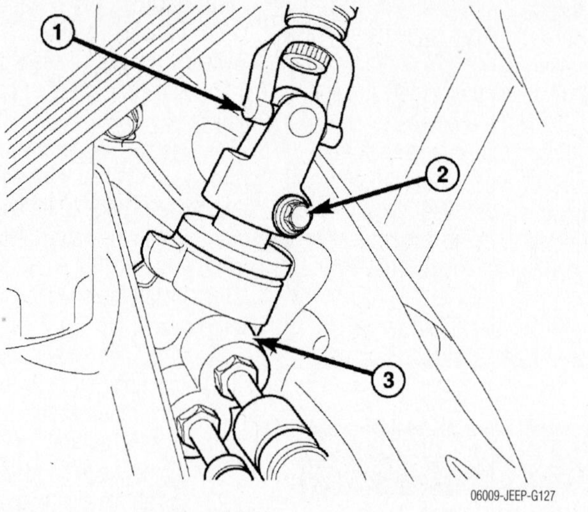

Remove the column (1) coupler shaft bolt (2) and remove the shaft from the gear (3)—2005 Grand Cherokee

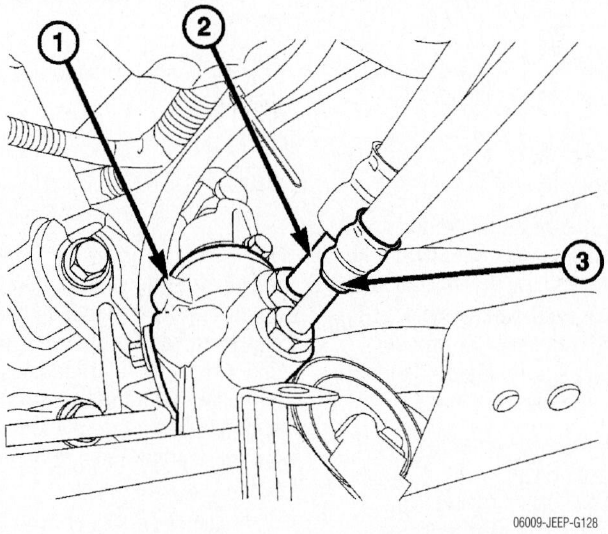

Remove the pressure line (2) and return line (3) at the gear (1)—2005 Grand Cherokee

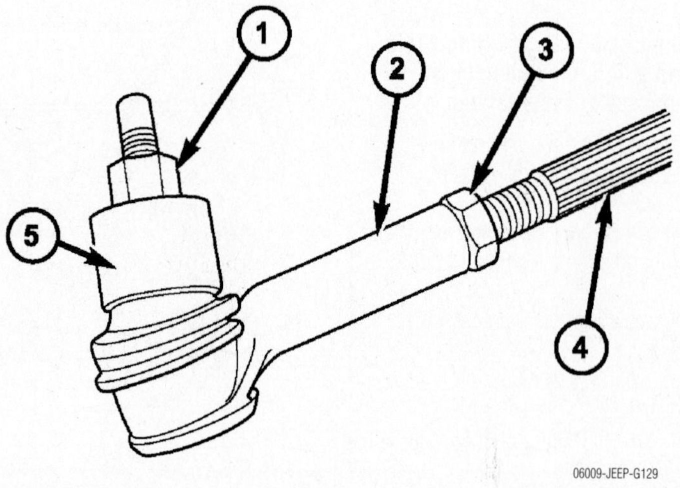

Tie rod end. (1) ball stud nut, (2) tie rod end, (3) jam nut, (4) tie rod, (5) ball joint—2005 Grand Cherokee

18. Support the axle with a lift/jack.
19. Remove bolts from left front axle bracket and cradle bracket bolt.
20. Remove the oil filter drip tray.
21. Mark and remove right control arm cam bolt.
22. Remove front bolt and rear bolt from right axle bracket frame mounts.
23. Remove bolt from left rear axle bracket frame mount.
24. Lower the axle and from vehicle.
25. Remove the lower coupler pinch bolt and slide the coupler off the gear.
26. Remove power steering pressure hose bracket.
27. Remove the power steering lines from the gear.
28. Remove the mounting bolts from the gear to the front cradle.
29. Remove the steering gear from the vehicle.

To install:

30. Transfer the tie rod ends to the new steering gear (if needed).
31. Install the steering gear to the vehicle.
32. Install the gear mounting bolts to the front cradle. Tighten the gear mounting bolts to 162 Nm (120 ft. lbs.)
33. Install the power steering lines to the gear.
34. Install the power steering pressure hose bracket.
35. Install the lower coupler pinch bolt and slide the coupler on to the gear.
36. Install the tie rod end to the knuckle and tighten the nuts.
37. Install right bracket to axle and tighten to 88 Nm (65 ft. lbs.).
38. Install left rear bracket to axle and tighten to 61 Nm (45 ft. lbs.).
39. Install left front bracket to axle and tighten to 61 Nm (45 ft. lbs.).
40. Raise axle up and align brackets with frame mounts.
41. Install frame mount bolts and tighten to 88 Nm (65 ft. lbs.).
42. Install halfshafts.
43. Install right front control arm cam bolt with marks aligned.
44. Install lower ball joint into lower control arms and tighten nuts.
45. Align clevis with knuckles and install clevis bolts.
46. Install stabilizer bar links to lower control arms and install bolts.
47. Install oil filter trough.
48. Install new halfshaft hub nuts and tighten to 136 Nm (100 ft. lbs.).
49. Install the propeller shaft.
50. Install the axle vent hose.
51. Fill the differential with gear lubricant.

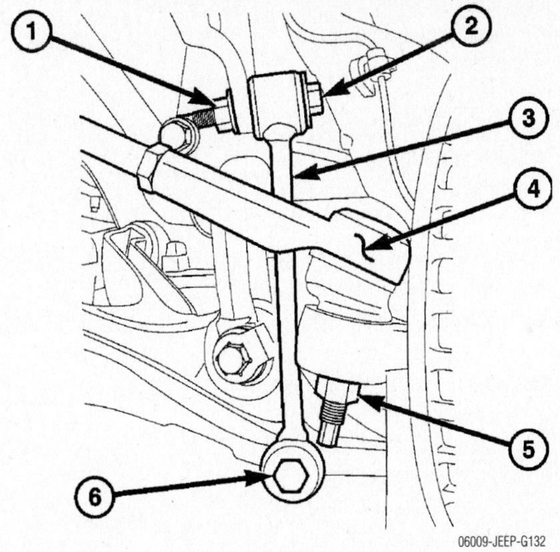

Stabilizer link upper attaching nut (1), upper attaching bolt (2), stabilizer link (3), tie rod end (4), tie rod ball stud nut (5), lower link attaching bolt (6)—2005 Grand Cherokee

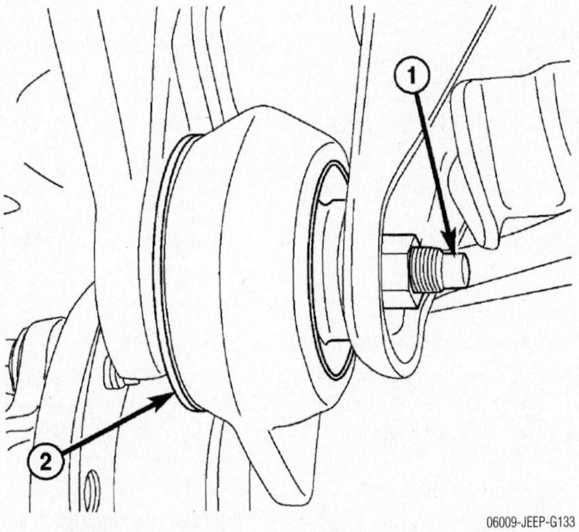

Remove bolt (1) from differential bushing (2)—2005 Grand Cherokee

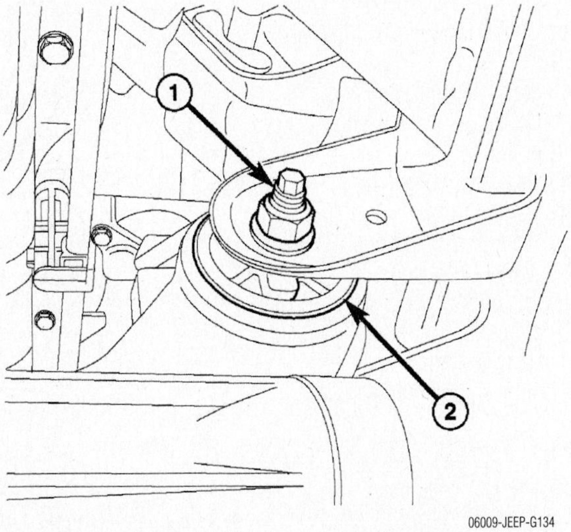

Remove bolt (1) from axle tube bushing (2)—2005 Grand Cherokee

52. Install the lower control arms.
53. Install the tire and wheel assembly.
54. Install the skid plate.
55. Lower the vehicle.
56. Unlock the steering wheel.
57. Fill the power steering fluid.
58. Align the front end.

Grand Cherokee

1. Before servicing the vehicle, refer to the precautions in the beginning of this section.
2. Place the front wheels in the straight ahead position with the steering wheel centered and locked with a steering wheel lock.
3. Drain or siphon the power steering system.
4. Remove the column coupler shaft bolt and remove the shaft from the gear.
5. Remove the oil drip tray.
6. Remove the pressure line at the gear.
7. Remove the return line at the gear.
8. Raise and support the vehicle.
9. Remove the front tires.
10. Loosen the tie rod end jam nuts.
11. Remove the outer tie rod end nut and separate the tie rod from the knuckle using the proper tool.
12. Remove the front splash shield.
13. With 4-wheel drive:
    a. With vehicle in neutral, position vehicle on hoist.
    b. Remove skid plate.
    c. Remove differential housing drain plug and drain fluid.
    d. Mark front driveshaft and pinion flange. Remove driveshaft from pinion flange.
    e. Remove halfshaft hub/bearing nuts.
    f. Remove wheel speed sensors from hub/bearings.
    g. Remove brake calipers bolts and remove calipers from caliper adapters.
    h. Remove stabilizer links bolts from lower control arms.
    i. Remove upper ball joint nuts and separate ball joints from knuckles with Remover 8677, or equivalent.
    j. Remove lower shock clevis nuts bolts from lower control arm.
    k. Lean the knuckles out and push halfshafts out of the hub/bearings.
    l. Pry halfshafts off the axles.
    m. Remove differential vent hose.
    n. Support axle with a lift/jack.
    o. Remove bolt from differential bushing.
    p. Remove bolt from axle tube bushing.
    q. Remove bolts from differential cover bracket bushing.

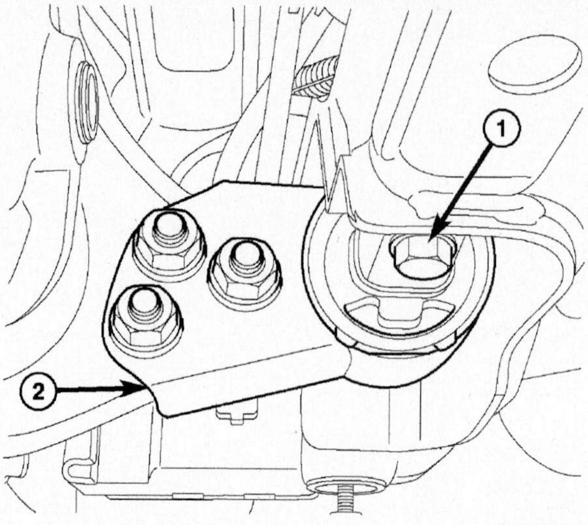

Remove bolts (1) from differential cover bracket (2) bushing—2005 Grand Cherokee

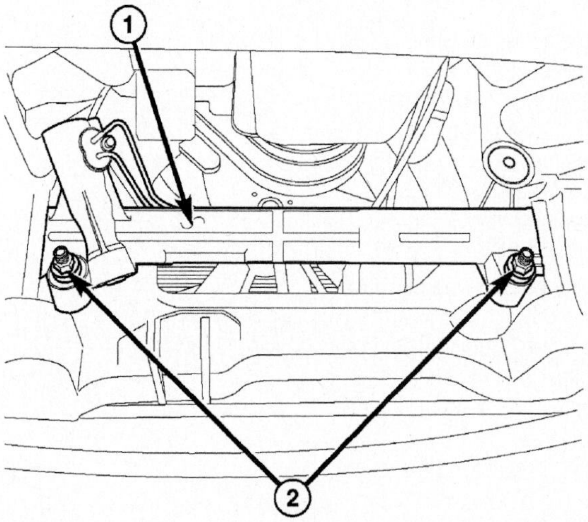

Steering gear mounting bolts (2)—2005 Grand Cherokee

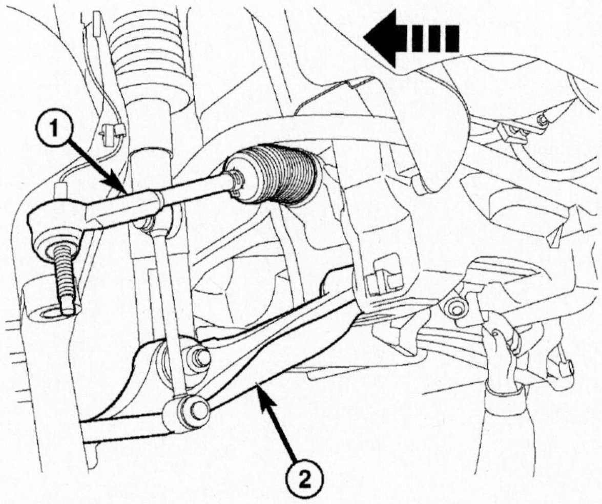

Move the gear (1) to the full right position to allow clearance over the control arm (2)—2005 Grand Cherokee

r. Lower axle and from vehicle.

14. Remove the two steering gear mounting bolts.

15. Move the gear to the full right position to allow clearance over the control arm then lower the gear down, then turn the gear to the full left position then to allow clearance then remove from the vehicle.

16. Remove the outer tie rod ends from the steering gear (if needed).

To install:

17. Install the outer tie rod ends.

18. Position the steering gear back into the vehicle the same way it was removed.

19. Install the steering gear mounting bolts. Tighten to 217 Nm (160 ft. lbs.).

20. Install the front splash shield.

21. Install the outer tie rod ends jam nuts loosely.

22. Install the outer tie rod ends to the knuckle and tighten the tie rod end nut to 95 Nm (70 ft. lbs.).

23. Install the pressure and return hoses to the steering gear and tighten to 28 Nm (21 ft. lbs.).

24. Install the oil filter drip tray.

25. Install the column coupler shaft into the lower coupling and install a new bolt. Tighten to 49 Nm (36 ft. lbs.).

26. With 4-wheel drive:

a. Raise axle to vehicle.

b. Install differential cover axle bracket bushing bolts and tighten bolts to 47 Nm (35 ft. lbs.)

c. Install axle tube bushing bolt and nut. Tighten nut to 95 Nm (70 ft. lbs.)

d. Install differential bushing bolt and nut. Tighten nut to 95 Nm (70 ft. lbs.)

e. Install differential vent hose.

f. Install halfshaft into axle and knuckles. Verify axles are engaged in the axle.

g. Install clevis bolt and nut on lower control arms and tighten to specification.

h. Install upper control arms on knuckles and tighten ball joint nuts to specifications.

i. Install tie rod ends and nuts on knuckles and tighten to specifications.

j. Install stabilizer bar links and bolts on lower control arms and tighten to specifications.

k. Install brake calipers on knuckles.

l. Install wheel speed sensor on hub/bearing.

m. Install halfshaft hub nuts and tighten to 135 Nm (100 ft. lbs.).

n. Install driveshaft on pinion flange with reference marks aligned.

o. Install drain plug and tighten to 60 Nm (44 ft. lbs.).

p. Fill differential housing and install fill plug. Tighten plug to 60 Nm (44 ft. lbs.).

q. Install skid plate.

27. Install the wheel and tire assembly.

28. Remove the support and lower the vehicle.

29. Remove the steering wheel lock.

30. Fill the power steering pump.

31. Set the toe.

Front Struts

REMOVAL & INSTALLATION

Liberty

LEFT SIDE

1. Before servicing the vehicle, refer to the precautions in the beginning of this section.

2. Remove or disconnect the following:
   - Negative battery cable
   - Battery
   - Power center
   - Battery tray
   - Battery temperature sensor
   - The 4 upper strut nuts
   - Left wheel
   - Lower clevis bracket bolt
   - Stabilizer link
   - Lower ball joint from the arm
   - Clevis bracket
   - Strut

3. Installation is the reverse of removal. Observe the following torques:
   - Upper strut nuts: 80 ft. lbs. (108 Nm)
   - Clevis bracket-to-strut bolt: 65 ft. lbs. (88 Nm)
   - Ball joint nut: 60 ft. lbs. (81 Nm)
   - Clevis bracket-to-arm bolt: 110 ft. lbs. (150 Nm)
   - Stabilizer link-to-arm bolt: 100 ft. lbs. (136 Nm)

RIGHT SIDE

1. Before servicing the vehicle, refer to the precautions in the beginning of this section.

2. Remove or disconnect the following:
   - Negative battery cable
   - Air box
   - Cruise control servo mounting nuts
   - The 4 upper strut nuts
   - Right wheel
   - Lower clevis bracket bolt
   - Stabilizer link
   - Lower ball joint from the arm
   - Clevis bracket

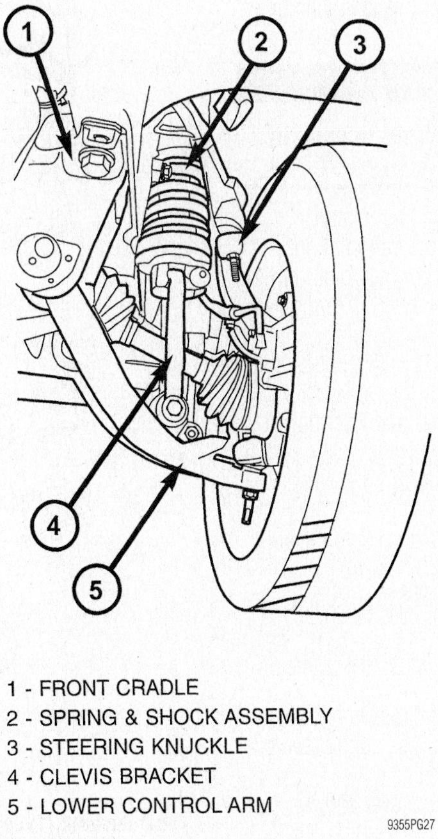

Strut and clevis assembly—Liberty

- Strut

3. Installation is the reverse of removal. Observe the following torques:
    - Upper strut nuts: 80 ft. lbs. (108 Nm)
    - Clevis bracket-to-strut bolt: 65 ft. lbs. (88 Nm)
    - Ball joint nut: 60 ft. lbs. (81 Nm)
    - Clevis bracket-to-arm bolt: 110 ft. lbs. (150 Nm)
    - Stabilizer link-to-arm bolt: 100 ft. lbs. (136 Nm)

2005 Grand Cherokee

RIGHT SIDE

1. Before servicing the vehicle, refer to the precautions in the beginning of this section.
2. Remove the air box cover and air intake hose.
3. Disconnect the cruise control servo electrical connector.
4. Remove the coolant reservoir mounting bolt and move the coolant reservoir off to the side.
5. Remove the four upper strut mounting nuts.
6. Raise and support the vehicle.
7. Remove the tire.
8. Remove the two brake caliper adapter bolts.
9. Support the brake caliper adaptor and caliper.

WARNING

Do not allow the caliper to hang by the brake hose.

10. Remove the disc brake rotor.
11. Remove the upper ball joint nut.
12. Separate the upper ball joint from the knuckle using special tool 8677, or equivalent.
13. Remove the lower clevis bolt at the lower control arm.
14. Remove the lower stabilizer bolt at the lower control arm.
15. Remove the strut from the vehicle.

To install:

16. Install the clevis bracket to the strut and tighten to 122 Nm (90 ft. lbs.).
17. Install the strut assembly to the vehicle.
18. Install the four upper strut nuts. Tighten to 95 Nm (70 ft. lbs.).
19. Install the coolant reservoir bolt.
20. Reconnect the cruise control servo wiring connector.
21. Install the air box cover and air intake hose.
22. Raise the vehicle up.

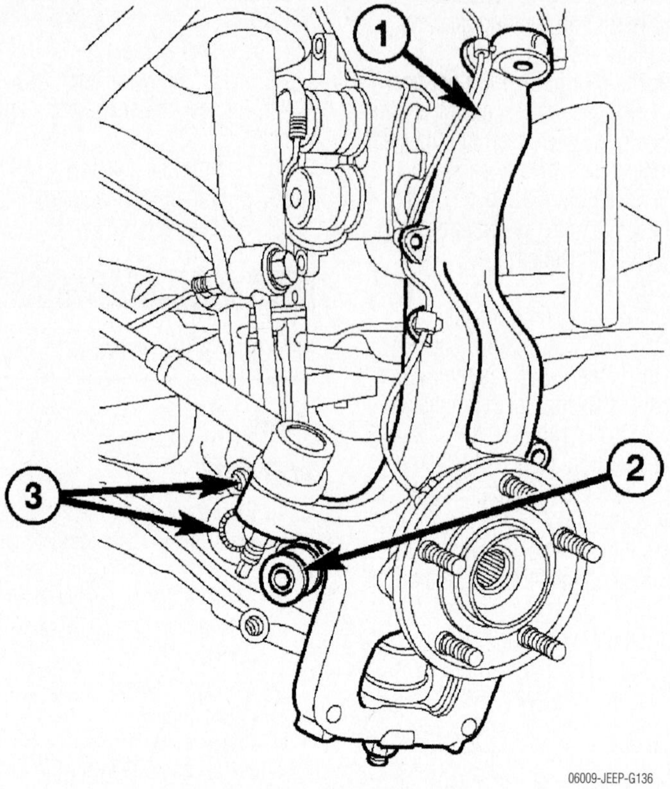

(1) knuckle, (2) lower clevis bolt, (3) lower stabilizer bolt—2005 Grand Cherokee

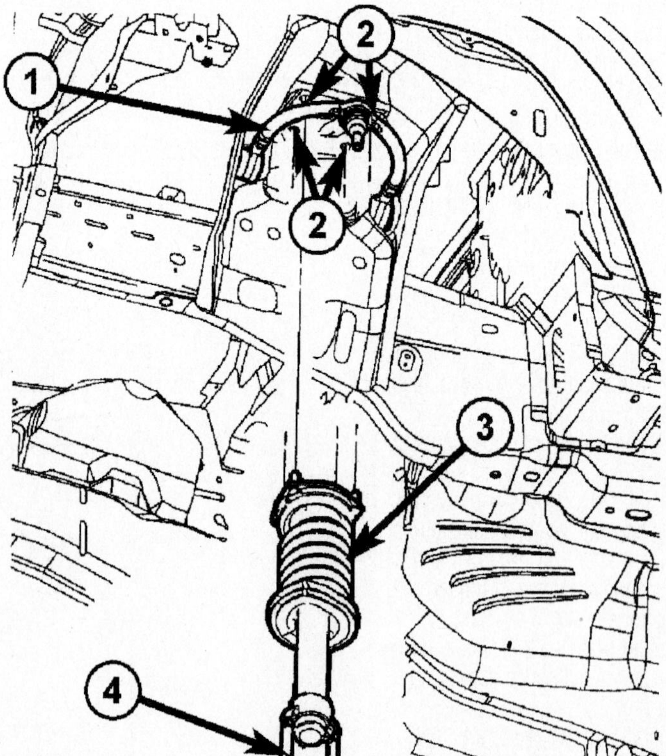

Strut removal. (1) upper control arm, (2) upper strut mounting holes, (3) strut, (4) strut clevis—2005 Grand Cherokee

23. Install the lower stabilizer bolt at the lower control arm. Torque to 85 ft. lbs. (115 Nm).
24. Install the lower clevis bolt at the lower control arm and tighten to 169 Nm (125 ft. lbs.).
25. Install the upper ball joint into the knuckle and tighten the nut to 75 Nm (55 ft. lbs.).
26. Install the disc brake rotor.
27. Install the caliper adaptor mounting bolts to 176 Nm (130 ft. lbs.).
28. Install the tire and wheel assembly.
29. Lower the vehicle.

LEFT SIDE

1. Before servicing the vehicle, refer to the precautions in the beginning of this section.
2. Remove the air box cover and air intake hose.
3. Remove the 3 PDC bracket nuts.
4. Move the PDC off to the side to access the four upper strut mount nuts.
5. Remove the four upper strut mount nuts.
6. Raise and support the vehicle.
7. Remove the tire.
8. Remove the two brake caliper adapter bolts.
9. Support the brake caliper adaptor and caliper.

⁂ WARNING

Do not allow the caliper to hang by the brake hose.

10. Remove the disc brake rotor.
11. Remove the upper ball joint nut.
12. Separate the upper ball joint from the knuckle using special tool 8677, or equivalent.
13. Remove the lower clevis bolt at the lower control arm.
14. Remove the lower stabilizer bolt at the lower control arm.
15. Remove the strut from the vehicle.
16. Remove the spring if necessary.

To install:

17. Install the clevis bracket to the strut and tighten to 122 Nm (90 ft. lbs.).
18. Install the strut assembly to the vehicle.
19. Install the four upper strut nuts. Tighten to 95 Nm (70 ft. lbs.).
20. Install the 3 PDC bracket nuts.
21. Raise the vehicle up.
22. Install the lower stabilizer bolt at the lower control arm and tighten to 115 Nm (85 ft. lbs.).
23. Install the lower clevis bolt at the lower control arm and tighten to 169 Nm (125 ft. lbs.).
24. Install the upper ball joint into the knuckle and tighten the nut to 75 Nm (55 ft. lbs.).
25. Install the disc brake rotor.
26. Install the caliper adaptor mounting bolts to 176 Nm (130 ft. lbs.).
27. Install the tire and wheel assembly.
28. Lower the vehicle.

Front Shock Absorber

REMOVAL & INSTALLATION

2002–04

1. Before servicing the vehicle, refer to the precautions in the beginning of this section.
2. Remove or disconnect the following:
   - Upper nut, washer and grommet from the upper stud
   - Lower fasteners
   - Shock absorber

To install:

3. Install or connect the following:
   - Shock absorber. Torque the lower fasteners to 17 ft. lbs. (23 Nm).
   - Upper grommet, washer, and nut to the stud. Torque it to 16 ft. lbs. (22 Nm).

2005 Wrangler

1. Before servicing the vehicle, refer to the precautions in the beginning of this section.
2. Remove the nut, retainer and grommet from the upper stud through engine compartment access hole.
3. Remove the lower nuts and bolts from the axle bracket and remove the shock absorber.

To install:

4. Position the lower retainer and grommet on the upper stud. Insert the shock absorber through the shock bracket hole.
5. Install the lower bolts and nuts. Tighten nuts to 28 Nm (250 inch lbs.).
6. Install the upper grommet and retainer on the stud and install the nut and tighten to 23 Nm (17 ft. lbs.).

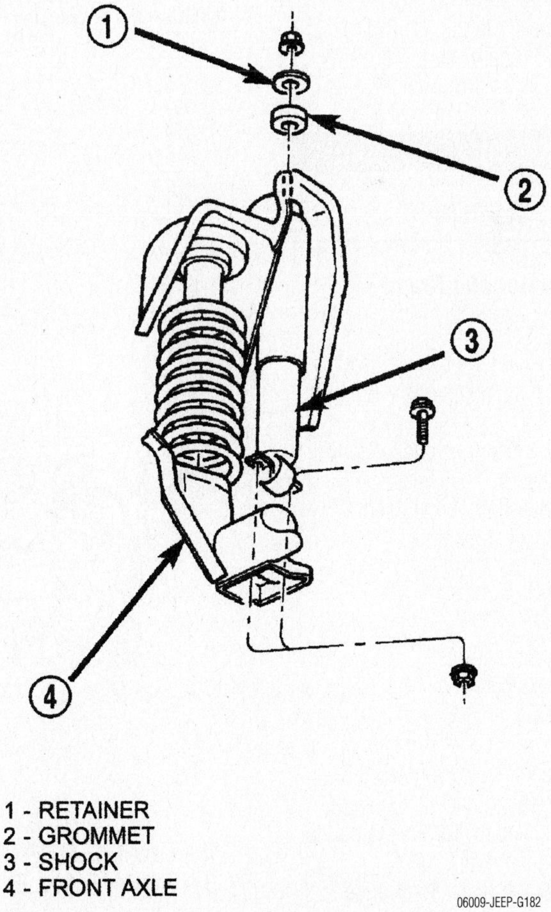

Coil spring and shock absorber—2005 Wrangler shown

Rear Shock Absorber

REMOVAL & INSTALLATION

1. Before servicing the vehicle, refer to the precautions in the beginning of this section.
2. Remove or disconnect the following:
   - Upper locknut and washer from the frame bracket stud, on the Grand Cherokee
   - Upper mounting bolts, on the Wrangler
   - Lower bolt, nut and washers from the axle shaft tube bracket
   - Shock absorber

To install:

3. Place the shock absorber upper end in position and tighten the fasteners to the following specifications:
   - Wrangler: 23 ft. lbs. (31 Nm).
   - Liberty and 2002–04 Grand Cherokee: 80 ft. lbs. (108 Nm).
   - 2005 Grand Cherokee: 70 ft. lbs. (95 Nm).
4. Place the shock absorber lower end in position and tighten the fasteners to the following specifications:
   - Wrangler: 74 ft. lbs. (100 Nm).
   - Grand Cherokee: 85 ft. lbs. (115 Nm).

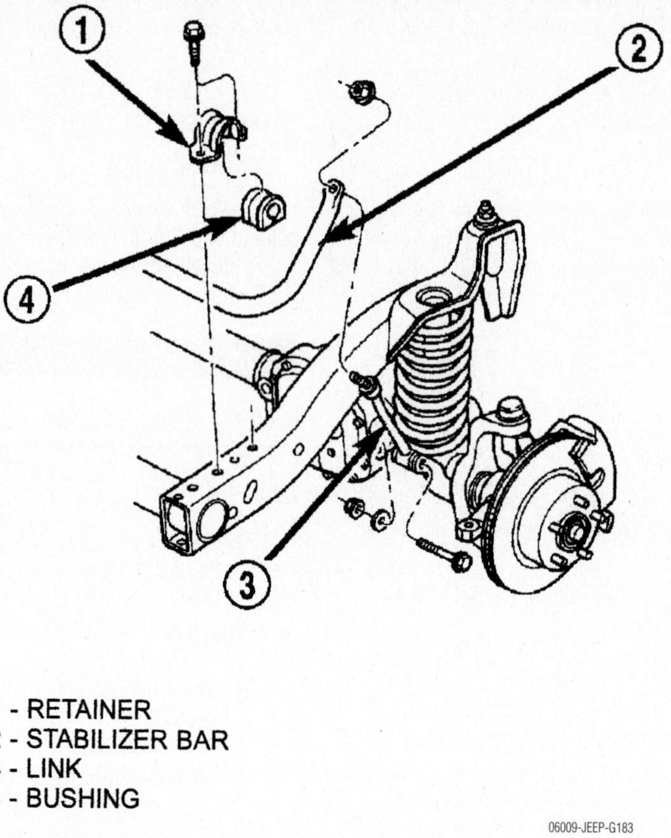

Stabilizer bar—Wrangler

Front Coil Spring

REMOVAL & INSTALLATION

2002–04 Wrangler and Grand Cherokee

1. Before servicing the vehicle, refer to the precautions in the beginning of this section.
2. Remove or disconnect the following:
   - Front wheels
   - Front driveshaft, if equipped
   - Lower suspension arm
   - Stabilizer bar links
   - Track bar
   - Drag link
   - Brake hose brackets
   - Spring retainers
   - Coil spring

To install:

3. Install or connect the following:
   - Coil spring
   - Spring retainers. Torque the bolt to 16 ft. lbs. (21 Nm).
   - Stabilizer bar links. Torque the bolts to 70 ft. lbs. (95 Nm).

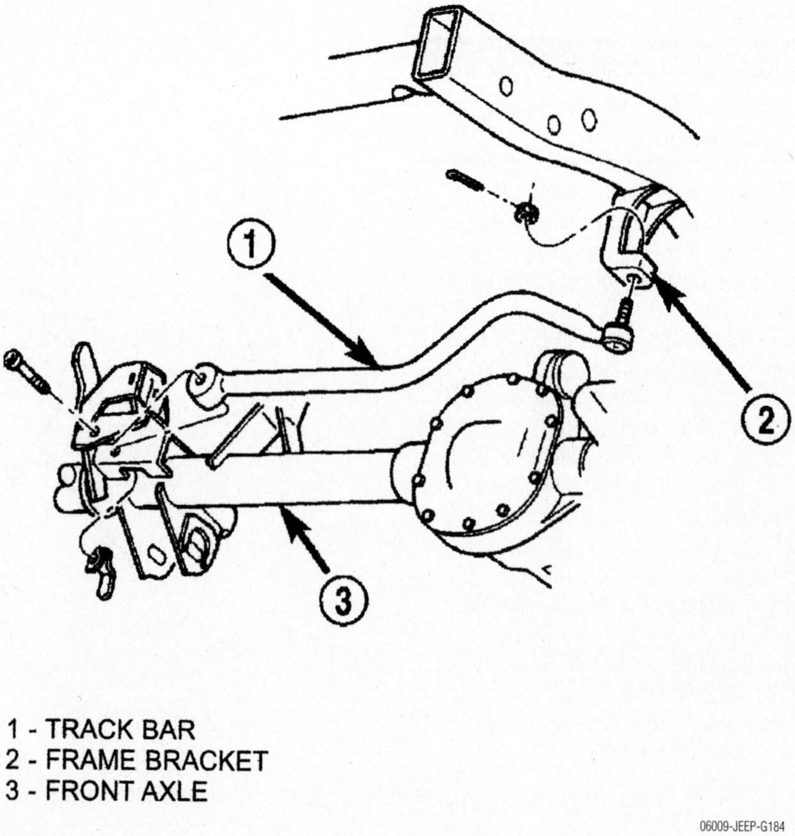

Track bar—Wrangler

- Track bar. Torque the bolt to 35 ft. lbs. (47 Nm).
- Front driveshaft to the axle
- Drag link
- Lower suspension arm. Torque the bolt to 133 ft. lbs. (180 Nm).
- Front driveshaft, if equipped
- Front wheels

2005 Wrangler

1. Before servicing the vehicle, refer to the precautions in the beginning of this section.
2. Raise and support the vehicle.
3. Remove the wheel and tire assemblies.
4. Position a hydraulic jack under the axle to support it.
5. Remove the front shocks at the lower mountings.
6. Remove the ABS wire mounting brackets at the axle, if equipped.
7. Remove lower suspension arms mounting nuts and bolts from the frame.
8. Remove the track bar from the axle bracket.
9. Remove the right side of the drag link from the right side knuckle.
10. Lower the axle until the spring is free from the upper mount.

➡Rotation of the spring and prying down slightly on the axle will aid in removal.

11. Remove the coil spring retainer clip and remove the spring.
12. Remove the upper spring isolator, if needed.
13. Pull jounce bumper out of mount, if needed.

To install:

14. Install jounce bumper into mount.
15. Install the spring isolator.

➡Rotation of the spring and prying down slightly on the axle will aid in installation.

16. Position the coil spring on the axle pad. It may be necessary to rotate the spring while installing.
17. Install the spring retainer clip and bolt. Tighten bolt to 21 Nm (16 ft. lbs.).
18. Raise the axle into position until the spring seats in the upper mount.
19. Install the shock at the axle.
20. Install the ABS wire mounting brackets at the axle, if equipped.
21. Install the track bar to the axle bracket. Torque to 40 ft. lbs. (47 Nm).
22. Install the lower suspension arms to the frame. Install mounting bolts and nuts finger tight.

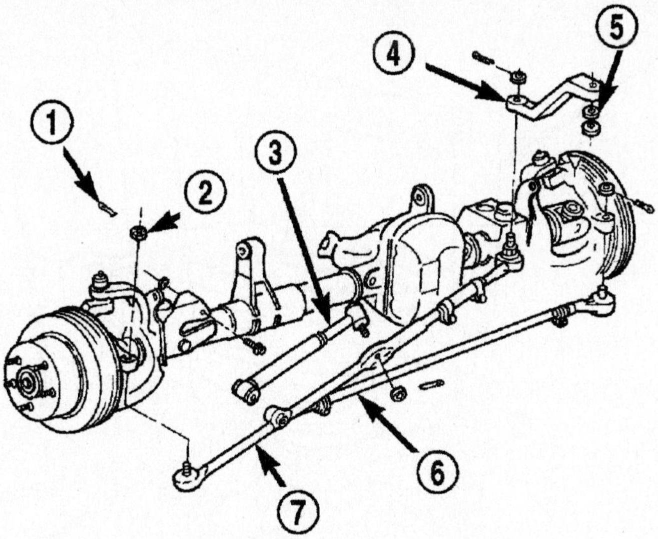

1 - COTTER PIN
2 - NUT
3 - DAMPENER
4 - PITMAN ARM
5 - WASHER
6 - TIE ROD
7 - DRAG LINK

06009-JEEP-G185

Steering linkage—Wrangler

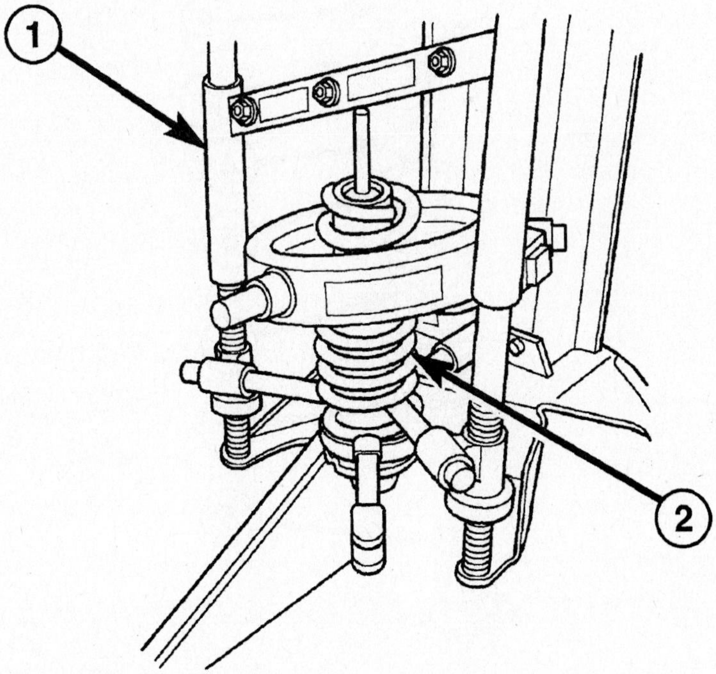

1 - SPRING COMPRESSOR
2 - SPRING

9355PG28

Front coil spring removal—Liberty

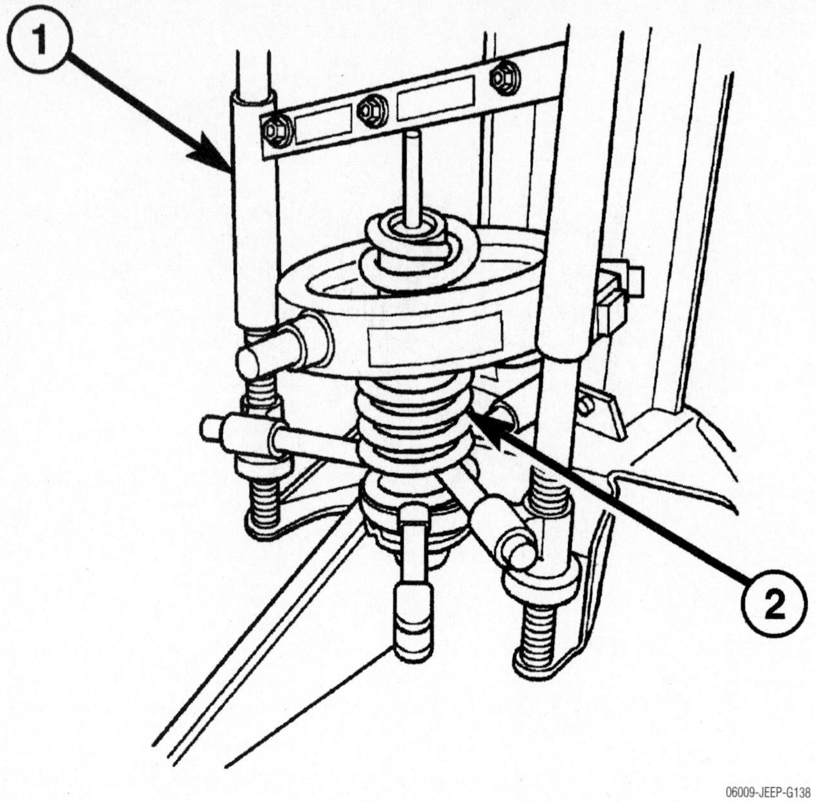

Strut installed in a spring compressor—2005 Grand Cherokee

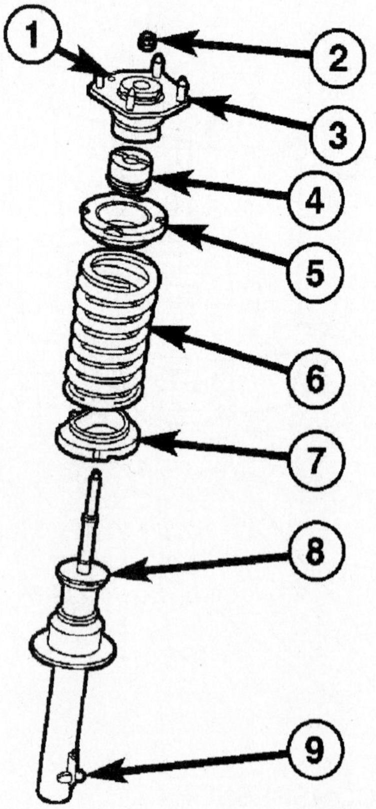

Strut components. (1) nub hole, (2) upper nut, (3) strut plate, (4) jounce bumper, (5) upper isolator, (6) spring, (7) lower isolator, (8) strut, (9) bracket—2005 Grand Cherokee

23. Install the drag link to the right side knuckle. Torque to 35 ft. lbs. (47 Nm). Use new cotter pins.
24. Remove the hydraulic jack from under the axle
25. Install the wheel and tire assemblies.
26. Remove the supports and lower the vehicle.
27. Tighten the lower suspension arms nuts to 115 Nm (85 ft. lbs.) at normal ride height with the vehicle weight.

Liberty

1. Before servicing the vehicle, refer to the precautions in the beginning of this section.
2. Remove the strut and place it in a Pentastar W7200 spring compressor, or equivalent.
3. Compress the spring, remove the strut mount nut and remove the strut from the compressor.
4. Installation is the reverse of removal. Torque the nut to 30 ft. lbs. (41 Nm).

2005 Grand Cherokee

1. Before servicing the vehicle, refer to the precautions in the beginning of this section.
2. Remove the strut.
3. Install the strut assembly in the Branick 7200T spring removal/installation tool or equivalent.
4. Compress the spring.
5. Remove the upper strut nut.

➡The nub in the upper strut mount must be 180° from the centerline of the lower bracket on the strut for proper installation.

6. Remove the strut.
7. Remove the strut upper mounting plate.
8. Remove and inspect the upper and lower spring isolators.

To install:

9. Compress the spring.
10. Install the lower isolator.
11. Position the strut into the coil spring. Make sure the jounce bumper is on the strut rod.
12. Install the upper isolator.

➡For proper orientation the nub hole in the upper strut plate must be 180° in a centerline from the bracket at the bottom of the strut.

13. Install the upper strut mounting plate.
14. Install the strut upper mounting nut. Tighten to 39 Nm (25 ft. lbs.).

15. Decompress the spring.
16. Remove the strut assembly from the spring compressor tool.
17. Install the strut assembly.

Rear Coil Springs

REMOVAL & INSTALLATION

2002–04 Wrangler and Grand Cherokee

1. Before servicing the vehicle, refer to the precautions in the beginning of this section.
2. Remove or disconnect the following:
    - Rear wheels
    - Stabilizer bar links
    - Shock absorbers
    - Track bar
    - Spring retainers
    - Coil springs

To install:

3. Install or connect the following:
    - Coil springs
    - Spring retainers
    - Track bar
    - Shock absorbers
    - Stabilizer bar links
    - Rear wheels

Liberty

1. Before servicing the vehicle, refer to the precautions in the beginning of this section.
2. Raise and support the vehicle. Position a hydraulic jack under the axle to support the axle.
3. Remove the shock absorber lower bolt from the axle bracket.
4. Lower the hydraulic jack and tilt the axle and remove the coil spring.
5. Remove and inspect the upper and lower spring isolators.

To install:

6. Install the upper isolator.
7. Install the lower isolator.
8. Pull down on the axle and position the coil spring in the lower isolator.

CAUTION

Ensure the spring is positioned on the lower isolator.

9. Raise the axle with the hydraulic jack.
10. Install the shock absorber to the axle bracket and tighten to 115 Nm (85 ft. lbs.).
11. Remove the supports and lower the vehicle.
12. Tighten the stabilizer bar links to 99 Nm (73 ft. lbs.).

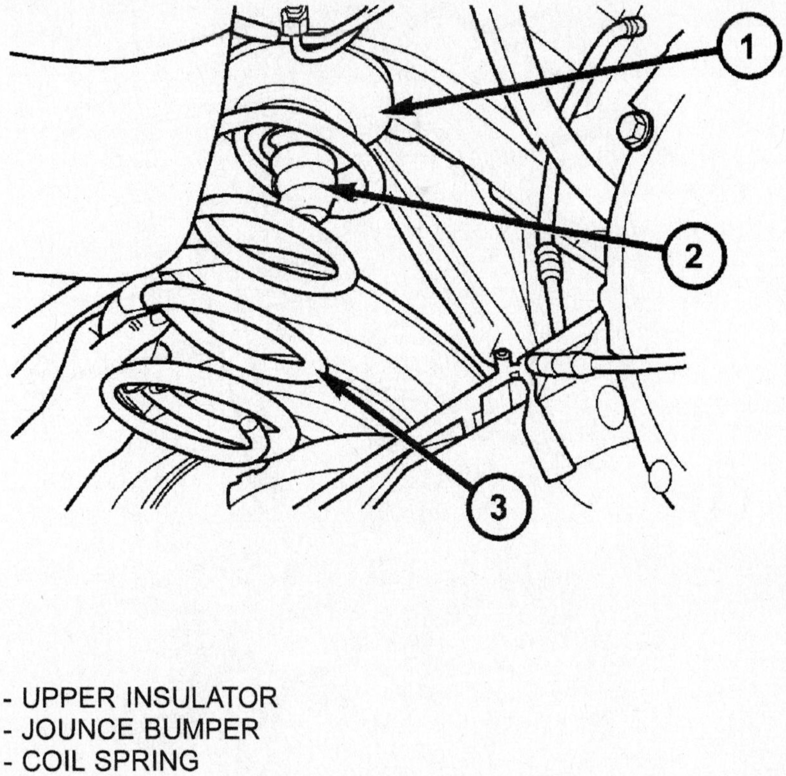

Rear coil spring—Liberty

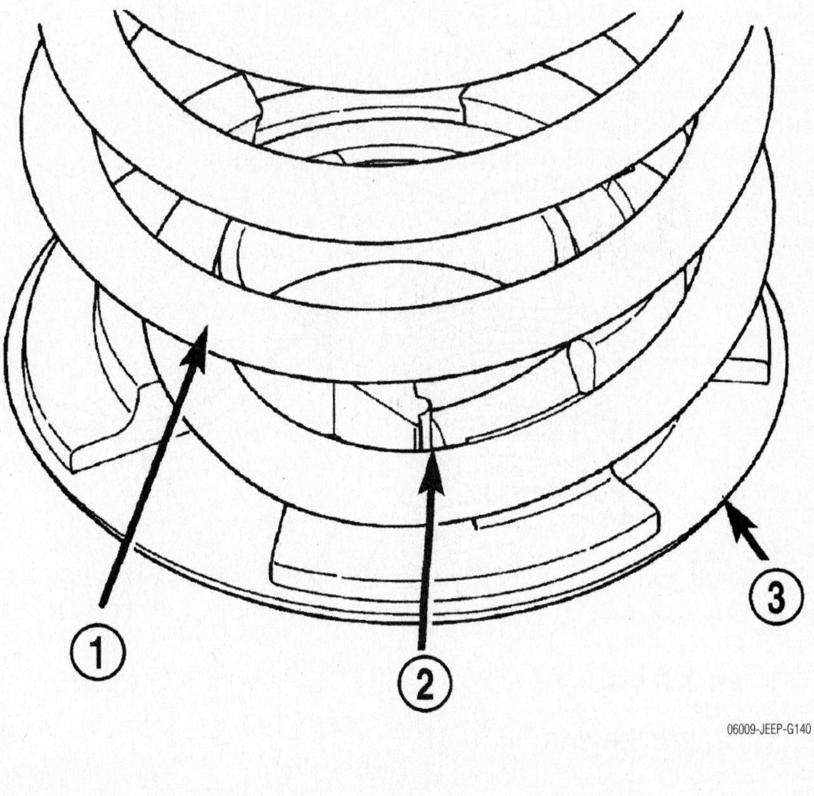

Proper rear spring positioning—2005 Grand Cherokee

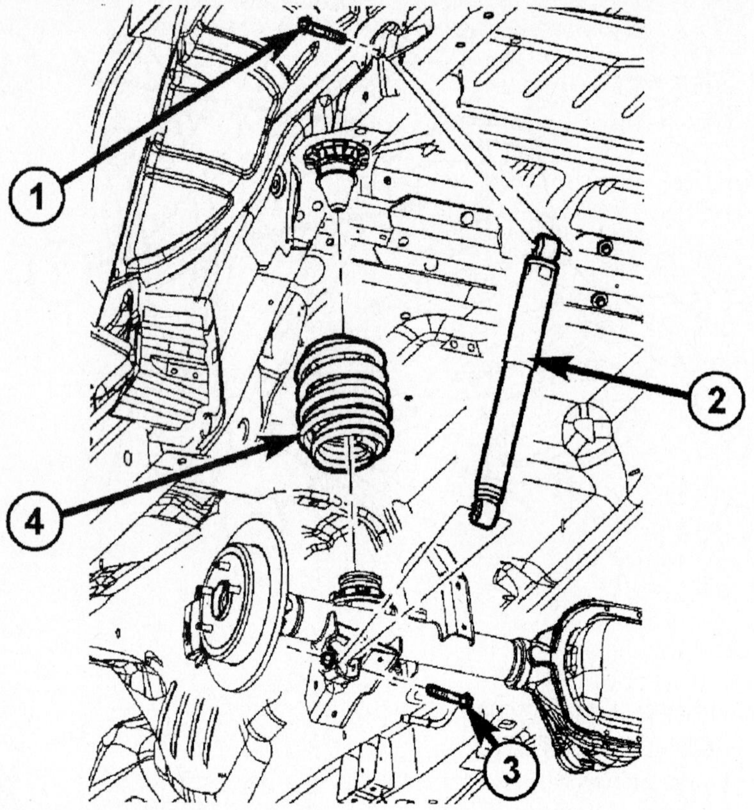

Rear spring and shock installation (1) upper shock bolt, (2) shock, (3) lower shock bolt, (4) spring—2005 Grand Cherokee

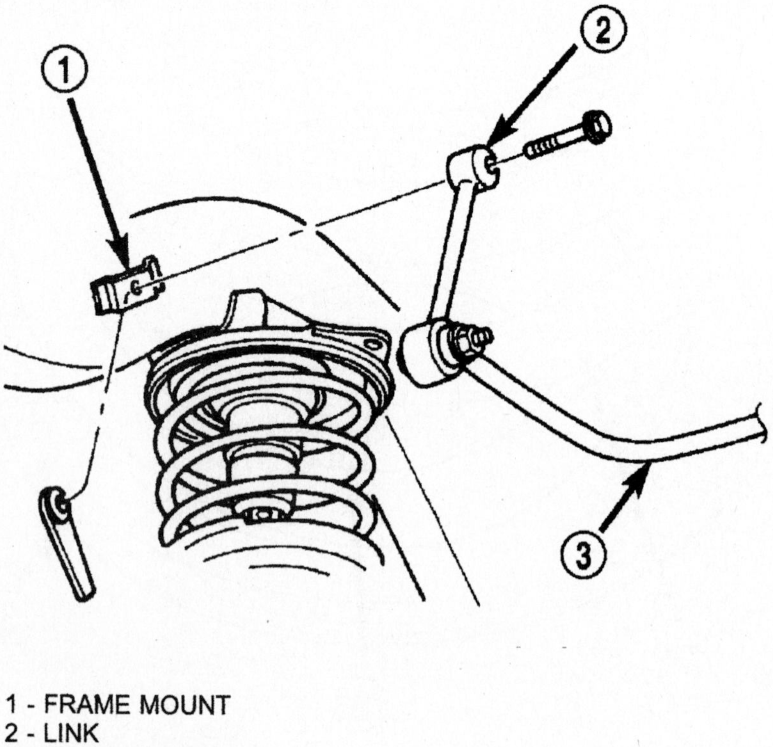

Rear stabilizer bar link—Wrangler

2005 Grand Cherokee

1. Before servicing the vehicle, refer to the precautions in the beginning of this section.
2. Raise and support the vehicle. Position a hydraulic jack under the axle to support the axle.
3. Remove the wheel and tire assemblies.
4. Remove the lower shock bolt from the axle bracket.
5. Remove the stabilizer bar link from the body rail.
6. Lower the hydraulic jack and tilt the axle and remove the coil spring.
7. Remove and inspect the spring isolators.

To install:

8. Install the upper isolator.
9. Install the lower isolator.
10. Pull down on the axle and position the coil spring in the lower isolator.
11. Raise the axle with the hydraulic jack.
12. Install the shock absorber to the axle bracket and tighten to 115 Nm (85 ft. lbs.).
13. Install the stabilizer bar link to the body rail, tighten to 102 Nm (75 ft. lbs.).
14. Install the wheel and tire assemblies.
15. Remove the supports and lower the vehicle.

2005 Wrangler

1. Before servicing the vehicle, refer to the precautions in the beginning of this section.
2. Raise and support the vehicle. Position a hydraulic jack under the axle to support it.
3. Disconnect the stabilizer bar links and shock absorbers from the axle brackets.
4. Disconnect the track bar from the frame rail bracket.
5. Lower the axle until the spring is free from the upper mount seat and remove the spring.

To install:

6. Position the coil spring on the axle pad isolator.
7. Raise the axle into position until the spring seats on the upper isolator.
8. Connect the stabilizer bar links and shock absorbers to the axle bracket. Connect the track bar to the frame rail bracket.
9. Remove the supports and lower the vehicle.
10. Observe the following torques:
    - Stabilizer bar links: 40 ft. lbs. (54 Nm)
    - Shock absorbers: 74 ft. lbs. (100 Nm)
    - Track bar: 74 ft. lbs. (100 Nm)

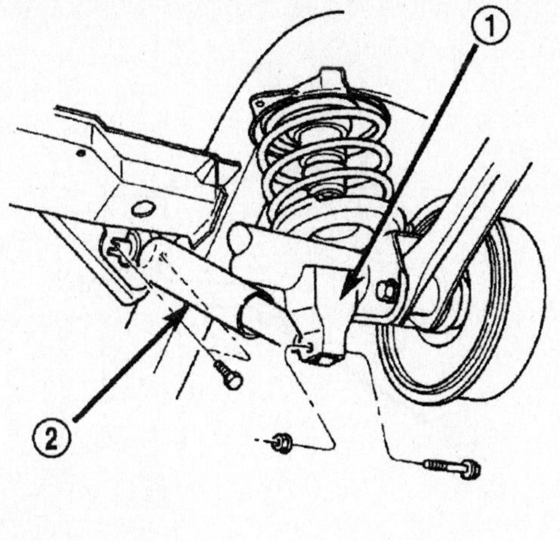

Rear spring and shock absorber—Wrangler

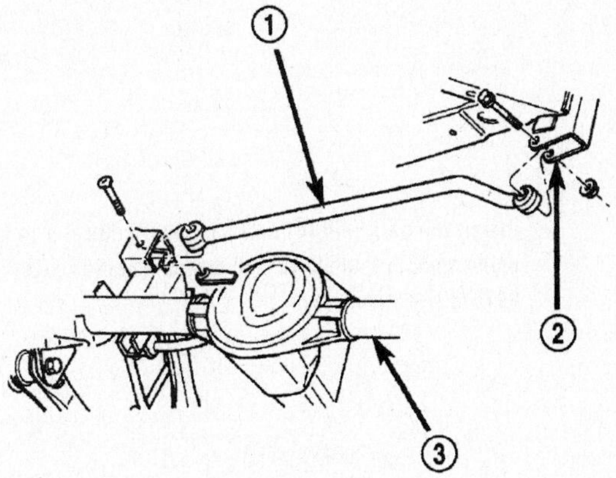

Rear track bar—Wrangler

Front Suspension Upper Ball Joint

REMOVAL & INSTALLATION

2002–04 Wrangler and Grand Cherokee

1. Before servicing the vehicle, refer to the precautions in the beginning of this section.
2. Remove or disconnect the following:
    - Front wheel
    - Disc brake caliper and rotor
    - Wheel bearing
    - Axle shaft
    - Steering knuckle
    - Upper ball joint

To install:

3. Install or connect the following:
    - Upper ball joint
    - Steering knuckle. Torque the nuts to 100 ft. lbs. (135 Nm).
    - Axle shaft
    - Wheel bearing
    - Disc brake caliper and rotor
    - Front wheel

Liberty

The upper ball joint is serviced as an assembly with the control arm.

2005 Grand Cherokee

1. Before servicing the vehicle, refer to the precautions in the beginning of this section.
2. Raise vehicle and support the axle.
3. Remove the tire and wheel.
4. Remove the upper ball joint retaining nut.
5. Separate the upper ball joint from the knuckle using special tool 8677.
6. Remove the wheel speed sensor wire from the upper control arm.
7. Move the knuckle out of the way to allow ball joint removal tool access.

➡Extreme pressure lubrication must be used on the threaded portions of the tool. This will increase the longevity of the tool and insure proper operation during the removal and installation process.

8. Press the ball joint from the upper control arm using special tools C-4212-F and 9652.

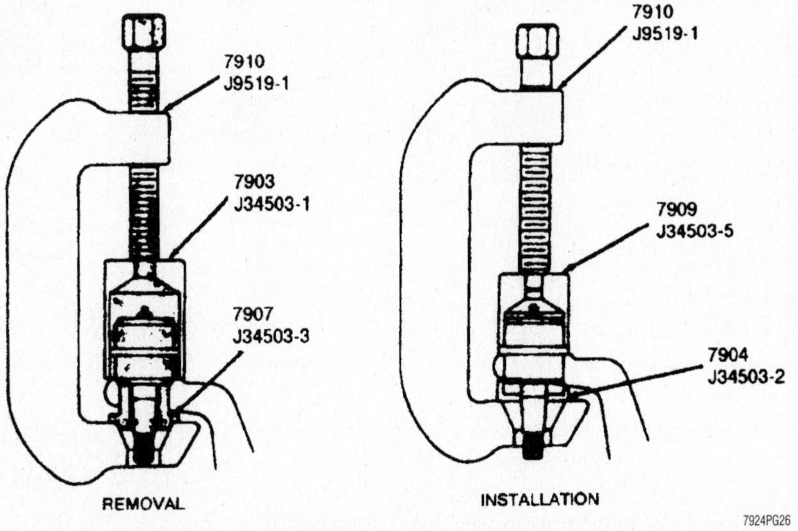

Upper ball joint removal and installation

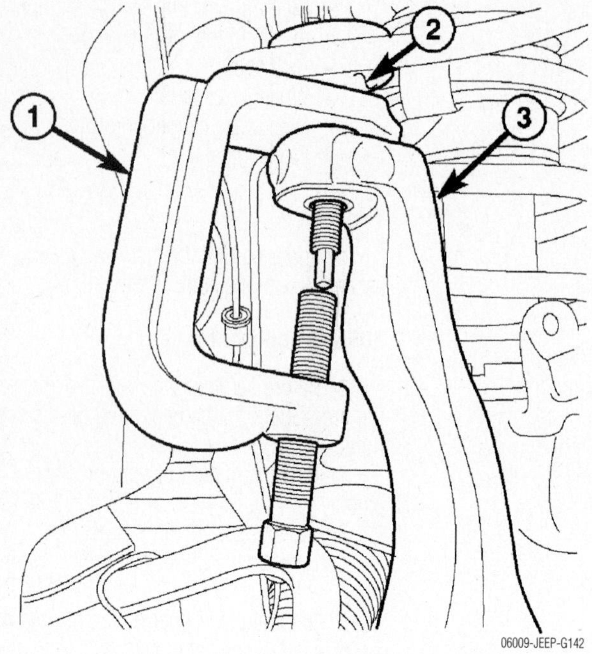

Separate the upper ball joint (2) from the knuckle (3) using special tool 8677 (1)

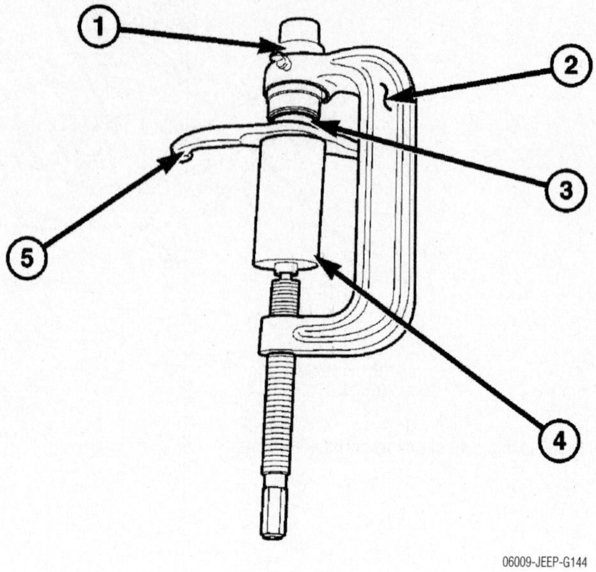

Install the ball joint (4) into the upper control arm (5) and press in using special tools C-4212-F (press) (2), 9652 (Driver) (1) and 8975-2 (Receiver) (4)

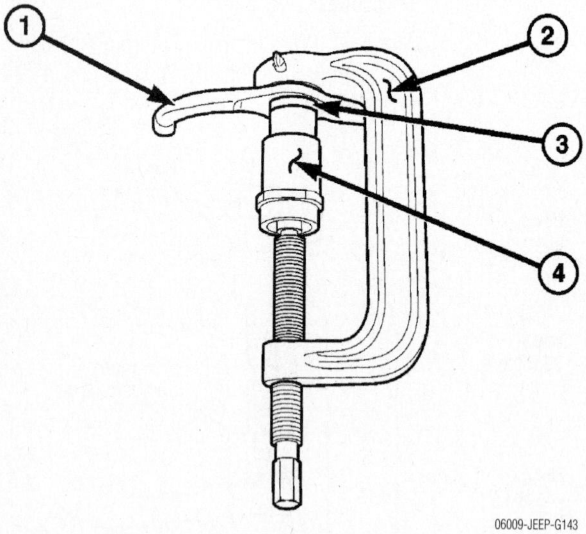

Press the ball joint from the upper control arm (1) using special tools C-4212-F (PRESS) (2) and 9652 (Driver) (3)

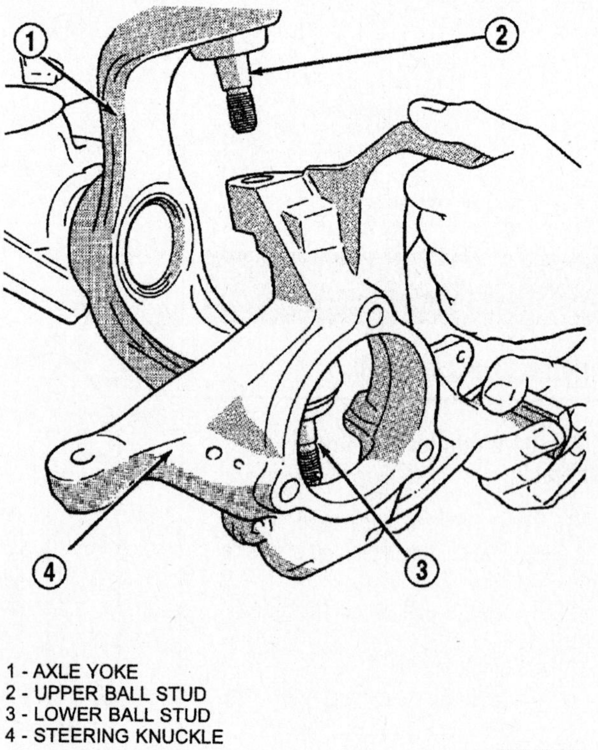

Knuckle removal—Wrangler

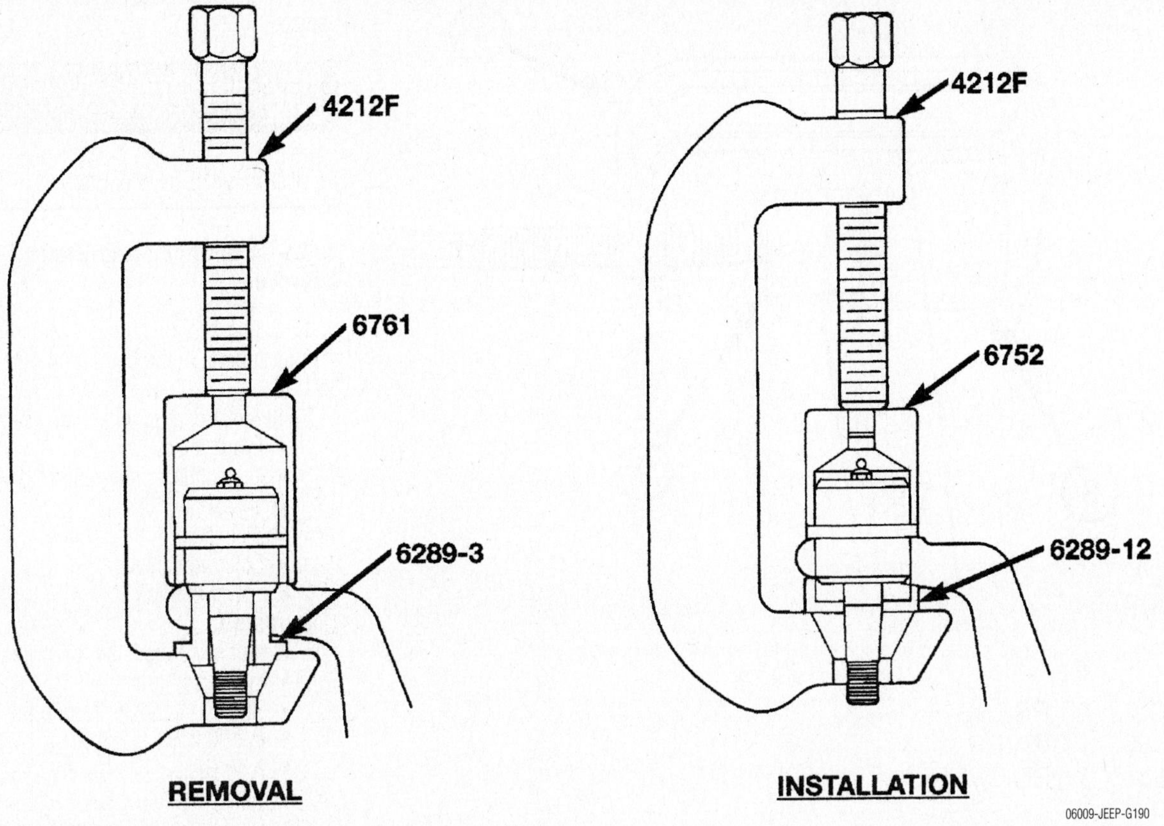

Upper ball joint removal/installation—Wrangler

To install:

➡Extreme pressure lubrication must be used on the threaded portions of the tool. This will increase the longevity of the tool and insure proper operation during the removal and installation process.

9. Install the ball joint into the upper control arm.
10. Install the upper ball joint into the knuckle.
11. Install the upper ball joint retaining nut and tighten to 75 Nm (55 ft. lbs.).
12. Install the wheel speed sensor wire to the upper control arm.
13. Install the tire and wheel.
14. Remove the supports and lower the vehicle.
15. Perform a wheel alignment.

2005 Wrangler

1. Before servicing the vehicle, refer to the precautions in the beginning of this section.
2. Remove hub/bearing and axle shaft.
3. Disconnect the tie-rod or drag link from the steering knuckle arm.
4. Remove the cotter pins from the upper and lower ball studs.
5. Remove the upper and lower ball stud nuts.
6. Using special tool C-4150A separate the ball joints from the steering knuckle. Remove knuckle from ball studs.

To install:

7. Position the steering knuckle on the ball studs.
8. Install and tighten the bottom retaining nut to 109 Nm (80 ft. lbs.) torque. Install new cotter pin.
9. Install and tighten the top retaining nut to 101 Nm (75 ft. lbs.) torque. Install new cotter pin.
10. Install the hub bearing and axle shaft.
11. Connect the tie-rod or drag link end to the steering knuckle arm. Torque to 35 ft. lbs. (47 Nm).

Rear Suspension Upper Ball Joint

REMOVAL & INSTALLATION

Liberty

1. Before servicing the vehicle, refer to the precautions in the beginning of this section.
2. Raise and support the vehicle.
3. Support the rear axle with a hydraulic jack.
4. Remove the ball joint pinch bolt from the top of the axle.
5. Separate the ball joint arm assembly from the differential housing by prying upwards.
6. Remove the upper suspension arm from the vehicle.
7. Secure the suspension arm in a vise.
8. Install special tools C-4212F (press), 8861-3 (driver) and 8861-2 (receiver).
9. Press out the old ball joint.

To install:

10. Install special tools C-4212F (press), 8861-1 (receiver) and 8861-2 (driver) with the ball joint in place.
11. Press the ball joint in the upper suspension arm.
12. Remove the upper suspension arm from the vise.
13. Reinstall the upper suspension arm.
14. Raise the rear axle with a hydraulic jack to align the ball joint with the differential housing bracket.
15. Insert the ball joint into the differential housing bracket.
16. Install the ball joint pinch bolt and tighten to 95 Nm (70 ft. lbs.).

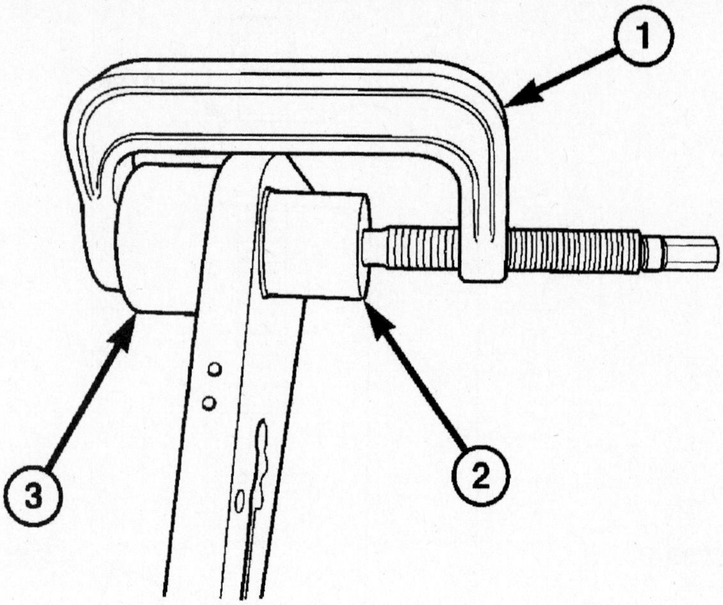

Rear upper ball joint removal—Liberty

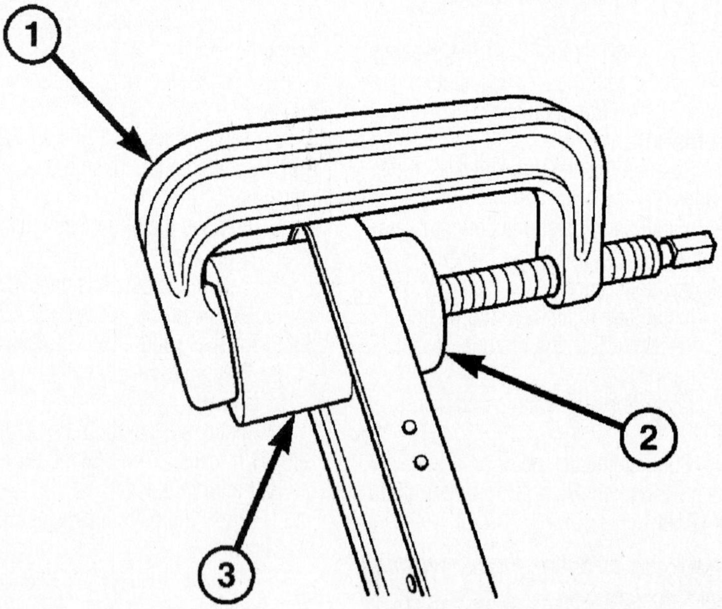

Rear upper ball joint installation—Liberty

17. Remove the supports and lower the vehicle.

Front Suspension Lower Ball Joint

REMOVAL & INSTALLATION

2002–04 Wrangler and Grand Cherokee

1. Before servicing the vehicle, refer to the precautions in the beginning of this section.
2. Remove or disconnect the following:
   - Front wheel
   - Disc brake caliper and rotor
   - Wheel bearing
   - Axle shaft
   - Steering knuckle
   - Lower ball joint

To install:

3. Install or connect the following:
   - Lower ball joint
   - Steering knuckle. Torque the nuts to 100 ft. lbs. (135 Nm).
   - Axle shaft
   - Wheel bearing
   - Disc brake caliper and rotor
   - Front wheel

Liberty

1. Before servicing the vehicle, refer to the precautions in the beginning of this section.
2. Remove or disconnect the following:
   - Front wheel
   - Lower clevis bracket bolt from the control arm
   - Stabilizer link at the control arm
   - Lower ball joint nut
   - Control arm from lower ball joint with tool C4150A, or equivalent
3. Installation is the reverse of removal. Torque the ball joint nut to 60 ft. lbs. (81 Nm); the stabilizer link bolt to 100 ft. lbs. (136 Nm); the lower clevis bracket bolt to 110 ft. lbs. (150 Nm).

2005 Grand Cherokee

1. Before servicing the vehicle, refer to the precautions in the beginning of this section.
2. Remove the tire and wheel assembly.
3. Remove the brake caliper and rotor.
4. Disconnect the tie rod from the steering knuckle using special tool C-3894-A.
5. Separate the upper ball joint from the knuckle using special tool 8677.

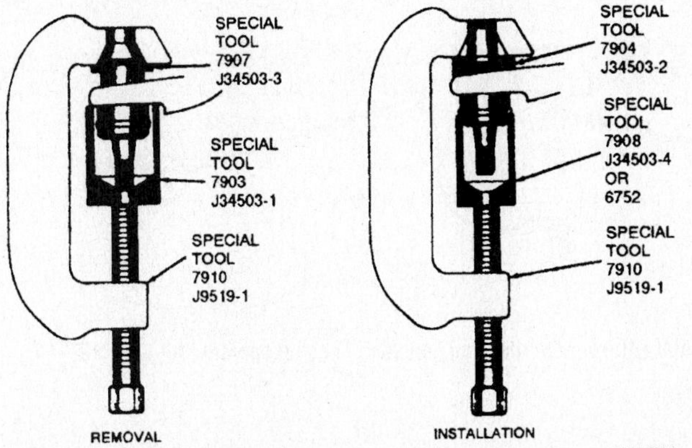

Lower ball joint removal and installation—except Liberty

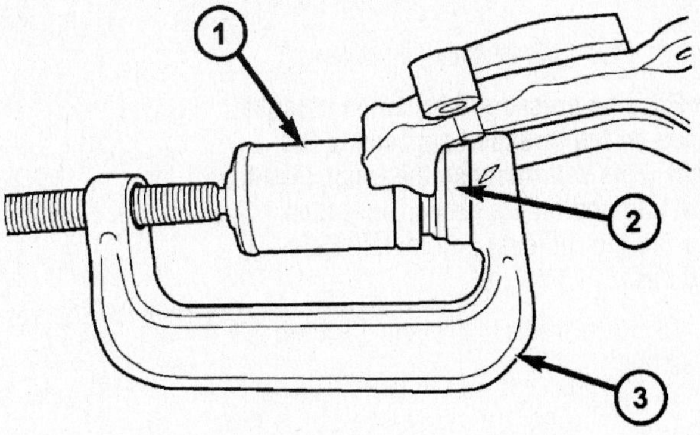

Front lower ball joint installation—Liberty

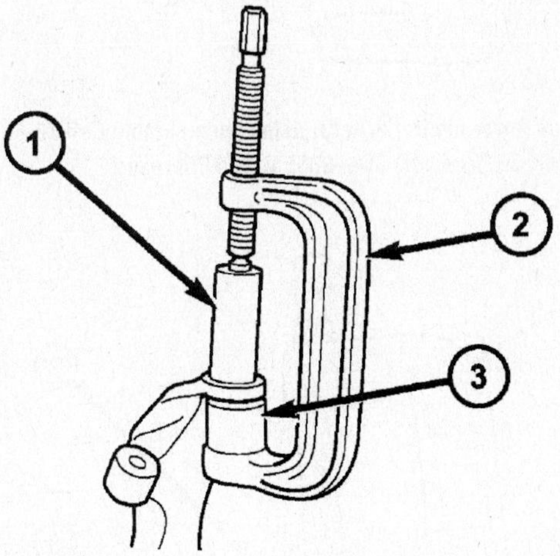

Front lower ball joint removal—Liberty

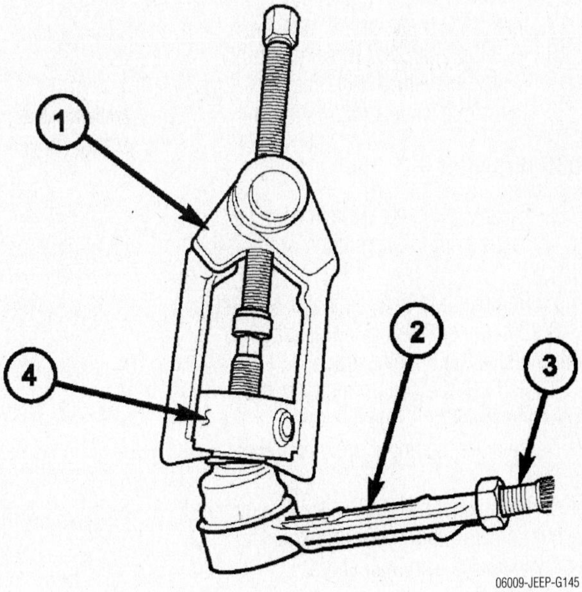

Disconnect the tie rod (1) from the steering knuckle using special tool C-3894-A (1)—2005 Grand Cherokee

6. Separate the lower ball joint from the steering knuckle using special tool 8677.
7. Remove the steering knuckle.
8. Move the halfshaft to the side and support the halfshaft out of the way (4X4 only).
9. Chisel out the ball joint stakes.

➡Extreme pressure lubrication must be used on the threaded portions of the tool. This will increase the longevity of the tool and insure proper operation during the removal and installation process.

10. Press the ball joint from the lower control arm.

To install:

➡Extreme pressure lubrication must be used on the threaded portions of the tool. This will increase the longevity of the tool and insure proper operation during the removal and installation process.

11. Install the ball joint into the control arm.
12. Stake the ball joint flange in four evenly spaced places around the ball joint flange, using a chisel and hammer.
13. Remove the support for the halfshaft and install into position (4X4 only).
14. Install the steering knuckle.
15. Install the tie rod end into the steering knuckle.
16. Install and tighten the halfshaft nut to 251 Nm (185 ft. lbs.) (if equipped).
17. Install the brake caliper and rotor.
18. Install the tire and wheel assembly.
19. Check the vehicle ride height
20. Perform a wheel alignment.

2005 Wrangler

1. Before servicing the vehicle, refer to the precautions in the beginning of this section.
2. Remove hub/bearing and axle shaft.
3. Disconnect the tie-rod or drag link from the steering knuckle arm.
4. Remove the cotter pins from the upper and lower ball studs.
5. Remove the upper and lower ball stud nuts.
6. Using special tool C-4150A separate the ball joints from the steering knuckle. Remove knuckle from ball studs.

To install:

7. Position the steering knuckle on the ball studs.
8. Install and tighten the bottom retaining nut to 109 Nm (80 ft. lbs.) torque. Install new cotter pin.

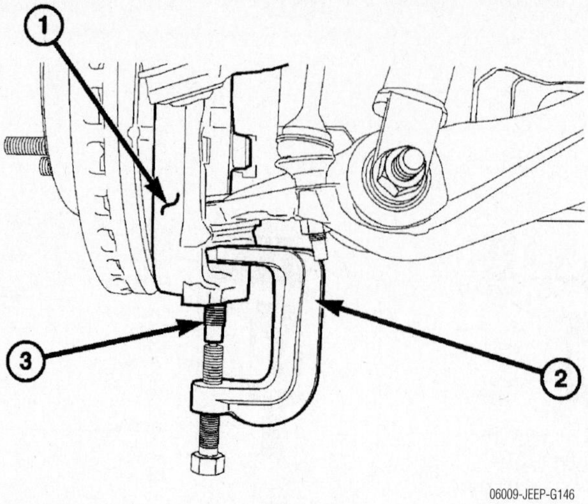

Separate the lower ball joint (3) from the steering knuckle (1) using special tool 8677 (2)—2005 Grand Cherokee

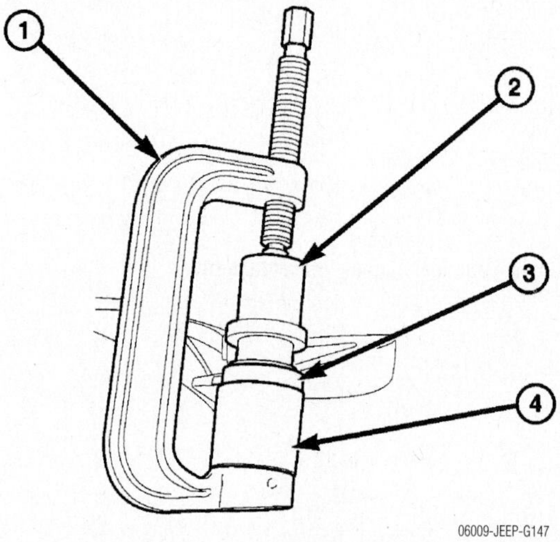

Press the ball joint from the lower control arm (3) using special tools C-4212-F (PRESS) (1), C-4212-3 (Driver) (2) and 9654-3 (Receiver) (4)—2005 Grand Cherokee

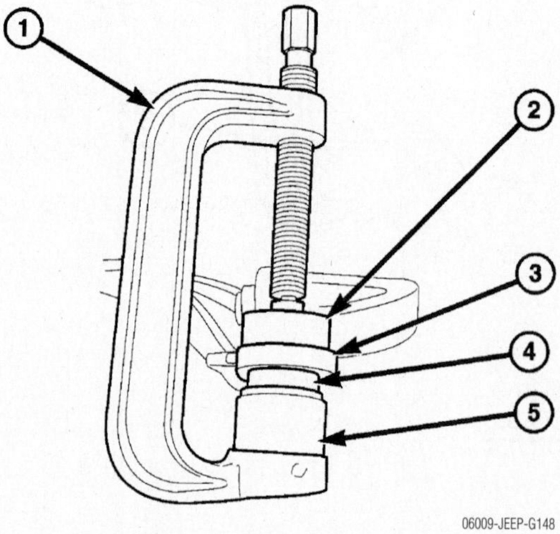

Install the ball joint (4) into the control arm (3) and press in using special tools C-4212-F (press) (1), 9654-1 (Driver) (5) and 9654-2 (Receiver) (2)—2005 Grand Cherokee

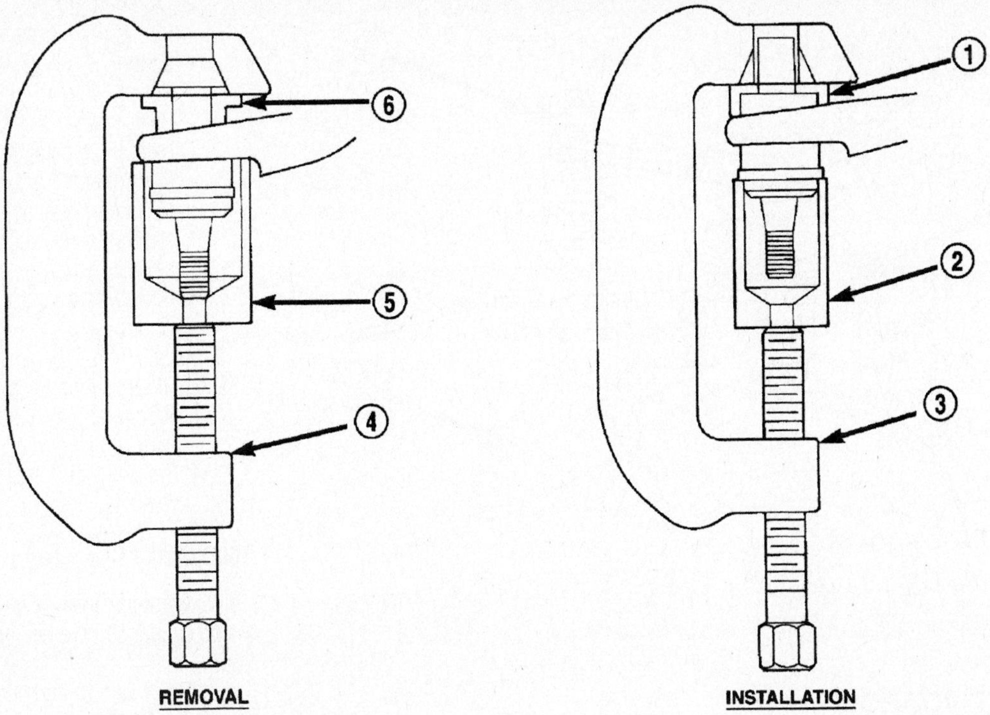

1 - SPECIAL TOOL 6289–12
2 - SPECIAL TOOL 6289–4
3 - SPECIAL TOOL 4212F
4 - SPECIAL TOOL 4212F
5 - SPECIAL TOOL 6289–1
6 - SPECIAL TOOL 6289–3

06009-JEEP-G191

Lower ball joint removal/installation—Wrangler

9. Install and tighten the top retaining nut to 101 Nm (75 ft. lbs.) torque. Install new cotter pin.
10. Install the hub bearing and axle shaft.
11. Connect the tie-rod or drag link end to the steering knuckle arm. Torque to 35 ft. lbs. (47 Nm).

Front Suspension Upper Control Arm

REMOVAL & INSTALLATION

2002–04 Wrangler and Grand Cherokee

1. Before servicing the vehicle, refer to the precautions in the beginning of this section.
2. Support the axle with a jackstand.
3. Unbolt and remove the upper control arm.

To install:

➡The weight of the vehicle must be supported by the springs before tightening the control arm fasteners.

4. Install the control arms.
5. Torque the axle fastener to 55 ft. lbs.

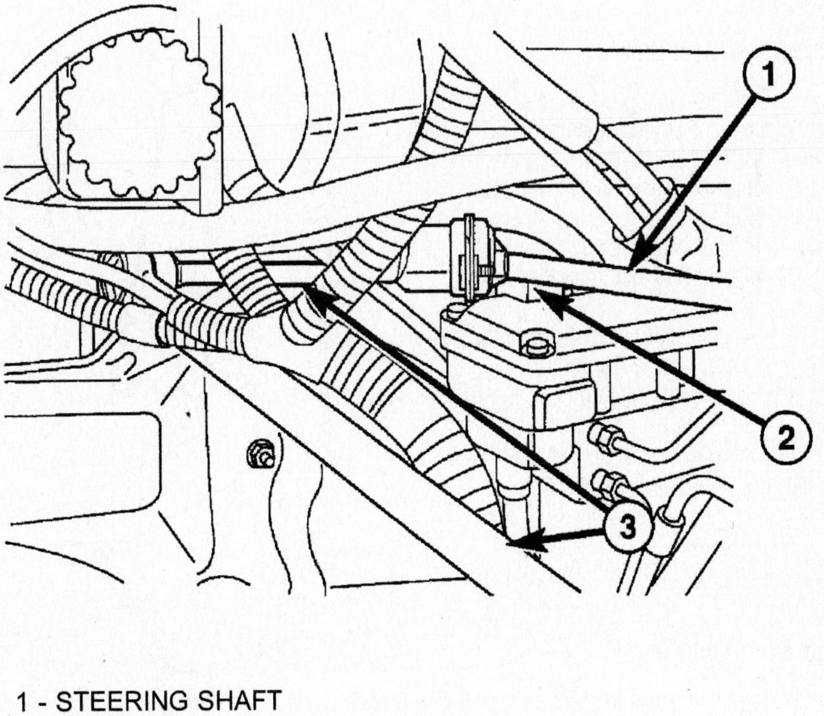

1 - STEERING SHAFT
2 - REAR BOLT
3 - RATCHET WITH AN EXTENSION

06009-JEEP-G71

Accessing the front left side upper control arm rear bolt—Liberty

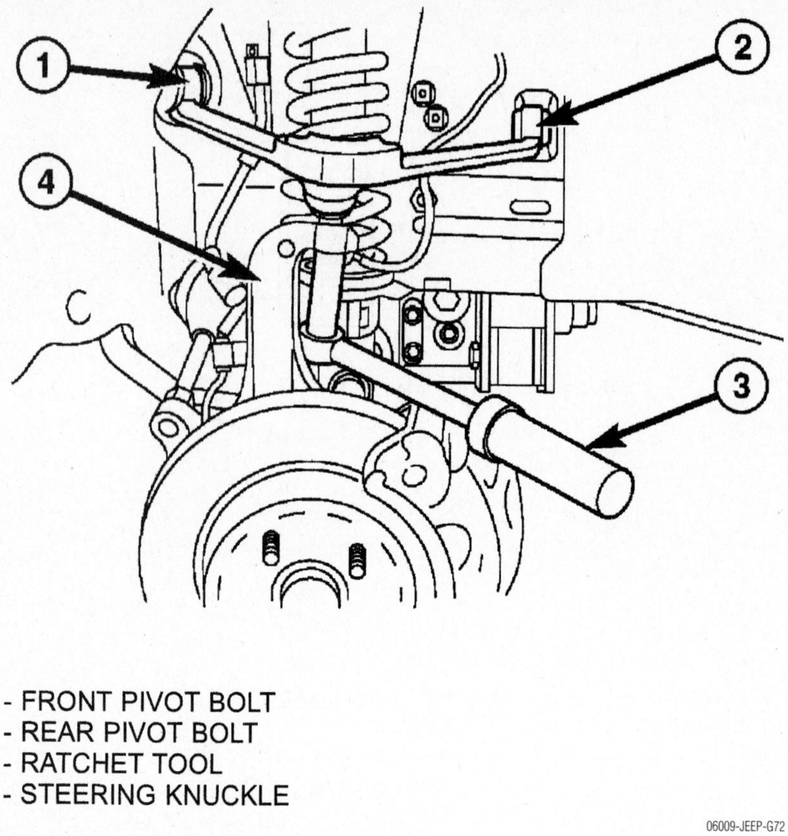

Front left side upper control arm removal—Liberty

(75 Nm) and the frame fastener to 66 ft. lbs. (90 Nm).

Liberty—Right Side

1. Before servicing the vehicle, refer to the precautions at the beginning of this section.
2. Remove or disconnect the following:
   - Wheel
   - Upper ball joint nut
   - Upper Ball joint from the knuckle
   - Air box
   - Cruise control servo mounting nuts
   - Upper arm rear bolt
   - Upper arm front bolt
   - Upper arm
3. Installation is the reverse of removal. Observe the following torques:
   - Front and rear bolts: 90 ft. lbs. (122 Nm)
   - Ball joint stud nut: 60 ft. lbs. (81 Nm)

Liberty—Left Side

1. Before servicing the vehicle, refer to the precautions at the beginning of this section.
2. Remove or disconnect the following:
   - Wheel
   - Ball joint nut
   - Ball joint from the knuckle
   - Battery
   - Power center
   - Battery tray
   - Battery temperature sensor
   - Control arm rear bolt, by using a ratchet and extension under the steering shaft, positioned by the power steering reservoir
   - Control arm front bolt
   - Control arm
3. Installation is the reverse of removal. Observe the following torques:
   - Front and rear bolts: 90 ft. lbs. (122 Nm)
   - Ball joint stud nut: 60 ft. lbs. (81 Nm)

2005 Grand Cherokee

1. Before servicing the vehicle, refer to the precautions in the beginning of this section.
2. Raise vehicle and support the axle.
3. Remove the tire and wheel.
4. Remove the inner fender well.
5. Remove the upper ball joint retaining nut.
6. Separate the upper ball joint from the knuckle using special tool 8677.

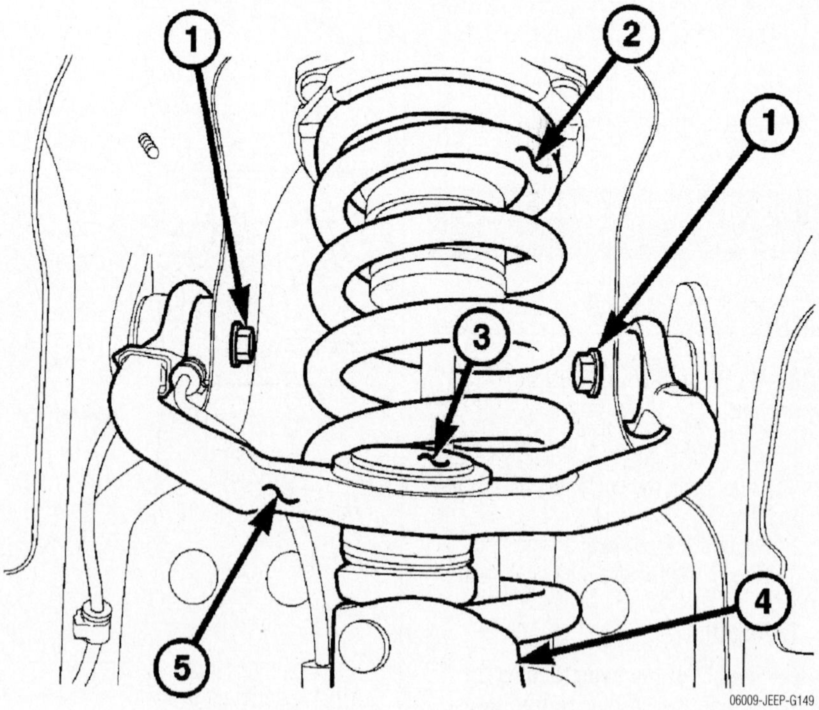

Upper control arm mounting. (1) control arm bolts, (2) spring, (3) ball joint, (4) knuckle, (5) upper control arm—2005 Grand Cherokee

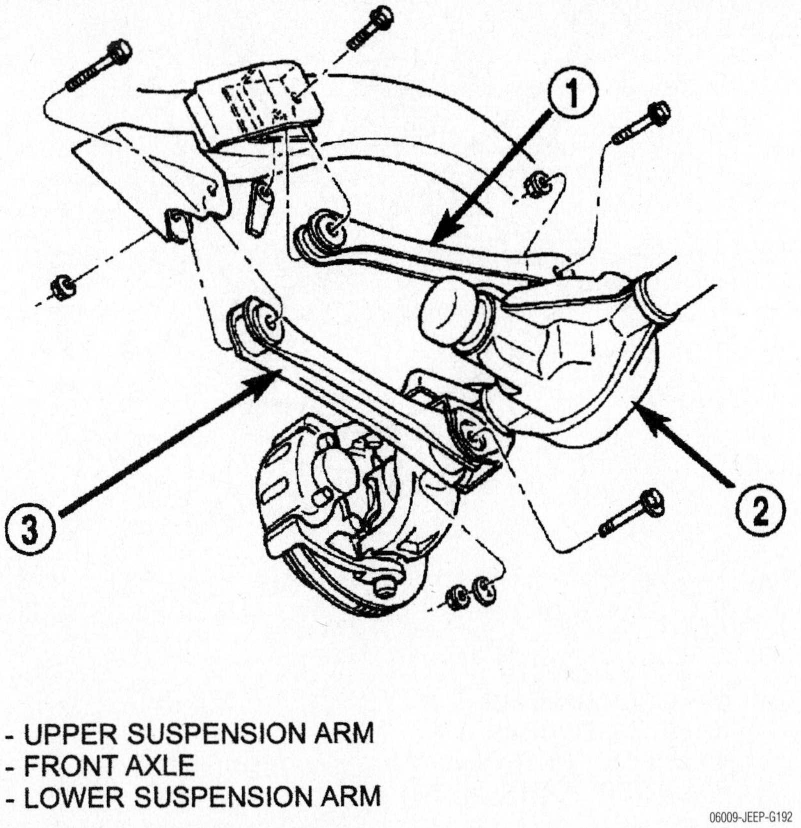

Upper and lower front control arms—Wrangler

7. Remove the wheel speed sensor wire from the upper control arm.
8. Remove the nut and bolt securing the upper control arm to the body.
9. Remove the upper control arm from the vehicle.

To install:

10. Install the upper control arm to the vehicle.
11. Install the nut and bolt securing the upper control arm to the body.
12. Install the wheel speed sensor wire to the upper control arm.
13. Install the upper ball joint into the knuckle.
14. Install the upper ball joint retaining nut and tighten the nut to 81 Nm (60 ft. lbs.).
15. Install the inner fender well.
16. Install the tire and wheel.
17. Remove the supports and lower the vehicle.
18. Perform a wheel alignment.

2005 Wrangler

1. Before servicing the vehicle, refer to the precautions in the beginning of this section.
2. Raise and support the vehicle.
3. Remove the upper suspension arm nut and bolt at the axle bracket.
4. Remove the nut and bolt at the frame rail and remove the upper suspension arm.

To install:

5. Position the upper suspension arm at the axle and frame rail.
6. Install the bolts and finger tighten the nuts.
7. Remove the supports and lower the vehicle.
8. Tighten the nut at the axle and frame brackets to 81 Nm (60 ft. lbs.).

CONTROL ARM BUSHING REPLACEMENT

The upper control arm bushings are serviced with the control arms as complete assemblies, with the exception of the front upper axle bushing, which may be replaced after removing the upper control arm.

Rear Suspension Upper Control Arm

REMOVAL & INSTALLATION

2002–05 Wrangler

1. Before servicing the vehicle, refer to the precautions in the beginning of this section.
2. Support the axle with a jackstand.
3. Remove or disconnect the following:
   - Parking brake cable and bracket
   - Wheel speed sensor wiring bracket, if equipped
   - Upper control arm

To install:

➡The weight of the vehicle must be supported by the springs before tightening the control arm fasteners.

4. Install or connect the following:
   - Upper control arm
   - Wheel speed sensor wiring bracket, if equipped
   - Parking brake cable and bracket
5. Torque the control arm fasteners to 55 ft. lbs. (75 Nm).

2002–04 Grand Cherokee

1. Before servicing the vehicle, refer to the precautions in the beginning of this section.
2. Support the axle with a jackstand.
3. Remove or disconnect the following:
   - Parking brake cable brackets
   - Brake hose brackets
   - Axle ball joint
   - Upper control arm

To install:

➡Use a new axle ball joint nut.

4. Install or connect the following:
   - Upper control arm. Torque the frame bracket bolts to 74 ft. lbs. (100 Nm) and the axle ball joint nut to 105 ft. lbs. (142 Nm).
   - Brake hose brackets
   - Parking brake cable brackets

2005 Grand Cherokee

LEFT SIDE

1. Before servicing the vehicle, refer to the precautions in the beginning of this section.
2. Raise and support the vehicle.
3. Support the rear axle.
4. Lower the fuel tank in order to gain access to the bolt.
5. Remove the upper suspension arm nut and bolt from the axle bracket.
6. Remove the nut and bolt from the frame rail and remove the upper suspension arm.

To install:

➡All torques should be done with vehicle on the ground with full vehicle weight.

7. Position the upper suspension arm in the frame rail bracket.

8. Install the mounting bolt and nut tighten to 129 Nm (95 ft. lbs.).

9. Position the upper suspension arm in the axle bracket.

10. Install the mounting bolt and nut tighten to 136 Nm (100 ft. lbs.).

11. Raise the fuel tank back into place and secure.

12. Remove the supports and lower the vehicle.

RIGHT SIDE

1. Before servicing the vehicle, refer to the precautions in the beginning of this section.
2. Raise and support the vehicle.
3. Support the rear axle.
4. Remove the upper suspension arm nut and bolt from the axle bracket.
5. Remove the nut and bolt from the frame rail and remove the upper suspension arm.

To install:

➡All torques should be done with vehicle on the ground with full vehicle weight.

6. Position the upper suspension arm in the frame rail bracket.
7. Install the mounting bolt and nut tighten to 129 Nm (95 ft. lbs.).
8. Position the upper suspension arm in the axle bracket.
9. Install the mounting bolt and nut tighten to 136 Nm (100 ft. lbs.).
10. Remove the supports and lower the vehicle.

Liberty

1. Before servicing the vehicle, refer to the precautions in the beginning of this section.
2. Raise and support the vehicle.
3. Support the rear axle with a hydraulic jack.
4. Remove the ball joint pinch bolt from the top of the differential housing bracket.
5. Remove partial nuts from the heat shield in order to lower the shield down enough to get the proper clearance to remove the right side bolt from the body.
6. Remove the upper suspension arm mounting bolts from the body and remove the arm.
7. Remove the support bracket mounting bolts if needed.

To install:

8. Position the upper suspension arm in the frame rail brackets.
9. Install the mounting bolts and tighten to 100 Nm (74 ft. lbs.).

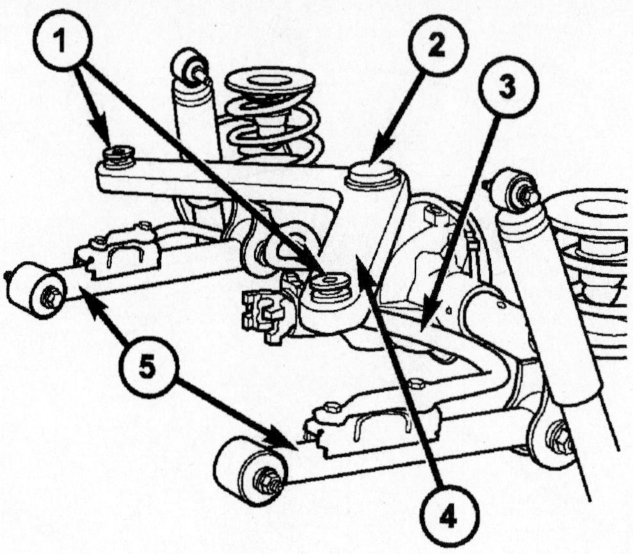

Rear suspension upper control arm—Liberty

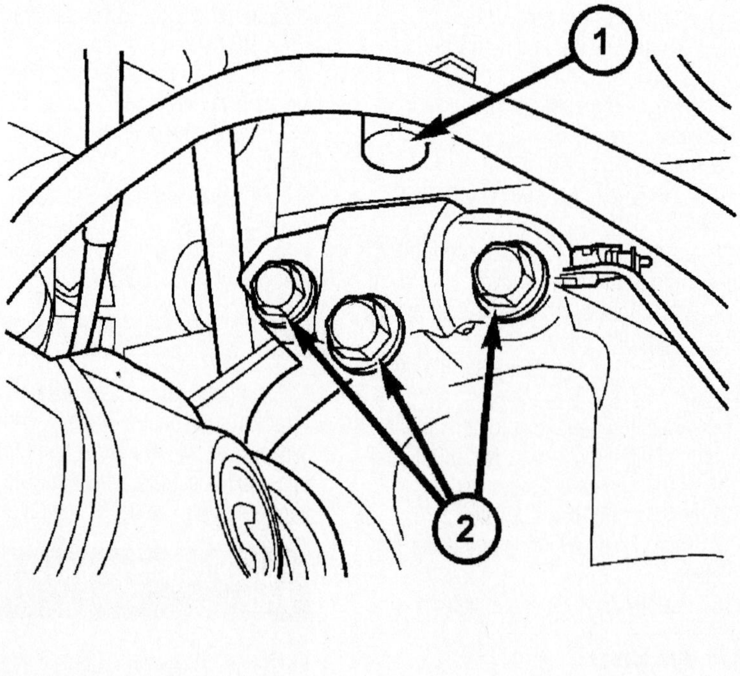

Rear suspension upper ball joint bracket—Liberty

10. Retighten the heat shield back into place.
11. Pull the arm down on the differential housing bracket and install the pinch bolt and nut. Tighten the nut to 95 Nm (70 ft. lbs.).
12. Remove the supports and lower the vehicle.

Front Suspension Lower Control Arms

REMOVAL & INSTALLATION

2002–04 Wrangler and Grand Cherokee

1. Before servicing the vehicle, refer to the precautions in the beginning of this section.
2. Support the axle with a jackstand.
3. Remove or disconnect the following:
   - Wheel speed sensor wiring, if equipped
   - Lower control arm

To install:

➡The weight of the vehicle must be supported by the springs before tightening the control arm fasteners.

4. Install or connect the following:
   - Lower control arm
   - Wheel speed sensor wiring, if equipped
5. Tighten the control arm bolts to the following specifications:
   - Wrangler: Axle fastener to 85 ft. lbs. (115 Nm) and frame bracket fastener to 130 ft. lbs. (176 Nm).
   - Grand Cherokee: Frame bracket bolt to 115 ft. lbs. (156 Nm) and axle bracket nut to 120 ft. lbs. (163 Nm).

Liberty

1. Before servicing the vehicle, refer to the precautions in the beginning of this section.
2. Remove or disconnect the following:
   - Front wheel
   - Lower clevis bracket bolt from the control arm
   - Stabilizer link at the control arm
   - Lower ball joint nut
   - Control arm from lower ball joint with tool C4150A

➡Matchmark the front and rear control arm pivot bolts.

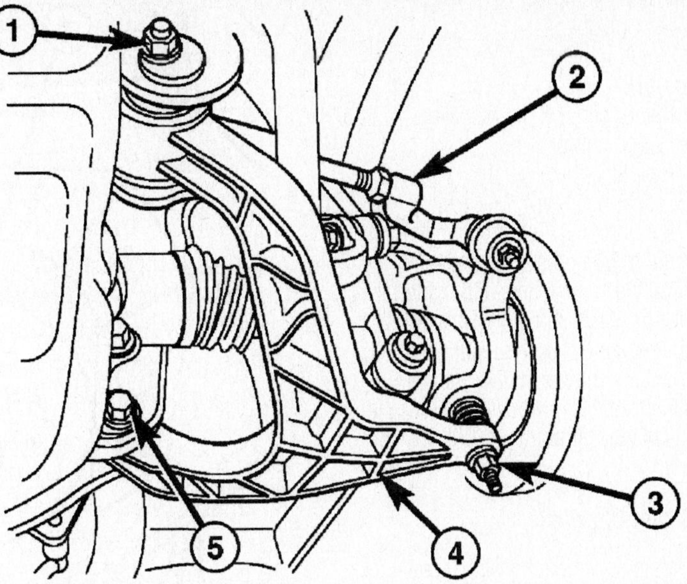

1 - FRONT CAM BOLT
2 - OUTER TIE ROD END
3 - LOWER BALL JOINT NUT
4 - LOWER CONTROL ARM
5 - REAR CAM BOLT

06009-JEEP-G78

Front lower control arm—Liberty

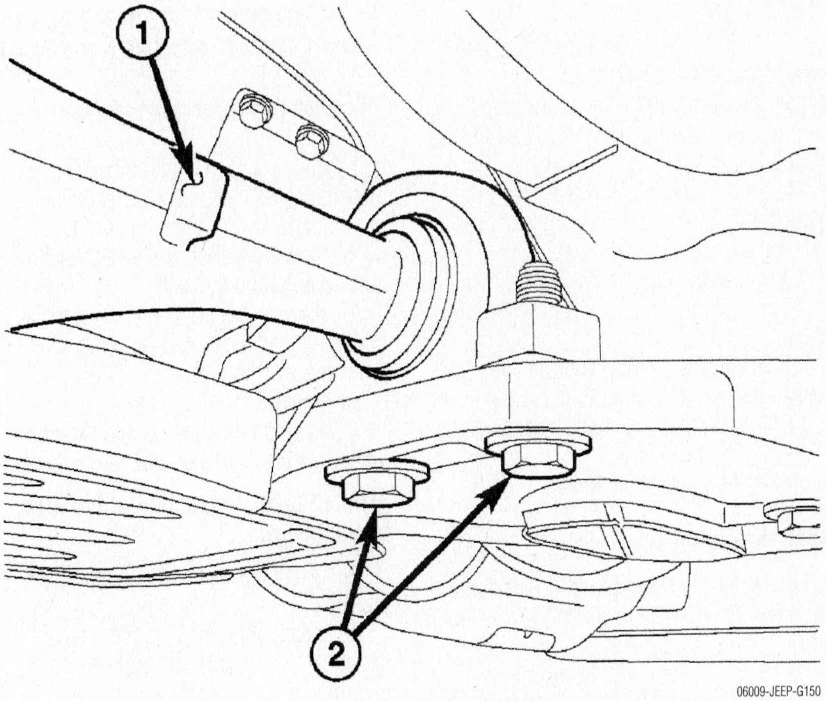

06009-JEEP-G150

Lower control arm (1) mounting bolts (2)—2005 Grand Cherokee

- Front pivot bolt
- Rear pivot bolt
- Control arm

To install:

3. Install or connect the following:
   - Lower control arm
   - Rear pivot bolt
   - Front pivot bolt
   - Ball joint nut. Torque the ball joint nut to 60 ft. lbs. (81 Nm)
4. Align the matchmarks and tighten the pivot bolts to 125 ft. lbs. (170 Nm).
5. The remainder of the installation is the reverse of removal. Torque the stabilizer link bolt to 100 ft. lbs. (136 Nm); the lower clevis bracket bolt to 110 ft. lbs. (150 Nm).

2005 Grand Cherokee

1. Before servicing the vehicle, refer to the precautions in the beginning of this section.
2. Raise and support the vehicle.
3. Remove the tire and wheel assembly.
4. Remove the steering knuckle.
5. Remove the shock clevis bracket from the lower control arm.
6. Remove the nut and bolt from the front of the lower control arm.
7. Remove the rear bolts and flag nuts from the lower control arm.
8. Remove the lower control arm from the vehicle.

To install:

9. Position the lower suspension arm into the frame rail bracket.
10. Install the rear bolts for the lower control arm to the frame, Tighten to 142 Nm (105 ft lbs).
11. Install the nut and bolt for the front of the lower control arm Tighten to 169 Nm (125 ft lbs).
12. Install the lower clevis bolt at the lower control arm and tighten to 88 Nm (65 ft. lbs.).
13. Install the steering knuckle and tighten the nut to 75 Nm (55 ft. lbs.).
14. Install the tire and wheel assembly.
15. Lower the vehicle.
16. Perform wheel alignment.

2005 Wrangler

1. Before servicing the vehicle, refer to the precautions in the beginning of this section.
2. Raise and support the vehicle.
3. If equipped with ABS brakes remove sensor wire from the inboard side of the arm.
4. If the vehicle is equipped with a cam bolt service package paint or scribe alignment marks on the cam adjusters and suspension arm for installation reference.

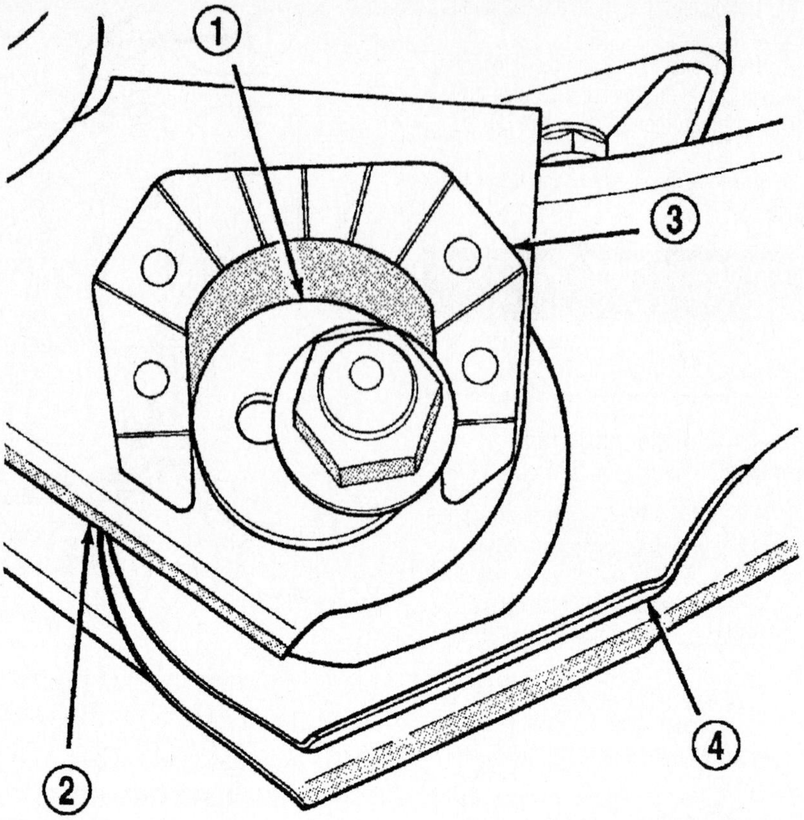

1 - ADJUSTMENT CAM
2 - AXLE BRACKET
3 - BRACKET REINFORCEMENT
4 - LOWER SUSPENSION ARM

06009-JEEP-G193

Cam bolt service package—Wrangler

5. Remove the lower suspension arm nut and bolt from the axle.
6. Remove the nut and bolt/cam bolt from the frame rail bracket and remove the lower suspension arm.

To install:

7. Position the lower suspension arm in the axle bracket and frame rail bracket.

➡Small holes in the side of the arm face inboard.

8. Install the rear bolt and nut finger tighten.
9. Install bolt/cam bolt and new nut finger tighten in the axle and align the reference marks.
10. If equipped with ABS brakes install sensor wire to the inboard side of the arm with new clips.
11. Lower the vehicle.
12. Tighten axle bracket nut to 176 Nm (130 ft. lbs.).
13. Tighten frame bracket nut to 176 Nm (130 ft. lbs.).

Rear Suspension Lower Control Arm

REMOVAL & INSTALLATION

Except 2005 Grand Cherokee

1. Before servicing the vehicle, refer to the precautions in the beginning of this section.
2. Support the axle with a jackstand.
3. On the Liberty, disconnect the stabilizer bar from the arm.
4. Unbolt and remove the lower control arm.

➡On the Liberty's right arm, it will be necessary to pry the exhaust pipe slightly to get to the frame rail-to-arm bolt.

To install:

➡The weight of the vehicle must be supported by the springs before tightening the control arm fasteners.

5. Install the lower control arm.
6. Tighten the lower control arm fasteners to the following specifications:
   - 2002–04 Wrangler: Both fasteners to 130 ft. lbs. (177 Nm)
   - 2005 Wrangler: Both fasteners to 150 ft. lbs. (203 Nm)
   - Liberty: Both fasteners to 120 ft. lbs. (163 Nm)
   - Grand Cherokee: Frame bracket nut to 115 ft. lbs. (156 Nm) and axle bracket nut to 120 ft. lbs. (163 Nm)

2005 Grand Cherokee

LEFT SIDE

1. Before servicing the vehicle, refer to the precautions in the beginning of this section.
2. Raise the vehicle and support the rear axle.
3. Remove the fuel tank.
4. Remove the lower suspension arm nut and bolt from the axle bracket.
5. Remove the nut and bolt from the frame rail and remove the lower suspension arm.

➡All torques should be done with vehicle on the ground with full vehicle weight.

To install:

6. Position the lower suspension arm in the frame rail.
7. Install the frame rail bracket bolt and nut. Tighten to176 Nm (130 ft. lbs.).
8. Position the lower suspension arm in the axle bracket.
9. Install the axle bracket bolt and nut. Tighten to 210 Nm (155 ft. lbs.).
10. Install the fuel tank.
11. Remove the supports and lower the vehicle.

RIGHT SIDE

1. Before servicing the vehicle, refer to the precautions in the beginning of this section.
2. Raise the vehicle and support the rear axle.
3. Remove the lower suspension arm nut and bolt from the axle bracket.
4. Remove the nut and bolt from the frame rail and remove the lower suspension arm.

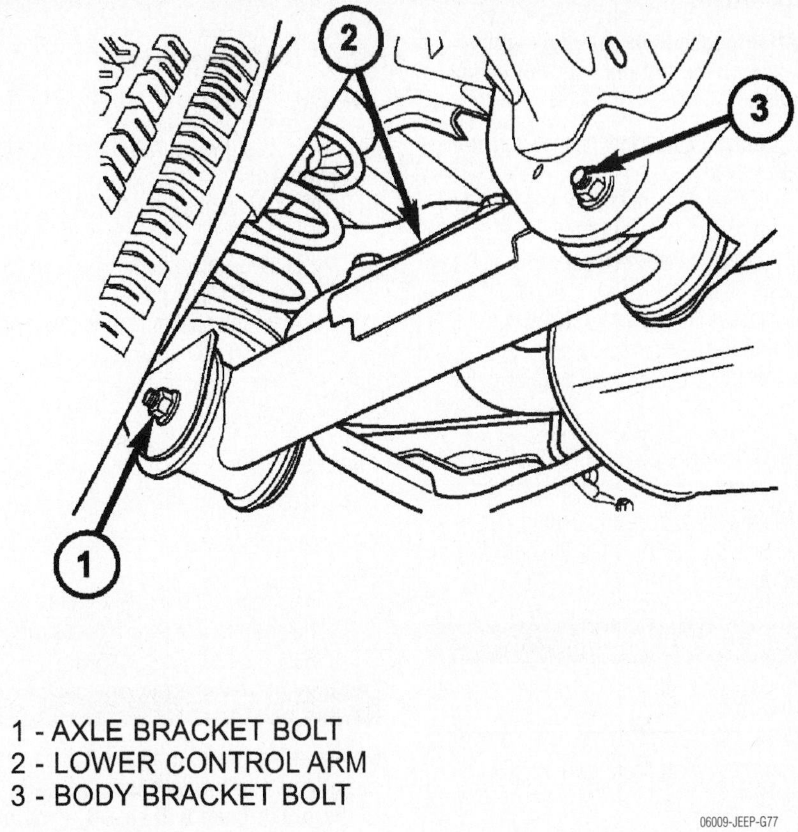

Rear lower control arm—Liberty

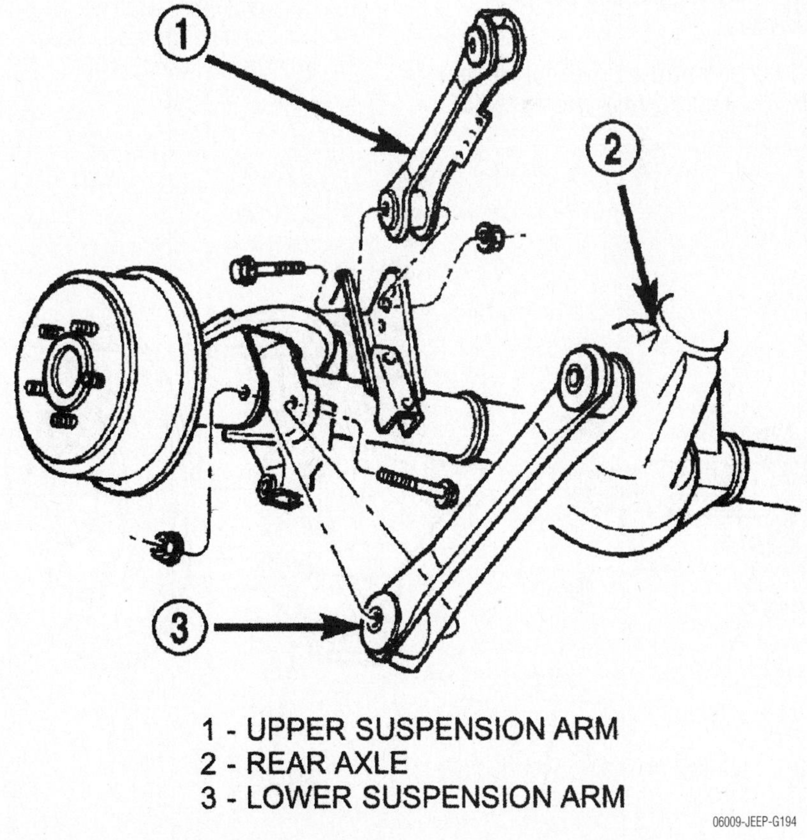

Rear upper and lower control arms—Wrangler

To install:

➡All torques should be done with vehicle on the ground with full vehicle weight.

5. Position the lower suspension arm in the frame rail.
6. Install the frame rail bracket bolt and nut tighten to 176 Nm (130 ft. lbs.).
7. Position the lower suspension arm in the axle bracket.
8. Install the axle bracket bolt and nut. Tighten to 210 Nm (155 ft. lbs.).
9. Remove the supports and lower the vehicle.

CONTROL ARM BUSHING REPLACEMENT

The lower control arm bushings are serviced with the control arms as complete assemblies.

Front Wheel Bearings

ADJUSTMENT

The front wheel bearings are not adjustable.

REMOVAL & INSTALLATION

2002–04 Wrangler and Grand Cherokee

1. Before servicing the vehicle, refer to the precautions in the beginning of this section.
2. Remove or disconnect the following:
   - Front wheel
   - Brake caliper and rotor
   - Wheel speed sensor and bracket
   - Axle stub shaft nut
   - Hub and wheel bearing assembly

To install:

3. Install or connect the following:
   - Hub assembly over the axle stub shaft. Tighten the hub bolts to 75 ft. lbs. (102 Nm)
   - Axle stub shaft nut. Torque the nut to 175 ft. lbs. (237 Nm)
   - Brake caliper and rotor
   - Front wheel

Liberty

1. Before servicing the vehicle, refer to the precautions in the beginning of this section.
2. Raise and support the vehicle.
3. Remove the tire and wheel assembly.
4. Remove the caliper adapter.

**** WARNING**

Never allow the disc brake caliper to hang from the brake hose. Damage to the brake hose will result. Provide a suitable support to hang the caliper securely.

5. Remove the disc brake rotor.
6. Remove the wheel speed sensor.
7. Remove the bracket securing the wheel speed sensor wire.
8. Remove the axle shaft nut, if equipped with four wheel drive.
9. Remove the three mounting bolts for the hub/bearing assembly.
10. Remove the hub/bearing.

To install:

11. Install the hub/bearing assembly to the vehicle.
12. Install the three mounting bolts for the hub/bearing. Tighten the bolt to 130 Nm (96 ft. lbs.).
13. Install the axle shaft nut. Tighten the nut to 135 Nm (100 ft. lbs.), if equipped with four wheel drive.
14. Install the bracket to the wheel speed sensor wire.
15. Install the wheel speed sensor to the hub. Tighten the bolt to 13.5 Nm (10 ft. lbs.).
16. Install the disc brake rotor.
17. Install the disc brake caliper adapter. Tighten the nut to 135 Nm (100 ft. lbs.).
18. Install the tire and wheel assembly.

2005 Wrangler

1. Before servicing the vehicle, refer to the precautions in the beginning of this section.
2. Raise and support the vehicle.
3. Remove the wheel and tire assembly.
4. Remove the brake caliper, rotor and ABS wheel speed sensor.
5. Remove the cotter pin, nut retainer and axle hub nut.
6. Remove the hub bearing mounting

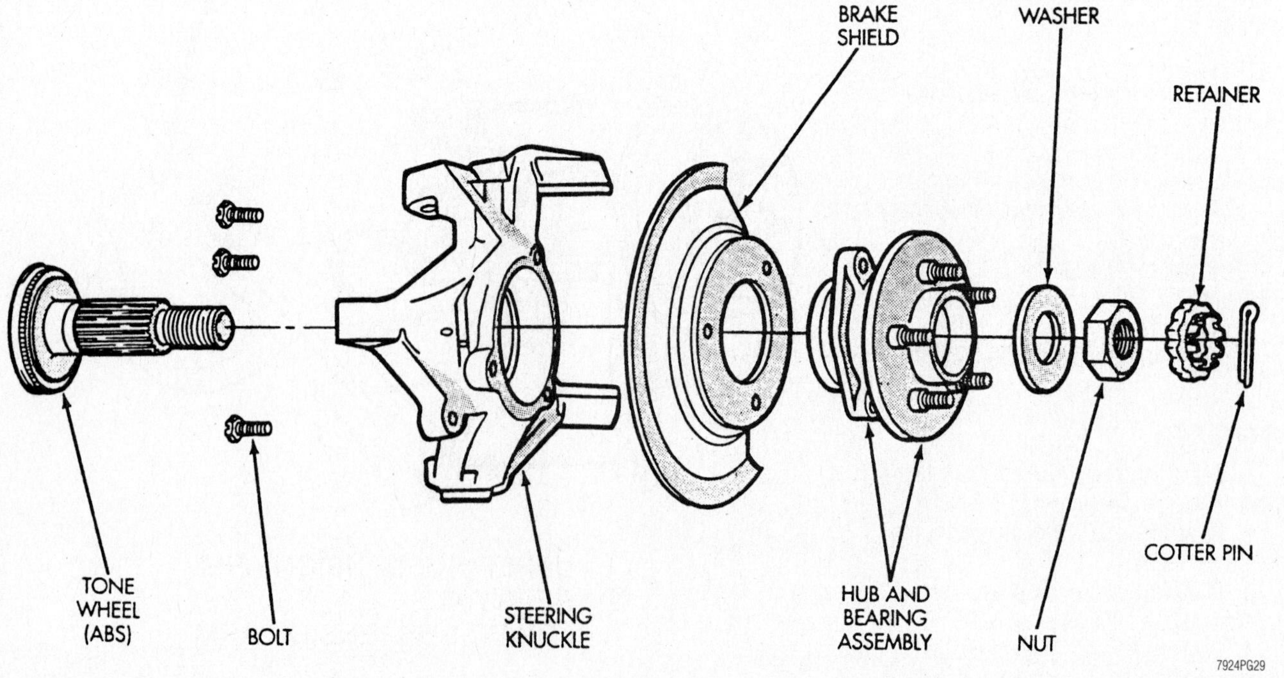

Exploded view of the hub assembly—2002–04 4WD models

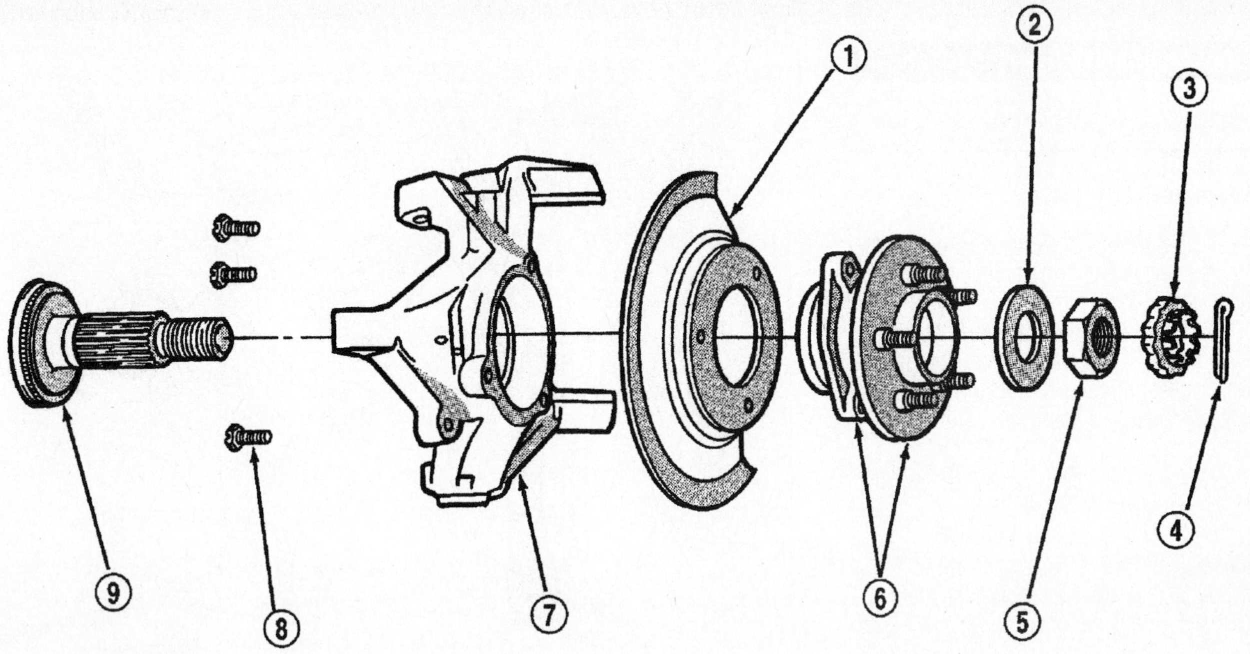

Hub/bearing and knuckle—Wrangler shown

bolts from the back of the steering knuckle. Remove hub bearing from the steering knuckle and off the axle shaft.

To install:

7. Install the hub bearing and brake dust shield to the knuckle.
8. Install the hub bearing to knuckle bolts and tighten to 102 Nm (75 ft. lbs.).
9. Install the hub washer and nut. Tighten the hub nut to 237 Nm (175 ft. lbs.). Install the nut retainer and a new cotter pin.
10. Install the brake rotor, caliper and ABS wheel speed sensor.
11. Install the wheel and tire assembly.
12. Remove support and lower the vehicle.

2005 Grand Cherokee

1. Before servicing the vehicle, refer to the precautions in the beginning of this section.
2. Raise and support the vehicle.
3. Remove the wheel and tire assembly.

**** WARNING**

Support the caliper. Do not let the caliper hang by the hose.

4. Remove the disc brake caliper.
5. Remove the brake caliper adaptor.
6. Remove and discard the O-ring and then remove the disc brake rotor.
7. Remove the wheel speed sensor nut.
8. Remove the wheel speed sensor.
9. Remove the 3 hub bearing mounting bolts from the back of the steering knuckle. Remove hub bearing from the steering knuckle.

To install:

10. Install the hub bearing to the knuckle.
11. Install the hub bearing to knuckle and the 3 bolts then tighten to 136 Nm (100 ft. lbs.).
12. Install the wheel speed sensor.
13. Install the wheel speed sensor nut.
14. Install the brake rotor.
15. Install the brake caliper adaptor.
16. Install the caliper
17. Install the wheel and tire assembly.
18. Remove the support and lower the vehicle.

BRAKES

Front Brake Caliper

REMOVAL & INSTALLATION

Wrangler

1. Before servicing the vehicle, refer to the precautions in the beginning of this section.
2. Drain ⅔ of the brake fluid from the front reservoir. Use the bleeder screw at the front outlet port to drain the fluid. If equipped with anti-lock brakes, relieve the system pressure.
3. Raise and safely support the vehicle.
4. Remove the wheels.
5. Place a C-clamp on the caliper so the solid end contacts the back of the caliper and screw end contacts the metal part of the outboard brake pad.
6. Tighten the clamp until the caliper moves far enough to force the piston to the bottom of the piston bore. This will back the brake pads off of the rotor surface to facilitate the removal and installation of the caliper assembly.
7. Remove the C-clamp.
8. Remove both of the mounting bolts and lift the caliper off the rotor.
9. If the caliper is being removed, it is necessary to disconnect the brake fluid hose. Clean the brake fluid hose-to-caliper connection thoroughly. Remove the hose-to-caliper bolt. Cap or tape the open ends to keep dirt out. Discard the copper gaskets.

To install:

⁂ WARNING

If new caliper bolts are being installed, or if the original reason for repair was a drag/pull condition, check caliper bolt length before proceeding. Bolts must not have a shank length greater than 67.6 mm (2.66 in.).

10. Connect the brake line to the caliper with new sealing washers and fitting bolt. Hand-tighten the fitting bolt.
11. Position the caliper into place over the rotor.
12. Coat the caliper mounting bolt with silicone grease and torque them to 11 ft. lbs. (15 Nm).
13. Position the brake line clear of all chassis components, untwisted and free of kinks. Torque the fitting bolt to 23 ft. lbs. (31 Nm).
14. Install the wheels.

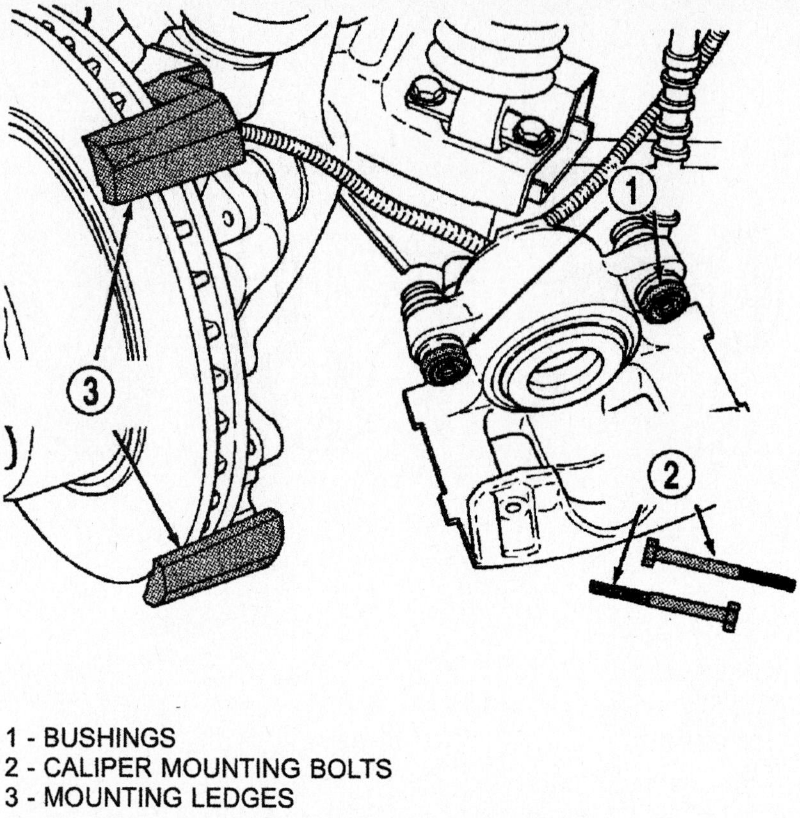

Caliper lubrication points—Wrangler shown

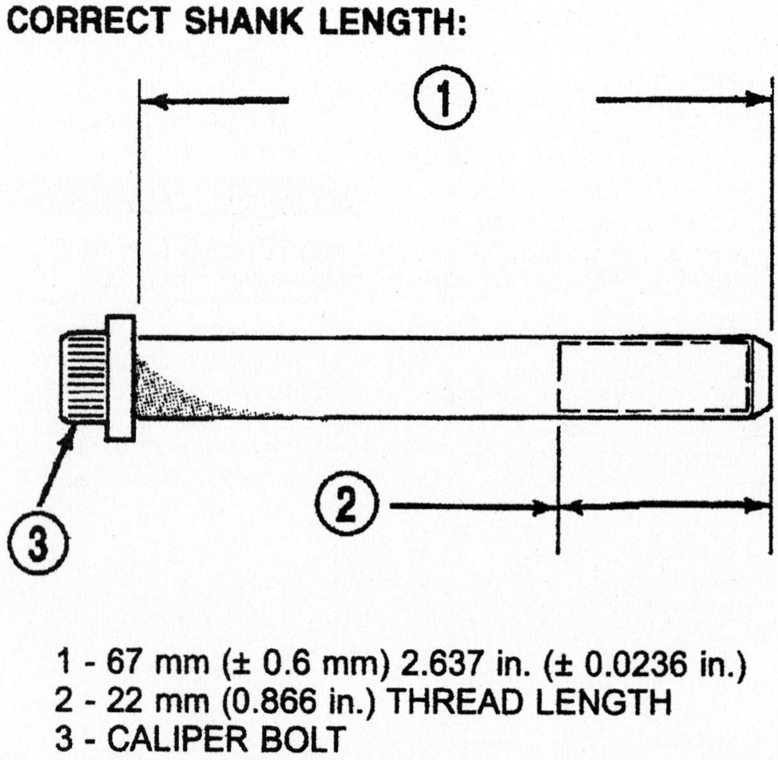

Caliper mounting bolt dimensions—Wrangler

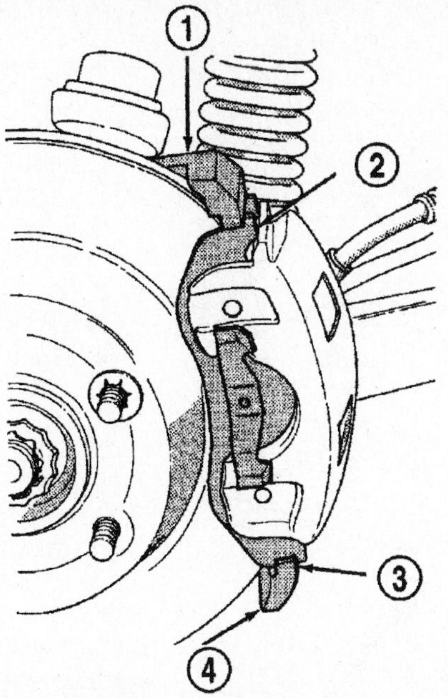

1 - TOP LEDGE
2 - BRAKESHOE TAB ON LEDGE OUTER SURFACE
3 - LEDGE SEATED IN BRAKESHOE NOTCH
4 - BOTTOM LEDGE

06009-JEEP-G198

Caliper installation—Wrangler

15. Fill the master cylinder with fluid and bleed the brake system.

16. Before driving the vehicle, pump the brakes several times to seat the pads.

2002 Grand Cherokee

1. Before servicing the vehicle, refer to the precautions in the beginning of this section.

2. Drain ⅔ of the brake fluid from the front reservoir. Use the bleeder screw at the front outlet port to drain the fluid. If equipped with anti-lock brakes, relieve the system pressure.

3. Raise and safely support the vehicle.

4. Remove the wheels.

5. Insert a small prybar through the caliper opening and pry the caliper (using the outboard brake pad) to bottom the pistons in the caliper bore.

➡This will back the brake pads off of the rotor surface to facilitate the removal and installation of the caliper assembly.

6. Remove the brake hose-to-caliper bolt, hose and washers.

7. Pry the caliper support spring out of the caliper.

8. Remove both caliper slide pin bushing caps and slide pins.

9. Lift the caliper from the anchor.

To install:

10. Position the caliper into place on the anchor.

11. Coat the caliper slide pins with silicone grease and torque them to 21–30 ft. lbs. (29–41 Nm). Install the slide pin bushing caps.

12. Install the caliper support spring in the top of the caliper under the anchor; then, install the other end into the lower caliper hole.

➡Hold the spring in the caliper hole with your thumb while prying the spring end out and under the anchor.

13. Using new gasket washers, install the brake line and torque the fitting bolt to 23 ft. lbs. (31 Nm).

14. Fill the master cylinder with fluid and bleed the brake system.

15. Before driving the vehicle, pump the brakes several times to seat the pads.

16. Install the wheels.

2003–2004 Grand Cherokee

1. Before servicing the vehicle, refer to the precautions in the beginning of this section.

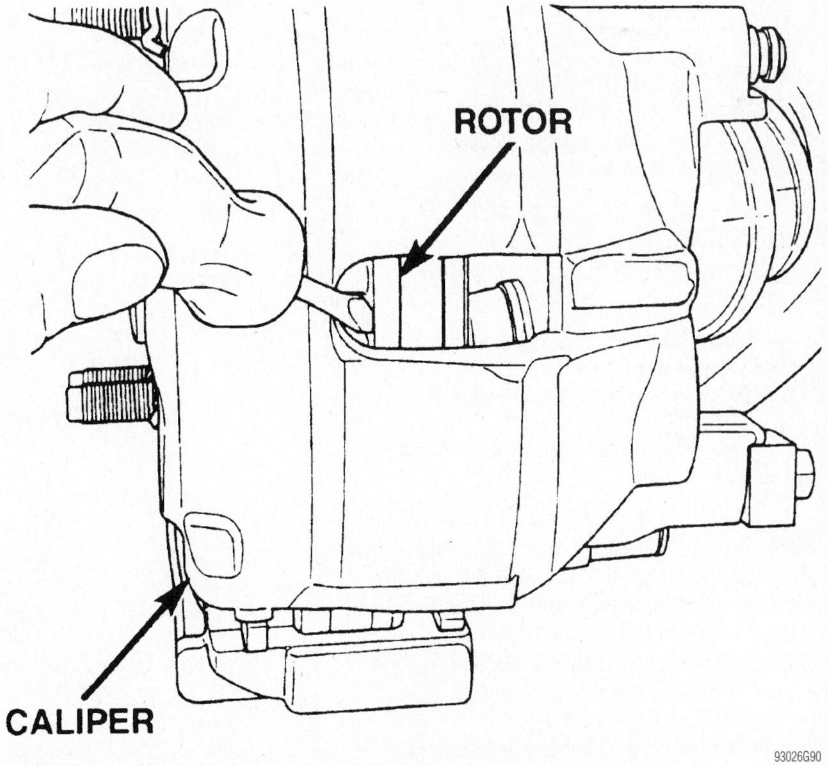

93026G90

Bottoming the pistons in the front caliper—Grand Cherokee

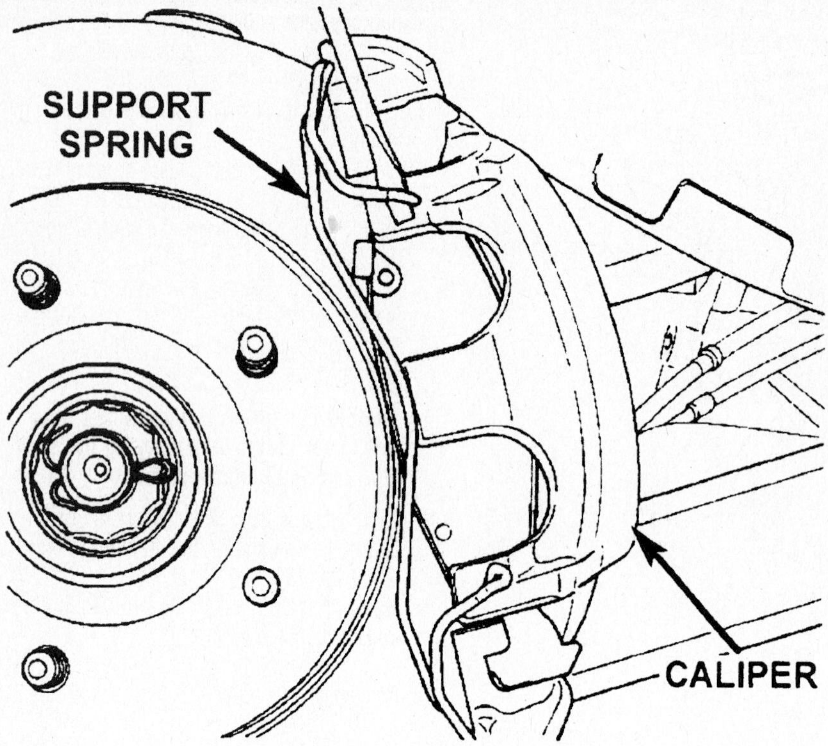

93026G91

View of the support springs on the front caliper—Grand Cherokee

2. Drain ⅔ of the brake fluid from the front reservoir. Use the bleeder screw at the front outlet port to drain the fluid. If equipped with anti-lock brakes, relieve the system pressure.

3. Raise and safely support the vehicle.

4. Remove the wheels.

5. Insert a small prybar through the caliper opening and pry the caliper (using the outboard brake pad) to bottom the pistons in the caliper bore.

➡This will back the brake pads off of the rotor surface to facilitate the removal and installation of the caliper assembly.

6. Remove the brake hose-to-caliper bolt, hose and washers.

7. Remove both caliper slide pin bushing caps and slide pins.

8. Lift the caliper from the anchor.

To install:

9. Position the caliper into place on the anchor.

10. Coat the caliper slide pins with silicone grease and torque them to 53 ft. lbs. (72 Nm). Install the slide pin bushing caps.

11. Using new gasket washers, install the brake line and torque the fitting bolt to 23 ft. lbs. (31 Nm).

12. Fill the master cylinder with fluid and bleed the brake system.

13. Before driving the vehicle, pump the brakes several times to seat the pads.

14. Install the wheels.

2005 Grand Cherokee

1. Before servicing the vehicle, refer to the precautions in the beginning of this section.

2. Raise and support vehicle.

1. Caliper
2. Caliper slide pin bolts
3. Anchor plate bolts
4. Anchor plate
5. Brake line
6. Banjo bolt

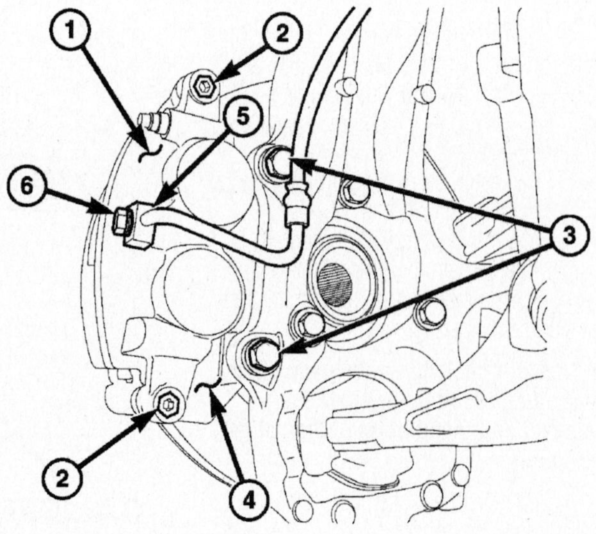

06009-JEEP-G151

Front caliper mounting—2005 Grand Cherokee

3. Remove front wheel and tire assembly.

4. Drain small amount of fluid from master cylinder brake reservoir with clean suction gun.

5. Bottom caliper pistons into the caliper by prying the caliper over.

6. Remove brake hose banjo bolt and gasket washers. Discard gasket washers.

7. Remove the caliper slide bolts.

8. Remove the caliper from the adapter.

To install:

9. Install the caliper on the adapter.

10. Caliper slide pins should be free from debris and lightly lubricated.

11. Install the caliper slide pin bolts and tighten to 72 Nm (53 ft. lbs.).

12. Gently lift one end of the slide pin boot to equalize air pressure, then release the boot and verify that the boot is fully covering the slide pin.

⁂ WARNING

Verify brake hose is not twisted or kinked before tightening banjo bolt.

13. Install brake hose to caliper with new copper washers and tighten banjo bolt to 31 Nm (23 ft. lbs.).

14. Fill and bleed brake system.

15. Install wheel and tire assemblies.

16. Remove supports and lower vehicle.

17. Verify brake fluid level.

Liberty

1. Before servicing the vehicle, refer to the precautions in the beginning of this section.

2. Drain ⅔ of the brake fluid from the

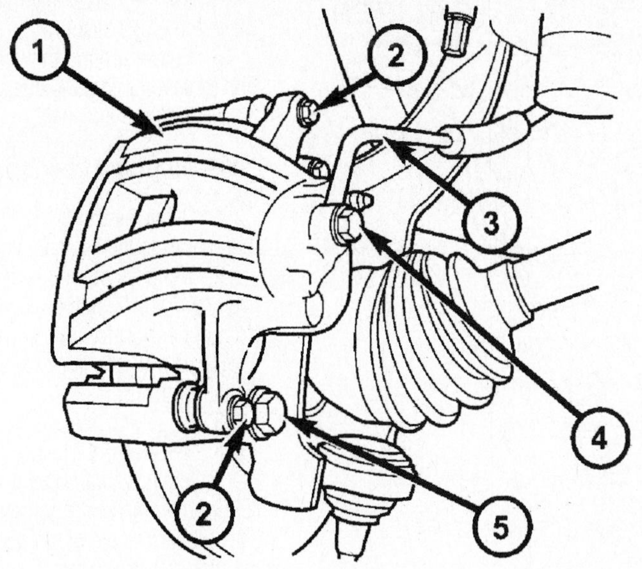

1 - DISC BRAKE CALIPER
2 - CALIPER SLIDE MOUNTING BOLTS
3 - BRAKE HOSE
4 - BANJO BOLT
5 - CALIPER ADAPTER MOUNTING BOLT

06009-JEEP-G79

Front caliper—Liberty

front reservoir. Use the bleeder screw at the front outlet port to drain the fluid. If equipped with anti-lock brakes, relieve the system pressure.

3. Raise and safely support the vehicle.
4. Remove the wheels.
5. Remove both of the mounting bolts and lift the caliper off the rotor.
6. If the caliper is being removed, it is necessary to disconnect the brake fluid hose. Clean the brake fluid hose-to-caliper connection thoroughly. Remove the hose-to-caliper bolt. Cap or tape the open ends to keep dirt out. Discard the copper gaskets.

To install:

7. Connect the brake line to the caliper with new sealing washers and fitting bolt. Hand-tighten the fitting bolt.
8. Position the caliper into place over the rotor.
9. Coat the caliper mounting bolts with silicone grease and torque them to 11 ft. lbs. (15 Nm).
10. Position the brake line clear of all chassis components, untwisted and free of kinks. Use new copper washers. Torque the fitting bolt to 23 ft. lbs. (31 Nm).
11. Install the wheels.
12. Fill the master cylinder with fluid and bleed the brake system.
13. Before driving the vehicle, pump the brakes several times to seat the pads.

Rear Brake Caliper

REMOVAL & INSTALLATION

Wrangler

1. Before servicing the vehicle, refer to the precautions in the beginning of this section.
2. Drain ⅔ of the brake fluid from the front reservoir. If equipped with anti-lock brakes, relieve the system pressure.
3. Raise and safely support the vehicle.
4. Remove the wheels.
5. Insert a small prybar through the caliper opening and pry the caliper (using the outboard brake pad) to bottom the piston in the caliper bore.

➡This will back the brake pads off of the rotor surface to facilitate the removal and installation of the caliper assembly.

6. Remove the brake hose-to-caliper bolt, hose and washers.
7. Remove both caliper slide pin bushing caps and slide pins.
8. Lift the caliper from the anchor.

To install:

9. Position the caliper into place on the anchor.
10. Coat the caliper slide pins with silicone grease and torque them to 11 ft. lbs. (15 Nm). Install the slide pin bushing caps.
11. Using new gasket washers, install the brake line and torque the fitting bolt to 23 ft. lbs. (31 Nm).
12. Fill the master cylinder with fluid and bleed the brake system.
13. Before driving the vehicle, pump the brakes several times to seat the pads.
14. Install the wheels.

2002 Grand Cherokee

1. Before servicing the vehicle, refer to the precautions in the beginning of this section.
2. Drain ⅔ of the brake fluid from the front reservoir. Use the bleeder screw at the front outlet port to drain the fluid. If equipped with anti-lock brakes, relieve the system pressure.
3. Raise and safely support the vehicle.
4. Remove the wheels.
5. Insert a small prybar through the caliper opening and pry the caliper (using the outboard brake pad) to bottom the piston in the caliper bore.

➡This will back the brake pads off of the rotor surface to facilitate the removal and installation of the caliper assembly.

6. Remove the brake hose-to-caliper bolt, hose and washers.
7. Pry the caliper support spring out of the caliper.
8. Remove both caliper slide pin bushing caps and slide pins.
9. Lift the caliper from the anchor.

To install:

10. Position the caliper into place on the anchor.
11. Coat the caliper slide pins with silicone grease and torque them to 26 ft. lbs. Install the slide pin bushing caps.
12. Install the caliper support spring in the top of the caliper under the anchor; then, install the other end into the lower caliper hole.

➡Hold the spring in the caliper hole with your thumb while prying the spring end out and under the anchor.

13. Using new gasket washers, install the brake line and torque the fitting bolt to 23 ft. lbs. (31 Nm).

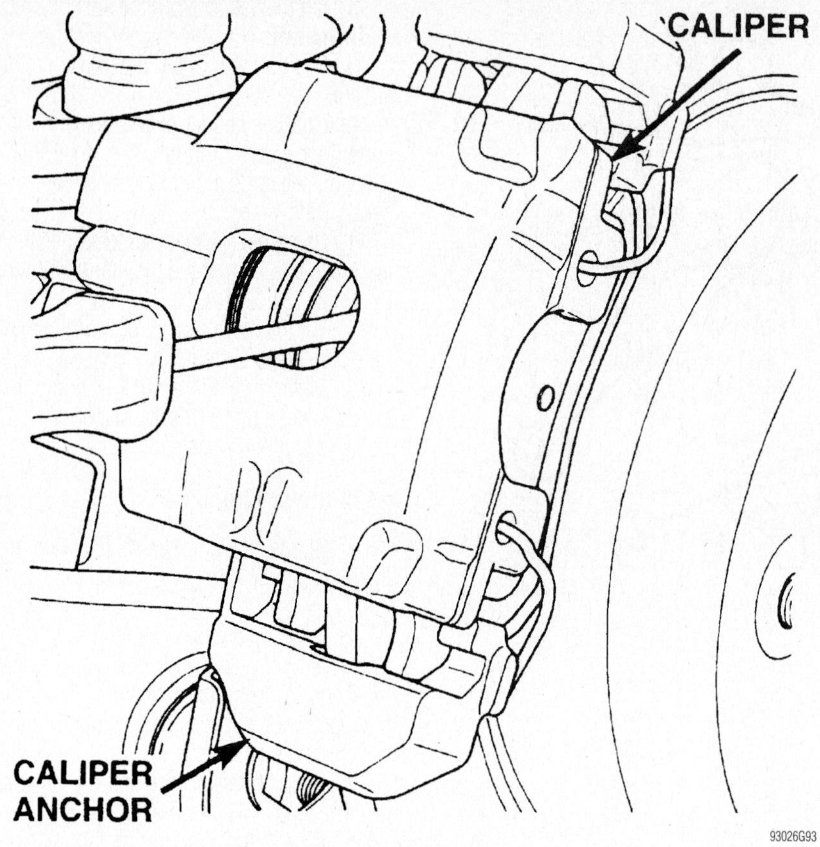

Bottoming the piston in the rear caliper—Grand Cherokee

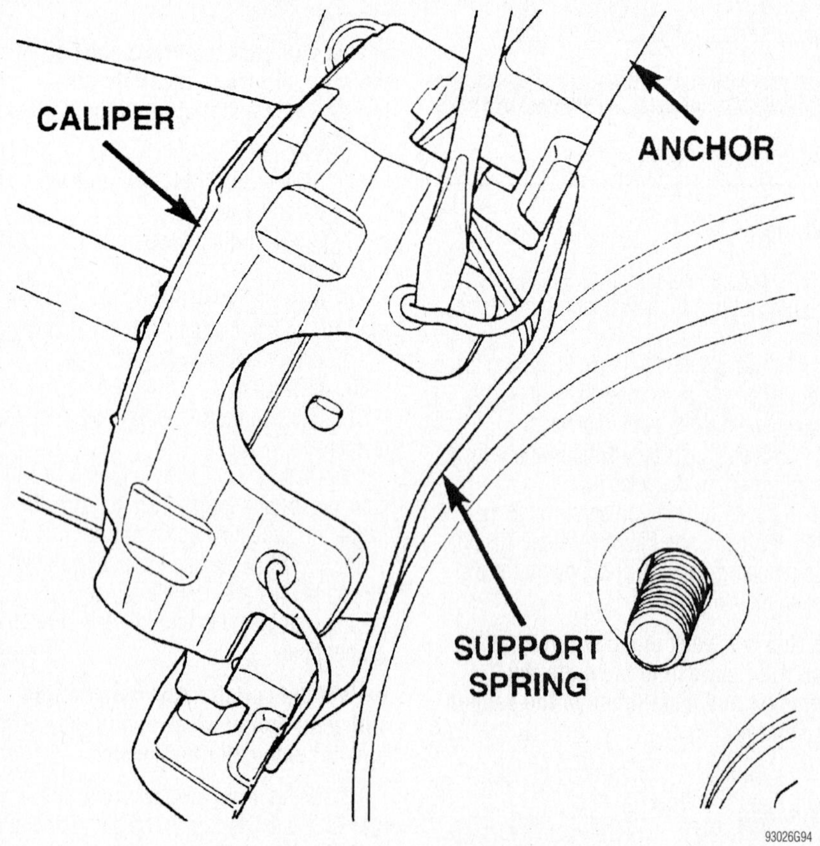

View of the support springs on the rear caliper—Grand Cherokee

14. Fill the master cylinder with fluid and bleed the brake system.
15. Before driving the vehicle, pump the brakes several times to seat the pads.
16. Install the wheels.

2003–05 Grand Cherokee

1. Before servicing the vehicle, refer to the precautions in the beginning of this section.
2. Drain ⅔ of the brake fluid from the front reservoir. Use the bleeder screw at the front outlet port to drain the fluid. If equipped with anti-lock brakes, relieve the system pressure.
3. Raise and safely support the vehicle.
4. Remove the wheels.
5. Insert a small prybar through the caliper opening and pry the caliper (using the outboard brake pad) to bottom the piston in the caliper bore.

➡This will back the brake pads off of the rotor surface to facilitate the removal and installation of the caliper assembly.

6. Remove the brake hose-to-caliper bolt, hose and washers.
7. Pry the caliper support spring out of the caliper.
8. Remove both caliper slide pin bushing caps and slide pins.
9. Lift the caliper from the anchor.

To install:

10. Position the caliper into place on the anchor.
11. Coat the caliper slide pins with silicone grease and torque them to 26 ft. lbs. (35 Nm). Install the slide pin bushing caps.
12. Install the caliper support spring in the top of the caliper under the anchor; then, install the other end into the lower caliper hole.

➡Hold the spring in the caliper hole with your thumb while prying the spring end out and under the anchor.

13. Using new gasket washers, install the brake line and torque the fitting bolt to 23 ft. lbs. (31 Nm).
14. Fill the master cylinder with fluid and bleed the brake system.
15. Before driving the vehicle, pump the brakes several times to seat the pads.
16. Install the wheels.

Liberty

1. Before servicing the vehicle, refer to the precautions in the beginning of this section.
2. Drain ⅔ of the brake fluid from the front reservoir. If equipped with anti-lock brakes, relieve the system pressure.

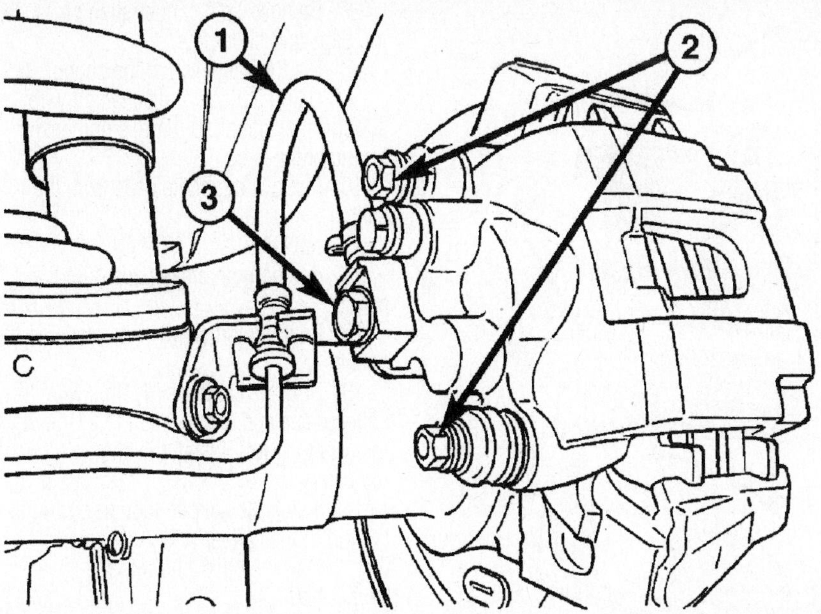

1 - BRAKE HOSE
2 - CALIPER MOUNTING BOLTS
3 - BANJO BOLT

06009-JEEP-G80

Rear brake caliper—Liberty

3. Raise and safely support the vehicle.
4. Remove the wheels.
5. Remove the brake hose-to-caliper bolt, hose and washers.
6. Remove both caliper slide pin bushing caps and slide pins.
7. Lift the caliper from the anchor.

To install:

8. Position the caliper into place on the anchor.
9. Coat the caliper slide pins with silicone grease and torque them to 18 ft. lbs. Install the slide pin bushing caps.
10. Using new gasket washers, install the brake line and torque the fitting bolt to 23 ft. lbs. (31 Nm).
11. Fill the master cylinder with fluid and bleed the brake system.
12. Before driving the vehicle, pump the brakes several times to seat the pads.
13. Install the wheels.

Front Disc Brake Pads

REMOVAL & INSTALLATION

Wrangler

1. Before servicing the vehicle, refer to the precautions in the beginning of this section.
2. Raise and support vehicle.
3. Remove wheel and tire assembly.
4. Remove caliper.
5. Pressing one end of outboard shoe inward to disengage shoe lug. Then rotate shoe upward until retainer spring clears caliper. Press opposite end of shoe inward to disengage shoe lug and rotate shoe up and out of caliper.
6. Grasp ends of inboard shoe and tilt shoe outward to release springs from caliper piston and remove shoe from caliper.

➡If original brake shoes will be used, keep them in sets left and right. They are not interchangeable.

7. Secure caliper to nearby suspension part with wire. Do not allow brake hose to support caliper weight.
8. Wipe caliper off with shop rags or towels.

To install:

9. Install inboard shoe in caliper and verify shoe retaining is fully seated into the piston.
10. Start one end of outboard shoe in caliper and rotate shoe downward into place. Verify shoe locating lugs and shoe spring are seated.

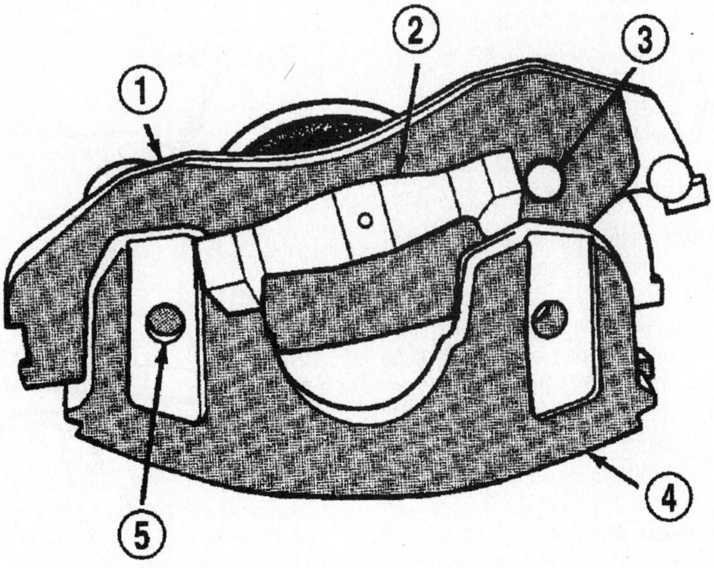

1 - OUTBOARD BRAKESHOE
2 - SHOE SPRING
3 - LOCATING LUG
4 - CALIPER
5 - LOCATING LUG

06009-JEEP-G199

Outboard front brake pad removal/installation—Wrangler

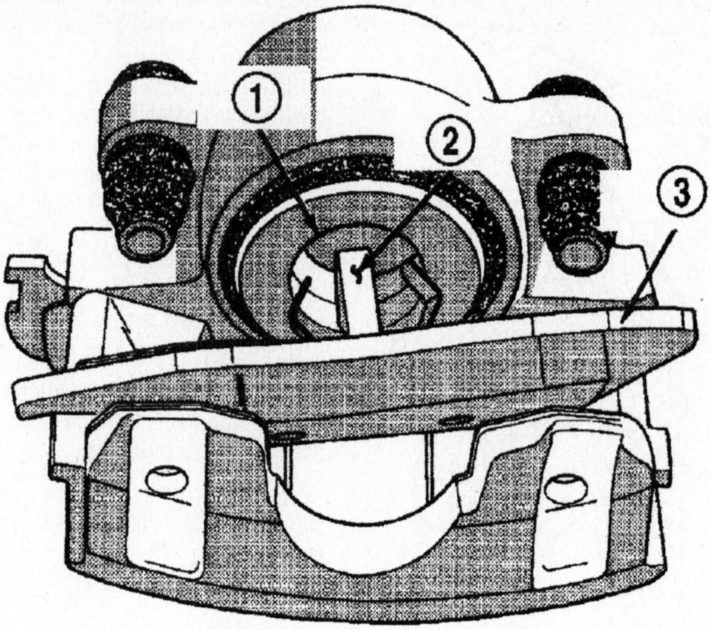

1 - CALIPER PISTON
2 - SHOE SPRINGS
3 - INBOARD BRAKESHOE

06009-JEEP-G200

Inboard front brake pad removal/installation—Wrangler

11. Install caliper.
12. Install wheel and tire assembly.
13. Remove support and lower vehicle.
14. Pump brake pedal until caliper pistons and brake shoes are seated.
15. Top off brake fluid level if necessary.

2002 Grand Cherokee

1. Before servicing the vehicle, refer to the precautions in the beginning of this section.
2. Drain ⅔ of the brake fluid from the front reservoir. Use the bleeder screw at the front outlet port to drain the fluid. If equipped with anti-lock brakes, relieve the system pressure.
3. Raise and safely support the vehicle.
4. Remove the wheels.
5. Insert a small prybar through the caliper opening and pry the caliper (using the outboard brake pad) to bottom the pistons in the caliper bore.

➡This will back the brake pads off of the rotor surface to facilitate the removal and installation of the caliper assembly.

6. Pry the caliper support spring out of the caliper.
7. Remove both caliper slide pin bushing caps and slide pins.
8. Lift the caliper from the anchor.
9. Using a piece of mechanic's wire, support the caliper so there is not tension on the brake hose.
10. Remove the brake pads from the caliper.

To install:

11. Position the brake pads onto the caliper.
12. Position the caliper into place on the anchor.
13. Coat the caliper slide pins with silicone grease and torque them to 21–30 ft. lbs. (29–41 Nm). Install the slide pin bushing caps.
14. Install the caliper support spring in the top of the caliper under the anchor; then, install the other end into the lower caliper hole.

➡Hold the spring in the caliper hole with your thumb while prying the spring end out and under the anchor.

15. Fill the master cylinder with fluid and bleed the brake system.
16. Before driving the vehicle, pump the brakes several times to seat the pads.
17. Install the wheels.

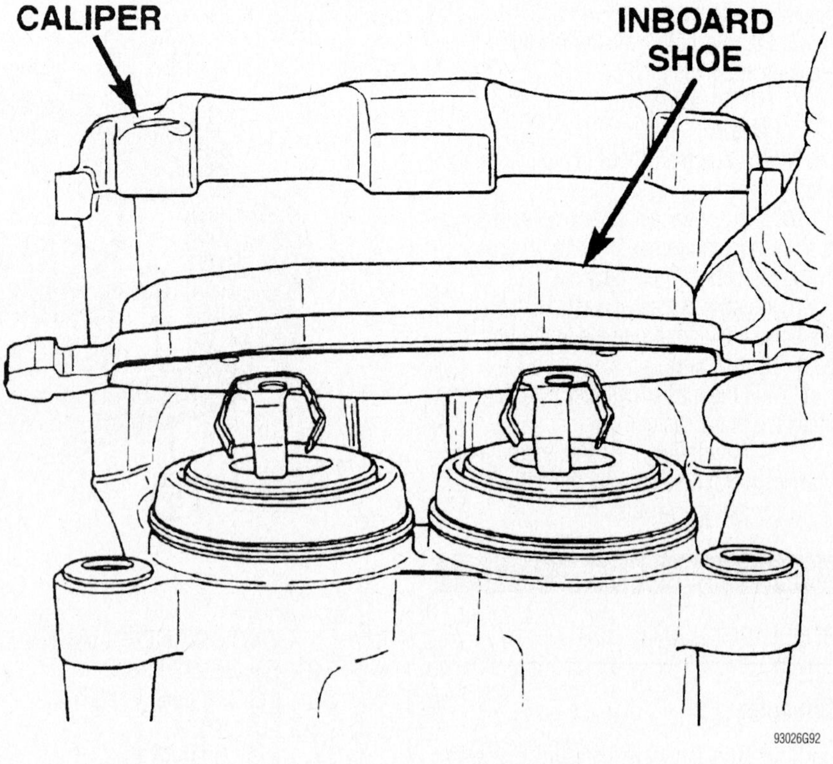

93026G92

Installing the inward brake pad on the front caliper pistons—Grand Cherokee

2003–05 Grand Cherokee

1. Before servicing the vehicle, refer to the precautions in the beginning of this section.
2. Drain ⅔ of the brake fluid from the front reservoir. Use the bleeder screw at the front outlet port to drain the fluid. If equipped with anti-lock brakes, relieve the system pressure.
3. Raise and safely support the vehicle.
4. Remove the wheels.
5. Insert a small prybar through the caliper opening and pry the caliper (using the outboard brake pad) to bottom the pistons in the caliper bore.

➡This will back the brake pads off of the rotor surface to facilitate the removal and installation of the caliper assembly.

6. Remove both caliper slide pin bushing caps and slide pins.
7. Lift the caliper from the anchor.
8. Using a piece of mechanic's wire, support the caliper so there is not tension on the brake hose.
9. Remove the brake pads from the caliper.

To install:

10. Position the brake pads onto the caliper.
11. Position the caliper into place on the anchor.
12. Coat the caliper slide pins with silicone grease and torque them to 53 ft. lbs. (72 Nm). Install the slide pin bushing caps.

➡Hold the spring in the caliper hole with your thumb while prying the spring end out and under the anchor.

13. Fill the master cylinder with fluid and bleed the brake system.
14. Before driving the vehicle, pump the brakes several times to seat the pads.
15. Install the wheels.

Liberty

1. Before servicing the vehicle, refer to the precautions in the beginning of this section.
2. Drain ⅔ of the brake fluid from the front reservoir. Use the bleeder screw at the front outlet port to drain the fluid. If equipped with anti-lock brakes, relieve the system pressure.
3. Raise and safely support the vehicle.
4. Remove the wheels.
5. Remove both caliper slide pin bushing caps and slide pins.
6. Lift the caliper from the anchor.
7. Using a piece of mechanic's wire, support the caliper so there is not tension on the brake hose.
8. Remove the brake pads from the caliper.

To install:

9. Position the brake pads onto the caliper.
10. Position the caliper into place on the anchor.
11. Coat the caliper slide pins with silicone grease and torque them to 11 ft. lbs. (15 Nm). Install the slide pin bushing caps.

➡Hold the spring in the caliper hole with your thumb while prying the spring end out and under the anchor.

12. Fill the master cylinder with fluid and bleed the brake system.
13. Before driving the vehicle, pump the brakes several times to seat the pads.
14. Install the wheels.

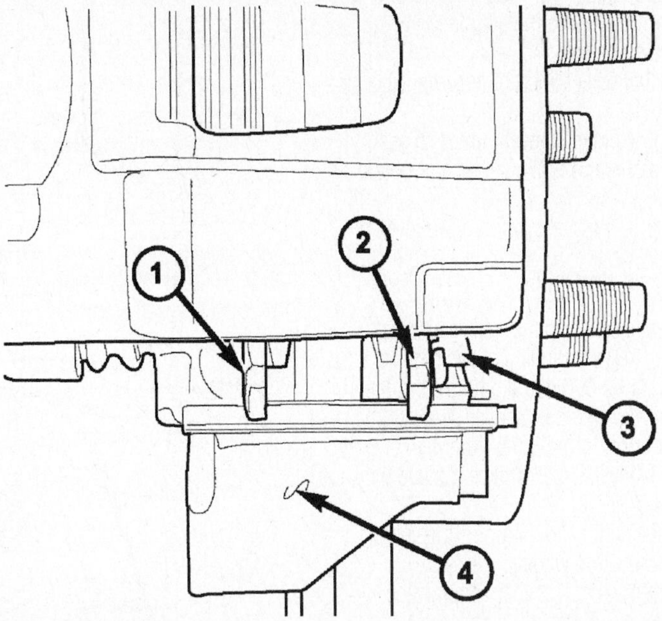

1 - INBOARD BRAKE PAD
2 - OUTBOARD BRAKE PAD
3 - ANTI-RATTLE SPRING CLIP
4 - CALIPER ADAPTER

06009-JEEP-G81

Front brake pad anti-rattle clip installation—Liberty

Rear Disc Brake Pads

REMOVAL & INSTALLATION

Wrangler

1. Before servicing the vehicle, refer to the precautions in the beginning of this section.
2. Raise and support vehicle.
3. Remove the wheel and tire assemblies.
4. Compress the caliper.
5. Remove the caliper.
6. Remove the caliper by tilting the top up and off the caliper adapter.

➡Do not allow brake hose to support caliper assembly.

7. Support and hang the caliper.
8. Remove the inboard brake pad from the caliper adapter.
9. Remove the outboard brake pad from the caliper adapter.

To install:

10. Bottom pistons in caliper bore with C-clamp. Place an old brake shoe between a C-clamp and caliper piston.
11. Clean caliper mounting adapter and anti-rattle springs.
12. Lubricate anti-rattle springs with Mopar brake grease, or equivalent.
13. Install anti-rattle springs.

➡Anti-rattle springs are not interchangeable.

14. Install inboard brake pad in adapter.
15. Install outboard brake pad in adapter.
16. Tilt the top of the caliper over rotor

and under adapter. Then push the bottom of the caliper down onto the adapter.

17. Install caliper.

18. Install wheel and tire assemblies and lower vehicle.

19. Apply brakes several times to seat caliper pistons and brake shoes and obtain firm pedal.

20. Top off master cylinder fluid level.

2002 Grand Cherokee

1. Before servicing the vehicle, refer to the precautions in the beginning of this section.
2. Drain ⅔ of the brake fluid from the front reservoir. Use the bleeder screw at the front outlet port to drain the fluid. If equipped with anti-lock brakes, relieve the system pressure.
3. Raise and safely support the vehicle.
4. Remove the wheels.
5. Insert a small prybar through the caliper opening and pry the caliper (using the outboard brake pad) to bottom the piston in the caliper bore.

➡This will back the brake pads off of the rotor surface to facilitate the removal and installation of the caliper assembly.

6. Pry the caliper support spring out of the caliper.
7. Remove both caliper slide pin bushing caps and slide pins.
8. Lift the caliper from the anchor.
9. Using a piece of mechanics wire, support the caliper so there is not tension on the brake hose.
10. Remove the brake pads from the caliper.

To install:

11. Position the brake pads onto the caliper.
12. Position the caliper into place on the anchor.
13. Coat the caliper slide pins with silicone grease and torque them to 21–30 ft. lbs. (29–41 Nm). Install the slide pin bushing caps.
14. Install the caliper support spring in the top of the caliper under the anchor; then, install the other end into the lower caliper hole.

➡Hold the spring in the caliper hole with your thumb while prying the spring end out and under the anchor.

15. Fill the master cylinder with fluid and bleed the brake system.
16. Before driving the vehicle, pump the brakes several times to seat the pads.
17. Install the wheels.

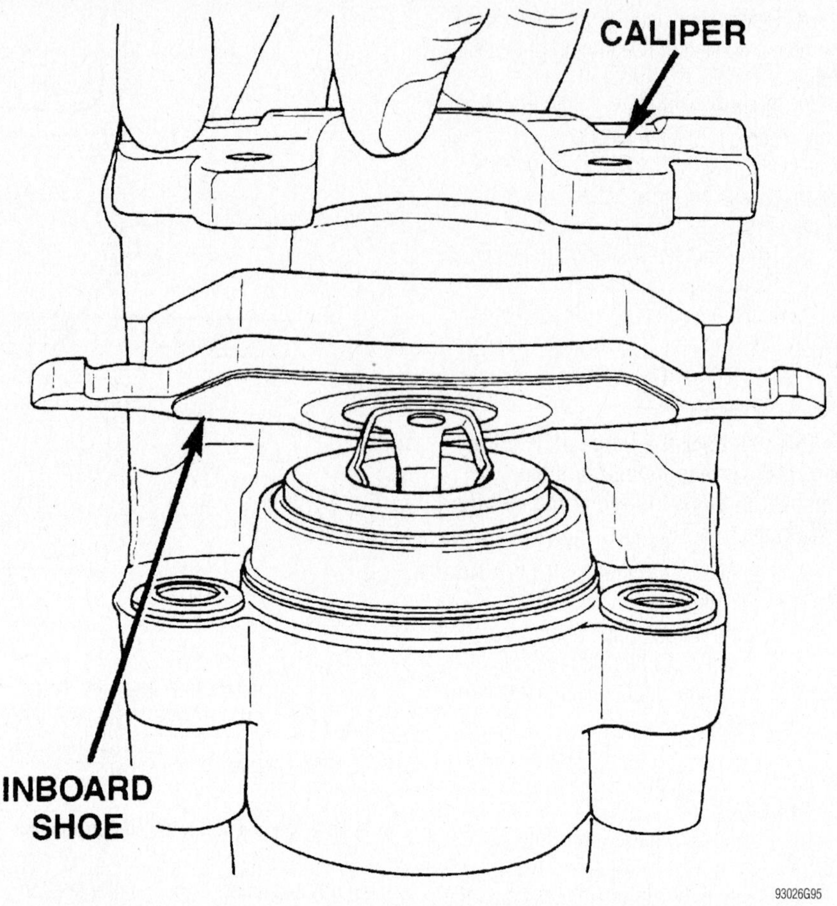

Installing the inward brake pad on the rear caliper piston—Grand Cherokee

2003–05 Grand Cherokee

1. Before servicing the vehicle, refer to the precautions in the beginning of this section.
2. Drain ⅔ of the brake fluid from the front reservoir. Use the bleeder screw at the front outlet port to drain the fluid. If equipped with anti-lock brakes, relieve the system pressure.
3. Raise and safely support the vehicle.
4. Remove the wheels.
5. Insert a small prybar through the caliper opening and pry the caliper (using the outboard brake pad) to bottom the piston in the caliper bore.

➡This will back the brake pads off of the rotor surface to facilitate the removal and installation of the caliper assembly.

6. Pry the caliper support spring out of the caliper.
7. Remove both caliper slide pin bushing caps and slide pins.
8. Lift the caliper from the anchor.
9. Using a piece of mechanics wire, support the caliper so there is not tension on the brake hose.
10. Remove the brake pads from the caliper.

To install:

11. Position the brake pads onto the caliper.
12. Position the caliper into place on the anchor.
13. Coat the caliper slide pins with silicone grease and torque them to 21–30 ft. lbs. (29–41 Nm). Install the slide pin bushing caps.
14. Install the caliper support spring in the top of the caliper under the anchor; then, install the other end into the lower caliper hole.

➡Hold the spring in the caliper hole with your thumb while prying the spring end out and under the anchor.

15. Fill the master cylinder with fluid and bleed the brake system.

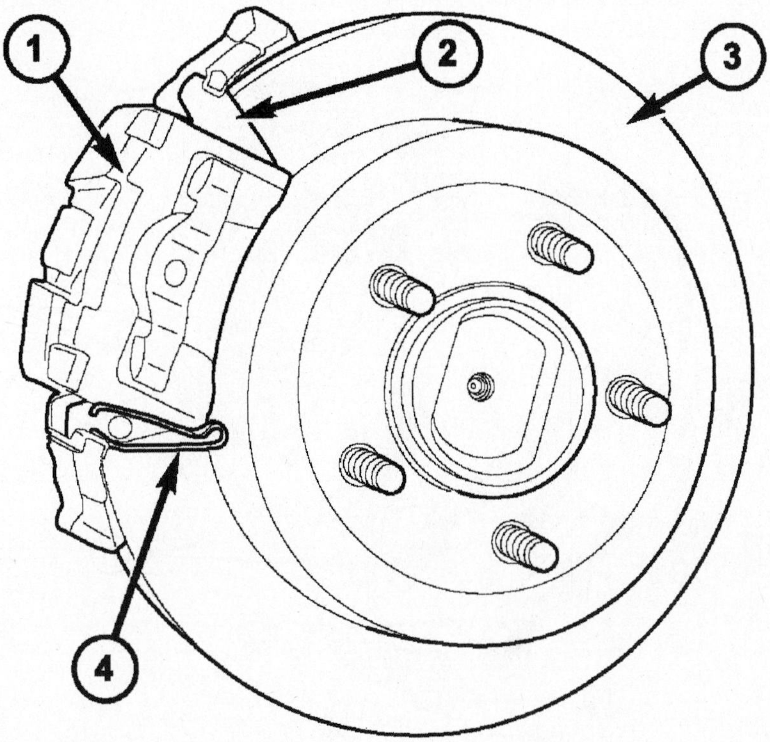

1 - DISC BRAKE CALIPER
2 - OUTBOARD BRAKE PAD
3 - DISC BRAKE ROTOR
4 - ANTI-RATTLE SPRING CLIP

06009-JEEP-G82

Rear brake pads—Liberty

16. Before driving the vehicle, pump the brakes several times to seat the pads.
17. Install the wheels.

Liberty

1. Before servicing the vehicle, refer to the precautions in the beginning of this section.
2. Drain ⅔ of the brake fluid from the front reservoir. Use the bleeder screw at the front outlet port to drain the fluid. If equipped with anti-lock brakes, relieve the system pressure.
3. Raise and safely support the vehicle.
4. Remove the wheels.
5. Pry the caliper support spring out of the caliper.
6. Remove both caliper slide pin bushing caps and slide pins.
7. Lift the caliper from the anchor.
8. Using a piece of mechanics wire, support the caliper so there is not tension on the brake hose.
9. Remove the brake pads from the caliper.

To install:

10. Position the brake pads onto the caliper.
11. Position the caliper into place on the anchor.
12. Coat the caliper slide pins with silicone grease and torque them to 18 ft. lbs. (25 Nm). Install the slide pin bushing caps.
13. Install the caliper support spring in the top of the caliper under the anchor; then, install the other end into the lower caliper hole.

➡Hold the spring in the caliper hole with your thumb while prying the spring end out and under the anchor.

14. Fill the master cylinder with fluid and bleed the brake system.
15. Before driving the vehicle, pump the brakes several times to seat the pads.
16. Install the wheels.

Brake Drum

REMOVAL & INSTALLATION

Wrangler

1. Before servicing the vehicle, refer to the precautions in the beginning of this section.
2. Raise and safely support the vehicle.
3. Remove the wheel.
4. Remove the spring nuts (if installed) from the lug bolts and remove the drum from the vehicle.

To install:

5. Ensure the contacting surfaces are clean and flat. Install the drum on the hub.
6. Adjust the brake shoes, if necessary.
7. Install the spring nuts on the lug bolts.
8. Install the wheel.

Brake Shoes

REMOVAL & INSTALLATION

Wrangler

1. Before servicing the vehicle, refer to the precautions in the beginning of this section.
2. Raise and safely support the vehicle.
3. Remove the wheel and brake drum.
4. Remove the U-clip and washer securing the adjuster cable to the parking brake lever.
5. Remove the primary and secondary return springs from the anchor pin.
6. Remove the hold-down springs, retainers and pins.
7. Install spring clamps on the wheel cylinders to hold the pistons in place.
8. Remove the adjuster lever, adjuster screw and spring.
9. Remove the adjuster cable and cable guide.
10. Remove the brake shoes and parking brake strut.
11. Disconnect the cable from the parking brake lever and remove the lever.

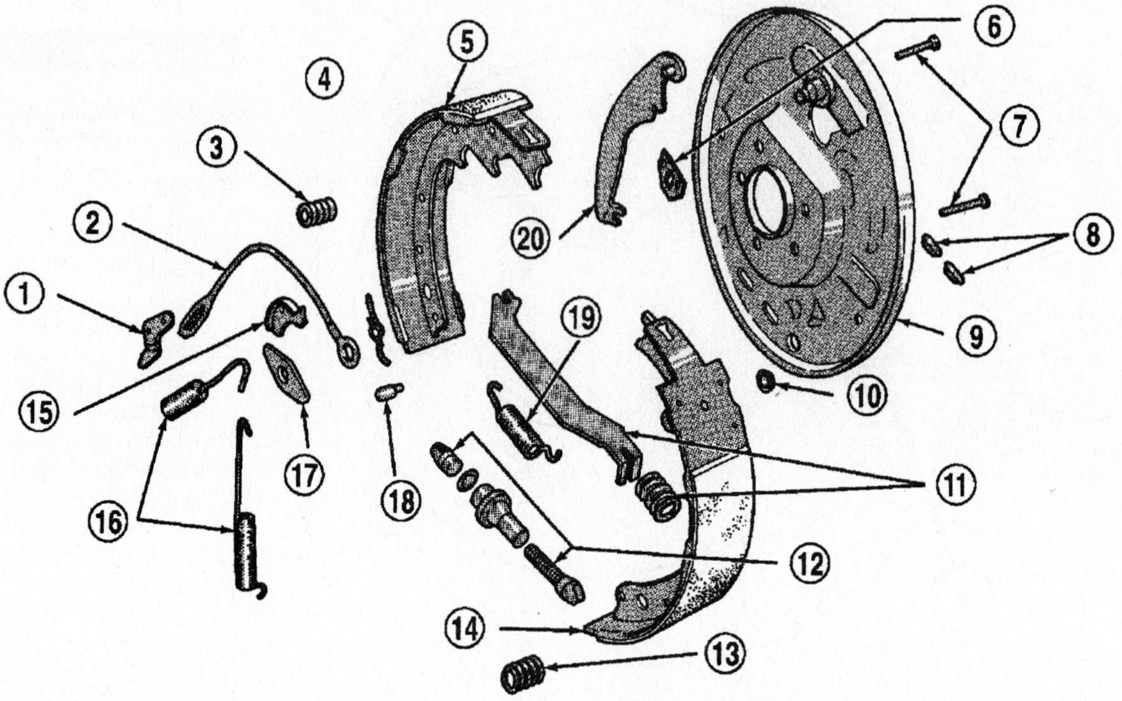

1 - ADJUSTER LEVER
2 - ADJUSTER CABLE
3 - HOLDDOWN SPRING AND RETAINERS
4 - ADJUSTER LEVER SPRING
5 - TRAILING SHOE
6 - CYLINDER-TO-SUPPORT SEAL
7 - HOLDDOWN PINS
8 - ACCESS PLUGS
9 - SUPPORT PLATE
10 - CABLE HOLE PLUG
11 - PARK BRAKE STRUT AND SPRING
12 - ADJUSTER SCREW ASSEMBLY
13 - HOLDDOWN SPRING AND RETAINERS
14 - LEADING SHOE
15 - CABLE GUIDE
16 - SHOE RETURN SPRINGS
17 - SHOE GUIDE PLATE
18 - PIN
19 - SHOE SPRING
20 - PARK BRAKE LEVER

06009-JEEP-G201

Drum brake components, right side shown—Wrangler

To install:

12. Clean the support plate with brake cleaner.
13. Apply multi-purpose grease to the brake shoe contact surfaces on the backing plate.
14. Lubricate the adjuster screw threads.
15. Attach the parking brake lever to the secondary brake shoe. Use a new washer and U-clip.
16. Remove the wheel cylinder clamps.
17. Attach the parking brake cable to the lever.
18. Install the brake shoes on the support plate. Secure the shoes with new hold-down springs, pins and retainers.
19. Install the parking brake strut and spring.
20. Install the guide plate and adjuster cable to the anchor pin.
21. Install the return springs.
22. Install the adjuster cable guide on the secondary shoe.
23. Install the adjuster screw, spring and lever. Connect to the adjuster cable.
24. Adjust the shoes to the drum. Install the drum.
25. Install the wheel/tire assemblies and lower the vehicle.
26. Verify a firm brake pedal before moving the vehicle.

Wheel Cylinders

REMOVAL & INSTALLATION

1. Before servicing the vehicle, refer to the precautions in the beginning of this section.
2. Remove wheel and tire assembly.
3. Remove brake drum.

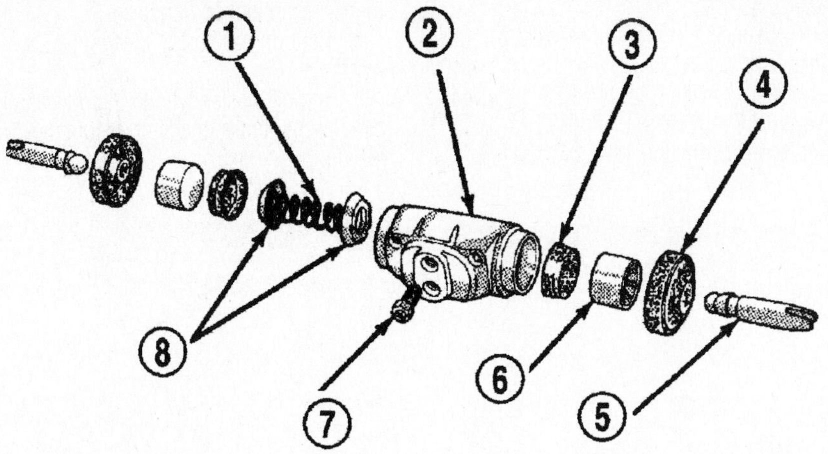

1 - SPRING
2 - CYLINDER
3 - PISTON CLIP
4 - BOOT
5 - PUSH ROD
6 - PISTON
7 - BLEED SCREW
8 - CUP EXPANDERS

06009-JEEP-G202

Wheel cylinder components

4. Remove wheel cylinder brake line.
5. Remove brake shoe return springs and move shoes out of engagement with cylinder push rods.
6. Remove cylinder attaching bolts and remove cylinder from support plate.

To install:

7. Apply bead of silicone sealer around cylinder mounting surface of support plate.
8. Install cylinder mounting bolts and tighten to 10 Nm (7 ft. lbs.).
9. Install brake line to cylinder and tighten to 16 Nm (12 ft. lbs.).
10. Install brake shoe return spring.
11. Install brake drum.
12. Install wheel and tire assembly.
13. Bleed base brake system.

Brake Hydraulic System

BLEEDING

Without ABS

1. Before servicing the vehicle, refer to the precautions in the beginning of this section.

➡Use brake fluid, or an equivalent quality fluid meeting SAE J1703-F and DOT 3 standards only. Use fresh, clean fluid from a sealed container at all times.

2. Remove reservoir filler caps and fill reservoir.
3. If calipers, or wheel cylinders were overhauled, open all caliper and wheel cylinder bleed screws. Then close each bleed screw as fluid starts to drip from it. Top off master cylinder reservoir once more before proceeding.
4. Attach one end of bleed hose to bleed screw and insert opposite end in glass container partially filled with brake fluid. Be sure end of bleed hose is immersed in fluid.
5. Open up bleeder, then have a helper press down the brake pedal. Once the pedal is down close the bleeder. Repeat bleeding until fluid stream is clear and free of bubbles. Then move to the next wheel.

With ABS

1. Before servicing the vehicle, refer to the precautions in the beginning of this section.

ABS system bleeding requires conventional bleeding methods plus use of the

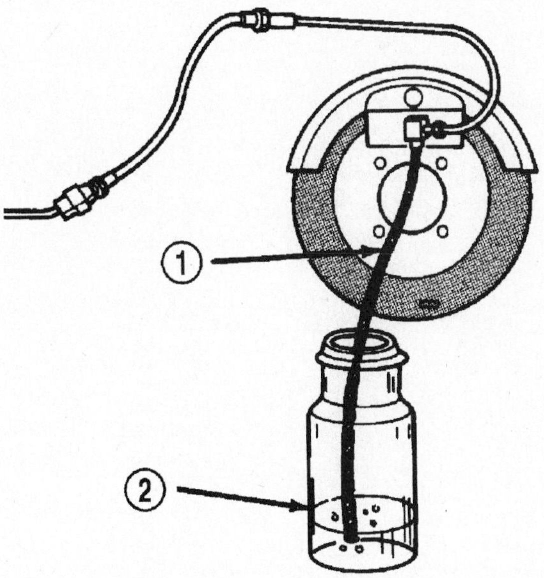

1 - BLEED HOSE
2 - FLUID CONTAINER PARTIALLY FILLED WITH FLUID

06009-JEEP-G203

Bleed hose setup

DRB scan tool. The procedure involves performing a base brake bleeding, followed by use of the scan tool to cycle and bleed the HCU pump and solenoids. A second base brake bleeding procedure is then required to remove any air remaining in the system.

2. Perform base brake bleeding (without ABS).
3. Connect scan tool to the Data Link Connector.
4. Select ANTILOCK BRAKES, followed by MISCELLANEOUS, then ABS BRAKES. Follow the instructions displayed. When scan tool displays TEST COMPLETE, disconnect scan tool and proceed.
5. Perform base brake bleeding a second time.
6. Top off master cylinder fluid level and verify proper brake operation before moving vehicle.

DODGE AND PLYMOUTH

8

Neon

SPECIFICATIONS AND MAINTENANCE CHARTS

ENGINE AND VEHICLE IDENTIFICATION

Engine						
Code ①	Liters	Cu. In. (cc)	Cyl.	Fuel Sys.	Engine Type	Eng. Mfg.
C	2.0	122 (1996)	4	MFI	SOHC	Chrysler
F	2.0	122 (1996)	4	MFI	SOHC	Chrysler
S	2.4	148 (2421)	4	MFI-Turbo	DOHC	Chrysler

Model Year	
Code ②	Year
2	2002
3	2003
4	2004
5	2005
6	2006

MFI: Multi-point Fuel Injection

SOHC: Single Overhead Camshaft

① 8th position of VIN

② 10th position of VIN

06009-NEON-C01

GENERAL ENGINE SPECIFICATIONS

All measurements are given in inches.

Year	Model	Engine Displacement Liters	Engine Series (ID/VIN)	Net Horsepower @ rpm	Net Torque @ rpm (ft. lbs.)	Bore x Stroke (in.)	Com-pression Ratio	Oil Pressure @ rpm
2002	Neon	2.0	C	132@6000	129@5000	3.44x3.26	9.8:1	25-80@3000
	Neon	2.0	F	150@4400	133@5500	3.44x3.26	9.8:1	25-80@3000
2003	Neon	2.0	C	132@6000	129@5000	3.44x3.26	9.8:1	25-80@3000
	Neon	2.0	F	150@4400	133@5500	3.44x3.26	9.8:1	25-80@3000
	Neon	2.4	S	230@5300	250@4400	3.44x3.97	8.1:1	25-80@3000
2004	Neon	2.0	C	132@6000	129@5000	3.44x3.26	9.8:1	25-80@3000
	Neon	2.0	F	150@4400	133@5500	3.44x3.26	9.8:1	25-80@3000
	Neon	2.4	S	230@5300	250@4400	3.44x3.97	8.1:1	25-80@3000
2005	Neon	2.0	C	132@6000	129@5000	3.44x3.26	9.3:1	25-80@3000
	Neon	2.4	S	230@5300	250@4400	3.44x3.97	8.1:1	25-80@3000

NA: Not available

06009-NEON-C02

ENGINE TUNE-UP SPECIFICATIONS

Year	Engine Displacement Liters	Engine ID/VIN	Spark Plugs Gap (in.)	Ignition Timing (deg.)		Fuel Pump (psi)	Idle Speed (rpm)		Valve Clearance	
				MT	AT		MT	AT	In.	Ex.
2002	2.0	C	0.035	①	①	48	②	②	HYD	HYD
	2.0	F	0.035	①	①	48	②	②	HYD	HYD
2003	2.0	C	0.035	①	①	48	②	②	HYD	HYD
	2.0	F	0.035	①	①	48	②	②	HYD	HYD
	2.4	S	0.035	—	①	48	②	②	HYD	HYD
2004	2.0	C	0.035	①	①	48	②	②	HYD	HYD
	2.0	F	0.035	①	①	48	②	②	HYD	HYD
	2.4	S	0.035	—	①	48	②	②	HYD	HYD
2005	2.0	C	0.035	①	①	48	②	②	HYD	HYD
	2.4	S	0.035	—	①	48	②	②	HYD	HYD

NOTE: The Vehicle Emission Control Information label often reflects specification changes made during production. The label figures must be used if they differ from those in this chart.

HYD : Hydraulic

① Refer to the Vehicle Emission Control Information label for correct timing specifications with a range of +/- 2 degrees.
Ignition timing cannot be adjusted. Base engine timing is set at TDC during assembly.

② Refer to Vehicle Emissions Control Information label for proper specification

06009-NEON-C03

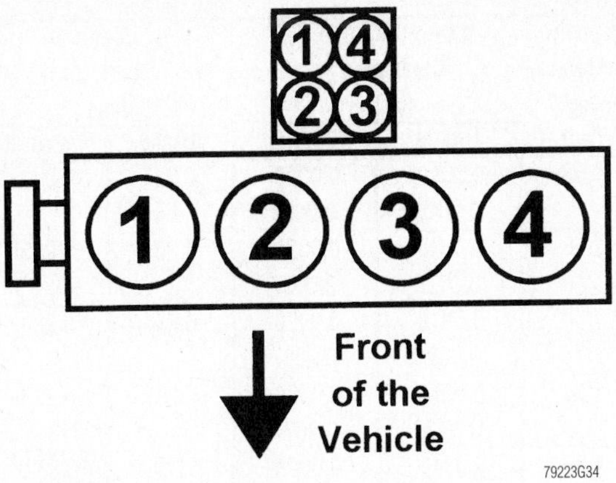

2.0L Engine
Firing Order: 1–3–4–2
Distributorless ignition system

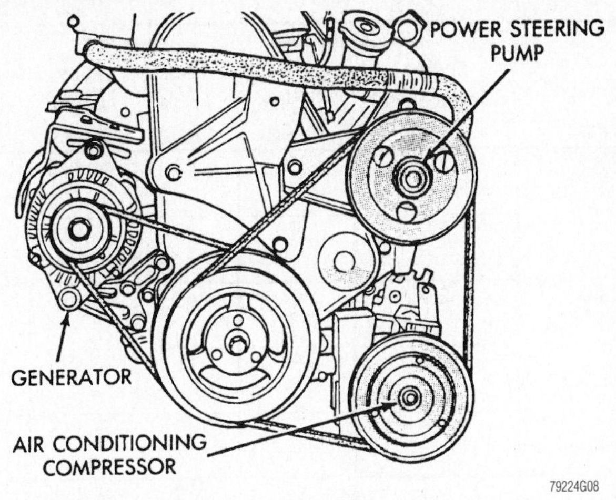

Accessory drive belt routing—2.0L engines

CAPACITIES

Year	Model	Engine Displacement Liters	Engine ID/VIN	Engine Oil with Filter (qts.)	Transmission (pts.) 5-Spd	Transmission (pts.) Auto.	Fuel Tank (gal.)	Cooling System (qts.)
2002	Neon	2.0	C	4.5	5.0-5.6	8.0	12.5	6.5
	Neon	2.0	F	4.5	5.0-5.6	8.0	12.5	6.5
2003	Neon	2.0	C	4.5	5.0-5.6	8.0	12.5	6.5
	Neon	2.0	F	4.5	5.0-5.6	8.0	12.5	6.5
	Neon	2.4	S	5.0	4.8-5.2 ①	—	12.5	6.5
2004	Neon	2.0	C	4.5	5.0-5.6	8.0	12.5	6.5
	Neon	2.0	F	4.5	5.0-5.6	8.0	12.5	6.5
	Neon	2.4	S	5.0	4.8-5.2 ①	—	12.5	6.5
2005	Neon	2.0	C	4.5	5.0-5.6	8.0	12.5	6.5
	Neon	2.4	S	5.0	4.8-5.2 ①	—	12.5	6.5

NOTE: All capacities are approximate. Add fluid gradually and ensure a proper fluid level is obtained.

① Includes 4 oz. Of Mopar Limited Slip Additive (P/N 04318060AB)

06009-NEON-C04

VALVE SPECIFICATIONS

Year	Engine Displacement Liters	Engine ID/VIN	Seat Angle (deg.)	Face Angle (deg.)	Spring Test Pressure (lbs. @ in.)	Spring Installed Height (in.)	Stem-to-Guide Clearance (in.) Intake	Stem-to-Guide Clearance (in.) Exhaust	Stem Diameter (in.) Intake	Stem Diameter (in.) Exhaust
2002	2.0	C	45	45-45.5	70@1.57	1.58	0.0018-0.0025	0.0029-0.0037	0.234	0.233
	2.0	F	45	45-45.5	72@1.57	1.58	0.0018-0.0025	0.0029-0.0037	0.234	0.233
2003	2.0	C	45	45-45.5	70@1.57	1.58	0.0018-0.0025	0.0029-0.0037	0.234	0.233
	2.0	F	45	45-45.5	72@1.57	1.58	0.0018-0.0025	0.0029-0.0037	0.234	0.233
	2.4	S	45	45-45.5	75@1.49	1.49	0.0018-0.0025	0.0029-0.0037	0.233-0.234	0.232-0.233
2004	2.0	C	45	45-45.5	70@1.57	1.58	0.0018-0.0025	0.0029-0.0037	0.234	0.233
	2.0	F	45	45-45.5	72@1.57	1.58	0.0018-0.0025	0.0029-0.0037	0.234	0.233
	2.4	S	45	45-45.5	75@1.49	1.49	0.0018-0.0025	0.0029-0.0037	0.233-0.234	0.232-0.233
2005	2.0	C	45	45-45.5	70@1.57	1.58	0.0018-0.0025	0.0029-0.0037	0.234	0.233
	2.4	S	45	45-45.5	75@1.49	1.49	0.0018-0.0025	0.0029-0.0037	0.233-0.234	0.232-0.233

06009-NEON-C05

CRANKSHAFT AND CONNECTING ROD SPECIFICATIONS

All measurements are given in inches.

Year	Engine Displacement Liters	Engine ID/VIN	Crankshaft Main Brg. Journal Dia.	Crankshaft Main Brg. Oil Clearance	Crankshaft Shaft End-play	Crankshaft Thrust on No.	Connecting Rod Journal Diameter	Connecting Rod Oil Clearance	Connecting Rod Side Clearance
2002	2.0	C	2.0469-2.0475	0.0008-0.0024	0.0035-0.0094	3	1.8894-1.8900	0.0010-0.0023	0.0050-0.0150
	2.0	F	2.0469-2.0475	0.0008-0.0024	0.0035-0.0094	3	1.8894-1.8900	0.0010-0.0023	0.005-0.0015
2003	2.0	C	2.0469-2.0475	0.0008-0.0024	0.0035-0.0094	3	1.8894-1.8900	0.0010-0.0023	0.0050-0.0150
	2.0	F	2.0469-2.0475	0.0008-0.0024	0.0035-0.0094	3	1.8894-1.8900	0.0010-0.0023	0.005-0.0015
	2.4	S	2.3620-2.3625	0.0007-0.0024	0.0035-0.0094	3	1.9680-1.9865	0.0009-0.0027	0.005-0.0015
2004	2.0	C	2.0469-2.0475	0.0008-0.0024	0.0035-0.0094	3	1.8894-1.8900	0.0010-0.0023	0.0050-0.0150
	2.0	F	2.0469-2.0475	0.0008-0.0024	0.0035-0.0094	3	1.8894-1.8900	0.0010-0.0023	0.005-0.0015
	2.4	S	2.3620-2.3625	0.0007-0.0024	0.0035-0.0094	3	1.9680-1.9865	0.0009-0.0027	0.005-0.0015
2005	2.0	C	2.0469-2.0475	0.0008-0.0024	0.0035-0.0094	3	1.8894-1.8900	0.0010-0.0023	0.0050-0.0150
	2.4	S	2.3620-2.3625	0.0007-0.0024	0.0035-0.0094	3	1.9680-1.9865	0.0009-0.0027	0.005-0.0015

06009-NEON-C06

PISTON AND RING SPECIFICATIONS

All measurements are given in inches.

Year	Engine Displacement Liters	Engine ID/VIN	Piston Clearance	Ring Gap Top Compression	Ring Gap Bottom Compression	Ring Gap Oil Control	Ring Side Clearance Top Compression	Ring Side Clearance Bottom Compression	Ring Side Clearance Oil Control
2002	2.0	C	0.0008-0.0020	0.0090-0.0200	0.0190-0.0310	0.009-0.026	0.0010-0.0026	0.0010-0.0026	0.0002-0.0070
	2.0	F	0.0008-0.0020	0.0090-0.0200	0.0190-0.0310	0.009-0.0260	0.0010-0.0026	0.0010-0.0026	0.0002-0.0070
2003	2.0	C	0.0008-0.0020	0.0090-0.0200	0.0190-0.0310	0.009-0.026	0.0010-0.0026	0.0010-0.0026	0.0002-0.0070
	2.0	F	0.0008-0.0020	0.0090-0.0200	0.0190-0.0310	0.009-0.0260	0.0010-0.0026	0.0010-0.0026	0.0002-0.0070
	2.4	S	0.0009-0.0022	0.0078-0.0157	0.0070-0.0015	0.0050-0.0250	0.0010-0.0030	0.0010-0.0030	0.0010-0.0060
2004	2.0	C	0.0008-0.0020	0.0090-0.0200	0.0190-0.0310	0.009-0.026	0.0010-0.0026	0.0010-0.0026	0.0002-0.0070
	2.0	F	0.0008-0.0020	0.0090-0.0200	0.0190-0.0310	0.009-0.0260	0.0010-0.0026	0.0010-0.0026	0.0002-0.0070
	2.4	S	0.0009-0.0022	0.0078-0.0157	0.0070-0.0015	0.0050-0.0250	0.0010-0.0030	0.0010-0.0030	0.0010-0.0060
2005	2.0	C	0.0008-0.0020	0.0090-0.0200	0.0190-0.0310	0.009-0.026	0.0010-0.0026	0.0010-0.0026	0.0002-0.0070
	2.4	S	0.0009-0.0022	0.0078-0.0157	0.0070-0.0015	0.0050-0.0250	0.0010-0.0030	0.0010-0.0030	0.0010-0.0060

06009-NEON-C07

TORQUE SPECIFICATIONS

All readings in ft. lbs.

Year	Engine Displacement Liters	Engine ID/VIN	Cylinder Head Bolts	Main Bearing Bolts	Rod Bearing Bolts	Crankshaft Damper Bolts	Flywheel Bolts	Manifold Intake	Manifold Exhaust	Spark Plugs	Oil Pan Drain Plug
2002	2.0	C	①	②	③	100	70	8.5	17	20	20
	2.0	F	①	②	③	100	70	8.5	17	20	20
2003	2.0	C	①	④	③	100	70	8.5	17	20	20
	2.0	F	①	④	③	100	70	8.5	17	20	20
	2.4	S	⑤	⑥	③	100	70	17	21	11-15	20
2004	2.0	C	①	④	③	100	70	8.5	17	20	20
	2.0	F	①	④	③	100	70	8.5	17	20	20
	2.4	S	⑤	⑥	③	100	70	17	21	11-15	20
2005	2.0	C	①	②	③	100	70	8.5	17	20	20
	2.4	S	⑤	⑥	③	100	70	17	21	11-15	20

① Step 1: All bolts 25 ft. lbs.
Step 2: bolts 1-6 50 ft. lbs.
Step 3: bolts 7-10 35 ft. lbs.
Step 4: bolts 1-6 50 ft. lbs.
Step 5: bolts 7-10 35 ft. lbs.
Step 6: all bolts plus 90 degrees

② M8: 25 ft. lbs.
M11: 60 ft. lbs.

③ Step 1: 20 ft. lbs.
Step 2: plus 90 degrees

④ Step 1:
M10 bolts: 15 ft. lbs.
M8 bolts: 11 ft. lbs.
Step 2:
M10 bolts: 30 ft. lbs.
Step 3:
M10 bolts: 30 ft. lbs.
Step 4:
M10 bolts: plus 90 degrees
M8 bolts: 22 ft. lbs.

⑤ Turbo models:
Step 1: 25 ft. lbs.
Step 2: 60 ft. lbs.
Step 3: 60 ft. lbs.
Step 4: all bolts plus 90 degrees
Non-turbo
Step 1: 25 ft. lbs.
Step 2: 50 ft. lbs.
Step 3: 50 ft. lbs.
Step 4: all bolts plus 90 degrees

⑥ M8 bolts: 21 ft. lbs.
M11 bolts: 55 ft. lbs.

06009-NEON-C08

WHEEL ALIGNMENT

			Caster		Camber		
Year	Model		Range (+/-Deg.)	Preferred Setting (Deg.)	Range (+/-Deg.)	Preferred Setting (Deg.)	Toe-in (in.)
2002	Neon	F	+1.00	+2.60	+0.40	0	0.05 +/- 0.10
		R	—	—	+0.40	-0.25	0.15 +/- 0.10
2003	Neon	F	+1.00	+2.60	+0.40	0	0.20 +/-0.20
		R	—	—	+0.40	-0.25	0.30 +/- 0.20
2004	Neon	F	+1.00	+2.60	+0.40	0	0.20 +/-0.20
		R	—	—	+0.40	-0.25	0.30 +/- 0.20
2005	Neon	F	+1.00	+2.60	+0.40	0	0.20 +/-0.20
		R	—	—	+0.40	-0.25	0.30 +/- 0.20

06009-NEON-C09

TIRE, WHEEL AND BALL JOINT SPECIFICATIONS

		OEM Tires		Tire Pressures (psi)		Wheel	Ball Joint	Lug Nut
Year	Model	Standard	Optional	Front	Rear	Size	Inspection	(ft. lbs.)
2002	Neon, base	P185/65R14	None	30	30	5.5-JJ	①	100
	Neon SE	P185/60R15	None	32	32	5.5-JJ	①	100
	Neon ES	P185/60R15	None	32	32	6-JJ	①	100
	Neon ACR	P185/60R15	None	32	32	6-JJ	①	100
	Neon RT	P195/50R16	None	32	32	6-JJ	①	100
2003	Neon, base	P185/65R14	None	30	30	5.5-JJ	①	100
	Neon SE	P185/60R15	None	32	32	5.5-JJ	①	100
	Neon ES	P185/60R15	None	32	32	6-JJ	①	100
	Neon ACR	P185/60R15	None	32	32	6-JJ	①	100
	Neon RT	P195/50R16	None	32	32	6-JJ	①	100
	Neon SRT-4	P205/50R17	None	32	32	6-JJ	①	100
2004	Neon, base	P185/65R14	None	30	30	5.5-JJ	①	100
	Neon SE	P185/60R15	None	32	32	5.5-JJ	①	100
	Neon ES	P185/60R15	None	32	32	6-JJ	①	100
	Neon ACR	P185/60R15	None	32	32	6-JJ	①	100
	Neon RT	P195/50R16	None	32	32	6-JJ	①	100
	Neon SRT-4	P205/50R17	None	32	32	6-JJ	①	100
2005	Neon, base	P185/65R14	None	30	30	5.5-JJ	①	100
	Neon SE	P185/60R15	None	32	32	5.5-JJ	①	100
	Neon ES	P185/60R15	None	32	32	6-JJ	①	100
	Neon ACR	P185/60R15	None	32	32	6-JJ	①	100
	Neon RT	P195/50R16	None	32	32	6-JJ	①	100
	Neon SRT-4	P205/50R17	None	32	32	6-JJ	①	100

OEM: Original Equipment Manufacturer

PSI: Pounds Per Square Inch

① Replace if any measurable movement is found.

06009-NEON-C10

BRAKE SPECIFICATIONS

All measurements in inches unless noted

			Brake Disc			Brake Drum Diameter				Brake Caliper	
Year	Model		Original Thickness	Minimum Thickness	Maximum Runout	Original Inside Diameter	Max. Wear Limit	Maximum Machine Diameter	Minimum Lining Thickness	Bracket Bolts (ft. lbs.)	Mounting Bolts (ft. lbs.)
2002	Neon	F	0.866	0.803	0.005	—	—	—	0.300	55	16
		R	0.354	0.285	0.005	7.88	NA	①	②	55	16
2003	Neon	F	0.866	0.803	0.005	—	—	—	0.300	55	16
		R	0.354	0.285	0.005	7.88	NA	①	②	55	16
2004	Neon	F	0.866	0.803	0.005	—	—	—	0.300	55	16
		R	0.354	0.285	0.005	7.88	NA	①	②	55	16
2005	Neon	F	0.866	0.803	0.005	—	—	—	0.300	55	16
		R	0.354	0.285	0.005	7.88	NA	①	②	55	16

NA: Not Available

F: Front

R: Rear

① Stamped on the outer edge of drum

② Disc brake pad total thickness: 0.281
Brake shoe lining thickness: 0.0625

06009-NEON-C11

SCHEDULED MAINTENANCE INTERVALS

2002 Dodge—Neon

TO BE SERVICED	TYPE OF SERVICE	VEHICLE MILEAGE INTERVAL (x1000)												
		7.5	15	22.5	30	37.5	45	52.5	60	67.5	75	82.5	90	97.5
Engine oil & filter	R	✓	✓	✓	✓	✓	✓	✓	✓	✓	✓	✓	✓	✓
Brake hoses	S/I	✓	✓	✓	✓	✓	✓	✓	✓	✓	✓	✓	✓	✓
Coolant level, hoses & clamps	S/I	✓	✓	✓	✓	✓	✓	✓	✓	✓	✓	✓	✓	✓
CV joints & front suspension components	S/I	✓	✓	✓	✓	✓	✓	✓	✓	✓	✓	✓	✓	✓
Exhaust system	S/I	✓	✓	✓	✓	✓	✓	✓	✓	✓	✓	✓	✓	✓
Manual transaxle oil	S/I	✓	✓	✓	✓	✓	✓	✓	✓	✓	✓	✓	✓	✓
Rotate tires	S/I	✓	✓	✓	✓	✓	✓	✓	✓	✓	✓	✓	✓	✓
Accessory drive belts	S/I		✓		✓		✓		✓		✓		✓	
Brake linings	S/I			✓			✓			✓			✓	
Air filter element	R				✓				✓				✓	
Spark plugs	R				✓				✓				✓	
Lubricate ball joints	S/I				✓				✓				✓	
Engine coolant	R						✓				✓			
PCV valve	S/I								✓				✓	
Ignition cables	R								✓					
Camshaft timing belt ①	R													

R: Replace S/I: Service or Inspect

① Camshaft timing belt: replace at 105,000 miles

FREQUENT OPERATION MAINTENANCE (SEVERE SERVICE)

If a vehicle is operated under any of the following conditions it is considered severe service:

- Extremely dusty areas

- 50% or more of the vehicle operation is in 32°C (90°F) or higher temperatures, or constant operation in temperatures below 0°C (32°F)

Prolonged idling (vehicle operation in stop and go traffic)

- Frequent short running periods (engine does not warm to normal operating temperatures)

- Police, taxi, delivery usage or trailer towing usage

Oil & oil filter change: change every 3000 miles

Rotate tires every 6000 miles

Brake linings: inspect every 12,000 miles

Air filter element: service or inspect every 15,000 miles

Automatic transaxle: change fluid & adjust bands every 15,000 miles

Manual transaxle fluid: replace every 15,000 miles

Engine coolant: replace at 36,000 miles and every 30,000 miles thereafter

06009-NEON-C12

SCHEDULED MAINTENANCE INTERVALS

2003-05 Dodge—Neon

TO BE SERVICED	TYPE OF SERVICE	VEHICLE MILEAGE INTERVAL (x1000)												
		6	12	18	24	30	36	42	48	54	60	66	72	78
Engine oil & filter	R	✓	✓	✓	✓	✓	✓	✓	✓	✓	✓	✓	✓	✓
Brake hoses	S/I	✓	✓	✓	✓	✓	✓	✓	✓	✓	✓	✓	✓	✓
Coolant level, hoses & clamps	S/I	✓	✓	✓	✓	✓	✓	✓	✓	✓	✓	✓	✓	✓
CV joints & front suspension components	S/I	✓	✓	✓	✓	✓	✓	✓	✓	✓	✓	✓	✓	✓
Exhaust system	S/I	✓	✓	✓	✓	✓	✓	✓	✓	✓	✓	✓	✓	✓
Manual transaxle oil	S/I	✓	✓	✓	✓	✓	✓	✓	✓	✓	✓	✓	✓	✓
Rotate tires	S/I	✓	✓	✓	✓	✓	✓	✓	✓	✓	✓	✓	✓	✓
Accessory drive belts	S/I		✓		✓		✓		✓		✓		✓	
Brake linings	S/I			✓			✓			✓			✓	
Air filter element	R				✓				✓				✓	
Spark plugs	R				✓				✓				✓	
Lubricate ball joints	S/I				✓				✓				✓	
Engine coolant	R						✓				✓			
PCV valve	S/I								✓				✓	
Ignition cables	R								✓					
Camshaft timing belt ①	R													

R: Replace S/I: Service or Inspect

① Camshaft timing belt: replace at 105,000 miles

FREQUENT OPERATION MAINTENANCE (SEVERE SERVICE)

If a vehicle is operated under any of the following conditions it is considered severe service:

- Extremely dusty areas

- 50% or more of the vehicle operation is in 32°C (90°F) or higher temperatures, or constant operation in temperatures below 0°C (32°F)

Prolonged idling (vehicle operation in stop and go traffic)

- Frequent short running periods (engine does not warm to normal operating temperatures)

- Police, taxi, delivery usage or trailer towing usage

Oil & oil filter change: change every 3000 miles

Rotate tires every 6000 miles

Brake linings: inspect every 12,000 miles

Air filter element: service or inspect every 15,000 miles

Automatic transaxle: change fluid & adjust bands every 15,000 miles

Manual transaxle fluid: replace every 15,000 miles

Engine coolant: replace at 36,000 miles and every 30,000 miles thereafter

06009-NEON-C13

PRECAUTIONS

Before servicing any vehicle, please be sure to read all of the following precautions, which deal with personal safety, prevention of component damage, and important points to take into consideration when servicing a motor vehicle:

- Never open, service or drain the radiator or cooling system when the engine is hot; serious burns can occur from the steam and hot coolant.
- Observe all applicable safety precautions when working around fuel. Whenever servicing the fuel system, always work in a well-ventilated area. Do not allow fuel spray or vapors to come in contact with a spark, open flame, or excessive heat (a hot drop light, for example). Keep a dry chemical fire extinguisher near the work area. Always keep fuel in a container specifically designed for fuel storage; also, always properly seal fuel containers to avoid the possibility of fire or explosion. Refer to the additional fuel system precautions later in this section.
- Fuel injection systems often remain pressurized, even after the engine has been turned **OFF**. The fuel system pressure must be relieved before disconnecting any fuel lines. Failure to do so may result in fire and/or personal injury.
- Brake fluid often contains polyglycol ethers and polyglycols. Avoid contact with the eyes and wash your hands thoroughly after handling brake fluid. If you do get brake fluid in your eyes, flush your eyes with clean, running water for 15 minutes. If eye irritation persists, or if you have taken brake fluid internally, IMMEDIATELY seek medical assistance.
- The EPA warns that prolonged contact with used engine oil may cause a number of skin disorders, including cancer! You should make every effort to minimize your exposure to used engine oil. Protective gloves should be worn when changing oil. Wash your hands and any other exposed skin areas as soon as possible after exposure to used engine oil. Soap and water, or waterless hand cleaner should be used.
- All new vehicles are now equipped with an air bag system, often referred to as a Supplemental Restraint System (SRS) or Supplemental Inflatable Restraint (SIR) system. The system must be disabled before performing service on or around system components, steering column, instrument panel components, wiring and sensors. Failure to follow safety and disabling procedures could result in accidental air bag deployment, possible personal injury and unnecessary system repairs.
- Always wear safety goggles when working with, or around, the air bag system. When carrying a non-deployed air bag, be sure the bag and trim cover are pointed away from your body. When placing a non-deployed air bag on a work surface, always face the bag and trim cover upward, away from the surface. This will reduce the motion of the module if it is accidentally deployed. Refer to the additional air bag system precautions later in this section.
- Clean, high quality brake fluid from a sealed container is essential to the safe and proper operation of the brake system. You should always buy the correct type of brake fluid for your vehicle. If the brake fluid becomes contaminated, completely flush the system with new fluid. Never reuse any brake fluid. Any brake fluid that is removed from the system should be discarded. Also, do not allow any brake fluid to come in contact with a painted surface; it will damage the paint.
- Never operate the engine without the proper amount and type of engine oil; doing so WILL result in severe engine damage.
- Timing belt maintenance is extremely important! Many models utilize an interference-type, non-freewheeling engine. If the timing belt breaks, the valves in the cylinder head may strike the pistons, causing potentially serious (also time-consuming and expensive) engine damage. Refer to the maintenance interval charts in the front of this manual for the recommended replacement interval for the timing belt, and to the timing belt section for belt replacement and inspection.
- Disconnecting the negative battery cable on some vehicles may interfere with the functions of the on-board computer system(s) and may require the computer to undergo a relearning process once the negative battery cable is reconnected.
- When servicing drum brakes, only disassemble and assemble one side at a time, leaving the remaining side intact for reference.
- Only an MVAC-trained, EPA-certified automotive technician should service the air conditioning system or its components.

ENGINE REPAIR

Alternator

REMOVAL & INSTALLATION

2.0L Engines

1. Before servicing the vehicle, refer to the precautions in the beginning of this section.
2. Remove or disconnect the following:
   - Negative battery cable
   - Alternator jam nut, loosen only
   - Alternator adjustment nut, loosen only
   - Accessory drive splash shield
   - Alternator lower bolt, loosen only
   - Alternator drive belt
   - Alternator field circuit wiring connector, push the **RED** locking tab to release
   - Battery positive terminal
   - Upper and lower bolts, move alternator off pivot bracket
   - Pivot bracket
   - Alternator

To install:

3. Install or connect the following:
   - Alternator
   - Pivot bracket and torque the bolts to 40 ft. lbs. (54 Nm)
   - Alternator, move it onto pivot bracket
   - Alternator field circuit wiring connector, push **RED** locking tab in until it snaps in place
   - Battery positive terminal
   - Alternator drive belt
   - Tension the drive belt to 100 lbs. (used) or 135 lbs. (new)
   - Alternator adjustment bolt
   - Alternator jam nut and torque nut to 40 ft. lbs. (54 Nm)
   - Alternator mount bolts and torque the bolts to 40 ft. lbs. (54 Nm)
   - Accessory drive splash shield
   - Negative battery cable

2.4L Engines

➡The alternator is located above the oil filter and axle shaft.

1. Before servicing the vehicle, refer to the precautions in the beginning of this section.
2. Remove or disconnect the following:
   - Negative battery cable
   - 2 bolts from the top of the heat shield on the alternator
   - Nut from the upper T-bolt adjustment bracket

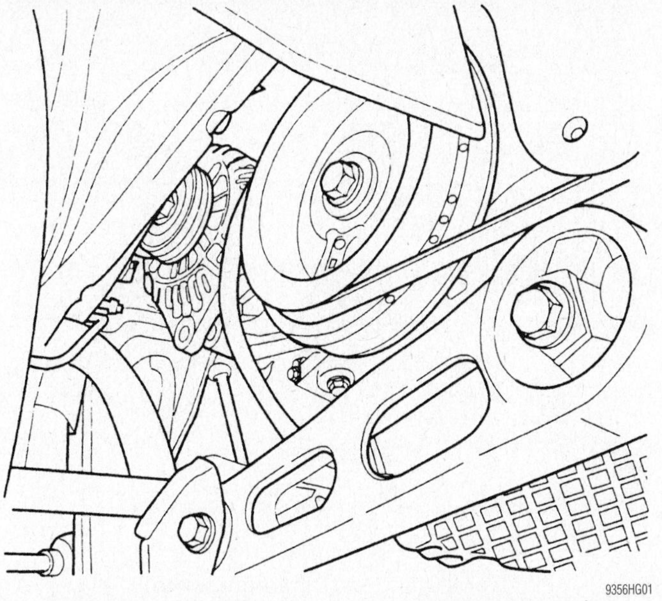

View of the alternator after removing the splash shield—2.0L engines

- Right front wheel
- Accessory drive splash shield
- Axle retaining nut
- Lower control arm from the steering knuckle
- 2 bolts for the axle shaft bearing support
- Axle shaft assembly. Put a pan under the transmission to catch the transmission fluid.
- Lower heat shield bolt
- Alternator heat shield
- Field circuit from the alternator
- B+ terminal nut and wire
- Loosen the accessory drive belt T-bolt
- Pencil strut
- Lower alternator pivot bolt
- Alternator belt
- Alternator from the lower mounting bracket and position aside
- Alternator lower mounting bracket

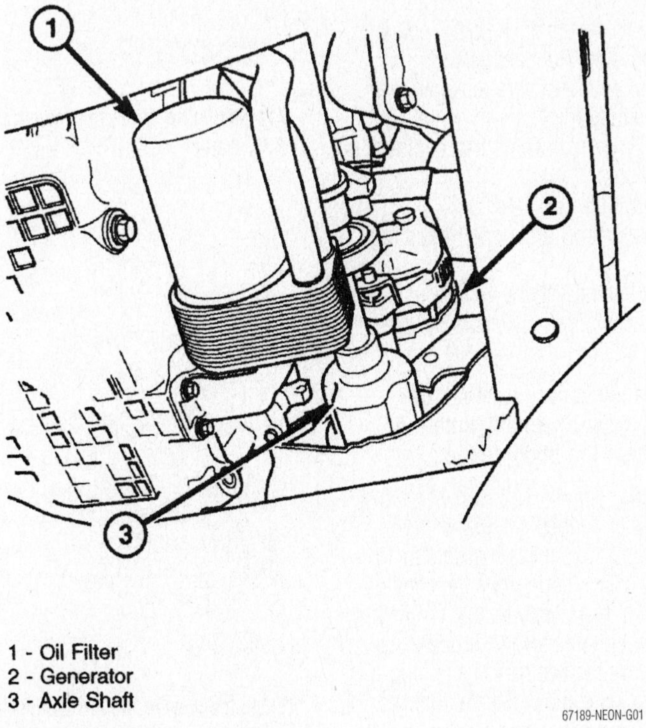

Alternator mounting—2.4L engine shown

- Alternator through the axle shaft hole

To install:

3. Install or connect the following:
   - Alternator into the vehicle through the axle shaft hole in the wheel well
   - Alternator on upper T-bolt and loosely install the nut
   - Lower mounting bracket to the block and tighten the bolts to 40 ft. lbs. (54 Nm)
   - Axle shaft into the steering knuckle
   - Lower control arm to the knuckle and secure with the bolt. Torque to 70 ft. lbs. (90 Nm)
   - Axle retaining nut and tighten to 180 ft. lbs. (240 Nm)
   - Alternator belt and adjust tension
   - Accessory belt T-bolt and tighten the nut to 40 ft. lbs. (54 Nm).
   - Lower generator pivot bolt and tighten to 40 ft. lbs. (54 Nm)
   - Pencil strut
   - Alternator heat shield
   - Lower heat shield bolt and tighten to 40 ft. lbs. (54 Nm)
   - Accessory drive splash shield
   - Right front wheel
   - Nut to the upper adjustment bracket and tighten to 18 ft. lbs. (25 Nm)
   - 2 heat shield bolts and tighten to 40 inch lbs. (4.5 Nm)
   - Negative battery cable

Ignition Timing

ADJUSTMENT

Ignition timing is controlled by the Powertrain Control Module (PCM). No adjustment is necessary or possible.

Engine Assembly

REMOVAL & INSTALLATION

2.0L Engines

1. Before servicing the vehicle, refer to the precautions in the beginning of this section.

➡After all components are installed on the engine, a DRB scan tool is necessary to perform the camshaft and crankshaft timing relearn procedure.

2. Properly recover the air conditioning system refrigerant.
3. Properly relieve the fuel system pressure.
4. Drain the engine oil.

5. Drain the cooling system.
6. Remove or disconnect the following:
   - Battery and battery tray
   - Air intake duct from the intake manifold
   - Throttle cables
   - Electrical connectors
   - Air cleaner assembly
   - Upper radiator hose
   - Fan module assembly
   - Lower radiator hose
   - Cooler lines if equipped with an automatic transaxle
   - Clutch cable or hydraulic line as necessary, if equipped with a manual transaxle
   - Shift linkage from the transaxle
   - Transaxle electrical connections
   - Engine wiring harness
   - Powertrain Control Module (PCM)
   - Positive cable from the Power Distribution Center (PDC)
   - Ground wires
   - Heater hoses
   - Brake booster vacuum hose
   - Coolant recovery hose
   - Accessory drive belts
   - Power steering pump and reservoir, move them aside
   - Air conditioning compressor, if equipped
7. Raise the vehicle.
   - Front wheels
   - Right inner splash shield
   - Halfshafts
   - Power steering cooler, if equipped, and position it aside
   - Downstream Oxygen (O_2S) sensor connector
   - Exhaust pipe from the manifold
   - Exhaust system isolators and move exhaust rearward
   - Lower engine torque strut
   - Alternator, lower bracket and upper mounting bolt
   - Structural collar
   - Flywheel or flexplate, as applicable
   - Bolt securing the power steering line to the engine block and reposition line.
8. Lower the vehicle.

9. Raise the vehicle enough to allow an Engine Dolly Tool 6135 and Cradle Tool 6710 to be placed under the engine.

10. Loosen the engine support posts in order to allow movement for positioning onto the engine locating holes and flange on the engine bedplate. Carefully lower the vehicle and position the cradle until the engine is resting on the support posts. Tighten the mounts to the cradle frame. This will keep the support posts from moving when removing or installing the engine and transaxle.

11. Install safety straps around the engine to the cradle; tighten the straps and lock them into position.

12. Raise the vehicle enough to see if the straps are tight enough to hold the cradle assembly to the engine.

13. Lower the vehicle so the weight of the engine and transaxle ONLY is on the cradle.

14. Remove the upper engine torque strut.

15. Remove the right and left engine/transaxle mount through-bolts.

16. Raise the vehicle slowly, until it is about 6 in. (15cm) above normal engine mount locations.

17. Remove the alternator.

18. Lower bracket and upper mounting bolt.

19. Continue raising the vehicle until the engine/transaxle assembly clears the compartment.

➡If may be necessary to move the engine/transaxle assembly with the cradle to clear the body flanges.

To install:

20. Install or connect the following:
    - Engine/transmission assembly to the vehicle and torque the engine mount bolts to 87 ft. lbs. (118 Nm)
    - Upper engine torque strut and remove the engine safety straps

21. Slowly raise the vehicle enough to remove the engine dolly and cradle.

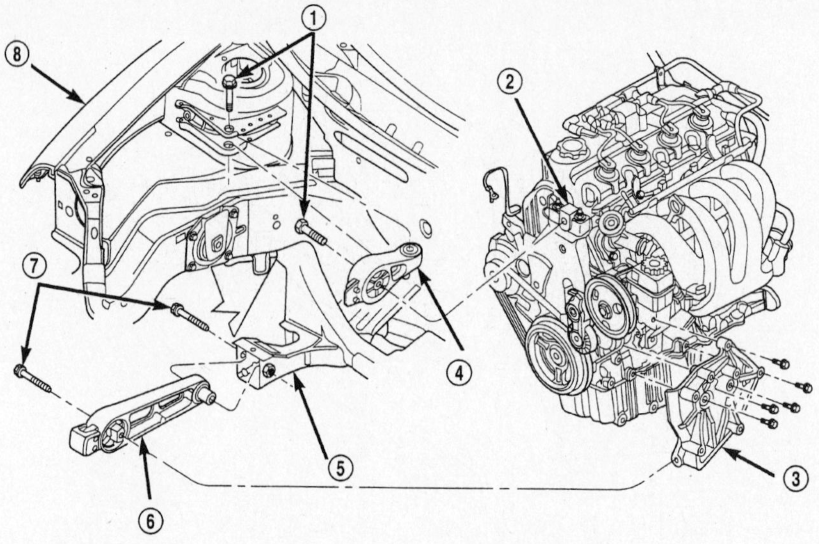

Exploded view of the torque struts and bracket—2.0L engine

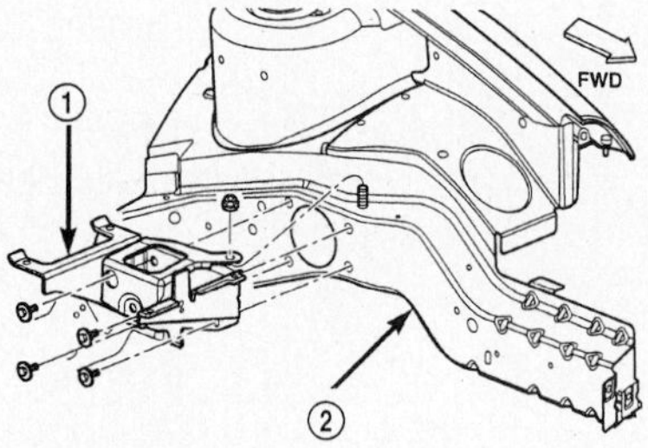

Exploded view of the left mount bracket—2.0L engine

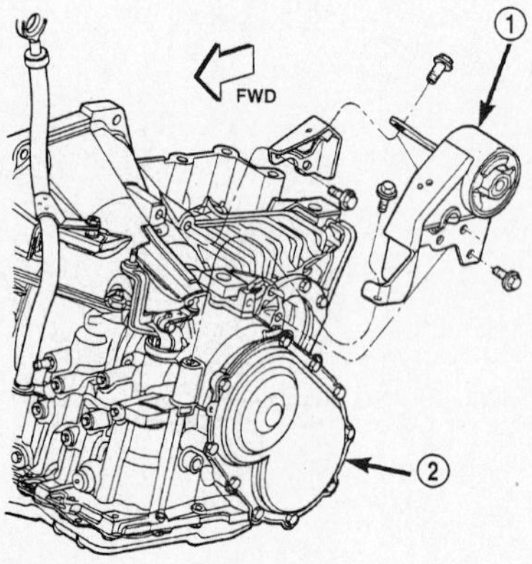

Exploded view of the left mount—Automatic transaxle—2.0L engine

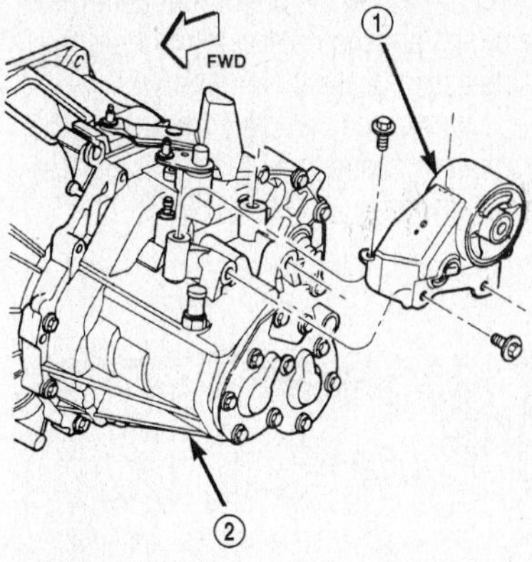

1 – MOUNT
2 – TRANSAXLE

9306EG16

Exploded view of the left mount—Manual transaxle 2.0L engine

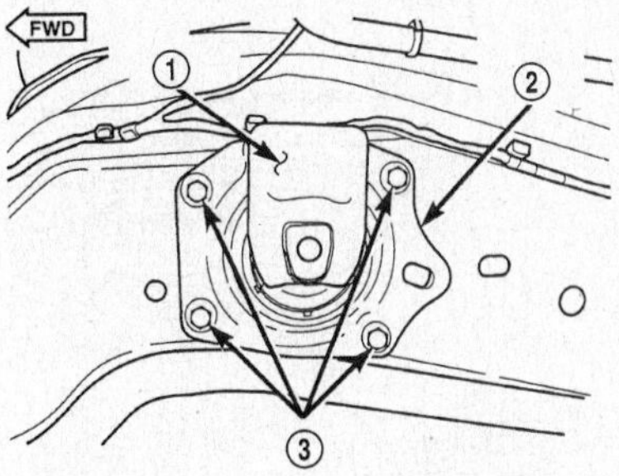

1 – SNUBBER PAD
2 – RIGHT ENGINE MOUNT
3 – BOLTS

9306EG17

Exploded view of the right mount—2.0L engine

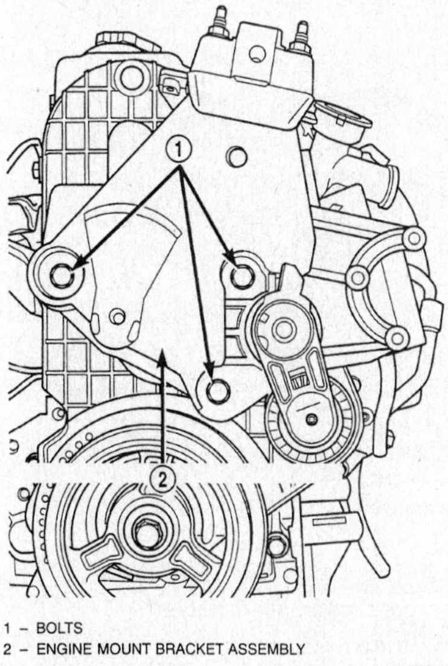

1 – BOLTS
2 – ENGINE MOUNT BRACKET ASSEMBLY

9306EG19

Exploded view of the right mount bracket—2.0L engine

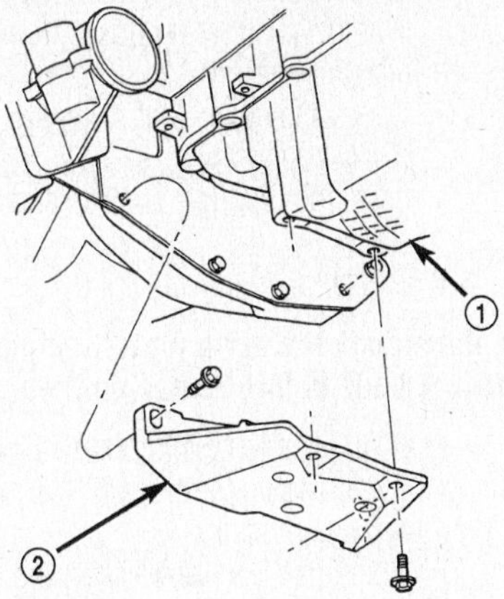

1 – OIL PAN
2 – STRUCTURAL COLLAR

9306EG18

Exploded view of the structural collar—2.0L engine

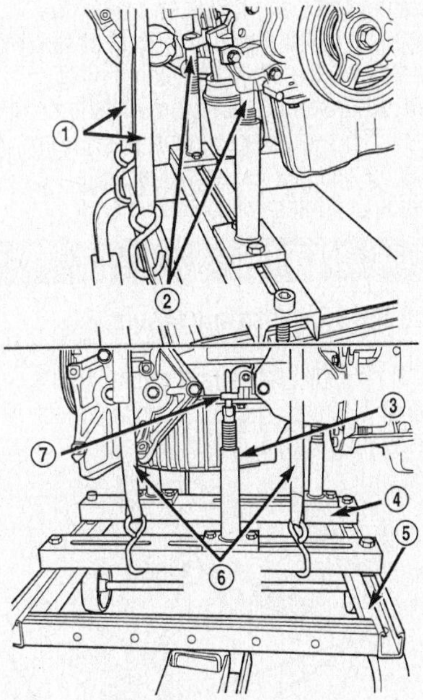

1 – SAFETY STRAPS
2 – PLACE REAR POSTS INTO LOCATING HOLES
3 – SPECIAL TOOL 6848
4 – SPECIAL TOOL 6710
5 – SPECIAL TOOL 6135
6 – SAFETY STRAPS
7 – PLACE FRONT POST UNDER BLOCK FLANGE

9306EG20

View of the engine cradle and dolly—2.0L engine

- Alternator, lower bracket and upper mounting bolt
- Halfshafts
- Flywheel or flexplate
- Bolt securing the power steering line to the engine block
- Power steering cooler screws, if equipped
- Dust shield
- Lateral bending brace

22. Install the structural collar and torque the bolts, in 3 steps, using the following procedure:
    a. Step 1: Collar-to-oil pan bolts to 30 inch lbs. (3 Nm).
    b. Step 2: Collar-to-transaxle bolts to 80 ft. lbs. (108 Nm).
    c. Step 3: Collar-to-oil pan bolts to 40 ft. lbs. (54 Nm).
23. Install or connect the following:
    - Lower engine torque strut and torque the through-bolts to 87 ft. lbs. (118 Nm)
    - Exhaust system isolators
    - Exhaust pipe to the manifold. Tighten the exhaust fasteners to 21 ft. lbs. (28 Nm)
    - Downstream O_2S sensor connector
    - A/C compressor, if equipped
    - Drive belts
    - Right inner splash shield
    - Front wheels, then lower vehicle
    - Power steering pump and reservoir

➡If equipped with a hydraulic clutch, it will not be necessary to bleed the system. The quick-connect fittings used on this system close immediately after disconnection and don't allow any fluid to leak out.

- Clutch cable/hydraulic line if equipped with a manual transaxle
- Automatic transaxle cooler lines, if equipped
- Shift linkage to the transaxle
- Transaxle electrical connectors
- Fuel lines
- Heater hoses
- Ground wires
- PCM
- Engine wire harness
- Lower radiator hose
- Fan module assembly
- Upper radiator hose
- Air cleaner assembly
- Electrical connectors
- Throttle cables
- Air intake duct to the intake manifold
- Battery and battery tray

24. Fill the engine with clean oil.
25. Fill the cooling system.
26. After all components are installed, perform the camshaft and crankshaft timing relearn procedure as follows:
    a. Connect a DRB scan tool to the DLC (located under the instrument panel, near the steering column).
    b. Turn the ignition switch **ON**, and access the "miscellaneous" screen.
    c. Select "re-learn cam/crank" option

and follow the directions on the scan tool screen.

27. If equipped with air conditioning, recharge the system.

2.4L Engine

1. Before servicing the vehicle, refer to the precautions in the beginning of this section.

➡After all components are installed on the engine, a DRB scan tool is necessary to perform the camshaft and crankshaft timing relearn procedure.

2. Properly recover the air conditioning system refrigerant.
3. Properly relieve the fuel system pressure.
4. Drain the engine oil.
5. Drain the cooling system.
6. Remove or disconnect the following:
   - Air cleaner housing
   - Battery cables, battery and battery tray
   - Throttle and speed control cables from the throttle body
   - Engine wiring harness from the Powertrain Control Module (PCM)
   - Positive cable from the Power Distribution Center (PDC) and ground wire from vehicle body
   - Bolts attaching PDC and set aside
   - Wiring connectors at lower battery tray support
   - Ground wire from the vehicle body-to-engine at the right side strut tower
   - Brake booster vacuum hose from intake manifold
   - Proportional purge hose from the throttle body
   - Coolant reserve/recovery hose from coolant outlet connector
   - Heater hoses
   - Upper and lower radiator hoses
   - Upper A/C line from A/C condenser
   - A/C lines at junction near upper torque strut
   - Clutch hydraulic line, using Special Tool 6638A
   - Transmission shift linkage
   - Transmission electrical connectors
   - Power steering hoses from radiator
   - Radiator fan electrical connector and cooling module assembly
   - Front wheels
   - Right inner splash shield
   - Halfshafts
   - Accessory drive belts
   - Alternator and support brackets
   - Charge air cooler hoses
   - Downstream Oxygen Sensor (O_2S) connector
   - Exhaust system from manifold
   - Both power steering hoses from steering gear
   - Upper and lower heat shields, elbow support bracket, turbocharger support bracket, and elbow
   - Structural collar

➡Matchmark the drive plate-to-clutch module bolts before removing them.

   - Drive plate to clutch module bolts
   - Lower engine torque strut
   - A/C compressor
   - Fluid lines from power steering pump and the pump

7. Raise vehicle enough to allow engine dolly, cradle, and posts, (Special Tools 6135, 6710, and 6848) to be installed under vehicle.
8. Loosen the engine support posts to allow movement for positioning onto engine locating holes and flange on the engine bedplate. Lower the vehicle and position the cradle until the engine is resting on the support posts. Tighten the mounts to cradle frame. This will keep support posts from moving when removing or installing the engine and transmission assembly.

✲✲ CAUTION

You MUST use safety straps.

9. Install safety straps around the engine to cradle. Tighten straps and lock them into position.
10. Raise vehicle enough to determine if the straps are secure enough to hold the cradle assembly to the engine.
11. Lower the vehicle so weight of the engine and transmission ONLY is on the cradle assembly.

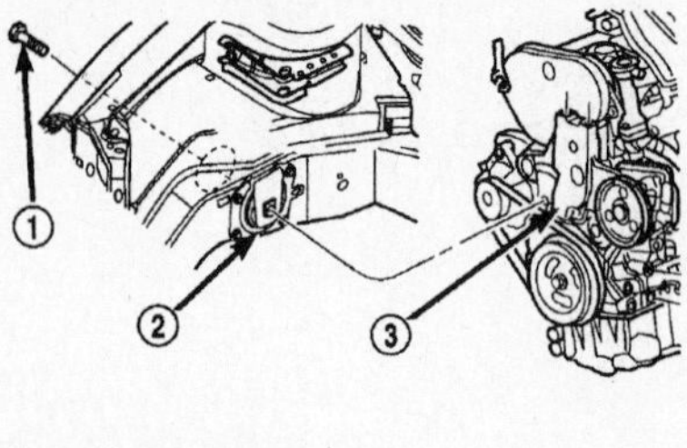

1 - BOLT
2 - RIGHT ENGINE MOUNT
3 - ENGINE MOUNT BRACKET

67189-NEON-G02

View of the right mount—2.4L engine

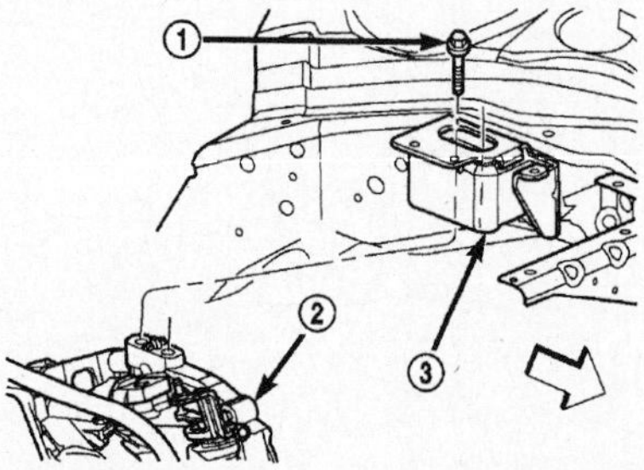

1 - BOLT
2 - TRANSAXLE
3 - LEFT MOUNT

67189-NEON-G03

View of the left mount—2.4L engine

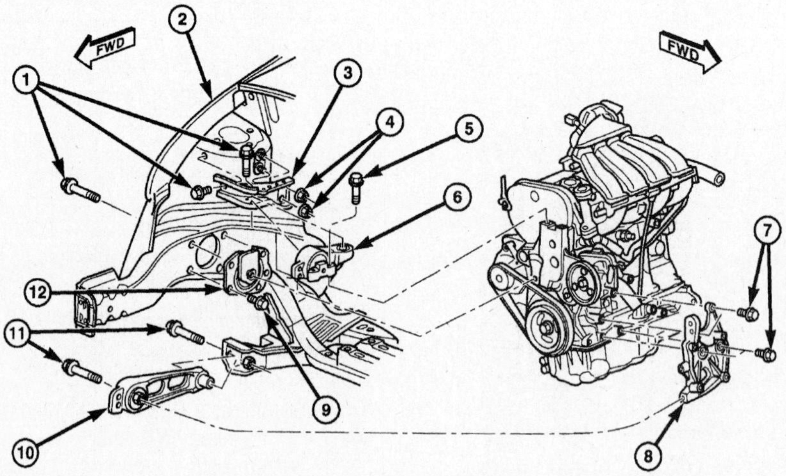

1 - BOLT
2 - RIGHT FENDER
3 - UPPER TORQUE STRUT BRACKET
4 - NUTS
5 - BOLT
6 - UPPER TORQUE STRUT
7 - BOLT
8 - LOWER TORQUE STRUT BRACKET
9 - BOLT
10 - LOWER TORQUE STRUT
11 - BOLT
12 - RIGHT ENGINE MOUNT

67189-NEON-G07

Exploded view of the torque struts—2.4L engine

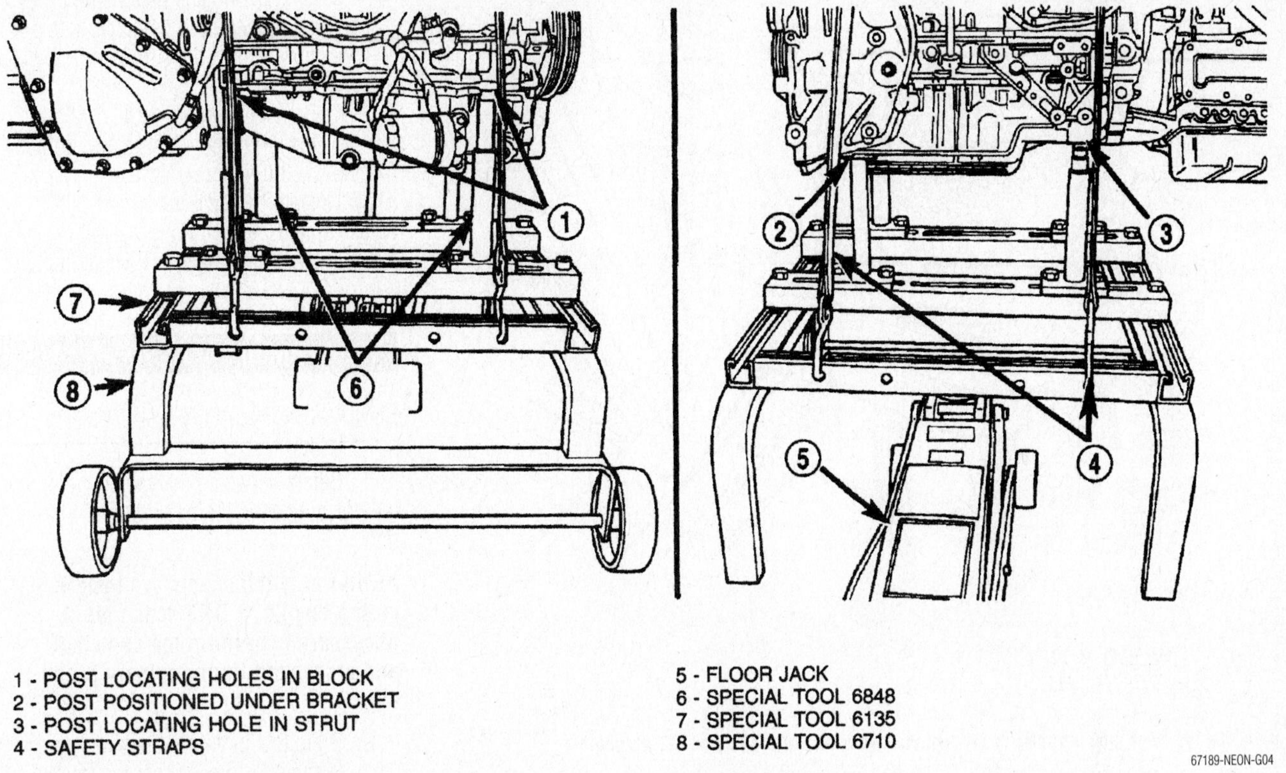

Proper positioning of the engine cradle—2.4L engine

12. Remove or disconnect the following:
- Upper engine torque strut
- Right mount through bolt and left mount attaching bolts

13. Raise the vehicle slowly until the engine/transaxle assembly clears the engine compartment. You may need to move the assembly with the cradle to allow for removal around body flanges.

To install:

14. Install or connect the following:
- Position the engine and transmission assembly under the vehicle and slowly lower the vehicle over the engine/transaxle assembly. Continue lowering vehicle until the engine/transaxle aligns to mounting locations.
- Mounting bolts at the right and left engine/transaxle mounts. Torque the bolts to 87 ft. lbs. (118 Nm)
- Upper engine torque strut. Remove the safety straps from engine/transaxle assembly. Slowly raise the vehicle enough to remove the engine dolly and cradle.
- Power steering pump
- Fluid lines to power steering pump
- A/C compressor
- Lower engine torque strut
- Drive plate to clutch module bolts
- Structural collar

WARNING

To prevent damaging the lower shield, you must install the upper shield first.

- Elbow, turbocharger support bracket, elbow support bracket, and upper and lower heat shields
- Power steering hoses to the steering gear
- Exhaust system to the manifold
- Downstream O_2S
- Charge air cooler hoses
- Alternator and mounting brackets
- Accessory drive belts
- Halfshafts
- Right inner splash shield
- Wheels
- Cooling module assembly, and connect the radiator fan electrical connector
- Power steering hoses to the radiator
- Clutch hydraulic line
- Transmission shift linkage and electrical connectors

➡You do not have to bleed the clutch. The quick-connect fittings seal closed immediately after disconnection and no air can get in the system

- A/C lines at junction near the upper torque strut
- Upper A/C line to the A/C condenser
- Upper and lower radiator hoses
- Fuel line and heater hoses
- Coolant reserve/recovery hose to the coolant outlet connector
- Brake booster vacuum hose to the intake manifold
- Proportional purge hose to the throttle body
- Ground straps, and connect all engine wiring
- PDC and retaining bolts
- Positive battery cable to the PDC and ground wire to the vehicle body
- Engine wiring harness to the PCM
- Throttle and speed control cables
- Battery tray, battery and cables
- Air cleaner housing assembly and clean air hose
- Oil filter. Fill engine with proper type and amount of oil

15. Fill power steering and engine cooling systems.
16. Evacuate and recharge A/C system.
17. Start the engine and run until operating temperature is reached. Check for leaks.
18. Perform torque strut adjustment procedure, as follows:

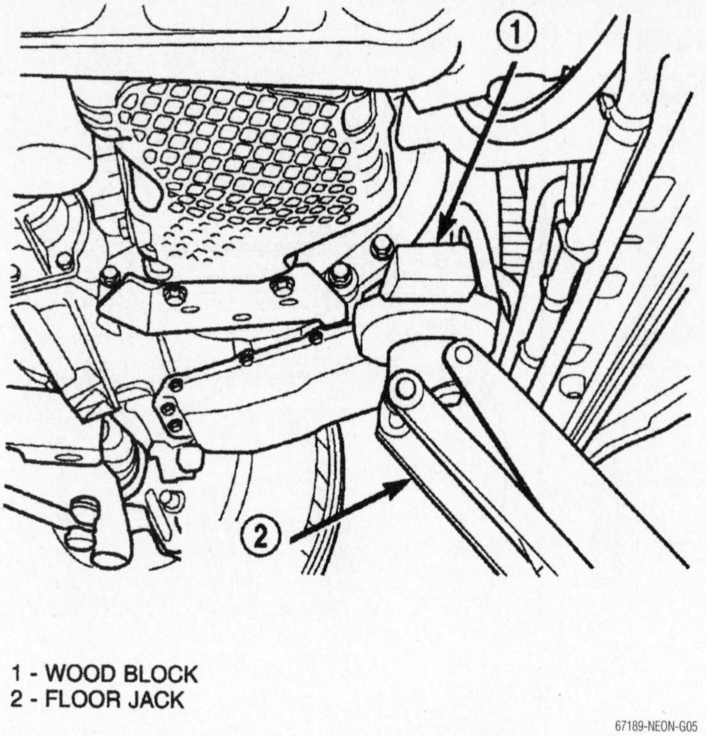

Position the floor jack properly to prevent upward lifting of the engine—2.4L engine

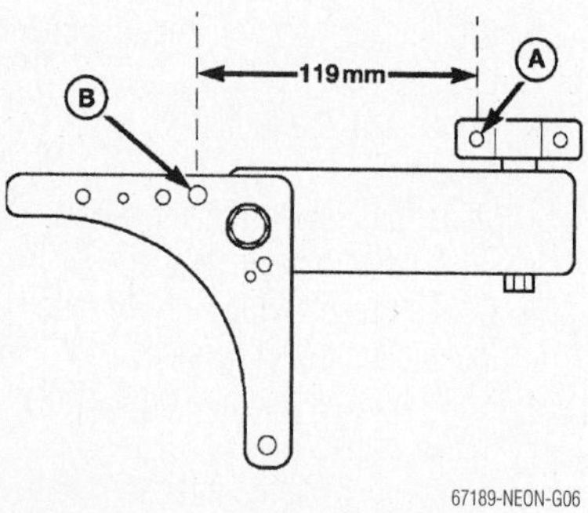

Measure the distance between the center of the rearmost attaching bolt on the engine mount bracket (point A) and the center of the hole on the shock tower bracket (point B)—2.4L engine

a. Remove the accessory drive belt splash shield.

b. Remove the pencil strut.

c. Loosen the upper and lower torque strut attaching bolt at the suspension crossmember and shock tower bracket.

d. Adjust the engine position by placing a floor jack on the forward edge of the transmission bell housing

e. With the engine supported, remove the upper and lower torque strut attachment bolt(s) at shock tower bracket and suspension crossmember. Make sure the torque struts are free to move within the shock tower bracket and crossmember. Reinstall the torque strut bolt(s), but do not tighten yet.

f. Apply upward force, allowing the upper engine to rotate rearward until the distance between the center of the rearmost attaching bolt on the engine mount bracket (point A) and the center of the hole on the shock tower bracket (point B) is 4.70 in. (119mm)

WARNING

Hold the engine in position with the jack until the upper and lower torque strut bolts are tightened.

g. With the engine held at the proper position, tighten both the upper and lower torque strut bolts to 85 ft. lbs. (115 Nm)

h. Remove the floor jack.

i. Install the pencil strut and tighten nuts to 43 ft. lbs. (58 Nm).

19. Install the accessory drive belt splash shield.

20. Adjust transmission linkage, if necessary.

Water Pump

REMOVAL & INSTALLATION

1. Before servicing the vehicle, refer to the precautions in the beginning of this section.

➡After all components are installed on the engine, a DRB scan tool is necessary to perform the camshaft and crankshaft timing relearn procedure.

2. Drain the cooling system.
3. Remove or disconnect the following:
   - Negative battery cable
   - Right inner splash shield
   - Accessory drive belts
   - Power steering pump and support the engine from the bottom with a jack
   - Upper and lower torque isolator struts
   - Right engine mount
   - Power steering pump bracket bolts, move the pump/bracket assembly aside

➡It is not necessary to disconnect the power steering lines.

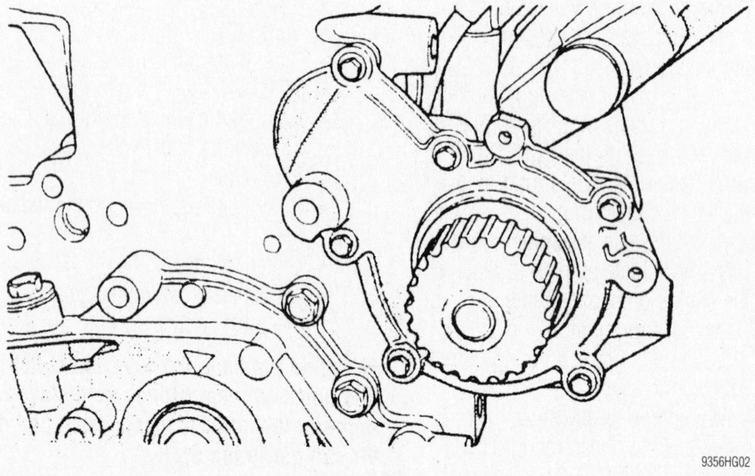

Water pump mounting—2002 Neon shown, other years similar

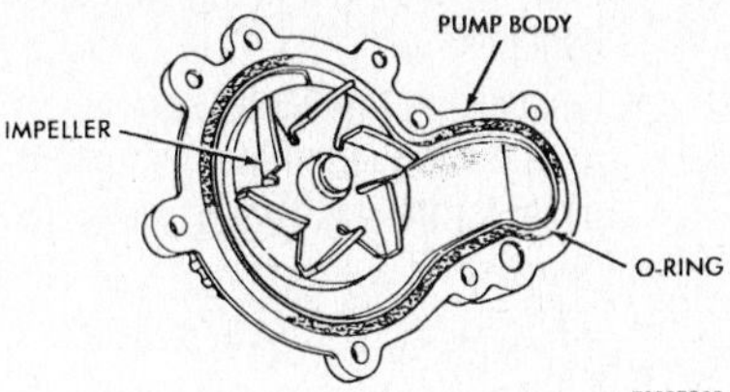

7922EG05

Be sure the O-ring is seated in the groove before installing the new pump

- Right engine mount bracket
- Timing belt and timing belt tensioner
- Camshaft sprocket(s)
- Inner (rear) timing belt cover
- Water pump and discard the O-ring

4. Clean all mating surfaces of any residual gasket material.

To install:

5. Install a new O-ring in the water pump groove.

➡Hold the O-ring in place with a few small dabs of silicone sealant.

WARNING

Before proceeding, be sure the O-ring gasket is properly seated in the water pump groove before tightening the screws. An improperly installed O-ring could cause a coolant leak.

6. Install or connect the following:
   - Water pump, torque the bolt to 105 inch lbs. (12 Nm). Use a pressure tester to pressurize the cooling system to 15 psi (103 kPa) and check the water pump shaft seal and O-ring for leaks.

➡Rotate the pump by hand to check for freedom of movement.

- Inner (rear) timing belt cover
- Camshaft sprocket
- Timing belt tensioner and timing belt
- Right engine mount bracket and engine mount
- Power steering pump
- Upper and lower torque isolator struts
- Accessory drive belts
- Negative battery cable

7. Refill the cooling system.
8. Use a DRB scan tool to perform the camshaft and crankshaft timing relearn procedure, as follows:
   a. Connect the scan tool to the DLC, located under the instrument panel near the steering column.
   b. Turn the ignition switch **ON**, and access the "miscellaneous" screen.
   c. Select the "re-learn cam/crank" option, then follow the instructions on the scan tool screen.

Heater Core

REMOVAL AND INSTALLATION

1. Before servicing the vehicle, refer to the precautions in the beginning of this section.
2. Disconnect the negative battery cable.
3. Discharge and recover the air conditioning system refrigerant.
4. Remove the instrument panel from the vehicle by removing or disconnecting the following:
   - Push the seats back all the way
   - Pry out the left and right A-pillar trim moldings and remove (using a trim stick tool)
   - Upper instrument panel cover
   - Pull up on the cluster bezel (carefully), and remove it from the vehicle

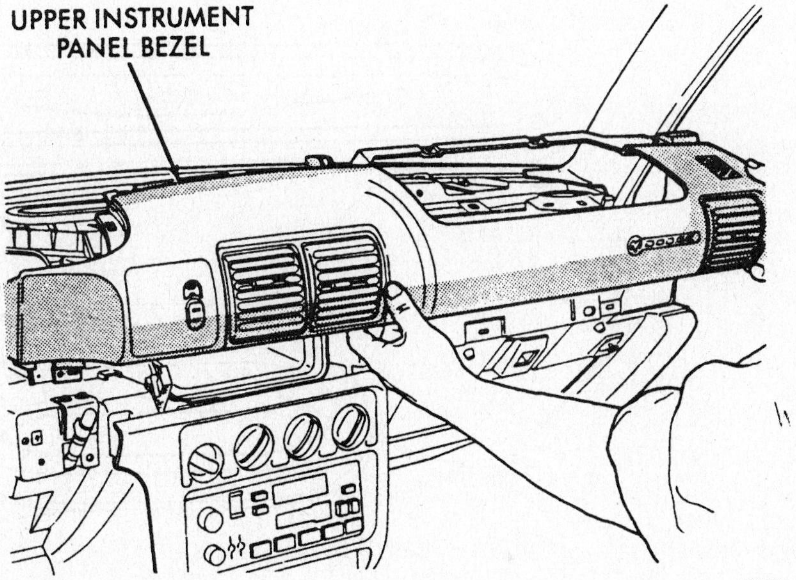

89716G11

Remove the right side upper instrument panel bezel

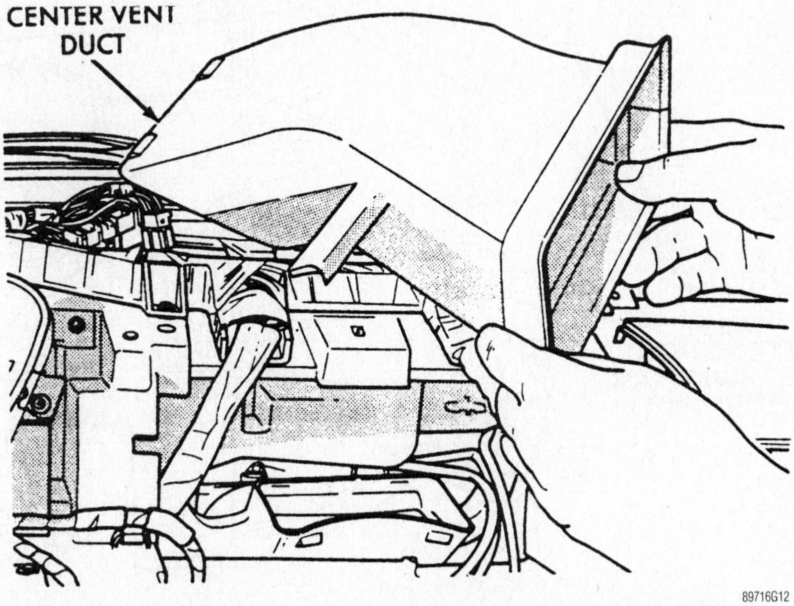

89716G12

You must remove the instrument panel center vent duct

- Pull the instrument panel cover rearward (carefully), and remove it from the vehicle

⁂ WARNING

Lock the steering wheel in the straight-ahead position; this will prevent damage to the clockspring

- Steering column as an assembly
- Left and right instrument panel end caps
- Center console
- Depress the Data Link Connector (DLC) sides and remove the DLC from the instrument panel reinforcement
- 4 bulkhead instrument panel screws
- 2 brake pedal support bracket bolts
- 2 center support mounting bolts
- Left and right A-pillar bolts; there are 2 on each side
- Antenna connector from the right side
- Left and right A-pillar door harness connectors
- 2 HVAC wiring harness connectors from the top right of the instrument panel
- Left side wiring harness connector from the top left of the instrument panel for the vanity and rear view mirrors
- Pull off the HVAC control head knobs
- 2 top front center bezel screws
- Using a trim stick, carefully pry out the instrument panel center bezel and remove it
- 2 HVAC control head screws

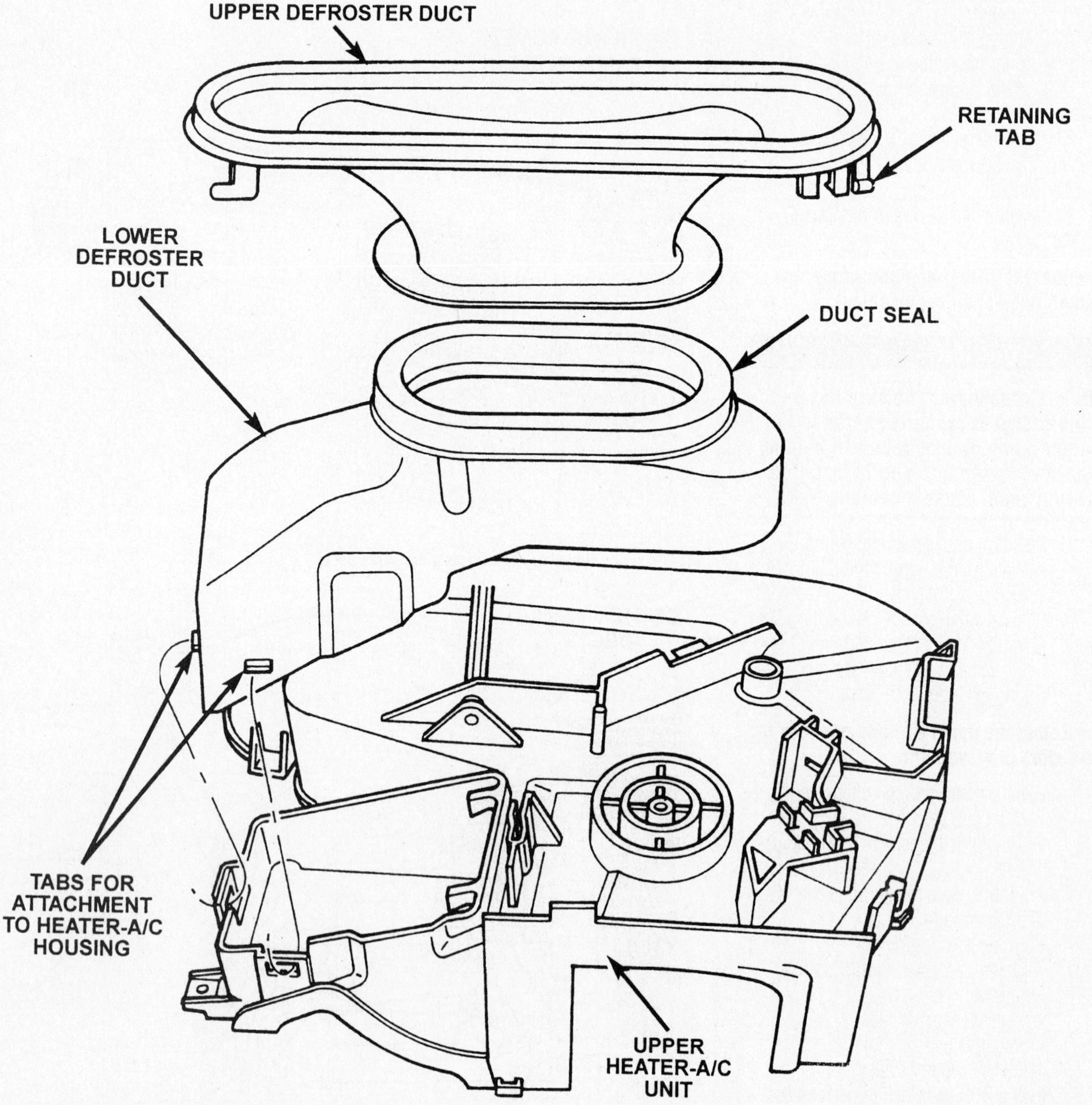

The upper defrost duct is secured with retaining tabs

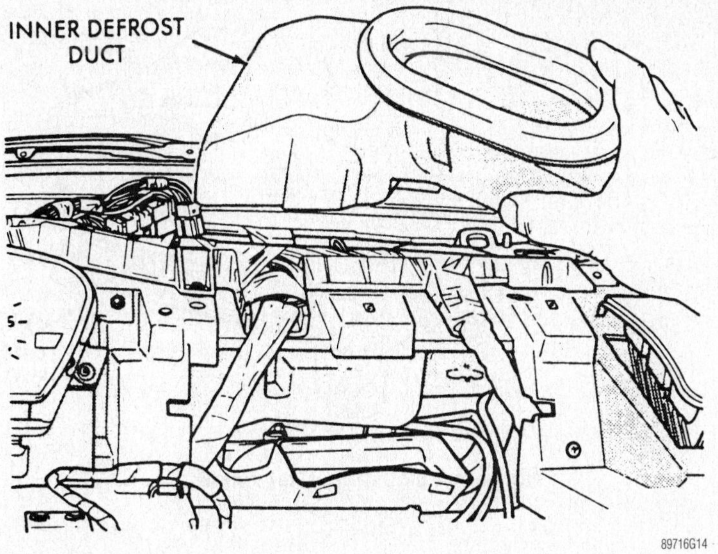

Remove the inner defrost duct

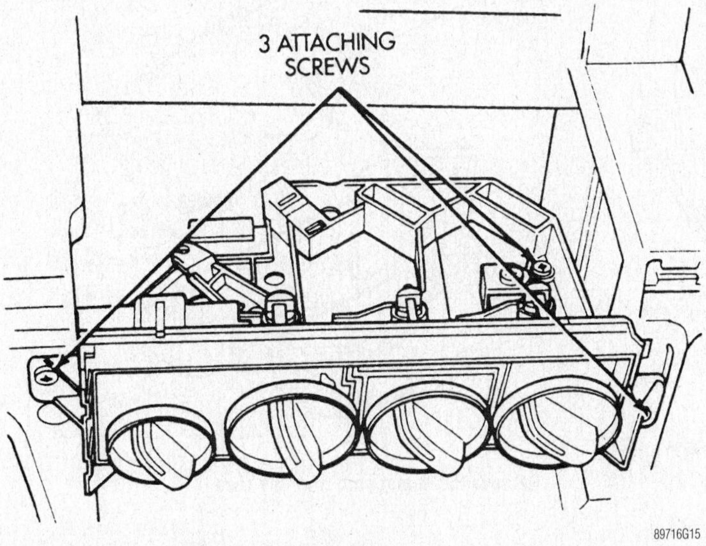

Exploded view of the control panel retainer locations

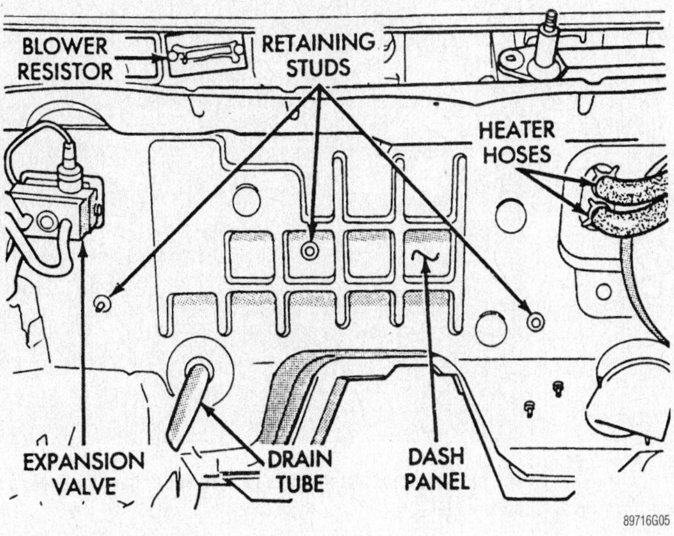

Location of the 3 dash panel retaining studs

- Instrument panel wiring harness connector
- Vacuum harness connector

5. Pull the HVAC control head out of the instrument panel, twist it 90 degrees and push it back through the opening; do not disconnect the control cables.
6. Remove or disconnect the following:
   - Air bag Control Module (ACM) from the center console
   - Parking Brake Warning Lamp Switch from the center console
   - Transmission Range Indicator Lamp from the center console
7. Using an assistant, pull the instrument panel rearward and remove it from the vehicle.
8. Drain the cooling system and remove the heater hoses at the dash panel. Place plugs in the heater core outlets to prevent coolant spillage during the unit housing removal.
9. Remove the suction line at the expansion valve. Place a piece of tape over the open refrigerant line to prevent moisture and/or dirt from entering the line.
10. Remove the expansion valve from the evaporator fitting to prevent moisture and/or dirt from entering the evaporator.
11. Remove or disconnect the following:
    - Rubber drain tube extension from the condensation drain tube
    - Vacuum harness from the power brake booster (if equipped)
    - Defroster duct
    - 3 retaining nuts located in the engine compartment, on the dash panel
    - Right side retaining screw
    - Remaining nut located on the dash panel stud
    - Blue 5-way wiring harness connector from the plenum
    - Heater/air conditioning assembly from the vehicle
    - Separate the air distribution outlet-to-parting case line foam seals
    - Evaporator lines and heater core tubes foam seals
    - Clips and screws that hold the unit to the housing
    - 4 retaining screws from the inlet air duct on the module
    - Air inlet and recirc door assembly from the module
    - Sensing switch from its harness
    - Upper-to-lower case retaining clips and screws
    - Separate the case halves, and remove the evaporator
    - Heater core

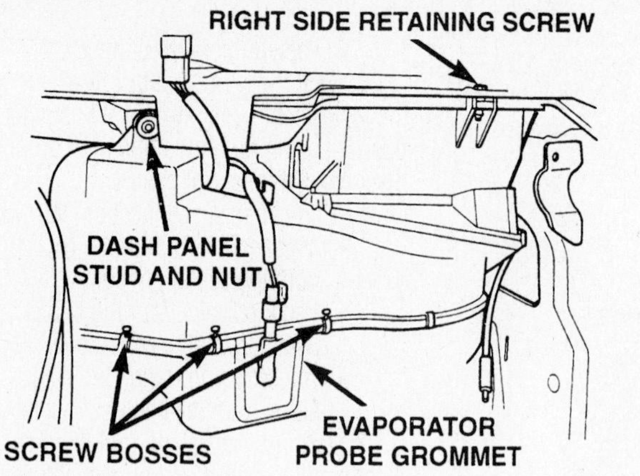

Unit housing retaining screw location

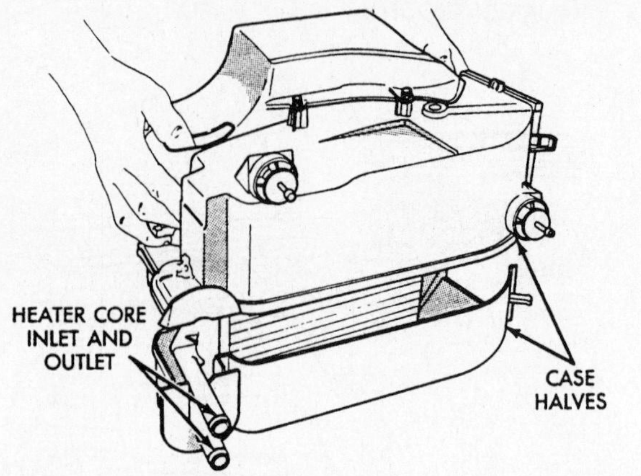

Then separate the 2 halves of the module

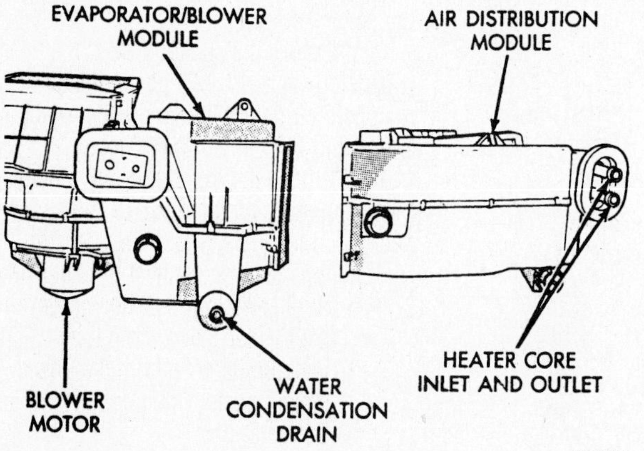

You must remove the retainers, then separate the air distribution module from the evaporator/blower module

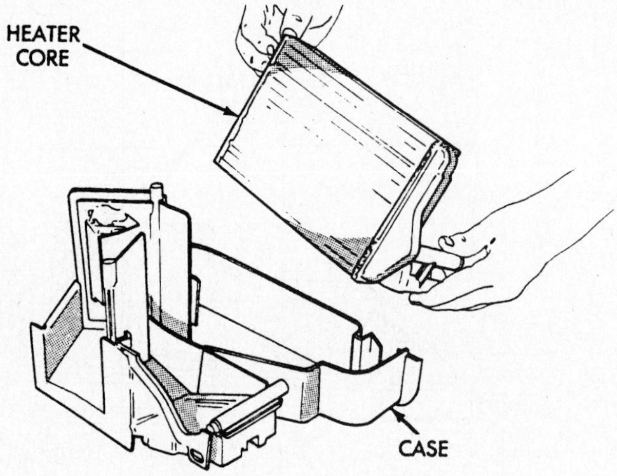

Remove the heater core from the case

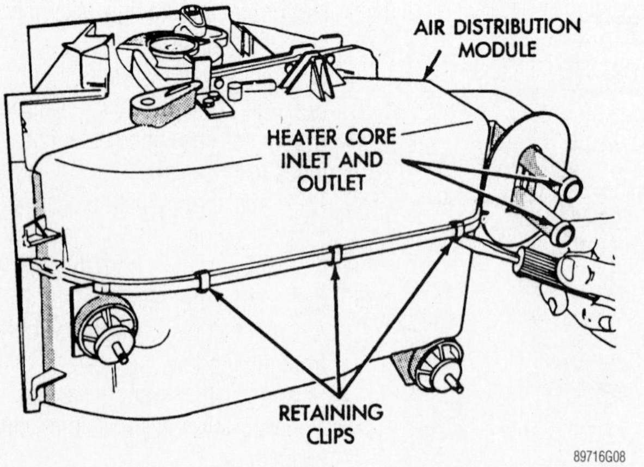

Remove the upper-to-lower housing retaining clips and screws

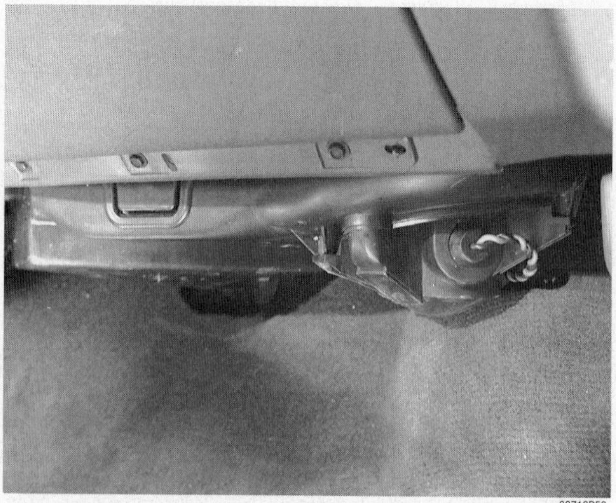

The blower motor assembly is located under the passenger's side dash panel

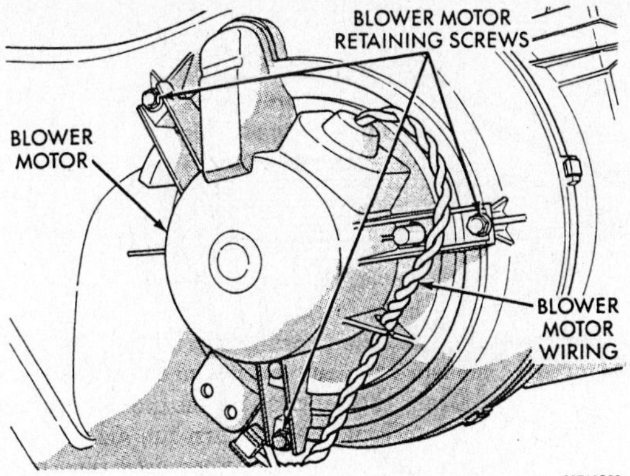

Blower motor mounting and location of the retaining screws and wiring vehicles with air conditioning

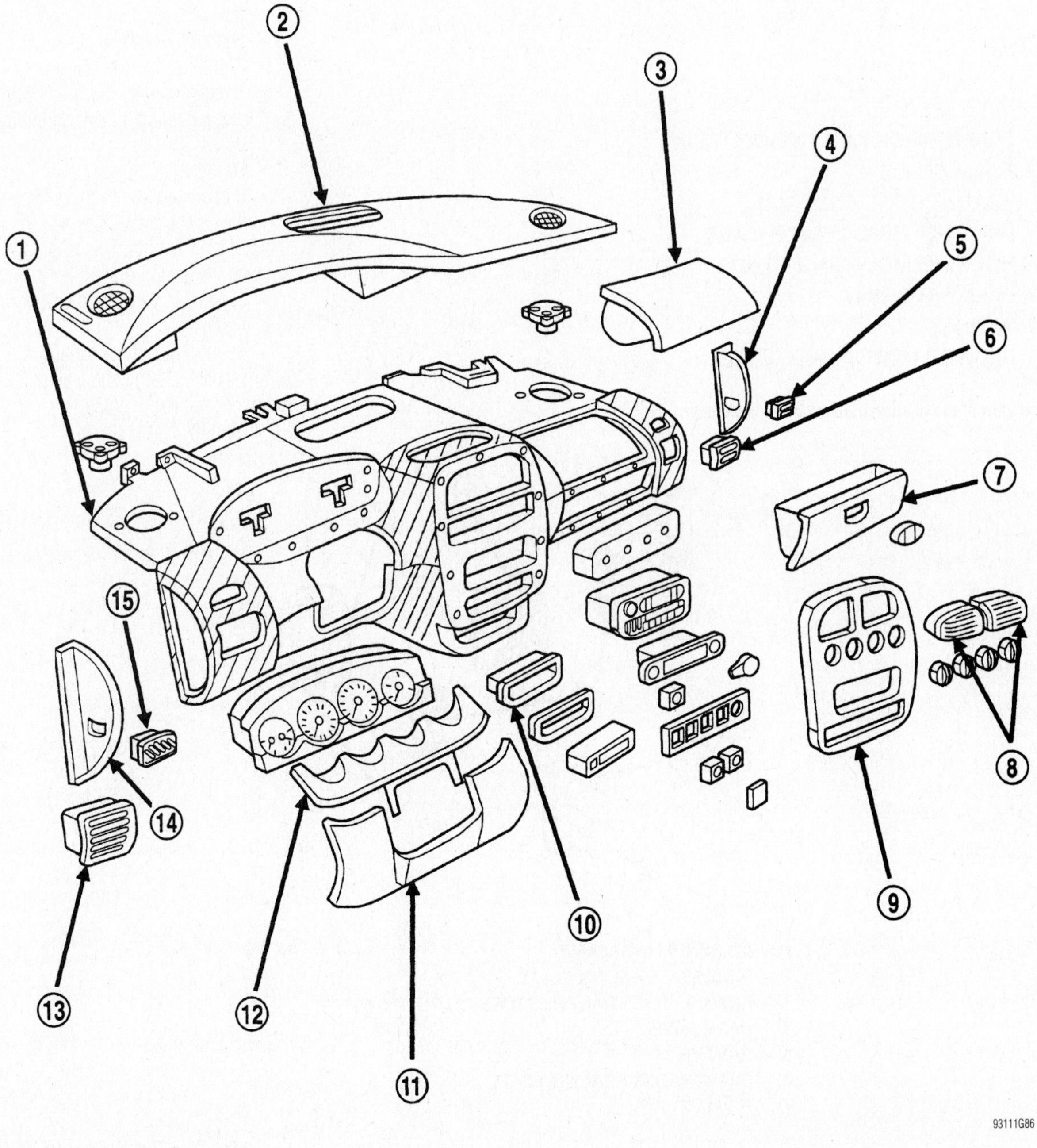

Exploded view of the instrument panel

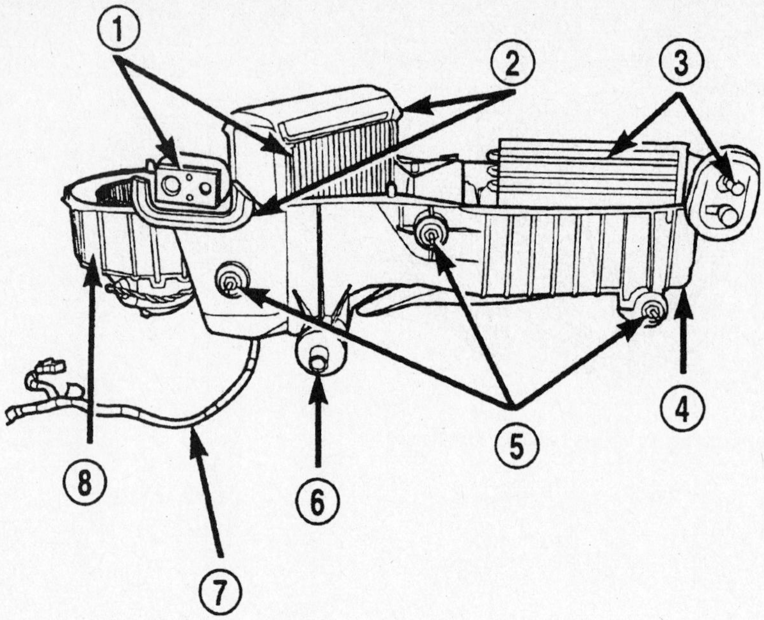

1 – EVAPORATOR AND CONNECTION
2 – FOAM SEALS
3 – HEATER CORE AND TUBES
4 – HVAC HOUSING LOWER CASE
5 – HOUSING MOUNTING STUDS
6 – HOUSING DRAIN
7 – WIRING
8 – BLOWER MOTOR AND WHEEL

93111G87

Sectional view of the heater/air conditioning assembly

To install:

12. Install or connect the following:
    - Heater core into the heater/air conditioning unit.
    - Evaporator and the upper-to-lower case retaining clips and screws
    - Sensing switch to its harness
    - Recirc door assembly to the module and the air inlet
    - Clips and screws that hold the unit to the housing
    - Evaporator lines and heater core tubes foam seals.
    - Air distribution outlet-to-parting case line foam seals
    - Assembly into the vehicle
    - Blue 5-way wiring harness connector to the plenum
    - Nut on the dash panel stud
    - Right side retaining screw
    - 3 retaining nuts located in the engine compartment, on the dash panel
    - Defroster duct
    - If equipped, the vacuum harness to the power brake booster
    - Rubber drain tube extension to the condensation drain tube
    - Expansion valve to the evaporator fitting
    - Suction line at the expansion valve
13. Evacuate, charge and leak-test the air conditioning system, if equipped.
14. Install the heater hoses.

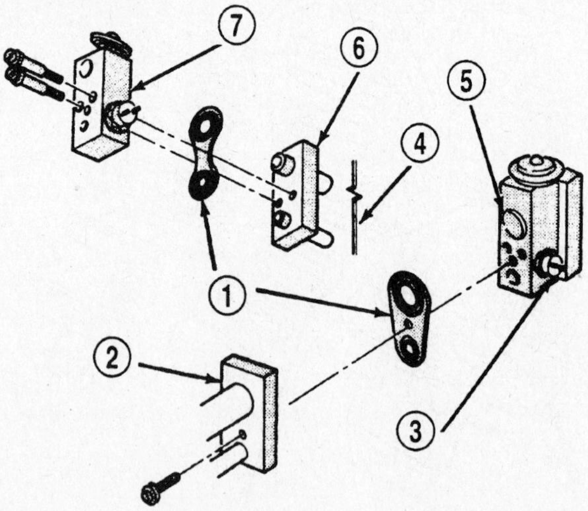

1 – ALUMINUM N-GASKET
2 – PLUMBING SEALING PLATE
3 – LOW/DIFFERENTIAL PRESSURE CUT-OFF SWITCH
4 – DASH PANEL
5 – H-VALVE
6 – EVAPORATOR SEALING PLATE
7 – H-VALVE

93111G88

Exploded view of the expansion valve

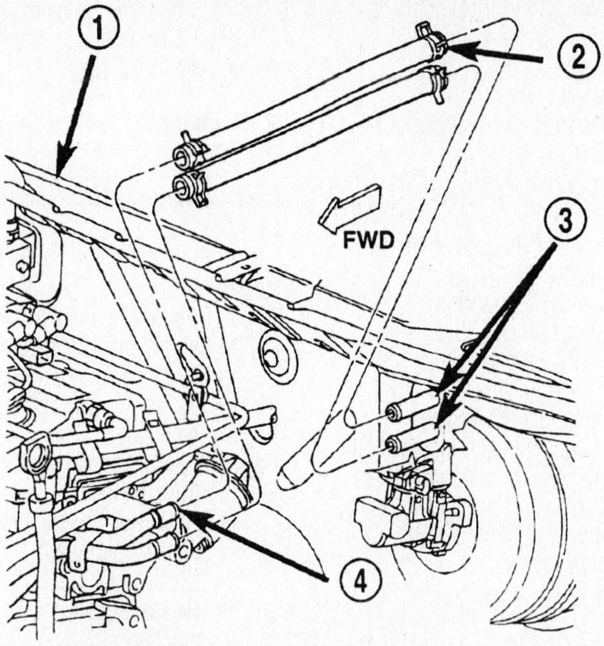

1 – COWL PANEL
2 – HEATER HOSE AND CLAMPS
3 – HEATER CORE TUBES
4 – HEATER HOSE SUPPLY AND RETURN TUBES
93111G89

View of the heater hoses and clamps

15. Refill the cooling system.
16. Install the instrument panel to the vehicle by performing the following procedure:
- Air bag Control Module (ACM)
- Parking Brake Warning Lamp Switch
- Transmission Range Indicator Lamp
- HVAC control head into the instrument panel
- Vacuum harness connector
- Instrument panel wiring harness connector
- 2 HVAC control head screws
- Instrument panel center bezel
- 2 top front center bezel screws
- HVAC control head knobs
- Left side wiring harness connector to the top left of the instrument panel for the vanity and rear view mirrors
- 2 HVAC wiring harness connectors to the top right of the instrument panel
- Left and right A-pillar door harness connectors
- Antenna connector to the right side
- Left and right A-pillar bolts; there are 2 on each side
- 2 center support mounting bolts
- 2 brake pedal support bracket bolts
- 4 bulkhead instrument panel screws
- Data Link Connector (DLC) to the instrument panel reinforcement
- Center console
- Left and right instrument panel end caps
- Steering column as an assembly
- Instrument panel cover
- Cluster bezel
- Upper instrument panel cover
- Left and right A-pillar trim moldings
- Move seats forward
- Negative battery cable

17. Operate the engine to normal operating temperatures; then, check the climate control operation and check for leaks.

Cylinder Head

REMOVAL & INSTALLATION

2.0L Engine

1. Before servicing the vehicle, refer to the precautions in the beginning of this section.

➡After all components are installed on the engine, a DRB scan tool is necessary to perform the camshaft and crankshaft timing relearn procedure.

2. Properly relieve the fuel system pressure.
3. Drain the cooling system.
4. Remove or disconnect the following:
- Negative battery cable
- Power steering/air conditioning drive belt
- Exhaust pipe from the manifold
- Right front wheel
- Right side splash shield
- Alternator drive belt
- Crankshaft damper
- Lower torque strut
- Upper torque strut
- Ground strap from the engine mount bracket
- Power steering hose support clip from the engine mount bracket
- Power steering pump, move it aside

5. Place a jack under the engine and support it.
- Right side engine mount-to-bracket through-bolt
- Lower engine mount bracket bolt

6. Slightly raise the engine.
- Upper engine mount bracket bolts
- Engine mount bracket

➡It may be necessary to raise and lower the engine until the bracket clears the engine components.

- Front timing belt cover

7. Rotate the crankshaft and align the timing marks.
- Timing belt and tensioner
- Camshaft sprocket
- Rear timing belt cover
- Fuel line from the fuel rail
- Coolant recovery container
- Ground wire and from the cylinder head
- Upper radiator hose
- Intake manifold
- Ignition coil electrical connector
- Coil pack
- Spark plug wires
- Crankcase Closed Ventilation (CCV) hose from the valve cover
- Cam sensor electrical connector
- Coolant Temperature (CT) sensor electrical connector
- Heater tube from the cylinder head
- Heater hose from the thermostat housing connector
- Cylinder head cover
- Cylinder head bolts, working from the center out
- Cylinder head and discard gasket

➡**The cylinder head bolts must be inspected before they can be reused. If the threads of bolts are stretched, they must be replaced. Check for thread stretching by holding a scale or other straightedge against the threads. If all the threads do not contact the scale, the bolts must be replaced.**

⁂ WARNING

Use only a plastic scraper to clean the mating surfaces. NEVER use metal, as this may gouge the surfaces and cause leaks!

8. Cover the combustion chambers, then use a plastic scraper to thoroughly and carefully clean the engine block and cylinder head mating surfaces.

To install:

9. Install the cylinder head with a new gasket after applying Mopar Gasket Sealant to both sides of the gasket.
10. Lubricate the cylinder head bolt threads. The 4 short bolts (164mm) are installed in positions 7, 8, 9 and 10.
11. Torque the cylinder head bolts, in the sequence, to:
    a. Step 1: All bolts to 25 ft. lbs. (34 Nm).
    b. Step 2: Bolts 1–6 to 50 ft. lbs. (68 Nm).
    c. Step 3: Bolts 7–10 35 ft. lbs. (49 Nm).
    d. Step 4: Bolts 1–6 to 50 ft. lbs. (68 Nm).
    e. Step 5: Bolts 7–10 35 ft. lbs. (49 Nm).
    f. Step 6: All bolts an additional ¼ turn using a torque angle wrench.
12. Install or connect the following:
    - Cylinder head cover
    - Heater hose to the thermostat housing connector
    - Heater tube to the cylinder head
    - CT sensor electrical connector
    - Cam sensor electrical connector
    - CCV hose to the valve cover
    - Coil pack
    - Spark plug wires
    - Ignition coil electrical connector
    - Intake manifold
    - Upper radiator hose
    - Ground wire to the cylinder head
    - Coolant recovery container
    - Fuel line to the fuel rail
    - Rear timing belt cover
    - Camshaft sprocket
    - Timing belt and tensioner
    - Front timing belt cover
    - Engine mount bracket
    - Upper engine mount bracket bolts
    - Lower engine mount bracket bolt
    - Right side engine mount-to-bracket through-bolt and torque it to 87 ft. lbs. (118 Nm)
13. Remove the jack from under the engine.
    - Power steering pump
    - Power steering hose support clip to the engine mount bracket
    - Ground strap to the engine mount bracket
    - Upper torque strut
    - Lower torque strut
    - Crankshaft damper and torque the bolt to 100 ft. lbs. (136 Nm)
    - Alternator drive belt
    - Right side splash shield
    - Right front wheel
    - Exhaust pipe to the manifold
    - Power steering/air conditioning drive belt
    - Negative battery cable
14. Refill the cooling system.
15. Use a DRB scan tool to perform the camshaft and crankshaft timing relearn procedure, as follows:
    a. Connect the scan tool to the DLC, located under the instrument panel near the steering column.
    b. Turn the ignition switch **ON**, and access the "miscellaneous" screen.
    c. Select the "re-learn cam/crank" option, then follow the instructions on the scan tool screen.

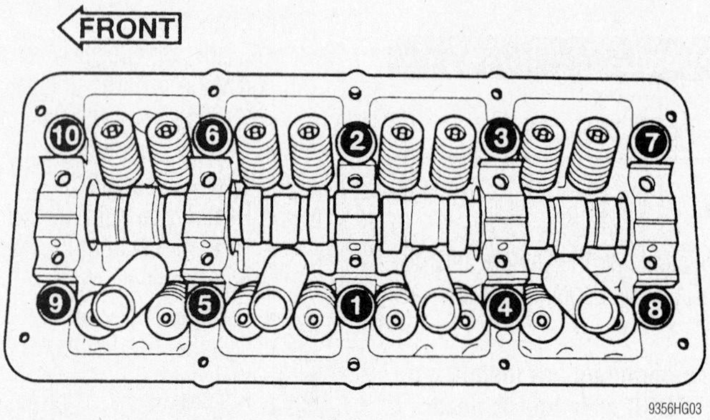

Cylinder head bolt torque sequence—2.0L engines

2.4L Engine

1. Before servicing the vehicle, refer to the precautions in the beginning of this section.
2. Properly relieve the fuel system pressure.
3. Drain the cooling system.
4. Remove or disconnect the following:
    - Negative battery cable
    - Air intake hose and air cleaner housing
    - Fuel supply line quick-connect from the fuel rail assembly
    - Heater tube support bracket from the cylinder head
    - Upper radiator hose
    - Heater hoses from the thermostat housing
    - Engine Coolant Temperature (ECT) sensor connector
    - Accessory drive belts
    - Exhaust pipe from the manifold
    - Turbocharger heat shields
    - Elbow support bracket
    - Turbocharger support bracket
    - Oil return tube and oil supply line
    - Coolant supply and return lines
    - Ignition coil wiring connector
    - Ignition coil and plug wires
    - Camshaft Position (CMP) sensor connector
    - Timing belt and camshaft sprockets
    - Timing belt idler pulley and rear timing belt cover
    - Cylinder head (valve) cover
    - Camshafts and rocker arms
    - Cylinder head bolts, in the reverse of the tightening sequence
    - Cylinder head and discard gasket

➡**The cylinder head bolts must be inspected before they can be reused. If the threads of bolts are stretched, they must be replaced. Check for thread stretching by holding a scale or other straightedge against the threads. If all the threads do not contact the scale, the bolts must be replaced.**

⁂ WARNING

Use only a plastic scraper to clean the mating surfaces. NEVER use metal, as this may gouge the surfaces and cause leaks!

5. Cover the combustion chambers, then use a plastic scraper to thoroughly and

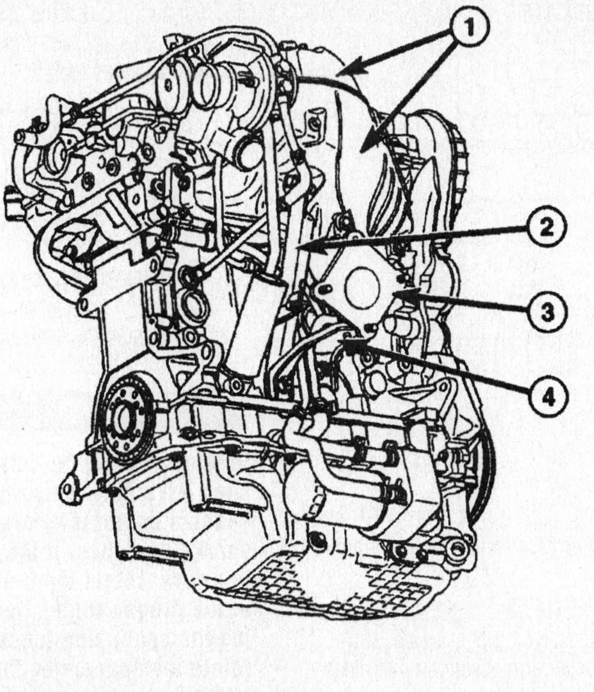

Turbocharger heat shield mounting—2.4L engine

carefully clean the engine block and cylinder head mating surfaces.

To install:

6. Install the cylinder head with a new gasket after applying Mopar Gasket Sealant to both sides of the gasket.
7. Lubricate the cylinder head bolt threads. The 4 short bolts (164mm) are installed in positions 7, 8, 9 and 10.
8. On turbo models, torque the cylinder head bolts, in the sequence, to:
   a. Step 1: All bolts to 25 ft. lbs. (34 Nm).
   b. Step 2: All bolts to 60 ft. lbs. (82 Nm).
   c. Step 3: All bolts to 60 ft. lbs. (82 Nm).
   d. Step 4: All bolts an additional 1/4 turn using a torque angle wrench.
9. On Non-turbo models, torque the cylinder head bolts, in the sequence, to:
   a. Step 1: 25 ft. lbs. (34 Nm).
   b. Step 2: 50 ft. lbs. (68 Nm).
   c. Step 3: 50 ft. lbs. (68 Nm).
   d. Step 4: All bolts an additional 1/4 turn using a torque angle wrench.
10. Install or connect the following:
   - Rocker arms and camshafts
   - Cylinder head (valve) cover
   - Rear timing belt cover and timing belt idler pulley
   - Camshaft sprockets and timing belt
   - CMP sensor connector
   - Ignition coil and plug wire
   - Ignition coil wiring connector
   - Coolant return and supply lines
   - Oil supply line and return tube
   - Exhaust elbow
   - Turbocharger support bracket
   - Elbow support bracket
   - Turbocharger heat shields
   - Exhaust pipe to the manifold. Tighten to 20 ft. lbs. (28 Nm)
   - Accessory drive belts
   - ECT sensor connector
   - Upper radiator hose
   - Heater hoses to the thermostat housing
   - Heater tube support bracket to the cylinder head
   - Lower intake manifold support bracket
   - Quick-connect fuel lines to the fuel rail assembly
   - Air intake hose and air cleaner housing
   - Negative battery cable
11. Refill the cooling system.

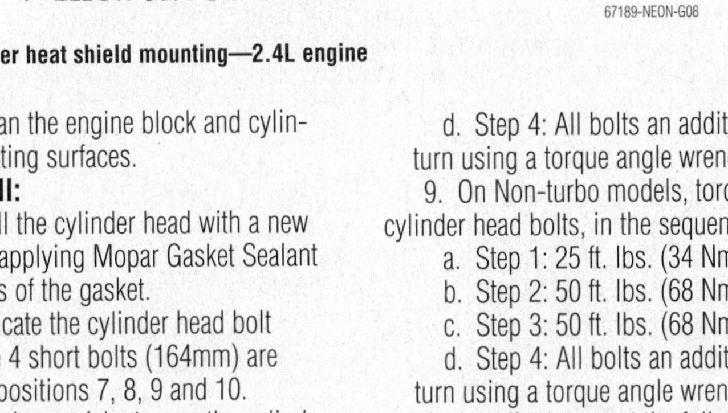

Cylinder head bolt torque sequence–2.4L engine

Rocker Arms/Shafts

REMOVAL & INSTALLATION

This procedure applies to Single Overhead Camshaft (SOHC) engines only. On Dual Overhead Camshaft (DOHC) engines, the valves are actuated directly by the camshafts and no rocker arms are used.

1. Before servicing the vehicle, refer to the precautions in the beginning of this section.
2. Remove or disconnect the following:
   - Negative battery cable
   - Cylinder head cover

➡Be sure to note the installed positions of the rocker arm shaft assemblies before removal.

   - Rocker arm shaft attaching fasteners, loosen them
   - Rocker arm shaft assembly from the cylinder head
   - Rocker arm assemblies, slide the rocker arms and spacers off the shaft

➡Be sure to keep the spacers and rocker arms in their original locations for installation.

3. Inspect the rocker arm for scoring, wear on the roller or damage to the rocker arm; replace any components showing damage. Check the location where the rocker arms mount to the shafts for wear or damage. Replace if damaged or worn. The

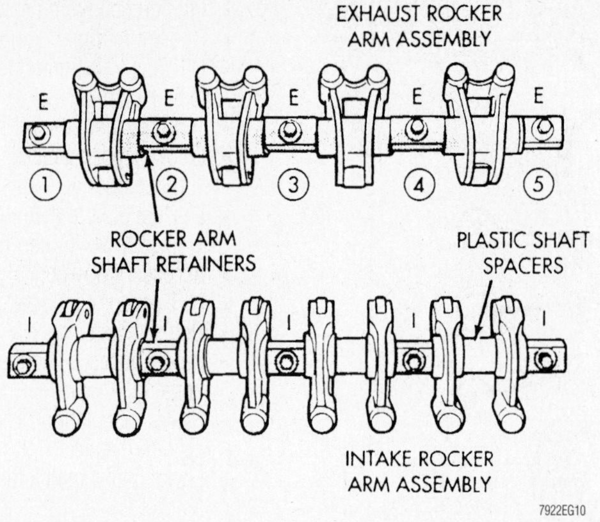

Be sure to note the rocker arm shaft component positions before disassembling the rocker arm shafts

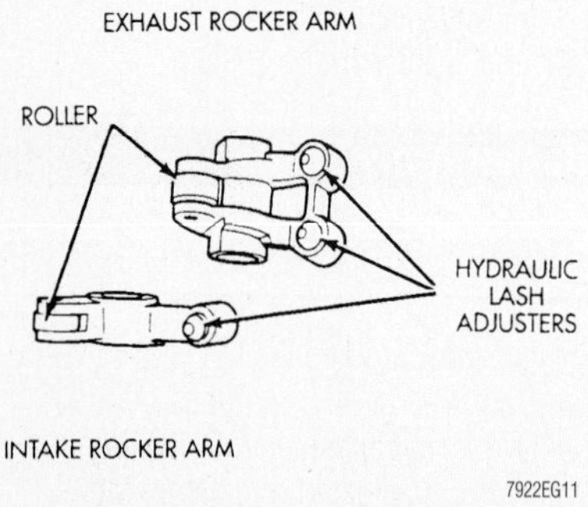

Disassembled view of an intake and exhaust rocker arm

rocker arm shaft is hollow and is used as a lubrication oil duct. Check the oil holes for clogs with a small piece of wire, and clean as required. Lubricate the rocker arms and spacers. Be sure to install in their original locations.

To install:

⁂ WARNING

Set the crankshaft to 3 notches before Top Dead Center (TDC) before installing the rocker arm shafts, otherwise valvetrain damage may occur when the mounting bolts are tightened.

4. Set the crankshaft sprocket to TDC by aligning the mark on the sprocket with the arrow on the oil pump housing, then back off to 3 notches before TDC.
5. Install the rocker arm/hydraulic lash adjuster assembly making sure that the adjusters are at least partially full of oil. This is indicated by little or no plunger travel when the lash adjuster is depressed. If there is excessive plunger travel, submerge the rocker arm assembly into clean engine oil and pump the plunger until the lash adjuster travel is taken up. If travel is not reduced, replace the assembly. The hydraulic lash adjuster and rocker arm are serviced as an assembly.
6. Install or connect the following:
   - Rocker arm and shaft assemblies. Position the rocker arm shafts with the **NOTCH FACING UP** and toward the timing belt side of the engine.
   - Retainers in their original positions on the exhaust and intake shafts. Tighten the bolts, in the sequence, to 21 ft. lbs. (28 Nm).

⁂ WARNING

When installing the intake rocker arm shaft assembly, be sure the plastic spacers do not interfere with the spark plug tubes. If the spacers do interfere, rotate them until they are at the proper angle. To avoid damaging the spark plug tubes, do not try to rotate the spacers by forcing the shaft down.

Be sure the notches in the rocker arm shafts point up and toward the timing belt side of the engine

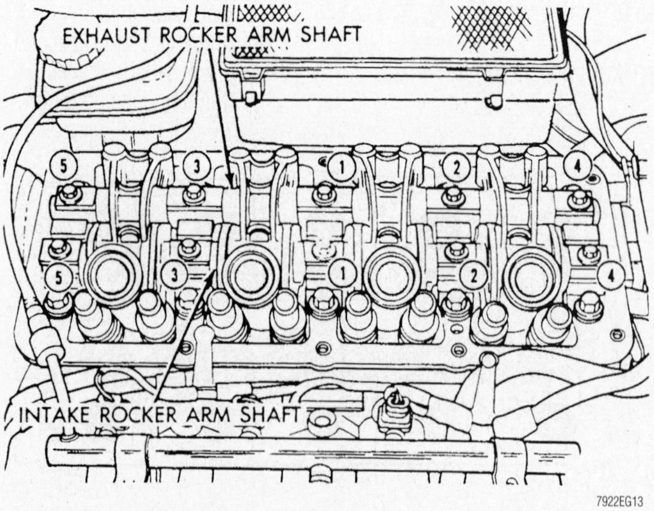

Rocker arm shaft assembly bolt tightening sequence

7. Install the rocker arm (valve) cover.
8. Connect the negative battery cable.

Intake Manifold

REMOVAL & INSTALLATION

2.0L SOHC Engine

EXCEPT HIGH OUTPUT ENGINES

1. Before servicing the vehicle, refer to the precautions in the beginning of this section.
2. Properly relieve the fuel system pressure.
3. Remove or disconnect the following:
   - Negative battery cable
   - Fresh air inlet duct from the air cleaner
   - Fuel supply line from the fuel tube assembly
   - Fuel injector cover
   - Fuel rail assembly
   - Brake booster vacuum hose
   - Positive Crankcase Ventilation (PCV) hose
   - Knock Sensor (KS) electrical connector
   - Starter wiring connectors
   - Intake manifold and discard the gasket
4. Clean all mating surfaces of any residual gasket material.

To install:

5. Install or connect the following:
   - New gaskets and seals
   - Intake manifold and torque the new retainers, in sequence, to 95 inch lbs. (11 Nm)
   - Fuel rail and torque the screws to 17 ft. lbs. (23 Nm)
   - PCV hose
   - Brake booster hose
6. Inspect the fuel line quick-connect fittings for damage and replace if necessary. Apply a small amount of clean engine oil to the fuel inlet tube.
   - Fuel supply hose to the fuel rail assembly. Ensure the connection is fastened securely by pulling on the connector.

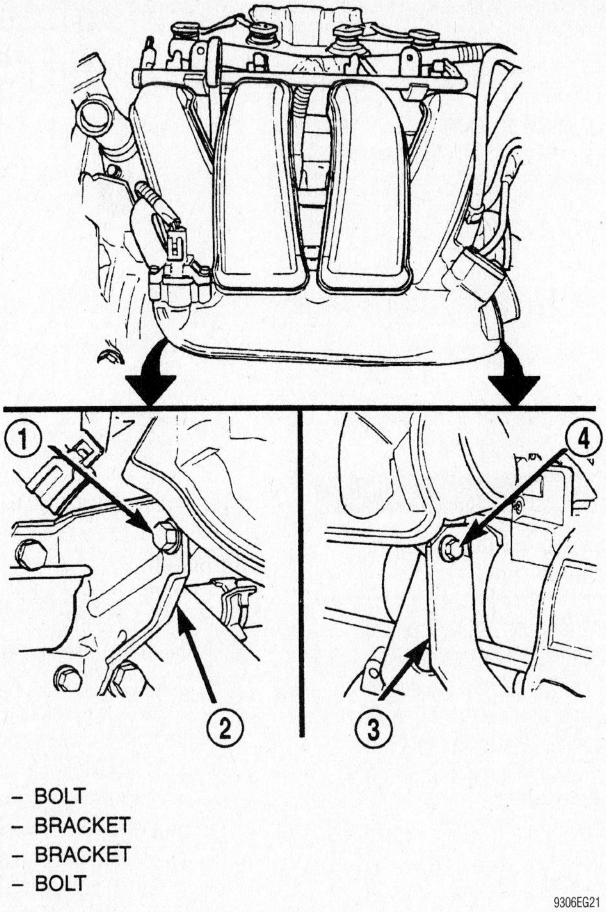

View of the intake manifold lower supports—SOHC engines

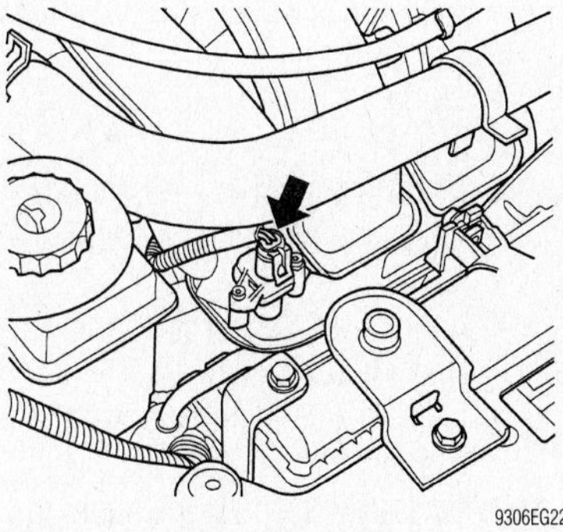

View of the Manifold Absolute Pressure (MAP) sensor—SOHC engines

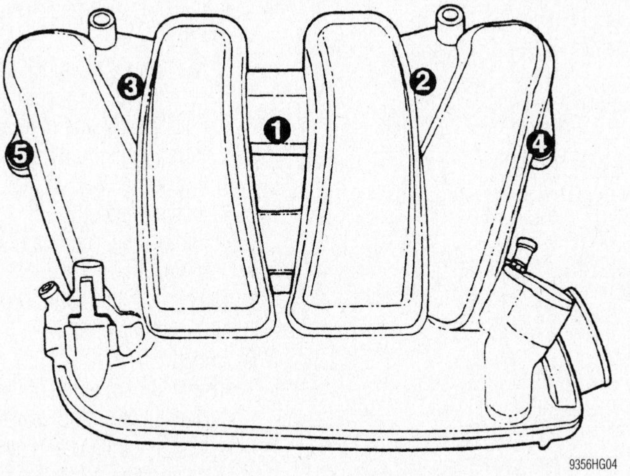

Intake manifold bolt torque sequence—2.0L (VIN C) SOHC engine

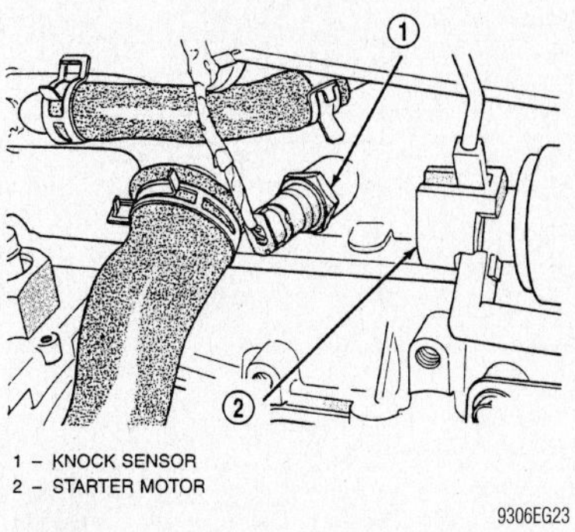

View of the Knock Sensor (KS)—SOHC engines

- Fuel injector cover
- MAP sensor wiring
- KS sensor connector
- Air inlet duct
- Negative battery cable

7. Start the vehicle, check for leaks and repair if necessary.

HIGH OUTPUT ENGINE

1. Before servicing the vehicle, refer to the precautions in the beginning of this section.
2. Properly relieve the fuel system pressure.
3. Remove or disconnect the following:
   - Negative battery cable

⁂ CAUTION

Wrap towels around the fitting to catch any spilled fuel.

   - Fuel supply line quick-connect from the fuel tube assembly
   - Fuel rail cover from the engine
   - Fuel injector connector and harness from the injectors and fuel rail
   - Positive Crankcase Ventilation (PCV) hose
   - Throttle body-to-intake manifold air duct
   - Lower intake manifold support bracket bolt
   - Oil dipstick
   - Manifold Absolute Pressure (MAP) sensor wiring
   - Manifold Tuning Valve (MTV) actuator wiring
   - Intake manifold-to-cylinder head bolts and move the manifold forward
   - Knock (KS) Sensor wiring
   - Brake booster hose
   - Intake manifold and discard the gasket
4. If necessary, separate the upper manifold from the lower manifold.

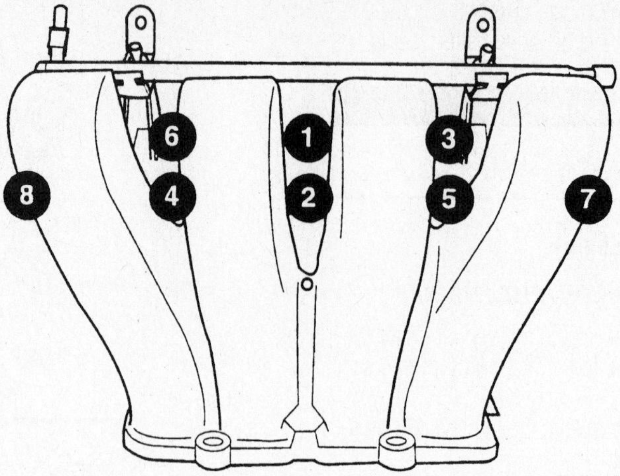

Upper intake manifold bolt torque sequence—High Output 2.0L (VIN F) SOHC engine

To install:

5. Clean the mating surfaces of any residual gasket material.
6. Install or connect the following:
   - Lower manifold to the upper manifold with a new gasket and torque the bolts to 105 inch lbs. (12 Nm)
   - Intake manifold on the cylinder head with a new gasket
   - Engine wiring harness between the middle intake runners
   - KS wiring connector
   - Brake booster hose and torque the intake manifold bolts, in sequence, to 105 inch lbs. (12 Nm)
   - Lower manifold support bracket bolt and torque the bolt to 21 ft. lbs. (28 Nm)
   - MAP sensor connector
   - MTV actuator connector
   - Oil dipstick
   - Inlet air duct to the manifold and the throttle body
   - PCV hose
   - Fuel injector connectors and harness to the fuel rail
   - Fuel rail cover
   - Fuel supply line
   - Negative battery cable
7. Start the vehicle, check for leaks and repair if necessary.

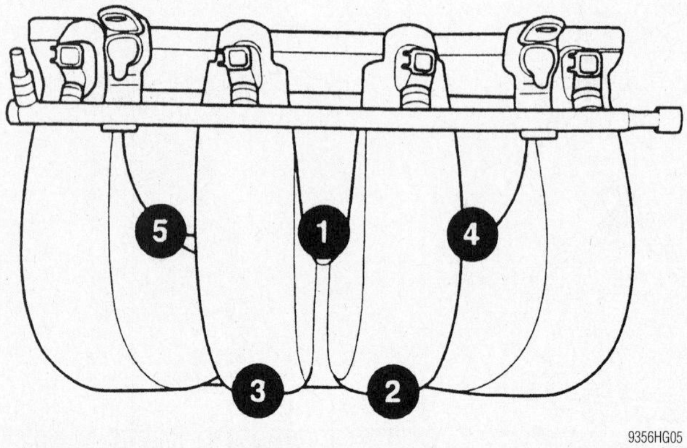

Lower intake manifold bolt torque sequence—High Output 2.0L (VIN F) SOHC engine

2.4L Engine

1. Before servicing the vehicle, refer to the precautions in the beginning of this section.
2. Properly relieve the fuel system pressure.
3. Remove or disconnect the following:
   - Negative battery cable
   - Fuel rail trim cover by pulling straight off the fuel rail studs
   - Charge air cooler-to-throttle body hose
   - Vacuum hoses from the throttle body and intake manifold
   - Throttle cable from the throttle body
   - Throttle Position Sensor (TPS) and Idle Air Control (IAC) motor electrical connectors
   - Intake manifold support bracket
   - Fuel injector electrical connectors
   - Injector wiring harness from fuel rail
   - Manifold Absolute Pressure (MAP) sensor electrical connector
   - Fuel supply line quick connect from the fuel rail assembly
   - Intake manifold fasteners
   - Intake manifold.
   - Intake manifold and discard the gasket
4. Clean all mating surfaces of any residual gasket material.

To install:

5. Install or connect the following:
   - New gaskets and seals
   - Intake manifold and torque the new retainers to 17 ft. lbs. (23 Nm)

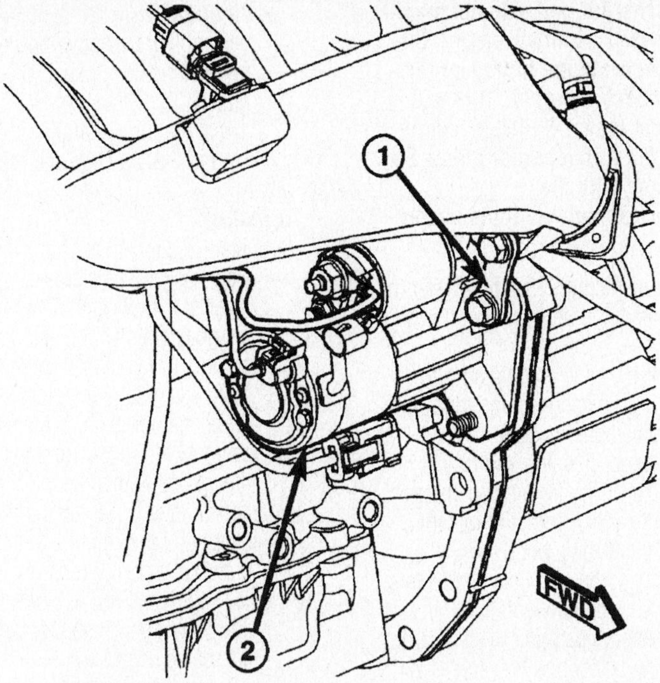

1 - INTAKE MANIFOLD SUPPORT BRACKET
2 - STARTER

67189-NEON-G10

Intake manifold support bracket—2.4L engine

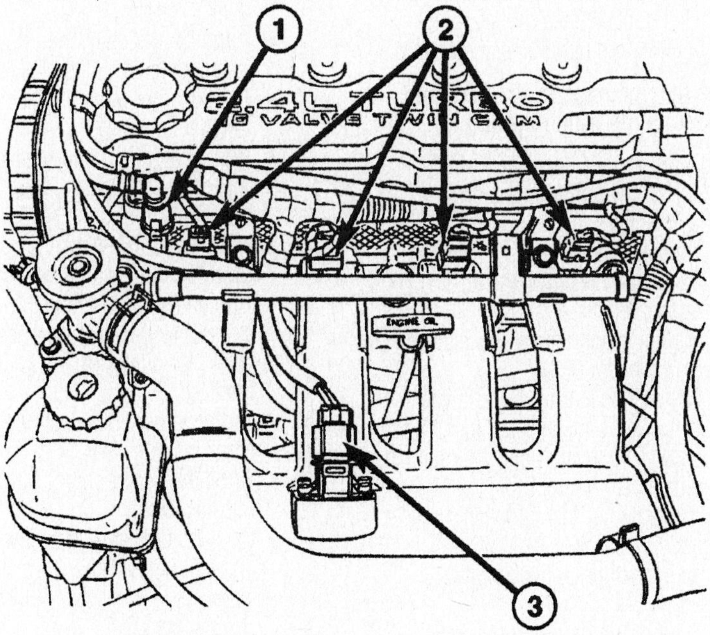

1 - FUEL SUPPLY LINE CONNECTION
2 - FUEL INJECTOR CONNECTORS
3 - MAP SENSOR

67189-NEON-G11

Location of some electrical connectors that need to be disconnected for intake manifold removal—2.4L engine

- Fuel rail assembly to the intake manifold, if removed. Tighten the retainers to screws to 17 ft. lbs. (23 Nm)
- Fuel supply line to fuel rail assembly. Make sure connection is secure by pulling on connector to insure it locked into position
- MAP sensor electrical connector
- Fuel injector electrical connectors
- Clip injector wiring harness to fuel rail
- Intake manifold support bracket. Torque the retainer to 17 ft. lbs. (23 Nm).
- TPS and IAC motor electrical connectors
- Throttle cable to the throttle body
- Vacuum hoses to the throttle body and intake manifold
- Charge air cooler to the throttle body hose
- Fuel rail trim cover by pushing it over the fuel rail studs
- Negative battery cable

Exhaust Manifold

REMOVAL & INSTALLATION

2.0L Engines

1. Before servicing the vehicle, refer to the precautions in the beginning of this section.
2. Remove or disconnect the following:
   - Negative battery cable
   - Wiring harness heat shield-to-exhaust manifold support bracket bolt, if equipped
   - Wiring harness heat shield-to-exhaust manifold, if equipped
   - Exhaust manifold support bracket bolt and bracket, if equipped
   - Exhaust flex joint-to-manifold flange fasteners. Move the exhaust system rearward to clear the flange studs
   - Make up air hose
   - Speed control vacuum reservoir, if equipped
   - Oxygen (O_2S) sensor connector and harness clip
   - Upper heat shield
   - Exhaust manifold bolts
   - Exhaust manifold and discard gasket

To install:

3. Install or connect the following:
   - Exhaust manifold with a new gasket and torque the bolts, in sequence, to 16 ft. lbs. (23 Nm)

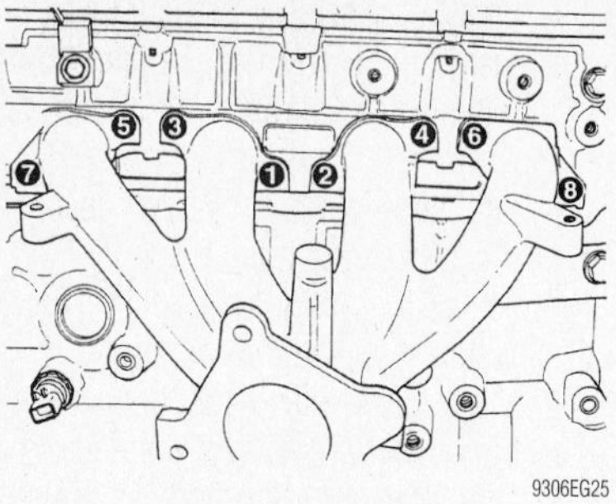

Exhaust manifold tightening sequence

- Upper heat shield, if ULEV equipped and torque bolts to 16 ft. lbs. (23 Nm)
- Upper and lower heat shields and torque the bolts to 95 inch lbs. (11 Nm)
- O_2S sensor connector and harness clip
- Cylinder head cover, if ULEV equipped
- Speed control vacuum reservoir, if equipped
- Make up air hose
- Exhaust manifold support bracket bolt, if LEV equipped. Torque the M10 bolt to 40 ft. lbs. (54 Nm), M12 bolt to 70 ft. lbs. (95 Nm) and the nut to 21 ft. lbs. (28 Nm).
- Cylinder head-to-exhaust manifold support bracket bolt, if equipped. Torque the bolts to 40 ft. lbs. (54 Nm).
- Wiring harness heat shield-to-exhaust manifold, if equipped
- Wiring harness heat shield-to-exhaust manifold support bracket bolt, if equipped
- Negative battery cable

2.4L Engines

On 2.4L turbocharged engines, the exhaust manifold is serviced as an assembly with the turbocharger. Refer to Turbocharger the removal and installation procedure.

Turbocharger Assembly

REMOVAL & INSTALLATION

2.4L Engine

⁂ WARNING

If the turbocharger is replaced due to a bearing failure, you must also replace the oil pressure feed line. The oil return tube should also be cleaned.

➡The turbocharger and exhaust manifold are serviced as an assembly. Do NOT try to separate the turbocharger from the exhaust manifold. This will cause exhaust leaks. Chrysler recommends that the turbocharger elbow be replaced along with the turbocharger/exhaust manifold assembly.

1. Before servicing the vehicle, refer to the precautions in the beginning of this section.
2. Drain the engine cooling system.
3. Remove or disconnect the following:
   - Negative battery cable
   - Air cleaner housing and lid
   - Clean air hose from turbocharger
   - Throttle and speed control cables from the throttle body
4. Disconnect the electrical connectors from the following components:
   - Inlet Air Temperature (IAT) Sensor
   - Manifold Absolute Pressure (MAP) Sensor
   - Idle Air Control (IAC) Motor
   - Throttle Position Sensor (TPS)
   - Ignition Coil Capacitor
   - Upstream Oxygen Sensor (O_2S)
   - Air inlet hose from the throttle body
   - Vacuum hoses from the throttle body and upper intake manifold
   - Upper intake manifold support bracket and upper intake manifold

➡Cover the openings in the manifold to prevent debris from entering.

   - Turbocharger upper heat shield
   - Oil supply line from the turbocharger
   - Coolant return line
   - Vacuum hoses from the turbocharger
   - Muffler ground strap and muffler
   - Downstream O_2S
   - Fasteners securing the catalytic converter to the exhaust manifold
   - Catalytic converter and intermediate pipe as an assembly
   - Turbocharger to charge air cooler hose assembly
   - Turbocharger elbow support bracket
   - Turbocharger support bracket
   - Oil return tube
   - Turbocharger coolant supply line
   - Turbocharger lower heat shield
   - Turbocharger elbow
   - Lower exhaust manifold fasteners that are accessible while vehicle raised
   - Upper exhaust manifold fasteners
   - Turbocharger/exhaust manifold assembly from above/between the engine and cowl panel. Discard the gasket
5. Inspect the turbocharger assembly for cracks and other damage and replace if necessary.

To install:

6. Install or connect the following:
   - New gasket. The stainless steel layer of the exhaust manifold gasket goes against the cylinder head, and the graphite layer of the gasket goes against the manifold surface.
   - Turbocharger/exhaust manifold assembly in place, from between the engine and brake master cylinder. Gradually tighten the fasteners, starting at the center and going outward in both directions to 21 ft. lbs. (28 Nm). You will need to raise and lower the vehicle for fastener access as necessary.
   - Alternator heat shield
   - Coolant return line, using new washers. Tighten the banjo fitting bolt to 27 ft. lbs. (37 Nm)
   - Turbocharger coolant supply line, using new washers. Tighten the banjo fitting bolt to 27 ft. lbs. (37 Nm). Torque the flared fitting to 23 ft. lbs. (31 Nm).
   - Turbocharger elbow and tighten the fasteners to 21 ft. lbs. (28 Nm)
   - Turbocharger upper heat shield and fasteners, loosely.
   - Turbocharger support bracket. Torque the M8 fasteners to 21 ft. lbs. (28 Nm) and the M10 fasteners to 40 ft. lbs. (54 Nm).
   - Oil return tube with a new gasket. Tighten the fasteners to 105 inch lbs. (12 Nm). Make sure heat shield for oil return line is properly installed.
   - Turbocharger elbow support bracket
   - Turbocharger-to-charge air cooler hose assembly
   - Catalytic converter with new gasket. Tighten the converter-to-manifold fasteners to 21 ft. lbs. (28 Nm).
   - Downstream O_2S
   - Exhaust extension pipe to catalytic converter
   - Power steering cooler fasteners
   - Oil supply line to the turbocharger with new washers. Tighten the banjo fitting bolt to 27 ft. lbs. (37 Nm).

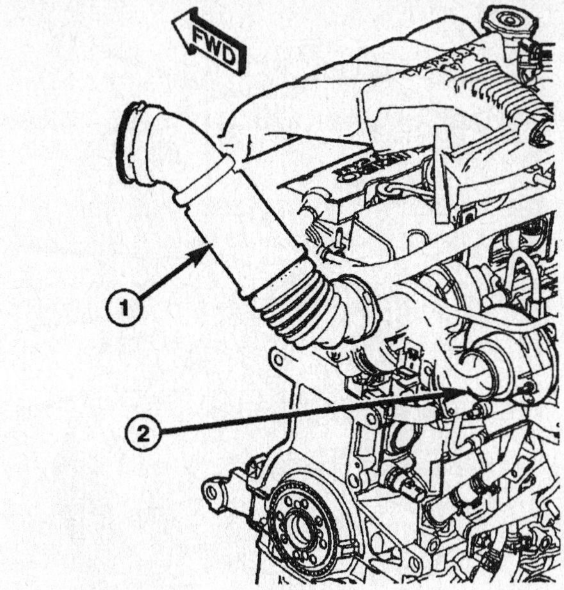

1 - CLEAN AIR HOSE
2 - TURBOCHARGER

67189-NEON-G12

Clean air hose—2.4L engine

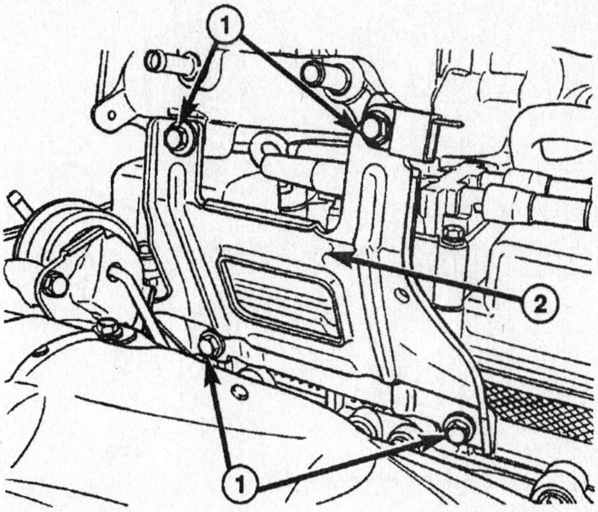

1 - FASTENERS
2 - UPPER INTAKE MANIFOLD SUPPORT BRACKET

67189-NEON-G14

Support bracket mounting—2.4L engine

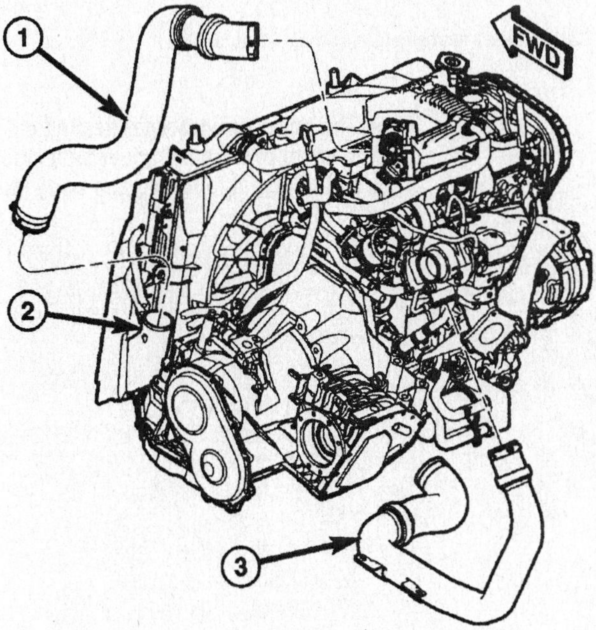

1 - HOSE - CHARGE AIR COOLER TO THROTTLE BODY
2 - CHARGE AIR COOLER
3 - HOSE - TURBOCHARGER TO CHARGE AIR COOLER

67189-NEON-G13

Charge air cooler hoses—2.4L engine

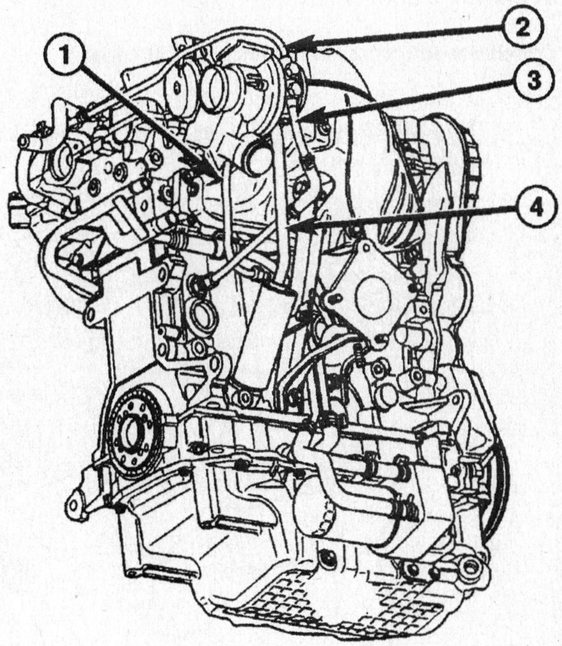

1 - OIL SUPPLY LINE
2 - COOLANT RETURN LINE
3 - COOLANT SUPPLY LINE
4 - OIL RETURN TUBE

67189-NEON-G15

Turbocharger lines and hoses—2.4L engine

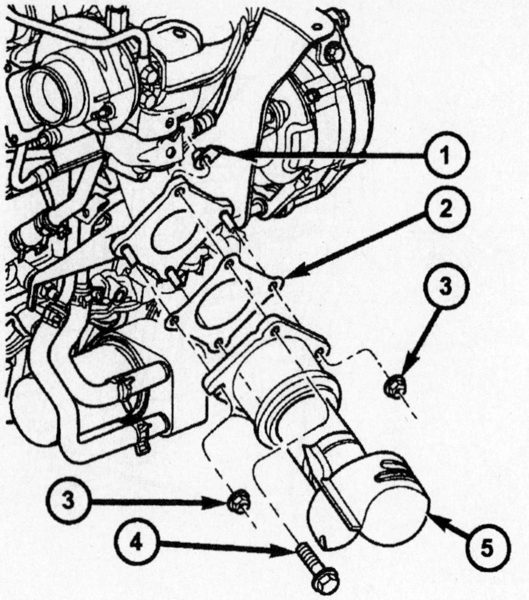

Converter-to-exhaust manifold attachment—2.4L engine

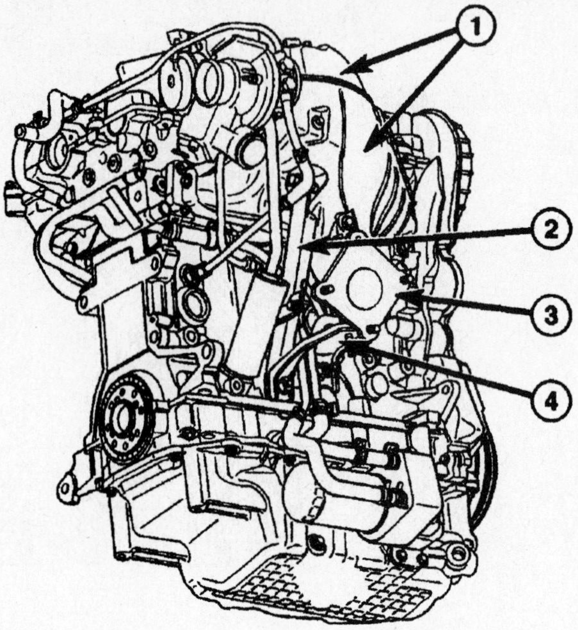

Turbocharger heat shield mounting—2.4L engine

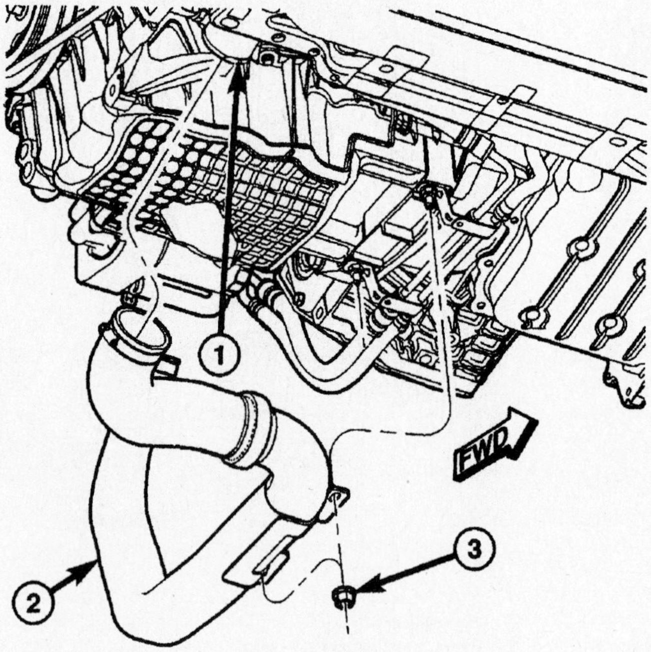

Charge air cooler hose—2.4L engine

➡The lower turbocharger heat shield tabs must overlap the upper heat shield tabs to prevent fatiguing and early failure.

- Turbocharger lower heat shield. Torque the upper and lower heat shield fasteners to 21 ft. lbs. (28 Nm)
- Vacuum hoses to turbocharger
- Upstream O_2S
- Ignition coil and electrical connector
- Electrical connector into bracket
- Clean air hose
- Negative battery cable

7. Fill the cooling system
8. Change the oil and filter.
9. Start the engine and check for exhaust leaks. Repair leaks as necessary.
10. Inspect the exhaust system for contact with the body panels. Make the necessary adjustments, if needed.

Front Crankshaft Seal

REMOVAL & INSTALLATION

2.0L Engines

1. Before servicing the vehicle, refer to the precautions in the beginning of this section.
2. Remove or disconnect the following:
   - Negative battery cable
   - Accessory drive belts

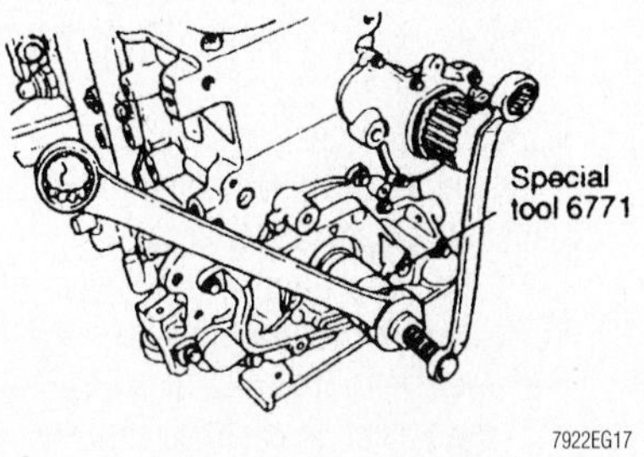

Removing the front crankshaft oil seal

   - Crankshaft damper, using Puller Tool 1026 and Insert Tool 6827-A
   - Timing belt
   - Crankshaft sprocket using special Tools 6793 and C-4685-C2
   - Crankshaft sprocket key
   - Crankshaft oil seal, using a Seal Puller Tool 6771

To install:

3. Install or connect the following:
   - New front crankshaft oil seal, using Crankshaft Installer Tool 6780–1
   - Crankshaft key
   - Crankshaft sprocket using Installer Tool 6792

➡Make sure the word FRONT on the crankshaft sprocket is facing outward.

   - Timing belt
   - Crankshaft damper using thrust bearing washer and bolt from Installer Tool 6792. Torque the damper bolt to 100 ft. lbs. (136 Nm).
   - Accessory drive belts
   - Negative battery cable

2.4L Engine

1. Before servicing the vehicle, refer to the precautions in the beginning of this section.
2. Remove or disconnect the following:
   - Negative battery cable

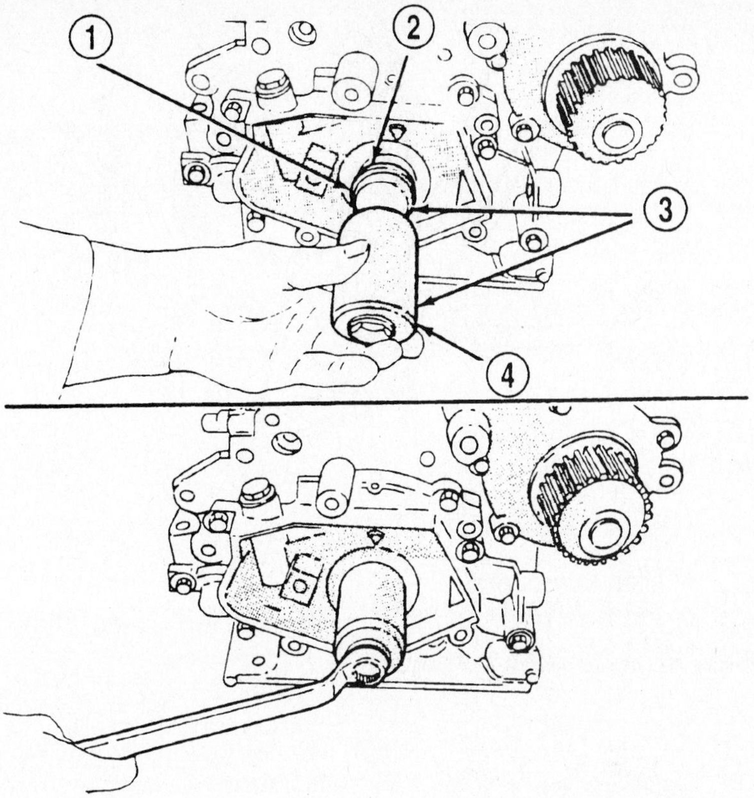

Installing a new seal, using Seal Installer 6780–1; proceed with caution if using substitute tools

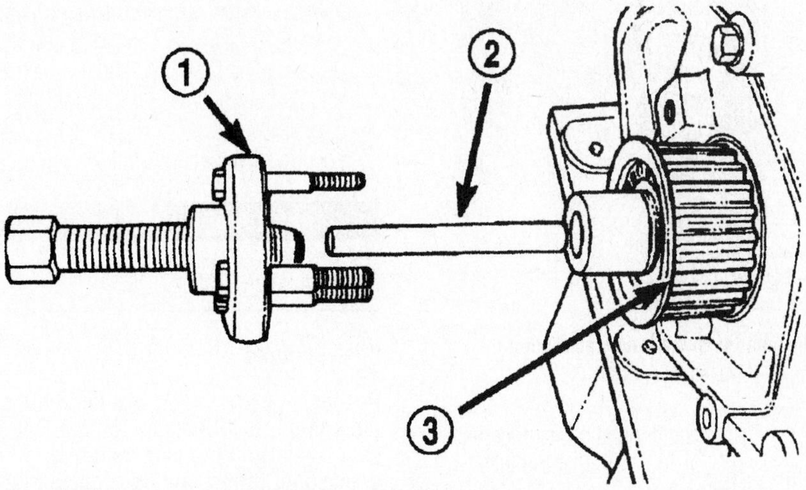

Removing the crankshaft sprocket—2.4L engine

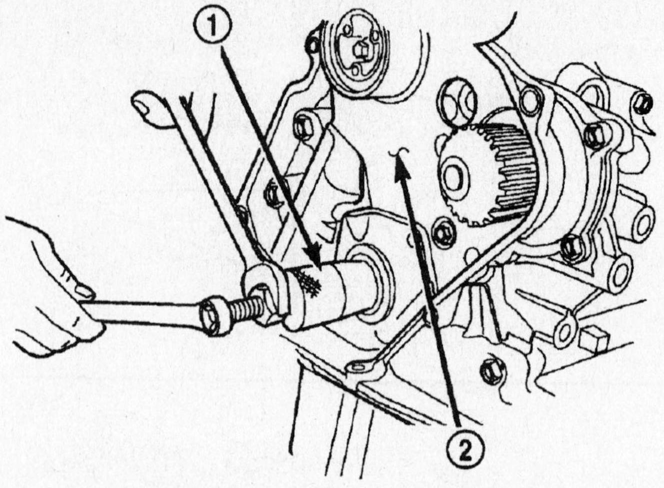

Front crankshaft oil seal removal—2.4L engine

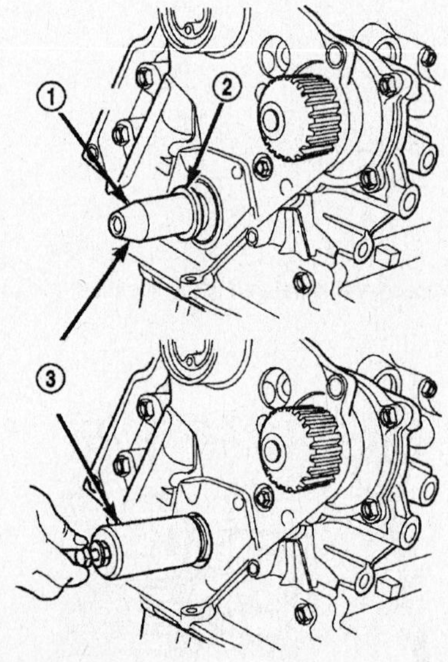

Installation of the front crankshaft oil seal—2.4L engine

- Crankshaft vibration damper with special Tools 1026 and 6287-A
- Timing belt
- Crankshaft sprocket with special Tools 6793 and insert C-4685-C2

⁂ WARNING

Be careful not to scratch or nick the seal surface or seal bore.

- Crankshaft oil seal with special Tool 6771

To install:

3. Install or connect the following:
   - Front crankshaft oil seal with special Tool 6780-1 until it is flush with the cover
   - Crankshaft sprocket with special Tool 6792

➡Make sure the word FRONT on the crankshaft sprocket is facing outward.

   - Timing belt
   - Crankshaft vibration damper using thrust bearing washer and bolt from Installer Tool 6792 and torque the damper bolt to 100 ft. lbs. (135 Nm)
   - Negative battery cable

Camshaft and Lifters

REMOVAL & INSTALLATION

2.0L Engines

➡After all components are installed on the engine, a DRB scan tool is necessary to perform the camshaft and crankshaft timing relearn procedure.

1. Before servicing the vehicle, refer to the precautions in the beginning of this section.
2. Relieve the fuel system pressure.
3. Drain the cooling system.
4. Remove or disconnect the following:
   - Battery and tray
   - Inlet Air Temperature (IAT) sensor electrical connector
   - Air intake duct from the intake manifold
   - Throttle cables
   - Throttle Position (TP) sensor
   - Idle Air Control (IAC) valve from the throttle body
   - Air make-up hose from the air cleaner
   - Air cleaner housing
   - Timing belt and tensioner
   - Camshaft sprocket
   - Rear timing belt cover
   - Cylinder head cover and mark the rocker arm shaft assemblies to ease installation
   - Rocker arm shaft assemblies
   - Engine Coolant Temperature (ECT) sensor electrical connector
   - Heater supply and return hoses
   - Heater tube support bracket
   - Power Distribution Center (PDC) screws aside to ease the camshaft removal
   - Camshaft Position (CMP) sensor and camshaft target magnet
   - Camshaft from the rear of the cylinder head

To install:

5. Lubricate the camshaft journals with clean engine oil.
6. Install or connect the following:
   - Camshaft and reposition the PDC and heater tubes
   - Camshaft target magnet and torque the screw to 30 inch lbs. (3 Nm)
   - CMP sensor
   - Front camshaft seal, if removed
   - Timing belt rear cover
   - Camshaft sprocket and torque to 85 ft. lbs. (115 Nm)
   - Timing belt and tensioner
   - Rocker arms in the proper order and torque to 21 ft. lbs. (28 Nm)
   - Cylinder head cover
   - Ignition coil and spark plug cables and torque the fasteners to 105 inch lbs. (12 Nm)
   - Heater tube bracket bolt
   - Heater tubes
   - ECT sensor electrical connector
   - PDC attaching screws
   - Battery and tray
   - Air cleaner housing
   - Throttle body connections
   - IAT sensor
7. Fill the cooling system to the proper level.
8. Use a DRB scan tool to perform the camshaft and crankshaft timing relearn procedure, as follows:

a. Connect the scan tool to the DLC (located under the instrument panel, near the steering column).
b. Turn the ignition switch **ON**, and access the "miscellaneous" screen.
c. Select the "re-learn cam/crank" option, then follow the instructions on the scan tool screen.

9. Start the engine and check for leaks. Run the engine with the radiator cap off so as the engine warms and the thermostat opens, coolant can be added to the radiator. Test drive vehicle to check for proper operation.

2.4L Engine

1. Before servicing the vehicle, refer to the precautions in the beginning of this section.
2. Remove or disconnect the following:
   - Cylinder head (valve) cover
   - Camshaft Position (CMP) sensor and camshaft target magnet
   - Timing belt
   - Camshaft sprockets and timing belt rear cover
   - Camshaft bearing caps, loosen the bearing caps in sequence, one camshaft at a time.

➡The bearing caps are identified for location. Remove the outside bearing caps first. If the bearing caps are difficult to remove, use a plastic hammer to gently tap the rear part of the camshaft.

   - Intake and exhaust camshafts

To install:

3. Before installation, clean the cylinder head and cover mating surfaces. Make certain that the rails are flat.
4. Lubricate the bearing journals, rocker arms and cam lobes with clean engine oil.

✻✻ WARNING

Make sure that NONE of the pistons are at Top Dead Center (TDC) when installing the camshafts.

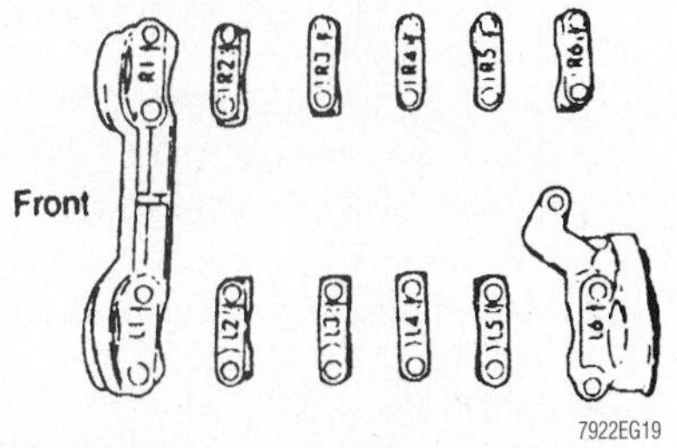

Identifying the camshaft bearing caps—2.4L engine

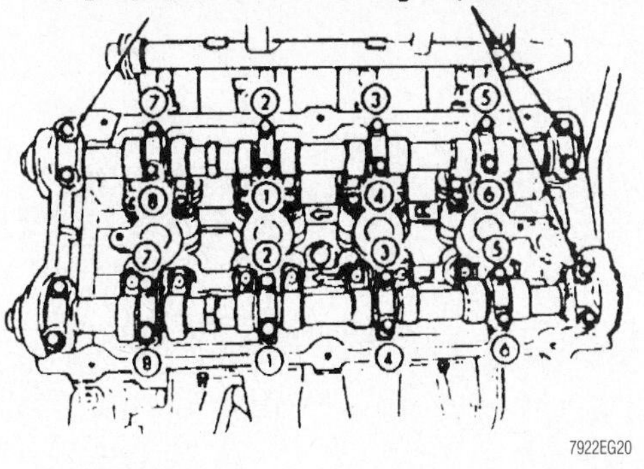

Bearing cap removal sequence—2.4L engine

5. Install or connect the following:
   - Camshafts on cylinder head bearing journals.
   - Right and left bearing caps No. 2–5 and No. 6. Torque the M6 fasteners, in sequence, to 105 inch lbs. (12 Nm)
   - Apply Mopar Gasket Maker to the No. 1 and No. 6 bearing caps. Install the caps and tighten the M8 fasteners to 21 ft. lbs. (28 Nm).

➡The bearing end caps must be installed before the seals can be installed.

   - Camshaft oil seals
   - Camshaft sprockets
   - Timing belt
   - Camshaft target magnet and CMP sensor
   - Cylinder head cover
   - Negative battery cable

6. Start the engine and check for proper operation and leaks.

Valve Lash

ADJUSTMENT

The engines in these vehicles do not require periodic valve lash adjustment.

Starter Motor

REMOVAL & INSTALLATION

2.0L Engines

1. Before servicing the vehicle, refer to the precautions in the beginning of this section.
2. Remove or disconnect the following:
   - Negative battery cable
   - Inlet hose from the intake manifold and reposition the air cleaner box, on high output engines only
   - Starter electrical connectors
   - Starter-to-engine bolts
   - Starter

To install:

3. Install or connect the following:
   - Starter and torque bolts to 40 ft. lbs. (54 Nm)
   - Starter electrical connectors
   - Intake manifold inlet hose on high output engines
   - Negative battery cable

2.4L Engine

1. Before servicing the vehicle, refer to the precautions in the beginning of this section.
2. Remove or disconnect the following:

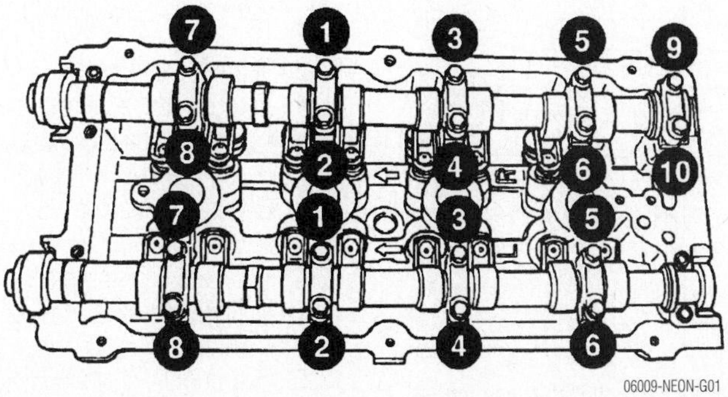

Bearing cap tightening sequence—2.4L engine

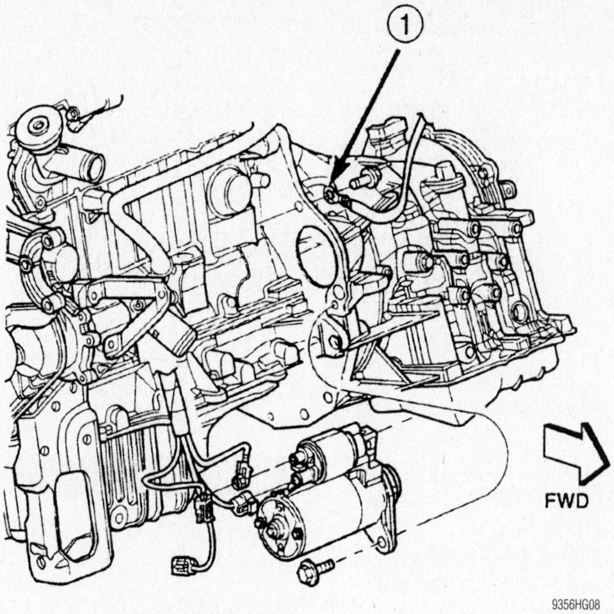

Starter mounting—2002 Neon shown, others similar

- Negative battery cable
- Air cleaner assembly
- Throttle body
- Upper mounting bolt and ground wire
- Positive battery cable from the starter
- Solenoid connector from the starter
- Middle starter mounting bolt
- Upper bolt for the intake manifold support (loosen only) and swing the bracket out of the way
- Starter lower mounting bolt and starter

To install:

3. Install or connect the following:
   - Starter and lower mounting bolt
   - Loosen the upper bolt for the intake manifold support bracket and swing bracket back into place
   - Middle starter mounting bolt
   - Tighten the intake manifold support bracket bolt
   - Tighten the starter mounting bolts to 40 ft. lbs. (54 Nm)
   - Solenoid connector to the starter
   - Positive battery cable to starter. Torque the captive nut to 89 inch lbs. (10 Nm)
   - Upper mounting bolt and ground wire, then tighten the starter mounting bolt to 40 ft. lbs. (54 Nm)
   - Throttle body
   - Air cleaner assembly
   - Negative battery cable

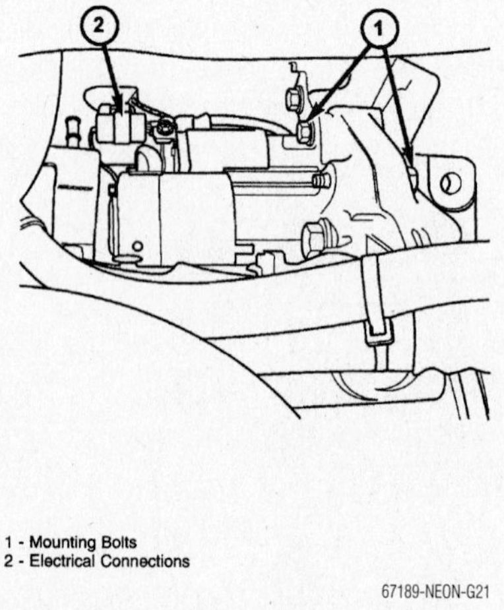

Starter mounting and connector location—2.4L engine

Oil Pan

REMOVAL & INSTALLATION

2.0L Engines

1. Before servicing the vehicle, refer to the precautions in the beginning of this section.
2. Raise and safely support the vehicle.
3. Drain the engine oil.
4. Remove or disconnect the following:
   - Negative battery cable
   - Oil filter
   - Oil filter adapter
   - Structural collar
   - Lateral bending brace
   - Transaxle lower dust cover
   - Oil pan
5. Thoroughly, clean the gasket mating surfaces.

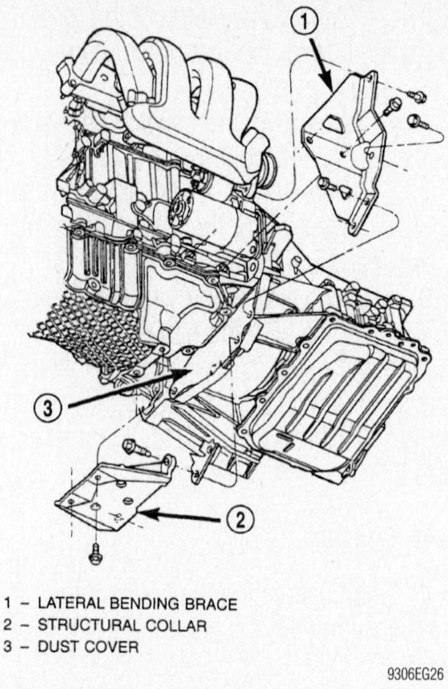

Exploded view of the bending brace, structural collar and dust cover

To install:

6. Apply silicone sealer to the oil pump-to-engine block parting line.
7. Install or connect the following:
   - New oil pan gasket
   - Oil pan and torque the screws to 105 inch lbs. (12 Nm)
   - Transaxle lower dust cover
   - Lateral bending brace
   - Structural collar
   - Oil filter adapter and torque the bolts to 60 ft. lbs. (80 Nm)
   - Oil filter
8. Fill the engine with clean oil
9. Start the engine, check for leaks and repair if necessary.

2.4L Engine

1. Before servicing the vehicle, refer to the precautions in the beginning of this section.
2. Raise and safely support the vehicle.
3. Drain the engine oil.
4. Remove or disconnect the following:
   - Negative battery cable
   - Oil filter
   - Accessory drive belt splash shield
   - Turbocharger-to-charge air cooler hose assembly
   - Oil cooler connector bolt. Do NOT disconnect the coolant lines from oil cooler, just position it aside.
   - Structural collar
   - Lower torque strut
   - Oil filter adapter and gasket
   - Oil pan and gasket
5. Thoroughly clean the gasket mating surfaces.

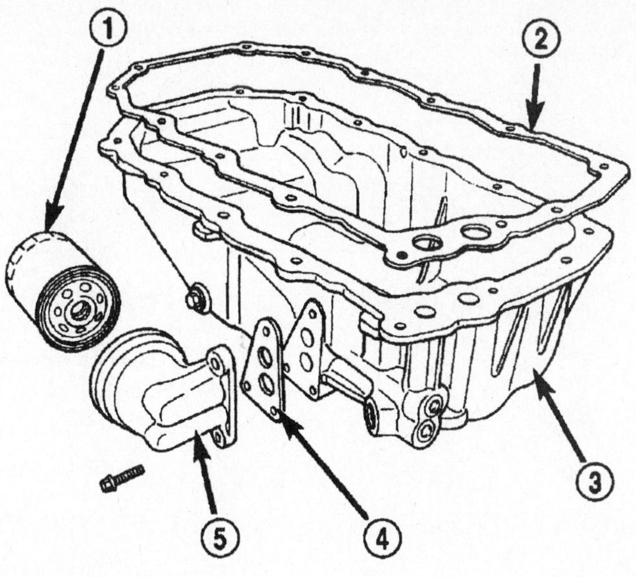

Exploded view of the oil pan—2.4L engine

To install:

6. Apply Mopar Engine RTV GEN II, or equivalent silicone sealer to the oil pump-to-engine block parting line.
7. Install or connect the following:
   - New oil pan gasket
   - Oil pan and torque the screws to 105 inch lbs. (12 Nm)
   - Oil filter adapter and gasket. Torque the screws to 105 inch lbs. (12 Nm)
   - Oil cooler seal. Lubricate seal and position oil cooler to oil filter adapter, aligning notch to tab.
   - Oil cooler connector bolt and tighten to 41 ft. lbs. (55 Nm).
   - Oil drain plug and oil filter
   - Structural collar
   - Lower torque strut
   - Turbocharger-to-charge air cooler hose assembly
8. Fill the engine with clean oil

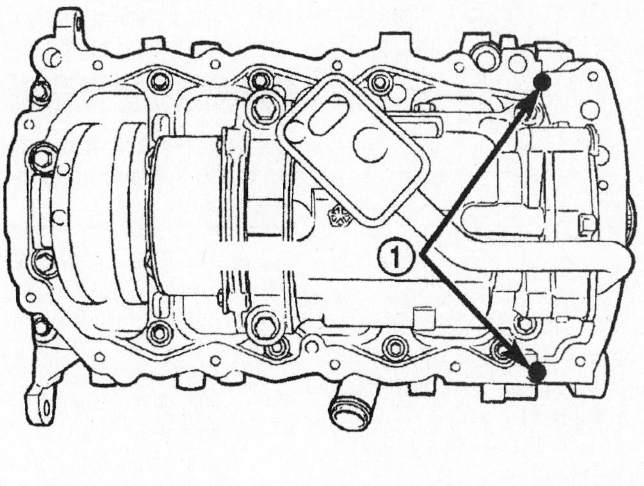

Oil pan sealant locations—2.4L engine

9. Start the engine, check for leaks and repair if necessary.

Oil Pump

REMOVAL & INSTALLATION

2.0L Engine

1. Before servicing the vehicle, refer to the precautions in the beginning of this section.
2. Drain the engine oil.
3. Remove or disconnect the following:
   - Negative battery cable
   - Crankshaft damper
   - Timing belt
   - Timing belt tensioner
   - Camshaft sprocket and rear timing belt cover
   - Oil pan
   - Crankshaft sprocket, using Tool 6795 and Insert Tool C-4685-C2
   - Oil pickup tube
   - Oil pump
   - Front crankshaft seal
   - Oil pump cover screws and lift the cover off
   - Oil pump rotors
4. Wash all parts in a solvent; then, inspect carefully for damage or wear, as follows:
   a. Inspect the mating surface of the oil pump should be smooth. Replace the pump cover, if scratched or grooved.
   b. Lay a straightedge across the pump cover surface. If a 0.003 in. (0.076mm) feeler gauge can be inserted between the cover and the straightedge, the cover should be replaced.
   c. Measure the thickness and diameter of the outer rotor. If the outer rotor thickness measures 0.301 in. (7.64mm) or less, or if the diameter is 3.148 in. (79.95mm) or less, replace the outer rotor.
   d. If the inner rotor measures 0.301 in. (7.64mm) or less, replace the inner rotor.
   e. Slide the outer rotor into the pump housing, press to one side with your fingers and measure the clearance between the rotor and the housing. If the measurement is 0.015 in. (0.39mm) or more, replace the housing only if the outer rotor is within specification.
   f. Install the inner rotor into the pump housing, If the clearance between the inner and outer rotors is 0.008 in. (0.203mm) or more, replace both rotors.
   g. Place a straightedge across the

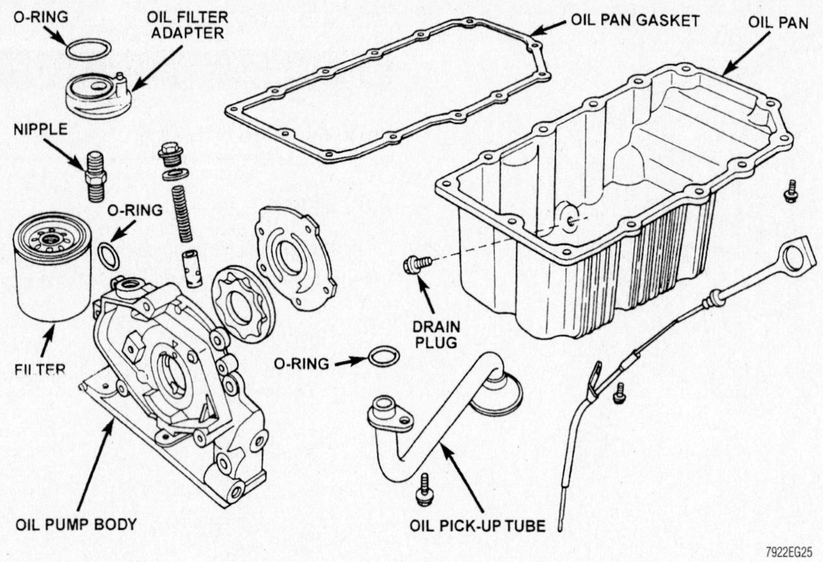

Exploded view of the oil pump and related component mounting—2.0L engine

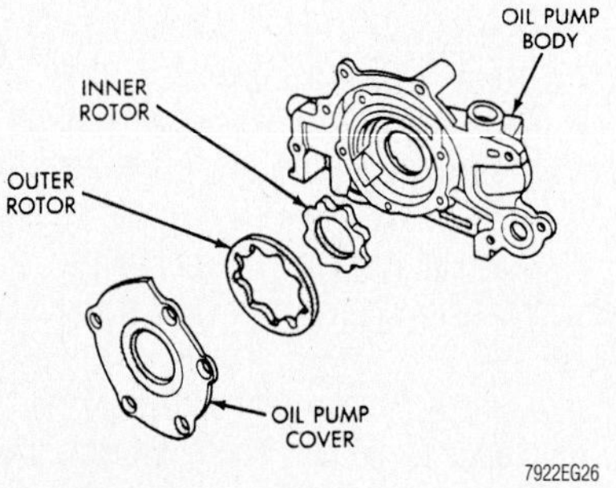

Exploded view of the oil pump assembly—2.0L engine

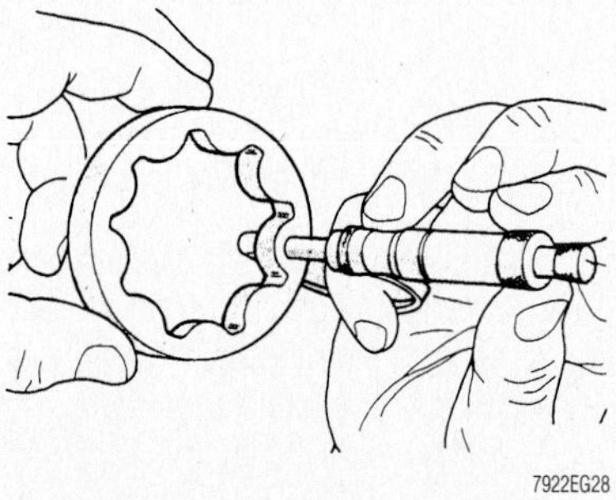

Use calipers to measure the outer rotor thickness—2.0L engine

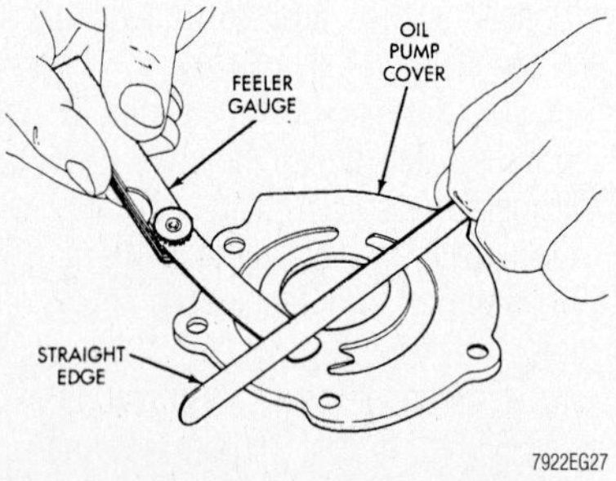

Use a feeler gauge and straightedge to check the oil pump cover for warpage—2.0L engine

and the inner rotor thickness—2.0L engine

face of the pump housing, between the bolt holes. If a feeler gauge of 0.004 in. (0.102mm) or more can be inserted between the rotors and the straightedge, replace the pump assembly.

h. Inspect the oil pressure relief valve plunger for scoring and free operation in its bore. Small marks may be removed with 400 grit wet or dry sandpaper.

i. The relief valve spring has a free length of about 2.39 in. (60.7mm) and should test between 18–19 lbs. (8.1–8.6 kg) when compressed to 1.60 in. (40.6mm). Replace the spring, if it falls outside of specifications.

j. If the oil pressure is low and the pump is within specifications, inspect for worn engine bearings or for other reasons for oil pressure loss.

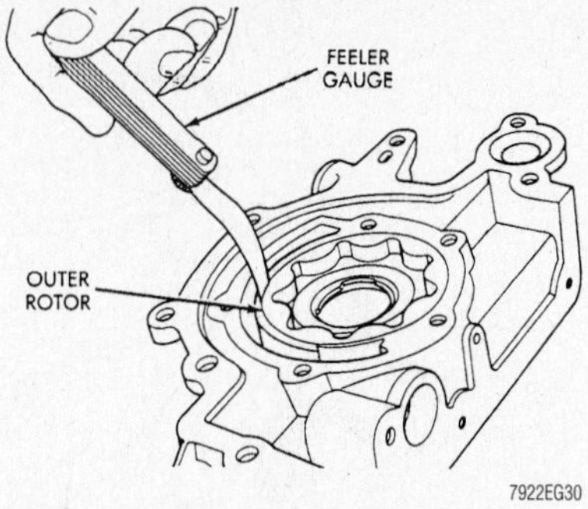

Measure the outer rotor clearance in the housing—2.0L engine

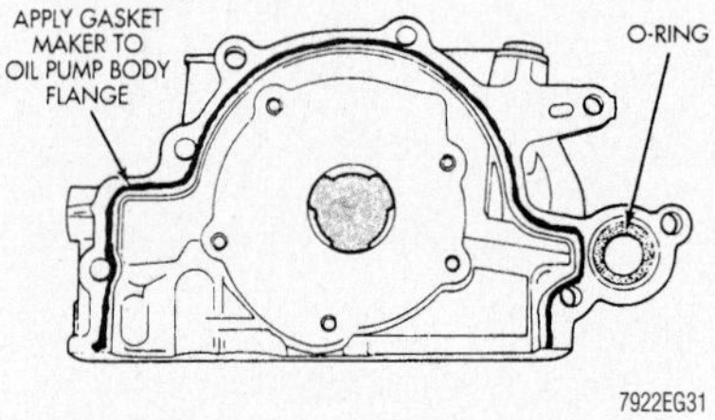

Apply a small amount of gasket maker to the pump body cover mounting surface—2.0L engine

To install:

5. Assemble the pump, using new parts as required, as follows:

a. Install the inner rotor with the chamfer facing the cast iron oil pump cover.

b. Apply Mopar gasket maker to the oil pump.

c. Install the oil ring into the oil pump body discharge passage.

6. Prime the oil pump before installation by filling the rotor cavity with engine oil.

7. Install or connect the following:

- Oil pump, align the rotor flats with the crankshaft flats. Torque the pump bolts to 21 ft. lbs. (28 Nm).

⁂ WARNING

The front crankshaft seal MUST be out of the pump to align or damage may result.

- New front crankshaft seal, using Seal Driver Tool 6780
- Crankshaft sprocket, using a Crankshaft Sprocket Installer Tool 6792
- Oil pump pickup tube
- Oil pan
- Rear timing belt cover and camshaft sprocket
- Timing belt tensioner
- Timing belt and front cover
- Crankshaft damper

- Accessory drive belts
- Negative battery cable

8. Fill the engine with clean oil.
9. Start the engine and check for leaks; then, recheck the fluid level and add as necessary.

2.4L Engine

1. Before servicing the vehicle, refer to the precautions in the beginning of this section.
2. Disconnect the negative battery cable.
3. Remove the timing belt.
4. Remove the timing belt rear cover.
5. Remove the oil pan.
6. Remove the crankshaft sprocket using tools 6793 and C-4685-C2.
7. Remove the crankshaft key.
8. Remove the oil pick-up tube.
9. Remove the oil pump and front crankshaft seal.
10. Disassemble the pump as follows:
    a. Remove the oil pump cover fasteners, and lift off the cover.
    b. Remove the pump rotors.
    c. Wash all parts in a suitable solvent and inspect carefully for damage or wear.
11. Inspect the pump components as follows:
    a. Inspect the mating surface of the oil pump. The surface should be smooth. Replace the pump cover if it is scratched or grooved.
    b. Lay a straightedge across the pump cover surface. If a 0.001 inch (0.025 mm) feeler gauge can be inserted between the cover and straight edge, the cover should be replaced.
    c. Measure the thickness and diameter of the outer rotor. If the outer rotor thickness measures 0.421 inch (10.699 mm) or less, or if the diameter is 3.382 inch (85.924 mm) or less, replace the outer rotor.
12. Assemble the pump as follows:
    a. Assemble the pump, using new parts as required. Install the inner rotor with the chamfer facing the cast iron oil pump cover.
    b. Prime the oil pump before installation by filling the rotor cavity with engine oil.
    c. Install the cover and tighten the fasteners to 118 inch lbs. (13 Nm).
13. Install the pump as follows:

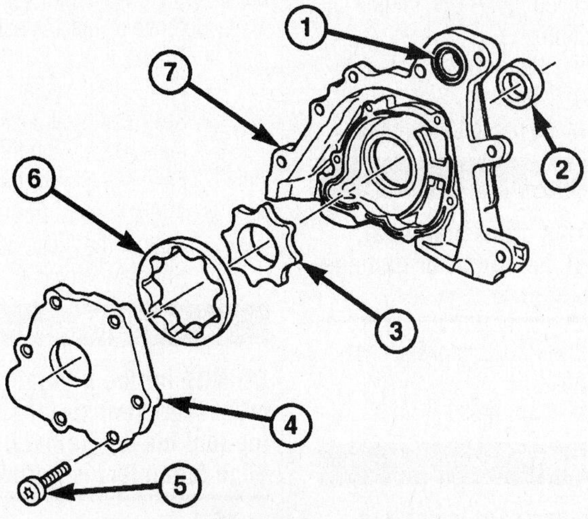

Exploded view of the oil pump assembly—2.4L engines

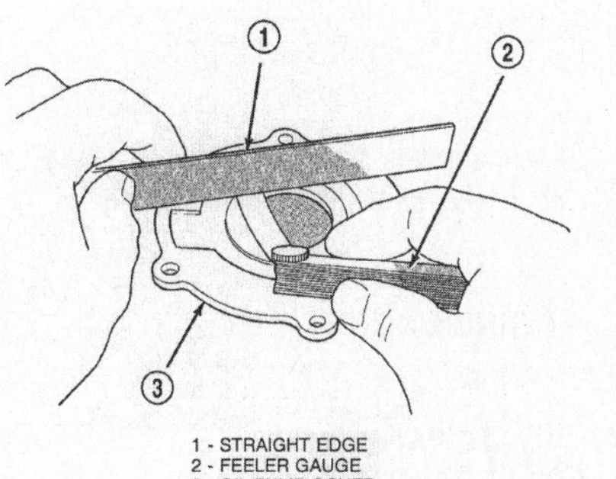

Check the oil pump cover flatness—2.4L engines

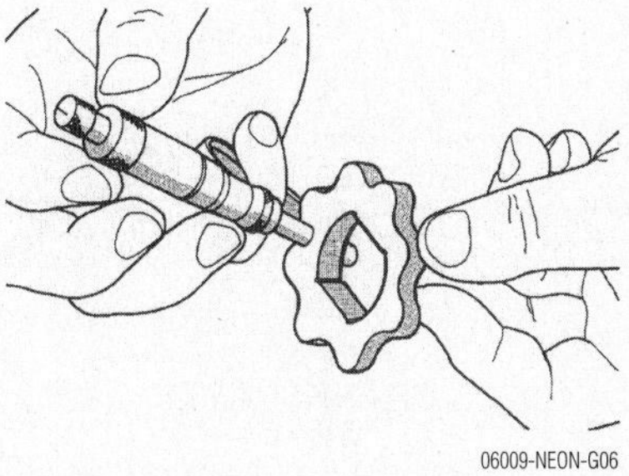

Check the oil pump inner rotor thickness—2.4L engines

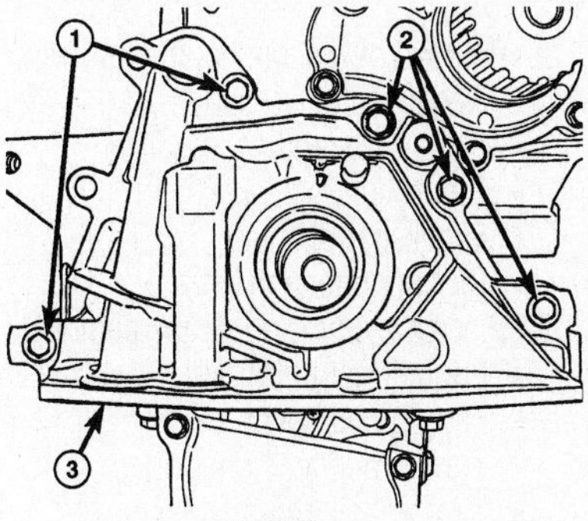

Oil pump retainer locations—2.4L engines

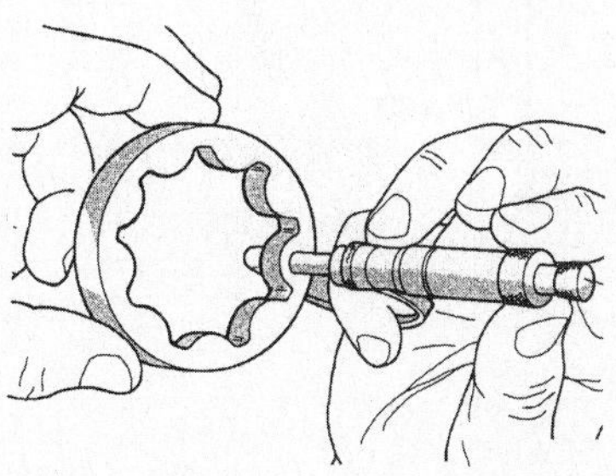

Check the oil pump outer rotor thickness—2.4L engines

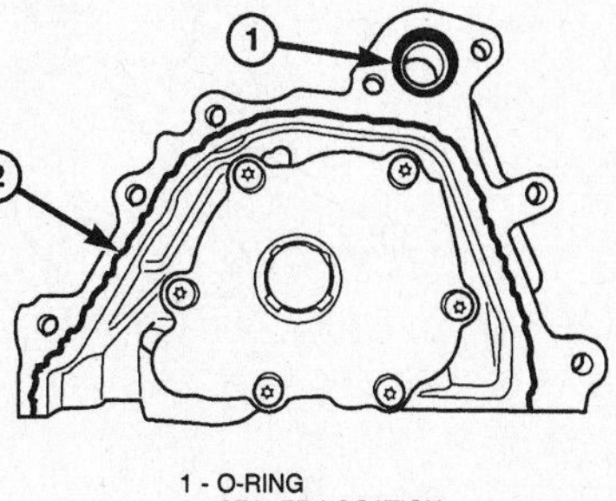

Apply gasket maker to the oil pump as shown and install the O-ring into oil pump body discharge passage —2.4L engines

a. Make sure all the surfaces are clean and free of oil and dirt.

b. Apply gasket maker to the oil pump as illustrated. Install the O-ring into oil pump body

discharge passage.

c. Prime the oil pump with engine oil before installation.

d. Align the oil pump rotor flats with the flats on crankshaft.

e. Install the oil pump to the block.

CAUTION

To align, the front crankshaft seal MUST be out of the pump, or damage may result.

f. Install a new front crankshaft seal using tool 6780.

g. Install the crankshaft key.

CAUTION

The crankshaft sprocket is set to a predetermined depth from the factory for correct timing belt tracking. If removed, use of tool 6792 is required to set the sprocket to its original installation depth. An incorrectly installed sprocket will result in timing belt and engine damage.

h. Install the crankshaft sprocket using tool 6792.

i. Install the oil pump pick-up tube.

j. Install the oil.

k. Install timing belt rear cover.

l. Install the timing.

14. Connect the negative battery cable.

Rear Main Seal

REMOVAL & INSTALLATION

1. Before servicing the vehicle, refer to the precautions in the beginning of this section.
2. Remove or disconnect the following:
   - Transaxle
   - Flexplate/flywheel
   - Rear main seal. Insert a seal remover between the dust lip and the metal case of the crankshaft seal. Angle the tool through the dust lip against the metal case of the seal. Pry out the seal.

WARNING

DO NOT let the prytool contact the crankshaft seal surface. Contact of the tool blade against the crankshaft edge (chamfer) is permitted.

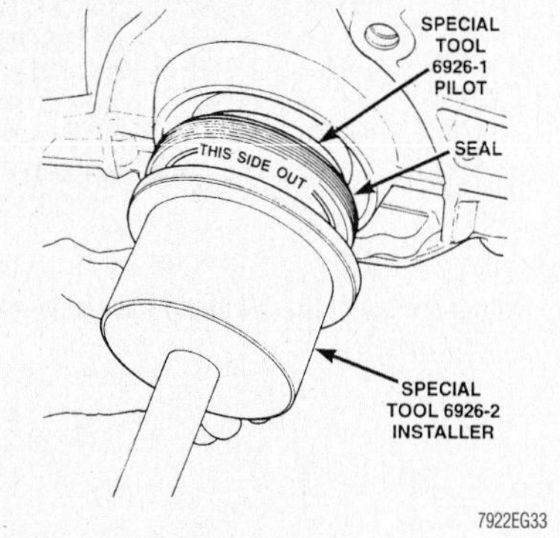

Place a proper size pilot tool with a magnetic base on the crankshaft

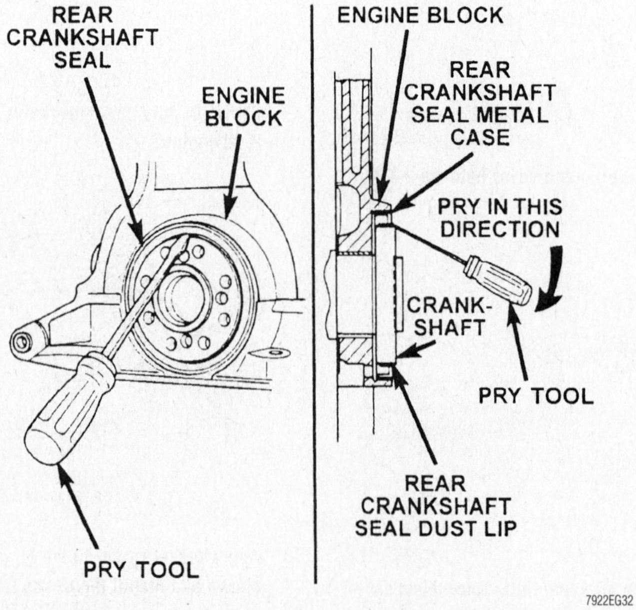

When prying the seal out, be sure to use the prytool at the proper angle

To install:

WARNING

If the crankshaft edge (chamfer) has any burrs or scratches on the, clean it up with 400 grit sand paper to prevent seal damage during installation of the new seal.

➡No lubrication is necessary when installing the seal.

3. Place Crankcase Seal Pilot Tool 6926–1 on the crankshaft; this is a pilot tool with a magnetic base.
4. Position the seal over the Pilot Tool; be sure the words THIS SIDE OUT on the seal can be read.

➡The pilot tool should stay on the crankshaft during installation of the seal. Be sure the seal lip faces the crankcase during installation.

WARNING

If the seal is driven in the block past flush, this may cause an oil leak.

5. Drive the seal into the block, using Crankshaft Seal Tool 6926-2 and Handle C-4171, until the tool bottoms out against the block.
6. Install the flexplate/flywheel. Apply Lock & Seal Adhesive to the bolt treads. Torque the bolts in a star pattern, to 70 ft. lbs. (95 Nm).
7. Install the transaxle.

Timing Belt

REMOVAL & INSTALLATION

2.0L (VIN C) Engine

1. Before servicing the vehicle, refer to the precautions in the beginning of this section.
2. Remove or disconnect the following:
   - Negative battery cable
   - Drive belts and accessories
   - Right inner splash-shield
   - Crankshaft damper
   - Right engine mount
3. Place a support under the engine.
   - Engine mount bracket
   - Timing belt cover
   - Timing belt tensioner bolts
   - Timing belt and the tensioner

➡When tensioner is removed from the engine it is necessary to compress the plunger into the tensioner body.

4. Place the tensioner in a soft-jawed vise to compress the tensioner.

5. After compressing the tensioner place a pin (a 5⁄64 in. Allen wrench will work) into the plunger side hole to retain the plunger until installation.

To install:

6. Set the crankshaft sprocket to Top Dead Center (TDC) by aligning the notch on the sprocket with the arrow on the oil pump housing, then back off the sprocket three notches before TDC.

7. Set the camshaft to align the timing marks.

8. Move the crankshaft to ½ notch before TDC.

9. Install the timing belt starting at the crankshaft, around the water pump, then around the camshaft last.

10. Move the crankshaft to TDC to take up the belt slack.

11. Reinstall the tensioner to the block but do not tighten it.

12. Using a torque wrench apply 250 inch lbs. (28 Nm) of torque to the tensioner pulley.

13. With torque being applied to the tensioner pulley, move the tensioner up against the tensioner bracket and tighten the fasteners to 275 inch lbs. (31 Nm).

14. Remove the tensioner plunger pin, the tension is correct when the plunger pin can be removed and replaced easily.

15. Rotate the crankshaft two revolutions and recheck the timing marks.

16. Install or connect the following:
- Timing belt cover
- Engine mount bracket
- Right engine mount

17. Remove the engine support.
- Crankshaft damper and tighten to 105 ft. lbs. (142 Nm)

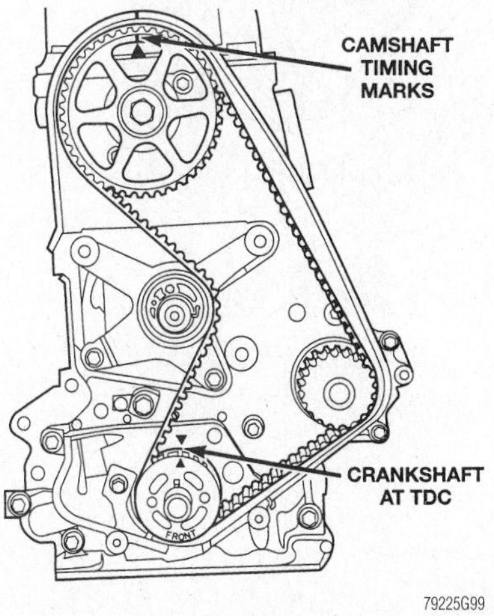

TDC alignment for timing belt installation—2.0L (VIN C) engine

- Drive belts and accessories
- Right inner splash-shield

18. Perform the crankshaft and camshaft relearn alignment procedure using the DRB scan tool or equivalent.

2.0L (VIN F) Engine

1. Before servicing the vehicle, refer to the precautions in the beginning of this section.

2. Remove or disconnect the following:
- Negative battery cable
- Engine undercover
- Engine mount bracket
- Drive belts
- Belt tensioner pulley
- Water pump pulleys
- Crankshaft pulley.
- Stud bolt from the engine support bracket, then remove the timing belt covers

3. Rotate the crankshaft clockwise to align the camshaft timing marks. Always turn the crankshaft in the forward direction only.

4. Loosen the tension pulley center bolt.

➡If the timing belt is to be reused, mark the direction of rotation on the flat side of the belt with an arrow.

5. Move the tension pulley towards the water pump and remove the timing belt.

6. Remove the crankshaft sprocket center bolt using special tool MB990767 to hold the crankshaft sprocket while removing the center bolt. Then, use MB998778 puller to remove the sprocket.

7. Mark the direction of rotation on the timing belt "B" with a arrow.

8. Loosen the center bolt on the tensioner and remove the belt.

9. To remove the camshaft sprocket, remove the cylinder head cover. Use a wrench to hold the hexagonal part of the camshaft and remove the sprocket mounting bolt.

⁂ WARNING

Do not rotate the camshafts or the crankshaft while the timing belt is removed.

To install:

10. Use a wrench to hold the camshaft, then install the sprocket and mounting bolt. Tighten the bolt(s) to 65 ft. lbs. (88 Nm).

11. Install the cylinder head cover.

12. Place the crankshaft sprocket on the crankshaft. Use tool MB990767 to hold the crankshaft sprocket while tightening the center bolt. Tighten the center bolt to 80–94 ft. lbs. (108–127 Nm).

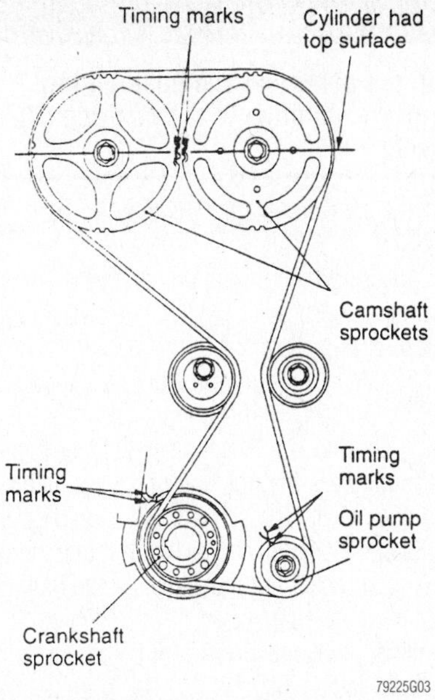

Notice the timing mark on the oil pump drive sprocket—2.0L (VIN F) engines

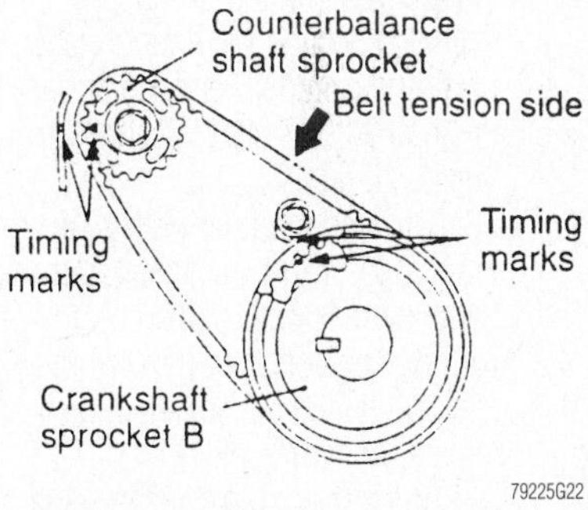

Timing belt "B" timing marks locations—2.0L (VIN F) engines

13. Align the timing marks on the crankshaft sprocket "B" and the balance shaft.

14. Install timing belt "B" on the sprockets. Position the center of the tensioner pulley to the left and above the center of the mounting bolt.

15. Push the pulley clockwise toward the crankshaft to apply tension to the belt and tighten the mounting bolt to 14 ft. lbs. (19 Nm). Do not let the pulley turn when tightening the bolt because it will cause excessive tension on the belt. The belt should deflect 0.20–0.28 in. (5–7mm) when finger pressure is applied between the pulleys.

16. Install the crankshaft sensing blade and the crankshaft sprocket. Apply engine oil to the mounting bolt and tighten the bolt to 80–94 ft. lbs. (108–127 Nm).

17. Use a press or vise to compress the auto tensioner pushrod. Insert a set pin when the holes are aligned.

⁂ WARNING

Do not compress the pushrod too quickly, damage to the pushrod can occur.

18. Install the auto tensioner on the engine.
19. Align the timing marks on the camshaft sprocket, crankshaft sprocket and the oil pump sprocket.
20. After aligning the mark on the oil pump sprocket, remove the cylinder block plug and insert a Phillips screwdriver in the hole to check the position of the counter balance shaft. The screwdriver should go in at least 2.36 in. or more. If not, rotate the oil pump sprocket once and realign the timing mark so the screwdriver goes in. Do not remove the screwdriver until the timing belt is installed.
21. Install or connect the following:
    - Timing belt on the intake camshaft and secure it with a clip
    - Timing belt on the exhaust camshaft. Align the timing marks with the cylinder head top surface using two wrenches. Secure the belt with another clip.
    - Belt around the idler pulley, oil pump sprocket, crankshaft sprocket and the tensioner pulley
22. Turn the tension pulley so the pin-holes are at the bottom. Press the pulley lightly against the timing belt.
23. Screw the special tool into the left engine support bracket until it contacts the tensioner arm, then screw the tool in a little more and remove the pushrod pin from the auto tensioner. Remove the special tool and tighten the center bolt to 35 ft. lbs. (48 Nm).
24. Turn the crankshaft ¼ turn counter-clockwise, then clockwise until the timing marks are aligned.
25. Loosen the center bolt. Install special tool MD998767 on the tension pulley. Turn the tension pulley counterclockwise with a torque of 2.6 ft. lbs. (3.5 Nm) and tighten the center bolt to 35 ft. lbs. (48 Nm). Do not let the tension pulley turn with the bolt.
26. Turn the crankshaft two revolutions to the right and align the timing marks. After 15 minutes, measure the protrusion of the pushrod on the auto tensioner. The standard measurement is 0.150–0.177 in. (3.8–4.5mm). If the protrusion is out of specification, loosen the tension pulley, apply the proper torque to the belt and retighten the center bolt.
27. Install or connect the following:
    - Crankshaft pulley. Tighten the mounting bolts to 18 ft. lbs. (25 Nm).
    - Water pump. Tighten the mounting bolts to 6.5 ft. lbs. (8.8 Nm).
    - Drive belts, then adjust
    - Engine mount bracket
    - Engine undercover
    - Negative battery cable

2.4L Engine

1. Before servicing the vehicle, refer to the precautions in the beginning of this section.
2. Remove or disconnect the following:
    - Negative battery cable
    - Upper torque strut attaching bolts and strut
    - Torque strut bracket from strut tower
    - Torque strut bracket from engine
    - Upper timing belt cover fasteners and cover
    - Right front wheel
    - Accessory drive belt splash shield
    - Accessory drive belts
    - Crankshaft vibration damper with special Tools 1026 and 6287-A
    - Lower torque strut
    - Exhaust system from the exhaust manifold
    - A/C pressure switch from the rear of the compressor housing

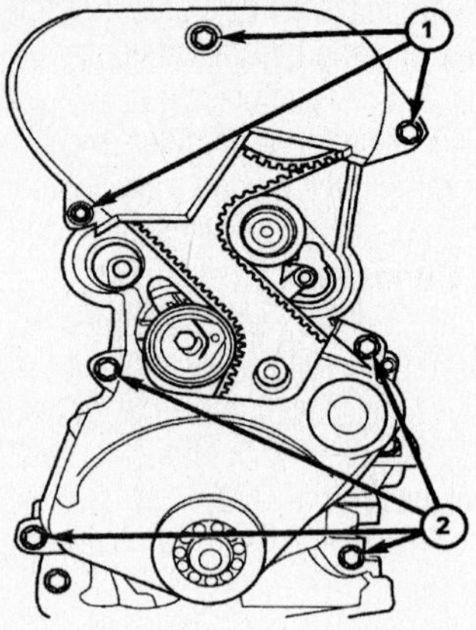

1 - UPPER COVER FASTENERS
2 - LOWER COVER FASTENERS

67189-NEON-G25

Upper and lower timing belt cover mounting—2.4L engine

3. Lower the vehicle and support the engine with a suitable jack.
    - Upper torque strut attaching bolts and remove strut
    - Power steering pump and bracket. Set the pump aside. Do not disconnect the fluid lines from pump.
    - Make sure the engine is properly supported, then remove the right engine mount through bolt

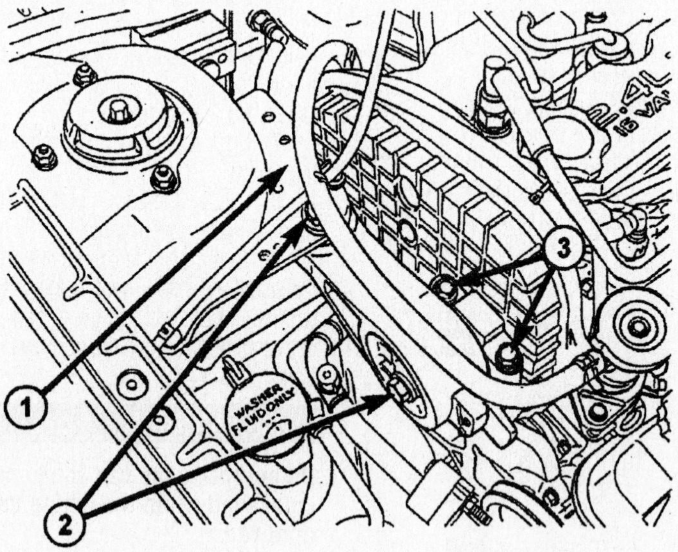

1 - TORQUE STRUT BRACKET
2 - TORQUE STRUT MOUNTING BOLTS
3 - ENGINE BRACKET MOUNTING BOLTS

67189-NEON-G24

Upper torque strut and bracket mounting—2.4L engine

- Raise the engine with jack until you can access the engine support bracket bolts
- Engine support bracket
- Lower timing belt cover fasteners and cover

4. Rotate the crankshaft until the TDC mark on oil pump housing aligns with the TDC mark on crankshaft sprocket (trailing edge of sprocket tooth)
5. Loosen the timing belt tensioner lock bolt.
6. Insert a 6mm Allen wrench into the hexagon opening located on the top plate of the belt tensioner pulley. Rotate the top plate clockwise until there is enough slack in timing belt to allow for removal.
7. Remove or disconnect the following:
   - Timing belt and camshaft sprockets.

⁂ WARNING

If the timing belt was damaged due to incorrect alignment, you must replace the belt tensioner pulley and bracket as an assembly.

- Crankshaft sprocket using Special Tool 6793 and insert C-4685-C2

To install:

➡The crankshaft sprocket is set to a pre-determined depth from the factory for correct timing belt tracking. If removed, use of Special Tool 6792 is required to set the sprocket to original installation depth. An incorrectly installed sprocket will result in timing belt and engine damage.

8. Install or connect the following:
   - Crankshaft sprocket using Special Tool 6792

⁂ WARNING

Do not use an impact wrench to tighten camshaft sprocket bolts. Damage to the camshaft-to-sprocket locating dowel pin may occur.

- Camshaft sprockets. Hold the sprockets with Special Tool 6847 while tightening center bolt to 85 ft. lbs. (115 Nm)

9. Set the crankshaft sprocket to TDC by aligning the sprocket with the arrow on

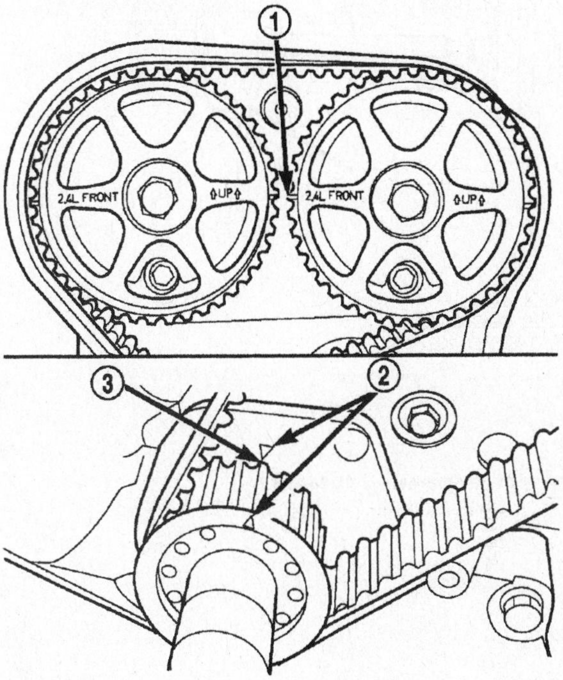

1 - CAMSHAFT TIMING MARKS
2 - CRANKSHAFT TDC MARKS
3 - TRAILING EDGE OF SPROCKET TOOTH

67189-NEON-G26

View of the proper timing alignment marks for timing belt removal—2.4L engine

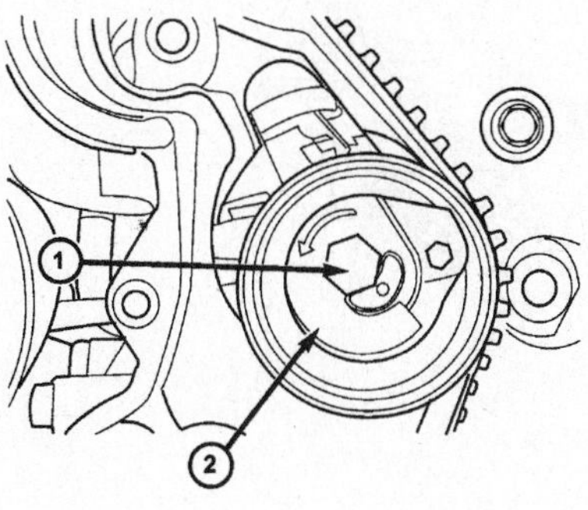

1 - LOCK BOLT
2 - TOP PLATE

67189-NEON-G27

Timing belt tensioner mounting—2.4L engine

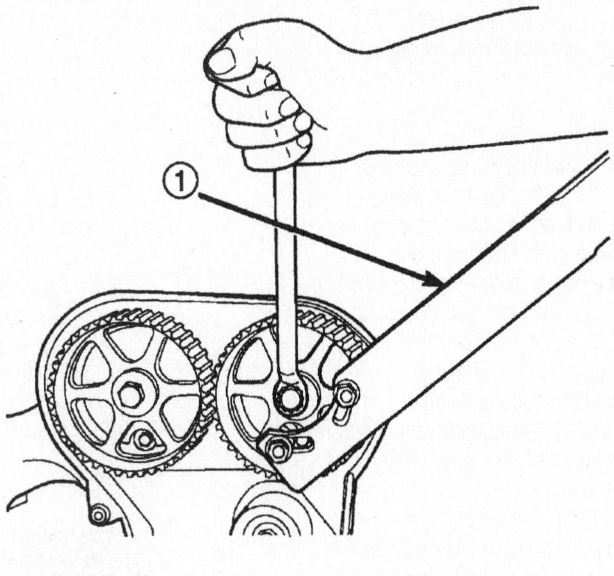

1 - SPECIAL TOOL 6847

67189-NEON-G28

Removal and installation of the camshaft sprockets—2.4L engine

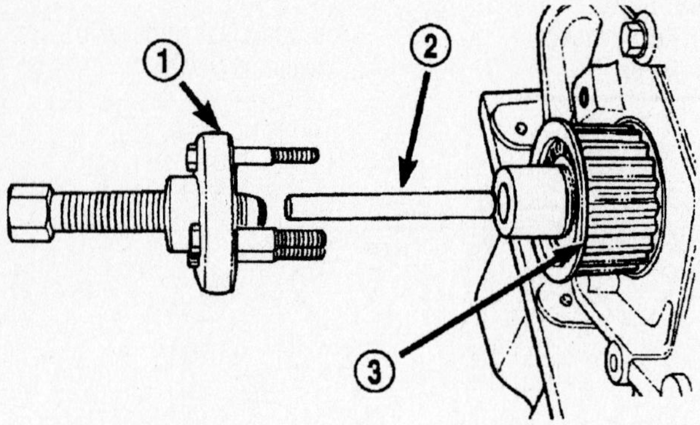

1 - SPECIAL TOOL 6793
2 - SPECIAL TOOL C-4685-C2
3 - CRANKSHAFT SPROCKET

67189-NEON-G29

Removal of the crankshaft sprocket—2.4L engine

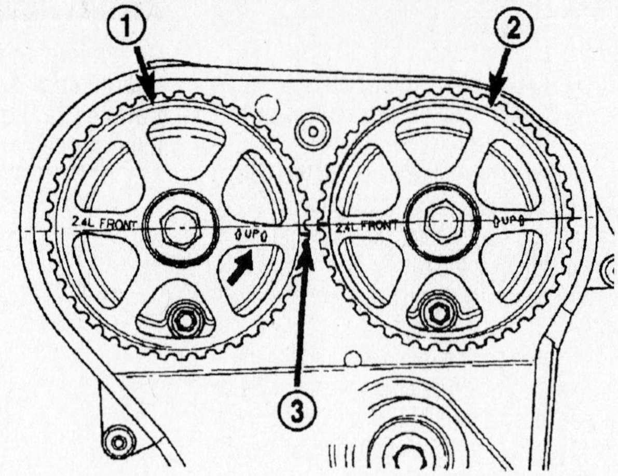

1 - CAMSHAFT SPROCKET-EXHAUST
2 - CAMSHAFT SPROCKET-INTAKE
3 - 1/2 NOTCH LOCATION

67189-NEON-G31

Make sure the marks are aligned before installing the timing belt—2.4L engine

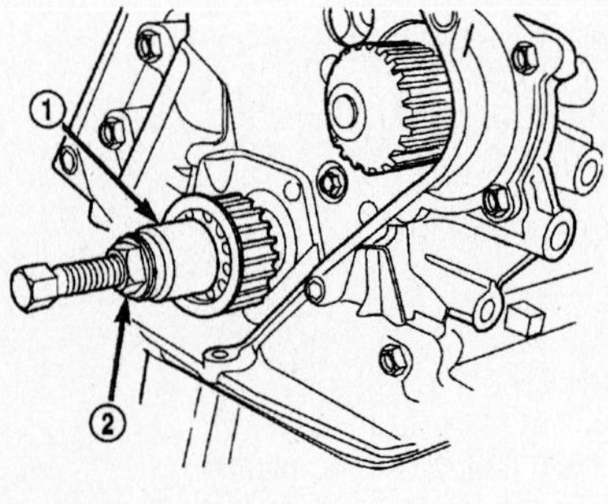

1 - SPECIAL TOOL 6792
2 - TIGHTEN NUT TO INSTALL

67189-NEON-G30

Installation of the crankshaft sprocket—2.4L engine

the oil pump housing. Set the camshafts timing marks so that the exhaust camshaft sprocket is a 1/2 notch below the intake camshaft sprocket. Make sure the arrows on the camshaft sprockets are facing up.

- Timing belt, starting at the crankshaft, go around the water pump sprocket, idler pulley, camshaft sprockets and then around the tensioner
- Move the exhaust camshaft sprocket counterclockwise to align the marks and take up any timing belt slack.

10. Insert a 6mm Allen wrench into the hexagon opening located on the top plate of the belt tensioner pulley. Rotate the top plate counterclockwise. The tensioner pulley will move against the belt and the tensioner setting notch will eventually start to move clock-

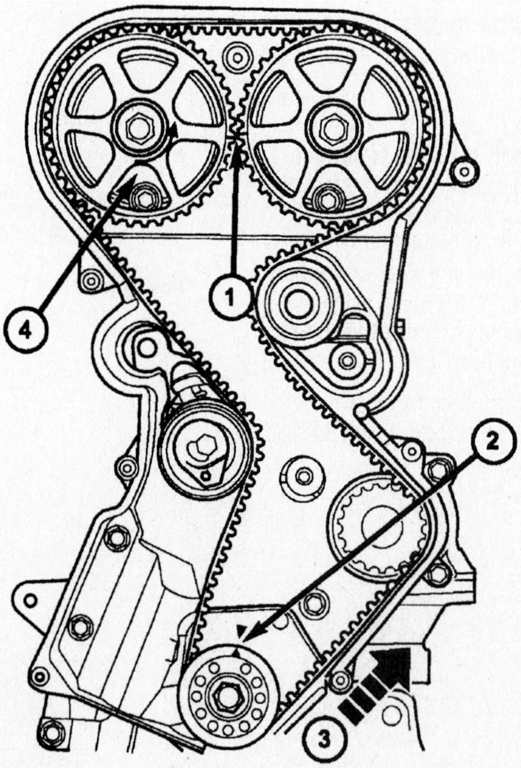

1 - CAMSHAFT TIMING MARKS 1/2 NOTCH LOCATION
2 - CRANKSHAFT AT TDC
3 - INSTALL BELT IN THIS DIRECTION
4 - ROTATE CAMSHAFT SPROCKET TO TAKE UP BELT SLACK

67189-NEON-G32

View of the timing belt properly installed—2.4L engine

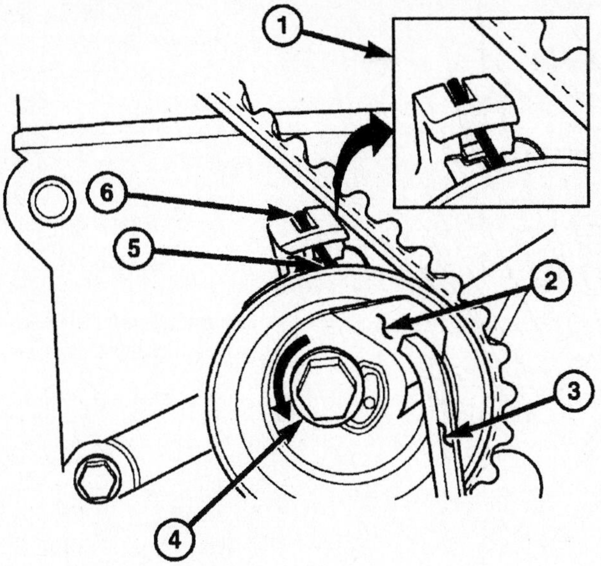

Timing belt tensioner adjustment—2.4L engine

wise. Watching the movement of the setting notch, continue rotating the top plate counterclockwise until the setting notch is aligned with the spring tang. Using the Allen wrench to prevent the top plate from moving, tighten the tensioner lock bolt to 18 ft. lbs. (25 Nm). The setting notch and spring tang should remain aligned after lock bolt is torqued.

11. Remove the Allen wrench and torque wrench.

WARNING

Repositioning the crankshaft to the TDC position must be done only during the clockwise rotation movement. If TDC is missed, rotate a further two revolutions until TDC is achieved. Do NOT rotate crankshaft counterclockwise as this will make verification of proper tensioner setting impossible.

12. Rotate the crankshaft clockwise 2 complete revolutions manually for seating of the timing belt, until the crankshaft is repositioned at the TDC position. Make sure that the camshaft and crankshaft timing marks are in proper position.

13. Check if the spring tang is within the tolerance window. If the spring tang is within the tolerance window, the installation process is complete and nothing further is required. If the spring tang is not within the tolerance window, repeat the previous 3 Steps.

14. Install or connect the following

- Lower timing belt cover and tighten the retainers to 50 inch lbs. (6 Nm)
- Engine support bracket. Make sure the power steering pump is properly located in mounting location on bracket. Tighten the mount bracket bolts to 45 ft. lbs. (61 Nm).

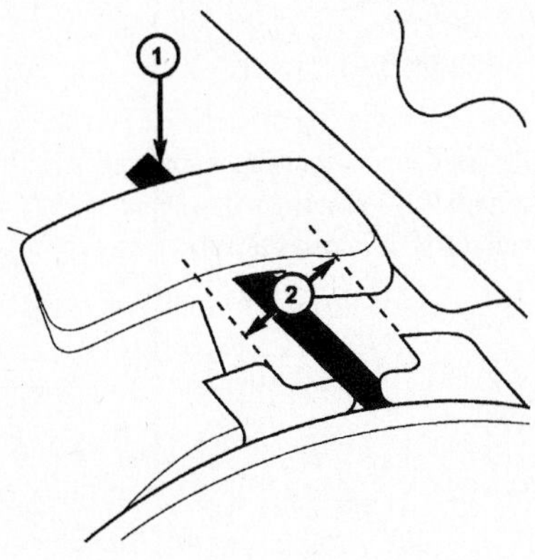

Make sure the spring tang is within the tolerance window—2.4L engine

- Lower engine into mounting position and right engine mount through bolt. Torque the bolt to 87 ft. lbs. (118 Nm)
- Power steering pump and bracket
- Upper torque strut attaching bolts
- Exhaust system to manifold
- A/C pressure switch connector
- Crankshaft vibration damper using thrust bearing washer and bolt from Installer Tool 6792 and torque the damper bolt to 100 ft. lbs. (135 Nm)
- Accessory drive belts
- Lower torque strut
- Accessory drive belt splash shield
- Right front wheel
- Upper timing belt cover and tighten fasteners to 50 inch lbs. (6 Nm)
- Torque strut bracket to engine
- Torque strut bracket to the strut tower
- Upper torque attaching bolts

15. Perform torque strut adjustment procedure, as follows:

a. Remove the accessory drive belt splash shield.

b. Remove the pencil strut.

c. Loosen the upper and lower torque strut attaching bolt at the suspension crossmember and shock tower bracket.

d. Adjust the engine position by placing a floor jack on the forward edge of the transmission bell housing

e. With the engine supported, remove the upper and lower torque strut attachment bolt(s) at shock tower bracket and suspension crossmember. Make sure the torque struts are free to move within the shock tower bracket and crossmember. Reinstall the torque strut bolt(s), but do not tighten yet.

f. Apply upward force, allowing the upper engine to rotate rearward until the distance between the center of the rearmost attaching bolt on the engine mount bracket (point A) and the center of the hole on the shock tower bracket (point B) is 4.70 in. (119mm)

WARNING

Hold the engine in position with the jack until the upper and lower torque strut bolts are tightened.

g. With the engine held at the proper position, tighten both the upper and lower torque strut bolts to 85 ft. lbs. (115 Nm)

h. Remove the floor jack.

i. Install the pencil strut and tighten nuts to 43 ft. lbs. (58 Nm).

16. Install the accessory drive belt splash shield.

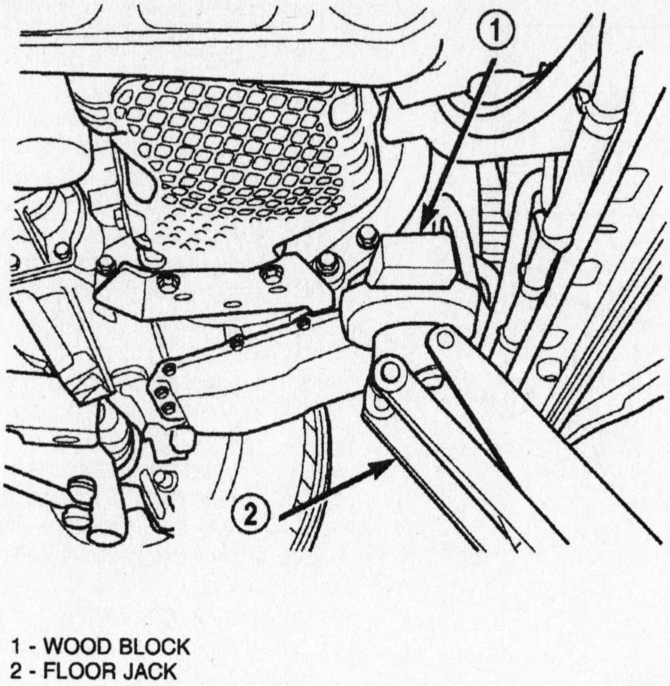

Position the floor jack properly to prevent upward lifting of the engine—2.4L engine

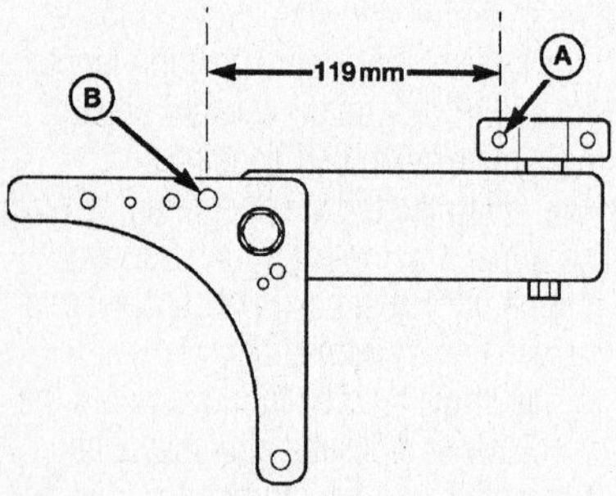

Measure the distance between the center of the rearmost attaching bolt on the engine mount bracket (point A) and the center of the hole on the shock tower bracket (point B)—2.4L engine

Piston and Ring

POSITIONING

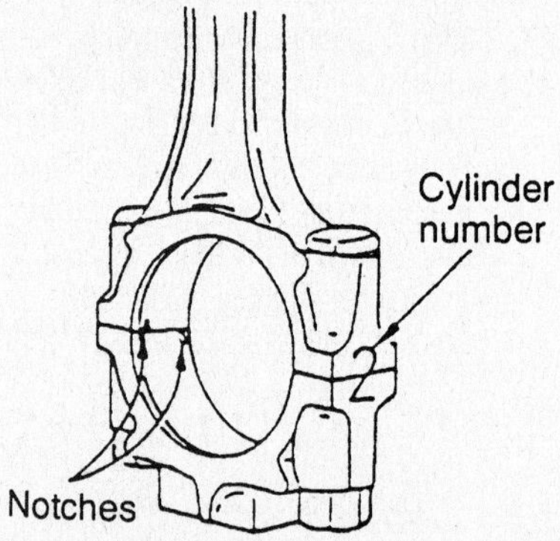

Chrysler engine connecting rod and cap installation—ensure to matchmark the cap and rod prior to disassembly

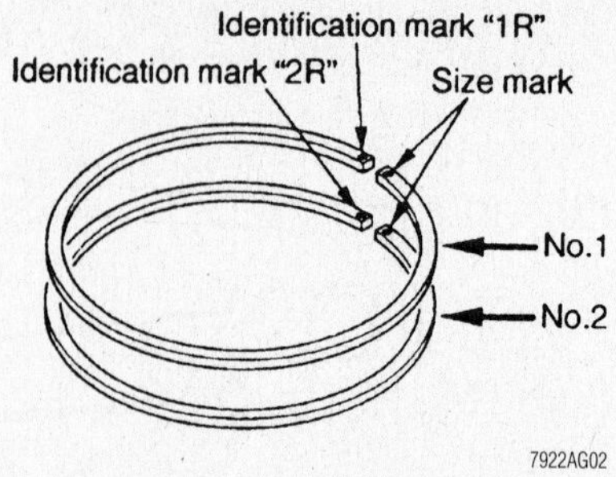

Common Chrysler piston ring identification mark locations

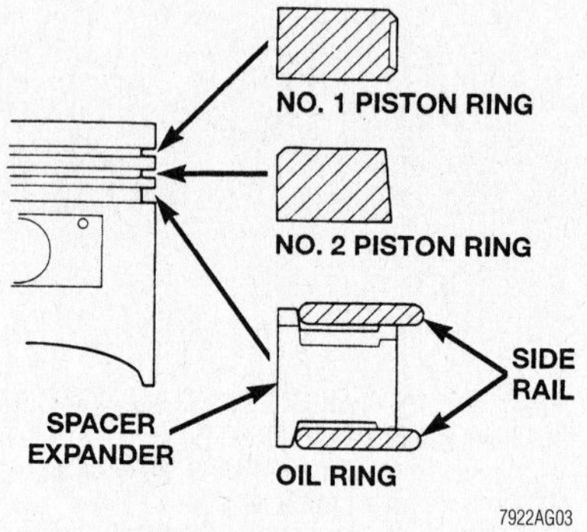

Piston ring orientation

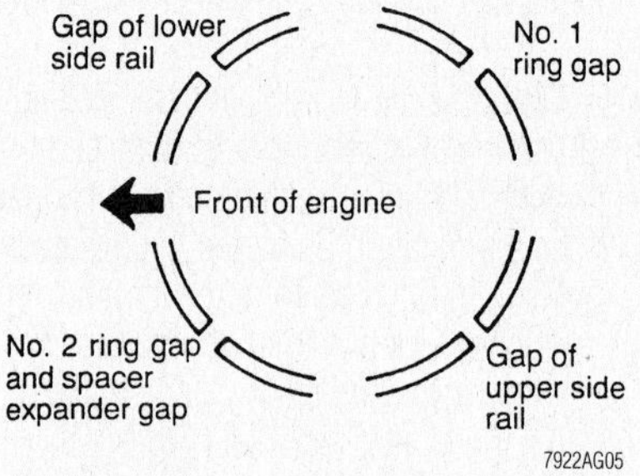

Piston ring end-gap spacing

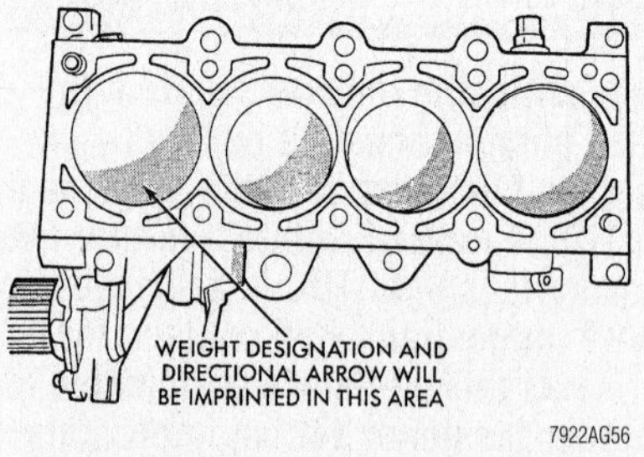

Piston positioning. The arrow or weight marking (L or H) must face the timing belt side of the engine—2.0L engines

FUEL SYSTEM

Fuel System Service Precautions

Safety is an important factor when servicing the fuel system. Failure to conduct maintenance and repairs in a safe manner may result in serious personal injury. Maintenance and testing of the vehicle's fuel system components can be accomplished safely and effectively by adhering to the following rules and guidelines:

- To avoid the possibility of fire and personal injury, always disconnect the negative battery cable unless the repair or test procedure requires that battery voltage be applied.
- Always relieve the fuel system pressure prior to disconnecting any fuel system component (injector, fuel rail, pressure regulator, etc.), fitting or fuel line connection. Exercise extreme caution whenever relieving fuel system pressure, to avoid exposing skin, face and eyes to fuel spray. Please be advised that fuel under pressure may penetrate the skin or any part of the body that it contacts.
- Always place a shop towel or cloth around the fitting or connection prior to loosening to absorb any excess fuel due to spillage. Ensure that all fuel spillage is quickly removed from engine surfaces. Ensure that all fuel soaked cloths or towels are deposited into a suitable waste container.
- Always keep a dry chemical (Class B) fire extinguisher near the work area.
- Do not allow fuel spray or fuel vapors to come into contact with a spark or open flame.
- Always use a back-up wrench when loosening and tightening fuel line connection fittings. This will prevent unnecessary stress and torsion to fuel line piping.
- Always replace worn fuel fitting O-rings. Do not substitute fuel hose where fuel pipe is installed.

Fuel System Pressure

RELIEVING

**** CAUTION**

Relieve the fuel system pressure before servicing any components of the fuel system. Service vehicles in well ventilated areas and avoid ignition sources. NEVER smoke while servicing the vehicle!

1. Before servicing the vehicle, refer to the precautions in the beginning of this section.
2. Remove the fuel pump relay from the Power Distribution Center (PDC). Location of the relay can be found on a label on the underside of the PDC cover.

➡Removal of the fuel pump relay may cause Diagnostic Trouble Code(s) (DTC's) to set. You must use a DRB scan tool to clear codes if necessary.

3. Start the engine and let run until it stalls.
4. Try to start the engine again until it will no longer run.
5. Turn the ignition key **OFF**.
6. Place a shop towel below the fuel line quick-connect fitting at the fuel rail.
7. Put the fuel pump relay back in the PDC.
8. The vehicle is now safe for servicing.

Fuel Filter

A combination fuel filter/pressure regulator assembly is used, which is located on the top of the fuel pump module.

REMOVAL & INSTALLATION

**** CAUTION**

Do not allow fuel spray or fuel vapors to come in contact with a spark or open flame. Keep a dry chemical fire extinguisher nearby. Never store fuel in an open container due to risk of fire or explosion.

1. Before servicing the vehicle, refer to the precautions in the beginning of this section.
2. Properly relieve the fuel system pressure.
3. Disconnect the negative battery cable.
4. Disconnect the quick-connect fuel supply line from the filter/regulator nipple.
5. Depress the locking spring tab, located on the side of the fuel filter/regulator, then rotate 90 degrees and pull out. Be sure the upper and lower O-rings are still on the filter assembly.

To install:

6. Lightly coat the filter O-rings with clean engine oil. Insert the filter into the opening in the fuel pump module, then align the 2 hold-down tabs with the flange.
7. While applying downward pressure, rotate the filter clockwise until the spring tab catches in the locating slot.

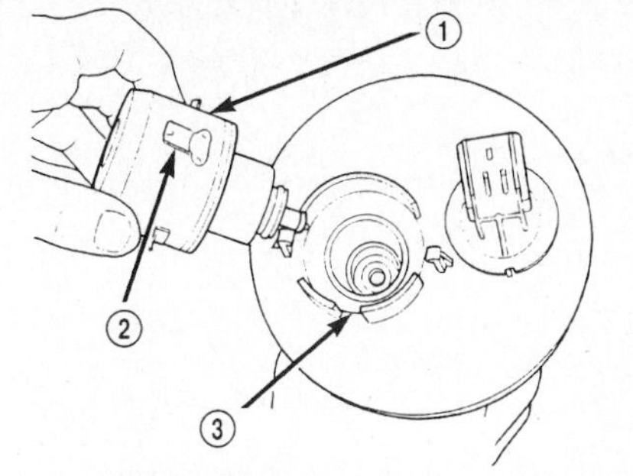

Depress the spring tab, then rotate and pull the fuel filter assembly out

8. Attach the fuel line to the filter/regulator assembly.
9. Connect the negative battery cable.

Fuel Pump

The fuel pump is integral with the pump module, which also contains the fuel reservoir, level sensor, inlet strainer and fuel pressure regulator. The inlet strainer, fuel pressure regulator and level sensor are the only serviceable items. If the fuel pump requires service, replace the entire fuel pump module.

REMOVAL & INSTALLATION

1. Before servicing the vehicle, refer to the precautions in the beginning of this section.
2. Properly relieve the fuel system pressure.
3. Drain the fuel tank.
4. Remove or disconnect the following:
   - Negative battery cable
   - Vapor line from the Evaporative Emissions (EVAP) canister tube
   - EVAP canister

**** WARNING**

The fuel reservoir of the fuel pump module does not empty out when the tank is drained. The fuel in the reservoir will spill out when the module is removed.

- Fuel line from the fuel pump module by depressing the quick-connect retainers with your thumb and forefinger
- Fuel pump module electrical connector by pushing down on the

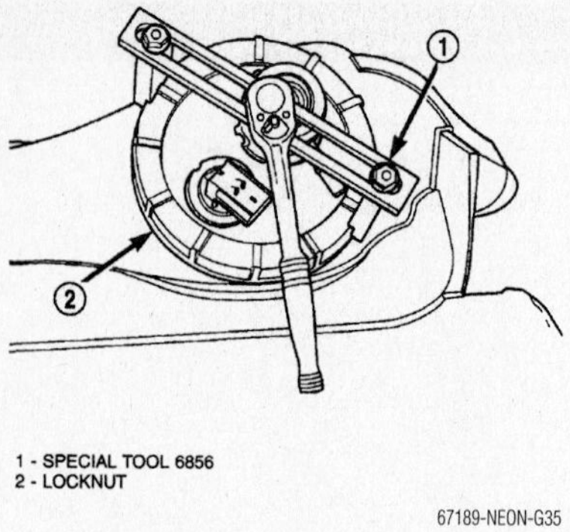

Fuel pump removal using the special tool—2004 vehicle shown

connector retainer and pulling the connector off of the module
- Fuel filler tube and filler vent tube from the fuel tank's filler hose

5. Support the fuel tank with a transmission jack.
   - Fuel tank strap bolts; then, lower the tank slightly for access to the module
   - Fuel pump module locknut using a ratchet and spanner wrench or special tool 6856, as necessary
   - Fuel pump and discard the O-ring seal

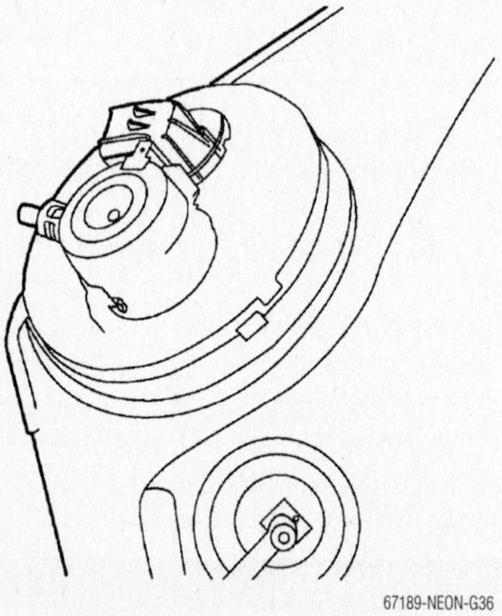

Be sure the alignment tab on the underside of the fuel pump module flange sits in the notch on the fuel tank

To install:

⁂ WARNING

Take care to avoid wiping dirt or debris into the fuel tank opening. Such contaminants may clog or damage the new fuel pump.

6. Wipe the fuel pump mounting flange and surrounding area of the tank clean.
7. Install or connect the following:
   - New O-ring seal in the tank opening
   - Fuel pump

➡Be sure the alignment tab on the underside of the fuel pump module flange sits in the notch on the fuel tank.

   - Fuel pump module locknut. Using the ratchet and spanner wrench, torque the locknut to 41 ft. lbs. (55 Nm).
   - Fuel tank
   - Fuel filler tube and filler vent tube to the fuel tank's filler hose
   - Fuel pump module electrical connector
   - Fuel line, to the fuel pump module
   - EVAP canister
   - Vapor line to the EVAP canister tube
   - Negative battery cable
8. Start the vehicle, check for leaks and repair if necessary.

Fuel Injector

REMOVAL & INSTALLATION

1. Before servicing the vehicle, refer to the precautions in the beginning of this section.
2. Release the fuel system pressure.
3. Remove or disconnect the following:
   - Negative battery cable
   - Fuel line(s) from fuel rail
   - Fuel injector electrical connectors
   - Fuel rail mounting screws
   - Fuel rail with the fuel injectors
   - Fuel injector retainers
   - Fuel injector(s) and discard the O-rings

To install:

4. Install or connect the following:
   - Fuel injector(s) using new O-rings

➡Lubricate the O-rings with clean engine oil.

   - Fuel injector retainers
   - Fuel rail with the fuel injectors and torque the bolts to 15–19 ft. lbs. (20–25 Nm)
   - Fuel injector electrical connectors
   - Fuel line(s) to fuel rail
   - Negative battery cable
5. Start the vehicle, check for leaks and repair if necessary.

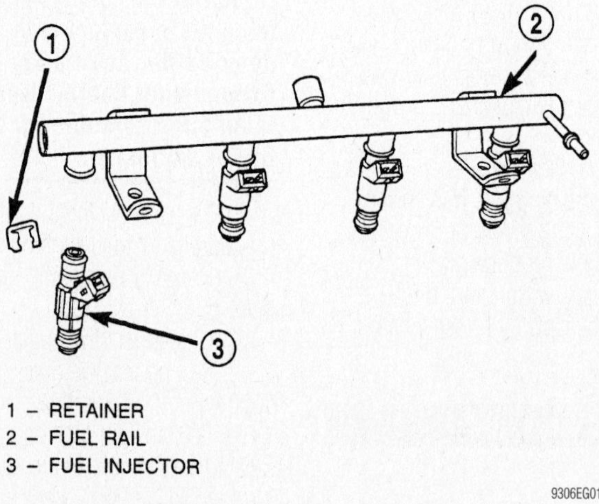

Exploded view of the fuel rail assembly

DRIVE TRAIN

Transaxle Assembly

REMOVAL & INSTALLATION

Manual Transaxle

T-350 TRANSAXLE

1. Before servicing the vehicle, refer to the precautions in the beginning of this section.
2. Drain the transaxle fluid.
3. Remove or disconnect the following:
   - Battery
   - Air cleaner/throttle body assembly
   - Proportional Purge Solenoid (PSS)
   - Crankcase vent hose
   - Throttle Position Sensor (TPS) and Idle Air Control (IAC) connectors
   - Throttle body air duct from the intake manifold
   - Accelerator cable
   - Cruise control cable, if equipped
   - Battery tray from the bracket
   - Ground cable from the battery tray bracket
   - Back-up light switch connector
   - Hydraulic line quick-connect fitting
   - Shift cable-to-bracket clips
   - Shift selector and crossover cables from the levers and move aside
   - Halfshafts
   - Structural collar
   - Left engine-to-transaxle lateral bending brace
   - Bell housing dust cover
   - Right engine-to-transaxle lateral bending brace
   - Starter
   - Driveplate-to-clutch module bolts
4. Place a jack and block of wood under the engine's oil pan and support the engine. Remove the transaxle upper mount through-bolt.
5. Lower the engine/transaxle assembly to provide clearance.
   - Transaxle-to-engine bolts, using an assistant to support the transaxle
   - Transaxle

To install:

6. If installing a new or replacement transaxle, transfer the upper mount to the new transaxle; then, torque the upper mount-to-transaxle bolts to 50 ft. lbs. (68 Nm).
7. Install or connect the following:
   - Transaxle
   - Transaxle-to-engine bolts, using an assistant to support the transaxle.

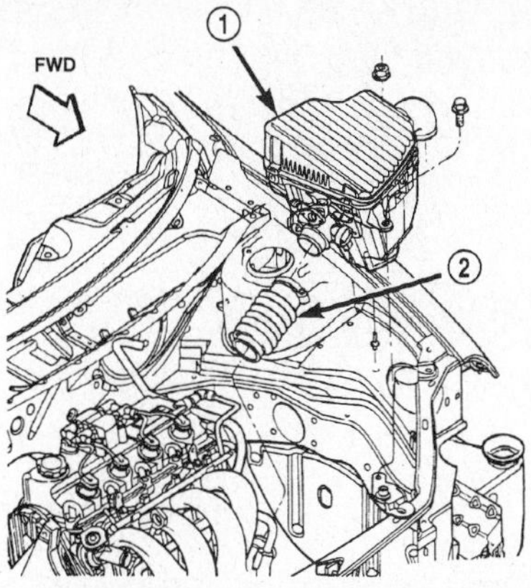

1 – AIR CLEANER ASSY.
2 – THROTTLE BODY DUCT

9306EG27

View of the air cleaner assembly and throttle body duct

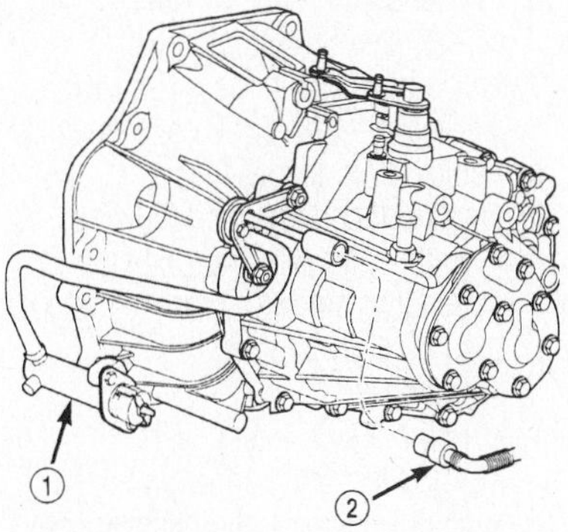

1 - SLAVE CYLINDER
2 - HYDRAULIC TUBE

9356HG10

View of the clutch slave cylinder connection —T-350 manual transaxle

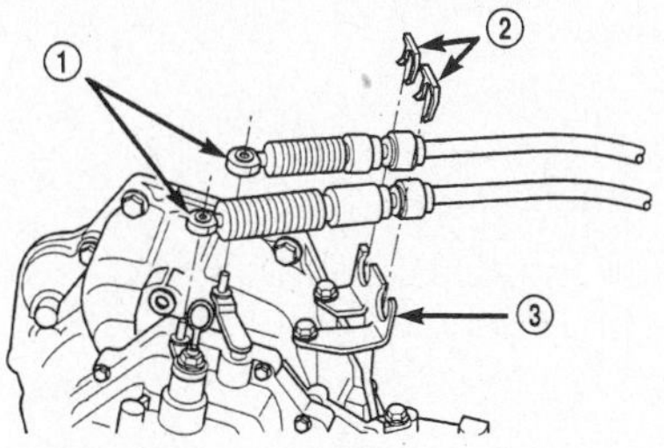

1 – SHIFT CABLES
2 – CLIPS
3 – BRACKET

9306EG29

View of the shift cables—T-350 manual transaxle

Torque the transaxle-to-engine bolts to 70 ft. lbs. (95 Nm).

8. Raise the engine/transaxle assembly until the upper mount aligns with the mount bracket hole. Torque the upper mount-to-mount bracket through-bolt to 80 ft. lbs. (108 Nm).
9. Remove the jack and wooden block.
   - Driveplate-to-clutch module bolts and torque to 65 ft. lbs. (88 Nm)
   - Starter and torque the bolts to 40 ft. lbs. (54 Nm)
   - Starter electrical connectors and torque the positive cable bolt to 89 inch lbs. (10 Nm)
   - Left engine-to-transaxle lateral bending brace and torque the bolts to 60 ft. lbs. (81 Nm)

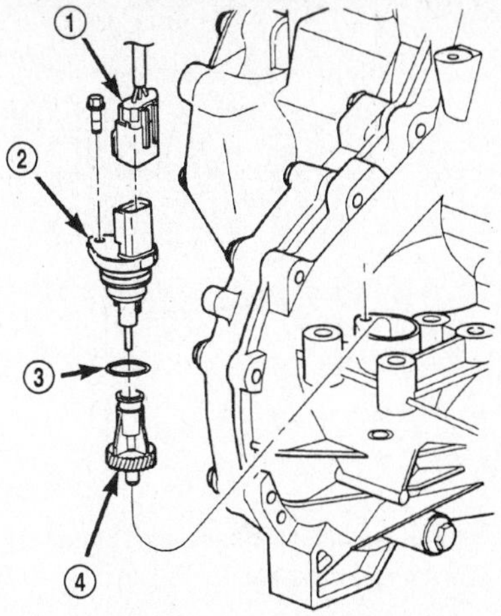

1 - CONNECTOR 3 - O-RING
2 - SENSOR 4 - SPEEDO PINION

9306EG30

View of the Vehicle Speed Sensor (VSS)—T-350 manual transaxle

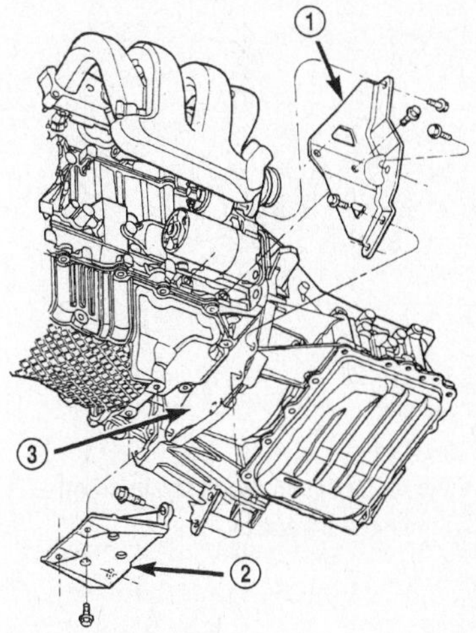

1 – LATERAL BENDING BRACE
2 – STRUCTURAL COLLAR
3 – DUST COVER

9306EG31

View of the left lateral bending brace and structural collar—T-350 manual transaxle

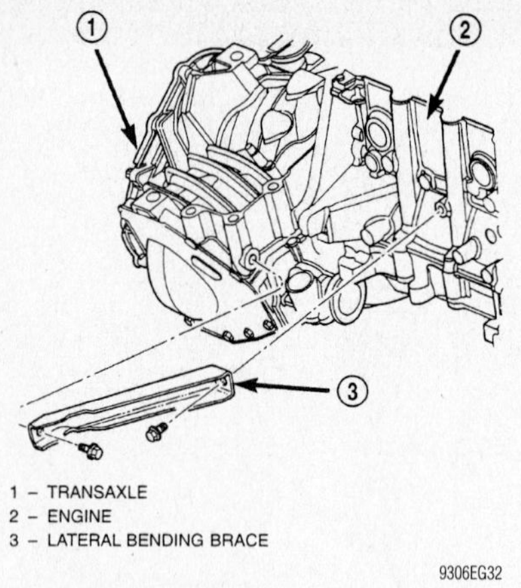

View of the right lateral bending brace—T-350 manual transaxle

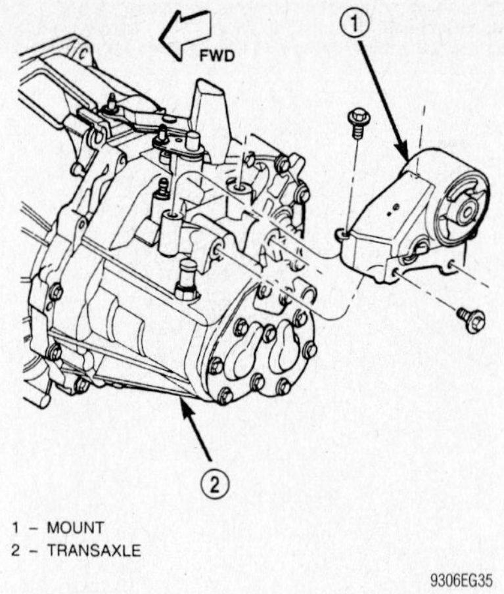

View of the upper mount—T-350 manual transaxle

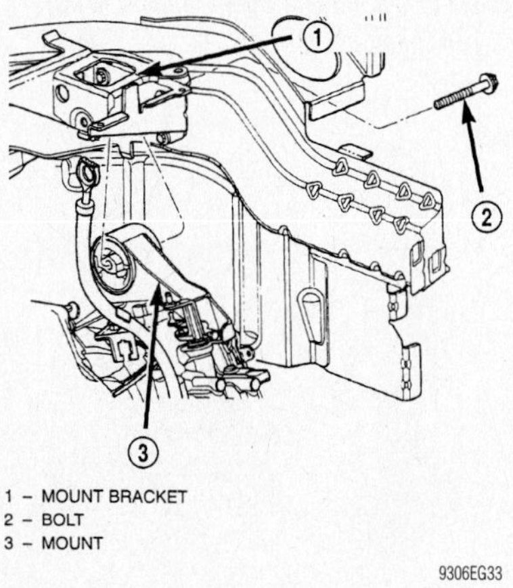

View of the upper mount through-bolt—T-350 manual transaxle

10. Install the structural collar and torque the bolts as follows:
    a. Step 1—Structural collar-to-oil pan bolts: 30 inch lbs. (3 Nm).
    b. Step 2—Structural collar-to-transaxle bolts: 80 ft. lbs. (108 Nm).
    c. Step 3—Structural collar-to-oil pan bolts: 40 ft. lbs. (54 Nm).
11. Install or connect the following:
    - Right engine-to-transaxle lateral bending brace and torque the bolts to 60 ft. lbs. (81 Nm)
    - Halfshafts
    - Bell housing dust cover
    - VSS electrical connector
    - Shift selector and crossover cables to the levers
    - Shift cable-to-bracket clips
    - Hydraulic line quick-connect fitting
    - Back-up light switch connector
    - Ground cable to the battery tray bracket
    - Battery tray to the bracket
    - Accelerator cable
    - Cruise control cable, if equipped
    - Air cleaner assembly and torque the nuts and bolts to 10 ft. lbs. (14 Nm)
    - Throttle body air duct to the intake manifold
    - TPS and IAC connectors
    - Crankcase vent hose to the throttle body
    - PSS to the throttle body
    - Battery
12. Fill the transaxle to the proper level.
13. Road test the vehicle and inspect for leaks.

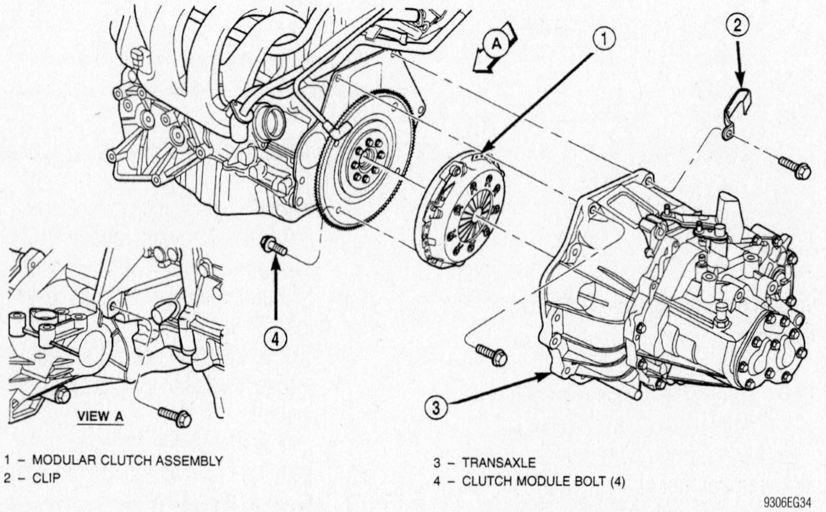

Exploded view of the transaxle and related components—T-350 manual transaxle

T-850 TRANSAXLE

1. Before servicing the vehicle, refer to the precautions in the beginning of this section.
2. Drain the transaxle fluid.
3. Remove or disconnect the following:
    - Turbocharger solenoid pack from the air cleaner assembly and position it aside
    - AAT sensor and remove air cleaner assembly
    - Throttle body inlet hose from the inter-cooler
    - Idle Air Control (IAC) and Throttle Position Sensor (TPS) connectors
    - Accelerator cable from throttle body linkage
    - Throttle body adapter
    - Battery and battery tray
    - Shift cables from the transaxle shift levers and bracket
    - Both front wheel/tire assemblies
    - Both Anti-lock Brake System (ABS) front wheel speed sensors
    - Halfshafts and intermediate shaft
    - Clutch hydraulic circuit quick-connect fitting, using Tool 6638A, or equivalent
4. Back-up lamp switch connector
    - Starter motor-to-transaxle case bolts
    - Front lateral bending brace
    - Turbocharger pipe and power steering lines from the structural collar
    - Structural collar
    - 4 modular clutch-to-engine drive-plate bolts. While removing the bolts, you will find one tight-tolerance (slotted) hole. When this bolt is removed, matchmark the drive-plate and modular clutch assembly at this location, and make sure to align marks upon reassembly.
5. Install and secure a suitable transmission jack.

- Left mount through-bolt

6. Lower the engine/transaxle assembly enough to gain access to left mount and bellhousing bolts.
   - Left mount-to-transaxle bolts and mount
7. Install a screw jack and wood block to support the engine after transaxle removal.
   - Transaxle bellhousing-to-block bolts, then remove the transaxle and modular clutch assembly.
8. If replacing the transmission, transfer the impact blocker, back-up lamp switch, vehicle speed sensor and gearshift cable bracket.

To install:

9. Install or connect the following:
   - Modular clutch to the input shaft assembly
   - Transaxle assembly to a suitable transmission jack, making sure it is secure
   - Transaxle into position. Torque the transaxle-to-block bolts to 80 ft. lbs. (108 Nm)
   - Upper mount/bracket to transaxle and tighten the bolts to 50 ft. lbs. (68 Nm)

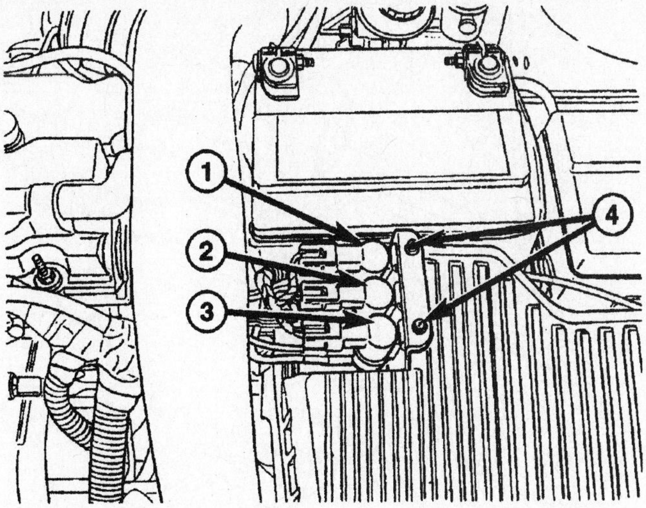

1 - TIP SOLENOID
2 - SURGE VALVE ACTUATOR SOLENOID
3 - WASTEGATE ACTUATOR SOLENOID
4 - BRACKET MOUNTING SCREWS

67189-NEON-G37

View of the turbocharger solenoids and mounting bracket—T-850 manual transaxle

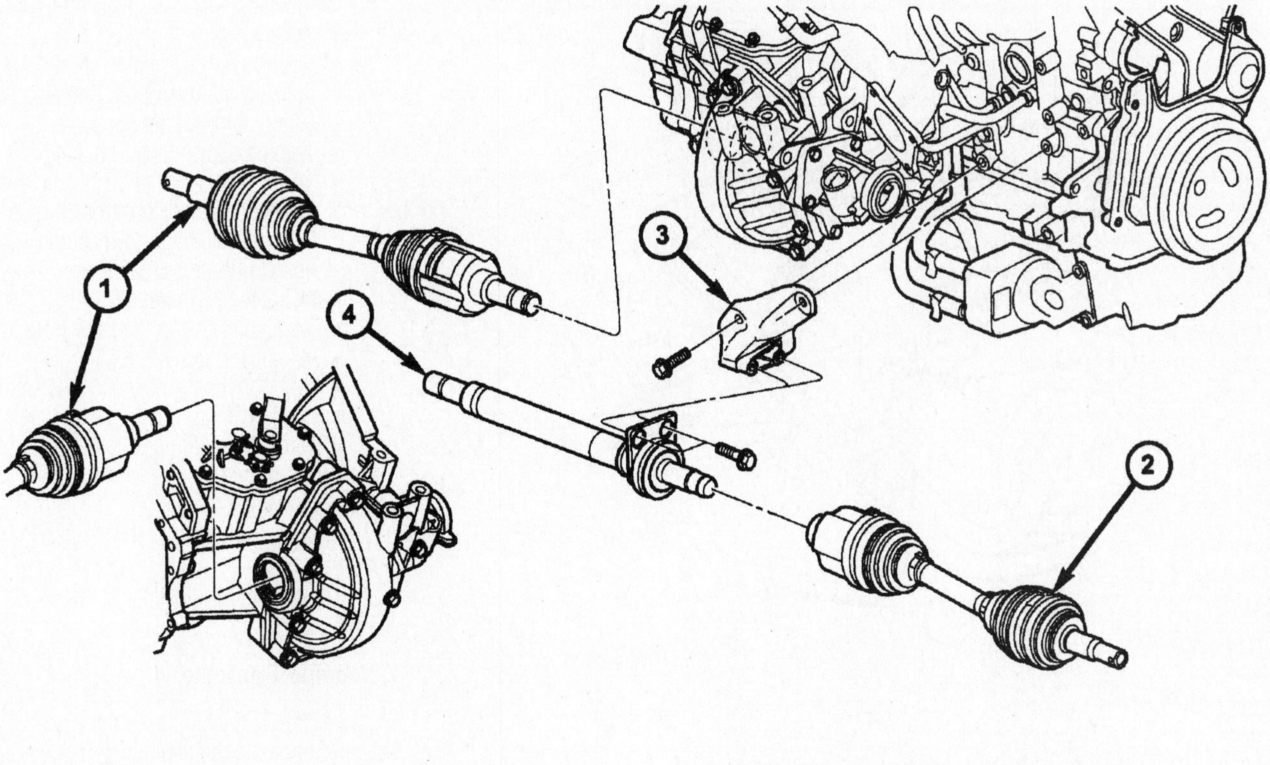

1 - HALFSHAFT - LEFT
2 - HALFSHAFT - RIGHT
3 - SUPPORT BRACKET - INTERMEDIATE SHAFT
4 - INTERMEDIATE SHAFT/BEARING ASSEMBLY

67189-NEON-G38

Exploded view of the halfshaft assemblies—2.4L turbo models

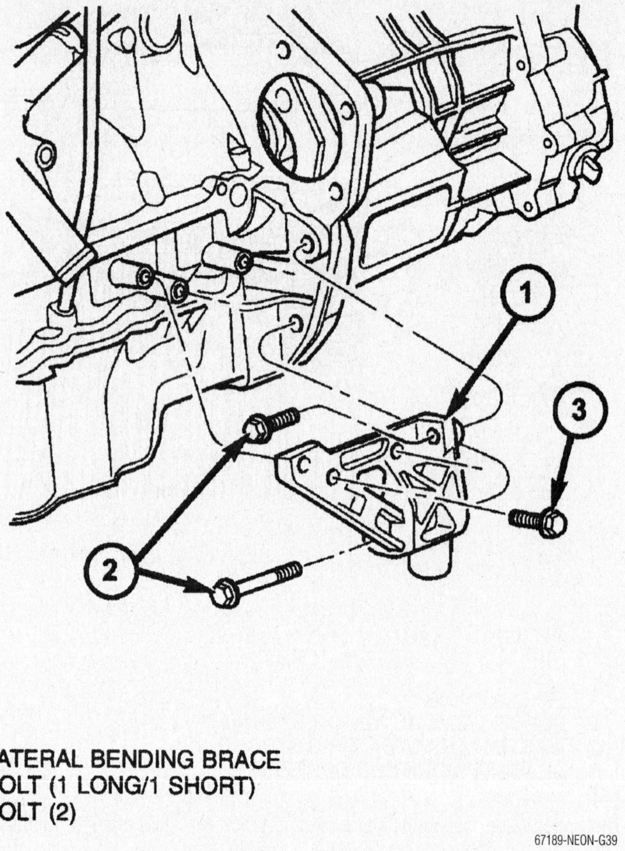

Lateral bending brace mounting—2.4L turbo models

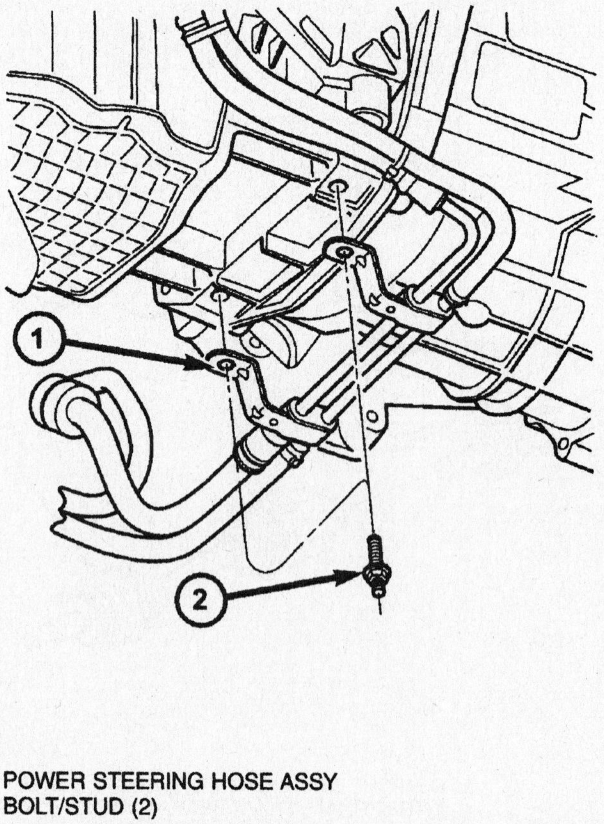

Structural collar/power steering hose retainers—2.4L turbo models

10. Raise the engine/transaxle assembly into position. Install and torque the through-bolt to 90 ft. lbs. (122 Nm).

- 4 modular clutch-to-driveplate bolts. Align the driveplate and modular clutch marks made during removal. Start with the tight-tolerance (slotted) hole, install and torque the bolts to 65 ft. lbs. (88 Nm).
- Structural collar and lateral bending brace. Finger-tighten all bolts, then tighten the structural collar horizontal bolts to 80 ft. lbs. (108 N). Torque the structural collar vertical bolts to 40 ft. lbs. (54 Nm) and the lateral bending brace bolts to 45 ft. lbs. (61 Nm).
- Intercooler pipe and power steering hoses into position and secure with nuts
- Clutch hydraulic plumbing to the slave cylinder quick-connect. You should hear a "click" when the connection is properly installed.
- Halfshafts and intermediate shaft
- ABS wheel speed sensors and torque the bolts to 105 inch lbs. (12 Nm)
- Gearshift cables to the transaxle bracket and secure with new retaining clips
- Battery tray
- Battery (with thermal wrap) and hold-down clamp/bolt. Connect battery temperature sensor.
- Throttle body adapter housing
- Accelerator cable while placing the throttle body into position. Torque the 2 throttle body-to-manifold bolts to 21 ft. lbs. (28 Nm).
- IAC and TPS connectors
- Intercooler-to-throttle body hose
- IAT sensor and TIP hose
- Air cleaner assembly and AAT sensor connector
- Turbo solenoid pack
- Air cleaner inlet tube to turbocharger

11. Fill the transaxle to the proper level.

12. Road test the vehicle and inspect for leaks.

Automatic Transaxle

1. Before servicing the vehicle, refer to the precautions in the beginning of this section.
2. Drain the transaxle fluid.
3. Remove or disconnect the following:

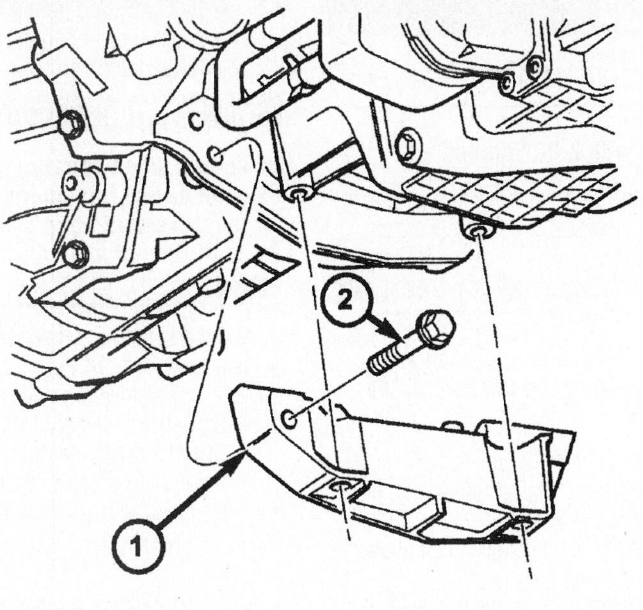

Structural collar mounting—2.4L turbo models

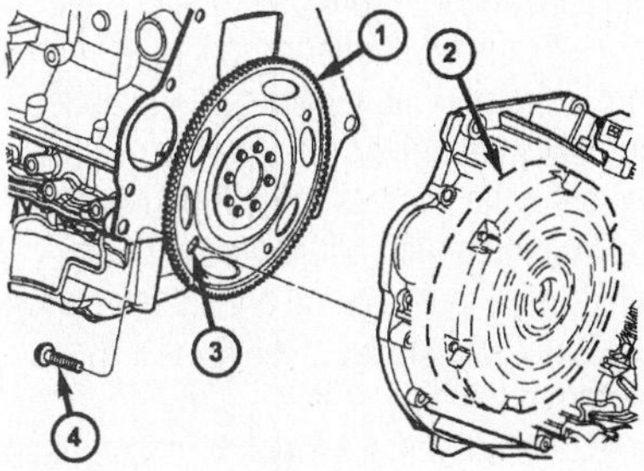

Modular clutch-to-driveplate mounting—T-850 manual transaxle

- Battery and battery tray

4. Remove the air cleaner/throttle body assembly by removing or disconnect the following:
   a. Proportional Purge Solenoid (PSS)
   b. Crankcase vent hose
   c. Throttle Position Sensor (TPS) and Idle Air Control (IAC) connectors
   d. Throttle body air duct from the intake manifold
   e. Accelerator cable
   f. Transaxle kickdown cable
   g. Cruise control cable, if equipped
   h. Air cleaner assembly
5. Remove or disconnect the following:

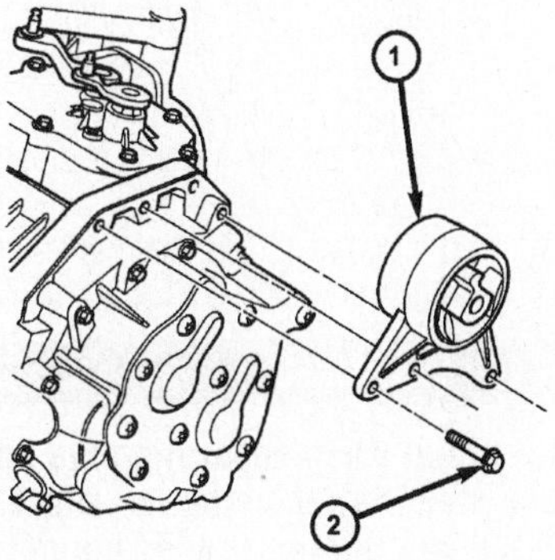

View of the transaxle upper mount and fasteners—T-850 manual transaxle

- Transaxle oil cooler lines. Use a cutter to cut the hoses off flush with the end of the fittings. You will need to install a service splice kit during installation.
- Input and output speed connectors
- Solenoid/pressure switch assembly connector
- Gearshift cable from manual valve
- Gearshift cable from upper mount bracket and position aside
- Upper starter mounting bolt
- Halfshafts
- Structural collar
- Left engine-to-transaxle lateral bending brace
- Starter
- Power steering cooler-to-cross-member fasteners and position the cooler aside
- Right lateral bending brace-to-transaxle bolt
- Torque converter dust shield
- Torque converter-to-driveplate bolts
- Both lower transaxle bellhousing-to-engine bolts

6. Using a jack and a block of wood, support the engine/transaxle assembly.
7. Remove the transaxle upper mount through-bolt.

➡The upper mount through-bolt may be accessed through the left side wheel house.

8. Lower the engine/transaxle assembly for clearance.
9. Remove or disconnect the following:
   - Transaxle-to-engine bolts, using an assistant to support the transaxle
   - Transaxle

To install:

10. Install transaxle and tighten the transaxle-to-engine bolts, using an assistant to support the transaxle. Torque the transaxle-to-engine bolts to 80 ft. lbs. (108 Nm).
11. Raise the engine/transaxle assembly until the upper mount aligns with the mount bracket hole. Torque the upper mount-to-mount bracket through-bolt to 87 ft. lbs. (118 Nm).
12. Remove the jack and wooden block.
13. Install or connect the following:
    - Torque converter-to-driveplate bolts and torque the bolts to 65 ft. lbs. (88 Nm)
    - Converter dust shield
    - Starter and starter electrical connections
    - Left lateral bending brace. Torque the brace-to-transaxle bolts to 80 ft. lbs. (108 Nm), the brace-to-engine bedplate bolt to 40 ft. lbs. (54 Nm) and the brace-to-intake manifold bolt to 21 ft. lbs. (28 Nm).
    - Structural collar
    - Power steering oil cooler
    - Halfshafts
    - Gearshift cable to mount. Secure the cable end to the manual valve lever
    - Solenoid/pressure switch assembly connector
    - Input and output speed connectors

- Cooler lines with service splice kit
- Air cleaner and throttle body assembly
- Battery tray and battery

14. Fill the transaxle and check for leaks.
15. Be sure the vehicle's back-up lights and speedometer are working properly.

Clutch

ADJUSTMENT

The clutches on these vehicles are self-adjusting. No manual adjustment is necessary or possible.

REMOVAL & INSTALLATION

1. Before servicing the vehicle, refer to the precautions in the beginning of this section.
2. Remove or disconnect the following:
   - Negative battery cable
   - Modular clutch-to-driveplate bolts and discard the bolts
   - Transaxle assembly
   - Modular clutch assembly from the transaxle input shaft

To install:

3. Install or connect the following:
   - Modular clutch assembly onto the transaxle input shaft
   - Transaxle assembly and torque the transaxle-to-engine bolts to 70 ft. lbs. (95 Nm)
   - New modular clutch-to-driveplate bolts and torque the bolts to 65 ft. lbs. (88 Nm)
   - Negative battery cable

Halfshafts

REMOVAL & INSTALLATION

Vehicles with 2.0L Engines

1. Before servicing the vehicle, refer to the precautions in the beginning of this section.
2. With the vehicle on the ground and brakes applied, loosen, but do not remove the stub axle-to-hub and bearing retaining nut.

➡The front hub and driveshaft are splined together and retained by the hub nut.

3. If equipped with Anti-lock Brake System (ABS), disconnect the front wheel speed sensor.
4. Remove or disconnect the following:
   - Front wheel
   - Front caliper-to-steering knuckle bolts
   - Caliper from the steering knuckle
5. Support the caliper out of the way by suspending it with a piece of wire from the strut. DO NOT allow the caliper to hang by the brake hose.
   - Rotor from the hub
   - Outer tie rod end-to-steering knuckle nut by holding the tie rod end stud with a 11/32 in. socket while loosening the nut
   - Tie rod end stud from the steering knuckle
   - Ball joint stud-to-steering knuckle nut and bolt

WARNING

Be careful when separating the ball joint stud from the steering knuckle, so the ball joint seal does not get damaged.

- Ball joint stud from the steering knuckle by prying down on the lower control arm

WARNING

Be careful when separating the inner CV-joint during this operation. Do not let the driveshaft hang by the inner CV-joint, the driveshaft must be supported.

6. Pull the steering knuckle assembly out and away from the outer CV-joint of the driveshaft assembly. Support the outer end of the driveshaft assembly.
7. Insert a prybar between the inner tripod joint and the transaxle case. Pry against the inner tripod joint until the joint retaining snapring is disengaged from the transaxle side gear.

➡Inner tripod joint removal is easier by applying outward pressure on the joint while hitting the punch with a hammer.

8. Remove the inner tripod joints from the transaxle side gears using a punch to dislodge the inner tripod joint retaining ring from the transaxle side gear. If removing the right side inner tripod joint, position the punch against the inner tripod joint. Hit the punch sharply with a hammer to dislodge the right inner joint from the side gear. If removing the left side inner tripod joint, position the punch in the groove of the inner tripod joint. Hit the punch sharply with a hammer to dislodge the left inner tripod joint from the side gear.
9. Hold the inner tripod joint and interconnecting shaft of the driveshaft assembly. Remove the inner tripod joint from the transaxle by pulling it straight out of the transaxle side gear and transaxle oil seal. When removing the tripod joint, do not let the spline or snapring drag across the sealing lip of the transaxle-to-tripod joint oil seal.

WARNING

The driveshaft, when installed, acts as a bolt which secures the front hub and bearing assembly. If the vehicle is to be supported or moved on its wheels with a driveshaft removed, install a proper-sized bolt and nut through the front hub. Tighten the bolt and nut to 135 ft. lbs. (183 Nm). This will ensure that the hub bearing cannot loosen.

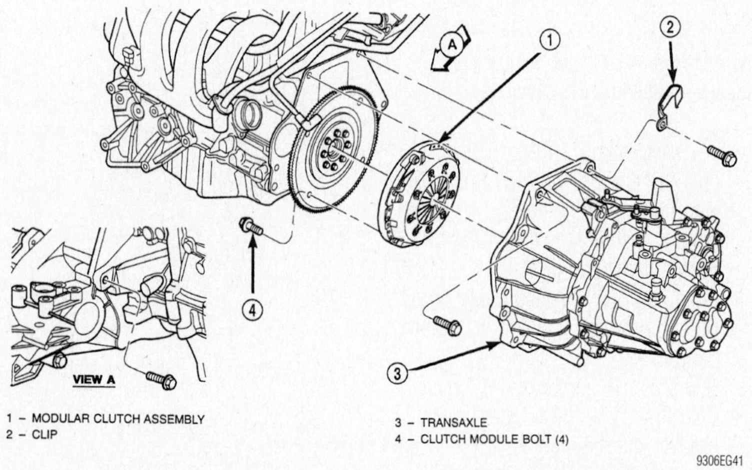

Exploded view of the manual transaxle, clutch module and related components

To install:

10. Thoroughly clean the spline and oil seal sealing surface on the tripod joint. Lightly lubricate the oil seal sealing surface on the tripod joint with fresh, clean transmission fluid.

11. Holding the driveshaft assembly by the tripod joint and interconnecting shaft, install the tripod joint into the transaxle side gear as far as possible by hand.

12. Carefully align the tripod joint with the transaxle side gears. Then, grasp the driveshaft interconnecting shaft and push the tripod joint into the transaxle side gear until fully seated. Be sure the snapring is fully engaged with the side gear by trying to remove the tripod joint from the transaxle by hand. If the snapring is fully seated with the side gear, the tripod joint will not be removable by hand.

13. Clean all debris and moisture out of the steering knuckle.

14. Be sure that the outer CV-joint, which fits into the steering knuckle, has no debris or moisture on it before installing into the steering knuckle.

15. Slide the driveshaft back into the front hub. Install the steering knuckle into the ball joint stud.

16. Install a NEW steering knuckle-to-ball joint stud bolt and nut. Tighten the nut and bolt to 70 ft. lbs. (95 Nm).

17. Insert the tie rod end into the steering knuckle. Start the tie rod end-to-steering knuckle nut onto the stud of the tie rod end. While holding the stud of the tie rod end stationary, tighten the nut. Then, using a crow's foot and 11/32 in socket, tighten the tie rod end nut to 45 ft. lbs. (61 Nm).

18. Install the rotor back onto the hub and bearing assembly.

19. Position the caliper on the steering knuckle. Slide the top of the caliper under the top abutment on the steering knuckle, then install the bottom of the caliper against the bottom abutment of the steering knuckle.

20. Install the caliper-to-knuckle bolts and tighten to 23 ft. lbs. (31 Nm).

21. On vehicles equipped with ABS, connect the front wheel speed sensor.

22. Clean all foreign matter from the threads of the outer CV-joint stub axle. Install hub nut and washer onto the threads of the stub axle and tighten the nut.

23. With the vehicle's brakes applied to prevent the axle shaft from turning, tighten the hub nut to 180 ft. lbs. (244 Nm).

24. Install the front wheel and tire assembly. Install the lug nuts and tighten to 100 ft. lbs. (135 Nm).

25. Check the transaxle fluid level, lowering the vehicle as necessary.

Vehicles with 2.4L Engines

1. Before servicing the vehicle, refer to the precautions in the beginning of this section.

2. Disconnect the negative battery cable.

3. With the vehicle on the ground and brakes applied, loosen, but do not remove the stub axle-to-hub and bearing retaining nut.

➡The front hub and driveshaft are splined together and retained by the hub nut.

4. If equipped with Anti-lock Brake System (ABS), disconnect the front wheel speed sensor.

5. Remove or disconnect the following:
   - Front wheel
   - Cotter pin, nut lock, and spring washer, and hub nut from the end of the outer CV-joint stub axle
   - Front wheel speed sensor and secure harness out of the way, if equipped with Anti-lock Brake System (ABS)
   - Nut and bolt securing the ball joint stud into steering knuckle

⁂ WARNING

Be careful when separating the ball joint stud from the steering knuckle to prevent damaging the components.

   - Ball joint stud from steering knuckle by prying down on the lower control arm

⁂ WARNING

Be careful not to separate the inner CV-joint during this operation. Do not let the halfshaft hang by inner CV-joint, it MUST be supported.

   - Halfshaft from the steering knuckle by pulling outward on knuckle while pressing in on the halfshaft. Support the outer end of halfshaft assembly.

➡If you are encountering difficulty separating the halfshaft from the hub, use puller 1026, do NOT hit the shaft with a hammer.

6. Support the outer end of the halfshaft assembly.

➡When the left halfshaft is removed from transaxle, some fluid may leak out.

➡It is easier to remove the inner tripod joint if you apply outward pressure on the joint as you strike the punch with a hammer. Do not pull on the interconnecting shaft to remove, as the inner joint will become separated.

7. Remove or disconnect the following:
   - Left halfshaft by applying outward pressure on joint by hand, dislodge the inner tripod joint from the differential side gear by striking outward with a punch. When removing the tripod joint and halfshaft, do not let the spline or snapring drag across axle seal.
   - Right halfshaft by sliding the inner tripod joint off the intermediate shaft. If necessary, use a punch to dislodge the joint from the intermediate shaft.
   - Intermediate shaft, if necessary, by removing the 2 intermediate shaft bearing-to-bracket bolts

To install:

⁂ WARNING

The driveshaft, when installed, acts as a bolt which secures the front hub and bearing assembly. If the vehicle is to be supported or moved on its wheels with a driveshaft removed, install a proper-sized bolt and nut through the front hub. Tighten the bolt and nut to 135 ft. lbs. (183 Nm). This will ensure that the hub bearing cannot loosen.

8. Install or connect the following:
   - Intermediate shaft/bearing, if removed. Torque the bearing-to-bracket bolts to 40 ft. lbs. (54 Nm).

9. On the left halfshaft, thoroughly clean the spline and oil seal sealing surface on the left tripod joint. Lightly lubricate the oil seal sealing surface on the tripod joint with fresh, clean transmission fluid.

10. Holding the driveshaft assembly by the tripod joint and interconnecting shaft, install the tripod joint into the transaxle side gear as far as possible by hand.

11. Carefully align the tripod joint with the transaxle side gears. Then, grasp the driveshaft interconnecting shaft and push the tripod joint into the transaxle side gear until fully seated. Be sure the snapring is fully engaged with the side gear by trying to remove the tripod joint from the transaxle by hand. If the snapring is fully seated with the side gear, the tripod joint will not be removable by hand.

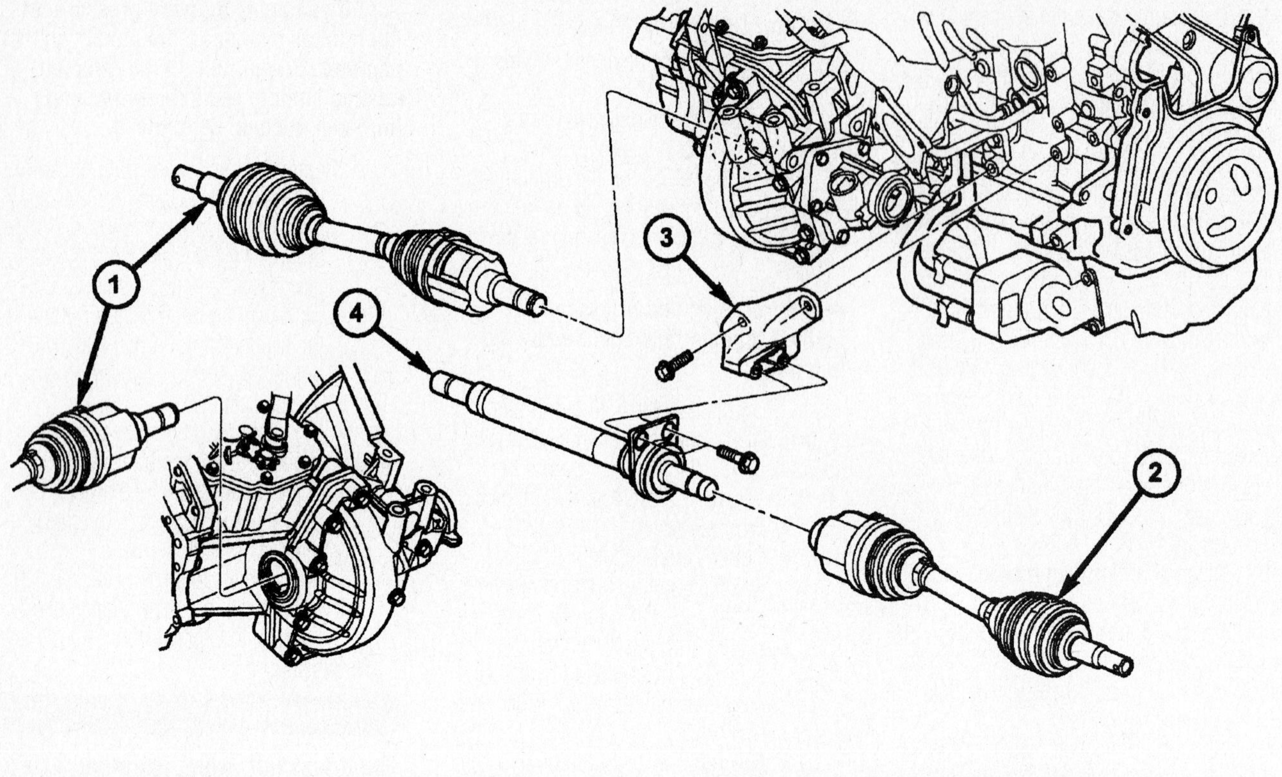

Exploded view of the halfshaft assemblies—2.4L engines

12. On the right halfshaft, thoroughly clean the right halfshaft tripod joint spline, and the intermediate shaft spline. While holding the halfshaft assembly by the tripod joint and interconnecting shaft, install tripod joint onto the intermediate shaft as far as possible by hand.

13. Slide the halfshaft back into front hub. Install the steering knuckle onto the ball joint stud

➡At this point, the outer joint will not seat completely into the front hub. The outer joint will be pulled into hub and seated when the hub nut is installed and torqued.

14. Install or connect the following:
- New steering knuckle-to-ball joint stud bolt and nut. Tighten to 70 ft. lbs. (95 Nm).

15. On vehicles equipped with ABS, connect the front wheel speed sensor.

16. Clean all foreign matter from the threads of the outer CV-joint stub axle. Install hub nut and washer onto the threads of the stub axle and tighten the nut.

17. With the vehicle's brakes applied to prevent the axle shaft from turning, tighten the hub nut to 180 ft. lbs. (244 Nm).
- Spring washer, nut lock, and cotter pin

18. Install the front wheel and tire assembly. Install the lug nuts and tighten to 100 ft. lbs. (135 Nm).

19. Check the transaxle fluid level, lowering the vehicle as necessary.

CV-Joints

OVERHAUL

Tri-Pod (Inner) Joint

1. Remove or disassemble the following:
- Halfshaft and place it in a soft-jawed vise
- Tri-pod joint boot clamps and the boot down the shaft

⁂ WARNING

When removing the spider joint, hold the rollers in place on the trunnions to keep the rollers and needle bearings in place.

- Spider/shaft assembly from the tri-pod housing
- Snapring from the shaft
- Spider assembly

➡If necessary, tap the spider assembly from the shaft with a brass drift.

⁂ WARNING

When removing the spider assembly, do not hit the outer bearings.

- Boot by sliding it off the shaft

⁂ WARNING

If any parts show excessive wear, replace the halfshaft assembly; the component parts are not serviceable.

To assemble:

⁂ WARNING

The Tri-pod sealing boots are made of 2 different types of material; silicon rubber (high temperature) which is soft and pliable or hytrel plastic (standard temperature) which is stiff and rigid. Be sure to replace the boot made of the correct material.

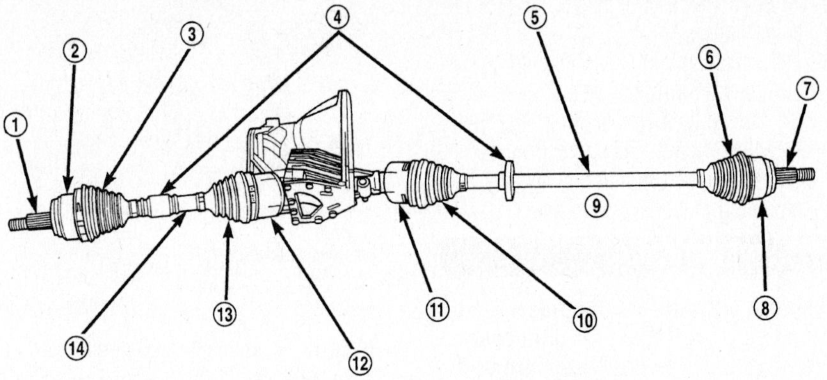

1 - STUB AXLE
2 - OUTER C/V JOINT
3 - OUTER C/V JOINT BOOT
4 - TUNED RUBBER DAMPER WEIGHT
5 - INTERCONNECTING SHAFT
6 - OUTER C/V JOINT BOOT
7 - STUB AXLE
8 - OUTER C/V JOINT
9 - RIGHT DRIVESHAFT
10 - INNER TRIPOD JOINT BOOT
11 - INNER TRIPOD JOINT
12 - INNER TRIPOD JOINT
13 - INNER TRIPOD JOINT BOOT
14 - INTERCONNECTING SHAFT LEFT DRIVESHAFT

9306EG02

View of the halfshaft assemblies—Typical

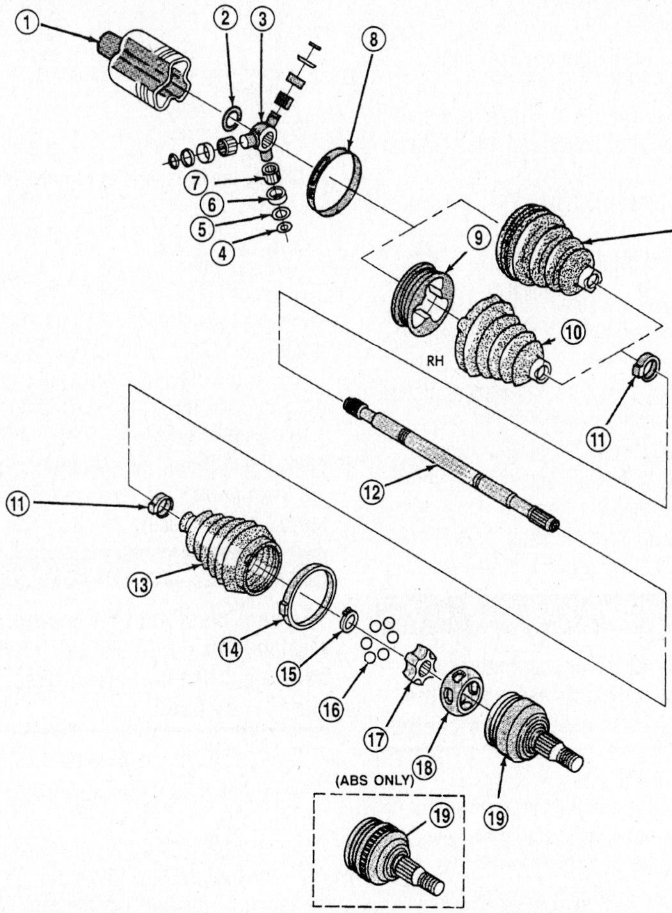

1 - HOUSING ASM, RETAINER &
2 - RING, SPACER
3 - SPIDER, TRIPOTD JOINT
4 - RING, RETAINING
5 - RETAINER, BALL & ROLLER
6 - BALL, TRIPOD JOINT
7 - ROLLER, NEEDLE
8 - CLAMP, SEAL RETAINING
9 - BUSHING, TRILOBAL TRIPOD
10 - SEAL, DRIVE AXLE INBOARD
11 - CLAMP, SEAL RETAINING
12 - SHAFT, AXLE (RH SHOWN, LH SIMILAR)
13 - SEAL, DRIVE AXLE OUTBOARD
14 - CLAMP, SEAL RETAINING
15 - RING, RACE RETAINING
16 - BALL, CHROME ALLOY
17 - RACE, C/V JOINT INNER
18 - CAGE, C/V JOINT
19 - RACE, C/V JOINT OUTER

9306EG03

Exploded view of the halfshaft assembly—Typical

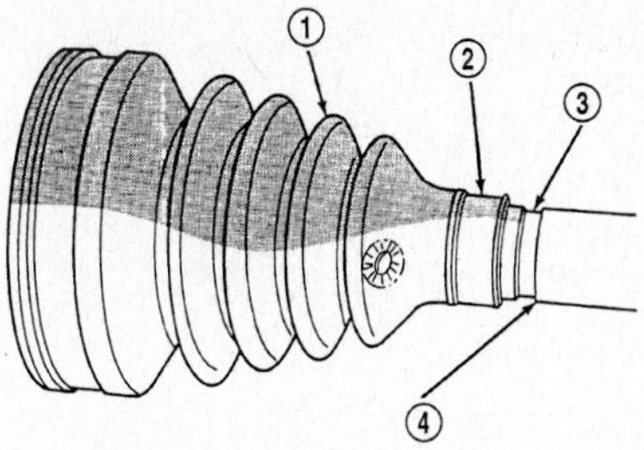

1 - SEALING BOOT
2 - RAISED BEAD IN THIS AREA OF SEALING BOOT
3 - GROOVE
4 - INTERCONNECTING SHAFT

9306EG04

Positioning the boot on the shaft—Tri-pod and Rzeppa joints

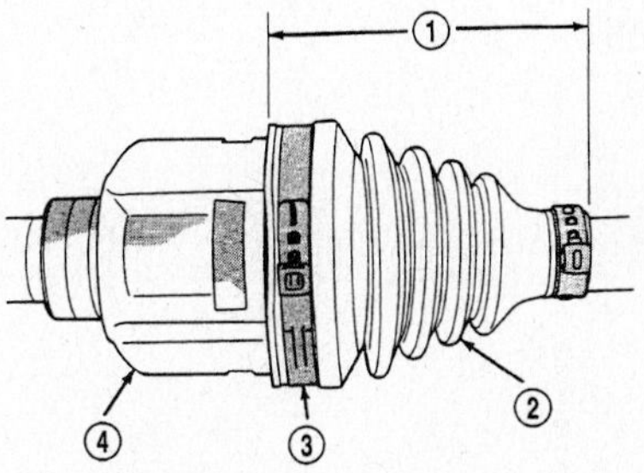

1 - 107 MILLIMETERS
2 - HYTREL SEALING BOOT
3 - SEALING BOOT CLAMP
4 - INNER TRIPOD JOINT

9306EG05

Adjusting the Tri-pod boot length—Hytrel plastic boot

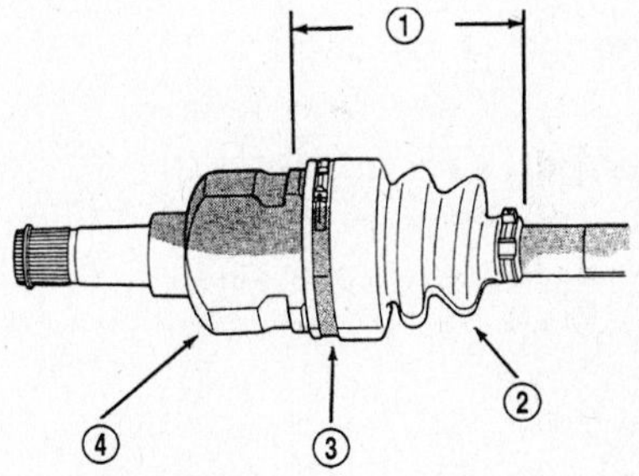

1 - 115 MILLIMETERS
2 - SILICONE SEALING BOOT
3 - CLAMP
4 - INNER TRIPOD JOINT

9306EG06

Adjusting the Tri-pod boot length—Silicone rubber boot

2. Install a new small boot clamp and slide it on the shaft.
3. Position the boot so that the raised bead on the inside the boot seal is in the shaft groove.
4. If installing a new boot, distribute ½ of the grease in the service package inside the tri-pod housing and the other ½ inside the boot.
5. Install or connect the following:
   - Spider assembly, face the chamfered side toward the shaft

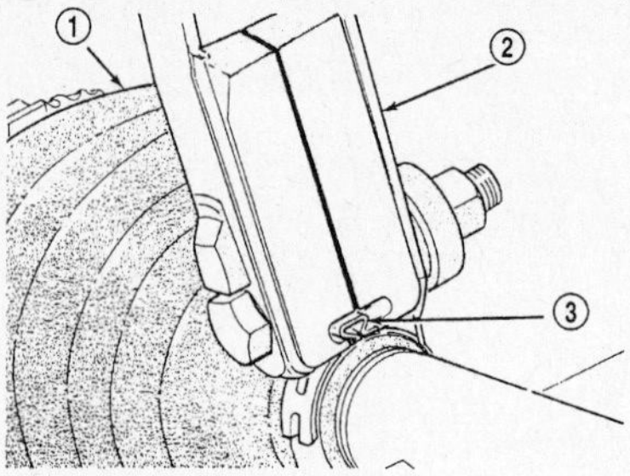

1 - SEALING BOOT
2 - SPECIAL TOOL C-4975
3 - CLAMP BRIDGE

9306BG13

Tightening the high profile boot clamp—Tri-pod joint (Hytrel plastic) boot and Rzeppa joint boots

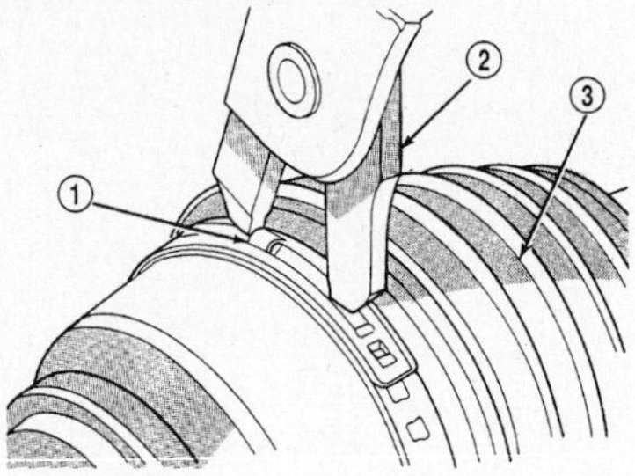

1 - CLAMP
2 - SPECIAL TOOL YA3050
3 - SEALING BOOT

9306EG07

Tightening the low profile boot clamp—Silicone rubber Tri-pod boot

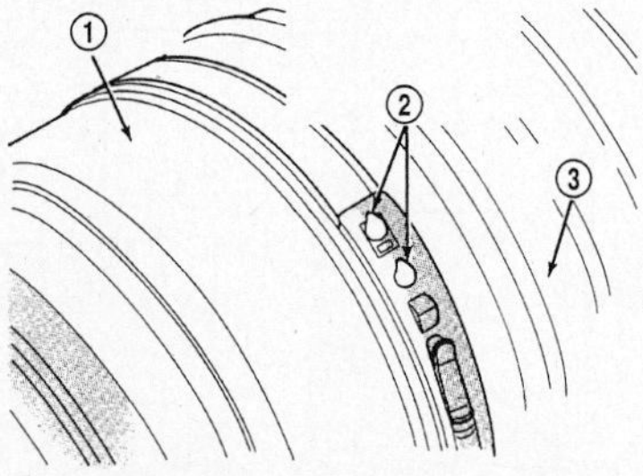

1 - INNER TRIPOD JOINT HOUSING
2 - TOP BANK OF CLAMP MUST BE RETAINED BY TABS AS SHOWN HERE TO CORRECTLY LATCH BOOT CLAMP
3 - SEALING BOOT

9306EG08

Secured the low profile boot clamp—Silicone rubber Tri-pod boot

WARNING

If necessary, tap the spider assembly onto the shaft using a brass drift; be careful not to hit the outer bearings.

- Snapring making sure it is fully seated in the groove
- Spider/shaft assembly into the tri-pod housing
- New inner boot clamp and position it evenly on the sealing boot

6. Using a trim stick, adjust the boot length to 107mm (hytrel plastic) or 115mm (silicone rubber).
7. If installing a high profile boot clamp, perform the following procedure:
    a. Using the Crimper Tool C-4975-A, place the tool over the clamp bridge, tighten the tool nut until the jaws are completely closed (face-to-face).

WARNING

The seal must not be dimpled, stretched or out of shape. If necessary, equalize the seal pressure and shape it by hand.

    b. Position the boot onto the tri-pod housing retaining groove and install the retaining clamp evenly on the boot.
    c. Using the Crimper Tool C-4975-A, place the tool over the clamp bridge, tighten the tool nut until the jaws are completely closed (face-to-face).
8. If installing a low profile latching type boot clamp, position Snap-On® Clamp Locking Tool YA3050 prongs in the clamp holes and squeeze the tool until the upper clamp band is latched behind the 2 tabs on the lower clamp band.
9. Install the halfshaft.

Rzeppa (Outer) Joint

1. Remove or disassemble the following:
    - Halfshaft and place it in a soft-jawed vise
    - Rzeppa joint boot clamps and slide the boot down the shaft
    - Rzeppa joint housing, by sharply hitting it with a soft-faced hammer to drive it off the shaft
    - Circlip from the shaft
    - Boot by sliding it off the shaft

WARNING

If any parts show excessive wear, replace the halfshaft assembly; the component parts are not serviceable.

To assemble:

2. Install or assemble the following:
    - New small boot clamp and slide it onto the shaft
    - Boot and slide it onto the shaft
    - Circlip, if removed
3. Position the boot so that the raised bead on the inside the boot seal is in the shaft groove.
    - Halfshaft hub nut onto the Rzeppa joint threaded shaft so it is flush with the end
    - Rzeppa joint, align the shaft splines and tap it onto the shaft with a soft-faced hammer so it locks on the circlip

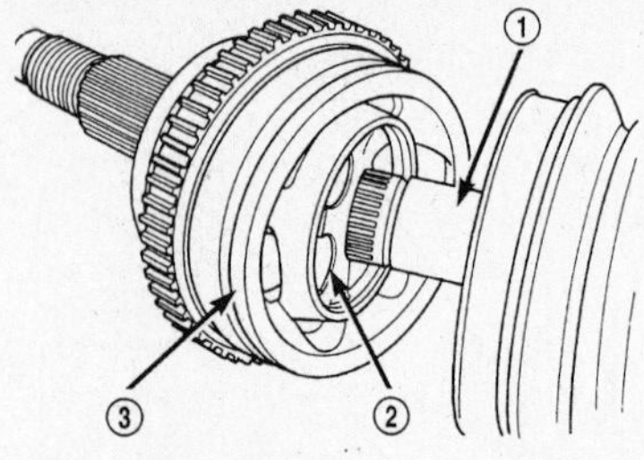

1 - INTERCONNECTING SHAFT
2 - CROSS
3 - OUTER C/V JOINT ASSEMBLY

9306EG09

Aligning the cross splines with the shaft splines—Rzeppa joint

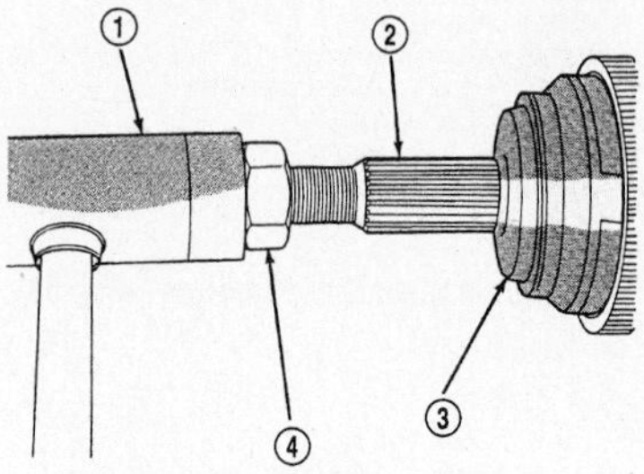

1 - SOFT FACED HAMMER
2 - STUB AXLE
3 - OUTER C/V JOINT
4 - NUT

9306EG10

Driving the Rzeppa joint onto the shaft

4. Distribute ½ of the grease in the service package inside the Rzeppa joint housing and the other ½ inside the boot.
5. Install the new small boot clamp and position it evenly on the sealing boot
6. Using the Crimper Tool C-4975-A, place the tool over the clamp bridge, tighten the tool nut until the jaws are completely closed (face-to-face).

WARNING

The seal must not be dimpled, stretched or out of shape. If necessary, equalize the seal pressure and shape it by hand.

7. Position the boot onto the Rzeppa housing retaining groove and install the retaining clamp evenly on the boot.
8. Using the Crimper Tool C-4975-A, place the tool over the clamp bridge, tighten the tool nut until the jaws are completely closed (face-to-face).

WARNING

The seal must not be dimpled, stretched or out of shape. If necessary, equalize the seal pressure and shape it by hand.

9. Install the halfshaft.

STEERING AND SUSPENSION

Air Bag

** CAUTION

Some vehicles are equipped with an air bag system. The system MUST BE disabled before performing service on or around system components, steering column, instrument panel components, wiring and sensors. Failure to follow safety and disabling procedures could result in accidental air bag deployment, possible personal injury and unnecessary system repairs.

PRECAUTIONS

Several precautions must be observed when handling the inflator module to avoid accidental deployment and possible personal injury:

- Never carry the inflator module by the wires or connector on the underside of the module.
- When carrying a live inflator module, hold securely with both hands, and ensure that the bag and trim cover are pointed away.
- Place the inflator module on a bench or other surface with the bag and trim cover facing up.
- With the inflator module on the bench, never place anything on or close to the module which may be thrown in the event of an accidental deployment.

DISARMING

Proper Supplemental Restraint System (SRS) disarming can be obtained by disconnecting and isolating the negative battery cable. Allow the air bag system capacitor at least 2 minutes to discharge before removing any air bag system components.

Rack and Pinion Steering Gear

These vehicles are designed and assembled using NET BUILD front suspension alignment settings. This means that the alignment settings are determined as the vehicle is designed by the location of the front suspension components in relation to the body. This is carried out when building the vehicle, by precisely locating the front crossmember to meter gauge holes located in the underbody of the vehicle. With this method of designing and building a vehicle, it is no longer possible to adjust a vehicle's front suspension alignment settings to the required specifications. As a result, whenever the crossmember is removed from a vehicle, it MUST be replaced in the same location on the body of the vehicle it was removed from. The front suspension toe settings can still be adjusted by the outer tie rod ends.

REMOVAL & INSTALLATION

1. Before servicing the vehicle, refer to the precautions in the beginning of this section.
2. Place the steering wheel in the straight-ahead position. Lock the steering wheel in place, using a steering wheel holder.

➡Locking the steering wheel keeps the clockspring in alignment position.

3. Working inside the passenger compartment, remove the steering column coupling retainer pin, pinch bolt nut/bolt and separate the couplings.
4. Remove or disconnect the following:
   - Both front wheels
   - Outer tie rods-to-steering knuckle nuts

➡Remove the tie rod nut by holding the stud securely.

   - Outer tie rod ends from the steering knuckle using Remover Tool MB991113
   - Tie rod heat shield
   - Power steering fluid pressure switch wiring connector by releasing the locking tab
   - Power steering fluid pressure hose from the steering gear
   - Power steering fluid return hose from the steering gear, if not equipped with a power steering fluid cooler
   - Power steering fluid cooler hose from the steering gear, if equipped with a power steering fluid cooler
   - Power steering fluid return hose from the routing clip C-clamps, if not equipped with a power steering fluid cooler
   - Power steering fluid pressure hose from the steering gear's routing clips
   - Power steering cooler hose from the steering gear's right routing clip, if equipped
   - Both power steering cooler screws from the front suspension crossmember, if equipped, and move the cooler aside

➡The screws are located behind the cooler and can be accessed from above.

   - Power steering cooler, if equipped, and move it aside
   - Engine torque strut-to-front suspension crossmember bolt from the right forward corner of the crossmember

5. Matchmark the front suspension crossmember-to-chassis location.

** WARNING

If the front suspension crossmember-to-chassis location is not matchmarked, the front wheel alignment setting will be lost.

6. Place a transmission jack under the front crossmember and support it.
7. Remove both front suspension crossmember-to-frame rail bolts, one located at each side.
8. Loosen both rear suspension crossmember-to-frame rail bolts, one located at each side, until they release from the threaded tapping plates in the bolt.

** WARNING

Do not completely remove the rear bolts for they are designed to disengage from the body threads and will stay within the lower control arm rear isolator bushing.

➡The threaded tapping plates allow the lower control arm to stay in place on the crossmember.

9. Using the transmission jack, lower the front suspension crossmember enough to allow the power steering gear to be removed form the rear of the crossmember. Use the jack to support the crossmember's weight.
10. Remove or disconnect the following:
   - Lower steering column coupling-to-power steering gear pinion shaft's roll pin, using a roll pin punch
   - Lower steering column coupling from the power steering column pinion shaft

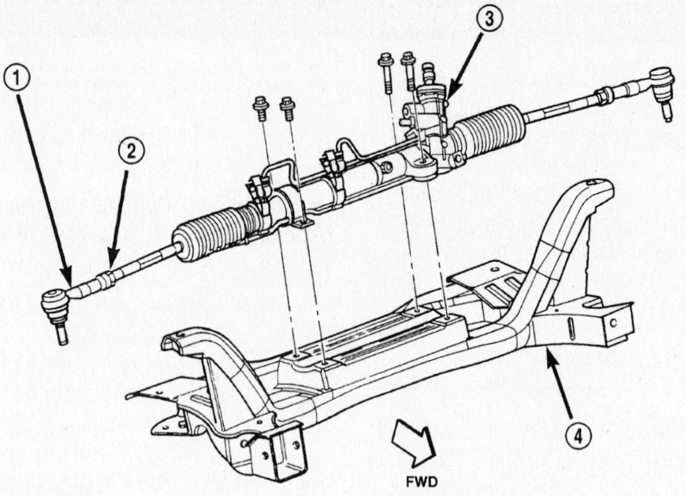

1 - OUTER TIE ROD
2 - JAM NUT
3 - STEERING GEAR
4 - FRONT SUSPENSION CROSSMEMBER

9306EG42

View of the power steering gear and crossmember

- Pinion shaft dash cover seal from the tabs cast into the power steering gear housing
- Power steering gear from the front suspension crossmember

To install:

11. Install or connect the following:
    - Power steering gear onto the front suspension crossmember and torque the bolts to 45 ft. lbs. (61 Nm)
    - Pinion shaft dash cover seal over the shaft and onto the power steering gear housing

➡Align the seal holes with the tabs cast into the power steering gear housing.

- Lower steering column coupling by aligning the coupling and steering gear pinion shaft flats

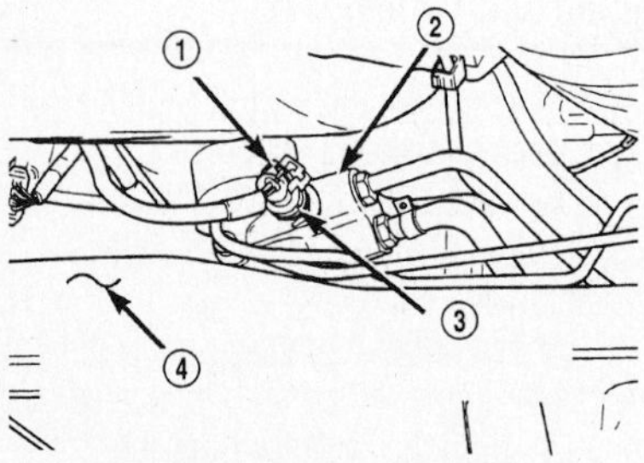

1 - WIRING HARNESS CONNECTOR
2 - POWER STEERING GEAR
3 - POWER STEERING FLUID PRESSURE SWITCH
4 - REAR OF FRONT SUSPENSION CROSSMEMBER

9306EG43

View of the power steering gear fluid pressure switch

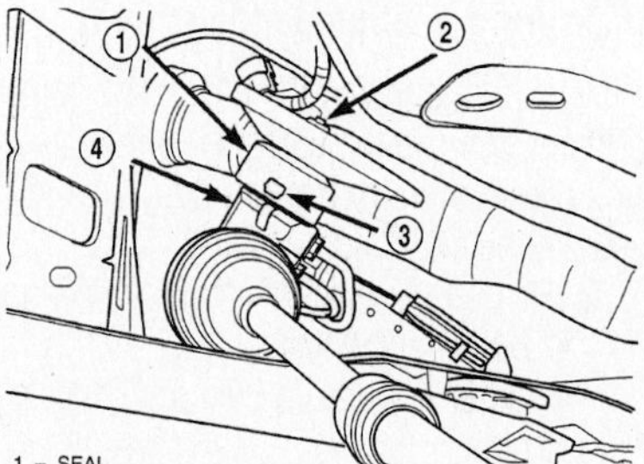

1 - SEAL
2 - PINION SHAFT
3 - TAB
4 - POWER STEERING GEAR

9306EG45

View of the pinion shaft dash cover seal

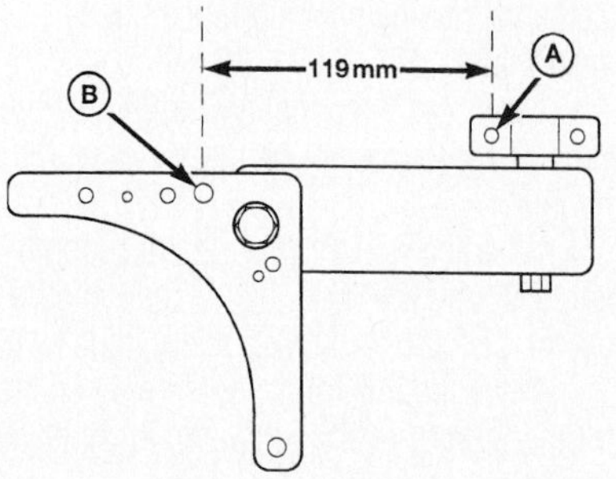

9306EG47

Measuring the engine torque bracket

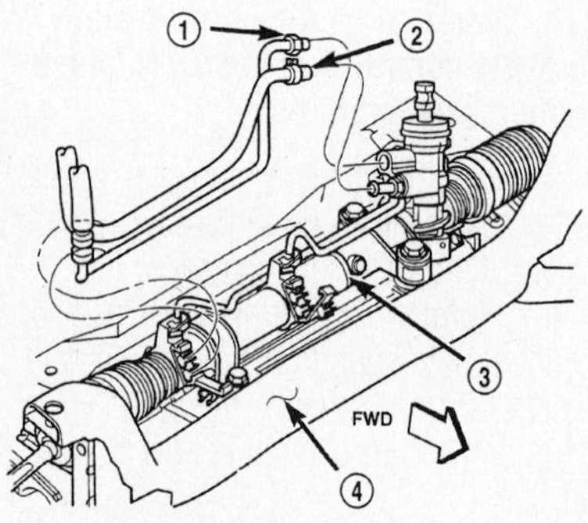

1 - PRESSURE HOSE TUBE NUT
2 - RETURN HOSE
3 - POWER STEERING GEAR
4 - FRONT SUSPENSION CROSSMEMBER

9306EG44

View of the power steering gear hoses and routing clips

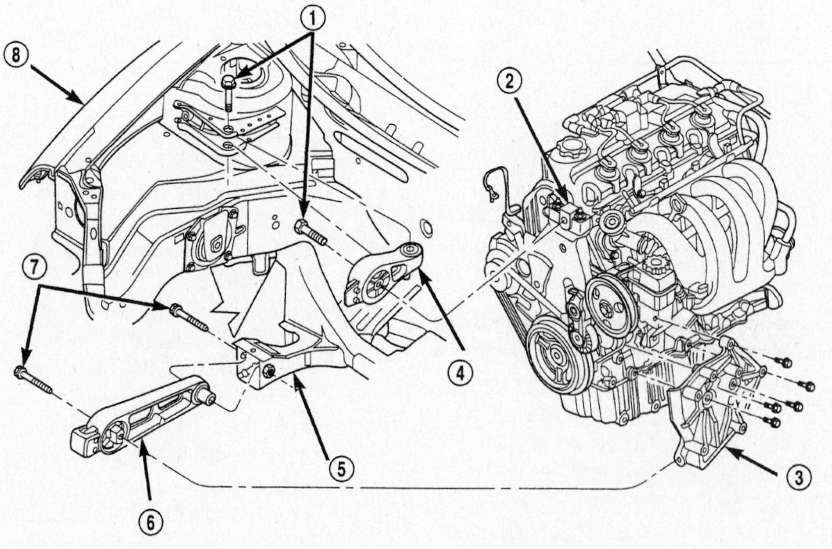

1 - BOLTS
2 - ENGINE MOUNT BRACKET
3 - TORQUE STRUT BRACKET
4 - UPPER TORQUE STRUT
5 - CROSSMEMBER
6 - LOWER TORQUE STRUT
7 - BOLTS
8 - RIGHT FENDER

9306EG46

View of the engine torque struts and related components

- Lower steering column coupling-to-pinion shaft's roll pin until it is centered

12. Center the power steering gear rack's travel.

- Front suspension crossmember/power steering gear assembly by raising it with the jack until is aligns with its matchmarks
- Lower steering column coupling, guide it through the dash panel hole as it is raised
- Both rear crossmember-to-tapping plate bolts
- Both front crossmember-to-frame rail bolts and torque them to 20 inch lbs. (2 Nm)

⁂ WARNING

Be sure to align the front suspension crossmember-to-chassis matchmarks; otherwise, the front wheel alignment setting will be lost.

- Once aligned, torque both rear crossmember-to-rear lower control arm bolts to 150 ft. lbs. (203 Nm) and both front crossmember bolts to 105 ft. lbs. (142 Nm)
- Engine torque strut to the right forward corner of the front suspension crossmember

13. Adjust the engine torque strut by performing the following procedure:

a. Loosen the upper torque strut at the shock tower bracket.

b. Position a floor jack on the forward edge of the bell housing to prevent the least amount of upward lifting of the engine.

c. Slowly, lift the assembly, allowing the engine to rotate rearward so the distance between center of the engine mount bracket's rearmost attaching stud (point A) and the center of the shock tower bracket's washer hose clip hole (point B) is 4.70 in. (119mm).

d. Torque the upper and lower torque strut bolts to 87 ft. lbs. (118 Nm).

e. Remove the floor jack.

14. Install or connect the following:

- Power steering hose-to-power steering gear using a new O-ring lubricated with power steering oil, if not equipped with a power steering cooler
- Power steering fluid cooler line-to-power steering gear, if equipped with a power steering cooler
- Power steering fluid return hose to the routing clip C-clamps, if not equipped with a power steering fluid cooler
- Power steering fluid pressure hose to the steering gear's routing clips
- Power steering cooler hose to the steering gear's right routing clip, if equipped
- Torque the power steering pressure hose-to-power steering gear nut to 25 ft. lbs. (34 Nm)
- Both power steering cooler screws to the front suspension crossmember, if equipped
- Power steering fluid pressure switch wiring connector be sure the locking tab is secure latched
- Tie rod heat shield, facing outboard
- Outer tie rod ends to the steering knuckle. Torque the nut, using a crowsfoot wrench, to 40 ft. lbs. (55 Nm), while holding the tie rod stationary
- Both front wheels and torque the lug nuts, in a crisscross pattern, to 100 ft. lbs. (135 Nm)
- Dash-to-lower coupling seal over the lower coupling's plastic collar

➡Verify that the seal's lip shows grease at the coupling's plastic collar contact.

- Steering column lower coupling-to-steering column upper coupling pinch bolt and torque the nut to 21 ft. lbs. (28 Nm)
- Pinch bolt retainer pin

15. Remove the steering wheel holder.
16. Fill and bleed the power steering system.
17. Check for leaks.
18. Check and/or adjust the front toe setting.

Strut

REMOVAL & INSTALLATION

Front

1. Before servicing the vehicle, refer to the precautions in the beginning of this section.
2. Remove the front wheels.
3. Mark each one right or left, as applicable, if both struts are being removed.
4. Remove the screw securing the ground strap to the rear of the strut, if equipped.
5. Remove the screw securing the ABS wheel speed sensor to the back of the strut, if equipped.

⁂ WARNING

The steering knuckle-to-strut assembly attaching bolts are serrated and must not be turned during removal.

6. Remove or disconnect the following:

- Steering knuckle nuts while holding the bolts stationary

➡If necessary, partially lower the vehicle for access to the upper mounting nuts.

- 3 upper strut mount-to-strut tower nuts
- Strut assembly

To install:

7. Install or connect the following:

- Strut assembly into the strut tower by aligning the 3 upper strut mount studs with the shock tower holes. Torque the 3 upper strut mount nut/washer assemblies to 25 ft. lbs. (34 Nm).

⁂ WARNING

The steering knuckle-to-strut assembly attaching bolts are serrated and must not be turned during installation.

- Lower end of the strut, in line with the upper end of the steering knuckle and align the mounting holes.
- Both strut-to-steering knuckle bolts and nuts. Torque both nuts, while holding the bolts stationary, to 40 ft. lbs. (53 Nm), plus an additional ¼ turn after the specified torque is met

➡The bolts should be installed with the nuts facing the front of the vehicle.

- ABS wheel speed sensor to the rearward ear of the strut, and secure with its mounting screw. Tighten the screw to 10 ft. lbs. (13 Nm)
- Hydraulic brake hose routing bracket and screw onto the strut damper bracket and torque the bracket bolts to 10 ft. lbs. (13 Nm), if necessary. If equipped with ABS, the hydraulic hose routing bracket is combined with the speed sensor cable routing bracket.
- Screw securing the ground strap to the rear of the strut, if equipped.
- Front wheels and torque the lug nuts, in a criss-cross pattern, to 100 ft. lbs. (135 Nm)

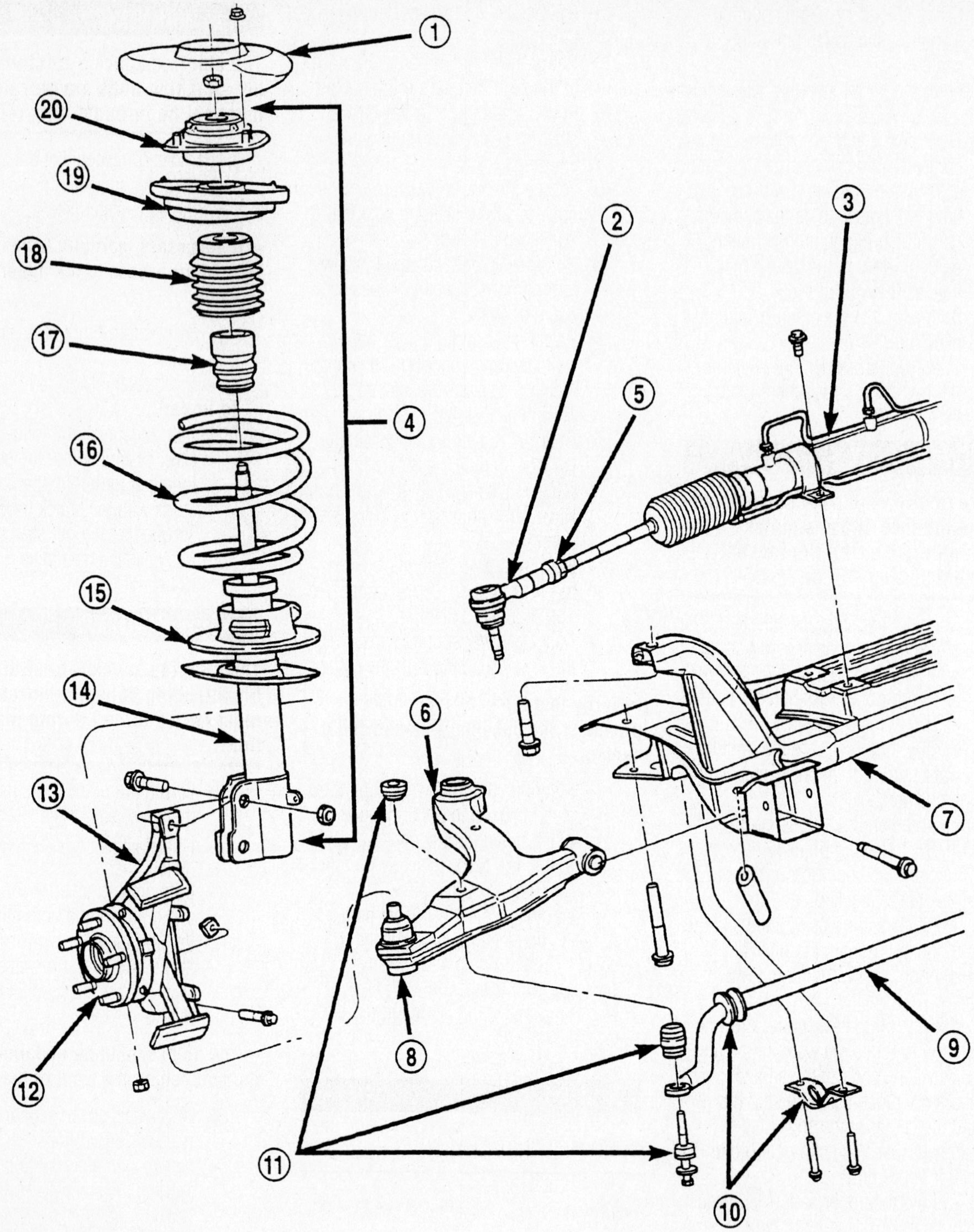

1 – VEHICLE STRUT TOWER
2 – OUTER TIE ROD
3 – STEERING GEAR
4 – STRUT ASSEMBLY
5 – JAM NUT
6 – LOWER CONTROL ARM
7 – CROSSMEMBER
8 – BALL JOINT
9 – STABILIZER BAR
10 – STABILIZER BAR CUSHION AND RETAINER
11 – STABILIZER BAR LINK
12 – HUB
13 – KNUCKLE
14 – STRUT
15 – LOWER SPRING ISOLATOR
16 – COIL SPRING
17 – JOUNCE BUMPER
18 – DUST SHIELD
19 – SPRING SEAT AND BEARING
20 – UPPER MOUNT

9306EG11

Exploded view of the front suspension

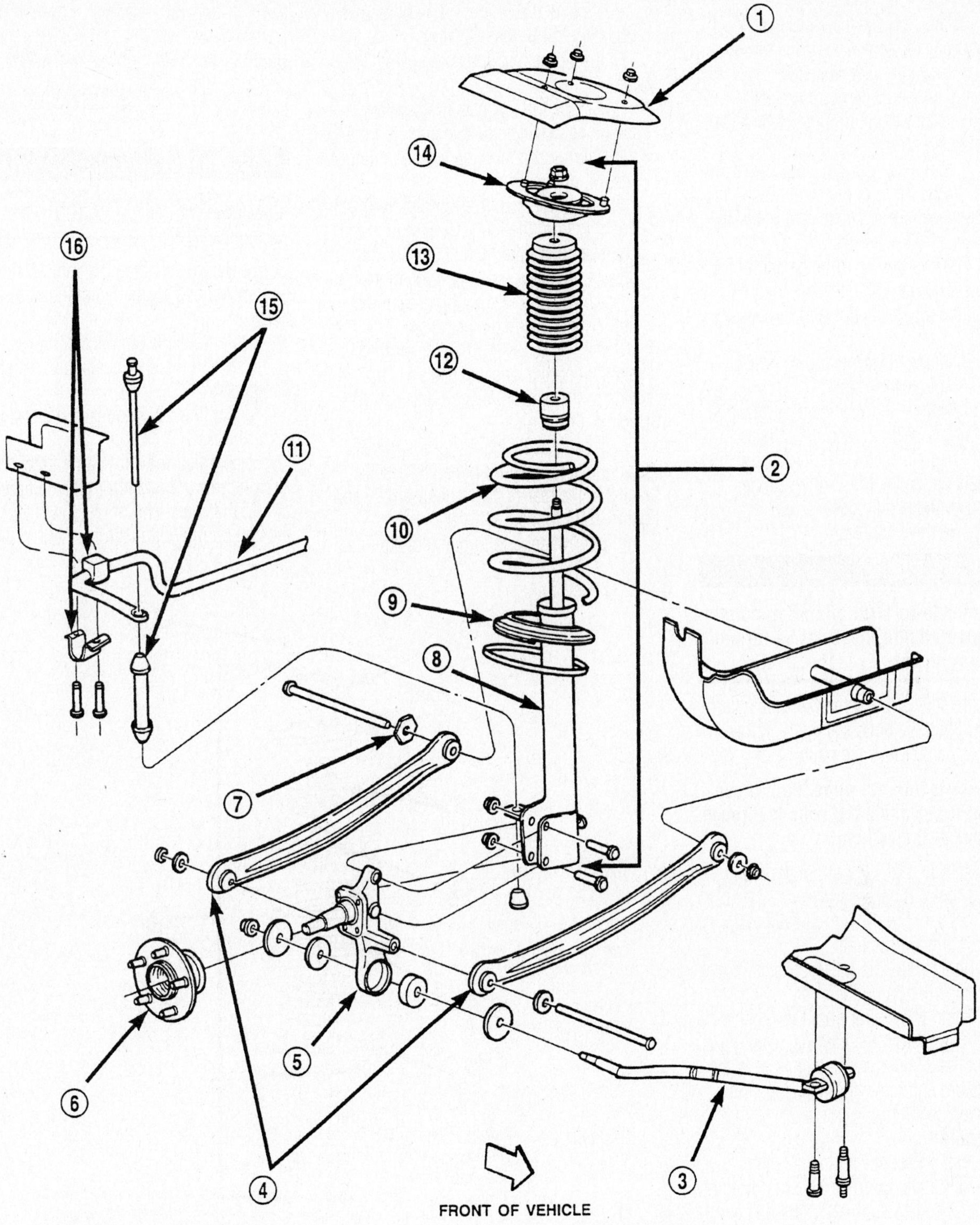

1 – VEHICLE STRUT TOWER
2 – STRUT ASSEMBLY
3 – TENSION STRUT
4 – LATERAL ARMS
5 – KNUCKLE
6 – HUB AND BEARING
7 – WHEEL ALIGNMENT ADJUSTMENT CAM
8 – STRUT
9 – LOWER SPRING ISOLATOR
10 – COIL SPRING
11 – STABILIZER BAR
12 – JOUNCE BUMPER
13 – DUST SHIELD
14 – UPPER MOUNT
15 – STABILIZER BAR LINK
16 – STABILIZER BAR CUSHION AND RETAINER

Exploded view of the rear suspension

Rear

1. Before servicing the vehicle, refer to the precautions in the beginning of this section.
2. Remove the rear wheel.
3. If equipped with drum brakes, remove the screw securing the brake hose bracket to the strut.
4. If equipped with Anti-lock Brake System (ABS), remove the screw securing the ABS wheel speed sensor bracket to the rear of the strut assembly.
5. Remove the nut from the end of the rear stabilizer bar link bolt. Pull the bolt out through the top of the link and remove the link.
6. If equipped with rear disc brakes, perform the following steps:
   a. Remove the 2 guide pins securing the caliper, then remove the caliper
   b. Use a piece of wire to suspend the caliper aside. Do NOT let the caliper hang by the brake hose.
   c. Remove the rotor.

WARNING

The knuckle-to-strut attaching bolts are serrated and must not be turned during removal.

7. Remove strut-to-rear knuckle nuts while holding the bolts stationary in the knuckle, then remove the bolts.

➡Access to the rear upper strut mount-to-strut tower attaching bolts is through the trunk of the vehicle.

8. Remove the carpet from the top of the strut tower, if necessary.
9. Loosen, but do not remove the 3 upper strut mounting nuts.
10. Remove the 3 strut-to-chassis mount nuts, while supporting the strut.
11. Remove the strut from the knuckle by sliding it away from the knuckle, lowering it between the 2 lateral arms, then tipping the top outward and removing it through the wheel well opening.

To install:

12. Install or connect the following:
    - Strut and torque the 3 strut mount-to-body nuts to 25 ft. lbs. (34 Nm)
    - Carpeting on top of the strut tower, then close the trunk lid
13. Align the holes in the strut clevis bracket on the lower end of the strut with the mounting holes in the knuckle.
    - 2 bolts attaching the strut to the rear knuckle and torque to 65 ft. lbs. (88 Nm)
14. If equipped with rear disc brakes, perform the following:
    a. Install the brake rotor and caliper.
    b. Install the caliper guide bolts and tighten to 16 ft. lbs. (22 Nm).
15. Install the stabilizer bar link, as follows:
    a. Place the link center sleeve and bushings between the eye in the end of the stabilizer bar and the link mounting bracket on the strut.
    b. Start the stabilizer bar link bolt with bushing from the top, down through the stabilizer bar, inner link bushings and sleeve, and strut link mounting bracket.
    c. Install a lower bushing, then the nut, but don't tighten the nut yet.
16. If equipped with ABS, install the screw holding the wheel speed sensor bracket to the rear of the strut and tighten to 10 ft. lbs. (13 Nm).
17. If equipped with rear drum brakes, install the screw securing the brake hose bracket to the rear of the strut assembly and tighten to 23 ft. lbs. (31 Nm).
18. Rear wheel and torque the lug nuts evenly, in sequence, to 100 ft. lbs. (135 Nm)
19. Lower the vehicle, then tighten the stabilizer bar link nut to 17 ft. lbs. (23 Nm).
20. Check the alignment and adjust, if necessary.

Coil Spring

REMOVAL AND INSTALLATION

1. Before servicing the vehicle, refer to the precautions in the beginning of this section.
2. Remove the strut assembly.
3. Install the strut assembly in a spring compressor.
4. Set the lower hooks first then the upper hooks.
5. Position the strut clevis bracket straight outward away from the compressor
6. Place a clamp on the lower end of the coil spring so the strut is held in place.

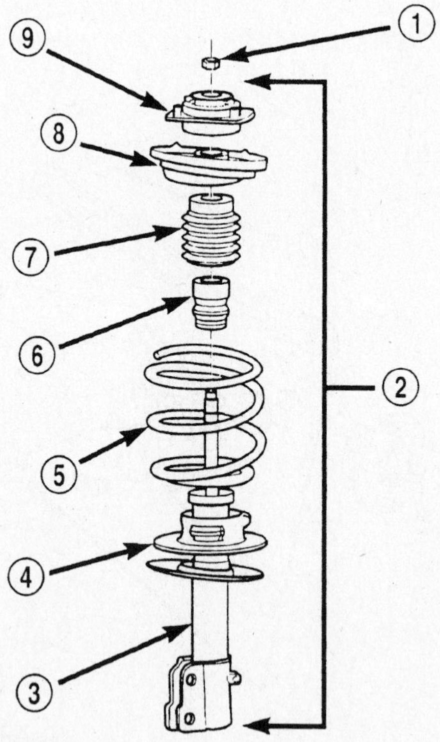

Exploded view of the strut assembly

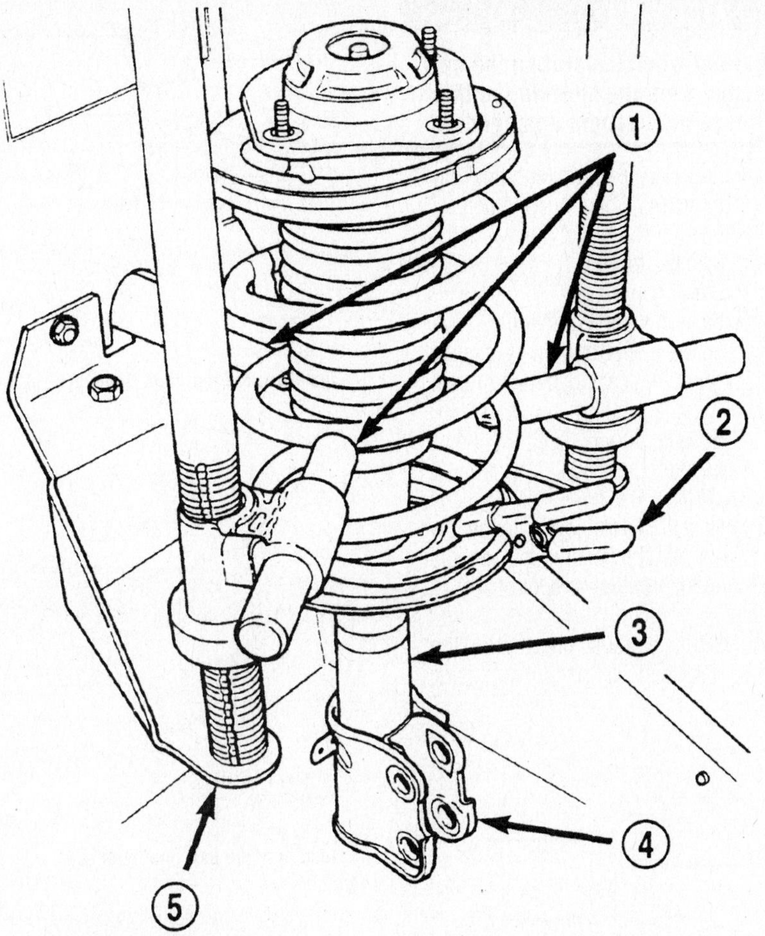

Assemble the coil spring as shown

7. Compress the spring until all tension is removed from the upper mount.

8. Install a strut nut socket, Special Tool 6864, on the shaft retaining nut.

9. Install socket on the hex end of the strut shaft.

10. Hold the shaft from turning and remove the nut from the shaft.

11. Remove or disconnect the following:
- Upper mount from the strut shaft
- Upper spring seat, bearing and upper spring isolator as an assembly
- Dust shield
- Jounce bumper
- Clamp from the bottom of the coil spring
- Strut assembly through the bottom of the coil spring
- Lower spring isolator
- Coil spring by releasing the tension on the compressor drive

To install:

12. Place the coil spring in a compressor and rotate the spring so that the end of the top coil is directly in back.

13. Slowly compress the coil spring until enough room is available for reassembly.

14. Install or connect the following:
- Lower spring isolator on the lower spring seat
- Strut through the bottom of the coil spring until the spring seat contacts the lower end of the coil spring

15. Rotate the strut, if necessary, until the clevis bracket is positioned straight outward and away from the compressor.
- Clamp on the lower end of the coil spring and strut
- Jounce bumper on the strut shaft with the small end facing down
- Dust shield on the strut shaft
- Upper spring isolator on the upper spring seat and bearing
- Upper spring seat and bearing on the top of the coil spring
- Strut upper mount over the strut shaft and on top of the upper spring seat and bearing
- Retaining nut on the strut shaft, loosely
- Strut nut socket Tool 6864 on the retaining nut
- Socket on the hex end of the shaft and torque the nut to 55 ft. lbs. (75 Nm)

16. Slowly release the tension from the from the coil spring and make certain that the coil spring is properly aligned.

17. Remove the clamp from the lower end of the coil spring and strut.

18. Remove the upper and lower hooks from the compressor tool and remove the tool.

19. Install the strut assembly.

Tension Strut

REMOVAL & INSTALLATION

1. Remove or disconnect the following:
- Rear wheels
- Tension strut-to-knuckle nut by holding the strut with a wrench
- Forward tension strut retainer
- Rear tension strut bayonet bushing
- Parking brake cable from tension strut bolt
- Tension strut from the chassis

To install:

2. Install or connect the following:
- Tension strut to the chassis and torque the bolts to 70 ft. lbs. (95 Nm)
- Parking brake cable to tension strut nut and torque the bolts to 21 ft. lbs. (28 Nm)

➡The mounting bolt with the stud on the head is installed on the inboard side.

- Rear tension strut bayonet bushing making sure that the stepped area faces toward the knuckle
- Tension strut retainer
- Tension strut-to-knuckle nut by holding the strut with a wrench and torque the nut to 70 ft. lbs. (95 Nm)

- Rear wheels and torque the lug nuts to 100 ft. lbs. (135 Nm)

3. Check and/or align the rear wheels.

Lower Ball Joint

REMOVAL & INSTALLATION

The front suspension ball joints operate with no free-play. The ball joints are replaceable ONLY as an assembly. Do not attempt any type of repair on the ball joint assembly. The ball joint is a press fit into the lower control arm with the joint stud retained in the steering knuckle by the clamp bolt. To check the ball joint, with the weight of the vehicle resting on the road wheels, grasp the grease fitting and without using any tools, attempt to move the grease fitting. If the ball joint is worn the grease fitting will move easily. If movement is noted, replacement of the ball joint is recommended.

1. Before servicing the vehicle, refer to the precautions in the beginning of this section.
2. Remove or disconnect the following:
   - Wheel
   - Steering knuckle-to-ball joint stud's pinch bolt and nut
   - Stabilizer bar-to-lower control arm links
3. Loosen, but do not remove the bolts holding the stabilizer bar retainers to the crossmember. Then, rotate the stabilizer bar and attaching links away from the lower control arms.

**** WARNING**

Pulling the steering knuckle outward after releasing the ball joint can separate the inner CV-joint.

4. Remove or disconnect the following:
   - Ball joint from the steering knuckle using a prybar

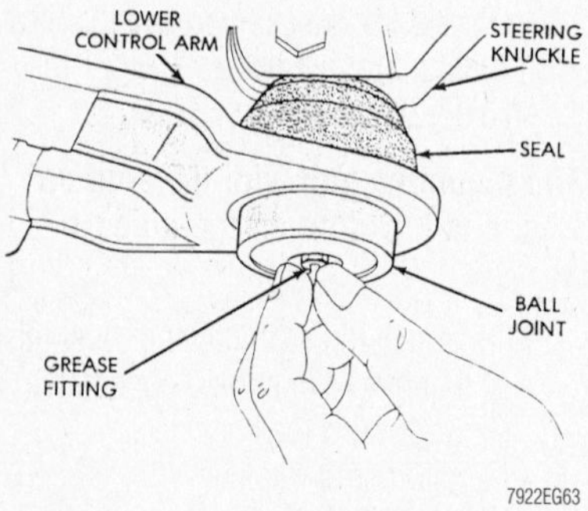

Wiggle the grease fitting with your fingers—if it moves, the ball joint should be replaced

**** WARNING**

Be careful when separating the ball joint stud from the knuckle, so the seal does not become damaged.

   - Front lower control arm bushing-to-crossmember nut and bolt
   - Rear lower control arm-to-crossmember bolt
   - Lower control arm
   - Ball joint using a prytool
5. Using a hydraulic press, press the ball joint from the lower control arm using tools:
   - Receiver Tool 6908-2
   - Adapter Tool 6804

To install:

6. Reinstall the ball joint into the lower control arm with the notch in the ball joint stud facing the front lower control arm bushing.
7. Using a hydraulic press, press the ball joint into the lower control arm using tools:

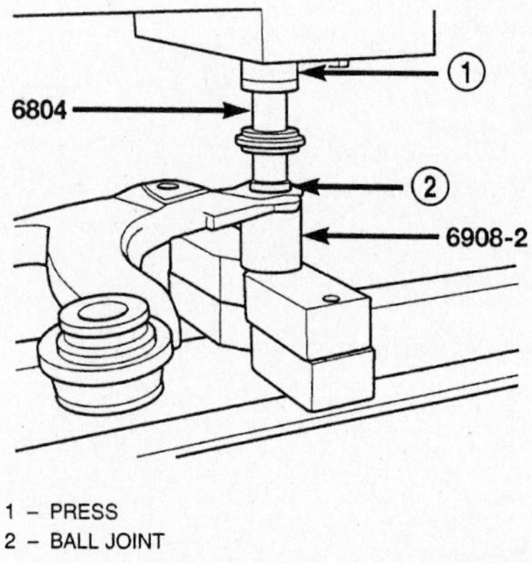

Removing the ball joint from the control arms

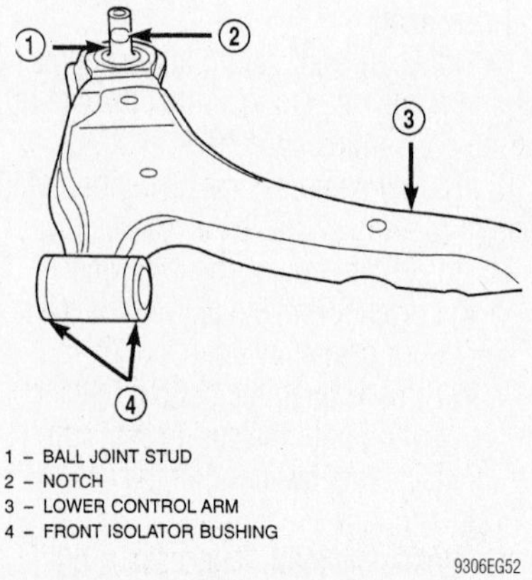

Aligning the ball joint stud notch to the control arms

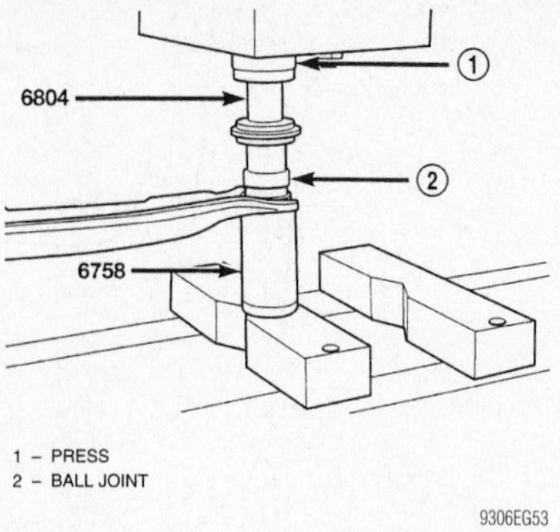

Installing the ball joint to the control arms

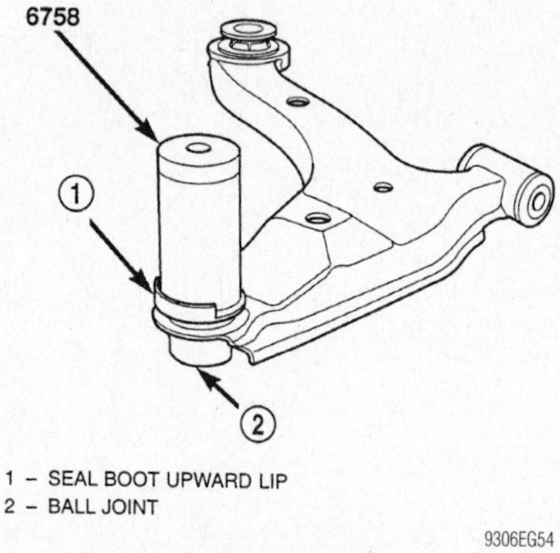

Installing the ball joint boot seal

   a. Receiver Tool 6758 and Adapter Tool 6804.

8. Install or connect the following:
   - Ball joint boot seal using a driver tool such as a large socket or suitable sized piece of pipe

**** WARNING**

Do not use a shop press that was used to install the ball joint, for the press exerts too much force.

   - Lower control arm into the front crossmember
   - Rear lower control arm-to-crossmember and frame rail bolt

➡DO NOT tighten the rear bolt at this time.

   - Front lower control arm-to-crossmember nut and bolt

9. Torque the lower control arm rear pivot bolt to 150 ft. lbs. (203 Nm) and the front pivot bolt to 120 ft. lbs. (163 Nm).
10. Install the ball joint stud into the steering knuckle. Torque the steering knuckle-to-ball joint stud pinch bolt and nut to 70 ft. lbs. (95 Nm).
11. Assemble the stabilizer bar-to-lower control arm link assemblies and bushings.

12. Rotate the stabilizer bar into position, installing the stabilizer bar links into the lower control arms. Install the top stabilizer bar link bushings and nuts. DO NOT tighten the link yet.

13. Install the wheel.

14. Lower the vehicle so the suspension is supporting the total weight of the vehicle.

15. Torque the stabilizer bar-to-lower control arm links to 17 ft. lbs. (23 Nm).

16. Torque the stabilizer bar bushing retainer-to-crossmember bolts to 21 ft. lbs. (28 Nm).

17. Check and/or adjust the toe, as necessary.

Wheel Bearings

ADJUSTMENT

Neons are equipped with sealed hub and bearing assemblies. The hub and bearing assembly is non-serviceable. If the assembly is damaged, the complete unit must be replaced.

REMOVAL & INSTALLATION

Front

1. Before servicing the vehicle, refer to the precautions in the beginning of this section.

2. Remove the steering knuckle and hub and bearing assembly.

3. Remove a wheel lug stud from the hub flange using a C-clamp and Adapter Tool 4150A.

4. Rotate the hub to align the removed lug stud with the notch in the bearing retainer plate.

5. Rotate the hub so the stud hole is facing away from the brake caliper's lower rail on the steering knuckle.

6. Install ½ of a Bearing Splitter Tool 1130, between the hub and the bearing retainer plate. The threaded hole in this ½ is to be aligned with the caliper rail on the steering knuckle.

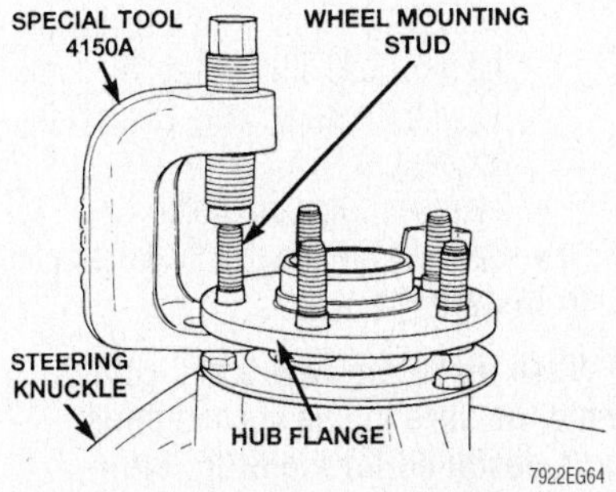

Use a proper C-clamp and adapter tool to press out one of the lug studs

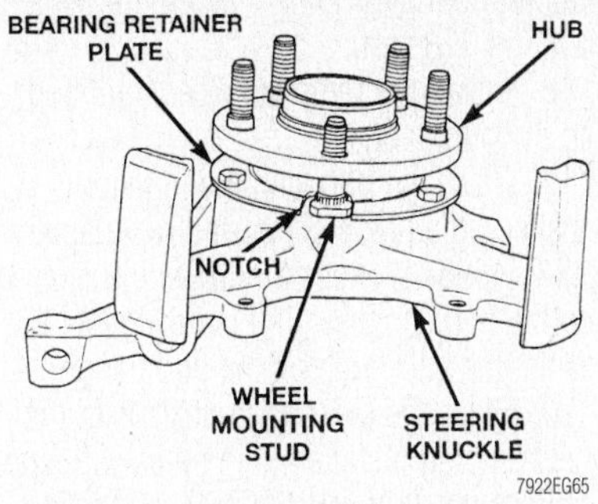

Rotate the hub in order to remove the lug stud

7. Install the remaining pieces of the bearing splitter on the steering knuckle. Hand-tighten the nuts to hold the splitter in place on the knuckle.

8. When the bearing splitter is installed, be sure the 3 bolts attaching the bearing retainer plate to the knuckle are contacting the bearing splitter. The bearing retainer plate should not support the knuckle or contact the splitter.

9. Place the steering knuckle in a hydraulic press, supported by the bearing splitter.

10. Position a driver on the small end of the hub. Using the press, remove the hub from the wheel bearing. The outer bearing race will come out of the wheel bearing when the hub is pressed out of the bearing.

11. Remove the bearing splitter tool from the knuckle.

12. Place the knuckle in a press supported by the press block. The blocks must not obstruct the bore in the steering knuckle so the wheel bearing can be pressed out of the knuckle. Place a driver on the outer race

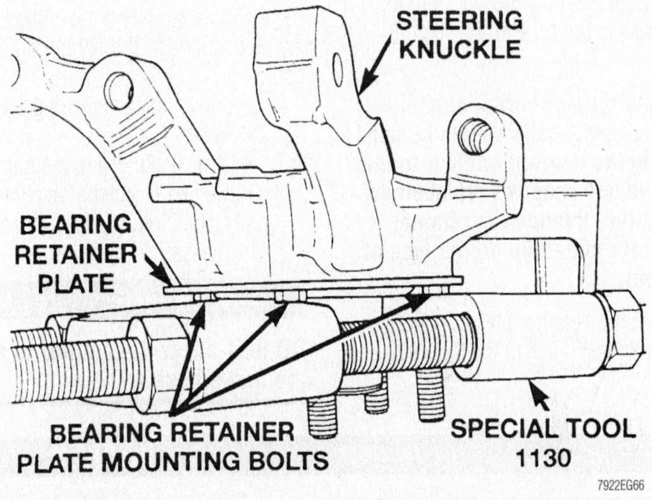

Proper installation of the bearing splitter

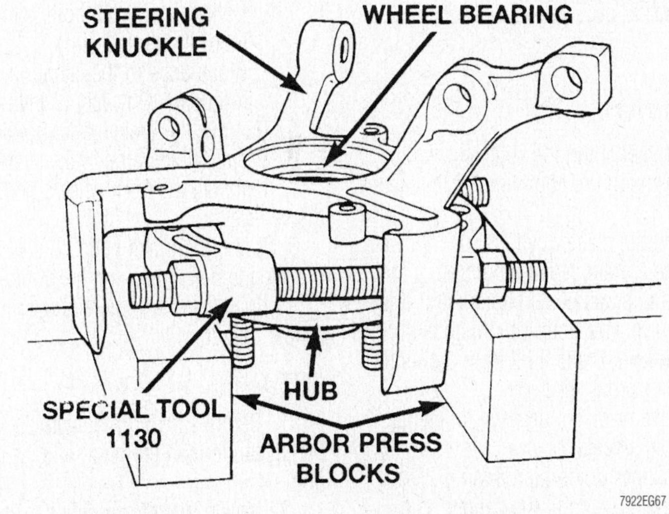

Properly support the steering knuckle for hub and bearing removal

of the wheel bearing, then press the bearing out of the knuckle.

13. Install the bearing splitter on the hub. The splitter is to be installed on the hub so it is between the flange of the hub and the bearing race on the hub. Place the hub, bearing race and splitter in a press. Use a driver to press the hub out of the bearing race.

To install:

14. Use clean, dry cloth to wipe and grease or dirt from the bore of the steering knuckle.

15. Clean the rust preventative from the replacement wheel bearing using a clean, dry towel.

16. Place the new wheel bearing into the bore of the steering knuckle. Be sure the bearing is placed squarely into the bore. Place the knuckle in a press with a receiver tool, C-4698–2 supporting the steering knuckle. Place a driver tool on the outer race of the wheel bearing. Press the wheel bearing into the steering knuckle until it is fully bottomed in the bore of the steering knuckle.

➡Only the original or original equipment replacement bolts should be used to mounting the bearing retainer to the knuckle. If a bolt requires replacement when installing the bearing retainer plate, be sure to get the proper type of replacement.

17. Install the bearing retainer plate on the steering knuckle. Install the 3 bearing retainer mounting bolts. Tighten the bolts to 21 ft. lbs. (28 Nm).

18. Install the removed wheel lug stud into the hub flange.

19. Place the hub with the lug stud installed, in a press supported by Adapter Tool C-4698–1. Press the wheel lug stud into the hub flange until it is fully seated against the back side on the hub flange.

20. Place the steering knuckle with the wheel bearing installed, in a press with special Receiver Tool MB-990799 supporting the inner race of the wheel bearing. Place the hub in the wheel bearing, making sure it is square with the bearing. Press the hub into the wheel bearing until it is fully bottomed in the wheel bearing.

21. Install the steering knuckle and the wheel.

22. Check and/or adjust the front alignment.

Rear

1. Before servicing the vehicle, refer to the precautions in the beginning of this section.

2. Remove or disconnect the following:
   - Wheel
   - Rear brake drum, if equipped
   - Caliper (suspend on a wire) and the rotor, if equipped with rear disc brakes

⁂ WARNING

DO NOT allow the caliper to hang by the brake hose.

   - Dust cap from the rear hub/bearing
   - Hub/bearing assembly-to-knuckle/spindle nut

➡Discard the hub nut and replace with a new one during installation.

   - Hub/bearing from the spindle by pulling it off the end of the spindle by hand

To install:

⁂ WARNING

The hub/bearing nut must be tightened to, but NOT over, its specified torque value. The proper specification is crucial to the life of the hub bearing.

3. Position the hub/bearing assembly on the rear spindle/knuckle. Install a NEW hub nut and tighten to 160 ft. lbs. (217 Nm).

4. Install or connect the following:
   - Dust cap and seat it using a soft face hammer to carefully tap it into place
   - Brake drum, if equipped with drum brakes
   - Rotor, if equipped with disc brakes
   - Caliper and 2 guide pin bolts, if equipped with disc brakes. Tighten the bolts to 16 ft. lbs. (22 Nm).

5. Install the wheel and tire assembly. Tighten the lug nuts in a crisscross pattern, to 100 ft. lbs. (135 Nm).

BRAKES

Brake Caliper

REMOVAL & INSTALLATION

Front

EXCEPT SRT-4

1. Before servicing the vehicle, refer to the precautions in the beginning of this section.

2. Isolate the master cylinder as follows:
   a. Use a brake pedal holding tool and depress the brake pedal past its first one inch of travel and hold it in this position. (This will keep brake fluid from draining from the master cylinder).

3. Remove or disconnect the following:
   - Front wheels
   - 2 caliper guide pin bolts
   - Brake hose from the caliper
   - Caliper from the steering spindle

To install:

4. Install or connect the following:
   - Caliper to the steering spindle
   - Caliper guide pin bolts and torque to 16 ft. lbs. (22 Nm)
   - Brake hose to the caliper and torque to 18 ft. lbs. (24 Nm)

5. Remove the brake pedal holding tool.

6. Properly bleed the brake system.

7. Reinstall the front wheels and lower the vehicle.

8. Pump the brake pedal to seat the front brake pads before moving the vehicle.

9. Road test the vehicle and check for proper operation.

SRT-4

1. Before servicing the vehicle, refer to the precautions in the beginning of this section.

2. Isolate the master cylinder as follows:
   a. Use a brake pedal holding tool and depress the brake pedal past its first one inch of travel and hold it in this position. (This will keep brake fluid from draining from the master cylinder).

3. Remove or disconnect the following:
   - Front wheels
   - Banjo bolt connecting the brake hose to the brake caliper. There are two washers (one on each side of the flex hose fitting) that will come off with the banjo bolt. Discard the washers.
   - 2 caliper guide pin bolts
   - Caliper from the disc brake adapter.

To install:

➡When installing new brake components, be sure to use correct parts. Parts designed for the BR3 Performance Brake System must not be mixed with other brake systems.

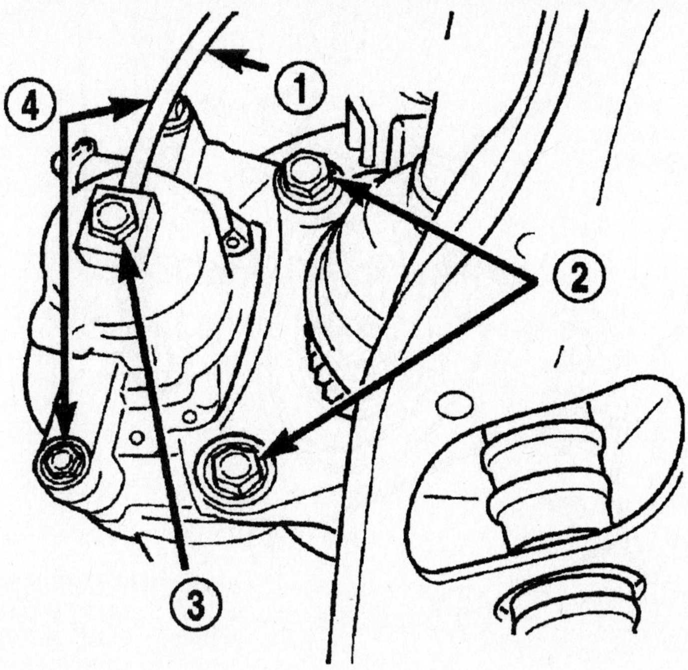

1 - BRAKE HOSE
2 - ADAPTER MOUNTING BOLTS
3 - BANJO BOLT
4 - CALIPER GUIDE PIN BOLTS

67189-NEON-G44

Front brake caliper mounting—SRT-4 models

4. If new pads were installed, completely retract the caliper piston back into the bore of the caliper. Use a C-clamp to retract the piston. Place a wood block over the piston before installing the C-clamp to avoid damaging the piston.

⁂ WARNING

Be careful when installing the caliper onto the disc brake adapter to avoid damaging the boots on the caliper guide pins.

5. Install or connect the following:
   - Caliper over the brake shoes on the brake caliper adapter. Make sure the springs on the shoes do not get caught in the hole formed into the center of the caliper housing.
   - Align the caliper guide pin bolt holes with the guide pins.
   - Caliper guide pin bolts and torque to 26 ft. lbs. (of 35 Nm)
   - Banjo bolt connecting the brake hose to the brake caliper, using new washers on each side of the hose fitting as the banjo bolt is guided through the fitting. Torque the banjo bolt to 18 ft. lbs. (24 Nm).
   - Tire and wheel. Torque the lug nuts to 100 ft. lbs. (135 Nm).
6. Remove the brake pedal holding tool.
7. Properly bleed the brake system.
8. Reinstall the front wheels and lower the vehicle.
9. Pump the brake pedal to seat the front brake pads before moving the vehicle.
10. Road test the vehicle and check for proper operation.

Rear

1. Before servicing the vehicle, refer to the precautions in the beginning of this section.
2. Isolate the master cylinder as follows:
   a. Use a brake pedal holding tool and depress the brake pedal past its first one inch of travel and hold it in this position. (This will keep brake fluid from draining from the master cylinder).
3. Remove or disconnect the following:
   - Rear wheels
   - Banjo bolt connecting the brake hose to the brake caliper. There are two washers (one on each side of the flex hose fitting) that will come

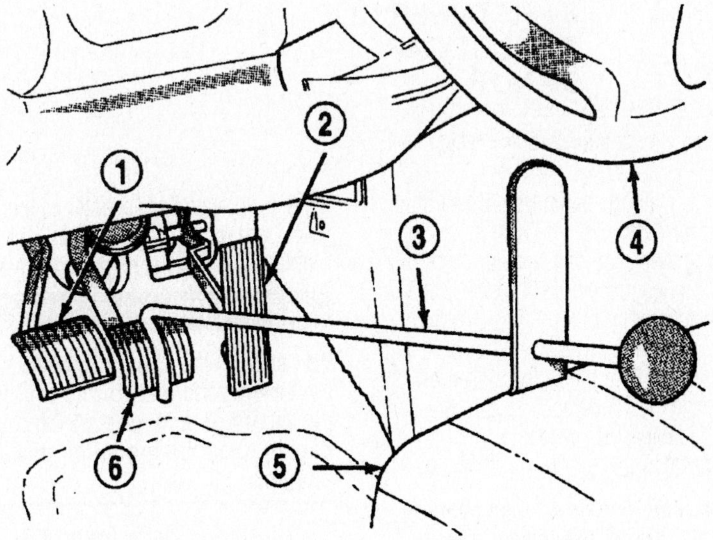

1 - CLUTCH PEDAL (IF EQUIPPED WITH MANUAL TRANSAXLE)
2 - THROTTLE PEDAL
3 - BRAKE PEDAL HOLDING TOOL
4 - STEERING WHEEL
5 - DRIVER'S SEAT
6 - BRAKE PEDAL

67189-NEON-G45

View of the brake pedal holding tool—all models

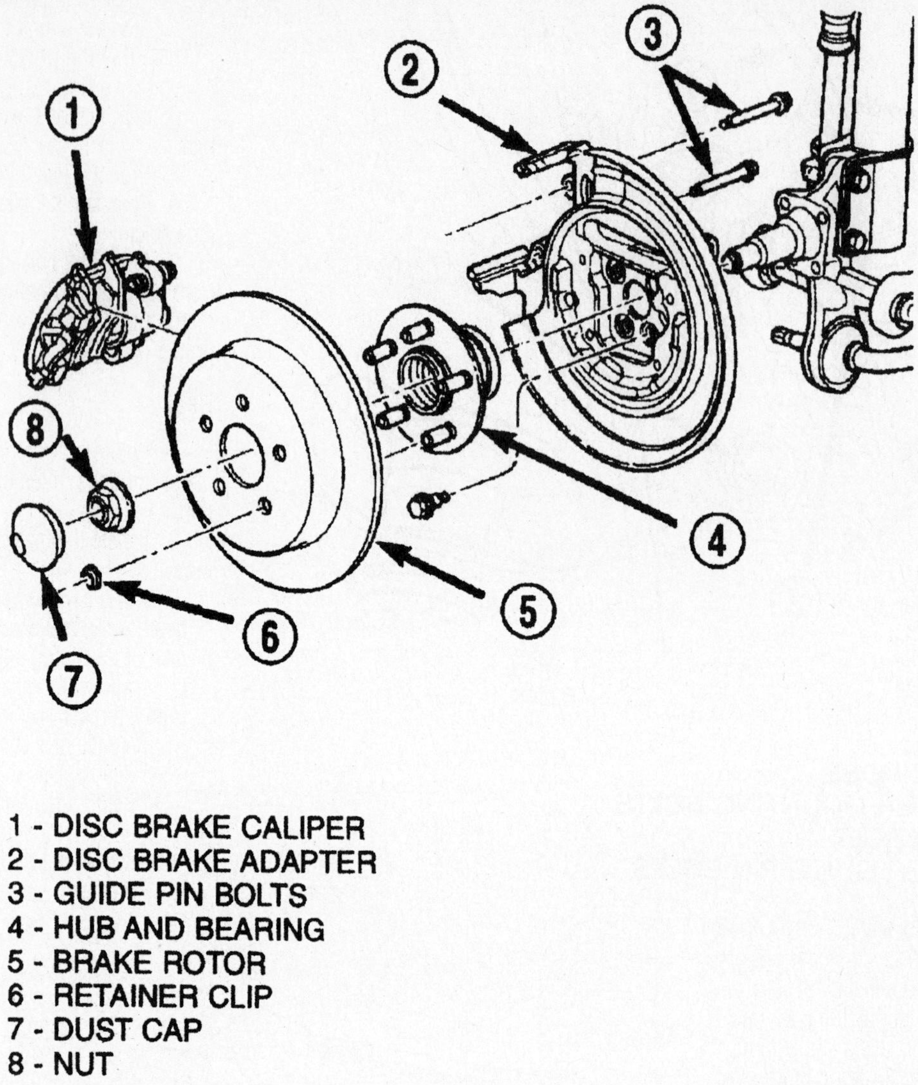

Exploded view of the rear brake caliper mounting

off with the banjo bolt. Discard the washers.

- 2 caliper guide pin bolts
- Caliper assembly from the brake adapter by first rotating the top of the caliper away from the rotor, and then lifting the caliper assembly off the machined abutment on the adapter

To install:

➡When installing new brake components, be sure to use correct parts. Parts designed for the BR3 Performance Brake System must not be mixed with other brake systems.

4. If new pad were installed, completely retract the caliper piston back into piston bore of the caliper.

5. Lubricate both adapter caliper slide abutments with a liberal amount of Mopar Multipurpose Lubricant, or an equivalent.

✻✻ WARNING

Be careful when installing the caliper assembly onto adapter so the guide pin bushings and sleeves do not get damaged by the mounting bosses on adapter.

6. Starting with the lower end, carefully lower the caliper and brake shoes over the brake rotor and catch the caliper's bottom edge behind the caliper slide abutment. Rotate the top of the caliper into mounting position on the adapter.

✻✻ WARNING

Be very careful not to cross thread the caliper guide pin bolts when you are installing them!

7. Install or connect the following:
   - Caliper guide pin bolts, then carefully torque them to 16 ft. lbs. (22 Nm)
   - Banjo bolt connecting the brake hose to the brake caliper. Install new washers, one on each side of the hose fitting as the banjo bolt is guided through the fitting. Torque the banjo bolt to 18 ft. lbs. (24 Nm).
   - Tire and wheel. Torque the lug nuts to 100 ft. lbs. (135 Nm).

8. Remove the brake pedal holding tool.
9. Properly bleed the brake system.
10. Reinstall the front wheels and lower the vehicle.
11. Pump the brake pedal to seat the front brake pads before moving the vehicle.
12. Road test the vehicle and check for proper operation.

Disc Brake Pads

REMOVAL & INSTALLATION

Front

EXCEPT SRT-4

1. Before servicing the vehicle, refer to the precautions in the beginning of this section.
2. Remove the front wheels.
3. Remove the 2 caliper to steering knuckle guide pin bolts.
4. Lift the caliper away from the steering knuckle by first rotating the free end of the caliper away from the steering knuckle. Then, slide the opposite end of the caliper out from under the machined end of the steering knuckle.
5. Support the caliper from the upper control arm to prevent the weight of the caliper from being supported by the brake flex hose that will damage the hose.
6. Remove the brake pads from the caliper. Remove the outboard brake pad by prying the pad retaining clip over the raised area on the caliper. Then, slide the pad down and off the caliper. Pull the inboard brake pad away from the piston until the retaining clip is free from the cavity in the piston.
7. If required, the rotor can be removed by pulling it straight off the wheel mounting studs.

To install:

8. Clean all parts well. Inspect the caliper for piston seal leaks (brake fluid in and around the boot area and inboard lining) and for any ruptures of the piston dust boot. If the boot is damaged or fluid leak is visible, disassemble the caliper and install a new seal and boot (and piston, if scored).
9. Inspect the caliper pin bushings. Replace if damaged, dry or brittle.
10. Completely compress the piston into the caliper using a large C-clamp or other suitable tool.
11. Lubricate the area on the steering knuckle where the caliper slides with high temperature grease.
12. Reinstall the rotor if removed.
13. Reinstall the brake pads into the caliper. Note that the inboard and outboard pads are different. Make sure the inboard brake shoe assembly is positioned squarely against the face of the caliper piston.

➡Be sure to remove the noise suppression gasket paper cover if the pads come so equipped.

14. Carefully position the caliper and brake shoe assemblies over the rotor by hooking the lower end of the caliper over the steering knuckle. Then, rotate the caliper into position at the top of the steering knuckle. Make sure the caliper guide pin bolts, bushings and sleeves are clear of the steering knuckle bosses.
15. Reinstall the caliper guide pin bolts and torque to 16 ft. lbs. (22 Nm).
16. Reinstall the wheels and torque the lug nuts to 100 ft. lbs. (135 Nm).
17. Pump the brake pedal until the brake pads are seated and a firm pedal is achieved before attempting to move the vehicle.
18. Road test the vehicle for proper operation.

SRT-4

1. Before servicing the vehicle, refer to the precautions in the beginning of this section.
2. Remove or disconnect the following:
   - Front wheels
   - 2 caliper to steering knuckle guide pin bolts
   - Caliper from the steering knuckle. Support the caliper from the upper control arm to prevent the weight of the caliper from being supported by the brake flex hose that will damage the hose.
   - Brake shoes from the disc brake caliper adapter
3. If required, the rotor can be removed by pulling it straight off the wheel mounting studs.

To install:

4. Clean all parts well. Inspect the caliper for piston seal leaks (brake fluid in and around the boot area and inboard lining) and for any ruptures of the piston dust boot. If the boot is damaged or fluid leak is visible, disassemble the caliper and install a new seal and boot (and piston, if scored).

➡The inboard brake shoes are not identical side-to-side. This is due to placement of the audible wear indicator on the end of each inboard shoe. Make sure that the audible wear indicators are placed toward the top when the inboard shoes are installed on each side of the vehicle.

5. Install or connect the following:
   - Brake shoes in the abutment shims

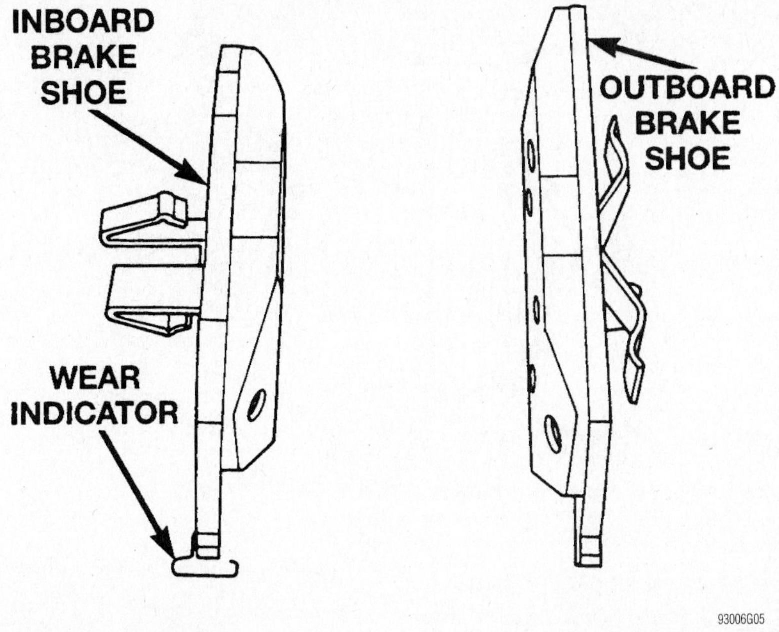

Disc brake pad identification

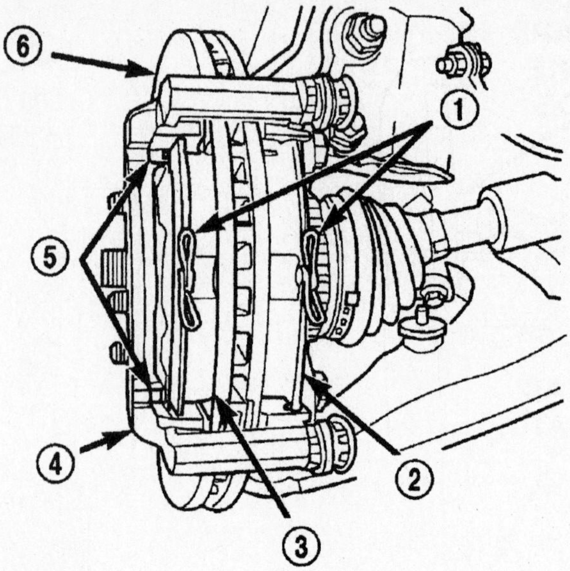

1 - SPRINGS
2 - INBOARD SHOE
3 - OUTBOARD SHOE
4 - DISC BRAKE ADAPTER
5 - ABUTMENT SHIMS
6 - BRAKE ROTOR

67189-NEON-G47

Install the brake shoes in the abutment shims clipped into the disc brake caliper adapter—SRT-4

clipped into the disc brake caliper adapter as shown in the accompanying figure

6. Place the shoe with the wear indicator attached on the inboard side
7. Completely retract the caliper piston back into the bore of the caliper.
8. Install the caliper over the brake shoes on the brake caliper adapter. Make sure the springs on the shoes do not get caught in the hole formed into the center of the caliper housing.
9. Align the caliper guide pin bolt holes with the guide pins. Install the caliper guide pin bolts and tighten them to 26 ft. lbs. (35 Nm)
10. Reinstall the wheels and torque the lug nuts to 100 ft. lbs. (135 Nm).
11. Pump the brake pedal until the brake pads are seated and a firm pedal is achieved before attempting to move the vehicle.
12. Road test the vehicle for proper operation.

Rear

1. Before servicing the vehicle, refer to the precautions in the beginning of this section.
2. Remove the rear wheels.
3. Remove the 2 caliper to steering knuckle guide pin bolts.
4. Lift the caliper away from the steering knuckle by first rotating the free end of the caliper away from the steering knuckle. Then, slide the opposite end of the caliper out from under the machined end of the steering knuckle.
5. Support the caliper from the upper control arm to prevent the weight of the caliper from being supported by the brake flex hose that will damage the hose.
6. Remove the brake pads from the caliper. Remove the outboard brake pad by prying the pad retaining clip over the raised area on the caliper. Then, slide the pad down and off the caliper. Pull the inboard brake pad away from the piston until the retaining clip is free from the cavity in the piston.
7. If required, the rotor can be removed by pulling it straight off the wheel mounting studs.

To install:

8. Clean all parts well. Inspect the caliper for piston seal leaks (brake fluid in and around the boot area and inboard lining) and for any ruptures of the piston dust boot. If the boot is damaged or fluid leak is visible, disassemble the caliper and install a new seal and boot (and piston, if scored).
9. Inspect the caliper pin bushings. Replace if damaged, dry or brittle.
10. Completely compress the piston into the caliper using a large C-clamp or other suitable tool.
11. Remove any protective paper from the noise suppression gasket on both inner and outer brake shoe assemblies (if equipped).
12. Install or connect the following:
   - Inboard brake shoe into the caliper piston by firmly pressing the shoe in with your thumbs. Make sure the inboard brake shoe is positioned squarely against the face of the caliper piston.
   - Slide the outboard brake shoe onto the caliper assembly. Make sure the retaining clip is squarely seated in the depressed areas on the caliper.
13. Lubricate both adapter caliper slide abutments with a liberal amount of Mopar Multipurpose Lubricant, or an equivalent.
14. Starting with the lower end, carefully lower the caliper and brake shoes over the brake rotor and catch the caliper's bottom edge behind the caliper slide abutment. Rotate the top of the caliper into mounting position on the adapter.

**** WARNING**

Be very careful not to cross thread the caliper guide pin bolts when they are installed.

15. Install or connect the following:
   - Caliper guide pin bolts and tighten to 16 ft. lbs. (22 Nm)
   - Rear tire and wheel assemblies. Torque the lug nuts to 100 ft. lbs. (135 Nm).
16. Pump the brake pedal until the brake pads are seated and a firm pedal is achieved before attempting to move the vehicle.
17. Road test the vehicle for proper operation.

Brake Drums

REMOVAL & INSTALLATION

1. Before servicing the vehicle, refer to the precautions in the beginning of this section.
2. Remove the rear wheels.
   a. Locate and remove the rubber plug from the top of the brake support plate (backing plate).

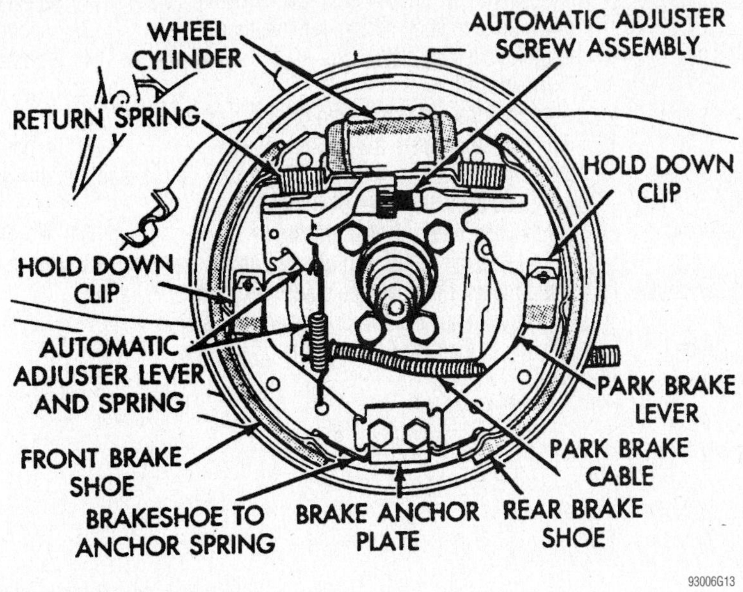

Kelsey Hayes rear brake assembly (left side shown)

b. Insert a small prying tool through the adjuster access hole and engage the teeth on the adjuster wheel.

c. Rotate the adjuster wheel so it is moved toward the front of the vehicle. This will back off the adjustment of the rear brake shoes.

d. Continue moving the adjuster wheel toward the front of the vehicle until it stops moving.

3. Remove the rear brake drum from the hub assembly.

To install:

4. Inspect the brake drums for cracks or signs of overheating. Measure the drum runout and diameter. If not to specification, resurface the drum. Runout should not exceed 0.006 inch (0.152mm). The diameter variation (oval shape) of the drum braking surface must not exceed either 0.0025 inch (0.0635mm) in 30° or 0.0035 inch (0.089mm) in 360°. All brake drums are marked with the maximum allowable brake drum diameter on the face of the drum.

5. Install the rear brake drum onto the hub assembly.

6. Reinstall the wheels and properly adjust the brakes.

7. Road test vehicle to check brake operation.

Brake Shoes

REMOVAL & INSTALLATION

1. Before servicing the vehicle, refer to the precautions in the beginning of this section.

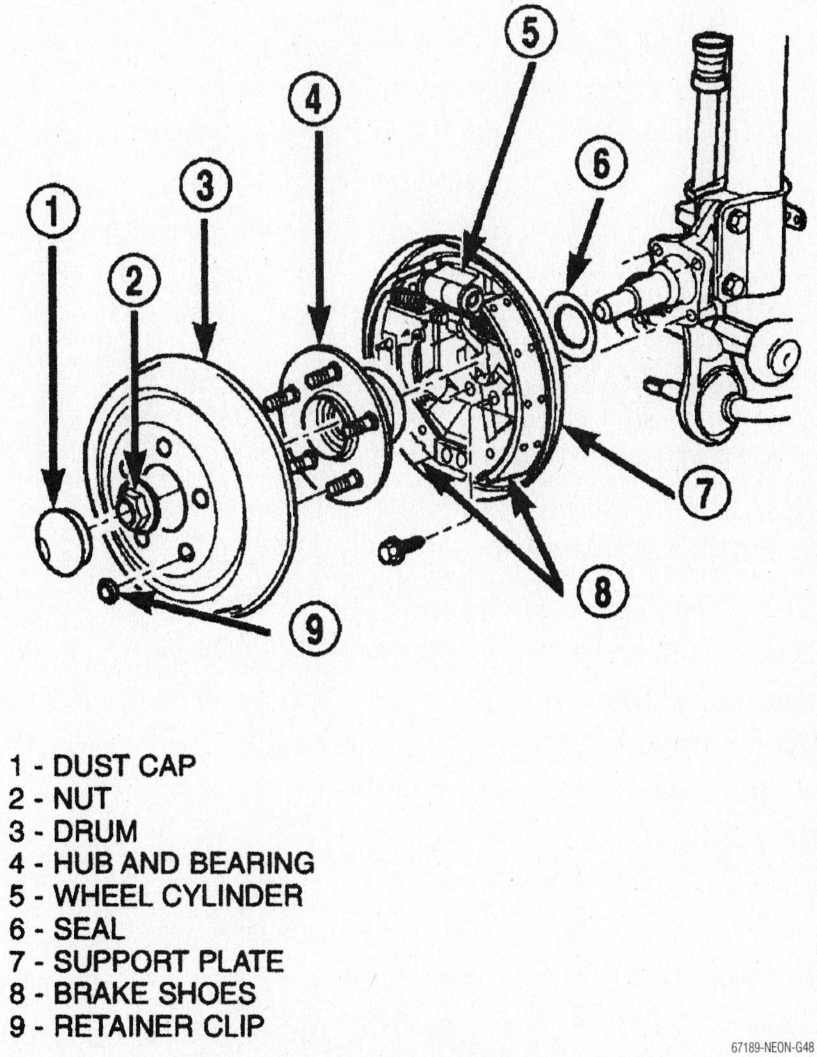

Exploded view of the drum brake assembly—2004 model shown

2. Remove the rear wheels.
3. Remove the drums.
4. Remove the automatic adjuster spring and lever.
5. Remove the hold-down clips and pins.
6. Rotate the automatic adjuster starwheel enough so both shoes move out far enough to be free of the wheel cylinder boots.
7. Disconnect the parking brake cable from the actuating lever.
8. Remove the lower shoe to shoe spring.
9. With the shoes held together by the upper shoe to shoe spring, remove them from the backing plate.

To install:

10. Thoroughly clean and dry the backing plate. To prepare the backing plate, lubricate the bosses, anchor pin and parking brake actuating lever pivot surface lightly with lithium based grease.
11. Remove, clean and dry all brake components. Lubricate the starwheel shaft threads with anti-seize lubricant and transfer all parts to their proper locations on the new shoes.
12. Reinstall the lower spring.
13. Reconnect the parking brake cable.
14. Reinstall the automatic adjuster lever and spring.
15. Adjust the starwheel.
16. Remove any grease from the linings and install the drum.
17. Reinstall the wheels and properly adjust the brakes.
18. Check for proper brake system operation.

CHRYSLER

Pacifica

9

SPECIFICATION CHARTS

ENGINE AND VEHICLE IDENTIFICATION

Engine						
Code ①	Liters (cc)	Cu. In.	Cyl.	Fuel Sys.	Engine Type	Eng. Mfg.
4	3.5 (3518)	215	V6	MFI	SOHC	Chrysler
L	3.8 (3785)	231	6	SMFI	OHV	Chrysler

Model Year	
Code ②	Year
2	2002
3	2003
4	2004
5	2005
6	2006

MFI: Multi-point Fuel Injection

SMFI: Sequential Multi-port Fuel Injection

SOHC: Single Overhead Camshaft

① 8th position of the Vehicle Identification Number (VIN)

② 10th position of VIN

06009-PACI-C01

GENERAL ENGINE SPECIFICATIONS

All measurements are given in inches.

Year	Model	Engine Displacement Liters (VIN)	Net Horsepower @ rpm	Net Torque @ rpm (ft. lbs.)	Bore x Stroke (in.)	Com-pression Ratio	Oil Pressure @ rpm
2003	Pacifica	3.5 (4)	250@6400	255@3950	3.78x3.18	10:01	45-105@3000
2004	Pacifica	3.5 (4)	250@6400	255@3950	3.78x3.18	10:01	45-105@3000
2005-06	Pacifica	3.5 (4)	250@6400	255@3950	3.78x3.18	10:01	45-105@3000
		3.8 (L)	210@5000	240@4000	3.79x3.42	9.6:1	30-80@3000

MFI: Multi-point Fuel Injection

06009-PACI-C02

ENGINE TUNE-UP SPECIFICATIONS

Year	Engine Displacement Liters (VIN)	Spark Plug Gap (in.)	Ignition Timing (deg.)	Fuel Pump (psi)	Idle Speed (rpm)	Valve Clearance Intake	Valve Clearance Exhaust
2003	3.5 (4)	0.048-0.053	①	58	②	HYD	HYD
2004	3.5 (4)	0.048-0.053	①	58	②	HYD	HYD
2005-06	3.5 (4)	0.048-0.053	①	58	②	HYD	HYD
	3.8 (L)	0.048-0.053	①	58	②	HYD	HYD

NOTE: The Vehicle Emission Control Information label often reflects specification changes made during production. The label figures must be used if they differ from those in this chart.

HYD: Hydraulic

① Ignition timing is regulated by the Powertrain Control Module (PCM), and cannot be adjusted.

② Idle speed is controled by the Powertrain Control Module (PCM), and cannot be adjusted.

06009-PACI-C03

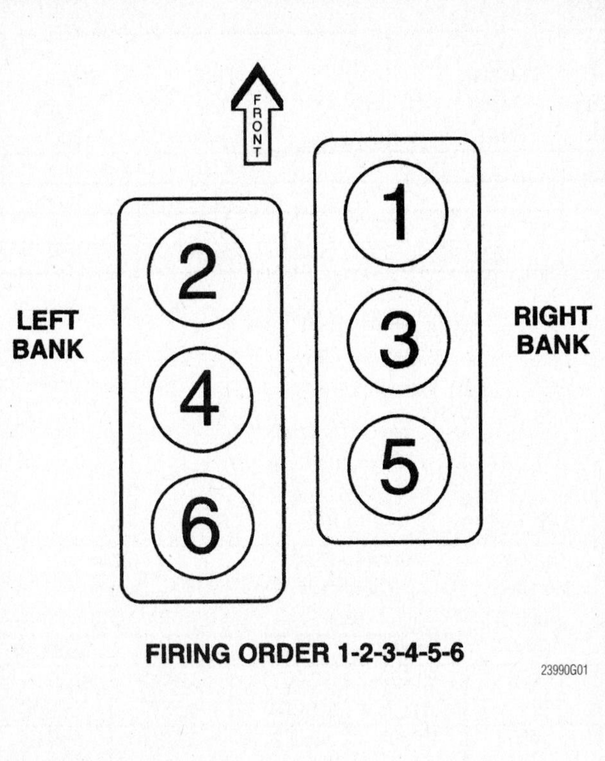

3.5L VIN 4 Engine
Firing Order: 1–2–3–4–5–6
Distributorless ignition system

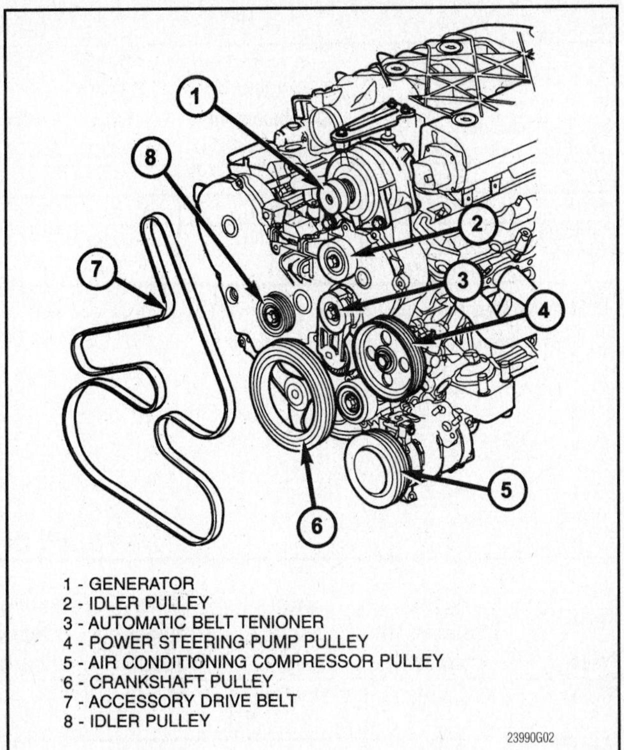

Accessory drive belt routing—3.5L engine

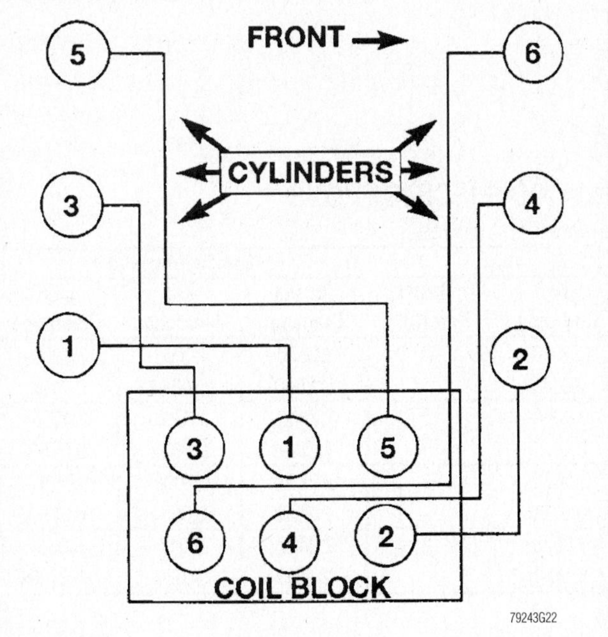

3.8L engines
Firing order: 1–2–3–4–5–6
Distributorless ignition system

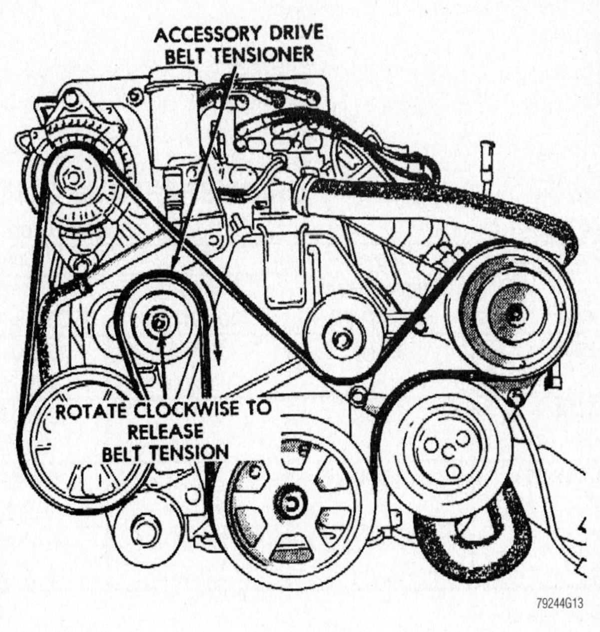

Accessory V-belt routing—3.8L engines

CAPACITIES

Year	Model	Engine Displacement Liters (VIN)	Engine Oil with Filter (qts.)	Auto. Transmission (pts.) ①	Transfer Case (qts.)	Front Drive Axle (pts.)	Fuel Tank (gal.)	Cooling System (qts.)
2003	Pacifica	3.5 (4)	5.5	19.8	—	2.0	23.0	10.5
2004	Pacifica	3.5 (4)	5.5	19.8	—	2.0	23.0	10.5
2005-06	Pacifica	3.5 (4)	5.5	19.8	—	2.0	23.0	10.5
	Pacifica	3.8 (L)	5.0	19.8	1.22	2.0	23.0	10.5

NOTE: All capacities are approximate. Add fluid gradually and ensure a proper fluid level is obtained.

① Overhaul fill capacity with torque converter empty
Estimated service fill: 8 pts.

06009-PACI-C04

VALVE SPECIFICATIONS

Year	Engine Displacement Liters (VIN)	Seat Angle (deg.)	Face Angle (deg.)	Spring Test Pressure (lbs. @ in.)	Spring Installed Height (in.)	Stem-to-Guide Clearance (in.)		Stem Diameter (in.)	
						Intake	Exhaust	Intake	Exhaust
2003	3.5 (4)	45-45.5	44.5-45	①	1.496	0.0009-0.0026	0.0020-0.0037	0.2730-0.2737	0.2719-0.2726
2004	3.5 (4)	45-45.5	44.5-45	①	1.496	0.0009-0.0026	0.0020-0.0037	0.2730-0.2737	0.2719-0.2726
2005-06	3.5 (4)	45-45.5	44.5-45	①	1.496	0.0009-0.0026	0.0020-0.0037	0.2730-0.2737	0.2719-0.2726
	3.8 (L)	45-45.5	45-45.5	②	1.61-1.68	0.0010-0.0025	0.0020-0.0037	0.2718-0.2725	0.2718-0.2725

① Intake: 69.5-80.5 lbs.@1.496 in. valve closed
Intake: 188.0-204.0 lbs.@1.1594 in. valve opened
Exhaust: 71-79 lbs.@1.4961 in. valve closed
Exhaust: 130-144 lbs.@1.239 in. valve opened

② Type A Valve Closed: 84.9-95.6 lbs. @ 1.65 in.
Type A Valve Open: 194.2-215.8 @ 1.65 in
Type B Valve Open: 197.9-216.3 @ 122 in.
Type B Valve Closed: 84.4-95.2 @ 1.65 in.

06009-PACI-C05

CRANKSHAFT AND CONNECTING ROD SPECIFICATIONS

All measurements are given in inches.

Year	Engine Displacement Liters (VIN)	Crankshaft				Connecting Rod		
		Main Brg. Journal Dia.	Main Brg. Oil Clearance	Shaft End-play	Thrust on No.	Journal Diameter	Oil Clearance	Side Clearance
2003	3.5 (4)	2.5190-2.5200	0.0004-0.0022	0.0040-0.0120	2	2.2830-2.2840	0.0008-0.0034	0.0050-0.0150
2004	3.5 (4)	2.5190-2.5200	0.0004-0.0022	0.0040-0.0120	2	2.2830-2.2840	0.0008-0.0034	0.0050-0.0150
2005-06	3.5 (4)	2.5190-2.5200	0.0004-0.0022	0.0040-0.0120	2	2.2830-2.2840	0.0008-0.0034	0.0050-0.0150
	3.8 (L)	2.5194-2.5202	0.0005-0.0022	0.0036-0.0095	2	2.2827-2.2837	0.017-0.020	0.005-0.0130

Max: Maximum

06009-PACI-C06

PISTON AND RING SPECIFICATIONS

All measurements are given in inches.

Year	Engine Displacement Liters (VIN)	Piston Clearance	Ring Gap Top Compression	Ring Gap Bottom Compression	Ring Gap Oil Control	Ring Side Clearance Top Compression	Ring Side Clearance Bottom Compression	Ring Side Clearance Oil Control
2003	3.5 (4)	0.0003-0.0018	0.008-0.014	0.0091-0.0197	0.010-0.030	0.0016-0.0031	0.0016-0.0031	0.0015-0.0073
2004	3.5 (4)	0.0003-0.0018	0.008-0.014	0.0091-0.0197	0.010-0.030	0.0016-0.0031	0.0016-0.0031	0.0015-0.0073
2005-06	3.5 (4)	0.0003-0.0018	0.008-0.014	0.0091-0.0197	0.010-0.030	0.0016-0.0031	0.0016-0.0031	0.0015-0.0073
	3.8 (L)	0.0002-0.0015	0.007-0.0150	0.011-0.0220	0.009-0.0300	0.0012-0.0027	0.0016-0.0033	0.0015-0.0078

06009-PACI-C07

TORQUE SPECIFICATIONS

All readings in ft. lbs.

Year	Engine Displacement Liters (VIN)	Cylinder Head Bolts	Main Bearing Bolts	Rod Bearing Bolts	Crankshaft Damper Bolts	Flywheel Bolts	Manifold Intake	Manifold Exhaust	Spark Plugs	Oil Pan Drain Plug
2003	3.5 (4)	①	②	③	70	70	④	⑤	20	20
2004	3.5 (4)	①	②	③	70	70	④	⑤	20	20
2005-06	3.5 (4)	①	②	③	70	70	④	⑤	20	20
	3.8 (L)	⑥	⑦	⑧	40	65	⑨	⑩	20	20

① Step 1: 45 ft. lbs.
Step 2: 65 ft. lbs.
Step 3: 65 ft. lbs.
Step 4: Plus 1/4 turn
Final torque should be over 90 ft. lbs.

② Main cap inside bolts: 15 ft. lbs. plus 1/4 turn
Main cap outside bolts: 20 ft. lbs. plus 1/4 turn
Main cap tie bolts: 250 inch lbs.

③ Step 1: 20 ft. lbs.
Step 2: Plus 1/4 turn

④ Upper 105 inch lbs.
Lower 250 inch lbs.

⑤ 200 inch lbs.

⑥ Step 1: 45 ft. lbs.
Step 2: 65 ft. lbs.
Step 3: 65 ft. lbs.
Step 4: Plus 90 degrees

⑦ Step 1: 30 ft. lbs.
Step 2: Plus 90 degrees

⑧ Step 1: 40 ft. lbs.
Step 2: Plus 90 degrees

⑨ Lower manifold: Step 1: 10 inch lbs.
Step 2: 200 inch lbs.
Step 3: 200 inch lbs.
Upper manifold: 105 inch lbs.

⑩ Refer to procedure for torque sequence and specifications

06009-PACI-C08

WHEEL ALIGNMENT

Year	Model		Caster Range (+/-Deg.)	Caster Preferred Setting (Deg.)	Camber Range (+/-Deg.)	Camber Preferred Setting (Deg.)	Toe-in (in.)
2003	Pacifica	F	+4.00 to + 5.00	+ 4.50	-0.60 to +0.20	-0.20	0.10 +/- 0.20
		R	—	—	-0.90 to +0.10	-0.40	0.10 +/- 0.30
2004	Pacifica	F	+4.00 to + 5.00	+ 4.50	-0.60 to +0.20	-0.20	0.10 +/- 0.20
		R	—	—	-0.90 to +0.10	-0.40	0.10 +/- 0.30
2005-06	Pacifica	F	+4.00 to + 5.00	+ 4.50	-0.60 to +0.20	-0.20	0.10 +/- 0.20
		R	—	—	-0.90 to +0.10	-0.40	0.10 +/- 0.30

06009-PACI-C09

TIRE, WHEEL AND BALL JOINT SPECIFICATIONS

		OEM Tires		Tire Pressures (psi)		Wheel	Ball Joint	Lug
Year	Model	Standard	Optional	Front	Rear	Size	Inspection	Nuts
2003	Pacifica	P235/65/R17	None	NA	NA	7-JJ	①	100
2004	Pacifica	P235/65/R17	None	NA	NA	7-JJ	①	100
2005-06	Pacifica	P235/65/R17	None	NA	NA	7-JJ	①	100

OEM: Original Equipment Manufacturer

PSI: Pounds Per Square Inch

NA: Not Available

① Do not lift car. Grasp the grease fitting and attempt to move or rotate. Replace if any movement is found.

06009-PACI-C10

BRAKE SPECIFICATIONS

All measurements in inches unless noted

			Brake Disc			Brake Drum Diameter			Minimum	Brake Caliper
			Original	Minimum	Maximum	Original Inside	Max. Wear	Maximum Machine	Lining	Mounting Bolts
Year	Model		Thickness	Thickness	Runout	Diameter	Limit	Diameter	Thickness	(ft. lbs.)
2003	Pacifica	F	1.107	1.040	0.0014	—	—	—	0.250	32
		R	0.556	0.492	0.0055	—	—	—	0.280	17
2004	Pacifica	F	1.107	1.040	0.0014	—	—	—	0.250	32
		R	0.556	0.492	0.0055	—	—	—	0.280	17
2005-06	Pacifica	F	1.107	1.040	0.0014	—	—	—	0.250	32
		R	0.556	0.492	0.0055	—	—	—	0.280	17

F: Front

R: Rear

06009-PACI-C11

SCHEDULED MAINTENANCE INTERVALS
Chrysler—Pacifica

TO BE SERVICED	TYPE OF SERVICE	VEHICLE MILEAGE INTERVAL (x1000)												
		7.5	15	22.5	30	37.5	45	52.5	60	67.5	75	82.5	90	97.5
Engine oil & filter	R	✓	✓	✓	✓	✓	✓	✓	✓	✓	✓	✓	✓	✓
Exhaust system	S/I	✓	✓	✓	✓	✓	✓	✓	✓	✓	✓	✓	✓	✓
Brake hoses	S/I	✓	✓	✓	✓	✓	✓	✓	✓	✓	✓	✓	✓	✓
CV joints & front suspension components	S/I	✓	✓	✓	✓	✓	✓	✓	✓	✓	✓	✓	✓	✓
Rotate tires	S/I	✓	✓	✓	✓	✓	✓	✓	✓	✓	✓	✓	✓	✓
Coolant level, hoses & clamps	S/I	✓	✓	✓	✓	✓	✓	✓	✓	✓	✓	✓	✓	✓
Accessory drive belts	S/I		✓		✓		✓		✓		✓		✓	
Brake linings	S/I		✓	✓				✓	✓				✓	
Spark plugs	R				✓				✓				✓	
Air filter element	R				✓				✓				✓	
Lubricate steering linkage & tie rod ends	S/I				✓				✓				✓	
Engine coolant	R						✓				✓			
PCV valve	S/I								✓				✓	
Ignition cables	R								✓					
Camshaft timing belt	R								✓					

R: Replace S/I: Service or Inspect

FREQUENT OPERATION MAINTENANCE (SEVERE SERVICE)

If a vehicle is operated under any of the following conditions it is considered severe service:

- Extremely dusty areas

- 50% or more of the vehicle operation is in 32°C (90°F) or higher temperatures, or constant operation in temperatures below 0°C (32°F)

Prolonged idling (vehicle operation in stop and go traffic)

- Frequent short running periods (engine does not warm to normal operating temperatures)

- Police, taxi, delivery usage or trailer towing usage

CV joints & front suspension components: check every 3000 miles

Oil & oil filter change: change every 3000 miles

Rotate tires: every 3000 miles

Brake linings: check every 9000 miles

Air filter element: change every 15,000 miles

Automatic transaxle fluid: change every 15,000 miles

Differential fluid: change every 15,000 miles

Tie rod ends & steering linkage: lubricate every 15,000 miles

PCV valve: check every 30,000 miles

06009-PACI-C12

PRECAUTIONS

Before servicing any vehicle, please be sure to read all of the following precautions, which deal with personal safety, prevention of component damage, and important points to take into consideration when servicing a motor vehicle:

- Never open, service or drain the radiator or cooling system when the engine is hot; serious burns can occur from the steam and hot coolant.
- Observe all applicable safety precautions when working around fuel. Whenever servicing the fuel system, always work in a well-ventilated area. Do not allow fuel spray or vapors to come in contact with a spark, open flame or excessive heat (a hot drop light, for example). Keep a dry chemical fire extinguisher near the work area. Always keep fuel in a container specifically designed for fuel storage; also, always properly seal fuel containers to avoid the possibility of fire or explosion. Refer to the additional fuel system precautions later in this section.
- Fuel injection systems often remain pressurized, even after the engine has been turned **OFF**. The fuel system pressure must be relieved before disconnecting any fuel lines. Failure to do so may result in fire and/or personal injury.
- Brake fluid often contains polyglycol ethers and polyglycols. Avoid contact with the eyes and wash your hands thoroughly after handling brake fluid. If you do get brake fluid in your eyes, flush your eyes with clean, running water for 15 minutes. If eye irritation persists, or if you have taken brake fluid internally, IMMEDIATELY seek medical assistance.
- The EPA warns that prolonged contact with used engine oil may cause a number of skin disorders, including cancer! You should make every effort to minimize your exposure to used engine oil. Protective gloves should be worn when changing oil. Wash your hands and any other exposed skin areas as soon as possible after exposure to used engine oil. Soap and water, or waterless hand cleaner should be used.
- All new vehicles are now equipped with an air bag system. The system must be disabled before performing service on or around system components, steering column, instrument panel components, wiring and sensors. Failure to follow safety and disabling procedures could result in accidental air bag deployment, possible personal injury and unnecessary system repairs.
- Always wear safety goggles when working with, or around, the air bag system. When carrying a non-deployed air bag, be sure the bag and trim cover are pointed away from your body. When placing a non-deployed air bag on a work surface, always face the bag and trim cover upward, away from the surface. This will reduce the motion of the module if it is accidentally deployed. Refer to the additional air bag system precautions later in this section.
- Clean, high quality brake fluid from a sealed container is essential to the safe and proper operation of the brake system. You should always buy the correct type of brake fluid for your vehicle. If the brake fluid becomes contaminated, completely flush the system with new fluid. Never reuse any brake fluid. Any brake fluid that is removed from the system should be discarded. Also, do not allow any brake fluid to come in contact with a painted surface; it will damage the paint.
- Never operate the engine without the proper amount and type of engine oil; doing so WILL result in severe engine damage.
- Timing belt maintenance is extremely important! Many models utilize an interference-type, non-freewheeling engine. If the timing belt breaks, the valves in the cylinder head may strike the pistons, causing potentially serious (also time-consuming and expensive) engine damage.
- Disconnecting the negative battery cable on some vehicles may interfere with the functions of the on-board computer system(s) and may require the computer to undergo a relearning process once the negative battery cable is reconnected.
- When servicing drum brakes, only disassemble and assemble one side at a time, leaving the remaining side intact for reference.
- Only an MVAC-trained, EPA-certified automotive technician should service the air conditioning system or its components.

ENGINE REPAIR

Alternator

REMOVAL & INSTALLATION

3.5L Engine

1. Before servicing the vehicle, refer to the precautions in the beginning of this section.
2. Remove or disconnect the following:
   - Negative battery cable
   - Engine cover
   - Drive belt
   - Alternator B+ terminal nut and wire
   - Alternator field circuit wiring by pushing the red locking tab
   - Upper mounting bracket
   - Lower mounting bolts and the alternator

To install:

3. Install or connect the following:
   - Alternator and lower mounting bolts, tighten to 40 ft. lbs. (54 Nm)
   - Upper mounting bracket and tighten the bolts 20 ft. lbs. (28 Nm)
   - Alternator field circuit wiring making sure to engage the red locking tab
   - Alternator B+ terminal nut and wire, tighten the nut to 110 inch lbs. (12 Nm)
   - Drive belt
   - Engine cover
   - Negative battery cable

3.8L Engines

1. Before servicing the vehicle, refer to the precautions in the beginning of this section.
2. Disconnect the negative battery cable.
3. Remove the air cleaner assembly.
4. Disconnect the alternator electrical connectors.
5. Remove the cover from the battery cable at the rear of the alternator.
6. Disconnect the positive battery cable from rear of the alternator.
7. Remove the accessory drive belt.
8. Remove the three mounting bolts for the power

Steering assembly and set the assembly aside.

9. Remove the idler pulley/lower mounting bolt.
10. Remove the alternator rear mounting bolt.
11. Remove the alternator upper mounting bolt.
12. Remove the alternator and idler pulley.

To install:

13. Attach the idler pulley to alternator and install the alternator/idler assembly in the vehicle.
14. Loosely install the alternator upper, rear and lower mounting bolts.
15. Tighten all mounting bolts to 40 ft. lbs. (54 Nm).
16. Install the power steering pump assembly and tighten the bolts to 200 inch lbs. (23 Nm).
17. Install the accessory drive belt.
18. Connect the battery cable to the alternator and tighten the nut to 110 inch lbs. (12 Nm).
19. Snap the cable cover into position.
20. Connect the alternator electrical connectors
21. Connect the negative battery cable.
22. Install the air cleaner assembly.

Ignition Timing

ADJUSTMENT

This model utilize a Distributorless Ignition System (DIS). It is a fixed ignition timing system, which means that basic ignition timing cannot be adjusted. All spark advance is permanently set by the Powertrain Control Module (PCM).

Engine Assembly

REMOVAL & INSTALLATION

3.5L Engines

1. Before servicing the vehicle, refer to the precautions in the beginning of this section.
2. Drain the engine oil.
3. Drain the engine coolant.
4. Properly relieve the fuel system pressure.
5. Remove or disconnect the following:
   - Negative battery cable
   - Hood
   - Cruise control servo
   - Radiator closure panel
   - Radiator core support
   - Upper radiator hose
   - Cooling fan
   - Air cleaner assembly
6. Recover the A/C system refrigerant using approved recycling equipment.
   - Throttle and cruise control cable from the throttle body
   - Cruise control and power brake booster vacuum hoses from the engine
   - Transmission wiring harness from the solenoid pack, input and output sensors and the range sensor connectors
   - Engine wiring harness grounds from the inner frame rail
   - Transmission shift cable
   - Transmission cooler lines
   - Engine block heater connector, if equipped
   - Coolant reservoir houses from the thermostat housing
   - Heater hoses from the heater core
   - Upper radiator hoses from the thermostat housing
   - Brake lines from the Hydraulic Control Unit (HCU)
   - A/C suction discharge hoses from the compressor and plug the openings
   - Lower radiator hose from the engine outlet
   - A/C clutch connection
   - Oil pressure sender and alternator connections
   - Fuel supply line from the fuel rail
   - Ignition harness connection at the intake manifold
   - Front wheel
   - Left inner fender well
   - Engine harness from the Powertrain Control Module (PCM)
   - Left wheel speed sensor and retainer
   - Left sway bar link from the strut
   - Left drive axle nut
   - Both left front steering knuckle pinch bolts
   - Left steering knuckle from the strut and support the knuckle
   - Negative battery cable from the transmission
   - Left fascia screws
   - Right inner fender well
   - Right fascia screws
   - Right wheel speed sensor and retainer
   - Right sway bar link from the strut
   - Right drive axle nut
   - Both right front steering knuckle pinch bolts
   - Right steering knuckle from the strut and support the knuckle
   - Fascia driving lamps, if equipped
   - Front fascia
   - Starter connectors and harness retainers
   - Oxygen (O_2S) sensor electrical wiring at the exhaust manifold and catalytic converter
   - Exhaust from the manifold
   - Driveshaft after marking the position at the front and rear of the shaft. This will help with installation and retain driveshaft balance.
   - Engine-to-transmission plate
   - Transmission inspection plate
   - Flex plate-to-torque converter bolt

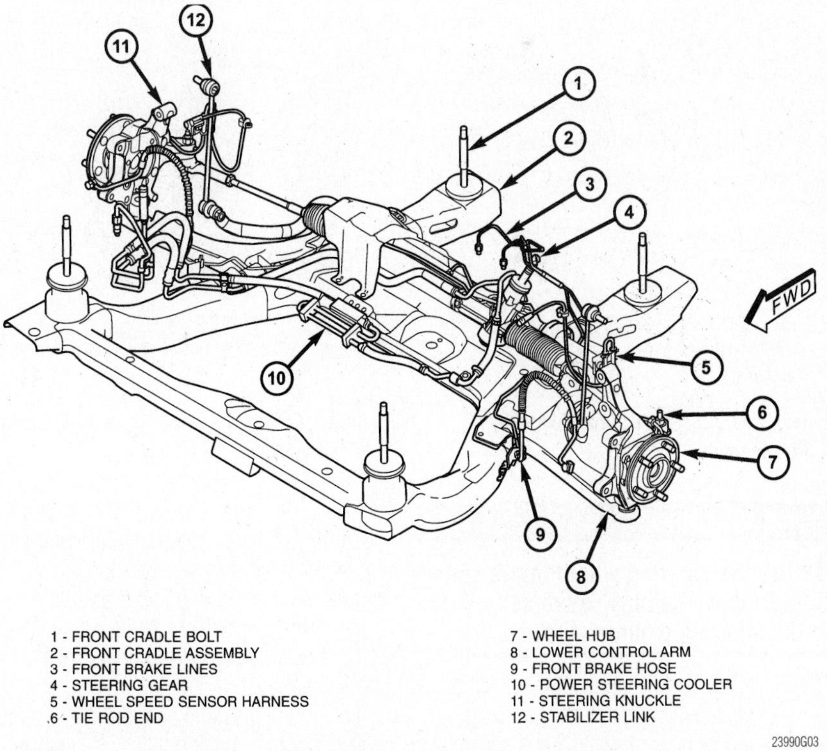

Exploded view of the front cradle assembly—3.5L engine

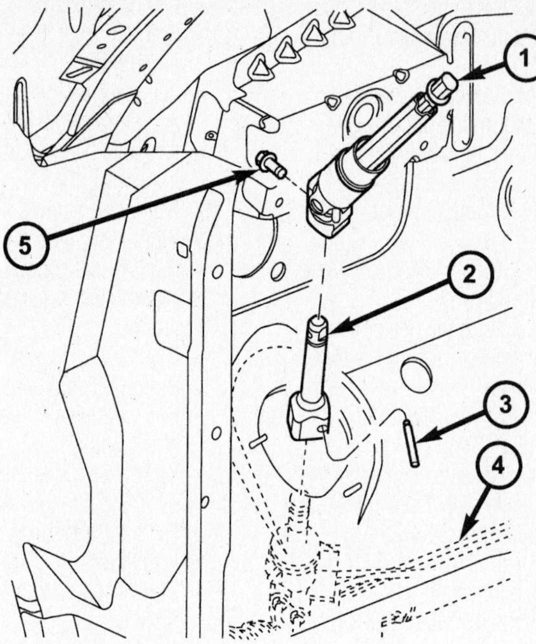

1 - LOWER STEERING COLUMN COUPLING ASSEMBLY
2 - INTERMEDIATE SHAFT TO STEERING GEAR
3 - PIN
4 - STEERING GEAR
5 - EXTENSION TO SHAFT BOLT

23990G04

Lower steering column coupling assembly—3.5L engine

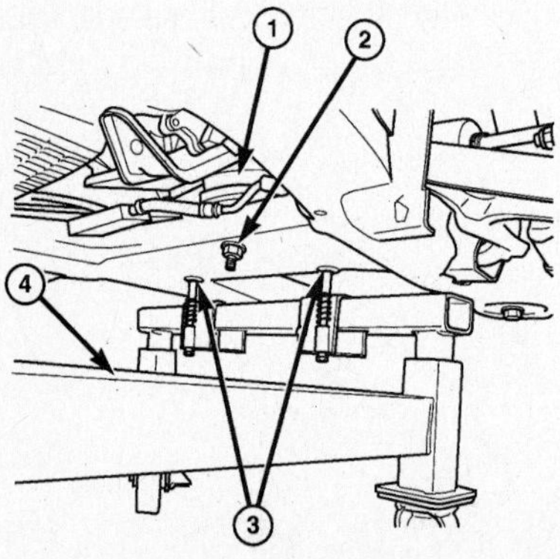

1 - RIGHT LOWER ENGINE MOUNT
2 - RIGHT LOWER ENGINE MOUNT NUT
3 - ALIGNMENT DOWEL ACCESS HOLES
4 - CRADLE SUPPORT FIXTURE

23990G05

Rear cradle support fixture mounting—3.5L engine

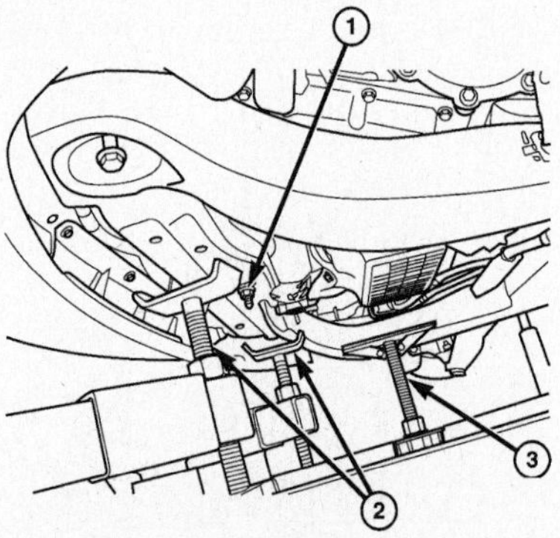

1 - LEFT LOWER ENGINE MOUNT NUT
2 - CRADLE SUPPORT FIXTURE ARMS
3 - CRADLE SUPPORT FIXTURE ENGINE STABILIZER

23990G06

Adjust the cradle support fixture to fit flush with the cradle—3.5L engine

⁂ WARNING

Secure the steering wheel to prevent it rotating which may cause damage to the steering column clock spring.

7. Remove or disconnect the following:
   - Lower steering column coupling pin using tool 6831-A and separate the union
   - Both lower engine mount bolts

8. Matchmark using paint the front cradle-to-body location and position the engine cradle support under the vehicle.

9. Remove or disconnect the following:

10. Lower the vehicle until it is just above the cradle

11. Align the cradle support dowels with cradle access holes and adjust the engine support fixture to fit flush with the oil pan and adjust the cradle support fixture support to fit flush with the cradle.

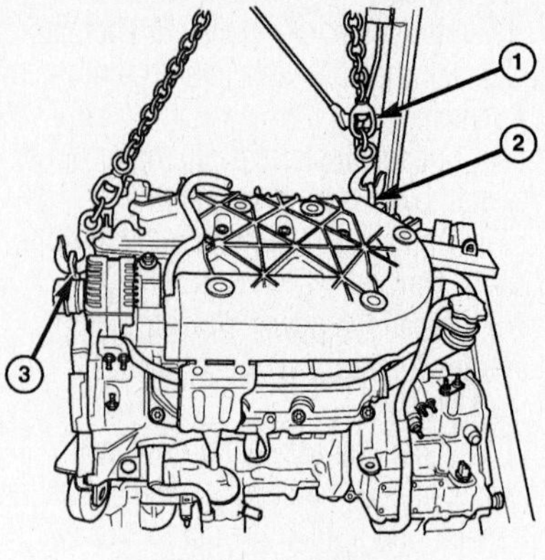

1 - SUITABLE LIFTING CHAIN
2 - 8537-7 PART OF KIT 8537-12
3 - 8537-15 PART OF KIT 8537-12

23990G07

Use suitable lift brackets and a hoist assembly to separate the engine from the transaxle—3.5L engine

12. Carefully lower the vehicle onto the cradle.

13. Remove the upper engine mount

14. Very **carefully** remove the front and rear cradle mounting bolts and raise the vehicle to separate the engine/transmission and cradle from the vehicle.

15. Connect lifting brackets to the engine, separate the Power Transfer Unit (PTU), if equipped.

16. Separate the engine from transmission using the lift brackets.

To install:

17. Installation is the reverse of removal, please note the following important specifications.

18. Remove or disconnect the following:
   - Cradle supporting bolts: 120 ft. lbs. (163 Nm)
   - Upper engine mount bolt to timing cover to 40 ft. lbs. (54 Nm) and the bolt to right rail to 50 ft. lbs. (68 Nm)
   - Drive axle-to-steering knuckle nut to 180 ft. lbs. (244 Nm)
   - Steering knuckle-to-strut bolts to 180 ft. lbs. (244 Nm)
   - Front speed sensor support bracket bolts to 105 inch lbs. (12 Nm)
   - Brake line support bracket bolts to 105 inch lbs. (12 Nm)
   - Flex plate-to-torque converter bolts to 55 ft. lbs. (75 Nm)
   - Engine-to-transmission support collar bolts to 40 ft. lbs. (54 Nm)

19. Fill the cooling system to the proper level.

20. Fill the engine with clean oil.

21. Bleed the brakes.

22. Recharge the A/C system
23. Check wheel alignment.
24. Start the vehicle, check for leaks and repair if necessary.

3.8L Engines

1. Before servicing the vehicle, refer to the precautions in the beginning of this section.
2. Drain the cooling system.
3. Drain the engine oil.
4. Relieve the fuel system pressure.
5. Recover the A/C refrigerant.
6. Secure the steering wheel in the straight ahead position.
7. Disconnect the CCV hose from the left cylinder head cover.
8. Remove the air cleaner cover and housing.
9. Disconnect the throttle and cruise control cables from the throttle body, separate the cables from the bracket and set to one side.
10. Disconnect the cruise control and power brake booster vacuum hoses and set aside.
11. Disconnect the upper radiator hose from the thermostat housing and the lower radiator hose from the radiator.
12. Disconnect the heater hoses and coolant reservoir hose from the intake manifold.
13. Disconnect the transmission controller and transmission sensor wire harness connectors.
14. Disconnect the brake controller wire harness connector.
15. Disconnect the transmission shift cable and remove the cable from the bracket.
16. Disconnect the transmission lines using tool 8875.
17. Disconnect the alternator electrical connectors.
18. Disconnect the fuel line from fuel rail
19. Disconnect the EVAP hose from the throttle body.
20. Disconnect the Throttle Position (TP), Idle Air Control (IAC) motor, Heated Oxygen (HOo2s), Camshaft Position (CMP), Manifold Absolute Pressure (MAP) sensors and the two engine wiring harness connectors. Remove the harness from the cover studs and set aside.
21. Disconnect the purge solenoid harness connector.
22. Remove engine wiring harness retainer from the left cylinder head cover and the oil dipstick tube.
23. Disconnect the A/C electrical connector, and both A/C lines from the compressor. Cover and seal all openings of hoses and compressor to avoid contamination.
24. Disconnect the oil pressure sending unit.
25. Disconnect the coolant hose at the engine oil cooler.
26. Raise and support the vehicle and remove the front wheels.
27 Remove the front fascia.
28. Disconnect the starter wiring and push pin.
29. Remove the flex plate inspection cover.
30. Rotate the engine at the crankshaft pulley bolt and remove the torque converter bolts.
31. Disconnect the rear Oo2s sensor and remove the exhaust system.
32. Separate the steering column intermediate shaft from the steering gear using tool #6831-A.
33. Disconnect the Knock (KS) sensor wiring harness connector and remove the Crankshaft Position (CKP) sensor.
34. Disconnect the in block heater, if equipped.
35. Disconnect the left wheel speed sensor and sensor harness push pin from the body, and set the harness aside.
36. Disconnect the left upper stabilizer link.
37. Disconnect the left strut to steering knuckle bolts and separate the union.
38. Disconnect the right wheel speed sensor and sensor harness push pin from the body, and set the harness aside.
39. Disconnect the right upper stabilizer link.
40. Disconnect the right strut to steering knuckle bolts and separate the union.
41. Disconnect the engine harness ground at the transmission.
42. Disconnect the rear transmission mount from the body.
43. Set up and adjust the cradle support fixture arms and dowels, refer to the accompanying illustrations for cradle support set up. Paint mark the cradle to body location at all four cradle bolt positions.

✲✲ CAUTION

Do not apply the total vehicle weight on the cradle support fixture.

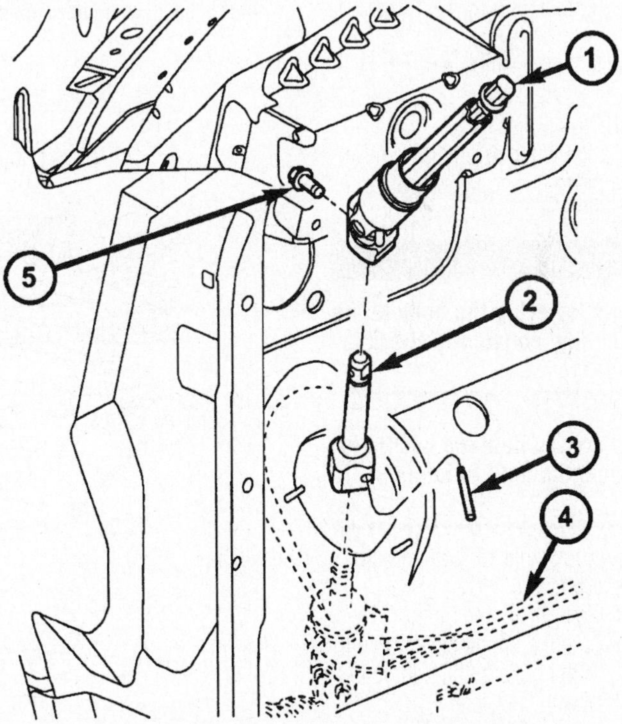

1 - LOWER STEERING COLUMN COUPLING ASSEMBLY
2 - INTERMEDIATE SHAFT TO STEERING GEAR
3 - PIN
4 - STEERING GEAR
5 - EXTENSION TO SHAFT BOLT

06009-PACI-G01

Exploded view of the steering column coupling

44. Lower the front cradle onto the table and remove all four front cradle mount bolts.

➡Use a brake pedal prop rod and depress the brake pedal to prevent excessive brake fluid loss when the brake system is open.

45. Disconnect both front brake lines at the brake controller and route aside.

46. Disconnect the right front engine mount.

47. Remove the left powertrain torque strut to frame bracket bolt.

⁂ CAUTION

While slowly separating the body from the cradle assembly, constant checks must be performed to assure proper positioning and that no damage occurs during separation.

48. Carefully raise the vehicle off of the engine and cradle assembly.

49. Connect an engine lift bracket, using tool # 8534-8 to the left cylinder head and tool # 8534-7 to the right cylinder head.

50. Separate the engine and transmission from the cradle using the lift brackets.

51. Separate the transmission from the engine.

To install:

52. Installation is the reverse of removal, keep in mind the following important steps and torque values.

⁂ CAUTION

While slowly lowering the body to the cradle assembly, constantly check to assure proper positioning and no damage occurs to other components until union is made.

53. Tighten the cradle supporting bolts to 120 ft. lbs. (163 Nm).

54. Install the right engine mount. tighten the bolts to the timing cover to 40 ft. lbs. 54 Nm) and the bolts to the right rail to 50 ft. lbs. (68 Nm).

55. Tighten the rear transmission torque strut bolts to 40 ft. lbs. (54 Nm).

56. Tighten the left front strut to steering knuckle bolts to 60 ft. lbs. (81 Nm) and then tighten an additional 90 degrees.

57. Tighten the left stabilizer link to 65 ft. lbs. (88 Nm).

58. Tighten the right front strut to steer-

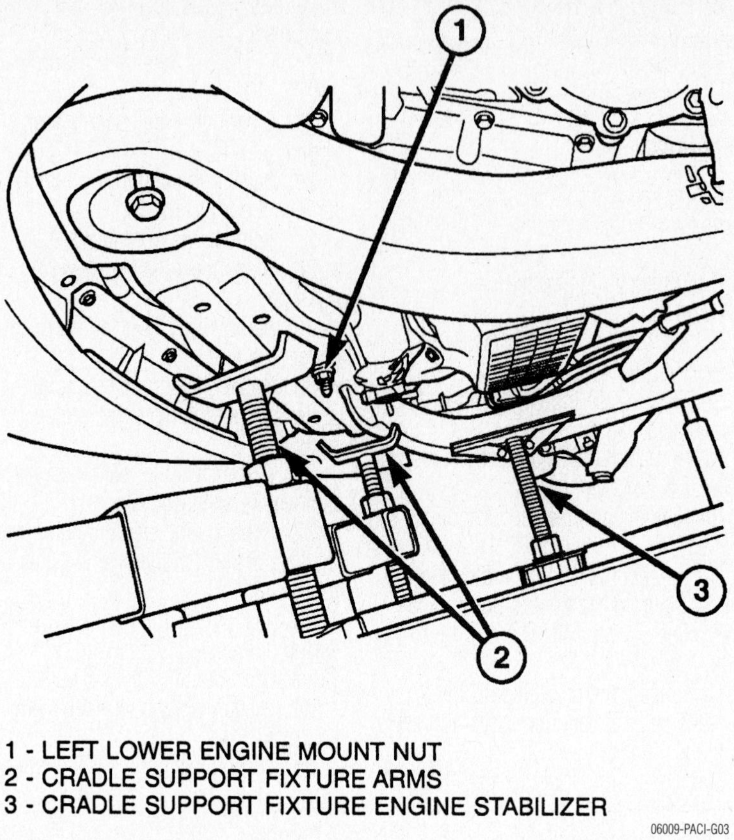

View of the front cradle support positioning

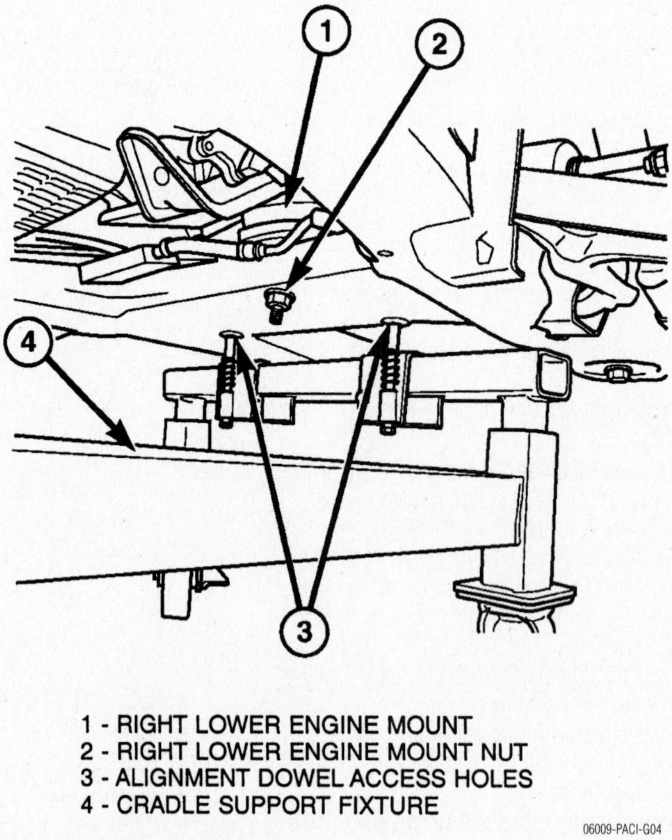

View of the rear cradle support positioning

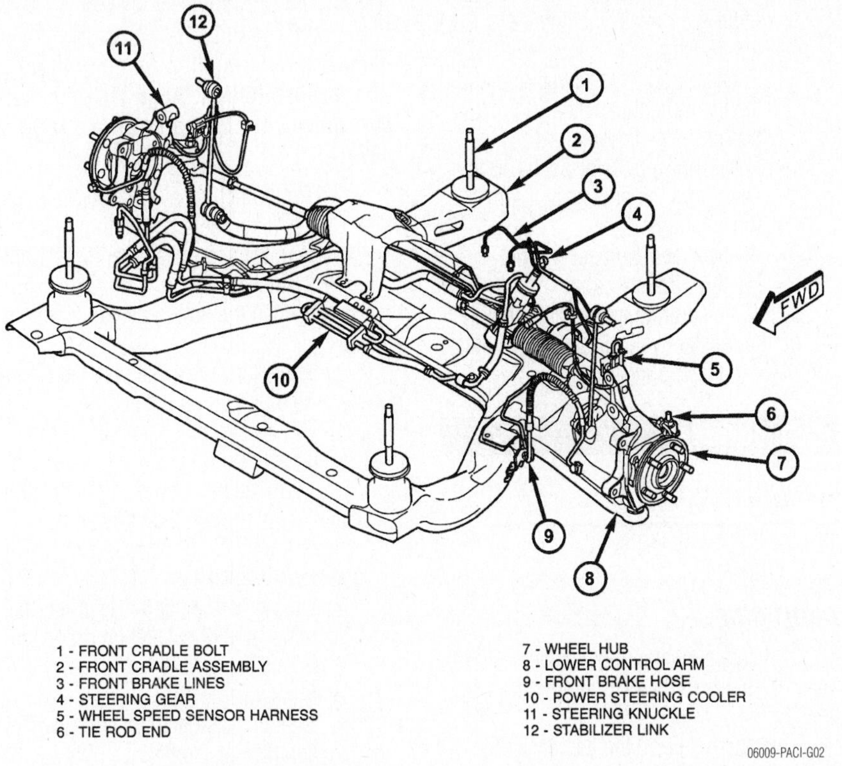

Exploded view of the front cradle assembly

ing knuckle bolts to 60 ft. lbs. (81 Nm) and then tighten an additional 90 degrees.

59. Tighten the right stabilizer link. Tighten to 65 ft. lbs. (88 Nm).
60. Tighten the torque converter bolts to 65 ft. lbs. (88 Nm).
61. Fill the cooling system.
62. Fill the engine with clean oil.
63. Recharge the A/C system.
64. Start the engine and check for leaks.

Water Pump

REMOVAL & INSTALLATION

3.5L Engines

The water pump has a die cast aluminum body and a stamped steel impeller. It bolts directly to the chain case cover using an O-ring for sealing. It is driven by the back side of the serpentine belt.

It is normal for a small amount of coolant to drip from the weep hole located on the water pump body (small black spot). If this condition exists, DO NOT replace the water pump. Only replace the water pump if a heavy deposit or steady flow of brown/green coolant is visible on the water pump body from the weep hole, which would indicate shaft seal failure. Before replacing the water pump, be sure to perform a thorough inspection. A defective pump will not be able to circulate heated coolant through the long heater hose.

1. Before servicing the vehicle, refer to the precautions in the beginning of this section.
2. Drain the cooling system.

✲✲ WARNING

Do not use pliers to open the plastic drain.

3. Remove or disconnect the following:
   - Negative battery cable
   - Coolant recovery cap and open the thermostat bleed valve
   - Timing belt

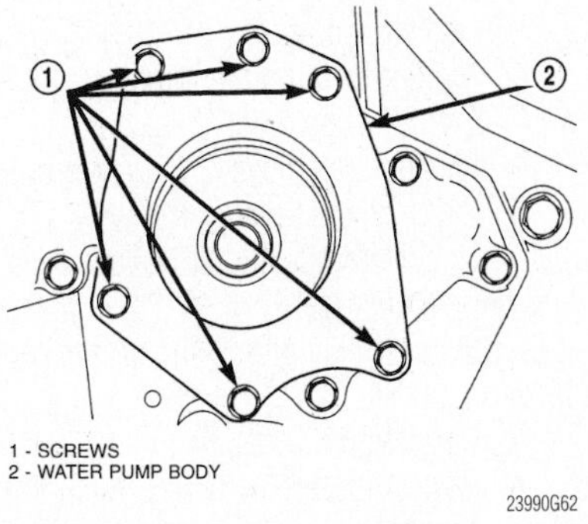

Location of the water pump mounting bolts—3.5L engine

➡It is good practice to turn the crankshaft until the No. 1 cylinder is at Top Dead Center (TDC) of its compression stroke (firing position).

   - Water pump mounting bolts and pump. Discard the O-ring seal.

4. Clean the gasket sealing surfaces, being careful not to scratch the aluminum surfaces.

To install:

5. Install or connect the following:
   - New O-ring and coat with dielectric grease prior to installation
   - Water pump with a new O-ring. Torque the water pump-to-engine bolts to 105 inch lbs. (12 Nm).

✲✲ WARNING

Rotate the pump and check for freedom of movement.

   - Timing belt

6. Fill the cooling system.
7. Reconnect the negative battery cable. Start the engine and allow it to reach normal operating temperatures.
8. Check the cooling system for leaks and correct coolant level.

Heater Core

REMOAVL & INSTALLATION

1. Before servicing the vehicle, refer to the precautions in the beginning of this section.
2. Disable the airbag system, by disconnecting the negative battery cable. Tape the end of the cable to avoid it grounding and wait 2 minutes for the system capacitor to discharge before performing any service.

➡If the vehicle is adjustable pedals, these must be removed prior to heater core removal.

3. Drain the cooling system.

✲✲ WARNING

Do not use pliers to open the plastic drain.

4. Remove or disconnect the following:
   - Negative battery cable if not already done to disable the air bag system
   - Silencer boot fasteners located around the base of the lower steering shaft from the dash panel so it can be moved aside
   - Brake lamp switch from the bracket

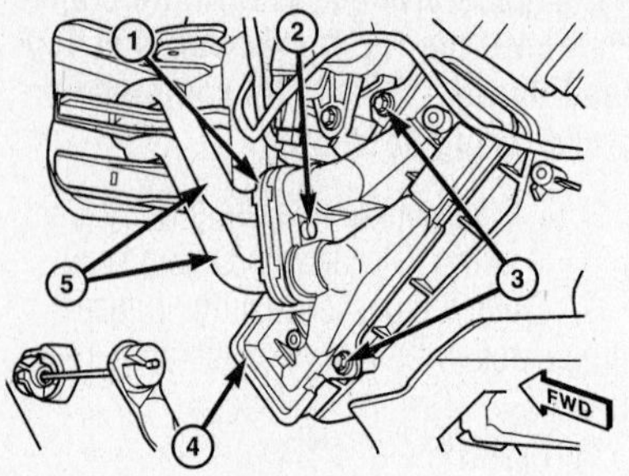

Heater core mounting and related components

- Power brake booster push rod from the pin on the brake pedal arm

5. Place plastic below the heater core inside the vehicle to protect against spills.
6. Remove or disconnect the following:
   - Screw that secures the heater core to tube sealing plate to the supply and return ports
   - Heater core tubes by pushing simultaneously towards the dash panel and disengaging the fittings and cap the tube fittings and heater core ports
   - Two screws that retain the heater core mounting plate to the distribution housing
7. Push the brake pedal downward and at the same time pull the accelerator pedal up to create enough clearance to remove the heater core.

To install:

8. Push the brake pedal downward and at the same time pull the accelerator pedal up to create enough clearance to install the heater core.
   - Two screws that retain the heater core mounting plate to the distribution housing and tighten to 17 inch lbs. (2 Nm)
9. Remove the plugs from the heater core tubes and if using the same heater core the core as well. Position the heater core tubes and sealer plate as a unit beneath the instrument panel.
10. Position both heater core tubes and the sealing plate at the same time to the heater core supply and return ports.
11. The tubes have a slot that must be indexed to a location tab within each of the core ports. Adjust the tube position so that the sealing plate fits flush against the supply and return ports which will ensure the tubes are correctly indexed.
12. Install or connect the following:
   - Screw that retains the heater core tube sealing plate to the supply and return ports and tighten to 27 inch lbs. (3 Nm)
   - Silencer under the driver side end if the instrument panel
   - Heater hoses to the tubes
   - Power brake booster push rod to the pin on the brake pedal arm
   - Brake lamp switch
   - Silencer boot and secure the
   - Negative battery cable
13. Fill the cooling system and operate for two thermostat cycles to assure elimination of air in the cooling system.

Cylinder Head

REMOVAL & INSTALLATION

3.5L Engines

RIGHT SIDE

1. Before servicing the vehicle, refer to the precautions in the beginning of this section.
2. Properly relieve the fuel system pressure.
3. Drain the cooling system.
4. Remove or disconnect the following:
   - Negative battery cable
   - Engine cover
   - Air cleaner assembly
   - Fuel line from the rail
   - Upper intake manifold
   - Lower intake manifold
   - Right front wheel
   - Right inner splash shield
   - Drive belt
   - Crankshaft damper using the proper puller
   - Lower drive belt idler pulley
   - Power steering bolts and set the pump aside
   - Lower timing belt cover bolts
   - Catalytic converter nuts
   - Oxygen (O_2S) sensor electrical wiring at the exhaust manifold and catalytic converter
   - Muffler-to-tail pipe union
   - Catalytic converter
   - Exhaust cross over pipe lower bolts
   - Right exhaust manifold
   - Upper drive belt idler pulley
   - Belt tensioner
5. Support the engine with jackstands and wood blocks.
   - Upper engine mount
   - Power steering reservoir bolts and set aside
   - All remaining outer timing belt cover bolts
6. Rotate the engine until it is at Top Dead center (TDC).

➡Mark the timing belt running direction for installation. Align the camshaft sprockets with the marks on the rear covers.

   - Timing belt tensioner and reset the tensioner. Refer to the timing belt removal and installation procedure in this section.
7. Pre-load the timing belt tensioner as follows:
   a. Place tensioner in a vise the same way it is mounted on the engine.
   b. Slowly compress the plunger into the tensioner body.
   c. Once the plunger is compressed, install a pin through the body and plunger to retain it in place until the tensioner is installed.
   - Timing belt. Refer to the timing belt removal and installation procedure in this section.
   - Right vale cover
   - Exhaust Gas Recirculation (EGR) and tube assembly
   - Right cylinder head cover
   - Right rocker arm assembly
   - Right rear camshaft thrust plate
   - Holding the camshaft gear, unfasten the right cam gear retaining bolt
8. Push the camshaft about 3.5 inches out of the back of the cylinder head and remove the cam gear.
   - Inner timing cover-to-cylinder head bolts
   - Cylinder head bolts in the reverse of tightening sequence. Refer to the cylinder head bolt tightening sequence illustration.
   - Cylinder head and gasket.

To install:

9. Thoroughly clean and dry the mating surfaces of the head and block.

⁂ WARNING

When cleaning the cylinder head and block mating surfaces, do not use a metal scraper because the soft aluminum surfaces could be cut or damaged. Instead, use a scraper made of wood or plastic.

10. Check the cylinder head for cracks, damage or engine coolant leakage. Check the head for flatness. End-to-end, the head should be within 0.002 in. (0.051mm) normally with 0.008 in. (0.203mm) the maximum allowed out of true. The resurface limit is 0.008 in. (0.203mm) maximum, the com-

bined total dimension of stock removal from the cylinder head, if any, and block top surface.

11. Place a new head gasket on the cylinder block locating dowels, being sure the gasket is on the correct side.

12. Inspect the cylinder head bolts for necking (stretching) by holding a straightedge against the threads of each bolt. If all of the threads are not contacting the scale, the bolt should be replaced.

⁂ WARNING

Due to the cylinder head bolt torque method used, it is imperative that the threads of the bolts be inspected for necking prior to installation. If the threads are necked down, the bolt should be replaced. Failure to do so may result in parts failure or damage. New bolts are always recommended.

13. Install the cylinder head into position on the engine block and over the dowels. Install the cylinder head bolts, lubricating the threads with clean engine oil prior to installation.

14. Torque the cylinder head bolts using the proper sequence as follows:

a. Step 1: torque in sequence to 45 ft. lbs. (61 Nm).

b. Step 2: torque in sequence to 65 ft. lbs. (88 Nm).

c. Step 3: torque in sequence to 65 ft. lbs. (88 Nm).

d. Step 4: torque in sequence an additional ¼ turn by using a torque angle meter.

➡Inspect the bolt torque after tightening. The torque should be over 90 ft. lbs. (122 Nm). If not, replace the cylinder head bolt.

15. Install or connect the following:

- Inner timing cover-to-cylinder head bolts and tighten to 40 ft. lbs. (54 Nm)
- Cam gear, holding the camshaft gear, fasten the right cam gear retaining bolt to 75 ft. lbs. (102 Nm) plus an additional ¼ turn
- Rear camshaft thrust plate

16. Rotate the camshaft gear until it is properly aligned and check the left camshaft gear and crankshaft timing alignment marks.

- Timing belt. Refer to the timing belt removal and installation procedure in this section.
- Timing belt outer cover
- Power steering reservoir
- Crankshaft damper
- Upper engine mount
- Belt tensioner
- Upper drive belt idler pulley
- Right exhaust manifold
- Exhaust cross over pipe
- Catalytic converter
- Muffler-to-tail pipe union
- O_2S sensor electrical wiring at the exhaust manifold and catalytic converter
- Right rocker arm assembly
- Right cylinder head cover
- EGR and tube assembly
- Lower drive belt idler pulley
- Right vale cover
- Lower intake manifold and fuel rail
- Upper intake manifold
- Fuel line to the rail
- Power steering pump
- Drive belt
- Air cleaner assembly
- Engine cover
- Right inner splash shield
- Right front wheel
- Negative battery cable

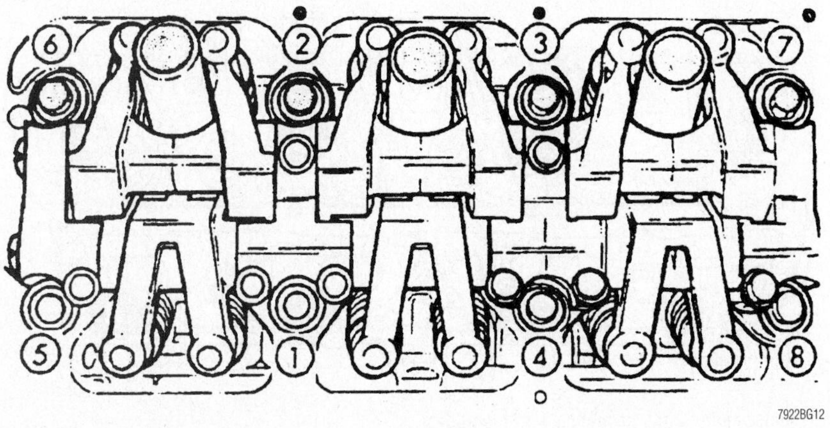

Cylinder head bolt tightening sequence—3.5L engine

LEFT SIDE

1. Before servicing the vehicle, refer to the precautions in the beginning of this section.

2. Properly relieve the fuel system pressure.

3. Drain the cooling system.

4. Remove or disconnect the following:

- Negative battery cable
- Engine cover
- Air cleaner assembly
- Radiator close out panel
- Radiator core support and fan assembly
- Fuel line from the rail
- Upper intake manifold
- Lower intake manifold
- Crankshaft damper
- Upper engine mount
- Upper drive belt idler pulley
- Belt tensioner
- Left exhaust manifold
- Exhaust cross over pipe
- Outer timing belt cover

5. Rotate the engine until it is at Top Dead center (TDC).

➡Mark the timing belt running direction for installation. Align the camshaft sprockets with the marks on the rear covers.

- Timing belt tensioner and reset the tensioner. Refer to the timing belt removal and installation procedure in this section.

6. Pre-load the timing belt tensioner as follows:

a. Place tensioner in a vise the same way it is mounted on the engine.

b. Slowly compress the plunger into the tensioner body.

c. Once the plunger is compressed, install a pin through the body and plunger to retain it in place until the tensioner is installed.

- Timing belt. Refer to the timing belt removal and installation procedure in this section.
- Left cylinder head cover
- Left rocker arm assembly
- Left rear camshaft thrust plate
- Holding the camshaft gear, unfasten the right cam gear retaining bolt

7. Push the camshaft about 3.5 inches out of the back of the cylinder head and remove the cam gear.

- Front timing belt housing-to-cylinder head bolts
- Cylinder head bolts in the reverse of tightening sequence. Refer to the

cylinder head bolt tightening sequence illustration.
- Cylinder head and gasket.

To install:

8. Thoroughly clean and dry the mating surfaces of the head and block.

⁂ WARNING

When cleaning the cylinder head and block mating surfaces, do not use a metal scraper because the soft aluminum surfaces could be cut or damaged. Instead, use a scraper made of wood or plastic.

9. Check the cylinder head for cracks, damage or engine coolant leakage. Check the head for flatness. End-to-end, the head should be within 0.002 in. (0.051mm) normally with 0.008 in. (0.203mm) the maximum allowed out of true. The resurface limit is 0.008 in. (0.203mm) maximum, the combined total dimension of stock removal from the cylinder head, if any, and block top surface.
10. Place a new head gasket on the cylinder block locating dowels, being sure the gasket is on the correct side.
11. Inspect the cylinder head bolts for necking (stretching) by holding a straightedge against the threads of each bolt. If all of the threads are not contacting the scale, the bolt should be replaced.

⁂ WARNING

Due to the cylinder head bolt torque method used, it is imperative that the threads of the bolts be inspected for necking prior to installation. If the threads are necked down, the bolt should be replaced. Failure to do so may result in parts failure or damage. New bolts are always recommended.

12. Install the cylinder head into position on the engine block and over the dowels. Install the cylinder head bolts, lubricating the threads with clean engine oil prior to installation.
13. Torque the cylinder head bolts using the proper sequence as follows:
    a. Step 1: torque in sequence to 45 ft. lbs. (61 Nm).
    b. Step 2: torque in sequence to 65 ft. lbs. (88 Nm).
    c. Step 3: torque in sequence to 65 ft. lbs. (88 Nm).
    d. Step 4: torque in sequence an additional ¼ turn by using a torque angle meter.

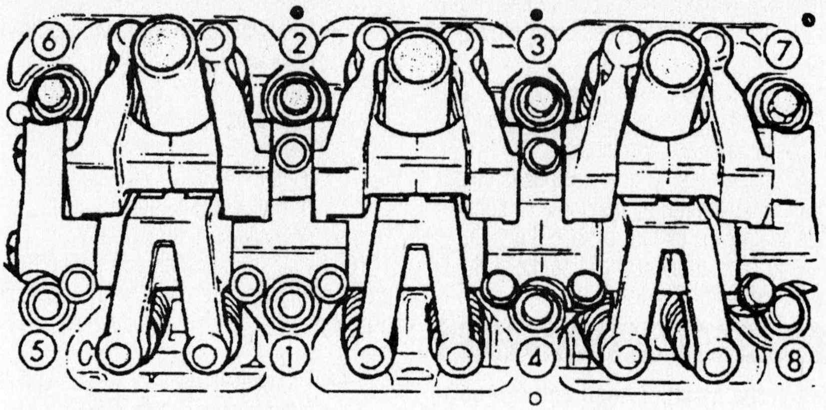

Cylinder head bolt tightening sequence—3.5L engine

➡Inspect the bolt torque after tightening. The torque should be over 90 ft. lbs. (122 Nm). If not, replace the cylinder head bolt.

14. Install or connect the following:
    - Inner timing cover-to-cylinder head bolts and tighten to 40 ft. lbs. (54 Nm)
    - Cam gear, holding the camshaft gear, fasten the right cam gear retaining bolt to 75 ft. lbs. (102 Nm) plus an additional ¼ turn
    - Rear camshaft thrust plate
15. Rotate the camshaft gear until it is properly aligned and check the left camshaft gear and crankshaft timing alignment marks.
    - Timing belt. Refer to the timing belt removal and installation procedure in this section.
    - Timing belt front cover
    - Crankshaft damper
    - Upper engine mount
    - Belt tensioner
    - Upper drive belt idler pulley
    - Left exhaust manifold
    - Exhaust cross over pipe
    - Left rocker arm assembly
    - Left cylinder head cover
    - Lower drive belt idler pulley
    - Lower intake manifold and fuel rail
    - Upper intake manifold
    - Fuel line to the rail
    - Radiator core support and fan assembly
    - Radiator close out panel
    - Drive belt
    - Air cleaner assembly
    - Engine cover
    - Negative battery cable

3.8L Engines

1. Before servicing the vehicle, refer to the precautions in the beginning of this section.
2. Drain the cooling system.
3. Relieve the fuel system pressure.
4. Remove or disconnect the following:
    - Negative battery cable
    - Upper and lower intake manifolds

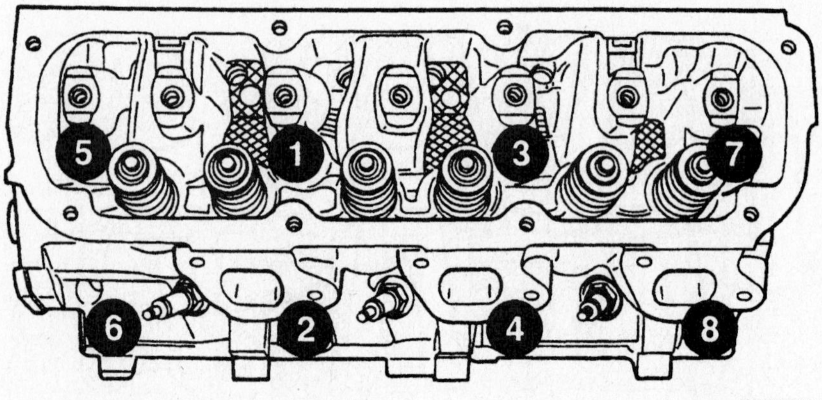

Cylinder head torque sequence—3.8L engines

✲✲ CAUTION

The intake manifold gaskets are made of a very thin metal and can cause injury if not properly handled.

- Valve covers.
- Spark plugs
- Dipstick tube
- Exhaust manifolds

➡Keep all valvetrain components in order for assembly

- Rocker arm and shaft assemblies
- Cylinder head bolts, heads and gaskets

5. Clean all gasket mating surfaces.

To install:

6. Examine the cylinder head bolts and replace any that have stretched.
7. Install the cylinder head with a new gasket. The left bank gasket has an L on it and is located at the front of the engine. The right gasket has an R stamped on it and is located at the rear of the engine. Tighten the bolts in sequence, as follows:
   a. Step 1: Tighten bolts 1–8 to 45 ft. lbs. (61 Nm).
   b. Step 2: Tighten bolts 1–8 to 65 ft. lbs. (88 Nm).
   c. Step 3: Tighten bolts 1–8 to 65 ft. lbs. (88 Nm).
   d. Step 4: Bolts 1–8 plus 90 degrees.
8. Check that the torque on bolts 1–8 has exceeded 90 ft. lbs. (122 Nm). If not, replace the bolt.
9. Install or connect the following:
   - Pushrods
   - Rocker arm and shaft assemblies
   - Valve covers
   - Exhaust manifolds
   - Dipstick tube with a new O–ring
   - Spark plugs
   - Intake manifolds
   - Negative battery cable
10. Fill the cooling system.
11. Start the engine and check for leaks.

Rocker Arm/Shafts

REMOVAL & INSTALLATION

3.5L Engines

1. Before servicing the vehicle, refer to the precautions in the beginning of this section.
2. Relieve the fuel system pressure.
3. Remove or disconnect the following:
   - Negative battery cable
   - Air cleaner assembly and the intake manifold plenum

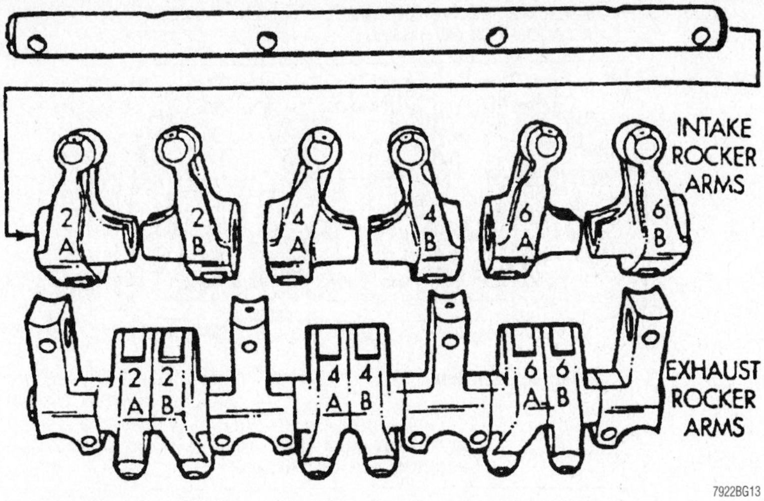

Left bank rocker arm and shaft identification—3.5L engine

➡Cover the lower intake manifold during service.

- Cylinder head covers
- Rocker arm assembly

4. Inspect the rocker arms for wear or damage. Inspect the roller for scuffing or wear. Replace assembly as necessary.

✲✲ WARNING

Do not remove the lash adjusters from the rocker arm assembly. The rocker arm and the adjuster are serviced as an assembly.

5. Identify the rocker arm assemblies and rocker arms and disassemble the shaft as follows:
   a. Thread a nut, washer and spacer onto a 4mm screw.
   b. Insert and tighten the 4mm screw into the dowel pin on the shaft.
   c. Loosen the nut on the screw. This will pull the dowel pin from the shaft support.
   d. Remove the rocker arms and pedestals, keeping them in order.
   e. Check the oil holes for restrictions with a small wire and clean as required.

To install:

6. Assemble the rocker shaft as follows:
   a. Install the rocker arms and pedestals onto the shaft, keeping them in the original order.
   b. Press the dowel pins into the pedestals until they bottom out in the pedestals.
7. Position the camshaft so that the timing mark on the right camshaft timing belt sprocket aligns with the timing mark on the

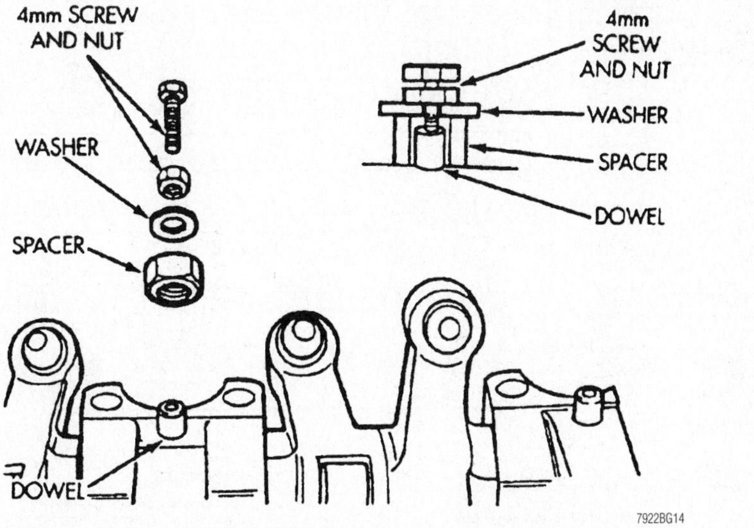

Remove the dowel pin using a 4mm screw, nut, spacer and washer installed into the pin—3.5L engine

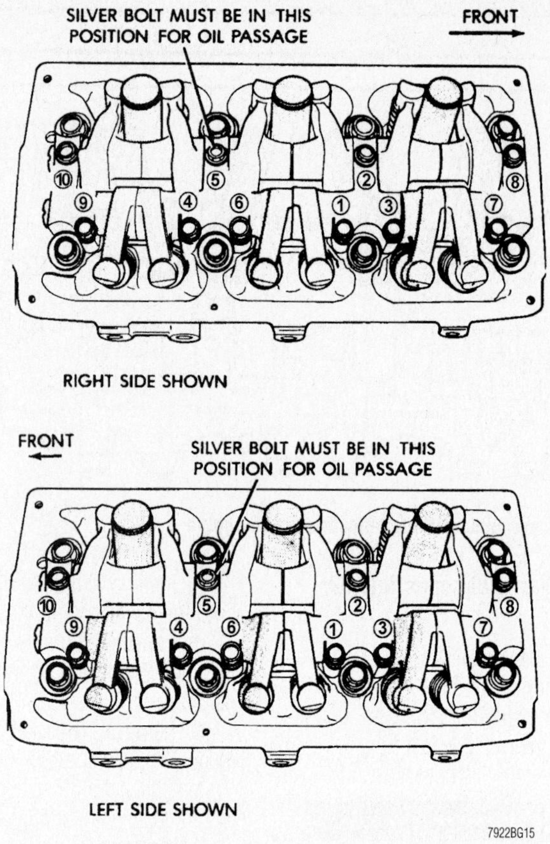

Proper torque sequence for the rocker arm and shaft assemblies—3.5L engine

rear timing belt cover and the timing mark on the left sprocket is 45 degrees from the mark on the rear timing belt cover. There will be no load on the shaft during installation. Install the rocker shafts so the identification marks are facing toward the front of the engine.

8. Install the oil feed bolt in the correct location on the rocker shaft retainer. Torque the bolts in proper sequence to 23 ft. lbs. (31 Nm).
9. Install or connect the following:
   - Valve covers and torque the bolts to 105 inch lbs. (12 Nm)

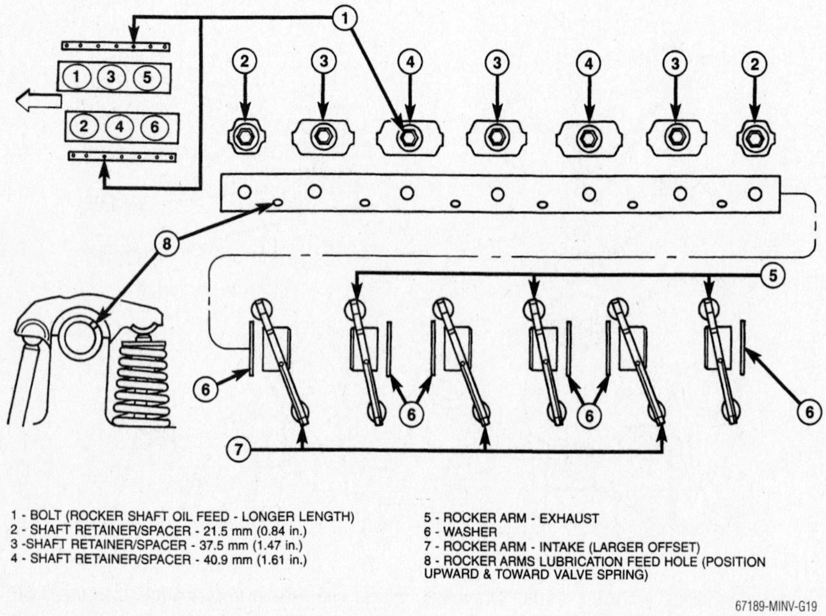

Rocker arm and shaft assembly—3.8L engines

- Intake manifold plenum
- Air cleaner assembly
- Negative battery cable

3.8L Engines

1. Before servicing the vehicle, refer to the precautions in the beginning of this section.
2. Remove or disconnect the following:
   - Negative battery cable
   - Valve covers
   - Rocker arm and shaft assemblies

➡Keep all valvetrain components in order for assembly.

To install:

➡Do not rotate the crankshaft during or immediately after rocker arm installation. Wait 20 minutes for the hydraulic lash adjusters to bleed down.

3. Install or connect the following:
   - Rocker arm and shaft assemblies. Tighten the bolts to several passes to 200 inch lbs. (23 Nm).
   - Valve covers
   - Negative battery cable
4. Start the engine and check for proper operation.

Intake Manifold

REMOVAL & INSTALLATION

3.5L Engines

UPPER

1. Before servicing the vehicle, refer to the precautions in the beginning of this section.
2. Properly relieve the fuel system pressure.
3. Drain the cooling system.
4. Remove or disconnect the following:
   - Negative battery cable
   - Air cleaner assembly
   - Engine cover from the top of the intake manifold
   - Accelerator and the speed control cable from the throttle lever
   - Two cable bracket bolts and position the bracket aside
   - Oil dipstick tube-to-upper intake bolt
5. Disconnect the following electrical connections:
   - Manifold Tuning Valve (MTV)
   - Sort Runner (SR) valve
   - Exhaust Gas Recirculation (EGR) valve
   - Throttle Position (TP) sensor

- Idle Air Control (IAC) motor
- Intake Air Temperature (IAT) sensor
- Manifold Absolute Pressure (MAP) sensor

6. Remove or disconnect the following:
   - Alternator support bracket
   - EGR tube
7. Disconnect the following vacuum lines:
   - Positive Crankcase Ventilation (PCV) valve
   - Evaporative Emissions (EVAP) solenoid
   - Brake booster
   - Front and rear manifold support brackets
   - Intake plenum mounting bolts
8. Remove the intake manifold plenum from the intake manifold.

➡Discard the old gasket. Cover the intake manifold openings with tape to keep debris from entering the engine.

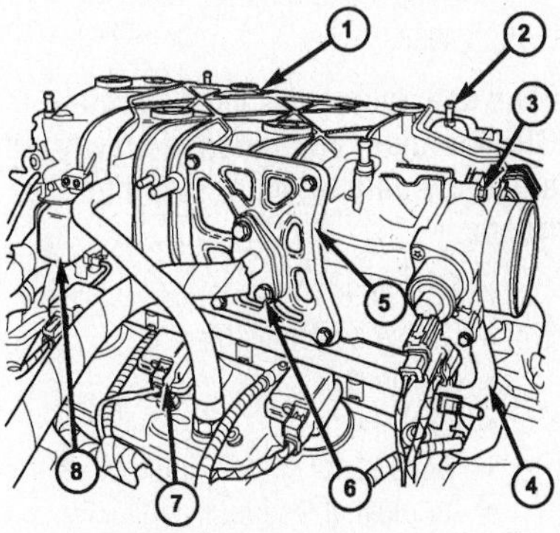

Upper intake manifold and related components—3.5L engine

To install:

9. Install or connect the following:
   - Intake manifold plenum with a new gasket in place. Torque the mounting bolts, working from the center outward, to 105 inch lbs. (12 Nm)

➡Do not overtighten bolts when working with light alloys.

   - Oil dipstick tube-to-upper intake bolt
   - EGR tube
10. Connect the following vacuum lines:
    - PCV valve
    - EVAP solenoid
    - Brake booster
11. Install or connect the following:
    - Alternator support bracket
12. Connect the following electrical connections:
    - (MTV valve
    - Sort Runner (SR) valve
    - EGR valve
    - TP sensor
    - IAC motor
    - IAT sensor
    - MAP sensor
13. Install or connect the following:
    - Cable bracket and bolts
    - Accelerator and the speed control cable to the throttle lever
    - Air cleaner assembly
    - Engine cover
    - Negative battery cable
14. Fill and bleed the cooling system.
15. Change the engine oil and filter.
16. Test run the engine, check for fuel and coolant leaks and verify correct engine operation.

LOWER

1. Before servicing the vehicle, refer to the precautions in the beginning of this section.
2. Remove or disconnect the following:
   - Upper radiator hose from the thermostat housing
   - Upper intake manifold
   - Power steering reservoir and position aside
   - Fuel injector and Coolant temperature (CTS) sensor electrical connections
   - Exhaust crossover bolt
   - Heater hose from the intake manifold
   - Coolant container hose at the thermostat housing
   - Fuel supply hose from the rail
   - Fuel rail bolts and rail with the injectors as an assembly
   - Lower intake manifold bolts and manifold

➡Clean all gasket mating surfaces and inspect for distortion with a good straightedge.

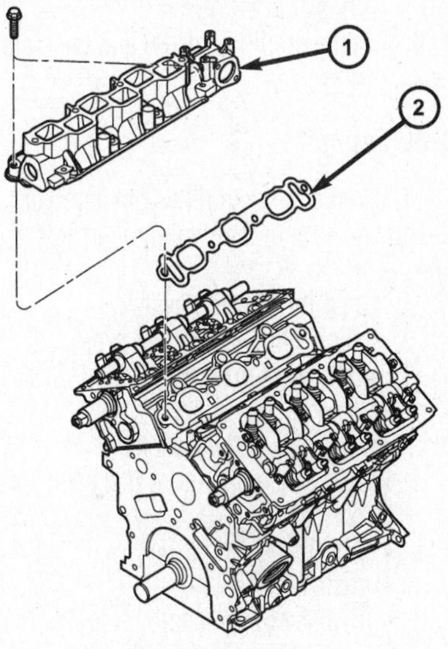

Lower intake manifold and gasket—3.5L engine

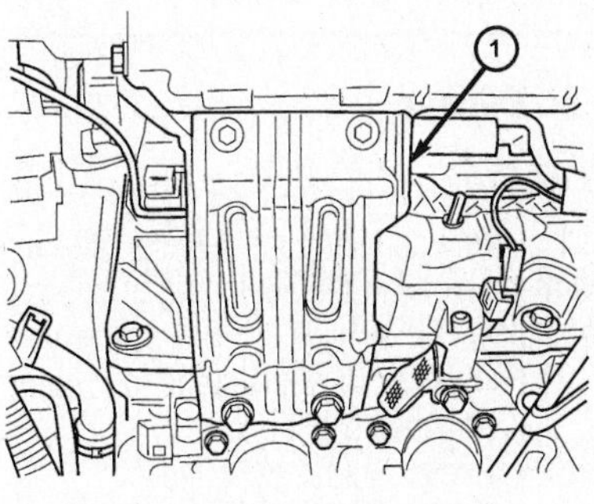

Upper intake manifold support bracket–left shown, right similar—3.5L engine

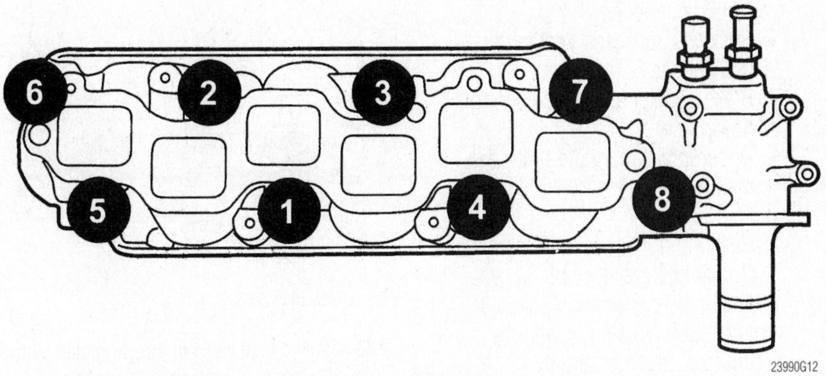

Lower intake manifold tightening sequence—3.5L engine

To install:

➡**Verify that all intake manifold and cylinder head sealing surfaces are clean.**

3. Install or connect the following:
   - Intake manifold gasket, then the lower manifold. Torque the bolts in the proper sequence to 21 ft. lbs. (28 Nm).
   - Fuel rail bolts and rail with the injectors as an assembly
   - Fuel supply hose to the rail
   - Heater hose to the intake manifold
   - Coolant container hose to the thermostat housing
   - Exhaust crossover bolt
   - Fuel injector and CTS sensor electrical connections
   - Power steering reservoir
   - Upper intake manifold
   - Upper radiator hose from the thermostat housing
4. Fill the cooling system and check for leaks.

3.8L Engines

1. Before servicing the vehicle, refer to the precautions in the beginning of this section.
2. Drain the cooling system.
3. Relieve the fuel system pressure.
4. Remove or disconnect the following:
   - Negative battery cable
   - Idle Air Temperature (IAT) sensor connector
   - Air cleaner and tube
   - Accelerator cable
   - Cruise control cable
   - EVAP hose from the throttle body
   - Exhaust Gas Recirculation (EGR) tube
   - Idle Air Control (IAC) motor and Throttle Position (TP) sensor connectors
   - Manifold Absolute Pressure (MAP) sensor connector
   - Positive Crankcase Ventilation (PCV) hose
   - Brake booster and cruise control hoses
   - Upper intake manifold bolts and manifold
   - Fuel line
   - Heater supply hose and Engine Coolant temperature (ECT) sensor
   - Fuel injector harness connectors
   - Fuel supply manifold with the injectors as an assembly
   - Upper radiator hose
   - Lower intake manifold bolts and manifold
   - Intake manifold seal retainer screws and the gasket

⁂ CAUTION

The intake manifold gasket is made of thin metal, handle with care to avoid injury.

5. Clean all gasket mating surfaces.

To install:

6. Install a ¼ inch bead of gasket maker onto each of the 4 manifold-to-cylinder head gasket corners.
7. Install the lower intake manifold gasket and tighten the seal retainer screws to 105 inch lbs. (12 Nm)

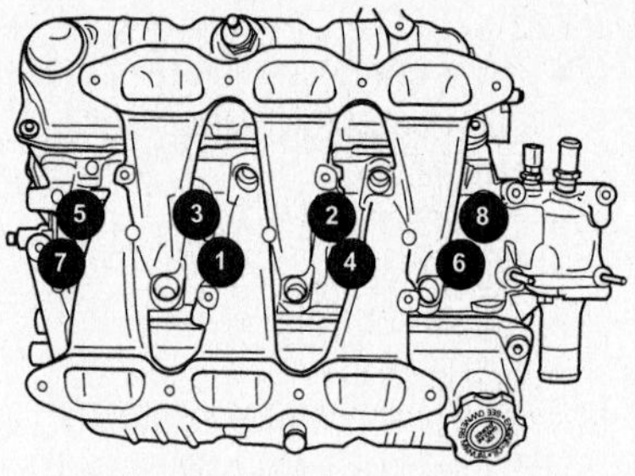

Lower intake manifold torque sequence—3.8L engines

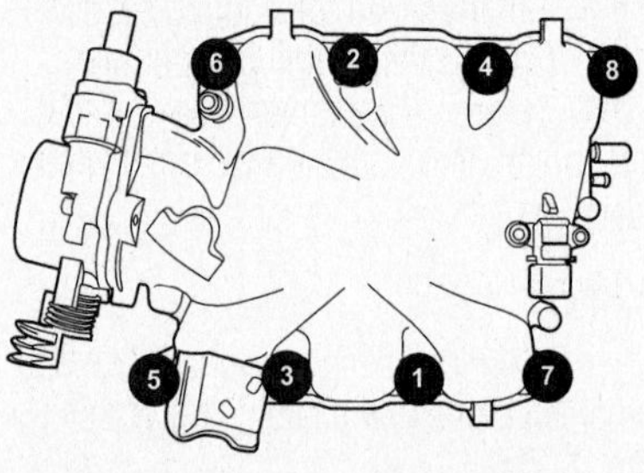

Upper intake manifold torque sequence—3.8L engines

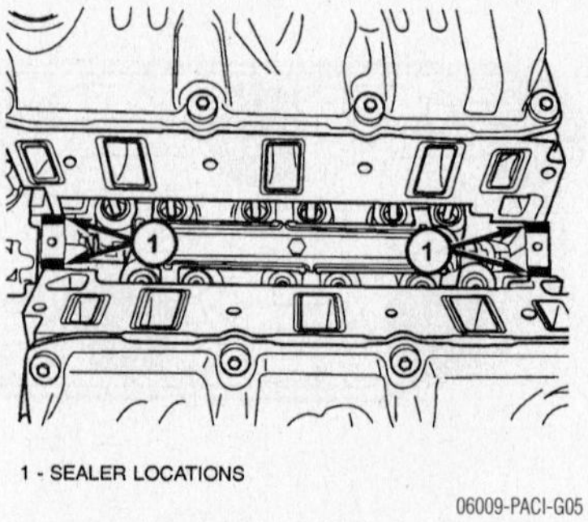

Install a ¼ inch bead of gasket maker onto each of the 4 manifold-to-cylinder head gasket corners—3.8L engines

8. Install the lower intake manifold and tighten the bolts in sequence, as follows:
   a. Step 1: 10 inch lbs. (1 Nm).
   b. Step 2: 17 ft. lbs. (22 Nm).
   c. Step 3: 17 ft. lbs. (22 Nm).
9. Install or connect the following:
   - Fuel supply manifold and injectors as an assembly
   - Fuel injector harness connectors
   - ECT sensor connector
   - Heater supply hose
   - Radiator hose
   - Fuel line
10. Place a new gasket in the channel and press lightly in place.
    - Upper intake manifold. Apply Mopar thread and lock adhesive (medium strength) to each upper intake manifold bolt. Tighten the bolts to 105 inch lbs. (12 Nm) in the sequence illustrated.
    - MAP sensor connector
    - Brake booster and cruise control hoses

➡**The special screws used to retain the EGR tube must be tightened slowly with hand tools only to avoid stripping the threads.**

   - EGR tube
   - Cruise control cable
   - Accelerator cable
   - EVAP hose
   - IAC and TP sensor connectors
   - Air cleaner and tube
   - IAT connector
   - PCV hose
   - Negative battery cable
11. Fill the cooling system.
12. Start the engine, check for leaks and repair if necessary.

Exhaust Manifold

REMOVAL & INSTALLATION

3.5L Engines

LEFT SIDE

1. Before servicing the vehicle, refer to the precautions in the beginning of this section.
2. Remove or disconnect the following:
   - Negative battery cable
   - Radiator close out panel
   - Radiator core support and cooling fan assembly
   - Oil dipstick tube bolt and move the tube aside

- Exhaust manifold crossover pipe bolts
- Exhaust manifold bolts and manifold

3. Inspect the manifold for damage or cracks. Check for distortion against a straight-edge or thickness gauge. Replace manifold if required.

4. Remove all traces of the old manifold gasket and clean both gasket mating surfaces.

To install:

- Exhaust manifold and gasket, tighten the bolts to 200 inch lbs. (23 Nm)
- Exhaust manifold crossover pipe and tighten the bolts to 275 inch lbs. (31 Nm)
- Oil dipstick tube and secure the bolt
- Radiator cooling fan assembly and core support

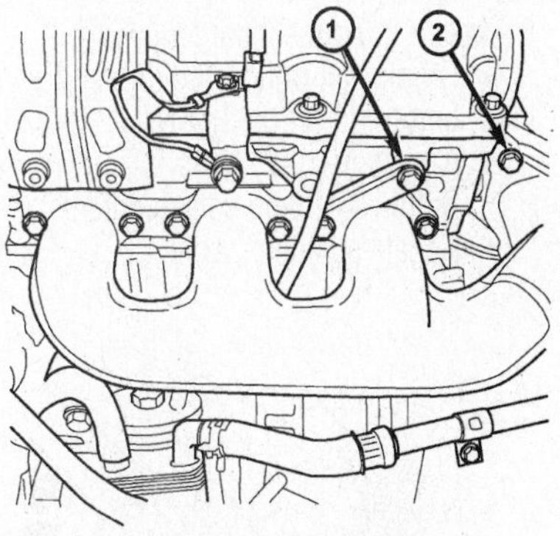

The oil dipstick tube and exhaust crossover pipe bolts must be removed prior to removing the manifold—3.5L engine

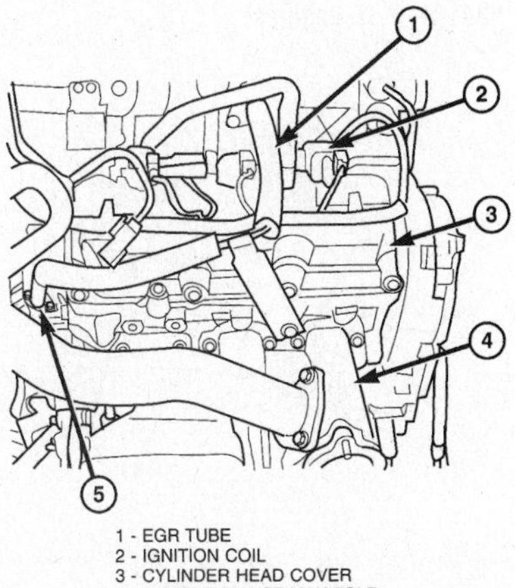

Exhaust Gas Recirculation (EGR) valve and tube assembly—3.5L engine

- Radiator close out panel
- Negative battery cable

RIGHT SIDE

1. Before servicing the vehicle, refer to the precautions in the beginning of this section.

2. Remove or disconnect the following:
   - Negative battery cable
   - Exhaust manifold crossover pipe bolts
   - Downstream Oxygen (O_2S) sensor electrical wiring
   - Exhaust manifold flange bolts and the hangers from the pipe assembly
   - Exhaust system from the vehicle
   - Exhaust Gas Recirculation (EGR) tube support bracket bolt
   - Lower manifold bolts
   - Manifold and gasket
   - O_2S from the manifold

3. Inspect the manifold for damage or cracks. Check for distortion against a straight-edge or thickness gauge. Replace manifold if required.

4. Remove all traces of the old manifold gasket and clean both gasket mating surfaces.

To install:

5. Install or connect the following:
   - Exhaust manifold and gasket, tighten the bolts to 200 inch lbs. (23 Nm)
   - Exhaust system to the vehicle
   - Exhaust manifold flange bolts and the hangers to the pipe assembly. tighten the flange bolts to 22 ft. lbs. (30 Nm).
   - Downstream O_2S sensor electrical wiring
   - EGR tube support bracket bolt
   - Exhaust manifold crossover pipe bolts and tighten to 275 inch lbs. (31 Nm)
   - O_2S to the manifold
   - Negative battery cable

3.8L Engines

RIGHT SIDE

1. Disconnect the negative battery cable.

2. Raise and safely support the vehicle.

3. Separate the exhaust pipe at the exhaust manifold.

4. Remove the bolts attaching the crossover pipe to exhaust manifold.

5. Remove the air cleaner housing.

6. Disconnect the Throttle Position (TP), Idle Air Control (IAC) sensor connectors and the purge vacuum hose from the throttle body.

7. Disconnect the upstream Oxygen (O_2S) sensor connector.

8. Disconnect the spark plug wires from the plugs.

9. Remove bolts attaching the exhaust manifold to the cylinder head and remove the manifold.

10. Discard gasket, if equipped and clean all mating surfaces.

11. Position the exhaust manifold on the cylinder head and install the bolts to the center runner (cyl # 3), and temporarily tighten to 25 inch lbs. (2.8 Nm).

➡Examine the crossover pipe bolts for damage caused due to heat or corrosion and replace using new OEM bolts if found to be defective.

12. Install a new gasket, attach the crossover pipe to the manifold and tighten the bolts to 30 ft. lbs. (41 Nm).

13. Install or connect the following:
    - Remaining manifold bolts. Torque the fasteners to 17 ft. lbs. (23 Nm).

14. Install and connect the upstream O_2S sensor.

15. Connect the spark plug.

16. Connect the IAC, TPS and the EVAP hose to the throttle body.

17. Install the air cleaner housing.

18. Attach the catalytic converter pipe to exhaust manifold using a new gasket and tighten bolts to 27 ft. lbs. (37 Nm).

19. Connect the negative battery cable.

20. Start the engine and check for leaks.

LEFT SIDE

1. Before servicing the vehicle, refer to the precautions in the beginning of this section.

2. Remove or disconnect the following:
   - Negative battery cable
   - Crossover pipe
   - Spark plug wires
   - Heat shield
   - Exhaust manifold bolt
   - Exhaust manifold and gasket

To install:

3. Position the exhaust manifold on the cylinder head and install the bolts to the center runner (cyl # 4), and temporarily tighten to 25 inch lbs. (2.8 Nm).

➡Examine the crossover pipe bolts for damage caused due to heat or corrosion and replace using new OEM bolts if found to be defective.

4. Install a new gasket, attach the crossover pipe to the manifold and tighten the bolts to 30 ft. lbs. (41 Nm).

5. Install or connect the following:

- Remaining manifold bolts. Torque the fasteners to 17 ft. lbs. (23 Nm).
- Heat shield
- Negative battery cable

6. Start the engine and check for leaks.

Front Crankshaft Seal

REMOVAL & INSTALLATION

3.5L Engines

Note that the timing belt must be removed from the vehicle to perform this service. Use care to be sure all valve timing marks are carefully aligned both before removing the belt and after belt installation and all service has been completed. It may be good practice to set the engine to Top Dead Center (TDC) No. 1 cylinder compression stroke (firing position) and aligning all timing marks before removing the timing belt. This serves as a reference for all work that follows.

1. Before servicing the vehicle, refer to the precautions at the beginning of this section.
2. Properly relieve the fuel system pressure.
3. Drain the cooling system.
4. Remove or disconnect the following:
   - Negative battery cable
   - Radiator and cooling fan module assembly
   - Accessory drive belts
   - Crankshaft damper bolt
   - Timing belt front cover

➡The sealer on the timing belt front cover may be reusable and should not be removed. Use silicone rubber adhesive sealant to replace any missing sealer.

   - Timing belt and tensioner. Refer to the appropriate procedure for removal and installation.
   - Crankshaft timing belt sprocket

5. Locate the small dowel pin in the crankshaft. With a small punch, carefully tap out the dowel from the end of the crankshaft.
6. Remove the crankshaft seal using tool 6341A, taking care not to nick the shaft seal surface or seal bore during removal.

To install:

7. Inspect the crankshaft seal lip surface for varnish and dirt. Polish the area using 400 grit sandpaper to remove varnish as necessary.
8. Install or connect the following:
   - Crankshaft seal using seal installer tool 6342
   - Rear lower timing belt cover
   - Dowel into the crankshaft so that it protrudes 0.047 in. (1.2mm)
   - Timing belt sprocket at the crankshaft using tool C-4685C1, thrust bearing, washer and 12mm bolt or an equivalent setup to pull the sprocket onto crankshaft. Do not hammer on the sprocket.
9. Verify that all valve timing marks are aligned.
   - Timing belt and tensioner using the recommended procedure
10. Rotate the crankshaft 2 complete turns and recheck the timing marks on the camshafts and crankshaft. The marks must align with their respective locations. If the marks do not align, repeat the timing belt installation procedure. When correct valve timing has been verified, install the timing belt covers.
    - Crankshaft damper. Hold the crankshaft damper, using tool L-3281, and torque the bolt to 85 ft. lbs. (115 Nm).
    - Accessory drive belts and adjust to the proper tension
    - Radiator and cooling fan assemblies
    - Negative battery cable
11. Fill and bleed the cooling system.
12. Start the vehicle, check for leaks and repair if necessary.

3.8L Engines

1. Before servicing the vehicle, refer to the precautions in the beginning of this section.
2. Remove or disconnect the following:
   - Negative battery cable
   - Right wheel and splash shield
   - Accessory drive belts
   - Crankshaft pulley
3. Place tool 6341A on the crankshaft nose and screw the tool into the seal until it is engaged being careful not to damage the crankshaft seal surface cover.
4. Turn the forcing screw on the tool to remove the seal

To install:

5. Position guide C–4992–2 on the crankshaft nose.
6. Position a new seal over the guide with the seal spring facing the engine cover.
7. Install the seal using tool C–42992–1 until the seal is flush with the cover and remove the tools.
8. Install or connect the following:
   - Crankshaft pulley and tighten the bolt to 40 ft. lbs. (54 Nm)
   - Accessory drive belts
   - Right wheel and splash shield
   - Negative battery cable

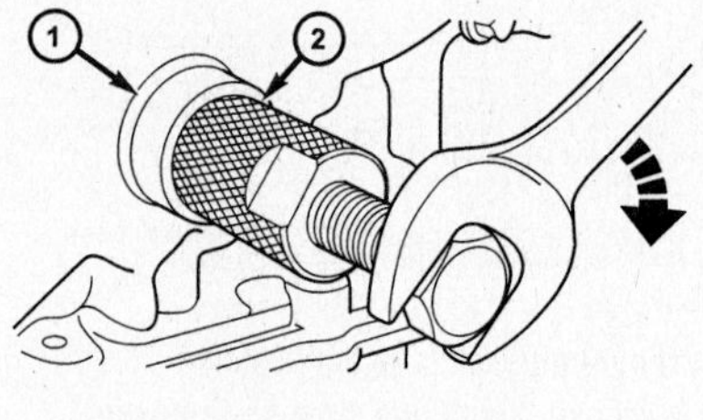

1 - SEAL
2 - SPECIAL TOOL 6341A

67189-MINV-G38

Removing the front seal using tool 6341A—3.8L engines

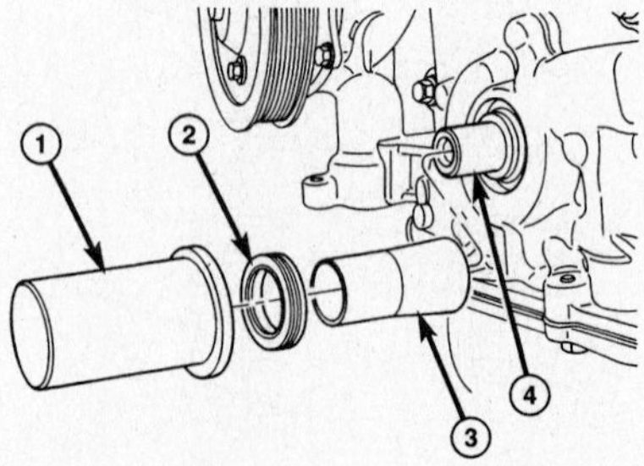

1 - SPECIAL TOOL C-4992-1
2 - SEAL
3 - SPECIAL TOOL C-4992-2
4 - CRANKSHAFT

67189-MINV-G35

Removing the front seal using tools C–4992–2 and C–42992–1—3.8L engines

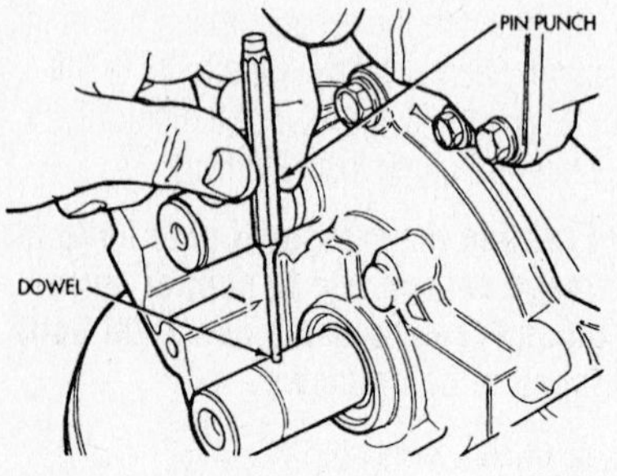

7922BG23

Removing the timing belt sprocket dowel pin from the crankshaft—3.5L engine

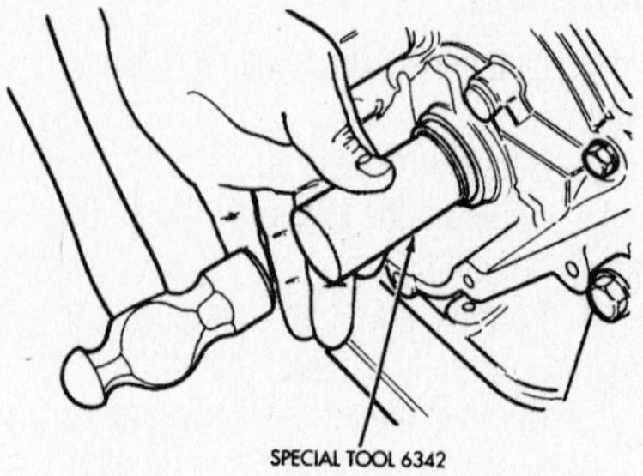

7922BG24

Installing the crankshaft oil seal—3.5L engine

Camshaft and Valve Lifters

REMOVAL & INSTALLATION

3.5L Engines

1. Before servicing the vehicle, refer to the precautions in the beginning of this section.

Camshafts are serviced from the rear of the cylinder head. Although the engine does not need to be removed for camshaft service, the cylinder head must be removed from the vehicle. Note too, that the camshaft sprockets have a D-shaped hole that allows it to rotate several degrees in each direction on its shaft.

2. Properly relieve the fuel system pressure.
3. Drain the cooling system.
4. Remove or disconnect the following:
   - Cylinder head bolts and cylinder head

➡Mark the rocker arm assembly to note component locations before disassembly.

5. Remove the rocker arm and shaft assemblies.
6. Remove the rear camshaft cover and O-ring.

⁂ WARNING

Carefully, remove the camshaft from the rear of the head taking care not to nick or scratch the journals.

7. Inspect camshaft journals for wear or damage. If wear is present, inspect the cylinder head for damage. Inspect the head

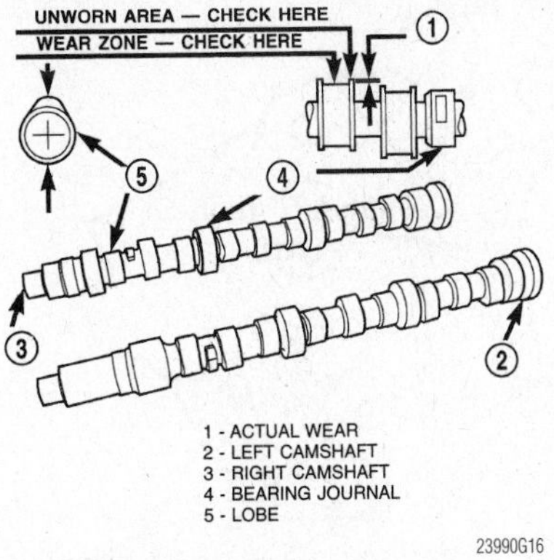

Measure the camshaft to ensure it meets specifications, if the camshaft is worn beyond the specifications it must be replaced—3.5L engine

oil holes for clogging. Replace the camshaft as required.

8. Measure the height of the cam using a micrometer. Measure in 2 places: the unworn area and in the wear zone. Subtract the figures to get cam wear. The standard specification is 0.001 in. (0.0254mm) with the wear limit being 0.010 in. (0.254mm). Replace the camshaft if it is worn beyond this specification.

To install:

9. Lubricate the camshaft journals and lobes with clean engine oil.
10. Install or connect the following:
    - Camshaft
    - Camshaft cover and O-ring and torque the bolts to 21 ft. lbs. (28 Nm)
    - Rocker arm assemblies
    - Cylinder head assembly
    - Negative battery cable
11. Fill and bleed the cooling system. An oil and filter change is recommended.
12. Start the vehicle, check for leaks and repair if necessary.

3.8L Engines

➡Keep all valvetrain components in order for assembly.

1. Before servicing the vehicle, refer to the precautions in the beginning of this section.
2. Remove the engine from the vehicle and mount it on a stand.
3. Remove or disconnect the following:
   - Valve covers
   - Rocker arm and shaft assemblies
   - Pushrods
   - Intake manifold
   - Cylinder heads
   - Yoke retainer
   - Aligning yokes
   - Hydraulic lifters
   - Oil pan
   - Oil pump pickup tube
   - Crankshaft pulley
   - Front cover
   - Timing chain and sprockets
   - Camshaft thrust plate
   - Camshaft

To install:

4. Install or connect the following:
   - Camshaft
   - Camshaft thrust plate. Torque the bolts to 105 inch lbs. (12 Nm).
   - Timing chain and sprockets. Torque the camshaft sprocket bolt to 40 ft. lbs. (54 Nm).
   - Front cover. Torque the bolts to 20 ft. lbs. (27 Nm).
   - Crankshaft pulley. Torque the bolt to 40 ft. lbs. (54 Nm).
   - Oil pump pickup tube
   - Oil pan

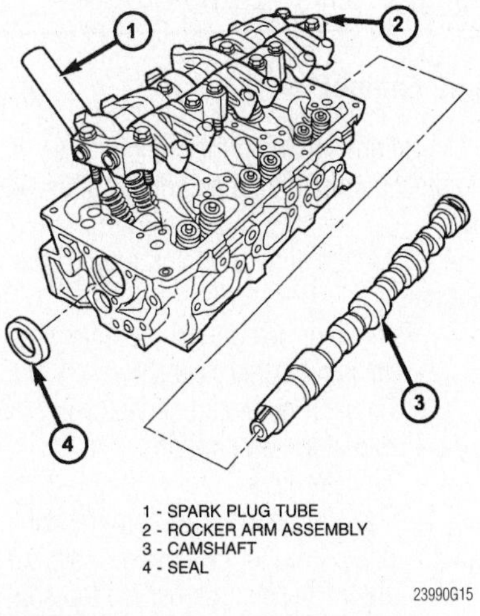

Exploded view of the camshaft assembly—3.5L engine

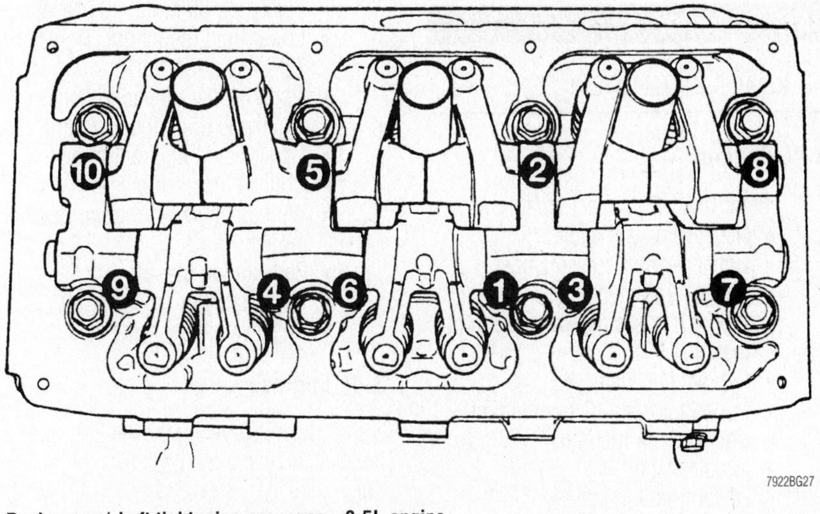

Rocker arm/shaft tightening sequence—3.5L engine

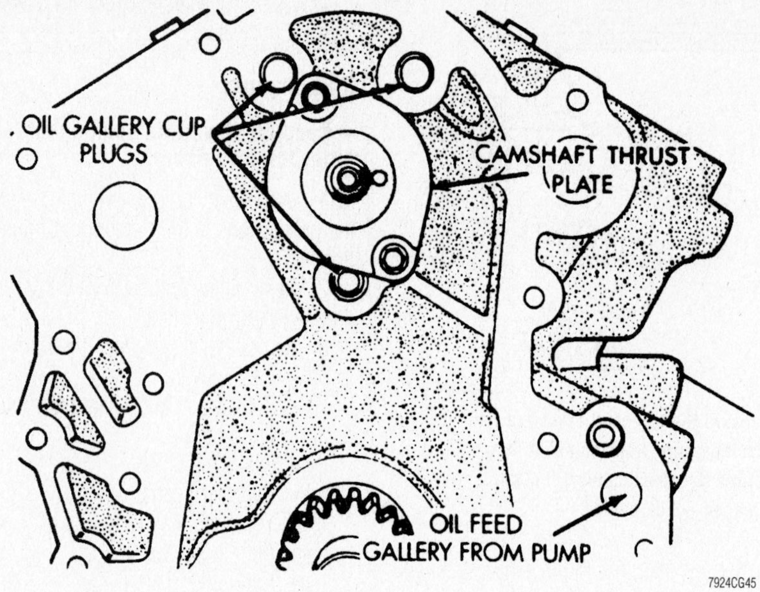

Remove the thrust plate and withdraw the camshaft from the engine—3.8L engines

- Hydraulic lifters
- Aligning yokes
- Yoke retainer. Torque the bolts to 105 inch lbs. (12 Nm).
- Cylinder heads
- Intake manifold
- Pushrods
- Rocker arm and shaft assemblies
- Valve covers

5. Install the engine to the vehicle.

Valve Lash

ADJUSTMENT

These engines use hydraulic roller lifters to take up the free-play in the valve train system, therefore no lash adjustments are necessary.

Starter Motor

REMOVAL & INSTALLATION

3.5L Engines

1. Remove or disconnect the following:
   - Negative battery cable
   - Upper radiator crossmember
   - Radiator fan module
   - Wiring harness clip from the front mount
   - Coolant line clamp bolt
   - Upper nut from the front mount
   - Upper starter mounting bolt
   - Middle starter bolt
   - Lower front mount nut
   - Lower bracket-to-transmission bolt
   - Lower bracket-to-engine block bolt
2. Move the front mount and bracket out of the way.
   - Positive battery cable from the starter
   - Solenoid connector from the starter
   - Lower starter bolt and the starter.

To install:

3. Install or connect the following:
   - Starter and lower bolt. Tighten to 35 ft. lbs. (47 Nm)
   - Solenoid connector to the starter
   - Positive battery cable to the starter
4. Move the front mount and bracket into position.
   - Middle starter bolt. Tighten to 35 ft. lbs. (47 Nm)
   - Lower bracket-to-engine block bolt. Tighten to 50 ft. lbs. (67 Nm)
   - Lower bracket-to-transmission bolt. Tighten to 50 ft. lbs. (67 Nm)
   - Lower front mount nut. Tighten to 75 ft. lbs. (101 Nm)
   - Upper starter mounting bolt. Tighten to 35 ft. lbs. (47 Nm)
   - Upper nut to the front mount
   - Coolant line clamp bolt
   - Wiring harness clip from the front mount
   - Radiator fan module
   - Upper radiator crossmember
   - Negative battery cable

3.8L Engines

1. Before servicing the vehicle, refer to the precautions in the beginning of this section.
2. Disconnect the negative battery cable.
3. Remove and relocate the speed control servo.
4. Unlock the electrical connector and disconnect the vacuum hose from servo.
5. Unlock and disconnect the electrical connectors from the cooling module.
6. Disconnect the wiring clips from cooling module.
7. Remove the retaining clip from the top of the cooling module.
8. Remove the two fasteners and remove fan module from the vehicle.
9. Remove the bolt from the oil cooler line bracket in the front motor mount.
10. Remove the battery positive cable from the starter.
11. Remove the lower, middle and upper starter bolt.
12. Place a piece of cardboard in front of the radiator core, to keep from damaging the radiator.
13. Remove the starter, and disconnect the solenoid wire from the starter.
14. Remove the starter.

To install:

15. Place a piece of cardboard in front of the radiator core, to keep from damaging the radiator.
16. Place the starter on the lower crossmember and connect the starter solenoid connector.
17. Install the dust shield on the nose of the starter, make sure of proper orientation.
18. Loosely install the starter mounting bolts.
19. Tighten the mounting bolts to 35 ft. lbs. (47 Nm).
20. Install the remaining components in the reverse order of removal.

Oil Pan

REMOVAL & INSTALLATION

3.5L Engines

1. Before servicing the vehicle, refer to the precautions in the beginning of this section.
2. Drain the engine oil and remove the oil filter.
3. Remove or disconnect the following:
   - Negative battery cable
   - Radiator close out panel
   - Radiator core support
   - Radiator fans
   - Top and bottom A/C compressor bolts. It is not necessary to evacuate the system. reposition the compressor.
   - A/C compressor bracket

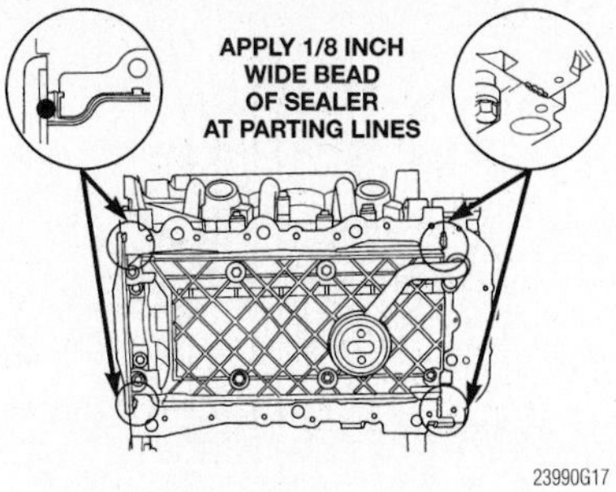

To ensure a proper seal, apply sealer as shown—3.5L engine

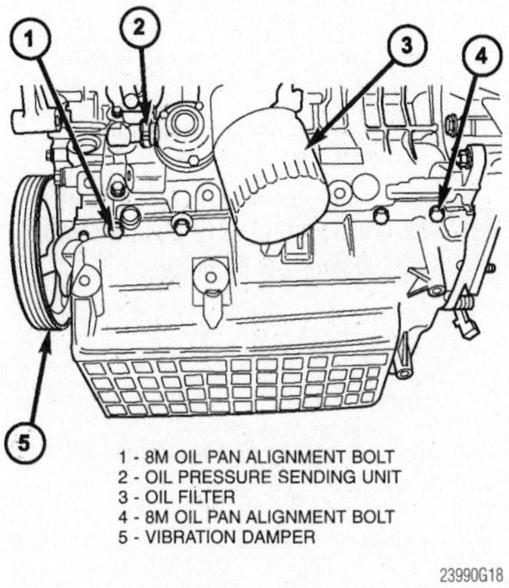

Location of the oil pan alignment bolts—3.5L engine

- Structural collar from the rear of the oil pan and transmission housing
- Inspection shield from between transmission and oil pan
- Dipstick and housing
- Oil pan mounting bolts, oil pan and gasket

4. Clean the oil pan and all gasket surfaces.

To install:

5. Apply a ⅛in. (3mm) bead of sealer at the parting line of the oil pump body and the rear seal retainer.
6. Install or connect the following:
   - Oil pan and torque the M8 nuts/bolts to 21 ft. lbs. (28 Nm) and the M6 nuts/bolts to 105 inch lbs. (12 Nm)
   - Dipstick and housing
   - Inspection shield from between transmission and oil pan
   - Structural collar to the rear of the oil pan and transmission housing
   - A/C compressor bracket and tighten the bolts to 40 ft. lbs. (54 Nm)
   - Top and bottom A/C compressor bolts to 250 inch lbs. (28 Nm)
   - Radiator fans
   - Radiator core support
   - Radiator close out panel
   - Negative battery cable

3.8L Engines

1. Before servicing the vehicle, refer to the precautions in the beginning of this section.
2. Disconnect the negative battery cable.
3. Remove and relocate the speed control servo.
4. Unlock the electrical connector and disconnect the vacuum hose from servo.
5. Unlock and disconnect the electrical connectors from the cooling module.
6. Disconnect the wiring clips from cooling module.
7. Remove the retaining clip from the top of the cooling module.
8. Remove the two fasteners and remove fan module from the vehicle.
9. Remove the bolt from the oil cooler line bracket in the front motor mount.
10. Remove the battery positive cable from the starter.
11. Remove the lower, middle and upper starter bolt.
12. Place a piece of cardboard in front of the radiator core, to keep from damaging the radiator.
13. Remove the starter, and disconnect the solenoid wire from the starter.
14. Remove the starter.

To install:

15. Place a piece of cardboard in front of the radiator core, to keep from damaging the radiator.
16. Place the starter on the lower crossmember and connect the starter solenoid connector.
17. Install the dust shield on the nose of the starter, make sure of proper orientation.
18. Loosely install the starter mounting bolts.
19. Tighten the mounting bolts to 35 ft. lbs. (47 Nm).
20. Install the remaining components in the reverse order of removal.
21. Drain the engine oil.
22. Disconnect the negative battery cable.
23. Remove the dipstick tube.
24. Remove the right engine mount heat shield.
25. Disconnect the in block heater, if equipped.
26. Remove the right engine mount structural bracket at the engine and transaxle.
27. Remove the flex plate inspection cover.
28. Remove oil pan bolts, oil pan and gasket.

To install:

29. Clean the sealing surfaces and apply a ⅛ inch bead of sealer at the parting line of the chain case cover and the rear seal retainer.
30. Position a new pan gasket on the oil pan.

(3) Install oil pan and tighten the fasteners to 105 inch lbs. (12 Nm).

31. Install the flex plate cover.
32. Install the right engine mount bracket.
33. Connect the in block heater, if equipped.
34. Install the right engine mount heat.
35. Install the dipstick tube.
36. Fill the crankcase to the correct level.
37. Start the engine, check for leaks and repair if necessary.

Oil Pump

REMOVAL & INSTALLATION

3.5L Engines

The timing belt must be removed to access the oil pump located behind the crankshaft drive sprocket. It is good practice to turn the crankshaft to Top Dead Center (TDC) No. 1 cylinder compression stroke (firing position) before starting disassembly. This should align all timing marks and be a good point of reference for all work to follow.

1. Before servicing the vehicle, refer to the precautions in the beginning of this section.
2. Drain the engine oil.
3. Drain the cooling system.
4. Remove or disconnect the following:
   - Negative battery cable
   - Timing belt. Refer to the appropriate procedure in this section.

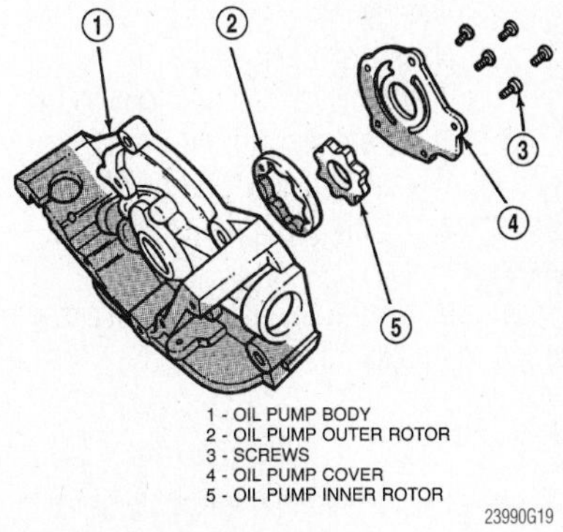

Exploded view of the oil pump assembly—3.5L engine

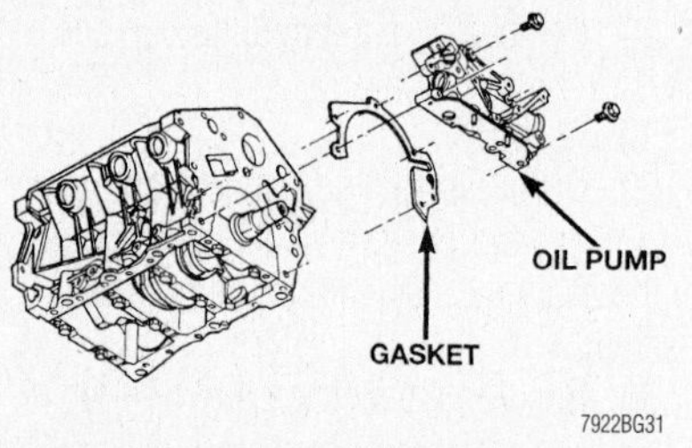

Prime the oil pump before installation, because a dry pump will wear prematurely and cause low oil pressure—3.5L engine

- Crankshaft sprocket using tool L-4407-A
- Oil pan
- Oil pump pickup tube
- Oil pump-to-engine screws and pump
- Oil pump rotors

5. Wash all parts in solvent and inspect carefully for damage or wear.

To install:

6. Clean all parts well. There should be no traces of old gasket/sealer on any components.
7. Assemble the oil pump with new parts as required.
8. Install the oil pump cover. Torque the fasteners to 108 inch lbs. (12 Nm).
9. Prime the oil pump prior to installation by filling the rotor cavity with clean engine oil.
10. Install the oil pump and tighten the oil pump-to-engine screws to 250 inch lbs. 28 Nm)
11. Install or connect the following:
    - Oil pump pickup tube using a new O–ring
    - Oil pan
    - Crankshaft sprocket using tool C-4685C1, thrust bearing, washer and 12mm bolt to draw the sprocket onto the crankshaft
    - Timing belt. Refer to the appropriate procedure for removal and installation.
    - Negative battery cable
12. Fill and bleed the cooling system.
13. Fill the engine with the correct amount of clean SAE 5W-30 or SAE 10W-30 engine oil only. Do not mix the two grades of oil.
14. Start the engine, check for leaks and proper oil pressure.

3.8L Engines

1. Before servicing the vehicle, refer to the precautions in the beginning of this section.
2. Drain the cooling system.

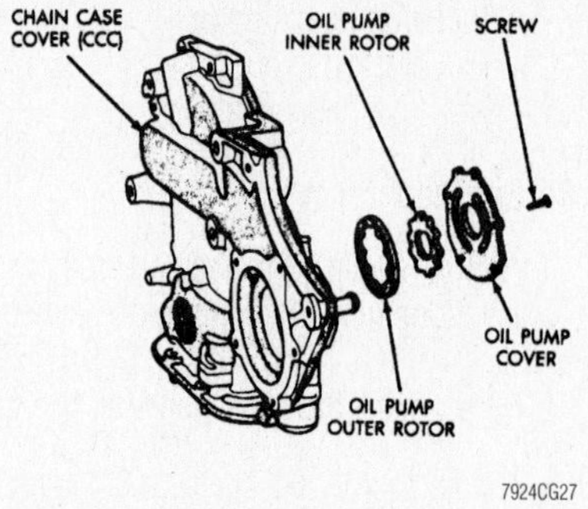

Exploded view of the oil pump assembly—3.8L engines

3. Drain the engine oil.
4. Remove or disconnect the following:
    - Negative battery cable
    - Oil pan
    - Timing chain cover
    - Oil pump from the case cover

To install:

5. Install or connect the following:
    - Oil pump to the front cover. Torque the cover screws to 105 inch lbs. (12 Nm).
    - Front cover. Torque the bolts to 20 ft. lbs. 927 Nm).
    - Oil pan
    - Negative battery cable
6. Fill the engine with clean oil.
7. Fill the cooling system.
8. Start the engine, check for leaks and repair if necessary.

Rear Main Seal

REMOVAL & INSTALLATION

3.5L Engines

1. Before servicing the vehicle, refer to the precautions in the beginning of this section.
2. Drain the transaxle fluid.
3. Remove the negative battery cable.
4. Remove the transaxle, inspection cover and flywheel/flexplate.
5. Using a small prytool, carefully pry out the rear oil seal. Be careful not to nick or damage the crankshaft flange seal surface or the retainer bore.

To install:

6. Place the Seal Pilot tool 6926-1 on the crankshaft.
7. Apply a light coating of engine oil to the entire circumference of the oil seal lip.
8. Place the seal over tool 6926-1 and use installer 6926-2 and handle C-4171 to drive the seal into place until it is flush with the housing surface.

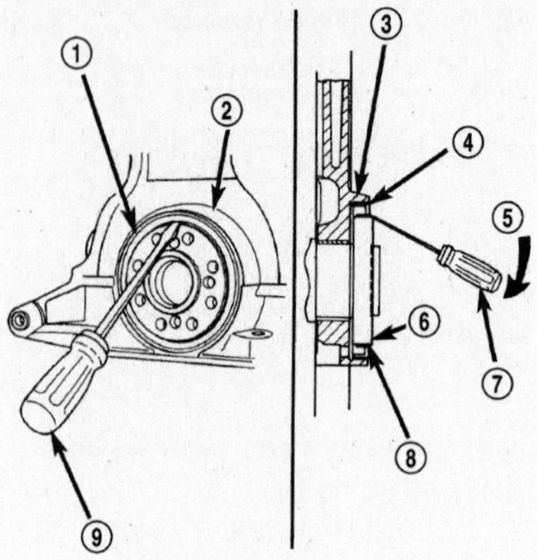

Use a suitable prytool to remove the rear oil seal as shown—3.5L engine

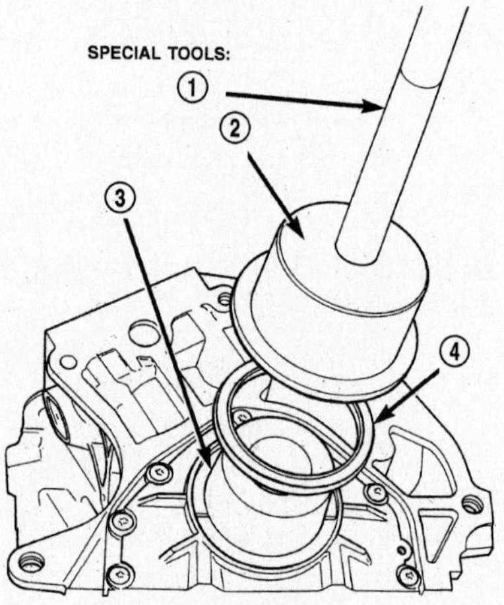

Installing the rear main oil seal—3.5L engine

9. Install the flexplate/flywheel and transaxle.
10. Connect the negative battery cable.
11. Fill the transaxle with the proper fluid.
12. Start the vehicle, check for leaks and repair if necessary.

3.8L Engines

1. Before servicing the vehicle, refer to the precautions at the beginning of this section.
2. Remove or disconnect the following:
    - Negative battery cable
    - Transaxle

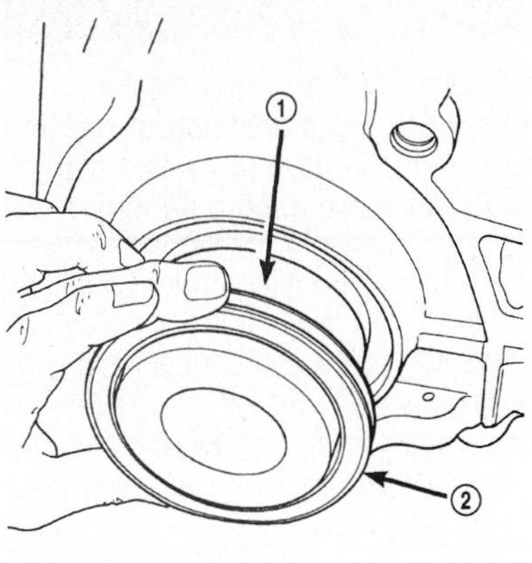

Install the rear main seal flush with the cylinder block surface—3.8L engines

- Flexplate
- Rear main seal

To install:

3. Place seal guide tool 6926–1 on the crankshaft and place the seal over the tool. Make sure the lip of the seal is facing towards the crankcase.
4. Install or connect the following:
   - Rear main seal flush with the cylinder block surface using the tools illustrated.
   - Flexplate
   - Transaxle
   - Negative battery cable
5. Run the engine and check for leaks.

Timing Chain, Sprockets, Front Cover and Seal

REMOVAL & INSTALLATION

3.8L Engines

1. Before servicing the vehicle, refer to the precautions in the beginning of this section.
2. Disconnect the negative battery cable.
3. Properly relieve the fuel system pressure.
4. Evacuate air conditioning system.
5. Drain and recycle the engine coolant.
6. Remove the air cleaner.
7. Loosen the water pump pulley bolts.
8. Remove the accessory drive belt and belt tensioner.
9. Remove the power steering pump and set the pump aside.
10. Remove the alternator.
11. Disconnect the air conditioning compressor clutch electrical connector and suction discharge hoses.
12. Remove the dipstick tube.
13. Drain the engine oil.
14. Remove right wheel and inner splash shield.
15. Remove the oil pan.
16. Disconnect the lower radiator hose and the oil cooler hose at the coolant cross over pipe.
17. Separate the power steering lines from the A/C bracket and bracket at the cradle.
18. Separate the air conditioning compressor from the engine and set aside.
19. Remove the crankshaft damper using tool # 8454 and insert # 8450.
20. Remove the coolant cross over pipe fasteners and pipe.
21. Remove the accessible lower timing cover fasteners through the wheel well.
22. Remove Camshaft Position (CMP) sensor from the timing chain cover.
23. Remove the water pump.
24. Support the engine with a floor jack and remove the upper engine mount assembly.
25. Remove the timing chain cover fasteners and cover.
26. Clean the timing cover mating surfaces.
27. Rotate the engine so that the timing marks are aligned.
28. Remove or disconnect the following:
    - Camshaft sprocket attaching bolt
    - Timing chain and camshaft sprocket
    - Crankshaft sprocket with special tools 8539, 5048–6 and 5048–1

To install:

29. Rotate the engine so the timing arrow is at the 12 o'clock position.
30. Lubricate the chain and sprockets with clean oil.
31. Hold the camshaft sprocket and chain and place the chain around the sprocket aligning the plated link with the dot on the sprocket. Position the timing arrow at the 6 o'clock position.
32. Place the chain around the crankshaft sprocket with the plated link lined up with the dot on the sprocket and install the camshaft sprocket.
33. Use a straight edge to check the timing alignment marks.
34. Align the timing chain colored links with the dots on the timing sprockets.
35. Tighten the camshaft sprocket bolt to 40 ft. lbs. (54 Nm).
36. Rotate the crankshaft 2 revolutions and verify the proper timing chain alignment, if not remove the components and reinstall as described above.

➡Do not use sealer on cover gasket

37. Position a new gasket on the timing cover. Adhere the new gasket to the chain case cover, making sure that the lower edge of the gasket is 0.020 inch (0.5 mm) beyond the lower edge of the cover.
38. Rotate the crankshaft so that the oil pump drive flats are in the vertical position.

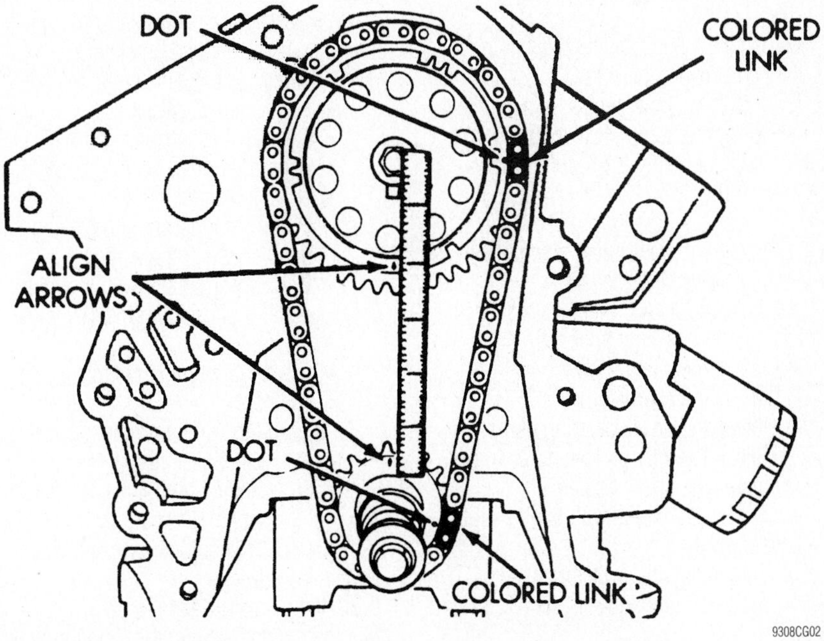

Timing mark alignment—3.8L engines

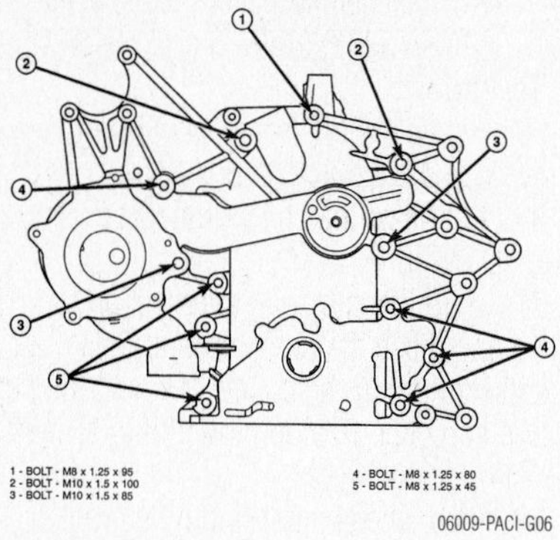

Timing chain cover bolt locations—3.8L engines

39. Position the oil pump inner rotor so the mating flats are in the same position as the crankshaft drive flats.

⁂ CAUTION

Make sure the oil pump is engaged on the crankshaft correctly or severe damage may result.

40. Install the timing cover.
41. Install the timing chain cover bolts. Tighten the M8 bolts to 20 ft. lbs. (27 Nm) and the M10 bolts to 40 ft. lbs. (54 Nm).
42. Install the crankshaft front oil seal.
43. Install the crankshaft damper.
44. Install the coolant cross over pipe.
45. Install the A/C compressor.
46. Position and install the power steering line hold downs at the cradle and A/C bracket.
47. Connect the coolant hoses to the coolant crossover pipe.
48. Install the inner fender splash shield.
49. Install oil pump pick-up tube with a new O-ring. Tighten the bolt to 250 inch lbs. (28 Nm).
50. Install the oil pan.
51. Install the water pump and pulley.
52. Install the CMP sensor.
53. Connect the A/C electrical connector and the suction, discharge hoses.
54. Install the dipstick tube.
55. Install the alternator.
56. Install the power steering pump.
57. Install the idler pulley and drive belt.
58. Install the upper engine.
59. Fill the crankcase with engine oil to the proper level.
60. Fill the cooling system.
61. Evacuate and recharge the A/C system.
62. Check and refill all fluid levels.
63. Connect the negative battery cable.

Timing Belt

REMOVAL & INSTALLATION

3.5L Engines

1. Before servicing the vehicle, refer to the precautions in the beginning of this section.

Use care when servicing a timing belt. Valve timing is absolutely critical to engine performance. If the valve timing marks on all drive sprockets are not properly aligned, engine damage will result. If only the belt and tensioner are being serviced, do not loosen the camshaft drive sprockets unless they are to be replaced. The sprockets have oversized openings and can be rotated several degrees in each direction on their shafts. This means the sprockets must be re-timed, requiring some special tools.

⁂ CAUTION

Fuel injection systems remain under pressure, even after the engine has been turned off. The fuel system pressure must be relieved before disconnecting any fuel lines. Failure to do so may result in fire and/or personal injury.

2. Loosen the valve train rocker assemblies before servicing the timing components.
3. Release the fuel system pressure using the recommended procedure.
4. Remove or disconnect the following:
   - Negative battery cable
   - Cylinder head covers
   - Accessory drive belts
   - Drive belt tensioner
   - Power steering pump bolts and position the pump aside
   - Crankshaft damper with a quality puller tool gripping the inside of the pulley
   - Lower front timing belt cover retainers
5. Support the engine with a floor jack.
   - Air cleaner assembly
   - Front engine mount
   - Fuel supply line from the rail
   - Upper timing belt cover bolts and the cover
6. If the timing belt is to be reused, mark the timing belt with the running direction for installation.

⁂ WARNING

Always align the timing marks always use the crankshaft to rotate the engine. Failure to do this will result in severe engine damage.

7. Rotate the engine clockwise until the crankshaft mark aligns with the Top Dead Center (TDC) mark on the oil pump housing and the camshaft sprocket timing marks align with the marks on the rear cover.
8. Remove the timing belt tensioner and the belt.
9. Pre-load the timing belt tensioner as follows:
   a. Place tensioner in a vise the same way it is mounted on the engine.
   b. Slowly compress the plunger into the tensioner body.
   c. Once the plunger is compressed, install a pin through the body and plunger to retain it in place until the tensioner is installed.

To install:

⁂ WARNING

If the camshafts have moved from the timing marks always rotate them towards the direction nearest the to the timing marks. Never turn the camshafts a full turn or sever engine damage will occur.

10. Align the crankshaft sprocket with the TDC mark on the oil pump.
11. Align the camshaft sprockets timing marks with the mark on the rear cover.
12. Install the belt starting at the crankshaft sprocket going in a counterclockwise direction. Install the belt around the last sprocket always maintaining tension on the belt as it is positioned around the tensioner pulley.

➡If the camshaft gears have been removed it is only necessary to have the camshaft gear retaining bolts installed to a snug torque at this time.

13. Holding the tensioner pulley against the belt, install the tensioner into the housing and tighten to 250 inch lbs. (28 Nm) making sure all the alignment marks remain aligned.

Pull the pin from the tensioner and allow the tensioner to extend to the pulley bracket.

14. Rotate the crankshaft sprocket 2 full turns and check the timing marks on the cam and crank shafts. Make sure the marks are aligned, if not repeat the procedure.
15. if the camshafts were removed using a dial indicator, position the number 1 piston at TDC.
16. Hold the camshaft sprocket hex with a

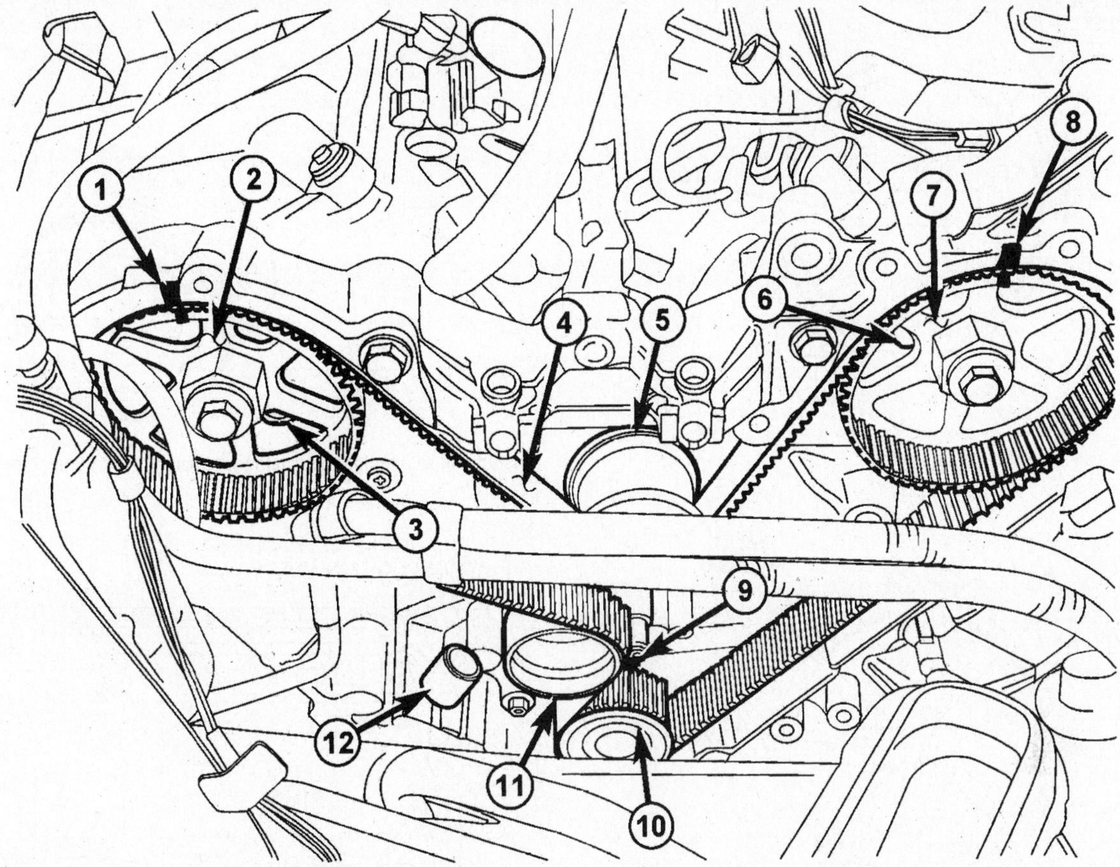

View of the timing belt alignment marks—3.5L engine

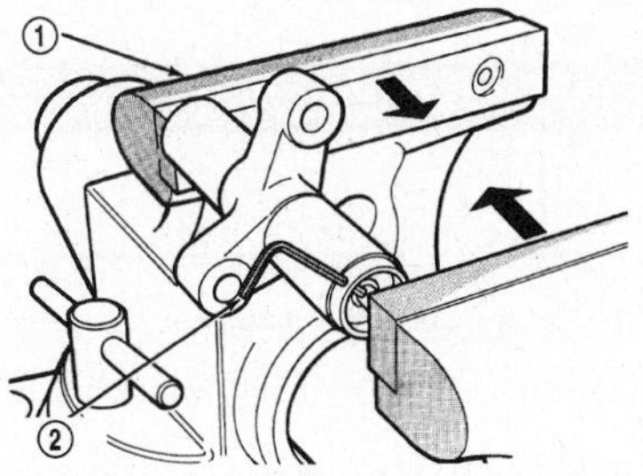

Pre-load the timing belt tensioner—3.5L engine

36mm wrench and tighten the right camshaft sprocket bolt to 75 ft. lbs. (102 Nm) plus an additional 90 degree turn. Tighten the left camshaft sprocket bolt to 85 ft. lbs. (115 Nm) plus an additional 90 degree turn.

17. Remove the dial indicator. Install the spark plug and tighten to 20 ft. lbs. (28 Nm).
18. Remove the camshaft alignment tools from the back of the cylinder heads and install the cam covers with new O-rings.
    - Upper timing belt cover and bolts
    - Fuel supply line to the rail
    - Front engine mount
    - Air cleaner assembly
    - Lower front timing belt cover retainers
    - Crankshaft damper
    - Power steering pump and bolts
    - Drive belt tensioner
    - Accessory drive belts
    - Cylinder head covers
    - Negative battery cable

Piston and Rings

POSITIONING

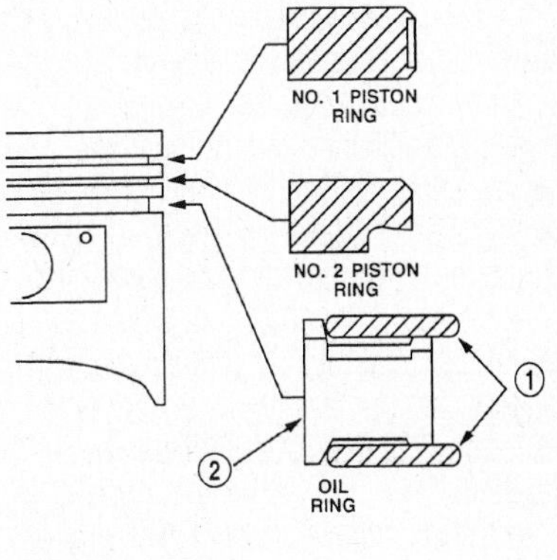

Cross-sectional view of the piston rings—3.5L engine

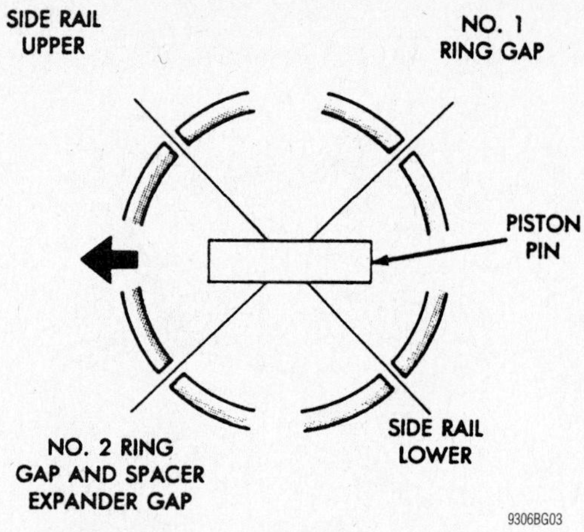

Piston ring gap positions—3.5L engine

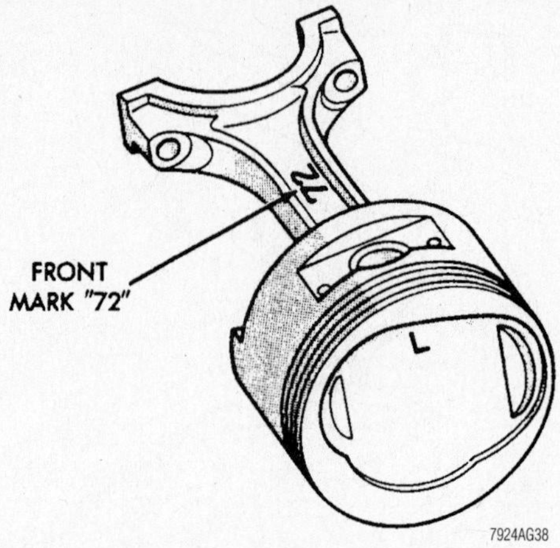

Piston and connecting rod front mark locations—3.8L engines

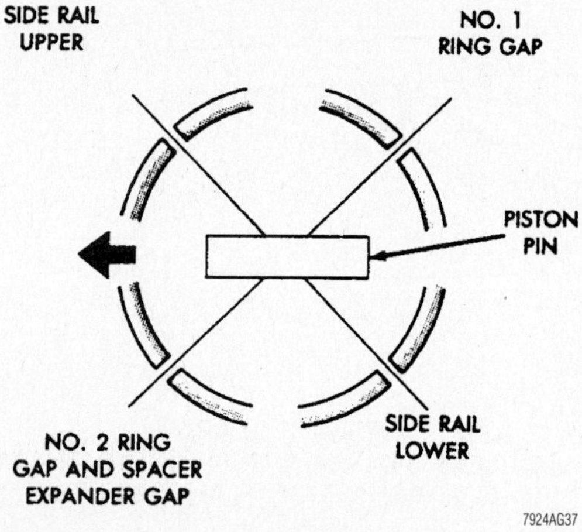

Piston ring end-gap spacing—3.8L engines

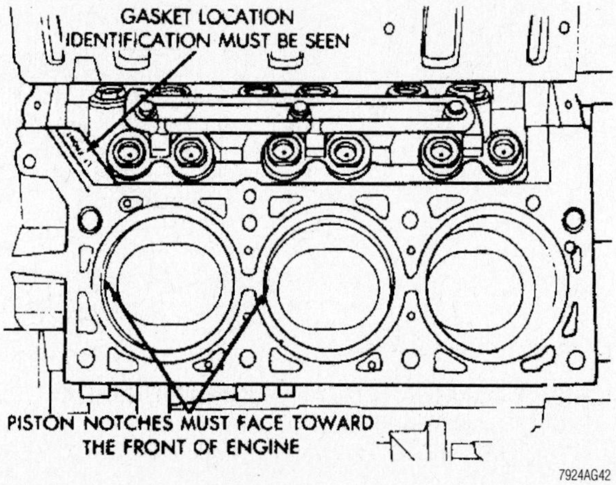

Piston-to-engine positioning mark locations—3.8L engines

FUEL SYSTEM

Fuel System Service Precautions

Safety is the most important factor when performing not only fuel system maintenance but any type of maintenance. Failure to conduct maintenance and repairs in a safe manner may result in serious personal injury or death. Maintenance and testing of the vehicle's fuel system components can be accomplished safely and effectively by adhering to the following rules and guidelines:

- To avoid the possibility of fire and personal injury, always disconnect the negative battery cable unless the repair or test procedure requires that battery voltage be applied.
- Always relieve the fuel system pressure prior to disconnecting any fuel system component (injector, fuel rail, pressure regulator, etc.), fitting or fuel line connection. Exercise extreme caution whenever relieving fuel system pressure, to avoid exposing skin, face and eyes to fuel spray. Please be advised that fuel under pressure may penetrate the skin or any part of the body that it contacts.
- Always place a shop towel or cloth around the fitting or connection prior to loosening to absorb any excess fuel due to spillage. Ensure that all fuel spillage (should it occur) is quickly removed from engine surfaces. Ensure that all fuel soaked cloths or towels are deposited into a suitable waste container.
- Always keep a dry chemical (Class B) fire extinguisher near the work area.
- Do not allow fuel spray or fuel vapors to come into contact with a spark or open flame.
- Always use a back-up wrench when loosening and tightening fuel line connection fittings. This will prevent unnecessary stress and torsion to fuel line piping.
- Always replace worn fuel fitting O-rings with new. Do not substitute fuel hose, where fuel pipe is installed.

Before servicing the vehicle, also make sure to refer to the precautions in the beginning of this section as well.

Fuel System Pressure

RELIEVING

1. Before servicing the vehicle, refer to the precautions in the beginning of this section.
2. At the Power Distribution Center (PDC), remove the Fuel Pump Relay.
3. Operate the engine until the engine stalls; then, continue restarting the engine until it will no longer run.
4. Turn the ignition switch **OFF**.

**** CAUTION**

The previous steps must be performed to relieve the high pressure fuel from the fuel rail. The following

steps must be performed to remove excess fuel from the fuel rail. Do not use the following steps to relieve high pressure for excessive fuel will be forced into a cylinder chamber.

5. Disconnect the electrical connector from any injector.
6. Connect 1 end of a jumper wire (with an alligator clip) to either injector terminal and the other end to the positive side of the battery.
7. Connect 1 end of a second jumper wire to the other injector terminal.

WARNING

Applying power to an injector for more than a few seconds will damage the injector.

8. Momentarily, touch the other end of the jumper wire to a ground for no more than a few seconds.
9. At the Power Distribution Center (PDC), install the fuel pump relay.

➡When the fuel pump relay is removed, 1 or more Diagnostic Trouble Codes (DTC's) may be stored in the Powertrain Control Module (PCM) memory. A DRB scan tool must by used to clear the DTC's.

Fuel Pump

REMOVAL & INSTALLATION

2003–04 Models

The in-tank fuel pump module contains the fuel pump and pressure regulator which adjusts fuel system pressure. Fuel pump voltage is supplied through the fuel pump relay.

The fuel pump is serviced as part of the fuel pump module. The fuel pump module is installed in the top of the fuel tank and contains the electric fuel pump, fuel pump reservoir, inlet strainer fuel gauge sending unit, fuel supply and return line connections and the pressure regulator. The inlet strainer, fuel pressure regulator and level sensor are the only serviceable items. If the fuel pump requires service, replace the fuel pump module.

1. Before servicing the vehicle, refer to the precautions in the beginning of this section.
2. Properly relieve the fuel system pressure.
3. Remove or disconnect the following:
   - Negative battery cable
4. Fold the drivers side rear seat and tilt it forward.
   - Two bolts for the rear seat
   - Seat electrical connection
   - Seat
   - Rear console scuff plate
   - Rear console electrical connector
   - Fourteen screws retaining the rear console
   - Screws from the cup holder pockets
   - Screw from the console pocket under the rubber mat
   - Left rear door sill plate
5. Cut the carpet along the mounting bracket for the center console. The cut does not need to exceed a foot in length. Pull the carpet back to access the pump cover.
6. Cut the seal for the pump access panel and remove the cover. Clean the sealer from the floor pan and cover. Use a vacuum cleaner to remove any debris before removing the pump lock ring.
7. Remove or disconnect the following:
   - Vapor line and electrical connection
   - Pump lock ring using a brass punch
   - Pump top
8. Drain the fuel tank.
   - Electrical connections from the bottom of the pump top and the passenger side level sensor assembly
   - Tabs attaching the return and supply lines
   - Return line from the pump using a small prytool to pry the tab back and tip the hose to one side, then pry the tab on the other side to release the hose
   - Fuel supply line from the top of the pump in the same manner as the return line
   - Fuel pump

To install:

9. Install or connect the following:
   - Fuel pump
   - Fuel supply line and return lines
   - Electrical connections to the bottom of the pump
   - Pump top, gasket and lock ring
   - Fuel line and electrical connection
   - Negative battery cable
10. Fill the fuel tank.
11. Use a DRB III to pressurize the fuel system and check for leaks. If no leaks detected, disconnect the negative battery cable.
12. Clean the access cover and floorpan.
13. Apply sealer to the outer edge of the access plate and the push pin holes.
14. Install or connect the following:
   - Access cover and the push pins into the holes

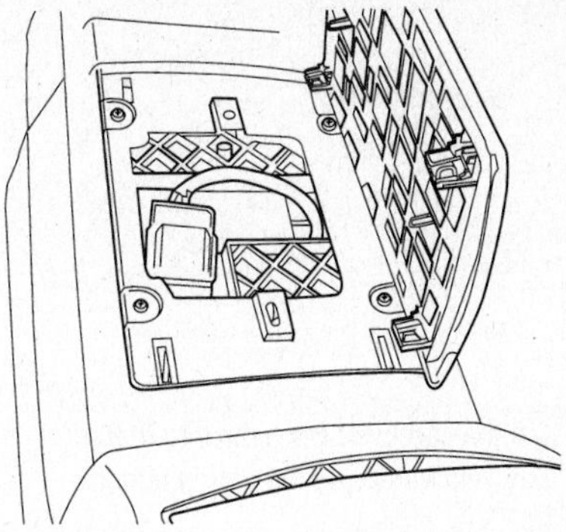
23990G24

Remove the rear console scuff plate—2003–04 models

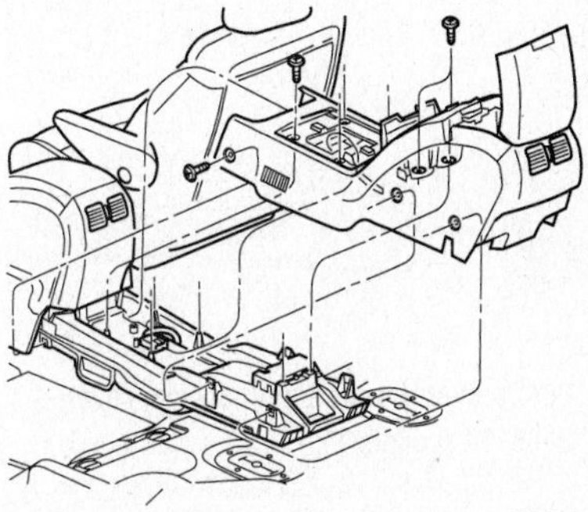
23990G25

Remove the rear console assembly—2003–04 models

23990G26

Remove the screws from the cup holder pockets—2003–04 models

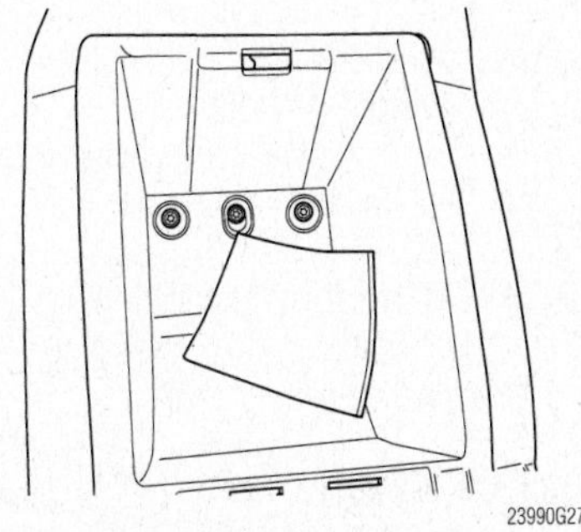
23990G27

Remove the screws from under the rubber mat in the console pocket—2003–04 models

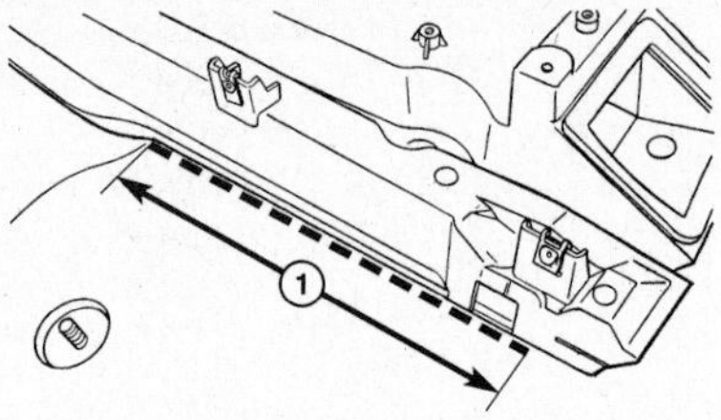

1 - Cut NO longer than 1 Foot

23990G28

Cut the carpet where indicated (dotted line), no longer than a foot in length—2003–04 models

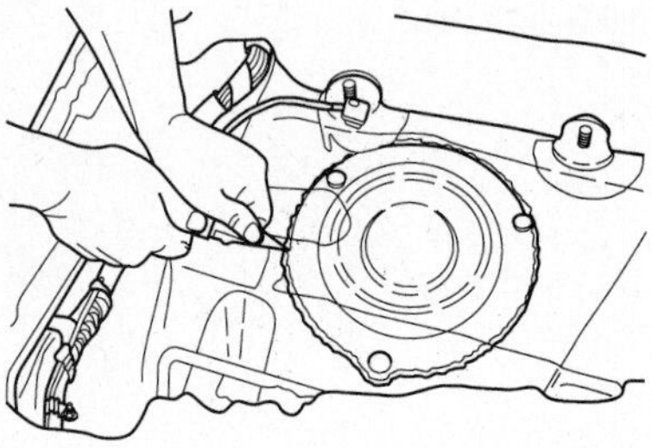

23990G29

Cut the seal for the pump access cover—2003–04 models

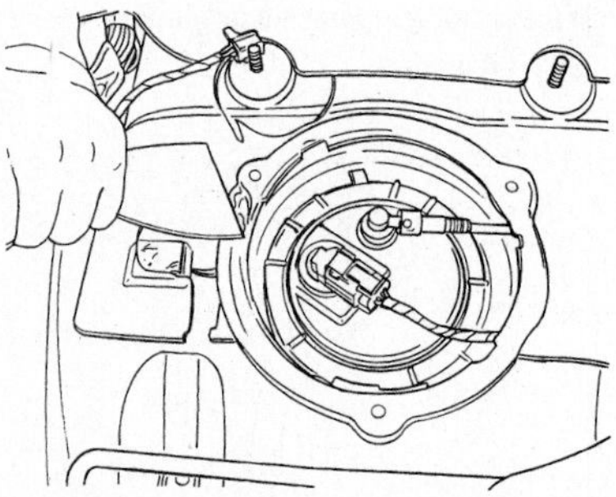

23990G30

Clean the sealer from the cover contact surfaces—2003–04 models

- Carpet and left rear door sill plate
- Screws to the console pocket under the rubber mat
- Screws to the cup holder pockets
- Fourteen screws retaining the rear console
- Rear console electrical connector
- Rear console scuff plate
- Seat
- Seat electrical connection
- Two bolts for the rear seat
- Negative battery cable

2005–06 Models

1. Before servicing the vehicle, refer to the precautions in the beginning of this section.

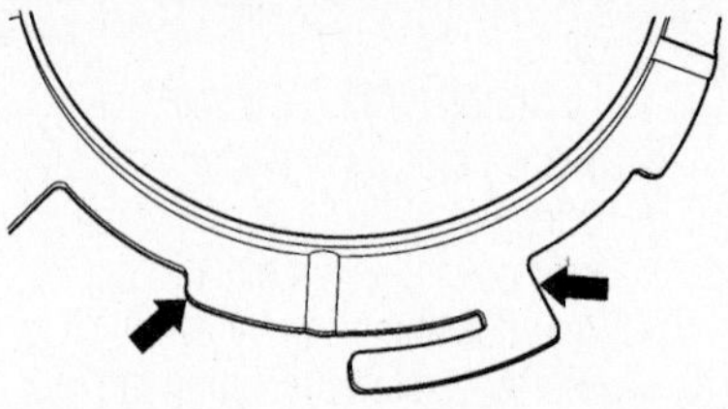

23990G31

Pump lock ring contact points, place the brass punch at these locations when removing/installing the ring—all models

23990G32

Apply sealer to the outer edge of the access plate and the push pin holes—2003–04 models

2. Properly relieve the fuel system pressure.
3. Disconnect the negative battery cable.
4. Drain the fuel tank.
5. Disconnect the exhaust system between the catalytic converter and the muffler.
6. Support the exhaust system with jack stands.
7. Remove muffler hanger rubber grommets and the center exhaust hanger rubber grommets.
8. Remove the rear portion of the exhaust system.
9. Disconnect the fuel tank electrical connector.
10. Disconnect the fuel line and EVAP line.
11. Disconnect the vapor recirculation tube.
12. Mark the driveshaft rear connection and disconnect the driveshaft rear connection.
13. Remove the driveshaft support/bearing.
14. Support the driveshaft with a jack.
15. Remove the left rear tire and splash shield.
16. Remove the fuel filler tube from the rubber hose at fuel tank.
17. Support the fuel tank with a transmission jack.
18. Remove the 2 fuel tank straps and lower fuel tank.
19. Drain any remaining fuel from the tank and remove the tank.
20. Clean around the fuel pump module lock ring prior to removal.
21. Disconnect the vapor line and electrical connector.
22. Remove the fuel pump module lock ring using a brass punch to remove the ¼ turn lock ring. Make contact only at the points illustrated.
23. Remove fuel pump module top.
24. Disconnect electrical connector.
25. Disconnect the wire connector to the passenger side level sensor before removing the reservoir from the tank.
26. Remove the return line from pump module. Use a small screwdriver to pry the tab back and tip the hose to one side. Pry the tab on other side back to release the hose.

CAUTION

Care should be taken to not crack/break the locking tabs or supply line fitting.

27. Remove the fuel line from the top of the pump module. Use a small screwdriver to pry the tab back and tip hose to one side. Pry the tab on the other side back to release the hose.
28. Tip fuel pump module on its side to remove the fuel from the reservoir.
29. Remove the fuel pump module.

To install:

30. Install the fuel pump module into the fuel tank.
31. Connect the siphon and fuel lines to fuel pump module.
32. Connect the fuel pump electrical connector to the bottom of the module top.
33. Connect the wire connector to the passenger side level sensor.
34. Install the fuel pump module top to the module.
35. Install the fuel pump module gasket and lock ring. Use a brass punch to install the ¼ turn lock ring.
36. Connect the vapor line and electrical connector.
37. Support the fuel tank with a transmission jack.
38. Raise fuel tank and install in the vehicle.
39. Install the 2 fuel tank straps.
40. Install the remaining components in the reverse order of removal.
41. Connect the negative battery cable.
42. Fill the fuel tank and use the DRBIIIt scan tool to pressurize the fuel system. Check for leaks.

Fuel Injector

REMOVAL & INSTALLATION

3.5L Engines

1. Before servicing the vehicle, refer to the precautions in the beginning of this section.
2. Relieve the fuel system pressure.
3. Remove or disconnect the following:
   - Negative battery cable
   - Intake manifold plenum and cover opening with a clean cloth
4. Tag the injectors to match them with their correct cylinder during installation.
5. Place a shop rag under the fuel rail's quick-connect fitting; then, squeeze the quick-connect fitting's retainer tabs together and pull the fitting assembly off of the fuel tube nipple.
   - Fuel injector electrical connectors
   - Fuel rail-to-engine bolts and fuel rail
   - Fuel injector-to-fuel rail retainer clips
   - Fuel injectors

To install:

6. Lubricate the injector O-rings with clean engine oil.
7. Install or connect the following:
   - Fuel injector and secure with retaining clips
   - Fuel rail onto cylinder head and press rail into place. Make sure that the injectors are fully seated.
   - Fuel rail-to-cylinder head bolts and torque bolts to 250 inch lbs. (28 Nm)
   - Fuel injector electrical connectors
   - Intake plenum
8. Lubricate the quick-connect fitting's O-rings with clean engine oil; then, push the connector together until the retainer seats and a click is heard.
9. Connect the negative battery cable.

3.8L Engines

1. Before servicing the vehicle, refer to the precautions in the beginning of this section.
2. Relieve the fuel system pressure.
3. Remove the air cleaner housing and inlet hose.
4. Remove the engine cover.
5. Disconnect the Throttle Position (TP) sensor connector.
6. Disconnect the Idle Air Control (IAC) valve connector.

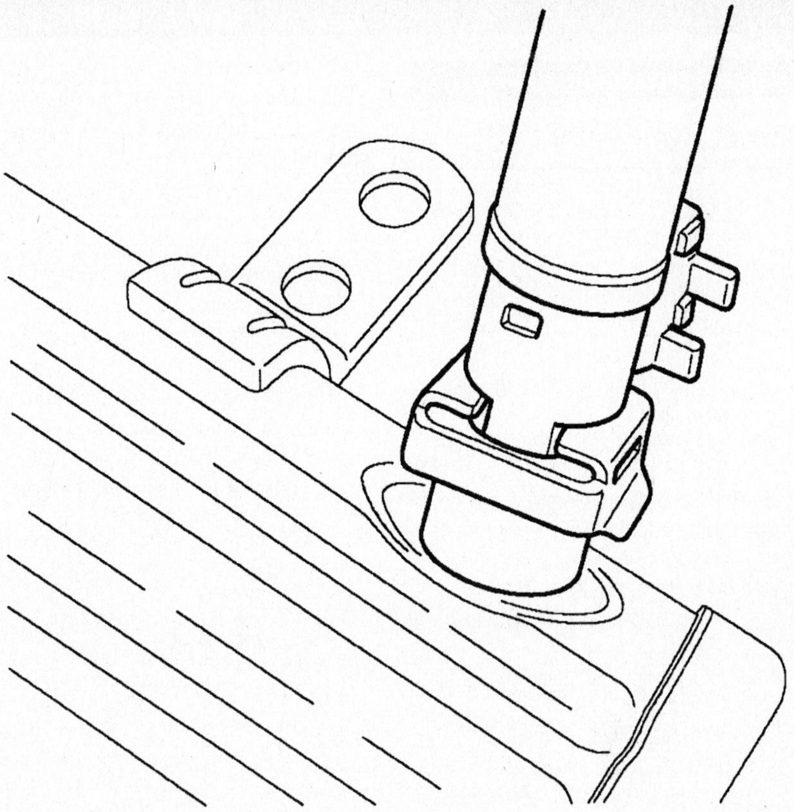

Proper positioning of the fuel injector in the fuel rail—3.8L engines

7. Disconnect the purge line.
8. Disconnect the throttle and cruise control cables, if equipped.
9. Disconnect the Manifold Absolute Pressure (MAP) sensor connector.
10. Disconnect the vacuum lines.
11. Remove the Exhaust Gas Recirculation (EGR) tube.
12. Remove the upper intake manifold.
13. Disconnect the injector electrical connectors.
14. If the injector connectors are not tagged with their cylinder number, tag them to identify the correct cylinder.
15. Remove the 4 mounting bolts from the fuel rail.
16. Disconnect the fuel supply line from the fuel rail.
17. Lift fuel rail straight up off of cylinder head.
18. Remove the retaining clips from fuel injectors at the fuel rail.
19. Remove the fuel injector from fuel rail.

To install:

20. Inspect and replace any damaged fuel injector O-rings.
21. Lightly lubricate the fuel injector O-rings with a couple drops of clean engine oil.
22. Install retaining clips on fuel.
23. Install the fuel injector locator flange on fuel rail.
24. Push the injectors into the fuel injector rail until the clips are in the correct position.
25. Position the fuel rail over the cylinder heads, and push the rail into place. Tighten the fuel rail mounting bolts to 250 inch lbs. (28 Nm).
26. Connect the electrical connectors to the fuel injectors.
27. Connect the fuel supply tube quick connect fitting to the fuel rail.
28. Install the upper intake manifold.
29. Connect and electrical connection for the MAP sensor.
30. Connect the vacuum lines.
31. Connect the electrical connectors to the TPS and IAC.
32. Connect the purge line.
33. Install all remaining components.
34. Connect the negative cable to battery.

DRIVE TRAIN

Transaxle Assembly

REMOVAL & INSTALLATION

1. Before servicing the vehicle, refer to the precautions in the beginning of this section.
2. Remove or disconnect the following:
   - Battery cables
   - Battery and tray
   - Gearshift cable from the manual valve lever
   - Oil cooler off of hose fitting connection.
3. Use a suitable cutting tool, cut the oil cooler hoses off flush at the fittings. A service splice kit will be required during installation. Cap the hoses.
   - Ground cable from the transaxle case
   - Input and output speed sensor connections
   - Crankshaft Position Sensor (CKP)
   - Transaxle range sensor and solenoid/pressure switch connections
   - Transaxle harness from retainers and position aside
   - Coolant bypass tube-to-engine and transaxle retainers
   - Front brake lines from the Hydraulic Control Unit (HCU) and position aside
   - Intermediate shaft extension from the steering gear using tool 6831A to remove the roll pin and then slide the shaft extension off the gear
4. Install a powertrain support fixture such as 8534B and adapter kit 8534-12.
5. Oil dipstick tube-to-cylinder head bolt
6. Install lift support bracket 8534-8 and use the dipstick tube bolt to secure it.
7. Remove the Coolant Temperature Sensor (CTS) at the thermostat housing. Remove the engine harness-to-cylinder head bolt and position the harness aside.
8. Install lift support bracket 8534-7 and bolt.
9. Assemble mounting bracket/sleeve assemblies 8534-2 to support tube 8534-1 and install on the vehicle allowing the brackets to rest on the inner fender ledges.
10. Assemble cross bar 8534-3, clamp 8534-5 and support leg 8534-4 to support tube allowing the support leg to rest on the radiator upper support.
11. Tighten the cross bar-to-support tube clamp 8534-5 as well as mounting bracket/sleeve 8534-3.
12. Install lift bracket assemblies as shown in the accompanying illustration.
13. Raise the vehicle and remove the front wheels.
14. Remove or disconnect the following:
   - ABS sensor connector and sensor brackets from the struts
   - Sway bar links from struts
   - Halfshafts
   - Engine front and rear mount-to-cradle nuts
   - Brake lines and brackets from the frame rails
   - Power steering hydraulic line from the bracket at the cradle located on the passenger side
   - Power steering oil cooler from the cradle

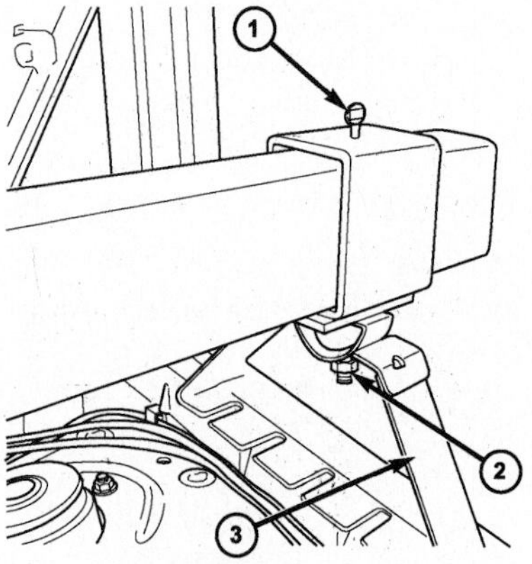

1 - THUMB SCREW
2 - PIVOT HEX NUT
3 - BRACKET/SLEEVE 8534-2

23990G33

Assemble the torque thumb screw and pivot nut bracket assembly as shown

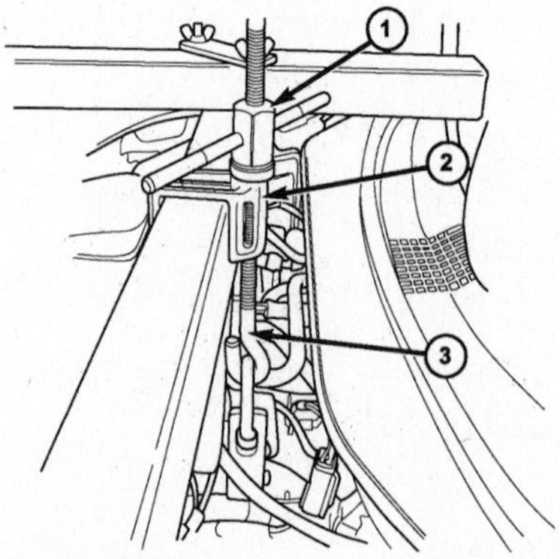

1 - T-HANDLE 8534-14
2 - LIFTING BRACKET 8534-13
3 - HOOK ASSEMBLY 8534-11

23990G34

Assemble the T–handle, bracket and hook assembly at the rear lift bracket as illustrated

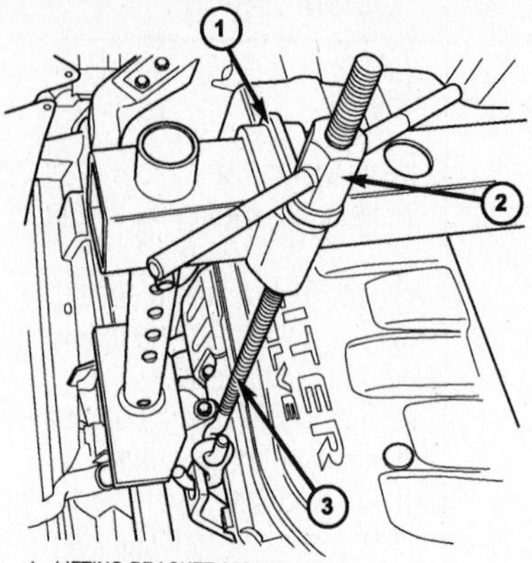

1 - LIFTING BRACKET 8534-13
2 - T-HANDLE 8534-14
3 - HOOK ASSEMBLY 8534-11

23990G35

Assemble the T–handle, bracket and hook assembly at the front lift bracket as shown

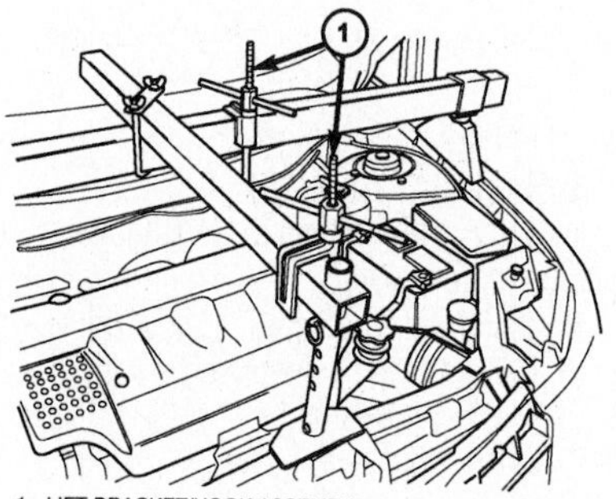

1 - LIFT BRACKET/HOOK ASSEMBLY

23990G36

The engine support system slack is taken up by using the T–handles

   - Power steering pressure and return lines from the steering gear and cap the lines
   - Transmission-to-cradle torque strut
15. Position drive line support table 8874 into position and lower the vehicle until the cradle and fixture engage as shown in the accompanying illustration.
16. Scribe alignment marks to reference cradle-to-body alignment to help during installation, then remove the four cradle-to-body bolts.
17. Slowly raise the vehicle and separate the cradle from the vehicle. Check the overhead fixture is secure on the fenders and radiator support.
18. Have an assistant guide the brake and power steering lines through as they remain attached.
19. Remove or disconnect the following:
   - Engine front mount/bracket
   - Starter motor

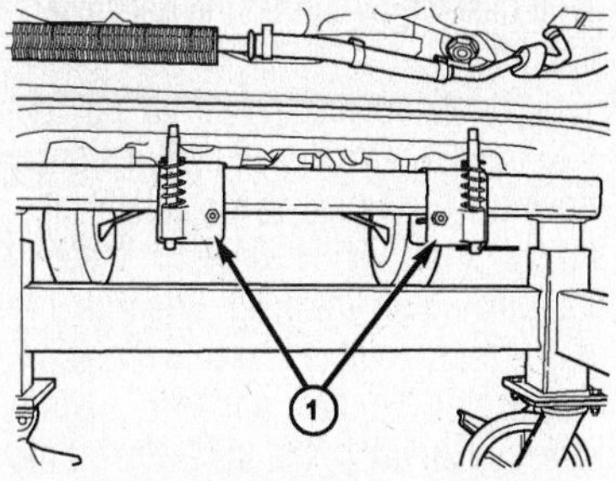

Position drive line support table 8874 into position and lower the vehicle until the cradle and fixture engage

20. On All Wheel Drive (AWD) models remove or disconnect the following:
- Propeller shaft
- Power Transfer Unit (PTU) rear mount bracket
- Oil pan-to-transaxle collar
- Heat shield
- PTU-to-transaxle upper bolts
- PTU-to-bracket lower bolts
- PTU

21. Remove or disconnect the following:
- Torque converter dust shield
- Four torque converter-to-flexplate bolts. Upon removing the bolts a tight tolerance (slotted) bolt will be found. Mark this location on the converter and flexplate to use as a reference during assembly.
- Four transaxle upper bellhousing bolts

22. Use the overhead fixture to lower the engine/transaxle assembly. and secure the transmission to a suitable jack.

23. Remove the two transaxle-to-engine lower bolts and remove the transaxle.

To install:

24. Install or connect the following:
- Transaxle
- Two transaxle-to-engine lower bolts and tighten to 70 ft. lbs. (95 Nm)

25. Remove the transmission jack.
- Four transaxle upper bellhousing bolts and tighten to 70 ft. lbs. (95 Nm)
- Four torque converter-to-flexplate bolts and tighten to 65 ft. lbs. (88 Nm)
- Torque converter dust shield

26. Loosely install the engine-to-transaxle collar.

27. On AWD models install or connect the following:
- PTU lower bracket and PTU
- PTU-to-bracket lower bolts and tighten to 21 ft. lbs. (28 Nm)

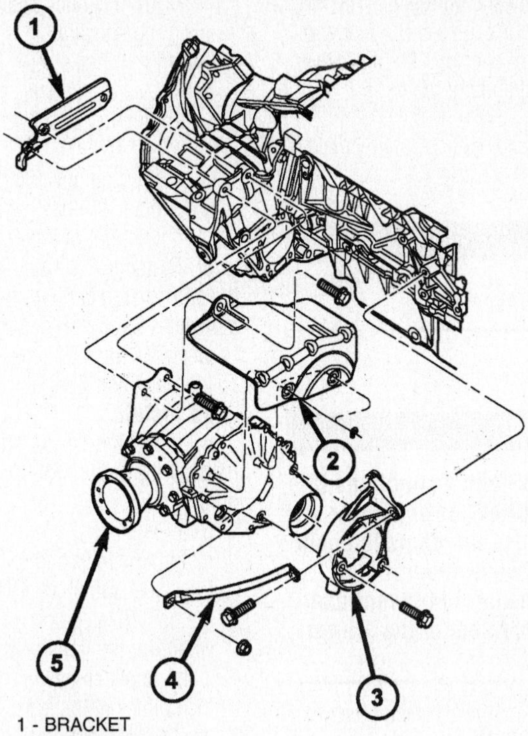

Exploded view of the Power Transfer Unit mounting and related components

- PTU-to-transaxle upper bolts and tighten to 40 ft. lbs. (54 Nm)
- PTU rear mount bracket and tighten the bolts to 40 ft. lbs. (54 Nm)
- Heat shield
- Propeller shaft

28. Install or connect the following:
- Starter motor
- Engine front mount/bracket and tighten the bolts to 50 ft. lbs. (67 Nm)

29. Align the scribe alignment marks made to reference cradle-to-body alignment during removal, then install the four cradle-to-body bolts and tighten to 120 ft. lbs. (162 Nm).

30. Route the brake and power steering lines.

31. Remove the overhead support fixture.

32. Install or connect the following:
- Engine front and rear mount-to-cradle nuts and tighten to 40 ft. lbs. (54 Nm)
- Steering shaft-to-gear coupler using tool 6831A
- Power steering pressure and return lines to the steering gear and tighten the fittings to 23 ft. lbs. (31 Nm)
- Power steering oil cooler and hose routing clip to the cradle
- Transmission-to-cradle torque strut and tighten to 40 ft. lbs. (54 Nm)
- Halfshafts
- Sway bar links to the struts and tighten the nuts to 65 ft. lbs. (88 Nm)
- ABS sensor brackets and electrical connectors
- Both wheels
- Front brake lines to the HCU
- Coolant bypass tube fasteners
- Solenoid/pressure switch and tighten the screw to 35 inch lbs. (4 Nm)
- Transaxle range sensor connection
- Input and output speed sensor connections
- Transaxle harness using the retainers
- CKP sensor
- Transaxle oil cooler splice kit using the kit instructions
- Gearshift cable to the manual valve lever
- Battery tray, battery and cables

33. Bleed the brakes.

34. Add power steering fluid.

35. Check and fill the transmission fluid to the proper level.

36. If equipped with AWD, remove the

PTU fill plug, check the fluid level. The PTU should have 2.1 pints of 75W-90 gear and axle lubricant. After checking the fluid lever, install the fill plug and tighten to 26 ft. lbs. (35 Nm).

37. Check and adjust the wheel alignment.

Halfshaft

REMOVAL & INSTALLATION

Front

✱✱ WARNING

Allowing the CV-joint assemblies to dangle unsupported, or pulling or pushing the ends, can damage boots or CV-joints. Always support both ends of the halfshaft to prevent damage or disengagement of the Tri-pot joint.

1. Before servicing the vehicle, refer to the precautions in the beginning of this section.
2. Remove or disconnect the following:
   - Negative battery cable
   - Hub and bearing-to-stub axle cotter pin and retainer nut
   - Front wheels
   - Front caliper assembly from steering knuckle
   - Front brake rotor from the hub
   - Steering knuckle-to-strut bolts
   - Steering knuckle from the strut bracket
   - Knuckle down and away from the outer CV-joint while pulling the joint out of the bearing assembly
3. Dislodge the inner Tri-pot joint from the stub shaft retaining snapring on the transaxle. To do this, insert a prybar between the transaxle case and the inner Tri-pot joint and pry on Tri-pot joint.

To install:

4. Install the Tri-pot joint side of the halfshaft by performing the following procedure:
   a. Replace the inner Tri-pot joint retaining circlip and O-ring seal on the transaxle stub shaft. These components are not reusable and must be replaced whenever the halfshaft is removed.
   b. Apply an even coat of grease on the splines of the inner Tri-pot joint, where the O-ring seats against the Tri-pot joint.
   c. Grasp the inner joint in 1 hand and interconnecting shaft in the other. Align the inner Tri-pot joint spline with the stub shaft spline on the transaxle. Use a rocking motion with the inner Tri-pot joint to get it past the circlip on the transaxle stub shaft.
   d. Continue pushing the Tri-pot joint onto transaxle stub shaft until it stops moving. The O-ring on the stub shaft should not be visible when the inner Tri-pot joint is fully installed. Check that the inner Tri-pot joint is locked in position by grasping the inner joint and pulling. If locked in position, the joint will not move on the stub shaft.
5. Install the outer CV-joint into the hub and bearing assembly.

✱✱ WARNING

The steering knuckle-to-strut assembly bolts are serrated and must not be turned during installation.. Also if the vehicle is equipped with eccentric strut assembly bolts, the eccentric bolt must be installed in the bottom (slotted) hole on the strut bract.

6. Install or connect the following:
   - Steering knuckle in the bracket of the damper assembly
   - Strut damper-to-steering knuckle bolts and tighten to 65 ft. lbs. (88 Nm) plus an additional ¼ turn
   - Rotor, caliper. Tighten the caliper adapter bolts to 125 ft. lbs. (169 Nm)

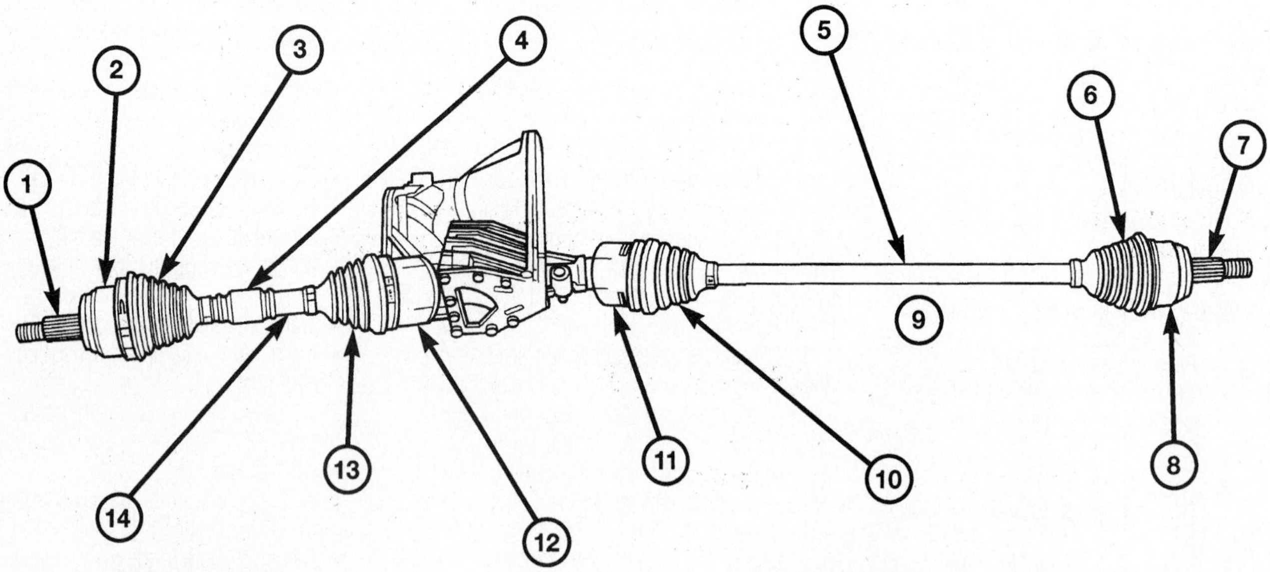

Exploded view of the front halfshaft and related components

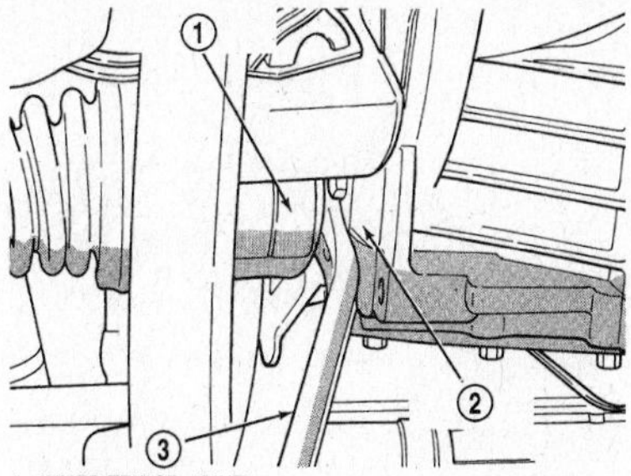

1 - INNER TRIPOD JOINT
2 - TRANSAXLE CASE
3 - PRY BAR

23990G40

Use a prybar to disengage the halfshaft from the transaxle

- Washer and bearing-to-stub axle bolt and hand-tighten.
- Wheels

7. Apply the brakes and tighten the bearing-to-stub axle bolt to 180 ft. lbs. (244 Nm) and install the cotter pin.
8. Road test the vehicle to check for noise or vibration.

Rear

WARNING

The rear suspension and drivetrain design requires this procedure to be performed on a drive on hoist as the front and rear suspension must be compressed to the vehicle ride height.

1. Before servicing the vehicle, refer to the precautions in the beginning of this section.
2. Remove or disconnect the following:
   - Negative battery cable
   - Wheel cover cap
   - Cotter pin, locknut and washer
   - Exhaust system center hanger at the propeller shaft center bearing
   - Exhaust system at the rear hanger and lower at least 10 inches (254mm) before securing with wire
3. Mark the propeller shaft and rear driveline module flanges for reference during installation.
   - 3 propeller shaft-to-driveline module bolts. Do not disconnect the shaft from the module as it will be disconnected upon lowering the module.
4. Support the driveline module with a transmission jack.
5. Using a suitable prytool, partially dislodge the halfshaft from the differential.
6. Install seal protector tool 9099 to protect the seal during disassembly.
7. Remove the 3 driveline-to-crossmember bolts.

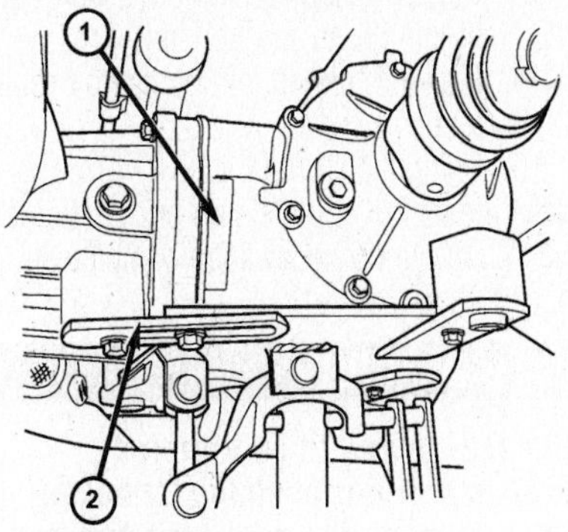

1 - DRIVELINE MODULE
2 - TRANSMISSION JACK

23990G41

Support the driveline module using a transmission jack

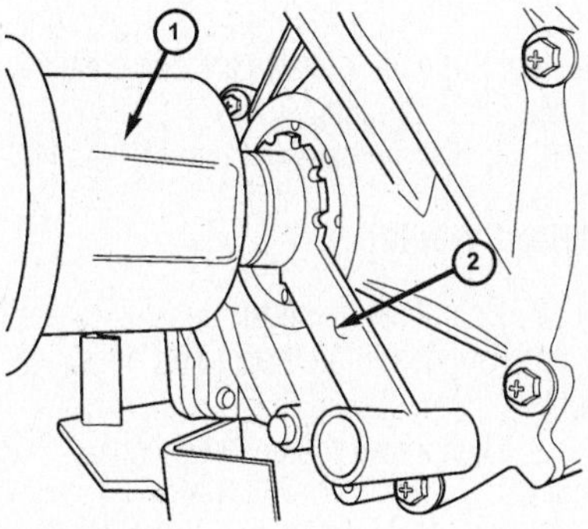

1 - HALFSHAFT
2 - SEAL PROTECTOR

23990G42

Use seal protector tool 9099 to protect against seal damage

8. Lower the driveline module enough to remove the shaft from the differential, making sure tool 9099 engages the seal.
9. Disconnect the propeller shaft from the differential and secure it to the exhaust system.
10. Remove the halfshaft from the bearing assembly.

To install:

11. Remove the halfshaft from the bearing assembly. Install the hub nut and washer and hand tighten.
12. Attach the shaft to the differential, using tool 9099.
13. Connect the propeller shaft to the module flange. Install but do not tighten the 3 bolts
14. Position the module and tighten the module-to-cradle bolts to 75 ft. lbs. (102 Nm).
15. Tighten the propeller shaft-to-module bolts to 40 ft. lbs. (54 Nm)
16. Tighten the halfshaft nut to 180 ft. lbs. (244 Nm).
17. Install or connect the following:
   - Washer, nut and new cotter pin
   - Wheel cover cap
   - Negative battery cable
18. Check and adjust the differential fluid.
19. Road test the vehicle to check for noise or vibration.

CV-Joints

OVERHAUL

Inner (Tri-pot) Joint

1. Before servicing the vehicle, refer to the precautions in the beginning of this section.
2. Disconnect the negative battery cable.
3. Remove the halfshaft and retaining clamps.
4. Slide the boot down the shaft away from the tri-pot housing.

➡When separating the spider joint from the tri-pot joint housing, hold the rollers in place on the trunnions to prevent the rollers and needle bearings from falling away.

5. Carefully, slide the shaft/spider assembly from the tri-pot housing.
6. Remove the spider assembly-to-shaft snapring; then, slide the spider assembly off the shaft.

WARNING

If necessary, tap the spider assembly off the shaft using a brass drift; be careful not to hit the outer bearings.

7. Slide the boot off the shaft.
8. Thoroughly, inspect all parts for signs of excessive wear; if necessary, replace the halfshaft.

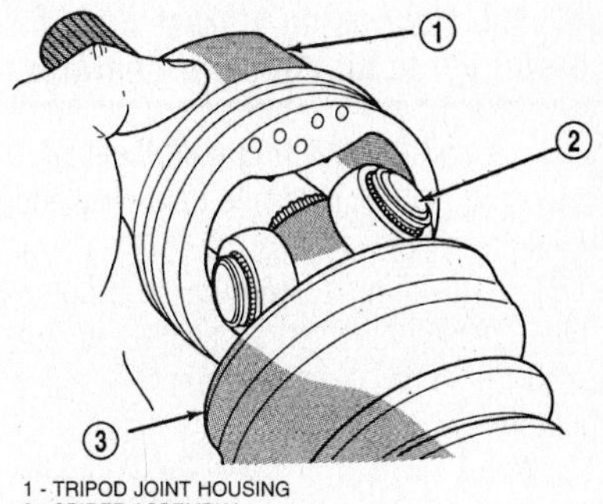

1 - TRIPOD JOINT HOUSING
2 - SPIDER ASSEMBLY
3 - SEALING BOOT

23990G43

Slide the boot down the shaft away from the tri-pot housing

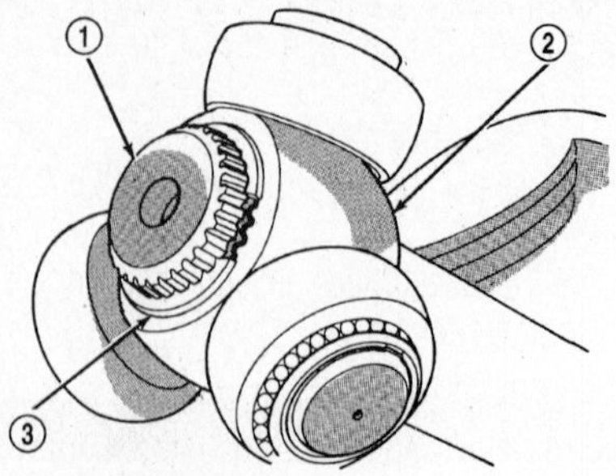

1 - INTERCONNECTING SHAFT
2 - SPIDER ASSEMBLY
3 - RETAINING SNAP-RING

23990G44

Remove the spider assembly-to-shaft snapring; then, slide the spider assembly off the shaft

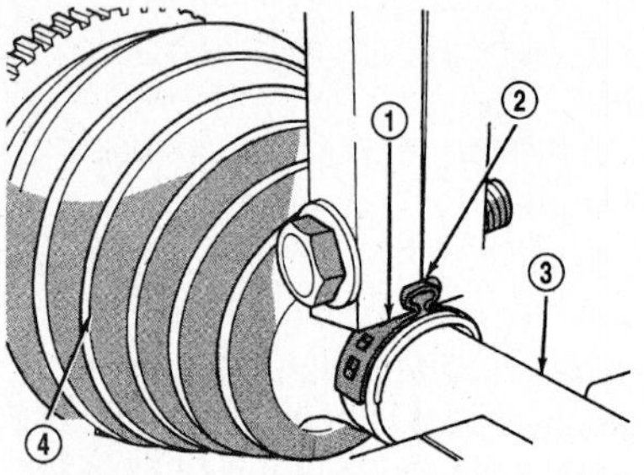

1 - CLAMP
2 - JAWS OF SPECIAL TOOL C-4975-A MUST BE CLOSED COMPLETELY TOGETHER HERE
3 - INTERCONNECTING SHAFT
4 - SEALING BOOT

23990G45

Securing the halfshaft boot clamp

➡Component parts are not serviceable and must be replaced as an assembly.

To install:

9. Slide the inner tri-pot boot clamp and boot onto the shaft; then, position the boot so that only the thinnest (sight) groove is visible on the shaft.
10. Install the spider assembly onto the shaft just far enough so that the snapring can be installed.

⁂ WARNING

If necessary, tap the spider assembly onto the shaft using a brass drift; be careful not to hit the outer bearings.

11. Install the snapring onto the shaft; make sure that the snapring is fully seated in the groove.
12. If installing a new boot, distribute ½ of the grease in the service package inside the tri-pot housing and the other ½ inside the boot.
13. Carefully, slide the spider assembly and shaft into the tri-pot housing.
14. Position the inner boot clamp evenly on the sealing boot.
15. Using the Crimper tool C-4975, place the tool over the clamp bridge, tighten the tool nut until the jaws are completely closed (face-to-face).

⁂ WARNING

The seal must not be dimpled, stretched or out of shape. If necessary, equalize the seal pressure and shape it by hand.

16. Position the boot onto the tri-pot housing retaining groove and install the retaining clamp evenly on the boot.
17. Using the Crimper tool C-4975, place the tool over the clamp bridge, tighten the tool nut until the jaws are completely closed (face-to-face).
18. Install the halfshaft into the vehicle.

Outer CV-Joint

1. Before servicing the vehicle, refer to the precautions in the beginning of this section.
2. Disconnect the negative battery cable.
3. Remove the halfshaft and the retaining clamps.
4. Slide the boot down the shaft away from the CV-joint housing.
5. Remove the grease to expose the CV-joint-to-shaft retaining ring.
6. Spread the snapring ears apart and slide the CV-joint assembly off of the shaft.
7. Slide the boot off the shaft.
8. Thoroughly, clean and inspect all parts for signs of excessive wear; if necessary, replace the halfshaft.

➡Component parts are not serviceable and must be replaced as an assembly.

To install:

9. Slide the outer CV-joint boot clamp and boot onto the shaft; then, position the boot so that only the thinnest (sight) groove is visible on the shaft.
10. Slide the outer CV-joint assembly on the shaft, spread the snapring ears, position the CV-joint and verify that the snapring is fully seated in the shaft groove.
11. If installing a new boot, distribute ½ of the grease in the service package into the CV-joint housing and the other ½inside the boot.

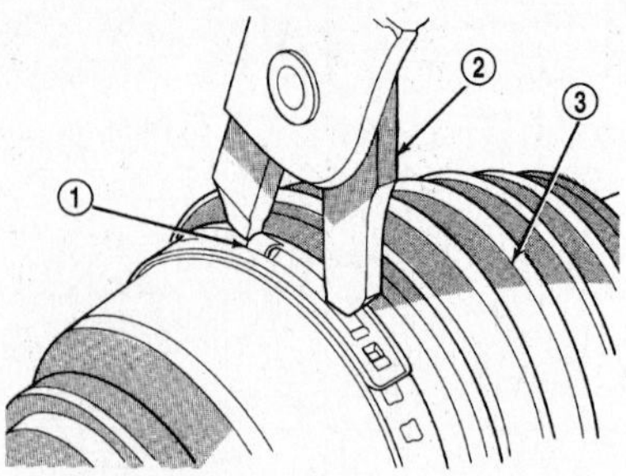

1 - CLAMP
2 - TOOL YA3050, OR EQUIVALENT
3 - SEALING BOOT

23990G46

The latch type boot clamp is tighten as illustrated....

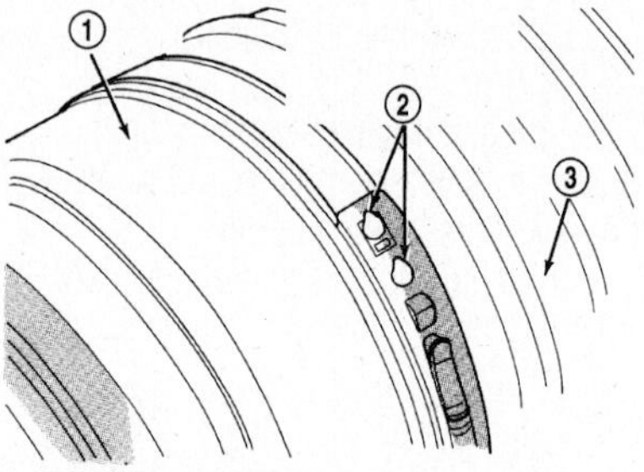

1 - INNER TRIPOD JOINT HOUSING
2 - TOP BAND OF CLAMP MUST BE RETAINED BY TABS AS SHOWN HERE TO CORRECTLY LATCH BOOT CLAMP
3 - SEALING BOOT

23990G47

....and should look like this when properly installed

12. Position the outer boot clamp evenly on the sealing boot.
13. Using the Crimper tool C-4975, place the tool over the clamp bridge, tighten the tool nut until the jaws are completely closed (face-to-face).

⁂ WARNING

The seal must not be dimpled, stretched or out of shape. If necessary, equalize the seal pressure and shape it by hand.

14. Position the boot onto the CV-joint housing retaining groove and install the retaining clamp evenly on the boot.
15. Using the Crimper tool C-4975, place the tool over the clamp bridge, tighten the tool nut until the jaws are completely closed (face-to-face).
16. Install the halfshaft into the vehicle.

STEERING AND SUSPENSION

Air Bag

CAUTION

Some vehicles are equipped with an air bag system. The system must be disabled before performing service on or around system components, steering column, instrument panel components, wiring and sensors. Failure to follow safety and disabling procedures could result in accidental air bag deployment, possible personal injury and unnecessary system repairs.

PRECAUTIONS

Several precautions must be observed when handling the inflator module to avoid accidental deployment and possible personal injury.

- Never carry the inflator module by the wires or connector on the underside of the module.
- When carrying a live inflator module, hold securely with both hands, and ensure that the bag and trim cover are pointed away.
- Place the inflator module on a bench or other surface with the bag and trim cover facing up.
- With the inflator module on the bench, never place anything on or close to the module which may be thrown in the event of an accidental deployment.

Before servicing the vehicle, also make sure to refer to the precautions in the beginning of this section as well.

DISARMING

CAUTION

The Air Bag system must be disarmed before repair and/or removal of any component in its immediate area including the air bag itself. Failure to do so may cause accidental deployment of the air bag, resulting in unnecessary system repairs and/or personal injury.

1. Disconnect the negative battery cable and isolate the cable using an appropriate insulator (wrap with quality electrical tape).
2. Allow the system capacitor to discharge for 2 minutes before starting any repair on any air bag system or related components. This will disable the air bag system.

CAUTION

Always wear safety goggles when working with or around the air bag system. When carrying a live air bag, be sure the bag and trim cover are pointed away from the body. In the unlikely event of an accidental deployment, the bag will, then deploy with minimal chance of injury. When placing a live air bag on a bench or other surface, always face the bag and trim cover up, away from the surface. This will reduce the motion of the module if it is accidentally deployed.

Power Rack and Pinion Steering Gear

REMOVAL & INSTALLATION

1. Before servicing the vehicle, refer to the precautions in the beginning of this section.
2. Turn the front wheels to the straight-ahead position.
3. Remove or disconnect the following:
   - Negative battery cable
   - Cap from the power steering fluid reservoir and siphon out as much fluid as possible
   - Front wheels
4. If transferring tie rod ends to a new gear, loosen the jam nuts.
   - Nut attaching tie rod end to the knuckle on both sides
   - Tie rod ends from the knuckle using tool C-3894-A
5. Remove the roll pin fastening intermediate shaft extension-to-power steering gear shaft as follows:
   a. Insert removal tool 631A through the roll pin attaching the intermediate shaft extension to the steering gear.
   b. Thread the knurled nut all the way onto the remover.
   c. Hold the remove head steady and turn the hex nut to pull the pin from the shafts and remove the tool.
6. Remove or disconnect the following:
   - Intermediate shaft extension off the gear shaft
   - Tube nut attaching return hose to the gear
   - Return hose from the gear port
   - Two gear mounting bolts and the gear.
7. If necessary, remove the tie rod ends counting how many turns it takes as this will aid during installation.

To install:

8. If removed, install the tie rod ends.
9. Install or connect the following:
   - Gear through the left wheel well being careful of the tie rod end shields

➡The drivers side gear mounting bolt is longer than the passenger side bolt.

   - Gear bolts after centering the gear over the mounting bosses over the holes in the cradle and tighten to 120 ft. lbs. (163 Nm)
10. Wipe the power steering hoses on the gear, the connections with a lint free cloth and replace the O-rings coated with clean power steering fluid.
   - Pressure and return hoses to the

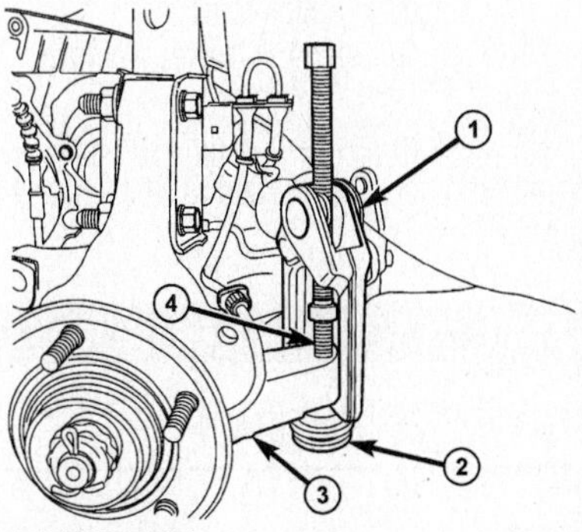

Use tool C-3894-A to separate the tie rod ends from the knuckle

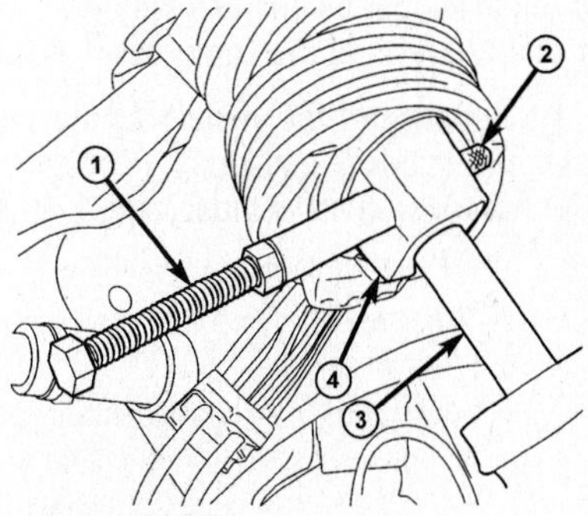

Insert removal tool 631A through the roll pin attaching the intermediate shaft extension to the steering gear

gear and tighten the tube nut to 275 inch lbs. (31 Nm)

11. Find the gear's center of travel and match the splines inside the intermediate shaft extension with those on the gear shaft, then slide the extension onto the shaft.

12. Install the roll pin as follows:

a. Position installer 6381-A onto the extension and gear shaft.

b. Slide the roll pin onto the remover shaft.

c. Thread the knurled nut all the way onto the end of the installer.

d. While holding the install head, turn the hex nut to pull the pin into the shafts. Once the pin is centered in the intermediate shaft extension, remove the tool.

13. Install or connect the following:

- Shaft cover over the top of the gear
- Outer tie rod ends to the knuckle using new nuts and tighten to 35 ft. lbs. (47 Nm), then an additional 180 degree (½turn
- Snug the jam nuts
- Wheels
- Negative battery cable

14. Fill and bleed the power steering system.

15. Perform a wheel alignment and tighten the tie rod jam nuts to 55 ft. lbs. (75 Nm).

Strut

REMOVAL & INSTALLATION

Front

1. Before servicing the vehicle, refer to the precautions in the beginning of this section.

➡Service of the coil spring requires the use of a coil spring compressor tool. It is required that 5 coils be captured within the jaws of the compressor tool.

➡Do not support the vehicle by placing supports under the suspension arms. The suspension arms must hang freely.

2. Remove or disconnect the following:

- Negative battery cable
- Front wheel(s)
- Speed sensor wiring harness mounting bracket from the strut, if equipped with Antilock Brake System (ABS)

➡When removing the nut from the stud of the stabilizer bar link, do not allow the stud to rotate in the socket by holding with an open ended wrench.

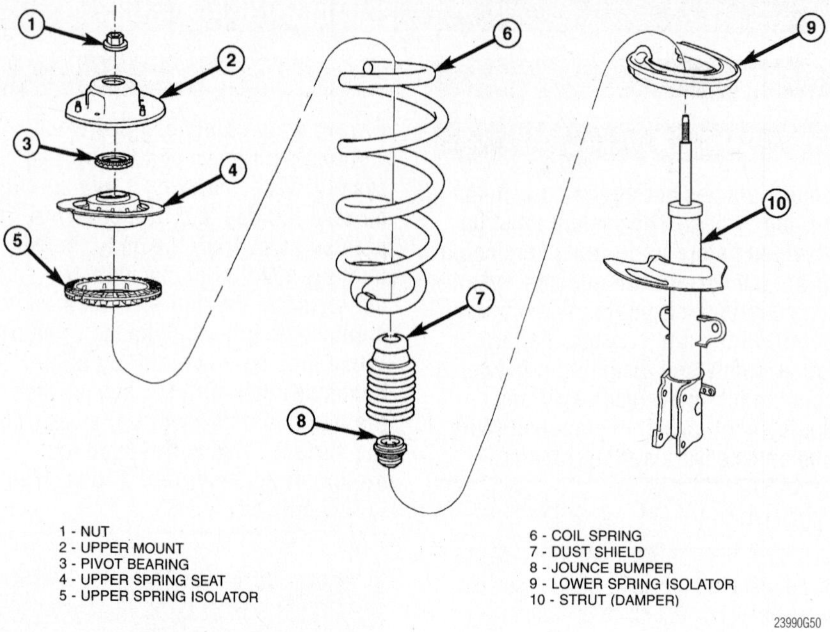

Exploded view of the front strut assembly

- Strut-to-steering knuckle bolts and nuts
- Stabilizer bar attaching link at the strut assembly

⁂ WARNING

The steering knuckle-to-strut assembly bolts are serrated and must not be turned during installation.

3. If removing the left strut assembly, unfasten the nuts attaching the coolant recovery bottle and position the bottle and hoses aside.

4. Remove the 3 strut assembly upper mount-to-shock tower mounting nuts and washers. Remove the strut from the vehicle.

5. Disassemble the strut by performing the following procedure:

a. Securely mount the strut assembly into a vise. Using paint, mark the strut unit, lower spring isolator, spring and upper strut mount for indexing of the parts at assembly.

b. Position the spring compressor tool onto the strut. Compress the coil spring until all load is off the upper strut mount assembly.

c. Install Strut Rod Socket tool 6864 on the strut shaft nut and a 10mm socket on the end of the strut shaft to prevent it from turning. Remove the strut shaft nut.

d. Remove the upper mount assembly, jounce bumper and seat bearing and dust shield as an assembly.

e. Remove the coil spring and compressor as an assembly from the strut. Remove the lower spring isolator from the strut assembly lower spring seat.

f. Inspect all components for abnormal wear, oil leakage or failure. Replace parts as required.

To install:

6. Assemble the strut by performing the following procedure:

a. Inspect the strut assembly for signs of leakage. Actual leakage will be a stream of fluid running down the side and dripping off the lower end of the strut. A slight amount of seepage between the strut rod and strut shaft seal is not unusual and does not affect performance of the strut assembly.

b. Install the lower spring isolator on the strut unit. Install the compressed coil spring onto the strut assembly aligning the paint marks made during removal.

c. Install the strut bearing into the bearing seat. The bearing must be installed into the seat with the notches on the bearings facing down.

d. Lower the seat bearing and dust shield onto the strut and spring assembly. Align the paint marks made during removal.

e. Install the jounce bumper and upper mount on the strut shaft, aligning the paint marks.

f. Install the strut mount-to-shaft retainer nut. Inspect all alignment marks made during removal and align as required. While holding the strut shaft from turning with a 10mm socket, torque the strut shaft nut to 85 ft. lbs. (115 Nm).

g. Equally loosen the spring compressor tool until all tension is released. Remove the spring compressor tool.

7. Install the front strut into the strut tower. Torque the 3 upper nuts to 250 inch lbs. (28 Nm).

8. If installing the left strut assembly, position the coolant recovery bottle and hoses. Fasten the bottle retainers.

WARNING

The steering knuckle-to-strut assembly bolts are serrated and must not be turned during installation.. Also if the vehicle is equipped with eccentric strut assembly bolts, the eccentric bolt must be installed in the bottom (slotted) hole on the strut bract.

➡The strut-to-steering are installed differently on each side. The left hand side bolts are installed from the vehicle rear to front and the right side bolts are installed from the vehicle front to rear.

9. Position the steering knuckle neck into the strut assembly. Install the strut assembly-to-steering knuckle bolts. Install the nuts onto the attaching bolts and torque to 60 ft. lbs. (81 Nm), then tighten an additional ¼ or 90 degrees turn.

➡When installing the nut to the stud of the stabilizer bar link, do not allow the stud to rotate in the socket by holding with an open ended wrench.

10. Install or connect the following:
- Stabilizer bar attaching link to the strut assembly, hand thread the nut then tighten to 65 ft. lbs. (88 Nm) using an open ended wrench on the machine surfaced to prevent from turning.
- Speed sensor wiring harness mounting bracket to the strut and tighten the retainers to 115 inch lbs. (13 Nm)
- Front wheel(s)
- Negative battery cable

11. Align the vehicle.

Rear

1. Before servicing the vehicle, refer to the precautions in the beginning of this section.

2. Remove or disconnect the following:
- Rear wheel

3. Position jack stands under the forward end of the engine cradle to support and stabilizer the vehicle.

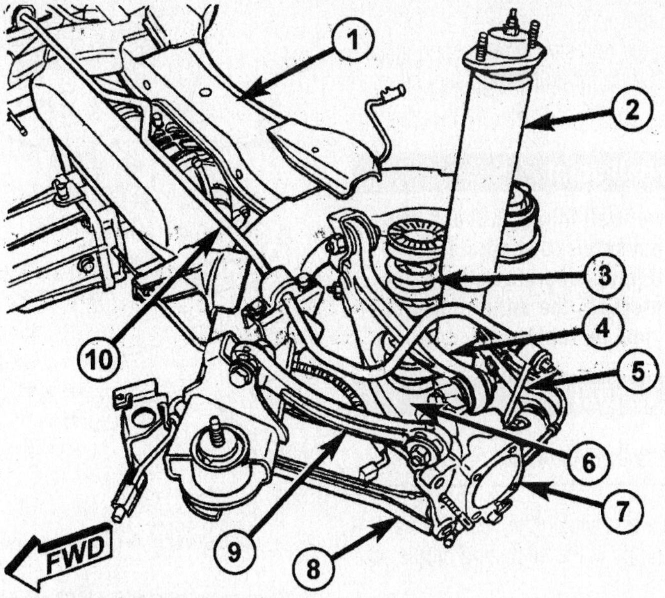

1 - CROSSMEMBER
2 - SHOCK ABSORBER (LOAD-LEVELING)
3 - COIL SPRING
4 - CAMBER LINK
5 - TOE LINK
6 - SPRING LINK
7 - KNUCKLE
8 - COMPRESSION LINK
9 - TENSION LINK
10 - STABILIZER BAR

23990G51

Exploded view of the rear suspension components

4. Place a transmission jack under the center of the rear suspension crossmember on models not equipped with All Wheel Drive (AWD) or rear driveline module, if equipped with AWD.

5. Remove or disconnect the following:
- Two shock absorber upper bolts
- Lower shock absorber bolts
- Front and rear crossmember bolts

6. Lower the jack slowly enough so to allow the top of the shock absorber to clear the body flange and remove the shock absorber by tipping it outwards and lifting the lower end out of the pocket in the spring link.

7. Remove the coil spring and isolator lower end first, if necessary. Never lower the jack any more than necessary for removal and installation.

To install:

8. Install or connect the following:
- Coil spring isolator on top of the sprint, then the spring top end first. Match the lower end coil against the abutment in the spring link.
- Shock absorber by setting the lower end into the pocket in the spring link, then tipping the top inwards. Hand start the upper mounting bolts

9. Raise the jack slowly guiding the spring and lower end of the shock absorber into position. Once the bolt holes are aligned, stop jacking. Install and hand tighten the bolts.

10. Raise the jack until the two crossmember bolts can be installed and tighten to 120 ft. lbs. (163 Nm).

11. Tighten the upper shock absorber bolts to 45 ft. lbs. (61 Nm) and remove the jack.

12. Install the wheels, position the vehicle on an alignment rack and raise the vehicle enough to access the lower shock absorber bolts. Tighten the bolts to 75 ft. lbs. (102 Nm).

13. Have the rear wheel toe set to specifications.

Coil Springs

REMOVAL & INSTALLATION

Front

Refer to the front strut removal and installation procedure for coil spring service information.

Rear

Refer to the rear strut removal and installation procedure for coil spring service information.

Lower Ball Joint

➡The lower ball joints on these vehicles are not serviced separately. The lower ball joints operate with no free-play. If defective, the entire lower control arm must be replaced.

Lower Control Arm

REMOVAL & INSTALLATION

1. Before servicing the vehicle, refer to the precautions in the beginning of this section.
2. Remove or disconnect the following:
   - Negative battery cable
   - Front wheels
   - Hub and bearing-to-stub axle cotter pin and retainer nut
   - Front caliper assembly from steering knuckle
   - Front brake rotor from the hub
   - Wheel Speed Sensor (WSS) harness connector and unclip the connector and retaining clip from the frame rail
   - Screw attaching the WSS routing bracket to the strut, open the routing clip at the knuckle and remove the cable
3. Push in the outer end of the halfshaft to disengage the splines from the hub splines.
   - Tie rod end nut from the steering knuckle by holding the tie rod while loosening the nut, then use tool C-

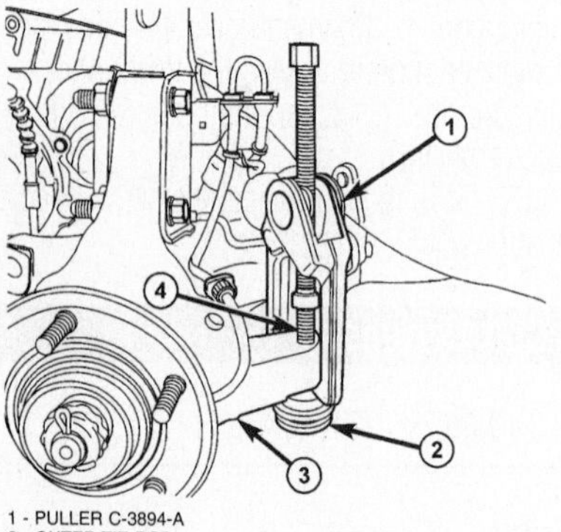

1 - PULLER C-3894-A
2 - OUTER TIE ROD
3 - STEERING KNUCKLE
4 - TIE ROD STUD

23990G48

Use tool C-3894-A to separate the tie rod ends from the knuckle

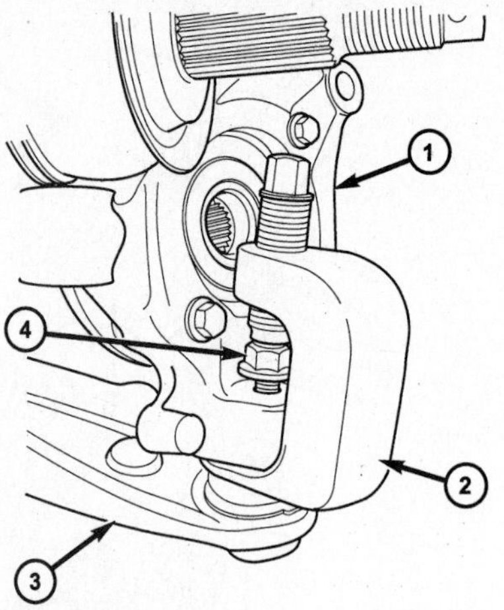

1 - ALUMINUM KNUCKLE
2 - SPECIAL TOOL C-4150A
3 - LOWER CONTROL ARM
4 - NUT INSTALLED ON BALL JOINT STEM

23990G52

Place removal tool C-4150A over the ball joint and stud and tighten the tool to release the stud. It may help to rotate the knuckle around so the inside of the knuckle faces outwards

   3894-A to separate the tie rod from the knuckle
   - Steering knuckle-to-strut bolts
   - Steering knuckle from the strut bracket
4. Tip the knuckle outward at the top and remove the halfshaft from the hum and suspend the halfshaft using wire.
   - Ball joint nut using an impact gun, then reinstall the nut until the top of the nut is even with the stud. this will prevent the stud from distorting while performing the next step.
5. Place removal tool C-4150A over the ball joint and stud and tighten the tool to release the stud. It may help to rotate the knuckle around so the inside of the knuckle faces outwards, then remove the tool.
   - Knuckle
   - Bolts attaching the lower control arm to the engine cradle and the control arm

To install:

➡When installing the lower control arm-to-cradle rear mounting bolts, make sure the flag on the left control arm nut are positioned upwards (above the control arm) and the right control arm nut are positioned downwards (below the control arm).

6. Install or connect the following:
   - New lower control arm and the control arm to cradle bolts from the rear as illustrated and hand tighten

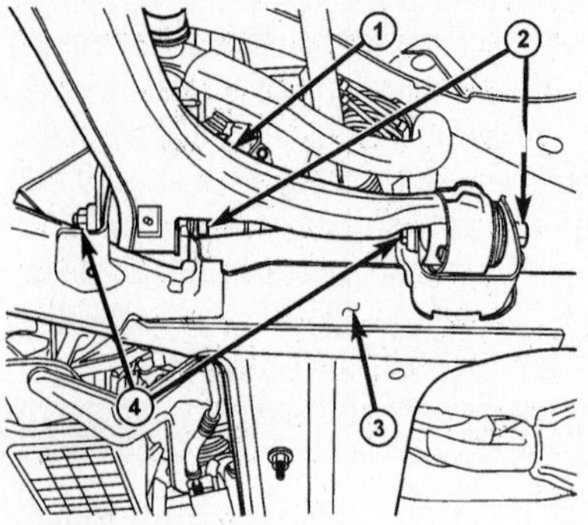

1 - LOWER CONTROL ARM
2 - MOUNTING BOLTS
3 - ENGINE CRADLE
4 - FLAG NUTS

23990G53

Exploded view of the lower control arm mounting, note the orientation of the bolts for installation purposes

7. Clean the ball joint stud and knuckle contact surfaces of grease and dirt to avoid damage.
   - Steering knuckle on the ball stud
   - New knuckle to ball joint nut, then tighten the nut while holding the stud with a hex wrench and tighten to 60 ft. lbs. (81 Nm) plus an additional 90 degrees (1/4) turn
   - Halfshaft into the hub assembly

⁂ WARNING

The steering knuckle-to-strut assembly bolts are serrated and must not be turned during installation.. Also if the vehicle is equipped with eccentric strut assembly bolts, the eccentric bolt must be installed in the bottom (slotted) hole on the strut bract.

➡The strut-to-steering are installed differently on each side. The left hand side bolts are installed from the vehicle rear to front and the right side bolts are installed from the vehicle front to rear.

8. Position the steering knuckle neck into the strut assembly. Install the strut assembly-to-steering knuckle bolts. Install the nuts onto the attaching bolts and torque to 60 ft. lbs. (81 Nm), then tighten an additional 1/4 or 90 degrees turn.
9. Clean the tie rod end and knuckle contact areas to prevent damage.
10. Install or connect the following:
    - Tie rod end to the knuckle and hand tighten the nut. While using a socket to hold the stud, tighten the nut to 35 ft. lbs. (47 Nm) plus an additional 180 degree (1/2 turn).

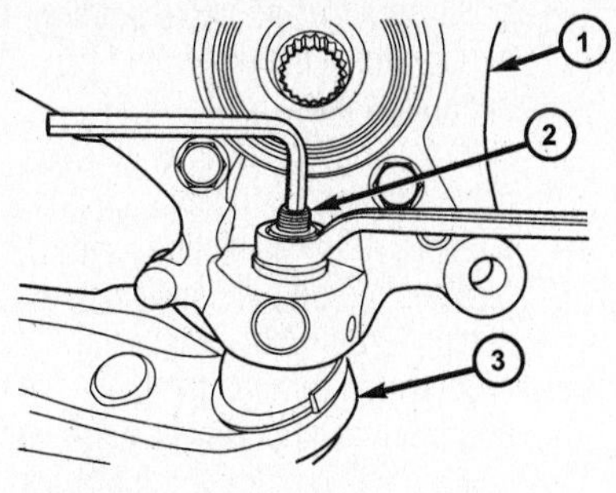

1 - KNUCKLE
2 - BALL JOINT STUD
3 - LOWER CONTROL ARM

23990G54

Tightening the ball joint nut

- WSS sensor clip, cable, routing clip and attach the connection
- Rotor and caliper

11. Have an assistant apply the brakes, install the washer, and hub nut. Tighten the nut to 180 ft. lbs. (244 Nm).
    - Spring washer, new hub nut and cotter pin
    - Wheels
    - Tighten the lower arm frame bolts to 120 ft. lbs. (163 Nm)
12. Check and adjust the brake fluid and brakes.
13. Align the vehicle.

CONTROL ARM BUSHING REPLACEMENT

The control arm bushings are an integral part of the assembly. If worn or damaged, the control arm assembly must be replaced.

Wheel Bearings

ADJUSTMENT

These front wheel drive vehicles are equipped with permanently sealed front and rear wheel bearings. There is no periodic lubrication or maintenance recommended for these units.

REMOVAL & INSTALLATION

Front

1. Before servicing the vehicle, refer to the precautions in the beginning of this section.
2. Remove or disconnect the following:
    - Negative battery cable
    - Front wheels
    - Hub and bearing-to-stub axle cotter pin and retainer nut
    - Front caliper assembly from steering knuckle
    - Front brake rotor from the hub
    - Wheel Speed Sensor (WSS) harness connector and unclip the connector and retaining clip from the frame rail
    - Screw attaching the WSS routing bracket to the strut, open the routing clip at the knuckle and remove the cable
3. Push in the outer end of the halfshaft to disengage the splines from the hub splines.
    - Four hub/bearing bolts from the rear of the knuckle
    - Hub/bearing with the WSS

To install:

4. Clean all mating and mounting surfaces prior to installation to prevent damage.
5. Install or connect the following:
    - Hub/bearing onto the halfshaft stub shaft and into the knuckle until it is squarely seated on the face of the knuckle.
    - Four hub/bearing bolts and using a crisscross pattern tighten progressively and equally to 45 ft. lbs. (65 Nm)
    - WSS sensor clip, cable, routing clip and attach the connection
    - Rotor and caliper
6. Have an assistant apply the brakes, install the washer, and hub nut. Tighten the nut to 180 ft. lbs. (244 Nm).
    - Spring washer , new hub nut and cotter pin
    - Wheels
7. Check and adjust the brake fluid and brakes.
8. Align the vehicle.

Rear

1. Before servicing the vehicle, refer to the precautions in the beginning of this section.
2. Remove or disconnect the following:
    - Rear wheel

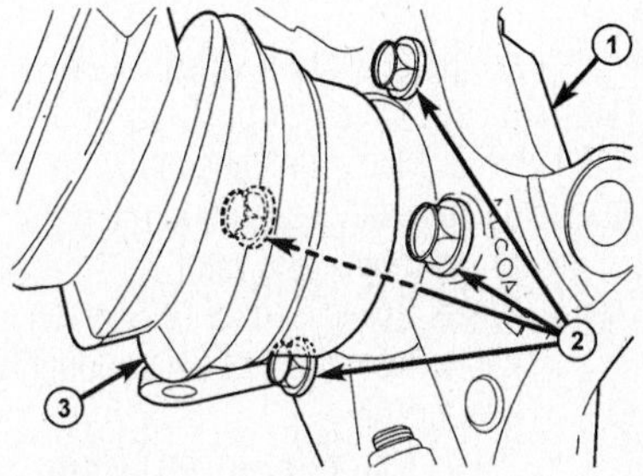

1 - KNUCKLE
2 - HUB AND BEARING MOUNTING BOLTS
3 - HALFSHAFT

23990G58

Location of the front hub/bearing mounting bolts

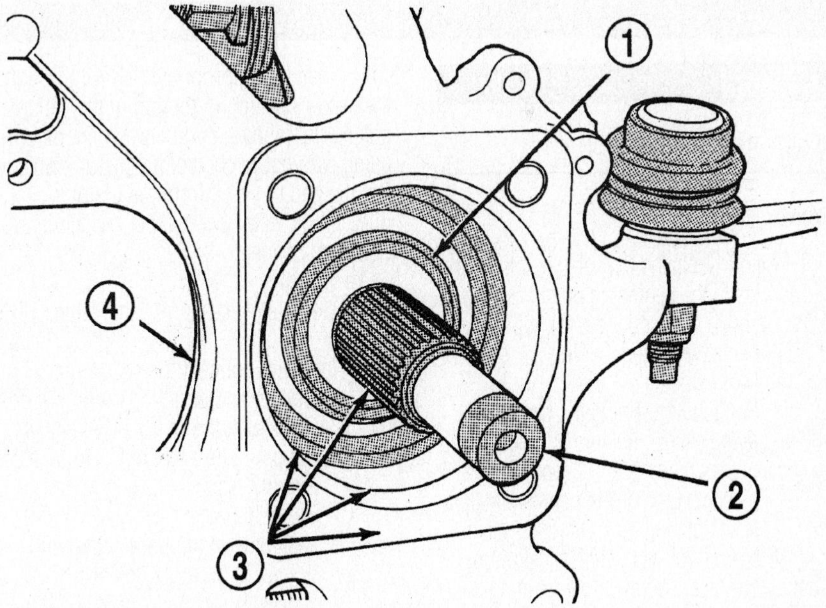

1 - HALFSHAFT OUTER C/V JOINT
2 - STUB SHAFT
3 - THESE SURFACES MUST BE CLEAN AND FREE OF NICKS BEFORE INSTALLING BEARING ASSEMBLY
4 - STEERING KNUCKLE

23990G59

Exploded view of the front hub/bearing assembly

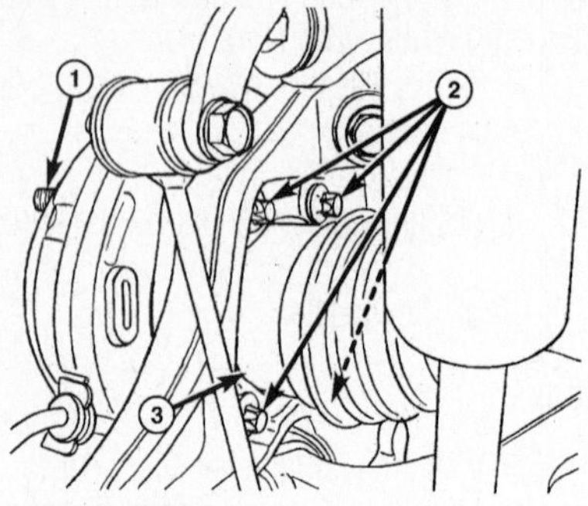

Location of the rear hub/bearing mounting bolts

3. If equipped with All Wheel Drive (AWD), remove the cotter pin, nut and spring washer. Have someone apply the brake pedal and remove the hub nut.
   - Brake caliper and rotor
   - Wheel Speed Sensor (WSS) connector from the spare tire mounting support
   - Two sensor cable routing clips along the toe link
   - Sensor cable from the bracket on the brake support
   - Loosen but do not remove the hub/bearing retaining bolts. Once loosened, push the bolts against the rear of the hub to keep the brake support in place when the hub is removed.
   - Hub and thread the WSS cable through the hole in the brake support plate as you remove the hub

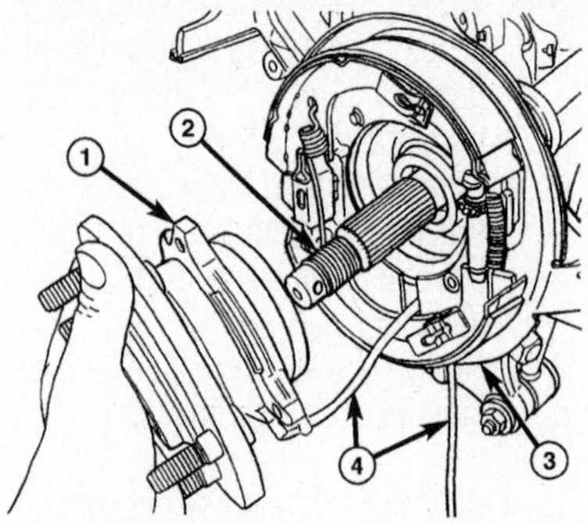

When removing/installing the rear hub/bearing assembly, thread the WSS cable through the hole in the brake support plate

To install:

4. Install or connect the following:
   - Hub/bearing bolts through the rear of the knuckle and parking brake support just enough to hold the support in place
   - Hub/bearing first feeding the WSS cable through the hole in the brake support. At the same time, slide the hub/bearing onto the halfshaft, if equipped with AWD. Place the hub/bearing through the brake support onto the knuckle and align the bolt mounting holes with bolts and place the WSS head at the bottom.
   - Four hub/bearing bolts and tighten to 60 ft. lbs. (81 Nm)
   - Sensor cable onto the bracket on the brake support and route the cable
   - Two sensor cable routing clips along the toe link
   - WSS connector to the spare tire mounting support
   - Brake caliper and rotor

5. If equipped with AWD, install the hub nut, have someone apply the brake and tighten the nut to 180 ft. lbs. (244 Nm). Install the spring washer, nut lock and cotter pin.
   - Wheels

6. Pump the brake pedal several times to ensure proper operation and road test the vehicle.

BRAKES

Caliper

REMOVAL & INSTALLATION

Front

1. Before servicing the vehicle, refer to the precautions in the beginning of this section.
2. Remove or disconnect the following:
   - Negative battery cable
   - Front wheels
   - Banjo bolt retaining the brake hose to the caliper. Be sure to plug the end of the brake hose or cover it with a plastic bag to prevent contamination from entering the hydraulic system.
3. Place a C-clamp over the caliper as illustrated, place a screw drive head against the outboard pad and hook against the rear of the caliper. Slowly tighten the screw drive and retract the caliper pistons into their bores and breaking the outboard brake pad from the caliper fingers.
4. Once the pad is free, slide the caliper in on the guides to provide clearance between the rotor and inboard pad.
5. Place an appropriate prytool through the center opening in the top of the caliper behind the inboard pad between the pistons using care not to contact the piston boots. Pry the pad to free it from the pistons.
6. Remove the two caliper guide pin bolts and caliper.

To install:

7. Completely compress the caliper pistons
8. Install or connect the following:
   - Caliper and align the guide pin bolt holes with the guide pins. Install the pin bolts and tighten to 32 ft. lbs. (43 Nm).
   - Banjo bolt with new washers on each side of the hose fitting and tighten to 35 ft. lbs. (47 Nm)
   - Wheels
   - Negative battery cable
9. Bleed the brake system and road test the vehicle.

Rear

1. Before servicing the vehicle, refer to the precautions in the beginning of this section.
2. Remove or disconnect the following:
   - Negative battery cable
   - Rear wheels
   - Banjo bolt retaining the brake hose to the caliper. Be sure to plug the end of the brake hose or cover it with a plastic bag to prevent contamination from entering the hydraulic system.
   - Caliper guide pin bolts
   - Caliper assembly from the brake adapter by rotating the bottom of the caliper away from the rotor, then lift the caliper with the pads away from the adapter abutment
   - Brake pads by pushing (outboard) or pulling (inboard) from the caliper fingers and piston

To install:

3. Completely compress the caliper piston
4. Install or connect the following:
   - Inboard pad clip against the piston cavity and press the pad until the clip is seated making sure the pad backing plate is flush against the piston

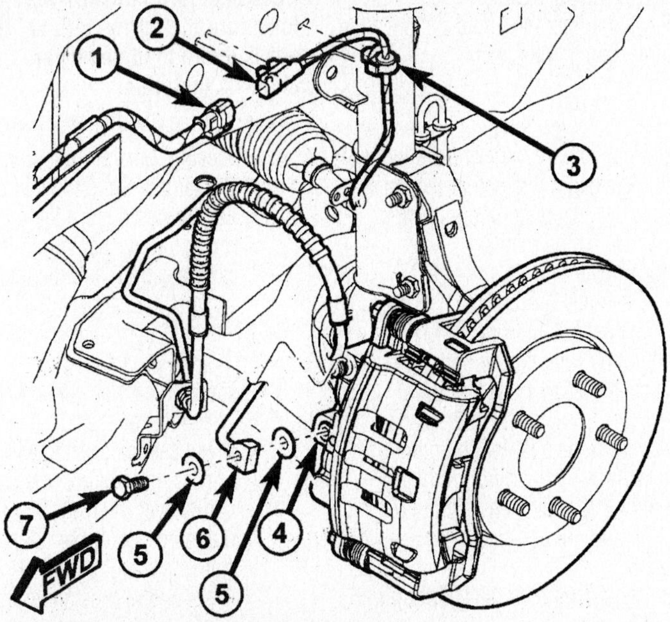

1 - WIRING HARNESS CONNECTOR
2 - WHEEL SPEED SENSOR CONNECTOR
3 - ROUTING CLIP
4 - PORT ON BRAKE CALIPER
5 - COPPER SEALING WASHER
6 - BRAKE HOSE BANJO FITTING
7 - BANJO BOLT

23990G57

Exploded view of the banjo bolt and related components

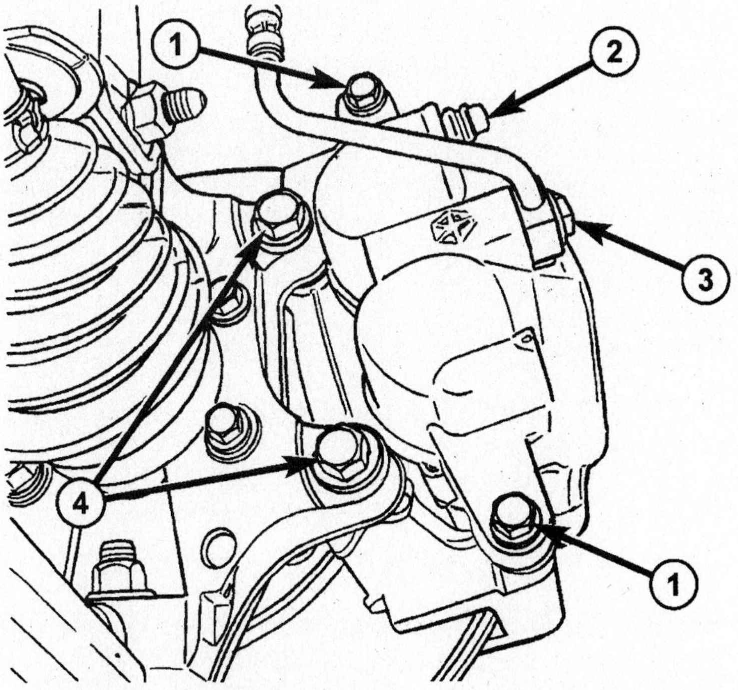

1 - CALIPER GUIDE PIN BOLTS
2 - BLEEDER SCREW
3 - BRAKE HOSE BANJO BOLT
4 - CALIPER ADAPTER MOUNTING BOLTS

23990G56

Exploded view of the caliper and adapter mounting

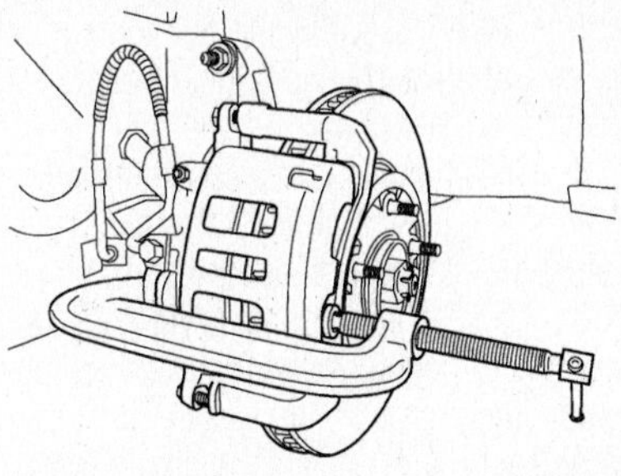

23990G55

Place a C-clamp over the caliper as illustrated

➡The outboard pads are side oriented, make sure the spring clip is installed so it is positioned downwards when the caliper is installed.

- Outboard pad making sure the locating pins are positioned against the ramps. Slide the pad onto the caliper and ensure the locating pins are squarely seated into the holes on the caliper and the pad is flush against the caliper fingers.

5. Make sure the abutment shims are in place on both slide abutments.
6. Retract the caliper guide pins to clear the caliper adapter bosses.
   - Brake caliper. Staring with the upper end, position the caliper and shoes over the rotor and align the outboard pad upper edge with the caliper slide abutment. Rotate the lower end of the caliper into position
   - Caliper guide pin bolts and tighten to 200 inch lbs. (23 Nm)
   - Banjo bolt with new washers on each side of the hose fitting and tighten to 35 ft. lbs. (47 Nm)
   - Wheels
   - Negative battery cable
7. Bleed the brake system and road test the vehicle.

Disc Brake Pads

REMOVAL & INSTALLATION

Front

1. Before servicing the vehicle, refer to the precautions in the beginning of this section.
2. Remove or disconnect the following:
   - Negative battery cable
   - Front wheels
3. Place a C-clamp over the caliper as illustrated, place a screw drive head against the outboard pad and hook against the rear

of the caliper. Slowly tighten the screw drive and retract the caliper pistons into their bores and breaking the outboard brake pad from the caliper fingers.

4. Once the pad is free, slide the caliper in on the guides to provide clearance between the rotor and inboard pad.

5. Place an appropriate prytool through the center opening in the top of the caliper behind the inboard pad between the pistons using care not to contact the piston boots. Pry the pad to free it from the pistons.

6. Remove the two caliper guide pin bolts and position the caliper aside. Remove the brake pads.

To install:

7. Completely compress the caliper pistons

8. Install or connect the following:
   - Brake pads with anti-rattle clips onto the adapter
   - Caliper and align the guide pin bolt holes with the guide pins. Install the pin bolts and tighten to 32 ft. lbs. (43 Nm).
   - Banjo bolt with new washers on each side of the hose fitting and tighten to 35 ft. lbs. (47 Nm)
   - Wheels
   - Negative battery cable

9. Check and adjust the brake system fluid level and road test the vehicle.

Rear

1. Before servicing the vehicle, refer to the precautions in the beginning of this section.

2. Remove or disconnect the following:
   - Negative battery cable
   - Rear wheels
   - Caliper guide pin bolts
   - Caliper assembly from the brake adapter by rotating the bottom of the caliper away from the rotor, then lift the caliper with the pads away from the adapter abutment
   - Brake pads by pushing (outboard) or pulling (inboard) from the caliper fingers and piston

To install:

3. Completely compress the caliper piston

4. Install or connect the following:
   - Inboard pad clip against the piston cavity and press the pad until the clip is seated making sure the pad backing plate is flush against the piston

➡The outboard pads are side oriented. Make sure the spring clip is installed so it is positioned downwards when the caliper is installed.

   - Outboard pad making sure the locating pins are positioned against the ramps. Slide the pad onto the caliper and ensure the locating pins are squarely seated into the holes on the caliper and the pad is flush against the caliper fingers.

5. Make sure the abutment shims are in place on both slide abutments.

6. Retract the caliper guide pins to clear the caliper adapter bosses.
   - Brake caliper. Staring with the upper end, position the caliper and shoes over the rotor and align the outboard pad upper edge with the caliper slide abutment. Rotate the lower end of the caliper into position
   - Caliper guide pin bolts and tighten to 200 inch lbs. (23 Nm)
   - Wheels
   - Negative battery cable

7. Check and adjust the brake system fluid level and road test the vehicle.

CHRYSLER

10

PT Cruiser

SPECIFICATION AND MAINTENANCE CHARTS

ENGINE AND VEHICLE IDENTIFICATION

Engine						
Code ①	Liters (cc)	Cu. In.	Cyl.	Fuel Sys.	Engine Type	Eng. Mfg.
B	2.4 (2429)	148	L4	SMFI	DOHC	Chrysler

Model Year	
Code ②	Year
2	2002
3	2003
4	2004
5	2005
6	2006

DOHC: Double Overhead Camshaft

① 8th position of VIN

② 10th position of VIN

06009-PTCR-C01

GENERAL ENGINE SPECIFICATIONS

Year	Model	Engine Displacement Liters (VIN)	Net Horsepower @ rpm	Net Torque @ rpm (ft. lbs.)	Bore x Stroke (in.)	Com-pression Ratio	Oil Pressure @ rpm
2002	PT Cruiser	2.4 (B)	150@5200	164@4000	3.44x3.98	9.4:1	25-80@3000
2003	PT Cruiser	2.4 (B)	150@5200	164@4000	3.44x3.98	9.4:1	25-80@3000
2004	PT Cruiser	2.4 (B)	150@5200	164@4000	3.44x3.98	9.4:1	25-80@3000
2005-06	PT Cruiser	2.4 (B)	150@5200	164@4000	3.44x3.98	9.4:1	25-80@3000

SMFI: Sequential Multi-port Fuel Injection

06009-PTCR-C02

ENGINE TUNE-UP SPECIFICATIONS

Year	Engine Displacement Liters (VIN)	Spark Plug Gap (in.)	Ignition Timing (deg.)	Fuel Pump (psi)	Idle Speed (rpm)	Valve Clearance	
						In.	Ex.
2002	2.4 (B)	0.048-0.053	①	49	②	HYD	HYD
2003	2.4 (B)	0.048-0.053	①	49	②	HYD	HYD
2004	2.4 (B)	0.048-0.053	①	49	②	HYD	HYD
2005-06	2.4 (B)	0.048-0.053	①	49	②	HYD	HYD

NOTE: The Vehicle Emission Control Information label often reflects specification changes made during production. The label figures must be used if they differ from those in this chart.

HYD: Hydraulic

① Ignition timing is regulated by the Powertrain Control Module (PCM), and cannot be adjusted.

② Idle speed is controled by the Powertrain Control Module (PCM), and cannot be adjusted.

06009-PTCR-C03

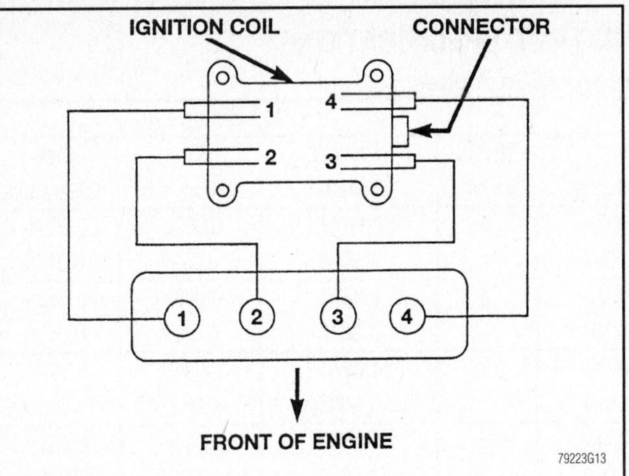

2.4L Engine
Firing order: 1–3–4–2
Distributor rotatation: Counterclockwise

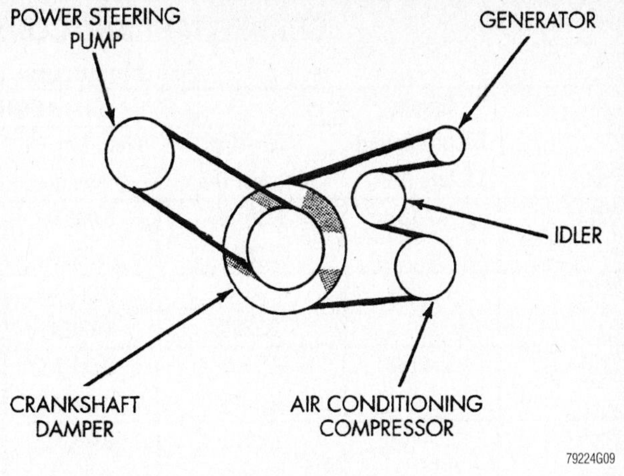

Accessory drive belt routing—2.4L engine

CAPACITIES

Year	Model	Engine Displacement Liters (VIN)	Engine Oil with Filter (qts.)	Automatic Transaxle (qts.)	Power Transfer Unit (qts.)	Rear Drive Axle (pts.)	Fuel Tank (gal.)	Cooling System (qts.)
2002	PT Cruiser	2.4 (B)	4.5	①	1.22	②	20.0	9.5
2003	PT Cruiser	2.4 (B)	4.5	①	1.22	②	20.0	9.5
2004	PT Cruiser	2.4 (B)	4.5	①	1.22	②	20.0	9.5
2005-06	PT Cruiser	2.4 (B)	4.5	①	1.22	②	20.0	9.5

NOTE: All capacities are approximate. Add fluid gradually and check to be sure a proper fluid level is obtained.

① 31TH overhaul fill capacity with torque converter empty: 8.5 qts.
41TE overhaul fill capacity with torque converter empty: 9.1 qts.

② Overrunning clutch: 0.75 pts.

06009-PTCR-C04

VALVE SPECIFICATIONS

Year	Engine Displacement Liters (VIN)	Seat Angle (deg.)	Face Angle (deg.)	Spring Test Pressure (lbs. @ in.)	Spring Installed Height (in.)	Stem-to-Guide Clearance (in.) Intake	Stem-to-Guide Clearance (in.) Exhaust	Stem Diameter (in.) Intake	Stem Diameter (in.) Exhaust
2002	2.4 (B)	45	44.5-45.0	129-143@ 1.17	1.50	0.0018-0.0025	0.0029-0.0037	0.2340	0.2330
2003	2.4 (B)	45	44.5-45.0	129-143@ 1.17	1.50	0.0018-0.0025	0.0029-0.0037	0.2340	0.2330
2004	2.4 (B)	45	44.5-45.0	129-143@ 1.17	1.50	0.0018-0.0025	0.0029-0.0037	0.2340	0.2330
2005-06	2.4 (B)	45	44.5-45.0	129-143@ 1.17	1.50	0.0018-0.0025	0.0029-0.0037	0.2340	0.2330

① Intake valve: 44.5 degrees
Exhaust valve: 45 degrees

06009-PTCR-C05

CRANKSHAFT AND CONNECTING ROD SPECIFICATIONS

All measurements are given in inches.

Year	Engine Displacement Liters (VIN)	Crankshaft Main Brg. Journal Dia.	Crankshaft Main Brg. Oil Clearance	Crankshaft Shaft End-play	Crankshaft Thrust on No.	Connecting Rod Journal Diameter	Connecting Rod Oil Clearance	Connecting Rod Side Clearance
2002	2.4 (B)	2.3610-2.3625	0.0007-0.0023	0.0035-0.0094	2	1.9670-1.9685	0.0009-0.0027	0.0051-0.0150
2003	2.4 (B)	2.3610-2.3625	0.0007-0.0023	0.0035-0.0094	2	1.9670-1.9685	0.0009-0.0027	0.0051-0.0150
2004	2.4 (B)	2.3610-2.3625	0.0007-0.0023	0.0035-0.0094	2	1.9670-1.9685	0.0009-0.0027	0.0051-0.0150
2005-06	2.4 (B)	2.3610-2.3625	0.0007-0.0023	0.0035-0.0094	2	1.9670-1.9685	0.0009-0.0027	0.0051-0.0150

06009-PTCR-C06

PISTON AND RING SPECIFICATIONS

All measurements are given in inches.

Year	Engine Displacement Liters (VIN)	Piston Clearance	Ring Gap Top Compression	Ring Gap Bottom Compression	Ring Gap Oil Control	Ring Side Clearance Top Compression	Ring Side Clearance Bottom Compression	Ring Side Clearance Oil Control
2002	2.4 (B)	0.0009-0.0022	0.0098-0.0200	0.0090-0.0180	0.0098-0.0250	0.0011-0.0031	0.0011-0.0031	0.0004-0.0070
2003	2.4 (B)	0.0009-0.0022	0.0098-0.0200	0.0090-0.0180	0.0098-0.0250	0.0011-0.0031	0.0011-0.0031	0.0004-0.0070
2004	2.4 (B)	0.0009-0.0022	0.0098-0.0200	0.0090-0.0180	0.0098-0.0250	0.0011-0.0031	0.0011-0.0031	0.0004-0.0070
2005-06	2.4 (B)	0.0009-0.0022	0.0098-0.0200	0.0090-0.0180	0.0098-0.0250	0.0011-0.0031	0.0011-0.0031	0.0004-0.0070

① Oil control ring side rails must be free to rotate after assembly

06009-PTCR-C07

TORQUE SPECIFICATIONS

All readings in ft. lbs.

Year	Engine Displacement Liters (VIN)	Cylinder Head Bolts	Main Bearing Bolts	Rod Bearing Bolts	Crankshaft Damper Bolts	Flywheel Bolts	Manifold Intake	Manifold Exhaust	Spark Plugs	Oil Pan Drain Plug
2002	2.4 (B)	①	②	③	100	70	20	17	20	20
2003	2.4 (B)	①	②	③	100	70	④	17	20	20
2004	2.4 (B)	①	②	③	100	70	④	17	20	20
2005-06	2.4 (B)	①	②	③	100	70	④	17	20	20

① Step 1: 25 ft. lbs.
Step 2: 50 ft. lbs.
Step 3: 50 ft. lbs.
Step 4: Plus 1/4 turn

② M8 bolts: 20 ft. lbs.
M11 bolts: 30 ft. lbs. plus 1/4 turn

③ Step 1: 20 ft. lbs.
Step 2: Plus 1/4 turn

④ Lower manifold: non turbo models 105 inch lbs.
Lower manifold: turbo models 250 inch lbs.
Upper manifold: non turbo models 105 inch lbs.
Upper manifold: turbo models 250 inch lbs.

06009-PTCR-C08

WHEEL ALIGNMENT SPECIFICATIONS

Year	Model		Caster Range (+/-Deg.)	Caster Preferred Setting (Deg.)	Camber Range (+/-Deg.)	Camber Preferred Setting (Deg.)	Toe-in (in.)
2002	PT Cruiser	F	+1.00	+2.45	+0.40	0.00	0.00 + 0.40
		R	—	—	+0.40	+0.20	+0.00 + 0.40
2003	PT Cruiser	F	+1.00	+2.45	+0.40	0.00	0.00 + 0.40
		R	—	—	+0.40	+0.20	+0.00 + 0.40
2004	PT Cruiser	F	+1.00	+2.45	+0.40	0.00	0.00 + 0.40
		R	—	—	+0.40	+0.20	+0.00 + 0.40
2005-06	PT Cruiser	F	+1.00	+2.45	+0.40	0.00	0.00 + 0.40
		R	—	—	+0.40	+0.20	+0.00 + 0.40

NA: Not Available

06009-PTCR-C09

TIRE, WHEEL AND BALL JOINT SPECIFICATIONS

Year	Model	OEM Tires Standard	OEM Tires Optional	Tire Pressures (psi) Front	Tire Pressures (psi) Rear	Wheel Size	Ball Joint Inspection	Lug Nuts
2002	PT Cruiser	P205/55R16	None	32	32	5.5-J	①	100
2003	PT Cruiser	P205/55R16	None	32	32	5.5-J	①	100
2004	PT Cruiser	P205/55R16	None	32	32	5.5-J	①	100
2005-06	PT Cruiser	P205/55R16	None	32	32	5.5-J	①	100

OEM: Original Equipment Manufacturer

PSI: Pounds Per Square Inch

① Replace if any measurable movement is found

06009-PTCR-C10

BRAKE SPECIFICATIONS

All measurements in inches unless noted

Year	Model		Brake Disc Original Thickness	Brake Disc Minimum Thickness	Brake Disc Maximum Run-out	Brake Drum Diameter Original Inside Diameter	Brake Drum Diameter Max. Wear Limit	Brake Drum Diameter Maximum Machine Diameter	Min. Lining Thick-ness	Caliper Guide Pin Bolts (ft. lbs.)
2002	PT Cruiser	F	0.902-0.909	0.803	0.005	—	—	—	0.313	26
		R	0.344-0.364	0.285	0.005	8.63-8.65	NA	NA	①	16
2003	PT Cruiser	F	0.902-0.909	0.803	0.005	—	—	—	0.313	26
		R	0.344-0.364	0.285	0.005	8.63-8.65	NA	NA	①	16
2004	PT Cruiser	F	0.902-0.909	0.803	0.005	—	—	—	0.313	26
		R	0.344-0.364	0.285	0.005	8.63-8.65	NA	NA	①	16
2005-06	PT Cruiser	F	0.902-0.909	0.803	0.005	—	—	—	0.313	26
		R	0.344-0.364	0.285	0.005	8.63-8.65	NA	NA	①	16

NA: Not Available

F: Front

R: Rear

① Rear Disc: 0.350 in.

Rear Bonded Shoes: 0.062 in.

Rear Riveted Shoes: 0.031 in.

06009-PTCR-C11

SCHEDULED MAINTENANCE INTERVALS
Chrysler—PT Cruiser

TO BE SERVICED	TYPE OF SERVICE	VEHICLE MILEAGE INTERVAL (x1000)												
		7.5	15	22.5	30	37.5	45	52.5	60	67.5	75	82.5	90	97.5
Engine oil & filter	R	✓	✓	✓	✓	✓	✓	✓	✓	✓	✓	✓	✓	✓
Brake hoses	S/I	✓	✓	✓	✓	✓	✓	✓	✓	✓	✓	✓	✓	✓
Coolant level, hoses & clamps	S/I	✓	✓	✓	✓	✓	✓	✓	✓	✓	✓	✓	✓	✓
CV joints & front suspension components	S/I	✓	✓	✓	✓	✓	✓	✓	✓	✓	✓	✓	✓	✓
Exhaust system	S/I	✓	✓	✓	✓	✓	✓	✓	✓	✓	✓	✓	✓	✓
Manual transaxle oil	S/I	✓	✓	✓	✓	✓	✓	✓	✓	✓	✓	✓	✓	✓
Rotate tires	S/I	✓	✓	✓	✓	✓	✓	✓	✓	✓	✓	✓	✓	✓
Accessory drive belts	S/I		✓		✓		✓		✓		✓		✓	
Brake linings	S/I			✓			✓			✓			✓	
Air filter element	R				✓				✓				✓	
Spark plugs	R				✓				✓				✓	
Lubricate ball joints	S/I				✓				✓				✓	
Engine coolant ①	R													
PCV valve	S/I								✓				✓	
Ignition cables	R								✓					
Camshaft timing belt ②	R													

R: Replace S/I: Service or Inspect

① Engine coolant: flush and replace at 100,000 miles.

② Camshaft timing belt: replace at 120,000 miles.

FREQUENT OPERATION MAINTENANCE (SEVERE SERVICE)

If a vehicle is operated under any of the following conditions it is considered severe service:

- Extremely dusty areas.

- 50% or more of the vehicle operation is in 32°C (90°F) or higher temperatures, or constant operation in temperatures below 0°C (32°F).

- Prolonged idling (vehicle operation in stop and go traffic).

- Frequent short running periods (engine does not warm to normal operating temperatures).

- Police, taxi, delivery usage or trailer towing usage.

Oil & oil filter change: change every 3000 miles.

Rotate tires every 6000 miles.

Brake linings: inspect every 12,000 miles.

Air filter element: service or inspect every 15,000 miles.

Automatic transaxle: change fluid & adjust bands every 15,000 miles.

Manual transaxle fluid: replace every 15,000 miles.

Engine coolant: replace at 36,000 miles and every 30,000 miles thereafter.

06009-PTCR-C12

PRECAUTIONS

Before servicing any vehicle, please be sure to read all of the following precautions, which deal with personal safety, prevention of component damage, and important points to take into consideration when servicing a motor vehicle:

- Never open, service or drain the radiator or cooling system when the engine is hot; serious burns can occur from the steam and hot coolant.
- Observe all applicable safety precautions when working around fuel. Whenever servicing the fuel system, always work in a well-ventilated area. Do not allow fuel spray or vapors to come in contact with a spark, open flame, or excessive heat (a hot drop light, for example). Keep a dry chemical fire extinguisher near the work area. Always keep fuel in a container specifically designed for fuel storage; also, always properly seal fuel containers to avoid the possibility of fire or explosion. Refer to the additional fuel system precautions later in this section.
- Fuel injection systems often remain pressurized, even after the engine has been turned **OFF**. The fuel system pressure must be relieved before disconnecting any fuel lines. Failure to do so may result in fire and/or personal injury.
- Brake fluid often contains polyglycol ethers and polyglycols. Avoid contact with the eyes and wash your hands thoroughly after handling brake fluid. If you do get brake fluid in your eyes, flush your eyes with clean, running water for 15 minutes. If eye irritation persists, or if you have taken brake fluid internally, IMMEDIATELY seek medical assistance.
- The EPA warns that prolonged contact with used engine oil may cause a number of skin disorders, including cancer! You should make every effort to minimize your exposure to used engine oil. Protective gloves should be worn when changing oil. Wash your hands and any other exposed skin areas as soon as possible after exposure to used engine oil. Soap and water, or waterless hand cleaner should be used.
- All new vehicles are now equipped with an air bag system, often referred to as a Supplemental Restraint System (SRS) or Supplemental Inflatable Restraint (SIR) system. The system must be disabled before performing service on or around system components, steering column, instrument panel components, wiring and sensors. Failure to follow safety and disabling procedures could result in accidental air bag deployment, possible personal injury and unnecessary system repairs.
- Always wear safety goggles when working with, or around, the air bag system. When carrying a non-deployed air bag, be sure the bag and trim cover are pointed away from your body. When placing a non-deployed air bag on a work surface, always face the bag and trim cover upward, away from the surface. This will reduce the motion of the module if it is accidentally deployed. Refer to the additional air bag system precautions later in this section.
- Clean, high quality brake fluid from a sealed container is essential to the safe and proper operation of the brake system. You should always buy the correct type of brake fluid for your vehicle. If the brake fluid becomes contaminated, completely flush the system with new fluid. Never reuse any brake fluid. Any brake fluid that is removed from the system should be discarded. Also, do not allow any brake fluid to come in contact with a painted surface; it will damage the paint.
- Never operate the engine without the proper amount and type of engine oil; doing so WILL result in severe engine damage.
- Timing belt maintenance is extremely important! Many models utilize an interference-type, non-freewheeling engine. If the timing belt breaks, the valves in the cylinder head may strike the pistons, causing potentially serious (also time-consuming and expensive) engine damage.
- Disconnecting the negative battery cable on some vehicles may interfere with the functions of the on-board computer system(s) and may require the computer to undergo a relearning process once the negative battery cable is reconnected.
- When servicing drum brakes, only disassemble and assemble one side at a time, leaving the remaining side intact for reference.
- Only an MVAC-trained, EPA-certified automotive technician should service the air conditioning system or its components.

ENGINE REPAIR

Distributor

The PT Cruiser uses a Direct Ignition System (DIS) and basic ignition timing is not adjustable.

Alternator

REMOVAL

Remove or disconnect the following:

- Negative battery cable
- Air cleaner lid
- Inlet Air Temperature (IAT) sensor and make-up hose
- Upper alternator adjustment lock nut, loosen only
- Right front wheel
- Splash shield
- Support bracket
- Lower pivot bolt, loosen only
- Drive belt T-bolt, loosen only
- Alternator wiring connectors
- Alternator belt
- Axle retaining nut
- Lower control arm from the steering knuckle
- Axle shaft
- Lower mounting bolt from the upper adjustment bracket
- Alternator

INSTALLATION

Install or connect the following:

- Alternator
- Alternator drive belt
- Axle shaft
- Lower control arm to the steering knuckle
- Lower ball joint nut and torque the nut to 70 ft. lbs. (95 Nm)
- Axle retaining nut and torque the nut to 120 ft. lbs. (163 Nm)
- Support bracket
- Alternator electrical connectors and torque the B+ terminal nut to 100 inch lbs. (11 Nm)
- Splash shield and right front wheel
- Right front wheel
- Upper pivot nut and torque to 40 ft. lbs. (54 Nm)
- Negative battery cable
- Air cleaner lid, IAT sensor and make up hose

Remove the lower pivot bolt from the alternator

Ignition Timing

ADJUSTMENT

Ignition timing is controlled by the Powertrain Control Module (PCM). No adjustment is necessary or possible.

Engine Assembly

REMOVAL & INSTALLATION

➡After all components are installed on the engine, a DRB scan tool is necessary to perform the camshaft and crankshaft timing relearn procedure.

1. Before servicing the vehicle, refer to the precautions in the beginning of this section.
2. Properly recover the air conditioning system refrigerant.
3. Properly relieve the fuel system pressure.
4. Drain the cooling system.
5. Drain the engine oil.
6. Remove or disconnect the following:
   - Battery and battery tray
   - Throttle and cruise control cables
   - Powertrain Control Module (PCM) wiring harness
   - Positive cable from the Power Distribution Center (PDC) and ground wire
   - Ground wire from the body to the engine
   - Brake booster vacuum hose
   - Proportional purge hoses from the intake manifold
   - Coolant recovery hose
   - Heater hoses
   - Upper radiator support crossmember
   - Upper and lower radiator hoses
   - A/C line from the condenser
   - A/C lines from the junction at the upper torque strut
   - Cooler lines, if equipped
   - Fan cooling module assembly
   - Transmission shift linkage and electrical connectors
   - Clutch hydraulic lines, if equipped
   - Power steering hoses from the radiator on turbo models
   - Radiator fan, radiator and condenser on 2003–06 models
   - Both front wheels and the right inner splash shield
   - Halfshafts
   - Accessory drive belts
   - Alternator and support brackets
   - Charge air cooler hoses on turbo models
   - Downstream Oxygen (O_2S) Sensor
   - Exhaust system from the manifold
   - Power steering pressure hose on non-turbo models
   - Power steering hoses from the steering gear on turbo models
   - Lower engine torque strut, if equipped
   - Upper and lower heat shields, elbow support bracket, turbocharger support bract and elbow if equipped with a turbocharger
   - Structural collar
   - Torque converter bolts
   - A/C compressor
   - Power steering return line
   - Power steering pump
7. Raise the vehicle enough to allow an engine dolly and cradle to be placed under the engine.
8. Loosen the engine support posts in order to allow movement for positioning onto the engine locating holes and flange on the engine bedplate. Lower the vehicle and position the cradle until the engine is resting on the support posts. Tighten the mounts to the cradle frame. This will keep the support posts from moving when removing or installing the engine and transaxle.
9. Install safety straps around the engine to the cradle; tighten the straps and lock them into position.
10. Raise the vehicle enough to see if the straps are tight enough to hold the cradle assembly to the engine.
11. Lower the vehicle so the weight of the engine and transaxle ONLY is on the cradle.
12. Remove the engine and transaxle mount through-bolts.
13. Raise the vehicle slowly, it might be necessary to move the engine/transaxle assembly with the cradle to allow removal around the body flanges.

To install:

14. Install or connect the following:
    - Engine/transaxle assembly by lowering the vehicle over the assembly
    - Engine and transaxle mounts and torque the bolts to 87 ft. lbs. (118 Nm)
    - Upper torque strut
15. Remove the support fixtures.
    - Alternator and support brackets
    - Halfshafts
16. Install the structural collar and torque the bolts, in 3 steps, using the following procedure:
    a. Step 1: Collar-to-oil pan bolts to 30 inch lbs. (3 Nm).
    b. Step 2: Collar-to-transmission bolts to 80 ft. lbs. (108 Nm).
    c. Step 3: Collar-to-oil pan bolts to 40 ft. lbs. (54 Nm).
17. Install or connect the following:
    - Upper and lower heat shields, elbow support bracket, turbocharger support bract and elbow if equipped with a turbocharger
    - Lower engine torque strut and

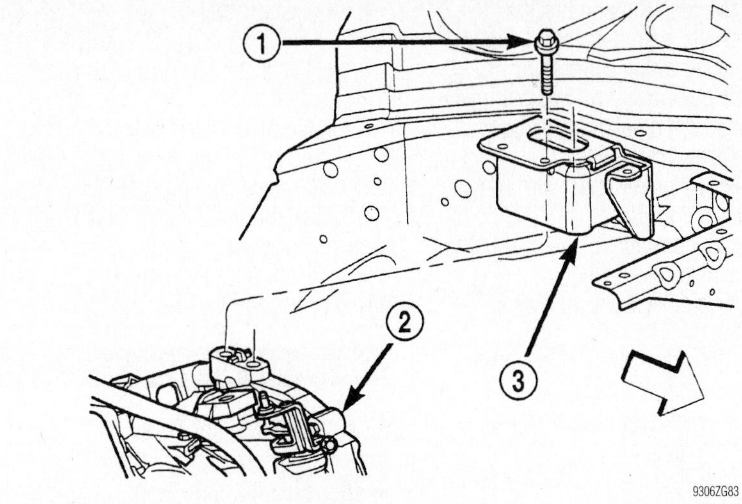

9306ZG83

Left engine mount location and bolt identification

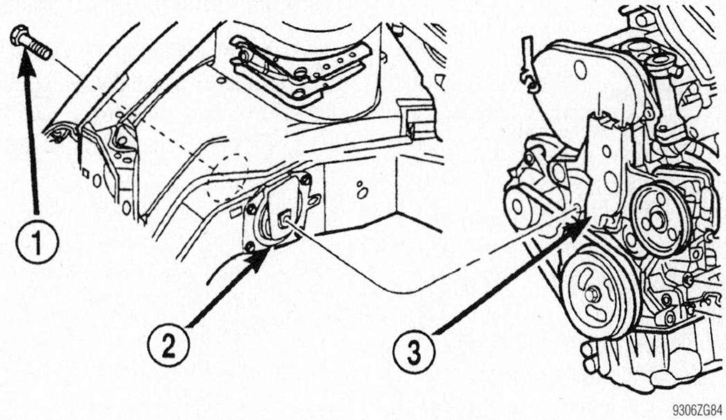

9306ZG84

Exploded view of the right mount through bolt

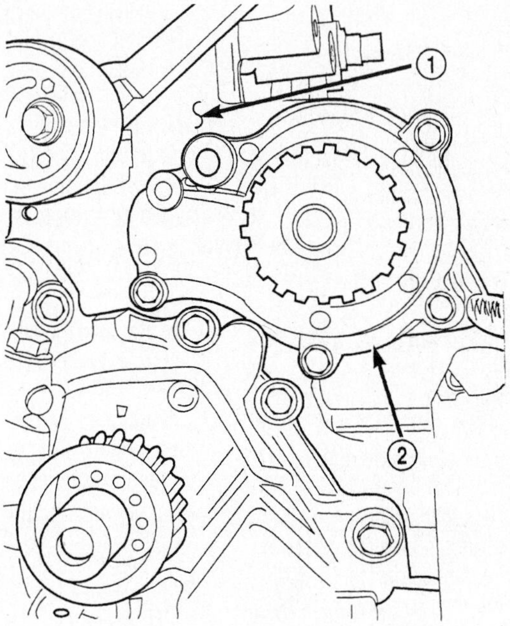

1 - CYLINDER BLOCK
2 - WATER PUMP

9306ZG85

Exploded view of the water pump

torque the through bolts to 87 ft. lbs. (118 Nm)

- Exhaust pipe to the manifold
- Downstream O_2S sensor connector
- Charge air cooler hoses on turbocharged models
- Power steering pressure hose to the steering gear
- A/C compressor
- Power steering pump
- Drive belts
- Right inner splash shield and both front wheels
- Power steering return hose
- Clutch hydraulic line, electrical connectors and shift linkage, if equipped
- Shift linkage and cooler lines, if equipped
- Fuel lines
- Heater hoses
- Engine wiring harness and ground strap
- Lower and upper radiator hoses
- Fan module assembly
- Throttle and speed control cables
- PDC, positive battery cable and ground strap
- PCM
- Battery and battery tray
- Battery cables

18. Perform the camshaft and crankshaft synchronization procedure.
19. Fill the engine with clean oil.
20. Fill the cooling system.
21. Recharge the A/C system.
22. After all components are installed, perform the camshaft and crankshaft timing relearn procedure as follows:
    a. Connect a DRB or equivalent, scan tool to the DLC (located under the instrument panel, near the steering column).
    b. Turn the ignition switch **ON**, and access the "miscellaneous" screen.
    c. Select "re-learn cam/crank" option and follow the directions on the scan tool screen.
23. Start the vehicle and check for leaks, repair if necessary.

Water Pump

REMOVAL & INSTALLATION

1. Before servicing the vehicle, refer to the precautions in the beginning of this section.
2. Drain the cooling system.
3. Remove or disconnect the following:
    - Negative battery cable
    - Timing belt

- Camshaft sprockets
- Rear timing belt cover
- Water pump and discard the O-ring

To install:

➡**Apply dielectric grease to the new O–ring before installation.**

4. Install or connect the following:
   - New O-ring in the water pump groove

⁂ WARNING

Before proceeding, be sure the O-ring gasket is properly seated in the water pump groove before tightening the screws. An improperly installed O-ring could cause a coolant leak.

- Water pump and torque the bolts to 105 inch lbs. (12 Nm)

➡**Rotate the pump by hand to check for freedom of movement.**

- Rear timing belt cover
- Camshaft sprockets
- Timing belt
- Negative battery cable

5. Fill the cooling system.
6. Start the vehicle and check for leaks, repair if necessary.

Heater Core

REMOVAL & INSTALLATION

1. Before servicing the vehicle, refer to the precautions in the beginning of this section.
2. Remove the A/C refrigerant by using approved equipment.
3. Drain the cooling system.
4. Disconnect the negative battery cable.
5. Remove the instrument panel by removing or disconnecting the following:
   - Left and right **A** pillar trim moldings using a trim stick
   - Front power window switch
   - Center bezel retaining screw
   - Heating, Ventilation and Air Conditioning (HVAC) control knobs
   - Center bezel using a trim stick
   - Top cover retaining screws and pull it rearward to remove
   - HVAC control unit from the instrument panel
   - Retaining screws from the upper and lower steering column shrouds
   - Left lower instrument panel bezel
   - Steering column wire connectors
   - Steering column
   - Brake pedal support bracket
   - Left side instrument panel end cap from the wiring connectors
   - Rear power window switch from the console and pull the parking handle all the way up
   - Auxiliary power outlet wire connector
   - Center console
   - Glove box assembly
   - Right side instrument panel end cap
   - 5 right side instrument panel wire connectors
   - Instrument panel from the vehicle, with the help of an assistant
6. Remove or disconnect the following:
   - Heater hoses from the heater core
   - Reposition the vehicle speed control servo, if equipped
   - Suction and liquid lines from the evaporator
   - Drain tube
   - A/C vacuum harness connector
   - Defroster duct from the heater—A/C unit housing
   - Heater—A/C unit housing from the vehicle
   - Heater core from the housing

To install:

7. Install or connect the following:
   - Heater core to the housing
   - Heater—A/C unit housing to the vehicle
   - Defroster duct to the housing
   - A/C vacuum harness
   - Drain tube
   - Suction and liquid lines to the evaporator
   - Reposition the vehicle speed control servo, if equipped
   - Heater hoses to the heater core
8. Install the instrument panel by installing or connecting the following:
   - Instrument panel
   - Right side instrument panel wire connectors
   - Right side instrument panel end cap
   - Glove box assembly
   - Center console
   - Auxiliary power outlet wire connector
   - Rear power window switch
   - Left side instrument panel end cap
   - Left side wire connectors
   - Brake pedal support bracket
   - Steering column with a new pinch bolt
   - Steering column wire connectors
   - Left lower instrument panel bezel
   - HVAC control unit
   - Center bezel
   - Front power window switch
   - Left and right **A** pillar trim moldings
   - Negative battery cable
9. Fill the cooling system.
10. Recharge the A/C system.
11. Start the vehicle and verify proper system operations.
12. Roadtest the vehicle and check for any rattles, repair if necessary.

Cylinder Head

REMOVAL & INSTALLATION

➡**After all components are installed on the engine, a DRB scan tool is necessary to perform the camshaft and crankshaft timing relearn procedure.**

1. Before servicing the vehicle, refer to the precautions in the beginning of this section.
2. Properly relieve the fuel system pressure.
3. Drain the cooling system.
4. Remove or disconnect the following:
   - Negative battery cable
   - Air cleaner inlet duct and air cleaner
   - Inlet Air Temperature (IAT) sensor and make-up air hose
   - Upper intake manifold
   - Dipstick tube fastener
   - Lower intake manifold support bracket on turbo models
   - Fuel supply line from the fuel rail
   - Heater tube support bracket
   - Upper radiator hose
   - Heater supply hoses
   - Accessory drive belt
   - Exhaust pipe from the manifold
5. On turbo models perform the following:
   a. Remove the turbocharger heat shields.
   b. Remove the elbow support bracket.
   c. Remove the turbocharger support bracket.
   d. Remove the oil return, oil supply, coolant return and coolant supply hoses.
6. Remove or disconnect the following:
   - Power steering pump, move it aside. DO NOT disconnect the fluid lines.
   - Ignition coil pack wiring connector
   - Ignition coil pack and bracket
   - Cam sensor electrical connector
   - Timing belt
   - Timing belt idler pulley
   - Camshaft sprocket

- Power steering pump reservoir and bracket, if necessary on non-turbo models
- Cylinder head cover
- Camshaft and cam followers
- Cylinder head bolts, working from the center outward
- Cylinder head

➡The cylinder head bolts must be inspected before they can be reused. If the threads of bolts are stretched, they must be replaced. Check for thread stretching by holding a scale or other straightedge against the threads. If all the threads do not contact the scale, the bolts must be replaced.

✲✲ WARNING

Use only a plastic scraper to clean the mating surfaces. NEVER use metal, as this may gouge the surfaces and cause leaks.

7. Cover the combustion chambers, then use a plastic scraper to thoroughly and carefully clean the engine block and cylinder head mating surfaces.

To install:

8. Install the cylinder head with a new gasket. Make certain that the part number on the new gasket is facing up.
9. Lubricate the cylinder head bolt threads with clean engine oil.
10. Torque the cylinder head bolts, in the following sequence:
    a. Step 1: 25 ft. lbs. (34 Nm).
    b. Step 2: 50 ft. lbs. (68 Nm).
    c. Step 3: 50 ft. lbs. (68 Nm).
    d. Step 4: An additional ¼ turn.
11. Install or connect the following:
    - Camshaft and cam follower assemblies
    - Cylinder head cover
    - Rear timing belt cover and pulley
    - Camshaft sprockets
    - Timing belt
    - Cam sensor wiring connector
    - Ignition coil and spark plug wires
    - Power steering pump reservoir and bracket, if removed
12. On turbo models perform the following:
    a. Install the oil return, oil supply, coolant return and coolant supply hoses.
    b. Install the elbow support bracket.
    c. Install the turbocharger heat shields.
    d. Install the turbocharger support bracket.

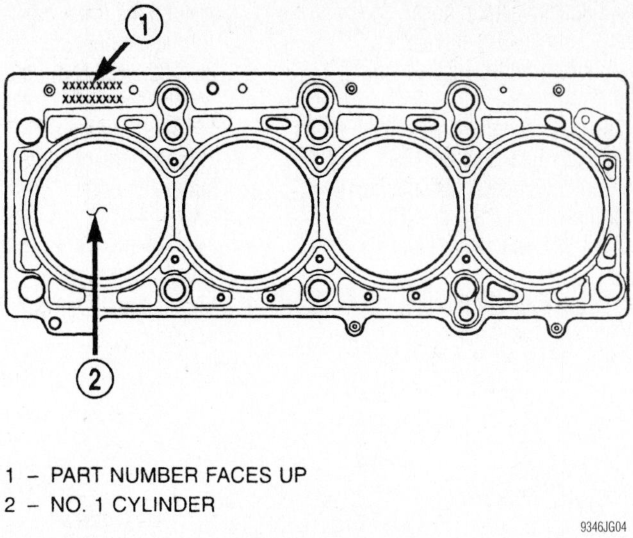

Install the new gasket with the part number facing upwards

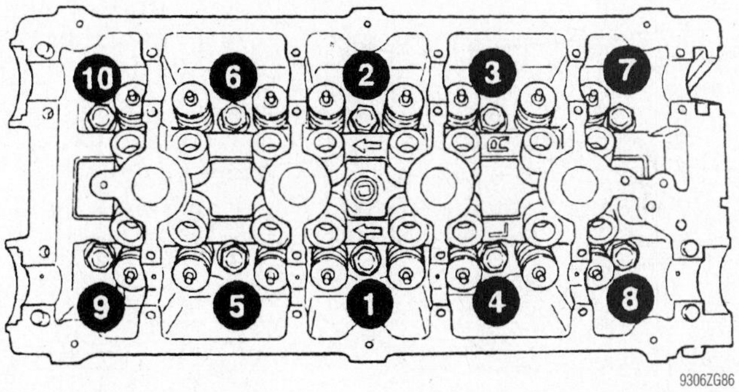

Cylinder head bolt torque sequence

13. Install or connect the following:
    - Exhaust pipe to the manifold
    - Accessory drive belts
    - Lower intake manifold
    - Upper radiator and heater supply hose
    - Heater support bracket
    - Lower intake manifold support bracket on turbo models
    - Dipstick tube fastener
    - Power brake vacuum hose to the intake manifold
    - Fuel supply line to the fuel rail
    - Vacuum lines and electrical wiring
    - Upper intake manifold
    - Air cleaner inlet duct and air cleaner
    - IAT sensor and make up hose
    - Negative battery cable
14. Fill the cooling system.
15. Turn the ignition switch **ON**, and access the "miscellaneous" screen.
16. Select the "re-learn cam/crank" option, then follow the instructions on the scan tool screen.

Intake Manifold

REMOVAL & INSTALLATION

Upper

NON-TURBO MODELS

1. Before servicing the vehicle, refer to the precautions in the beginning of this section.
2. Properly relieve the fuel system pressure.
3. Remove or disconnect the following:
    - Negative battery cable
    - Inlet Air Temperature (IAT) sensor and air hose
    - Engine cover
    - Throttle and cruise control cables from the throttle lever bracket
    - Manifold Absolute Pressure (MAP) sensor
    - Idle Air Control (IAC) motor electrical connector
    - Throttle Position (TPS) sensor wiring connector
    - Proportional purge hoses

- Brake booster vacuum hose
- Positive Crankcase Ventilation (PCV) hose from the intake manifold
- Throttle body support bracket bolt
- Exhaust Gas Recirculation (EGR) tube from the upper intake manifold
- Upper intake manifold

4. Clean the mating surfaces.

To install:

5. Install or connect the following:
   - New gaskets and seals
   - Intake manifold on the EGR tube then on the lower intake manifold. Torque the bolts, in sequence, to 20 ft. lbs. (28 Nm) on 2002 models or 105 inch lbs. (12 Nm) on 2003–06 models.
   - Throttle body support bracket and torque the bolt to 28 ft. lbs. (20 Nm)
   - EGR retainer plate and torque the smaller bolt to 95 inch lbs. (11 Nm) and the large bolt to 28 ft. lbs. (21 Nm)
   - PCV hose to the intake manifold
   - MAP sensor electrical connector
   - Proportional purge hoses
   - Brake booster hose
   - IAC motor and TPS connectors
   - Throttle and speed control cables
   - Air cleaner assembly
   - Engine cover
   - IAT sensor
   - Negative battery cable

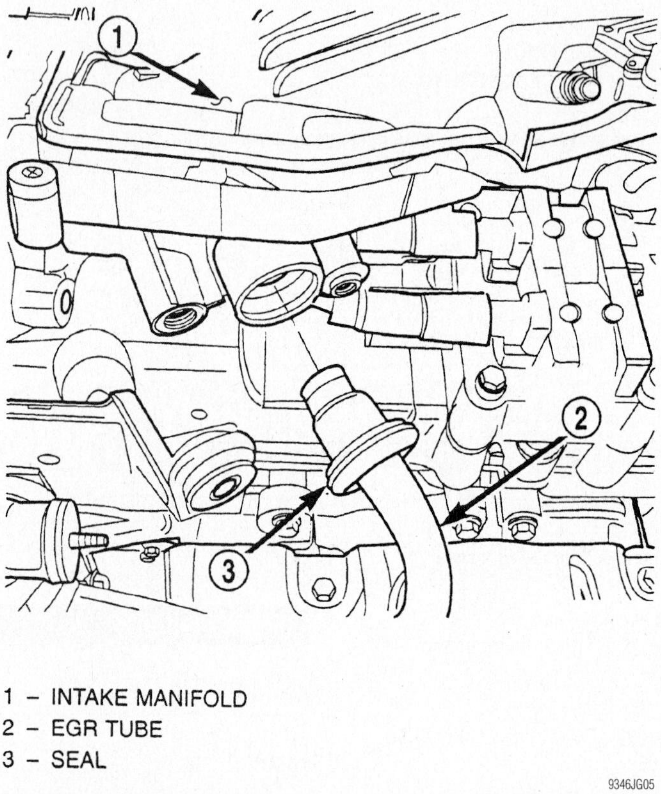

Remove the EGR tube from the upper intake manifold

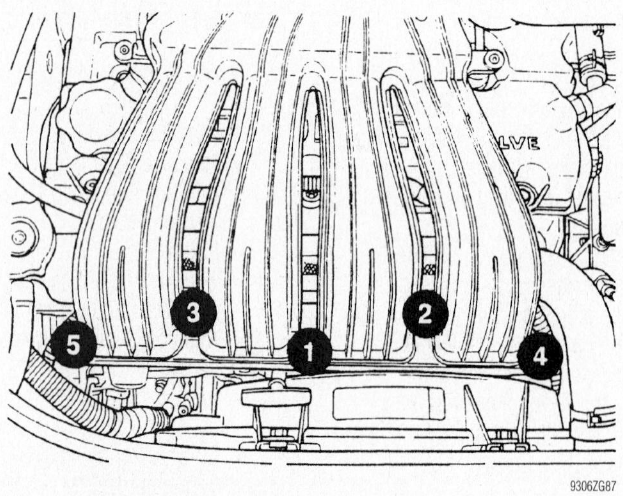

Upper intake manifold bolt torque sequence—non-turbo models

TURBO MODELS

➡The engines in these vehicles are equipped with either an aluminum or plastic manifold.

1. Before servicing the vehicle, refer to the precautions in the beginning of this section.
2. Properly relieve the fuel system pressure.
3. Remove or disconnect the following:
   - Negative battery cable
   - Inlet Air Temperature (IAT) sensor
   - Manifold Absolute Pressure (MAP) sensor
   - Throttle inlet pressure hose
   - Charge air cooler hose from the throttle body
   - Idle Air Control (IAC) motor electrical connector
   - Throttle control shield
   - Throttle and cruise control cables from the throttle lever bracket
   - Throttle cable bracket
   - Brake booster vacuum hose
   - Positive Crankcase Ventilation (PCV) hose from the intake manifold
   - Purge solenoid hose from the throttle body
   - Upper intake manifold support bracket and manifold

➡Cover the lower intake manifold to avoid dirt and other objects from entering.

4. Clean the mating surfaces.

To install:

5. Remove the cover from the lower intake manifold.
6. Install or connect the following:
   - New gasket
   - Intake manifold on the lower intake manifold. If equipped with an aluminum manifold, torque the bolts in sequence to 250 inch lbs. (28 Nm). If equipped with a plastic manifold, torque the bolts in sequence to 105 inch lbs. (12 Nm).
   - Upper intake manifold support bracket Torque the retainers to 250 inch lbs. (28 Nm).

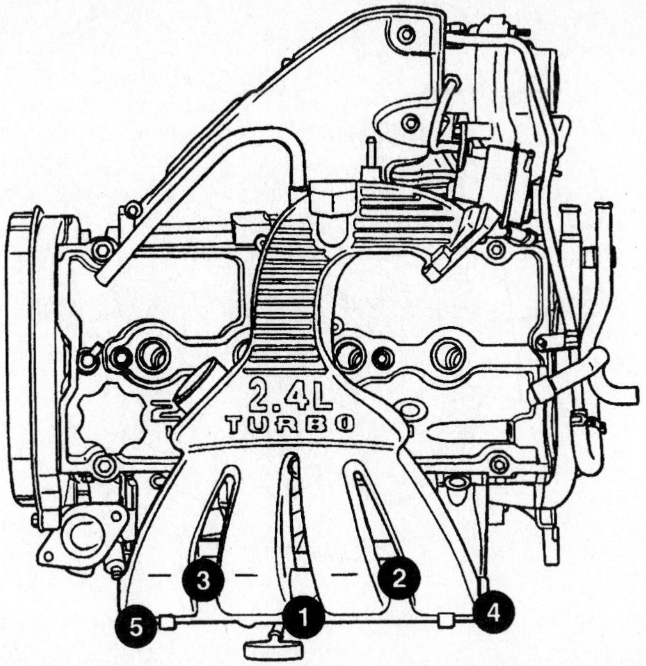

67189-PTCR-G05

Upper intake manifold bolt torque sequence—turbo models

- Purge solenoid hose to the throttle body
- Brake booster vacuum hose
- PCV hose
- Throttle cable bracket and tighten the screws to 105 inch lbs. (12 Nm)
- Throttle and cruise control cables to the throttle lever bracket
- Throttle control shield
- IAC motor electrical connector
- MAP sensor
- Charge air cooler hose to the throttle body
- IAT sensor
- Throttle inlet pressure hose
- Negative battery cable

Lower

➡The engines in these vehicles are equipped with either an aluminum or plastic manifold.

1. Before servicing the vehicle, refer to the precautions in the beginning of this section.
2. Properly relieve the fuel system pressure.
3. Drain the coolant system.
4. Remove or disconnect the following:
   - Negative battery cable
   - Inlet Air Temperature (IAT) sensor and make-up hose
   - Air cleaner
   - Upper intake manifold
   - Upper radiator hose and coolant outlet connector
   - Fuel supply line quick-connect from the fuel rail
   - Fuel injector wiring harness
   - Oil dipstick tube from the lower intake manifold
   - Intake manifold and discard the gaskets and seals
5. Thoroughly clean the gasket mating surfaces.

To install:

6. Install or connect the following:
   - New seals on non-turbo models
   - Intake manifold and torque the bolts in sequence to 105 inch lbs. (12 Nm) on non-turbo models
   - New gaskets on turbo models
   - Intake manifold. If equipped with an aluminum manifold, torque the bolts in sequence to 250 inch lbs. (28 Nm). If equipped with a plastic manifold, torque the bolts in sequence to 150 inch lbs. (12 Nm) on turbo models.

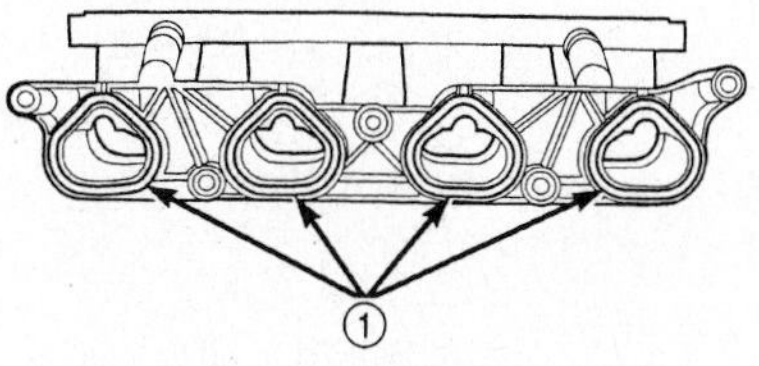

9346JG06

Install new seals on the lower intake manifold–non-turbo models

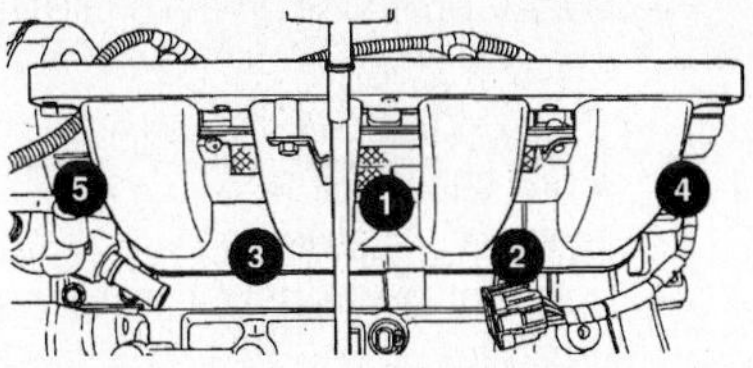

67189-PTCR-G06

Lower intake manifold bolt torque sequence

   - Lower intake manifold support bracket retainers on turbocharged models to 40 ft. lbs. (54 Nm)
   - Fuel injector wiring harness
   - Fuel supply line quick-connect to the fuel tube assembly
   - Oil dipstick tube to the lower intake manifold
   - Upper radiator hose
   - Upper intake manifold
   - IAT sensor
   - Air cleaner assembly
   - Negative battery cable
7. Fill the coolant system.
8. Pressurize the fuel system.
9. Start the vehicle and check for leaks, repair if necessary.

Turbocharger

REMOVAL & INSTALLATION

1. Before servicing the vehicle, refer to the precautions in the beginning of this section.

➡The exhaust manifold on turbocharged models is removed as an assembly with the turbocharger. Do not remove the turbocharger from the manifold always replace as a complete assembly.

⁂ CAUTION

If the turbocharger is being replaced due to bearing failure, the oil pressure feed line has to be replaced and the return tube should be cleaned.

2. Properly relieve the fuel system pressure.
3. Remove or disconnect the following:
   - Negative battery cable
   - Air cleaner housing
   - Clean air hose from the turbocharger
   - Throttle and cruise control cables from the throttle lever bracket
   - Inlet Air Temperature (IAT) sensor
   - Manifold Absolute Pressure (MAP) sensor

- Idle Air Control (IAC) motor electrical connector
- Throttle Position (TPS) sensor wiring connector
- Ignition coil capacitor
- Upstream Oxygen (O_2S) sensor connector
- Air inlet hose from the throttle body
- Vacuum hoses from the throttle body and upper intake manifold
- Upper intake manifold support bracket and manifold

➡Cover the lower intake manifold to avoid dirt and other objects from entering.

- Turbocharger lower heat shield
- Oil supply line from the turbocharger
- Coolant return line
- Vacuum hoses
- Muffler ground strap
- Downstream O_2S sensor connector
- Catalytic converter and intermediate pipe as an assembly
- Turbocharger to charge air cooler hose assembly
- Turbocharger, elbow support and support brackets
- Oil return tube
- Turbocharger coolant supply line and upper heat shield
- Turbocharger elbow
- Lower exhaust manifold fasteners from below, then lower the vehicle and remove the upper manifold fasteners
- Turbocharger/manifold assembly from between the engine and cowl panel.
- Discard the gasket

4. Clean the mating surfaces.
5. Install or connect the following:
   - New gasket. Use no sealer when installing the gasket.
   - Turbocharger/manifold assembly between the engine and cowl panel. Tighten the fasteners working from the center out in progressing in both directions to 259 inch lbs. (28 Nm).
   - Turbocharger elbow. Tighten the fasteners to 259 inch lbs. (28 Nm).
   - Turbocharger upper heat shield. Tighten the fasteners to 259 inch lbs. (28 Nm).
   - Coolant supply line using new washers. Tighten the banjo bolt to 22 ft. lbs. (30 Nm) and the flared fitting to 23 ft. lbs. (31 Nm).
   - New oil return tube gasket, return tub and tighten to 105 inch lbs. (12 Nm).

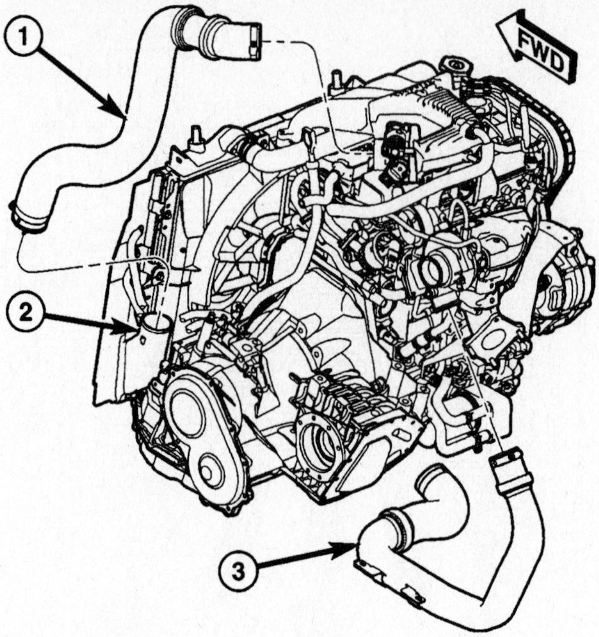

1 - HOSE - CHARGE AIR COOLER TO THROTTLE BODY
2 - CHARGE AIR COOLER
3 - HOSE - TURBOCHARGER TO CHARGE AIR COOLER

67189-PTCR-G01

Charge air cooler hoses—Turbo models

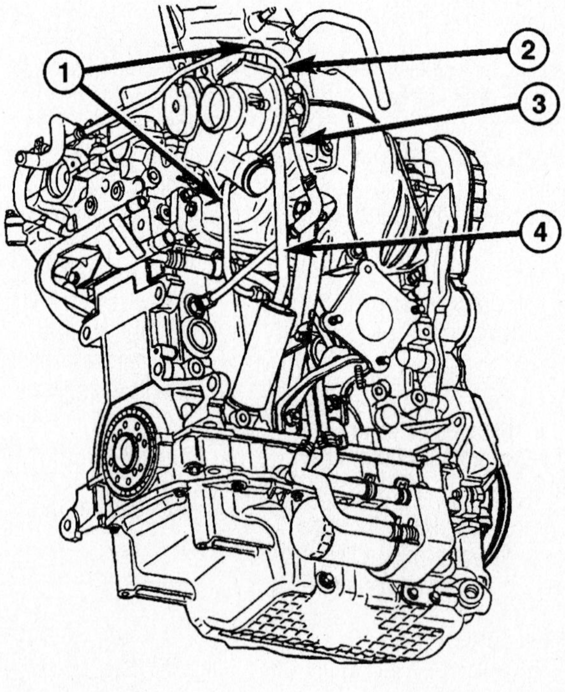

1 - OIL SUPPLY LINE
2 - COOLANT RETURN LINE
3 - COOLANT SUPPLY LINE
4 - OIL RETURN TUBE

67189-PTCR-G02

Turbocharger line and hose locations—Turbo models

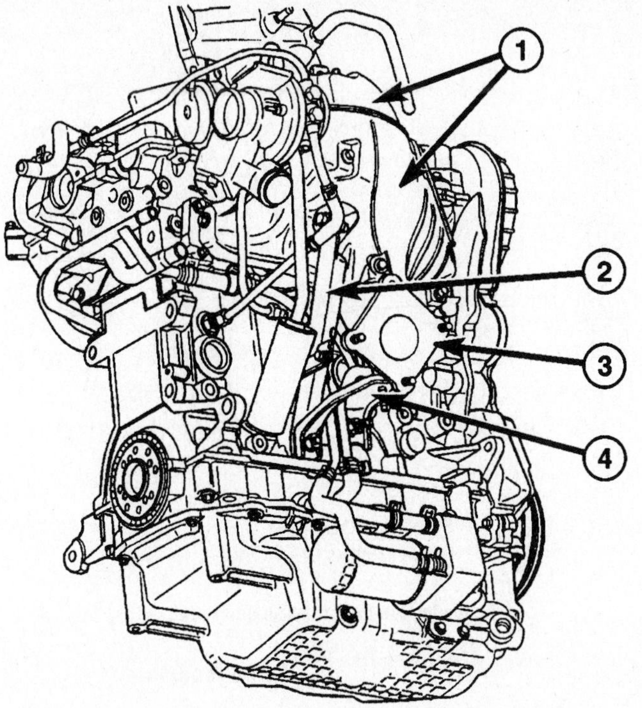

1 - UPPER/LOWER HEAT SHIELDS
2 - TURBOCHARGER SUPPORT BRACKET
3 - ELBOW
4 - ELBOW SUPPORT BRACKET

67189-PTCR-G03

Turbocharger bracket and heat shield locations—Turbo models

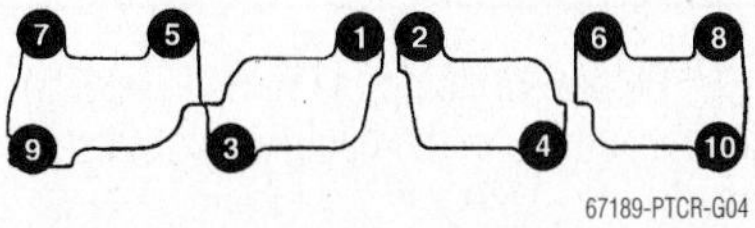

Exhaust manifold torque sequence—Turbo models

make sure the heat shield for the oil return line is installed properly.

- Turbocharger support bracket. Tighten the M8 fasteners to 259 inch lbs. (28 Nm) and the M10 fasteners to 40 ft. lbs. (54 Nm).
- Turbocharger elbow support bracket
- Turbocharger to charge air cooler hose assembly
- Catalytic converter and intermediate pipe as an assembly
- Muffler ground strap
- Downstream O_2S sensor connector
- Vacuum hose
- Coolant return line using new washers. Tighten the banjo bolt to 22 ft. lbs. (30 Nm)
- Oil supply line and tighten the flared fitting to 23 ft. lbs. (31 Nm)

➡The lower heat shield tabs must overlap the upper heat shield to prevent fatigue and premature union failure.

- Turbocharger lower heat shield. Tighten the fasteners to 259 inch lbs. (28 Nm).

6. Remove the cover placed on the lower intake manifold.
   - Upper intake manifold and support bracket
   - Vacuum hoses to the throttle body and upper intake manifold
   - Air inlet hose to the throttle body
   - IAT sensor connector
   - MAP sensor
   - TPS sensor wiring connector
   - IAC motor electrical connector
   - Ignition coil capacitor
   - Upstream O_2S sensor connector
   - Throttle and cruise control cables to the throttle lever bracket
   - Clean air hose to the turbocharger
   - Air cleaner housing
   - Negative battery cable
7. Fill the cooling system.
8. Chain the oil and filter.
9. Start the vehicle and check for exhaust system leaks and also check the system is not contacting any body panels. Adjust the system as necessary to avoid panel contact.

Exhaust Manifold

REMOVAL & INSTALLATION

Non Turbo Models

1. Before servicing the vehicle, refer to the precautions in the beginning of this section.
2. Remove or disconnect the following:
   - Negative battery cable
   - Air cleaner assembly and bracket
   - Throttle and speed control cables
   - Manifold Absolute Pressure (MAP) sensor electrical connector
   - Power steering reservoir, move it aside. DO NOT disconnect the fluid lines.
   - Coolant recovery bottle
   - Exhaust manifold upper heat shield
   - Exhaust pipe from the manifold
   - Engine wiring heat shield
   - Manifold support bracket
   - Lower exhaust manifold heat shield
   - Upstream Heated Oxygen (HO_2S) sensor connector
   - Exhaust manifold and discard the gasket
3. Thoroughly clean the mating surfaces.

To install:

4. Install or connect the following:
   - New gasket
   - Exhaust manifold and torque the bolts, in sequence, to 17 ft. lbs. (23 Nm)
   - Alternator bracket bolt, if loosened
   - Exhaust manifold heat shields and torque the bolts to 105 inch lbs. (12 Nm)
   - Exhaust manifold support bracket
   - Engine wiring heat shield
   - Upstream HO_2Ssensor wiring connector
   - Exhaust pipe to the manifold and torque the bolts to 21 ft. lbs. (28 Nm)
   - Coolant recovery bottle
   - Power steering pump reservoir
   - MAP sensor connector
   - Throttle and speed control cables
   - Air cleaner bracket and assembly
   - Negative battery cable
5. Start the vehicle and check for leaks, repair if necessary.

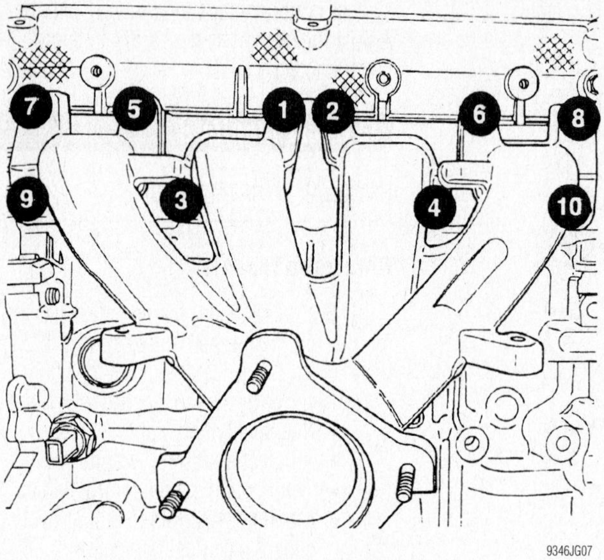

Torque the exhaust manifold bolts in sequence

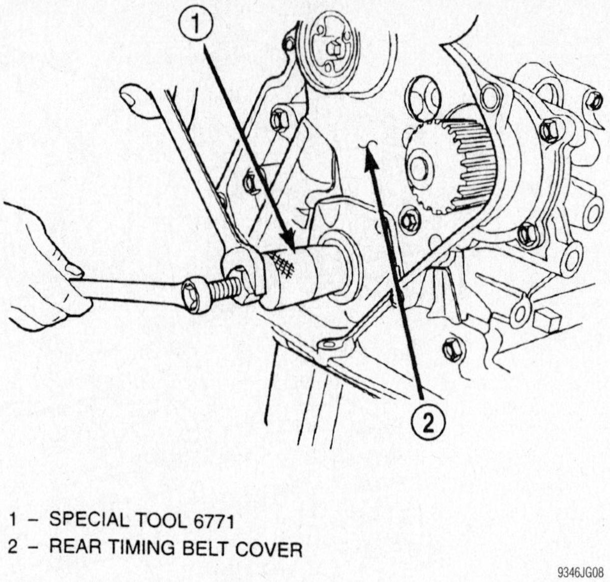

Removing the front crankshaft oil seal

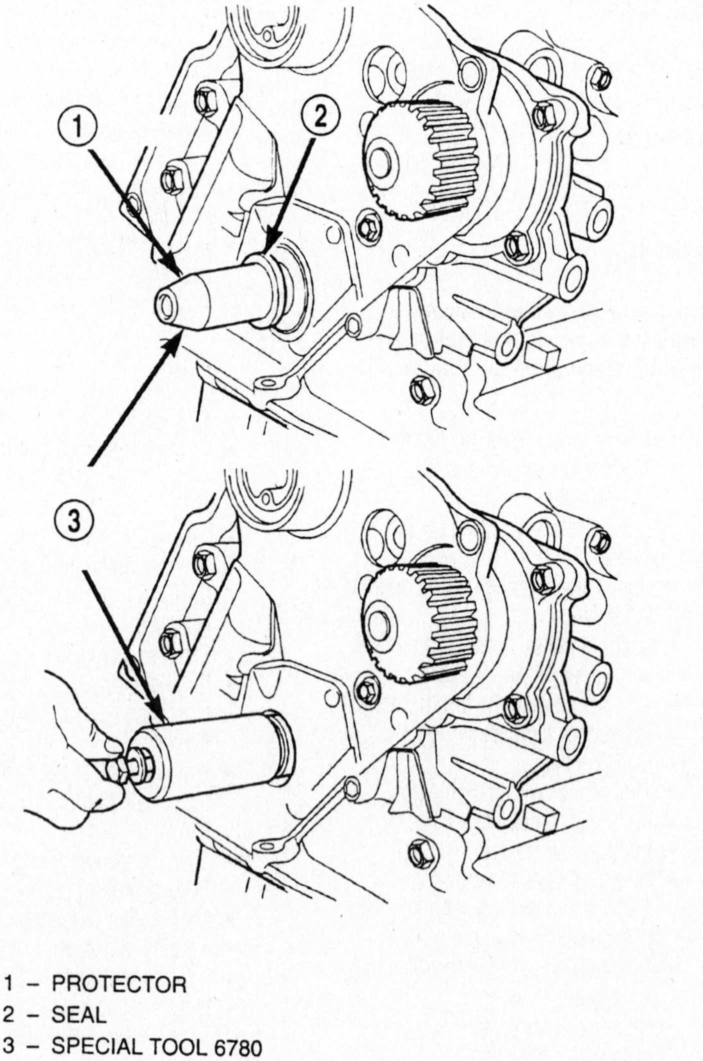

Installing a new seal using Seal Installer 6780-1; proceed with caution if using substitute tools

Turbo Models

The exhaust manifold on turbocharged models is removed as an assembly with the turbocharger. Refer to the turbocharger removal and installation procedure in this manual for manifold removal.

Front Crankshaft Seal

REMOVAL & INSTALLATION

1. Before servicing the vehicle, refer to the precautions in the beginning of this section.
2. Remove or disconnect the following:
   - Negative battery cable
   - Crankshaft damper bolt
   - Crankshaft damper, using Puller Tool 1026 and Insert Tool 6827-A
   - Timing belt
   - Crankshaft sprocket, using Tool 6793 and Insert Tool C-4685-C2
   - Crankshaft oil seal, using a Seal Puller Tool 6771

➡Be careful not to damage the seal surface of the cover.

To install:

3. Install or connect the following:
   - New front crankshaft oil seal with the seal spring facing the engine, using Crankshaft Installer Tool 6780
   - Crankshaft sprocket with special T 6792

➡Make sure the word FRONT on the crankshaft sprocket is facing outward.

- Timing belt
- Crankshaft damper using thrust bearing washer and bolt from Installer Tool 6792. Torque the damper bolt to 105 ft. lbs. (142 Nm).
- Negative battery cable

4. Start the vehicle and check for leaks, repair if necessary.

Camshaft and Lifters

REMOVAL & INSTALLATION

➡After all components are installed on the engine, a DRB scan tool is necessary to perform the camshaft and crankshaft timing relearn procedure.

1. Before servicing the vehicle, refer to the precautions in the beginning of this section.
2. Relieve the fuel system pressure.
3. Remove or disconnect the following:
   - Negative battery cable
   - Cylinder head cover
   - Camshaft Position (CMP) sensor and target magnet
   - Timing belt
   - Camshaft sprocket
   - Rear timing belt cover
   - Loosen the camshaft bearing caps in sequence from the rear of the cylinder head
   - Camshafts

To install:

Make certain that the pistons are **NOT** at Top Dead Center (TDC) before installing the camshafts.

4. Install the camshafts and cam followers, lubricate the bearing journals thoroughly.
5. Install the left and right camshaft bearing caps, No's 2–5 and right side No. 6. Torque these fasteners to 105 inch lbs. (12 Nm). Apply Mopar® gasket maker to No.1 and left side No. 6. Torque these fasteners to 18 ft. lbs. (24 Nm) on 2002 models or 250 ft. lbs. (28 Nm) on 2003–06 models.
6. Install or connect the following:
   - Camshaft seals
   - Rear timing belt cover
   - Camshaft sprockets and torque the bolt to 85 ft. lbs. (115 Nm)
   - Timing belt
   - Target magnet
   - CMP sensor and torque the screws to 85 inch lbs. (9.6 Nm)
   - Cylinder head cover
   - Negative battery cable

➡An oil and filter change are recommended.

7. Use a DRB scan tool to perform the camshaft and crankshaft timing relearn procedure, as follows:
   a. Connect the scan tool to the DLC (located under the instrument panel, near the steering column).
   b. Turn the ignition switch **ON** and access the "miscellaneous" screen.
   c. Select the "re-learn cam/crank" option, then follow the instructions on the scan tool screen.
8. Start the engine and check for leaks. Run the engine with the radiator cap off so as the engine warms and the thermostat opens, coolant can be added to the radiator. Test drive vehicle to check for proper operation.

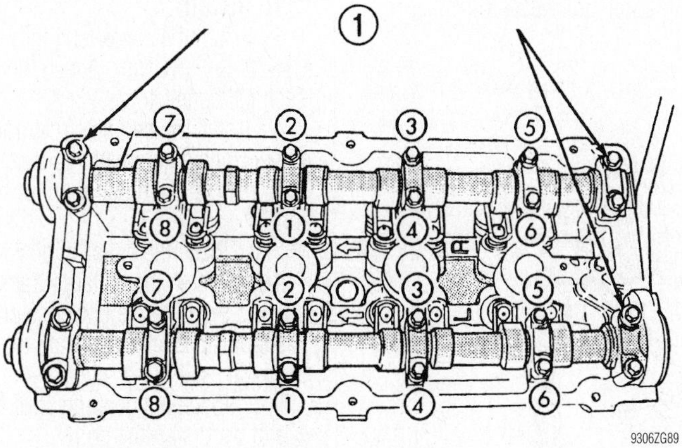

Remove the camshaft bearing caps in sequence

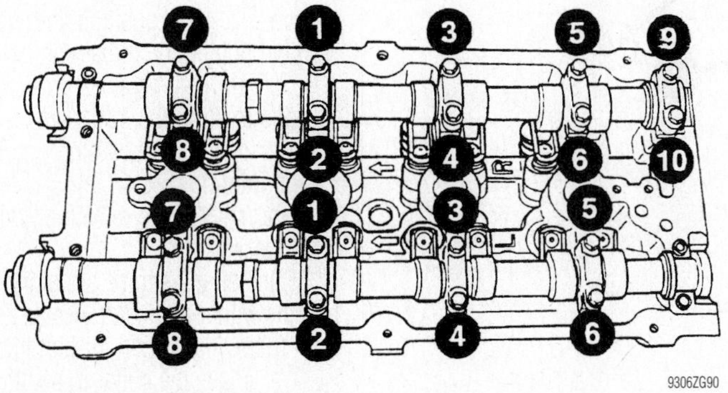

Camshaft bearing cap tightening sequence

Valve Lash

ADJUSTMENT

The engines in these vehicles do not require periodic valve lash adjustment.

Starter Motor

REMOVAL & INSTALLATION

Non Turbo Models

1. Before servicing the vehicle, refer to the precautions in the beginning of this section.
2. Remove or disconnect the following:
   - Negative battery cable
   - Air cleaner box cover
   - Engine structural collar
   - Starter electrical connectors
   - Starter

To install:

3. Install the starter and torque the bolts to 40 ft. lbs. (54 Nm). Attach the starter electrical connectors.
4. Install the engine structural collar on models with a automatic transaxle as follows, refer to the illustration for bolt locations:
   a. Place the collar in position and hand tighten the collar to transaxle bolt.
   b. Position the power steering hose support bracket and install the collar to oil pan bolt.
   c. Place the bending strut in place and hand start the bolt.
   d. Install the bolt through the strut and collar and hand tighten.

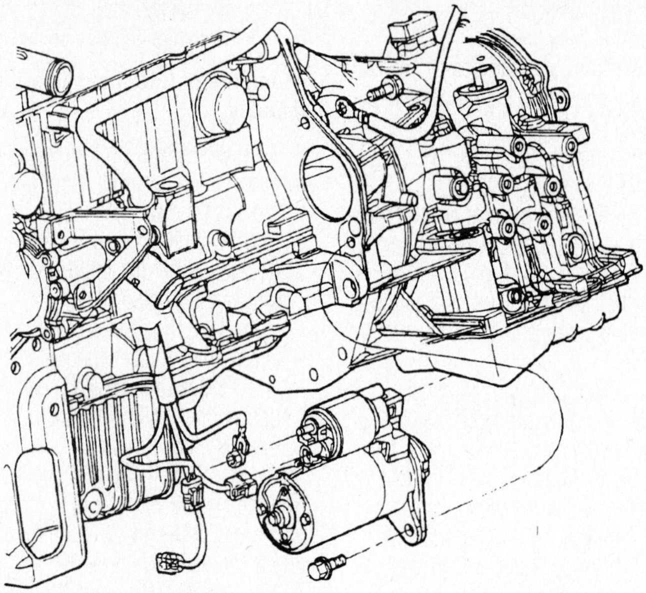

9306ZG91

Removal of the starter motor mounting

e. Place the power steering hose support bracket in position and install and hand tighten the remaining bolts.

f. Tighten the collar-to-transmission bolts to 75 ft. lbs. (101 Nm).

g. Install the bolts through the strut and into the block.

h. Tighten the remaining bolts to 45 ft. lbs. (61 Nm).

5. Install the engine structural collar on models with a manual transaxle as follows, refer to the illustration for bolt locations:

a. Place the collar in position and hand tighten the collar to transaxle b olt.

b. Position the power steering hose support bracket and install the collar to oil pan bolt.

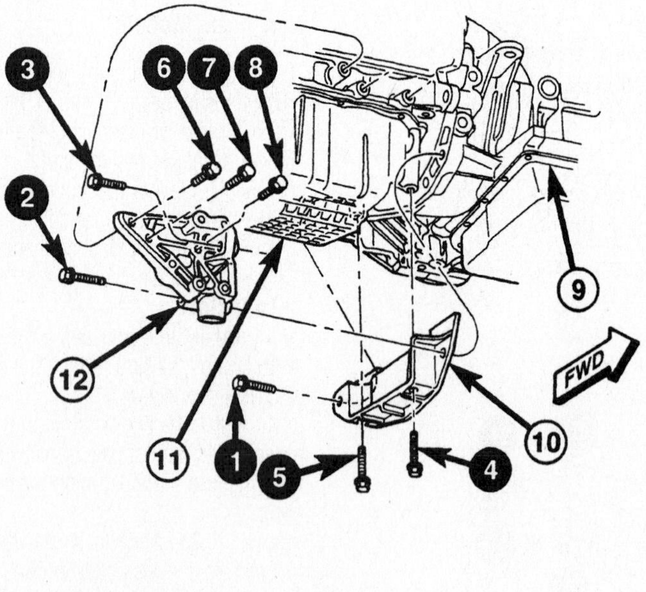

1–8 – BOLT TIGHTENING SEQUENCE
9 – TRANSAXLE
10 – COLLAR
11 – OIL PAN
12 – STRUT

67189-PTCR-G07

Structural collar assembly—automatic transaxle

c. Position the clutch slave cylinder into position and hand start the bolts.

d. Tighten bolts (1) to 75 ft. lbs. (101 Nm).

e. Tighten bolts (2 and 5) to 45 ft. lbs. (61 Nm).

f. Tighten bolts (3 and 4) to 20 ft. lbs. (28 Nm).

6. Install the air cleaner box cover and connect the negative battery cable.

Turbo Models

1. Before servicing the vehicle, refer to the precautions in the beginning of this section.
2. Remove or disconnect the following:
   - Negative battery cable
   - Air cleaner box cover
   - Upper starter bolt by pushing the inner cooler up and out of the way
   - Inner cooler lower hose from the inner cooler
   - Nuts retaining the inner cooler tube
   - Studs retaining the power steering lines
   - Position the power steering lines aside
   - Engine structural collar
   - Starter electrical connectors
   - Starter

To install:

3. Install the starter and torque the bolts to 40 ft. lbs. (54 Nm). Attach the starter electrical connectors.
4. Install the engine structural collar on models with a automatic transaxle as follows, refer to the illustration for bolt locations:

a. Place the collar in position and hand tighten the collar to transaxle bolt.

b. Position the power steering hose support bracket and install the collar to oil pan bolt.

c. Place the bending strut in place and hand start the bolt.

d. Install the bolt through the strut and collar and hand tighten.

e. Place the power steering hose support bracket in position and install and hand tighten the remaining bolts.

f. Tighten the collar-to-transmission bolts to 75 ft. lbs. (101 Nm).

g. Install the bolts through the strut and into the block.

h. Tighten the remaining bolts to 45 ft. lbs. (61 Nm).

5. Install the engine structural collar on models with a manual transaxle as follows, refer to the illustration for bolt locations:

a. Place the collar in position and hand tighten the collar to transaxle bolt.

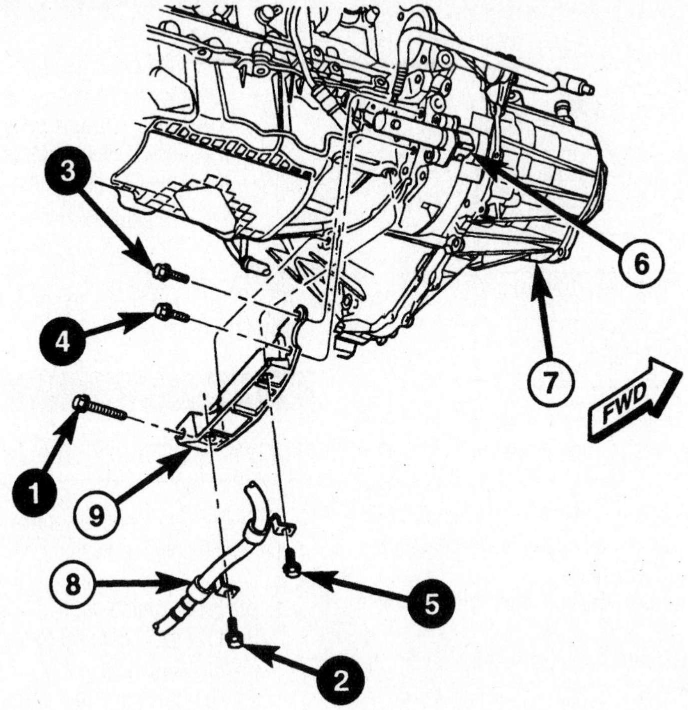

1–5 – BOLT TIGHTENING SEQUENCE
6 – HYDRAULIC CLUTCH SLAVE CYLINDER
7 – TRANSAXLE
8 – POWER STEERING HOSE
9 – COLLAR

67189-PTCR-G08

Structural collar assembly—manual transaxle

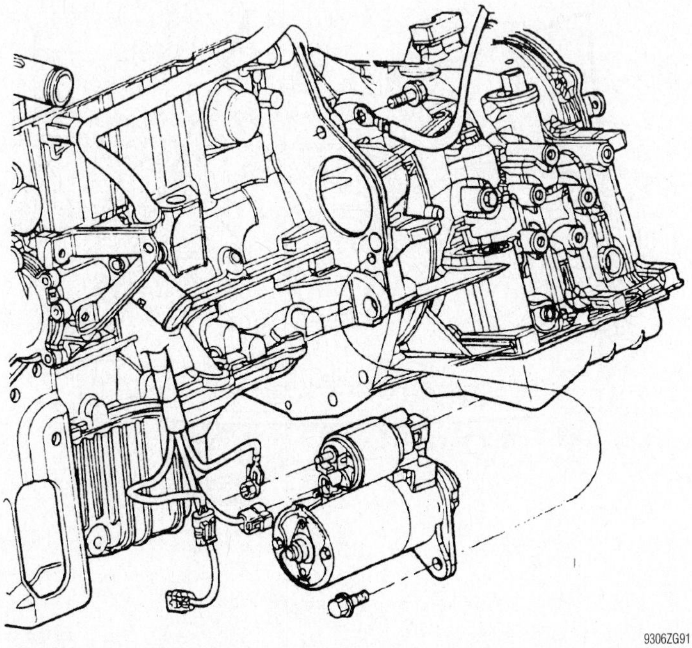

9306ZG91

Removal of the starter motor mounting

b. Position the power steering hose support bracket and install the collar to oil pan bolt.

c. Position the clutch slave cylinder into position and hand start the bolts.

d. Tighten bolts (1) to 75 ft. lbs. (101 Nm).

e. Tighten bolts (2 and 5) to 45 ft. lbs. (61 Nm).

f. Tighten bolts (3 and 4) to 20 ft. lbs. (28 Nm).

6. Install or connect the following:
   - Power steering lines aside
   - Studs retaining the power steering lines and tighten to 45 ft. lbs. (61 Nm)
   - Nuts retaining the inner cooler tube
   - Inner cooler lower hose to the inner cooler
   - Upper starter bolt by pushing the inner cooler up and out of the way. Tighten to 40 ft. lbs. (54 Nm).
   - Air cleaner box cover
   - Negative battery cable

Oil Pan

REMOVAL & INSTALLATION

1. Before servicing the vehicle, refer to the precautions in the beginning of this section.
2. Drain the engine oil and remove the oil filter.
3. Support the powertrain assembly.
4. Remove or disconnect the following:
   - Negative battery cable
   - Right inner splash shield
   - Turbocharger-to-charge air cooler hose assembly, if equipped
   - Oil cooler connector bolt, if equipped with a turbocharger. Do not disconnect the coolant lines from the oil cooler and reposition the cooler.
   - Engine structural collar
   - Lower torque strut
   - Oil filter adapter

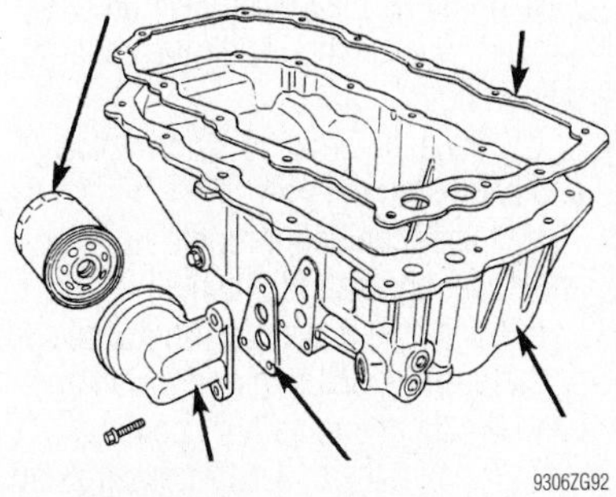

9306ZG92

Remove the oil filter adapter

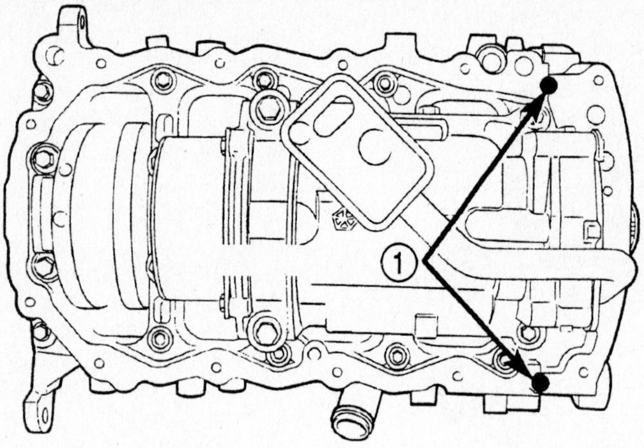

Silicone sealer application locations

- Oil pan and gasket

5. Thoroughly clean all gasket mating surfaces.

To install:

6. Apply silicone sealer to the oil pump-to-engine block parting line.
7. Install or connect the following:
   - New gasket on the oil pan
   - Oil pan and torque the bolts to 105 inch lbs. (12 Nm)
   - Oil filter and adapter and torque the screws to 105 inch lbs. (12 Nm)
8. If equipped with a turbocharger, replace the oil cooler seal. Lubricate the seal with and place the oil cooler-to-oil filter adapter in position making sure to align the notch on the tab. Install the oil cooler connector bolt and tighten to 41 ft. lbs. (55 Nm).

WARNING

Follow the proper tightening sequence for the structural collar or damage to the collar or oil pan may occur.

9. Install the engine structural collar on models with a automatic transaxle as follows, refer to the illustration for bolt locations:
   a. Place the collar in position and hand tighten the collar to transaxle bolt.
   b. Position the power steering hose support bracket and install the collar to oil pan bolt.
   c. Place the bending strut in place and hand start the bolt.
   d. Install the bolt through the strut and collar and hand tighten.
   e. Place the power steering hose support bracket in position and install and hand tighten the remaining bolts.
   f. Tighten the collar-to-transmission bolts to 75 ft. lbs. (101 Nm).
   g. Install the bolts through the strut and into the block.
   h. Tighten the remaining bolts to 45 ft. lbs. (61 Nm).
10. Install the engine structural collar on models with a manual transaxle as follows, refer to the illustration for bolt locations:
   a. Place the collar in position and hand tighten the collar to transaxle bolt.
   b. Position the power steering hose support bracket and install the collar to oil pan bolt.
   c. Position the clutch slave cylinder into position and hand start the bolts.
   d. Tighten bolts (1) to 75 ft. lbs. (101 Nm).
   e. Tighten bolts (2 and 5) to 45 ft. lbs. (61 Nm).
   f. Tighten bolts (3 and 4) to 20 ft. lbs. (28 Nm).
11. Install or connect the following:
   - Lower torque strut
   - Turbocharger-to-charge air cooler hose assembly, if equipped
   - Right inner splash shield
   - Negative battery cable
12. Fill the engine with clean oil and a new filter.
13. Start the vehicle and check for leaks, repair if necessary.

Oil Pump

REMOVAL & INSTALLATION

1. Before servicing the vehicle, refer to the precautions in the beginning of this section.
2. Drain the engine oil.
3. Remove or disconnect the following:
   - Negative battery cable
   - Timing belt and rear cover
   - Oil pan
   - Crankshaft sprocket, using Tool 6793 and Insert Tool C-4685-C2
   - Crankshaft key
   - Oil pickup tube
   - Oil pump

To install:

4. Wash all parts in a solvent; then, inspect carefully for damage or wear, as follows:
   a. Inspect the mating surface of the

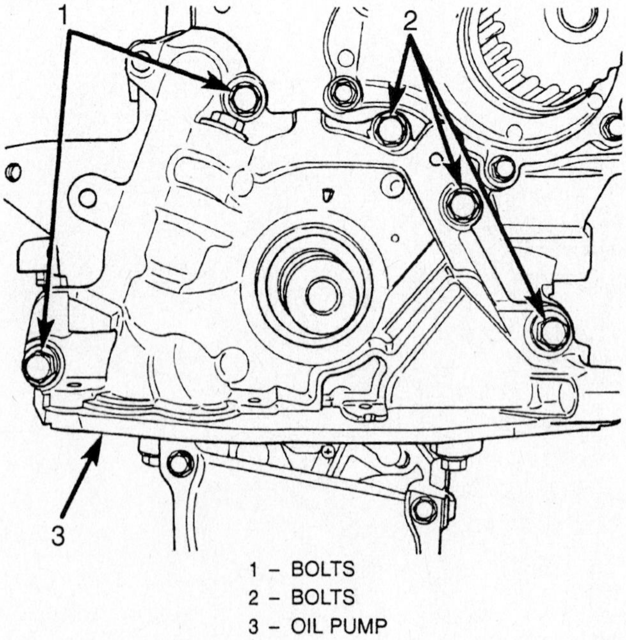

Exploded view of the oil pump mounting bolts

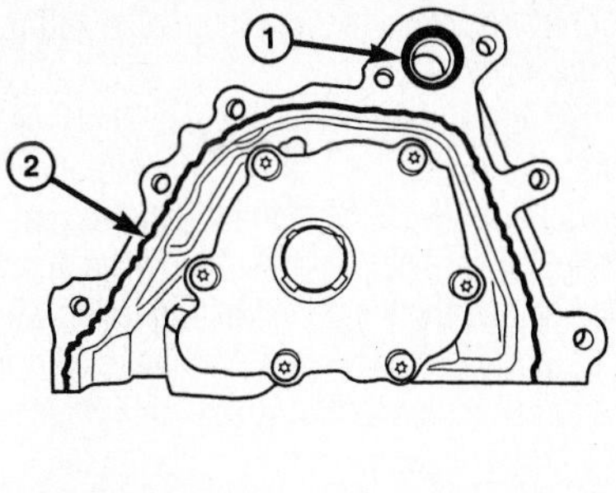

1 - O-RING
2 - SEALER LOCATION
67189-PTCR-G10

Apply a small amount of gasket maker to the pump body cover mounting surface

oil pump should be smooth. Replace the pump cover, if scratched or grooved.

b. Apply Mopar® gasket maker to the oil pump.

c. Install the O-ring into the oil pump body discharge passage.

5. Prime the oil pump before installation by filling the rotor cavity with engine oil.

6. Install or connect the following:

- Oil pump, align the rotor flats with the crankshaft flats and torque the bolts to 21 ft. lbs. (28 Nm).

WARNING

The front crankshaft seal MUST be out of the pump to align or damage may result.

- New front crankshaft seal, using Seal Driver Tool 6780
- Crankshaft key
- Crankshaft sprocket, using a Crankshaft Sprocket Installer Tool 6792
- Oil pump pickup tube
- Oil pan
- Rear timing belt cover
- Timing belt
- Negative battery cable

7. Fill the engine with clean oil.

8. Start the engine and check for leaks; repair if necessary.

Timing Belt

REMOVAL & INSTALLATION

1. Before servicing the vehicle, refer to the precautions in the beginning of this section.

2. Remove the A/C refrigerant using approved equipment.

3. Remove or disconnect the following to remove the upper timing cover:

- Negative battery cable

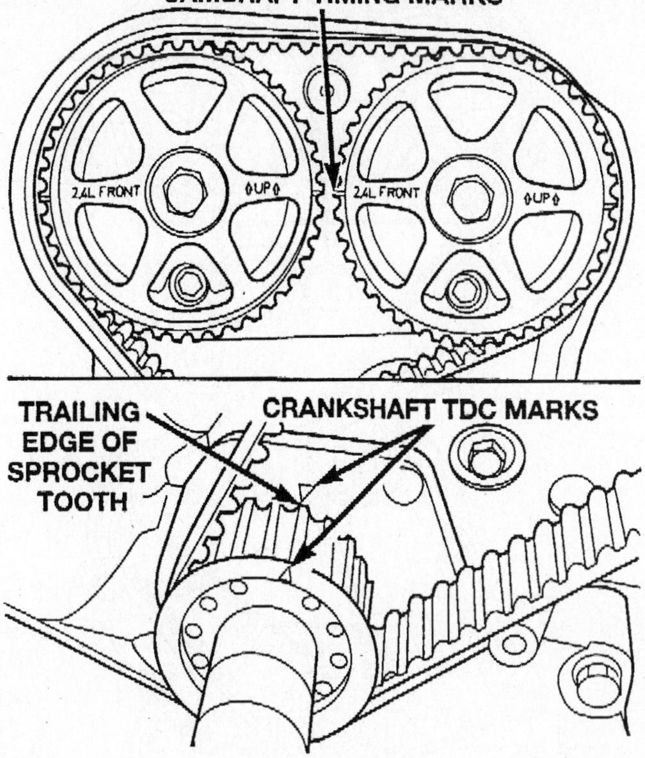

79245G14

Camshaft and crankshaft alignment marks—2.4L engine

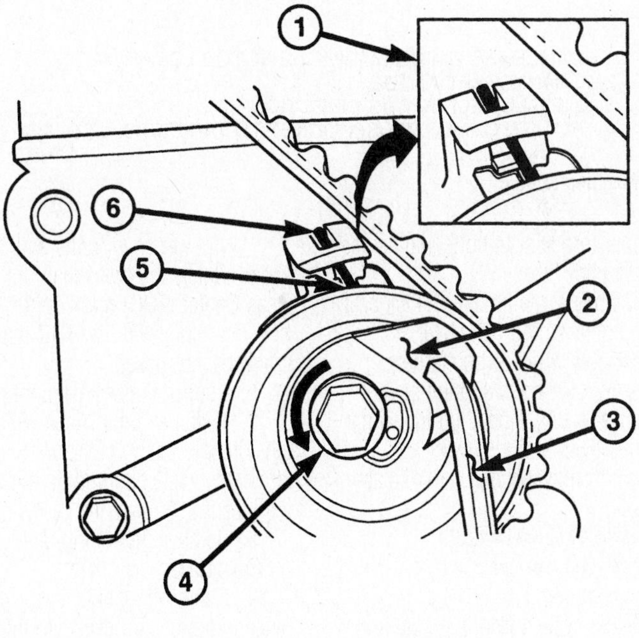

1 - ALIGN SETTING NOTCH WITH SPRING TANG
2 - TOP PLATE
3 - 6mm ALLEN WRENCH
4 - LOCK BOLT
5 - SETTING NOTCH
6 - SPRING TANG

67189-PTCR-G11

Install the belt tensioner

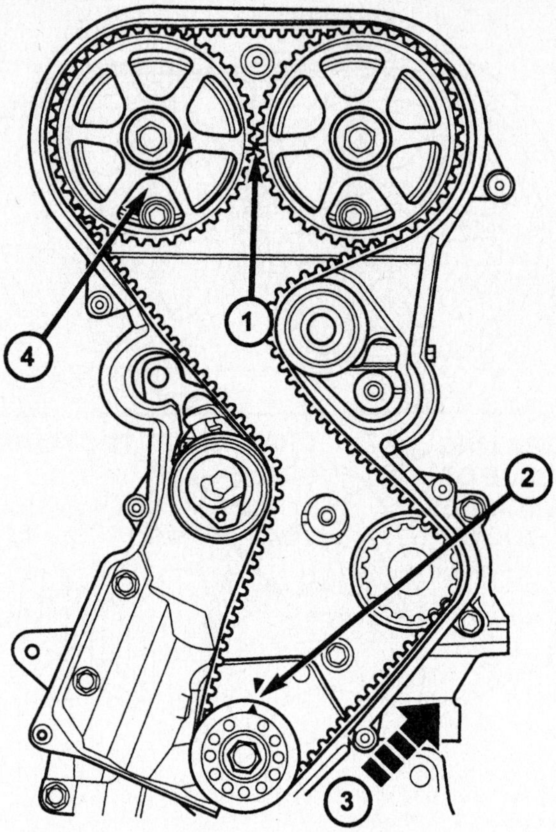

1 - CAMSHAFT TIMING MARKS 1/2 NOTCH LOCATION
2 - CRANKSHAFT AT TDC
3 - INSTALL BELT IN THIS DIRECTION
4 - ROTATE CAMSHAFT SPROCKET TO TAKE UP BELT SLACK

67189-PTCR-G12

Proper timing belt routing

- Upper torque strut bolts and set the strut aside
- A/C lines at the junction block near the upper timing belt cover, if equipped with a turbocharger
- Upper cover bolts and the cover

4. Remove or disconnect the following to remove the lower timing cover:

- Right front wheel and inner splash shield
- Accessory drive belts
- Crankshaft damper
- Lower torque strut
- Exhaust system from the manifold
- A/C compressor switch
- Upper torque strut and bracket
- Upper radiator support crossmember
- Power steering pump and bracket without disconnecting the lines
- Right engine mount through bolt after supporting the engine
- Engine support bracket
- Lower cover bolts and the cover

5. Rotate the crankshaft until the Top Dead Center (TDC) mark on the oil pump housing aligns with the TDC mark on the crankshaft sprocket.
6. Loosen the belt tensioner lock bolt.
7. Install an 6mm Allen wrench into the tensioner. Rotate the tensioner counterclockwise while pushing on the wrench until it slides into the locking hole.
8. Remove the timing belt.

To install:

9. Set the crankshaft sprocket at TDC by aligning the sprocket with the arrow on the oil pump housing.
10. Set the camshafts timing marks so that the exhaust camshaft sprocket is a ½notch below the intake camshaft sprocket.
11. Install the timing belt by starting at the crankshaft. Go around the water pump sprocket, idler pulley, camshaft sprockets and the tensioner.
12. Move the exhaust camshaft sprocket counterclockwise to align the marks and to remove any slack.
13. Insert a 6mm Allen wrench into the tensioner opening on the top plate of the tensioner pulley. Rotate the top plate counterclockwise. The pulley will move against the belt and the tensioner setting notch will start to move clockwise. Continue to move the top plate counterclockwise until the setting notch is aligned with the spring tang. Using the Allen wrench, to prevent the top plate from moving, tighten the tensioner lock bolt to 220 inch lbs. (25 Nm). Make sure the setting notch and spring tang are still aligned after the lock nut is tighten. If not repeat the procedure.
14. Remove the wrench from the belt tensioner.

➡When repositioning the crankshaft to the TDC position, this must be done during the clockwise rotation movement. If the TDC is missed, rotate a further two full turns until TDC is reached. Do not rotate the crankshaft counterclockwise as this will result in improper tensioner settings.

15. Rotate the crankshaft two full revolutions and verify that the TDC marks are properly aligned.
16. Check the spring tang is within the tolerance window, if not repeat the previous two steps.
17. Install or connect the following:

- Lower timing belt cover and torque the bolts to 50 inch lbs. (6 Nm).
- Upper timing belt cover and torque the bolts to 50 inch lbs. (6 Nm).
- A/C lines at the junction block near the upper timing belt cover, if equipped with a turbocharger
- Right engine support bracket and reposition the power steering pump. Torque the bracket bolts to 45 ft. lbs. (61 Nm).
- Right engine mount through bolt and torque it to 87 ft. lbs. (118 Nm)
- Upper radiator support crossmember
- Torque strut bracket
- Upper torque strut
- A/C lines and pressure switch
- Exhaust system to the manifold
- Crankshaft damper
- Accessory drive belts
- Lower torque strut
- Right splash shield and wheel
- Negative battery cable

18. Recharge the A/C system.
19. Perform the camshaft/crankshaft synchronization procedure.

Rear Main Seal

REMOVAL & INSTALLATION

1. Before servicing the vehicle, refer to the precautions in the beginning of this section.
2. Remove or disconnect the following:
   - Transmission
   - Flexplate/flywheel
   - Rear main seal
3. Insert a seal remover between the dust lip and the metal case of the crankshaft seal. Angle the tool through the dust lip against the metal case of the seal. Pry out the seal.

**** WARNING**

DO NOT let the prytool contact the crankshaft seal surface. Contact of the tool blade against the crankshaft edge (chamfer) is permitted.

To install:

**** WARNING**

If the crankshaft edge (chamfer) has any burrs or scratches on the, clean it up with 400 grit sand paper to prevent seal damage during installation of the new seal.

➡No lubrication is necessary when installing the seal.

4. Place Crankcase Seal Pilot Tool 6926-1 on the crankshaft; this is a pilot tool with a magnetic base
5. Position the seal over the pilot Tool; be sure the words THIS SIDE OUT on the seal can be read.

➡The pilot tool should stay on the crankshaft during installation of the seal. Be sure the seal lip faces the crankcase during installation.

**** WARNING**

If the seal is driven in the block past flush, this may cause an oil leak.

6. Drive the seal into the block, using Crankshaft Seal Tool 6926-2 and handle C-4171, until the tool bottoms out against the block.
7. Install or connect the following:
   - Flexplate/flywheel. Apply Lock & Seal Adhesive to the bolt treads and torque the bolts in a star pattern, to 70 ft. lbs. (95 Nm).
   - Transmission
   - Negative battery cable
8. Start the vehicle and check for leaks, repair if necessary.

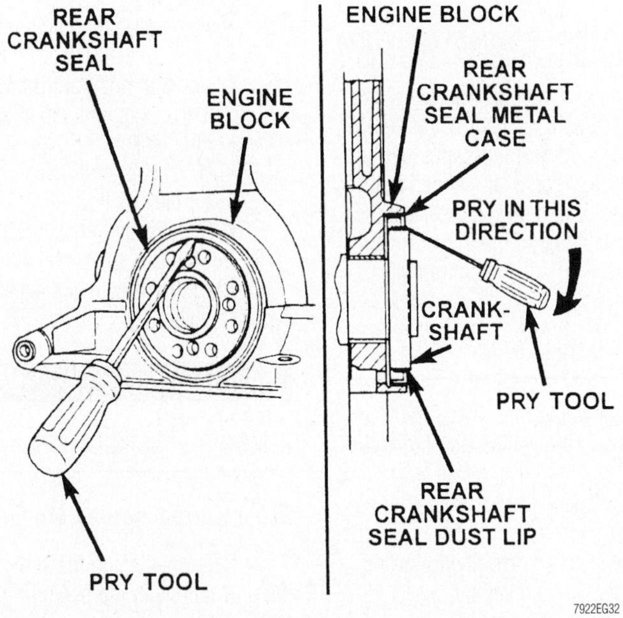

When prying the seal out, be sure to use the prytool at the proper angle

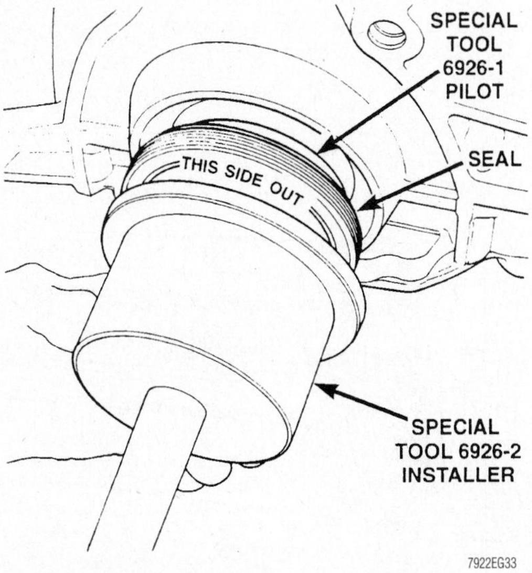

Place a proper size pilot tool with a magnetic base on the crankshaft

Piston and Ring

POSITIONING

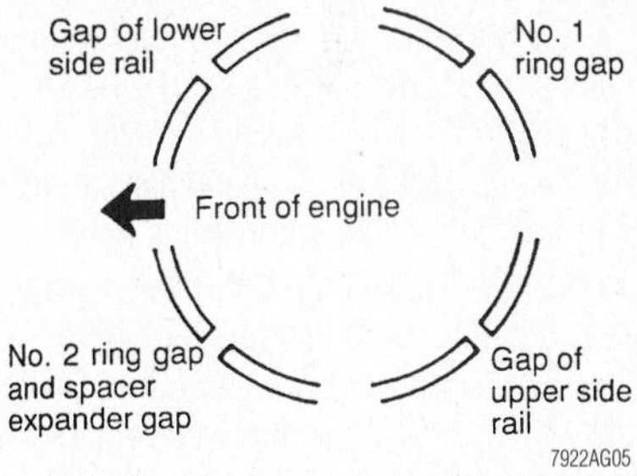

Piston ring end-gap spacing—2.4L engine

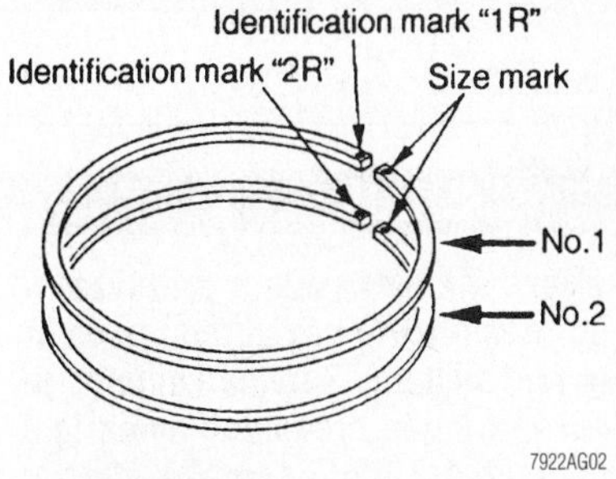

Common Chrysler piston ring identification mark locations

FUEL SYSTEM

Fuel System Service Precautions

Safety is an important factor when servicing the fuel system. Failure to conduct maintenance and repairs in a safe manner may result in serious personal injury. Maintenance and testing of the vehicle's fuel system components can be accomplished safely and effectively by adhering to the following rules and guidelines:

- To avoid the possibility of fire and personal injury, always disconnect the negative battery cable unless the repair or test procedure requires that battery voltage be applied.
- Always relieve the fuel system pressure prior to disconnecting any fuel system component (injector, fuel rail, pressure regulator, etc.), fitting or fuel line connection. Exercise extreme caution whenever relieving fuel system pressure, to avoid exposing skin, face and eyes to fuel spray. Please be advised that fuel under pressure may penetrate the skin or any part of the body that it contacts.
- Always place a shop towel or cloth around the fitting or connection prior to loosening to absorb any excess fuel due to spillage. Ensure that all fuel spillage is quickly removed from engine surfaces. Ensure that all fuel soaked cloths or towels are deposited into a suitable waste container.
- Always keep a dry chemical (Class B) fire extinguisher near the work area.
- Do not allow fuel spray or fuel vapors to come into contact with a spark or open flame.
- Always use a back-up wrench when loosening and tightening fuel line connection fittings. This will prevent unnecessary stress and torsion to fuel line piping.
- Always replace worn fuel fitting O-rings. Do not substitute fuel hose where fuel pipe is installed.

Fuel System Pressure

RELIEVING

✱✱ CAUTION

Relieve the fuel system pressure before servicing any components of the fuel system. Service vehicles in well ventilated areas and avoid ignition sources. NEVER smoke while servicing the vehicle!

1. Before servicing the vehicle, refer to the precautions in the beginning of this section.
2. Remove the negative battery cable.
3. Remove the fuel pump relay from the Power Distribution Center (PDC).
4. Start and run the engine until it stalls.
5. Turn the ignition key to the OFF position.

Fuel Filter

The fuel filter is part of the fuel pump module located in the fuel tank. Refer to the fuel pump module procedure for the fuel filter.

Fuel Pump

The fuel pump is integral with the pump module, which also contains the fuel reservoir, level sensor, inlet strainer and fuel pressure regulator. The inlet strainer, fuel pressure regulator and level sensor are the only serviceable items. If the fuel pump requires service, replace the entire fuel pump module.

REMOVAL & INSTALLATION

1. Before servicing the vehicle, refer to the precautions in the beginning of this section.
2. Properly relieve the fuel system pressure.
3. Remove or disconnect the following:
   - Negative battery cable
   - Air cleaner lid
   - Inlet Air Temperature (IAT) sensor and make up air hose
   - Fuel tank
   - Fuel pump module and seal from the tank
   - Locknut to release the fuel pump module, using a Ring Spanner Tool No. 6856
   - Fuel filter from the fuel pump module
   - Fuel pump and seal from the fuel tank

To install:

4. Install or connect the following:
   - Fuel filter to the fuel pump module
   - Fuel filter lines to the fuel pump module
   - Fuel pump module in the tank
   - Locknut while holding the fuel pump in position. Using Tool 6856, torque the nut to 56 ft. lbs. (75 Nm).
   - Fuel tank
   - IAT sensor and make-up hose
   - Air cleaner lid
   - Negative battery cable
5. Fill the fuel tank.
6. Start the vehicle and check for leaks, repair if necessary.

Fuel Injector

REMOVAL & INSTALLATION

Except Turbocharged Models

1. Before servicing the vehicle, refer to the precautions in the beginning of this section.

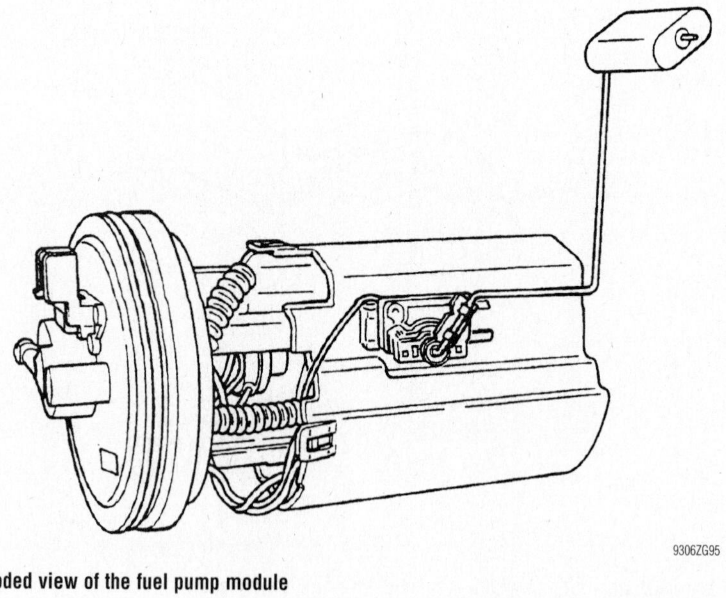

Exploded view of the fuel pump module

2. Release the fuel system pressure.
3. Remove or disconnect the following:
   - Negative battery cable
   - Remove the air cleaner lid
   - Inlet Air Temperature (IAT) sensor and the make-up hose
   - Engine cover or throttle control shield, if equipped
   - Fuel supply tube from fuel rail
   - Intake manifold
   - Fuel injector electrical connectors
   - Fuel rail with the fuel injectors
   - Fuel injector(s) and discard the O-rings

To install:

➡**Lubricate the O-rings with clean engine oil.**

4. Install or connect the following:
   - Fuel injector(s) to the fuel rail using new O-rings
   - Fuel injector nozzles into the intake manifold and torque the fuel rail bolts to 8 ft. lbs. (12 Nm)
   - Fuel injector electrical connectors
   - Fuel supply tube to fuel rail

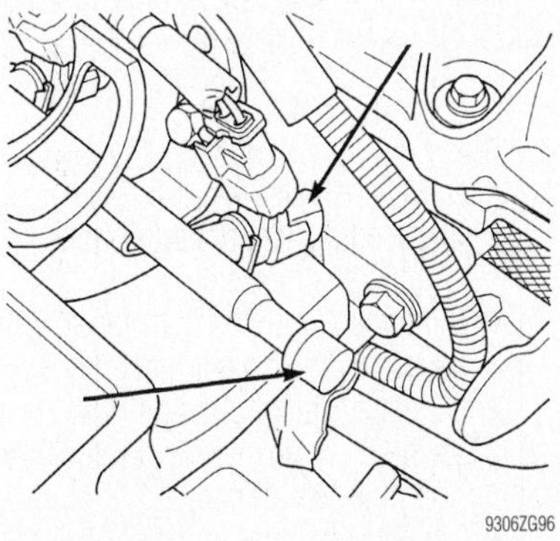

Remove the fuel rail and injectors as an assembly

   - Intake manifold
   - Engine cover or throttle control shield, if equipped
   - IAT sensor and make-up hose
   - Air cleaner lid
   - Negative battery cable

5. Start the vehicle and check for leaks, repair if necessary.

Turbocharged Models

1. Before servicing the vehicle, refer to the precautions in the beginning of this section.
2. Release the fuel system pressure.
3. Drain the cooling system.
4. Remove or disconnect the following:
   - Negative battery cable
   - Throttle body inlet hose from the throttle body
   - Purge hose from the throttle body
   - Electrical connections from the throttle body
   - Throttle control shield
   - Accelerator and cruise control cable from the throttle body
   - Manifold Absolute Pressure (MAP) sensor connector
   - Vacuum lines from the rear of the intake manifold
   - 5 bolts from the front and 2 bolts from the rear of the intake manifold
   - Intake manifold and cover the lower manifold to avoid contamination
   - Upper radiator hose up and out of the way
   - 2 small hoses from the thermostat housing
   - 2 bolts from the thermostat housing and rotate the assembly up and out of the way
   - Fuel lines from the rail
   - Fuel injector electrical connectors and wire from the rail
   - Fuel rail with the fuel injectors
   - Fuel injector(s) and discard the O-rings

To install:

➡**Lubricate the O-rings with clean engine oil.**

5. Install or connect the following:
   - Fuel injector(s) to the fuel rail using new O-rings
   - Fuel injector nozzles into the intake manifold and torque the fuel rail bolts to 170 inch lbs. (19 Nm)
   - Fuel injector electrical connectors and wiring to the fuel rail
   - Fuel line to fuel rail
   - Rotate the thermostat assembly into position and tighten the 2 bolts
   - 2 small hoses to the thermostat housing
   - Upper radiator hose
   - Intake manifold
   - 5 bolts front and 2 rear bolts on the intake manifold
   - Vacuum lines to the rear of the intake manifold
   - MAP sensor connector
   - Accelerator and cruise control cable to the throttle body
   - Throttle control shield
   - Electrical connections to the throttle body
   - Purge hose to the throttle body
   - Throttle body inlet hose to the throttle body
   - Negative battery cable
6. Fill the cooling system.
7. Start the vehicle and check for leaks, repair if necessary.

DRIVE TRAIN

Transaxle Assembly

REMOVAL & INSTALLATION

Manual

T350 TRANSAXLE

1. Before servicing the vehicle, refer to the precautions in the beginning of this section.
2. Drain the transaxle fluid.
3. Remove or disconnect the following:
   - Battery cables
   - Battery and tray
   - Air cleaner
   - Back-up lamp switch wiring from the transaxle
   - Shift selector and crossover cable and move them out of the way
   - Vehicle Speed Sensor (VSS) wire
   - Clutch master cylinder hydraulic tube from the slave cylinder
   - Both halfshafts
   - Bell housing dust cover
   - Power steering hose from the structural collar
   - Left side engine-to-transaxle lateral bending brace and structural collar
   - Right side engine-to-transaxle lateral bending brace
   - Starter
   - Driveplate-to-clutch module bolts and support the engine at the oil pan
   - Transaxle upper mount bolts and lower the powertrain assembly onto the support
   - Transaxle-to-engine mounting bolts
   - Transaxle
   - Clutch module from the input shaft
   - Slave cylinder
   - Upper transaxle mount

To install:

4. Install or connect the following:
   - Clutch module to the transaxle input shaft and place the transaxle into position

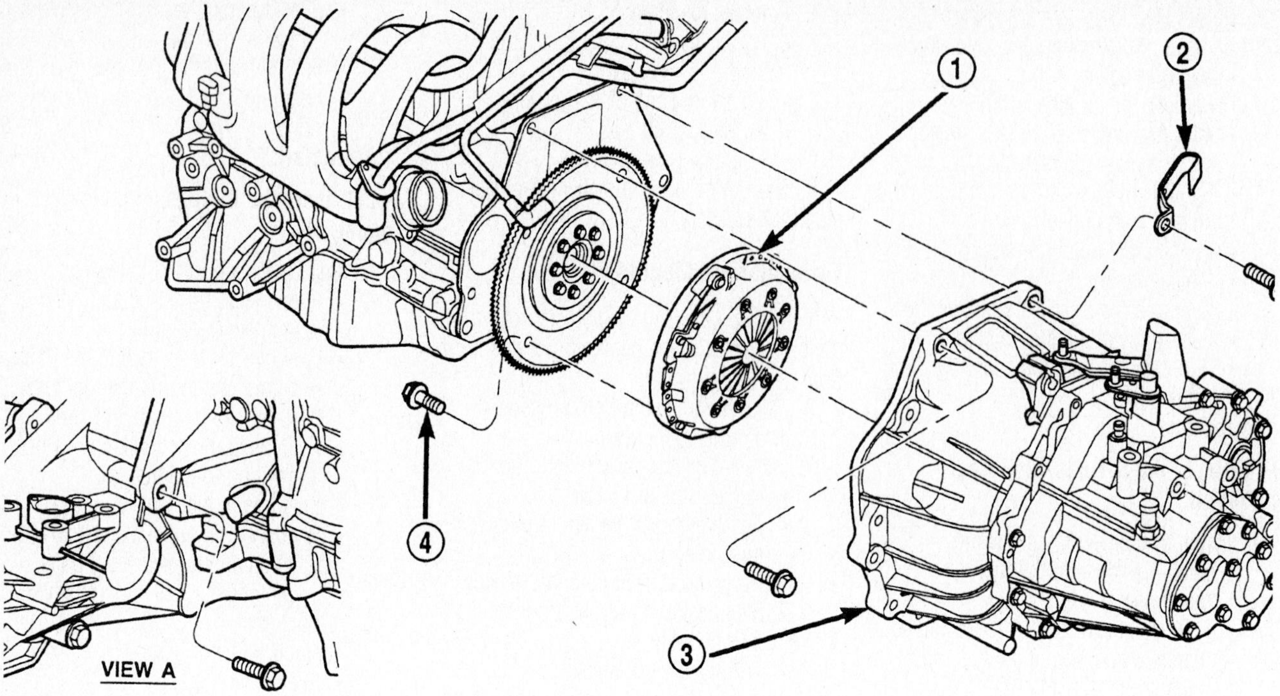

Remove the clutch module from the transaxle assembly

- Transaxle-to-engine mounting bolts and torque them to 80 ft. lbs. (108 Nm)
- Upper transaxle mount and torque the bolts to 45 ft. lbs. (62 Nm)
- Driveplate-to-clutch module bolts and torque to 65 ft. lbs. (88 Nm)
- Starter
- Starter electrical connectors
- Bell housing dust cover
- Left side engine-to-transaxle bending brace and structural brace. Torque the bolts to 60 ft. lbs. (81 Nm).
- Right lateral bending brace and torque the bolts to 60 ft. lbs. (81 Nm)
- Power steering hose to the structural collar
- Both halfshafts
- Clutch master cylinder tube to the slave cylinder
- VSS electrical connector
- Shift crossover and selector cables to the shift lever
- Cables to the bracket and install a new retainer clip
- Back-up lamp switch connector
- Battery tray
- Battery and cables
- Air cleaner assembly

5. Fill the transaxle fluid to the proper level.

6. Be sure the vehicle's back-up lights and speedometer are functioning properly.

7. Start the vehicle and check for leaks, repair if necessary.

G288 TRANSAXLE

1. Before servicing the vehicle, refer to the precautions in the beginning of this section.
2. Drain the transaxle fluid.
3. Remove or disconnect the following:
   - Negative battery cable
   - Air cleaner
   - Power Distribution Center (PDC) from the bracket
   - Air cleaner/PDC mounting bracket
   - Gearshift cable from the shift mechanism
   - Gearshift cable bracket
   - Back-up lamp switch wiring from the transaxle
   - Vehicle Speed Sensor (VSS) wire
   - Clutch master cylinder hydraulic tube from the slave cylinder
   - Upper transaxle bell housing bolts
   - Transaxle drain plug, drain the fluid and reinstall the plug
   - Both halfshafts
   - Intermediate shaft/bearing bolts and the shaft assembly
   - 2 intercooler connector pipe-to-oil pan bolts and position aside
   - Oil pan-to-bell housing bolts
4. Place a screw jack with a piece of wood on top onto the oil pan.
   - 2 transaxle upper mount-to-bracket bolts and lower the engine onto the wood and screw jack
   - Transaxle upper mount bracket
   - Attach a transaxle jack to the transaxle
   - Starter lower bolt and ground cable
   - 4 modular clutch drive-to-drive plate bolts. While removing the bolts one tolerance (slotted) drive plate hole will be seen. When this bolt is removed, mark the drive plate and clutch assembly alignment to aid during reinstallation for proper alignment
   - Remaining transaxle-to-engine bolts
   - Transaxle

To install:

5. Install or connect the following:
   - Transaxle into position
   - Accessible transaxle-to-engine

mounting bolts and torque them to 80 ft. lbs. (108 Nm)

- 4 modular clutch-to-drive plate bolts, align the drive plate and clutch assembly marks made during removal, start with the slotted hole first and tighten the bolts to 65 ft. lbs. (88 Nm)
- Upper mount bracket
- raise the transaxle using the jack until the mount bracket is aligned and install the bolts
- Starter motor bolt and ground cable
- Lower bellhousing bolt
- 2 intercooler connector pipe-to-oil pan bolts
- Remove the drain plug and fill the transaxle with the correct type and amount of fluid. Install the plug and tighten to 35 ft. lbs. (47 Nm).
- Intermediate/bearing assembly and tighten the bolts
- Both halfshafts
- Clutch master cylinder tube to the slave cylinder
- VSS electrical connector
- Gearshift cable bracket and tighten the bolts
- Gearshift crossover and selector cables and secure using new clips
- Back-up lamp switch connector
- PDC/air cleaner bracket
- PDC
- Air cleaner assembly
- Negative battery cable

6. Be sure the vehicle's back-up lights and speedometer are functioning properly.
7. Start the vehicle and check for leaks, repair if necessary.

Automatic

1. Before servicing the vehicle, refer to the precautions in the beginning of this section.
2. Drain the transaxle fluid.
3. Remove or disconnect the following:
   - Battery cables
   - Air cleaner assembly
   - Battery and tray
   - Upper starter-to-transaxle bell housing bolt
   - Transaxle dipstick and tube
   - Gearshift cable end from the transaxle shift lever
   - Gearshift cable bracket bolt from the transaxle
   - Transaxle oil cooler lines. Plug the lines to prevent contamination.
   - Input and output speed sensor electrical connectors
   - Transaxle range sensor connector
   - Solenoid/pressure switch assembly connector
   - Both front wheels
   - Left front splash shield
   - Both halfshafts
   - Power steering hose from the structural collar
   - Left and right side engine-to-transaxle lateral brace and structural collar
   - Starter motor electrical connectors
   - Starter
   - Gearshift cable bracket
   - Driveplate-to-torque converter bolts and support the powertrain assembly
   - Transaxle upper mount to bracket bolts and lower the powertrain assembly
   - Transaxle

To install:

4. Install or connect the following:
   - Transaxle and torque the bolts to 80 ft. lbs. (105 Nm) while supporting the transaxle with a jack
   - Mount-to-transaxle bracket. Torque the bolts to 50 ft. lbs. (68 Nm) and remove the jack.
   - Driveplate-to-torque converter bolts and torque the bolts to 65 ft. lbs. (88 Nm)
   - Starter and hand tighten the bolts
   - Dipstick tube and secure the bracket to the transaxle. Torque the upper starter bolt to 40 ft. lbs. (54 Nm).
   - Cable bracket to the bell housing and torque the bolt to 45 ft. lbs. (61 Nm)
   - Starter lower bolt and torque the bolt to 40 ft. lbs. (54 Nm)

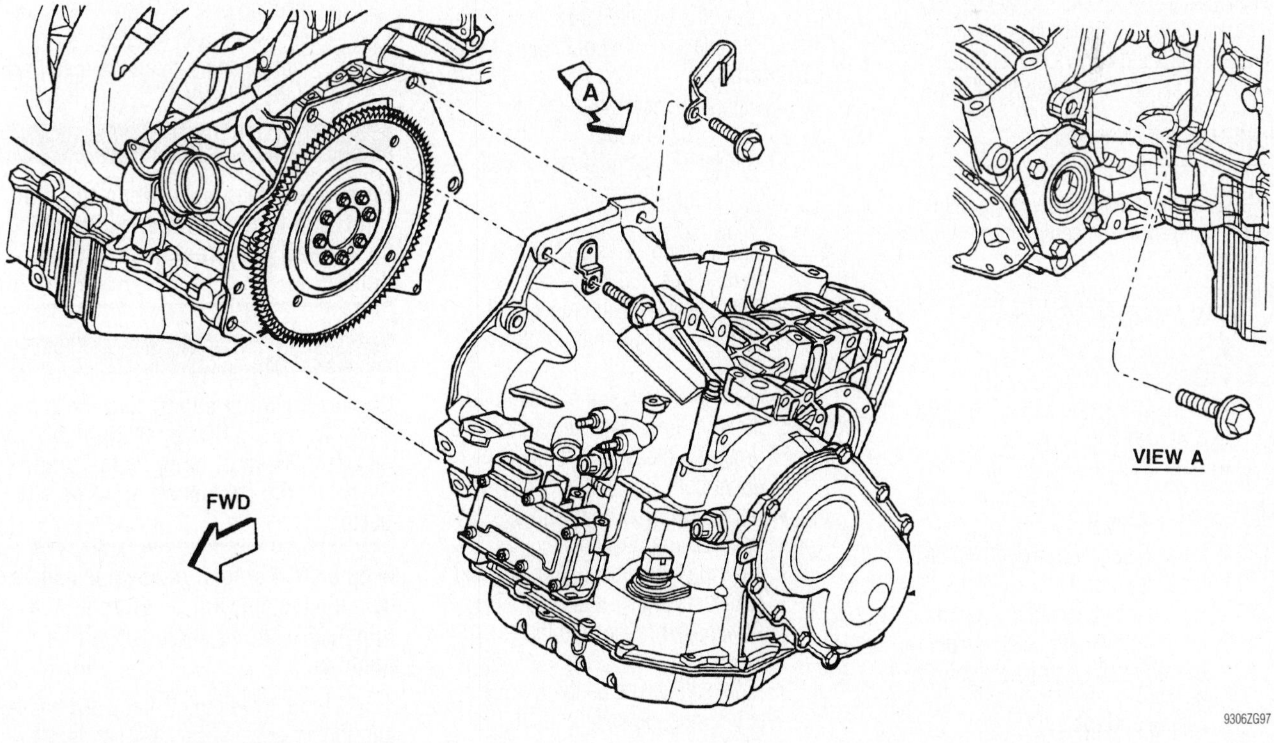

Exploded view of automatic transaxle removal

- Starter electrical connections
- Bell housing dust cover
- Lower dust shield screws and torque the bell housing cover bolts to 108 inch lbs. (12 Nm)
- Right side lateral bending brace and torque the bolts to 60 ft. lbs. (81 Nm)
- Both halfshafts
- Left side splash shield and both wheels
- Cooler lines to the transaxle and secure with constant tension clamps
- Solenoid/pressure switch assembly
- Transmission range sensor connector
- Input/output sensor connectors
- Gearshift cable to the bracket and connect it to the manual valve lever
- Battery tray and battery
- Air cleaner assembly
- VSS wiring
- Battery cables

5. Fill the transaxle fluid to the proper level.
6. Be sure the vehicle's back-up lights and speedometer are working properly.

Clutch

ADJUSTMENT

This vehicle utilizes a modular clutch assembly located between the engine and the transaxle. The modular clutch is serviced as an assembly and is self-adjusting. The self-adjusting feature of the clutch relies on a sensor ring and adjuster ring.

REMOVAL & INSTALLATION

1. Before servicing the vehicle, refer to the precautions in the beginning of this section.
2. Remove or disconnect the following:
   - Negative battery cable
   - Air cleaner assembly
   - Battery and tray
   - Back-up lamp electrical connector
   - Shift cable-to-bracket clips
   - Shift lever and crossover cable from the levers. Move the cables out of the way.
   - Vehicle Speed Sensor (VSS) electrical connector
   - Clutch master cylinder tube from the slave cylinder using a clutch hydraulic quick connect Tool 6638A
   - Both halfshafts

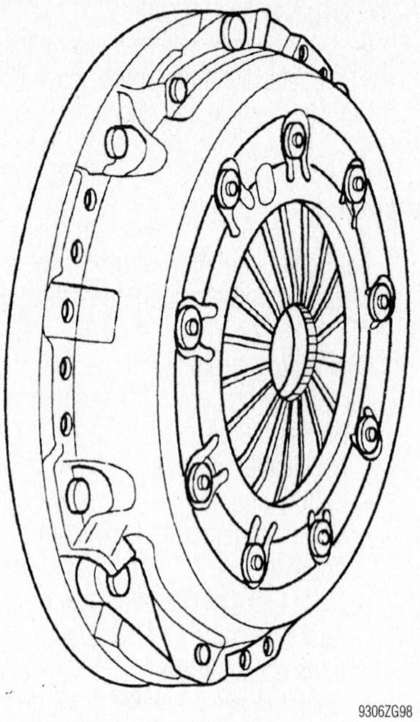

Exploded view of the modular clutch assembly

   - Power steering hose from the structural collar
   - Left side lateral bending brace and structural collar
   - Bell housing dust cover
   - Right side lateral bending brace
   - Starter
   - Driveplate-to-clutch module bolts and support the engine at the oil pan
   - Transaxle upper mount bolts and lower the engine/transaxle assembly
   - Modular clutch from the input shaft

To install:

3. Install or connect the following:
   - Clutch module to the input shaft
   - Transaxle-to-engine mount bolts and torque them to 80 ft. lbs. (108 Nm)
   - Upper mount bolts and torque them to 65 ft. lbs. (88 Nm)
   - Driveplate-to-clutch module bolts and torque them to 65 ft. lbs. (88 Nm)
   - Starter and torque the bolts to 40 ft. lbs. (54 Nm). Make certain that the ground cable is fastened to the upper bolt.
   - Starter electrical connectors
   - Bell housing dust cover
   - Left side lateral bending brace and structural collar
   - Power steering hose to the structural collar
   - Right side lateral bending brace and torque the bolts to 60 ft. lbs. (81 Nm)
   - Both halfshafts
   - Clutch master cylinder tube to the slave cylinder
   - VSS electrical connector
   - Back-up lamp switch electrical connector
   - Air cleaner assembly
   - Battery and tray
   - Battery cables
4. Fill the transmission to the proper level.
5. Road test the vehicle and check for proper clutch operation.
6. Check the fluid level and adjust if needed.

Halfshafts

REMOVAL & INSTALLATION

1. Before servicing the vehicle, refer to the precautions in the beginning of this section.
2. Place the transaxle in the **P** position, for automatic and neutral for manual.
3. Remove or disconnect the following:
   - Negative battery cable
   - Front wheel
   - Cotter pin, locknut and spring washer from the end of the outer Constant Velocity (CV) joint stub axle
   - Driveshaft to hub and bearing nut
   - Front wheel speed sensor, if equipped
   - Steering knuckle from the ball joint
   - Driveshaft from the steering knuckle and support the outer end of the driveshaft

✱✱ WARNING

Be careful when separating the ball joint stud from the steering knuckle, so the ball joint seal does not get damaged.

✱✱ WARNING

Be careful when separating the inner CV-joint during this operation. Do not let the driveshaft hang by the inner CV-joint, the driveshaft must be supported.

➡Inner Tri-Pot joint removal is easier by applying outward pressure on the joint while hitting the punch with a hammer.

4. Inner Tri-Pot joints from the transmission side gears using a punch to dis-

lodge the inner Tri-Pot joint retaining ring from the transmission side gear. If removing the right side inner Tri-Pot joint, position the punch against the inner Tri-Pot joint. Hit the punch sharply with a hammer to dislodge the right inner joint from the side gear. If removing the left side inner Tri-Pot joint, position the punch in the groove of the inner Tri-Pot joint. Hit the punch sharply with a hammer to dislodge the left inner Tri-Pot joint from the side gear.

5. Hold the inner Tri-Pot joint and interconnecting shaft of the driveshaft assembly. Remove the inner Tri-Pot joint from the transaxle by pulling it straight out of the transaxle side gear and transmission oil seal. When removing the Tri-Pot joint, do not let the spline or snapring drag across the sealing lip of the transaxle-to-Tri-Pot joint oil seal.

**** WARNING**

The driveshaft, when installed, acts as a bolt which secures the front hub and bearing assembly. If the vehicle is to be supported or moved on its wheels with a driveshaft removed, install a proper-sized bolt and nut through the front hub. Tighten the bolt and nut to 135 ft. lbs. (183 Nm). This will ensure that the hub bearing cannot loosen.

To install:

6. Thoroughly clean the spline and oil seal sealing surface on the Tri-Pot joint. Lightly lubricate the oil seal sealing surface on the Tri-Pot joint with fresh, clean transmission fluid.

7. Holding the driveshaft assembly by the Tri-Pot joint and interconnecting shaft, install the Tri-Pot joint into the transaxle side gear as far as possible by hand

8. Align the Tri-Pot joint with the transmission side gears, grasp the driveshaft interconnecting shaft and push the Tri-Pot joint into the transaxle side gear until fully seated. Be sure the snapring is fully engaged with the side gear by trying to remove the Tri-Pot joint from the transaxle by hand. If the snapring is fully seated with the side gear, the Tri-Pot joint will not be removable by hand.

9. Install or connect the following:
   - Driveshaft back into the front hub
   - Steering knuckle into the ball joint stud
   - New steering knuckle-to-ball joint stud bolt and nut. Torque the nut and bolt to 70 ft. lbs. (95 Nm).
   - Washer and hub nut to the stub axle and torque the nut to 180 ft. lbs. (244 Nm)
   - Spring washer, locknut and cotter pin
   - VSS, if equipped
   - Front wheel
   - Negative battery cable

10. Check the transaxle fluid and adjust if needed.

CV-Joints

OVERHAUL

Tri-Pot (Inner) Joint

1. Before servicing the vehicle, refer to the precautions in the beginning of this section.
2. Remove the negative battery cable.
3. Remove the halfshaft.
4. Remove the tri-pot joint boot clamps and slide the boot down the shaft.

**** WARNING**

When removing the spider joint, hold the rollers in place on the trunnions to keep the rollers and needle bearings in place.

5. Slide the interconnecting shaft and spider assembly from the tri-pot housing.
6. Remove the snapring from the shaft.
7. Remove the spider assembly.

➡If necessary, tap the spider assembly from the shaft with a brass drift.

**** WARNING**

When removing the spider assembly, do not hit the outer bearings.

8. Remove the boot by sliding it off the shaft.

**** WARNING**

If any parts show excessive wear, replace the halfshaft assembly; the component parts are not serviceable.

To install:

**** WARNING**

The Tri-pot sealing boots are made of 2 different types of material, silicon rubber (high temperature) which is soft and pliable or hytrel plastic (standard temperature) which is stiff and rigid. Be sure to replace the boot made of the correct material.

9. Install a new small boot clamp and slide it on the shaft.
10. Install the boot and slide it on the shaft.
11. Position the boot so that the raised bead on the inside the boot seal is in the shaft groove.

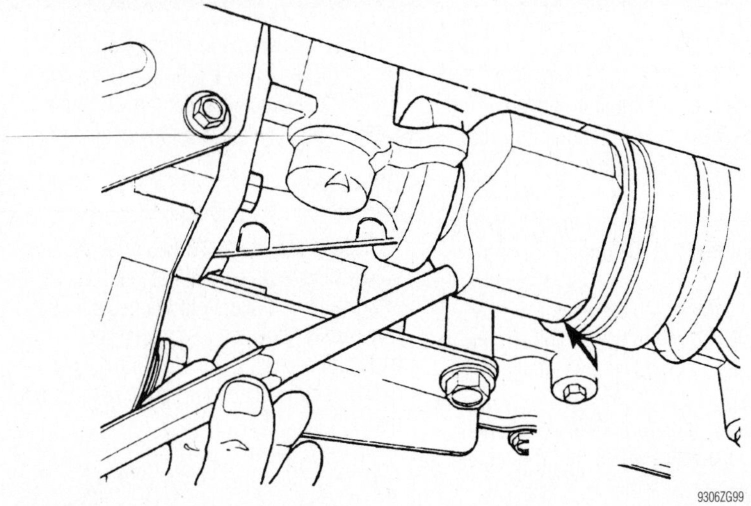

93062G99

Remove the Tri-Pot joint from the transaxle

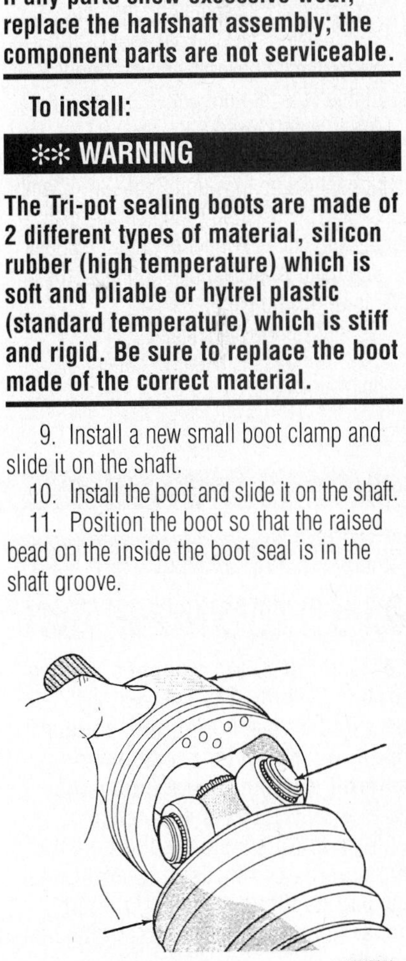

93062G0A

Install the Tri-Pot housing on to the spider assembly

12. Install the spider assembly, face the chamfered side toward the shaft.
13. Install the snapring making sure it is fully seated in the groove.
14. Install the spider/shaft assembly into the tri-pot housing.
15. Install a new inner boot clamp and position it evenly on the sealing boot.
16. Using a trim stick, adjust the boot length to 115mm (hytrel plastic) or 115mm (silicone rubber).
17. If installing a high profile boot clamp, perform the following procedure:
  a. Using the Crimper Tool C-4975-A, place the tool over the clamp bridge, tighten the tool nut until the jaws are completely closed (face-to-face).

✲✲ WARNING

The seal must not be dimpled, stretched or out of shape. If necessary, equalize the seal pressure and shape it by hand.

  b. Position the boot onto the tri-pot housing retaining groove and install the retaining clamp evenly on the boot.
  c. Using the Crimper Tool C-4975-A, place the tool over the clamp bridge, tighten the tool nut until the jaws are completely closed (face-to-face).
18. If installing a low profile latching type boot clamp, position Snap-On® Clamp Locking Tool YA3050 prongs in the clamp holes and squeeze the tool until the upper clamp band is latched behind the 2 tabs on the lower clamp band.
19. Install the halfshaft.
20. Connect the negative battery cable.

Outer Joint

1. Before servicing the vehicle, refer to the precautions in the beginning of this section.
2. Remove the halfshaft.
3. Remove the clamps from the CV-joint boot and discard.
4. Remove the boot from the CV-joint housing and slide it down the interconnecting shaft.
5. Remove the outer CV-joint from the interconnecting shaft by sharply hitting it with a soft-faced hammer to drive it off the shaft.
6. Remove the circlip from the shaft.
7. Remove the CV-joint by sliding it off the shaft.

✲✲ WARNING

If any parts show excessive wear, replace the halfshaft assembly; the component parts are not serviceable.

To install:

8. Install a new small boot clamp and slide it onto the shaft.

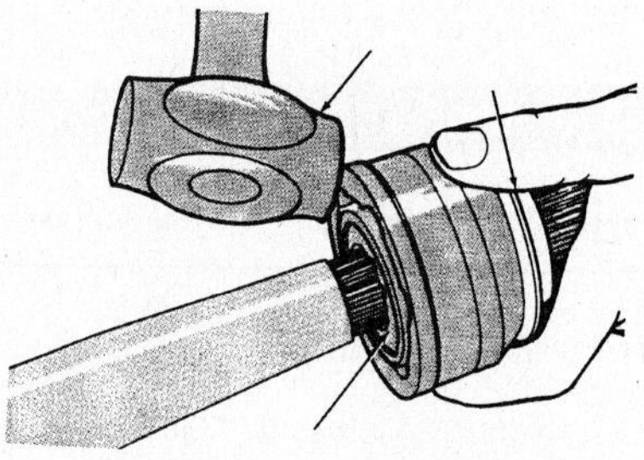
9306ZG0B

Remove the outer C/V joint from the interconnecting shaft

9. Install the boot and slide it onto the shaft.
10. Install the circlip.
11. Position the boot so that the raised bead on the inside the boot seal is in the shaft groove.
12. Install the halfshaft hub nut onto the joint threaded shaft so it is flush with the end.
13. Align the shaft splines and tap it onto the shaft with a soft-faced hammer so it locks on the circlip.
14. Distribute ½ of the grease in the service package inside the joint housing and the other ½ inside the boot.
15. Install a new small boot clamp and position it evenly on the sealing boot.

✲✲ CAUTION

Clamp the boot to the shaft using the Crimper Tool C-4975-A, place the tool over the clamp bridge, tighten the tool nut until the jaws are completely closed (face-to-face).

✲✲ WARNING

The seal must not be dimpled, stretched or out of shape. If necessary, equalize the seal pressure and shape it by hand.

16. Position the boot onto the retaining groove and install the retaining clamp evenly on the boot.
17. Clamp the boot to the outer CV-joint housing using the Crimper Tool C-4975-A, place the tool over the clamp bridge, tighten the tool nut until the jaws are completely closed (face-to-face).
18. Install the halfshaft

STEERING AND SUSPENSION

Air Bag

✲✲ CAUTION

Some vehicles are equipped with an air bag system. The system MUST BE disabled before performing service on or around system components, steering column, instrument panel components, wiring and sensors. Failure to follow safety and disabling procedures could result in accidental air bag deployment, possible personal injury and unnecessary system repairs.

PRECAUTIONS

Several precautions must be observed when handling the inflator module to avoid accidental deployment and possible personal injury:

1. Never carry the inflator module by the wires or connector on the underside of the module.
2. When carrying a live inflator module, hold securely with both hands, and ensure that the bag and trim cover are pointed away.
3. Place the inflator module on a bench or other surface with the bag and trim cover facing up.
4. With the inflator module on the bench, never place anything on or close to the module which may be thrown in the the event of an accidental deployment.

DISARMING

Proper Supplemental Restraint System (SRS) disarming can be obtained by disconnecting and isolating the negative battery cable. Allow the air bag system capacitor at least 2 minutes to discharge before removing any air bag system components.

Rack and Pinion Steering Gear

REMOVAL & INSTALLATION

1. Before servicing the vehicle, refer to the precautions in the beginning of this section.
2. Place the steering wheel in the straight-ahead position. Lock the steering wheel in place, using a steering wheel holder.

➡Locking the steering wheel keeps the clockspring in alignment position.

3. Remove or disconnect the following:
   - Silencer pad from below the knee blocker panel
   - Knee blocker
   - Steering column coupling retainer pin, pinch bolt nut/bolt and separate the couplings
   - Both front wheels
   - Outer tie rod-to-steering knuckle nuts
   - Outer tie rod ends from the steering knuckle using remover Tool MB991113
   - Tie rod heat shield
   - Power steering fluid pressure switch wiring connector by releasing the locking tab
   - Power steering fluid pressure hose from the steering gear
   - Power steering fluid return hose from the steering gear, if not equipped with a power steering fluid cooler
   - Power steering fluid cooler hose from the steering gear, if equipped with a power steering fluid cooler
   - Power steering fluid return hose from the routing clip C-clamps, if not equipped with a power steering fluid cooler
   - Power steering fluid pressure hose from the steering gear's routing clips
   - Power steering cooler hose from the steering gear's right routing clip, if equipped
   - Both power steering cooler screws from the front suspension crossmember, if equipped, and move the cooler aside
   - Drive belt splash shield
   - Engine torque strut-to-front suspension crossmember bolt from the right forward corner of the crossmember
4. Matchmark the front suspension crossmember-to-chassis location.

✻✻ WARNING

If the front suspension crossmember-to-chassis location is not matchmarked, the front wheel alignment setting will be lost.

5. Place a transmission jack under the front crossmember and support it.
6. Remove the front suspension crossmember-to-frame rail bolts, one located at each side.
7. Loosen both rear suspension crossmember-to-frame rail bolts, one located at each side, until they release from the threaded tapping plates in the bolt.

✻✻ WARNING

Do not completely remove the rear bolts for they are designed to disengage from the body threads and will stay within the lower control arm rear isolator bushing.

➡The threaded tapping plates allow the lower control arm to stay in place on the crossmember.

8. Using the transmission jack, lower the front suspension crossmember enough to allow the power steering gear to be removed form the rear of the crossmember. Use the jack to support the crossmember's weight.
9. Remove or disconnect the following:
   - Lower steering column coupling-to-power steering gear pinion shaft's roll pin, using a roll pin punch
   - Lower steering column coupling from the power steering column pinion shaft
   - Pinion shaft dash cover seal from the tabs cast into the power steering gear housing
   - Power steering gear from the front suspension crossmember

To install:

10. Install or connect the following:
   - Power steering gear onto the front suspension crossmember and torque the bolts to 45 ft. lbs. (61 Nm)
   - Pinion shaft dash cover seal over the shaft and onto the power steering gear housing. Align the seal holes with the tabs cast into the power steering gear housing.
   - Lower steering column coupling by aligning the coupling and steering gear pinion shaft flats
   - Lower steering column coupling-to-pinion shaft's roll pin until it is centered. Center the power steering gear rack's travel.
   - Front suspension crossmember/power steering gear assembly by raising it with the jack until is aligns with its matchmarks
   - Lower steering column coupling, guide it through the dash panel hole as it is raised

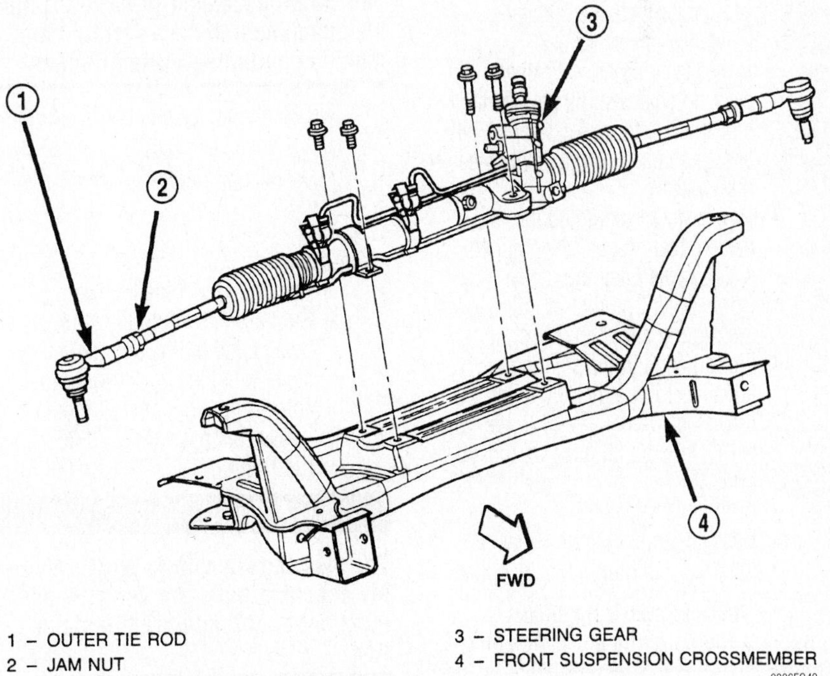

View of the power steering gear and crossmember

- Both rear crossmember-to-tapping plate bolts
- Both front crossmember-to-frame rail bolts and torque the 4 bolts to 20 inch lbs. (2 Nm)

⁂ WARNING

Be sure to align the front suspension crossmember-to-chassis matchmarks; otherwise, the front wheel alignment setting will be lost.

11. Once aligned, torque both rear crossmember-to-rear lower control arm bolts to 185 ft. lbs. (250 Nm) and both front crossmember bolts to 113 ft. lbs. (153 Nm).
12. Install the engine torque strut to the right forward corner of the front suspension crossmember.
13. Adjust the engine torque strut by performing the following procedure:
    a. Loosen the upper torque strut at the shock tower bracket.
    b. Position a floor jack on the forward edge of the bell housing to prevent the least amount of upward lifting of the engine.
    c. Slowly, lift the assembly, allowing the engine to rotate rearward so the distance between center of the engine mount bracket's rearmost attaching the and the center of the shock tower bracket's washer hose clip hole is 4.70 in. (119mm).
    d. Torque the upper and lower torque strut bolts to 87 ft. lbs. (118 Nm).
    e. Remove the floor jack.
14. Install or connect the following:
    - Drive belt splash shield
    - Power steering hose-to-power steering gear using a new O-ring lubricated with power steering oil, if not equipped with a power steering cooler
    - Power steering fluid cooler line-to-power steering gear, if equipped with a power steering cooler
    - Power steering fluid return hose to the routing clip C-clamps
    - Tie rod ends to the steering knuckle. Torque the nut, using a crowsfoot wrench, to 40 ft. lbs. (55 Nm), while holding the tie rod stationary.
    - Both front wheels
    - Dash-to-lower coupling seal over the lower coupling's plastic collar

➡Verify that the seal's lip shows grease at the coupling's plastic collar contact.

- Steering column lower coupling-to-steering column upper coupling pinch bolt and torque the nut to 21 ft. lbs. (28 Nm)
- Pinch bolt retainer pin
- Knee blocker and silencer pad

15. Remove the steering wheel holder.
16. Fill and bleed the power steering system.
17. Start the vehicle and check for leaks, repair if necessary.
18. Check and/or adjust the front toe setting.

Strut

REMOVAL & INSTALLATION

Front

1. Before servicing the vehicle, refer to the precautions in the beginning of this section.
2. Install or connect the following:
    - Negative battery cable
    - Front wheels
    - Mark each one right or left, as applicable, if both struts are being removed
    - Ground strap from the rear of the strut
    - Anti-lock Brake System (ABS) wheel speed sensor from the strut, if equipped

⁂ WARNING

The steering knuckle-to-strut assembly attaching bolts are serrated and must not be turned during removal.

- Steering knuckle nuts while holding the bolts stationary
- 3 upper strut mount-to-strut tower nuts
- Strut assembly

To install:

3. Install or connect the following:
    - Strut assembly into the strut tower by aligning the 3 upper strut mount studs with the shock tower holes. Torque the 3 upper strut mount nut/washer assemblies to 25 ft. lbs. (34 Nm).

⁂ WARNING

The steering knuckle-to-strut assembly attaching bolts are serrated and must not be turned during installation.

- Steering knuckle nuts while holding the bolts stationary
- Steering knuckle arm and position it into the strut assembly by aligning the strut assembly-to-steering knuckle holes
- Both strut-to-steering knuckle bolts. On 2002–04 models, torque both bolts to 40 ft. lbs. (53 Nm), plus an additional 90 degrees after the specified torque is met. On 2005–06 models, torque both bolts to 120 ft. lbs. (163 Nm).

➡The bolts should be installed with the nuts facing the front of the vehicle.

- ABS wheel sensor to the rear of the strut and torque the screw to 120 inch lbs. (13 Nm)
- Ground strap to the rear of the strut and torque the screw to 120 inch lbs. (13 Nm)
- Front wheel
- Negative battery cable

Shock Absorber

REMOVAL & INSTALLATION

Rear

1. Before servicing the vehicle, refer to the precautions in the beginning of this section.
2. Remove the rear wheel and position a transmission jack under the center of the axle. Raise the jack enough to support the axle.
3. Remove or disconnect the following:
    - Shock absorber lower mounting bolt from the axle
    - Upper mounting bolt
    - Shock absorber

To install:

4. Install or connect the following:
    - Shock absorber eye to the body bracket. Hand tighten the upper mounting bolt.
    - Lower the jack and install the lower mounting bolt through the axle flange and shock absorber. Torque the bolt to 65 ft. lbs. (88 Nm) on 2002–04 models or 65 ft. lbs. (88 Nm) on 2005–06 models. Torque the upper mounting bolt to 73 ft. lbs. (99 Nm).
    - Rear wheel
    - Negative battery cable

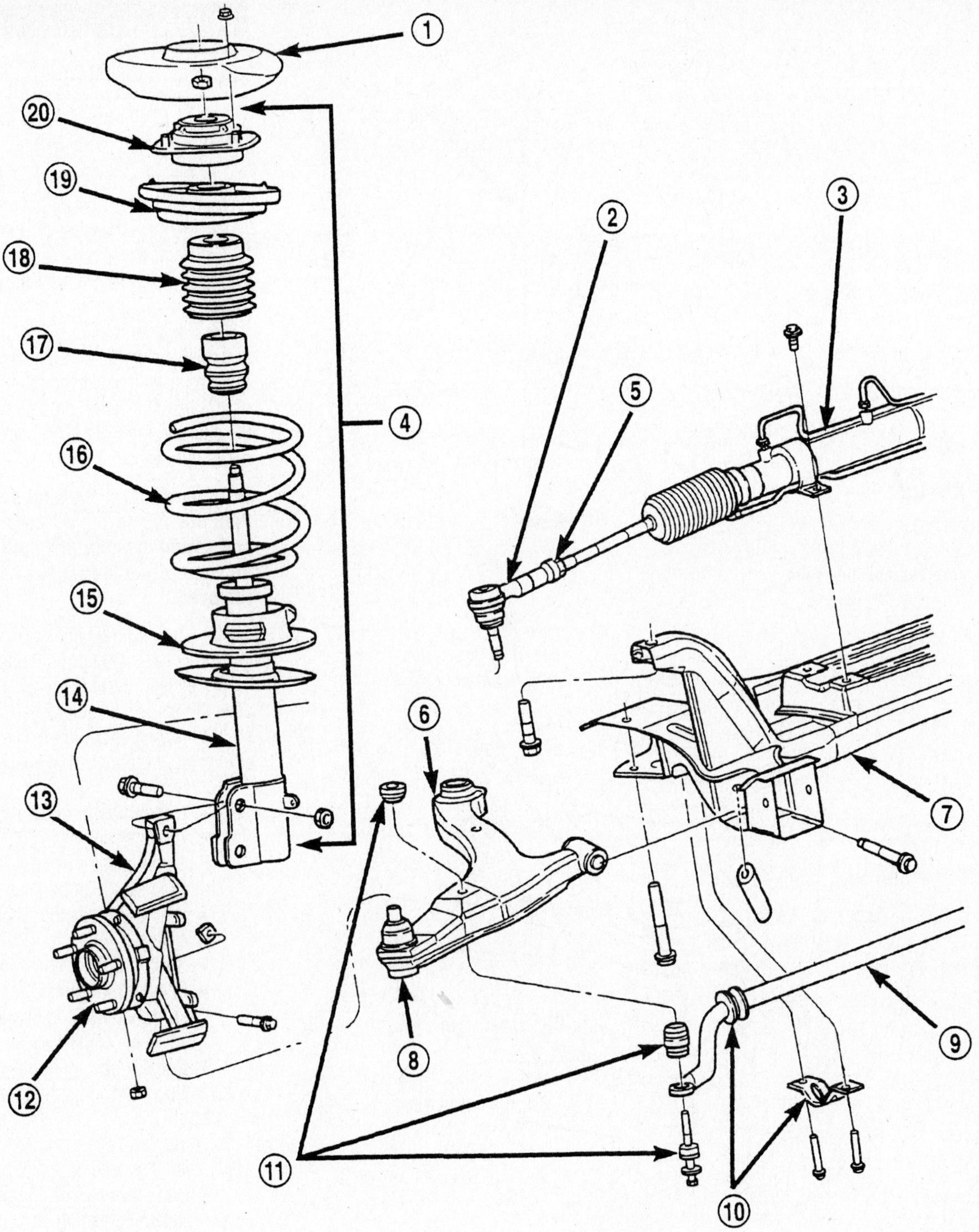

1 – VEHICLE STRUT TOWER
2 – OUTER TIE ROD
3 – STEERING GEAR
4 – STRUT ASSEMBLY
5 – JAM NUT
6 – LOWER CONTROL ARM
7 – CROSSMEMBER
8 – BALL JOINT
9 – STABILIZER BAR
10 – STABILIZER BAR CUSHION AND RETAINER
11 – STABILIZER BAR LINK
12 – HUB
13 – KNUCKLE
14 – STRUT
15 – LOWER SPRING ISOLATOR
16 – COIL SPRING
17 – JOUNCE BUMPER
18 – DUST SHIELD
19 – SPRING SEAT AND BEARING
20 – UPPER MOUNT

Exploded view of the front suspension

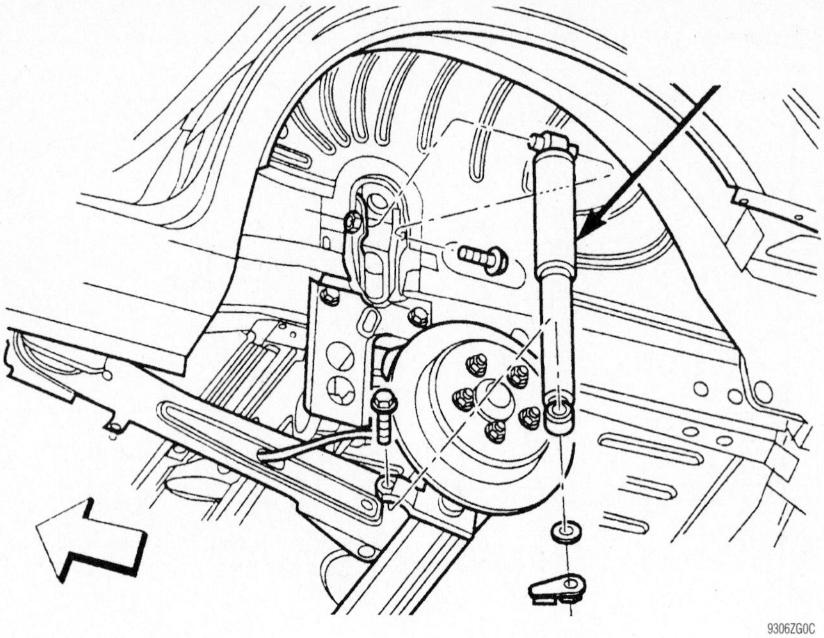

Shock absorber mounting bolts

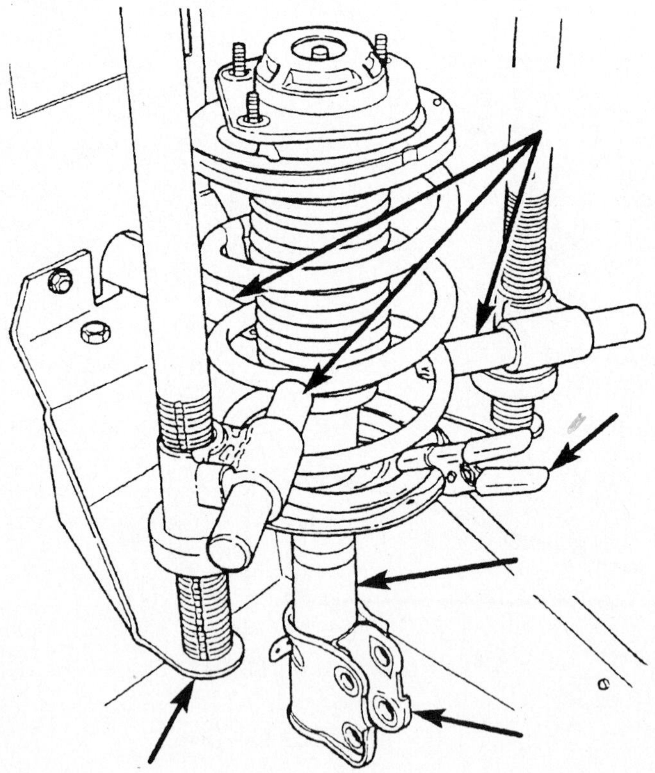

Coil spring mounted in the coil spring compressor tool

Coil Spring

REMOVAL & INSTALLATION

Front

1. Before servicing the vehicle, refer to the precautions in the beginning of this section.
2. Remove the wheel.
3. Remove the strut assembly and place it in a Strut Spring Compressor Tool W-7200.
4. Set the lower hooks then the upper hooks. Position the clevis bracket straight outward away from the compressor tool
5. Install a clamp on the lower end of the coil spring to secure the strut when the nut is removed.

➡Do not remove the strut shaft nut until the coil spring is compressed. The coil spring is under pressure and must be compressed before the shaft nut is removed.

6. Compress the coil spring until all tension is removed from the upper mount.
7. Install a Strut Nut Socket Tool 6864 once the spring is compressed.
8. Install a socket on the hex end of the strut shaft and remove the nut.
9. Remove or disconnect the following:
   - Upper mount from the strut shaft
   - Upper spring seat, bearing and upper isolator as an assembly
   - Dust shield and jounce bumper
   - Clamp from the bottom of the coil spring
   - Strut through the bottom of the coil
   - Release the tension from the coil spring by backing off the compressor drive completely
   - Coil spring
10. Inspect the coil spring for any signs of damage

To install:

11. Install or connect the following:
    - Coil spring in the compressor. Rotate the spring so that the end of the top coil is directly in the front.
    - Slowly compress the coil until enough room is available to install the strut
    - Lower spring isolator on the lower spring seat
    - Strut through the bottom of the of coil spring until the lower spring seat contacts the lower end of the coil spring. Rotate the strut until the clevis bracket is positioned straight outward away from the compressor.

- Clamp on the lower end of the coil spring and strut
- Jounce bumper on the strut shaft with the smaller end pointing downward
- Dust shield until the bottom of the shield snaps on to the retainer
- Upper spring isolator
- Upper spring seat and bearing on top of the coil spring. Position the notch formed into the edge of the upper seat straight out away from the compressor.
- Strut upper mount over the strut shaft and onto the top of the upper spring seat and bearing. Position the mount so that the third mounting stud is inward toward the compressor.
- Retaining nut on the strut shaft, loosely

12. Install the strut nut socket on the strut shaft retaining nut. Install a socket on the hex end of the shaft. Secure the strut shaft and torque the nut to 55 ft. lbs. (75 Nm).

13. Slowly release the tension from the coil spring by backing off the compressor completely.

14. Remove the clamp from the bottom of the coil spring and strut. Push back the spring compressor upper and lower hooks and remove the strut from the compressor.

15. Install the strut assembly to the vehicle.

Rear

1. Before servicing the vehicle, refer to the precautions in the beginning of this section.
2. Remove or disconnect the following:
   - Both rear wheels
   - Watts link bell crank from the center of the axle
   - Sway bar cushion retainers
   - Sway bar from the rear axle and place a jack under the rear axle
   - Shock absorber and lower the jack
   - Coil springs and rubber isolators

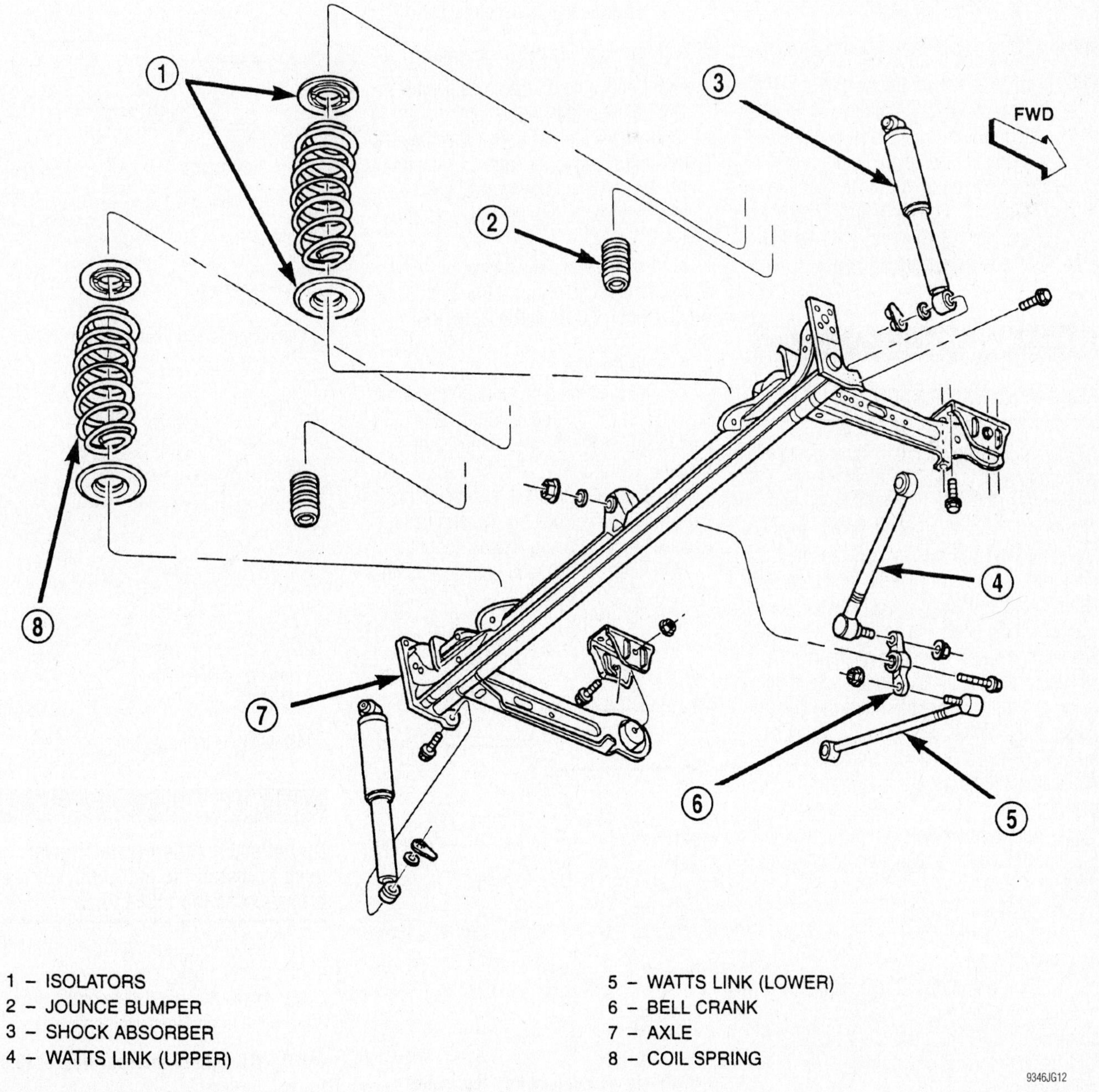

Exploded view of the rear suspension

To install:

3. Install or connect the following:
   - Rubber isolator on each end of the coil spring and wrap the finger around the coil
   - Coil springs on top of the axle spring perches and make certain that the upper coils end near the outboard sides of the vehicle and not at 180 degrees of that location
   - Coil springs into the spring mounting brackets
   - Washer and nut on the lower mounting bolts and torque the bolts to 65 ft. lbs. (88 Nm) on 2002–04 models or 65 ft. lbs. (88 Nm) on 2005–06 models.
   - Lower end of the sway bar retainers in the slots at the back of the axle
   - Mounting bolt through the cushion retainer and torque to 40 ft. lbs. (54 Nm) on 2002–04 models or 45 ft. lbs. (61 Nm) on 2005–06 models
   - Watts link bell crank to the center of the axle and torque the bolts to 90 ft. lbs. (122 Nm) on 2002–04 models or 110 ft. lbs. (149 Nm) on 2005–06 models
   - Rear wheels

Lower Ball Joint

REMOVAL & INSTALLATION

The front suspension ball joints operate with no free-play. The ball joints are replaceable ONLY as an assembly. Do not attempt any type of repair on the ball joint assembly. The ball joint is a press fit into the lower control arm with the joint stud retained in the steering knuckle by the clamp bolt. To check the ball joint, with the weight of the vehicle resting on the road wheels, grasp the grease fitting and without using any tools, attempt to move the grease fitting. If the ball joint is worn the grease fitting will move easily. If movement is noted, replacement of the ball joint is recommended.

1. Before servicing the vehicle, refer to the precautions in the beginning of this section.
2. Remove the wheel.
3. Remove the stabilizer bar-to-lower control arm links.
4. Loosen, but do not remove the bolts holding the stabilizer bar retainers to the crossmember. Then, rotate the stabilizer bar and attaching links away from the lower control arms.

**** WARNING**

Pulling the steering knuckle outward after releasing the ball joint can separate the inner CV-joint.

5. Remove the steering knuckle-to-ball joint stud's pinch bolt and nut.
6. Remove the ball joint from the steering knuckle using a prybar.

**** WARNING**

Be careful when separating the ball joint stud from the knuckle, so the seal does not become damaged.

7. If removing the right lower control arm, perform the following steps:
   a. Remove the drive belt splash shield.
   b. Remove the pencil strut from the right front corner of the crossmember
   c. Remove the engine torque strut.
8. Remove the pivot bolts attaching the lower control arm to the front crossmember.
9. Remove the lower control arm.
10. Remove the ball joint using a pry tool.
11. Using a hydraulic press, press the ball joint from the lower control arm using Receiver tool 6908-2 and Adapter Tool 6804.

To install:

12. Reinstall the ball joint into the lower control arm with the notch in the ball joint stud facing the front lower control arm bushing.
13. Using a hydraulic press, press the ball joint into the lower control arm using Receiver Tool 6758 and Adapter tool 6804.
14. Install the ball joint boot seal using a driver tool such as a large socket or suitable sized piece of pipe.

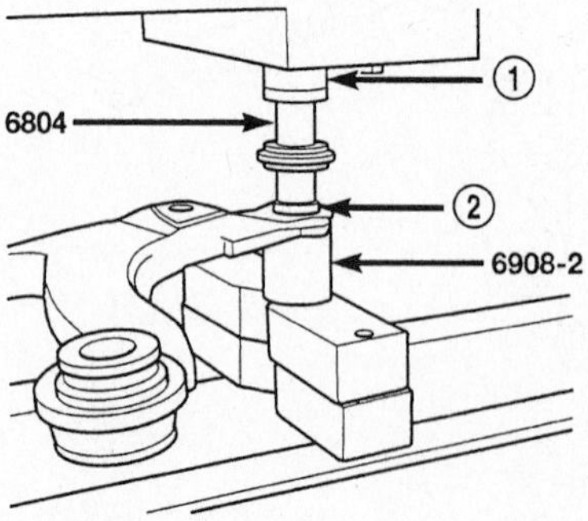

1 - PRESS
2 - BALL JOINT

9306EG51

Removing the ball joint from the control arm

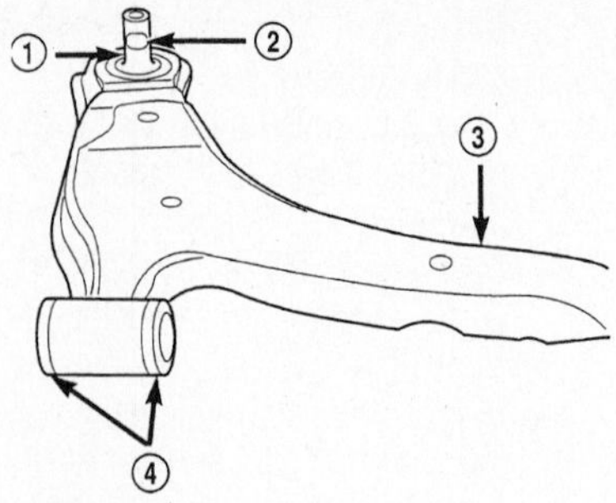

1 - BALL JOINT STUD
2 - NOTCH
3 - LOWER CONTROL ARM
4 - FRONT ISOLATOR BUSHING

9306EG52

Aligning the ball joint stud notch to the control arm

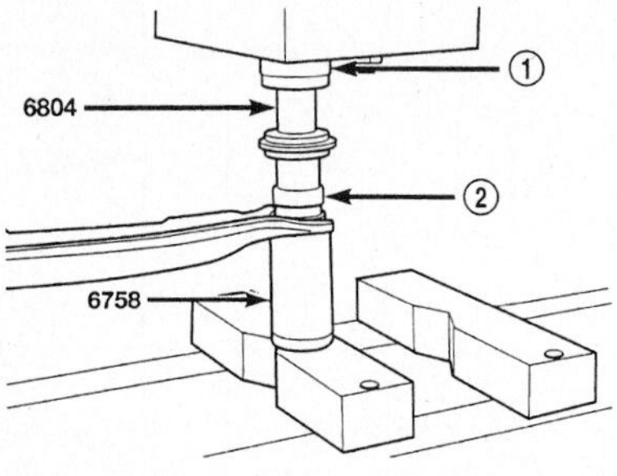

1 - PRESS
2 - BALL JOINT

9306EG53

Installing the ball joint to the control arm

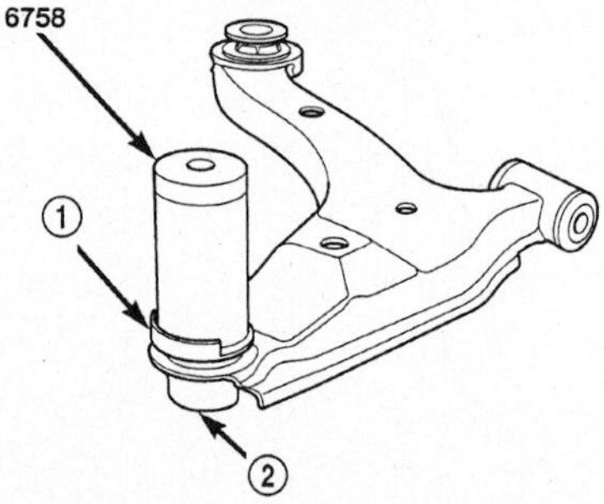

1 - SEAL BOOT UPWARD LIP
2 - BALL JOINT

9306EG54

Installing the ball joint boot seal

**** WARNING**

Do not use a shop press that was used to install the ball joint, for the press exerts too much force.

15. Install the lower control arm into the front crossmember.
16. Install the rear lower control arm-to-crossmember and frame rail bolt.

➡DO NOT tighten the rear bolt at this time.

17. Install the front lower control arm-to-crossmember nut and bolt.
18. Torque the lower control arm to rear pivot bolt to 185 ft. lbs. (250 Nm) and the front pivot bolt 120 ft. lbs. (163 Nm).
19. Install the ball joint stud into the steering knuckle. Torque the steering knuckle-to-ball joint stud pinch bolt/nut to 70 ft. lbs. (95 Nm).
20. If the right side lower control arm has been service, install the following:
21. Install the engine torque strut.
22. Install the pencil strut to the right front corner of the crossmember and torque the nuts to 43 ft. lbs. (58 Nm).
23. Install the drive belt splash shield.
24. Install the front fascia-to-reinforcement screws.
25. Install the stabilizer bar-to-lower control arm link assemblies and bushings.
26. Rotate the stabilizer bar into position, installing the stabilizer bar links into the lower control arms.
27. Install the top stabilizer bar link bushings and nuts. DO NOT tighten the link yet.
28. Install the wheel.
29. Lower the vehicle so the suspension is supporting the total weight of the vehicle.
30. Torque the stabilizer bar-to-lower control arm links to 21 ft. lbs. (28 Nm).
31. Torque the stabilizer bar bushing retainer-to-crossmember bolts to 21 ft. lbs. (28 Nm).
32. Check and/or adjust the toe, as necessary.

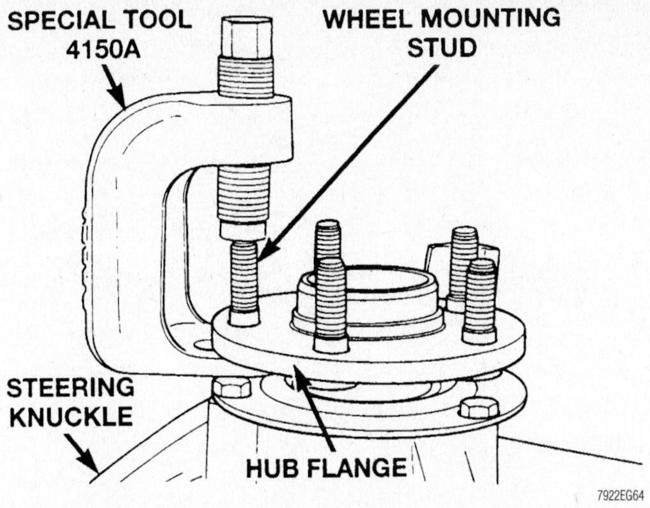

Use a proper C-clamp and adapter tool to press out one of the lug studs

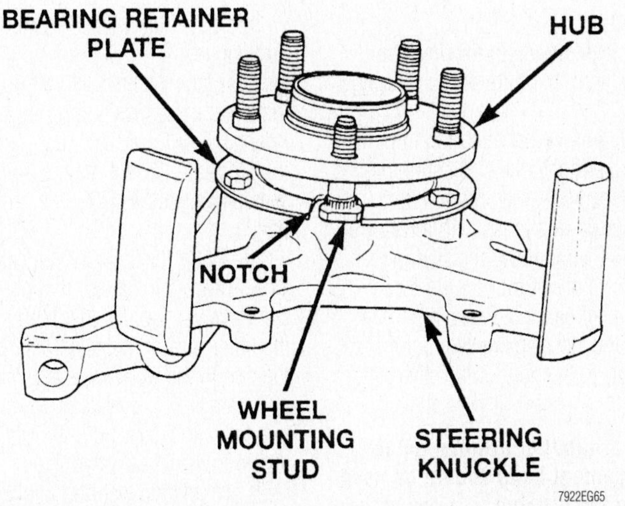

Rotate the hub in order to remove the lug stud

Wheel Bearings

ADJUSTMENT

The PT Cruiser is equipped with a sealed hub and bearing assemblies. The hub and bearing assembly is non-serviceable. If the assembly is damaged, the complete unit must be replaced.

REMOVAL & INSTALLATION

Front

1. Before servicing the vehicle, refer to the precautions in the beginning of this section.
2. Remove the steering knuckle and hub and bearing assembly.
3. Remove the wheel lug stud from the hub flange, using a C-clamp and Adapter Tool 4150A.
4. Rotate the hub to align the removed lug stud with the notch in the bearing retainer plate.
5. Rotate the hub so the stud hole is facing away from the brake caliper's lower rail on the steering knuckle.
6. Install ½ of a Bearing Splitter Tool 1130, between the hub and the bearing retainer plate. The threaded hole in this ½ is to be aligned with the caliper rail on the steering knuckle.
7. Install the remaining pieces of the bearing splitter on the steering knuckle. Hand-tighten the nuts to hold the splitter in place on the knuckle.
8. When the bearing splitter is installed, be sure the 3 bolts attaching the bearing retainer plate to the knuckle are contacting the bearing splitter. The bearing retainer plate should not support the knuckle or contact the splitter.
9. Place the steering knuckle in a hydraulic press, supported by the bearing splitter.
10. Position a driver on the small end of the hub. Using the press, remove the hub from the wheel bearing. The outer bearing race will come out of the wheel bearing when the hub is pressed out of the bearing.
11. Remove the bearing splitter tool from the knuckle.
12. Remove the 3 bolts mounting the bearing retainer plate to the steering knuckle.
13. Place the knuckle in a press supported by the press block. The blocks must not obstruct the bore in the steering knuckle so the wheel bearing can be pressed out of the knuckle. Place a driver on the outer race of the wheel bearing, then press the bearing out of the knuckle.
14. Install the bearing splitter on the hub. The splitter is to be installed on the hub so it is between the flange of the hub and the bearing race on the hub. Place the hub, bearing race and splitter in a press. Use a driver to press the hub out of the bearing race.

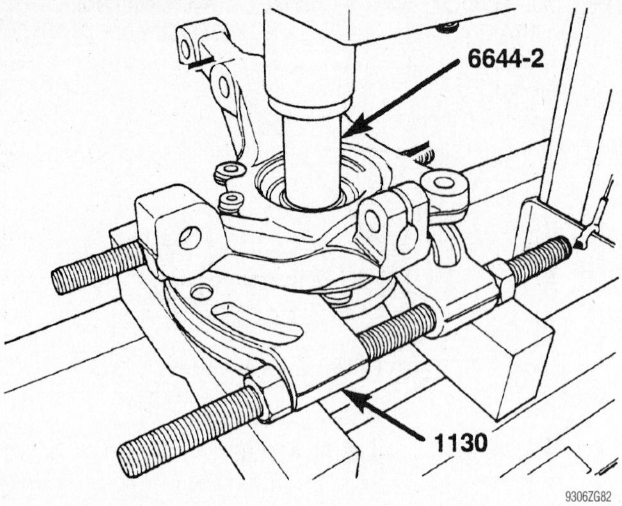

Properly support the steering knuckle for hub and bearing removal

To install:

15. Use clean, dry cloth to wipe and grease or dirt from the bore of the steering knuckle.

16. Install a new wheel bearing into the bore of the steering knuckle. Be sure the bearing is placed squarely into the bore. Place the knuckle in a press with a Receiver Tool, C-4698-2 supporting the steering knuckle. Place a driver tool on the outer race of the wheel bearing. Press the wheel bearing into the steering knuckle until it is fully bottomed in the bore of the steering knuckle.

➡Only the original or original equipment replacement bolts should be used to mounting the bearing retainer to the knuckle. If a bolt requires replacement when installing the bearing retainer plate, be sure to get the proper type of replacement.

17. Install the bearing retainer plate on the steering knuckle. Install the 3 bearing retainer mounting bolts. Tighten the bolts to 21 ft. lbs. (28 Nm).

18. Install the wheel lug stud into the hub flange.

19. Place the hub with the lug stud installed, in a press supported by an Adapter Tool C-4698-1 and press the wheel lug stud into the hub flange until it is fully seated against the back side on the hub flange.

20. Place the steering knuckle with the wheel bearing installed, in a press with special Receiver Tool MB-990799 supporting the inner race of the wheel bearing. Place the hub in the wheel bearing, making sure it is square with the bearing. Press the hub into the wheel bearing until it is fully bottomed in the wheel bearing.

21. Install the steering knuckle and the wheel.

22. Check and/or adjust the front alignment.

Rear

1. Before servicing the vehicle, refer to the precautions in the beginning of this section.

2. Remove or disconnect the following:
   - Wheel
   - Rear brake drum, if equipped
   - Caliper (suspend on a wire) and the rotor, if equipped with rear disc brakes

✻✻ WARNING

DO NOT allow the caliper to hang by the brake hose.

   - Dust cap from the rear hub/bearing
   - Hub/bearing assembly-to-knuckle/spindle nut

➡Discard the hub nut and replace with a new one during installation.

   - Hub/bearing from the spindle by pulling it off the end of the spindle by hand

To install:

✻✻ WARNING

The hub/bearing nut must be tightened to, but NOT over, its specified torque value. The proper specification is crucial to the life of the hub bearing.

3. Position the hub/bearing assembly on the rear spindle/knuckle. Install a NEW hub nut and tighten to 160 ft. lbs. (217 Nm)

4. Install or connect the following:
   - Dust cap and seat it using a soft face hammer to carefully tap it into place
   - Brake drum, if equipped
   - Rotor, if equipped
   - Caliper and 2 guide pin bolts, if equipped with disc brakes. Torque the bolts to 16 ft. lbs. (22 Nm).

BRAKES

Caliper

REMOVAL & INSTALLATION

Front

1. Before servicing the vehicle, refer to the precautions in the beginning of this section.

2. Isolate the master cylinder as follows:
   a. Use a brake pedal holding tool and depress the brake pedal past its first one inch of travel and hold it in this position. (This will keep brake fluid from draining from the master cylinder).

3. Remove or disconnect the following:
   - Front wheels
   - 2 caliper guide pin bolts
   - Brake hose from the caliper
   - Caliper from the steering spindle

To install:

4. Install or connect the following:
   - Brake hose to the caliper and torque to 210 inch lbs. (24 Nm)
   - Caliper to the steering spindle
   - Caliper guide pin bolts and torque to 26 ft. lbs. (35 Nm)

5. Properly bleed the brake system.
   - Front wheels and lower the vehicle.

6. Pump the brake pedal to seat the front brake pads before moving the vehicle.

7. Road test the vehicle and check for proper operation.

Rear

1. Before servicing the vehicle, refer to the precautions in the beginning of this section.

2. Isolate the master cylinder as follows:
a. Use a brake pedal holding tool and depress the brake pedal past its first one inch of travel and hold it in this position. (This will keep brake fluid from draining from the master cylinder).
3. Remove or disconnect the following:
- Front wheels
- 2 caliper guide pin bolts
- Brake hose from the caliper
- Caliper from the steering spindle

To install:

4. Install or connect the following:
- Brake hose to the caliper and torque to 210 inch lbs. (24 Nm)
- Caliper to the steering spindle
- Caliper guide pin bolts and torque to 192 inch lbs. (22 Nm)
5. Properly bleed the brake system.
- Front wheels and lower the vehicle.
6. Pump the brake pedal to seat the front brake pads before moving the vehicle.
7. Road test the vehicle and check for proper operation.

Disc Brake Pads

REMOVAL & INSTALLATION

1. Before servicing the vehicle, refer to the precautions in the beginning of this section.
2. Remove or disconnect the following:
- Front wheels
- 2 caliper to steering knuckle guide pin bolts
- Caliper away from the steering knuckle by first rotating the free end of the caliper away from the steering knuckle. Then, slide the opposite end of the caliper out from under the machined end of the steering knuckle.
3. Support the caliper from the upper control arm to prevent the weight of the caliper from being supported by the brake flex hose that will damage the hose.
- Brake pads from the caliper
- Outboard brake pad by prying the pad retaining clip over the raised area on the caliper. Then, slide the pad down and off the caliper
- Inboard brake pad by pulling it away from the piston until the retaining clip is free from the cavity in the piston.
- Rotor, if necessary, by pulling it straight off the wheel mounting studs

To install:

4. Clean all parts well. Inspect the caliper for piston seal leaks (brake fluid in and around the boot area and inboard lining) and for any ruptures of the piston dust boot. If the boot is damaged or fluid leak is visible, disassemble the caliper and install a new seal and boot (and piston, if scored).
5. Inspect the caliper pin bushings. Replace if damaged, dry or brittle.
6. Completely compress the piston into the caliper using a large C-clamp or other suitable tool.
7. Lubricate the area on the steering knuckle where the caliper slides with high temperature grease.
8. Install or connect the following:
- Rotor if removed
- Brake pads into the caliper. Note that the inboard and outboard pads are different. Make sure the inboard brake shoe assembly is positioned squarely against the face of the caliper piston.

➡Be sure to remove the noise suppression gasket paper cover if the pads come so equipped.

- Caliper and brake shoe assemblies over the rotor by hooking the lower end of the caliper over the steering knuckle. Then, rotate the caliper into position at the top of the steering knuckle. Make sure the caliper guide pin bolts, bushings and sleeves are clear of the steering knuckle bosses.
- Caliper guide pin bolts and torque to 26 ft. lbs. (35 Nm) on the front caliper or 192 inch lbs. (22 Nm) on the rear caliper
- Wheels and torque the mounting bolts to 100 ft. lbs. (135 Nm)
9. Pump the brake pedal until the brake pads are seated and a firm pedal is achieved before attempting to move the vehicle.
10. Road test the vehicle for proper operation.

Brake Drums

REMOVAL & INSTALLATION

1. Before servicing the vehicle, refer to the precautions in the beginning of this section.
2. Remove or disconnect the following:
- Rear wheels
- Rubber plug from the top of the brake support plate (backing plate)
3. Insert a small prying tool through the adjuster access hole and engage the teeth on the adjuster wheel.
4. Rotate the adjuster wheel so it is moved toward the front of the vehicle. This will back off the adjustment of the rear brake shoes. Continue moving the adjuster wheel toward the front of the vehicle until it stops moving.
- Rear brake drum from the hub assembly

To install:

5. Inspect the brake drums for cracks or signs of overheating. Measure the drum runout and diameter. If not to specification, resurface the drum. Runout should not exceed 0.006 inch (0.152mm). The diameter variation (oval shape) of the drum braking surface must not exceed either 0.0025 inch (0.0635mm) in 30° or 0.0035 inch (0.089mm) in 360°. All brake drums are marked with the maximum allowable brake drum diameter on the face of the drum.
6. Install or connect the following:

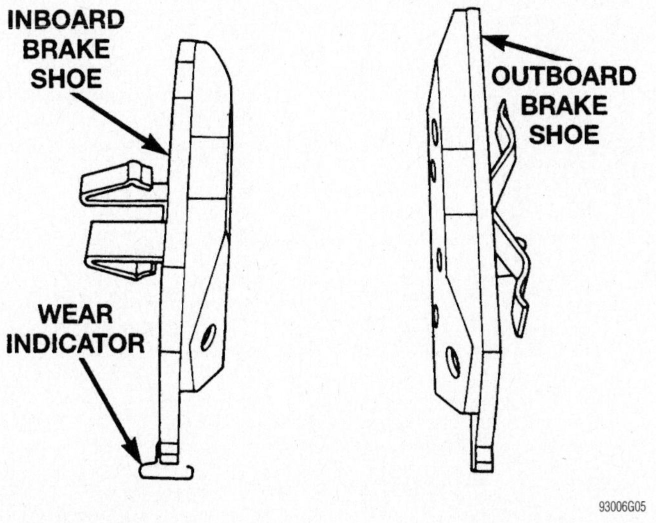

Disc brake pad identification

- Rear brake drum onto the hub assembly
- Wheels and properly adjust the brakes

7. Road test vehicle to check brake operation.

Brake Shoes

REMOVAL & INSTALLATION

1. Before servicing the vehicle, refer to the precautions in the beginning of this section.
2. Remove or disconnect the following:
   - Rear wheels
   - Drums
   - Automatic adjuster spring and lever
   - Hold-down clips and pins
3. Rotate the automatic adjuster starwheel enough so both shoes move out far enough to be free of the wheel cylinder boots.
   - Parking brake cable from the actuating lever
   - Lower shoe-to-shoe spring
   - Shoes from the backing plate, holding them together by the upper shoe-to-shoe spring

To install:

4. Thoroughly clean and dry the backing plate. To prepare the backing plate, lubricate the bosses, anchor pin and parking brake actuating lever pivot surface lightly with lithium based grease.
5. Remove, clean and dry all brake components. Lubricate the starwheel shaft threads with anti-seize lubricant and transfer all parts to their proper locations on the new shoes.
6. Install or connect the following:
   - Lower spring
   - Parking brake cable
   - Automatic adjuster lever and spring
7. Adjust the starwheel.
8. Remove any grease from the linings and install the drum.
   - Wheels and properly adjust the brakes
9. Check for proper brake system operation.

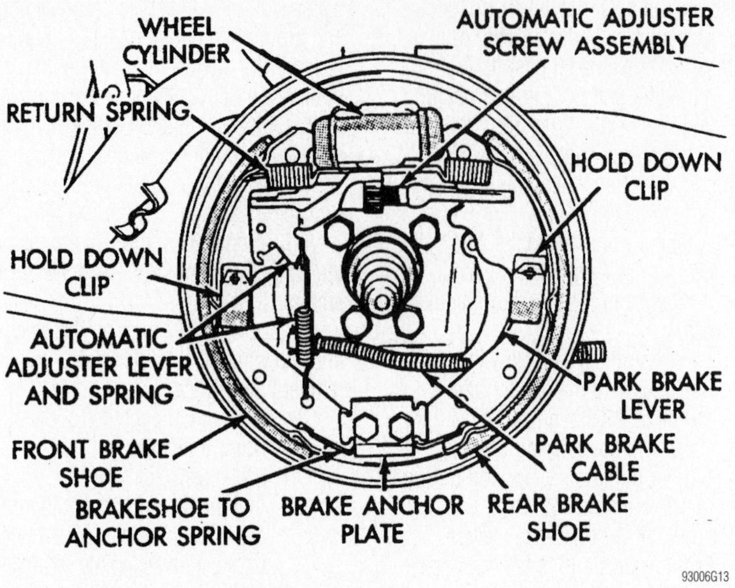

Kelsey Hayes rear brake assembly (left side shown)

DODGE

11

Ram Trucks • Ram Vans

SPECIFICATIONS AND MAINTENANCE CHARTS

ENGINE AND VEHICLE IDENTIFICATION

Engine						
Code ①	Liters (cc)	Cu. In.	Cyl.	Fuel Sys.	Engine Type	Eng. Mfg.
5	5.9 (5899)	360	8	SMFI	OHV	Chrysler
6	5.9 (5882)	359	6	DSL-24V Turbo	OHV	Cummins
7	5.9 (5882)	359	6	DSL-24V Turbo	OHV	Cummins
C	5.9 (5882)	359	6	DSL-24V Turbo	OHV	Cummins
D	5.7 (5653)	345	8	SMPI	OHV	Chrysler
K	3.7 (3701)	226	6	MPI	OHV	Chrysler
N	4.7 (4701)	287	8	MPI	OHV	Chrysler
P	4.7 (4701)	287	8	Flex	OHV	Chrysler
W	8.0 (7994)	488	10	SMFI	OHV	Chrysler
X	3.9 (3916)	238	6	SMFI	OHV	Chrysler
Y	5.2 (5208)	318	8	SMFI	OHV	Chrysler
Z	5.9 (5899)	360	8	SMFI	OHV	Chrysler

Model Year	
Code ②	Year
2	2002
3	2003
4	2004
5	2005
6	2006

OHV: Overhead Valve

DSL-24V: Diesel with 24-valve cylinder head

SMFI: Sequential Multi-port Fuel Injection

① 8th position of VIN

② 10th position of VIN

06009-RAMT-C01

GENERAL ENGINE SPECIFICATIONS

Year	Model	Engine Displacement Liters	Engine Series VIN	Net Horsepower @ rpm	Net Torque @ rpm (ft. lbs.)	Bore x Stroke (in.)	Com-pression Ratio	Oil Pressure @ rpm
2002	Ram Truck 1500	3.9	X	175@4800	220@3200	3.91x3.31	9.1:1	30-80@3000
		5.2	Y	220@4400	300@3200	3.91x3.31	9.1:1	30-80@3000
		5.9	Z	230@4000	330@3250	4.00x3.58	9.1:1	30-80@3000
	Ram Truck 2500	5.9	6	①	②	4.02x4.72	16.5	30@2500
		5.9	7	245@2700	505@1600	4.02x4.72	17.0:1	30@2500
		5.9	Z	230@4000	330@3250	4.00x3.58	9.1:1	30-80@3000
		8.0	W	300@4000	450@2400	4.00x3.58	8.4:1	50-60@3000
	Ram Truck 3500	5.9	6	①	②	4.02x4.72	16.5	30@2500
		5.9	7	245@2700	505@1600	4.02x4.72	17.0:1	30@2500
		5.9	5	230@4000	330@2800	4.00x3.58	8.9:1	30-80@3000
		8.0	W	300@4000	450@2400	4.00x3.58	8.4:1	50-60@3000
	Ram Van 1500	3.9	X	175@4800	220@3200	3.91x3.31	9.1:1	30-80@3000
		5.2	Y	220@4400	300@3200	3.91x3.31	9.1:1	30-80@3000
	Ram Van 2500	3.9	X	175@4800	220@3200	3.91x3.31	9.1:1	30-80@3000
		5.2	Y	220@4400	300@3200	3.91x3.31	9.1:1	30-80@3000
		5.9	Z	230@4000	330@3250	4.00x3.58	9.1:1	30-80@3000
	Ram Van 3500	5.2	Y	220@4400	300@3200	3.91x3.31	9.1:1	30-80@3000
		5.9	Z	230@4000	330@3250	4.00x3.58	9.1:1	30-80@3000
2003	Ram Truck 1500	3.7	K	210@5200	225@4200	3.66x3.40	9.1:1	25-110@3000
		4.7	N	235@4800	295@3200	3.66x3.40	9.3:1	25-110@3000
		5.7	D	345@5400	375@4200	3.91x3.58	9.6:1	25-110@3000
		5.9	Z	230@4000	330@3250	4.00x3.58	9.1:1	30-80@3000
	Ram Truck 2500	5.9	6	①	②	4.02x4.72	16.5	30@2500
		5.9	C	305@2900	555@1400	4.02x4.72	17.2:1	30@2500
		5.7	D	345@5400	375@4200	3.91x3.58	9.6:1	25-110@3000
		8.0	W	300@4000	450@2400	4.00x3.58	8.4:1	50-60@3000
	Ram Truck 3500	5.9	6	①	②	4.02x4.72	16.5	30@2500
		5.9	C	305@2900	555@1400	4.02x4.72	17.2:1	30@2500
		5.7	D	345@5400	375@4200	3.91x3.58	9.6:1	25-110@3000
		8.0	W	300@4000	450@2400	4.00x3.58	8.4:1	50-60@3000
	Ram Van 1500	3.9	X	175@4800	220@3200	3.91x3.31	9.1:1	30-80@3000
		5.2	Y	220@4400	300@3200	3.91x3.31	9.1:1	30-80@3000
	Ram Van 2500	3.9	X	175@4800	220@3200	3.91x3.31	9.1:1	30-80@3000
		5.2	Y	220@4400	300@3200	3.91x3.31	9.1:1	30-80@3000
		5.9	Z	230@4000	330@3250	4.00x3.58	9.1:1	30-80@3000
	Ram Van 3500	5.2	Y	220@4400	300@3200	3.91x3.31	9.1:1	30-80@3000
		5.9	Z	230@4000	330@3250	4.00x3.58	9.1:1	30-80@3000
2004	Ram Truck 1500	3.7	K	210@5200	225@4200	3.66x3.40	9.1:1	25-110@3000
		4.7	N	235@4800	295@3200	3.66x3.40	9.3:1	25-110@3000
		5.7	D	345@5400	375@4200	3.91x3.58	9.6:1	25-110@3000
	Ram Truck 2500	5.9	6	①	②	4.02x4.72	16.5	30@2500
		5.9	C	305@2900	555@1400	4.02x4.72	17.2:1	30@2500
		5.7	D	345@5400	375@4200	3.91x3.58	9.6:1	25-110@3000
	Ram Truck 3500	5.9	6	①	②	4.02x4.72	16.5	30@2500
		5.9	C	305@2900	555@1400	4.02x4.72	17.2:1	30@2500
		5.7	D	345@5400	375@4200	3.91x3.58	9.6:1	25-110@3000

06009-RAMT-C02

GENERAL ENGINE SPECIFICATIONS

Year	Model	Engine Displacement Liters	Engine Series VIN	Net Horsepower @ rpm	Net Torque @ rpm (ft. lbs.)	Bore x Stroke (in.)	Com-pression Ratio	Oil Pressure @ rpm
2005	Ram Truck 1500	3.7	K	211@5200	236@4000	3.66x3.40	9.6:1	25-110@3000
		4.7	N	235@4800	295@3200	3.66x3.40	9.0:1	35-105@3000
		5.7	D	345@5400	375@4200	3.91x3.58	9.6:1	25-110@3000
	Ram Truck 2500	5.9	C	325@2900	600@1400	4.02x4.72	17.2:1	30@2500
		5.7	D	345@5400	375@4200	3.91x3.58	9.6:1	25-110@3000
	Ram Truck 3500	5.9	C	325@2900	600@1400	4.02x4.72	17.2:1	30@2500
		5.7	D	345@5400	375@4200	3.91x3.58	9.6:1	25-110@3000

① AT: 215@2700rpm
MT: 235@2700rpm

② AT: 420@1600rpm
MT: 460@1600rpm

06009-RAMT-C03

GASOLINE ENGINE TUNE-UP SPECIFICATIONS

Year	Engine Displacement Liters	Engine VIN	Spark Plug Gap (in.)	Ignition Timing (deg.)	Fuel Pump (psi)	Idle Speed (rpm)	Valve Clearance	
							Intake	Exhaust
2002	3.9	X	0.040	①	44.2-54.2	②	HYD	HYD
	5.2	Y	0.040	①	44.2-54.2	②	HYD	HYD
	5.9	5	0.040	①	44.2-54.2	②	HYD	HYD
	5.9	Z	0.040	①	44.2-54.2	②	HYD	HYD
	8.0	W	0.045	①	44.2-54.2	②	HYD	HYD
2003	3.7	K	0.042	①	44-54	②	HYD	HYD
	3.9	X	0.040	①	44.2-54.2	②	HYD	HYD
	4.7	N	0.040	①	47-51	②	HYD	HYD
	5.2	Y	0.040	①	44.2-54.2	②	HYD	HYD
	5.7	D	0.045	①	49.0-49.4	②	HYD	HYD
	5.9	Z	0.040	①	44.2-54.2	②	HYD	HYD
	8.0	W	0.045	①	44.2-54.2	②	HYD	HYD
2004	3.7	K	0.042	①	44-54	②	HYD	HYD
	4.7	N	0.040	①	47-51	②	HYD	HYD
	5.7	D	0.045	①	49.0-49.4	②	HYD	HYD
2005	3.7	K	0.042	①	NA	②	HYD	HYD
	4.7	N	0.040	①	NA	②	HYD	HYD
	5.7	D	0.045	①	NA	②	HYD	HYD

NOTE: The Vehicle Emission Control Information (VECI) label often reflects specification changes made during production.
The label figures must be used if they differ from those in this chart.
HYD: Hydraulic

① Ignition timing is controlled by the PCM and is not adjustable.

② Idle speed is controlled by the PCM and is not adjustable

NA: Not available

06009-RAMT-C04

DIESEL ENGINE TUNE-UP SPECIFICATIONS

Year	Engine Displacement Liters	Engine VIN	Valve Clearance Intake (in.)	Valve Clearance Exhaust (in.)	Intake Valve Opens (deg.)	Injection Pump Setting (deg.)	Injection Nozzle Pressure (psi) New	Injection Nozzle Pressure (psi) Used	Idle Speed (rpm)	Cranking Compression Pressure (psi)
2002	5.9	6	0.006-0.015	0.0015-0.0300	NA	①	4250-4750	NA	②	NA
	5.9	7	0.006-0.015	0.0015-0.0300	NA	①	4250-4750	NA	②	NA
2003	5.9	6	0.006-0.015	0.0015-0.0300	NA	①	4250-4750	NA	②	NA
	5.9	C	0.006-0.015	0.0015-0.0300	NA	①	4250-4750	NA	②	NA
2004	5.9	6	0.006-0.015	0.0015-0.0300	NA	①	4250-4750	NA	②	NA
	5.9	C	0.006-0.015	0.0015-0.0300	NA	①	4250-4750	NA	②	NA
2005	5.9	C	0.006-0.015	0.0015-0.0300	NA	①	4250-4750	NA	②	NA

NOTE: The Vehicle Emission Control Information (VECI) label often reflects specification changes made during production.
The label figures must be used if they differ from those in this chart

NA: Not Available

① Federal models with manual transmissions: 13.5 degrees BTDC
Except Federal models with manual transmissions: 14.0 degrees BTDC

② Automatic transmission: 750-800 rpm
Manual transmission: 780 rpm

③ All fuel adjustments are made by the ECM

06009-RAMT-C05

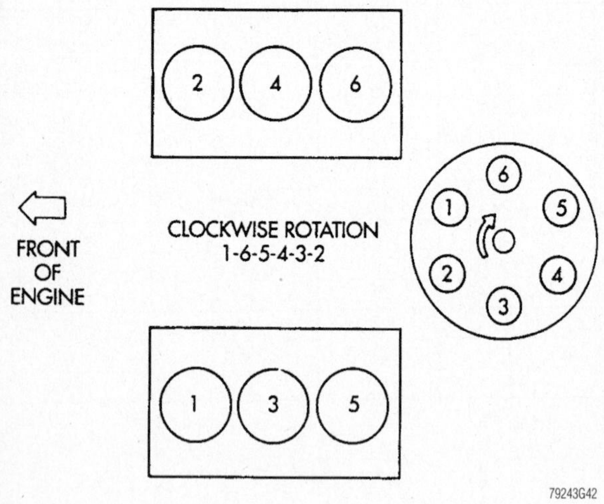

3.9L engine
Firing Order: 1–6–5–4–3–2
Distributor rotation: Clockwise

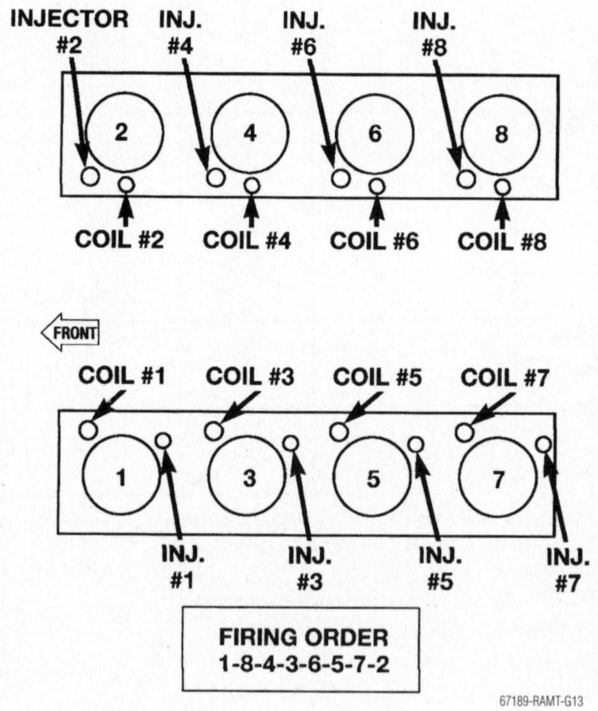

4.7L engine firing order
Distributorless ignition

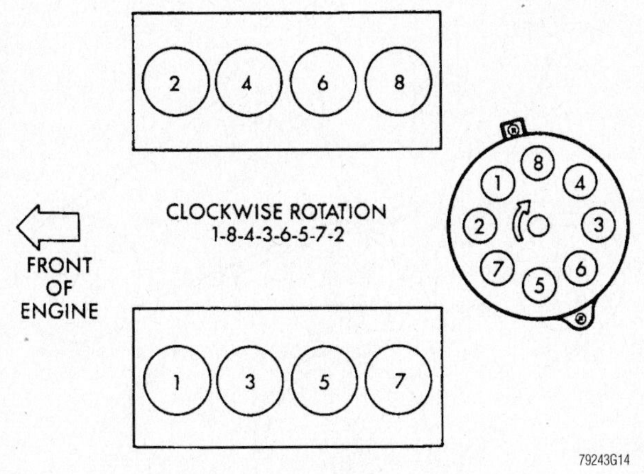

5.2L and 5.9L engine
Firing Order: 1–8–4–3–6–5–7–2
Distributor rotation: Clockwise

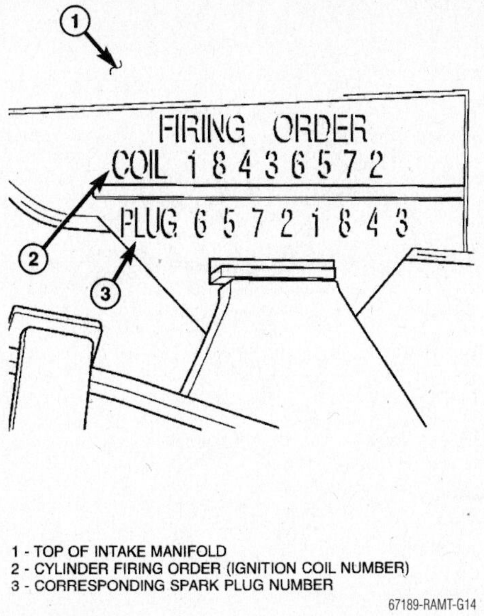

5.7L engine firing order
Distributorless ignition

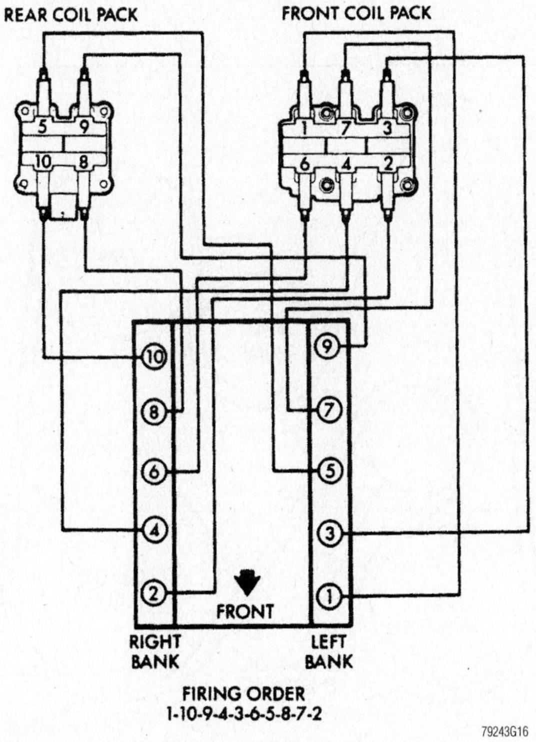

8.0L engine
Firing Order: 1–10–9–4–3–6–5–8–7–2
Distributorless ignition system

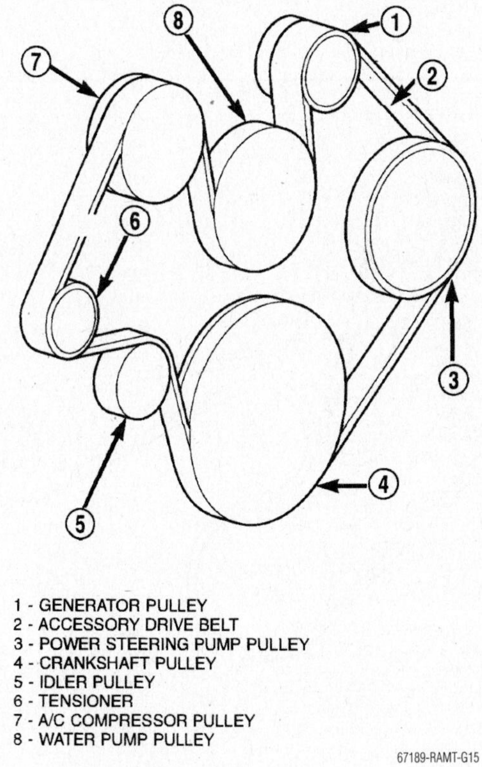

Accessory serpentine belt routing—3.7L and 4.7L gasoline engines

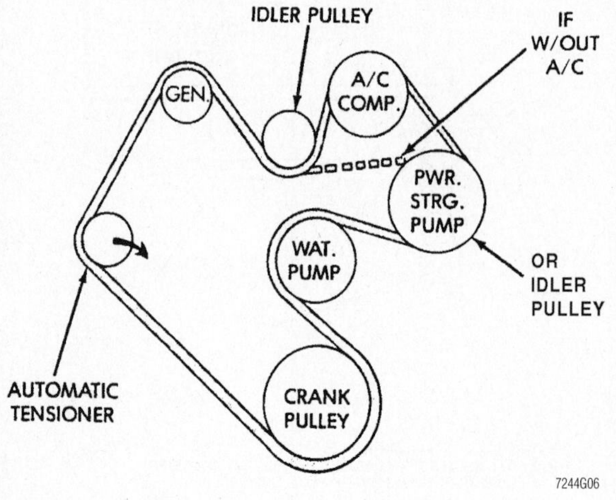

Accessory serpentine belt routing—3.9L, 5.2L and 5.9L LDC gasoline engines

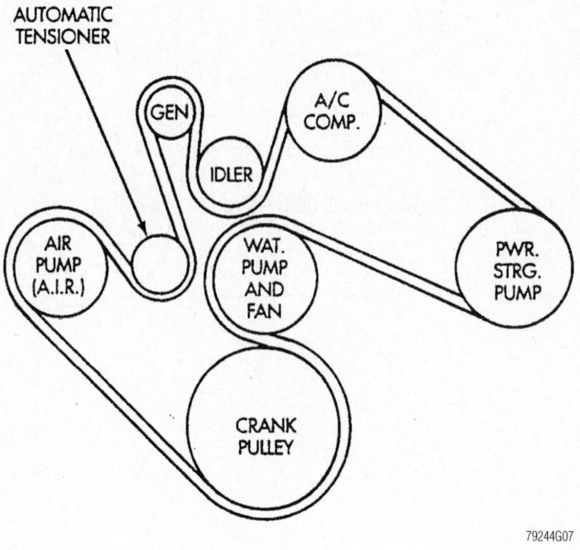

Accessory serpentine belt routing—5.9L HDC and 8.0L gasoline engines with A/C

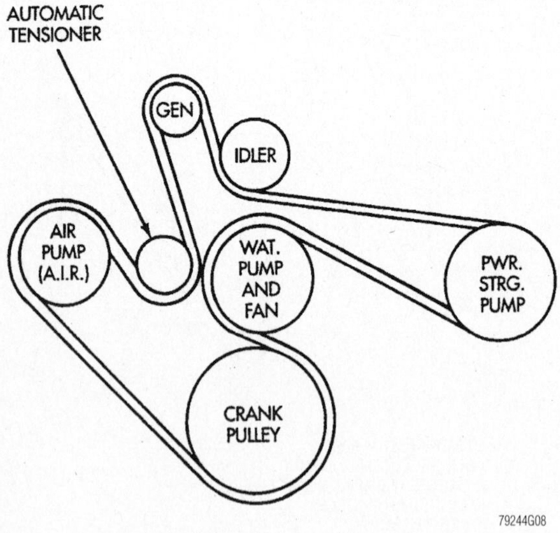

Accessory serpentine belt routing—5.9L HDC and 8.0L gasoline engines without A/C

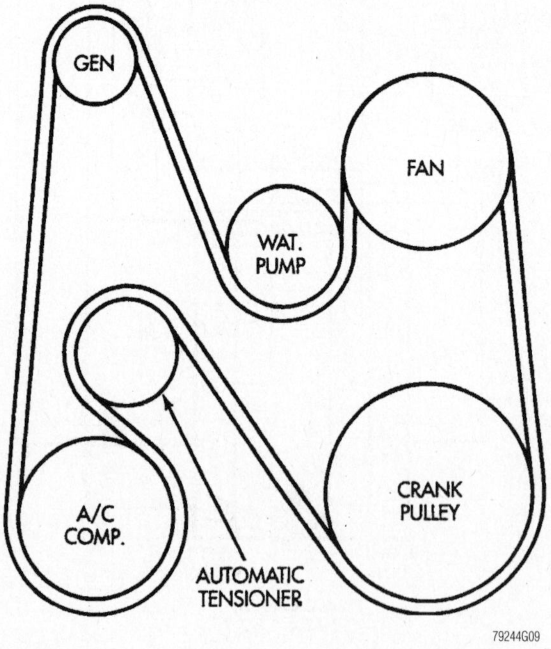

Accessory serpentine belt routing—2001-2003 5.9L Diesel engines with A/C

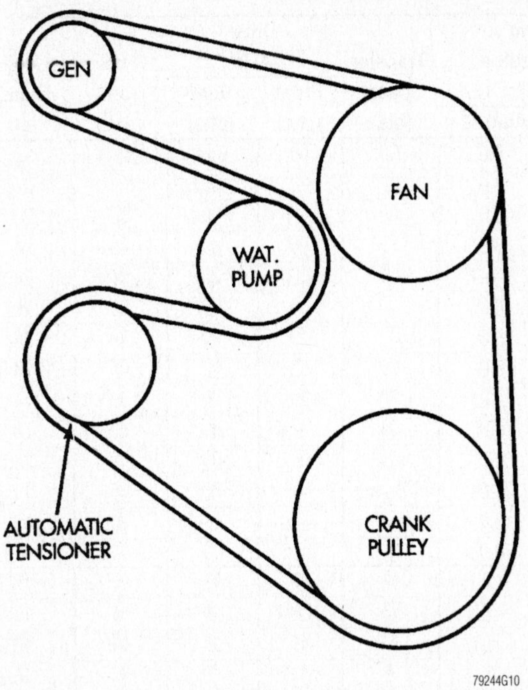

Accessory serpentine belt routing—2001-2003 5.9L Diesel engines without A/C

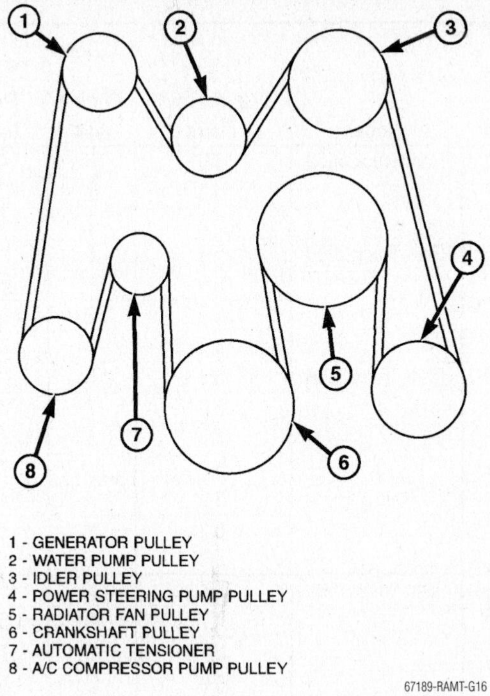

Accessory serpentine belt routing—2004 5.9L Diesel engines with A/C

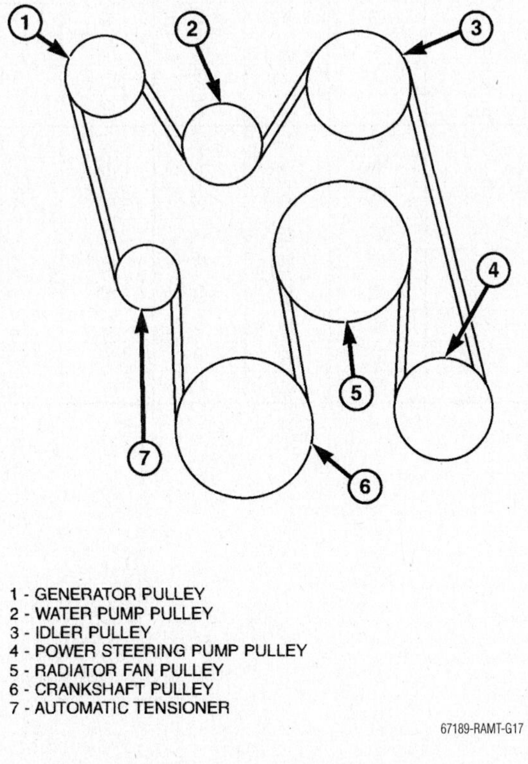

Accessory serpentine belt routing—2004 5.9L Diesel engines without A/C

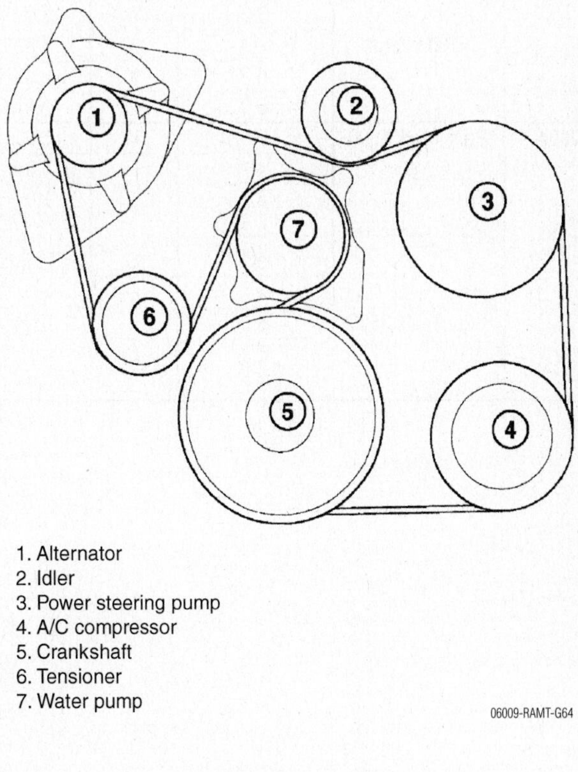

Accessory serpentine belt routing—5.7L engines

CAPACITIES

Year	Model	Engine Displacement Liters	Engine VIN	Oil with Filter (qts.)	Engine Transmission (pts.)		Transfer Case (pts.)	Drive Axle		Fuel Tank (gal.)	Cooling System (qts.)
					Manual	Auto. ①		Front (pts.)	Rear (pts.)		
2002	Ram Truck 1500	3.9	X	4.5	②	8.0	③	④	⑤	⑥	20.0
		5.2	Y	5.0	②	8.0	③	④	⑤	⑤	20.0
		5.9	Z	5.0	—	8.0	③	④	⑤	⑤	20.0
	Ram Truck 2500	5.9	6	11.0	②	8.0	③	④	⑤	⑥	24.0
		5.9	7	11.0	②	—	③	④	⑤	⑤	24.0
		5.9	Z	5.0	②	8.0	③	④	⑤	⑤	20.0
		8.0	W	7.0	②	8.0	③	④	⑤	⑥	26.0
	Ram Truck 3500	5.9	6	11.0	②	8.0	③	④	⑤	⑤	24.0
		5.9	7	11.0	②	—	③	④	⑤	⑥	24.0
		5.9	5	5.0	②	8.0	③	④	⑤	⑤	20.0
		8.0	W	7.0	②	8.0	③	④	⑤	⑤	26.0
	Ram Van	3.9	X	4.0	—	8.0	—	—	⑦	⑧	14.5
		5.2	Y	5.0	—	8.0	—	—	⑦	⑧	16.5
		5.9	Z	5.0	—	8.0	—	—	⑦	⑧	⑨
2003	Ram Truck 1500	3.7	K	5.0	⑩	⑪	⑫	⑬	⑭	⑮	16.2
		4.7	N	6.0	⑩	⑪	⑫	⑬	⑭	⑮	16.2
		5.7	D	7.0	⑩	⑪	⑫	⑬	⑭	⑮	16.2
		5.9	Z	5.0	⑩	⑪	⑫	⑬	⑭	⑮	16.3
	Ram Truck 2500	5.9	6	11.0	⑩	⑪	⑫	⑬	⑭	⑮	29.5
		5.9	C	11.0	⑩	⑪	⑫	⑬	⑭	⑮	29.5
		5.7	D	7.0	⑩	⑪	⑫	⑬	⑭	⑮	16.2
		8.0	W	7.0	⑩	⑪	⑫	⑬	⑭	⑮	24.3
	Ram Truck 3500	5.9	6	11.0	⑩	⑪	⑫	⑬	⑭	⑮	29.5
		5.9	C	11.0	⑩	⑪	⑫	⑬	⑭	⑮	29.5
		5.7	D	7.0	⑩	⑪	⑫	⑬	⑭	⑮	16.2
		8.0	W	7.0	⑩	⑪	⑫	⑬	⑭	⑮	24.3
	Ram Van	3.9	X	4.0	—	8.0	—	—	⑦	⑧	14.5
		5.2	Y	5.0	—	8.0	—	—	⑦	⑧	16.5
		5.9	Z	5.0	—	8.0	—	—	⑦	⑧	⑨
2004	Ram Truck 1500	3.7	K	5.0	⑩	⑪	⑫	⑬	⑭	⑮	16.2
		4.7	N	6.0	⑩	⑪	⑫	⑬	⑭	⑮	16.2
		5.7	D	7.0	⑩	⑪	⑫	⑬	⑭	⑮	16.2
	Ram Truck 2500	5.9	6	12.0	⑩	⑪	⑫	⑬	⑭	⑮	29.5
		5.9	C	12.0	⑩	⑪	⑫	⑬	⑭	⑮	29.5
		5.7	D	7.0	⑩	⑪	⑫	⑬	⑭	⑮	16.2
	Ram Truck 3500	5.9	6	12.0	⑩	⑪	⑫	⑬	⑭	⑮	29.5
		5.9	C	12.0	⑩	⑪	⑫	⑬	⑭	⑮	29.5
		5.7	D	7.0	⑩	⑪	⑫	⑬	⑭	⑮	16.2

06009-RAMT-C06

CAPACITIES

Year	Model	Engine Displacement Liters	Engine VIN	Oil with Filter (qts.)	Engine Transmission (pts.)		Transfer Case (pts.)	Drive Axle		Fuel Tank (gal.)	Cooling System (qts.)
					Manual	Auto. ①		Front (pts.)	Rear (pts.)		
2005	Ram Truck 1500	3.7	K	5.0	⑰	⑰	⑫	⑬	⑭	⑯	16.2
		4.7	N	6.0	⑰	⑰	⑫	⑬	⑭	⑯	16.2
		5.7	D	7.0	⑰	⑰	⑫	⑬	⑭	⑯	16.2
	Ram Truck 2500	5.9	C	12.0	⑰	⑰	⑫	⑬	⑭	⑯	29.5
		5.7	D	7.0	⑰	⑰	⑫	⑬	⑭	⑯	16.2
	Ram Truck 3500	5.9	C	12.0	⑰	⑰	⑫	⑬	⑭	⑯	29.5
		5.7	D	7.0	⑰	⑰	⑫	⑬	⑭	⑯	16.2

NOTE: All capacities are approximate. Add fluid gradually and check to be sure a proper fluid level is obtained.

① For fluid drain and filter replacement only.

② NV3500: 4.2 pts.
NV4500: 8.0 pts.
NV4500 HD: 8.0 pts.
NV5600: 9.5 pts.

③ NV231HD: 2.5 pts.
NV241: 4.6 pts.
NV241HD: 6.5 pts.

④ 216FBI: 4.8 pts.
248FBI: 8.5 pts.

⑤ 9.25 in.: 4.5 pts.
248RBI (2wd) 6.3 pts.
248RBI (4wd): 7.0 pts.
267RBI (2wd): 7.0 pts.
267RBI (4wd): 7.5 pts.
286RBI (2wd): 6.8 pts.
286RBI (4wd): 10.1 pts.

⑥ 1500 series w/6.5 ft. box: 26.0
2500 series w/6.5 ft. box: 34.0
All others: 35.0

⑦ 8.25 in. 4.4 pts.
9.25 in.: 4.5 pts.
248RBI: 6.25 pts.

⑧ 109 in. wheebase: 31.0
132 in. wheelbase: 35.0

⑨ With rear heater: 16.0

⑩ NV3500 2wd: 4.8 pts.
NV3500 4wd: 4.2 pts.
NV4500: 8.0 pts.
NV5600: 9.5 pts.

⑪ 46RE: 8.0 pts.
45RFE and 545RFE
2wd: 11.0 pts.
4wd: 13.0 pts.

⑫ NV241 GEN II: 3.4 pts.
NV243: 3.4 pts.
NV 271: 4.0 pts.
NV 273: 4.0 pts.

⑬ C205F: 3.5 pts.
9.25AA: 4.75 pts.

⑭ 9.25: 4.9 pts. W/Trac-Lok + 4 oz. Additive
10.5AA: 4.75 pts.
11.5AA: 7.65 pts.

⑮ Short bed: 26.0
Long bed: 35.0

⑯ 1500 Short bed: 26.0
2500/3500 Short bed: 34.0
All Long bed: 35.0

⑰ 48RE: 8.0 pts.
45RFE/545RFE:
4x2: 11.0 pts.
4x4: 13.0 pts.

⑰ Getrag 238: 8.0 pts.
G56: 12.0 pts.
NV4500: 8.0 pts.
NV5600: 9.5 pts.

06009-RAMT-C07

VALVE SPECIFICATIONS

Year	Engine Displacement Liters	Engine VIN	Seat Angle (deg.)	Face Angle (deg.)	Spring Test Pressure (lbs. @ in.)	Spring Installed Height (in.)	Stem-to-Guide Clearance (in.)		Stem Diameter (in.)	
							Intake	Exhaust	Intake	Exhaust
2002	3.9	X	44.25-44.75	43.25-43.75	200@1.21	1.64	0.001-0.003	0.001-0.003	0.311-0.312	0.311-0.312
	5.2	Y	44.25-44.75	43.25-43.75	200@1.21	1.64	0.001-0.003	0.001-0.003	0.311-0.312	0.311-0.312
	5.9	Z	44.25-44.75	43.25-43.75	200@1.21	1.64	0.001-0.003	0.002-0.004	0.372-0.373	0.371-0.372
	5.9	6	①	①	76.4@1.39	1.39	0.002	0.002	0.274-0.276	0.274-0.276
	5.9	7	①	①	76.4@1.39	1.39	0.002	0.002	0.274-0.276	0.274-0.276
	8.0	W	44.5	45.0	200@1.212	1.64	0.001-0.003	0.001-0.003	0.311-0.312	0.311-0.312
	5.9	5	44.25-44.75	43.25-43.75	200@1.21	1.64	0.001-0.003	0.002-0.004	0.372-0.373	0.371-0.372
2003	3.7	K	44.5-45	45-45.5	221-242@1.107	1.619	0.0008-0.0028	0.0019-0.0039	0.2729-0.2739	0.2717-0.2728
	3.9	X	44.25-44.75	43.25-43.75	200@1.21	1.64	0.001-0.003	0.001-0.003	0.311-0.312	0.311-0.312
	4.7	N	44.5-45	45-45.5	176.2-192.4@1.1532	1.602	0.0011-0.0017	0.0029	0.2728-0.2739	0.2717-0.2728
	5.2	Y	44.25-44.75	43.25-43.75	200@1.21	1.64	0.001-0.003	0.001-0.003	0.311-0.312	0.311-0.312
	5.7	D	44.5-45	45-45.5	95@1.81	1.81	0.0008-0.0025	0.0019-0.0037	0.3120-0.3130	0.3110-0.3120
	5.9	Z	44.25-44.75	43.25-43.75	200@1.21	1.64	0.001-0.003	0.002-0.004	0.372-0.373	0.371-0.372
	5.9	6	①	①	76.4@1.39	1.39	0.002	0.002	0.274-0.276	0.274-0.276
	5.9	C	①	①	76.4@1.39	1.39	0.002	0.002	0.274-0.276	0.274-0.276
	8.0	W	44.5	45.0	200@1.212	1.64	0.001-0.003	0.001-0.003	0.311-0.312	0.311-0.312
2004	3.7	K	44.5-45	45-45.5	221-234@1.107	1.579	0.0008-0.0028	0.0019-0.0039	0.2729-0.2739	0.2717-0.2728
	4.7	N	44.5-45	45-45.5	176.2-192.4@1.1532	1.579	0.0008-0.0028	0.0019-0.0039	0.2728-0.2739	0.2717-0.2728
	5.7	D	44.5-45	45-45.5	95@1.81	1.81	0.0008-0.0025	0.0019-0.0037	0.3120-0.3130	0.3110-0.3120
	5.9	6	①	①	76.4@1.39	1.39	0.002	0.002	0.274-0.276	0.274-0.276
	5.9	C	①	①	76.4@1.39	1.39	0.002	0.002	0.274-0.276	0.274-0.276

06009-RAMT-C08

VALVE SPECIFICATIONS

Year	Engine Displacement Liters	Engine VIN	Seat Angle (deg.)	Face Angle (deg.)	Spring Test Pressure (lbs. @ in.)	Spring Installed Height (in.)	Stem-to-Guide Clearance (in.) Intake	Stem-to-Guide Clearance (in.) Exhaust	Stem Diameter (in.) Intake	Stem Diameter (in.) Exhaust
2005	3.7	K	44.5-45	45-45.5	213-233@ 1.107	1.579	0.0008-0.0028	0.0019-0.0039	0.2729-0.2739	0.2717-0.2728
	4.7	N	44.5-45	45-45.5	174.4-195.5 @1.137	1.579	0.0008-0.0028	0.0019-0.0039	0.2729-0.2739	0.2717-0.2728
	5.7	D	44.5-45	45-45.5	242@1.322	1.81	0.0008-0.0025	0.0019-0.0037	0.3120-0.3130	0.3110-0.3120
	5.9	C	①	①	76.4@1.39	1.39	0.002	0.002	0.274-0.276	0.274-0.276

① Intake: 30 degrees
Exhaust: 45 degrees

06009-RAMT-C09

CRANKSHAFT AND CONNECTING ROD SPECIFICATIONS

All measurements are given in inches.

Year	Engine Displacement Liters	Engine VIN	Crankshaft Main Brg. Journal Dia.	Crankshaft Main Brg. Oil Clearance	Crankshaft Shaft End-play	Crankshaft Thrust on No.	Connecting Rod Journal Diameter	Connecting Rod Oil Clearance	Connecting Rod Side Clearance
2002	3.9	X	2.4995-2.5005	①	0.0020-0.0070	2	2.1240-2.1250	0.0005-0.0022	0.0060-0.0140
	5.2	Y	2.4995-2.5005	①	0.0020-0.0070	2	2.1240-2.1250	0.0005-0.0022	0.0060-0.0140
	5.9	6	3.2662-3.2682	0.0037-0.0057	0.0040-0.0017	2	2.7150-2.7170	0.0035	0.0040-0.0120
	5.9	7	3.2662-3.2682	0.0037-0.0057	0.0040-0.0017	2	2.7150-2.7170	0.0035	0.0040-0.0120
	5.9	5	2.8095-2.8105	①	0.0020-0.0070	2	2.1240-2.1250	0.0005-0.0022	0.0060-0.0140
	5.9	Z	2.8095-2.8105	①	0.0020-0.0070	2	2.1240-2.1250	0.0005-0.0022	0.0060-0.0140
	8.0	W	2.9995-3.0005	0.0002-0.0023	0.0030-0.0120	2	2.1240-2.1250	0.0002-0.0029	0.0100-0.0180
2003	3.7	K	2.4996-2.5005	0.0020-0.0034	0.0021-0.0112	2	2.2794-2.2797	0.0004-0.0019	0.0040-0.0138
	3.9	X	2.4995-2.5005	①	0.0020-0.0070	2	2.1240-2.1250	0.0005-0.0022	0.0060-0.0140
	4.7	N	2.4996-2.5005	0.0002-0.0013	0.0021-0.0112	2	2.0076-2.0082	0.0004-0.0019	0.0040-0.0138
	5.2	Y	2.4995-2.5005	①	0.0020-0.0070	2	2.1240-2.1250	0.0005-0.0022	0.0060-0.0140
	5.7	D	2.5585-2.5595	0.0009-0.0020	0.0020-0.0110	3	2.1250-2.1260	0.0007-0.0023	0.0030-0.0137
	5.9	6	3.2662-3.2682	0.0037-0.0057	0.0040-0.0017	2	2.7150-2.7170	0.0035	0.0040-0.0120
	5.9	C	3.2662-3.2682	0.0037-0.0057	0.0040-0.0017	2	2.7150-2.7170	0.0035	0.0040-0.0120
	5.9	Z	2.8095-2.8105	①	0.0020-0.0070	2	2.1240-2.1250	0.0005-0.0022	0.0060-0.0140
	8.0	W	2.9995-3.0005	0.0002-0.0023	0.0030-0.0120	2	2.1240-2.1250	0.0002-0.0029	0.0100-0.0180
2004	3.7	K	2.4996-2.5005	0.0020-0.0034	0.0021-0.0112	2	2.2794-2.2797	0.0004-0.0022	0.0040-0.0138
	4.7	N	2.4996-2.5005	0.0008-0.0021	0.0021-0.0112	3	2.0076-2.0082	0.0006-0.0022	0.0040-0.0138
	5.7	D	2.5585-2.5595	0.0009-0.0020	0.0020-0.0110	3	2.1250-2.1260	0.0007-0.0023	0.0030-0.0137
	5.9	6	3.2662-3.2682	0.0037-0.0057	0.0040-0.0017	2	2.7150-2.7170	0.0035	0.0040-0.0120
	5.9	C	3.2662-3.2682	0.0037-0.0057	0.0040-0.0017	2	2.7150-2.7170	0.0035	0.0040-0.0120

06009-RAMT-C10

CRANKSHAFT AND CONNECTING ROD SPECIFICATIONS

All measurements are given in inches.

	Engine		Crankshaft				Connecting Rod		
Year	Displacement Liters	Engine VIN	Main Brg. Journal Dia.	Main Brg. Oil Clearance	Shaft End-play	Thrust on No.	Journal Diameter	Oil Clearance	Side Clearance
2005	3.7	K	2.4996-2.5005	0.0020-0.0034	0.0021-0.0112	2	2.2794-2.2797	0.0004-0.0022	0.0040-0.0138
	4.7	N	2.4996-2.5005	0.0008-0.0021	0.0021-0.0112	3	2.0076-2.0082	0.0006-0.0022	0.0040-0.0138
	5.7	D	2.5585-2.5595	0.0009-0.0020	0.0020-0.0110	3	2.1250-2.1260	0.0007-0.0023	0.0030-0.0137
	5.9	C	3.2662-3.2682	0.0037-0.0057	0.0040-0.0017	6	2.7190-2.7200	0.0035	0.0040-0.0130

① No.1: 0.0005-0.0015
All others: 0.0005-0.0020

06009-RAMT-C11

PISTON AND RING SPECIFICATIONS

All measurements are given in inches.

Year	Engine Displacement Liters	Engine VIN	Piston Clearance	Ring Gap			Ring Side Clearance		
				Top Compression	Bottom Compression	Oil Control	Top Compression	Bottom Compression	Oil Control
2002	3.9	X	0.0005-0.0015	0.0100-0.0200	0.0100-0.0200	0.0100-0.0500	0.0015-0.0030	0.0015-0.0030	0.0020-0.0080
	5.2	Y	0.0005-0.0015	0.0100-0.0200	0.0100-0.0200	0.0100-0.0500	0.0015-0.0030	0.0015-0.0030	0.0020-0.0080
	5.9	6	NA	0.0138-0.0178	0.0335-0.0453	0.0098-0.0217	0.0037	0.0037	0.0033
	5.9	7	NA	0.0138-0.0178	0.0335-0.0453	0.0098-0.0217	0.0037	0.0037	0.0033
	5.9	5	0.0005-0.0015	0.0120-0.0220	0.0220-0.0310	0.0150-0.0550	0.0016-0.0033	0.0016-0.0033	0.0020-0.0080
	5.9	Z	0.0005-0.0015	0.0120-0.0220	0.0220-0.0310	0.0150-0.0550	0.0016-0.0033	0.0016-0.0033	0.0020-0.0080
	8.0	W	0.0005-0.0015	0.0100-0.0200	0.0100-0.0200	0.0150-0.0550	0.0029-0.0038	0.0029-0.0038	0.1020-0.1080
2003	3.7	K	0.0014	0.0146-0.0249	0.0146-0.0249	0.0100-0.0300	0.0020-0.0037	0.0016-0.0031	0.0007-0.0091
	3.9	X	0.0005-0.0015	0.0100-0.0200	0.0100-0.0200	0.0100-0.0500	0.0015-0.0030	0.0015-0.0030	0.0020-0.0080
	4.7	N	0.0008-0.0020	0.0146-0.0249	0.0146-0.0249	0.0100-0.0500	0.0020-0.0041	0.0016-0.0032	0.0007-0.0091
	5.2	Y	0.0005-0.0015	0.0100-0.0200	0.0100-0.0200	0.0100-0.0500	0.0015-0.0030	0.0015-0.0030	0.0020-0.0080
	5.7	D	0.0008-0.0019	0.0090-0.0149	0.0137-0.0236	0.0059-0.0259	0.0007-0.0026	0.0007-0.0022	0.0007-0.0091
	5.9	6	NA	0.0138-0.0178	0.0335-0.0453	0.0098-0.0217	0.0037	0.0037	0.0033
	5.9	C	NA	0.0138-0.0178	0.0335-0.0453	0.0098-0.0217	0.0037	0.0037	0.0033
	5.9	Z	0.0005-0.0015	0.0120-0.0220	0.0220-0.0310	0.0150-0.0550	0.0016-0.0033	0.0016-0.0033	0.0020-0.0080
	8.0	W	0.0005-0.0015	0.0100-0.0200	0.0100-0.0200	0.0150-0.0550	0.0029-0.0038	0.0029-0.0038	0.1020-0.1080
2004	3.7	K	0.0014	0.0079-0.0142	0.0146-0.0249	0.0100-0.0300	0.0020-0.0037	0.0016-0.0031	0.0007-0.0091
	4.7	N	0.0008-0.0020	0.0079-0.0142	0.0146-0.0249	0.0100-0.0300	0.0020-0.0037	0.0016-0.0032	0.0007-0.0091
	5.7	D	0.0008-0.0019	0.0090-0.0149	0.0137-0.0236	0.0059-0.0259	0.0007-0.0026	0.0007-0.0022	0.0007-0.0091
	5.9	6	NA	0.0138-0.0178	0.0335-0.0453	0.0098-0.0217	0.0037	0.0037	0.0033
	5.9	C	NA	0.0138-0.0178	0.0335-0.0453	0.0098-0.0217	0.0037	0.0037	0.0033
2005	3.7	K	0.0014	0.0079-0.0142	0.0146-0.0249	0.0099-0.0300	0.0020-0.0037	0.0016-0.0031	0.0007-0.0091
	4.7	N	0.0008-0.0019	0.0079-0.0142	0.0146-0.0249	0.0100-0.0300	0.0020-0.0037	0.0016-0.0032	0.0007-0.0091
	5.7	D	0.0008-0.0019	0.0090-0.0149	0.0137-0.0236	0.0059-0.0259	0.0007-0.0026	0.0007-0.0022	0.0007-0.0091
	5.9	C	NA	0.0100-0.0140	0.0330-0.0450	0.0100-0.0210	N/A	0.0018-0.0037	0.0016-0.0033

06009-RAMT-C12

TORQUE SPECIFICATIONS

All readings in ft. lbs.

Year	Engine Displacement Liters	Engine VIN	Cylinder Head Bolts	Main Bearing Bolts	Rod Bearing Bolts	Crankshaft Damper Bolts	Flywheel Bolts	Manifold Intake	Manifold Exhaust	Spark Plugs	Oil Pan Drain Plug
2002	3.9	X	①	85	45	18	55	②	25	30	③
	5.2	Y	①	85	45	18	55	②	25	30	③
	5.9	6	④	⑤	⑥	92	⑦	18	32	—	③
	5.9	7	④	⑤	⑥	92	⑦	18	32	—	③
	5.9	5	①	85	45	18	55	②	25	30	③
	5.9	Z	①	85	45	18	55	②	25	30	③
	8.0	W	⑧	⑨	45	230	55	⑩	16	30	③
2003	3.7	K	⑪	⑫	⑬	130	45	9	18	27	③
	3.9	X	①	85	45	18	55	②	25	30	③
	4.7	N	⑭	⑮	⑯	130	45	9	18	27	③
	5.2	Y	①	85	45	18	55	②	25	30	③
	5.7	D	⑰	⑱	⑲	90	70	⑪	18	18	③
	5.9	C	④	⑤	⑥	92	⑦	18	32	—	③
	5.9	6	④	⑤	⑥	92	⑦	18	32	—	③
	5.9	Z	①	85	45	18	55	②	25	30	③
	8.0	W	⑧	⑨	45	230	55	⑩	16	30	③
2004	3.7	K	⑪	⑫	⑬	130	45	9	18	27	③
	4.7	N	⑭	⑮	⑬	130	45	9	18	27	③
	5.7	D	⑰	⑱	⑲	130	55	⑪	18	18	③
	5.9	C	④	⑤	⑥	92	⑦	18	32	—	③
	5.9	6	④	⑤	⑥	92	⑦	18	32	—	③
2005	3.7	K	⑪	⑫	⑬	130	45	12	18	27	③
	4.7	N	⑪	⑮	⑬	130	45	12	18	27	③
	5.7	D	⑰	⑱	⑲	130	55	⑪	18	18	③
	5.9	C	⑳	㉑	㉒	92	⑦	18	32	—	③

① Step 1: 50 ft. lbs.
Step 2: 105 ft. lbs.

② Step 1: 48 inch lbs.
Step 2: 84 inch lbs.
Step 3: Recheck 84 inch lbs.

③ Gas engines:25 ft. lbs.
Diesel engine: 37 ft. lbs.

④ Step 1: 59 ft. lbs.
Step 2: 77 ft. lbs.
Step 3: Recheck at 77 ft. lbs.
Step 4: All bolts an additional 1/4 turn (90 degrees)

⑤ Step 1: 45 ft. lbs.
Step 2: 60 ft. lbs.
Step 3: additional 1/4 turn (90 degrees)

⑥ Step 1: 26 ft. lbs.
Step 2: 51 ft. lbs.
Step 3: 73 ft. lbs.

⑦ Manual transmission: 101 ft. lbs.
Automatic transmission: 32 ft. lbs.

⑧ Step 1: 43 ft. lbs.
Step 2: 105 ft. lbs.

⑨ Step 1: 20 ft. lbs.
Step 2: 85 ft. lbs.

⑩ Upper: 16 ft. lbs.
Lower: 40 ft. lbs.

⑪ See procedure

⑫ Bed plate: see procedure

⑬ 20 ft. lbs. plus 90 degrees

⑭ M11 bolts: 60 ft. lbs.
M8 bolts: 250 inch lbs.

⑮ Step 1: Bolts 1-10 to 25 inch lbs.
Step 2: Bolts 1-10 plus 90 degrees
Step 3: Bolts A-L to 40 ft. lbs.
Step 4: Bolts A1-A6 to 20 ft. lbs.

⑯ 15 ft. lbs. plus 110 degrees

⑰ M8
Step 1: 15 ft. lbs.
Step 2: 25 ft. lbs.
M12
Step 1: 25 ft. lbs.
Step 2: 40 ft. lbs.
Step 3: plus 90 degree turn

⑱ M12 bolts: 12 ft. lbs. plus 90 degrees
M8 bolts: 13 ft. lbs.

⑲ 15 ft. lbs. Plus 90 degrees

⑳ Step 1: 52 ft. lbs.
Step 2: Back off 360 degrees
Step 3: 77 ft. lbs.
Step 4: Recheck 77 ft. lbs.
Step 5: Plus 90 degrees

㉑ Step 1: 37 ft. lbs.
Step 2: 59 ft. lbs.
Step 3: Plus 90 degrees

㉒ Step 1: 22 ft. lbs.
Step 2: 44 ft. lbs.
Step 3: Plus 60 degrees

06009-RAMT-C13

WHEEL ALIGNMENT

				Caster		Camber		
Year	Model	GVW	Wheel Base (in.)	Range (+/-Deg.)	Preferred Setting (Deg.)	Range (+/-Deg.)	Preferred Setting (Deg.)	Toe-in (in.)
2002	4x2 Pickup	6,400	118.7	1.00	+3.66	0.50	+0.50	0.10+/-0.10
	4x2 Pickup	6,400	134.7	1.00	+3.89	0.50	+0.50	0.10+/-0.10
	4x2 Pickup	6,400	138.7	1.00	+3.99	0.50	+0.50	0.10+/-0.10
	4x2 Pickup	6,400	154.7	1.00	+4.17	0.50	+0.50	0.10+/-0.10
	4x2 Pickup	8,800	134.7	1.00	+3.53	0.50	+0.50	0.10+/-0.10
	4x2 Pickup	8,800	138.7	1.00	+3.59	0.50	+0.50	0.10+/-0.10
	4x2 Pickup	8,800	154.7	1.00	+3.78	0.50	+0.50	0.10+/-0.10
	4x2 Pickup	10,500	134.7	1.00	+3.33	0.50	+0.50	0.10+/-0.10
	4x2 Pickup	10,500	154.7	1.00	+3.58	0.50	+0.50	0.10+/-0.10
	4x4 Pickup	6,400	118.7	1.00	+2.86	0.50	NA	0.10+/-0.10
	4x4 Pickup	6,400	134.7	1.00	+3.04	0.50	NA	0.10+/-0.10
	4x4 Pickup	6,400	138.7	1.00	+3.19	0.50	NA	0.10+/-0.10
	4x4 Pickup	6,400	154.7	1.00	+3.37	0.50	NA	0.10+/-0.10
	4x4 Pickup	8,800	134.7	1.00	+2.68	0.50	NA	0.10+/-0.10
	4x4 Pickup	8,800	138.7	1.00	+2.74	0.50	NA	0.10+/-0.10
	4x4 Pickup	8,800	154.7	1.00	+2.88	0.50	NA	0.10+/-0.10
	4x4 Pickup	10,500	134.7	1.00	+2.48	0.50	NA	0.10+/-0.10
	4x4 Pickup	10,500	154.7	1.00	+2.63	0.50	NA	0.10+/-0.10
	Van	All	All	1.50	+2.75	0.30	0	0.25+/-0.25
2003	1500 4x2	—	120.5	0.75	+4.0	0.50	0	0.10+/-0.10
	1500 4x2	—	140.5	0.75	+4.2	0.50	0	0.10+/-0.10
	1500 4x2	—	160.5	0.75	+4.4	0.50	0	0.10+/-0.10
	1500 4x4	—	120.5	0.75	+4.2	0.50	0	0.10+/-0.10
	1500 4x4	—	140.5	0.75	+4.4	0.50	0	0.10+/-0.10
	1500 4x4	—	160.5	0.75	+4.6	0.50	0	0.10+/-0.10
	2500 & 3500 4x2	—	140	0.75	+4.0	0.50	0	0.10+/-0.05
	2500 & 3500 4x2	—	160	0.75	+4.3	0.50	0	0.10+/-0.05
	2500 & 3500 4x4	—	140	0.75	+4.5	0.25	+0.25	0.10+/-0.05
	2500 & 3500 4x4	—	160	0.75	+4.7	0.25	0.25	0.10+/-0.05
	Van	All	All	1.00	+3.5	0.30	0	0.25+/-0.25
2004	1500 4x2	—	120.5	0.75	+4.0	0.50	0	0.10+/-0.10
	1500 4x2	—	140.5	0.75	+4.2	0.50	0	0.10+/-0.10
	1500 4x2	—	160.5	0.75	+4.4	0.50	0	0.10+/-0.10
	1500 4x4	—	120.5	0.75	+4.2	0.50	0	0.10+/-0.10
	1500 4x4	—	140.5	0.75	+4.4	0.50	0	0.10+/-0.10
	1500 4x4	—	160.5	0.75	+4.6	0.50	0	0.10+/-0.10
	2500 & 3500 4x2	—	140	0.75	+4.0	0.50	0	0.10+/-0.05
	2500 & 3500 4x2	—	160	0.75	+4.3	0.50	0	0.10+/-0.05
	2500 & 3500 4x4	—	140	0.75	+4.5	0.25	+0.25	0.10+/-0.05
	2500 & 3500 4x4	—	160	0.75	+4.7	0.25	0.25	0.10+/-0.05
2005	1500 4x2	—	120.5	0.75	+4.0	0.50	0	0.10+/-0.10
	1500 4x2	—	140.5	0.75	+4.2	0.50	0	0.10+/-0.10
	1500 4x2	—	160.5	0.75	+4.4	0.50	0	0.10+/-0.10
	1500 4x4	—	120.5	0.75	+4.2	0.50	0	0.10+/-0.10
	1500 4x4	—	140.5	0.75	+4.4	0.50	0	0.10+/-0.10
	1500 4x4	—	160.5	0.75	+4.6	0.50	0	0.10+/-0.10
	2500 & 3500 4x2	—	140	0.75	+4.0	0.50	0	0.10+/-0.05
	2500 & 3500 4x2	—	160	0.75	+4.3	0.50	0	0.10+/-0.05
	2500 & 3500 4x4	—	140	0.75	+4.5	0.25	0.25	0.10+/-0.05
	2500 & 3500 4x4	—	160	0.75	+4.7	0.25	0.25	0.10+/-0.05

NA: Not adjustable

06009-RAMT-C14

TIRE AND WHEEL SPECIFICATIONS

		OEM Tires		Tire Pressures (psi)		Wheel	Wheel Lug
Year	Model	Standard	Optional	Front	Rear	Size	Nut Torque
2002	1500 PU	P265/70R17	P275/55R20	②	②	Std: 8-J Opt: 9-J	①
	2500 ST PU	LT245/75R16E	LT265/75R16E	②	②	Std: 7.5-J Opt: 8J	①
	2500 SLT PU	LT265/75R16E	None	②	②	8-J	①
	3500 PU	LT235/85R16E	None	②	②	6-J	①
	1500 Cargo Van	P235/75R15	None	②	②	6.5-J	①
	2500 Cargo Van	LT225/75R16D	None	②	②	6.5-J	①
	3500 Cargo Van	LT225/75R16E	None	②	②	6.5-J	①
	1500 Ram Wagon	P235/75R15	None	②	②	6.5-J	①
	2500 Ram Wagon	LT225/75R16D	None	②	②	6.5-J	①
	3500 Ram Wagon	LT245/75R16E	None	②	②	6.5-J	①
2003	1500 PU	P265/70R17	P275/55R20	②	②	Std: 8-J Opt: 9-J	①
	2500 ST PU	LT245/75R16E	LT265/75R16E	②	②	Std: 7.5-J Opt: 8J	①
	2500 SLT PU	LT265/75R16E	None	②	②	8-J	①
	3500 PU	LT235/85R16E	None	②	②	6-J	①
	1500 Cargo Van	P235/75R15	None	②	②	6.5-J	①
	2500 Cargo Van	LT225/75R16D	None	②	②	6.5-J	①
	3500 Cargo Van	LT225/75R16E	None	②	②	6.5-J	①
	1500 Ram Wagon	P235/75R15	None	②	②	6.5-J	①
	2500 Ram Wagon	LT225/75R16D	None	②	②	6.5-J	①
	3500 Ram Wagon	LT245/75R16E	None	②	②	6.5-J	①
2004	1500 PU	P265/70R17	P275/55R20	②	②	Std: 8-J Opt: 9-J	①
	2500 ST PU	LT245/75R16E	LT265/75R16E	②	②	Std: 7.5-J Opt: 8J	①
	2500 SLT PU	LT265/75R16E	None	②	②	8-J	①
	3500 PU	LT235/85R16E	None	②	②	6-J	①
2005	1500 Laramie	P265/70R17	None	②	②	NA	95
	1500 ST/SLT	P245/70R17	P265/70R17	②	②	NA	95
	2500 ST	LT245/70R17E	None	②	②	NA	145
	2500 SLT/Laramie	LT265/70R17E	None	②	②	NA	145
	2500 Power Wagon	LT285/70R17D	None	②	②	NA	145
	3500	LT235/80R17E	LT265/70R17E	②	②	NA	155

OEM: Original Equipment Manufacturer

PSI: Pounds Per Square Inch

STD: Standard

OPT: Optional

① 5 stud wheel: 95 ft. lbs.
8 stud wheel: 135 ft. lbs.
8 stud dual wheel: 145 ft. lbs.

② See sticker on door

06009-RAMT-C15

BRAKE SPECIFICATIONS

All measurements in inches unless noted

			Brake Disc			Brake Drum			Minimum Lining Thickness		Brake Caliper	
Year	Model		Original Thickness	Minimum Thickness	Maximum Run-out	Original Inside Diameter	Max. Wear Limit	Max. Machine Diameter	Front	Rear	Bracket Bolts (ft. lbs.)	Mounting Bolts (ft. lbs.)
2002	B1500 Van		1.26	1.181	0.004	11.03	①	①	②	③	110	15
	B2500 Van		1.26	1.181	0.004	11.03	①	①	②	③	110	15
	B3500 Van		1.26	1.181	0.004	12.125	①	①	②	③	110	15
	Ram 1500 Pick-up		1.18	1.117	0.005	11.00	11.09	11.06	②	③	④	24
	Ram 2500 Pick-up	F	1.50	1.334	0.005	—	—	—	②	—	④	24
		R	1.18	1.117	0.005	—	—	—	—	NA	—	24
	Ram 3500 Pick-up	F	1.50	1.334	0.005	—	—	—	②	—	④	24
		R	1.18	1.117	0.005	—	—	—	—	NA	—	24
2003		NA	NA	NA	NA	NA	NA	NA	NA	NA	⑤	⑤
2004	Ram Pick-up LD	F	1.10	1.039	0.005	—	—	—	NA	—	④	24
		R	0.86	1.117	0.005	—	—	—	—	NA	—	11
	Ram Pick-up HD	F	1.39	1.334	0.005	—	—	—	NA	—	④	24
		R	1.18	1.117	0.005	—	—	—	—	NA	—	11
2005	Ram Pick-up LD	F	NA	NA	0.005	—	—	—	NA	—	④	24
		R	NA	NA	0.005	—	—	—	—	NA	—	11
	Ram Pick-up HD	F	NA	NA	0.005	—	—	—	NA	—	④	24
		R	NA	NA	0.005	—	—	—	—	NA	—	11

NA: Not Available

① Maximum allowable drum diameter, either from wear or machining is stamped on the drum

② Riveted brake pads: 0.0625 in.
Bonded brake pads: 0.1875 in.

③ Riveted brake shoes: 0.031 in.
Bonded brake shoes: 0.0625 in.

④ LD adapter: 130 ft. lbs.
HD adapter: 210 ft. lbs.

⑤ See procedures in the text

06009-RAMT-C16

SCHEDULED MAINTENANCE INTERVALS

2002-05 RAM PICKUP (Exc. 8.0L)

TO BE SERVICED	TYPE OF SERVICE	VEHICLE MILEAGE INTERVAL (x1000)														
		7.5	15	22.5	30	37.5	45	52.5	60	67.5	75	82.5	90	97.5	100	105
Engine oil & filter	R	✓	✓	✓	✓	✓	✓	✓	✓	✓	✓	✓	✓	✓		✓
Engine coolant & hoses	I	✓	✓	✓	✓	✓	✓	✓	✓	✓	✓	✓	✓	✓		✓
Brake hoses	I	✓	✓	✓	✓	✓	✓	✓	✓	✓	✓	✓	✓	✓		✓
Steering linkage	L	✓	✓	✓	✓	✓	✓	✓	✓	✓	✓	✓	✓	✓		✓
Manual trans. fluid level	I	✓	✓	✓	✓	✓	✓	✓	✓	✓	✓	✓	✓	✓		✓
Exhaust system	I	✓	✓	✓	✓	✓	✓	✓	✓	✓	✓	✓	✓	✓		✓
Brake linings	I			✓			✓			✓			✓			
Front wheel bearings (2wd)	S/I			✓			✓			✓			✓			
Ball joints	L			✓			✓			✓			✓			
Air cleaner element	R				✓				✓				✓			
Spark plugs	R				✓				✓				✓			
Transfer case fluid	R					✓					✓					
Rear drum brakes	Adj	✓	✓	✓	✓	✓	✓	✓	✓	✓	✓	✓	✓	✓		✓
Tires	Rotate	✓	✓	✓	✓	✓	✓	✓	✓	✓	✓	✓	✓	✓		✓
Engine coolant ①	R							✓								✓
Spark plug wires	R								✓							
PCV valve	S/I								✓							
Drive belt tensioner	S/I										✓					
Automatic trans. fluid	R														✓	
Automatic trans. bands	Adj														✓	

R: Replace S/I: Service or Inspect Adj: Adjust L: Lubricate

① Replace every 36 months, regardless of mileage.

FREQUENT OPERATION MAINTENANCE (SEVERE SERVICE)

If a vehicle is operated under any of the following conditions it is considered severe service:

- Extremely dusty areas.

- 50% or more of the vehicle operation is in 32°C (90°F) or higher temperatures, or constant operation in temperatures below 0°C (32°F).

- Prolonged idling (vehicle operation in stop and go traffic.

- Frequent short running periods (engine does not warm to normal operating temperatures).

- Police, taxi, delivery usage or trailer towing usage.

Oil & oil filter change: change every 3000 miles.

Air filter/air pump air filter: change every 24,000 miles.

Engine coolant level, hoses & clamps: check every 6,000 miles.

Exhaust system: check every 6000 miles.

Drive belts: check every 18,000 miles; replace every 24,000 miles.

Crankcase inlet air filter (6 & 8 cyl.): clean every 24,000 miles.

Oxygen sensor: replace every 82,500 miles.

Automatic transmission fluid, filter & bands: change & adjust every 12,000 miles.

Steering linkage: lubricate every 6000 miles.

Rear axle fluid: change every 12,000 miles.

06009-RAMT-C17

SCHEDULED MAINTENANCE INTERVALS

2002-03 Ram Van

		VEHICLE MILEAGE INTERVAL (x1000)														
TO BE SERVICED	TYPE OF SERVICE	7.5	15	22.5	30	37.5	45	52.5	60	67.5	75	82.5	90	97.5	100	105
Engine oil & filter	R	✓	✓	✓	✓	✓	✓	✓	✓	✓	✓	✓	✓	✓		✓
Engine coolant & hoses	I	✓	✓	✓	✓	✓	✓	✓	✓	✓	✓	✓	✓	✓		✓
Brake hoses	I	✓	✓	✓	✓	✓	✓	✓	✓	✓	✓	✓	✓	✓		✓
Steering linkage	L		✓		✓		✓		✓		✓		✓		✓	
Exhaust system	I	✓	✓	✓	✓	✓	✓	✓	✓	✓	✓	✓	✓	✓		✓
Brake linings	I			✓			✓			✓			✓			✓
Front wheel bearings	S/I			✓			✓			✓			✓			✓
Ball joints	L			✓			✓			✓			✓			✓
Air cleaner element	R				✓				✓				✓			
Spark plugs	R				✓				✓				✓			
Tires	Rotate	✓	✓	✓	✓	✓	✓	✓	✓	✓	✓	✓	✓	✓		✓
Engine coolant ①	R						✓						✓			
Spark plug wires	R								✓							
PCV valve	S/I								✓							
Drive belt tensioner	S/I								✓							
Automatic trans. fluid	R														✓	
Automatic trans. bands	Adj														✓	

R: Replace S/I: Service or Inspect Adj: Adjust L: Lubricate

① Replace every 36 months, regardless of mileage.

FREQUENT OPERATION MAINTENANCE (SEVERE SERVICE)

If a vehicle is operated under any of the following conditions it is considered severe service:

- **Extremely dusty areas.**
- **50% or more of the vehicle operation is in 32°C (90°F) or higher temperatures, or constant operation in temperatures below 0°C (32°F).**
- **Prolonged idling (vehicle operation in stop and go traffic.**
- **Frequent short running periods (engine does not warm to normal operating temperatures).**
- **Police, taxi, delivery usage or trailer towing usage.**

Oil & oil filter change: change every 3000 miles.

Air filter/air pump air filter: change every 24,000 miles.

Engine coolant level, hoses & clamps: check every 6,000 miles.

Exhaust system: check every 6000 miles.

Drive belts: check every 18,000 miles; replace every 24,000 miles.

Crankcase inlet air filter (6 & 8 cyl.): clean every 24,000 miles.

Oxygen sensor: replace every 82,500 miles.

Automatic transmission fluid, filter & bands: change & adjust every 12,000 miles.

Steering linkage: lubricate every 6000 miles.

Rear axle fluid: change every 12,000 miles.

06009-RAMT-C18

SCHEDULED MAINTENANCE INTERVALS

2002-05 Medium Duty (California 8.0L 2500 & 3500 only)

TO BE SERVICED	TYPE OF SERVICE	VEHICLE MILEAGE INTERVAL (x1000)														
		6	12	18	24	30	36	42	48	54	60	66	72	78	84	90
Engine oil and filter	R	✓	✓	✓	✓	✓	✓	✓	✓	✓	✓	✓	✓	✓	✓	✓
Brake linings	I			✓			✓			✓			✓			✓
Front wheel bearings (2wd)	S/I			✓			✓			✓			✓			✓
Auto trans fluid and filter	R				✓				✓				✓			
Auto trans bands	Adj				✓				✓				✓			
Air cleaner element	R					✓					✓					✓
Spark plugs	R					✓					✓					✓
Exhaust system	I	✓	✓	✓	✓	✓	✓	✓	✓	✓	✓	✓	✓	✓	✓	✓
Brake hoses	I	✓	✓	✓	✓	✓	✓	✓	✓	✓	✓	✓	✓	✓	✓	✓
Rear drum brakes	Adj	✓	✓	✓	✓	✓	✓	✓	✓	✓	✓	✓	✓	✓	✓	✓
Tires	Rotate	✓	✓	✓	✓	✓	✓	✓	✓	✓	✓	✓	✓	✓	✓	✓
Steering linkage	L	✓	✓	✓	✓	✓	✓	✓	✓	✓	✓	✓	✓	✓	✓	✓
Transfer case fluid	R						✓						✓			
Engine coolant ①	R						✓						✓			
Spark plug wires	R										✓					

R: Replace S/I: Service or Inspect Adj: Adjust L: Lubricate

① Replace every 36 months, regardless of mileage.

FREQUENT OPERATION MAINTENANCE (SEVERE SERVICE)

If a vehicle is operated under any of the following conditions it is considered severe service:

- **Extremely dusty areas.**
- **50% or more of the vehicle operation is in 32°C (90°F) or higher temperatures, or constant operation in temperatures below 0°C (32°F).**
- **Prolonged idling (vehicle operation in stop and go traffic.**
- **Frequent short running periods (engine does not warm to normal operating temperatures).**
- **Police, taxi, delivery usage or trailer towing usage.**

Oil & oil filter change: change every 3000 miles.

Air filter/air pump air filter: change every 24,000 miles.

Engine coolant level, hoses & clamps: check every 6,000 miles.

Exhaust system: check every 6000 miles.

Drive belts: check every 18,000 miles; replace every 24,000 miles.

Crankcase inlet air filter (6 & 8 cyl.): clean every 24,000 miles.

Oxygen sensor: replace every 82,500 miles.

Automatic transmission fluid, filter & bands: change & adjust every 12,000 miles.

Steering linkage: lubricate every 6000 miles.

Rear axle fluid: change every 12,000 miles.

06009-RAMT-C19

SCHEDULED MAINTENANCE INTERVALS

2002-05 Heavy Duty (Federal 2500 8.0L and 3500 5.9L & 8.0L)

TO BE SERVICED	TYPE OF SERVICE	VEHICLE MILEAGE INTERVAL (x1000)														
		6	12	18	24	30	36	42	48	54	60	66	72	78	82.5	84
Engine oil and filter	R	✓	✓	✓	✓	✓	✓	✓	✓	✓	✓	✓	✓	✓		✓
Air cleaner element (8.0L)	R		✓				✓				✓					✓
Brake linings	I			✓			✓			✓			✓			
Front wheel bearings (4wd)	S/I			✓			✓			✓			✓			
Air cleaner element (5.9L)	R				✓				✓				✓			
Air pump filter	R				✓				✓				✓			
Crankcase inlet air filter (5.9L)	C/L				✓				✓				✓			
Auto trans fluid & filter	R				✓				✓				✓			
Auto trans bands	Adj				✓				✓				✓			
Front wheel bearings (2wd)	S/I				✓				✓				✓			
Spark plugs	R					✓					✓					
Transfer case fluid	R						✓						✓			
Exhaust system	I	✓	✓	✓	✓	✓	✓	✓	✓	✓	✓	✓	✓	✓		✓
Brake hoses	I	✓	✓	✓	✓	✓	✓	✓	✓	✓	✓	✓	✓	✓		✓
Tires	Rotate	✓	✓	✓	✓	✓	✓	✓	✓	✓	✓	✓	✓	✓		✓
Steering linkage	L	✓	✓	✓	✓	✓	✓	✓	✓	✓	✓	✓	✓	✓		✓
Engine coolant ①	R						✓						✓			
Spark plug wires	R										✓					
PCV valve (5.9L)	R										✓					
Distributor cap & rotor (5.9L)	R										✓					
Oxygen sensor (5.9L)	R														✓	

R: Replace S/I: Service or Inspect Adj: Adjust C/L: Clean & lubricate

① Replace every 36 months, regardless of mileage.

FREQUENT OPERATION MAINTENANCE (SEVERE SERVICE)

If a vehicle is operated under any of the following conditions it is considered severe service:

- Extremely dusty areas.

- 50% or more of the vehicle operation is in 32°C (90°F) or higher temperatures, or constant operation in temperatures below 0°C (32°F).

- Prolonged idling (vehicle operation in stop and go traffic.

- Frequent short running periods (engine does not warm to normal operating temperatures).

- Police, taxi, delivery usage or trailer towing usage.

Oil & oil filter change: change every 3000 miles.

Air filter/air pump air filter: change every 24,000 miles.

Engine coolant level, hoses & clamps: check every 6,000 miles.

Exhaust system: check every 6000 miles.

Drive belts: check every 18,000 miles; replace every 24,000 miles.

Crankcase inlet air filter (6 & 8 cyl.): clean every 24,000 miles.

Oxygen sensor: replace every 82,500 miles.

Automatic transmission fluid, filter & bands: change & adjust every 12,000 miles.

Steering linkage: lubricate every 6000 miles.

Rear axle fluid: change every 12,000 miles.

06009-RAMT-C20

SCHEDULED MAINTENANCE INTERVALS

2005 Turbo Diesel

TO BE SERVICED	TYPE OF SERVICE	VEHICLE MILEAGE INTERVAL (x1000)													
		7.5	15	22.5	30	37.5	45	52.5	60	67.5	75	82.5	90	97.5	100
Engine oil & filter	R		✓		✓		✓		✓		✓		✓		
Crankcase breather canister	D	✓	✓	✓	✓	✓	✓	✓	✓	✓	✓	✓	✓	✓	
Water pump weep hole	I		✓		✓		✓		✓		✓		✓		
Fuel filter	R		✓		✓		✓		✓		✓		✓		
Fuel sensor	C		✓		✓		✓		✓		✓		✓		
Drive belts	I			✓			✓			✓			✓		
Brake linings	I			✓			✓			✓			✓		
Fan hub	I				✓				✓				✓		
Damper	I				✓				✓				✓		
Accessory drive belt	S/I			✓			✓			✓			✓		
Transfer case fluid	I				✓				✓				✓		
Auto trans & filter	R														✓
Auto trans bands	Adj				✓				✓				✓		
Front wheel bearings (2WD)	S/I				✓				✓				✓		
Engine coolant ①	R				✓				✓				✓		
Exhaust system	I	✓	✓	✓	✓	✓	✓	✓	✓	✓	✓	✓	✓	✓	
Steering linkage	L	✓	✓	✓	✓	✓	✓	✓	✓	✓	✓	✓	✓	✓	
Rear drum brakes	Adj	✓	✓	✓	✓	✓	✓	✓	✓	✓	✓	✓	✓	✓	
Tires	Rotate	✓	✓	✓	✓	✓	✓	✓	✓	✓	✓	✓	✓	✓	

R: Replace S/I: Service or Inspect D: Drain C: Clean Adj: Adjust

① Replace every 36 months, regardless of mileage

FREQUENT OPERATION MAINTENANCE (SEVERE SERVICE)

If a vehicle is operated under any of the following conditions it is considered severe service:

- Extremely dusty areas.

- 50% or more of the vehicle operation is in 32°C (90°F) or higher temperatures, or constant operation in temperatures below 0°C (32°F).

- Prolonged idling (vehicle operation in stop and go traffic.

- Frequent short running periods (engine does not warm to normal operating temperatures).

- Police, taxi, delivery usage or trailer towing usage.

Oil & oil filter change: change every 7500 miles.

Air filter/air pump air filter: change every 24,000 miles.

Engine coolant level, hoses & clamps: check every 6,000 miles.

Brake linings: check every 15,000 miles

Exhaust system: inspect every 6000 miles.

Drive belts: check every 18,000 miles; replace every 24,000 miles.

Crankcase inlet air filter (6 & 8 cyl.): clean every 24,000 miles.

Oxygen sensor: replace every 82,500 miles.

Automatic transmission fluid, filter & bands: change & adjust every 30,000 miles.

Manual transmission fluid: change every 75,000 miles

Steering linkage: lubricate every 6000 miles.

Rear axle fluid: change every 15,000 miles.

Valve lash: adjust every 150,000 miles.

06009-RAMT-C21

PRECAUTIONS

Before servicing any vehicle, please be sure to read all of the following precautions, which deal with personal safety, prevention of component damage, and important points to take into consideration when servicing a motor vehicle:

- Never open, service or drain the radiator or cooling system when the engine is hot; serious burns can occur from the steam and hot coolant.
- Observe all applicable safety precautions when working around fuel. Whenever servicing the fuel system, always work in a well-ventilated area. Do not allow fuel spray or vapors to come in contact with a spark, open flame, or excessive heat (a hot drop light, for example). Keep a dry chemical fire extinguisher near the work area. Always keep fuel in a container specifically designed for fuel storage; also, always properly seal fuel containers to avoid the possibility of fire or explosion. Refer to the additional fuel system precautions later in this section.
- Fuel injection systems often remain pressurized, even after the engine has been turned **OFF**. The fuel system pressure must be relieved before disconnecting any fuel lines. Failure to do so may result in fire and/or personal injury.
- Brake fluid often contains polyglycol ethers and polyglycols. Avoid contact with the eyes and wash your hands thoroughly after handling brake fluid. If you do get brake fluid in your eyes, flush your eyes with clean, running water for 15 minutes. If eye irritation persists, or if you have taken brake fluid internally, IMMEDIATELY seek medical assistance.
- The EPA warns that prolonged contact with used engine oil may cause a number of skin disorders, including cancer! You should make every effort to minimize your exposure to used engine oil. Protective gloves should be worn when changing oil. Wash your hands and any other exposed skin areas as soon as possible after exposure to used engine oil. Soap and water, or waterless hand cleaner should be used.
- All new vehicles are now equipped with an air bag system, often referred to as a Supplemental Restraint System (SRS) or Supplemental Inflatable Restraint (SIR) system. The system must be disabled before performing service on or around system components, steering column, instrument panel components, wiring and sensors. Failure to follow safety and disabling procedures could result in accidental air bag deployment, possible personal injury and unnecessary system repairs.
- Always wear safety goggles when working with, or around, the air bag system. When carrying a non-deployed air bag, be sure the bag and trim cover are pointed away from your body. When placing a non-deployed air bag on a work surface, always face the bag and trim cover upward, away from the surface. This will reduce the motion of the module if it is accidentally deployed. Refer to the additional air bag system precautions later in this section.
- Clean, high quality brake fluid from a sealed container is essential to the safe and proper operation of the brake system. You should always buy the correct type of brake fluid for your vehicle. If the brake fluid becomes contaminated, completely flush the system with new fluid. Never reuse any brake fluid. Any brake fluid that is removed from the system should be discarded. Also, do not allow any brake fluid to come in contact with a painted surface; it will damage the paint.
- Never operate the engine without the proper amount and type of engine oil; doing so WILL result in severe engine damage.
- Timing belt maintenance is extremely important! Many models utilize an interference-type, non-freewheeling engine. If the timing belt breaks, the valves in the cylinder head may strike the pistons, causing potentially serious (also time-consuming and expensive) engine damage. Refer to the maintenance interval charts in the front of this manual for the recommended replacement interval for the timing belt, and to the timing belt section for belt replacement and inspection.
- Disconnecting the negative battery cable on some vehicles may interfere with the functions of the on-board computer system(s) and may require the computer to undergo a relearning process once the negative battery cable is reconnected.
- When servicing drum brakes, only disassemble and assemble one side at a time, leaving the remaining side intact for reference.
- Only an MVAC-trained, EPA-certified automotive technician should service the air conditioning system or its components.

GASOLINE ENGINE REPAIR

➡Disconnecting the negative battery cable on some vehicles may interfere with the functions of the on board computer system. The computer may undergo a relearning process once the negative battery cable is reconnected.

Distributor

➡The 3.7L, 4.7L and 5.7L engines do not use a distributor.

REMOVAL

3.9L, 5.2L and 5.9L Engines

1. Before servicing the vehicle, refer to the precautions in the beginning of this section.
2. Remove or disconnect the following:
   - Negative battery cable
   - Air cleaner tube
   - Distributor cap
   - Camshaft Position (CMP) sensor connector
3. Matchmark the distributor housing and the rotor.
4. Matchmark the distributor housing and the intake manifold.
5. Remove the distributor.

INSTALLATION

Timing Not Disturbed

3.9L, 5.2L AND 5.9L ENGINES

1. Before servicing the vehicle, refer to the precautions in the beginning of this section.

➡The rotor will rotate clockwise as the gears engage.

2. Position the rotor slightly counterclockwise of the matchmark made during removal.
3. Install the distributor.
4. Align the distributor housing and intake manifold matchmarks and check that the distributor housing matchmark and rotor are also aligned.
5. Install or connect the following:
   - Distributor housing clamp and bolt. Tighten the bolt to 17 ft. lbs. (23 Nm).
   - CMP sensor connector
   - Distributor cap
   - Air cleaner tube
   - Negative battery cable

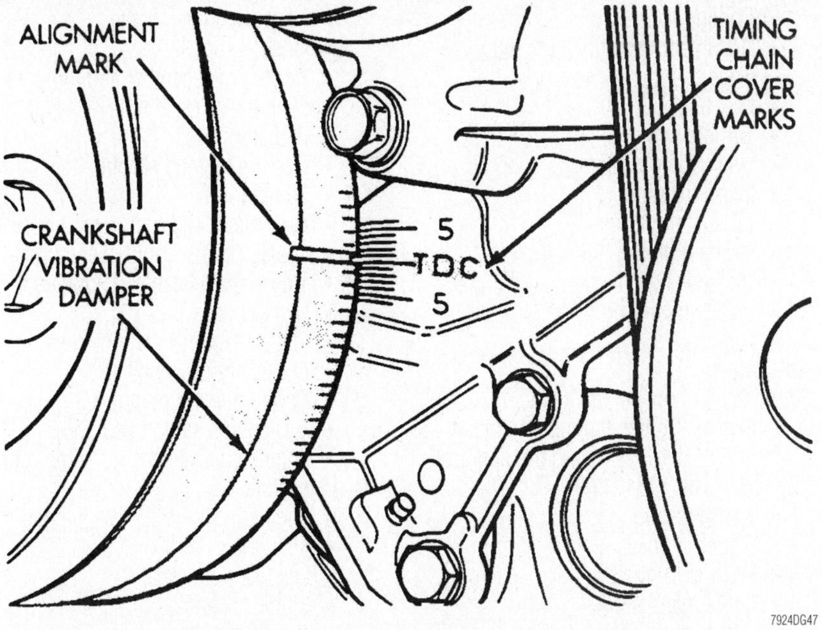

Crankshaft pulley and timing chain cover marks aligned at Top Dead Center (TDC)

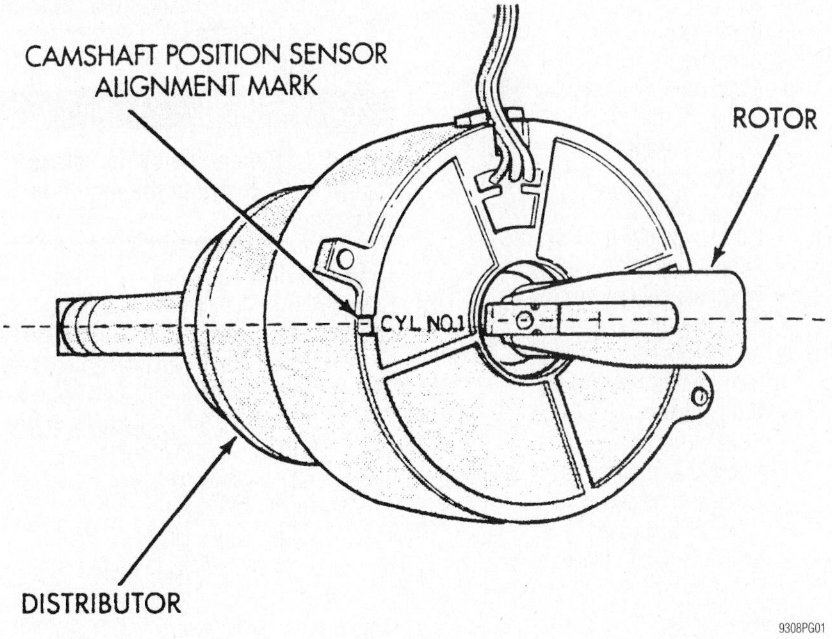

Distributor rotor alignment—3.9L, 5.2L and 5.9L engines

Timing Disturbed

3.9L, 5.2L AND 5.9L ENGINES

1. Before servicing the vehicle, refer to the precautions in the beginning of this section.
2. Set the engine at Top Dead Center (TDC) of the No. 1 cylinder compression stroke.
3. Install the distributor, clamp and bolt.
4. Rotate the distributor so that the rotor aligns with the No. 1 cylinder mark on the Camshaft Position (CMP) sensor. Tighten the clamp bolt to 17 ft. lbs. (23 Nm).
5. Install or connect the following:
   - CMP sensor connector
   - Distributor cap
   - Air cleaner tube
   - Negative battery cable
6. The final distributor position must be set with a scan tool. Follow the instructions supplied by the scan tool manufacturer.

Alternator

REMOVAL

3.9L, 5.2L, 5.9L and 8.0L engines

1. Before servicing the vehicle, refer to the precautions in the beginning of this section.
2. Remove or disconnect the following:
   - Negative battery cable
   - Accessory drive belt
   - Alternator harness connectors
   - Alternator

3.7L, 4.7L and 5.7L Engines

1. Before servicing the vehicle, refer to the precautions in the beginning of this section.
2. Remove or disconnect the following:
   - Negative battery cable
   - Accessory drive belt
   - Alternator harness connectors
   - Support bracket nuts and bolt
   - Mounting bolts and alternator

➡The 3.7L and 4.7L engines have 1 vertical and 2 horizontal bolts.

INSTALLATION

3.9L, 5.2L, 5.9L and 8.0L engines

1. Before servicing the vehicle, refer to the precautions in the beginning of this section.
2. Install the alternator and tighten the bolts to 30 ft. lbs. (41 Nm).
3. Install or connect the following:
   - Alternator harness connectors
   - Accessory drive belt
   - Negative battery cable

3.7L and 4.7L engines

1. Before servicing the vehicle, refer to the precautions in the beginning of this section.
2. Install the alternator and tighten the bolts to the following specifications:
   - 2003-2004 3.7L engines: Tighten the horizontal bolts to 42 ft. lbs. (57 Nm), then the vertical bolt to 29 ft. lbs. (40 Nm)
   - 2005 3.7L engines: Vertical bolt and long horizontal bolt to 40 ft. lbs. (55 Nm), short horizontal bolt to 55 ft. lbs. (74 Nm)
   - 4.7L engines: Vertical bolt and long horizontal bolt to 41 ft. lbs. (56 Nm), short horizontal bolt to 55 ft. lbs. (74 Nm)

3. Install or connect the following:
   - Alternator harness connectors
   - Accessory drive belt
   - Negative battery cable

5.7L engines

1. Before servicing the vehicle, refer to the precautions in the beginning of this section.
2. Install or connect the following:
   - Alternator and tighten the bolts to 30 ft. lbs. (41 Nm)
   - Support bracket and tighten the nuts and bolt to 30 ft. lbs. (41 Nm)
   - Alternator harness connectors
   - Accessory drive belt
   - Negative battery cable

Ignition Timing

ADJUSTMENT

The ignition timing is controlled by the Powertrain Control Module (PCM) and is not adjustable.

Engine Assembly

REMOVAL & INSTALLATION

Ram Van

1. Before servicing the vehicle, refer to the precautions in the beginning of this section.
2. Drain the cooling system.
3. Recover the A/C refrigerant, if equipped.
4. Drain the engine oil.
5. Relieve the fuel system pressure.
6. Remove or disconnect the following:
   - Negative battery cable
   - Hood
   - Grille
   - Radiator support brace
   - Inside engine cover
   - Air cleaner assembly
   - Transmission oil cooler, if equipped
   - Radiator hoses
   - Heater hoses
   - Radiator and fan shroud
   - A/C condenser
   - Accessory drive belt
   - Power steering pump
   - Air injection pump
   - Intake manifold vacuum lines
   - Washer solvent bottle
   - A/C compressor, if equipped
   - Throttle linkage
   - Engine control sensor harness connectors
   - Alternator
   - Cooling fan
   - Distributor cap and spark plug wires
   - Fuel line
   - Throttle body
   - Fuel supply manifold
   - Intake manifold
   - Exhaust front pipe
   - Starter motor
   - Transmission
   - Engine mounts
   - Engine

To install:

7. Install or connect the following:
   - Engine. Tighten the mount bolts to 30 ft. lbs. (41 Nm) and the nuts to 75 ft. lbs. (101 Nm).
   - Transmission
   - Starter motor
   - Exhaust front pipe
   - Intake manifold
   - Fuel supply manifold
   - Throttle body
   - Fuel line
   - Distributor cap and spark plug wires
   - Cooling fan
   - Alternator
   - Engine control sensor harness connectors
   - Throttle linkage
   - A/C compressor, if equipped
   - Washer solvent bottle
   - Intake manifold vacuum lines
   - Air injection pump
   - Power steering pump
   - Accessory drive belt
   - A/C condenser
   - Radiator and fan shroud
   - Heater hoses
   - Radiator hoses
   - Transmission oil cooler, if equipped
   - Air cleaner assembly
   - Inside engine cover
   - Radiator support brace
   - Grille
   - Hood
   - Negative battery cable
8. Fill the crankcase to the correct level.
9. Fill the cooling system.
10. Start the engine and check for leaks.

Ram Truck

3.9L, 5.2L, 5.9L AND 8.0L ENGINES

1. Before servicing the vehicle, refer to the precautions in the beginning of this section.
2. Drain the cooling system.
3. Drain the engine oil.
4. Relieve the fuel system pressure.
5. Remove or disconnect the following:
   - Negative battery cable
   - Hood
   - Upper crossmember and top core support
   - Radiator hoses
   - Cooling fan and shroud
   - Radiator
   - Accelerator cable
   - Cruise control cable, if equipped
   - Transmission cable, if equipped
   - Heater hoses
   - Intake manifold vacuum lines
   - Accessory drive belt
   - Power steering pump
   - A/C compressor, if equipped
   - Engine control sensor harness connectors
   - Engine block heater, if equipped
   - Fuel line
   - Exhaust front pipe
   - Starter motor
   - Torque converter, if equipped
   - Transmission oil cooler lines, if equipped
   - Engine mounts
   - Transmission flange bolts. Support the transmission.
   - Engine

⁂ WARNING

Do not lift the engine by the intake manifold. Damage to the manifold may result.

To install:

6. Install or connect the following:
   - Engine. Tighten the engine mount bolts to 70 ft. lbs. (95 Nm).
   - Transmission flange bolts. Tighten the bolts to 40–45 ft. lbs. (54–61 Nm).
   - Transmission oil cooler lines, if equipped
   - Torque converter, if equipped. Tighten the bolts to 23 ft. lbs. (31 Nm).
   - Starter motor
   - Exhaust front pipe
   - Fuel line
   - Engine block heater, if equipped
   - Engine control sensor harness connectors
   - A/C compressor, if equipped
   - Power steering pump
   - Accessory drive belt
   - Intake manifold vacuum lines
   - Heater hoses
   - Transmission cable, if equipped
   - Cruise control cable, if equipped
   - Accelerator cable

- Radiator
- Cooling fan and shroud
- Radiator hoses
- Upper crossmember and top core support
- Hood
- Negative battery cable

7. Fill the crankcase to the correct level.
8. Fill the cooling system.
9. Start the engine and check for leaks.

3.7L ENGINE

1. Before servicing the vehicle, refer to the precautions in the beginning of this section.
2. Properly relieve the fuel system pressure.
3. Drain the cooling system.
4. Drain the engine oil.
5. Remove or disconnect the following:
   - Negative battery cable
   - Hood
   - Air cleaner assembly
   - Radiator
   - Electric and mechanical fan assemblies
   - A/C compressor, if equipped, and secure it out of the way with the lines attached. DO NOT DISCHARGE!
   - Power steering pump, with the lines attached
   - Alternator
   - Coolant bottle
   - Heater hoses
   - Accelerator and speed control cables
   - Lower and upper radiator hoses
   - Engine ground straps
   - Intake Air Temperature (IAT) sensor
   - Fuel injection wiring connectors
   - Throttle Position (TP) sensor
   - Idle Air Control (IAC) motor
   - Oil pressure sender connector
   - Engine Coolant Temperature (ECT) sensor
   - Manifold Absolute Pressure (MAP) sensor
   - Camshaft position sensor
   - Ignition coil wiring connector
   - Crankshaft Position (CKP) sensor
   - Coil pack
   - Fuel rail
   - PCV hose
   - Vacuum hoses from the intake manifold
   - Knock sensor connectors
   - Oil dipstick tube
   - Intake manifold
   - Heated Oxygen (HO_2S) sensor connector
   - Block heater connector
   - Front driveshaft at the differential
   - Starter
   - Structural cover
   - With a manual transmission, remove the transmission
   - Torque converter bolts and matchmark the converter
   - Automatic transmission-to-engine bolts
   - Exhaust front pipes
   - Left and right engine mounts
6. Place a support stand under the transmission.
7. Install an engine lift plate
8. Lift the engine out of the vehicle.

To install:

9. If equipped with a manual transmission, install the transmission
10. Lower the engine and install the mounts. Don't tighten the bolts yet.
11. If equipped with an automatic transmission, perform the following steps:
    a. Align the torque converter housing to the engine.
    b. Torque the bolts to 30 ft. lbs. (41 Nm).
12. Install or connect the following:
    - Install the torque converter to flexplate bolts. Torque the bolts to 50 ft. lbs. (68 Nm)
    - Torque the through bolts to 45 ft. lbs. (61 Nm)
    - Engine ground strap
    - Starter motor
    - CKP sensor
    - Block heater cable
    - Structural cover

✻✻ WARNING

The structural cover must be held tightly against the engine and bellhousing during tightening. The torque for all bolts is 40 ft. lbs. (54 Nm); the bolts must be tightened in the order shown.

- Exhaust pipes. New flange clamps MUST be used!
- HO_2S sensor connectors
- KS sensors
- Intake Manifold
- Dipstick tube
- Vacuum hoses to the intake manifold
- PCV and breather hoses
- Fuel rail
- Ignition coil
- IAT sensor
- Fuel injector connectors
- TP sensor
- IAC motor
- Oil pressure sender
- ECT sensor electrical connector
- MAP sensor
- CMP sensor
- Radiator hoses
- Cruise control cable, if equipped
- Throttle cable
- Heater hoses
- Coolant bottle
- Power steering pump
- Alternator
- A/C compressor
- Radiator
- Fan assemblies

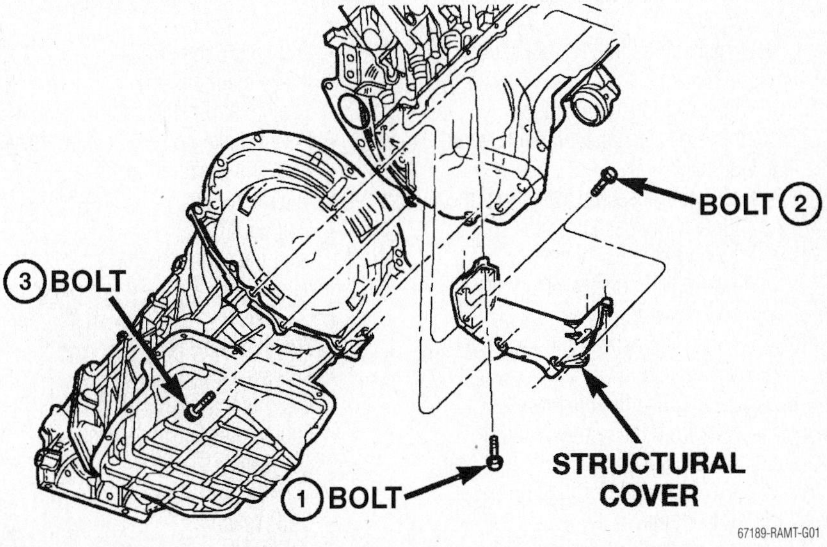

Tighten the structural cover bolts in this order—3.7L, 4.7L and 5.7L engines

- Air cleaner assembly
- Negative battery cable

13. Fill and bleed the power steering system.
14. Fill the engine with clean oil.
15. Start the engine and check for leaks, repair if necessary.

2005 3.7L ENGINE

1. Before servicing the vehicle, refer to the precautions in the beginning of this section.
2. Drain the engine oil.
3. Drain the cooling system.
4. Relieve fuel system pressure.
5. Remove or disconnect the following:
   - Negative battery cable
   - Radiator support bracket
   - Radiator fan
   - Accessory drive belt
   - A/C compressor and secure away from the engine with lines attached
   - Alternator
   - Power steering pump and secure away from the engine with lines attached
   - Heater hoses
   - Accelerator and speed control cables
   - Radiator hoses
   - Radiator
   - Engine ground straps
6. Remove the engine wiring harness from the following locations:
   - Intake air temperature (IAT) sensor
   - Fuel injectors
   - Throttle position (TPS) switch
   - Idle air control (IAC) motor
   - Engine oil pressure switch
   - Engine coolant temperature (ECT) sensor
   - Manifold absolute pressure (MAP) sensor
   - Camshaft position (CMP) sensor
   - Coil pack
   - Crankshaft position sensor
7. Remove or disconnect the following:
   - Coil pack
   - Fuel rail and secure away from the engine with lines attached
   - PCV and breather hoses
   - Power brake booster vacuum hose
   - Knock sensors
   - Engine oil dipstick tube assembly
   - Intake manifold
8. Attach engine lifting fixture 8247 using the bolts from the removed intake manifold and fuel rail.
9. Secure the engine wiring harnesses away from the engine.
10. Raise and support the vehicle.
11. Remove or disconnect the following:
    - Oxygen (O_2) sensor wiring
    - Crankshaft position sensor
    - Engine block heater power cable, if equipped
    - Front driveshaft from the differential
    - Starter
    - Left and right exhaust pipes from the manifold and crossover
    - Structural cover
12. Matchmark and remove the torque converter bolts, if equipped.
13. Remove the transmission bellhousing to engine bolts.
14. Remove the left and right engine mount thru bolts.
15. Lower the vehicle and support the transmission with a suitable jack.
16. Attach a suitable lifting device to the engine and remove the engine from the vehicle.

To install:

17. Lower the engine into the vehicle.
18. Align the engine with the left and right engine mounts, but do not tighten.
19. Raise and support the vehicle.
20. Install the transmission bellhousing to engine bolts and tighten to 30 ft. lbs. (41 Nm).
21. Tighten the engine mount bolts.
22. Install the torque converter bolts, if equipped.
23. Install or connect the following:
    - Starter
    - Crankshaft position sensor
    - Engine block heater power cable, if equipped
    - Structural cover
    - Left and right exhaust pipes using new clamps
    - O_2 sensors
24. Lower the vehicle and remove the engine lift plate.
25. Install or connect the following:
    - Knock sensors
    - Engine ground straps
    - Intake manifold
    - Engine oil dipstick tube assembly
    - Power brake booster vacuum hose
    - PCV and breather hoses
    - Fuel rails
    - Coil pack
26. Install the engine wiring harness to the following points:
    - IAT sensor
    - Fuel injectors
    - TPS switch
    - IAC motor
    - Engine oil pressure switch
    - ECT sensor
    - MAP sensor
    - CMP sensor
    - Coil pack
    - Crankshaft position sensor
27. Install or connect the following:
    - Radiator and radiator hoses
    - Accelerator and speed control cables
    - Heater hoses
    - Power steering pump
    - Alternator
    - A/C compressor
    - Accessory drive belt
    - Radiator fan and radiator support bracket
    - Air intake assembly
    - Negative battery cable
28. Fill the engine with oil.
29. Fill the cooling system to the correct level.
30. Start the engine and check for leaks.

4.7L ENGINE

1. Before servicing the vehicle, refer to the precautions in the beginning of this section.
2. Drain the cooling system and engine oil.
3. Remove or disconnect the following:
   - Negative battery cable
   - Battery and tray
   - Exhaust crossover pipe
   - On 4wd, the axle vent tube
   - Left and right engine mount through bolts
   - On 4wd, the left and right engine mount bracket locknuts
   - Ground straps
   - CKP sensor
   - On 4wd, the axle isolator bracket
   - Structural cover
   - Starter
   - Torque converter bolts
   - Transmission-to-engine bolts
   - Engine block heater
   - Resonator and air inlet
   - Throttle and speed control cables
   - Crankcase breathers
   - A/C compressor
   - Shroud and fan assemblies
   - Transmission cooler lines
   - Radiator hoses
   - Radiator
   - Alternator
   - Heater hoses
   - Engine harness
   - Vacuum lines
   - Fuel system pressure
   - Fuel line at the rail
   - Power steering pump
4. Install lifting eyes and take up the weight of the engine with a crane.
5. Support the transmission with a jack.
6. Remove the engine.

To install:

7. Installation is the reverse of removal. Observe the following:

- Left and right engine mount through bolts: 2wd 70 ft. lbs. (95 Nm); 4wd 75 ft. lbs. (102 Nm)
- On 4wd, the bracket locknuts: 30 ft. lbs. (41 Nm)
- Transmission-to-engine bolts: 30 ft. lbs. (41 Nm)

⁂ WARNING

The structural cover must be held tightly against the engine and bell-housing during tightening. The torque for all bolts is 40 ft. lbs. (54 Nm); the bolts must be tightened in the order shown.

2005 4.7L ENGINE

1. Before servicing the vehicle, refer to the precautions in the beginning of this section.
2. Drain the cooling system.
3. Drain the engine oil.
4. Relieve the fuel system pressure.
5. Discharge the A/C system.
6. Raise and support the vehicle.
7. Remove or disconnect the following:
   - Negative battery cable
   - Exhaust crossover pipe
   - Axle vent tube from engine mount—4x4 models only
   - Retaining bolt from left and right engine mounts
   - Locknut from left and right engine mount brackets—4x4 models only
   - Engine ground straps
   - Crankshaft position sensor
   - Axle isolator bracket from engine, transmission and axle—4x4 models with equipped with automatic transmission only
   - Structural cover
   - Starter
   - Torque converter bolts, if equipped
   - Transmission to engine mounting bolts
   - Engine block heater power cable, if equipped

   lower
   - Throttle body resonator assembly
   - Air intake hose
   - Accelerator and speed control cables
   - Crankcase breather tubes
   - A/C compressor
   - Engine fan assembly
   - Accessory drive belt
   - Transmission oil cooler lines
   - Radiator hoses
   - Alternator
   - Heater hoses
8. Disconnect the engine wiring harness at the following points:
   - Intake air temperature (IAT) sensor
   - Fuel injectors
   - Throttle position (TPS) switch
   - Idle air control (IAC) motor
   - Engine oil pressure switch
   - Engine coolant temperature (ECT) sensor
   - Manifold absolute pressure (MAP) sensor
   - Camshaft position (CMP) sensor
   - Coil pack
9. Remove or disconnect the following:
   - Vacuum lines
   - Power steering pump and place securely aside
10. Install Special Tools 8400 lifting studs into the cylinder heads.
11. Install Engine Lifting Fixture 8437 as follows:
    - Holding the lifting fixture at a slight angle, slide the large bore in the front plate over the hex portion of the lifting stud.
    - Position the two remaining fixture arms onto the two Special Tools 8400 Lifting Studs, in the cylinder heads.
    - Pull forward and upward on the lifting fixture so that the lifting stud rest in the slotted area below the large bore.
    - Secure the lifting fixture to the three studs using three 7/16—14 N/C locknuts.
    - Make sure the lifting loop in the lifting fixture is in the last hole (closest to the throttle body) to minimize the angle of engine during removal.
12. Disconnect the body group straps.
13. Place a suitable jack under the transmission.
14. Remove the engine from the vehicle.

To install:

15. Lower the engine into the vehicle and install the left and right engine mount bolts. Tighten the nuts as follows:
    a. 4x2 models—70 ft. lbs. (95 Nm)
    b. 4x4 models—75 ft. lbs. (102 Nm)

    and tighten locknuts to 30 ft. lbs. (41 Nm)
16. Remove the Engine Lifting Fixture 8347 and Special Tools 8400 Lifting Studs.
17. Install or connect the following:
    - Body ground straps
    - Power steering pump
    - Vacuum lines
18. Connect the wiring harness at the following points:
    - IAT sensor
    - IAC motor
    - Fuel injectors
    - TPS switch
    - Engine oil pressure switch
    - ECT switch
    - MAP sensor
    - CMP sensor
    - Coil pack
19. Install or connect the following:
    - Heater hoses
    - Alternator
    - Radiator hoses
    - Transmission oil cooler
    - Accessory drive belt
    - A/C compressor
    - Crankcase breather tubes
    - Throttle body resonator assembly
    - Air intake hose
    - Transmission to engine mounting bolts. Tighten to 30 ft. lbs. (41 Nm).
    - Torque converter bolts, if equipped
    - Crankshaft position sensor
    - Axle isolator bracket to engine, transmission and axle—4x4 models with equipped with automatic transmission only
    - Starter
    - Structural cover
    - Exhaust crossover pipe
    - Engine block heater power cable, if equipped
    - Axle vent tube to left side engine mount—4x4 models only
    - Negative battery cable
20. Fill the cooling system to the correct level.
21. Fill the engine with oil.
22. Start the engine and check for leaks.

5.7L ENGINE

1. Before servicing the vehicle, refer to the precautions in the beginning of this section.
2. Drain the cooling system.
3. Drain the engine oil.
4. Relieve the fuel system pressure.
5. Remove or disconnect the following:
   - Negative battery cable
   - Hood
   - Air cleaner and resonator
   - Engine fan
   - Accessory drive belt
   - Fan shroud
   - Radiator
   - Upper crossmember and top core support
   - A/C compressor, if equipped, and secure aside leaving the lines attached

- Alternator
- Intake manifold and IAFM as an assembly
- Heater hoses
- Power steering pump, and secure aside leaving the lines attached
- Fuel supply line
- Engine front mount thru-bolt nuts
- Right side axle retaining bolts
- Transmission oil cooler lines, if equipped
- Exhaust pipes at the manifolds
- Starter
- Structural dust cover and transmission inspection cover
- Torque converter-to-flexplate bolts, if equipped
- Transmission flange bolts. Support the transmission.

6. Install engine lifting fixture 8984

➡**Do not use the adapter.**

7. Separate the engine from the transmission and remove engine from the vehicle.

To install:

8. Lower the engine into the vehicle and align the engine with the transmission.
   a. Manual transmission—Align clutch disc assembly if disturbed. Install transmission input shaft into clutch disc while mating engine and transmission surfaces. Install two transmission to engine block mounting bolts finger tight.
   b. Automatic transmission—Mate engine and transmission and install two transmission to engine block mounting bolts finger tight.
9. Position the thru-bolt into the support cushion brackets.
10. Lower the engine assembly until engine mount through bolts rest in mount perches.
11. Install remaining transmission flange bolts. Tighten to 40 ft. lbs. (54 Nm).
12. Install the engine mount bolts. Tighten to 70 ft. lbs. (95 Nm).
13. Install or connect the following:
    - Torque converter-to-flexplate bolts, if equipped
    - Structural dust cover and transmission inspection cover
    - Starter
    - Exhaust pipes
    - Transmission oil cooler lines, if equipped
    - Torque converter, if equipped. Tighten the bolts to 23 ft. lbs. (31 Nm).
    - Structural dust cover and transmission inspection cover
    - Starter motor
14. Remove the engine lifting fixture 8984.
15. Install or connect the following:
    - Fuel line
    - Power steering pump
    - Heater hoses
    - Intake manifold and IAFM as an assembly
    - Alternator
    - A/C compressor, if equipped
    - Accessory drive belt
    - Upper crossmember and top core support
    - Radiator
    - Engine fan
    - Air cleaner and resonator
    - Hood
    - Negative battery cable
16. Fill the crankcase to the correct level.
17. Fill the cooling system.
18. Start the engine and check for leaks.

Water Pump

REMOVAL & INSTALLATION

3.7L and 4.7L Engines

1. Before servicing the vehicle, refer to the precautions in the beginning of this section.
2. Drain the cooling system.
3. Remove or disconnect the following:
   - Negative battery cable
   - Fan and clutch assembly from the pump
   - Radiator fan shroud and fan assembly. If you're reusing the fan clutch, keep it upright to avoid silicone fluid loss!
   - Accessory drive belt
   - Lower radiator hose
   - Water pump mounting bolts
   - Water pump and gasket

To install:

4. Install or connect the following:
   - Water pump with new gasket. Tighten bolts in sequence to 43 ft. lbs. (58 Nm).
   - Lower radiator hose
   - Accessory drive belt
   - Fan and clutch assembly
   - Negative battery cable
5. Fill the cooling system to the correct level.
6. Start the engine and check for leaks.

5.7L Engines

1. Before servicing the vehicle, refer to the precautions in the beginning of this section.

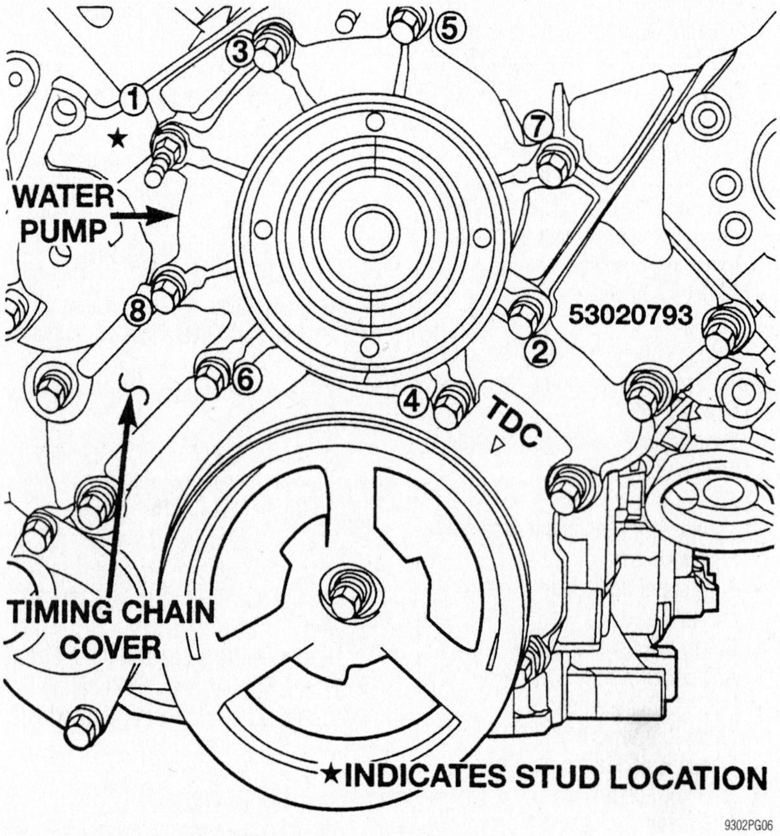

Water pump tightening sequence—3.7L and 4.7L engines

2. Drain the cooling system.
3. Remove or disconnect the following:
   - Negative battery cable
   - Accessory drive belt
   - Fan clutch assembly
   - Coolant recovery bottle
   - Washer bottle wiring and hose
   - Fan shroud assembly
   - A/C compressor and alternator brace
   - Idler pulleys
   - Belt tensioner
   - Radiator hoses
   - Heater hoses
   - Water pump

To install:

4. Install or connect the following:
   - Water pump and tighten the bolts to 18 ft. lbs. (24 Nm)
   - Heater hoses
   - Radiator hoses
   - Idler pulleys
   - A/C compressor and alternator brace and tighten the bolts to 21 ft. lbs. (28 Nm)
   - Fan shroud assembly
   - Washer bottle wiring and hose
   - Coolant recovery bottle
   - Fan clutch assembly
   - Accessory drive belt
   - Negative battery cable
5. Fill the cooling system.
6. Start the engine and check for leaks.

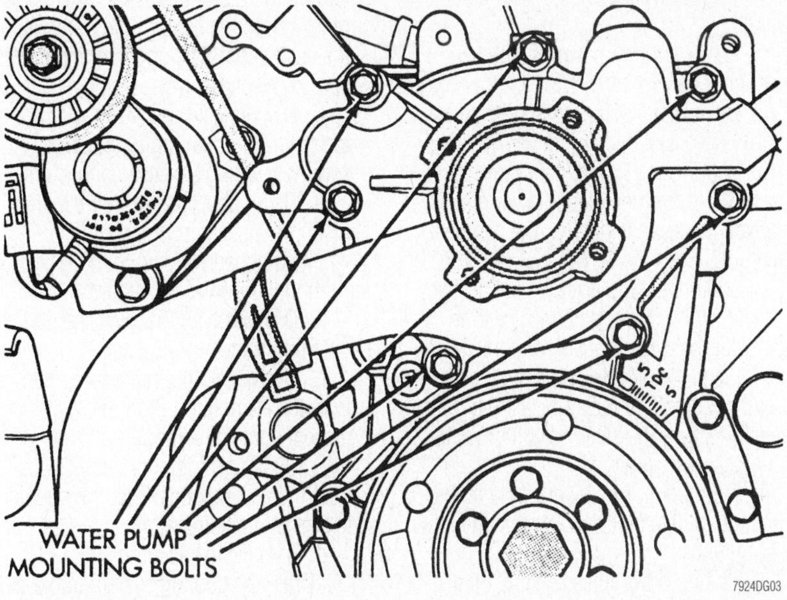

Water pump mounting bolt locations—3.9L, 5.2L and 5.9L engines, 8.0L engine is similar

3.9L, 5.2L and 5.9L Engines

1. Before servicing the vehicle, refer to the precautions in the beginning of this section.
2. Drain the cooling system.
3. Remove or disconnect the following:
   - Negative battery cable
   - Engine cooling fan and shroud

➡Do not store the fan clutch assembly horizontally, silicone may leak into the bearing grease and cause contamination.

   - Accessory drive belt
   - Water pump pulley
   - Lower radiator hose
   - Heater hose and tube
   - Bypass hose
   - Water pump

To install:

4. Install or connect the following:
   - Water pump, using a new gasket. Tighten the bolts to 30 ft. lbs. (40 Nm).
   - Bypass hose
   - Heater hose and tube. Use a new O-ring seal.
   - Lower radiator hose
   - Water pump pulley. Tighten the bolts to 20 ft. lbs. (27 Nm).
   - Accessory drive belt
   - Engine cooling fan and shroud
   - Negative battery cable
5. Fill the cooling system.
6. Start the engine and check for leaks.

8.0L Engine

1. Before servicing the vehicle, refer to the precautions in the beginning of this section.
2. Drain the cooling system.
3. Remove or disconnect the following:
   - Negative battery cable
   - Washer solvent bottle
   - Upper radiator hose

➡The 8.0L engine is equipped with a fan clutch that threads directly onto the water pump shaft. This fan clutch is equipped with right-hand threads.

➡Do not store the fan clutch assembly horizontally, silicone may leak into the bearing grease and cause contamination.

   - Engine cooling fan and shroud
   - Accessory drive belt
   - Water pump pulley
   - Lower radiator hose
   - Heater hose
   - Bypass hose
   - Water pump

To install:

4. Install or connect the following:
   - Water pump. Use a new O-ring seal and tighten the bolts to 30 ft. lbs. (40 Nm).
   - Bypass hose
   - Heater hose
   - Lower radiator hose
   - Water pump pulley. Tighten the bolts to 16 ft. lbs. (22 Nm).
   - Accessory drive belt
   - Engine cooling fan and shroud
   - Upper radiator hose
   - Washer solvent bottle
   - Negative battery cable
5. Fill the cooling system.
6. Start the engine and check for leaks.

Heater Core

REMOVAL & INSTALLATION

Dodge Ram Pick-Up—2005 Models

1. Disconnect the negative battery cable.

⁂ CAUTION

After disconnecting the negative battery cable, wait 2 minutes for the driver's/passenger's air bag system capacitor to discharge before attempting to do any work around the steering column or instrument panel.

2. On diesel models, remove the passenger side battery and battery tray.
3. Remove the instrument panel by performing the following procedures:
   a. Remove the air bag control module

(ACM) and bracket from the floor panel tunnel.

b. Remove the trim from both cowl side inner panels.

c. Remove the steering column-to-instrument panel opening cover.

d. Remove the 2 hood latch release handle-to-instrument panel lower reinforcement screws and lower the handle to the floor.

e. Disconnect the driver's side air bag module wiring harness connector from the lower instrument panel reinforcement.

f. Place the wheels in the straight-ahead position and lock the steering wheel. Remove the steering column without disassembling it.

g. From under the driver's side of the instrument, disconnect or remove the following items:

- Parking brake release handle linkage rod from the parking brake mechanism located on the left cowl side inner panel.
- Instrument panel wiring harness connector from the parking brake switch located on the parking brake mechanism.
- Three wiring harness connectors (body wiring harness, headlight, dash) from the 3 junction block connector receptacles located closest to the dash panel.
- Head light/dash-to-instrument panel bulkhead wiring harness connector screw and disconnect the connector.
- Instrument panel-to-door wiring harness connector located directly below the bulkhead wiring harness connector.
- Infinity sound system wiring harness connector (if equipped), located at the outboard side of the instrument panel bulkhead connector.
- Stop light switch electrical connector.
- Vacuum harness connector located near the left side of the heater/air conditioning housing.

h. Under the passenger's side of the instrument panel, disconnect the radio antenna coaxial cable connector.

i. Loosen both sides of the instrument panel cowl side roll-down bracket screws about ½ inch (13mm).

j. Remove the 5 upper instrument panel-to-upper dash panel screws; remove the center screw last.

k. Roll down the instrument panel and install a temporary hook in the center hole on top of the panel. Attach the other end to the center hole in the top of the dash panel. The opening should be approximately 18 inches (46cm).

l. Disconnect the instrument panel-to-heater/air conditioning housing assembly wiring harness connectors.

m. Using an assistant, remove the instrument panel from the vehicle.

4. If equipped with air conditioning, perform the following procedure:

a. Discharge and recover the air conditioning system refrigerant.

b. Disconnect the refrigerant lines from the evaporator. Plug the openings to prevent contamination.

c. Disconnect the refrigerant lines from the accumulator. Plug the openings to prevent contamination.

d. Remove the accumulator.

5. Drain the cooling system into a clean container for reuse.

6. Disconnect the heater hoses from the heater core tubes.

7. Remove the PCM from the dash panel and move it aside. Do not disconnect the PCM harness connector.

8. In the engine compartment, remove the heater/air conditioning housing assembly-to-chassis nuts.

9. In the passenger's compartment, remove the heater/air conditioning housing-to-dash panel nuts.

10. Pull the heater/air conditioning housing assembly rearward far enough to clear the studs and air conditioning drain tube holes.

11. Remove the heater/air conditioning housing assembly from the vehicle.

12. Remove the upper-to-lower heater/air conditioning housing screws and remove the upper housing.

13. Remove the heater core from the lower housing.

To install:

14. Install the heater core in the lower housing.

15. Install the upper housing and the upper-to-lower heater/air conditioning housing screws.

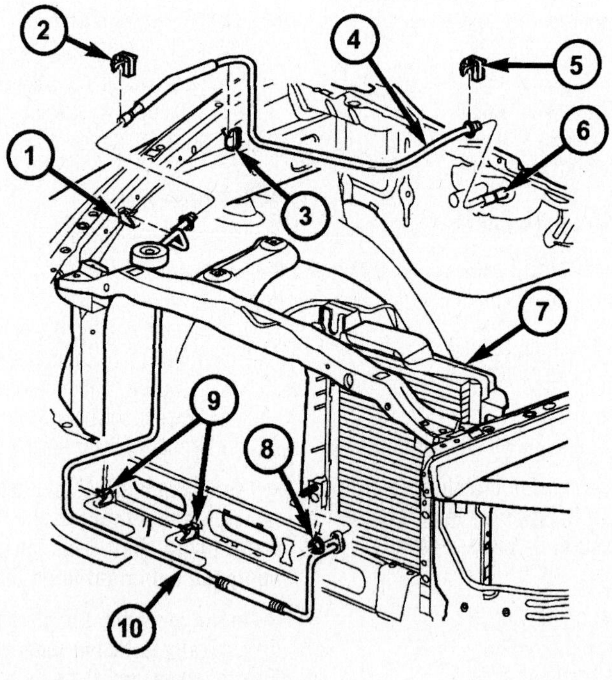

Exploded view of refrigerant line—Dodge Ram Pick-Up

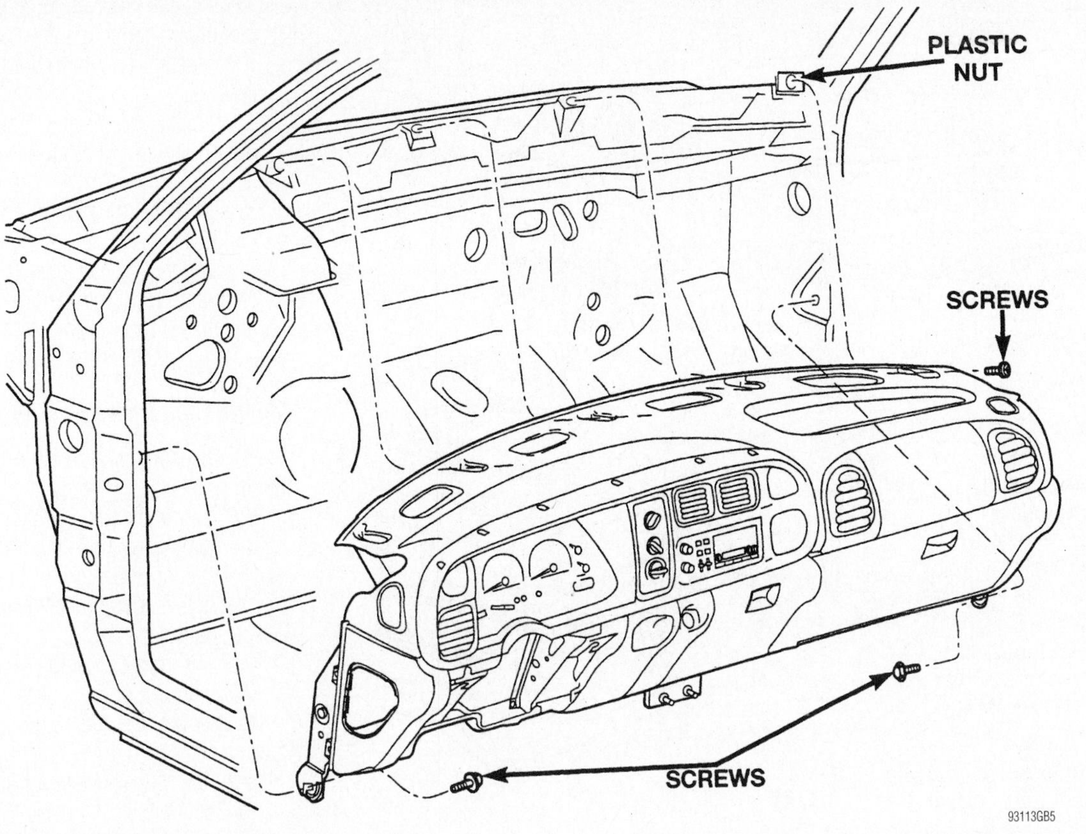

View of the instrument panel—Dodge Ram Pick-Up

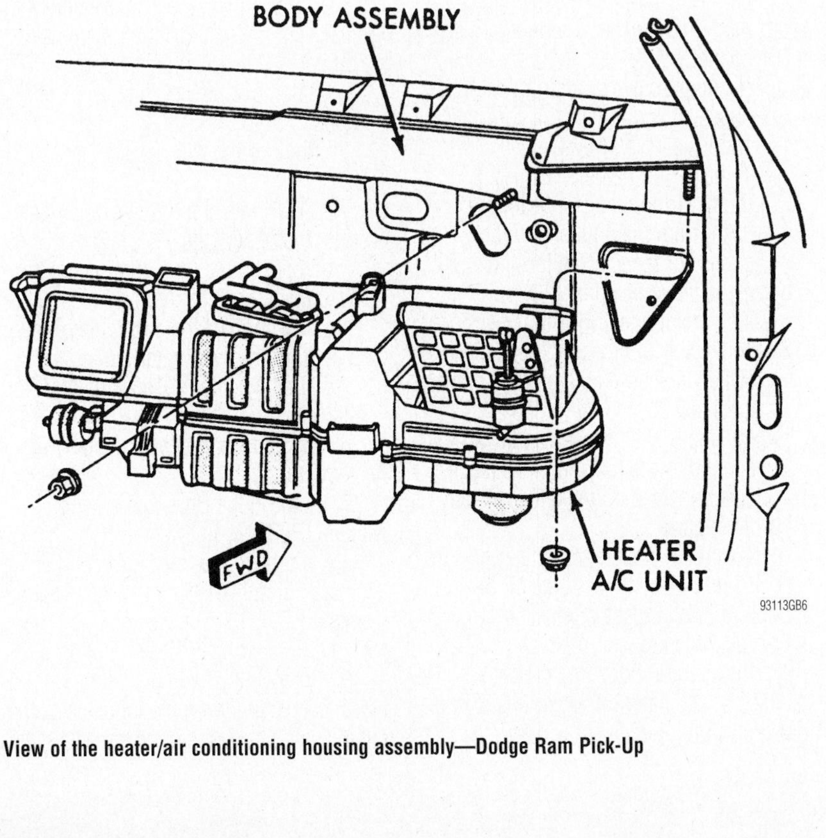

View of the heater/air conditioning housing assembly—Dodge Ram Pick-Up

16. Install the heater/air conditioning housing assembly to the vehicle.
17. Push the heater/air conditioning housing assembly forward far enough to engage the studs and air conditioning drain tube holes.
18. In the passenger's compartment, install the heater/air conditioning housing-to-dash panel nuts.
19. In the engine compartment, install the heater/air conditioning housing assembly-to-chassis nuts.
20. Install the PCM to the dash panel.
21. Connect the heater hoses to the heater core tubes.
22. Refill the cooling system.
23. If equipped with air conditioning, perform the following procedure:
    a. Install the accumulator.
    b. Connect the refrigerant lines to the accumulator.
    c. Connect the refrigerant lines from the evaporator.
    d. Evacuate and charge the air conditioning system refrigerant.
24. Install the instrument panel by performing the following procedures:
    a. Using an assistant, install the instrument panel to the vehicle.
    b. Connect the instrument panel-to-

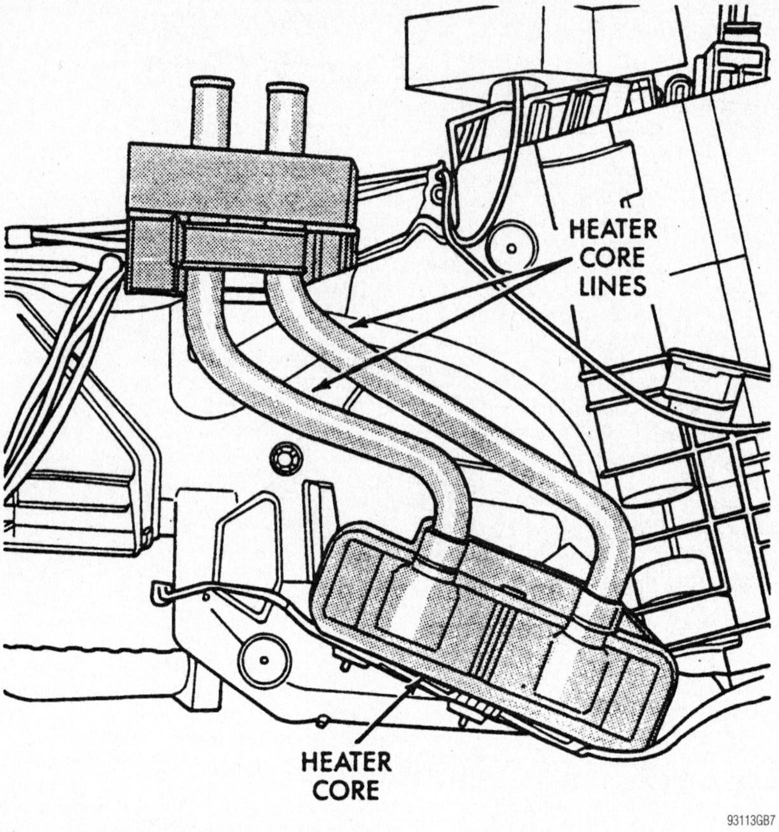

View of the heater core—Dodge Ram Pick-Up

heater/air conditioning housing assembly wiring harness connectors.

c. Roll-up the instrument panel and install a temporary hook in the center hole on top of the panel and the top of the dash panel. The opening should be approximately 18 inches (46cm).

d. Install the 5 upper instrument panel-to-upper dash panel screws; install the center screw first.

e. Install both sides of the instrument panel cowl side roll-down bracket screws to about ½ inch (13mm).

f. Under the passenger's side of the instrument panel, connect the radio antenna coaxial cable connector.

g. Under the driver's side of the instrument, connect or install the following items:

- Vacuum harness connector located near the left side of the heater/air conditioning housing.
- Stoplight switch electrical connector.
- Infinity sound system wiring harness connector (if equipped), located at the outboard side of the instrument panel bulkhead connector.
- Infinity sound system wiring harness connector (if equipped), located at the outboard side of the instrument panel bulkhead connector.
- Instrument panel-to-door wiring harness connector located directly below the bulkhead wiring harness connector.
- Headlight/dash-to-instrument panel bulkhead wiring harness connector and install the screw.
- Three wiring harness connectors (body wiring harness, headlight, dash) from the 3 junction block connector receptacles located closest to the dash panel.
- Instrument panel wiring harness connector to the parking brake switch located on the parking brake mechanism.
- Parking brake release handle linkage rod to the parking brake mechanism located on the left cowl side inner panel.

h. Install the steering column.

i. Connect the driver's side air bag module wiring harness connector to the lower instrument panel reinforcement.

j. Install the 2 hood latch release handle and the handle-to-instrument panel lower reinforcement screws.

k. Install the steering column-to-instrument panel opening cover.

l. Install the trim to both cowl side inner panels.

m. Install the air bag control module (ACM) and bracket to the floor panel tunnel.

25. On diesel models, install the battery tray and battery.
26. Connect the negative battery cable.
27. Run the engine to normal operating temperatures; then, check the climate control operation and check for leaks.

Dodge Ram Pick-up—2005 Models

1. Discharge and recover the air conditioning system refrigerant.
2. Drain the engine cooling system.
3. Remove or disconnect the following:
   - Negative battery cable
   - Passenger side battery—Diesel engines only
   - Air intake assembly—Gasoline engines only
4. Remove the A/C liquid lines as follows:

a. Disconnect the front A/C liquid line from condenser.

b. Disconnect the rear A/C liquid line from the evaporator.

c. Release the rear A/C liquid line from body retaining clips.

d. Remove the retaining clip that secures the front section of the liquid line to the rear section.

e. Disconnect the front section of the line from the rear section using the proper A/C line disconnect tool.

f. Remove the A/C liquid line sections from the vehicle.

5. Remove the A/C accumulator as follows:

a. Disconnect the A/C suction line from the outlet tube.

b. Loosen the bolts that secure the accumulator to the support bracket.

c. Disconnect the accumulator inlet tube from the evaporator outlet tube.

d. Remove the A/C accumulator from the vehicle.

6. Disconnect the heater hoses from the heater core.
7. Remove the Powertrain Control Module (PCM) as follows:

a. Remove the PCM cover.

b. Carefully unplug the electrical connectors.

c. Remove the PCM mounting bolts.

8. Remove the two nuts behind the PCM that secure the HVAC housing to the engine compartment side of the dash panel.

9. Remove the instrument panel as follows:
   a. Remove or disconnect the following:
   - Left A-pillar trim
   - Instrument panel top cover
   - Headliner wiring harness
   - Left instrument panel side cover using trim stick C-4755 or equivalent
   - Left cowl trim panel
   - Steering column
   - Steering column support bracket mounting bolts
   - Parking brake release handle actuator
   - Instrument panel wiring harness connector from the bulkhead wiring harness
   - Instrument panel side bezel
   - Left side instrument panel mounting bolts
   - Floor console, if equipped
   - Air bag control module cover, if equipped
   - Air bag control module electrical connector, if equipped
   - Instrument panel-to-floor mounting bolts
   - Right cowl trim cover
   - Right side wiring harnesses and radio antenna
   - Bolt below the glove box securing the panel to the HVAC housing
   - Right instrument panel side cover
   - Right A-pillar trim
   - Right side instrument panel mounting bolts
   - Instrument panel-to-cowl mounting bolts
   - Instrument panel

10. Remove the mounting bolts that secure the HVAC housing to the floor bracket and dash panel.
11. Remove the HVAC housing.
12. Remove the foam seal from the heater core tubes.
13. If equipped with the dual zone heating-A/C system, remove the linkage rod to gain access to the heater core.
14. Remove the two screws that secure the heater core tube bracket to the HVAC housing and remove the bracket.
15. Carefully pull the heater core out of the front of the HVAC housing.

To install:

16. Carefully install the heater core into the front of the HVAC housing.
17. Position the heater core tube bracket onto the HVAC housing.
18. Install the heater core bracket to the HVAC housing and tighten to 20 inch lbs. (2.2 Nm)

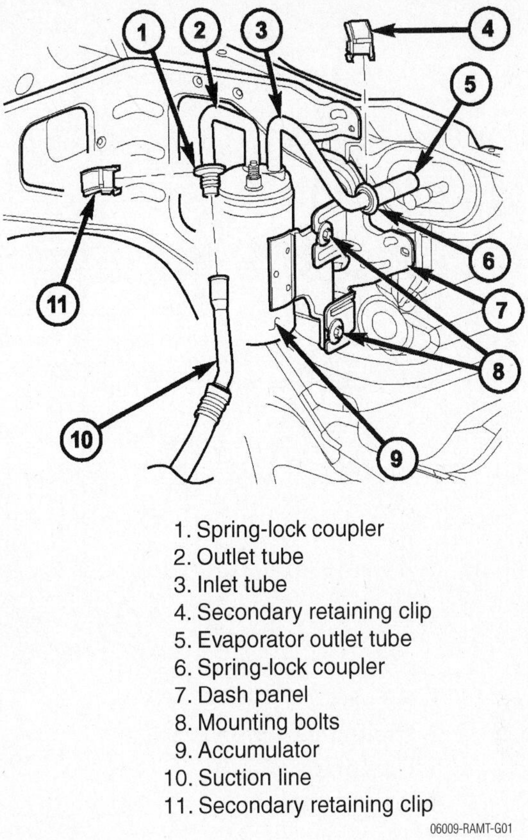

View of the A/C accumulator

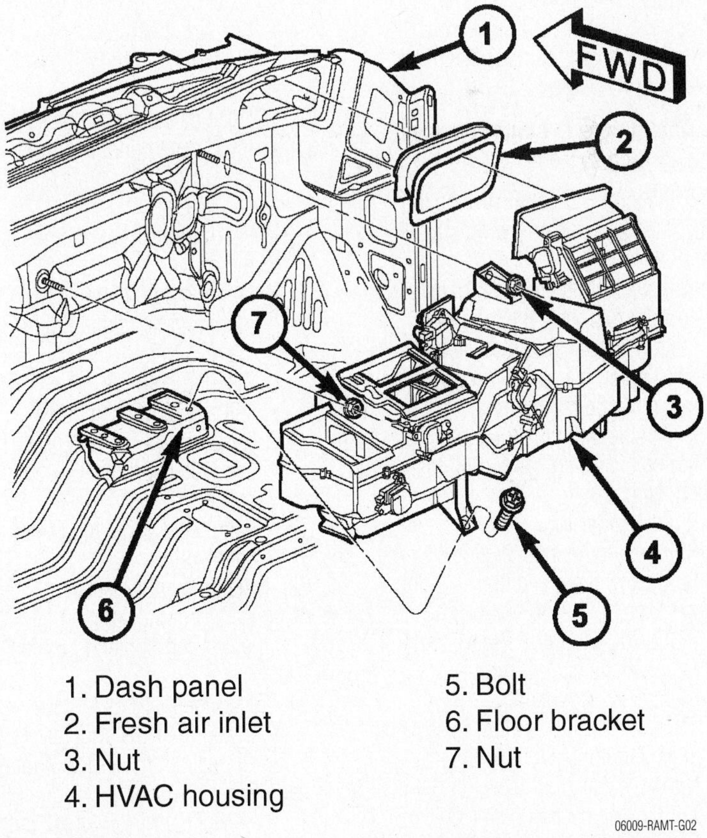

Removing the HVAC housing.

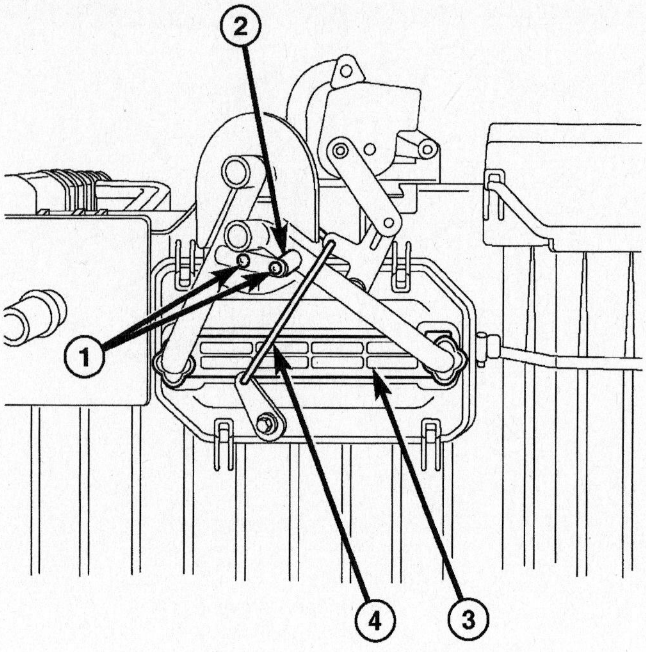

1. Screws
2. Heater core tube bracket
3. Heater core
4. Linkage rod

Removing the heater core.

19. If equipped with the dual zone heating-A/C system, install the linkage rod.
20. Install the foam seal onto the heater core tubes.

➡If the heater core is being replaced, flush the cooling system.

21. Position the HVAC housing on the mounting studs and tighten the nuts to 55 inch lbs. (6.2 Nm)
22. Install the instrument panel as follows:
    a. Install or connect the following:
    - Instrument panel into the vehicle and tighten the panel-to-cowl mounting bolts to 9 ft. lbs. (12 Nm)
    - Right side instrument panel mounting bolts and tighten to 9 ft. lbs. (12 Nm)
    - Right A-pillar trim
    - Right instrument panel side cover
    - Bolt below the glove box securing the panel to the HVAC housing
    - Right side wiring harnesses and radio antenna connection
    - Right cowl trim panel
    - Instrument panel-to-floor mounting bolts and tighten to 9 ft. lbs. (12 Nm)
    - Air bag control module electrical connection and cover, if equipped
    - Floor console, if equipped
    - Left side instrument panel mounting bolts and tighten to 9 ft. lbs. (12 Nm)
    - Left side instrument panel side bezel
    - Instrument panel wiring harness connector from the bulkhead wiring harness
    - Parking brake release handle actuator
    - Steering column support bracket mounting bolts and tighten to 10 ft. lbs. (14 Nm)
    - Steering column
    - Left side cowl trim panel
    - Headliner wiring harness
    - Left side instrument panel side cover
    - Instrument panel top cover
    - Left A-pillar
23. Install the nuts that secure the HVAC housing to the engine compartment side of the dash panel and tighten to 55 inch lbs. (6.2 Nm).
24. Install the PCM as follows:
    a. Install the PCM mounting bolts.
    b. Connect the electrical harnesses and install the cover.
25. Connect the heater hoses to the heater core tubes.
26. Install the A/C accumulator as follows:
    a. Position the A/C accumulator on the mounting bracket and loosely install the mounting bolts.
    b. Connect the inlet tube to the evaporator outlet tube using the retaining clip.
    c. Tighten the accumulator mounting bolts to 40 inch lbs. (4.5 Nm).
    d. Connect the A/C suction line to the accumulator outlet tube.
27. Install the A/C liquid lines as follows:
    a. Position the front and rear sections in the engine compartment.
    b. Connect the front section of the line to the rear section using the secondary retaining clip.
    c. Engage the lines into the body retaining clips.
    d. Connect the front section line to the outlet port of the A/C condenser.
    e. Connect the rear section line to the evaporator inlet tube.
28. Install the passenger side battery—Diesel Engine models
29. Install the air intake assembly—Gasoline Engine models
30. Connect the negative battery cable.
31. Fill the cooling system to the correct level.
32. Evacuate and charge the refrigerant system.
33. Start the engine and check for leaks.

Dodge Ram Van/Wagon

FRONT HEATER

1. Disconnect the negative battery cable.

CAUTION

After disconnecting the negative battery cable, wait 2 minutes for the driver's/passenger's air bag system capacitor to discharge before attempting to do any work around the steering column or instrument

2. If equipped with air conditioning, perform the following procedure:
    a. Discharge and recover the air conditioning system refrigerant.
    b. Disconnect the suction line jumper from the evaporator
    c. Remove the filter/drier and bracket-to-heater/air conditioning housing and cowl panel screws; then, move the filter/drier, bracket and refrigerant lines toward the center of the vehicle.
    d. Disconnect the wire harness connector from the fin sensing cycling clutch switch.

e. Disconnect the wire harness connectors from the blower motor resistor and the high speed blower motor relay.

3. Cover the alternator to protect it from coolant spillage.

4. Drain the cooling system into a clean container for reuse.

5. Disconnect the heater hoses from the heater core tubes.

6. Remove the right headlight assembly, the grille panel, the right cowl grille support panel and the right radiator core support assembly.

7. Remove both lower heater/air conditioning housing flange-to-blower housing screws.

8. In the passenger's compartment, perform the following procedures:

a. From the top of the distribution duct, disconnect the blend-air door motor link.

b. Reach through the glove box opening and remove the heater/air conditioning housing-to-dash panel stamped nuts (2) and screw (1). The fasteners are located below the distribution duct.

c. Reach through the glove box opening; then, remove the 2 heater/air conditioning housing-to-dash panel stamped nuts. The nuts are located above the distribution duct.

9. In the engine compartment, remove the heater/air conditioning housing-to-dash stamped nut.

10. Carefully, separate the lower heater/air conditioning housing from the blower housing.

11. Pull the heater/air conditioning housing from the dash until the blend-air door link is clear of the dash panel hole.

12. Remove the heater/air conditioning housing assembly from the vehicle and place it on a bench with the top cover facing upward.

13. Disassemble the heater/air conditioning housing assembly by performing the following procedure:

a. Remove the blend-air door pivot shaft nut.

b. Remove the blend-air door link-to-rear mounting flange boot.

c. Remove the blend-air door link, the boot and the lever as a unit.

d. Remove the high speed blower motor relay mounting bracket-to-heater/air conditioning housing screw and remove the relay and bracket.

e. Remove the top cover-to-heater/air conditioning housing screws and the cover.

14. Remove the heater core tube support bracket-to-mounting boss screw and remove the heater core from the heater/air conditioning housing assembly.

To install:

15. Install the heater core to the heater/air conditioning housing assembly and heater core tube support bracket-to-mounting boss screw, then, tighten the screw to 20 inch lbs. (2.2 Nm).

16. Assemble the heater/air conditioning housing assembly by performing the following procedure:

a. Install the top cover and the cover-to-heater/air conditioning housing screws.

b. Install the high speed blower motor relay and mounting bracket, then install the relay and bracket -to-heater/air conditioning housing screw.

c. Install the blend-air door link, the boot and the lever as a unit.

d. Install the blend-air door link-to-rear mounting flange boot.

e. Install the blend-air door pivot shaft nut.

17. Install the heater/air conditioning housing assembly into the vehicle.

18. Push the heater/air conditioning housing into the dash until the blend-air door link falls into the dash panel hole.

19. Carefully, assemble the lower heater/air conditioning housing to the blower housing.

20. In the engine compartment, install the heater/air conditioning housing-to-dash stamped nut.

21. In the passenger's compartment, perform the following procedures:

a. Reach through the glove box opening; then, install the 2 heater/air conditioning housing-to-dash panel stamped nuts. The nuts are located above the distribution duct.

b. Reach through the glove box opening; then, install the heater/air conditioning housing-to-dash panel stamped nuts (2) and screw (1). The fasteners are located below the distribution duct.

c. At the top of the distribution duct, connect the blend-air door motor link.

22. Install both lower heater/air conditioning housing flange-to-blower housing screws.

23. Install the right radiator core support assembly, the right cowl grille support panel, the grille panel and the right headlight assembly.

24. Connect the heater hoses to the heater core tubes.

25. Refill the cooling system.

26. Uncover the alternator.

27. If equipped with air conditioning, perform the following procedure:

a. Connect the wire harness connectors to the blower motor resistor and the high-speed blower motor relay.

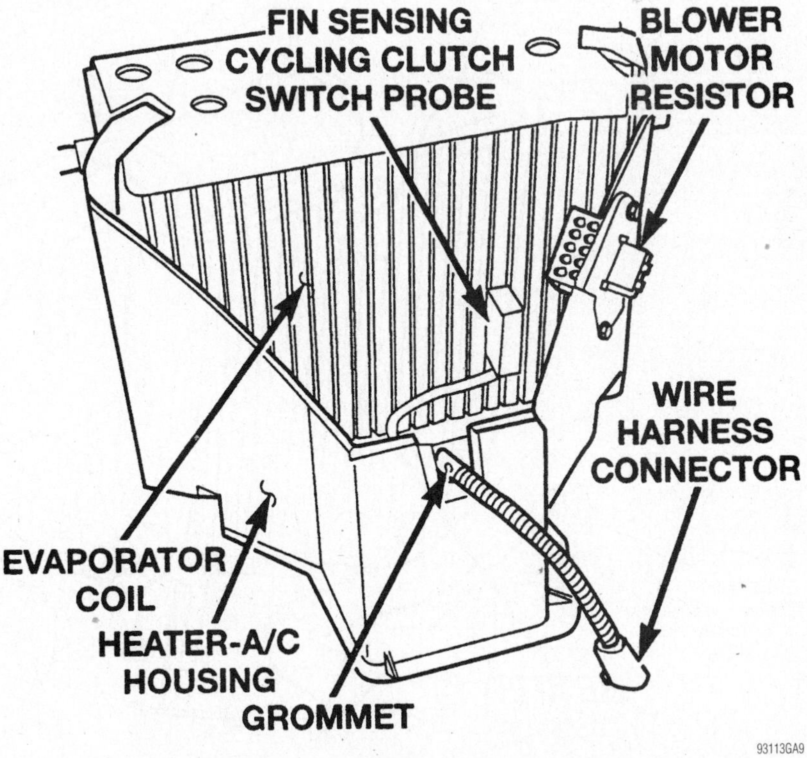

View of the air conditioning fin sensing cycling clutch switch—Dodge Ram Van/Wagon

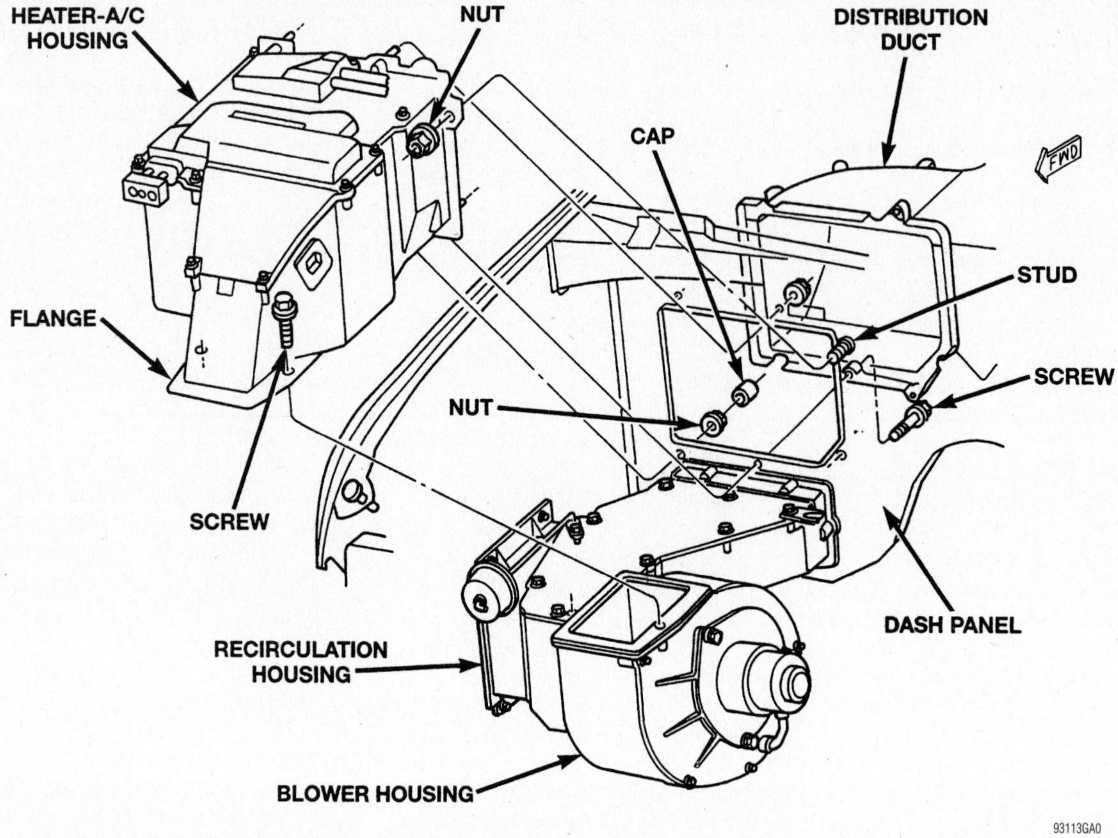

Exploded view of the heater/air conditioning housing and related components—Dodge Ram Van/Wagon

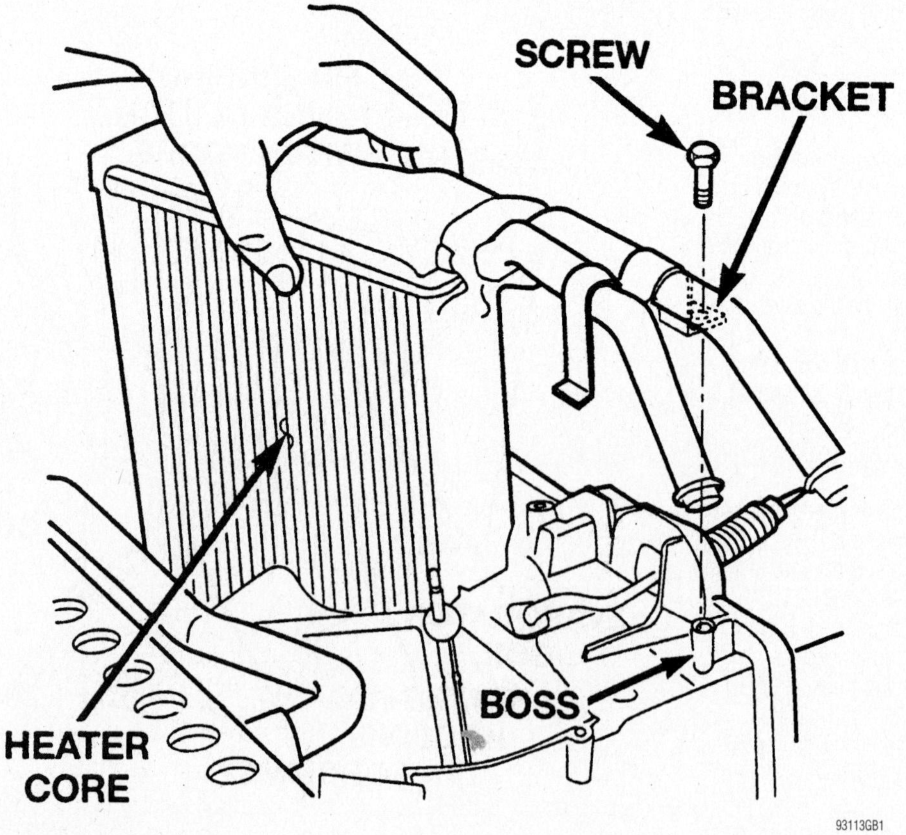

View of the heater core—Dodge Ram Van/Wagon

b. Connect the wire harness connector to the fin sensing cycling clutch switch.

c. Install the filter/drier and bracket-to-heater/air conditioning housing and cowl panel screws.

d. Connect the suction line jumper to the evaporator

e. Evacuate and charge the air conditioning system refrigerant.

28. Connect the negative battery cable.

29. Run the engine to normal operating temperatures; then, check the climate control operation and check for leaks.

REAR AUXILIARY HEATER

The combination coil is used only with the optional rear heater/air conditioning housing assembly.

1. Disconnect the negative battery cable.
2. Discharge and recover the air conditioning system refrigerant.
3. Drain the cooling system into a clean container for reuse.
4. If equipped, remove the rear bench seat.
5. Raise and safely support the rear of the vehicle.
6. Disconnect the underbody plumbing from the rear heater/air conditioning housing plumbing connections. Plug all of the openings to prevent contamination.
7. Remove the rear heater/air conditioning housing-to-underbody panel screw.
8. Lower the vehicle.
9. Remove the 3 cover-to-rear heater/air conditioning housing screws and the cover.
10. From the rear heater/air conditioning housing assembly, perform the following procedures:
    a. Remove the vertical duct.
    b. Remove the horizontal duct.
    c. From behind the unit, remove the ground wire eyelet-to-left side panel strainer screw.
11. Disassemble the rear heater/air conditioning housing assembly by performing the following procedure:
    a. Lift both relay wiring harness connectors upward and disconnect them from the mounting tabs located at the rear of the housing.
    b. Disconnect the wiring connectors from the blower motor and the rear mode control motor.
    c. Disconnect the blower motor cooling tube from the lower housing nipple.
    d. Remove the control cable from the water valve.
    e. Remove the rear inboard corner lower housing-to-upper housing clip.
    f. Remove the upper-to-lower housing screws.
    g. Remove the upper housing.
12. Remove the combination coil from the lower heater/air conditioning housing.

To install:

13. Install the combination coil to the lower heater/air conditioning housing.
14. Assemble the rear heater/air conditioning housing assembly by performing the following procedure:
    a. Install the upper housing.
    b. Install the upper-to-lower housing screws and torque to 20 inch lbs. (2.2 Nm).
    c. Install the rear inboard corner lower housing-to-upper housing clip.
    d. Install the control cable to the water valve.

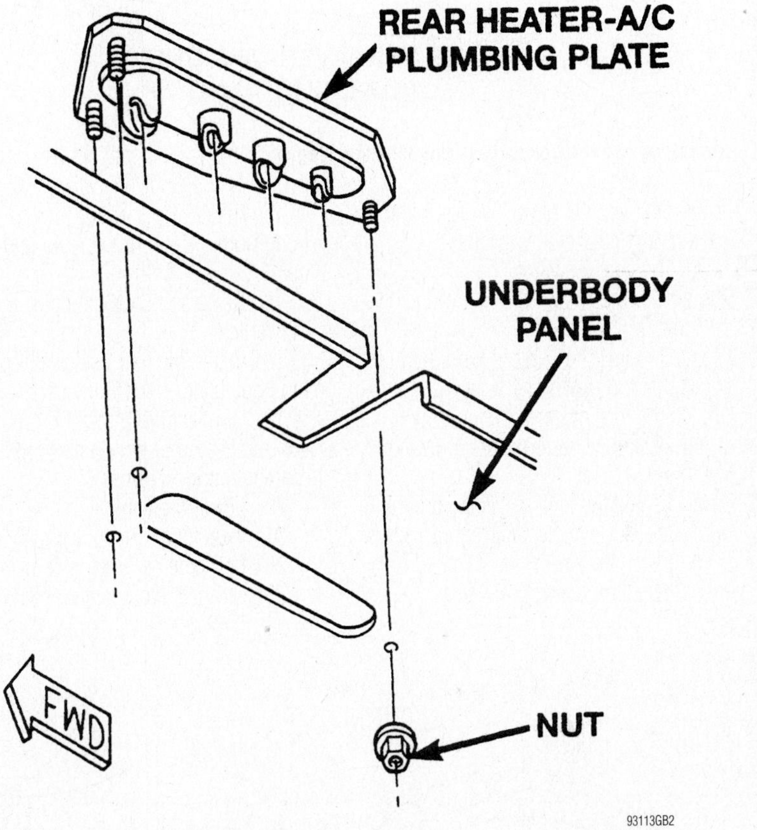

View of the rear heater/air conditioning housing assembly plumbing plate—Dodge Ram Van/Wagon

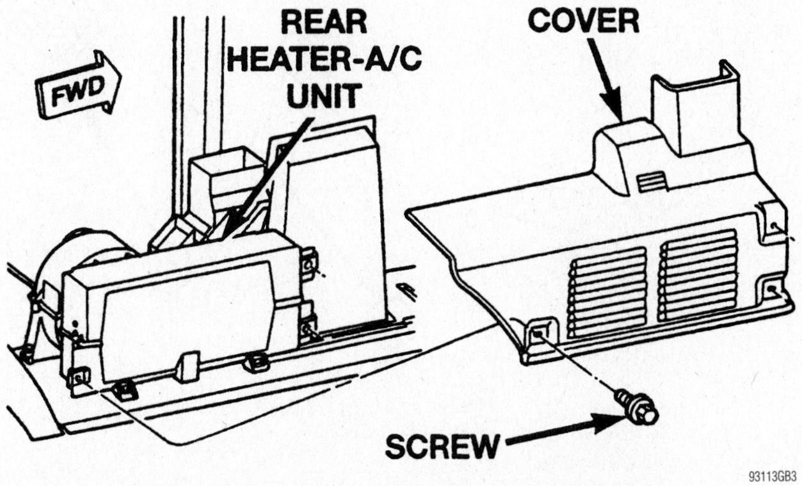

View of the rear heater/air conditioning housing assembly—Dodge Ram Van/Wagon

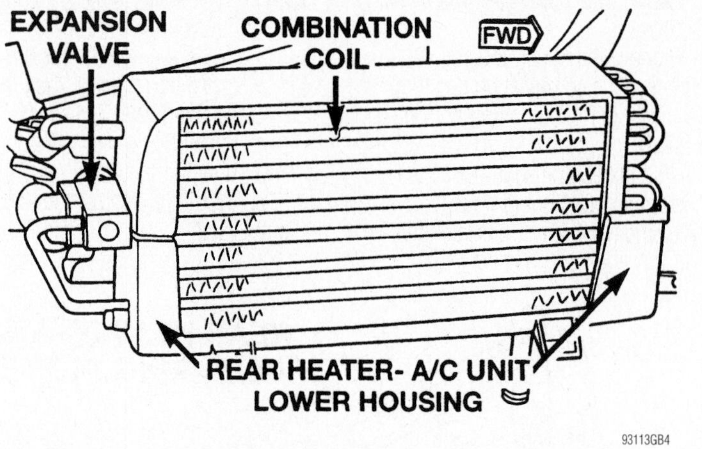

View of the combination coil—Dodge Ram Van/Wagon

e. Connect the blower motor cooling tube to the lower housing nipple.

f. Connect the wiring connectors to the blower motor and the rear mode control motor.

g. Connect them to the mounting tabs located at the rear of the housing.

15. At the rear heater/air conditioning housing assembly, perform the following procedures:

a. Behind the unit, install the ground wire eyelet-to-left side panel strainer screw.

b. Install the horizontal duct.

c. Install the vertical duct.

16. Install the 3 cover-to-rear heater/air conditioning housing screws and the cover.

17. Raise and safely support the rear of the vehicle.

18. Install the rear heater/air conditioning housing-to-underbody panel screw.

19. Connect the underbody plumbing to the rear heater/air conditioning housing plumbing connections.

20. Lower the vehicle.

21. If equipped, install the rear bench seat.

22. Refill the cooling system.

23. Connect the negative battery cable.

24. Evacuate and charge the air conditioning system refrigerant.

25. Run the engine to normal operating temperatures; then, check the climate control operation and check for leaks.

Cylinder Head

REMOVAL & INSTALLATION

3.7L Engine

LEFT SIDE

1. Before servicing the vehicle, refer to the precautions in the beginning of this section.
2. Drain the cooling system.
3. Properly relieve the fuel system pressure.
4. Remove or disconnect the following:
   - Negative battery cable
   - Exhaust Y-pipe
   - Intake manifold
   - Brake booster and master cylinder
   - Cylinder head cover
   - Engine cooling fan and shroud
   - Accessory drive belt
   - Power steering pump
5. Rotate the crankshaft so that the crankshaft timing mark aligns with the Top

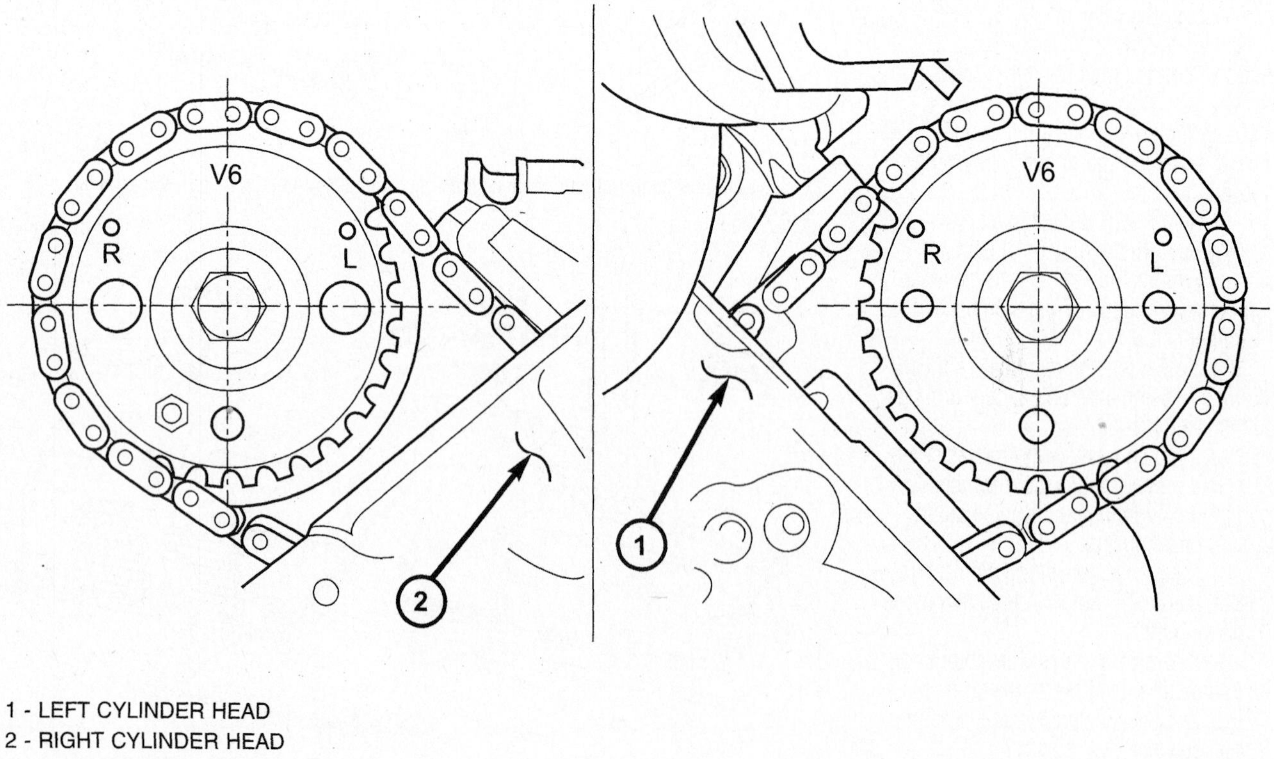

Camshaft sprocket timing marks—3.7L

Dead Center (TDC) mark on the front cover, and the **V6** marks on the camshaft sprockets are at 12 o'clock as shown.

- Crankshaft damper
- Front cover

6. Lock the secondary timing chain to the idler sprocket with Timing Chain Locking tool 8429.
7. Matchmark the secondary timing chain one link on each side of the V6 mark to the camshaft sprocket.

- Left secondary timing chain tensioner
- Cylinder head access plug
- Secondary timing chain guide
- Camshaft sprocket
- Cylinder head

➡The cylinder head is retained by twelve bolts. Four of the bolts are smaller and are at the front of the head.

To install:

8. Check the cylinder head bolts for signs of stretching and replace as necessary.
9. Lubricate the threads of the 11mm bolts with clean engine oil.
10. Coat the threads of the 8mm bolts with Mopar® Lock and Seal Adhesive.
11. Install the cylinder heads. Use new gaskets and tighten the bolts, in sequence, as follows:
   a. Step 1: Bolts 1–8 to 20 ft. lbs. (27 Nm)
   b. Step 2: Bolts 1–8 verify torque without loosening
   c. Step 3: Bolts 9–12 to 10 ft. lbs. (14 Nm)
   d. Step 4: Bolts 1–8 plus 90°
   e. Step 5: Bolts 1–8 plus 90° again
   f. Step 6: Bolts 9–12 to 19 ft. lbs. (26 Nm)
12. Install or connect the following:
   - Camshaft sprocket. Align the secondary chain matchmarks and tighten the bolt to 90 ft. lbs. (122 Nm).
   - Secondary timing chain guide
   - Cylinder head access plug
   - Secondary timing chain tensioner. Refer to the timing chain procedure in this section.
13. Remove the Timing Chain Locking tool.
14. Install or connect the following:
   - Front cover
   - Crankshaft damper. Torque the bolt to 130 ft. lbs. (175 Nm).
   - Power steering pump
   - Accessory drive belt
   - Engine cooling fan and shroud

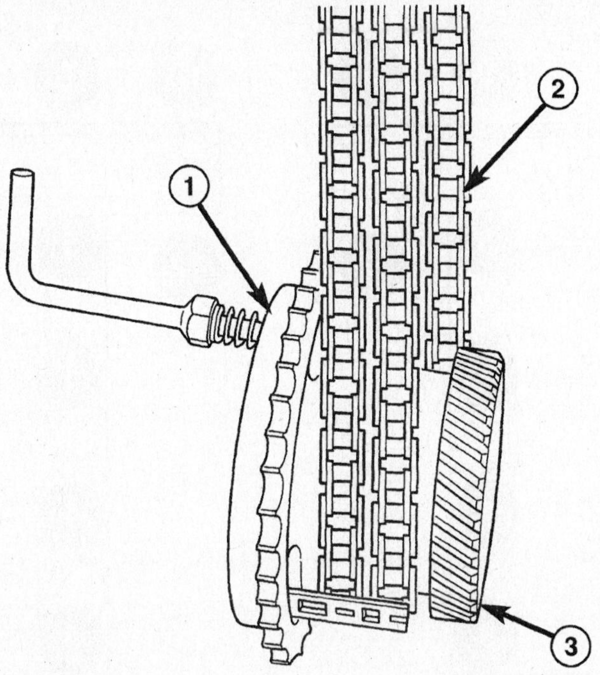

Camshaft locking tool—3.7L

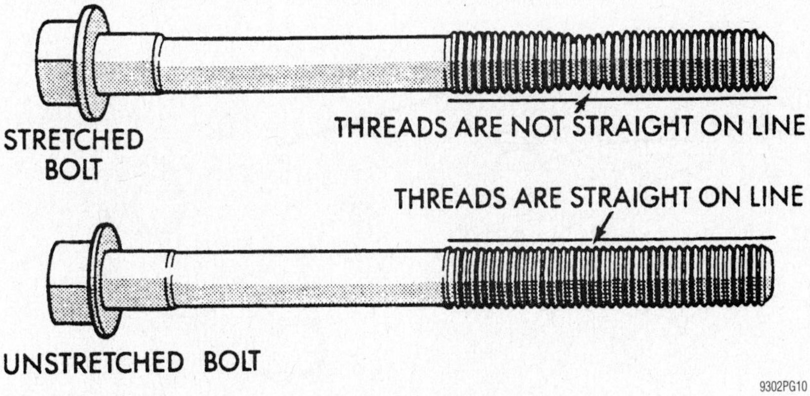

Examine the head bolts for signs of stretching—3.7L engine

   - Cover
   - Intake manifold
   - Exhaust Y-pipe
   - Negative battery cable

15. Fill and bleed the cooling system.
16. Start the engine, check for leaks and repair if necessary.

RIGHT SIDE

1. Before servicing the vehicle, refer to the precautions in the beginning of this section.
2. Drain the cooling system.
3. Properly relieve the fuel system pressure.
4. Remove or disconnect the following:
   - Negative battery cable
   - Exhaust Y-pipe
   - Intake manifold
   - Valve cover
   - Engine cooling fan and shroud
   - Accessory drive belt
   - Oil fill housing
   - Power steering pump
5. Rotate the crankshaft so that the crankshaft timing mark aligns with the Top

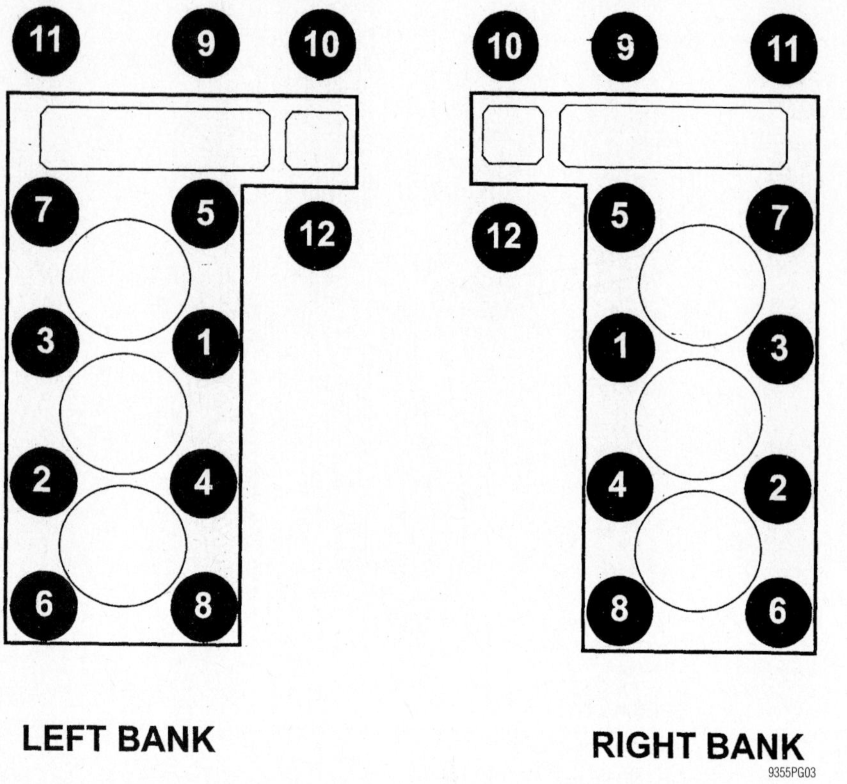

Cylinder head bolt torque sequence—3.7L

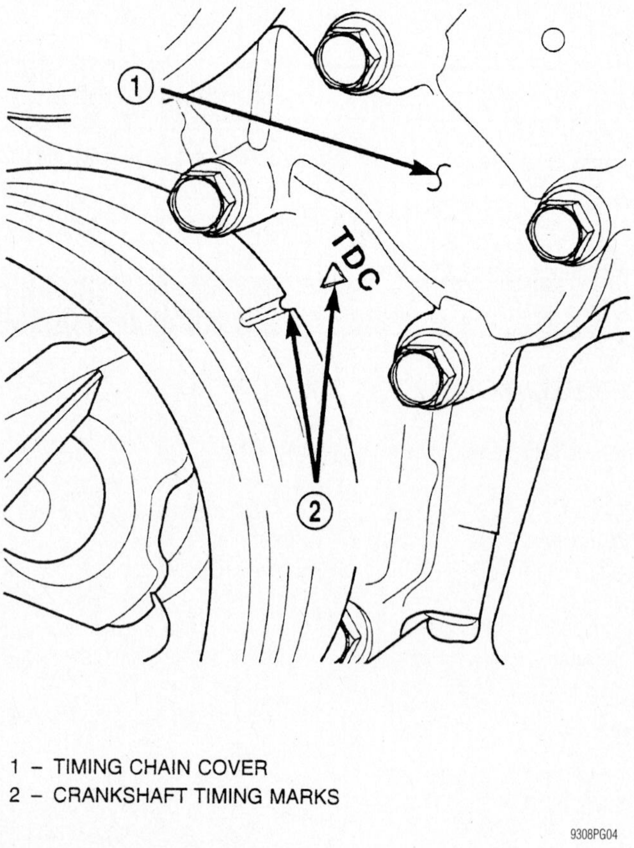

Crankshaft timing marks—3.7L engine

Dead Center (TDC) mark on the front cover, and the **V6** marks on the camshaft sprockets are at 12 o'clock as shown.

6. Remove or disconnect the following:
   - Crankshaft damper
   - Front cover
7. Lock the secondary timing chains to the idler sprocket with Timing Chain Locking tool 8429.
8. Matchmark the secondary timing chains to the camshaft sprockets.
9. Remove or disconnect the following:
   - Secondary timing chain tensioners
   - Cylinder head access plugs
   - Secondary timing chain guides
   - Camshaft sprockets

✱✱ CAUTION

The right side camshaft sprocket and camshaft sensor target wheel are serviced as an assembly. The nut on the sprocket should not be removed.

   - Cylinder head and gasket

➡The cylinder head is retained by eight 11mm bolts and four 8mm bolts.

To install:

10. Check the cylinder head bolts for signs of stretching and replace as necessary.
11. Lubricate the threads of the 11mm bolts with clean engine oil.
12. Coat the threads of the 8mm bolts with Mopar® Lock and Seal Adhesive.
13. Install the cylinder heads. Use new gaskets and tighten the bolts, in sequence, as follows:
    a. Step 1: Bolts 1–8 to 20 ft. lbs. (27 Nm)
    b. Step 2: Bolts 1–8 Verify torque without loosening
    c. Step 3: Bolts 9–12 to 10 ft. lbs. (14 Nm)
    d. Step 4: Bolts 1–8 plus 90°
    e. Step 4: Bolts 1–8 plus 90° again
    f. Step 5: Bolts 9–12 to 19 ft. lbs. (26 Nm)
14. Install or connect the following:
    - Camshaft sprockets. Align the secondary chain matchmarks and tighten the bolts to 90 ft. lbs. (122 Nm).
    - Secondary timing chain guides
    - Cylinder head access plugs
    - Secondary timing chain tensioners. Refer to the timing chain procedure in this section.
15. Remove the Timing Chain Locking Tool 8429.
16. Install or connect the following:

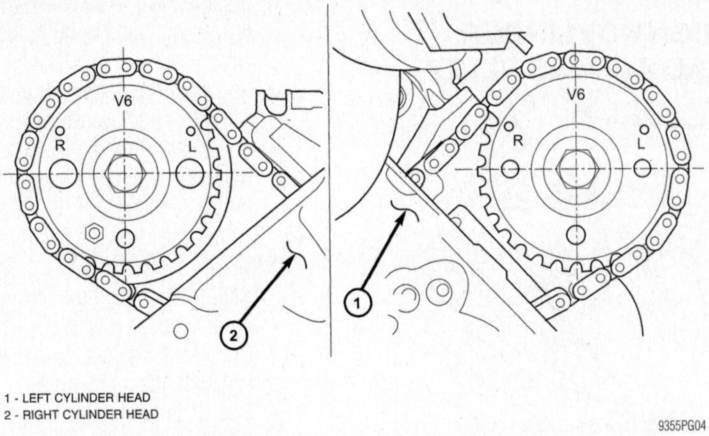

Camshaft positioning—3.7L engine

- Front cover
- Crankshaft damper. Torque the bolt to 130 ft. lbs. (175 Nm).
- Rocker arms
- Power steering pump
- Oil fill housing
- Accessory drive belt
- Engine cooling fan and shroud
- Valve covers
- Intake manifold
- Exhaust Y-pipe
- Negative battery cable

17. Fill and bleed the cooling system.
18. Start the engine, check for leaks and repair if necessary.

4.7L Engine

1. Before servicing the vehicle, refer to the precautions in the beginning of this section.
2. Drain the cooling system.
3. Properly relieve the fuel system pressure.
4. Remove or disconnect the following:
   - Negative battery cable
   - Exhaust Y-pipe
   - Intake manifold
   - Brake master cylinder and booster assembly
   - Cylinder head cover
   - Engine cooling fan and shroud
   - Accessory drive belt
   - Oil fill housing
   - Power steering pump
5. Rotate the crankshaft so that the crankshaft timing mark aligns with the Top Dead Center (TDC) mark on the front cover, and the **V8** marks on the camshaft sprockets are at 12 o'clock as shown.
6. Remove or disconnect the following:
   - Crankshaft damper
   - Front cover
7. Lock the secondary timing chains to the idler sprocket with Timing Chain Locking tool 8515.
8. Matchmark the secondary timing chains to the camshaft sprockets.
9. Remove or disconnect the following:
   - Secondary timing chain tensioners
   - Cylinder head access plugs
   - Secondary timing chain guides
   - Camshaft sprockets
   - Cylinder heads

➡**Each cylinder head is retained by ten 11mm bolts and four 8mm bolts.**

To install:

10. Check the cylinder head bolts for signs of stretching and replace as necessary.

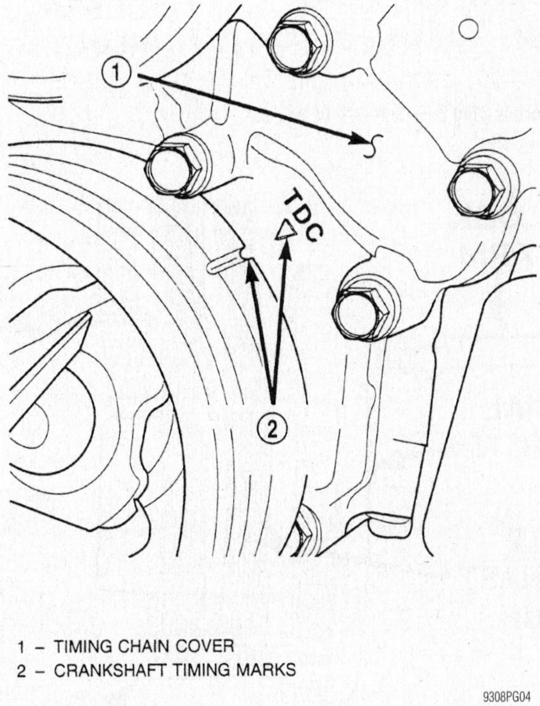

Crankshaft timing marks—4.7L engine

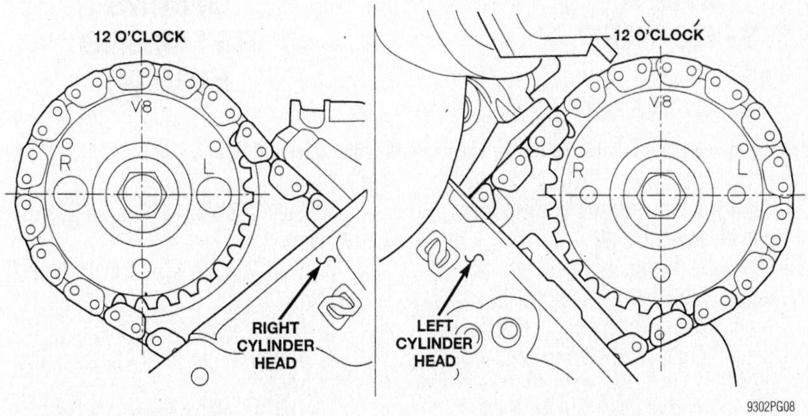

Camshaft positioning—4.7L engine

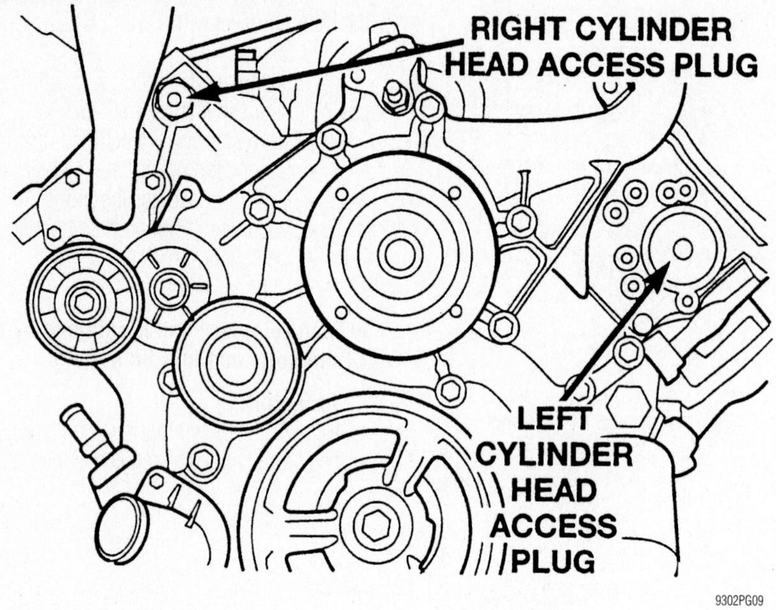

Cylinder head access plug locations—4.7L engine

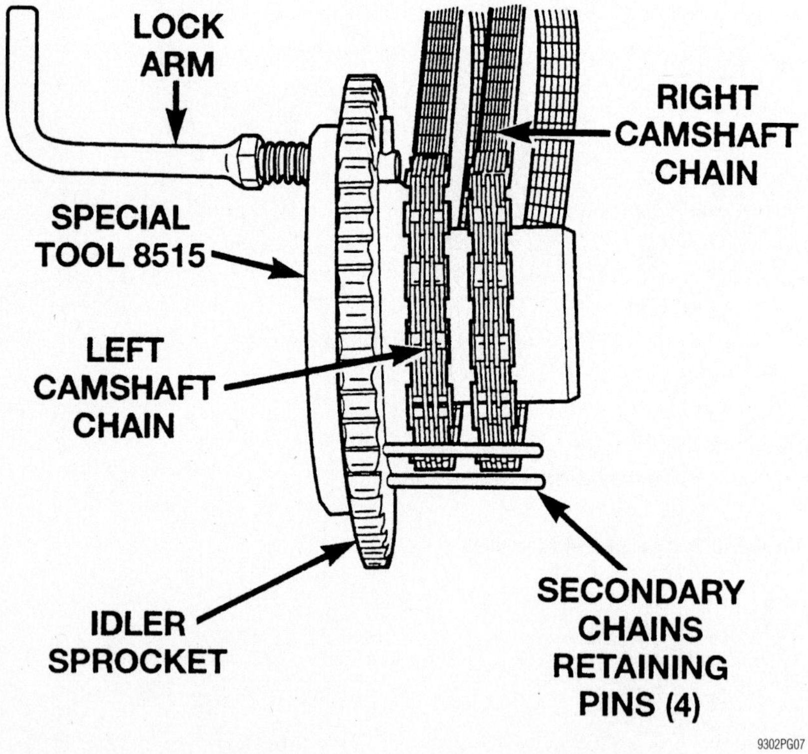

Use the special tool to lock the timing chains on the idler gear—4.7L engine

11. Lubricate the threads of the 11mm bolts with clean engine oil.
12. Coat the threads of the 8mm bolts with Mopar® Lock and Seal Adhesive.
13. Install the cylinder heads. Use new gaskets and tighten the bolts, in sequence, as follows:
    a. Step 1: Bolts 1–10 to 15 ft. lbs. (20 Nm)
    b. Step 2: Bolts 1–10 to 35 ft. lbs. (47 Nm)
    c. Step 3: Bolts 11–14 to 18 ft. lbs. (25 Nm)
    d. Step 4: Bolts 1–10 plus 90°
    e. Step 5: Bolts 11–14 to 22 ft. lbs. (30 Nm)
14. Install or connect the following:
    - Camshaft sprockets. Align the secondary chain matchmarks and tighten the bolts to 90 ft. lbs. (122 Nm).
    - Secondary timing chain guides
    - Cylinder head access plugs
    - Secondary timing chain tensioners. Refer to the timing chain procedure in this section.
15. Remove the Timing Chain Locking tool 8515.
16. Install or connect the following:
    - Front cover
    - Crankshaft damper. Torque the bolt to 130 ft. lbs. (175 Nm).
    - Power steering pump
    - Oil fill housing
    - Accessory drive belt
    - Engine cooling fan and shroud
    - Valve covers
    - Intake manifold
    - Exhaust Y-pipe
    - Negative battery cable
17. Fill and bleed the cooling system.
18. Start the engine, check for leaks and repair if necessary.

5.7L Engine

2002-2004 MODELS

1. Before servicing the vehicle, refer to the precautions in the beginning of this section.
2. Drain the cooling system.
3. Relieve the fuel system pressure.
4. Remove or disconnect the following:
    - Negative battery cable
    - Air cleaner resonator and ducts
    - Accessory drive belt
    - Alternator
    - Closed crankcase ventilation (CCV) system
    - Fuel supply line
    - EVAP control system
    - Heater hoses
    - Cylinder head covers
    - Intake manifold and throttle body
    - Rocker arms and pushrods
    - Cylinder head bolts
    - Cylinder heads and gasket

To install:

➡The head gaskets are not interchangeable. They are marked "L" and "R".

5. Install the cylinder heads. Use new gaskets and tighten the bolts, in sequence, as follows:
    a. Step 1: 12 mm bolts–25 ft. lbs. (34 Nm); 8mm bolts–15 ft. lbs. (20 Nm)
    b. Step 2: 12mm bolts–40 ft. lbs. (54 Nm); 8mm bolts retorque–15 ft. lbs. (20 Nm)

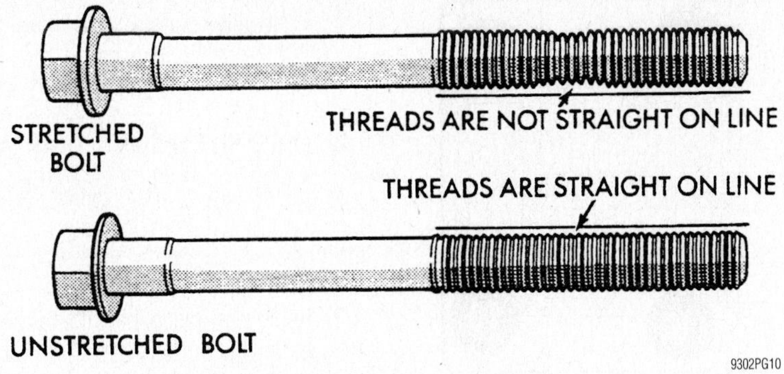

9302PG10

Examine the head bolts for signs of stretching—4.7L engine

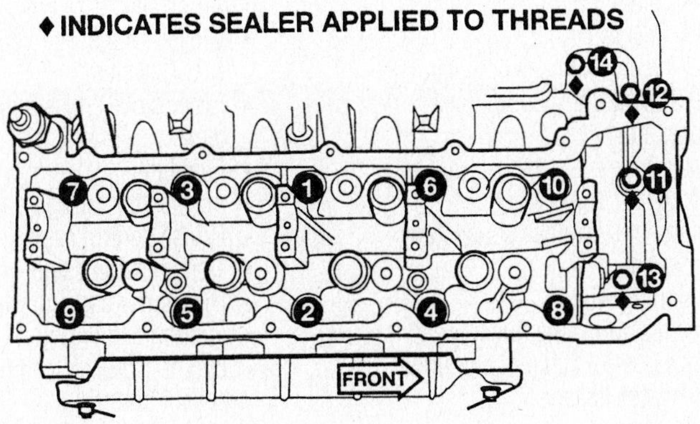

9302PG11

Cylinder head torque sequence—4.7L engine

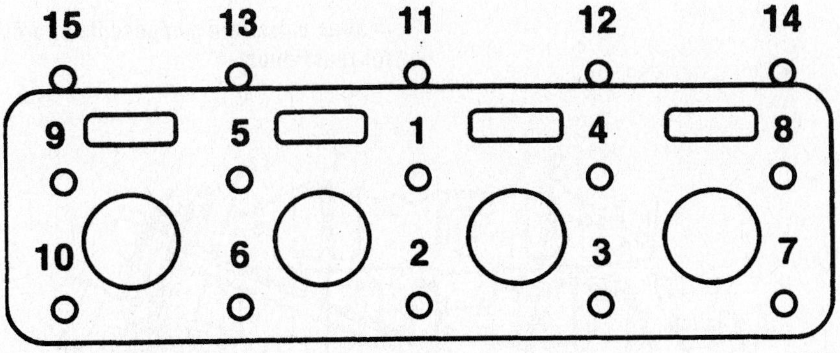

2399PG03

Cylinder head torque sequence—5.7L engine

c. Step 3: 12mm bolts–plus 90°; 8mm bolts–25 ft. lbs. (34 Nm)

6. Install or connect the following:
   - Rocker arms and pushrods
   - Intake manifold
   - Heater hoses
   - Alternator
   - Accessory drive belt
   - Cylinder head covers. Torque the studs and bolts to 70 inch lbs.
   - Air cleaner resonator and ducts
   - Negative battery cable

2005 MODELS

1. Before servicing the vehicle, refer to the precautions in the beginning of this section.
2. Drain the cooling system.
3. Relieve the fuel system pressure.
4. Remove or disconnect the following:
   - Negative battery cable
   - Air intake assembly
   - Closed Crankcase Ventilation (CCV) system
   - Exhaust Y-pipe
   - EVAP control system
   - Heater hoses
   - Power steering pump
   - Cylinder head cover bolts, loosing in sequence shown
   - Intake manifold and throttle body
   - Rocker arms and pushrods
   - Cylinder head bolts, loosening in sequence shown
   - Cylinder head and gasket

To install:

5. Install the cylinder head with new gasket.

➡The head gaskets are not interchangeable. They are marked "L" and "R" to indicate left and right sides and "TOP" to indicate which side goes up.

6. Tighten the cylinder head bolts as follows:
   a. Step 1: 12 mm bolts–25 ft. lbs. (34 Nm); 8mm bolts–15 ft. lbs. (20 Nm)
   b. Step 2: 12mm bolts–40 ft. lbs. (54 Nm); 8mm bolts retorque–15 ft. lbs. (20 Nm)
   c. Step 3: 12mm bolts–plus 90°; 8mm bolts–25 ft. lbs. (34 Nm)
7. Install or connect the following:
   - Rocker arms and pushrods
   - Heater hoses
   - Fuel supply line
   - Power steering pump
   - Accessory drive belt
   - Cylinder head covers, tightening bolts in sequence shown.
   - EVAP control system
   - Air intake assembly
   - Fill the cooling system to the correct level
   - Negative battery cable
8. Start the engine and check for leaks.

3.9L Engine

1. Before servicing the vehicle, refer to the precautions in the beginning of this section.
2. Relieve the fuel pressure.
3. Drain the cooling system.
4. Remove or disconnect the following:
   - Negative battery cable
   - Accessory drive belt
   - Alternator
   - A/C compressor, if equipped
   - Alternator and A/C compressor bracket
   - Air injection pump, if equipped
   - Closed Crankcase Ventilation (CCV) system
   - Air cleaner and hose
   - Fuel line
   - Accelerator linkage

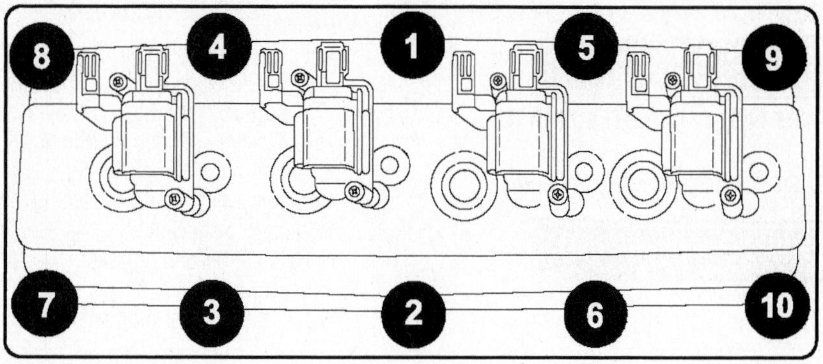

Cylinder head cover bolt sequence—5.7L engine

- Cruise control cable, if equipped
- Transmission cable, if equipped
- Spark plug wires
- Distributor
- Ignition coil harness connectors
- Engine Coolant Temperature (ECT) sensor connector
- Heater hoses
- Bypass hose
- Intake manifold vacuum lines
- Fuel injector harness connectors
- Valve covers
- Intake manifold
- Exhaust front pipe
- Exhaust manifolds

➡Keep all valvetrain components in order for assembly.

- Rocker arms
- Pushrods
- Cylinder heads

To install:

⁂ WARNING

Position the crankshaft so that no piston is at Top Dead Center (TDC) prior to installing the cylinder heads. Do not rotate the crankshaft during or immediately after rocker arm installation. Wait 5 minutes for the hydraulic lash adjusters to bleed down.

5. Install the cylinder heads with new gaskets. Tighten the bolts in sequence as follows:
   a. Step 1: 50 ft. lbs. (68 Nm).
   b. Step 2: 105 ft. lbs. (143 Nm).
   c. Step 3: 105 ft. lbs. (143 Nm).
6. Install or connect the following:
   - Pushrods in their original locations
   - Rocker arms in their original locations. Tighten the bolts to 21 ft. lbs. (28 Nm).
   - Exhaust manifolds
   - Exhaust front pipe
   - Intake manifold
   - Valve covers
   - Fuel injector harness connectors
   - Intake manifold vacuum lines
   - Bypass hose
   - Heater hoses
   - ECT sensor connector
   - Ignition coil harness connectors
   - Distributor
   - Spark plug wires
   - Transmission cable, if equipped
   - Cruise control cable, if equipped
   - Accelerator linkage
   - Fuel line
   - Air cleaner and hose
   - CCV system
   - Air injection pump, if equipped
   - Alternator and A/C compressor bracket
   - Alternator
   - A/C compressor
   - Accessory drive belt
   - Negative battery cable
7. Fill the cooling system.
8. Start the engine and check for leaks.

5.2L and 5.9L Engines

1. Before servicing the vehicle, refer to the precautions in the beginning of this section.
2. Drain the cooling system.
3. Remove or disconnect the following:
   - Negative battery cable
   - Accessory drive belt
   - Alternator
   - A/C compressor, if equipped
   - Air injection pump, if equipped
   - Air cleaner assembly
   - Closed Crankcase Ventilation (CCV) system
   - Evaporative emissions control system
   - Fuel line
   - Accelerator linkage
   - Cruise control cable
   - Transmission cable
   - Distributor cap and wires
   - Ignition coil wiring
   - Engine Coolant Temperature (ECT) sensor connector
   - Heater hoses
   - Bypass hose
   - Upper radiator hose
   - Intake manifold
   - Valve covers

➡Keep valvetrain components in order for reassembly.

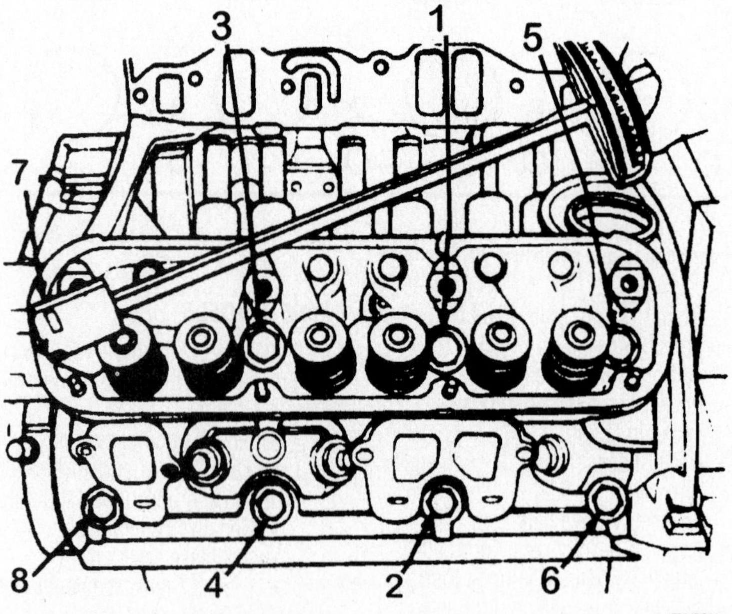

Cylinder head torque sequence—3.9L engine

- Rocker arms
- Pushrods
- Exhaust manifolds
- Spark plugs
- Cylinder heads

To install:

WARNING

Position the crankshaft so that no piston is at Top Dead Center (TDC) prior to installing the cylinder heads. Do not rotate the crankshaft during or immediately after rocker arm installation. Wait 5 minutes for the hydraulic lash adjusters to bleed down.

4. Install the cylinder heads. Use new gaskets and tighten the bolts in sequence as follows:
   a. Step 1: 50 ft. lbs. (68 Nm).
   b. Step 2: 105 ft. lbs. (143 Nm).
   c. Step 3: 105 ft. lbs. (143 Nm).
5. Install or connect the following:
   - Spark plugs
   - Exhaust manifolds
   - Pushrods and rocker arms in their original positions
   - Valve covers
   - Intake manifold
   - Upper radiator hose
   - Bypass hose
   - Heater hoses
   - ECT sensor connector
   - Ignition coil wiring
   - Distributor cap and wires
   - Transmission cable
   - Cruise control cable
   - Accelerator linkage
   - Fuel line
   - Evaporative emissions control system
   - CCV system
   - Air cleaner assembly
   - A/C compressor, if equipped
   - Air injection pump, if equipped
   - Alternator
   - Accessory drive belt
   - Negative battery cable
6. Fill the cooling system.
7. Start the engine and check for leaks.

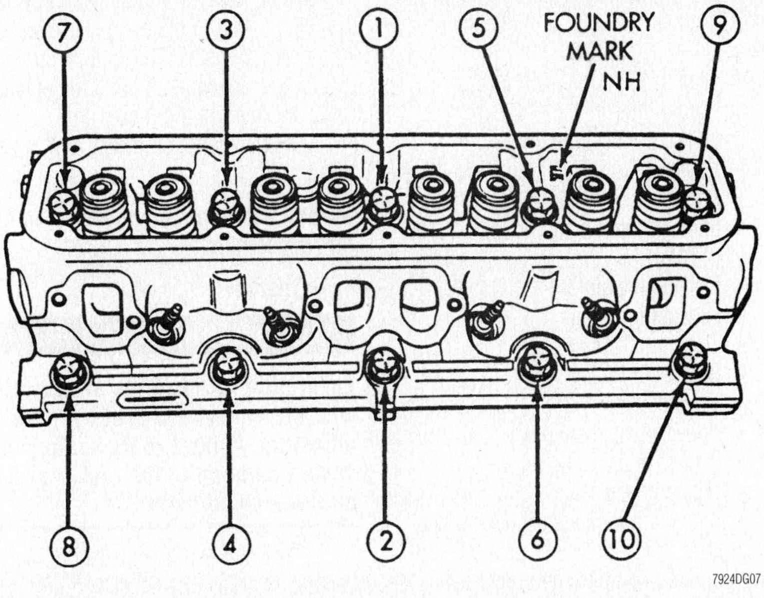

Cylinder head torque sequence—5.2L and 5.9L engines

8.0L Engine

1. Before servicing the vehicle, refer to the precautions in the beginning of this section.
2. Drain the cooling system.
3. Relieve the fuel system pressure.
4. Remove or disconnect the following:
   - Negative battery cable
   - Spark plug wires and heat shields
   - Accessory drive belt
   - Alternator
   - A/C compressor, if equipped
   - Alternator and A/C compressor bracket
   - Air injection pump, if equipped
   - Closed Crankcase Ventilation (CCV) system
   - Evaporative Emissions (EVAP) canister vacuum lines
   - Air cleaner and hose
   - Fuel line
   - Accelerator linkage
   - Cruise control cable, if equipped
   - Transmission cable, if equipped
   - Ignition coil pack and bracket
   - Temperature gauge sender connector
   - Heater hoses
   - Bypass hose
   - Upper intake manifold
   - Valve covers
   - Exhaust Gas Recirculation (EGR) tube
   - Lower intake manifold
   - Exhaust front pipe
   - Exhaust manifolds

➡**Keep all valvetrain components in order for assembly.**

- Rocker arms
- Pushrods
- Cylinder heads

To install:

WARNING

Position the crankshaft so that no piston is at Top Dead Center (TDC) prior to installing the cylinder heads. Do not rotate the crankshaft during or immediately after rocker arm installation. Wait 5 minutes for the hydraulic lash adjusters to bleed down.

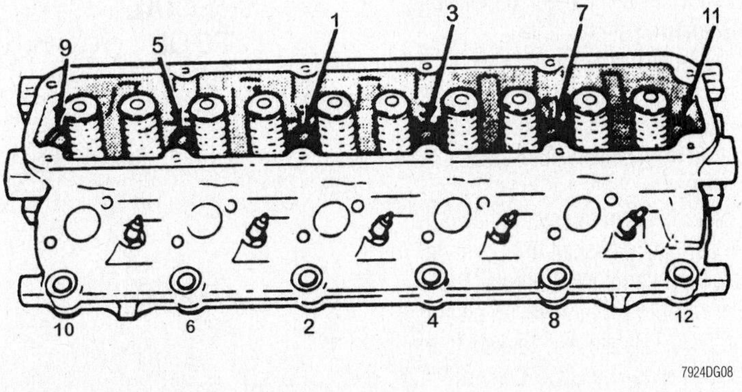

Cylinder head torque sequence—8.0L engine

5. Install the cylinder heads with new gaskets. Tighten the bolts in sequence as follows:
   a. Step 1: 43 ft. lbs. (58 Nm).
   b. Step 2: 105 ft. lbs. (143 Nm).
6. Install or connect the following:
   - Pushrods in their original locations
   - Rocker arms in their original locations. Tighten the bolts to 21 ft. lbs. (28 Nm).
   - Exhaust manifolds
   - Exhaust front pipe
   - Lower intake manifold
   - EGR tube
   - Valve covers
   - Upper intake manifold
   - Bypass hose
   - Heater hoses
   - Temperature gauge sender connector
   - Ignition coil pack and bracket
   - Transmission cable, if equipped
   - Cruise control cable, if equipped
   - Accelerator linkage
   - Fuel line
   - Air cleaner and hose
   - EVAP canister vacuum lines
   - CCV system
   - Air injection pump, if equipped
   - Alternator and A/C compressor bracket
   - A/C compressor, if equipped
   - Alternator
   - Accessory drive belt
   - Spark plug wires and heat shields
   - Negative battery cable
7. Fill the cooling system.
8. Start the engine and check for leaks.

Rocker Arms/Shafts

REMOVAL & INSTALLATION

3.7L Engines

1. Before servicing the vehicle, refer to the precautions in the beginning of this section.
2. Remove or disconnect the following:
   - Negative battery cable
   - Cylinder head cover
3. For rocker arm removal on cylinder No. 4, Rotate the crankshaft until cylinder No. 1 is at Bottom Dead Center (BDC) intake stroke.
4. For rocker arm removal on cylinder No. 1, Rotate the crankshaft until cylinder No. 1 is at BDC combustion stroke.
5. For rocker arm removal on cylinders No. 3 and No. 5, Rotate the crankshaft until cylinder No. 1 is at Top Dead Center (TDC) exhaust stroke.
6. For rocker arm removal on cylinders No. 2 and No. 6, Rotate the crankshaft until cylinder No. 1 is at TDC ignition stroke.
7. Use special tool 8516 to depress the valve and remove the rocker arm.
8. Repeat for each rocker arm to be serviced.

➡Keep valvetrain components in order for reassembly.

To install:

⁂ CAUTION

Ensure the rocker arms are installed with the concave pocket over the lash adjusters. Failure to do so may cause severe damage to the rocker arms and/or lash adjusters.

9. For rocker arm installation on cylinders No. 4, rotate the crankshaft until cylinder No. 1 is at BDC intake stroke.
10. For rocker arm installation on cylinder No. 1, rotate the crankshaft until cylinder No. 1 is at BDC combustion stroke.
11. For rocker arm installation on cylinders No. 3 and No. 5, rotate the crankshaft until cylinder No. 1 is at TDC exhaust stroke.
12. For rocker arm installation on cylinders No. 2 and No. 6, rotate the crankshaft until cylinder No. 1 is at TDC ignition stroke.
13. Using special tool 8516 press downward on the valve spring and install rocker arm.
14. Repeat for each rocker arm to be installed.
15. Install or connect the following:
   - Cylinder head cover
   - Negative battery cable

4.7L Engine

1. Before servicing the vehicle, refer to the precautions in the beginning of this section.
2. Remove or disconnect the following:
   - Negative battery cable
   - Cylinder head cover
3. For rocker arm removal on cylinders No. 3 and No. 5, rotate the crankshaft until cylinder #1 is at Top Dead Center (TDC) exhaust stroke.
4. For rocker arm removal on cylinders No. 2 and No. 8, rotate the crankshaft until cylinder #1 is at TDC compression stroke.
5. For rocker arm removal on cylinders No. 4 and No. 6, rotate the crankshaft until cylinder No .3 is at TDC compression stroke.
6. For rocker arm removal on cylinders

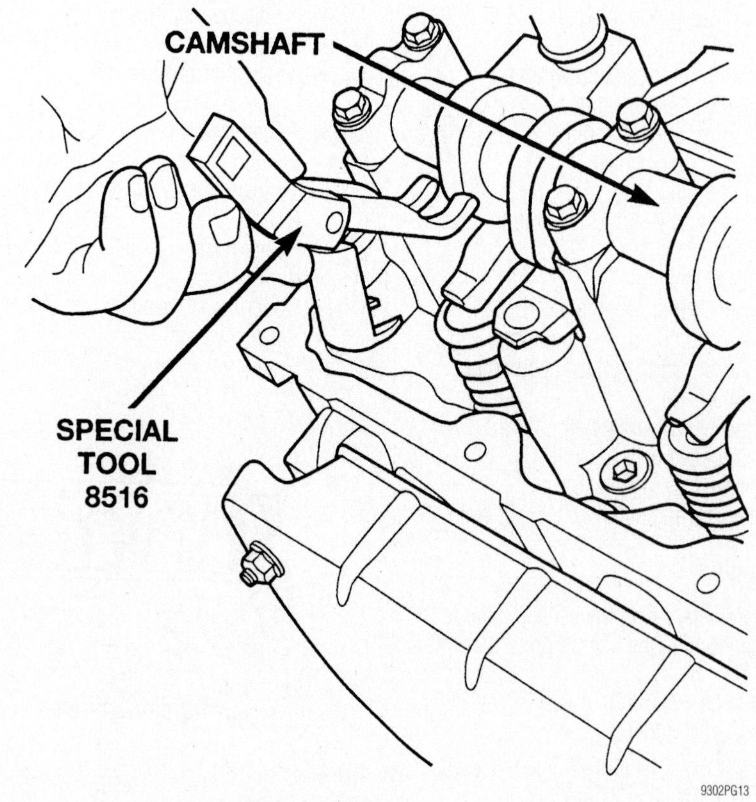

Rocker arm service—3.7L and 4.7L engines

No. 1 and No. 7, rotate the crankshaft until cylinder No. 2 is at TDC compression stroke.

7. Use special tool 8516 to depress the valve and remove the rocker arm.

8. Repeat for each rocker arm to be serviced.

➡**Keep valvetrain components in order for reassembly.**

To install:

CAUTION

Ensure the rocker arms are installed with the concave pocket over the lash adjusters. Failure to do so may cause severe damage to the rocker arms and/or lash adjusters.

9. For rocker arm installation on cylinders No. 3 and No. 5, rotate the crankshaft until cylinder No. 1 is at TDC exhaust stroke.

10. For rocker arm installation on cylinders No. 2 and No. 8, rotate the crankshaft until cylinder No. 1 is at TDC compression stroke.

11. For rocker arm installation on cylinders No. 4 and No. 6, rotate the crankshaft until cylinder No. 3 is at TDC compression stroke.

12. For rocker arm installation on cylinders No. 1 and No. 7, rotate the crankshaft until cylinder No. 2 is at TDC compression stroke.

13. Using special tool 8516 press downward on the valve spring and install rocker arm.

14. Repeat for each rocker arm to be installed.

15. Install or connect the following:
- Cylinder head cover
- Negative battery cable

3.9L Engine

1. Before servicing the vehicle, refer to the precautions in the beginning of this section.

2. Remove or disconnect the following:
- Negative battery cable
- Valve covers
- Rocker arms

➡**Keep all valvetrain components in order for assembly.**

To install:

3. Rotate the crankshaft so that the **V6** mark on the crankshaft damper aligns with the timing mark on the front cover. The **V6** mark is located 147 degrees **AFTER** Top Dead Center (TDC).

4. Install the rocker arms in their original positions and tighten the bolts to 21 ft. lbs. (28 Nm).

CAUTION

Do not rotate the crankshaft during or immediately after rocker arm installation. Wait 5 minutes for the hydraulic lash adjusters to bleed down.

5. Install or connect the following:
- Valve covers
- Negative battery cable

5.7L Engine

2002-2004 MODELS

1. Before servicing the vehicle, refer to the precautions in the beginning of this section.

2. Remove or disconnect the following:
- Negative battery cable
- Cylinder head cover

3. Loosen the rocker shaft bolts as follows: center, center left, center right, left, right.

➡**The rocker shaft assemblies are not interchangeable. The intake side is marked "I". The shorter pushrods are used on the intake side.**

To install:

4. Install or connect the following:
- Pushrods
- Rocker shaft assemblies. Torque the shafts to 16 ft. lbs. (22 Nm) in this sequence: center, center right, center left, right, left.

WARNING

Do not rotate the crankshaft during or immediately after rocker shaft installation. Allow the roller tappets to leakdown for about 5 minutes.

- Cylinder head covers. Torque the fasteners to 70 inch lbs.

2005 MODELS

1. Before servicing the vehicle, refer to the precautions in the beginning of this section.

2. Disconnect the negative battery cable.

3. Remove the cylinder head cover.

4. Install pushrod retaining plate 9070

5. Loosen the rocker shafts using the sequence shown.

6. Remove the rocker shaft assemblies and pushrods.

CAUTION

Note location for reassembly. The rocker shaft assemblies are not interchangeable between intake and exhaust. The intake rocker arms are marked with an "I". The shorter pushrods are for the intake side.

To install:

7. Install the following:
- Pushrods
- Pushrod retaining plate 9070

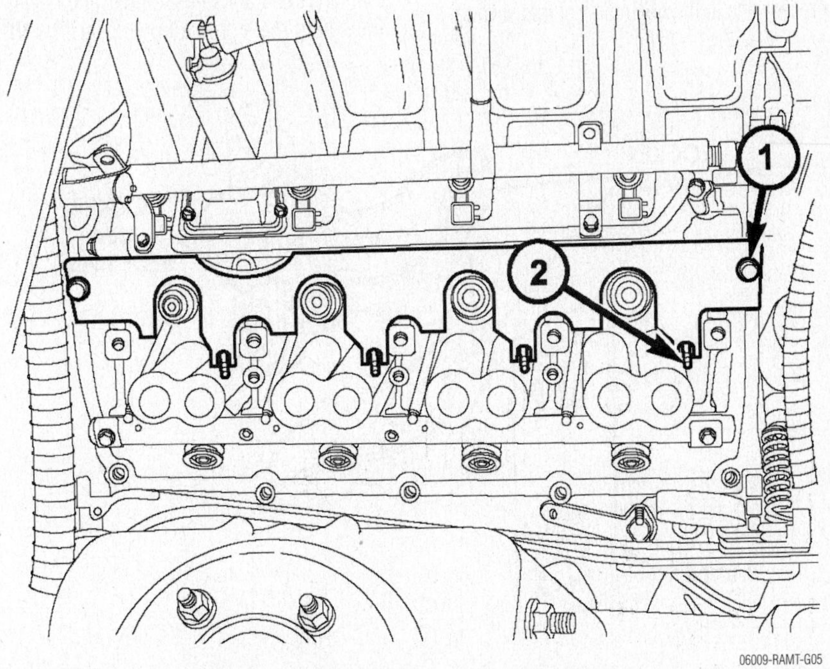

Installing the pushrod retaining plate 9070

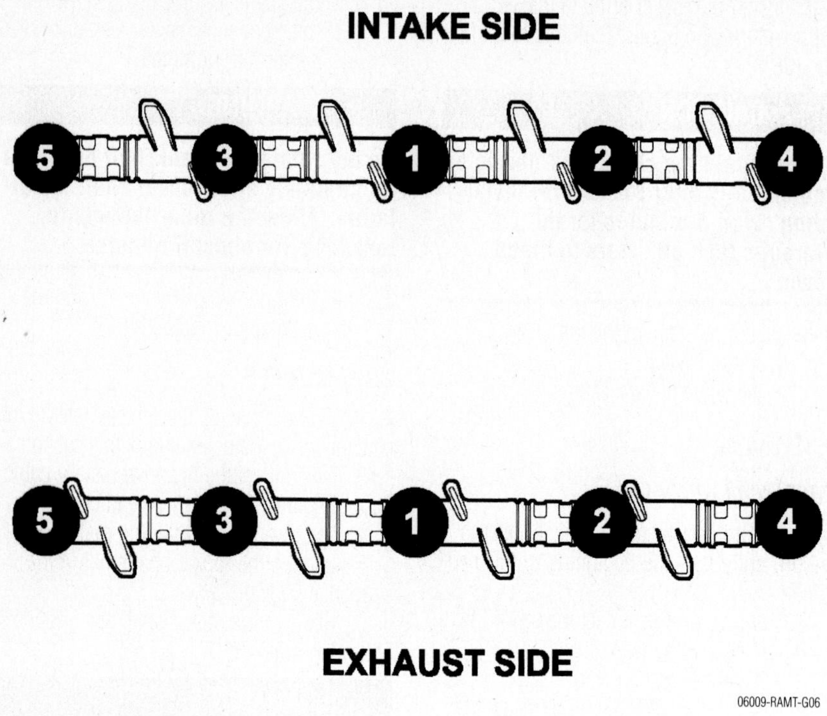

Rocker arm removal order

- Rocker shaft assemblies. Torque the shaft bolts 195 inch lbs (22 Nm).

8. Remove the pushrod retaining plate.
9. Install or connect the following:
   - Cylinder head cover
   - Negative battery cable

5.2L and 5.9L Engines

1. Before servicing the vehicle, refer to the precautions in the beginning of this section.
2. Remove or disconnect the following:
   - Negative battery cable
   - Valve covers
   - Rocker arms

➡**Keep valvetrain components in order for reassembly.**

To install:

3. Rotate the crankshaft so that the **V8** mark on the crankshaft damper aligns with the timing mark on the front cover. The **V8** mark is located 147 degrees **AFTER** Top Dead Center (TDC).
4. Install the rocker arms in their original positions and tighten the bolts to 21 ft. lbs. (28 Nm).

⁂ CAUTION

Do not rotate the crankshaft during or immediately after rocker arm installation. Wait 5 minutes for the hydraulic lash adjusters to bleed down.

5. Install or connect the following:
   - Valve covers
   - Negative battery cable

8.0L Engine

1. Before servicing the vehicle, refer to the precautions in the beginning of this section.
2. Remove or disconnect the following:
   - Negative battery cable
   - Valve covers
   - Rocker arms

➡**Keep valvetrain components in order for reassembly.**

To install:

⁂ CAUTION

When installing the rocker arms, ensure that the piston is not at Top Dead Center. Tighten the rocker arm bolts slowly. Do not rotate the crankshaft during or immediately after rocker arm installation. Wait 5 minutes for the hydraulic lash adjusters to bleed down.

3. Install or connect the following:
   - Rocker arms in their original positions. Tighten the bolts to 21 ft. lbs. (28 Nm).
   - Valve covers
   - Negative battery cable

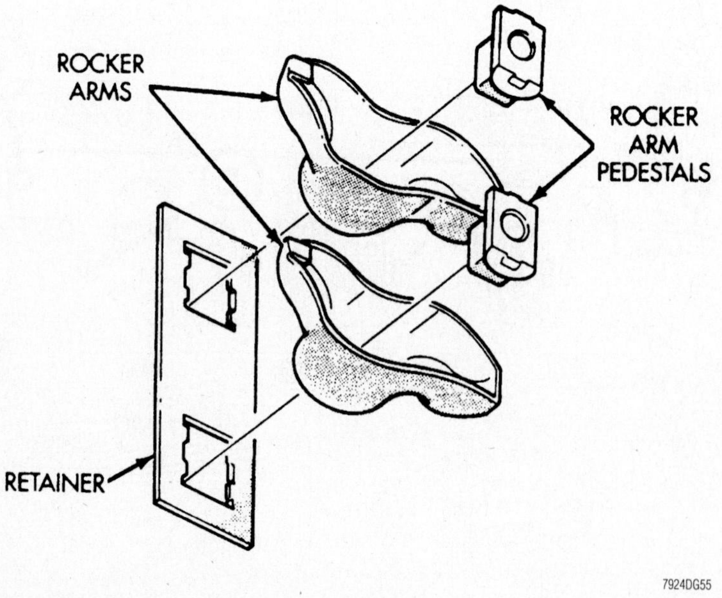

Exploded view of the rocker arm assembly—8.0L engine

Intake Manifold

REMOVAL & INSTALLATION

3.7L and 4.7L Engines

1. Before servicing the vehicle, refer to the precautions in the beginning of this section.
2. Drain the cooling system.
3. Relieve fuel system pressure.
4. Remove or disconnect the following:
   - Negative battery cable

- Air cleaner assembly
- Accelerator cable and cruise control cable
- Manifold Absolute Pressure (MAP) sensor connector
- Intake Air Temperature (IAT) sensor connector
- Throttle Position (TPS) Sensor connector
- Coolant Temperature (CTS) Sensor
- Idle Air Control (IAC) valve connector
- Positive Crankcase Ventilation (PCV) valve and hose
- Canister purge vacuum line
- Brake booster vacuum line
- Cruise control servo hose
- Alternator connections
- A/C compressor connections
- Left and right radio suppressor straps
- Engine ground straps
- Ignition coil towers
- Oil dipstick tube
- Fuel rail
- Throttle body assembly and mounting bracket
- Heater hoses from engine front cover and heater core
- Heater hoses and tubes and intake manifold
- CTS sensor
- Intake manifold. Remove the fasteners in reverse of the tightening sequence.

To install:

5. Install or connect the following:
   - Intake manifold using new gaskets. Torque the bolts, in sequence, to 105 inch lbs. (12 Nm).
   - Left and right radio suppressor straps
   - Throttle body assembly and mounting bracket
   - Accelerator and cruise control cables
   - Fuel rail
   - Ignition coil towers
   - Heater hoses and tubes on intake manifold
   - Heater hoses on front engine cover and heater core
   - MAP sensor connector
   - IAT sensor connector
   - TPS sensor connector
   - CTS sensor connector
   - IAC valve connector
   - PCV hose
   - Oil dipstick tube
   - Engine ground straps
   - Alternator connections
   - A/C compressor connections

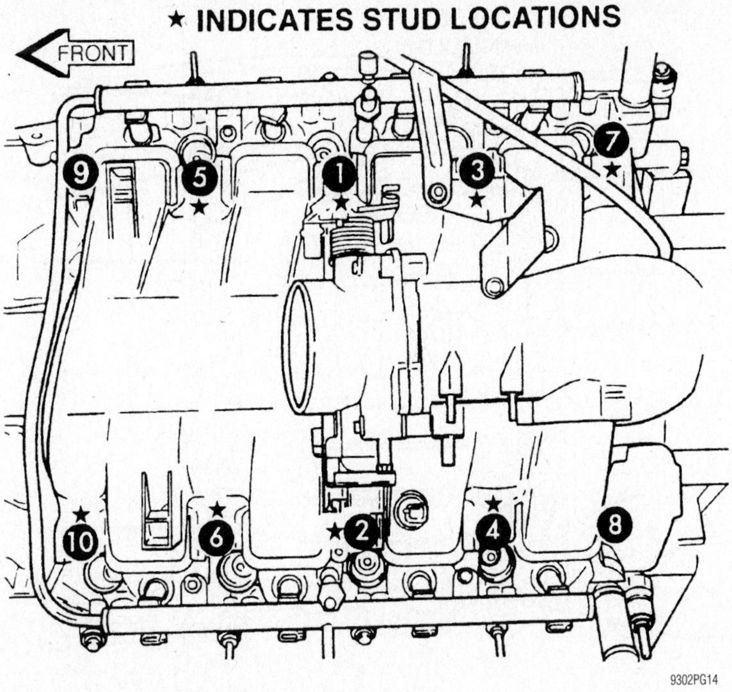

Intake manifold torque sequence—3.7L & 4.7L engines

   - Canister purge vacuum line
   - Brake booster vacuum line
   - Cruise control servo hose
   - Cruise control cable
   - Air cleaner assembly
   - Negative battery cable
6. Fill and bleed the cooling system.
7. Start the engine and check for leaks.

3.9L Engine

1. Before servicing the vehicle, refer to the precautions in the beginning of this section.
2. Drain the cooling system.
3. Remove or disconnect the following:
   - Negative battery cable
   - Accessory drive belt
   - Alternator
   - A/C compressor
   - Alternator and A/C compressor bracket
   - Air cleaner assembly
   - Fuel line
   - Fuel supply manifold
   - Accelerator cable
   - Transmission cable
   - Cruise control cable
   - Distributor cap and wires
   - Ignition coil wiring
   - Engine Coolant Temperature (ECT) sensor connector
   - Heater hose
   - Upper radiator hose
   - Bypass hose
   - Closed Crankcase Ventilation (CCV) system
   - Evaporative emissions system
   - Intake manifold

To install:

4. Install the intake manifold. Use a new gasket and tighten the bolts in sequence as follows:
   a. Step 1: 48 inch lbs.
   b. Step 2: 84 inch lbs.
   c. Step 3: Recheck 84 inch lbs.
5. Install or connect the following:
   - Evaporative emissions system
   - CCV system
   - Bypass hose
   - Upper radiator hose
   - Heater hose
   - ECT sensor connector
   - Ignition coil wiring
   - Distributor cap and wires
   - Cruise control cable
   - Transmission cable
   - Accelerator cable
   - Fuel supply manifold
   - Fuel line
   - Air cleaner assembly
   - Alternator and A/C compressor bracket
   - A/C compressor
   - Alternator
   - Accessory drive belt
   - Negative battery cable
6. Fill the cooling system.
7. Start the engine and check for leaks.

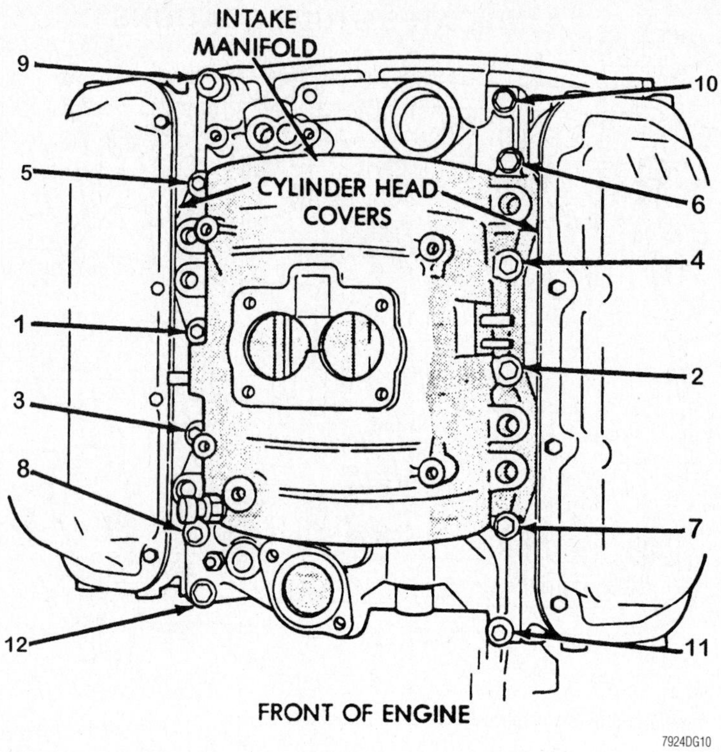

Intake manifold torque sequence—3.9L engine

5.2L and 5.9L Engines

1. Before servicing the vehicle, refer to the precautions in the beginning of this section.
2. Drain the cooling system.
3. Remove or disconnect the following:
   - Negative battery cable
   - Accessory drive belt
   - Alternator
   - A/C compressor
   - Alternator and A/C compressor bracket
   - Air cleaner assembly
   - Fuel line
   - Fuel supply manifold
   - Accelerator cable
   - Transmission cable
   - Cruise control cable
   - Distributor cap and wires
   - Ignition coil wiring
   - Engine Coolant Temperature (ECT) sensor connector
   - Heater hose
   - Upper radiator hose
   - Bypass hose
   - Closed Crankcase Ventilation (CCV) system
   - Evaporative emissions system
   - Intake manifold

To install:

4. Install the intake manifold. Use a new gasket and tighten the bolts in sequence as follows:
   a. Step 1: 48 inch lbs.
   b. Step 2: 84 inch lbs.
   c. Step 3: Recheck 84 inch lbs.
   d. Step 1: Bolts 1–4 to 72 inch lbs. (8 Nm) using 12 inch lb. (1.4 Nm) increments.
5. Install or connect the following:
   - Evaporative emissions system
   - CCV system
   - Bypass hose
   - Upper radiator hose
   - Heater hose
   - ECT sensor connector
   - Ignition coil wiring
   - Distributor cap and wires
   - Cruise control cable
   - Transmission cable
   - Accelerator cable
   - Fuel supply manifold
   - Fuel line
   - Air cleaner assembly
   - Alternator and A/C compressor bracket
   - A/C compressor
   - Alternator
   - Accessory drive belt
   - Negative battery cable
6. Fill the cooling system.
7. Start the engine and check for leaks.

5.7L Engines

1. Before servicing the vehicle, refer to the precautions in the beginning of this section.
2. Drain the cooling system.
3. Relieve the fuel system pressure.
4. Remove or disconnect the following:
   - Negative battery cable
   - Air cleaner assembly
   - Accessory drive belt
   - Manifold Absolute Pressure (MAP) sensor connector
   - Intake Air Temperature (IAT) sensor connector
   - Throttle Position Sensor (TPS) connector
   - Coolant Temperature Sensor (CTS) connector
   - Brake booster hose

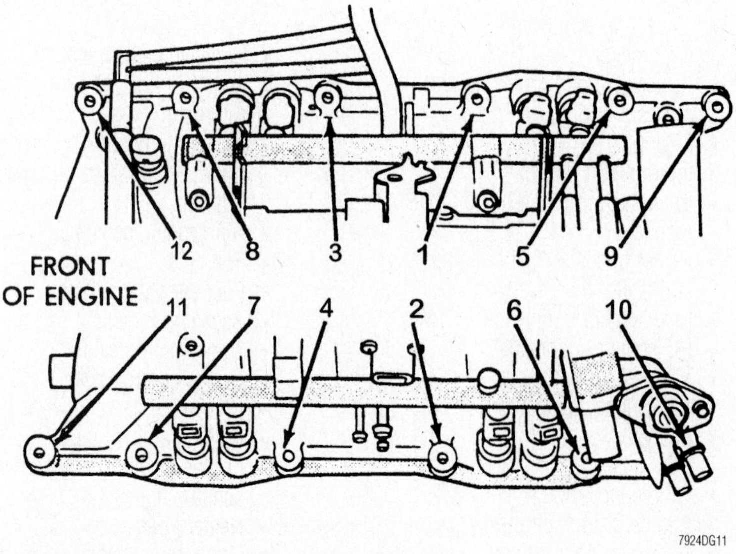

Intake manifold torque sequence—5.2L and 5.9L engines

- Positive Crankcase Ventilation (PCV) hose
- Alternator
- A/C compressor and secure aside leaving lines attached
- Intake manifold bolts, in a criss-cross pattern from the outside to the center
- Intake manifold and IAFM as an assembly

To install:

5. Position new intake manifold seals.
6. Install the intake manifold and IAFM assembly. Tighten the bolts in sequence from the center outwards in a criss-cross pattern, to 105 inch lbs. (12 Nm).
7. Install or connect the following:
   - MAP connector
   - IAT connector
   - TPS connector
   - CTS connector
   - Alternator
   - A/C compressor
   - Accessory drive belt
   - Brake booster hose
   - PCV hose
   - Air cleaner assembly
   - Negative battery cable

8.0L Engines

1. Before servicing the vehicle, refer to the precautions in the beginning of this section.
2. Drain the cooling system.
3. Relieve the fuel system pressure.
   - Negative battery cable
   - Accessory drive belt
   - Alternator and brace
   - A/C compressor and brace
   - Air cleaner housing
   - Fuel line
   - Accelerator linkage
   - cruise control cable, if equipped
   - Transmission cable, if equipped
   - Ignition coil pack and spark plug wires
   - Intake manifold vacuum lines
   - Engine control sensor harness connectors
   - Heater hoses
   - Bypass hose
   - Evaporative Emissions (EVAP) system
   - Closed Crankcase Ventilation (CCV) system
   - Throttle body
   - Upper intake manifold
   - Lower intake manifold

To install:

4. Install or connect the following:
   - Lower intake manifold. Tighten the bolts in sequence to 40 ft. lbs. (54 Nm).
   - Upper intake manifold. Tighten the bolts in sequence to 16 ft. lbs. (22 Nm).
   - Throttle body. Tighten the bolts to 17 ft. lbs. (23 Nm).
   - CCV system
   - EVAP system
   - Bypass hose
   - Heater hoses
   - Engine control sensor harness connectors
   - Intake manifold vacuum lines
   - Ignition coil pack and spark plug wires
   - Transmission cable, if equipped
   - cruise control cable, if equipped
   - Accelerator linkage
   - Fuel line
   - Air cleaner housing
   - A/C compressor and brace
   - Alternator and brace
   - Accessory drive belt
   - Negative battery cable
5. Fill the cooling system.
6. Start the engine and check for leaks.

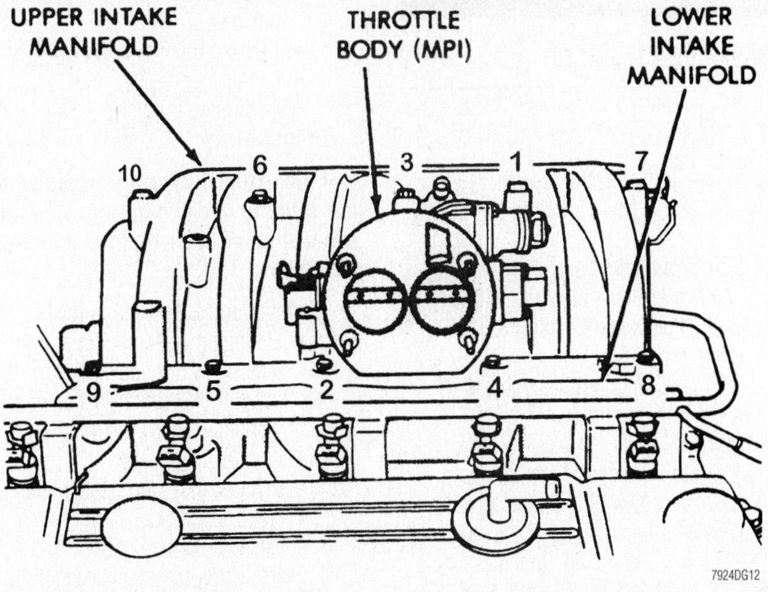

Upper intake manifold torque sequence—8.0L engine

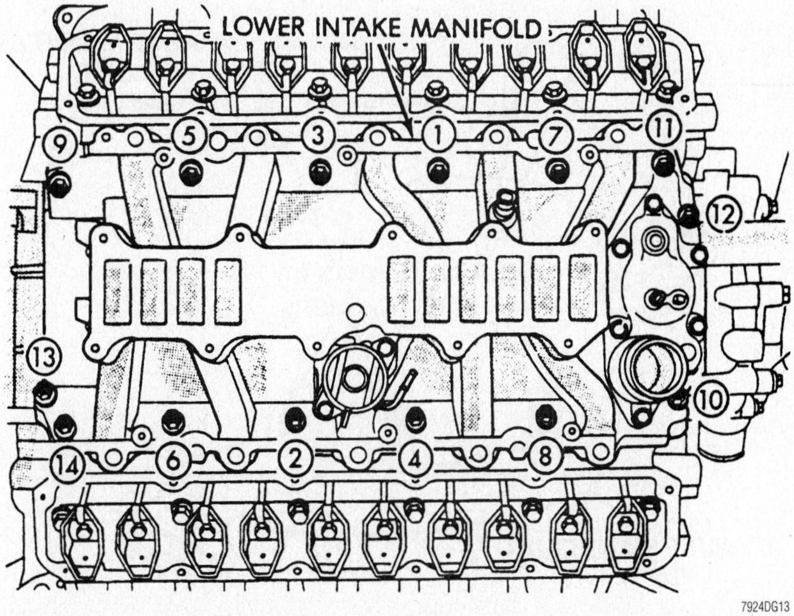

Lower intake manifold torque sequence—8.0L engine

Exhaust Manifold

REMOVAL & INSTALLATION

3.7L Engines

1. Before servicing the vehicle, refer to the precautions in the beginning of this section.
2. Drain the cooling system.
3. Remove or disconnect the following:
    - Negative battery cable
    - Exhaust Y-pipe
    - Exhaust manifold heat shield
    - Exhaust manifold

To install:

4. Position the exhaust manifold and new gasket onto mounting studs. Install conical washers and nuts on these studs.
5. Install the remaining conical washers. Tighten the bolts to 18 ft. lbs. (25 Nm), starting with the inner bolts and working outward.
6. Install or connect the following:
    - Exhaust heat shields
    - Exhaust Y-pipe to the exhaust manifold and tighten the bolts and nuts to 25 ft. lbs. (34 Nm).
    - Negative battery cable

3.9L Engine

1. Before servicing the vehicle, refer to the precautions in the beginning of this section.
2. Remove or disconnect the following:
    - Negative battery cable
    - Heated Oxygen (HO_2S) sensor connectors
    - Exhaust manifold heat shields
    - Exhaust front pipe
    - Exhaust manifolds

To install:

➡If the exhaust manifold studs came out with the nuts when removing the exhaust manifolds, replace them with new studs.

3. Install or connect the following:
    - Exhaust manifolds. Tighten the fasteners to 25 ft. lbs. (34 Nm), starting with the center fasteners and working out to the ends.
    - Exhaust front pipe
    - Exhaust manifold heat shields
    - HO_2S sensor connectors
    - Negative battery cable
4. Start the engine and check for leaks.

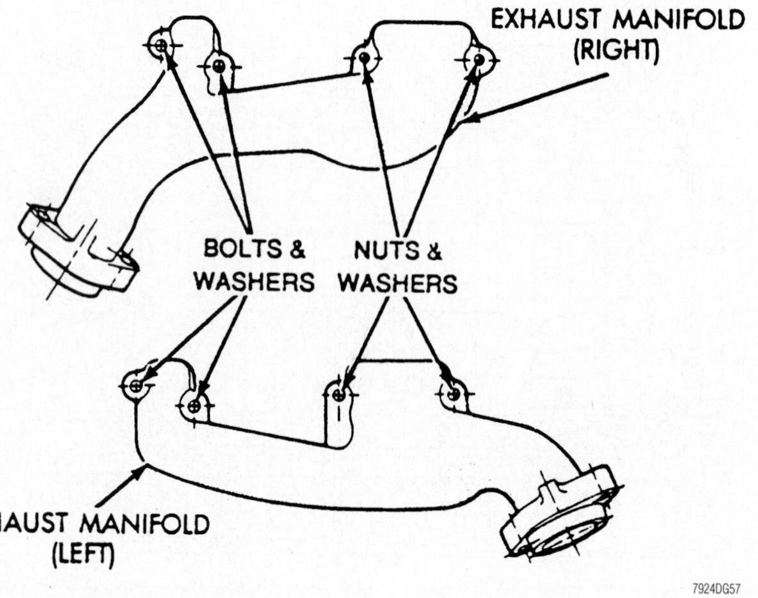

Stud and bolt locations—3.9L engine

4.7L Engines

RIGHT SIDE

1. Before servicing the vehicle, refer to the precautions in the beginning of this section.
2. Drain the cooling system.
3. Remove or disconnect the following:
    - Negative battery cable
    - Air cleaner assembly
    - Accessory drive belt
    - A/C compressor
    - A/C accumulator support bracket fastener
    - Heater hoses at engine
    - Exhaust manifold heat shield
    - Exhaust Y-pipe from manifold
4. Remove the fasteners attaching the starter. Secure the starter aside.
5. Remove the exhaust manifold fasteners.
6. Remove the exhaust manifold and gasket.

➡The manifold is removed from below the engine compartment.

To install:

7. Install the exhaust manifold and gasket.
8. Install the exhaust manifold fasters.

➡Do not tighten until all fasteners are in place.

9. Tighten all bolts to 18 ft. lbs. (25 Nm) starting in the center and working outward.
10. Install the exhaust manifold heat shield. Tighten to 72 inch lbs. and then back off the nuts 45°

⁂ CAUTION

Do not overtighten!

11. Install or connect the following:
    - Starter
    - Exhaust Y-pipe
    - Heater hoses
    - A/C accumulator fastener
    - A/C compressor
    - Accessory drive belt
    - Air intake assembly
    - Negative battery cable
12. Fill the cooling system.
13. Start the engine and check for leaks.

LEFT SIDE

1. Before servicing the vehicle, refer to the precautions in the beginning of this section.
2. Remove or disconnect the following:
    - Negative battery cable
    - Exhaust Y-pipe at manifold
    - Exhaust manifold heat shield
    - Exhaust manifold retaining bolts
    - Exhaust manifold and gasket

➡The manifold is removed from below the engine compartment.

To install:

3. Install the exhaust manifold and gasket.
4. Install the exhaust manifold fasters.

➡Do not tighten until all fasteners are in place.

5. Tighten all bolts to 18 ft. lbs. (25 Nm) starting in the center and working outward.

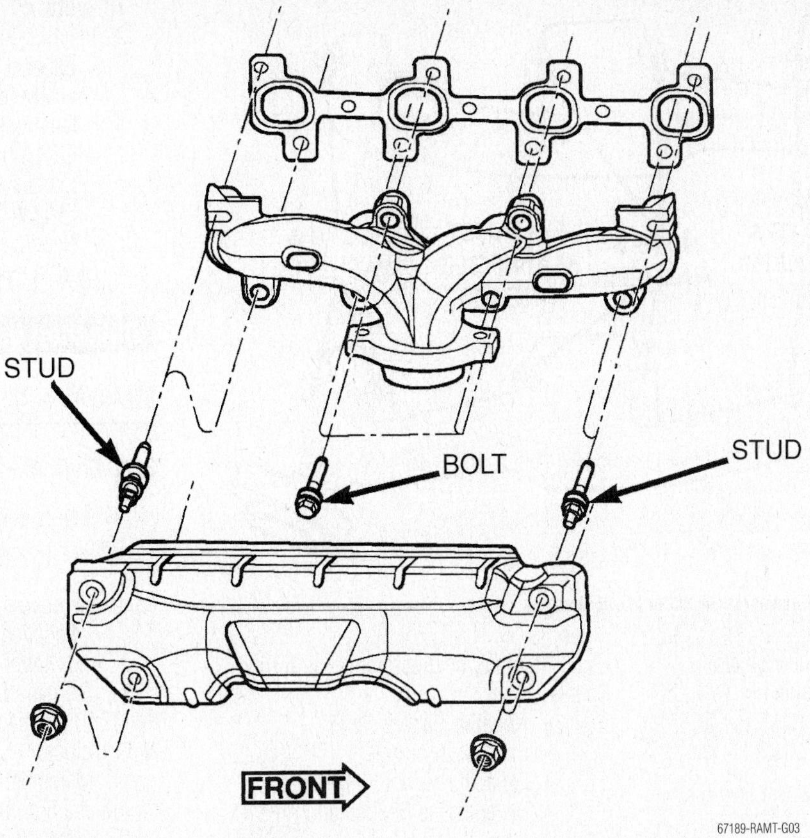

Right side exhaust manifold—4.7L engine, left side and 3.7L engine similar

6. Install the exhaust manifold heat shield. Tighten to 72 inch lbs. and then back off the nuts 45°

⁂ CAUTION

Do not overtighten!

7. Connect the exhaust Y-pipe to the manifold.
8. Connect the negative battery cable.
9. Start the engine and check for leaks.

5.7L Engines

1. Before servicing the vehicle, refer to the precautions in the beginning of this section.
2. Drain the cooling system.
3. Remove or disconnect the following:
   - Battery
   - Exhaust pipe-to-manifold bolts
4. Install engine lifting fixture #8534. Loosen the engine mount bolts and raise the engine enough to provide clearance for manifold removal.
5. Remove the exhaust manifold heat shields.
6. Remove the exhaust manifold bolts in sequence as shown.
7. Remove the exhaust manifold.

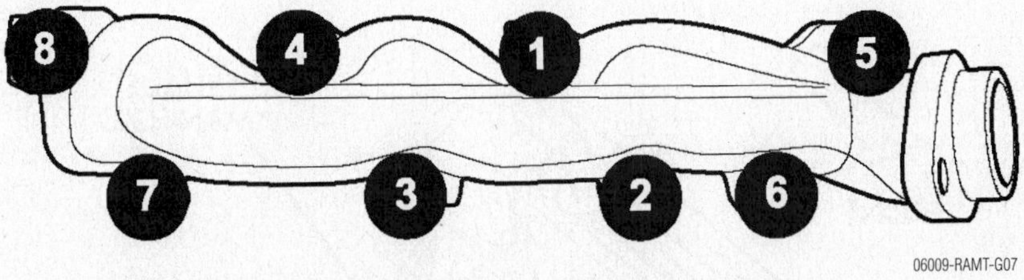

5.7L exhaust manifold torque sequence

To install:

8. Install or connect the following:
   - Exhaust manifold, using new gasket. Tighten the bolts to 18 ft. lbs. (25 Nm) in sequence as shown.
   - Exhaust manifold heat shield. Tighten nuts to 11 ft. lbs. (15 Nm).
9. Lower the engine and install the right and left mount through bolts. Remove the engine lifting fixture.
10. Install or connect the following:
    - Exhaust pipe-to-manifold bolts
    - Negative battery

5.2L and 5.9L Engines

1. Before servicing the vehicle, refer to the precautions in the beginning of this section.
2. Remove or disconnect the following:
   - Negative battery cable
   - Exhaust manifold heat shields
   - Exhaust Gas Recirculation (EGR) tube
   - Exhaust Y-pipe
   - Exhaust manifolds

To install:

➡If the exhaust manifold studs came out with the nuts when removing the exhaust manifolds, replace them with new studs.

3. Install or connect the following:
   - Exhaust manifolds. Tighten the fasteners to 25 ft. lbs. (27 Nm), starting with the center nuts and work out to the ends.
   - Exhaust Y-pipe
   - EGR tube

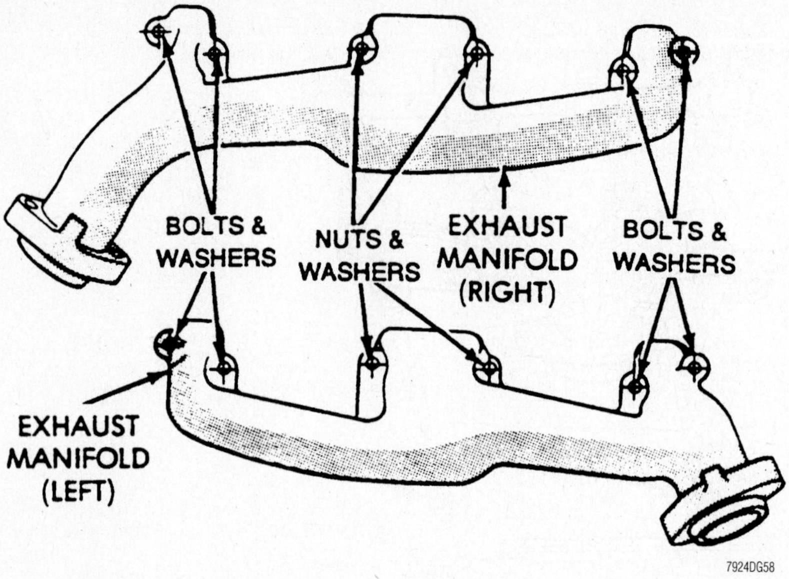

Exhaust manifold fastener locations—5.2L and 5.9L engines

- Exhaust manifold heat shields
- Negative battery cable

4. Fill the cooling system.
5. Start the engine and check for leaks.

8.0L Engine

1. Before servicing the vehicle, refer to the precautions in the beginning of this section.
2. Remove or disconnect the following:
   - Negative battery cable
   - Exhaust front pipe
   - Exhaust manifold heat shields
   - Exhaust Gas Recirculation (EGR) tube
   - Oil dipstick tube bracket
   - Exhaust manifolds

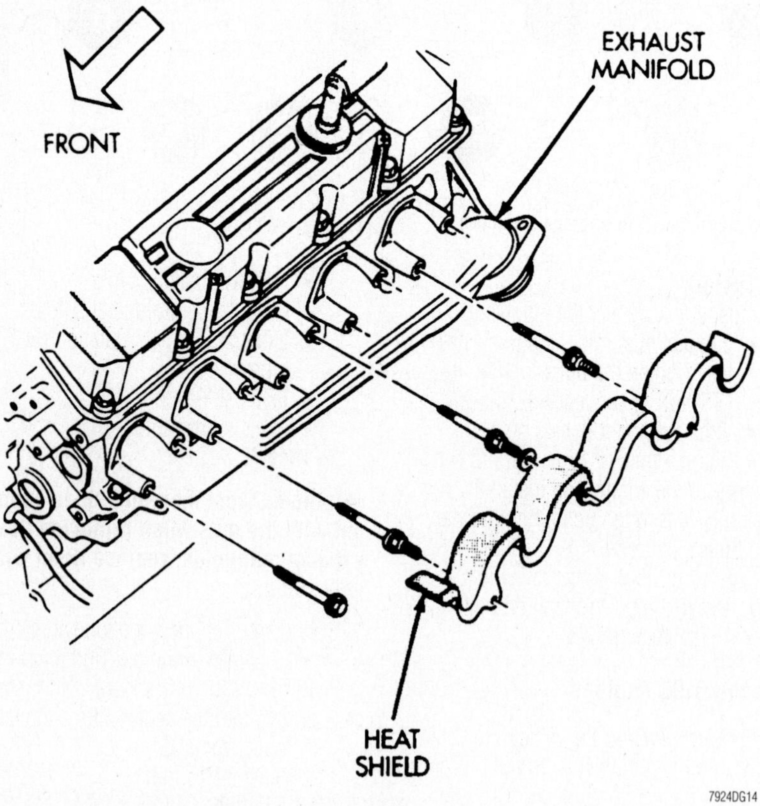

Exhaust manifold fastener locations—8.0L engine

To install:

3. Install or connect the following:
   - Exhaust manifolds. Tighten the fasteners to 16 ft. lbs. (22 Nm).
   - Oil dipstick tube bracket
   - EGR tube
   - Exhaust manifold heat shields
   - Exhaust front pipe
   - Negative battery cable
4. Start the engine and check for leaks.

Camshaft and Valve Lifters

REMOVAL & INSTALLATION

3.7L Engines

1. Before servicing the vehicle, refer to the precautions in the beginning of this section.
2. Remove or disconnect the following:
   - Negative battery cable
   - Cylinder head cover
3. Set cylinder No. 1 at Top Dead Center (TDC) and camshaft sprocket "V6" mark at 12 o'clock.
4. Matchmark a link on the secondary timing chain on both side of the "V6" mark on the camshaft sprocket.
5. Loosen the camshaft sprocket retaining bolt. Leave the bolt snug against the sprocket.
6. Position Special Tool 8379 timing chain wedge between the timing chain strands, tap the tool to securely wedge the timing chain against the tensioner arm and guide.
7. Remove the camshaft position sensor.
8. Hold the camshaft with Special Tool 8428 Camshaft Wrench, while removing the camshaft sprocket bolt and sprocket.
9. Using Special Tool 8428 Camshaft Wrench, gently allow the camshaft to rotate 5° clockwise until the camshaft is in the neutral position (no valve load).
10. Starting at the outside working inward, loosen the camshaft bearing cap retaining bolts 1/2 turn at a time. Repeat until all load is off the bearing caps.

➡When the camshaft is removed the rocker arms may slide downward, mark the rocker arms before removing camshaft.

11. Remove the camshaft bearing caps and the camshaft.

To install:

12. Lubricate the camshaft journals with clean engine oil.

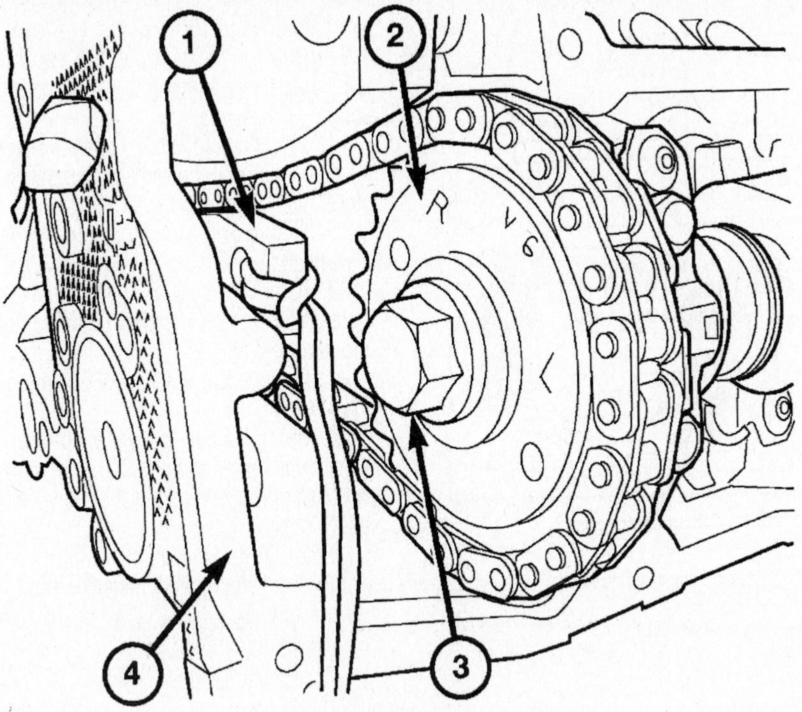

Installing the timing chain wedge—3.7L engine

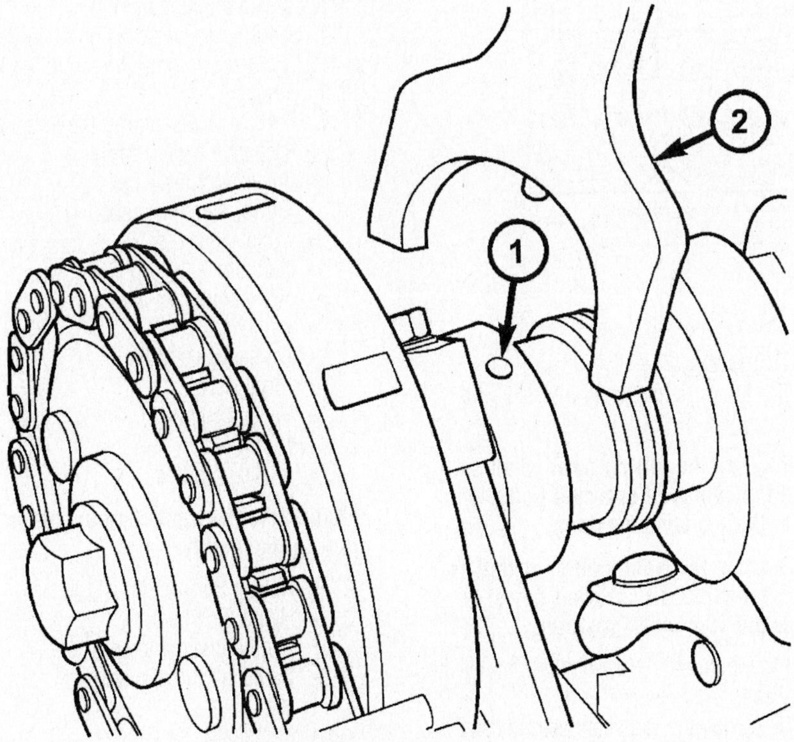

Hold the camshaft with Special Tool 8428 Camshaft wrench—3.7L engine

➡Position the right side camshaft so that the camshaft sprocket dowel is near the 10 o'clock position.

13. Position the camshaft into the cylinder head.
14. Install the camshaft bearing caps, hand tighten the retaining bolts.

➡Caps should be installed so that the stamped numbers on the caps are in numerical order, (1 thru 4) from the front to the rear of the engine. All caps should be installed so that the stamped arrows on the caps point toward the front of the engine.

15. Working in ½ turn increments, tighten the bearing cap retaining bolts starting with the middle cap working outward.
16. Torque the camshaft bearing cap retaining bolts to 100 inch lbs. (11 Nm).
17. Position the camshaft drive gear into the timing chain aligning the "V6" mark between the two marked chain links.
18. Using Special Tool 8428 Camshaft Wrench, rotate the camshaft until the camshaft sprocket dowel is aligned with the slot in the camshaft sprocket. Install the sprocket onto the camshaft.
19. Remove excess oil from camshaft sprocket bolt, then install the camshaft sprocket retaining bolt and hand tighten.
20. Remove timing chain wedge special tool 8379.
21. Using Special Tool 6958 spanner wrench with adapter pins 8346, torque the camshaft sprocket retaining bolt to 90 ft. lbs. (122 Nm)
22. Install or connect the following:
    - Camshaft position sensor
    - Cylinder head cover
    - Negative battery cable

4.7L Engines

1. Before servicing the vehicle, refer to the precautions in the beginning of this section.
2. Remove or disconnect the following:
    - Negative battery cable
    - Remove cylinder head cover
3. Set engine to TDC cylinder #1, camshaft sprocket V8 marks at the 12 o'clock position.
4. Matchmark a link on the secondary timing chain on both side of the "V8" mark on the camshaft sprocket.
5. Loosen the camshaft sprocket retaining bolt. Leave the bolt snug against the sprocket.
6. Position Special Tool 8350 timing chain wedge between the timing chain strands, tap the tool to securely wedge the

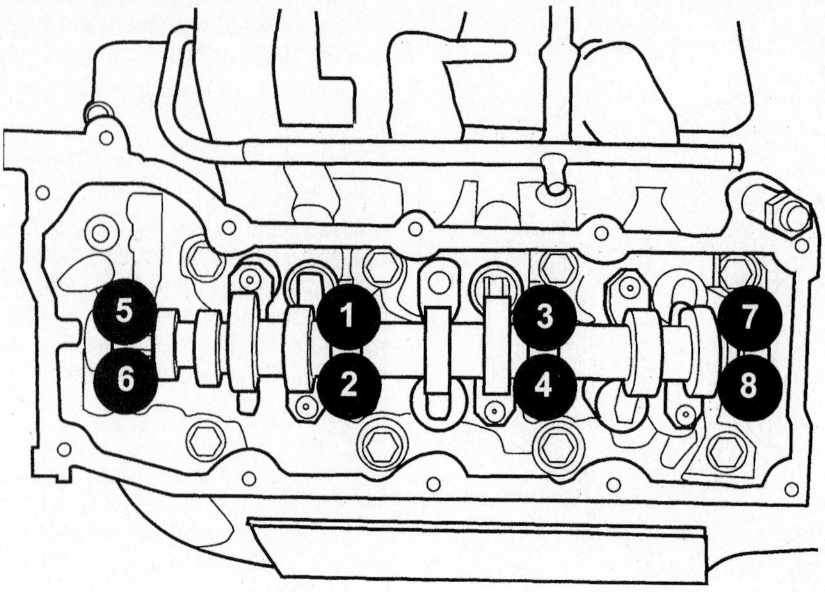

Camshaft bearing cap bolt tightening sequence—3.7L

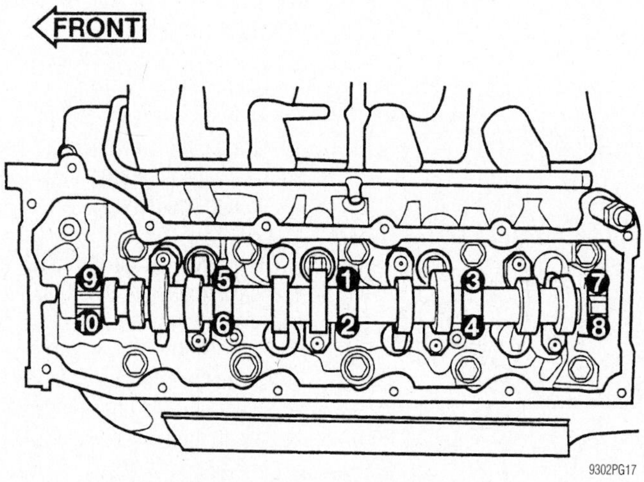

Camshaft bearing cap bolt tightening sequence—4.7L engine

timing chain against the tensioner arm and guide.

7. Remove the camshaft position sensor, right side only.

8. Hold the camshaft with adjustable pliers while removing the camshaft sprocket bolt and sprocket.

9. Using the pliers, gently allow the camshaft to rotate the specified amount until the camshaft is in the neutral position.

a. Left side camshaft: 15°clockwise

b. Right side camshaft: 45° clockwise

10. Starting at the outside working inward, loosen the camshaft bearing cap retaining bolts 1/2 turn at a time. Repeat until all load is off the bearing caps.

11. Remove the camshaft bearing caps.

12. Remove the camshaft.

To install:

13. Lubricate camshaft journals with clean engine oil.

➡Position the left side camshaft so that the camshaft sprocket dowel is near the 1 o'clock position.

➡Position the right side camshaft so that the camshaft sprocket dowel is near the 10 o'clock position.

14. Position the camshaft into the cylinder head.

15. Install the camshaft bearing caps, hand tighten the retaining bolts.

16. Working in ½ turn increments, tighten the bearing cap retaining bolts starting with the middle cap working outward.

17. Torque the camshaft bearing cap retaining bolts to 100 inch lbs. (11 Nm).

18. Position the camshaft drive gear into the timing chain aligning the "V8" mark between the matchmarks.

19. Using the adjustable pliers, rotate the camshaft until the camshaft sprocket dowel is aligned with the slot in the camshaft sprocket. Install the sprocket onto the camshaft.

20. Remove excess oil from bolt, then install the camshaft sprocket retaining bolt and hand tighten.

21. Remove Special Tool 8350 timing chain wedge.

22. Using Special Tool 6958 spanner wrench with adapter pins 8346, torque the camshaft sprocket retaining bolt to 90 ft. lbs. (122 Nm).

23. Install or connect the following:

- Camshaft position sensor, right side only
- Cylinder head cover
- Negative battery cable

3.9L, 5.2L and 5.9L Engines

1. Before servicing the vehicle, refer to the precautions in the beginning of this section.

2. Drain the cooling system.

3. Recover the A/C refrigerant, if equipped with air conditioning.

4. Set the crankshaft to Top Dead Center (TDC) of the compression stroke for the No. 1 cylinder.

5. Remove or disconnect the following:

- Negative battery cable
- Accessory drive belt
- Power steering pump
- Water pump
- Radiator
- A/C condenser
- Grille
- Crankshaft damper
- Front cover
- Valve covers
- Distributor
- Intake manifold

➡Keep all valvetrain components in order for assembly.

- Rocker arms and pushrods
- Hydraulic lifters
- Timing chain and sprockets
- Camshaft thrust plate and chain oil tab
- Camshaft

To install:

6. Install or connect the following:

- Camshaft
- Camshaft Holding Tool C-3509

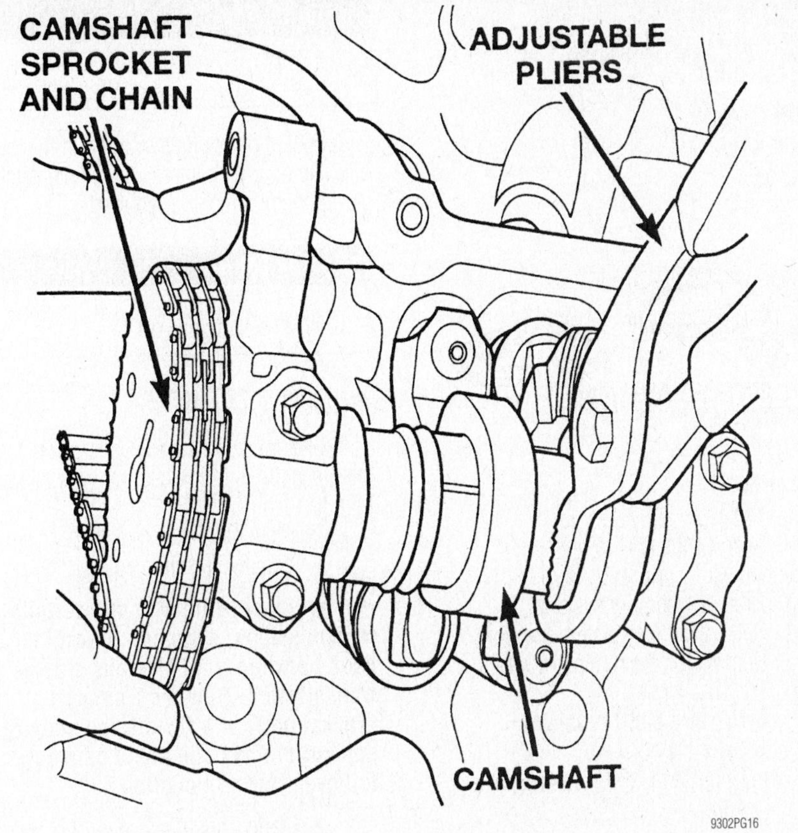

Turn the camshaft with pliers, if needed, to align the dowel in the sprocket—4.7L engine

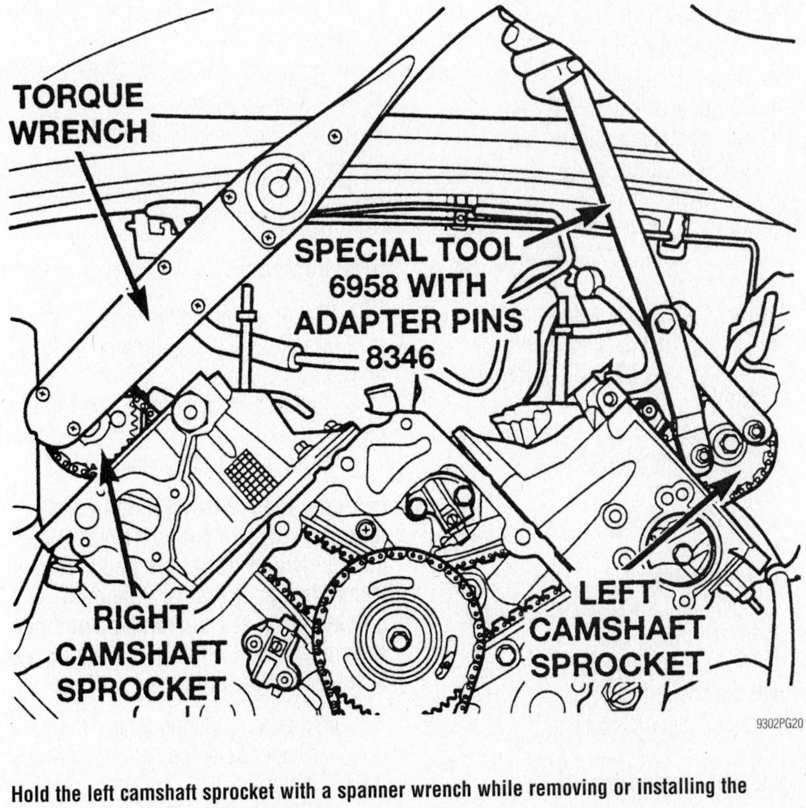

Hold the left camshaft sprocket with a spanner wrench while removing or installing the camshaft sprocket bolts—4.7L engine

- Camshaft thrust plate and chain oil tab. Tighten the bolts to 18 ft. lbs. (24 Nm).
- Timing chain and sprockets. Tighten the camshaft sprocket bolt to 50 ft. lbs. (68 Nm).

7. Remove the camshaft holding tool.
8. Install or connect the following:
   - Hydraulic lifters in their original positions
   - Rocker arms and pushrods in their original positions
   - Intake manifold
   - Distributor
   - Valve covers
   - Front cover
   - Crankshaft damper
   - Grille
   - A/C condenser
   - Radiator
   - Water pump
   - Power steering pump
   - Accessory drive belt
   - Negative battery cable
9. Fill the cooling system.
10. Recharge the A/C system, if equipped.
11. Start the engine and check for leaks.

5.7L Engines

1. Before servicing the vehicle, refer to the precautions in the beginning of this section.
2. Drain the engine oil.
3. Drain the cooling system.
4. Recover the A/C refrigerant, if equipped with air conditioning.
5. Set the crankshaft to Top Dead Center (TDC) of the compression stroke for the No. 1 cylinder.
6. Remove or disconnect the following:
   - Negative battery cable
   - Rear cam bearing core plug
   - Air intake assembly
   - Accessory drive belt
   - Alternator
   - A/C compressor
   - Radiator
   - Intake manifold
   - Cylinder head covers
   - Cylinder heads
   - Oil pan
   - Timing chain front cover
   - Oil pickup tube
   - Oil pump
   - Timing chain and sprockets
   - Camshaft thrust plate
   - Hydraulic lifters
   - Camshaft

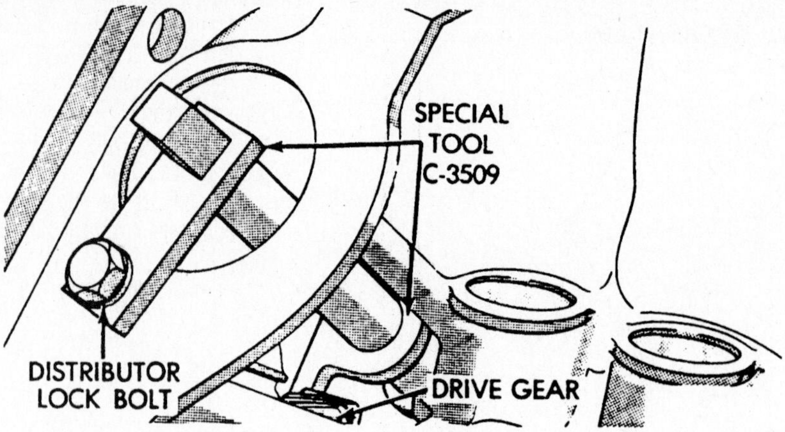

Camshaft holding tool C-3509—3.9L, 5.2L and 5.9L engines

To install:

7. Install a new core hole plug at the rear of camshaft, using suitable flat-faced tool. The plug must be fully seated on the cylinder block shoulder.
8. Install or connect the following:
   - Camshaft
   - Camshaft thrust plate. Tighten the bolts to 21 ft. lbs. (28 Nm).
   - Timing chain and sprockets

➡If camshaft end play is not within limits, a new thrust plate must be installed.

   - Oil pump
   - Oil pickup tube

➡Lifters must be replaced in their original positions. They must be replaced if camshaft is replaced.

   - Hydraulic lifters
   - Cylinder heads
   - Pushrods
   - Rocker arms
   - Timing case front cover
   - Oil pan
   - Cylinder head covers
   - Intake manifold
   - A/C compressor
   - Alternator
   - Accessory drive belt
   - Radiator
   - Air intake assembly
   - Negative battery cable
9. Refill the engine with oil.
10. Fill the cooling system to the correct level.
11. Recharge the A/C system, if equipped.
12. Start the engine and check for leaks.

8.0L Engine

1. Before servicing the vehicle, refer to the precautions in the beginning of this section.
2. Recover the A/C refrigerant, if equipped with air conditioning.
3. Drain the cooling system.
4. Relieve the fuel system pressure.
5. Remove or disconnect the following:
   - Negative battery cable
   - Valve covers
   - Upper and lower intake manifolds

➡Keep all valvetrain components in order for assembly.

   - Rocker arms
   - Pushrods
   - Cylinder heads
   - Valve lifters
   - Crankshaft pulley
   - Front cover
   - Timing chain and sprockets
   - Camshaft thrust plate
   - Camshaft

To install:

6. Install or connect the following:
   - Camshaft
   - Camshaft thrust plate
   - Timing chain and sprockets. Tighten the bolt to 55 ft. lbs. (75 Nm).
   - Front cover
   - Crankshaft pulley
   - Valve lifters
   - Cylinder heads
   - Pushrods
   - Rocker arms
   - Upper and lower intake manifolds
   - Valve covers
   - Negative battery cable
7. Fill the cooling system.
8. Recharge the A/C system, if equipped.
9. Start the engine and check for leaks.

Valve Lash

ADJUSTMENT

All gasoline engines covered in this section use hydraulic lifters. No maintenance or periodic adjustment is required.

Starter Motor

REMOVAL & INSTALLATION

3.7L and 4.7L Engines

1. Before servicing the vehicle, refer to the precautions in the beginning of this section.
2. Remove or disconnect the following:
   - Negative battery cable

➡If equipped with 4WD and certain transmissions, a support bracket is used between front axle and side of transmission. Remove 2 support bracket bolts at transmission and pry support bracket slightly to gain access to lower starter mounting bolt.

   - Starter mounting bolts
   - Starter solenoid harness connections
   - Starter

To install:

3. Connect the starter solenoid wiring connectors.
4. Install the starter and torque the bolts to 40 ft. lbs. (54 Nm).
5. Install the support bracket bolts, if equipped.
6. Install the negative battery cable and check for proper operation.

5.7L Engines

1. Before servicing the vehicle, refer to the precautions in the beginning of this section.
2. Remove or disconnect the following:
   - Negative battery cable

➡If equipped with 4WD and certain transmissions, a support bracket is used between front axle and side of transmission. Remove 2 support bracket bolts at transmission and pry support bracket slightly to gain access to lower starter mounting bolt.

   - Starter mounting bolts
   - Starter solenoid harness connections
   - Starter

To install:

3. Connect the starter solenoid wiring connectors.
4. Install the starter and torque the bolts to 50 ft. lbs. (68 Nm).
5. Install the support bracket bolts, if equipped.
6. Install the negative battery cable and check for proper operation.

3.9L, 5.2L, 5.9L and 8.0L Engines

1. Before servicing the vehicle, refer to the precautions in the beginning of this section.
2. Remove or disconnect the following:
   - Negative battery cable
   - Starter mounting bolts
   - Starter solenoid harness connections
   - Starter

To install:

3. Connect the starter solenoid wiring connectors.
4. Install the starter and tighten the bolts to 50 ft. lbs. (68 Nm).
5. Install the negative battery cable and check for proper operation.

Oil Pan

REMOVAL & INSTALLATION

3.7L, 4.7L and 2002-2004 5.7L Engines

1. Before servicing the vehicle, refer to the precautions in the beginning of this section.
2. Drain the engine oil.
3. Install engine lifting fixture 8534, but do not raise the engine at this time.
4. Loosen, but don't remove, the left and right mount through-bolts.
5. Remove or disconnect the following:
   - Negative battery cable
   - Structural dust cover, if equipped
   - Front crossmember
6. Raise the engine just enough for clearance.

➡Don't pry on the pan. The gasket is integral with the windage tray, and doesn't come out with the pan.

7. Remove or disconnect the following:
   - Oil pan mounting bolts
   - Oil pan

➡The double ended studs must be installed in their original locations.

To install:

8. Replace oil pan gasket if necessary.
9. Install the oil pan. Tighten the bolts, in sequence, to 105 inch lbs. (12 Nm).

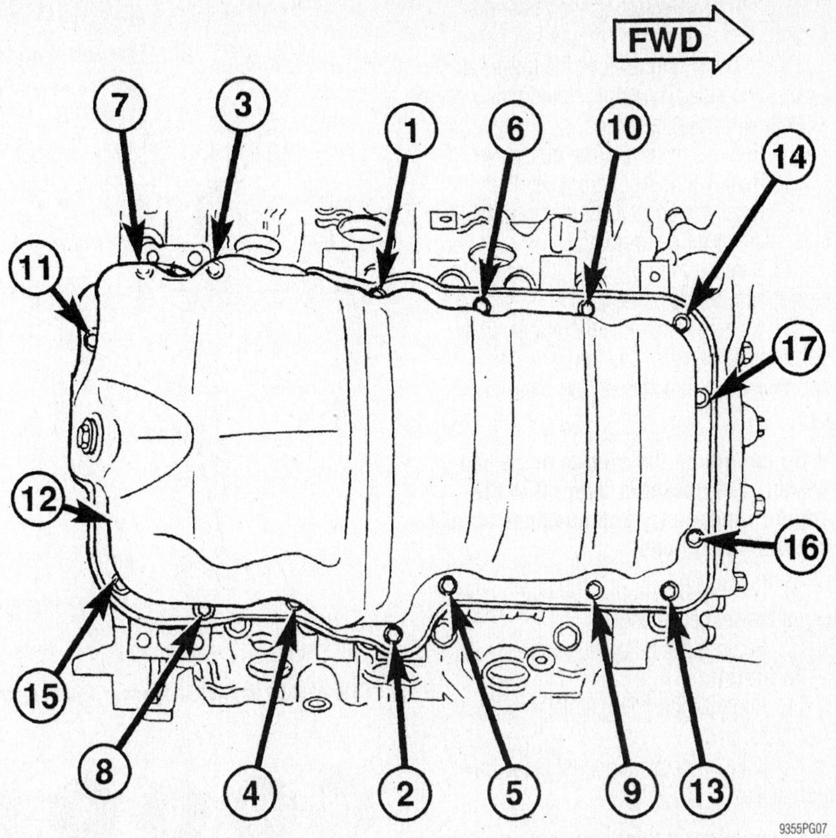

Oil pan bolt torque sequence—3.7L

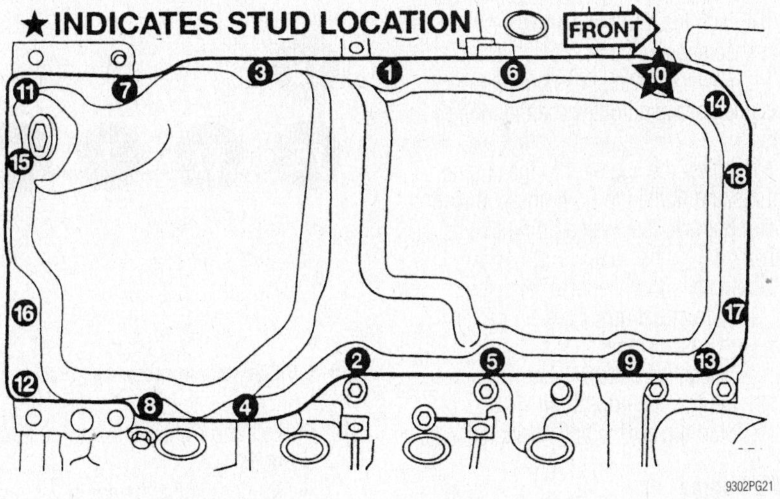

Oil pan mounting bolt tightening sequence—4.7L engine

10. Lower the engine into the engine mounts and tighten to 50 ft. lbs. (68 Nm).
11. Remove the engine lifting fixture 8534.
12. Install or connect the following:
    - Structural dust cover, if equipped
    - Front crossmember
    - Negative battery cable
13. Fill the engine with oil.
14. Start the engine and check for leaks.

2005 5.7L Engines

1. Before servicing the vehicle, refer to the precautions in the beginning of this section.
2. Drain the engine oil.
3. Install engine lifting fixture 8534 but do not raise the engine.
   a. 2WD vehicles: Remove both engine mount to frame through bolts.

b. 4WD vehicles: Loosen both left and right side engine mount bracket to mount bolts, and remove the lower bolt on each side. The upper hole in the engine bracket is slotted.

4. Remove the structural dust cover.
5. Remove the front crossmember.
6. Raise the engine slightly to provide clearance for oil pan removal.
7. Remove the nuts from the transmission mount to crossmember.
8. Using a transmission jack or other suitable device, raise the transmission approximately 3 inches off the crossmember.

➡Do not pry on the oil pan or oil pan gasket. The gasket is integral to the engine windage try and does not come out with the oil pan.

9. Remove the oil pan mounting bolts using the sequence shown.
10. Remove the oil pan.

To install:

11. Apply Mopar® Engine RTV to the 4 T-joints shown.
12. Install a new integral windage tray and gasket.

➡A new integral windage tray and gasket assembly must be installed when the oil pan is removed.

13. Install the oil pan. Tighten the bolts to 105 inch lbs. (12 Nm) using the sequence shown.
14. Lower the transmission onto the crossmember and tighten the mounting nuts.
15. Lower the engine into the engine mounts and tighten to 50 ft. lbs. (68 Nm).
16. Remove the engine lifting fixture 8534.
17. Install or connect the following:
    - Structural dust cover
    - Crossmember
    - Negative battery cable
18. Fill the engine with oil
19. Start the engine and check for leaks.

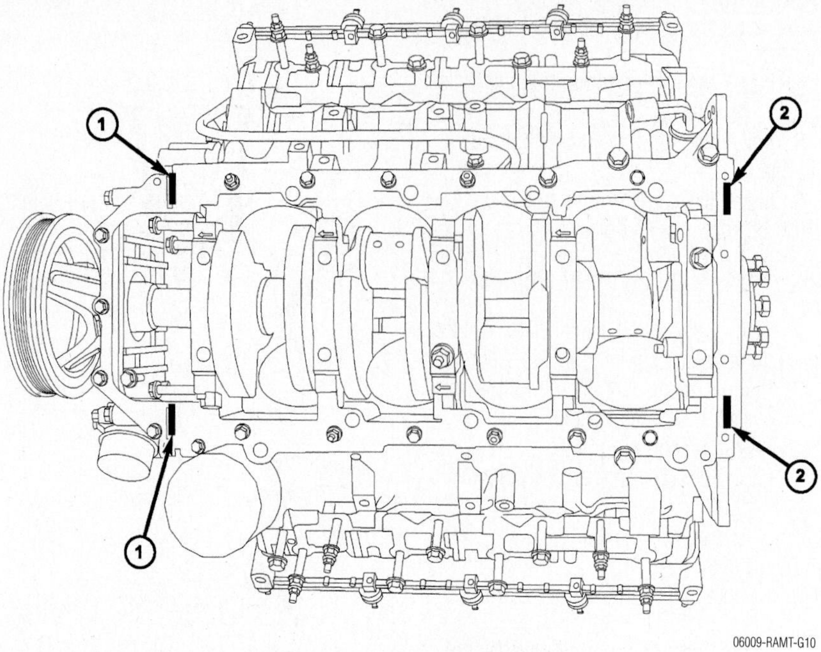

RTV must be applied to the four T-joint locations—2005 5.7L engine

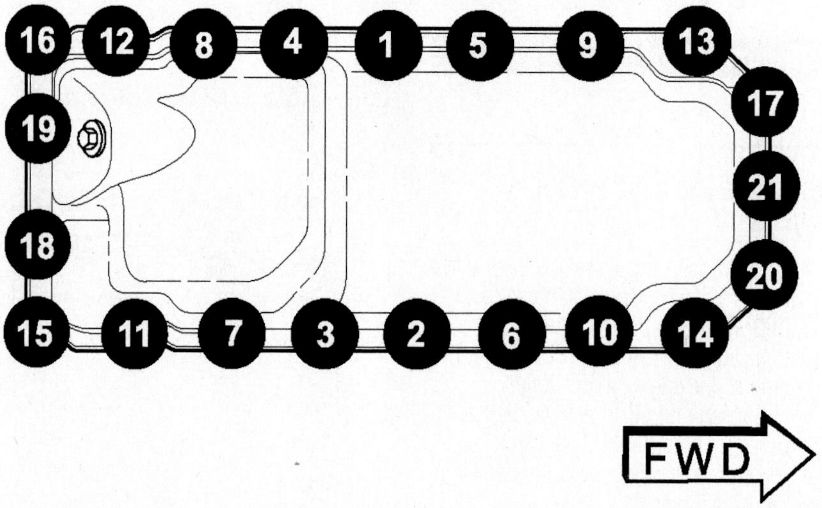

5.7L Oil pan torque sequence

3.9L Engine

2-WHEEL DRIVE MODELS

1. Before servicing the vehicle, refer to the precautions in the beginning of this section.
2. Drain the engine oil.
3. Remove or disconnect the following:
    - Negative battery cable
    - Distributor cap
    - Oil dipstick
    - Exhaust front pipe
    - Flywheel access panel, if equipped
    - Left and right motor mount through bolts
    - Oil pan. Raise the engine as necessary for clearance.

To install:

4. Fabricate 4 alignment dowels from 1½ inch x ¼ inch bolts. Cut the heads off the bolts and cut a slot into the top of the dowel to allow installation/removal with a screwdriver.
5. Install or connect the following:
    - Alignment dowels
    - Oil pan. Replace the dowels with bolts and tighten all bolts to 17 ft. lbs. (23 Nm).
    - Left and right motor mount through bolts
    - Flywheel access panel, if equipped
    - Exhaust front pipe
    - Oil dipstick
    - Distributor cap
    - Negative battery cable
6. Fill the crankcase to the correct level.
7. Start the engine and check for leaks.

4-WHEEL DRIVE MODELS

1. Before servicing the vehicle, refer to the precautions in the beginning of this section.
2. Drain the engine oil.

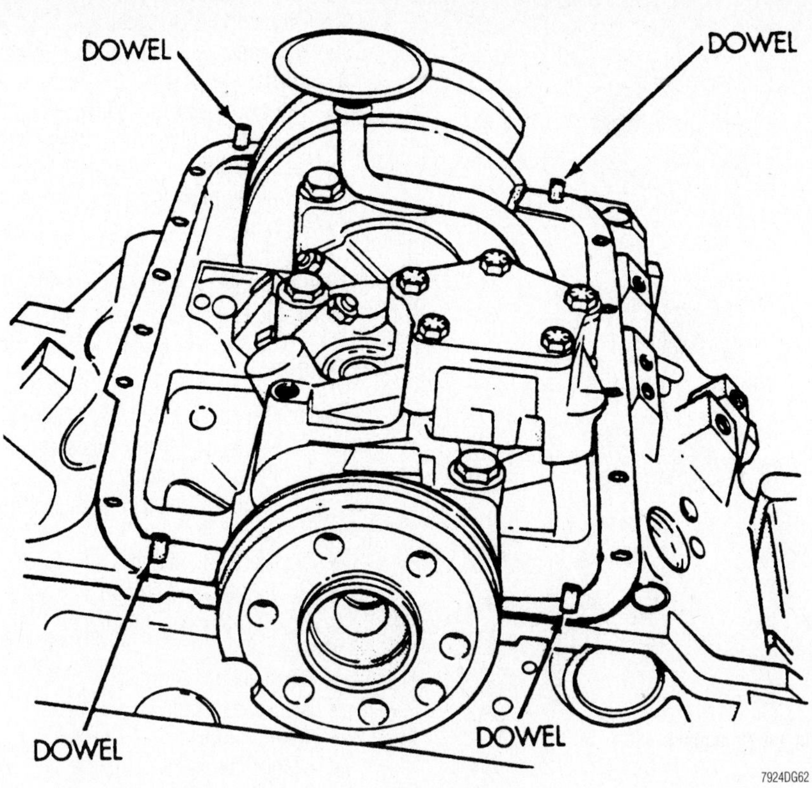

Oil pan alignment dowel placement—3.9L engine shown

3. Remove or disconnect the following:
   - Negative battery cable
   - Engine oil dipstick
   - Front axle. Support the engine
   - Exhaust front pipe
   - Flywheel access panel, if equipped
   - Oil pan

To install:

4. Fabricate 4 alignment dowels from 1½ inch x ¼ inch bolts. Cut the heads off the bolts and cut a slot into the top of the dowel to allow installation/removal with a screwdriver.
5. Install or connect the following:
   - Alignment dowels
   - Oil pan. Replace the dowels with bolts and tighten all bolts to 17 ft. lbs. (23 Nm).
   - Flywheel access panel, if equipped
   - Exhaust front pipe
   - Front axle
   - Engine oil dipstick
   - Negative battery cable
6. Fill the crankcase to the correct level.
7. Start the engine and check for leaks.

5.2L and 5.9L Engines

1. Before servicing the vehicle, refer to the precautions in the beginning of this section.
2. Drain the engine oil.
3. Remove or disconnect the following:
   - Oil filter

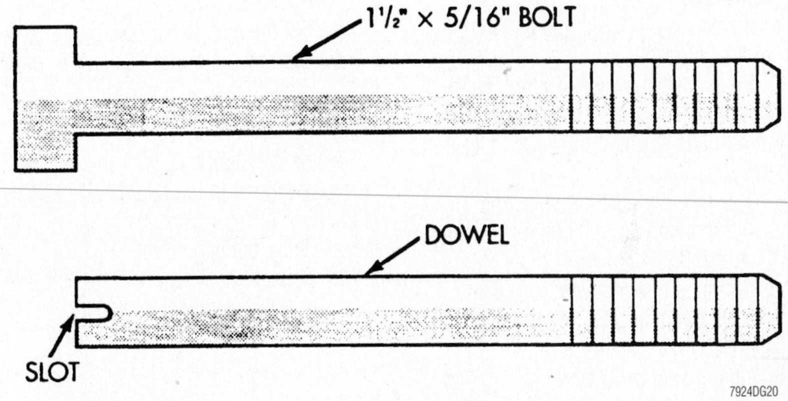

Oil pan alignment dowels—3.9L, 5.2L, 5.9L and 8.0L engines

   - Starter motor
   - Cooler lines
   - Oil level sensor connector
   - Heated Oxygen (HO_2S) sensor connector
   - Exhaust Y-pipe
   - Oil pan

To install:

4. Install or connect the following:
   - Oil pan, using a new gasket. Tighten the bolts to 18 ft. lbs. (24 Nm).
   - Exhaust Y-pipe
   - Heated Oxygen (HO_2S) sensor connector
   - Oil level sensor connector
   - Cooler lines
   - Starter motor
   - Oil filter
5. Fill the crankcase.
6. Start the engine and check for leaks.

8.0L Engine

1. Before servicing the vehicle, refer to the precautions in the beginning of this section.
2. Drain the engine oil.
3. Remove or disconnect the following:
   - Negative battery cable
   - Engine oil dipstick
   - Left transmission brace
   - Oil pan

To install:

4. Install or connect the following:
   - Oil pan. Tighten the 5⁄16 bolts to 12 ft. lbs. (16 Nm) and the ¼ bolts to 96 inch lbs. (11 Nm).
   - Left transmission brace
   - Engine oil dipstick
   - Negative battery cable
5. Fill the crankcase.
6. Start the engine and check for leaks.

Oil Pump

REMOVAL & INSTALLATION

3.7L and 4.7L Engines

1. Before servicing the vehicle, refer to the precautions in the beginning of this section.
2. Drain the engine oil.
3. Remove or disconnect the following:
   - Negative battery cable
   - Oil pan and pick-up tube
   - Timing chain cover
   - Timing chains and tensioners

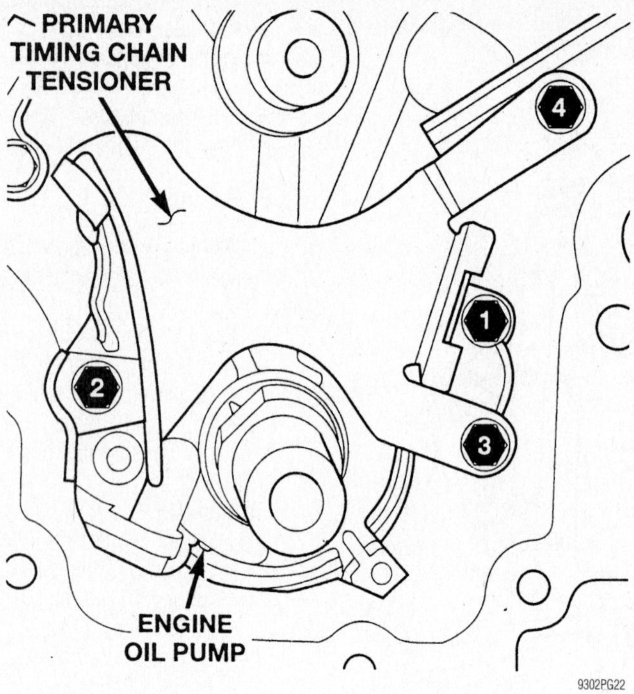

Oil pump and chain tensioner torque sequence—3.7L and 4.7L engines

- Oil pump mounting bolts
- Primary timing chain tensioner
- Oil pump

To install:

4. Install the oil pump and install one oil pump retaining bolt.
5. Position the primary timing chain tensioner and install the three remaining mounting bolts. Tighten the bolts to 21 ft. lbs. (28 Nm).
6. Install or connect the following:
   - Secondary timing chain tensioners and timing chains
   - Timing chain cover
   - Oil pan and pick-up tube
   - Negative battery cable
7. Fill the engine with oil.
8. Start the engine and check for leaks.

5.7L Engine

1. Before servicing the vehicle, refer to the precautions in the beginning of this section.
2. Drain the engine oil.
3. Remove or disconnect the following:
   - Negative battery cable
   - Oil pan
   - Oil pump pick-up tube
   - Timing chain cover
   - Oil pump mounting bolts
   - Oil pump

To install:

4. Install or connect the following:
   - Oil pump. Tighten the bolts to 21 ft. lbs. (28 Nm).
   - Timing chain cover
   - Oil pump pick-up tube
   - Oil pan
   - Negative battery cable
5. Fill the engine with oil.
6. Start the engine and check for leaks.

3.9L, 5.2L and 5.9L Engines

1. Before servicing the vehicle, refer to the precautions in the beginning of this section.
2. Drain the engine oil.
3. Remove or disconnect the following:
   - Negative battery cable
   - Oil pan
   - Oil pump pick-up tube
   - Oil pump

To install:

4. Install or connect the following:
   - Oil pump. Tighten the bolts to 30 ft. lbs. (41 Nm).
   - Oil pump pick-up tube
   - Oil pan
   - Negative battery cable
5. Fill the crankcase to the correct level.
6. Start the engine and check for leaks.

8.0L Engine

1. Before servicing the vehicle, refer to the precautions in the beginning of this section.
2. Drain the cooling system.
3. Remove or disconnect the following:
   - Negative battery cable
   - Accessory drive belt
   - Cooling fan and shroud
   - A/C compressor, if equipped
   - Alternator
   - Air injection pump
   - Bracket assembly
   - Water pump
   - Crankshaft pulley
   - Front oil pan bolts
   - Front cover
   - Oil pump pressure relief valve
   - Oil pump cover
   - Oil pump rotors

To install:

4. Install or connect the following:
   - Oil pump rotors
   - Oil pump cover. Tighten the bolts to 10 ft. lbs. (14 Nm).
   - Oil pump pressure relief valve. Tighten the plug to 15 ft. lbs. (20 Nm).
   - Front cover. Tighten the bolts to 35 ft. lbs. (47 Nm).

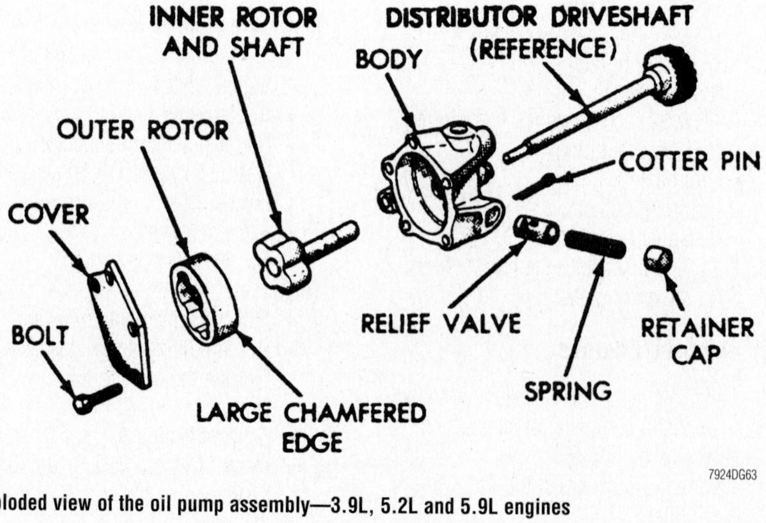

Exploded view of the oil pump assembly—3.9L, 5.2L and 5.9L engines

- Front oil pan bolts. Tighten the bolts to 12 ft. lbs. (16 Nm).
- Crankshaft pulley. Tighten the bolt to 135 ft. lbs. (183 Nm).
- Water pump
- Bracket assembly
- Air injection pump
- Alternator
- A/C compressor, if equipped
- Cooling fan and shroud
- Accessory drive belt
- Negative battery cable

5. Fill the cooling system.
6. Start the engine and check for leaks.

Rear Main Seal

REMOVAL & INSTALLATION

3.7L, 4.7L and 5.7L Engines

1. Before servicing the vehicle, refer to the precautions in the beginning of this section.
2. Remove or disconnect the following:
   - Transmission
   - Flexplate
3. Thread Oil Seal Remover 8506 into the rear main seal as far as possible and remove the rear main seal.

To install:

4. Install or connect the following:
5. Install the Seal Guide 8349-2 onto the crankshaft rear face.
6. Position the crankshaft rear main seal on the seal guide.
7. Using the Crankshaft Rear Oil Seal Installer 8349 and Driver Handle C-4171; tap the seal into place until the installer is flush with the cylinder block.
8. Install the flexplate. Torque the bolts in the sequence shown to 45 ft. lbs. (60 Nm) on auto transmission, or 70 ft. lbs. (95 Nm) for manual transmission.
9. Transmission the transmission
10. Start the engine and check for leaks.

3.9L, 5.2L and 5.9L Engines

1. Before servicing the vehicle, refer to the precautions in the beginning of this section.
2. Drain the engine oil.
3. Remove or disconnect the following:
   - Oil pan
   - Oil pump
   - Rear main bearing cap
4. Loosen the other main bearing cap bolts for clearance and remove the rear main seal halfs.

To install:

5. Install or connect the following:
   - New upper seal half to the cylinder block

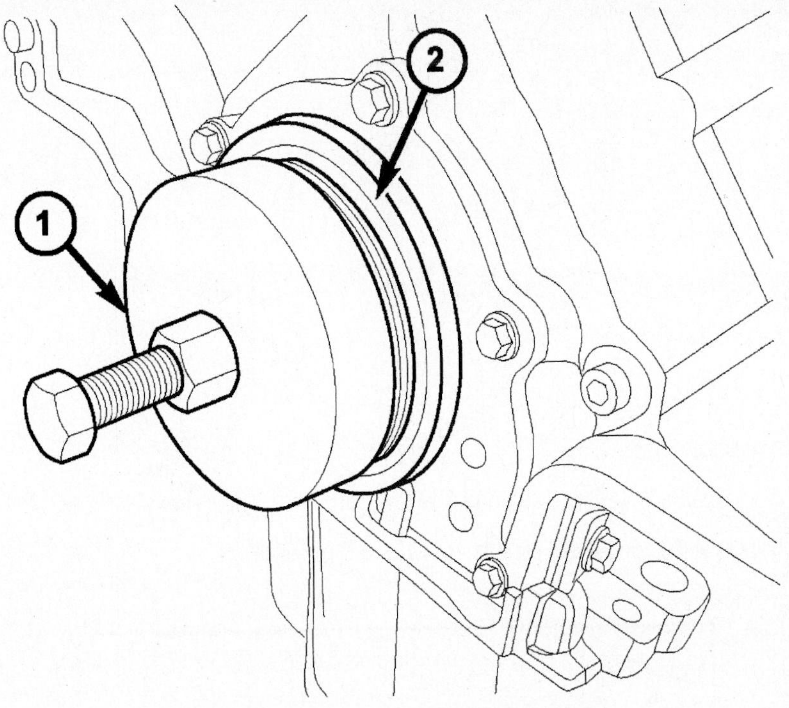

Remove the seal using Oil Seal Remover 8506

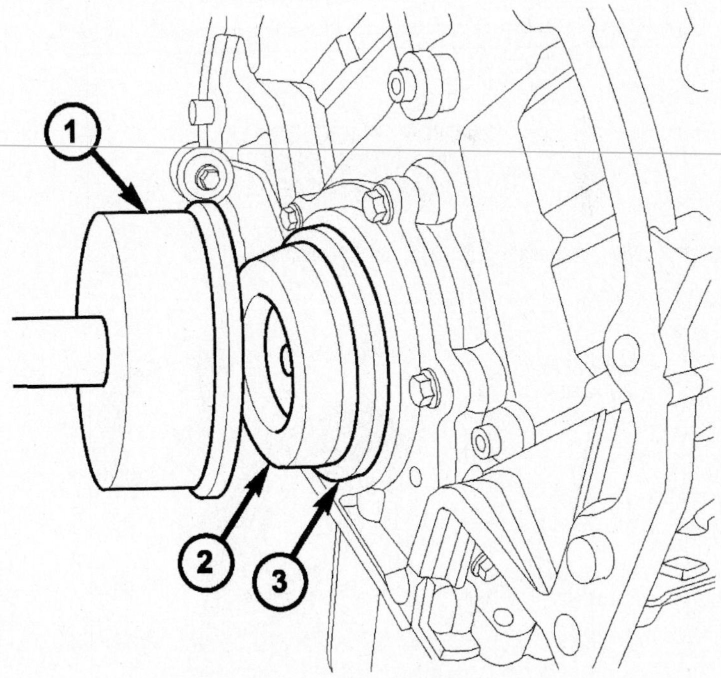

Install the seal using Rear seal installer 8349

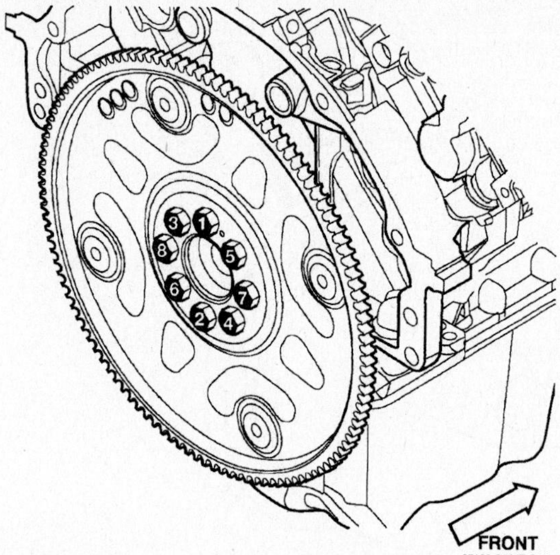

Flexplate bolt tightening sequence—3.7L, 4.7L and 5.7L engines

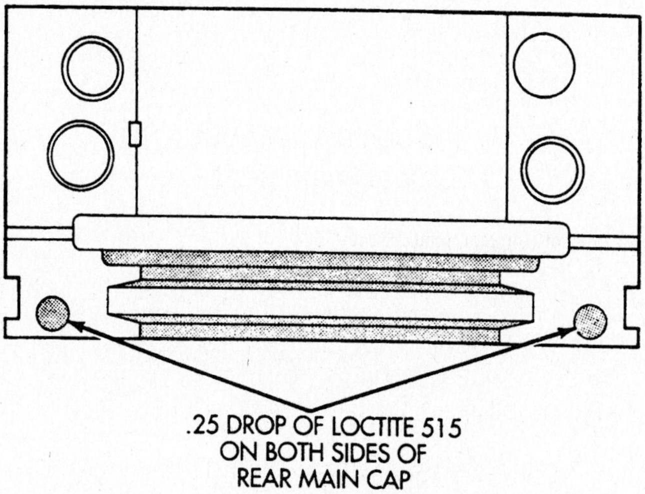

Sealant application locations —3.9L, 5.2L and 5.9L engines

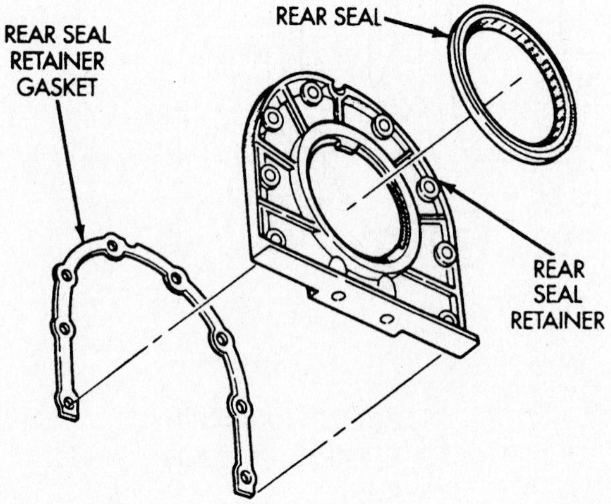

Exploded view of the rear oil seal and retainer—8.0L engine

- New lower seal half to the bearing cap

6. Apply sealant to the rear main bearing cap.
7. Install or connect the following:
   - Rear main bearing cap. Tighten **all** main bearing cap bolts to 85 ft. lbs. (115 Nm).
   - Oil pump and oil pan
8. Fill the engine.
9. Start the engine and check for leaks.

8.0L Engine

1. Before servicing the vehicle, refer to the precautions in the beginning of this section.
2. Remove or disconnect the following:
   - Transmission
   - Clutch, if equipped
   - Flywheel
   - Rear main seal

To install:

3. Install the rear main seal so that it is flush with the cylinder block.
4. Install or connect the following:
   - Flywheel. Tighten the bolts to 55 ft. lbs. (75 Nm).
   - Clutch, if equipped
   - Transmission
5. Start the engine and check for leaks.

Timing Chain, Sprockets, Front Cover and Seal

REMOVAL & INSTALLATION

3.7L and 4.7L Engines

1. Before servicing the vehicle, refer to the precautions in the beginning of this section.

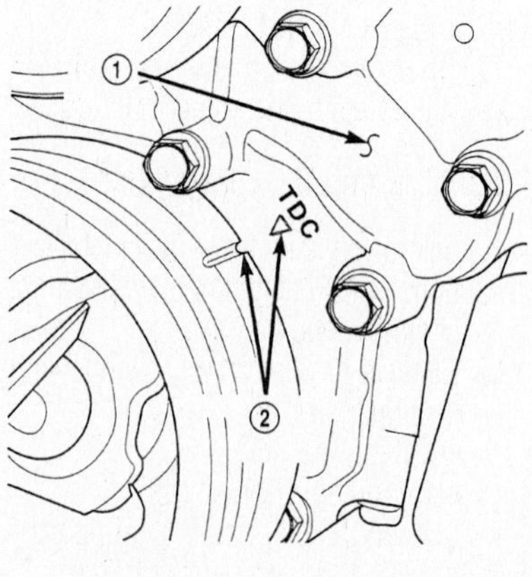

Crankshaft timing marks—3.7L and 4.7L engines

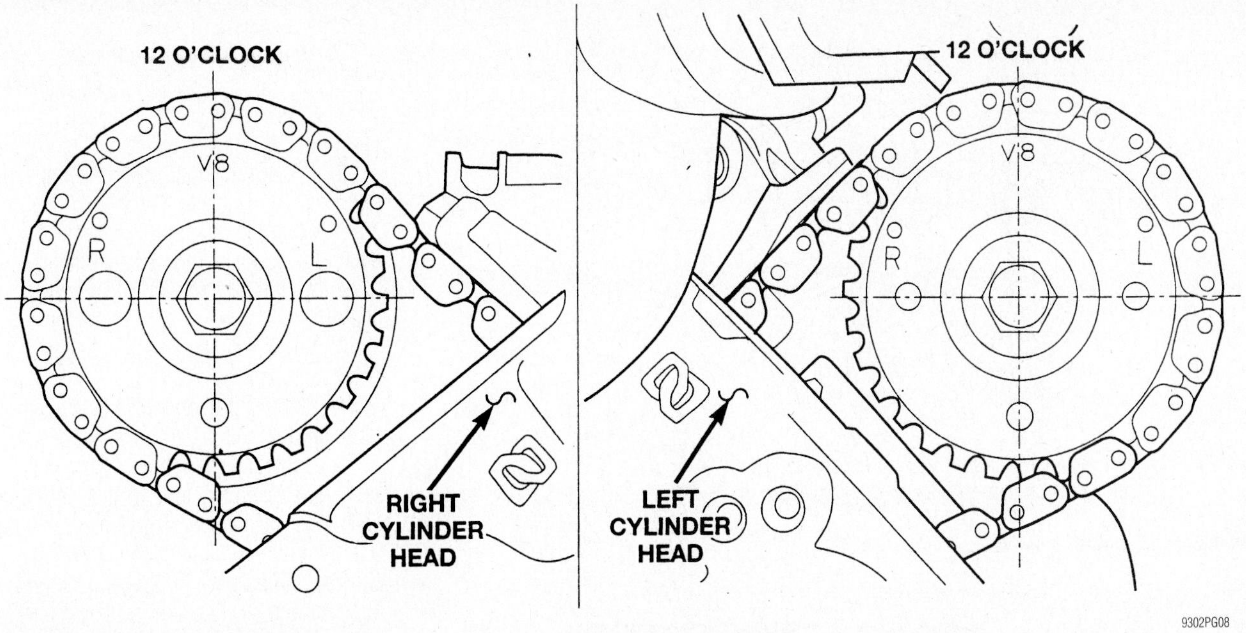

Camshaft positioning—4.7L engine

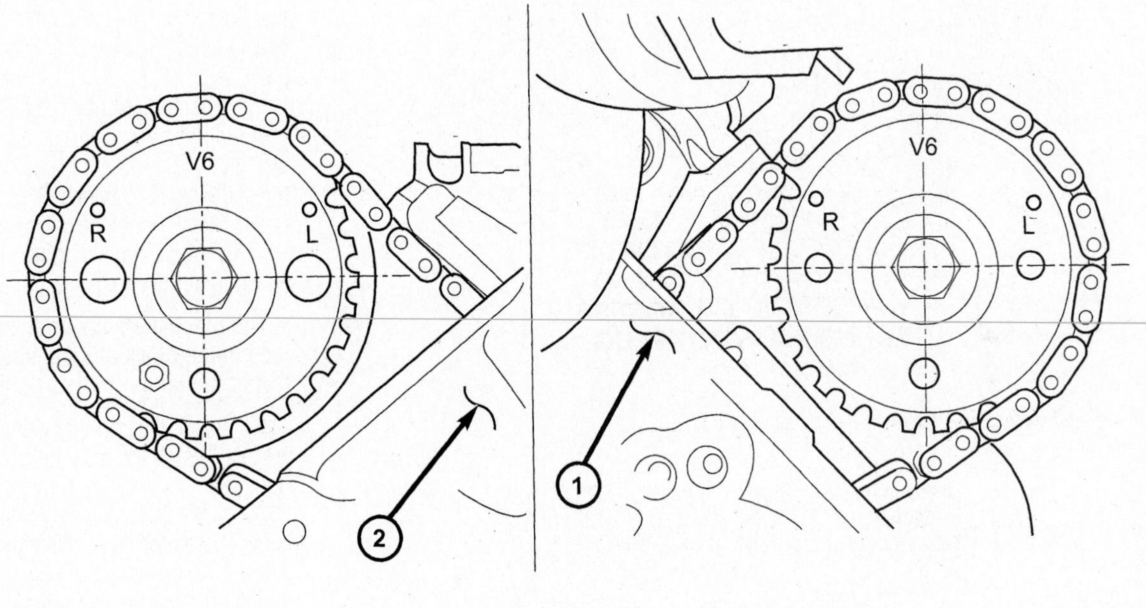

Camshaft sprocket timing marks—3.7L

2. Drain the cooling system.
3. Remove or disconnect the following:
- Negative battery cable
- Cylinder head covers
- Radiator fan shroud
- Accessory drive belt

4. Rotate the crankshaft so that the crankshaft timing mark aligns with the Top Dead Center (TDC) mark on the front cover, and the **V6** or **V8** marks on the camshaft sprockets are at 12 o'clock.
5. Remove or disconnect the following:
- Power steering pump
- Cylinder head access plugs to access chain guide fasteners
- Oil fill housing to access right side tensioner arm fastener
- Crankshaft damper
- Timing chain front cover
- Collapse and pin primary chain tensioner
- Secondary chain tensioners
- Camshaft position sensor
- Camshaft sprocket bolts

6. While holding the left camshaft steel tube with Special Tool 8428 Camshaft Wrench, remove the left camshaft sprocket.

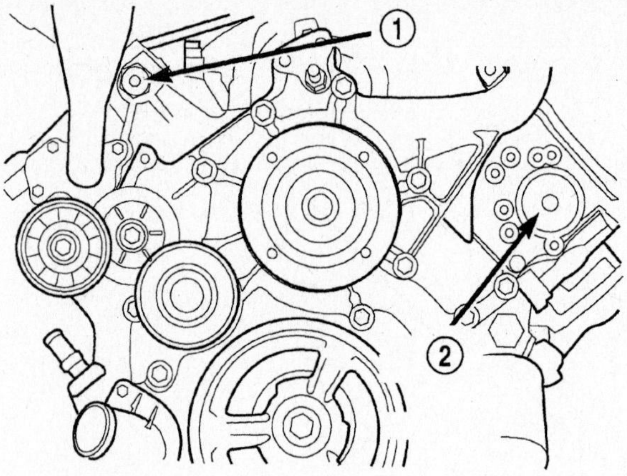

1 - RIGHT CYLINDER HEAD ACCESS PLUG
2 - LEFT CYLINDER HEAD ACCESS PLUG

9355PG11

Cylinder head access plugs—3.7L

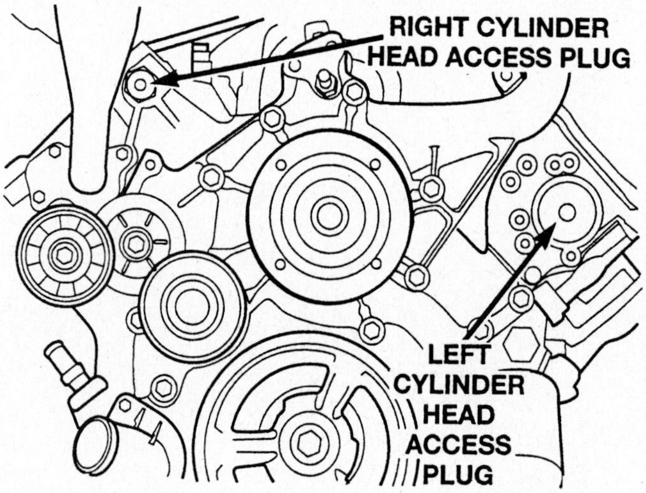

9302PG09

Cylinder head access plug locations—4.7L engine

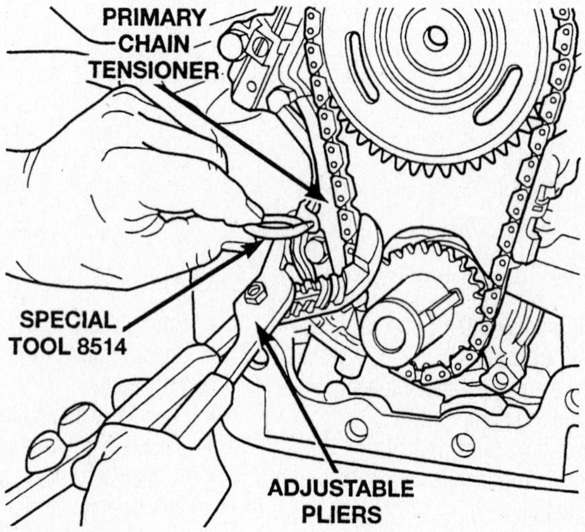

9302PG23

Compress and lock the primary chain tensioner—4.7L engine

Slowly rotate the camshaft approximately 5 degrees clockwise to a neutral position.

7. While holding the right camshaft steel tube with Special Tool 8428 Camshaft Wrench, remove the right camshaft sprocket.
8. Remove idler sprocket assembly bolt.
9. Slide the idler sprocket assembly and crank sprocket forward simultaneously to remove the primary and secondary chains.
10. Remove both pivoting tensioner arms and chain guides.
11. Remove primary chain tensioner.

To install:

12. Using a vise, lightly compress the secondary chain tensioner piston until the piston step is flush with the tensioner body. Using a pin or suitable tool, release ratchet pawl by pulling pawl back against spring force through access hole on side of tensioner. While continuing to hold pawl back, Push ratchet device to approximately 0.08 inches (2 mm) from the tensioner body. Install Special Tool 8514 Insert Lock Pin into hole on front of tensioner. Slowly open vise to transfer piston spring force to lock pin.
13. Position primary chain tensioner over oil pump and insert bolts into lower two holes on tensioner bracket. Tighten bolts to 28 Nm (250 inch lbs.).
14. Install right side chain tensioner arm. Install Torx® bolt. Tighten Torx® bolt to 28 Nm (250 inch lbs.).

➡The black bolts fasten the guide to the engine block and the silver bolts fasten the guide to the cylinder head.

15. Install the secondary timing chain guides. Tighten the bolts to 21 ft. lbs. (28 Nm).
16. Install the secondary timing chains to the idler sprocket so that the double plated links on each chain are visible through the slots in the primary idler sprocket.
17. Lock the secondary timing chains to the idler sprocket with Timing Chain Locking tool as shown.
18. Align the primary chain double plated links with the idler sprocket timing mark and the single plated link with the crankshaft sprocket timing mark.
19. Lubricate idler shaft and bushings.
20. Install all chains, crankshaft sprocket, and idler sprocket as an assembly. After guiding both secondary chains through the block and cylinder head openings, affix chains with a elastic strap or

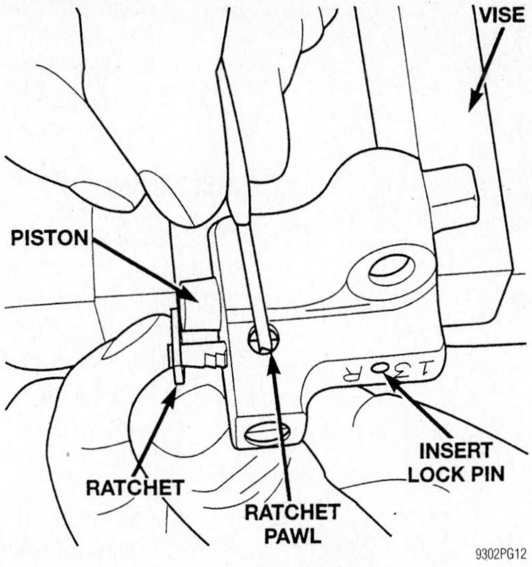

Secondary timing chain tensioner preparation—3.7L and 4.7L engines

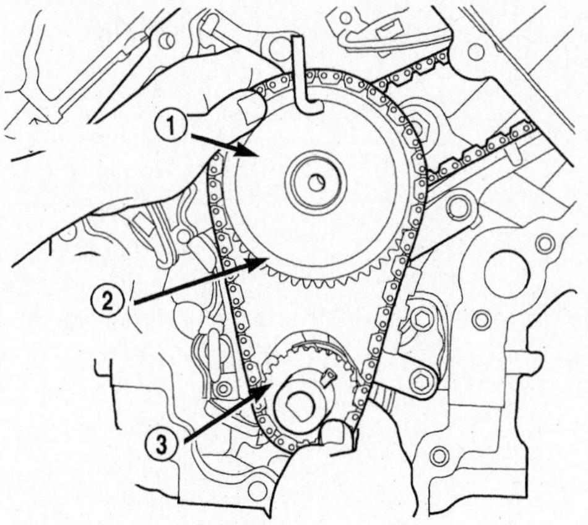

Installing the idler gear and timing chains—3.7L and 4.7L engines

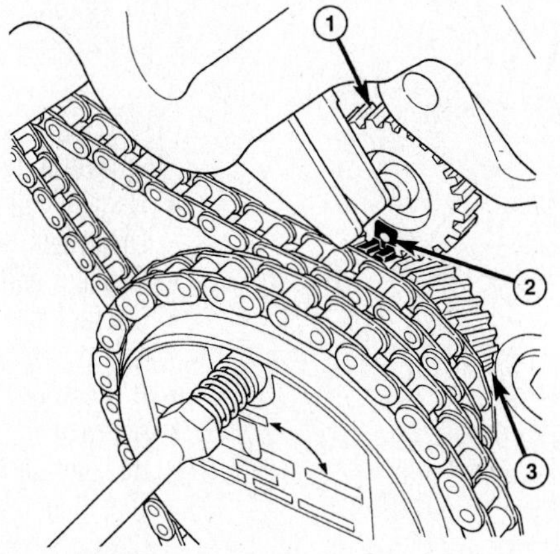

Counterbalance shaft timing marks—3.7L

equivalent. This will maintain tension on chains to aid in installation.

21. Align the timing mark on the idler sprocket gear to the timing mark on the counterbalance shaft drive gear, then seat idler sprocket fully. Before installing idler sprocket bolt, lubricate washer with oil, and tighten idler sprocket assembly retaining bolt to 25 ft. lbs. (34 Nm).

22. Align the dot at the **L** mark on the left sprocket with the plated link on the left chain and the dot at the **R** mark on the right sprocket with the plated link on the right chain.

⁑ CAUTION

Remove excess oil from the camshaft sprocket bolt. Failure to do so can result in over-torque of bolt resulting in bolt failure.

23. Remove timing chain locking tool, then attach both sprockets to camshafts. Install sprocket bolts, but do not tighten.

24. Verify that all plated links are aligned with the marks on all sprockets and the timing marks on camshaft sprockets are at the 12 o'clock position.

25. Install the secondary chain tensioners. Tighten the bolts to 21 ft. lbs. (28 Nm).

26. Remove the primary and secondary timing chain tensioner locking pins.

27. Using Special Tool 6958, Spanner with Adaptor Pins 8346, tighten left and right camshaft sprocket bolts to 90 ft. lbs. (122 Nm).

28. Rotate engine two full revolutions. Verify timing marks are at the follow locations:

- Primary chain idler sprocket dot is at 12 o'clock
- Primary chain crankshaft sprocket dot is at 6 o'clock
- Secondary chain camshaft sprocket timing marks are at 12 o'clock
- Counterbalance shaft drive gear dot is aligned to the idler sprocket gear dot

29. Install or connect the following:

- Front timing chain cover. Tighten the bolts, in sequence, to 40 ft. lbs. (54 Nm).
- Crankshaft damper. Tighten the bolt to 130 ft. lbs. (175 Nm).
- Cylinder head covers
- Right cylinder head access plug. Tighten to 60 ft. lbs. (81 Nm).
- Oil fill housing
- Left cylinder head access plug. Tighten to 60 ft. lbs. (81 Nm).
- Power steering pump

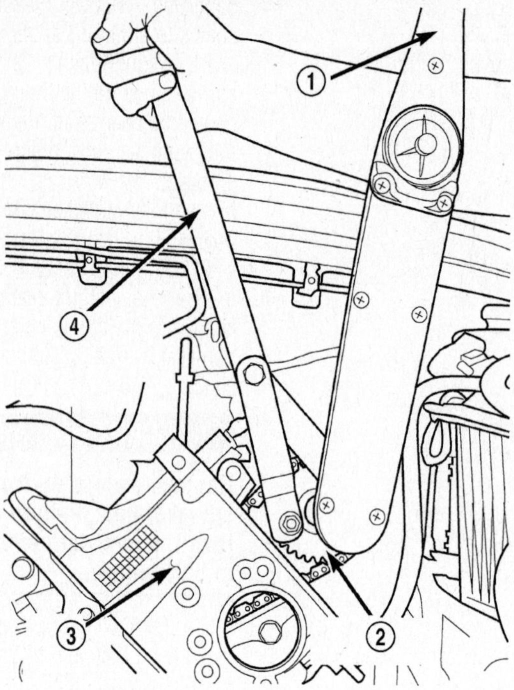

1 - TORQUE WRENCH
2 - CAMSHAFT SPROCKET
3 - LEFT CYLINDER HEAD
4 - SPECIAL TOOL 6958 SPANNER WITH ADAPTER PINS 8346

9355PG14

Tightening the left side camshaft sprocket—3.7L and 4.7L engines

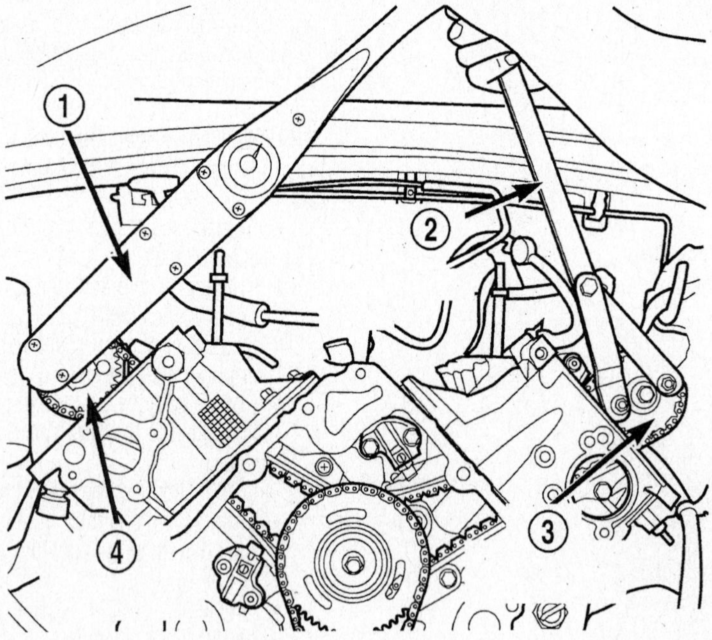

1 - TORQUE WRENCH
2 - SPECIAL TOOL 6958 WITH ADAPTER PINS 8346
3 - LEFT CAMSHAFT SPROCKET
4 - RIGHT CAMSHAFT SPROCKET

9355PG15

Tightening the right side camshaft sprocket—3.7L and 4.7L engines

- Accessory drive belt
- Radiator fan shroud
- Negative battery cable

30. Fill and bleed the cooling system.
31. Start the engine and check for leaks.

3.9L Engine

1. Before servicing the vehicle, refer to the precautions in the beginning of this section.
2. Drain the cooling system.
3. Remove or disconnect the following:
   - Negative battery cable
   - Accessory drive belt
   - Radiator
   - Cooling fan
   - Water pump
   - Crankshaft pulley
   - Front crankshaft seal
   - Front cover
   - Timing chain and gears

To install:

4. Install or connect the following:
   - Timing chain and gears. Align the timing marks and tighten the camshaft sprocket bolt to 35 ft. lbs. (47 Nm).
   - Front cover. Tighten the bolts to 30 ft. lbs. (41 Nm).
   - Front crankshaft seal
   - Crankshaft pulley. Tighten the bolt to 135 ft. lbs. (183 Nm).
   - Water pump
   - Cooling fan
   - Radiator
   - Accessory drive belt
   - Negative battery cable
5. Fill the cooling system.
6. Start the engine and check for leaks.

5.7L Engines

1. Before servicing the vehicle, refer to the precautions in the beginning of this section.
2. Drain the cooling system.
3. Remove or disconnect the following:
   - Negative battery cable
   - Drive belt
   - Radiator and cooling fan
   - Coolant and washer bottles
   - Fan shroud
   - A/C compressor
   - Alternator
   - Radiator and heater hoses
   - Tensioner and idler pulleys
   - Crankshaft damper
   - Power steering pump
   - Oil pan and pickup tube
   - Timing chain cover
4. Re-install the damper bolt finger tight.

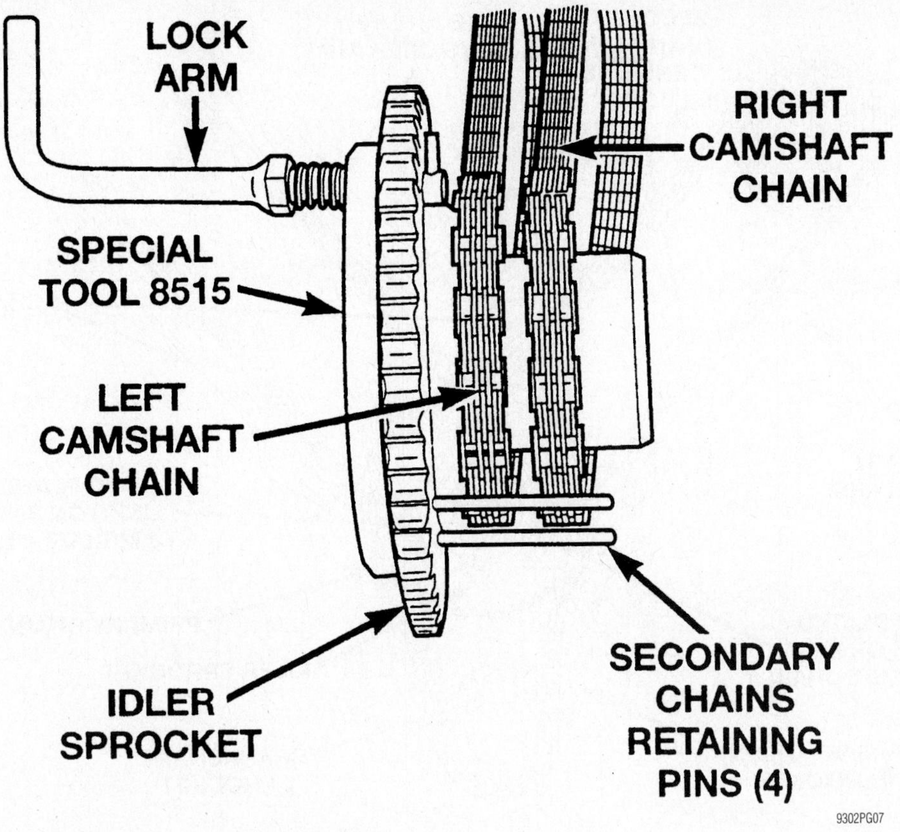

Use the Timing Chain Locking tool to lock the timing chains on the idler gear—4.7L engine

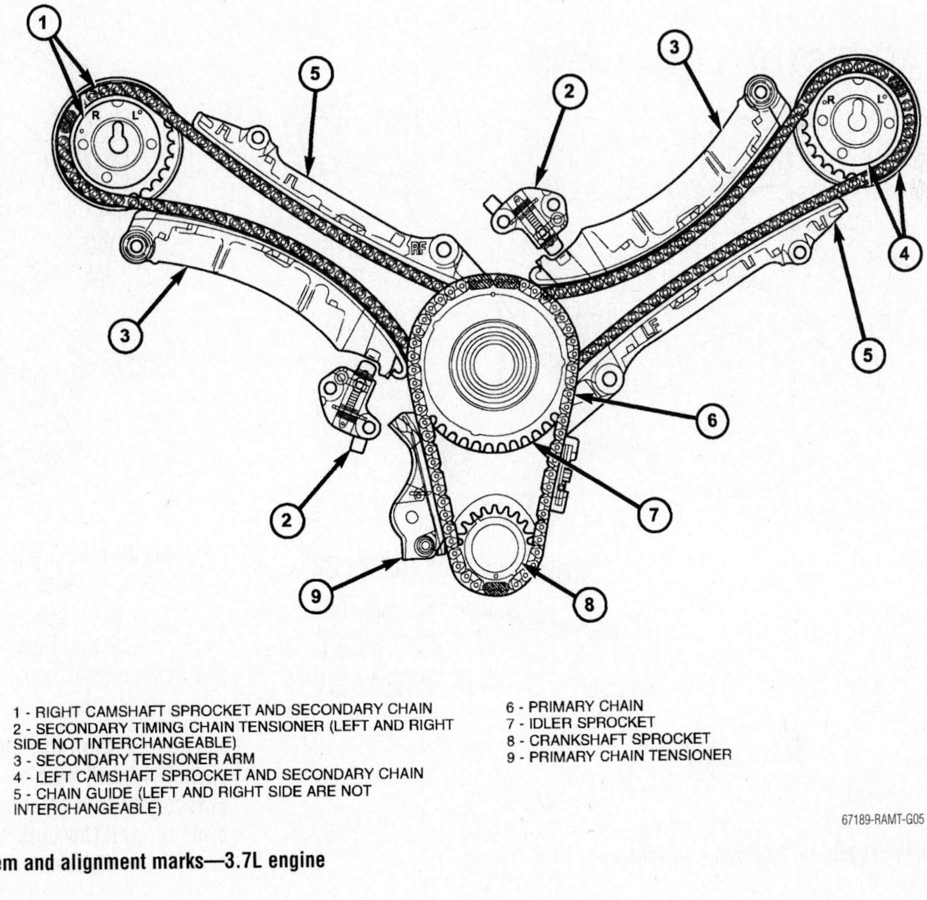

Timing chain system and alignment marks—3.7L engine

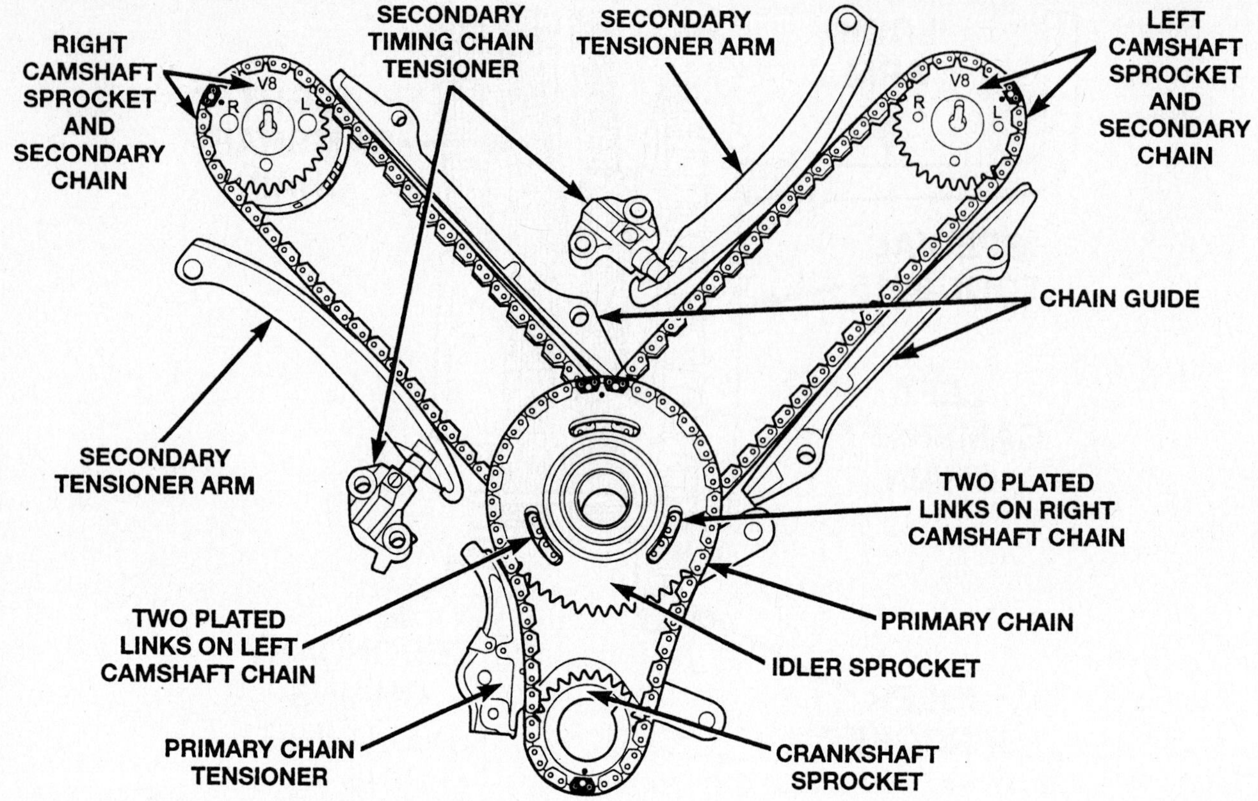

Timing chain system and alignment marks—4.7L engine

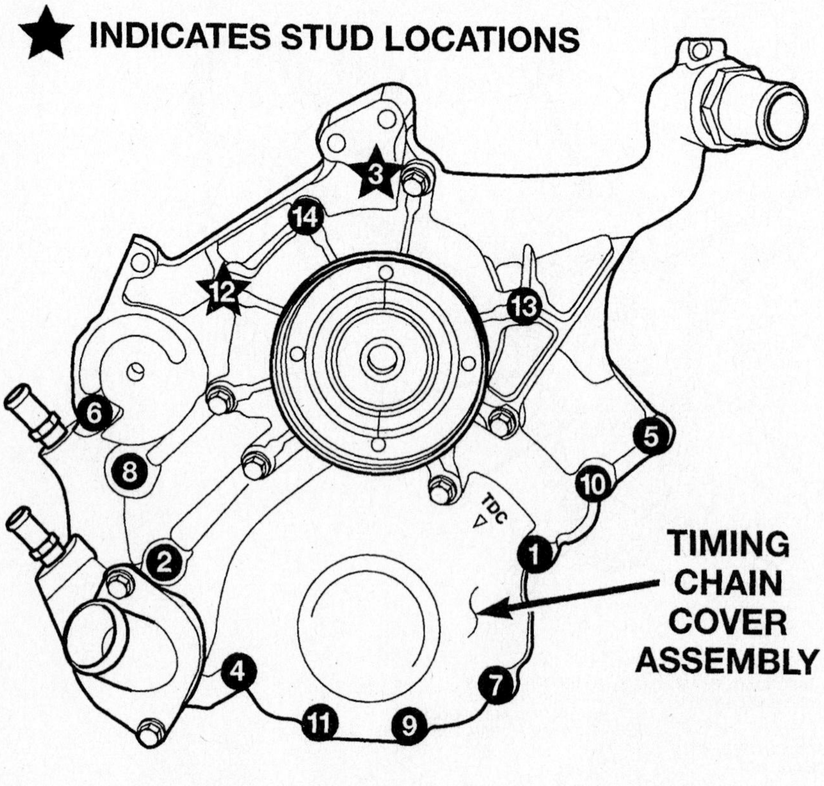

Timing cover bolt torque sequence—3.7L and 4.7L engines

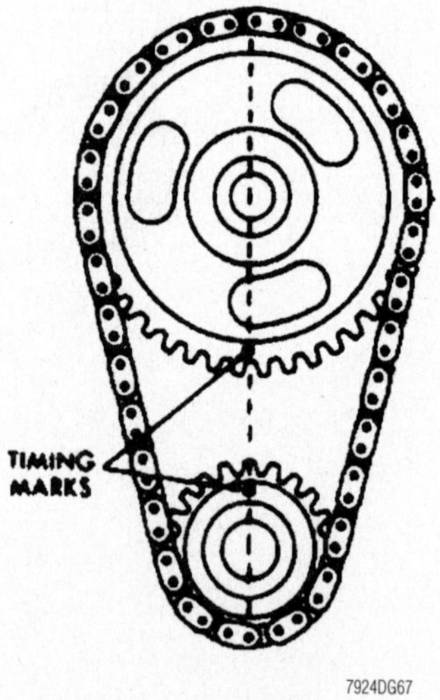

Timing chain alignment marks—3.9L, 5.2L, 5.9L and 8.0L engines

5. Rotate the crankshaft so that the camshaft sprocket and crankshaft sprocket timing marks are aligned.

➡The camshaft pin and slot in the cam sprocket must be a 12 o'clock, the crankshaft keyway must be at 2 o'clock, and the dots or paint on the crank sprocket must be at 6 o'clock.

6. Remove the oil pump.
7. Pin back the tensioner shoe.
8. Remove the timing chain and sprockets.

To install:

9. With the timing marks aligned, wrap the chain around the sprockets. The chain must be installed with the single plated link aligned with the dot or paint on the cam sprocket. The dot or paint on the crank sprocket should be aligned between the 2 plated links.
10. Install the assembly and torque the cam sprocket bolt to 90 ft. lbs. (122 Nm).
11. Unpin the tensioner and verify the timing mark alignment.
12. Install or connect the following:
    - Timing cover. Torque all fasteners to 21 ft. lbs. (28 Nm). Torque the large lifting lug to 40 ft. lbs. (55 Nm).
    - Oil pan and pickup tube
    - Power steering pump
    - Crankshaft damper
    - Tensioner and idler pulleys
    - Radiator and heater hoses
    - Alternator
    - A/C compressor
    - Fan shroud
    - Coolant and washer bottles
    - Radiator fan
    - Drive belt
    - Negative battery cable

5.2L and 5.9L Engines

1. Before servicing the vehicle, refer to the precautions in the beginning of this section.
2. Drain the cooling system.
3. Remove or disconnect the following:
    - Negative battery cable
    - Accessory drive belt
    - Cooling fan and shroud
    - Water pump
    - Power steering pump
    - Crankshaft damper
    - Front crankshaft seal
    - Front cover
4. Rotate the crankshaft so that the camshaft sprocket and crankshaft sprocket timing marks are aligned.
5. Remove the timing chain and sprockets.

To install:

6. Install the timing chain and sprockets with the timing marks aligned. Tighten the camshaft sprocket bolt to 50 ft. lbs. (68 Nm).
7. Install or connect the following:
    - Front cover. Tighten the cover bolts

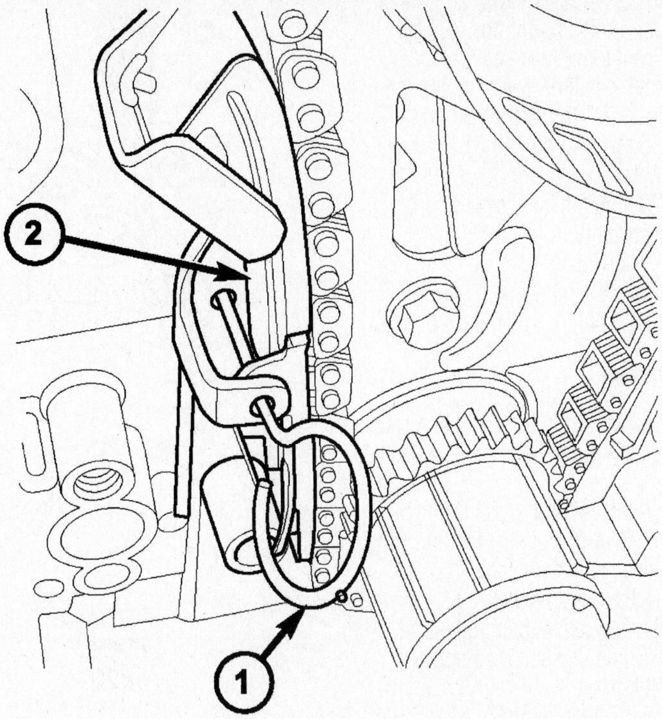

Pinning back the tensioner shoe

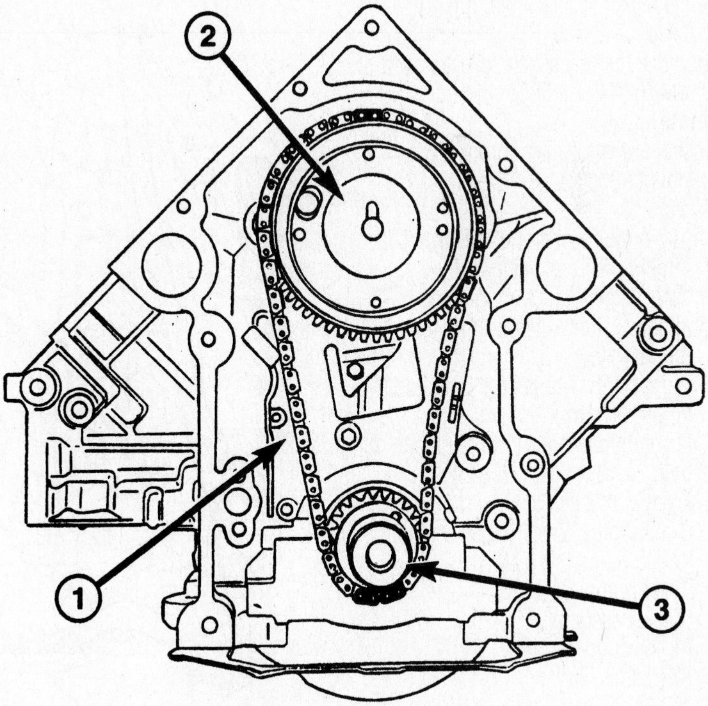

Timing mark alignment—5.7L engine

to 30 ft. lbs. (41 Nm) and the oil pan bolts to 18 ft. lbs. (24 Nm).
- Front crankshaft seal
- Crankshaft damper. Tighten the bolt to 135 ft. lbs. (183 Nm).
- Power steering pump
- Water pump
- Cooling fan and shroud
- Accessory drive belt
- Negative battery cable

8. Fill the cooling system.
9. Start the engine and check for leaks.

8.0L Engine

1. Before servicing the vehicle, refer to the precautions in the beginning of this section.
2. Drain the cooling system.
3. Remove or disconnect the following:
   - Negative battery cable
   - Accessory drive belt
   - Cooling fan and shroud
   - A/C compressor, if equipped
   - Alternator
   - Air injection pump
   - Bracket assembly
   - Water pump
   - Crankshaft pulley
   - Front crankshaft seal
   - Front oil pan bolts
   - Front cover
   - Camshaft sprocket and timing chain
4. Remove the crankshaft sprocket with Special tools 6444 and 6820.

To install:

5. Install the crankshaft sprocket with Special tools C-3688, C-3718 and MD990799.
6. Install or connect the following:
   - Timing chain and camshaft sprocket with the timing marks aligned. Tighten the bolt to 45 ft. lbs. (61 Nm).
   - Front cover. Tighten the bolts to 35 ft. lbs. (47 Nm).
   - Front oil pan bolts. Tighten the bolts to 12 ft. lbs. (16 Nm).
   - Front crankshaft seal
   - Crankshaft pulley. Tighten the bolt to 135 ft. lbs. (183 Nm).
   - Water pump
   - Bracket assembly
   - Air injection pump
   - Alternator
   - A/C compressor, if equipped
   - Cooling fan and shroud
   - Accessory drive belt
   - Negative battery cable
7. Fill the cooling system.
8. Start the engine and check for leaks.

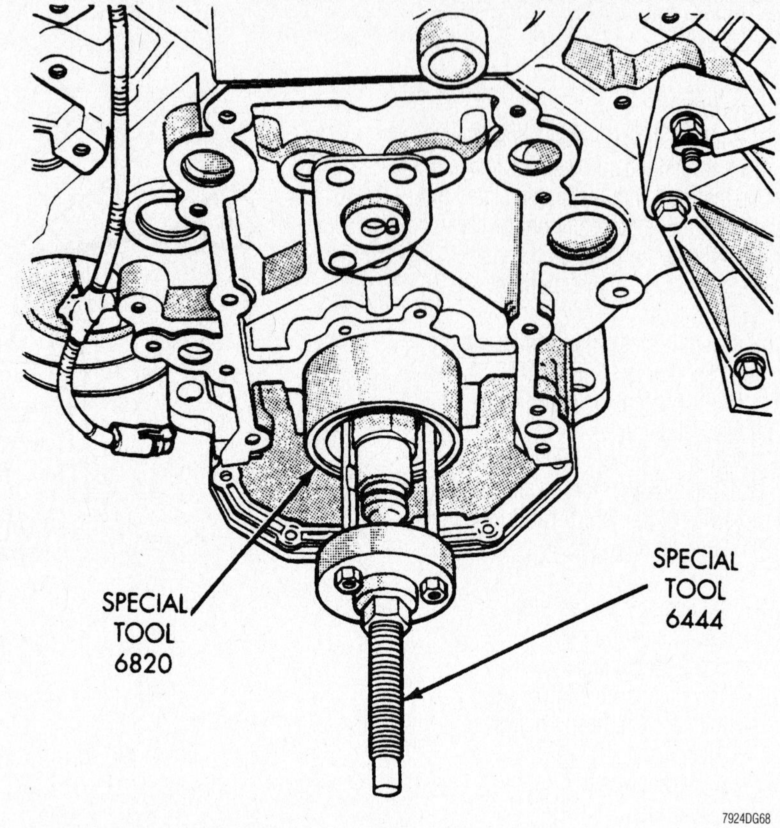

Crankshaft sprocket removal tools—8.0L engine

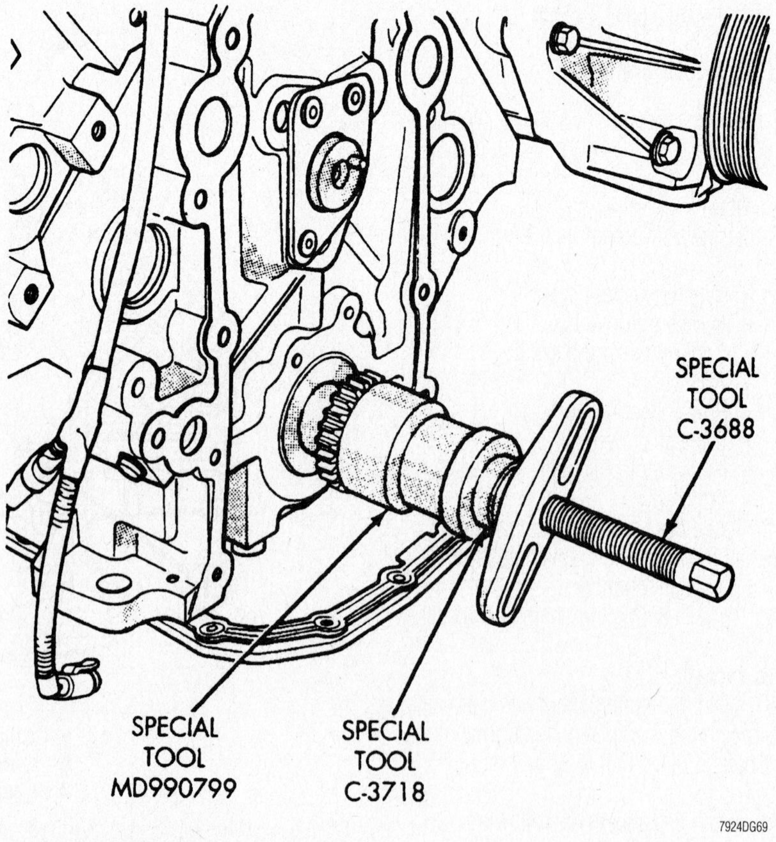

Crankshaft installation tools—8.0L engine

Piston and Ring

POSITIONING

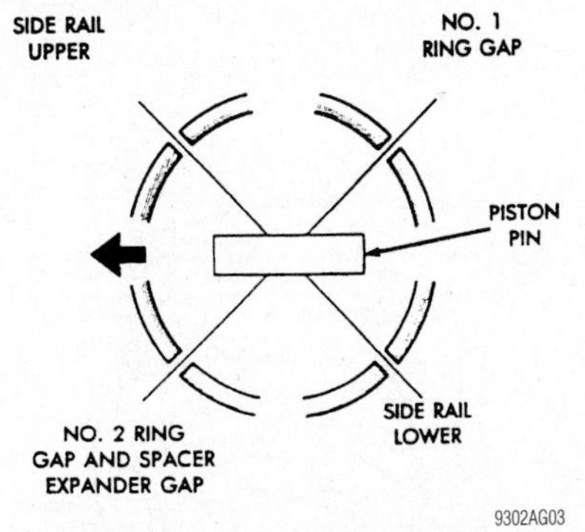

Piston ring end-gap spacing. Position raised "F" on piston toward front of engine—3.7L and 4.7L engines

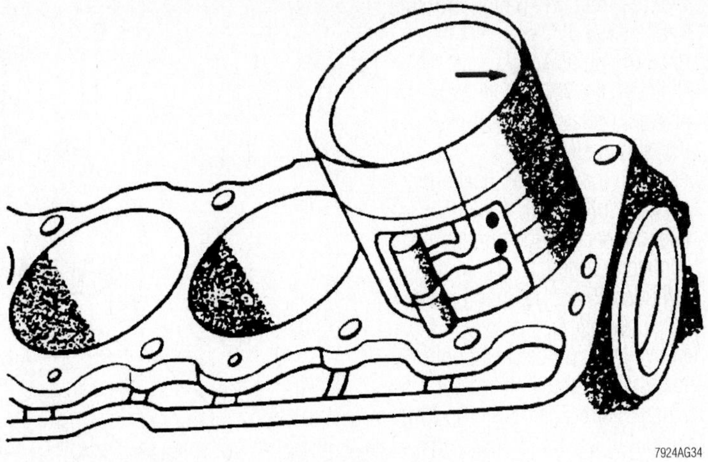

Piston to engine positioning—3.9L, 5.2L, 5.9L and 8.0L engines

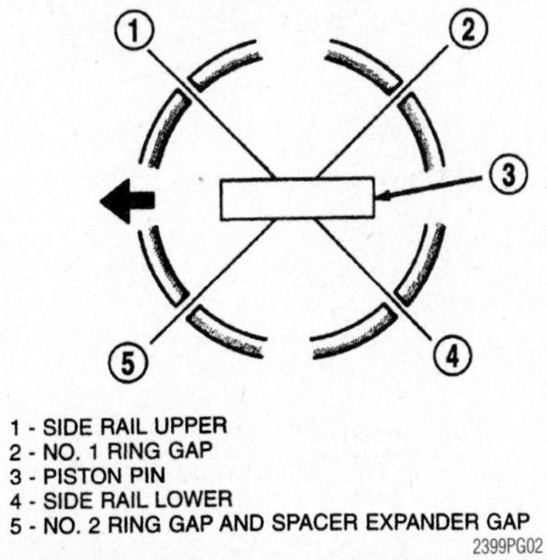

Piston ring end-gap spacing—5.7L engines

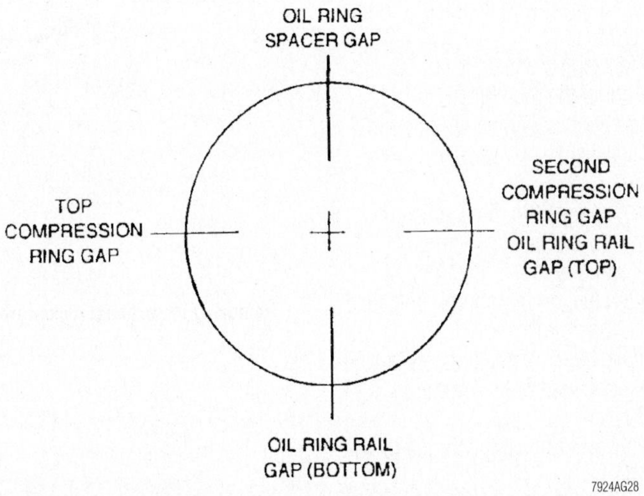

Piston ring end-gap spacing—3.9L, 5.2L, 5.9L and 8.0L engines

DIESEL ENGINE REPAIR

Engine Assembly

REMOVAL & INSTALLATION

1. Before servicing the vehicle, refer to the precautions in the beginning of this section.
2. Drain the cooling system.
3. Drain the engine oil.
4. Recover the A/C refrigerant, if equipped.
5. Remove or disconnect the following:
   - Both negative battery cables
   - Engine grid heater harness at grid heater relays
   - Electrical connections from rear of alternator
   - Radiator fan and fan drive
   - Radiator fan shroud
   - Upper radiator hose
   - Coolant recovery bottle hose
   - Heater hoses
   - Transmission
   - Transfer case, if equipped
   - Exhaust pipe from turbocharger extension pipe
   - Engine harness to vehicle harness connectors
   - Starter motor
   - Flywheel/flexplate
   - Transmission adapter
   - A/C suction/discharge hose from the rear of the A/C compressor, if equipped
   - Lower radiator hose
   - Transmission oil cooler lines from in front of radiator using Special Tool #6931, if equipped
   - Radiator
   - A/C condenser refrigerant lines, if equipped
   - Charge air cooler piping
   - Charge air cooler
   - A/C condenser, if equipped
   - Damper and speed indicator ring
   - Fan shroud brackets
   - Engine block heater connector
   - A/C compressor and pressure sensor electrical connectors
   - Battery ground cables from the engine block
   - Power steering pump
   - ECM power connector
   - ECM ground wire from the hydroform screw
   - Fuel supply and return hoses
   - Cylinder head cover
   - Wire harnesses from the injectors
   - Cylinder head cover carrier gasket
   - #6 fuel line shield bolts and rotate shield out of the way

- Cylinder #5 and #6 high pressure fuel lines
- Rear engine lift bracket
- Cylinder #4, #5, and #6 intake and exhaust rocker arms (1), pedestals (2), and push tubes. Note the original location for re-assembly.
- Fuel connector tube nut and fuel connector tube
- Cylinder #5 and #6 fuel injector
- Rocker housing

6. Remove two cylinder head bolts as shown and install Engine Lifting Tool # 9009. Torque bolts to 77ft. lbs. (105 Nm).
7. Loosen but do not remove engine mount through bolts and nuts.
8. Disconnect hood support struts and position hood out of the way.
9. Attach a chain with two hooks to the engine lift brackets.
10. While keeping engine level, lift straight up out of the mounts.
11. Rotate nose of engine upward and pull out of chassis.

To install:

12. Lower engine into the engine compartment and install the engine mount through bolts and nuts.
13. Tighten the mount through bolts and nuts to 65 ft. lbs. (88 Nm).
14. Remove the engine lifting device and reinstall cylinder head bolts. Torque to specifications.
15. Install rocker housing. Torque to 18 ft. lbs. (24 Nm).
16. Replace injector o-ring and sealing washer on injectors #5 and #6 and install injectors.
17. Install rear engine lift bracket. Torque to 57 ft. lbs. (77 Nm).
18. Install #5 and #6 high pressure fuel lines. Torque fuel line fittings to 22 ft. lbs. (30 Nm).
19. Install push tubes, rocker arms and pedestals for cylinders #4, #5, and #6. Torque the mounting bolts to (27 ft. lbs. (36 Nm).
20. Reset valve lash on cylinders #4, #5, and #6. Torque adjusting nuts to 18 ft. lbs. (24 Nm).
21. Install or connect the following:
    - Injector harness nuts to injectors. Torque to 11 inch lbs. (1.25 Nm).
    - Injector wiring harness connectors to injector connections on cylinder head cover carrier gasket
    - Cylinder head cover
    - Breather tube and lube oil drain tube to cylinder head cover
    - Fuel supply and return hoses
    - ECM ground to hydroform screw
    - ECM power connector

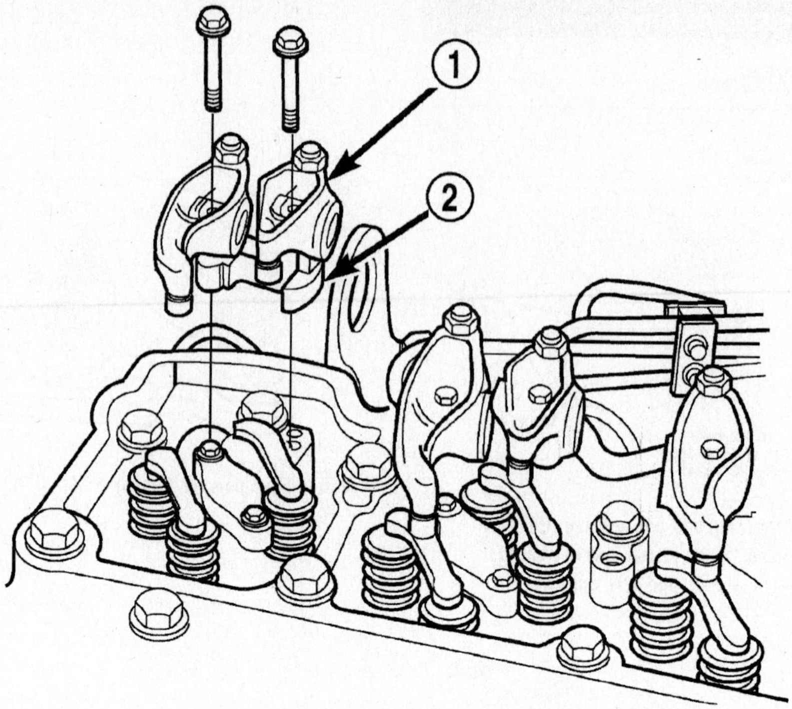

Removing the exhaust rocker arms

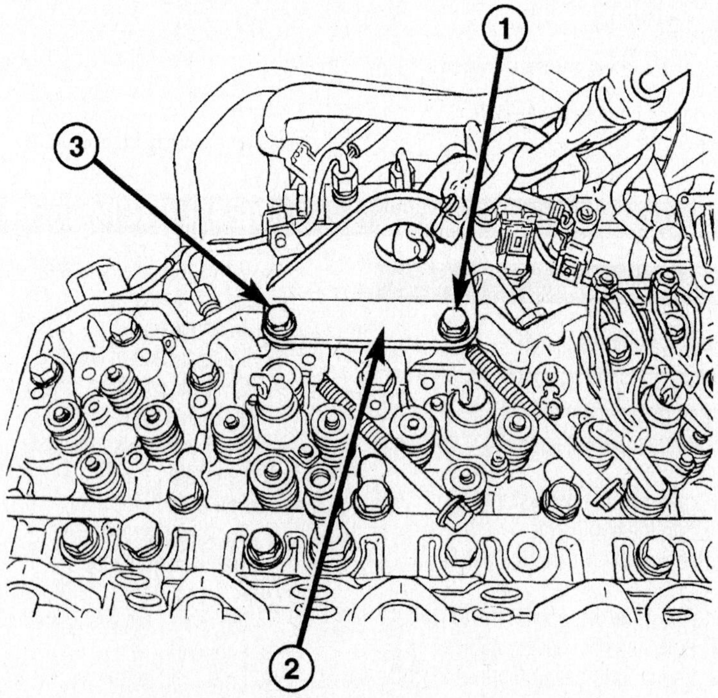

Install engine lifting fixture 9009

- Power steering pump
- Damper and speed indicator ring. Torque to 30 ft. lbs. (40 Nm) plus 60°.
- Lower fan shroud bracket. Torque to 18 ft. lbs. (24 Nm).
- Upper fan shroud brackets, but do not tighten fasteners at this time
- Engine block heater connection
- A/C compressor and pressure sensor connectors
- Charge air cooler and tighten the mounting bolts to 17 inch lbs. (2 Nm)
- Charge air cooler tubes. Torque all clamps to 72 inch lbs. (8 Nm).
- A/C condenser
- A/C refrigerant lines to the A/C condenser
- Radiator upper support panel
- Radiator
- Transmission quick-connect oil cooler lines, if equipped
- A/C compressor suction/discharge hose, if equipped
- Lower radiator hose
- Negative battery cables to the engine block on both sides
- Transmission adapter with a new camshaft rectangular ring seal. Torque to 57 ft. lbs. (77 Nm).
- Flywheel/flexplate
- Starter
- Engine to vehicle harness connectors
- Transmission
- Transfer case, if equipped
- Exhaust pipe to the turbocharger elbow
- Transmission auxiliary oil cooler lines, if equipped
- Heater hoses
- Radiator fan and fan drive
- Radiator fan shroud
- Coolant recovery bottle
- Upper radiator hose
- Grid heater harness at grid heater relays
- Electrical connections to rear of alternator

22. Refill the cooling system.
23. Fill the engine with oil to the correct level.
24. Start the engine and check for leaks.

Water Pump

REMOVAL & INSTALLATION

1. Before servicing the vehicle, refer to the precautions in the beginning of this section.

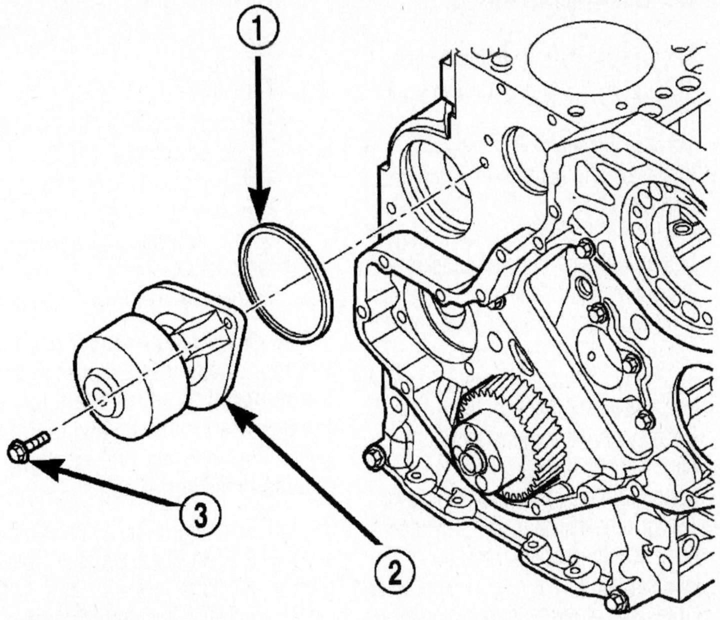

Exploded view of the water pump mounting—5.9L diesel engine

2. Drain the cooling system.
3. Remove or disconnect the following:
   - Negative battery cables
   - Accessory drive belt
   - Water pump

To install:

4. Install or connect the following:
   - Water pump with new O-ring seal. Tighten the bolts to 18 ft. lbs. (24 Nm).
   - Accessory drive belt
   - Negative battery cables
5. Fill the cooling system.
6. Start the engine and check for leaks.

Glow Plugs

REMOVAL & INSTALLATION

The 5.9L diesel engine uses an intake manifold air heater instead of glow plugs to preheat the air for improved starting ability. The heater element is located within the intake manifold top cover. Refer to the intake manifold removal and installation procedure to service the intake manifold air heater.

Cylinder Head

REMOVAL & INSTALLATION

1. Before servicing the vehicle, refer to the precautions in the beginning of this section.
2. Drain the cooling system.
3. Drain the engine oil.
4. Remove or disconnect the following:
   - Negative battery cables
   - Exhaust pipe from turbocharger elbow
   - Turbocharger drain tube bolts at turbocharger
   - Air inlet temperature/pressure sensor
   - Air intake assembly
   - Heater hoses
   - Turbocharger oil supply line at the turbocharger end
   - Exhaust manifold-to-cylinder head bolts, spacers, heat shield, retention straps, and cab heater plumbing
   - Exhaust manifold and turbocharger from the vehicle as an assembly
   - Radiator fan assembly and shroud
   - Accessory drive belt
   - Radiator fan support from cylinder block
   - Upper alternator bolt, loosen lower alternator bolt and rotate alternator away from cylinder head
   - Upper radiator hose from the thermostat housing
   - Intake Air Temperature sensor connector
   - Manifold Air Pressure sensor connector

- Coolant Temperature sensor connectors
- Engine harness to cylinder head attaching bolts and P-clips at front of head
- Throttle linkage cover
- Intake air grid heater wires from the grid heater
- Engine oil dipstick tube mounting bolt at fuel filter housing bracket and inlet air connection
- Charge air cooler-to-air inlet housing pipe
- Engine wiring harness mounting bolt and wiring harness push-in fastener from air inlet housing
- Air inlet housing and intake grid heater from the intake manifold cover
- Two grid heater harness-to-cylinder head attaching bolts at front of cylinder head

5. Remove the high pressure pump to fuel rail fuel line as follows:
   a. Loosen fuel line nuts at fuel pump and at fuel rail.
   b. Use a back-up wrench on the fitting at the fuel pump to keep it from loosening.
6. Remove the fuel rail to cylinder head fuel lines as follows:
   a. Loosen No. 6 high pressure fuel line shield and position out of way.
   b. Loosen the fuel line nuts at the fuel rail and at the cylinder head. Use a back-up wrench on HPC nut.
   c. Remove the fuel line bracket bolts at the intake manifold cover.
7. Remove the engine lift bracket from the rear of the cylinder head.
8. Remove the fuel rail as follows:
   a. Remove fuel rail pressure sensor connector.
   b. Remove banjo fitting at pressure limiting valve.
   c. Remove fuel rail bolts and fuel rail.
9. Remove the low pressure lines as follows:
   a. Remove the fuel drain banjo fitting on the front side of fuel filter housing.
   b. Remove the fuel drain banjo fitting on rear side of fuel filter housing.
   c. Remove the fuel drain line support bracket on rear side of filter housing.
   d. Disconnect fuel drain hose.
   e. Remove banjo fitting at bottom of fuel filter housing.
   f. Disconnect fuel supply hose at fuel filter.
10. Remove or disconnect the following:
    - Fuel heater and water-in-fuel sensor.
    - Fuel filter assembly-to-cylinder assembly
    - Wiring harness P-clip from cylinder head (located behind filter housing)
    - Cylinder head cover
    - Rocker housing injector harness connectors
    - Injector harness nuts from injectors
    - Rocker arms, cross heads and push rods. Matchmark each component so they can be

installed in their original positions.

➡The #5 cylinder exhaust and the #6 cylinder intake and exhaust push rods are removed by lifting them up and through the provided cowl panel access holes. Remove the rubber plugs to expose these relief holes.

- Fuel return line and banjo bolt at the rear of the cylinder head.
- Fuel injectors
- Rocker housing bolts and rocker housing and gasket

11. Reinstall the engine lift bracket at the rear of cylinder head. Torque to 57 ft. lbs. (77 Nm).
12. Remove the 26 cylinder head-to-block bolts.
13. Attach an engine lift crane to engine lift brackets and lift cylinder head out of the vehicle.
14. Remove the cylinder head gasket.

To install:

➡Check the cylinder head bolt length. If length exceeds 5.2 inches (132.1 mm), the bolt must be replaced.

15. Install a new gasket with the part number side up.
16. Lower the cylinder head onto the engine. Tighten the bolts as follow:
    a. Step 1: 52 ft. lbs. (70 Nm).
    b. Step 2: Loosen each bolt in sequence 360°
    c. Step 3: 77 ft. lbs. (104 Nm).
    d. Step 4: Retighten to 77 ft. lbs. (80 Nm).
    e. Step 5: Tighten an additional 90°
17. Install or connect the following:
    - Push rods into their original locations
    - Rocker housing and bolts. Tighten to 18 ft. lbs. (24 Nm)
    - Fuel injectors
    - Crossheads in their original locations
    - Rocker arms and pedestals in their original locations. Tighten the bolts to 27 ft. lbs. (36 Nm).
    - Cylinder head cover gasket onto rocker housing
    - Injector harness nuts. Tighten to 11 inch lbs. (1.25 Nm).

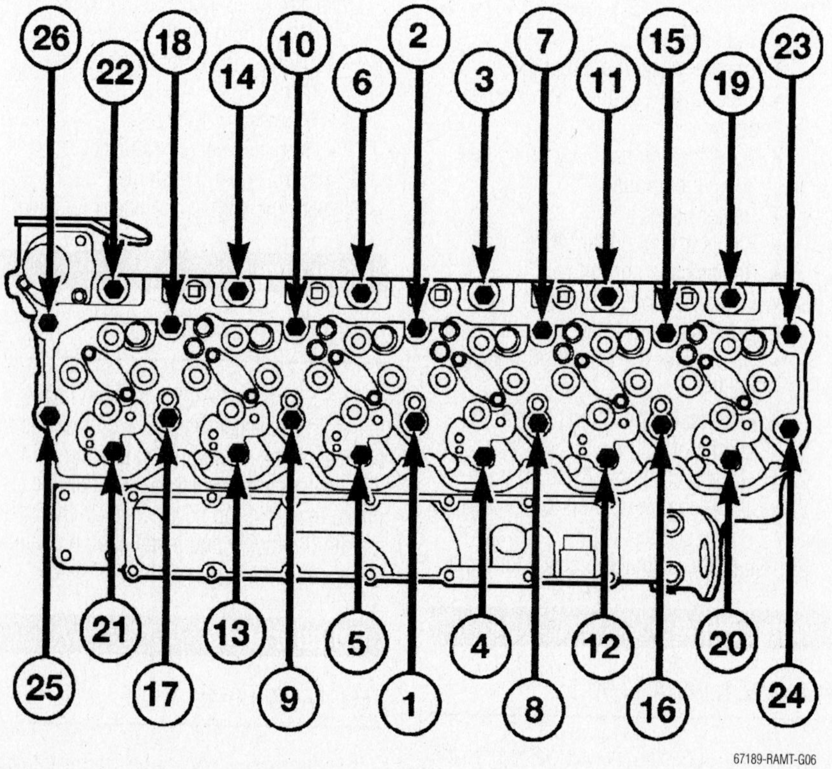

Cylinder head torque sequence—5.9L diesel engine

- Injector harness connectors at cylinder head cover gasket.
- Wiring harness P-clip to cylinder head behind filter housing and tighten to 18 ft. lbs. (24 Nm)
- IAT/MAP sensor connector
- Fuel filter housing assembly. Tighten the mounting bolts to 18 ft. lbs. (24 Nm)
- WIF sensor and fuel heater

18. Install the fuel rail and high pressure fuel lines as follows:
   a. Hand tighten fuel rail bolts.
   b. Hand tighten fuel drain line to pressure limiting valve. Hand tighten banjo bolt at fuel filter housing.
   c. Hand tighten fuel rail-to-cylinder head high pressure fuel lines.
   d. Install fuel line brace bolts-to-intake manifold finger tight for fuel lines #2 thru #5.
   e. Install fuel line brace nut to rear lift bracket stud, finger tight, for fuel line #6.
   f. Hand tighten fuel pump to fuel rail line.
   g. Torque fuel line nuts at cylinder head to 22 ft. lbs. (30 Nm).
   h. Torque fuel line nuts at fuel rail to 22 ft. lbs. (30 Nm).
   i. Using a back up wrench, torque fuel pump to fuel rail line to 27 ft. lbs. (37 Nm).
   j. Torque fuel pump to fuel rail line to 37 Nm (27 ft. lbs.) at fuel rail.
   k. Torque fuel drain banjo bolt at pressure limiting valve and front of fuel filter housing to 18 ft. lbs. (24 Nm).
   l. Torque fuel line brace bolts to 18 ft. lbs. (24 Nm).
   m. Torque fuel line brace nut to 18 ft. lbs. (24 Nm).
   n. Torque rail bolts to 18 ft. lbs. (24 Nm).
   o. Connect fuel pressure sensor.
19. Reposition fuel line #6 shield and torque to 32 ft. lbs. (43 Nm).
20. Install the fuel filter to injection pump low pressure line. Torque banjo bolts to 18 ft. lbs. (24 Nm).
21. Hand tighten the fuel return line at back of cylinder head.
22. Hand tighten the fuel return line at filter housing
23. Tighten the banjo connections at cylinder head and fuel filter housing to 18 ft. lbs. (24 Nm).
24. Install or connect the following:
   - Bracket to rear of filter housing. Tighten to 18 ft. lbs. (24 Nm).
   - Intake grid heater and air inlet housing using new gasket. Tighten bolts to 18 ft. lbs. (24 Nm).
   - Wiring harness P-clip. Push on clip to air inlet housing.
   - Engine oil dipstick tube at fuel filter housing and at air inlet housing
   - Charge air cooler-to-air inlet housing duct assembly. Tighten all clamps to 100 inch lbs. (11 Nm).
   - Intake grid heater wires
   - Engine harness to front of cylinder head
   - Engine coolant temperature sensor connector
   - Radiator upper hose to thermostat housing
25. Rotate alternator into position and tighten the mounting bolts.
26. Install or connect the following:
   - Wiring harness push-on clip below bracket
   - Wiring harness P-clip to top of bracket
   - Radiator fan support and torque to 24 ft. lbs. (32 Nm)
   - Radiator fan assembly and shroud.
   - Accessory drive belt
   - Exhaust manifold/turbocharger assembly, using new gaskets. Torque bolts to 32 ft. lbs. (43 Nm).
   - Exhaust manifold heat shield. Tighten to 18 ft. lbs. (24 Nm).
   - Exhaust bolt retention straps across cylinders 5 & 6
   - Turbocharger oil drain tube using new gasket. Tighten to 18 ft. lbs. (24 Nm).
27. Perform the turbocharger pre-lube procedure.
28. Install or connect the following:
   - Turbocharger oil supply line. Tighten to 18 ft. lbs. (24 Nm).
   - Air intake assembly
   - Air inlet temperature/pressure sensor
   - Exhaust pipe to turbocharger elbow. Tighten bolts to 100 inch lbs. (11 Nm).
29. Fill the engine with coolant to the correct level.
30. Start the engine and check for leaks.

Rocker Arms/Shafts

REMOVAL & INSTALLATION

1. Before servicing the vehicle, refer to the precautions in the beginning of this section.
2. Remove or disconnect the following:
   - Negative battery cables
   - Cylinder head cover

✲✲ WARNING

The sockets may fall out of the rocker arms as the rocker arms are lifted from the cylinder head. Do not drop the sockets into the engine.

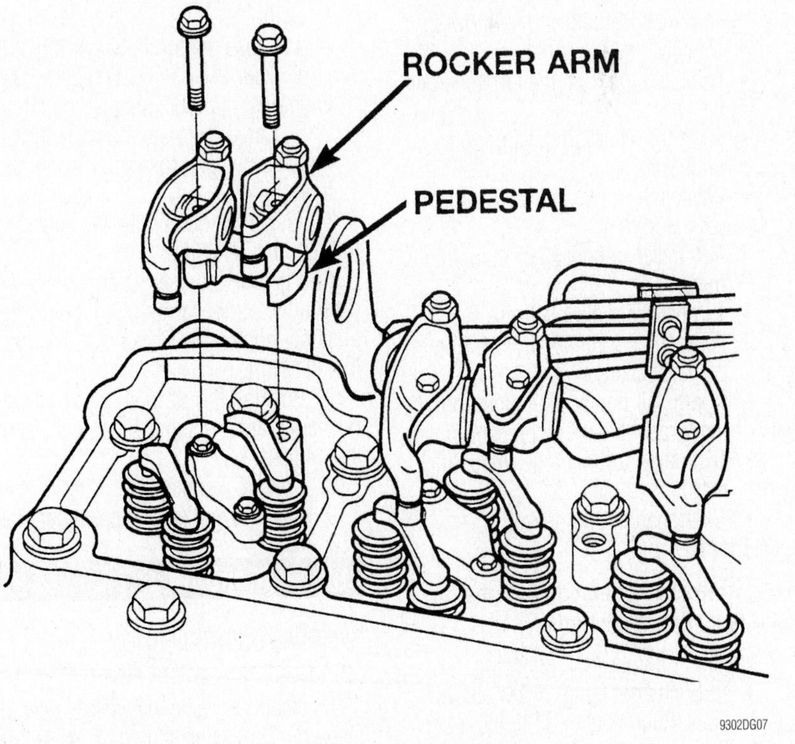

Exploded view of the rocker arm mounting—5.9L diesel engine

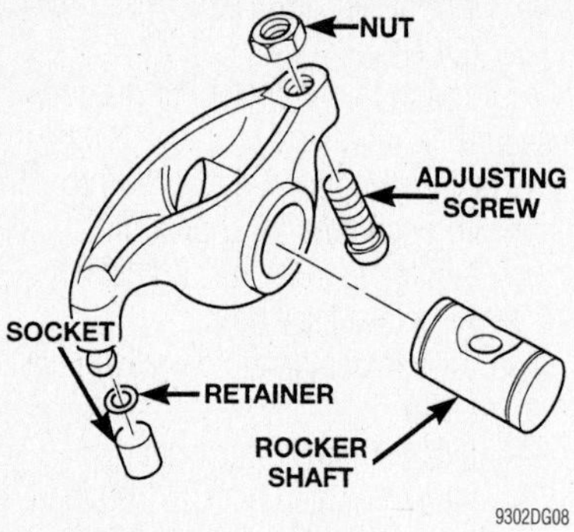

Exploded view of the rocker arm—5.9 L diesel engine

➡**Keep all valvetrain components in order for assembly.**

- Rocker arms and pedestal assemblies

To install:

3. Install or connect the following:
   - Rocker arm and pedestal assemblies. Tighten the bolts to 27 ft. lbs. (36 Nm).
   - Valve cover. Tighten the bolts to 18 ft. lbs. (24 Nm).
   - Negative battery cables

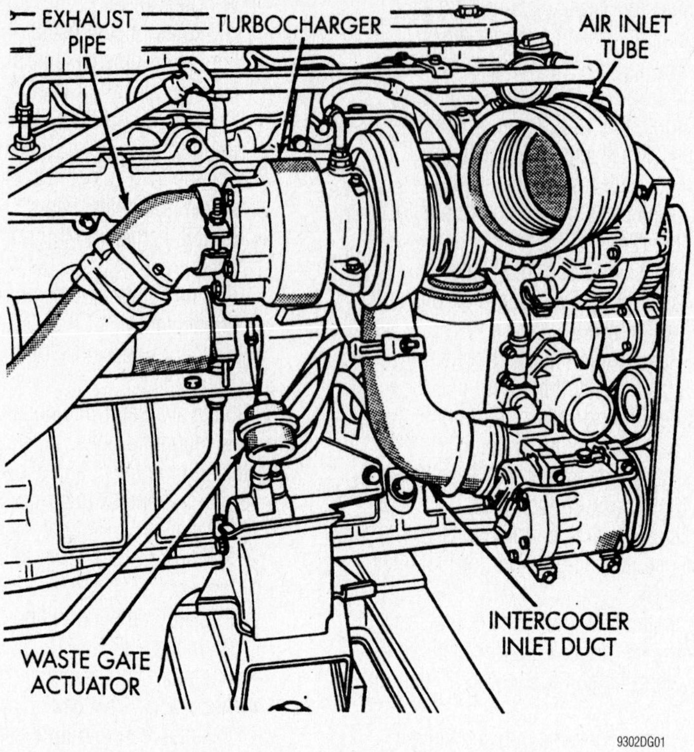

Turbocharger and related components—5.9L diesel engine

Turbocharger

REMOVAL & INSTALLATION

1. Before servicing the vehicle, refer to the precautions in the beginning of this section.
2. Remove or disconnect the following:
   - Negative battery cables
   - Exhaust pipe from the turbocharger elbow
   - Bolts from the turbocharger oil drain tube
   - Wastegate command valve connector
   - Two lower turbocharger mounting nuts
   - Turbocharger air inlet hose
   - Turbocharger oil supply line from the turbocharger
   - Charge air cooler inlet pipe from the turbocharger compressor outlet
   - Two upper turbocharger mounting nuts
   - Turbocharger and gasket

To install:

➡**Use anti-seize compound on the turbocharger mounting studs.**

3. Install or connect the following:
   - Turbocharger using a new gasket. Tighten the nuts to 32 ft. lbs. (43 Nm).
   - Oil drain tube to the turbocharger using new gasket. Tighten the bolts to 18 ft. lbs. 24 Nm).
   - Exhaust pipe to the turbocharger and tighten the bolts to 100 inch lbs. (11 Nm).
   - Wastegate command valve connector
4. Pre-lube the turbocharger as follows:
   a. Pour 2-3 oz. (50-60 cc) of clean engine oil in the oil supply line fitting.
   b. Carefully rotate the turbocharger impeller by hand to distribute the oil thoroughly.
5. Install the oil supply line and tighten to 18 ft. lbs. (24 Nm)
6. Position the charge air cooler inlet pipe to the turbocharger. With the clamp in position, tighten the clamp nut to 95 inch lbs. (11 Nm).
7. Position the air inlet hose to the turbocharger and tighten the clamp to 35 inch lbs. (4 Nm).
8. Connect the negative battery cables.
9. Start the engine and check for leaks.

Intake Manifold

REMOVAL & INSTALLATION

1. Before servicing the vehicle, refer to the precautions in the beginning of this section.
2. Remove or disconnect the following:
   - Negative battery cables
   - Charge air cooler outlet tube from the air inlet housing
   - Engine oil dipstick tube
   - Air grid heater power cables at the cable mounting studs
   - Engine wiring harness connections at the air inlet housing elbow
   - Air inlet housing from top of the heater elements
   - Intake air grid heater
   - High pressure fuel lines and fuel rail
   - Manifold air temperature/pressure sensor connector
   - Remaining intake manifold cover-to-cylinder head bolts
   - Intake manifold cover and gasket

To install:

➡**Use liquid Teflon® sealer on the intake manifold cover bolts.**

3. Install or connect the following:
   - Intake manifold cover, using new gasket. Install the cover-to-cylinder head bolts that do not hold down the fuel rail and tighten to 18 ft. lbs. (24 Nm).
   - High pressure rail and fuel lines
   - Intake air grid heater and air inlet housing using new gaskets
   - Manifold air temperature/pressure sensor connector

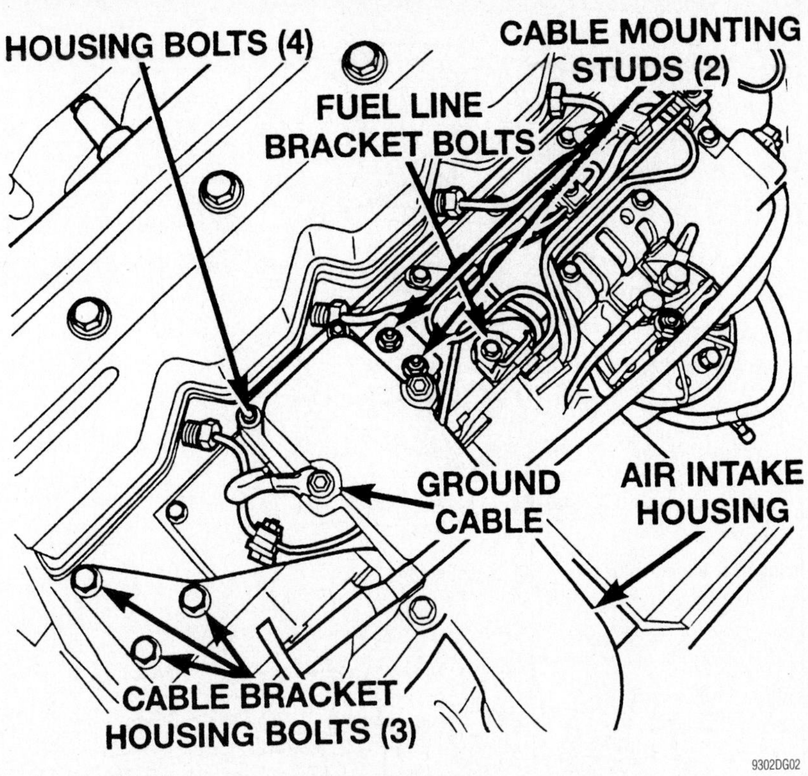

Intake air heater electrical connections—5.9L diesel engine

- Air grid heater power supply cables and tighten to 120 inch lbs. (14 Nm)
- Engine oil dipstick tube
- Charge air cooler outlet tube onto the air inlet housing. Tighten the clamps to 100 inch lbs. (11 Nm)
- Engine wire harness to the air inlet housing elbow
- Negative battery cables

4. Start the engine and check for leaks.

Exhaust Manifold

REMOVAL & INSTALLATION

1. Before servicing the vehicle, refer to the precautions in the beginning of this section.
2. Remove or disconnect the following:
   - Negative battery cables
   - Turbocharger
   - Cab heater supply and return lines
   - Heat shield
   - Exhaust manifold

To install:

➡Use anti-seize compound on the exhaust manifold bolts.

3. Install or connect the following:
   - Exhaust manifold. Tighten the bolts in sequence to 32 ft. lbs. (43 Nm).
   - Cab heater supply and return lines
   - Turbocharger
   - Negative battery cables
4. Start the engine and check for leaks.

Camshaft and Valve Lifters

REMOVAL & INSTALLATION

1. Before servicing the vehicle, refer to the precautions in the beginning of this section.
2. Recover the A/C refrigerant, if equipped.
3. Drain the cooling system.
4. Remove or disconnect the following:
   - Negative battery cables
   - Radiator fan assembly and shroud
   - Radiator
   - Automatic transmission oil cooler lines, if equipped
   - Upper radiator support panel
   - Charge air cooler
   - A/C condenser, if equipped
   - Accessory drive belt
   - Accessory drive belt tensioner

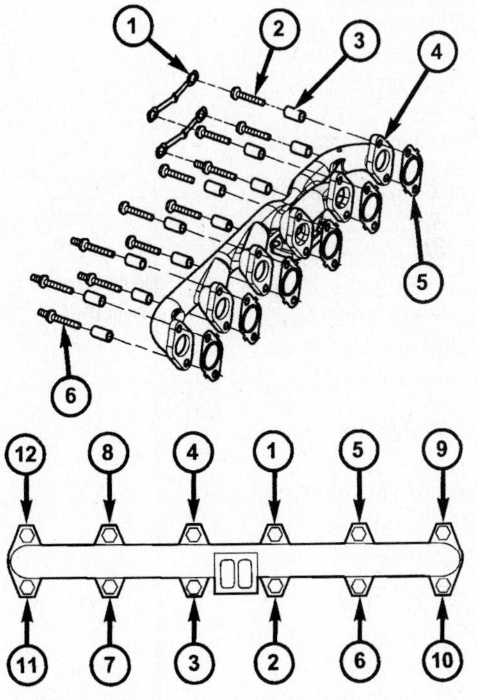

Exhaust manifold torque sequence—5.9L diesel engine

- Fan support/hub assembly
- Vibration damper
- Front cover

5. Using Special Tool 7471-B Crankshaft Barring Tool, rotate the crankshaft to align the timing marks on the crankshaft and the camshaft gears.

6. Remove the cylinder head cover.

7. Matchmark and remove the rocker arms, cross heads and push rods.

➡The #5 cylinder intake and the #6 cylinder intake and exhaust pushrods are removed by lifting them up and through the provided cowl panel access holes. Remove the rubber plugs to expose these relief holes.

8. Raise the valve lifters as follows, using the wooden dowel rods provided with the Miller Tool 8502 Tappet Replacement Kit.

a. Insert the slotted end of the dowel rod into the valve lifter. The dowel rods for the rear two cylinders will have to be cut for cowl panel clearance. Press firmly to ensure that it is seated in the valve lifter.

b. Raise the dowel rod to bring the valve lifter to the top of its travel, and wrap a rubber band around the dowel rods to prevent the lifters from dropping into the crankcase.

c. Repeat this procedure for the remaining cylinders.

9. Verify that the camshaft timing marks are lined up with the crankshaft mark.

10. Remove the bolts from the thrust plate.

11. Remove the engine mount bolts

12. Install engine support fixture special tool # 8534 and steel bracket/wing nut special tool #8534A. Raise the engine enough to allow for camshaft removal.

13. Remove the camshaft, gear and thrust plate.

14. Remove the valve lifters as follows, using the trough provided with the Miller Tool 8502 Tappet Replacement Kit.

a. Insert the trough the full length of the camshaft bore. Make sure the cap end goes in first and the open side faces up.

b. Remove only one tappet at a time. Remove rubber band from one cylinder pair and

c. Attach valve lifter dowel not being removed to the next cylinder pair. Raise dowel rod (disengage from lifter) and allow lifter to fall into trough.

d. Carefully remove trough (do not rotate) and lifter. If the lifter is not being

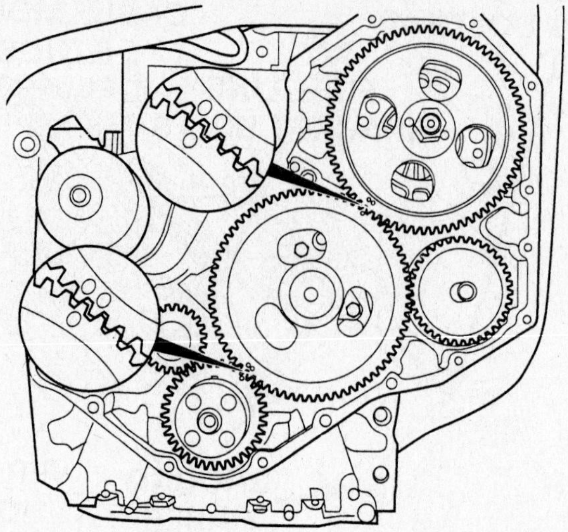
9302DG06

Timing gear alignment marks—5.9L diesel engine

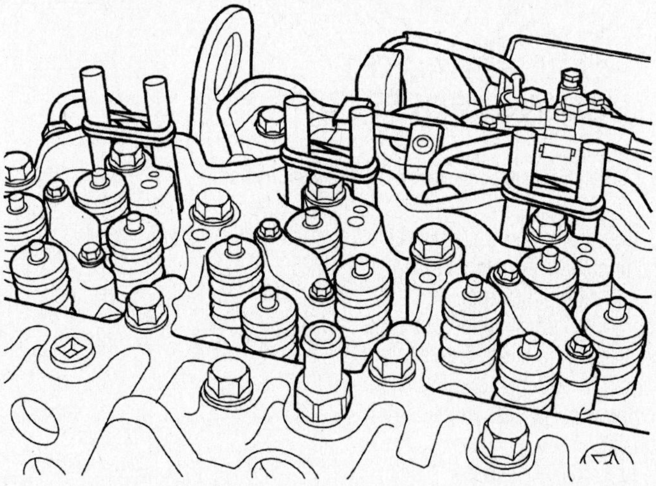
06009-RAMT-G18

Use dowel tools and rubber bands to hold the lifters up in the bore while removing the camshaft—5.9L diesel engine.

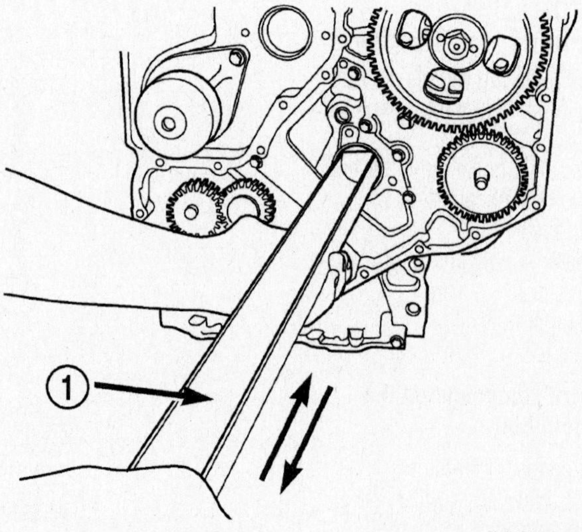

06009-RAMT-G19

Insert the trough to remove lifters.

replaced, mark it so it can be installed in its original location.

e. Reinstall trough and repeat procedure on remaining lifters.

To install:

15. Install the valve lifters as follows, using the installation tools provided with the Miller Tool 8502 Tappet Replacement Kit.

a. Insert the trough the full length of the camshaft bore. Make sure the cap end goes in first and the open side faces up.

b. Lower the lifter installation tool through the push rod hole and into the trough.

c. Retrieve the lifter installation tool using the hooked rod provided with the tool kit

d. Lubricate the lifter with clean engine oil or suitable equivalent and install the lifter to the installation tool.

e. Pull the lifter up and into position. If difficulty is experienced getting the lifter to make the turn into the valve lifter bore, wiggle the trough while gently pulling up on the lifter.

f. With the valve lifter in place, rotate the trough one half turn so the open side is down (toward crankshaft).

g. Remove the tappet installation tool from the tappet.

h. Reinstall a dowel rod and secure the rod with a rubber band.

i. Rotate the trough one half turn and repeat the procedure for the remaining lifters.

16. Install the camshaft and thrust plate. Align the timing marks and tighten the bolts to 18 ft. lbs. (24 Nm).

17. Remove the wooden dowels and rubber bands from valve lifters.

18. Lubricate the push rods with engine oil and install in their original location.

19. Lubricate the valve tips with engine oil and install the crossheads in their original locations.

20. Lubricate the crossheads and push rod sockets with engine oil and install the rocker arms and pedestals in their original locations. Tighten bolts to 27 ft. lbs. (36 Nm).

21. Verify valve lash adjustment

22. Install or connect the following:

- Cylinder head cover
- Gear housing cover
- Vibration damper
- Fan support/hub assembly
- Accessory drive belt tensioner. Tighten to 32 ft. lbs. (43 Nm).
- Accessory drive belt
- Charge air cooler

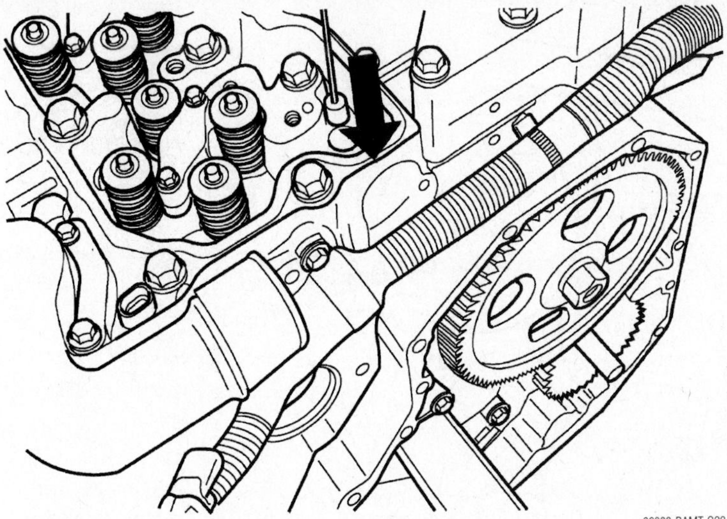

06009-RAMT-G20

Lower the lifter installation tool through the pushrod hole.

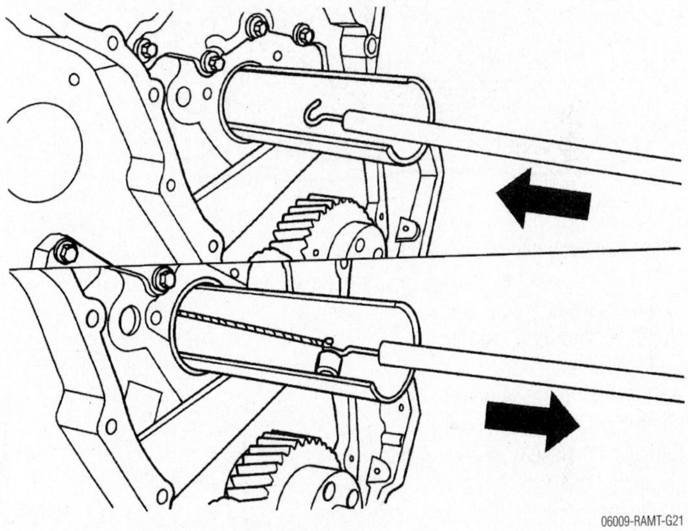

06009-RAMT-G21

Insert the hooked rod to retrieve the installation tool.

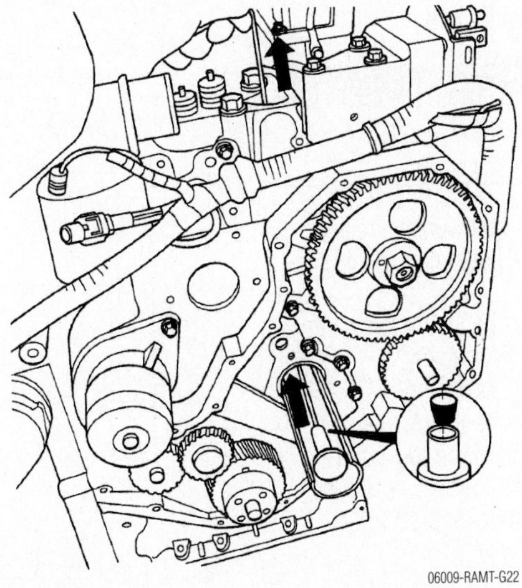

06009-RAMT-G22

Pulling the lifter into position.

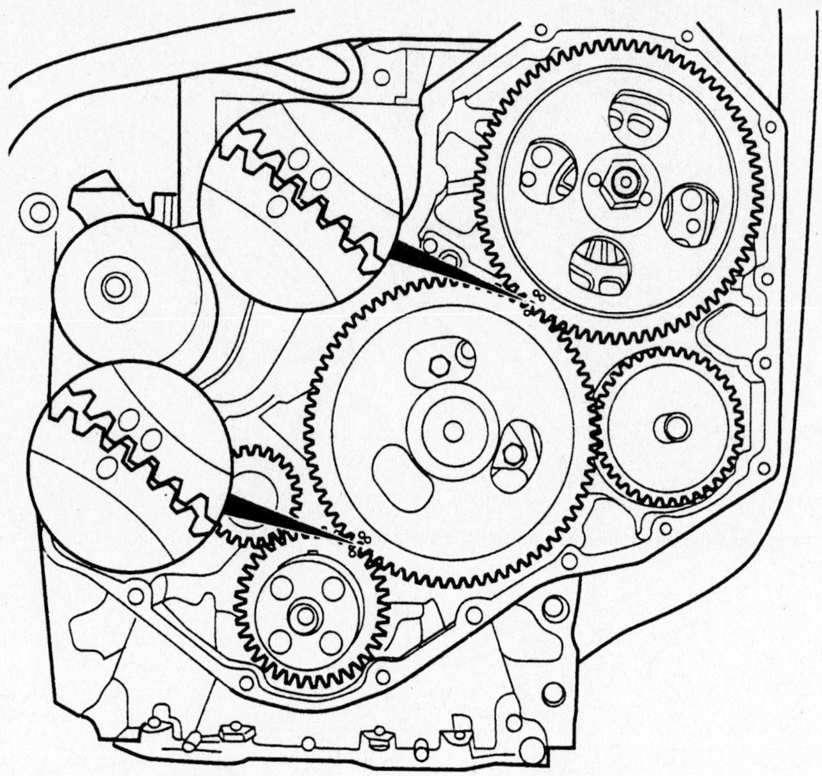

9302DG06

Timing gear alignment marks—5.9L diesel engine

- A/C condenser, if equipped
- Upper radiator support panel
- Radiator
- Automatic transmission auxiliary oil cooler lines, if equipped
- Radiator fan assembly and shroud
- Negative battery cables

23. Fill the cooling system to the correct level.
24. Charge the A/C system with refrigerant.
25. Start engine and check for leaks.

Valve Lash

ADJUSTMENT

1. Before servicing the vehicle, refer to the precautions in the beginning of this section.
2. Remove or disconnect the following:
   - Negative battery cables
   - Cylinder head cover
3. Position the gear as shown and measure the clearance of the indicated valves. No adjustment is necessary if the lash falls within the following specifications:
   a. Intake—0.006–0.015 inch (0.152–0.381mm).
   b. Exhaust—0.015–0.030 inch (0.381–0.762mm).
4. If adjustment is necessary:
   a. Loosen the lock nut on the rocker arms and turn the adjusting screw until the desired lash is obtained.
   - Intake—0.010 in (0.254 mm)
   - Exhaust—0.020 in (0.508 mm)

   b. Tighten the lock nut to 18 ft. lbs. (24 Nm) and recheck the valve lash.
5. Install or connect the following:
   - Cylinder head cover
   - Negative battery cables

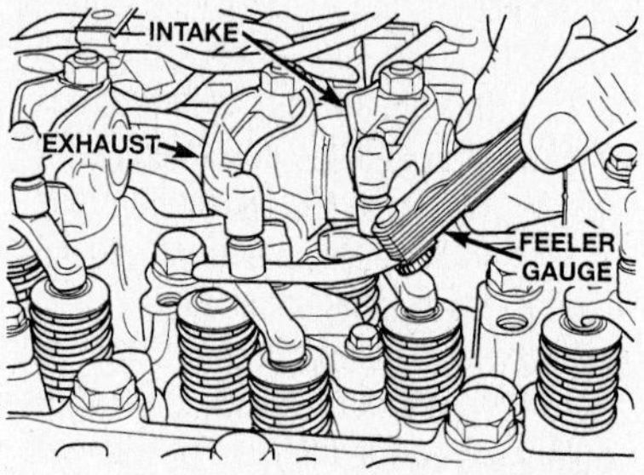

9302DG04

Use a feeler gauge to measure the valve lash—diesel engine

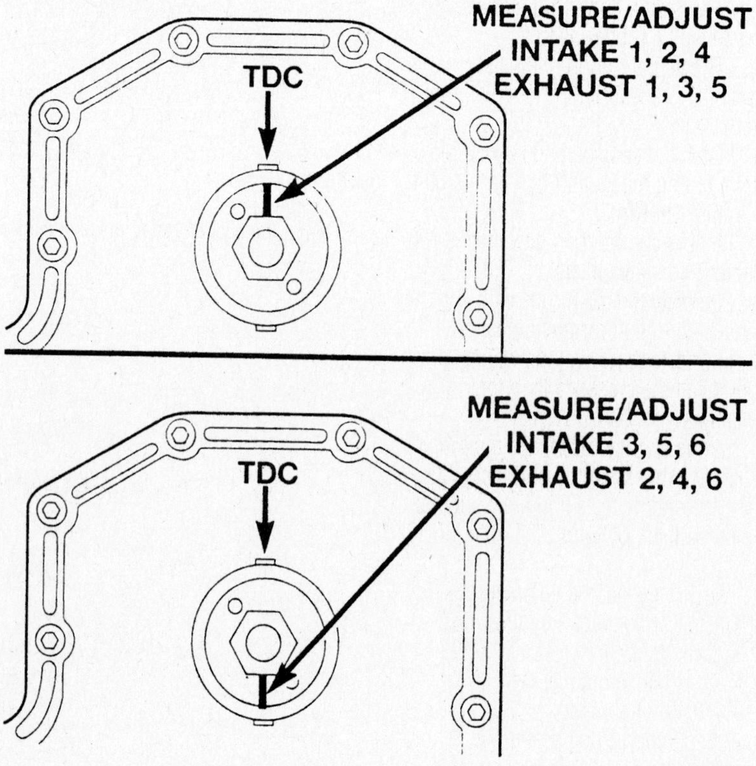

9302DG03

Adjust the specified valves when the mark on the pump gear is in either of the 2 positions—diesel engines

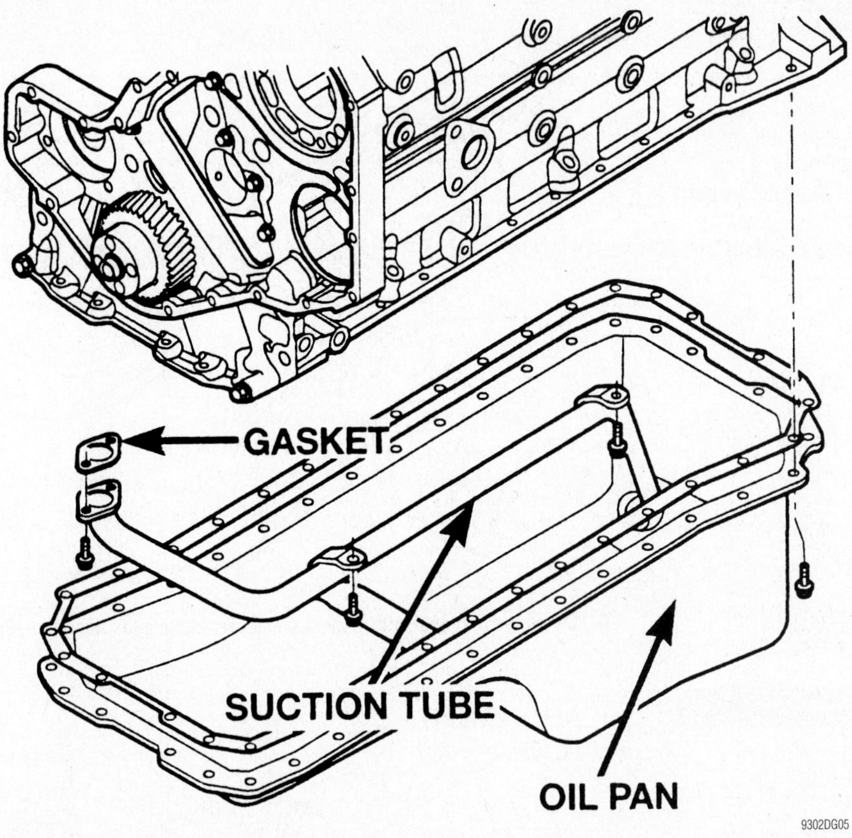

Exploded view of the oil pan mounting—diesel engines

Oil Pan

REMOVAL & INSTALLATION

1. Before servicing the vehicle, refer to the precautions in the beginning of this section.
2. Drain the engine oil.
3. Remove or disconnect the following:
   - Negative battery cables
   - Starter motor
   - Transmission
   - Transfer case, if equipped
   - Flywheel and transmission adapter plate.
   - Oil pan bolts
   - Oil pump suction tube
   - Oil pan

To install:

4. Install or connect the following:
   - Oil pan
   - Oil pump suction tube. Tighten the bolts to 18 ft. lbs. (24 Nm).
   - Oil pan bolts. Tighten the bolts to 21 ft. lbs. (28 Nm).
   - Flywheel and transmission adapter plate.
   - Transfer case, if equipped
   - Transmission
   - Starter motor
5. Fill the crankcase to the correct level.

Oil Pump

REMOVAL & INSTALLATION

1. Before servicing the vehicle, refer to the precautions in the beginning of this section.
2. Drain the cooling system.
3. Remove or disconnect the following:
   - Negative battery cables
   - Radiator fan assembly and shroud
   - Accessory drive belt
   - Radiator
   - Vibration damper

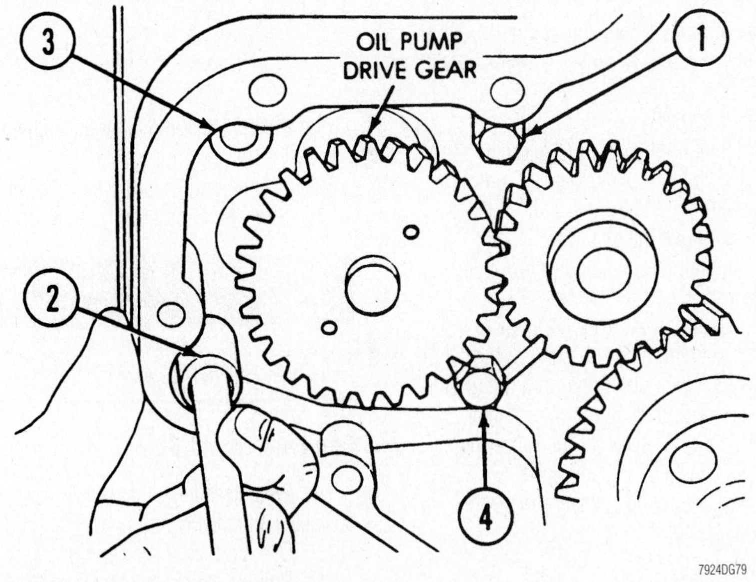

Oil pump torque sequence—5.9L diesel engine

- Power steering pump
- Accessory drive belt tensioner
- Front cover
- Oil pump

To install:

➡When the pump is correctly installed, the flange on the pump does not touch the block; the back plate on the pump seats against the bottom of the bore.

4. Install the oil pump. Tighten the bolts in sequence as follows:
   a. Step 1: 44 inch lbs. (5 Nm).
   b. Step 2: 18 ft. lbs. (24 Nm).
5. Install or connect the following:
   - Front cover
   - Vibration damper
   - Accessory drive belt tensioner
   - Accessory drive belt
   - Radiator
   - Radiator fan assembly and shroud
   - Negative battery cables

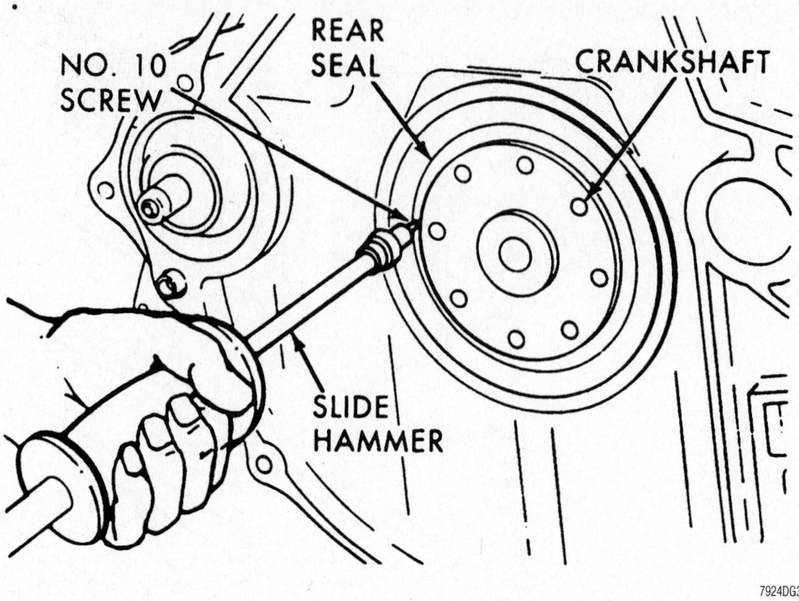

Removing the rear seal with a sheet metal screw and slide hammer—5.9L diesel engine

Rear Main Seal

REMOVAL & INSTALLATION

1. Before servicing the vehicle, refer to the precautions in the beginning of this section.
2. Remove or disconnect the following:
   - Negative battery cables
   - Transmission
   - Clutch and pressure plate, if equipped
   - Flywheel
3. Drill holes 180° apart into the seal.

➡Be careful not to contact the drill against the crankshaft.

4. Install #10 sheet metal screws in the drilled holes and remove the rear seal with a slide hammer.

To install:

5. Install the seal pilot and new seal, provided in the replacement kit, onto the crankshaft.
6. Remove the seal pilot.
7. Install the installation tool over crankshaft.
8. Using a ball peen hammer or suitable equivalent, strike the tool at the 12, 3, 6, and 9 o'clock positions until the alignment tool bottoms out on the retainer.
9. Install or connect the following:
   - Flywheel
   - Clutch and pressure plate, if equipped
   - Transmission
   - Negative battery cables

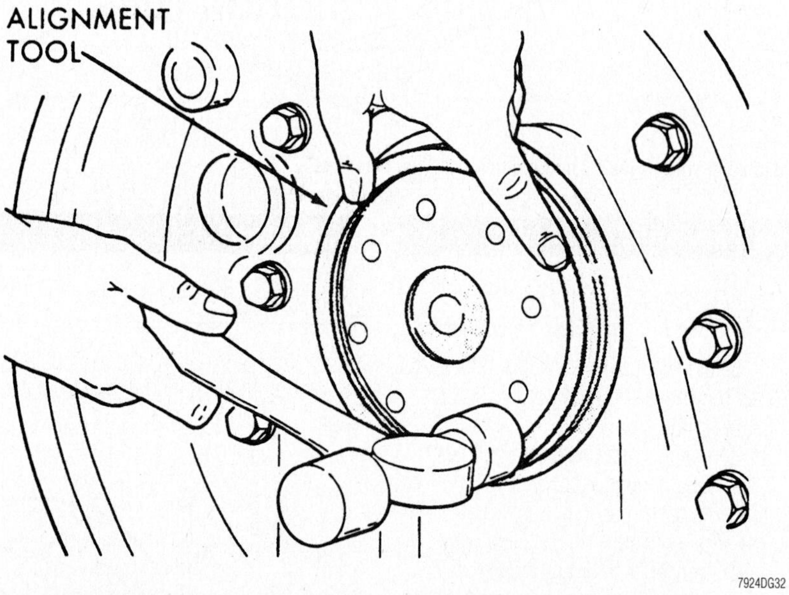

Place the alignment tool on the seal and tap the seal into place—5.9L diesel engine

10. Fill the engine oil to the correct level.
11. Start the engine and check for leaks.

Timing Gears, Front Cover and Seal

REMOVAL & INSTALLATION

Front Cover and Seal

1. Before servicing the vehicle, refer to the precautions in the beginning of this section.
2. Drain the cooling system.
3. Remove or disconnect the following:
   - Negative battery cables
   - Upper radiator hose
   - Cooling fan assembly and shroud
   - Accessory drive belt
   - Cooling fan support
   - Crankshaft pulley
   - Fan shroud brackets
   - Power steering pump
   - Accessory drive belt tensioner
4. Remove the cover-to-housing bolts and gently pry the cover away from the housing.
5. Remove or disconnect the following:
   - Fuel injection pump
   - Camshaft position sensor connector
   - Engine speed sensor

- Camshaft
- Six front oil pan fasteners
- Front cover fasteners

6. Slide a feeler gauge, or suitable equivalent, between the cover and oil pan gasket to break the seal.
7. Remove the front cover and gasket.

To install:

➡If the oil pan gasket is damaged, it must be replaced.

8. Install or connect the following:
   - Front crankshaft seal
   - Front cover. Tighten the bolts to 18 ft. lbs. (24 Nm).
   - Camshaft position sensor
   - Engine speed sensor
   - Fuel injection pump
   - Camshaft
   - Crankshaft pulley
   - Fan shroud brackets
   - Cooling fan support. Tighten the bolts to 24 ft. lbs. (32 Nm).
   - Power steering pump
   - Accessory drive belt tensioner. Tighten the bolts to 32 ft. lbs. (43 Nm).
   - Accessory drive belt
   - Cooling fan assembly and shroud
   - Upper radiator hose
   - Negative battery cables
9. Fill the cooling system to the correct level.
10. Start the engine and check for leaks.

Timing Gears

1. Before servicing the vehicle, refer to the precautions in the beginning of this section.
2. Remove or disconnect the following:
   - Negative battery cables
   - Accessory drive belt
   - Cooling fan and shroud
   - Belt tensioner
   - Oil fill tube and adapter
   - Crankshaft pulley
   - Front cover
   - Camshaft
3. Press the camshaft out of the timing gear.

To install:

4. Install the camshaft key.
5. Heat the timing gear in an oven to 350°F (177°C) for 45 minutes.

➡The camshaft gear will be permanently distorted if overheated. Do not exceed 350°F (177°C).

6. Install the timing gear to the camshaft with the timing marks facing out and the gear seated on the camshaft shoulder.
7. Install or connect the following:

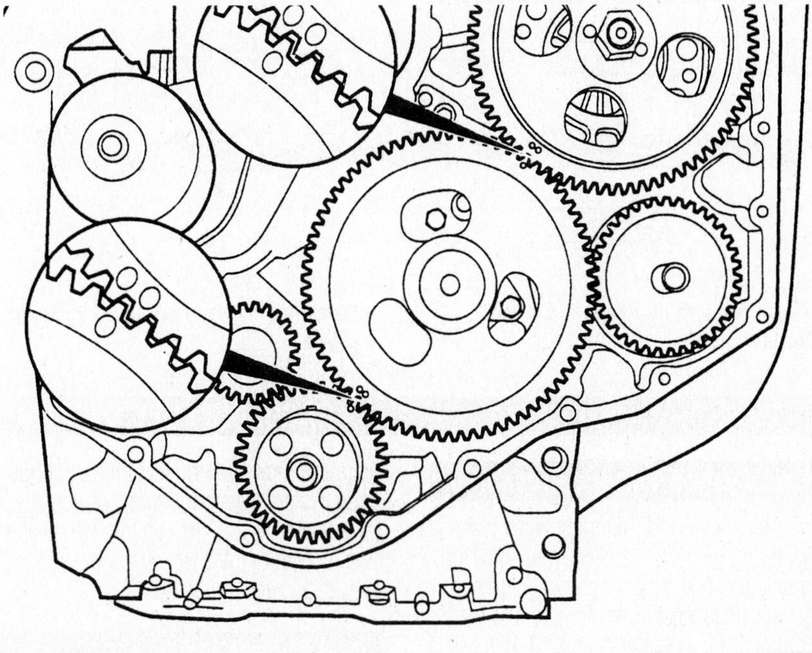

9302DG06

Timing gear alignment marks—5.9L diesel engine

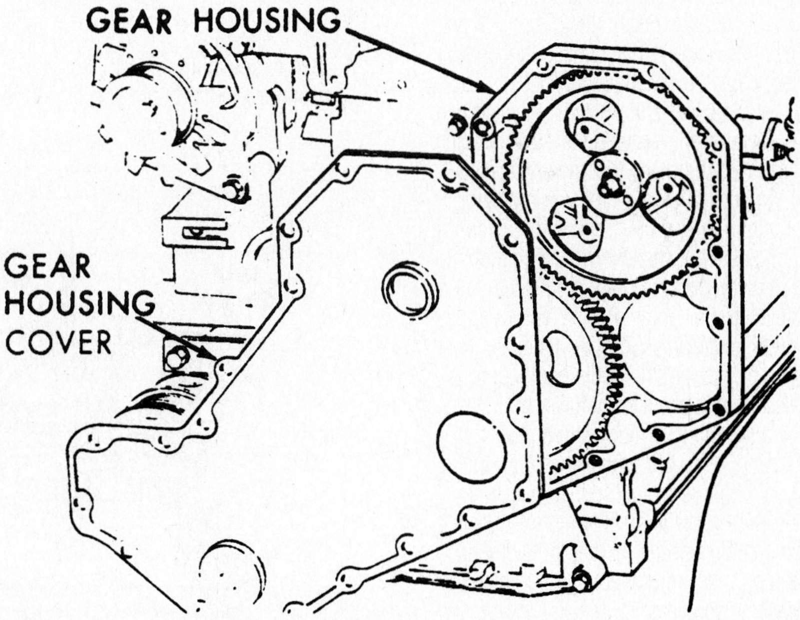

7924DG81

Remove the gear cover to replace the timing gears—5.9L diesel engine

- Camshaft with the timing marks aligned
- Front cover. Tighten the bolts to 18 ft. lbs. (24 Nm).
- Crankshaft pulley. Tighten the bolt to 92 ft. lbs. (125 Nm).
- Oil fill tube and adapter. Tighten the mounting bolts to 32 ft. lbs. (43 Nm).
- Belt tensioner. Tighten the mounting bolts to 32 ft. lbs. (43 Nm).
- Cooling fan and shroud
- Accessory drive belt
- Negative battery cables

Piston and Ring

POSITIONING

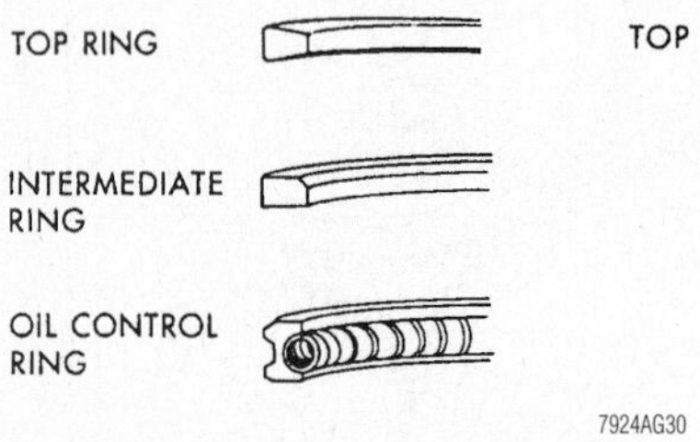

Piston ring identification—5.9L diesel engine

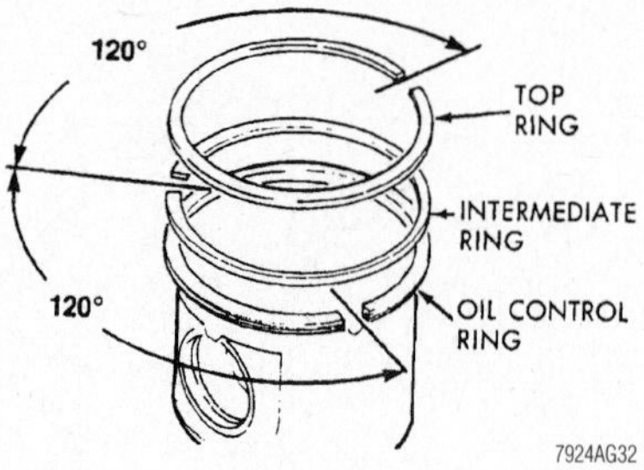

Piston ring end gap spacing—5.9L diesel engine

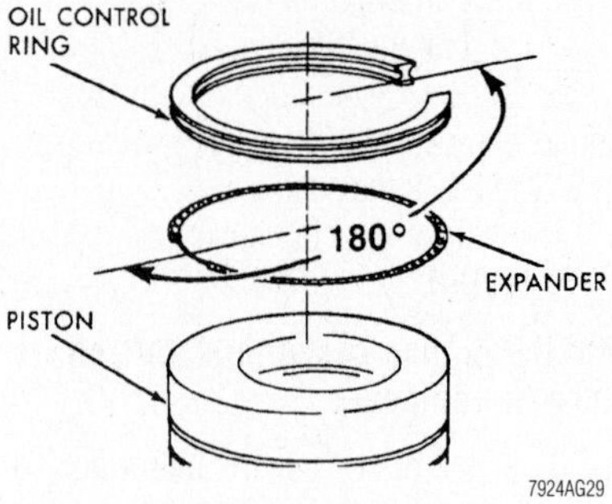

Oil control ring-to-spacer end gap spacing—5.9L diesel engine

GASOLINE FUEL SYSTEM

Fuel System Service Precautions

Safety is the most important factor when performing not only fuel system maintenance but any type of maintenance. Failure to conduct maintenance and repairs in a safe manner may result in serious personal injury or death. Maintenance and testing of the vehicle's fuel system components can be accomplished safely and effectively by adhering to the following rules and guidelines.

- To avoid the possibility of fire and personal injury, always disconnect the negative battery cable unless the repair or test procedure requires that battery voltage be applied.
- Always relieve the fuel system pressure prior to detaching any fuel system component (injector, fuel rail, pressure regulator, etc.), fitting or fuel line connection. Exercise extreme caution whenever relieving fuel system pressure to avoid exposing skin, face and eyes to fuel spray. Please be advised that fuel under pressure may penetrate the skin or any part of the body that it contacts.
- Always place a shop towel or cloth around the fitting or connection prior to loosening to absorb any excess fuel due to spillage. Ensure that all fuel spillage (should it occur) is quickly removed from engine surfaces. Ensure that all fuel soaked cloths or towels are deposited into a suitable waste container.
- Always keep a dry chemical (Class B) fire extinguisher near the work area.
- Do not allow fuel spray or fuel vapors to come into contact with a spark or open flame.
- Always use a back-up wrench when loosening and tightening fuel line connection fittings. This will prevent unnecessary stress and torsion to fuel line piping.
- Always replace worn fuel fitting O-rings with new. Do not substitute fuel hose or equivalent, where fuel pipe is installed.

Before servicing the vehicle, also make sure to refer to the precautions in the beginning of this section as well.

Fuel System Pressure

RELIEVING

2002

1. Before servicing the vehicle, refer to the precautions in the beginning of this section.
2. Disconnect the negative battery cable.
3. Remove the fuel tank filler cap to release any fuel tank pressure.
4. Unscrew the plastic cap from the pressure test port on the fuel rail. On the 8.0L engine, the test port is found at the front of the engine.
5. Obtain a fuel pressure gauge/hose from a Fuel Pressure Gauge tool set No. 5069, or equivalent. Remove the gauge, then place the gauge end of the hose into a suitable gasoline container.
6. Place a shop towel under the test port.
7. Screw the other end of the hose onto

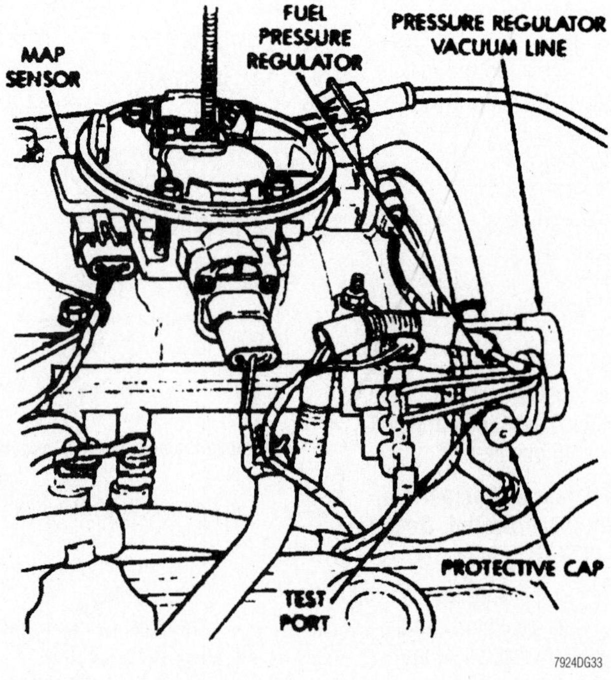

Fuel pressure test port—3.9L engine

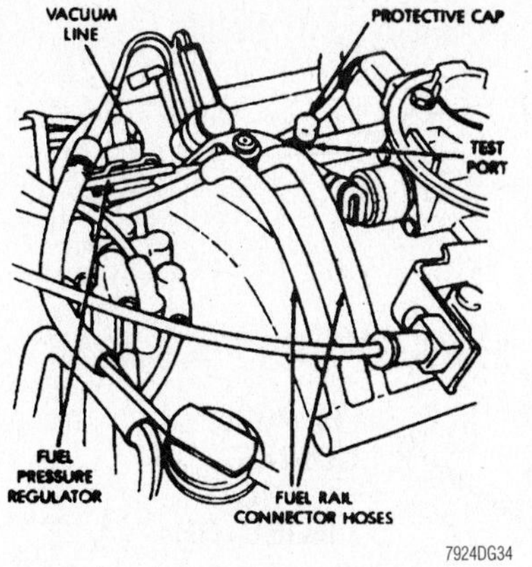

Fuel pressure test port—5.2L engine, 5.9L engine is similar

the fuel pressure port to relieve the pressure.

8. When the pressure has been relieved, remove the hose and cap the port.

2003–05

1. Before servicing the vehicle, refer to the precautions in the beginning of this section.
2. Disconnect the negative battery cable.
3. Remove the fuel tank filler cap to release any fuel tank pressure.
4. Remove the fuel pump relay from the PDC.
5. Start and run the engine until it stops.
6. Unplug the connector from any injector and connect a jumper wire from either injector terminal to the positive battery terminal. Connect another jumper wire to the other terminal and momentarily touch the other end to the negative battery terminal.

**** WARNING**

Just touch the jumper to the battery. Powering the injector for more than a few seconds will permanently damage it.

7. Place a rag below the quick-disconnect coupling at the fuel rail and disconnect it.

Fuel Filter

REMOVAL & INSTALLATION

1. Before servicing the vehicle, refer to the precautions in the beginning of this section.
2. Relieve the fuel system pressure.
3. Remove or disconnect the following:
   - Negative battery cable

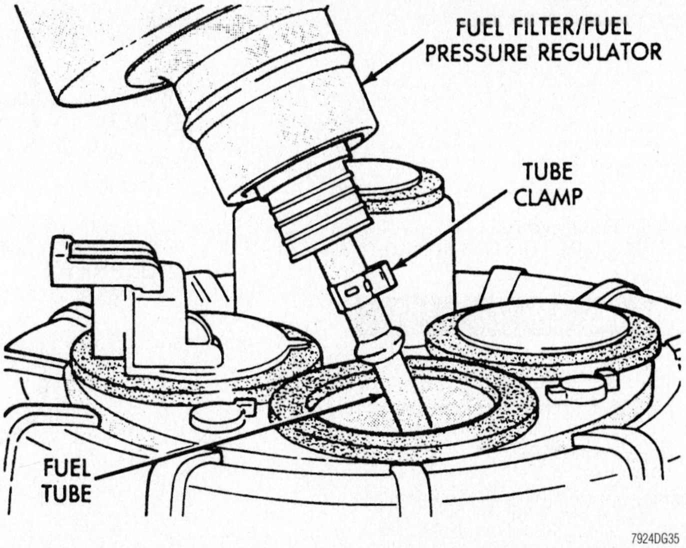

Pull and twist the filter/regulator to remove it from the top of the fuel pump module

   - Fuel tank
   - Fuel pump module locknut, using Special Tool 6856
   - Fuel pump module
4. Pull the filter/regulator out of the rubber grommet. Cut the hose clamp and remove the fuel line.

To install:

5. Install the filter/regulator with a new clamp and push it into the rubber grommet.
6. Install or connect the following:
   - Fuel pump module
   - Fuel pump module locknut, using Special Tool 6856
   - Fuel tank
   - Negative battery cable
7. Start the engine and check for leaks.

Fuel Pump

REMOVAL & INSTALLATION

1. Before servicing the vehicle, refer to the precautions in the beginning of this section.
2. Relieve the fuel system pressure and drain the fuel tank.
3. Remove or disconnect the following:
   - Negative battery cable
   - Fuel pump module harness connector
   - Fuel line
   - Fuel tank
   - Fuel pump module locknut, using Special Tool 6856
   - Fuel pump module

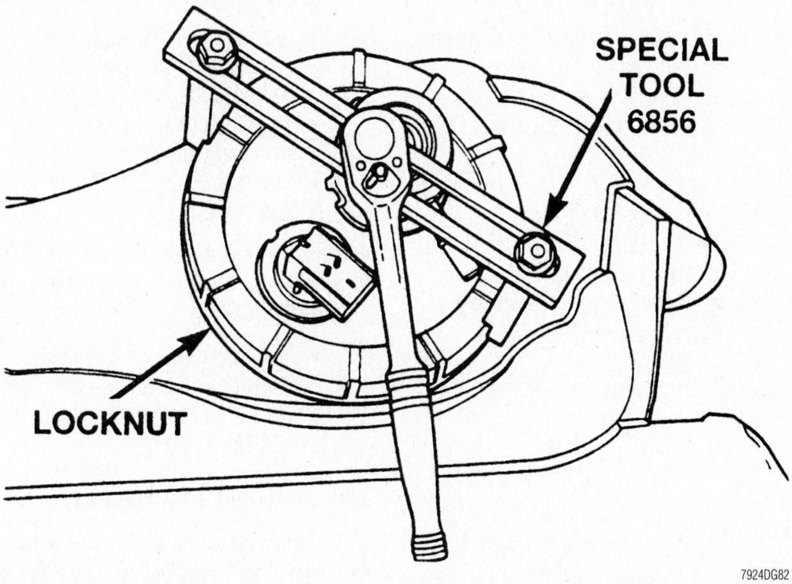

Fuel pump module locknut removal

To install:

4. Install or connect the following:
   - Fuel pump module
   - Fuel pump module locknut, using Special Tool 6856
   - Fuel tank
   - Fuel line
   - Fuel pump module harness connector
   - Negative battery cable
5. Start the engine and check for leaks.

Fuel Injector

REMOVAL & INSTALLATION

3.7L and 4.7L Engines

1. Before servicing the vehicle, refer to the precautions in the beginning of this section.

➡The left and right fuel rails are replaced as an assembly. Do not attempt to separate rail halves at connector tubes.

2. Relieve fuel system pressure.
3. Remove or disconnect the following:
   - Negative battery cable
   - Air intake assembly
   - Fuel line
   - Throttle body vacuum lines and electrical connectors
   - Fuel injector connectors
   - Manifold Absolute Pressure (MAP) sensor connector
   - Intake Air Temperature (IAT) sensor connector
   - Ignition coils
   - Fuel rail with injectors attached
   - Fuel injectors

To install:

4. Install or connect the following:
   - Fuel injectors to the fuel rail, using new O-rings
   - Fuel rail with injectors attached. Tighten the bolts to 20 ft. lbs. (27 Nm).
   - Ignition coils
   - IAT sensor connector
   - MAP sensor connector
   - Fuel injector connectors
   - Throttle body vacuum lines and electrical connectors
   - Fuel line
   - Alternator wiring connectors
   - Air intake assembly
   - Negative battery cable
5. Start the engine and check for leaks.

5.7L Engines

1. Before servicing the vehicle, refer to the precautions in the beginning of this section.

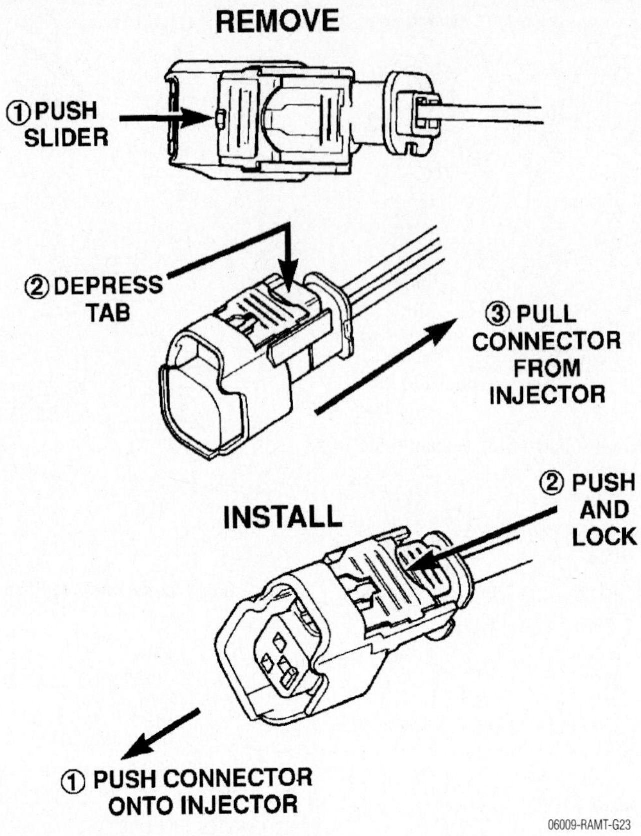

Fuel Injector connectors—3.7L and 4.7L Engines

2. Relieve fuel system pressure.
3. Remove or disconnect the following:
   - Negative battery cable
   - Air intake assembly
   - Spark plug wires at the plugs and coils

➡Do not remove the wires from the routing tray.

   - Routing tray
   - Ignition coil connectors
   - Injector wires and matchmark original locations
   - Throttle body vacuum lines and electrical connectors
   - Fuel rail mounting bolts and clamps
   - Fuel rail with injectors attached

➡Rock the left end of the rail until the injectors just clear, then rock the right side until clear.

   - Fuel injectors

4. Installation is the reverse of removal. Torque the fuel rail mounting bolts to 100 inch lbs. (11 Nm).

3.9L, 5.2L and 5.9L Engines

1. Before servicing the vehicle, refer to the precautions in the beginning of this section.
2. Relieve fuel system pressure.
3. Remove or disconnect the following:
   - Negative battery cable
   - Air intake tube
   - Throttle body
   - A/C compressor bracket
   - Fuel injector connectors
   - Fuel line
   - Fuel supply manifold with injectors
   - Fuel injectors

To install:

4. Install or connect the following:
   - Fuel injectors with new O-ring seals
   - Fuel supply manifold with injectors. Tighten the bolts to 17 ft. lbs. (23 Nm).
   - Fuel line
   - Fuel injector connectors
   - A/C compressor bracket
   - Throttle body
   - Air intake tube
   - Negative battery cable
5. Start the engine and check for leaks.

8.0L Engine

1. Before servicing the vehicle, refer to the precautions in the beginning of this section.

2. Relieve the fuel system pressure.
3. Remove or disconnect the following:
   - Negative battery cable
   - Air cleaner housing and tube
   - Throttle body
   - Ignition coil pack and bracket
   - Upper intake manifold
   - Fuel injector harness connectors
   - Fuel line
   - Fuel supply manifold with injectors attached
   - Fuel injector retainer clips
   - Fuel injectors

To install:

4. Install or connect the following:
   - Fuel injectors with new O-ring seals
   - Fuel injector retainer clips
   - Fuel supply manifold. Tighten the bolts to 11 ft. lbs. (15 Nm).
   - Fuel line
   - Fuel injector harness connectors
   - Upper intake manifold
   - Ignition coil pack and bracket
   - Throttle body
   - Air cleaner housing and tube
   - Negative battery cable
5. Start the engine and check for leaks.

DIESEL FUEL SYSTEM

Fuel System Service Precautions

Safety is the most important factor when performing not only fuel system maintenance, but any type of maintenance. Failure to conduct maintenance and repairs in a safe manner may result in serious personal injury or death. Maintenance and testing of the vehicle's fuel system components can be accomplished safely and effectively by adhering to the following rules and guidelines.

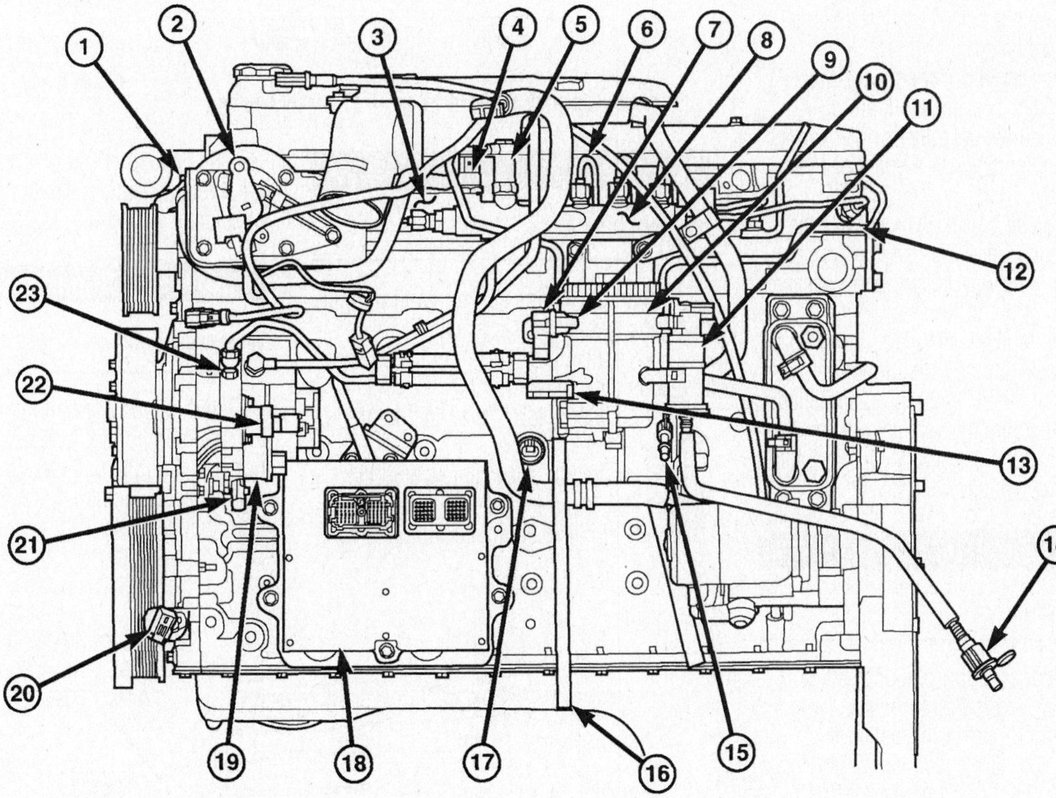

1 - ENGINE COOLANT TEMPERATURE (ECT) SENSOR
2 - THROTTLE LEVER BELLCRANK AND APPS (ACCELERATOR PEDAL POSITION SENSOR)
3 - INTAKE MANIFOLD AIR HEATER/ELEMENTS
4 - FUEL PRESSURE SENSOR
5 - FUEL PRESSURE LIMITING VALVE
6 - HIGH-PRESSURE FUEL LINES
7 - FUEL HEATER
8 - HIGH-PRESSURE FUEL RAIL
9 - FUEL HEATER TEMPERATURE SENSOR (THERMOSTAT)
10 - FUEL FILTER/WATER SEPARATOR
11 - FUEL TRANSFER (LIFT) PUMP
12 - FUEL DRAIN MANIFOLD (CYLINDER HEAD FUEL RETURN LINE)
13 - DRAIN VALVE
14 - FUEL SUPPLY LINE (LOW-PRESSURE, TO ENGINE)
15 - FUEL RETURN LINE CONNECTION (TO FUEL TANK)
16 - FUEL DRAIN TUBE
17 - OIL PRESSURE SWITCH
18 - ENGINE CONTROL MODULE (ECM)
19 - FUEL INJECTION PUMP
20 - CRANKSHAFT POSITION (ENGINE SPEED) SENSOR
21 - CAMSHAFT POSITION SENSOR (CMP)
22 - FUEL CONTROL ACTUATOR (FCA)
23 - CASCADE OVERFLOW VALVE

67189-RAMT-G08

5.9L Diesel fuel system components—2004 shown, other years similar

• To avoid the possibility of fire and personal injury, always disconnect the negative battery cable unless the repair or test procedure requires that battery voltage be applied.

• Always relieve the fuel system pressure prior to disengaging any fuel system component (injector, fuel rail, pressure regulator, etc.), fitting or fuel line connection. Exercise extreme caution whenever relieving fuel system pressure, to avoid exposing skin, face and eyes to fuel spray. Please be advised that fuel under pressure may penetrate the skin or any part of the body that it contacts.

• Always place a shop towel or cloth around the fitting or connection prior to loosening to absorb any excess fuel due to spillage. Ensure that all fuel spillage (should it occur) is quickly removed from engine surfaces. Ensure that all fuel soaked cloths or towels are deposited into a suitable waste container.

• Always keep a dry chemical (Class B) fire extinguisher near the work area.

• Do not allow fuel spray or fuel vapors to come into contact with a spark or open flame.

• Always use a back-up wrench when loosening and tightening fuel line connection fittings. This will prevent unnecessary stress and torsion to fuel line piping.

• Always replace worn fuel fitting O-rings with new. Do not substitute fuel hose or equivalent, where fuel pipe is installed.

Before servicing the vehicle, also make sure to refer to the precautions in the beginning of this section as well.

Fuel System

BLEEDING AIR

The fuel transfer (lift) pump is self-priming: When the key is first turned on (without cranking engine), the pump operates for approximately 1 to 2 second and then shuts off. The pump will also operate for up to 25 seconds after the starter is quickly engaged, and then disengaged without allowing the engine to start. The pump shuts off immediately if the key is on and the engine stops running.

1. Turn key to **CRANK** position and quickly release key to **ON** position before engine starts. This will operate fuel transfer pump for approximately 25 seconds.
2. Crank engine. If the engine does not start after 25 seconds, turn key to **OFF** position, and leave it off for at least 5 seconds. Repeat previous step until engine starts.
3. Fuel system priming is now complete.

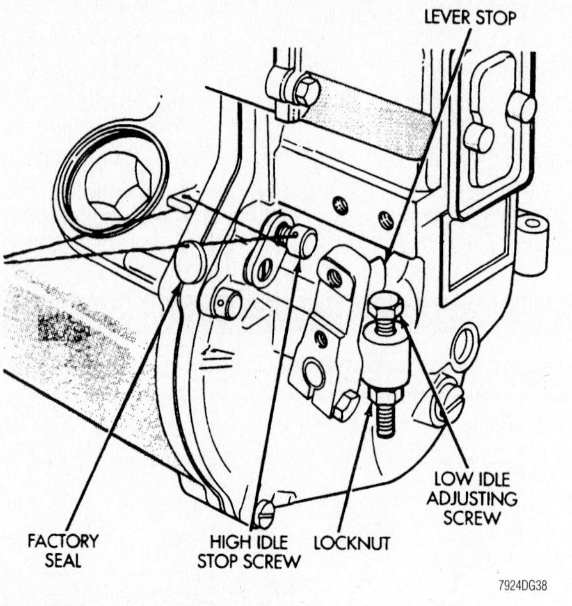

Idle speed adjusting screw location—5.9L diesel engine

Idle Speed

ADJUSTMENT

1. Start the engine and run until normal operating temperature is reached.
2. An optical tachometer must be used to read engine speed.
3. If equipped, turn the air conditioning **ON**.
4. Turn the idle speed screw until the desired idle speed is obtained. The specification for a vehicle equipped with automatic transmission is 700 rpm. The specification for a vehicle equipped with manual transmission is 750 rpm.

Fuel Water Separator/Filter

DRAINING WATER

Filtration and separation of water from the fuel is important for trouble-free operation and long life of the fuel system. Regular maintenance, including draining moisture from the fuel/water separator filter is essen-

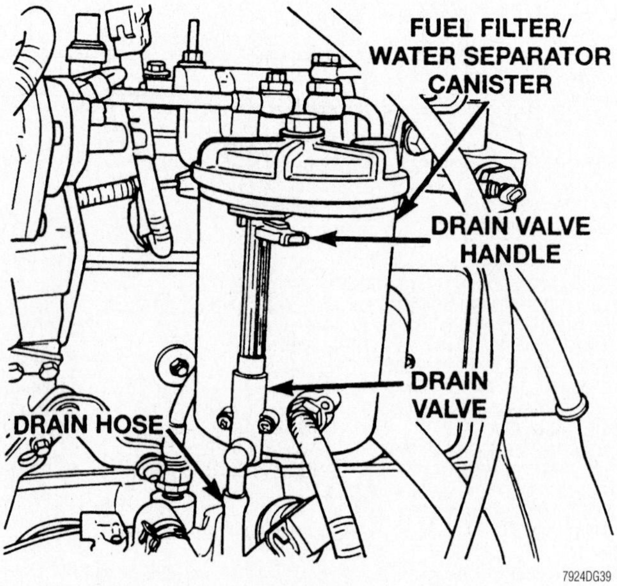

Fuel filter/water separator assembly and related components

tial to keep water out of the fuel pump. To remove the collected water, unscrew the drain at the bottom of the Water-In-Filter (WIF) assembly located at the bottom of the filter separator.

REMOVAL & INSTALLATION

1. Before servicing the vehicle, refer to the precautions in the beginning of this section.
2. Remove or disconnect the following:
   - Negative battery cables
   - Water In Filter (WIF) sensor connector
   - Separator filter assembly

To install:

3. Install or connect the following:
   - Separator filter assembly with a new O-ring seal
   - WIF sensor connector
   - Negative battery cables
4. Bleed air from the system.
5. Start the engine and check for leaks.

Diesel Injection Pump

REMOVAL & INSTALLATION

2001–2003

➡The Bosch VE lever is indexed to the shaft during pump calibration. Do not remove it from the pump during removal.

1. Before servicing the vehicle, refer to the precautions in the beginning of this section.
2. Remove or disconnect the following:
   - Negative battery cables
   - Throttle linkage and bracket
   - Fuel drain manifold
   - Injection pump supply line
   - High pressure lines
   - Fuel air control tube
   - Fuel shut off valve connector
   - Pump support bracket
   - Oil fill tube and adapter
3. Place a shop towel in the gear cover opening in a position that will prevent the nut and washer from falling into the gear housing. Remove the gear retaining nut and washer.
4. Turn the engine until the keyway on the fuel pump shaft is pointing approximately in the 6 o'clock position.
5. Locate Top Dead Center (TDC) for cylinder No. 1 by turning the engine slowly while pushing in on the TDC pin. Stop turning the engine as soon as the pin engages with the gear timing hole. Disengage the pin after locating TDC and remove the turning equipment.
6. Loosen the lockscrew, remove the special washer from the injection pump and wire it to the line above it so it will not get misplaced. Retighten the lockscrew to 22 ft. lbs. (30 Nm) to lock the driveshaft.
7. Press the pump drive gear from the driveshaft.

➡Be careful not to drop the drive gear key into the front cover when removing or installing the pump. If it does drop in, it must be removed before proceeding.

8. Remove the 3 mounting nuts and remove the injection pump from the vehicle.
9. Remove the gasket and clean the mounting surface.

To install:

➡The shaft of a new or reconditioned pump is locked so the key aligns with the drive gear keyway with cylinder No. 1 at TDC.

10. Install the pump and finger-tighten the mounting nuts; the pump must be free to move in the slots.
11. Install the pump drive gear, washer and nut to the driveshaft. The pump will rotate slightly because of gear helix and clearance. This is acceptable providing the pump is free to move on the flange slots and the crankshaft does not move. Tighten the nut to 11–15 ft. lbs. (15–20 Nm). Do not overtighten.
12. If installing the original pump, rotate the pump to align the original timing marks and tighten the mounting nuts to 18 ft. lbs. (24 Nm).
13. If installing a replacement pump, take up gear lash by rotating the pump counterclockwise toward the cylinder head, and tighten the mounting nuts to 18 ft. lbs. (24 Nm). Permanently mark the new injection pump flange to match the mark on the gear housing.
14. Loosen the lockscrew and install the special washer under the lockscrew; tighten to 13 ft. lbs. (18 Nm). Disengage the TDC pin.
15. Install or connect the following:
   - Pump support bracket. Tighten the pump drive gear nut to 48 ft. lbs. (65 Nm).
   - Oil fill tube and adapter
   - Fuel shut off valve connector
   - Fuel air control tube
   - High pressure lines
   - Injection pump supply line

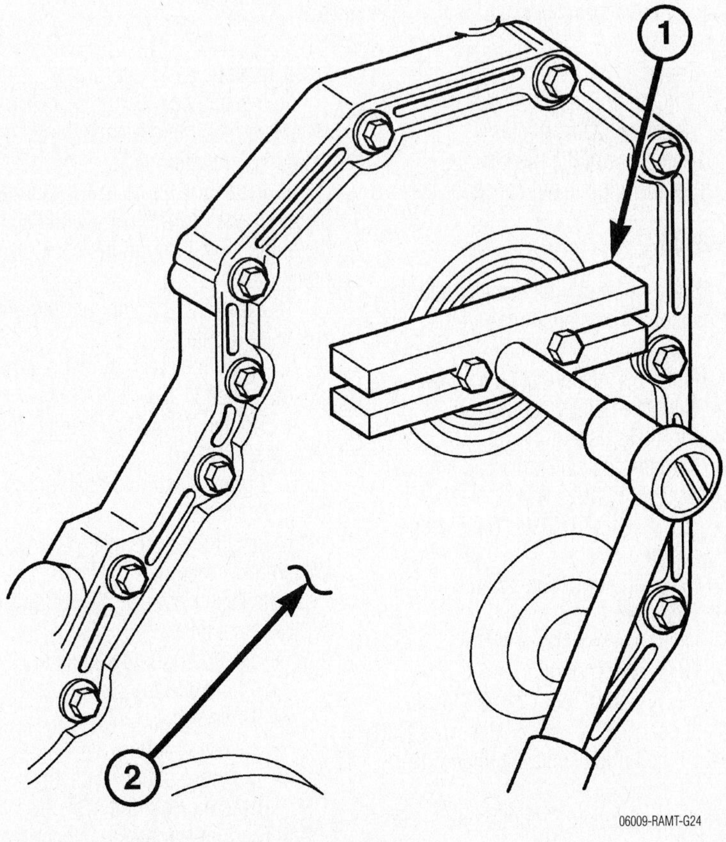

Removing the drive gear using Special Tool C3428B.

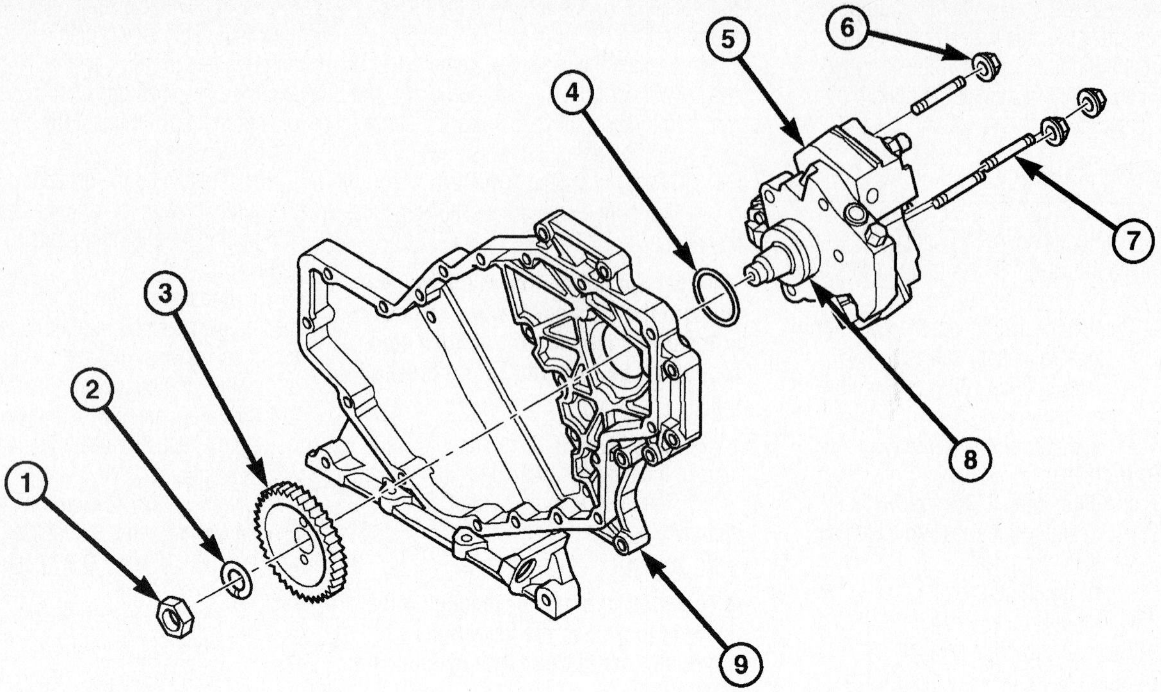

1 - PUMP DRIVE GEAR NUT
2 - WASHER
3 - PUMP DRIVE GEAR
4 - RUBBER O-RING
5 - FUEL INJECTION PUMP
6 - PUMP MOUNTING NUTS (3)
7 - PUMP MOUNTING STUDS (3)
8 - O-RING MACHINED GROOVE
9 - FRONT TIMING GEAR HOUSING

67189-RAMT-G09

Exploded view of diesel injection pump—2004 models

- Fuel drain manifold
- Throttle linkage and bracket
- Negative battery cables

16. Bleed air from the system.
17. Start the engine and check for leaks.

2004–2005

1. Before servicing the vehicle, refer to the precautions in the beginning of this section.
2. Remove or disconnect the following:
   - Negative battery cables
   - Air intake assembly
   - Radiator fan assembly and shroud
   - Accessory drive belt
   - Fuel Control Actuator (FCA) connector
   - Injection pump supply line
   - Fuel lines
   - Drive gear access cover
   - Drive gear mounting nut
3. Using Special Tool C3428B, or equivalent gear puller, remove the drive gear and leave it hanging within the timing gear cover.
4. Remove 3 pump mounting nuts and remove the pump.

To install:

5. Install a new O-ring coated with clean engine oil into the machined groove at the pump mounting area.
6. Install the pump to the mounting flange on the gear housing while aligning the pump shaft through the back of the pump gear.
7. Install the 3 pump mounting nuts and finger tighten.
8. Install the shaft washer and nut and tighten by hand.
9. Tighten the pump mounting nuts to 18 ft. lbs. (25 Nm).
10. Tighten the pump shaft nut to 77 ft. lbs. (105 Nm).
11. Install the drive gear cover and tighten to 71 inch lbs. (8 Nm).
12. Install or connect the following:
    - Fuel lines
    - Injection pump supply line
    - Pump wiring harness
    - Drive belt
    - Intake air tube
    - Negative battery cables
13. Bleed air from the system.
14. Start the engine and check for leaks.

Fuel Injectors

REMOVAL & INSTALLATION

2001–2003

1. Before servicing the vehicle, refer to the precautions in the beginning of this section.
2. Remove or disconnect the following:
   - Negative battery cables
   - High pressure fuel lines
   - Fuel drain manifold
   - Fuel injector. Hold the injector with a backing wrench while loosening the mounting nut.

To install:

3. Use a new copper washer and coat the backing nut threads with anti-seize compound.
4. Install or connect the following:
   - Fuel injector. Tighten the mounting nut to 44 ft. lbs. (60 Nm).
   - Fuel drain manifold. Use new copper gaskets and tighten the fitting screws to 84 inch lbs. (9 Nm).
   - High pressure fuel lines

- Negative battery cables

5. Bleed the fuel system.
6. Start the engine and check for leaks.

2004

1. Before servicing the vehicle, refer to the precautions in the beginning of this section.
2. Remove or disconnect the following:
   - Negative battery cables
   - Cylinder head cover
   - Fuel line connector
   - Wiring harness retainer nuts
3. Use high pressure tool 9015 and remove high pressure connectors.
4. Remove the necessary exhaust rocker arms.
5. Disconnect the injector solenoid wire nuts from top of injectors.
6. Remove injector hold-down clamp bolts.
7. Remove the fuel injector using tool 9010 as follows:
   a. Remove rocker housing bolt.
   b. Install lower half of mounting stud to center of rocker housing bridge.
   c. Install upper half of mounting stud to lower half.
   d. Place tool handle to mounting stud and install nut loosely.

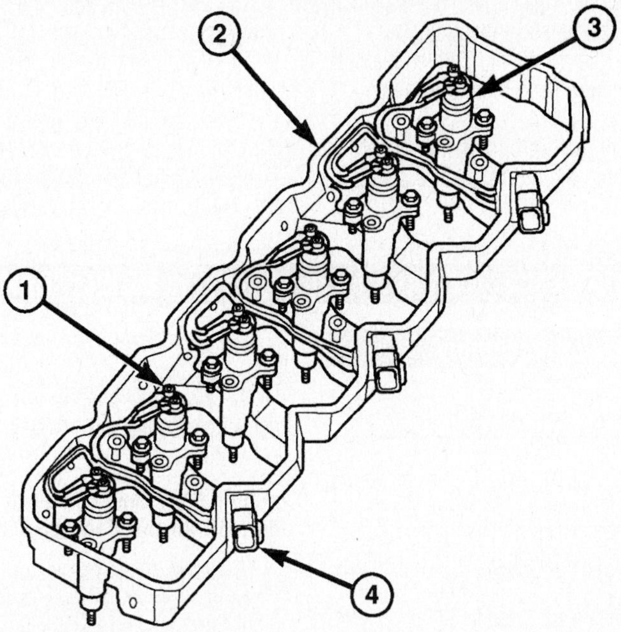

1 - SOLENOID CONNECTIONS
2 - ROCKER HOUSING
3 - FUEL INJECTOR
4 - PASSTHROUGH CONNECTOR

67189-RAMT-G11

Fuel injector mounting—2004

   e. Place lower part of clamshells to sides of injector.
   f. Place upper part of clamshells onto tool handle head.
   g. Slide retainer sleeve over pivoting handle.
   h. Depress handle downward to force injector up and out of cylinder head.
8. Remove and discard the sealing washer from the injector.

To install:

9. Use a new sealing washer and O-ring on the injector.

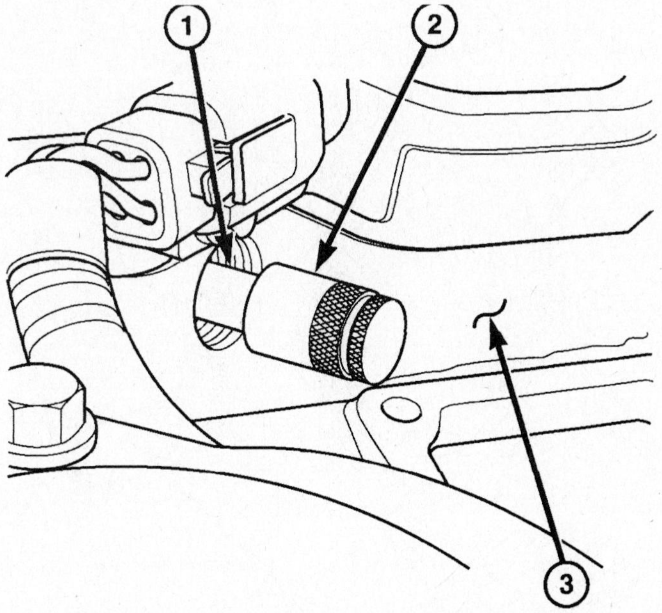

1 - CONNECTOR TUBE
2 - TOOL #9015
3 - CYLINDER HEAD (LEFT SIDE)

67189-RAMT-G10

Using tool 9015 to remove high pressure connectors—2004

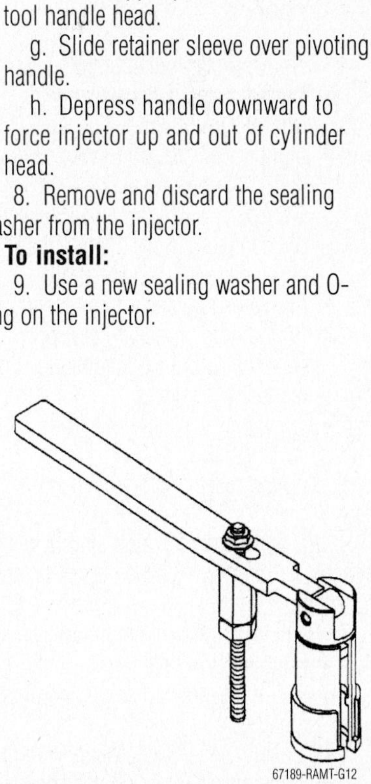

67189-RAMT-G12

Fuel injector removal tool 9010—2004

10. Install the injector with the male connector port facing intake manifold.
11. Fuel injector clamp. Tighten the bolt to 44 inch lbs. (5 Nm), then loosen the bolts.
12. Install the high pressure connector and tighten the nut to 11 ft. lbs. (15 Nm).
13. Alternately tighten the injector hold down bolts to 89 inch lbs. (10 Nm).
14. Retighten the high pressure retaining nut to 37 ft. lbs. (50 Nm).
15. Install solenoid wires and nuts. Tighten nuts to 11 inch lbs. (1.25 Nm).
16. Install or connect the following:
   - Exhaust rocker arms
   - Check valve lash
   - Fuel line connector
   - Wiring harness retaining nuts
   - Valve cover
   - Negative battery cables
17. Bleed the fuel system.
18. Start the engine and check for leaks.

DRIVE TRAIN

Transmission Assembly

REMOVAL & INSTALLATION

Automatic

1. Before servicing the vehicle, refer to the precautions in the beginning of this section.
2. Remove or disconnect the following:
   - Negative battery cable
   - Skid plates, if equipped
   - Rear driveshaft
   - Front driveshaft, if necessary
   - Engine to transmission structural dust cover
   - Exhaust support bracket
   - Exhaust front pipe
   - Starter motor
3. Rotate crankshaft in clockwise direction until converter bolts are accessible. Then remove bolts one at a time. Rotate crankshaft with socket wrench on dampener bolt.
4. Remove or disconnect the following:
   - Output speed sensor connector from the output speed sensor
   - Input speed sensor connector from the input speed sensor
   - Transmission solenoid/TRS assembly connector from the transmission solenoid/TRS assembly
   - Line pressure sensor connector from the line pressure sensor
   - Gearshift cable from transmission manual valve lever
   - Transmission vent hose from the transmission
5. Support rear of engine with safety stand or jack.
6. Raise the transmission slightly with service jack to relieve load on crossmember and supports.
7. Remove or disconnect the following:
   - Bolts securing rear support and cushion to transmission and crossmember
   - Transfer case, if necessary
   - Transmission fluid cooler lines at transmission fittings and clips
   - Remaining converter housing bolts
8. Work the transmission and torque converter assembly rearward off engine block dowels.

➡Hold the torque converter in place during transmission removal.

9. Lower the transmission and remove assembly from under the vehicle.
10. Slide the torque converter out of the transmission.

To install:

11. Slide the torque converter onto the transmission and secure with a C-clamp.
12. Position transmission on jack and secure it with chains.
13. Apply a light coating of Mopar® High Temp Grease to the torque converter hub pocket in the rear pocket of the engine's crankshaft.
14. Raise the transmission into the vehicle.

➡Install two bolts to attach the transmission to the engine.

15. Install or connect the following:
   - Remaining torque converter housing to engine bolts. Tighten to 50 ft. lbs. (68 Nm).
   - Transfer case, if equipped
   - Transmission rear support and tighten bolts to 35 ft. lbs. (47 Nm)
   - Transmission crossmember. Tighten the bracket to crossmember bolts to 35 ft. lbs. (47 Nm) and bracket to rear support bolt to 50 ft. lbs. (68 Nm).

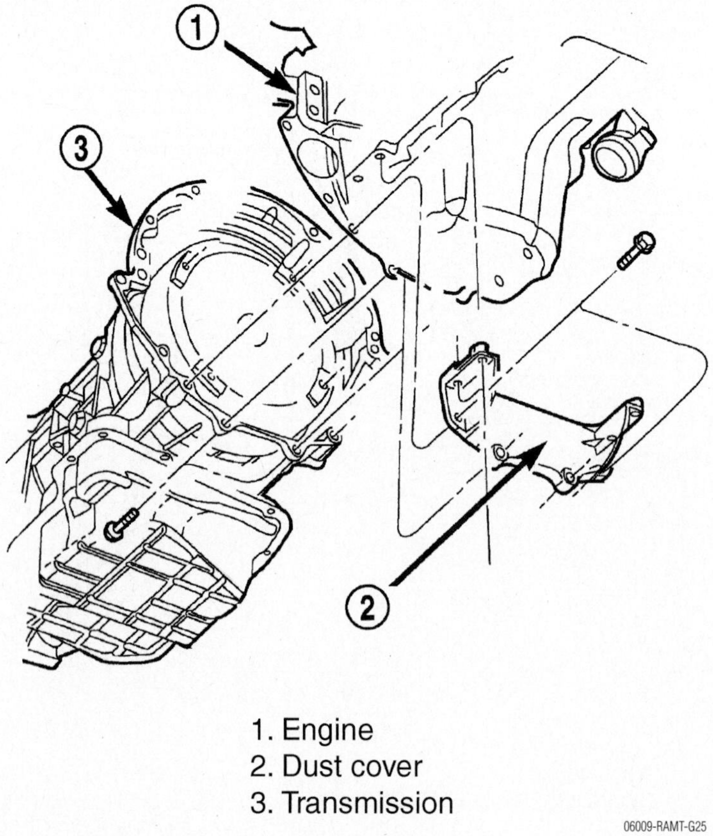

Removing the engine to transmission structural dust cover.

- Gearshift cable to transmission
- Wires to solenoid and pressure switch assembly connector
- Wires to input speed sensor
- Wires to output speed sensor
- Wires to line pressure sensor
- Torque converter-to-driveplate bolts. Tighten bolts to 270 inch lbs. (31 Nm)
- Starter motor
- Exhaust components, if necessary
- Structural dust cover
- Driveshaft(s)
- Skid plates, if equipped

16. Fill transmission to the correct level.
17. Start the engine and check for leaks.

Manual

1. Before servicing the vehicle, refer to the precautions in the beginning of this section.
2. Remove or disconnect the following:
   - Negative battery cable
   - Shift boot
   - Shift tower and isolator plate
   - Skid plate, if equipped
   - Driveshaft
   - Exhaust front pipe
   - Speed sensor connector
   - Back-up light switch connector
3. If equipped with four wheel drive:
   a. Remove the transfer case linkage.
   b. Remove the transfer case mounting nuts.
   c. Slide the transfer case rearward until the input gear clears the transmission mainshaft. Lower the transfer case from the vehicle.
4. Remove or disconnect the following:
   - Rear crossmember
   - Clutch slave cylinder bolts and secure cylinder aside
   - Starter, 5.7L only
   - Transmission wiring harnesses
   - Transmission to clutch housing bolts
5. Lower the transmission from the vehicle.

To install:

6. Raise the transmission into the vehicle until the input shaft is centered in the release bearing and clutch disc hub.
7. Work the transmission forward until it is seated against the clutch housing.
8. If equipped with four wheel drive:
   a. Align the transfer case input gear with the transmission mainshaft.
   b. Work the transfer case forward until seated flush on the adapter.
   c. Tighten the transfer case nuts as follows:
      - 3/8 studs: 30-35 ft. lbs. (41-47 Nm)
      - 5/16 studs: 22-30 ft. lbs. (30-41 Nm)
   d. Install the transfer case linkage.
9. Install or connect the following:
   - Transmission to clutch housing bolts and tighten to 80 ft. lbs. (108 Nm)
   - Rear crossmember
   - Clutch slave cylinder
   - Speed sensor connector
   - Back-up light switch connector
   - Starter, 5.7L only
   - Exhaust front pipe
   - Driveshaft
   - Skid plate, if equipped
   - Shift tower and isolator plate
   - Shift boot
   - Negative battery cable

Clutch

REMOVAL & INSTALLATION

1. Before servicing the vehicle, refer to the precautions in the beginning of this section.
2. Remove or disconnect the following:
   - Negative battery cable
   - Transfer case, if equipped
   - Transmission
   - Pressure plate. Loosen the bolts evenly in ½ turn steps.
   - Clutch disc

To install:

3. Install or connect the following:
   - Clutch disc and pressure plate. Tighten the pressure plate bolts evenly in ½ turns to 37 ft. lbs. (50 Nm).
   - Transmission
   - Transfer case, if equipped
   - Negative battery cable

Hydraulic Clutch System

BLEEDING

The system is self-bleeding. Press the clutch pedal repeatedly to release air from the fluid. The air will be vented from the reservoir.

Transfer Case Assembly

REMOVAL & INSTALLATION

1. Before servicing the vehicle, refer to the precautions in the beginning of this section.
2. Shift the transfer case into **N**.
3. Drain the transfer case fluid.

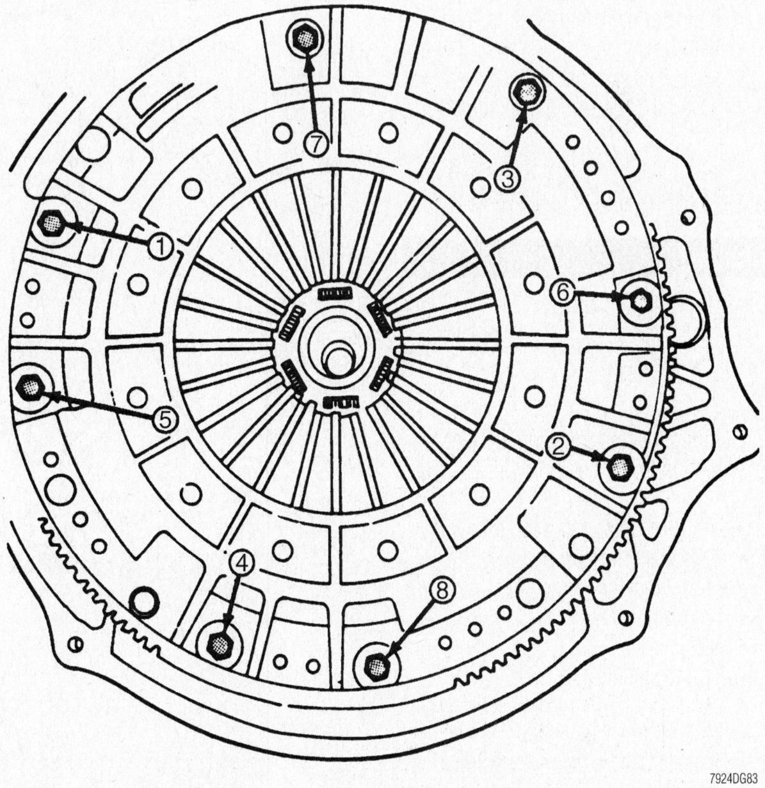

Pressure plate torque sequence

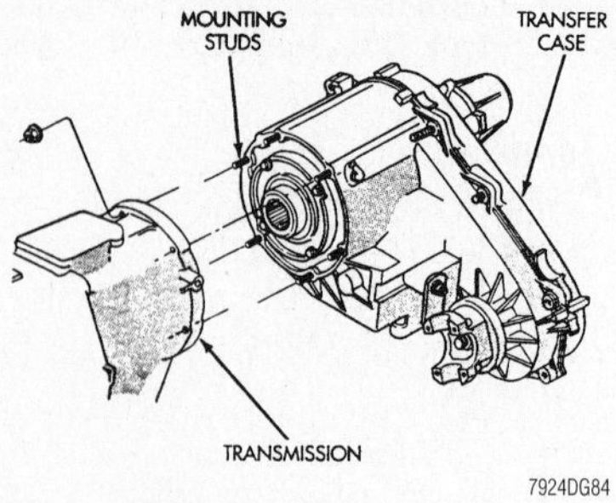

Typical transfer case mounting

4. Remove or disconnect the following:
   - Skid plate, if equipped
   - Vent house
   - Indicator switch connector
   - Shift linkage
   - Front and rear driveshafts
   - Transmission mount and crossmember. Support the transmission.
   - Vehicle Speed (VSS) sensor connector
   - Transfer case attaching nuts
   - Transfer case

To install:

5. Install or connect the following:
   - Transfer case. Tighten the nuts to 26 ft. lbs. (35 Nm).
   - VSS sensor connector
   - Transmission mount and crossmember
   - Front and rear driveshafts
   - Shift linkage
   - Indicator switch connector
   - Vent hose
   - Skid plate, if equipped
6. Fill the transfer case to the correct level.
7. Adjust the shift lever if necessary.
8. Start then engine and check for leaks.

Halfshaft

REMOVAL & INSTALLATION

2002

1. Before servicing the vehicle, refer to the precautions in the beginning of this section.
2. Remove or disconnect the following:
   - Skid plate, if equipped
   - Front wheel
   - Split pin
   - Nut lock
   - Spring washer
   - Hub nut
   - Brake caliper and rotor
   - Wheel speed sensor, if equipped
   - Wheel bearing and hub assembly

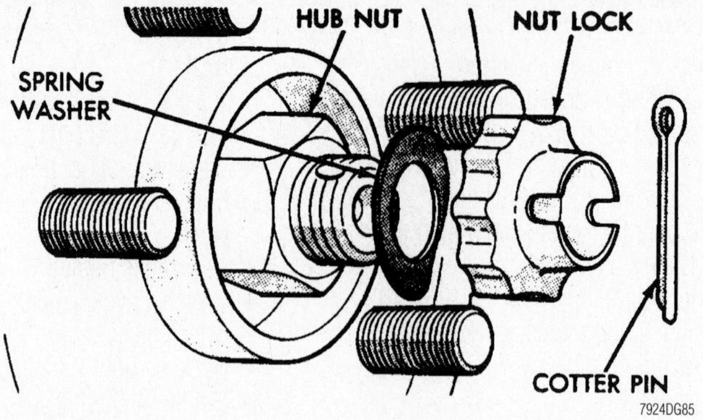

To separate the halfshaft from the hub, remove the cotter pin, nut lock and spring washer from the axle shaft—2002 models

3. Pry the inner tripod joint out of the differential and remove the axle halfshaft.

To install:

4. Install the axle halfshaft so that the snapring is felt to seat in the joint housing groove.
5. Install or connect the following:
   - Wheel bearing and hub assembly
   - Wheel speed sensor, if equipped
   - Brake caliper and rotor
   - Hub nut. Tighten the nut to 180 ft. lbs. (244 Nm).
   - Spring washer
   - Nut lock
   - Split pin
   - Front wheel
   - Skid plate, if equipped

2003–2004

1. Before servicing the vehicle, refer to the precautions in the beginning of this section.
2. Remove or disconnect the following:
   - Skid plate, if equipped
   - Front wheel
   - Hub nut

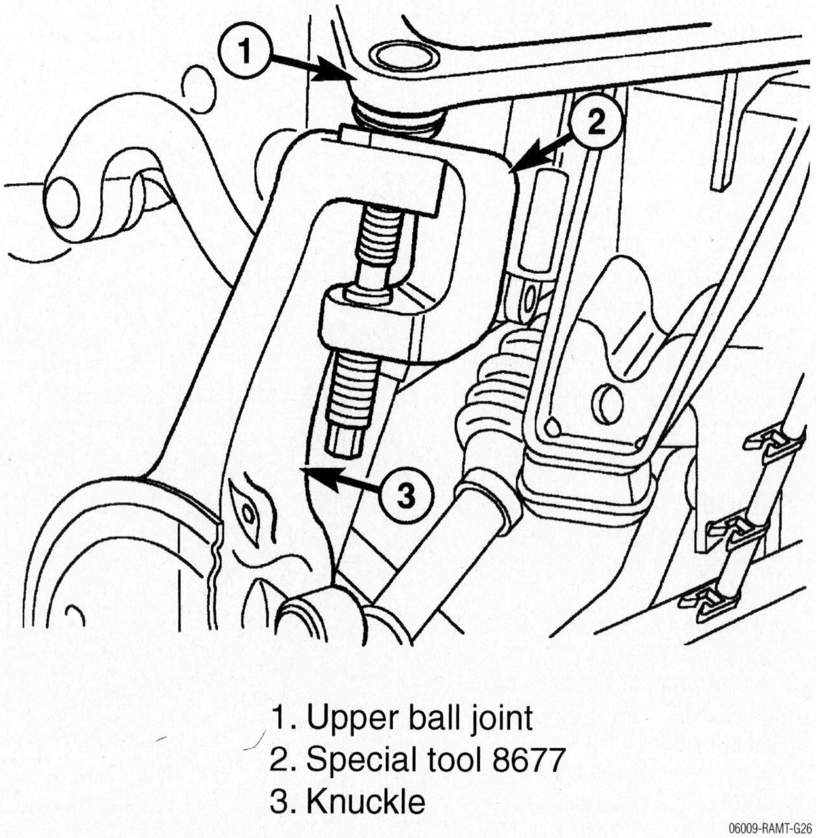

Removing the upper ball joint using Special tool 8677.

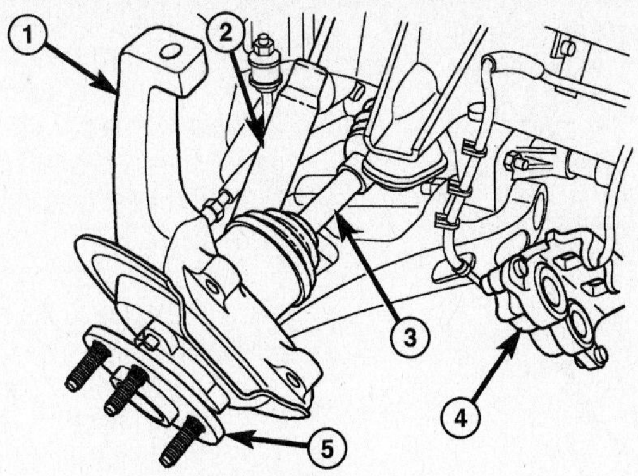

1. Knuckle
2. Shock
3. Halfshaft
4. Caliper
5. Hub flange

06009-RAMT-G27

Removing the halfshaft.

- Brake caliper and rotor

3. Unload the suspension with a jack.
4. Remove or disconnect the following:
   - Lower shock absorber bolt
   - Upper ball joint using Special Tool 8677
5. Pry the inner joint out of the differential and remove the axle halfshaft.
6. Disengage inner C/V joint from axle shaft with two pry bars between the C/V housing and axle housing.
7. Tilt the knuckle out and push half shaft out of the knuckle.

To install:

8. Install the axle halfshaft so that the snapring is felt to seat in the joint housing groove.
9. Install or connect the following:
   - Upper ball joint
   - Lower shock absorber bolt
   - Brake caliper and rotor
   - Hub nut. Tighten the nut to 185 ft. lbs. (251 Nm).
   - Front wheel
   - Skid plate, if equipped

CV-Joints

OVERHAUL

Outer CV-Joint

1. Before servicing the vehicle, refer to the precautions in the beginning of this section.
2. Clamp shaft in a vise (with soft jaws) and support C/V joint.
3. Remove clamps with a cut-off wheel or grinder.
4. Slide the boot down the shaft.
5. Remove lubricant to expose the C/V joint snap ring.
6. Spread snap ring and slide the joint off the shaft.
7. Slide boot off the shaft and discard old boot.
8. Mark alignment marks on the inner race/hub, bearing cage and housing.
9. Clamp C/V joint in a vertical position in a soft jawed vise.
10. Press down one side of the bearing cage to gain access to the ball at the opposite side.

➡If joint is tight, use a hammer and brass drift to loosen the bearing hub. Do not contact the bearing cage with the drift.

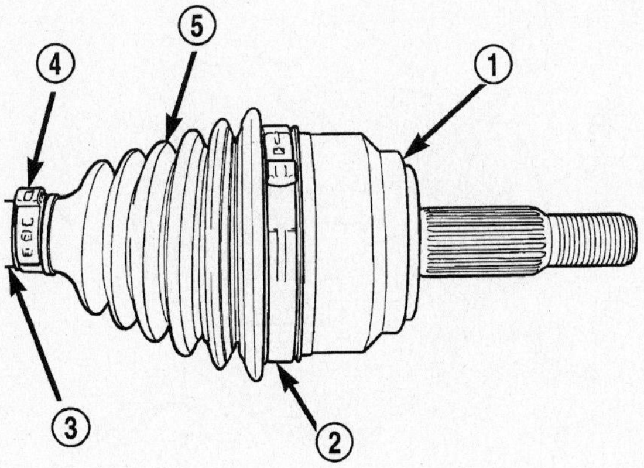

1. Hub
2. Clamp
3. Shaft
4. Clamp
5. Boot

06009-RAMT-G28

Outer CV-joint components

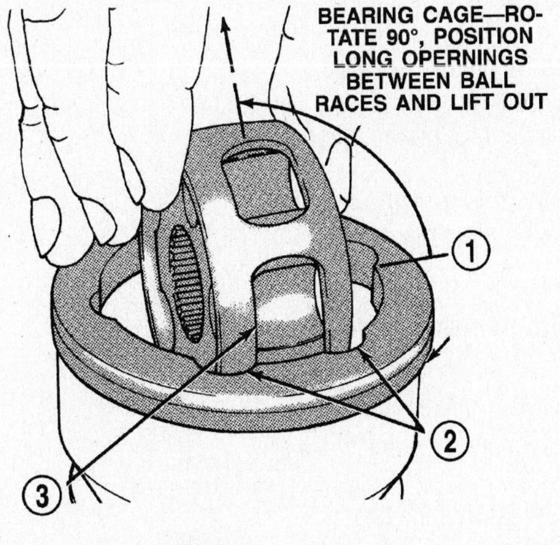

1. Housing
2. Inner race
3. Bearing

06009-RAMT-G29

Press down one side of bearing cage to remove.

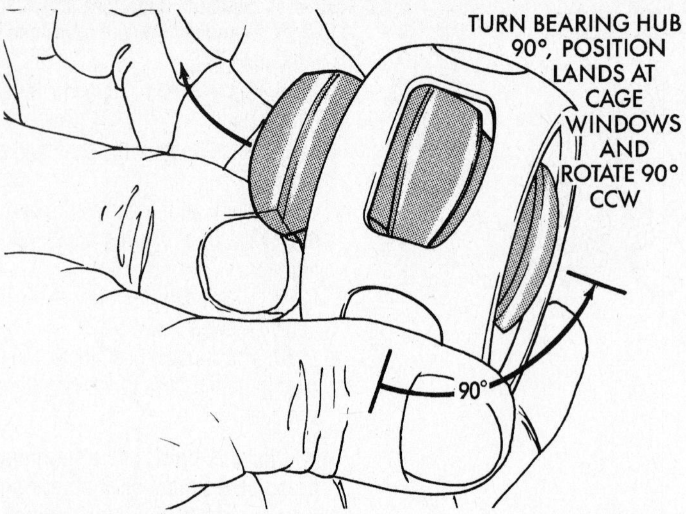

Rotate the bearing 90 degrees.

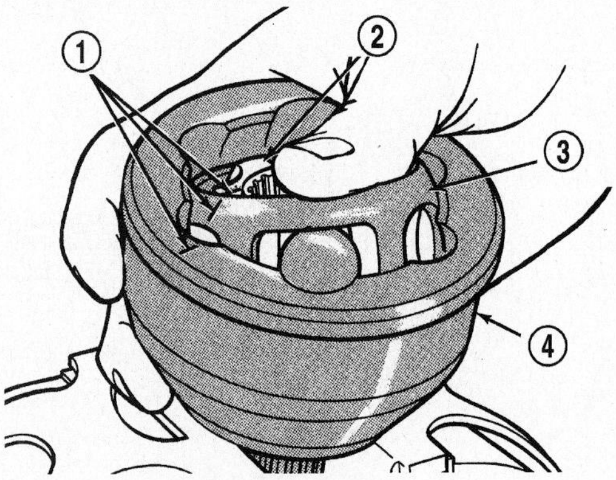

1. Alignment marks
2. Inner race
3. Bearing cage
4. Housing

Installing the bearing cage using the alignment marks.

11. Remove the ball from the bearing cage.
12. Repeat step above until all six balls are removed from the bearing cage.
13. Lift cage and inner race upward and out from the housing
14. Turn inner race 90° in the cage and rotate the inner race/hub out of the cage.

To install:

➡If C/V joint is worn, replace entire C/V joint and boot.

15. Align the inner race cage and housing according to the alignment reference marks.
16. Insert the inner race into the cage and rotate race into the cage.
17. Rotate the inner race/hub in the cage.
18. Insert cage into the housing. Rotate the cage 90° into the housing so the large bearing hub counterbore is facing outwards.
19. Apply the grease supplied with the joint/boot to the ball races. Spread the grease equally between all the races.
20. Tilt inner race/hub and cage and install the balls.
21. Place new clamps onto the new boot and slide boot onto the shaft to its original position.
22. Apply the rest of the grease to the C/V joint and boot.
23. Install the joint onto the shaft. Push the joint onto the shaft until the snap ring seats in the groove.

➡Pull on the joint to verify the span ring has engaged.

24. Position the boot on the joint in it's original position.

➡Verify boot is not twisted and remove any excess air.

25. Secure both boot clamps with Clamp Installer C-4975A. Place tool on clamp bridge and tighten tool until jaws of the tool are closed.
26. Install the axle halfshaft.

Inner Tripod Joint

1. Remove the axle shaft.
2. Clamp the axle shaft in a vise and support C/V joint.
3. Remove the clamps with a cut-off wheel or grinder.
4. Remove the housing from the half shaft and slide boot down the shaft.
5. Remove housing bushing from the housing.
6. Remove tripod snap ring.

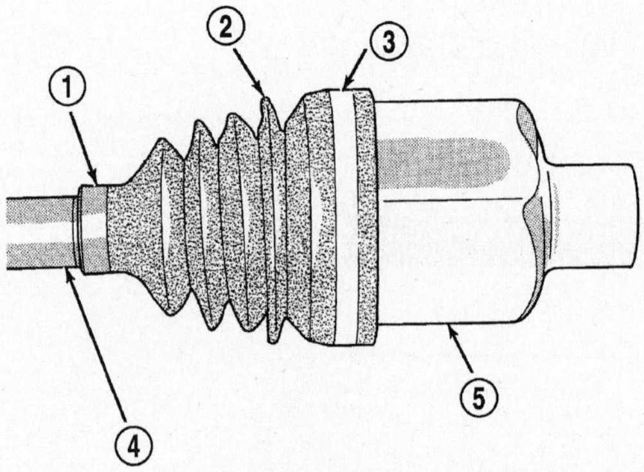

1. Clamp
2. Boot
3. Clamp
4. Shaft
5. Housing

Inner CV-joint components

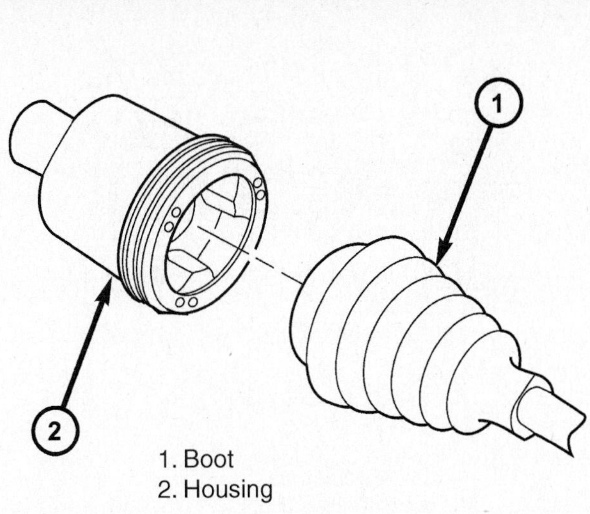

Removing the housing from halfshaft.

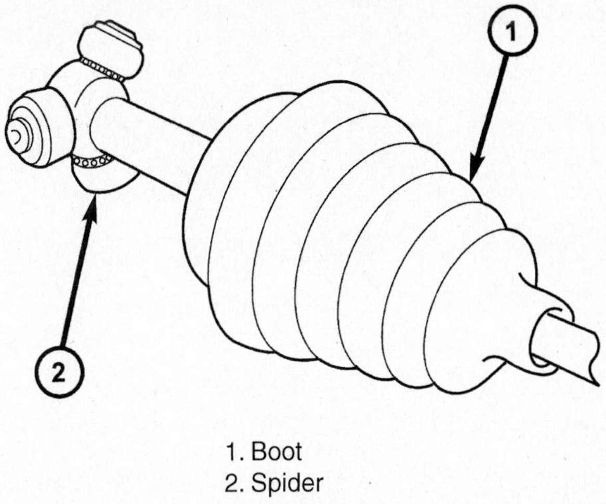

Slide the new boot down the half shaft.

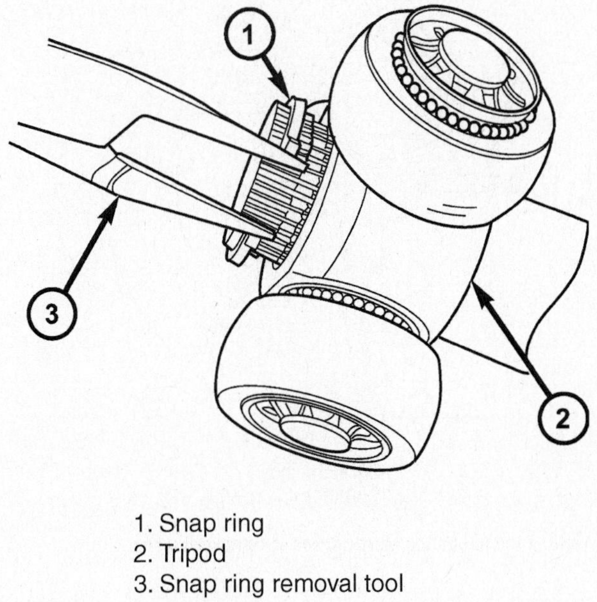

Removing the snap ring from the tripod.

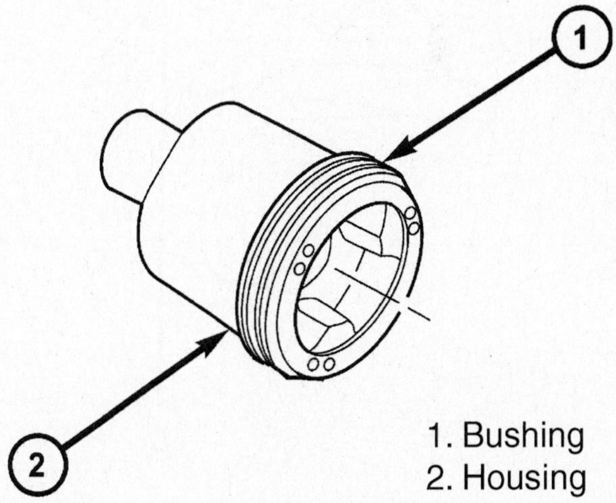

Installing a new bushing onto the housing.

7. Remove tripod and boot from the half shaft.

To install:

8. Clean and inspect C/V components for excessive wear and damage. Replace the tripod as a unit only if necessary.
9. Slide new boot down the half shaft.
10. Install tripod and tripod snap ring on the half shaft.
11. Pack grease supplied with the joint/boot into the housing and boot.
12. Coat tripod with supplied grease
13. Install new bushing onto the housing.
14. Insert the tripod and shaft in the housing.
15. Position the boot on the housing and shaft in it's original position.

➡Verify boot is not twisted and remove any excess air.

16. Measure the distance from the end of the housing to the end of the boot on the shaft. This measurement should be 10.25 in. (260 mm).

➡If measurement is not correct, allow more or less air into the boot.

17. Secure both boot clamps with Clamp Installer C-4975A. Place tool on clamp bridge and tighten tool until the jaws of the tool are closed.
18. Install the axle halfshaft.

Axle Shaft, Bearing and Seal

REMOVAL & INSTALLATION

Front

1. Before servicing the vehicle, refer to the precautions in the beginning of this section.
2. Remove or disconnect the following:
   - Brake caliper
   - Brake rotor
   - ABS wheel speed sensor, if equipped
   - Axle shaft cotter pin
   - Hub nut and washer

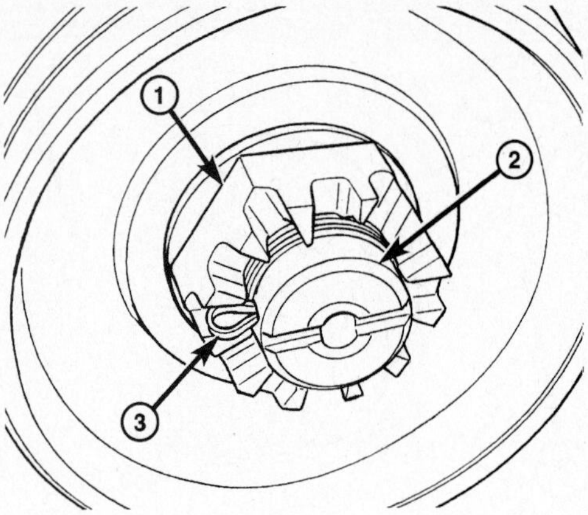

1. Hub nut
2. Axle shaft
3. Cotter pin

06009-RAMT-G37

Removing the axle shaft cotter pin.

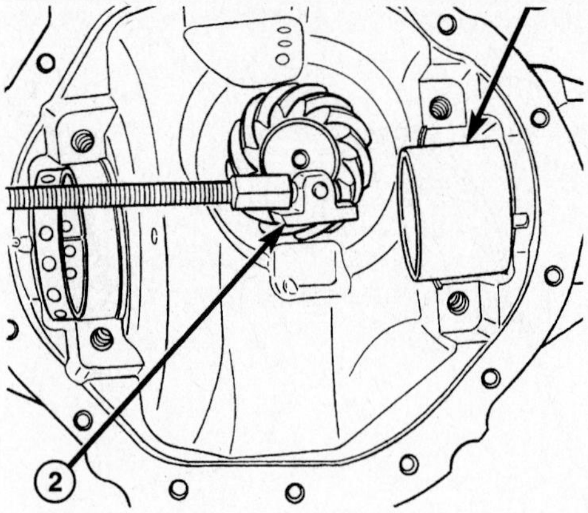

1. Special tool 8498
2. Extractor foot 6310-9

06009-RAMT-G40

Install Special Tool 6310 extractor with extractor foot through the receiver and seal.

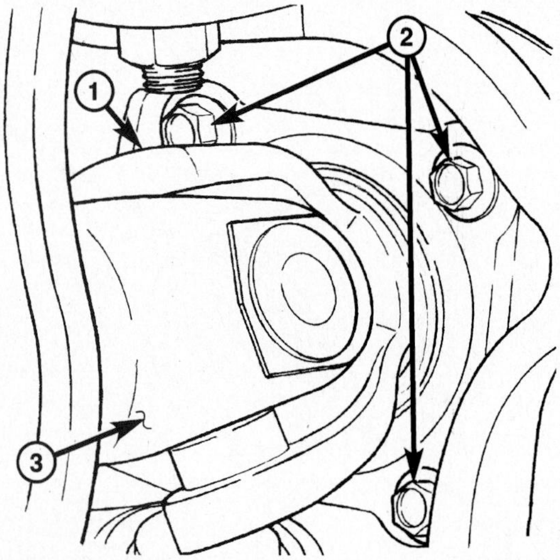

06009-RAMT-G38

Hub bearing bolt locations.

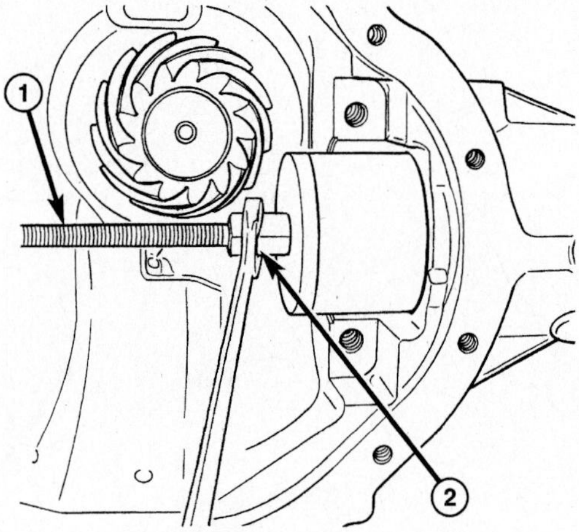

1. Extractor rod
2. Extractor nut 6310-7

06009-RAMT-G41

Tighten the nut on the extractor rod to remove the seal.

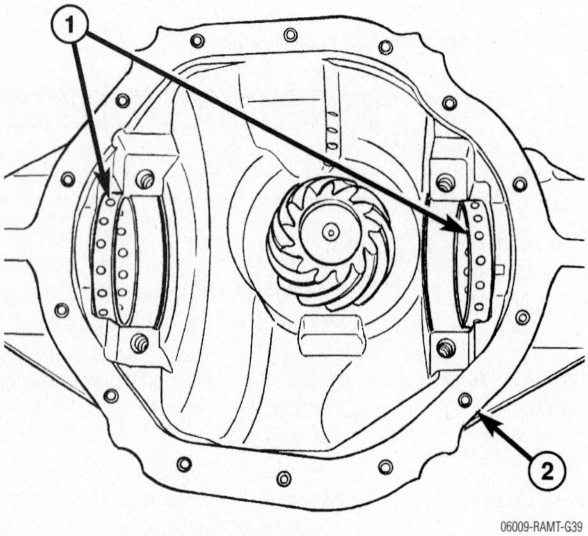

06009-RAMT-G39

Remove the differential bearing adjusters.

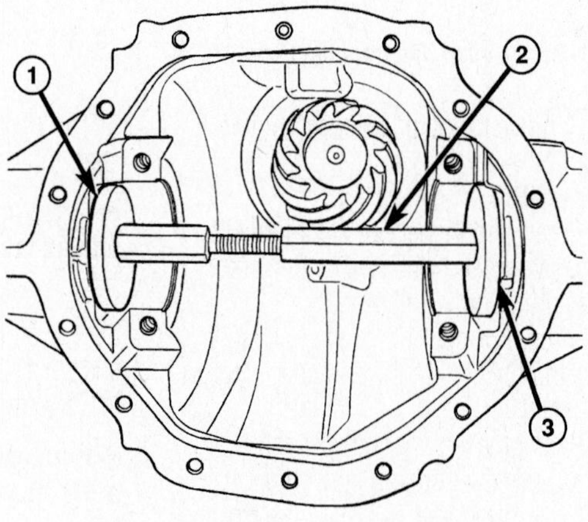

1. Installer cup 8885-2
2. Turnbuckle 8885-1
3. Installer cup 8885-2

06009-RAMT-G42

Use install cups and turnbuckle to install the axle seals.

- Hub bearing bolts from the back of the steering knuckle
- Hub bearing
- Axle shaft
- Differential
- Differential bearing adjusters

3. Install Special Tool 8498 Receiver into the adjuster bore.
4. Install Extractor Rod 6310 with Extractor Foot 6310-9 through the receiver and the axle seal.
5. Tighten nut on the extractor rod and pull the seal out and into the receiver.

To install:

6. Install axle seal on Installer Cups 8885-2 and position cups with seals into the housing.

➡Seal are installed with the axle guide facing outward.

7. Install Turnbuckle 8885-1 into the installer cups and expand the turnbuckle until the seals bottom out in the housing.
8. Install or connect the following:
   - Differential into the axle housing
   - Axle shaft through the steering knuckle and into the differential side gears
   - Hub bearing in the knuckle
   - Hub bearing bolts and tighten to 149 ft. lbs. (202 Nm)
   - ABS wheel sensor, if equipped
9. Install the axle washer and nut. Tighten nut to 132 ft. lbs. (179 Nm).
10. Rotate the axle several times to seat the wheel bearing.
11. Tighten the axle nut to a final torque of 263 ft. lbs. (356 Nm).
12. Align the nut to the next cotter pin hole and install a new cotter pin.
13. Fill the differential with fluid to the correct level.

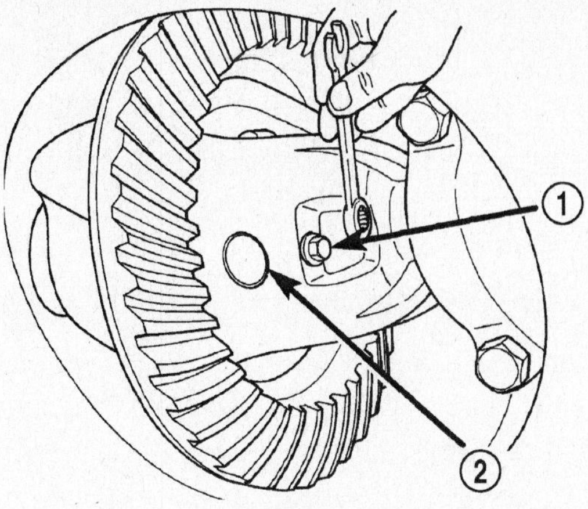

1. Lock screw
2. Pinion mate shaft

06009-RAMT-G43

Removing the lock screw and pinion mate shaft.

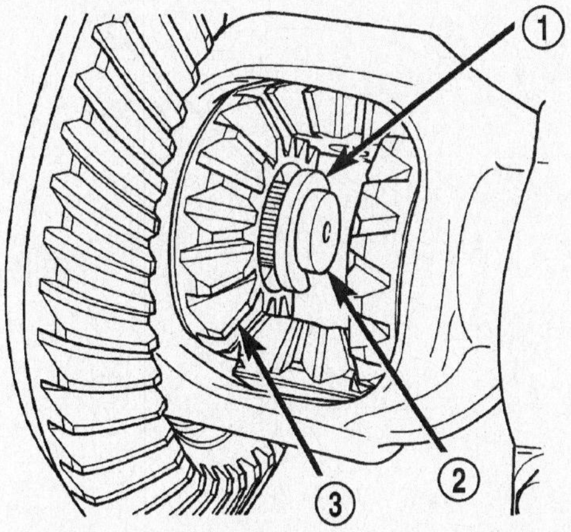

1. C-clip
2. Axle shaft
3. Side gear

06009-RAMT-G44

Removing the axle shaft C-clip.

Rear

9¼ AXLE

1. Before servicing the vehicle, refer to the precautions in the beginning of this section.
2. Remove or disconnect the following:
   - Brake caliper and adapter
   - Brake rotor
   - Differential housing cover
3. Rotate differential case so pinion mate shaft lock screw is accessible. Remove lock screw and pinion mate shaft from differential case.
4. Push the axle shaft inward and remove the C-clip from the axle shaft.
5. Remove the axle shaft carefully to prevent damage to the shaft bearing and seal in the axle tube.
6. Remove the axle shaft seal from the axle tube with a small pry bar.
7. Remove the axle shaft bearing with Special Tool 6310 and Foot 6310-9.

To install:

8. Install the axle shaft bearing with Special Tool C-4198 Installer and Special Tool C-4171 Handle. Drive bearing in until tool contacts the axle tube.

➡Bearing is installed with the bearing part number against the installer.

9. Coat lip of the new axle seal with axle lubricant and install with Special Tool C-4076-B Installer and Special Tool C-4735-1 Handle.
10. Install axle shaft and engage into side gear splines.
11. Insert C-clip in end of axle shaft then push axle shaft outward to seat C-lock in side gear.
12. Insert pinion shaft into differential case and through thrust washers and differential pinions.
13. Align hole in pinion mate shaft with hole in the differential case and install lock screw with Loctite® on the threads. Tighten lock screw to 8 ft. lbs. (11 Nm).
14. Install or connect the following:
   - Differential cover
   - Brake rotor
   - Caliper adapter and caliper
15. Fill the differential with fluid to the correct level.

10½ AND 11½ AXLE

1. Before servicing the vehicle, refer to the precautions in the beginning of this section.

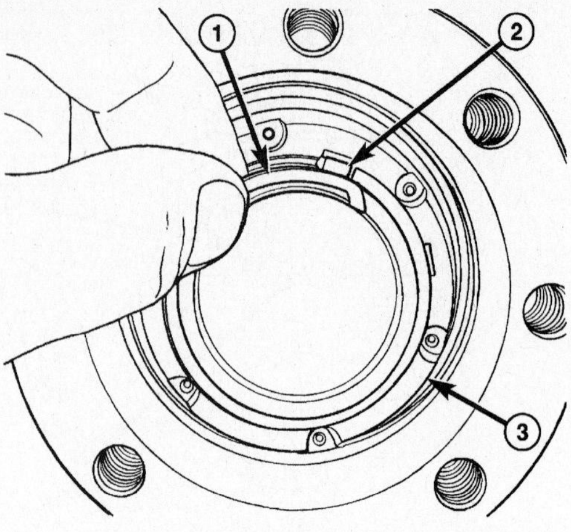

Remove the retainer ring from axle shaft tube.

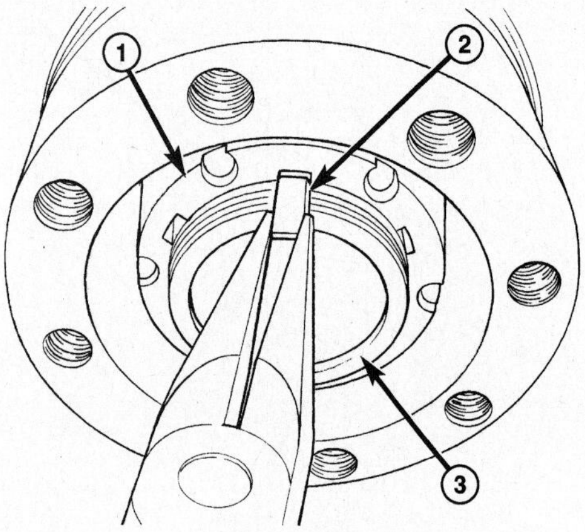

Removing the hub bearing nut locking.

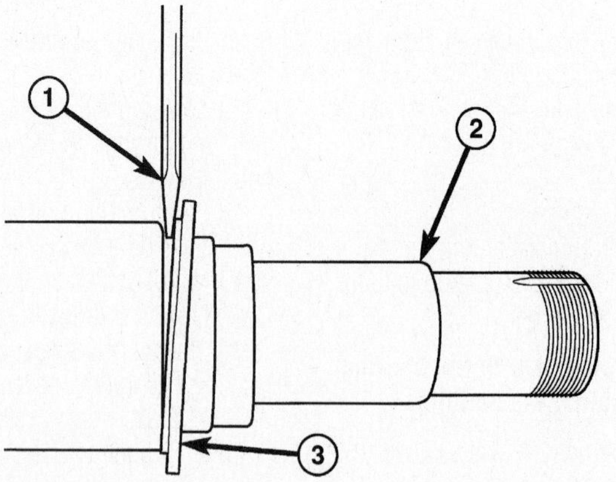

Prying off the bearing seal.

2. Remove the rear brake components.
3. Remove the axle shaft flange bolts.
4. Slide the axle shaft out of the axle tube.
5. Remove the retainer ring from the axle shaft tube.
6. Remove the hub bearing nut with Special Tool 8954.
7. Remove the hub and bearings from the axle.
8. Pry out hub bearing seal from the back of the hub.
9. Remove the rear bearing.
10. Remove the hub bearing cups with a hammer and drift.

To install:

11. Install the outer hub bearing cup with Special Tool 8961 Installer and Special Tool C-4171 Handle.
12. Install the inner hub bearing cup with Special Tool 8962 Installer and Special Tool C-4171 Handle.
13. Pack the bearings with the appropriate grease.
14. Install the rear bearing and install new a grease seal with Special Tool 8963 Installer and Special Tool C-4171 Handle.
15. Slide hub on the axle tube and install front bearing into the hub.
16. Install hub bearing nut with Special Tool 8954 Socket and tighten to 22 ft. lbs. (30 Nm) while rotating the hub.
17. Back off nut about 30° and align next hub nut key slot with axle tube key slot and install locking key.
18. Install retainer ring with ring end in the key slot.
19. Install new axle shaft gasket and install axle shaft.
20. Install the axle shaft flange bolts and tighten to 95 ft. lbs. (129 Nm).
21. Install the rear brake components.
22. Fill the differential with fluid to the correct level.

Pinion Seal

REMOVAL & INSTALLATION

Front

1. Before servicing the vehicle, refer to the precautions in the beginning of this section.
2. Matchmark and remove the drive shaft.
3. Remove the hub bearings and axle shafts.
4. Rotate the pinion a few times and record the bearing preload.
5. Hold the pinion flange with Special

Tool 8979 Flange wrench and remove the pinion flange nut and washer.

6. Remove the pinion flange with Special Tool 8992 Pinion Flange Puller.

7. Remove the pinion shaft seal with a pry tool or slide hammer mounted screw.

To install:

8. Install a new pinion seal with Special Tool 8882 Installer and Special Tool C-4171 Handle.

9. Lightly tap the pinion flange onto the pinion until a few threads are showing.

10. Install flange washer and new pinion nut.

11. Hold flange with Special Tool 8979 Flange Wrench and tighten pinion nut until pinion end play is taken up.

12. Rotate the pinion several times to seat bearings.

13. Measure bearing preload with an inch pound torque wrench and compare it to recorded measurement. Tighten pinion nut in small increments, until bearing preload is 3-5 inch lbs. (0.40-0.57 Nm) greater than recorded measurement.

14. Rotate pinion several times and then verify bearing preload again.

15. Install the axle shafts and hub bearings.

16. Install the drive shaft with reference marks aligned.

Rear

9¼ AXLE

1. Before servicing the vehicle, refer to the precautions in the beginning of this section.

2. Matchmark and remove the drive shaft from the companion flange.

3. Remove the brake rotors to prevent any drag.

4. Rotate companion flange three or four times and record the bearing preload.

5. Matchmark the companion flange to the pinion shaft.

6. Install two bolts into the companion flange threaded holes, 180° apart. Position Special Tool 6719A Holder against the companion flange and install and tighten two bolts and washers into the remaining holes.

7. Hold the companion flange with Special Tool 6719A Holder and remove pinion nut and washer.

8. Remove companion flange with Special Tool C-452 Remover.

9. Remove pinion seal with pry tool or slide-hammer mounted screw.

To install:

10. Install a new pinion seal with Special Tool C-3860-A Installer and Special Tool C-4171 Handle.

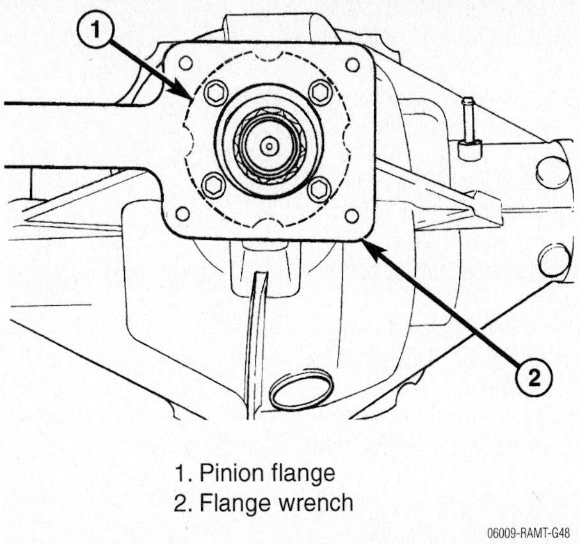

Use the 8979 Flange Wrench to hold the pinion while removing the flange nut and washer.

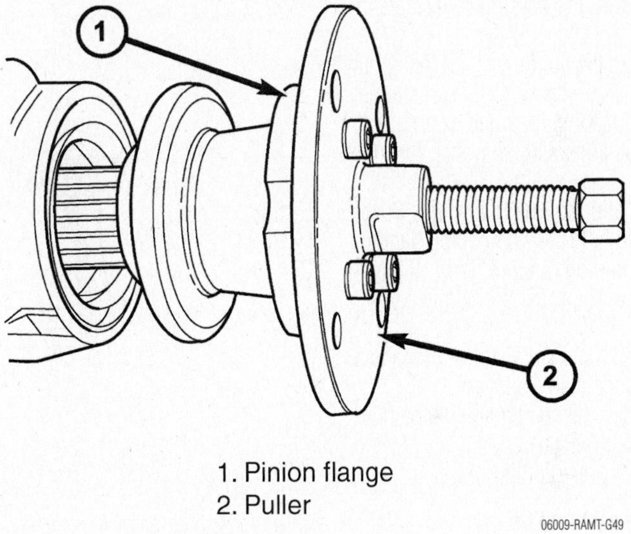

Flange puller 8992 installed on the pinion flange.

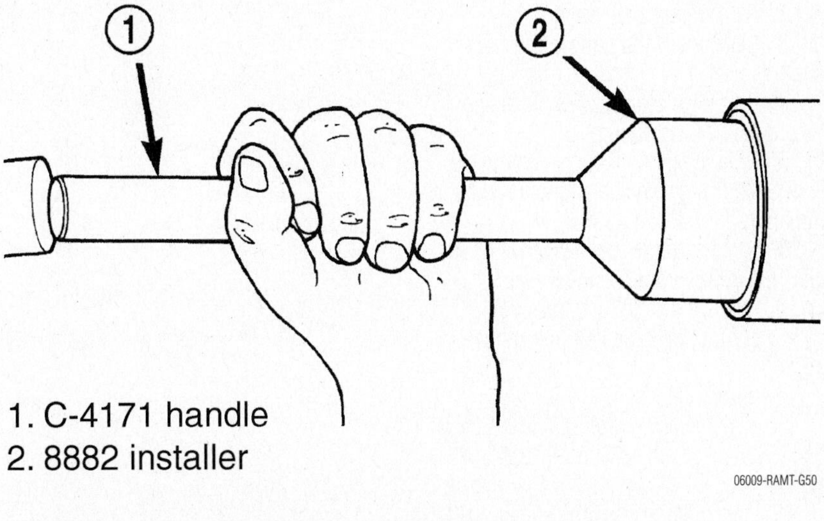

Installing the pinion seal using the 8882 installation tool.

11. Install the companion flange on the end of the shaft with the reference marks aligned.

12. Install two bolts into the threaded holes in the companion flange, 180° apart.

13. Position Special Tool 6719 Holder against the companion flange and install a bolt and washer into one of the remaining threaded holes. Tighten the bolts so holder is held to the flange.

14. Install companion flange on pinion shaft with Special Tool C-3718 Installer and Special Tool 6719 Holder.

15. Install pinion washer and a new pinion nut. The convex side of the washer must face outward.

16. Hold companion flange with Special Tool 6719 Holder and tighten the pinion nut with a torque set to 210 ft. lbs. (285 Nm). Rotate pinion several revolutions to ensure the bearing rollers are seated.

17. Rotate the pinion with an inch pound torque wrench. Rotating torque should be equal to the reading recorded during removal plus an additional 5 inch lbs. (0.56 Nm).

18. If rotating torque is low, use Special Tool 6719 to hold the companion flange and tighten pinion nut in 5 ft. lb. (6.8 Nm) increments until proper preload is achieved.

19. Install the drive shaft with the installation reference marks aligned.

20. Tighten the companion flange bolts to 80 ft. lbs. (108 Nm).

21. Install brake rotors.

10½ AND 11½ AXLE

1. Before servicing the vehicle, refer to the precautions in the beginning of this section.

2. Remove the axle shafts.

3. Matchmark and remove the drive shaft.

4. Rotate the pinion a few times and record the bearing preload.

5. Hold the pinion flange with Special Tool 8979 Flange wrench and remove the pinion flange nut and washer.

6. Remove the pinion flange with Special Tool 8992 Pinion Flange Puller.

7. Remove the pinion shaft seal with a pry tool or slide hammer mounted screw.

To install:

8. Install a new pinion seal with Special Tool 8896 Installer and Special Tool C-4171 Handle.

9. Lightly tap the pinion flange onto the pinion until a few threads are showing.

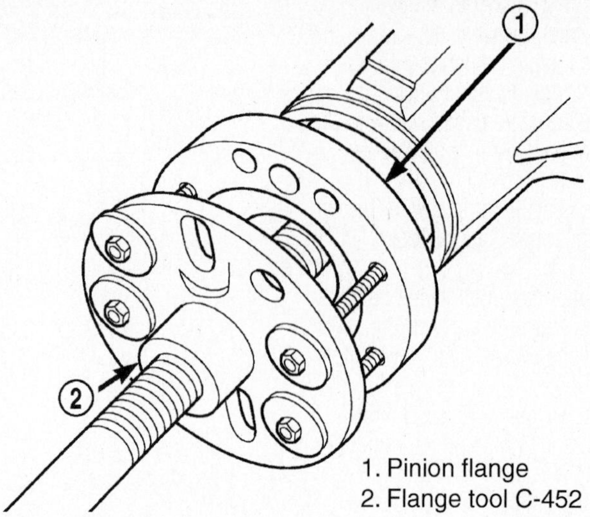

Removing companion flange with Special Tool C-452.

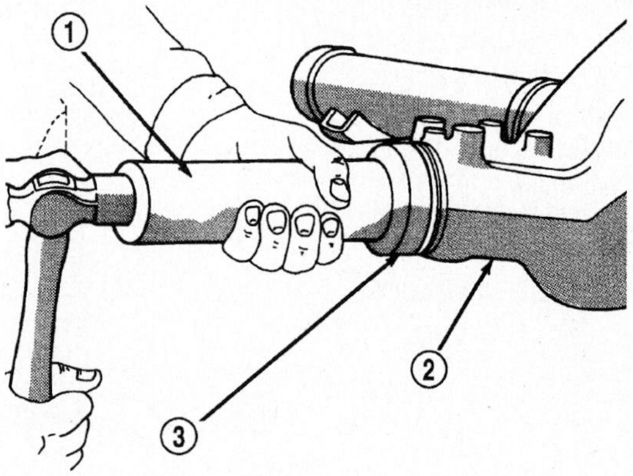

Installing the new pinion seal—9¼ rear axle.

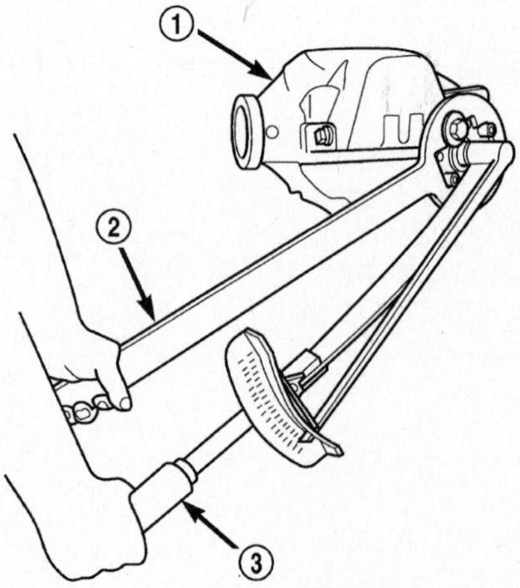

Check the pinion bearing preload.

10. Install flange washer and new pinion nut.

11. Hold flange with Special Tool 8979 Flange Wrench and tighten pinion nut until pinion end play is taken up.

12. Rotate the pinion several times to seat bearings.

13. Measure bearing preload with an inch pound torque wrench and compare it to recorded measurement. Tighten pinion nut in small increments, until bearing preload is 3-5 inch lbs. (0.40-0.57 Nm) greater than recorded measurement.

14. Rotate pinion several times and then verify bearing preload again.

15. Install the axle shafts and drive shaft with reference marks aligned

STEERING AND SUSPENSION

Air Bag

⁂ CAUTION

Some vehicles are equipped with an air bag system. The system must be disarmed before performing service on, or around, system components, the steering column, instrument panel components, wiring and sensors. Failure to follow the safety precautions and the disarming procedure could result in accidental air bag deployment, possible injury and unnecessary system repairs.

PRECAUTIONS

Several precautions must be observed when handling the inflator module to avoid accidental deployment and possible personal injury.

- Never carry the inflator module by the wires or connector on the underside of the module.
- When carrying a live inflator module, hold securely with both hands, and ensure that the bag and trim cover are pointed away.
- Place the inflator module on a bench or other surface with the bag and trim cover facing up.
- With the inflator module on the bench, never place anything on or close to the module which may be thrown in the event of an accidental deployment.

Before servicing the vehicle, also make sure to refer to the precautions in the beginning of this section as well.

DISARMING

1. Disconnect and isolate the negative battery cable. Wait 2 minutes for the system capacitor to discharge before performing any service.
2. When repairs are completed, connect the negative battery cable.

Recirculating Ball Power Steering Gear

REMOVAL & INSTALLATION

1. Before servicing the vehicle, refer to the precautions in the beginning of this section.
2. Drain the power steering fluid from the reservoir.
3. Remove or disconnect the following:
   - Power steering pressure and return lines
   - Coupler pinch bolt
   - Intermediate shaft
   - Pitman arm
   - Steering gear mounting bolts
   - Steering gear

To install:

4. Install the steering gear. Tighten the mounting bolts to 145 ft. lbs. (196 Nm).
5. Align the intermediate shaft on the steering gear. Install the pinch bolt and tighten to 36 ft. lbs. (49 Nm).
6. Install or connect the following:
   - Pitman arm. Tighten the nut to 225 ft. lbs. (305 Nm).
   - Power steering pressure and return lines and tighten to 23 ft. lbs. (31 Nm)
7. Fill the power steering fluid reservoir.
8. Start the engine and check for leaks.

Rack and Pinion Steering Gear

REMOVAL & INSTALLATION

1. Before servicing the vehicle, refer to the precautions in the beginning of this section.
2. Drain the power steering fluid from the reservoir.
3. Remove or disconnect the following:
   - Front wheels
   - Skid plate, if equipped
   - Steering shaft coupler. Discard the pinch bolt.
   - Power steering hoses
   - Outer tie rod end nuts
4. Separate the tie rod ends from the knuckles using Special Tool 8677.
5. Remove the steering gear mounting bolts.
6. Remove the steering gear.

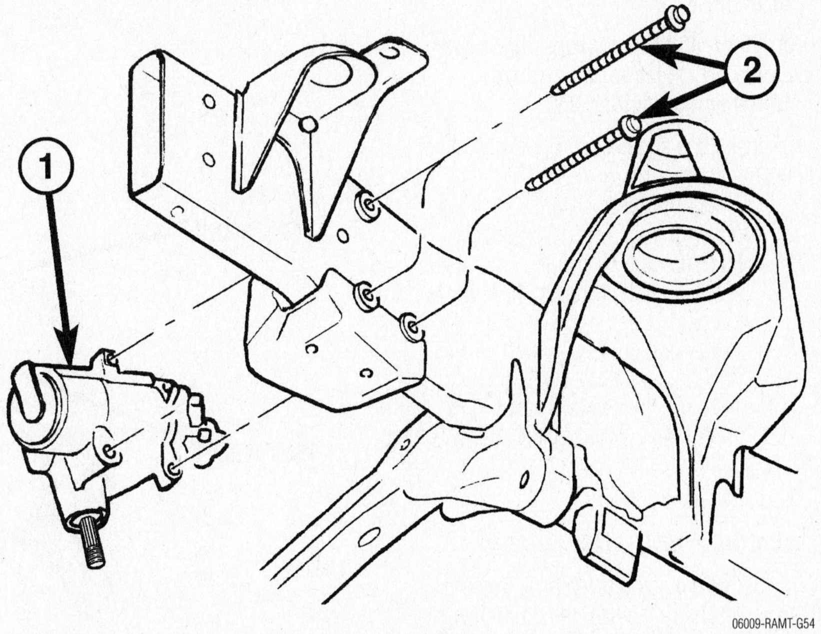

Typical recirculating ball power steering gear mounting.

To install:

7. Install the steering gear. Tighten the bolts as follows:
   a. 4x2 HD Models: 185 ft. lbs. (251 Nm)
   b. All other models: 235 ft. lbs. (319 Nm)
8. Slide the shaft coupler onto the gear. Install a new pinch bolt and tighten to 36 ft. lbs. (49 Nm).
9. Install or connect the following:
   - Tie rod ends into the steering knuckles and tighten the nuts to 45 ft. lbs. (61 Nm)
   - Power steering hoses. Tighten the pressure line fitting to 23 ft. lbs. (32 Nm); return line to 52 ft. lbs. (71 Nm).
   - Skid plate, if equipped
   - Front wheels
10. Fill the power steering fluid reservoir to the correct level.

Shock Absorber

REMOVAL & INSTALLATION

Front

INDEPENDENT FRONT SUSPENSION

1. Before servicing the vehicle, refer to the precautions in the beginning of this section.
2. Remove or disconnect the following:
   - Front wheel
   - Upper mounting bolt
   - Lower mounting bolts
   - Shock absorber

To install:

➡On 2003–05 2WD pickups, the upper nut must be replaced, or if-reused, installed using Loctite® 242.

3. Install the shock absorber. Tighten the bolts as follows:
   a. Upper bolt or nut:
   - All 2WD and 2002 4WD Models: 25 ft. lbs. (34 Nm)
   - 2003–05 4WD Models: 40 ft. lbs. (54 Nm)

   b. Lower bolt or nut:
   - All 2WD and 2002 4WD Models: 25 ft. lbs. (34 Nm)
   - 2003–2005 4WD Models: 100 ft. lbs. (135 Nm)
4. Install the front wheel.

LINK/COIL FRONT SUSPENSION

1. Before servicing the vehicle, refer to the precautions in the beginning of this section.

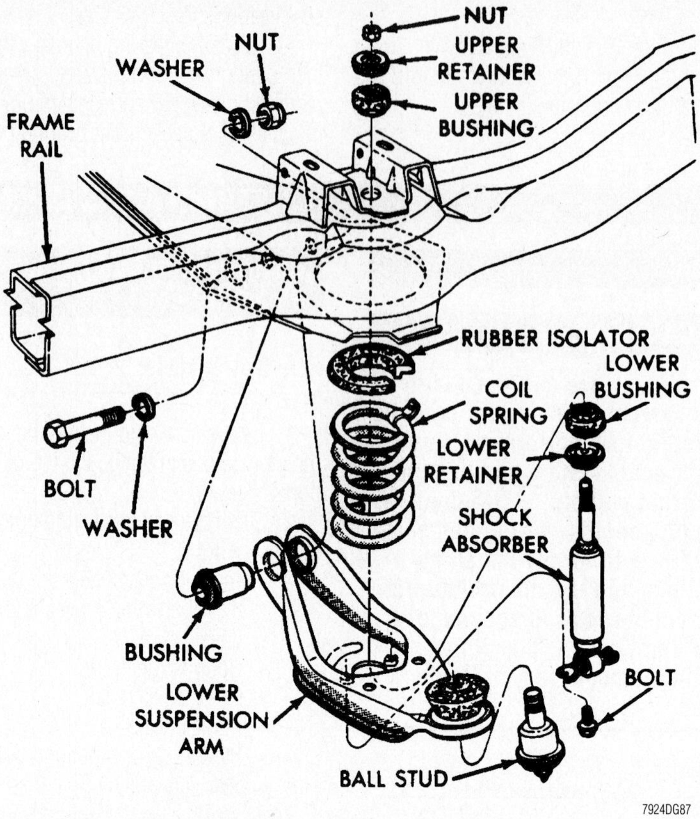

Front shock absorber mounting—Ram Van models

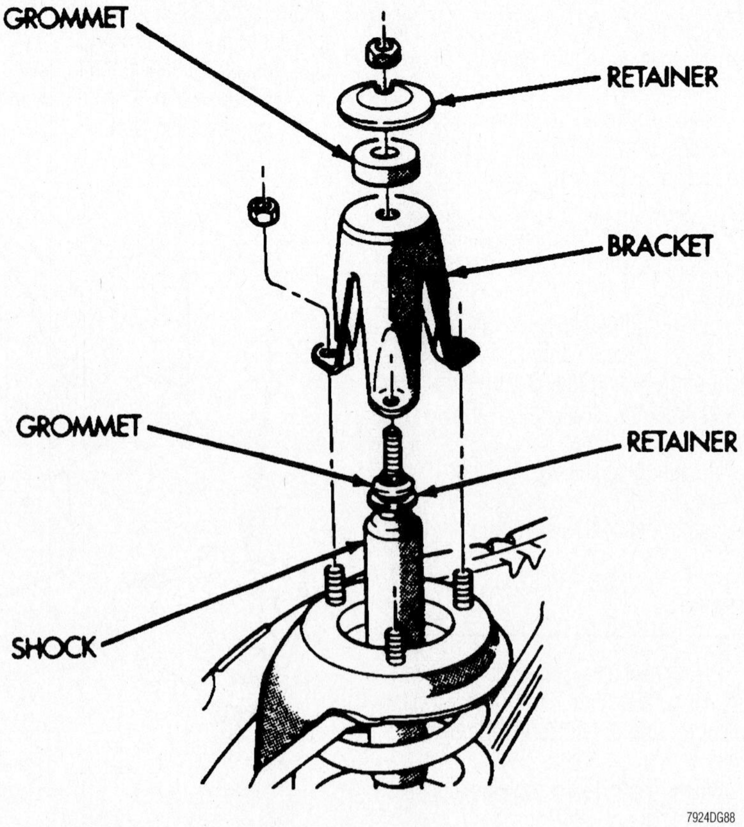

Upper shock absorber mounting—2002 Ram Truck models

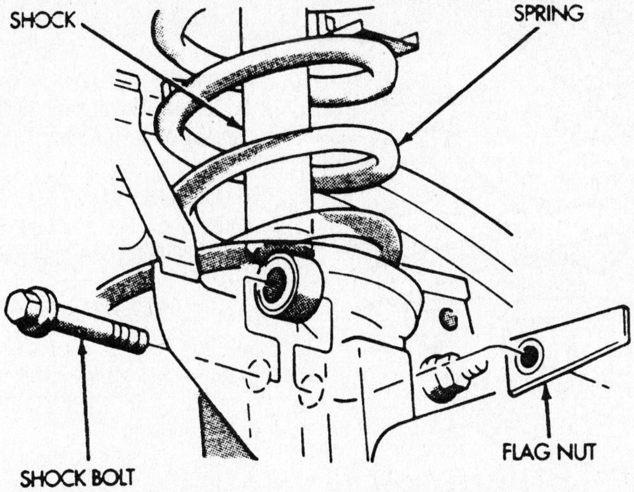

Lower shock absorber mounting—2002 Ram Truck models

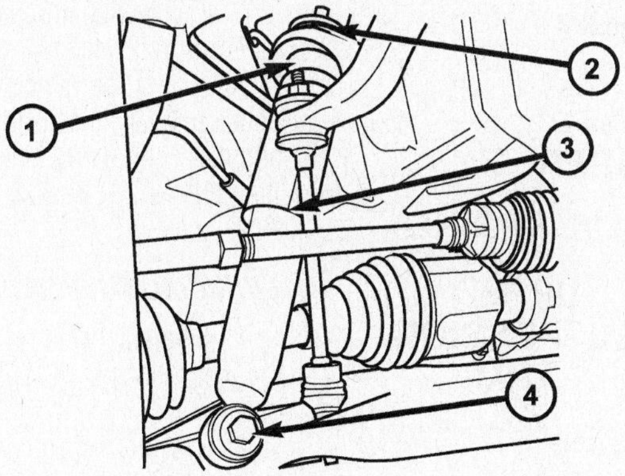

Typical shock absorber mounting—2003-05 Ram Truck models

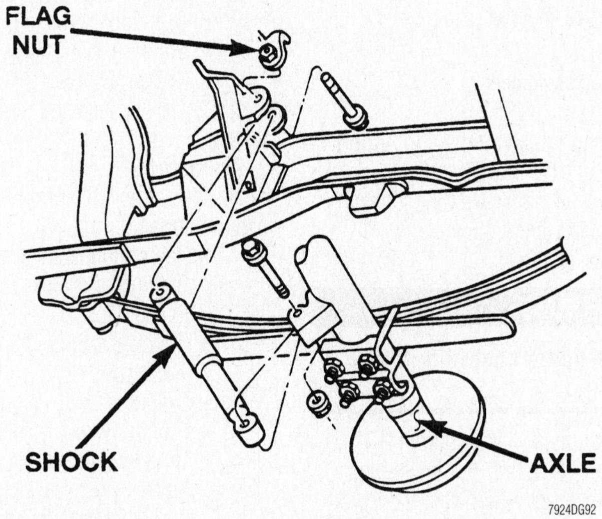

Rear shock absorber mounting—Ram Truck and Van models

2. Remove or disconnect the following:
   - Upper mounting nut
   - 3 upper mounting nuts from the upper shock bracket
   - Lower mounting bolt
   - Shock absorber

To install:

3. Install or connect the following:
   - Shock absorber. Tighten the lower bolt to 100 ft. lbs. (135 Nm); upper bracket nuts to 55 ft. lbs. (75 Nm); Upper shock nut to 40 ft. lbs. (54 Nm)

Rear

1. Before servicing the vehicle, refer to the precautions in the beginning of this section.
2. Support the axle.
3. Remove or disconnect the following:
   - Upper bolt
   - Lower bolt
   - Shock absorber

To install:

4. Install the bolts through the brackets and shock and tighten them as follows:
   - Tighten the upper bolt to 70 ft. lbs. (95 Nm) and the lower bolt to 100 ft. lbs. (136 Nm) on 2002 models; 100 ft. lbs. (135 Nm) for upper and lower on 2003–05 models.

Coil Spring

REMOVAL & INSTALLATION

2002

1. Before servicing the vehicle, refer to the precautions in the beginning of this section.
2. Support the lower control arm on a floor jack.
3. Remove or disconnect the following:
   - Front wheel
   - Brake caliper and rotor
   - Outer tie rod end
   - Stabilizer bar link
   - Lower ball joint
   - Shock absorber
4. Lower the jack and remove the coil spring.

To install:

5. Install the coil spring and raise the control arm into position.
6. Install or connect the following:
   - Shock absorber
   - Lower ball joint. Tighten the nut to 95 ft. lbs. (129 Nm).
   - Stabilizer bar link. Tighten the nut to 27 ft. lbs. (37 Nm).

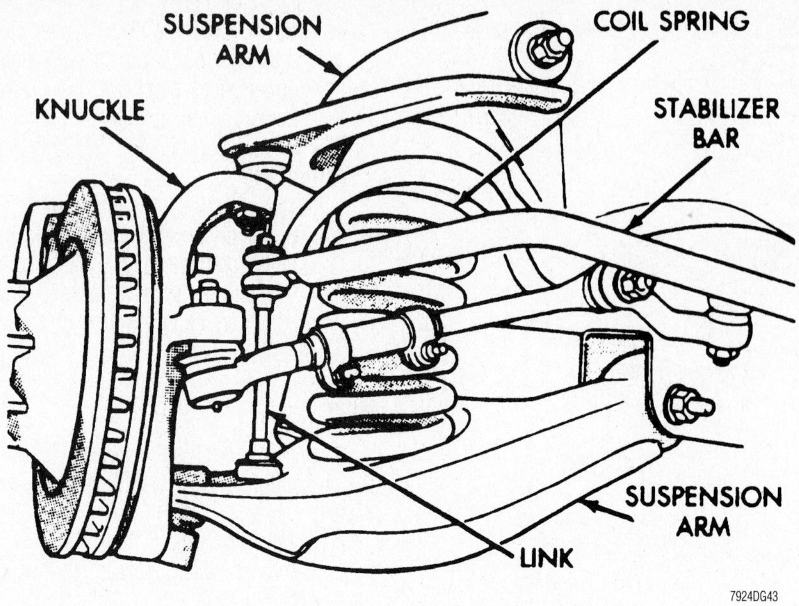

Independent front suspension components—2WD Ram Truck and Van models

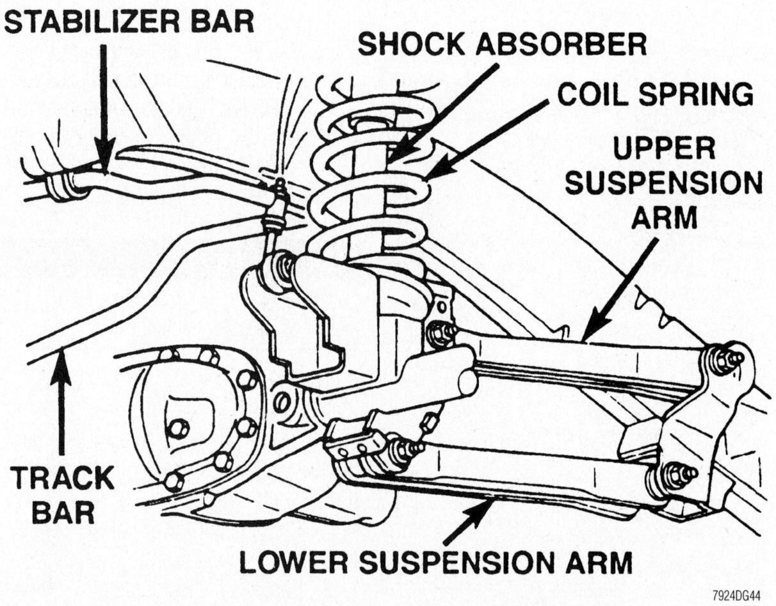

Link and coil front suspension components—4WD Ram Truck and Van models

- Outer tie rod end. Tighten the nut to 65 ft. lbs. (88 Nm).
- Brake caliper and rotor
- Front wheel

2003–2005

INDEPENDENT FRONT SUSPENSION

1. Before servicing the vehicle, refer to the precautions in the beginning of this section.
2. Support the lower control arm on a floor jack.
3. Remove or disconnect the following:
   - Front wheel
   - Shock absorber
4. Compress the spring using Spring Compressor DD-1278 or equivalent.
5. Remove or disconnect the following:
   - Stabilizer bar link
   - Lower ball joint

⁂ WARNING

Support the upper control arm and knuckle.

   - Lower control arm support
6. Loosen the tension on the spring compressor tool slowly allowing the lower suspension arm to pivot downward.
7. Remove the spring compressor tool.
8. Remove the coil spring and isolator pad.

To install:

9. Install the coil spring with the isolator pad and raise the control arm into position.
10. Compress the spring using Spring Compressor DD-1278 or equivalent.
11. Install the Lower ball joint. Tighten the nut as follows:
    a. 1500 Series: 38 ft. lbs. (52 Nm), plus 90°
    b. HD 4x2 Series: 100 ft. lbs. (135 Nm)
    c. All other models: 1500 Series: 38 ft. lbs. (52 Nm)
12. Remove the spring compressor
13. Install or connect the following:
    - Shock absorber
    - Stabilizer bar link. Tighten the nut to 27 ft. lbs. (37 Nm).
    - Lower control arm support
    - Front wheel
14. Lower the vehicle and allow the suspension to take the weight. Tighten the front and rear control arm pivot bolts to 150 ft. lbs. (204 Nm) (LD); 210 ft. lbs. (285 Nm) HD

LINK/COIL SUSPENSION

1. Before servicing the vehicle, refer to the precautions in the beginning of this section.
2. Support the axle on a floor jack.
3. Matchmark the lower arm adjuster and axle bracket.
4. Remove or disconnect the following:
   - Front wheel
   - Upper control arm and loosen the lower arm bolts
   - Front drive shaft, 4x4 models only
   - Track bar from the frame rail bracket
   - Drag link from the pitman arm
   - Stabilizer bar link
   - Shock absorber
5. Lower the jack and remove the coil spring.

To install:

6. Install the coil spring and raise the axle into position.
7. Install or connect the following:
   - Stabilizer bar link. Tighten the nut to 45 ft. lbs. (61 Nm).
   - Shock absorber
   - Track bar
   - Drag link
   - Front drive shaft, 4x4 models only
   - Upper control arm and lower arm bolts. Upper arm nuts to 110 ft. lbs. (149 Nm); lower arm nuts to 160 ft. lbs. (217 Nm)

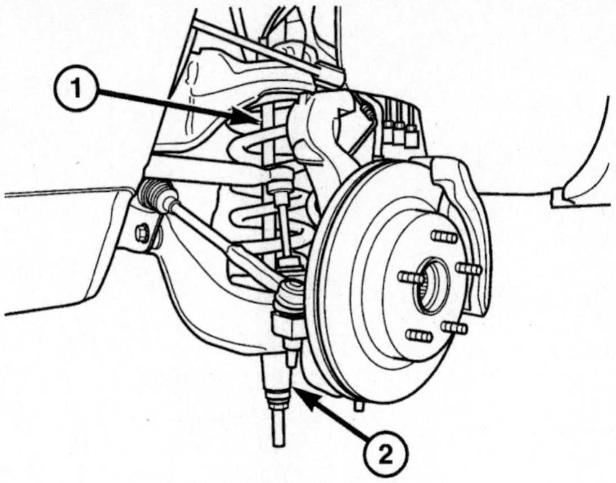

Compressing the spring using compress tool DD-1278.

Torsion Bar

REMOVAL & INSTALLATION

1. Before servicing the vehicle, refer to the precautions in the beginning of this section.
2. Loosen the adjustment bolt to remove spring load. Note the number of turns for installation.
3. Install Special Tool 8686 or equivalent to the anchor arm and the cross member.
4. Increase the tension on the anchor arm tool until the load is removed from the adjustment bolt and the adjuster nut.
5. Turn the adjustment bolt counterclockwise to remove the bolt and the adjuster nut.
6. Remove the Special Tool 8686 or equivalent, allowing the torsion bar to unload.
7. Remove the torsion bar and anchor.
8. Separate the anchor from the torsion bar.

To install:

9. Install the torsion bar ends into the anchor and suspension arm.
10. Position the anchor in the frame crossmember.
11. Install Special Tool 8686 or equivalent to the anchor arm and the cross member.
12. Increase the tension on the anchor in order to load the torsion bar.
13. Install the adjustment bolt and the adjuster nut.
14. Turn adjustment bolt clockwise the recorded amount of turns.
15. Remove Special Tool 8686 from the torsion bar crossmember.
16. Lower the vehicle and align the wheels as necessary.

Leaf Spring

REMOVAL & INSTALLATION

1. Before servicing the vehicle, refer to the precautions in the beginning of this section.
2. Support the vehicle at the frame rails.
3. Support the rear axle with a jack.
4. Remove or disconnect the following:
   - Rear wheel
   - Stabilizer bar link
   - Axle U-bolts
   - Spring bracket
   - Leaf spring

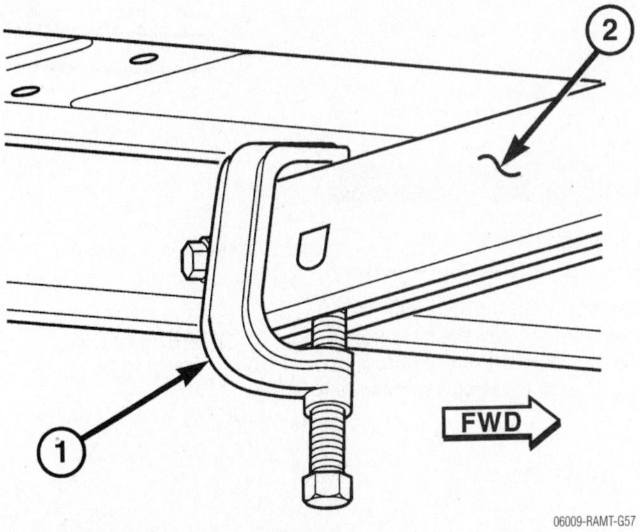

Using special tool 8686 to unload the torsion bar.

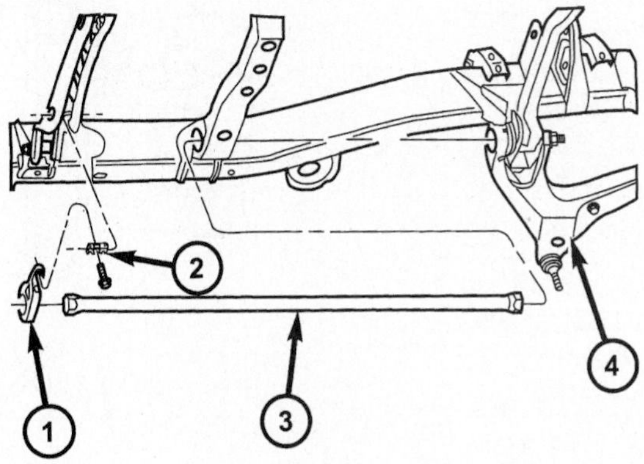

Install the Torsion bar components.

To install:

➡The weight of the vehicle must be supported by the springs when the spring eye and stabilizer bar fasteners are tightened.

5. Install or connect the following:
   - Leaf spring
   - Spring bracket
   - Axle U-bolts. Tighten the nuts to 52 ft. lbs. (70 Nm) on 2002 models; 110 ft. lbs. (149 Nm) on 2003–05 models.
   - Stabilizer bar link
   - Rear wheel
6. On 2002 models, tighten the front spring eye bolt and nut to 115 ft. lbs. (156 Nm). Tighten the rear spring eye bolt and nut to 80 ft. lbs. (108 Nm). Tighten the stabilizer bar nuts 55 ft. lbs. (74 Nm).
7. On 2003–05 LD models, torque the front and rear pivot bolt nuts to 120 ft. lbs. (163 Nm).
8. On 2003–05 HD models, torque the front and rear pivot bolt nuts to 170 ft. lbs. (230 Nm).

Upper Ball Joint

REMOVAL & INSTALLATION

➡The upper ball joint on 2003–05 independent front suspensions, is not replaceable. If the ball joint is defective, the upper arm must be replaced.

2002

2-WHEEL DRIVE

1. Before servicing the vehicle, refer to the precautions in the beginning of this section.
2. Support the lower control arm.
3. Remove the front wheel.
4. Remove the stud nut and separate the ball joint from the steering knuckle.
5. Unscrew the ball joint from the control arm with tool C-3561.

To install:

6. Thread the ball joint into the upper control arm and tighten it to 125 ft. lbs. (169 Nm). Tighten the upper ball stud nut to 135 ft. lbs. (183 Nm).
7. Install the front wheel.

4-WHEEL DRIVE

1. Before servicing the vehicle, refer to the precautions in the beginning of this section.
2. Remove or disconnect the following:
   - Front wheel
   - Brake caliper and rotor

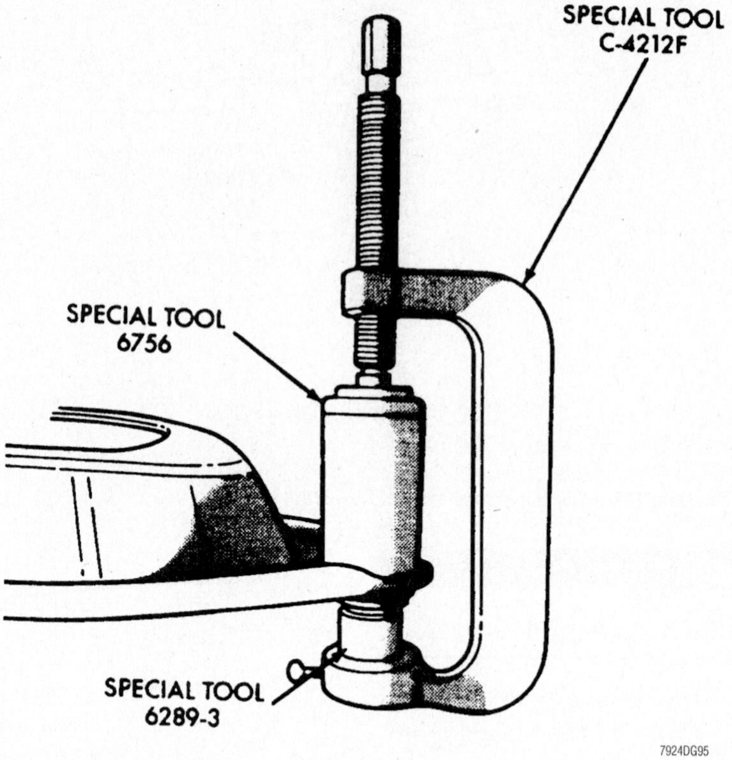

Upper ball joint removal— Ram Truck and Ram Van

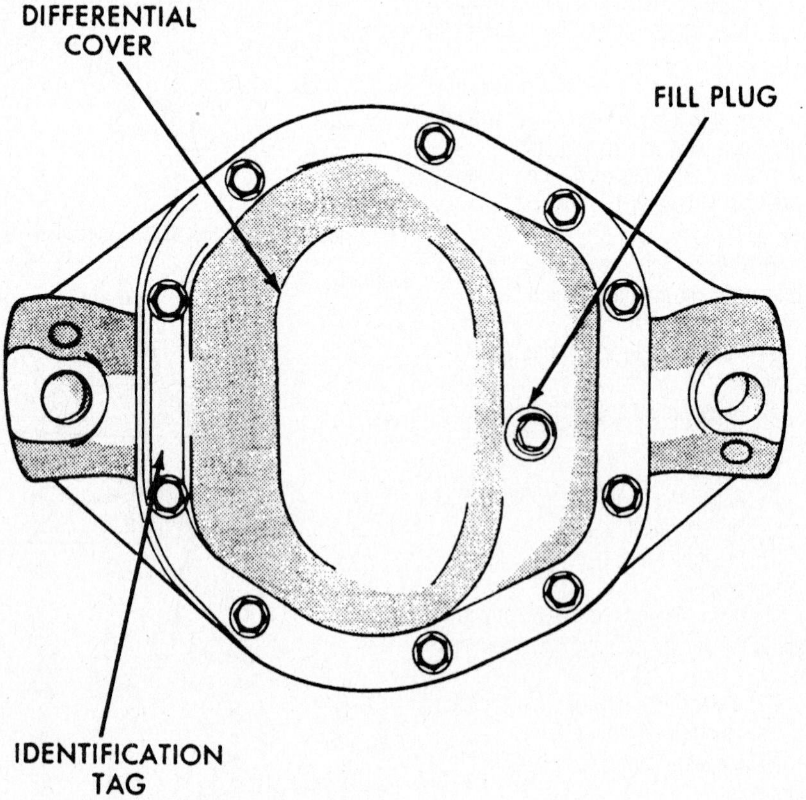

Axle identification—Model 44 front axle differential cover

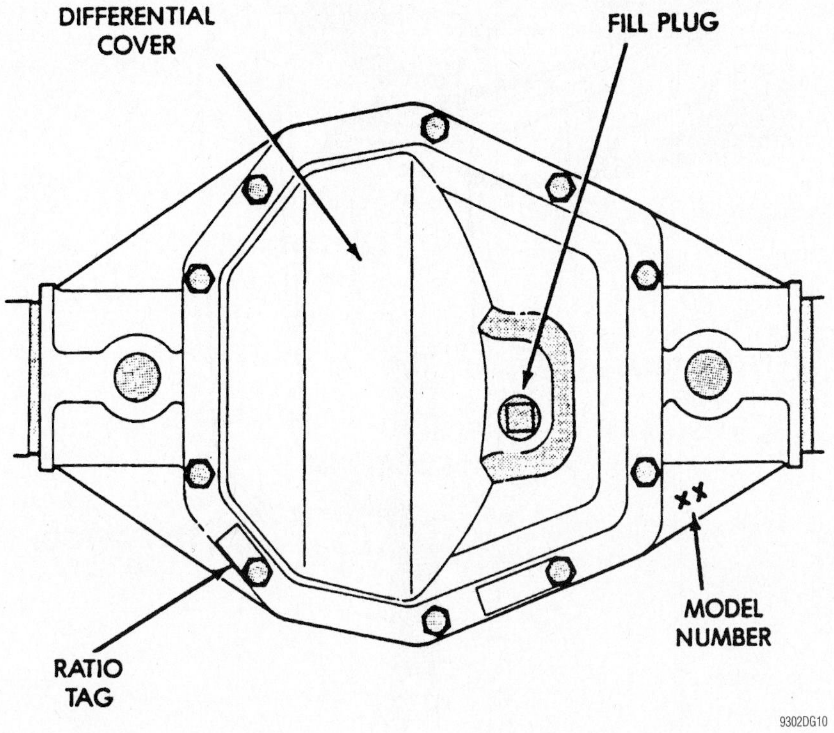

Axle identification—Model 60 front axle differential cover

- Hub retainer
- Axle shaft
- Outer tie rod ends
- Upper and lower ball joint stud nuts

3. If equipped with a Model 44 front axle, use a brass drift and hammer to separate the steering knuckle from the axle tube yoke. Use tool C-4169 to remove the sleeve from the upper yoke arm.

4. Remove the snapring from the ball joint. Install the knuckle in a vise and use tools D-150-1, D-150–3 and C-4212-L to remove the ball joint from the knuckle.

5. If equipped with a Model 60 front axle, remove the bolts from the knuckle lower cap. Dislodge the cap from the steering knuckle and axle tube yoke. Remove the steering knuckle. Use tool D-192 to remove the upper socket pin from the axle tube upper arm bore. Remove the seal.

To install:

6. If equipped with a Model 44 front axle, use tools C-4212-L and C-4288 to force the upper ball joint into the steering knuckle. Install the snapring and install a new rubber boot. Thread the replacement sleeve into the upper yoke bore so that 2 threads are exposed at the top of the yoke. Position the knuckle on the axle tube yoke

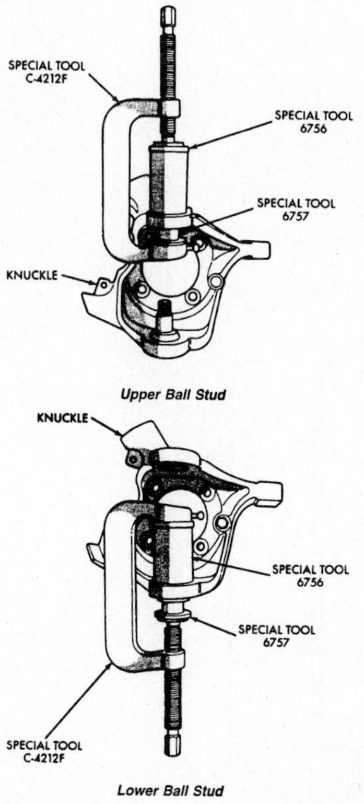

Typical ball joint removal using the special tools—248 FBI axle shown

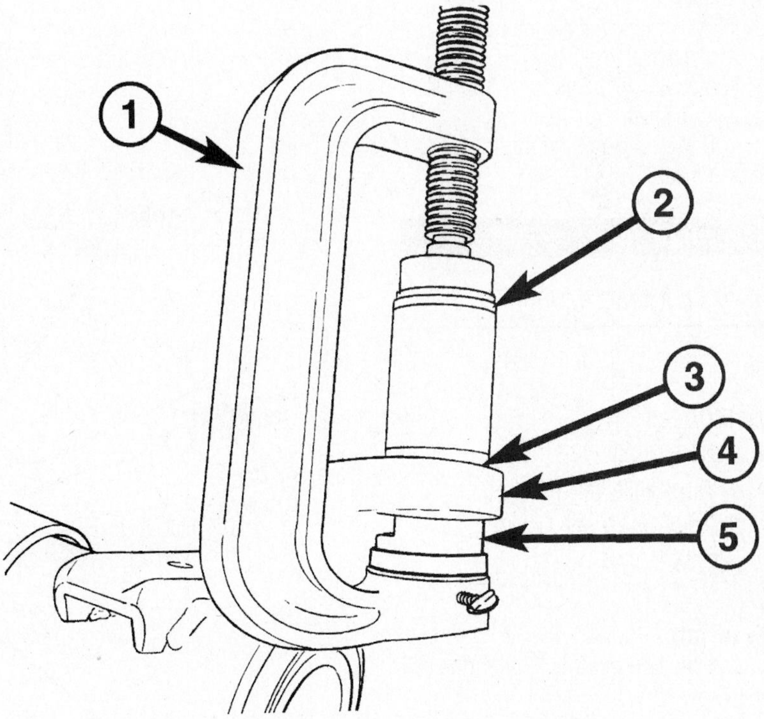

Installing the upper ball joint using the installation tools.

and install a new lower ball stud nut, then tighten to 80 ft. lbs. (108 Nm). Using the special socket, tighten the sleeve to 40 ft. lbs. (54 Nm). Install the upper ball stud nut and tighten to 100 ft. lbs. (136 Nm) and install a new cotter pin.

7. If equipped with a Model 60 front axle, use tool D-192 to install the upper socket pin in the axle tube upper arm bore. Install a new seal. Tighten to 500–600 ft. lbs. (668–813 Nm). Position the knuckle over the socket pin. Fill the lower socket cavity with grease. Install the lower cap and tighten the bolts to 80 ft. lbs. (108 Nm).

8. Install or connect the following:
- Outer tie rod ends
- Axle shaft
- Hub retainer
- Brake caliper and rotor
- Front wheel

2003–2005 Link/Coil Suspension

1. Before servicing the vehicle, refer to the precautions in the beginning of this section.
2. Remove or disconnect the following:
   - Wheel
   - Brake caliper and rotor
   - Axle shaft
   - Upper Ball Joint, using tool 6761 and 8445-3 with C-4212-F

To install:

3. Install or connect the following:
   - Upper ball joint, using driver 8445-2, receiver 8975-5 and tool C-4212-F
   - Axle shaft
   - Brake caliper and rotor
   - Wheel

Lower Ball Joint

REMOVAL & INSTALLATION

2002

2WD MODELS

- Front wheel
- Shock absorber
- Stabilizer bar link
- Coil spring

1. Press the ball joint out of the control arm.

To install:

2. Use the remover tool to press the ball joint into the arm.
3. Install or connect the following:
   - Coil spring. Tighten 11⁄16 lower ball joint nuts to 135 ft. lbs. (183 Nm). Tighten 3⁄4 nuts to 175 ft. lbs. (237 Nm).

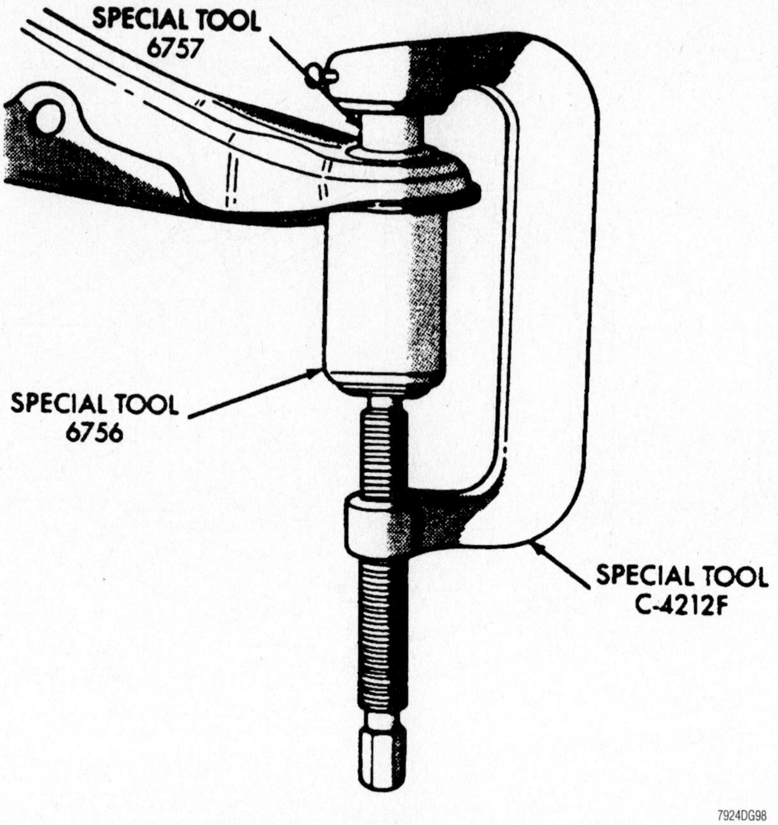

Lower ball joint removal—2WD Ram Truck and Van models

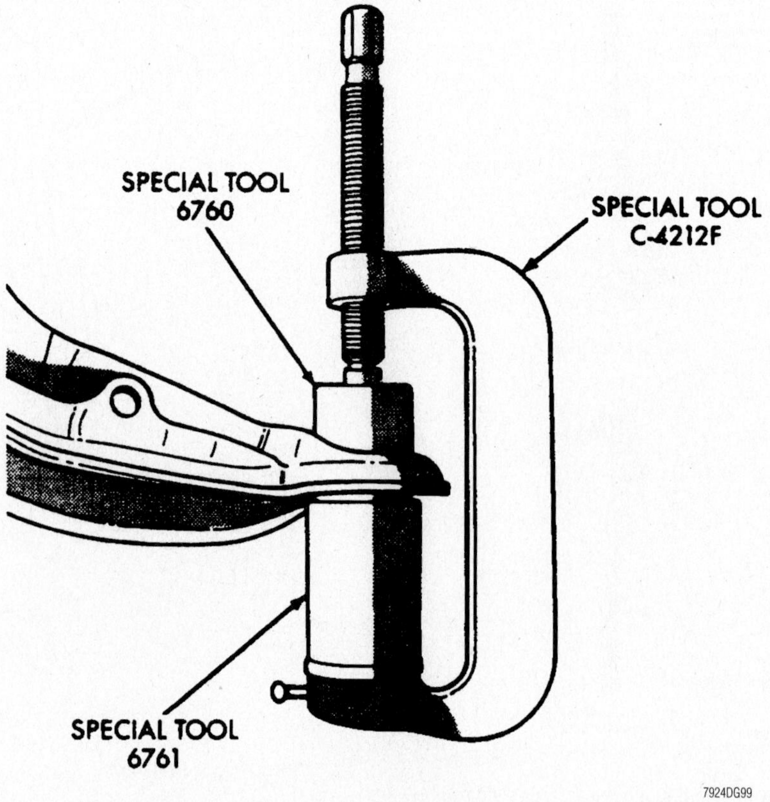

Lower ball joint installation—2WD Ram Truck and Van models

- Stabilizer bar link
- Shock absorber
- Front wheel

4WD MODELS

1. Before servicing the vehicle, refer to the precautions in the beginning of this section.
2. Remove or disconnect the following:
   - Front wheel
   - Brake caliper and rotor
   - Hub retainer
   - Axle shaft
   - Outer tie rod ends
   - Upper and lower ball joint stud nuts
3. If equipped with a Model 44 front axle, use a brass drift and hammer to separate the steering knuckle from the axle tube yoke.
4. Remove the snapring from the ball joint. Install the knuckle in a vise and use tools D-150-1, D-150–3 and C-4212-L to remove the ball joint from the knuckle.
5. If equipped with a Dana 60 front axle, use tools C-4212-L, C-4366–1 and C-4366-2 (or equivalents) to remove the lower ball joint.

To install:

6. If equipped with a Model 44 front axle, use tools C-4212-L and C-4288 to force the lower ball joint into the steering knuckle. Install the snapring and install a new rubber boot. Position the knuckle on the axle tube yoke and install a new lower ball stud nut. Tighten to 80 ft. lbs. (108 Nm). Install the upper ball stud nut and tighten to 100 ft. lbs. (136 Nm).
7. If equipped with a Model 60 front axle, use tools C-4212-L, C-4366-3 and C-4366-4 to install the seal and lower bearing cup into the axle tube yoke lower bore. Reposition the tools and install the lower bearing and seal into the bore. Position the knuckle over the socket pin. Fill the lower socket cavity with grease. Install the lower cap and tighten the bolts to 80 ft. lbs. (108 Nm).
8. Install or connect the following:
   - Outer tie rod ends
   - Axle shaft
   - Hub retainer
   - Brake caliper and rotor
   - Front wheel

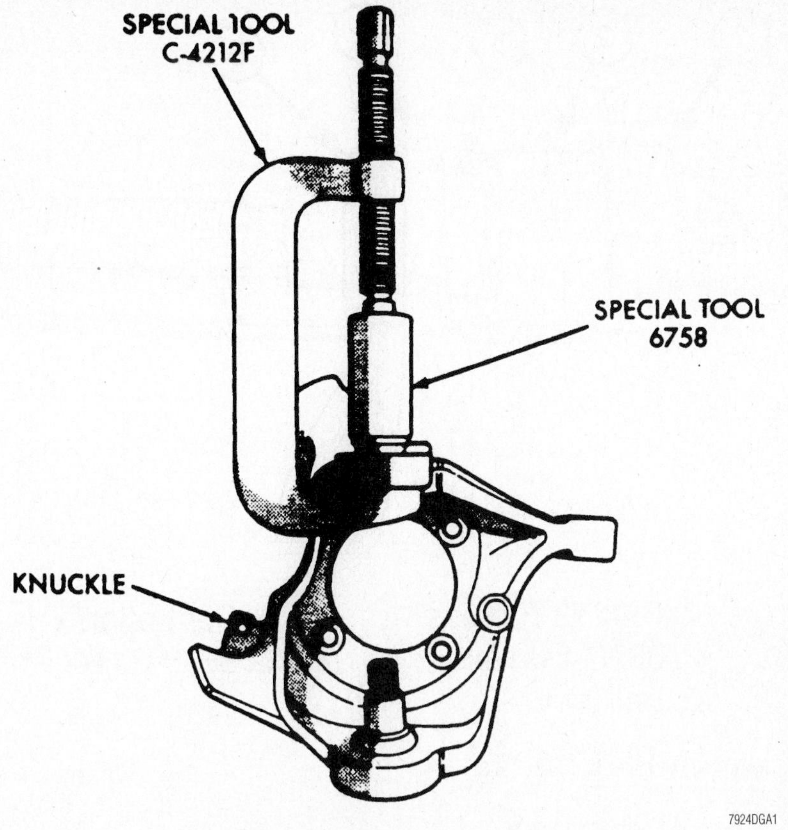

Lower ball joint installation—2002 4WD Ram Truck and Van models

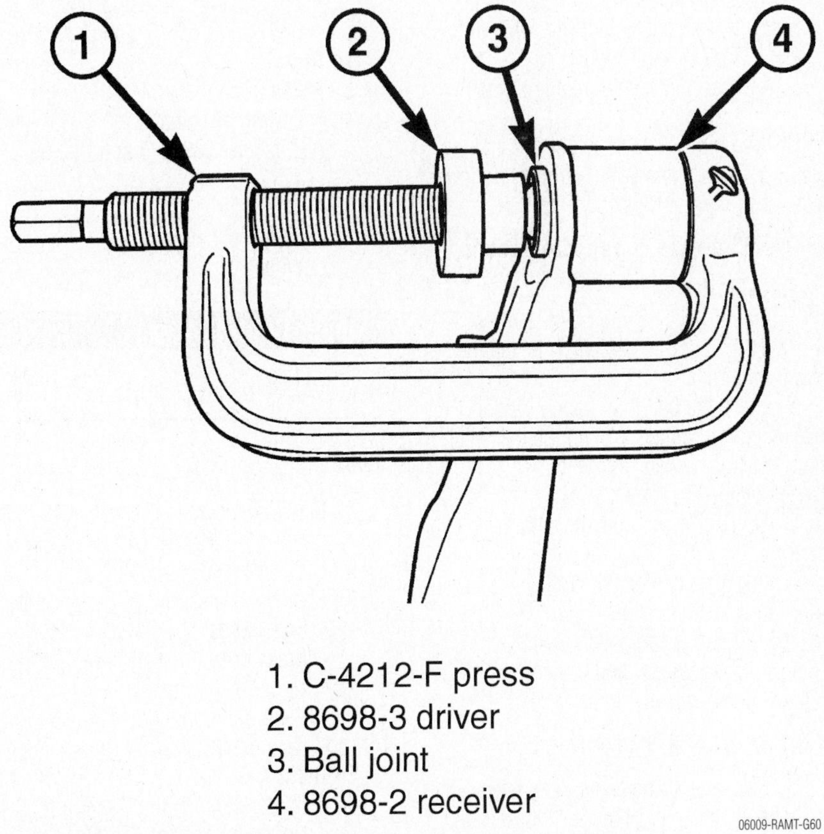

Lower ball joint removal tools—IFS

2003–2005

INDEPENDENT FRONT SUSPENSION

1. Before servicing the vehicle, refer to the precautions in the beginning of this section.
2. Remove or disconnect the following:
   - Front wheel

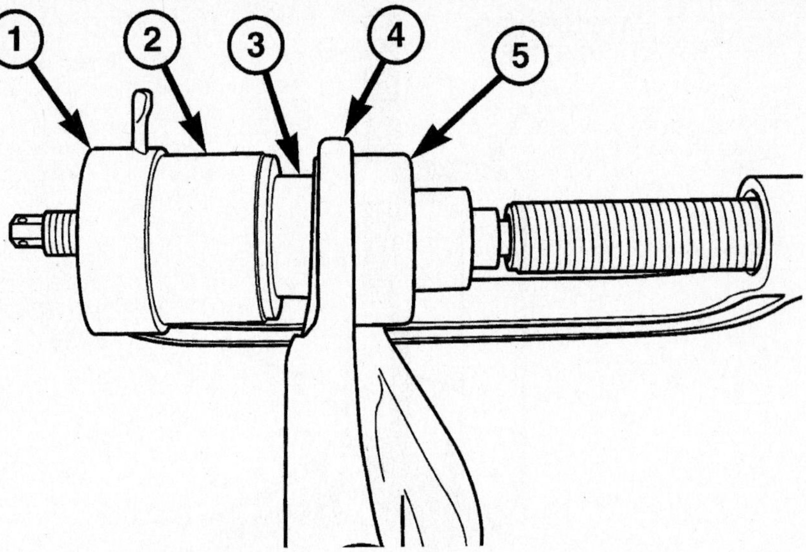

1. C-4212-F press
2. 8698-1 driver
3. Ball joint
4. Control arm
5. 8698-3 receiver

06009-RAMT-G61

Installing the lower ball joint—IFS

- Brake caliper and rotor
- Outer tie rod ends from the steering knuckle
- Lower ball joint nut from the steering knuckle using Special Tool 8677 or equivalent
- Snapring from the ball joint (HD 4x2 only)

3. Use tools C-4212-F, 8698-2 and 8698-3 (or equivalents) to remove the lower ball joint.

➡Use EP grease on the tool threads.

To install:

4. Using Special Tools C-4212-F, 8698-1 and 8698-3 (or equivalents), install the lower ball joint.
5. Stake the ball joint flange in four evenly spaced places around the ball joint flange, using a chisel and hammer (LD Series only).
6. On HD 4x2 models, install a new snapring
7. Install or connect the following:
   - Knuckle.
   - Outer tie rod ends
   - Brake caliper and rotor
   - Front wheel

LINK/COIL SUSPENSION

1. Before servicing the vehicle, refer to the precautions in the beginning of this section.
2. Remove or disconnect the following:
   - Wheel
   - Brake caliper and rotor
   - Axle shaft
   - Lower ball joint, using Special Tool 6761 and 8445-3 with C-4212-F

To install:

3. Install or connect the following:
   - Lower ball joint, using driver 8445-2, receiver 8975-5 and tool C-4212-F.
   - Axle shaft
   - Brake caliper and rotor
   - Wheel

Upper Control Arm

REMOVAL & INSTALLATION

2002

1. Before servicing the vehicle, refer to the precautions in the beginning of this section.
2. Support the lower control arm.
3. Remove or disconnect the following:
   - Front wheel
   - Brake hose brackets
   - Upper ball joint
   - Pivot mounting nuts
   - Upper control arm

To install:

4. Install or connect the following:
   - Upper control arm. Tighten the pivot nuts to 155 ft. lbs. (210 Nm).
   - Upper ball joint. Tighten the nut to 60 ft. lbs. (81 Nm).
   - Brake hose brackets
   - Front wheel
5. Check the wheel alignment and adjust as necessary.

2003–2005

INDEPENDENT FRONT SUSPENSION

1. Before servicing the vehicle, refer to the precautions in the beginning of this section.
2. Support the lower control arm.
3. Remove or disconnect the following:
   - Front wheel
   - Upper ball joint
   - Pivot mounting nuts
   - Upper control arm

To install:

4. Install the upper control arm. Tighten the pivot nuts as follows:
   a. LD Series: 97 ft. lbs. (132 Nm)
   b. HD 4x2: 125 ft. lbs. (170 Nm)
5. Install or connect the following:
   - Upper ball joint
   - Front wheel

LINK/COIL SUSPENSION—LEFT SIDE

1. Before servicing the vehicle, refer to the precautions in the beginning of this section.
2. Support the lower control arm.
3. Remove or disconnect the following:
   - Front wheel
   - Nut and bolt at the axle bracket
   - Nut and bolt at the frame rail
   - Upper control arm

To install:

➡New nuts and bolts should be used when servicing the suspension arms.

4. Position the control arm and install all fasteners finger-tight.
5. Install the front wheel.
6. Lower the vehicle to load the suspension.
7. Tighten all the fasteners to 120 ft. lbs. (163 Nm).

LINK/COIL SUSPENSION—RIGHT SIDE

1. Before servicing the vehicle, refer to the precautions in the beginning of this section.
2. Support the lower control arm.
3. Remove or disconnect the following:
   - Front wheel
   - Exhaust system at the manifolds
   - Exhaust mounts at the muffler
4. Support the transmission
5. Remove or disconnect the following:

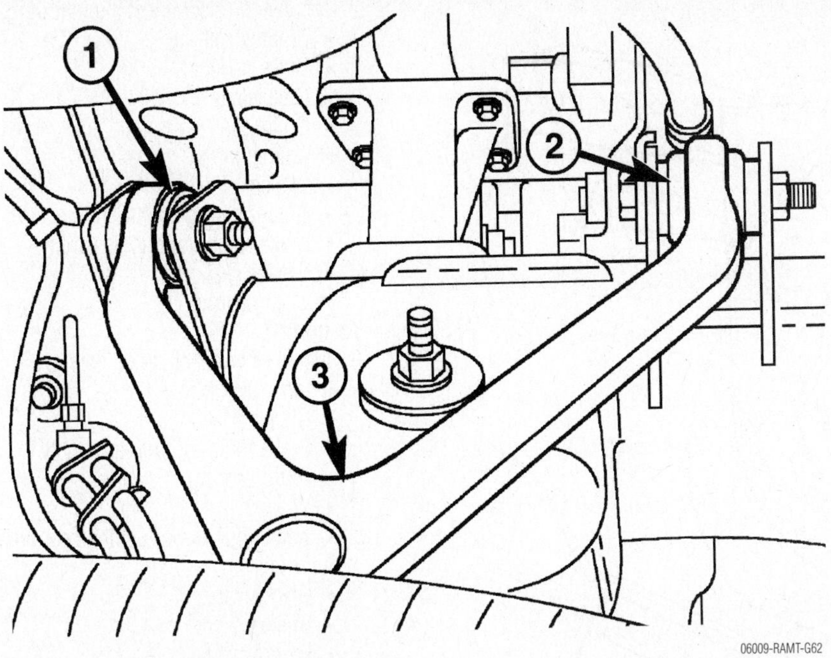

Location of the control arm pivot mounting nuts

- Transmission crossmember
- Nut and bolt at the axle bracket
- Nut and bolt at the frame rail
- Upper control arm

To install:

6. Position the control arm and install all fasteners finger-tight.
7. Install or connect the following:
   - Exhaust mounts at the muffler
   - Exhaust system at the manifolds
   - Transmission crossmember
8. Install the front wheel. Lower the vehicle to load the suspension.
9. Tighten all the fasteners to 120 ft. lbs. (163 Nm).

CONTROL ARM BUSHING REPLACEMENT

The control arm bushings are serviced with the control arm as an assembly.

Lower Control Arm

REMOVAL & INSTALLATION

2002

1. Before servicing the vehicle, refer to the precautions in the beginning of this section.
2. Support the lower control arm with a floor jack.
3. Remove or disconnect the following:
   - Front wheel
   - Brake caliper and rotor
   - Stabilizer bar link
   - Lower ball joint
   - Coil spring
   - Crossmember nuts
   - Lower control arm

To install:

4. Install or connect the following:
   - Lower control arm. Tighten the crossmember nuts to 145 ft. lbs. (196 Nm).
   - Coil spring
   - Lower ball joint. Tighten the nut to 135 ft. lbs. (183 Nm).
   - Stabilizer bar link
   - Brake caliper and rotor
   - Front wheel

2003–2005

INDEPENDENT FRONT SUSPENSION—2WD

1. Before servicing the vehicle, refer to the precautions in the beginning of this section.
2. Support the lower control arm with a floor jack.
3. Remove or disconnect the following:
   - Front wheel
   - Brake caliper and rotor
   - Shock
4. Install Spring Compressor DD-1278 and tighten the spring compressor nut against bell-shaped adapter finger tight then loosen 1/2 turn.
5. Remove or disconnect the following:
   - Stabilizer bar link
   - Lower ball joint nut
6. Support the upper control arm and steering knuckle out of the way.
7. Tighten the spring compressor tool to allow clearance for the lower ball joint to be removed out of the knuckle.
8. Loosen the tension on the spring compressor tool slowly allowing the lower suspension arm to pivot downward.
9. Remove or disconnect the following:
   - Spring compressor tool
   - Coil spring and isolator pad
   - Lower control arm front and rear pivot bolts
   - Lower control arm

To install:

10. Install the lower control arm and pivot bolts finger tight.
11. Install the coil spring into the frame pocket
12. Install the Spring Compressor DD-1278 and tighten the tool nut to compress the coil spring.
13. Remove the support from the upper control arm and steering knuckle.
14. Install the lower ball joint
15. Support the lower control arm at the outboard side of the lower control arm to support vehicle weight.
16. Remove the spring compressor tool.
17. Install the shock absorber and stabilizer link.
18. Remove the lower control arm support and install the wheel.
19. Lower the vehicle to the floor with vehicle weight and tighten the front and rear control arm pivot bolts as follows:
    a. LD Series: 150 ft. lbs. (204 Nm)
    b. HD 4x2: 210 ft. lbs. (285 Nm)
20. Align the vehicle.

COIL SPRING SUSPENSION—4WD

1. Before servicing the vehicle, refer to the precautions in the beginning of this section.
2. Remove or disconnect the following:
   - Front wheel
   - Upper ball joint from the knuckle
   - Halfshaft
   - Torsion bar
   - Shock absorber lower bolt
   - Stabilizer bar link
   - Lower ball joint from the knuckle
   - Lower control arm

To install:

3. Install or connect the following:
   - Lower control arm. Tighten bolts finger-tight.
   - Lower ball joint
   - Torsion bar
   - Shock absorber lower bolt. Torque it to 100 ft. lbs. (135 Nm).
   - Halfshaft

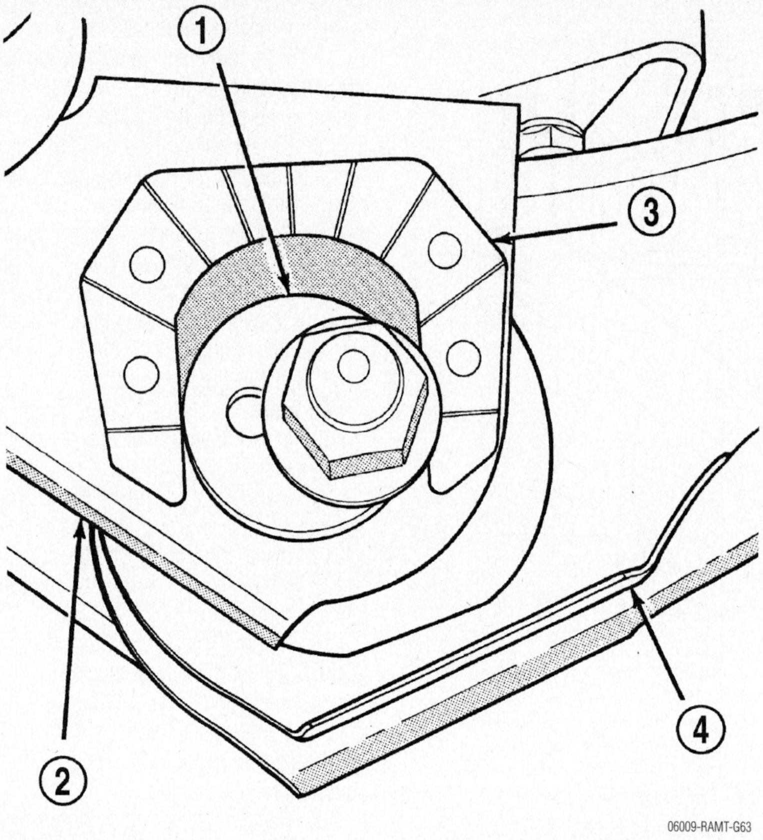

Matchmark cam adjusters for reinstallation.

- Upper ball joint
- Stabilizer bar link

4. Tighten the control arm pivot bolts to 150 ft. lbs. (204 Nm).
   - Front wheel

COIL/LINK SUSPENSION

1. Before servicing the vehicle, refer to the precautions in the beginning of this section.
2. Paint or scribe matchmarks on the cam adjusters and suspension arm.
3. Remove or disconnect the following:
   - Lower arm nut, cam and cam bolt
   - Nut and bolt from the frame rail
   - Lower arm

To install:

4. Install or connect the following:
   - Lower control arm. Tighten bolts finger-tight. Align the reference marks.
5. Lower the truck to load the suspension
6. Tighten the cam nuts to 160 ft. lbs. (217 Nm).

CONTROL ARM BUSHING REPLACEMENT

The control arm bushings are serviced with the control arm as an assembly.

Front Wheel Bearings

ADJUSTMENT

2002 Pickups; 2002–03 Vans

1. Before servicing the vehicle, refer to the precautions in the beginning of this section.
2. Tighten the wheel bearing nut to 30–40 ft. lbs. (41–54 Nm) while turning the rotor.
3. Loosen the wheel bearing adjusting nut completely.
4. Tighten the nut finger-tight.
5. Check the wheel bearing end-play. The specification is 0.001–0.003 inch (0.025–0.076mm).
6. Install the nut lock and cotter pin.

2003–2005 Pickups

These are unitized hub/bearing assemblies. No adjustment is possible.

REMOVAL & INSTALLATION

2002

2 WHEEL DRIVE

1. Before servicing the vehicle, refer to the precautions in the beginning of this section.
2. Remove or disconnect the following:
   - Front wheel
   - Brake caliper
   - Grease cap
   - Cotter pin
   - Adjusting nut
   - Washer
   - Outer bearing
   - Brake rotor and hub assembly
   - Grease seal
   - Inner bearing

To install:

3. Install or connect the following:
   - Inner bearing
   - Grease seal
   - Brake rotor and hub assembly
   - Outer bearing
   - Washer
   - Adjusting nut. Adjust the bearings.
   - Cotter pin
   - Grease cap
   - Brake caliper
   - Front wheel

4 WHEEL DRIVE

1. Before servicing the vehicle, refer to the precautions in the beginning of this section.
2. Remove or disconnect the following:
   - Front wheel
   - Brake caliper and rotor
   - Hub retainer nut
   - Hub and bearing assembly

To install:

3. Install or connect the following:
   - Hub and bearing assembly. Tighten the bolts to 125 ft. lbs. (170 Nm).
   - Hub retainer nut. Tighten the nut to 175 ft. lbs. (237 Nm).
   - Brake caliper and rotor
   - Front wheel

2003–2005 Pickups

INDEPENDENT FRONT SUSPENSION—2 WHEEL DRIVE

1. Before servicing the vehicle, refer to the precautions in the beginning of this section.
2. Remove or disconnect the following:
   - Front wheel
   - Brake caliper and rotor
   - ABS wheel speed sensor, if equipped
   - Hub/bearing mounting bolts
   - Hub/bearing assembly

To install:

3. Install the hub/bearing assembly. Tighten the bolts as follows:
   a. LD Series: 120 ft. lbs. (163 Nm)
   b. HD Series: 130 ft. lbs. (176 Nm)
4. Install or connect the following:

- Wheel speed sensor
- Brake caliper and rotor
- Front wheel

INDEPENDENT FRONT SUSPENSION—4 WHEEL DRIVE

1. Before servicing the vehicle, refer to the precautions in the beginning of this section.
2. Remove or disconnect the following:
   - Front wheel
   - Brake caliper and rotor
   - ABS wheel speed sensor, if equipped
   - Halfshaft nut
   - Tie rod end from the knuckle
   - Upper ball joint from the knuckle
3. Pull down on the steering knuckle to separate the halfshaft from the hub/bearing assembly.
4. Remove the three hub/bearing mounting bolts from the steering knuckle.
5. Remove the hub/bearing assembly.

To install:

6. Install or connect the following:
   - Hub and bearing assembly. Tighten the bolts to 120 ft. lbs. (163 Nm).
   - Upper ball joint to the knuckle
   - Tie rod end to the knuckle. Tighten the nut to 45 ft. lbs. (61 Nm) plus 90°.
   - Halfshaft nut. Torque to 185 ft. lbs. (251 Nm).
   - ABS wheel speed sensor, if equipped
   - Brake caliper and rotor
   - Front wheel

COIL/LINK SUSPENSION

1. Before servicing the vehicle, refer to the precautions in the beginning of this section.
2. Remove or disconnect the following:
   - Front wheel
   - Hub extension from the rotor, if equipped
   - Brake caliper
   - Hub nut
   - ABS wheel speed sensor wiring, if equipped
3. Back off each hub/bearing mounting bolt ¼ inch each. Then, tap the bolts with a mallet to loosen the hub/bearing assembly.

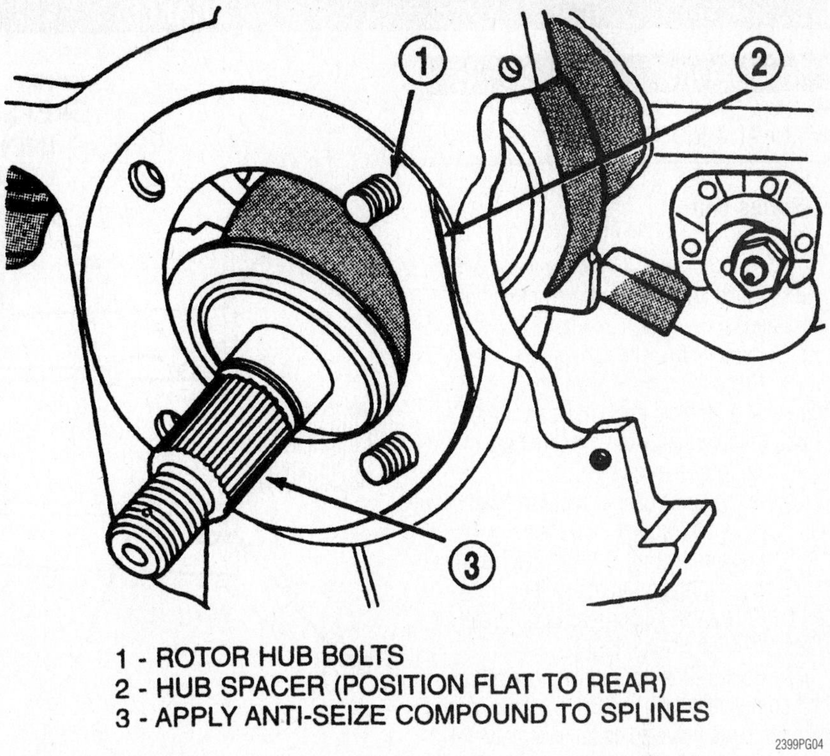

Hub bolt positioning

4. Remove the hub/bearing assembly, wheel studs/extension studs, rotor, shield and spacer.
5. Remove the wheel speed sensor from the bearing assembly, if equipped.
6. Press out the wheel studs/hub extension studs and separate the rotor from the hub/bearing assembly.

To install:

7. Install the wheel speed sensor in the hub bearing, if equipped.
8. Position the rotor on the hub/bearing.
9. Press the wheel studs/hub extension studs through the back side of the rotor and through the hub/bearing flange.
10. Apply a liberal amount of anti-seize compound on the splines of the front halfshaft.
11. Insert the 2 rearmost, top and bottom rotor hub bolts in the knuckle, so they extend from the front face as shown.
12. Position the hub spacer.
13. Position the brake shield on the bolts just installed on the knuckle.

➡If equipped with wheel speed sensor, the shield must be positioned on the hub bearing.

14. Align the rotor hub with the drive shaft and start the shaft into the rotor hub splines.
15. Align the bolt holes in the flange with the installed bolts. Thread the bolts onto the flange far enough to hold the unit in place.
16. Install the remaining bolts. Tighten all hub/bearing bolts to 149 ft. lbs. (202 Nm).
17. Install the washer and shaft nut. Tighten the nut to 132 ft. lbs. (179 Nm). Rotate the axle 5 to 10 times to seat the bearings. Tighten the nut to 263 ft. lbs. (356 Nm). Install a new cotter pin, advancing the nut for alignment.
18. Install or connect the following:
    - ABS wheel speed sensor wiring, if equipped
    - Brake caliper
    - Hub extension
    - Front wheel

BRAKES

Brake Caliper

REMOVAL & INSTALLATION

B-Series Van

FRONT

1. Raise and support the front end on jackstands.
2. Remove the wheels.
3. Press the caliper piston back into the bore with a suitable prytool. Use a large C-clamp to drive the piston into the bore of additional force is required.
4. Remove the caliper mounting bolts with a 3/8 in. hex wrench or socket.
5. Loosen the bolt that secures the front brake hose fitting bolt in the caliper.
6. Rotate the caliper rearward off the rotor and out from its mount.
7. Remove the front brake hose fitting bolt completely, then remove the caliper with the pads installed as an assembly. Take care not to drip fluid onto the pad surfaces.
8. Cover the open end of the front brake hose fitting to prevent dirt entry.

To install:

9. Clean the caliper and steering knuckle sliding surfaces with a wire brush. Then, apply a coat of Mopar® multi-mileage grease or equivalent.
10. Lubricate the caliper mounting bolts, collars, bushings and bores with Dow 111® or GE 661® silicone grease or equivalent.
11. Install the caliper over the rotor and seat it in its original position until flush.
12. Install the mounting bolts by hand, then tighten them to 38 ft. lbs. (51 Nm) for 2001–02 models; 24 ft. lbs. (33 Nm) for 2003–04 models.
13. Install the wheels.
14. Lower the vehicle.
15. Pump the brakes several times to seat the pads.

REAR

1. Before servicing the vehicle, refer to the precautions in the beginning of this section.
2. Compress the piston.
3. Remove or disconnect the following:
   - Caliper pin bolts
   - Banjo bolt
   - Caliper
4. Installation is the reverse of removal. Use new copper washers. On 2001–02 models, torque the caliper pin bolts to 25 ft. lbs. (33Nm) and the banjo bolt to 28 ft. lbs. (38Nm). On 2003–04 models, torque the

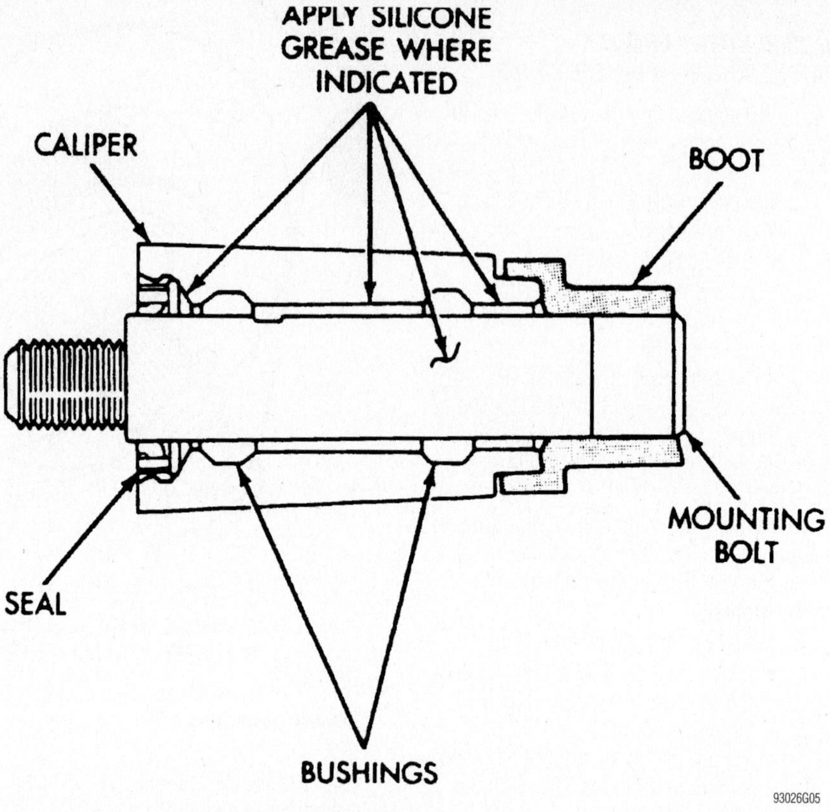

Mounting bolt lubrication points—Ram Pick-Up 75mm caliper

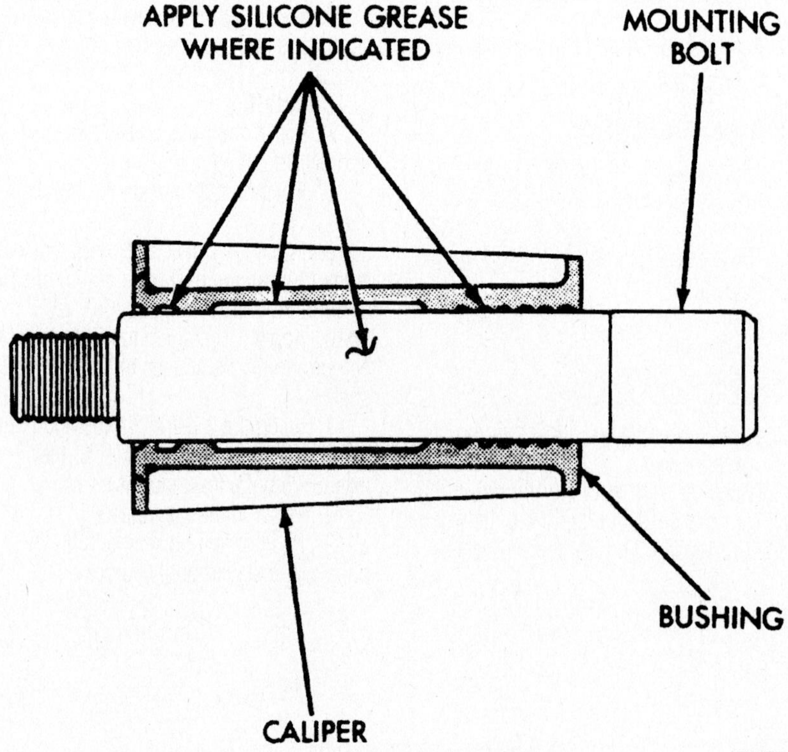

Mounting bolt lubrication points—Ram Pick-Up 80 or 86mm caliper pin bolts to 24 ft. lbs. (32 Nm) and the banjo bolt to 21 ft. lbs. (28 Nm).

Ram Pick-Up

FRONT

1. Raise and support the front end on jackstands.
2. Remove the wheels.
3. Press the caliper piston back into the bore with a suitable prytool. Use a large C-clamp to drive the piston into the bore of additional force is required.
4. Remove the caliper mounting bolts with a ⅜ in. hex wrench or socket.
5. Loosen the bolt that secures the front brake hose fitting bolt in the caliper if caliper is to be removed from the vehicle.
6. Rotate the caliper rearward off the rotor and out from its mount.
7. Remove the front brake hose fitting bolt completely, then remove the caliper with the pads installed as an assembly. Take care not to drip fluid onto the pad surfaces.
8. Cover the open end of the front brake hose fitting to prevent dirt entry.

To install:

9. Clean the caliper and steering knuckle sliding surfaces with a wire brush. Then, apply a coat of Mopar® multi-mileage grease or equivalent.
10. Lubricate the caliper mounting bolts, collars, bushings and bores with Dow 111® or GE 661® silicone grease or equivalent.
11. Install the caliper over the rotor and seat it in its original position until flush.
12. Install the mounting bolts by hand, then tighten them to 38 ft. lbs. (51 Nm) on 2002 models; 24 ft. lbs. (32 Nm) on 2003–05 models.
13. Attach the brake hose, using new washers, and torque the bolt to 20 ft. lbs. (27 Nm).
14. Install the wheels.
15. Lower the vehicle.
16. Pump the brakes several times to seat the pads.

REAR

1. Before servicing the vehicle, refer to the precautions in the beginning of this section.
2. Compress the piston.
3. Remove or disconnect the following:
   - Rear wheel
   - Caliper slide bolts
   - Brake hose banjo bolt if caliper is being replaced
   - Caliper
4. Installation is the reverse of removal. Use new copper washers. On 2002 models, torque the caliper pin bolts to 25 ft. lbs. (33Nm) and the banjo bolt to 28 ft. lbs. (38Nm). On 2003–04 models, torque the pin bolts to 22 ft. lbs. (30 Nm) and the banjo bolt to 245 inch lbs. (27 Nm).

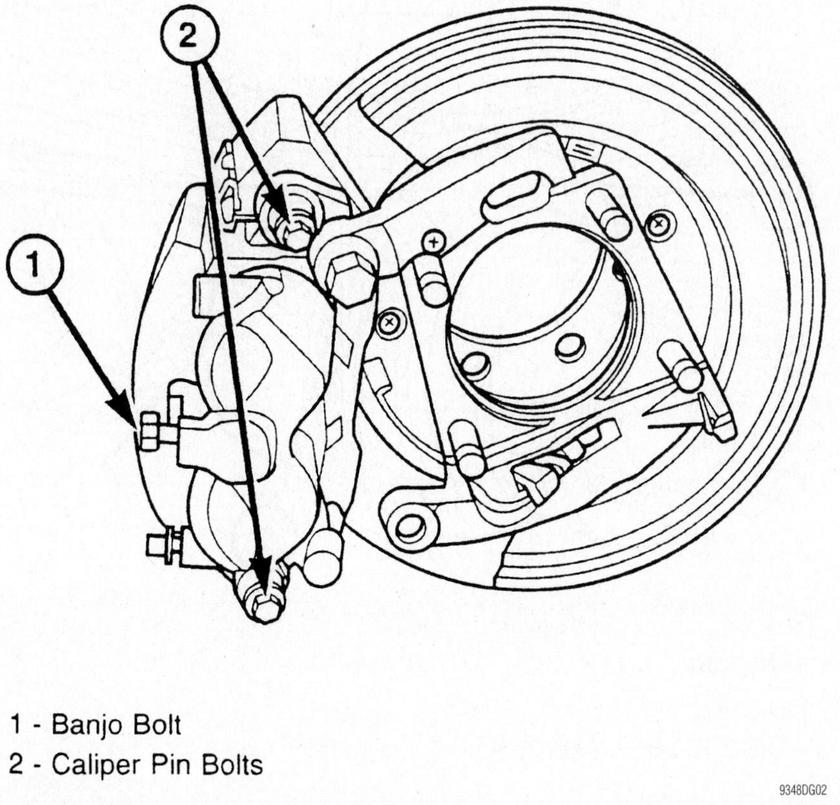

Rear brake caliper

Disc Brake Pads

REMOVAL & INSTALLATION

B-Series Van

FRONT

1. Raise and support the front end on jackstands.
2. Remove the wheels.
3. Press the caliper piston back into the bore with a suitable prytool. Use a large C-clamp to drive the piston into the bore of additional force is required.
4. Remove the caliper mounting bolts with a ⅜ in. hex wrench or socket.
5. Rotate the caliper rearward off the rotor and out from its mount.
6. Set the caliper on a crate or sturdy box, then remove the inboard and outboard brake pads. The inboard pad has a spring clip that holds it in the caliper. Tilt this pad out at the top to unseat the clip. The outboard pad has a retaining spring that secures it in the caliper. Unseat 1 spring end and rotate the pad out of the caliper.
7. Secure the caliper to a chassis or suspension component with a sturdy wire. Do not let it hang from the hose.

To install:

8. Clean the caliper and steering knuckle sliding surfaces with a wire brush. Then, apply a coat of Mopar® multi-mileage grease or equivalent.

➡If there is minor rust or corrosion on the pins, first polish them with a crocus cloth. If they are severely rusted, replace them.

9. Lubricate the caliper mounting bolts, collars, bushings and bores with Dow 111® or GE 661® silicone grease or equivalent.
10. Install the inboard pad and its spring clip.
11. Install the outboard brake pad.
12. Install the caliper over the rotor and seat it until flush in its original position.
13. Install the mounting bolts by hand, then tighten them to 38 ft. lbs. (51 Nm) for 2001–02 models; 24 ft. lbs. (33 Nm) for 2003–05 models.
14. Install the wheels.
15. Lower the vehicle.
16. Pump the brakes several times to seat the pads.

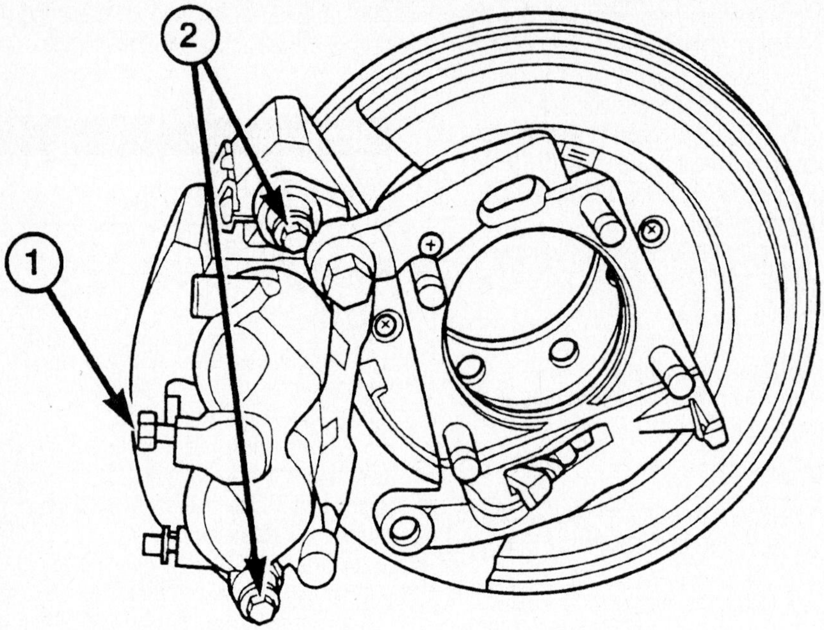

1 - Banjo Bolt
2 - Caliper Pin Bolts

9348DG02

Rear brake caliper assembly in position

REAR

1. Before servicing the vehicle, refer to the precautions in the beginning of this section.
2. Compress the piston.
3. Remove or disconnect the following:
   - Caliper pin bolts
   - Caliper
   - Inboard and outboard pads
   - Anti-rattle springs

➡The springs are not interchangeable.

4. Installation is the reverse of removal. Use brake grease on the springs.

Ram Pick-Up

FRONT

1. Raise and support the front end on jackstands.
2. Remove the wheels.
3. Press the caliper piston back into the bore with a suitable prytool. Use a large C-clamp to drive the piston into the bore of additional force is required.
4. Remove the caliper mounting bolts and rotate the caliper rearward off the rotor and out from its mount.
5. Secure the caliper to a chassis or suspension component with a sturdy wire. Do not let it hang from the hose.
6. Remove or disconnect the following:
   - Inboard brake pad
   - Outboard brake pad
   - Top anti-rattle spring
   - Bottom anti-rattle spring

To install:

7. Clean the caliper and steering knuckle sliding surfaces with a wire brush. Then, apply a coat of Mopar® multi-mileage grease or equivalent.

➡If there is minor rust or corrosion on the pins, first polish them with a crocus cloth. If they are severely rusted, replace them.

8. Lubricate the caliper mounting bolts, collars, bushings and bores with Dow 111® or GE 661® silicone grease or equivalent.

➡NOTE: Anti-rattle springs are not interchangeable.

9. Install new top anti-rattle springs.
10. Install new bottom anti-rattle springs.
11. Install the inboard pad in the adapter.
12. Install the outboard brake pad in the adapter.
13. Install the caliper over the rotor and seat it until flush in its original position.

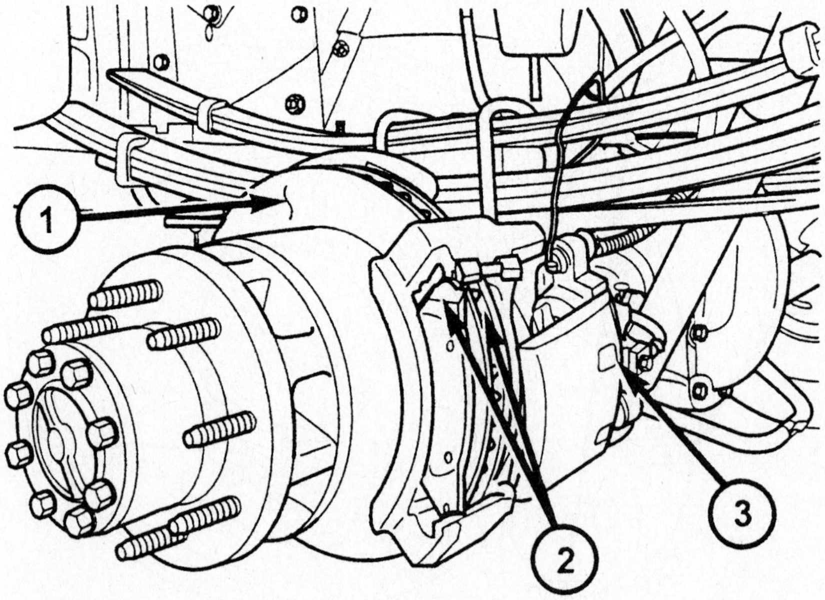

1 - ROTOR
2 - BRAKE SHOES
3 - DISC BRAKE CALIPER

9348DG01

Rear brake caliper assembly in position

14. Install the mounting bolts by hand, then tighten them to 38 ft. lbs. (51 Nm) for 2002 models; 24 ft. lbs. (33 Nm) for 2003–05 models.
15. Install the wheels.
16. Lower the vehicle.
17. Pump the brakes several times to seat the pads.

REAR

1. Before servicing the vehicle, refer to the precautions in the beginning of this section.
2. Compress the piston.
3. Remove or disconnect the following:
   - Rear wheel
   - Caliper pin bolts
   - Caliper
   - Inboard and outboard pads
   - Anti-rattle springs

➡The springs are not interchangeable.

4. Installation is the reverse of removal. On 2002 models, torque the caliper pin bolts to 25 ft. lbs. (33Nm). On 2003–05 models, torque the pin bolts to 22 ft. lbs. (30 Nm).

Rear Disc Rotor

REMOVAL & INSTALLATION

2002

1. Before servicing the vehicle, refer to the precautions in the beginning of this section.
2. Compress the piston.
3. Remove or disconnect the following:
   - Caliper pin bolts
   - Caliper
   - With dual wheels, the axle shaft
   - 2500, the rotor
   - 3500, the hub/rotor assembly
4. Installation is the reverse of removal. Torque the rotor-to-hub bolts to 95 ft. lbs. (128Nm).

2003–2005

1. Before servicing the vehicle, refer to the precautions in the beginning of this section.
2. Remove or disconnect the following:
   - Wheel
   - Caliper
   - Caliper adapter bolts
   - Rotor
3. Installation is the reverse of removal. Torque the caliper adapter bolts as follows:
   a. LD Series: 100 ft. lbs. (135 Nm)
   b. HD Series: 145 ft. lbs. (197 Nm)

Brake Drums

REMOVAL & INSTALLATION

B-Series Van

CHRYSLER SERVO TYPE WITH SINGLE ANCHOR

1. Raise and safely support the truck.
2. Remove the plug from the brake adjustment access hole.
3. Insert a thin bladed screwdriver through the adjusting hole and hold the adjusting lever away from the starwheel.
4. Release the brake by prying down against the starwheel with a brake spoon.

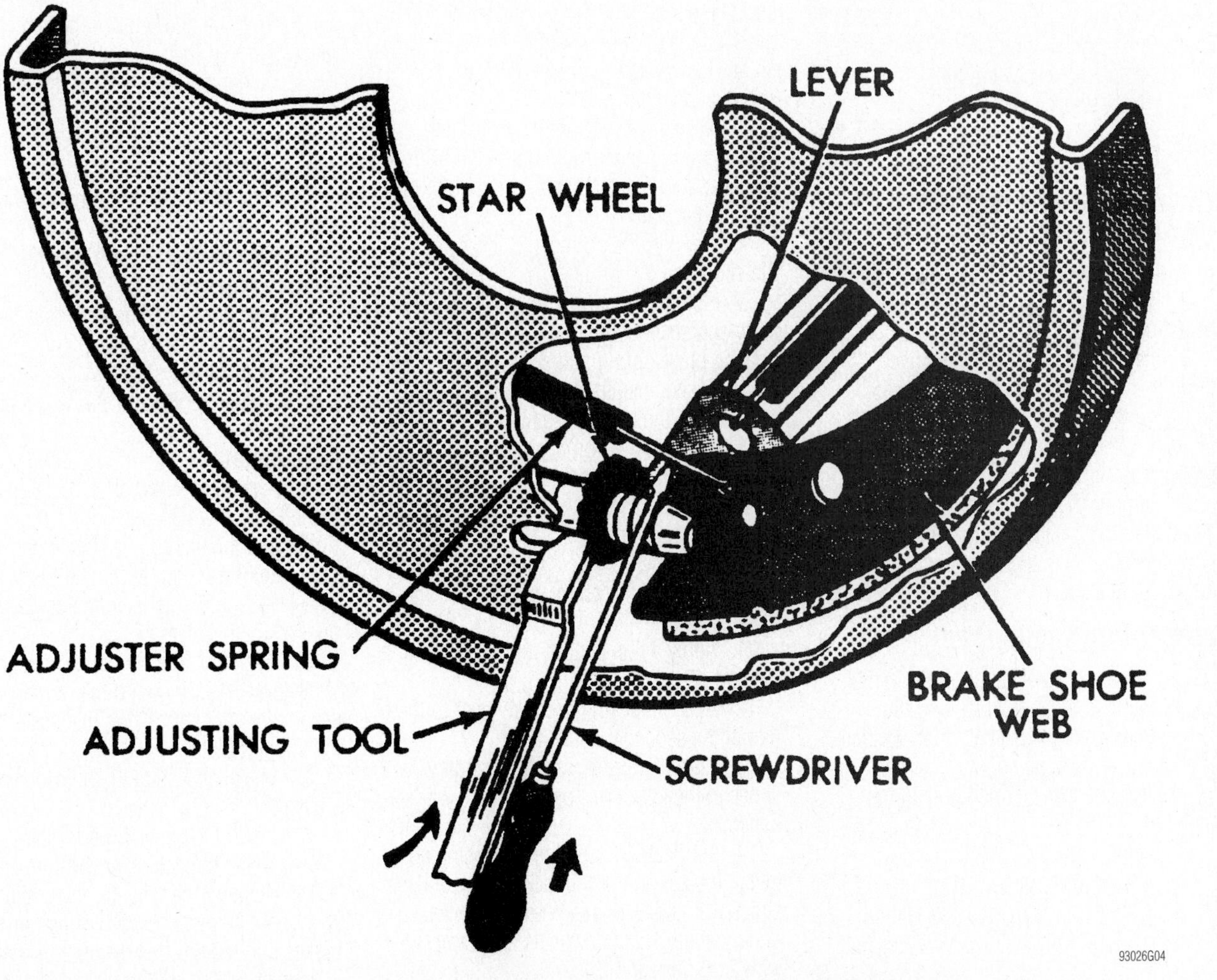

Use a lever releasing tool to depress the adjuster lever while turning the starwheel with a prytool—Bendix brakes

5. Remove the rear wheel and clips from the wheel studs. Remove the brake drum.
6. Installation is the reverse of removal. Adjust the brakes.

BENDIX DUO-SERVO TYPE

1. Raise and safely support the vehicle.
2. Remove the rear wheel and tire.
3. Remove the axle shaft nuts, washers and cones. If the cones do not readily release, rap the axle shaft sharply in the center.
4. Remove the axle shaft.
5. Remove the outer hub nut.
6. Straighten the lockwasher tab and remove it along with the inner nut and bearing.
7. Carefully remove the drum.

To install:

8. Position the drum on the axle housing.
9. Install the bearing and inner nut. While rotating the wheel and tire, tighten the adjusting nut until a slight drag is felt.
10. Back off the adjusting nut 1/6 turn so that the wheel rotates freely without excessive end-play.
11. Install the lockrings and nut. Place a new gasket on the hub and install the axle shaft, cones, lockwashers and nuts.
12. Install the wheel and tire.
13. Road-test the vehicle.

Ram Pick-Up

CHRYSLER SERVO TYPE WITH SINGLE ANCHOR

1. Raise and safely support the truck.
2. Remove the plug from the brake adjustment access hole.
3. Insert a thin bladed screwdriver through the adjusting hole and hold the adjusting lever away from the starwheel.
4. Release the brake by prying down against the starwheel with a brake spoon.
5. Remove the rear wheel and clips from the wheel studs. Remove the brake drum.
6. Installation is the reverse of removal. Adjust the brakes.

BENDIX DUO-SERVO TYPE

1. Raise and safely support the vehicle.
2. Remove the rear wheel and tire.
3. Remove the axle shaft nuts, washers and cones. If the cones do not readily release, rap the axle shaft sharply in the center.
4. Remove the axle shaft.
5. Remove the outer hub nut.
6. Straighten the lockwasher tab and remove it along with the inner nut and bearing.
7. Carefully remove the drum.

To install:

8. Position the drum on the axle housing.
9. Install the bearing and inner nut. While rotating the wheel and tire, tighten the adjusting nut until a slight drag is felt.
10. Back off the adjusting nut 1/6 turn so that the wheel rotates freely without excessive end-play.
11. Install the lockrings and nut. Place a new gasket on the hub and install the axle shaft, cones, lockwashers and nuts.
12. Install the wheel and tire.
13. Road-test the vehicle.

Brake Shoes

REMOVAL & INSTALLATION

B-Series Van

SERVO TYPE WITH SINGLE ANCHOR

1. Raise and support the vehicle.
2. Remove the rear wheel, drum retaining clips and the brake drum.
3. Remove the brake shoe return springs, noting how the secondary spring overlaps the primary spring.
4. Remove the brake shoe retainer, springs and nails.
5. Disconnect the automatic adjuster cable from the anchor and unhook it from the lever. Remove the cable, cable guide, and anchor plate.
6. Remove the spring and lever from the shoe web.
7. Spread the anchor ends of the primary and secondary shoes and remove the parking brake spring and strut.
8. Disconnect the parking brake cable and remove the brake assembly.
9. Remove the primary and secondary brake shoe assemblies and the star adjuster as an assembly. Block the wheel cylinders to retain the pistons.

To install:

10. Measure the drum as described in this section.
11. Apply a thin coat of lubricant to the support platforms.
12. Attach the parking brake lever to the rear of the secondary shoe.
13. Place the primary and secondary shoes in their relative positions on a workbench.
14. Lubricate the adjuster screw threads. Install it between the primary and secondary shoes with the star wheel next to the secondary shoe. The star wheels are stamped with an **L** (left) and **R** (right).
15. Overlap the ends of the primary and second brake shoes and install the adjusting spring and lever at the anchor end.
16. Hold the shoes in position and install the parking brake cable into the lever.
17. Install the parking brake strut and spring between the parking brake lever and primary shoe.
18. Place the brake shoes on the support and install the retainer nails and springs.
19. Install the anchor pin plate.
20. Install the eye of the adjusting cable over the anchor pin and install the return spring between the anchor pin and primary shoe.
21. Install the cable guide in the secondary shoe and install the secondary return spring. Be sure that the primary spring overlaps the secondary spring.
22. Position the adjusting cable in the groove of the cable guide and engage the hook of the cable in the adjusting lever.
23. Install the brake drum and retaining clips. Install the wheel and tire.
24. Adjust the brakes and road-test the truck.

BENDIX DUO-SERVO TYPE

1. Unhook and remove the adjusting lever return spring.
2. Remove the lever from the lever pivot pin.
3. Unhook the adjuster lever from the adjuster cable.
4. Unhook the upper shoe-to-shoe spring.
5. Unhook and remove the shoe hold-down springs.
6. Disconnect the parking brake cable from the parking brake lever.
7. Remove the shoes with the lower shoe-to-shoe spring and star wheel as an assembly.

To install:

8. The pivot screw and adjusting nut on the left side have left-hand threads and right-hand threads on the right side.
9. Lubricate and assemble the star wheel assembly. Lubricate the guide pads on the support plates.
10. Assemble the star wheel, lower shoe-to-shoe spring, and the primary and secondary shoes. Position this assembly on the support plate.
11. Install and hook the hold-down springs.
12. Install the upper shoe-to-shoe spring.
13. Install the cable and retaining clip.
14. Position the adjuster lever return spring on the pivot (green springs on left brakes and red springs on right brakes).
15. Install the adjuster lever. Route the adjuster cable and connect it to the adjuster.

16. Install the brake drum and adjust the brakes.

Ram Pick-Up

SERVO TYPE WITH SINGLE ANCHOR

1. Raise and support the vehicle.
2. Remove the rear wheel, drum retaining clips and the brake drum.
3. Remove the brake shoe return springs, noting how the secondary spring overlaps the primary spring.
4. Remove the brake shoe retainer, springs and nails.
5. Disconnect the automatic adjuster cable from the anchor and unhook it from the lever. Remove the cable, cable guide, and anchor plate.
6. Remove the spring and lever from the shoe web.
7. Spread the anchor ends of the primary and secondary shoes and remove the parking brake spring and strut.
8. Disconnect the parking brake cable and remove the brake assembly.
9. Remove the primary and secondary brake shoe assemblies and the star adjuster as an assembly. Block the wheel cylinders to retain the pistons.

To install:

10. Measure the drum as described in this section.
11. Apply a thin coat of lubricant to the support platforms.
12. Attach the parking brake lever to the rear of the secondary shoe.
13. Place the primary and secondary shoes in their relative positions on a workbench.
14. Lubricate the adjuster screw threads. Install it between the primary and secondary shoes with the star wheel next to the secondary shoe. The star wheels are stamped with an **L** (left) and **R** (right).
15. Overlap the ends of the primary and second brake shoes and install the adjusting spring and lever at the anchor end.
16. Hold the shoes in position and install the parking brake cable into the lever.
17. Install the parking brake strut and spring between the parking brake lever and primary shoe.
18. Place the brake shoes on the support and install the retainer nails and springs.
19. Install the anchor pin plate.
20. Install the eye of the adjusting cable over the anchor pin and install the return spring between the anchor pin and primary shoe.
21. Install the cable guide in the secondary shoe and install the secondary return spring. Be sure that the primary spring overlaps the secondary spring.
22. Position the adjusting cable in the groove of the cable guide and engage the hook of the cable in the adjusting lever.
23. Install the brake drum and retaining clips. Install the wheel and tire.
24. Adjust the brakes and road-test the truck.

BENDIX DUO-SERVO TYPE

1. Unhook and remove the adjusting lever return spring.
2. Remove the lever from the lever pivot pin.
3. Unhook the adjuster lever from the adjuster cable.
4. Unhook the upper shoe-to-shoe spring.
5. Unhook and remove the shoe hold-down springs.
6. Disconnect the parking brake cable from the parking brake lever.
7. Remove the shoes with the lower shoe-to-shoe spring and star wheel as an assembly.

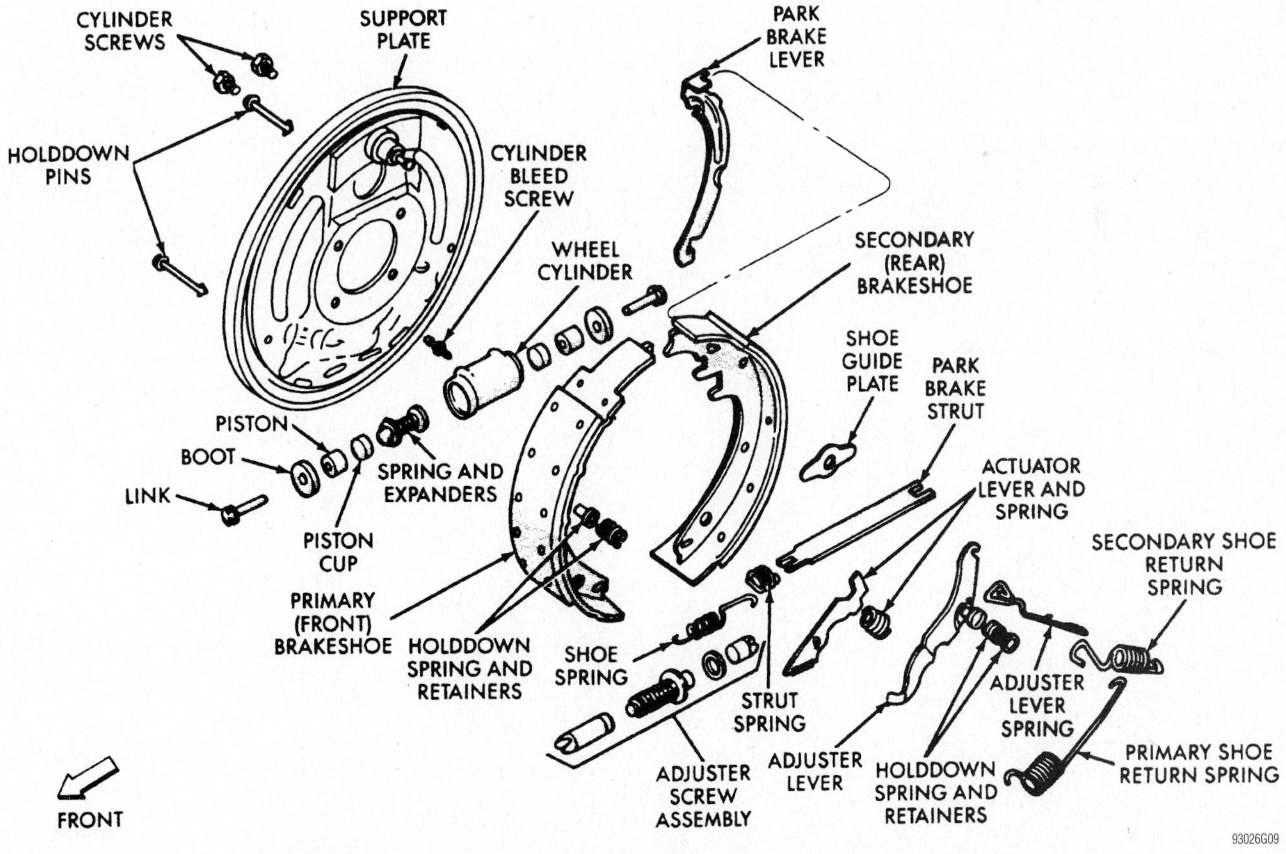

Exploded view of the rear brake components

To install:

8. The pivot screw and adjusting nut on the left side have left-hand threads and right-hand threads on the right side.
9. Lubricate and assemble the star wheel assembly. Lubricate the guide pads on the support plates.
10. Assemble the star wheel, lower shoe-to-shoe spring, and the primary and secondary shoes. Position this assembly on the support plate.
11. Install and hook the hold-down springs.
12. Install the upper shoe-to-shoe spring.
13. Install the cable and retaining clip.
14. Position the adjuster lever return spring on the pivot (green springs on left brakes and red springs on right brakes).
15. Install the adjuster lever. Route the adjuster cable and connect it to the adjuster.
16. Install the brake drum and adjust the brakes.

CHRYSLER AND DODGE

12

Sebring Convertible & Sedan • Stratus

SPECIFICATIONS AND MAINTENANCE CHARTS

ENGINE AND VEHICLE IDENTIFICATION

Engine						
Code ①	Liters (cc)	Cu. In.	Cyl.	Fuel Sys.	Engine Type	Eng. Mfg.
S	2.4 (2429)	148	I4	MFI	DOHC	DaimlerChrysler
J (PZEV)	2.4 (2429)	148	I4	MFI	DOHC	DaimlerChrysler
R	2.7 (2736)	167	V6	MFI	DOHC	DaimlerChrysler

Model Year	
Code ②	Year
2	2002
3	2003
4	2004
5	2005
6	2006

MFI: Multi-point Fuel Injection

DOHC: Double Overhead Camshaft

PZEV: Partially Zero Emissions Vehicle (California)

① 8th position of VIN

② 10th position of VIN

06009-SEBR-C01

GENERAL ENGINE SPECIFICATIONS

All measurements are given in inches.

Year	Model	Engine Displacement Liters	Engine Series VIN	Net Horsepower @ rpm	Net Torque @ rpm (ft. lbs.)	Bore x Stroke (in.)	Com-pression Ratio	Oil Pressure @ rpm
2002	Sebring Convertible	2.4	S	150@5200	167@4000	3.44x3.98	9.4:1	25-80@3000
	Sebring Convertible	2.7	R	200@5800	190@4850	3.38x3.09	9.7:1	45-105@3000
	Sebring Sedan	2.4	S	150@5200	167@4000	3.44x3.98	9.4:1	25-80@3000
	Sebring Sedan	2.7	R	200@5900	192@4300	3.39x3.09	9.7:1	45-105@3000
	Stratus Sedan	2.4	S	150@5200	167@4000	3.44x3.98	9.4:1	25-80@3000
	Stratus Sedan	2.7	R	200@5900	192@4300	3.39x3.09	9.7:1	45-105@3000
2003	Sebring Convertible	2.4	S	150@5200	167@4000	3.44x3.98	9.4:1	25-80@3000
	Sebring Convertible	2.7	R	200@5800	190@4850	3.38x3.09	9.7:1	45-105@3000
	Sebring Sedan	2.4	S	150@5200	167@4000	3.44x3.98	9.4:1	25-80@3000
	Sebring Sedan	2.7	R	200@5900	192@4300	3.39x3.09	9.7:1	45-105@3000
	Stratus Sedan	2.4	S	150@5200	167@4000	3.44x3.98	9.4:1	25-80@3000
	Stratus Sedan	2.7	R	200@5900	192@4300	3.39x3.09	9.7:1	45-105@3000
2004	Sebring Convertible	2.4	S	150@5200	167@4000	3.44x3.98	9.4:1	25-80@3000
	Sebring Convertible	2.7	R	200@5800	190@4850	3.38x3.09	9.7:1	45-105@3000
	Sebring Sedan	2.4	S	150@5200	167@4000	3.44x3.98	9.4:1	25-80@3000
	Sebring Sedan	2.7	R	200@5900	192@4300	3.39x3.09	9.7:1	45-105@3000
	Stratus Sedan	2.4	S	150@5200	167@4000	3.44x3.98	9.4:1	25-80@3000
	Stratus Sedan	2.7	R	200@5900	192@4300	3.39x3.09	9.7:1	45-105@3000
2005	Sebring Convertible	2.4	S, J	150@5200	167@4000	3.44x3.98	9.4:1	25-80@3000
	Sebring Convertible	2.7	R	200@5800	190@4850	3.38x3.09	9.7:1	45-105@3000
	Sebring Sedan	2.4	S	150@5200	167@4000	3.44x3.98	9.4:1	25-80@3000
	Sebring Sedan	2.7	R	200@5900	192@4300	3.39x3.09	9.7:1	45-105@3000
	Stratus Sedan	2.4	S	150@5200	167@4000	3.44x3.98	9.4:1	25-80@3000
	Stratus Sedan	2.7	R	200@5900	192@4300	3.39x3.09	9.7:1	45-105@3000

06009-SEBR-C02

GASOLINE ENGINE TUNE-UP SPECIFICATIONS

Year	Engine Displacement Liters	Engine VIN	Spark Plug Gap (in.)	Ignition Timing (deg.) MT	Ignition Timing (deg.) AT	Fuel Pump (psi) ①	Idle Speed (rpm) MT	Idle Speed (rpm) AT	Valve Clearance In.	Valve Clearance Ex.
2002	2.4	S	0.050	—	①	58	②	②	HYD	HYD
	2.7	R	0.048-0.053	①	①	58	②	②	HYD	HYD
2003	2.4	S	0.050	—	①	58	②	②	HYD	HYD
	2.7	R	0.048-0.053	①	①	58	②	②	HYD	HYD
2004	2.4	S	0.050	—	①	58	②	②	HYD	HYD
	2.7	R	0.048-0.053	①	①	58	②	②	HYD	HYD
2005	2.4	S, J	0.048-0.053	—	①	58	②	②	HYD	HYD
	2.7	R	0.048-0.053	①	①	58	②	②	HYD	HYD

NOTE: The Vehicle Emission Control Information label often reflects specification changes made during production.
The label figures must be used if they differ from those in this chart.

HYD: Hydraulic

① Ignition timing cannot be adjusted. Base engine timing is set at TDC during assembly.

② Refer to the Vehicle Emission Control Information label for correct specifications.

06009-SEBR-C03

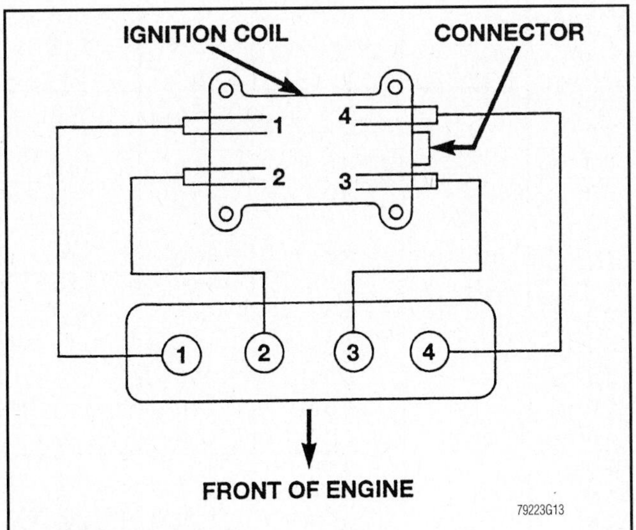

2.4L Engines
Firing Order: 1–3–4–2
Distributorless ignition system

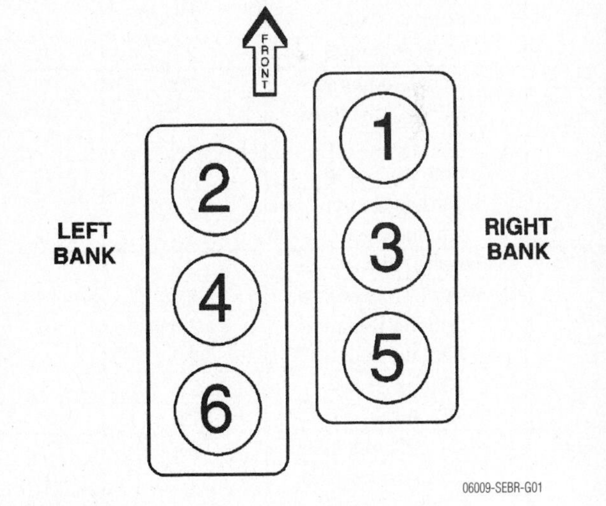

2.7L Engines
Firing order: 1–2–3–4–5–6
Distributorless ignition system. One coil per cylinder

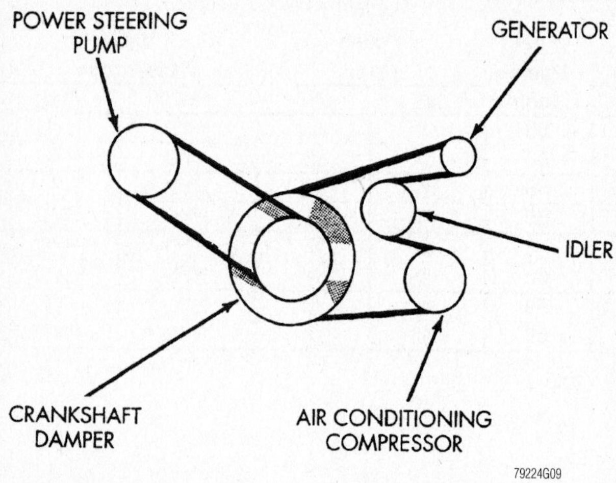

Accessory drive belt routing—2.7L engine

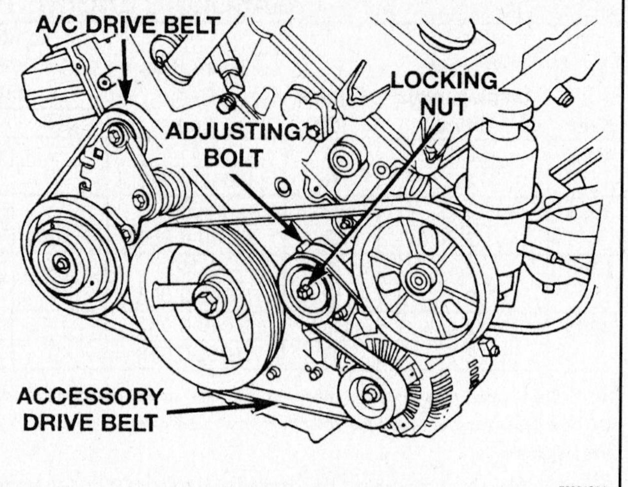

Accessory drive belt routing—2.4L engine

CAPACITIES

Year	Model	Engine Displacement Liters	Engine VIN	Engine Oil with Filter (qts.)	Transmission (pts.) MT	Transmission (pts.) Auto.	Fuel Tank (gal.)	Cooling System (qts.)
2002	Sebring Convertible	2.4	S	5.0	—	8.0 ①	16.0	10.5
	Sebring Convertible	2.7	R	5.0	—	8.0 ①	16.0	9.5
	Sebring Sedan	2.4	S	5.0	—	8.0 ①	16.0	10.5
	Sebring Sedan	2.7	R	5.0	—	8.0 ①	16.0	9.5
	Stratus Sedan	2.4	S	5.0	—	8.0 ①	16.0	10.5
	Stratus Sedan	2.7	R	5.0	5.0-5.6	8.0 ①	16.0	9.5
2003	Sebring Convertible	2.4	S	5.0	—	8.0 ①	16.0	10.5
	Sebring Convertible	2.7	R	5.0	—	8.0 ①	16.0	9.5
	Sebring Sedan	2.4	S	5.0	—	8.0 ①	16.0	10.5
	Sebring Sedan	2.7	R	5.0	—	8.0 ①	16.0	9.5
	Stratus Sedan	2.4	S	5.0	—	8.0 ①	16.0	10.5
	Stratus Sedan	2.7	R	5.0	5.0-5.6	8.0 ①	16.0	9.5
2004	Sebring Convertible	2.4	S	5.0	—	8.0 ①	16.0	10.5
	Sebring Convertible	2.7	R	5.0	—	8.0 ①	16.0	9.5
	Sebring Sedan	2.4	S	5.0	—	8.0 ①	16.0	10.5
	Sebring Sedan	2.7	R	5.0	—	8.0 ①	16.0	9.5
	Stratus Sedan	2.4	S	5.0	—	8.0 ①	16.0	10.5
	Stratus Sedan	2.7	R	5.0	5.0-5.6	8.0 ①	16.0	9.5
2005	Sebring Convertible	2.4	S, J	5.0	—	8.0 ①	16.0	8.0
	Sebring Convertible	2.7	R	5.0	—	8.0 ①	16.0	9.5
	Sebring Sedan	2.4	S, J	5.0	—	8.0 ①	16.0	8.0
	Sebring Sedan	2.7	R	5.0	—	8.0 ①	16.0	9.5
	Stratus Sedan	2.4	S, J	5.0	—	8.0 ①	16.0	8.0
	Stratus Sedan	2.7	R	5.0	5.0-5.6	8.0 ①	16.0	9.5

NOTE: All capacities are approximate. Add fluid gradually and ensure a proper fluid level is obtained.

① Overhaul fill capacity with torque converter empty: 18.4 pts.

06009-SEBR-C04

VALVE SPECIFICATIONS

Year	Engine Displacemen Liters	Engine VIN	Seat Angle (deg.)	Face Angle (deg.)	Spring Test Pressure (lbs. @ in.)	Spring Installed Height (in.)	Stem-to-Guide Clearance (in.)		Stem Diameter (in.)	
							Intake	Exhaust	Intake	Exhaust
2002	2.4	S	45	44.5-45	76@1.50	1.496	0.0018-0.0025	0.0029-0.0037	0.2337-0.2344	0.2326-0.2333
	2.7	R	44.5-45	44.5-45.5	148@1.14	1.496	0.0009-0.0026	0.0020-0.0037	0.2337-0.2344	0.2326-0.2333
2003	2.4	S	45	44.5-45	76@1.50	1.496	0.0018-0.0025	0.0029-0.0037	0.2337-0.2344	0.2326-0.2333
	2.7	R	44.5-45	44.5-45.5	148@1.14	1.496	0.0009-0.0026	0.0020-0.0037	0.2337-0.2344	0.2326-0.2333
2004	2.4	S	45	44.5-45	76@1.50	1.496	0.0018-0.0025	0.0029-0.0037	0.2337-0.2344	0.2326-0.2333
	2.7	R	44.5-45	44.5-45.5	148@1.14	1.496	0.0009-0.0026	0.0020-0.0037	0.2337-0.2344	0.2326-0.2333
2005	2.4	S, J	45	44.5-45	76@1.50	1.496	0.0018-0.0025	0.0029-0.0037	0.2337-0.2344	0.2326-0.2333
	2.7	R	44.5-45	44.5-45.5	148@1.14	1.496	0.0009-0.0026	0.0020-0.0037	0.2337-0.2344	0.2326-0.2333

06009-SEBR-C05

CRANKSHAFT AND CONNECTING ROD SPECIFICATIONS

All measurements are given in inches.

Year	Engine Displacement Liters	Engine VIN	Crankshaft				Connecting Rod		
			Main Brg. Journal Dia.	Main Brg. Oil Clearance	Shaft End-play	Thrust on No.	Journal Diameter	Oil Clearance	Side Clearance
2002	2.4	S	2.3610-2.3625	0.0007-0.0023	0.0035-0.0094	3	1.9670-1.9685	0.0009-0.0027	0.0051-0.0150
	2.7	R	2.4997-2.5004	0.0014-0.0021	0.017 max.	3	2.1067-2.1060	0.0010-0.0026	0.0052-0.0150
2003	2.4	S	2.3610-2.3625	0.0007-0.0023	0.0035-0.0094	3	1.9670-1.9685	0.0009-0.0027	0.0051-0.0150
	2.7	R	2.4997-2.5004	0.0014-0.0021	0.017 max.	3	2.1067-2.1060	0.0010-0.0026	0.0052-0.0150
2004	2.4	S	2.3610-2.3625	0.0007-0.0023	0.0035-0.0094	3	1.9670-1.9685	0.0009-0.0027	0.0051-0.0150
	2.7	R	2.4997-2.5004	0.0014-0.0021	0.017 max.	3	2.1067-2.1060	0.0010-0.0026	0.0052-0.0150
2005	2.4	S, J	2.3610-2.3625	0.0007-0.0023	0.0035-0.0094	3	1.9670-1.9685	0.0009-0.0027	0.0051-0.0150
	2.7	R	2.4997-2.5004	0.0014-0.0021	0.017 max.	3	2.1067-2.1060	0.0010-0.0026	0.0052-0.0150

06009-SEBR-C06

PISTON AND RING SPECIFICATIONS

All measurements are given in inches.

Year	Engine Displacement Liters	Engine VIN	Piston Clearance	Ring Gap Top Compression	Ring Gap Bottom Compression	Ring Gap Oil Control	Ring Side Clearance Top Compression	Ring Side Clearance Bottom Compression	Ring Side Clearance Oil Control
2002	2.4	S	0.0009-0.0022	0.010-0.020	0.009-0.018	0.010-0.025	0.0011-0.0031	0.0011-0.0031	0.0004-0.0070
	2.7	R	0.0003-0.0016	0.008-0.014	0.0146-0.0249	0.010-0.030	0.0013-0.0032	0.0016-0.0031	0.0022-0.0080
2003	2.4	S	0.0009-0.0022	0.010-0.020	0.009-0.018	0.010-0.025	0.0011-0.0031	0.0011-0.0031	0.0004-0.0070
	2.7	R	0.0003-0.0016	0.008-0.014	0.0146-0.0249	0.010-0.030	0.0013-0.0032	0.0016-0.0031	0.0022-0.0080
2004	2.4	S	0.0009-0.0022	0.010-0.020	0.009-0.018	0.010-0.025	0.0011-0.0031	0.0011-0.0031	0.0004-0.0070
	2.7	R	0.0003-0.0016	0.008-0.014	0.0146-0.0249	0.010-0.030	0.0013-0.0032	0.0016-0.0031	0.0022-0.0080
2005	2.4	S, J	0.0009-0.0022	0.010-0.020	0.009-0.018	0.010-0.025	0.0011-0.0031	0.0011-0.0031	0.0004-0.0070
	2.7	R	0.0003-0.0016	0.008-0.014	0.0146-0.0249	0.010-0.030	0.0013-0.0032	0.0016-0.0031	0.0022-0.0080

06009-SEBR-C07

TORQUE SPECIFICATIONS

All readings in ft. lbs.

Year	Engine Displacement Liters	Engine VIN	Cylinder Head Bolts	Main Bearing Bolts	Rod Bearing Bolts	Crankshaft Damper Bolts	Flywheel Bolts	Manifold Intake	Manifold Exhaust	Spark Plug	Oil Pan Drain Plug
2002	2.4	S	①	②	③	100	70	21	17	13	20
	2.7	R	④	⑤	③	125	—	9	17	15	20
2003	2.4	S	①	②	③	100	70	21	17	13	20
	2.7	R	④	⑤	③	125	—	9	17	15	20
2004	2.4	S	①	②	③	100	70	21	17	13	20
	2.7	R	④	⑤	③	125	—	9	17	15	20
2005	2.4	S, J	⑥	②	③	100	70	21	17	13	20
	2.7	R	④	⑤	③	125	—	9	17	15	20

① Step 1: 25 ft. lbs.
Step 2: 50 ft. lbs.
Step 3: 50 ft. lbs.
Step 4: + 1/4 turn

② M8 bolts: 21 ft. lbs.
M11 bolts: 55 ft. lbs.

③ Step 1: 20 ft. lbs.
Step 2: Plus 1/4 turn

④ See text for illustration:
Step 1: bolts 1-8 35 ft. lbs.
Step 2: bolts 1-8 55 ft. lbs.
Step 3: bolts 1-8 55 ft. lbs.
Step 4: bolts 1-8 Plus 1/4 turn
Step 5: bolts 9-11 21 ft. lbs.

⑤ Main cap inside bolts: 15 ft. lbs. Plus 1/4 turn
Main cap outside bolts: 20 ft. lbs. Plus 1/4 turn
Main cap tie bolts: 21 ft. lbs.

⑥ Non PZEC engines:
Step 1: 25 ft. lbs
Step 2: 50 ft. lbs.
Step 3: 50 ft. lbs.
Step 4 + 1/4 turn
PZEC engines:
Step 1: 25 ft. lbs
Step 2: 60 ft. lbs.
Step 3: 60 ft. lbs.
Step 4: + 1/4 turn

06009-SEBR-C08

WHEEL ALIGNMENT

Year	Model		Caster Range (+/-Deg.)	Caster Preferred Setting (Deg.)	Camber Range (+/-Deg.)	Camber Preferred Setting (Deg.)	Toe-in (in.)
2002	All	F	1.00	+3.31	0.59	0	0.06 +/- 0.10
		R	—	—	0.38	-0.19	0.06 +/- 0.10
2003	All	F	1.00	+3.31	0.59	0	0.06 +/- 0.10
		R	—	—	0.38	-0.19	0.06 +/- 0.10
2004	All	F	1.00	+3.31	0.59	0	0.06 +/- 0.10
		R	—	—	0.38	-0.19	0.06 +/- 0.10
2005	All	F	1.00	+3.31	-0.9 to +0.3	-0.30	0.12 +/- 0.10
		R	—	—	0.50	-0.50	0.06 +/- 0.10

06009-SEBR-C09

TIRE, WHEEL AND BALL JOINT SPECIFICATIONS

Year	Model	OEM Tires Standard	OEM Tires Optional	Tire Pressures (psi) Front	Tire Pressures (psi) Rear	Wheel Size	Ball Joint Inspection	Lug Nut Torque (ft. lbs.)
2002	Sebring Conv.	P205/65R15	P205/55R16	32	32	6-JJ	①	100
	Sebring Sedan	P205/65TR15	P205/60TR16	32	32	6-JJ	①	100
	Stratus Sedan	P205/65TR15	P205/60TR16 ②	31	31	6-JJ	①	100
2003	Sebring Conv.	P205/65R15	P205/55R16	32	32	6-JJ	①	100
	Sebring Sedan	P205/65TR15	P205/60TR16	32	32	6-JJ	①	100
	Stratus Sedan	P205/65TR15	P205/60TR16 ②	31	31	6-JJ	①	100
2004	Sebring Conv.	P205/65R15	P205/55R16	32	32	6-JJ	①	100
	Sebring Sedan	P205/65TR15	P205/60TR16	32	32	6-JJ	①	100
	Stratus Sedan	P205/65TR15	P205/60TR16 ②	31	31	6-JJ	①	100
2005	Sebring Conv.	P205/65R15	P205/60R16	③	③	6.5	④	100
	Sebring Sedan	P205/60R16	none	③	③	6.5	④	100
	Stratus SXT Sedan	P205/60R16	none	③	③	6.5	④	100
	Stratus R/T Sedan	P215/50R17	none	③	③	7.0	④	100

OEM: Original Equipment Manufacturer

PSI: Pounds Per Square Inch

① Replace if any measurable movement is found

② P215/50VR17 tires are available on the ES models with the Performance Handling Group

③ See placard on vehicle

④ Lower: 1.5mm movement max.
Upper: Replace if any movement is noted

06009-SEBR-C10

BRAKE SPECIFICATIONS

All measurements in inches unless noted

						Brake Drum Diameter				Brake Caliper
			Brake Disc			Original	Max.	Maximum	Minimum	Guide
Year	Model		Original Thickness	Minimum Thickness	Maximum Run-out	Inside Diameter	Wear Limit	Machine Diameter	Lining Thickness	Pin Bolts (ft. lbs.)
2002	Sebring Convertible	F	0.911	0.843	0.005	—	—	—	①	26
		R	0.400	0.330	0.005	9.80	②	②	0.040	16
	Sebring Sedan	F	0.911	0.843	0.005	—	—	—	①	26
		R	0.360	0.285	0.005	NA	②	②	③	16
	Stratus	F	0.911	0.843	0.003	—	—	—	①	26
		R	0.360	0.285	0.005	NA	②	②	③	16
2003	Sebring Convertible	F	0.911	0.843	0.005	—	—	—	①	26
		R	0.400	0.330	0.005	9.80	②	②	0.040	16
	Sebring Sedan	F	0.911	0.843	0.005	—	—	—	①	26
		R	0.360	0.285	0.005	NA	②	②	③	16
	Stratus	F	0.911	0.843	0.003	—	—	—	①	26
		R	0.360	0.285	0.005	NA	②	②	③	16
2004	Sebring Convertible	F	0.911	0.843	0.005	—	—	—	①	26
		R	0.400	0.330	0.005	9.80	②	②	0.040	16
	Sebring Sedan	F	0.911	0.843	0.005	—	—	—	①	26
		R	0.360	0.285	0.005	NA	②	②	③	16
	Stratus	F	0.911	0.843	0.003	—	—	—	①	26
		R	0.360	0.285	0.005	NA	②	②	③	16
2005	Sebring Convertible	F	0.911	0.843	0.003	—	—	—	NA	26
		R	0.360	0.285	0.005	NA	②	②	NA	16
	Sebring Sedan	F	0.911	0.843	0.003	—	—	—	NA	26
		R	0.360	0.285	0.005	NA	②	②	NA	16
	Stratus Sedan	F	0.911	0.843	0.003	—	—	—	NA	26
		R	0.360	0.285	0.005	NA	②	②	NA	16

NA: Information not available

① Maximum diameter is stamped on drum

② Disc brake lining and backing total thickness: 9/32 inch
Drum brake shoe lining and backing total thickness: 1/8 inch

③ Disc brake lining and backing: 3/8 inch

06009-SEBR-C11

SCHEDULED MAINTENANCE INTERVALS

2002-04 Chrysler Sebring Sedan & Convertible, Dodge Stratus

		VEHICLE MILEAGE INTERVAL (x1000)												
TO BE SERVICED	TYPE OF SERVICE	7.5	15	22.5	30	37.5	45	52.5	60	67.5	75	82.5	90	97.5
Engine oil & filter	R	✓	✓	✓	✓	✓	✓	✓	✓	✓	✓	✓	✓	✓
Brake hoses	S/I	✓	✓	✓	✓	✓	✓	✓	✓	✓	✓	✓	✓	✓
Coolant level, hoses & clamps	S/I	✓	✓	✓	✓	✓	✓	✓	✓	✓	✓	✓	✓	✓
CV joints & front suspension components	S/I	✓	✓	✓	✓	✓	✓	✓	✓	✓	✓	✓	✓	✓
Exhaust system	S/I	✓	✓	✓	✓	✓	✓	✓	✓	✓	✓	✓	✓	✓
Rotate tires	S/I	✓	✓	✓	✓	✓	✓	✓	✓	✓	✓	✓	✓	✓
Accessory drive belts	S/I		✓		✓		✓		✓		✓		✓	
Brake linings	S/I			✓			✓			✓			✓	
Air filter element	R				✓				✓				✓	
Spark plugs ① ②	R													
Lubricate front & rear ball joints	S/I				✓				✓				✓	
Engine coolant	R						✓				✓			
PCV valve	S/I								✓				✓	
Ignition cables ② ③	R													
Camshaft timing belt ④	R													

R: Replace S/I: Service or Inspect

① 4-cylinder: every 30,000 miles.

② 6-cylinder: 100,000 miles.

③ 4-cylinder: 60,000 miles.

④ Replace at 105,000 miles for normal service; replace at 102,000 miles for severe service

FREQUENT OPERATION MAINTENANCE (SEVERE SERVICE)

If a vehicle is operated under any of the following conditions it is considered severe service:

- Extremely dusty areas.

- 50% or more of the vehicle operation is in 32°C (90°F) or higher temperatures, or constant operation in temperatures below 0°C (32°F).

- Prolonged idling (vehicle operation in stop and go traffic).

- Frequent short running periods (engine does not warm to normal operating temperatures).

- Police, taxi, delivery usage or trailer towing usage.

Oil & oil filter change: change every 3000 miles.

Rotate tires every 6000 miles.

Brake linings: check every 12,000 miles.

Air filter element: change every 15,000 miles.

Automatic transaxle fluid: service or inspect every 15,000 miles.

PCV valve: check every 30,000 miles.

Engine coolant, replace at 36,000, 51,000 & 81,000 miles.

06009-SEBR-C12

SCHEDULED MAINTENANCE INTERVALS
2005 Chrysler Sebring Sedan and Convertible, Dodge Stratus Sedan

TO BE SERVICED	TYPE OF SERVICE	VEHICLE MILEAGE INTERVAL (x1000)												
		3	6	9	12	15	18	21	24	27	30	33	36	39
Engine oil & filter	R	✓	✓	✓	✓	✓	✓	✓	✓	✓	✓	✓	✓	✓
Tires	Rotate		✓		✓		✓		✓		✓		✓	
Brake linings	I				✓				✓				✓	
Engine air filter	S/I	✓	✓	✓	✓	✓	✓	✓	✓	✓	✓	✓	✓	✓
Engine air filter	R										✓			
Spark plugs (2.4L, exc. PZEC engine)	R										✓			
PCV valve	S/I										✓			
Power steering belt (2.4L)	I/R										✓			
Auto. trans. fluid and filter	R	Every 60,000 miles												
Spark plug wires (2.4L, exc. PZEC engine)	R	Every 60,000 miles												
Accessory drive belts (2.7L)	R	Every 60,000 miles												
Engine coolant	R	Every 60,000 miles												
Timing belt (2.4L)	R	Every 90,000 miles												
Spark plugs (2.7L)	R	Every 100,000 miles												
Accessory drive belts (2.7L)	S/I	Every 105,000 miles												

R: Replace S: Service I: Inspect Adj: Adjust I/R: Inspect and replace if necessary

The above schedule is to be used if you drive under any of the following conditions:

Driving in temperatures under 32 degrees F
Stop and go traffic
Extensive engine idling
Driving in dusty conditions
Frequent trips under 10 miles
More than 50 % of your driving is in hot weather (90 deg. F) above 50 miles per hour
Trailer towing
Taxi, police or delivery service

If none of these conditions is met, double the maintenance intervals

06009-SEBR-C13

PRECAUTIONS

Before servicing any vehicle, please be sure to read all of the following precautions. The following precautions deal with personal safety, preventing of component damage, and important points to take into consideration when servicing a motor vehicle:

- Never open, service or drain the radiator or cooling system when the engine is hot; serious burns can occur from the steam and hot coolant.
- Observe all applicable safety precautions when working around fuel. Whenever servicing the fuel system, always work in a well-ventilated area. Do not allow fuel spray or vapors to come in contact with a spark, open flame or excessive heat (a hot drop light, for example). Keep a dry chemical fire extinguisher near the work area. Always keep fuel in a container specifically designed for fuel storage; also, always properly seal fuel containers to avoid the possibility of fire or explosion. Refer to the additional fuel system precautions later in this section.
- Fuel injection systems often remain pressurized, even after the engine has been turned **OFF**. The fuel system pressure must be relieved before disconnecting any fuel lines. Failure to do so may result in fire and/or personal injury.
- Brake fluid often contains polyglycol ethers and polyglycols. Avoid contact with the eyes and wash your hands thoroughly after handling brake fluid. If you do get brake fluid in your eyes, flush your eyes with clean, running water for 15 minutes. If eye irritation persists, or if you have taken brake fluid internally, IMMEDIATELY seek medical assistance.
- The EPA warns that prolonged contact with used engine oil may cause a number of skin disorders, including cancer! You should make every effort to minimize your exposure to used engine oil. Protective gloves should be worn when changing the oil. Wash your hands and any other exposed skin areas as soon as possible after exposure to used engine oil. Soap and water, or waterless hand cleaner should be used.
- All vehicles are equipped with an air bag system, often referred to as a Supplemental Restraint System (SRS) or as a Supplemental Inflatable Restraint (SIR) system. The system must be disabled before performing service on or around system components, steering column, instrument panel components, wiring and sensors. Failure to follow safety and disabling procedures could result in accidental air bag deployment, possible personal injury and unnecessary system repairs.
- Always wear safety goggles when working with, or around, the air bag system. When carrying a non-deployed air bag, be sure the bag and trim cover are pointed away from your body. When placing a non-deployed air bag on a work surface, always face the bag and trim cover upward, away from the surface. This will reduce the motion of the module if it is accidentally deployed.
- Clean, high quality brake fluid from a sealed container is essential for the safe and proper operation of the brake system. You should always buy the grade of fluid recommended for your vehicle. If the brake fluid becomes contaminated, drain and flush the system, then refill the master cylinder with new fluid. Never reuse any brake fluid. Any brake fluid that is removed from the system should be discarded. Also, do not allow any brake fluid to come in contact with a painted surface; it will damage the paint.
- Never operate the engine without the proper amount and type of engine oil; doing so WILL result in severe engine damage.
- Timing belt maintenance is extremely important! Many models may utilize an interference-type, non-free-wheeling engine. If the timing belt breaks, the valves in the cylinder head may strike the pistons, causing potentially serious (also time-consuming and expensive) engine damage. Refer to the maintenance interval charts in the front of this manual for the recommended replacement interval for the timing belt, and to the timing belt section for belt replacement and inspection.
- Disconnecting the negative battery cable on some vehicles may interfere with the functions of the on board computer system(s) and may require the computer to undergo a relearning process once the negative battery cable is reconnected.
- When servicing drum brakes, only disassemble and assemble one side at a time, leaving the remaining side intact for reference.
- Only an MVAC-trained, EPA-certified, automotive technician should service the air conditioning system or its components.

ENGINE REPAIR

Distributor

These vehicles are equipped with distributorless ignition systems.

Alternator

REMOVAL & INSTALLATION

2.4L Engine

1. Before servicing the vehicle, refer to the precautions in the beginning of this section.
2. Remove or disconnect the following:
   - Negative battery cable from the shock tower
   - Accessory belt cover
   - Alternator electrical connectors
   - Accessory belt splash shield
   - Accessory belt
   - Manifold Absolute Pressure (MAP) sensor from the intake manifold
   - Alternator

To install:

3. Install or connect the following:
   - Alternator and torque the bolts to 33 ft. lbs. (44 Nm)
   - MAP sensor
   - Drive belt and splash shield
   - Alternator electrical connectors
   - Accessory belt cover
   - Negative battery cable

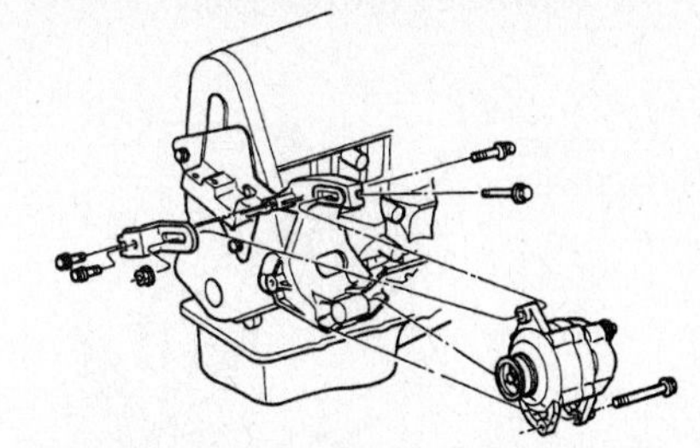

9306FG05

Exploded view of the alternator and related components—2.4L engine

2.7L Engine

1. Before servicing the vehicle, refer to the precautions in the beginning of this section.

2. Remove or disconnect the following:
   - Negative battery cable from the shock tower
   - Accessory belt splash shield and loosen the belt
   - Alternator electrical connectors
   - A/C pressure switch and clutch electrical connectors
   - Engine oil dipstick
   - Alternator

To install:

3. Install or connect the following:
   - Alternator and torque the bolts to 40 ft. lbs. (54 Nm)
   - Engine oil dipstick tube
   - A/C pressure switch and clutch electrical connectors
   - Alternator electrical connectors
   - Drive belt and splash shield
   - Negative battery cable

Ignition Timing

ADJUSTMENT

These engines use a fixed ignition system. The Powertrain Control Module (PCM) regulates the ignition timing. Basic ignition timing is not adjustable.

Engine Assembly

REMOVAL & INSTALLATION

2.4L Engine

1. Before servicing the vehicle, refer to the precautions in the beginning of this section.
2. Properly relieve the fuel system pressure.
3. Drain the engine oil.
4. Drain the cooling system.
5. Recover the A/C refrigerant.

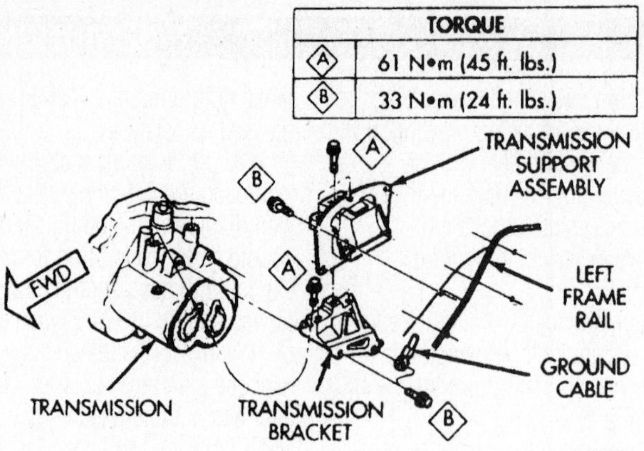

Exploded view of the left side engine mount—Type 1

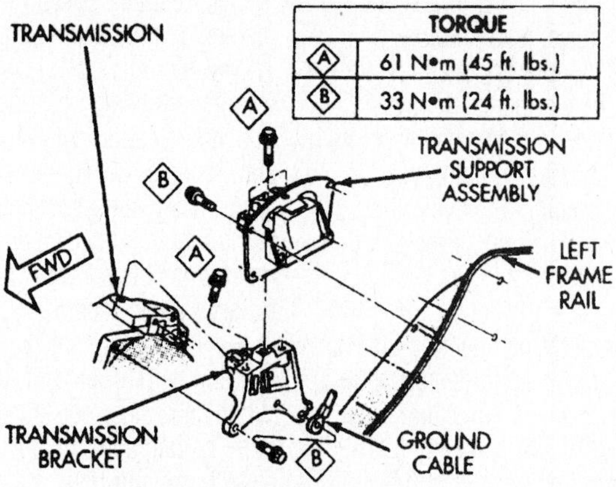

Exploded view of the left side engine mount—Type 2

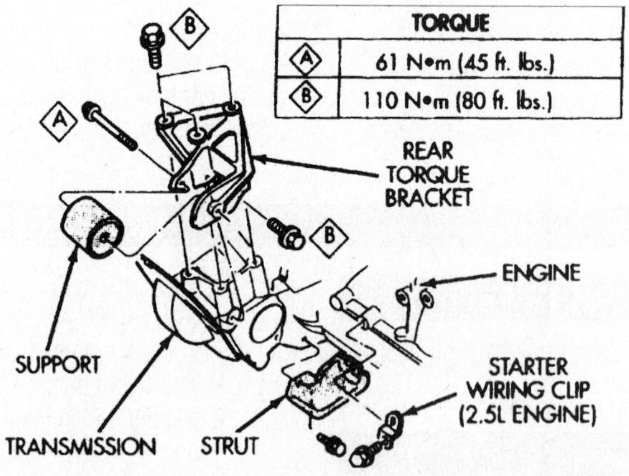

Exploded view of the rear engine mounting torque bracket—2.4L engines

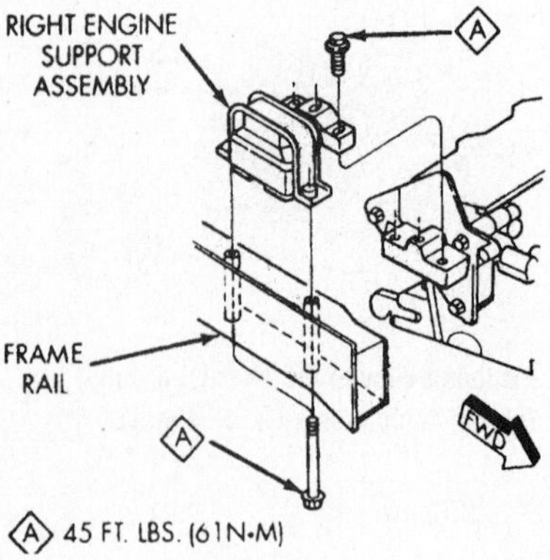

Exploded view of the right side engine mount—all engines

6. Remove or disconnect the following:
   - Negative battery cable
   - Throttle body air inlet hose
   - Air cleaner assembly
   - Upper radiator crossmember
   - Upper and lower radiator hoses
   - Transmission cooler lines
   - A/C lines at the condenser
   - Radiator, fan module and A/C condenser
   - Transmission electrical harness connectors
   - Transmission shift cable
   - Engine electrical harness from the Powertrain Control Module (PCM) and bulkhead

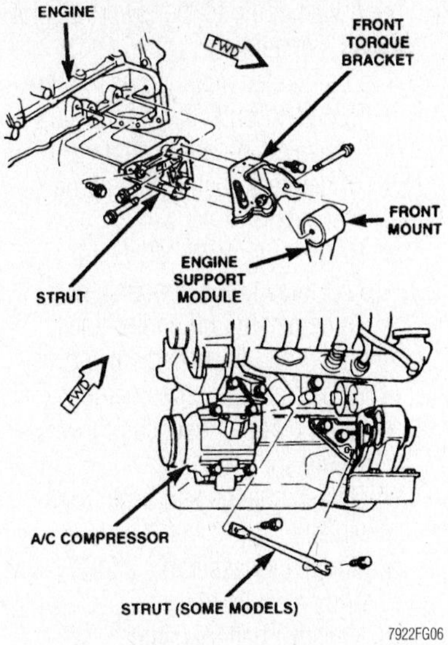

Exploded view of the front engine mounting torque bracket—2.4L engines

- Both front wheels and splash shields
- Both halfshafts
- Drive belts
- Power steering pump from the bracket and move it aside
- Heater return hose from the right front frame rail
- A/C compressor electrical connectors
- Exhaust pipe from the manifold
- Through bolts from the front and rear engine mounts
- Rear mount bracket from the transmission
- Structural collar and torque reaction bracket
- Torque converter bolts after matchmarking them
- Positive battery cable from the battery and the Power Distribution Center (PDC)
- Ground cable from the left side transmission mount bracket
- Throttle and speed control cables
- Coolant overflow hose
- Heater hose from the thermostat housing
- Engine ground straps
- Brake booster and vacuum purge hoses
- Fuel lines from the rail
- Intake manifold
- Alternator
- A/C suction line from the compressor and place an engine dolly and cradle in position

7. Loosen the cradle engine mounts and lower the engine/transmission assembly onto the dolly.

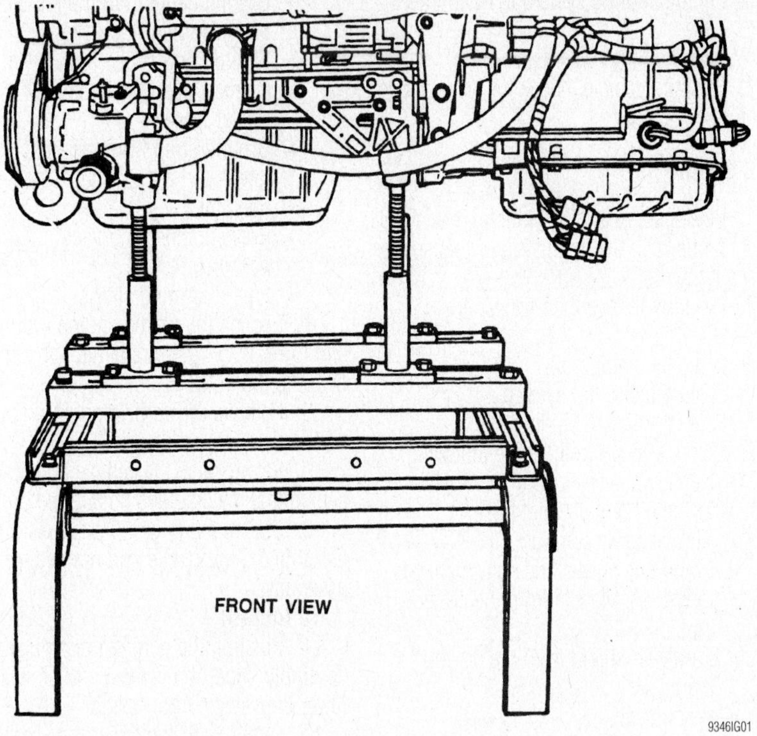

Place the engine/transmission assembly on the cradle as shown

8. Remove the vertical engine mount bolts
9. Slowly raise the vehicle and make certain that the cradle is positioned properly.
10. Secure the engine/transmission assembly to the cradle and remove the assembly.

To install:

11. Position the engine/transmission assembly under the vehicle and slowly lower the vehicle into position.
12. Continue to the lower the vehicle until the right side engine mount and left side transaxle mount are properly aligned.
13. Install or connect the following:
    - Mounting bolts and torque them to 45 ft. lbs. (61 Nm). Raise the vehicle and remove the cradle
    - A/C compressor and suction line
    - Alternator
    - Intake manifold and torque the bolts to 105 inch lbs. (12 Nm)
    - Alternator electrical connectors
    - Fuel line to the rail
    - Brake booster and vacuum purge hoses
    - Engine ground straps
    - Heater hose to the thermostat housing
    - Coolant overflow hose
    - Throttle and speed control cables
    - Ground cable to the left side transmission mount bracket
    - Positive battery cable to the PDC
    - Torque converter bolts
    - Structural collar and the torque reaction bracket
    - Rear mount bracket to the transmission
    - Front and rear engine mount through bolts. Torque the bolts to 45 ft. lbs. (61 Nm)
    - Exhaust pipe to the manifold and torque the bolts to 21 ft. lbs. (28 Nm)
    - A/C compressor electrical connectors
    - Heater return hose to the right front frame rail
    - Power steering pump to the bracket
    - Drive belts
    - Both halfshafts
    - Splash shields and front wheels
    - Engine electrical harness to the PCM and the bulkhead connectors
    - Transmission shift cable
    - Transmission electrical connectors
    - Radiator, fan module and A/C condenser
    - A/C lines to the condenser
    - Transmission cooler lines
    - Upper and lower radiator hoses
    - Upper radiator crossmember
    - Air cleaner assembly
    - Throttle body air inlet hose
    - Negative battery cable
14. Fill the engine with clean oil.

15. Fill the cooling system to the proper level.
16. Recharge the A/C system.
17. Start the vehicle, check for leaks and repair if necessary.

2.7L Engine

1. Before servicing the vehicle, refer to the precautions in the beginning of this section.
2. Properly relieve the fuel system pressure.
3. Drain the engine oil.
4. Drain the cooling system.
5. Recover the A/C refrigerant.
6. Remove or disconnect the following:
   - Negative battery cable
   - Throttle body air inlet hose
   - Air cleaner assembly
   - Both front wheels and splash shields
   - Lower front fascia from the crossmember
   - Lower air shield from the crossmember
   - Front fascia
   - Upper radiator crossmember
   - Upper and lower radiator hoses
   - Transmission oil cooler lines
   - A/C lines from the condenser
   - Radiator, fan and A/C condenser
   - Transmission electrical harness connectors
   - Transmission shift cable
   - Engine electrical harness from the Powertrain Control Module (PCM) and bulkhead connectors
   - Antilock Brake System (ABS) brake module from the lower radiator crossmember
   - Brake line from the lower radiator crossmember
   - Both halfshafts
   - Front engine mount through bolt
   - Front engine mount from the lower radiator crossmember
   - Lower radiator crossmember
   - Drive belts
   - Power steering pump and bracket. Do not disconnect the lines
   - Heater return hose from the right front frame rail
   - A/C compressor electrical connectors and reposition the compressor
   - Structural collar
   - Exhaust system cross under pipe
   - Rear engine mount and transaxle bracket
   - Torque converter housing cover and match mark the bolts
   - Positive battery cable from the Power Distribution Center (PDC)
   - Ground cable from the left side transaxle mount bracket
   - Throttle and speed control cables
   - Coolant overflow hose
   - Heater hose
   - Ground strap from the right shock tower
   - Fuel lines
   - Brake booster and vacuum purge hoses
   - Engine ground straps
7. Loosen the cradle engine mounts and lower the engine/transmission assembly onto the dolly.
8. Remove the vertical engine mount bolts
9. Slowly raise the vehicle and make certain that the cradle is positioned properly.
10. Secure the engine/transmission assembly to the cradle and remove the assembly.

To install:

11. Position the engine/transmission assembly under the vehicle and slowly lower the vehicle into position.
12. Continue to the lower the vehicle until the right side engine mount and left side transaxle mount are properly aligned.
13. Install or connect the following:
    - Mounting bolts and torque them to 45 ft. lbs. (61 Nm). Raise the vehicle and remove the cradle
    - Engine ground straps
    - Brake booster and vacuum purge hoses
    - Fuel lines
    - Ground strap to the right shock tower
    - Heater hose
    - Coolant overflow hose
    - Throttle and speed control cables
    - Ground cable to the left transaxle mount bracket
    - Positive battery cable to the PDC
    - Torque converter bolts and cover
    - Rear engine mount and transaxle bracket
    - Exhaust system cross under pipe
    - Structural collar
    - A/C compressor to the bracket
    - A/C compressor clutch electrical connectors
    - Heater return hose to the right side frame rail
    - Power steering pump and bracket
    - Drive belts
    - Lower radiator crossmember
    - Front engine mount to the lower radiator crossmember
    - Both halfshafts
    - Brake line to the lower radiator crossmember
    - ABS module to the lower radiator crossmember
    - Engine electrical harness to the PCM and bulkhead connectors
    - Transmission shift cable
    - Transmission electrical harness connectors
    - Radiator, fan and A/C condenser
    - A/C lines to the condenser
    - Transmission oil cooler lines
    - Upper and lower radiator hoses
    - Upper radiator crossmember
    - Front bumper fascia and lower air shield
    - Splash shields and both front wheels
    - Air cleaner assembly and throttle body air inlet hose
    - Negative battery cable
14. Fill the engine with clean oil and replace the filter.
15. Fill the cooling system to the proper level.
16. Recharge the A/C system with the proper refrigerant.
17. Start the vehicle, check for leaks and repair if necessary.

Water Pump

REMOVAL & INSTALLATION

2.4L Engine

This engine uses a die-cast aluminum body water pump with a stamped steel impeller. The water pump bolts directly to the block. Cylinder block-to-water pump sealing is provided by a large rubber O-ring. The water pump is driven by the timing belt, that must be removed to service the water pump.

1. Before servicing the vehicle, refer to the precautions in the beginning of this section.
2. Disconnect the negative battery cable from the left shock tower.

➡The ground cable is equipped with an insulator grommet which should be placed on the stud to prevent the negative battery cable from accidentally grounding.

➡This procedure requires removing the engine timing belt and the autotensioner. The factory specifies that the timing marks should always be aligned before removing the timing belt. Set the engine at Top Dead Center (TDC) on No. 1 compression stroke. This should align all timing marks on the crankshaft sprocket and both camshaft sprockets.

3. Drain the cooling system.
4. Remove or disconnect the following:
   - Right inner splash shield
   - Accessory drive belts and support the engine
   - Right motor mount
   - Power steering pump bracket bolts and move the pump/bracket assembly aside

➡**Do not disconnect the power steering fluid lines.**

   - Right engine mount bracket
   - Timing belt front covers
   - Timing belt tensioner and timing belt by loosening the tensioner screws

⁂ WARNING

With the timing belt removed, DO NOT rotate the camshaft or crankshaft or damage to the engine could occur.

   - Camshaft sprockets

➡**Do not allow the camshafts to turn when the camshaft sprockets are being removed.**

   - Rear timing belt cover
   - Water pump

To install:

5. Thoroughly clean all sealing surfaces. Replace the water pump if there are any cracks, signs of coolant leakage from the shaft seal, loose or rough turning bearings, damaged impeller or sprocket or sprocket flange loose or damaged.
6. Install or connect the following:
   - New rubber O-ring into the water pump

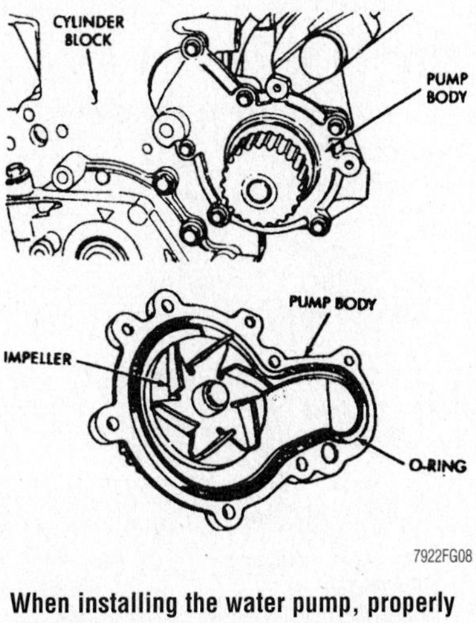

When installing the water pump, properly install the O-ring to ensure a tight seal—2.4L engines

⁂ WARNING

Be sure the O-ring is properly seated in the water pump groove before tightening the screws. An improperly located O-ring may cause damage to the O-ring and cause a coolant leak.

   - Water pump. Torque the bolts to 105 inch lbs. (12 Nm)

7. Pressurize the cooling system to 15 psi (103.4 kPa) and check for leaks. If okay, release the pressure and continue the engine assembly process.
8. Install or connect the following:
   - Rear timing belt cover
   - Camshaft sprockets. Torque the bolts to 75 ft. lbs. (101 Nm)

➡**DO NOT allow the camshafts to turn while the sprocket bolts are being tightened to maintain timing mark alignment.**

⁂ WARNING

Do not attempt to compress the tensioner plunger with the tensioner assembly installed in the engine. This will cause damage to the tensioner and other related components. The tensioner MUST be compressed in a vise.

   - Timing belt tensioner and timing belt. Properly tension the timing belt.
   - Front upper and lower timing belt covers
   - Right engine mount bracket and engine mount
   - Crankshaft damper. Torque the center bolt to 105 ft. lbs. (142 Nm)
   - Right inner splash shield
   - Power steering pump bracket and power steering pump. Torque the bracket mounting bolts to 40 ft. lbs. (54 Nm)
   - Accessory drive belts. Properly tension the drive belts

9. Refill and bleed the cooling system.
10. Start the engine and check for proper operation.
11. Check and top off cooling system, if necessary.

2.7L Engine

1. Before servicing the vehicle, refer to the precautions in the beginning of this section.
2. Drain the cooling system.
3. Remove or disconnect the following:
   - Negative battery cable
   - Timing chain cover
   - Timing chain and guides
   - Water pump and discard the gasket

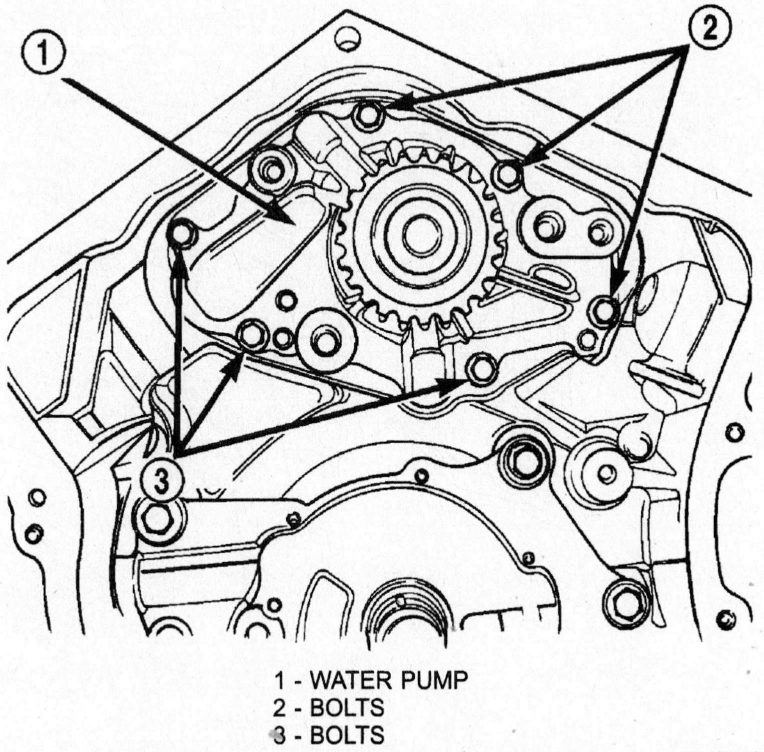

Water pump—2.7L engine

To install:

4. Clean all mating surfaces of any residual gasket material.
5. Install or connect the following:
   - Water pump with a new gasket and torque the bolts to 105 inch lbs. (12 Nm)
   - Timing chain guides and chain
   - Timing chain cover
   - Negative battery cable
6. Fill the cooling system to the proper level.
7. Start the vehicle, check for leaks and repair if necessary.

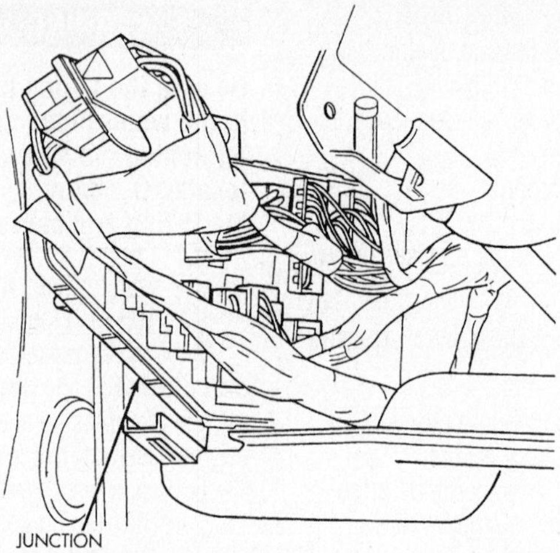

Junction block location—Stratus and Sebring convertible

Heater Core

REMOVAL & INSTALLATION

2002–04

1. Disconnect the negative battery cable.
2. Drain the cooling system into a clean container for reuse.
3. From the center of the instrument panel, grasp and pull to remove the radio/control module bezel.
4. At the right side of the instrument panel, remove the side trim.
5. Remove or disconnect the following:
   - 2 lower right side support beam screws
   - Instrument panel support-to-A-pillar bolt

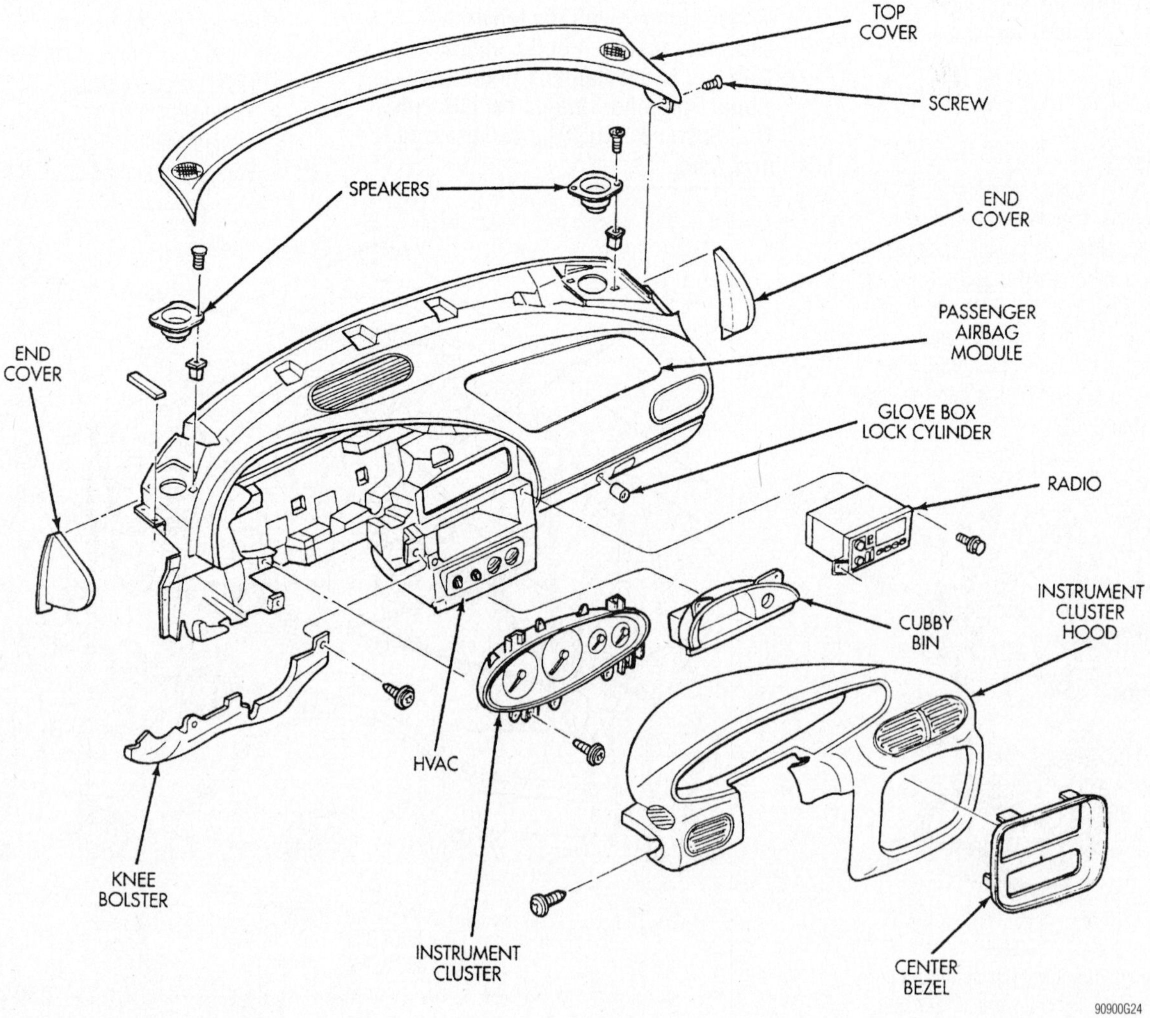

Exploded view of the instrument panel assembly

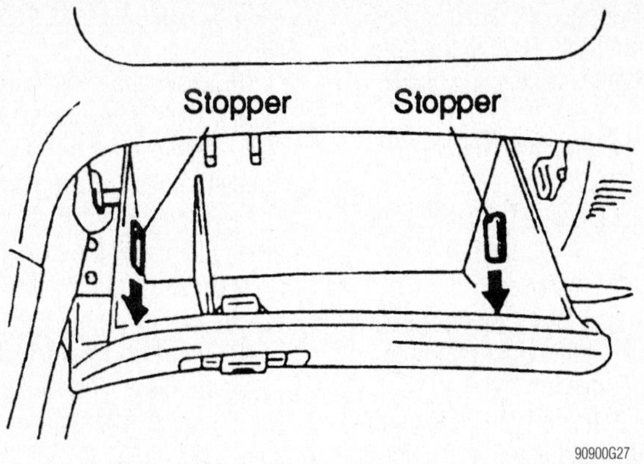

Unlock the glove box door stoppers by pushing in the direction shown—Stratus and Sebring convertible

- On left side of the instrument panel, remove the side trim
- Upper instrument panel bezel
- Lower knee bolster
- Console-to-instrument panel screws
- Gearshift knob and the shifter bezel
- Rear console screws and the rear console half
- Front console screws and the front console half
- Right side instrument panel support strut
- Heater hoses from the heater core tubes
- Heater core cover-to-heater/air conditioning housing assembly screws and the cover
- Heater core from the heater/air conditioning housing assembly

To install:

6. Install or connect the following:
   - Heater core to the heater/air conditioning housing assembly
   - Heater core cover and the cover-to-heater/air conditioning housing assembly screws
   - Heater hoses to the heater core tubes
   - Right side instrument panel support strut
   - Front console half and the front console screws
   - Rear console half and the rear console screws
   - Shifter bezel and the gearshift knob
   - Console-to-instrument panel screws
   - Lower knee bolster
   - Upper instrument panel bezel

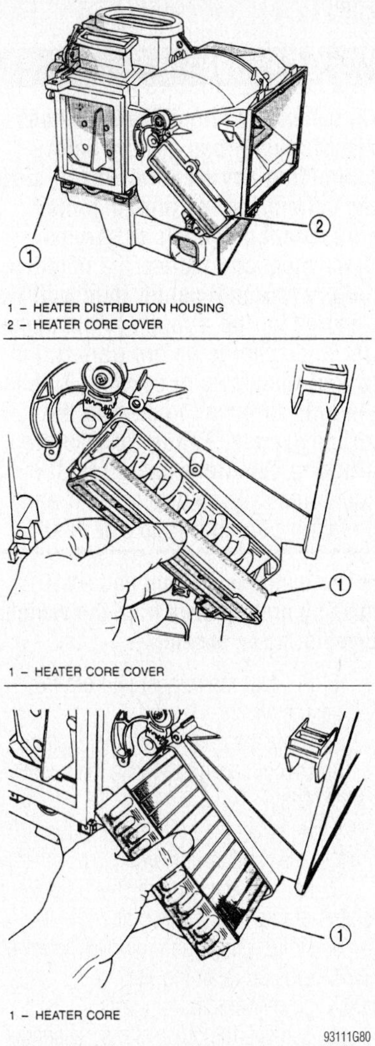

View of the heater housing assembly and the heater core—Stratus

   - On left side of the instrument panel, install the side trim
   - Instrument panel support-to-A-pillar bolt
   - 2 lower right side support beam screws
   - Right side of the instrument panel, install the side trim
   - On center of the instrument panel, install the radio/control module bezel
7. Refill the cooling system.
8. Connect the negative battery cable.
9. Operate the engine to normal operating temperatures; then, check the climate control operation and check for leaks.

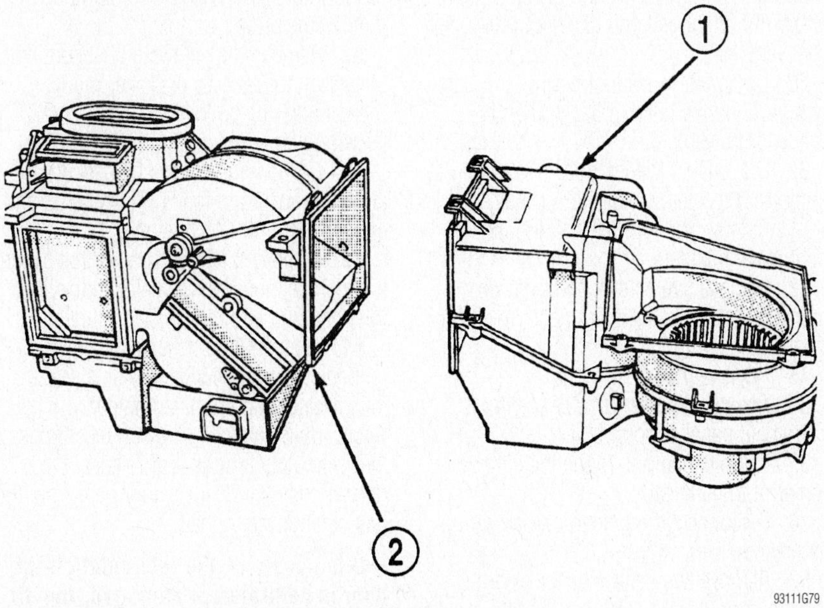

View of the heater/air conditioning housing assembly—Stratus

2005

CAUTION

On vehicles equipped with airbags, disable the airbag system before attempting any steering wheel, steering column, or instrument panel component diagnosis or service. Disconnect and isolate the negative battery (ground) cable, then wait two minutes for the airbag system capacitor to discharge before performing further diagnosis or service. This is the only sure way to disable the airbag system. Failure to take the proper precautions could result in accidental airbag deployment and possible personal injury or death.

➡The instrument panel and HVAC housing are removed from the vehicle together as an assembly.

1. Recover the refrigerant from the refrigerant system.
2. Drain the engine cooling system.
3. Disconnect and isolate the negative battery cable.
4. Remove one bolt retaining the A/C lines to the right strut tower.
5. Remove the two top HVAC housing retaining nuts in engine compartment.
6. Hoist vehicle and remove the lower HVAC housing retaining nut.
7. Lower vehicle from hoist.
8. Remove the driver's and passenger's seats:
   a. Position seat far enough forward to gain access to rear mount bolts on floor.
   b. Remove bolts holding rear of seat track to floor.
   c. Slide seat rearward.
   d. Remove bolts attaching front of the seat track to floor kick up.
   e. Disconnect negative battery cable, if electrical seat.
   f. Disconnect front seat wire harness connector from body harness connector.

CAUTION

Caution should be taken not to handle seat by adjuster release bar when removing spring loaded seat from vehicle.

   g. Remove front seat from vehicle.
9. Disconnect and isolate the battery negative remote cable.
10. Remove one screw to shift knob and transmission range indicator bezel from floor console.
11. Remove floor center console. Remove two mounting screws in the front and two mounting screws inside console storage bin.
12. Remove the amplifier retaining fasteners.
13. Disconnect the wire harness connector.
14. Remove the amplifier.
15. Using a trim stick or equivalent, gently pry out on the left and right rear floor heat ducts and remove.
16. Disconnect the console wiring connectors and unclip the push in fasteners.
17. Using a trim stick or equivalent, gently pry up on and remove the right sill plate.
18. Remove the two screws to the right cowl kick panel and remove from vehicle.
19. Using a trim stick or equivalent, gently pry out on the right end cover and remove from vehicle.
20. Using a trim stick or equivalent, gently pry out on the right A-pillar trim and remove from vehicle.
21. Open the glove box. Pinch in on the sides and drop to the floor. Pull down to disengage hinges.
22. Disconnect the right door harness connectors located within the A-pillar.
23. Remove the four right A-pillar instrument panel retainers.
24. Remove the right silencer.
25. Remove the one right side, center instrument panel support bolt.
26. Using a trim stick or equivalent, gently pry up on and remove the left sill plate.
27. Remove the two screws to the left cowl kick panel and remove from vehicle.
28. Using a trim stick or equivalent, gently pry out on the left end cover and remove from vehicle.
29. Using a trim stick or equivalent, gently pry out on the left A-pillar trim and remove from vehicle.
30. Disconnect the left door harness connectors located within the A-pillar.
31. Remove the four left A-pillar instrument panel retainers.
32. Remove five screws and one wire connector to left lower instrument panel trim bezel.
33. Remove the left silencer.
34. Remove the one left side, center instrument panel support bolt.
35. Remove one bolt to the left side support strut to bulkhead.
36. Disconnect the junction block wiring connectors.
37. Remove two screws to the steering column shrouds and remove.
38. Disconnect all the column wiring and unclip the harness from the steering column.
39. Disconnect the transmission shift interlock and wire connector.
40. Remove the four steering column retaining bolts and drop column to floor.
41. Using a trim stick or equivalent, pry up on the rear of the top cover and pull rearward as you lift it off of instrument panel.
42. Remove instrument panel pad.
43. Remove seven bolts and five screws attaching instrument panel to fence line and cowl side.
44. With the help of an assistant, lift up instrument panel and move rearward to remove. Tilt the panel rearward to prevent the coolant from leaking out. If replacing the instrument panel, transfer all non-damaged parts and wiring harness to the new instrument panel assembly.
45. Disconnect HVAC wire harness from the instrument panel wire harness connector at the left side of the HVAC housing.
46. Remove the two nuts and two bolts that secure the HVAC housing to the instrument panel.
47. Separate the HVAC housing from the instrument panel.
48. Disconnect the HVAC wire harness connectors from the blower motor, blower motor resistor or power module (depending on application), evaporator temperature sensor and each actuator.
49. Remove the two screws that secure the HVAC harness connector to the HVAC housing and remove the HVAC wire harness.
50. Remove the screws that secure the blower motor to the HVAC housing and remove the blower motor.
51. Remove the screws that secure the blower motor resistor or power module (depending on application) to the HVAC housing and remove it from the housing.
52. Remove the screws that secure the air inlet housing to the HVAC housing and remove the air inlet housing.
53. Remove the three screws that secure the recirculation door actuator to the air inlet housing and remove the actuator.
54. Remove the screw that secures the recirculation door actuator gear to the air inlet housing and remove the gear.
55. Using a small screwdriver, depress the recirculation door gear release and remove the recirculation door gear from the recirculation door shaft.

➡If the seals on the recirculation-air door is deformed or damaged, the air door must be replaced.

56. Tilt the gear end of the recirculation-air door up and pull the door out of the air inlet housing.

➡If either foam seal is deformed or damaged, it must be replaced.

57. Carefully, remove the foam seals from the heater core and the evaporator tube mounting flange of the HVAC housing.

58. Remove the screws that secure the upper and lower HVAC housing halves together and separate the housing halves.

59. Using a flat blade pry tool, disengage the locking tab on the access plate of the evaporator temperature sensor. Turn the access plate clockwise one quarter turn and push the access plate inside of the HVAC housing, then orientate the plate in such a way that it can be removed from the housing.

60. Remove the evaporator temperature sensor probe from the A/C evaporator and remove the sensor.

61. Lift the A/C evaporator out of the bottom half of the HVAC housing.

62. Lift the heater core out of the bottom half of the HVAC housing.

To install:

63. Install the heater core into the bottom half of the HVAC housing.

64. Align the blend and mode door shafts with the holes in the HVAC housing and install the top half of the housing to the bottom half of the housing.

65. Install the screws that secure the two halves of the HVAC housing together. Tighten the screws to 2 Nm (17 inch lbs.).

66. Install the foam seals on the heater core and evaporator coil tube mounting flange.

67. Install the recirculation-air door into the air inlet housing.

68. Install the recirculation door gear onto the recirculation door shaft.

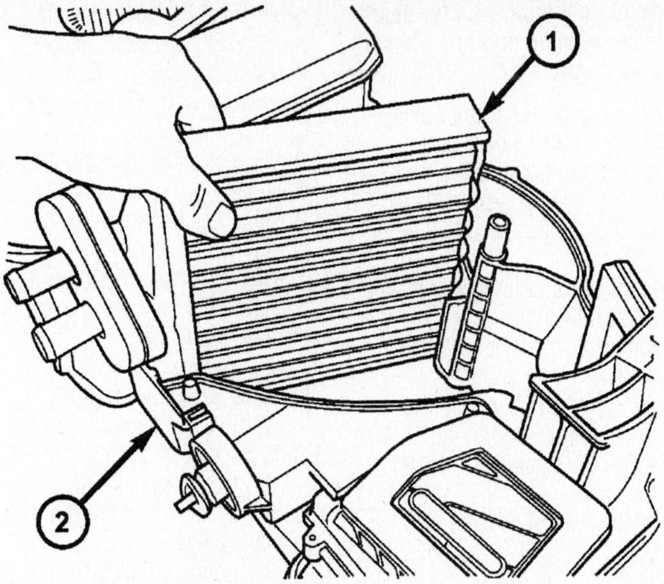

Heater core

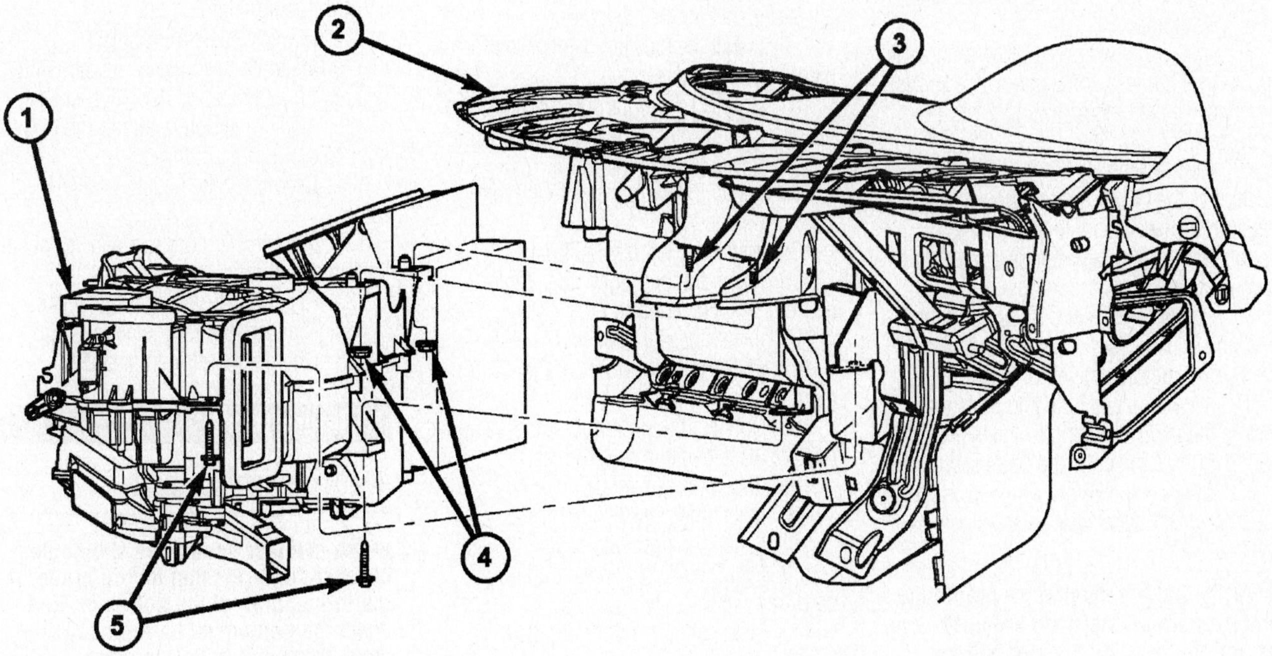

HVAC housing

➡The recirculation door actuator gear and door gear must be properly aligned to operate correctly.

69. Align and install the recirculation door actuator gear with the door gear.
70. Install the screw that secure the recirculation door actuator gear to the air inlet housing. Tighten the screw to 2 Nm (17 inch lbs.).
71. Align the recirculation door actuator spline with the spline in the recirculation door gear and install the recirculation door actuator.
72. Install the three screws that secure the recirculation door actuator to the HVAC housing. Tighten the screws to 2 Nm (17 inch lbs.).
73. Install the air inlet housing onto the HVAC housing.
74. Install the screws that secure the air inlet housing to the HVAC housing. Tighten the screws to 2 Nm (17 inch lbs.).
75. If removed, install the blower motor resistor or power module (depending on application). Install and tighten the screws to 2 Nm (17 inch lbs.).
76. If removed, install the blower motor. Install and tighten the screws to 2 Nm (17 inch lbs.).

➡Make sure that all HVAC wire harness leads are routed through all HVAC housing wiring retainers.

77. Install the HVAC wire harness onto the HVAC housing.
78. Install the two screws that secure the HVAC wire harness connector to the HVAC housing. Tighten the screws to 2 Nm (17 inch lbs.).
79. Connect the HVAC wire harness connectors to the blower motor, blower motor resistor or power module (depending on application), evaporator temperature sensor and each actuator.
80. Position the HVAC housing to the instrument panel. Make sure that all air ducts are properly aligned or connected.
81. Install the two nuts and bolts that secure the HVAC housing to the instrument panel. Tighten the nuts and bolts securely.
82. Connect the HVAC wire harness to the instrument panel wire harness connector.

➡Verify that the evaporator drain tube and evaporator seals are properly seated to the HVAC housing before installing the instrument panel and HVAC Housing into the vehicle.

83. With the help of an assistant, place the instrument panel in the vehicle, carefully guiding the HVAC housing studs through the bulkhead and into position.
84. Install the seven bolts and five screws attaching instrument panel to fence line and cowl side.
85. Place instrument panel pad into position.
86. Place instrument panel top cover in position on vehicle.
87. Push top cover forward to engage hooks to hold front of top cover to instrument panel.
88. Engage clips to hold rear edge of top cover to instrument panel.
89. Pull top cover rearward as you firmly push down and snap into place.
90. Install A-pillar trim.
91. Install the four steering column retaining bolts.
92. Connect the transmission shift interlock and wire connector.
93. Connect all the column wiring and clip the harness to the steering column.
94. Install two screws to the steering column shrouds.
95. Connect the junction block wiring connectors.
96. Install one bolt to the left side support strut to bulkhead.
97. Install five screws and one wire connector to left lower instrument panel trim bezel.
98. Install the one left side, center instrument panel support bolt.
99. Install the left silencer.
100. Install the four left A-pillar instrument panel retainers.
101. Connect the right door harness connectors located within the A-pillar.
102. Install the left end cover.
103. Install the two screws to the left cowl trim panel.
104. Install the left door sill plate.
105. Install the one right side, center instrument panel support bolt.
106. Install the right silencer.
107. Install the four right A-pillar instrument panel retainers.
108. Connect the right door harness connectors located within the A-pillar.
109. Install the glove box.
110. Install the right A-pillar trim.
111. Install the right end cover.
112. Install the two screws to the right cowl trim panel.
113. Install the right door sill plate.
114. Connect the console wiring connectors and clip the push in fasteners in.
115. Install the left and right rear floor heat ducts.
116. Connect the two harness connectors to the amplifier.
117. Connect wire harness connector to amplifier.
118. Install amplifier. Tighten the forward screw to 3 Nm (26.5 inch lbs.) and the rear nut to 2 Nm (18 inch lbs.).
119. Install floor center console. Install two mounting screws in the front and two mounting screws inside console storage bin.
120. Install the transmission range indicator bezel and one screw to shift knob.
121. Connect the battery negative remote cable.

⁂ CAUTION

Do not handle by seat adjuster release bar.

➡Ensure that the seat tracks are rearward. Ensure the seat tracks on the right and left are in equal latch positions.

122. Place front seat in position in vehicle.
123. Connect front seat wire harness connector to body harness connector.
124. Install bolts to attach front of seat track to the floor kick up. Tighten bolts to 61 Nm (45 ft. lbs.) torque.
125. Slide seat forward and install bolts attaching rear of seat track to the floor. Tighten bolts to 61 Nm (45 ft. lbs.) torque.
126. Verify front seat operation.
127. Hoist vehicle and install the lower HVAC housing retaining nut.
128. Lower vehicle from hoist.
129. Install the two top HVAC housing retaining nuts in engine compartment.
130. Install one bolt retaining the A/C lines to the right strut tower.
131. Reconnect the negative battery cable.
132. If the heater core is being replaced, flush the cooling system.
133. Refill the engine cooling system.
134. Evacuate the refrigerant system.
135. Charge the refrigerant system.

Cylinder Head

REMOVAL & INSTALLATION

➡The cylinder head bolts should be checked for stretching before reuse. If the thread area of the bolt is necked-down the bolts must be replaced with new. New head bolts are recommended.

2.4L Engine

This engine uses a Dual Over Head Camshaft (DOHC) 4-valves per cylinder

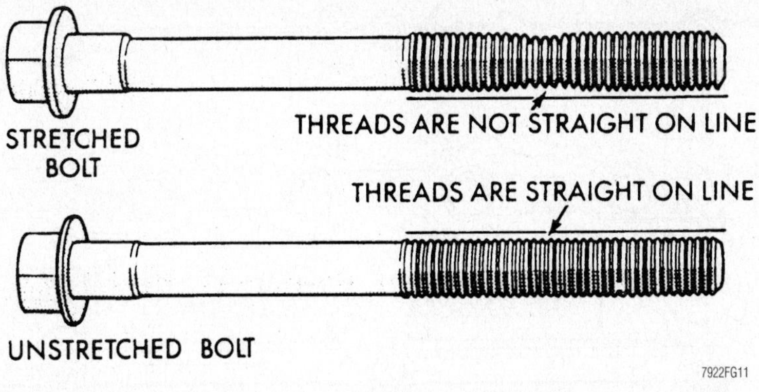

Before reusing old cylinder head bolts, check them for necking (stretching)—2.4L engine

cross flow aluminum cylinder head. The valves are actuated by roller cam followers which pivot on stationary hydraulic valve adjusters. Care must be taken to be sure all valve timing marks align after cylinder head and valvetrain service.

1. Before servicing the vehicle, refer to the precautions in the beginning of this section.
2. Relieve the fuel system pressure.
3. Drain the cooling system.
4. Disconnect the negative battery cable from the left shock tower.

➡The ground cable is equipped with an insulator grommet which should be placed on the stud to prevent the negative battery cable from accidentally grounding.

5. Remove or disconnect the following:
   - Air cleaner assembly
   - All vacuum lines, electrical connectors and fuel lines from the throttle body
   - Throttle linkage
   - Accessory drive belts
   - Power brake vacuum hose from the intake manifold
   - Exhaust pipe from the exhaust manifold
   - Power steering pump and move it aside. Do not disconnect the fluid lines
   - Spark plug wires
   - Coil pack electrical connector
   - Coil pack with the spark plug wires
   - Cam sensor and fuel injector electrical connectors
   - Timing belt covers
   - Timing belt
   - Camshaft sprockets
   - Timing belt idler pulley
   - Rear timing belt cover
   - Cylinder head cover
   - Ground strap
6. Identify the camshafts if they are to be reused for later installation. The camshafts are not interchangeable.
7. Remove or disconnect the following:
   - Camshaft bearing cap bolts in sequence
   - Camshafts

➡Refer to the camshaft removal and installation procedure for correct bolt removal/installation sequence.

   - Camshaft followers

➡Any components that are to be reused must be installed in their original locations. Use care to identify and mark the positions of any removed valvetrain components so they may be reinstalled correctly.

   - Intake and exhaust manifolds
   - Cylinder head bolts
   - Cylinder head

➡Be careful not to damage the aluminum gasket surfaces.

8. Remove all gasket material from the cylinder head and engine block. Be careful not to gouge or scratch the sealing surface of the aluminum head. The cylinder head should be checked for flatness using a good straightedge and feeler gauges. The cylinder head must be flat within 0.004 in. (0.1mm).
9. Inspect the camshaft bearing oil feed holes in the cylinder head for clogging. Inspect the camshaft bearing journals for wear or scoring. Check the cam surface for abnormal wear and damage. A visible worn groove in the roller path or on the cam lobes is cause for replacement. Valve service may be performed at this time.

To install:

10. Thoroughly clean all parts.

➡The cylinder head bolts are stretch-type. New cylinder head bolts are recommended.

11. Thoroughly clean all sealing surfaces.
12. Install or connect the following:
    - New cylinder head gasket
    - Cylinder head
13. Before installing the bolts, the threads should be oiled with clean engine oil.
14. With non-PZEV engines, torque the cylinder head bolts in sequence, using the following 4 Steps:
    - Tighten all bolts to 25 ft. lbs. (34 Nm).
    - Tighten all bolts to 50 ft. lbs. (68 Nm).
    - Tighten all bolts again to 50 ft. lbs. (68 Nm).
    - Tighten all bolts and additional 1/4 turn.

➡Do not use a torque wrench for the 4th step.

15. On PZEV (Partially Zero Emissions Vehicle) equipped engines tighten the cylinder head bolts in the sequence shown, using the 4 step torque turn method:
    - Tighten all bolts to 34 Nm (25 ft. lbs.)
    - Tighten all bolts to 82 Nm (60 ft. lbs.)
    - Tighten all bolts to 82 Nm (60 ft. lbs.)

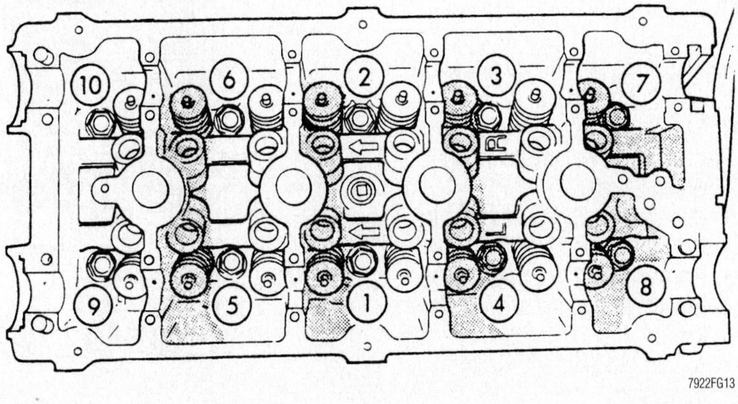

Cylinder head bolt torque sequence—2.4L engine

➡**Do not use a torque wrench for the fourth step.**

- Turn all bolts an additional ¼ turn

16. Check the camshaft end-play using the recommended procedure, then install the camshaft.
17. Apply Mopar Gasket Maker sealer to the No. 1 and No. 6 bearing caps. Install the bearing caps and tighten the M8 fasteners to 21 ft. lbs. (28 Nm). The end caps must be installed before the seals may be installed.
18. Apply a light coating of clean engine oil to the lip of the new camshaft seal. Install the camshaft seal until it fits flush with the cylinder head.
19. Install or connect the following:
    - Camshaft sprockets, if removed
    - Rear timing belt cover
    - Timing belt and properly align the timing marks
    - Timing belt cover

⁂ WARNING

Verify that all timing marks are correct. If the timing belt or sprockets are incorrectly installed, engine damage will occur. Take time to be sure all timing marks are correctly aligned.

- Intake and exhaust manifolds
- New cylinder head cover gasket
- Cylinder head cover

⁂ WARNING

DO NOT allow oil or solvents to contact the timing belt as they can deteriorate the rubber and cause tooth skipping.

➡**Apply Mopar Silicone Rubber Adhesive Sealant at the camshaft cap corners and at the top edge of the ½ round seal.**

20. Torque the cylinder head cover fasteners, in sequence, using the following Steps:
    a. Step 1: 40 inch lbs. (4.5 Nm).
    b. Step 2: 80 inch lbs. (9 Nm).
    c. Step 3: 105 inch lbs. (12 Nm).
21. Install or connect the following:
    - Ground strap
    - Coil pack and spark plug wiring
    - Cam sensor and fuel injector wiring
    - Power steering pump assembly
    - Exhaust pipe to the exhaust manifold
    - All vacuum lines and electrical connectors

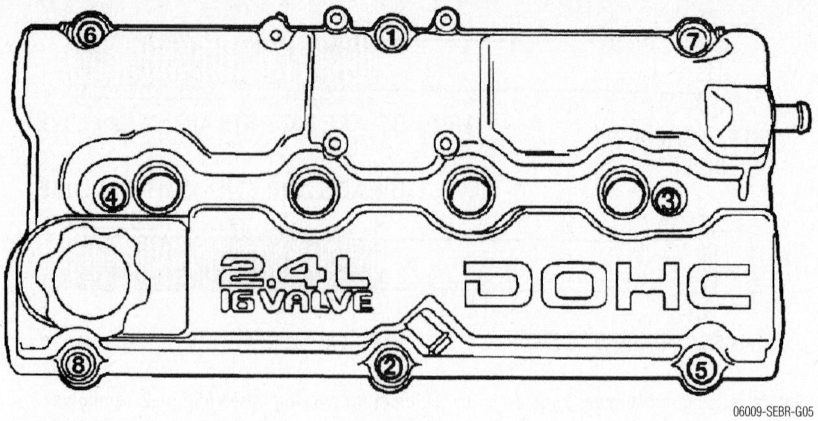

Cylinder head cover torque sequence—2.4L engine

- Throttle linkage and fuel lines
- Accessory drive belts and adjust them
- Negative battery cable

22. Refill the cooling system.

➡**An oil and filter change is recommended since coolant can enter the oil system when a head is removed.**

23. Connect the remaining air ducting. Test run vehicle. Check for leaks and for proper operation.

2.7L Engine

1. Before servicing the vehicle, refer to the precautions in the beginning of this section.
2. Relieve the fuel system pressure.
3. Drain the cooling system.
4. Remove or disconnect the following:
    - Negative battery cable
    - Drive belts
    - Vibration damper
    - Exhaust cross under pipe
    - Catalytic converter
    - Upper and lower intake manifolds
    - Rocker arm covers
    - Water outlet connector and rotate the crankshaft until the sprocket timing mark aligns with the timing mark on the oil pump housing
    - Primary timing chain
    - Camshaft bearing caps gradually
    - Camshafts and valve train components
    - Cylinder head and discard the gasket

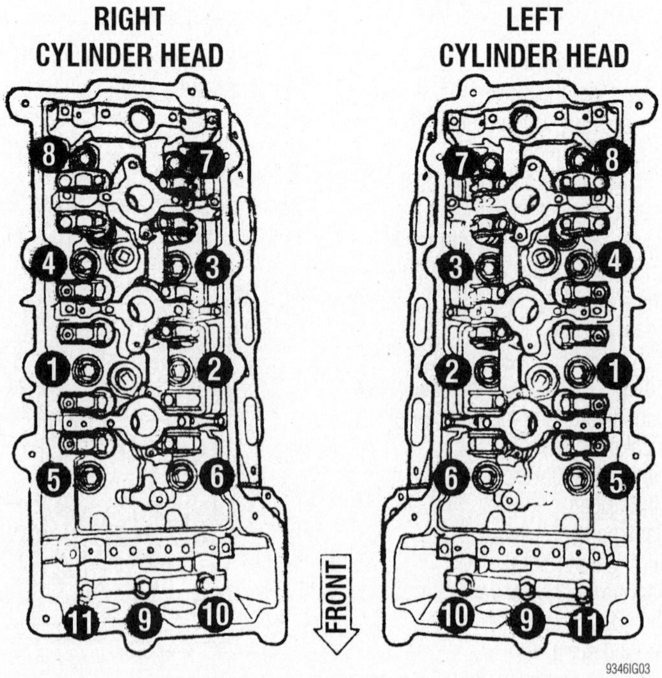

Cylinder head bolt torque sequence—2.7L engine

To install:

5. Clean all mating surfaces of any residual gasket material.
6. Lubricate the threads of the cylinder head bolts with clean engine oil.
7. Install or connect the following:
   - Cylinder head with a new gasket
8. Torque the bolts, in sequence, using the following procedure:
   a. Step 1: Bolts 1–8: 35 ft. lbs. (48 Nm).
   b. Step 2: Bolts 1–8: 55 ft. lbs. (75 Nm).
   c. Step 3: Bolts 1–8: 55 ft. lbs. (75 Nm).
   d. Step 4: Bolts 1–8: An additional 90 degrees. Do not use a torque wrench for this step.
   e. Step 5: Bolts 9–11: 21 ft. lbs. (28 Nm).
   - Valve train components and the camshafts
   - Timing chain and sprockets
   - Water outlet connector
   - Rocker arm covers
   - Timing chain cover
   - Crankshaft vibration damper
   - Lower and upper intake manifolds
   - Catalytic converter
   - Exhaust cross under pipe
   - Drive belts
   - Negative battery cable
9. Fill the cooling system to the proper level.
10. Start the vehicle, check for leaks and repair if necessary.

Rocker Arm/Shaft

REMOVAL & INSTALLATION

2.7L Engine

1. Before servicing the vehicle, refer to the precautions in the beginning of this section.
2. Relieve the fuel system pressure using the recommended procedure.
3. Remove or disconnect the following:
   - Negative battery cable
   - Rocker arm cover and rotate the engine until the cam lobe is on its base circle
4. Using special tools 8215—A and 8216—A, depress the valve spring enough to release the tension on the rocker arm.
5. Remove the rocker arm from the cylinder head.

To install:

6. Lubricate the rocker arms with clean engine oil.
7. Depress the valve spring far enough to install the rocker arm.
8. Install or connect the following:
   - Rocker arm into its original position over the valve and lash adjuster.
   - Rocker arm cover
   - Negative battery cable

Intake Manifold

REMOVAL & INSTALLATION

2.4L Engine

The intake manifold is a long branch design made of cast aluminum. It is attached to the cylinder head with 8 fasteners.

1. Before servicing the vehicle, refer to the precautions in the beginning of this section.
2. Disconnect the negative battery cable from the left shock tower.

➡The ground cable is equipped with an insulator grommet which should be placed on the stud to prevent the negative battery cable from accidentally grounding.

3. Relieve the fuel system pressure using the recommended procedure.
4. Remove the air inlet resonator as follows:
   - Both air inlet resonator-to-intake manifold bolts
   - Resonator-to-throttle body screw, loosen it
   - Air inlet resonator-to-air inlet tube clamp, loosen it
   - Resonator
5. Remove or disconnect the following:
   - Fuel supply line quick-disconnect at the fuel tube assembly

➡Squeeze the retainer tabs together and pull the fuel tube/quick-disconnect fitting assembly from the fuel tube nipple. The retainer will remain on the fuel tube. Use shop towels to catch any dripping fuel.

   - Fuel rail

⁂ WARNING

Use care when handling the fuel injectors. Do not set them on their tips. Cover the fuel injector openings after fuel rail removal.

   - Accelerator, kickdown and speed control cables, from the throttle lever and bracket
   - Idle Air Control (IAC) motor and Throttle Position (TP) sensor electrical connections
   - Throttle body vacuum hoses
   - Manifold Absolute Pressure (MAP) sensor and Intake Air Temperature (IAT) sensor electrical connections
   - Vapor and brake booster hoses
   - Knock sensor electrical connector
   - Wiring harness from the intake manifold tab
   - Transaxle-to-throttle body support bracket fasteners at the throttle body and loosen the fastener at the transaxle end

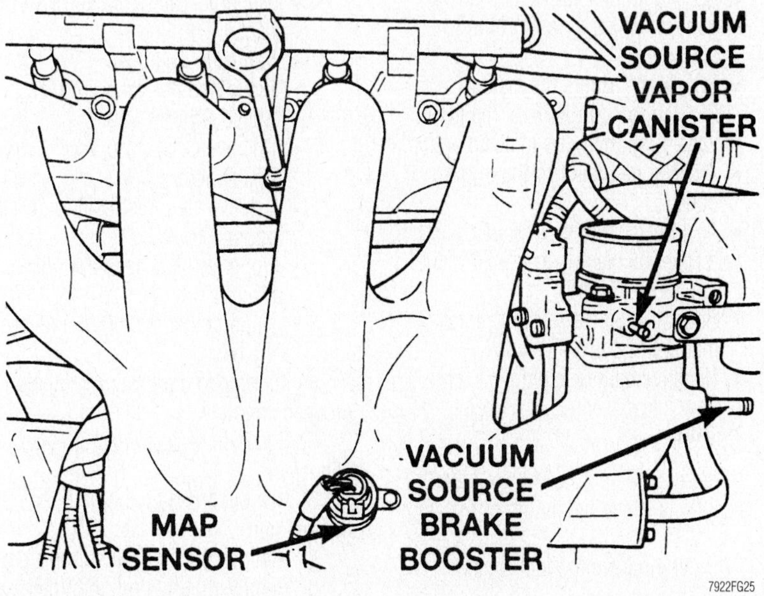

MAP sensor location—2.4L engine

- Throttle body
- Exhaust Gas Recirculation (EGR) tube
- Intake manifold support bracket
- Intake manifold

To install:

6. Thoroughly clean all parts. Clean all sealing surfaces
7. Install or connect the following:
   - New gasket
   - Intake manifold. Torque the fasteners to 17 ft. lbs. (23 Nm), in correct sequence, starting at the center and working outward

➡Make sure the fuel injector openings clean.

   - Fuel rail assembly to the intake manifold and torque the screws to 17 ft. lbs. (23 Nm)
   - Positive Crankcase Ventilation (PCV) and brake booster hoses
8. Lubricate the fuel tube with engine oil.
9. Install or connect the following:
   - Fuel supply line to the fuel rail. Pull on the connector to insure it is locked into position
   - Throttle body and torque the fasteners to 17 ft. lbs. (23 Nm)
   - Transaxle-to-throttle body support bracket. Torque the fasteners to 105 inch lbs. (12 Nm) at the throttle body first; then, at the transaxle
   - MAP and IAT electrical connectors
   - Knock sensor electrical connector
   - Wiring harness to the intake manifold tab
   - IAC motor and TP sensor electrical connectors
   - Throttle body vacuum hoses
   - Accelerator, kickdown and speed control cables to the throttle lever and bracket
   - EGR tube. Torque the fasteners to 95 inch lbs. (11 Nm), at the EGR valve first; then, the intake manifold
   - Air inlet resonator to the throttle body
   - Air inlet tube to the resonator and torque the clamps to 20–30 inch lbs. (2.5–3.5 Nm)
   - Both air inlet resonator-to-intake manifold bolts
   - Negative battery cable

2.7L Engine

1. Before servicing the vehicle, refer to the precautions in the beginning of this section.
2. Properly relieve the fuel system pressure.

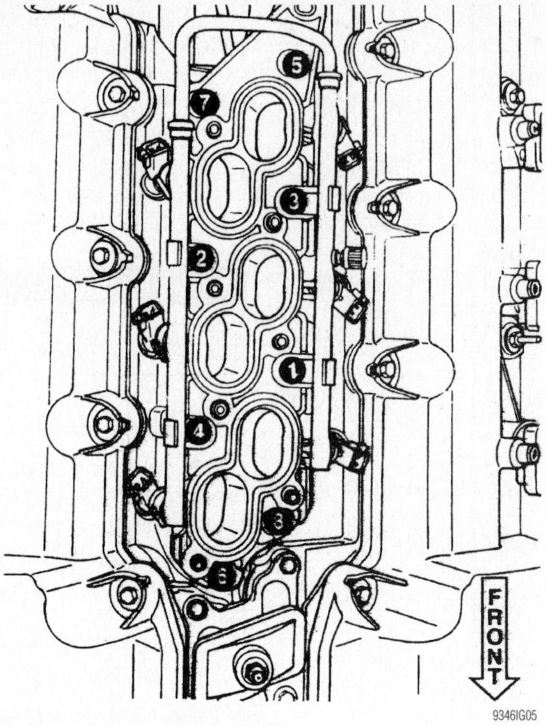

Intake manifold bolt torque sequence 2.7L engine

3. Remove or disconnect the following:
   - Negative battery cable
   - Throttle body air inlet hose
   - Air cleaner assembly
   - Throttle cable shield
   - Throttle and speed control cables
   - Throttle cable bracket
   - Manifold Absolute Pressure (MAP) sensor connector
   - Throttle Position Sensor (TPS) connector
   - Idle Air Control (IAC) motor connector
   - Manifold Tuning Valve (MTV) connector
   - Vapor purge hose
   - Brake booster vacuum hose
   - Speed Control Servo
   - Positive Crankcase Ventilation (PCV) hose
   - Exhaust Gas Recirculation (EGR) upper tube
   - Upper throttle body support bracket
   - Upper intake manifold and discard the gasket
4. To remove the lower intake manifold proceed as follows:
5. Remove the fuel injector electrical connectors.
6. Remove the fuel supply hose from the fuel rail.
7. Remove the fuel rail support bracket.
8. Remove the fuel rail and injectors as an assembly.
9. Remove the lower intake manifold and discard the gasket.

To install:

10. Clean all mating surfaces of any residual gasket material.
11. Install or connect the following:
    - Lower intake manifold with a new gasket
    - Fuel rail and injectors and torque the lower intake manifold bolts, in sequence, to 105 inch lbs. (12 Nm)
    - Fuel supply hose to the fuel rail
    - Fuel rail support bracket to the throttle body
    - Fuel injector electrical connectors
    - Upper intake manifold with a new gasket on the lower intake manifold and torque the bolts, in sequence, to 105 inch lbs. (12 Nm)
    - EGR upper tube
    - Speed control servo
    - PCV, brake booster and vapor purge hoses
    - MAP sensor electrical connector
    - TPS sensor electrical connector
    - IAC motor electrical connector
    - MTV electrical connector
    - Throttle cable bracket
    - Throttle and speed control cables
    - Throttle cable shield
    - Throttle body air inlet hose
    - Air cleaner assembly
    - Negative battery cable
12. Start the vehicle, check for leaks and repair if necessary.

Exhaust Manifold

REMOVAL & INSTALLATION

2.4L Engine

NON-PZEV ENGINES

1. Before servicing the vehicle, refer to the precautions in the beginning of this section.
2. Disconnect the negative battery cable from the left shock tower.

➡The ground cable is equipped with an insulator grommet which should be placed on the stud to prevent the negative battery cable from accidentally grounding.

3. Remove or disconnect the following:
   - Exhaust pipe from the exhaust manifold
   - Exhaust manifold heat shield
   - Heated Oxygen Sensor (HO_2S), if necessary
   - Exhaust manifold and discard the gasket

To install:

4. Thoroughly clean all parts. Clean all sealing surfaces of the manifold and cylinder head. Check the manifold gasket surface for flatness with a straightedge and feeler gauge. The surface must be flat within 0.006 in. (0.15mm) per foot (30cm) of manifold length. Inspect the manifold for cracks or distortion. Replace if necessary.
5. Install or connect the following:
   - New gasket
   - Exhaust manifold. Torque the bolts, starting at the center and working outward, to 17 ft. lbs. (23 Nm)
   - HO_2S sensor
   - Heat shield
   - Exhaust pipe and torque the fasteners to 21 ft. lbs. (28 Nm)
   - Negative battery cable

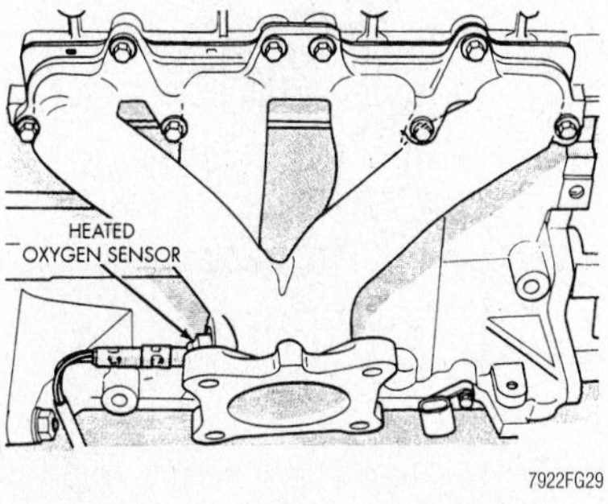

Be careful not to damage the oxygen sensor when servicing the manifold—2.4L engine

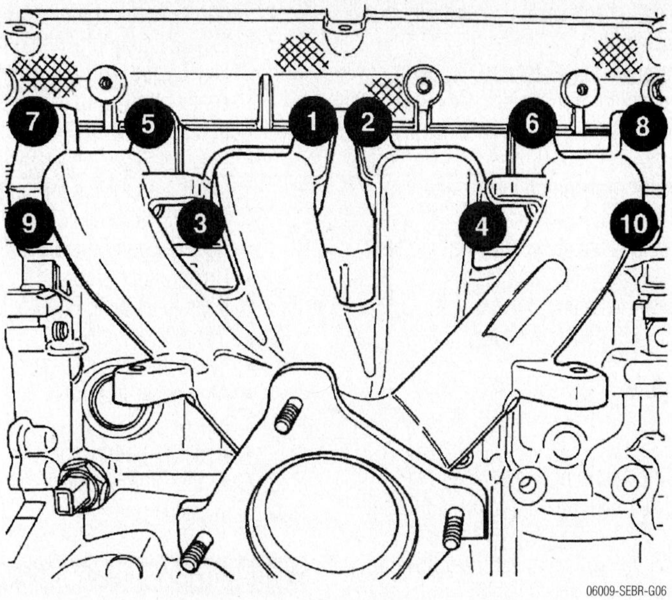

Exhaust manifold torque sequence—2.4L non-PZEV engines

6. Start the engine and allow it to idle while inspecting the manifold for exhaust leaks.

PZEV ENGINES

1. Disconnect negative battery cable.
2. Raise vehicle on hoist.
3. Remove complete exhaust system.
4. Remove rear engine mount and transaxle bracket.
5. Disconnect connector tube from check valve. Squeeze connector and pull.
6. Remove exhaust manifold heat shield.
7. Disconnect oxygen sensor electrical connector.
8. Remove exhaust manifold retaining fasteners and remove exhaust manifold.
9. Remove and discard manifold gasket.

To install:

10. Install new exhaust manifold gasket. DO NOT APPLY SEALER.
11. Position exhaust manifold in place. Tighten fasteners in sequence shown to 23 Nm (200 inch lbs.).
12. Install exhaust manifold heat shield.

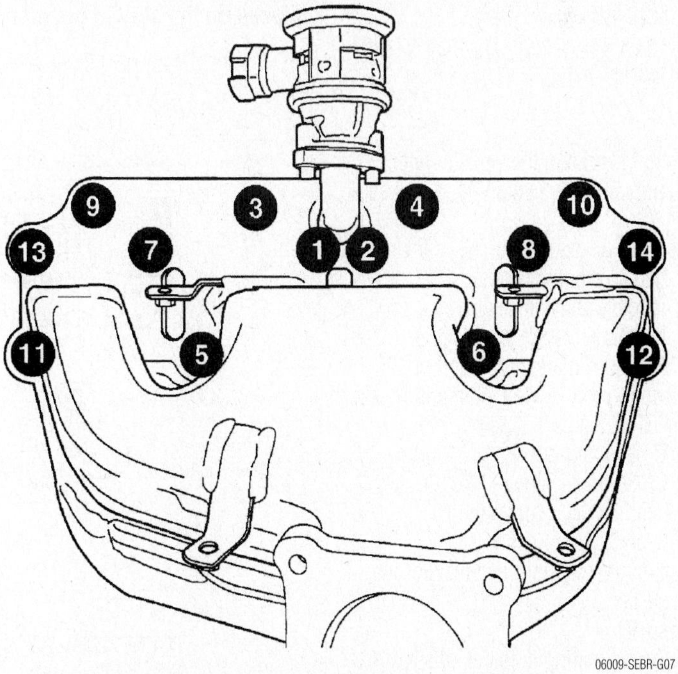

Exhaust manifold torque sequence—2.4L PZEV engines

13. Install check valve tighten bolts to 12 Nm (105 inch lbs.).
14. Connect connector tube to check valve by pushing fitting over check valve.
15. Connect oxygen sensor electrical connector.
16. Install rear engine mount and transaxle bracket.
17. Install exhaust system. Tighten fasteners to 28 Nm (250 inch lbs.).
18. Lower vehicle.
19. Connect negative battery cable.

2.7L Engine

FRONT

1. Before servicing the vehicle, refer to the precautions in the beginning of this section.
2. Remove or disconnect the following:
   - Negative battery cable
   - Cross under pipe
   - Front catalytic converter
   - Exhaust manifold and discard the gasket

To install:

3. Clean all mating surfaces of any residual gasket material.
4. Install or connect the following:
   - Exhaust manifold with a new gasket. Torque the bolts, starting in the center and working outward, to 17 ft. lbs. (23 Nm)
   - Front catalytic converter and heat shield and torque the bolts to 21 ft. lbs. (28 Nm)
   - Cross under pipe and torque the bolts to 21 ft. lbs. (28 Nm)
   - Negative battery cable
5. Start the vehicle, check for leaks and repair if necessary.

REAR

1. Before servicing the vehicle, refer to the precautions in the beginning of this section.
2. Remove or disconnect the following:
   - Negative battery cable
   - Throttle body air inlet hose
   - Air cleaner assembly
   - Exhaust Gas Recirculation (EGR) tube from the exhaust manifold and EGR valve
   - Exhaust system
   - Cross under pipe
   - Rear catalytic converter
   - Rear Oxygen Sensor (O_2S)
   - Exhaust manifold and discard the gasket

To install:

3. Clean all mating surfaces of any residual gasket material.
4. Install or connect the following:
   - Exhaust manifold with a new gasket. Torque the bolts, starting in the center and working outward, to 17 ft. lbs. (23 Nm)
   - Rear catalytic converter and heat shield and torque the bolts to 21 ft. lbs. (28 Nm)
   - Rear Oxygen Sensor (O_2S)
   - Cross under pipe and torque the bolts to 21 ft. lbs. (28 Nm)
   - Exhaust system
   - EGR tube with new gaskets
   - Air cleaner assembly and air intake hose
   - Negative battery cable
5. Start the vehicle, check for leaks and repair if necessary.

Front Crankshaft Seal

REMOVAL & INSTALLATION

2.4L Engine

The timing belt must be removed for this procedure. Use care that all timing marks are aligned after installation or the engine will be damaged.

1. Drain the engine oil.
2. Before servicing the vehicle, refer to the precautions in the beginning of this section.
3. Disconnect the negative battery cable from the left shock tower.

➡The ground cable is equipped with an insulator grommet which should be placed on the stud to prevent the negative battery cable from accidentally grounding.

4. Remove or disconnect the following:
   - Accessory drive belts
   - Crankshaft damper/pulley using a jaw puller tool
   - Timing belt
   - Crankshaft timing belt sprocket using a gear/sprocket puller

➡Be careful not to nick the seal surface of the crankshaft or the seal bore.

   - Front crankshaft seal using Seal Removal Tool No. 6771

➡Be careful not to damage the seal contact area of the crankshaft.

To install:

5. Lubricate the new oil seal lip with engine oil.
6. Install or connect the following:
   - New crankshaft oil seal with the spring facing inward using Oil Seal Installer Tool 6780–1, until it is flush with the front cover
   - Crankshaft timing belt sprocket using Tool No. 6792

➡Be sure the word "FRONT" on the timing belt sprocket is facing outward.

   - Timing belt and timing belt cover
   - Crankshaft damper/pulley using the thrust bearing/washer and 12M-1.75 x 150mm bolt from special Tool No. 6792. Torque the bolt to 105 ft. lbs. (142 Nm)
   - Accessory drive belts and adjust the tension

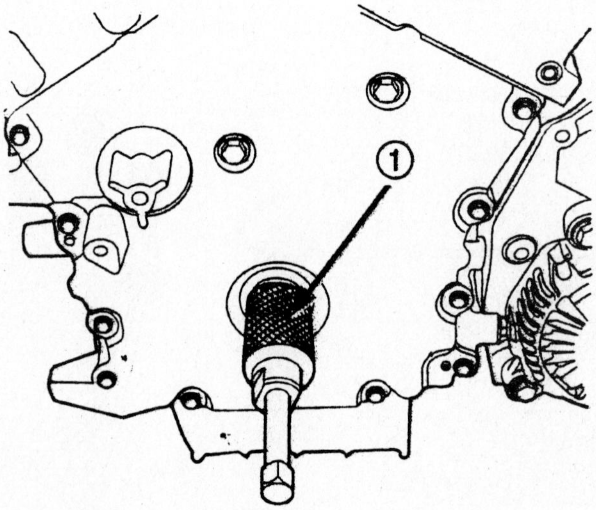

Remove the front crankshaft seal with special tool 6771

- Negative battery cable

⁂ WARNING

Operating the engine without the proper amount and type of engine oil will result in severe engine damage.

7. Fill the engine with clean oil.
8. Start the engine and check for leaks.

2.7L Engine

1. Before servicing the vehicle, refer to the precautions in the beginning of this section.
2. Remove or disconnect the following:
   - Negative battery cable
   - Crankshaft damper
   - Front crankshaft seal with special Tool 6771

To install:

3. Install or connect the following:
   - New seal with special Tools 6780–2, 8179 and 6780–1
   - Crankshaft damper
   - Negative battery cable

Camshaft

REMOVAL & INSTALLATION

2.4L Engine

This engine uses a DOHC, 4-valves per cylinder, cross-flow aluminum cylinder head. The valves are actuated by roller cam followers which pivot on stationary hydraulic valve adjusters. Care must be taken to ensure all valve timing marks align after cylinder head and valvetrain service.

1. Before servicing the vehicle, refer to the precautions in the beginning of this section.
2. Disconnect the negative battery cable from the left shock tower.

➡The ground cable is equipped with an insulator grommet which should be placed on the stud to prevent the negative battery cable from accidentally grounding.

3. Relieve the fuel system pressure using the recommended procedure.
4. Remove or disconnect the following:
   - Spark plugs cables
   - Ignition coil pack with the spark plug cables
   - Cylinder head cover and discard the gasket
   - Ground strap
   - Timing belt covers and timing belt
   - Camshaft sprockets by holding them with Tool 6847, while removing the bolt
5. Take note that the camshaft bearing caps are numbered for correct location during installation. Remove the outer bearing caps first.
6. Loosen, but do not remove, the camshaft bearing cap retaining fasteners in the correct sequence, inside working outward. Perform this step on one camshaft at a time.

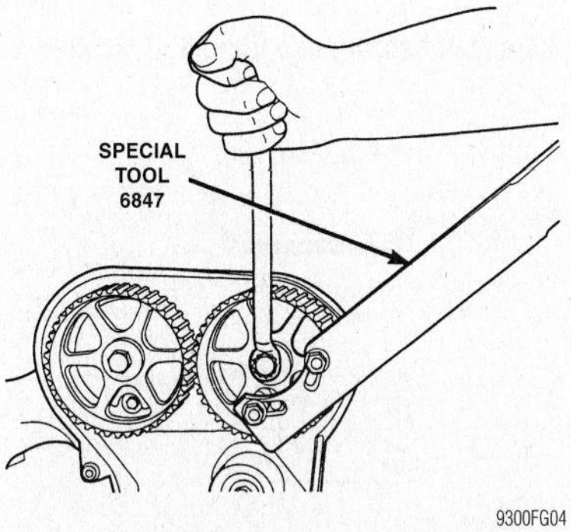

Use special tool 6847 to hold the camshaft sprocket while removing or installing the center bolt—2.4L engine

7. Identify the camshafts, if they are to be reused for later installation. The camshafts are not interchangeable.
8. Remove or disconnect the following:
   - Camshaft bearing caps
   - Camshafts
   - Camshaft followers
9. Any components that are to be reused must be installed in their original locations. Use care to identify and mark the positions of any removed valvetrain components so they may be reinstalled correctly.
10. Inspect the camshaft bearing oil feed holes in the cylinder head for clogging. Inspect the camshaft bearing journals for wear or scoring. Check the cam surface for abnormal wear and damage. A visible worn groove in the roller path or on the cam lobes is cause for replacement.

To install:

11. Thoroughly clean all camshaft and related parts.
12. Inspect the camshaft end-play using the following procedure:
    a. Lubricate the camshaft journals and install the camshaft **WITHOUT** the cam follower assemblies. Install the rear

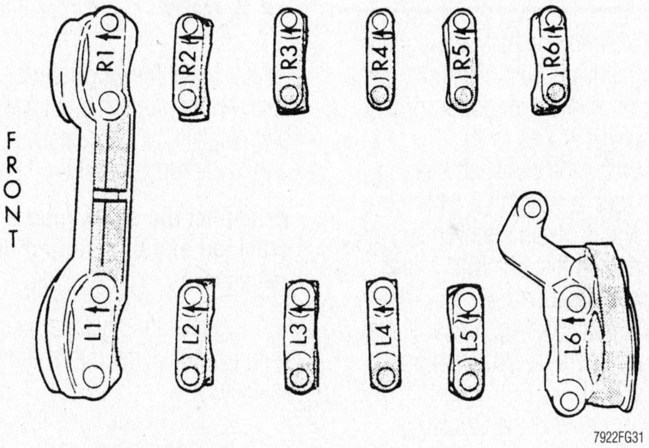

Camshaft bearing cap identification—2.4L engine

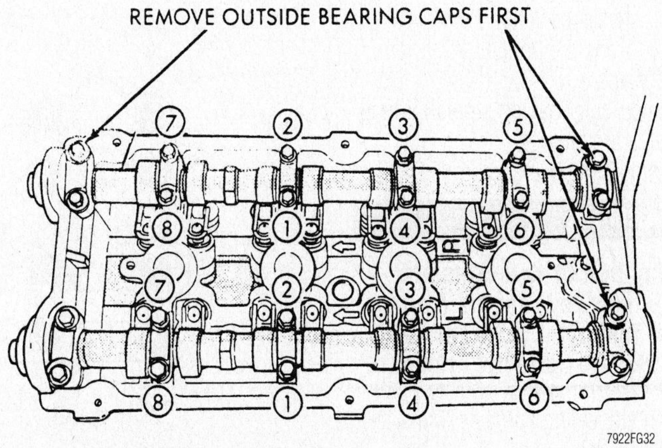

Camshaft bearing cap bolt removal sequence—2.4L engine

cam caps and tighten to 21 ft. lbs. (28 Nm).

b. Push the camshaft rearward as far as it will go.

c. Adjust a dial indicator to rest against the front of the camshaft (the sprocket end). Zero the indicator.

d. Move the camshaft forward as far as it will go. Read the dial indicator. End-play specification is 0.002–0.010 in. (0.05–0.15mm).

e. If excessive end-play is present, inspect the cylinder head and camshaft for wear; replace if necessary.

13. If the fit and condition of the camshafts are acceptable, remove the camshafts for installation of the cam followers.

14. The hydraulic valve lash adjusters are inside the roller cam followers. Be sure they are clean, well lubricated with engine oil and properly positioned. Install the cam followers in their original positions on the hydraulic adjuster and valve stem.

⁂ WARNING

Be sure NONE of the pistons are at Top Dead Center (TDC) when installing the camshafts.

15. Lubricate the camshaft bearing journals and cam followers with clean engine oil

16. Install or connect the following:

- Camshafts
- Right/left-side camshaft bearing caps Nos. 2 through 5 and right-side No. 6. Torque the M6 fasteners to 105 inch lbs. (12 Nm) in the correct sequence.

17. Apply Mopar® Gasket Maker sealer to the No. 1 and left-side No. 6 bearing caps.

18. Install or connect the following:

- Bearing caps and torque the M8 fasteners to 21 ft. lbs. (28 Nm)
- Camshaft end seals
- Camshaft sprockets, if removed, and torque the bolts to 75 ft. lbs. (101 Nm)
- Timing belt, making sure all timing marks are aligned
- Timing belt covers

⁂ WARNING

If the timing belt or sprockets are incorrectly installed, engine damage will occur. Take time to be sure all timing marks are correctly aligned.

19. Clean all sealing surfaces. Make certain the rails are flat.

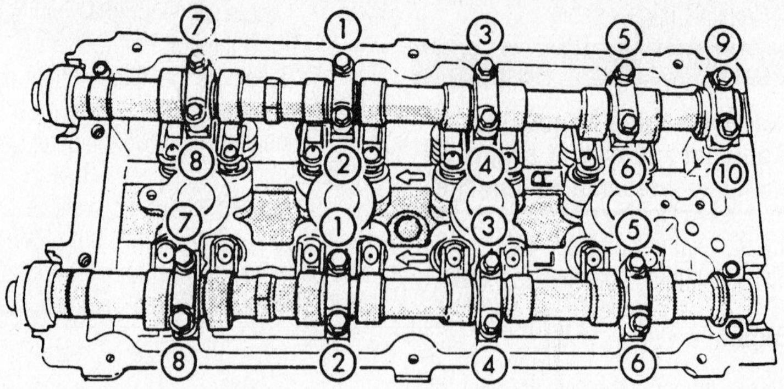

7922FG33

Camshaft bearing cap tightening sequence—2.4L engine

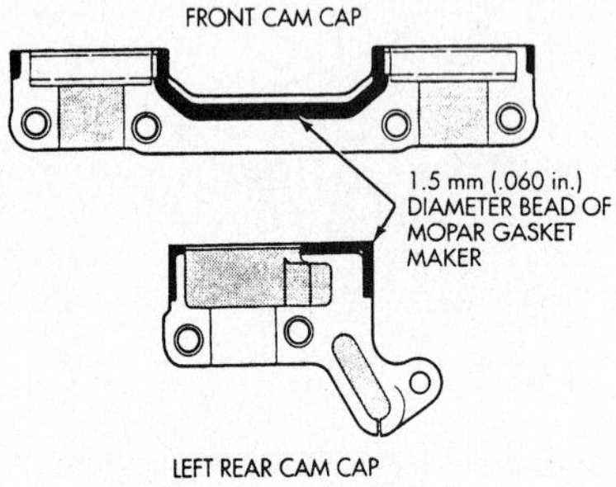

7922FG34

Apply sealer as shown to prevent oil leakage from the camshaft bearing end caps—2.4L engine

20. Install new cylinder head cover gaskets. Apply Mopar® Silicone Rubber Adhesive Sealant, at the camshaft cap corners and at the top edge of the ½ round seal.

➡Inspect the spark plug well seals for cracking and/or swelling and replace if necessary.

21. Install the cylinder head cover and torque the fasteners, in sequence, using the following:

a. Step 1: 40 inch lbs. (4.5 Nm).

b. Step 2: 80 inch lbs. (9 Nm).

c. Step 3: 105 inch lbs. (12 Nm).

22. Install or connect the following:

- Ignition coil pack and torque the fasteners to 105 inch lbs. (12 Nm)
- Spark plug cables
- Ground strap
- All vacuum lines and wiring
- Negative battery cable

➡An oil and filter change is recommended.

23. Test run vehicle. Check for leaks and for proper operation.

2.7L Engine

1. Before servicing the vehicle, refer to the precautions in the beginning of this section.

2. Remove or disconnect the following:

- Negative battery
- Timing chain and secondary chain tensioner bolts
- Camshaft bearing caps
- Camshafts, secondary chain and tensioner as an assembly
- Tensioner and chain from the camshafts

To install:

3. Install or connect the following:

- Camshaft chain on the cams and

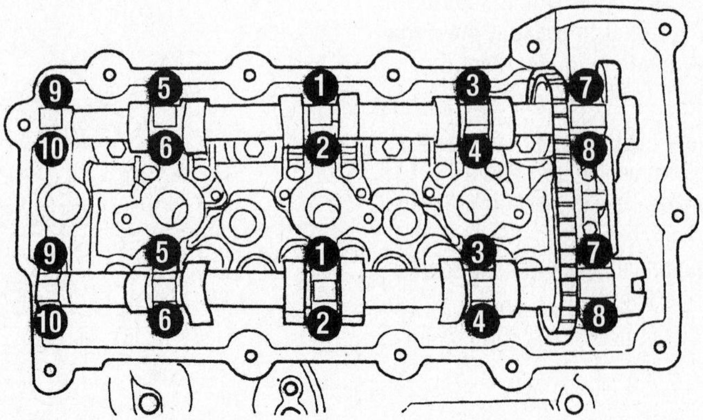

9346IG07

Tighten the camshaft bearing caps in the proper sequence–2.7L Engine

make certain that the plated links are facing forward

4. Compress the chain tensioner as follows:
   a. Place the tensioner in a soft jaw vise.
   b. Slowly compress the tensioner until the fabricated lock pin can be installed in the locking holes.
   - Tensioner between the camshafts and chain. Rotate the cams until the plated links and dots are in the 12 o'clock position
   - Cams to the cylinder head
   - Camshaft bearing caps. Slowly, and in sequence, torque the bolts to 105 inch lbs. (12 Nm)
   - Secondary chain tensioner bolts and torque them to 105 inch lbs. (12 Nm). Remove the locking pin from the tensioner
   - Primary timing chain
   - Negative battery cable

Valve Lash

ADJUSTMENT

2.4L and 2.7L Engines

The valves are actuated by roller cam followers which pivot on stationary Hydraulic Lash Adjusters (HLAs). The HLAs are precision units installed in machined openings of the cam follower. Valve clearance adjustments are not performed.

Starter Motor

REMOVAL & INSTALLATION

2.4L Engine

1. Before servicing the vehicle, refer to the precautions in the beginning of this section.
2. Remove or disconnect the following:
   - Negative battery cable
   - Air cleaner assembly
   - Starter cover
   - Starter electrical connectors
   - Starter

To install:

3. Install or connect the following:
   - Starter and torque the bolt to 23 ft. lbs. (30 Nm)
   - Electrical connectors to the starter
   - Starter cover and torque the bolt to 44 inch lbs. (5 Nm)
   - Air cleaner assembly
   - Negative battery cable

2.7L Engine

1. Before servicing the vehicle, refer to the precautions in the beginning of this section.
2. Remove or disconnect the following:
   - Negative battery cable
   - Oxygen Sensor (O_2S) electrical connector
   - Front engine mount through bolt
   - Starter electrical connectors
   - Starter

To install:

3. Install or connect the following:
   - Starter and hand tighten the lower bolt
   - Starter electrical connectors and torque the starter bolts to 40 ft. lbs. (54 Nm)
   - Engine mount through bolt and torque it to 45 ft. lbs. (61 Nm)
   - O_2 sensor and torque it to 20 ft. lbs. (27 Nm)
   - Negative battery cable

Oil Pan

REMOVAL & INSTALLATION

2.4L Engine

2002–04

1. Disconnect negative battery cable.

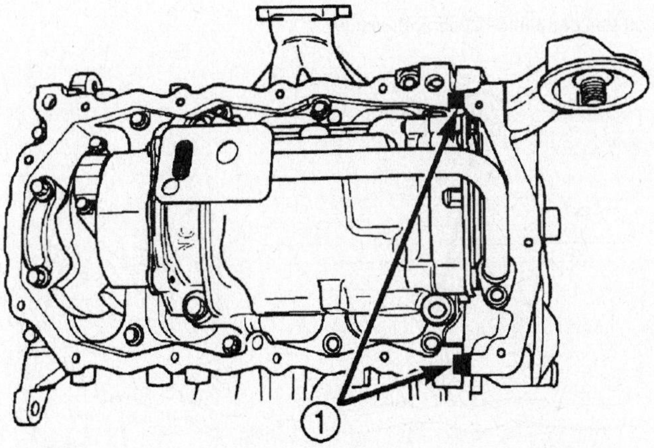

1 - PLACE A 3 MM (1/8 INCH.) BEAD OF SEALANT AT THE PARTING LINE OF THE OIL PUMP BODY TO ENGINE BLOCK
06009-SEBR-G56

Oil pan sealing—2002–04 2.4L engine

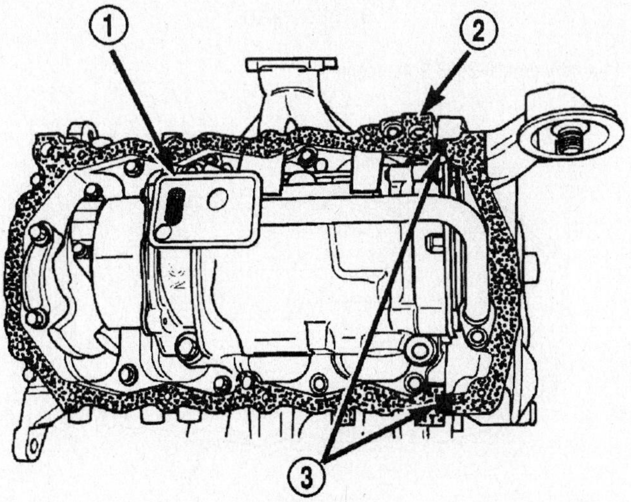

1 - OIL PICK-UP TUBE
2 - OIL PAN GASKET
3 - SEALER
06009-SEBR-G57

Oil pan gasket installation—2002–04 2.4L engine

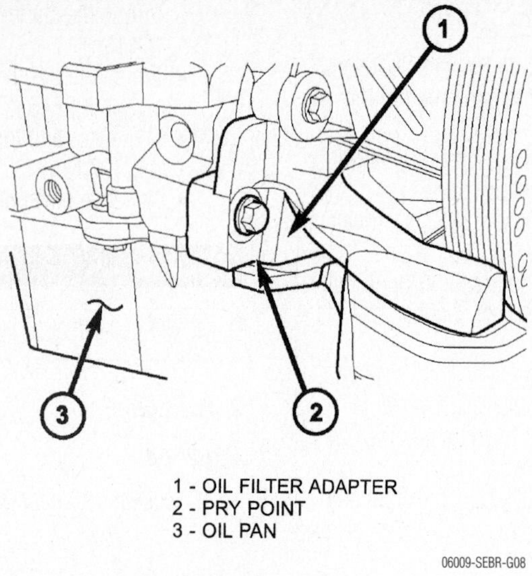

Right side oil pan pry point—2005 2.4L engine

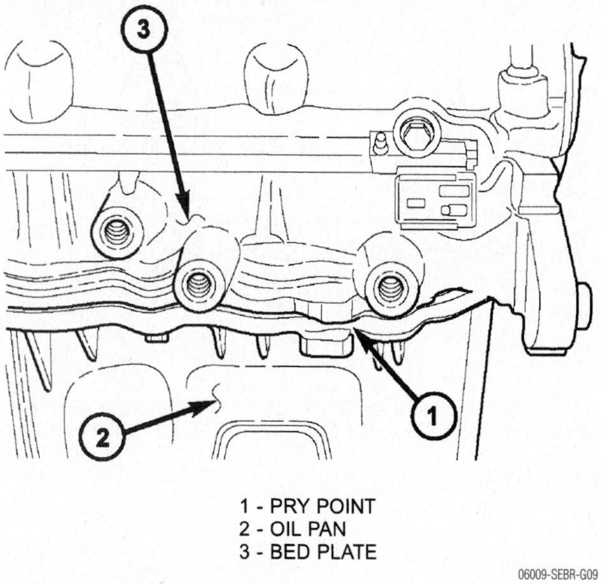

Left side oil pan pry point—2005 2.4L engine

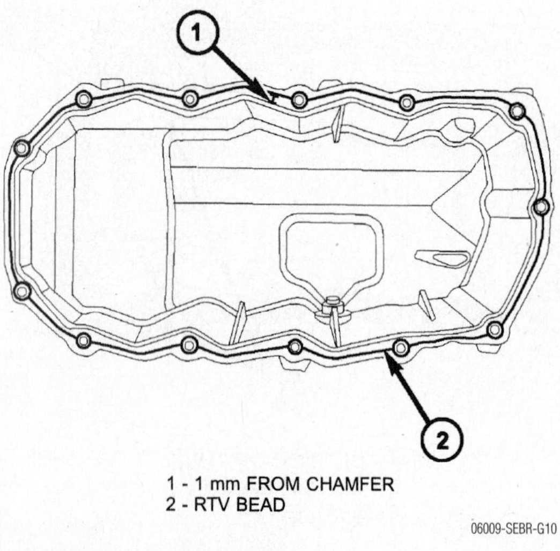

RTV application for the 2005 2.4L engine oil pan

2. Raise vehicle on hoist and drain engine oil.
3. Remove structural collar and torque reaction bracket.
4. Remove bolts attaching oil pan.
5. Remove oil pan.
6. Clean oil pan and all gasket surfaces.

To install:

7. Apply Mopar® Engine RTV GEN II, or equivalent, at the oil pump to engine block parting line.
8. Install the oil pan gasket to the block.
9. Install pan and tighten the screws to 12 Nm (105 inch lbs.).
10. Install structural collar and torque reaction bracket.
11. Lower vehicle and fill engine crankcase with proper oil to correct level.
12. Connect negative battery cable.

2005

1. Disconnect negative battery cable.
2. Raise vehicle on hoist and drain engine oil.
3. Remove structural collar.
4. Remove air conditioning compressor bracket to oil pan bolt.

➡The oil pan is sealed with RTV and may need to be pried apart.

5. Remove bolts attaching oil pan.

➡Prior to removing the oil pan, score RTV with a razor blade between the oil pan and the bed plate.

6. Insert pry bar at locations shown and gently pry oil pan away from the engine.

To install:

➡You must assemble oil pan to bed plate before RTV "skins over". If RTV skins before assembly, parts must be cleaned and a new bead of RTV applied.

7. Clean oil pan and block surfaces. Clean cured RTV from oil pan and bed plate with a plastic or brass scraper.
8. Clean all sealing surfaces with MOPAR® Brake Parts Cleaner, or equivalent.
9. Inspect surfaces for damage, replace as needed.
10. Apply a 3–4 mm diameter bead of MOPAR® Engine RTV, or equivalent, around the perimeter of the oil pan flange, 1 mm away from the chamfer on the inner edge of the oil pan.
11. Position oil pan on bedplate.
12. Install oil pan attaching bolts and tighten to 12 Nm (105 inch lbs.).
13. Install air conditioning compressor bracket to oil pan bolt.

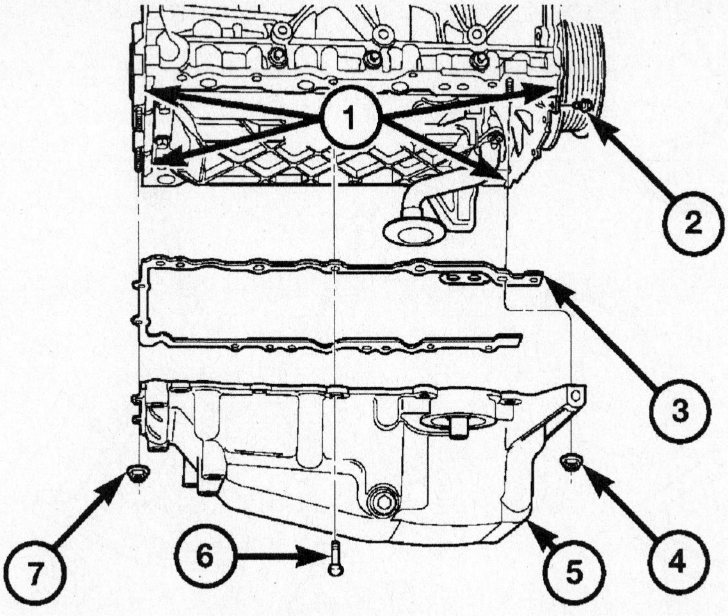

1 - SEALER LOCATION
2 - BOLT-M6
3 - GASKET
4 - NUT-M6
5 - OIL PAN
6 - BOLT-M8
7 - NUT-M6

06009-SEBR-G11

Oil pan installation—2.7L engine

14. Install structural collar.
15. Lower vehicle and fill engine crankcase with proper oil to correct level.
16. Connect negative battery cable.
17. Start engine and check for leaks.

2.7L Engine

1. Before servicing the vehicle, refer to the precautions in the beginning of this section.
2. Drain the engine oil.
3. Remove the engine support module as follows:
   - Negative battery cable
   - Oil dipstick and tube
   - Structural collar
   - Exhaust cross under pipe
   - Torque converter cover
   - Lower bolt from the A/C compressor
   - Oil pan and discard the gasket

To install:

4. Clean all mating surfaces of any residual gasket material.
   - Oil pan with a new gasket
5. Torque the oil pan bolts as follows:
   a. Timing chain cover to oil pan bolts and torque them to 105 inch lbs. (12 Nm).
   b. Oil pan bolts to 21 ft. lbs. (28 Nm).
   c. Oil pan nuts to 105 inch lbs. (12 Nm).
   d. A/C compressor bolt to 21 ft. lbs. (28 Nm).
   - Torque converter cover
   - Exhaust cross under pipe
   - Structural collar
   - Oil dipstick and tube
   - Negative battery cable
6. Fill the engine with clean oil.
7. Start the vehicle, check for leaks and repair if necessary.

Oil Pump

REMOVAL & INSTALLATION

2.4L Engine

The oil drawn up through the pick-up tube is pressurized by the pump and routed through the full flow filter to the main oil galley running the length of the cylinder block. The oil pick-up, pump and check valve provide oil flow to the main oil gallery. A vertical hole at the No. 5 bulkhead routes pressurized oil through a restrictor up past a cylinder head bolt to an oil galley running the length of the cylinder head. The camshaft journals are slotted to allow pressurized oil to pass into the bearing cap cavities. Small holes in the bearing caps direct oil to the camshaft lobes.

1. Before servicing the vehicle, refer to the precautions in the beginning of this section.
2. Drain the engine oil.
3. Remove the engine support module as follows:
   - Negative battery cable
   - Crankshaft damper
   - Timing belt and tensioner
   - Camshaft sprocket
   - Rear timing belt cover
   - Oil pan
   - Oil pick up tube
   - Oil pump and discard the gasket
4. Clean all mating surfaces of any residual gasket material.

To install:

5. Prime the oil pump before installation.
6. Install or connect the following:
   - Oil pump with the flats aligned with the crankshaft flats and torque the bolts to 21 ft. lbs. (28 Nm)
   - New front crankshaft seal
   - Crankshaft sprocket
   - Oil pick up tube and oil pan
   - Camshaft sprocket
   - Rear timing belt cover
   - Timing belt and tensioner
   - Front timing belt cover
   - Crankcase damper
   - Negative battery cable
7. Fill the engine with clean oil.
8. Start the vehicle, check for leaks and repair if necessary.

2.7L Engine

1. Before servicing the vehicle, refer to the precautions in the beginning of this section.
2. Drain the engine oil.
3. Remove the engine support module as follows:
   - Negative battery cable
   - Crankshaft damper
   - Timing chain cover
   - Timing chain and sprockets
   - Oil pan
   - Oil pick up tube and O-ring and make certain that the crankshaft is at 60 degrees After Top Dead Center (ATDC)
   - Oil pump

To install:

4. Prime the pump before installation by filling the rotor cavity with clean engine oil.

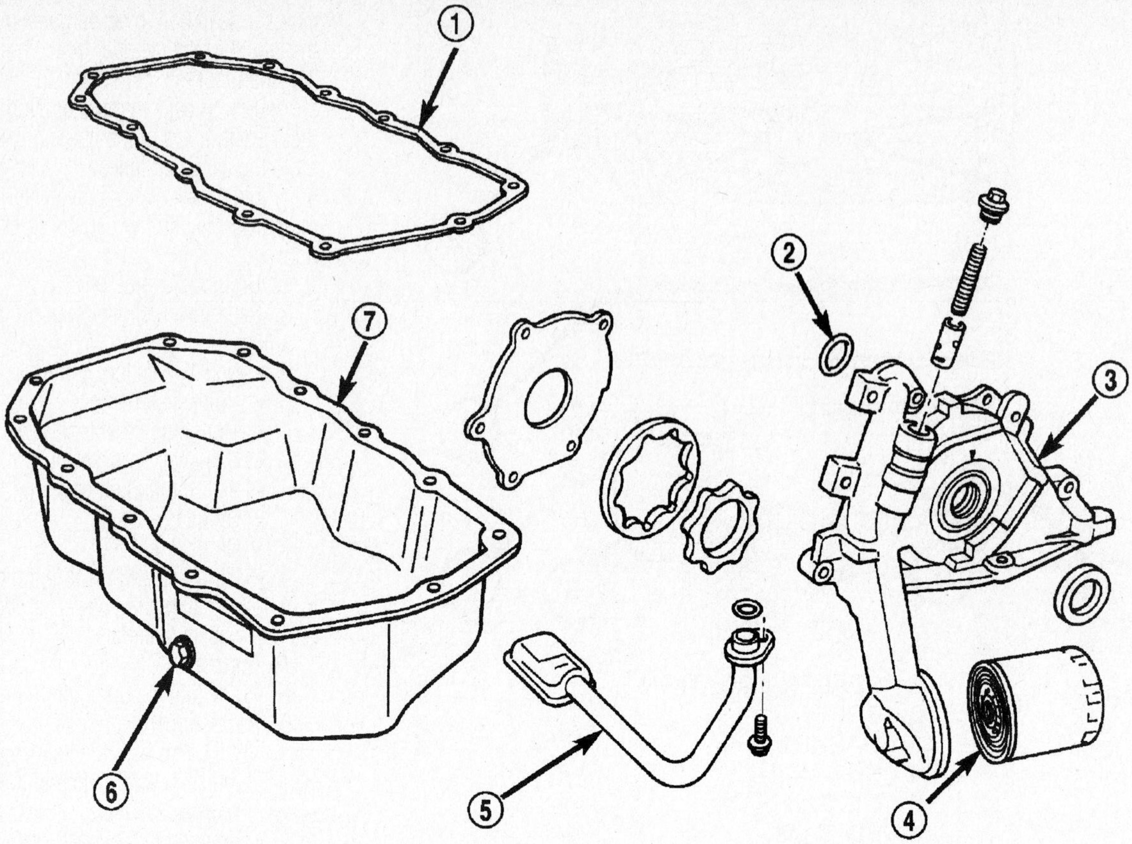

1 - OIL PAN GASKET
2 - O-RING
3 - OIL PUMP BODY
4 - FILTER
5 - OIL PICK-UP TUBE
6 - DRAIN PLUG
7 - OIL PAN

06009-SEBR-G12

Oil pump and related parts—2.4L engine

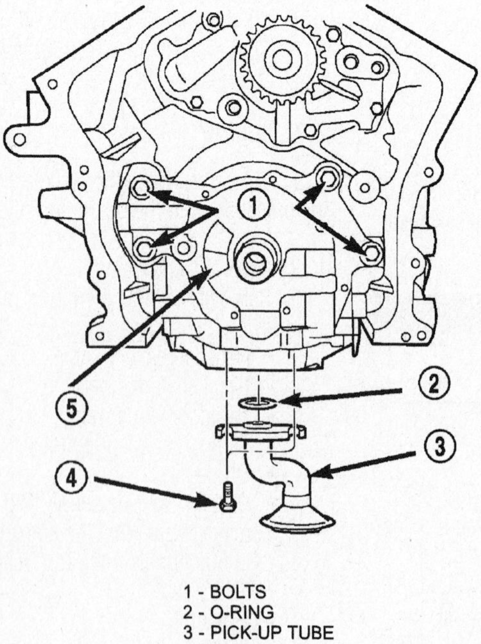

1 - BOLTS
2 - O-RING
3 - PICK-UP TUBE
4 - BOLT
5 - OIL PUMP

06009-SEBR-G13

Oil pump and pick-up tube—2.7L engine

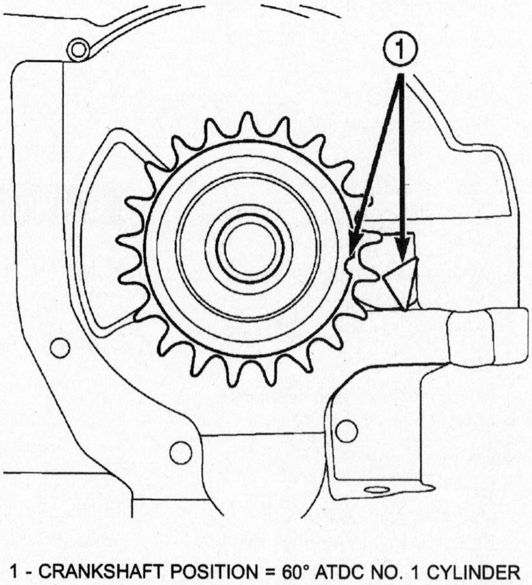

1 - CRANKSHAFT POSITION = 60° ATDC NO. 1 CYLINDER

06009-SEBR-G14

Crankshaft positioned at 60 degrees ATDC—2.7L engine

5. Install or connect the following:
   - Oil pump over the crankshaft and torque the bolts to 21 ft. lbs. (28 Nm)
   - Oil pick up tube with a new O-ring and torque the bolt to 21 ft. lbs. (28 Nm)
   - Oil pan with a new gasket
   - Timing chain and sprockets
   - Timing chain and cover
   - Crankshaft damper
   - Negative battery cable
6. Fill the engine with clean oil.
7. Start the vehicle, check for leaks and repair if necessary.

Rear Main Seal

REMOVAL & INSTALLATION

1. Before servicing the vehicle, refer to the precautions in the beginning of this section.
2. Remove or disconnect the following:
   - Negative battery cable from the left shock tower.

➡The ground cable is equipped with an insulator grommet which should be placed on the stud to prevent the negative battery cable from accidentally grounding.

   - Transaxle
   - Flexplate/flywheel
   - Rear crankshaft oil seal using a flat-bladed prying tool

To install:

➡When installing the seal there is no need to lubricate the sealing surface.

3. Install or connect the following:
   - New seal using a suitable driver or seal installer
   - Flexplate/flywheel and torque the bolts to 70 ft. lbs. (95 Nm)
   - Transaxle
   - Negative battery cable
4. Start the engine and check for leaks.

Balance Shaft Chain, Sprockets And Cover

REMOVAL & INSTALLATION

2.4L Engine

1. Before servicing the vehicle, refer to the precautions in the beginning of this section.
2. Drain the engine oil.
3. Remove or disconnect the following:
   - Oil pan and discard the gasket
   - Chain cover, guide and tensioner
   - Gear cover and balance shaft gears
   - Balance shaft gear and crankshaft chain sprocket
   - Carrier gear cover and balance shafts

BALANCE SHAFT TIMING AND INSTALLATION

➡During installation of the balance shafts to crankshaft, the proper timing procedure must be followed to prevent internal engine damage.

1. Install or connect the following:
2. Install the balance shafts and carrier on the crankshaft. Torque the bolts to 40 ft. lbs. (54 Nm)
3. Turn the balance shafts until both key ways are pointing up. Install the short hub drive gear on the sprocket driven shaft.
4. Make certain that the key ways and the gear timing marks are properly aligned.
5. Install the gear cover and torque the double ended stud and washer to 105 inch lbs. (12 Nm)
6. Align the flat of the balance shaft drive sprocket to the flat on the crankshaft
7. Install the balance shaft drive sprocket to the crankshaft with special tool 6052.
8. Turn the crankshaft until the No. 1 cylinder is at Top Dead Center (TDC).
9. Make certain that the timing marks on the chain sprocket are aligned with the parting line on the left side main bearing cap.
10. Install the chain over the crankshaft and be sure the plated link is over the No. 1 cylinder timing mark on the balance shaft crankshaft sprocket.
11. Place the balance shaft sprocket into the timing chain and align the sprocket dot to the lower plate link on the chain.
12. Slide the balance shaft sprocket on the nose of the balance shaft with the key ways pointing at the 12 o'clock position.
13. When properly timed, install the balance shaft bolts and torque to 21 ft. lbs. (28 Nm)
14. To tighten the chain use the following procedure:

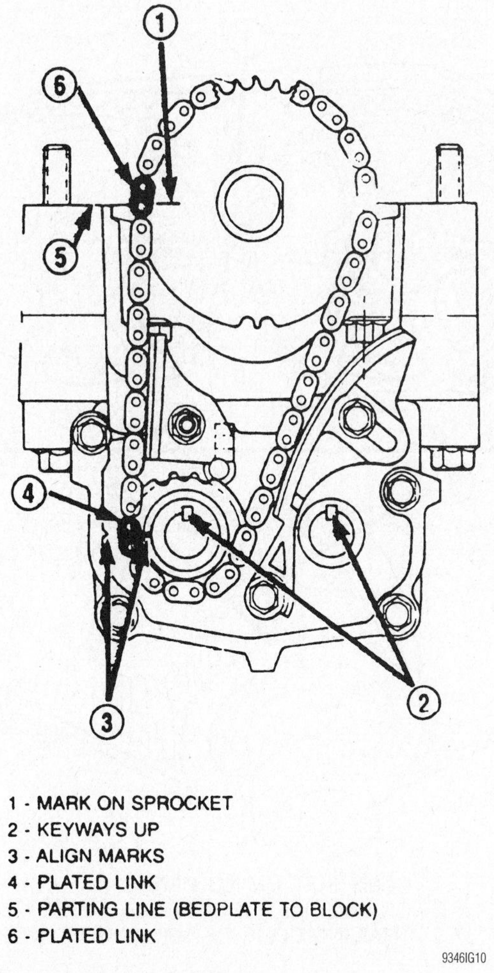

Align the key ways and the gear timing marks

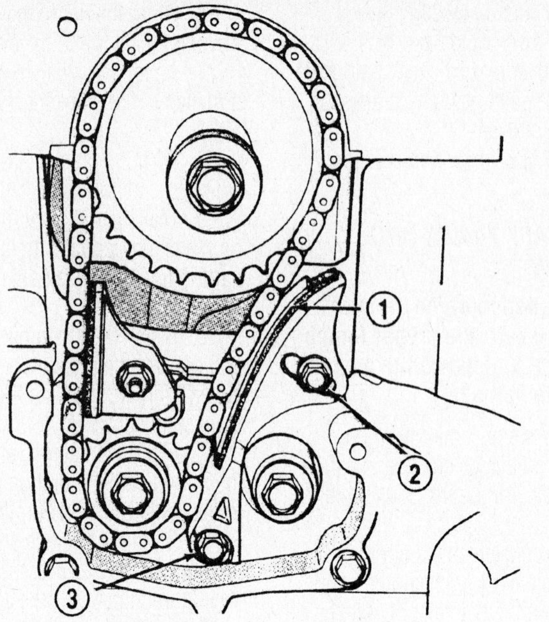

Properly adjust the chain tension

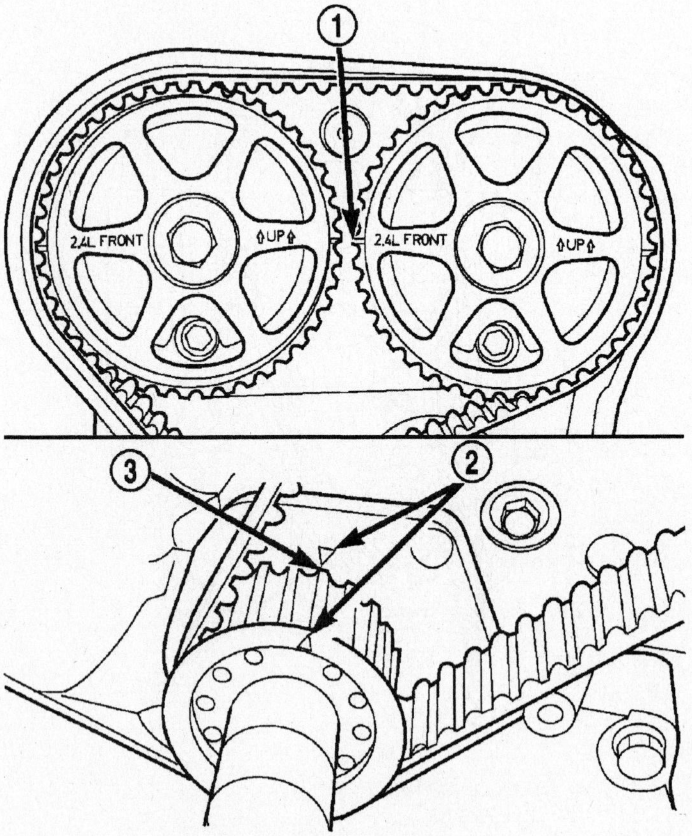

Crankshaft and camshaft timing marks—2.4L engine

a. Loosely install the chain tensioner.

b. Position the guide on the double ended stud and make certain that the tab on the guide fits into the slot on the gear cover.

c. Install and tighten the nut/washer assembly to 105 inch lbs. (12 Nm).

d. Place a 1 mm thick by 70 mm long shim between the tensioner and chain.

e. With a load applied, torque the top tensioner bolt to 105 inch lbs. (12 Nm).

f. Torque the bottom pivot bolt to 105 inch lbs. (12 Nm).

g. Remove the shim.

15. Install the carrier covers and torque the bolts to 105 inch lbs. (12 Nm).
16. Install the oil pan.
17. Fill the engine with clean oil.
18. Start the vehicle, check for leaks and repair if necessary.

Timing Belt

REMOVAL & INSTALLATION

2.4L Engine

1. Before servicing the vehicle, refer to the precautions in the beginning of this section.
2. Remove the A/C refrigerant using approved equipment.
3. Remove or disconnect the following:
   - Negative battery cable
   - Right front wheel and inner splash shield
   - Accessory drive belts
   - Crankshaft damper
   - Lower torque strut
   - Exhaust system from the manifold
   - A/C compressor switch
   - Upper torque strut and bracket
   - Upper radiator support crossmember
   - Power steering pump and bracket without disconnecting the lines
   - Right engine mount through bolt after supporting the engine
   - Engine support bracket
4. Rotate the crankshaft until the Top Dead Center (TDC) mark on the oil pump housing aligns with the TDC mark on the crankshaft sprocket.
5. Install an allen wrench into the tensioner. Rotate the tensioner counterclockwise while pushing on the wrench until it slides into the locking hole.
6. Remove the timing belt.

To install:

7. Set the crankshaft sprocket at TDC by aligning the sprocket with the arrow on the oil pump housing.

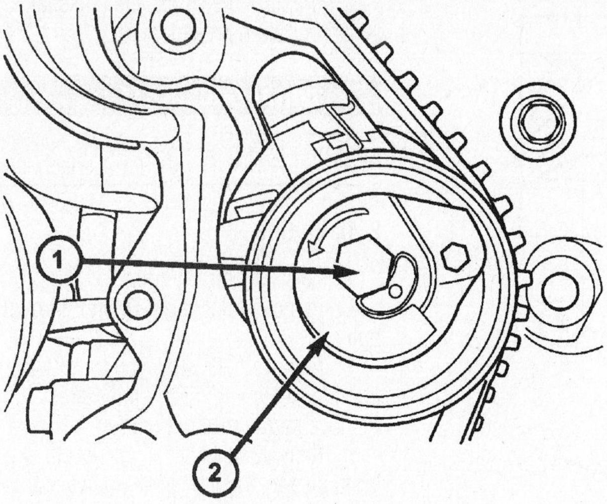

1 - LOCK BOLT
2 - TOP PLATE

06009-SEBR-G16

Timing belt tensioner—2.4L engine

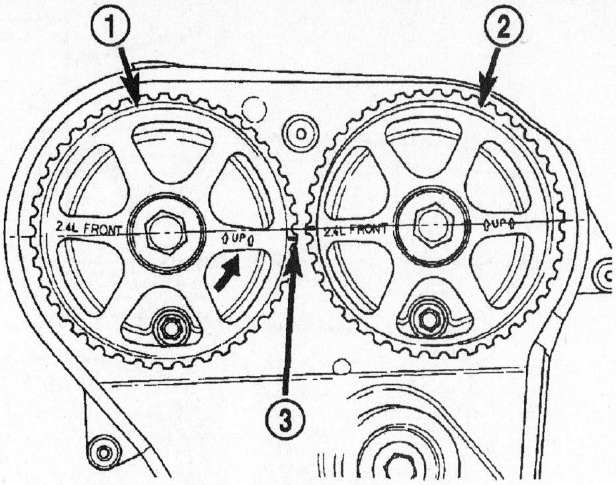

1 - CAMSHAFT SPROCKET-EXHAUST
2 - CAMSHAFT SPROCKET-INTAKE
3 - 1/2 NOTCH LOCATION

06009-SEBR-G17

Camshaft sprocket alignment—2.4L engine

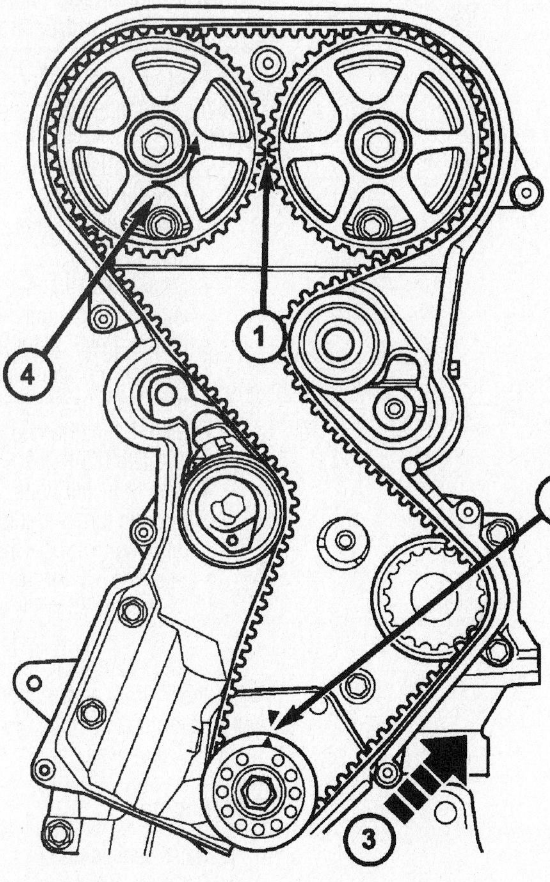

1 - CAMSHAFT TIMING MARKS 1/2 NOTCH LOCATION
2 - CRANKSHAFT AT TDC
3 - INSTALL BELT IN THIS DIRECTION
4 - ROTATE CAMSHAFT SPROCKET TO TAKE UP BELT SLACK

06009-SEBR-G18

Timing belt installation—2.4L engine

8. Set the camshafts timing marks so that the exhaust camshaft sprocket is a ½ notch below the intake camshaft sprocket.
9. Install the timing belt by starting at the crankshaft. Go around the water pump sprocket idler pulley, camshaft sprockets and the tensioner.
10. Move the exhaust camshaft sprocket counterclockwise to align the marks and to remove any slack.
11. Remove the wrench from the belt tensioner.
12. Rotate the crankshaft two full revolutions and verify that the TDC marks are properly aligned.
13. Install or connect the following:
    - Lower timing belt cover and torque the bolts to 40 inch (4.5 Nm)
    - Upper timing belt cover and torque the bolts to 40 inch lbs. (4.5 Nm)
    - Right engine support bracket and reposition the power steering pump. Torque the bracket bolts to 45 ft. lbs. (61 Nm).
    - Right engine mount through bolt and torque it to 87 ft. lbs. (118 Nm)
    - Upper radiator support crossmember
    - Torque strut bracket
    - Upper torque strut
    - A/C lines and pressure switch
    - Exhaust system to the manifold
    - Crankshaft damper
    - Accessory drive belts
    - Lower torque strut
    - Right splash shield and wheel
    - Negative battery cable
14. Recharge the A/C system.

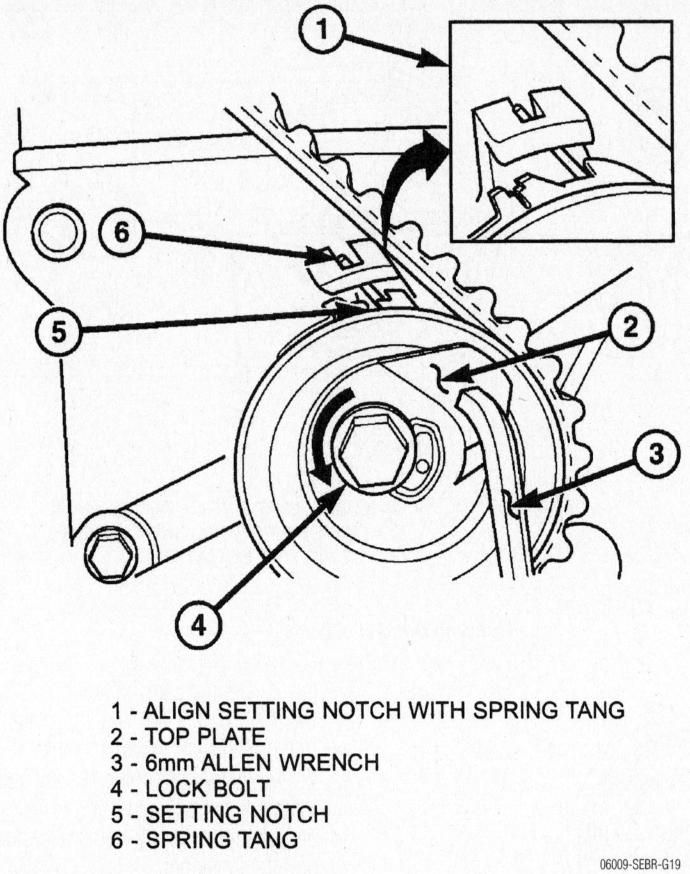

Timing belt tension adjustment—2.4L engine

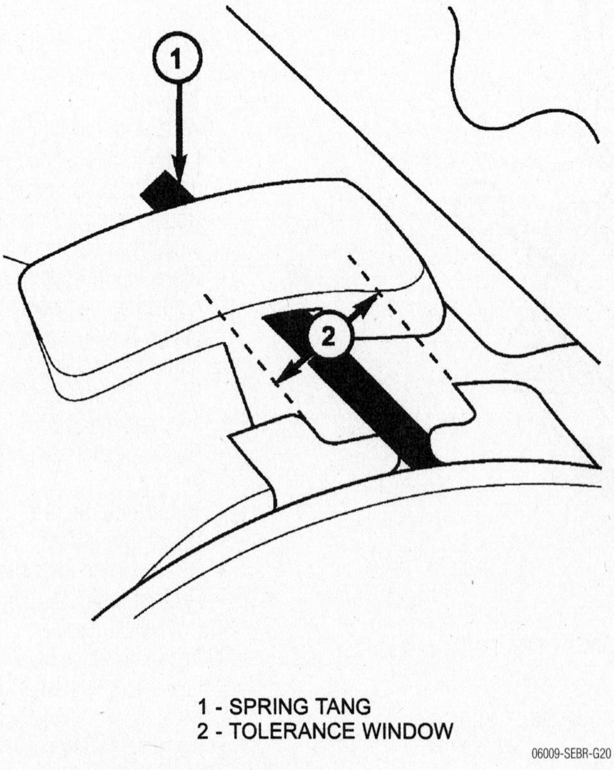

Timing belt tension verification—2.4L engine

15. Perform the camshaft/crankshaft synchronization procedure.

Structural Collar

REMOVAL & INSTALLATION

2.4L Engine

1. Raise vehicle on hoist.
2. Remove front engine mount through bolt.
3. Remove bolts attaching torque reaction bracket to engine and transaxle. Remove torque reaction bracket.
4. Remove bolts attaching structural collar to oil pan and transaxle. Remove structural collar.

To install:

⁂ WARNING

Torque procedure for structural collar and torque reaction bracket must be followed or damage could occur to oil pan, collar, and/or bracket.

5. Perform the following steps for installing structural collar and torque reaction bracket:
   - Step 1: Place collar into position between transaxle and oil pan. Install collar to transaxle bolt hand tight only.
   - Step 2: Install collar to oil pan bolts and hand tight only.
   - Step 3: Position torque reaction bracket in place. Install attaching bolts and, hand tight only.
   - Step 4: Final torque bolts 1–3 to 101 Nm (75 ft. lbs.).
   - Step 5: Install bolts 6 and 7 through torque reduction bracket into engine block, hand tight only.
   - Step 6: Final torque bolts 4 and 5 to 45 Nm (35 ft. lbs.).
   - Step 7: Final torque bolts 6 and 7 to 61 Nm (45 ft. lbs.).
6. Install front engine mount through bolt and torque to 61 Nm (45 ft. lbs.).
7. Lower vehicle.

2.7L Engine

1. Raise vehicle on hoist.
2. Remove bolts attaching structural collar to oil pan and transmission housing.
3. Remove collar.

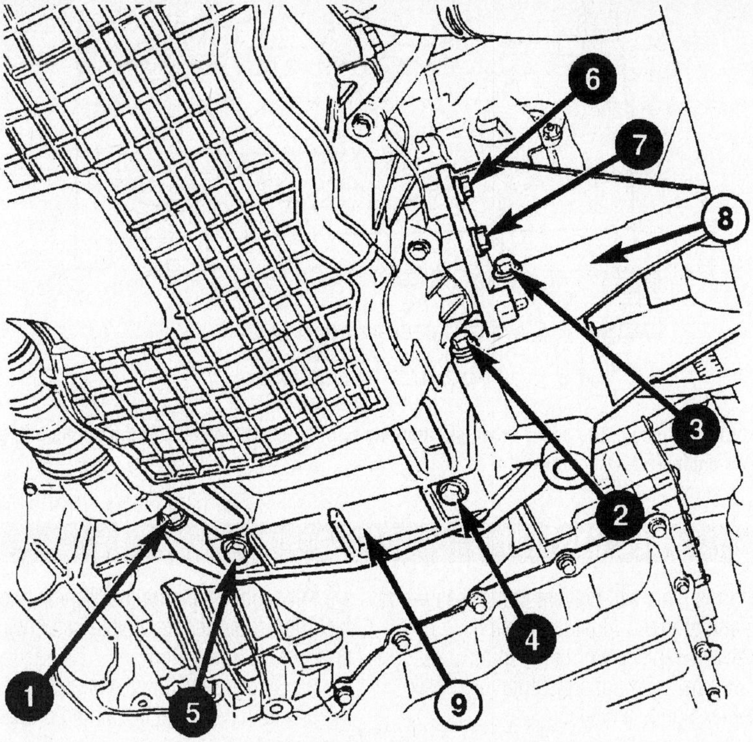

Structural collar and torque reaction bracket—2.4L engine

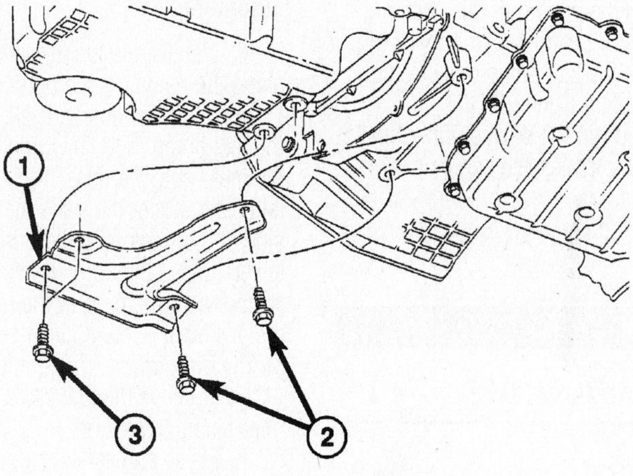

Structural collar installation—2.7L engine

To install:

⁂ WARNING

The collar must be tightened using this service procedure, as damage to transaxle case and/or oil pan may occur.

4. Position structural collar on oil pan and transaxle.
5. Loosely install all bolts.
6. Tighten the collar to oil pan bolts to 55 Nm (40 ft. lbs.).
7. Tighten collar to transaxle bolts to 55 Nm (40 ft. lbs.).
8. Lower vehicle.

Piston and Ring

POSITIONING

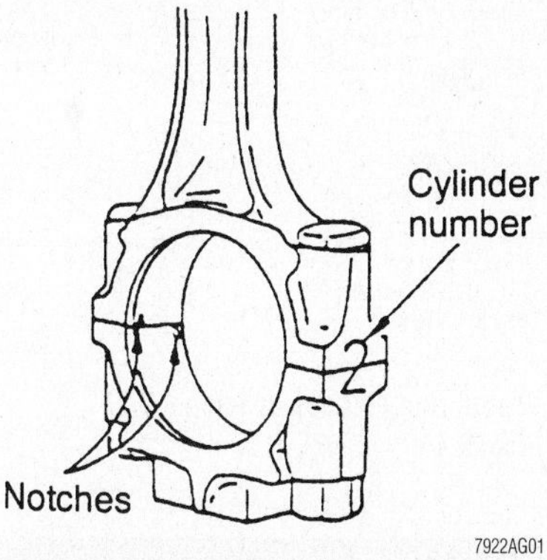

Engine connecting rod and cap installation—ensure to matchmark the cap and rod prior to disassembly

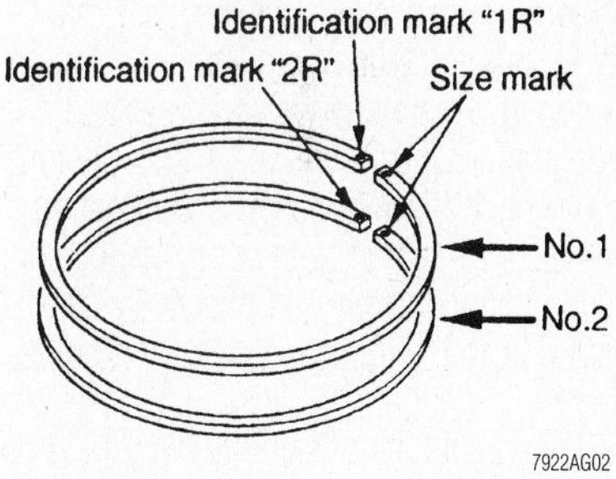

Piston ring identification mark locations

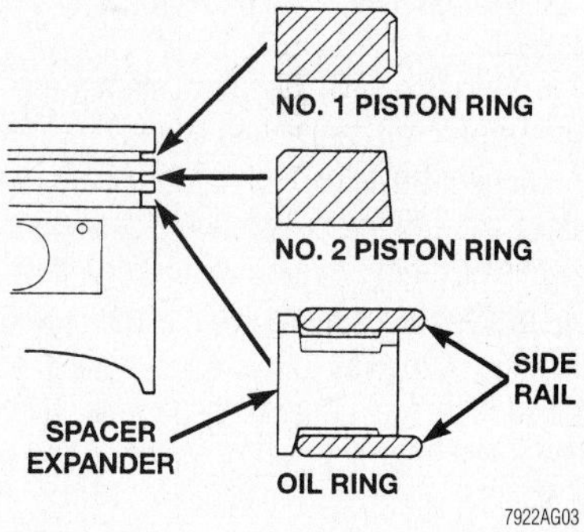

Piston ring orientation—2.4L and 2.7L engines

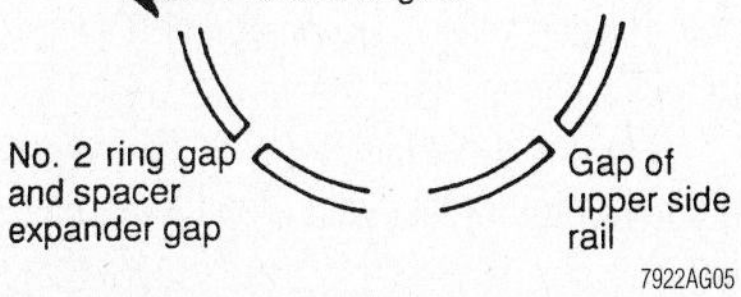

Piston ring end-gap spacing—2.4L engines

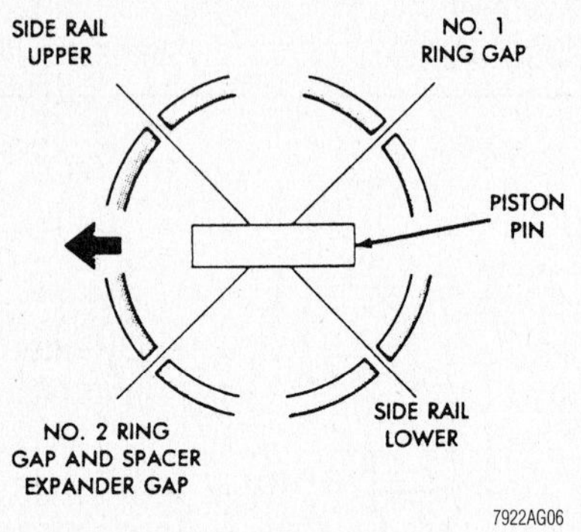

Piston ring end-gap spacing—2.7L engines

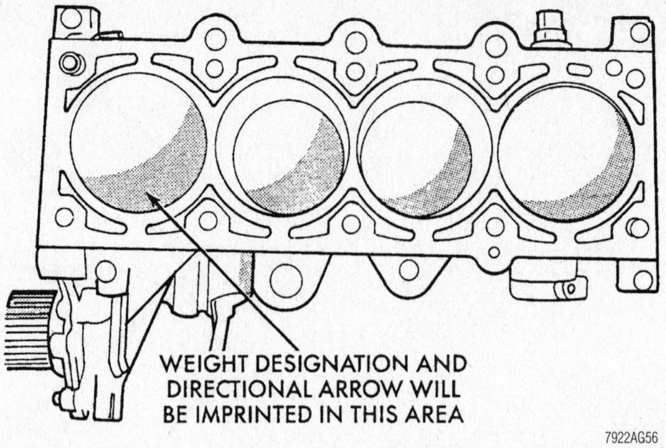

Piston positioning. The arrow or weight marking (L or H) must face toward the timing belt side of the engine—2.4L engines

FUEL SYSTEM

Fuel System Service Precautions

Safety is the most important factor when performing not only fuel system maintenance but any type of maintenance. Failure to conduct maintenance and repairs in a safe manner may result in serious personal injury or death. Maintenance and testing of the vehicle's fuel system components can be accomplished safely and effectively by adhering to the following rules and guidelines.

- To avoid the possibility of fire and personal injury, always disconnect the negative battery cable unless the repair or test procedure requires that battery voltage be applied.
- Always relieve the fuel system pressure prior to disconnecting any fuel system component (injector, fuel rail, pressure regulator, etc.), fitting or fuel line connection. Exercise extreme caution whenever relieving fuel system pressure to avoid exposing skin, face and eyes to fuel spray. Please be advised that fuel under pressure may penetrate the skin or any part of the body that it contacts.
- Always place a shop towel or cloth around the fitting or connection prior to loosening to absorb any excess fuel due to spillage. Ensure that all fuel spillage (should it occur) is quickly removed from engine surfaces. Ensure that all fuel soaked cloths or towels are deposited into a suitable waste container.
- Always keep a dry chemical (Class B) fire extinguisher near the work area.
- Do not allow fuel spray or fuel vapors to come into contact with a spark or open flame.
- Always use a back-up wrench when loosening and tightening fuel line connection fittings. This will prevent unnecessary stress and torsion to fuel line piping.
- Always replace worn fuel fitting O-rings with new. Do not substitute fuel hose where fuel pipe is installed.

Fuel System Pressure

RELIEVING

1. Before servicing the vehicle, refer to the fuel system precautions and to the precautions in the beginning of this section.
2. Remove or disconnect the following:
3. Remove the fuel pump relay from the Power Distribution Center (PDC).
4. Start and run the engine until it stalls.
5. Attempt to restart the engine until it no longer runs.
6. Turn the ignition key to the off position.

Fuel Filter

REMOVAL & INSTALLATION

Frame Mounted

2002 MODELS

The 2002 fuel delivery system contains a replaceable inline filter. The fuel filter mounts to the frame above the rear of the fuel tank. The fuel tank assembly must be loosened and lowered slightly to access the filter. The inlet and outlet tubes are permanently attached to the filter. Please note that the fuel system pressure must be relieved before servicing fuel system components. In addition, quick-disconnect fittings are used on fuel line connections. When the tubes are fully connected, the locking ears and the fuel tube shoulder are visible in the windows of the connector.

1. Before servicing the vehicle, refer to the precautions in the beginning of this section.
2. Disconnect the negative battery cable from the left shock tower.

➡The ground cable is equipped with an insulator grommet which should be placed on the stud to prevent the negative battery cable from accidentally grounding.

3. Relieve the fuel system pressure using the recommended procedure.
4. From inside the trunk, disconnect the fuel pump module wiring jumper from the main body harness. The 4-pin connector is located under the trunk mat on the left side of the trunk near the base of the shock tower. Locate the body grommet for the jumper near the base of the rear seat. Push the grommet out and feed the jumper completely through the hole in the body.
5. Remove the fuel cap slowly to release tank pressure.
6. Raise and safely support the vehicle.
7. Locate the drain plug on the bottom left of the fuel tank. Place an approved fuel

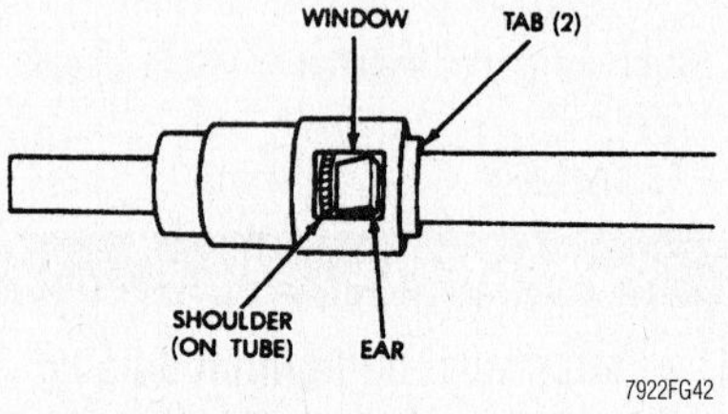

Be sure the shoulder and locking tabs are visible through the window of the connector when attached

container with a capacity of at least 16 gallons, under the drain plug. Remove the plug and drain the fuel tank. When finished draining, install the plug since there will be 1–2 gallons of fuel remaining. Tighten the drain plug to 32 inch lbs. (3.6 Nm).

8. Remove or disconnect the following:
   - Driver's side fuel tank strap
   - Passenger's side fuel tank strap loosen it

➡Allow the tank to lower until the fuel tank neck touches the rear suspension crossmember.

⁂ CAUTION

Wrap shop towels around the fuel hoses to catch any gasoline spillage.

   - Fuel line quick-disconnect fittings from the fuel pump module. Squeeze fuel filter hose connector releasing tabs and detach the fuel lines
   - Fuel filter

To install:

➡The fuel supply (to filter) tube and the return tube (to fuel pump module) are permanently attached to the fuel filter. The ends of the fuel supply and return tubes have different size quick-disconnect fittings. The large quick-disconnect fitting attaches to the large nipple (supply side) on the fuel pump module. The smaller quick-disconnect fitting attaches the small nipple (return side) on the fuel pump module.

9. Lubricate the fuel filter nipples with engine oil.
10. Install or connect the following:
   - Fuel tubes
   - Fuel tank raise it into position
   - Fuel tank straps and torque the bolts to 17 ft. lbs. (23 Nm)

➡Be sure the fuel pump module electrical harness grommet is installed in the body as the tank is raised into position.

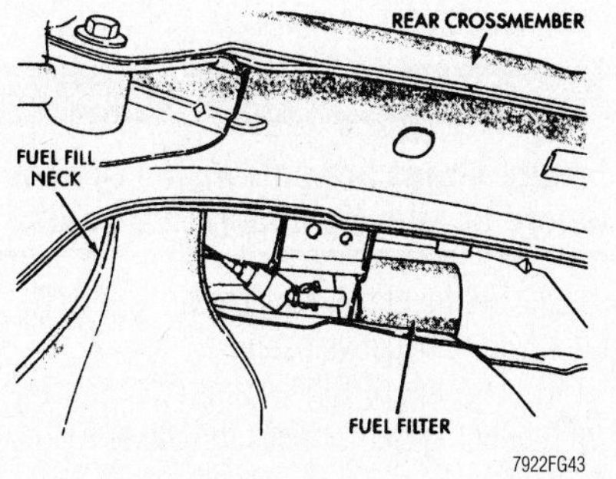

Fuel filter mounting location

   - Fuel pump module connector
   - Negative battery cable
11. Refill the fuel tank.

Inlet Filter

1. Before servicing the vehicle, refer to the precautions in the beginning of this section.
2. Properly relieve the fuel system pressure.
3. Remove or disconnect the following:
   - Negative battery cable
   - Fuel pump module
   - Locking tabs on the fuel pump
   - Strainer and O-ring
   - Inlet filter from the assembly

To install:

4. Install or connect the following:
   - Inlet filter
   - New O-ring and strainer
   - Locking tabs on the fuel pump module
   - Fuel pump Module
   - Negative battery cable
5. Start the vehicle, check for leaks and repair if necessary.

Fuel Pump

REMOVAL & INSTALLATION

2002 Models

The in-tank fuel pump module contains the fuel pump and pressure regulator which adjusts fuel system pressure to approximately 49 psi (337.8 kPa). Voltage to the fuel pump is supplied through the fuel pump relay.

The fuel pump is serviced as part of the fuel pump module. The fuel pump module is installed in the top of the fuel tank and contains the electric fuel pump, fuel pump reservoir, inlet strainer fuel gauge sending unit, fuel supply and return line connections and the pressure regulator. The inlet strainer, fuel pressure regulator and level sensor are the only serviceable items. If the fuel pump requires service, replace the fuel pump module.

1. Before servicing the vehicle, refer to the precautions in the beginning of this section.
2. Disconnect the negative battery cable from the left shock tower.

➡The ground cable is equipped with an insulator grommet which should be placed on the stud to prevent the negative battery cable from accidentally grounding.

3. Remove the fuel filler cap and relieve the fuel system pressure using the recommended procedure.
4. Remove or disconnect the following:
   - Fuel pump harness electrical connector on the left side of the trunk near the base of the shock tower
   - Fuel pump harness grommet and push the harness through the hole
5. Raise the vehicle and drain the fuel tank.
6. Remove or disconnect the following:
   - Fuel tank inlet and from the filler hose
   - Fuel tank straps. Support with a jack prior to loosening the straps.
7. Carefully lower the tank.
8. Clean the top of the tank to remove any loose dirt.
9. Remove or disconnect the following:
   - Fuel lines from the fuel pump module
   - Fuel pump locknut using Spanner wrench tool 6856

⁂ CAUTION

The fuel reservoir of the fuel pump module does not empty out when the tank is drained. The fuel in the reservoir may spill out when the module is removed.

   - Fuel pump
   - Fuel tank O-ring and discard it

To install:

10. Thoroughly clean all parts. Wipe the seal area of the tank clean.
11. Install or connect the following:
   - New O-ring onto the fuel tank
   - Fuel pump module in the tank

➡Be sure the alignment tab on the underside of the pump module flange sits in the corresponding notch in the fuel tank.

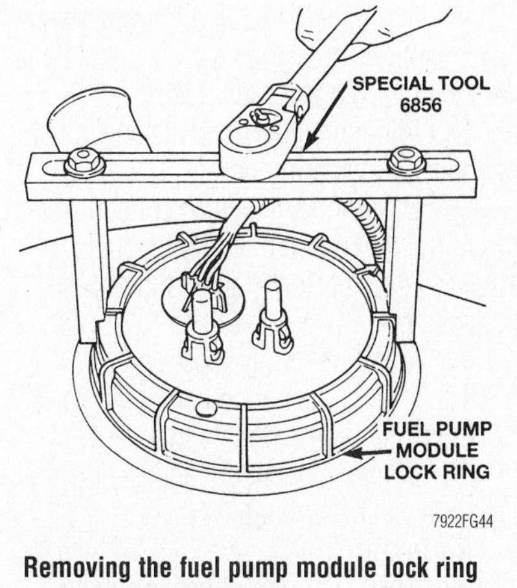

Removing the fuel pump module lock ring using special tool 6856

- Locking ring and torque it to 41 ft. lbs. (55 Nm), using Spanner Wrench Tool 6856
- Fuel tank assembly
- Negative battery cable

12. Refill the fuel tank.
13. Turn the ignition switch **ON** to pressurize the system. Check the fuel system for leaks.

2003–05 Models

1. Remove fuel filler cap and perform fuel system pressure release procedure.
2. Disconnect negative cable from battery.
3. Drain fuel tank dry into holding tank or a properly labeled gasoline safety container.
4. Raise vehicle on hoist and support.
5. Use a transmission jack to support fuel tank. Remove bolts from fuel tank straps.
6. Lower tank slightly.
7. Disconnect the fuel filler vent tube.
8. Disconnect fuel line and vapor line.
9. Disconnect the vacuum line from the LDP.
10. Disconnect fuel filler tube by loosing the clamp and removing hose.
11. Slide fuel pump module electrical connector lock to unlock. The fuel pump module electrical connector has a retainer that locks it in place.
12. Push down on connector retainer and pull connector off module.
13. Lower tank from vehicle.

⁂ CAUTION

The fuel reservoir of the fuel pump module does not empty out when the tank is drained. The fuel in the reservoir will spill out when the module is removed.

14. Disconnect fuel line from fuel pump module by depressing quick connect retainers with thumb and forefinger.
15. Slide fuel pump module electrical connector lock to unlock.
16. Disconnect the electrical connection from the fuel pump module, by pushing down on connector retainer and pulling connector off of module.
17. Mark fuel pump location.
18. Use a brass punch and hammer to remove fuel pump module lock ring.
19. Remove fuel pump and O-ring seal from tank. Discard old seal.

To install:

20. Wipe seal area of tank clean and place a new seal in position in the tank opening.
21. Position fuel pump in the tank.
22. Position the lockring over the fuel pump module.
23. Tighten the lockring using a brass punch and hammer.
24. Connect the electrical connector.
25. Connect the fuel line to the fuel pump.
26. Position fuel tank on transmission jack.
27. Raise tank into position and carefully work filler tube onto the fuel tank. A light coating of clean engine oil on the tube end may be used to aid assembly. Connect fuel filler tube hose and tighten clamp.
28. Connect fuel pump/module electrical connector. Place retainer in locked position.
29. Connect the EVAP, vapor, and vacuum lines.
30. Lubricate the fuel supply line with clean 30 weight engine oil, install the quick connect fuel fitting.
31. Attach filler vent line to filler tube. Pull on connector to make sure of connection.
32. Place NVLD air filter under right rear tank strap before tightening.
33. Tighten strap bolts to 54 Nm (40 ft. lbs.) torque. Remove transmission jack.
34. Tighten filler hose clamp to 3.6 Nm (32 inch lbs.).

⁂ WARNING

Ensure straps are not twisted or bent before or after tightening strap nuts.

35. Fill fuel tank, replace cap, and connect battery negative cable.
36. Use the DRBIII® scan tool to pressurize the fuel system. Check for leaks.

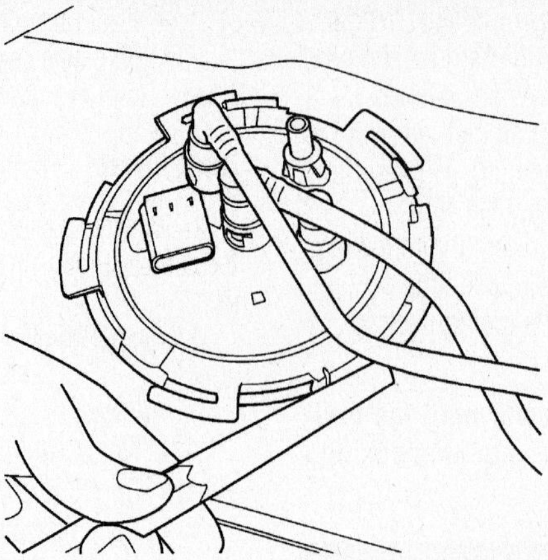

06009-SEBR-G23

Use a brass punch and hammer to remove fuel pump module lock ring—2003–05 models

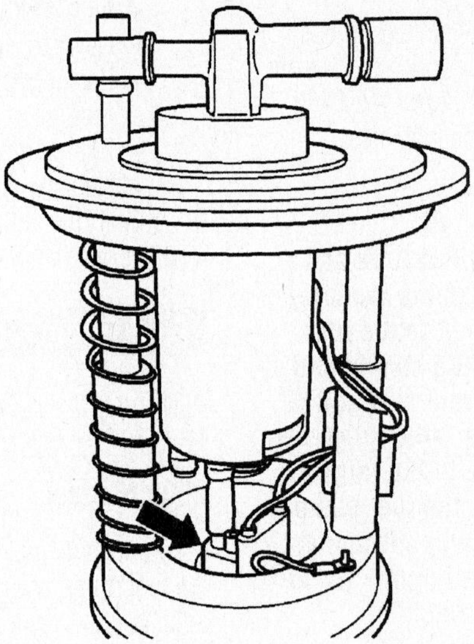

06009-SEBR-G24

Fuel pump location

Fuel Injector

REMOVAL & INSTALLATION

2.4L Engine

1. Perform fuel system pressure release procedure before servicing or starting repairs.
2. Disconnect negative cable from battery.
3. Remove wiring harness from fuel rail support bracket.
4. Remove fuel rail support bracket.
5. Disconnect the wiring connectors for fuel injectors harness.
6. Remove wiring harness from fuel rail brackets.
7. Disconnect the connectors from the fuel injectors.
8. Remove throttle and speed control cables from the fuel rail.
9. Remove fuel hose quick connect fitting from the chassis tube. Place a shop towel under the connections to absorb any fuel spilled from the fitting.

⁂ CAUTION

Wrap a shop towel around hoses to catch any gasoline spillage.

10. Remove fuel rail attaching bolts.
11. Remove fuel rail. Be careful not to damage the injector O-rings upon removal from their ports.
12. Position fuel rail assembly so that the fuel injectors are easily accessible.
13. Remove injector clip and pull injector out of fuel rail.
14. Check injector O-ring for damage. If O-ring is damaged, it must be replaced. Replace the injector clip if it is damaged.
15. Repeat for remaining injectors.

To install:

16. Before installing an injector the rubber O-ring must be lubricated with a drop of clean engine oil to aid in installation.
17. Install injector clip by sliding open end into the top slot of the injector. The edge of the receiver cup will slide into the side slots of clip.
18. Install injector top end into fuel rail receiver cap. Be careful not to damage O-ring during installation.
19. Repeat steps for remaining injectors.
20. Ensure injector holes are clean. Replace O-rings if damaged.
21. Lubricate injector O-rings with a drop of clean engine oil to ease installation.
22. Put the tip of each injector into their ports. Push the assembly into place until the injectors are seated in the ports.

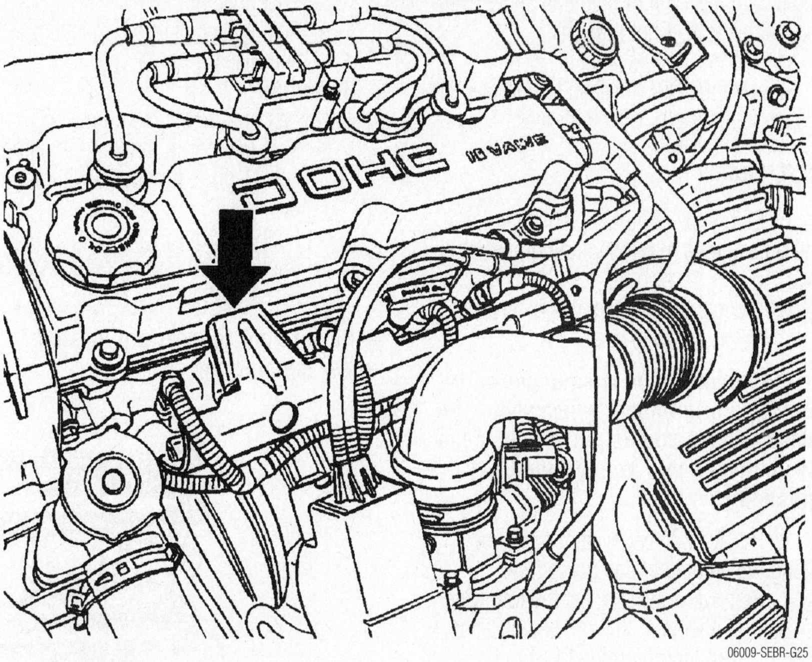

Fuel rail and bracket—2.4L engine

Fuel injector and rail—2.4L engine

23. Install the 2 fuel rail mounting bolts to intake manifold. Tighten bolts to 12 Nm (105 inch lbs.) torque.
24. Install the fuel rail mounting bolt to cylinder head. Tighten bolt to 22 Nm (195 inch lbs.) torque
25. Connect the connectors to the fuel injectors.
26. Install fuel rail support bracket and loosely install all bolts.
27. Tighten three fasteners to intake manifold to 28 Nm (250 inch lbs.) torque.
28. Tighten two bolts to fuel rail to 8.5 Nm (75 inch lbs.) torque.
29. Install wiring harness to brackets.

30. Connect the throttle and speed control cables to fuel rail bracket.
31. Connect negative battery cable.
32. Use the DRBIII® scan tool to pressurize the fuel system. Check for leaks.

2.7L Engine

1. Before servicing the vehicle, refer to the precautions in the beginning of this section.
2. Disconnect the negative battery cable from the left shock tower.

➡The ground cable is equipped with an insulator grommet which should be placed on the stud to prevent the negative battery cable from accidentally grounding.

3. Relieve the fuel system pressure using the recommended procedure.
4. Remove or disconnect the following:
   - Fuel supply line quick quick-connect fitting from the fuel rail
   - Upper intake manifold plenum and discard the gasket
   - Fuel injector electrical connectors
   - Fuel rail from the intake manifold

➡Cover the intake manifold openings to keep dirt from entering the system.

   - Fuel injector clip
   - Fuel injector(s) from the fuel rail and discard the O-rings

To install:

5. Install or connect the following:
   - Fuel injector to the fuel rail using new O-rings lubricated with engine oil
   - Fuel injector clip
   - Fuel rail and torque the bolts to 8 ft. lbs. (12 Nm)

➡Be sure the spacers are located under the fuel rail

   - Fuel injector electrical connectors
   - Fuel supply line quick quick-connect fitting to the fuel rail
   - Upper intake manifold plenum using a new gasket and torque the bolts to 13 ft. lbs. (18 Nm)
   - Throttle cables
   - Sensor electrical connectors
   - Negative battery cable
6. Use a Diagnostic Readout Box (DRB) scan tool Automatic Shutdown (ASD) fuel system test to pressurize the fuel system. Check for leaks.

View of the fuel rail assembly—2.7L engine

DRIVE TRAIN

Transaxle Assembly

REMOVAL & INSTALLATION

Manual

2002–04

⁂ WARNING

If the vehicle is going to be rolled on its wheels while the transaxle is out of the vehicle, obtain 2 outer CV-joints to install in the hubs. If the vehicle is rolled without the proper torque applied to the front wheel bearings, the bearings will no longer be usable.

1. Before servicing the vehicle, refer to the precautions in the beginning of this section.
2. Remove or disconnect the following:
   - Negative battery cable from the left shock tower.

➡The ground cable is equipped with an insulator grommet which should be placed on the stud to prevent the negative battery cable from accidentally grounding.

   - Air cleaner and intake hoses
   - Clutch housing vent cap
   - Clutch cable from the bell housing

⁂ WARNING

Using equal force, pry up on both sides of the shifter cable isolator bushings to avoid damaging the cable isolator bushing.

   - Selector lever and crossover cables from the transaxle
   - Shift cable mounting bracket from the transaxle
   - Accelerator cables from the throttle body
   - Accelerator cable bracket from the throttle body
   - Upper starter-to-throttle body support bracket bolt
   - Upper bell housing-to-throttle body support bracket stud nut
   - Throttle body support bracket
   - Left transaxle mount upper bolts
   - Upper bell housing bolts
   - Vehicle Speed Sensor (VSS)
   - Back-up light electrical connector from the transaxle
3. Install an engine support fixture tool and support the engine.
4. Remove or disconnect the following:
   - Front wheels
   - Halfshafts
   - Lower splash shield/battery cover from the left side
   - Lower bracket bolts, from the left transaxle mount
   - Engine-to-lower crossbar bolts
   - Front steel engine mount bracket
   - 3 front aluminum engine mount bracket bolts
   - Starter
   - Rear transaxle mount bracket
   - Transaxle-to-rear lateral bending strut from engine/transaxle assembly
   - Transaxle bell housing cover
5. Using a transaxle jack, support the transaxle.
6. Rotate the engine clockwise to gain access to the driveplate clutch bolts.
   - Driveplate clutch bolts
   - Lower engine-to-transaxle bolts
   - Transaxle

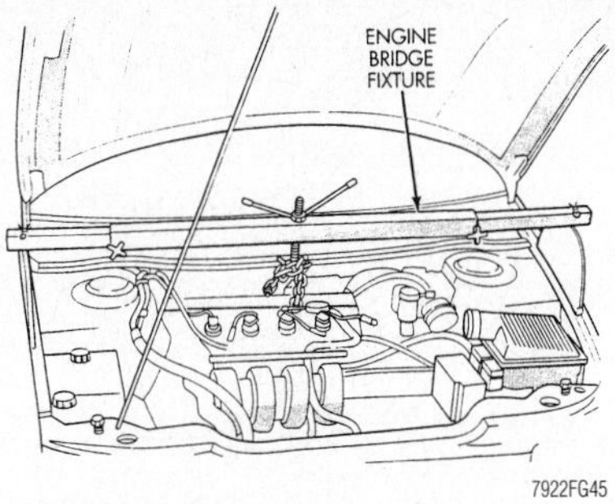

Secure the engine assembly with an appropriate support fixture—2002–04

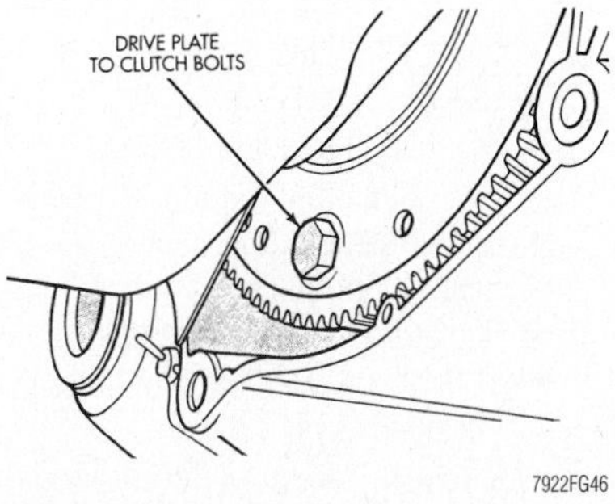

Remove the driveplate clutch bolts—rotate the engine clockwise to advance to the next bolt—2002–04

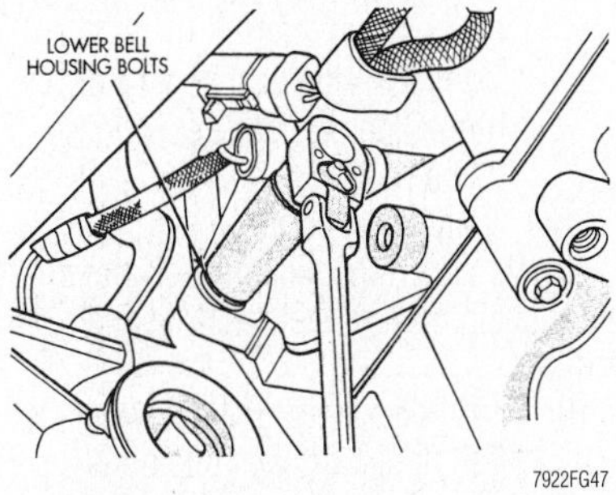

Removing the lower engine-to-transaxle mounting bolts—2002–04

7. To prepare the vehicle for rolling, support the engine with a suitable support or reinstall the front motor mount to the engine. Then, reinstall the ball joints to the steering knuckle and install the retaining bolt. Install the obtained outer CV-joints to the hubs, install the washers and tighten the axle nuts to 180 ft. lbs. (244 Nm). The vehicle may now be safely rolled.

To install:

8. Install or connect the following:
   - Transaxle and torque the transaxle-to-engine bolts to 70 ft. lbs. (95 Nm)
   - Driveplate clutch bolts
   - Lower engine-to-transaxle bolts and torque the bolts to 70 ft. lbs. (95 Nm)
   - Transaxle bell housing cover and torque bolts to 9 ft. lbs. (12 Nm)
   - Transaxle-to-rear lateral bending strut to the engine/transaxle assembly and torque the bolt to 40 ft. lbs. (54 Nm)
   - Rear transaxle mount bracket and torque the bolts to 40 ft. lbs. (54 Nm)
   - Starter
   - 3 front aluminum engine mount bracket bolts and torque the bolts to 40 ft. lbs. (54 Nm)
   - Front steel engine mount bracket and torque the bolt to 45 ft. lbs. (61 Nm)
   - Engine-to-lower crossbar bolts
   - Lower bracket bolts to the left transaxle mount and torque the bolts to 40 ft. lbs. (54 Nm)
   - Lower splash shield/battery cover to the left side
   - Halfshafts
   - Front wheels
9. Remove engine support fixture tool.
10. Install or connect the following:
    - Back-up light electrical connector
    - VSS sensor
    - Upper bell housing bolts and torque the bolts to 70 ft. lbs. (95 Nm)
    - Left transaxle mount upper bolts and torque the bolts to 40 ft. lbs. (54 Nm)
    - Throttle body support bracket
    - Upper bell housing-to-throttle body support bracket stud nut and torque the nut to 32 ft. lbs. (43 Nm)
    - Upper starter-to-throttle body support bracket bolt
    - Accelerator cable bracket to the throttle body
    - Accelerator cables to the throttle body

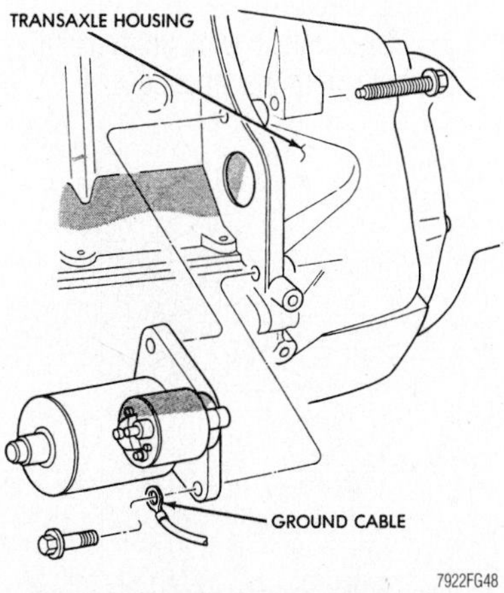

Be sure to connect the ground cable at the starter—2002–04

   - Shift cable mounting bracket to the transaxle
   - Selector lever and crossover cables to the transaxle
   - Clutch cable to the bell housing
   - Clutch housing vent cap
   - Air cleaner and intake hoses
   - Negative battery cable

11. Check to be sure that all fasteners are tightened and connections made.
12. Refill the transaxle.
13. Check the transaxle for proper operation. Be sure the reverse lights turn **ON** when in reverse.

2005

1. Raise hood.
2. Disconnect battery negative cable.
3. Remove air cleaner assembly.
4. Disconnect Vehicle Speed Sensor/Back-up Lamp Switch harness connector.
5. Remove clutch release access cap.
6. Disconnect clutch release cable.
7. Disconnect crossover and selector cables from transaxle.
8. Remove gearshift cable retainer clips from bracket and position gearshift cables out of way.
9. Remove starter motor and secure out of way.
10. Remove three rear mount bracket-to-transaxle bolts.
11. Raise vehicle on hoist.
12. Remove halfshafts.
13. Remove rear mount bracket-to-transaxle horizontal bolt.
14. Remove rear mount through-bolt.
15. Remove three rear mount-to-crossmember bolts.
16. Remove rear mount and bracket.
17. Remove front mount through bolt.
18. Remove engine front pencil strut brace.
19. Remove front mount bracket from engine/transaxle.
20. Remove structural collar.
21. Remove bellhousing dust cover.
22. Remove left wheel opening splash shield.
23. Remove four modular clutch-to-drive plate bolts. While removing bolts, one tight-tolerance (slotted) drive plate hole will be encountered. When this bolt is removed, mark driveplate and modular clutch assembly at this location, and be sure to align marks upon reassembly.
24. Position screw jack and wood block to oil pan.
25. Remove transaxle upper mount through bolt.
26. Lower engine transaxle assembly.

27. Remove transaxle upper mount.
28. Obtain helper and transmission jack.
29. Remove transaxle-to-engine block bolts.
30. Remove transaxle from engine.

To install:

31. Install modular clutch assembly to transaxle input shaft.
32. Using a helper and transmission jack, install transaxle to engine.
33. Install and torque transaxle-to-engine bolts to 95 Nm (70 ft. lbs.).
34. Install transaxle upper mount bracket and torque bolts to 65 Nm (48 ft. lbs.).
35. Raise engine/transaxle assembly into position with screw jack.
36. Install and torque transaxle upper mount-to-bracket bolts to 61 Nm (45 ft. lbs.).
37. Remove screw jack and wood block.
38. Install four modular clutch-to-drive-plate bolts. Align drive plate and modular clutch alignment marks placed upon disassembly. Start with tight-tolerance (slotted) hole, install and torque bolts to 88 Nm (65 ft. lbs.) torque.
39. Install bellhousing dust cover.
40. Install structural collar.
41. Install left wheel house splash shield.
42. Install front mount and bracket.
43. Install rear mount to crossmember.
44. Install rear mount bracket into position and loosely install horizontal bolt.
45. Install halfshafts.
46. Lower vehicle.
47. Install three rear mount bracket-to-transaxle and torque to 110 Nm (80 ft. lbs.).
48. Raise vehicle.
49. Torque rear mount bracket-to-transaxle horizontal bolt to 110 Nm (80 ft. lbs.).
50. Install rear mount through-bolt and torque to 61 Nm (45 ft. lbs.).
51. Install starter motor.
52. Install gearshift cables to bracket and install new retaining clips.
53. Connect gearshift crossover and selector cable to crossover and selector levers.
54. Install clutch release cable to release lever and case. Install cap.
55. Connect Vehicle Speed Sensor and Back-up Lamp Switch harness.
56. Install air cleaner assembly.
57. Connect battery negative cables.
58. Fill transaxle with suitable amount of fluid.

Automatic

2002–04

**** WARNING**

If the vehicle is going to be rolled on its wheels while the transaxle is out of the vehicle, obtain 2 outer CV-joints to install to the hubs. If the vehicle is rolled without the proper torque applied to the front wheel bearings, the bearings will no longer be usable.

1. Before servicing the vehicle, refer to the precautions in the beginning of this section.
2. Drain the transaxle.
3. Disconnect the negative battery cable from the left shock tower.

➡The ground cable is equipped with an insulator grommet which should be placed on the stud to prevent the negative battery cable from accidentally grounding.

4. Remove or disconnect the following:
   - Air cleaner duct
   - Transmission Control Module (TCM)
   - Solenoid pack electrical connector
   - Dipstick tube from the transaxle
   - Transaxle cooler lines
   - Shift lever cable from the transaxle lever
5. Using an engine support fixture tool, support the engine assembly.
6. Remove or disconnect the following:
   - Upper transaxle mount top bolts from the left side
   - Front wheels
   - Lower splash shields from both sides
   - Exhaust pipe from the exhaust manifold
   - Upper transaxle mount remaining bolts from the left side

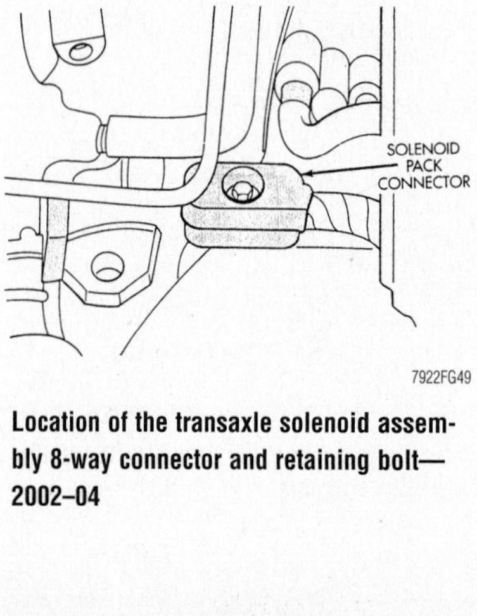

Location of the transaxle solenoid assembly 8-way connector and retaining bolt—2002–04

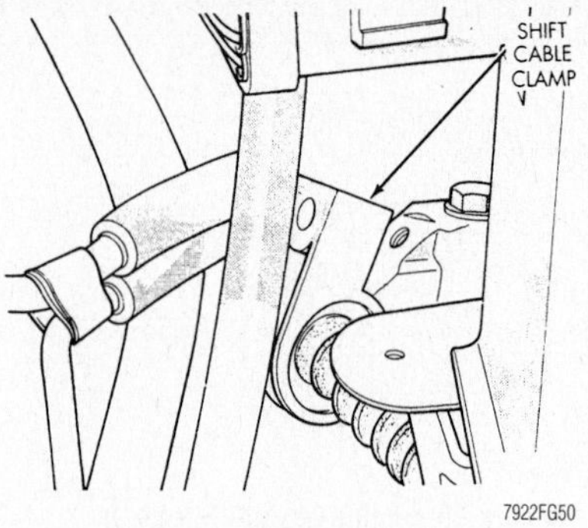

Removing the shift cable and clamp—2002–04

   - Engine oil filter
   - Starter
   - Front engine mount bracket
   - Rear mount bracket through-bolt
   - Centermember bolts
   - Rear mount bracket
   - Radiator lower crossmember
   - Both lateral bending strut brackets
   - Flexplate cover
7. Matchmark the converter-to-flexplate location. Rotate the crankshaft clockwise to align the converter bolts.
8. Install or connect the following:
   - Converter-to-flexplate bolts
   - Crankshaft Position (CKP) sensor, if equipped
   - Transaxle electrical connectors
   - Right side steering gear and K-frame bolts
   - Sway bar mounts
9. Using a transmission jack and a safety chain, secure the transaxle and support it.
10. Install or connect the following:
    - Bell housing-to-engine bolts
    - Transaxle, by moving the K-frame rearward
11. To prepare the vehicle for rolling, support the engine with a suitable support or reinstall the front motor mount to the engine. Then, reinstall the ball joints to the steering knuckle and install the retaining bolt. Install the obtained outer CV-joints to the hubs, install the washers and tighten the axle nuts to 180 ft. lbs. (244 Nm). The vehicle may now be safely rolled.

To install:

12. Install or connect the following:
    - Transaxle, by moving the K-frame rearward
    - Bell housing-to-engine bolts and torque the bolts to 70 ft. lbs. (95 Nm)
    - Sway bar mounts
    - Right side steering gear and K-frame bolts

- Transaxle electrical connectors
- CKP sensor, if equipped

13. Align the converter-to-flexplate matchmark. Rotate the crankshaft clockwise to align the converter bolts.
14. Install or connect the following:
    - Torque converter and torque the converter-to-flexplate bolts to 55 ft. lbs. (74 Nm)
    - Flexplate cover and torque the bolts to 108 inch lbs. (12 Nm)
    - Both lateral bending strut brackets and torque the bolts to 45 ft. lbs. (61 Nm)
    - Radiator's lower crossmember and torque the bolts to 45 ft. lbs. (61 Nm)
    - Rear mount bracket
    - Centermember bolts and torque the bolts to 45 ft. lbs. (61 Nm)
    - Rear mount bracket through-bolt and torque the through-bolt to 45 ft. lbs. (61 Nm)
    - Front engine mount bracket and torque the bolts to 24 ft. lbs. (33 Nm)
    - Starter
    - New engine oil filter
    - Upper transaxle mount remaining bolts to the left side
    - Exhaust pipe to the exhaust manifold
    - Lower splash shields to both sides
    - Front wheels
    - Upper transaxle mount top bolts, to the left side
    - Shift lever cable to the transaxle lever and torque the nut to 14 ft. lbs. (19 Nm)
    - Transaxle cooler lines
    - Dipstick tube to the transaxle
    - Solenoid pack electrical connector
    - TCM
    - Air cleaner duct
    - Negative battery cable
15. Adjust the gearshift and throttle cables.
16. Refill the transaxle.
17. Check the transaxle for proper operation. Be sure the back-up lights and speedometer are working properly.

2005

➡If transaxle assembly is being replaced or overhauled (clutch and/or seal replacement), it is necessary to perform the TCM Quick Learn Procedure.

1. Disconnect battery cable.
2. Remove air cleaner assembly.
3. Remove dipstick tube. Plug hole to prevent debris from entering transaxle.
4. Using Tool 8875A, disconnect the transmission cooler line quick-connect fittings.
5. Disconnect Solenoid/Pressure Switch Assy. connector.
6. Disconnect Transmission Range Sensor connector.
7. Disconnect Input Speed Sensor connector.
8. Disconnect Output Speed Sensor connector.
9. Disconnect shift cable from manual valve lever and bracket.
10. Disconnect Crankshaft Position Sensor.
11. Remove throttle body support bracket.
12. Disconnect Oxygen Sensor harness retainer from transaxle case.
13. Remove rear mount bracket-to-transaxle case bolts.
14. Remove starter upper bracket-to-block bolt.
15. Raise vehicle.
16. Remove halfshafts.
17. Remove rear mount bracket lower bolt.
18. Remove rear mount through-bolt.
19. Remove rear mount-to-cross member bolts.
20. Remove rear mount and bracket.
21. Remove front mount through-bolt.
22. Remove front mount to radiator lower crossmember bolts.
23. Remove front mount bracket to block and transaxle.
24. Remove front mount bracket and mount.
25. Remove starter lower bolt and starter motor assembly.
26. Remove converter dust shield.
27. Remove torque converter bolts.
28. Support engine/transaxle with screw jack and wood block on engine oil pan.
29. Remove left mount bracket-to-transaxle bolts.
30. Carefully lower engine/transaxle assembly to gain access to and remove transaxle-to-engine bolts.
31. With aid of helper or transmission jack, remove transaxle assembly from vehicle.

To install:

32. Install transaxle to engine. Install and torque transaxle-to-engine bolts to 95 Nm (70 ft. lbs.).
33. Install upper mount to transaxle.
34. Raise engine/transaxle assembly into position and install through-bolt. Torque through-bolt to 70 Nm (55 ft. lbs.).
35. Remove transmission jack and screw jack.
36. Install torque converter bolts.
37. Install torque converter dust shield.
38. Install starter motor assembly and lower bolt.
39. Install front mount/bracket assembly.
40. Install rear mount/bracket assembly.
41. Install halfshafts.
42. Lower vehicle.
43. Install starter upper bracket-to-block bolt.
44. Connect oxygen sensor harness connector.
45. Install throttle body support bracket (if equipped).
46. Connect crankshaft position sensor connector (if equipped).
47. Connect gearshift cable to manual valve lever.
48. Connect output speed sensor connector.
49. Connect input speed sensor connector.
50. Connect transmission range sensor connector.
51. Connect solenoid/pressure switch assembly connector.
52. Connect oil cooler lines to the transaxle. An audible 'click' should be heard. Verify connection by pulling outward.
53. Install dipstick tube.
54. Install air cleaner assembly.
55. Connect battery negative cable.

Clutch

ADJUSTMENT

Free-Play

The manual transaxle clutch release system has a unique self-adjusting mechanism to compensate for clutch disc wear. This adjuster mechanism is located with the clutch cable assembly. The preload spring maintains tension on the cable. This tension keeps the clutch release bearing continuously loaded against the fingers of the clutch cover assembly. No manual adjustment is necessary.

When servicing this vehicle or if removing and installing the clutch cable, do not pull on the clutch cable housing to remove it from the dash panel. Damage to the cable self-adjuster may occur.

To check the function of the adjuster mechanism, use the following procedure:

1. With slight pressure, pull the clutch release lever end of the cable to draw the cable taut.
2. Push the clutch cable housing toward the dash panel. With less than 25 lbs. (11

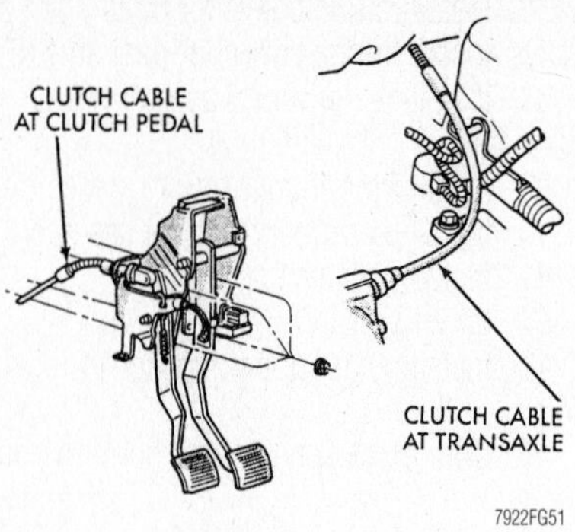

Clutch cable routing

kg) of effort, the cable housing should move 1.2–2.0 in. (30–50mm). This indicates proper adjuster mechanism function.

3. If the cable does not adjust, determine if the mechanism is properly seated on the bracket.

REMOVAL & INSTALLATION

➡The transaxle assembly must be removed to service the clutch assembly.

1. Before servicing the vehicle, refer to the precautions in the beginning of this section.
2. Remove or disconnect the following:
   - Negative battery cable from the left shock tower

➡The ground cable is equipped with an insulator grommet which should be placed on the stud to prevent the negative battery cable from accidentally grounding.

   - Starter
   - Rear and front transaxle support brackets
   - Clutch inspection cover
   - Modular clutch-to-flywheel bolts
   - Transaxle assembly with the clutch as an assembly
   - Clutch assembly from the transaxle input shaft

To install:

3. Clean all parts well. Inspect for oil leakage through the engine rear crankshaft oil seal and transaxle input shaft seal. If leakage is noted, it should be corrected at this time.
4. Examine the throwout or clutch release bearing. It is pre-lubricated and sealed and should not be washed in solvent. The bearing should turn smoothly when held in the hand with a light thrust load. A light drag caused by the lubricant fill is normal. If the bearing is noisy, rough or dry, replace the complete bearing assembly. In most cases where a clutch is being serviced, the complete clutch assembly and release bearing are usually replaced together.
5. Check the condition of the stud pivot spring clips on the back side of the clutch fork. If the clips are broken or distorted, replace the clutch fork. The pivot ball pocket in the fork is Teflon® coated and should be installed **WITHOUT** any lubricant such as grease which will break down the Teflon® coating. Be sure the ball stud and fork pocket are clean of contamination and dirt. When assembling the fork to the bearing, the small pegs on the bearing must go over the fork arms.
6. Check the flywheel for cracks, glazing or grooves. If any of these conditions exist, machine (reface) or replace the flywheel to prevent clutch chatter and premature clutch wear.

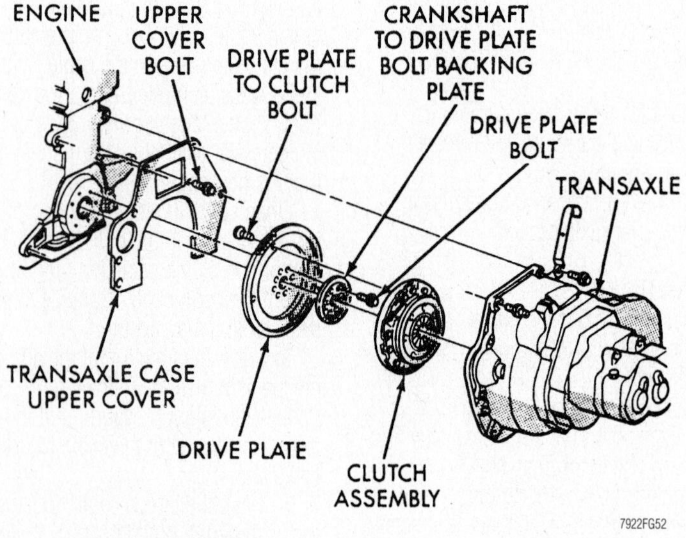

Exploded view of the clutch assembly

➡The manual transaxle is equipped with a reverse brake. It functions as a synchronizer, but only if the vehicle is not moving. When the clutch pedal is depressed to the floor and held for 3 seconds, and the transaxle shifts to reverse, no gear clash should be present. If there is, the input shaft should be checked. When the transaxle is removed for clutch service, check the input clutch shaft, clutch disc splines and release bearing for dry rust. If present, clean rust off and apply a light coat of high temperature bearing grease to the input shaft splines. Apply grease on the input shaft splines only where the clutch disc slides. Verify that the clutch disc slides freely along the input shaft splines.

7. Install or connect the following:
   - Modular clutch assembly onto the transaxle input shaft
   - Transaxle assembly
   - New clutch-to-driveplate (flywheel) bolts. Tighten the bolts, in a criss-cross pattern, a few turns at a time to 55 ft. lbs. (75 Nm)
   - Clutch inspection cover
   - Transaxle lower support brackets
   - Starter
   - Negative battery cable
8. Road test the vehicle to check for proper clutch operation.

Halfshafts

REMOVAL & INSTALLATION

1. Raise vehicle on hoist.
2. Remove front tire and wheel assembly.
3. Remove caliper adapter-to-knuckle bolts.
4. Remove caliper/adapter assembly and support with mechanic's wire or equivalent. Do not support assembly by the flexible brake hose.
5. Remove the brake rotor.
6. Remove the tie rod-to-steering knuckle nut. Remove nut from tie rod end by holding tie rod end stud with an $^{11}/_{32}$ socket and loosen and remove nut.
7. Separate tie rod end from steering knuckle arm, using Tool MB-991113, or equivalent.
8. Remove the halfshaft-to-hub and bearing retaining nut.
9. Remove cotter pin and castle nut from stud of lower ball joint at the steering knuckle.

⁂ WARNING

No tool is to be inserted between the steering knuckle and the lower ball joint to separate stud of lower ball joint from the steering knuckle. The steering knuckle is to be separated from the stud of the ball joint only using the procedure as described below.

10. Turn steering knuckle so the front of the steering knuckle is facing as far outboard in the wheel well as possible. Using a hammer strike steering knuckle boss until steering knuckle separates from stud of lower ball joint. When striking steering knuckle, care MUST be taken not to hit lower control arm or ball joint grease seal.

➡Care must be taken not to separate the inner CV-joint during this operation. Do not allow driveshaft to hang by inner CV-joint. Driveshaft must be supported.

11. Separate halfshaft from steering knuckle. If difficulty is encountered separating halfshaft from knuckle, DO NOT strike halfshaft with a hammer to break free. Install Puller 1026 as shown in to press shaft out of hub/bearing assembly.

➡When inserting pry bar between inner tripod joint and transaxle case, care must be used to ensure no damage occurs to oil seal in transaxle case.

12. Support the outer end of the halfshaft assembly. Insert a pry bar between inner tripod joint and transaxle case. Pry against inner tripod joint, until tripod joint retaining snapring is disengaged from transaxle side gear.

13. Hold inner tripod joint and interconnecting shaft of driveshaft assembly. Remove inner tripod joint from transaxle, by pulling it straight out of transaxle side gear and transaxle oil seal.

➡When removing tripod joint, do not let spline or snapring drag across sealing lip of the transaxle to tripod joint oil seal.

To install:

14. Thoroughly clean spline and oil seal sealing surface, on tripod joint. Lightly lubricate oil seal sealing surface on tripod joint with fresh clean transmission lubricant.

15. Holding halfshaft assembly by tripod joint and interconnecting shaft, install tripod joint into transaxle side gear as far as possible by hand.

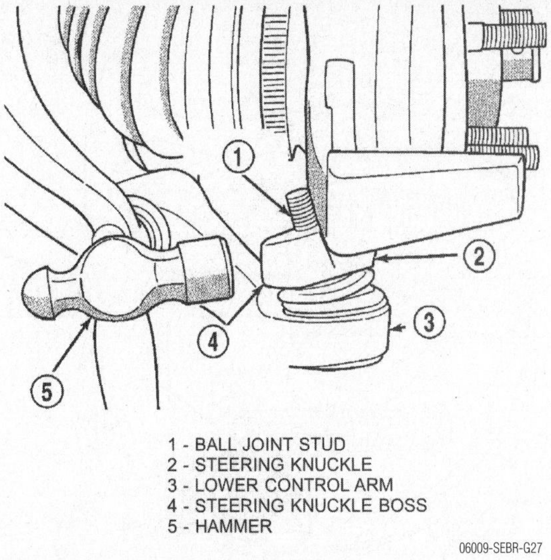

Separating the ball joint stud from the steering knuckle

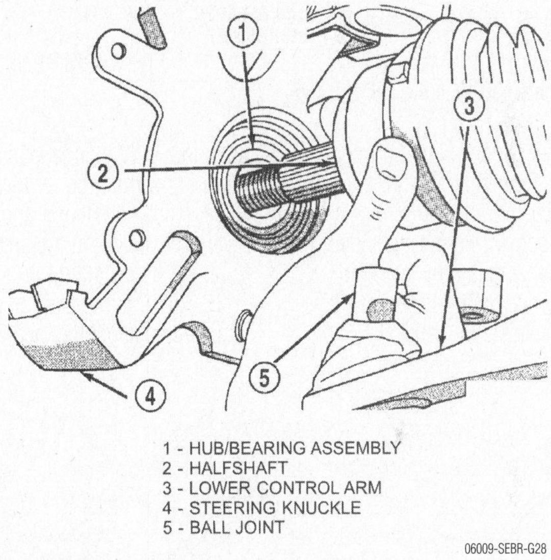

Separating the halfshaft from the knuckle

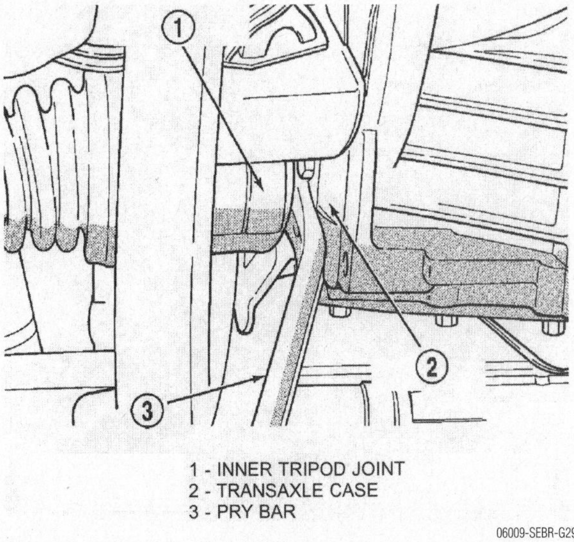

Disengaging the inner tripod joint from the transaxle

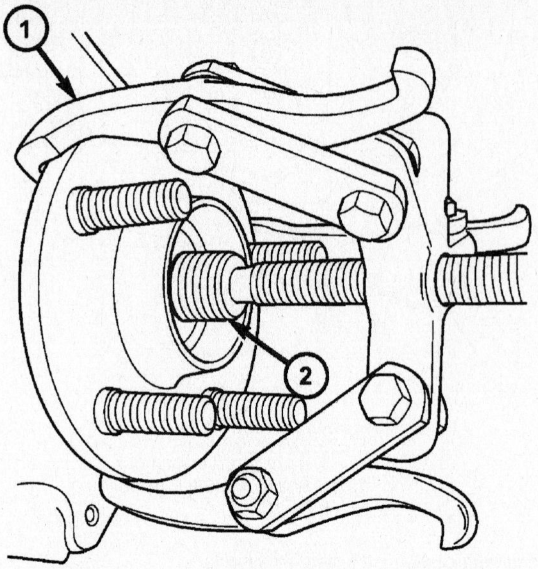

Separating the halfshaft from the hub/bearing

16. Grasp inner tripod joint and interconnecting shaft. Forcefully push the tripod joint into side gear of transaxle, until snapring is engaged with transaxle side gear. Test that snapring is fully engaged with side gear by attempting to remove tripod joint from transaxle by hand. If snapring is fully engaged with side gear, tripod joint will not be removable by hand.

17. Clean all debris and moisture out of steering knuckle, in the area were outer CV-joint will be installed into steering knuckle.

18. Ensure that front of outer CV-joint which fits against the face of the hub and bearing is free of debris and moisture before installing outer CV-joint into hub and bearing assembly.

19. Slide halfshaft back into front hub. Then install steering knuckle onto the lower control arm ball joint stud.

20. Install the steering knuckle to ball joint stud castle nut. Tighten the castle nut to 95 Nm (70 ft. lbs.).

21. Install tie rod end into the steering knuckle. Start tie rod end to steering knuckle nut onto stud of tie rod end. While holding stud of tie rod end stationary, tighten tie rod end to steering knuckle nut. Using a crowfoot and 11⁄32 socket, tighten the nut to 55 Nm (41 ft. lbs.).

22. Install brake rotor to hub.

23. Install caliper/adapter bracket assembly on steering knuckle. Install and torque caliper adapter-to-knuckle bolts to 88 Nm (65 ft. lbs.).

24. Clean all foreign matter from the threads of the outer CV-joint stub axle. Install hub nut onto threads of halfshaft. With vehicle brakes applied to keep braking disc from turning, tighten hub nut to 150 Nm (110 ft. lbs.).

25. Install front wheel and tire assembly. Install front wheel lug nuts and tighten in the correct sequence. Tighten lug nuts to 135 Nm (100 ft. lbs.).

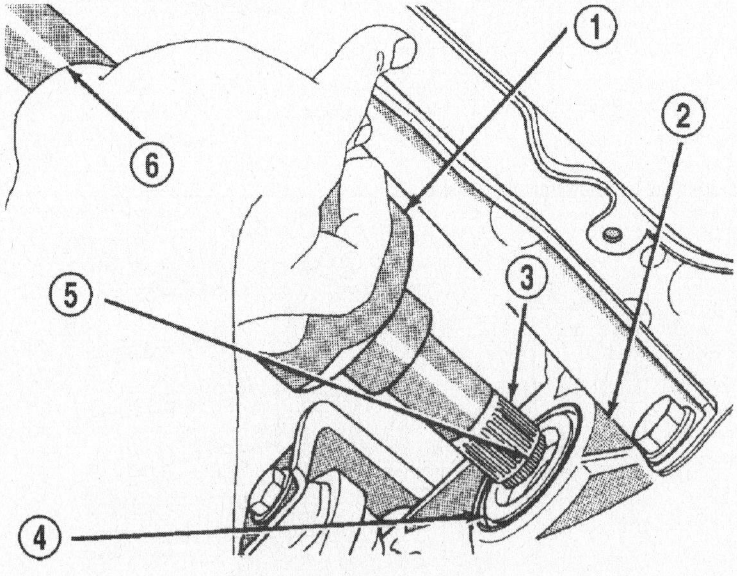

Tripod joint removal from the transaxle

CV-Joints

OVERHAUL

Inner Tri-Pod Joint

1. Remove or disassemble the following:
   - Negative battery cable
   - Halfshaft
   - Large and small boot retaining clamps
2. Slide the boot down the shaft away from the tri-pod housing.

➡When separating the spider joint from the tri-pod joint housing, hold the rollers in place on the trunnions to prevent the rollers and needle bearings from falling away.

3. Carefully slide the shaft/spider assembly from the tri-pod housing.
4. Remove the spider assembly-to-shaft snapring; then, slide the spider assembly off the shaft.

⁂ WARNING

If necessary, tap the spider assembly off the shaft using a brass drift; be careful not to hit the outer bearings.

5. Slide the boot off the shaft.
6. Thoroughly inspect all parts for signs of excessive wear. If necessary, replace the halfshaft.

➡**Component parts are not serviceable and must be replaced as an assembly.**

To assemble:

⁂ WARNING

The Tri-pod sealing boots are made of 2 different types of material; silicon rubber (high temperature) which is soft and pliable or Hytrel plastic (standard temperature) which is stiff and rigid. Be sure to replace the boot made of the correct material.

7. Slide the inner tri-pod boot clamp and boot onto the shaft; then, position the boot so that only the thinnest (sight) groove is visible on the shaft.
8. Install the spider assembly onto the shaft just far enough so that the snapring can be installed.

⁂ WARNING

If necessary, tap the spider assembly onto the shaft using a brass drift; be careful not to hit the outer bearings.

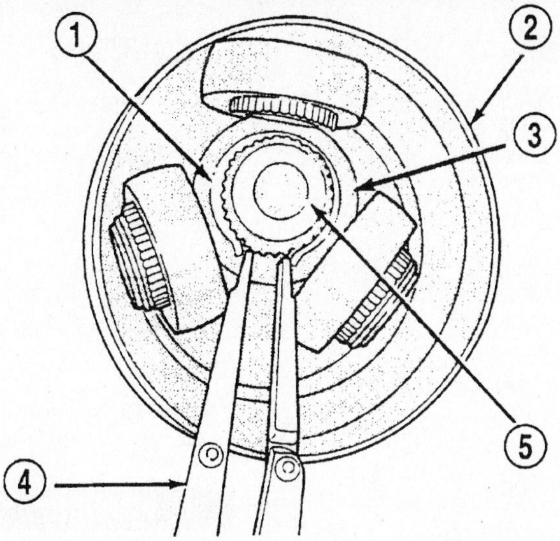

1 – SNAP RING
2 – SEALING BOOT
3 – SPIDER ASSEMBLY
4 – SNAP RING PLIERS
5 – INTERCONNECTING SHAFT

9306BG11

View of the halfshaft inner tri-pod joint and snapring

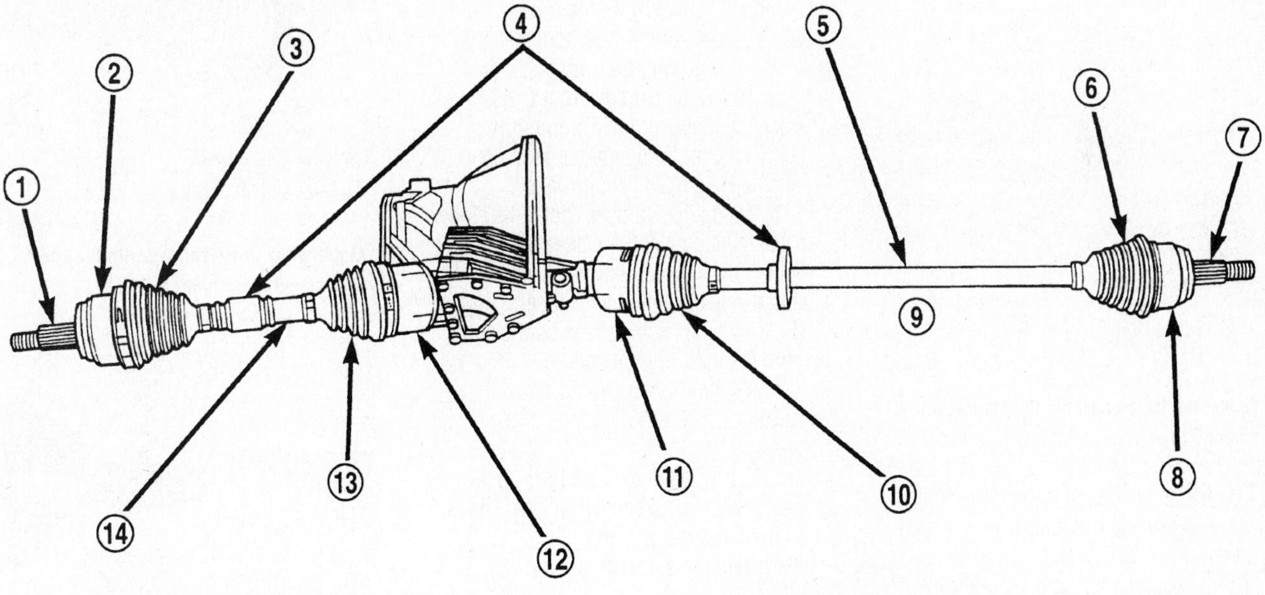

1 – STUB AXLE
2 – OUTER C/V JOINT
3 – OUTER C/V JOINT BOOT
4 – TUNED RUBBER DAMPER WEIGHT
5 – INTERCONNECTING SHAFT
6 – OUTER C/V JOINT BOOT
7 – STUB AXLE
8 – OUTER C/V JOINT
9 – RIGHT DRIVESHAFT
10 – INNER TRIPOD JOINT BOOT
11 – INNER TRIPOD JOINT
12 – INNER TRIPOD JOINT
13 – INNER TRIPOD JOINT BOOT
14 – INTERCONNECTING SHAFT LEFT DRIVESHAFT

9306EG02

View of the halfshaft assemblies—Typical

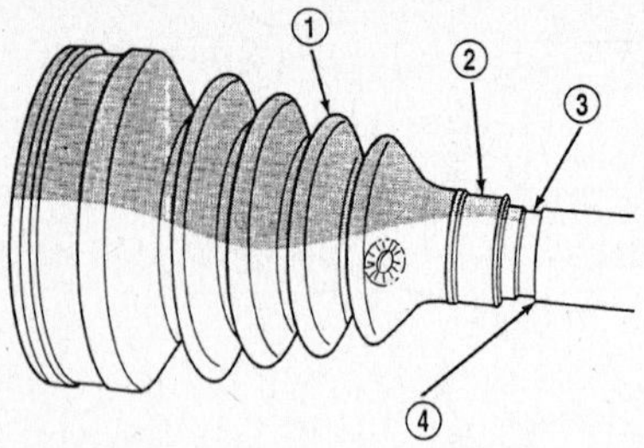

1 – SEALING BOOT
2 – RAISED BEAD IN THIS AREA OF SEALING BOOT
3 – GROOVE
4 – INTERCONNECTING SHAFT

9306EG04

View of the halfshaft boot and shaft

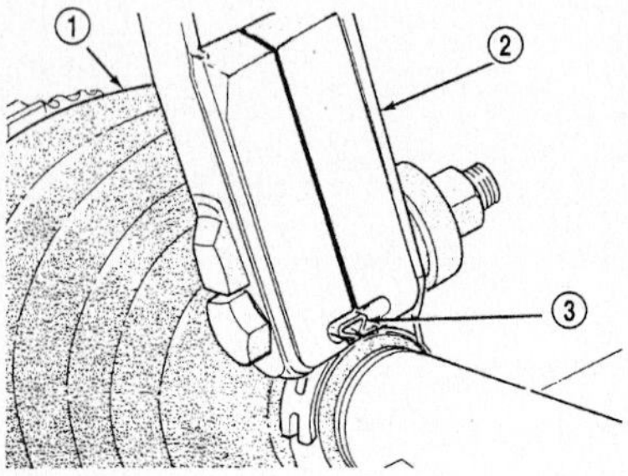

1 – SEALING BOOT
2 – SPECIAL TOOL C-4975
3 – CLAMP BRIDGE

9306BG13

Securing the halfshaft boot clamp

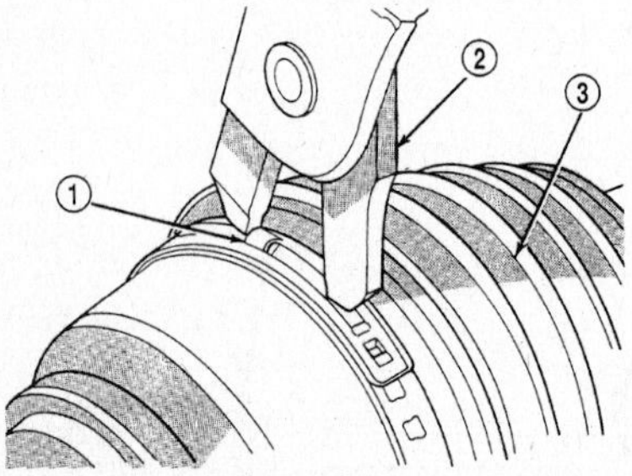

1 – CLAMP
2 – SPECIAL TOOL YA3050
3 – SEALING BOOT

9306EG07

Tightening the low profile boot clamp—Silicone rubber Tri-pod boot

9. Install the snapring onto the shaft; make sure that the snapring is fully seated in the groove.

10. If installing a new boot, distribute ½ of the grease in the service package inside the tri-pod housing and the other ½ inside the boot.

11. Carefully, slide the spider assembly and shaft into the tri-pod housing.

12. Position the inner boot clamp evenly on the sealing boot.

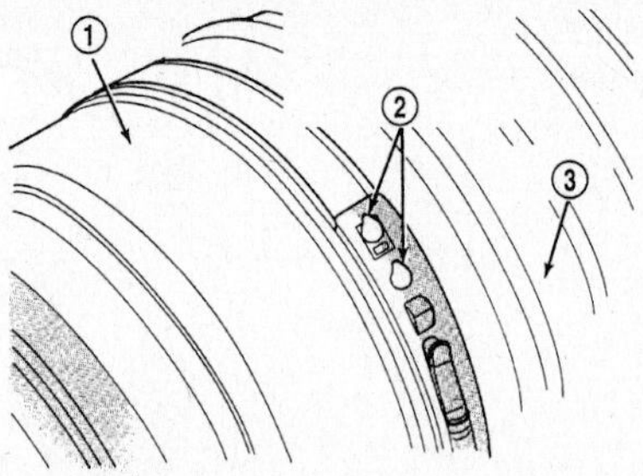

1 – INNER TRIPOD JOINT HOUSING
2 – TOP BANK OF CLAMP MUST BE RETAINED BY TABS AS SHOWN HERE TO CORRECTLY LATCH BOOT CLAMP
3 – SEALING BOOT

9306EG08

Secured the low profile boot clamp—Silicone rubber Tri-pod boot

⁂ WARNING

The seal must not be dimpled, stretched or out of shape. If necessary, use a trim stick to seat and equalize the seal pressure and shape it by hand.

➡If using a Hytrel (hard plastic) boot, be sure the stick is inserted between the soft rubber insert and the tri-pod housing, not between the hard plastic sealing boot and the soft rubber insert.

13. Position the boot onto the tri-pod housing retaining groove and install the retaining clamp evenly on the boot.

14. If using a crimp type boot clamp, perform the following procedure:
   a. Using the Crimper Tool C-4975-A, place the tool over the clamp bridge, tighten the tool nut until the jaws are completely closed (face-to-face).
   b. Using the Crimper Tool C-4975-A, place the tool over the clamp bridge, tighten the tool nut until the jaws are completely closed (face-to-face).

15. If using a latching type boot clamp, perform the following procedure:
   a. Position Snap-On® Clamp Locking Tool YA3050 prongs in the clamp holes.
   b. Squeeze the tool until the upper clamp band is latched behind the 2 tabs on the lower clamp band.

16. Install the halfshaft into the vehicle.

Rzeppa (Outer) Joint

1. Remove or disassemble the following:
   - Halfshaft and place it in a soft-jawed vise
   - Rzeppa joint boot clamps and slide the boot down the shaft
   - Rzeppa joint housing by sharply hitting it with a soft-faced hammer to drive it off the shaft
   - Circlip from the shaft
   - Boot by sliding it off the shaft

⁂ WARNING

If any parts show excessive wear, replace the halfshaft assembly; the component parts are not serviceable.

To assemble:

2. Install or assemble the following:
   - New small boot clamp and slide it onto the shaft
   - Boot and slide it onto the shaft
   - Circlip, if removed

3. Position the boot so that the raised bead on the inside the boot seal is in the shaft groove.

4. Install or assemble the following:
   - Halfshaft hub nut onto the Rzeppa joint threaded shaft so it is flush with the end
   - Rzeppa joint and align the shaft splines and tap it onto the shaft

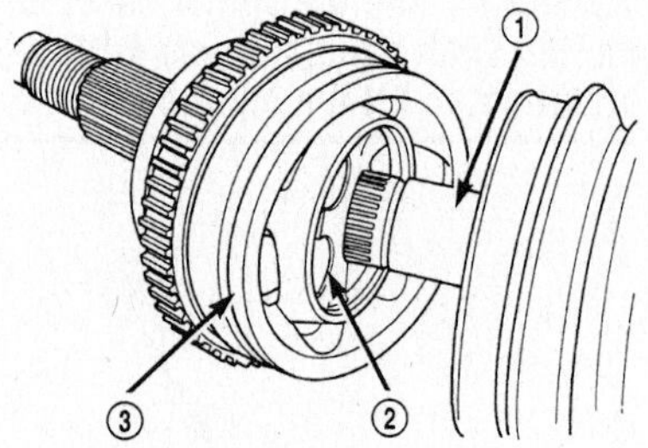

1 – INTERCONNECTING SHAFT
2 – CROSS
3 – OUTER C/V JOINT ASSEMBLY

9306EG09

Aligning the cross splines with the shaft splines—Rzeppa joint

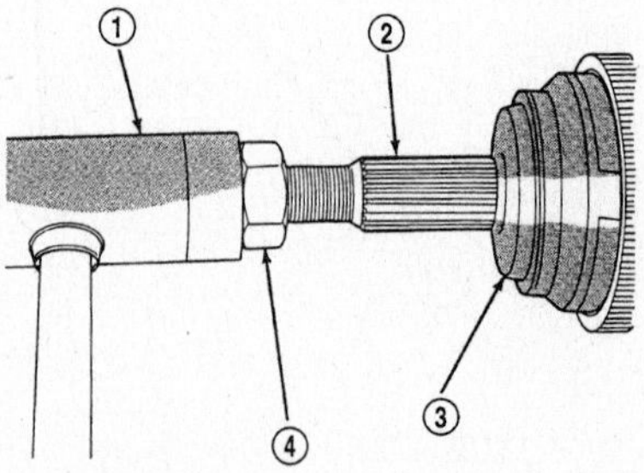

1 – SOFT FACED HAMMER
2 – STUB AXLE
3 – OUTER C/V JOINT
4 – NUT

9306EG10

Driving the Rzeppa joint onto the shaft

with a soft-faced hammer so it locks on the circlip

5. Distribute ½ of the grease in the service package inside the Rzeppa joint housing and the other ½ inside the boot.

6. Install or connect the following:

- New small boot clamp and position it evenly on the sealing boot

7. Using the Crimper Tool C-4975-A, place the tool over the clamp bridge, tighten the tool nut until the jaws are completely closed (face-to-face).

**** WARNING**

The seal must not be dimpled, stretched or out of shape. If necessary, equalize the seal pressure and shape it by hand.

8. Position the boot onto the Rzeppa housing retaining groove and install the retaining clamp evenly on the boot.

9. Using the Crimper Tool C-4975-A, place the tool over the clamp bridge, tighten the tool nut until the jaws are completely closed (face-to-face).

**** WARNING**

The seal must not be dimpled, stretched or out of shape. If necessary, equalize the seal pressure and shape it by hand.

10. Install the halfshaft into the vehicle.

STEERING AND SUSPENSION

Air Bag

**** CAUTION**

Some vehicles are equipped with an air bag system, also known as the Supplemental Inflatable Restraint (SIR) or Supplemental Restraint System (SRS). The system must be disabled before performing service on or around system components, steering column, instrument panel components, wiring and sensors. Failure to follow safety and disabling procedures could result in accidental air bag deployment, possible personal injury and unnecessary system repairs.

PRECAUTIONS

Several precautions must be observed when handling the inflator module to avoid accidental deployment and possible personal injury.

➡Before servicing the vehicle, also refer to the precautions in the beginning of this section.

- Never carry the inflator module by the wires or connector on the underside of the module.
- When carrying a live inflator module, hold securely with both hands, and ensure that the bag and trim cover are pointed away.
- Place the inflator module on a bench or other surface with the bag and trim cover facing up.
- With the inflator module on the bench, never place anything on or close to the module which may be thrown in the event of accidental deployment.

DISARMING

This air bag system is a sensitive, complex, electromechanical unit. Proper SRS (also called Supplemental Inflatable Restraint, or SIR, or air bag system) disarming can be obtained by disconnecting and isolating the negative battery cable (wrapping the battery cable end with electrical tape is a good method for isolating it). Failure to disconnect the battery could result in accidental air bag deployment and possible personal injury. Before beginning service work, allow the system capacitor 2 minutes to discharge after disconnecting and isolating the negative battery cable.

Do not place an intact undeployed airbag face down on a solid surface. The airbag will propel into the air if accidentally deployed and could result in personal injury. When carrying or handling an undeployed airbag, the trim side of the airbag should be pointing towards the body to minimize possibility of personal injury or death if accidental deployment occurs.

Power Rack and Pinion Steering Gear

REMOVAL & INSTALLATION

1. Before servicing the vehicle, refer to the precautions in the beginning of this section.

2. Remove or disconnect the following:

- Negative battery cable from the left shock tower

➡The ground cable is equipped with an insulator grommet which should be placed on the stud to prevent the negative battery cable from accidentally grounding.

- Intermediate shaft coupler pin bolt retaining pin
- Intermediate shaft coupler pin bolt
- Front wheels
- Tie rod ends by holding the tie rod end stud with a 11/32 in. socket and loosen the retaining nut
- Tie rod end from the steering knuckle

➡Before removing the front suspension crossmember from the vehicle scribe the front suspension crossmember and the vehicle body. This must be done to retain the proper alignment. The caster and camber are not adjustable.

3. Scribe a line on the body and on the crossmember on all 4 sides.

4. Remove or disconnect the following:

- Stabilizer
- 3 anti-lock brake controller-to-crossmember bolts and secure it to the chassis, if equipped with Anti-lock Brake System (ABS)
- Strut clevis from the lower control arm
- Both engine support bracket-to-crossmember bolts
- Engine support bracket-to-transaxle mounting bracket bolt

5. Place a lifting device under the front suspension crossmember.

6. Remove or disconnect the following:

- 8 crossmember-to-chassis bolts

7. Lower the lifting device enough to gain access to the steering rack.

8. Remove or disconnect the following:

- Power steering lines and drain the fluid
- Power steering pressure switch electrical connector
- Solenoid control module electrical connector, if equipped with speed proportional steering
- Both steering rack isolator bolts
- Both steering rack saddle bracket bolts

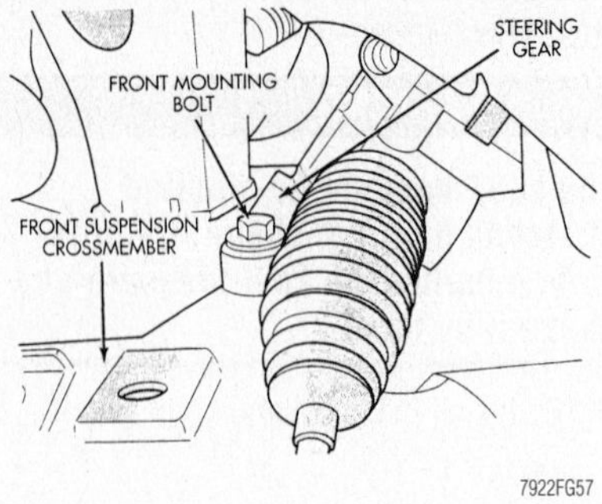

Steering gear front mounting bolt location

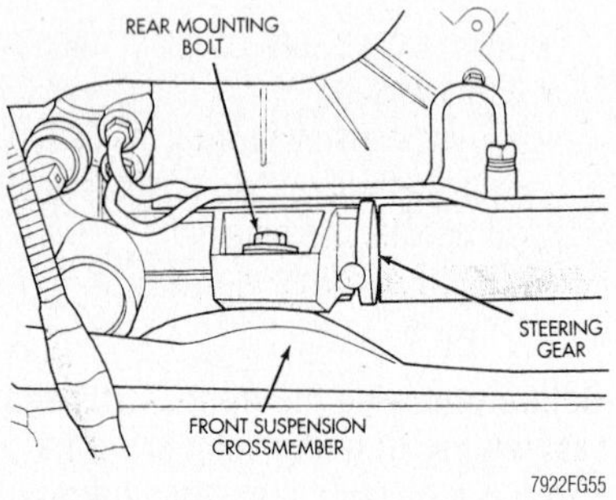

Steering gear rear mounting bolt location

- Steering rack

To install:

9. Install or connect the following:
   - Steering rack
   - Isolator and saddle bracket and torque the bolts to 50 ft. lbs. (68 Nm)
   - Power steering pressure and return lines and torque the lines to 23 ft. lbs. (31 Nm)
   - Crossmember, install the rear bolts first and torque the bolts to 20 inch lbs. (2 Nm)
10. Using a soft faced hammer tap the crossmember into position.

➡Be sure to align the scribed marks on the crossmember.

11. Install or connect the following:
    - Crossmember, starting with the rear bolts and torque the bolts to 120 ft. lbs. (163 Nm)
    - Both engine support bracket-to-crossmember bolts
    - Engine support bracket-to-transaxle bracket bolt and torque the 3 bolts to 55 ft. lbs. (75 Nm)
    - Power steering pressure switch
    - ABS control unit and torque the bolts to 21 ft. lbs. (28 Nm)
    - Heat shield on the tie rod ends
    - Shock clevis to the lower control arm
    - Tie rod ends and torque the nuts to 45 ft. lbs. (61 Nm)
    - Both stabilizer clamps
    - Strut clevis bolt and torque the bolt to 68 ft. lbs. (92 Nm)
    - Wheels and torque the lug nuts to 100 ft. lbs. (135 Nm)
    - Intermediate shaft pin bolt and retaining pin and torque the pin bolt to 20 ft. lbs. (27 Nm)
    - Negative battery cable
12. Refill the power steering system.
13. Start the engine and allow it to run for a few minutes.
14. Shut **OFF** the engine and check the power steering fluid.
15. Add power steering fluid if necessary.
16. Raise the front wheels off the ground.
17. Start the engine and turn the wheel from stop-to-stop to bleed any air from the system.
18. Check the fluid level and add, if necessary.
19. Check and adjust the alignment.

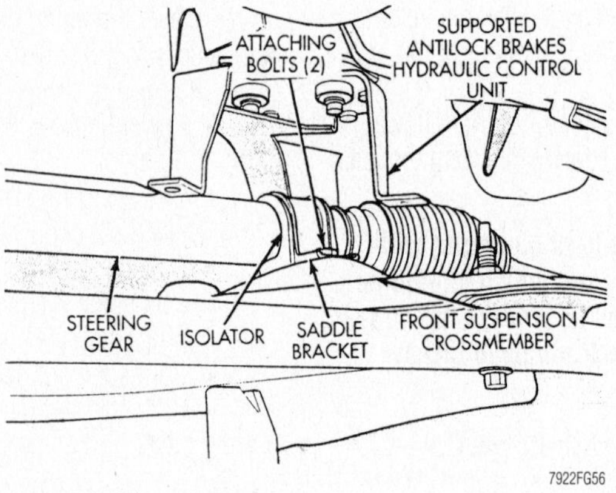

Check the condition of the isolator—if worn or oil soaked, replace

Strut

REMOVAL & INSTALLATION

Front

1. Loosen wheel nuts.
2. Raise vehicle.
3. Remove the wheel and tire assembly.
4. If both shock assemblies are removed, mark the shock assemblies right and left according to which side of the vehicle they were removed from.
5. Remove the wheel speed sensor cable routing bracket from the steering knuckle.
6. Remove the nut from the upper ball joint stud.
7. Remove the upper ball joint stud from the steering knuckle using a puller. Pull steering knuckle outward and position toward the rear of the front wheel opening.
8. Remove pinch bolt attaching shock absorber clevis to shock absorber.
9. Remove the nut and through-bolt attaching the shock absorber clevis to the lower control arm.
10. Remove the clevis from the shock absorber by carefully tapping the clevis off the shock absorber using a soft (brass) drift.
11. Remove the 4 bolts attaching the shock absorber/upper control arm mounting bracket to the shock tower of the vehicle.
12. Remove the shock assembly from the vehicle. The shock assembly is removed out through the front area of the front wheel well.

To install:

13. Install the shock assembly, with the clevis removed, into shock tower. Aligning the 2 locating pins and the 4 mounting holes on the upper control arm shock absorber mount with the 4 holes in shock tower. Install the 4 upper control arm mount to shock tower mounting bolts. Tighten the 4 bolts to a torque of 95 Nm (70 ft. lbs.).
14. Install the clevis on the shock absorber. Clevis is installed by tapping it onto the fluid reservoir of the shock absorber using a soft (brass) drift until fully seated against locating tab on shock absorber. Orientation tab on locating tab must be positioned in the split of the clevis.
15. Install the pinch bolt retaining the shock clevis to the shock absorber. Tighten the pinch bolt to a torque of 88 Nm (65 ft. lbs.)
16. Install the clevis bracket to lower control arm through-bolt. Do not tighten the through-bolt at this time.
17. Install upper ball joint into steering

knuckle. Install nut on ball joint stud. Tighten nut to a torque of 27 Nm (20 ft. lbs.).

18. Install the routing bracket for the wheel speed sensor cable on the steering knuckle. Install and securely tighten the routing bracket attaching bolt.

WARNING

When supporting lower control arm with jackstand, do not position jack stand under the ball joint cap on the lower control arm. Position in area of lower control arm shown.

19. Lower vehicle to the ground with a jackstand positioned under the lower control arm. Continue to lower vehicle so the total weight of the vehicle is supported by the jack stand and lower control arm.

20. Tighten the shock absorber clevis to lower control arm bushing through-bolt nut to a torque of 88 Nm (65 ft. lbs.).

21. Raise the vehicle, then remove the jack stand.

22. Install the tire and wheel assembly. Progressively tighten the wheel mounting nuts in a crisscross sequence until all nuts are torqued to half specification. Then repeat the tightening sequence to the full specified torque of 135 Nm (100 ft. lbs.).

23. Lower the vehicle.

Rear

1. Before servicing the vehicle, refer to the precautions in the beginning of this section.
2. Remove or disconnect the following:
   - Carpet, pull it back from the rear strut tower
   - Plastic cover from the top of the strut tower
   - Both strut assembly-to-chassis nuts
   - Rear wheel
   - Splash shield from the upper strut mount
   - Strut-to-rear knuckle bolt

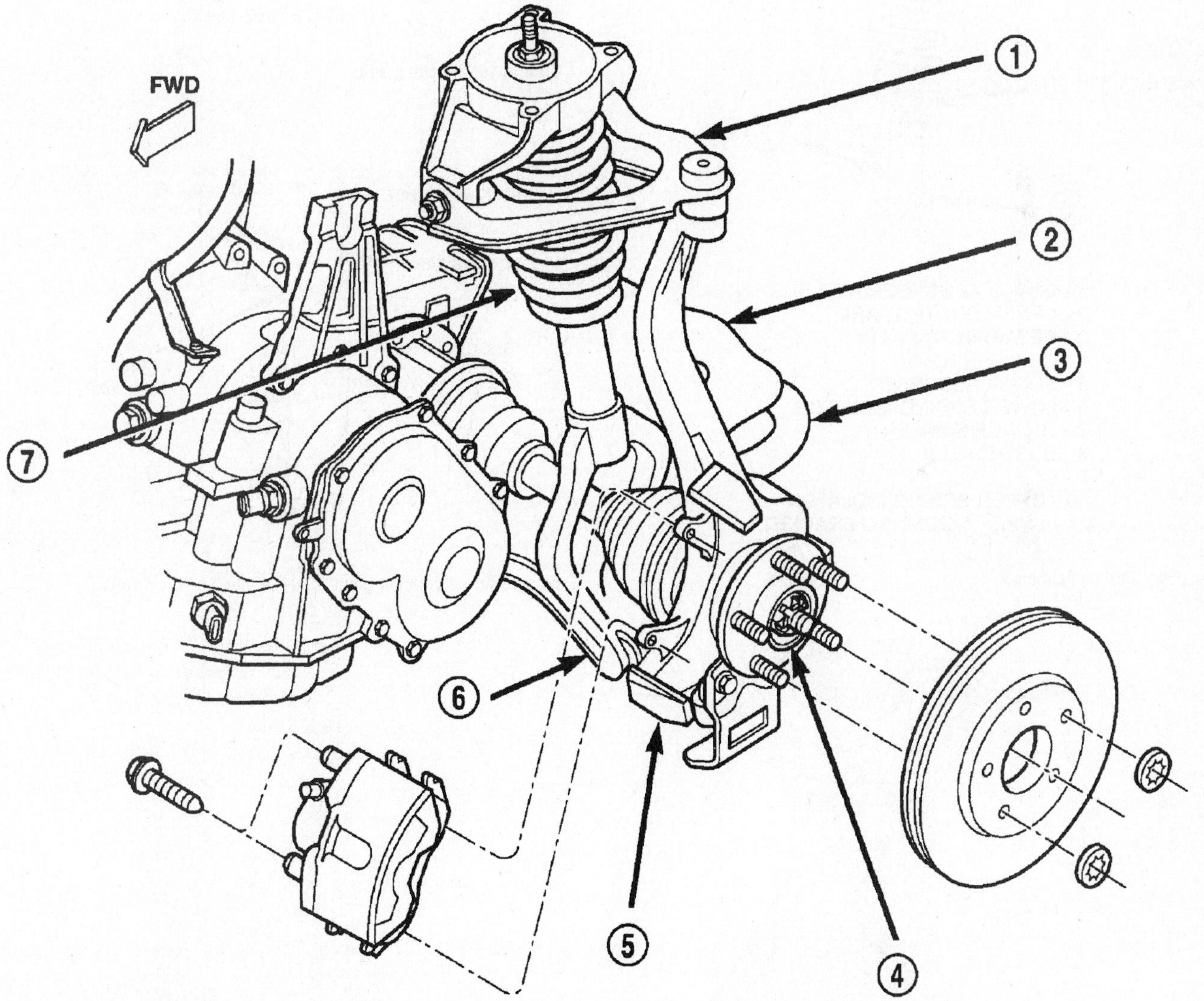

1 - UPPER CONTROL ARM
2 - FRONT SUSPENSION CROSSMEMBER
3 - STABILIZER BAR
4 - HUB AND BEARING
5 - STEERING KNUCKLE
6 - LOWER CONTROL ARM
7 - SHOCK ASSEMBLY

06009-SEBR-G32

Front suspension components

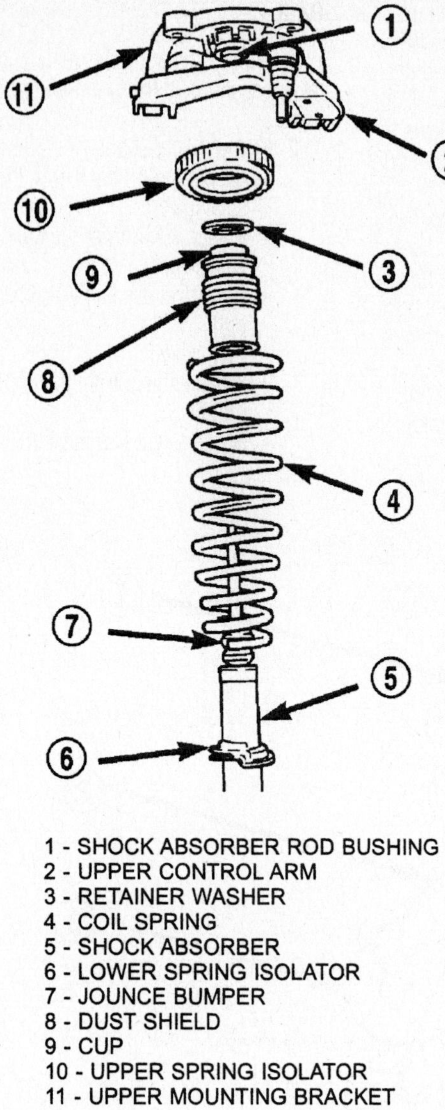

Front strut components

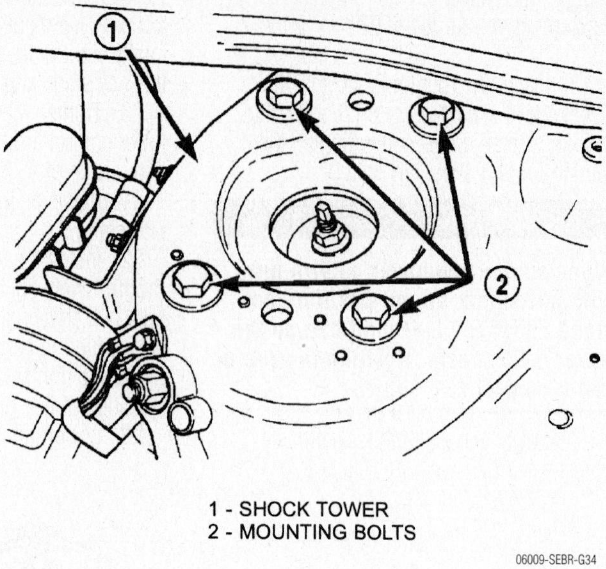

Strut upper attaching bolts

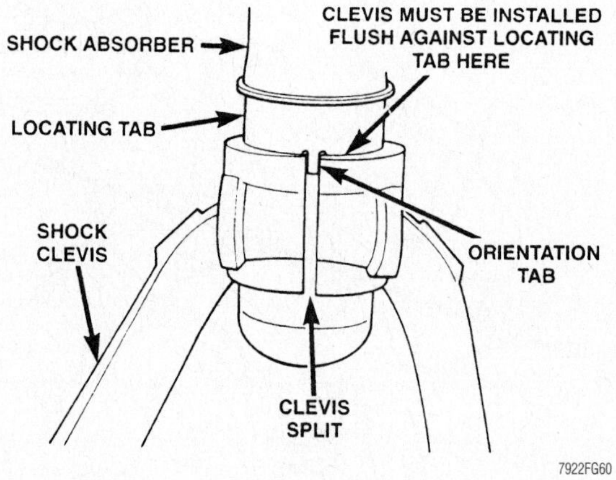

Be sure the orientation tab is situated into the clevis split

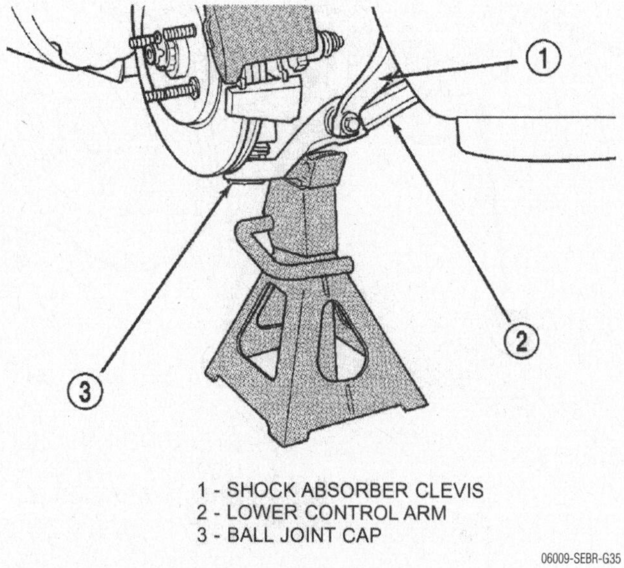

Proper jackstand placement under the lower control arm

- Strut by pushing the rear suspension downward and tilting the top of the strut outward

To install:

3. Install or connect the following:
   - Strut by pushing the rear suspen-

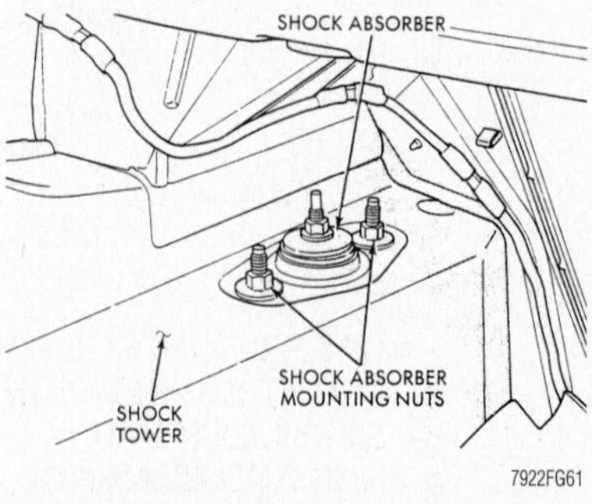

Access the rear strut upper mounting from inside the trunk

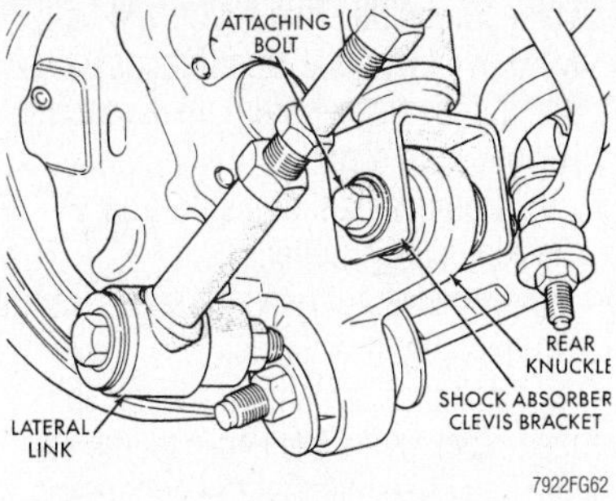

View of the rear strut lower mounting

sion downward and inserting the top of the strut into the vehicle

- Strut-to-rear knuckle bolt and torque the bolt to 70 ft. lbs. (95 Nm)
- Strut upper mounting nuts and torque the nuts to 25 ft. lbs. (34 Nm) for 2002–04; 40 ft. lbs. (54 Nm) for 2005.
- Strut top cover
- Rear wheel

Coil Spring

REMOVAL & INSTALLATION

Front

1. Before servicing the vehicle, refer to the precautions in the beginning of this section.
2. Disassemble as follows:
3. Remove the strut/shock assembly and place it in a suitable compressor tool
4. Set the lower hooks and install the clamp on the lower end of the spring.
5. Rotate the spring so the ball joint sits directly below the front upper hook.
6. Compress the spring until all tension is removed.
7. Remove the retaining nut and washer.
8. Remove the clamp from the bottom of the spring.
9. Remove the strut/shock assembly.
10. Remove the lower spring, jounce bumper and cup.
11. Remove the dust boot and lower bushing washer.
12. Release the tension from the spring.
13. Remove the spring from the compressor tool.

To assemble:

14. Install the upper isolator on the mounting bracket.
15. Install the sleeve into the lower isolator bushing.
16. Install the bushing and sleeve into the bottom of the upper mounting bracket.
17. Install the upper isolator bushing into the center of the upper mounting bracket.
18. Install the lower end of the spring in the compressor tool supported by the hooks.
19. Install the upper mounting bracket on top of the coil spring matching the spring to the isolator on the upper mounting bracket.
20. Install the control arm ball joint directly below the front upper hook.
21. Compress the spring.
22. Install the lower spring isolator on the spring seat.
23. Install the jounce bumper with the pointed end downward.
24. Install the collar with the undercut side facing down.
25. Install the dust shield and cup
26. Install the lower bushing retainer washer
27. Install the strut/shock assembly through the coil spring.
28. When properly aligned, install the upper mounting nut and torque the nut to 40 ft. lbs. (55 Nm).
29. Slowly release the tension from the compressor tool.
30. Install the strut/shock assembly in the vehicle.

Rear

1. Before servicing the vehicle, refer to the precautions in the beginning of this section.
2. Disassemble as follows:
3. Remove the rear shock from the vehicle.
4. Install the assembly in a spring compressor tool.
5. Remove the jam nut, washer and upper bushing.
6. Remove the upper bracket assembly and spring pad.
7. Remove the collar and bushing.
8. Remove the cup, bump rubber and dust cover.
9. Remove the coil spring

To assemble:

10. Install the coil spring to the shock absorber using compressor Tools MB991237 and MB991239.
11. Install the dust cover and bump rubber.
12. Install the cup and bushing
13. Install the collar, supper spring pad and bracket assembly.
14. Install the upper bushing and washer.
15. When properly aligned, install the jam nut and torque to 17 ft. lbs. (23 Nm).
16. Remove the compressor tools and install the shock absorber.

Upper Ball Joint

REMOVAL & INSTALLATION

The upper ball joint is an integrated part of the upper control arm assembly, and cannot be serviced separately. A worn or damaged ball joint requires replacement of upper control arm assembly.

Lower Ball Joint

REMOVAL & INSTALLATION

On all vehicles, the ball joint cannot be serviced separately. If the ball joint is defective it will require replacement of the lower control arm.

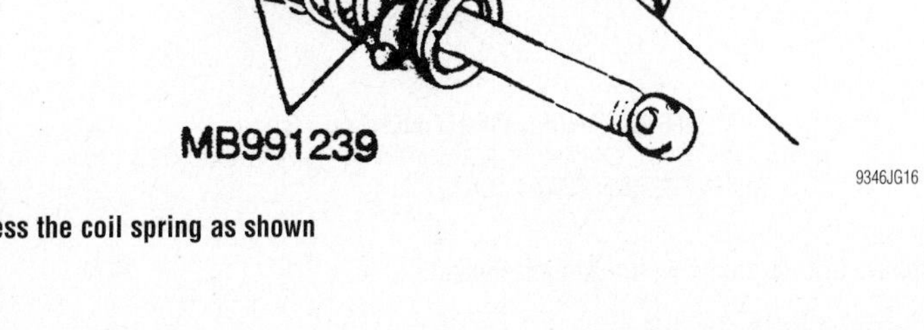

Compress the coil spring as shown

Upper Control Arm

REMOVAL & INSTALLATION

Front

1. Remove the front strut assembly from the vehicle.
2. Disassemble the strut assembly until the upper mounting bracket is removed from the coil spring.
3. Remove the 2 bolts attaching the upper control arm to the bushings in the upper mounting bracket.
4. Remove the upper control arm from the mounting bracket.

To install:

5. Install the upper control arm on the upper (strut absorber/upper control arm) mounting bracket.
6. Install the 2 bolts attaching the upper control arm to the bushings in the mounting bracket.

WARNING

For clearance reasons the control arm mounting bolts must be installed from center, so the heads are toward the coil spring when it is installed. Otherwise the bolts may rub the coil spring, damaging it.

7. Install the control arm mounting bolt nuts. Position the control arm at a 90 degree angle to the mounting bracket and tighten the bolts to a torque of 90 Nm (66 ft. lbs.).
8. Reinstall the upper mounting bracket and control arm on the coil spring. Reassemble the front strut assembly.
9. Reinstall the front strut assembly on the vehicle.

Rear

1. Raise vehicle.
2. Remove both rear wheel and tire assemblies from the vehicle.
3. Remove the shock absorber clevis bracket to rear knuckle attaching bolt and nut on both sides of the vehicle.
4. Remove muffler support bracket from rear frame rail.
5. Remove the rear exhaust pipe hanger bracket from the rear suspension crossmember. Let exhaust system drop down as far as possible.
6. If vehicle is equipped with antilock brakes, remove routing clips for wheel speed sensor cable from brackets on both upper control arms.
7. If vehicle is equipped with antilock brakes, remove bolts and wheel speed sensor heads from both rear knuckles.
8. Remove both rear disc brake calipers from knuckles and hang out of way using wire or cord. Do not allow calipers to hang by brake hoses.
9. Release brake tubing from clips on top of crossmember.
10. On the side of the vehicle requiring control arm removal, separate the control arm ball joint from the rear knuckle using following procedure:
    a. Back off nut retaining upper control arm ball joint to knuckle until nut is even with end of stud. This action will help avoid damaging the stud threads when the stud is released from the knuckle in the following step.
    b. Remove ball joint stud from knuckle using a puller.
    c. Remove nut retaining upper control arm ball joint to knuckle.
11. Position a transmission jack and wooden block under the center of the rear suspension crossmember to support and lower crossmember during removal.
12. Remove the 4 bolts attaching rear suspension crossmember to rear frame rails.

WARNING

When lowering rear suspension crossmember do not put a strain on the rear brake flex hoses.

13. Lower the rear suspension crossmember far enough to access the upper control arm pivot bar to crossmember attaching bolts.
14. Remove the 2 bolts attaching the upper control arm to the rear suspension crossmember.
15. Remove the upper control arm from the rear suspension crossmember.
16. Transfer any required components to the replacement control arm.

To install:

17. Align the pivot bar on the upper control arm with the mounting holes in the rear suspension crossmember. Install the pivot bar attaching bolts. Tighten the 2 pivot bar attaching bolts to a torque of 108 Nm (80 ft. lbs.).
18. Using transmission jack, raise rear suspension crossmember up to the rear frame rails and loosely install the 4 attaching bolts.
19. Position an appropriate size drift into the positioning hole in each side of rear suspension crossmember and crossmember locating holes in frame rails of the vehicle. This is required to properly position rear suspension crossmember to the body of the vehicle. Tighten the 4 crossmember to frame rail attaching bolts to 108 Nm (80 ft. lbs.). Remove drifts from rear suspension crossmember.
20. Remove transmission jack supporting rear suspension crossmember.
21. Install upper ball joint stud in knuckle. Install and tighten the ball joint stud nut to a torque of 27 Nm (20 ft. lbs.).
22. Install the brake tubing above the rear suspension crossmember into the routing clips on top of the crossmember.

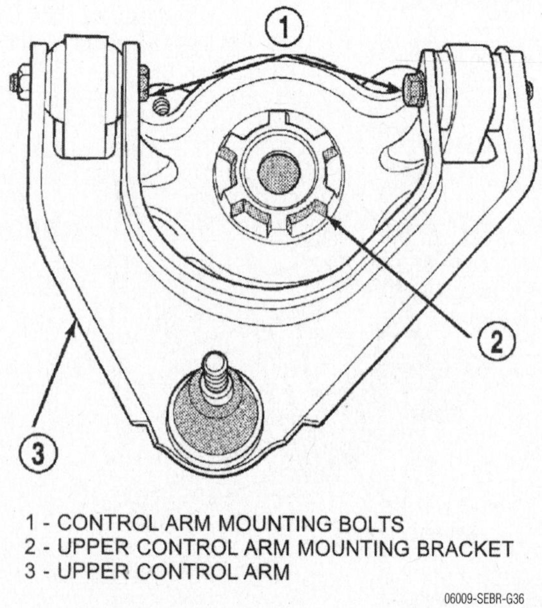

Upper control arm-to-mounting bracket attachment

23. Install rear brake calipers onto rear knuckles.

24. If vehicle is equipped with antilock brakes, attach wheel speed sensor heads to both rear knuckles. Tighten sensor mounting bolts to a torque of 8 Nm (75 inch lbs.).

25. If vehicle is equipped with antilock brakes, install routing clips for wheel speed sensor cables onto brackets on both upper control arms. Securely tighten routing clip attaching bolts.

26. Install muffler support bracket on rear frame rail.

27. Install rear exhaust pipe hanger on rear suspension crossmember.

28. Install the shock absorber clevis brackets on the rear knuckles. Tighten the shock absorber mounting bolts to a torque of 95 Nm (70 ft. lbs.).

29. Install wheel and tire assembly on vehicle. Progressively tighten the wheel mounting nuts in crisscross sequence until all nuts are torqued to half specification. Then repeat the tightening sequence to the full specified torque of 135 Nm (100 ft. lbs.).

30. Lower vehicle.

31. Check and reset if required, rear wheel camber and toe to preferred specifications.

Lower Control Arm

REMOVAL & INSTALLATION

Front

1. Raise vehicle.
2. Remove tire and wheel assembly.

➡Removing the outer tie rod from the steering knuckle allows the steering knuckle to be turned further. This allows better access to the steering knuckle when striking it to remove the ball joint stud from the steering knuckle.

3. Remove nut attaching the outer tie rod to steering knuckle. Remove nut from tie rod end by holding tie rod stud with a socket while loosening and removing nut with wrench.

4. Remove outer tie rod from steering knuckle.

5. Remove cotter pin and castle nut from lower ball joint stud.

⁂ WARNING

No tool is to be inserted between the steering knuckle and the lower ball joint to separate the lower ball joint from the steering knuckle. The steering knuckle is to be separated from the ball joint only using the procedure as described below.

⁂ WARNING

When striking the steering knuckle, do not hit the heat shield covering the ball joint grease seal. Bending the heat shield against the ball joint grease seal will cause the grease seal to fail.

6. Turn steering knuckle so front of steering knuckle is facing as far outboard in the wheel opening as possible. Using a hammer, strike steering knuckle boss until steering knuckle separates from lower ball joint. When striking steering knuckle, care MUST be taken not to hit lower control arm or ball joint grease seal.

⁂ WARNING

Pulling the steering knuckle outward from the vehicle after releasing it from the ball joint, can separate driveshaft inner CV-joint, thus damaging it.

7. Remove shock absorber clevis to lower control arm bushing, nut and through-bolt. Separate clevis from lower control arm.

8. Remove nut attaching stabilizer bar link to lower control arm. When removing nut, hold stud of stabilizer bar link from turning by inserting an Allen wrench in end of stud.

9. Remove bolts attaching closest stabilizer bar bushing clamp to front suspension crossmember and body of vehicle.

10. Lower that side of stabilizer bar away from lower control arm and body of vehicle.

11. Remove nut and bolt attaching rear of lower control arm to front suspension crossmember.

12. Remove nut and bolt attaching the front of lower control arm to front suspension crossmember.

⁂ WARNING

When removing lower control arm from crossmember care must be taken to prevent hitting lower ball joint seal against steering knuckle, causing damage to the ball joint seal.

13. Remove front of lower control arm from suspension crossmember first, then remove rear of lower control arm from suspension crossmember. When removing rear of lower control arm from crossmember, keep control arm as level as possible. This will keep rear bushing from binding on crossmember making it easier to remove control arm from crossmember.

To install:

14. Position rear of lower control arm

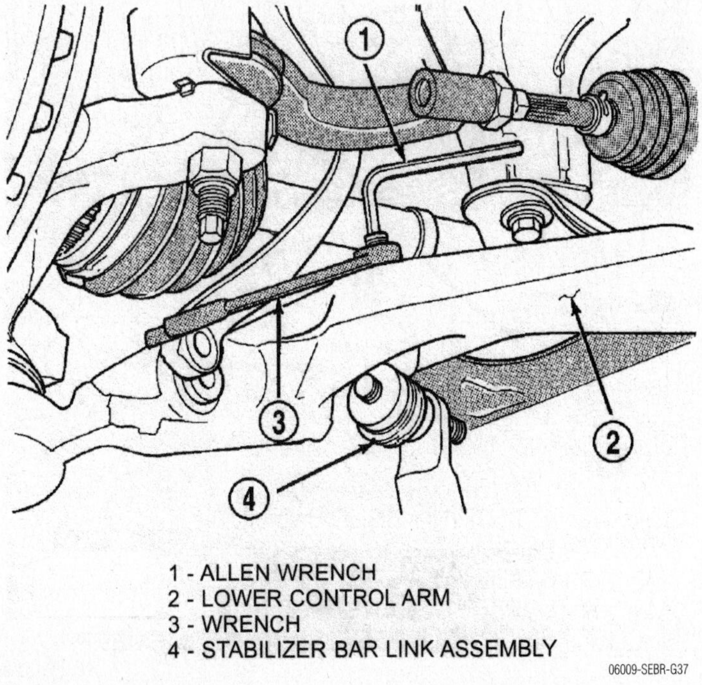

Removing nut from stabilizer link

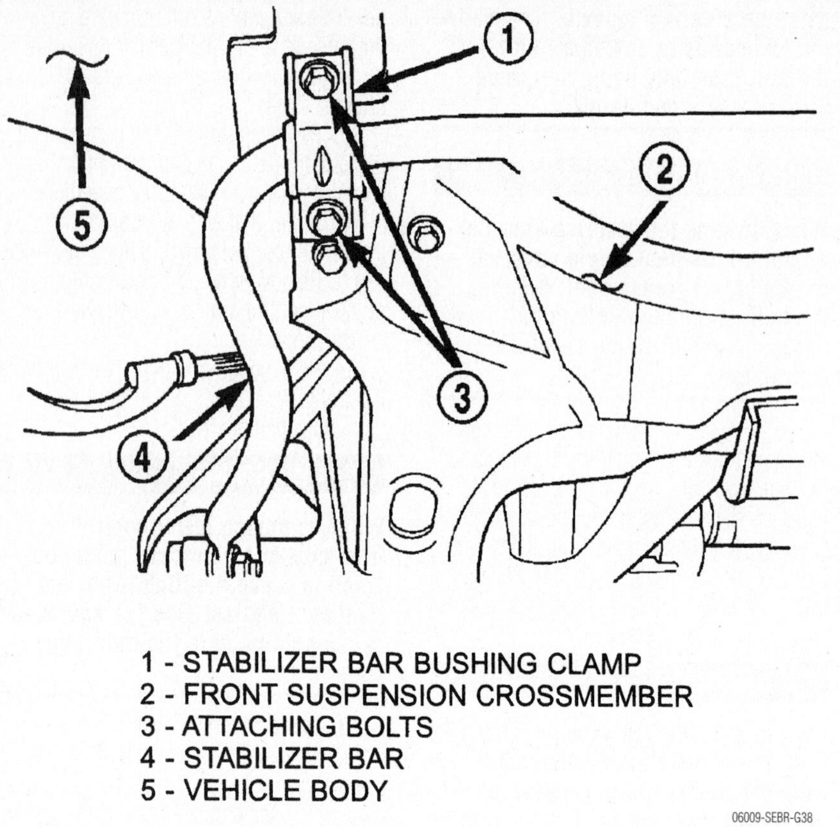

Stabilizer bar clamp attachment

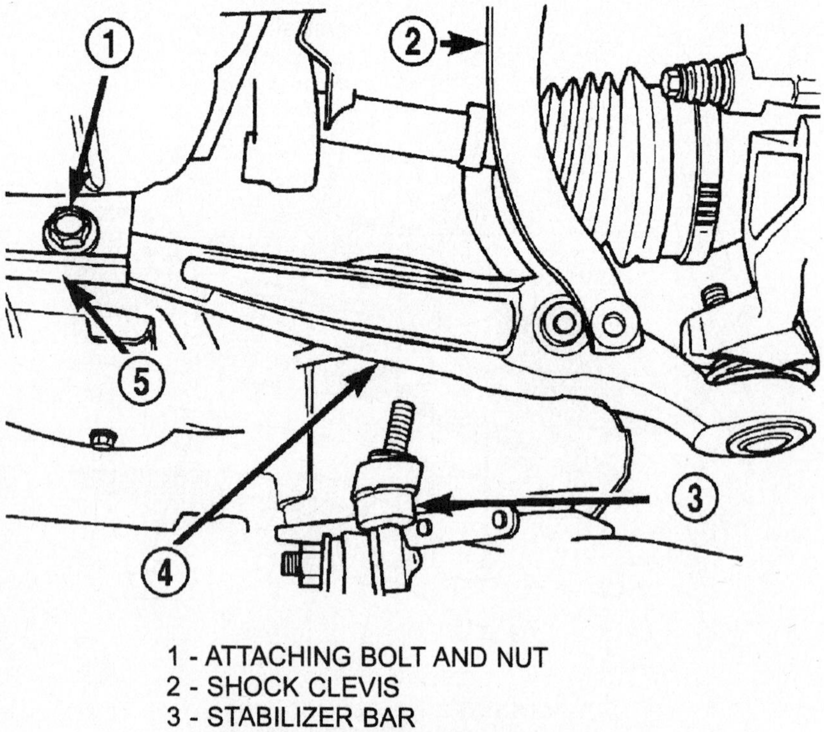

Front of lower control arm-to-crossmember attachment

into front suspension crossmember first, then install front of lower control arm in front suspension crossmember. Install bolts and nuts attaching front and rear of lower control arm to front suspension crossmember. Do not tighten front attaching bolt at this time. Tighten lower control arm rear attaching nut and bolt to 95 Nm (70 ft. lbs.) torque.

15. Install lower control arm ball joint stud into steering knuckle. Install the steering knuckle to ball joint stud castle nut. Do not tighten nut at this time.

⁂ WARNING

When installing tie rod on steering knuckle, the heat shield must be installed. If heat shield is not installed, tie rod seal boot can fail due to excessive heat from brake rotor.

16. Install outer tie rod stud into steering knuckle. Start outer tie rod-to-steering knuckle attaching nut onto tie rod stud. While holding stud of tie rod end stationary, tighten nut using a crowfoot and 11⁄32 socket. Tighten nut to 55 Nm (40 ft. lbs.) torque.

17. Tighten lower control arm ball joint castle nut to 74 Nm (55 ft. lbs.) torque. Install cotter pin.

18. Position stabilizer bar link into its lower control arm mounting hole.

19. Align stabilizer bar bushing clamp with mounting holes in front suspension crossmember and body of vehicle. Install and tighten bushing clamp mounting bolts to 61 Nm (45 ft. lbs.) torque.

20. Install and tighten stabilizer bar link-to-lower control arm nut to 105 Nm (77 ft. lbs.) torque. When tightening nut, hold stud of link from turning using an Allen wrench.

21. Align shock absorber clevis with bushing in lower control arm and install through-bolt and nut. Do not tighten nut at this time.

⁂ WARNING

When supporting lower control arm with jackstand, do not position jack stand under the ball joint cap on the lower control arm. Position stand in area of lower control arm shown.

22. Position a jack stand under lower control arm, so that when arm is lowered onto it, vehicle will be at curb height.

23. Lower vehicle with jack stand positioned under lower control arm. Continue to lower vehicle until total weight of that corner

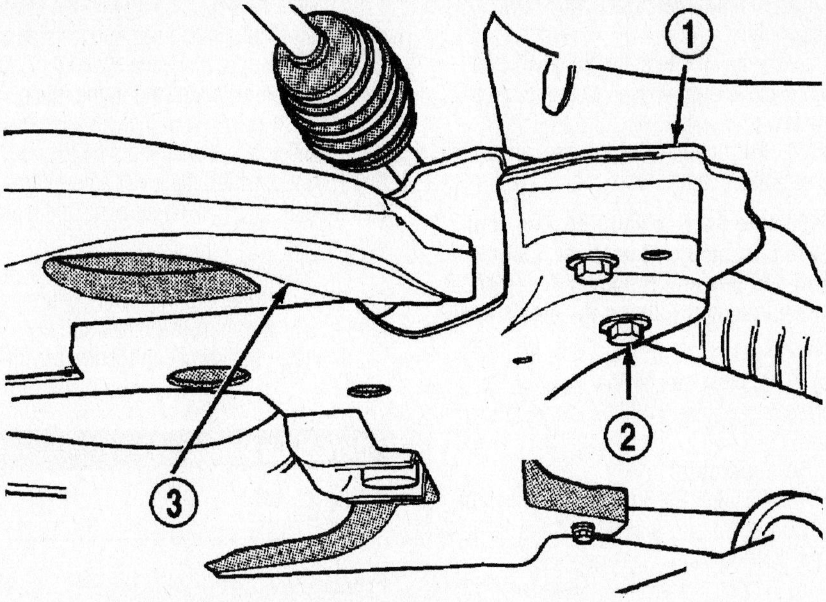

1 - FRONT SUSPENSION CROSSMEMBER
2 - LOWER CONTROL ARM ATTACHING BOLT AND NUT
3 - LOWER CONTROL ARM

06009-SEBR-G40

Rear of lower control arm-to-crossmember attachment

of vehicle is supported by jack stand and lower control arm.

⁂ WARNING

When tightening the through-bolt and nut, do not turn the bolt in the clevis. The serrations on the bolt and the hole in the clevis will be damaged.

24. With the vehicle's suspension at curb height, tighten the clevis-to-lower control arm bushing through-bolt nut to 88 Nm (65 ft. lbs.) torque.
25. Tighten front lower control arm nut and bolt to 142 Nm (104 ft. lbs.) torque.
26. Raise vehicle and remove jack stand from under lower control arm.
27. Install wheel and tire assembly. Install wheel mounting (lug) nuts and progressively tighten in proper sequence to 135 Nm (100 ft. lbs.) torque.
28. Lower vehicle.
29. Check wheel alignment specifications and set front toe to preferred specifications.

CONTROL ARM BUSHING REPLACEMENT

Front Isolator Bushing

1. Before servicing the vehicle, refer to the precautions in the beginning of this section.
2. Remove the lower control arm assembly
3. Install the Bushing Remover Tool 6602-5 and Bushing Receiver Tool MB-990799 on Special Tool C-4212-F.
4. Position the lower control arm on the assembled removal tools.

➡Be sure the Bushing Receiver Tool MB-990799 is square on the lower control arm and Bushing Remover Tool 6602-5 is positioned correctly on the isolator bushing.

5. Tighten Special Tool C-4212-F to press the bushing from the lower control arm.

To assemble:

6. Position the Bushing Installer Tool 6876 onto the screw portion of Special Tool C-4212-F.
7. Start the new bushing into the lower

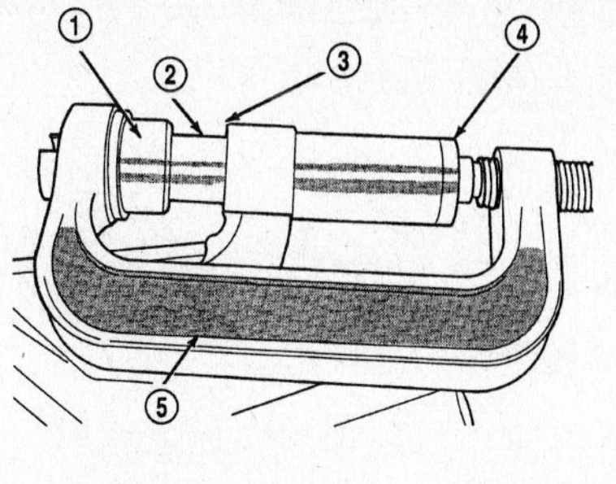

1 – SPECIAL TOOL 6876
2 – ISOLATOR BUSHING
3 – MACHINED SURFACE SIDE OF LOWER CONTROL ARM
4 – SPECIAL TOOL 6758
5 – SPECIAL TOOL C-4212-F

9306FG10

Installing the front isolator bushing to the lower control arm

control arm hole machined surface side by hand, making sure it is square with its mounting hole.

8. Assemble Special Tools 6758, 6876 and C-4212-F; then, position the lower control arm on the assembly, making sure that the tools are aligned.
9. Tighten Special Tool C-4212-F to press the bushing into the lower control arm until it is flush on the machined surface.
10. Install the lower control arm assembly.

Rear Isolator Bushing

1. Before servicing the vehicle, refer to the precautions in the beginning of this section.
2. Remove the lower control arm assembly
3. Install the Bushing Remover Tool 6756 and Bushing Receiver Tool C-4366-2 on Special Tool C-4212-F.
4. Position the lower control arm on the assembled removal tools.

➡Be sure the Bushing Receiver Tool C-4366-2 is square on the lower control arm and Bushing Remover Tool 6756 is positioned correctly on the isolator bushing.

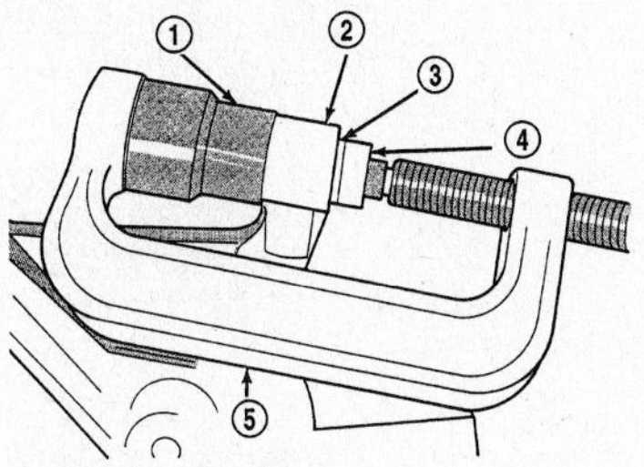

1 – SPECIAL TOOL MB-990799
2 – LOWER CONTROL ARM
3 – FRONT ISOLATOR BUSHING
4 – SPECIAL TOOL 6602-5
5 – SPECIAL TOOL C-4212-F

9306FG09

Removing the front isolator bushing from the lower control arm

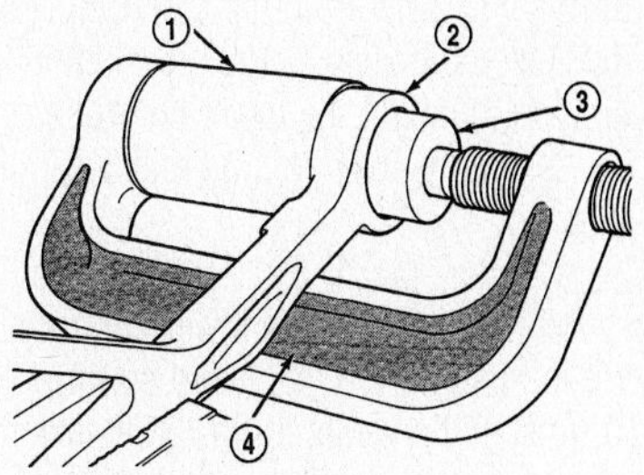

1 – SPECIAL TOOL C-4366-2
2 – LOWER CONTROL ARM
3 – SPECIAL TOOL 6756
4 – SPECIAL TOOL C-4212-F

9306FG11

Removing the rear isolator bushing from the lower control arm

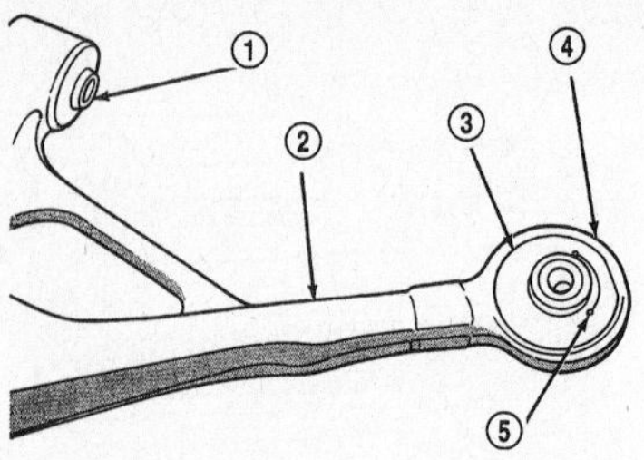

1 – FRONT ISOLATOR BUSHING
2 – LOWER CONTROL ARM
3 – REAR ISOLATOR BUSHING
4 – MACHINED SURFACE
5 – VOID IN BUSHING IN THIS DIRECTION

9306FG12

Positioning the rear isolator bushing to the lower control arm

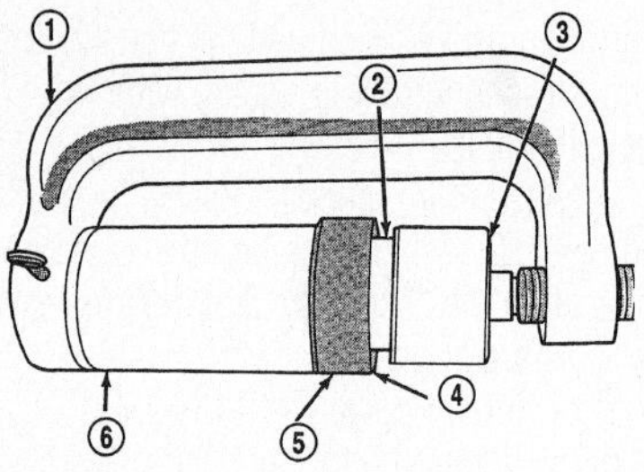

1 – SPECIAL TOOL C-4212-F
2 – REAR BUSHING
3 – SPECIAL TOOL 6760
4 – MACHINED SURFACE ON LOWER CONTROL ARM
5 – LOWER CONTROL ARM
6 – SPECIAL TOOL 6756

9306FG13

Installing the rear isolator bushing to the lower control arm

5. Tighten Special Tool C-4212-F to press the bushing from the lower control arm.

To assemble:

6. Position the Bushing Installer Tool 6760 onto the screw portion of special Tool C-4212-F.
7. Start the new bushing into the lower control arm hole's machined surface side by hand, making sure it is square with its mounting hole with the void in the rubber portion facing away from the ball joint.
8. Assemble special Tools 6760 and 6756; then, position the lower control arm on the assembly, making sure that the tools are aligned.
9. Tighten special Tool C-4212-F to press the bushing into the lower control arm until it is flush on the machined surface.
10. Install the lower control arm assembly.

Clevis Bushing

1. Before servicing the vehicle, refer to the precautions in the beginning of this section.
2. Remove the lower control arm assembly
3. Install the Bushing Remover Tool 6877 and Bushing Receiver Tool 6876 on special Tool C-4212-F.
4. Position the lower control arm on the assembled removal tools.

➡**Be sure the Bushing Receiver Tool 6876 is square on the lower control arm and Bushing Remover Tool 6877 is positioned correctly on the clevis bushing.**

5. Tighten special Tool C-4212-F to press the bushing from the lower control arm.

To assemble:

6. Position the Bushing Installer Tool 6877 onto the screw portion of special Tool C-4212-F.

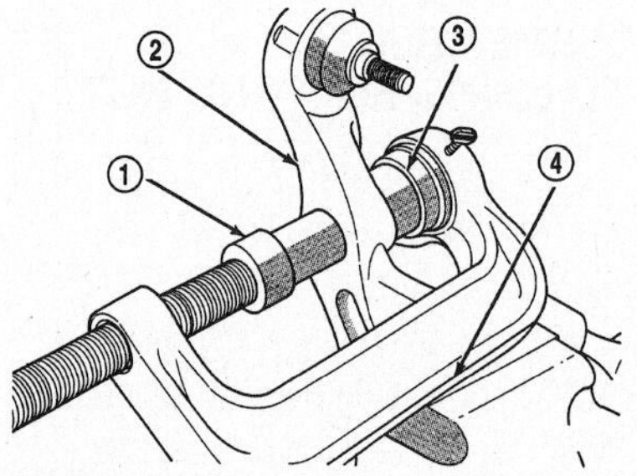

1 – SPECIAL TOOL 6877
2 – LOWER CONTROL ARM
3 – SPECIAL TOOL 6876
4 – SPECIAL TOOL C-4212-F

9306FG14

Removing the clevis bushing from the lower control arm

7. Start the new bushing into the lower control arm hole's machined surface side by hand, making sure it is square with its mounting hole with the void in the rubber portion facing away from the ball joint.
8. Assemble special Tools 6876 and 6877; then, position the lower control arm on the assembly, making sure that the tools are aligned.
9. Tighten special Tool C-4212-F to press the bushing into the lower control arm until it is flush on the machined surface.
10. Install the lower control arm assembly.

Wheel Bearings

ADJUSTMENT

Front

The front hub wheel bearing is designed for the life of the vehicle and requires no type of adjustment or periodic maintenance. The bearing is a sealed unit with the wheel hub and can only be removed and/or replaced as one unit.

Rear

The rear hub and wheel bearing assembly is designed for the life of the vehicle and requires no type of adjustment or periodic maintenance. The bearing is a sealed unit with the wheel hub and can only be removed and/or replaced as one unit.

The following procedure may be used for evaluation of bearing condition:

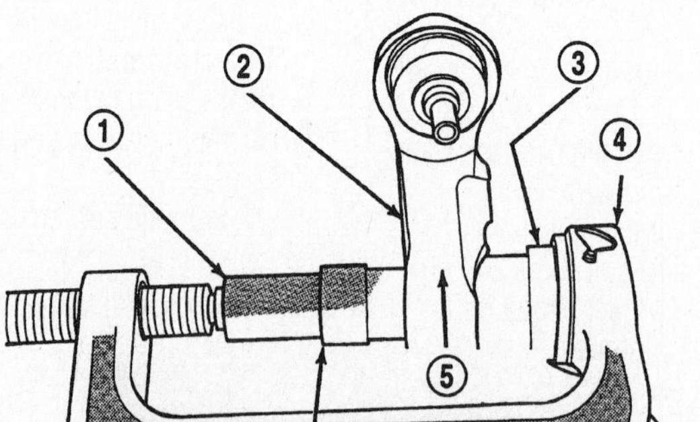

1 – SPECIAL TOOL 6877
2 – MACHINED SURFACE SIDE OF LOWER CONTROL ARM
3 – SPECIAL TOOL 6876
4 – SPECIAL TOOL C-4212-F
5 – LOWER CONTROL ARM
6 – CLEVIS BUSHING

9306FG15

Installing the clevis bushing to the lower control arm

1. Raise and safely support the vehicle.
2. Remove the rear wheels and brake drums.
3. Turn the hub flange carefully. Excessive roughness, lateral play or resistance to rotation may indicate dirt intrusion or bearing failure.
4. If the rear wheel bearings exhibit the conditions during inspection, the hub and bearing assembly should be replaced.
5. Damaged bearing seals and resulting excessive grease loss may also require bearing replacement. Moderate grease loss from the bearing is considered normal and should not require replacement of the hub and bearing assembly.

REMOVAL & INSTALLATION

Front

The front wheel bearing used on this vehicle is a bolt-in type wheel bearing.

The wheel bearing is serviced separately from the front steering knuckle and front hub assembly. Retention of the front wheel bearing into the steering knuckle is by means of 3 bolts installed from the rear of the steering knuckle. The 3 bolts attach the hub/bearing to the front surface of the steering knuckle. Removal and installation of the hub/bearing assembly from the steering knuckle must be done with the steering knuckle removed from the vehicle.

The face of the outer CV-joint has a metal bearing shield pressed on it. This design deters direct water splash on the bearing seal while allowing any water that gets in to run out the bottom of the steering knuckle. It is important to thoroughly clean the outer CV-joint and the wheel bearing area in the steering knuckle before it is assembled after servicing the front wheel bearing or driveshaft.

At no time when servicing this vehicle, can a sheet metal screw, bolt or other metal fastener be installed in the shock tower to take the place of an original plastic clip. Also, NO holes can be drilled into the front shock tower for the installation of any metal fasteners into the shock tower. Because of the minimum clearance in this area installation of metal fasteners could damage the coil spring's protective coating and lead to corrosion failure of the spring. If a plastic clip is missing, lost or broken during servicing a vehicle, replace only with the equivalent part listed in the Mopar parts catalog.

1. Before servicing the vehicle, refer to the precautions in the beginning of this section.
2. Remove or disconnect the following:
   - Front wheel
   - Steering knuckle assembly
   - 3 hub/bearing assembly-to-steering knuckle bolts
   - Hub/bearing assembly from the steering knuckle. If necessary, tap the bearing out, using a soft-faced hammer

➡The wheel bearing is transferable to a replacement steering knuckle if the bearing is found in serviceable condition.

To install:

3. Clean all parts well. Thoroughly, clean all the hub/bearing assembly mounting surfaces on the steering knuckle.

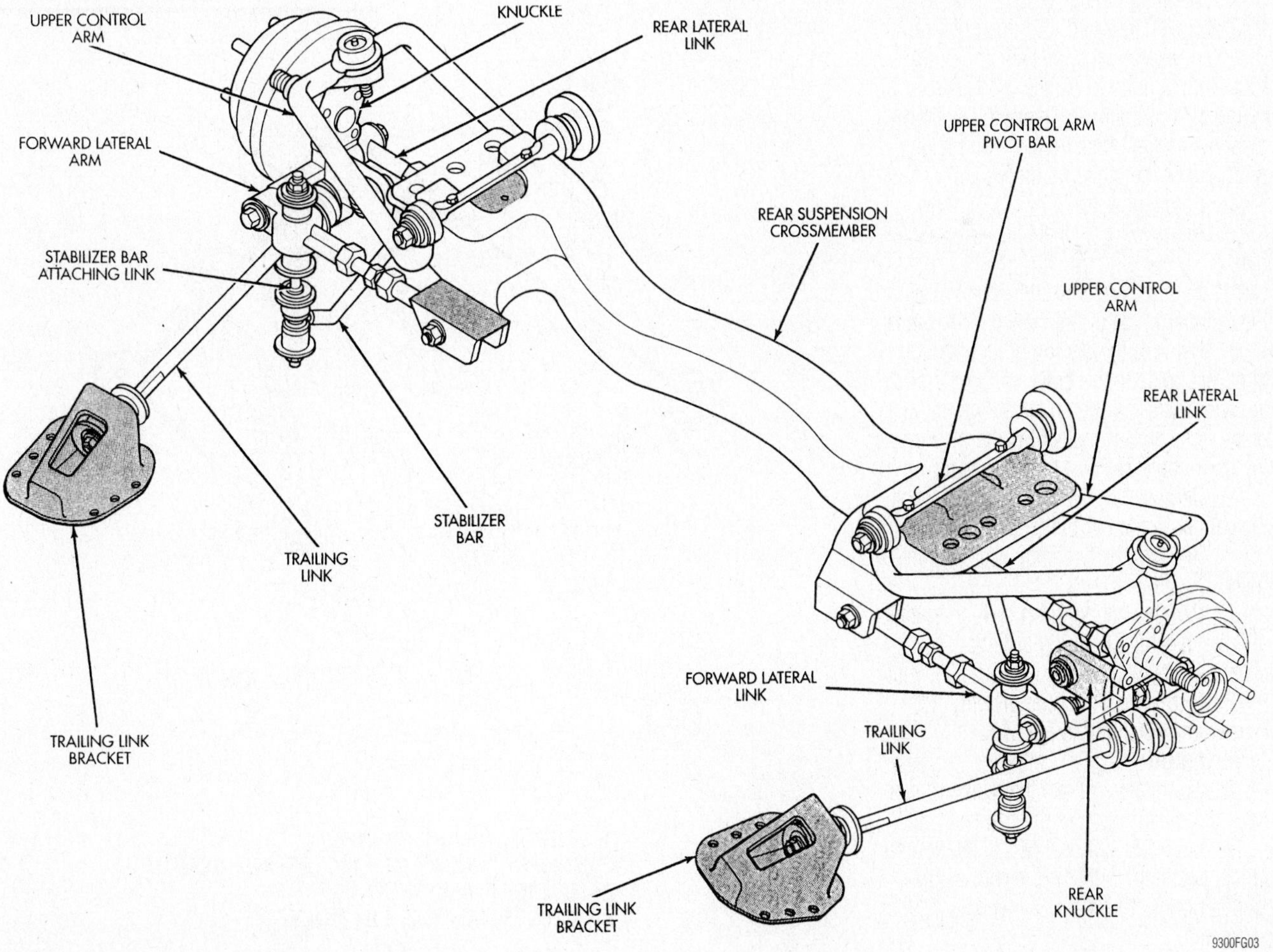

Independent rear suspension component identification

4. Install or connect the following:
   - Hub/bearing assembly and torque the bolts to 80 ft. lbs. (110 Nm)
   - Steering knuckle
   - Front wheel

Rear

All vehicles are equipped with permanently lubricated and sealed for life rear wheel bearings. There is no periodic lubrication or maintenance recommended for these units.

To evaluate the condition of the rear wheel bearings, remove the wheel and brake drum or rotor and rotate the flanged outer ring of the hub. Excessive roughness or resistance to rotation may indicate dirt intrusion or wheel bearing failure. If the rear wheel bearings exhibit these conditions during inspection, the hub and bearing assembly should be replaced. Damaged bearing seals and resulting excessive grease loss may also require bearing replacement. Moderate grease loss from the bearing is considered normal and should not require replacement of the hub and bearing assembly. If service requires removal for inspection or replacement of the rear wheel bearing and hub assembly, use the following procedure.

1. Before servicing the vehicle, refer to the precautions in the beginning of this section.
2. Remove or disconnect the following:
   - Rear wheel
   - Brake drum, if equipped with drum brakes
   - Brake caliper and rotor, if equipped with disc brakes
   - Hub dust cap
   - Hub nut and discard it
   - Hub/bearing assembly by pulling straight off the spindle

To install:

3. Install or connect the following:
   - New bearing on the spindle
   - Hub using a new nut and torque the nut to 185 ft. lbs. (250 Nm)
   - Dust cap
   - Brake drum or rotor and caliper
   - Rear wheel and torque the lug nuts to 100 ft. lbs. (135 Nm)

BRAKES

Brake Caliper

REMOVAL & INSTALLATION

Front

1. Depress the brake pedal past its first inch of travel and hold it in this position using a brake pedal depressor (holding) tool. This is done to isolate the master cylinder from the brake hydraulic system disallowing the brake fluid to completely drain out of the brake fluid reservoir.
2. Raise the vehicle.
3. Remove front wheel and tire assembly.
4. Remove the anti-rattle spring from the outboard side of the caliper and adapter.
5. Remove the banjo bolt connecting the brake hose to the brake caliper. There are two washers (one on each side of the brake hose fitting) that will come off with the banjo bolt. Discard these washers.
6. Remove the two caps in place over the caliper guide pin bolts.
7. Remove the two caliper guide pin bolts.
8. Remove caliper from caliper adapter and brake rotor. The outboard shoe will probably stay with the caliper adapter while the inboard shoe will come off with the caliper as the caliper is removed.

To install:

9. Completely retract the caliper piston back into piston bore of the caliper.
10. Lubricate both caliper adapter abutments where the shoes slide with a sufficient amount of Mopar® Silicone Grease, or equivalent.

➡If shoes are being transferred or newly installed, make note that the inboard brake shoes are stamped L or R for left or right side of the vehicle. The P-slot or void on piston insulator must be positioned upward when brake caliper is mounted.

⁂ WARNING

Use care when installing the brake caliper assembly onto the adapter, so that the seals on the caliper guide pin bushings do not get damaged by the adapter bosses.

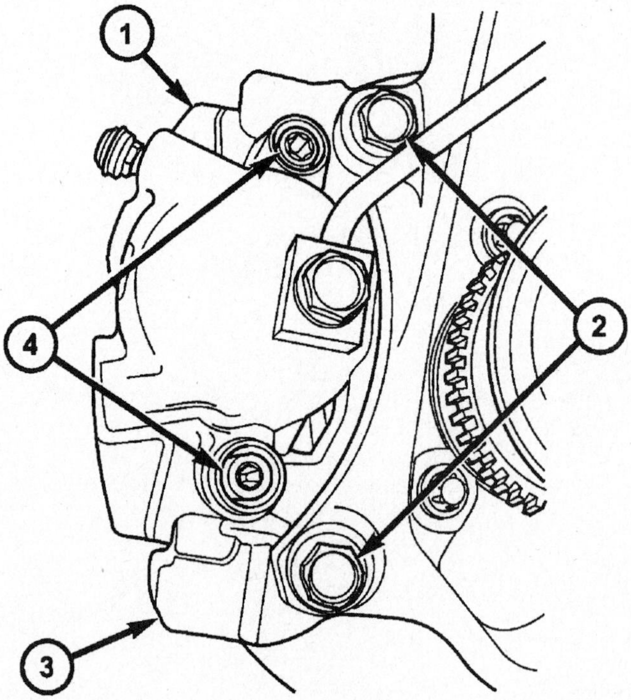

Front caliper mounting—2005 models shown

11. Carefully position the brake caliper (with inboard shoe) over the brake rotor, outboard shoe and adapter.

12. Install the caliper guide pin bolts and tighten to a torque of 35 Nm (26 ft. lbs.).

➡**Extreme caution should be taken not to cross thread the caliper guide pin bolts.**

13. Install the caps over the ends of the guide pin bolts.

14. Install the anti-rattle spring on the outboard side of the caliper. Start the clip into the holes on the caliper, then stretch the clip legs past the abutments on the caliper adapter.

⁂ WARNING

When connecting the brake hose to the caliper, install new brake hose to caliper special washers.

15. Install the brake hose on the caliper. To do this, first place one NEW special fitting washer on each side of the hose fitting, then slide the banjo bolt through the fitting. Next, thread the banjo bolt into the threaded port on the rear of the brake caliper. Tighten the banjo bolt to a torque of 35 Nm (26 ft. lbs.).

16. Install the wheel and tire assembly. Tighten the wheel mounting stud nuts in proper sequence until all nuts are torqued to half specification, then repeat the tightening sequence to the full specified torque of 135 Nm (100 ft. lbs.).

17. Lower the vehicle.

18. Remove the brake pedal depressor (holding) tool.

19. Bleed the hydraulic brake circuit to the brake caliper.

20. Road test the vehicle and make several stops to wear off any foreign material on the brakes and to seat the brake shoe linings.

Rear

1. Depress the brake pedal past its first inch of travel and hold it in this position using a brake pedal depressor (holding) tool. This is done to isolate the master cylinder from the brake hydraulic system disallowing the brake fluid to completely drain out of the brake fluid reservoir.

2. Raise vehicle.

3. Remove rear wheel and tire from vehicle.

4. Remove the banjo bolt connecting the brake hose to the brake caliper. There are two washers (one on each side of the brake hose fitting) that will come off with the banjo bolt. Discard these washers.

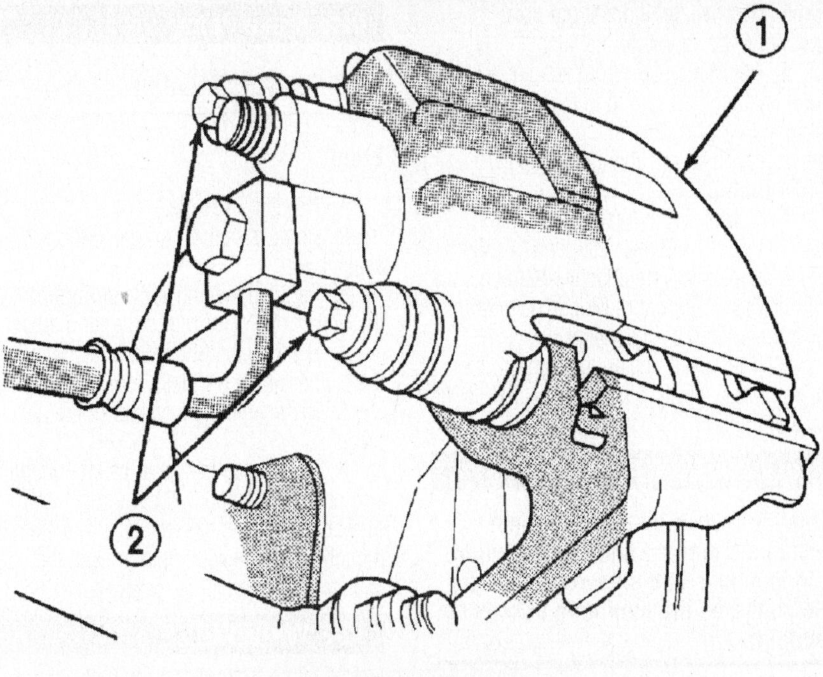

Rear caliper mounting—2005 models

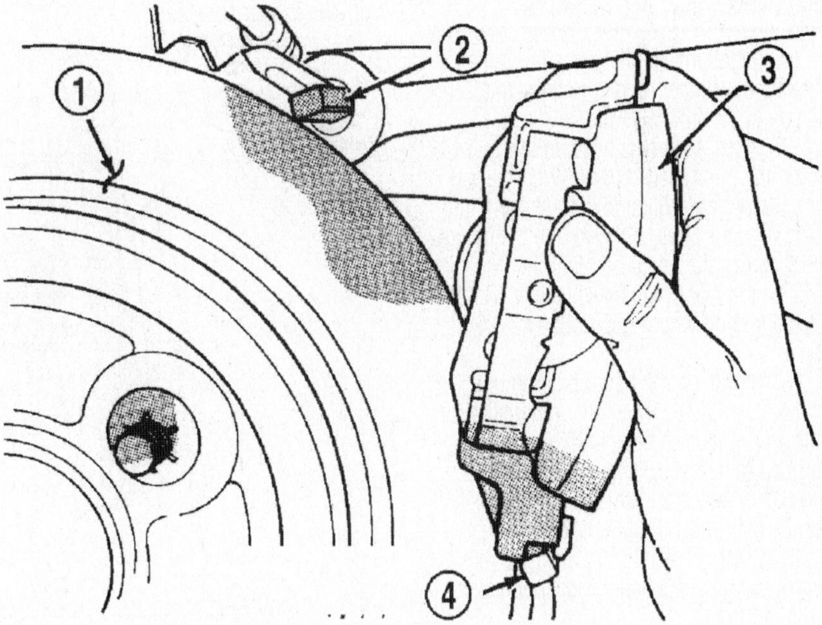

Rear caliper removal—2005 models

5. Remove the 2 guide pin bolts mounting the caliper to the adapter.

6. Remove caliper from adapter and rotor by first rotating top of caliper away from the adapter, then lifting the caliper off lower machined abutment on adapter.

To install:

7. Completely retract caliper piston back into piston bore of caliper assembly.

8. Lubricate both adapter abutments with a liberal amount of Mopar® Multipurpose Lubricant, or equivalent.

9. If removed, install the rotor on the hub making sure it is squarely seated on the face of the hub.

**** WARNING**

Use care when installing caliper assembly onto adapter so the guide pin bushings and sleeves do not get damaged by the mounting bosses on adapter.

10. Carefully lower caliper and brake shoes over rotor reversing the removal procedure. Make sure that the caliper guide pin bolts, bushings and sleeves are clear of the adapter bosses.

**** WARNING**

Extreme caution should be taken not to cross thread the caliper guide pin bolts when they are installed.

11. Install the caliper guide pin bolts. Tighten the caliper guide pin bolts to a torque of 22 Nm (16 ft. lbs.).

12. Install the brake hose on the caliper. To do this, first place one NEW special fitting washer on each side of the hose fitting, then slide the banjo bolt through the fitting. Next, thread the banjo bolt into the threaded port on the rear of the brake caliper. Tighten the banjo bolt to a torque of 35 Nm (26 ft. lbs.).

13. Install wheels and tire assembly.

14. Tighten the wheel mounting stud nuts in proper sequence until all nuts are torqued to half specification. Then repeat the tightening sequence to the full specified torque of 135 Nm (100 ft. lbs.).

15. Lower vehicle.

16. Remove the brake pedal depressor (holding) tool.

17. Bleed the hydraulic brake system.

18. Road test the vehicle and make several stops to wear off any foreign material on the brakes and to seat the brake pads.

Disc Brake Pads

REMOVAL & INSTALLATION

Front

1. Raise the vehicle.

2. Remove both front wheel and tire assemblies.

3. Remove the anti-rattle spring from the outboard side of the caliper and adapter.

4. Remove the two caps in place over the caliper guide pin bolts.

5. Remove the two caliper guide pin bolts.

6. Remove caliper from caliper adapter and brake rotor. The outboard shoe will probably stay with the caliper adapter while the inboard shoe will come off with the caliper as the caliper is removed.

**** WARNING**

Supporting weight of caliper by the flexible brake fluid hose can damage the hose.

7. Using wire or cord, hang the caliper from the front strut assembly. Support the caliper firmly to prevent weight of caliper from being supported by the brake fluid hose.

8. Remove the outboard brake shoe from the caliper adapter.

9. Pull the inboard brake shoe away from the caliper piston until the retaining clip on shoe is free from the cavity in the caliper piston.

10. Repeat the above procedure on other side of the vehicle.

To install:

➡There may be more than one lining material released. Make sure proper linings are being installed.

11. Begin on one side of the vehicle or the other.

12. Completely retract the caliper piston back into its bore in the brake caliper (This is required for caliper installation on the brake rotor with new brake shoes installed).

13. If applied, remove the protective paper from the noise suppression gasket on the rear of both the inner and outer brake shoe assemblies.

➡Inboard brake shoes are stamped L or R for left or right side of the vehicle. The P-slot or void on piston insulator must be positioned upward when brake caliper is mounted.

14. Install the new inboard brake shoe into the caliper piston by firmly pressing its retaining clip into the piston bore. Be sure the inboard brake shoe is positioned squarely against the face of the caliper piston.

15. Lubricate both adapter abutments

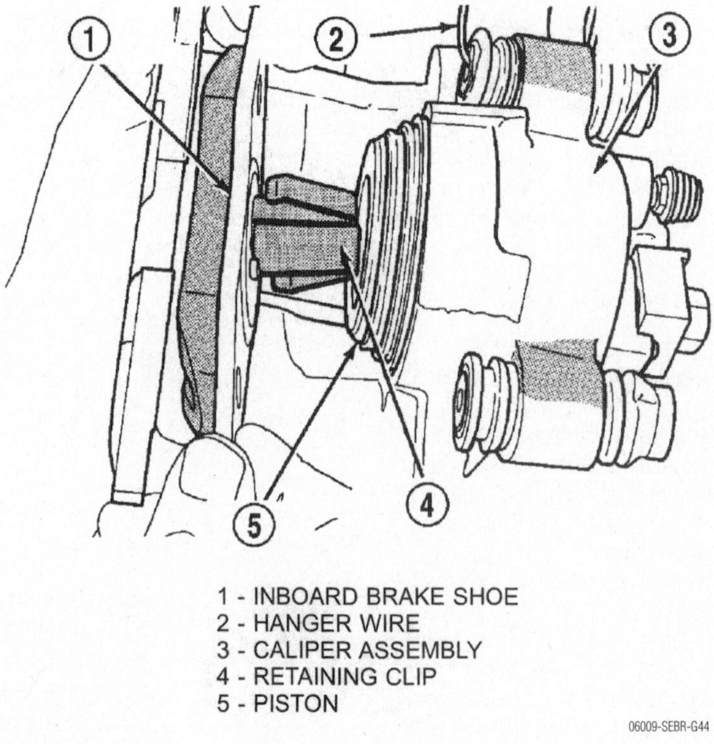

Removing inboard pad

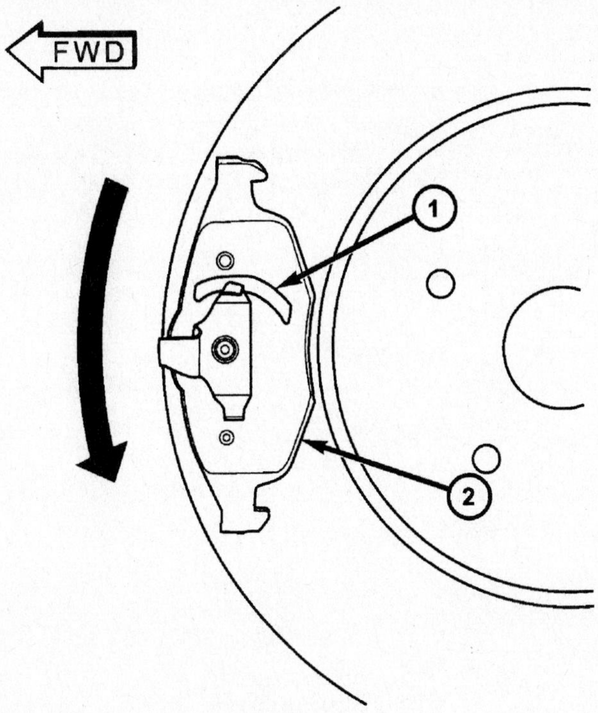

Inboard shoe orientation

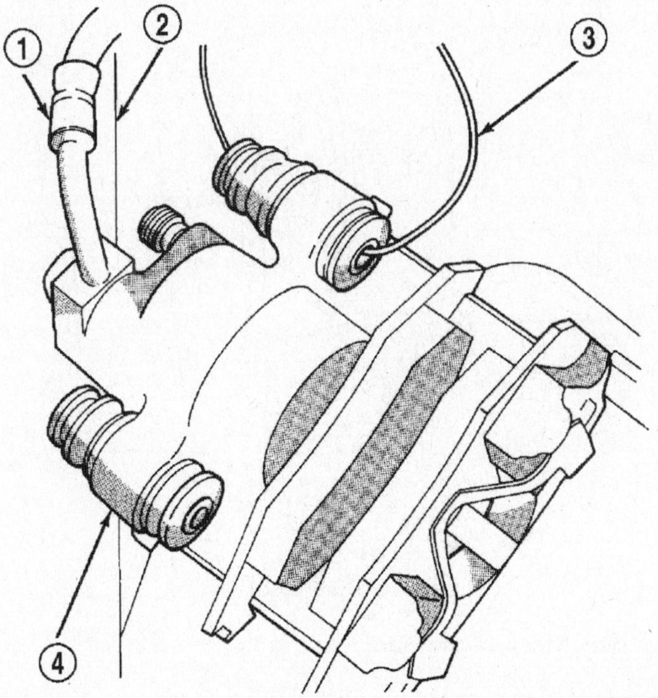

Hanging the rear caliper

where the shoes slide with a small amount of Mopar® Silicone grease, or equivalent.

16. Slide the new outboard brake shoe into the caliper adapter with the lining up against the outside of the brake rotor.

⁂ WARNING

Use care when installing the caliper assembly onto the caliper adapter, so the caliper guide pin bushings do not get damaged by the adapter bosses.

17. Carefully position the brake caliper over the brake rotor and adapter.

18. Install the caliper guide pin bolts and tighten to a torque of 35 Nm (26 ft. lbs.). Extreme caution should be taken not to cross thread the caliper guide pin bolts.

19. Install the caps over the caliper guide pin bolts.

20. Install the new caliper hold-down spring (anti-rattle clip) on the outboard side of the caliper. Start the spring into the holes on the caliper, then stretch the clip legs past the abutments on the caliper adapter.

21. Repeat the above procedure on other side of the vehicle.

22. Install the wheel and tire assemblies. Tighten the wheel mounting nuts in proper sequence until all nuts are torqued to half specification, then repeat the tightening sequence to the full specified torque of 135 Nm (100 ft. lbs.).

23. Lower vehicle.

24. Pump the brake pedal several times. This will set the shoes to the brake rotor.

25. Check and adjust brake fluid level as necessary.

26. Road test the vehicle and make several stops to wear off any foreign material on the brakes and to seat the brake shoes.

Rear

1. Raise vehicle.
2. Remove rear wheels and tires from vehicle.
3. Remove the 2 guide pin bolts mounting the caliper to the adapter.
4. Remove caliper from adapter and rotor by first rotating the top of the caliper away from adapter, and then lifting the caliper off the lower machined abutment on adapter.
5. Support caliper from rear strut to prevent weight of caliper from damaging the flexible brake hose.
6. Remove rear rotor from hub/bearing. Then inspect drum-in-hat parking brake shoes and parking brake braking surface on rotor for any signs of excessive wear or

damage. Replace parking brake shoes if required.

7. Remove outboard brake pad from caliper by prying brake pad retaining clip over raised area on caliper. Then slide brake pad down and off the caliper.

8. Pull inboard brake pad away from caliper piston, until retaining clip is free from cavity in piston.

To install:

9. Completely retract the caliper piston back into the piston bore of the caliper. This is required for caliper installation when new brake pad assemblies are installed on caliper.

10. Lubricate both adapter abutments with a liberal amount of Mopar® Multipurpose Lubricant, or equivalent.

11. Install rear rotor on hub making sure it is squarely seated on face of hub.

12. Remove protective paper from noise suppression gasket on both inner and outer brake pad assemblies (if equipped).

13. Install new inboard brake pad into caliper piston by firmly pressing it into bore of piston using thumbs. Be sure inboard brake shoe is positioned squarely against face of piston.

14. Slide new outboard brake pad onto the caliper. Be sure retaining clip is squarely seated in the depressed areas on the caliper.

**** WARNING**

Use care when installing caliper assembly onto adapter, so the guide pin bushings and sleeves do not get damaged by the mounting bosses on adapter.

15. Carefully lower caliper and brake shoes over rotor reversing the required removal procedure. Make sure that caliper guide pin bolts, bushings and sleeves are clear of the adapter bosses.

**** WARNING**

Extreme caution should be taken not to cross thread the caliper guide pin bolts when they are installed.

16. Install caliper guide pin bolts into adapter. Tighten the guide pin bolts to a torque of 22 Nm (16 ft. lbs.).

17. Install the wheels and tires.

18. Tighten the wheel mounting stud nuts in proper sequence until all nuts are torqued to half specification. Then repeat the tightening sequence to the full specified torque of 135 Nm (100 ft. lbs.).

19. Lower vehicle.

20. Check brake fluid level.

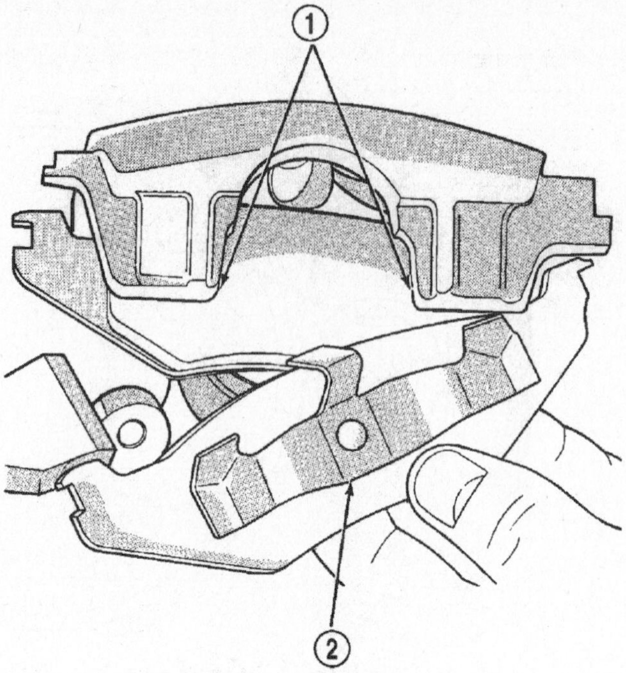

Removing outboard brake pad—rear brakes

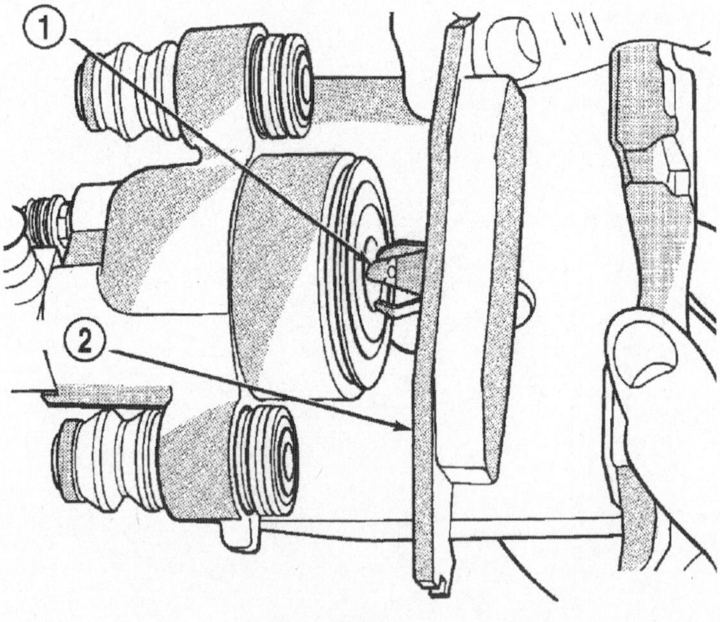

Removing inboard brake pad—rear brakes

Brake Drums

REMOVAL & INSTALLATION

1. Raise vehicle.
2. Remove rear tire and wheel assembly.
3. Slide brake drum off mounted wheel studs of hub and remove from vehicle. If drum does not come off, further brake clearance can be obtained by backing off brake adjuster screw using following steps:
   a. Fabricate pawl release tool like one

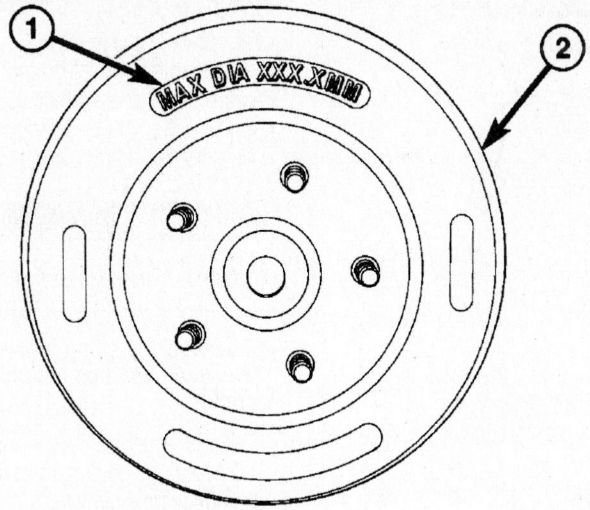

1 - MAXIMUM DIAMETER IDENTIFICATION 230.6 MM (9.079 IN.)
2 - BRAKE DRUM

06009-SEBR-G51

Drum maximum diameter stamping

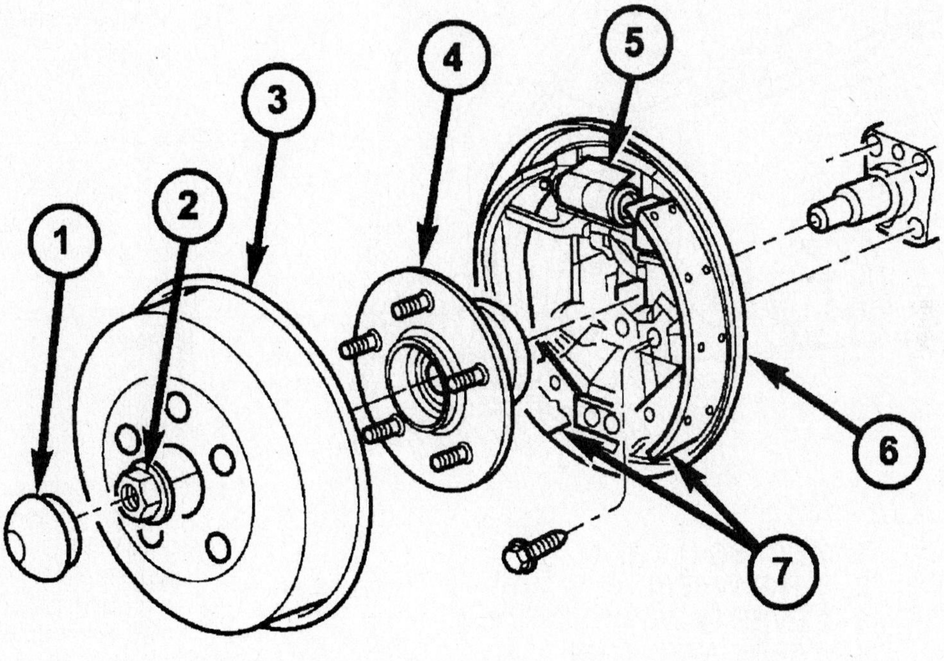

1 - DUST CAP
2 - NUT
3 - DRUM
4 - HUB AND BEARING
5 - WHEEL CYLINDER
6 - SUPPORT PLATE
7 - BRAKE SHOES

06009-SEBR-G52

Brake drum removal

shown using a coat hanger or heavy mechanic's wire.

b. Remove plug from rear of support plate below wheel cylinder.

c. Insert fabricated tool through access hole in support plate, under adjuster, against lever pawl. Pawl is attached to and pivots from leading shoe. When inserting tool through hole, directing tool somewhat forward (toward leading shoe) assures better contact with pawl.

d. While pushing on pawl with fabricated tool to disengage it from adjuster star-wheel teeth, rotate star-wheel downward to back off adjustment using a small screwdriver or brake adjuster tool.

e. Once adjuster screw is backed off a sufficient amount, drum should slide off wheel studs.

f. Remove tools and reinstall plug in support plate.

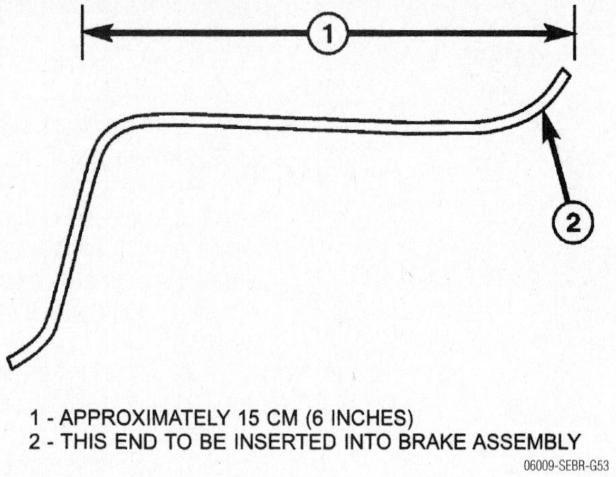

Fabricated tool

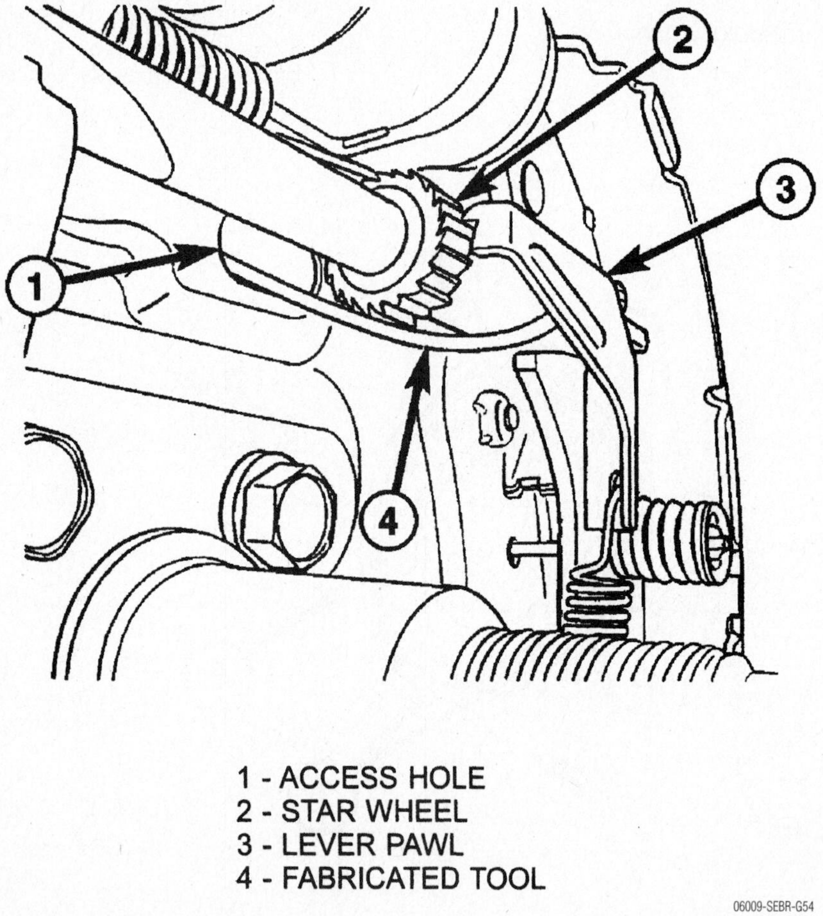

Fabricated tool positioned against pawl

To install:

➡**Before installing drum, inspect brake shoe linings for wear, alignment, and contamination. Repair or replace as necessary. If rust or any foreign material is present on hub, drum or wheel mating surfaces, wet wire brush these areas to remove prior to assembly of parts.**

4. Properly remove any buildup formed along outer edge of drum's machined braking surface.
5. Adjust brake shoes to drum diameter using a brake shoe gauge.
6. Install brake drum on rear hub and bearing.
7. Install tire and wheel assembly. Tighten wheel mounting nuts to 135 Nm (100 ft. lbs.) torque.
8. Lower vehicle.
9. Road test vehicle stopping in both forward and reverse directions. Automatic adjuster will continue to adjust brakes as necessary during road test.

Brake Shoes

REMOVAL & INSTALLATION

➡**Make sure parking brake is in "released" position before raising vehicle.**

1. Raise vehicle.
2. Remove both rear tire and wheel assemblies from vehicle.
3. Remove brake drums from both sides of vehicle.
4. Remove hub and bearing dust cap.
5. Remove hub and bearing retaining nut, then slide hub and bearing off spindle.
6. Compress cable return spring, then remove parking brake cable from parking brake lever.
7. Compress and remove shoe hold-down springs.
8. Remove lower return spring.
9. Remove both shoes and remaining parts as an assembly from support plate.
10. Remove adjuster spring from leading shoe and lever pawl.
11. Remove lever pawl from pivot on leading shoe and slide out from under adjuster.
12. Remove adjuster and upper spring.

To install:

13. Lubricate six shoe contact areas on support plate and anchor using Mopar® Brake Lubricant or equivalent.
14. Before installing shoes on vehicle, perform the following on the bench:
    a. Place one leading and one trailing shoe on bench (trailing shoe has parking brake lever attached to it).
    b. Install adjuster and upper spring.
    c. Slide lever pawl under adjuster and onto pivot on leading shoe.
    d. Install adjuster spring between leading shoe and lever pawl.
15. Install pre-assembled brake shoe assembly on brake support plate.
16. Install lower return spring.
17. Install both shoe hold-down pins from rear, through support plate and shoes. Compress and install shoe hold-down springs on pins.
18. Compress parking brake cable return spring, then install cable on parking brake

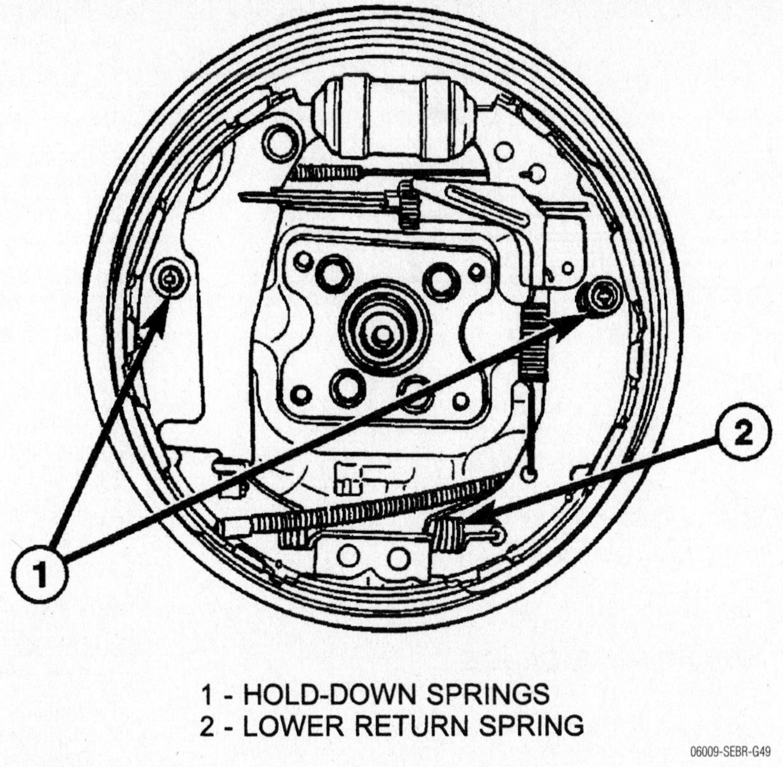

Hold-down and lower return springs

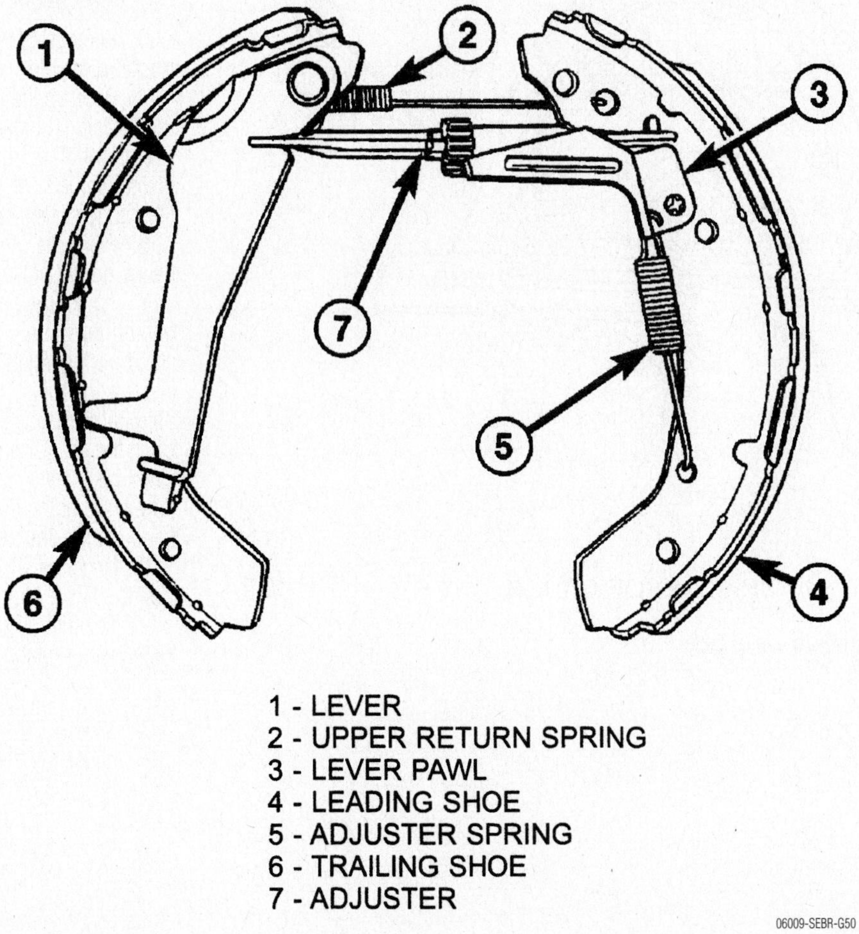

Brake shoes assembled—right side shown

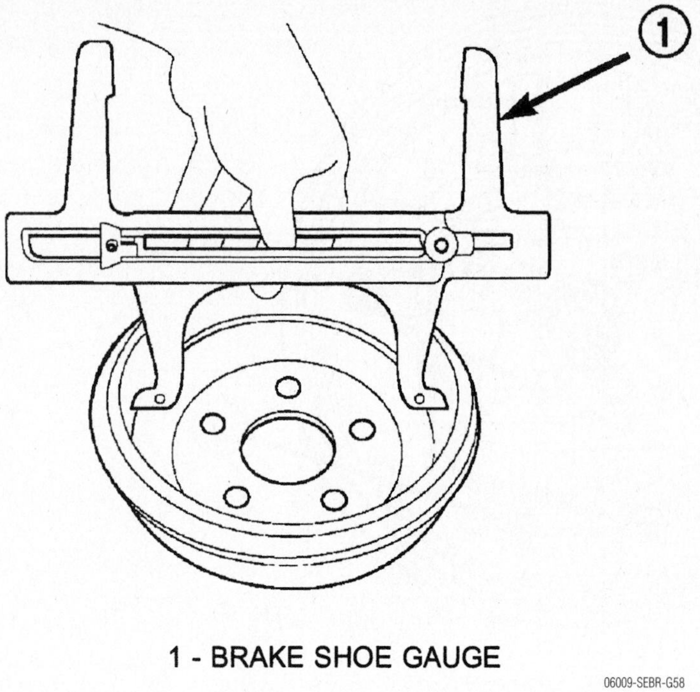

Brake drum inner diameter

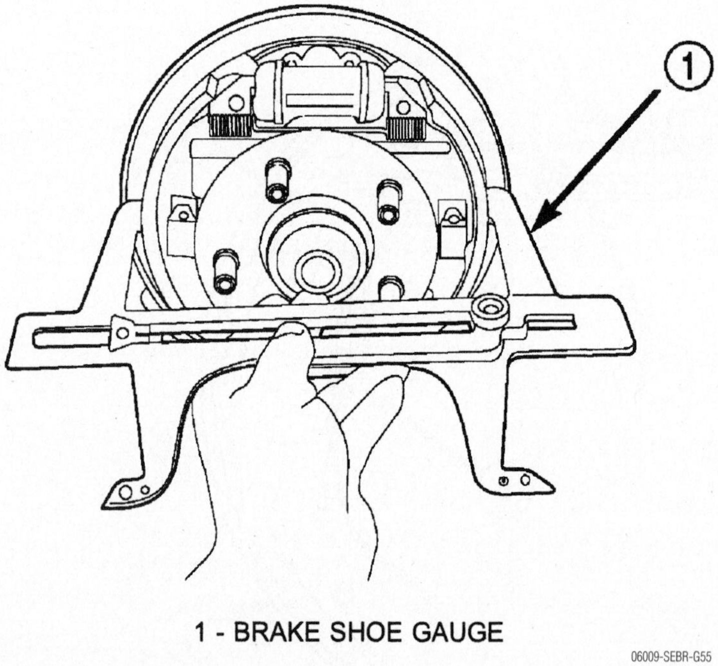

Measuring brake shoes at their widest point

lever. Release spring guiding it beneath retaining tab on lever.

19. Insert cable into cable guide attached to anchor retaining plate.
20. Slide hub and bearing onto spindle. Install retaining nut and tighten to 250 Nm (185 ft. lbs.) torque.
21. Install hub and bearing dust cap.
22. Adjust brake shoes to drum diameter using brake shoe gauge.
23. Install brake drum.
24. Install both tire and wheel assemblies. Tighten wheel mounting nuts to 135 Nm (100 ft. lbs.) torque.
25. Slowly rotate wheels and verify that brake drum lightly drags on shoes.
26. Lower vehicle.
27. Road test vehicle stopping in both forward and reverse directions. Automatic adjuster will continue to adjust brakes as necessary during road test.

ADJUSTMENT

1. Verify the parking brake lever is in the fully released position.
2. Raise the vehicle.

➡Perform the following steps on each rear drum brake assembly.

3. Remove the tire and wheel assembly from the vehicle.
4. Remove the brake drum.
5. Using a brake shoe gauge, Special Tool C-3919, or equivalent, measure the inside diameter of the brake drum. Tighten the gauge set-screw at this measurement.
6. Place the other side of the brake shoe gauge on the brake shoes as shown.
7. Adjust the shoe diameter to the setting on the gauge. To adjust the shoe diameter, turn the adjuster screw starwheel using a screwdriver inserted through the adjusting hole in the rear of the shoe support plate. Once the tip of the screwdriver contacts the star-wheel teeth, move the handle of tool downward using the support plate as a pivot to adjust the shoes outward.
8. Once the shoe diameter is set, reinstall the brake drum.
9. Turn the drum. A slight drag should be felt while rotating the drum.

CHRYSLER AND DODGE

13

Sebring Coupe • Stratus Coupe

SPECIFICATIONS AND MAINTENANCE CHARTS

ENGINE AND VEHICLE IDENTIFICATION

Engine							Model Year	
Code ①	Liters (cc)	Cu. In.	Cyl.	Fuel Sys.	Engine Type	Eng. Mfg.	Code ②	Year
G	2.4 (2350)	143	4	MFI	SOHC	Mitsubishi	2	2002
H	3.0 (2972)	181	6	MFI	DOHC	Mitsubishi	3	2003
							4	2004
							5	2005
							6	2006

MFI: Multi-port Fuel Injection

DOHC: Double Overhead Camshaft

SOHC: Single Overhead Camshaft

① 8th position of the Vehicle Identification Number (VIN).

② 10th position of VIN.

06009-COUP-C01

GENERAL ENGINE SPECIFICATIONS

All measurements are given in inches.

Year	Model	Engine Displacement Liters	Engine Series VIN	Net Horsepower @ rpm	Net Torque @ rpm (ft. lbs.)	Bore x Stroke (in.)	Com-pression Ratio	Oil Pressure @ rpm
2002	Sebring Coupe	2.4	G	140@5500	158@4000	3.40 x 3.94	9.5:1	43-100@3500
		3.0	H	200@5500	205@4000	3.59 x 2.99	9.0:1	43-100@3500
	Stratus Coupe	2.4	G	140@5500	158@4000	3.40 x 3.94	9.5:1	43-100@3500
		3.0	H	200@5500	205@4000	3.59 x 2.99	9.0:1	43-100@3500
2003	Sebring Coupe	2.4	G	140@5500	158@4000	3.40 x 3.94	9.5:1	43-100@3500
		3.0	H	200@5500	205@4000	3.59 x 2.99	9.0:1	43-100@3500
	Stratus Coupe	2.4	G	140@5500	158@4000	3.40 x 3.94	9.5:1	43-100@3500
		3.0	H	200@5500	205@4000	3.59 x 2.99	9.0:1	43-100@3500
2004	Sebring Coupe	2.4	G	140@5500	158@4000	3.40 x 3.94	9.5:1	43-100@3500
		3.0	H	200@5500	205@4000	3.59 x 2.99	9.0:1	43-100@3500
	Stratus Coupe	2.4	G	140@5500	158@4000	3.40 x 3.94	9.5:1	43-100@3500
		3.0	H	200@5500	205@4000	3.59 x 2.99	9.0:1	43-100@3500
2005	Sebring Coupe	2.4	G	140@5500	158@4000	3.40 x 3.94	9.5:1	43-100@3500
		3.0	H	200@5500	205@4000	3.59 x 2.99	9.0:1	43-100@3500
	Stratus Coupe	2.4	G	140@5500	158@4000	3.40 x 3.94	9.5:1	43-100@3500
		3.0	H	200@5500	205@4000	3.59 x 2.99	9.0:1	43-100@3500

06009-COUP-C02

ENGINE TUNE-UP SPECIFICATIONS

Year	Engine Displacement Liters	Engine VIN	Spark Plug Gap (in.)	Ignition Timing (deg.)	Fuel Pump (psi)	Idle Speed (rpm) MT	Idle Speed (rpm) AT	Valve Clearance In.	Valve Clearance Ex.
2002	2.4	G	0.039-0.043	①	47-50	700	700	HYD	HYD
	3.0	H	0.039-0.043	①	47-50	700	700	HYD	HYD
2003	2.4	G	0.039-0.043	①	47-50	700	700	HYD	HYD
	3.0	H	0.039-0.043	①	47-50	700	700	HYD	HYD
2004	2.4	G	0.039-0.043	①	47-50	700	700	HYD	HYD
	3.0	H	0.028-0.031	①	47-50	700	700	HYD	HYD
2005	2.4	G	0.039-0.043	①	49-51	700	700	HYD	HYD
	3.0	H	0.028-0.031	①	49-51	700	700	HYD	HYD

NOTE: The Vehicle Emission Control Information label often reflects specification changes made during production.
The label figures must be used if they differ from those in this chart.

HYD: Hydraulic

① Basic ignition timing is not adjustable.

06009-COUP-C03

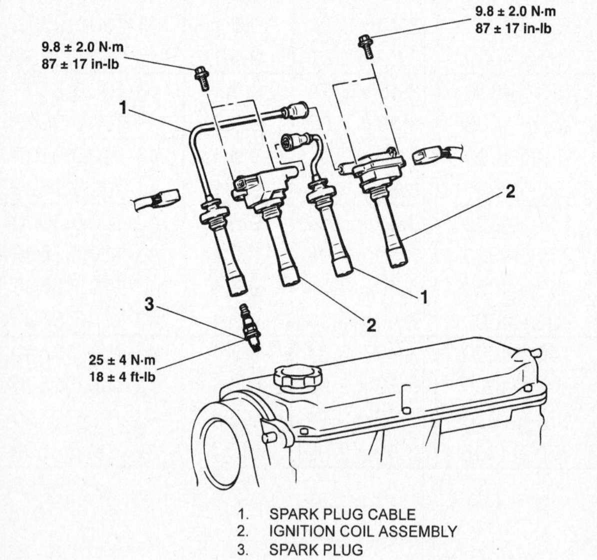

1. SPARK PLUG CABLE
2. IGNITION COIL ASSEMBLY
3. SPARK PLUG

06009-COUP-G01

2.4L engine
Firing order: 1-3-4-2
Distributorless ignition. Coil on 2 and 4

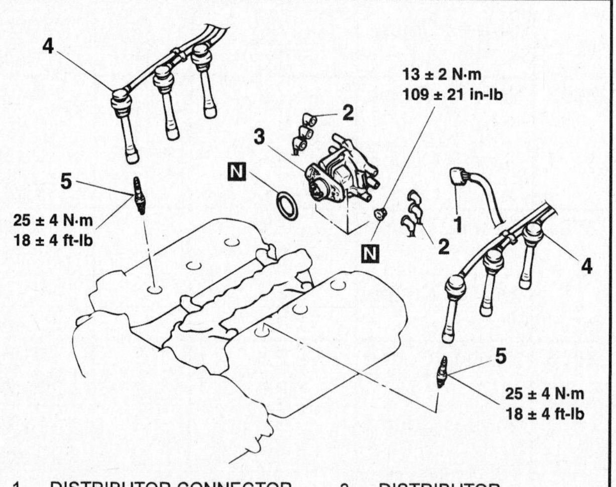

1. DISTRIBUTOR CONNECTOR
2. SPARK PLUG CABLE CONNECTION
3. DISTRIBUTOR
4. SPARK PLUG CABLE
5. SPARK PLUG

06009-COUP-G02

3.0L engine
Firing order: 1-2-3-4-5-6
Distributor ignition. Coil in distributor

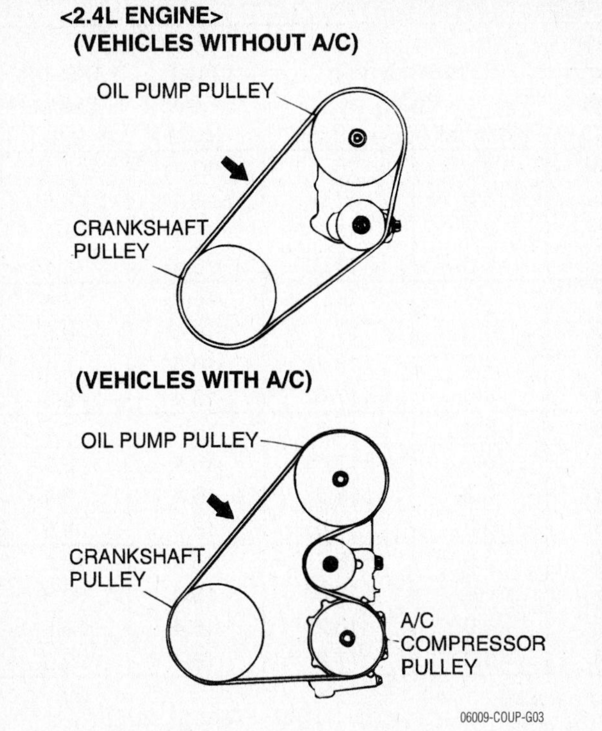

Accessory drive belt routing, part 1—2.4L engine

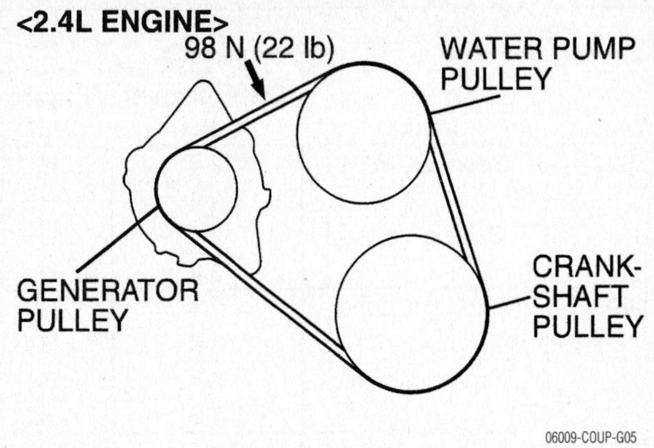

Accessory drive belt routing, part 2—2.4L engine

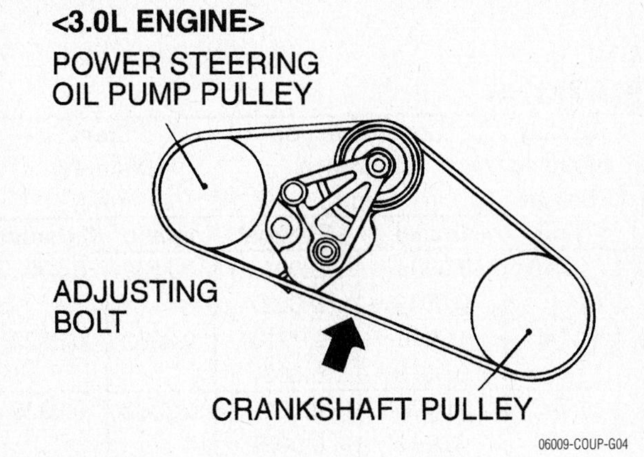

Accessory drive belt routing, part 1—3.0L engine

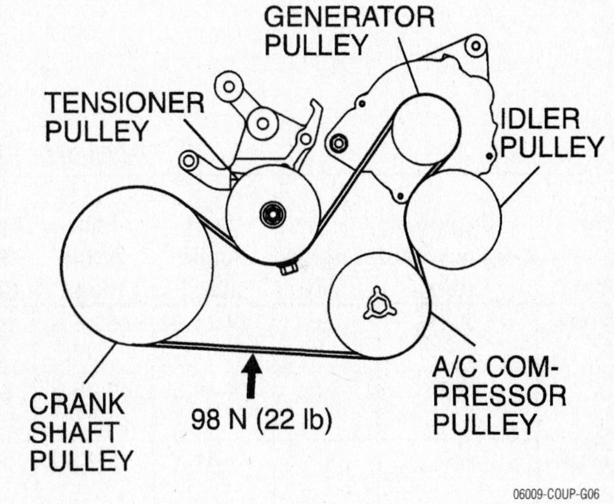

Accessory drive belt routing, part 2—3.0L engine

CAPACITIES

Year	Model	Engine Displacement Liters	Engine VIN	Engine Oil with Filter (qts.)	Transmission (pts.) Manual	Transmission (pts.) Auto.①	Fuel Tank (gal.)	Cooling System (qts.)
2002	Sebring Coupe	2.4	G	4.5	4.6	16.2	16.4	7.4
		3.0	H	4.5	6.0	17.0	16.4	8.5
	Stratus Coupe	2.4	G	4.5	4.6	16.2	16.4	7.4
		3.0	H	4.5	6.0	17.0	16.4	8.5
2003	Sebring Coupe	2.4	G	4.5	4.6	16.2	16.4	7.4
		3.0	H	4.5	6.0	17.0	16.4	8.5
	Stratus Coupe	2.4	G	4.5	4.6	16.2	16.4	7.4
		3.0	H	4.5	6.0	17.0	16.4	8.5
2004	Sebring Coupe	2.4	G	4.8	4.6	16.2	16.4	7.4
		3.0	H	4.8	6.0	17.0	16.4	8.5
	Stratus Coupe	2.4	G	4.8	4.6	16.2	16.4	7.4
		3.0	H	4.8	6.0	17.0	16.4	8.5
2005	Sebring Coupe	2.4	G	4.5	4.6	16.2	16.4	8.5
		3.0	H	4.5	6.0	17.8	16.4	9.6
	Stratus Coupe	2.4	G	4.5	4.6	16.2	16.4	8.5
		3.0	H	4.5	6.0	17.8	16.4	9.6

NOTE: All capacities are approximate. Add fluid gradually and ensure a proper fluid level is obtained.

① Overhaul fill capacity with torque converter empty

06009-COUP-C04

VALVE SPECIFICATIONS

Year	Engine Displacement Liters	Engine VIN	Seat Angle (deg.)	Face Angle (deg.)	Spring Test Pressure (lbs. @ in.)	Spring Installed Height (in.)	Stem-to-Guide Clearance (in.) Intake	Stem-to-Guide Clearance (in.) Exhaust	Stem Diameter (in.) Intake	Stem Diameter (in.) Exhaust
2002	2.4	G	44-44.5	45-45.5	60@1.740	1.740	0.0008-0.0019	0.0012 0.0027	0.2362	0.2323
	3.0	H	44-44.5	45-45.5	60@1.740	1.740	0.0008-0.0019	0.0016-0.0027	0.2362	0.2323
2003	2.4	G	44-44.5	45-45.5	60@1.740	1.740	0.0008-0.0019	0.0012 0.0027	0.2362	0.2323
	3.0	H	44-44.5	45-45.5	60@1.740	1.740	0.0008-0.0019	0.0016-0.0027	0.2362	0.2323
2004	2.4	G	44-44.5	45-45.5	60@1.740	1.740	0.0008-0.0019	0.0012 0.0027	0.2362	0.2323
	3.0	H	44-44.5	45-45.5	60@1.740	1.740	0.0008-0.0019	0.0016-0.0027	0.2362	0.2323
2005	2.4	G	44-44.5	45-45.5	44@1.740	1.740	0.0008-0.0019	0.0012 0.0027	0.2362	0.2323
	3.0	H	44-44.5	45-45.5	44@1.740	1.740	0.0008-0.0019	0.0016-0.0027	0.2362	0.2323

06009-COUP-C05

CRANKSHAFT AND CONNECTING ROD SPECIFICATIONS

All measurements are given in inches.

Year	Engine Displacement Liters	Engine VIN	Crankshaft Main Brg. Journal Dia.	Crankshaft Main Brg. Oil Clearance	Crankshaft Shaft End-play	Crankshaft Thrust on No.	Connecting Rod Journal Diameter	Connecting Rod Oil Clearance	Connecting Rod Side Clearance
2002	2.4	G	2.2441	0.0008-0.0015	0.0020-0.0090	3	1.7716	0.0008-0.0020	0.0040-0.0090
	3.0	H	2.3622	0.0008-0.0015	0.0020-0.0090	3	1.9685	0.0008-0.0019	0.0030-0.0090
2003	2.4	G	2.2441	0.0008-0.0015	0.0020-0.0090	3	1.7716	0.0008-0.0020	0.0040-0.0090
	3.0	H	2.3622	0.0008-0.0015	0.0020-0.0090	3	1.9685	0.0008-0.0019	0.0030-0.0090
2004	2.4	G	2.2441	0.0008-0.0015	0.0020-0.0090	3	1.7716	0.0008-0.0020	0.0040-0.0090
	3.0	H	2.3622	0.0008-0.0015	0.0020-0.0090	3	1.9685	0.0008-0.0019	0.0030-0.0090
2005	2.4	G	2.2441	0.0008-0.0015	0.0020-0.0090	3	1.7716	0.0008-0.0019	0.0040-0.0090
	3.0	H	2.3622	0.0008-0.0015	0.0020-0.0090	3	1.9685	0.0008-0.0019	0.0030-0.0090

06009-COUP-C06

PISTON AND RING SPECIFICATIONS

All measurements are given in inches.

Year	Engine Displacement Liters	Engine VIN	Piston Clearance	Ring Gap Top Compression	Ring Gap Bottom Compression	Ring Gap Oil Control	Ring Side Clearance Top Compression	Ring Side Clearance Bottom Compression	Ring Side Clearance Oil Control
2002	2.4	G	0.0008-0.0015	0.010-0.013	0.016-0.021	0.004-0.015	0.0008-0.0023	0.0012-0.0023	NA
	3.0	H	0.0008-0.0015	0.012-0.017	0.018-0.023	0.008-0.023	0.0012-0.0027	0.0008-0.0023	NA
2003	2.4	G	0.0008-0.0015	0.010-0.013	0.016-0.021	0.004-0.015	0.0008-0.0023	0.0012-0.0023	NA
	3.0	H	0.0008-0.0015	0.012-0.017	0.018-0.023	0.008-0.023	0.0012-0.0027	0.0008-0.0023	NA
2004	2.4	G	0.0008-0.0015	0.010-0.013	0.016-0.021	0.004-0.015	0.0008-0.0023	0.0012-0.0023	NA
	3.0	H	0.0008-0.0015	0.012-0.017	0.018-0.023	0.008-0.023	0.0012-0.0027	0.0008-0.0023	NA
2005	2.4	G	0.0008-0.0015	0.010-0.013	0.016-0.021	0.004-0.015	0.0008-0.0023	0.0012-0.0023	NA
	3.0	H	0.0008-0.0015	0.012-0.017	0.018-0.023	0.008-0.023	0.0012-0.0027	0.0008-0.0023	NA

NA: Information not available

06009-COUP-C07

TORQUE SPECIFICATIONS

All readings in ft. lbs.

Year	Engine Displacement Liters	Engine VIN	Cylinder Head Bolts	Main Bearing Bolts	Rod Bearing Bolts	Crankshaft Damper Bolts	Flywheel Bolts	Manifold Intake	Manifold Exhaust	Spark Plugs	Oil Pan Drain Plug
2002	2.4	G	①	18 ②	14 ②	87	98	14	⑥	18	29
	3.0	H	③	69	38	134	55	④	33	18	29
2003	2.4	G	①	18 ②	14 ②	87	98	14	⑥	18	29
	3.0	H	③	69	38	134	55	④	33	18	29
2004	2.4	G	①	18 ②	14 ②	87	98	14	⑥	18	29
	3.0	H	③	69	38	134	55	④	33	18	29
2005	2.4	G	①	18 ②	14 ②	123 ⑤	98	15	⑥	18	29
	3.0	H	③	69	20 ②	137	55	⑦	33	18	29

① Step 1: 55-61 ft. lbs.
Step 2: Loosen all bolts
Step 3: 14-16 ft. lbs.
Step 4: 1/4 turn
Step 5: 1/4 turn

② Plus 1/4 turn

③ Step 1: 80 ft. lbs.
Step 2: Loosen all bolts
Step 3: 80 ft. lbs.

④ Step 1: Right bank nuts to 56 inch lbs.
Step 2: Left bank nuts to 16 ft. lbs.
Step 3: Right bank nuts to 16 ft. lbs.
Step 4: Left bank nuts to 16 ft. lbs.
Step 5: Right bank nuts to 16 ft. lbs.

⑤ Sprocket bolt

⑥ M8 bolts: 26 ft. lbs.
M10 bolts: 33 ft. lbs.

⑦ Upper: 13 ft. lbs.
Lower: 16 ft. lbs.

06009-COUP-C08

WHEEL ALIGNMENT

Year	Model		Caster Range (+/-Deg.)	Caster Preferred Setting (Deg.)	Camber Range (+/-Deg.)	Camber Preferred Setting (Deg.)	Toe-in (in.)
2002	ALL	F	0.50	+3.00	0.50	0	0 +/- 0.12
		R	—	—	0.50	-1.34	0.13 +/- 0.13
2003	ALL	F	0.50	+3.00	0.50	0	0 +/- 0.12
		R	—	—	0.50	-1.34	0.13 +/- 0.13
2004	ALL	F	0.50	+3.00	0.50	0	0 +/- 0.12
		R	—	—	0.50	-1.34	0.13 +/- 0.13
2005	ALL	F	0.50	+3.00	0.50	0	0 +/- 0.12
		R	—	—	0.50	-1.34	0.12 +/- 0.12

06009-COUP-C09

TIRE, WHEEL AND BALL JOINT SPECIFICATIONS

Year	Model	OEM Tires		Tire Pressures (psi)		Wheel	Ball Joint	Lug Nuts
		Standard	Optional	Front	Rear	Size	Inspection	(ft. lbs.)
2002	Sebring LX	P205/60HR16	None	32	29	6JJ	①	73
	Sebring Lxi	P215/50HR17	None	32	29	6.5JJ	①	73
	Stratus SE	P205/60HR16	None	32	29	6JJ	①	73
	Stratus R/T	P215/50HR17	None	32	29	6.5JJ	①	73
2003	Sebring LX	P205/60HR16	None	32	29	6JJ	①	73
	Sebring Lxi	P215/50HR17	None	32	29	6.5JJ	①	73
	Stratus SE	P205/60HR16	None	32	29	6JJ	①	73
	Stratus R/T	P215/50HR17	None	32	29	6.5JJ	①	73
2004	Sebring LX	P205/60HR16	None	32	29	6JJ	①	73
	Sebring Lxi	P215/50HR17	None	32	29	6.5JJ	①	73
	Stratus SE	P205/60HR16	None	32	29	6JJ	①	73
	Stratus R/T	P215/50HR17	None	32	29	6.5JJ	①	73
2005	Sebring LX	P205/60HR16	None	②	②	6JJ	①	73
	Sebring Lxi	P215/50HR17	None	②	②	6.5JJ	①	73
	Stratus SE	P205/60HR16	None	②	②	6JJ	①	73
	Stratus R/T	P215/50HR17	None	②	②	6.5JJ	①	73

OEM: Original Equipment Manufacturer

PSI: Pounds Per Square Inch

STD: Standard

OPT: Optional

① Ball joint rotating torque: 22-54 inch lbs. If torque falls outside that range, replace it.

② See placard on vehicle

06009-COUP-C10

BRAKE SPECIFICATIONS

All measurements in inches unless noted

			Brake Disc			Brake Drum Diameter		Min.	Brake Caliper	
Year	Model		Original Thickness	Minimum Thickness	Maximum Run-out	Original Inside Diameter	Maximum Machine Diameter	Lining Thick-ness	Bracket Bolts (ft. lbs.)	Mounting Bolts (ft. lbs.)
2002	Sebring	F	①	②	0.002	—	—	0.080	74	28
	Coupe	R	0.394	0.331	0.003	9.00	—	③	④	32
	Stratus	F	①	②	0.002	—	—	0.080	74	28
	Coupe	R	0.394	0.331	0.003	9.00	—	③	④	32
2003	Sebring	F	①	②	0.002	—	—	0.080	74	28
	Coupe	R	0.394	0.331	0.003	9.00	—	③	④	32
	Stratus	F	①	②	0.002	—	—	0.080	74	28
	Coupe	R	0.394	0.331	0.003	9.00	—	③	④	32
2004	Sebring	F	①	②	0.002	—	—	0.080	74	⑤
	Coupe	R	0.394	0.331	0.003	9.00	—	③	④	32
	Stratus	F	①	②	0.002	—	—	0.080	74	⑤
	Coupe	R	0.394	0.330	0.003	9.00	—	③	④	32
2005	Sebring	F	①	②	0.002	—	—	0.080	74	⑤
	Coupe	R	0.394	0.331	0.003	9.00	—	③	④	32
	Stratus	F	①	②	0.002	—	—	0.080	74	⑤
	Coupe	R	0.394	0.331	0.003	9.00	—	③	④	32

F: Front

R: Rear

① 2.4L: 0.944
3.0L: 1.024

② 2.4L: 0.882
3.0L: 0.961

③ With rear disc: 0.080 in.
With rear drum: 0.040

④ Flange bolt: 44 ft. lbs.
Bolt and washer: 41 ft. lbs.

⑤ 2.4L: Guide pin and lock pin: 54 ft. lbs.
3.0L: 28 ft. Lbs.

06009-COUP-C11

SCHEDULED MAINTENANCE INTERVALS

2002 Chrysler Sebring Coupe, Dodge Stratus Coupe

TO BE SERVICED	TYPE OF SERVICE	VEHICLE MILEAGE INTERVAL (x1000)												
		7.5	15	22.5	30	37.5	45	52.5	60	67.5	75	82.5	90	97.5
Engine oil & filter	R	✓	✓	✓	✓	✓	✓	✓	✓	✓	✓	✓	✓	✓
Coolant level, hoses & clamps	S/I	✓	✓	✓	✓	✓	✓	✓	✓	✓	✓	✓	✓	✓
Rotate tires	S/I	✓	✓	✓	✓	✓	✓	✓	✓	✓	✓	✓	✓	✓
Automatic transaxle fluid level	S/I		✓		✓		✓		✓		✓		✓	
Brake hoses & disc brake pads	S/I		✓		✓		✓		✓		✓		✓	
Driveshaft boots & front suspension components	S/I		✓		✓		✓		✓		✓		✓	
Air filter element	R				✓				✓				✓	
Engine coolant	R				✓				✓				✓	
Spark plugs (3.0L)	R				✓				✓				✓	
Spark plugs (2.4L) ①	R				✓				✓				✓	
Accessory drive belts	S/I				✓				✓				✓	
Ball joints & steering linkage seals	S/I				✓				✓				✓	
Exhaust system	S/I				✓				✓				✓	
Fuel hoses	S/I				✓				✓				✓	
Manual transaxle oil	S/I				✓				✓				✓	
PCV valve	S/I				✓				✓				✓	
Rear drum brake lining & rear wheel cylinders	S/I				✓				✓				✓	
Camshaft timing belt ②	R								✓					
Ignition cables	R								✓					
Distributor cap & rotor	S/I								✓					
EVAP system	S/I								✓					
Fuel system	S/I								✓					

R: Replace S/I: Service or Inspect

① Spark plugs: replace every 100,000 miles.

② If equipped with the 3.0L engine, replace at 100.000 miles

FREQUENT OPERATION MAINTENANCE (SEVERE SERVICE)

If a vehicle is operated under any of the following conditions it is considered severe service:

- Extremely dusty areas.

- 50% or more of the vehicle operation is in 32°C (90°F) or higher temperatures, or constant operation in temperatures below 0°C (32°F).

- Prolonged idling (vehicle operation in stop and go traffic).

- Frequent short running periods (engine does not warm to normal operating temperatures).

- Police, taxi, delivery usage or trailer towing usage.

Oil & oil filter change: change every 3000 miles.

Disc brake pads: check every 6000 miles.

Air filter element: change every 15,000 miles.

Automatic transaxle fluid: change every 15,000 miles.

Rear drum brake linings & rear wheel cylinders: check every 15,000 miles.

Spark plugs: change every 15,000 miles.

06009-COUP-C12

SCHEDULED MAINTENANCE INTERVALS

2003 Chrysler Sebring Coupe, Dodge Stratus Coupe

		VEHICLE MILEAGE INTERVAL (x1000)												
TO BE SERVICED	TYPE OF SERVICE	6	12	18	24	30	36	42	48	54	60	66	72	78
Engine oil & filter	R	✓	✓	✓	✓	✓	✓	✓	✓	✓	✓	✓	✓	✓
Coolant level, hoses & clamps	S/I	✓	✓	✓	✓	✓	✓	✓	✓	✓	✓	✓	✓	✓
Rotate tires	S/I	✓	✓	✓	✓	✓	✓	✓	✓	✓	✓	✓	✓	✓
Automatic transaxle fluid level	S/I		✓		✓		✓		✓		✓		✓	
Brake hoses & disc brake pads	S/I		✓		✓		✓		✓		✓		✓	
Driveshaft boots & front suspension components	S/I		✓		✓		✓		✓		✓		✓	
Air filter element	R				✓				✓				✓	
Engine coolant	R				✓				✓				✓	
Spark plugs (3.0L)	R				✓				✓				✓	
Spark plugs (2.4L) ①	R				✓				✓				✓	
Accessory drive belts	S/I				✓				✓				✓	
Ball joints & steering linkage seals	S/I				✓				✓				✓	
Exhaust system	S/I				✓				✓				✓	
Fuel hoses	S/I				✓				✓				✓	
Manual transaxle oil	S/I				✓				✓				✓	
PCV valve	S/I				✓				✓				✓	
Rear drum brake lining & rear wheel cylinders	S/I				✓				✓				✓	
Camshaft timing belt ②	R								✓					
Ignition cables	R								✓					
Distributor cap & rotor	S/I								✓					
EVAP system	S/I								✓					
Fuel system	S/I								✓					

R: Replace S/I: Service or Inspect

① Platinum spark plugs: replace every 100,000 miles.

② If equipped with the 3.0L engine, replace at 100.000 miles

FREQUENT OPERATION MAINTENANCE (SEVERE SERVICE)

If a vehicle is operated under any of the following conditions it is considered severe service:

- Extremely dusty areas.

- 50% or more of the vehicle operation is in 32°C (90°F) or higher temperatures, or constant operation in temperatures below 0°C (32°F).

- Prolonged idling (vehicle operation in stop and go traffic).

- Frequent short running periods (engine does not warm to normal operating temperatures).

- Police, taxi, delivery usage or trailer towing usage.

Oil & oil filter change: change every 3000 miles.

Disc brake pads: check every 6000 miles.

Air filter element: change every 15,000 miles.

Automatic transaxle fluid: change every 15,000 miles.

Rear drum brake linings & rear wheel cylinders: check every 15,000 miles.

Spark plugs: change every 15,000 miles.

06009-COUP-C13

SCHEDULED MAINTENANCE INTERVALS

2004-05 Chrysler Sebring Coupe, Dodge Stratus Coupe

TO BE SERVICED	TYPE OF SERVICE	VEHICLE MILEAGE INTERVAL (x1000)												
		6	12	18	24	30	36	42	48	54	60	66	72	78
Engine oil & filter	R	✓	✓	✓	✓	✓	✓	✓	✓	✓	✓	✓	✓	✓
Coolant level, hoses & clamps	S/I					✓					✓			
Rotate tires	S/I	✓	✓	✓	✓	✓	✓	✓	✓	✓	✓	✓	✓	✓
Automatic transaxle fluid level	S/I		✓		✓		✓		✓		✓		✓	
Brake hoses & disc brake pads	S/I		✓		✓		✓		✓		✓		✓	
Driveshaft boots & front suspension components	S/I		✓		✓		✓		✓		✓		✓	
Air filter element	R					✓					✓			
Engine coolant	R										✓			
Spark plugs (standard)	R					✓					✓			
Spark plugs (Iridium tipped)	R	Every 102,000 miles												
Accessory drive belts	S/I					✓					✓			
Ball joints & steering linkage seals	S/I					✓					✓			
Exhaust system	S/I					✓					✓			
Fuel system connections	I										✓			
Manual transaxle oil	S/I					✓					✓			
Air bag system	I	Every 10 years												
Timing belt	R										✓			
Distributor cap & rotor	S/I					✓					✓			
EVAP system	I										✓			
Fuel hoses	I					✓					✓			

R: Replace   S/I: Service or Inspect

FREQUENT OPERATION MAINTENANCE (SEVERE SERVICE)

If a vehicle is operated under any of the following conditions it is considered severe service:

- Extremely dusty areas.

- 50% or more of the vehicle operation is in 32°C (90°F) or higher temperatures, or constant operation in temperatures below 0°C (32°F).

- Prolonged idling (vehicle operation in stop and go traffic).

- Frequent short trips of less than 10 miles.

- Police, taxi, delivery usage or trailer towing usage.

Engine oil & oil filter change: change every 3000 miles.

Disc brake pads, rotors and calipers: check every 6000 miles.

Air filter element: change every 15,000 miles.

Manual transaxle fluid: change every 30,000 miles

Automatic transaxle fluid: change every 30,000 miles.

Rear drum brake linings & rear wheel cylinders: check every 15,000 miles.

Spark plugs (standard): change every 15,000 miles.

Tire rotation: every 3,000 miles

06009-COUP-C14

PRECAUTIONS

Before servicing any vehicle, please be sure to read all of the following precautions, which deal with personal safety, prevention of component damage, and important points to take into consideration when servicing a motor vehicle:

- Never open, service or drain the radiator or cooling system when the engine is hot; serious burns can occur from the steam and hot coolant.
- Observe all applicable safety precautions when working around fuel. Whenever servicing the fuel system, always work in a well-ventilated area. Do not allow fuel spray or vapors to come in contact with a spark, open flame or excessive heat (a hot drop light, for example). Keep a dry chemical fire extinguisher near the work area. Always keep fuel in a container specifically designed for fuel storage; also, always properly seal fuel containers to avoid the possibility of fire or explosion. Refer to the additional fuel system precautions later in this section.
- Fuel injection systems often remain pressurized, even after the engine has been turned **OFF**. The fuel system pressure must be relieved before disconnecting any fuel lines. Failure to do so may result in fire and/or personal injury.
- Brake fluid often contains polyglycol ethers and polyglycols. Avoid contact with the eyes and wash your hands thoroughly after handling brake fluid. If you do get brake fluid in your eyes, flush your eyes with clean, running water for 15 minutes. If eye irritation persists, or if you have taken brake fluid internally, IMMEDIATELY seek medical assistance.
- The EPA warns that prolonged contact with used engine oil may cause a number of skin disorders, including cancer! You should make every effort to minimize your exposure to used engine oil. Protective gloves should be worn when changing oil. Wash your hands and any other exposed skin areas as soon as possible after exposure to used engine oil. Soap and water, or waterless hand cleaner should be used.
- All new vehicles are now equipped with an air bag system, often referred to as a Supplemental Restraint System (SRS) or Supplemental Inflatable Restraint (SIR) system. The system must be disabled before performing service on or around system components, steering column, instrument panel components, wiring and sensors. Failure to follow safety and disabling procedures could result in accidental air bag deployment, possible personal injury and unnecessary system repairs.
- Always wear safety goggles when working with, or around, the air bag system. When carrying a non-deployed air bag, be sure the bag and trim cover are pointed away from your body. When placing a non-deployed air bag on a work surface, always face the bag and trim cover upward, away from the surface. This will reduce the motion of the module if it is accidentally deployed. Refer to the additional air bag system precautions later in this section.
- Clean, high quality brake fluid from a sealed container is essential to the safe and proper operation of the brake system. You should always buy the correct type of brake fluid for your vehicle. If the brake fluid becomes contaminated, completely flush the system with new fluid. Never reuse any brake fluid. Any brake fluid that is removed from the system should be discarded. Also, do not allow any brake fluid to come in contact with a painted surface; it will damage the paint.
- Never operate the engine without the proper amount and type of engine oil; doing so WILL result in severe engine damage.
- Timing belt maintenance is extremely important! Many models utilize an interference-type, non-freewheeling engine. If the timing belt breaks, the valves in the cylinder head may strike the pistons, causing potentially serious (also time-consuming and expensive) engine damage.
- Disconnecting the negative battery cable on some vehicles may interfere with the functions of the on-board computer system(s) and may require the computer to undergo a relearning process once the negative battery cable is reconnected.
- When servicing drum brakes, only disassemble and assemble one side at a time, leaving the remaining side intact for reference.
- Only an MVAC-trained, EPA-certified automotive technician should service the air conditioning system or its components.

ENGINE REPAIR

Alternator

REMOVAL

2.4L Engine

1. Before servicing the vehicle, refer to the precautions in the beginning of this section.
2. Remove or disconnect the following:
   - Negative battery cable
   - Engine under cover
   - Engine mount bracket
   - Oil pressure hose and tube clamp bolts
   - Oil return tube clamp bolt
   - Power steering and A/C drive belt
   - Alternator drive belt
   - Water pump pulley
   - Alternator connector

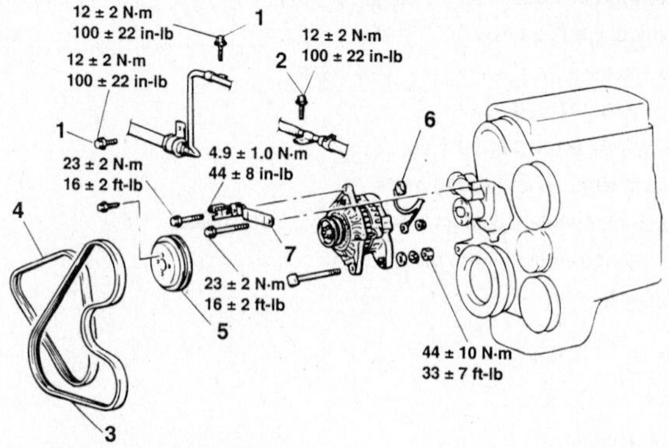

1. OIL PRESSURE HOSE AND TUBE ASSEMBLY CLAMP BOLTS
2. OIL RETURN TUBE ASSEMBLY CLAMP BOLT
3. DRIVE BELT (POWER STEERING, A/C)
4. DRIVE BELT (GENERATOR)
5. WATER PUMP PULLEY
6. GENERATOR CONNECTOR
7. GENERATOR BRACE

9356CG01

Exploded view of the alternator and related components—2.4L engine

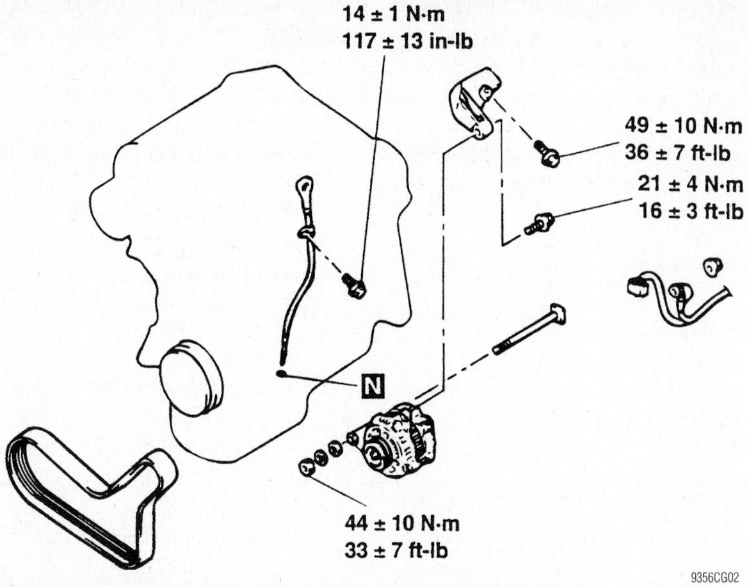

Exploded view of the alternator and related components—3.0L engine

- Alternator brace
- Alternator bolt and nut
- Alternator

3.0L Engine

1. Before servicing the vehicle, refer to the precautions in the beginning of this section.
2. Remove or disconnect the following:
   - Negative battery cable
   - Engine under cover
   - Engine mount bracket
   - Drive belt
   - Alternator belt
   - Oil level gauge unit
   - Alternator electrical connectors
   - Alternator brace
   - Alternator

INSTALLATION

2.4L Engine

1. Before servicing the vehicle, refer to the precautions in the beginning of this section.
2. Install or connect the following:
   - Alternator
   - Alternator bolt and nut. Tighten to 26–40 ft. lbs. (34–54 Nm).
   - Alternator brace
   - Alternator connector
   - Water pump pulley
   - Alternator drive belt
   - Power steering and A/C drive belt
   - Oil return tube clamp bolt
   - Oil pressure hose and tube clamp bolts
   - Engine mount bracket
3. Adjust the belt tension.
   - Negative battery cable
   - Under cover

3.0L Engine

1. Before servicing the vehicle, refer to the precautions in the beginning of this section.
2. Install or connect the following:
   - Alternator and torque the bolts to 33 ft. lbs. (44 Nm)
   - Alternator brace and torque the bolts to 36 ft. lbs. (49 Nm)
   - Oil level gauge unit and torque the bolt to 10 ft. lbs. (15 Nm)
   - Alternator belt
   - Drive belt
   - Engine mount bracket
3. Adjust the belt tension.
   - Negative battery cable
   - Under cover

Ignition Timing

Ignition timing is controlled by the Powertrain Control Module (PCM) and is not adjustable.

Engine Assembly

REMOVAL & INSTALLATION

2.4L Engine

1. Before servicing the vehicle, refer to the precautions in the beginning of this section.
2. Properly relieve the fuel system pressure.
3. Drain the engine oil.
4. Drain the cooling system.
5. Recover the A/C refrigerant.
6. Remove or disconnect the following:
   - Hood
   - Negative battery cable
   - Strut tower bar
   - Air cleaner assembly
   - Radiator and recovery tank
   - Front exhaust pipe
   - Accelerator cable connection
   - Purge hose connection
   - Brake booster vacuum hose connection
   - Ignition coil connector
   - Injector connectors
   - Manifold differential pressure sensor connector
   - Throttle Position Sensor (TP) connection
   - Heated Oxygen Sensor (HO_2S) connection
   - Capacitor connection
   - Engine Coolant Temperature (ECT) connector
   - Camshaft Position (CMP) sensor connector
   - Knock Sensor (KS) connector
   - Engine coolant temperature gauge unit connector
   - Idle Air Control (IAC) motor connector
   - Evaporative emission purge solenoid valve
   - Exhaust Gas Recirculation (EGR) solenoid valve connector
   - High pressure and fuel return fuel hoses
   - Pressure hose connection
   - Oil dipstick and tube
   - Heater hose
   - Alternator connector
   - Oil pressure switch connector
   - Drive belts
   - Crankshaft Position (CKP) sensor connector
   - Power steering pressure switch connector
7. Unbolt the power steering pump and bracket, and place aside, as an assembly with the hose attached.

WARNING

Do NOT disconnect the A/C compressor refrigerant lines!

   - A/C compressor connector. Unbolt and place the A/C compressor aside with the lines attached.
8. Remove the transaxle. Do NOT

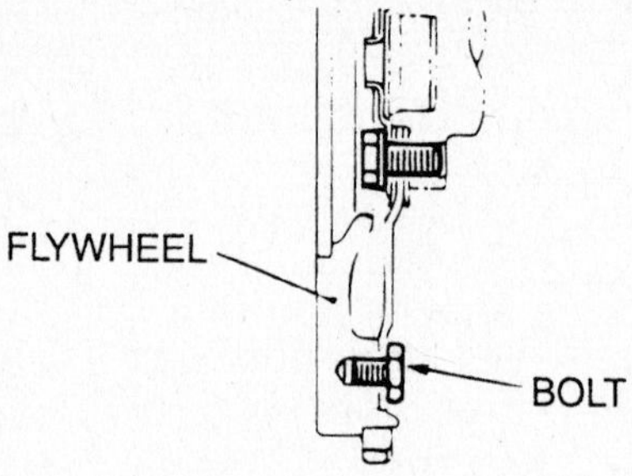

Do not remove the flywheel bolt shown, or you will damage the flywheel

remove the flywheel mounting bolt shown in the accompanying figure. Removal of this bolt will throw the flywheel out of balance and damage it.

9. Remove the engine mount bracket, as follows:
   a. Support the engine with a suitable jack.
   b. Remove the special tools that were attached during transaxle removal
   c. Hold the engine assembly with a chain block or similar tool.
   d. Place a jack against the oil pan (with a block of wood in between), the jack up the engine so that the weight of the engine is no longer on the engine mount, then remove the engine mount bracket.
10. Remove the engine mount stopper.
11. Make sure that all cables, hoses and electrical connections are disconnected; slowly raise the engine up and out of the engine compartment.

To install:

12. Install or connect the following:
   - Engine assembly
   - Engine mount stopper and torque the bolts to 60 ft. lbs. (81 Nm). Clamp the mount stopper so that the arrow points in the direction shown in the accompanying figure.

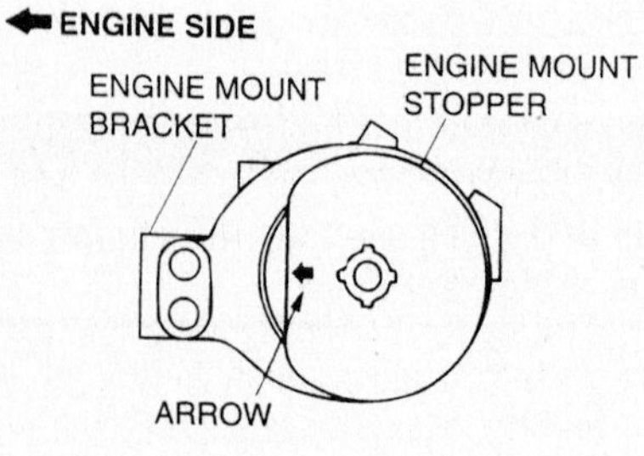

Correct installation of the engine mount stopper—2.4L engine

   - Engine mount bracket and torque the bolts to 64 ft. lbs. (86 Nm). Remove the engine lifting device
   - Transaxle
   - A/C compressor and connector
   - Power steering pump and bracket
   - Power steering pressure switch connector
   - CKP sensor connector
   - Drive belts
   - Oil pressure switch connector
   - Alternator connector
   - Heater hose
   - Pressure hose
   - Oil dipstick tube and dipstick
   - Pressure hose connector
   - Fuel return hose
   - High pressure fuel hose. Apply a small quantity of oil to the hose O-ring and then installing using a twisting motion. Make sure that the O-ring does not get twisted or clamped during installation.
   - EGR solenoid valve connector
   - EVAP purge solenoid valve
   - IAC motor connector
   - ECT gauge unit connector
   - KS connector
   - CMP sensor connector
   - ECT sensor connector
   - Capacitor connector
   - HO_2S connector
   - TP sensor connector
   - Manifold differential pressure sensor connector
   - Injector connector
   - Ignition coil connector
   - Vacuum hose connection
   - Brake booster vacuum hose
   - Purge hose
   - Accelerator cable
   - Front exhaust pipe
   - Radiator reservoir tank and radiator
   - Air cleaner
   - Strut tower bar
   - Negative battery cable
   - Hood and adjust as needed
13. Fill the engine with clean oil and replace the filter.
14. Fill the cooling system to the proper level.
15. Recharge the A/C system with the proper refrigerant.
16. Start the vehicle, check for leaks and repair if necessary.

3.0L Engine

1. Before servicing the vehicle, refer to the precautions in the beginning of this section.
2. Properly relieve the fuel system pressure.
3. Drain the engine oil.
4. Drain the cooling system.
5. Recover the A/C refrigerant.
6. Remove or disconnect the following:
   - Hood
   - Negative battery cable
   - Strut tower bar
   - Air cleaner
   - Radiator reservoir tank
   - Front exhaust pipe
   - Throttle cable
   - Manifold differential pressure sensor connector
   - Control wiring harness and power steering wiring harness combination connector
   - Exhaust Gas Recirculation (EGR) valve connector
   - Evaporative (EVAP) Emission purge solenoid valve connector
   - Knock Sensor (KS) connector
   - Crankshaft Position (CKP) sensor connector
   - Right bank Heated Oxygen Sensor (HO_2S)
   - Fuel injector connector
   - Distributor connector
   - Control wiring/injector wiring harness combination connector
   - Throttle Position Sensor (TPS) connector
   - Idle Air Control (IAC) motor connector
   - Ground wire
   - Engine Coolant Temperature (ECT) gauge unit
   - ECT sensor connector
   - Left bank HO_2S
   - Brake booster vacuum hose
   - Fuel supply and return hoses
   - Heater hoses
   - Starter electrical connectors
   - Oil Pressure switch
   - Alternator wiring
   - Drive belts
   - A/C compressor
   - Power steering pressure switch
   - Power steering oil pump. Do not disconnect the hose.
   - Engine mount stay and install an engine Lifting Tool, such as MZ203827
   - Engine mount bracket and stopper
   - Engine assembly

To install:

7. Install or connect the following:
   - Engine assembly
   - Engine mount bracket and stopper and torque the bolts to 60 ft. lbs. (81 Nm)

- Engine mount stay and torque the bolts to 26 ft. lbs. (35 Nm). Remove the engine lifting device.
- Power steering oil pump and torque the bolts to 31 ft. lbs. (42 Nm)
- Power steering pressure switch
- A/C compressor
- Drive belts
- Alternator wiring
- Oil pressure switch
- Starter electrical connectors
- Heater hoses
- Fuel supply and return lines
- Brake booster vacuum hose
- Left bank HO_2S
- ECT sensor connector
- ECT gauge unit
- Ground wire
- IAC motor electrical connector
- TPS connector
- Control wiring/injector wiring harness combination connector
- Distributor connector
- Fuel injector connector
- Right bank HO_2S
- CKP sensor connector
- KS connector
- EVAP solenoid valve connector
- EGR valve connector
- Control wiring harness and power steering harness combination connector
- Manifold differential Pressure sensor connector
- Throttle cable
- Front exhaust pipe
- Radiator reservoir tank
- Air cleaner
- Strut tower bar
- Negative battery cable
- Hood and adjust as needed

8. Fill the engine with clean oil and replace the filter.
9. Fill the cooling system to the proper level.
10. Recharge the A/C system with the proper refrigerant.
11. Start the vehicle, check for leaks and repair if necessary.

Water Pump

REMOVAL & INSTALLATION

2.4L Engine

1. Before servicing the vehicle, refer to the precautions in the beginning of this section.
2. Drain the cooling system.
3. Remove or disconnect the following:

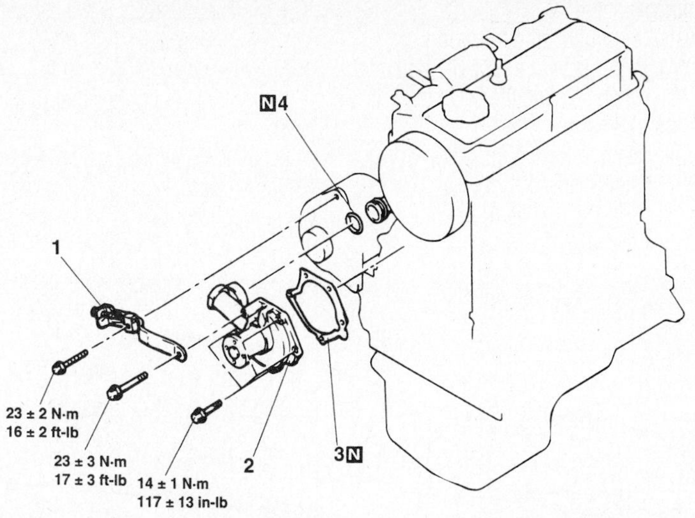

1. GENERATOR BRACE
2. WATER PUMP ASSEMBLY
3. WATER PUMP GASKET
4. O-RING

9356CG05

Exploded view of the water pump—2.4L engine

- Negative battery cable
- Timing belt tensioner pulley
- Alternator brace
- Water pump and discard the gasket and O-ring

To install:

4. Clean all mating surfaces of any residual gasket material
5. Install or connect the following:
   - New O-ring gasket
   - Water pump and torque the bolts to 117 inch lbs. (14 Nm)
   - Alternator brace and torque the bolts to 17 ft. lbs. (23 Nm)
   - Timing belt tensioner pulley
   - Negative battery cable
6. Fill the cooling system to the proper level.
7. Start the vehicle, check for leaks and repair if necessary.

3.0L Engine

1. Before servicing the vehicle, refer to the precautions in the beginning of this section.
2. Drain the cooling system.
3. Remove or disconnect the following:
   - Negative battery cable
   - Timing belt
   - Thermostat and bracket
   - Water pump and discard the gasket and O-ring

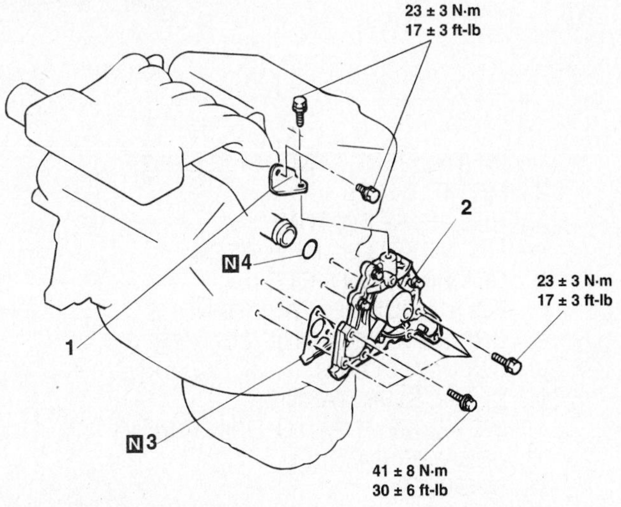

1. BRACKET
2. WATER PUMP ASSEMBLY
3. WATER PUMP GASKET
4. O-RING

06009-COUP-G07

Water pump installation—3.0L engine

To install:

4. Install or connect the following:
   - Water pump with a new O-ring and gasket and torque the bolts to 17 ft. lbs. (23 Nm)
   - Bracket and torque the bolts to 17 ft. lbs. (23 Nm)
   - Thermostat
   - Timing belt
   - Negative battery cable
5. Fill the cooling system to the proper level.
6. Start the vehicle, check for leaks and repair if necessary.

Heater Core

REMOVAL & INSTALLATION

1. Before servicing the vehicle, refer to the precautions in the beginning of this section.

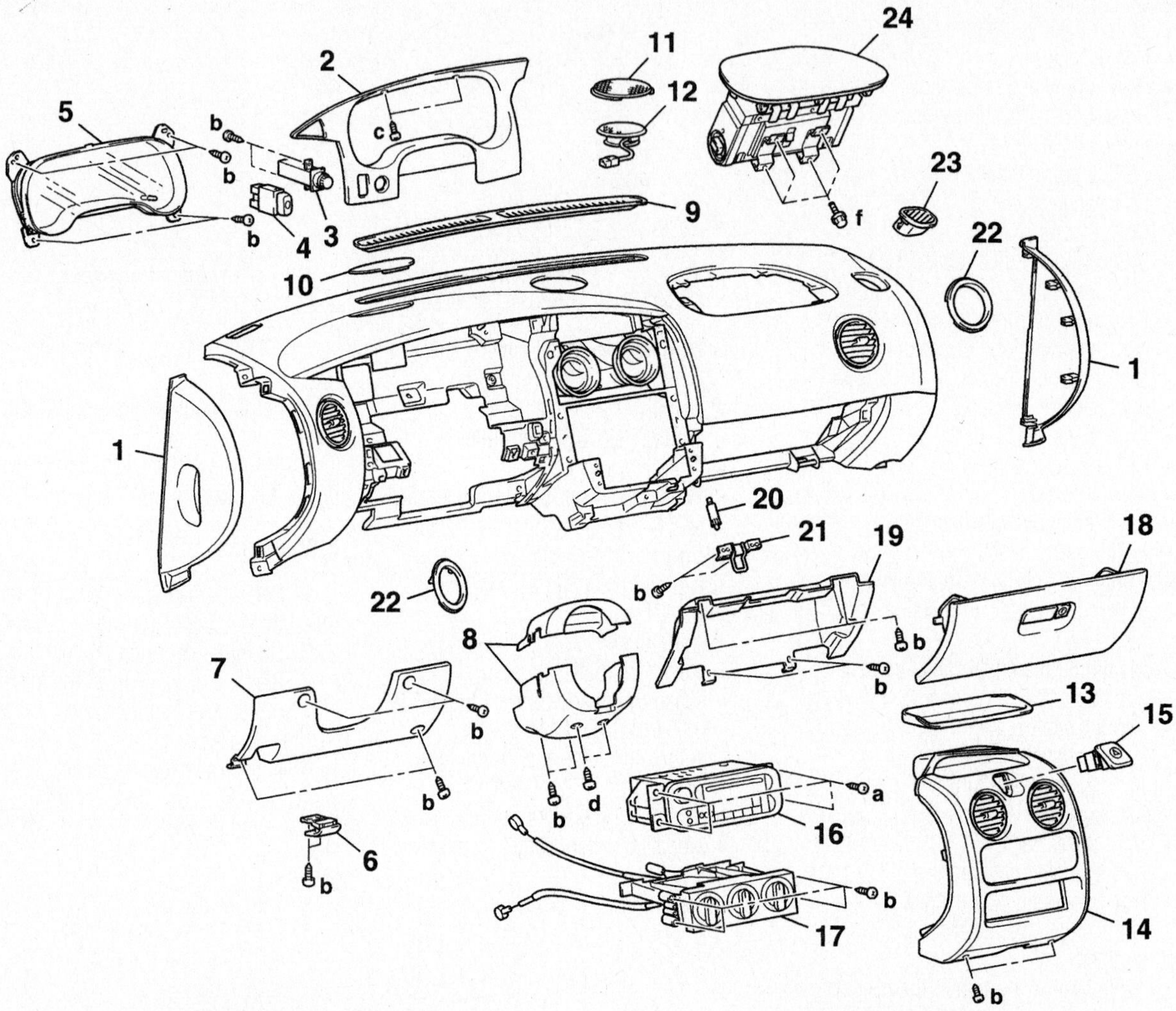

1. INSTRUMENT PANEL SIDE COVER
2. METER BEZEL
3. RHEOSTAT SWITCH
4. FRONT FOG LIGHT SWITCH
5. COMBINATION METER
6. FOOD LOCK RELEASE HANDLE
7. DRIVER'S SIDE UNDER COVER
8. COLUMN COVER
9. DEFROSTER GARNISH
10. INSTRUMENT PANEL UPPER PLUG
11. SPEAKER GARNISH
12. SPEAKER
13. RUBBER MAT
14. CENTER PANEL ASSEMBLY
15. HAZARD SWITCH
16. RADIO AND TAPE PLAYER ASSEMBLY
17. HEATER CONTROL ASSEMBLY
18. GLOVE BOX ASSEMBLY
19. GLOVE BOX CASE
20. GLOVE BOX LIGHT SWITCH ASSEMBLY
21. GLOVE BOX STRIKER
22. AIR OUTLET BEZEL
23. SIDE DEFROSTER GRILLE RH
24. PASSENGER'S SIDE AIR BAG MODULE ASSEMBLY

06009-COUP-G08

Instrument panel

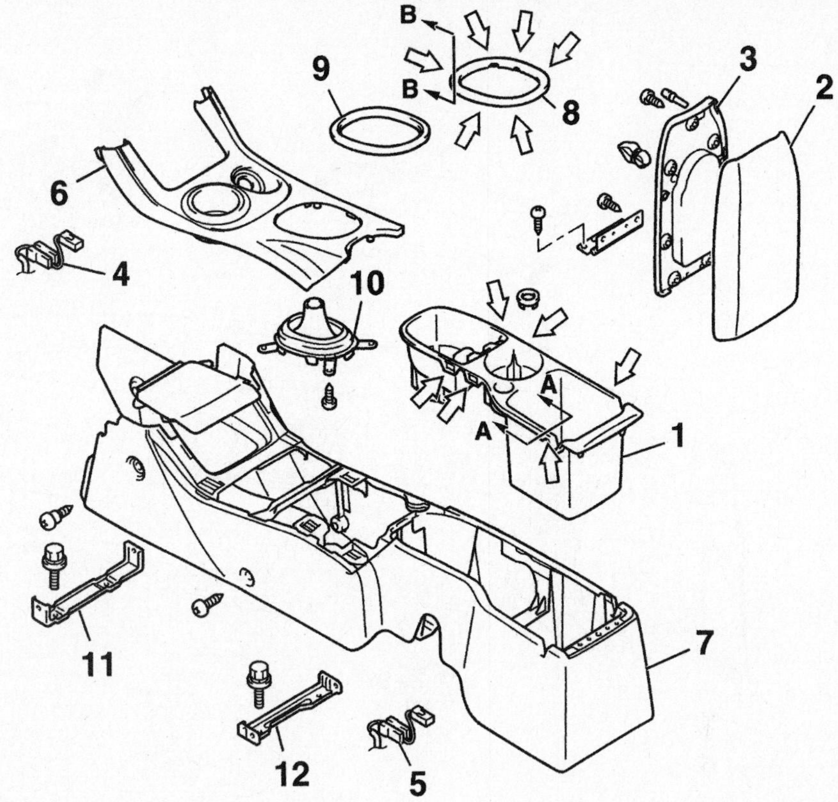

NOTE
⇦ : CLAW POSITIONS

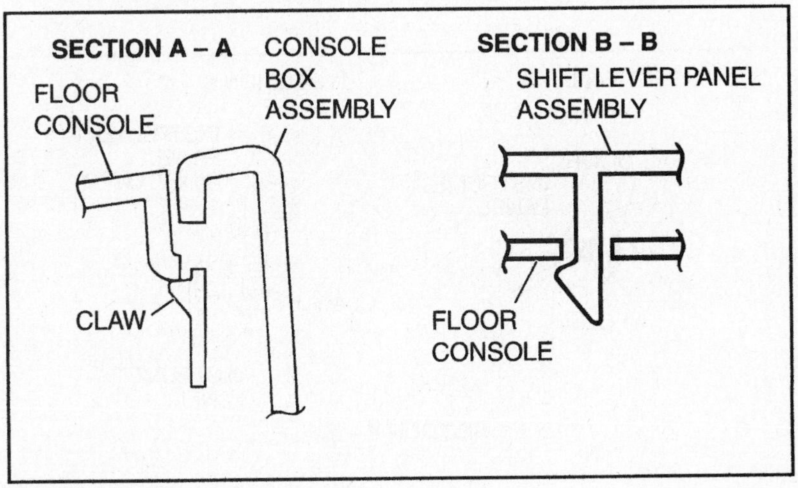

1. CONSOLE BOX ASSEMBLY
2. CONSOLE LID
3. CONSOLE LID COVER
4. ACCESSORY SOCKET HARNESS (FRONT)
5. ACCESSORY SOCKET HARNESS (REAR)
6. FLOOR CONSOLE INDICATOR PANEL
7. FLOOR CONSOLE
8. SHIFT LEVER PANEL ASSEMBLY <M/T>
9. GARNISH <A/T>
10. SHIFT LEVER BOOT <M/T>
11. FLOOR CONSOLE BRACKET A
12. FLOOR CONSOLE BRACKET B

06009-COUP-G09

Floor console

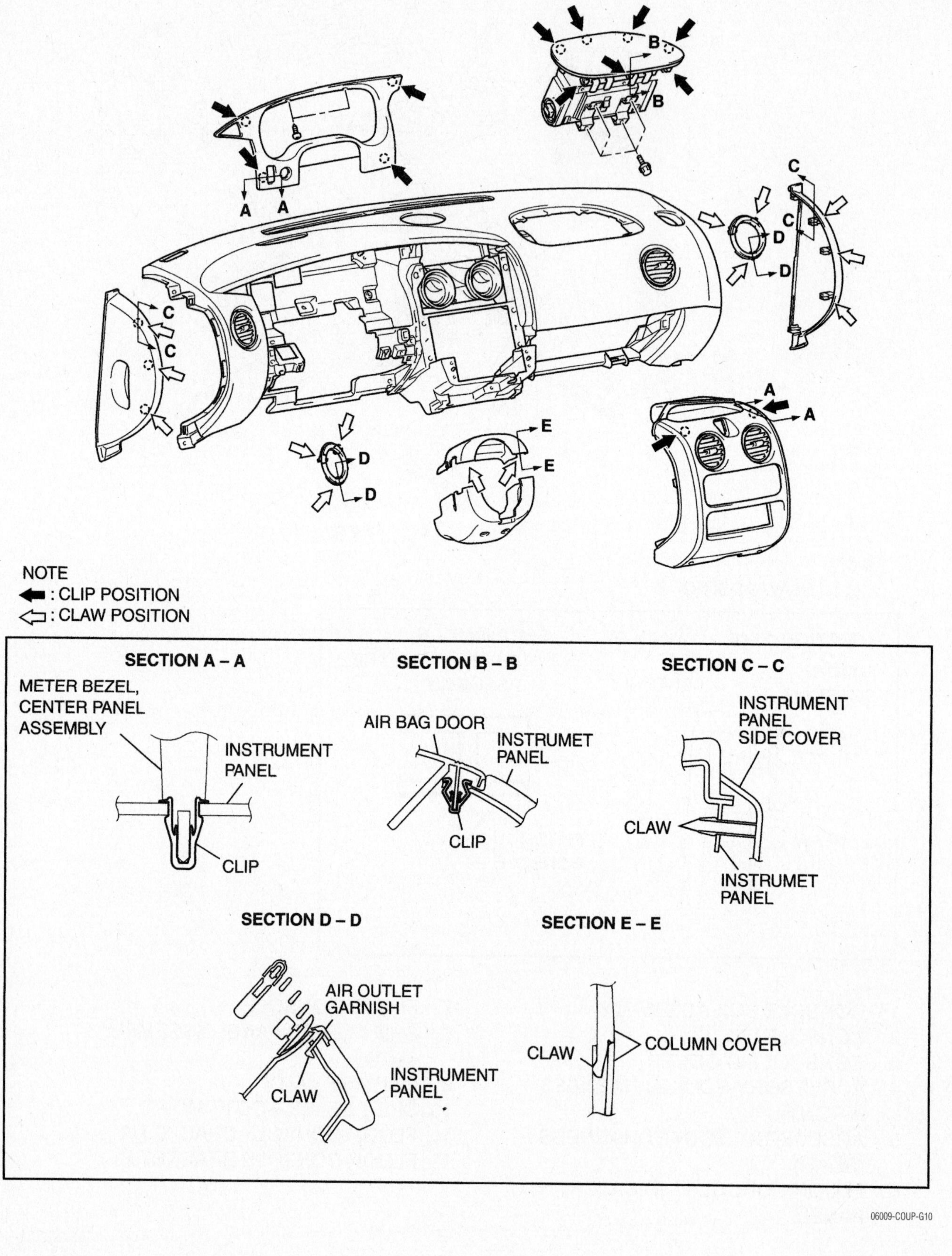

Clip positions

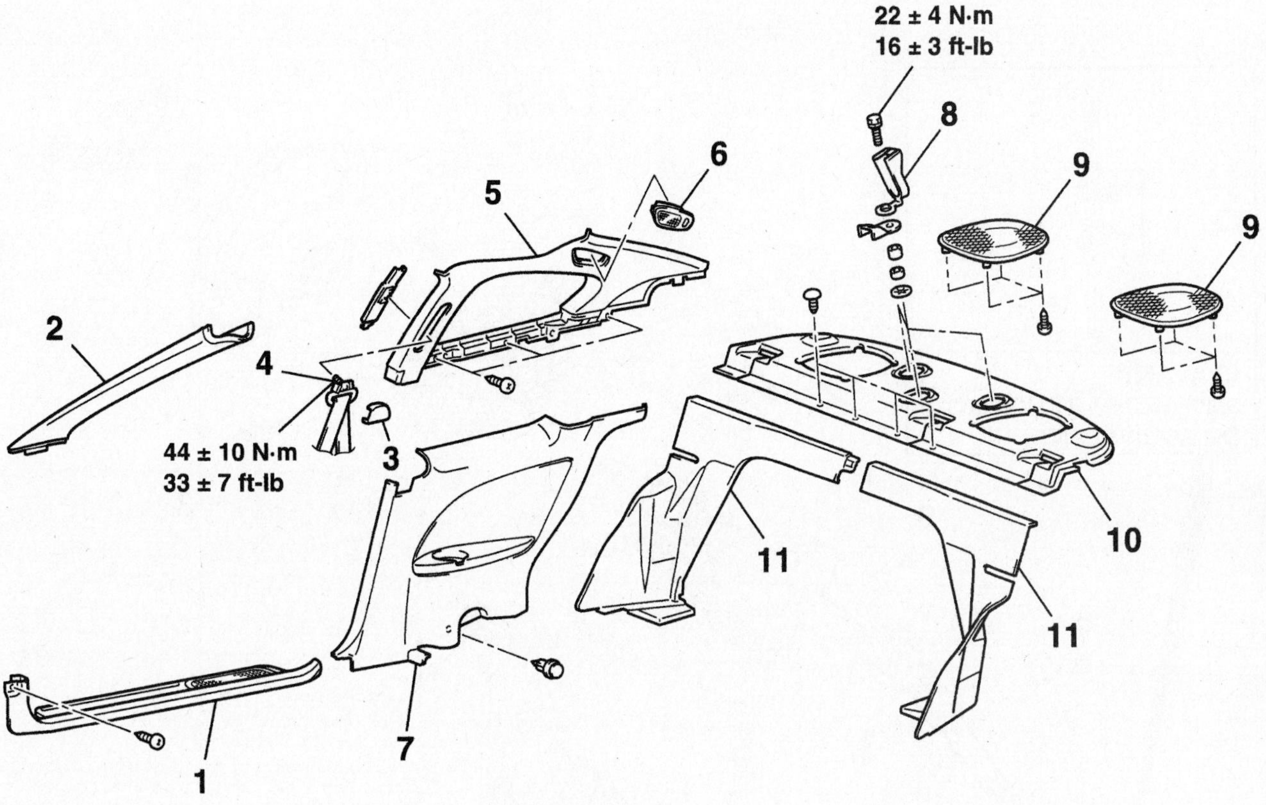

1. FRONT SCUFF PLATE
2. FRONT PILLAR TRIM
3. SEAT BELT SASH GUIDE COVER
4. FRONT SEAT BELT CONNECTION
5. QUARTER TRIM, UPPER
6. REAR PERSONAL LAMP ASSEMBLY
7. QUARTER TRIM, LOWER
8. UPPER TETHER ANCHOR
9. SPEAKER GARNISH
10. SHELF TRIM
11. FRONT SHELF TRIM

06009-COUP-G11

Trim plates

2. See the accompanying illustrations and remove:
   - Driver's side under cover
   - Center panel
   - Glove box
   - Joint duct
   - Air bag
   - Steering wheel
   - Steering shaft
   - Front pillar trim
   - A/C controller
   - Radio and tape player
   - Heater controller
   - Instrument panel assembly
3. Discharge and safely capture the refrigerant.
4. Disconnect the refrigerant lines at the evaporator.
5. Remove the expansion valve. Discard the O-rings.
6. Remove the evaporator.
7. Remove the A/C sensor.
8. Drain the cooling system
9. Disconnect the heater hoses at the core.
10. See the accompanying illustrations and remove:
    - Front deck crossmember
    - Floor console
    - Instrument panel center reinforcement
11. Remove the HVAC unit.
12. Remove the heater core.

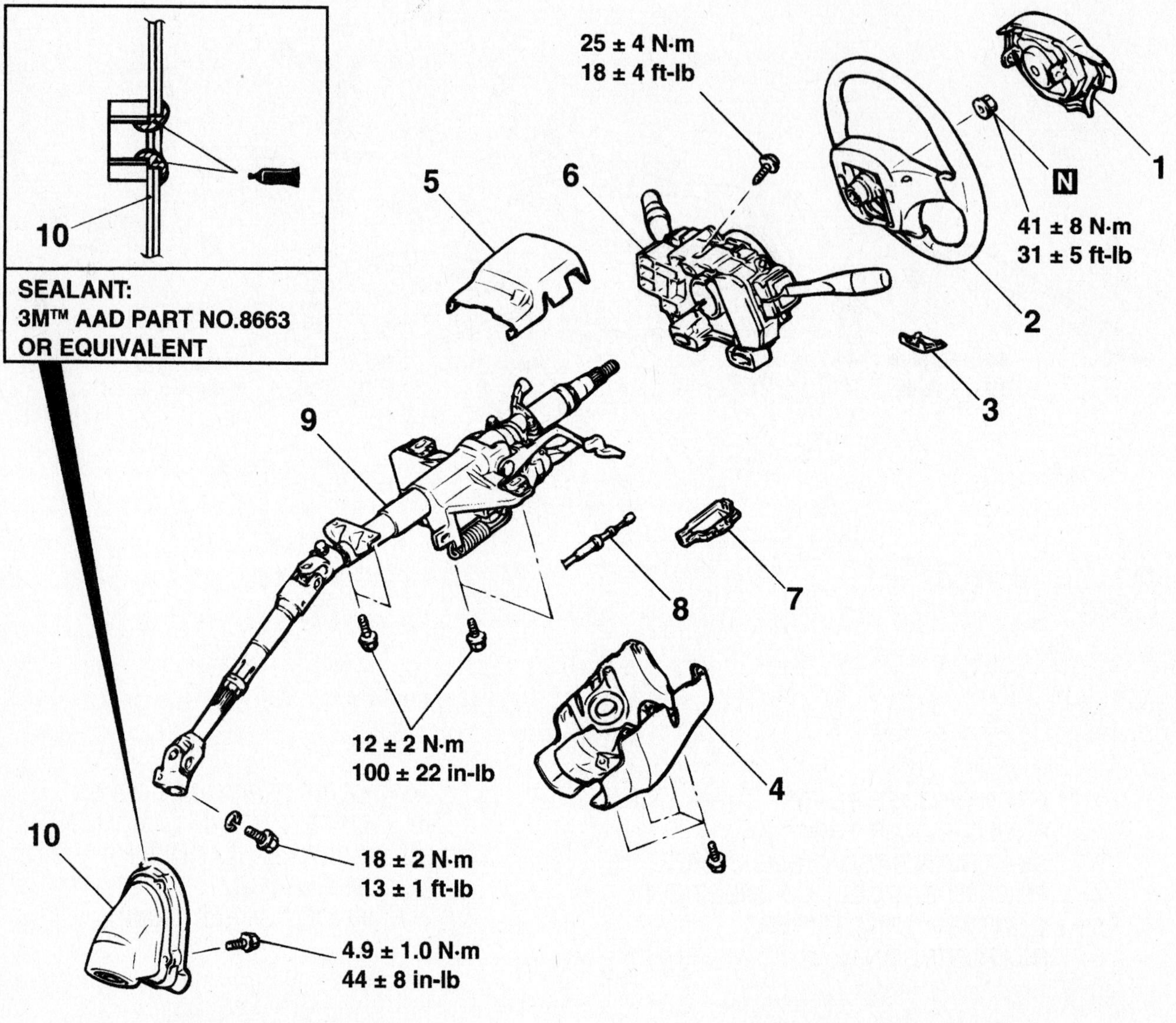

1. AIR BAG MODULE
2. STEERING WHEEL
3. COVER
- INSTRUMENT PANEL UNDER COVER
4. LOWER COLUMN COVER
5. UPPER COLUMN COVER
6. CLOCK SPRING AND COLUMN SWITCH ASSEMBLY
7. COVER <A/T>
8. KEY INTERLOCK CABLE <A/T>
9. STEERING SHAFT ASSEMBLY
10. STEERING COVER ASSEMBLY

06009-COUP-G12

Steering wheel and shaft

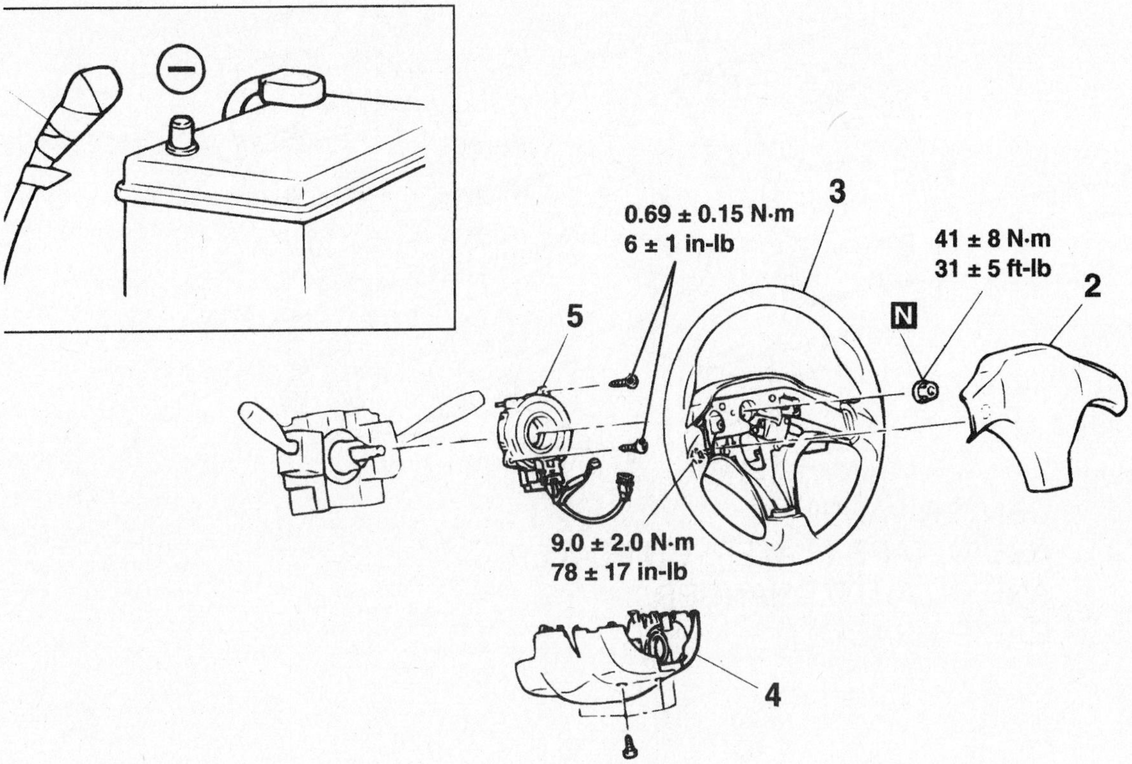

AIR BAG MODULE REMOVAL STEPS

1. NEGATIVE (–) BATTERY CABLE CONNECTION
2. AIR BAG MODULE (DRIVER'S SIDE)

CLOCK SPRING REMOVAL STEPS

1. NEGATIVE (–) BATTERY CABLE CONNECTION
2. AIR BAG MODULE (DRIVER'S SIDE)
3. STEERING WHEEL
4. COLUMN COVER LOWER
5. CLOCK SPRING

AIR BAG MODULE INSTALLATION STEPS

- PRE-INSTALLATION INSPECTION

2. AIR BAG MODULE (DRIVER'S SIDE)
1. NEGATIVE (–) BATTERY CABLE CONNECTION

- POST-INSTALLATION INSPECTION

CLOCK SPRING INSTALLATION STEPS

- PRE-INSTALLATION INSPECTION

5. CLOCK SPRING
4. COLUMN COVER LOWER
3. STEERING WHEEL
2. AIR BAG MODULE (DRIVER'S SIDE)
1. NEGATIVE (–) BATTERY CABLE CONNECTION

- POST-INSTALLATION INSPECTION

Required Special Tools:

- MB990502: Scan Tool (MUT-II)
- MB990803: Steering Wheel Puller

06009-COUP-G13

Driver's side air bag and clock spring

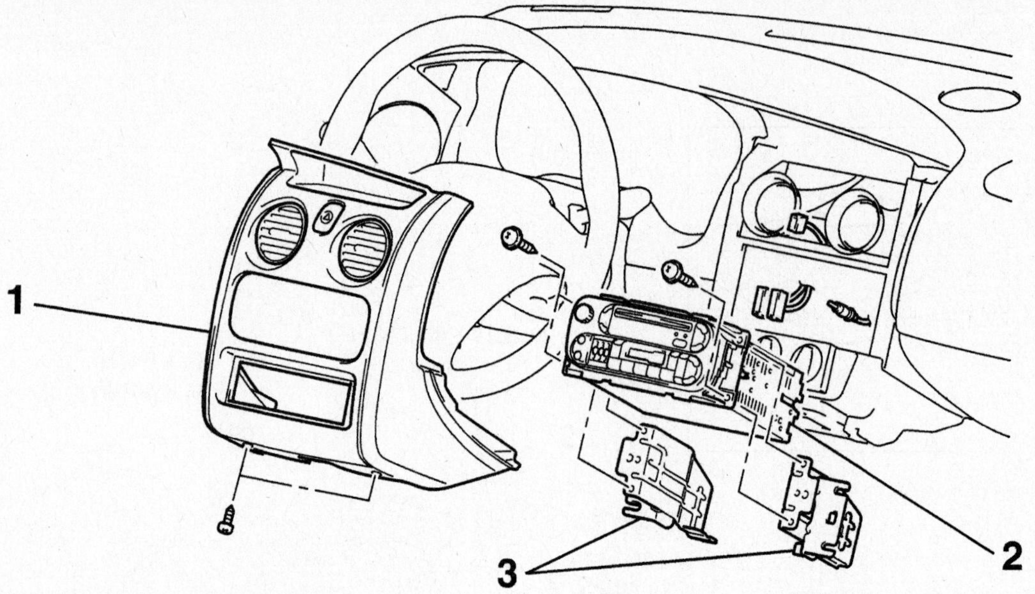

1. CENTER PANEL
2. RADIO, TAPE PLAYER, CD PLAYER AND CD AUTO CHANGER
3. RADIO BRACKET

06009-COUP-G14

Audio unit

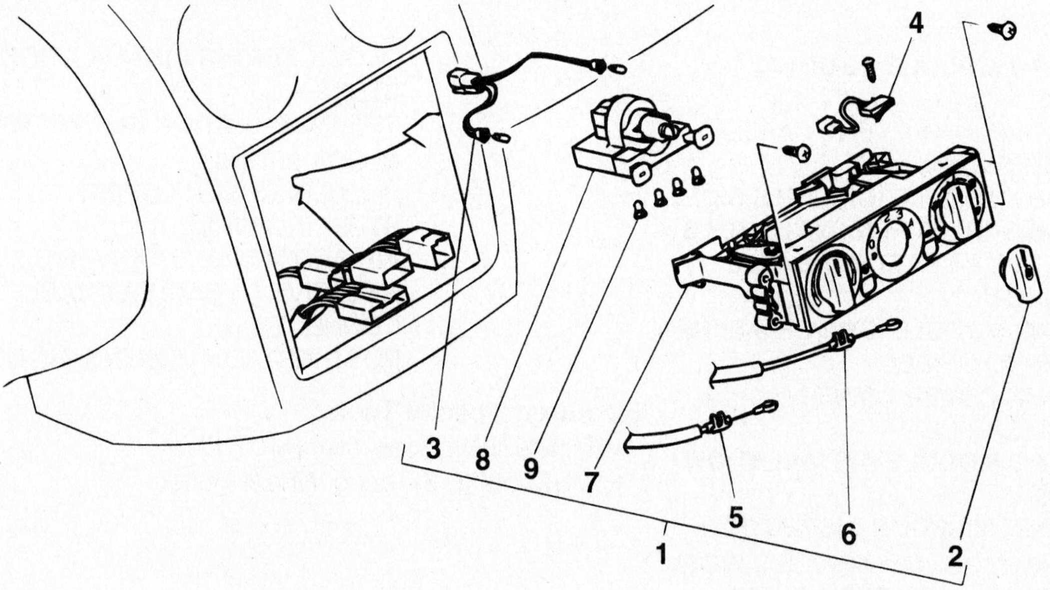

1. HEATER CONTROL ASSEMBLY
2. KNOB ASSEMBLY
3. BULB HARNESS
4. DEFROSTER SWITCH
5. AIR MIX DAMPER LEVER CABLE
6. AIR OUTLET CHANGEOVER DAMPER LINK CABLE
7. HEATER CONTROL PANEL
8. BLOWER SWITCH ASSEMBLY
9. BULB

06009-COUP-G15

Heater controller

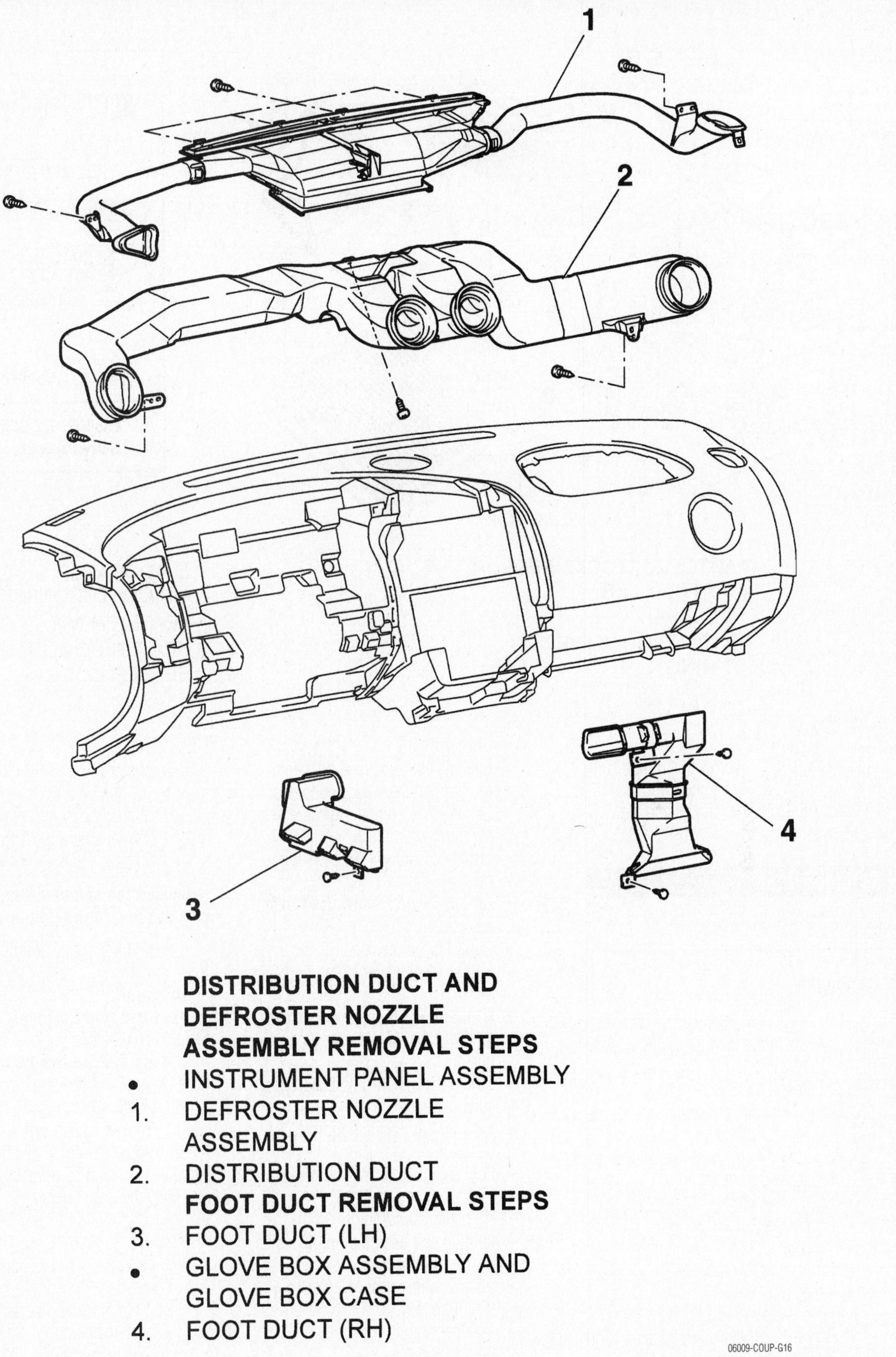

Air duct system

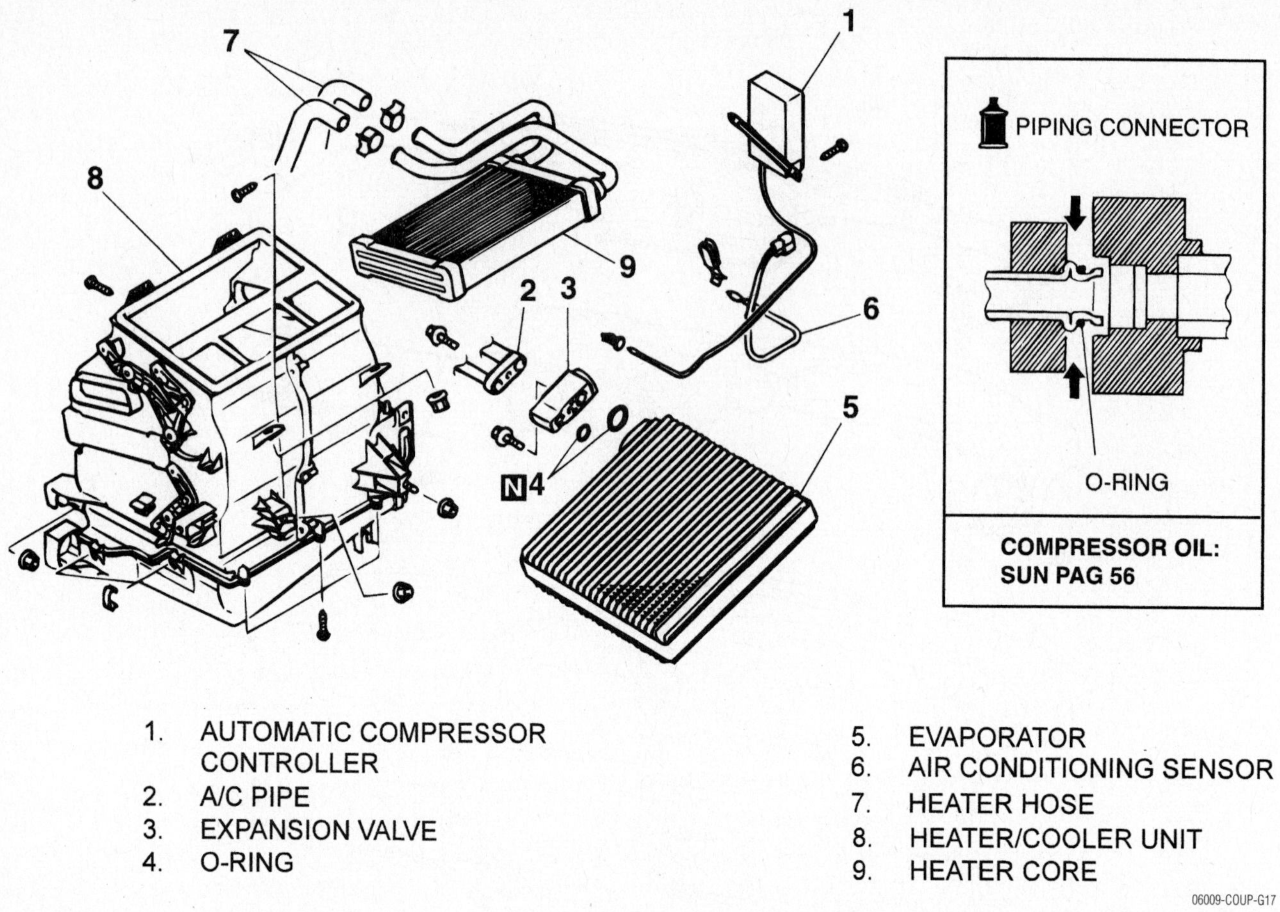

HVAC unit

13. Installation is the reverse of removal. Always use new O-rings coated with clean refrigerant oil. Evacuate, charge and leak test the system.

Cylinder Head

REMOVAL & INSTALLATION

2.4L Engine

1. Before servicing the vehicle, refer to the precautions in the beginning of this section.
2. Relieve the fuel system pressure.
3. Drain the cooling system.
4. Drain the engine oil.
5. Remove or disconnect the following:
   - Negative battery cable
   - Strut tower bar
   - Air cleaner assembly
   - Thermostat case
   - Front exhaust pipe
   - Accelerator cable
   - Purge hose
   - Brake booster vacuum hose
   - Ignition coil
   - Fuel injector electrical connector
   - Ignition failure sensor connector
   - Manifold differential pressure sensor connector
   - Throttle Position (TP) sensor connector
   - Heated Oxygen Sensor (HO_2S) connector
   - Capacitor connector
   - Engine Coolant Temperature (ECT) sensor
   - Camshaft Position (CMP) sensor
   - Idle Air Control (IAC) motor connector
   - Evaporative (EVAP) emission purge solenoid valve
   - Exhaust Gas Recirculation (EGR) valve
   - Fuel supply and return hoses
   - Pressure hose
   - Oil dipstick and guide
   - Ignition coil and spark plug cables
   - Upper radiator hose
   - Positive Crankcase Ventilation (PCV) hose
   - Breather hose
   - Rocker arm cover
   - Spark plug guide oil seal
   - Water hose
   - Timing belt
   - Power steering pressure switch
   - Power steering pump and bracket
   - Exhaust manifold bracket

➡Loosen the cylinder head bolts, in 2 or 3 passes, in the proper sequence.

   - Cylinder head and discard the gasket

To install:

6. Clean all mating surfaces of any residual gasket material.
7. Lubricate the threads of the cylinder head bolts with clean engine oil.
8. Install or connect the following:
   - Cylinder head with a new gasket
9. Torque the bolts, at a 90 degree angle with special Tool MB991654, using the following procedure:
   a. Step 1: 58 ft. lbs. (79 Nm).
   b. Step 2: Fully loosen all the bolts.
   c. Step 3: 15 ft. lbs. (20 Nm).
   d. Step 4: An additional 90 degrees.
   e. Step 5: An additional 90 degrees.
   - Exhaust manifold bracket and torque the bolts to 26 ft. lbs. (35 Nm)

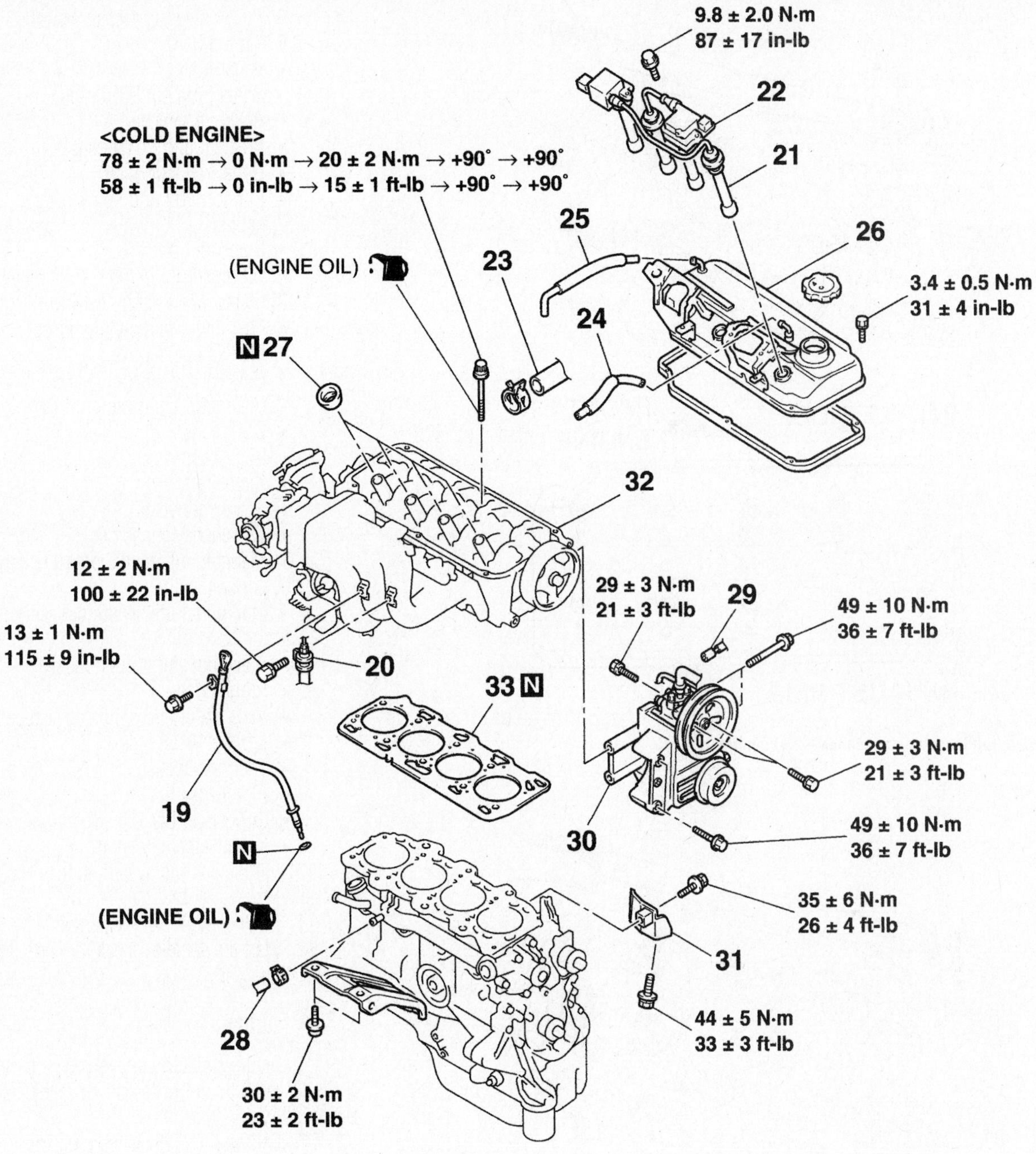

19. OIL DIPSTICK AND DIPSTICK GUIDE
20. PRESSURE HOSE CONNECTION
21. SPARK PLUG CABLE
22. IGNITION COIL
23. RADIATOR UPPER HOSE CONNECTION
24. PCV HOSE CONNECTION
25. BREATHER HOSE CONNECTION
26. ROCKER COVER
27. SPARK PLUG GUIDE OIL SEAL
28. WATER HOSE CONNECTION
- TIMING BELT
29. POWER STEERING PRESSURE SWITCH
30. POWER STEERING OIL PUMP AND BRACKET ASSEMBLY
31. EXHAUST MANIFOLD BRACKET
32. CYLINDER HEAD ASSEMBLY
33. CYLINDER HEAD GASKET

06009-COUP-G18

Cylinder head and related parts—2.4L engine

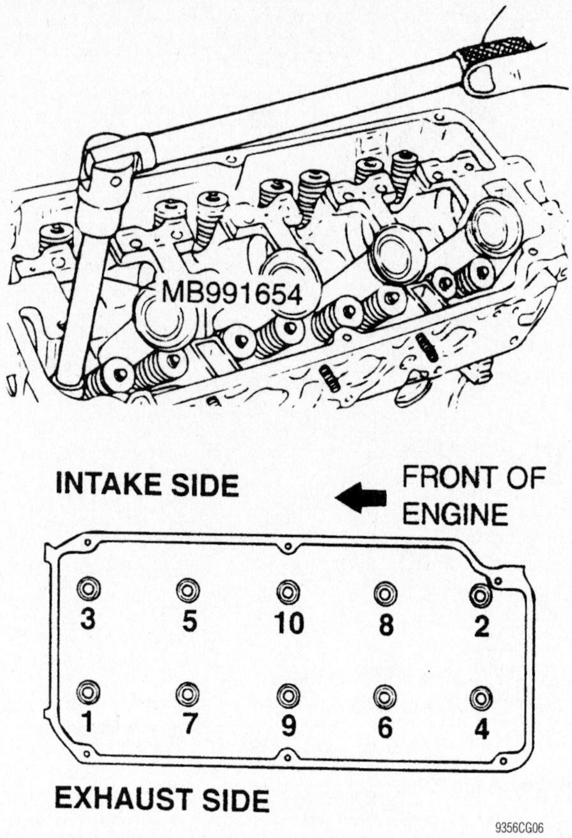

Cylinder head bolt loosening sequence—2.4L engine

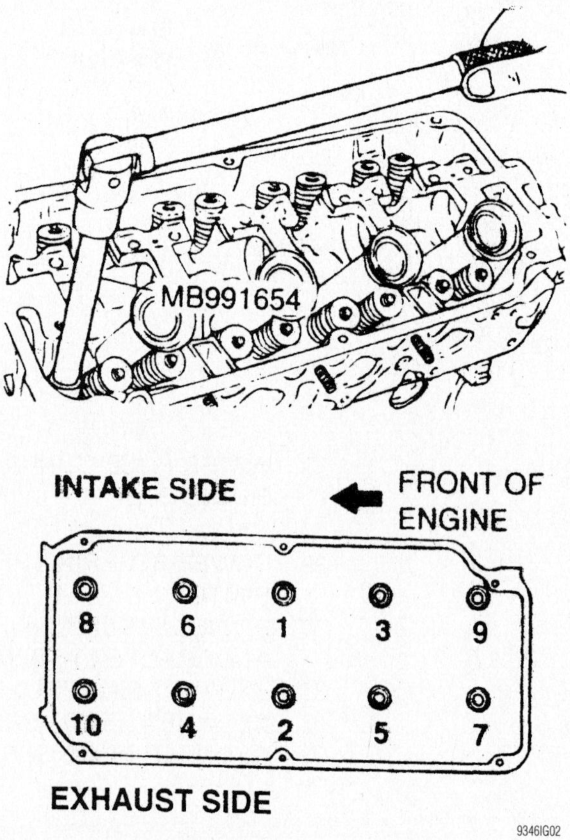

Cylinder head bolt torque sequence—2.4L engine

- Power steering pump and bracket
- Power steering pressure switch
- Timing belt
- Water hose
- Spark plug guide oil seal
- Rocker arm cover
- Breather and PCV hoses
- Upper radiator hose
- Ignition coil and spark plug cables
- Pressure hose
- Oil dipstick and tube. Torque the bolt to 177 inch lbs. (14 Nm).
- Fuel return and supply hoses
- EGR solenoid valve
- EVAP purge solenoid valve
- IAC motor connector
- CMP sensor
- ECT sensor
- Capacitor
- HO_2S connector
- TP sensor connector
- Manifold differential pressure sensor connector
- Ignition failure sensor connector
- Ignition coil connector
- Brake booster vacuum hose
- Purge hose
- Accelerator and adjust as needed
- Front exhaust pipe
- Thermostat case
- Air cleaner
- Strut tower bar
- Negative battery cable

10. Fill the cooling system to the proper level.
11. Fill the engine with clean oil.
12. Start the vehicle, check for leaks and repair if necessary.

3.0L Engine

1. Before servicing the vehicle, refer to the precautions in the beginning of this section.
2. Relieve the fuel system pressure.
3. Drain the cooling system.
4. Remove or disconnect the following:
   - Negative battery cable
   - Timing belt
   - Alternator
   - Intake manifold
   - Exhaust manifold
   - Water inlet pipe
   - Breather and blow by hoses
   - Positive Crankcase Ventilation (PCV) hose
   - Spark plug cables
   - Rocker arm cover
   - Timing belt rear cover
   - Cylinder head and discard the gasket

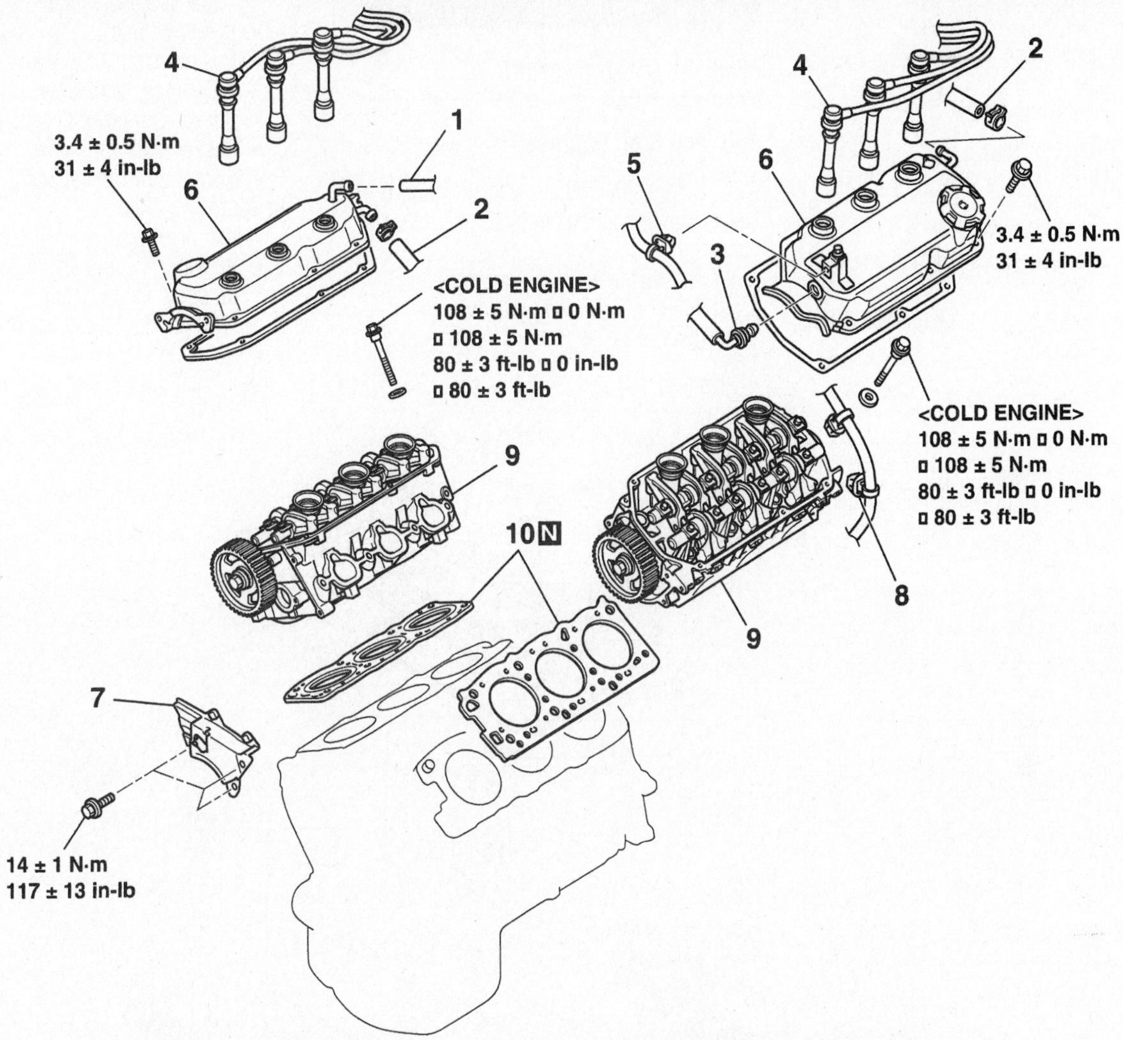

1. BREATHER HOSE
2. BLOW-BY HOSE
3. PCV HOSE
4. SPARK PLUG CABLE
5. BATTERY CABLE CONNECTION
6. ROCKER COVER
7. TIMING BELT REAR COVER
8. BATTERY CABLE AND HARNESS BRACKET
9. CYLINDER HEAD ASSEMBLY
10. CYLINDER HEAD GASKET

06009-COUP-G19

Cylinder heads and related parts—3.0L engine

To install:

5. Clean all mating surfaces of any residual gasket material.
6. Lubricate the threads of the cylinder head bolts with clean engine oil.
7. Install or connect the following:
   - Cylinder head with a new gasket
8. Torque the bolts, in sequence, as follows:
   a. Step 1: 80 ft. lbs. (108 Nm).
   b. Step 2: Loosen all bolts in sequence.
   c. Step 3: 80 ft. lbs. (108 Nm).
9. Install or connect the following:
   - Rear timing belt cover
   - Rocker arm cover

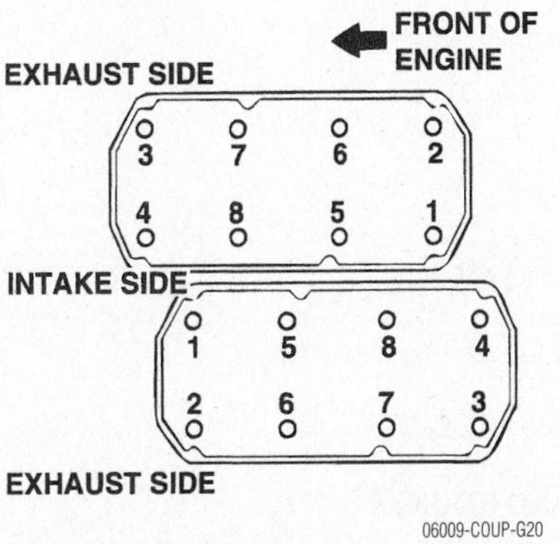

06009-COUP-G20

Cylinder head bolt loosening sequence—3.0L engine

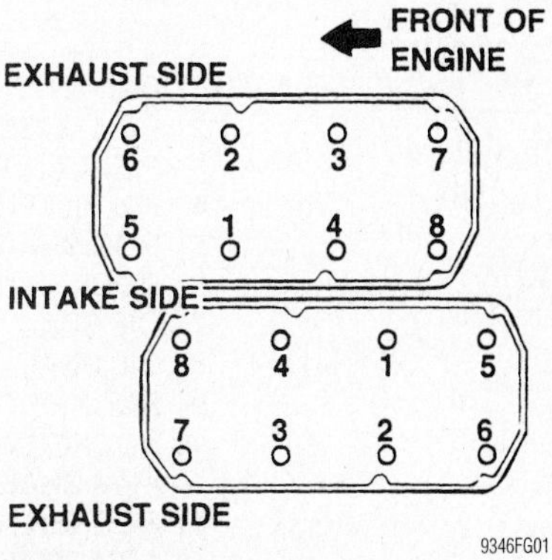

9346FG01

Cylinder head bolt torque sequence—3.0L engine

- Spark plug cables
- PCV and blow by hoses
- Breather hose
- Water inlet pipe
- Exhaust manifold
- Intake manifold
- Alternator
- Timing belt
- Negative battery cable

10. Fill the cooling system to the proper level.
11. Start the vehicle, check for leaks and repair if necessary.

Rocker Arm/Shafts

REMOVAL & INSTALLATION

2.4L and 3.0L Engines

1. Before servicing the vehicle, refer to the precautions in the beginning of this section.
2. Relieve the fuel system pressure using the recommended procedure.
3. Remove or disconnect the following:
   - Negative battery cable
   - Breather hose
   - Positive Crankcase Ventilation (PCV) hose and valve
   - Rocker arm cover
   - Rocker shaft cap
   - Rocker arms and shafts

To install:

4. Install or connect the following:
   - Rocker arms and shafts. When properly aligned, torque the rocker shaft cap to 23 ft. lbs. (31 Nm).
   - Oil seals

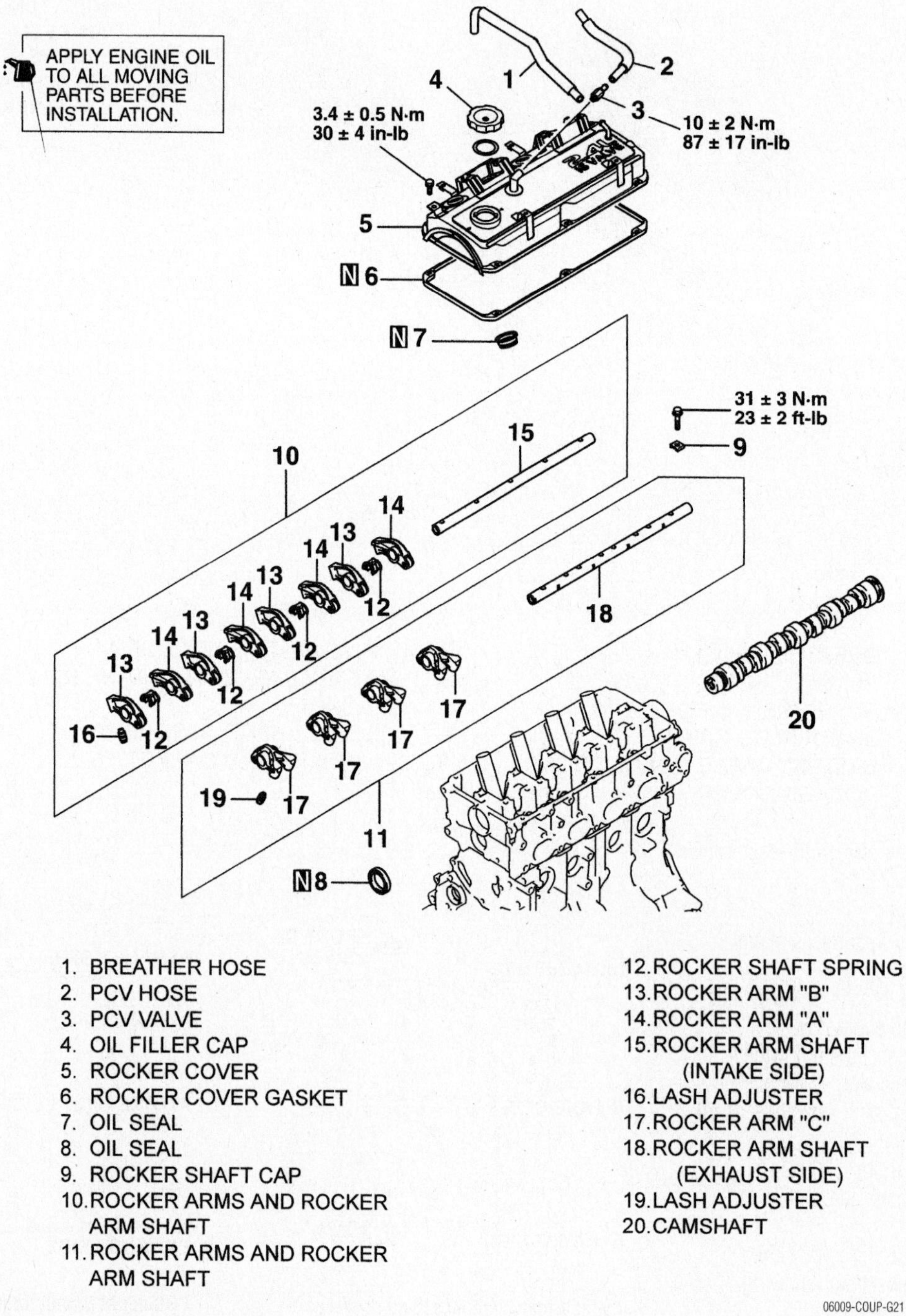

Rocker arm assemblies and related parts—2.4L engine

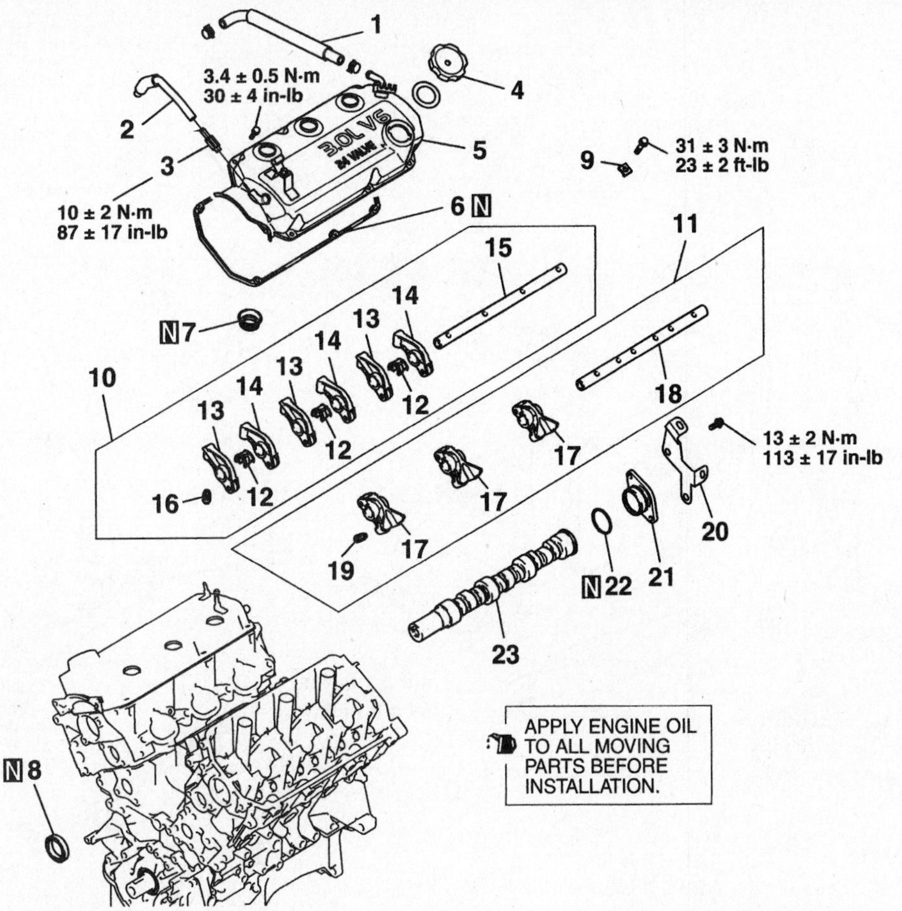

1. BREATHER HOSE
2. POSITIVE CRANKCASE VENTILATION HOSE
3. POSITIVE CRANKCASE VENTILATION VALVE
4. OIL FILLER CAP
5. ROCKER COVER
6. ROCKER COVER GASKET
7. OIL SEAL
8. CAMSHAFT OIL SEAL
9. ROCKER SHAFT CAP
10. ROCKER ARMS AND SHAFT
11. ROCKER ARMS AND SHAFT
12. ROCKER SHAFT SPRING
13. ROCKER ARM A
14. ROCKER ARM B
15. ROCKER ARM SHAFT
16. LASH ADJUSTER
17. ROCKER ARM C
18. ROCKER ARM SHAFT
19. LASH ADJUSTER
20. HARNESS BRACKET
21. THRUST CASE <LEFT BANK ONLY>
22. O-RING <LEFT BANK ONLY>
23. CAMSHAFT

06009-COUP-G22

Rocker arm assemblies and related parts—3.0L engine

- Rocker arm cover with a new gasket
- PCV valve with a new gasket
- PCV hose
- Breather hose
- Negative battery cable

Intake Manifold

REMOVAL & INSTALLATION

2.4L Engine

The intake manifold is a long branch design made of cast aluminum. It is attached to the cylinder head with 8 fasteners.

1. Before servicing the vehicle, refer to the precautions in the beginning of this section.
2. Drain the cooling system.
3. Properly relieve the fuel system pressure.
4. Remove or disconnect the following:
   - Negative battery cable
   - Air cleaner assembly
   - Throttle body assembly
   - Thermostat case assembly
   - Purge hose connection
   - Brake booster vacuum hose
   - Ignition coil connector
   - Fuel injector electrical connectors
   - Manifold differential pressure sensor connector
   - Evaporative Emission (EVAP) purge solenoid valve connector
   - Exhaust Gas Recirculation (EGR) solenoid valve connector
   - High pressure fuel line and fuel return hose
   - Oil dipstick and tube
   - Positive Crankcase Ventilation (PCV) hose

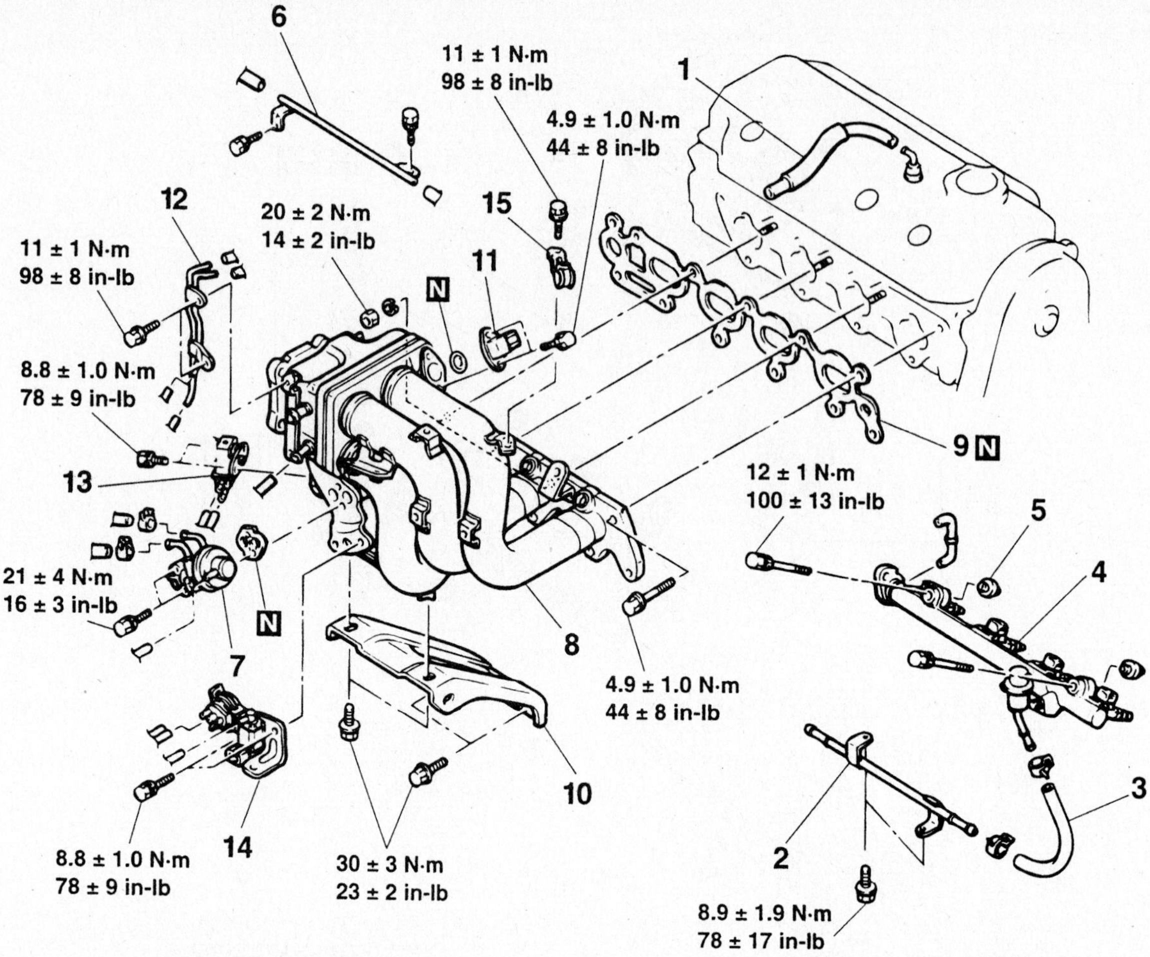

1. PCV HOSE
2. FUEL RETURN PIPE
3. FUEL HOSE
4. FUEL RAIL, INJECTOR AND FUEL PRESSURE REGULATOR
5. INSULATOR
6. VACUUM PIPE
7. EGR VALVE
8. INTAKE MANIFOLD
9. INTAKE MANIFOLD GASKET
10. INTAKE MANIFOLD STAY
11. MANIFOLD DIFFERENTIAL PRESSURE SENSOR
12. VACUUM PIPE
13. EVAPORATIVE EMISSION PURGE SOLENOID VALVE
14. EGR SOLENOID VALVE AND VACUUM CONTROL VALVE
15. ACCELERATOR CABLE CLAMP

9356CG07

Exploded view of the intake manifold—2.4L engine

- Fuel return pipe
- Fuel hose
- Fuel rail assembly, including the injectors and pressure regulator
- Insulator and vacuum pipe
- EGR valve
- Intake manifold retainers, manifold and gasket

5. If replacing the manifold, remove the following and transfer to the new manifold:
   - Intake manifold stay
   - Manifold differential pressure sensor
   - Vacuum pipe
   - Evaporative Emission (EVAP) purge solenoid valve
   - EGR solenoid valve and vacuum control valve
   - Accelerator cable clamp

To install:

6. Install or connect the following:
   - Intake manifold with a new gasket and torque the bolts, in sequence, to 14 ft. lbs. (19 Nm)
   - EGR valve
   - Vacuum pipe and insulator
   - Fuel rail assembly
   - Fuel hose and return pipe
   - PCV hose
   - Oil dipstick tube and dipstick
   - Fuel return hose
   - High pressure fuel hose. Apply a small quantity of oil to the hose O-ring and then installing using a twisting motion. Make sure that the O-ring does not get twisted or clamped during installation.
   - EGR solenoid valve connector
   - EVAP purge solenoid valve connector
   - Manifold differential pressure sensor connector

- Injector connector
- Ignition coil connector
- Brake booster vacuum hose connection
- Purge hose connection
- Thermostat case assembly
- Throttle body assembly
- Air cleaner assembly
- Negative battery cable

7. Fill the cooling system to the proper level.
8. Start the vehicle, check for leaks and repair if necessary.

3.0L Engine

1. Before servicing the vehicle, refer to the precautions in the beginning of this section.
2. Properly relieve the fuel system pressure.
3. Remove or disconnect the following:
   - Negative battery cable
   - Intake manifold plenum
   - Fuel injector connector
   - Fuel supply and return hoses
   - Vacuum hose
   - Fuel rail, injectors and pressure regulator as an assembly
   - Insulators
   - Positive Crankcase Ventilation (PCV) hose
   - Right front upper timing belt cover and bracket
   - Intake manifold and discard the gasket

To install:

4. Clean all mating surfaces of any residual gasket material.
5. Install or connect the following:

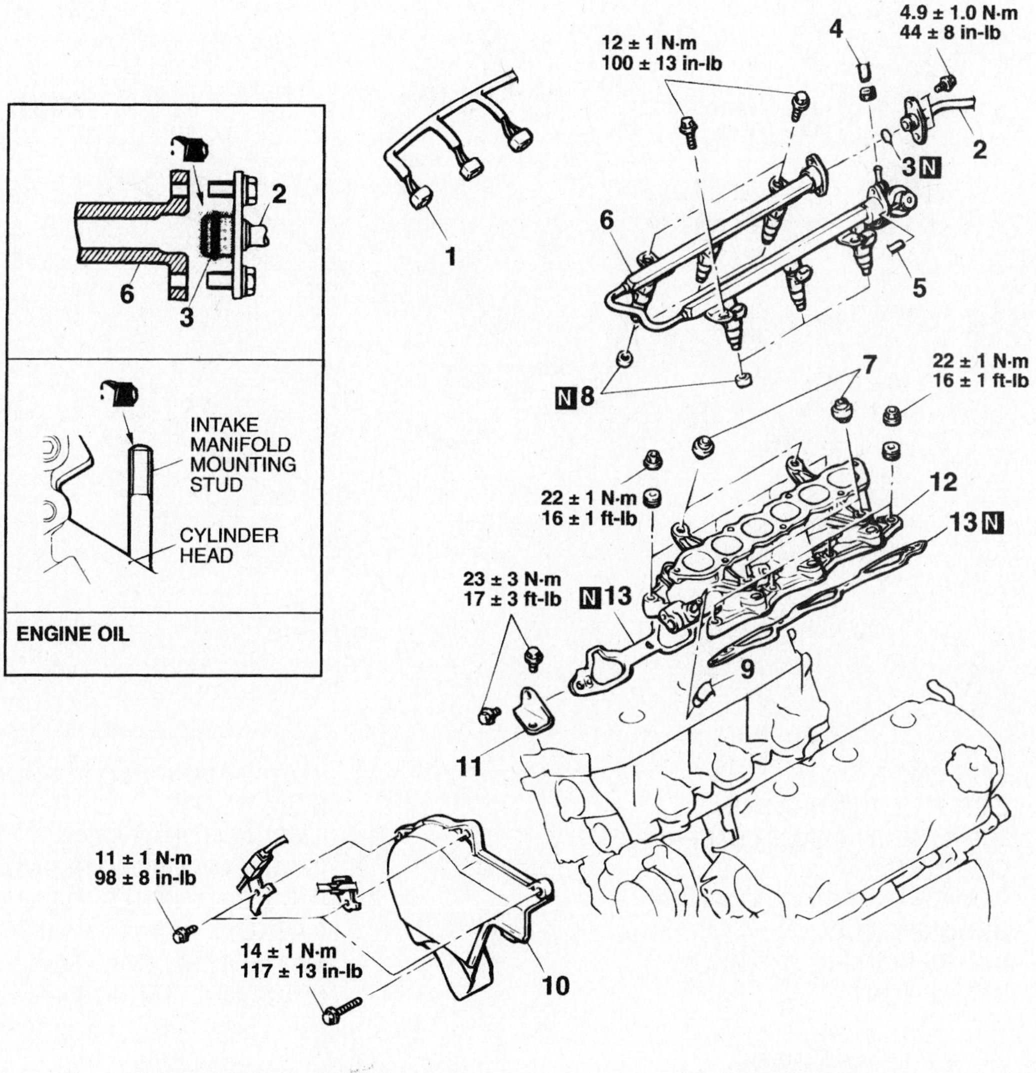

1. INJECTOR CONNECTOR
2. FUEL HIGH-PRESSURE HOSE CONNECTION
3. O-RING
4. FUEL RETURN HOSE CONNECTION
5. VACUUM HOSE CONNECTION
6. FUEL RAIL, INJECTOR AND FUEL PRESSURE REGULATOR
7. INSULATORS
8. INSULATORS
9. PCV HOSE CONNECTION
10. TIMING BELT FRONT UPPER COVER, RIGHT
11. BRACKET
12. INTAKE MANIFOLD
13. INTAKE MANIFOLD GASKET

06009-COUP-G24

Intake manifold plenum—3.0L engine

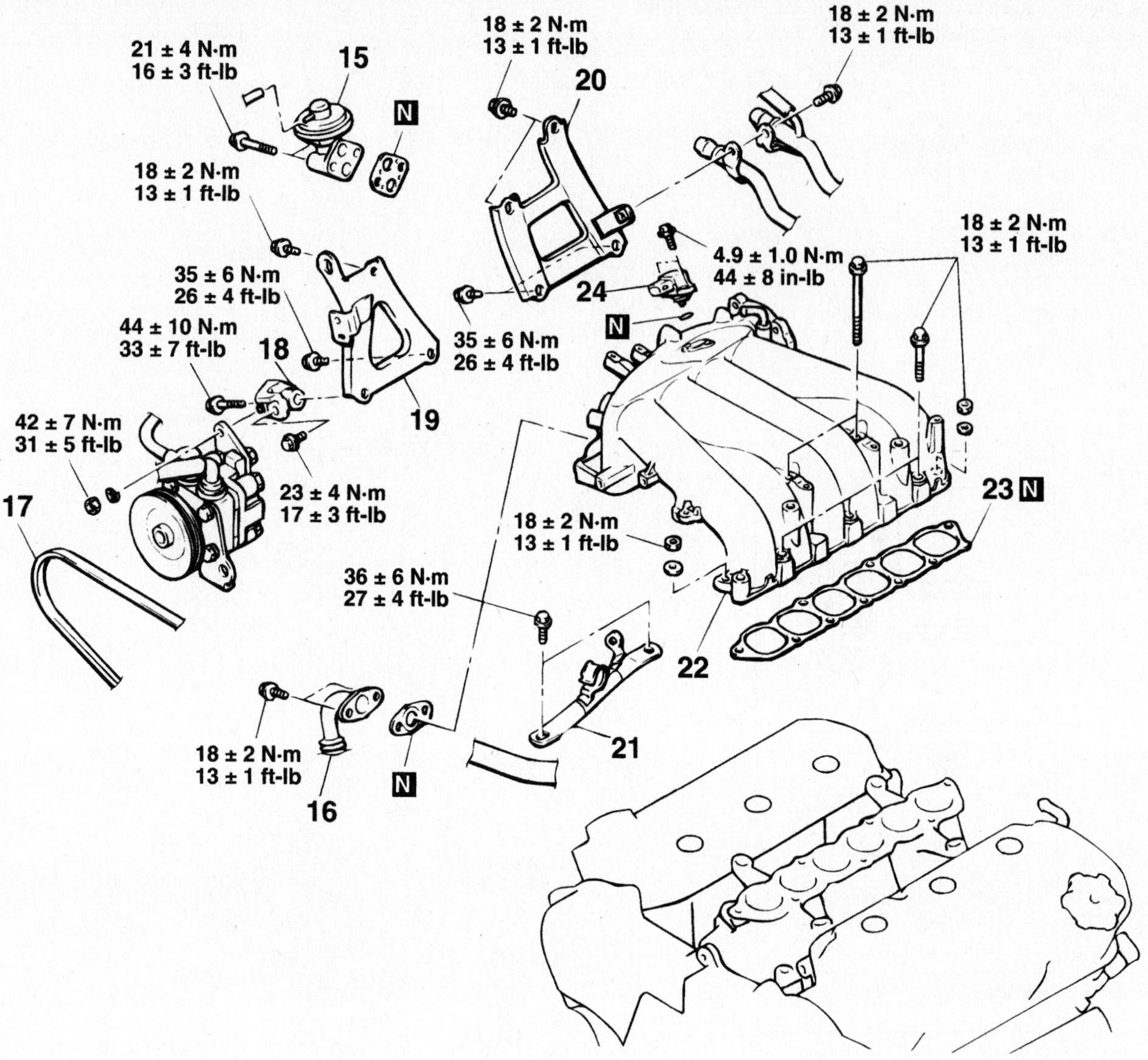

15. EGR VALVE
16. EGR PIPE CONNECTION
17. DRIVE BELT <POWER STEERING OIL PUMP>
18. POWER STEERING OIL PUMP BRACKET STAY
19. INTAKE MANIFOLD PLENUM STAY, FRONT
20. INTAKE MANIFOLD PLENUM STAY, REAR
21. ENGINE MOUNT STAY
22. INTAKE MANIFOLD PLENUM
23. INTAKE MANIFOLD PLENUM GASKET
24. MANIFOLD ABSOLUTE PRESSURE SENSOR

06009-COUP-G23

Intake manifold and related parts—3.0L engine

- Intake manifold with a new gasket. Torque the nuts as follows:
  a. Right bank to 62 inch lbs. (7 Nm).
  b. Left bank to 16 ft. lbs. (22 Nm).
  c. Right bank to 16 ft. lbs. (22 Nm).
  d. Left bank to 16 ft. lbs. (22 Nm).
  e. Right bank to 16 ft. lbs. (22 Nm).
- Right front upper timing belt cover and bracket and torque the bolts to 10 ft. lbs. (15 Nm)
- PCV hose
- Fuel rail assembly and torque the bolts to 102 inch lbs. (12 Nm)
- Vacuum hose
- Fuel supply and return hoses
- Fuel injector electrical connector
- Negative battery cable

6. Start the vehicle, check for leaks and repair if necessary.

Exhaust Manifold

REMOVAL & INSTALLATION

2.4L Engine

1. Before servicing the vehicle, refer to the precautions in the beginning of this section.

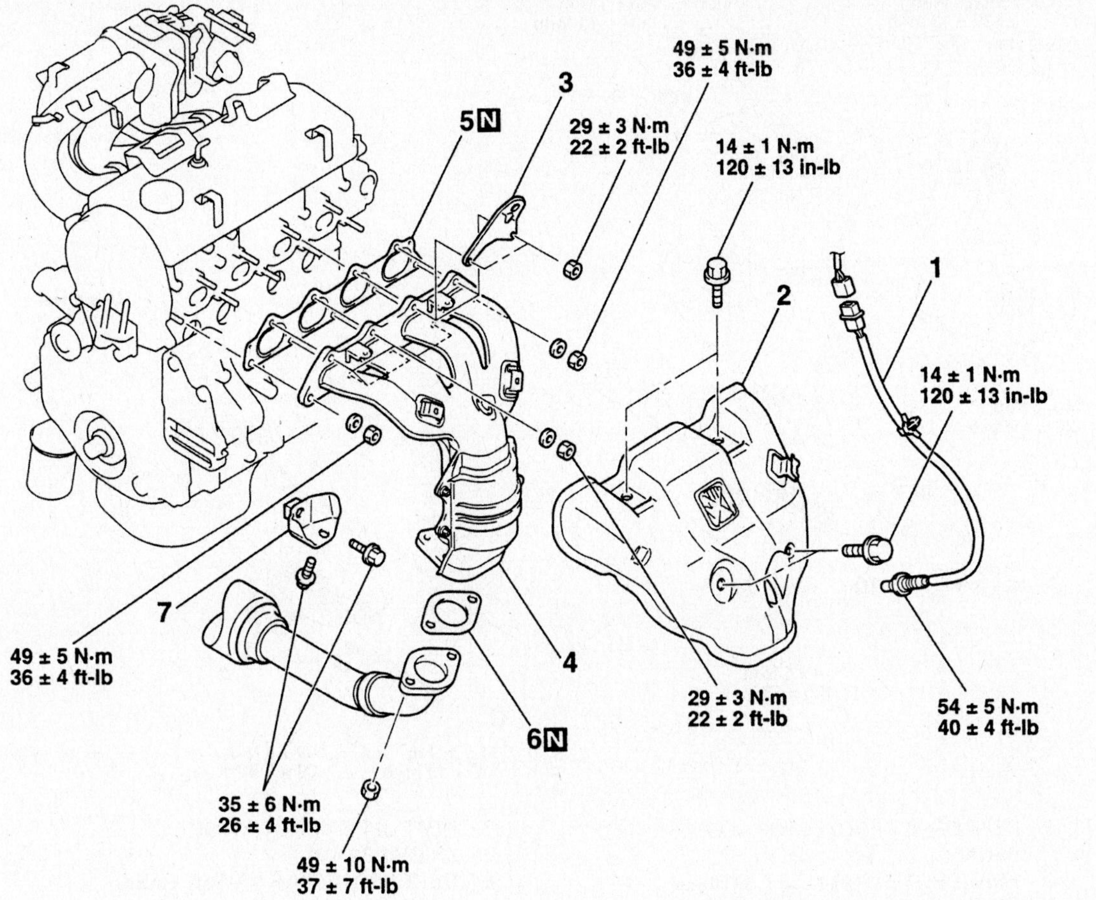

Exploded view of the exhaust manifold—2.4L engine

2. Remove or disconnect the following:
   - Negative battery cable
   - Front Heated Oxygen Sensor (HO_2S), using a special tool
   - Exhaust manifold heat shield
   - Exhaust pipe from the exhaust manifold
   - Exhaust manifold and discard the gaskets
   - Exhaust manifold bracket, if necessary

To install:

3. Thoroughly clean all parts. Clean all sealing surfaces of the manifold and cylinder head. Check the manifold gasket surface for flatness with a straightedge and feeler gauge. The surface must be flat within 0.006 in. (0.15mm) per foot (30cm) of manifold length. Inspect the manifold for cracks or distortion. Replace if necessary.
4. Install or connect the following:
   - Exhaust manifold bracket, if removed. Tighten the bolts to 26 ft. lbs. (35 Nm).
   - New gaskets
   - Exhaust manifold. Torque the M8 nuts to 22 ft. lbs.(29 Nm) and the M10 bolts to 36 ft. lbs.(49 Nm).
   - Exhaust pipe to the exhaust manifold. Tighten the nuts to 37 ft. lbs. (49 Nm).
   - Exhaust manifold heat shield
   - HO_2S
   - Negative battery cable
5. Start the engine and allow it to idle while inspecting the manifold for exhaust leaks.

3.0L Engine

1. Before servicing the vehicle, refer to the precautions in the beginning of this section.
2. Remove or disconnect the following:
   - Battery and tray
   - Front exhaust pipe
   - Air cleaner assembly
   - Oil dipstick guide
   - Strut tower bar
   - Upper and lower heat shields
   - Exhaust Gas Recirculation (EGR) pipe, left side only
   - Exhaust manifold and discard the gasket

To install:

3. Clean all mating surfaces of any residual gasket material.
4. Install or connect the following:
   - Exhaust manifold with a new gasket and torque the bolts to 33 ft. lbs. (44 Nm)
   - EGR pipe, if removed. Torque the bolt to 13 ft. lbs. (18 Nm).
   - Upper and lower heat shields and

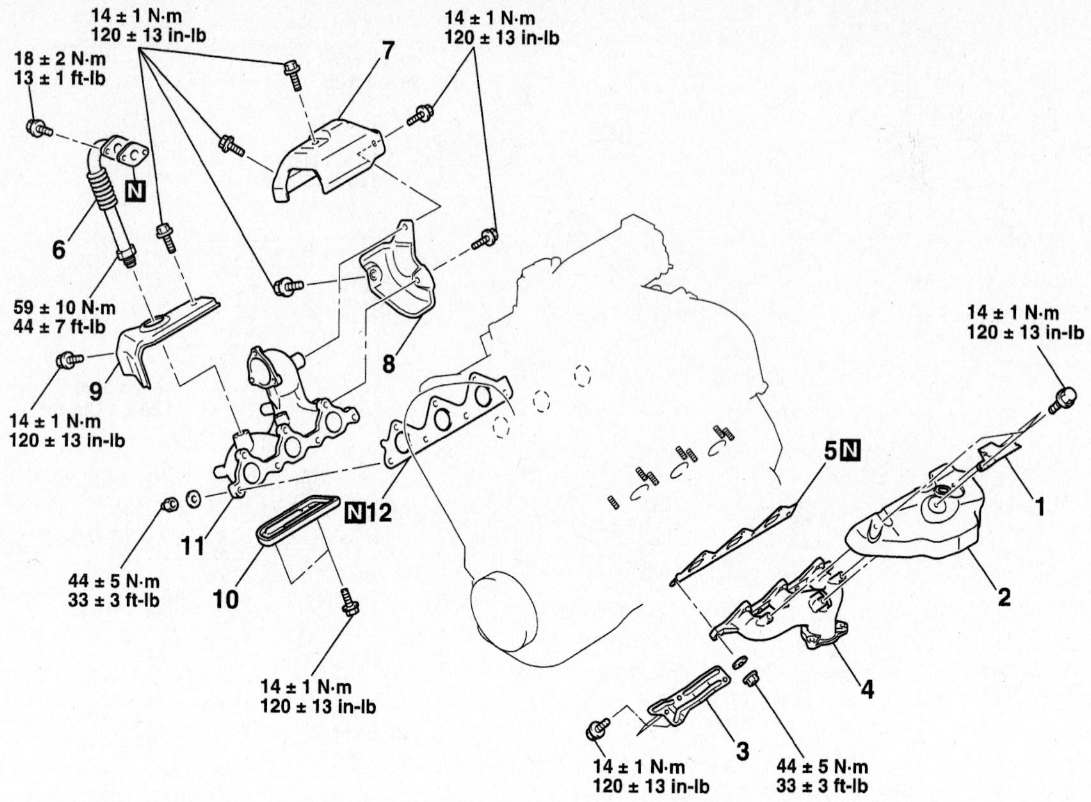

1. HEAT UPPER PROTECTOR <LEFT BANK>
2. HEAT PROTECTOR <LEFT BANK>
3. HEAT LOWER PROTECTOR <LEFT BANK>
4. EXHAUST MANIFOLD <LEFT BANK>
5. EXHAUST MANIFOLD GASKET <LEFT BANK>
6. EGR PIPE
7. HEAT UPPER PROTECTOR <RIGHT BANK>
8. HEAT PROTECTOR <RIGHT BANK>
9. HEAT FRONT PROTECTOR <RIGHT BANK>
10. HEAT LOWER PROTECTOR <RIGHT BANK>
11. EXHAUST MANIFOLD <RIGHT BANK>
12. EXHAUST MANIFOLD GASKET <RIGHT BANK>

06009-COUP-G25

Exhaust manifold and related parts—3.0L engine

torque the bolts to 10 ft. lbs. (15 Nm)

- Strut tower bar
- Oil dipstick guide
- Air cleaner assembly
- Front exhaust pipe
- Battery and tray

5. Start the vehicle, check for leaks and repair if necessary.

Front Crankshaft Seal

REMOVAL & INSTALLATION

2.4L Engine

The timing belt must be removed for this procedure. Use care that all timing marks are aligned after installation or the engine will be damaged.

1. Drain the engine oil.
2. Before servicing the vehicle, refer to the precautions in the beginning of this section.
3. Disconnect the negative battery cable.
4. Remove or disconnect the following:
   - Timing belt
   - Crankshaft Position (CKP) sensor
   - Crankshaft timing belt sprocket using a gear/sprocket puller
   - Crankshaft sensing blade
   - Timing belt "B"
   - Crankshaft sprocket "B"

➡Be careful not to nick the seal surface of the crankshaft or the seal bore.

- Key
- Front crankshaft seal using a suitable spanner tool

➡Be careful not to damage the seal contact area of the crankshaft.

To install:

5. Lubricate the new oil seal lip with engine oil.
6. Install or connect the following:
   - New crankshaft oil seal using Oil Seal Installer Tool MD998375, until it is flush with the front cover
   - Crankshaft timing belt sprocket using Tool No. 6792
   - Key

➡To prevent the crankshaft bolt from loosening, clean the mating surfaces to the crankshaft, sprocket "B", crankshaft sensing blade and crankshaft at the positions shown in the accompanying figure.

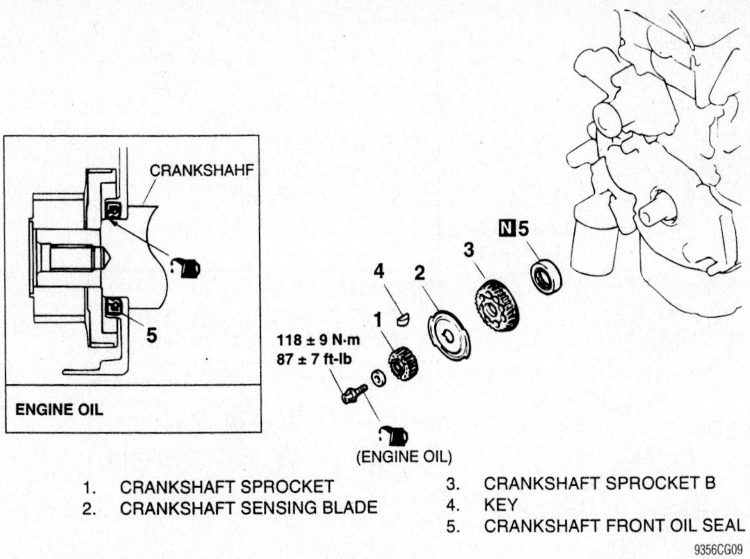

Exploded view of the front crankshaft oil seal—2.4L engine

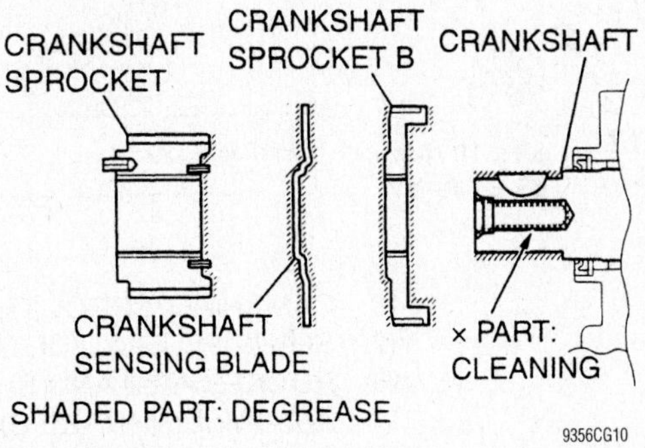

View of the proper installation and mating order cleaning surfaces—2.4L engine

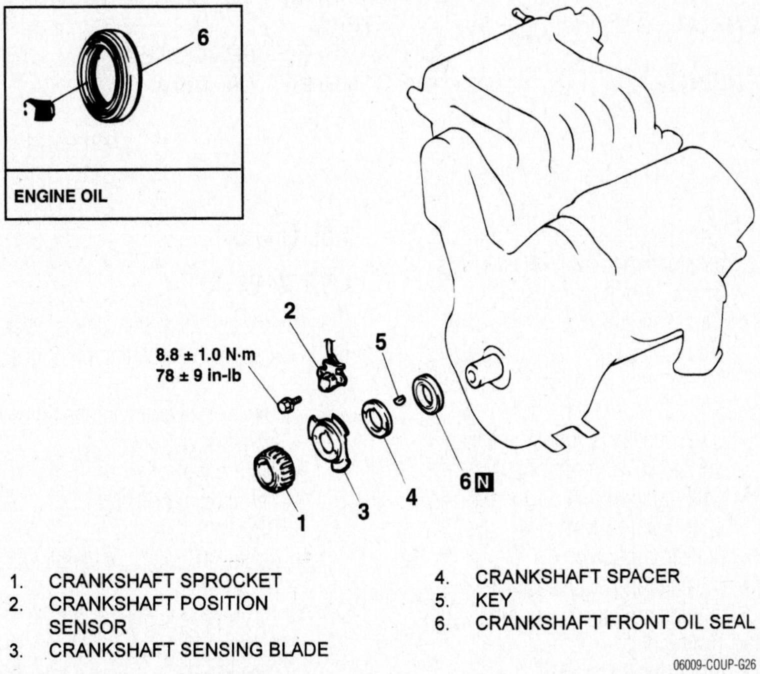

Crankshaft front seal and related parts—3.0L engine

- Crankshaft sprocket "B"
- Timing belt "B"
- Crankshaft sensing blade, so it faces the direction shown by the figure
- Crankshaft sprocket

7. Apply a very small amount of engine oil to the seating surface and thread of the crankshaft bolt
8. Use the crankshaft pulley mounting bolt to secure the special spanner tool
9. Tighten the crankshaft sprocket bolt to 87 ft. lbs. (118 Nm).
   - CKP sensor
   - Timing belt
   - Negative battery cable

WARNING

Operating the engine without the proper amount and type of engine oil will result in severe engine damage.

10. Fill the engine with clean oil.
11. Start the engine and check for leaks.

3.0L Engine

1. Before servicing the vehicle, refer to the precautions in the beginning of this section.
2. Remove or disconnect the following:
   - Negative battery cable
   - Timing belt
   - Crankshaft sprocket
   - Crankshaft Position (CKP) sensor
   - Crankshaft sensing blade
   - Crankshaft spacer and key
   - Front oil seal

To install:

3. Lubricate the oil seal lip with clean engine oil.
4. Install or connect the following:
   - New oil seal with Special Tool MD998717
   - Crankshaft key and spacer
   - Crankshaft sensing blade
   - CKP sensor
   - Crankshaft sprocket
   - Timing belt
   - Negative battery cable

Camshaft and Valve Lifters

REMOVAL & INSTALLATION

2.4L Engine

1. Before servicing the vehicle, refer to the precautions in the beginning of this section.
2. Remove or disconnect the following:
   - Negative battery
   - Air cleaner assembly

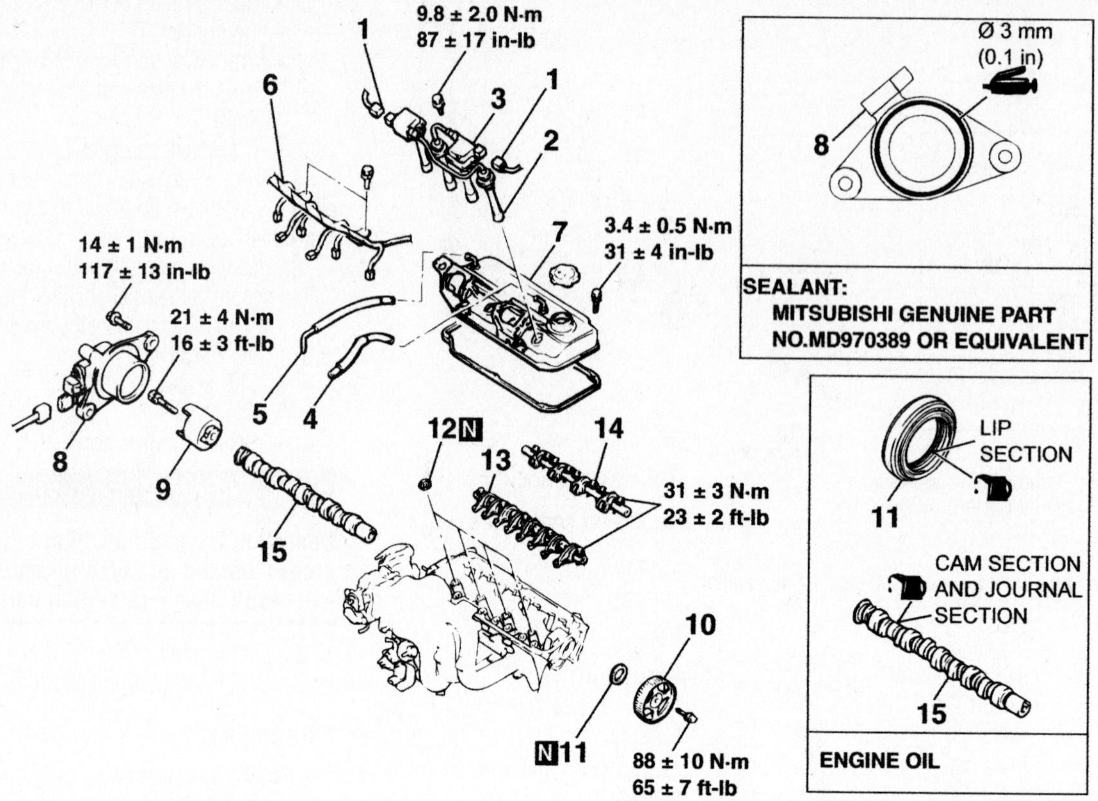

1. IGNITION COIL CONNECTOR
2. SPARK PLUG CABLE
3. IGNITION COIL
4. PCV HOSE
5. BREATHER HOSE
6. CONTROL HARNESS CONNECTION
7. ROCKER COVER
8. CAMSHAFT POSITION SENSOR SUPPORT
9. CAMSHAFT POSITION SENSING CYLINDER
10. CAMSHAFT SPROCKET
12. SPARK PLUG GUIDE OIL SEAL
13. ROCKER ARM AND SHAFT ASSEMBLY (INTAKE SIDE)
14. ROCKER ARM AND SHAFT ASSEMBLY (EXHAUST SIDE)
15. CAMSHAFT

CAMSHAFT OIL SEAL REMOVAL STEPS

10. CAMSHAFT SPROCKET
11. CAMSHAFT OIL SEAL

06009-COUP-G27

Camshaft and related parts—2.4L engine

- Timing belt
- Ignition coils and spark plug cables
- Positive Crankcase Ventilation (PCV) hose
- Breather hose
- Rocker arm cover
- Camshaft Position (CMP) sensor support
- CMP sensing cylinder
- Camshaft sprocket
- Spark plug guide oil seal
- Rocker arm and shaft assembly
- Camshaft

To install:

3. Install or connect the following:
   - Camshaft
   - Rocker arm and shaft assemblies in there proper position. Torque the assemblies to 23 ft. lbs. (31 Nm).
   - Spark plug guide oil seals
   - Camshaft sprocket with special Tools MB990767 and MD998719. Torque the bolt to 65 ft. lbs. (88 Nm).
   - CMP sensing cylinder and torque the bolt to 16 ft. lbs. (21 Nm)
   - CMP support and torque the bolt to 10 ft. lbs. (15 Nm)
   - Rocker arm cover
   - Breather and PCV hoses
   - Ignition coil and spark plug cables
   - Timing belt
   - Air cleaner assembly
   - Negative battery cable

3.0L Engine

LEFT BANK

1. Before servicing the vehicle, refer to the precautions in the beginning of this section.
2. Remove or disconnect the following:
   - Negative battery
   - Timing belt
   - Thermostat housing
   - Blow-by hose
   - Positive Crankcase Ventilation (PCV) hose
   - Spark plug cables
   - Rocker arm cover
   - Rocker arm and shaft
   - Camshaft sprocket

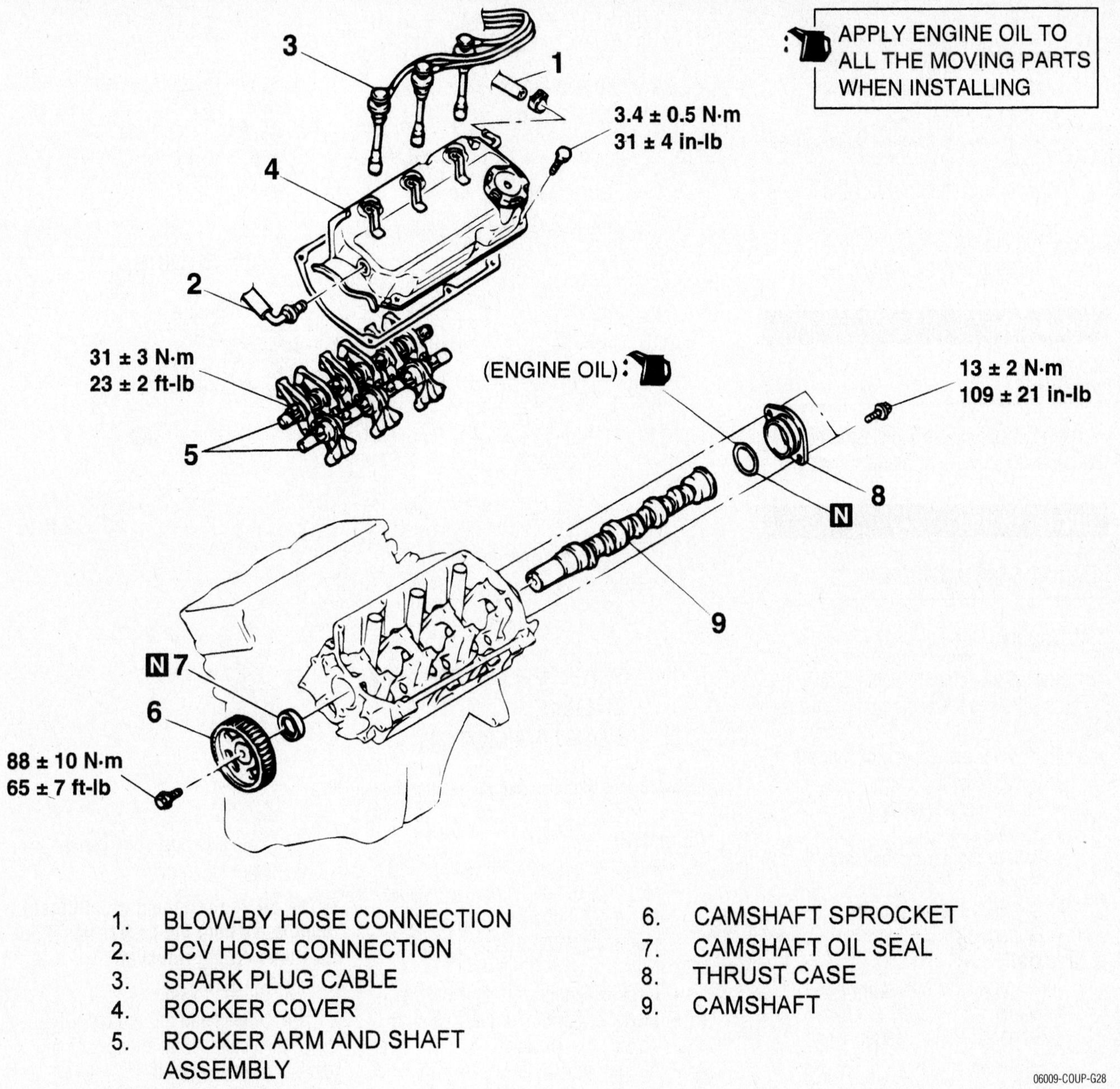

Camshafts and related parts—3.0L engine

- Thrust case
- Camshaft

To install:

3. Install or connect the following:
   - Camshaft
   - Thrust case and torque the bolt to 109 inch lbs. (13 Nm)
   - Camshaft sprocket and torque the bolt to 65 ft. lbs. (88 Nm)
   - Rocker arm and shaft assembly and torque to 23 ft. lbs. (31 Nm)
   - Rocker arm cover and torque the bolts to 35 inch lbs. (4 Nm)
   - Spark plug cables
   - PCV and blow-by hoses
   - Thermostat housing
   - Timing belt
   - Negative battery cable

RIGHT BANK

1. Before servicing the vehicle, refer to the precautions in the beginning of this section.
2. Remove or disconnect the following:
   - Negative battery
   - Timing belt
   - Intake manifold plenum
   - Breather hose
   - Blow-by hose
   - Spark plug cables
   - Rocker arm cover
   - Rocker arm and shaft assembly with Tool MD998443
   - Distributor
   - Camshaft sprocket with Tool MB990767 and MD998715
   - Camshaft

To install:

3. Install or connect the following:
   - Camshaft and a new oil seal, if removed
   - Camshaft sprocket and torque the bolt to 65 ft. lbs. (88 Nm)
   - Distributor
4. Rotate the camshaft until the dowel

pin on the front end is properly positioned.

5. Install or connect the following:
   - Rocker arm cover and torque the bolts to 35 inch lbs. (4 Nm)
   - Spark plug cables
   - Blow-by and breather hose connections
   - Intake manifold plenum and torque the bolts to 13 ft. lbs. (18 Nm)
   - Timing belt
   - Negative battery cable

Valve Lash

ADJUSTMENT

The engines in these vehicles do not require periodic valve lash adjustment.

Starter Motor

REMOVAL & INSTALLATION

2.4L Engine

1. Before servicing the vehicle, refer to the precautions in the beginning of this section.
2. Remove or disconnect the following:
   - Negative battery cable
   - Air cleaner assembly
   - Starter cover
   - Battery cable from the starter
   - Solenoid connector
   - Starter-to-transaxle bolts
   - Starter

To install:

3. Install or connect the following:
   - Starter
   - Starter-to-transaxle bolt(s) and torque the bolt to 23 ft. lbs. (30 Nm)
   - Electrical connectors to the starter
   - Starter cover and tighten the bolt to 44 inch lbs. (4.9 Nm)
   - Air cleaner assembly
   - Negative battery cable to the shock tower

3.0L Engine

1. Before servicing the vehicle, refer to the precautions in the beginning of this section.
2. Remove or disconnect the following:
   - Negative battery cable from the shock tower
   - Air cleaner resonator
   - Starter motor cover
3. Starter electrical connectors
   - Starter assembly

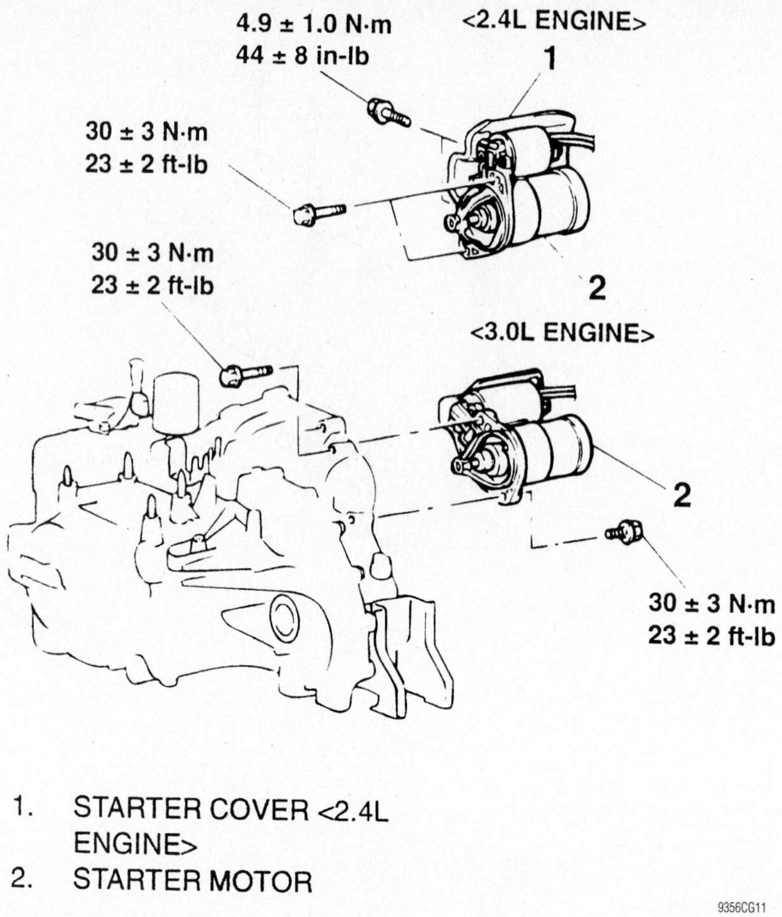

Exploded view of the starter and related components

To install:

4. Install or connect the following:
   - Starter assembly. Torque the starter bolt to 23 ft. lbs. (30 Nm) and the rear bracket bolt to 8 inch lbs. (1 Nm).
   - Electrical connectors to the starter
   - Starter cover and torque the bolt to 15 inch lbs. (6 Nm)
   - Air cleaner resonator
   - Negative battery cable to the shock tower

Oil Pan

REMOVAL & INSTALLATION

2.4L Engine

1. Before servicing the vehicle, refer to the precautions in the beginning of this section.
2. Drain the engine oil. Install a new drain plug gasket.
3. Remove or disconnect the following:
   - Negative battery cable
   - Oil dipstick
   - Front exhaust pipe
   - Transaxle dust shield/bellhousing cover

➡The oil pan retainers are different lengths, so note the proper installed locations during removal.

   - Oil pan retainers
   - Oil pan and discard the gasket

4. Clean all mating surfaces of any residual gasket material.

To install:

⁂ WARNING

Make sure to install the oil pan within 5 minutes of applying the sealant!

5. Using a suitable gasket sealant apply a 0.2 in. (4mm) bead at the oil pump-to-engine block parting line.
6. Install or connect the following:
   - Oil pan with a new gasket. Torque the bolts to the specifications shown in the accompanying figure.
   - Transaxle dust shield/bellhousing cover. Torque the bolts to the specifications shown in the accompanying figure.

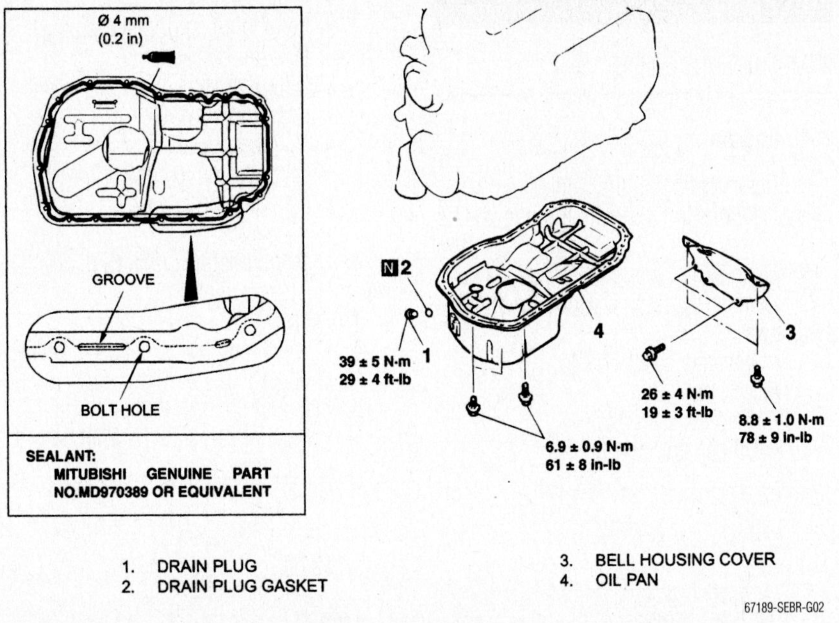

Exploded view of the oil pan and related components—2.4L engine

- Front exhaust pipe
- Oil dipstick
- Negative battery cable

7. Fill the engine with clean oil.

⁂ WARNING

After installing the oil pan, you MUST wait at least one hour before starting the engine.

- Start the vehicle, check for leaks and repair if necessary.

3.0L Engine

1. Before servicing the vehicle, refer to the precautions in the beginning of this section.
2. Drain the engine oil.
3. Remove the engine support module as follows:
   - Negative battery cable
   - Front exhaust pipe

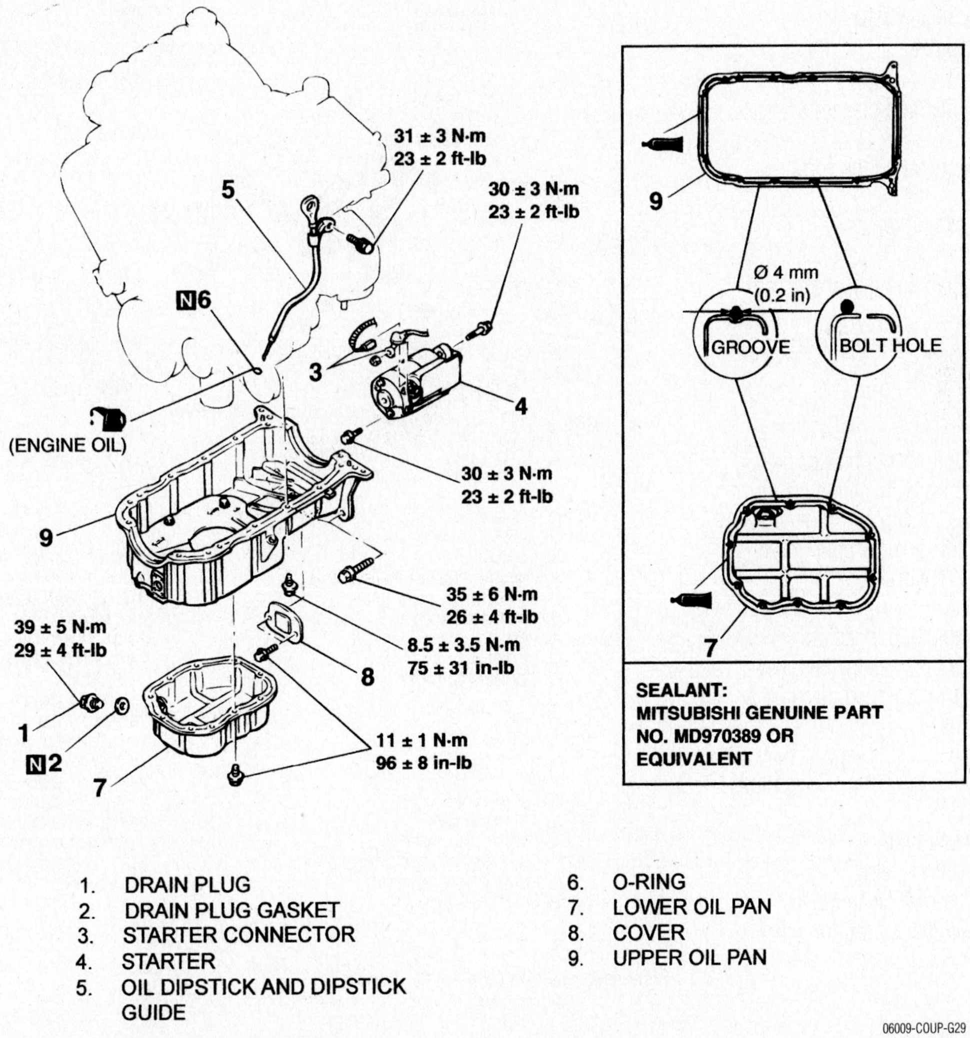

Upper and lower oil pans and related parts—3.0L engine

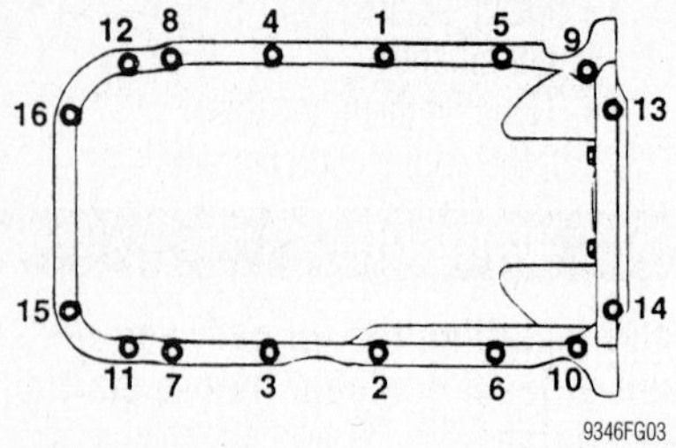

Tighten the upper oil pan bolts in sequence—3.0L engine

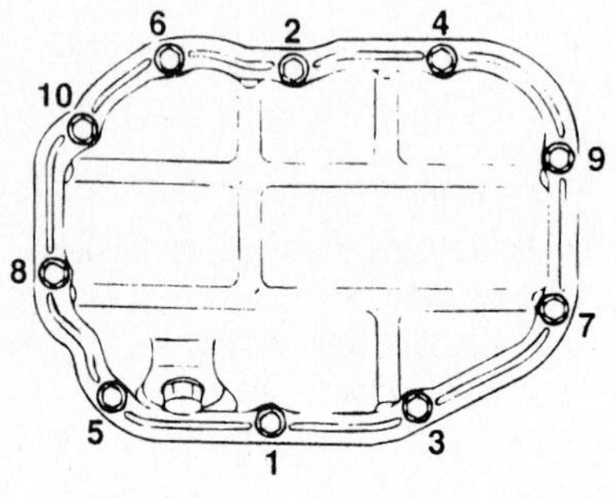

Tighten the lower oil pan bolts in sequence—3.0L engine

- Lower oil pan and discard the gasket
- Starter electrical connectors
- Starter
- Oil dipstick and guide
- Cover
- Upper oil pan and discard the gasket

4. Clean all mating surfaces of any residual gasket material.

To install:

5. Install or connect the following:
   - Upper oil pan with a new gasket. Torque the bolts, in sequence, to 53 inch lbs. (6 Nm).
   - Cover and torque the bolts to 96 inch lbs. (11 Nm)
   - Starter
   - Starter electrical connectors
   - Oil dipstick and guide with a new O-ring and torque the bolt to 36 ft. lbs. (48 Nm)
   - Lower oil pan with a new gasket and torque the bolts to 96 inch lbs. (11 Nm)
   - Front exhaust pipe
   - Negative battery cable
6. Fill the engine with clean oil.
   - Start the vehicle, check for leaks and repair if necessary.

Oil Pump

REMOVAL & INSTALLATION

2.4L Engine

1. Before servicing the vehicle, refer to the precautions in the beginning of this section.
2. Drain the engine oil.
3. Remove the engine support module as follows:
   - Negative battery cable
   - Oil filter
   - Oil pressure switch
   - Oil pan and screen
   - Flange bolt
   - Relief plug, spring and plunger
   - Oil filter bracket
   - Oil pump case
   - Front case gasket
   - Oil pump cover
   - Oil pump driven gear
   - Oil pump drive gear
   - Oil pump
4. Clean all mating surfaces of any residual gasket material.

To install:

5. Prime the oil pump before installation.
6. Install or connect the following:
   - New oil seal into the front case
   - Oil pump

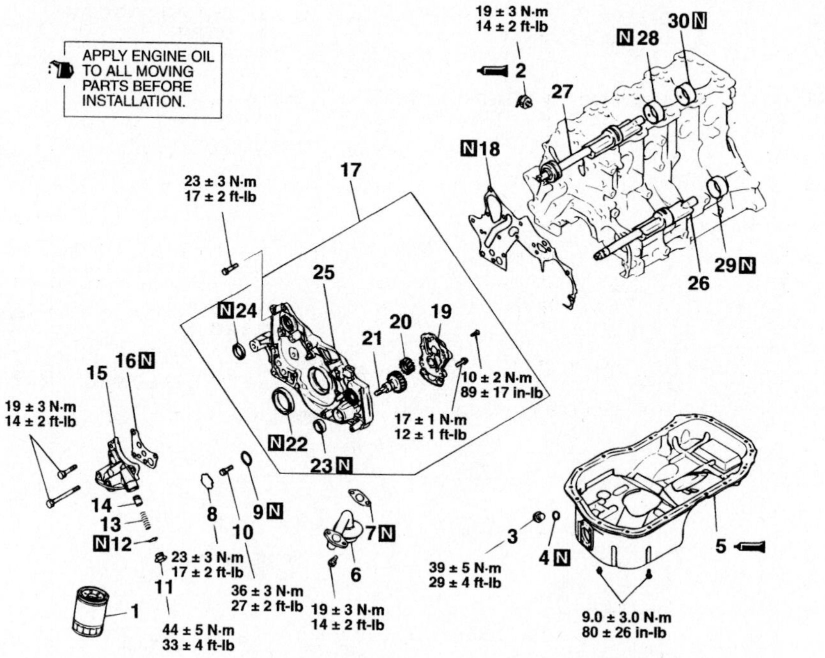

1. OIL FILTER
2. ENGINE OIL PRESSURE SWITCH
3. DRAIN PLUG
4. DRAIN PLUG GASKET
5. OIL PAN
6. OIL SCREEN
7. OIL SCREEN GASKET
8. PLUG
9. O-RING
10. FLANGE BOLT
11. RELIEF PLUG
12. GASKET
13. RELIEF SPRING
14. RELIEF PLUNGER
15. OIL FILTER BRACKET
16. OIL FILTER BRACKET GASKET
17. OIL PUMP CASE ASSEMBLY
18. FRONT CASE GASKET
19. OIL PUMP COVER
20. OIL PUMP DRIVEN GEAR
21. OIL PUMP DRIVE GEAR
22. CRANKSHAFT FRONT OIL SEAL
23. OIL PUMP OIL SEAL
24. COUNTERBALANCE SHAFT OIL SEAL
25. FRONT CASE
26. COUNTERBALANCE SHAFT, LEFT
27. COUNTERBALANCE SHAFT, RIGHT
28. COUNTERBALANCE SHAFT, FRONT BEARING
29. COUNTERBALANCE SHAFT, REAR BEARING, LEFT
30. COUNTERBALANCE SHAFT, REAR BEARING, RIGHT

06009-COUP-G30

Front case and oil pump—2.4L engine

- Oil pump drive gear
- Oil pump driven gear
- Oil pump cover
- Front case gasket
- Oil pump case
- Oil filter bracket
- Relief plunger, spring, gasket and plug
- Flange bolt
- Oil pan and screen
- Oil pressure switch
- New oil filter
- Negative battery cable

7. Fill the engine with clean oil.
8. Start the vehicle, check for leaks and repair if necessary.

3.0L Engine

1. Before servicing the vehicle, refer to the precautions in the beginning of this section.
2. Drain the engine oil.
3. Remove the engine support module as follows:
   - Negative battery cable

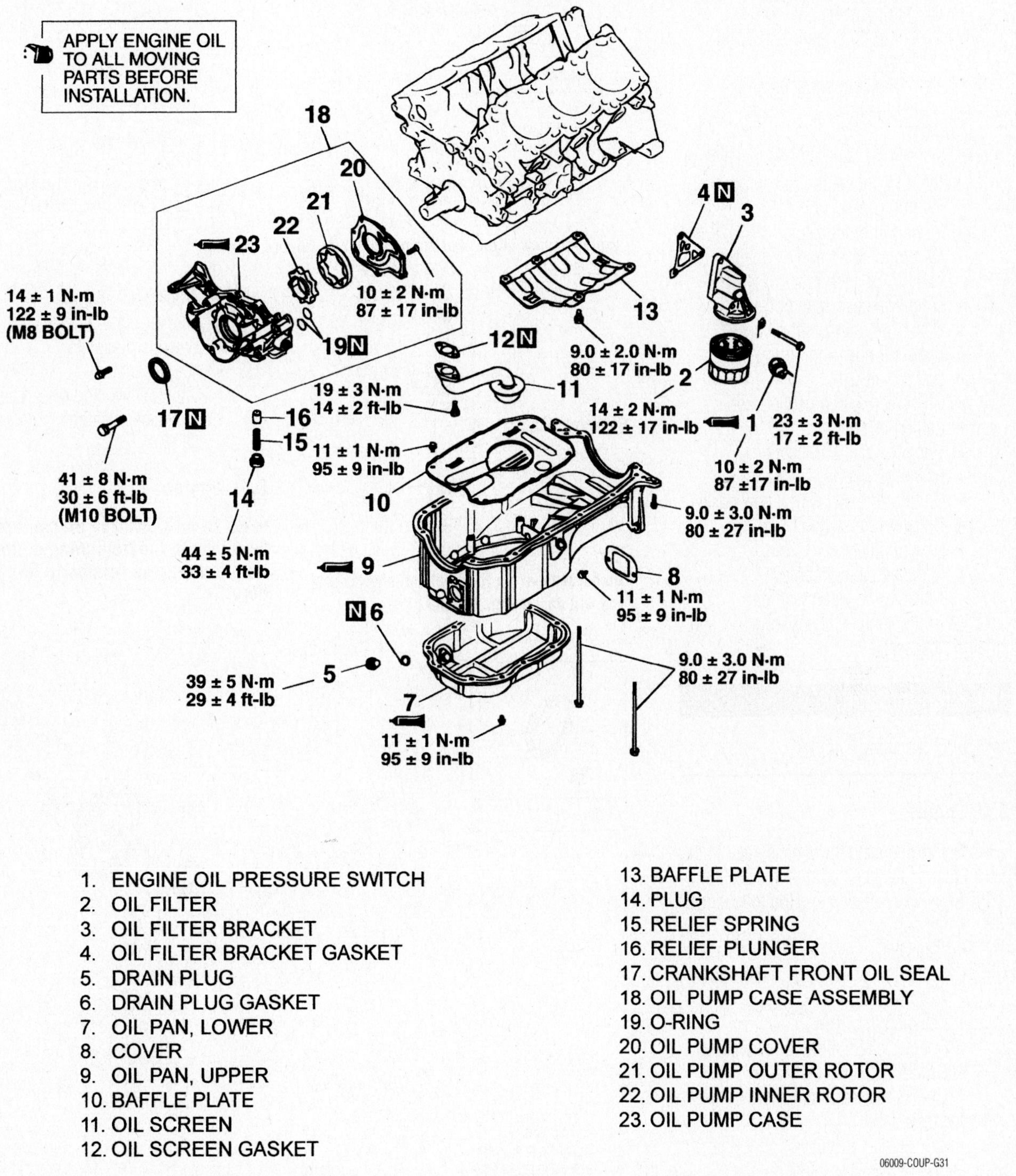

Oil pump and related parts—3.0L engine

- Oil filter
- Oil pressure switch
- Oil filter bracket
- Lower and upper oil pan
- Baffle plate
- Oil screen and gasket
- Baffle plate
- Relief spring and plunger
- Front crankshaft oil seal
- Oil pump case
- Oil pump cover
- Oil pump outer and inner rotors
- Oil pump case

4. Clean all mating surfaces of any residual gasket material.

To install:

5. Prime the oil pump before installation.
6. Install or connect the following:
   - Oil pump case
   - Inner and outer rotors
   - Oil pump cover and torque the bolts to 89 inch lbs. (10 Nm)
   - Oil pump case assembly
   - New front crankshaft oil seal
   - Relief plunger and spring
   - Baffle plate
   - Oil screen with a new gasket
   - Baffle plate
   - Upper oil pan and cover
   - Lower oil pan
   - Oil filter bracket with a new gasket
   - Oil pressure switch
   - New oil filter
   - Negative battery cable
7. Fill the engine with clean oil.
8. Start the vehicle, check for leaks and repair if necessary.

Rear Main Seal

REMOVAL & INSTALLATION

2.4L Engine

1. Before servicing the vehicle, refer to the precautions in the beginning of this section.
2. Remove or disconnect the following:
   - Oil pan
   - Transaxle
   - Clutch cover and disc, if equipped
   - Adapter plate
   - Flexplate/flywheel
   - Adapter plate, if equipped with M/T
   - Crankshaft bushing
   - Crankshaft rear main seal

To install:

3. Apply a small amount of oil to the inside of the rear main seal lip.
4. Install or connect the following:
   - Rear main seal, using the proper installation tools
   - Crankshaft bushing
   - Adapter plate, if removed
   - Flywheel/flexplate
   - Adapter plate
   - Clutch cover and disc if equipped
   - Transaxle assembly
   - Oil pan

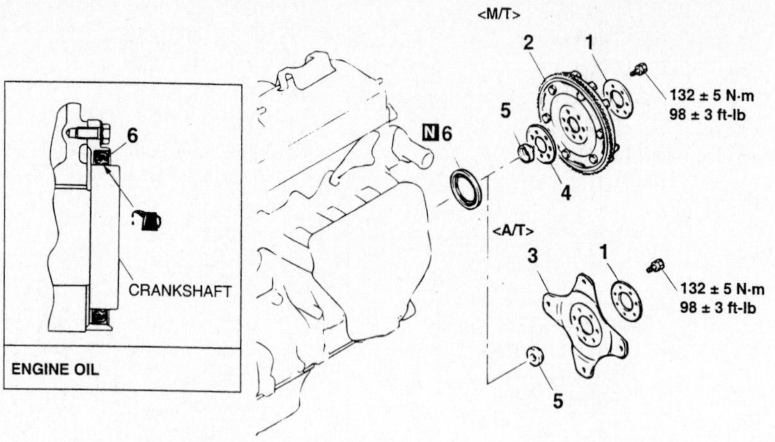

1. ADAPTER PLATE
2. FLYWHEEL <M/T>
3. DRIVE PLATE <A/T>
4. ADAPTER PLATE <M/T>
5. CRANKSHAFT BUSHING
6. CRANKSHAFT REAR OIL SEAL

9356CG12

Exploded view of the rear main seal—2.4L engine shown

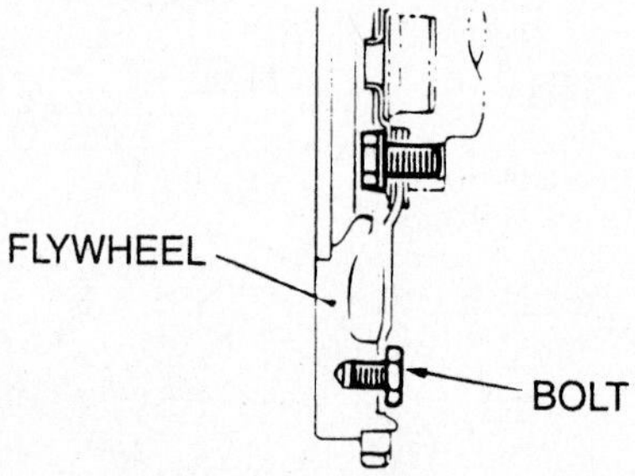

9356CG03

Do not remove the flywheel bolt shown, or you will damage the flywheel

3.0L Engine

➡Be sure to observe all cautions and warnings in the beginning of the section that may be related to this procedure.

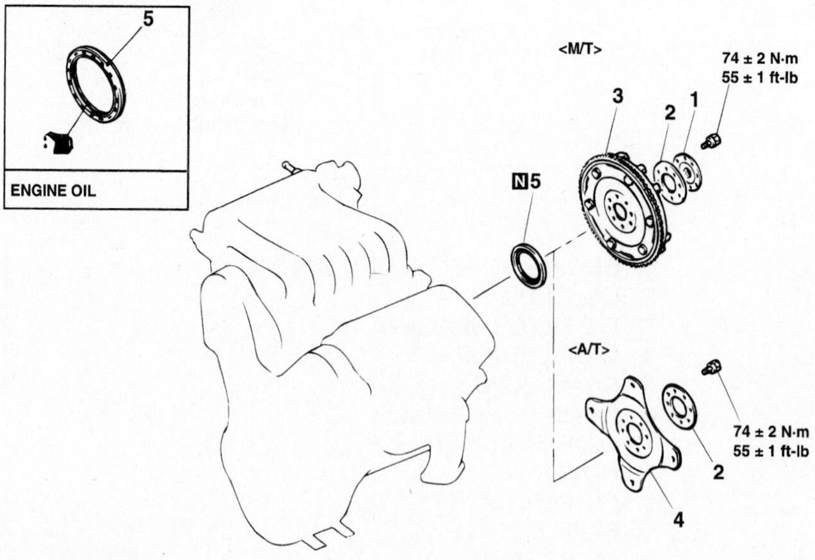

1. PLATE <M/T>
2. ADAPTER PLATE
3. FLYWHEEL <M/T>
4. DRIVE PLATE <A/T>
5. CRANKSHAFT REAR OIL SEAL

06009-COUP-G32

Rear main seal and related parts—3.0L engine

1. Remove or disconnect the following:
   - Transaxle
   - Flexplate/flywheel
   - Rear crankshaft oil seal

➡Pry the oil seal from the housing using a suitable flat bladed prying tool.

To install:

➡When installing the new seal there is no need to lubricate sealing surface.

2. Install or connect the following:
   - Oil seal into housing using a suitable installation tool
   - Flexplate/flywheel and torque the bolts to 68 ft. lbs. (92 Nm)
   - Transaxle

Timing Belt

REMOVAL & INSTALLATION

2.4L Engine

1. Before servicing the vehicle, refer to the precautions in the beginning of this section.
2. Remove or disconnect the following:
   - Negative battery cable
   - Right inner splash-shield
   - Radiator reservoir tank
   - Accessory drive belts
   - Water pump pulley
   - Crankshaft damper/pulley
3. Place a suitable floor jack under the vehicle to support the engine.
   - Engine mount insulator mounting bolt
   - Engine mount bracket
   - Engine mount stopper
   - Timing belt upper and lower covers

➡Do not rotate the crankshaft or the camshafts after the timing belt has been removed. Damage to the valve components may occur. Before removing the timing belt, always align the timing marks.

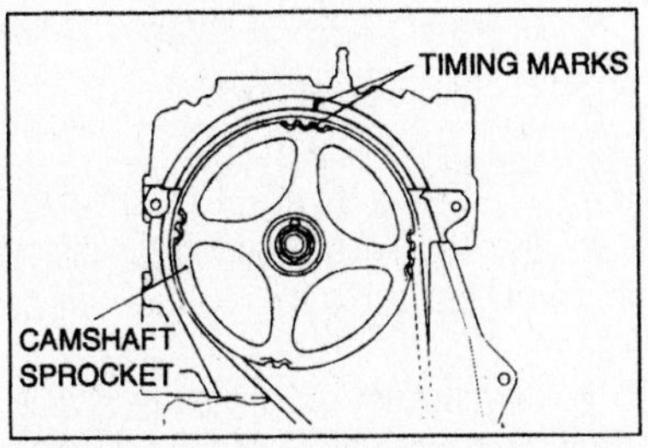

67189-SEBR-G03

Turn the crankshaft in the forward direction only to align the timing marks—2.4L engine

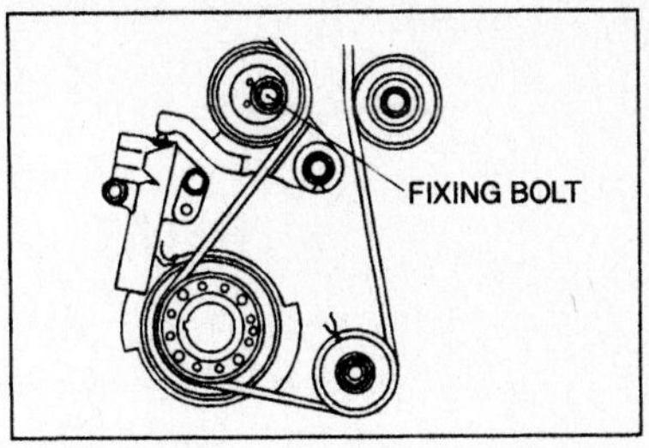

67189-SEBR-G04

Location of the tension pulley fixing bolt—2.4L engine

4. Align the timing marks of the timing belt sprockets to the timing marks on the rear timing belt cover and oil pump cover.

➡If the timing belt is to be reused, draw an arrow indicating the direction of rotation on the back of the belt for reinstallation.

5. Loosen the timing belt tension pulley fixing bolt, then move the tension pulley to the water pump side.
6. Remove or disconnect the following:
   - Timing belt
   - Tensioner pulley
   - Auto-tensioner
   - Camshaft timing belt sprockets
   - Crankshaft timing belt sprocket using special removal tool No. 6793

To install:

7. Install or connect the following:
   - Crankshaft timing belt sprocket onto the crankshaft, using special tool No. 6792
   - Camshaft sprockets onto the camshafts
   - Camshaft sprocket bolts, and torque to 75 ft. lbs. (101 Nm)
8. Apply 22–44 lbs. (98–196 N) of force to the pushrod of the auto tensioner by pushing it against a metal object (ex. engine block). Measure the movement of the pushrod, as shown in the accompanying figure. Standard value is within 0.04 in. (1mm). A is the length when it is free (not pressed). B is the length when it is pressed. A minus B equals movement. If the measurement is outside the standard, replace the tensioner.
9. Place the tensioner in a vise and slowly compress the pushrod until pin hole A of the pushrod and pinhole B of the tensioner cylinder are aligned. Take care not to damage the pushrod. When the holes are aligned, insert the set pin.

➡When replacing the tensioner with a new part, it will come with a pin.

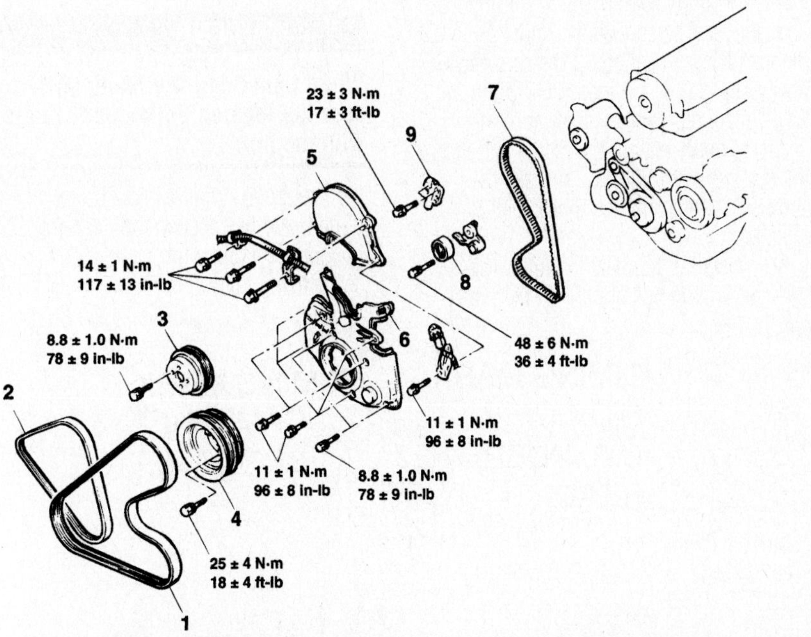

1. DRIVE BELT (POWER STEERING OIL PUMP AND A/C COMPRESSOR)
2. DRIVE BELT (GENERATOR)
3. WATER PUMP PULLEY
4. CRANKSHAFT PULLEY
5. TIMING BELT UPPER COVER ASSEMBLY
6. TIMING BELT LOWER COVER ASSEMBLY
- TIMING BELT TENSION ADJUSTMENT
7. TIMING BELT
8. TENSIONER PULLEY
9. AUTO-TENSIONER

67189-SEBR-G07

Exploded view of the timing belt and related components—2.4L engine

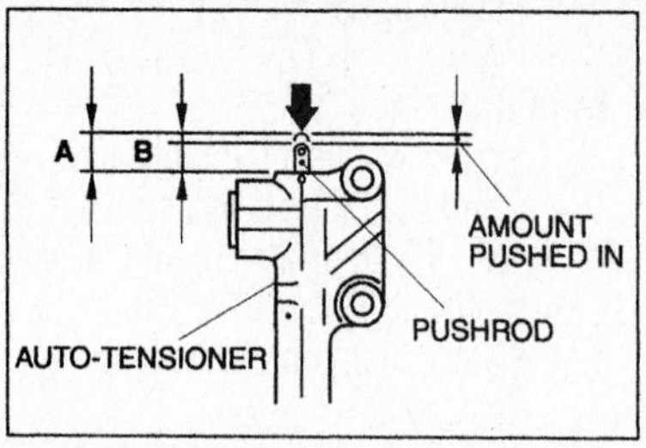

Measuring the auto tensioner—2.4L and 3.0L engines

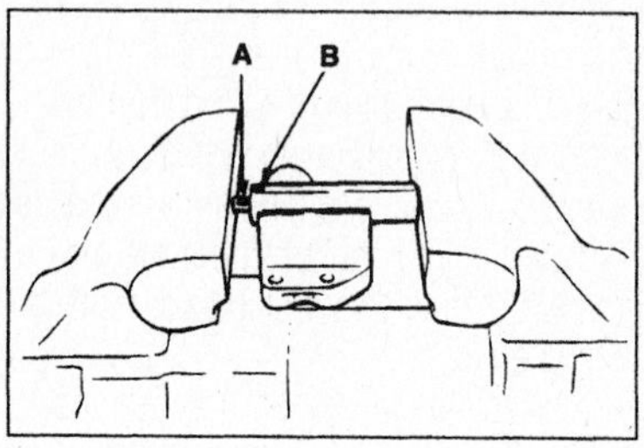

Compress the pushrod until holes A & B are aligned—2.4L engine

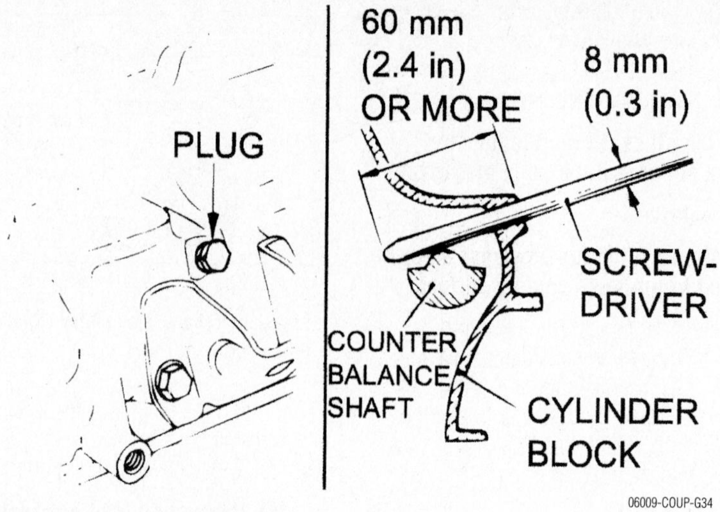

Screwdriver installed—2.4L engine

10. Align the timing marks on the camshaft sprocket, crankshaft sprocket and oil pump sprocket.
11. Remove the cylinder block plug and insert a Phillips head 0.3 inch (8mm) screwdriver. The screwdriver must go in 2.4 inches (60mm) or more. If the screwdriver only goes in 0.8–1.0 inch (20–25mm) hitting the counterbalance shaft, turn the sprocket once, realign the marks and insert the screwdriver again. Do not take the screwdriver out until the timing belt is installed.
12. Install or connect the following:
   - Timing belt starting at the crankshaft, oil pump sprocket and camshaft sprocket, in that order, so there is no slack.
13. Set the tension pulley so that the pin holes are at the bottom, press the tension pulley lightly against the timing belt and tighten the fixing bolt.
14. Adjust belt tension as follows:
   a. After turning the crankshaft ¼ revolution counterclockwise, turn it in the clockwise direction until the timing marks are aligned.

WARNING

When tightening the fixing bolt, make sure the tension pulley does not turn with the bolt.

   b. Loosen the tension pulley fixing bolt, and use special tool MD998767 or equivalent and a torque wrench to tighten the fixing bolt to 32–40 ft. lbs. (42–54 Nm) while applying tension of 31 inch lbs. (3.5 Nm) to the timing belt.
   c. Remove the set pin from the tensioner.
   d. Turn the crankshaft 2 revolutions clockwise so that the timing marks are aligned. After leaving it for 15 minutes, measure the amount of protrusion of the auto-tensioner. It should be 0.15–0.18 in. (3.8–5.5 Nm). If the measurement is outside the specifications, repeat steps A.-D.
   e. Make sure the timing marks of each sprocket are still aligned.
   - Front and rear timing belt covers
   - Engine mount stopper
   - Engine mount bracket
   - Engine mount insulator mounting bolt and tighten to 60 ft. lbs. (81 Nm)
15. Remove the floor jack from under the vehicle.
   - Crankshaft damper and tighten to 105 ft. lbs. (142 Nm)
   - Water pump pulley
   - Accessory drive belts
   - Radiator reservoir tank
   - Right inner splash-shield
16. Properly fill the cooling system.
17. Connect the negative battery cable.
18. Check for leaks and proper engine and cooling system operation.

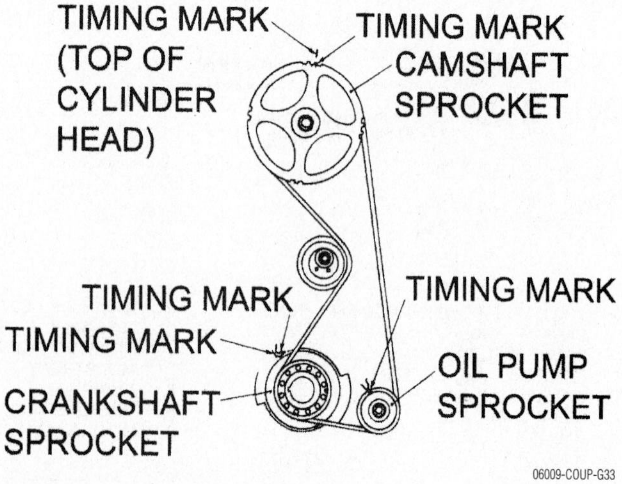

Timing mark alignment—2.4L engine

3.0L Engine

1. Before servicing the vehicle, refer to the precautions in the beginning of this section.
2. Place a floor jack under the engine oil pan, with a block of wood in between, and raise the engine so that the weight of the engine is no longer being applied to the engine support bracket.

- Upper engine mount. Spraying lubricant, slowly remove the reamer (alignment) bolt and remaining bolts and remove the engine support bracket.

➡The reamer bolt is sometimes heat-seized on the engine support bracket.

3. Remove or disconnect the following:
   - Negative battery cable
   - Alternator
   - Accessory drive belts
   - Crankshaft bolt and pulley, using Crankshaft Holding Tools MB990767 and MB998754/ MD998715
   - Heated oxygen sensor connection, if necessary
   - Power steering pump with the hose attached and position it aside
   - Power steering pump bracket
   - Tensioner pulley assembly
   - Upper right front timing belt cover
   - Upper left front timing belt cover

➡If the timing belt is to be reused, draw an arrow indicating the direction of rotation on the back of the belt for reinstallation.

4. Align the timing marks by turning the crankshaft with MD998769 Crankshaft Turning tool. Loosen the center bolt on the timing belt tensioner pulley and remove the belt.

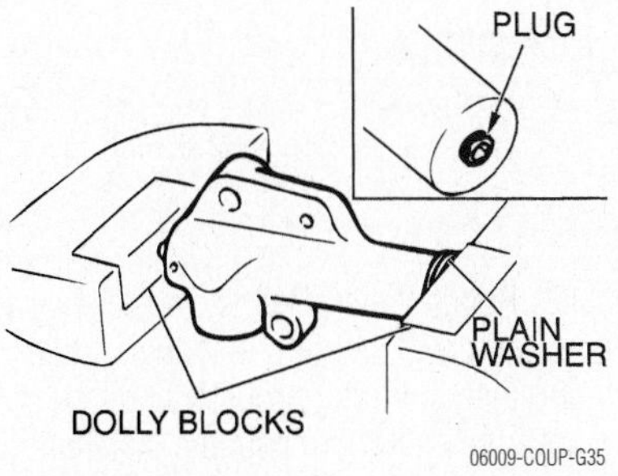

Compressing the plunger—3.0L engine

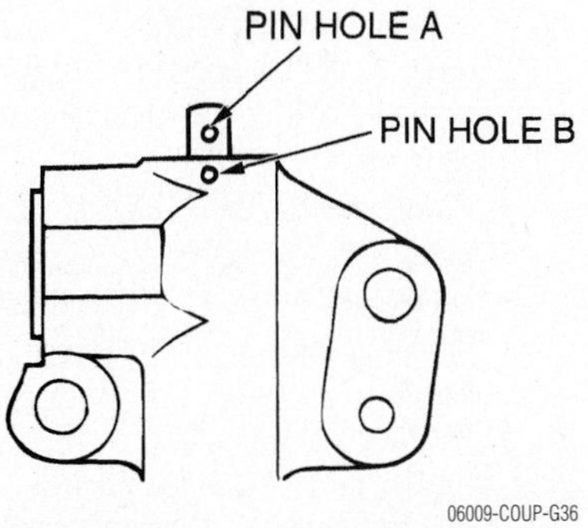

Compress the plunger until the holes line up—3.0L engine

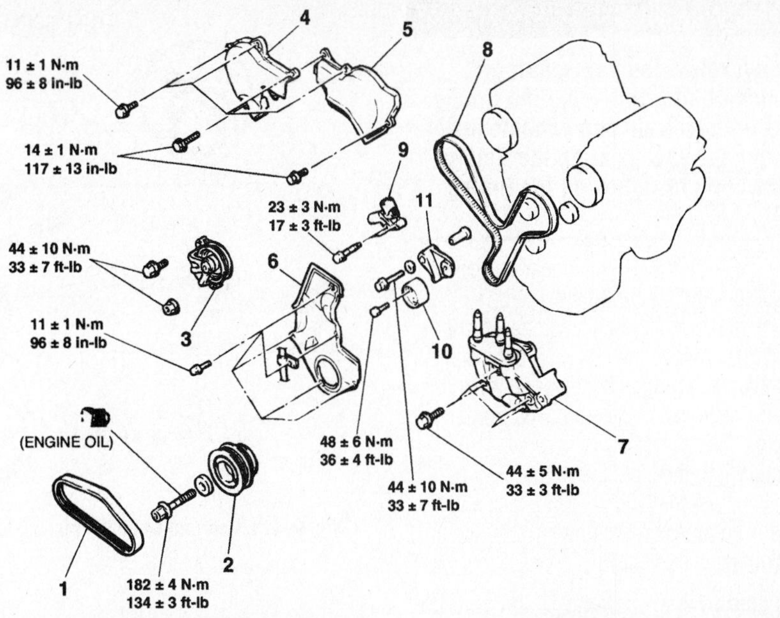

1. DRIVE BELT (POWER STEERING OIL PUMP)
2. CRANKSHAFT PULLEY
3. TENSIONER PULLEY ASSEMBLY (POWER STEERING OIL PUMP)
4. TIMING BELT FRONT UPPER COVER, RIGHT
5. TIMING BELT FRONT UPPER COVER, LEFT
6. TIMING BELT FRONT LOWER COVER
7. ENGINE SUPPORT BRACKET, RIGHT
8. TIMING BELT
9. AUTO-TENSIONER
10. TENSIONER PULLEY
11. TENSIONER ARM

67189-SEBR-G08

Exploded view of the timing belt and related components—3.0L engine

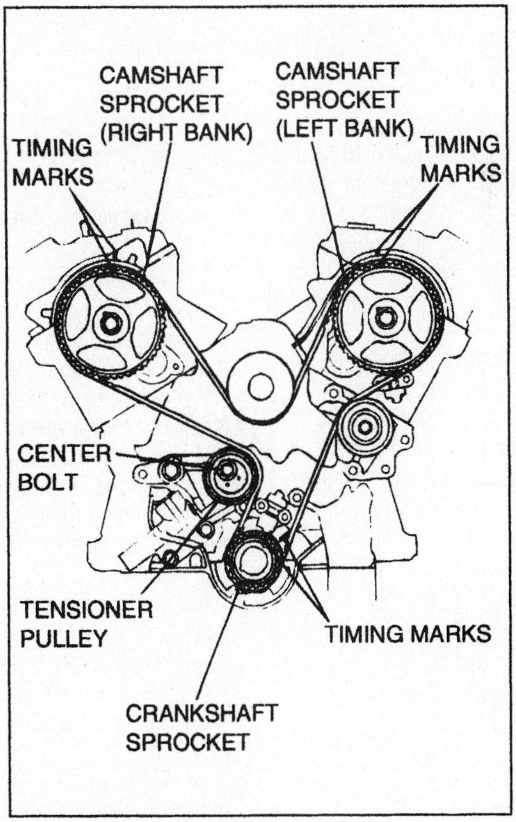

Alignment of the timing marks—3.0L engine

⁂ WARNING

Do not rotate the crankshaft or camshaft after removing the timing belt or valvetrain components may be damaged. Always align the timing marks before removing the timing belt.

5. Check the belt tensioner for leaks and check the pushrod for cracks.
6. If the timing belt tensioner is to be replaced, remove the retaining bolts and remove the timing belt tensioner. When the timing belt tensioner is removed from the engine, it is necessary to compress the plunger into the tensioner body.
7. Place the tensioner in a vise and slowly compress the plunger. Take care not to damage the pushrod.

➡Position the tensioner in the vise the same way it will be installed on the engine. This is to ensure proper pin orientation for when the tensioner is installed on the engine.

8. When the plunger is compressed into the tensioner body, install a pin through the body and plunger to hold the plunger in place until the tensioner is installed.

To install:

9. Install the timing belt tensioner and tighten the retaining bolts to 17 ft. lbs. (24 Nm), but do not remove the pin at this time.
10. Check that all timing marks are still aligned.
11. Use bulldog clips (large paper binder clips) or other suitable tool to secure the timing belt and to prevent it from slacking. Install the timing belt. Starting at the crankshaft, go around the idler pulley, then the front camshaft sprocket, the water pump pulley, the rear camshaft sprocket and the tensioner pulley.
12. Be sure the belt is tight between the crankshaft and front camshaft sprocket, between the camshaft sprockets and the water pump. Gently raise the tensioner pulley, so that the belt does not sag, and temporarily tighten the center bolt.

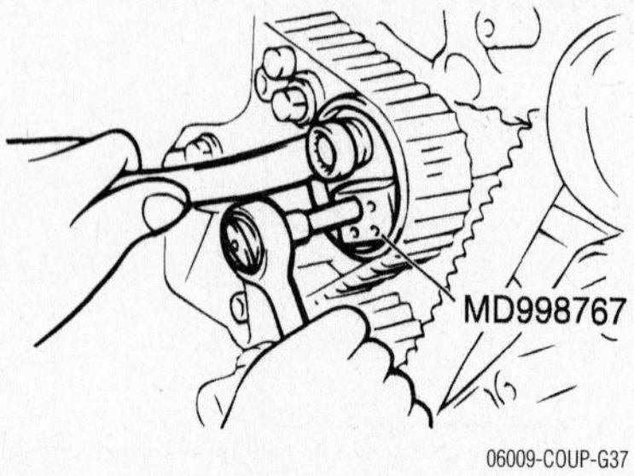

Using tool MD998767—3.0L engine

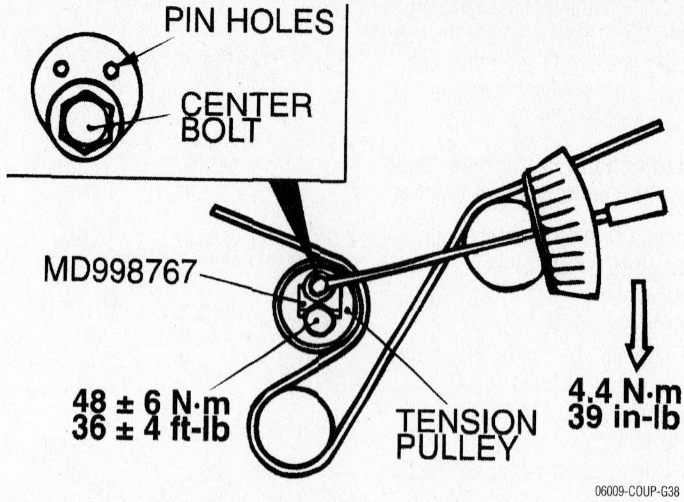

Pretension is correct when the pin can be removed and installed easily—3.0L engine

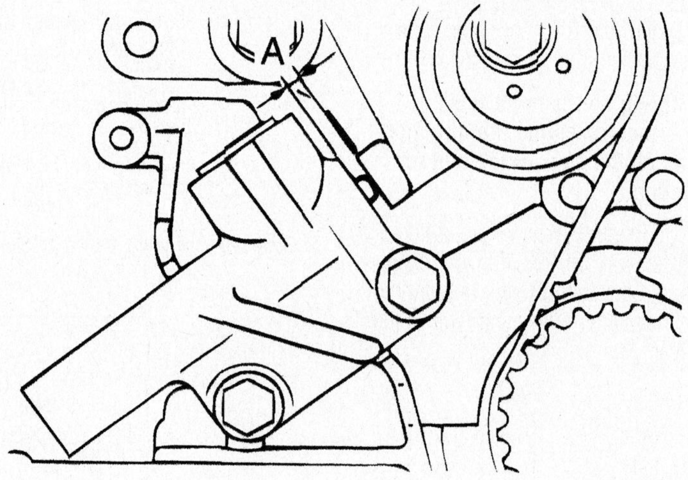

Measuring pushrod extension—3.0L engine

13. Move the crankshaft ¼ turn counterclockwise, then turn it clockwise to the position where the timing marks are aligned.
14. Loosen the center bolt of the tensioner pulley. Using MD998767 tensioner tool, and a torque wrench apply 3.3 ft. lbs. (4.4 Nm) tensional torque to the timing belt and tighten the center bolt to 35 ft. lbs. (48 Nm). When tightening the bolt, be sure that the tensioner pulley shaft does not rotate with the bolt.
15. Remove the tensioner plunger pin. Pretension is correct when the pin can be removed and installed easily. If the pin cannot be easily removed and installed it is still satisfactory as long as it is within its standard value.
16. Check that the tensioner pushrod is within the standard value. When the tensioner is engaged the pushrod should measure 0.149–0.177 in. (3.8–4.5mm).
17. Rotate the crankshaft two revolutions and check the timing marks. If the timing marks are not properly aligned remove the belt and repeat the installation steps.
18. Install or connect the following:
   - Timing belt covers
   - Engine mounting bracket
19. Lower the engine enough to install the engine mount onto bracket and remove the floor jack.
   - Power steering pump bracket and pump
   - Crankshaft pulley and tighten the retaining bolt to 13 ft. lbs. (18 Nm)
   - Accessory drive belts
20. Properly fill the cooling system.
21. Connect the negative battery cable.
22. Check for leaks and proper engine and cooling system operation.

Piston and Ring

POSITIONING

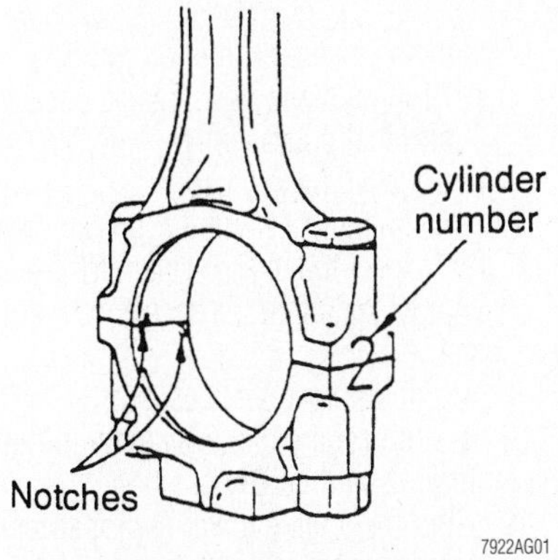

Connecting rod and cap installation—ensure to matchmark the cap and rod prior to disassembly

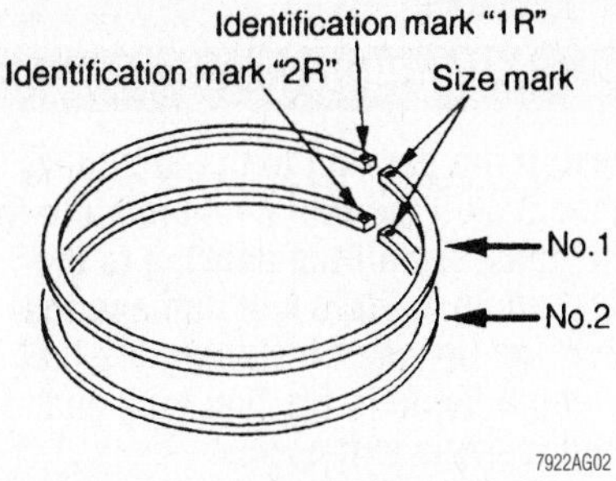

Piston ring identification mark locations

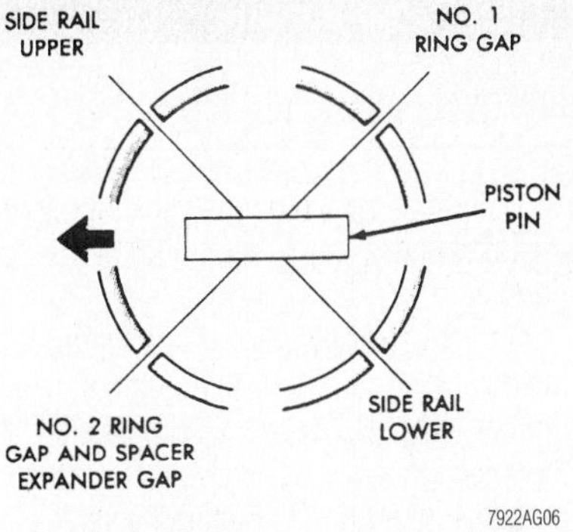

Piston ring end-gap spacing—3.0L engine

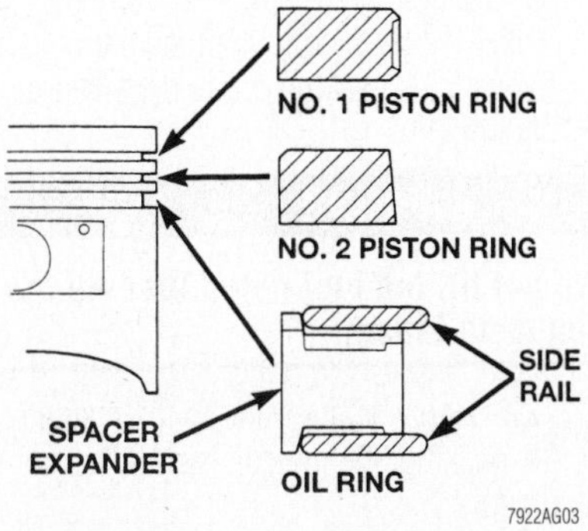

Piston ring orientation—2.4L engine

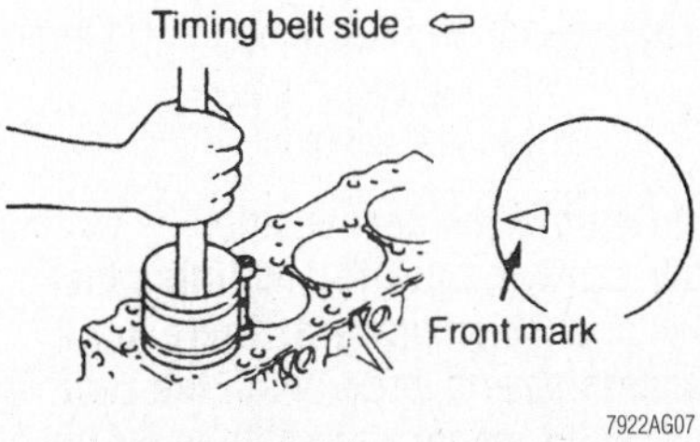

Piston positioning. The small arrows on the crown of the pistons must point toward the front of the engine

FUEL SYSTEM

Fuel System Service Precautions

Safety is the most important factor when performing not only fuel system maintenance but any type of maintenance. Failure to conduct maintenance and repairs in a safe manner may result in serious personal injury or death. Maintenance and testing of the vehicle's fuel system components can be accomplished safely and effectively by adhering to the following rules and guidelines.

- To avoid the possibility of fire and personal injury, always disconnect the negative battery cable unless the repair or test procedure requires that battery voltage be applied.
- Always relieve the fuel system pressure before disconnecting any fuel system component (injector, fuel rail, pressure regulator, etc.), fitting or fuel line connection. Exercise extreme caution whenever relieving fuel system pressure, to avoid exposing skin, face and eyes to fuel spray. Please be advised that fuel under pressure may penetrate the skin or any part of the body that it contacts.
- Always place a shop towel or cloth around the fitting or connection prior to loosening to absorb any excess fuel due to spillage. Ensure that all fuel spillage (should it occur) is quickly removed from engine surfaces. Ensure that all fuel soaked cloths or towels are deposited into a suitable waste container.
- Always keep a dry chemical (Class B) fire extinguisher near the work area.
- Do not allow fuel spray or fuel vapors to come into contact with a spark or open flame.
- Always use a back-up wrench when loosening and tightening fuel line connection fittings. This will prevent unnecessary stress and torsion to fuel line piping.
- Always replace worn fuel fitting O-rings with new. Do not substitute fuel hose, where fuel pipe is installed.

Fuel System Pressure

RELIEVING

1. Before servicing the vehicle, refer to the precautions in the beginning of this section.
2. Remove the fuel filler cap to release fuel tank pressure.
3. Remove the driver's side instrument panel side cover to expose the fuel box.
4. Remove the fuel pump relay.
5. Start the vehicle and allow it to run until it stalls from lack of fuel. Turn the key to the **OFF** position.

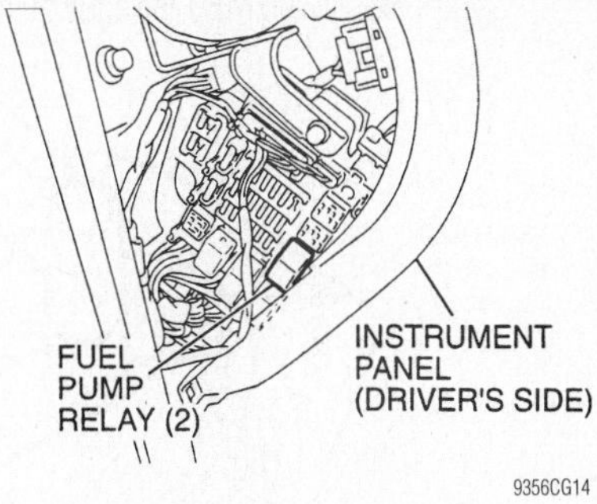

Location of the fuel pump relay

6. Disconnect the negative battery cable, then reconnect the fuel pump relay and install the cover.
7. Install the fuel filler cap.

CAUTION

Always wrap shop towels around a fitting that is being disconnected to absorb residual fuel in the lines.

Fuel Filter

REMOVAL & INSTALLATION

The fuel filter is part of the fuel pump module.

Fuel Pump Module

REMOVAL & INSTALLATION

1. Before servicing the vehicle, refer to the precautions in the beginning of this section.

Do not use conventional fuel filters, hoses or clamps when servicing fuel injection systems. They are not compatible with the injection system and could fail, causing personal injury or damage to the vehicle. Use only hoses and clamps specifically designed for fuel injection.

2. Relieve the fuel system pressure.
3. Remove or disconnect the following:
   - Negative battery cable
   - Rear seat cushion

➡Remove the seat cushion by pulling the stopper outward and lifting the lower cushion upward. There are 2 access covers underneath the seat. The panel on the far right side is for the fuel pump.

   - Access cover/protector
   - Fuel pump wiring
   - Return hose and high pressure fuel hose
   - Fuel pump module, using special Tool No. MB991480, or equivalent

To install:

⁂ WARNING

Install the packing to the fuel tank, then install the fuel pump module in the tank. Install the packing to the fuel pump module will damage the packing lip when installing the fuel pump module to the fuel tank and fuel leakage will occur.

4. Perform the following:
   a. Check that the fuel tank is not damaged or deformed, then securely install the packing to the fuel tank. Replace the packing if it is damaged or deformed.

⁂ WARNING

Do not tilt the fuel pump module during installation.

   b. Apply soapy water to the inside of the packing, then install the fuel pump module.

⁂ WARNING

When tightening, be careful not to let the fuel pump module turn together with the cap. If the mating marks are misaligned, the float may measure a remaining amount of fuel incorrectly, causing the low fuel warning light to malfunction.

   c. Use special Tool No. MB991480 to align the mating marks on the fuel tank and fuel pump module, then tighten the cap.
5. Install or connect the following:
   - High pressure hose, return hose and fuel pump wiring
   - Negative battery cable
6. Check the fuel pump for proper pressure and inspect the entire system for leaks.
7. Apply sealant to the access cover and install the cover.
8. Install the rear seat cushion.
9. Pressurize the fuel system by turning the ignition key to the **ON** position.
10. Start the engine to verify proper fuel pump performance.

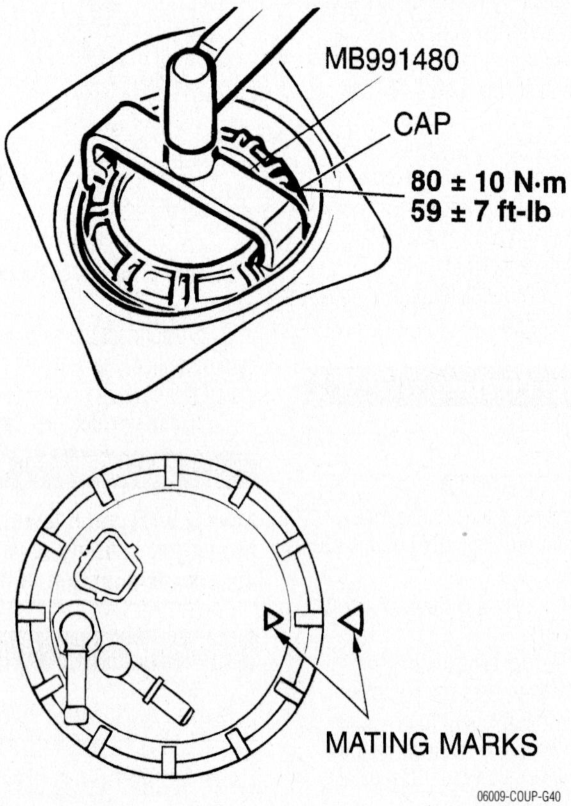

Fuel pump module installation

Fuel Injector

REMOVAL & INSTALLATION

2.4L Engine

1. Before servicing the vehicle, refer to the precautions in the beginning of this section.
2. Relieve the fuel system pressure.
3. Drain the engine coolant.
4. Remove or disconnect the following:
   - Air cleaner assembly
   - Accelerator cable connection
   - Throttle Position (TP) sensor connector
   - Idle Air Control (IAC) motor connector
   - Vacuum hose connector(s)
   - Water hose
   - Throttle body stay
   - Throttle body and gasket
   - Positive Crankcase Ventilation (PCV) hose connection
   - Ignition coil connector
   - Fuel injector connectors
   - Manifold Differential Pressure (MDP) sensor
   - Fuel lines
   - Vacuum hose connections
   - Fuel pressure regulator and O-ring
   - Fuel rail with the injectors attached
   - Insulators
   - Injectors, O-rings and grommets

To install:

5. Install or connect the following
   - Grommets, O-rings and injectors to the fuel rail
   - Insulators
   - Fuel rail assembly

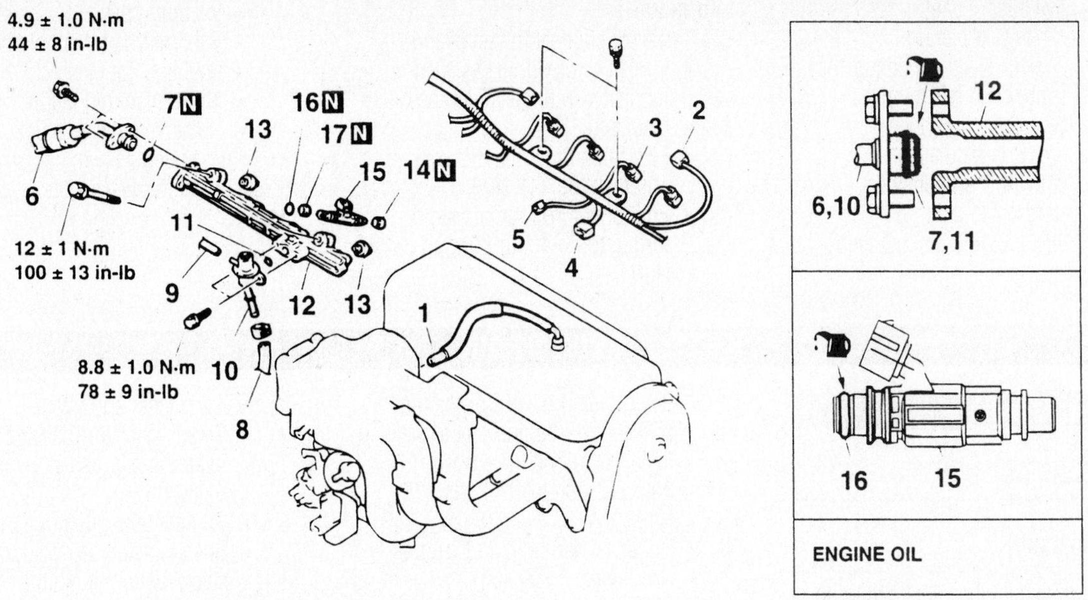

1. PCV HOSE CONNECTION
2. IGNITION COIL CONNECTOR
3. INJECTOR CONNECTOR
4. MANIFOLD DIFFERENTIAL PRESSURESENSOR CONNECTOR
5. HIGH-PRESSURE FUEL HOSE CONNECTION
6. O-RING
7. FUEL HOSE CONNECTION
8. VACUUM HOSE CONNECTION
9. FUEL PRESSURE REGULATOR
10. O-RING
11. FUEL RAIL
12. INSULATORS
13. INSULATORS
14. INJECTORS
15. O-RINGS
16. GROMMETS

9356CG16

Exploded view of the fuel rail, injectors and related components—2.4L engine

- Vacuum hose connections
- Fuel lines
- MDP sensor
- Injector connections
- Ignition coil connector
- PCV hose
- Throttle body gasket and throttle body
- Throttle body stay
- Water hose
- Vacuum hose connector(s)
- IAC motor connector
- TP connector
- Accelerator cable connection
- Air cleaner assembly
- Negative battery cable.

6. Start the vehicle, check for leaks and repair if necessary.

3.0L Engine

1. Before servicing the vehicle, refer to the precautions in the beginning of this section.
2. Properly relieve the fuel system pressure.
3. Drain the engine coolant.
4. Remove or disconnect the following:

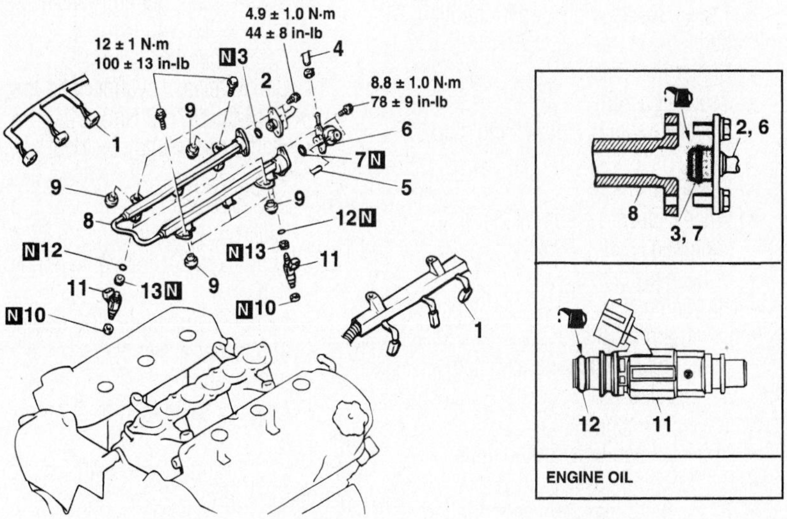

1. INJECTOR CONNECTOR
2. FUEL HIGH-PRESSURE HOSE CONNECTION
3. O-RING
4. FUEL RETURN HOSE CONNECTION
5. VACUUM HOSE CONNECTION
6. FUEL PRESSURE REGULATOR
7. O-RING
8. FUEL RAIL
9. INSULATORS
10. INSULATORS
11. INJECTORS
12. O-RING
13. GROMMETS

06009-COUP-G41

Exploded view of the fuel rail, injectors and related components—3.0L engine

- Negative battery cable
- Intake manifold plenum
- Fuel injector electrical connector
- Fuel supply and return hoses
- Vacuum hose
- Fuel pressure regulator
- Fuel rail with the injectors
- Insulators
- Fuel injectors and discard the O-rings

To install:

5. Install or connect the following:
   - Fuel injector with a new O-ring lubricated with clean engine oil
   - Insulators
   - Fuel rail and torque the bolts to 100 inch lbs. (12 Nm)
   - Fuel pressure regulator and torque the bolts to 80 inch lbs. (9 Nm)
   - Vacuum hose
   - Fuel return and supply hoses
   - Fuel injector electrical connector
   - Intake manifold plenum
   - Negative battery cable
6. Fill the engine with coolant, start the vehicle, check for leaks and repair if necessary.

DRIVE TRAIN

Transaxle Assembly

REMOVAL & INSTALLATION

Manual Transaxle

1. Before servicing the vehicle, refer to the precautions in the beginning of this section.
2. Drain the transaxle fluid.
3. Remove or disconnect the following:
   - Battery and tray
   - Air cleaner assembly
   - Front exhaust pipe, 3.0L engine
   - Shift and selection cable connections
   - Back-up light switch connection
   - Vehicle Speed Sensor (VSS) connector
   - Starter
   - Clutch release cylinder connector
   - Transaxle assembly upper coupling bolts
   - Center member
   - Rear roll stopper
   - Transaxle mount bracket and stopper
   - Stabilizer link
   - Wheel Speed Sensor (WSS), if equipped
   - Brake hose clamp
   - Tie rod end
   - Lower arm
   - Clutch release bearing engagement, 3.0L engine
   - Driveshaft connection
   - Upper oil pan bolt, 3.0L engine
   - Bell housing cover
   - Transaxle assembly lower coupling bolts
   - Transaxle assembly

To install:

4. Install or connect the following:
   - Transaxle assembly. Torque the lower coupling bolts to 36 ft. lbs. (48 Nm) on the 2.4L engine and to 52 ft. lbs. (71 Nm) for the 3.0L engine.
   - Bell housing cover. Torque the cover—to—engine bolt to 80 inch lbs. (9 Nm) and the cover—to—transaxle bolt to 19 ft. lbs. (26 Nm).
   - Upper oil pan bolt, 3.0L engine
   - Driveshaft connection
   - Clutch release bearing engagement
   - Lower arm and torque the nut to 80 ft. lbs. (108 Nm)
   - Tie rod end and torque the nut to 21 ft. lbs. (28 Nm)
   - WSS, if equipped
   - Stabilizer link and torque the nut to 33 ft. lbs. (45 Nm)
   - Transaxle mount bracket and torque the nut to 42 ft. lbs. (57 Nm)
   - Transaxle mount stopper and torque the nut to 60 ft. lbs. (81 Nm)
   - Rear roll stopper and torque the nut to 33 ft. lbs. (45 Nm)
   - Center member
   - Clutch release cylinder and torque to 13 ft. lbs. (18 Nm)
   - Starter motor and electrical connectors
   - VSS connector
   - Back-up light switch connector
   - Shift and selection cable connections
   - Front exhaust pipe, 3.0L engine
   - Air cleaner assembly
   - Battery and tray
5. Fill the transaxle fluid to the proper level.
6. Start the vehicle, check for leaks and repair if necessary.

Automatic Transaxle

1. Before servicing the vehicle, refer to the precautions in the beginning of this section.
2. Drain the transaxle fluid.
3. Remove or disconnect the following:
   - Battery and battery tray
   - Battery brace
   - Air cleaner and intake hoses
   - Shifter lever-to-transaxle nut, cable retaining clip and cable from the transaxle
   - Shifter cable mounting bracket
   - Speedometer, solenoid, neutral safety switch (inhibitor switch), the pulse generator, kickdown servo switch and oil temperature sensor electrical connectors
   - Transaxle oil cooler lines
   - Transaxle dipstick and tube
   - Starter and using Special Tool 7137 or C-4852, support the engine assembly
   - Rear roll stopper mounting bracket
   - Transaxle mount bracket
   - Upper transaxle mounting bolts
   - Front wheels
   - Left side undercover
   - Tie rod end from the steering knuckle
   - Stabilizer bar link from the damper fork
   - Damper fork from the lateral lower control arm
   - Lateral lower arm and compression lower arm, lower ball joints from the steering knuckle
   - Halfshafts from the transaxle and secure aside
   - Bell housing cover
   - Engine front roll stopper through-bolt
   - Center member
   - Flexplate-to-torque converter bolts. Rotate the crankshaft to bring the bolts into a position for removal, one at a time.

➡To make installation easier, use chalk or paint to make matchmarks on the torque converter and flexplate. These marks will be used at assembly to realign the assembly, keeping these parts in balance.

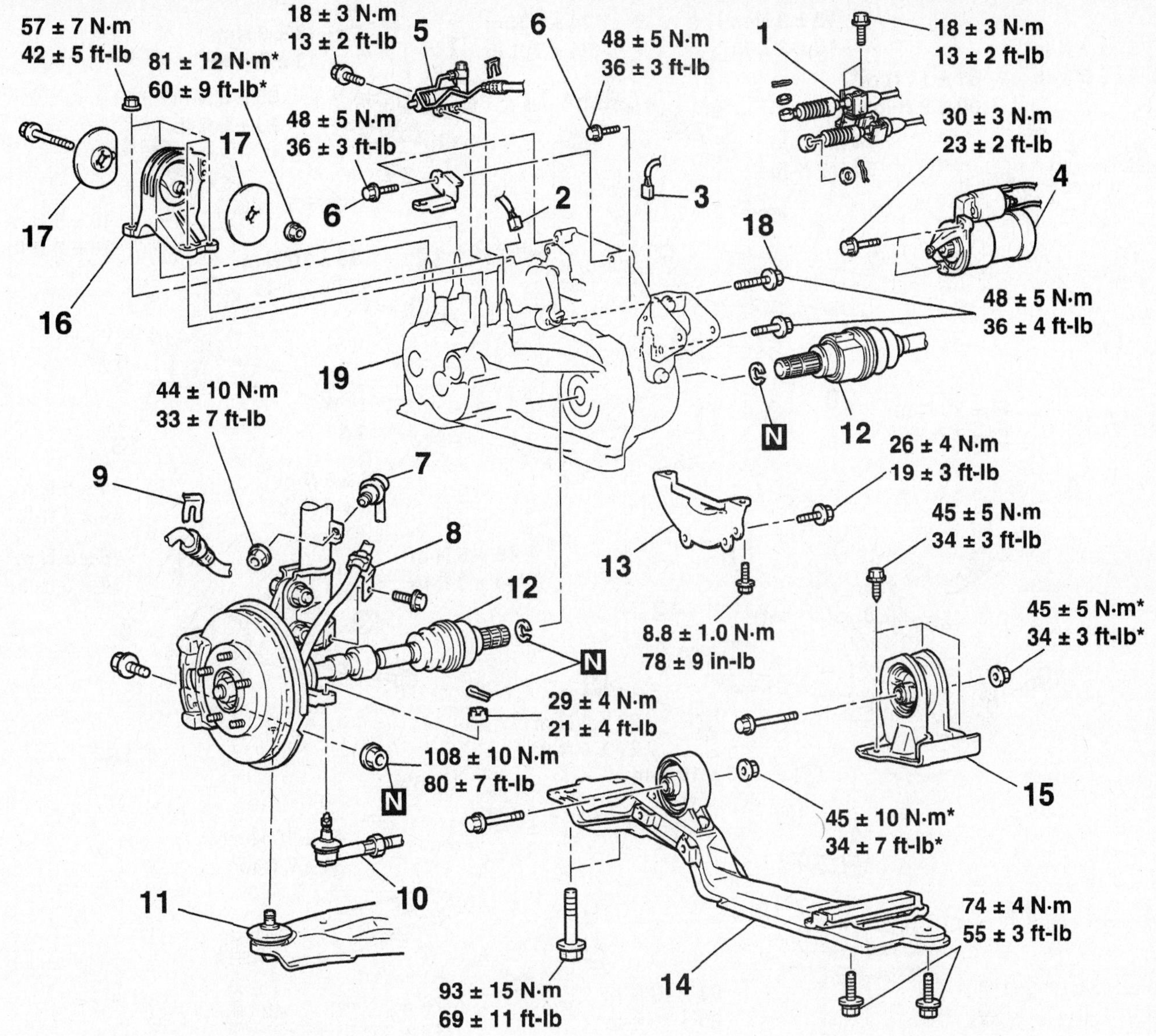

1. SHIFT CABLE AND SELECT CABLE CONNECTION
2. BACKUP LIGHT SWITCH CONNECTOR
3. VEHICLE SPEED SENSOR CONNECTOR
4. STARTER MOTOR
5. CLUTCH RELEASE CYLINDER CONNECTION
6. TRANSAXLE ASSEMBLY UPPER PART COUPLING BOLTS
7. STABILIZER LINK CONNECTION <STRUT SIDE>
8. WHEEL SPEED SENSOR CABLE CONNECTION <VEHICLES WITH ABS>
9. BRAKE HOSE CLAMP
10. TIE ROD END CONNECTION
11. LOWER ARM CONNECTION
12. DRIVESHAFT CONNECTION
13. BELL HOUSING COVER
14. CENTERMEMBER ASSEMBLY
15. REAR ROLL STOPPER
16. TRANSAXLE MOUNT BRACKET
17. TRANSAXLE MOUNT STOPPER
18. TRANSAXLE ASSEMBLY LOWER PART COUPLING BOLTS
19. TRANSAXLE ASSEMBLY

06009-COUP-G42

Manual transaxle and related parts—with 2.4L engine

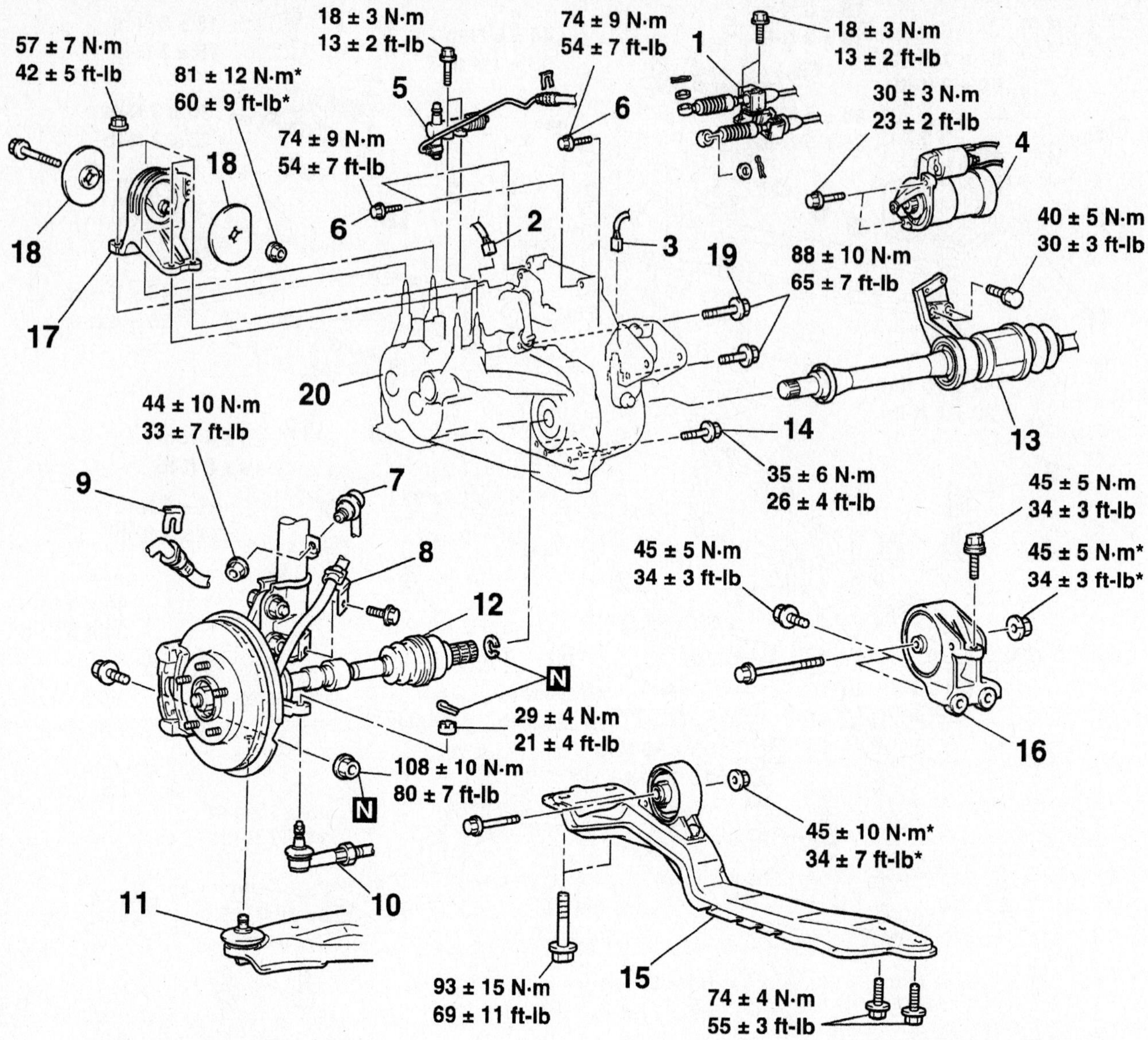

1. SHIFT CABLE AND SELECT CABLE CONNECTION
2. BACKUP LIGHT SWITCH CONNECTOR
3. VEHICLE SPEED SENSOR CONNECTOR
4. STARTER MOTOR
5. CLUTCH RELEASE CYLINDER CONNECTION
6. TRANSAXLE ASSEMBLY UPPER PART COUPLING BOLTS
7. STABILIZER LINK CONNECTION <STRUT SIDE>
8. WHEEL SPEED SENSOR CABLE CONNECTION <VEHICLES WITH ABS>
9. BRAKE HOSE CLAMP
10. TIE ROD END CONNECTION
11. LOWER ARM CONNECTION
12. DRIVESHAFT CONNECTION
13. DRIVESHAFT AND INNER SHAFT CONNECTION
14. UPPER OIL PAN CONNECTING BOLT
15. CENTERMEMBER ASSEMBLY
16. REAR ROLL STOPPER
17. TRANSAXLE MOUNT BRACKET
18. TRANSAXLE MOUNT STOPPER
19. TRANSAXLE ASSEMBLY LOWER PART COUPLING BOLTS
20. TRANSAXLE ASSEMBLY

06009-COUP-G43

Manual transaxle and related parts—with 3.0L engine

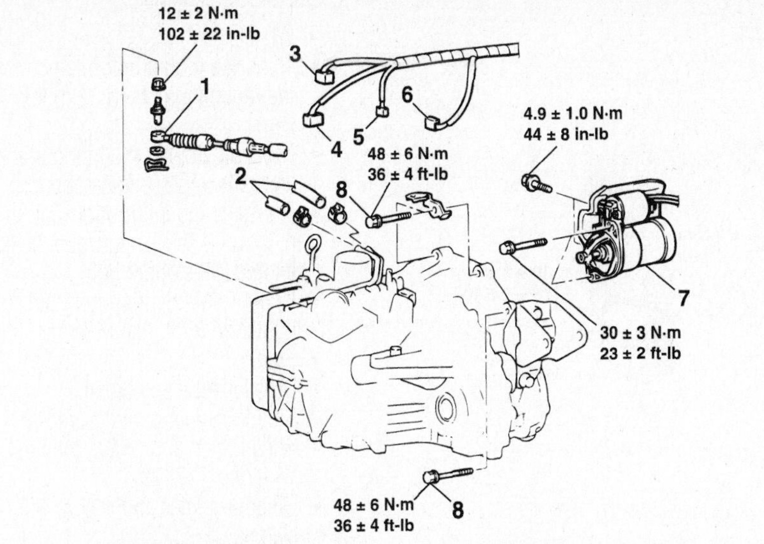

1. TRANSAXLE CONTROL CABLE CONNECTION
2. TRANSAXLE OIL COOLER HOSES CONNECTION
3. TRANSMISSION RANGE SWITCH CONNECTOR
4. A/T CONTROL SOLENOID VALVE ASSEMBLY CONNECTOR
5. INPUT SHAFT SPEED SENSOR CONNECTOR
6. OUTPUT SHAFT SPEED SENSOR CONNECTOR
7. STARTER MOTOR
8. TRANSAXLE UPPER CONNECTING BOLTS

06009-COUP-G44

Automatic transaxle—with 2.4L engine

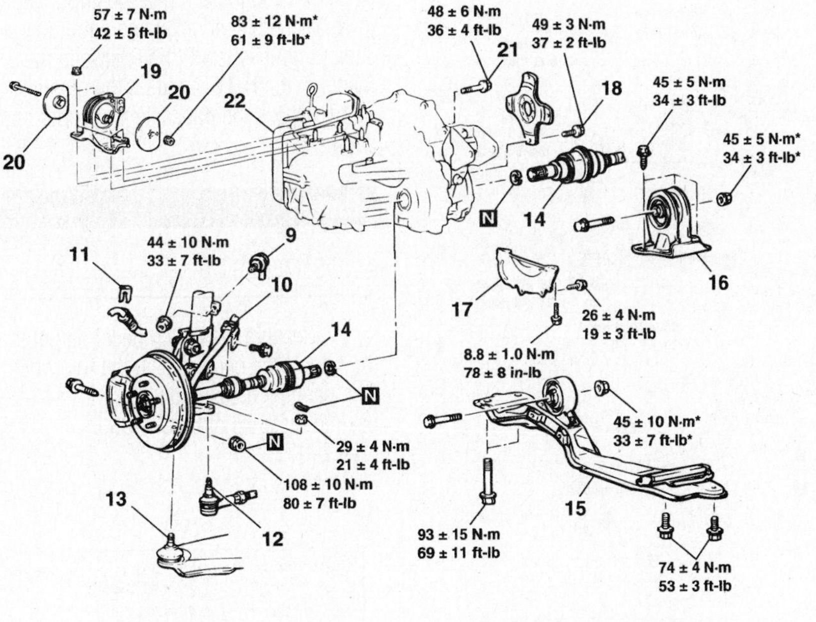

9. STABILIZER LINK CONNECTION <STRUT SIDE>
10. WHEEL SPEED SENSOR CABLE CONNECTION <VEHICLES WITH ABS>
11. BRAKE HOSE CLAMP
12. TIE ROD END CONNECTION
13. LOWER ARM CONNECTION
14. DRIVE SHAFT CONNECTION
15. CENTERMEMBER ASSEMBLY
16. REAR ROLL STOPPER
17. BELL HOUSING COVER
18. DRIVE PLATE BOLTS
19. TRANSAXLE MOUNT BRACKET
20. TRANSAXLE MOUNT STOPPER
21. TRANSAXLE LOWER CONNECTING BOLTS
22. TRANSAXLE ASSEMBLY

06009-COUP-G45

Automatic transaxle mounting—with 2.4L engine

⁂ WARNING

After removing the bolts, push the torque converter toward the transaxle. This will prevent the converter from remaining in contact with the engine, possibly damaging the converter.

4. Support the transaxle using a transmission jack (at the side of the case, NOT at the pan), and remove the transaxle lower coupling bolt.

➡The coupling bolt is inserted from the engine side into the transaxle and is located just above the halfshaft opening.

5. Slide the transaxle rearward and carefully lower it from the vehicle.

To install:

6. After the torque converter has been mounted on the transaxle, install the transaxle assembly to the engine. Install the mounting bolts and tighten to 70 ft. lbs. (95 Nm).
7. Align the balance matchmarks and torque the torque converter-to-flexplate bolts to 55 ft. lbs. (75 Nm).
8. Install or connect the following:
   - Bellhousing cover and torque the bolts to 108 inch lbs. (12 Nm)
   - Center member and torque the front mounting bolts to 65 ft. lbs. (88 Nm) and the rear bolts to 51–58 ft. lbs. (69–78 Nm).
   - Front engine roll stopper through-bolt and lightly tighten. Once the full weight of the engine is on the mounts, torque the bolt to 42 ft. lbs. (56 Nm).
   - Halfshafts using new circlips

⁂ WARNING

When installing the halfshaft, keep the inboard joint straight in relation to the axle to avoid damaging the oil seal lip of the transaxle with the splined part of the halfshaft.

   - Tie rod and ball joints to the steering knuckle. Torque the ball joint self-locking nuts to 48 ft. lbs. (65 Nm), the tie rod end nut to 21 ft. lbs. (28 Nm) and secure with a new cotter pin.
   - Damper fork to the lower control arm and torque the through-bolt to 65 ft. lbs. (88 Nm)
   - Stabilizer link to the damper fork and torque the self-locking nut to 29 ft. lbs. (39 Nm)

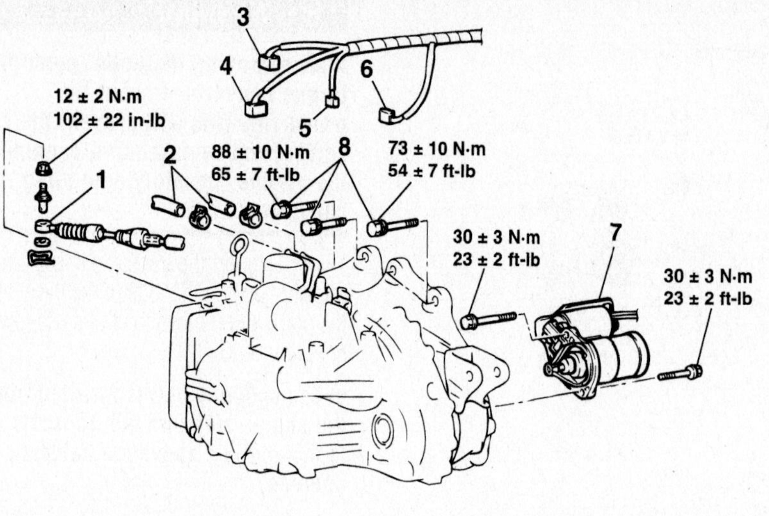

1. TRANSAXLE CONTROL CABLE CONNECTION
2. TRANSAXLE OIL COOLER HOSES CONNECTION
3. TRANSMISSION RANGE SWITCH CONNECTOR
4. A/T CONTROL SOLENOID VALVE ASSEMBLY CONNECTOR
5. INPUT SHAFT SPEED SENSOR CONNECTOR
6. OUTPUT SHAFT SPEED SENSOR CONNECTOR
7. STARTER MOTOR
8. TRANSAXLE UPPER CONNECTING BOLTS

06009-COUP-G46

Automatic transaxle—with 3.0L engine

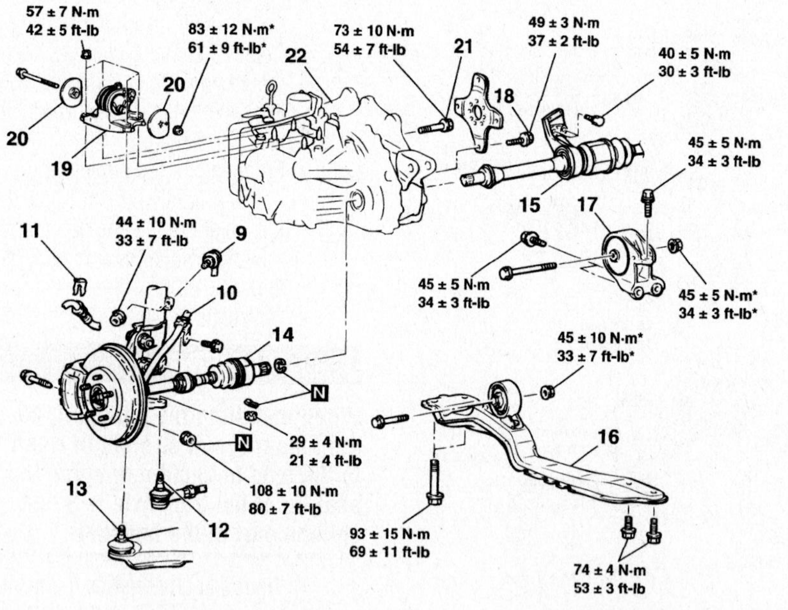

9. STABILIZER LINK CONNECTION <STRUT SIDE>
10. WHEEL SPEED SENSOR CABLE CONNECTION <VEHICLES WITH ABS>
11. BRAKE HOSE CLAMP
12. TIE ROD END CONNECTION
13. LOWER ARM CONNECTION
14. DRIVE SHAFT CONNECTION
15. DRIVESHAFT AND INNER SHAFT CONNECTION
16. CENTERMEMBER ASSEMBLY
17. REAR ROLL STOPPER
18. DRIVE PLATE BOLTS
19. TRANSAXLE MOUNT BRACKET
20. TRANSAXLE MOUNT STOPPER
• ENGINE ASSEMBLY SUPPORT
21. TRANSAXLE LOWER CONNECTING BOLTS
22. TRANSAXLE ASSEMBLY

06009-COUP-G47

Automatic transaxle mounting—with 3.0L engine

- Left side undercover
- Wheels
- Transaxle mount bracket and torque the mounting nuts to 32 ft. lbs. (43 Nm)
- Rear roll stopper mounting bracket and remove the engine support
- Transaxle mount through-bolt to 51 ft. lbs. (69 Nm) and torque the front engine roll stopper bolt
- Upper transaxle mounting bolts and tighten to 35 ft. lbs. (48 Nm)
- Starter
- Dipstick tube and dipstick
- Shifter cable mounting bracket
- Shifter lever and torque the nut to 14 ft. lbs. (19 Nm)
- Oil cooler lines and secure with clamps
- Speedometer, solenoid, neutral safety switch (inhibitor switch), the pulse generator, kickdown servo switch and oil temperature sensor electrical connectors
- Air cleaner and air intake hose
- Battery tray and battery

9. Refill the transaxle with MOPAR® ATF PLUS transmission fluid. Start the engine and allow it to idle for 2 minutes. Apply the parking brake and move the selector through each gear position, ending in **N**. Recheck fluid level and add if necessary. Fluid level should be between the marks in the **HOT** range on the dipstick.

10. Check the transaxle for proper operation.

Clutch

ADJUSTMENTS

1. Measure the clutch pedal height from the face of the pedal pad to the floorboard. Compare the measured value with the desired distance of 6.88–7.00 in. (174.7–177.7mm).

2. Measure the clutch pedal clevis pin

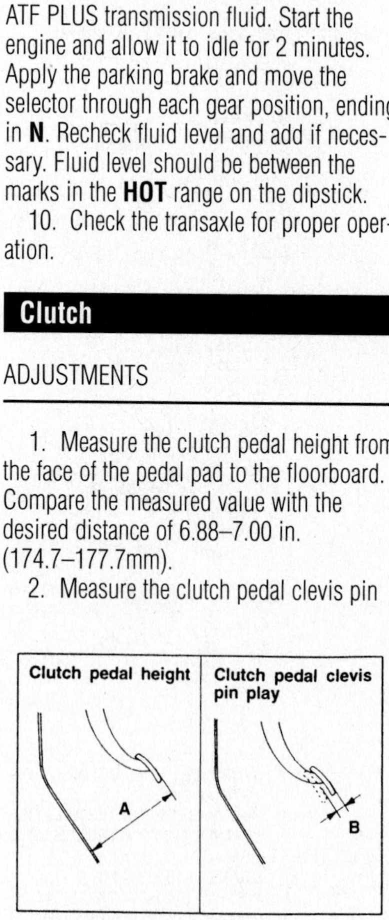

9300CG01

Clutch pedal height and free-play adjustment measurements

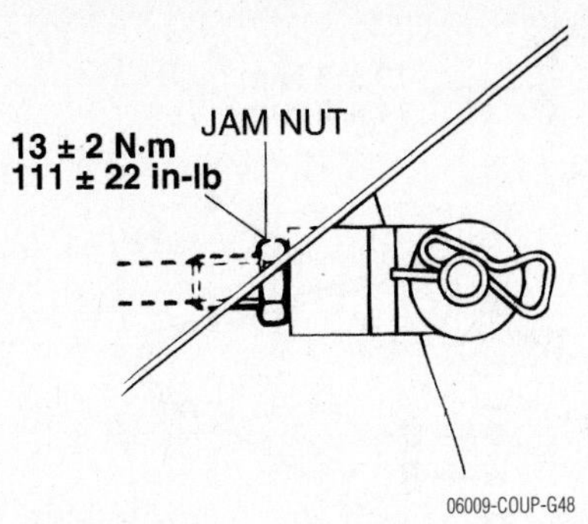

Jam nut

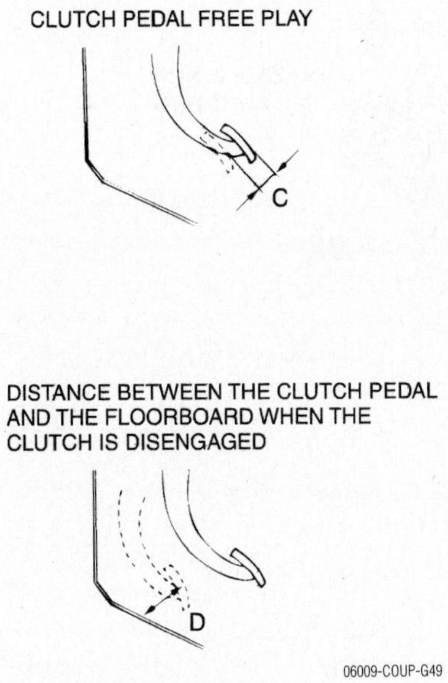

Free play and disengagement measurements

play at the face of the pedal pad. Press the pedal lightly until resistance is met, and measure this distance. The clutch pedal clevis pin play should be within 0.040–0.120 in. (1–3mm).

3. If the clutch pedal height or clevis pin play, is not within the standard values, loosen the jam nut and adjust the pedal height to the standard value pushing the clutch pedal position switch.

➡Do not push in the master cylinder pushrod at this time.

4. If the clutch pedal play is not within the standard value, loosen the jam nut and move the pushrod to adjust.

5. After the adjustments, confirm that the clutch pedal free play (measured at the face of the pedal pad) and the distance between the clutch pedal (the face of the pedal pad) and the floorboard when the clutch is disengaged are within the standard value ranges:

- Standard value (C): 6–13 mm (0.2–0.5 inch)
- Standard value (D):
- 2.4L engine: 95.4 mm (3.8 inch) or more
- 3.0L engine: 89.5 mm (3.5 inch) or more

6. If the measured free play and distance do not agree with the standard value ranges, it is probably the result of either air in the hydraulic system or a faulty master cylinder or clutch. Bleed the air, or disassemble and inspect the master cylinder or clutch.

7. Reinstall the carpet, etc.

REMOVAL & INSTALLATION

1. Before servicing the vehicle, refer to the precautions in the beginning of this section.
2. Remove or disconnect the following:
   - Negative battery cable
   - Clutch fluid line bracket
   - Insulator, washer and clutch tube
   - Union, union bolt and gasket
   - Valve and spring, 3.0L engine
   - Clutch release cylinder
   - Clutch cover and disc
   - Release fork shaft, 3.0L engine
   - Left support spring, 3.0L engine
   - Release fork
   - Clutch release bearing
   - Release fork boot and fulcrum, 2.4L engine
   - Release fork boot and maintenance cover, 3.0L engine

To install:

3. Install or connect the following:
   - Release fork boot and maintenance cover, 3.0L engine
   - Release fork boot and fulcrum and torque the fulcrum to 26 ft. lbs. (35 Nm)
   - Clutch release bearing
   - Release fork
   - Left support spring and release fork shaft, 3.0L engine
   - Clutch disc and cover and torque the bolts to 14 ft. lbs. (19 Nm)
   - Clutch release cylinder
   - Valve and spring, 3.0L engine
   - Union, bolt and gasket and torque the bolt to 17 ft. lbs. (23 Nm)

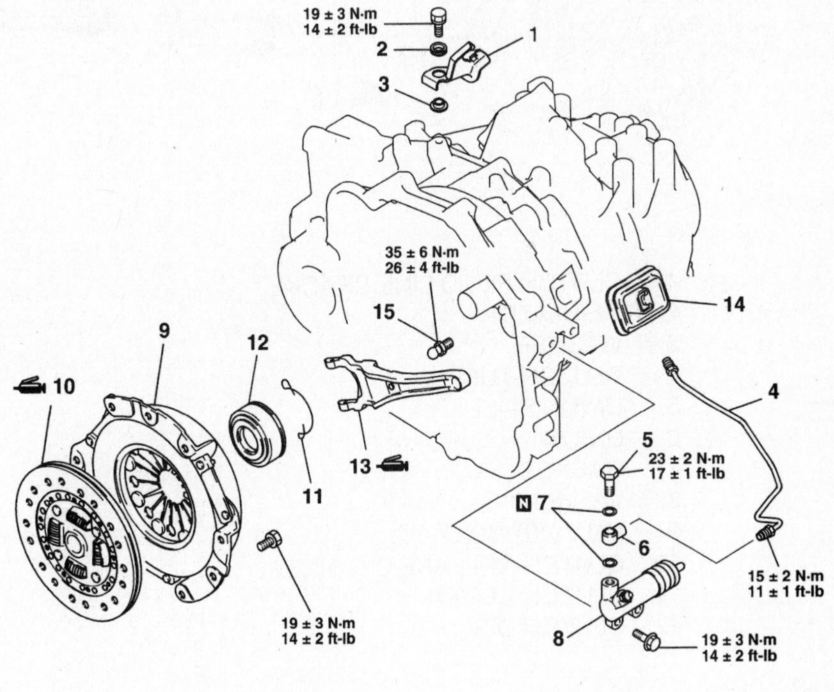

Clutch and related parts—with 2.4L engine

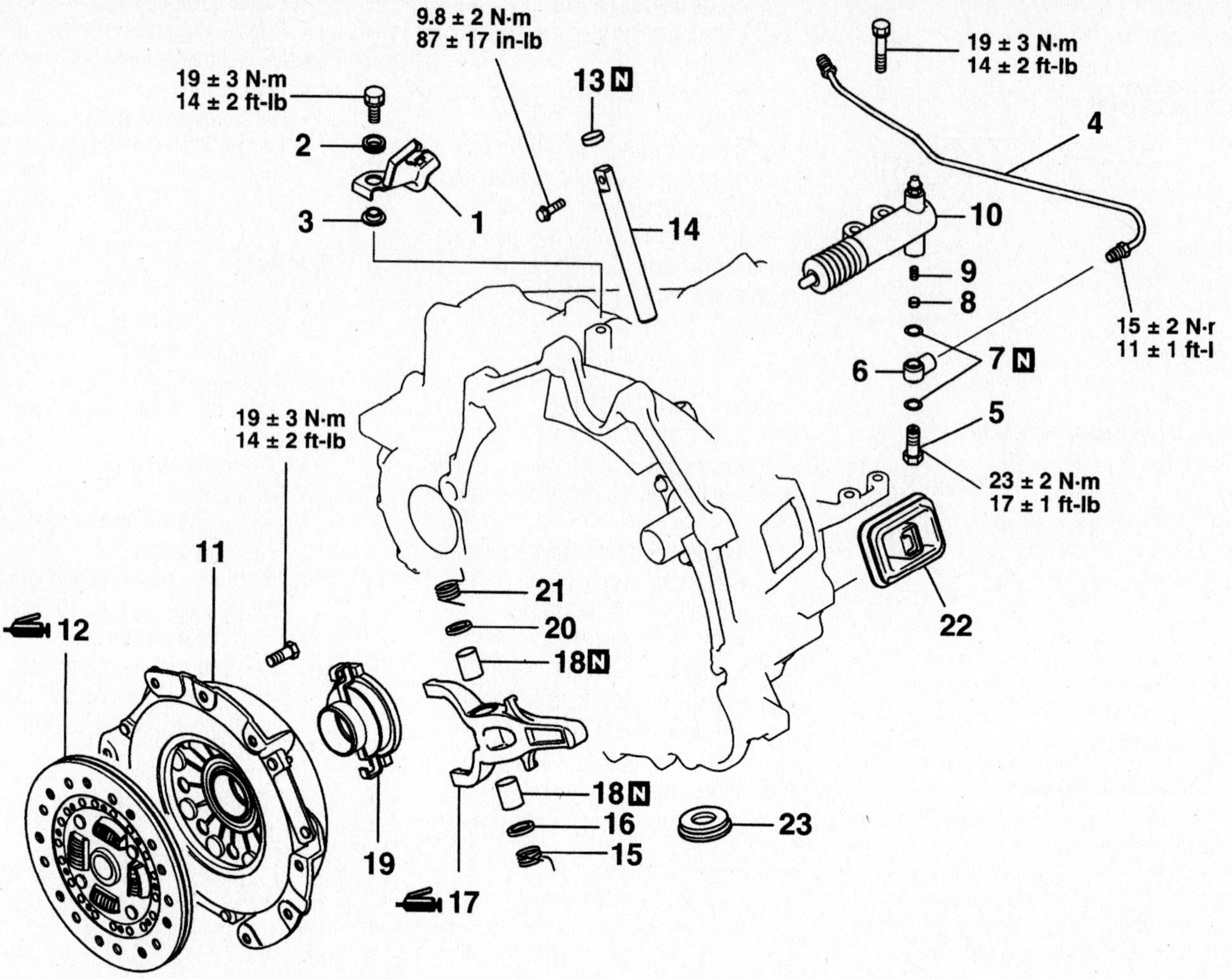

1.	CLUTCH FLUID LINE BRACKET	13.	SEALING CAP
2.	INSULATOR	14.	RELEASE FORK SHAFT
3.	WASHER	15.	SUPPORT SPRING (L)
4.	CLUTCH TUBE	16.	PACKING
5.	UNION BOLT	17.	RELEASE FORK
6.	UNION	18.	BUSHING
7.	GASKET	19.	CLUTCH RELEASE BEARING
8.	VALVE	20.	PACKING
9.	VALVE SPRING	21.	SUPPORT SPRING (R)
10.	CLUTCH RELEASE CYLINDER	22.	RELEASE FORK BOOT
11.	CLUTCH COVER	23.	MAINTENANCE HOLE COVER
12.	CLUTCH DISC		

06009-COUP-G51

Clutch and related parts—with 3.0L engine

- Insulator, washer and clutch tube and torque the fastener to 11 ft. lbs. (15 Nm)
- Fluid line bracket and torque the bolt to 14 ft. lbs. (19 Nm)
- Negative battery cable

4. Check and top off the fluid level, if necessary.

Hydraulic Clutch System

BLEEDING

1. Before servicing the vehicle, refer to the precautions in the beginning of this section.

2. Fill the reservoir with clean DOT 3 or DOT 4 brake fluid.

3. Loosen the bleed screw, then have the clutch pedal pressed to the floor.

4. Tighten the bleed screw, then release the clutch pedal.

5. Repeat the procedure until the fluid is free of air bubbles.

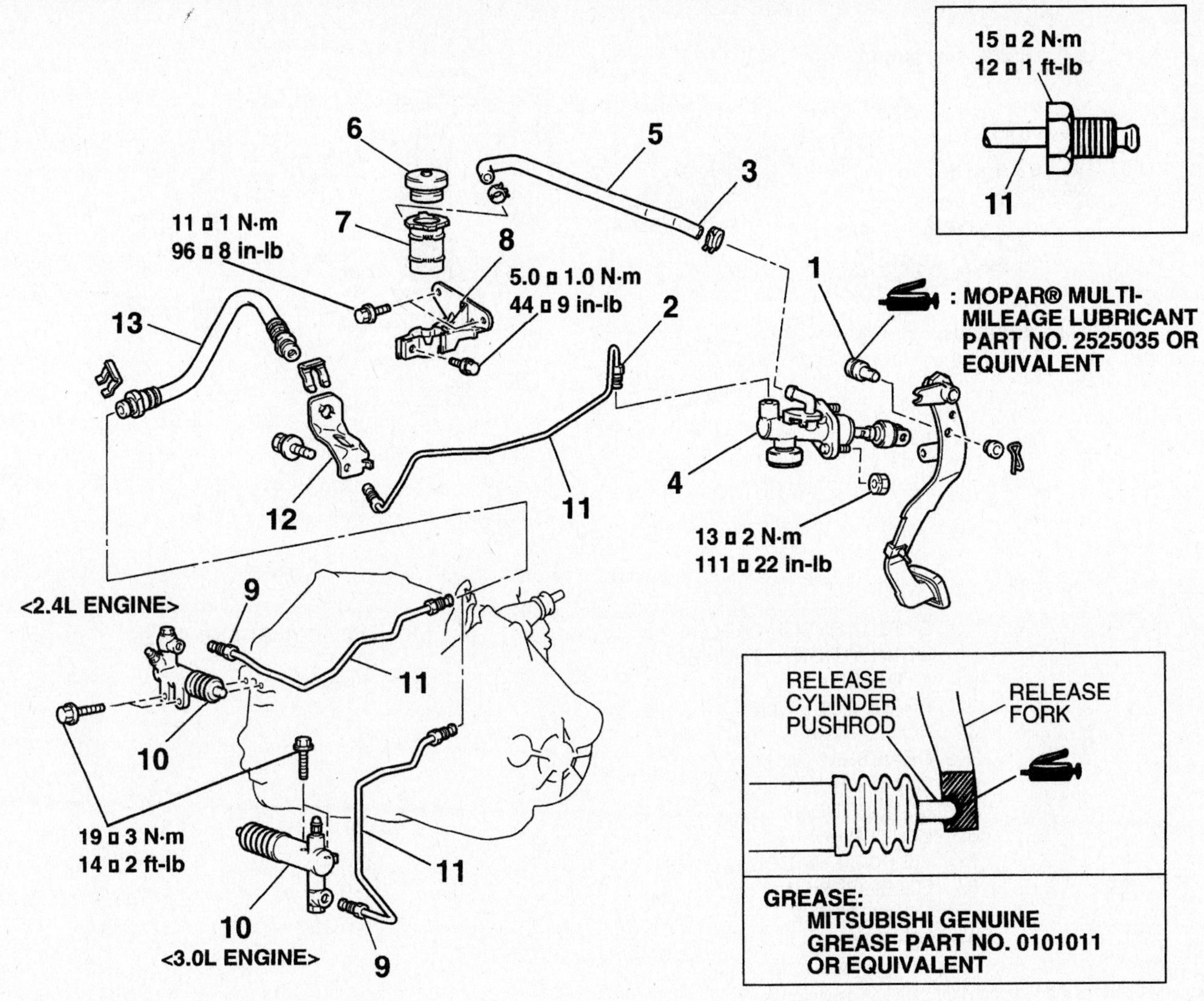

1. CLEVIS PIN
2. CLUTCH TUBE CONNECTION
3. RESERVOIR HOSE CONNECTION
4. CLUTCH MASTER CYLINDER
5. RESERVOIR HOSE
6. RESERVOIR CAP
7. RESERVOIR TANK
8. RESERVOIR BRACKET
9. CLUTCH TUBE CONNECTION
10. CLUTCH RELEASE CYLINDER
11. CLUTCH TUBE
12. BRACKET
13. CLUTCH HOSE

06009-COUP-G52

Clutch hydraulic system

➡It is suggested that a hose be attached to the bleed screw with the other end immersed in a container at least half full of brake fluid during the bleeding operation. Do not allow the reservoir to run out of fluid during bleeding.

6. Refill the reservoir with clean brake fluid.
7. Check the clutch for proper operation.

Halfshaft

REMOVAL & INSTALLATION

1. Before servicing the vehicle, refer to the precautions in the beginning of this section.
2. Remove or disconnect the following:
   - Negative battery cable
   - Front wheel
   - Vehicle Speed Sensor (VSS) cable, if equipped
   - Brake hose clip and cotter pin
   - Driveshaft nut
   - Lower arm ball joint connection
   - Tie rod end
   - Stabilizer link
   - Driveshaft and circlip

To install:

3. Install or connect the following:

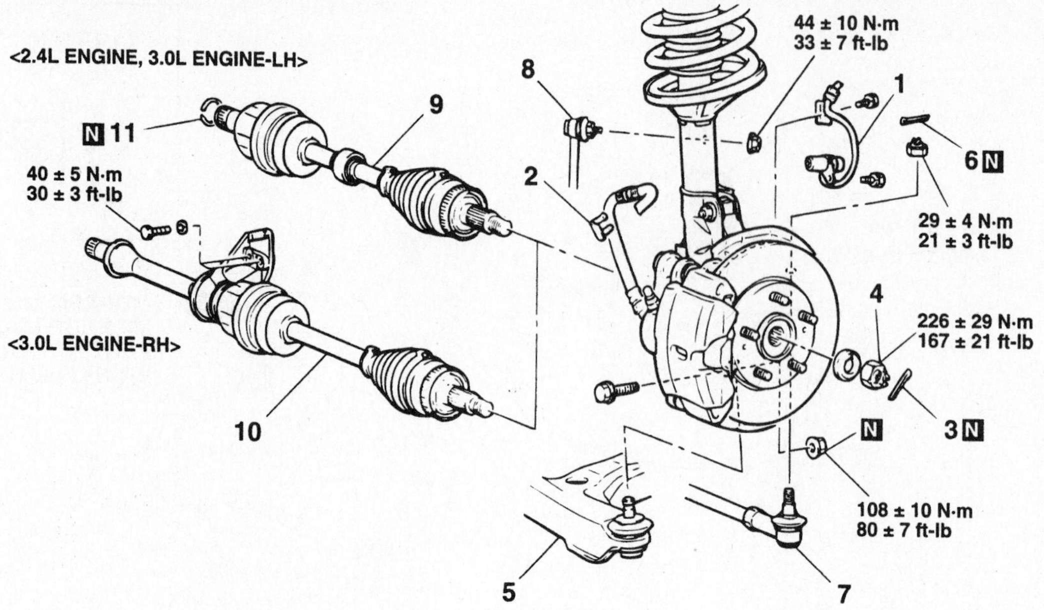

1. SPEED SENSOR CABLE CONNECTION <VEHICLES WITH ABS>
2. BRAKE HOSE CLIP
3. COTTER PIN
4. DRIVESHAFT NUT
5. LOWER ARM BALL JOINT CONNECTION
6. COTTER PIN
7. TIE ROD END CONNECTION
8. STABILIZER LINK CONNECTION
9. DRIVESHAFT
10. DRIVESHAFT AND INNER SHAFT
11. CIRCLIP

9356CG18

Exploded view of the left and right halfshaft assemblies

- Driveshaft
- Stabilizer link and torque the nut to 33 ft. lbs. (44 Nm)
- Tie rod end and torque the nut to 21 ft. lbs. (28 Nm)
- New cotter pin to the tie rod end
- Lower arm ball joint
- Driveshaft nut and torque it to a maximum of 167 ft. lbs. (226 Nm)
- New cotter pin to the driveshaft nut
- Brake hose clip
- VSS cable, if equipped
- Front wheel
- Negative battery cable

CV-Joints

OVERHAUL

WARNING

The Birfield joint assembly, located on the wheel side of the halfshaft, is not to be disassembled; repair of this joint is only by replacement of the halfshaft.

Tri-Pot Joint

1. Before servicing the vehicle, refer to the precautions in the beginning of this section.

2. Remove halfshaft and place it in a soft jawed vise.

3. Disassemble or remove:

- Tri-pot boot bands
- Tri-pot case for left halfshaft or tri-pot case/inner shaft assembly for right halfshaft

➡Wipe the grease from the tri-pot case.

- Halfshaft snapring
- Tri-pot spider assembly

WARNING

Do not disassemble the spider assembly.

- Tri-pot boot

WARNING

If the boot is to be reused, wrap plastic tape around the shaft splines to protect the boot from damage.

4. If working with the right halfshaft and it is necessary to replace the tri-pot case, press the case from the inner shaft assembly.

To install:

5. If installing a new tri-pot case onto the right halfshaft, perform the following procedure:

a. Lubricate the inner shaft splines with Multi-Mileage Grease No. 2525035.

b. Press the inner shaft assembly into tri-pot case.

c. Secure the tri-pot case with special Tool MB991248 on a hydraulic press with the case facing upward.

d. Position a new seal plate in the center of the tri-pot case.

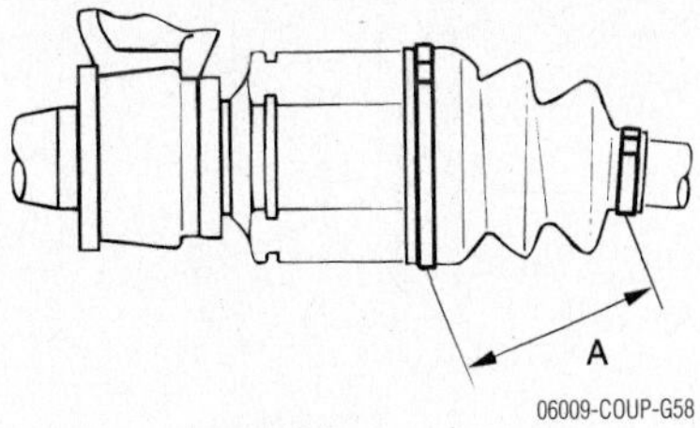

Boot band installation measurement

e. Place a 1.18 in. (30mm) pipe on the seal plate and press the seal into the tri-pot case.

⁂ WARNING

Wrap plastic tape around the shaft splines to protect the boot from damage.

6. Install a new small boot clamp onto the halfshaft followed by the tri-pot boot.
7. Apply the specified repair kit grease between the spider axle and the roller.
8. Install the tri-pot spider assembly onto the shaft from the spline beveled section direction and secure with the snapring.
9. Distribute a portion of the 4.23 oz. (120 g) specified repair kit grease into the tri-pot case, insert the spider assembly and add the remaining grease.
10. Install the tri-pot boot by performing the following procedure:
   a. Position the boots large end on the tri-pot case.
   b. Position the large and small boot clamps onto the boot.
   c. Adjust the boot so that the bands are spaced at 3.3 for 2.4L engines; 3.1 in. for 3.0L engines.
   d. Tighten the band securely.

Center Bearing

1. Before servicing the vehicle, refer to the precautions in the beginning of this section.
2. Remove the right halfshaft and place it in a soft-jawed vise.
3. Remove the tri-pot spider assembly from the tri-pot case.
4. Remove the tri-pot case by performing the following procedure:
   a. Position the inner shaft/tri-pot case assembly on a hydraulic press supported by Tool MB991248 with the tri-pot case facing upward.
   b. Position a bar inside the tri-pot case, on the end of the inner shaft and press the case from the inner shaft assembly.
5. Place the inner shaft assembly in a soft-jawed vise.

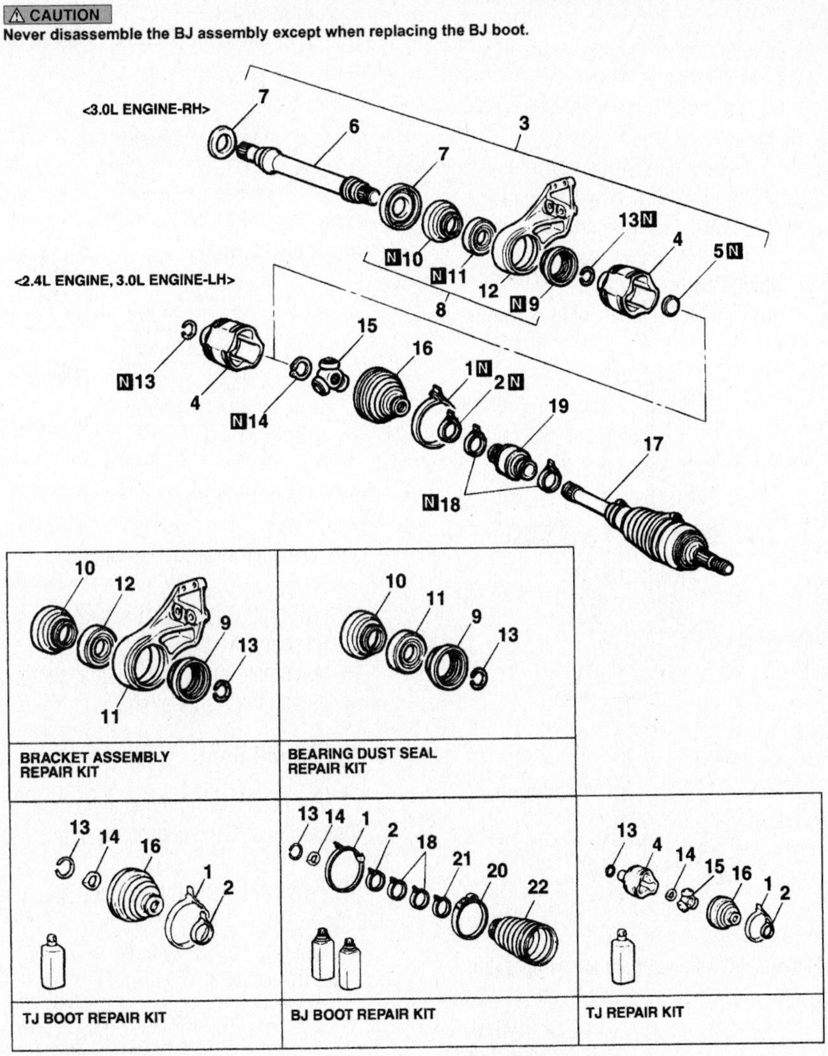

1. TJ BOOT BAND (LARGE)
2. TJ BOOT BAND (SMALL)
3. TJ CASE INNER SHAFT ASSEMBLY
4. TJ CASE
5. SEAL PLATE
6. INNER SHAFT
7. DUST COVER
8. BRACKET ASSEMBLY
9. DUST SEAL OUTER
10. DUST SEAL INNER
11. CENTER BEARING
12. CENTER BEARING BRACKET
13. CIRCLIP
14. SNAP RING
15. SPIDER ASSEMBLY
16. TJ BOOT
17. BJ ASSEMBLY
18. DAMPER BAND <2.4L ENGINE, 3.0L ENGINE-LH>
19. DYNAMIC DAMPER <2.4L ENGINE, 3.0L ENGINE-LH>
20. BJ BOOT BAND (LARGE)
21. BJ BOOT BAND (SMALL)
22. BJ BOOT

NOTE:
BJ: Birfield Joint
TJ: Tripod Joint

67189-SEBR-G11

Exploded view of the halfshaft assemblies

6. Using a wheel puller, press the center bearing bracket from the inner shaft.
7. Using a hydraulic press, Installer Adapter Tool MB990932 and Snap-in Bar Tool MB990938, press the center bearing and inner dust seal from the center bearing bracket.

To install:

8. Using a hydraulic press, Installer Adapter Tool MB990932 and Snap-in Bar Tool MB990938, press the center bearing into the center bearing bracket.
9. Using a hydraulic press, Installer Adapter Tool MB990933 and Snap-in Bar Tool MB990938, press the inner duct seal into the center bearing bracket.
10. Using a hydraulic press, Installer Adapter Tool MB990931 and Snap-in Bar Tool MB990938, press the outer duct seal into the center bearing bracket.
11. Using a hydraulic press and Adapter Tool MB991172, press the inner shaft into the center bearing bracket.
12. Install the tri-pot case onto the right

halfshaft by performing the following procedure:

a. Lubricate the inner shaft splines with Multi-Mileage Grease No. 2525035.

b. Press the inner shaft assembly into tri-pot case.

c. Secure the tri-pot case with special Tool MB991248 on a hydraulic press with the case facing upward.

d. Position a new seal plate in the center of the tri-pot case.

e. Place a 1.18 in. (30mm) pipe on the seal plate and press the seal into the tri-pot case.

13. Assemble the tri-pot joint assembly.

14. Install the right halfshaft.

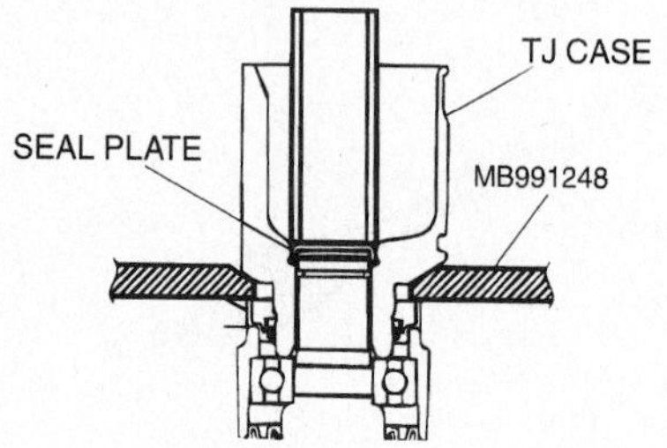

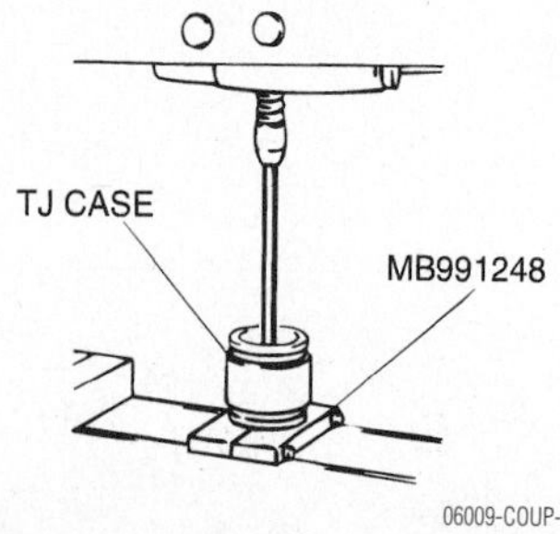

Inner shaft removal from the tri-pot joint

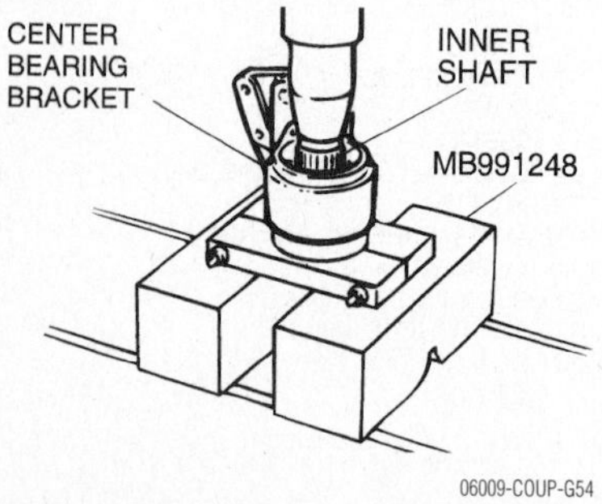

Inner shaft removal from the center bearing bracket

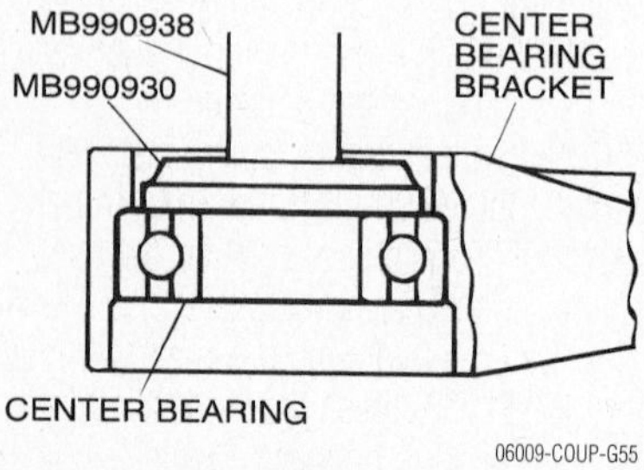

Center bearing removal

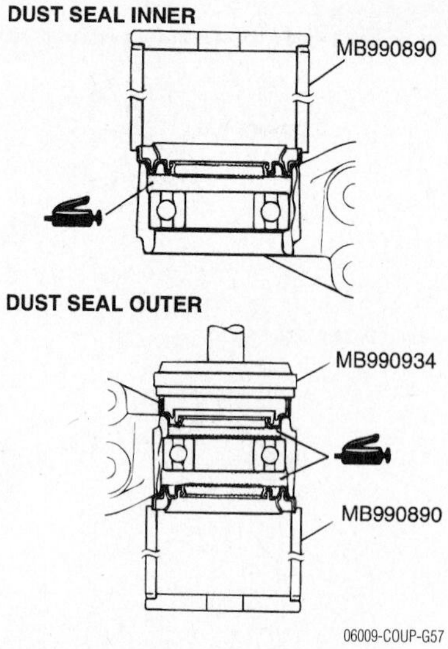

Inner and outer dust seal installation

Birfield Joint Boot

1. Before servicing the vehicle, refer to the precautions in the beginning of this section.

2. Remove halfshaft and place it in a soft jawed vise.

3. Disassemble or remove:
   - Tri-pot joint
   - Dynamic damper, if equipped
   - Birfield joint boot bands
   - Birfield joint boot

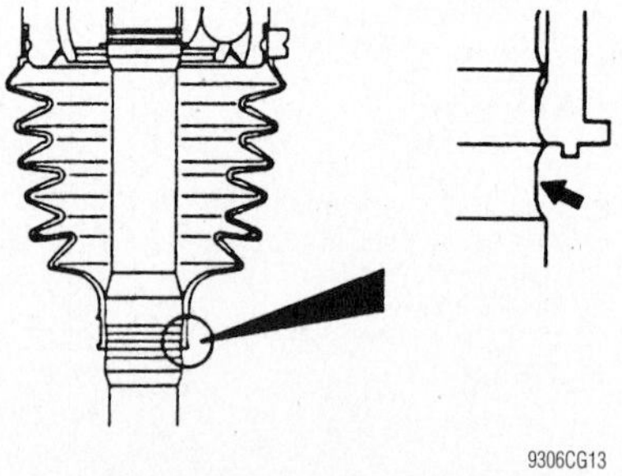

Positioning the Birfield joint boot's small diameter

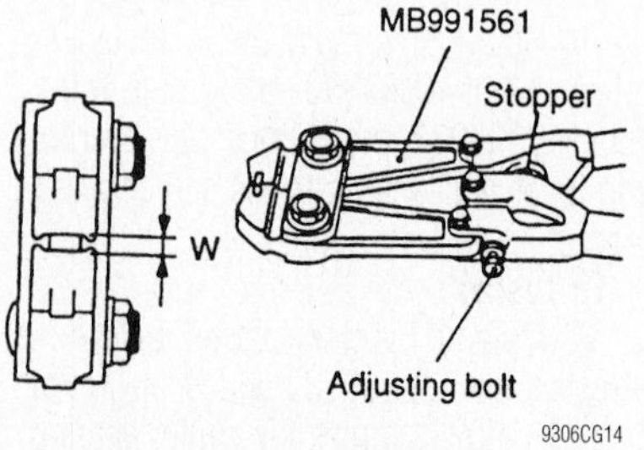

View of the band crimper tool

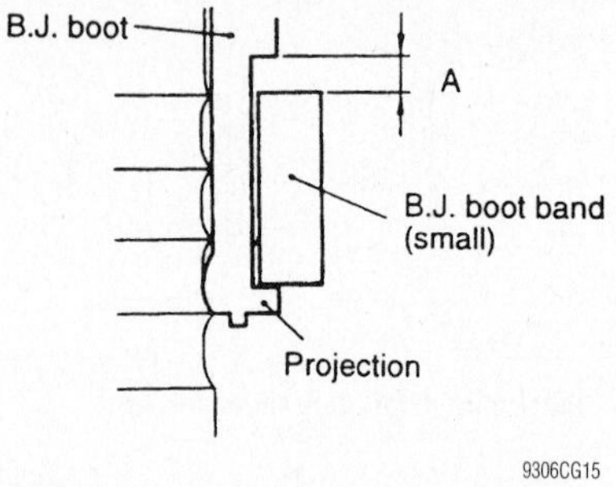

Positioning the Birfield joint boot's small band

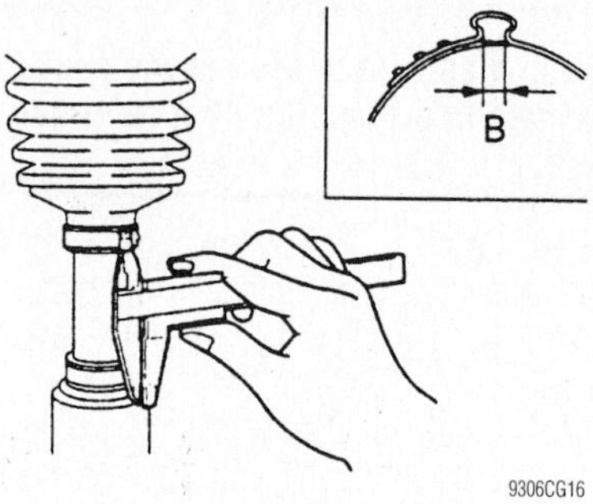

Measuring the small band's crimp

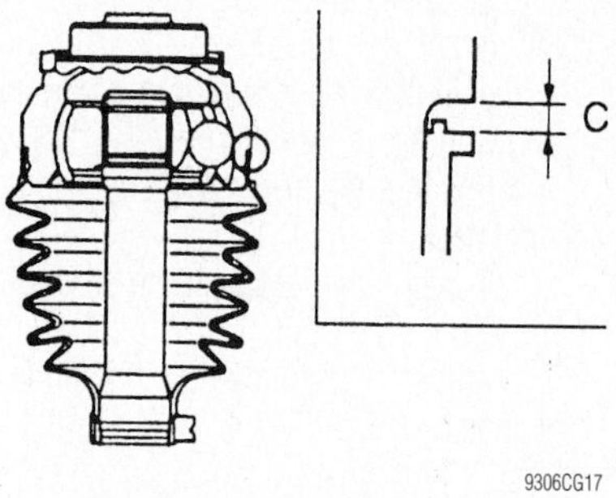

Positioning the Birfield joint boot's large diameter

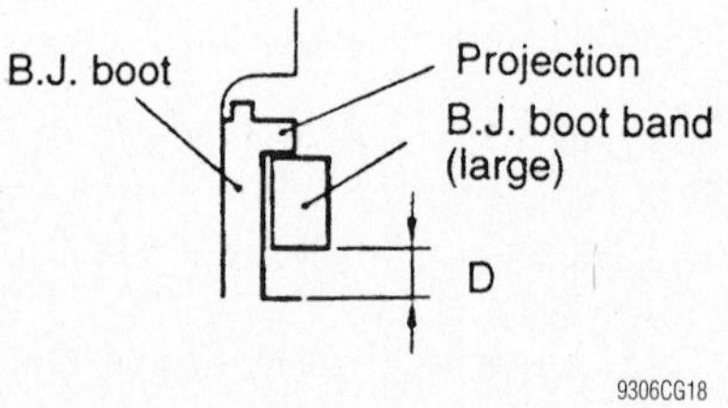

Positioning the Birfield joint boot's large band

To install:

4. Assemble or install:
   - Birfield joint boot
   - Birfield joint boot small band

5. Place the halfshaft in a soft jawed

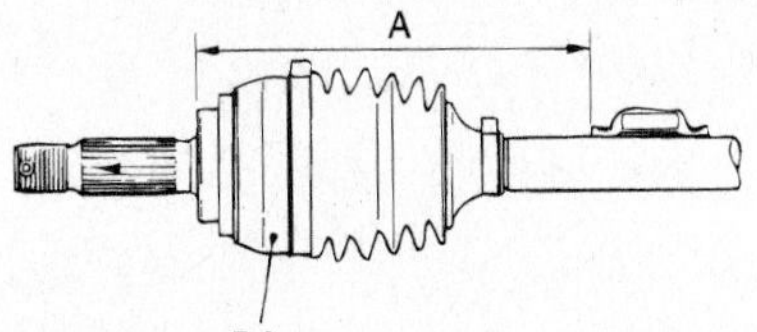

Dynamic damper installation. Dimension A should be 9.5 in. on the left side; 10.0 in. on the right side.

vise so that the Birfield joint is standing vertically.

6. Position the boots small diameter so that only 1 groove is exposed on the shaft.

7. Adjust the Band Crimper Tool MB991561 so that the jaw opening is 0.114 in. (2.9mm).

8. Position the small band so there is clearance between the boot and the bands mating edge.

9. Using a Band Crimper Tool MB991561, crimp the small band until the internal measurement of the crimp is 0.094–0.110 in. (2.4–2.8mm).

**** WARNING**

If the crimp dimension is not correct, remove the band install a new one.

10. Using 5.47 oz. (155g) specified amount of grease in the repair kit, pack the Birfield boot.

11. Position the boot so 0.004–0.061 in. (0.1–1.55mm) of clearance exists between the boots large diameter end and the Birfield joint housing shoulder.

12. Adjust the Band Crimper Tool MB991561 so that the jaw opening is 0.126 in. (3.2mm).

13. Install and position the large boot band so that it rests against the boot(s) projection and a gap exists between the clamp and the boot.

14. Using a Band Crimper Tool MB991561, crimp the small band until the internal measurement of the crimp is 0.094–0.110 in. (2.4–2.8mm).

**** WARNING**

If the crimp dimension is not correct, remove the band install a new one.

15. If equipped, install the dynamic damper by performing the following procedure:

a. Slide the damper onto the halfshaft with new bands.

b. Position the damper so the distance from the front of the Birfield joint to the front edge of the damper assembly is 14.60–14.84 in. (371–377mm) for the right halfshaft or 7.52–7.76 in. (191–197mm) for the left halfshaft.

c. Tighten and secure the damper bands.

16. Install the tri-pot joint.

STEERING AND SUSPENSION

Air Bag

**** CAUTION**

Some vehicles are equipped with an air bag system, also known as the Supplemental Inflatable Restraint (SIR) system or Supplemental Restraint System (SRS). The system must be disabled before performing service on or around system components, steering column, instrument panel components, wiring and sensors. Failure to follow safety and disabling procedures could result in accidental air bag deployment, possible personal injury and unnecessary system repairs.

PRECAUTIONS

Several precautions must be observed when handling the inflator module to avoid accidental deployment and possible personal injury. Along with the precautions in the beginning of this section, observe the following:

1. Never carry the inflator module by the wires or connector on the underside of the module.

2. When carrying a live inflator module, hold securely with both hands, and ensure that the bag and trim cover are pointed away.

3. Place the inflator module on a bench or other surface with the bag and trim cover facing up.

4. With the inflator module on the bench, never place anything on or close to the module which may be thrown in the event of an accidental deployment.

5. Do not attempt to repair any of the air bag system wiring harness connectors. If any of the connectors or wires are faulty, replace that harness.

6. Air bag components should not be subjected to heat over 200°F (93°C). Remove the Supplemental Restraint System Air Bag Control Unit (SRS-ECU), the air bag modules themselves and the clock spring before drying or baking the vehicle after painting.

7. After air bag system service, check the SRS warning light operation to be sure that the system functions properly.

8. Make certain that the ignition switch is in the **OFF** position when a scan tool is connected or disconnected.

DO NOT use any electrical test equipment on or near any SRS components except those specified by Chrysler corporation:

9. Use a digital multi-meter for which the maximum test current is 2 milliamp (mA) or less at the minimum range of resistance measurement for use with the Chrysler SRS Check Harness when checking the SRS electrical circuitry.

10. Chrysler Special Tool MB991613 SRS Check Harness acts like a "break-out box" for checking SRS wiring. There are other factory special tool wiring adapters that are available and may be used.

11. DRB III scan tool for reading and erasing air bag diagnostic codes.

NEVER ATTEMPT TO REPAIR THE FOLLOWING COMPONENTS:

12. Air Bag Control Unit (SRS-ECU)

13. Air Bag Modules

14. If any of these components are diagnosed as faulty, they should only be replaced.

DISARMING

The front system consists of 2 air bag modules, one located in the center of the steering wheel and another located above the glove box, which contains the folded air bag and an inflator unit. The air bag Electronic Control Unit (SRS-ECU) located under the floor console assembly monitors the system and which contains a safing G sensor and analog G sensor. An SRS warning light is located on the instrument panel which indicates the status of the air bag system. A clock spring interconnection is located within the steering column.

To deploy the air bags, the SRS-ECU must respond to the output signal from the analog G sensor and the safing G sensor must be ON. The SRS-ECU, then causes the air bag modules to ignite and deploy.

Service technicians should use care when working around any vehicle equipped with an air bag system, to avoid injury to the technician by inadvertent deployment of the air bag or to the driver by rendering the air bag system inoperative.

The SRS-ECU not only controls the air bag system, it can provide diagnostic information. The SRS-ECU monitors the air bag system and stores data concerning any detected faults in the system. When the ignition key is turned to the **ON** or **START** position, the SRS warning light should illuminate for about 7 seconds, then turn off. That indicates that the SRS system is in operating condition. If the SRS warning light does not illuminate as described or stays on for more than 7 seconds or if the SRS light illuminates while driving, immediate inspection is required. If the vehicle's SRS warning light is in any of these 3 conditions, the SRS system must be inspected, diagnosed and serviced.

To avoid injury from accidental deployment of the air bag during vehicle servicing, refer to all service precautions.

**** CAUTION**

The Air Bag system must be disarmed before removing many components. Failure to do so may cause accidental deployment of the air bag, resulting in unnecessary system repairs and/or personal injury.

1. Disarm the air bag system using the following procedure:
   a. Position the front wheels in the straight-ahead position and place the key in the **LOCK** position. Remove the key from the ignition lock cylinder.
   b. Disconnect the negative battery cable and insulate the cable end with high-quality electrical tape or similar non-conductive wrapping.
   c. Wait at least one minute before working on the vehicle. The air bag system is designed to retain enough voltage to deploy the air bag for a short period of time even after the battery has been disconnected.

Power Rack and Pinion Steering Gear

REMOVAL

1. Before servicing the vehicle, refer to the precautions in the beginning of this section.
2. Drain the power steering fluid.

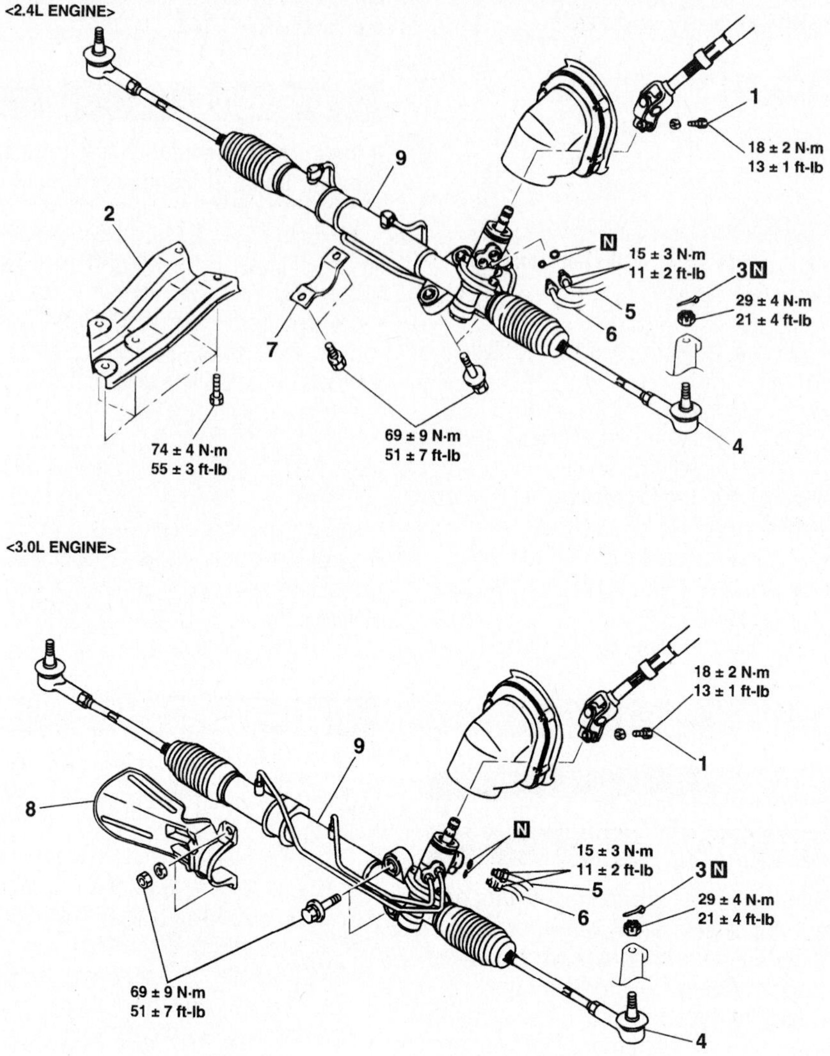

1. STEERING SHAFT ASSEMBLY AND GEAR BOX CONNECTING BOLT
2. STAY <2.4L ENGINE>
3. COTTER PIN
4. TIE ROD END AND KNUCKLE CONNECTION
5. RETURN HOSE CONNECTION
6. PRESSURE TUBE CONNECTION
7. CYLINDER CLAMP <2.4L ENGINE>
8. CYLINDER CLAMP ASSEMBLY <3.0L ENGINE>
9. GEAR BOX ASSEMBLY

06009-COUP-G59

Exploded view of the steering gear assemblies

3. Remove or disconnect the following:
   - Negative battery cable from the left shock tower
   - Front exhaust pipe
   - Center member
   - Sway bar, 2.4L engine only
   - Steering shaft assembly
   - Stay, 2.4L engine only
   - Tie rod end from the steering knuckle
   - Return hose and pressure tube connections
   - Cylinder clamp
   - Steering gear from the vehicle

INSPECTION

Gear Box Total Pinion Torque Check

**** WARNING**

When holding the steering gear box assembly in a vice, secure its mounting positions. If it is secured in any other place, the gear housing may become deformed or damaged.

1. Using special tool MB990228 or MB991006, rotate the pinion gear at the rate

of one rotation in approximately 4 to 6 seconds to check the total pinion torque.

- Standard value: 0.8–1.9 Nm (6.9–16.5 in-lb)
- Change in torque: 0.7 Nm (6.1 inch lbs.) or less

➡**When measuring, remove the bellows from the rack housing.**

2. Measure the pinion torque through the whole stroke of the rack. If the measured value is not within the standard range, first adjust the rack support cover, and then check the total pinion torque again.
3. If the total pinion torque cannot be adjusted to within the standard range by adjusting the rack support cover, check the rack support cover, rack support spring, rack support and replace any parts if necessary.

Tie Rod Swing Resistance Check

1. Give 10 hard swings to the tie rod.
2. Measure the tie rod swing resistance with a spring balance. Standard value: 4.0–18.6 N (17.8–82.7 lbs.) [1.0–4.9 Nm (8.7–43.4 inch lbs.)]
3. If the measured value exceeds the standard value, replace tie rod.
4. If the measured value is below the standard value, the tie rod can be re-used if it swings smoothly without excessive play.

INSTALLATION

1. Install or connect the following:
   - Steering gear to the vehicle
   - Cylinder clamp and torque the bolts to 51 ft. lbs. (69 Nm)
   - Pressure tube and return hoses and torque to 11 ft. lbs. (15 Nm)
   - Tie rod ends to the steering knuckle and torque the nut to 21 ft. lbs. (29 Nm)
   - Stay and torque the bolts to 55 ft. lbs. (74 Nm), 2.4L engine only
   - Steering shaft assembly and torque the bolt to 13 ft. lbs. (18 Nm)
   - Sway bar, 2.4L engine only
   - Center member
   - Front exhaust pipe, 2.4L engine only
   - Negative battery cable
2. Fill and bleed the power steering system.
3. Start the vehicle, check for leaks and repair if necessary.
4. Perform a front wheel alignment.

Strut

REMOVAL & INSTALLATION

Front

1. Before servicing the vehicle, refer to the precautions in the beginning of this section.
2. Remove or disconnect the following:
   - Negative battery cable
   - Front wheel

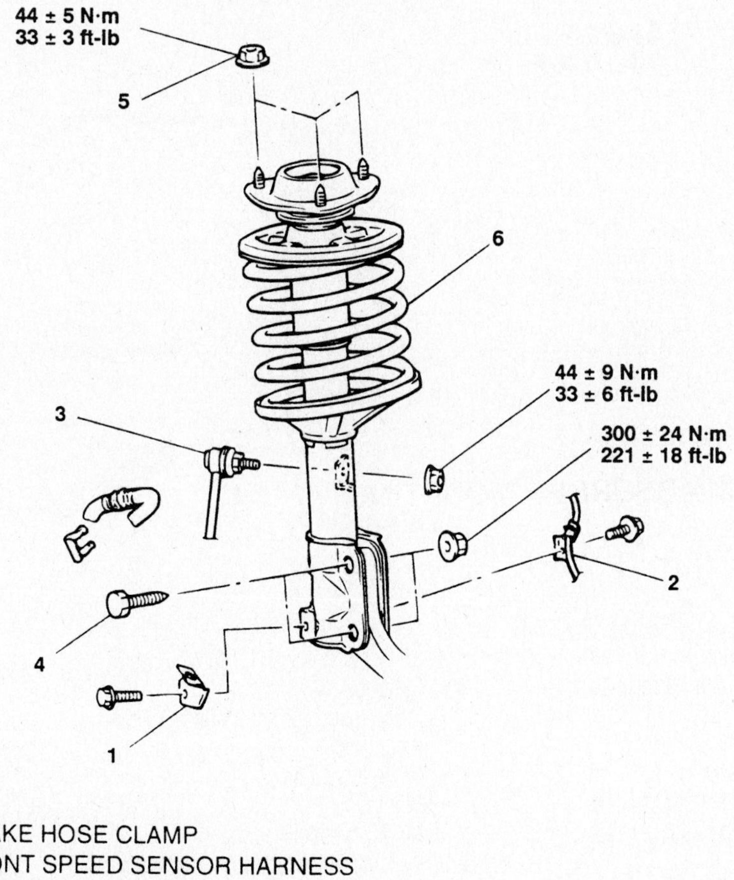

1. BRAKE HOSE CLAMP
2. FRONT SPEED SENSOR HARNESS CLAMP <VEHICLES WITH ABS>
3. STABILIZER LINK
4. BOLTS
5. NUT
6. STRUT ASSEMBLY

9346FG08

Front strut assembly

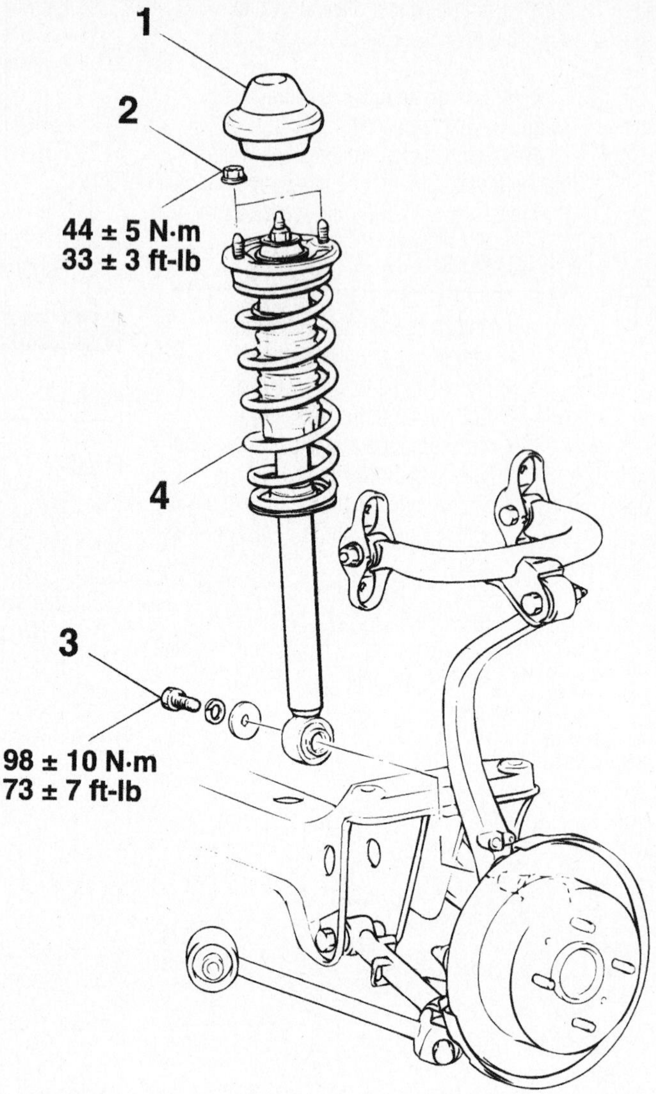

Rear strut assembly

- Brake hose clamp
- Front speed sensor harness clamp, if equipped
- Sway bar link
- Lower control arm
- Strut from the steering knuckle
- Upper mounting bolts
- Strut assembly

To install:

3. Install or connect the following:
   - Strut assembly and torque the upper bolts to 33 ft. lbs. (44 Nm)
   - Strut assembly to the lower control arm and steering knuckle. Torque the bolts to 221 ft. lbs. (300 Nm).
   - Sway bar link and torque the nut to 33 ft. lbs. (44 Nm)
   - Front speed sensor harness clamp, if equipped
   - Brake hose clamp
   - Front wheel
   - Negative battery cable

4. Check and adjust the front end alignment, if needed.

Rear

1. Before servicing the vehicle, refer to the precautions in the beginning of this section.
2. Remove or disconnect the following:
   - Negative battery cable
   - Rear shelf trim
   - Cover from the upper shock tower
   - Upper mounting nuts
   - Shock from the rear steering knuckle
   - Shock assembly

To install:

3. Install or connect the following:
   - Shock absorber and torque the shock to steering knuckle bolt to 73 ft. lbs. (98 Nm)
   - Upper shock nuts and torque to 32 ft. lbs. (45 Nm)

- Upper tower cover
- Rear shelf trim
- Negative battery cable

Coil Spring

REMOVAL & INSTALLATION

Front

1. Before servicing the vehicle, refer to the precautions in the beginning of this section.
2. Remove the strut/shock assembly and place it in a suitable compressor tool.
3. Install Special Tool MB991176 to secure the strut and remove the dust cover and jam nut.
4. Remove the strut insulator.
5. Remove the upper spring seat and pad.
6. Remove the bump rubber and dust cover.
7. Remove the coil spring.
8. Remove the lower spring pad.

To assemble:

9. Install the lower spring pad and coil spring.
10. Install the dust cover and bump rubber.
11. Install the upper spring pad and seat.
12. Carefully compress the spring and install the insulator and jam nut.
13. When properly aligned, torque the jam nut to 47 ft. lbs. (64 Nm).
14. Install the dust cover and remove the compressor tools.
15. Install the strut/shock to the vehicle.

Rear

1. Before servicing the vehicle, refer to the precautions in the beginning of this section.
2. Disassemble as follows:
3. Remove the rear shock from the vehicle.
4. Install the assembly in a spring compressor tool.
5. Remove the jam nut, washer and upper bushing.
6. Remove the upper bracket assembly and spring pad.
7. Remove the collar and bushing.

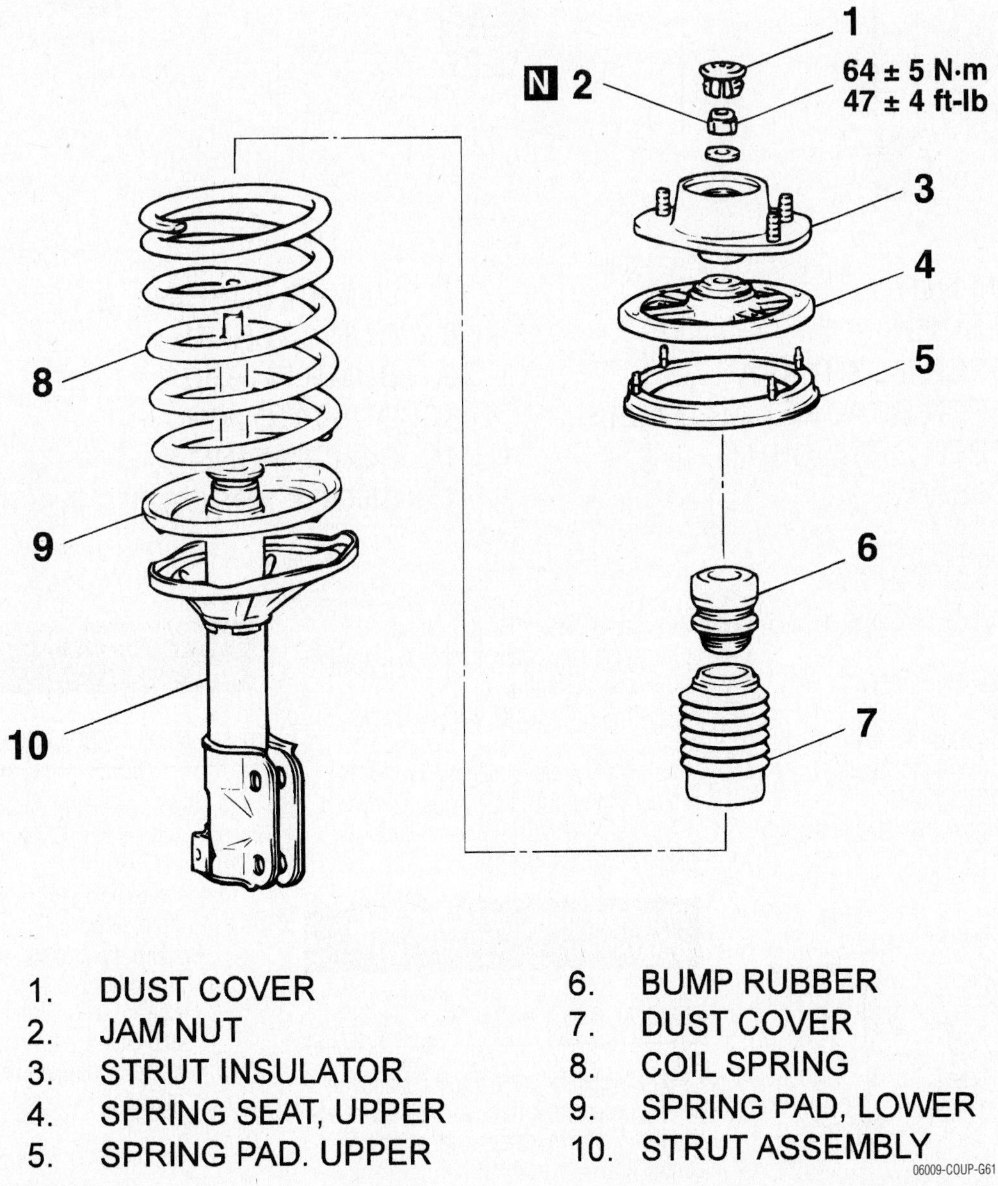

Front strut exploded view

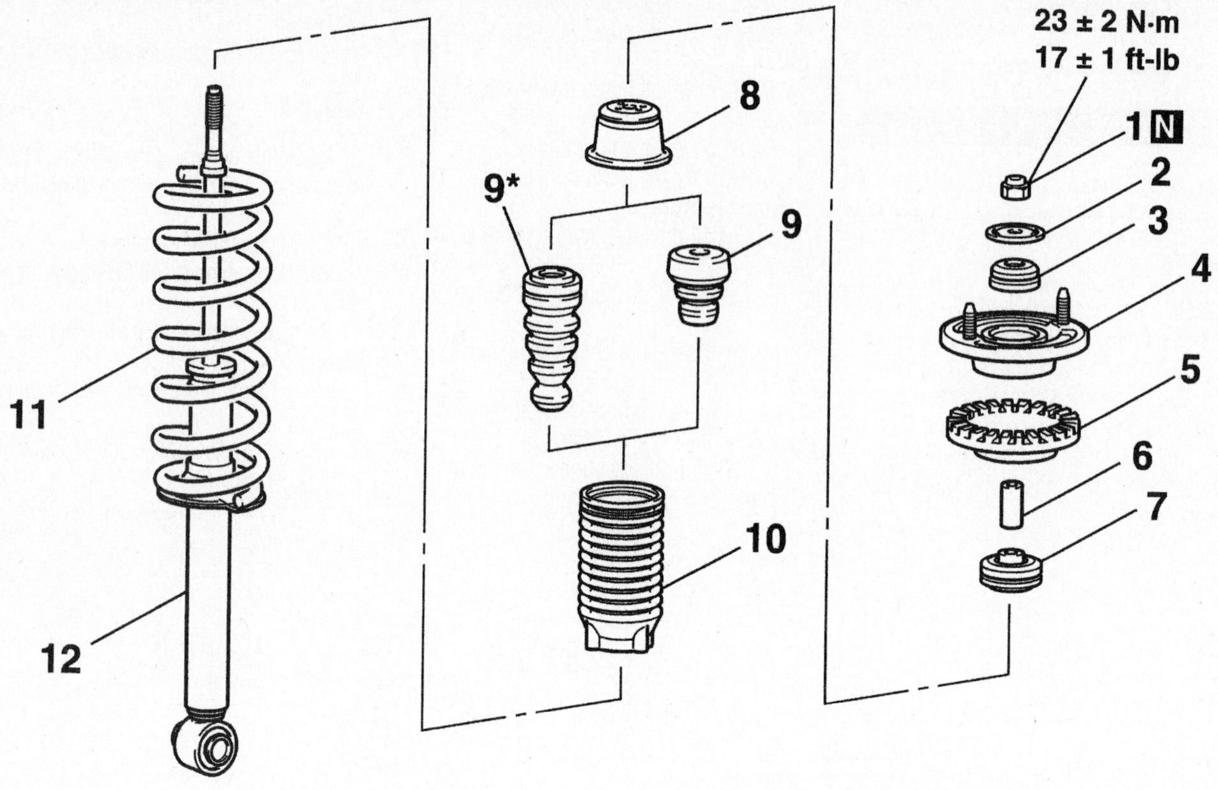

1. JAM NUT
2. WASHER
3. UPPER BUSHING A
4. UPPER BRACKET ASSEMBLY
5. UPPER SPRING PAD
6. COLLAR
7. UPPER BUSHING B
8. CUP ASSEMBLY
9. BUMP RUBBER
10. DUST COVER
11. COIL SPRING
12. SHOCK ABSORBER ASSEMBLY

06009-COUP-G62

Rear strut exploded view

8. Remove the cup, bump rubber and dust cover.
9. Remove the coil spring

To assemble:

10. Install the coil spring to the shock absorber using Compressor Tools MB991237 and MB991239.
11. Install the dust cover and bump rubber.

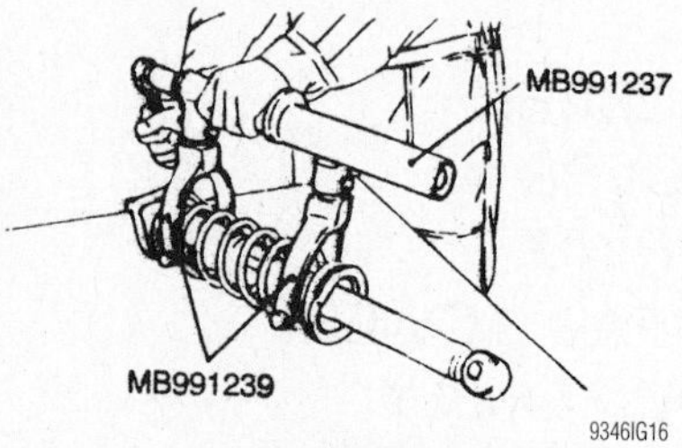

Compress the coil spring as shown

12. Install the cup and bushing
13. Install the collar, supper spring pad and bracket assembly.
14. Install the upper bushing and washer.
15. When properly aligned, install the jam nut and torque to 17 ft. lbs. (23 Nm).
16. Remove the compressor tools and install the shock absorber.

Front Suspension Lower Ball Joint

REMOVAL & INSTALLATION

On all vehicles, the ball joint cannot be serviced separately. If the ball joint is defective it will require replacement of the lower control arm.

Front Suspension Lower Control Arm

REMOVAL & INSTALLATION

1. Before servicing the vehicle, refer to the precautions in the beginning of this section.
2. Remove or disconnect the following:
   - Front wheel
   - Lower control arm from the steering knuckle
   - Lower control arm mounting bolt and clamp
   - Lower control arm
3. If replacing the bushing, perform the following:
   a. Apply soapy water between the shaft and old bushing and pry the old bushing out.

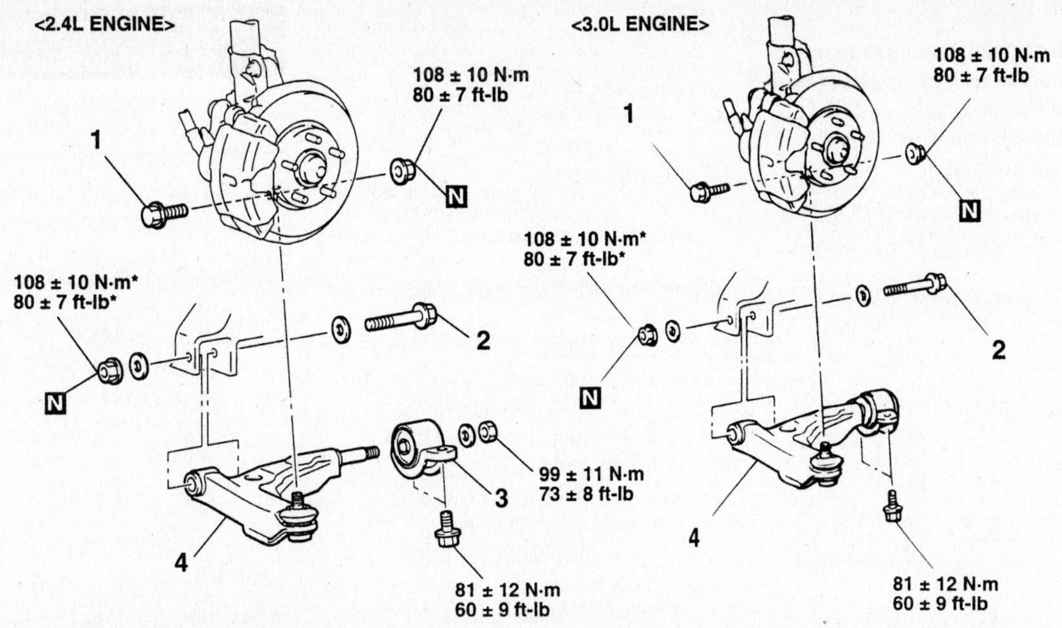

1. LOWER ARM AND KNUCKLE CONNECTION
2. LOWER ARM MOUNTING BOLT
3. LOWER ARM CLAMP <2.4L ENGINE>
4. LOWER ARM

9356CG19

Exploded view of the front lower control arm and related components—2002 models

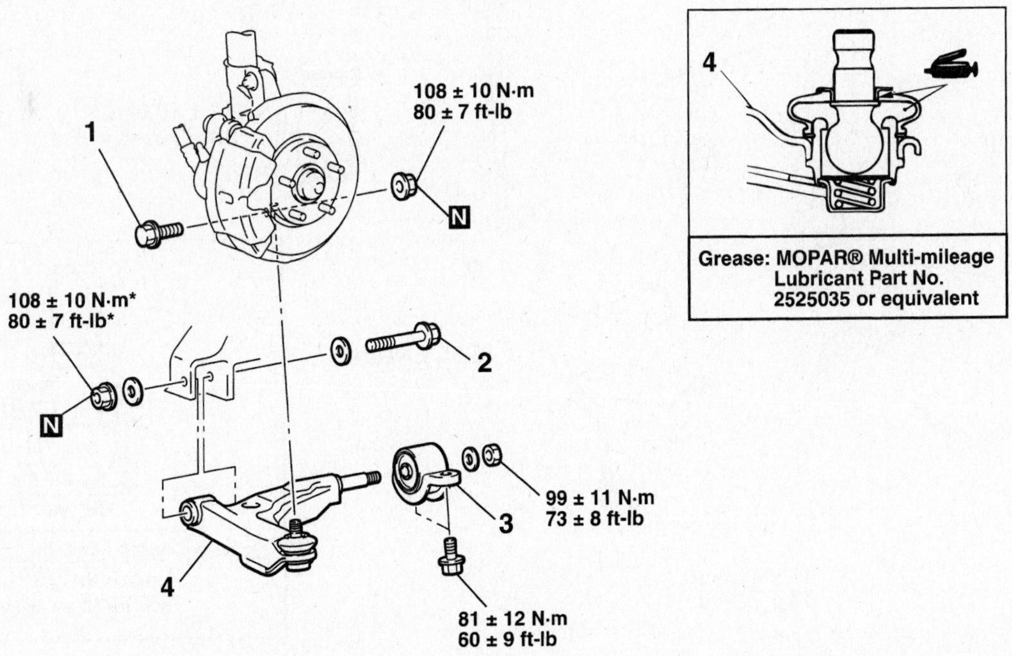

1. LOWER ARM AND KNUCKLE CONNECTION
2. LOWER ARM MOUNTING BOLT
3. LOWER ARM CLAMP
4. LOWER ARM

06009-COUP-G63

Lower control arm and related parts—all 2003–05 models

To install:

➡See the accompanying illustrations. Fasteners marked with an asterisk (*) should be temporarily tightened, then, fully tightened with the suspension loaded (on the ground).

4. Apply soapy water to the shaft and new bushing and install the new bushing, if removed.
5. Install or connect the following:

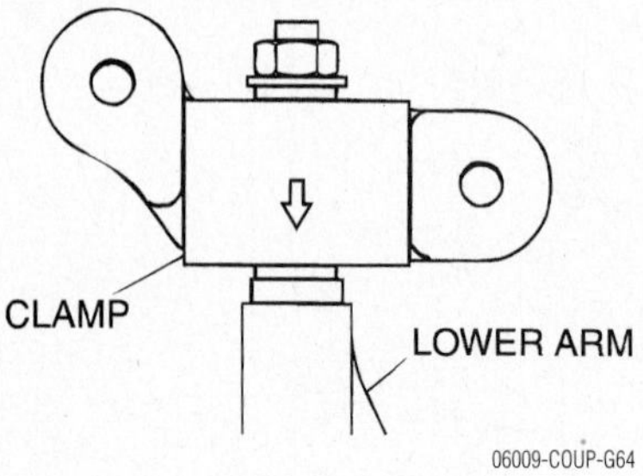

Observe the arrow mark on the clamp

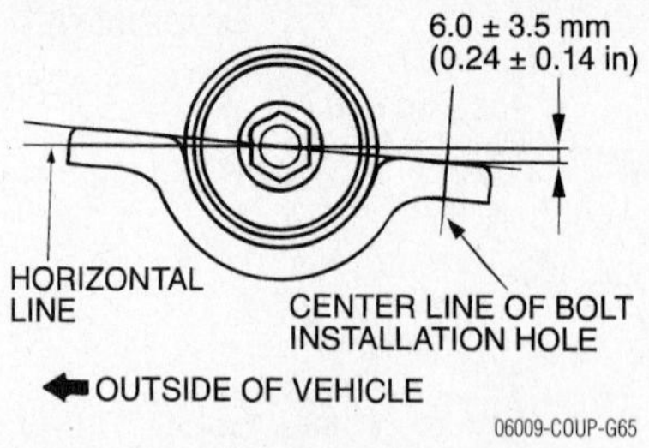

Install the clamp at the angle shown

- Lower control arm
- Lower control arm clamp (position as shown) and torque the clamp bolt to 60 ft. lbs. (81 Nm); the arm-to-clamp nut to 73 ft. lbs.
- Lower control arm to the steering knuckle and torque the bolts to 80 ft. lbs. (108 Nm)
- Front wheel

6. Check and adjust the front toe, if needed.

Rear Suspension Lower Control Arm and Toe Control Arm

REMOVAL & INSTALLATION

1. Before servicing the vehicle, refer to the precautions in the beginning of this section.
2. Remove or disconnect the following:
   - Rear wheel
   - Stabilizer link
   - Wheel Speed Sensor (WSS), if equipped
   - Lower control arm assembly from knuckle
   - Lower control arm
   - Toe control arm from knuckle
   - To control arm from frame

To install:

➡See the accompanying illustrations. Fasteners marked with an asterisk (*) should be temporarily tightened, then,

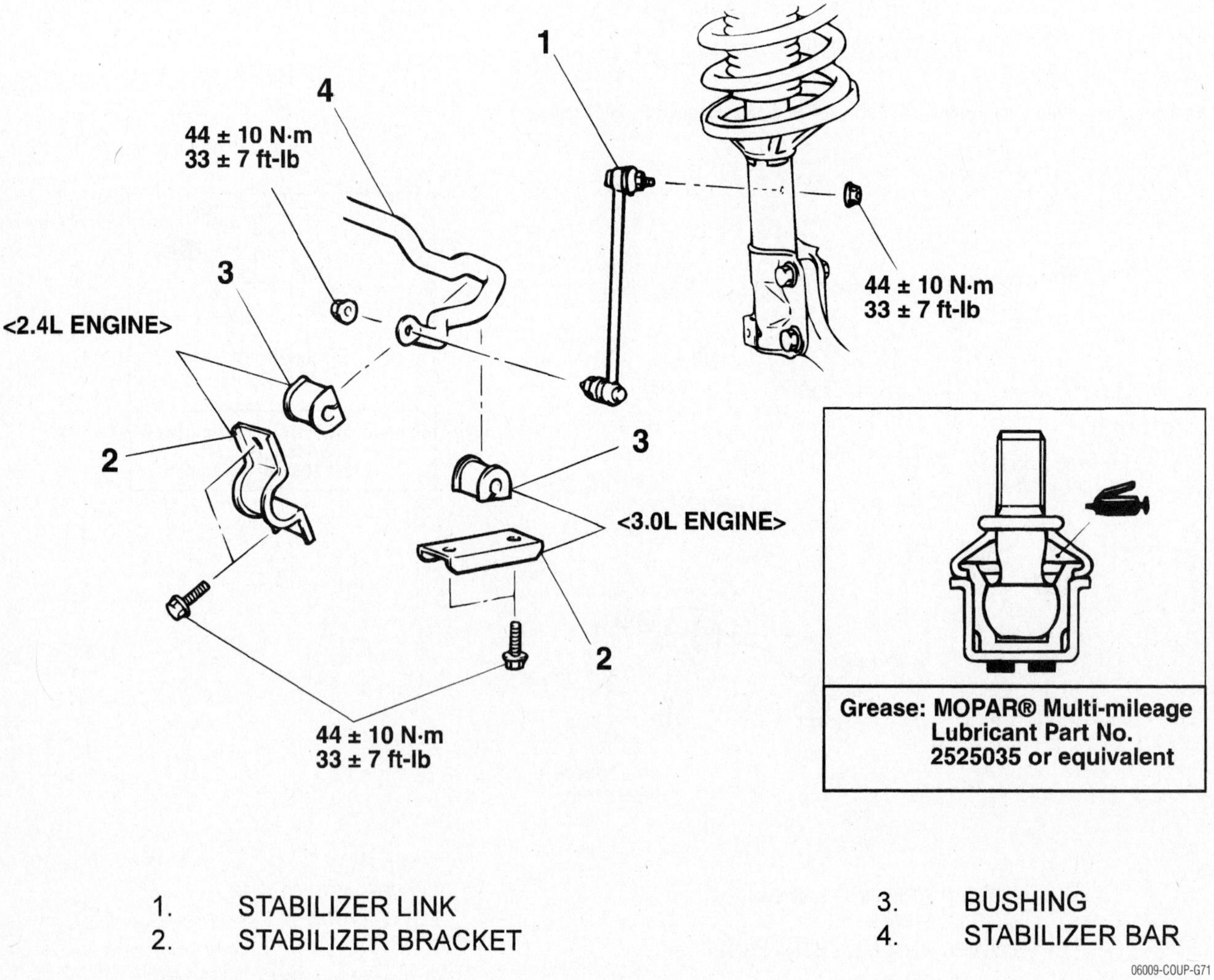

Front stabilizer bar and related parts

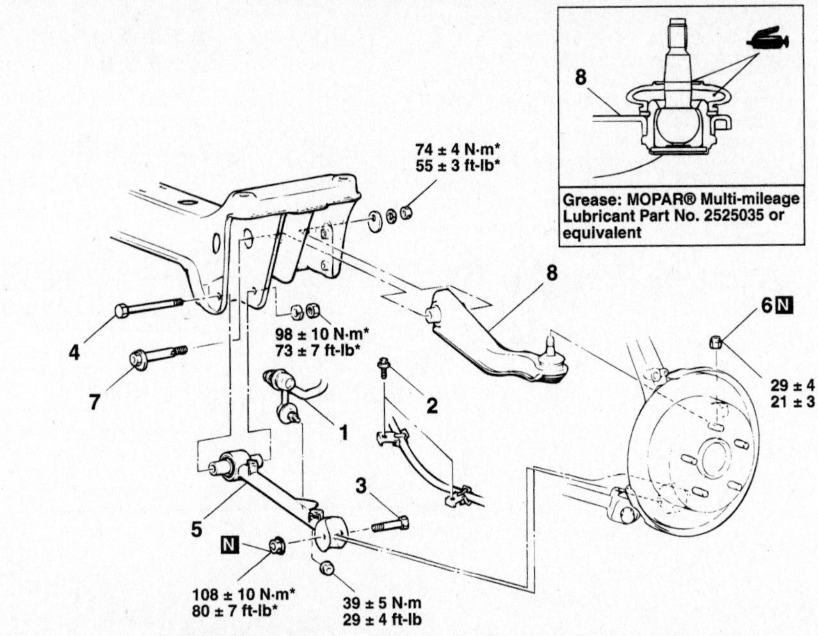

1. STABILIZER LINK CONNECTION
2. WHEEL SPEED SENSOR MOUNTING BOLTS
3. LOWER ARM ASSEMBLY AND KNUCKLE CONNECTING BOLT
4. LOWER ARM ASSEMBLY MOUNTING BOLT
5. LOWER ARM ASSEMBLY
6. TOE CONTROL ARM AND KNUCKLE CONNECTION
7. TOE CONTROL ARM ASSEMBLY MOUNTING BOLT
8. TOE CONTROL ARM ASSEMBLY

06009-COUP-G66

Rear suspension lower control arm and toe control arm

fully tightened with the suspension loaded (on the ground).

3. Install or connect the following:
   - To control arm-to-frame. Torque the bolt/nut to 55 ft. lbs. (74 Nm)
   - To control arm-to-knuckle and torque the nut to 21 ft. lbs. (29 Nm)
   - Lower control arm-to-frame and torque the bolt/nut to 73 ft. lbs. (98 Nm)
   - Lower control arm to the knuckle and torque the bolt to 80 ft. lbs. (108 Nm)
   - WSS, if equipped
   - Stabilizer link and torque the bolt to 29 ft. lbs. (39 Nm)
   - Rear wheel

Rear Suspension Upper Control Arm

REMOVAL & INSTALLATION

➡See the accompanying illustrations. Fasteners marked with an asterisk (*) should be temporarily tightened, then, fully tightened with the suspension loaded (on the ground).

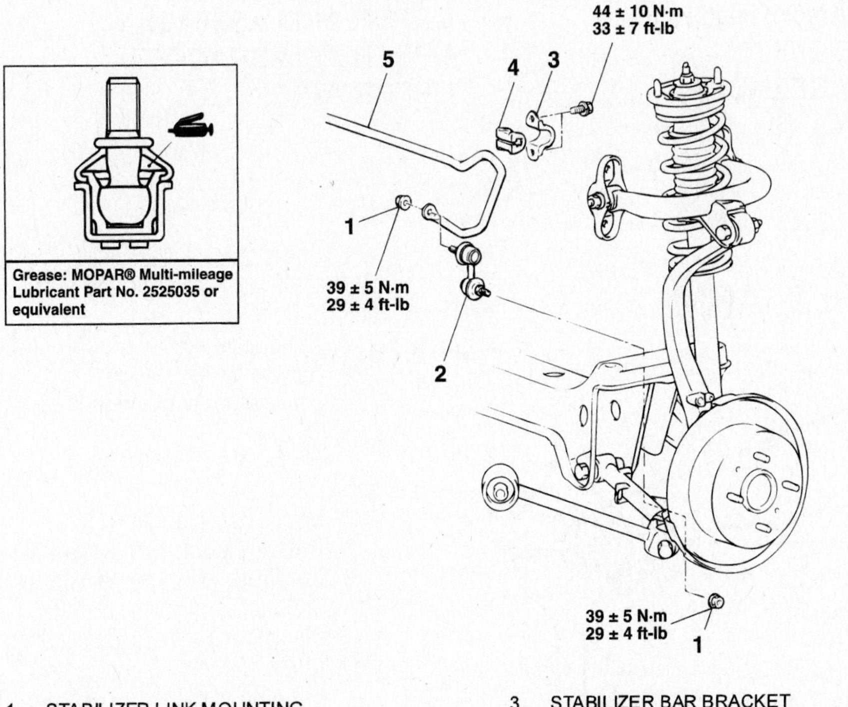

1. STABILIZER LINK MOUNTING NUTS
2. STABILIZER LINK
3. STABILIZER BAR BRACKET
4. BUSHING
5. STABILIZER BAR

06009-COUP-G72

Rear stabilizer bar and related parts

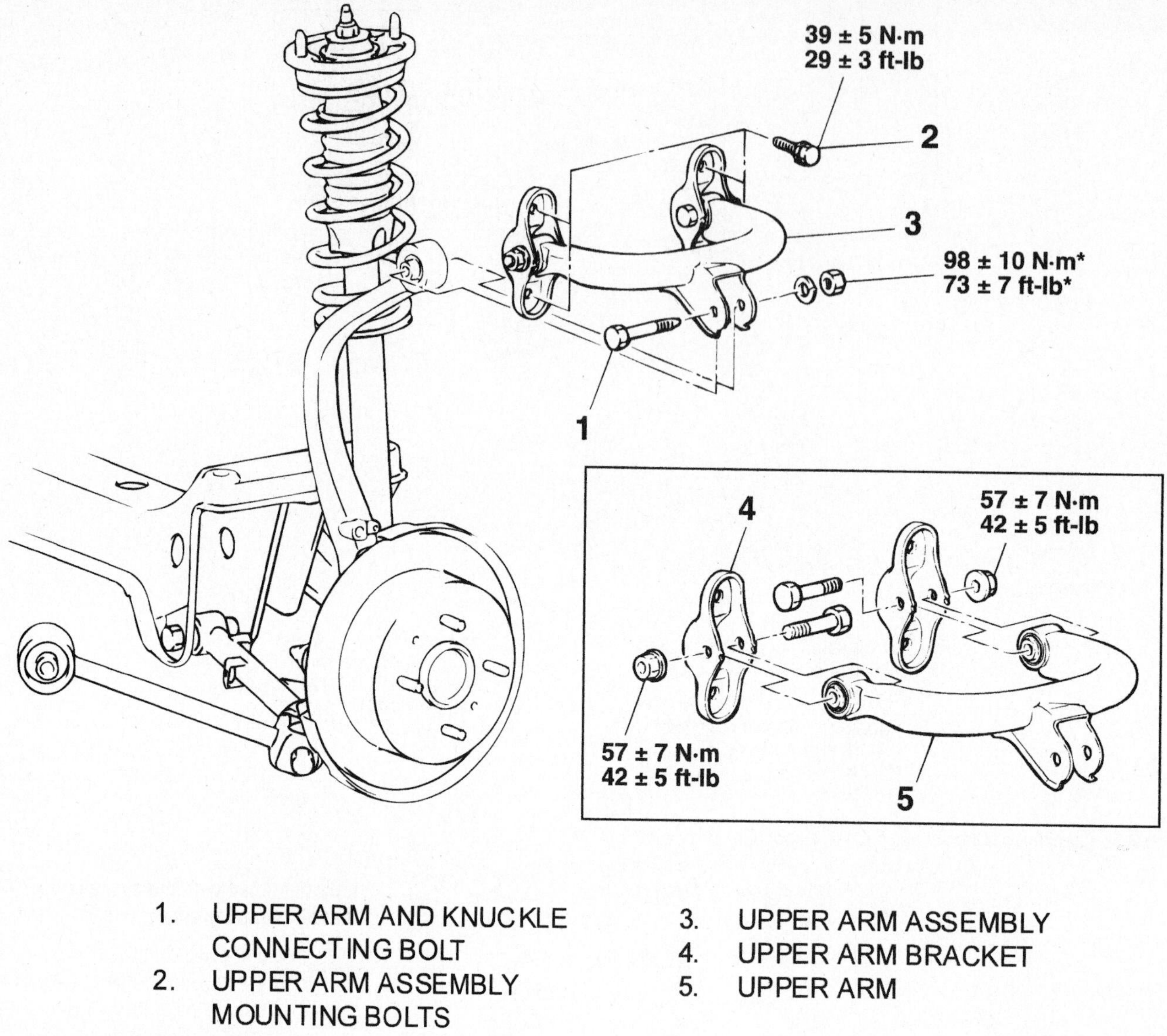

Rear suspension upper control arm

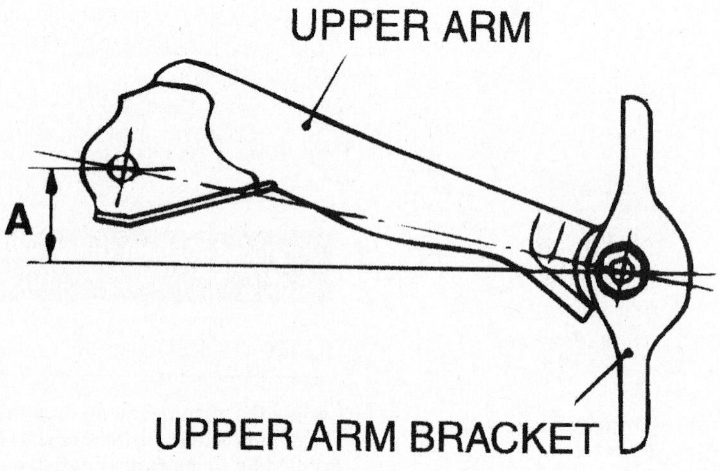

Upper arm bracket installation dimension

1. Before servicing the vehicle, refer to the precautions in the beginning of this section.
2. Remove the wheel.
3. Support the suspension.
4. Remove the upper arm-to-knuckle bolt.
5. Remove the upper arm mounting bolts.
6. Remove the upper arm.
7. Remove the bracket from the arm.
8. Installation is the reverse of removal. See the accompanying illustrations for installation dimensions and angles. Observe the following torques:
   - Arm-to-bracket bolts: 42 ft. lbs. (57 Nm)
   - Mounting bolts: 29 ft. lbs. (39 Nm)
   - Arm-to-knuckle: 73 ft. lbs. (98 Nm)

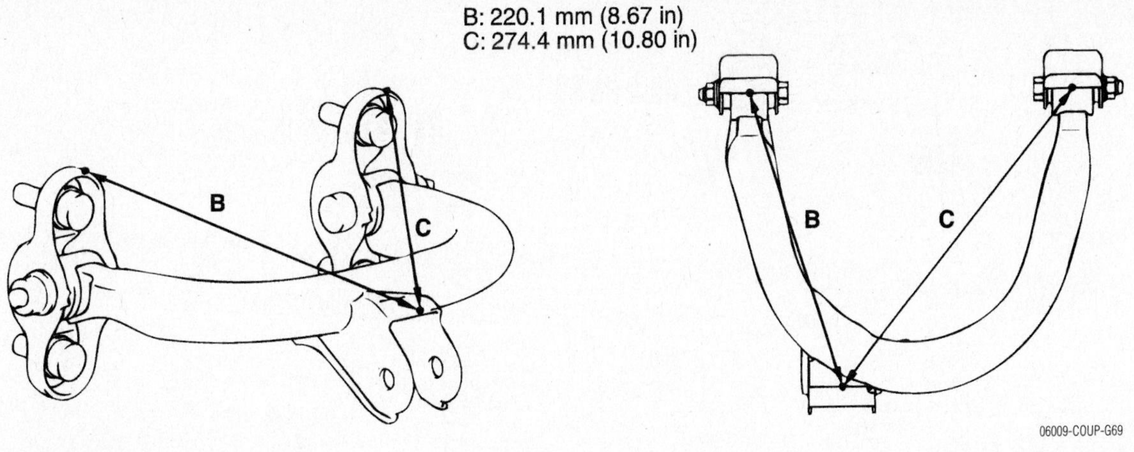

Upper arm installation angle

Rear Suspension Trailing Arm

REMOVAL & INSTALLATION

➡See the accompanying illustrations. Fasteners marked with an asterisk (*) should be temporarily tightened, then, fully tightened with the suspension loaded (on the ground).

1. Before servicing the vehicle, refer to the precautions in the beginning of this section.
2. Remove the wheel.
3. Support the suspension.
4. Remove the trailing arm-to-knuckle bolt.
5. Remove the grommet.
6. Remove the trailing arm mounting bolt.
7. Remove stoppers A and B.

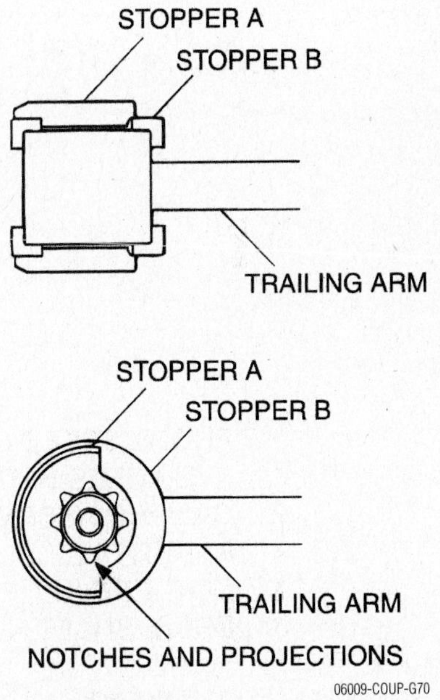

Stopper installation

8. Remove the arm.
9. Installation is the reverse of removal. See the accompanying illustration for stopper installation. Observe the following torques:
   - Arm-to-knuckle bolt: 94 ft. lbs. (128 Nm)
   - Mounting bolt: 109 ft. lbs. (147 Nm)

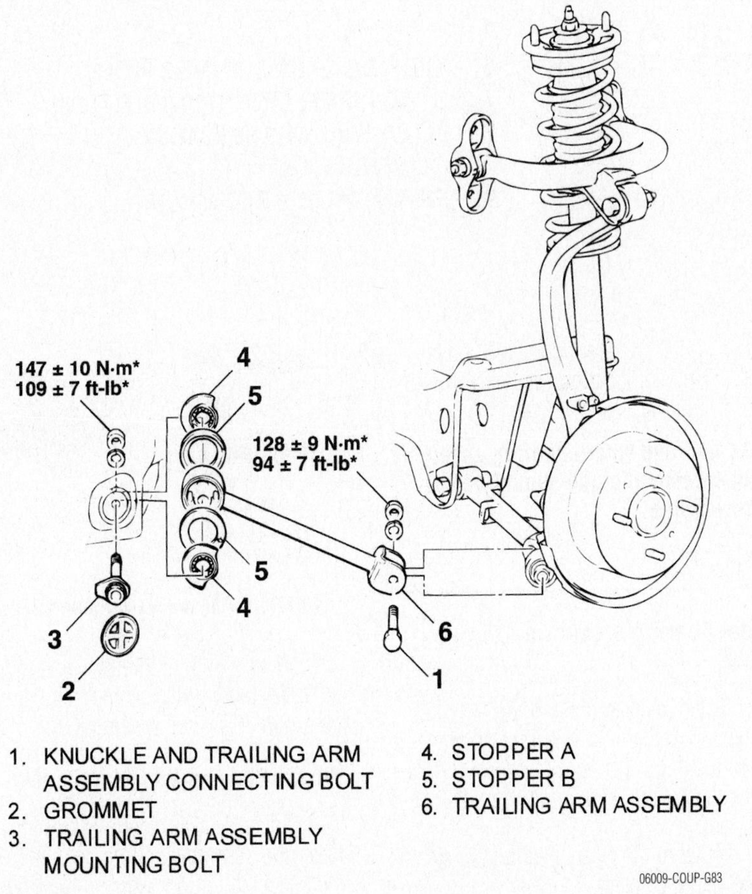

Trailing arm

Front Wheel Hub/Bearing

REMOVAL

1. Before servicing the vehicle, refer to the precautions in the beginning of this section.
2. Remove the cotter pin, halfshaft nut and washer.

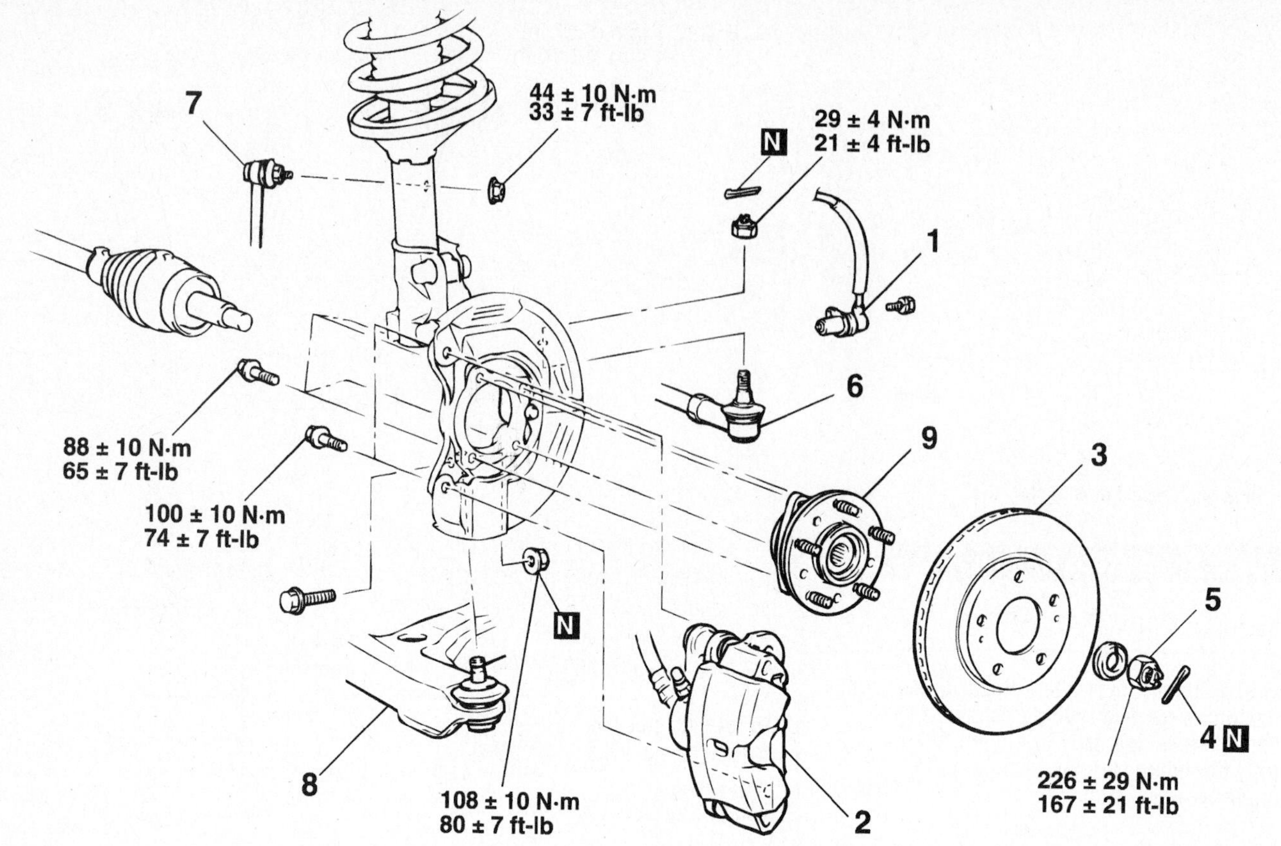

1. FRONT SPEED SENSOR <VEHICLES WITH ABS>
2. CALIPER ASSEMBLY
3. BRAKE DISC
4. COTTER PIN
5. DRIVE SHAFT NUT
6. TIE ROD END CONNECTION
7. STABILIZER LINK CONNECTION
8. LOWER ARM ASSEMBLY CONNECTION
9. FRONT HUB ASSEMBLY

06009-COUP-G73

Front hub/bearing and related parts

3. Remove or disconnect the following:
   - Front wheel
   - Vehicle Speed Sensor (VSS), if equipped
   - Caliper and brake pads; then, support the caliper out of the way using wire
   - Brake rotor from the hub assembly
   - Cotter pin and halfshaft nut
   - Tie rod end from knuckle
   - Stabilizer link
   - Lower arm from the steering

⁂ WARNING

Use of improper methods of joint separation can result in damage to joint, leading to possible failure. Never use wedge-type tools or the ball joint can be damaged.

   - 4 hub-to-steering knuckle bolts
   - Hub and bearing assembly from the knuckle

➡The hub and wheel bearing assembly is not serviceable and should not be disassembled.

INSPECTION

Wheel Bearing Breakaway Torque Check

1. Install special tools MB990998 and MB990326 to the front hub assembly and tighten the nut to the specified torque. Tightening torque: 226 +/- 29 Nm (167 +/- 21 ft. lbs.)
2. Measure the wheel bearing breakaway torque with the special tools. Limit: 1.0 Nm (9 inch lbs.) or less

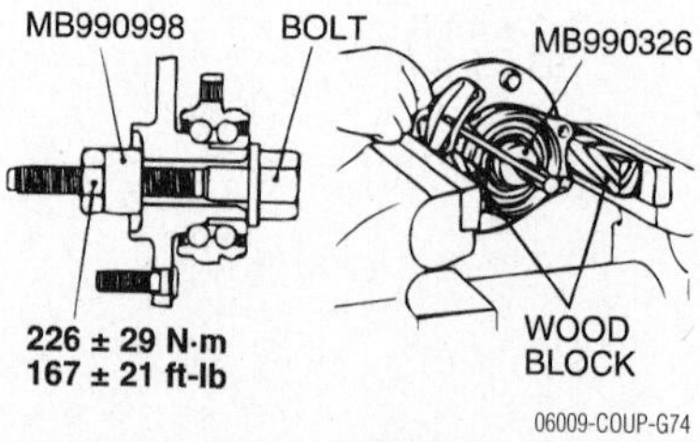

06009-COUP-G74

Wheel bearing breakaway torque check

3. Wheel bearing breakaway torque must be under the limit value and there should be no roughness when rotating the hub.

Wheel Bearing End Play Check

1. Install special tool MB990998 to the front hub assembly and tighten the nut to the specified torque. Tightening torque: 226 +/- 29 Nm (167 +/- 21 ft. lbs.)

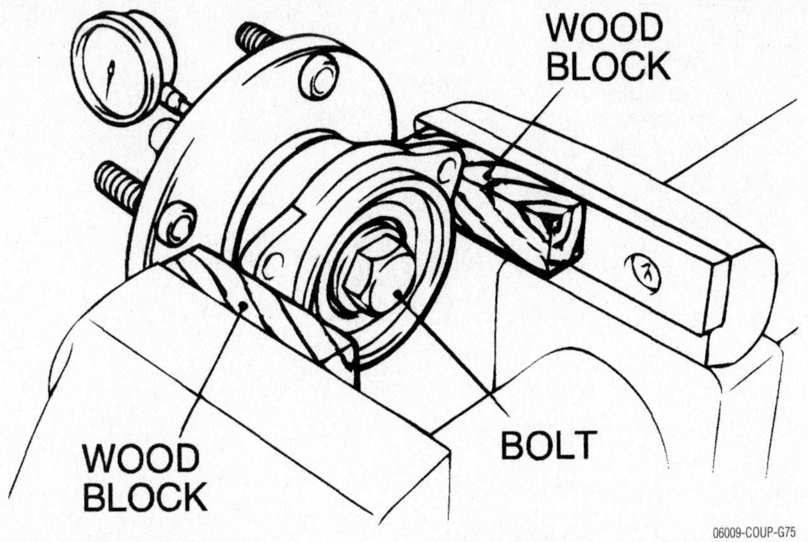

Wheel bearing end play check

2. Measure the play in the hub axial direction. Limit: 0.05 mm (0.002 inch)

3. If the play exceeds the limit, replace the front hub assembly.

INSTALLATION

1. Install or connect the following:
   - Hub to the steering knuckle and torque the bolts to 65 ft. lbs. (88 Nm)
   - Ball joint to the steering knuckle and torque the self-locking nut to 21 ft. lbs. (28 Nm)

2. Position the rotor on the hub. Install a couple of lug nuts and lightly tighten to hold the rotor on the hub.
   - Caliper holder and place the brake pads in the holder. Slide the caliper over the brake pads and install the guide pins. Once the caliper is secured, the lug nuts can be removed.
   - VSS, if equipped
   - Front wheel
   - New cotter pin and bend to secure

3. Examine the driveshaft (halfshaft) hub washer. Locate the chamfered side. This side is installed outward, away from the hub. Install the washer and hub nut. Tighten the axle nut with the brakes applied. Torque the nut to 167 ft. lbs. (229 Nm). Install a new cotter pin.

⁂ WARNING

Pump the brake pedal until hard, before attempting to move the vehicle.

Rear Wheel Hub/Bearing

REMOVAL & INSTALLATION

The rear hub and wheel bearing assembly is designed for the life of the vehicle and requires no type of adjustment or periodic maintenance. The bearing is a sealed unit with the wheel hub and can only be removed and/or replaced as one unit.

The following procedure may be used for evaluation of bearing condition:

1. Raise and safely support the vehicle.
2. Remove the rear wheels and brake drums.
3. Turn the hub flange carefully. Excessive roughness, lateral play or resistance to rotation may indicate dirt intrusion or bearing failure.
4. If the rear wheel bearings exhibit these conditions during inspection, the hub and bearing assembly should be replaced.
5. Damaged bearing seals and resulting excessive grease loss may also require bearing replacement. Moderate grease loss from the bearing is considered normal and should not require replacement of the hub and bearing assembly.

With Drum Brakes

1. Before servicing the vehicle, refer to the precautions in the beginning of this section.
2. Remove or disconnect the following:
   - Rear wheel
   - Vehicle Speed Sensor (VSS), if equipped with Anti-lock Brake System (ABS)
   - Brake drum
   - 4 hub to the knuckle bolts
   - Hub and bearing assembly from the knuckle

➡The hub assembly is not serviceable and should not be disassembled.

3. If replacing the hub, use special Socket MB991248 and a press, to remove the wheel sensor rotor from the hub.

To install:

4. Press the wheel sensor rotor onto the hub.
5. Install or connect the following:
   - Hub to the knuckle and torque the bolts to 53 ft. lbs. (71 Nm)
   - Brake drum
   - VSS, if equipped with ABS
   - Rear wheel and lower the vehicle

With Disc Brakes

1. Before servicing the vehicle, refer to the precautions in the beginning of this section.
2. Remove or disconnect the following:
   - Rear wheel
   - Vehicle Speed Sensor (VSS), if equipped with Anti-lock Brake System (ABS)
   - Caliper and brake pads; then, support the caliper out of the way using wire
   - Brake rotor
3. Remove the parking brake shoes as follows:
   a. Upper shoe-to-anchor springs.
   b. Lower shoe-to-shoe spring.
   c. Brake shoe hold-down springs.
   d. Parking brake cable from the actuating lever.
4. Remove the 4 hub-to-knuckle bolts.
5. Remove the hub and bearing assembly from the knuckle.

➡The hub assembly is not serviceable and should not be disassembled.

6. If replacing the hub, use special Socket MB991248 and a press, to remove the wheel sensor rotor from the hub.

To install:

7. Press the wheel sensor rotor onto the hub.
8. Install the hub to the knuckle and torque the bolts to 53 ft. lbs. (71 Nm)
9. Install the parking brake shoes.
10. Position the rotor on the hub. Install a couple of lug nuts and lightly tighten to hold rotor on hub.
11. Install the caliper holder and place brake pads in holder. Slide the caliper over brake pads and install guide pins. Once caliper is secured, lug nuts can be removed.
12. Install the VSS, if equipped
13. Install the rear wheel.

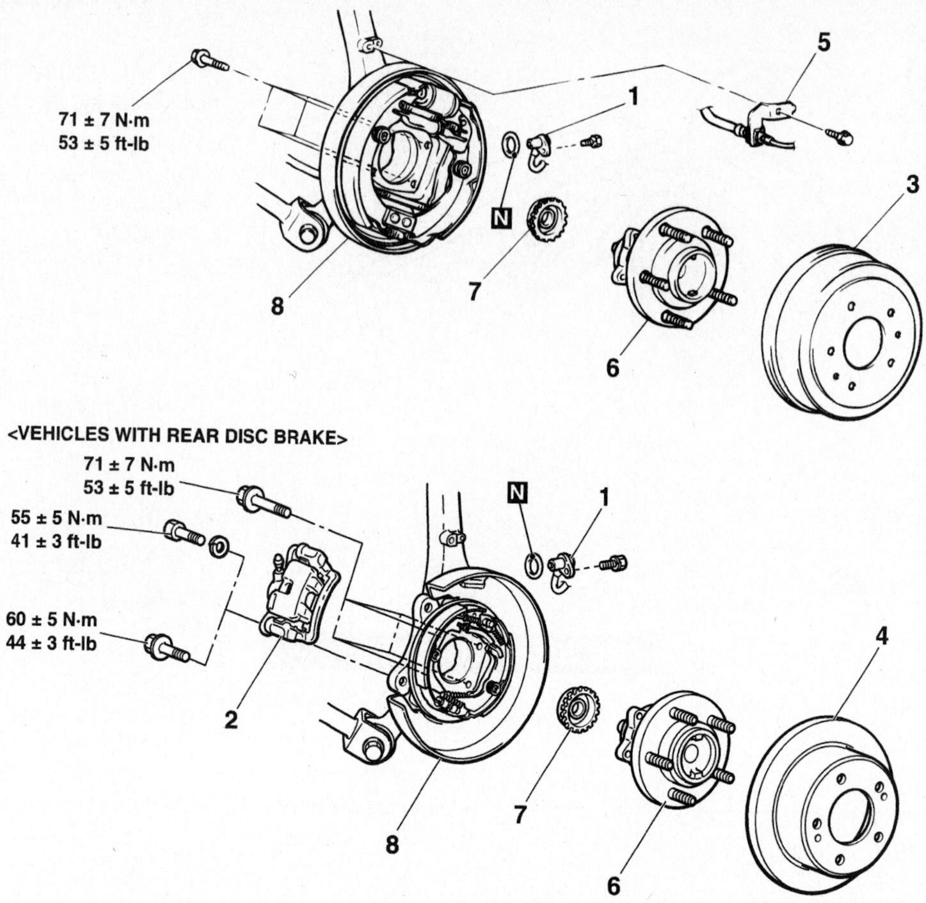

1. REAR WHEEL-SPEED SENSOR <VEHICLES WITH ABS>
2. CALIPER ASSEMBLY
3. BRAKE DRUM
4. BRAKE DISC
5. BRAKE HOSE INSTALLATION BRACKET
6. REAR HUB ASSEMBLY
7. ABS ROTOR <VEHICLES WITH ABS>
8. BACKING PLATE

06009-COUP-G76

Exploded view of the rear hub assembly and related parts

FRONT BRAKES

Brake Caliper

REMOVAL & INSTALLATION

1. Before servicing the vehicle, refer to the precautions in the beginning of this section.
2. Remove or disconnect the following:
   - About half of the brake fluid from the master cylinder
   - Wheel assembly
3. Position a C-clamp, or other suitable tool, over the caliper. Smoothly apply pressure, forcing the caliper piston into the caliper bore until it bottoms. Remove the C-clamp, if used.
   - Brake hose attaching bolt and hose from the caliper, if the caliper is to be completely removed from the vehicle. Plug the hose to prevent fluid contamination or loss.
   - Caliper mounting bolts and lift the caliper off of the support bracket
   - Caliper from the vehicle. If the caliper is only removed for access to other components, support the caliper, with the brake hose attached, so that there is no strain on the brake hose.

To install:

4. Install or connect the following:
   - Caliper to the steering knuckle. Lubricate and install the mounting bolts. See the accompanying illustrations for bolt torques.
   - Brake line hose to the caliper, if removed, and torque the inlet fitting bolt to 22 ft. lbs. (29 Nm)
   - Master cylinder with fresh brake fluid and, if the brake hose was removed, bleed the brake system.
   - Wheel assembly
5. Depress the brake pedal 3–4 times to seat the brake linings and to restore pressure in the system.

⁂ CAUTION

Do not move the vehicle until a firm pedal is obtained.

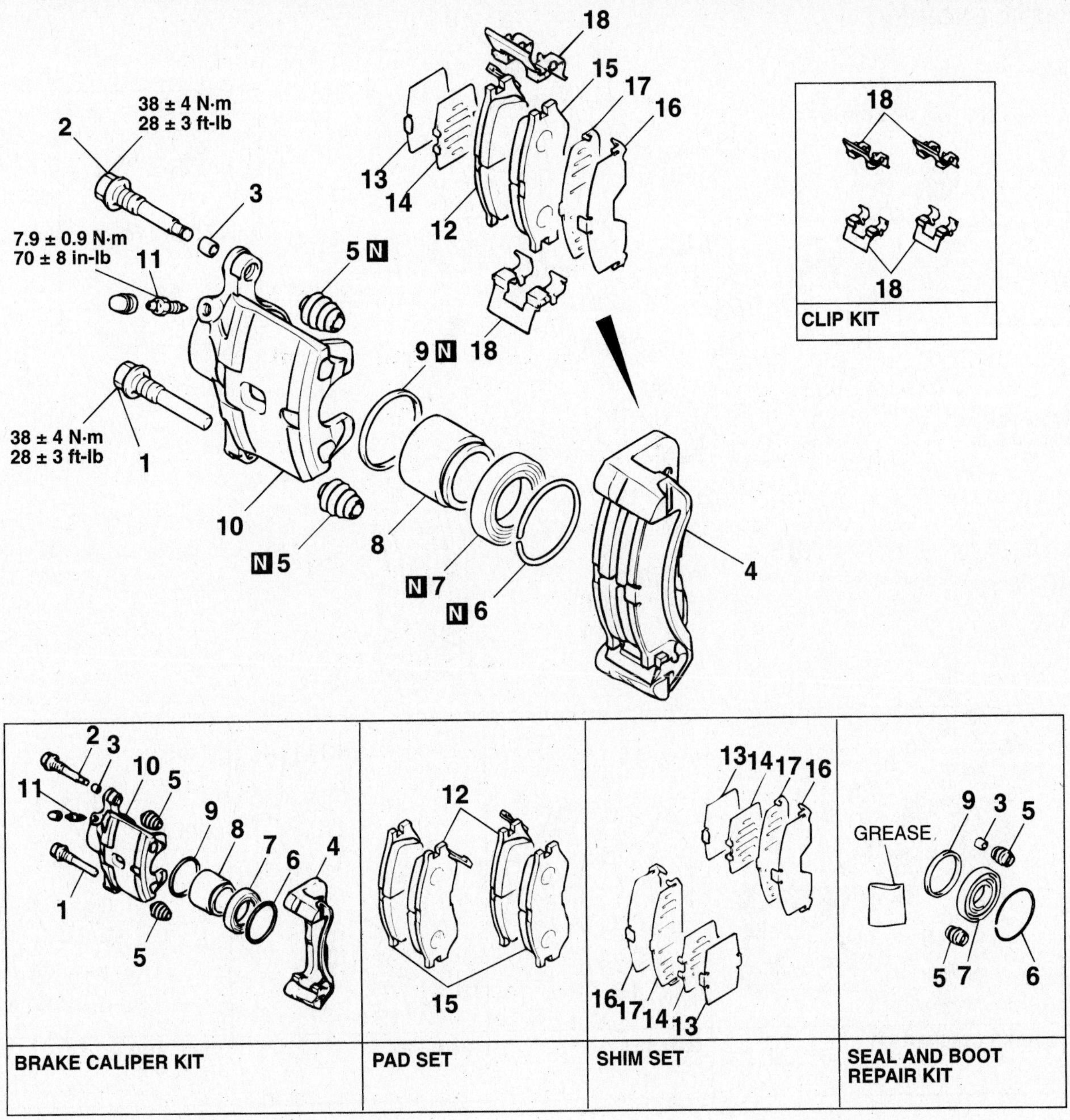

CALIPER ASSEMBLY DISASSEMBLY STEPS

1. GUIDE PIN
2. LOCK PIN
3. BUSHING
4. CALIPER SUPPORT, PAD, CLIP AND SHIM ASSEMBLY
5. PIN BOOT
6. BOOT RING
7. PISTON BOOT
8. PISTON
9. PISTON SEAL
10. CALIPER BODY
11. BLEEDER SCREW

PAD ASSEMBLY DISASSEMBLY STEPS

1. GUIDE PIN
2. LOCK PIN
3. BUSHING
4. CALIPER SUPPORT, PAD, CLIP AND SHIM ASSEMBLY
12. PAD AND WEAR INDICATOR ASSEMBLY
13. INNER SHIM B
14. INNER SHIM A
15. PAD ASSEMBLY
16. OUTER SHIM B
17. OUTER SHIM A
18. CLIP

06009-COUP-G77

Front brake caliper exploded view—2002–03 models

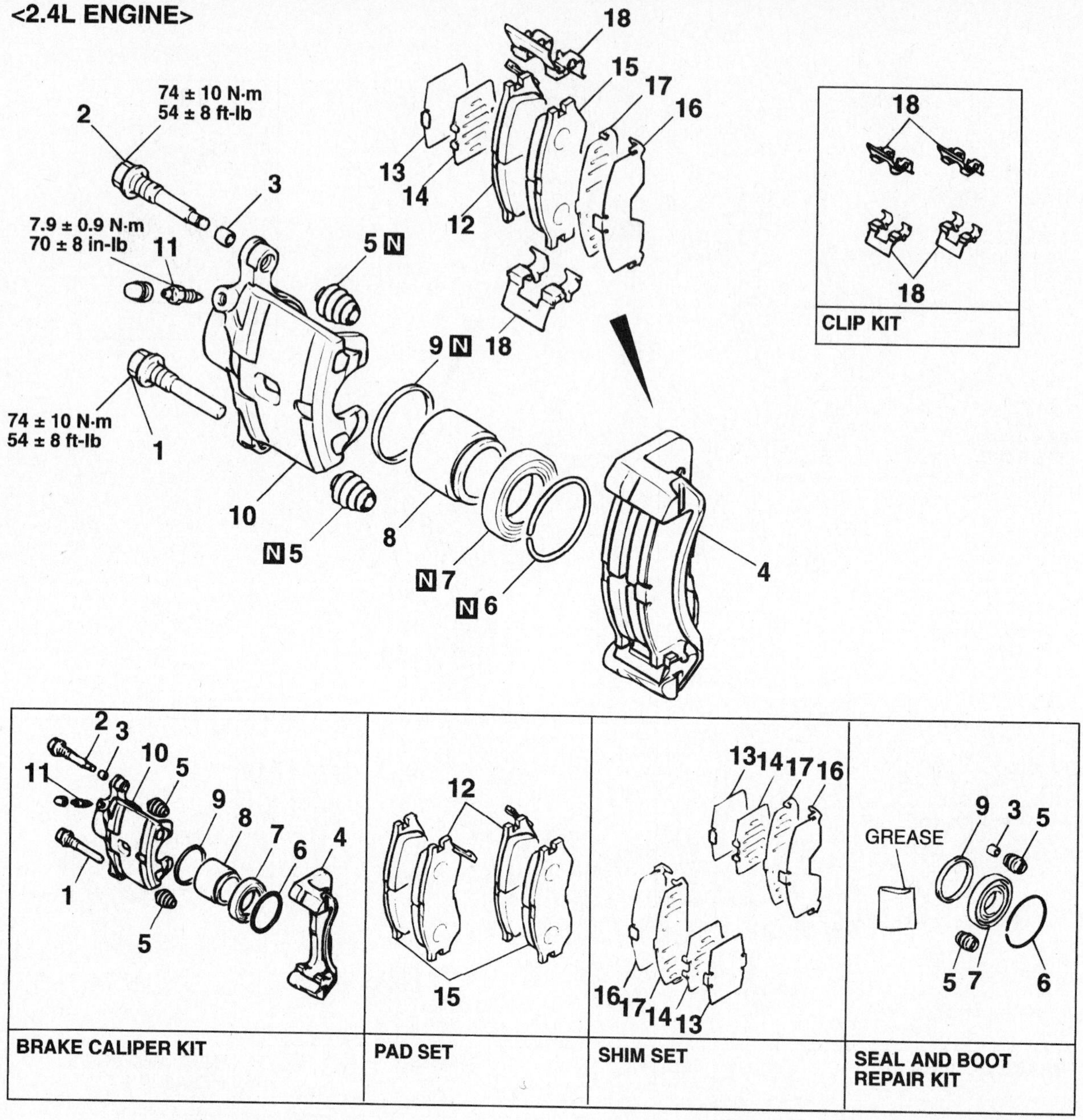

CALIPER ASSEMBLY DISASSEMBLY STEPS

1. GUIDE PIN
2. LOCK PIN
3. BUSHING
4. CALIPER SUPPORT, PAD, CLIP AND SHIM ASSEMBLY
5. PIN BOOT
6. BOOT RING
7. PISTON BOOT
8. PISTON
9. PISTON SEAL
10. CALIPER BODY
11. BLEEDER SCREW

PAD ASSEMBLY DISASSEMBLY STEPS

1. GUIDE PIN
2. LOCK PIN
3. BUSHING
4. CALIPER SUPPORT, PAD, CLIP AND SHIM ASSEMBLY
12. PAD AND WEAR INDICATOR ASSEMBLY
13. INNER SHIM B
14. INNER SHIM A
15. PAD ASSEMBLY
16. OUTER SHIM B
17. OUTER SHIM A
18. CLIP

06009-COUP-G78

Front brake caliper exploded view—2004–05 models with the 2.4L engine

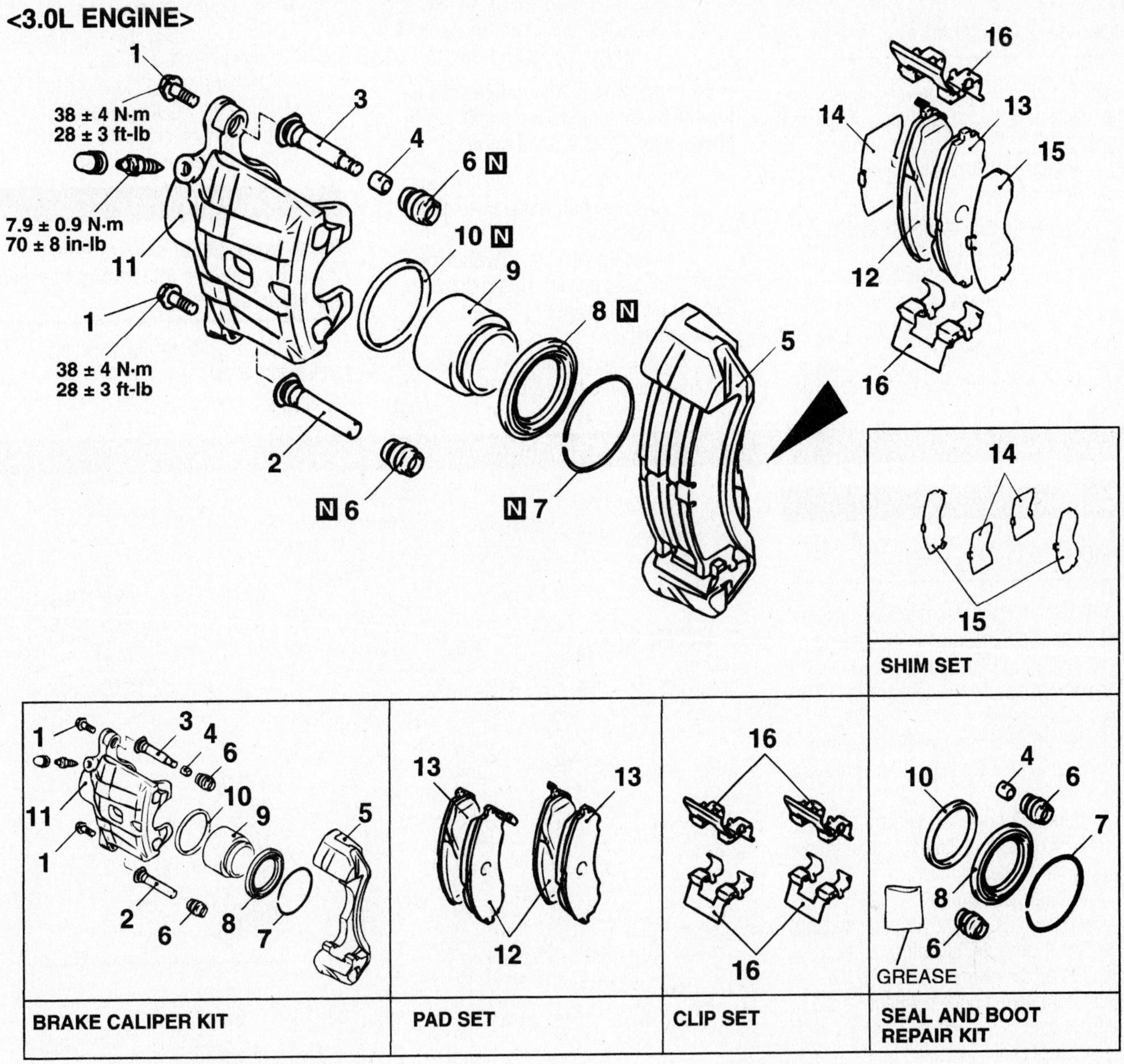

CALIPER ASSEMBLY DISASSEMBLY STEPS

1. PIN BOLT
2. GUIDE PIN
3. LOCK PIN
4. BUSHING
5. CALIPER SUPPORT, PAD, CLIP AND SHIM ASSEMBLY
6. PIN BOOT
7. BOOT RING
8. PISTON BOOT
9. PISTON
10. PISTON SEAL
11. CALIPER BODY

PAD ASSEMBLY DISASSEMBLY STEPS

1. PIN BOLT
2. GUIDE PIN
3. LOCK PIN
4. BUSHING
5. CALIPER SUPPORT, PAD, CLIP AND SHIM ASSEMBLY
12. PAD AND WEAR INDICATOR ASSEMBLY
13. PAD ASSEMBLY
14. INNER SHIM
15. OUTER SHIM
16. PAD CLIP

06009-COUP-G79

Front brake caliper exploded view—2004–05 models with the 3.0L engine

Disc Brake Pads

REMOVAL & INSTALLATION

1. Before servicing the vehicle, refer to the precautions in the beginning of this section.
2. Remove or disconnect the following:
   - Some of the brake fluid from the master cylinder reservoir. The reservoir should be no more than ½ full. When the pistons are depressed into the calipers, excess fluid will flow up into the reservoir.
   - Tire and wheel
   - Caliper guide and lock pins and lift the caliper assembly from the caliper support. Tie the caliper out of the way using wire. Do not allow the caliper to hang by the brake line.

➡On some models the caliper can be flipped up by leaving the upper pin in place, using it as a pivot point.

   - Brake pads, spring clip and shims. Take note of positioning to aid installation.
3. Install the wheel lug nuts onto the studs and lightly tighten. This is done to hold the disc on the hub.

To install:

4. Use a large C-clamp to compress the piston(s) back into the caliper bore.
5. Install or connect the following:
   - Brake pads, shims and spring clip onto the caliper support, after lubricating the slide points
   - Caliper over the brake pads
   - Lubricate caliper guide and lock pins in their original positions.
   - Tire and wheel

⁂ CAUTION

Pump brake pedal several times, until firm, before attempting to move the vehicle.

6. Road test the vehicle and check brakes for proper operation.

REAR BRAKES

Caliper

REMOVAL & INSTALLATION

Unlike many rear disc brake designs, this system does not incorporate the parking brake system, into the rear brake caliper. The rear brake system is serviced the same as the front system.

1. Before servicing the vehicle, refer to the precautions in the beginning of this section.
2. Remove or disconnect the following:
   - About half of the brake fluid from the master cylinder
   - Wheel assembly
3. Position a C-clamp, or other suitable tool, over the caliper. Smoothly apply pressure, forcing the caliper piston into the caliper bore until it bottoms. Remove the C-clamp, if used.
   - Brake hose attaching bolt and hose from the caliper, if the caliper is to be completely removed from the vehicle. Plug the hose to prevent fluid contamination or loss.
   - Caliper mounting bolts and lift the caliper off of the support bracket
   - Caliper from the vehicle. If the caliper is only removed for access to other components, support the caliper, with the brake hose attached, so that there is no strain on the brake hose.

To install:

4. Install or connect the following:
   - Caliper on the support bracket, lubricate and install the mounting bolts. See the accompanying illustrations for the torque values.
   - Brake line hose to the caliper, if

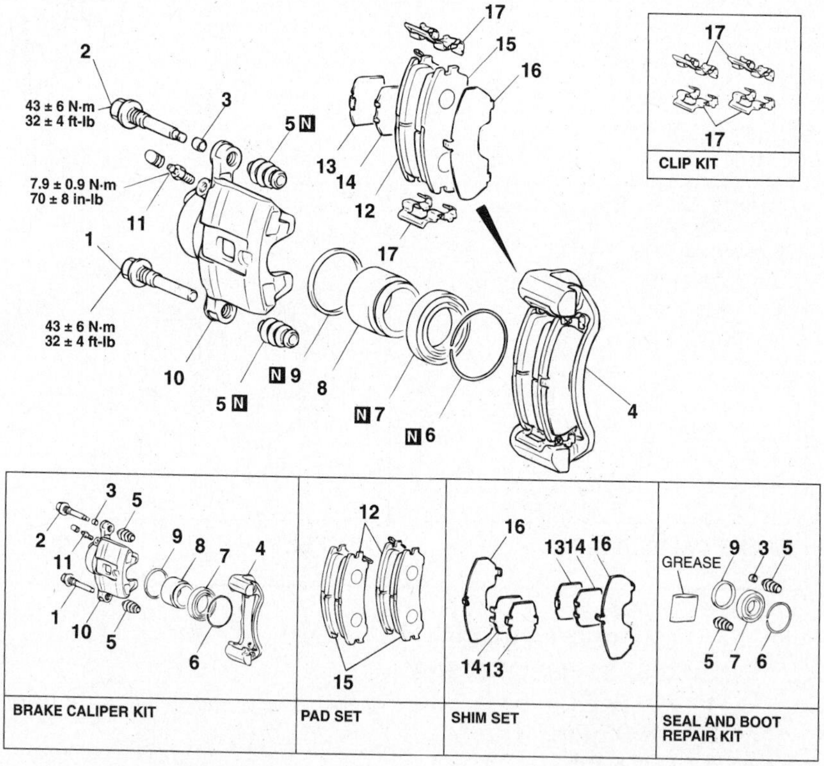

CALIPER ASSEMBLY DISASSEMBLY STEPS
1. GUIDE PIN
2. LOCK PIN
3. BUSHING
4. CALIPER SUPPORT, PAD, CLIP AND SHIM ASSEMBLY
5. BOOT
6. BOOT RING
7. PISTON BOOT
8. PISTON
9. PISTON SEAL
10. CALIPER BODY
11. BLEEDER SCREW

PAD ASSEMBLY DISASSEMBLY STEPS
1. GUIDE PIN
2. LOCK PIN
3. BUSHING
4. CALIPER SUPPORT, PAD, CLIP AND SHIM ASSEMBLY
12. PAD AND WEAR INDICATOR ASSEMBLY
13. INNER SHIM B
14. INNER SHIM A
15. PAD ASSEMBLY
16. OUTER SHIM
17. CLIP

06009-COUP-G80

Rear caliper exploded view

removed, and torque the inlet fitting bolt to 11 ft. lbs. (15 Nm)
- Master cylinder with fresh brake fluid and, if the brake hose was removed, bleed the brake system
- Wheel assembly

5. Depress the brake pedal 3–4 times to seat the brake linings and to restore pressure in the system.

CAUTION

Do not move the vehicle until a firm pedal is obtained.

Disc Brake Pads

REMOVAL & INSTALLATION

Unlike many rear disc brake designs, this system does not incorporate the parking brake system, into the rear brake caliper, therefore, the rear brake system is serviced the same as the front system.

1. Before servicing the vehicle, refer to the precautions in the beginning of this section.
2. Remove or disconnect the following:
   - Some of the brake fluid from the master cylinder reservoir. The reservoir should be no more than ½ full. When the pistons are depressed into the calipers, excess fluid will flow up into the reservoir.
   - Tire and wheel
   - Caliper guide and lock pins and lift the caliper assembly from the caliper support. Tie the caliper out of the way using wire. Do not allow the caliper to hang by the brake hose.

➡On some models, the caliper can be flipped up by leaving the upper pin in place, using it as a pivot point.

   - Brake pads, spring clip and shims. Take note of positioning to aid installation.

3. Install the wheel lug nuts onto the studs and lightly tighten. This is done to hold the brake disc on the hub.

To install:

4. Use a large C-clamp to compress the piston(s) back into the caliper bore.
5. Install or connect the following:
   - Brake pads, shims and spring clip, after lubricating the slide points, onto the caliper support
   - Caliper over the brake pads
   - Lubricated caliper guide and lock pins in their original positions. See the accompanying illustration for the torque values.
   - Tire and wheel

CAUTION

Pump brake pedal several times, until firm, before attempting to move the vehicle.

6. Road test the vehicle and check brakes for proper operation.

Brake Drums

REMOVAL & INSTALLATION

1. Before servicing the vehicle, refer to the precautions in the beginning of this section.
2. Remove or disconnect the following:
   - Wheel assembly
   - Brake drum detent (retaining) screw
   - Drum from the axle
3. If difficulty is encountered in removing the drum, perform the following:
   a. Verify the parking brake is released.
   b. Loosen the parking brake cable.
   c. Remove the access hole plug from the backing plate and move the parking brake lever until the lever stop rests on the brake shoe.
   d. If necessary, turn the adjuster so that it draws in to allow more clearance between the shoe and drum.

To install:

4. Inspect all parts. Check the drum for cracks or excessive wear.
5. Turn the adjuster until it is drawn all the way in to the stop. Check that the adjuster turns freely. The nut must NOT lock at the end of the adjuster. Check that the parking brake lever stops are against the edge of the shoe web.
6. Install or connect the following:
   - Brake drum and detent screw
   - Wheel assembly
7. Apply the foot brake at least 10 times until clicking of the adjustment actuator can no longer be heard. This procedure will automatically adjust the clearance between the shoe and drum.

Brake Shoes

REMOVAL & INSTALLATION

1. Before servicing the vehicle, refer to the precautions in the beginning of this section.

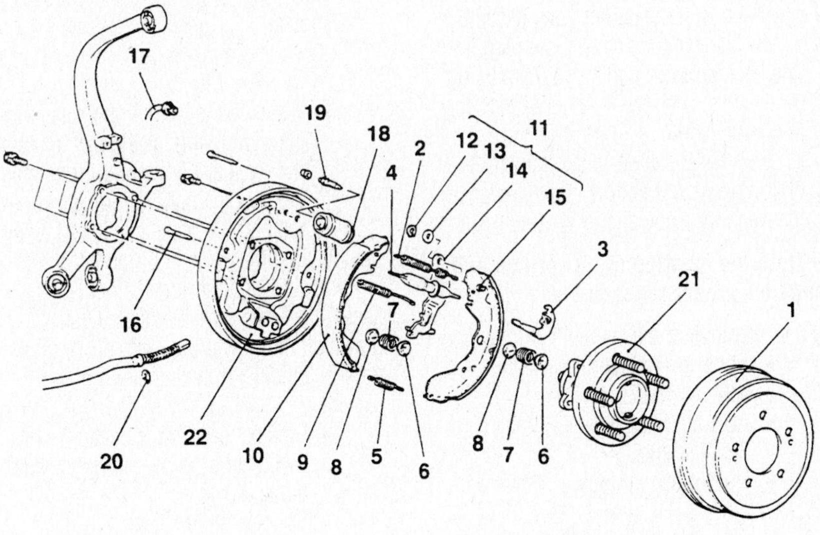

1. Brake drum
2. Shoe-to-lever spring
3. Adjuster lever
4. Auto adjuster assembly
5. Retainer spring
6. Shoe hold-down cup
7. Shoe hold-down spring
8. Shoe hold-down cup
9. Shoe-to-shoe spring
10. Shoe and lining assembly
11. Shoe and lever assembly
12. Retainer
13. Wave washer
14. Parking lever
15. Shoe and lining assembly
16. Shoe hold-down pin
17. Brake pipe connection
18. Wheel cylinder
19. Bleeder screw
20. Snap ring
21. Rear hub assembly
22. Backing plate

93006G10

Exploded view of the drum brake assembly

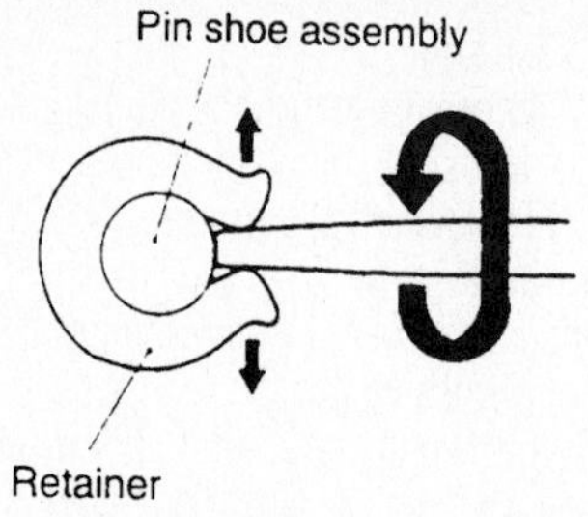

93006G11

Opening the retainer clip

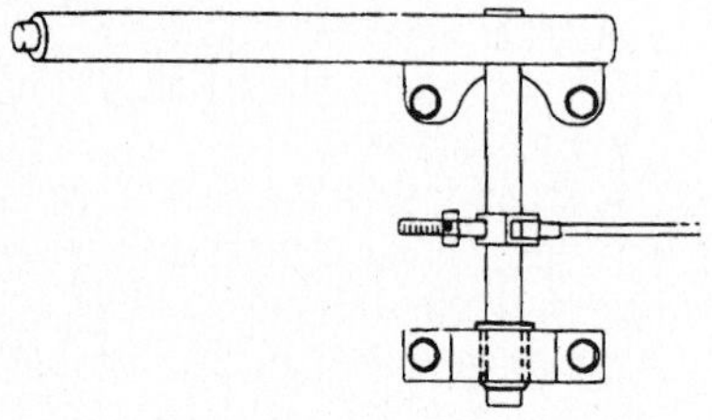

93006G12

Adjusting nut and nut holder

2. Remove or disconnect the following:
   - Wheel assembly
   - Brake drum detent (retaining) screw
   - Drum from the axle
3. If difficulty is encountered in removing the drum:
   a. Verify the parking brake is released.
   b. Loosen the parking brake cable.
   c. Remove the access hole plug from the backing plate and move the parking brake lever until the lever stop rests on the brake shoe.
   d. If necessary, turn the adjuster so that it draws in to allow more clearance between the shoe and drum.

➡Note the location of all springs and clips for proper reassembly.

4. Remove or disconnect the following:
   - Shoe-to-lever spring and the adjuster lever
   - Auto adjuster assembly
   - Retainer spring
   - Hold-down springs, washers and pins
   - Shoe-to-shoe spring
   - Brake shoes from the backing plate
5. Using a flat-tipped tool, open up the parking brake lever retaining clip.
   - Clip and washer from the pin on the shoe assembly and the shoe from the lever assembly.

To install:

6. Thoroughly clean and dry the backing plate. Lubricate the backing plate at the brake shoe contact points.

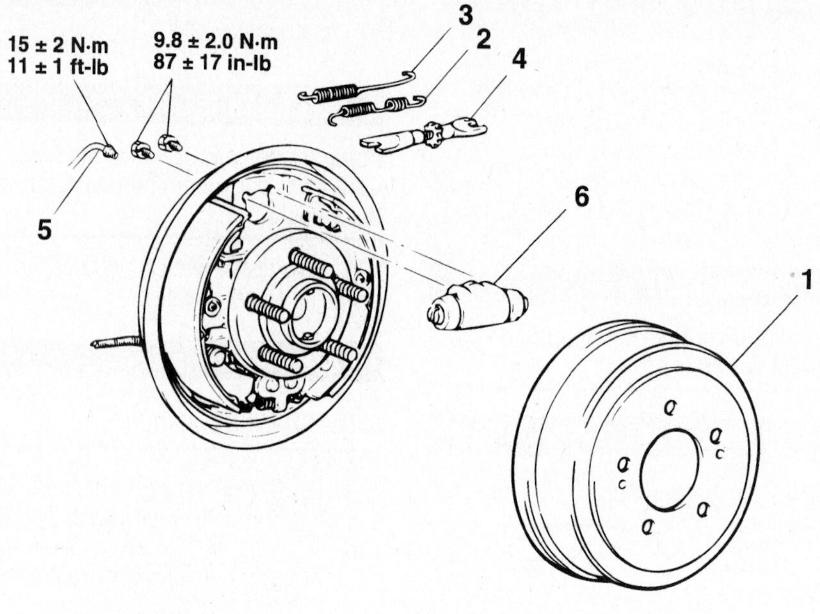

1. BRAKE DRUM
2. SHOE-TO-LEVER SPRING
3. SHOE-TO-SHOE SPRING
4. AUTO ADJUSTER ASSEMBLY
5. CONNECTION FOR THE BRAKE PIPE
6. WHEEL CYLINDER

06009-COUP-G82

Wheel cylinder installation

7. Lubricate backing plate bosses, anchor pin, and parking brake actuating mechanism with a lithium-based grease.
8. Install or connect the following:
   - Parking brake lever assembly on the lever pin
   - Wave washer and a new retaining clip. Use pliers or the like to install the retainer on the pin. If removed, connect the parking brake lever to the parking brake cable and verify that the cable is properly routed.
9. Clean and lubricate the adjuster assembly. Make sure the nut-adjuster is drawn all the way to the stop, but the nut must NOT lock firmly at the end of the assembly.
   - Brake shoes on the backing plate with the hold-down springs, washers and pins
   - Shoe-to-shoe spring
   - Retainer spring
   - Auto adjuster assembly, the adjuster lever and the shoe-to-lever spring
10. Pre-adjust the shoes so the drum slides on with a light drag and install the brake drum.
11. Adjust the rear brake shoes and install the rear wheels.
12. Adjust the parking brake cable.
13. Check for proper brake operation.

Wheel Cylinder

REMOVAL & INSTALLATION

1. Before servicing the vehicle, refer to the precautions in the beginning of this section.
2. Remove the drum.
3. Remove the shoe-to-lever spring.
4. Remove the shoe-to-shoe spring.
5. Remove the auto-adjuster assembly.
6. Disconnect the brake pipe at the cylinder
7. Remove the wheel cylinder bolts and wheel cylinder.
8. Installation is the reverse of removal. Torque the bolts to 87 inch lbs. (10 Nm).
9. Bleed the brakes.

Parking Brake Unit

REMOVAL & INSTALLATION

With Disc Brakes

1. Before servicing the vehicle, refer to the precautions in the beginning of this section.
2. Remove the caliper.
3. Remove the rotor.

4. Remove the shoe-to-anchor spring.
5. Remove the adjuster screw spring.
6. Remove the adjuster.
7. Remove the strut.
8. Remove the strut return spring.
9. Remove the show hold-down cup.
10. Remove the hold-down spring.
11. Remove the hold-down pin.
12. Remove the shoes.
13. Remove the clip.
14. Installation is the reverse of removal.

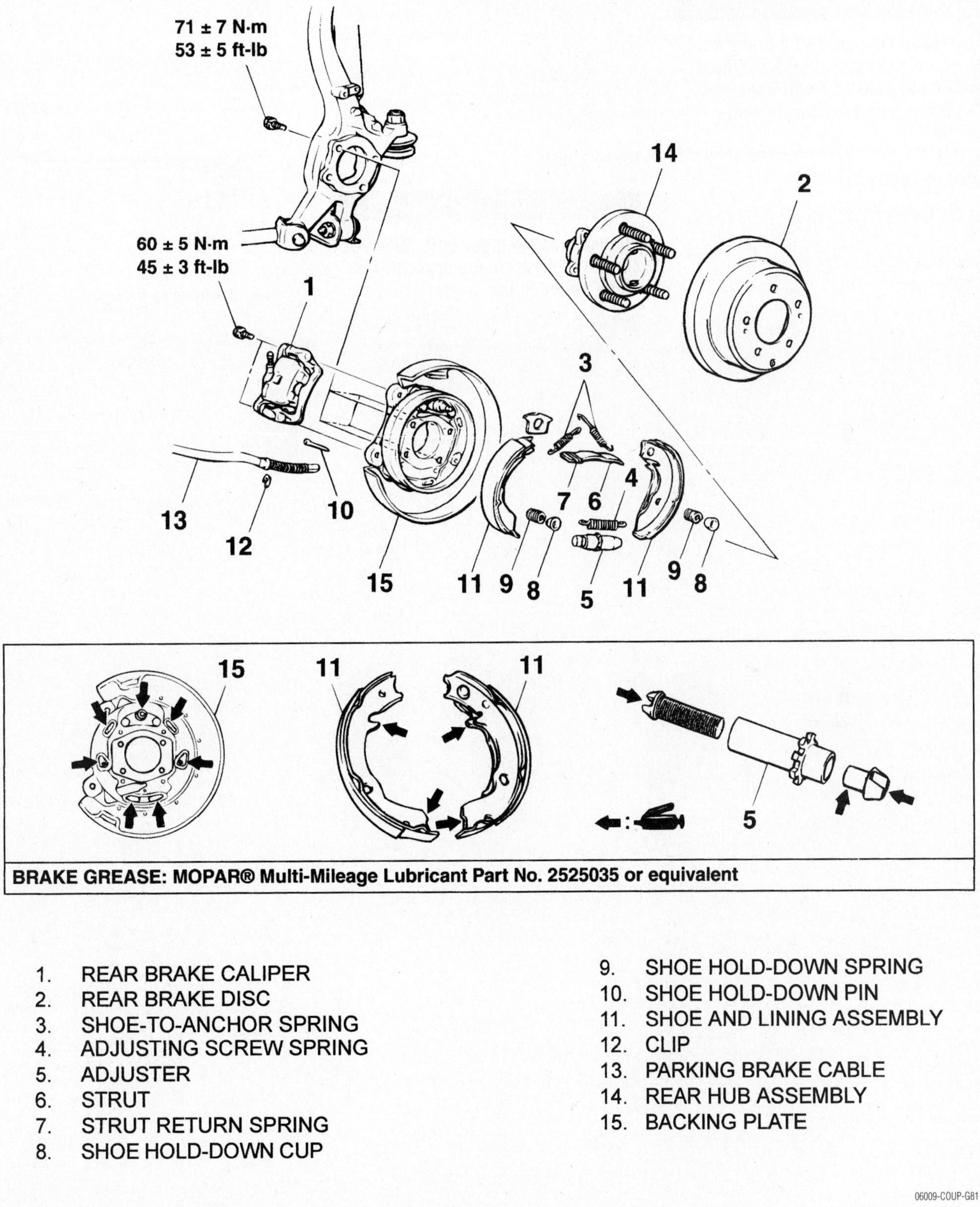

Parking brake unit with disc brakes

BRAKE HYDRAULIC SYSTEM

Brake Master Cylinder and Lines

BLEEDING

**** CAUTION**

Use only DOT 3 or DOT 4 brake fluid. Never mix the specified brake fluid with other fluid as it will influence the braking performance significantly.

Master Cylinder

The master cylinder used has no check valve, so if bleeding is carried out by the following procedure, bleeding of air from the brake pipeline will become easier.

(When brake fluid is not contained in the master cylinder.)

1. Fill the reserve tank with brake fluid.
2. Keep the brake pedal depressed.
3. Have another person cover the master cylinder outlet with a finger.
4. With the outlet still closed, release the brake pedal.
5. Repeat steps 2–4 three or four times to fill the inside of the master cylinder with brake fluid.

Brake Lines

**** CAUTION**

For vehicles equipped with ABS, be sure to filter/strain the brake fluid being added to the master cylinder reservoir tank. Debris may damage the hydraulic unit.

Start the engine and bleed the air in the sequence shown in the accompanying illustration.

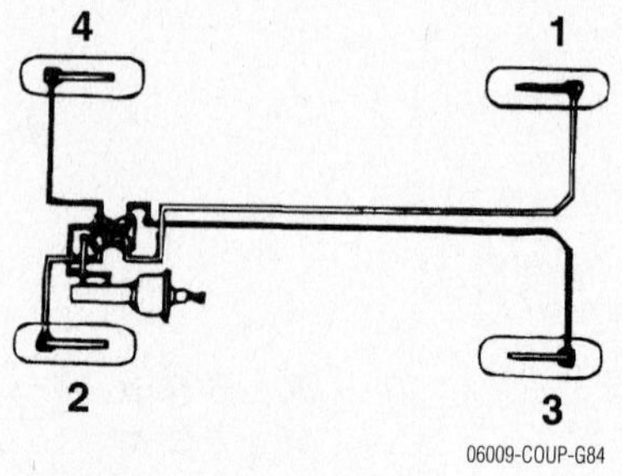

Brake line bleeding sequence

DODGE

14

SRT-10 Pick-up

SPECIFICATIONS AND MAINTENANCE CHARTS

ENGINE AND VEHICLE IDENTIFICATION

Engine						
Code ①	Liters (cc)	Cu. In.	Cyl.	Fuel Sys.	Engine Type	Eng. Mfg.
H	8.3 (8275)	505	10	SMFI	OHV	Chrysler

Model Year	
Code ②	Year
4	2004
5	2005

SMFI: Sequential Multi-port Fuel Injection

① 8th position of VIN

② 10th position of VIN

06009-SRT1-C01

GENERAL ENGINE SPECIFICATIONS

Year	Model	Engine Displacement Liters	Engine Series VIN	Net Horsepower @ rpm	Net Torque @ rpm (ft. lbs.)	Bore x Stroke (in.)	Com-pression Ratio	Oil Pressure @ rpm
2004	SRT-10	8.3	H	501@5600	525@4100	4.03x3.96	9.6:1	45-75@3000
2005	SRT-10	8.3	H	501@5600	525@4100	4.03x3.96	9.6:1	45-75@3000

06009-SRT1-C02

GASOLINE ENGINE TUNE-UP SPECIFICATIONS

Year	Engine Displacement Liters	Engine VIN	Spark Plug Gap (in.)	Ignition Timing (deg.)	Fuel Pump (psi)	Idle Speed (rpm)	Valve Clearance	
							Intake	Exhaust
2004	8.3	H	0.033-0.038	①	56-60	②	HYD	HYD
2005	8.3	H	0.033-0.038	①	56-60	②	HYD	HYD

NOTE: The Vehicle Emission Control Information (VECI) label often reflects specification changes made during production. The label figures must be used if they differ from those in this chart.

HYD: Hydraulic

① Ignition timing is controlled by the PCM and is not adjustable.

② Idle speed is controlled by the PCM and is not adjustable

06009-SRT1-C03

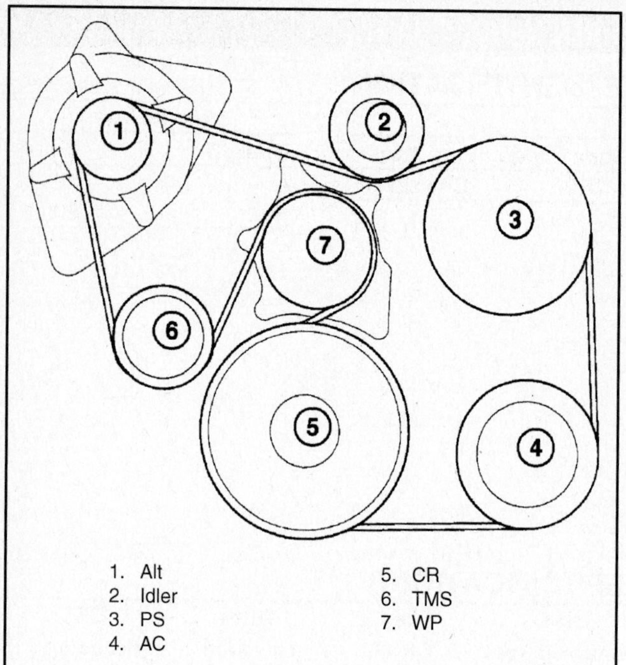

Accessory drive belt routing: (1) Alternator, (2) Idler, (3) Power Steering Pump, (4) A/C Compressor, (5) Crankshaft Pulley, (6) Tensioner, (7) Water Pump

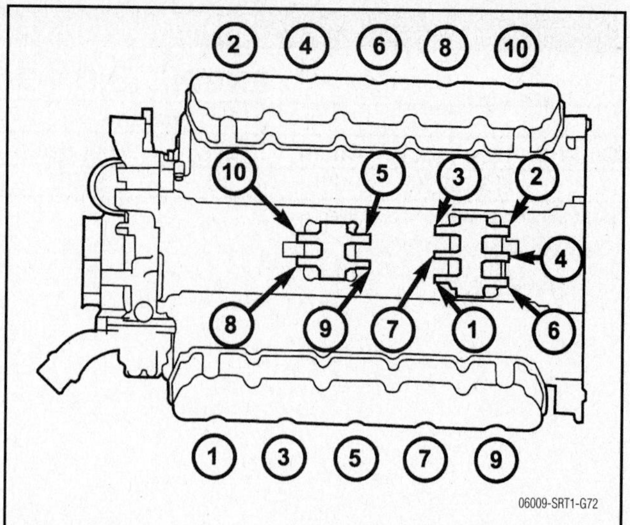

Firing order

CAPACITIES

Year	Model	Engine Displacement Liters	Engine VIN	Oil with Filter (qts.)	Transmission (pts.)		Drive Axle (pts.)	Fuel Tank (gal.)	Cooling System (qts.)
					Manual	Auto. ①			
2004	SRT-10	8.3	H	8.8	4.8	—	2.8 ②	26.0	18.0
2005	SRT-10	8.3	H	8.8	4.8	8.0	2.8 ②	26.0	18.0

NOTE: All capacities are approximate. Add fluid gradually and check to be sure a proper fluid level is obtained.

① For fluid drain and filter replacement only.

② With Trac-Lok: add 4 0z. Of limited slip additive

06009-SRT1-C04

VALVE SPECIFICATIONS

Year	Engine Displ. Liters	Engine VIN	Seat Angle (deg.)	Face Angle (deg.)	Spring Test Pressure (lbs. @ in.)	Spring Installed Height (in.)	Stem-to-Guide Clearance (in.)		Stem Diameter (in.)	
							Intake	Exhaust	Intake	Exhaust
2004	8.3	H	44-44.5	45-45.5	280@1.207	1.750	0.001-0.003	0.001-0.003	0.312-0.331	0.312-0.331
2005	8.3	H	44-44.5	45-45.5	280@1.207	1.750	0.001-0.003	0.001-0.003	0.312-0.331	0.312-0.331

06009-SRT1-C05

CRANKSHAFT AND CONNECTING ROD SPECIFICATIONS

All measurements are given in inches.

Year	Engine Displacement Liters	Engine VIN	Crankshaft				Connecting Rod		
			Main Brg. Journal Dia.	Main Brg. Oil Clearance	Shaft End-play	Thrust on No.	Journal Diameter	Oil Clearance	Side Clearance
2004	8.3	H	3.0013-3.0018	0.0010-0.0020	0.0020-0.0070	3	2.1256-2.1263	0.0002-0.0029	0.010-0.0180
2005	8.3	H	3.0013-3.0018	0.0010-0.0020	0.0020-0.0070	3	2.1256-2.1263	0.0002-0.0029	0.010-0.0180

06009-SRT1-C06

PISTON AND RING SPECIFICATIONS

All measurements are given in inches.

Year	Engine Displacement Liters	Engine VIN	Piston Clearance	Ring Gap Top Compression	Ring Gap Bottom Compression	Ring Gap Oil Control	Ring Side Clearance Top Compression	Ring Side Clearance Bottom Compression	Ring Side Clearance Oil Control
2004	8.3	H	0.0010-0.0022	0.0080-0.0140	0.0220-0.0331	0.010-0.030	0.0016-0.0033	0.0016-0.0033	0.0018-0.0079
2005	8.3	H	0.0010-0.0022	0.0080-0.0140	0.0220-0.0331	0.010-0.030	0.0016-0.0033	0.0016-0.0033	0.0018-0.0079

06009-SRT1-C07

TORQUE SPECIFICATIONS

All readings in ft. lbs.

Year	Engine Displacement Liters	Engine VIN	Cylinder Head Bolts	Main Bearing Bolts	Rod Bearing Bolts	Crankshaft Damper Bolts	Flywheel Bolts	Manifold Intake	Manifold Exhaust	Spark Plugs	Oil Pan Drain Plug
2004	8.3	H	①	②	50	130	55	③	16	20	25
2005	8.3	H	①	②	50	130	55	③	16	20	25

① See the illustration in the text
Bolts 1-12
Step 1: 35 ft. lbs.
Step 2: 90 ft. lbs.
Bolts A-H: 95 inch lbs.

② See the illustration below
Bolts 1-12: 15 ft. lbs.
Bolts A1: 45 ft. lbs.
Bolts A2: 85 ft. lbs.
Bolts A3: 40 ft. lbs.

③ 95 inch lbs.

06009-SRT1-C08

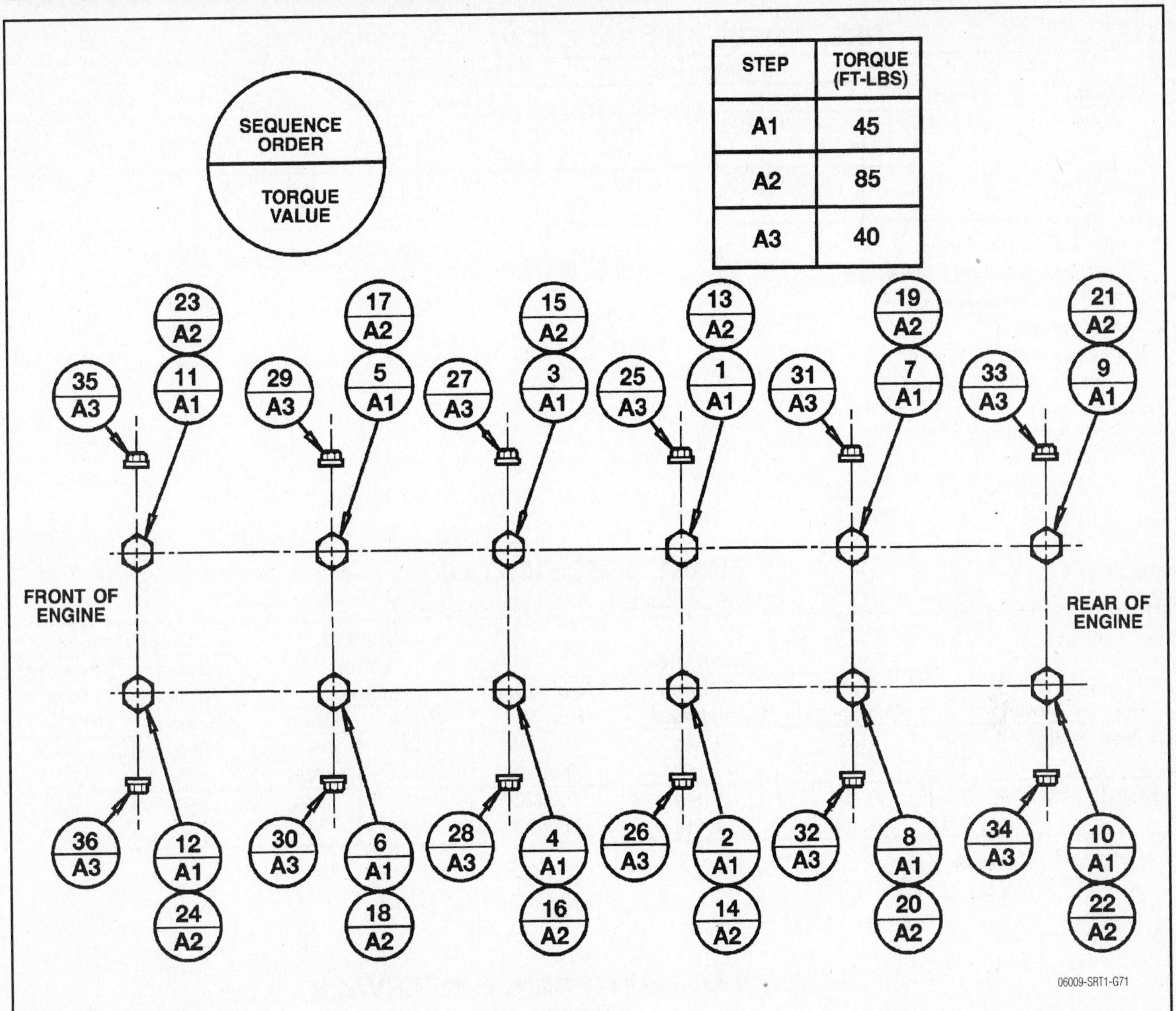

Main bearing torque sequence

WHEEL ALIGNMENT

Year	Model	Caster Range (+/-Deg.)	Caster Preferred Setting (Deg.)	Camber Range (+/-Deg.)	Camber Preferred Setting (Deg.)	Toe-in (in.)
2004	SRT-10	0.50	+5.75	0.30	0	0.24+/-0.16
2005	SRT-10	0.50	+5.75	0.30	0	0.24+/-0.16

06009-SRT1-C11

TIRE, WHEEL AND BALL JOINT SPECIFICATIONS

Year	Model	OEM Tires		Tire Pressures (psi)		Wheel Size	Wheel Lug Nut Torque
		Standard	Optional	Front	Rear		
2004	SRT-10	305/40ZR22	None	①	①	10	95
2005	SRT-10	305/40ZR22	None	①	①	10	95

NA: Information not available
OEM: Original Equipment Manufacturer
PSI: Pounds Per Square Inch
STD: Standard
OPT: Optional
① See placard on the vehicle

06009-SRT1-C10

BRAKE SPECIFICATIONS

All measurements in inches unless noted

Year	Model		Brake Disc Original Thickness	Brake Disc Minimum Thickness	Brake Disc Maximum Run-out	Minimum Lining Thickness Front	Minimum Lining Thickness Rear	Brake Caliper Bracket Bolts (ft. lbs.)	Brake Caliper Mounting Pins (ft. lbs.)
2004	SRT-10	F	1.397	1.344	0.0020	NA	NA	250	24
		R	1.180	1.118	0.0020	NA	NA	145	22
2005	SRT-10	F	1.397	1.344	0.0020	NA	NA	250	24
		R	1.180	1.118	0.0020	NA	NA	145	22

NA: Information not available

06009-SRT1-C09

SCHEDULED MAINTENANCE INTERVALS

SRT-10

TO BE SERVICED	TYPE OF SERVICE	VEHICLE MILEAGE INTERVAL (x1000) 3	6	9	12	15	18	21	24	27	30	33	36	39
Engine oil & filter	R	✓	✓	✓	✓	✓	✓	✓	✓	✓	✓	✓	✓	✓
Auto. trans. fluid and filter ①	R					✓					✓			
Auto. trans. Bands ①	Adj.					✓					✓			
Tires	Rotate		✓		✓		✓		✓		✓		✓	
Rear axle fluid	R					✓					✓			
Brake linings	I				✓				✓				✓	
Engine air filter	S/I					✓					✓			
Spark plugs	R										✓			
PCV valve	S/I										✓			
Engine coolant	R	Every 5 years, regardless of mileage												
Spark plug cables	R	Every 60,000 miles												
Accessory drive belt	I/R	Every 75,000 miles												
Power steering fluid	R	Every 100,000 miles												

R: Replace S: Service I: Inspect Adj: Adjust
① After 105,000 miles, perform the service every 15,000 miles

06009-SRT1-C12

PRECAUTIONS

Before servicing any vehicle, please be sure to read all of the following precautions, which deal with personal safety, prevention of component damage, and important points to take into consideration when servicing a motor vehicle:

- Never open, service or drain the radiator or cooling system when the engine is hot; serious burns can occur from the steam and hot coolant.
- Observe all applicable safety precautions when working around fuel. Whenever servicing the fuel system, always work in a well-ventilated area. Do not allow fuel spray or vapors to come in contact with a spark, open flame, or excessive heat (a hot drop light, for example). Keep a dry chemical fire extinguisher near the work area. Always keep fuel in a container specifically designed for fuel storage; also, always properly seal fuel containers to avoid the possibility of fire or explosion. Refer to the additional fuel system precautions later in this section.
- Fuel injection systems often remain pressurized, even after the engine has been turned **OFF**. The fuel system pressure must be relieved before disconnecting any fuel lines. Failure to do so may result in fire and/or personal injury.
- Brake fluid often contains polyglycol ethers and polyglycols. Avoid contact with the eyes and wash your hands thoroughly after handling brake fluid. If you do get brake fluid in your eyes, flush your eyes with clean, running water for 15 minutes. If eye irritation persists, or if you have taken brake fluid internally, IMMEDIATELY seek medical assistance.
- The EPA warns that prolonged contact with used engine oil may cause a number of skin disorders, including cancer! You should make every effort to minimize your exposure to used engine oil. Protective gloves should be worn when changing oil. Wash your hands and any other exposed skin areas as soon as possible after exposure to used engine oil. Soap and water, or waterless hand cleaner should be used.
- All new vehicles are now equipped with an air bag system, often referred to as a Supplemental Restraint System (SRS) or Supplemental Inflatable Restraint (SIR) system. The system must be disabled before performing service on or around system components, steering column, instrument panel components, wiring and sensors. Failure to follow safety and disabling procedures could result in accidental air bag deployment, possible personal injury and unnecessary system repairs.
- Always wear safety goggles when working with, or around, the air bag system. When carrying a non-deployed air bag, be sure the bag and trim cover are pointed away from your body. When placing a non-deployed air bag on a work surface, always face the bag and trim cover upward, away from the surface. This will reduce the motion of the module if it is accidentally deployed. Refer to the additional air bag system precautions later in this section.
- Clean, high quality brake fluid from a sealed container is essential to the safe and proper operation of the brake system. You should always buy the correct type of brake fluid for your vehicle. If the brake fluid becomes contaminated, completely flush the system with new fluid. Never reuse any brake fluid. Any brake fluid that is removed from the system should be discarded. Also, do not allow any brake fluid to come in contact with a painted surface; it will damage the paint.
- Never operate the engine without the proper amount and type of engine oil; doing so WILL result in severe engine damage.
- Timing belt maintenance is extremely important! Many models utilize an interference-type, non-freewheeling engine. If the timing belt breaks, the valves in the cylinder head may strike the pistons, causing potentially serious (also time-consuming and expensive) engine damage.
- Disconnecting the negative battery cable on some vehicles may interfere with the functions of the on-board computer system(s) and may require the computer to undergo a relearning process once the negative battery cable is reconnected.
- When servicing drum brakes, only disassemble and assemble one side at a time, leaving the remaining side intact for reference.

ENGINE REPAIR

Alternator

REMOVAL & INSTALLATION

1. Disconnect the negative battery cable.
2. Remove air cleaner assembly.
3. Remove the accessory drive belt.
4. Disconnect the field connector.
5. Disconnect the battery positive cable from the alternator. First disconnect the B+ cap on the wiring harness side.
6. Remove the upper mounting bolt.
7. Remove the lower mounting bolt and nut.
8. Remove alternator.

To install:

9. Install alternator.
10. Install lower mounting bolt and nut.
11. Install upper mounting bolt.

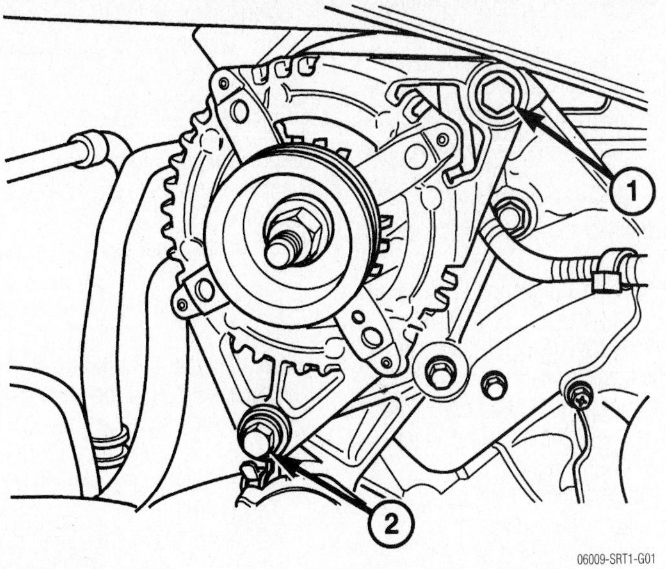

Alternator mounting

12. Tighten bolts to 55 Nm (41 ft. lbs).
13. Connect battery positive cable to alternator, and install B+ stud cap.
14. Connect field connector.
15. Install accessory drive belt.
16. Install air cleaner assembly.
17. Connect negative battery cable, tighten nut to 14.2 Nm (125 inch. lbs.).

Engine Assembly

REMOVAL & INSTALLATION

1. Perform fuel pressure release procedure.
2. Disconnect negative battery cable.
3. Disconnect throttle body air inlet duct, remove air cleaner cover.
4. Raise and support the vehicle.
5. Remove the lower engine shield.
6. Drain cooling system.
7. Paint mark and remove the driveshaft.
8. Remove the transmission.
9. Remove the starter.
10. Disconnect the front exhaust pipe flange to exhaust manifold fasteners.
11. Disconnect the hydraulic cooling fan lines and capture any spillage.
12. Drain engine oil and remove the oil filter.
13. Separate the air conditioning hose from the fan shroud.
14. Disconnect the lower radiator hose.
15. Disconnect the power steering line support bracket at the lower left of the radiator.
16. Loosen both engine mount through bolts.
17. Separate the ground strap from the floor board, above the right exhaust flange.
18. Lower the vehicle.
19. Discharge air conditioning system.
20. Remove the upper radiator hose.
21. Disconnect windshield washer hose at the splice.
22. Remove radiator core support.
23. Disconnect the cooling fan harness connector.
24. Remove the radiator and radiator fan assembly.
25. Remove the A/C condenser, cap and position the hoses aside.
26. Disconnect the heater hoses at the top front cover and at the lower coolant housing and position aside.
27. Disconnect the oil cooler hoses from the engine and position aside.
28. Remove accessory drive belt.
29. Disconnect and remove alternator.

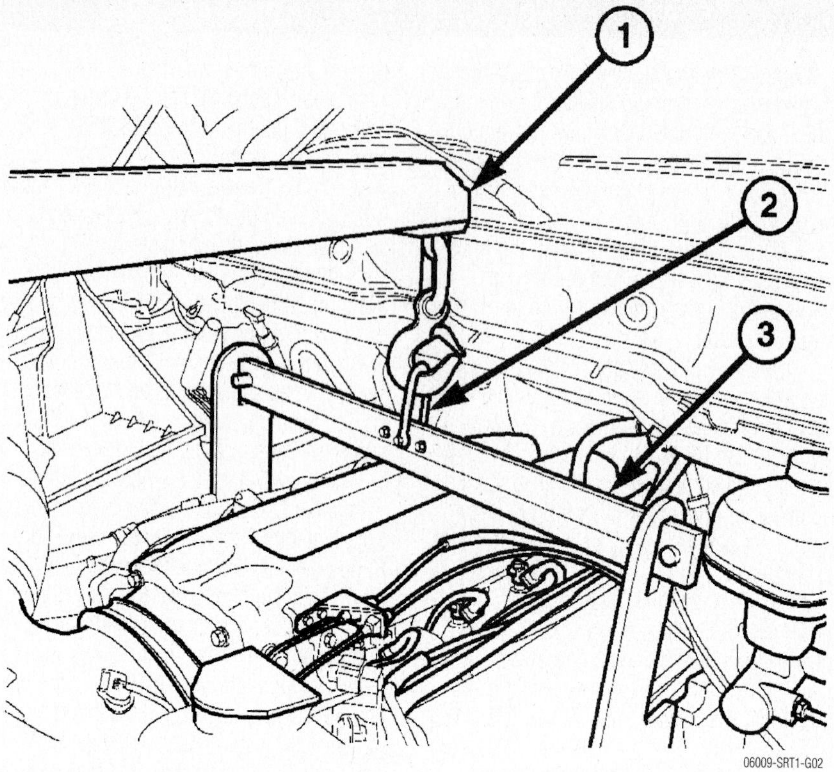

Engine crane (1), lifting eye (2) and lift bar (3)

30. Remove power steering pump mounting bolts and set pump aside.
31. Disconnect A/C lines and electrical connector from A/C compressor. Plug holes in A/C compressor to prevent debris from entering A/C compressor.
32. Disconnect brake booster vacuum hose and fuel line to fuel rail.
33. Disconnect the cruise control, accelerator cable and fuel vapor purge harness, then set aside.
34. Disconnect the engine harness connectors on top of the intake manifold.
35. Disconnect the spark plug wires from the spark plugs and set the plug wires on top of the engine.
36. Disconnect the ground strap at the thermostat housing and pull the electrical harness underneath the intake manifold and set aside.
37. Install Special Tool 9363 (Engine Lifting Brackets), or equivalent, and to the exhaust manifold heat shield studs where indicated.
38. Insert engine lifting bar through the access holes in the engine lifting brackets and secure with the bar retaining pins.
39. Remove engine with an engine hoist.

To install:

40. Attach engine hoist and lower the engine into engine compartment, aligning the left and right engine mounts
41. Install left and right engine mount insulator bolts and nuts. Tighten nuts to 102 Nm (75 ft. lbs.).
42. Remove engine hoist and engine lifting bar and brackets.
43. Install the exhaust manifold heat shield nuts and tighten to 6 Nm (50 in. lbs.).
44. Connect the ground strap to the thermostat housing and route the electrical and purge harness underneath the throttle body and attach the connectors.
45. Connect the fuel vapor purge lines.
46. Properly route and connect the spark plug wires.
47. Connect the engine electrical harnesses.
48. Connect the brake booster vacuum hose, fuel line, cruise control and accelerator cables.
49. Install the A/C lines at the A/C compressor.
50. Connect A/C clutch electrical connector.
51. Install the power steering pump.
52. Install the alternator.
53. Install the accessory drive belt.
54. Connect the heater hoses to the engine.
55. Install the radiator and cooling fan assembly.
56. Install the A/C condenser.
57. Connect the A/C lines to the condenser.

58. Connect the cooling fan electrical connector.

59. Connect the windshield washer hose at the splice.

60. Install the coolant recovery container.

61. Install the upper radiator hose.

62. Install the throttle body air inlet duct and the air cleaner element cover.

63. Raise and support the vehicle.

64. Install transmission.

65. Install transmission bolts and tighten bolts to:

- 7⁄16 in. transmission bolts and washers: 68 Nm (50 ft. lbs.)
- Transmission to top of engine: 41 Nm (30 ft. lbs.)
- Transmission to oil pan: 30 Nm (22 ft. lbs.)

66. Install drive shaft.

67. Install the ground strap to the floor pan above the right exhaust manifold flange.

68. Connect the front exhaust pipes to the exhaust manifolds and tighten the fasteners to 34 Nm (25 ft. lbs.).

69. Connect oil cooler lines and connect the oil pressure sensor and oil temperature sensor.

70. Install the power steering line support bracket at the radiator.

71. Install lower radiator hose.

72. Connect the cooling fan hydraulic lines.

73. Connect the A/C line to the fan shroud.

74. Install under body shield.

75. Lower vehicle.

76. Fill engine crankcase with the proper oil to the correct level.

77. Evacuate and recharge the air conditioning.

78. Fill the cooling system.

79. Fill power steering to proper level and purge the system.

80. Connect the negative battery cable.

81. Start the engine and run until operating temperature is obtained.

82. Turn engine off and inspect for leaks.

83. Recheck all fluid levels, fill as required.

Water Pump

REMOVAL & INSTALLATION

1. Disconnect negative battery cable from battery.

2. Drain cooling system.

➡**Do not waste reusable coolant. If solution is clean, drain coolant into a clean container for reuse.**

3. Remove windshield washer reservoir tank from radiator fan shroud.

4. Remove the four fan shroud mounting bolts at the radiator. Do not attempt to remove shroud from vehicle at this time.

⁂ WARNING

Constant tension hose clamps are used on most cooling system hoses. When removing or installing, use only tools designed for servicing this type of clamp, such as special clamp tool number 6093. Always wear safety glasses when servicing constant tension clamps.

➡**A number or letter is stamped into the tongue of constant tension clamps. If replacement is necessary, use only an original equipment clamp with a matching number or letter.**

5. Remove radiator upper hose at radiator.

6. Remove accessory drive belt.

7. Remove the radiator lower hose at water pump.

8. Remove heater hose at water pump fitting.

9. Remove the seven water pump mounting bolts.

10. Loosen the clamp at the water pump end of bypass hose. Slip the bypass hose from the water pump while removing pump from vehicle. Do not remove the clamp from the bypass hose.

11. Discard the water pump-to-timing chain/case cover o-ring seal.

12. Remove the heater hose fitting from water pump if pump replacement is necessary. Note position (direction) of fitting before removal. Fitting must be re-installed to same position.

⁂ WARNING

Do not pry the water pump at timing chain case/cover. The machined surfaces may be damaged resulting in leaks.

13. Clean the gasket mating surface. Use caution not to damage the gasket sealing surface.

14. Inspect the water pump assembly for cracks in the housing, water leaks from shaft seal, worn bearing or impeller rubbing either the pump body or timing chain case/cover.

To install:

15. If water pump is being replaced, install the heater hose fitting to the pump. Tighten fitting to 16 Nm (144 inch lbs.) torque. After fitting has been torqued, posi-

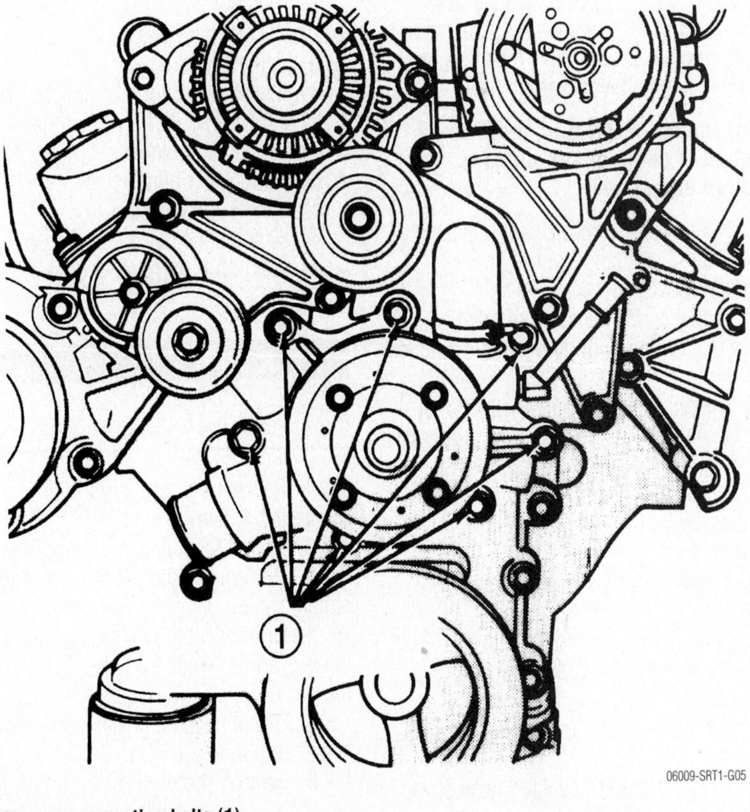

Water pump mounting bolts (1)

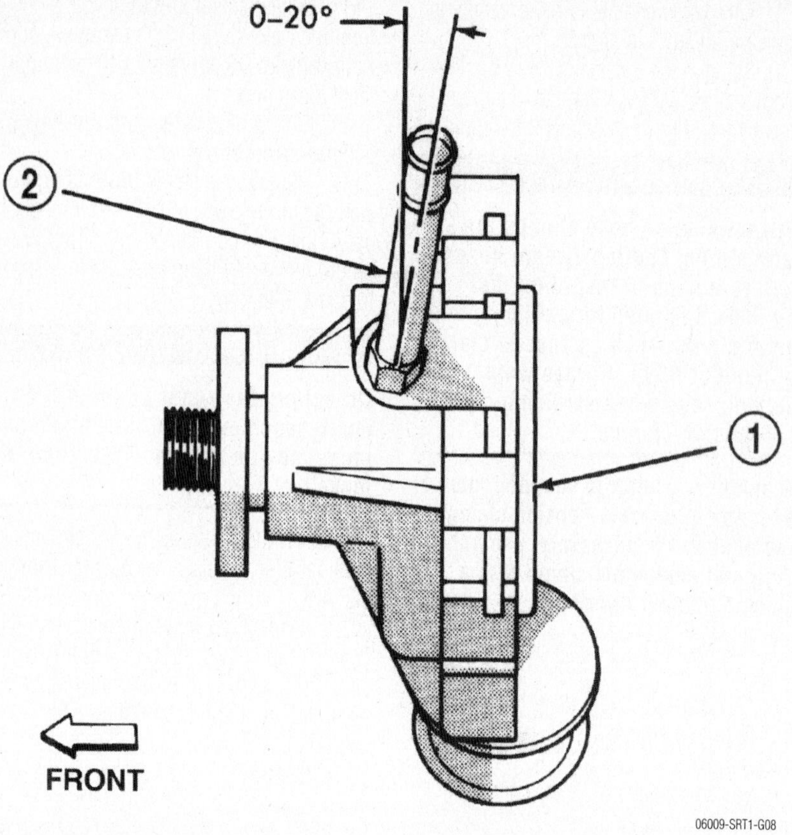

If water pump is being replaced, install the heater hose fitting to the pump. Tighten fitting to 16 Nm (144 inch lbs.) torque. After fitting has been torqued, position fitting as shown

tion fitting as shown. When positioning fitting, do not back off (rotate counterclockwise). Use a sealant on the fitting such as Mopar® Thread Sealant With Teflon. Refer to the directions on the package.

➡This heater hose fitting must be installed to pump before pump is installed to engine.

16. Clean the o-ring mating surfaces at rear of water pump and front of timing chain/case cover.
17. Apply a small amount of petroleum jelly to o-ring. This will help retain o-ring to water pump.
18. Install water pump to engine as follows: Guide water pump fitting into bypass hose as pump is being installed. Install water pump bolts. Tighten water pump mounting bolts to 40 Nm (30 ft. lbs.) torque.
19. Position bypass hose clamp to bypass hose.
20. Spin water pump to be sure that pump impeller does not rub against timing chain case/cover.
21. Connect radiator lower hose to water pump.
22. Connect heater hose and hose clamp to heater hose fitting.
23. Install drive belt.
24. Position fan shroud and fan blade/viscous fan drive assembly to vehicle as a complete unit.
25. Install fan shroud to radiator. Tighten bolts to 6 Nm (50 inch lbs.) torque.
26. Install fan blade/viscous fan drive assembly to water pump shaft.
27. Fill cooling system.
28. Connect negative battery cable.
29. Start and warm the engine. Check for leaks.

Heater Core

REMOVAL & INSTALLATION

1. Recover the refrigerant from the refrigerant system.
2. Drain the engine cooling system.
3. Disconnect and isolate the negative battery cable.
4. Disconnect the A/C liquid line and the A/C accumulator from the A/C evaporator.
5. Disconnect the heater hoses from the heater core tubes.
6. Remove the powertrain control module (PCM) to gain access to the two nuts that secure the HVAC housing to the engine compartment side of the dash panel and remove the nuts

➡The PCM is located in the engine compartment attached to the dash panel. To avoid possible voltage spike damage to the PCM, ignition key must be off, and negative battery cable must be disconnected before unplugging PCM connectors.

a. Disconnect negative battery cable at battery.
b. Remove cover over electrical connectors. Cover snaps onto PCM.
c. Carefully unplug the three 32-way connectors (four 38-way connectors if equipped with NGC) from PCM.
d. Remove three PCM mounting bolts and remove PCM from vehicle.

7. Disconnect and isolate the negative battery cable.
8. Remove the left a-pillar trim.
9. Remove the instrument panel top cover.
10. Disconnect the headliner wire harness connector located at the a-pillar.
11. Using a trim stick C-4755 or equivalent, from the notch on the bottom, remove the left instrument panel side cover.
12. Remove the left cowl trim panel.
13. Position the front wheels straight ahead.
14. Disconnect the negative (ground) cable from the battery.
15. Remove the two switches from the steering wheel.
16. From the underside of the steering wheel, remove the two screws that secure the driver airbag to the steering wheel armature.

⁂ WARNING

Do not pull on the horn switch feed pigtail wire to disengage the connector from the driver airbag housing or to disconnect the horn switch to steering wheel wire harness connection. Improper pulling on this pigtail wire or connection can result in damage to the horn switch membrane or feed circuit.

17. Pull the driver airbag away from the steering wheel far enough to access the three electrical connections on the back of the airbag housing.
18. Disconnect the steering wheel wire harness connector for the horn switch from the horn switch feed pigtail wire connector, which is located on the back of the driver airbag housing.

WARNING

Do not pull on the clockspring pigtail wires or pry on the connector insulator to disengage the connector from the driver airbag inflator connector receptacle. Improper removal of these pigtail wires and their connector insulators can result in damage to the airbag circuits or connector insulators.

19. The clockspring driver airbag pigtail wire connectors are secured by integral latches to the airbag inflator connector receptacles, which are located on the back of the driver airbag housing. Depress the latches on each side of the connector insulator and pull the insulators straight out from the airbag inflator to disconnect them from the connector receptacles.
20. Remove the driver airbag from the steering wheel.
21. Remove the steering wheel with special tool CJ98-1 or an appropriate steering wheel puller.

WARNING

Ensure the puller bolts are fully engaged into the steering wheel and not into the clockspring, before attempting to remove the wheel. Failure to do so may damage the steering wheel/clockspring.

22. Remove the steering column opening cover.
23. Remove the tilt lever.
24. Remove the column shrouds.
25. Remove the clock spring.
26. Disconnect the wiring harness to the column.
27. Remove the shift cable from the column shift lever actuator.
28. Release the shift cable from the column bracket and remove it from the bracket.
29. Remove the SKIM module in order to disconnect the electrical connector.
30. Remove the upper steering shaft coupler bolt.
31. Separate the shaft from the coupler.
32. Remove the brake light switch and discard.
33. Remove the four steering column mounting nuts.
34. Remove the steering column assembly from the vehicle.
35. Remove the two bolts that secure the steering column support bracket to the instrument panel.
36. Remove the park brake release handle actuator rod.
37. Disconnect the instrument panel wire harness connector located above the brake pedal from the bulkhead wire harness connector.
38. Using a trim stick C-4755 or equivalent, from the notch on the bottom, remove the left instrument panel side cover.
39. Remove the left cowl trim panel.
40. Remove the two screws that secure the lower driver side instrument panel bezel to the instrument panel.
41. Using a trim stick C-4755 or equivalent, disengage the retaining clips that secure the top of the lower driver side instrument panel bezel to the instrument panel.
42. Tilt the top of the lower driver side instrument panel bezel downward and, If equipped, disconnect the wire harness connector from the adjustable pedal switch.
43. Remove the lower driver side instrument panel bezel from the lower retaining hooks and remove the bezel.
44. Remove the three bolts that secure the left side of the instrument panel to the dash panel.
45. If equipped, remove the 4WD gear shift boot.
46. If equipped with a manual transmission, remove the transmission gear shift lever extension.
47. Using a trim stick C-4755 or equivalent, disengage the retaining tabs that secure the manual transmission gear shift boot or storage bin (depending on application) to the floor console and remove the boot or storage bin.
48. Remove the three inserts from the floor console.
49. Remove the three bolts that secure the floor console to the floor panel.
50. Lift up the rear of the floor console to clear the gear shift lever, if equipped.
51. Slide the floor console rearward and remove it from the instrument panel.
52. Using a trim stick or another suitable wide flat-bladed tool, gently pry each side of the ACM cover away from the instrument panel at each side of the center bracket on the floor panel transmission tunnel far enough to disengage the two snap clip retainers from the instrument panel receptacles.
53. Remove the ACM cover from the instrument panel.
54. Remove the two bolts that secure the instrument panel to the center of the floor panel.
55. Using a trim stick C-4755 or equivalent, disengage the retaining tabs of the

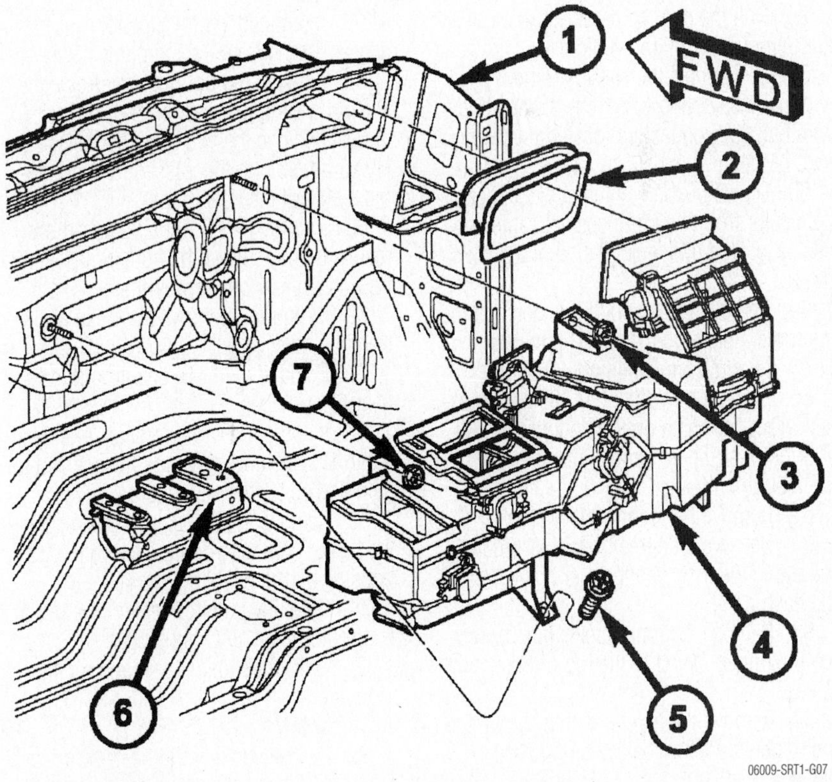

HVAC housing removal. (1) Dash panel, (2) air inlet, (3) Retaining nut, (4) Housing, (5) Retaining bolt, (6) Floor bracket, (7) Retaining nut

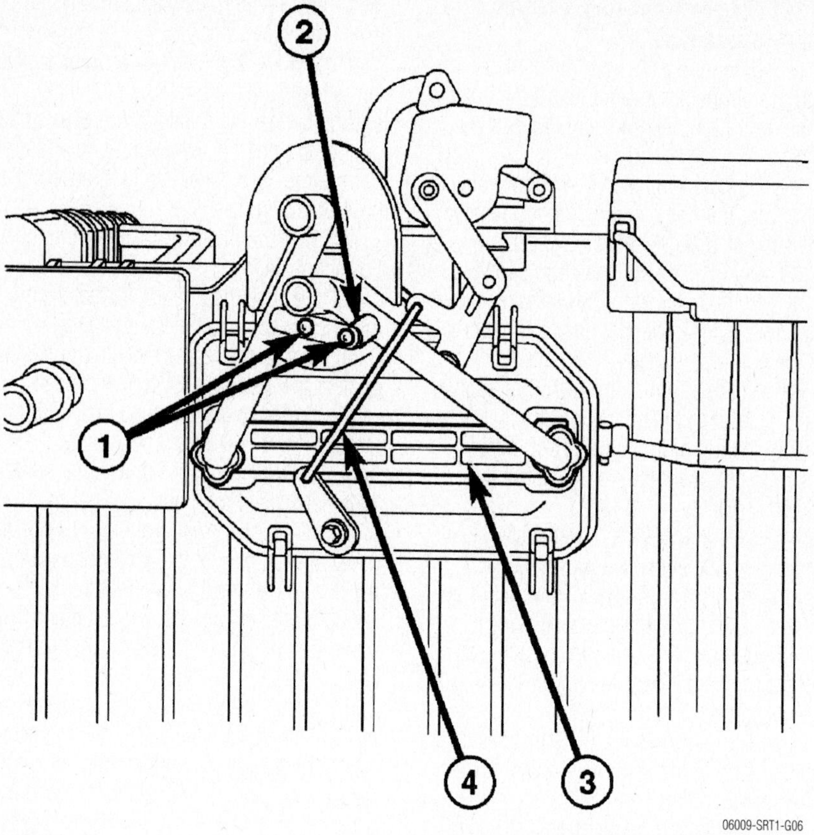

Heater core

cowl trim panel from the retainer clips in the door sill.

56. Pull the cowl trim panel rearward and remove it from the vehicle.

57. Disconnect the two instrument panel wire harness connectors from the two body wire harness connectors located on the right side of the cowl panel.

58. Disconnect the antenna coaxial cable connector from the radio coaxial cable connector located on the right side of the cowl panel.

59. Remove the one bolt that secures the instrument panel to the HVAC housing below the glove box opening.

60. Using a trim stick C-4755 or equivalent, from the notch on the bottom, remove the right instrument panel side cover from the instrument panel.

61. Remove the right a-pillar trim.

62. Remove the three bolts that secure the right side instrument panel bracket to the dash panel.

63. Remove the four screws that secure the instrument panel to the top of the cowl panel.

64. Remove the two bolts that secure the instrument panel to the top of the cowl panel.

65. With the help of an assistant, lift the instrument panel up and off of the cowl panel and remove the instrument panel from the vehicle.

66. If required, remove the four plastic screws inserts from the top of the cowl panel.

67. Remove the bolt that secures the HVAC housing to the floor bracket.

68. Remove the two nuts that secure the HVAC housing to the passenger compartment side of the dash panel.

69. Pull the HVAC housing assembly rearward and remove the housing assembly from the passenger compartment.

70. If required, remove the fresh air inlet from the dash panel.

71. Remove the foam seal from the heater core tubes.

72. If equipped with the dual zone heating-A/C system, remove the linkage rod to gain access to the heater core.

73. Remove the two screws that secure the heater core tube bracket to the HVAC housing and remove the bracket.

74. Carefully pull the heater core out of the front of the HVAC housing.

75. Inspect all foam seals and replace as required.

To install:

76. Carefully install the heater core into the front of the HVAC housing.

77. Position the heater core tube bracket onto the HVAC housing.

78. Install the two screws that secures the heater core bracket to the HVAC housing. Tighten the screws to 2.2 Nm (20 inch lbs.).

79. If equipped with the dual zone heating-A/C system, install the linkage rod.

80. Install the foam seal onto the heater core tubes.

➡If the heater core is being replaced, flush the cooling system.

81. If removed, install the fresh air inlet onto the dash panel.

82. Position the HVAC housing assembly into the passenger compartment with the mounting studs and the condensate drain tube in their proper locations in the dash panel.

83. Install the two nuts that secure the HVAC housing to the passenger compartment side of the dash panel. Tighten the nuts to 6.2 Nm (55 inch lbs.).

84. Install the bolt that secures the HVAC housing to the floor bracket. Tighten the bolt to 6.2 Nm (55 inch lbs.).

85. If removed, install the four plastic screw inserts into the top of the cowl panel.

86. With the help of an assistant, position the instrument panel into the vehicle and install the right side guide pin and the left side guide hook to the sides of the cowl panel.

87. Install the two bolts that secure the instrument panel to the top of the cowl panel. Tighten the bolts to 12 Nm (9 ft. lbs.).

88. Install the four screws that secure the instrument panel to the top of the cowl panel. Tighten the screws to 2.2 Nm (20 inch lbs.).

89. Install the three bolts that secure the right side instrument panel bracket to the cowl panel. Tighten the bolts to 12 Nm (9 ft. lbs.).

90. Install the right a-pillar trim.

91. Install the right instrument panel side cover.

92. Install the one bolt that secures the instrument panel to the HVAC housing below the glove
box opening. Tighten the bolt securely.

93. Connect the radio coaxial cable connector to the antenna coaxial cable connector located on the right side of the cowl panel.

94. Connect the two instrument panel wire harness connectors to the two body wire harness connectors located on the right side of the cowl panel.

95. Install the right cowl trim panel.

96. Install the two bolts that secure the instrument panel to the center of the floor

panel. Tighten the bolts to 12 Nm (9 ft. lbs.).

97. Connect the air bag control module electrical connector.

98. Install the air bag control module cover, if equipped.

99. Install the floor console, if equipped.

100.Install the three bolts that secure the left side of the instrument panel to the dash panel. Tighten the bolts to 12 Nm (9 ft. lbs.).

100. Install the instrument panel driver's side bezel.

102. Install the left cowl trim panel.

103. Connect the instrument panel wire harness connector located above the brake pedal to the bulkhead wire harness connector.

104. Install the park brake release handle actuator rod.

105. Install the two bolts that secure the steering column support bracket to the instrument panel. Tighten the bolts to 14 Nm (10 ft. lbs.).

106. Install the steering column.

107. Install the left cowl trim cover.

108. Install the left instrument panel side cover.

109. Connect the headliner wire harness connector located at the a-pillar.

110. Install the instrument panel top cover.

111. Install the left a-pillar trim.

112. Reconnect the negative battery cable.

113. Install the two nuts that secure the HVAC housing to the engine compartment side of the dash panel. Tighten the nuts to 6.2 Nm (55 inch lbs.).

114. Install the powertrain control module (PCM).

115. Connect the heater hoses to the heater core tubes.

116. Connect the A/C liquid line and the A/C accumulator to the A/C evaporator.

117. Reconnect the negative battery cable.

118. If the heater core is being replaced, flush the cooling system.

119. Refill the engine cooling system.

120. Evacuate the refrigerant system.

121. Charge the refrigerant system.

Cylinder Head

REMOVAL & INSTALLATION

➡Care must be taken to protect the intake manifold and cylinder head covers powder coating from scrapes and abrasions. This procedure covers either the left or right cylinder head.

➡The aluminum alloy cylinder head is held in place by 12½ inch bolts and 8¼ inch bolts.

1. Release the fuel pressure.
2. Disconnect negative battery cable.
3. Disconnect the throttle body air inlet duct, IAT sensor, CCV hose and remove the air cleaner cover.
4. Drain the cooling system.
5. Raise and support the vehicle.
6. Disconnect the front exhaust pipe to exhaust manifold connection.
7. Lower the vehicle.
8. Remove intake manifold.
9. Disconnect the spark plug wires by pulling on the boot straight out in line with the spark plug. Dislodge spark plug retainers from the cylinder head cover studs and position the spark plug wires aside.
10. Disconnect the oil level indicator tube and set aside.
11. Disconnect CCV hoses from cylinder head covers.
12. Remove the cylinder head covers.
13. Remove exhaust manifold heat shields.
14. Remove exhaust manifold(s) and gasket(s).

➡If rocker arms and push rods are to be reused, identify each component to ensure installation in original locations.

15. Remove the rocker arm and pedestal assemblies.
16. Remove push rods.
17. Remove the 12 head bolts and 8 cylinder head tappet gallery bolts from cylinder head(s).
18. Remove cylinder head(s) and gasket(s).
19. Before cleaning, check for leaks, damage and cracks.
20. Clean cylinder head and oil passages.
21. Check cylinder head for flatness.
22. Inspect all surfaces with a straightedge if there is any reason to suspect leakage. If out-of-flatness exceeds 0.019 mm (0.00075 in.) times the span length in inches in any direction, either replace head or lightly machine the head surface. As an example, if a 12.0 inch span is 1.0 mm (0.004 in.) out-of-flat, allowable is 12 x 0.019 mm (0.00075 in.) equals 0.22 mm (0.009 in.). This amount of out-of-flat is acceptable. Maximum of 0.2 mm (0.008 in.) for grinding is permitted.

➡This is a combined total dimension of stock removal from cylinder head surface.

To install:

➡This procedure covers either the left or right cylinder head.

23. Clean all surfaces of engine block and cylinder heads.

➡Remove all gasket material from cylinder head and block using a plastic scraper only. DO NOT use a metal scraper, as damage to sealing surface may occur.

➡NOTE: A multi-layer steel head gasket is used.

24. Install new gasket(s) on the engine block. Assure all coolant passages and bolt holes align properly.

➡Cylinder head bolts are coated, DO NOT oil.

25. Install cylinder head(s) over dowel pins. Install cylinder head bolts 1–12. Tighten cylinder head bolts 1–12 in the sequence shown in using a two step torque sequence:
   a. First to 47 Nm (35 ft. lbs.).
   b. Second to 122 Nm (90 ft. lbs.).
26. After cylinder head bolts 1–12 have

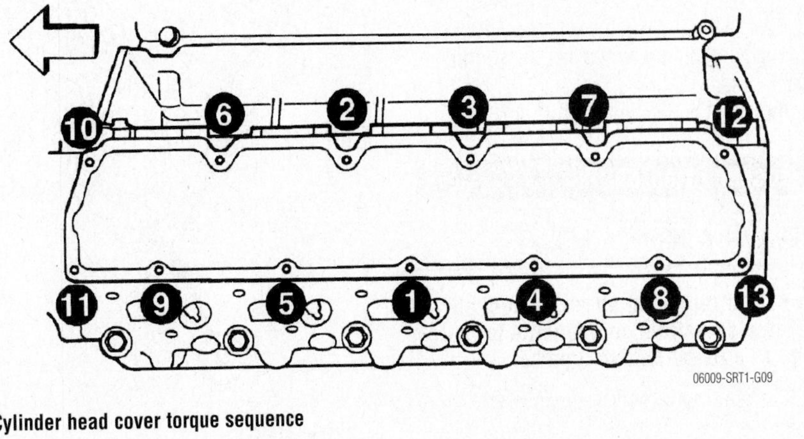

Cylinder head cover torque sequence

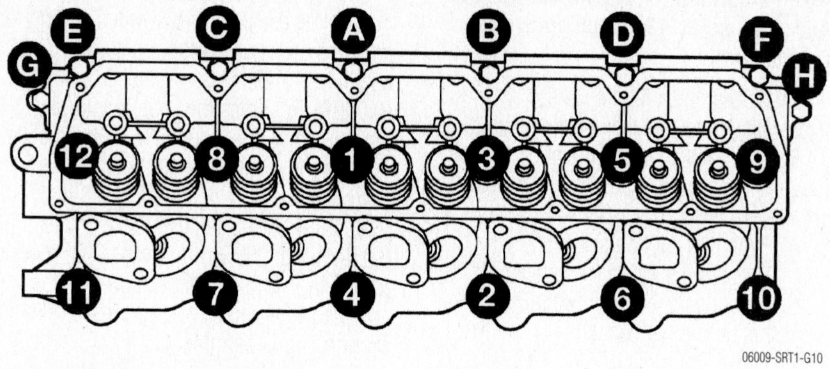

Cylinder head bolt torque sequence

been tightened to specifications, install and tighten cylinder head tappet gallery bolts A-H in sequence shown to 11 Nm (95 inch lbs.).

27. Inspect the push rods. Replace worn or bent push rods as necessary.
28. Install pushrods. Install rocker arms and pedestal assemblies into their original positions. Ensure push rods are seated properly under each rocker arm. Tighten rocker arm bolts to 35 Nm (26 ft. lbs.).
29. Install exhaust manifold(s) and gasket(s).
30. Install exhaust manifold heat shields.
31. Install cylinder head covers.
32. Position and install the oil level indicator tube. Tighten fastener to 11 Nm (95 inch lbs.).
33. Connect CCV hoses to cylinder head covers.
34. Install the spark plug wires and plug wire retainers.
35. Install intake manifold.
36. Raise and support the vehicle.
37. Connect the exhaust system to the exhaust manifolds.
38. Drain the engine oil.
39. Lower the vehicle.
40. Fill engine oil to the proper level.
41. Fill the cooling system.
42. Connect negative battery cable.
43. Install the air cleaner cover, throttle inlet duct and connect the IAT sensor and CCV hose.
44. Start engine and inspect for leaks.

Rocker Arms/Shafts

REMOVAL & INSTALLATION

➡Before replacing parts, inspect all related valvetrain components for damage to prevent engine misfire.

1. Remove cylinder head cover(s).
2. Remove two rocker pedestal bolts per each cylinder.
3. Remove rocker arm assemblies as a pair.
4. If rocker arm assemblies are disassembled for cleaning or replacement, reassemble rocker arms in their original position.

To install:

➡Before replacing parts, inspect all related valvetrain components for damage to prevent engine misfire.

5. Install rocker arm and pedestal assemblies into position. Ensure push rods are seated properly under each rocker arm. Tighten rocker arm bolts to 35 Nm (26 ft. lbs.).

⁂ WARNING

The rocker arm pedestal bolts should be torqued down slowly. Allow 20 minutes tappet bleed down time after installation of the rocker bolts before engine operation.

6. Install cylinder head cover(s).

Intake Manifold

REMOVAL & INSTALLATION

➡Care must be taken to protect the intake manifold powder coating from scrapes and abrasions.

1. Perform the fuel system pressure release procedure, before attempting any repairs.
2. Disconnect negative battery cable.
3. Disconnect throttle body air inlet duct, IAT sensor, CCV hose and remove air cleaner cover.
4. Disconnect throttle cable from throttle cam. Remove throttle cable from bracket and set aside.

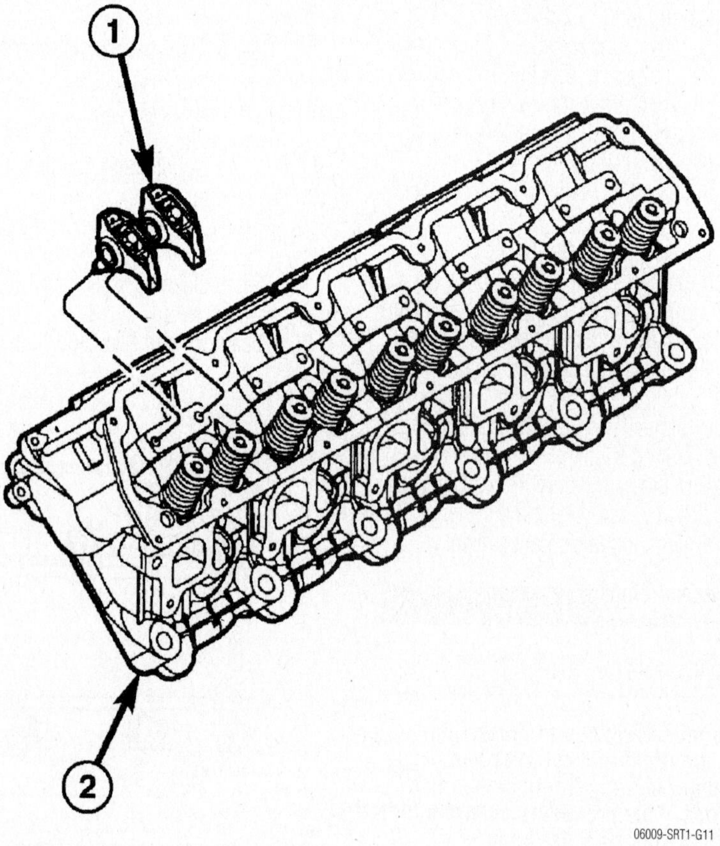

Rocker arms

5. Disconnect speed control cable from throttle cam and set aside.
6. Disconnect brake booster vacuum hose from the intake manifold and set aside.

WARNING

Wrap a shop towel around fuel Line to catch any gasoline spillage.

7. Disconnect fuel line quick connect fitting from fuel rail.
8. Disconnect the following electrical connectors:
    - Left cylinder head fuel injectors
    - MAP sensor
    - Engine Coolant Temperature (ECT) Sensor
    - Idle Speed Control Sensor (ISC)
    - Throttle Position Sensor (TPS)
    - Right cylinder head fuel injectors
9. Carefully separate the fuel injector harness push pins from the intake manifold.
10. Disconnect the purge solenoid hose from beneath the throttle body.
11. Disconnect CCV hoses from left and right cylinder head covers.

CAUTION

Care must be taken when removing the No. nine cylinder inside intake manifold bolt. Failure to do so may result in personal injury or possible damage.

12. Remove the intake manifold bolts and intake manifold.
13. Cover the intake manifold openings with a suitable cover when servicing to prevent any debris from entering cylinder heads.
14. Inspect manifold for cracks or other damage.
15. Inspect manifold sealing surfaces for damage.

To install:

CAUTION

When cleaning the intake manifold and cylinder head mating surfaces, do not use a metal scraper because the surfaces could be cut or ground. Instead, use a wooden or plastic scraper.

➡Care must be taken to protect the intake manifold powder coat from possible scrapes and abrasions.

16. Clean gasket surfaces and remove covering from intake manifold openings.
17. Carefully install the intake manifold gaskets and alignment pins.

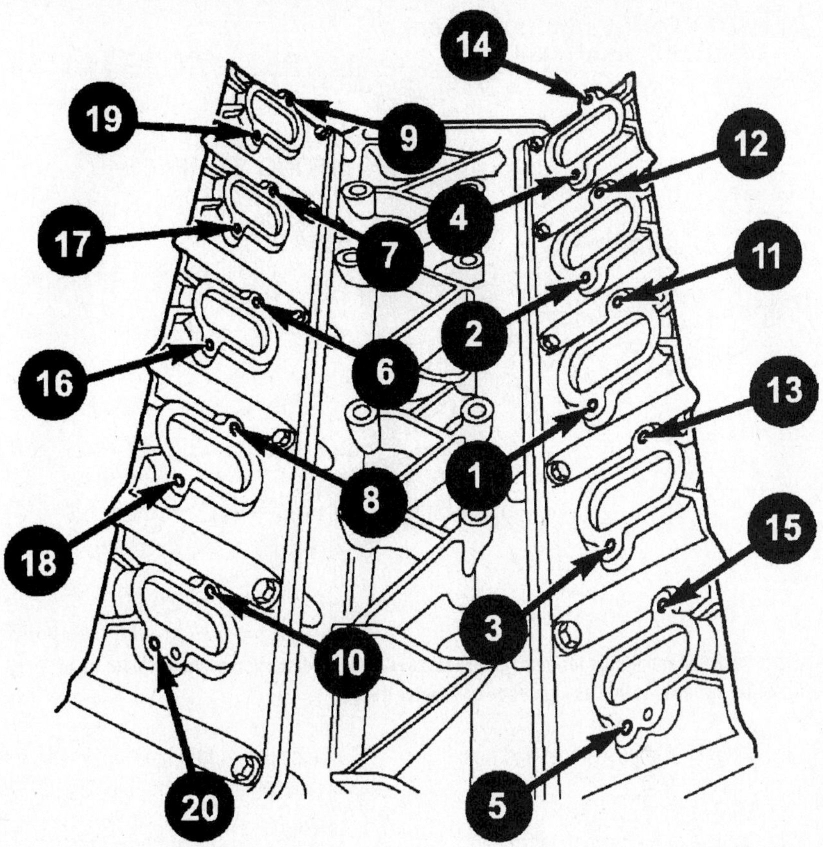

Intake manifold bolt torque sequence

18. Set the intake manifold in position on the engine. Install intake manifold bolts and tighten initially to 1 Nm (10 inch lbs.). Tighten bolts starting at the center and working outward to 11 Nm (95 inch lbs.).
19. Connect CCV hoses to left and right cylinder head covers.
20. Connect fuel line to fuel rail. Pull backward on the quick connect fitting to ensure complete insertion.
21. Connect the brake booster vacuum hose.
22. Connect speed control cable to throttle linkage.
23. Connect the throttle cable housing to the bracket and then attach the cable to the throttle linkage
24. Attach the purge solenoid hose to the underside of the throttle body housing.
25. Reconnect the following electrical connectors:
    - Left cylinder head fuel injectors
    - Engine Coolant Temperature (ECT) Sensor
    - MAP sensor
    - Idle Speed Control
    - Throttle Position Sensor
    - Right cylinder head fuel injectors
26. Install air cleaner cover assembly, connect throttle body air inlet hose, CCV hose and connect the Intake Air Temperature Sensor.
27. Connect negative battery cable.
28. With the Scan Tool, use the ASD Fuel System Test to pressurize fuel system to check for leaks.

CAUTION

When using the ASD Fuel System Test, the Auto Shutdown (ASD) relay will remain energized for 7 minutes or until the ignition switch is turned to the OFF position, or Stop All Test is selected.

Exhaust Manifold

REMOVAL & INSTALLATION

➡Care must be taken to protect the cylinder head covers powder coating from scrapes and abrasions. This procedure covers either the left or right exhaust manifold.

1. Disconnect negative battery cable.
2. When servicing the right exhaust manifold it is necessary to remove the

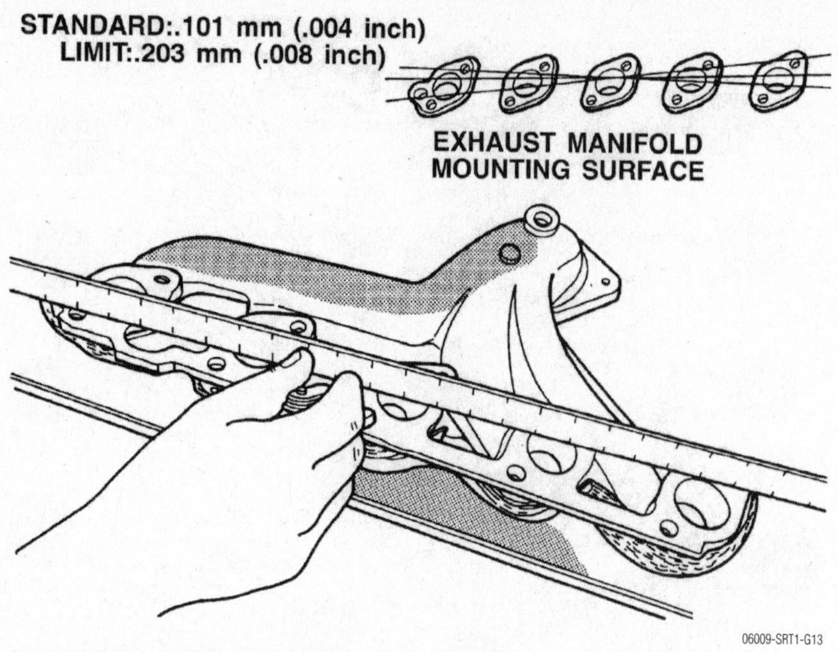

Inspect exhaust manifolds for damage or cracks. Check for distortion of the cylinder head mounting surfaces with a straightedge and feeler gauge

upper oil dip stick tube fastener and position the tube aside.

3. Raise vehicle on hoist.
4. Loosen and remove the fasteners from the exhaust pipe to exhaust manifold.
5. Separate exhaust pipe from the exhaust manifold.
6. Lower the vehicle.
7. Disconnect the spark plug wires from the spark plugs and set aside.
8. Remove manifold heat shield from exhaust manifold.
9. Note the location of exhaust manifold studs and bolts for proper installation.
10. Remove studs and bolts attaching manifold to cylinder head and remove manifold.
11. Inspect exhaust manifolds for damage or cracks. Check for distortion of the cylinder head mounting surfaces with a straightedge and feeler gauge.

To install:

➡Care must be taken to protect the cylinder head covers powder coating from scrapes and abrasions. This procedure covers either the left or right exhaust manifold.

12. Install new gasket and install exhaust manifold Tighten attaching fasteners to 23 Nm (200 inch lbs.).
13. When installing right exhaust manifold, properly position the oil level indicator tube, install fastener and tighten to the cylinder head cover specification.
14. Install manifold heat shield and tighten attaching fasteners to 6 Nm (50 inch lbs.).
15. Install the spark plug wires.
16. Raise vehicle on hoist.
17. Connect exhaust pipe to exhaust manifold. Tighten fasteners to 34 Nm (25 ft. lbs.).
18. Lower vehicle.
19. Connect negative battery cable.

Camshaft and Valve Lifters

REMOVAL & INSTALLATION

1. Perform the fuel pressure relief procedure.
2. Disconnect the negative battery cables.
3. Raise and support the vehicle.
4. Remove the front engine shield.
5. Drain the cooling system.
6. Drain the power steering fluid.
7. Remove the oil pan.
8. Lower the vehicle.
9. Remove air intake tube between intake manifold and air filter assembly.
10. Unclip A/C lines from radiator fan shroud, and position out of the way.
11. Remove coolant recover bottle and coolant container tube.
12. Remove upper radiator hose from radiator and position out of the way.
13. Raise vehicle on hoist.
14. Disconnect the high and low pressure hydraulic hoses at the hydraulic fan motor.
15. Disconnect solenoid electrical connector.
16. Lower vehicle.
17. Remove the four fan shroud retaining bolts.
18. Remove the fan motor and shroud assembly from the vehicle.
19. Remove the power steering cooler mounting bolts and pushpins. Position the power steering cooler out of the way.
20. Lower vehicle.
21. Remove the two radiator upper mounting bolts.
22. Lift radiator straight up and out of engine compartment.

➡The bottom of the radiator is equipped with two alignment dowels that fit into holes in the lower radiator support panel. Rubber biscuits (insulators) are installed to these dowels. Take care not to damage cooling fins or tubes on the radiator and air conditioning condenser when removing.

23. Remove the radiator and radiator fan assembly.
24. Remove intake manifold.
25. Remove cylinder head covers.
26. Remove timing chain cover.
27. Remove rocker arm assemblies.

➡Identify each push rod to be replaced in its original location.

28. Remove pushrods.
29. Remove cylinder heads.
30. Remove hydraulic tappets.
31. Remove timing chain and camshaft sprocket.
32. Remove camshaft thrust plate.
33. Slowly remove camshaft. Use care not to damage camshaft bearing surfaces with the cam lobes.
34. Remove cylinder head.

➡If tappets (3) are to be reused, identify tappet location to ensure installation in original location.

35. Remove tappet aligning yoke bolt (1).
36. Remove tappets (3) from bores.

⁂ WARNING

Do not disassemble tappets.

To install:

37. Lubricate tappets.
38. Install tappets in their original positions with oil holes facing up.

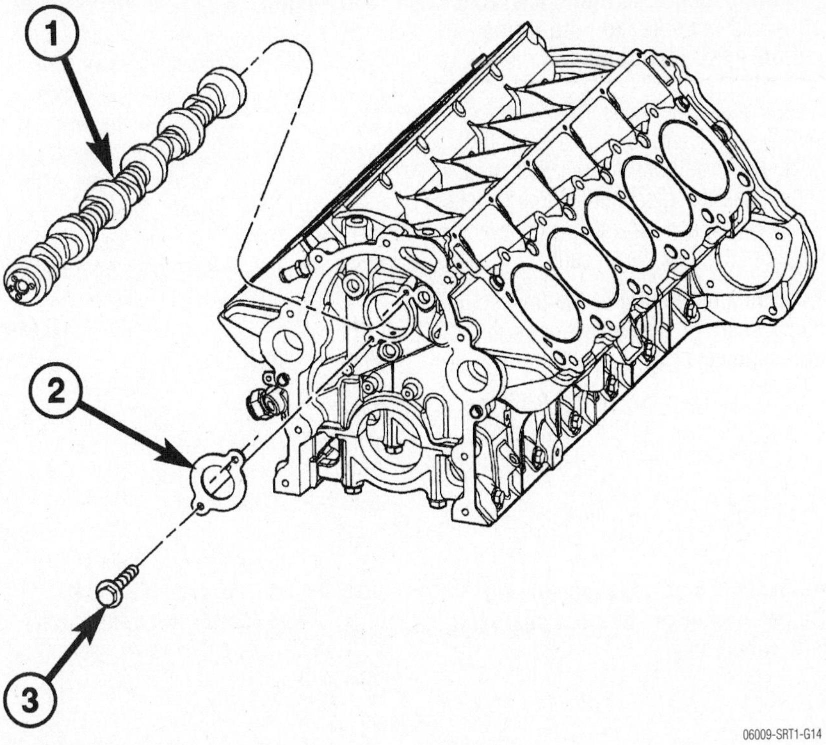

Camshaft (1) removal along with thrust plate (2) and bolt (3)

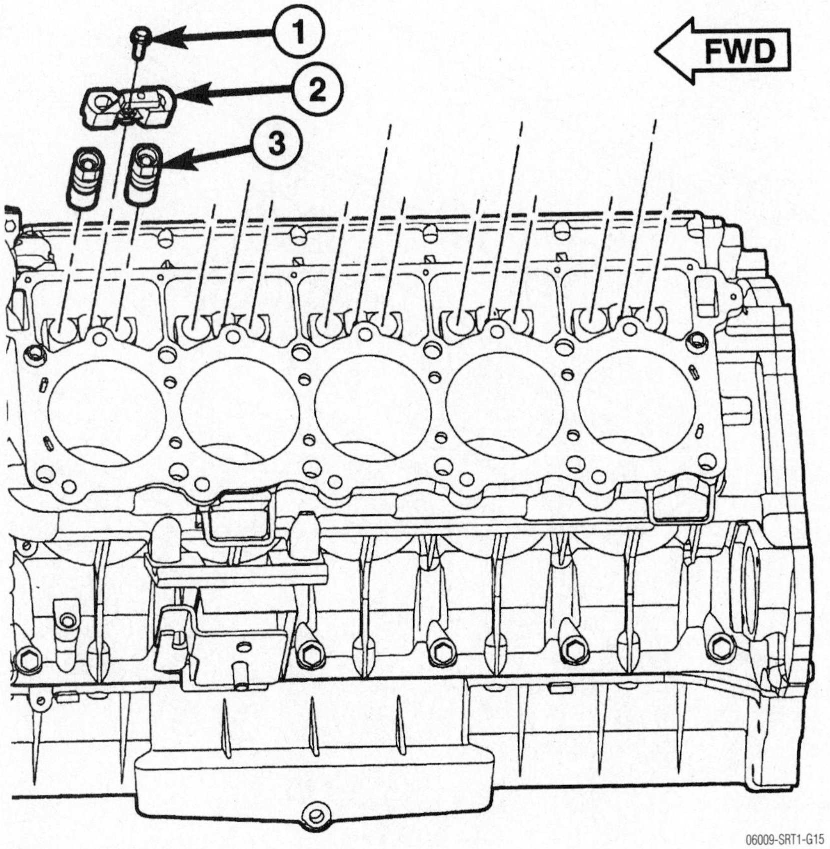

Hydraulic tappets—(1) is the oil holes

⁂ WARNING

The "handle" on the tappet aligning yoke must be installed in the outboard position; i.e. toward the cylinders.

39. With roller tappets in proper position, install aligning yoke assemblies. Apply Mopar® Lock & Seal Adhesive (Medium Strength Threadlocker) to the screws and tighten to 11 Nm (95 inch lbs.).
40. Install cylinder heads.
41. Lubricate camshaft lobes and camshaft bearing journals.
42. Slowly install camshaft. Use care not to damage camshaft bearing surfaces with the cam lobes.
43. Install camshaft thrust plate bolts. Tighten to 24 Nm (210 inch lbs.).
44. Install timing chain and camshaft sprocket.
45. Measure camshaft end play. End Play should measure 0.0127–0.381 mm (0.005–0.015 in.) 0.381 mm (0.015 in. Max.). If not within limits, install a new thrust plate.

➡When camshaft is replaced, all of the tappets must be replaced. Each hydraulic tappet must be installed with oil hole facing up.

46. Install timing chain cover.
47. Install push rods in their original location.
48. Install rocker arms.
49. Install intake manifold.
50. Install cylinder head covers.
51. Install the radiator and fan assembly.
52. Raise and support the vehicle.
53. Install the oil pan.
54. Install the front engine shield.
55. Lower the vehicle.
56. Fill and purge air from the power steering.
57. Fill and purge air from the cooling system.
58. Fill engine crankcase with the proper oil to the correct level.
59. Connect the negative battery cable.
60. Start the vehicle and inspect for leaks.

⁂ WARNING

To prevent damage to valve mechanism, engine must not be run above fast idle until all hydraulic tappets have filled with oil and have become quiet.

Oil Pan

REMOVAL & INSTALLATION

1. Disconnect the negative battery cable.
2. Disconnect engine oil indicator tube fastener from the right cylinder head cover.
3. Install engine lifting fixture tool 9363 and support tool.
4. Raise vehicle on hoist.
5. Remove the lower engine shield.
6. Drain engine oil.
7. Remove the crossmember bolts.

WARNING

Do not use any flame or plasma cutting equipment to cut the frame in this procedure. This is due to the inaccurate nature of the cut-line and the fact that the high temperatures achieved during flame or plasma cutting will change the metal characteristics and may weaken the frame and/or repair location.

8. Remove the welds using a grinder or equivalent and remove the crossmember.
9. Remove flywheel inspection cover.
10. Loosen engine mount through bolts.
11. Wiggle and separate the oil level indicator tube from the oil pan.

➡Careful positioning of the jack stand is necessary to prevent damage to other components.

12. Using lifting fixture tool 9363 and support tool, raise the front of the engine.
13. Remove transmission to oil pan fasteners.
14. Remove oil pan bolts and remove oil pan.

➡Clean and inspect oil pan gasket. The pan gasket can be used again if in good condition.

To install:

15. Clean gasket surfaces. Inspect oil pan gasket. Replace as necessary.
16. Apply a ⅛ inch bead of Mopar Engine RTV at the two joints between the cylinder block to timing chain case cover and the two joints between the rear main seal retainer and engine block.
17. Position oil pan and gasket to engine block, seat the oil level indicator tube in the oil pan and hand start all fasteners.
18. Tighten the transmission to oil pan bolts in finger tight.
19. Torque 5⁄16-18 oil pan bolts to 23 Nm (200 inch lbs.). Torque ¼-20 oil pan bolts to 11 Nm (95 inch lbs.).
20. Torque transmission to oil pan bolts to 23 Nm (200 inch lbs.).
21. Lower the engine.
22. Install and tighten engine mount through bolts to 102 Nm (75 ft. lbs.).
23. Install flywheel inspection cover.
24. Install the crossmember bolts. Tighten the bolts to 61 Nm (45 ft. lbs.).

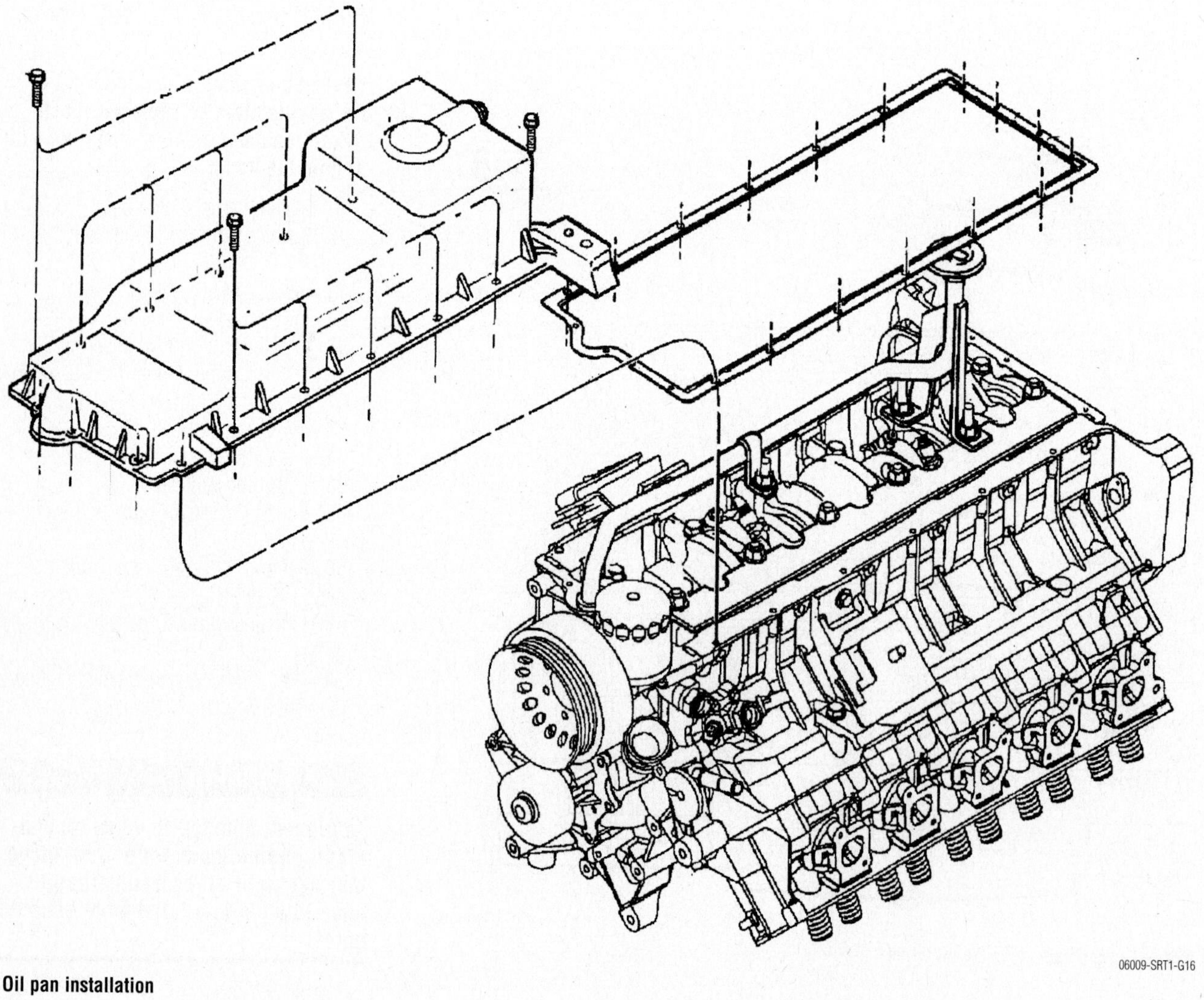

Oil pan installation

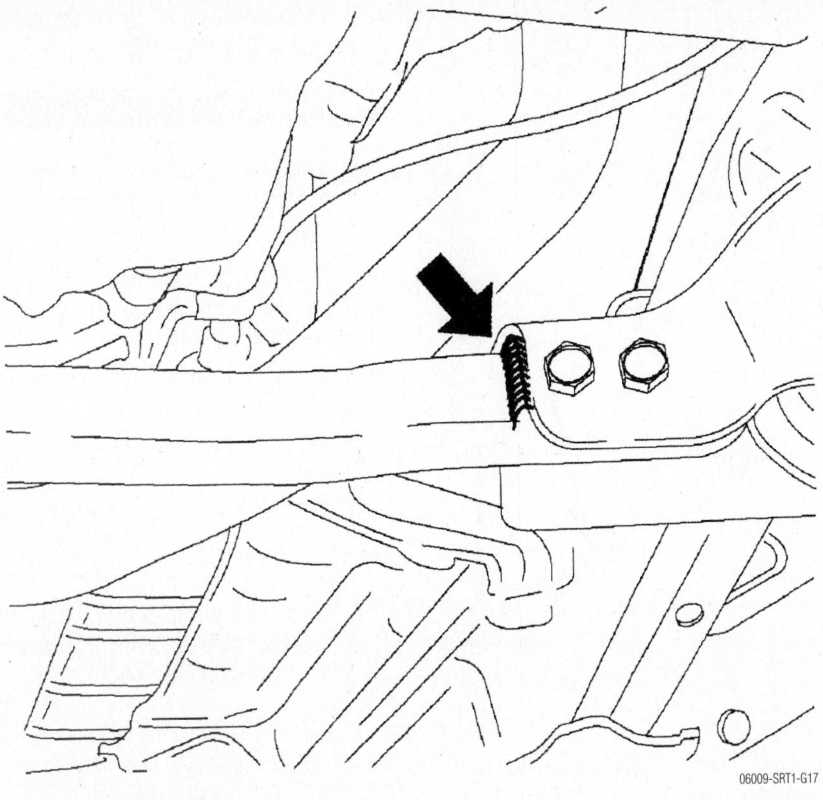

Front crossmember weld

25. Remove all OEM e-coat within 51 mm (2.0 in.) of the weld on both the crossmember and the frame.

WARNING

Shield the surrounding area and components from exposure to the welding spatter and heat.

➡Gas Metal Arc Welding Process: Butt joints; apply two layers (passes) of weld metal. First pass should only fill approximately ½ the thickness. Vertical position welds, maintain electrode wire at leading edge of weld puddle while traveling down hill to produce maximum penetration into the sleeve. These techniques work for FCAW as well.

26. Apply new welds to the crossmember.

➡Any burned surface coatings will need to be removed prior to application of corrosion preventive coatings.

27. Dress the welded area and apply corrosion resistant coatings.
   a. Apply etch-primer to the repair area.
   b. Apply a durable top coat to the outside of the repair area.
28. Lower vehicle.
29. Install the oil level indicator tube fastener.
30. Fill the crankcase with the correct oil to proper level.
31. Start the engine and inspect for leaks.

Oil Pump

REMOVAL

➡The oil pump is serviced with the timing chain cover.

1. Remove timing chain cover.
2. Disassemble oil pump.
3. Inspect oil pump parts.

DISASSEMBLY

1. Remove oil pump cover screws and lift off cover.
2. Remove pump rotors.
3. Remove oil pressure relief valve and spring.
4. Wash all parts in a suitable solvent and inspect carefully for damage or wear.

CLEANING

Clean all parts thoroughly. Mating surface of the chain case cover should be smooth. Replace pump cover if scratched or grooved.

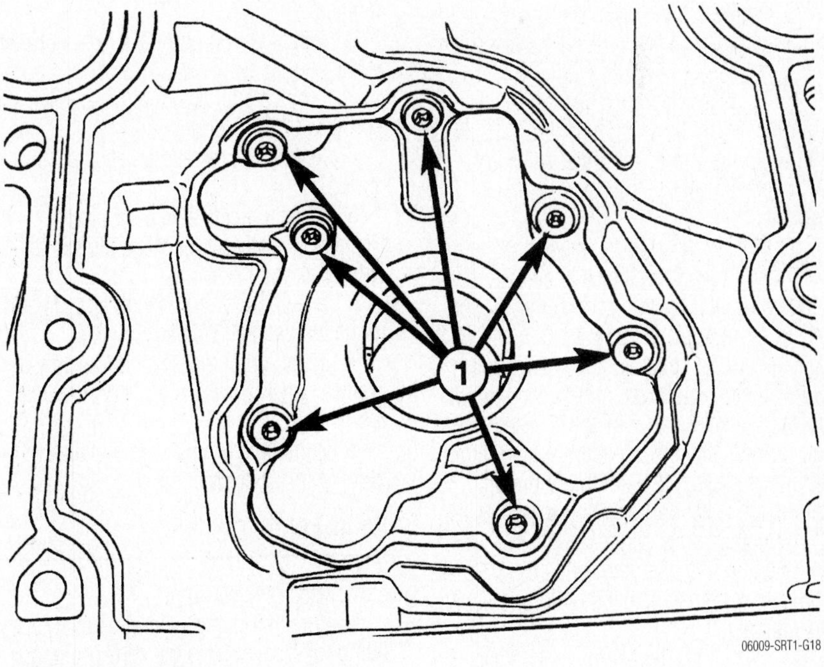

Oil pump cover screws

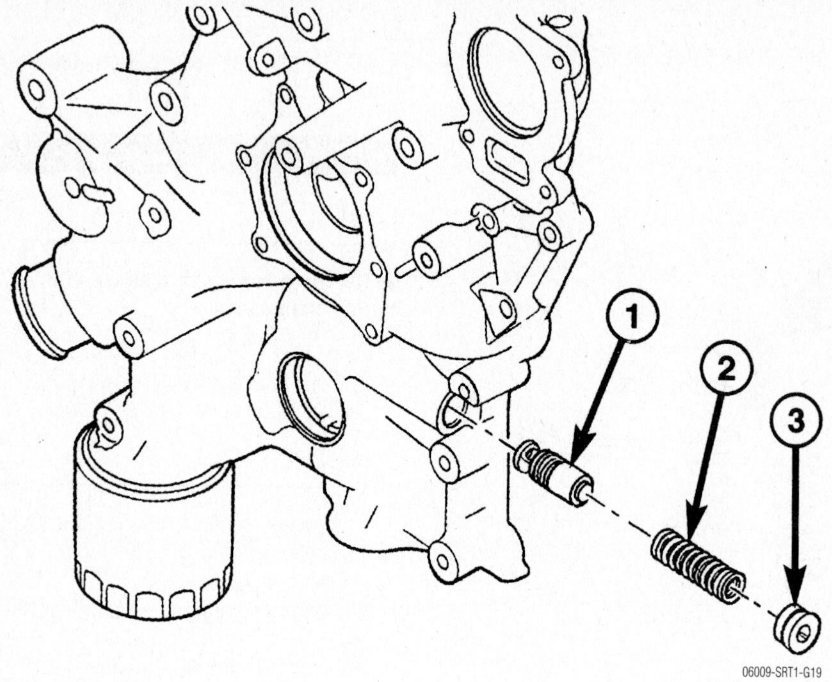

Relief valve and spring

INSPECTION

1. Lay a straightedge across the pump cover surface. If a 0.076 mm (0.003 in.) feeler gauge can be inserted between cover and straight edge, cover should be replaced.
2. Measure thickness and diameter of rotors. If the outer rotor diameter is 82.45 mm (3.246 in.) or less, replace rotor set and.
3. If either rotor thickness measures 18.92 mm (0.744 in.) or less, replace rotor set.
4. Measure clearance between outer rotor and case. If measurement is 0.19 mm (0.007 in.) or more, and outer rotor is within specification, replace chain case cover.
5. Measure clearance between the outer rotor and inner rotor. If clearance between inner and outer rotors is 0.150 mm (0.006 in.) or more, replace rotor set.
6. Place a straightedge across the face of the chain case cover, in between the bolt holes. If a feeler gauge of 0.077 mm (0.003 in.) or more can be inserted between rotors and the straightedge, and the rotors are within specification, replace the case cover.
7. Inspect oil pressure relief valve for scoring and free operation in its bore. Small marks may be removed with 400 grit wet or dry sandpaper.
8. The relief valve spring has a free length of approximately 68.7 mm (2.70 in.) and should test between 10.2 kg and 12.4 kg (22.5 and 27.5 lbs.) when compressed to 57 mm (2.24 in.). Replace spring that fails to meet specifications.
9. If oil pressure is low and pump is within specifications, inspect for worn engine bearings or other reasons for oil pressure loss.

ASSEMBLY

1. Assemble pump using new parts as required.
2. Install spring onto oil pressure relief valve.
3. Lubricate oil pressure relief valve and spring with clean engine oil.
4. Install oil pressure relief valve and spring.
5. Inspect relief valve plug o-ring. Replace as necessary. Torque plug to 34 Nm (25 ft. lbs.)
6. Install pump rotors with chamfer side to the inside of chain case cover.
7. Install oil pump cover and screws. Tighten cover screws to 23 Nm (200 inch lbs.).
8. Prime oil pump before installation by filling rotor cavity with engine oil.

INSTALLATION

1. Assemble oil pump.
2. Prime oil pump before installation by filling rotor cavity with the correct engine oil.
3. Install timing chain cover.
4. Install NEW oil filter.
5. Fill the crankcase with the correct engine oil to the proper level.

Rear Main Seal

REMOVAL & INSTALLATION

1. Remove the transmission.
2. Remove the clutch and flywheel.

➡The oil seal and retainer are serviced as an assembly.

3. Remove the seal retainer bolts from the cylinder block and oil pan.
4. Remove the oil seal and retainer assembly.

To install:

5. Clean the engine block and retainer. Make sure the surfaces are clean and free of oil.
6. Install Special Tool 9060 (seal guide) onto the end of crankshaft. Align the hole in the tool with the dowel pin on the crankshaft.
7. Carefully position the retainer/oil seal on the block. Tighten the retainer bolts to 11 Nm (95 inch lbs.).
8. Install the flywheel and clutch.
9. Install the transmission.

Flywheel

REMOVAL & INSTALLATION

1. Remove transmission clutch housing and clutch assembly.
2. Remove flywheel bolts and remove flywheel.

To install:

3. Clean flywheel bolts and apply thread sealer.
4. Install flywheel on engine and install flywheel bolts.
5. Tighten flywheel bolts to 74 Nm (55 ft. lbs.).
6. Install clutch assembly, clutch housing and transmission.

Crankshaft Front Seal

REMOVAL & INSTALLATION

1. Disconnect the air inlet duct, IAT sensor, CCV hose at the throttle body, and remove the air cleaner housing cover.
2. Remove accessory drive belt.
3. Remove crankshaft damper
4. Use Special Tool C-4679A to remove crankshaft oil seal. Be careful not to damage the crankshaft seal surface of cover.

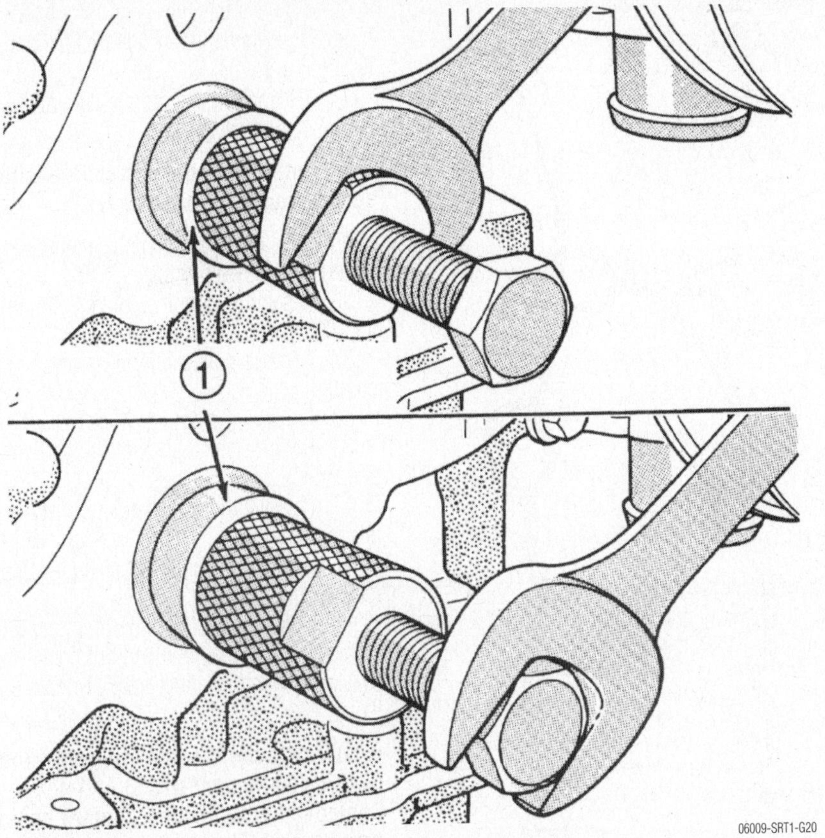

Front seal removal using the special tool (1)

To install:

5. Install the new seal by using Special Tool MD 998306.
6. Place seal into opening with seal spring towards the inside of engine. Install seal until it bottoms out in timing chain cover.
7. Install crankshaft damper.
8. Install accessory drive belt.
9. Install air cleaner cover connecting the CVV hose, IAT sensor, and air duct to the throttle body housing.

Timing Chain, Sprockets, and Front Cover

REMOVAL & INSTALLATION

1. Disconnect negative battery cable.
2. Disconnect throttle body air inlet duct at the throttle body, CCV hose, disconnect the IAT sensor and remove air cleaner housing cover.
3. Remove the oil indicator tube fastener at the right cylinder head cover.
4. Raise and support the vehicle.
5. Remove the lower engine shield.
6. Drain cooling system.
7. Drain engine oil and remove oil filter.
8. Drain power steering.
9. Disconnect the cooling fan return hose at radiator fan.
10. Disconnect the radiator fan pressure hose to power steering rack.
11. Disconnect the power steering hose to power steering pump.
12. Separate the A/C hose from the radiator fan module.
13. Remove the lower radiator hoses.
14. Remove the front frame cross member.
15. Remove the flywheel inspection cover.

➡Capture and properly dispose of any fuel spillage.

16. Disconnect oil cooler line at the timing cover using Special Tool 9005.
17. Separate the oil cooler coolant hose from the timing cover.
18. Wiggle oil indicator tube free of oil pan.
19. Loosen engine mounts.

➡Care must be taken not to damage the A/C compressor when placing the jack stand and raising the engine.

20. Raise and support the engine with a jack stand at the A/C bracket.
21. Remove the engine oil pan.
22. Lower the engine and remove the jackstand.
23. Remove the oil pump pick up tube.
24. Remove accessory drive belt.
25. Remove A/C compressor mounting bolts. Reposition A/C compressor. Do not disconnect A/C lines.
26. Disconnect the power steering line support bracket at the radiator.
27. Lower the vehicle.
28. Remove the upper radiator hose and coolant recovery container.
29. Remove the cooling fan module.
30. Remove the upper idler pulley.
31. Remove power steering pump mounting bolts. Reposition power steering pump aside. Do not disconnect power steering lines.

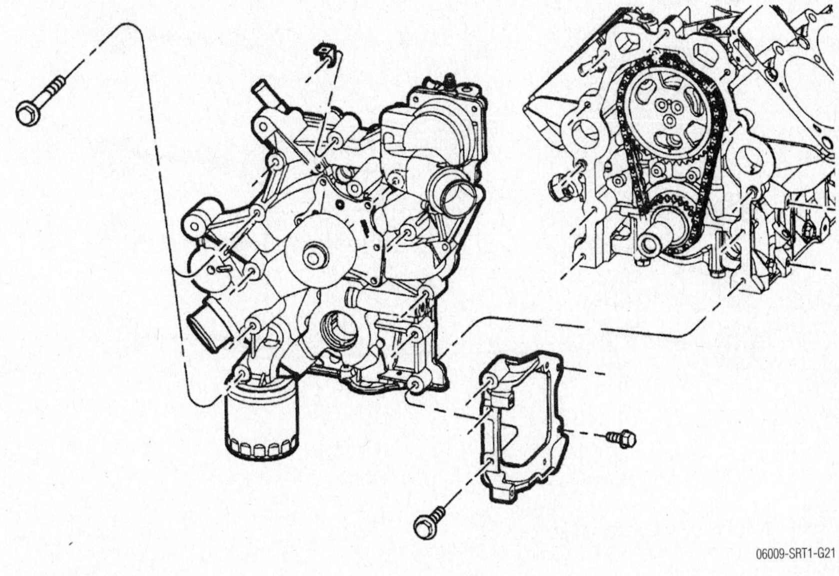

Front cover removal

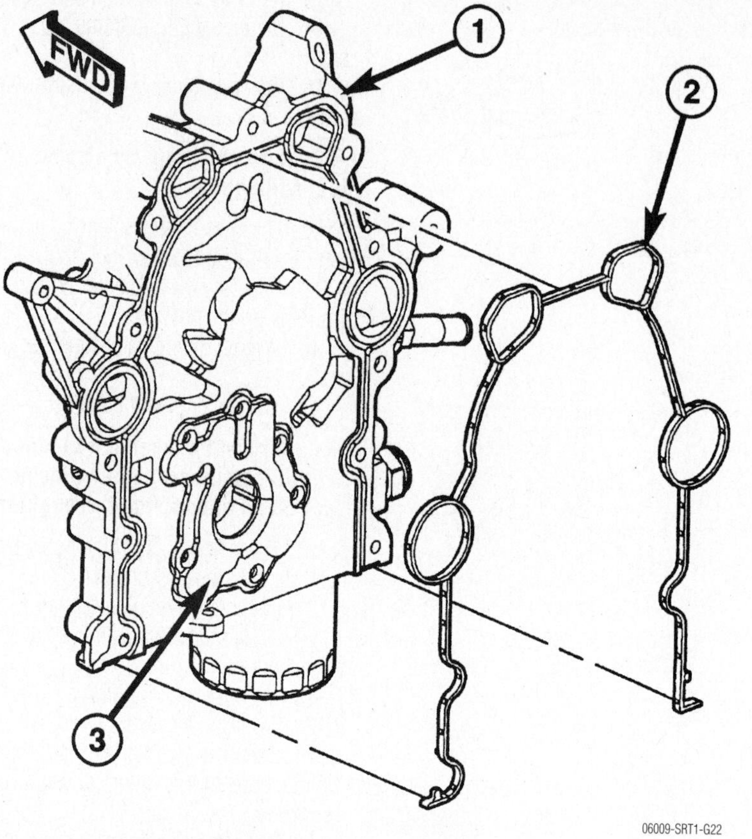

Front cover (1) installation, (2) is the gasket, (3) is the oil pump

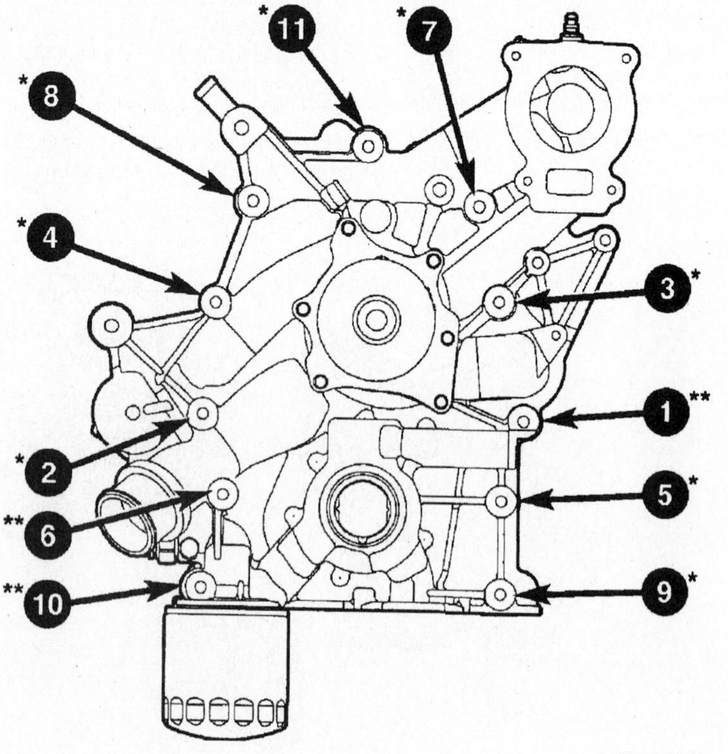

Front cover fastener torque sequence

32. Remove the alternator.
33. Disconnect the heater hose at the timing cover.
34. Disconnect the ground wire at the thermostat housing.

➡Do Not remove the camshaft position sensor from the timing cover.

35. Disconnect the camshaft sensor wiring harness connector.
36. Disconnect engine coolant temperature (ECT) sensor connector.
37. Remove crankshaft pulley and damper.
38. Remove the timing cover.
39. Rotate crankshaft until timing marks are aligned.
40. Remove the camshaft sprocket attaching bolts.
41. Remove the timing chain with camshaft sprocket.
42. Install Special Tool 8194 into end of crankshaft. Lubricate the end of Special Tool 8194 with wheel bearing grease or equivalent.

➡Lubricate the threads of Special Tool 5048 using Mopar® Nickel Anti-seize Compound or equivalent, before beginning crankshaft sprocket removal.

43. Using Special Tools 9056 and 5048, remove crankshaft sprocket.

To install:

44. Position a new crankshaft sprocket on the crankshaft with timing mark facing out. Align crankshaft keyway with slot in crankshaft sprocket.

➡Lubricate the threads of Special Tool 9055 using Mopar® Nickel Anti-seize Compound or equivalent, before beginning crankshaft sprocket installation.

45. Install the crankshaft sprocket using Special Tool 9055.
46. Rotate the crankshaft sprocket so the timing mark is at the 12 o'clock position.
47. Place the timing chain around the camshaft sprocket and place the timing mark to the 6 o'clock position.
48. Place the timing chain around crankshaft sprocket and install the camshaft sprocket into position.
49. Using straight edge, check alignment of timing marks.
50. Install the camshaft sprocket bolts. Tighten the bolts to 23 Nm (200 inch lbs.).
51. Rotate crankshaft 2 revolutions. Timing marks should line up. If timing marks do not line up, remove cam sprocket and realign.

52. Measure camshaft end play. If not within specifications, install new thrust plate.

⁂ WARNING

Do not use a metal scraper to clean aluminum sealing surfaces, as damage to surfaces may occur. Use only a plastic scraper and Mopar® Brake Parts Cleaner or the equivalent.

53. Be sure mating surfaces of chain case cover and cylinder block are clean and free from burrs. Crankshaft oil seal must be removed to insure correct oil pump engagement.
54. Install the timing chain cover gasket.
55. Position the oil pump inner rotor mating flats in the same position as the crankshaft drive flats.
56. Install the timing chain cover onto crankshaft. Ensure the oil pump is engaged on the crankshaft correctly or severe damage may result.
57. Hand start bolts No.1 and No.2.
58. Loosely install A/C compressor mounting bracket.
59. Hand start remaining timing chain cover bolts.
60. Torque timing chain cover bolts in the sequence shown in to 23 Nm (200 inch lbs.).
61. Torque A/C compressor mounting bracket bolts to 23 Nm (200 inch lbs.).
62. Install crankshaft oil seal.
63. Raise vehicle on hoist.
64. Install oil pump pick up tube using a new o-ring. Torque 1⁄4-20 bolt to 11 Nm (95 inch lbs.). Torque 5⁄16-18 nuts to 23 Nm (200 inch lbs.).
65. Support the engine with a jackstand.
66. Install oil pan and drain plug.
67. Remove the jackstand.
68. Tighten the engine mount to frame nuts to 102 Nm (75 ft. lbs.).
69. Connect the oil cooler coolant hose to the timing cover.
70. Connect the oil cooler line.
71. Install the lower radiator hose.
72. Install the flywheel inspection cover.
73. Install the front frame cross member.
74. Lower vehicle.
75. Position the compressor and install A/C compressor mounting bolts. Torque bolts to 23 Nm (200 inch lbs.).
76. Connect heater hose at timing cover.
77. Install crankshaft damper and pulley.
78. Install alternator. Connect electrical connectors.
79. Install the idler pulley.
80. Connect the camshaft position sensor wiring harness connector.
81. Connect engine coolant temperature (ECT) sensor wiring harness connector.
82. Install power steering pump mounting bolts.
83. Install accessory drive belt.
84. Install the radiator fan assembly.
85. Raise and support the vehicle.
86. Connect the hydraulic lines to the fan assembly and steering rack then connect the A/C hose to the fan shroud.
87. Install a new oil filter.
88. Lower the vehicle.
89. Install upper radiator hose and connect the ground wire at the thermostat housing.
90. Install air cleaner housing cover assembly, connect throttle body air inlet duct, IAT sensor and CCV hose.
91. Fill and purge air from the power steering system.
92. Fill engine crankcase with proper oil to correct level.
93. Fill and purge air from the cooling system.
94. Connect negative battery cable.
95. Start the vehicle and inspect for leaks.
96. Install the lower engine shield.

Piston and Ring

POSITIONING

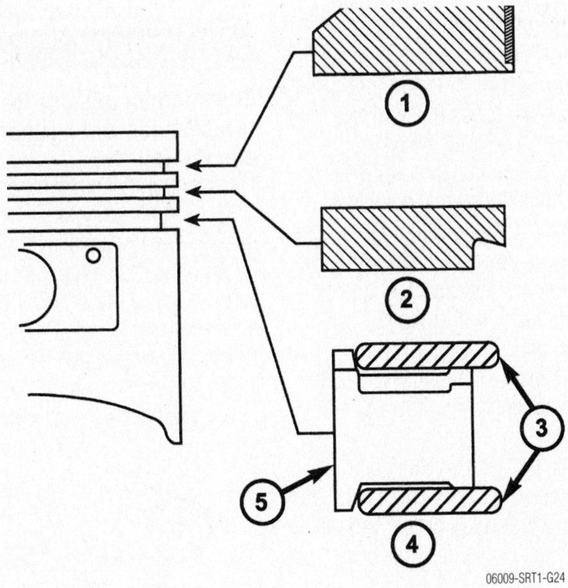

Piston ring installation

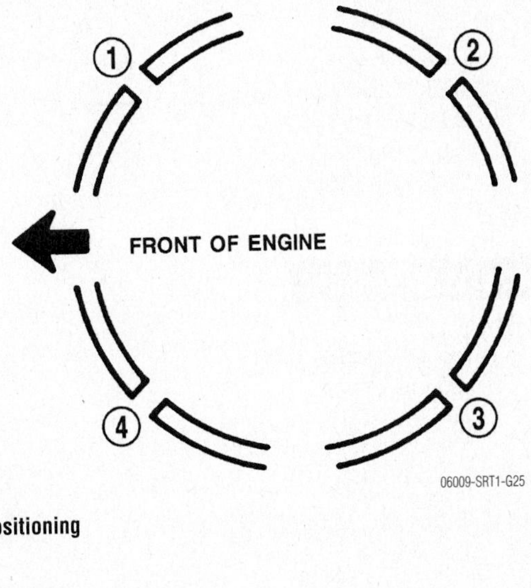

Piston ring gap positioning

FUEL SYSTEM

Fuel System Service Precautions

Safety is the most important factor when performing not only fuel system maintenance but any type of maintenance. Failure to conduct maintenance and repairs in a safe manner may result in serious personal injury or death. Maintenance and testing of the vehicle's fuel system components can be accomplished safely and effectively by adhering to the following rules and guidelines.

- To avoid the possibility of fire and personal injury, always disconnect the negative battery cable unless the repair or test procedure requires that battery voltage be applied.
- Always relieve the fuel system pressure prior to detaching any fuel system component (injector, fuel rail, pressure regulator, etc.), fitting or fuel line connection. Exercise extreme caution whenever relieving fuel system pressure to avoid exposing skin, face and eyes to fuel spray. Please be advised that fuel under pressure may penetrate the skin or any part of the body that it contacts.
- Always place a shop towel or cloth around the fitting or connection prior to loosening to absorb any excess fuel due to spillage. Ensure that all fuel spillage (should it occur) is quickly removed from engine surfaces. Ensure that all fuel soaked cloths or towels are deposited into a suitable waste container.
- Always keep a dry chemical (Class B) fire extinguisher near the work area.
- Do not allow fuel spray or fuel vapors to come into contact with a spark or open flame.
- Always use a back-up wrench when loosening and tightening fuel line connection fittings. This will prevent unnecessary stress and torsion to fuel line piping.
- Always replace worn fuel fitting O-rings with new. Do not substitute fuel hose or equivalent, where fuel pipe is installed.

Before servicing the vehicle, also make sure to refer to the precautions in the beginning of this section as well.

Relieving Fuel System Pressure

1. Remove fuel fill cap.
2. Remove fuel pump relay from Power Distribution Center (PDC). For location of relay, refer to label on underside of PDC cover.
3. Start and run engine until it stalls.
4. Attempt restarting engine until it will no longer run.
5. Turn ignition key to OFF position.

** CAUTION

Steps 1, 2, 3 and 4 must be performed to relieve high pressure fuel from within fuel rail. Do not attempt to use following steps to relieve this pressure as excessive fuel will be forced into a cylinder chamber.

6. Unplug connector from any fuel injector.
7. Attach one end of a jumper wire with alligator clips (18 gauge or smaller) to either injector terminal.
8. Connect other end of jumper wire to positive side of battery.
9. Connect one end of a second jumper wire to remaining injector terminal.

** CAUTION

Powering an injector for more than a few seconds will permanently damage the injector.

10. Momentarily touch other end of jumper wire to negative terminal of battery for no more than a few seconds.
11. Place a rag or towel below fuel line quick-connect fitting at fuel rail.
12. Disconnect quick-connect fitting at fuel rail. Refer to Quick-Connect Fittings.
13. Return fuel pump relay to PDC.
14. One or more Diagnostic Trouble Codes (DTCs) may have been stored in PCM memory due to fuel pump relay removal. The DRBT scan tool must be used to erase a DTC.

Fuel Filter

REMOVAL & INSTALLATION

1. Disconnect and separate fuel vent line from fuel fill bezel.
2. Disconnect electrical connector at top of fuel pump module. To disconnect electrical connector:
   a. Push upward on red colored tab to unlock.
   b. Push on black colored tab while removing connector.
3. Disconnect necessary emission vent lines from leak pump and EVAP canister.
4. Disconnect fuel supply line from fuel pump module.
5. Disconnect necessary emission vent lines from check valves and/or at top of tank.
6. Loosen clamps and disconnect rubber fuel hoses at tank fittings.
7. Support tank with a hydraulic jack.
8. Remove two fuel tank strap nuts and remove both tank support straps.
9. Continue to lower tank for removal.

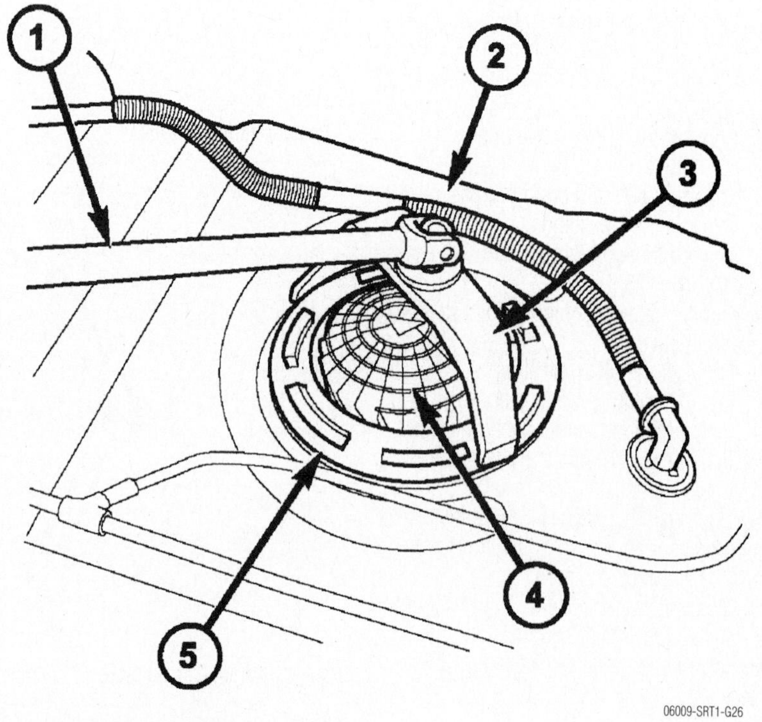

Removing the module lockring

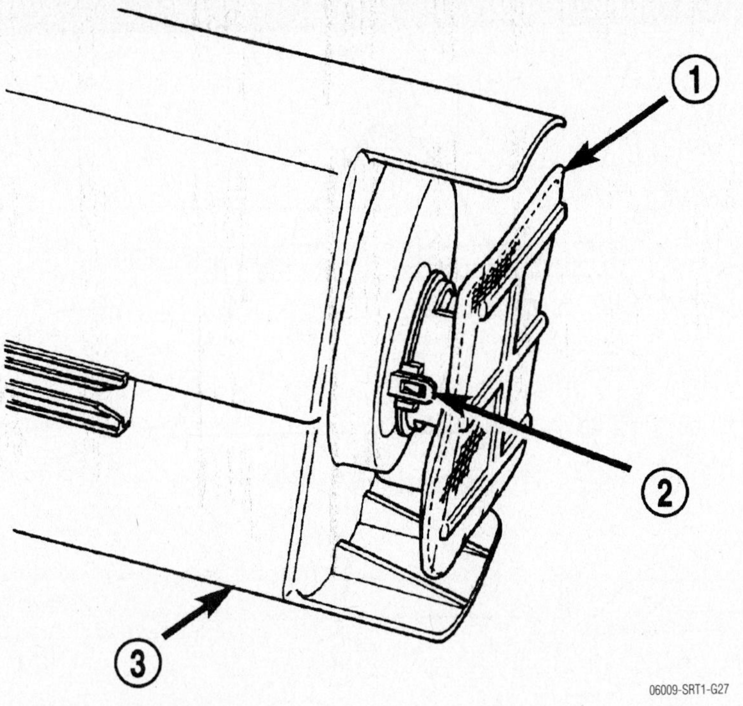

Fuel filter, (2) Lock tabs, (3) Module

10. Note rotational position of module before attempting removal. An indexing arrow is located on top of module for this purpose.

11. Position Special Tool 9340 into notches on outside edge of lockring.

12. Install ½ inch drive breaker bar to tool 9340.

13. Rotate breaker bar counter-clockwise to remove lockring.

14. Remove lockring. The module will spring up slightly when lockring is removed.

15. Remove module from fuel tank. Be careful not to bend float arm while removing.

16. Remove filter by carefully prying 2 lock tabs at bottom of module with 2 screwdrivers. Filter is snapped to module.

17. Clean bottom of pump module.

To install:

18. Snap new filter to bottom of module. Be sure o-ring is in correct position.

19. Using a new seal (gasket), position fuel pump module into opening in fuel tank.

20. Position lockring over top of fuel pump module.

21. Rotate module until embossed alignment arrow points to center alignment mark. This step must be performed to prevent float from contacting side of fuel tank. Also be sure fuel fitting on top of pump module is pointed to drivers side of vehicle.

22. Install Special Tool 9340 to lockring.

23. Install ½ inch drive breaker into Special Tool 9340.

24. Tighten lockring (clockwise) until all seven notches have engaged.

25. Disconnect clamps and remove rubber fuel fill hose and fuel vent hose at fuel fill tube. Install these two hoses to two fuel tank fittings. Rotate hoses until paint marks on hoses line up with alignment marks. Tighten both clamps.

26. Position fuel tank to hydraulic jack.

27. Raise tank until positioned near body.

28. Continue raising tank until positioned snug to body.

29. Install and position both tank support straps. Install two fuel tank strap nuts and tighten. Torque to 30 ft. lbs. (41 Nm).

➡Tighten rear strap nut first.

30. Connect rubber fill and vent hoses to fuel fill tube and tighten clamps.

31. Connect electrical connector to top of fuel pump module.

32. Connect necessary emission vent lines to leak pump and EVAP canister.

33. Connect fuel supply line to fuel pump module.

34. Connect necessary emission vent lines from check valves and/or at top of tank.

35. Connect fuel vent line to fuel fill bezel.

36. Fill fuel tank with fuel.

37. Start engine and check for fuel leaks near top of module.

Fuel Pump

REMOVAL & INSTALLATION

See Fuel Filter, above.

Fuel Injectors

REMOVAL & INSTALLATION

1. Release fuel system pressure.

2. Remove the battery cover and disconnect negative battery cable.

3. Remove the air cleaner assembly.

4. Disconnect the electrical connector to the MAP sensor and Coolant Temperature sensor.

5. Disconnect the electrical connector to the TPS and Idle Air Control.

6. Remove the wiring harness from the wiring clips under the throttle body.

7. Disconnect the electrical connector from the fuel injectors.

8. Disconnect the fuel line quick connector.

9. Remove the bolts for the fuel rail.

10. Pull fuel rail and injectors straight up and out of the intake manifold.

11. Move the fuel rail forward and out from under the intake manifold and throttle body.

12. Remove the fuel injector from the fuel rail.

To install:

13. Install the fuel injectors to the fuel rail.

14. Install fuel rail under throttle body.

15. Apply a light coating of clean engine oil to the O-ring on the nozzle end of each injector.

16. Insert fuel injector nozzles into openings in intake manifold. Seat the injectors in place. Tighten fuel rail bolts to 12 Nm (105 inch lbs.).

17. Attach electrical connectors to fuel injectors.

18. Connect the electrical connector to the MAP sensor and Coolant Temperature sensor.

19. Connect the electrical connector to the TPS and Idle Air Control.

20. Install the wiring harness to the wiring clips under the throttle body.

21. Connect fuel supply tube to fuel rail.

22. Install the negative battery cable and install the battery cover.

23. Install the air cleaner assembly.

24. Use the DRBIIIT scan tool ASD Fuel System Test to pressurize the fuel system. Check for leaks.

DRIVE TRAIN

Manual Transmission

REMOVAL & INSTALLATION

1. With vehicle in neutral, position it on a hoist.
2. Mark installation reference marks on the propeller shaft, pinion yoke and transmission output shaft.
3. Remove propeller shaft clamp bolts and remove shaft.
4. Remove shift cover bolts.
5. Remove hydraulic clutch line quick connector at the transmission with Clutch Line Disconnect Tool 6638 or 6638A to disconnect.
6. Remove reverse lockout solenoid connector.
7. Remove fuel line clip from top plate stud.
8. Remove back up lamp switch connector.
9. Position transmission jack and secure to transmission.
10. Remove transmission to clutch housing bolts.
11. Remove transmission mount bolts.
12. Remove transmission crossmember bolts.
13. Slide transmission back to clear the bell housing. Then lower transmission.

To install:

14. Slide transmission back into the clutch and seat against clutch housing.
15. Install transmission bolts to clutch housing bolts and tighten to and tighten bolts to 35 Nm (26 ft. lbs.)
16. Install transmission crossmember and bolts/nuts. Tighten nuts to 61 Nm (45 ft. lbs.).
17. Install transmission mount bolts and tighten to 41 Nm (30 ft. lbs.).
18. Remove transmission jack.
19. Install back up lamp switch connector.
20. Install fuel line to top plate stud.
21. Install reverse lockout solenoid harness connector.
22. Install hydraulic clutch quick connector to the slave cylinder.
23. Clean transmission shifter tower sealing surfaces and apply a bead of Mopar® Gasket Maker.
24. Install shifter tower and tighten bolts to 18 Nm (156 inch lbs.).
25. Install propeller shaft with reference marks aligned.

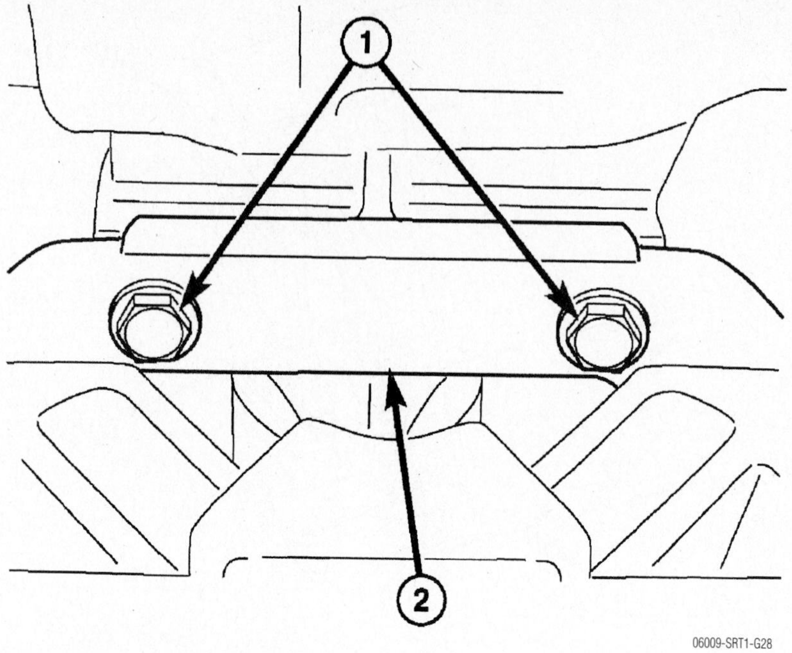

Manual transmission mount (2) and bolts (1)

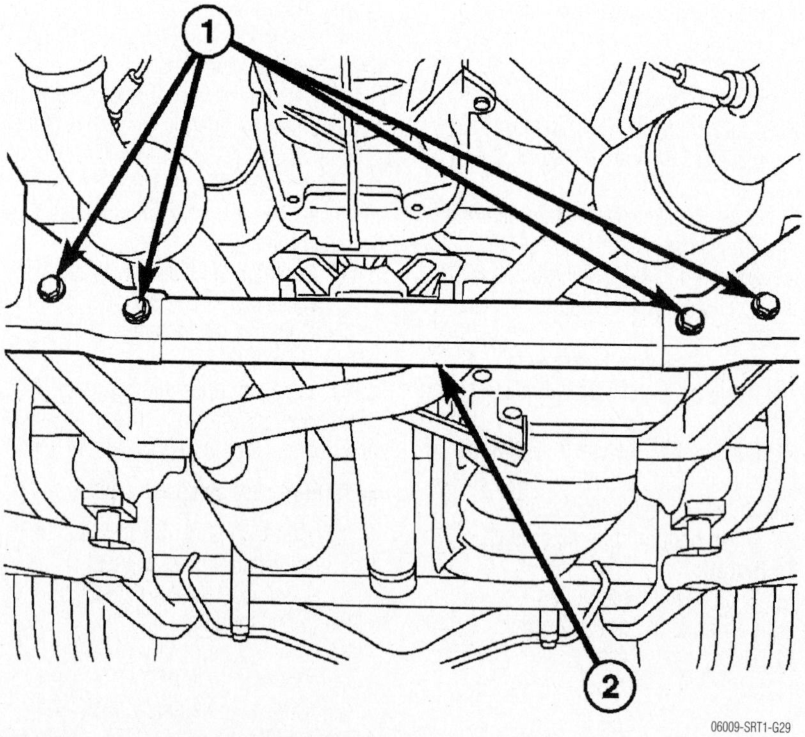

Manual transmission crossmember (2) and bolts (1)

⁂ WARNING

The drive shaft strap bolts must be replaced or cleaned and Loctite® 242 applied. Replace drive shaft joint straps whenever they are removed. The straps slightly deform when torqued in place. Reuse of the straps may result in reduced clamp load on the universal joint cap.

Automatic Transmission

REMOVAL & INSTALLATION

➡The overdrive unit can be removed and serviced separately. It is not necessary to remove the entire transmission assembly to perform overdrive unit repairs.

1. Disconnect battery negative cable.
2. Raise vehicle.
3. Disconnect and lower or remove any necessary exhaust components.
4. Remove engine-to-transmission struts.
5. Remove starter motor.
6. Disconnect and remove the crankshaft position sensor. Retain the sensor attaching bolts.
7. If transmission is being removed for overhaul, remove transmission oil pan, drain fluid and reinstall pan.
8. Remove torque converter access cover.
9. Rotate crankshaft in clockwise direction until converter bolts are accessible. Then remove bolts one at a time. Rotate crankshaft with socket wrench on dampener bolt.
10. Mark propeller shaft and axle yokes for assembly alignment. Then disconnect and remove propeller shaft.
11. Disconnect wires from the transmission range sensor and transmission solenoid connector.
12. Disconnect gearshift cable from the transmission manual lever.
13. Disconnect throttle valve cable from transmission bracket and throttle valve lever.
14. Support rear of engine with safety stand or jack.
15. Raise transmission slightly with service jack to relieve load on crossmember and supports.
16. Remove the nuts securing rear support to the transmission crossmember.
17. Remove the bolts holding the rear support to the transmission remove the rear support.
18. Remove bolts attaching crossmember to frame and remove crossmember.
19. Remove fill tube bracket bolts and pull tube out of transmission. Retain fill tube seal.
20. Disconnect fluid cooler lines at transmission.
21. Carefully work transmission and torque converter assembly rearward off engine block dowels.
22. Lower transmission and remove assembly from under the vehicle.
23. To remove torque converter, remove C-clamp from edge of bell housing and carefully slide torque converter out of the transmission.
24. Remove all bolts holding the transmission to the engine adapter.
25. Carefully work transmission and torque converter assembly rearward off engine block dowels.
26. Lower transmission and remove assembly from under the vehicle.
27. To remove torque converter, remove C-clamp from edge of bell housing and carefully slide torque converter out of the transmission.

To install:

28. Check torque converter hub inner and outer diameters and hub drive notches for sharp edges burrs, scratches, or nicks. Polish the hub and notches with 320/400 grit paper and crocus cloth if necessary. The hub must be smooth to avoid damaging pump seal at installation.
29. Lubricate pocket in the rear oil pump seal lip with transmission fluid.
30. Lubricate converter pilot hub of the crankshaft with a light coating of Mopar® High Temp Grease.
31. Align and install converter in oil pump.
32. Carefully insert converter in oil pump. Then rotate converter back and forth until fully seated in pump gears.
33. Check converter seating with steel scale and straightedge. Surface of converter lugs should be 19mm (0.75 in.) to rear of straightedge when converter is fully seated.
34. Temporarily secure converter with C-clamp.
35. Position transmission on jack and secure it with chains.
36. Check condition of converter driveplate. Replace the plate if cracked, distorted or damaged.

➡Be sure transmission dowel pins are seated in engine block and protrude far enough to hold transmission in alignment.

37. Raise transmission and align converter with drive plate and converter housing with engine block.
38. Move transmission forward. Then raise, lower or tilt transmission to align converter housing with engine block dowels.
39. Carefully work transmission forward and over engine block dowels until converter hub is seated in crankshaft.
40. Install bolts attaching transmission to engine.
41. Install rear support.
42. Install the rear transmission crossmember.
43. Lower transmission onto crossmember and install bolts attaching transmission mount to crossmember.
44. Remove engine support fixture.
45. Install crankshaft position sensor.
46. Connect gearshift cable and throttle cable to transmission.
47. Connect cooler lines to transmission.
48. Connect wires to the transmission range sensor and transmission solenoid connector. Be sure the transmission harnesses are properly routed.

⁂ WARNING

It is essential that correct length bolts be used to attach the converter to the driveplate. Bolts that are too long will damage the clutch surface inside the converter.

49. Install torque converter-to-driveplate bolts.
50. Install converter housing access cover.
51. Install starter motor and cooler line bracket.
52. Install transmission fill tube. Install new seal on tube before installation.
53. Install any exhaust components previously removed.
54. Align and connect propeller shaft.
55. Adjust gearshift cable and throttle valve cable, if necessary.
56. Lower vehicle.
57. Fill transmission with Mopar® ATF +4, Automatic Transmission fluid.

Clutch

REMOVAL & INSTALLATION

1. Remove transmission.
2. Remove clutch housing bolts and remove clutch housing from engine.
3. Mark pressure plate and flywheel to maintain their position when installing clutch assembly.
4. Insert Clutch Disc Aligning Tool through the clutch disc hub to prevent the clutch disc from falling and damaging the facings.
5. Loosen pressure plate bolts one or two turns at a time, in a crisscross pattern. This will release spring pressure evenly and avoid clutch cover damage.
6. Remove pressure plate assembly and disc.
7. Clean flywheel face with crocus cloth or 400-600 grade sandpaper, then wipe the

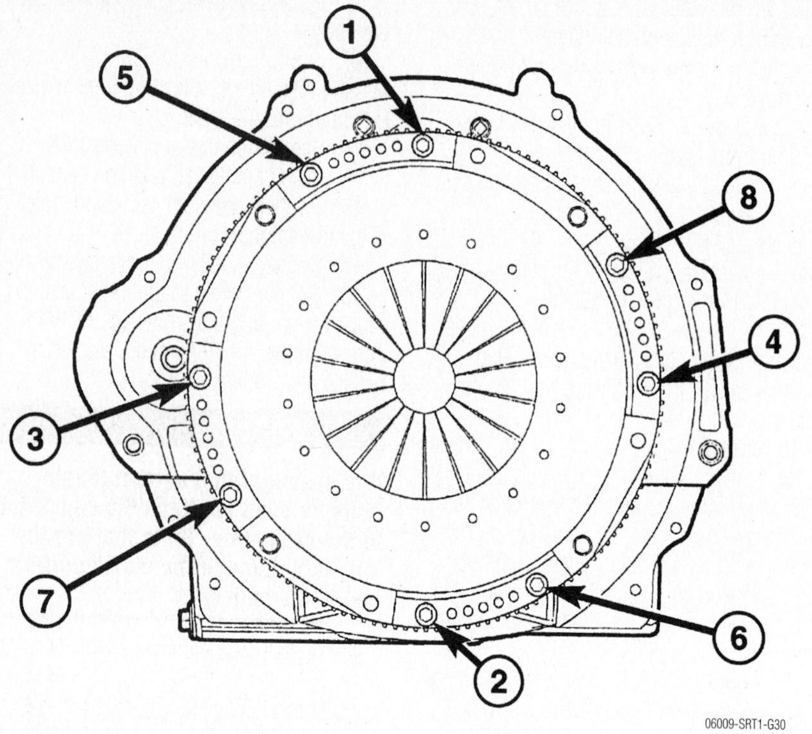

Pressure plate torque sequence

surface with mineral spirits. Wipe the friction surface of the pressure plate with mineral spirits. If the surface is severely scored, heat checked or warped, replace the flywheel.

➡Never machine the flywheel face. If flywheel surface is bad the flywheel must be replaced.

To install:

➡Clutch disc is marked flywheel side.

8. Mount clutch assembly on flywheel, being careful to properly align dowels and alignment marks made before removal. If new clutch or flywheel is installed, align cover balance spot as close as possible to flywheel balance orange spot. Apply pressure to the alignment tool. Lightly tighten the clutch cover bolts enough to hold the disc in position.
9. Tighten pressure plate bolts a few turns at a time in a crisscross pattern until bolts are all seated. Then tighten bolts sequence to 30 Nm (23 ft. lbs.). Remove clutch disc alignment tool.
10. Install bellhousing on engine.
11. Install transmission.

Hydraulic Clutch System

REMOVAL & INSTALLATION

1. Remove nuts attaching slave cylinder to studs on clutch housing.

➡The hydraulic linkage has a quick disconnect at the slave cylinder. This fitting should never be disconnected or tampered with. Once the hydraulic line is connected to the slave cylinder, it should never be disconnected.

2. Raise and support vehicle.
3. Remove heat shield over hydraulic line.
4. Remove slave cylinder from clutch housing.
5. Remove plastic clip securing the hydraulic line to the dash panel from the lower dash panel flange.
6. Remove plastic clip securing hydraulic line to the dash panel from the upper dash panel stud.
7. Lower vehicle.
8. Disconnect clutch pedal interlock switch connector.
9. Remove clutch master cylinder rod pin.
10. Verify that cap on clutch master cylinder reservoir is tight. This will avoid spillage during removal.
11. Remove clutch master cylinder nuts holding the to the dash panel.
12. Remove clutch cylinders, reservoir and connecting lines from vehicle.

To install:

13. Position cylinders and connecting line in vehicle engine compartment. Position clutch hydraulic line against the dash panel and behind all engine hoses and wiring.
14. Apply a light coating of grease to the inside diameter of the master cylinder push rod eye.
15. Install clutch master cylinder on dash panel and tighten clutch master cylinder nuts to 28 Nm (21 ft. lbs.).
16. Install clutch master cylinder push rod pin.
17. Connect clutch pedal position interlock switch connector.
18. Install plastic clip securing hydraulic

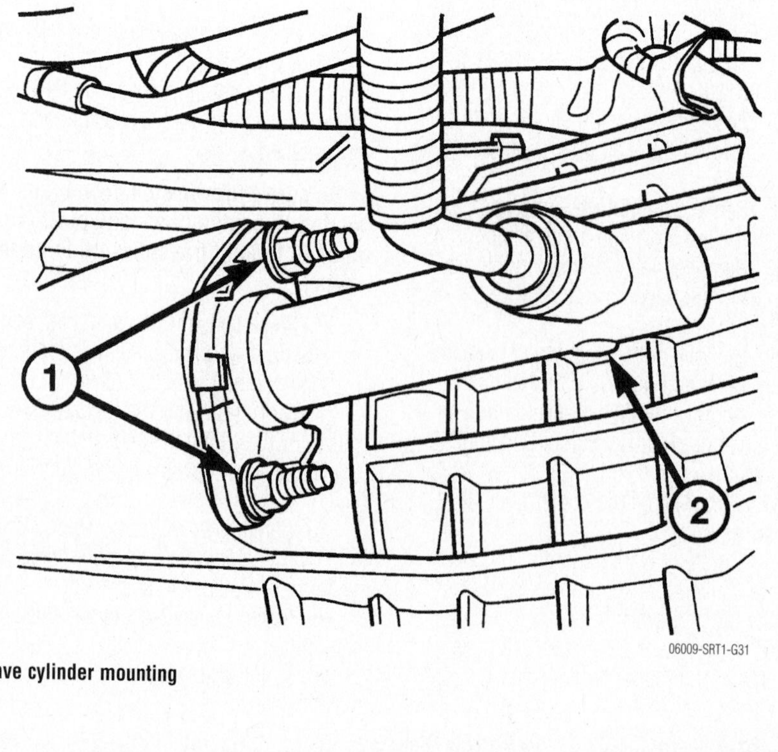

Slave cylinder mounting

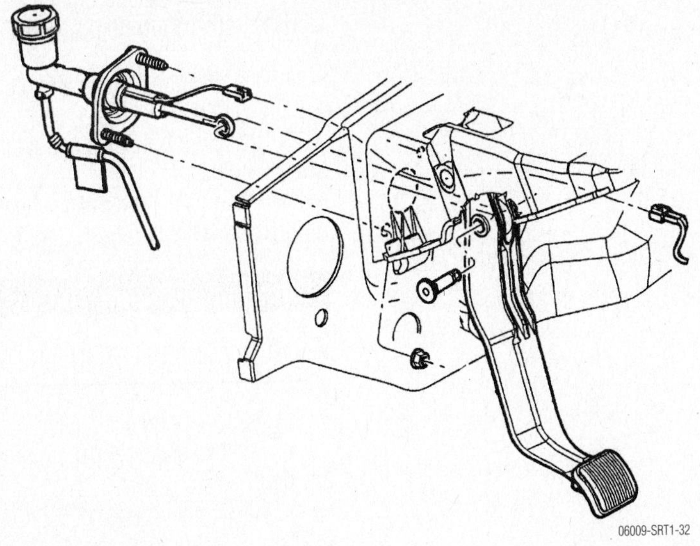

Master cylinder mounting

line to the dash panel into the lower dash panel flange.

19. Install plastic clip securing hydraulic line to the dash panel onto the upper dash panel stud.
20. Raise vehicle.
21. Install slave cylinder and verify cylinder rod is properly seated in release lever.
22. Install and tighten slave cylinder nuts to 23 Nm (17 ft. lbs.).
23. If new clutch linkage is being installed, connect the clutch hydraulic line to the clutch slave cylinder.

➡Once the clutch hydraulic line is connected to the slave cylinder, it should never be disconnected.

24. Install heat shield over hydraulic line.
25. Operate linkage several times to verify proper operation.

Rear Axle Shaft, Bearing and Seal

REMOVAL & INSTALLATION

1. With vehicle in neutral, position vehicle on hoist.
2. Remove brake caliper adapter with caliper and remove rotor.
3. Remove differential housing cover and drain lubricant.
4. Remove pinion mate shaft lock screw.
5. Remove pinion mate shaft from differential case.
6. Push axle shaft inward and remove axle shaft C-lock.
7. Remove axle shaft.
8. Remove axle seal with pry bar.
9. Position bearing Receiver 9345 on axle tube.
10. Insert bearing Remover 6310 with Foot 6310-9 through receiver and bearing.
11. Tighten Remove 6310 nut to pull bearing into the receiver.

To install

12. Remove any old sealer/burrs from axle tube.
13. Install axle shaft bearing with an installer. Drive bearing in until tool contacts the axle tube.

➡Bearing is installed with the bearing part number against the installer.

14. Coat new axle seal lip with axle lubricant and install with a seal driver.
15. Lubricate bearing bore and seal lip with gear lubricant.

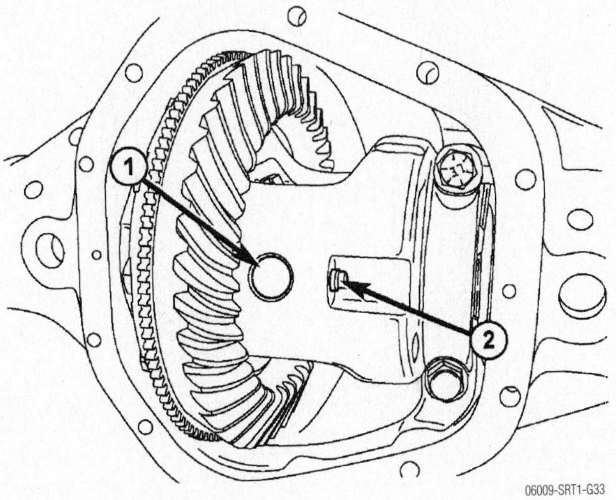

Pinion mate shaft (1) and lock screw (2)

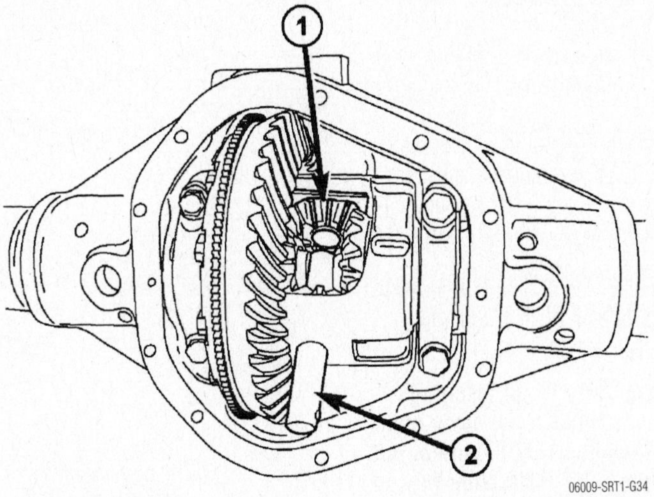

Remove the pinion mate shaft

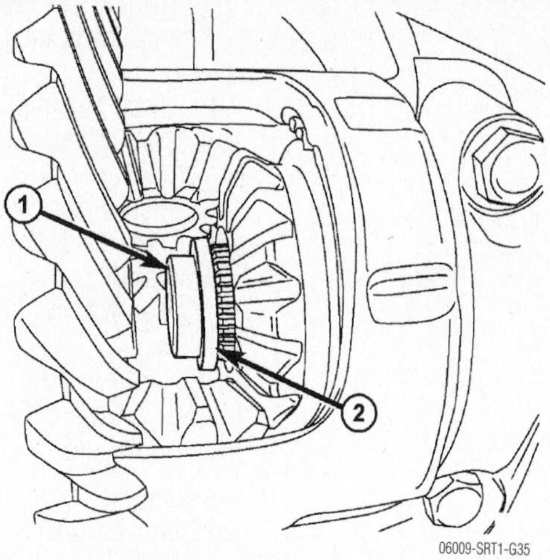

Push the axle shaft (1) inward and remove the C-lock (2)

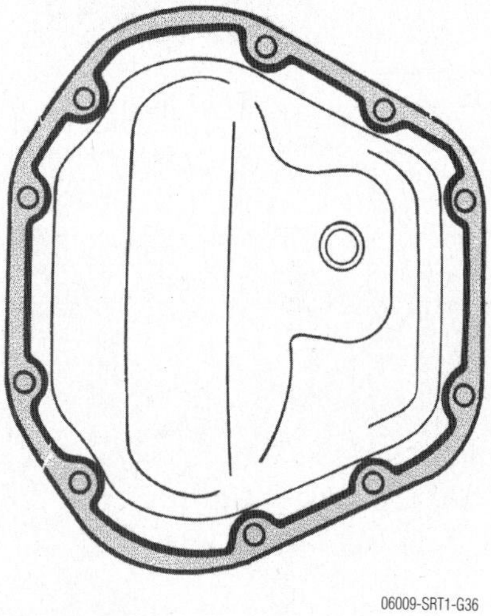

RTV application on differential cover

16. Insert axle shaft through seal, bearing and engage into side gear splines.
17. Insert C-lock in end of axle shaft then push axle shaft outward to seat C-lock in side gear.
18. Insert pinion mate shaft into differential case and through thrust washer and pinion gear.
19. Align hole in pinion mate shaft with hole in differential case and install lock screw with Loctite® 242 or equivalent. Tighten lock screw to 27 Nm (20 ft. lbs.).
20. Apply a bead of orange Mopar Axle RTV sealant or equivalent to the housing cover.

➡If housing cover is not installed within 3 to 5 minutes, the cover must be cleaned and new RTV applied. Failure to follow these instruction will result in a leak.

21. Install cover and tighten bolts in a criss-cross pattern to 48 Nm (35 ft. lbs.).
22. Fill differential lubricant to bottom of the fill plug hole.
23. Install fill hole plug and tighten to 27 Nm (20 ft. lbs.).
24. Install brake rotor and caliper adapter with caliper.

Differential Pinion Seal

REMOVAL & INSTALLATION

1. Remove propeller shaft.
2. Remove brake calipers to prevent any drag.
3. Rotate pinion yoke three or four times.
4. Record pinion gear rotating torque with an inch pound torque wrench.
5. Hold yoke with Yoke Holder 6719A and remove pinion shaft nut and washer.
6. Remove pinion yoke with Remover C-452.
7. Remove pinion seal with pry tool or slide-hammer mounted screw.

To install:

8. Coat lip of pinion seal with gear lubricant.
9. Install new pinion seal with Installer 9348 and Handle C-4171.
10. Install yoke on pinion shaft with Installer C-3718 and Yoke Holder 6719A.
11. Install pinion yoke washer with the concave surface against the yoke end.
12. Install new pinion nut.

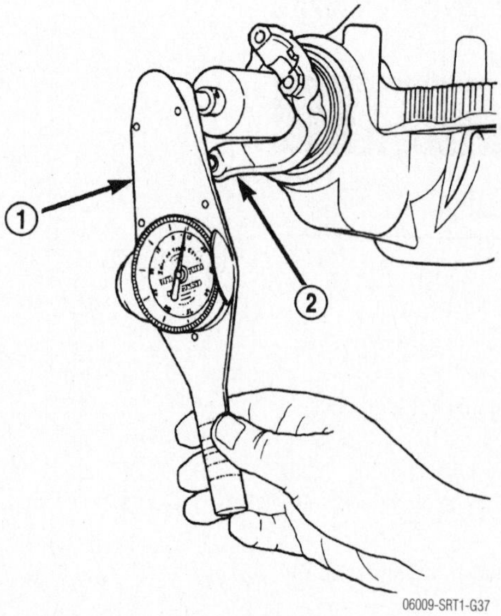

Measuring pinion gear rotating torque. (1) Torque wrench, (2) Yoke

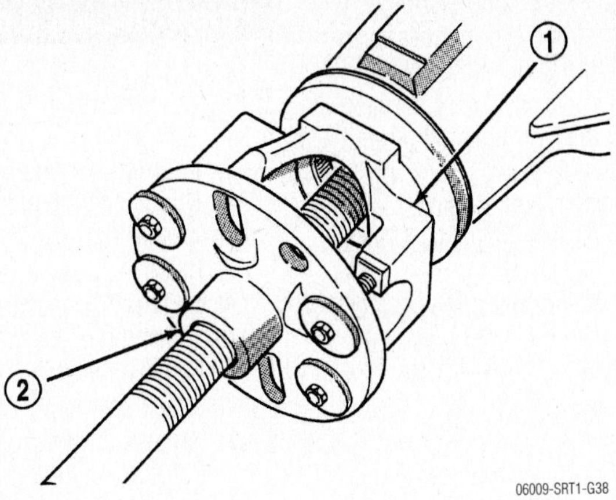

Removing the yoke

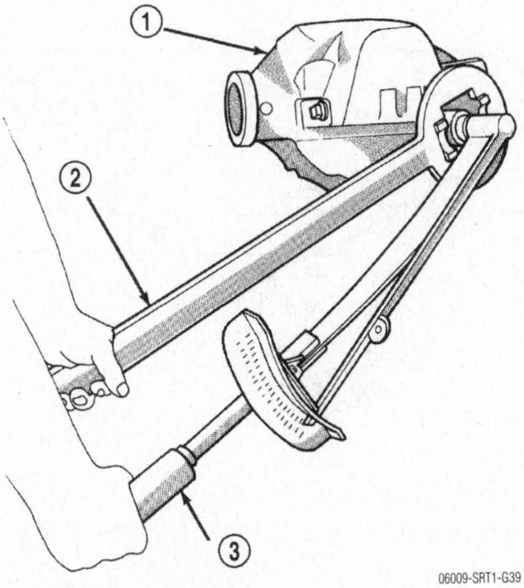

Holding the yoke and tightening the pinion nut.

✱✱ WARNING

Never exceed minimum tightening torque when installing the pinion yoke retaining nut at this point. Failure to heed caution may result in damage.

13. Hold pinion yoke with Yoke Holder 6719A and tighten nut with torque wrench to 292 Nm (215 ft. lbs.). Rotate pinion shaft several revolutions to ensure bearing rollers are seated.
14. Rotate pinion with an inch pound torque wrench. Rotating torque should be equal to reading recorded during removal, plus 0.56 Nm (5 inch lbs.).

✱✱ WARNING

Never loosen pinion nut to decrease pinion rotating torque and never exceed specified preload torque. Failure to follow these instruction will result in damage.

15. If rotating torque is low, use Yoke Holder 6719 to hold the pinion yoke and tighten the pinion shaft nut in 6.8 Nm (60 inch lbs.) increments until proper rotating torque is achieved.

➡The bearing rotating torque should be constant during a complete revolution of the pinion. If the rotating torque varies, this indicates a binding condition.

16. Install brake components.
17. Install propeller shaft.

STEERING AND SUSPENSION

Air Bag

PRECAUTIONS

To avoid personal injury or death, on vehicles equipped with airbags, disable the supplemental restraint system before attempting any steering wheel, steering column, airbag, seat belt tensioner, impact sensor, or instrument panel component diagnosis or service. Disconnect and isolate the battery negative (ground) cable, then wait two minutes for the system capacitor to discharge before performing further diagnosis or service. This is the only sure way to disable the supplemental restraint system. Failure to take the proper precautions could result in accidental airbag deployment.

Before performing any welding operations disconnect and isolate the battery negative (ground) cable and disconnect all wire harness connectors from the Airbag Control Module (ACM). Failure to take the proper precautions could result in accidental airbag deployment and other possible damage to the supplemental restraint system circuits and components.

Do not attempt to dismantle an airbag unit or tamper with its inflator. Do not puncture, incinerate, or bring into contact with electricity. Do not store at temperatures exceeding 93° C (200° F). An airbag inflator unit may contain sodium azide and potassium nitrate. These materials are poisonous and extremely flammable. Contact with acid, water, or heavy metals may produce harmful and irritating gases (sodium hydroxide is formed in the presence of moisture) or combustible compounds. An airbag inflator unit may also contain a gas canister pressurized to over 2500 psi.

At no time should any source of electricity be permitted near the inflator on the back of a non-deployed airbag or seat belt tensioner. When carrying a non-deployed airbag, the trim cover or airbag cushion side of the unit should be pointed away from the body to minimize injury in the event of an accidental deployment. If the airbag unit is placed on a bench or any other surface, the trim cover or airbag cushion side of the unit should be face up to minimize movement in the event of an accidental deployment.

DISARMING THE SYSTEM

Disconnect and isolate the battery negative (ground) cable, then wait two minutes for the system capacitor to discharge before performing further diagnosis or service. This is the only sure way to disable the supplemental restraint system. Failure to take the proper precautions could result in accidental airbag deployment.

Power Steering Pump

REMOVAL & INSTALLATION

1. Drain and siphon the power steering fluid from the reservoir.
2. Remove the serpentine belt.

⁂ WARNING

Do not remove the fitting on the pump that the high pressure hose screws into. The fitting may come loose unless it is backed up using another wrench. If the fitting does come loose, it must be retightened before continuing. (57–67Nm, 40–50 ft. lbs.) If this fitting comes out of the pump body, the internal spring and valve parts will fall out of the pump and they cannot be reinstalled properly. If this occurs the pump needs to be replaced with a new pump.

3. Disconnect the return hose (3).
4. Disconnect the pressure hose (4).
5. Access to remove the three bolts securing the pump to the cylinder head can be gained through the pulley holes (1).

To install:

6. Align the pump with the mounting holes in the left cylinder head.
7. Install 3 pump mounting bolts through the pulley access holes (1). Tighten the bolts to 28 Nm (21 ft. lbs.).
8. Reconnect the pressure line (4) and return hose (3) to the pump and reservoir (2). Tighten the pressure line to 37 Nm (27 ft. lbs.).
9. Install the serpentine drive belt.
10. Fill the power steering pump.

Power steering pump mounting

Power Steering Gear

REMOVAL & INSTALLATION

1. Lock the steering wheel.
2. Drain and siphon the power steering fluid from the reservoir.
3. Raise the vehicle.
4. Remove and discard the steering coupler pinch bolt.
5. Remove the power steering hoses from the rack & pinion.
6. Remove the tire and wheel assembly.
7. Remove the tie rod end nuts and separate tie rod ends from the knuckles with a tie rod end separator.

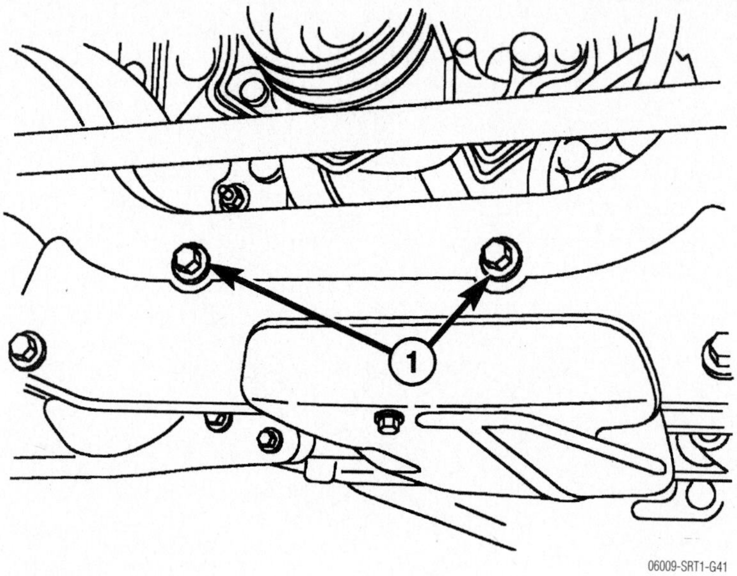

Steering gear mounting bolts (1)

8. Remove the rack & pinion mounting bolts.
9. Remove the rack & pinion from the vehicle.

To install:

➡Before installing gear inspect bushings and replace if worn or damaged. In the frame there is two holes for the mounting of the steering gear one is slotted and one is round. When tightening the gear to specifications make sure to tighten the mounting bolt with the hole first to avoid movement of the steering gear.

10. Install the gear on the front crossmember and tighten the mounting bolts to 319 Nm (235 ft. lbs.)
11. Slide the shaft coupler onto the gear. Install new pinch bolt and tighten to 49 Nm (36 ft. lbs.).

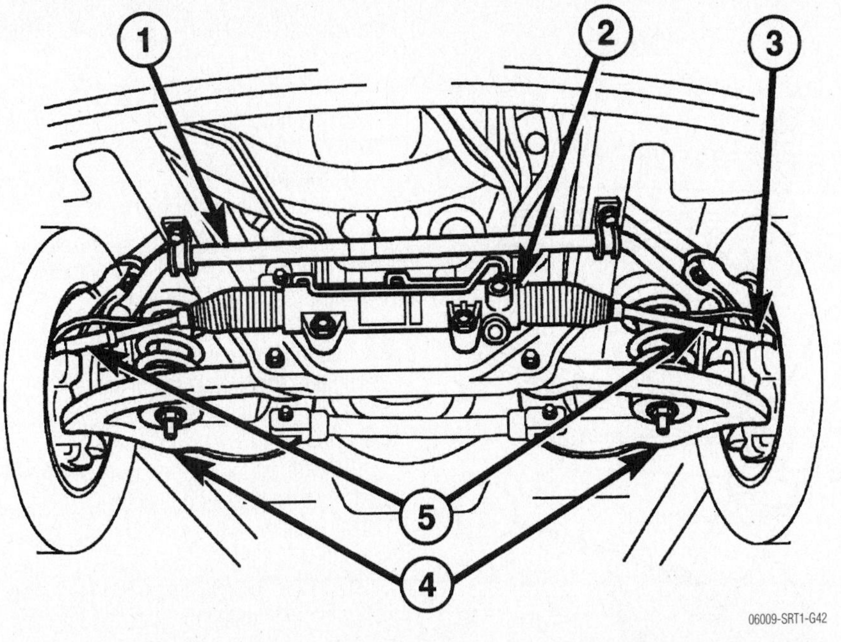

Steering gear installed. (1) Sway bar, (2) Steering gear, (3 & 5) Tie rods/ends, (4) Control arms

12. Clean and dry the tie rod end studs and the knuckle tapers.
13. Install the tie rod ends into the steering knuckles and tighten the nuts to 61 Nm (45 ft. lbs.) then an additional 90°.
14. Install the pressure power steering hose to the steering gear and tighten to 32 Nm (23 ft. lbs.).
15. Install the return power steering hose to the steering gear and tighten to 71 Nm (52 ft. lbs.).
16. Install the tire and wheel assembly.
17. Remove the support and lower the vehicle.
18. Fill the system with fluid.
19. Adjust the toe position.

Shock Absorber

REMOVAL & INSTALLATION

Front

1. Raise and support vehicle.
2. Support the lower control arm outboard end.
3. Remove the upper shock absorber nut, retainer and grommet.
4. Remove the lower nuts and remove the shock absorber.

To install:

➡Upper shock nut must be replaced or use Mopar Lock 'N Seal or Loctite® 242 on existing nut.

5. Install the lower retainer and grommet on the shock absorber stud. Insert the shock absorber through the frame bracket hole.
6. Install the lower nuts and tighten the nuts to 35 Nm (25 ft. lbs.).
7. Install the upper grommet, retainer and new nut or use Mopar Lock 'N Seal or Loctite® 242 on existing nut, on the shock absorber stud. Tighten nut to 54 Nm (40 ft. lbs.).
8. Remove the support from the lower control arm outboard end.
9. Lower the vehicle.

Rear

1. Raise vehicle and support the axle.

➡The rear upper shock attachment uses a flag nut. Do not use an air tool to remove the bolt, the flag may rotate into the bottom of the bed and cause damage. Use a wrench to hold the nut when loosening.

2. Remove the upper shock bolt and nut.
3. Remove the lower shock bolt and nut.
4. Remove the rear shock absorber from the vehicle.

To install:

5. Position the shock absorber in the brackets.
6. Install the bolts through the brackets and the shock. Install the flag nut on the top bolt and nut on lower bolt.
7. Tighten the upper and lower bolt/nuts Tighten to 135 Nm (100 ft. lbs.)
8. Remove the support and lower the vehicle.

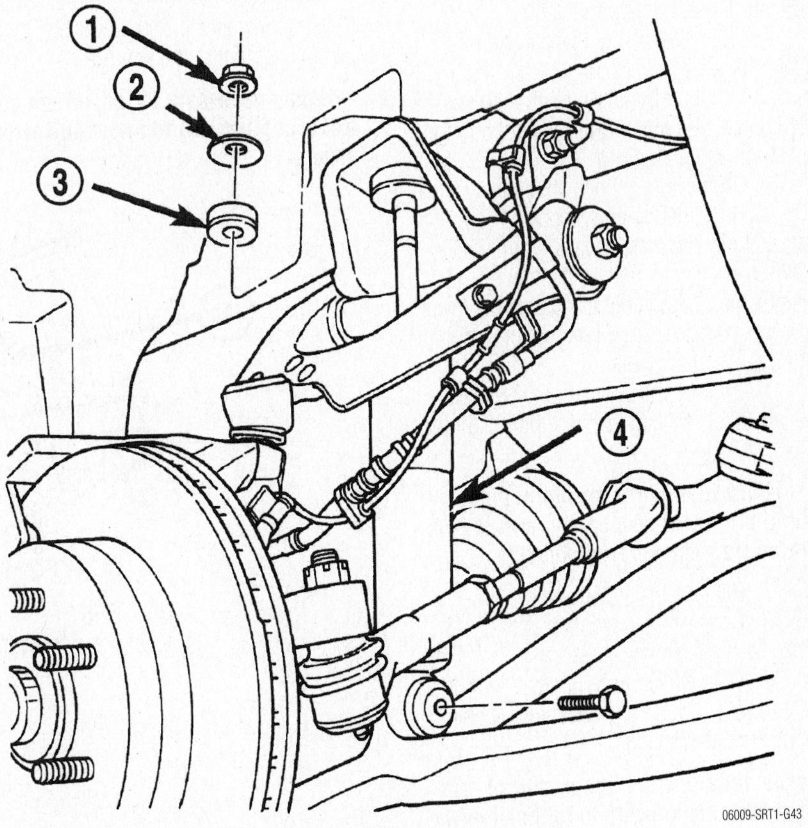

Front shock absorber mounting. (1) Nut, (2) Retainer, (3) Grommet

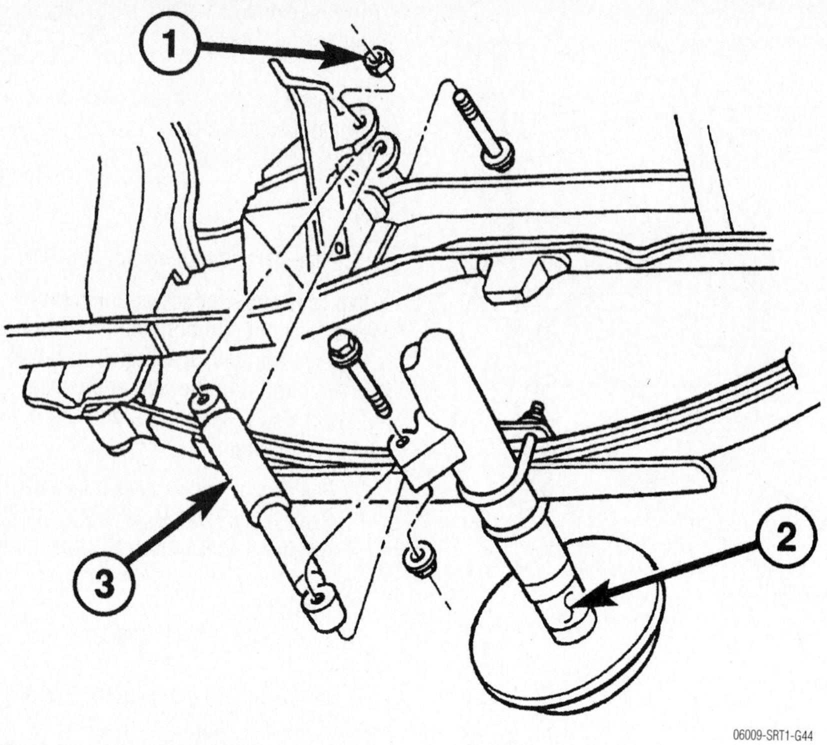

Rear shock absorber mounting. (1) Nut, (2) Axle, (3) Shock absorber

Coil Spring

REMOVAL & INSTALLATION

1. Raise and support vehicle.
2. Remove the front wheel and tire assembly.
3. Support the lower control arm at the outboard side of the lower control arm to support vehicle weight.
4. Remove the shock absorber.
5. Install a spring compressor up through the lower suspension arm, coil spring and shock hole in the frame. The bell-shaped adapter goes against the lower suspension arm. Install the nut on top of the tool at the shock hole.
6. Tighten the spring compressor nut against bell-shaped adapter finger tight then loosen ½ turn.

➡This will hold the spring in place until the lower suspension arm is separated from the steering knuckle.

7. Remove the stabilizer link.
8. Remove the lower ball joint nut at the steering knuckle.
9. Install a ball joint remover tool on the lower ball joint and separate the ball joint from the knuckle.

➡Do not allow the upper control arm and steering knuckle to rebound downwards they must be supported.

10. Support the upper control arm and steering knuckle out of the way.
11. Remove the lower control arm support.
12. Tighten the spring compressor tool to allow clearance for the lower ball joint to be removed out of the knuckle.

➡It may necessary to loosen the control arm pivot bolt to allow downward swing.

13. Loosen the tension on the spring compressor tool slowly allowing the lower suspension arm to pivot downward.
14. Remove the spring compressor tool.
15. Remove coil spring and isolator pad from the vehicle.

To install:

16. Tape the isolator pad to the top of the coil spring. Position the spring in the lower suspension arm well. Be sure that the coil spring is seated in the well.
17. Install a spring compressor up through the lower suspension arm, coil spring and shock hole in the frame.
18. Tighten the tool nut to compress the coil spring.
19. Remove the support from the upper control arm and steering knuckle.
20. Position the lower ball joint into the steering knuckle.
21. Install the retaining nut on the lower ball joint and tighten to 52 Nm (38 ft. lbs.) an additional 90° turn is required.
22. Remove the spring compressor tool.
23. Support the lower control arm at the outboard side of the lower control arm to support vehicle weight.
24. Install the shock absorber.
25. Install the stabilizer link.
26. Remove the lower control arm support.
27. Install the wheel and tire assembly and lower the vehicle.
28. Lower the vehicle to the floor with vehicle weight and Tighten the front and rear control arm pivot bolts if loosened to 204 Nm (150 ft. lbs.)
29. Perform a wheel alignment.

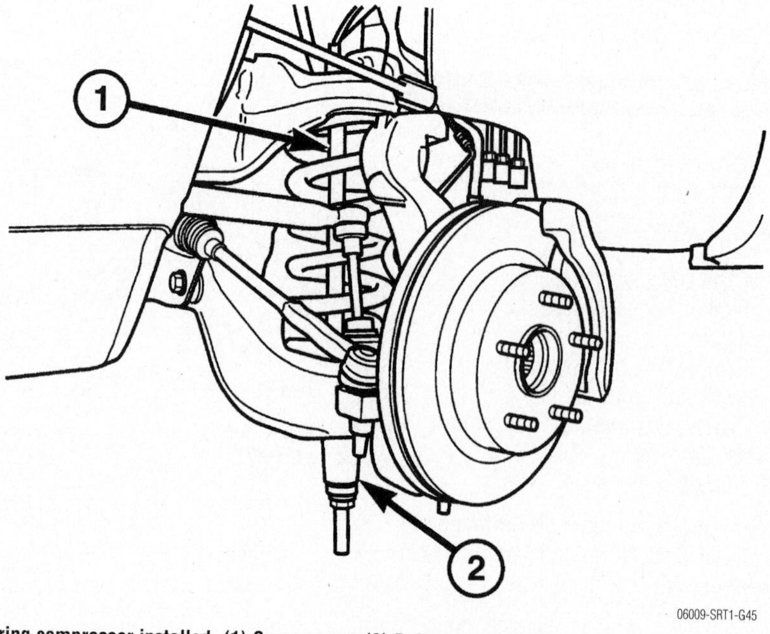

Spring compressor installed. (1) Compressor, (2) Bell-shaped adapter

Leaf Springs

REMOVAL & INSTALLATION

1. Raise and support the vehicle.
2. Support the axle with a suitable holding fixture.
3. Remove the nuts, spring clamp bolts and the plate that attach the spring to the axle.
4. Remove the nuts and bolts from the spring front and rear shackle.
5. Remove the spring from the vehicle.

To install:

6. Position spring on axle shaft tube so spring center bolt is inserted into the locating hole in the axle tube.
7. Align the front of the spring with the bolt hole in the front bracket. Install the eye pivot bolt and nut.
8. Align the rear of the spring into the shackle and install the bolt and nut.
9. Tighten the spring front and rear eye pivot bolt snug do not torque.
10. Install the spring clamp bolts, plate and the retaining nuts.
11. Remove the holding fixture for the rear axle.
12. Remove the supports and lower the vehicle so that the weight is being supported by the tires.
13. Tighten the spring clamp retaining nuts to 149 Nm (110 ft. lbs.).
14. Tighten the spring front and rear pivot bolt nuts to 163 Nm (120 ft. lbs.).

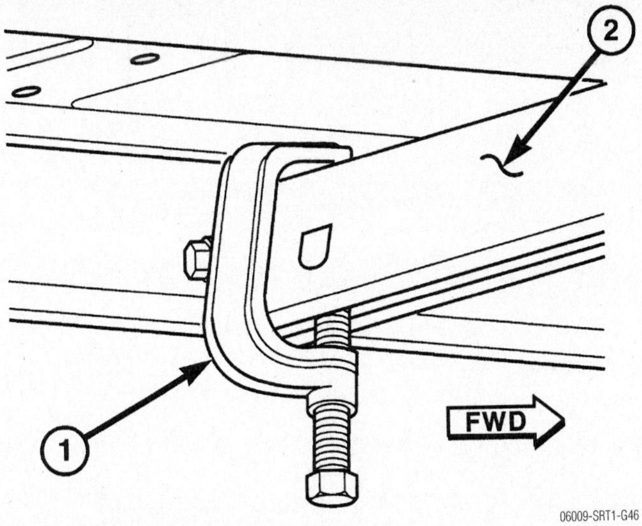

Special tool 8686 (1) installed on the crossmember (2)

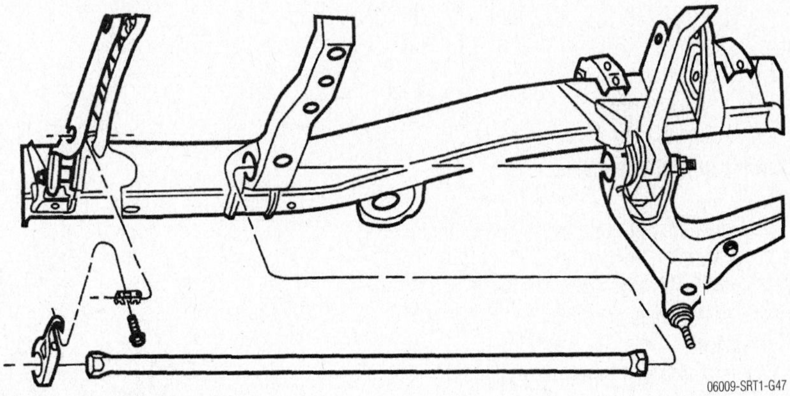

Torsion bar mounting

Torsion Bar

REMOVAL & INSTALLATION

➡The left and right side torsion bars are NOT interchangeable. The bars are identified and stamped R or L, for right or left. The bars do not have a front or rear end and can be installed with either end facing forward.

1. Raise and support the vehicle with the front suspension hanging.

➡Count and record the number of turns for installation reference.

2. Mark the adjustment bolt setting.
3. Install Special Tool 8686 to the anchor arm and the cross member.
4. Increase the tension on the anchor arm tool until the load is removed from the adjustment bolt and the adjuster nut.
5. Turn the adjustment bolt counterclockwise to remove the bolt and the adjuster nut.
6. Remove the Special Tool 8686, allowing the torsion bar to unload.
7. Remove torsion bar and anchor. Remove anchor from torsion bar.
8. Remove all foreign material from torsion bar mounting in anchor and suspension arm.
9. Inspect adjustment bolt, bearing and swivel for damage.

To install:

10. Insert torsion bar ends into anchor and suspension arm.
11. Position the anchor in the frame crossmember.
12. Install Special Tool 8686 to the anchor and the crossmember.
13. Increase the tension on the anchor in order to load the torsion bar.
14. Install the adjustment bolt and the adjuster nut.
15. Turn adjustment bolt clockwise the recorded amount of turns.
16. Remove tool 8686 from the torsion bar crossmember.
17. Lower vehicle and adjust the front suspension height.
18. Perform a wheel alignment.

Upper Ball Joint

REMOVAL & INSTALLATION

The upper ball joint is integral with the control arm. If the ball joint is defective, the whole arm must be replaced.

Lower Ball Joint

REMOVAL & INSTALLATION

1. Remove the tire and wheel assembly.
2. Remove the brake caliper and rotor.
3. Disconnect the tie rod from the steering knuckle.
4. Remove the lower ball joint nut and separate the lower ball joint from the steering knuckle.

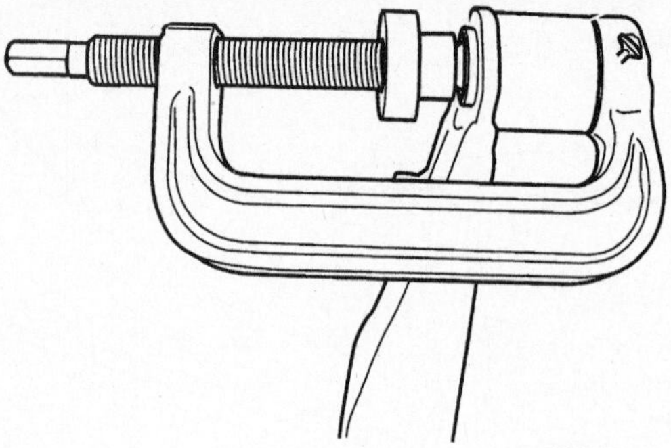

06009-SRT1-G48

Lower ball joint removal

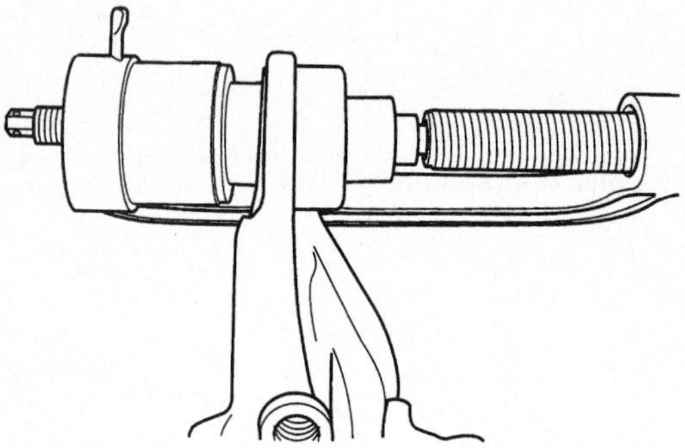

06009-SRT1-G49

Lower ball joint installation

5. Press the ball joint from the lower control arm using special tools C-4212-F (press), 8698-2 (receiver) and 8698-3 (driver), or equivalents.

To install:

6. Install the ball joint into the control arm and press in using special tools C-4212-F (press), 8698-1 (driver) and 8698-3 (receiver), or equivalent.
7. Install the ball joint boot.
8. Stake the ball joint flange in four evenly spaced places around the ball joint flange, using a chisel and hammer.
9. Install the steering knuckle. Torque the lower ball joint stud nut to 38 ft. lbs. (52 Nm) plus 90 degrees.
10. Install the tie rod end into the steering knuckle. Torque to 45 ft. lbs. (61 Nm) plus 90 degrees.
11. Install the brake caliper and rotor.
12. Install the tire and wheel assembly.
13. Check the vehicle ride height.
14. Perform a wheel alignment.

Upper Control Arm

REMOVAL & INSTALLATION

1. Raise and support vehicle.
2. Remove wheel and tire assembly.
3. Remove the nut from upper ball joint.
4. Separate upper ball joint from the steering knuckle.

➡When installing the tool to separate the ball joint, be careful not to damage the ball joint seal.

5. Remove the control arm pivot bolts and remove control arm.

To install:

6. Position the control arm into the frame brackets. Install bolts and tighten to 132 Nm (97 ft. lbs.).
7. Insert the ball joint in steering knuckle and tighten the upper ball joint nut to 54 Nm (40 ft. lbs.) plus 90 degrees.
8. Install the wheel and tire assembly.
9. Remove the support and lower vehicle.
10. Perform a wheel alignment.

Lower Control Arm

REMOVAL AND & INSTALLATION

1. Raise and support the vehicle.
2. Remove the tire and wheel assembly.
3. Support the lower control arm at the outboard side of the lower control arm to support vehicle weight.

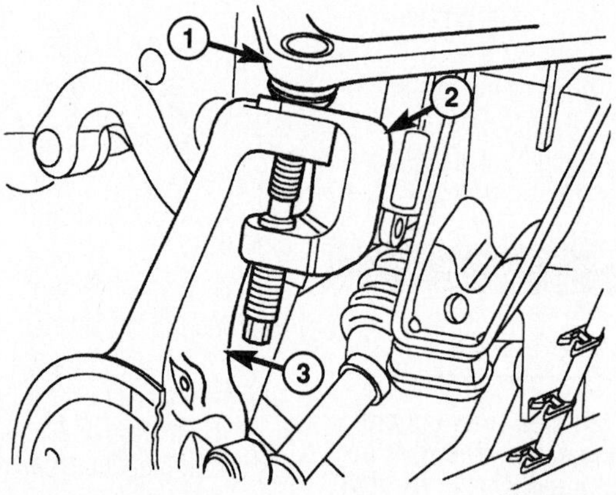

06009-SRT1-G50

Separating upper ball joint from the knuckle. (1) Control arm, (2) Tool, (3) Knuckle

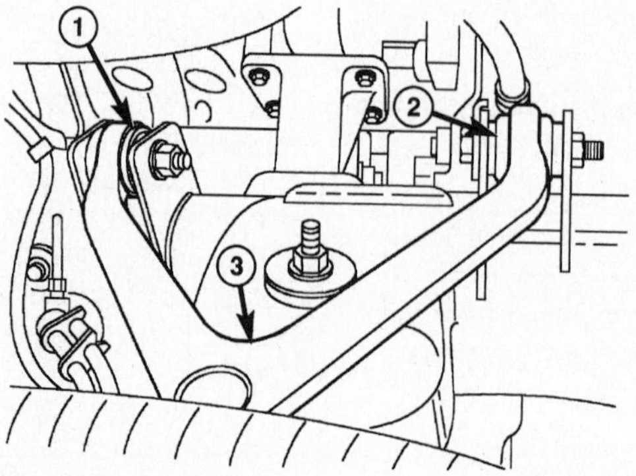

06009-SRT1-G51

Upper control arm pivot bolts (1 & 2), control arm (3)

4. Remove the shock.

5. Install a spring compressor up through the lower suspension arm, coil spring and shock hole in the frame. The bell-shaped adapter goes against the lower suspension arm. Install the nut on top of the tool at the shock hole.

6. Tighten the spring compressor nut against bell-shaped adapter finger tight then loosen ½ turn.

➡This will hold the spring in place until the lower suspension arm is separated from the steering knuckle.

7. Remove the stabilizer link.

8. Remove the lower ball joint nut at the steering knuckle.

9. Install a ball joint remover tool on the lower ball joint and separate the ball joint from the knuckle.

➡Do not allow the upper control arm and steering knuckle to rebound downwards they must be supported.

10. Support the upper control arm and steering knuckle out of the way.

11. Remove the lower control arm support.

12. Tighten the spring compressor tool to allow clearance for the lower ball joint to be removed out of the knuckle.

13. Loosen the tension on the spring compressor tool slowly allowing the lower suspension arm to pivot downward.

14. Remove the spring compressor tool.

15. Remove coil spring and isolator pad from the vehicle.

16. Remove the front and rear pivot bolts.

17. Remove the lower control arm.

To install:

18. Install the lower control arm into place on the vehicle.

19. Install the front and rear control arm pivot bolts finger tight.

20. Install the coil spring into the frame pocket.

21. Install a spring compressor up through the lower suspension arm, coil spring and shock hole in the frame.

22. Tighten the tool nut to compress the coil spring.

23. Remove the support from the upper control arm and steering knuckle.

24. Position the lower ball joint into the steering knuckle.

25. Install the retaining nut on the lower ball joint and tighten to 52 Nm (38 ft. lbs), an additional 90 degree turn is required.

26. Support the lower control arm at the outboard side of the lower control arm to support vehicle weight.

27. Remove the spring compressor tool.

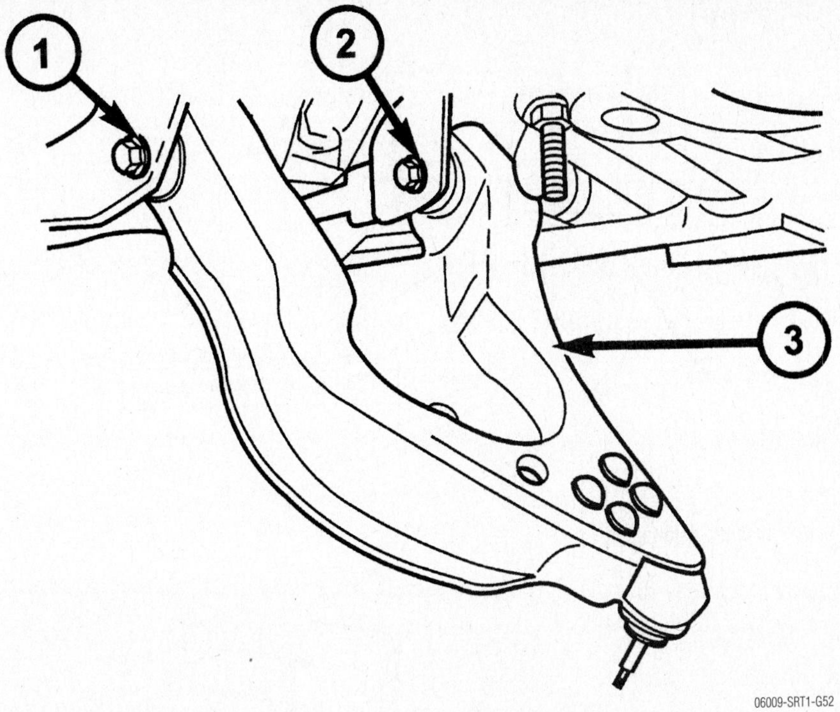

Lower control arm removal. (1 & 2) Pivot bolts, (3) control arm

28. Install the shock absorber.

29. Install the stabilizer link.

30. Remove the lower control arm support.

31. Install the wheel and tire assembly and lower the vehicle.

32. Lower the vehicle to the floor with vehicle weight and Tighten the front and rear control arm pivot bolts to 204 Nm (150 ft. lbs.).

33. Perform a wheel alignment.

LOWER CONTROL ARM BUSHING REPLACEMENT

Large Bushing

1. Install bushing remover tools 8836-2 (receiver 4), 8836-4 (spacer 1) and 8836-5 (driver 2) with the threaded rod 8839 (3) and the bearing as shown for replacement of the large bushing.

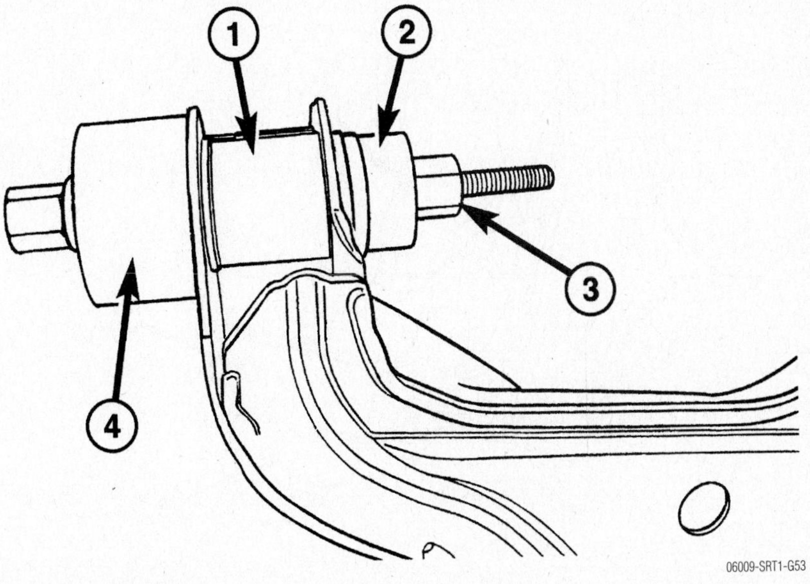

Large bushing removal

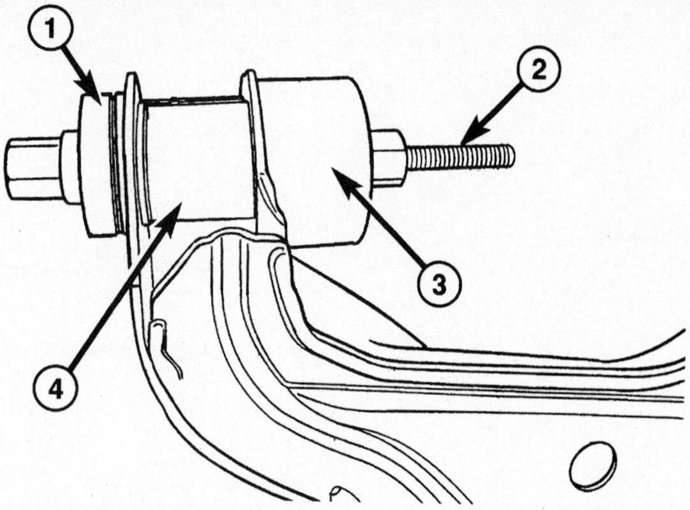

Large bushing installation

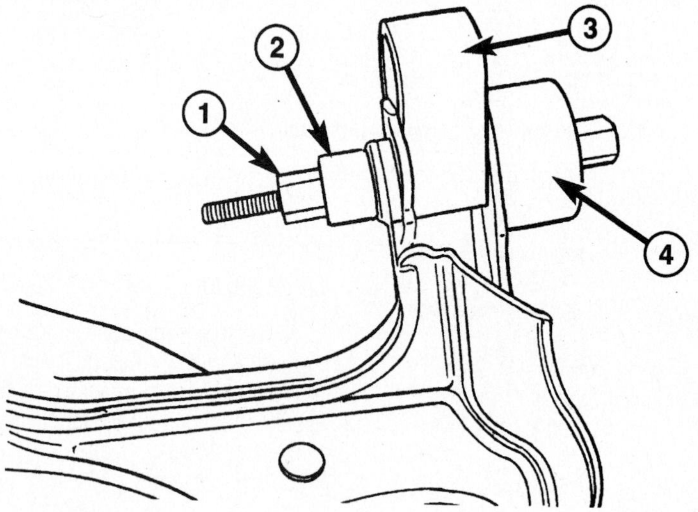

Small bushing removal

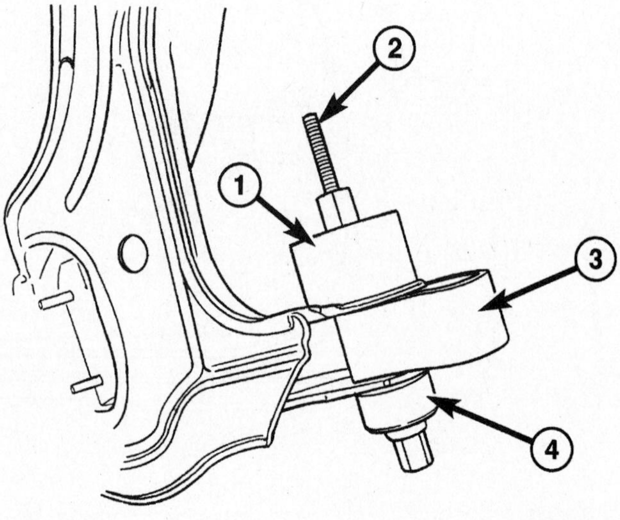

Small bushing installation

➡Extreme pressure lubrication must be used on the threaded portions of the tool. This will increase the longevity of the tool and insure proper operation during the removal and installation process.

2. Install the new bushing into the lower control arm using special tools 8836-2 (receiver 3), 8836-1 (driver 1), 8836-4 (spacer 4) with the bearing and the threaded rod (8839 2).

Small Bushing

1. Install the bushing tool 8836-6 (driver 2), 8836-3 (spacer 3) and 8836-2 (receiver 4) with the threaded rod 8839 (1) and the bearing as shown for the replacement of the small bushing.
2. Install the small bushings into the lower control arm using tools 8836-7 (driver 4), 8836-2 (receiver 1), 8836-3 (spacer 3) and the bearing with the threaded rod (8839 2).
3. Remove the control arm from the vise.
4. Install the lower control arm.
5. Reset the vehicle ride height.
6. Perform a wheel alignment.

Knuckle

REMOVAL & INSTALLATION

1. Raise and support the vehicle.
2. Remove the wheel and tire assembly.
3. Remove the brake caliper, rotor, shield and ABS wheel speed sensor if equipped.
4. Remove the tie rod end nut. Separate the tie rod from the knuckle.

➡When installing the remover to separate the ball joint, be careful not to damage the ball joint seal.

5. Remove the lower ball joint nut. Separate the ball joint from the knuckle and remove the knuckle.
6. Remove the upper ball joint nut. Separate the ball joint from the knuckle.
7. Remove the hub/bearing from the steering knuckle.

To install:

➡The ball joint stud tapers must be clean and dry before installing the knuckle. Clean the stud tapers with mineral spirits to remove dirt and grease.

➡When installing hub/bearing with ABS brakes, position the speed sensor opening towards the front of the vehicle.

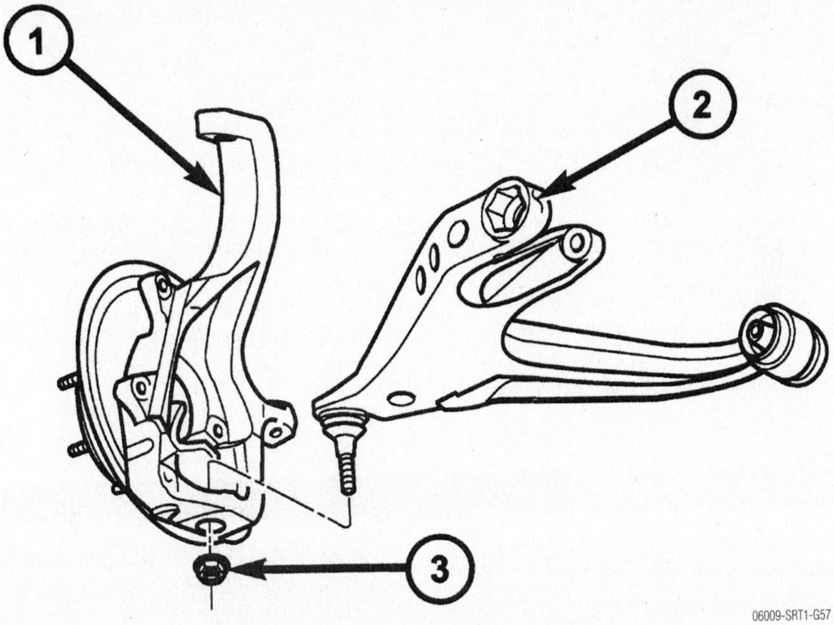

Knuckle installation. (1) knuckle, (2) lower control arm, (3) lower ball joint nut

8. Install the hub/bearing to the steering knuckle and tighten the bolts to 163 Nm (120 ft. lbs.).
9. Install the knuckle onto the upper and lower ball joints.
10. Install the upper and lower ball joint nuts. Tighten the upper ball joint nut to 54 Nm (40 ft. lbs.) plus 90 degrees, and the lower ball joint nut to 52 Nm (38 ft. lbs.) plus 90 degrees.
11. Remove the hydraulic jack from the lower suspension arm.
12. Install the tie rod end and tighten the nut to 61 Nm (45 ft. lbs.).
13. Install the ABS wheel speed sensor if equipped and brake shield, rotor and caliper.
14. Install the wheel and tire assembly.
15. Remove the support and lower the vehicle.
16. Perform a wheel alignment.

Stabilizer Bar

REMOVAL & INSTALLATION

Front

➡To service the stabilizer bar the vehicle should be on a drive on hoist. The vehicle suspension must be at curb height for stabilizer bar installation.

1. Remove the stabilizer bar link upper nuts and remove the retainers and grommets.
2. Remove the stabilizer bar retainer, bolts and retainers from the frame crossmember and remove the bar.
3. If necessary, remove the bushings from the stabilizer bar.

➡Do not cut the old bushings off the stabilizer bar use a mixture of soapy water in order to aid in sliding the bushing off.

To install:

4. If removed, Install the bushings on the stabilizer bar using a mixture of soapy water or equivalent in order to slide the bushing over the bar with ease.

➡Do not cut the new bushing for installation.

5. Position the stabilizer bar on the frame crossmember brackets and install the bracket and bolts finger-tight.

➡Check the alignment of the bar to ensure there is no interference with the either frame rail or chassis component. Spacing should be equal on both sides.

6. Install the stabilizer bar to the stabilizer link and install the grommets and retainers.
7. Install the nuts to the stabilizer link and tighten to 27 Nm (20 ft. lbs.).
8. Tighten the brackets to the frame to 61 Nm (45 ft. lbs.).

Rear

1. Raise and support vehicle.
2. Remove lower links mounting nuts.
3. Remove stabilizer bar retainer bolts and retainers at the axle.
4. Remove stabilizer bar and remove bushings.

To install:

5. Install stabilizer bar bushings.
6. Install the stabilizer bar and center it with equal spacing on both sides.
7. Install stabilizer bar retainers. Tighten the bolts to 75 Nm (55 ft. lbs.).
8. Install the park brake cable bracket to the retainers.
9. Install the links on the stabilizer bar. Install mounting bolts and nuts.
10. Remove support and lower vehicle.
11. Tighten stabilizer link lower nuts to 111 Nm (82 ft. lbs.).

Power Hop Damper

REMOVAL & INSTALLATION

1. Raise and support the vehicle.
2. Remove the damper nuts and bolts.
3. Remove the damper bracket from the differential cover if necessary.
4. Remove the from the vehicle.

To install:

5. Install the damper bracket to the rear differential cover if removed. Tighten the bolts (5) to 41 Nm (30 ft. lbs.).
6. Compress the damper for easier installation.
7. Install the damper to the mounting brackets.
8. Install the damper nuts and bolts at the axle end. Tighten the bolts to 102 Nm (75 ft. lbs.).
9. Decompress the damper.
10. Install the damper nut and bolt to the frame. Tighten the bolt to 102 Nm (75 ft. lbs.).
11. Lower the vehicle.

Front Wheel Bearings

ADJUSTMENT

The hub and bearing are an integral unit. No adjustment is possible.

REMOVAL & INSTALLATION

1. Raise and support the vehicle.
2. Remove the wheel and tire assembly.
3. Remove the brake caliper and rotor.
4. Remove the ABS wheel speed sensor if equipped.

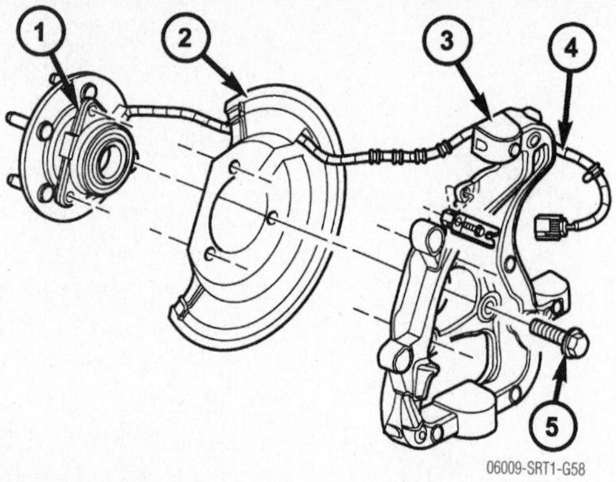

Front hub and bearing assembly. (1) hub/bearing, (2) dust shield (not on all SRT-10 models), (3) knuckle, (4) ABS wire, (5) hub bolts

5. Remove the three hub/bearing mounting bolts from the steering knuckle.
6. Slide the hub/bearing out of the steering knuckle.

To install

7. Install the hub/bearing into the steering knuckle and tighten the bolts to 163 Nm (120 ft. lbs.).
8. Install the brake rotor and caliper.
9. Install the ABS wheel speed sensor if equipped.
10. Install the wheel and tire assembly.
11. Remove the support and lower vehicle.

BRAKES

Brake Caliper

REMOVAL & INSTALLATION

Front

**** WARNING**

Never allow the disc brake caliper to hang from the brake hose. Damage to the brake hose will result. Provide a suitable support to hang the caliper securely.

1. Raise and support the vehicle.
2. Remove the tire and wheel assembly.
3. Compress the disc brake caliper.
4. Remove the banjo bolt and discard the copper washers.
5. Remove the caliper mounting bolts.
6. Remove the disc brake caliper.

To install:

➡Install a new copper washers on the banjo bolt when installing.

7. Install the disc brake pads, anti-rattle springs and pad retaining pin.
8. Install the disc brake caliper.

**** CAUTION**

Verify brake hose is not twisted or kinked before tightening fitting bolt.

9. Install the brake hose banjo bolt with new copper washers to the caliper. Tighten to 27 Nm (20 ft. lbs.)
10. Install the caliper mounting bolts. Tighten to 32 Nm (24 ft. lbs.).
11. Bleed the base brake system.
12. Install the tire and wheel assembly.
13. Lower the vehicle.

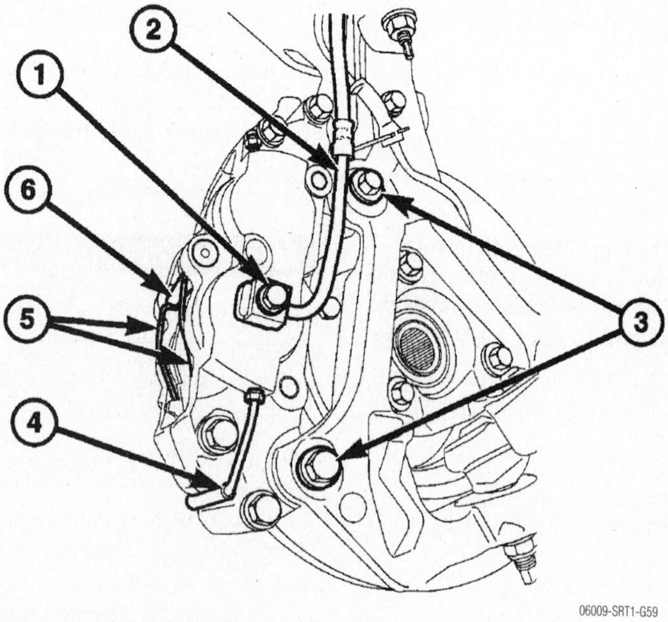

Front caliper. (1) banjo bolt, (2) brake line, (3) caliper mounting bolts, (4) fluid line, (5) anti-rattle springs, (6) pad retaining pin

Rear

1. Raise and support the vehicle.
2. Remove the tire and wheel assembly.
3. Compress the disc brake caliper using tool C4212F, or equivalent.
4. Remove the caliper pin bolts.
5. Remove the banjo bolt and discard the copper washer.

**** WARNING**

Never allow the disc brake caliper to hang from the brake hose. Damage to the brake hose with result. Provide a suitable support to hang the caliper securely.

6. Remove the rear disc brake caliper.

To install:

➡Install a new copper washers on the banjo bolt when installing

7. Install the rear disc brake caliper.
8. Install the banjo bolt with new copper washers to the caliper. Tighten to 27 Nm (20 ft. lbs.)
9. Install the caliper pin bolts. Tighten to 30 Nm (22 ft. lbs.)
10. Bleed the base brake system.
11. Install the tire and wheel assembly.
12. Lower the vehicle.

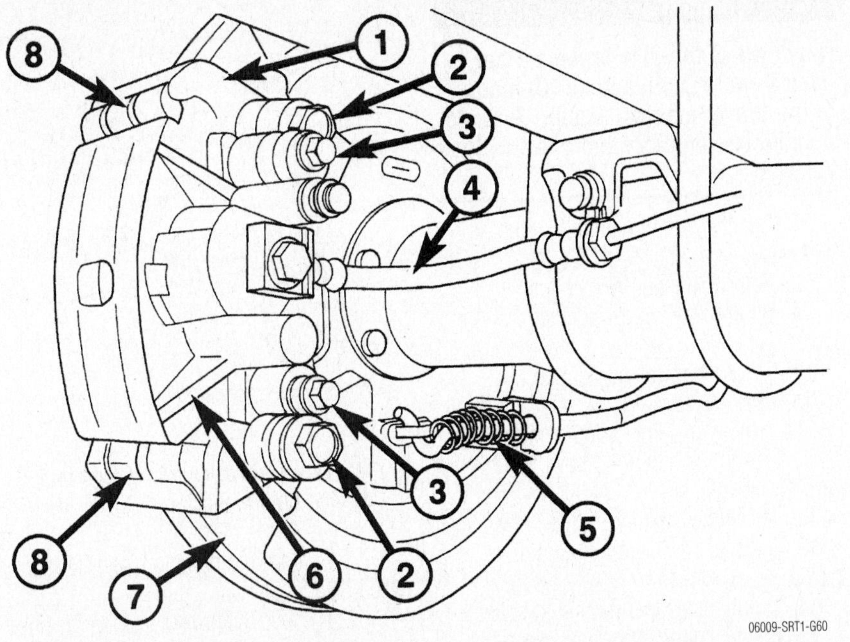

Rear caliper. (1) adapter, (2) adapter bolts, (3) caliper mounting bolts, (4) brake line, (5) parking brake spring, (6) caliper, (7) rotor, (8) anti-rattle springs

Disc Brake Pads

REMOVAL AND INSTALLATION

Front

1. Raise and support vehicle.
2. Remove the wheel and tire assemblies.
3. Compress the caliper.
4. Remove the caliper.
5. Remove the caliper by tilting the top up and off the caliper adapter.

➡Do not allow brake hose to support caliper assembly.

6. Support and hang the caliper.
7. Remove the inboard brake shoe from the caliper adapter.
8. Remove the outboard brake shoe from the caliper adapter.

➡Anti-rattle springs are not interchangeable.

9. Remove the top anti-rattle springs from the caliper adapter.
10. Remove the bottom anti-rattle springs from the caliper adapter.

To install:

11. Bottom pistons in caliper bore with C-clamp. Place an old brake shoe between a C-clamp and caliper piston.
12. Clean caliper mounting adapter and anti-rattle springs.
13. Lubricate anti-rattle springs with Mopar brake grease, or equivalent.
14. Install new top anti-rattle springs.
15. Install new bottom anti-rattle springs.
16. Install inboard brake shoe in adapter.
17. Install outboard brake shoe in adapter.
18. Tilt the top of the caliper over rotor and under adapter. Then push the bottom of the caliper down onto the adapter.
19. Install caliper.
20. Install wheel and tire assemblies and lower vehicle.
21. Apply brakes several times to seat caliper pistons and brake shoes and obtain firm pedal.
22. Top off master cylinder fluid level.

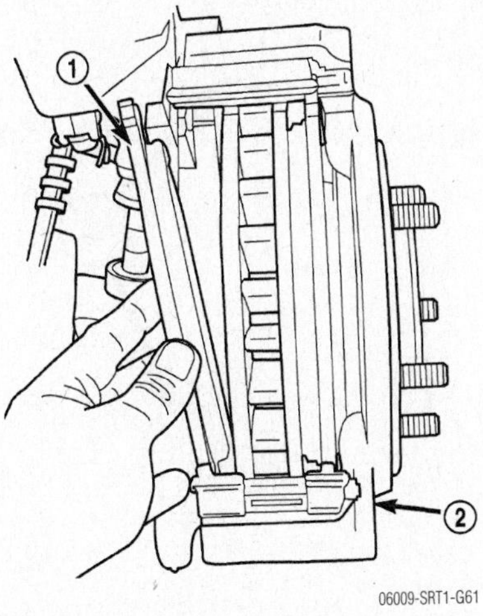

Front inboard pad removal. (1) pad, (2) adapter

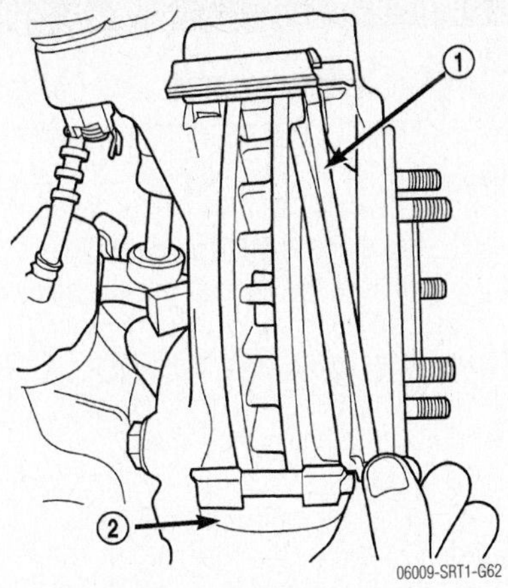

Front outboard pad removal

Rear

1. Raise and support the vehicle.
2. Remove the rear wheel and tire assemblies.
3. Compress the caliper.
4. Remove caliper slide bolts.

➡Do not allow brake hose to support caliper assembly.

5. Remove the caliper, and then tilt the top up and off the caliper adapter.
6. Remove inboard brake shoe from the caliper adapter.
7. Remove outboard brake shoe from caliper adapter.

➡Anti-rattle springs are not interchangeable.

8. Remove the top anti-rattle spring from the caliper adapter.

To install:

9. Clean caliper mounting adapter and anti-rattle springs.
10. Lubricate anti-rattle springs with Mopar brake grease.
11. Install new top anti-rattle spring.
12. Install new bottom anti-rattle spring.
13. Install inboard brake shoe in adapter.
14. Install outboard brake shoe in adapter.
15. Tilt the bottom of the caliper over rotor and under adapter. Then push the top of the caliper down onto the adapter.
16. Install caliper.
17. Install wheel and tire assemblies and lower vehicle.
18. Apply brakes several times to seat caliper pistons and brake shoes and obtain firm pedal.
19. Top off master cylinder fluid level.

Caliper Adapter

REMOVAL & INSTALLATION

Front

1. Raise and support the vehicle.
2. Remove the tire and wheel assembly.
3. Remove the disc brake caliper.
4. Remove the bolts securing the caliper adapter to the steering knuckle.
5. Remove the caliper adapter.

To install:

6. Install the caliper adapter to the steering knuckle.
7. Install the caliper adapter mounting bolts and tighten to 176 Nm (130 ft. lbs.).
8. Install the disc brake caliper.
9. Install the tire and wheel assembly.
10. Remove the support and lower the vehicle.

Rear

1. Raise and support the vehicle.
2. Remove the wheel and tire assembly.
3. Drain a small amount of fluid from master cylinder brake reservoir with a clean suction gun.
4. Bottom the caliper pistons into the caliper by prying the caliper over.
5. Remove the caliper slide bolts.
6. Remove the disc brake caliper from the mount.

⁂ WARNING

Never allow the disc brake caliper to hang from the brake hose. Damage to the brake hose will result. Provide a suitable support to hang the caliper securely.

7. Remove the inboard and outboard brake pads.
8. Remove the anti-rattle clips.
9. Remove the caliper adapter mounting bolts.
10. Remove the caliper adapter.

To install:

11. Install the caliper adapter mounting bolts. Tighten the mounting bolts to 135 Nm (100 ft. lbs).
12. Install the anti-rattle clips.
13. Install the inboard and outboard pads.
14. Install the caliper mounting bolts.
15. Install the tire and wheel assembly.

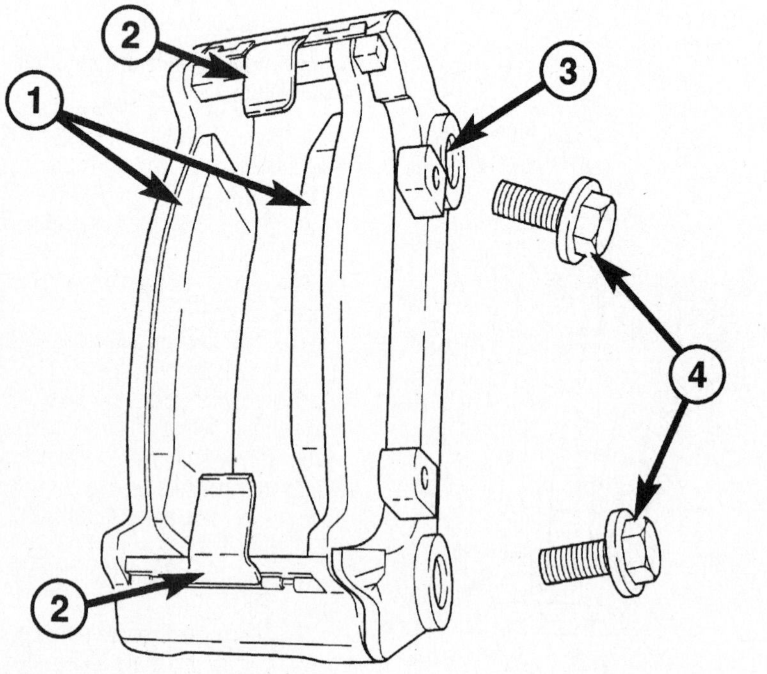

Front caliper adapter (rear similar). (1) pads, (2) anti-rattle clips, (3) adapter, (4) adapter mounting bolts

Rear Caliper Adapter Mount

REMOVAL & INSTALLATION

1. Remove wheel and tire assembly.
2. Remove the disc brake caliper.
3. Remove the caliper adapter.
4. Remove the rotor.
5. Remove the axle shaft.
6. Remove the park brake shoes.

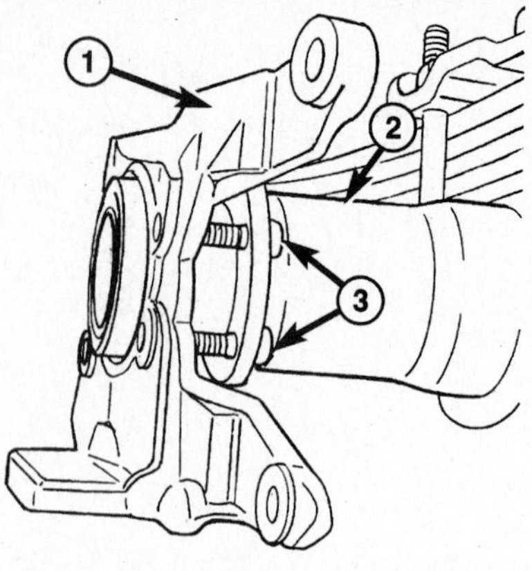

Rear caliper adapter mount. (1) mount, (2) axle, (3) mount bolts

7. Remove the parking brake cable from the brake lever.
8. Remove the bolts attaching the support plate to the axle and remove the support plate.
9. Remove the caliper adapter mount from the axle housing.

To install:

10. Install the caliper adapter mount on the axle housing.
11. Install support plate on axle flange. Tighten attaching bolts to 68 Nm (50 ft. lbs.).
12. Install parking brake cable in the brake lever.
13. Install the park brake shoes.
14. Install axle shaft.
15. Adjust brake shoes to drum with brake gauge.
16. Install the rotor.
17. Install the caliper adapter.
18. Install the caliper.
19. Install wheel and tire assembly.

Parking Brake Shoes

REMOVAL

1. Raise and support the vehicle.
2. Remove the tire and wheel assembly.
3. Remove the disc brake caliper.
4. Remove the disc brake rotor.
5. Lockout the parking brake cable.
6. Disengage the park brake cable from behind the rotor assembly to allow easier disassembly of the park brake shoes.
7. Remove the axle shaft.
8. Disassemble the rear park brake shoes.

INSPECTION

Riveted brake shoes should be replaced when worn to within 0.78 mm (1⁄32 in.) of the rivet heads. Bonded lining should be replaced when worn to a thickness of 1.6 mm 1⁄16 in.).

Examine the lining contact pattern to determine if the shoes are bent or the drum is tapered. The lining should exhibit contact across its entire width. Shoes exhibiting contact only on one side should be replaced and the drum checked for runout or taper. Inspect the adjuster screw assembly. Replace the assembly if the star wheel or threads are damaged, or the components are severely rusted or corroded.

Discard the brake springs and retainer components if worn, distorted or collapsed. Also replace the springs if a brake drag condition had occurred. Overheating will distort and weaken the springs.

Inspect the brake shoe contact pads on the support plate, replace the support plate if any of the pads are worn or rusted through. Also replace the plate if it is bent or distorted.

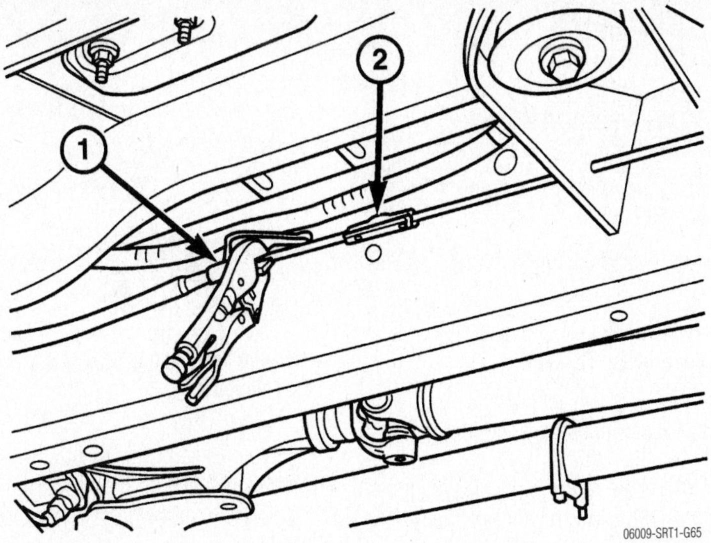

Lockout the parking brake cable. (1) locking pliers, (2) cable

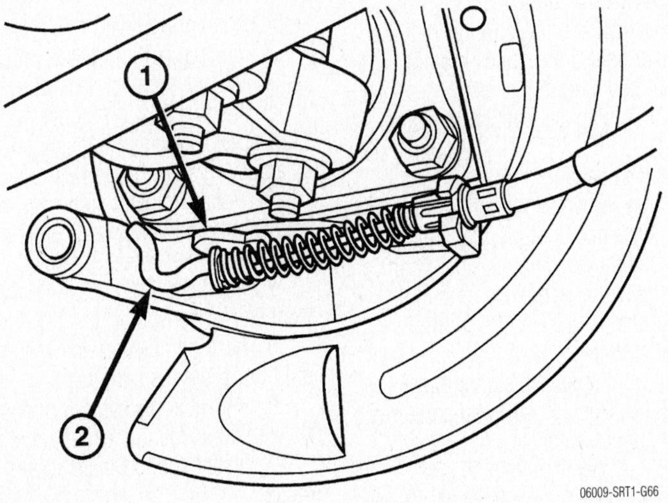

Disengage the cable (2) from the lever (1)

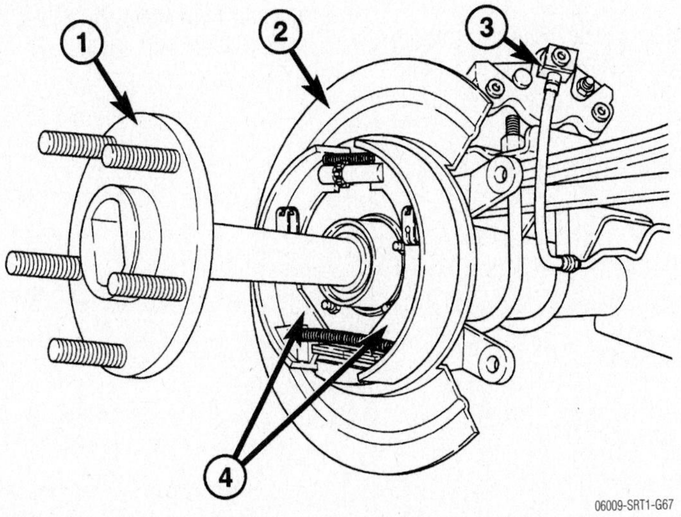

Remove the axle shaft. (1) axle shaft, (2) dust shield, (3) caliper, (4) shoes

INSTALLATION

➡On a new vehicle or after parking brake lining replacement, it is recommended that the parking brake system be conditioned prior to use. This is done by making one stop from 25 mph on dry pavement or concrete using light to moderate force on the parking brake foot pedal.

1. Reassemble the rear park brake shoes.
2. Install the axle shaft.
3. Install the park brake cable to the lever behind the support plate.
4. Unlock the park brake cable.
5. Install the disc brake rotor.
6. Install the disc brake caliper.
7. Adjust the rear brake shoes.
8. Install the tire and wheel assembly.
9. Lower the vehicle.

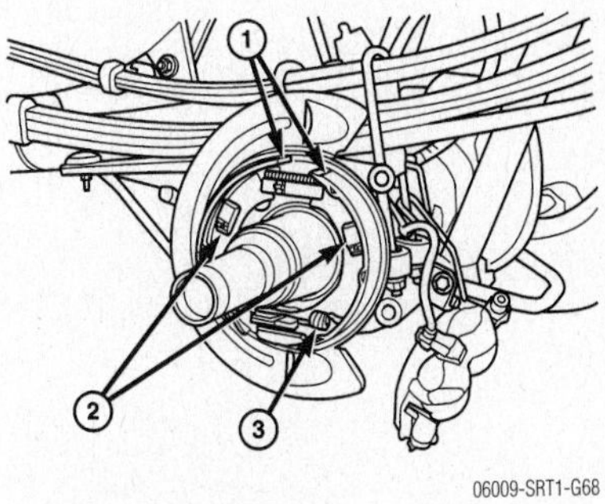

Install the shoes. (1) shoes, (2) hold-down clips, (3) lower spring

ADJUSTMENTS

With Brake Gauge

➡Before adjusting the park brake shoes be sure that the park brake pedal is in the fully released position. If park brake pedal is not in the fully released position, the park brake shoes can not be accurately adjusted.

1. Raise vehicle.
2. Remove tire and wheel.
3. Remove disc brake caliper from caliper adapter.
4. Remove rotor from the axle shaft.

➡When measuring the brake drum diameter, the diameter should be measured in the center of the area in which the park brake shoes contact the surface of the brake drum.

5. Using Brake Shoe Gauge, Special Tool C-3919, or equivalent, accurately measure the inside diameter of the park brake drum portion of the rotor.
6. Using a ruler that reads in 64th of an inch, accurately read the measurement of the inside diameter of the park brake drum from the special tool.
7. Reduce the inside diameter measurement of the brake drum that was taken using Special Tool C-3919 by 1⁄64 of an inch. Reset Gauge, Brake Shoe, Special Tool C-3919 or the equivalent used, so that the outside measurement jaws are set to the reduced measurement.
8. Place Gauge, Brake Shoe, Special Tool C-3919, or equivalent over the park brake shoes. The special tool must be located diagonally across at the top of one shoe and bottom of opposite shoe (widest point) of the park brake shoes.
9. Using the star wheel adjuster, adjust the park brake shoes until the lining on the park brake shoes just touches the jaws on the special tool.

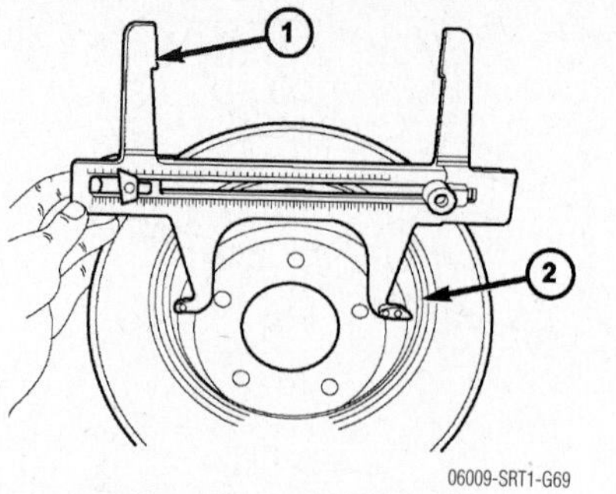

Measure the drum inner diameter. (1) brake shoe gauge, (2) drum

10. Repeat step 8 above and measure shoes in both directions.
11. Install brake rotor on the axle shaft.
12. Rotate rotor to verify that the park brake shoes are not dragging on the brake drum. If park brake shoes are dragging, remove rotor and back off star wheel adjuster one notch and recheck for brake shoe drag against drum. Continue with the previous step until brake shoes are not dragging on brake drum.
13. Install disc brake caliper on caliper adapter.
14. Install wheel and tire.
15. Tighten the wheel mounting nuts in the proper sequence until all nuts are torqued to half the specified torque. Then repeat the tightening sequence to the full specified torque of 129 Nm (95 ft. lbs.).
16. Lower vehicle.
17. Apply and release the park brake pedal one time. This will seat and correctly adjust the park brake cables.
18. Road test the vehicle to ensure proper function of the vehicle's brake system.

With Adjusting Tool

Adjustment can be made with a standard brake gauge or with adjusting tool. Adjust-

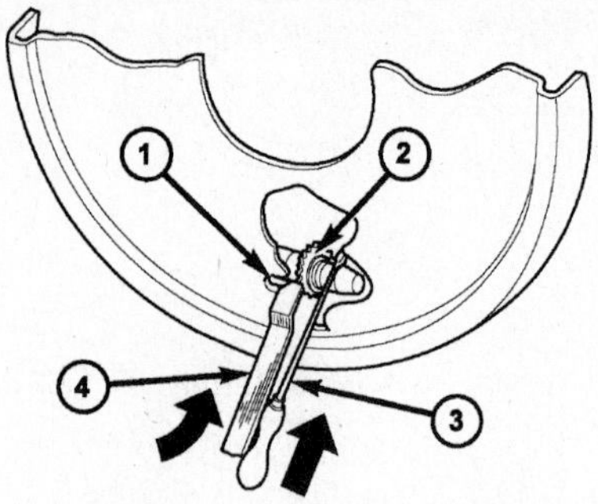

Adjusting the parking brake with a brake adjusting tool

ment is performed with the complete brake assembly installed on the backing plate.

1. Be sure parking brake lever is fully released.
2. Raise vehicle so rear wheels can be rotated freely.
3. Remove plug from each access hole in brake support plates.
4. Loosen parking brake cable adjustment nut until there is slack in front cable.
5. Insert adjusting tool through support plate access hole and engage tool in teeth of adjusting screw star wheel.
6. Rotate adjuster screw star wheel (move tool handle upward) until slight drag can be felt when wheel is rotated.
7. Push and hold adjuster lever away from star wheel with thin screwdriver.
8. Back off adjuster screw star wheel until brake drag is eliminated.
9. Repeat adjustment at opposite wheel. Be sure adjustment is equal at both wheels.
10. Install support plate access hole plugs.
11. Adjust parking brake cable and lower vehicle.
12. Depress park brake pedal and make sure park brakes hold the vehicle stationary.
13. Release park brake pedal.

GLOSSARY

ABS: Anti-lock braking system. An electro-mechanical braking system which is designed to minimize or prevent wheel lock-up during braking.

ABSOLUTE PRESSURE: Atmospheric (barometric) pressure plus the pressure gauge reading.

ACCELERATOR PUMP: A small pump located in the carburetor that feeds fuel into the air/fuel mixture during acceleration.

ACCUMULATOR: A device that controls shift quality by cushioning the shock of hydraulic oil pressure being applied to a clutch or band.

ACTUATING MECHANISM: The mechanical output devices of a hydraulic system, for example, clutch pistons and band servos.

ACTUATOR: The output component of a hydraulic or electronic system.

ADVANCE: Setting the ignition timing so that spark occurs earlier before the piston reaches top dead center (TDC).

ADAPTIVE MEMORY (ADAPTIVE STRATEGY): The learning ability of the TCM or PCM to redefine its decision-making process to provide optimum shift quality.

AFTER TOP DEAD CENTER (ATDC): The point after the piston reaches the top of its travel on the compression stroke.

AIR BAG: Device on the inside of the car designed to inflate on impact of crash, protecting the occupants of the car.

AIR CHARGE TEMPERATURE (ACT) SENSOR: The temperature of the airflow into the engine is measured by an ACT sensor, usually located in the lower intake manifold or air cleaner.

AIR CLEANER: An assembly consisting of a housing, filter and any connecting ductwork. The filter element is made up of a porous paper, sometimes with a wire mesh screening, and is designed to prevent airborne particles from entering the engine through the carburetor or throttle body.

AIR INJECTION: One method of reducing harmful exhaust emissions by injecting air into each of the exhaust ports of an engine. The fresh air entering the hot exhaust manifold causes any remaining fuel to be burned before it can exit the tailpipe.

AIR PUMP: An emission control device that supplies fresh air to the exhaust manifold to aid in more completely burning exhaust gases.

AIR/FUEL RATIO: The ratio of air-to-gasoline by weight in the fuel mixture drawn into the engine.

ALDL (assembly line diagnostic link): Electrical connector for scanning ECM/PCM/TCM input and output devices.

ALIGNMENT RACK: A special drive-on vehicle lift apparatus/measuring device used to adjust a vehicle's toe, caster and camber angles.

ALL WHEEL DRIVE: Term used to describe a full time four wheel drive system or any other vehicle drive system that continuously delivers power to all four wheels. This system is found primarily on station wagon vehicles and SUVs not utilized for significant off road use.

ALTERNATING CURRENT (AC): Electric current that flows first in one direction, then in the opposite direction, continually reversing flow.

ALTERNATOR: A device which produces AC (alternating current) which is converted to DC (direct current) to charge the car battery.

AMMETER: An instrument, calibrated in amperes, used to measure the flow of an electrical current in a circuit. Ammeters are always connected in series with the circuit being tested.

AMPERAGE: The total amount of current (amperes) flowing in a circuit.

AMPLIFIER: A device used in an electrical circuit to increase the voltage of an output signal.

AMP/HR. RATING (BATTERY): Measurement of the ability of a battery to deliver a stated amount of current for a stated period of time. The higher the amp/hr. rating, the better the battery.

AMPERE: The rate of flow of electrical current present when one volt of electrical pressure is applied against one ohm of electrical resistance.

ANALOG COMPUTER: Any microprocessor that uses similar (analogous) electrical signals to make its calculations.

ANODIZED: A special coating applied to the surface of aluminum valves for extended service life.

ANTIFREEZE: A substance (ethylene or propylene glycol) added to the coolant to prevent freezing in cold weather.

ANTI-FOAM AGENTS: Minimize fluid foaming from the whipping action encountered in the converter and planetary action.

ANTI-WEAR AGENTS: Zinc agents that control wear on the gears, bushings, and thrust washers.

ANTI-LOCK BRAKING SYSTEM: A supplementary system to the base hydraulic system that prevents sustained lock-up of the wheels during braking as well as automatically controlling wheel slip.

ANTI-ROLL BAR: See stabilizer bar.

ARC: A flow of electricity through the air between two electrodes or contact points that produces a spark.

ARMATURE: A laminated, soft iron core wrapped by a wire that converts electrical energy to mechanical energy as in a motor or relay. When rotated in a magnetic field, it changes mechanical energy into electrical energy as in a generator.

ATDC: After Top Dead Center.

ATF: Automatic transmission fluid.

ATMOSPHERIC PRESSURE: The pressure on the Earth's surface caused by the weight of the air in the atmosphere. At sea level, this pressure is 14.7 psi at 32°F (101 kPa at 0°C).

ATOMIZATION: The breaking down of a liquid into a fine mist that can be suspended in air.

AUXILIARY ADD-ON COOLER: A supplemental transmission fluid cooling device that is installed in series with the heat exchanger (cooler), located inside the radiator, to provide additional support to cool the hot fluid leaving the torque converter.

AUXILIARY PRESSURE: An added fluid pressure that is introduced into a regulator or balanced valve system to control valve movement. The auxiliary pressure itself can be either a fixed or a variable value. (See balanced valve; regulator valve.)

AWD: All wheel drive.

AXIAL FORCE: A side or end thrust force acting in or along the same plane as the power flow.

AXIAL PLAY: Movement parallel to a shaft or bearing bore.

AXLE CAPACITY: The maximum load-carrying capacity of the axle itself, as specified by the manufacturer. This is usually a higher number than the GAWR.

AXLE RATIO: This is a number (3.07:1, 4.56:1, for example) expressing the ratio between driveshaft revolutions and wheel revolutions. A low numerical ratio allows the engine to work easier because it doesn't have to turn as fast. A high numerical ratio means that the engine has to turn more rpm's to move the wheels through the same number of turns.

BACKFIRE: The sudden combustion of gases in the intake or exhaust system that results in a loud explosion.

BACKLASH: The clearance or play between two parts, such as meshed gears.

BACKPRESSURE: Restrictions in the exhaust system that slow the exit of exhaust gases from the combustion chamber.

BAKELITE®: A heat resistant, plastic insulator material commonly used in printed circuit boards and transistorized components.

BALANCED VALVE: A valve that is positioned by opposing auxiliary hydraulic pressures and/or spring force. Examples include mainline regulator, throttle, and governor valves. (See regulator valve.)

BAND: A flexible ring of steel with an inner lining of friction material. When tightened around the outside of a drum, a planetary member is held stationary to the transmission/transaxle case.

BALL BEARING: A bearing made up of hardened inner and outer races between which hardened steel balls roll.

BALL JOINT: A ball and matching socket connecting suspension components (steering knuckle to lower control arms). It permits rotating movement in any direction between the components that are joined.

BARO (BAROMETRIC PRESSURE SENSOR): Measures the change in the intake manifold pressure caused by changes in altitude.

BAROMETRIC MANIFOLD ABSOLUTE PRESSURE (BMAP) SENSOR: Operates similarly to a conventional MAP sensor; reads intake manifold pressure and is also responsible for determining altitude and barometric pressure prior to engine operation.

BAROMETRIC PRESSURE: (See atmospheric pressure.)

BALLAST RESISTOR: A resistor in the primary ignition circuit that lowers voltage after the engine is started to reduce wear on ignition components.

BATTERY: A direct current electrical storage unit, consisting of the basic active materials of lead and sulfuric acid, which converts chemical energy into electrical energy. Used to provide current for the operation of the starter as well as other equipment, such as the radio, lighting, etc.

BEAD: The portion of a tire that holds it on the rim.

BEARING: A friction reducing, supportive device usually located between a stationary part and a moving part.

BEFORE TOP DEAD CENTER (BTDC): The point just before the piston reaches the top of its travel on the compression stroke.

BELTED TIRE: Tire construction similar to bias-ply tires, but using two or more layers of reinforced belts between body plies and the tread.

BEZEL: Piece of metal surrounding radio, headlights, gauges or similar components; sometimes used to hold the glass face of a gauge in the dash.

BIAS-PLY TIRE: Tire construction, using body ply reinforcing cords which run at alternating angles to the center line of the tread.

BI-METAL TEMPERATURE SENSOR: Any sensor or switch made of two dissimilar types of metal that bend when heated or cooled due to the different expansion rates of the alloys. These types of sensors usually function as an on/off switch.

BLOCK: See Engine Block.

BLOW-BY: Combustion gases, composed of water vapor and unburned fuel, that leak past the piston rings into the crankcase during normal engine operation. These gases are removed by the PCV system to prevent the buildup of harmful acids in the crankcase.

BOOK TIME: See Labor Time.

BOOK VALUE: The average value of a car, widely used to determine trade-in and resale value.

BOOST VALVE: Used at the base of the regulator valve to increase mainline pressure.

BORE: Diameter of a cylinder.

BRAKE CALIPER: The housing that fits over the brake disc. The caliper holds the brake pads, which are pressed against the discs by the caliper pistons when the brake pedal is depressed.

BRAKE HORSEPOWER (BHP): The actual horsepower available at the engine flywheel as measured by a dynamometer.

BRAKE FADE: Loss of braking power, usually caused by excessive heat after repeated brake applications.

BRAKE HORSEPOWER: Usable horsepower of an engine measured at the crankshaft.

BRAKE PAD: A brake shoe and lining assembly used with disc brakes.

BRAKE PROPORTIONING VALVE: A valve on the master cylinder which restricts hydraulic brake pressure to the wheels to a specified amount, preventing wheel lock-up.

BREAKAWAY: Often used by Chrysler to identify first-gear operation in D and 2 ranges. In these ranges, first-gear operation depends on a one-way roller clutch that holds on acceleration and releases (breaks away) on deceleration, resulting in a freewheeling coast-down condition.

BRAKE SHOE: The backing for the brake lining. The term is, however, usually applied to the assembly of the brake backing and lining.

BREAKER POINTS: A set of points inside the distributor, operated by a cam, which make and break the ignition circuit.

BRINNELLING: A wear pattern identified by a series of indentations at regular intervals. This condition is caused by a lack of lube, overload situations, and/or vibrations.

BTDC: Before Top Dead Center.

BUMP: Sudden and forceful apply of a clutch or band.

BUSHING: A liner, usually removable, for a bearing; an anti-friction liner used in place of a bearing.

CALIFORNIA ENGINE: An engine certified by the EPA for use in California only; conforms to more stringent emission regulations than Federal engine.

CALIPER: A hydraulically activated device in a disc brake system, which is mounted straddling the brake rotor (disc). The caliper contains at least one piston and two brake pads. Hydraulic pressure on the piston(s) forces the pads against the rotor.

CAPACITY: The quantity of electricity that can be delivered from a unit, as from a battery in ampere-hours, or output, as from a generator.

CAMBER: One of the factors of wheel alignment. Viewed from the front of the car, it is the inward or outward tilt of the wheel. The top of the tire will lean outward (positive camber) or inward (negative camber).

CAMSHAFT: A shaft in the engine on which are the lobes (cams) which operate the valves. The camshaft is driven by the crankshaft, via a belt, chain or gears, at one half the crankshaft speed.

CAPACITOR: A device which stores an electrical charge.

CARBON MONOXIDE (CO): A colorless, odorless gas given off as a normal byproduct of combustion. It is poisonous and extremely dangerous in confined areas, building up slowly to toxic levels without warning if adequate ventilation is not available.

CARBURETOR: A device, usually mounted on the intake manifold of an engine, which mixes the air and fuel in the proper proportion to allow even combustion.

CASTER: The forward or rearward tilt of an imaginary line drawn through the upper ball joint and the center of the wheel. Viewed from the sides, positive caster (forward tilt) lends directional stability, while negative caster (rearward tilt) produces instability.

CATALYTIC CONVERTER: A device installed in the exhaust system, like a muffler, that converts harmful byproducts of combustion into carbon dioxide and water vapor by means of a heat-producing chemical reaction.

CENTRIFUGAL ADVANCE: A mechanical method of advancing the spark timing by using flyweights in the distributor that react to centrifugal force generated by the distributor shaft rotation.

CENTRIFUGAL FORCE: The outward pull of a revolving object, away from the center of revolution. Centrifugal force increases with the speed of rotation.

CETANE RATING: A measure of the ignition value of diesel fuel. The higher the cetane rating, the better the fuel. Diesel fuel cetane rating is roughly comparable to gasoline octane rating.

CHECK VALVE: Any one-way valve installed to permit the flow of air, fuel or vacuum in one direction only.

CHOKE: The valve/plate that restricts the amount of air entering an engine on the induction stroke, thereby enriching the air/fuel ratio.

CHUGGLE: Bucking or jerking condition that may be engine related and may be most noticeable when converter clutch is engaged; similar to the feel of towing a trailer.

CIRCLIP: A split steel snapring that fits into a groove to hold various parts in place.

CIRCUIT BREAKER: A switch which protects an electrical circuit from overload by opening the circuit when the current flow exceeds a pre-determined level. Some circuit breakers must be reset manually, while most reset automatically.

CIRCUIT: Any unbroken path through which an electrical current can flow. Also used to describe fuel flow in some instances.

CIRCUIT, BYPASS: Another circuit in parallel with the major circuit through which power is diverted.

CIRCUIT, CLOSED: An electrical circuit in which there is no interruption of current flow.

CIRCUIT, GROUND: The non-insulated portion of a complete circuit used as a common potential point. In automotive circuits, the ground is composed of metal parts, such as the engine, body sheet metal, and frame and is usually a negative potential.

CIRCUIT, HOT: That portion of a circuit not at ground potential. The hot circuit is usually insulated and is connected to the positive side of the battery.

CIRCUIT, OPEN: A break or lack of contact in an electrical circuit, either intentional (switch) or unintentional (bad connection or broken wire).

CIRCUIT, PARALLEL: A circuit having two or more paths for current flow with common positive and negative tie points. The same voltage is applied to each load device or parallel branch.

CIRCUIT, SERIES: An electrical system in which separate parts are connected end to end, using one wire, to form a single path for current to flow.

CIRCUIT, SHORT: A circuit that is accidentally completed in an electrical path for which it was not intended.

CLAMPING (ISOLATION) DIODES: Diodes positioned in a circuit to prevent self-induction from damaging electronic components.

CLEARCOAT: A transparent layer which, when sprayed over a vehicle's paint job, adds gloss and depth as well as an additional protective coating to the finish.

CLUTCH: Part of the power train used to connect/disconnect power to the rear wheels.

CLUTCH, FLUID: The same as a fluid coupling. A fluid clutch or coupling performs the same function as a friction clutch by utilizing fluid friction and inertia as opposed to solid friction used by a friction clutch. (See fluid coupling.)

CLUTCH, FRICTION: A coupling device that provides a means of smooth and positive engagement and disengagement of engine torque to the vehicle powertrain. Transmission of power through the clutch is accomplished by bringing one or more rotating drive members into contact with complementing driven members.

COAST: Vehicle deceleration caused by engine braking conditions.

COEFFICIENT OF FRICTION: The amount of surface tension between two contacting surfaces; identified by a scientifically calculated number.

COIL: Part of the ignition system that boosts the relatively low voltage supplied by the car's electrical system to the high voltage required to fire the spark plugs.

COMBINATION MANIFOLD: An assembly which includes both the intake and exhaust manifolds in one casting.

COMBINATION VALVE: A device used in some fuel systems that routes fuel vapors to a charcoal storage canister instead of venting them into the atmosphere. The valve relieves fuel tank pressure and allows fresh air into the tank as the fuel level drops to prevent a vapor lock situation.

COMBUSTION CHAMBER: The part of the engine in the cylinder head where combustion takes place.

COMPOUND GEAR: A gear consisting of two or more simple gears with a common shaft.

COMPOUND PLANETARY: A gearset that has more than the three elements found in a simple gearset and is constructed by combining members of two planetary gearsets to create additional gear ratio possibilities.

COMPRESSION CHECK: A test involving removing each spark plug and inserting a gauge. When the engine is cranked, the gauge will record a pressure reading in the individual cylinder. General operating condition can be determined from a compression check.

COMPRESSION RATIO: The ratio of the volume between the piston and cylinder head when the piston is at the bottom of its stroke (bottom dead center) and when the piston is at the top of its stroke (top dead center).

COMPUTER: An electronic control module that correlates input data according to prearranged engineered instructions; used for the management of an actuator system or systems.

CONDENSER: An electrical device which acts to store an electrical charge, preventing voltage surges.

2. A radiator-like device in the air conditioning system in which refrigerant gas condenses into a liquid, giving off heat.

CONDUCTOR: Any material through which an electrical current can be transmitted easily.

CONNECTING ROD: The connecting link between the crankshaft and piston.

CONSTANT VELOCITY JOINT: Type of universal joint in a halfshaft assembly in which the output shaft turns at a constant angular velocity without variation, provided that the speed of the input shaft is constant.

CONTINUITY: Continuous or complete circuit. Can be checked with an ohmmeter.

CONTROL ARM: The upper or lower suspension components which are mounted on the frame and support the ball joints and steering knuckles.

CONVENTIONAL IGNITION: Ignition system which uses breaker points.

CONVERTER: (See torque converter.)

CONVERTER LOCKUP: The switching from hydrodynamic to direct mechanical drive, usually through the application of a friction element called the converter clutch.

COOLANT: Mixture of water and anti-freeze circulated through the engine to carry off heat produced by the engine.

CORROSION INHIBITOR: An inhibitor in ATF that prevents corrosion of bushings, thrust washers, and oil cooler brazed joints.

COUNTERSHAFT: An intermediate shaft which is rotated by a mainshaft and transmits, in turn, that rotation to a working part.

COUPLING PHASE: Occurs when the torque converter is operating at its greatest hydraulic efficiency. The speed differential between the impeller and the turbine is at its minimum. At this point, the stator freewheels, and there is no torque multiplication.

CRANKCASE: The lower part of an engine in which the crankshaft and related parts operate.

CRANKSHAFT: Engine component (connected to pistons by connecting rods) which converts the reciprocating (up and down) motion of pistons to rotary motion used to turn the driveshaft.

CURB WEIGHT: The weight of a vehicle without passengers or payload, but including all fluids (oil, gas, coolant, etc.) and other equipment specified as standard.

CURRENT: The flow (or rate) of electrons moving through a circuit. Current is measured in amperes (amp).

CURRENT FLOW CONVENTIONAL: Current flows through a circuit from the positive terminal of the source to the negative terminal (plus to minus).

CURRENT FLOW, ELECTRON: Current or electrons flow from the negative terminal of the source, through the circuit, to the positive terminal (minus to plus).

CV-JOINT: Constant velocity joint.

CYCLIC VIBRATIONS: The off-center movement of a rotating object that is affected by its initial balance, speed of rotation, and working angles.

CYLINDER BLOCK: See engine block.

CYLINDER HEAD: The detachable portion of the engine, usually fastened to the top of the cylinder block and containing all or most of the combustion chambers. On overhead valve engines, it contains the valves and their operating parts. On overhead cam engines, it contains the camshaft as well.

CYLINDER: In an engine, the round hole in the engine block in which the piston(s) ride.

DATA LINK CONNECTOR (DLC): Current acronym/term applied to the federally mandated, diagnostic junction connector that is used to monitor ECM/PC/TCM inputs, processing strategies, and outputs including diagnostic trouble codes (DTCs).

DEAD CENTER: The extreme top or bottom of the piston stroke.

DECELERATION BUMP: When referring to a torque converter clutch in the applied position, a sudden release of the accelerator pedal causes a forceful reversal of power through the drivetrain (engine braking), just prior to the apply plate actually being released.

DELAYED (LATE OR EXTENDED): Condition where shift is expected but does not occur for a period of time, for example, where clutch or band engagement does not occur as quickly as expected during part throttle or wide open throttle apply of accelerator or when manually downshifting to a lower range.

DETENT: A spring-loaded plunger, pin, ball, or pawl used as a holding device on a ratchet wheel or shaft. In automatic transmissions, a detent mechanism is used for locking the manual valve in place.

DETENT DOWNSHIFT: (See kickdown.)

DETERGENT: An additive in engine oil to improve its operating characteristics.

DETONATION: An unwanted explosion of the air/fuel mixture in the combustion chamber caused by excess heat and compression, advanced timing, or an overly lean mixture. Also referred to as "ping".

DEXRON®: A brand of automatic transmission fluid.

DIAGNOSTIC TROUBLE CODES (DTCs): A digital display from the control module memory that identifies the input, processor, or output device circuit that is related to the powertrain emission/driveability malfunction detected. Diagnostic trouble codes can be read by the MIL to flash any codes or by using a handheld scanner.

DIAPHRAGM: A thin, flexible wall separating two cavities, such as in a vacuum advance unit.

DIESELING: The engine continues to run after the car is shut off; caused by fuel continuing to be burned in the combustion chamber.

DIFFERENTIAL: A geared assembly which allows the transmission of motion between drive axles, giving one axle the ability to rotate faster than the other, as in cornering.

DIFFERENTIAL AREAS: When opposing faces of a spool valve are acted upon by the same pressure but their areas differ in size, the face with the larger area produces the differential force and valve movement. (See spool valve.)

DIFFERENTIAL FORCE: (See differential areas)

DIGITAL READOUT: A display of numbers or a combination of numbers and letters.

DIGITAL VOLT OHMMETER: An electronic diagnostic tool used to measure voltage, ohms and amps as well as several other functions, with the readings displayed on a digital screen in tenths, hundredths and thousandths.

DIODE: An electrical device that will allow current to flow in one direction only.

DIRECT CURRENT (DC): Electrical current that flows in one direction only.

DIRECT DRIVE: The gear ratio is 1:1, with no change occurring in the torque and speed input/output relationship.

DISC BRAKE: A hydraulic braking assembly consisting of a brake disc, or rotor, mounted on an axle shaft, and a caliper assembly containing, usually two brake pads which are activated by hydraulic pressure. The pads are forced against the sides of the disc, creating friction which slows the vehicle.

DISPERSANTS: Suspend dirt and prevent sludge buildup in a liquid, such as engine oil.

DOUBLE BUMP (DOUBLE FEEL): Two sudden and forceful applies of a clutch or band.

DISPLACEMENT: The total volume of air that is displaced by all pistons as the engine turns through one complete revolution.

DISTRIBUTOR: A mechanically driven device on an engine which is responsible for electrically firing the spark plug at a pre-determined point of the piston stroke.

DOHC: Double overhead camshaft.

DOUBLE OVERHEAD CAMSHAFT: The engine utilizes two camshafts mounted in one cylinder head. One camshaft operates the exhaust valves, while the other operates the intake valves.

DOWEL PIN: A pin, inserted in mating holes in two different parts allowing those parts to maintain a fixed relationship.

DRIVELINE: The drive connection between the transmission and the drive wheels.

DRIVE TRAIN: The components that transmit the flow of power from the engine to the wheels. The components include the clutch, transmission, drive-shafts (or axle shafts in front wheel drive), U-joints and differential.

DRUM BRAKE: A braking system which consists of two brake shoes and one or two wheel cylinders, mounted on a fixed backing plate, and a brake drum, mounted on an axle, which revolves around the assembly.

DRY CHARGED BATTERY: Battery to which electrolyte is added when the battery is placed in service.

DVOM: Digital volt ohmmeter

DWELL: The rate, measured in degrees of shaft rotation, at which an electrical circuit cycles on and off.

DYNAMIC: An application in which there is rotating or reciprocating motion between the parts.

EARLY: Condition where shift occurs before vehicle has reached proper speed, which tends to labor engine after upshift.

EBCM: See Electronic Control Unit (ECU).

ECM: See Electronic Control Unit (ECU).

ECU: Electronic control unit.

ELECTRODE: Conductor (positive or negative) of electric current.

ELECTROLYSIS: A surface etching or bonding of current conducting transmission/transaxle components that may occur when grounding straps are missing or in poor condition.

ELECTROLYTE: A solution of water and sulfuric acid used to activate the battery. Electrolyte is extremely corrosive.

ELECTROMAGNET: A coil that produces a magnetic field when current flows through its windings.

ELECTROMAGNETIC INDUCTION: A method to create (generate) current flow through the use of magnetism.

ELECTROMAGNETISM: The effects surrounding the relationship between electricity and magnetism.

ELECTROMOTIVE FORCE (EMF): The force or pressure (voltage) that causes current movement in an electrical circuit.

ELECTRONIC CONTROL UNIT: A digital computer that controls engine (and sometimes transmission, brake or other vehicle system) functions based on data received from various sensors. Examples used by some manufacturers include Electronic Brake Control Module (EBCM), Engine Control Module (ECM), Powertrain Control Module (PCM) or Vehicle Control Module (VCM).

ELECTRONIC IGNITION: A system in which the timing and firing of the spark plugs is controlled by an electronic control unit, usually called a module. These systems have no points or condenser.

ELECTRONIC PRESSURE CONTROL (EPC) SOLENOID: A specially designed solenoid containing a spool valve and spring assembly to control fluid mainline pressure. A variable current flow, controlled by the ECM/PCM, varies the internal force of the solenoid on the spool valve and resulting mainline pressure. (See variable force solenoid.)

ELECTRONICS: Miniaturized electrical circuits utilizing semiconductors, solid-state devices, and printed circuits. Electronic circuits utilize small amounts of power.

ELECTRONIFICATION: The application of electronic circuitry to a mechanical device. Regarding automatic transmissions, electrification is incorporated into converter clutch lockup, shift scheduling, and line pressure control systems.

ELECTROSTATIC DISCHARGE (ESD): An unwanted, high-voltage electrical current released by an individual who has taken on a static charge of electricity. Electronic components can be easily damaged by ESD.

ELEMENT: A device within a hydrodynamic drive unit designed with a set of blades to direct fluid flow.

ENAMEL: Type of paint that dries to a smooth, glossy finish.

END BUMP (END FEEL OR SLIP BUMP): Firmer feel at end of shift when compared with feel at start of shift.

END-PLAY: The clearance/gap between two components that allows for expansion of the parts as they warm up, to prevent binding and to allow space for lubrication.

ENERGY: The ability or capacity to do work.

ENGINE: The primary motor or power apparatus of a vehicle, which converts liquid or gas fuel into mechanical energy.

ENGINE BLOCK: The basic engine casting containing the cylinders, the crankshaft main bearings, as well as machined surfaces for the mounting of other components such as the cylinder head, oil pan, transmission, etc.

ENGINE BRAKING: Use of engine to slow vehicle by manually downshifting during zero-throttle coast down.

ENGINE CONTROL MODULE (ECM): Manages the engine and incorporates output control over the torque converter clutch solenoid. (Note: Current designation for the ECM in late model vehicles is PCM.)

ENGINE COOLANT TEMPERATURE (ECT) SENSOR: Prevents converter clutch engagement with a cold engine; also used for shift timing and shift quality.

EP LUBRICANT: EP (extreme pressure) lubricants are specially formulated for use with gears involving heavy loads (transmissions, differentials, etc.).

ETHYL: A substance added to gasoline to improve its resistance to knock, by slowing down the rate of combustion.

ETHYLENE GLYCOL: The base substance of antifreeze.

EXHAUST MANIFOLD: A set of cast passages or pipes which conduct exhaust gases from the engine.

FAIL-SAFE (BACKUP) CONTROL: A substitute value used by the PCM/TCM to replace a faulty signal from an input sensor. The temporary value allows the vehicle to continue to be operated.

FAST IDLE: The speed of the engine when the choke is on. Fast idle speeds engine warm-up.

FEDERAL ENGINE: An engine certified by the EPA for use in any of the 49 states (except California).

FEEDBACK: A circuit malfunction whereby current can find another path to feed load devices.

FEELER GAUGE: A blade, usually metal, of precisely predetermined thickness, used to measure the clearance between two parts.

FILAMENT: The part of a bulb that glows; the filament creates high resistance to current flow and actually glows from the resulting heat.

FINAL DRIVE: An essential part of the axle drive assembly where final gear reduction takes place in the powertrain. In RWD applications and north-south FWD applications, it must also change the power flow direction to the axle shaft by ninety degrees. (Also see axle ratio).

FIRING ORDER: The order in which combustion occurs in the cylinders of an engine. Also the order in which spark is distributed to the plugs by the distributor.

FIRM: A noticeable quick apply of a clutch or band that is considered normal with medium to heavy throttle shift; should not be confused with harsh or rough.

FLAME FRONT: The term used to describe certain aspects of the fuel explosion in the cylinders. The flame front should move in a controlled pattern across the cylinder, rather than simply exploding immediately.

FLARE (SLIPPING): A quick increase in engine rpm accompanied by momentary loss of torque; generally occurs during shift.

FLAT ENGINE: Engine design in which the pistons are horizontally opposed. Porsche, Subaru and some old VW are common examples of flat engines.

FLAT RATE: A dealership term referring to the amount of money paid to a technician for a repair or diagnostic service based on that particular service versus dealership's labor time (NOT based on the actual time the technician spent on the job).

FLAT SPOT: A point during acceleration when the engine seems to lose power for an instant.

FLOODING: The presence of too much fuel in the intake manifold and combustion chamber which prevents the air/fuel mixture from firing, thereby causing a no-start situation.

FLUID: A fluid can be either liquid or gas. In hydraulics, a liquid is used for transmitting force or motion.

FLUID COUPLING: The simplest form of hydrodynamic drive, the fluid coupling consists of two look-alike members with straight radial varies referred to as the impeller (pump) and the turbine. Input torque is always equal to the output torque.

FLUID DRIVE: Either a fluid coupling or a fluid torque converter. (See hydrodynamic drive units.)

FLUID TORQUE CONVERTER: A hydrodynamic drive that has the ability to act both as a torque multiplier and fluid coupling. (See hydrodynamic drive units; torque converter.)

FLUID VISCOSITY: The resistance of a liquid to flow. A cold fluid (oil) has greater viscosity and flows more slowly than a hot fluid (oil).

FLYWHEEL: A heavy disc of metal attached to the rear of the crankshaft. It smoothes the firing impulses of the engine and keeps the crankshaft turning during periods when no firing takes place. The starter also engages the flywheel to start the engine.

FOOT POUND (ft. lbs., lbs. ft. or sometimes, ft. lb.): The amount of energy or work needed to raise an item weighing one pound, a distance of one foot.

FREEZE PLUG: A plug in the engine block which will be pushed out if the coolant freezes. Sometimes called expansion plugs, they protect the block from cracking should the coolant freeze.

FRICTION: The resistance that occurs between contacting surfaces. This relationship is expressed by a ratio called the coefficient of friction (CL).

FRICTION, COEFFICIENT OF: The amount of surface tension between two contacting surfaces; expressed by a scientifically calculated number.

FRONT END ALIGNMENT: A service to set caster, camber and toe-in to the correct specifications. This will ensure that the car steers and handles properly and that the tires wear properly.

FRICTION MODIFIER: Changes the coefficient of friction of the fluid between the mating steel and composition clutch/band surfaces during the engagement process and allows for a certain amount of intentional slipping for a good "shift-feel".

FRONTAL AREA: The total frontal area of a vehicle exposed to air flow.

FUEL FILTER: A component of the fuel system containing a porous paper element used to prevent any impurities from entering the engine through the fuel system. It usually takes the form of a canister-like housing, mounted in-line with the fuel hose, located anywhere on a vehicle between the fuel tank and engine.

FUEL INJECTION: A system replacing the carburetor that sprays fuel into the cylinder through nozzles. The amount of fuel can be more precisely controlled with fuel injection.

FULL FLOATING AXLE: An axle in which the axle housing extends through the wheel giving bearing support on the outside of the housing. The front axle of a four-wheel drive vehicle is usually a full floating axle, as are the rear axles of many larger (1 ton and over) pick-ups and vans.

FULL-TIME FOUR-WHEEL DRIVE: A four-wheel drive system that continuously delivers power to all four wheels. A differential between the front and rear driveshafts permits variations in axle speeds to control gear wind-up without damage.

FULL THROTTLE DETENT DOWNSHIFT: A quick apply of accelerator pedal to its full travel, forcing a downshift.

FUSE: A protective device in a circuit which prevents circuit overload by breaking the circuit when a specific amperage is present. The device is constructed around a strip or wire of a lower amperage rating than the circuit it is designed to protect. When an amperage higher than that stamped on the fuse is present in the circuit, the strip or wire melts, opening the circuit.

FUSIBLE LINK: A piece of wire in a wiring harness that performs the same job as a fuse. If overloaded, the fusible link will melt and interrupt the circuit.

FWD: Front wheel drive.

GAWR: (Gross axle weight rating) the total maximum weight an axle is designed to carry.

GCW: (Gross combined weight) total combined weight of a tow vehicle and trailer.

GARAGE SHIFT: initial engagement feel of transmission, neutral to reverse or neutral to a forward drive.

GARAGE SHIFT FEEL: A quick check of the engagement quality and responsiveness of reverse and forward gears. This test is done with the vehicle stationary.

GEAR: A toothed mechanical device that acts as a rotating lever to transmit power or turning effort from one shaft to another. (See gear ratio.)

GEAR RATIO: A ratio expressing the number of turns a smaller gear will make to turn a larger gear through one revolution. The ratio is found by dividing the number of teeth on the smaller gear into the number of teeth on the larger gear.

GEARBOX: Transmission

GEAR REDUCTION: Torque is multiplied and speed decreased by the factor of the gear ratio. For example, a 3:1 gear ratio changes an input torque of 180 ft. lbs. and an input speed of 2700 rpm to 540 Ft. lbs. and 900 rpm, respectively. (No account is taken of frictional losses, which are always present.)

GEARTRAIN: A succession of intermeshing gears that form an assembly and provide for one or more torque changes as the power input is transmitted to the power output.

GEL COAT: A thin coat of plastic resin covering fiberglass body panels.

GENERATOR: A device which produces direct current (DC) necessary to charge the battery.

GOVERNOR: A device that senses vehicle speed and generates a hydraulic oil pressure. As vehicle speed increases, governor oil pressure rises.

GROUND CIRCUIT: (See circuit, ground.)

GROUND SIDE SWITCHING: The electrical/electronic circuit control switch is located after the circuit load.

GVWR: (Gross vehicle weight rating) total maximum weight a vehicle is designed to carry including the weight of the vehicle, passengers, equipment, gas, oil, etc.

HALOGEN: A special type of lamp known for its quality of brilliant white light. Originally used for fog lights and driving lights.

HARD CODES: DTCs that are present at the time of testing; also called continuous or current codes.

HARSH(ROUGH): An apply of a clutch or band that is more noticeable than a firm one; considered undesirable at any throttle position.

HEADER TANK: An expansion tank for the radiator coolant. It can be located remotely or built into the radiator.

HEAT RANGE: A term used to describe the ability of a spark plug to carry away heat. Plugs with longer nosed insulators take longer to carry heat off effectively.

HEAT RISER: A flapper in the exhaust manifold that is closed when the engine is cold, causing hot exhaust gases to heat the intake manifold providing better cold engine operation. A thermostatic spring opens the flapper when the engine warms up.

HEAVY THROTTLE: Approximately three-fourths of accelerator pedal travel.

HEMI: A name given an engine using hemispherical combustion chambers.

HERTZ (HZ): The international unit of frequency equal to one cycle per second (10,000 Hertz equals 10,000 cycles per second).

HIGH-IMPEDANCE DVOM (DIGITAL VOLT-OHMMETER): This styled device provides a built-in resistance value and is capable of limiting circuit current flow to safe milliamp levels.

HIGH RESISTANCE: Often refers to a circuit where there is an excessive amount of opposition to normal current flow.

HORSEPOWER: A measurement of the amount of work; one horsepower is the amount of work necessary to lift 33,000 lbs. one foot in one minute. Brake horsepower (bhp) is the horsepower delivered by an engine on a dynamometer. Net horsepower is the power remaining (measured at the flywheel of the engine) that can be used to turn the wheels after power is consumed through friction and running the engine accessories (water pump, alternator, air pump, fan etc.)

HOT CIRCUIT: (See circuit, hot; hot lead.)

HOT LEAD: A wire or conductor in the power side of the circuit. (See circuit, hot.)

HOT SIDE SWITCHING: The electrical/electronic circuit control switch is located before the circuit load.

HUB: The center part of a wheel or gear.

HUNTING (BUSYNESS): Repeating quick series of up-shifts and down-shifts that causes noticeable change in engine rpm, for example, as in a 4-3-4 shift pattern.

HYDRAULICS: The use of liquid under pressure to transfer force of motion.

HYDROCARBON (HC): Any chemical compound made up of hydrogen and carbon. A major pollutant formed by the engine as a by-product of combustion.

HYDRODYNAMIC DRIVE UNITS: Devices that transmit power solely by the action of a kinetic fluid flow in a closed recirculating path. An impeller energizes the fluid and discharges the high-speed jet stream into the turbine for power output.

HYDROMETER: An instrument used to measure the specific gravity of a solution.

HYDROPLANING: A phenomenon of driving when water builds up under the tire tread, causing it to lose contact with the road. Slowing down will usually restore normal tire contact with the road.

HYPOID GEARSET: The drive pinion gear may be placed below or above the centerline of the driven gear; often used as a final drive gearset.

IDLE MIXTURE: The mixture of air and fuel (usually about 14:1) being fed to the cylinders. The idle mixture screw(s) are sometimes adjusted as part of a tune-up.

IDLER ARM: Component of the steering linkage which is a geometric duplicate of the steering gear arm. It supports the right side of the center steering link.

IMPELLER: Often called a pump, the impeller is the power input (drive) member of a hydrodynamic drive. As part of the torque converter cover, it acts as a centrifugal pump and puts the fluid in motion.

INCH POUND (inch lbs.; sometimes in. lb. or in. lbs.): One twelfth of a foot pound.

INDUCTANCE: The force that produces voltage when a conductor is passed through a magnetic field.

INDUCTION: A means of transferring electrical energy in the form of a magnetic field. Principle used in the ignition coil to increase voltage.

INITIAL FEEL: A distinct firmer feel at start of shift when compared with feel at finish of shift.

INJECTOR: A device which receives metered fuel under relatively low pressure and is activated to inject the fuel into the engine under relatively high pressure at a predetermined time.

INPUT: In an automatic transmission, the source of power from the engine is absorbed by the torque converter, which provides the power input into the transmission. The turbine drives the input(turbine)shaft.

INPUT SHAFT: The shaft to which torque is applied, usually carrying the driving gear or gears.

INTAKE MANIFOLD: A casting of passages or pipes used to conduct air or a fuel/air mixture to the cylinders.

INTERNAL GEAR: The ring-like outer gear of a planetary gearset with the gear teeth cut on the inside of the ring to provide a mesh with the planet pinions.

ISOLATION (CLAMPING) DIODES: Diodes positioned in a circuit to prevent self-induction from damaging electronic components.

IX ROTARY GEAR PUMP: Contains two rotating members, one shaped with internal gear teeth and the other with external gear teeth. As the gears separate, the fluid fills the gaps between gear teeth, is pulled across a crescent-shaped divider, and then is forced to flow through the outlet as the gears mesh.

IX ROTARY LOBE PUMP: Sometimes referred to as a gerotor type pump. Two rotating members, one shaped with internal lobes and the other with external lobes, separate and then mesh to cause fluid to flow.

JOURNAL: The bearing surface within which a shaft operates.

JUMPER CABLES: Two heavy duty wires with large alligator clips used to provide power from a charged battery to a discharged battery mounted in a vehicle.

JUMPSTART: Utilizing the sufficiently charged battery of one vehicle to start the engine of another vehicle with a discharged battery by the use of jumper cables.

KEY: A small block usually fitted in a notch between a shaft and a hub to prevent slippage of the two parts.

KICKDOWN: Detent downshift system; either linkage, cable, or electrically controlled.

KILO: A prefix used in the metric system to indicate one thousand.

KNOCK: Noise which results from the spontaneous ignition of a portion of the air-fuel mixture in the engine cylinder caused by overly advanced ignition timing or use of incorrectly low octane fuel for that engine.

KNOCK SENSOR: An input device that responds to spark knock, caused by over advanced ignition timing.

LABOR TIME: A specific amount of time required to perform a certain repair or diagnostic service as defined by a vehicle or after-market manufacturer .

LACQUER: A quick-drying automotive paint.

LATE: Shift that occurs when engine is at higher than normal rpm for given amount of throttle.

LIGHT-EMITTING DIODE (LED): A semiconductor diode that emits light as electrical current flows through it; used in some electronic display devices to emit a red or other color light.

LIGHT THROTTLE: Approximately one-fourth of accelerator pedal travel.

LIMITED SLIP: A type of differential which transfers driving force to the wheel with the best traction.

LIMP-IN MODE: Electrical shutdown of the transmission/ transaxle output solenoids, allowing only forward and reverse gears that are hydraulically energized by the manual valve. This permits the vehicle to be driven to a service facility for repair.

LIP SEAL: Molded synthetic rubber seal designed with an outer sealing edge (lip) that points into the fluid containing area to be sealed. This type of seal is used where rotational and axial forces are present.

LITHIUM-BASE GREASE: Chassis and wheel bearing grease using lithium as a base. Not compatible with sodium-base grease.

LOAD DEVICE: A circuit's resistance that converts the electrical energy into light, sound, heat, or mechanical movement.

LOAD RANGE: Indicates the number of plies at which a tire is rated. Load range B equals four-ply rating; C equals six-ply rating; and, D equals an eight-ply rating.

LOAD TORQUE: The amount of output torque needed from the transmission/transaxle to overcome the vehicle load.

LOCKING HUBS: Accessories used on part-time four-wheel drive systems that allow the front wheels to be disengaged from the drive train when four-wheel drive is not being used. When four-wheel drive is desired, the hubs are engaged, locking the wheels to the drive train.

LOCKUP CONVERTER: A torque converter that operates hydraulically and mechanically. When an internal apply plate (lockup plate) clamps to the torque converter cover, hydraulic slippage is eliminated.

LOCK RING: See Circlip or Snapring

MAGNET: Any body with the property of attracting iron or steel.

MAGNETIC FIELD: The area surrounding the poles of a magnet that is affected by its attraction or repulsion forces.

MAIN LINE PRESSURE: Often called control pressure or line pressure, it refers to the pressure of the oil leaving the pump and is controlled by the pressure regulator valve.

MALFUNCTION INDICATOR LAMP (MIL): Previously known as a check engine light, the dash-mounted MIL illuminates and signals the driver that an emission or driveability problem with the powertrain has been detected by the ECM/PCM. When this occurs, at least one diagnostic trouble code (DTC) has been stored into the control module memory.

MANIFOLD ABSOLUTE PRESSURE (MAP) SENSOR: Reads the amount of air pressure (vacuum) in the engine's intake manifold system; its signal is used to analyze engine load conditions.

MANIFOLD VACUUM: Low pressure in an engine intake manifold formed just below the throttle plates. Manifold vacuum is highest at idle and drops under acceleration.

MANIFOLD: A casting of passages or set of pipes which connect the cylinders to an inlet or outlet source.

MANUAL LEVER POSITION SWITCH (MLPS): A mechanical switching unit that is typically mounted externally to the transmission/transaxle to inform the PCM/ECM which gear range the driver has selected.

MANUAL VALVE: Located inside the transmission/transaxle, it is directly connected to the driver's shift lever. The position of the manual valve determines which hydraulic circuits will be charged with oil pressure and the operating mode of the transmission.

MANUAL VALVE LEVER POSITION SENSOR (MVLPS): The input from this device tells the TCM what gear range was selected.

MASS AIR FLOW (MAF) SENSOR: Measures the airflow into the engine.

MASTER CYLINDER: The primary fluid pressurizing device in a hydraulic system. In automotive use, it is found in brake and hydraulic clutch systems and is pedal activated, either directly or, in a power brake system, through the power booster.

MacPherson STRUT: A suspension component combining a shock absorber and spring in one unit.

MEDIUM THROTTLE: Approximately one-half of accelerator pedal travel.

MEGA: A metric prefix indicating one million.

MEMBER: An independent component of a hydrodynamic unit such as an impeller, a stator, or a turbine. It may have one or more elements.

MERCON: A fluid developed by Ford Motor Company in 1988. It contains a friction modifier and closely resembles operating characteristics of Dexron.

METAL SEALING RINGS: Made from cast iron or aluminum, their primary application is with dynamic components involving pressure sealing circuits of rotating members. These rings are designed with either butt or hook lock end joints.

METER (ANALOG): A linear-style meter representing data as lengths; a needle-style instrument interfacing with logical numerical increments. This style of electrical meter uses relatively low impedance internal resistance and cannot be used for testing electronic circuitry.

METER (DIGITAL): Uses numbers as a direct readout to show values. Most meters of this style use high impedance internal resistance and must be used for testing low current electronic circuitry.

MICRO: A metric prefix indicating one-millionth (0.000001).

MILLI: A metric prefix indicating one-thousandth (0.001).

MINIMUM THROTTLE: The least amount of throttle opening required for upshift; normally close to zero throttle.

MISFIRE: Condition occurring when the fuel mixture in a cylinder fails to ignite, causing the engine to run roughly.

MODULE: Electronic control unit, amplifier or igniter of solid state or integrated design which controls the current flow in the ignition primary circuit based on input from the pick-up coil. When the module opens the primary circuit, high secondary voltage is induced in the coil.

MODULATED: In an electronic-hydraulic converter clutch system (or shift valve system), the term modulated refers to the pulsing of a solenoid, at a variable rate. This action controls the buildup of oil pressure in the hydraulic circuit to allow a controlled amount of clutch slippage.

MODULATED CONVERTER CLUTCH CONTROL (MCCC): A pulse width duty cycle valve that controls the converter lockup apply pressure and maximizes smoother transitions between lock and unlock conditions.

MODULATOR PRESSURE (THROTTLE PRESSURE): A hydraulic signal oil pressure relating to the amount of engine load, based on either the amount of throttle plate opening or engine vacuum.

MODULATOR VALVE: A regulator valve that is controlled by engine vacuum, providing a hydraulic pressure that varies in relation to engine torque. The hydraulic torque signal functions to delay the shift pattern and provide a line pressure boost. (See throttle valve.)

MOTOR: An electromagnetic device used to convert electrical energy into mechanical energy.

MULTIPLE-DISC CLUTCH: A grouping of steel and friction lined plates that, when compressed together by hydraulic pressure acting upon a piston, lock or unlock a planetary member.

MULTI-WEIGHT: Type of oil that provides adequate lubrication at both high and low temperatures.

MUSHY: Same as soft; slow and drawn out clutch apply with very little shift feel.

MUTUAL INDUCTION: The generation of current from one wire circuit to another by movement of the magnetic field surrounding a current-carrying circuit as its ampere flow increases or decreases.

NEEDLE BEARING: A bearing which consists of a number (usually a large number) of long, thin rollers.

NITROGEN OXIDE (NOx): One of the three basic pollutants found in the exhaust emission of an internal combustion engine. The amount of NOx usually varies in an inverse proportion to the amount of HC and CO.

NONPOSITIVE SEALING: A sealing method that allows some minor leakage, which normally assists in lubrication.

O2 SENSOR: Located in the engine's exhaust system, it is an input device to the ECM/PCM for managing the fuel delivery and ignition system. A scanner can be used to observe the fluctuating voltage readings produced by an 02 sensor as the oxygen content of the exhaust is analyzed.

O-RING SEAL: Molded synthetic rubber seal designed with a circular cross-section. This type of seal is used primarily in static applications.

OBD II (ON-BOARD DIAGNOSTICS, SECOND GENERATION): Refers to the federal law mandating tighter control of 1996 and newer vehicle emissions, active monitoring of related devices, and standardization of terminology, data link connectors, and other technician concerns.

OCTANE RATING: A number, indicating the quality of gasoline based on its ability to resist knock. The higher the number, the better the quality. Higher compression engines require higher octane gas.

OEM: Original Equipment Manufactured. OEM equipment is that furnished standard by the manufacturer.

OFFSET: The distance between the vertical center of the wheel and the mounting surface at the lugs. Offset is positive if the center is outside the lug circle; negative offset puts the center line inside the lug circle.

OHM'S LAW: A law of electricity that states the relationship between voltage, current, and resistance. Volts = amperes x ohms

OHM: The unit used to measure the resistance of conductor-to-electrical flow. One ohm is the amount of resistance that limits current flow to one ampere in a circuit with one volt of pressure.

OHMMETER: An instrument used for measuring the resistance, in ohms, in an electrical circuit.

ONE-WAY CLUTCH: A mechanical clutch of roller or sprag design that resists torque or transmits power in one direction only. It is used to either hold or drive a planetary member.

ONE-WAY ROLLER CLUTCH: A mechanical device that transmits or holds torque in one direction only.

OPEN CIRCUIT: A break or lack of contact in an electrical circuit, either intentional (switch) or unintentional (bad connection or broken wire).

ORIFICE: Located in hydraulic oil circuits, it acts as a restriction. It slows down fluid flow to either create back pressure or delay pressure buildup downstream.

OSCILLOSCOPE: A piece of test equipment that shows electric impulses as a pattern on a screen. Engine performance can be analyzed by interpreting these patterns.

OUTPUT SHAFT: The shaft which transmits torque from a device, such as a transmission.

OUTPUT SPEED SENSOR (OSS): Identifies transmission/transaxle output shaft speed for shift timing and may be used to calculate TCC slip; often functions as the VSS (vehicle speed sensor).

OVERDRIVE: (1.) A device attached to or incorporated in a transmission/transaxle that allows the engine to turn less than one full revolution for every complete revolution of the wheels. The net effect is to reduce engine rpm, thereby using less fuel. A typical overdrive gear ratio would be .87:1, instead of the normal 1:1 in high gear. (2.) A gear assembly which produces more shaft revolutions than that transmitted to it.

OVERDRIVE PLANETARY GEARSET: A single planetary gearset designed to provide a direct drive and overdrive ratio. When coupled to a three-speed transmission/transaxle configuration, a four-speed/overdrive unit is present.

OVERHEAD CAMSHAFT (OHC): An engine configuration in which the camshaft is mounted on top of the cylinder head and operates the valve either directly or by means of rocker arms.

OVERHEAD VALVE (OHV): An engine configuration in which all of the valves are located in the cylinder head and the camshaft is located in the cylinder block. The camshaft operates the valves via lifters and pushrods.

OVERRUNCLUTCH: Another name for a one-way mechanical clutch. Applies to both roller and sprag designs.

OVERSTEER: The tendency of some vehicles, when steering into a turn, to over-respond or steer more than required, which could result in excessive slip of the rear wheels. Opposite of under-steer.

OXIDATION STABILIZERS: Absorb and dissipate heat. Automatic transmission fluid has high resistance to varnish and sludge buildup that occurs from excessive heat that is generated primarily in the torque converter. Local temperatures as high as 6000F (3150C) can occur at the clutch plates during engagement, and this heat must be absorbed and dissipated. If the fluid cannot withstand the heat, it burns or oxidizes, resulting in an almost immediate destruction of friction materials, clogged filter screen and hydraulic passages, and sticky valves.

OXIDES OF NITROGEN: See nitrogen oxide (NOx).

OXYGEN SENSOR: Used with a feedback system to sense the presence of oxygen in the exhaust gas and signal the computer which can use the voltage signal to determine engine operating efficiency and adjust the air/fuel ratio.

PARALLEL CIRCUIT: (See circuit, parallel.)

PARTS WASHER: A basin or tub, usually with a built-in pump mechanism and hose used for circulating chemical solvent for the purpose of cleaning greasy, oily and dirty components.

PART-TIME FOUR WHEEL DRIVE: A system that is normally in the two wheel drive mode and only runs in four-wheel drive when the system is manually engaged because more traction is desired. Two or four wheel drive is normally selected by a lever to engage the front axle, but if locking hubs are used, these must also be manually engaged in the Lock position. Otherwise, the front axle will not drive the front wheels.

PASSIVE RESTRAINT: Safety systems such as air bags or automatic seat belts which operate with no action required on the part of the driver or passenger. Mandated by Federal regulations on all vehicles sold in the U.S. after 1990.

PAYLOAD: The weight the vehicle is capable of carrying in addition to its own weight. Payload includes weight of the driver, passengers and cargo, but not coolant, fuel, lubricant, spare tire, etc.

PCM: Powertrain control module.

PCV VALVE: A valve usually located in the rocker cover that vents crankcase vapors back into the engine to be reburned.

PERCOLATION: A condition in which the fuel actually "boils," due to excessive heat. Percolation prevents proper atomization of the fuel causing rough running.

PICK-UP COIL: The coil in which voltage is induced in an electronic ignition.

PING: A metallic rattling sound produced by the engine during acceleration. It is usually due to incorrect ignition timing or a poor grade of gasoline.

PINION: The smaller of two gears. The rear axle pinion drives the ring gear which transmits motion to the axle shafts.

PINION GEAR: The smallest gear in a drive gear assembly.

PISTON: A disc or cup that fits in a cylinder bore and is free to move. In hydraulics, it provides the means of converting hydraulic pressure into a usable force. Examples of piston applications are found in servo, clutch, and accumulator units.

PISTON RING: An open-ended ring which fits into a groove on the outer diameter of the piston. Its chief function is to form a seal between the piston and cylinder wall. Most automotive pistons have three rings: two for compression sealing; one for oil sealing.

PITMAN ARM: A lever which transmits steering force from the steering gear to the steering linkage.

PLANET CARRIER: A basic member of a planetary gear assembly that carries the pinion gears.

PLANET PINIONS: Gears housed in a planet carrier that are in constant mesh with the sun gear and internal gear. Because they have their own independent rotating centers, the pinions are capable of rotating around the sun gear or the inside of the internal gear.

PLANETARY GEAR RATIO: The reduction or overdrive ratio developed by a planetary gearset.

PLANETARY GEARSET: In its simplest form, it is made up of a basic assembly group containing a sun gear, internal gear, and planet carrier. The gears are always in constant mesh and offer a wide range of gear ratio possibilities.

PLANETARY GEARSET (COMPOUND): Two planetary gearsets combined together.

PLANETARY GEARSET (SIMPLE): An assembly of gears in constant mesh consisting of a sun gear, several pinion gears mounted in a carrier, and a ring gear. It provides gear ratio and direction changes, in addition to a direct drive and a neutral.

PLY RATING: A. rating given a tire which indicates strength (but not necessarily actual plies). A two-ply/four-ply rating has only two plies, but the strength of a four-ply tire.

POLARITY: Indication (positive or negative) of the two poles of a battery.

PORT: An opening for fluid intake or exhaust.

POSITIVE SEALING: A sealing method that completely prevents leakage.

POTENTIAL: Electrical force measured in volts; sometimes used interchangeably with voltage.

POWER: The ability to do work per unit of time, as expressed in horsepower; one horsepower equals 33,000 ft. lbs. of work per minute, or 550 ft. lbs. of work per second.

POWER FLOW: The systematic flow or transmission of power through the gears, from the input shaft to the output shaft.

POWER-TO-WEIGHT RATIO: Ratio of horsepower to weight of car.

POWERTRAIN: See Drivetrain.

POWERTRAIN CONTROL MODULE (PCM): Current designation for the engine control module (ECM). In many cases, late model vehicle control units manage the engine as well as the transmission. In other settings, the PCM controls the engine and is interfaced with a TCM to control transmission functions.

Ppm: Parts per million; unit used to measure exhaust emissions.

PREIGNITION: Early ignition of fuel in the cylinder, sometimes due to glowing carbon deposits in the combustion chamber. Preignition can be damaging since combustion takes place prematurely.

PRELOAD: A predetermined load placed on a bearing during assembly or by adjustment.

PRESS FIT: The mating of two parts under pressure, due to the inner diameter of one being smaller than the outer diameter of the other, or vice versa; an interference fit.

PRESSURE: The amount of force exerted upon a surface area.

PRESSURE CONTROL SOLENOID (PCS): An output device that provides a boost oil pressure to the mainline regulator valve to control line pressure. Its operation is determined by the amount of current sent from the PCM.

PRESSURE GAUGE: An instrument used for measuring the fluid pressure in a hydraulic circuit.

PRESSURE REGULATOR VALVE: In automatic transmissions, its purpose is to regulate the pressure of the pump output and supply the basic fluid pressure necessary to operate the transmission. The regulated fluid pressure may be referred to as mainline pressure, line pressure, or control pressure.

PRESSURE SWITCH ASSEMBLY (PSA): Mounted inside the transmission, it is a grouping of oil pressure switches that inputs to the PCM when certain hydraulic passages are charged with oil pressure.

PRESSURE PLATE: A spring-loaded plate (part of the clutch) that transmits power to the driven (friction) plate when the clutch is engaged.

PRIMARY CIRCUIT: The low voltage side of the ignition system which consists of the ignition switch, ballast resistor or resistance wire, bypass, coil, electronic control unit and pick-up coil as well as the connecting wires and harnesses.

PROFILE: Term used for tire measurement (tire series), which is the ratio of tire height to tread width.

PROM (PROGRAMMABLE READ-ONLY MEMORY): The heart of the computer that compares input data and makes the engineered program or strategy decisions about when to trigger the appropriate output based on stored computer instructions.

PULSE GENERATOR: A two-wire pickup sensor used to produce a fluctuating electrical signal. This changing signal is read by the controller to determine the speed of the object and can be used to measure transmission/transaxle input speed, output speed, and vehicle speed.

PSI: Pounds per square inch; a measurement of pressure.

PULSE WIDTH DUTY CYCLE SOLENOID (PULSE WIDTH MODULATED SOLENOID): A computer-controlled solenoid that turns on and off at a variable rate producing a modulated oil pressure; often referred to as a pulse width modulated (PWM) solenoid. Employed in many electronic automatic transmissions and transaxles, these solenoids are used to manage shift control and converter clutch hydraulic circuits.

PUSHROD: A steel rod between the hydraulic valve lifter and the valve rocker arm in overhead valve (OHV) engines.

PUMP: A mechanical device designed to create fluid flow and pressure buildup in a hydraulic system.

QUARTER PANEL: General term used to refer to a rear fender. Quarter panel is the area from the rear door opening to the tail light area and from rear wheel well to the base of the trunk and roof-line.

RACE: The surface on the inner or outer ring of a bearing on which the balls, needles or rollers move.

RACK AND PINION: A type of automotive steering system using a pinion gear attached to the end of the steering shaft. The pinion meshes with a long rack attached to the steering linkage.

RADIAL TIRE: Tire design which uses body cords running at right angles to the center line of the tire. Two or more belts are used to give tread strength. Radials can be identified by their characteristic sidewall bulge.

RADIATOR: Part of the cooling system for a water-cooled engine, mounted in the front of the vehicle and connected to the engine with rubber hoses. Through the radiator, excess combustion heat is dissipated into the atmosphere through forced convection using a water and glycol based mixture that circulates through, and cools, the engine.

RANGE REFERENCE AND CLUTCH/BAND APPLY CHART: A guide that shows the application of clutches and bands for each gear, within the selector range positions. These charts are extremely useful for understanding how the unit operates and for diagnosing malfunctions.

RAVIGNEAUX GEARSET: A compound planetary gearset that features matched dual planetary pinions (sets of two) mounted in a single planet carrier. Two sun gears and one ring mesh with the carrier pinions.

REACTION MEMBER: The stationary planetary member, in a planetary gearset, that is grounded to the transmission/transaxle case through the use of friction and wedging devices known as bands, disc clutches, and one-way clutches.

REACTION PRESSURE: The fluid pressure that moves a spool valve against an opposing force or forces; the area on which the opposing force acts. The opposing force can be a spring or a combination of spring force and auxiliary hydraulic force.

REACTOR, TORQUE CONVERTER: The reaction member of a fluid torque converter, more commonly called a stator. (See stator.)

REAR MAIN OIL SEAL: A synthetic or rope-type seal that prevents oil from leaking out of the engine past the rear main crankshaft bearing.

RECIRCULATING BALL: Type of steering system in which recirculating steel balls occupy the area between the nut and worm wheel, causing a reduction in friction.

RECTIFIER: A device (used primarily in alternators) that permits electrical current to flow in one direction only.

REDUCTION: (See gear reduction.)

REGULATOR VALVE: A valve that changes the pressure of the oil in a hydraulic circuit as the oil passes through the valve by bleeding off (or exhausting) some of the volume of oil supplied to the valve.

REFRIGERANT 12 (R-12) or 134 (R-134): The generic name of the refrigerant used in automotive air conditioning systems.

REGULATOR: A device which maintains the amperage and/or voltage levels of a circuit at predetermined values.

RELAY: A switch which automatically opens and/or closes a circuit.

RELAY VALVE: A valve that directs flow and pressure. Relay valves simply connect or disconnect interrelated passages without restricting the fluid flow or changing the pressure.

RELIEF VALVE: A spring-loaded, pressure-operated valve that limits oil pressure buildup in a hydraulic circuit to a predetermined maximum value.

RELUCTOR: A wheel that rotates inside the distributor and triggers the release of voltage in an electronic ignition.

RESERVOIR: The storage area for fluid in a hydraulic system; often called a sump.

RESIN: A liquid plastic used in body work.

RESIDUAL MAGNETISM: The magnetic strength stored in a material after a magnetizing field has been removed.

RESISTANCE: The opposition to the flow of current through a circuit or electrical device, and is measured in ohms. Resistance is equal to the voltage divided by the amperage.

RESISTOR SPARK PLUG: A spark plug using a resistor to shorten the spark duration. This suppresses radio interference and lengthens plug life.

RESISTOR: A device, usually made of wire, which offers a preset amount of resistance in an electrical circuit.

RESULTANT FORCE: The single effective directional thrust of the fluid force on the turbine produced by the vortex and rotary forces acting in different planes.

RETARD: Set the ignition timing so that spark occurs later (fewer degrees before TDC).

RHEOSTAT: A device for regulating a current by means of a variable resistance.

RING GEAR: The name given to a ring-shaped gear attached to a differential case, or affixed to a flywheel or as part of a planetary gear set.

ROADLOAD: grade.

ROCKER ARM: A lever which rotates around a shaft pushing down (opening) the valve with an end when the other end is pushed up by the pushrod. Spring pressure will later close the valve.

ROCKER PANEL: The body panel below the doors between the wheel opening.

ROLLER BEARING: A bearing made up of hardened inner and outer races between which hardened steel rollers move.

ROLLER CLUTCH: A type of one-way clutch design using rollers and springs mounted within an inner and outer cam race assembly.

ROTARY FLOW: The path of the fluid trapped between the blades of the members as they revolve with the rotation of the torque converter cover (rotational inertia).

ROTOR: (1.) The disc-shaped part of a disc brake assembly, upon which the brake pads bear; also called, brake disc. (2.) The device mounted atop the distributor shaft, which passes current to the distributor cap tower contacts.

ROTARY ENGINE: See Wankel engine.

RPM: Revolutions per minute (usually indicates engine speed).

RTV: A gasket making compound that cures as it is exposed to the atmosphere. It is used between surfaces that are not perfectly machined to one another, leaving a slight gap that the RTV fills and in which it hardens. The letters RTV represent room temperature vulcanizing.

RUN-ON: Condition when the engine continues to run, even when the key is turned off. See dieseling.

SEALED BEAM: A automotive headlight. The lens, reflector and filament from a single unit.

SEATBELT INTERLOCK: A system whereby the car cannot be started unless the seatbelt is buckled.

SECONDARY CIRCUIT: The high voltage side of the ignition system, usually above 20,000 volts. The secondary includes the ignition coil, coil wire, distributor cap and rotor, spark plug wires and spark plugs.

SELF-INDUCTION: The generation of voltage in a current-carrying wire by changing the amount of current flowing within that wire.

SEMI-CONDUCTOR: A material (silicon or germanium) that is neither a good conductor nor an insulator; used in diodes and transistors.

SEMI-FLOATING AXLE: In this design, a wheel is attached to the axle shaft, which takes both drive and cornering loads. Almost all solid axle passenger cars and light trucks use this design.

SENDING UNIT: A mechanical, electrical, hydraulic or electromagnetic device which transmits information to a gauge.

SENSOR: Any device designed to measure engine operating conditions or ambient pressures and temperatures. Usually electronic in nature and designed to send a voltage signal to an on-board computer, some sensors may operate as a simple on/off switch or they may provide a variable voltage signal (like a potentiometer) as conditions or measured parameters change.

SERIES CIRCUIT: (See circuit, series.)

SERPENTINE BELT: An accessory drive belt, with small multiple v-ribs, routed around most or all of the engine-powered accessories such as the alternator and power steering pump. Usually both the front and the back side of the belt comes into contact with various pulleys.

SERVO: In an automatic transmission, it is a piston in a cylinder assembly that converts hydraulic pressure into mechanical force and movement; used for the application of the bands and clutches.

SHIFT BUSYNESS: When referring to a torque converter clutch, it is the frequent apply and release of the clutch plate due to uncommon driving conditions.

SHIFT VALVE: Classified as a relay valve, it triggers the automatic shift in response to a governor and a throttle signal by directing fluid to the appropriate band and clutch apply combination to cause the shift to occur.

SHIM: Spacers of precise, predetermined thickness used between parts to establish a proper working relationship.

SHIMMY: Vibration (sometimes violent) in the front end caused by misaligned front end, out of balance tires or worn suspension components.

SHORT CIRCUIT: An electrical malfunction where current takes the path of least resistance to ground (usually through damaged insulation). Current flow is excessive from low resistance resulting in a blown fuse.

SHUDDER: Repeated jerking or stick-slip sensation, similar to chuggle but more severe and rapid in nature, that may be most noticeable during certain ranges of vehicle speed; also used to define condition after converter clutch engagement.

SIMPSON GEARSET: A compound planetary gear train that integrates two simple planetary gearsets referred to as the front planetary and the rear planetary.

SINGLE OVERHEAD CAMSHAFT: See overhead camshaft.

SKIDPLATE: A metal plate attached to the underside of the body to protect the fuel tank, transfer case or other vulnerable parts from damage.

SLAVE CYLINDER: In automotive use, a device in the hydraulic clutch system which is activated by hydraulic force, disengaging the clutch.

SLIPPING: Noticeable increase in engine rpm without vehicle speed increase; usually occurs during or after initial clutch or band engagement.

SLUDGE: Thick, black deposits in engine formed from dirt, oil, water, etc. It is usually formed in engines when oil changes are neglected.

SNAP RING: A circular retaining clip used inside or outside a shaft or part to secure a shaft, such as a floating wrist pin.

SOFT: Slow, almost unnoticeable clutch apply with very little shift feel.

SOFTCODES: DTCs that have been set into the PCM memory but are not present at the time of testing; often referred to as history or intermittent codes.

SOHC: Single overhead camshaft.

SOLENOID: An electrically operated, magnetic switching device.

SPALLING: A wear pattern identified by metal chips flaking off the hardened surface. This condition is caused by foreign particles, overloading situations, and/or normal wear.

SPARK PLUG: A device screwed into the combustion chamber of a spark ignition engine. The basic construction is a conductive core inside of a ceramic insulator, mounted in an outer conductive base. An electrical charge from the spark plug wire travels along the conductive core and jumps a preset air gap to a grounding point or points at the end of the conductive base. The resultant spark ignites the fuel/air mixture in the combustion chamber.

SPECIFIC GRAVITY (BATTERY): The relative weight of liquid (battery electrolyte) as compared to the weight of an equal volume of water.

SPLINES: Ridges machined or cast onto the outer diameter of a shaft or inner diameter of a bore to enable parts to mate without rotation.

SPLIT TORQUE DRIVE: In a torque converter, it refers to parallel paths of torque transmission, one of which is mechanical and the other hydraulic.

SPONGY PEDAL: A soft or spongy feeling when the brake pedal is depressed. It is usually due to air in the brake lines.

SPOOLVALVE: A precision-machined, cylindrically shaped valve made up of lands and grooves. Depending on its position in the valve bore, various inter-connecting hydraulic circuit passages are either opened or closed.

SPRAG CLUTCH: A type of one-way clutch design using cams or con-toured-shaped sprags between inner and outer races. (See one-way clutch.)

SPRUNG WEIGHT: The weight of a car supported by the springs.

SQUARE-CUT SEAL: Molded synthetic rubber seal designed with a square- or rectangular-shaped cross-section. This type of seal is used for both dynamic and static applications.

SRS: Supplemental restraint system

STABILIZER (SWAY) BAR: A bar linking both sides of the suspension. It resists sway on turns by taking some of added load from one wheel and putting it on the other.

STAGE: The number of turbine sets separated by a stator. A turbine set may be made up of one or more turbine members. A three-element converter is clas-sified as a single stage.

STALL: In fluid drive transmission/transaxle applications, stall refers to engine rpm with the transmission/transaxle engaged and the vehicle stationary; throttle valve can be in any position between closed and wide open.

STALL SPEED: In fluid drive transmission/transaxle applications, stall speed refers to the maximum engine rpm with the transmission/transaxle engaged and vehicle stationary, when the throttle valve is wide open. (See stall; stall test.)

STALL TEST: A procedure recommended by many manufacturers to help determine the integrity of an engine, the torque converter stator, and certain clutch and band combinations. With the shift lever in each of the forward and reverse positions and with the brakes firmly applied, the accelerator pedal is momentarily pressed to the wide open throttle (WOT) position. The engine rpm reading at full throttle can provide clues for diagnosing the condition of the items listed above.

STALL TORQUE: The maximum design or engineered torque ratio of a fluid torque converter, produced under stall speed conditions. (See stall speed.)

STARTER: A high-torque electric motor used for the purpose of starting the engine, typically through a high ratio geared drive connected to the flywheel ring gear.

STATIC: A sealing application in which the parts being sealed do not move in relation to each other.

STATOR (REACTOR): The reaction member of a fluid torque converter that changes the direction of the fluid as it leaves the turbine to enter the impeller vanes. During the torque multiplication phase, this action assists the impeller's rotary force and results in an increase in torque.

STEERING GEOMETRY: Combination of various angles of suspension components (caster, camber, toe-in); roughly equivalent to front end alignment.

STRAIGHT WEIGHT: Term designating motor oil as suitable for use within a narrow range of temperatures. Outside the narrow temperature range its flow characteristics will not adequately lubricate.

STROKE: The distance the piston travels from bottom dead center to top dead center.

SUBSTITUTION: Replacing one part suspected of a defect with a like part of known quality.

SUMP: The storage vessel or reservoir that provides a ready source of fluid to the pump. In an automatic transmission, the sump is the oil pan. All fluid eventually returns to the sump for recycling into the hydraulic system.

SUN GEAR: In a planetary gearset, it is the center gear that meshes with a cluster of planet pinions.

SUPERCHARGER: An air pump driven mechanically by the engine through belts, chains, shafts or gears from the crankshaft. Two general types of super-charger are the positive displacement and centrifugal type, which pump air in direct relationship to the speed of the engine.

SUPPLEMENTAL RESTRAINT SYSTEM: See air bag.

SURGE: Repeating engine-related feeling of acceleration and deceleration that is less intense than chuggle.

SWITCH: A device used to open, close, or redirect the current in an electri-cal circuit.

SYNCHROMESH: A manual transmission/transaxle that is equipped with devices (synchronizers) that match the gear speeds so that the transmission/transaxle can be downshifted without clashing gears.

SYNTHETIC OIL: Non-petroleum based oil.

TACHOMETER: A device used to measure the rotary speed of an engine, shaft, gear, etc., usually in rotations per minute.

TDC: Top dead center. The exact top of the piston's stroke.

TEFLON SEALING RINGS: Teflon is a soft, durable, plastic-like material that is resistant to heat and provides excellent sealing. These rings are designed with either scarf-cut joints or as one-piece rings. Teflon sealing rings have replaced many metal ring applications.

TERMINAL: A device attached to the end of a wire or cable to make an elec-trical connection.

TEST LIGHT, CIRCUIT-POWERED: Uses available circuit voltage to test circuit continuity.

TEST LIGHT, SELF-POWERED: Uses its own battery source to test circuit continuity.

THERMISTOR: A special resistor used to measure fluid temperature; it decreases its resistance with increases in temperature.

THERMOSTAT: A valve, located in the cooling system of an engine, which is closed when cold and opens gradually in response to engine heating, con-trolling the temperature of the coolant and rate of coolant flow.

THERMOSTATIC ELEMENT: A heat-sensitive, spring-type device that con-trols a drain port from the upper sump area to the lower sump. When the transaxle fluid reaches operating temperature, the port is closed and the upper sump fills, thus reducing the fluid level in the lower sump.

THROTTLE POSITION (TP) SENSOR: Reads the degree of throttle open-ing; its signal is used to analyze engine load conditions. The ECM/PCM decides to apply the TCC, or to disengage it for coast or load conditions that need a converter torque boost.

THROTTLE PRESSURE/MODULATOR PRESSURE: A hydraulic signal oil pressure relating to the amount of engine load, based on either the amount of throttle plate opening or engine vacuum.

THROTTLE VALVE: A regulating or balanced valve that is controlled mechanically by throttle linkage or engine vacuum. It sends a hydraulic signal to the shift valve body to control shift timing and shift quality. (See balanced valve; modulator valve.)

THROW-OUT BEARING: As the clutch pedal is depressed, the throwout bearing moves against the spring fingers of the pressure plate, forcing the pres-sure plate to disengage from the driven disc.

TIE ROD: A rod connecting the steering arms. Tie rods have threaded ends that are used to adjust toe-in.

TIE-UP: Condition where two opposing clutches are attempting to apply at same time, causing engine to labor with noticeable loss of engine rpm.

TIMING BELT: A square-toothed, reinforced rubber belt that is driven by the crankshaft and operates the camshaft.

TIMING CHAIN: A roller chain that is driven by the crankshaft and operates the camshaft.

TIRE ROTATION: Moving the tires from one position to another to make the tires wear evenly.

TOE-IN (OUT): A term comparing the extreme front and rear of the front tires. Closer together at the front is toe-in; farther apart at the front is toe-out.

TOP DEAD CENTER (TDC): The point at which the piston reaches the top of its travel on the compression stroke.

TORQUE: Measurement of turning or twisting force, expressed as foot-pounds or inch-pounds.

TORQUE CONVERTER: A turbine used to transmit power from a driving member to a driven member via hydraulic action, providing changes in drive ratio and torque. In automotive use, it links the driveplate at the rear of the engine to the automatic transmission.

TORQUE CONVERTER CLUTCH: The apply plate (lockup plate) assembly used for mechanical power flow through the converter.

TORQUE PHASE: Sometimes referred to as slip phase or stall phase, torque multiplication occurs when the turbine is turning at a slower speed than the impeller, and the stator is reactionary (stationary). This sequence generates a boost in output torque.

TORQUE RATING (STALL TORQUE): The maximum torque multiplication that occurs during stall conditions, with the engine at wide open throttle (WOT) and zero turbine speed.

TORQUE RATIO: An expression of the gear ratio factor on torque effect. A 3:1 gear ratio or 3:1 torque ratio increases the torque input by the ratio factor of 3. Input torque (100 ft. lbs.) x 3 = output torque (300 ft. lbs.)

TRACTION: The amount of usable tractive effort before the drive wheels slip on the road contact surface.

TORSION BAR SUSPENSION: Long rods of spring steel which take the place of springs. One end of the bar is anchored and the other arm (attached to the suspension) is free to twist. The bars' resistance to twisting causes springing action.

TRACK: Distance between the centers of the tires where they contact the ground.

TRACTION CONTROL: A control system that prevents the spinning of a vehicle's drive wheels when excess power is applied.

TRACTIVE EFFORT: The amount of force available to the drive wheels, to move the vehicle.

TRANSAXLE: A single housing containing the transmission and differential. Transaxles are usually found on front engine/front wheel drive or rear engine/rear wheel drive cars.

TRANSDUCER: A device that changes energy from one form to another. For example, a transducer in a microphone changes sound energy to electrical energy. In automotive air-conditioning controls used in automatic temperature systems, a transducer changes an electrical signal to a vacuum signal, which operates mechanical doors.

TRANSMISSION: A powertrain component designed to modify torque and speed developed by the engine; also provides direct drive, reverse, and neutral.

TRANSMISSION CONTROL MODULE (TCM): Manages transmission functions. These vary according to the manufacturer's product design but may include converter clutch operation, electronic shift scheduling, and mainline pressure.

TRANSMISSION FLUID TEMPERATURE (TFT) SENSOR: Originally called a transmission oil temperature (TOT) sensor, this input device to the ECM/PCM senses the fluid temperature and provides a resistance value. It operates on the thermistor principle.

TRANSMISSION INPUT SPEED (TIS) SENSOR: Measures turbine shaft (input shaft) rpm's and compares to engine rpm's to determine torque converter slip. When compared to the transmission output speed sensor or VSS, gear ratio and clutch engagement timing can be determined.

TRANSMISSION OIL TEMPERATURE (TOT) SENSOR: (See transmission fluid temperature (TFT) sensor.)

TRANSMISSION RANGE SELECTOR (TRS) SWITCH: Tells the module which gear shift position the driver has chosen.

TRANSFER CASE: A gearbox driven from the transmission that delivers power to both front and rear driveshafts in a four-wheel drive system. Transfer cases usually have a high and low range set of gears, used depending on how much pulling power is needed.

TRANSISTOR: A semi-conductor component which can be actuated by a small voltage to perform an electrical switching function.

TREAD WEAR INDICATOR: Bars molded into the tire at right angles to the tread that appear as horizontal bars when 1⁄16 in. of tread remains.

TREAD WEAR PATTERN: The pattern of wear on tires which can be "read" to diagnose problems in the front suspension.

TUNE-UP: A regular maintenance function, usually associated with the replacement and adjustment of parts and components in the electrical and fuel systems of a vehicle for the purpose of attaining optimum performance.

TURBINE: The output (driven) member of a fluid coupling or fluid torque converter. It is splined to the input (turbine) shaft of the transmission.

TURBOCHARGER: An exhaust driven pump which compresses intake air and forces it into the combustion chambers at higher than atmospheric pressures. The increased air pressure allows more fuel to be burned and results in increased horsepower being produced.

TURBULENCE: The interference of molecules of a fluid (or vapor) with each other in a fluid flow.

TYPE F: Transmission fluid developed and used by Ford Motor Company up to 1982. This fluid type provides a high coefficient of friction.

TYPE 7176: The preferred choice of transmission fluid for Chrysler automatic transmissions and transaxles. Developed in 1986, it closely resembles Dexron and Mercon. Type 7176 is the recommended service fill fluid for all Chrysler products utilizing a lockup torque converter dating back to 1978.

U-JOINT (UNIVERSAL JOINT): A flexible coupling in the drive train that allows the driveshafts or axle shafts to operate at different angles and still transmit rotary power.

UNDERSTEER: The tendency of a car to continue straight ahead while negotiating a turn.

UNIT BODY: Design in which the car body acts as the frame.

UNLEADED FUEL: Fuel which contains no lead (a common gasoline additive). The presence of lead in fuel will destroy the functioning elements of a catalytic converter, making it useless.

UNSPRUNG WEIGHT: The weight of car components not supported by the springs (wheels, tires, brakes, rear axle, control arms, etc.).

UPSHIFT: A shift that results in a decrease in torque ratio and an increase in speed.

VACUUM: A negative pressure; any pressure less than atmospheric pressure.

VACUUM ADVANCE: A device which advances the ignition timing in response to increased engine vacuum.

VACUUM GAUGE: An instrument used for measuring the existing vacuum in a vacuum circuit or chamber. The unit of measure is inches (of mercury in a barometer).

VACUUM MODULATOR: Generates a hydraulic oil pressure in response to the amount of engine vacuum.

VALVES: Devices that can open or close fluid passages in a hydraulic system and are used for directing fluid flow and controlling pressure.

VALVE BODY ASSEMBLY: The main hydraulic control assembly of the transmission/transaxle that contains numerous valves, check balls, and other components to control the distribution of pressurized oil throughout the transmission.

VALVE CLEARANCE: The measured gap between the end of the valve stem and the rocker arm, cam lobe or follower that activates the valve.

VALVE GUIDES: The guide through which the stem of the valve passes. The guide is designed to keep the valve in proper alignment.

VALVE LASH (clearance): The operating clearance in the valve train.

VALVE TRAIN: The system that operates intake and exhaust valves, consisting of camshaft, valves and springs, lifters, pushrods and rocker arms.

VAPOR LOCK: Boiling of the fuel in the fuel lines due to excess heat. This will interfere with the flow of fuel in the lines and can completely stop the flow. Vapor lock normally only occurs in hot weather.

VARIABLE DISPLACEMENT (VARIABLE CAPACITY) VANE PUMP: Slipper-type vanes, mounted in a revolving rotor and contained within the bore of a movable slide, capture and then force fluid to flow. Movement of the slide to various positions changes the size of the vane chambers and the amount of fluid flow. **Note:** GM refers to this pump design as variable displacement, and Ford terms it variable capacity.

VARIABLE FORCE SOLENOID (VFS): Commonly referred to as the electronic pressure control (EPC) solenoid, it replaces the cable/linkage style of TV system control and is integrated with a spool valve and spring assembly to control pressure. A variable computer-controlled current flow varies the internal force of the solenoid on the spool valve and resulting control pressure.

VARIABLE ORIFICE THERMAL VALVE: Temperature-sensitive hydraulic oil control device that adjusts the size of a circuit path opening. By altering the size of the opening, the oil flow rate is adapted for cold to hot oil viscosity changes.

VARNISH: Term applied to the residue formed when gasoline gets old and stale.

VCM: See Electronic Control Unit (ECU).

VEHICLE SPEED SENSOR (VSS): Provides an electrical signal to the computer module, measuring vehicle speed, and affects the torque converter clutch engagement and release.

VESPEL SEALING RINGS: Hard plastic material that produces excellent sealing in dynamic settings. These rings are found in late versions of the 4T60 and in all 4T60-E and 4T80-E transaxles.

VISCOSITY: The ability of a fluid to flow. The lower the viscosity rating, the easier the fluid will flow. 10 weight motor oil will flow much easier than 40 weight motor oil.

VISCOSITY INDEX IMPROVERS: Keeps the viscosity nearly constant with changes in temperature. This is especially important at low temperatures, when the oil needs to be thin to aid in shifting and for cold-weather starting. Yet it must not be so thin that at high temperatures it will cause excessive hydraulic leakage so that pumps are unable to maintain the proper pressures.

VISCOUS CLUTCH: A specially designed torque converter clutch apply plate that, through the use of a silicon fluid, clamps smoothly and absorbs torsional vibrations.

VOLT: Unit used to measure the force or pressure of electricity. It is defined as the pressure needed to move one amp through the resistance of one ohm.

VOLTAGE: The electrical pressure that causes current to flow. Voltage is measured in volts (V).

VOLTAGE, APPLIED: The actual voltage read at a given point in a circuit. It equals the available voltage of the power supply minus the losses in the circuit up to that point.

VOLTAGE DROP: The voltage lost or used in a circuit by normal loads such as a motor or lamp or by abnormal loads such as a poor (high-resistance) lead or terminal connection.

VOLTAGE REGULATOR: A device that controls the current output of the alternator or generator.

VOLTMETER: An instrument used for measuring electrical force in units called volts. Voltmeters are always connected parallel with the circuit being tested.

VORTEX FLOW: The crosswise or circulatory flow of oil between the blades of the members caused by the centrifugal pumping action of the impeller.

WANKEL ENGINE: An engine which uses no pistons. In place of pistons, triangular-shaped rotors revolve in specially shaped housings.

WATER PUMP: A belt driven component of the cooling system that mounts on the engine, circulating the coolant under pressure.

WATT: The unit for measuring electrical power. One watt is the product of one ampere and one volt (watts equals amps times volts). Wattage is the horsepower of electricity (746 watts equal one horsepower).

WHEEL ALIGNMENT: Inclusive term to describe the front end geometry (caster, camber, toe-in/out).

WHEEL CYLINDER: Found in the automotive drum brake assembly, it is a device, actuated by hydraulic pressure, which, through internal pistons, pushes the brake shoes outward against the drums.

WHEEL WEIGHT: Small weights attached to the wheel to balance the wheel and tire assembly. Out-of-balance tires quickly wear out and also give erratic handling when installed on the front.

WHEELBASE: Distance between the center of front wheels and the center of rear wheels.

WIDE OPEN THROTTLE (WOT): Full travel of accelerator pedal.

WORK: The force exerted to move a mass or object. Work involves motion; if a force is exerted and no motion takes place, no work is done. Work per unit of time is called power. Work = force x distance = ft. lbs. 33,000 ft. lbs. in one minute = 1 horsepower

ZERO-THROTTLE COAST DOWN: A full release of accelerator pedal while vehicle is in motion and in drive range.

Commonly Used Abbreviations

2

2WD Two Wheel Drive

4

4WD Four Wheel Drive

A

A/C Air Conditioning
ABDC After Bottom Dead Center
ABS Anti-lock Brakes
AC Alternating Current
ACL Air cleaner
ACT Air Charge Temperature
AIR Secondary Air Injection
ALCL Assembly Line Communications Link
ALDL Assembly Line Diagnostic Link
AT Automatic Transaxle/Transmission
ATDC After Top Dead Center
ATF Automatic Transmission Fluid
ATS Air Temperature Sensor
AWD All Wheel Drive

B

BAP Barometric Absolute Pressure
BARO Barometric Pressure
BBDC Before Bottom Dead Center
BCM Body Control Module
BDC Bottom Dead Center
BPT Backpressure Transducer
BTDC Before Top Dead Center
BVSV Bimetallic Vacuum Switching Valve

C

CAC Charge Air Cooler
CARB California Air Resources Board
CAT Catalytic Converter
CCC Computer Command Control
CCCC Computer Controlled Catalytic Converter
CCCI Computer Controlled Coil Ignition
CCD Computer Controlled Dwell
CDI Capacitor Discharge Ignition
CEC Computerized Engine Control
CFI Continuous Fuel Injection
CIS Continuous Injection System
CIS-E Continuous Injection System - Electronic
CKP Crankshaft Position
CL Closed Loop
CMP Camshaft Position
CPP Clutch Pedal Position
CTOX Continuous Trap Oxidizer System
CTP Closed Throttle Position
CVC Constant Vacuum Control
CYL Cylinder

D

DBC Dual Bed Catalyst
DC Direct Current
DFI Direct Fuel Injection
DIS Distributorless Ignition System
DLC Data Link Connector
DMM Digital Multimeter
DOHC Double Overhead Camshaft
DRB Diagnostic Readout Box
DTC Diagnostic Trouble Code
DTM Diagnostic Test Mode
DVOM Digital Volt/Ohmmeter

E

EBCM Electronic Brake Control Module
ECM Engine Control Module
ECT Engine Coolant Temperature
ECU Engine Control Unit or Electronic Control Unit
EDIS Electronic Distributorless Ignition System
EEC Electronic Engine Control
EEPROM Electrically Erasable Programmable Read Only Memory
EFE Early Fuel Evaporation
EGR Exhaust Gas Recirculation
EGRT Exhaust Gas Recirculation Temperature
EGRVC EGR Valve Control
EPROM Erasable Programmable Read Only Memory
EVAP Evaporative Emissions
EVP EGR Valve Position

F

FBC Feedback Carburetor
FEEPROM Flash Electrically Erasable Programmable Read Only Memory
FF Flexible Fuel
FI Fuel Injection
FT Fuel Trim
FWD Front Wheel Drive

G

GND Ground

H

HAC High Altitude Compensation
HEGO Heated Exhaust Gas Oxygen sensor
HEI High Energy Ignition
HO2 Sensor Heated Oxygen Sensor

I

IAC Idle Air Control
IAT Intake Air Temperature
ICM Ignition Control Module
IFI Indirect Fuel Injection
IFS Inertia Fuel Shutoff
ISC Idle Speed Control
IVSV Idle Vacuum Switching Valve

Commonly Used Abbreviations

K

KOEO Key On, Engine Off
KOER Key ON, Engine Running
KS Knock Sensor

M

MAF Mass Air Flow
MAP Manifold Absolute Pressure
MAT Manifold Air Temperature
MC Mixture Control
MDP Manifold Differential Pressure
MFI Multiport Fuel Injection
MIL Malfunction Indicator Lamp or Maintenance
MST Manifold Surface Temperature
MVZ Manifold Vacuum Zone

N

NVRAM Nonvolatile Random Access Memory

O

O2 Sensor Oxygen Sensor
OBD On-Board Diagnostic
OC Oxidation Catalyst
OHC Overhead Camshaft
OL Open Loop

P

P/S Power Steering
PAIR Pulsed Secondary Air Injection
PCM Powertrain Control Module
PCS Purge Control Solenoid
PCV Positive Crankcase Ventilation
PIP Profile Ignition Pick-up
PNP Park/Neutral Position
PROM Programmable Read Only Memory
PSP Power Steering Pressure
PTO Power Take-Off
PTOX Periodic Trap Oxidizer System

R

RABS Rear Anti-lock Brake System
RAM Random Access Memory
ROM Read Only Memory
RPM Revolutions Per Minute
RWAL Rear Wheel Anti-lock Brakes
RWD Rear Wheel Drive

S

SBC Single Bed Converter
SBEC Single Board Engine Controller
SC Supercharger
SCB Supercharger Bypass
SFI Sequential Multiport Fuel Injection
SIR Supplemental Inflatible Restraint
SOHC Single Overhead Camshaft
SPL Smoke Puff Limiter
SPOUT Spark Output
SRI Service Reminder Indicator
SRS Supplemental Restraint System
SRT System Readiness Test
SSI Solid State Ignition
ST Scan Tool
STO Self-Test Output

T

TAC Thermostatic Air Clearner
TBI Throttle Body Fuel Injection
TC Turbocharger
TCC Torque Converter Clutch
TCM Transmission Control Module
TDC Top Dead Center
TFI Thick Film Ignition
TP Throttle Position
TR Sensor Transaxle/Transmission Range Sensor
TVV Thermal Vacuum Valve
TWC Three-way Catalytic Converter

V

VAF Volume Air Flow, or Vane Air Flow
VAPS Variable Assist Power Steering
VRV Vacuum Regulator Valve
VSS Vehicle Speed Sensor
VSV Vacuum Switching Valve

W

WOT Wide Open Throttle
WU-TWC Warm Up Three-way Catalytic Converter

ENGLISH TO METRIC CONVERSION: TORQUE

To convert foot-pounds (ft. lbs.) to Newton-meters (Nm), multiply the number of ft. lbs. by 1.36
To convert Newton-meters (Nm) to foot-pounds (ft. lbs.), multiply the number of Nm by 0.7376

ft. lbs.	Nm	ft. lbs.	Nm	ft. lbs.	Nm	ft. lbs.	Nm
0.1	0.1	34	46.2	76	103.4	118	160.5
0.2	0.3	35	47.6	77	104.7	119	161.8
0.3	0.4	36	49.0	78	106.1	120	163.2
0.4	0.5	37	50.3	79	107.4	121	164.6
0.5	0.7	38	51.7	80	108.8	122	165.9
0.6	0.8	39	53.0	81	110.2	123	167.3
0.7	1.0	40	54.4	82	111.5	124	168.6
0.8	1.1	41	55.8	83	112.9	125	170.0
0.9	1.2	42	57.1	84	114.2	126	171.4
1	1.4	43	58.5	85	115.6	127	172.7
2	2.7	44	59.8	86	117.0	128	174.1
3	4.1	45	61.2	87	118.3	129	175.4
4	5.4	46	62.6	88	119.7	130	176.8
5	6.8	47	63.9	89	121.0	131	178.2
6	8.2	48	65.3	90	122.4	132	179.5
7	9.5	49	66.6	91	123.8	133	180.9
8	10.9	50	68.0	92	125.1	134	182.2
9	12.2	51	69.4	93	126.5	135	183.6
10	13.6	52	70.7	94	127.8	136	185.0
11	15.0	53	72.1	95	129.2	137	186.3
12	16.3	54	73.4	96	130.6	138	187.7
13	17.7	55	74.8	97	131.9	139	189.0
14	19.0	56	76.2	98	133.3	140	190.4
15	20.4	57	77.5	99	134.6	141	191.8
16	21.8	58	78.9	100	136.0	142	193.1
17	23.1	59	80.2	101	137.4	143	194.5
18	24.5	60	81.6	102	138.7	144	195.8
19	25.8	61	83.0	103	140.1	145	197.2
20	27.2	62	84.3	104	141.4	146	198.6
21	28.6	63	85.7	105	142.8	147	199.9
22	29.9	64	87.0	106	144.2	148	201.3
23	31.3	65	88.4	107	145.5	149	202.6
24	32.6	66	89.8	108	146.9	150	204.0
25	34.0	67	91.1	109	148.2	151	205.4
26	35.4	68	92.5	110	149.6	152	206.7
27	36.7	69	93.8	111	151.0	153	208.1
28	38.1	70	95.2	112	152.3	154	209.4
29	39.4	71	96.6	113	153.7	155	210.8
30	40.8	72	97.9	114	155.0	156	212.2
31	42.2	73	99.3	115	156.4	157	213.5
32	43.5	74	100.6	116	157.8	158	214.9
33	44.9	75	102.0	117	159.1	159	216.2

METRIC TO ENGLISH CONVERSION: TORQUE

To convert foot-pounds (ft. lbs.) to Newton-meters (Nm), multiply the number of ft. lbs. by 1.36

To convert Newton-meters (Nm) to foot-pounds (ft. lbs.), multiply the number of Nm by 0.7376

Nm	ft. lbs.	Nm	ft. lbs.	Nm	ft. lbs.	Nm	ft. lbs.	Nm	ft. lbs.
0.1	0.1	34	25.0	76	55.9	118	86.8	160	117.6
0.2	0.1	35	25.7	77	56.6	119	87.5	161	118.4
0.3	0.2	36	26.5	78	57.4	120	88.2	162	119.1
0.4	0.3	37	27.2	79	58.1	121	89.0	163	119.9
0.5	0.4	38	27.9	80	58.8	122	89.7	164	120.6
0.6	0.4	39	28.7	81	59.6	123	90.4	165	121.3
0.7	0.5	40	29.4	82	60.3	124	91.2	166	122.1
0.8	0.6	41	30.1	83	61.0	125	91.9	167	122.8
0.9	0.7	42	30.9	84	61.8	126	92.6	168	123.5
1	0.7	43	31.6	85	62.5	127	93.4	169	124.3
2	1.5	44	32.4	86	63.2	128	94.1	170	125.0
3	2.2	45	33.1	87	64.0	129	94.9	171	125.7
4	2.9	46	33.8	88	64.7	130	95.6	172	126.5
5	3.7	47	34.6	89	65.4	131	96.3	173	127.2
6	4.4	48	35.3	90	66.2	132	97.1	174	127.9
7	5.1	49	36.0	91	66.9	133	97.8	175	128.7
8	5.9	50	36.8	92	67.6	134	98.5	176	129.4
9	6.6	51	37.5	93	68.4	135	99.3	177	130.1
10	7.4	52	38.2	94	69.1	136	100.0	178	130.9
11	8.1	53	39.0	95	69.9	137	100.7	179	131.6
12	8.8	54	39.7	96	70.6	138	101.5	180	132.4
13	9.6	55	40.4	97	71.3	139	102.2	181	133.1
14	10.3	56	41.2	98	72.1	140	102.9	182	133.8
15	11.0	57	41.9	99	72.8	141	103.7	183	134.6
16	11.8	58	42.6	100	73.5	142	104.4	184	135.3
17	12.5	59	43.4	101	74.3	143	105.1	185	136.0
18	13.2	60	44.1	102	75.0	144	105.9	186	136.8
19	14.0	61	44.9	103	75.7	145	106.6	187	137.5
20	14.7	62	45.6	104	76.5	146	107.4	188	138.2
21	15.4	63	46.3	105	77.2	147	108.1	189	139.0
22	16.2	64	47.1	106	77.9	148	108.8	190	139.7
23	16.9	65	47.8	107	78.7	149	109.6	191	140.4
24	17.6	66	48.5	108	79.4	150	110.3	192	141.2
25	18.4	67	49.3	109	80.1	151	111.0	193	141.9
26	19.1	68	50.0	110	80.9	152	111.8	194	142.6
27	19.9	69	50.7	111	81.6	153	112.5	195	143.4
28	20.6	70	51.5	112	82.4	154	113.2	196	144.1
29	21.3	71	52.2	113	83.1	155	114.0	197	144.9
30	22.1	72	52.9	114	83.8	156	114.7	198	145.6
31	22.8	73	53.7	115	84.6	157	115.4	199	146.3
32	23.5	74	54.4	116	85.3	158	116.2	200	147.1
33	24.3	75	55.1	117	86.0	159	116.9	201	147.8

ENGLISH/METRIC CONVERSION: TEMPERATURE

To convert Fahrenheit (F°) to Celsius (C°), take F° temperature and subtract 32, multiply the result by 5 and divide the result by 9
To convert Celsius (C°) to Fahrenheit (F°), take C° temperature and multiply it by 9, divide the result by 5 and add 32

F°	C°	F°	C°	C°	F°	C°	F°
-40	-40.0	150	65.6	-38	-36.4	46	114.8
-35	-37.2	155	68.3	-36	-32.8	48	118.4
-30	-34.4	160	71.1	-34	-29.2	50	122
-25	-31.7	165	73.9	-32	-25.6	52	125.6
-20	-28.9	170	76.7	-30	-22	54	129.2
-15	-26.1	175	79.4	-28	-18.4	56	132.8
-10	-23.3	180	82.2	-26	-14.8	58	136.4
-5	-20.6	185	85.0	-24	-11.2	60	140
0	-17.8	190	87.8	-22	-7.6	62	143.6
1	-17.2	195	90.6	-20	-4	64	147.2
2	-16.7	200	93.3	-18	-0.4	66	150.8
3	-16.1	205	96.1	-16	3.2	68	154.4
4	-15.6	210	98.9	-14	6.8	70	158
5	-15.0	212	100.0	-12	10.4	72	161.6
10	-12.2	215	101.7	-10	14	74	165.2
15	-9.4	220	104.4	-8	17.6	76	168.8
20	-6.7	225	107.2	-6	21.2	78	172.4
25	-3.9	230	110.0	-4	24.8	80	176
30	-1.1	235	112.8	-2	28.4	82	179.6
35	1.7	240	115.6	0	32	84	183.2
40	4.4	245	118.3	2	35.6	86	186.8
45	7.2	250	121.1	4	39.2	88	190.4
50	10.0	255	123.9	6	42.8	90	194
55	12.8	260	126.7	8	46.4	92	197.6
60	15.6	265	129.4	10	50	94	201.2
65	18.3	270	132.2	12	53.6	96	204.8
70	21.1	275	135.0	14	57.2	98	208.4
75	23.9	280	137.8	16	60.8	100	212
80	26.7	285	140.6	18	64.4	102	215.6
85	29.4	290	143.3	20	68	104	219.2
90	32.2	295	146.1	22	71.6	106	222.8
95	35.0	300	148.9	24	75.2	108	226.4
100	37.8	305	151.7	26	78.8	110	230
105	40.6	310	154.4	28	82.4	112	233.6
110	43.3	315	157.2	30	86	114	237.2
115	46.1	320	160.0	32	89.6	116	240.8
120	48.9	325	162.8	34	93.2	118	244.4
125	51.7	330	165.6	36	96.8	120	248
130	54.4	335	168.3	38	100.4	122	251.6
135	57.2	340	171.1	40	104	124	255.2
140	60.0	345	173.9	42	107.6	126	258.8
145	62.8	350	176.7	44	111.2	128	262.4

LENGTH CONVERSION

To convert inches (in.) to millimeters (mm), multiply the number of inches by 25.4
To convert millimeters (mm) to inches (in.), multiply the number of millimeters by 0.04

Inches	Millimeters	Inches	Millimeters	Inches	Millimeters	Inches	Millimeters
0.0001	0.00254	0.005	0.1270	0.09	2.286	4	101.6
0.0002	0.00508	0.006	0.1524	0.1	2.54	5	127.0
0.0003	0.00762	0.007	0.1778	0.2	5.08	6	152.4
0.0004	0.01016	0.008	0.2032	0.3	7.62	7	177.8
0.0005	0.01270	0.009	0.2286	0.4	10.16	8	203.2
0.0006	0.01524	0.01	0.254	0.5	12.70	9	228.6
0.0007	0.01778	0.02	0.508	0.6	15.24	10	254.0
0.0008	0.02032	0.03	0.762	0.7	17.78	11	279.4
0.0009	0.02286	0.04	1.016	0.8	20.32	12	304.8
0.001	0.0254	0.05	1.270	0.9	22.86	13	330.2
0.002	0.0508	0.06	1.524	1	25.4	14	355.6
0.003	0.0762	0.07	1.778	2	50.8	15	381.0
0.004	0.1016	0.08	2.032	3	76.2	16	406.4

ENGLISH/METRIC CONVERSION: LENGTH

To convert inches (in.) to millimeters (mm), multiply the number of inches by 25.4
To convert millimeters (mm) to inches (in.), multiply the number of millimeters by 0.04

Inches		Millimeters	Inches		Millimeters	Inches		Millimeters
Fraction	Decimal	Decimal	Fraction	Decimal	Decimal	Fraction	Decimal	Decimal
1/64	0.016	0.397	11/32	0.344	8.731	11/16	0.688	17.463
1/32	0.031	0.794	23/64	0.359	9.128	45/64	0.703	17.859
3/64	0.047	1.191	3/8	0.375	9.525	23/32	0.719	18.256
1/16	0.063	1.588	25/64	0.391	9.922	47/64	0.734	18.653
5/64	0.078	1.984	13/32	0.406	10.319	3/4	0.750	19.050
3/32	0.094	2.381	27/64	0.422	10.716	49/64	0.766	19.447
7/64	0.109	2.778	7/16	0.438	11.113	25/32	0.781	19.844
1/8	0.125	3.175	29/64	0.453	11.509	51/64	0.797	20.241
9/64	0.141	3.572	15/32	0.469	11.906	13/16	0.813	20.638
5/32	0.156	3.969	31/64	0.484	12.303	53/64	0.828	21.034
11/64	0.172	4.366	1/2	0.500	12.700	27/32	0.844	21.431
3/16	0.188	4.763	33/64	0.516	13.097	55/64	0.859	21.828
13/64	0.203	5.159	17/32	0.531	13.494	7/8	0.875	22.225
7/32	0.219	5.556	35/64	0.547	13.891	57/64	0.891	22.622
15/64	0.234	5.953	9/16	0.563	14.288	29/32	0.906	23.019
1/4	0.250	6.350	37/64	0.578	14.684	59/64	0.922	23.416
17/64	0.266	6.747	19/32	0.594	15.081	15/16	0.938	23.813
9/32	0.281	7.144	39/64	0.609	15.478	61/64	0.953	24.209
19/64	0.297	7.541	5/8	0.625	15.875	31/32	0.969	24.606
5/16	0.313	7.938	41/64	0.641	16.272	63/64	0.984	25.003
21/64	0.328	8.334	21/32	0.656	16.669	1/1	1.000	25.400
			43/64	0.672	17.066			

Domestic Manual ISBN 1-4180-0606-8/Part No. 130606
Import Manual ISBN 1-4180-1537-7/Part No. 131537

Chilton has added so much to its labor guide manual that we've had to put it in two volumes! We've added hundreds of new labor operations—including maintenance services and electronic system diagnosis—to the *Chilton® 2006 Import and Domestic Labor Guide* manuals. All labor times for 1981 through 2006 vehicles consider the real world environment in which technicians work: worn, rusted or dirty components, being serviced with tools commonly used in the aftermarket. Chilton labor times are accepted by most insurance and extended warranty companies. Vehicle makes and models conform to current Automotive Aftermarket Industry Association standards.

Labor Guide Manual Benefits:

- parts terminology is more standardized across different OEMs to simplify reference.
- a total of more than 2,500 pages of updated Chilton labor times appear in these two volumes
- make sure your students have this latest edition because our experts have updated hundreds of labor times for earlier models

Hardcover Manuals are 8 7/8" x 11", ©2006

Labor Guide CD-ROM Benefits:

- easy-to-use software to create and print professional-quality estimates and invoices
- three user-defined levels of labor rates correspond to different types of job scenarios, for "real-world" application
- functions as a database of aftermarket labor times for monitoring warranty and insurance claims
- software keeps track of customers and prior estimates for time-saving recall
- customizable application allows service writers to add labor operations and times, and parts companies to add labor times to existing parts ordering systems

CD ISBN 1-4180-0605-X/Part No. 130605
©2006

Previous Year Editions:
Chilton 2005 Labor Guide Manual, ISBN 1-4018-7412-6/Part No. 27412
Chilton 2005 Labor Guide CD-ROM, ISBN 1-4018-7818-0/Part No. 27818

FOR CUSTOMER SUPPORT CALL **1-800-477-3692**

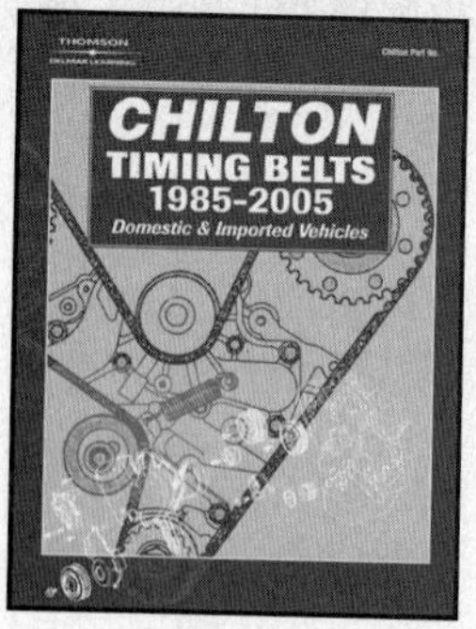

Chilton Timing Belts, 1985-2005

Chilton

ISBN 1-4018-9880-7/Part No. 129880

Timing belt procedures can represent increased profits for automotive repair shops and service stations, and this manual contains all the information automotive technicians need to properly service timing belts on domestic and imported cars, vans, and light trucks through 2005 models. Clear, straightforward procedures, illustrations, and specifications help to communicate 20 years of vehicle applications for fast, accurate inspection, replacement, and tensioning of timing belts. Readers will learn step-by-step how to perform key procedures both quickly and safely, while learning the correct labor time to charge for the service. OEM-recommended replacement intervals for proper maintenance of customer's vehicles are also featured.

BENEFITS

- detailed illustrations clearly demonstrate important concepts, such as how to correctly align camshaft and crankshaft timing marks, and how to simplify serpentine belt installation
- readers are made aware of potential hazards and time-wasting practices that can impede safe and profitable service procedures
- special tools are identified so that completing the service is as easy and quick as possible

544 pp, 8 1/2" x 11", softcover, ©2006

The *Chilton® 2006 Diagnostic Service Manuals* provide technicians with the critical diagnostic information they need to accurately identify and solve engine performance problems. Clear explanations, specifications, and illustrations help technicians diagnose second generation on-board diagnostic (OBD-II) systems. Chilton Diagnostic Service Manuals, when used with an engine analyzer, scan tool, or lab scope, allow diagnosticians to understand functions of engine performance components and systems, simplify testing procedures, and diagnose trouble codes.

Diagnostic Service Manual Benefits:

- provide training information in addition to reference material
- explain engine performance components and system operation
- function as exceptional diagnostic companions when analyzing automotive drive-train performance problems
- provide a comprehensive list of trouble code titles, conditions, and possible causes
- reduce diagnostic and repair time using expert testing procedures and trouble-shooting hints

2006 Editions

Chilton 2006 DaimlerChrysler Diagnostic Service Manual
ISBN 1-4180-2118-0/Part No. 132118
Chilton 2006 Ford Diagnostic Service Manual
ISBN 1-4180-2119-9/Part No. 132119
Chilton 2006 General Motors Diagnostic Service Manual
ISBN 1-4180-2120-2/Part No. 132120
Chilton 2006 Asian Diagnostic Service Manual, Volume 1
ISBN 1-4180-2913-0/Part No. 132913
Chilton 2006 Asian Diagnostic Service Manual, Volume 2
ISBN 1-4180-2914-9/Part No. 132914
Chilton 2006 Asian Diagnostic Service Manual, Volume 3
ISBN 1-4180-2915-7/Part No. 132915
Chilton 2006 European Diagnostic Service Manual
ISBN 1-4180-2924-6/Part No. 132924
Manuals are 8 1/2" x 11", ©2006

2005 Editions

Chilton 2005 General Motors Diagnostic Service Manual
ISBN 1-4180-0552-5/Part No. 130552
Chilton 2005 Chrysler Diagnostic Service Manual
ISBN 1-4180-0550-9/Part No. 130550
Chilton 2005 Ford Diagnostic Service Manual
ISBN 1-4180-0551-7/Part No. 130551
Chilton 2005 Asian Diagnostic Service Manual
ISBN 1-4180-0553-3/Part No. 130553
Manuals are 8 1/2" x 11", ©2005

The *Chilton® Perennial Editions* contain repair and maintenance information for popular mechanical systems that may not be available elsewhere. They offer a wide range of repair information on cars, trucks, vans, and SUVs dating back to the early 1960s, and as current as 2002. Information for 1993 and later model years includes scheduled maintenance interval charts.

Benefits:

- covers the most common vehicle models found in the repair aftermarket today
- gain quick understanding of systems using exploded-view illustrations, diagrams, and charts
- simplify tough jobs with easy-to-follow removal and installation instructions for heater core and other components
- obtain complete coverage of repair procedures from drive train to chassis and associated components

Auto Repair Manual, 1998-2002, 1,426 pages
ISBN 0-8019-9362-8/Part No. 9362
Auto Repair Manual, 1993-1997, 2,064 pages
ISBN 0-8019-7919-6/Part No. 7919
Auto Repair Manual, 1988-1992, 1,284 pages
ISBN 0-8019-7906-4/Part No. 7906
Auto Repair Manual, 1980-1987, 1,344 pages
ISBN 0-8019-7670-7/Part No. 7670

Import Car Repair Manual, 1998-2002, 1,792 pps
ISBN 0-8019-9363-6/Part No. 9363
Import Car Repair Manual, 1993-1997, 2,080 pps
ISBN 0-8019-7920-X/Part No. 7920
Import Car Repair Manual, 1988-1992, 1,632 pages
ISBN 0-8019-7907-2/Part No. 7907
Import Car Repair Manual, 1980-1987, 1,488 pages
ISBN 0-8019-7672-3/Part No. 7672

Truck & Van Repair Manual, 1998-2002, 1,408 pages
ISBN 0-8019-9364-4/Part No. 9364
Truck & Van Repair Manual, 1993-1997, 2,096 pages
ISBN 0-8019-7921-8/Part No. 7921
Truck & Van Repair Manual, 1991-1995, 1,664 pages
ISBN 0-8019-7911-0/Part No. 7911
Truck & Van Repair Manual, 1986-1990, 1,536 pages
ISBN 0-8019-7902-1/Part No. 7902
Truck & Van Repair Manual, 1979-1986, 1,440 pages
ISBN 0-8019-7655-3/Part No. 7655

SUV Repair Manual, 1998-2002, 1,292 pages
ISBN 0-8019-9365-2/Part No. 9365

Hardcover manuals are 8 1/2" x 11".

Chilton Collector's Editions—*Reference Manuals for Vintage Vehicles*
Auto Repair Manual, 1964-1971, ISBN 0-8019-5974-8/Part No. 5974,
Truck & Van Repair Manual, 1961-1971, ISBN 0-8019-6198-X/Part No. 6198
Truck & Van Repair Manual, 1971-1978, ISBN 0-8019-7012-1/Part No. 7012